FUNDAMENTAL CONSTANTS

Constant	Symbol	Value		
			Power of 10	Units
Speed of light	c	2.997 924 58*	10^{8}	$m\ s^{-1}$
Elementary charge	e	1.602 176 565	10^{-19}	C
Planck's constant	h	6.626 069 57	10^{-34}	J s
	$\hbar = h/2\pi$	1.054 571 726	10^{-34}	J s
Boltzmann's constant	k	1.380 6488	10^{-23}	$J\ K^{-1}$
Avogadro's constant	N_A	6.022 141 29	10^{23}	mol^{-1}
Gas constant	$R = N_A k$	8.314 4621		$J\ K^{-1}\ mol^{-1}$
Faraday's constant	$F = N_A e$	9.648 533 65	10^{4}	$C\ mol^{-1}$
Mass				
Electron	m_e	9.109 382 91	10^{-31}	kg
Proton	m_p	1.672 621 777	10^{-27}	kg
Neutron	m_n	1.674 927 351	10^{-27}	kg
Atomic mass constant	m_u	1.660 538 921	10^{-27}	kg
Vacuum permeability	μ_0	4π*	10^{-7}	$J\ s^2\ C^{-2}\ m^{-1}$
Vacuum permittivity	$\varepsilon_0 = 1/\mu_0 c^2$	8.854 187 817	10^{-12}	$J^{-1}\ C^2\ m^{-1}$
	$4\pi\varepsilon_0$	1.112 650 056	10^{-10}	$J^{-1}\ C^2\ m^{-1}$
Bohr magneton	$\mu_B = e\hbar/2m_e$	9.274 009 68	10^{-24}	$J\ T^{-1}$
Nuclear magneton	$\mu_N = e\hbar/2m_p$	5.050 783 53	10^{-27}	$J\ T^{-1}$
Proton magnetic moment	μ_p	1.410 606 743	10^{-26}	$J\ T^{-1}$
g-Value of electron	g_e	2.002 319 304		
Magnetogyric ratio				
Electron	$\gamma_e = -g_e e/2m_e$	−1.001 159 652	10^{10}	$C\ kg^{-1}$
Proton	$\gamma_p = 2\mu_p/\hbar$	2.675 222 004	10^{8}	$C\ kg^{-1}$
Bohr radius	$a_0 = 4\pi\varepsilon_0\hbar^2/e^2 m_e$	5.291 772 109	10^{-11}	m
Rydberg constant	$\tilde{R}_\infty = m_e e^4/8h^3 c\varepsilon_0^2$	1.097 373 157	10^{5}	cm^{-1}
	$hc\tilde{R}_\infty/e$	13.605 692 53		eV
Fine-structure constant	$\alpha = \mu_0 e^2 c/2h$	7.297 352 5698	10^{-3}	
	α^{-1}	1.370 359 990 74	10^{2}	
Second radiation constant	$c_2 = hc/k$	1.438 777 0	10^{-2}	m K
Stefan–Boltzmann constant	$\sigma = 2\pi^5 k^4/15h^3c^2$	5.670 373	10^{-8}	$W\ m^{-2}\ K^{-4}$
Standard acceleration of free fall	g	9.806 65*		$m\ s^{-2}$
Gravitational constant	G	6.673 84	10^{-11}	$N\ m^2\ kg^{-2}$

* Exact value. For current values of the constants, see the National Institute of Standards and Technology (NIST) website.

PHYSICAL CHEMISTRY

Thermodynamics, Structure, and Change

Tenth edition

Peter Atkins
Fellow of Lincoln College,
University of Oxford,
Oxford, UK

Julio de Paula
Professor of Chemistry,
Lewis & Clark College,
Portland, Oregon, USA

W. H. Freeman and Company
New York

Publisher: Jessica Fiorillo
Associate Director of Marketing: Debbie Clare
Associate Editor: Heidi Bamatter
Media Acquisitions Editor: Dave Quinn
Marketing Assistant: Samantha Zimbler

Library of Congress Control Number: 2013939968

Physical Chemistry: Thermodynamics, Structure, and Change, Tenth Edition

ISBN-13: 978-1-4292-9019-7
ISBN-10: 1-4292-9019-6

Published in Great Britain by Oxford University Press

This edition has been authorized by Oxford University Press for sales in the United States and Canada only and not export therefrom.

Fifth printing

W. H. Freeman and Company
41 Madison Avenue
New York, NY 10010

www.whfreeman.com

Printed in China

PREFACE

This new edition is the product of a thorough revision of content and its presentation. Our goal is to make the book even more accessible to students and useful to instructors by enhancing its flexibility. We hope that both categories of user will perceive and enjoy the renewed vitality of the text and the presentation of this demanding but engaging subject.

The text is still divided into three parts, but each chapter is now presented as a series of short and more readily mastered *Topics*. This new structure allows the instructor to tailor the text within the time constraints of the course as omissions will be easier to make, emphases satisfied more readily, and the trajectory through the subject modified more easily. For instance, it is now easier to approach the material either from a 'quantum first' or a 'thermodynamics first' perspective because it is no longer necessary to take a linear path through chapters. Instead, students and instructors can match the choice of Topics to their learning objectives. We have been very careful not to presuppose or impose a particular sequence, except where it is demanded by common sense.

We open with a *Foundations* chapter, which reviews basic concepts of chemistry and physics used through the text. Part 1 now carries the title *Thermodynamics*. New to this edition is coverage of ternary phase diagrams, which are important in applications of physical chemistry to engineering and materials science. Part 2 (*Structure*) continues to cover quantum theory, atomic and molecular structure, spectroscopy, molecular assemblies, and statistical thermodynamics. Part 3 (*Change*) has lost a chapter dedicated to catalysis, but not the material. Enzyme-catalysed reactions are now in Chapter 20, and heterogeneous catalysis is now part of a new Chapter 22 focused on surface structure and processes.

As always, we have paid special attention to helping students navigate and master this material. Each chapter opens with a brief summary of its Topics. Then each Topic begins with three questions: 'Why do you need to know this material?', 'What is the key idea?', and 'What do you need to know already?'. The answers to the third question point to other Topics that we consider appropriate to have studied or at least to refer to as background to the current Topic. The *Checklists* at the end of each Topic are useful distillations of the most important concepts and equations that appear in the exposition.

We continue to develop strategies to make mathematics, which is so central to the development of physical chemistry, accessible to students. In addition to associating *Mathematical background* sections with appropriate chapters, we give more help with the development of equations: we motivate them, justify them, and comment on the steps taken to derive them. We also added a new feature: *The chemist's toolkit*, which offers quick and immediate help on a concept from mathematics or physics.

This edition has more worked *Examples*, which require students to organize their thoughts about how to proceed with complex calculations, and more *Brief illustrations*, which show how to use an equation or deploy a concept in a straightforward way. Both have *Self-tests* to enable students to assess their grasp of the material. We have structured the end-of-chapter *Discussion questions*, *Exercises*, and *Problems* to match the grouping of the Topics, but have added Topic- and Chapter-crossing *Integrated activities* to show that several Topics are often necessary to solve a single problem. The *Resource section* has been restructured and augmented by the addition of a list of integrals that are used (and referred to) throughout the text.

We are, of course, alert to the development of electronic resources and have made a special effort in this edition to encourage the use of web-based tools, which are identified in the *Using the book* section that follows this preface. Important among these tools are *Impact* sections, which provide examples of how the material in the chapters is applied in such diverse areas as biochemistry, medicine, environmental science, and materials science.

Overall, we have taken this opportunity to refresh the text thoroughly, making it even more flexible, helpful, and up to date. As ever, we hope that you will contact us with your suggestions for its continued improvement.

PWA, Oxford
JdeP, Portland

USING THE BOOK

For the tenth edition of *Physical Chemistry: Thermodynamics, Structure, and Change* we have tailored the text even more closely to the needs of students. First, the material within each chapter has been reorganized into discrete topics to improve accessibility, clarity, and flexibility. Second, in addition to the variety of learning features already present, we have significantly enhanced the mathematics support by adding new Chemist's toolkit boxes, and checklists of key concepts at the end of each topic.

Organizing the information

➤ Innovative new structure

Each chapter has been reorganized into short topics, making the text more readable for students and more flexible for instructors. Each topic opens with a comment on why it is important, a statement of the key idea, and a brief summary of the background needed to understand the topic.

> ➤ Why do you need to know this material?
>
> Because chemistry is about matter and the changes that it can undergo, both physically and chemically, the properties of matter underlie the entire discussion in this book.
>
> ➤ What is the key idea?
>
> The bulk properties of matter are related to the identities

➤ Notes on good practice

Our *Notes on good practice* will help you avoid making common mistakes. They encourage conformity to the international language of science by setting out the conventions and procedures adopted by the International Union of Pure and Applied Chemistry (IUPAC).

> applicable only to perfect gases (and other idealized systems) are labelled, as here, with a number in blue.
>
> *A note on good practice* Although the term 'ideal gas' is almost universally used in place of 'perfect gas', there are reasons for preferring the latter term. In an ideal system the interactions between molecules in a mixture are all the same. In a perfect gas not only are the interactions all the same but they are in fact zero. Few, though, make this useful distinction.
>
> Equation A.5, the **perfect gas equation**, is a summary of three empirical conclusions, namely Boyle's law ($p \propto 1/V$ at

➤ Resource section

The comprehensive *Resource section* at the end of the book contains a table of integrals, data tables, a summary of conventions about units, and character tables. Short extracts of these tables often appear in the topics themselves, principally to give an idea of the typical values of the physical quantities we are introducing.

> RESOURCE SECTION
>
> Contents
>
>

➤ Checklist of concepts

A *Checklist of key concepts* is provided at the end of each topic so that you can tick off those concepts which you feel you have mastered.

Checklist of concepts

- ☐ 1. The **entropy** acts as a signpost of spontaneous change.
- ☐ 2. Entropy change is defined in terms of heat transactions (the **Clausius definition**).
- ☐ 3. The **Boltzmann formula** defines absolute entropies in terms of the number of ways of achieving a configuration.

Presenting the mathematics

➤ Justifications

Mathematical development is an intrinsic part of physical chemistry, and to achieve full understanding you need to see how a particular expression is obtained and if any assumptions have been made. The *Justifications* are set off from the text to let you adjust the level of detail to meet your current needs and make it easier to review material.

Justification 3A.1 Heating accompanying reversible adiabatic expansion

This *Justification* is based on two features of the cycle. One feature is that the two temperatures T_h and T_c in eqn 3A.7 lie on the same adiabat in Fig. 3A.7. The second feature is that the energy transferred as heat during the two isothermal stages are

$$q_h = nRT_h \ln\frac{V_B}{V_A} \qquad q_c = nRT_c \ln\frac{V_D}{V_C}$$

We now show that the two volume ratios are related in a very simple way. From the relation between temperature and volume for reversible adiabatic processes (VT^c=constant, Topic 2D):

➤ Chemist's toolkits

New to the tenth edition, the *Chemist's toolkits* are succinct reminders of the mathematical concepts and techniques that you will need in order to understand a particular derivation being described in the main text.

The chemist's toolkit A.1 Quantities and units

The result of a measurement is a **physical quantity** that is reported as a numerical multiple of a unit:

physical quantity = numerical value × unit

It follows that units may be treated like algebraic quantities and may be multiplied, divided, and cancelled. Thus, the expression (physical quantity)/unit is the numerical value (a dimensionless quantity) of the measurement in the specified

➤ Mathematical backgrounds

There are six *Mathematical background* sections dispersed throughout the text. They cover in detail the main mathematical concepts that you need to understand in order to be able to master physical chemistry. Each one is located at the end of the chapter to which it is most relevant.

Mathematical background 1 Differentiat

Two of the most important mathematical techniques in the physical sciences are differentiation and integration. They occur throughout the subject, and it is essential to be aware of the procedures involved.

MB1.1 Differentiation: definitions

Differentiation is concerned with the slopes of functions, such as the rate of change of a variable with time. The formal definition of the **derivative**, df/dx, of a function $f(x)$ is

➤ Annotated equations and equation labels

We have annotated many equations to help you follow how they are developed. An annotation can take you across the equals sign: it is a reminder of the substitution used, an approximation made, the terms that have been assumed constant, the integral used, and so on. An annotation can also be a reminder of the significance of an individual term in an expression. We sometimes color a collection of numbers or symbols to show how they carry from one line to the next. Many of the equations are labelled to highlight their significance.

$$w = -nRT\int_{V_i}^{V_f}\frac{dV}{V} \overset{\text{Integral A.2}}{=} -nRT\ln\frac{V_f}{V_i}$$

Perfect gas, reversible, isothermal — Work of expansion (2A.9)

➤ Checklists of equations

You don't have to memorize every equation in the text. A checklist at the end of each topic summarizes the most important equations and the conditions under which they apply.

Checklist of equations

Property	Equation
Compression factor	$Z = V_m/V_m^\circ$
Virial equation of state	$pV_m = RT(1+B/V_m+C/V_m^3+\cdots)$
van der Waals equation of state	$p = nRT/(V-nb) - a(n/V)^2$
Reduced variables	$X_r = X_m/X_c$

Setting up and solving problems

➤ Brief illustrations

A *Brief illustration* shows you how to use equations or concepts that have just been introduced in the text. They help you to learn how to use data, manipulate units correctly, and become familiar with the magnitudes of properties. They are all accompanied by a Self-test question which you can use to monitor your progress.

Brief illustration 1C.5 Corresponding states

The critical constants of argon and carbon dioxide are given in Table 1C.2. Suppose argon is at 23 atm and 200 K, its reduced pressure and temperature are then

$$p_r = \frac{23\,\text{atm}}{48.0\,\text{atm}} = 0.48 \qquad T_r = \frac{200\,\text{K}}{150.7\,\text{K}} = 1.33$$

For carbon dioxide to be in a corresponding state, its pressure and temperature would need to be

$$p = 0.48\times(72.9\,\text{atm}) = 35\,\text{atm} \qquad T = 1.33\times 304.2\,\text{K} = 405\,\text{K}$$

Self-test 1C.6 What would be the corresponding state of ammonia?

Answer: 53 atm, 539 K

➤ Worked examples

Worked *Examples* are more detailed illustrations of the application of the material, which require you to assemble and develop concepts and equations. We provide a suggested method for solving the problem and then implement it to reach the answer. Worked examples are also accompanied by *Self-test* questions.

Example 3A.2 Calculating the entropy change for a composite process

Calculate the entropy change when argon at 25 °C and 1.00 bar in a container of volume 0.500 dm^3 is allowed to expand to 1.000 dm^3 and is simultaneously heated to 100 °C.

Method As remarked in the text, use reversible isothermal expansion to the final volume, followed by reversible heating at constant volume to the final temperature. The entropy change in the first step is given by eqn 3A.16 and that of the second step, provided C_V is independent of temperature, by eqn 3A.20 (with C_V in place of C_p). In each case we need to

➤ Discussion questions

Discussion questions appear at the end of every chapter, where they are organized by topic. These questions are designed to encourage you to reflect on the material you have just read, and to view it conceptually.

➤ Exercises and Problems

Exercises and *Problems* are also provided at the end of every chapter, and organized by topic. They prompt you to test your understanding of the topics in that chapter. Exercises are designed as relatively straightforward numerical tests whereas the problems are more challenging. The Exercises come in related pairs, with final numerical answers available on the Book Companion Site for the 'a' questions. Final numerical answers to the odd-numbered problems are also available on the Book Companion Site.

TOPIC 3A Entropy

Discussion questions

3A.1 The evolution of life requires the organization of a very large number of molecules into biological cells. Does the formation of living organisms violate the Second Law of thermodynamics? State your conclusion clearly and present detailed arguments to support it.

3A.2 Discuss the significance of the terms 'dispersal' and 'disorder' in the context of the Second Law.

Exercises

3A.1(a) During a hypothetical process, the entropy of a system increases by 125 J K^{-1} while the entropy of the surroundings decreases by 125 J K^{-1}. Is the process spontaneous?
3A.1(b) During a hypothetical process, the entropy of a system increases by 105 J K^{-1} while the entropy of the surroundings decreases by 95 J K^{-1}. Is the process spontaneous?

3A.2(a) A certain ideal heat engine uses water at the triple point as the hot source and an organic liquid as the cold sink. It withdraws 10.00 kJ of heat from the hot source and generates 3.00 kJ of work. What is the temperature of the organic liquid?
3A.2(b) A certain ideal heat engine uses water at the triple point as the hot source and an organic liquid as the cold sink. It withdraws 2.71 kJ of heat from the hot source and generates 0.71 kJ of work. What is the temperature of the organic liquid?

➤ Integrated activities

At the end of most chapters, you will find questions that cross several topics and chapters, and are designed to help you use your knowledge creatively in a variety of ways. Some of the questions refer to the Living Graphs on the Book Companion Site, which you will find helpful for answering them.

➤ Solutions manuals

Two solutions manuals have been written by Charles Trapp, Marshall Cady, and Carmen Giunta to accompany this book.

The *Student Solutions Manual* (ISBN 1-4641-2449-3) provides full solutions to the 'a' exercises and to the odd-numbered problems.

The *Instructor's Solutions Manual* provides full solutions to the 'b' exercises and to the even-numbered problems (available to download from the Book Companion Site for registered adopters of the book only).

BOOK COMPANION SITE

The Book Companion Site to accompany *Physical Chemistry: Thermodynamics, Structure, and Change*, tenth edition provides a number of useful teaching and learning resources for students and instructors.

The site can be accessed at:

http://www.whfreeman.com/pchem10e/

Instructor resources are available only to registered adopters of the textbook. To register, simply visit **http://www.whfreeman.com/pchem10e/** and follow the appropriate links.

Student resources are openly available to all, without registration.

Materials on the Book Companion Site include:

'Impact' sections

'Impact' sections show how physical chemistry is applied in a variety of modern contexts. New for this edition, the Impacts are linked from the text by QR code images. Alternatively, visit the URL displayed next to the QR code image.

Group theory tables

Comprehensive group theory tables are available to download.

Figures and tables from the book

Instructors can find the artwork and tables from the book in ready-to-download format. These may be used for lectures without charge (but not for commercial purposes without specific permission).

Molecular modeling problems

PDFs containing molecular modeling problems can be downloaded, designed for use with the Spartan Student™ software. However they can also be completed using any modeling software that allows Hartree-Fock, density functional, and MP2 calculations.

Living graphs

These interactive graphs can be used to explore how a property changes as various parameters are changed. Living graphs are sometimes referred to in the Integrated activities at the end of a chapter.

ACKNOWLEDGEMENTS

A book as extensive as this could not have been written without significant input from many individuals. We would like to reiterate our thanks to the hundreds of people who contributed to the first nine editions. Many people gave their advice based on the ninth edition, and others, including students, reviewed the draft chapters for the tenth edition as they emerged. We wish to express our gratitude to the following colleagues:

Oleg Antzutkin, *Luleå University of Technology*
Mu-Hyun Baik, *Indiana University — Bloomington*
Maria G. Benavides, *University of Houston — Downtown*
Joseph A. Bentley, *Delta State University*
Maria Bohorquez, *Drake University*
Gary D. Branum, *Friends University*
Gary S. Buckley, *Cameron University*
Eleanor Campbell, *University of Edinburgh*
Lin X. Chen, *Northwestern University*
Gregory Dicinoski, *University of Tasmania*
Niels Engholm Henriksen, *Technical University of Denmark*
Walter C. Ermler, *University of Texas at San Antonio*
Alexander Y. Fadeev, *Seton Hall University*
Beth S. Guiton, *University of Kentucky*
Patrick M. Hare, *Northern Kentucky University*
Grant Hill, *University of Glasgow*
Ann Hopper, *Dublin Institute of Technology*
Garth Jones, *University of East Anglia*
George A. Kaminsky, *Worcester Polytechnic Institute*
Dan Killelea, *Loyola University of Chicago*
Richard Lavrich, *College of Charleston*
Yao Lin, *University of Connecticut*
Tony Masiello, *California State University — East Bay*
Lida Latifzadeh Masoudipour, *California State University — Dominquez Hills*
Christine McCreary, *University of Pittsburgh at Greensburg*
Ricardo B. Metz, *University of Massachusetts Amherst*
Maria Pacheco, *Buffalo State College*
Sid Parrish, Jr., *Newberry College*
Nessima Salhi, *Uppsala University*
Michael Schuder, *Carroll University*
Paul G. Seybold, *Wright State University*
John W. Shriver, *University of Alabama Huntsville*
Jens Spanget-Larsen, *Roskilde University*
Stefan Tsonchev, *Northeastern Illinois University*
A. L. M. van de Ven, *Eindhoven University of Technology*
Darren Walsh, *University of Nottingham*
Nicolas Winter, *Dominican University*
Georgene Wittig, *Carnegie Mellon University*
Daniel Zeroka, *Lehigh University*

Because we prepared this edition at the same time as its sister volume, *Physical Chemistry: Quanta, matter, and change*, it goes without saying that our colleague on that book, Ron Friedman, has had an unconscious but considerable impact on this text too, and we cannot thank him enough for his contribution to this book. Our warm thanks also go to Charles Trapp, Carmen Giunta, and Marshall Cady who once again have produced the *Solutions manuals* that accompany this book and whose comments led us to make a number of improvements. Kerry Karukstis contributed helpfully to the Impacts that are now on the web.

Last, but by no means least, we would also like to thank our two commissioning editors, Jonathan Crowe of Oxford University Press and Jessica Fiorillo of W. H. Freeman & Co., and their teams for their encouragement, patience, advice, and assistance.

FULL CONTENTS

TABLES

CHEMIST'S TOOLKITS

Foundations

Chemistry is the science of matter and the changes it can undergo. Physical chemistry is the branch of chemistry that establishes and develops the principles of the subject in terms of the underlying concepts of physics and the language of mathematics. It provides the basis for developing new spectroscopic techniques and their interpretation, for understanding the structures of molecules and the details of their electron distributions, and for relating the bulk properties of matter to their constituent atoms. Physical chemistry also provides a window on to the world of chemical reactions, and allows us to understand in detail how they take place.

A Matter

Throughout the text we draw on a number of concepts that should already be familiar from introductory chemistry, such as the 'nuclear model' of the atom, 'Lewis structures' of molecules, and the 'perfect gas equation'. This Topic reviews these and other concepts of chemistry that appear at many stages of the presentation.

B Energy

Because physical chemistry lies at the interface between physics and chemistry, we also need to review some of the concepts from elementary physics that we need to draw on in the text. This Topic begins with a brief summary of 'classical mechanics', our starting point for discussion of the motion and energy of particles. Then it reviews concepts of 'thermodynamics' that should already be part of your chemical vocabulary. Finally, we introduce the 'Boltzmann distribution' and the 'equipartition theorem', which help to establish connections between the bulk and molecular properties of matter.

C Waves

This Topic describes waves, with a focus on 'harmonic waves', which form the basis for the classical description of electromagnetic radiation. The classical ideas of motion, energy, and waves in this Topic and Topic B are expanded with the principles of quantum mechanics (Chapter 7), setting the stage for the treatment of electrons, atoms, and molecules. Quantum mechanics underlies the discussion of chemical structure and chemical change, and is the basis of many techniques of investigation.

A Matter

Contents

➤ **Why do you need to know this material?**

Because chemistry is about matter and the changes that it can undergo, both physically and chemically, the properties of matter underlie the entire discussion in this book.

➤ **What is the key idea?**

The bulk properties of matter are related to the identities and arrangements of atoms and molecules in a sample.

➤ **What do you need to know already?**

This Topic reviews material commonly covered in introductory chemistry.

The presentation of physical chemistry in this text is based on the experimentally verified fact that matter consists of atoms. In this Topic, which is a review of elementary concepts and language widely used in chemistry, we begin to make connections between atomic, molecular, and bulk properties. Most of the material is developed in greater detail later in the text.

A.1 Atoms

The atom of an element is characterized by its **atomic number**, Z, which is the number of protons in its nucleus. The number of neutrons in a nucleus is variable to a small extent, and the **nucleon number** (which is also commonly called the *mass number*), A, is the total number of protons and neutrons in the nucleus. Protons and neutrons are collectively called **nucleons**. Atoms of the same atomic number but different nucleon number are the **isotopes** of the element.

(a) The nuclear model

According to the **nuclear model**, an atom of atomic number Z consists of a nucleus of charge $+Ze$ surrounded by Z electrons each of charge $-e$ (e is the fundamental charge: see inside the front cover for its value and the values of the other fundamental constants). These electrons occupy **atomic orbitals**, which are regions of space where they are most likely to be found, with no more than two electrons in any one orbital. The atomic orbitals are arranged in **shells** around the nucleus, each shell being characterized by the **principal quantum number**, $n=1, 2, \ldots$. A shell consists of n^2 individual orbitals, which are grouped together into n **subshells**; these subshells, and the orbitals they contain, are denoted s, p, d, and f. For all neutral atoms other than hydrogen, the subshells of a given shell have slightly different energies.

(b) The periodic table

The sequential occupation of the orbitals in successive shells results in periodic similarities in the **electronic configurations**, the specification of the occupied orbitals, of atoms when they are arranged in order of their atomic number. This periodicity of structure accounts for the formulation of the **periodic table** (see the inside the back cover). The vertical columns of the periodic table are called **groups** and (in the modern convention) numbered from 1 to 18. Successive rows of the periodic table are called **periods**, the number of the period being equal to the principal quantum number of the **valence shell**, the outermost shell of the atom.

Some of the groups also have familiar names: Group 1 consists of the **alkali metals**, Group 2 (more specifically, calcium, strontium, and barium) of the **alkaline earth metals**, Group 17 of the **halogens**, and Group 18 of the **noble gases**. Broadly speaking, the elements towards the left of the periodic table are **metals** and those towards the right are **non-metals**; the two classes of substance meet at a diagonal line running from boron to polonium, which constitute the **metalloids**, with properties intermediate between those of metals and non-metals.

The periodic table is divided into s, p, d, and f **blocks**, according to the subshell that is last to be occupied in the formulation of the electronic configuration of the atom. The members of the d block (specifically the members of Groups 3–11 in the d block) are also known as the **transition metals**; those of the f block (which is not divided into numbered groups) are sometimes called the **inner transition metals**. The upper row of the f block (Period 6) consists of the **lanthanoids** (still commonly the 'lanthanides') and the lower row (Period 7) consists of the **actinoids** (still commonly the 'actinides').

(c) Ions

A monatomic **ion** is an electrically charged atom. When an atom gains one or more electrons it becomes a negatively charged **anion**; when it loses one or more electrons it becomes a positively charged **cation**. The charge number of an ion is called the **oxidation number** of the element in that state (thus, the oxidation number of magnesium in Mg^{2+} is +2 and that of oxygen in O^{2-} is −2). It is appropriate, but not always done, to distinguish between the oxidation number and the **oxidation state**, the latter being the physical state of the atom with a specified oxidation number. Thus, the oxidation number *of* magnesium is +2 when it is present as Mg^{2+}, and it is present *in* the oxidation state Mg^{2+}.

The elements form ions that are characteristic of their location in the periodic table: metallic elements typically form cations by losing the electrons of their outermost shell and acquiring the electronic configuration of the preceding noble gas atom. Nonmetals typically form anions by gaining electrons and attaining the electronic configuration of the following noble gas atom.

A.2 Molecules

A **chemical bond** is the link between atoms. Compounds that contain a metallic element typically, but far from universally, form **ionic compounds** that consist of cations and anions in a crystalline array. The 'chemical bonds' in an ionic compound are due to the Coulombic interactions between all the ions in the crystal and it is inappropriate to refer to a bond between a specific pair of neighbouring ions. The smallest unit of an ionic compound is called a **formula unit**. Thus $NaNO_3$, consisting of a Na^+ cation and a NO_3^- anion, is the formula unit of sodium nitrate. Compounds that do not contain a metallic element typically form **covalent compounds** consisting of discrete molecules. In this case, the bonds between the atoms of a molecule are **covalent**, meaning that they consist of shared pairs of electrons.

A note on good practice Some chemists use the term 'molecule' to denote the smallest unit of a compound with the composition of the bulk material regardless of whether it is an ionic or covalent compound and thus speak of 'a molecule of NaCl'. We use the term 'molecule' to denote a discrete covalently bonded entity (as in H_2O); for an ionic compound we use 'formula unit'.

(a) Lewis structures

The pattern of bonds between neighbouring atoms is displayed by drawing a **Lewis structure**, in which bonds are shown as lines and **lone pairs** of electrons, pairs of valence electrons that are not used in bonding, are shown as dots. Lewis structures are constructed by allowing each atom to share electrons until it has acquired an **octet** of eight electrons (for hydrogen, a *duplet* of two electrons). A shared pair of electrons is a **single bond**, two shared pairs constitute a **double bond**, and three shared pairs constitute a **triple bond**. Atoms of elements of Period 3 and later can accommodate more than eight electrons in their valence shell and 'expand their octet' to become **hypervalent**, that is, form more bonds than the octet rule would allow (for example, SF_6), or form more bonds to a small number of atoms (see *Brief illustration* A.1). When more than one Lewis structure can be written for a given arrangement of atoms, it is supposed that **resonance**, a blending of the structures, may occur and distribute multiple-bond character over the molecule (for example, the two Kekulé structures of benzene). Examples of these aspects of Lewis structures are shown in Fig. A.1.

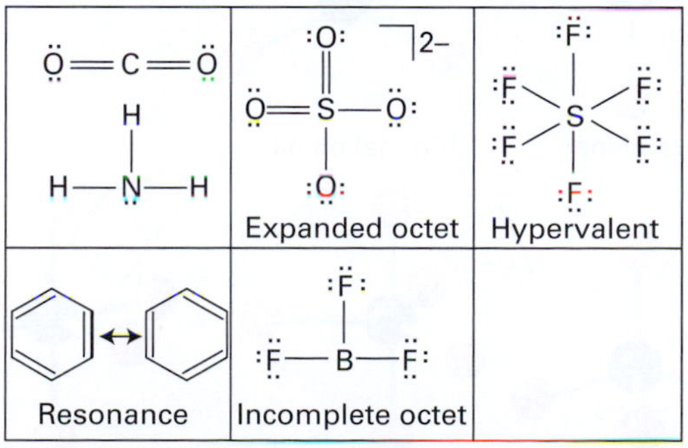

Figure A.1 Examples of Lewis structures.

Brief illustration A.1 Octet expansion

Octet expansion is also encountered in species that do not necessarily require it, but which, if it is permitted, may acquire a lower energy. Thus, of the structures (**1a**) and (**1b**) of the SO_4^{2-} ion, the second has a lower energy than the first. The actual structure of the ion is a resonance hybrid of both structures (together with analogous structures with double bonds in different locations), but the latter structure makes the dominant contribution.

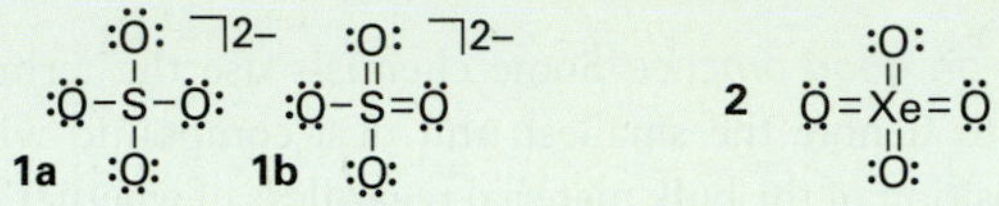

1a **1b** **2**

Self-test A.1 Draw the Lewis structure for XeO_4.

Answer: See 2

(b) VSEPR theory

Except in the simplest cases, a Lewis structure does not express the three-dimensional structure of a molecule. The simplest approach to the prediction of molecular shape is **valence-shell electron pair repulsion theory** (VSEPR theory). In this approach, the regions of high electron density, as represented by bonds—whether single or multiple—and lone pairs, take up orientations around the central atom that maximize their separations. Then the position of the attached atoms (not the lone pairs) is noted and used to classify the shape of the molecule. Thus, four regions of electron density adopt a tetrahedral arrangement; if an atom is at each of these locations (as in CH_4), then the molecule is tetrahedral; if there is an atom at only three of these locations (as in NH_3), then the molecule is trigonal pyramidal, and so on. The names of the various shapes that are commonly found are shown in Fig. A.2. In a refinement of the theory, lone pairs are assumed to repel bonding pairs more strongly than bonding pairs repel each other. The shape a molecule then adopts, if it is not determined fully by symmetry, is such as to minimize repulsions from lone pairs.

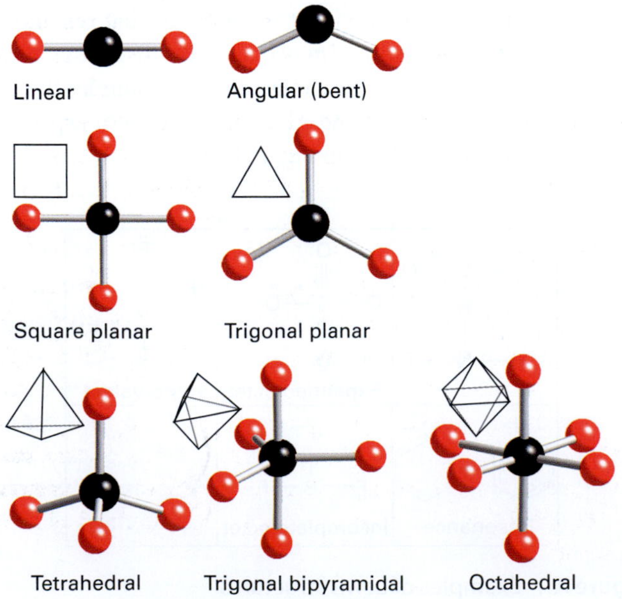

Figure A.2 The shapes of molecules that result from application of VSEPR theory.

Brief illustration A.2 Molecular shapes

In SF_4 the lone pair adopts an equatorial position and the two axial S–F bonds bend away from it slightly, to give a bent see-saw shaped molecule (Fig. A.3).

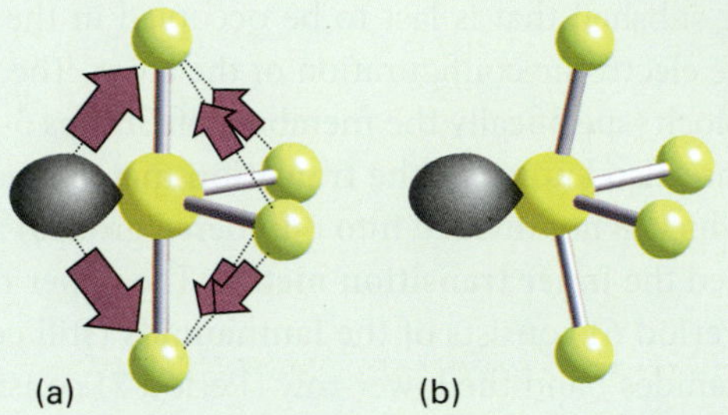

Figure A.3 (a) In SF_4 the lone pair adopts an equatorial position. (b) The two axial S–F bonds bend away from it slightly, to give a bent see-saw shaped molecule.

Self-test A.2 Predict the shape of the SO_3^{2-} ion.

Answer: Trigonal pyramid

(c) Polar bonds

Covalent bonds may be **polar**, or correspond to an unequal sharing of the electron pair, with the result that one atom has a partial positive charge (denoted δ+) and the other a partial negative charge (δ–). The ability of an atom to attract electrons to itself when part of a molecule is measured by the **electronegativity**, χ (chi), of the element. The juxtaposition of equal and opposite partial charges constitutes an **electric dipole**. If those charges are $+Q$ and $-Q$ and they are separated by a distance d, the magnitude of the **electric dipole moment**, μ, is

$$\mu = Qd \qquad \text{Definition} \qquad \text{Magnitude of the electric dipole moment} \qquad (A.1)$$

Brief illustration A.3 Nonpolar molecules with polar bonds

Whether or not a molecule as a whole is polar depends on the arrangement of its bonds, for in highly symmetrical molecules there may be no net dipole. Thus, although the linear CO_2 molecule (which is structurally OCO) has polar CO bonds, their effects cancel and the molecule as a whole is nonpolar.

Self-test A.3 Is NH_3 polar?

Answer: Yes

A.3 Bulk matter

Bulk matter consists of large numbers of atoms, molecules, or ions. Its physical state may be solid, liquid, or gas:

A **solid** is a form of matter that adopts and maintains a shape that is independent of the container it occupies.

A **liquid** is a form of matter that adopts the shape of the part of the container it occupies (in a gravitational field, the lower part) and is separated from the unoccupied part of the container by a definite surface.

A **gas** is a form of matter that immediately fills any container it occupies.

A liquid and a solid are examples of a **condensed state** of matter. A liquid and a gas are examples of a **fluid** form of matter: they flow in response to forces (such as gravity) that are applied.

(a) Properties of bulk matter

The state of a bulk sample of matter is defined by specifying the values of various properties. Among them are:

The **mass**, m, a measure of the quantity of matter present (unit: 1 kilogram, 1 kg).

The **volume**, V, a measure of the quantity of space the sample occupies (unit: 1 cubic metre, 1 m^3).

The **amount of substance**, n, a measure of the number of specified entities (atoms, molecules, or formula units) present (unit: 1 mole, 1 mol).

Brief illustration A.4 Volume units

Volume is also expressed as submultiples of 1 m^3, such as cubic decimetres (1 dm^3 = 10^{-3} m^3) and cubic centimetres (1 cm^3 = 10^{-6} m^3). It is also common to encounter the non-SI unit litre (1 L = 1 dm^3) and its submultiple the millilitre (1 mL = 1 cm^3). To carry out simple unit conversions, simply replace the fraction of the unit (such as 1 cm) by its definition (in this case, 10^{-2} m). Thus, to convert 100 cm^3 to cubic decimetres (litres), use 1 cm = 10^{-1} dm, in which case 100 cm^3 = 100 (10^{-1} dm)^3, which is the same as 0.100 dm^3.

Self-test A.4 Express a volume of 100 mm^3 in units of cm^3.

Answer: 0.100 cm^3

An **extensive property** of bulk matter is a property that depends on the amount of substance present in the sample; an **intensive property** is a property that is independent of the amount of substance. The volume is extensive; the mass density, ρ (rho), with

$$\rho = \frac{m}{V} \qquad \text{Mass density} \qquad (A.2)$$

is intensive.

The **amount of substance**, n (colloquially, 'the number of moles'), is a measure of the number of specified entities present in the sample. 'Amount of substance' is the official name of the quantity; it is commonly simplified to 'chemical amount' or simply 'amount'. The unit 1 mol is currently defined as the number of carbon atoms in exactly 12 g of carbon-12. (In 2011 the decision was taken to replace this definition, but the change has not yet, in 2014, been implemented.) The number of entities per mole is called **Avogadro's constant**, N_A; the currently accepted value is 6.022×10^{23} mol^{-1} (note that N_A is a constant with units, not a pure number).

The **molar mass of a substance**, M (units: formally kilograms per mole but commonly grams per mole, g mol^{-1}) is the mass per mole of its atoms, its molecules, or its formula units. The amount of substance of specified entities in a sample can readily be calculated from its mass, by noting that

$$n = \frac{m}{M} \qquad \text{Amount of substance} \qquad (A.3)$$

A note on good practice Be careful to distinguish atomic or molecular mass (the mass of a single atom or molecule; units kg) from molar mass (the mass per mole of atoms or molecules; units kg mol^{-1}). *Relative* molecular masses of atoms and molecules, $M_r = m/m_u$, where m is the mass of the atom or molecule and m_u is the atomic mass constant (see inside front cover), are still widely called 'atomic weights' and 'molecular weights' even though they are dimensionless quantities and not weights (the gravitational force exerted on an object).

A sample of matter may be subjected to a **pressure**, p (unit: 1 pascal, Pa; 1 Pa = 1 kg m^{-1} s^{-2}), which is defined as the force, F, it is subjected to divided by the area, A, to which that force is applied. A sample of gas exerts a pressure on the walls of its container because the molecules of gas are in ceaseless, random motion, and exert a force when they strike the walls. The frequency of the collisions is normally so great that the force, and therefore the pressure, is perceived as being steady.

Although 1 pascal is the SI unit of pressure (*The chemist's toolkit* A.1), it is also common to express pressure in bar (1 bar = 10^5 Pa) or atmospheres (1 atm = 101 325 Pa exactly), both of which correspond to typical atmospheric pressure. Because many physical properties depend on the pressure acting on a sample, it is appropriate to select a certain value of the pressure to report their values. The **standard pressure** for reporting physical quantities is currently defined as $p^{\ominus} = 1$ bar exactly.

The chemist's toolkit A.1 Quantities and units

The result of a measurement is a **physical quantity** that is reported as a numerical multiple of a unit:

$$\text{physical quantity} = \text{numerical value} \times \text{unit}$$

It follows that units may be treated like algebraic quantities and may be multiplied, divided, and cancelled. Thus, the expression (physical quantity)/unit is the numerical value (a dimensionless quantity) of the measurement in the specified units. For instance, the mass m of an object could be reported as $m=2.5\,\text{kg}$ or $m/\text{kg}=2.5$. See Table A.1 in the *Resource section* for a list of units. Although it is good practice to use only SI units, there will be occasions where accepted practice is so deeply rooted that physical quantities are expressed using other, non-SI units. By international convention, all physical quantities are represented by oblique (sloping) symbols; all units are roman (upright).

Units may be modified by a prefix that denotes a factor of a power of 10. Among the most common SI prefixes are those listed in Table A.2 in the *Resource section*. Examples of the use of these prefixes are:

$$1\,\text{nm}=10^{-9}\,\text{m} \qquad 1\,\text{ps}=10^{-12}\,\text{s} \qquad 1\,\mu\text{mol}=10^{-6}\,\text{mol}$$

Powers of units apply to the prefix as well as the unit they modify. For example, $1\,\text{cm}^3=1\,(\text{cm})^3$, and $(10^{-2}\,\text{m})^3=10^{-6}\,\text{m}^3$. Note that $1\,\text{cm}^3$ does not mean $1\,\text{c}(\text{m}^3)$. When carrying out numerical calculations, it is usually safest to write out the numerical value of an observable in scientific notation (as $n.nnn\times 10^n$).

There are seven SI base units, which are listed in Table A.3 in the *Resource section*. All other physical quantities may be expressed as combinations of these base units (see Table A.4 in the *Resource section*). *Molar concentration* (more formally, but very rarely, *amount of substance concentration*) for example, which is an amount of substance divided by the volume it occupies, can be expressed using the derived units of $\text{mol}\,\text{dm}^{-3}$ as a combination of the base units for amount of substance and length. A number of these derived combinations of units have special names and symbols and we highlight them as they arise.

To specify the state of a sample fully it is also necessary to give its **temperature**, T. The temperature is formally a property that determines in which direction energy will flow as heat when two samples are placed in contact through thermally conducting walls: energy flows from the sample with the higher temperature to the sample with the lower temperature. The symbol T is used to denote the **thermodynamic temperature** which is an absolute scale with $T=0$ as the lowest point. Temperatures above $T=0$ are then most commonly expressed by using the **Kelvin scale**, in which the gradations of temperature are expressed as multiples of the unit 1 kelvin (1 K). The Kelvin scale is currently defined by setting the triple point of water (the temperature at which ice, liquid water, and water vapour are in mutual equilibrium) at exactly 273.16 K (as for certain other units, a decision has been taken to revise this definition, but it has not yet, in 2014, been implemented). The freezing point of water (the melting point of ice) at 1 atm is then found experimentally to lie 0.01 K below the triple point, so the freezing point of water is 273.15 K. The Kelvin scale is unsuitable for everyday measurements of temperature, and it is common to use the **Celsius scale**, which is defined in terms of the Kelvin scale as

$$\theta/{}^\circ\text{C} = T/\text{K} - 273.15 \qquad \textit{Definition} \quad \text{Celsius scale} \qquad \text{(A.4)}$$

Thus, the freezing point of water is 0 °C and its boiling point (at 1 atm) is found to be 100 °C (more precisely 99.974 °C). Note that in this text T invariably denotes the thermodynamic (absolute) temperature and that temperatures on the Celsius scale are denoted θ (theta).

A note on good practice Note that we write $T=0$, not $T=0\,\text{K}$. General statements in science should be expressed without reference to a specific set of units. Moreover, because T (unlike θ) is absolute, the lowest point is 0 regardless of the scale used to express higher temperatures (such as the Kelvin scale). Similarly, we write $m=0$, not $m=0\,\text{kg}$ and $l=0$, not $l=0\,\text{m}$.

(b) The perfect gas equation

The properties that define the state of a system are not in general independent of one another. The most important example of a relation between them is provided by the idealized fluid known as a **perfect gas** (also, commonly, an 'ideal gas'):

$$pV = nRT \qquad \text{Perfect gas equation} \qquad \text{(A.5)}$$

Here R is the **gas constant**, a universal constant (in the sense of being independent of the chemical identity of the gas) with the value $8.3145\,\text{J}\,\text{K}^{-1}\,\text{mol}^{-1}$. Throughout this text, equations applicable only to perfect gases (and other idealized systems) are labelled, as here, with a number in blue.

A note on good practice Although the term 'ideal gas' is almost universally used in place of 'perfect gas', there are reasons for preferring the latter term. In an ideal system the interactions between molecules in a mixture are all the same. In a perfect gas not only are the interactions all the same but they are in fact zero. Few, though, make this useful distinction.

Equation A.5, the **perfect gas equation**, is a summary of three empirical conclusions, namely Boyle's law ($p\propto 1/V$ at constant temperature and amount), Charles's law ($p\propto T$ at constant volume and amount), and Avogadro's principle ($V\propto n$ at constant temperature and pressure).

Example A.1 Using the perfect gas equation

Calculate the pressure in kilopascals exerted by 1.25 g of nitrogen gas in a flask of volume 250 cm^3 at 20 °C.

Method To use eqn A.5, we need to know the amount of molecules (in moles) in the sample, which we can obtain from the mass and the molar mass (by using eqn A.3) and to convert the temperature to the Kelvin scale (by using eqn A.4).

Answer The amount of N_2 molecules (of molar mass 28.02 g mol^{-1}) present is

$$n(\mathrm{N_2}) = \frac{m}{M(\mathrm{N_2})} = \frac{1.25\,\mathrm{g}}{28.02\,\mathrm{g\,mol^{-1}}} = \frac{1.25}{28.02}\,\mathrm{mol}$$

The temperature of the sample is

$$T/\mathrm{K} = 20 + 273.15, \text{ so } T = (20 + 273.15)\,\mathrm{K}$$

Therefore, after rewriting eqn A.5 as $p = nRT/V$,

$$p = \frac{\overbrace{(1.25/28.02)\,\mathrm{mol}}^{n} \times \overbrace{(8.3145\,\mathrm{J\,K^{-1}\,mol^{-1}})}^{R} \times \overbrace{(20+273.15)\,\mathrm{K}}^{T}}{\underbrace{(2.50\times10^{-4})\,\mathrm{m^3}}_{V}}$$

$$= \frac{(1.25/28.02)\times(8.3145)\times(20+273.15)}{2.50\times10^{-4}}\,\frac{\mathrm{J}}{\mathrm{m^3}}$$

$$\overset{1\,\mathrm{J\,m^{-3}} = 1\,\mathrm{Pa}}{=} 4.35\times10^5\,\mathrm{Pa} = 435\,\mathrm{kPa}$$

A note on good practice It is best to postpone a numerical calculation to the last possible stage, and carry it out in a single step. This procedure avoids rounding errors. When we judge it appropriate to show an intermediate result without committing ourselves to a number of significant figures, we write it as *n.nnn*....

Self-test A.5 Calculate the pressure exerted by 1.22 g of carbon dioxide confined in a flask of volume 500 dm^3 (5.00 × 10^2 dm^3) at 37 °C.

Answer: 143 Pa

All gases obey the perfect gas equation ever more closely as the pressure is reduced towards zero. That is, eqn A.5 is an example of a **limiting law**, a law that becomes increasingly valid in a particular limit, in this case as the pressure is reduced to zero. In practice, normal atmospheric pressure at sea level (about 1 atm) is already low enough for most gases to behave almost perfectly, and unless stated otherwise, we assume in this text that the gases we encounter behave perfectly and obey eqn A.5.

A mixture of perfect gases behaves like a single perfect gas. According to **Dalton's law**, the total pressure of such a mixture is the sum of the pressures to which each gas would give rise if it occupied the container alone:

$$p = p_A + p_B + \cdots \qquad \text{Dalton's law} \qquad \text{(A.6)}$$

Each pressure, p_J, can be calculated from the perfect gas equation in the form $p_J = n_J RT/V$.

Checklist of concepts

- ☐ 1. In the **nuclear model** of an atom negatively charged electrons occupy atomic orbitals which are arranged in shells around a positively charged nucleus.
- ☐ 2. The **periodic table** highlights similarities in electronic configurations of atoms, which in turn lead to similarities in their physical and chemical properties.
- ☐ 3. **Covalent compounds** consist of discrete molecules in which atoms are linked by covalent bonds.
- ☐ 4. **Ionic compounds** consist of cations and anions in a crystalline array.
- ☐ 5. **Lewis structures** are useful models of the pattern of bonding in molecules.
- ☐ 6. The **valence-shell electron pair repulsion theory** (VSEPR theory) is used to predict the three-dimensional shapes of molecules from their Lewis structures.
- ☐ 7. The electrons in **polar covalent bonds** are shared unequally between the bonded nuclei.
- ☐ 8. The physical states of bulk matter are solid, liquid, or gas.
- ☐ 9. The state of a sample of bulk matter is defined by specifying its properties, such as mass, volume, amount, pressure, and temperature.
- ☐ 10. The **perfect gas equation** is a relation between the pressure, volume, amount, and temperature of an idealized gas.
- ☐ 11. A **limiting law** is a law that becomes increasingly valid in a particular limit.

Checklist of equations

Property	Equation	Comment	Equation number
Electric dipole moment	$\mu = Qd$	μ is the magnitude of the moment	A.1
Mass density	$\rho = m/V$	Intensive property	A.2
Amount of substance	$n = m/M$	Extensive property	A.3
Celsius scale	$\theta/°C = T/K - 273.15$	Temperature is an intensive property; 273.15 is exact.	A.4
Perfect gas equation	$pV = nRT$		A.5
Dalton's law	$p = p_A + p_B + \cdots$		A.6

B Energy

Contents

➤ Why do you need to know this material?

Energy is the central unifying concept of physical chemistry, and you need to gain insight into how electrons, atoms, and molecules gain, store, and lose energy.

➤ What is the key idea?

Energy, the capacity to do work, is restricted to discrete values in electrons, atoms, and molecules.

➤ What do you need to know already?

You need to review the laws of motion and principles of electrostatics normally covered in introductory physics and concepts of thermodynamics normally covered in introductory chemistry.

Much of chemistry is concerned with transfers and transformations of energy, and from the outset it is appropriate to define this familiar quantity precisely. We begin here by reviewing **classical mechanics**, which was formulated by Isaac Newton in the seventeenth century, and establishes the vocabulary used to describe the motion and energy of particles. These classical ideas prepare us for **quantum mechanics**, the more fundamental theory formulated in the twentieth century for the study of small particles, such as electrons, atoms, and molecules. We develop the concepts of quantum mechanics throughout the text. Here we begin to see why it is needed as a foundation for understanding atomic and molecular structure.

B.1 Force

Molecules are built from atoms and atoms are built from subatomic particles. To understand their structures we need to know how these bodies move under the influence of the forces they experience.

(a) Momentum

'Translation' is the motion of a particle through space. The **velocity**, $\boldsymbol{v}$, of a particle is the rate of change of its position $\boldsymbol{r}$:

$$\boldsymbol{v}=\frac{\mathrm{d}\boldsymbol{r}}{\mathrm{d}t}$$ *Definition* Velocity (B.1)

For motion confined to a single dimension, we would write $v_x = \mathrm{d}x/\mathrm{d}t$. The velocity and position are vectors, with both direction and magnitude (vectors and their manipulation are treated in detail in *Mathematical background* 5). The magnitude of the velocity is the **speed**, v. The **linear momentum**, $\boldsymbol{p}$, of a particle of mass m is related to its velocity, $\boldsymbol{v}$, by

$$\boldsymbol{p}=m\boldsymbol{v}$$ *Definition* Linear momentum (B.2)

Like the velocity vector, the linear momentum vector points in the direction of travel of the particle (Fig. B.1); its magnitude is denoted p.

The description of rotation is very similar to that of translation. The rotational motion of a particle about a central point is described by its **angular momentum**, $\boldsymbol{J}$. The angular

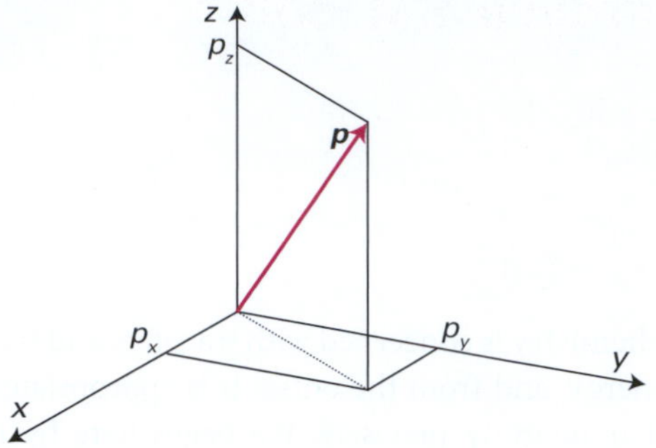

Figure B.1 The linear momentum $\boldsymbol{p}$ is denoted by a vector of magnitude p and an orientation that corresponds to the direction of motion.

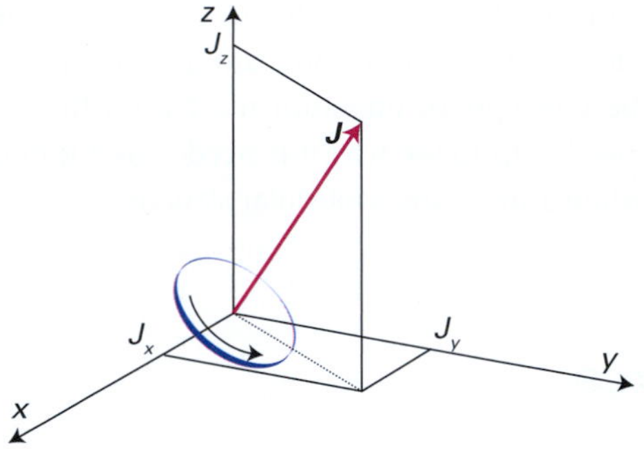

Figure B.2 The angular momentum $\boldsymbol{J}$ of a particle is represented by a vector along the axis of rotation and perpendicular to the plane of rotation. The length of the vector denotes the magnitude J of the angular momentum. The direction of motion is clockwise to an observer looking in the direction of the vector.

momentum is a vector: its magnitude gives the rate at which a particle circulates and its direction indicates the axis of rotation (Fig. B.2). The magnitude of the angular momentum, J, is

$$J = I\omega \qquad \text{Angular momentum} \qquad \text{(B.3)}$$

where ω is the **angular velocity** of the body, its rate of change of angular position (in radians per second), and I is the **moment of inertia**, a measure of its resistance to rotational acceleration. For a point particle of mass m moving in a circle of radius r, the moment of inertia about the axis of rotation is

$$I = mr^2 \qquad \textit{Point particle} \qquad \text{Moment of inertia} \qquad \text{(B.4)}$$

Brief illustration B.1 The moment of inertia

There are two possible axes of rotation in a $C^{16}O_2$ molecule, each passing through the C atom and perpendicular to the axis of the molecule and to each other. Each O atom is at a distance R from the axis of rotation, where R is the length of a CO bond, 116 pm. The mass of each ^{16}O atom is $16.00m_u$, where $m_u = 1.660\,54\times10^{-27}$ is the atomic mass constant. The C atom is stationary (it lies on the axis of rotation) and does not contribute to the moment of inertia. Therefore, the moment of inertia of the molecule around the rotation axis is

$$I = 2m(^{16}O)R^2 = 2\times\left(\overbrace{16.00\times\overbrace{1.660\,54\times10^{-27}\,\text{kg}}^{m_u}}^{m(^{16}O)}\right)\times\left(\overbrace{1.16\times10^{-10}\,\text{m}}^{R}\right)^2$$

$$= 7.15\times10^{-46}\,\text{kg}\,\text{m}^2$$

Note that the units of moments of inertia are kilograms-metre squared ($\text{kg}\,\text{m}^2$).

Self-test B.1 The moment of inertia for rotation of a hydrogen molecule, $^1H_2$, about an axis perpendicular to its bond is $4.61\times10^{-48}\,\text{kg}\,\text{m}^2$. What is the bond length of H_2?

Answer: 74.14 pm

(b) Newton's second law of motion

According to **Newton's second law of motion**, *the rate of change of momentum is equal to the force acting on the particle*:

$$\frac{d\boldsymbol{p}}{dt} = \boldsymbol{F} \qquad \text{Newton's second law of motion} \qquad \text{(B.5a)}$$

For motion confined to one dimension, we would write $dp_x/dt = F_x$. Equation B.5a may be taken as the definition of force. The SI units of force are newtons (N), with

$$1\,\text{N} = 1\,\text{kg}\,\text{m}\,\text{s}^{-2}$$

Because $\boldsymbol{p} = m(d\boldsymbol{r}/dt)$, it is sometimes more convenient to write eqn B.5a as

$$m\boldsymbol{a} = \boldsymbol{F} \qquad \boldsymbol{a} = \frac{d^2\boldsymbol{r}}{dt^2} \qquad \textit{Alternative form} \qquad \text{Newton's second law of motion} \qquad \text{(B.5b)}$$

where $\boldsymbol{a}$ is the **acceleration** of the particle, its rate of change of velocity. It follows that if we know the force acting everywhere and at all times, then solving eqn B.5 will give the **trajectory**, the position and momentum of the particle at each instant.

Brief illustration B.2 Newton's second law of motion

A *harmonic oscillator* consists of a particle that experiences a 'Hooke's law' restoring force, one that is proportional to its displacement from equilibrium. An example is a particle of mass m attached to a spring or an atom attached to another by a chemical bond. For a one-dimensional system, $F_x=-k_f x$, where the constant of proportionality k_f is called the *force constant*. Equation B.5b becomes

$$m\frac{d^2x}{dt^2}=-k_f x$$

(Techniques of differentiation are reviewed in *Mathematical background* 1 following Chapter 1.) If $x=0$ at $t=0$, a solution (as may be verified by substitution) is

$$x(t)=A\sin(2\pi\nu t) \qquad \nu=\frac{1}{2\pi}\left(\frac{k_f}{m}\right)^{1/2}$$

This solution shows that the position of the particle varies *harmonically* (that is, as a sine function) with a frequency ν, and that the frequency is high for light particles (m small) attached to stiff springs (k_f large).

Self-test B.2 How does the momentum of the oscillator vary with time?

Answer: $p=2\pi\nu Am\cos(2\pi\nu t)$

To accelerate a rotation it is necessary to apply a **torque**, T, a twisting force. Newton's equation is then

$$\frac{dJ}{dt}=T \qquad \textit{Definition} \quad \text{Torque} \quad \text{(B.6)}$$

The analogous roles of m and I, of v and ω, and of p and J in the translational and rotational cases respectively should be remembered because they provide a ready way of constructing and recalling equations. These analogies are summarized in Table B.1.

Table B.1 Analogies between translation and rotation

Translation		Rotation	
Property	Significance	Property	Significance
Mass, m	Resistance to the effect of a force	Moment of inertia, I	Resistance to the effect of a torque
Speed, v	Rate of change of position	Angular velocity, ω	Rate of change of angle
Magnitude of linear momentum, p	$p=mv$	Magnitude of angular momentum, J	$J=I\omega$
Translational kinetic energy, E_k	$E_k=\frac{1}{2}mv^2$ $=p^2/2m$	Rotational kinetic energy, E_k	$E_k=\frac{1}{2}I\omega^2$ $=J^2/2I$
Equation of motion	$dp/dt=F$	Equation of motion	$dJ/dt=T$

B.2 Energy: a first look

Before defining the term 'energy', we need to develop another familiar concept, that of 'work', more formally. Then we preview the uses of these concepts in chemistry.

(a) Work

Work, w, is done in order to achieve motion against an opposing force. For an infinitesimal displacement through $d\boldsymbol{s}$ (a vector), the work done is

$$dw=-\boldsymbol{F}\cdot d\boldsymbol{s} \qquad \textit{Definition} \quad \text{Work} \quad \text{(B.7a)}$$

where $\boldsymbol{F}\cdot d\boldsymbol{s}$ is the 'scalar product' of the vectors $\boldsymbol{F}$ and $d\boldsymbol{s}$:

$$\boldsymbol{F}\cdot d\boldsymbol{s}=F_x dx+F_y dy+F_z dz \qquad \textit{Definition} \quad \text{Scalar product} \quad \text{(B.7b)}$$

For motion in one dimension, we write $dw=-F_x dx$. The total work done along a path is the integral of this expression, allowing for the possibility that F changes in direction and magnitude at each point of the path. With force in newtons and distance in metres, the units of work are joules (J), with

$$1\,J=1\,N\,m=1\,kg\,m^2\,s^{-2}$$

Brief illustration B.3 The work of stretching a bond

The work needed to stretch a chemical bond that behaves like a spring through an infinitesimal distance dx is

$$dw=-F_x dx=-(-k_f x)dx=k_f x dx$$

The total work needed to stretch the bond from zero displacement ($x=0$) at its equilibrium length R_e to a length R, corresponding to a displacement $x=R-R_e$, is

$$w=\int_0^{R-R_e} k_f x dx=k_f\int_0^{R-R_e} x dx=\tfrac{1}{2}k_f(R-R_e)^2$$

We see that the work required increases as the square of the displacement: it takes four times as much work to stretch a bond through 20 pm as it does to stretch the same bond through 10 pm.

Self-test B.3 The force constant of the H–H bond is about $575\,N\,m^{-1}$. How much work is needed to stretch this bond by 10 pm?

Answer: 28.8 zJ

(b) The definition of energy

Energy is the capacity to do work. The SI unit of energy is the same as that of work, namely the joule. The rate of supply of energy is called the **power** (P), and is expressed in watts (W):

$$1\,\text{W} = 1\,\text{J s}^{-1}$$

Calories (cal) and kilocalories (kcal) are still encountered in the chemical literature. The calorie is now defined in terms of the joule, with 1 cal = 4.184 J (exactly). Caution needs to be exercised as there are several different kinds of calorie. The 'thermochemical calorie', cal_{15}, is the energy required to raise the temperature of 1 g of water at 15 °C by 1 °C and the 'dietary Calorie' is 1 kcal.

A particle may possess two kinds of energy, kinetic energy and potential energy. The **kinetic energy**, E_k, of a body is the energy the body possesses as a result of its motion. For a body of mass m travelling at a speed v,

$$E_k = \tfrac{1}{2}mv^2 \qquad \textit{Definition} \quad \text{Kinetic energy} \qquad \text{(B.8)}$$

It follows from Newton's second law that if a particle of mass m is initially stationary and is subjected to a constant force F for a time τ, then its speed increases from zero to $F\tau/m$ and therefore its kinetic energy increases from zero to

$$E_k = \frac{F^2\tau^2}{2m} \qquad \text{(B.9)}$$

The energy of the particle remains at this value after the force ceases to act. Because the magnitude of the applied force, F, and the time, τ, for which it acts may be varied at will, eqn B.9 implies that the energy of the particle may be increased to any value.

The **potential energy**, E_p or V, of a body is the energy it possesses as a result of its position. Because (in the absence of losses) the work that a particle can do when it is stationary in a given location is equal to the work that had to be done to bring it there, we can use the one-dimensional version of eqn B.7 to write $\mathrm{d}V = -F_x\mathrm{d}x$, and therefore

$$F_x = -\frac{\mathrm{d}V}{\mathrm{d}x} \qquad \textit{Definition} \quad \text{Potential energy} \qquad \text{(B.10)}$$

No universal expression for the potential energy can be given because it depends on the type of force the body experiences. For a particle of mass m at an altitude h close to the surface of the Earth, the gravitational potential energy is

$$V(h) = V(0) + mgh \qquad \text{Gravitational potential energy} \qquad \text{(B.11)}$$

where g is the **acceleration of free fall** (g depends on location, but its 'standard value' is close to 9.81 m s^{-2}). The zero of potential energy is arbitrary. For a particle close to the surface of the Earth, it is common to set $V(0) = 0$.

The **total energy** of a particle is the sum of its kinetic and potential energies:

$$E = E_k + E_p, \text{or } E = E_k + V \qquad \textit{Definition} \quad \text{Total energy} \qquad \text{(B.12)}$$

We make use of the apparently universal law of nature that *energy is conserved*; that is, energy can neither be created nor destroyed. Although energy can be transferred from one location to another and transformed from one form to another, the total energy is constant. In terms of the linear momentum, the total energy of a particle is

$$E = \frac{p^2}{2m} + V \qquad \text{(B.13)}$$

This expression may be used in place of Newton's second law to calculate the trajectory of a particle.

Brief illustration B.4 The trajectory of a particle

Consider an argon atom free to move in one direction (along the x-axis) in a region where $V = 0$ (so the energy is independent of position). Because $v = \mathrm{d}x/\mathrm{d}t$, it follows from eqns B.1 and B.8 that $\mathrm{d}x/\mathrm{d}t = (2E_k/m)^{1/2}$. As may be verified by substitution, a solution of this differential equation is

$$x(t) = x(0) + \left(\frac{2E_k}{m}\right)^{1/2} t$$

The linear momentum is

$$p(t) = mv(t) = m\frac{\mathrm{d}x}{\mathrm{d}t} = (2mE_k)^{1/2}$$

and is a constant. Hence, if we know the initial position and momentum, we can predict all later positions and momenta exactly.

Self-test B.4 Consider an atom of mass m moving along the x direction with an initial position x_1 and initial speed v_1. If the atom moves for a time interval Δt in a region where the potential energy varies as $V(x)$, what is its speed v_2 at position x_2?

Answer: $v_2 = v_1 \left|\mathrm{d}V(x)/\mathrm{d}x\right|_{x_1} \Delta t/m$

(c) The Coulomb potential energy

One of the most important kinds of potential energy in chemistry is the **Coulomb potential energy** between two electric charges. The Coulomb potential energy is equal to the work that must be done to bring up a charge from infinity to a distance r from a second charge. For a point charge Q_1 at a distance r in a vacuum from another point charge Q_2, their potential energy is

$$V(r)=\frac{Q_1Q_2}{4\pi\varepsilon_0 r} \qquad \textit{Definition} \quad \text{Coulomb potential energy} \qquad \text{(B.14)}$$

Charge is expressed in coulombs (C), often as a multiple of the fundamental charge, e. Thus, the charge of an electron is $-e$ and that of a proton is $+e$; the charge of an ion is ze, with z the **charge number** (positive for cations, negative for anions). The constant ε_0 (epsilon zero) is the **vacuum permittivity**, a fundamental constant with the value $8.854\times10^{-12}\,C^2\,J^{-1}\,m^{-1}$. It is conventional (as in eqn B.14) to set the potential energy equal to zero at infinite separation of charges. Then two opposite charges have a negative potential energy at finite separations whereas two like charges have a positive potential energy.

Brief illustration B.5 The Coulomb potential energy

The Coulomb potential energy resulting from the electrostatic interaction between a positively charged sodium cation, Na^+, and a negatively charged chloride anion, Cl^-, at a distance of 0.280 nm, which is the separation between ions in the lattice of a sodium chloride crystal, is

$$V=\frac{\overbrace{(-1.602\times10^{-19}\,C)}^{Q(Cl^-)}\times\overbrace{(1.602\times10^{-19}\,C)}^{Q(Na^+)}}{4\pi\times\underbrace{(8.854\times10^{-12}\,C^2\,J^{-1}\,m^{-1})}_{\varepsilon_0}\times\underbrace{(0.280\times10^{-9}\,m)}_{r}}$$

$$=-8.24\times10^{-19}\,J$$

This value is equivalent to a molar energy of

$$V\times N_A=(-8.24\times10^{-19}\,J)\times(6.022\times10^{23}\,mol^{-1})=-496\,kJ\,mol^{-1}$$

A note on good practice: Write units at *every* stage of a calculation and do not simply attach them to a final numerical value. Also, it is often sensible to express all numerical quantities in scientific notation using exponential format rather than SI prefixes to denote powers of ten.

Self-test B.5: The centres of neighbouring cations and anions in magnesium oxide crystals are separated by 0.21 nm. Determine the molar Coulomb potential energy resulting from the electrostatic interaction between a Mg^{2+} and an O^{2-} ion in such a crystal.

Answer: $2600\,kJ\,mol^{-1}$

In a medium other than a vacuum, the potential energy of interaction between two charges is reduced, and the vacuum permittivity is replaced by the **permittivity**, ε, of the medium. The permittivity is commonly expressed as a multiple of the vacuum permittivity:

$$\varepsilon=\varepsilon_r\varepsilon_0 \qquad \textit{Definition} \quad \text{Permittivity} \qquad \text{(B.15)}$$

with ε_r the dimensionless **relative permittivity** (formerly, the *dielectric constant*). This reduction in potential energy can be substantial: the relative permittivity of water at 25 °C is 80, so the reduction in potential energy for a given pair of charges at a fixed difference (with sufficient space between them for the water molecules to behave as a fluid) is by nearly two orders of magnitude.

Care should be taken to distinguish *potential energy* from *potential*. The potential energy of a charge Q_1 in the presence of another charge Q_2 can be expressed in terms of the **Coulomb potential**, ϕ (phi):

$$V(r)=Q_1\phi(r) \qquad \phi(r)=\frac{Q_2}{4\pi\varepsilon_0 r} \qquad \textit{Definition} \quad \text{Coulomb potential} \qquad \text{(B.16)}$$

The units of potential are joules per coulomb, $J\,C^{-1}$, so when ϕ is multiplied by a charge in coulombs, the result is in joules. The combination joules per coulomb occurs widely and is called a volt (V):

$$1\,V=1\,J\,C^{-1}$$

If there are several charges Q_2, Q_3, … present in the system, the total potential experienced by the charge Q_1 is the sum of the potential generated by each charge:

$$\phi=\phi_2+\phi_3+\cdots \qquad \text{(B.17)}$$

Just as the potential energy of a charge Q_1 can be written $V=Q_1\phi$, so the magnitude of the force on Q_1 can be written $F=Q_1\mathcal{E}$, where $\mathcal{E}$ is the magnitude of the **electric field strength** (units: volts per metre, $V\,m^{-1}$) arising from Q_2 or from some more general charge distribution. The electric field strength (which, like the force, is actually a vector quantity) is the negative gradient of the electric potential. In one dimension, we write the magnitude of the electric field strength as

$$\mathcal{E}=-\frac{d\phi}{dx} \qquad \text{Electric field strength} \qquad \text{(B.18)}$$

The language we have just developed inspires an important alternative energy unit, the **electronvolt** (eV): 1 eV is defined as the kinetic energy acquired when an electron is accelerated from rest through a potential difference of 1 V. The relation between electronvolts and joules is

$$1\,eV=1.602\times10^{-19}\,J$$

Many processes in chemistry involve energies of a few electronvolts. For example, to remove an electron from a sodium atom requires about 5 eV.

A particularly important way of supplying energy in chemistry (as in the everyday world) is by passing an electric current through a resistance. An **electric current** (I) is defined as the rate of supply of charge, $I = \mathrm{d}Q/\mathrm{d}t$, and is measured in *amperes* (A):

$$1\,\mathrm{A} = 1\,\mathrm{C\,s^{-1}}$$

If a charge Q is transferred from a region of potential ϕ_i, where its potential energy is $Q\phi_i$, to where the potential is ϕ_f and its potential energy is $Q\phi_f$, and therefore through a potential difference $\Delta\phi = \phi_f - \phi_i$, the change in potential energy is $Q\Delta\phi$. The rate at which the energy changes is $(\mathrm{d}Q/\mathrm{d}t)\Delta\phi$, or $I\Delta\phi$. The power is therefore

$$P = I\Delta\phi \qquad \text{Electrical power} \qquad \text{(B.19)}$$

With current in amperes and the potential difference in volts, the power is in watts. The total energy, E, supplied in an interval Δt is the power (the rate of energy supply) multiplied by the duration of the interval:

$$E = P\Delta t = I\Delta\phi\Delta t \qquad \text{(B.20)}$$

The energy is obtained in joules with the current in amperes, the potential difference in volts, and the time in seconds.

(d) Thermodynamics

The systematic discussion of the transfer and transformation of energy in bulk matter is called **thermodynamics**. This subtle subject is treated in detail in the text, but it will be familiar from introductory chemistry that there are two central concepts, the **internal energy**, U (units: joules, J), and the **entropy**, S (units: joules per kelvin, $\mathrm{J\,K^{-1}}$).

The internal energy is the total energy of a system. The **First Law of thermodynamics** states that the internal energy is constant in a system isolated from external influences. The internal energy of a sample of matter increases as its temperature is raised, and we write

$$\Delta U = C\Delta T \qquad \text{Change in internal energy} \qquad \text{(B.21)}$$

where ΔU is the change in internal energy when the temperature of the sample is raised by ΔT. The constant C is called the **heat capacity**, C (units: joules per kelvin, $\mathrm{J\,K^{-1}}$), of the sample. If the heat capacity is large, a small increase in temperature results in a large increase in internal energy. This remark can be expressed in a physically more significant way by inverting it: if the heat capacity is large, then even a large transfer of energy into the system leads to only a small rise in temperature. The heat capacity is an extensive property, and values for a substance are commonly reported as the **molar heat capacity**, $C_m = C/n$ (units: joules per kelvin per mole, $\mathrm{J\,K^{-1}\,mol^{-1}}$) or the **specific heat capacity**, $C_s = C/m$ (units: joules per kelvin per gram, $\mathrm{J\,K^{-1}\,g^{-1}}$), both of which are intensive properties.

Thermodynamic properties are often best discussed in terms of infinitesimal changes, in which case we would write eqn B.21 as $\mathrm{d}U = C\mathrm{d}T$. When this expression is written in the form

$$C = \frac{\mathrm{d}U}{\mathrm{d}T} \qquad \textit{Definition} \quad \text{Heat capacity} \qquad \text{(B.22)}$$

we see that the heat capacity can be interpreted as the slope of the plot of the internal energy of a sample against the temperature.

As will also be familiar from introductory chemistry and will be explained in detail later, for systems maintained at constant pressure it is usually more convenient to modify the internal energy by adding to it the quantity pV, and introducing the **enthalpy**, H (units: joules, J):

$$H = U + pV \qquad \textit{Definition} \quad \text{Enthalpy} \qquad \text{(B.23)}$$

The enthalpy, an extensive property, greatly simplifies the discussion of chemical reactions, in part because changes in enthalpy can be identified with the energy transferred as heat from a system maintained at constant pressure (as in common laboratory experiments).

Brief illustration B.6 The relation between U and H

The internal energy and enthalpy of a perfect gas, for which $pV = nRT$, are related by

$$H = U + nRT$$

Division by n and rearrangement gives

$$H_m - U_m = RT$$

where H_m and U_m are the molar enthalpy and the molar internal energy, respectively. We see that the difference between H_m and U_m increases with temperature.

Self-test B.6 By how much does the molar enthalpy of oxygen gas differ from its molar internal energy at 298 K?

Answer: $2.48\,\mathrm{kJ\,mol^{-1}}$

The **entropy**, S, is a measure of the *quality* of the energy of a system. If the energy is distributed over many modes of motion (for example, the rotational, vibrational, and translational motions for the particles that comprise the system), then the entropy is high. If the energy is distributed over only a small number of modes of motion, then the entropy is low. The **Second Law of thermodynamics** states that any spontaneous (that is, natural) change in an isolated system is accompanied by an increase in the entropy of the system. This tendency is commonly expressed by saying that the natural direction of change is accompanied by dispersal of energy from a localized region or its conversion to a less organized form.

The entropy of a system and its surroundings is of the greatest importance in chemistry because it enables us to identify the spontaneous direction of a chemical reaction and to identify the composition at which the reaction is at **equilibrium**. In a state of *dynamic* equilibrium, which is the character of all chemical equilibria, the forward and reverse reactions are occurring at the same rate and there is no net tendency to change in either direction. However, to use the entropy to identify this state we need to consider both the system and its surroundings. This task can be simplified if the reaction is taking place at constant temperature and pressure, for then it is possible to identify the state of equilibrium as the state at which the **Gibbs energy**, G (units: joules, J), of the system has reached a minimum. The Gibbs energy is defined as

$$G = H - TS \qquad \text{Definition} \quad \text{Gibbs energy} \qquad \text{(B.24)}$$

and is of the greatest importance in chemical thermodynamics. The Gibbs energy, which informally is called the 'free energy', is a measure of the energy stored in a system that is free to do useful work, such as driving electrons through a circuit or causing a reaction to be driven in its nonspontaneous (unnatural) direction.

B.3 The relation between molecular and bulk properties

The energy of a molecule, atom, or subatomic particle that is confined to a region of space is **quantized**, or restricted to certain discrete values. These permitted energies are called **energy levels**. The values of the permitted energies depend on the characteristics of the particle (for instance, its mass) and the extent of the region to which it is confined. The quantization of energy is most important—in the sense that the allowed energies are widest apart—for particles of small mass confined to small regions of space. Consequently, quantization is very important for electrons in atoms and molecules, but usually unimportant for macroscopic bodies, for which the separation of translational energy levels of particles in containers of macroscopic dimensions is so small that for all practical purposes their translational motion is unquantized and can be varied virtually continuously.

The energy of a molecule other than its unquantized translational motion arises mostly from three modes of motion: rotation of the molecule as a whole, distortion of the molecule through vibration of its atoms, and the motion of electrons around nuclei. Quantization becomes increasingly important as we change focus from rotational to vibrational and then to electronic motion. The separation of rotational energy levels (in small molecules, about 10^{-21} J or 1 zJ, corresponding to about 0.6 kJ mol^{-1}) is smaller than that of vibrational energy levels

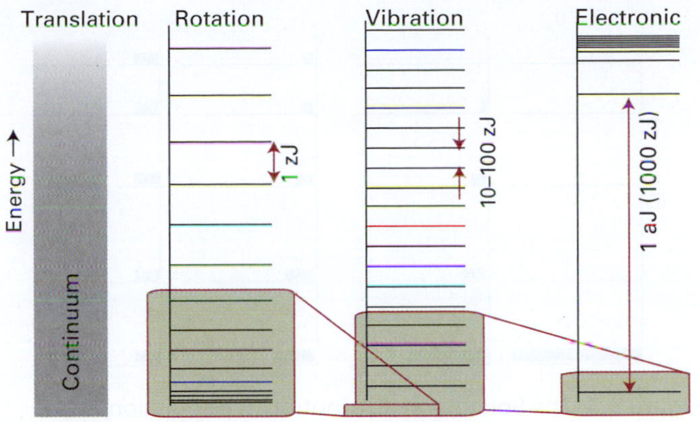

Figure B.3 The energy level separations typical of four types of system. (1 zJ = 10^{-21} J; in molar terms, 1 zJ is equivalent to about 0.6 kJ mol^{-1}.)

(about 10–100 zJ, or 6–60 kJ mol^{-1}), which itself is smaller than that of electronic energy levels (about 10^{-18} J or 1 aJ, where a is another uncommon but useful SI prefix, standing for atto, 10^{-18}, corresponding to about 600 kJ mol^{-1}). Figure B.3 depicts these typical energy level separations.

(a) The Boltzmann distribution

The continuous thermal agitation that the molecules experience in a sample at $T > 0$ ensures that they are distributed over the available energy levels. One particular molecule may be in a state corresponding to a low energy level at one instant, and then be excited into a high energy state a moment later. Although we cannot keep track of the state of a single molecule, we can speak of the *average* numbers of molecules in each state; even though individual molecules may be changing their states as a result of collisions, the average number in each state is constant (provided the temperature remains the same).

The average number of molecules in a state is called the **population** of the state. Only the lowest energy state is occupied at $T = 0$. Raising the temperature excites some molecules into higher energy states, and more and more states become accessible as the temperature is raised further (Fig. B.4). The formula for calculating the relative populations of states of various energies is called the **Boltzmann distribution** and was derived by the Austrian scientist Ludwig Boltzmann towards the end of the nineteenth century. This formula gives the ratio of the numbers of particles in states with energies ε_i and ε_j as

$$\frac{N_i}{N_j} = e^{-(\varepsilon_i - \varepsilon_j)/kT} \qquad \text{Boltzmann distribution} \qquad \text{(B.25a)}$$

where k is **Boltzmann's constant**, a fundamental constant with the value $k = 1.381 \times 10^{-23}$ J K^{-1}. In chemical applications it is common to use not the individual energies but energies per mole of molecules, E_i, with $E_i = N_A\varepsilon_i$, where N_A is Avogadro's

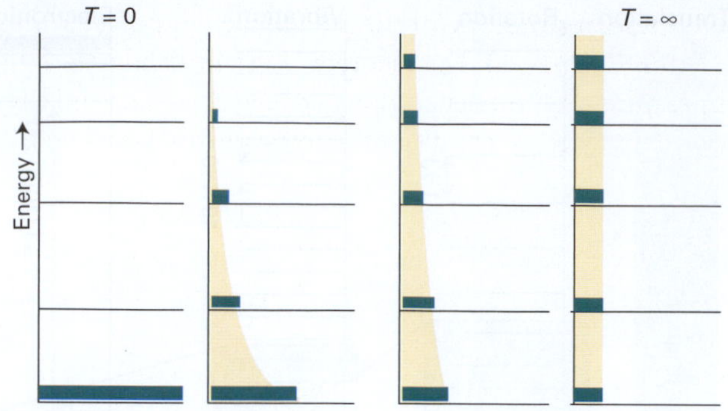

Figure B.4 The Boltzmann distribution of populations for a system of five energy levels as the temperature is raised from zero to infinity.

constant. When both the numerator and denominator in the exponential are multiplied by N_A, eqn B.25a becomes

$$\frac{N_i}{N_j}=e^{-(E_i-E_j)/RT} \quad \textit{Alternative form} \quad \text{Boltzmann distribution} \quad \text{(B.25b)}$$

where $R=N_Ak$. We see that k is often disguised in 'molar' form as the gas constant. The Boltzmann distribution provides the crucial link for expressing the macroscopic properties of matter in terms of microscopic behaviour.

Brief illustration B.7 Relative populations

Methyl cyclohexane molecules may exist in one of two conformations, with the methyl group in either an equatorial or axial position. The equatorial form is lower in energy with the axial form being 6.0 kJ mol⁻¹ higher in energy. At a temperature of 300 K, this difference in energy implies that the relative populations of molecules in the axial and equatorial states is

$$\frac{N_a}{N_e}=e^{-(E_a-E_e)/RT}=e^{-(6.0\times10^3\,\text{J mol}^{-1})/(8.3145\,\text{J K}^{-1}\,\text{mol}^{-1}\times300\,\text{K})}=0.090$$

where E_a and E_e are molar energies. The number of molecules in an axial conformation is therefore just 9 per cent of those in the equatorial conformation.

Self-test B.7 Determine the temperature at which the relative proportion of molecules in axial and equatorial conformations in a sample of methyl cyclohexane is 0.30 or 30 per cent.

Answer: 600 K

The important features of the Boltzmann distribution to bear in mind are:

- The distribution of populations is an exponential function of energy and temperature.
- At a high temperature more energy levels are occupied than at a low temperature.
- More levels are significantly populated if they are close together in comparison with kT (like rotational and translational states), than if they are far apart (like vibrational and electronic states).

Physical interpretation

Figure B.5 summarizes the form of the Boltzmann distribution for some typical sets of energy levels. The peculiar shape of the population of rotational levels stems from the fact that eqn B.25 applies to *individual states*, and for molecular rotation quantum theory shows that the number of rotational states corresponding to a given energy level—broadly speaking, the number of planes of rotation—increases with energy; therefore, although the population of each *state* decreases with energy, the population of the *levels* goes through a maximum.

One of the simplest examples of the relation between microscopic and bulk properties is provided by **kinetic molecular theory**, a model of a perfect gas. In this model, it is assumed that the molecules, imagined as particles of negligible size, are in ceaseless, random motion and do not interact except during their brief collisions. Different speeds correspond to different energies, so the Boltzmann formula can be used to predict the proportions of molecules having a specific speed at a particular temperature. The expression giving the fraction of molecules that have a particular speed is called the **Maxwell–Boltzmann distribution** and has the features summarized in Fig. B.6. The Maxwell–Boltzmann distribution can be used to show that the average speed, v_{mean}, of the molecules depends on the temperature and their molar mass as

$$v_{\text{mean}}=\left(\frac{8RT}{\pi M}\right)^{1/2} \quad \textit{Perfect gas} \quad \text{Average speed of molecules} \quad \text{(B.26)}$$

Thus, the average speed is high for light molecules at high temperatures. The distribution itself gives more information. For instance, the tail towards high speeds is longer at high temperatures than at low, which indicates that at high temperatures more molecules in a sample have speeds much higher than average.

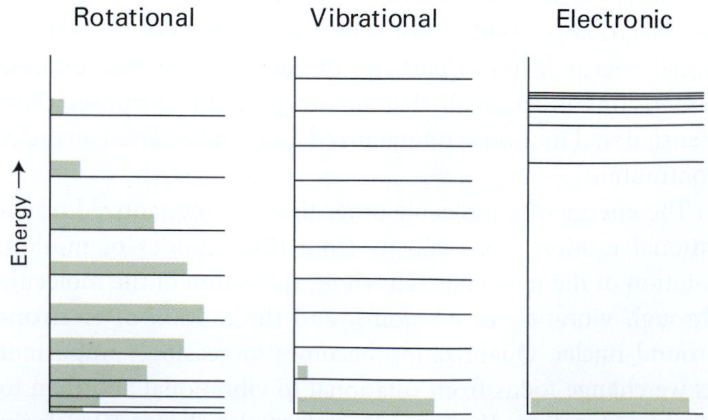

Figure B.5 The Boltzmann distribution of populations for rotational, vibrational, and electronic energy levels at room temperature.

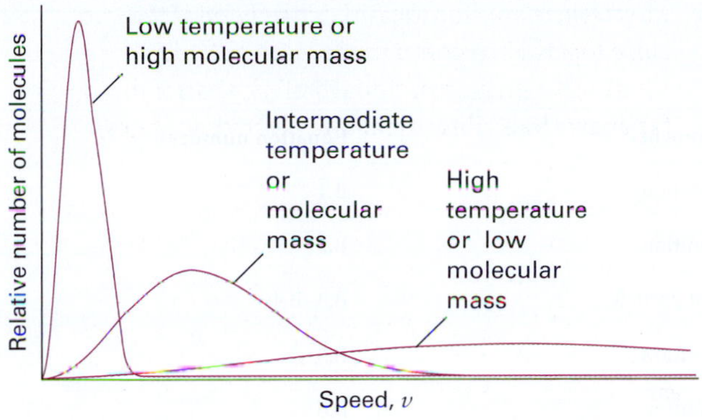

Figure B.6 The (Maxwell–Boltzmann) distribution of molecular speeds with temperature and molar mass. Note that the most probable speed (corresponding to the peak of the distribution) increases with temperature and with decreasing molar mass, and simultaneously the distribution becomes broader.

(b) Equipartition

Although the Boltzmann distribution can be used to calculate the average energy associated with each mode of motion of an atom or molecule in a sample at a given temperature, there is a much simpler shortcut. When the temperature is so high that many energy levels are occupied, we can use the **equipartition theorem**:

> For a sample at thermal equilibrium the average value of each quadratic contribution to the energy is $\frac{1}{2}kT$.

By a 'quadratic contribution' we mean a term that is proportional to the square of the momentum (as in the expression for the kinetic energy, $E_k = p^2/2m$) or the displacement from an equilibrium position (as for the potential energy of a harmonic oscillator, $E_p = \frac{1}{2}k_f x^2$). The theorem is strictly valid only at high temperatures or if the separation between energy levels is small because under these conditions many states are populated. The equipartition theorem is most reliable for translational and rotational modes of motion. The separation between vibrational and electronic states is typically greater than for rotation or translation, and so the equipartition theorem is unreliable for these types of motion.

Brief illustration B.8 Average molecular energies

An atom or molecule may move in three dimensions and its translational kinetic energy is therefore the sum of three quadratic contributions

$$E_{\text{trans}} = \tfrac{1}{2}mv_x^2 + \tfrac{1}{2}mv_y^2 + \tfrac{1}{2}mv_z^2$$

The equipartition theorem predicts that the average energy for each of these quadratic contributions is $\frac{1}{2}kT$. Thus, the average kinetic energy is $E_{\text{trans}} = 3\times\frac{1}{2}kT = \frac{3}{2}kT$. The molar translational energy is thus $E_{\text{trans,m}} = \frac{3}{2}kT \times N_A = \frac{3}{2}RT$. At 300 K

$$E_{\text{trans,m}} = \tfrac{3}{2}\times(8.3145\,\text{J K}^{-1}\,\text{mol}^{-1})\times(300\,\text{K}) = 3700\,\text{J mol}^{-1}$$
$$= 3.7\,\text{kJ mol}^{-1}$$

Self-test B.8 A linear molecule may rotate about two axes in space, each of which counts as a quadratic contribution. Calculate the rotational contribution to the molar energy of a collection of linear molecules at 500 K.

Answer: $4.2\,\text{kJ mol}^{-1}$

Checklist of concepts

- ☐ 1. **Newton's second law of motion** states that the rate of change of momentum is equal to the force acting on the particle.
- ☐ 2. **Work** is done in order to achieve motion against an opposing force.
- ☐ 3. **Energy** is the capacity to do work.
- ☐ 4. The **kinetic energy** of a particle is the energy it possesses as a result of its motion.
- ☐ 5. The **potential energy** of a particle is the energy it possesses as a result of its position.
- ☐ 6. The total energy of a particle is the sum of its kinetic and potential energies.
- ☐ 7. The **Coulomb potential energy** between two charges separated by a distance r varies as $1/r$.
- ☐ 8. The **First Law of thermodynamics** states that the internal energy is constant in a system isolated from external influences.
- ☐ 9. The **Second Law of thermodynamics** states that any spontaneous change in an isolated system is accompanied by an increase in the entropy of the system.
- ☐ 10. **Equilibrium** is the state at which the **Gibbs energy** of the system has reached a minimum.
- ☐ 11. The energy levels of confined particles are quantized.
- ☐ 12. The **Boltzmann distribution** is a formula for calculating the relative populations of states of various energies.
- ☐ 13. The **equipartition theorem** states that for a sample at thermal equilibrium the average value of each quadratic contribution to the energy is $\frac{1}{2}kT$.

Checklist of equations

Property	Equation	Comment	Equation number
Velocity	$v = \mathrm{d}r/\mathrm{d}t$	Definition	B.1
Linear momentum	$p = mv$	Definition	B.2
Angular momentum	$J = I\omega$, $I = mr^2$	Point particle	B.3–B.4
Force	$F = ma = \mathrm{d}p/\mathrm{d}t$	Definition	B.5
Torque	$T = \mathrm{d}J/\mathrm{d}t$	Definition	B.6
Work	$\mathrm{d}w = -\boldsymbol{F}\cdot\mathrm{d}\boldsymbol{s}$	Definition	B.7
Kinetic energy	$E_k = \frac{1}{2}mv^2$	Definition	B.8
Potential energy and force	$F_x = -\mathrm{d}V/\mathrm{d}x$	One dimension	B.10
Coulomb potential energy	$V(r) = Q_1Q_2/4\pi\varepsilon_0 r$	Vacuum	B.14
Coulomb potential	$\phi = Q_2/4\pi\varepsilon_0 r$	Vacuum	B.16
Electric field strength	$\mathcal{E} = -\mathrm{d}\phi/\mathrm{d}x$	One dimension	B.18
Electrical power	$P = I\Delta\phi$	I is the current	B.19
Heat capacity	$C = \mathrm{d}U/\mathrm{d}T$	U is the internal energy	B.22
Enthalpy	$H = U + pV$	Definition	B.23
Gibbs energy	$G = H - TS$	Definition	B.24
Boltzmann distribution	$N_i/N_j = \mathrm{e}^{-(\varepsilon_i - \varepsilon_j)/kT}$		B.25a
Average speed of molecules	$v_{\text{mean}} = (8RT/\pi M)^{1/2}$	Perfect gas	B.26

C Waves

Contents

➤ Why do you need to know this material?

Several important investigative techniques in physical chemistry, such as spectroscopy and X-ray diffraction, involve electromagnetic radiation, a wavelike electromagnetic disturbance. We shall also see that the properties of waves are central to the quantum mechanical description of electrons in atoms and molecules. To prepare for those discussions, we need to understand the mathematical description of waves.

➤ What is the key idea?

A wave is a disturbance that propagates through space with a displacement that can be expressed as a harmonic function.

➤ What do you need to know already?

You need to be familiar with the properties of harmonic (sine and cosine) functions.

A **wave** is an oscillatory disturbance that travels through space. Examples of such disturbances include the collective motion of water molecules in ocean waves and of gas particles in sound waves. A **harmonic wave** is a wave with a displacement that can be expressed as a sine or cosine function.

C.1 Harmonic waves

A harmonic wave is characterized by a **wavelength**, λ (lambda), the distance between the neighbouring peaks of the wave, and its **frequency**, ν (nu), the number of times per second at which its displacement at a fixed point returns to its original value (Fig. C.1). The frequency is measured in *hertz*, where $1\,\text{Hz} = 1\,\text{s}^{-1}$. The wavelength and frequency are related by

$$\lambda\nu = v \quad \text{Relation between frequency and wavelength} \quad \text{(C.1)}$$

where v is the speed of propagation of the wave.

First, consider the snapshot of a harmonic wave at $t=0$. The displacement $\psi(x,t)$ varies with position x as

$$\psi(x,0) = A\cos\{(2\pi/\lambda)x + \phi\} \quad \text{Harmonic wave at } t=0 \quad \text{(C.2a)}$$

where A is the **amplitude** of the wave, the maximum height of the wave, and ϕ is the **phase** of the wave, the shift in the location of the peak from $x=0$ and which may lie between $-\pi$ and π (Fig. C.2). As time advances, the peaks migrate along the x-axis (the direction of propagation), and at any later instant the displacement is

$$\psi(x,t) = A\cos\{(2\pi/\lambda)x - 2\pi\nu t + \phi\} \quad \text{Harmonic wave at } t>0 \quad \text{(C.2b)}$$

A given wave can also be expressed as a sine function with the same argument but with ϕ replaced by $\phi + \frac{1}{2}\pi$.

If two waves, in the same region of space, with the same wavelength, have different phases then the resultant wave, the sum of the two, will have either enhanced or diminished amplitude. If the phases differ by $\pm\pi$ (so the peaks of one wave coincide with the troughs of the other), then the resultant wave, the sum of the two, will have a diminished amplitude. This effect is called **destructive interference.** If the phases of the two waves

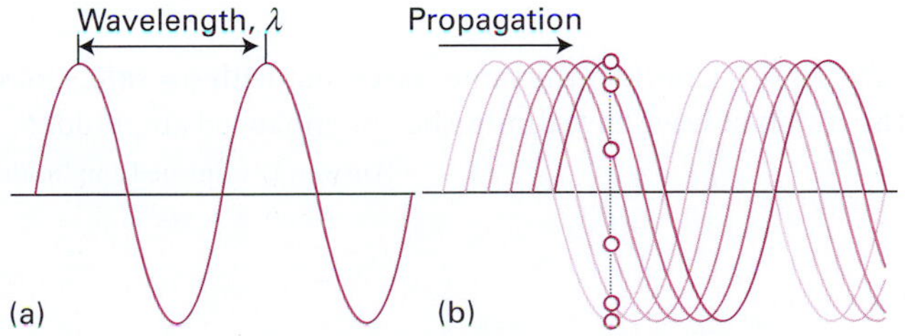

Figure C.1 (a) The wavelength, λ, of a wave is the peak-to-peak distance. (b) The wave is shown travelling to the right at a speed v. At a given location, the instantaneous amplitude of the wave changes through a complete cycle (the six dots show half a cycle) as it passes a given point. The frequency, ν, is the number of cycles per second that occur at a given point. Wavelength and frequency are related by $\lambda\nu = v$.

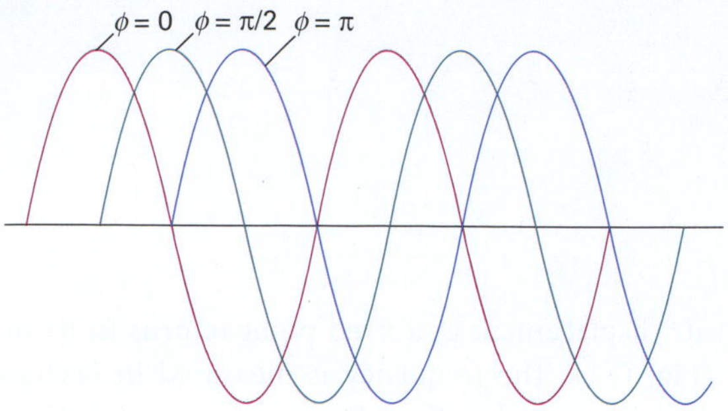

Figure C.2 The phase ϕ of a wave specifies the relative location of its peaks.

are the same (coincident peaks), the resultant has an enhanced amplitude. This effect is called **constructive interference.**

Brief illustration C.1 Resultant waves

To gain insight into cases in which the phase difference is a value other than $\pm\pi$, consider the addition of the waves $f(x) = \cos(2\pi x/\lambda)$ and $g(x) = \cos\{(2\pi x/\lambda) + \phi\}$. Figure C.3 shows plots of $f(x)$, $g(x)$, and $f(x) + g(x)$ against x/λ for $\phi = \pi/3$. The resultant wave has a greater amplitude than either $f(x)$ or $g(x)$, and has peaks between the peaks of $f(x)$ and $g(x)$.

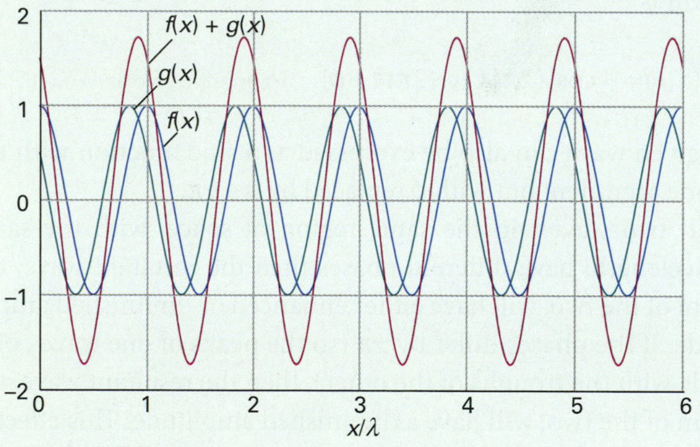

Figure C.3 Interference between the waves discussed in *Brief illustration* C.1.

Self-test C.1 Consider the same waves, but with $\phi = 3\pi/4$. Does the resultant wave have diminished or enhanced amplitude?

Answer: Diminished amplitude

C.2 The electromagnetic field

Light is a form of electromagnetic radiation. In classical physics, electromagnetic radiation is understood in terms of the **electromagnetic field**, an oscillating electric and magnetic disturbance that spreads as a harmonic wave through space. An **electric field** acts on charged particles (whether stationary or moving) and a **magnetic field** acts only on moving charged particles.

The wavelength and frequency of an electromagnetic wave in a vacuum are related by

$$\lambda\nu = c \qquad \textit{Electromagnetic wave in a vacuum} \qquad \text{Relation between frequency and wavelength} \qquad \text{(C.3)}$$

where $c = 2.997\,924\,58 \times 10^8\,\mathrm{m\,s^{-1}}$ (which we shall normally quote as $2.998 \times 10^8\,\mathrm{m\,s^{-1}}$) is the speed of light in a vacuum. When the wave is passing through a medium (even air), its speed is reduced to c' and, although the frequency remains unchanged, its wavelength is reduced accordingly. The reduced speed of light in a medium is normally expressed in terms of the **refractive index**, n_r, of the medium, where

$$n_r = \frac{c}{c'} \qquad \text{Refractive index} \qquad \text{(C.4)}$$

The refractive index depends on the frequency of the light, and for visible light typically increases with frequency. It also depends on the physical state of the medium. For yellow light in water at 25 °C, $n_r = 1.3$, so the wavelength is reduced by 30 per cent.

The classification of the electromagnetic field according to its frequency and wavelength is summarized in Fig. C.4. It is often desirable to express the characteristics of an electromagnetic wave by giving its **wavenumber**, $\tilde{\nu}$ (nu tilde), where

$$\tilde{\nu} = \frac{\nu}{c} = \frac{1}{\lambda} \qquad \textit{Electromagnetic radiation} \qquad \text{Wavenumber} \qquad \text{(C.5)}$$

A wavenumber can be interpreted as the number of complete wavelengths in a given length (of vacuum). Wavenumbers are normally reported in reciprocal centimetres ($\mathrm{cm^{-1}}$), so a wavenumber of $5\,\mathrm{cm^{-1}}$ indicates that there are 5 complete wavelengths in 1 cm.

Brief illustration C.2 Wavenumbers

The wavenumber of electromagnetic radiation of wavelength 660 nm is

$$\tilde{\nu} = \frac{1}{\lambda} = \frac{1}{660 \times 10^{-9}\,\mathrm{m}} = 1.5 \times 10^6\,\mathrm{m^{-1}} = 15\,000\,\mathrm{cm^{-1}}$$

You can avoid errors in converting between units of $\mathrm{m^{-1}}$ and $\mathrm{cm^{-1}}$ by remembering that wavenumber represents the number of wavelengths in a given distance. Thus, a wavenumber expressed as the number of waves per centimetre and hence in units of $\mathrm{cm^{-1}}$ must be 100 times less than the equivalent quantity expressed per metre in units of $\mathrm{m^{-1}}$.

Self-test C.2 Calculate the wavenumber and frequency of red light, of wavelength 710 nm.

Answer: $\tilde{\nu} = 1.41 \times 10^6\,\mathrm{m^{-1}} = 1.41 \times 10^4\,\mathrm{cm^{-1}}$, $\nu = 422\,\mathrm{THz}$ ($1\,\mathrm{THz} = 10^{12}\,\mathrm{s^{-1}}$)

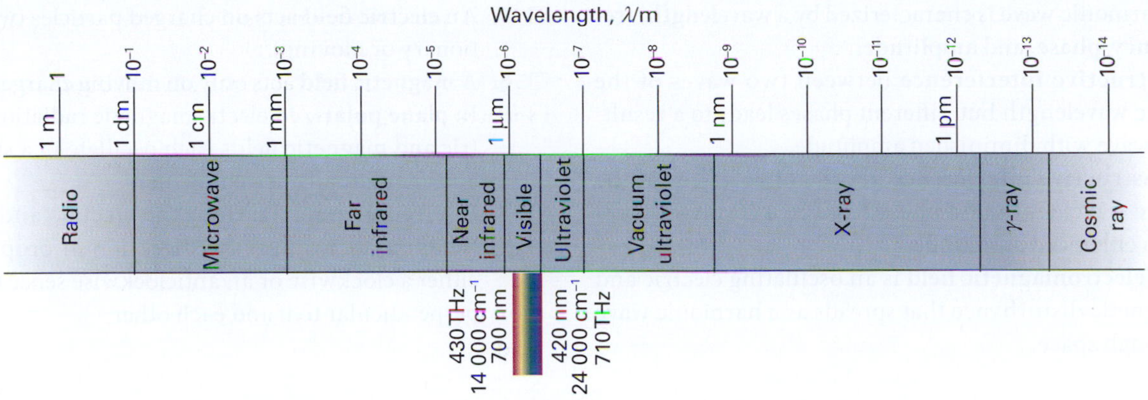

Figure C.4 The electromagnetic spectrum and its classification into regions (the boundaries are not precise).

The functions that describe the oscillating electric field, $\mathcal{E}(x,t)$, and magnetic field, $\mathcal{B}(x,t)$, travelling along the x-direction with wavelength λ and frequency ν are

$$\mathcal{E}(x,t) = \mathcal{E}_0 \cos\{(2\pi/\lambda)x - 2\pi\nu t + \phi\} \quad \text{(C.6a)}$$

Electromagnetic radiation Electric field

$$\mathcal{B}(x,t) = \mathcal{B}_0 \cos\{(2\pi/\lambda)x - 2\pi\nu t + \phi\} \quad \text{(C.6b)}$$

Electromagnetic radiation Magnetic field

where $\mathcal{E}_0$ and $\mathcal{B}_0$ are the amplitudes of the electric and magnetic fields, respectively, and ϕ is the phase of the wave. In this case the amplitude is a vector quantity, because the electric and magnetic fields have direction as well as amplitude. The magnetic field is perpendicular to the electric field and both are perpendicular to the propagation direction (Fig. C.5). According to classical electromagnetic theory, the **intensity** of electromagnetic radiation, a measure of the energy associated with the wave, is proportional to the square of the amplitude of the wave.

Equation C.6 describes electromagnetic radiation that is **plane polarized**; it is so called because the electric and magnetic fields each oscillate in a single plane. The plane of polarization may be orientated in any direction around the direction of propagation. An alternative mode of polarization is **circular polarization**, in which the electric and magnetic fields rotate around the direction of propagation in either a clockwise or an anticlockwise sense but remain perpendicular to it and to each other (Fig. C.6).

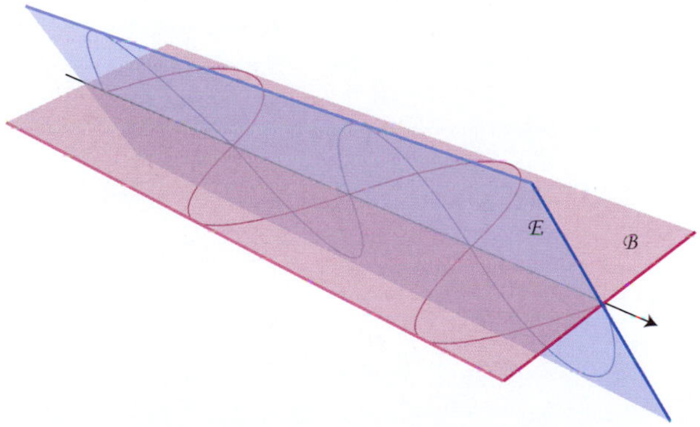

Figure C.5 In a plane polarized wave, the electric and magnetic fields oscillate in orthogonal planes and are perpendicular to the direction of propagation.

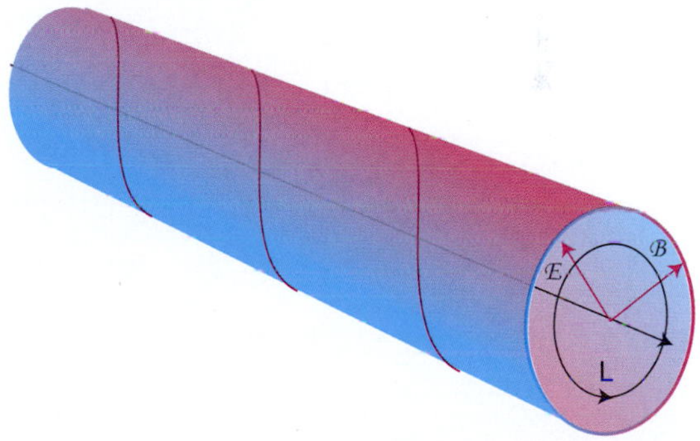

Figure C.6 In a circularly polarized wave, the electric and magnetic fields rotate around the direction of propagation but remain perpendicular to one another. The illustration also defines 'right' and 'left-handed' polarizations ('left-handed' polarization is shown as L).

Checklist of concepts

☐ 1. A **wave** is an oscillatory disturbance that travels through space.

☐ 2. A **harmonic wave** is a wave with a displacement that can be expressed as a sine or cosine function.

- ☐ 3. A harmonic wave is characterized by a **wavelength, frequency, phase**, and **amplitude**.
- ☐ 4. **Destructive interference** between two waves of the same wavelength but different phases leads to a resultant wave with diminished amplitude.
- ☐ 5. **Constructive interference** between two waves of the same wavelength and phase leads to a resultant wave with enhanced amplitude.
- ☐ 6. The **electromagnetic field** is an oscillating electric and magnetic disturbance that spreads as a harmonic wave through space.
- ☐ 7. An **electric field** acts on charged particles (whether stationary or moving).
- ☐ 8. A **magnetic field** acts only on moving charged particles.
- ☐ 9. In **plane polarized** electromagnetic radiation, the electric and magnetic fields each oscillate in a single plane and are mutually perpendicular.
- ☐ 10. In **circular polarization**, the electric and magnetic fields rotate around the direction of propagation in either a clockwise or an anticlockwise sense but remain perpendicular to it and each other.

Checklist of equations

Property	Equation	Comment	Equation number
Relation between the frequency and wavelength	$\lambda\nu = v$	For electromagnetic radiation in a vacuum, $v = c$	C.1
Refractive index	$n_r = c/c'$	Definition; $n_r \geq 1$	C.4
Wavenumber	$\tilde{\nu} = \nu/c = 1/\lambda$	Electromagnetic radiation	C.5

FOUNDATIONS

TOPIC A Matter

Discussion questions

A.1 Summarize the features of the nuclear model of the atom. Define the terms atomic number, nucleon number, and mass number.

A.2 Where in the periodic table are metals, non-metals, transition metals, lanthanoids, and actinoids found?

A.3 Summarize what is meant by a single bond and a multiple bond.

A.4 Summarize the principal concepts of the VSEPR theory of molecular shape.

A.5 Compare and contrast the properties of the solid, liquid, and gas states of matter.

Exercises

A.1(a) Express the typical ground-state electron configuration of an atom of an element in (i) Group 2, (ii) Group 7, (iii) Group 15 of the periodic table.
A.1(b) Express the typical ground-state electron configuration of an atom of an element in (i) Group 3, (ii) Group 5, (iii) Group 13 of the periodic table.

A.2(a) Identify the oxidation numbers of the elements in (i) $MgCl_2$, (ii) FeO, (iii) Hg_2Cl_2.
A.2(b) Identify the oxidation numbers of the elements in (i) CaH_2, (ii) CaC_2, (iii) LiN_3.

A.3(a) Identify a molecule with a (i) single, (ii) double, (iii) triple bond between a carbon and a nitrogen atom.
A.3(b) Identify a molecule with (i) one, (i) two, (iii) three lone pairs on the central atom.

A.4(a) Draw the Lewis (electron dot) structures of (i) SO_3^{2-}, (ii) XeF_4, (iii) P_4.
A.4(b) Draw the Lewis (electron dot) structures of (i) O_3, (ii) ClF_3^+, (iii) N_3^-.

A.5(a) Identify three compounds with an incomplete octet.
A.5(b) Identify four hypervalent compounds.

A.6(a) Use VSEPR theory to predict the structures of (i) PCl_3, (ii) PCl_5, (iii) XeF_2, (iv) XeF_4.
A.6(b) Use VSEPR theory to predict the structures of (i) H_2O_2, (ii) FSO_3^-, (iii) KrF_2, (iv) PCl_4^+.

A.7(a) Identify the polarities (by attaching partial charges δ+ and δ−) of the bonds (i) C–Cl, (ii) P–H, (iii) N–O.
A.7(b) Identify the polarities (by attaching partial charges δ+ and δ−) of the bonds (i) C–H, (ii) P–S, (iii) N–Cl.

A.8(a) State whether you expect the following molecules to be polar or nonpolar: (i) CO_2, (ii) SO_2, (iii) N_2O, (iv) SF_4.
A.8(b) State whether you expect the following molecules to be polar or nonpolar: (i) O_3, (ii) XeF_2, (iii) NO_2, (iv) C_6H_{14}.

A.9(a) Arrange the molecules in Exercise A.8(a) by increasing dipole moment.
A.9(b) Arrange the molecules in Exercise A.8(b) by increasing dipole moment.

A.10(a) Classify the following properties as extensive or intensive: (i) mass, (ii) mass density, (iii) temperature, (iv) number density.
A.10(b) Classify the following properties as extensive or intensive: (i) pressure, (ii) specific heat capacity, (iii) weight, (iv) molality.

A.11(a) Calculate (i) the amount of C_2H_5OH (in moles) and (ii) the number of molecules present in 25.0 g of ethanol.
A.11(b) Calculate (i) the amount of $C_6H_{12}O_6$ (in moles) and (ii) the number of molecules present in 5.0 g of glucose.

A.12(a) Calculate (i) the mass, (ii) the weight on the surface of the Earth (where $g = 9.81\ \mathrm{m\,s^{-2}}$) of 10.0 mol H_2O(l).
A.12(b) Calculate (i) the mass, (ii) the weight on the surface of Mars (where $g = 3.72\ \mathrm{m\,s^{-2}}$) of 10.0 mol C_6H_6(l).

A.13(a) Calculate the pressure exerted by a person of mass 65 kg standing (on the surface of the Earth) on shoes with soles of area 150 cm^2.
A.13(b) Calculate the pressure exerted by a person of mass 60 kg standing (on the surface of the Earth) on shoes with stiletto heels of area 2 cm^2 (assume that the weight is entirely on the heels).

A.14(a) Express the pressure calculated in Exercise A.13(a) in atmospheres.
A.14(b) Express the pressure calculated in Exercise A.13(b) in atmospheres.

A.15(a) Express a pressure of 1.45 atm in (i) pascal, (ii) bar.
A.15(b) Express a pressure of 222 atm in (i) pascal, (ii) bar.

A.16(a) Convert blood temperature, 37.0 °C, to the Kelvin scale.
A.16(b) Convert the boiling point of oxygen, 90.18 K, to the Celsius scale.

A.17(a) Equation A.4 is a relation between the Kelvin and Celsius scales. Devise the corresponding equation relating the Fahrenheit and Celsius scales and use it to express the boiling point of ethanol (78.5 °C) in degrees Fahrenheit.
A.17(b) The Rankine scale is a version of the thermodynamic temperature scale in which the degrees (°R) are the same size as degrees Fahrenheit. Derive an expression relating the Rankine and Kelvin scales and express the freezing point of water in degrees Rankine.

A.18(a) A sample of hydrogen gas was found to have a pressure of 110 kPa when the temperature was 20.0 °C. What can its pressure be expected to be when the temperature is 7.0 °C?
A.18(b) A sample of 325 mg of neon occupies 2.00 dm^3 at 20.0 °C. Use the perfect gas law to calculate the pressure of the gas.

A.19(a) At 500 °C and 93.2 kPa, the mass density of sulfur vapour is 3.710 kg m^{-3}. What is the molecular formula of sulfur under these conditions?
A.19(b) At 100 °C and 16.0 kPa, the mass density of phosphorus vapour is 0.6388 kg m^{-3}. What is the molecular formula of phosphorus under these conditions?

A.20(a) Calculate the pressure exerted by 22 g of ethane behaving as a perfect gas when confined to 1000 cm^3 at 25.0 °C.
A.20(b) Calculate the pressure exerted by 7.05 g of oxygen behaving as a perfect gas when confined to 100 cm^3 at 100.0 °C.

A.21(a) A vessel of volume 10.0 dm^3 contains 2.0 mol H_2 and 1.0 mol N_2 at 5.0 °C. Calculate the partial pressure of each component and their total pressure.
A.21(b) A vessel of volume 100 cm^3 contains 0.25 mol O_2 and 0.034 mol CO_2 at 10.0 °C. Calculate the partial pressure of each component and their total pressure.

TOPIC B Energy

Discussion questions

B.1 What is energy?

B.2 Distinguish between kinetic and potential energy.

B.3 State the Second Law of thermodynamics. Can the entropy of the system that is not isolated from its surroundings decrease during a spontaneous process?

B.4 What is meant by quantization of energy? In what circumstances are the effects of quantization most important for microscopic systems?

B.5 What are the assumptions of the kinetic molecular theory?

B.6 What are the main features of the Maxwell–Boltzmann distribution of speeds?

Exercises

B.1(a) A particle of mass 1.0 g is released near the surface of the Earth, where the acceleration of free fall is $g = 9.81\ \mathrm{m\,s^{-2}}$. What will be its speed and kinetic energy after (i) 1.0 s, (ii) 3.0 s. Ignore air resistance.
B.1(b) The same particle in Exercise B.1(a) is released near the surface of Mars, where the acceleration of free fall is $g = 3.72\ \mathrm{m\,s^{-2}}$. What will be its speed and kinetic energy after (i) 1.0 s, (ii) 3.0 s. Ignore air resistance.

B.2(a) An ion of charge ze moving through water is subject to an electric field of strength $\mathcal{E}$ which exerts a force $ze\mathcal{E}$, but it also experiences a frictional drag proportional to its speed s and equal to $6\pi\eta Rs$, where R is its radius and η (eta) is the viscosity of the medium. What will be its terminal velocity?
B.2(b) A particle descending through a viscous medium experiences a frictional drag proportional to its speed s and equal to $6\pi\eta Rs$, where R is its radius and η (eta) is the viscosity of the medium. If the acceleration of free fall is denoted g, what will be the terminal velocity of a sphere of radius R and mass density ρ?

B.3(a) Confirm that the general solution of the harmonic oscillator equation of motion $(m\mathrm{d}^2x/\mathrm{d}t^2 = -k_f x)$ is $x(t) = A\sin\omega t + B\cos\omega t$ with $\omega = (k_f/m)^{1/2}$.
B.3(b) Consider a harmonic oscillator with $B = 0$ (in the notation of Exercise B.3(a)); relate the total energy at any instant to its maximum displacement amplitude.

B.4(a) The force constant of a C–H bond is about $450\ \mathrm{N\,m^{-1}}$. How much work is needed to stretch the bond by (i) 10 pm, (ii) 20 pm?
B.4(b) The force constant of the H–H bond is about $510\ \mathrm{N\,m^{-1}}$. How much work is needed to stretch the bond by 20 pm?

B.5(a) An electron is accelerated in an electron microscope from rest through a potential difference $\Delta\phi = 100\ \mathrm{kV}$ and acquires an energy of $e\Delta\phi$. What is its final speed? What is its energy in electronvolts (eV)?
B.5(b) A $C_6H_4^{2+}$ ion is accelerated in a mass spectrometer from rest through a potential difference $\Delta\phi = 20\ \mathrm{kV}$ and acquires an energy of $e\Delta\phi$. What is its final speed? What is its energy in electronvolts (eV)?

B.6(a) Calculate the work that must be done in order to remove a Na^+ ion from 200 pm away from a Cl^- ion to infinity (in a vacuum). What work would be needed if the separation took place in water?
B.6(b) Calculate the work that must be done in order to remove an Mg^{2+} ion from 250 pm away from an O^{2-} ion to infinity (in a vacuum). What work would be needed if the separation took place in water?

B.7(a) Calculate the electric potential due to the nuclei at a point in a LiH molecule located at 200 pm from the Li nucleus and 150 pm from the H nucleus.
B.7(b) Plot the electric potential due to the nuclei at a point in a Na^+Cl^- ion pair located on a line half way between the nuclei (the internuclear separation is 283 pm) as the point approaches from infinity and ends at the mid-point between the nuclei.

B.8(a) An electric heater is immersed in a flask containing 200 g of water, and a current of 2.23 A from a 15.0 V supply is passed for 12.0 minutes. How much energy is supplied to the water? Estimate the rise in temperature (for water, $C = 75.3\ \mathrm{J\,K^{-1}\,mol^{-1}}$).
B.8(b) An electric heater is immersed in a flask containing 150 g of ethanol, and a current of 1.12 A from a 12.5 V supply is passed for 172 s. How much energy is supplied to the ethanol? Estimate the rise in temperature (for ethanol, $C = 111.5\ \mathrm{J\,K^{-1}\,mol^{-1}}$).

B.9(a) The heat capacity of a sample of iron was $3.67\ \mathrm{J\,K^{-1}}$. By how much would its temperature rise if 100 J of energy were transferred to it as heat?
B.9(b) The heat capacity of a sample of water was $5.77\ \mathrm{J\,K^{-1}}$. By how much would its temperature rise if 50.0 kJ of energy were transferred to it as heat?

B.10(a) The molar heat capacity of lead is $26.44\ \mathrm{J\,K^{-1}\,mol^{-1}}$. How much energy must be supplied (by heating) to 100 g of lead to increase its temperature by 10.0 °C?
B.10(b) The molar heat capacity of water is $75.2\ \mathrm{J\,K^{-1}\,mol^{-1}}$. How much energy must be supplied by heating to 10.0 g of water to increase its temperature by 10.0 °C?

B.11(a) The molar heat capacity of ethanol is $111.46\ \mathrm{J\,K^{-1}\,mol^{-1}}$. What is its specific heat capacity?
B.11(b) The molar heat capacity of sodium is $28.24\ \mathrm{J\,K^{-1}\,mol^{-1}}$. What is its specific heat capacity?

B.12(a) The specific heat capacity of water is $4.18\ \mathrm{J\,K^{-1}\,g^{-1}}$. What is its molar heat capacity?
B.12(b) The specific heat capacity of copper is $0.384\ \mathrm{J\,K^{-1}\,g^{-1}}$. What is its molar heat capacity?

B.13(a) By how much does the molar enthalpy of hydrogen gas differ from its molar internal energy at 1000 °C? Assume perfect gas behaviour.
B.13(b) The mass density of water is $0.997\ \mathrm{g\,cm^{-3}}$. By how much does the molar enthalpy of water differ from its molar internal energy at 298 K?

B.14(a) Which do you expect to have the greater entropy at 298 K and 1 bar, liquid water or water vapour?
B.14(b) Which do you expect to have the greater entropy at 0 °C and 1 atm, liquid water or ice?

B.15(a) Which do you expect to have the greater entropy, 100 g of iron at 300 K or 3000 K?
B.15(b) Which do you expect to have the greater entropy, 100 g of water at 0 °C or 100 °C?

B.16(a) Give three examples of a system that is in dynamic equilibrium.
B.16(b) Give three examples of a system that is in static equilibrium.

B.17(a) Suppose two states differ in energy by 1.0 eV (electronvolts, see inside the front cover); what is the ratio of their populations at (a) 300 K, (b) 3000 K?
B.17(b) Suppose two states differ in energy by 2.0 eV (electronvolts, see inside the front cover); what is the ratio of their populations at (a) 200 K, (b) 2000 K?

B.18(a) Suppose two states differ in energy by 1.0 eV, what can be said about their populations when $T = 0$?
B.18(b) Suppose two states differ in energy by 1.0 eV, what can be said about their populations when the temperature is infinite?

B.19(a) A typical vibrational excitation energy of a molecule corresponds to a wavenumber of 2500 cm^{-1} (convert to an energy separation by multiplying by hc; see *Foundations* C). Would you expect to find molecules in excited vibrational states at room temperature (20 °C)?
B.19(b) A typical rotational excitation energy of a molecule corresponds to a frequency of about 10 GHz (convert to an energy separation by multiplying by h; see *Foundations* C). Would you expect to find gas-phase molecules in excited rotational states at room temperature (20 °C)?

B.20(a) Suggest a reason why most molecules survive for long periods at room temperature.
B.20(b) Suggest a reason why the rates of chemical reactions typically increase with increasing temperature.

B.21(a) Calculate the relative mean speeds of N_2 molecules in air at 0 °C and 40 °C.
B.21(b) Calculate the relative mean speeds of CO_2 molecules in air at 20 °C and 30 °C.

B.22(a) Calculate the relative mean speeds of N_2 and CO_2 molecules in air.
B.22(b) Calculate the relative mean speeds of Hg_2 and H_2 molecules in a gaseous mixture.

B.23(a) Use the equipartition theorem to calculate the contribution of translational motion to the internal energy of 5.0 g of argon at 25 °C.
B.23(b) Use the equipartition theorem to calculate the contribution of translational motion to the internal energy of 10.0 g of helium at 30 °C.

B.24(a) Use the equipartition theorem to calculate the contribution to the total internal energy of a sample of 10.0 g of (i) carbon dioxide, (ii) methane at 20 °C; take into account translation and rotation but not vibration.
B.24(b) Use the equipartition theorem to calculate the contribution to the total internal energy of a sample of 10.0 g of lead at 20 °C, taking into account the vibrations of the atoms.

B.25(a) Use the equipartition theorem to compute the molar heat capacity of argon.
B.25(b) Use the equipartition theorem to compute the molar heat capacity of helium.

B.26(a) Use the equipartition theorem to estimate the heat capacity of (i) carbon dioxide, (ii) methane.
B.26(b) Use the equipartition theorem to estimate the heat capacity of (i) water vapour, (ii) lead.

TOPIC C Waves

Discussion questions

C.1 How many types of wave motion can you identify?

C.2 What is the wave nature of the sound of a sudden 'bang'?

Exercises

C.1(a) What is the speed of light in water if the refractive index of the latter is 1.33?
C.1(b) What is the speed of light in benzene if the refractive index of the latter is 1.52?

C.2(a) The wavenumber of a typical vibrational transition of a hydrocarbon is 2500 cm^{-1}. Calculate the corresponding wavelength and frequency.
C.2(b) The wavenumber of a typical vibrational transition of an O–H bond is 3600 cm^{-1}. Calculate the corresponding wavelength and frequency.

Integrated activities

F.1 In Topic 1B we show that for a perfect gas the fraction of molecules that have a speed in the range v to $v+dv$ is $f(v)dv$, where

$$f(v)=4\pi\left(\frac{M}{2\pi RT}\right)^{3/2} v^2 e^{-Mv^2/2RT}$$

is the Maxwell–Boltzmann distribution (eqn 1B.4). Use this expression and mathematical software, a spreadsheet, or the *Living graphs* on the web site of this book for the following exercises:

(a) Refer to the graph in Fig. B.6. Plot different distributions by keeping the molar mass constant at 100 g mol^{-1} and varying the temperature of the sample between 200 K and 2000 K.
(b) Evaluate numerically the fraction of molecules with speeds in the range 100 m s^{-1} to 200 m s^{-1} at 300 K and 1000 K.

F.2 Based on your observations from Problem F.1, provide a molecular interpretation of temperature.

PART ONE

Thermodynamics

Part 1 of the text develops the concepts of thermodynamics, the science of the transformations of energy. Thermodynamics provides a powerful way to discuss equilibria and the direction of natural change in chemistry. Its concepts apply to both physical change, such as fusion and vaporization, and chemical change, including electrochemistry. We see that through the concepts of energy, enthalpy, entropy, Gibbs energy, and the chemical potential it is possible to obtain a unified view of these core features of chemistry and to treat equilibria quantitatively.

The chapters in Part 1 deal with the bulk properties of matter; those of Part 2 show how these properties stem from the behaviour of individual atoms.

CHAPTER 1

The properties of gases

A **gas** is a form of matter that fills whatever container it occupies. This chapter establishes the properties of gases that will be used throughout the text.

1A The perfect gas

The chapter begins with an account of an idealized version of a gas, a 'perfect gas', and shows how its equation of state may be assembled from the experimental observations summarized by Boyle's law, Charles's law, and Avogadro's principle.

1B The kinetic model

One central feature of physical chemistry is its role in building models of molecular behaviour that seek to explain observed phenomena. A prime example of this procedure is the development of a molecular model of a perfect gas in terms of a collection of molecules (or atoms) in ceaseless, essentially random motion. This model is the basis of 'kinetic molecular theory'. As well as accounting for the gas laws, this theory can be used to predict the average speed at which molecules move in a gas, and that speed's dependence on temperature. In combination with the Boltzmann distribution (*Foundations* B), the kinetic theory can also be used to predict the spread of molecular speeds and its dependence on molecular mass and temperature.

1C Real gases

The perfect gas is an excellent starting point for the discussion of properties of all gases, and its properties are invoked throughout the chapters on thermodynamics that follow this chapter. However, actual gases, 'real gases', have properties that differ from those of perfect gases, and we need to be able to interpret these deviations and build the effects of molecular attractions and repulsions into our model. The discussion of real gases is another example of how initially primitive models in physical chemistry are elaborated to take into account more detailed observations.

What is the impact of this material?

The perfect gas law and the kinetic theory can be applied to the study of phenomena confined to a reaction vessel or encompassing an entire planet or star. We have identified two applications. In *Impact* I1.1 we see how the gas laws are used in the discussion of meteorological phenomena—the weather. In *Impact* I1.2 we examine how the kinetic model of gases has a surprising application: to the discussion of dense stellar media, such as the interior of the Sun.

To read more about the impact of this material, scan the QR code, or go to bcs.whfreeman.com/webpub/chemistry/pchem10e/impact/pchem-1-1.html

1A The perfect gas

Contents

➤ **Why do you need to know this material?**

Equations related to perfect gases provide the basis for the development of many equations in thermodynamics. The perfect gas law is also a good first approximation for accounting for the properties of real gases.

➤ **What is the key idea?**

The perfect gas law, which is based on a series of empirical observations, is a limiting law that is obeyed increasingly well as the pressure of a gas tends to zero.

➤ **What do you need to know already?**

You need to be aware of the concepts of pressure and temperature introduced in *Foundations* A.

In molecular terms, a gas consists of a collection of molecules that are in ceaseless motion and which interact significantly with one another only when they collide. The properties of gases were among the first to be established quantitatively (largely during the seventeenth and eighteenth centuries) when the technological requirements of travel in balloons stimulated their investigation.

1A.1 Variables of state

The **physical state** of a sample of a substance, its physical condition, is defined by its physical properties. Two samples of the same substance that have the same physical properties are in the same state. The variables needed to specify the state of a system are the amount of substance it contains, n, the volume it occupies, V, the pressure, p, and the temperature, T.

(a) Pressure

The origin of the force exerted by a gas is the incessant battering of the molecules on the walls of its container. The collisions are so numerous that they exert an effectively steady force, which is experienced as a steady pressure. The SI unit of pressure, the *pascal* (Pa, $1\,\text{Pa}=1\,\text{N m}^{-2}$) is introduced in *Foundations* A. As discussed there, several other units are still widely used (Table 1A.1). A pressure of 1 bar is the **standard pressure** for reporting data; we denote it $p^{\ominus}$.

If two gases are in separate containers that share a common movable wall (a 'piston', Fig. 1A.1), the gas that has the higher pressure will tend to compress (reduce the volume of) the gas that has lower pressure. The pressure of the high-pressure gas will fall as it expands and that of the low-pressure gas will rise as it is compressed. There will come a stage when the two pressures are equal and the wall has no further tendency to move. This condition of equality of pressure on either side of a movable wall is a state of **mechanical equilibrium** between the two gases. The pressure of a gas is therefore an indication of whether a container that contains the gas will be in mechanical equilibrium with another gas with which it shares a movable wall.

Table 1A.1 Pressure units*

Name	Symbol	Value
pascal	1 Pa	**1 N m^{-2}, 1 kg m^{-1} s^{-2}**
bar	1 bar	**10^5 Pa**
atmosphere	1 atm	**101.325 kPa**
torr	1 Torr	**(101 325/760) Pa**=133.32… Pa
millimetres of mercury	1 mmHg	133.322… Pa
pounds per square inch	1 psi	6.894 757… kPa

* Values in bold are exact.

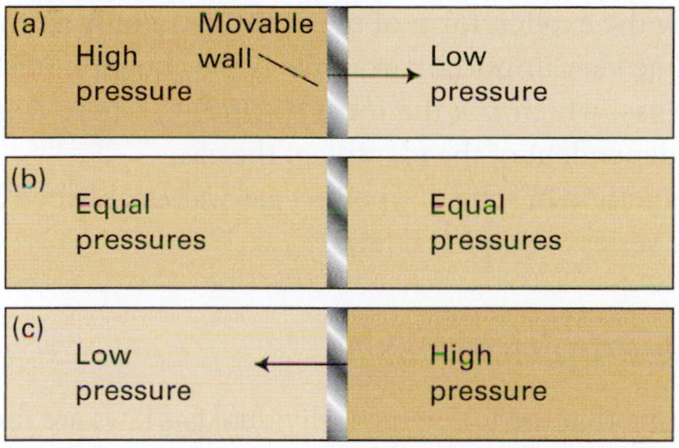

Figure 1A.1 When a region of high pressure is separated from a region of low pressure by a movable wall, the wall will be pushed into one region or the other, as in (a) and (c). However, if the two pressures are identical, the wall will not move (b). The latter condition is one of mechanical equilibrium between the two regions.

The pressure exerted by the atmosphere is measured with a *barometer*. The original version of a barometer (which was invented by Torricelli, a student of Galileo) was an inverted tube of mercury sealed at the upper end. When the column of mercury is in mechanical equilibrium with the atmosphere, the pressure at its base is equal to that exerted by the atmosphere. It follows that the height of the mercury column is proportional to the external pressure.

Example 1A.1 Calculating the pressure exerted by a column of liquid

Derive an equation for the pressure at the base of a column of liquid of mass density ρ (rho) and height h at the surface of the Earth. The pressure exerted by a column of liquid is commonly called the 'hydrostatic pressure'.

Method According to *Foundations* A, the pressure is the force, F, divided by the area, A, to which the force is applied: $p=F/A$. For a mass m subject to a gravitational field at the surface of the earth, $F=mg$, where g is the acceleration of free fall. To calculate F we need to know the mass m of the column of liquid, which is its mass density, ρ, multiplied by its volume, V: $m=\rho V$. The first step, therefore, is to calculate the volume of a cylindrical column of liquid.

Answer Let the column have cross-sectional area A, then its volume is Ah and its mass is $m=\rho Ah$. The force the column of this mass exerts at its base is

$$F=mg=\rho Ahg$$

The pressure at the base of the column is therefore

$$p=\frac{F}{A}=\frac{\rho Agh}{A}=\rho gh \qquad \text{Hydrostatic pressure} \qquad (1A.1)$$

Note that the hydrostatic pressure is independent of the shape and cross-sectional area of the column. The mass of the column of a given height increases as the area, but so does the area on which the force acts, so the two cancel.

Self-test 1A.1 Derive an expression for the pressure at the base of a column of liquid of length l held at an angle θ (theta) to the vertical (**1**).

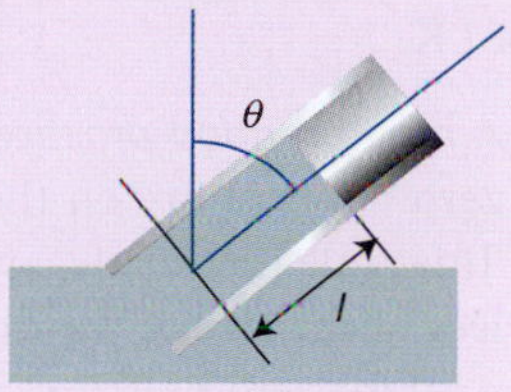

1

Answer: $p=\rho gl\cos\theta$

The pressure of a sample of gas inside a container is measured by using a pressure gauge, which is a device with properties that respond to the pressure. For instance, a *Bayard–Alpert pressure gauge* is based on the ionization of the molecules present in the gas and the resulting current of ions is interpreted in terms of the pressure. In a *capacitance manometer*, the deflection of a diaphragm relative to a fixed electrode is monitored through its effect on the capacitance of the arrangement. Certain semiconductors also respond to pressure and are used as transducers in solid-state pressure gauges.

(b) Temperature

The concept of temperature is introduced in *Foundations* A. In the early days of thermometry (and still in laboratory practice today), temperatures were related to the length of a column of liquid, and the difference in lengths shown when the thermometer was first in contact with melting ice and then with boiling water was divided into 100 steps called 'degrees', the lower point being labelled 0. This procedure led to the **Celsius scale** of temperature. In this text, temperatures on the Celsius scale are denoted θ (theta) and expressed in *degrees Celsius* (°C). However, because different liquids expand to different extents, and do not always expand uniformly over a given range, thermometers constructed from different materials showed different numerical values of the temperature between their fixed points. The pressure of a gas, however, can be used to construct a **perfect-gas temperature scale** that is independent of the identity of the gas. The perfect-gas scale turns out to be identical to the **thermodynamic temperature scale** introduced in Topic 3A, so we shall use the latter term from now on to avoid a proliferation of names.

On the thermodynamic temperature scale, temperatures are denoted T and are normally reported in *kelvins* (K; not °K). Thermodynamic and Celsius temperatures are related by the exact expression

$$T/\mathrm{K} = \theta/^\circ\mathrm{C} + 273.15$$ Definition of Celsius scale (1A.2)

This relation is the current definition of the Celsius scale in terms of the more fundamental Kelvin scale. It implies that a difference in temperature of 1 °C is equivalent to a difference of 1 K.

A note on good practice We write $T=0$, not $T=0\,\mathrm{K}$ for the zero temperature on the thermodynamic temperature scale. This scale is absolute, and the lowest temperature is 0 regardless of the size of the divisions on the scale (just as we write $p=0$ for zero pressure, regardless of the size of the units we adopt, such as bar or pascal). However, we write 0 °C because the Celsius scale is not absolute.

Brief illustration 1A.1 Temperature conversion

To express 25.00 °C as a temperature in kelvins, we use eqn 1A.4 to write

$$T/\mathrm{K} = (25.00\ ^\circ\mathrm{C})/^\circ\mathrm{C} + 273.15 = 25.00 + 273.15 = 298.15$$

Note how the units (in this case, °C) are cancelled like numbers. This is the procedure called 'quantity calculus' in which a physical quantity (such as the temperature) is the product of a numerical value (25.00) and a unit (1 °C); see *The chemist's toolkit* A.1 of *Foundations*. Multiplication of both sides by the unit K then gives $T = 298.15$ K.

A note on good practice When the units need to be specified in an equation, the approved procedure, which avoids any ambiguity, is to write (physical quantity)/units, which is a dimensionless number, just as (25.00 °C)/°C = 25.00 in this illustration. Units may be multiplied and cancelled just like numbers.

1A.2 Equations of state

Although in principle the state of a pure substance is specified by giving the values of n, V, p, and T, it has been established experimentally that it is sufficient to specify only three of these variables, for then the fourth variable is fixed. That is, it is an experimental fact that each substance is described by an **equation of state**, an equation that interrelates these four variables.

The general form of an equation of state is

$$p = f(T, V, n)$$ General form of an equation of state (1A.3)

This equation tells us that if we know the values of n, T, and V for a particular substance, then the pressure has a fixed value. Each substance is described by its own equation of state, but we know the explicit form of the equation in only a few special cases. One very important example is the equation of state of a 'perfect gas', which has the form $p = nRT/V$, where R is a constant independent of the identity of the gas.

The equation of state of a perfect gas was established by combining a series of empirical laws.

(a) The empirical basis

We assume that the following individual gas laws are familiar:

Boyle's law: $pV = \text{constant, at constant } n, T$ (1A.4a)

Charles's law: $V = \text{constant} \times T, \text{ at constant } n, p$ (1A.4b)

$p = \text{constant} \times T, \text{ at constant } n, V$ (1A.4c)

Avogadro's principle:

$V = \text{constant} \times n \text{ at constant } p, T$ (1A.4d)

Boyle's and Charles's laws are examples of a **limiting law**, a law that is strictly true only in a certain limit, in this case $p \to 0$. For example, if it is found empirically that the volume of a substance fits an expression $V = aT + bp + cp^2$, then in the limit of $p \to 0$, $V = aT$. Throughout this text, equations valid in this limiting sense are labelled with a blue equation number, as in these expressions. Although these relations are strictly true only at $p = 0$, they are reasonably reliable at normal pressures ($p \approx 1$ bar) and are used widely throughout chemistry.

Avogadro's principle is commonly expressed in the form 'equal volumes of gases at the same temperature and pressure contain the same numbers of molecules'. It is a principle rather than a law (a summary of experience) because it depends on the validity of a model, in this case the existence of molecules. Despite there now being no doubt about the existence of molecules, it is still a model-based principle rather than a law.

Figure 1A.2 depicts the variation of the pressure of a sample of gas as the volume is changed. Each of the curves in the

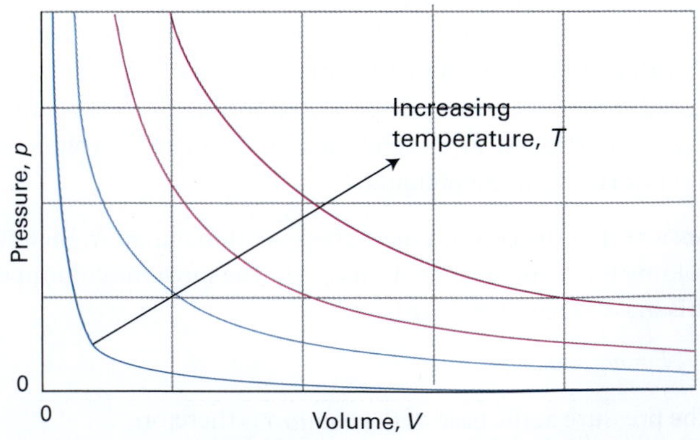

Figure 1A.2 The pressure–volume dependence of a fixed amount of perfect gas at different temperatures. Each curve is a hyperbola (pV = constant) and is called an isotherm.

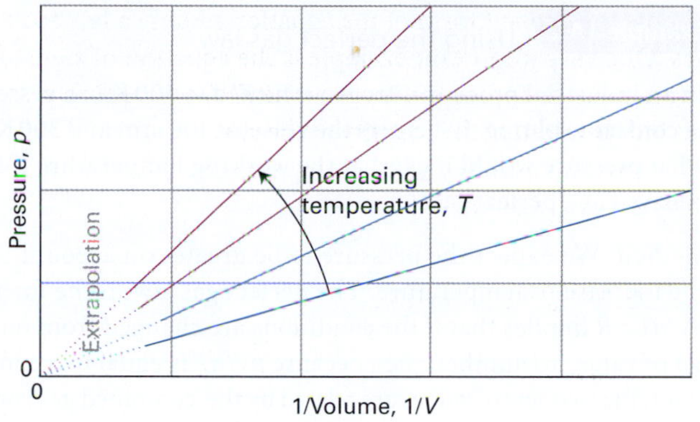

Figure 1A.3 Straight lines are obtained when the pressure is plotted against $1/V$ at constant temperature.

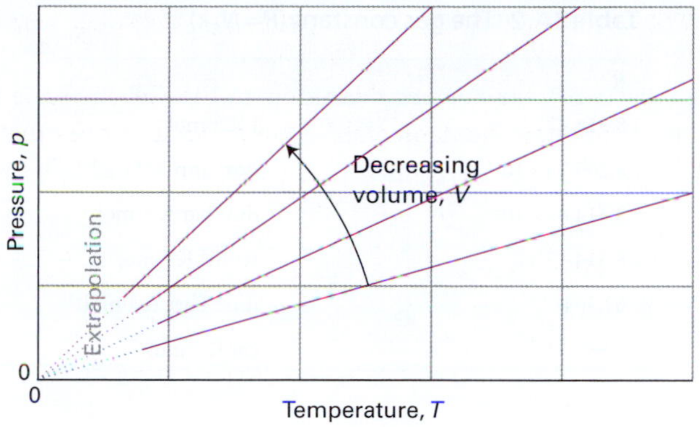

Figure 1A.5 The pressure also varies linearly with the temperature at constant volume, and extrapolates to zero at $T=0$ (−273 °C).

graph corresponds to a single temperature and hence is called an **isotherm**. According to Boyle's law, the isotherms of gases are hyperbolas (a curve obtained by plotting y against x with xy=constant, or y=constant/x). An alternative depiction, a plot of pressure against 1/volume, is shown in Fig. 1A.3. The linear variation of volume with temperature summarized by Charles's law is illustrated in Fig. 1A.4. The lines in this illustration are examples of **isobars**, or lines showing the variation of properties at constant pressure. Figure 1A.5 illustrates the linear variation of pressure with temperature. The lines in this diagram are **isochores**, or lines showing the variation of properties at constant volume.

A note on good practice To test the validity of a relation between two quantities, it is best to plot them in such a way that they should give a straight line, for deviations from a straight line are much easier to detect than deviations from a curve. The development of expressions that, when plotted, give a straight line is a very important and common procedure in physical chemistry.

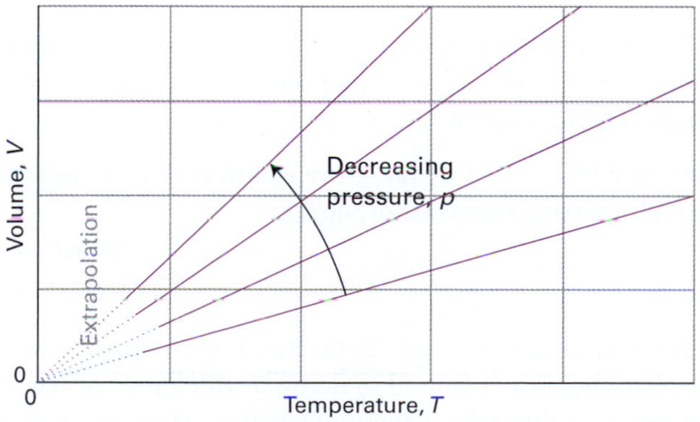

Figure 1A.4 The variation of the volume of a fixed amount of gas with the temperature at constant pressure. Note that in each case the isobars extrapolate to zero volume at $T=0$, or $\theta=-273$ °C.

The empirical observations summarized by eqn 1A.5 can be combined into a single expression:

$$pV = \text{constant} \times nT$$

This expression is consistent with Boyle's law (pV=constant) when n and T are constant, with both forms of Charles's law ($p \propto T$, $V \propto T$) when n and either V or p are held constant, and with Avogadro's principle ($V \propto n$) when p and T are constant. The constant of proportionality, which is found experimentally to be the same for all gases, is denoted R and called the (molar) **gas constant**. The resulting expression

$$pV = nRT \quad \text{Perfect gas law} \quad (1A.5)$$

is the **perfect gas law** (or *perfect gas equation of state*). It is the approximate equation of state of any gas, and becomes increasingly exact as the pressure of the gas approaches zero. A gas that obeys eqn 1A.5 exactly under all conditions is called a **perfect gas** (or *ideal gas*). A **real gas**, an actual gas, behaves more like a perfect gas the lower the pressure, and is described exactly by eqn 1A.5 in the limit of $p \rightarrow 0$. The gas constant R can be determined by evaluating $R=pV/nT$ for a gas in the limit of zero pressure (to guarantee that it is behaving perfectly). However, a more accurate value can be obtained by measuring the speed of sound in a low-pressure gas (argon is used in practice), for the speed of sound depends on the value of R and extrapolating its value to zero pressure. Another route to its value is to recognize (as explained in *Foundations* B) that it is related to Boltzmann's constant, k, by

$$R = N_A k \quad \text{The (molar) gas constant} \quad (1A.6)$$

where N_A is Avogadro's constant. There are currently (in 2014) plans to use this relation as the sole route to R, with defined values of N_A and k. Table 1A.2 lists the values of R in a variety of units.

A note on good practice Despite 'ideal gas' being the more common term, we prefer 'perfect gas'. As explained in Topic 5A, in an 'ideal mixture' of A and B, the AA, BB, and AB

Table 1A.2 The gas constant ($R = N_A k$)

R	
8.314 47	J K^{-1} mol^{-1}
8.205 74 × 10^{-2}	dm^3 atm K^{-1} mol^{-1}
8.314 47 × 10^{-2}	dm^3 bar K^{-1} mol^{-1}
8.314 47	Pa m^3 K^{-1} mol^{-1}
62.364	dm^3 Torr K^{-1} mol^{-1}
1.987 21	cal K^{-1} mol^{-1}

interactions are all the same but not necessarily zero. In a perfect gas, not only are the interactions all the same, they are also zero.

The surface in Fig. 1A.6 is a plot of the pressure of a fixed amount of perfect gas against its volume and thermodynamic temperature as given by eqn 1A.5. The surface depicts the only possible states of a perfect gas: the gas cannot exist in states that do not correspond to points on the surface. The graphs in Figs. 1A.2 and 1A.4 correspond to the sections through the surface (Fig. 1A.7).

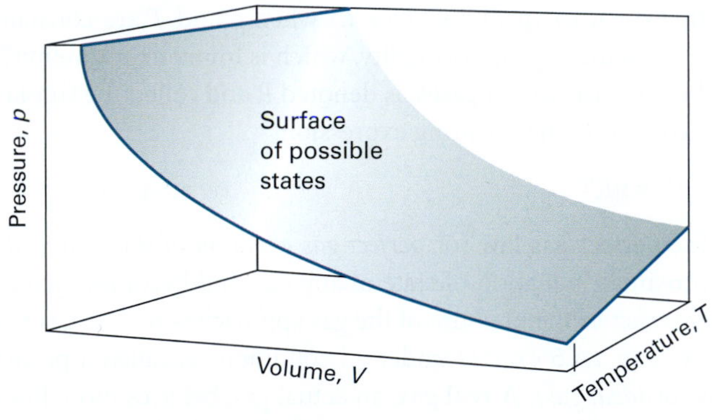

Figure 1A.6 A region of the p,V,T surface of a fixed amount of perfect gas. The points forming the surface represent the only states of the gas that can exist.

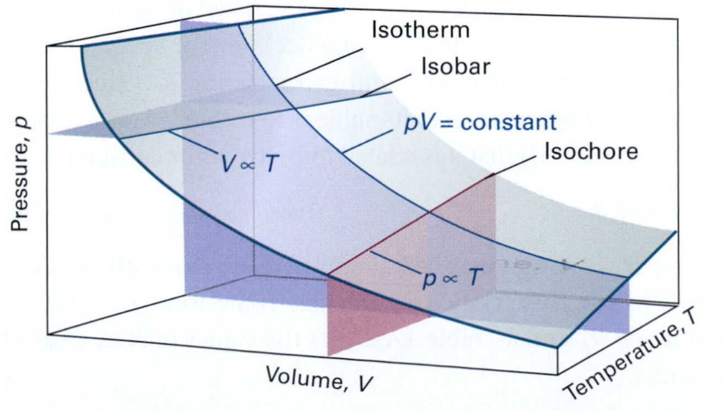

Figure 1A.7 Sections through the surface shown in Fig. 1A.6 at constant temperature give the isotherms shown in Fig. 1A.2, the isobars shown in Fig. 1A.4, and the isochores shown in Fig. 1A.5.

Example 1A.2 Using the perfect gas law

In an industrial process, nitrogen is heated to 500 K in a vessel of constant volume. If it enters the vessel at 100 atm and 300 K, what pressure would it exert at the working temperature if it behaved as a perfect gas?

Method We expect the pressure to be greater on account of the increase in temperature. The perfect gas law in the form $pV/nT = R$ implies that if the conditions are changed from one set of values to another, then because pV/nT is equal to a constant, the two sets of values are related by the 'combined gas law'

$$\frac{p_1V_1}{n_1T_1} = \frac{p_2V_2}{n_2T_2} \qquad \text{Combined gas law} \qquad (1A.7)$$

This expression is easily rearranged to give the unknown quantity (in this case p_2) in terms of the known. The known and unknown data are summarized as follows:

	n	p	V	T
Initial	Same	100	Same	300
Final	Same	?	Same	500

Answer Cancellation of the volumes (because $V_1 = V_2$) and amounts (because $n_1 = n_2$) on each side of the combined gas law results in

$$\frac{p_1}{T_1} = \frac{p_2}{T_2}$$

which can be rearranged into

$$p_2 = \frac{T_2}{T_1} \times p_1$$

Substitution of the data then gives

$$p_2 = \frac{500\ \text{K}}{300\ \text{K}} \times (100\ \text{atm}) = 167\ \text{atm}$$

Experiment shows that the pressure is actually 183 atm under these conditions, so the assumption that the gas is perfect leads to a 10 per cent error.

Self-test 1A.2 What temperature would result in the same sample exerting a pressure of 300 atm?

Answer: 900 K

The perfect gas law is of the greatest importance in physical chemistry because it is used to derive a wide range of relations that are used throughout thermodynamics. However, it is also of considerable practical utility for calculating the properties of a gas under a variety of conditions. For instance, the molar volume, $V_m = V/n$, of a perfect gas under the conditions called **standard ambient temperature and pressure** (SATP), which means

298.15 K and 1 bar (that is, exactly 10^5 Pa), is easily calculated from $V_m = RT/p$ to be 24.789 dm^3 mol^{-1}. An earlier definition, **standard temperature and pressure** (STP), was 0 °C and 1 atm; at STP, the molar volume of a perfect gas is 22.414 dm^3 mol^{-1}.

The molecular explanation of Boyle's law is that if a sample of gas is compressed to half its volume, then twice as many molecules strike the walls in a given period of time than before it was compressed. As a result, the average force exerted on the walls is doubled. Hence, when the volume is halved the pressure of the gas is doubled, and pV is a constant. Boyle's law applies to all gases regardless of their chemical identity (provided the pressure is low) because at low pressures the average separation of molecules is so great that they exert no influence on one another and hence travel independently. The molecular explanation of Charles's law lies in the fact that raising the temperature of a gas increases the average speed of its molecules. The molecules collide with the walls more frequently and with greater impact. Therefore they exert a greater pressure on the walls of the container. For a quantitative account of these relations, see Topic 1B.

(b) Mixtures of gases

When dealing with gaseous mixtures, we often need to know the contribution that each component makes to the total pressure of the sample. The **partial pressure**, p_J, of a gas J in a mixture (any gas, not just a perfect gas), is defined as

$$p_J = x_J p \qquad \textit{Definition} \quad \text{Partial pressure} \qquad (1A.8)$$

where x_J is the **mole fraction** of the component J, the amount of J expressed as a fraction of the total amount of molecules, n, in the sample:

$$x_J = \frac{n_J}{n} \qquad n = n_A + n_B + \cdots \qquad \textit{Definition} \quad \text{Mole fraction} \qquad (1A.9)$$

When no J molecules are present, $x_J = 0$; when only J molecules are present, $x_J = 1$. It follows from the definition of x_J that, whatever the composition of the mixture, $x_A + x_B + \ldots = 1$ and therefore that the sum of the partial pressures is equal to the total pressure:

$$p_A + p_B + \cdots = (x_A + x_B + \cdots)p = p \qquad (1A.10)$$

This relation is true for both real and perfect gases.

When all the gases are perfect, the partial pressure as defined in eqn 1A.9 is also the pressure that each gas would exert if it occupied the same container alone at the same temperature. The latter is the original meaning of 'partial pressure'. That identification was the basis of the original formulation of **Dalton's law**:

> The pressure exerted by a mixture of gases is the sum of the pressures that each one would exert if it occupied the container alone. — Dalton's law

Now, however, the relation between partial pressure (as defined in eqn 1A.8) and total pressure (as given by eqn 1A.10) is true for all gases and the identification of partial pressure with the pressure that the gas would exert on its own is valid only for a perfect gas.

Example 1A.3 Calculating partial pressures

The mass percentage composition of dry air at sea level is approximately N_2: 75.5; O_2: 23.2; Ar: 1.3. What is the partial pressure of each component when the total pressure is 1.20 atm?

Method We expect species with a high mole fraction to have a proportionally high partial pressure. Partial pressures are defined by eqn 1A.8. To use the equation, we need the mole fractions of the components. To calculate mole fractions, which are defined by eqn 1A.9, we use the fact that the amount of molecules J of molar mass M_J in a sample of mass m_J is $n_J = m_J/M_J$. The mole fractions are independent of the total mass of the sample, so we can choose the latter to be exactly 100 g (which makes the conversion from mass percentages very easy). Thus, the mass of N_2 present is 75.5 per cent of 100 g, which is 75.5 g.

Answer The amounts of each type of molecule present in 100 g of air, in which the masses of N_2, O_2, and Ar are 75.5 g, 23.2 g, and 1.3 g, respectively, are

$$n(N_2) = \frac{75.5\,\text{g}}{28.02\,\text{g mol}^{-1}} = \frac{75.5}{28.02}\,\text{mol} = 2.69\,\text{mol}$$

$$n(O_2) = \frac{23.2\,\text{g}}{32.00\,\text{g mol}^{-1}} = \frac{23.2}{32.00}\,\text{mol} = 0.725\,\text{mol}$$

$$n(\text{Ar}) = \frac{1.3\,\text{g}}{39.95\,\text{g mol}^{-1}} = \frac{1.3}{39.95}\,\text{mol} = 0.033\,\text{mol}$$

The total is 3.45 mol. The mole fractions are obtained by dividing each of the above amounts by 3.45 mol and the partial pressures are then obtained by multiplying the mole fraction by the total pressure (1.20 atm):

	N_2	O_2	Ar
Mole fraction:	0.780	0.210	0.0096
Partial pressure/atm:	0.936	0.252	0.012

We have not had to assume that the gases are perfect: partial pressures are defined as $p_J = x_J p$ for any kind of gas.

Self-test 1A.3 When carbon dioxide is taken into account, the mass percentages are 75.52 (N_2), 23.15 (O_2), 1.28 (Ar), and 0.046 (CO_2). What are the partial pressures when the total pressure is 0.900 atm?

Answer: 0.703, 0.189, 0.0084, 0.00027 atm

Checklist of concepts

- ☐ 1. The **physical state** of a sample of a substance, its physical condition, is defined by its physical properties.
- ☐ 2. **Mechanical equilibrium** is the condition of equality of pressure on either side of a shared movable wall.
- ☐ 3. An **equation of state** is an equation that interrelates the variables that define the state of a substance.
- ☐ 4. Boyle's and Charles's laws are examples of a **limiting law**, a law that is strictly true only in a certain limit, in this case $p \to 0$.
- ☐ 5. An **isotherm** is a line in a graph that corresponds to a single temperature.
- ☐ 6. An **isobar** is a line in a graph that corresponds to a single pressure.
- ☐ 7. An **isochore** is a line in a graph that corresponds to a single volume.
- ☐ 8. A **perfect gas** is a gas that obeys the perfect gas law under all conditions.
- ☐ 9. **Dalton's law** states that the pressure exerted by a mixture of (perfect) gases is the sum of the pressures that each one would exert if it occupied the container alone.

Checklist of equations

Property	Equation	Comment	Equation number
Relation between temperature scales	$T/\text{K} = \theta/°\text{C} + 273.15$	273.15 is exact	1A.2
Equation of state	$p = f(n,V,T)$		1A.3
Perfect gas law	$pV = nRT$	Valid for real gases in the limit $p \to 0$	1A.5
Partial pressure	$p_J = x_J p$	Valid for all gases	1A.8

1B The kinetic model

Contents

➤ **Why do you need to know this material?**

This material illustrates an important skill in science: the ability to extract quantitative information from a qualitative model. Moreover, the model is used in the discussion of the transport properties of gases (Topic 19A), reaction rates in gases (Topic 20F), and catalysis (Topic 22C).

➤ **What is the key idea?**

A gas consists of molecules of negligible size in ceaseless random motion and obeying the laws of classical mechanics in their collisions.

➤ **What do you need to know already?**

You need to be aware of Newton's second law of motion, that the acceleration of a body is proportional to the force acting on it, and the conservation of linear momentum.

In the **kinetic theory** of gases (which is sometimes called the *kinetic-molecular theory*, KMT) it is assumed that the only contribution to the energy of the gas is from the kinetic energies of the molecules. The kinetic model is one of the most remarkable—and arguably most beautiful—models in physical chemistry, for from a set of very slender assumptions, powerful quantitative conclusions can be reached.

1B.1 The model

The kinetic model is based on three assumptions:

1. The gas consists of molecules of mass m in ceaseless random motion obeying the laws of classical mechanics.
2. The size of the molecules is negligible, in the sense that their diameters are much smaller than the average distance travelled between collisions.
3. The molecules interact only through brief elastic collisions.

An **elastic collision** is a collision in which the total translational kinetic energy of the molecules is conserved.

(a) Pressure and molecular speeds

From the very economical assumptions of the kinetic model, we show in the following *Justification* that the pressure and volume of the gas are related by

$$pV = \tfrac{1}{3}nMv_{\rm rms}^2$$ *Perfect gas* Pressure (1B.1)

where $M = mN_A$, the molar mass of the molecules of mass m, and $v_{\rm rms}$ is the square root of the mean of the squares of the speeds, v, of the molecules:

$$v_{\rm rms} = \langle v^2\rangle^{1/2}$$ *Definition* Root-mean-square speed (1B.2)

Justification 1.1B The pressure of a gas according to the kinetic model

Consider the arrangement in Fig. 1B.1. When a particle of mass m that is travelling with a component of velocity v_x parallel to the x-axis collides with the wall on the right and is reflected, its linear momentum changes from mv_x before the collision to $-mv_x$ after the collision (when it is travelling in the opposite direction). The x-component of momentum therefore changes by $2mv_x$ on each collision (the y- and z-components are unchanged). Many molecules collide with the wall in an interval Δt, and the total change of momentum is the product of the change in momentum of each molecule multiplied

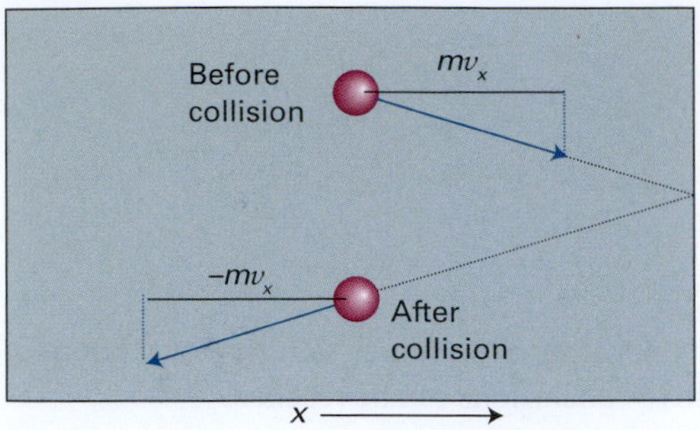

Figure 1B.1 The pressure of a gas arises from the impact of its molecules on the walls. In an elastic collision of a molecule with a wall perpendicular to the *x*-axis, the *x*-component of velocity is reversed but the *y*- and *z*-components are unchanged.

by the number of molecules that reach the wall during the interval.

Because a molecule with velocity component v_x can travel a distance $v_x\Delta t$ along the x-axis in an interval Δt, all the molecules within a distance $v_x\Delta t$ of the wall will strike it if they are travelling towards it (Fig. 1B.2). It follows that if the wall has area A, then all the particles in a volume $A\times v_x\Delta t$ will reach the wall (if they are travelling towards it). The number density of particles is nN_A/V, where n is the total amount of molecules in the container of volume V and N_A is Avogadro's constant, so the number of molecules in the volume $Av_x\Delta t$ is $(nN_A/V)\times Av_x\Delta t$.

At any instant, half the particles are moving to the right and half are moving to the left. Therefore, the average number of collisions with the wall during the interval Δt is $\frac{1}{2}nN_A Av_x\Delta t/V$. The total momentum change in that interval is the product of this number and the change $2mv_x$:

$$\text{Momentum change} = \frac{nN_A Av_x\Delta t}{2V}\times 2mv_x = \frac{n\overbrace{mN_A}^{M} Av_x^2\Delta t}{V} = \frac{nMAv_x^2\Delta t}{V}$$

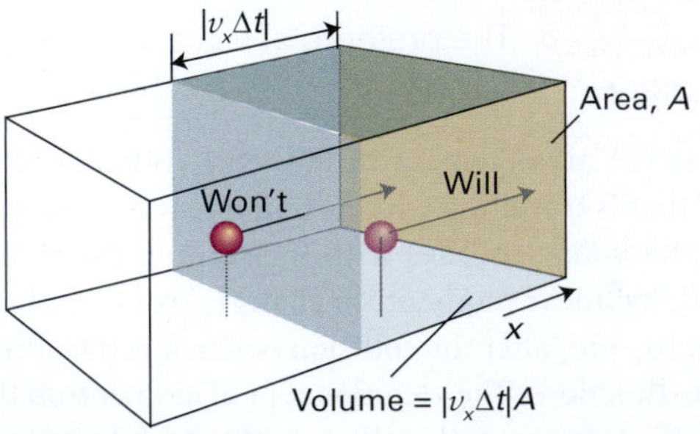

Figure 1B.2 A molecule will reach the wall on the right within an interval Δt if it is within a distance $v_x\Delta t$ of the wall and travelling to the right.

Next, to find the force, we calculate the rate of change of momentum, which is this change of momentum divided by the interval Δt during which it occurs:

$$\text{Rate of change of momentum} = \frac{nMAv_x^2}{V}$$

This rate of change of momentum is equal to the force (by Newton's second law of motion). It follows that the pressure, the force divided by the area, is

$$\text{Pressure} = \frac{nMv_x^2}{V}$$

Not all the molecules travel with the same velocity, so the detected pressure, p, is the average (denoted $\langle\ldots\rangle$) of the quantity just calculated:

$$p = \frac{nM\langle v_x^2\rangle}{V}$$

This expression already resembles the perfect gas equation of state.

To write an expression for the pressure in terms of the root-mean-square speed, v_{rms}, we begin by writing the speed of a single molecule, v, as $v^2 = v_x^2 + v_y^2 + v_z^2$. Because the root-mean-square speed is defined as $v_{rms} = \langle v^2\rangle^{1/2}$, it follows that

$$v_{rms}^2 = \langle v^2\rangle = \langle v_x^2\rangle + \langle v_y^2\rangle + \langle v_z^2\rangle$$

However, because the molecules are moving randomly, all three averages are the same. It follows that $v_{rms}^2 = \langle 3v_x^2\rangle$. Equation 1B.1 follows immediately by substituting $\langle v_x^2\rangle = \frac{1}{3}\langle v_{rms}^2\rangle$ into $p = nM\langle v_x^2\rangle/V$.

Equation 1B.1 is one of the key results of the kinetic model. We see that, if the root-mean-square speed of the molecules depends only on the temperature, then at constant temperature

$$pV = \text{constant}$$

which is the content of Boyle's law. Moreover, for eqn 1B.1 to be the equation of state of a perfect gas, its right-hand side must be equal to nRT. It follows that the root-mean-square speed of the molecules in a gas at a temperature T must be

$$v_{rms} = \left(\frac{3RT}{M}\right)^{1/2} \qquad \textit{Perfect gas} \quad \text{RMS speed} \quad (1B.3)$$

Brief illustration 1B.1 Molecular speeds

For N_2 molecules at 25 °C, we use $M = 28.02\ \text{g mol}^{-1}$, then

$$v_{rms} = \left\{\frac{3\times(8.3145\ \text{J K}^{-1}\ \text{mol}^{-1})\times(298\ \text{K})}{0.02802\ \text{kg mol}^{-1}}\right\}^{1/2} = 515\ \text{m s}^{-1}$$

Shortly we shall encounter the mean speed, v_{mean}, and the most probable speed v_{mp}; they are, respectively,

$$v_{\text{mean}}=\left(\frac{8}{3\pi}\right)^{1/2} v_{\text{rms}}=0.921\ldots\times(515\,\text{m s}^{-1})=475\,\text{m s}^{-1}$$

$$v_{\text{mp}}=\left(\frac{2}{3}\right)^{1/2} v_{\text{rms}}=0.816\ldots\times(515\,\text{m s}^{-1})=420\,\text{m s}^{-1}$$

Self-test 1B.1 Evaluate the root-mean-square speed of H_2 molecules at 25 °C.

Answer: $1.92\,\text{km s}^{-1}$

(b) The Maxwell–Boltzmann distribution of speeds

Equation 1B.2 is an expression for the mean square speed of molecules. However, in an actual gas the speeds of individual molecules span a wide range, and the collisions in the gas continually redistribute the speeds among the molecules. Before a collision, a molecule may be travelling rapidly, but after a collision it may be accelerated to a very high speed, only to be slowed again by the next collision. The fraction of molecules that have speeds in the range v to $v+dv$ is proportional to the width of the range, and is written $f(v)dv$, where $f(v)$ is called the **distribution of speeds**. Note that, in common with other distribution functions, $f(v)$ acquires physical significance only after it is multiplied by the range of speeds of interest. In the following *Justification* we show that the fraction of molecules that have a speed in the range v to $v+dv$ is $f(v)dv$, where

$$f(v)=4\pi\left(\frac{M}{2\pi RT}\right)^{3/2} v^2 e^{-Mv^2/2RT} \qquad \text{Perfect gas} \qquad \text{Maxwell–Boltzmann distribution} \qquad (1B.4)$$

The function $f(v)$ is called the **Maxwell–Boltzmann distribution of speeds**.

Justification 1B.2 The Maxwell–Boltzmann distribution of speeds

The Boltzmann distribution (*Foundations* B) implies that the fraction of molecules with velocity components v_x, v_y, and v_z is proportional to an exponential function of their kinetic energy: $f(v)=Ke^{-\varepsilon/kT}$, where K is a constant of proportionality. The kinetic energy is

$$\varepsilon=\tfrac{1}{2}mv_x^2+\tfrac{1}{2}mv_y^2+\tfrac{1}{2}mv_z^2$$

Therefore, we can use the relation $a^{x+y+z}=a^xa^ya^z$ to write

$$f(v)=Ke^{-(mv_x^2+mv_y^2+mv_z^2)/2kT}=Ke^{-mv_x^2/2kT}e^{-mv_y^2/2kT}e^{-mv_z^2/2kT}$$

The distribution factorizes into three terms, and we can write $f(v)=f(v_x)f(v_y)f(v_z)$ and $K=K_xK_yK_z$, with

$$f(v_x)=K_xe^{-mv_x^2/2kT}$$

and likewise for the other two axes.

To determine the constant K_x, we note that a molecule must have a velocity component somewhere in the range $-\infty<v_x<\infty$, so

$$\int_{-\infty}^{\infty} f(v_x)dv_x=1$$

Substitution of the expression for $f(v_x)$ then gives

$$1=K_x\int_{-\infty}^{\infty} e^{-mv_x^2/2kT}dv_x \overset{\text{Integral G.1}}{=} K_x\left(\frac{2\pi kT}{m}\right)^{1/2}$$

Therefore, $K_x=(m/2\pi kT)^{1/2}$ and at this stage we can write

$$f(v_x)=\left(\frac{m}{2\pi kT}\right)^{1/2} e^{-mv_x^2/2kT} \qquad (1B.5)$$

The probability that a molecule has a velocity in the range v_x to v_x+dv_x, v_y to v_y+dv_y, v_z to v_z+dv_z, is therefore

$$f(v_x)f(v_y)f(v_z)=\left(\frac{m}{2\pi kT}\right)^{3/2} e^{-mv_x^2/2kT}e^{-mv_y^2/2kT}e^{-mv_z^2/2kT}\times dv_xdv_ydv_z$$

$$=\left(\frac{m}{2\pi kT}\right)^{3/2} e^{-mv^2/2kT}dv_xdv_ydv_z$$

where $v^2=v_x^2+v_y^2+v_z^2$.

To evaluate the probability that the molecules have a speed in the range v to $v+dv$ regardless of direction we think of the three velocity components as defining three coordinates in 'velocity space', with the same properties as ordinary space except that the coordinates are labelled (v_x,v_y,v_z) instead of (x,y,z). Just as the volume element in ordinary space is $dxdydz$, so the volume element in velocity space is $dv_xdv_ydv_z$. The sum of all the volume elements in ordinary space that lie at a distance r from the centre is the volume of a spherical shell of radius r and thickness dr. That volume is the product of its surface area, $4\pi r^2$, and its thickness dr, and is therefore $4\pi r^2dr$. Similarly, the analogous volume in velocity space is the volume of a shell of radius v and thickness dv, namely $4\pi v^2dv$ (Fig. 1B.3). Now, because $f(v_x)f(v_y)f(v_z)$, the term in blue in the last equation, depends only on v^2, and has the same value everywhere in a shell of radius v, the total probability of the molecules possessing a speed in the range v to $v+dv$ is the product of the term in blue and the volume of the

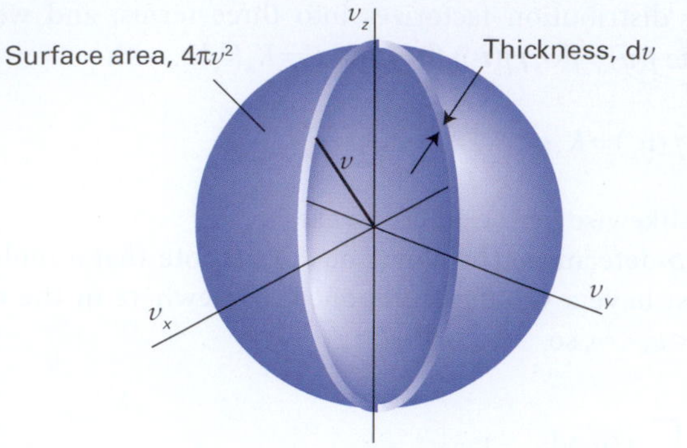

Figure 1B.3 To evaluate the probability that a molecule has a speed in the range v to $v+\mathrm{d}v$, we evaluate the total probability that the molecule will have a speed that is anywhere on the surface of a sphere of radius $v=(v_x^2+v_y^2+v_z^2)^{1/2}$ by summing the probabilities that it is in a volume element $\mathrm{d}v_x\mathrm{d}v_y\mathrm{d}v_z$ at a distance v from the origin.

shell of radius v and thickness dv. If this probability is written $f(v)\mathrm{d}v$, it follows that

$$f(v)\mathrm{d}v = 4\pi v^2 \mathrm{d}v\left(\frac{m}{2\pi kT}\right)^{3/2} \mathrm{e}^{-mv^2/2kT}$$

and $f(v)$ itself, after minor rearrangement, is

$$f(v) = 4\pi\left(\frac{m}{2\pi kT}\right)^{3/2} v^2\mathrm{e}^{-mv^2/2kT}$$

Because $m/k=M/R$, this expression is eqn 1B.4.

The important features of the Maxwell–Boltzmann distribution are as follows (and are shown pictorially in Fig. 1B.4):

Physical interpretation

- Equation 1B.4 includes a decaying exponential function (more specifically, a Gaussian function). Its presence implies that the fraction of molecules with very high speeds will be very small because e^{-x^2} becomes very small when x is large.
- The factor $M/2RT$ multiplying v^2 in the exponent is large when the molar mass, M, is large, so the exponential factor goes most rapidly towards zero when M is large. That is, heavy molecules are unlikely to be found with very high speeds.
- The opposite is true when the temperature, T, is high: then the factor $M/2RT$ in the exponent is small, so the exponential factor falls towards zero relatively slowly as v increases. In other words, a greater fraction of the molecules can be expected to have high speeds at high temperatures than at low temperatures.

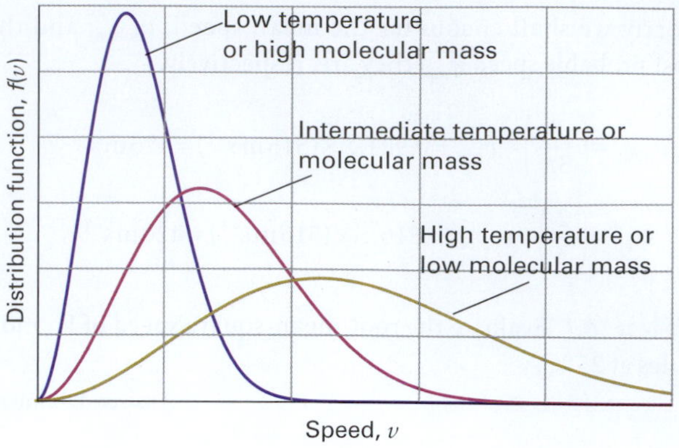

Figure 1B.4 The distribution of molecular speeds with temperature and molar mass. Note that the most probable speed (corresponding to the peak of the distribution) increases with temperature and with decreasing molar mass, and simultaneously the distribution becomes broader.

- A factor v^2 (the term before the e) multiplies the exponential. This factor goes to zero as v goes to zero, so the fraction of molecules with very low speeds will also be very small whatever their mass.
- The remaining factors (the term in parentheses in eqn 1B.4 and the 4π) simply ensure that, when we sum the fractions over the entire range of speeds from zero to infinity, then we get 1.

The Maxwell distribution has been verified experimentally. For example, molecular speeds can be measured directly with a velocity selector (Fig. 1B.5). The spinning discs have slits that permit the passage of only those molecules moving through them at the appropriate speed, and the number of molecules can be determined by collecting them at a detector.

(c) Mean values

Once we have the Maxwell–Boltzmann distribution, we can calculate the mean value of any power of the speed by evaluating the appropriate integral. For instance, to evaluate the

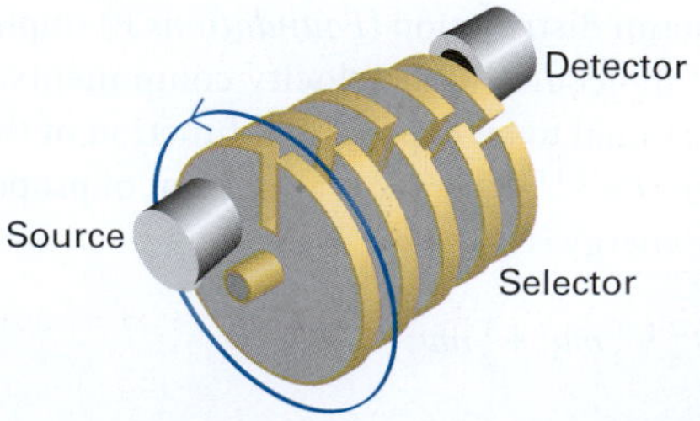

Figure 1B.5 A velocity selector. Only molecules travelling at speeds within a narrow range pass through the succession of slits as they rotate into position.

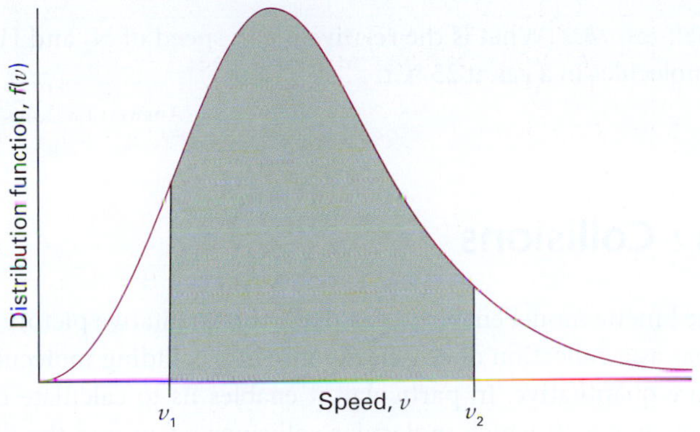

Figure 1B.6 To calculate the probability that a molecule will have a speed in the range v_1 to v_2, we integrate the distribution between those two limits; the integral is equal to the area of the curve between the limits, as shown shaded here.

fraction of molecules in the range v_1 to v_2 we evaluate the integral:

$$F(v_1,v_2)=\int_{v_1}^{v_2} f(v)\mathrm{d}v \qquad (1B.6)$$

This integral is the area under the graph of f as a function of v and, except in special cases, has to be evaluated numerically by using mathematical software (Fig. 1B.6). To evaluate the average value of v^n we calculate

$$\langle v^n\rangle=\int_0^{\infty} v^n f(v)\mathrm{d}v \qquad (1B.7)$$

In particular, integration with $n=2$ results in eqn 1B.3 for the mean square speed $\langle v^2\rangle$ of the molecules at a temperature T. We can conclude that the root-mean-square speed of the molecules of a gas is proportional to the square root of the temperature and inversely proportional to the square root of the molar mass. That is, the higher the temperature, the higher the root-mean-square speed of the molecules, and, at a given temperature, heavy molecules travel more slowly than light molecules. Sound waves are pressure waves, and for them to propagate the molecules of the gas must move to form regions of high and low pressure. Therefore, we should expect the root-mean-square speeds of molecules to be comparable to the speed of sound in air (340 m s^{-1}). As we have seen, the root-mean-square speed of N_2 molecules, for instance, is 515 m s^{-1} at 298 K.

Example 1B.1 Calculating the mean speed of molecules in a gas

Calculate the mean speed, v_{mean}, of N_2 molecules in air at 25 °C.

Method The mean speed is obtained by evaluating the integral

$$v_{\mathrm{mean}}=\int_0^{\infty} v f(v)\mathrm{d}v$$

with $f(v)$ given in eqn 1B.4. Either use mathematical software or use the standard integrals in the *Resource section*.

Answer The integral required is

$$v_{\mathrm{mean}}=4\pi\left(\frac{M}{2\pi RT}\right)^{3/2}\int_0^{\infty} v^3\mathrm{e}^{-mv^2/2kT}\mathrm{d}v$$

$$\overset{\text{Integral G.4}}{=} 4\pi\left(\frac{M}{2\pi RT}\right)^{3/2}\times\frac{1}{2}\left(\frac{2RT}{M}\right)^{1/2}=\left(\frac{8RT}{\pi M}\right)^{1/2}$$

Substitution of the data then gives

$$v_{\mathrm{mean}}=\left(\frac{8\times(8.3145\,\mathrm{J\,K^{-1}\,mol^{-1}})\times(298\,\mathrm{K})}{\pi\times(28.02\times10^{-3}\,\mathrm{kg\,mol^{-1}})}\right)^{1/2}=475\,\mathrm{m\,s^{-1}}$$

We have used 1 J = 1 kg m^2 s^{-2} (the difference from the earlier value of 474 is due to rounding effects in that calculation; this value is more accurate).

Self-test 1B.2 Evaluate the root-mean-square speed of the molecules by integration. Use mathematical software or use a standard integral in the *Resource section*.

Answer: $v_{\mathrm{rms}}=(3RT/M)^{1/2}=515\,\mathrm{m\,s^{-1}}$

As shown in *Example* 1B.1, we can use the Maxwell–Boltzmann distribution to evaluate the **mean speed**, v_{mean}, of the molecules in a gas:

$$v_{\mathrm{mean}}=\left(\frac{8RT}{\pi M}\right)^{1/2}=\left(\frac{8}{3\pi}\right)^{1/2}v_{\mathrm{rms}} \qquad \textit{Perfect gas} \quad \text{Mean speed} \qquad (1B.8)$$

We can identify the **most probable speed**, v_{mp}, from the location of the peak of the distribution:

$$v_{\mathrm{mp}}=\left(\frac{2RT}{M}\right)^{1/2}=\left(\frac{2}{3}\right)^{1/2}v_{\mathrm{rms}} \qquad \textit{Perfect gas} \quad \text{Most probable speed} \qquad (1B.9)$$

The location of the peak of the distribution is found by differentiating $f(v)$ with respect to v and looking for the value of v at which the derivative is zero (other than at $v=0$ and $v=\infty$); see Problem 1B.3. Figure 1B.7 summarizes these results and some numerical values were calculated in *Brief illustration* 1B.1.

The **mean relative speed**, v_{rel}, the mean speed with which one molecule approaches another of the same kind, can also be calculated from the distribution:

$$v_{\mathrm{rel}}=2^{1/2}v_{\mathrm{mean}} \qquad \textit{Perfect gas, identical molecules} \quad \text{Mean relative speed} \qquad (1B.10a)$$

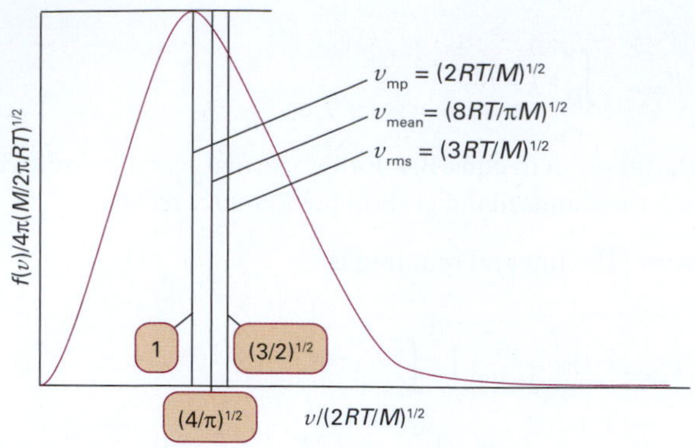

Figure 1B.7 A summary of the conclusions that can be deduced form the Maxwell distribution for molecules of molar mass M at a temperature T: v_{mp} is the most probable speed, v_{mean} is the mean speed, and v_{rms} is the root-mean-square speed.

This result is much harder to derive, but the diagram in Fig. 1B.8 should help to show that it is plausible. For the relative mean speed of two dissimilar molecules of masses m_A and m_B:

$$v_{rel} = \left(\frac{8kT}{\pi\mu}\right)^{1/2} \qquad \mu = \frac{m_A m_B}{m_A + m_B}$$ *Perfect gas* Mean relative speed (1B.10b)

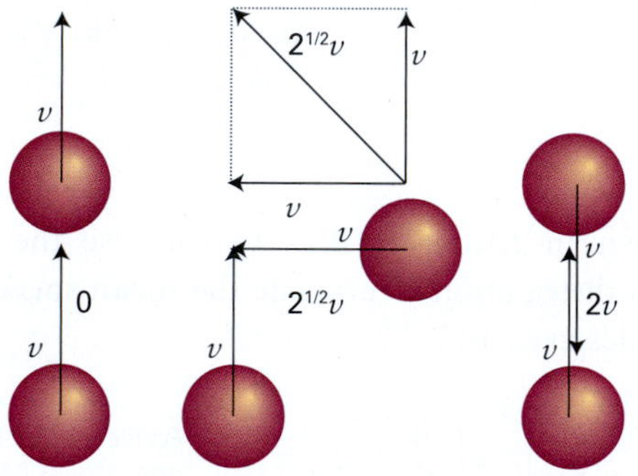

Figure 1B.8 A simplified version of the argument to show that the mean relative speed of molecules in a gas is related to the mean speed. When the molecules are moving in the same direction, the mean relative speed is zero; it is $2v$ when the molecules are approaching each other. A typical mean direction of approach is from the side, and the mean speed of approach is then $2^{1/2}v$. The last direction of approach is the most characteristic, so the mean speed of approach can be expected to be about $2^{1/2}v$. This value is confirmed by more detailed calculation.

Brief illustration 1B.2 Relative molecular speeds

We have already seen (in *Brief illustration* 1B.1) that the rms speed of N_2 molecules at 25 °C is 515 m s^{-1}. It follows from eqn 1B.10a that their relative mean speed is

$$v_{rel} = 2^{1/2} \times (515\,\mathrm{m\,s^{-1}}) = 728\,\mathrm{m\,s^{-1}}$$

Self-test 1B.3 What is the relative mean speed of N_2 and H_2 molecules in a gas at 25 °C?

Answer: 1.83 km s^{-1}

1B.2 Collisions

The kinetic model enables us to make the qualitative picture of a gas as a collection of ceaselessly moving, colliding molecules more quantitative. In particular, it enables us to calculate the frequency with which molecular collisions occur and the distance a molecule travels on average between collisions.

(a) The collision frequency

Although the kinetic-molecular theory assumes that the molecules are point-like, we can count a 'hit' whenever the centres of two molecules come within a distance d of each other, where d, the collision diameter, is of the order of the actual diameters of the molecules (for impenetrable hard spheres d is the diameter). As we show in the following *Justification*, we can use the kinetic model to deduce that the **collision frequency**, z, the number of collisions made by one molecule divided by the time interval during which the collisions are counted, when there are N molecules in a volume V is

$$z = \sigma v_{rel} \mathcal{N}$$ *Perfect gas* Collision frequency (1B.11a)

with $\mathcal{N} = N/V$, the number density, and v_{rel} given by eqn 1B.10. The area $\sigma = \pi d^2$ is called the **collision cross-section** of the molecules. Some typical collision cross-sections are given in Table 1B.1. In terms of the pressure (as is also shown in the following *Justification*),

$$z = \frac{\sigma v_{rel} p}{kT}$$ *Perfect gas* Collision frequency (1B.11b)

Table 1B.1* Collision cross-sections, σ/nm^2

	σ/nm^2
C_6H_6	0.88
CO_2	0.52
He	0.21
N_2	0.43

* More values are given in the *Resource section*.

Justification 1B.3 The collision frequency according to the kinetic model

Consider the positions of all the molecules except one to be frozen. Then note what happens as one mobile molecule travels through the gas with a mean relative speed v_{rel} for a

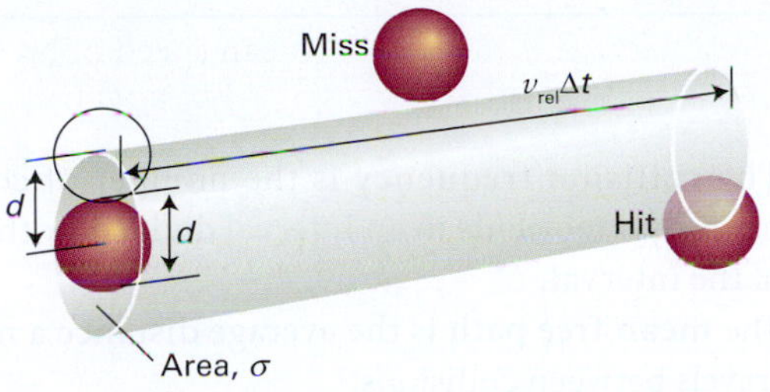

Figure 1B.9 The calculation of the collision frequency and the mean free path in the kinetic theory of gases.

time Δt. In doing so it sweeps out a 'collision tube' of cross-sectional area $\sigma=\pi d^2$ and length $v_{rel}\Delta t$ and therefore of volume $\sigma v_{rel}\Delta t$ (Fig. 1B.9). The number of stationary molecules with centres inside the collision tube is given by the volume of the tube multiplied by the number density $\mathcal{N}=N/V$, and is $\mathcal{N}\sigma v_{rel}\Delta t$. The number of hits scored in the interval Δt is equal to this number, so the number of collisions divided by the time interval is $\mathcal{N}\sigma v_{rel}$, which is eqn 1B.11a. The expression in terms of the pressure of the gas is obtained by using the perfect gas equation to write

$$\mathcal{N}=\frac{N}{V}=\frac{nN_A}{V}=\frac{nN_A}{nRT/p}=\frac{p}{kT}$$

Equation 1B.11a shows that, at constant volume, the collision frequency increases with increasing temperature. Equation 1B.11b shows that, at constant temperature, the collision frequency is proportional to the pressure. Such a proportionality is plausible because the greater the pressure, the greater the number density of molecules in the sample, and the rate at which they encounter one another is greater even though their average speed remains the same.

Brief illustration 1B.3 Molecular collisions

For an N_2 molecule in a sample at 1.00 atm (101 kPa) and 25 °C, from *Brief illustration* 1B.2 we know that $v_{rel}=728\,\mathrm{m\,s^{-1}}$. Therefore, from eqn 1B.11b, and taking $\sigma=0.45\,\mathrm{nm^2}$ (corresponding to $0.45\times10^{-18}\,\mathrm{m^2}$) from Table 1B.1,

$$z=\frac{(0.43\times10^{-18}\,\mathrm{m^2})\times(728\,\mathrm{m\,s^{-1}})\times(1.01\times10^5\,\mathrm{Pa})}{(1.381\times10^{-23}\,\mathrm{J\,K^{-1}})\times(298\,\mathrm{K})}$$
$$=7.7\times10^9\,\mathrm{s^{-1}}$$

so a given molecule collides about 8×10^9 times each second. We are beginning to appreciate the timescale of events in gases.

Self-test 1B.4 Evaluate the collision frequency between H_2 molecules in a gas under the same conditions.

Answer: $4.1\times10^9\,\mathrm{s^{-1}}$

(b) The mean free path

Once we have the collision frequency, we can calculate the **mean free path**, λ (lambda), the average distance a molecule travels between collisions. If a molecule collides with a frequency z, it spends a time $1/z$ in free flight between collisions, and therefore travels a distance $(1/z)v_{rel}$. It follows that the mean free path is

$$\lambda=\frac{v_{rel}}{z} \qquad \textit{Perfect gas} \quad \text{Mean free path} \qquad (1B.12)$$

Substitution of the expression for z in eqn 1B.11b gives

$$\lambda=\frac{kT}{\sigma p} \qquad \textit{Perfect gas} \quad \text{Mean free path} \qquad (1B.13)$$

Doubling the pressure reduces the mean free path by half.

Brief illustration 1B.4 The mean free path

In *Brief illustration* 1B.2 we noted that $v_{rel}=728\,\mathrm{m\,s^{-1}}$ for N_2 molecules at 25 °C, and in *Brief illustration* 1B.3 that $z=7.7\times10^9\,\mathrm{s^{-1}}$ when the pressure is 1.00 atm. Under these circumstances, the mean free path of N_2 molecules is

$$\lambda=\frac{728\,\mathrm{m\,s^{-1}}}{7.7\times10^9\,\mathrm{s^{-1}}}=9.5\times10^{-8}\,\mathrm{m}$$

or 95 nm, about 10^3 molecular diameters.

Self-test 1B.5 Evaluate the mean free path of benzene molecules at 25 °C in a sample where the pressure is 0.10 atm.

Answer: 460 nm

Although the temperature appears in eqn 1B.13, in a sample of constant volume, the pressure is proportional to T, so T/p remains constant when the temperature is increased. Therefore, the mean free path is independent of the temperature in a sample of gas in a container of fixed volume: the distance between collisions is determined by the number of molecules present in the given volume, not by the speed at which they travel.

In summary, a typical gas (N_2 or O_2) at 1 atm and 25 °C can be thought of as a collection of molecules travelling with a mean speed of about $500\,\mathrm{m\,s^{-1}}$. Each molecule makes a collision within about 1 ns, and between collisions it travels about 10^3 molecular diameters. The kinetic model of gases is valid and the gas behaves nearly perfectly if the diameter of the molecules is much smaller than the mean free path ($d\ll\lambda$), for then the molecules spend most of their time far from one another.

Checklist of concepts

- ☐ 1. The **kinetic model** of a gas considers only the contribution to the energy from the kinetic energies of the molecules.
- ☐ 2. Important results from the model include expressions for the pressure and the **root-mean-square speed**.
- ☐ 3. The **Maxwell–Boltzmann distribution of speeds** gives the fraction of molecules that have speeds in a specified range.
- ☐ 4. The **collision frequency** is the number of collisions made by a molecule in an interval divided by the length of the interval.
- ☐ 5. The **mean free path** is the average distance a molecule travels between collisions.

Checklist of equations

Property	Equation	Comment	Equation number
Pressure of a perfect gas from the kinetic model	$pV=\frac{1}{3}nMv_{\text{rms}}^2$	Kinetic model	1B.1
Maxwell–Boltzmann distribution of speeds	$f(v)=4\pi(M/2\pi RT)^{3/2}v^2e^{-Mv^2/2RT}$		1B.4
Root-mean-square speed in a perfect gas	$v_{\text{rms}}=(3RT/M)^{1/2}$		1B.3
Mean speed in a perfect gas	$v_{\text{mean}}=(8RT/\pi M)^{1/2}$		1B.8
Most probable speed in a perfect gas	$v_{\text{mp}}=(2RT/M)^{1/2}$		1B.9
Mean relative speed in a perfect gas	$v_{\text{rel}}=(8kT/\pi\mu)^{1/2}$ $\mu=m_A m_B/(m_A+m_B)$		1B.10
The collision frequency in a perfect gas	$z=\sigma v_{\text{rel}}p/kT$, $\sigma=\pi d^2$		1B.11
Mean free path in a perfect gas	$\lambda=v_{\text{rel}}/z$		1B.12

1C Real gases

Contents

➤ Why do you need to know this material?

Actual gases, so-called 'real gases', differ from perfect gases and it is important to be able to discuss their properties. Moreover, the deviations from perfect behaviour give insight into the nature of the interactions between molecules. Accounting for these interactions is also an introduction to the technique of model building in physical chemistry.

➤ What is the key idea?

Attractions and repulsions between gas molecules account for modifications to the isotherms of a gas and account for critical behaviour.

➤ What do you need to know already?

This Topic builds on and extends the discussion of perfect gases in Topic 1A. The principal mathematical technique employed is differentiation to identify a point of inflexion of a curve.

Real gases do not obey the perfect gas law exactly except in the limit of $p \to 0$. Deviations from the law are particularly important at high pressures and low temperatures, especially when a gas is on the point of condensing to liquid.

1C.1 Deviations from perfect behaviour

Real gases show deviations from the perfect gas law because molecules interact with one another. A point to keep in mind is that repulsive forces between molecules assist expansion and attractive forces assist compression.

Repulsive forces are significant only when molecules are almost in contact: they are short-range interactions, even on a scale measured in molecular diameters (Fig. 1C.1). Because they are short-range interactions, repulsions can be expected to be important only when the average separation of the molecules is small. This is the case at high pressure, when many molecules occupy a small volume. On the other hand, attractive intermolecular forces have a relatively long range and are effective over several molecular diameters. They are important when the molecules are fairly close together but not necessarily touching (at the intermediate separations in Fig. 1C.1).

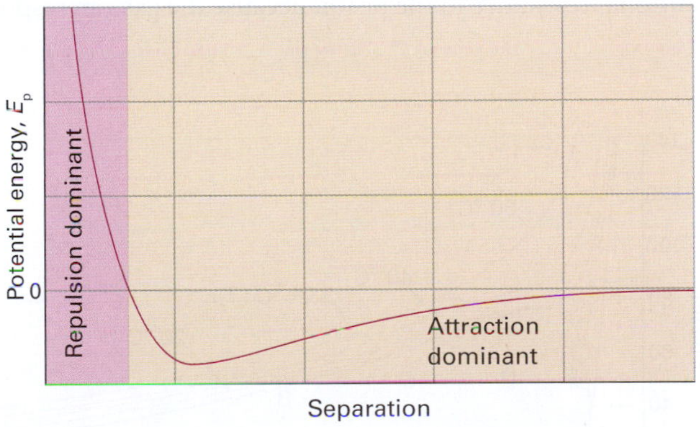

Figure 1C.1 The variation of the potential energy of two molecules on their separation. High positive potential energy (at very small separations) indicates that the interactions between them are strongly repulsive at these distances. At intermediate separations, where the potential energy is negative, the attractive interactions dominate. At large separations (on the right) the potential energy is zero and there is no interaction between the molecules.

Attractive forces are ineffective when the molecules are far apart (well to the right in Fig. 1C.1). Intermolecular forces are also important when the temperature is so low that the molecules travel with such low mean speeds that they can be captured by one another.

The consequences of these interactions are shown by shapes of experimental isotherms (Fig. 1C.2). At low pressures, when the sample occupies a large volume, the molecules are so far apart for most of the time that the intermolecular forces play no significant role, and the gas behaves virtually perfectly. At moderate pressures, when the average separation of the molecules is only a few molecular diameters, the attractive forces dominate the repulsive forces. In this case, the gas can be expected to be more compressible than a perfect gas because the forces help to draw the molecules together. At high pressures, when the average separation of the molecules is small, the repulsive forces dominate and the gas can be expected to be less compressible because now the forces help to drive the molecules apart.

Consider what happens when we compress (reduce the volume of) a sample of gas initially in the state marked A in Fig. 1C.2 at constant temperature by pushing in a piston. Near A, the pressure of the gas rises in approximate agreement with Boyle's law. Serious deviations from that law begin to appear when the volume has been reduced to B.

At C (which corresponds to about 60 atm for carbon dioxide), all similarity to perfect behaviour is lost, for suddenly the piston slides in without any further rise in pressure: this stage is represented by the horizontal line CDE. Examination of the contents of the vessel shows that just to the left of C a liquid appears, and there are two phases separated by a sharply defined surface. As the volume is decreased from C through D to E, the amount of liquid increases. There is no additional resistance to the piston because the gas can respond by condensing. The pressure corresponding to the line CDE, when both liquid and vapour are present in equilibrium, is called the **vapour pressure** of the liquid at the temperature of the experiment.

At E, the sample is entirely liquid and the piston rests on its surface. Any further reduction of volume requires the exertion of considerable pressure, as is indicated by the sharply rising line to the left of E. Even a small reduction of volume from E to F requires a great increase in pressure.

(a) The compression factor

As a first step in making these observations quantitative we introduce the **compression factor**, Z, the ratio of the measured molar volume of a gas, $V_m = V/n$, to the molar volume of a perfect gas, V_m°, at the same pressure and temperature:

$$Z = \frac{V_m}{V_m^\circ} \qquad \textit{Definition} \quad \text{Compression factor} \qquad (1C.1)$$

Because the molar volume of a perfect gas is equal to RT/p, an equivalent expression is $Z = RT/pV_m^\circ$, which we can write as

$$pV_m = RTZ \qquad (1C.2)$$

Because for a perfect gas $Z=1$ under all conditions, deviation of Z from 1 is a measure of departure from perfect behaviour.

Some experimental values of Z are plotted in Fig. 1C.3. At very low pressures, all the gases shown have $Z \approx 1$ and behave nearly perfectly. At high pressures, all the gases have $Z > 1$, signifying that they have a larger molar volume than a perfect gas. Repulsive forces are now dominant. At intermediate pressures, most gases have $Z < 1$, indicating that the attractive forces are reducing the molar volume relative to that of a perfect gas.

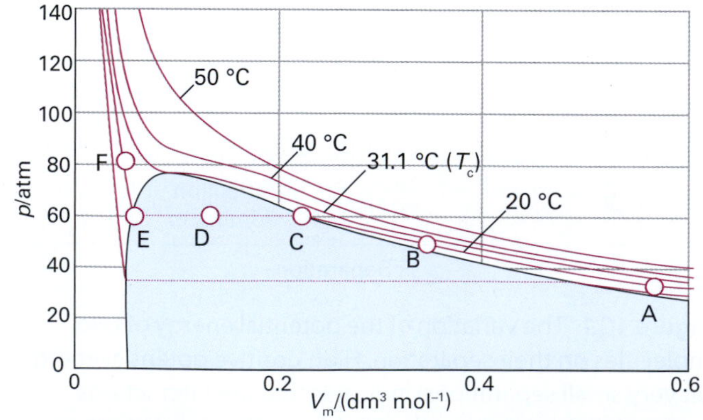

Figure 1C.2 Experimental isotherms of carbon dioxide at several temperatures. The 'critical isotherm', the isotherm at the critical temperature, is at 31.1 °C.

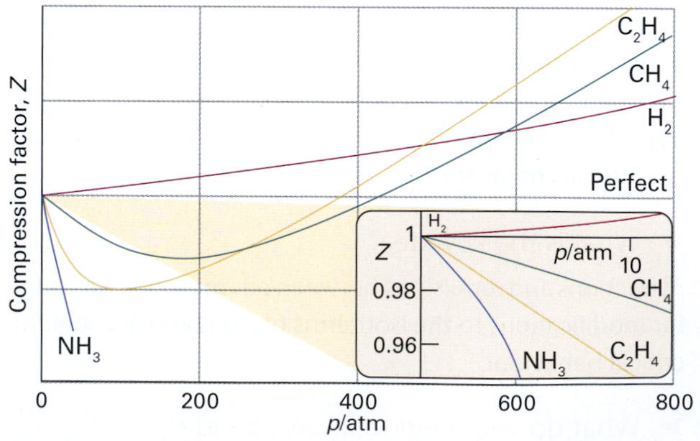

Figure 1C.3 The variation of the compression factor, Z, with pressure for several gases at 0 °C. A perfect gas has $Z=1$ at all pressures. Notice that, although the curves approach 1 as $p \to 0$, they do so with different slopes.

Brief illustration 1C.1 The compression factor

The molar volume of a perfect gas at 500 K and 100 bar is $V_m^\circ = 0.416\,dm^3\,mol^{-1}$. The molar volume of carbon dioxide under the same conditions is $V_m = 0.366\,dm^3\,mol^{-1}$. It follows that at 500 K

$$Z = \frac{0.366\,dm^3\,mol^{-1}}{0.416\,dm^3\,mol^{-1}} = 0.880$$

The fact that $Z < 1$ indicates that attractive forces dominate repulsive forces under these conditions.

Self-test 1C.1 The mean molar volume of air at 60 bar and 400 K is $0.9474\,dm^3\,mol^{-1}$. Are attractions or repulsions dominant?

Answer: Repulsions

(b) Virial coefficients

Now we relate Z to the experimental isotherms in Fig. 1C.2. At large molar volumes and high temperatures the real-gas isotherms do not differ greatly from perfect-gas isotherms. The small differences suggest that the perfect gas law $pV_m = RT$ is in fact the first term in an expression of the form

$$pV_m = RT(1 + B'p + C'p^2 + \cdots) \quad (1C.3a)$$

This expression is an example of a common procedure in physical chemistry, in which a simple law that is known to be a good first approximation (in this case $pV_m = RT$) is treated as the first term in a series in powers of a variable (in this case p). A more convenient expansion for many applications is

$$pV_m = RT\left(1 + \frac{B}{V_m} + \frac{C}{V_m^2} + \cdots\right) \quad \text{Virial equation of state} \quad (1C.3b)$$

These two expressions are two versions of the **virial equation of state**.[1] By comparing the expression with eqn 1C.2 we see that the term in parentheses in eqn 1C.3b is just the compression factor, Z.

The coefficients B, C, …, which depend on the temperature, are the second, third, ... **virial coefficients** (Table 1C.1); the first virial coefficient is 1. The third virial coefficient, C, is usually less important than the second coefficient, B, in the sense that at typical molar volumes $C/V_m^2 \ll B/V_m$. The values of the virial coefficients of a gas are determined from measurements of its compression factor.

[1] The name comes from the Latin word for force. The coefficients are sometimes denoted B_2, B_3, ….

Table 1C.1* Second virial coefficients, $B/(cm^3\,mol^{-1})$

	Temperature	
	273 K	600 K
Ar	−21.7	11.9
CO_2	−149.7	−12.4
N_2	−10.5	21.7
Xe	−153.7	−19.6

* More values are given in the *Resource section*.

Brief illustration 1C.2 The virial equation of state

To use eqn 1C.3b (up to the B term), to calculate the pressure exerted at 100 K by 0.104 mol O_2(g) in a vessel of volume 0.225 dm³, we begin by calculating the molar volume:

$$V_m = \frac{V}{n_{O_2}} = \frac{0.225\,dm^3}{0.104\,mol} = 2.16\,dm^3\,mol^{-1} = 2.16\times10^{-3}\,m^3\,mol^{-1}$$

Then, by using the value of B found in Table 1C.1 of the *Resource section*,

$$p = \frac{RT}{V_m}\left(1 + \frac{B}{V_m}\right)$$
$$= \frac{(8.3145\,J\,mol^{-1}\,K^{-1})\times(100\,K)}{2.16\times10^{-3}\,m^3\,mol^{-1}}\left(1 - \frac{1.975\times10^{-4}\,m^3\,mol^{-1}}{2.16\times10^{-3}\,m^3\,mol^{-1}}\right)$$
$$= 3.50\times10^5\,Pa, \text{ or } 350\,kPa$$

where we have used $1\,Pa = 1\,J\,m^{-3}$. The perfect gas equation of state would give the calculated pressure as 385 kPa, or 10 per cent higher than the value calculated by using the virial equation of state. The deviation is significant because under these conditions $B/V_m \approx 0.1$ which is not negligible relative to 1.

Self-test 1C.2 What pressure would 4.56 g of nitrogen gas in a vessel of volume 2.25 dm³ exert at 273 K if it obeyed the virial equation of state?

Answer: 104 kPa

An important point is that although the equation of state of a real gas may coincide with the perfect gas law as $p \to 0$, not all its properties necessarily coincide with those of a perfect gas in that limit. Consider, for example, the value of dZ/dp, the slope of the graph of compression factor against pressure. For a perfect gas $dZ/dp = 0$ (because $Z = 1$ at all pressures), but for a real gas from eqn 1C.3a we obtain

$$\frac{dZ}{dp} = B' + 2pC' + \cdots \to B' \text{ as } p \to 0 \quad (1C.4a)$$

However, B' is not necessarily zero, so the slope of Z with respect to p does not necessarily approach 0 (the perfect gas

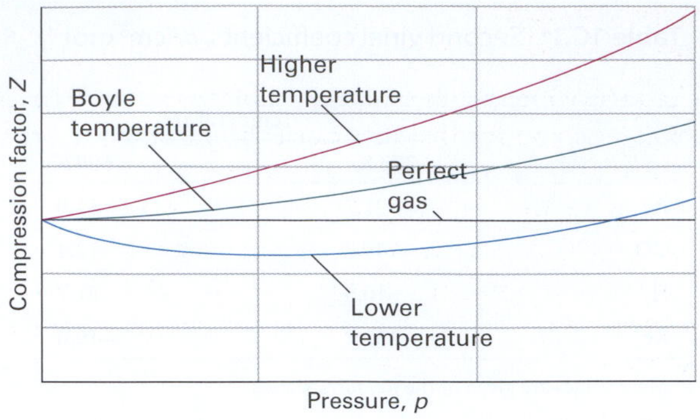

Figure 1C.4 The compression factor, Z, approaches 1 at low pressures, but does so with different slopes. For a perfect gas, the slope is zero, but real gases may have either positive or negative slopes, and the slope may vary with temperature. At the Boyle temperature, the slope is zero and the gas behaves perfectly over a wider range of conditions than at other temperatures.

value), as we can see in Fig. 1C.4. Because several physical properties of gases depend on derivatives, the properties of real gases do not always coincide with the perfect gas values at low pressures. By a similar argument,

$$\frac{dZ}{d(1/V_m)} \to B \text{ as } V_m \to \infty \qquad \text{(1C.4b)}$$

Because the virial coefficients depend on the temperature, there may be a temperature at which $Z \to 1$ with zero slope at low pressure or high molar volume (as in Fig. 1C.4). At this temperature, which is called the **Boyle temperature**, T_B, the properties of the real gas do coincide with those of a perfect gas as $p \to 0$. According to eqn 1C.4b, Z has zero slope as $p \to 0$ if $B=0$, so we can conclude that $B=0$ at the Boyle temperature. It then follows from eqn 1C.3 that $pV_m \approx RT_B$ over a more extended range of pressures than at other temperatures because the first term after 1 (that is, B/V_m) in the virial equation is zero and C/V_m^2 and higher terms are negligibly small. For helium $T_B = 22.64$ K; for air $T_B = 346.8$ K; more values are given in Table 1C.2.

(c) Critical constants

The isotherm at the temperature T_c (304.19 K, or 31.04 °C for CO_2) plays a special role in the theory of the states of matter.

Table 1C.2* Critical constants of gases

	p_c/atm	V_c/(cm^3 mol^{-1})	T_c/K	Z_c	T_B/K
Ar	48.0	75.3	150.7	0.292	411.5
CO_2	72.9	94.0	304.2	0.274	714.8
He	2.26	57.8	5.2	0.305	22.64
O_2	50.14	78.0	154.8	0.308	405.9

* More values are given in the *Resource section*.

An isotherm slightly below T_c behaves as we have already described: at a certain pressure, a liquid condenses from the gas and is distinguishable from it by the presence of a visible surface. If, however, the compression takes place at T_c itself, then a surface separating two phases does not appear and the volumes at each end of the horizontal part of the isotherm have merged to a single point, the **critical point** of the gas. The temperature, pressure, and molar volume at the critical point are called the **critical temperature**, T_c, **critical pressure**, p_c, and **critical molar volume**, V_c, of the substance. Collectively, p_c, V_c, and T_c are the **critical constants** of a substance (Table 1C.2).

At and above T_c, the sample has a single phase which occupies the entire volume of the container. Such a phase is, by definition, a gas. Hence, the liquid phase of a substance does not form above the critical temperature. The single phase that fills the entire volume when $T > T_c$ may be much denser that we normally consider typical of gases, and the name **supercritical fluid** is preferred.

Brief illustration 1C.3 The critical temperature

The critical temperature of oxygen signifies that it is impossible to produce liquid oxygen by compression alone if its temperature is greater than 155 K. To liquefy oxygen—to obtain a fluid phase that does not occupy the entire volume—the temperature must first be lowered to below 155 K, and then the gas compressed isothermally.

Self-test 1C.3 Under which conditions can liquid nitrogen be formed by the application of pressure?

Answer: At $T < 126$ K

1C.2 The van der Waals equation

We can draw conclusions from the virial equations of state only by inserting specific values of the coefficients. It is often useful to have a broader, if less precise, view of all gases. Therefore, we introduce the approximate equation of state suggested by J.D. van der Waals in 1873. This equation is an excellent example of an expression that can be obtained by thinking scientifically about a mathematically complicated but physically simple problem; that is, it is a good example of 'model building'.

(a) Formulation of the equation

The **van der Waals equation** is

$$p = \frac{nRT}{V - nb} - a\frac{n^2}{V^2} \qquad \text{Van der Waals equation of state} \qquad \text{(1C.5a)}$$

and a derivation is given in the following *Justification*. The equation is often written in terms of the molar volume $V_m = V/n$ as

$$p = \frac{RT}{V_m - b} - \frac{a}{V_m^2} \qquad (1C.5b)$$

The constants a and b are called the **van der Waals coefficients**. As can be understood from the following *Justification*, a represents the strength of attractive interactions and b that of the repulsive interactions between the molecules. They are characteristic of each gas but independent of the temperature (Table 1C.3). Although a and b are not precisely defined molecular properties, they correlate with physical properties such as critical temperature, vapour pressure, and enthalpy of vaporization that reflect the strength of intermolecular interactions. Correlations have also been sought where intermolecular forces might play a role. For example, the potency of certain general anaesthetics shows a correlation in the sense that a higher activity is observed with lower values of a (Fig. 1C.5).

Table 1C.3* van der Waals coefficients

	$a/(\text{atm dm}^6\,\text{mol}^{-2})$	$b/(10^{-2}\,\text{dm}^3\,\text{mol}^{-1})$
Ar	1.337	3.20
CO_2	3.610	4.29
He	0.0341	2.38
Xe	4.137	5.16

* More values are given in the *Resource section*.

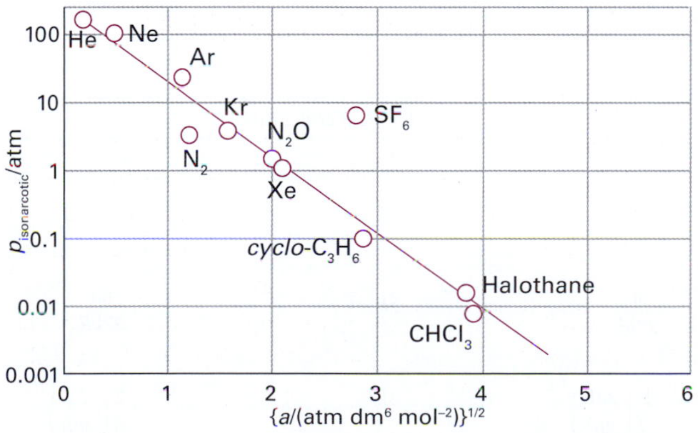

Figure 1C.5 The correlation of the effectiveness of a gas as an anaesthetic and the van der Waals parameter a. (Based on R.J. Wulf and R.M. Featherstone, *Anesthesiology* **18**, 97 (1957).) The isonarcotic pressure is the pressure required to bring about the same degree of anaesthesia.

Justification 1C.1 The van der Waals equation of state

The repulsive interactions between molecules are taken into account by supposing that they cause the molecules to behave as small but impenetrable spheres. The non-zero volume of the molecules implies that instead of moving in a volume V they are restricted to a smaller volume $V-nb$, where nb is approximately the total volume taken up by the molecules themselves. This argument suggests that the perfect gas law $p=nRT/V$ should be replaced by

$$p = \frac{nRT}{V-nb}$$

when repulsions are significant. To calculate the excluded volume we note that the closest distance of two hard-sphere molecules of radius r, and volume $V_{\text{molecule}} = \frac{4}{3}\pi r^3$, is $2r$, so the volume excluded is $\frac{4}{3}\pi(2r)^3$ or $8V_{\text{molecule}}$. The volume excluded per molecule is one-half this volume, or $4V_{\text{molecule}}$, so $b \approx 4V_{\text{molecule}}N_A$.

The pressure depends on both the frequency of collisions with the walls and the force of each collision. Both the frequency of the collisions and their force are reduced by the attractive interaction, which act with a strength proportional to the molar concentration, n/V, of molecules in the sample. Therefore, because both the frequency and the force of the collisions are reduced by the attractive interactions, the pressure is reduced in proportion to the square of this concentration. If the reduction of pressure is written as $a(n/V)^2$, where a is a positive constant characteristic of each gas, the combined effect of the repulsive and attractive forces is the van der Waals equation of state as expressed in eqn 1C.5.

In this *Justification* we have built the van der Waals equation using vague arguments about the volumes of molecules and the effects of forces. The equation can be derived in other ways, but the present method has the advantage that it shows how to derive the form of an equation out of general ideas. The derivation also has the advantage of keeping imprecise the significance of the coefficients a and b: they are much better regarded as empirical parameters that represent attractions and repulsions, respectively, rather than as precisely defined molecular properties.

Example 1C.1 Using the van der Waals equation to estimate a molar volume

Estimate the molar volume of CO_2 at 500 K and 100 atm by treating it as a van der Waals gas.

Method We need to find an expression for the molar volume by solving the van der Waals equation, eqn 1C.5b. To do so, we multiply both sides of the equation by $(V_m - b)V_m^2$, to obtain

$$(V_m - b)V_m^2\, p = RTV_m^2 - (V_m - b)a$$

Then, after division by p, collect powers of V_m to obtain

$$V_m^3 - \left(b + \frac{RT}{p}\right)V_m^2 + \left(\frac{a}{p}\right)V_m = 0ab/p$$

Although closed expressions for the roots of a cubic equation can be given, they are very complicated. Unless analytical solutions are essential, it is usually more expedient to solve such equations with commercial software; graphing calculators can also be used to help identify the acceptable root.

Answer According to Table 1C.3, $a = 3.592\,\mathrm{dm^6\,atm\,mol^{-2}}$ and $b = 4.267 \times 10^{-2}\,\mathrm{dm^3\,mol^{-1}}$. Under the stated conditions, $RT/p = 0.410\,\mathrm{dm^3\,mol^{-1}}$. The coefficients in the equation for V_m are therefore

$$b + RT/p = 0.453\,\mathrm{dm^3\,mol^{-1}}$$
$$a/p = 3.61 \times 10^{-2}(\mathrm{dm^3\,mol^{-1}})^2$$
$$ab/p = 1.55 \times 10^{-3}(\mathrm{dm^3\,mol^{-1}})^3$$

Therefore, on writing $x = V_m/(\mathrm{dm^3\,mol^{-1}})$, the equation to solve is

$$x^3 - 0.453x^2 + (3.61 \times 10^{-2})x - (1.55 \times 10^{-3}) = 0$$

The acceptable root is $x = 0.366$ (Fig. 1C.6), which implies that $V_m = 0.366\,\mathrm{dm^3\,mol^{-1}}$. For a perfect gas under these conditions, the molar volume is $0.410\,\mathrm{dm^3\,mol^{-1}}$.

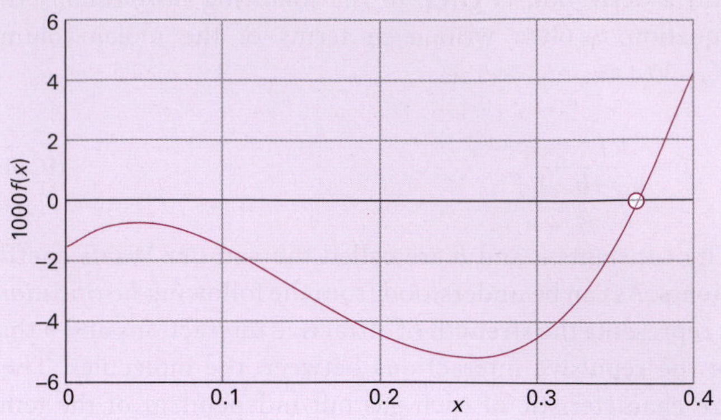

Figure 1C.6 The graphical solution of the cubic equation for *V* in *Example* 1C.1.

Self-test 1C.4 Calculate the molar volume of argon at 100 °C and 100 atm on the assumption that it is a van der Waals gas.

Answer: $0.298\,\mathrm{dm^3\,mol^{-1}}$

(b) The features of the equation

We now examine to what extent the van der Waals equation predicts the behaviour of real gases. It is too optimistic to expect a single, simple expression to be the true equation of state of all substances, and accurate work on gases must resort to the virial equation, use tabulated values of the coefficients at various temperatures, and analyse the systems numerically. The advantage of the van der Waals equation, however, is that it is analytical (that is, expressed symbolically) and allows us to draw some general conclusions about real gases. When the equation fails we must use one of the other equations of state that have been proposed (some are listed in Table 1C.4), invent a new one, or go back to the virial equation.

Table 1C.4 Selected equations of state

			Critical constants		
	Equation	Reduced form*	p_c	V_c	T_c
Perfect gas	$p = \dfrac{nRT}{V}$				
van der Waals	$p = \dfrac{nRT}{V - nb} - \dfrac{n^2a}{V^2}$	$p_r = \dfrac{8T_r}{3V_r - 1} - \dfrac{3}{V_r^2}$	$\dfrac{a}{27b^2}$	$3b$	$\dfrac{8a}{27bR}$
Berthelot	$p = \dfrac{nRT}{V - nb} - \dfrac{n^2a}{TV^2}$	$p_r = \dfrac{8T_r}{3V_r - 1} - \dfrac{3}{T_rV_r^2}$	$\dfrac{1}{12}\left(\dfrac{2aR}{3b^3}\right)^{1/2}$	$3b$	$\dfrac{2}{3}\left(\dfrac{2a}{3bR}\right)^{1/2}$
Dieterici	$p = \dfrac{nRT\mathrm{e}^{-aRTV/n}}{V - nb}$	$p_r = \dfrac{T_r\mathrm{e}^{2(1-1/T_rV_r)}}{2V_r - 1}$	$\dfrac{a}{4\mathrm{e}^2b^2}$	$2b$	$\dfrac{a}{4bR}$
Virial	$p = \dfrac{nRT}{V}\left\{1 + \dfrac{nB(T)}{V} + \dfrac{n^2C(T)}{V^2} + \cdots\right\}$				

* Reduced variables are defined in Section 1C.2(c). Equations of state are sometimes expressed in terms of the molar volume, $V_m = V/n$.

That having been said, we can begin to judge the reliability of the equation by comparing the isotherms it predicts with the experimental isotherms in Fig. 1C.2. Some calculated isotherms are shown in Fig. 1C.7 and Fig. 1C.8. Apart from the oscillations below the critical temperature, they do resemble experimental isotherms quite well. The oscillations, the **van der Waals' loops**, are unrealistic because they suggest that under some conditions an increase of pressure results in an increase of volume. Therefore they are replaced by horizontal lines drawn so the loops define equal areas above and below the lines: this procedure is called the **Maxwell construction** (**1**). The van der Waals coefficients, such as those in Table 1C.3, are found by fitting the calculated curves to the experimental curves.

1

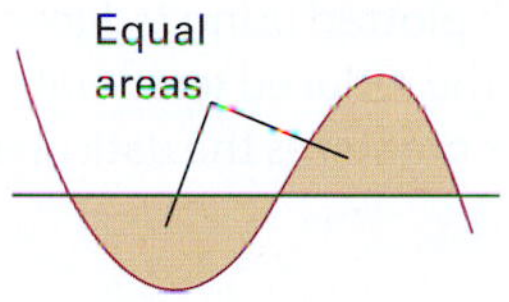

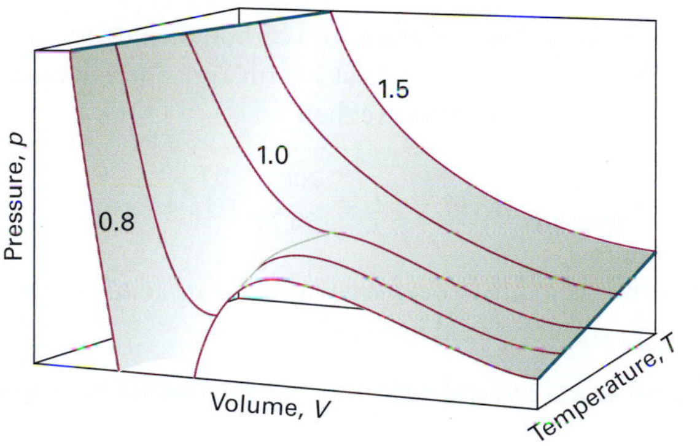

Figure 1C.7 The surface of possible states allowed by the van der Waals equation. Compare this surface with that shown in Fig. 1C.8.

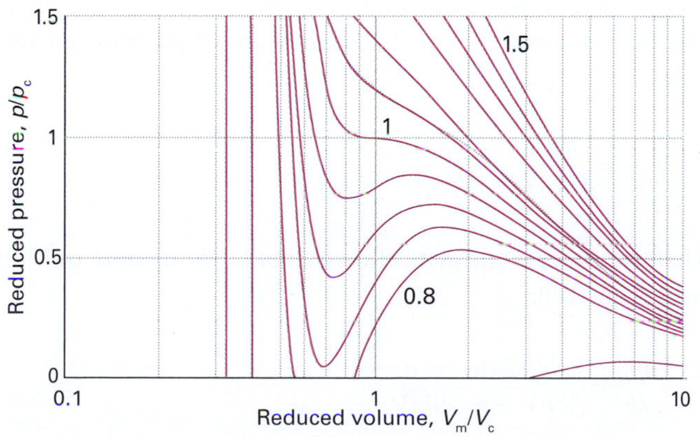

Figure 1C.8 van der Waals isotherms at several values of T/T_c. Compare these curves with those in Fig. 1C.2. The van der Waals loops are normally replaced by horizontal straight lines. The critical isotherm is the isotherm for $T/T_c = 1$.

The principal features of the van der Waals equation can be summarized as follows.

1. Perfect gas isotherms are obtained at high temperatures and large molar volumes.

When the temperature is high, RT may be so large that the first term in eqn 1C.5b greatly exceeds the second. Furthermore, if the molar volume is large in the sense $V_m \gg b$, then the denominator $V_m - b \approx V_m$. Under these conditions, the equation reduces to $p = RT/V_m$, the perfect gas equation.

2. Liquids and gases coexist when the attractive and repulsive effects are in balance.

The van der Waals loops occur when both terms in eqn 1C.5b have similar magnitudes. The first term arises from the kinetic energy of the molecules and their repulsive interactions; the second represents the effect of the attractive interactions.

3. The critical constants are related to the van der Waals coefficients.

For $T < T_c$, the calculated isotherms oscillate, and each one passes through a minimum followed by a maximum. These extrema converge as $T \to T_c$ and coincide at $T = T_c$; at the critical point the curve has a flat inflexion (**2**). From the properties of curves, we know that an inflexion of this type occurs when both the first and second derivatives are zero. Hence, we can find the critical constants by calculating these derivatives and setting them equal to zero at the critical point:

2

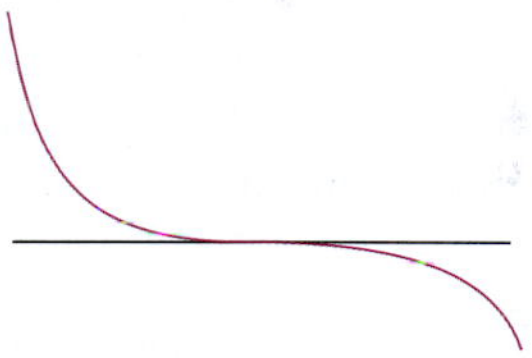

$$\frac{dp}{dV_m} = -\frac{RT}{(V_m - b)^2} + \frac{2a}{V_m^3} = 0$$

$$\frac{d^2p}{dV_m^2} = \frac{2RT}{(V_m - b)^3} - \frac{6a}{V_m^4} = 0$$

The solutions of these two equations (and using eqn 1C.5b to calculate p_c from V_c and T_c) are

$$V_c = 3b \qquad p_c = \frac{a}{27b^2} \qquad T_c = \frac{8a}{27Rb} \qquad (1C.6)$$

These relations provide an alternative route to the determination of a and b from the values of the critical constants. They can be tested by noting that the **critical compression factor**, Z_c, is predicted to be equal to

$$Z_c = \frac{p_cV_c}{RT_c} = \frac{3}{8} \qquad (1C.7)$$

for all gases that are described by the van der Waals equation near the critical point. We see from Table 1C.2 that although $Z_c < \frac{3}{8} = 0.375$, it is approximately constant (at 0.3) and the discrepancy is reasonably small.

Brief illustration 1C.4 Criteria for perfect gas behaviour

For benzene $a = 18.57$ atm dm^6 mol^{-2} (1.882 Pa m^6 mol^{-2}) and $b = 0.1193$ dm^3 mol^{-1} (1.193×10^{-4} m^3 mol^{-1}); its normal boiling point is 353 K. Treated as a perfect gas at $T = 400$ K and $p = 1.0$ atm, benzene vapour has a molar volume of $V_m = RT/p = 33$ dm mol^{-1}, so the criterion $V_m \gg b$ for perfect gas behaviour is satisfied. It follows that $a/V_m^2 \approx 0.017$ atm, which is 1.7 per cent of 1.0 atm. Therefore, we can expect benzene vapour to deviate only slightly from perfect gas behaviour at this temperature and pressure.

Self-test 1C.5 Can argon gas be treated as a perfect gas at 400 K and 3.0 atm?

Answer: Yes

(c) The principle of corresponding states

An important general technique in science for comparing the properties of objects is to choose a related fundamental property of the same kind and to set up a relative scale on that basis. We have seen that the critical constants are characteristic properties of gases, so it may be that a scale can be set up by using them as yardsticks. We therefore introduce the dimensionless **reduced variables** of a gas by dividing the actual variable by the corresponding critical constant:

$$V_r = \frac{V_m}{V_c} \quad p_r = \frac{p}{p_c} \quad T_r = \frac{T}{T_c} \qquad \text{Definition} \quad \text{Reduced variables} \qquad (1C.8)$$

If the reduced pressure of a gas is given, we can easily calculate its actual pressure by using $p = p_r p_c$, and likewise for the volume and temperature. van der Waals, who first tried this procedure, hoped that gases confined to the same reduced volume, V_r, at the same reduced temperature, T_r, would exert the same reduced pressure, p_r. The hope was largely fulfilled (Fig. 1C.9). The illustration shows the dependence of the compression factor on the reduced pressure for a variety of gases at various reduced temperatures. The success of the procedure is strikingly clear: compare this graph with Fig. 1C.3, where similar data are plotted without using reduced variables. The observation that real gases at the same reduced volume and reduced temperature exert the same reduced pressure is called the **principle of corresponding states**. The principle is only an approximation. It works best for gases composed of spherical molecules; it fails, sometimes badly, when the molecules are non-spherical or polar.

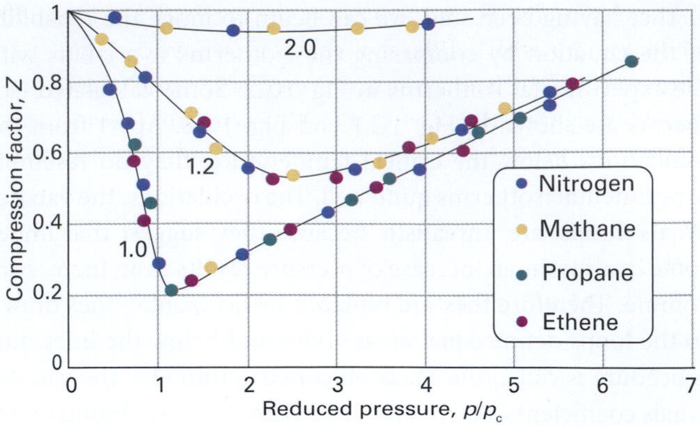

Figure 1C.9 The compression factors of four of the gases shown in Fig. 1C.3 plotted using reduced variables. The curves are labelled with the reduced temperature $T_r = T/T_c$. The use of reduced variables organizes the data on to single curves.

Brief illustration 1C.5 Corresponding states

The critical constants of argon and carbon dioxide are given in Table 1C.2. Suppose argon is at 23 atm and 200 K, its reduced pressure and temperature are then

$$p_r = \frac{23\,\text{atm}}{48.0\,\text{atm}} = 0.48 \qquad T_r = \frac{200\,\text{K}}{150.7\,\text{K}} = 1.33$$

For carbon dioxide to be in a corresponding state, its pressure and temperature would need to be

$$p = 0.48 \times (72.9\,\text{atm}) = 35\,\text{atm} \qquad T = 1.33 \times 304.2\,\text{K} = 405\,\text{K}$$

Self-test 1C.6 What would be the corresponding state of ammonia?

Answer: 53 atm, 539 K

The van der Waals equation sheds some light on the principle. First, we express eqn 1C.5b in terms of the reduced variables, which gives

$$p_r p_c = \frac{RT_r T_c}{V_r V_c - b} - \frac{a}{V_r^2 V_c^2}$$

Then we express the critical constants in terms of a and b by using eqn 1C.8:

$$\frac{ap_r}{27b^2} = \frac{8aT_r/27b}{3bpV_r - b} - \frac{a}{9b^2 V_r^2}$$

which can be reorganized into

$$p_r = \frac{8T_r}{3V_r - 1} - \frac{3}{V_r^2} \qquad (1C.9)$$

This equation has the same form as the original, but the coefficients a and b, which differ from gas to gas, have disappeared. It follows that if the isotherms are plotted in terms of the reduced variables (as we did in fact in Fig. 1C.8 without drawing attention to the fact), then the same curves are obtained whatever the gas. This is precisely the content of the principle of corresponding states, so the van der Waals equation is compatible with it.

Looking for too much significance in this apparent triumph is mistaken, because other equations of state also accommodate the principle (like those in Table 1C.4). In fact, all we need are two parameters playing the roles of a and b, for then the equation can always be manipulated into reduced form. The observation that real gases obey the principle approximately amounts to saying that the effects of the attractive and repulsive interactions can each be approximated in terms of a single parameter. The importance of the principle is then not so much its theoretical interpretation but the way that it enables the properties of a range of gases to be coordinated on to a single diagram (for example, Fig. 1C.9 instead of Fig. 1C.3).

Checklist of concepts

☐ 1. The extent of deviations from perfect behaviour is summarized by introducing the **compression factor**.

☐ 2. The **virial equation** is an empirical extension of the perfect gas equation that summarizes the behaviour of real gases over a range of conditions.

☐ 3. The isotherms of a real gas introduce the concepts of **vapour pressure** and **critical behaviour**.

☐ 4. A gas can be liquefied by pressure alone only if its temperature is at or below its **critical temperature**.

☐ 5. The **van der Waals equation** is a model equation of state for a real gas expressed in terms of two parameters, one (a) corresponding to molecular attractions and the other (b) to molecular repulsions.

☐ 6. The van der Waals equation captures the general features of the behaviour of real gases, including their critical behaviour.

☐ 7. The properties of real gases are coordinated by expressing their equations of state in terms of **reduced variables**.

Checklist of equations

Property	Equation	Comment	Equation number
Compression factor	$Z = V_m/V_m^\circ$	Definition	1C.1
Virial equation of state	$pV_m = RT(1+B/V_m+C/V_m^2+\cdots)$	B, C depend on temperature	1C.3
van der Waals equation of state	$p = nRT/(V-nb) - a(n/V)^2$	a parameterizes attractions, b parameterizes repulsions	1C.5
Reduced variables	$X_r = X/X_c$	$X = p$, V_m, or T	1C.8

CHAPTER 1 The properties of gases

TOPIC 1A The perfect gas

Discussion questions

1A.1 Explain how the perfect gas equation of state arises by combination of Boyle's law, Charles's law, and Avogadro's principle.

1A.2 Explain the term 'partial pressure' and explain why Dalton's law is a limiting law.

Exercises

1A.1(a) Could 131 g of xenon gas in a vessel of volume 1.0 dm^3 exert a pressure of 20 atm at 25 °C if it behaved as a perfect gas? If not, what pressure would it exert?

1A.1(b) Could 25 g of argon gas in a vessel of volume 1.5 dm^3 exert a pressure of 2.0 bar at 30 °C if it behaved as a perfect gas? If not, what pressure would it exert?

1A.2(a) A perfect gas undergoes isothermal compression, which reduces its volume by 2.20 dm^3. The final pressure and volume of the gas are 5.04 bar and 4.65 dm^3, respectively. Calculate the original pressure of the gas in (i) bar, (ii) atm.

1A.2(b) A perfect gas undergoes isothermal compression, which reduces its volume by 1.80 dm^3. The final pressure and volume of the gas are 1.97 bar and 2.14 dm^3, respectively. Calculate the original pressure of the gas in (i) bar, (ii) torr.

1A.3(a) A car tyre (i.e. an automobile tire) was inflated to a pressure of 24 lb in^{-2} (1.00 atm = 14.7 lb in^{-2}) on a winter's day when the temperature was −5 °C. What pressure will be found, assuming no leaks have occurred and that the volume is constant, on a subsequent summer's day when the temperature is 35 °C? What complications should be taken into account in practice?

1A.3(b) A sample of hydrogen gas was found to have a pressure of 125 kPa when the temperature was 23 °C. What can its pressure be expected to be when the temperature is 11 °C?

1A.4(a) A sample of 255 mg of neon occupies 3.00 dm^3 at 122 K. Use the perfect gas law to calculate the pressure of the gas.

1A.4(b) A homeowner uses 4.00×10^3 m^3 of natural gas in a year to heat a home. Assume that natural gas is all methane, CH_4, and that methane is a perfect gas for the conditions of this problem, which are 1.00 atm and 20 °C. What is the mass of gas used?

1A.5(a) A diving bell has an air space of 3.0 m^3 when on the deck of a boat. What is the volume of the air space when the bell has been lowered to a depth of 50 m? Take the mean density of sea water to be 1.025 g cm^{-3} and assume that the temperature is the same as on the surface.

1A.5(b) What pressure difference must be generated across the length of a 15 cm vertical drinking straw in order to drink a water-like liquid of density 1.0 g cm^{-3}?

1A.6(a) A manometer consists of a U-shaped tube containing a liquid. One side is connected to the apparatus and the other is open to the atmosphere. The pressure inside the apparatus is then determined from the difference in heights of the liquid. Suppose the liquid is water, the external pressure is 770 Torr, and the open side is 10.0 cm lower than the side connected to the apparatus. What is the pressure in the apparatus? (The density of water at 25 °C is 0.997 07 g cm^{-3}.)

1.A6(b) A manometer like that described in Exercise 1.6(a) contained mercury in place of water. Suppose the external pressure is 760 Torr, and the open side is 10.0 cm higher than the side connected to the apparatus. What is the pressure in the apparatus? (The density of mercury at 25 °C is 13.55 g cm^{-3}.)

1A.7(a) In an attempt to determine an accurate value of the gas constant, *R*, a student heated a container of volume 20.000 dm^3 filled with 0.251 32 g of helium gas to 500 °C and measured the pressure as 206.402 cm of water in a manometer at 25 °C. Calculate the value of *R* from these data. (The density of water at 25 °C is 0.99707 g cm^{-3}; the construction of a manometer is described in Exercise 1.6(a).)

1A.7(b) The following data have been obtained for oxygen gas at 273.15 K. Calculate the best value of the gas constant *R* from them and the best value of the molar mass of O_2. The density of oxygen at 273.15 K is 13.55 g cm^{-3}.

p/atm	0.750 000	0.500 000	0.250 000
$V_m/(dm^3\,mol^{-1})$	29.9649	44.8090	89.6384

1A.8(a) At 500 °C and 93.2 kPa, the mass density of sulfur vapour is 3.710 kg m^{-3}. What is the molecular formula of sulfur under these conditions?

1A.8(b) At 100 °C and 16.0 kPa, the mass density of phosphorus vapour is 0.6388 kg m^{-3}. What is the molecular formula of phosphorus under these conditions?

1A.9(a) Calculate the mass of water vapour present in a room of volume 400 m^3 that contains air at 27 °C on a day when the relative humidity is 60 per cent.

1A.9(b) Calculate the mass of water vapour present in a room of volume 250 m^3 that contains air at 23 °C on a day when the relative humidity is 53 per cent.

1A.10(a) Given that the density of air at 0.987 bar and 27 °C is 1.146 kg m^{-3}, calculate the mole fraction and partial pressure of nitrogen and oxygen assuming that (i) air consists only of these two gases, (ii) air also contains 1.0 mole per cent Ar.

1A.10(b) A gas mixture consists of 320 mg of methane, 175 mg of argon, and 225 mg of neon. The partial pressure of neon at 300 K is 8.87 kPa. Calculate (i) the volume and (ii) the total pressure of the mixture.

1A.11(a) The density of a gaseous compound was found to be 1.23 kg m^{-3} at 330 K and 20 kPa. What is the molar mass of the compound?

1A.11(b) In an experiment to measure the molar mass of a gas, 250 cm^3 of the gas was confined in a glass vessel. The pressure was 152 Torr at 298 K, and after correcting for buoyancy effects, the mass of the gas was 33.5 mg. What is the molar mass of the gas?

1A.12(a) The densities of air at −85 °C, 0 °C, and 100 °C are 1.877 g dm^{-3}, 1.294 g dm^{-3}, and 0.946 g dm^{-3}, respectively. From these data, and assuming that air obeys Charles's law, determine a value for the absolute zero of temperature in degrees Celsius.

1A.12(b) A certain sample of a gas has a volume of 20.00 dm^3 at 0 °C and 1.000 atm. A plot of the experimental data of its volume against the Celsius temperature, *θ*, at constant *p*, gives a straight line of slope 0.0741 $dm^3\,°C^{-1}$. From these data determine the absolute zero of temperature in degrees Celsius.

1A.13(a) A vessel of volume 22.4 dm^3 contains 2.0 mol H_2 and 1.0 mol N_2 at 273.15 K. Calculate (i) the mole fractions of each component, (ii) their partial pressures, and (iii) their total pressure.

1A.13(b) A vessel of volume 22.4 dm^3 contains 1.5 mol H_2 and 2.5 mol N_2 at 273.15 K. Calculate (i) the mole fractions of each component, (ii) their partial pressures, and (iii) their total pressure.

Problems

1A.1 Recent communication with the inhabitants of Neptune have revealed that they have a Celsius-type temperature scale, but based on the melting point (0 °N) and boiling point (100 °N) of their most common substance, hydrogen. Further communications have revealed that the Neptunians know about perfect gas behaviour and they find that in the limit of zero pressure, the value of pV is 28 dm^3 atm at 0 °N and 40 dm^3 atm at 100 °N. What is the value of the absolute zero of temperature on their temperature scale?

1A.2 Deduce the relation between the pressure and mass density, ρ, of a perfect gas of molar mass M. Confirm graphically, using the following data on dimethyl ether at 25 °C, that perfect behaviour is reached at low pressures and find the molar mass of the gas.

p/kPa	12.223	25.20	36.97	60.37	85.23	101.3
ρ/(kg m^{-3})	0.225	0.456	0.664	1.062	1.468	1.734

1A.3 Charles's law is sometimes expressed in the form $V=V_0(1+\alpha\theta)$, where θ is the Celsius temperature, is a constant, and V_0 is the volume of the sample at 0 °C. The following values for have been reported for nitrogen at 0 °C:

p/Torr	749.7	599.6	333.1	98.6
$10^3\alpha$/°C^{-1}	3.6717	3.6697	3.6665	3.6643

For these data calculate the best value for the absolute zero of temperature on the Celsius scale.

1A.4 The molar mass of a newly synthesized fluorocarbon was measured in a gas microbalance. This device consists of a glass bulb forming one end of a beam, the whole surrounded by a closed container. The beam is pivoted, and the balance point is attained by raising the pressure of gas in the container, so increasing the buoyancy of the enclosed bulb. In one experiment, the balance point was reached when the fluorocarbon pressure was 327.10 Torr; for the same setting of the pivot, a balance was reached when CHF_3 (M=70.014 g mol^{-1}) was introduced at 423.22 Torr. A repeat of the experiment with a different setting of the pivot required a pressure of 293.22 Torr of the fluorocarbon and 427.22 Torr of the CHF_3. What is the molar mass of the fluorocarbon? Suggest a molecular formula.

1A.5 A constant-volume perfect gas thermometer indicates a pressure of 6.69 kPa at the triple point temperature of water (273.16 K). (a) What change of pressure indicates a change of 1.00 K at this temperature? (b) What pressure indicates a temperature of 100.00 °C? (c) What change of pressure indicates a change of 1.00 K at the latter temperature?

1A.6 A vessel of volume 22.4 dm^3 contains 2.0 mol H_2 and 1.0 mol N_2 at 273.15 K initially. All the H_2 reacted with sufficient N_2 to form NH_3. Calculate the partial pressures and the total pressure of the final mixture.

1A.7 Atmospheric pollution is a problem that has received much attention. Not all pollution, however, is from industrial sources. Volcanic eruptions can be a significant source of air pollution. The Kilauea volcano in Hawaii emits 200–300 t of SO_2 per day. If this gas is emitted at 800 °C and 1.0 atm, what volume of gas is emitted?

1A.8 Ozone is a trace atmospheric gas which plays an important role in screening the Earth from harmful ultraviolet radiation, and the abundance of ozone is commonly reported in *Dobson units*. One Dobson unit is the thickness, in thousandths of a centimetre, of a column of gas if it were collected as a pure gas at 1.00 atm and 0 °C. What amount of O_3 (in moles) is found in a column of atmosphere with a cross-sectional area of 1.00 dm^2 if the abundance is 250 Dobson units (a typical mid-latitude value)? In the seasonal Antarctic ozone hole, the column abundance drops below 100 Dobson units; how many moles of O_3 are found in such a column of air above a 1.00 dm^2 area? Most atmospheric ozone is found between 10 and 50 km above the surface of the earth. If that ozone is spread uniformly through this portion of the atmosphere, what is the average molar concentration corresponding to (a) 250 Dobson units, (b) 100 Dobson units?

1A.9 The barometric formula (see *Impact* 1.1) relates the pressure of a gas of molar mass M at an altitude h to its pressure p_0 at sea level. Derive this relation by showing that the change in pressure dp for an infinitesimal change in altitude dh where the density is ρ is $dp=-\rho g dh$. Remember that ρ depends on the pressure. Evaluate (a) the pressure difference between the top and bottom of a laboratory vessel of height 15 cm, and (b) the external atmospheric pressure at a typical cruising altitude of an aircraft (11 km) when the pressure at ground level is 1.0 atm.

1A.10 Balloons are still used to deploy sensors that monitor meteorological phenomena and the chemistry of the atmosphere. It is possible to investigate some of the technicalities of ballooning by using the perfect gas law. Suppose your balloon has a radius of 3.0 m and that it is spherical. (a) What amount of H_2 (in moles) is needed to inflate it to 1.0 atm in an ambient temperature of 25 °C at sea level? (b) What mass can the balloon lift at sea level, where the density of air is 1.22 kg m^{-3}? (c) What would be the payload if He were used instead of H_2?

1A.11‡ The preceding problem is most readily solved with the use of Archimedes principle, which states that the lifting force is equal to the difference between the weight of the displaced air and the weight of the balloon. Prove Archimedes principle for the atmosphere from the barometric formula. *Hint*: Assume a simple shape for the balloon, perhaps a right circular cylinder of cross-sectional area A and height h.

1A.12‡ Chlorofluorocarbons such as CCl_3F and CCl_2F_2 have been linked to ozone depletion in Antarctica. As of 1994, these gases were found in quantities of 261 and 509 parts per trillion (10^{12}) by volume (World Resources Institute, *World resources* 1996–97). Compute the molar concentration of these gases under conditions typical of (a) the mid-latitude troposphere (10 °C and 1.0 atm) and (b) the Antarctic stratosphere (200 K and 0.050 atm).

1A.13‡ The composition of the atmosphere is approximately 80 per cent nitrogen and 20 per cent oxygen by mass. At what height above the surface of the Earth would the atmosphere become 90 per cent nitrogen and 10 per cent oxygen by mass? Assume that the temperature of the atmosphere is constant at 25 °C. What is the pressure of the atmosphere at that height?

‡ These problems were supplied by Charles Trapp and Carmen Giunta.

TOPIC 1B The kinetic model

Discussion questions

1B.1 Specify and analyse critically the assumptions that underlie the kinetic model of gases.

1B.2 Provide molecular interpretations for the dependencies of the mean free path on the temperature, pressure, and size of gas molecules.

Exercises

1B.1(a) Determine the ratios of (i) the mean speeds, (ii) the mean translational kinetic energies of H_2 molecules and Hg atoms at 20 °C.
1B.1(b) Determine the ratios of (i) the mean speeds, (ii) the mean kinetic energies of He atoms and Hg atoms at 25 °C.

1B.2(a) Calculate the root mean square speeds of H_2 and O_2 molecules at 20 °C.
1B.2(b) Calculate the root mean square speeds of CO_2 molecules and He atoms at 20 °C.

1B.3(a) Use the Maxwell–Boltzmann distribution of speeds to estimate the fraction of N_2 molecules at 400 K that have speeds in the range 200 to 210 m s^{-1}.
1B.3(b) Use the Maxwell–Boltzmann distribution of speeds to estimate the fraction of CO_2 molecules at 400 K that have speeds in the range 400 to 405 m s^{-1}.

1B.4(a) Calculate the most probable speed, the mean speed, and the mean relative speed of CO_2 molecules in air at 20 °C.
1B.4(b) Calculate the most probable speed, the mean speed, and the mean relative speed of H_2 molecules in air at 20 °C.

1B.5(a) Assume that air consists of N_2 molecules with a collision diameter of 395 pm. Calculate (i) the mean speed of the molecules, (ii) the mean free path, (iii) the collision frequency in air at 1.0 atm and 25 °C.
1B.5(b) The best laboratory vacuum pump can generate a vacuum of about 1 nTorr. At 25 °C and assuming that air consists of N_2 molecules with a collision diameter of 395 pm, calculate (i) the mean speed of the molecules, (ii) the mean free path, (iii) the collision frequency in the gas.

1B.6(a) At what pressure does the mean free path of argon at 20 °C become comparable to the diameter of a 100 cm^3 vessel that contains it? Take $\sigma = 0.36$ nm^2.
1B.6(b) At what pressure does the mean free path of argon at 20 °C become comparable to 10 times the diameters of the atoms themselves?

1B.7(a) At an altitude of 20 km the temperature is 217 K and the pressure 0.050 atm. What is the mean free path of N_2 molecules? ($\sigma = 0.43$ nm^2).
1B.7(b) At an altitude of 15 km the temperature is 217 K and the pressure 12.1 kPa. What is the mean free path of N_2 molecules? ($\sigma = 0.43$ nm^2).

Problems

1B.1 A rotating slotted-disc apparatus like that in Fig. 1B.5 consists of five coaxial 5.0 cm diameter disks separated by 1.0 cm, the slots in their rims being displaced by 2.0° between neighbours. The relative intensities, I, of the detected beam of Kr atoms for two different temperatures and at a series of rotation rates were as follows:

ν/Hz	20	40	80	100	120
I (40 K)	0.846	0.513	0.069	0.015	0.002
I (100 K)	0.592	0.485	0.217	0.119	0.057

Find the distributions of molecular velocities, $f(v_x)$, at these temperatures, and check that they conform to the theoretical prediction for a one-dimensional system.

1B.2 A Knudsen cell was used to determine the vapour pressure of germanium at 1000 °C. During an interval of 7200 s the mass loss through a hole of radius 0.50 mm amounted to 43 µg. What is the vapour pressure of germanium at 1000 °C? Assume the gas to be monatomic.

1B.3 Start from the Maxwell–Boltzmann distribution and derive an expression for the most probable speed of a gas of molecules at a temperature T. Go on to demonstrate the validity of the equipartition conclusion that the average translational kinetic energy of molecules free to move in three dimensions is $\frac{3}{2}kT$.

1B.4 Consider molecules that are confined to move in a plane (a two-dimensional gas). Calculate the distribution of speeds and determine the mean speed of the molecules at a temperature T.

1B.5 A specially constructed velocity-selector accepts a beam of molecules from an oven at a temperature T but blocks the passage of molecules with a speed greater than the mean. What is the mean speed of the emerging beam, relative to the initial value, treated as a one-dimensional problem?

1B.6 What, according to the Maxwell–Boltzmann distribution, is the proportion of gas molecules having (a) more than, (b) less than the root mean square speed? (c) What are the proportions having speeds greater and smaller than the mean speed?

1B.7 Calculate the fractions of molecules in a gas that have a speed in a range Δv at the speed nv_{mp} relative to those in the same range at v_m itself? This calculation can be used to estimate the fraction of very energetic molecules (which is important for reactions). Evaluate the ratio for $n=3$ and $n=4$.

1B.8 Derive an expression for $\langle v^n \rangle^{1/n}$ from the Maxwell–Boltzmann distribution of speeds. You will need standard integrals given in the *Resource section*.

1B.9 Calculate the escape velocity (the minimum initial velocity that will take an object to infinity) from the surface of a planet of radius R. What is the value for (a) the Earth, $R = 6.37 \times 10^6$ m, $g = 9.81$ m s^{-2}, (b) Mars, $R = 3.38 \times 10^6$ m, $m_{Mars}/m_{Earth} = 0.108$. At what temperatures do H_2, He, and O_2 molecules have mean speeds equal to their escape speeds? What proportion of the molecules have enough speed to escape when the temperature is (a) 240 K, (b) 1500 K? Calculations of this kind are very important in considering the composition of planetary atmospheres.

1B.10 The principal components of the atmosphere of the Earth are diatomic molecules, which can rotate as well as translate. Given that the translational kinetic energy density of the atmosphere is 0.15 J cm^{-3}, what is the total kinetic energy density, including rotation? The average rotational energy of a linear molecule is kT.

1B.11 Plot different Maxwell–Boltzmann speed distributions by keeping the molar mass constant at 100 g mol^{-1} and varying the temperature of the sample between 200 K and 2000 K.

1B.12 Evaluate numerically the fraction of molecules with speeds in the range 100 m s^{-1} to 200 m s^{-1} at 300 K and 1000 K.

TOPIC 1C Real gases

Discussion questions

1C.1 Explain how the compression factor varies with pressure and temperature and describe how it reveals information about intermolecular interactions in real gases.

1C.2 What is the significance of the critical constants?

1C.3 Describe the formulation of the van der Waals equation and suggest a rationale for one other equation of state in Table 1C.6.

1C.4 Explain how the van der Waals equation accounts for critical behaviour.

Exercises

1C.1(a) Calculate the pressure exerted by 1.0 mol C_2H_6 behaving as a van der Waals gas when it is confined under the following conditions: (i) at 273.15 K in 22.414 dm^3, (ii) at 1000 K in 100 cm^3. Use the data in Table 1C.3.
1C.1(b) Calculate the pressure exerted by 1.0 mol H_2S behaving as a van der Waals gas when it is confined under the following conditions: (i) at 273.15 K in 22.414 dm^3, (ii) at 500 K in 150 cm^3. Use the data in Table 1C.3.

1C.2(a) Express the van der Waals parameters $a=0.751$ atm dm^6 mol^{-2} and $b=0.0226$ dm^3 mol^{-1} in SI base units.
1C.2(b) Express the van der Waals parameters $a=1.32$ atm dm^6 mol^{-2} and $b=0.0436$ dm^3 mol^{-1} in SI base units.

1C.3(a) A gas at 250 K and 15 atm has a molar volume 12 per cent smaller than that calculated from the perfect gas law. Calculate (i) the compression factor under these conditions and (ii) the molar volume of the gas. Which are dominating in the sample, the attractive or the repulsive forces?
1C.3(b) A gas at 350 K and 12 atm has a molar volume 12 per cent larger than that calculated from the perfect gas law. Calculate (i) the compression factor under these conditions and (ii) the molar volume of the gas. Which are dominating in the sample, the attractive or the repulsive forces?

1C.4(a) In an industrial process, nitrogen is heated to 500 K at a constant volume of 1.000 m^3. The gas enters the container at 300 K and 100 atm. The mass of the gas is 92.4 kg. Use the van der Waals equation to determine the approximate pressure of the gas at its working temperature of 500 K. For nitrogen, $a=1.352$ dm^6 atm mol^{-2}, $b=0.0387$ dm^3 mol^{-1}.
1C.4(b) Cylinders of compressed gas are typically filled to a pressure of 200 bar. For oxygen, what would be the molar volume at this pressure and 25 °C based on (i) the perfect gas equation, (ii) the van der Waals equation? For oxygen, $a=1.364$ dm^6 atm mol^{-2}, $b=3.19\times10^{-2}$ dm^3 mol^{-1}.

1C.5(a) Suppose that 10.0 mol C_2H_6(g) is confined to 4.860 dm^3 at 27 °C. Predict the pressure exerted by the ethane from (i) the perfect gas and (ii) the van der Waals equations of state. Calculate the compression factor based on these calculations. For ethane, $a=5.507$ dm^6 atm mol^{-2}, $b=0.0651$ dm^3 mol^{-1}.

1C.5(b) At 300 K and 20 atm, the compression factor of a gas is 0.86. Calculate (i) the volume occupied by 8.2 mmol of the gas under these conditions and (ii) an approximate value of the second virial coefficient B at 300 K.

1C.6(a) The critical constants of methane are $p_c=45.6$ atm, $V_c=98.7$ cm^3 mol^{-1}, and $T_c=190.6$ K. Calculate the van der Waals parameters of the gas and estimate the radius of the molecules.
1C.6(b) The critical constants of ethane are $p_c=48.20$ atm, $V_c=148$ cm^3 mol^{-1}, and $T_c=305.4$ K. Calculate the van der Waals parameters of the gas and estimate the radius of the molecules.

1C.7(a) Use the van der Waals parameters for chlorine in Table 1C.3 of the *Resource section* to calculate approximate values of (i) the Boyle temperature of chlorine and (ii) the radius of a Cl_2 molecule regarded as a sphere.
1C.7(b) Use the van der Waals parameters for hydrogen sulfide in Table 1C.3 of the *Resource section* to calculate approximate values of (i) the Boyle temperature of the gas and (ii) the radius of a H_2S molecule regarded as a sphere.

1C.8(a) Suggest the pressure and temperature at which 1.0 mol of (i) NH_3, (ii) Xe, (iii) He will be in states that correspond to 1.0 mol H_2 at 1.0 atm and 25 °C.
1C.8(b) Suggest the pressure and temperature at which 1.0 mol of (i) H_2S, (ii) CO_2, (iii) Ar will be in states that correspond to 1.0 mol N_2 at 1.0 atm and 25 °C.

1C.9(a) A certain gas obeys the van der Waals equation with $a=0.50$ m^6 Pa mol^{-2}. Its volume is found to be 5.00×10^{-4} m^3 mol^{-1} at 273 K and 3.0 MPa. From this information calculate the van der Waals constant b. What is the compression factor for this gas at the prevailing temperature and pressure?
1C.9(b) A certain gas obeys the van der Waals equation with $a=0.76$ m^6 Pa mol^{-2}. Its volume is found to be 4.00×10^{-4} m^3 mol^{-1} at 288 K and 4.0 MPa. From this information calculate the van der Waals constant b. What is the compression factor for this gas at the prevailing temperature and pressure?

Problems

1C.1 Calculate the molar volume of chlorine gas at 350 K and 2.30 atm using (a) the perfect gas law and (b) the van der Waals equation. Use the answer to (a) to calculate a first approximation to the correction term for attraction and then use successive approximations to obtain a numerical answer for part (b).

1C.2 At 273 K measurements on argon gave $B=-21.7$ cm^3 mol^{-1} and $C=1200$ cm^6 mol^{-2}, where B and C are the second and third virial coefficients in the expansion of Z in powers of $1/V_m$. Assuming that the perfect gas law holds sufficiently well for the estimation of the second and third terms of the expansion, calculate the compression factor of argon at 100 atm and 273 K. From your result, estimate the molar volume of argon under these conditions.

1C.3 Calculate the volume occupied by 1.00 mol N_2 using the van der Waals equation in the form of a virial expansion at (a) its critical temperature, (b) its Boyle temperature, and (c) its inversion temperature. Assume that the pressure is 10 atm throughout. At what temperature is the gas most perfect? Use the following data: $T_c=126.3$ K, $a=1.352$ dm^6 atm mol^{-2}, $b=0.0387$ dm^3 mol^{-1}.

1C.4‡ The second virial coefficient of methane can be approximated by the empirical equation $B(T)=a+be^{-c/T^2}$, where $a=-0.1993$ bar^{-1}, $b=0.2002$ bar^{-1}, and $c=1131$ K^2 with 300 K < T < 600 K. What is the Boyle temperature of methane?

1C.5 The mass density of water vapour at 327.6 atm and 776.4 K is 133.2 kg m^{-3}. Given that for water $T_c=647.4$ K, $p_c=21.3$ atm, $a=5.464$ dm^6 atm mol^{-2}, $b=0.03049$ dm^3 mol^{-1}, and $M=18.02$ g mol^{-1}, calculate (a) the molar volume. Then calculate the compression factor (b) from the data, (c) from the virial expansion of the van der Waals equation.

1C.6 The critical volume and critical pressure of a certain gas are 160 cm^3 mol^{-1} and 40 atm, respectively. Estimate the critical temperature by assuming that the gas obeys the Berthelot equation of state. Estimate the radii of the gas molecules on the assumption that they are spheres.

1C.7 Estimate the coefficients a and b in the Dieterici equation of state from the critical constants of xenon. Calculate the pressure exerted by 1.0 mol Xe when it is confined to 1.0 dm^3 at 25 °C.

1C.8 Show that the van der Waals equation leads to values of $Z<1$ and $Z>1$, and identify the conditions for which these values are obtained.

1C.9 Express the van der Waals equation of state as a virial expansion in powers of $1/V_m$ and obtain expressions for B and C in terms of the parameters a and b. The expansion you will need is $(1-x)^{-1}=1+x+x^2+\ldots$. Measurements on argon gave $B=-21.7$ cm^3 mol^{-1} and $C=1200$ cm^6 mol^{-2} for the virial coefficients at 273 K. What are the values of a and b in the corresponding van der Waals equation of state?

1C.10‡ Derive the relation between the critical constants and the Dieterici equation parameters. Show that $Z_c=2e^{-2}$ and derive the reduced form of the Dieterici equation of state. Compare the van der Waals and Dieterici predictions of the critical compression factor. Which is closer to typical experimental values?

1C.11 A scientist proposed the following equation of state:

$$p=\frac{RT}{V_m}-\frac{B}{V_m^2}+\frac{C}{V_m^3}$$

Show that the equation leads to critical behaviour. Find the critical constants of the gas in terms of B and C and an expression for the critical compression factor.

1C.12 Equations 1C.3a and 1C.3b are expansions in p and $1/V_m$, respectively. Find the relation between B, C and B', C'.

1C.13 The second virial coefficient B' can be obtained from measurements of the density ρ of a gas at a series of pressures. Show that the graph of p/ρ against p should be a straight line with slope proportional to B'. Use the data on dimethyl ether in Problem 1A.2 to find the values of B' and B at 25 °C.

1C.14 The equation of state of a certain gas is given by $p=RT/V_m+(a+bT)/V_m^2$, where a and b are constants. Find $(\partial V/\partial T)_p$.

1C.15 The following equations of state are occasionally used for approximate calculations on gases: (gas A) $pV_m=RT(1+b/V_m)$, (gas B) $p(V_m-b)=RT$. Assuming that there were gases that actually obeyed these equations of state, would it be possible to liquefy either gas A or B? Would they have a critical temperature? Explain your answer.

1C.16 Derive an expression for the compression factor of a gas that obeys the equation of state $p(V-nb)=nRT$, where b and R are constants. If the pressure and temperature are such that $V_m=10b$, what is the numerical value of the compression factor?

1C.17‡ The discovery of the element argon by Lord Rayleigh and Sir William Ramsay had its origins in Rayleigh's measurements of the density of nitrogen with an eye toward accurate determination of its molar mass. Rayleigh prepared some samples of nitrogen by chemical reaction of nitrogen-containing compounds; under his standard conditions, a glass globe filled with this 'chemical nitrogen' had a mass of 2.2990 g. He prepared other samples by removing oxygen, carbon dioxide, and water vapor from atmospheric air; under the same conditions, this 'atmospheric nitrogen' had a mass of 2.3102 g (Lord Rayleigh, *Royal Institution Proceedings* **14**, 524 (1895)). With the hindsight of knowing accurate values for the molar masses of nitrogen and argon, compute the mole fraction of argon in the latter sample on the assumption that the former was pure nitrogen and the latter a mixture of nitrogen and argon.

1C.18‡ A substance as elementary and well known as argon still receives research attention. Stewart and Jacobsen have published a review of thermodynamic properties of argon (R.B. Stewart and R.T. Jacobsen, *J. Phys. Chem. Ref. Data* **18**, 639 (1989)) which included the following 300 K isotherm.

p/MPa	0.4000	0.5000	0.6000	0.8000	1.000
V_m/(dm^3 mol^{-1})	6.2208	4.9736	4.1423	3.1031	2.4795
p/MPa	1.500	2.000	2.500	3.000	4.000
V_m/(dm^3 mol^{-1})	1.6483	1.2328	0.98357	0.81746	0.60998

(a) Compute the second virial coefficient, B, at this temperature. (b) Use non-linear curve-fitting software to compute the third virial coefficient, C, at this temperature.

1C.19 Use mathematical software, a spreadsheet, or the *Living graphs* on the web site for this book to: (a) Explore how the pressure of 1.5 mol CO_2(g) varies with volume as it is compressed at (a) 273 K, (b) 373 K from 30 dm^3 to 15 dm^3. (c) Plot the data as p against $1/V$.

1C.20 Calculate the molar volume of chlorine gas on the basis of the van der Waals equation of state at 250 K and 150 kPa and calculate the percentage difference from the value predicted by the perfect gas equation.

1C.21 Is there a set of conditions at which the compression factor of a van der Waals gas passes through a minimum? If so, how does the location and value of the minimum value of Z depend on the coefficients a and b?

Mathematical background 1 Differentiation and integration

Two of the most important mathematical techniques in the physical sciences are differentiation and integration. They occur throughout the subject, and it is essential to be aware of the procedures involved.

MB1.1 Differentiation: definitions

Differentiation is concerned with the slopes of functions, such as the rate of change of a variable with time. The formal definition of the **derivative**, df/dx, of a function $f(x)$ is

$$\frac{df}{dx}=\lim_{\delta x\to 0}\frac{f(x+\delta x)-f(x)}{\delta x}$$ *Definition* First derivative (MB1.1)

As shown in Fig. MB1.1, the derivative can be interpreted as the slope of the tangent to the graph of $f(x)$. A positive first derivative indicates that the function slopes upwards (as x increases), and a negative first derivative indicates the opposite. It is sometimes convenient to denote the first derivative as $f'(x)$. The **second derivative**, d^2f/dx^2, of a function is the derivative of the first derivative (here denoted f'):

$$\frac{d^2f}{dx^2}=\lim_{\delta x\to 0}\frac{f'(x+\delta x)-f'(x)}{\delta x}$$ *Definition* Second derivative (MB1.2)

It is sometimes convenient to denote the second derivative f''. As shown in Fig. MB1.1, the second derivative of a function can be interpreted as an indication of the sharpness of the curvature of the function. A positive second derivative indicates that the function is ∪ shaped, and a negative second derivative indicates that it is ∩ shaped.

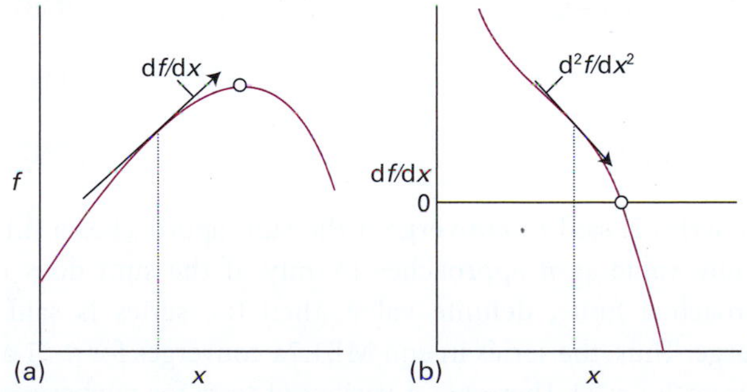

Figure MB1.1 (a) The first derivative of a function is equal to the slope of the tangent to the graph of the function at that point. The small circle indicates the extremum (in this case, maximum) of the function, where the slope is zero. (b) The second derivative of the same function is the slope of the tangent to a graph of the first derivative of the function. It can be interpreted as an indication of the curvature of the function at that point.

The derivatives of some common functions are as follows:

$$\frac{d}{dx}x^n=nx^{n-1}$$ (MB1.3a)

$$\frac{d}{dx}e^{ax}=ae^{ax}$$ (MB1.3b)

$$\frac{d}{dx}\sin ax=a\cos ax \qquad \frac{d}{dx}\cos ax=-a\sin ax$$ (MB1.3c)

$$\frac{d}{dx}\ln ax=\frac{1}{x}$$ (MB1.3d)

When a function depends on more than one variable, we need the concept of a **partial derivative**, $\partial f/\partial x$. Note the change from d to ∂: partial derivatives are dealt with at length in *Mathematical background* 2; all we need know at this stage is that they signify that all variables other than the stated variable are regarded as constant when evaluating the derivative.

Brief illustration MB1.1 Partial derivatives

Suppose we are told that f is a function of two variables, and specifically $f=4x^2y^3$. Then, to evaluate the partial derivative of f with respect to x, we regard y as a constant (just like the 4), and obtain

$$\frac{\partial f}{\partial x}=\frac{\partial}{\partial x}(4x^2y^3)=4y^3\frac{\partial}{\partial x}x^2=8xy^3$$

Similarly, to evaluate the partial derivative of f with respect to y, we regard x as a constant (again, like the 4), and obtain

$$\frac{\partial f}{\partial y}=\frac{\partial}{\partial y}(4x^2y^3)=4x^2\frac{\partial}{\partial y}y^3=12x^2y^2$$

MB1.2 Differentiation: manipulations

It follows from the definition of the derivative that a variety of combinations of functions can be differentiated by using the following rules:

$$\frac{d}{dx}(u+v)=\frac{du}{dx}+\frac{dv}{dx}$$ (MB1.4a)

$$\frac{d}{dx}uv=u\frac{dv}{dx}+v\frac{du}{dx}$$ (MB1.4b)

$$\frac{d}{dx}\frac{u}{v}=\frac{1}{v}\frac{du}{dx}-\frac{u}{v^2}\frac{dv}{dx}$$ (MB1.4c)

Brief illustration MB1.2 Derivatives

To differentiate the function $f=\sin^2 ax/x^2$ use eqn MB1.4 to write

$$\begin{aligned}\frac{\mathrm{d}}{\mathrm{d}x}\frac{\sin^2 ax}{x^2}&=\frac{\mathrm{d}}{\mathrm{d}x}\left(\frac{\sin ax}{x}\right)\left(\frac{\sin ax}{x}\right)=2\left(\frac{\sin ax}{x}\right)\frac{\mathrm{d}}{\mathrm{d}x}\left(\frac{\sin ax}{x}\right)\\&=2\left(\frac{\sin ax}{x}\right)\left\{\frac{1}{x}\frac{\mathrm{d}}{\mathrm{d}x}\sin ax+\sin ax\frac{\mathrm{d}}{\mathrm{d}x}\frac{1}{x}\right\}\\&=2\left\{\frac{a}{x^2}\sin ax\cos ax-\frac{\sin^2 ax}{x^3}\right\}\end{aligned}$$

The function and this first derivative are plotted in Fig. MB1.2.

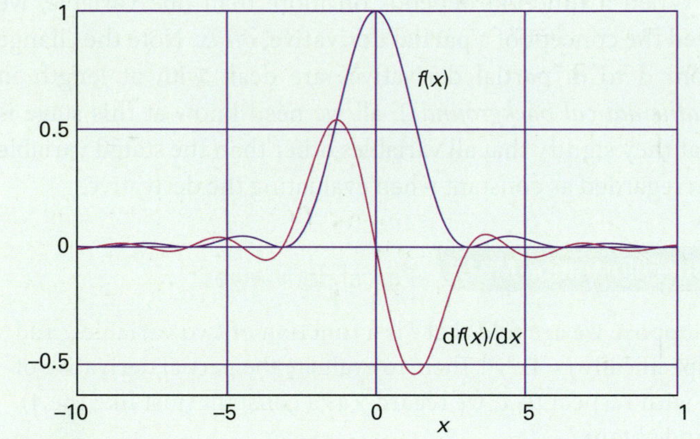

Figure MB1.2 The function considered in *Brief illustration* MB1.2 and its first derivative.

MB1.3 Series expansions

One application of differentiation is to the development of power series for functions. The **Taylor series** for a function $f(x)$ in the vicinity of $x=a$ is

$$\begin{aligned}f(x)&=f(a)+\left(\frac{\mathrm{d}f}{\mathrm{d}x}\right)_a(x-a)+\frac{1}{2!}\left(\frac{\mathrm{d}^2 f}{\mathrm{d}x^2}\right)_a(x-a)^2+\cdots\\&=\sum_{n=0}^{\infty}\frac{1}{n!}\left(\frac{\mathrm{d}^n f}{\mathrm{d}x^n}\right)_a(x-a)^n\end{aligned}\qquad \text{Taylor series} \quad \text{(MB1.5)}$$

where the notation $(\ldots)_a$ means that the derivative is evaluated at $x=a$ and $n!$ denotes a **factorial** given by

$$n!=n(n-1)(n-2)\ldots 1,\quad 0!=1 \qquad \text{Factorial} \quad \text{(MB1.6)}$$

The **Maclaurin series** for a function is a special case of the Taylor series in which $a=0$.

Brief illustration MB1.3 Series expansion

To evaluate the expansion of $\cos x$ around $x=0$ we note that

$$\left(\frac{\mathrm{d}}{\mathrm{d}x}\cos x\right)_0=(-\sin x)_0=0 \qquad \left(\frac{\mathrm{d}^2}{\mathrm{d}x^2}\cos x\right)_0=(-\cos x)_0=-1$$

and in general

$$\left(\frac{\mathrm{d}^n}{\mathrm{d}x^n}\cos x\right)_0=\begin{cases}0 \text{ for } n \text{ odd}\\(-1)^{n/2} \text{ for } n \text{ even}\end{cases}$$

Therefore,

$$\cos x=\sum_{n \text{ even}}^{\infty}\frac{(-1)^{n/2}}{n!}x^n=1-\tfrac{1}{2}x^2+\tfrac{1}{24}x^4-\cdots$$

The following Taylor series (specifically, Maclaurin series) are used at various stages in the text:

$$(1+x)^{-1}=1-x+x^2-\cdots=\sum_{n=0}^{\infty}(-1)^n x^n \qquad \text{(MB1.7a)}$$

$$e^x=1+x+\tfrac{1}{2}x^2+\cdots=\sum_{n=0}^{\infty}\frac{x^n}{n!} \qquad \text{(MB1.7b)}$$

$$\ln(1+x)=x-\tfrac{1}{2}x^2+\tfrac{1}{3}x^3-\cdots=\sum_{n=1}^{\infty}(-1)^{n+1}\frac{x^n}{n} \qquad \text{(MB1.7c)}$$

Taylor series are used to simplify calculations, for when $x\ll 1$ it is possible, to a good approximation, to terminate the series after one or two terms. Thus, provided $x\ll 1$ we can write

$$(1+x)^{-1}\approx 1-x \qquad \text{(MB1.8a)}$$

$$e^x\approx 1+x \qquad \text{(MB1.8b)}$$

$$\ln(1+x)\approx x \qquad \text{(MB1.8c)}$$

A series is said to **converge** if the sum approaches a finite, definite value as n approaches infinity. If the sum does not approach a finite, definite value, then the series is said to **diverge**. Thus, the series in eqn MB1.7a converges for $x<1$ and diverges for $x\geq 1$.There are a variety of tests for convergence, which are explained in mathematics texts.

MB1.4 Integration: definitions

Integration (which formally is the inverse of differentiation) is concerned with the areas under curves. The **integral** of a

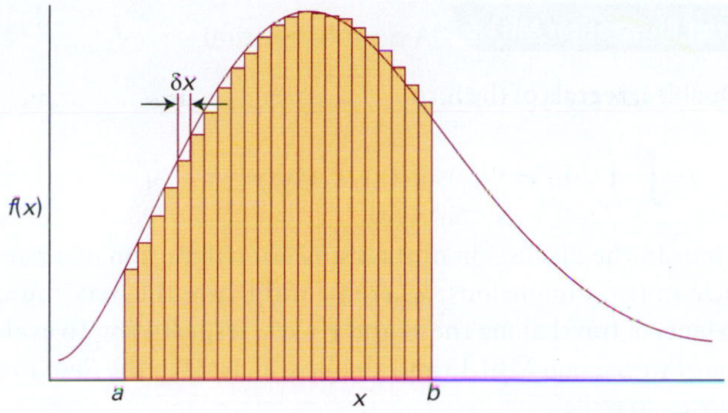

Figure MB1.3 A definite integral is evaluated by forming the product of the value of the function at each point and the increment δx, with δx→0, and then summing the products f(x)δx for all values of x between the limits a and b. It follows that the value of the integral is the area under the curve between the two limits.

function $f(x)$, which is denoted $\int f\,dx$ (the symbol $\int$ is an elongated S denoting a sum), between the two values $x=a$ and $x=b$ is defined by imagining the x axis as divided into strips of width δx and evaluating the following sum:

$$\int_a^b f(x)dx = \lim_{\delta x\to 0}\sum_i f(x_i)\delta x \quad \text{Definition} \quad \text{Integration} \quad \text{(MB1.9)}$$

As can be appreciated from Fig. MB1.3, the integral is the area under the curve between the limits a and b. The function to be integrated is called the **integrand**. It is an astonishing mathematical fact that the integral of a function is the inverse of the differential of that function in the sense that if we differentiate f and then integrate the resulting function, then we obtain the original function f (to within a constant). The function in eqn MB1.9 with the limits specified is called a **definite integral**. If it is written without the limits specified, then we have an **indefinite integral**. If the result of carrying out an indefinite integration is $g(x)+C$, where C is a constant, the following notation is used to evaluate the corresponding definite integral:

$$\begin{aligned}I &= \int_a^b f(x)dx = \{g(x)+C\}\Big|_a^b = \{g(b)+C\}-\{g(a)+C\} \\ &= g(b)-g(a)\end{aligned} \quad \text{Definite integral} \quad \text{(MB1.10)}$$

Note that the constant of integration disappears. The definite and indefinite integrals encountered in this text are listed in the *Resource section*.

MB1.5 Integration: manipulations

When an indefinite integral is not in the form of one of those listed in the *Resource section* it is sometimes possible to transform it into one of the forms by using integration techniques such as:

Substitution. Introduce a variable u related to the independent variable x (for example, an algebraic relation such as $u=x^2-1$ or a trigonometric relation such as $u=\sin x$). Express the differential dx in terms of du (for these substitutions, $du=2x\,dx$ and $du=\cos x\,dx$, respectively). Then transform the original integral written in terms of x into an integral in terms of u upon which, in some cases, a standard form such as one of those listed in the *Resource section* can be used.

Brief illustration MB1.4 Integration by substitution

To evaluate the indefinite integral $\int\cos^2 x\sin x\,dx$ we make the substitution $u=\cos x$. It follows that $du/dx=-\sin x$, and therefore that $\sin x\,dx=-du$. The integral is therefore

$$\int\cos^2 x\sin x\,dx = -\int u^2\,du = -\tfrac{1}{3}u^3+C = -\tfrac{1}{3}\cos^3 x+C$$

To evaluate the corresponding definite integral, we have to convert the limits on x into limits on u. Thus, if the limits are $x=0$ and $x=\pi$, the limits become $u=\cos 0=1$ and $u=\cos\pi=-1$:

$$\int_0^\pi \cos^2 x\sin x\,dx = -\int_1^{-1} u^2\,du = \left\{-\tfrac{1}{3}u^3+C\right\}\Big|_1^{-1} = \frac{2}{3}$$

Integration by parts. For two functions $f(x)$ and $g(x)$:

$$\int f\frac{dg}{dx}dx = fg - \int g\frac{df}{dx}dx \quad \text{Integration by parts} \quad \text{(MB1.11a)}$$

which may be abbreviated as:

$$\int f dg = fg - \int g df \quad \text{(MB1.11b)}$$

Brief illustration MB1.5 Integration by parts

Integrals over xe^{-ax} and their analogues occur commonly in the discussion of atomic structure and spectra. They may be integrated by parts, as in the following:

$$\begin{aligned}\int_0^\infty \overbrace{x}^{f}\,\overbrace{e^{-ax}}^{dg/dx}dx &= \overbrace{x}^{f}\overbrace{\frac{e^{-ax}}{-a}}^{g}\Bigg|_0^\infty - \int_0^\infty \overbrace{\frac{e^{-ax}}{-a}}^{g}\,\overbrace{1}^{df/dx}\,dx \\ &= -\frac{xe^{-ax}}{a}\Bigg|_0^\infty + \frac{1}{a}\int_0^\infty e^{-ax}dx = 0 - \frac{e^{-ax}}{a^2}\Bigg|_0^\infty \\ &= \frac{1}{a^2}\end{aligned}$$

MB1.6 Multiple integrals

A function may depend on more than one variable, in which case we may need to integrate over both the variables:

$$I=\int_a^b\int_c^d f(x,y)\mathrm{d}x\mathrm{d}y \tag{MB1.12}$$

We (but not everyone) adopt the convention that a and b are the limits of the variable x and c and d are the limits for y (as depicted by the colours in this instance). This procedure is simple if the function is a product of functions of each variable and of the form $f(x,y)=X(x)Y(y)$. In this case, the double integral is just a product of each integral:

$$I=\int_a^b\int_c^d X(x)Y(y)\mathrm{d}x\mathrm{d}y=\int_a^b X(x)\mathrm{d}x\int_c^d Y(y)\mathrm{d}y \tag{MB1.13}$$

Brief illustration MB1.6 A double integral

Double integrals of the form

$$I=\int_0^{L_1}\int_0^{L_2}\sin^2(\pi x/L_1)\sin^2(\pi y/L_2)\mathrm{d}x\mathrm{d}y$$

occur in the discussion of the translational motion of a particle in two dimensions, where L_1 and L_2 are the maximum extents of travel along the x- and y-axes, respectively. To evaluate I we use eqn MB1.13 and an integral listed in the *Resource section* to write

$$\begin{aligned}I &\overset{\text{Integral T.2}}{=} \int_0^{L_1}\sin^2(\pi x/L_1)\mathrm{d}x\int_0^{L_2}\sin^2(\pi y/L_2)\mathrm{d}y\\ &=\left\{\tfrac{1}{2}x-\frac{\sin(2\pi x/L_1)}{4\pi/L_1}+C\right\}\Bigg|_0^{L_1}\left\{\tfrac{1}{2}y-\frac{\sin(2\pi y/L_2)}{4\pi/L_2}+C\right\}\Bigg|_0^{L_2}\\ &=\tfrac{1}{4}L_1L_2\end{aligned}$$

CHAPTER 2

The First Law

The release of energy can be used to provide heat when a fuel burns in a furnace, to produce mechanical work when a fuel burns in an engine, and to generate electrical work when a chemical reaction pumps electrons through a circuit. In chemistry, we encounter reactions that can be harnessed to provide heat and work, reactions that liberate energy that is unused but which give products we require, and reactions that constitute the processes of life. **Thermodynamics**, the study of the transformations of energy, enables us to discuss all these matters quantitatively and to make useful predictions.

2A Internal energy

First, we examine the ways in which a system can exchange energy with its surroundings in terms of the work it may do or have done on it or the heat that it may produce or absorb. These considerations lead to the definition of the 'internal energy', the total energy of a system, and the formulation of the 'First Law' of thermodynamics, which states that the internal energy of an isolated system is constant.

2B Enthalpy

The second major concept of the chapter is 'enthalpy', which is a very useful book-keeping property for keeping track of the heat output (or requirements) of physical processes and chemical reactions that take place at constant pressure. Experimentally, changes in internal energy or enthalpy may be measured by techniques known collectively as 'calorimetry'.

2C Thermochemistry

'Thermochemistry' is the study of heat transactions during chemical reactions. We describe both computational and experimental methods for the determination of enthalpy changes associated with both physical and chemical changes.

2D State functions and exact differentials

We also begin to unfold some of the power of thermodynamics by showing how to establish relations between different properties of a system. We see that one very useful aspect of thermodynamics is that a property can be measured indirectly by measuring others and then combining their values. The relations we derive also enable us to discuss the liquefaction of gases and to establish the relation between the heat capacities of a substance under different conditions.

2E Adiabatic changes

'Adiabatic' processes occur without transfer of energy as heat. We focus on adiabatic changes involving perfect gases because they figure prominently in our presentation of thermodynamics.

What is the impact of this material?

Concepts of thermochemistry apply to the chemical reactions associated with the conversion of food into energy in organisms, and so form a basis for the discussion of bioenergetics. In *Impact* I2.1, we explore some of the thermochemical calculations related to the metabolism of fats, carbohydrates, and proteins.

To read more about the impact of this material, scan the QR code, or go to bcs.whfreeman.com/webpub/chemistry/pchem10e/impact/pchem-2-1.html

2A Internal energy

Contents

➤ Why do you need to know this material?

The First Law of thermodynamics is the foundation of the discussion of the role of energy in chemistry. Wherever we are interested in the generation or use of energy in physical transformations or chemical reactions, lying in the background are the concepts introduced by the First Law.

➤ What is the key idea?

The total energy of an isolated system is constant.

➤ What do you need to know already?

This Topic makes use of the discussion of the properties of gases (Topic 1A), particularly the perfect gas law. It builds on the definition of work given in *Foundations* B.

For the purposes of thermodynamics, the universe is divided into two parts, the system and its surroundings. The **system** is the part of the world in which we have a special interest. It may be a reaction vessel, an engine, an electrochemical cell, a biological cell, and so on. The **surroundings** comprise the region outside the system and are where we make our measurements. The type of system depends on the characteristics of the boundary that divides it from the surroundings (Fig. 2A.1). If matter can be transferred through the boundary between the system and its surroundings the system is classified as **open**. If matter cannot pass through the boundary the system is classified as **closed**. Both open and closed systems can exchange energy with their surroundings. For example, a closed system can expand and thereby raise a weight in the surroundings; a closed system may also transfer energy to the surroundings if they are at a lower temperature. An **isolated system** is a closed system that has neither mechanical nor thermal contact with its surroundings.

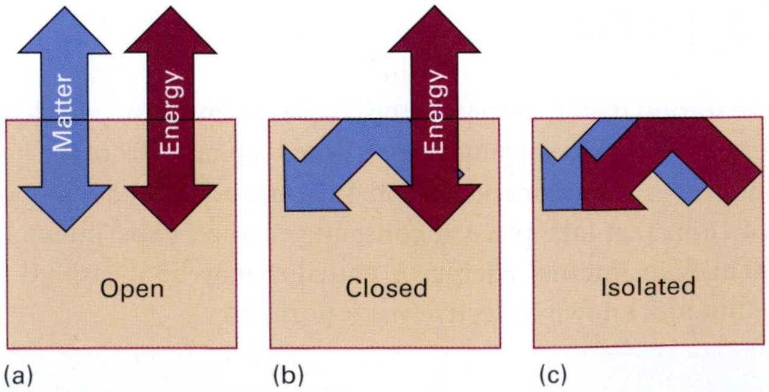

Figure 2A.1 (a) An open system can exchange matter and energy with its surroundings. (b) A closed system can exchange energy with its surroundings, but it cannot exchange matter. (c) An isolated system can exchange neither energy nor matter with its surroundings.

2A.1 Work, heat, and energy

Although thermodynamics deals with observations on bulk systems, it is immeasurably enriched by understanding the molecular origins of these observations. In each case we shall set out the bulk observations on which thermodynamics is based and then describe their molecular interpretations.

(a) Operational definitions

The fundamental physical property in thermodynamics is work: **work** is done to achieve motion against an opposing force. A simple example is the process of raising a weight against the pull of gravity. A process does work if in principle it can be harnessed to raise a weight somewhere in the surroundings. An example of doing work is the expansion of a gas that pushes out a piston: the motion of the piston can in principle be used to raise a weight. A chemical reaction that drives an electric current through a resistance also does work, because the same current could be passed through a motor and used to raise a weight.

The **energy** of a system is its capacity to do work. When work is done on an otherwise isolated system (for instance, by compressing a gas or winding a spring), the capacity of the system to do work is increased; in other words, the energy of the system is increased. When the system does work (i.e. when the piston moves out or the spring unwinds), the energy of the system is reduced and it can do less work than before.

Experiments have shown that the energy of a system may be changed by means other than work itself. When the energy of a system changes as a result of a temperature difference between the system and its surroundings we say that energy has been transferred as **heat**. When a heater is immersed in a beaker of water (the system), the capacity of the system to do work increases because hot water can be used to do more work than the same amount of cold water. Not all boundaries permit the transfer of energy even though there is a temperature difference between the system and its surroundings. Boundaries that do permit the transfer of energy as heat are called **diathermic**; those that do not are called **adiabatic**.

An **exothermic process** is a process that releases energy as heat into its surroundings. All combustion reactions are exothermic. An **endothermic process** is a process in which energy is acquired from its surroundings as heat. An example of an endothermic process is the vaporization of water. To avoid a lot of awkward language, we say that in an exothermic process energy is transferred 'as heat' to the surroundings and in an endothermic process energy is transferred 'as heat' from the surroundings into the system. However, it must never be forgotten that heat is a process (the transfer of energy as a result of a temperature difference), not an entity. An endothermic process in a diathermic container results in energy flowing into the system as heat to restore the temperature to that of the surroundings. An exothermic process in a similar diathermic container results in a release of energy as heat into the surroundings. When an endothermic process takes place in an adiabatic container, it results in a lowering of temperature of the system; an exothermic process results in a rise of temperature. These features are summarized in Fig. 2A.2.

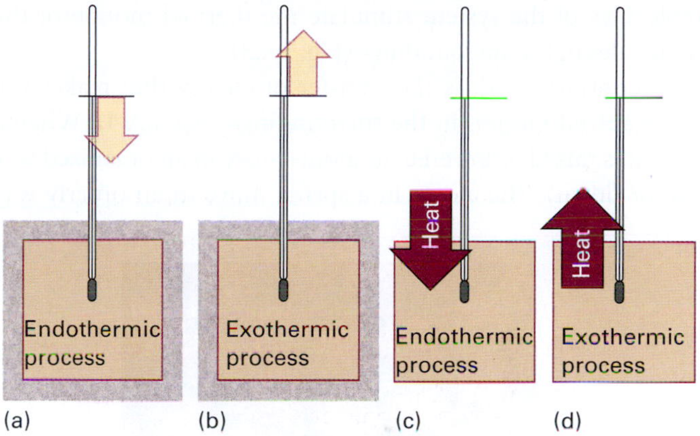

Figure 2A.2 (a) When an endothermic process occurs in an adiabatic system, the temperature falls; (b) if the process is exothermic, then the temperature rises. (c) When an endothermic process occurs in a diathermic container, energy enters as heat from the surroundings, and the system remains at the same temperature. (d) If the process is exothermic, then energy leaves as heat, and the process is isothermal.

Brief illustration 2A.1 Combustions in adiabatic and diathermic containers

Combustions are chemical reactions in which substances react with oxygen, normally with a flame. An example is the combustion of methane gas, $CH_4(g)$:

$$CH_4(g)+2O_2(g)\rightarrow CO_2(g)+2H_2O(l)$$

All combustions are exothermic. Although the temperature typically rises in the course of the combustion, if we wait long enough, the system returns to the temperature of its surroundings so we can speak of a combustion 'at 25 °C', for instance. If the combustion takes place in an adiabatic container, the energy released as heat remains inside the container and results in a permanent rise in temperature.

Self-test 2A.1 How may the isothermal expansion of a gas be achieved?

Answer: Immerse the system in a water bath

(b) The molecular interpretation of heat and work

In molecular terms, heating is the transfer of energy that makes use of disorderly, apparently random, molecular motion in the surroundings. The disorderly motion of molecules is called **thermal motion**. The thermal motion of the molecules in the hot surroundings stimulates the molecules in the cooler system to move more vigorously and, as a result, the energy of the system is increased. When a system heats its surroundings, molecules of the system stimulate the thermal motion of the molecules in the surroundings (Fig. 2A.3).

In contrast, work is the transfer of energy that makes use of organized motion in the surroundings (Fig. 2A.4). When a weight is raised or lowered, its atoms move in an organized way (up or down). The atoms in a spring move in an orderly way when it is wound; the electrons in an electric current move in the same direction. When a system does work it causes atoms or electrons in its surroundings to move in an organized way. Likewise, when work is done on a system, molecules in the surroundings are used to transfer energy to it in an organized way, as the atoms in a weight are lowered or a current of electrons is passed.

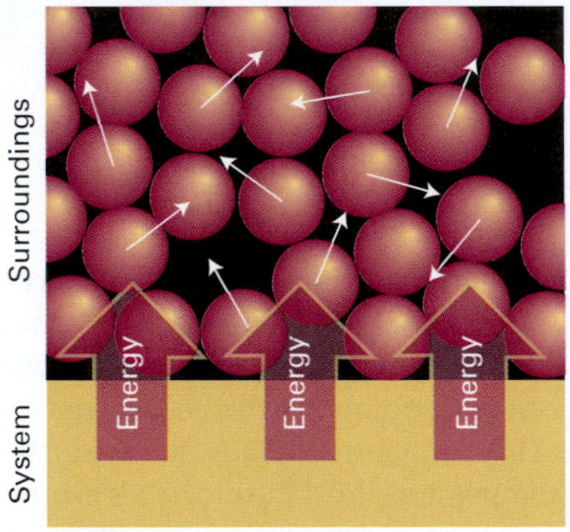

Figure 2A.3 When energy is transferred to the surroundings as heat, the transfer stimulates random motion of the atoms in the surroundings. Transfer of energy from the surroundings to the system makes use of random motion (thermal motion) in the surroundings.

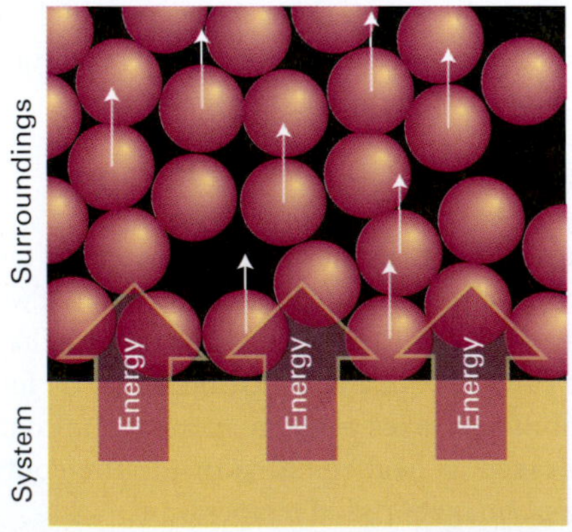

Figure 2A.4 When a system does work, it stimulates orderly motion in the surroundings. For instance, the atoms shown here may be part of a weight that is being raised. The ordered motion of the atoms in a falling weight does work on the system.

The distinction between work and heat is made in the surroundings. The fact that a falling weight may stimulate thermal motion in the system is irrelevant to the distinction between heat and work: work is identified as energy transfer making use of the organized motion of atoms in the surroundings, and heat is identified as energy transfer making use of thermal motion in the surroundings. In the adiabatic compression of a gas, for instance, work is done on the system as the atoms of the compressing weight descend in an orderly way, but the effect of the incoming piston is to accelerate the gas molecules to higher average speeds. Because collisions between molecules quickly randomize their directions, the orderly motion of the atoms of the weight is in effect stimulating thermal motion in the gas. We observe the falling weight, the orderly descent of its atoms, and report that work is being done even though it is stimulating thermal motion.

2A.2 The definition of internal energy

In thermodynamics, the total energy of a system is called its **internal energy**, U. The internal energy is the total kinetic and potential energy of the constituents (the atoms, ions, or molecules) of the system. It does not include the kinetic energy arising from the motion of the system as a whole, such as its kinetic energy as it accompanies the Earth on its orbit round the Sun. That is, the internal energy is the energy 'internal' to the system. We denote by ΔU the change in internal energy when a system changes from an initial state i with internal energy U_i to a final state f of internal energy U_f:

$$\Delta U = U_f - U_i \qquad (2A.1)$$

Throughout thermodynamics, we use the convention that $\Delta X = X_f - X_i$, where X is a property (a 'state function') of the system.

The internal energy is a **state function** in the sense that its value depends only on the current state of the system and is independent of how that state has been prepared. In other words, internal energy is a function of the properties that determine the current state of the system. Changing any one of the state variables, such as the pressure, results in a change in internal energy. That the internal energy is a state function has consequences of the greatest importance, as we shall start to unfold in Topic 2D.

The internal energy is an extensive property of a system (a property that depends on the amount of substance present, *Foundations* A) and is measures in joules ($1\,\text{J}=1\,\text{kg m}^2\,\text{s}^{-2}$). The molar internal energy, U_m, is the internal energy divided by the amount of substance in a system, $U_\text{m}=U/n$; it is an intensive property (a property independent of the amount of substance) and commonly reported in kilojoules per mole (kJ mol^{-1}).

(a) Molecular interpretation of internal energy

A molecule has a certain number of motional degrees of freedom, such as the ability to translate (the motion of its centre of mass through space), rotate around its centre of mass, or vibrate (as its bond lengths and angles change, leaving its centre of mass unmoved). Many physical and chemical properties depend on the energy associated with each of these modes of motion. For example, a chemical bond might break if a lot of energy becomes concentrated in it, for instance as vigorous vibration.

The 'equipartition theorem' of classical mechanics introduced in *Foundations* B can be used to predict the contributions of each mode of motion of a molecule to the total energy of a collection of non-interacting molecules (that is, of a perfect gas, and providing quantum effects can be ignored). For translation and rotational modes the contribution of a mode is proportional to the temperature, so the internal energy of a sample increases as the temperature is raised.

Brief illustration 2A.2 The internal energy of a perfect gas

In *Foundations* B it is shown that the mean energy of a molecule due to its translational motion is $\frac{3}{2}kT$ and therefore to the molar energy of a collection the contribution is $\frac{3}{2}RT$. Therefore, considering only the translational contribution to internal energy,

$$U_\text{m}(T)=U_\text{m}(0)+\tfrac{3}{2}N_\text{A}kT=U_\text{m}(0)+\tfrac{3}{2}RT$$

where $U_\text{m}(0)$, the internal energy at $T=0$, can be greater than zero (see, for example, Chapter 8). At 25 °C, $RT=2.48\,\text{kJ mol}^{-1}$, so the translational motion contributes $3.72\,\text{kJ mol}^{-1}$ to the molar internal energy of gases.

Self-test 2A.2 Calculate the molar internal energy of carbon dioxide at 25 °C, taking into account its translational and rotational degrees of freedom.

Answer: $U_\text{m}(T)=U_\text{m}(0)+\frac{5}{2}RT$

The contribution to the internal energy of a collection of perfect gas molecules is independent of the volume occupied by the molecules: there are no intermolecular interactions in a perfect gas, so the distance between the molecules has no effect on the energy. That is, *the internal energy of a perfect gas is independent of the volume it occupies.*

The internal energy of interacting molecules in condensed phases also has a contribution from the potential energy of their interaction, but no simple expressions can be written down in general. Nevertheless, it remains true that as the temperature of a system is raised, the internal energy increases as the various modes of motion become more highly excited.

(b) The formulation of the First Law

It has been found experimentally that the internal energy of a system may be changed either by doing work on the system or by heating it. Whereas we may know how the energy transfer has occurred (because we can see if a weight has been raised or lowered in the surroundings, indicating transfer of energy by doing work, or if ice has melted in the surroundings, indicating transfer of energy as heat), the system is blind to the mode employed. *Heat and work are equivalent ways of changing a system's internal energy.* A system is like a bank: it accepts deposits in either currency, but stores its reserves as internal energy. It is also found experimentally that if a system is isolated from its surroundings, then no change in internal energy takes place. This summary of observations is now known as the **First Law of thermodynamics** and is expressed as follows:

The internal energy of an isolated system is constant.

First Law of thermodynamics

We cannot use a system to do work, leave it isolated, and then come back expecting to find it restored to its original state with the same capacity for doing work. The experimental evidence for this observation is that no 'perpetual motion machine', a machine that does work without consuming fuel or using some other source of energy, has ever been built.

These remarks may be summarized as follows. If we write w for the work done on a system, q for the energy transferred as heat to a system, and ΔU for the resulting change in internal energy, then it follows that

$$\Delta U=q+w \qquad \text{Mathematical statement of the First Law} \qquad (2\text{A}.2)$$

Equation 2A.2 summarizes the equivalence of heat and work and the fact that the internal energy is constant in an isolated system (for which $q=0$ and $w=0$). The equation states that the change in internal energy of a closed system is equal to the energy that passes through its boundary as heat or work. It employs the 'acquisitive convention', in which w and q are positive if energy is transferred to the system as work or heat and are negative if energy is lost from the system. In other words, we view the flow of energy as work or heat from the system's perspective.

Brief illustration 2A.3 Changes in internal energy

If an electric motor produced 15 kJ of energy each second as mechanical work and lost 2 kJ as heat to the surroundings, then the change in the internal energy of the motor each second is $\Delta U = -2\text{ kJ} - 15\text{ kJ} = -17\text{ kJ}$. Suppose that, when a spring was wound, 100 J of work was done on it but 15 J escaped to the surroundings as heat. The change in internal energy of the spring is $\Delta U = 100\text{ J} - 15\text{ J} = +85\text{ J}$.

A note on good practice Always include the sign of ΔU (and of ΔX in general), even if it is positive.

Self-test 2A.3 A generator does work on an electric heater by forcing an electric current through it. Suppose 1 kJ of work is done on the heater and it heats its surroundings by 1 kJ. What is the change in internal energy of the heater?

Answer: 0

2A.3 Expansion work

The way is opened to powerful methods of calculation by switching attention to infinitesimal changes of state (such as infinitesimal change in temperature) and infinitesimal changes in the internal energy dU. Then, if the work done on a system is dw and the energy supplied to it as heat is dq, in place of eqn 2A.2 we have

$$dU = dq + dw \tag{2A.3}$$

To use this expression we must be able to relate dq and dw to events taking place in the surroundings.

We begin by discussing **expansion work**, the work arising from a change in volume. This type of work includes the work done by a gas as it expands and drives back the atmosphere. Many chemical reactions result in the generation of gases (for instance, the thermal decomposition of calcium carbonate or the combustion of octane), and the thermodynamic characteristics of the reaction depend on the work that must be done to make room for the gas it has produced. The term 'expansion work' also includes work associated with negative changes of volume, that is, compression.

(a) The general expression for work

The calculation of expansion work starts from the definition used in physics, which states that the work required to move an object a distance dz against an opposing force of magnitude $|F|$ is

$$dw = -|F|dz \qquad \text{Definition} \quad \text{Work done} \tag{2A.4}$$

The negative sign tells us that, when the system moves an object against an opposing force of magnitude $|F|$, and there are no other changes, then the internal energy of the system doing the work will decrease. That is, if dz is positive (motion to positive z), dw is negative, and the internal energy decreases (dU in eqn 2A.3 is negative provided that $dq=0$).

Now consider the arrangement shown in Fig. 2A.5, in which one wall of a system is a massless, frictionless, rigid, perfectly fitting piston of area A. If the external pressure is p_{ex}, the magnitude of the force acting on the outer face of the piston is $|F| = p_{ex}A$. When the system expands through a distance dz against an external pressure p_{ex}, it follows that the work done is $dw = -p_{ex}Adz$. The quantity Adz is the change in volume, dV, in the course of the expansion. Therefore, the work done when the system expands by dV against a pressure p_{ex} is

$$dw = -p_{ex}dV \qquad \text{Expansion work} \tag{2A.5a}$$

To obtain the total work done when the volume changes from an initial value V_i to a final value V_f we integrate this expression between the initial and final volumes:

$$w = -\int_{V_i}^{V_f} p_{ex}dV \tag{2A.5b}$$

The force acting on the piston, $p_{ex}A$, is equivalent to the force arising from a weight that is raised as the system expands. If the system is compressed instead, then the same weight is lowered in the surroundings and eqn 2A.5b can still be used, but now $V_f < V_i$. It is important to note that it is still the external pressure that determines the magnitude of the work. This somewhat perplexing conclusion seems to be inconsistent with the fact that the gas *inside* the container is opposing the compression. However, when a gas is compressed, the ability of the *surroundings* to do work is diminished by an amount determined by the weight that is lowered, and it is this energy that is transferred into the system.

Other types of work (for example, electrical work), which we shall call either **non-expansion work** or **additional work**, have

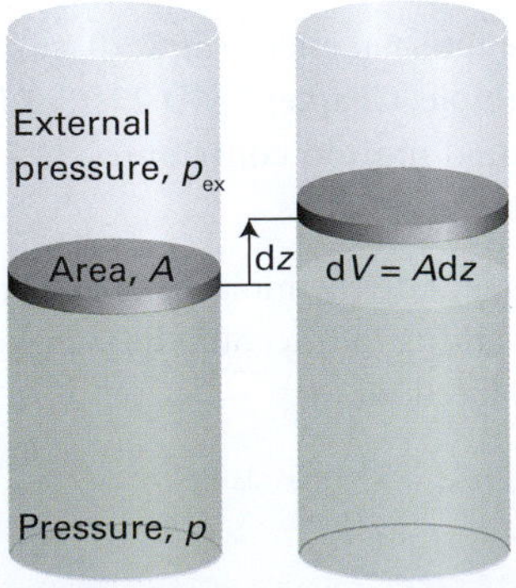

Figure 2A.5 When a piston of area A moves out through a distance dz, it sweeps out a volume $dV = Adz$. The external pressure p_{ex} is equivalent to a weight pressing on the piston, and the magnitude of the force opposing expansion is $|F| = p_{ex}A$.

Table 2A.1 Varieties of work*

Type of work	dw	Comments	Units†
Expansion	$-p_{ex}dV$	p_{ex} is the external pressure dV is the change in volume	Pa m^3
Surface expansion	$\gamma d\sigma$	γ is the surface tension $d\sigma$ is the change in area	N m^{-1} m^2
Extension	fdl	f is the tension dl is the change in length	N m
Electrical	ϕdQ	ϕ is the electric potential dQ is the change in charge	V C
	$Qd\phi$	$d\phi$ is the potential difference Q is the charge transferred	V C

* In general, the work done on a system can be expressed in the form $dw=-|F|dz$, where $|F|$ is the magnitude of a 'generalized force' and dz is a 'generalized displacement'.

† For work in joules (J). Note that 1 N m = 1 J and 1 V C = 1 J.

analogous expressions, with each one the product of an intensive factor (the pressure, for instance) and an extensive factor (the change in volume). Some are collected in Table 2A.1. For the present we continue with the work associated with changing the volume, the expansion work, and see what we can extract from eqn 2A.5b.

Brief illustration 2A.4 The work of extension

To establish an expression for the work of stretching an elastomer, a polymer that can stretch and contract, to an extension l given that the force opposing extension is proportional to the displacement from the resting state of the elastomer we write $|F|=k_f x$, where k_f is a constant and x is the displacement. It then follows from eqn 2A.4 that for an infinitesimal displacement from x to $x+dx$, $dw=-k_f x dx$. For the overall work of displacement from $x=0$ to the final extension l,

$$w=-\int_0^l k_f x\,dx=-\tfrac{1}{2}k_f l^2$$

Self-test 2A.4 Suppose the restoring force weakens as the elastomer is stretched, and $k_f(x)=a-bx^{1/2}$. Evaluate the work of extension to l.

Answer: $w=-\tfrac{1}{2}al^2+\tfrac{2}{5}bl^{5/2}$

(b) Expansion against constant pressure

Suppose that the external pressure is constant throughout the expansion. For example, the piston may be pressed on by the atmosphere, which exerts the same pressure throughout the expansion. A chemical example of this condition is the expansion of a gas formed in a chemical reaction in a container that can expand. We can evaluate eqn 2A.5b by taking the constant p_{ex} outside the integral:

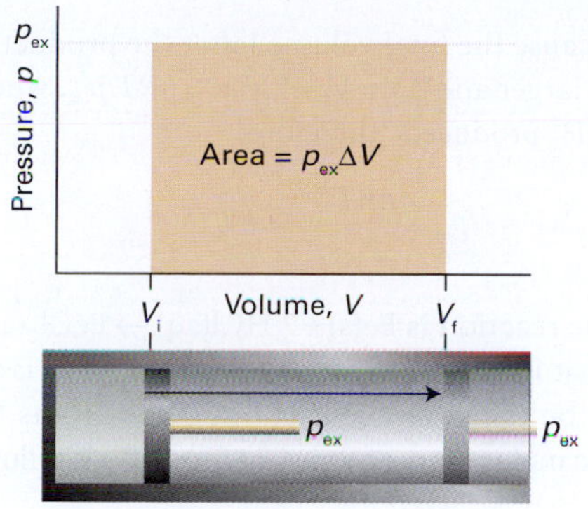

Figure 2A.6 The work done by a gas when it expands against a constant external pressure, p_{ex}, is equal to the shaded area in this example of an indicator diagram.

$$w=-p_{ex}\int_{V_i}^{V_f} dV=-p_{ex}(V_f-V_i)$$

Therefore, if we write the change in volume as $\Delta V=V_f-V_i$,

$$w=-p_{ex}\Delta V \quad \textit{Constant external pressure} \quad \text{Expansion work} \qquad (2A.6)$$

This result is illustrated graphically in Fig. 2A.6, which makes use of the fact that an integral can be interpreted as an area. The magnitude of w, denoted $|w|$, is equal to the area beneath the horizontal line at $p=p_{ex}$ lying between the initial and final volumes. A pV-graph used to illustrate expansion work is called an **indicator diagram**; James Watt first used one to indicate aspects of the operation of his steam engine.

Free expansion is expansion against zero opposing force. It occurs when $p_{ex}=0$. According to eqn 2A.6,

$$w=0 \quad \text{Work of free expansion} \qquad (2A.7)$$

That is, no work is done when a system expands freely. Expansion of this kind occurs when a gas expands into a vacuum.

Example 2A.1 Calculating the work of gas production

Calculate the work done when 50 g of iron reacts with hydrochloric acid to produce $FeCl_2(aq)$ and hydrogen in (a) a closed vessel of fixed volume, (b) an open beaker at 25 °C.

Method We need to judge the magnitude of the volume change and then to decide how the process occurs. If there is no change in volume, there is no expansion work however the process takes place. If the system expands against a constant external pressure, the work can be calculated from eqn 2A.6. A general feature of processes in which a condensed phase changes into a gas is that the volume of the former may usually be neglected relative to that of the gas it forms.

Answer In (a) the volume cannot change, so no expansion work is done and $w=0$. In (b) the gas drives back the atmosphere and therefore $w=-p_{ex}\Delta V$. We can neglect the initial volume because the final volume (after the production of gas) is so much larger and $\Delta V = V_f - V_i \approx V_f = nRT/p_{ex}$, where n is the amount of H_2 produced. Therefore,

$$w = -p_{ex}\Delta V \approx -p_{ex} \times \frac{nRT}{p_{ex}} = -nRT$$

Because the reaction is $Fe(s) + 2\,HCl(aq) \rightarrow FeCl_2(aq) + H_2(g)$, we know that 1 mol H_2 is generated when 1 mol Fe is consumed, and n can be taken as the amount of Fe atoms that react. Because the molar mass of Fe is 55.85 g mol^{-1}, it follows that

$$w = -\frac{50\,\text{g}}{55.85\,\text{g mol}^{-1}} \times (8.3145\,\text{J K}^{-1}\,\text{mol}^{-1}) \times (298\,\text{K})$$
$$\approx -2.2\,\text{kJ}$$

The system (the reaction mixture) does 2.2 kJ of work driving back the atmosphere. Note that (for this perfect gas system) the magnitude of the external pressure does not affect the final result: the lower the pressure, the larger the volume occupied by the gas, so the effects cancel.

Self-test 2A.5 Calculate the expansion work done when 50 g of water is electrolysed under constant pressure at 25 °C.

Answer: −10 kJ

(c) Reversible expansion

A **reversible change** in thermodynamics is a change that can be reversed by an infinitesimal modification of a variable. The key word 'infinitesimal' sharpens the everyday meaning of the word 'reversible' as something that can change direction. One example of reversibility that we have encountered already is the thermal equilibrium of two systems with the same temperature. The transfer of energy as heat between the two is reversible because, if the temperature of either system is lowered infinitesimally, then energy flows into the system with the lower temperature. If the temperature of either system at thermal equilibrium is raised infinitesimally, then energy flows out of the hotter system. There is obviously a very close relationship between reversibility and equilibrium: systems at equilibrium are poised to undergo reversible change.

Suppose a gas is confined by a piston and that the external pressure, p_{ex}, is set equal to the pressure, p, of the confined gas. Such a system is in mechanical equilibrium with its surroundings because an infinitesimal change in the external pressure in either direction causes changes in volume in opposite directions. If the external pressure is reduced infinitesimally, the gas expands slightly. If the external pressure is increased infinitesimally, the gas contracts slightly. In either case the change is reversible in the thermodynamic sense. If, on the other hand, the external pressure differs measurably from the internal pressure, then changing p_{ex} infinitesimally will not decrease it below the pressure of the gas, so will not change the direction of the process. Such a system is not in mechanical equilibrium with its surroundings and the expansion is thermodynamically irreversible.

To achieve reversible expansion we set p_{ex} equal to p at each stage of the expansion. In practice, this equalization could be achieved by gradually removing weights from the piston so that the downward force due to the weights always matches the changing upward force due to the pressure of the gas. When we set $p_{ex} = p$, eqn 2A.5a becomes

$$dw = -p_{ex}dV = -pdV \qquad \text{Reversible expansion work} \qquad (2A.8a)$$

Although the pressure inside the system appears in this expression for the work, it does so only because p_{ex} has been set equal to p to ensure reversibility. The total work of reversible expansion from an initial volume V_i to a final volume V_f is therefore

$$w = -\int_{V_i}^{V_f} p\,dV \qquad (2A.8b)$$

The integral can be evaluated once we know how the pressure of the confined gas depends on its volume. Equation 2A.8b is the link with the material covered in the Topics of Chapter 1 for, if we know the equation of state of the gas, then we can express p in terms of V and evaluate the integral.

(d) Isothermal reversible expansion

Consider the isothermal, reversible expansion of a perfect gas. The expansion is made isothermal by keeping the system in thermal contact with its surroundings (which may be a constant-temperature bath). Because the equation of state is $pV = nRT$, we know that at each stage $p = nRT/V$, with V the volume at that stage of the expansion. The temperature T is constant in an isothermal expansion, so (together with n and R) it may be taken outside the integral. It follows that the work of reversible isothermal expansion of a perfect gas from V_i to V_f at a temperature T is

$$w = -nRT\int_{V_i}^{V_f} \frac{dV}{V} \overset{\text{Integral A.2}}{=} -nRT\ln\frac{V_f}{V_i}$$

Perfect gas, reversible, isothermal Work of expansion (2A.9)

Brief illustration 2A.5 The work of isothermal reversible expansion

When a sample of 1.00 mol Ar, regarded here as a perfect gas, undergoes an isothermal reversible expansion at 20.0 °C from 10.0 dm^3 to 30.0 dm^3 the work done is

$$w = -(1.00\,\text{mol}) \times (8.3145\,\text{J K}^{-1}\,\text{mol}^{-1}) \times (293.2\,\text{K})\ln\frac{30.0\,\text{dm}^3}{10.0\,\text{dm}^3}$$
$$= -2.68\,\text{kJ}$$

Self-test 2A.6 Suppose that attractions are important between gas molecules, and the equation of state is $p=nRT/V - n^2a/V^2$. Derive an expression for the reversible, isothermal expansion of this gas. Is more or less work done *on the surroundings* when it expands (compared with a perfect gas)?

Answer: $w=-nRT\ln(V_f/V_i)-n^2a(1/V_f-1/V_i)$; less

When the final volume is greater than the initial volume, as in an expansion, the logarithm in eqn 2A.9 is positive and hence $w<0$. In this case, the system has done work on the surroundings and there is a corresponding negative contribution to its internal energy. (Note the cautious language: we shall see later that there is a compensating influx of energy as heat, so overall the internal energy is constant for the isothermal expansion of a perfect gas.) The equations also show that more work is done for a given change of volume when the temperature is increased: at a higher temperature the greater pressure of the confined gas needs a higher opposing pressure to ensure reversibility and the work done is correspondingly greater.

We can express the result of the calculation as an indicator diagram, for the magnitude of the work done is equal to the area under the isotherm $p=nRT/V$ (Fig. 2A.7). Superimposed on the diagram is the rectangular area obtained for irreversible expansion against constant external pressure fixed at the same final value as that reached in the reversible expansion. More work is obtained when the expansion is reversible (the area is greater) because matching the external pressure to the internal pressure at each stage of the process ensures that none of the system's pushing power is wasted. We cannot obtain more work than for the reversible process because increasing the external pressure even infinitesimally at any stage results in compression. We may infer from this discussion that, because some pushing power is wasted when $p>p_{ex}$, the maximum work available from a system operating between specified initial and final states and passing along a specified path is obtained when the change takes place reversibly.

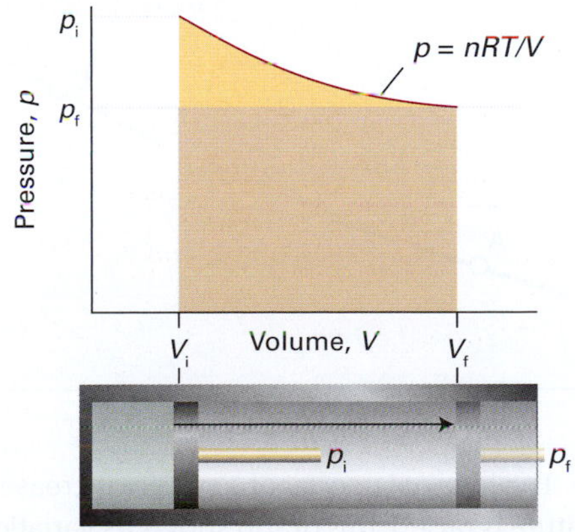

Figure 2A.7 The work done by a perfect gas when it expands reversibly and isothermally is equal to the area under the isotherm $p=nRT/V$. The work done during the irreversible expansion against the same final pressure is equal to the rectangular area shown slightly darker. Note that the reversible work is greater than the irreversible work.

We have introduced the connection between reversibility and maximum work for the special case of a perfect gas undergoing expansion. In Topic 3A we see that it applies to all substances and to all kinds of work.

2A.4 Heat transactions

In general, the change in internal energy of a system is

$$dU=dq+dw_{exp}+dw_e \tag{2A.10}$$

where dw_e is work in addition (e for 'extra') to the expansion work, dw_{exp}. For instance, dw_e might be the electrical work of driving a current through a circuit. A system kept at constant volume can do no expansion work, so $dw_{exp}=0$. If the system is also incapable of doing any other kind of work (if it is not, for instance, an electrochemical cell connected to an electric motor), then $dw_e=0$ too. Under these circumstances:

$$dU=dq \quad \text{Heat transferred at constant volume} \tag{2A.11a}$$

We express this relation by writing $dU=dq_V$, where the subscript implies a change at constant volume. For a measurable change between states i and f along a path at constant volume,

$$\overbrace{\int_i^f dU}^{U_f-U_i}=\overbrace{\int_i^f dq}^{q_V}$$

which we summarize as

$$\Delta U=q_V \tag{2A.11b}$$

Note that we do not write the integral over dq as Δq because q, unlike U, is not a state function. It follows that, by measuring the energy supplied to a constant-volume system as heat ($q_V>0$) or released from it as heat ($q_V<0$) when it undergoes a change of state, we are in fact measuring the change in its internal energy.

(a) Calorimetry

Calorimetry is the study of the transfer of energy as heat during physical and chemical processes. A **calorimeter** is a device for measuring energy transferred as heat. The most common device for measuring q_V (and therefore ΔU) is an **adiabatic bomb calorimeter** (Fig. 2A.8). The process we wish to study—which may be a chemical reaction—is initiated inside a constant-volume container, the 'bomb'. The bomb is immersed in

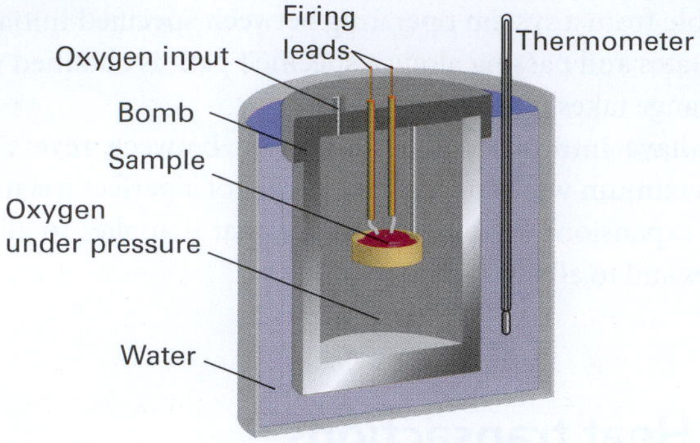

Figure 2A.8 A constant-volume bomb calorimeter. The 'bomb' is the central vessel, which is strong enough to withstand high pressures. The calorimeter (for which the heat capacity must be known) is the entire assembly shown here. To ensure adiabaticity, the calorimeter is immersed in a water bath with a temperature continuously readjusted to that of the calorimeter at each stage of the combustion.

a stirred water bath, and the whole device is the calorimeter. The calorimeter is also immersed in an outer water bath. The water in the calorimeter and of the outer bath are both monitored and adjusted to the same temperature. This arrangement ensures that there is no net loss of heat from the calorimeter to the surroundings (the bath) and hence that the calorimeter is adiabatic.

The change in temperature, ΔT, of the calorimeter is proportional to the energy that the reaction releases or absorbs as heat. Therefore, by measuring ΔT we can determine q_V and hence find ΔU. The conversion of ΔT to q_V is best achieved by calibrating the calorimeter using a process of known energy output and determining the **calorimeter constant**, the constant C in the relation

$$q = C\Delta T \tag{2A.12}$$

The calorimeter constant may be measured electrically by passing a constant current, I, from a source of known potential difference, $\Delta\phi$, through a heater for a known period of time, t, for then

$$q = It\Delta\phi \tag{2A.13}$$

Electrical charge is measured in *coulombs*, C. The motion of charge gives rise to an electric current, I, measured in coulombs per second, or *amperes*, A, where $1\ \mathrm{A} = 1\ \mathrm{C\ s^{-1}}$. If a constant current I flows through a potential difference $\Delta\phi$ (measured in volts, V), the total energy supplied in an interval t is $It\Delta\phi$. Because $1\ \mathrm{A\,V\,s} = 1\ (\mathrm{C\,s^{-1}})\ \mathrm{V\,s} = 1\ \mathrm{C\,V} = 1\ \mathrm{J}$, the energy is obtained in joules with the current in amperes, the potential difference in volts, and the time in seconds.

Brief illustration 2A.6 Electrical heating

If a current of 10.0 A from a 12 V supply is passed for 300 s, then from eqn 2A.13 the energy supplied as heat is

$$q = (10.0\,\mathrm{A}) \times (12\,\mathrm{V}) \times (300\,\mathrm{s}) = 3.6 \times 10^4\,\mathrm{A\,V\,s} = 36\,\mathrm{kJ}$$

because $1\ \mathrm{A\ V\ s} = 1\ \mathrm{J}$. If the observed rise in temperature is 5.5 K, then the calorimeter constant is $C = (36\,\mathrm{kJ})/(5.5\,\mathrm{K}) = 6.5\,\mathrm{kJ\,K^{-1}}$.

Self-test 2A.7 What is the value of the calorimeter constant if the temperature rises by 4.8 °C when a current of 8.6 A from an 11 V supply is passed for 280 s?

Answer: $5.5\,\mathrm{kJ\,K^{-1}}$

Alternatively, C may be determined by burning a known mass of substance (benzoic acid is often used) that has a known heat output. With C known, it is simple to interpret an observed temperature rise as a release of heat.

(b) Heat capacity

The internal energy of a system increases when its temperature is raised. The increase depends on the conditions under which the heating takes place and for the present we suppose that the system has a constant volume. For example, it may be a gas in a container of fixed volume. If the internal energy is plotted against temperature, then a curve like that in Fig. 2A.9 may be obtained. The slope of the tangent to the curve at any temperature is called the **heat capacity** of the system at that temperature. The **heat capacity at constant volume** is denoted C_V and is defined formally as

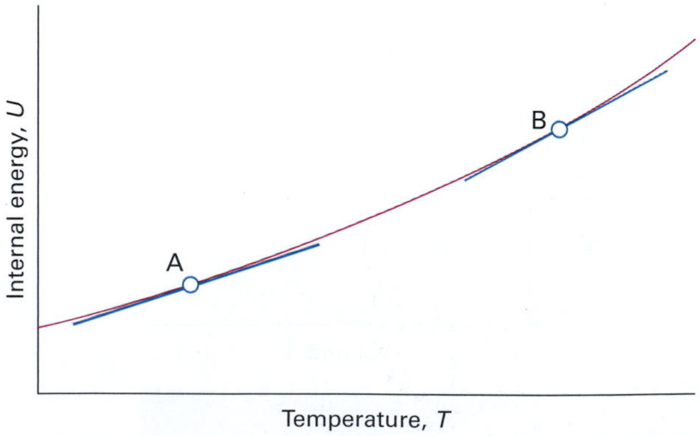

Figure 2A.9 The internal energy of a system increases as the temperature is raised; this graph shows its variation as the system is heated at constant volume. The slope of the tangent to the curve at any temperature is the heat capacity at constant volume at that temperature. Note that, for the system illustrated, the heat capacity is greater at B than at A.

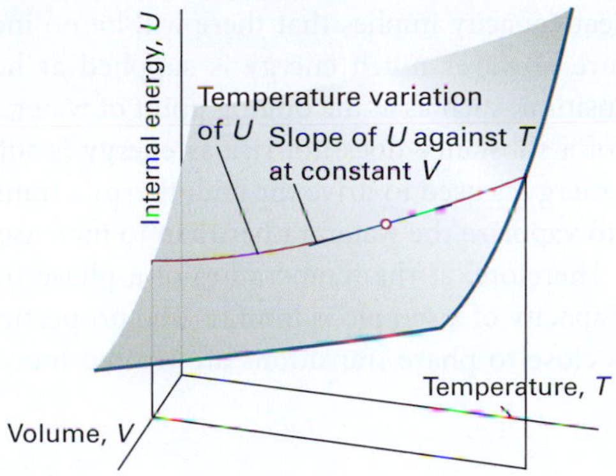

Figure 2A.10 The internal energy of a system varies with volume and temperature, perhaps as shown here by the surface. The variation of the internal energy with temperature at one particular constant volume is illustrated by the curve drawn parallel to T. The slope of this curve at any point is the partial derivative $(\partial U/\partial T)_V$.

$$C_V = \left(\frac{\partial U}{\partial T}\right)_V \quad \textit{Definition} \quad \text{Heat capacity at constant volume} \quad (2A.14)$$

Partial derivatives are reviewed in *Mathematical background* 2 following this chapter. The internal energy varies with the temperature and the volume of the sample, but here we are interested only in its variation with the temperature, the volume being held constant (Fig. 2A.10).

Brief illustration 2A.7 Heat capacity

The heat capacity of a monatomic perfect gas can be calculated by inserting the expression for the internal energy derived in *Brief illustration* 2A.2 where we saw that $U_m(T)=U_m(0)+\frac{3}{2}RT$, so from eqn 2A.14

$$C_{V,m} = \frac{\partial}{\partial T}\{U_m(0)+\tfrac{3}{2}RT\} = \tfrac{3}{2}R$$

The numerical value is 12.47 J K^{-1} mol^{-1}.

Self-test 2A.8 Estimate the molar constant-volume heat capacity of carbon dioxide.

Answer: $\frac{5}{2}R = 21\,\text{J K}^{-1}\,\text{mol}^{-1}$

Heat capacities are extensive properties: 100 g of water, for instance, has 100 times the heat capacity of 1 g of water (and therefore requires 100 times the energy as heat to bring about the same rise in temperature). The **molar heat capacity at constant volume**, $C_{V,m}=C_V/n$, is the heat capacity per mole of substance, and is an intensive property (all molar quantities are intensive). Typical values of $C_{V,m}$ for polyatomic gases are close to 25 J K^{-1} mol^{-1}. For certain applications it is useful to know the **specific heat capacity** (more informally, the 'specific heat') of a substance, which is the heat capacity of the sample divided by the mass, usually in grams: $C_{V,s}=C_V/m$. The specific heat capacity of water at room temperature is close to 4.2 J K^{-1} g^{-1}. In general, heat capacities depend on the temperature and decrease at low temperatures. However, over small ranges of temperature at and above room temperature, the variation is quite small and for approximate calculations heat capacities can be treated as almost independent of temperature.

The heat capacity is used to relate a change in internal energy to a change in temperature of a constant-volume system. It follows from eqn 2A.14 that

$$dU = C_V dT \quad \textit{Constant volume} \quad (2A.15a)$$

That is, at constant volume, an infinitesimal change in temperature brings about an infinitesimal change in internal energy, and the constant of proportionality is C_V. If the heat capacity is independent of temperature over the range of temperatures of interest, then

$$\Delta U = \int_{T_1}^{T_2} C_V dT = C_V \int_{T_1}^{T_2} dT = C_V \overbrace{(T_2 - T_1)}^{\Delta T}$$

and a measurable change of temperature, ΔT, brings about a measurable change in internal energy, ΔU, where

$$\Delta U = C_V \Delta T \quad \textit{Constant volume} \quad (2A.15b)$$

Because a change in internal energy can be identified with the heat supplied at constant volume (eqn 2A.11b), the last equation can also be written

$$q_V = C_V \Delta T \quad (2A.16)$$

This relation provides a simple way of measuring the heat capacity of a sample: a measured quantity of energy is transferred as heat to the sample (electrically, for example), and the resulting increase in temperature is monitored. The ratio of the energy transferred as heat to the temperature rise it causes ($q_V/\Delta T$) is the constant-volume heat capacity of the sample.

Brief illustration 2A.8 The determination of a heat capacity

Suppose a 55 W electric heater immersed in a gas in a constant-volume adiabatic container was on for 120 s and it was found that the temperature of the gas rose by 5.0 °C (an increase equivalent to 5.0 K). The heat supplied is (55 W)×(120 s) = 6.6 kJ (we have used 1 J = 1 W s), so the heat capacity of the sample is

$$C_V = \frac{6.6\text{ kJ}}{5.0\text{ K}} = 1.3\text{ kJ K}^{-1}$$

Self-test 2A.9 When 229 J of energy is supplied as heat to 3.0 mol of a gas at constant volume, the temperature of the gas increases by 2.55 °C. Calculate C_V and the molar heat capacity at constant volume.

Answer: $89.8\,\mathrm{J\,K^{-1}}$, $29.9\,\mathrm{J\,K^{-1}\,mol^{-1}}$

A large heat capacity implies that, for a given quantity of energy transferred as heat, there will be only a small increase in temperature (the sample has a large capacity for heat). An infinite heat capacity implies that there will be no increase in temperature however much energy is supplied as heat. At a phase transition, such as at the boiling point of water, the temperature of a substance does not rise as energy is supplied as heat: the energy is used to drive the endothermic transition, in this case to vaporize the water, rather than to increase its temperature. Therefore, at the temperature of a phase transition, the heat capacity of a sample is infinite. The properties of heat capacities close to phase transitions are treated more fully in Topic 4B.

Checklist of concepts

☐ 1. **Work** is done to achieve motion against an opposing force
☐ 2. **Energy** is the capacity to do work.
☐ 3. **Heating** is the transfer of energy that makes use of disorderly molecular motion.
☐ 4. Work is the transfer of energy that makes use of organized motion.
☐ 5. **Internal energy**, the total energy of a system, is a state function.
☐ 6. The **equipartition theorem** can be used to estimate the contribution to the internal energy of classical modes of motion.
☐ 7. The **First Law** states that the internal energy of an isolated system is constant.
☐ 8. Free expansion (expansion against zero pressure) does no work.
☐ 9. To achieve **reversible expansion**, the external pressure is matched at every stage to the pressure of the system.
☐ 10. The energy transferred as heat at constant volume is equal to the change in internal energy of the system.
☐ 11. **Calorimetry** is the measurement of heat transactions.

Checklist of equations

Property	Equation	Comment	Equation number
First Law of thermodynamics	$\Delta U = q + w$	Acquisitive convention	2A.2
Work of expansion	$\mathrm{d}w = -p_{\mathrm{ex}}\mathrm{d}V$		2A.5a
Work of expansion against a constant external pressure	$w = -p_{\mathrm{ex}}\Delta V$	$p_{\mathrm{ex}} = 0$ corresponds to free expansion	2A.6
Reversible work of expansion of a gas	$w = -nRT\ln(V_f/V_i)$	Isothermal, perfect gas	2A.9
Internal energy change	$\Delta U = q_V$	Constant volume, no other forms of work	2A.11b
Electrical heating	$q = It\Delta\phi$		2A.13
Heat capacity at constant volume	$C_V = (\partial U/\partial T)_V$	Definition	2A.14

2B Enthalpy

Contents

➤ Why do you need to know this material?

The concept of enthalpy is central to many thermodynamic discussions about processes taking place under conditions of constant pressure, such as the discussion of the heat requirements or output of physical transformations and chemical reactions.

➤ What is the key idea?

A change in enthalpy is equal to the energy transferred as heat at constant pressure.

➤ What do you need to know already?

This Topic makes use of the discussion of internal energy (Topic 2A) and draws on some aspects of perfect gases (Topic 1A).

The change in internal energy is not equal to the energy transferred as heat when the system is free to change its volume, such as when it is able to expand or contract under conditions of constant pressure. Under these circumstances some of the energy supplied as heat to the system is returned to the surroundings as expansion work (Fig. 2B.1), so dU is less than dq. In this case the energy supplied as heat at constant pressure is equal to the change in another thermodynamic property of the system, the enthalpy.

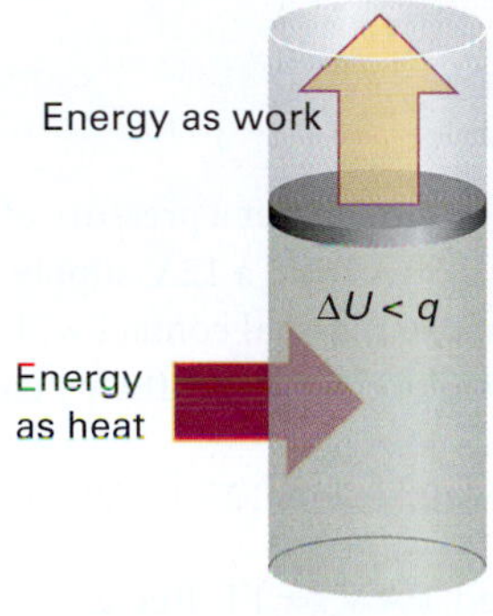

Figure 2B.1 When a system is subjected to constant pressure and is free to change its volume, some of the energy supplied as heat may escape back into the surroundings as work. In such a case, the change in internal energy is smaller than the energy supplied as heat.

2B.1 The definition of enthalpy

The **enthalpy**, H, is defined as

$$H = U + pV \qquad \text{Definition} \quad \text{Enthalpy} \qquad (2B.1)$$

where p is the pressure of the system and V is its volume. Because U, p, and V are all state functions, the enthalpy is a state function too. As is true of any state function, the change in enthalpy, ΔH, between any pair of initial and final states is independent of the path between them.

(a) Enthalpy change and heat transfer

Although the definition of enthalpy may appear arbitrary, it has important implications for thermochemistry. For instance, we show in the following *Justification* that eqn 2B.1 implies that *the change in enthalpy is equal to the energy supplied as heat at constant pressure* (provided the system does no additional work):

$$\mathrm{d}H = \mathrm{d}q_p \qquad \text{Heat transferred at constant pressure} \qquad (2B.2a)$$

For a measurable change between states i and f along a path at constant pressure, we write

$$\overbrace{\int_{\mathrm{i}}^{\mathrm{f}} \mathrm{d}H}^{H_\mathrm{f}-H_\mathrm{i}} = \overbrace{\int_{\mathrm{i}}^{\mathrm{f}} \mathrm{d}q_p}^{q_p}$$

and summarize the result as

$$\Delta H = q_p \tag{2B.2b}$$

Note that we do not write the integral over dq as Δq because q, unlike H, is not a state function.

Brief illustration 2B.1 A change in enthalpy

Water is heated to boiling under a pressure of 1.0 atm. When an electric current of 0.50 A from a 12 V supply is passed for 300 s through a resistance in thermal contact with it, it is found that 0.798 g of water is vaporized. The enthalpy change is

$$\Delta H = q_p = It\Delta\phi = (0.50\,\text{A})\times(12\,\text{V})\times(300\,\text{s}) = (0.50\times12\times300)\,\text{J}$$

Here we have used 1 A V s = 1 J. Because 0.798 g of water is $(0.798\,\text{g})/(18.02\,\text{g mol}^{-1}) = (0.798/18.02)$ mol H_2O, the enthalpy of vaporization per mole of H_2O is

$$\Delta H_\text{m} = \frac{(0.50\times12\times300)\,\text{J}}{(0.798/18.02)\,\text{mol}} = +41\,\text{kJ mol}^{-1}$$

Self-test 2B.1 The molar enthalpy of vaporization of benzene at its boiling point (353.25 K) is 30.8 kJ mol^{-1}. For how long would the same 12 V source need to supply a 0.50 A current in order to vaporize a 10 g sample?

Answer: 6.6×10^2 s

Justification 2B.1 The relation $\Delta H = q_p$

For a general infinitesimal change in the state of the system, U changes to $U + dU$, p changes to $p + dp$, and V changes to $V + dV$, so from the definition in eqn 2B.1, H changes from $U + pV$ to

$$\begin{aligned} H + dH &= (U + dU) + (p + dp)(V + dV) \\ &= U + dU + pV + p dV + V dp + dp dV \end{aligned}$$

The last term is the product of two infinitesimally small quantities and can therefore be neglected. As a result, after recognizing $U + pV = H$ on the right (in blue), we find that H changes to

$$H + dH = H + dU + p dV + V dp$$

and hence that

$$dH = dU + p dV + V dp$$

If we now substitute $dU = dq + dw$ into this expression, we get

$$dH = dq + dw + p dV + V dp$$

If the system is in mechanical equilibrium with its surroundings at a pressure p and does only expansion work, we can write $dw = -p dV$ and obtain

$$dH = dq + V dp$$

Now we impose the condition that the heating occurs at constant pressure by writing $dp = 0$. Then

$$dH = dq \text{ (at constant pressure, no additional work)}$$

as in eqn 2B.2a. Equation 2B.2b then follows, as explained in the text.

(b) Calorimetry

The process of measuring heat transactions between a system and its surroundings is called **calorimetry**. An enthalpy change can be measured calorimetrically by monitoring the temperature change that accompanies a physical or chemical change occurring at constant pressure. A calorimeter for studying processes at constant pressure is called an **isobaric calorimeter**. A simple example is a thermally insulated vessel open to the atmosphere: the heat released in the reaction is monitored by measuring the change in temperature of the contents. For a combustion reaction an **adiabatic flame calorimeter** may be used to measure ΔT when a given amount of substance burns in a supply of oxygen (Fig. 2B.2).

Another route to ΔH is to measure the internal energy change by using a bomb calorimeter, and then to convert ΔU to ΔH. Because solids and liquids have small molar volumes, for them pV_m is so small that the molar enthalpy and molar internal energy are almost identical ($H_\text{m} = U_\text{m} + pV_\text{m} \approx U_\text{m}$). Consequently, if a process involves only solids or liquids, the values of ΔH and ΔU are almost identical. Physically, such processes are accompanied by a very small change in volume; the system does negligible work on the surroundings when the process occurs, so the energy supplied as heat stays entirely

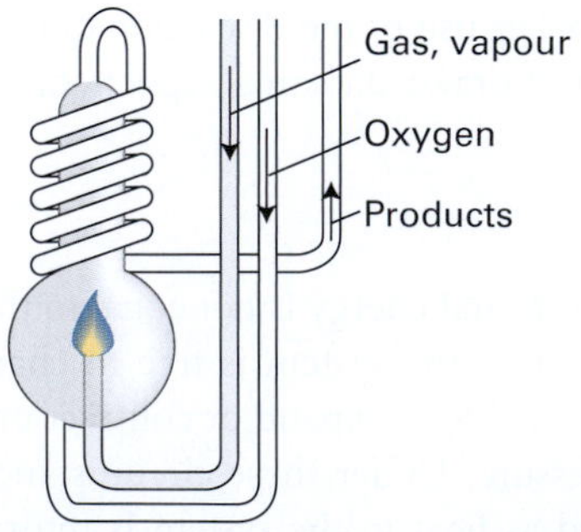

Figure 2B.2 A constant-pressure flame calorimeter consists of this component immersed in a stirred water bath. Combustion occurs as a known amount of reactant is passed through to fuel the flame, and the rise of temperature is monitored.

within the system. The most sophisticated way to measure enthalpy changes, however, is to use a *differential scanning calorimeter* (DSC), as explained in Topic 2C. Changes in enthalpy and internal energy may also be measured by non-calorimetric methods (see Topic 6C).

Example 2B.1 Relating ΔH and ΔU

The change in molar internal energy when $CaCO_3(s)$ as calcite converts to another form, aragonite, is $+0.21\,kJ\,mol^{-1}$. Calculate the difference between the molar enthalpy and internal energy changes when the pressure is 1.0 bar given that the densities of the polymorphs are $2.71\,g\,cm^{-3}$ (calcite) and $2.93\,g\,cm^{-3}$ (aragonite).

Method The starting point for the calculation is the relation between the enthalpy of a substance and its internal energy (eqn 2B.1). The difference between the two quantities can be expressed in terms of the pressure and the difference of their molar volumes, and the latter can be calculated from their molar masses, M, and their mass densities, ρ, by using $\rho = M/V_m$.

Answer The change in enthalpy when the transition occurs is

$$\begin{aligned}\Delta H_m &= H_m(\text{aragonite}) - H_m(\text{calcite})\\ &= \{U_m(a) + pV_m(a)\} - \{U_m(c) + pV_m(c)\}\\ &= \Delta U_m + p\{V_m(a) - V_m(c)\}\end{aligned}$$

where a denotes aragonite and c calcite. It follows by substituting $V_m = M/\rho$ that

$$\Delta H_m - \Delta U_m = pM\left(\frac{1}{\rho(a)} - \frac{1}{\rho(c)}\right)$$

Substitution of the data, using $M = 100.09\,g\,mol^{-1}$, gives

$$\begin{aligned}\Delta H_m - \Delta U_m &= (1.0\times10^5\,Pa)\times(100.09\,g\,mol^{-1})\\ &\quad\times\left(\frac{1}{2.93\,g\,cm^{-3}} - \frac{1}{2.71\,g\,cm^{-3}}\right)\\ &= -2.8\times10^5\,Pa\,cm^3\,mol^{-1} = -0.28\,Pa\,m^3\,mol^{-1}\end{aligned}$$

Hence (because $1\,Pa\,m^3 = 1\,J$), $\Delta H_m - \Delta U_m = -0.28\,J\,mol^{-1}$, which is only 0.1 per cent of the value of ΔU_m. We see that it is usually justifiable to ignore the difference between the molar enthalpy and internal energy of condensed phases, except at very high pressures, when $p\Delta V_m$ is no longer negligible.

Self-test 2B.2 Calculate the difference between ΔH and ΔU when 1.0 mol Sn(s, grey) of density $5.75\,g\,cm^{-3}$ changes to Sn(s, white) of density $7.31\,g\,cm^{-3}$ at 10.0 bar. At 298 K, $\Delta H = +2.1\,kJ$.

Answer: $\Delta H - \Delta U = -4.4\,J$

In contrast to processes involving condensed phases, the values of the changes in internal energy and enthalpy may differ significantly for processes involving gases. Thus, the enthalpy of a perfect gas is related to its internal energy by using $pV = nRT$ in the definition of H:

$$H = U + pV = U + nRT \tag{2B.3}$$

This relation implies that the change of enthalpy in a reaction that produces or consumes gas under isothermal conditions is

$$\Delta H = \Delta U + \Delta n_g RT \qquad \textit{Perfect gas, isothermal} \qquad \text{Relation between } \Delta H \text{ and } \Delta U \tag{2B.4}$$

where Δn_g is the change in the amount of gas molecules in the reaction.

Brief illustration 2B.2 Processes involving gases

In the reaction $2\,H_2(g) + O_2(g) \rightarrow 2\,H_2O(l)$, 3 mol of gas-phase molecules are replaced by 2 mol of liquid-phase molecules, so $\Delta n_g = -3\,mol$. Therefore, at 298 K, when $RT = 2.5\,kJ\,mol^{-1}$, the enthalpy and internal energy changes taking place in the system are related by

$$\Delta H_m - \Delta U_m = (-3\,mol)\times RT \approx -7.4\,kJ\,mol^{-1}$$

Note that the difference is expressed in kilojoules, not joules as in *Example* 2B.2. The enthalpy change is smaller (in this case, less negative) than the change in internal energy because, although heat escapes from the system when the reaction occurs, the system contracts when the liquid is formed, so energy is restored to it from the surroundings.

Self-test 2B.3 Calculate the value of $\Delta H_m - \Delta U_m$ for the reaction $N_2(g) + 3\,H_2(g) \rightarrow 2\,NH_3(g)$.

Answer: $-5.0\,kJ\,mol^{-1}$

2B.2 The variation of enthalpy with temperature

The enthalpy of a substance increases as its temperature is raised. The relation between the increase in enthalpy and the increase in temperature depends on the conditions (for example, constant pressure or constant volume).

(a) Heat capacity at constant pressure

The most important condition is constant pressure, and the slope of the tangent to a plot of enthalpy against temperature at constant pressure is called the **heat capacity at constant**

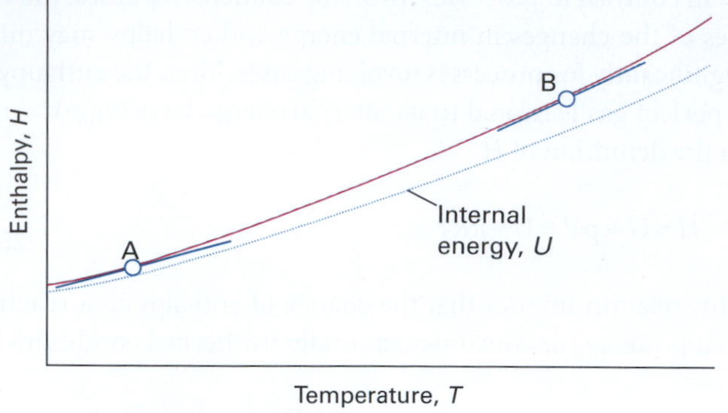

Figure 2B.3 The constant-pressure heat capacity at a particular temperature is the slope of the tangent to a curve of the enthalpy of a system plotted against temperature (at constant pressure). For gases, at a given temperature the slope of enthalpy versus temperature is steeper than that of internal energy versus temperature, and $C_{p,m}$ is larger than $C_{V,m}$.

pressure (or *isobaric heat capacity*), C_p, at a given temperature (Fig. 2B.3). More formally:

$$C_p = \left(\frac{\partial H}{\partial T}\right)_p \quad \textit{Definition} \quad \text{Heat capacity at constant pressure} \quad (2B.5)$$

The heat capacity at constant pressure is the analogue of the heat capacity at constant volume (Topic 1A) and is an extensive property. The **molar heat capacity at constant pressure**, $C_{p,m}$, is the heat capacity per mole of substance; it is an intensive property.

The heat capacity at constant pressure is used to relate the change in enthalpy to a change in temperature. For infinitesimal changes of temperature,

$$dH = C_p dT \quad \text{(at constant pressure)} \quad (2B.6a)$$

If the heat capacity is constant over the range of temperatures of interest, then for a measurable increase in temperature

$$\Delta H = \int_{T_1}^{T_2} C_p dT = C_p \int_{T_1}^{T_2} dT = C_p \overbrace{(T_2 - T_1)}^{\Delta T}$$

which we can summarize as

$$\Delta H = C_p \Delta T \quad \text{(at constant pressure)} \quad (2B.6b)$$

Because a change in enthalpy can be equated with the energy supplied as heat at constant pressure, the practical form of the latter equation is

$$q_p = C_p \Delta T \quad (2B.7)$$

Table 2B.1* Temperature variation of molar heat capacities, $C_{p,m}/(\mathrm{J\,K^{-1}\,mol^{-1}}) = a + bT + c/T^2$

	a	$b/(10^{-3}\,\mathrm{K^{-1}})$	$c/(10^5\,\mathrm{K^2})$
C(s, graphite)	16.86	4.77	−8.54
CO_2(g)	44.22	8.79	−8.62
H_2O(l)	75.29	0	0
N_2(g)	28.58	3.77	−0.50

* More values are given in the *Resource section*.

This expression shows us how to measure the heat capacity of a sample: a measured quantity of energy is supplied as heat under conditions of constant pressure (as in a sample exposed to the atmosphere and free to expand), and the temperature rise is monitored.

The variation of heat capacity with temperature can sometimes be ignored if the temperature range is small; this approximation is highly accurate for a monatomic perfect gas (for instance, one of the noble gases at low pressure). However, when it is necessary to take the variation into account, a convenient approximate empirical expression is

$$C_{p,m} = a + bT + \frac{c}{T^2} \quad (2B.8)$$

The empirical parameters a, b, and c are independent of temperature (Table 2B.1) and are found by fitting this expression to experimental data.

Example 2B.2 Evaluating an increase in enthalpy with temperature

What is the change in molar enthalpy of N_2 when it is heated from 25 °C to 100 °C? Use the heat capacity information in Table 2B.1.

Method The heat capacity of N_2 changes with temperature, so we cannot use eqn 2B.6b (which assumes that the heat capacity of the substance is constant). Therefore, we must use eqn 2B.6a, substitute eqn 2B.8 for the temperature dependence of the heat capacity, and integrate the resulting expression from 25 °C (298 K) to 100 °C (373 K).

Answer For convenience, we denote the two temperatures T_1 (298 K) and T_2 (373 K). The relation we require is

$$\int_{H_m(T_1)}^{H_m(T_2)} dH_m = \int_{T_1}^{T_2} \left(a + bT + \frac{c}{T^2}\right) dT$$

After using Integral A.1 in the *Resource section*, it follows that

$$H_m(T_2) - H_m(T_1) = a(T_2 - T_1) + \tfrac{1}{2}b(T_2^2 - T_1^2) - c\left(\frac{1}{T_2} - \frac{1}{T_1}\right)$$

Substitution of the numerical data results in

$$H_m(373\,K) = H_m(298\,K) + 2.20\,kJ\,mol^{-1}$$

If we had assumed a constant heat capacity of $29.14\,J\,K^{-1}\,mol^{-1}$ (the value given by eqn 2B.8 for $T = 298\,K$), we would have found that the two enthalpies differed by $2.19\,kJ\,mol^{-1}$.

Self-test 2B.4 At very low temperatures the heat capacity of a solid is proportional to T^3, and we can write $C_{p,m} = aT^3$. What is the change in enthalpy of such a substance when it is heated from 0 to a temperature T (with T close to 0)?

Answer: $\Delta H_m = \frac{1}{4}aT^4$

(b) The relation between heat capacities

Most systems expand when heated at constant pressure. Such systems do work on the surroundings and therefore some of the energy supplied to them as heat escapes back to the surroundings. As a result, the temperature of the system rises less than when the heating occurs at constant volume. A smaller increase in temperature implies a larger heat capacity, so we conclude that in most cases the heat capacity at constant pressure of a system is larger than its heat capacity at constant volume. We show in Topic 2D that there is a simple relation between the two heat capacities of a perfect gas:

$$C_p - C_V = nR \qquad \textit{Perfect gas} \qquad \text{Relation between heat capacities} \qquad (2B.9)$$

It follows that the molar heat capacity of a perfect gas is about $8\,J\,K^{-1}\,mol^{-1}$ larger at constant pressure than at constant volume. Because the molar constant-volume heat capacity of a monatomic gas is about $\frac{3}{2}R = 12\,J\,K^{-1}\,mol^{-1}$, the difference is highly significant and must be taken into account.

Checklist of concepts

- ☐ 1. Energy transferred as heat at constant pressure is equal to the change in **enthalpy** of a system.
- ☐ 2. Enthalpy changes are measured in a constant-pressure calorimeter.
- ☐ 3. The **heat capacity at constant pressure** is equal to the slope of enthalpy with temperature.

Checklist of equations

Property	Equation	Comment	Equation number
Enthalpy	$H = U + pV$	Definition	2B.1
Heat transfer at constant pressure	$dH = dq_p$, $\Delta H = q_p$	No additional work	2B.2
Relation between ΔH and ΔU	$\Delta H = \Delta U + \Delta n_g RT$	Molar volumes of the participating condensed phases are negligible; isothermal process	2B.4
Heat capacity at constant pressure	$C_p = (\partial H/\partial T)_p$	Definition	2B.5
Relation between heat capacities	$C_p - C_V = nR$	Perfect gas	2B.9

2C Thermochemistry

Contents

➤ Why do you need to know this material?

Thermochemistry is one of the principal applications of thermodynamics in chemistry, for thermochemical data provide a way of assessing the heat output of chemical reactions, including those involved in the consumption of fuels and foods. The data are also used widely in other chemical applications of thermodynamics.

➤ What is the key idea?

Reaction enthalpies can be combined to provide data on other reactions of interest.

➤ What do you need to know already?

You need to be aware of the definition of enthalpy and its status as a state function (Topic 2B). The material on temperature dependence of reaction enthalpies makes use of information on heat capacity (Topic 2B).

The study of the energy transferred as heat during the course of chemical reactions is called **thermochemistry**. Thermochemistry is a branch of thermodynamics because a reaction vessel and its contents form a system, and chemical reactions result in the exchange of energy between the system and the surroundings. Thus we can use calorimetry to measure the energy supplied or discarded as heat by a reaction, and can identify q with a change in internal energy if the reaction occurs at constant volume or with a change in enthalpy if the reaction occurs at constant pressure. Conversely, if we know ΔU or ΔH for a reaction, we can predict the heat the reaction can produce.

As pointed out in Topic 2A, a process that releases energy as heat into the surroundings is classified as exothermic and one that absorbs energy as heat from the surroundings is classified as endothermic. Because the release of heat at constant pressure signifies a decrease in the enthalpy of a system, it follows that an exothermic process is one for which $\Delta H<0$. Conversely, because the absorption of heat results in an increase in enthalpy, an endothermic process has $\Delta H>0$:

$$\text{exothermic process}: \Delta H<0 \quad \text{endothermic process}: \Delta H>0$$

2C.1 Standard enthalpy changes

Changes in enthalpy are normally reported for processes taking place under a set of standard conditions. In most of our discussions we shall consider the **standard enthalpy change**, $\Delta H^{\ominus}$, the change in enthalpy for a process in which the initial and final substances are in their standard states:

> The **standard state** of a substance at a specified temperature is its pure form at 1 bar. *Specification of standard state*

For example, the standard state of liquid ethanol at 298 K is pure liquid ethanol at 298 K and 1 bar; the standard state of solid iron at 500 K is pure iron at 500 K and 1 bar. The definition of standard state is more sophisticated for solutions (Topic 5E). The standard enthalpy change for a reaction or a physical process is the difference between the products in their standard states and the reactants in their standard states, all at the same specified temperature.

As an example of a standard enthalpy change, the *standard enthalpy of vaporization*, $\Delta_{vap}H^{\ominus}$, is the enthalpy change per mole of molecules when a pure liquid at 1 bar vaporizes to a gas at 1 bar, as in

$$H_2O(l) \rightarrow H_2O(g) \qquad \Delta_{vap}H^{\ominus}(373\,K) = +40.66\,kJ\,mol^{-1}$$

As implied by the examples, standard enthalpies may be reported for any temperature. However, the conventional temperature for reporting thermodynamic data is 298.15 K. Unless otherwise mentioned or indicated by attaching the temperature to $\Delta H^{\ominus}$, all thermodynamic data in this text are for this conventional temperature.

A note on good practice The attachment of the name of the transition to the symbol Δ, as in $\Delta_{vap}H$, is the current convention. However, the older convention, ΔH_{vap}, is still widely used. The current convention is more logical because the subscript identifies the type of change, not the physical observable related to the change.

(a) Enthalpies of physical change

The standard enthalpy change that accompanies a change of physical state is called the **standard enthalpy of transition** and is denoted $\Delta_{trs}H^{\ominus}$ (Table 2C.1). The **standard enthalpy of vaporization**, $\Delta_{vap}H^{\ominus}$, is one example. Another is the **standard enthalpy of fusion**, $\Delta_{fus}H^{\ominus}$, the standard enthalpy change accompanying the conversion of a solid to a liquid, as in

$$H_2O(s) \rightarrow H_2O(l) \qquad \Delta_{fus}H^{\ominus}(273\,K) = +6.01\,kJ\,mol^{-1}$$

Table 2C.1* Standard enthalpies of fusion and vaporization at the transition temperature, $\Delta_{trs}H^{\ominus}$/(kJ mol^{-1})

	T_f/K	Fusion	T_b/K	Vaporization
Ar	83.81	1.188	87.29	6.506
C_6H_6	278.61	10.59	353.2	30.8
H_2O	273.15	6.008	373.15	40.656 (44.016 at 298 K)
He	3.5	0.021	4.22	0.084

* More values are given in the *Resource section*.

Table 2C.2 Enthalpies of transition

Transition	Process	Symbol*
Transition	Phase α → phase β	$\Delta_{trs}H$
Fusion	s → l	$\Delta_{fus}H$
Vaporization	l → g	$\Delta_{vap}H$
Sublimation	s → g	$\Delta_{sub}H$
Mixing	Pure → mixture	$\Delta_{mix}H$
Solution	Solute → solution	$\Delta_{sol}H$
Hydration	$X^{\pm}(g) \rightarrow X^{\pm}(aq)$	$\Delta_{hyd}H$
Atomization	Species(s, l, g) → atoms(g)	$\Delta_{at}H$
Ionization	$X(g) \rightarrow X^+(g) + e^-(g)$	$\Delta_{ion}H$
Electron gain	$X(g) + e^-(g) \rightarrow X^-(g)$	$\Delta_{eg}H$
Reaction	Reactants → products	Δ_rH
Combustion	Compound(s, l, g) + $O_2(g) \rightarrow CO_2(g)$, H_2O(l, g)	Δ_cH
Formation	Elements → compound	Δ_fH
Activation	Reactants → activated complex	$\Delta^{\ddagger}H$

* IUPAC recommendations. In common usage, the transition subscript is often attached to ΔH, as in ΔH_{trs}.

As in this case, it is sometimes convenient to know the standard enthalpy change at the transition temperature as well as at the conventional temperature of 298 K. The different types of enthalpies encountered in thermochemistry are summarized in Table 2C.2. We meet them again in various locations throughout the text.

Because enthalpy is a state function, a change in enthalpy is independent of the path between the two states. This feature is of great importance in thermochemistry, for it implies that the same value of $\Delta H^{\ominus}$ will be obtained however the change is brought about between the same initial and final states. For example, we can picture the conversion of a solid to a vapour either as occurring by sublimation (the direct conversion from solid to vapour),

$$H_2O(s) \rightarrow H_2O(g) \qquad \Delta_{sub}H^{\ominus}$$

or as occurring in two steps, first fusion (melting) and then vaporization of the resulting liquid:

$$\begin{array}{lll} & H_2O(s) \rightarrow H_2O(l) & \Delta_{fus}H^{\ominus} \\ & H_2O(l) \rightarrow H_2O(g) & \Delta_{vap}H^{\ominus} \\ \text{Overall:} & H_2O(s) \rightarrow H_2O(g) & \Delta_{fus}H^{\ominus} + \Delta_{vap}H^{\ominus} \end{array}$$

Because the overall result of the indirect path is the same as that of the direct path, the overall enthalpy change is the same in each case (**1**), and we can conclude that (for processes occurring at the same temperature)

$$\Delta_{sub}H^{\ominus} = \Delta_{fus}H^{\ominus} + \Delta_{vap}H^{\ominus} \qquad (2C.1)$$

An immediate conclusion is that, because all enthalpies of fusion are positive, the enthalpy of sublimation of a substance is greater than its enthalpy of vaporization (at a given temperature).

1
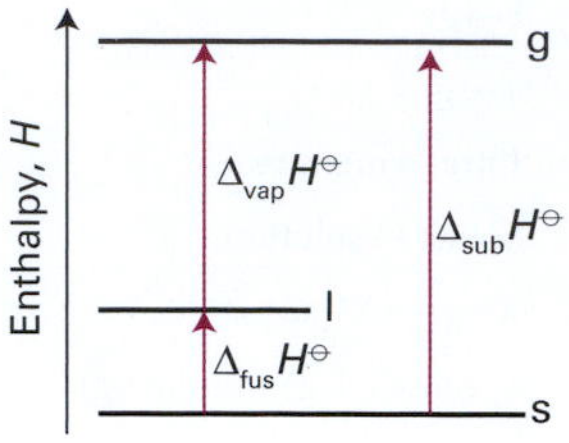

Another consequence of H being a state function is that the standard enthalpy changes of a forward process and its reverse differ in sign (**2**):

$$\Delta H^{\ominus}(A \rightarrow B) = -\Delta H^{\ominus}(B \rightarrow A) \quad (2C.2)$$

For instance, because the enthalpy of vaporization of water is +44 kJ mol^{-1} at 298 K, its enthalpy of condensation at that temperature is −44 kJ mol^{-1}.

2
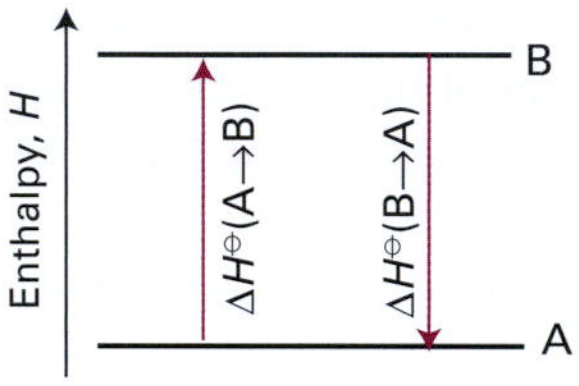

The vaporization of a solid often involves a large increase in energy, especially when the solid is ionic and the strong Coulombic interaction of the ions must be overcome in a process such as

$$MX(s) \rightarrow M^{+}(g) + X^{-}(g)$$

The **lattice enthalpy**, ΔH_L, is the change in standard molar enthalpy for this process. The lattice enthalpy is equal to the lattice internal energy at $T=0$; at normal temperatures they differ by only a few kilojoules per mole, and the difference is normally neglected.

Experimental values of the lattice enthalpy are obtained by using a **Born–Haber cycle**, a closed path of transformations starting and ending at the same point, one step of which is the formation of the solid compound from a gas of widely separated ions.

Brief illustration 2C.1 A Born–Haber cycle

A typical Born–Haber cycle, for potassium chloride, is shown in Fig. 2C.1.

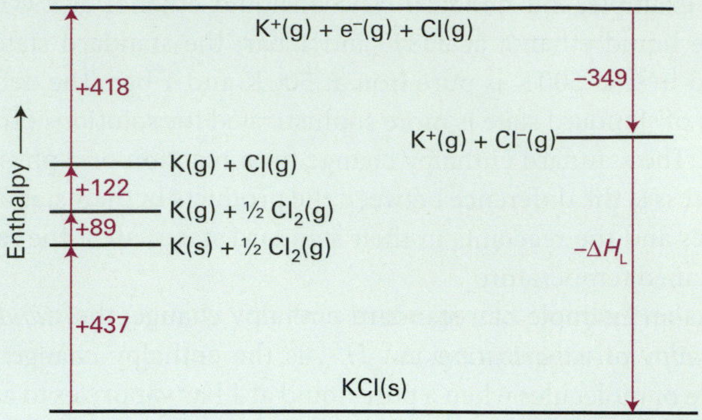

Figure 2C.1 The Born–Haber cycle for KCl at 298 K. Enthalpy changes are in kilojoules per mole.

It consists of the following steps (for convenience, starting at the elements):

	$\Delta H^{\ominus}$/(kJ mol^{-1})	
1. Sublimation of K(s)	+89	[dissociation enthalpy of K(s)]
2. Dissociation of $\frac{1}{2}Cl_2(g)$	+122	[$\frac{1}{2}$ × dissociation enthalpy of $Cl_2(g)$]
3. Ionization of K(g)	+418	[ionization enthalpy of K(g)]
4. Electron attachment to Cl(g)	−349	[electron gain enthalpy of Cl(g)]
5. Formation of solid from gas	$-\Delta H_L$/(kJ mol^{-1})	
6. Decomposition of compound	+437	[negative of enthalpy of formation of KCl(s)]

Because the sum of these enthalpy changes is equal to zero, we can infer from

$$89 + 122 + 418 - 349 - \Delta H_L/(\text{kJ mol}^{-1}) + 437 = 0$$

that $\Delta H_L = +717$ kJ mol^{-1}.

Self-test 2C.1 Assemble a similar cycle for the lattice enthalpy of magnesium chloride.

Answer: 2523 kJ mol^{-1}

Lattice enthalpies obtained in the same way as in *Brief illustration* 2C.1 are listed in Table 2C.3. They are large when the ions are highly charged and small, for then they are close together and attract each other strongly. We examine the quantitative relation between lattice enthalpy and structure in Topic 18B.

(b) Enthalpies of chemical change

Now we consider enthalpy changes that accompany chemical reactions. There are two ways of reporting the change in enthalpy that accompanies a chemical reaction. One is to write

Table 2C.3* Lattice enthalpies at 298 K, $\Delta H_L/(\text{kJ mol}^{-1})$

NaF	787
NaBr	751
MgO	3850
MgS	3406

* More values are given in the *Resource section*.

the **thermochemical equation**, a combination of a chemical equation and the corresponding change in standard enthalpy:

$$CH_4(g)+2\,O_2(g)\rightarrow CO_2(g)+2\,H_2O(g)\quad \Delta H^\ominus=-890\,\text{kJ}$$

$\Delta H^\ominus$ is the change in enthalpy when reactants in their standard states change to products in their standard states:

Pure, separate reactants in their standard states
→ pure, separate products in their standard states

Except in the case of ionic reactions in solution, the enthalpy changes accompanying mixing and separation are insignificant in comparison with the contribution from the reaction itself. For the combustion of methane, the standard value refers to the reaction in which 1 mol CH_4 in the form of pure methane gas at 1 bar reacts completely with 2 mol O_2 in the form of pure oxygen gas to produce 1 mol CO_2 as pure carbon dioxide at 1 bar and 2 mol H_2O as pure liquid water at 1 bar; the numerical value is for the reaction at 298.15 K.

Alternatively, we write the chemical equation and then report the **standard reaction enthalpy**, $\Delta_r H^\ominus$ (or 'standard enthalpy of reaction'). Thus, for the combustion of methane, we write

$$CH_4(g)+2\,O_2(g)\rightarrow CO_2(g)+2\,H_2O(g)$$
$$\Delta_r H^\ominus=-890\,\text{kJ mol}^{-1}$$

For a reaction of the form 2 A+B→3 C+D the standard reaction enthalpy would be

$$\Delta_r H^\ominus=\{3H_m^\ominus(C)+H_m^\ominus(D)\}-\{2H_m^\ominus(A)+H_m^\ominus(B)\}$$

where $H_m^\ominus(J)$ is the standard molar enthalpy of species J at the temperature of interest. Note how the 'per mole' of $\Delta_r H^\ominus$ comes directly from the fact that molar enthalpies appear in this expression. We interpret the 'per mole' by noting the stoichiometric coefficients in the chemical equation. In this case, 'per mole' in $\Delta_r H^\ominus$ means 'per 2 mol A', 'per mole B', 'per 3 mol C', or 'per mol D'. In general,

$$\Delta_r H^\ominus=\sum_{\text{Products}}\nu H_m^\ominus-\sum_{\text{Reactants}}\nu H_m^\ominus \quad \textit{Definition} \quad \text{Standard reaction enthalpy} \quad (2C.3)$$

where in each case the molar enthalpies of the species are multiplied by their (dimensionless and positive) stoichiometric coefficients, ν. This formal definition is of little practical value because the absolute values of the standard molar enthalpies are unknown: we see in Section 2C.2a how that problem is overcome.

Some standard reaction enthalpies have special names and a particular significance. For instance, the **standard enthalpy of combustion**, $\Delta_c H^\ominus$, is the standard reaction enthalpy for the complete oxidation of an organic compound to CO_2 gas and liquid H_2O if the compound contains C, H, and O, and to N_2 gas if N is also present.

Brief illustration 2C.2 Enthalpy of combustion

The combustion of glucose is

$$C_6H_{12}O_6(s)+6\,O_2(g)\rightarrow 6\,CO_2(g)+6\,H_2O(l)$$
$$\Delta_c H^\ominus=-2808\,\text{kJ mol}^{-1}$$

The value quoted shows that 2808 kJ of heat is released when 1 mol $C_6H_{12}O_6$ burns under standard conditions (at 298 K). More values are given in Table 2C.4.

Self-test 2C.2 Predict the heat output of the combustion of 1.0 dm³ of octane at 298 K. Its mass density is 0.703 g cm⁻³.

Answer: 34 MJ

Table 2C.4* Standard enthalpies of formation ($\Delta_f H^\ominus$) and combustion ($\Delta_c H^\ominus$) of organic compounds at 298 K

	$\Delta_f H^\ominus/(\text{kJ mol}^{-1})$	$\Delta_c H^\ominus/(\text{kJ mol}^{-1})$
Benzene, $C_6H_6(l)$	+49.0	−3268
Ethane, $C_2H_6(g)$	−84.7	−1560
Glucose, $C_6H_{12}O_6(s)$	−1274	−2808
Methane, $CH_4(g)$	−74.8	−890
Methanol, $CH_3OH(l)$	−238.7	−721

* More values are given in the *Resource section*.

(c) Hess's law

Standard enthalpies of individual reactions can be combined to obtain the enthalpy of another reaction. This application of the First Law is called **Hess's law**:

> The standard enthalpy of an overall reaction is the sum of the standard enthalpies of the individual reactions into which a reaction may be divided. *Hess's law*

The individual steps need not be realizable in practice: they may be hypothetical reactions, the only requirement being that their chemical equations should balance. The thermodynamic basis of the law is the path-independence of the value of $\Delta_r H^\ominus$ and the implication that we may take the specified reactants, pass through any (possibly hypothetical) set of reactions to the specified products, and overall obtain the same change of enthalpy. The importance of Hess's law is that information about a reaction of interest, which may be difficult to determine directly, can be assembled from information on other reactions.

Example 2C.1 Using Hess's law

The standard reaction enthalpy for the hydrogenation of propene,

$$CH_2{=}CHCH_3(g)+H_2(g)\rightarrow CH_3CH_2CH_3(g)$$

is −124 kJ mol⁻¹. The standard reaction enthalpy for the combustion of propane,

$$CH_3CH_2CH_3(g)+5\,O_2(g)\rightarrow 3\,CO_2(g)+4\,H_2O(l)$$

is −2220 kJ mol⁻¹. The standard reaction enthalpy for the formation of water,

$$H_2(g)+\tfrac{1}{2}O_2(g)\rightarrow H_2O(l)$$

is −286 kJ mol⁻¹. Calculate the standard enthalpy of combustion of propene.

Method The skill to develop is the ability to assemble a given thermochemical equation from others. Add or subtract the reactions given, together with any others needed, so as to reproduce the reaction required. Then add or subtract the reaction enthalpies in the same way.

Answer The combustion reaction we require is

$$C_3H_6(g)+\tfrac{9}{2}O_2(g)\rightarrow 3\,CO_2(g)+3\,H_2O(l)$$

This reaction can be recreated from the following sum:

	$\Delta_r H^\ominus/(\text{kJ mol}^{-1})$
$C_3H_6(g)+H_2(g)\rightarrow C_3H_8(g)$	−124
$C_3H_8(g)+5\,O_2(g)\rightarrow 3\,CO_2(g)+4\,H_2O(l)$	−2220
$H_2O(l)\rightarrow H_2(g)+\frac{1}{2}O_2(g)$	+286
$C_3H_6(g)+\frac{9}{2}O_2(g)\rightarrow 3\,CO_2(g)+3\,H_2O(l)$	−2058

Self-test 2C.3 Calculate the enthalpy of hydrogenation of benzene from its enthalpy of combustion and the enthalpy of combustion of cyclohexane.

Answer: −206 kJ mol⁻¹

2C.2 Standard enthalpies of formation

The **standard enthalpy of formation**, $\Delta_f H^\ominus$, of a substance is the standard reaction enthalpy for the formation of the compound from its elements in their reference states:

The **reference state** of an element is its most stable state at the specified temperature and 1 bar. *Specification of reference state*

For example, at 298 K the reference state of nitrogen is a gas of N_2 molecules, that of mercury is liquid mercury, that of carbon is graphite, and that of tin is the white (metallic) form. There is one exception to this general prescription of reference states: the reference state of phosphorus is taken to be white phosphorus despite this allotrope not being the most stable form but simply the more reproducible form of the element. Standard enthalpies of formation are expressed as enthalpies per mole of molecules or (for ionic substances) formula units of the compound. The standard enthalpy of formation of liquid benzene at 298 K, for example, refers to the reaction

$$6\,C(s,\text{graphite})+3\,H_2(g)\rightarrow C_6H_6(l)$$

and is +49.0 kJ mol⁻¹. The standard enthalpies of formation of elements in their reference states are zero at all temperatures because they are the enthalpies of such 'null' reactions as $N_2(g)\rightarrow N_2(g)$. Some enthalpies of formation are listed in Tables 2C.5 and 2C.6.

The standard enthalpy of formation of ions in solution poses a special problem because it is impossible to prepare a solution of cations alone or of anions alone. This problem is solved by

Table 2C.5* Standard enthalpies of formation of inorganic compounds at 298 K, $\Delta_f H^\ominus/(\text{kJ mol}^{-1})$

	$\Delta_f H^\ominus/(\text{kJ mol}^{-1})$
$H_2O(l)$	−285.83
$H_2O(g)$	−241.82
$NH_3(g)$	−46.11
$N_2H_4(l)$	+50.63
$NO_2(g)$	+33.18
$N_2O_4(g)$	+9.16
NaCl(s)	−411.15
KCl(s)	−436.75

* More values are given in the *Resource section.*

Table 2C.6* Standard enthalpies of formation of organic compounds at 298 K, $\Delta_f H^\ominus/(\text{kJ mol}^{-1})$

	$\Delta_f H^\ominus/(\text{kJ mol}^{-1})$
$CH_4(g)$	−74.81
$C_6H_6(l)$	+49.0
$C_6H_{12}(l)$	−156
$CH_3OH(l)$	−238.66
$CH_3CH_2OH(l)$	−277.69

* More values are given in the *Resource section.*

defining one ion, conventionally the hydrogen ion, to have zero standard enthalpy of formation at all temperatures:

$$\Delta_f H^{\ominus}(H^+,aq)=0 \qquad \textit{Convention} \qquad \text{Ions in solution} \qquad (2C.4)$$

Brief illustration 2C.3 Enthalpies of formation of ions in solution

If the enthalpy of formation of HBr(aq) is found to be -122 kJ mol^{-1}, then the whole of that value is ascribed to the formation of $Br^-(aq)$, and we write $\Delta_f H^{\ominus}(Br^-,aq)=-122$ kJ mol^{-1}. That value may then be combined with, for instance, the enthalpy of formation of AgBr(aq) to determine the value of $\Delta_f H^{\ominus}(Ag^+,aq)$, and so on. In essence, this definition adjusts the actual values of the enthalpies of formation of ions by a fixed amount, which is chosen so that the standard value for one of them, $H^+(aq)$, has the value zero.

Self-test 2C.4 Determine the value of $\Delta_f H^{\ominus}(Ag^+,aq)$; the standard enthalpy of formation of AgBr(aq) is -17 kJ mol^{-1}.

Answer: $+105$ kJ mol^{-1}

(a) The reaction enthalpy in terms of enthalpies of formation

Conceptually, we can regard a reaction as proceeding by decomposing the reactants into their elements and then forming those elements into the products. The value of $\Delta_r H^{\ominus}$ for the overall reaction is the sum of these 'unforming' and forming enthalpies. Because 'unforming' is the reverse of forming, the enthalpy of an unforming step is the negative of the enthalpy of formation (**3**).

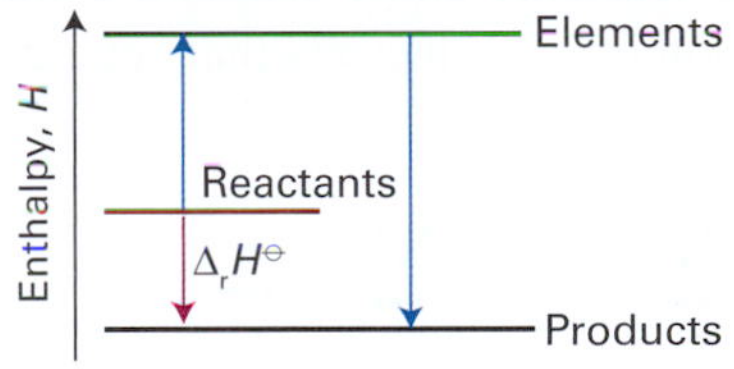

3

Hence, in the enthalpies of formation of substances, we have enough information to calculate the enthalpy of any reaction by using

$$\Delta_r H^{\ominus} = \sum_{\text{Products}} \nu\Delta_f H^{\ominus} - \sum_{\text{Reactants}} \nu\Delta_f H^{\ominus} \qquad \textit{Practical implementation} \qquad \text{Standard reaction enthalpy} \qquad (2C.5a)$$

where in each case the enthalpies of formation of the species that occur are multiplied by their stoichiometric coefficients. This procedure is the practical implementation of the formal definition in eqn 2C.3. A more sophisticated way of expressing the same result is to introduce the **stoichiometric numbers** ν_J (as distinct from the stoichiometric coefficients), which are positive for products and negative for reactants. Then we can write

$$\Delta_r H^{\ominus} = \sum_J \nu_J \Delta_f H^{\ominus}(J) \qquad (2C.5b)$$

Stoichiometric *numbers*, which have a sign, are denoted ν_J or $\nu(J)$. Stoichiometric *coefficients*, which are all positive, are denoted simply ν (with no subscript).

Brief illustration 2C.4 Enthalpies of formation

According to eqn 2C.5a, the standard enthalpy of the reaction $2\,HN_3(l)+2\,NO(g)\rightarrow H_2O_2(l)+4\,N_2(g)$ is calculated as follows:

$$\begin{aligned}\Delta_r H^{\ominus} &= \{\Delta_f H^{\ominus}(H_2O_2,l)+4\Delta_f H^{\ominus}(N_2,g)\}\\ &\quad -\{2\Delta_f H^{\ominus}(HN_3,l)+2\Delta_f H^{\ominus}(NO,g)\}\\ &=\{-187.78+4(0)\}\,\text{kJ mol}^{-1}-\{2(264.0)\\ &\quad +2(90.25)\}\,\text{kJ mol}^{-1}\\ &=-896.3\,\text{kJ mol}^{-1}\end{aligned}$$

To use eqn 2C.5b, we identify $\nu(HN_3)=-2$, $\nu(NO)=-2$, $\nu(H_2O_2)=+1$, and $\nu(N_2)=+4$, and then write

$$\begin{aligned}\Delta_r H^{\ominus} &= \Delta_f H^{\ominus}(H_2O_2,l)+4\Delta_f H^{\ominus}(N_2,g)-2\Delta_f H^{\ominus}(HN_3,l)\\ &\quad -2\Delta_f H^{\ominus}(NO,g)\}\end{aligned}$$

which gives the same result.

Self-test 2C.5 Evaluate the standard enthalpy of the reaction $C(graphite)+H_2O(g)\rightarrow CO(g)+H_2(g)$.

Answer: $+131.29$ kJ mol^{-1}

(b) Enthalpies of formation and molecular modelling

We have seen how to construct standard reaction enthalpies by combining standard enthalpies of formation. The question that now arises is whether we can construct standard enthalpies of formation from a knowledge of the chemical constitution of the species. The short answer is that there is no thermodynamically exact way of expressing enthalpies of formation in terms of contributions from individual atoms and bonds. In the past, approximate procedures based on **mean bond enthalpies**, $\Delta H(A{-}B)$, the average enthalpy change associated with the breaking of a specific A—B bond,

$$A{-}B(g)\rightarrow A(g)+B(g) \qquad \Delta H(A{-}B)$$

have been used. However, this procedure is notoriously unreliable, in part because the $\Delta H(A{-}B)$ are average values for a series of related compounds. Nor does the approach distinguish between geometrical isomers, where the same atoms and bonds may be present but experimentally the enthalpies of formation might be significantly different.

Computer-aided molecular modelling has largely displaced this more primitive approach. Commercial software packages use the principles developed in Topic 10E to calculate the standard enthalpy of formation of a molecule drawn on the computer screen. These techniques can be applied to different conformations of the same molecule. In the case of methylcyclohexane, for instance, the calculated conformational energy difference ranges from 5.9 to 7.9 kJ mol^{-1}, with the equatorial conformer having the lower standard enthalpy of formation. These estimates compare favourably with the experimental value of 7.5 kJ mol^{-1}. However, good agreement between calculated and experimental values is relatively rare. Computational methods almost always predict correctly which conformer is more stable but do not always predict the correct magnitude of the conformational energy difference. The most reliable technique for the determination of enthalpies of formation remains calorimetry, typically by using enthalpies of combustion.

Brief illustration 2C.5 Molecular modelling

Each software package has its own procedures; the general approach, though, is the same in most cases: the structure of the molecule is specified and the nature of the calculation selected. When the procedure is applied to the axial and equatorial isomers of methylcyclohexane, a typical value for the standard enthalpy of formation of equatorial isomer in the gas phase is −183 kJ mol^{-1} (using the AM1 semi-empirical procedure) whereas that for the axial isomer is −177 kJ mol^{-1}, a difference of 6 kJ mol^{-1}. The experimental difference is 7.5 kJ mol^{-1}.

Self-test 2C.6 If you have access to modelling software, repeat this calculation for the two isomers of cyclohexanol.

Answer: Using AM1: eq: −345 kJ mol^{-1}; ax: −349 kJ mol^{-1}

2C.3 The temperature dependence of reaction enthalpies

The standard enthalpies of many important reactions have been measured at different temperatures. However, in the absence of this information, standard reaction enthalpies at different temperatures may be calculated from heat capacities and the reaction enthalpy at some other temperature (Fig. 2C.2). In many cases heat capacity data are more accurate than reaction enthalpies. Therefore, providing the information is available, the procedure we are about to describe is more accurate than the direct measurement of a reaction enthalpy at an elevated temperature.

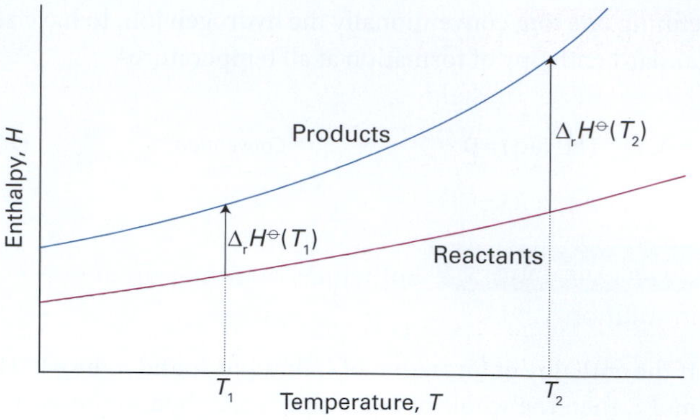

Figure 2C.2 When the temperature is increased, the enthalpy of the products and the reactants both increase, but may do so to different extents. In each case, the change in enthalpy depends on the heat capacities of the substances. The change in reaction enthalpy reflects the difference in the changes of the enthalpies.

It follows from eqn 2B.6a ($\mathrm{d}H = C_p \mathrm{d}T$) that, when a substance is heated from T_1 to T_2, its enthalpy changes from $H(T_1)$ to

$$H(T_2) = H(T_2) + \int_{T_1}^{T_2} C_p \mathrm{d}T \qquad (2C.6)$$

(We have assumed that no phase transition takes place in the temperature range of interest.) Because this equation applies to each substance in the reaction, the standard reaction enthalpy changes from $\Delta_r H^{\ominus}(T_1)$ to

$$\Delta_r H^{\ominus}(T_2) = \Delta_r H^{\ominus}(T_2) + \int_{T_1}^{T_2} \Delta_r C_p^{\ominus} \mathrm{d}T \qquad \text{Kirchhoff's law} \qquad (2C.7a)$$

where $\Delta_r C_p^{\ominus}$ is the difference of the molar heat capacities of products and reactants under standard conditions weighted by the stoichiometric coefficients that appear in the chemical equation:

$$\Delta_r C_{p,m}^{\ominus} = \sum_{\text{Products}} \nu C_{p,m}^{\ominus} - \sum_{\text{Reactants}} \nu C_{p,m}^{\ominus} \qquad (2C.7b)$$

or, in the notation of eqn 2C.5b,

$$\Delta_r C_{p,m}^{\ominus} = \sum_{\mathrm{J}} \nu_{\mathrm{J}} C_{p,m}^{\ominus}(\mathrm{J}) \qquad (2C.7c)$$

Equation 2C.7a is known as **Kirchhoff's law**. It is normally a good approximation to assume that $\Delta_r C_p^{\ominus}$ is independent of the temperature, at least over reasonably limited ranges. Although the individual heat capacities may vary, their difference varies less significantly. In some cases the temperature dependence of heat capacities is taken into account by using eqn 2B.8.

Example 2C.2 Using Kirchhoff's law

The standard enthalpy of formation of $H_2O(g)$ at 298 K is −241.82 kJ mol⁻¹. Estimate its value at 100 °C given the following values of the molar heat capacities at constant pressure: $H_2O(g)$: 33.58 J K⁻¹ mol⁻¹; $H_2(g)$: 28.84 J K⁻¹ mol⁻¹; $O_2(g)$: 29.37 J⁻¹ mol⁻¹. Assume that the heat capacities are independent of temperature.

Method When $\Delta_r C_p^\ominus$ is independent of temperature in the range T_1 to T_2, the integral in eqn 2C.7a evaluates to $(T_2-T_1)\Delta_r C_p^\ominus$. Therefore,

$$\Delta_r H^\ominus(T_2)=\Delta_r H^\ominus(T_1)+(T_2-T_1)\Delta_r C_p^\ominus$$

To proceed, write the chemical equation, identify the stoichiometric coefficients, and calculate $\Delta_r C_p^\ominus$ from the data.

Answer The reaction is $H_2(g)+\frac{1}{2}O_2(g)\rightarrow H_2O(g)$, so

$$\Delta_r C_p^\ominus = C_{p,m}^\ominus(H_2O,g)-\left\{C_{p,m}^\ominus(H_2,g)+\tfrac{1}{2}C_{p,m}^\ominus(O_2,g)\right\}$$
$$=-9.94\,J\,K^{-1}\,mol^{-1}$$

It then follows that

$$\Delta_r H^\ominus(373\,K)=-241.82\,kJ\,mol^{-1}+(75\,K)\times(-9.94\,J\,K^{-1}\,mol^{-1})=-242.6\,kJ\,mol^{-1}$$

Self-test 2C.7 Estimate the standard enthalpy of formation of cyclohexane, $C_6H_{12}(l)$, at 400 K from the data in Table 2C.6.

Answer: −163 kJ mol⁻¹

2C.4 Experimental techniques

The classic tool of thermochemistry is the calorimeter, as summarized in Topic 2B. However, technological advances have been made that allow measurements to be made on samples with mass as little as a few milligrams. We describe two of them here.

(a) Differential scanning calorimetry

A **differential scanning calorimeter** (DSC) measures the energy transferred as heat to or from a sample at constant pressure during a physical or chemical change. The term 'differential' refers to the fact that the behaviour of the sample is compared to that of a reference material that does not undergo a physical or chemical change during the analysis. The term 'scanning' refers to the fact that the temperatures of the sample and reference material are increased, or scanned, during the analysis.

A DSC consists of two small compartments that are heated electrically at a constant rate. The temperature, T, at time t during a linear scan is $T=T_0+\alpha t$, where T_0 is the initial temperature and α is the scan rate. A computer controls the electrical power supply that maintains the same temperature in the sample and reference compartments throughout the analysis (Fig. 2C.3).

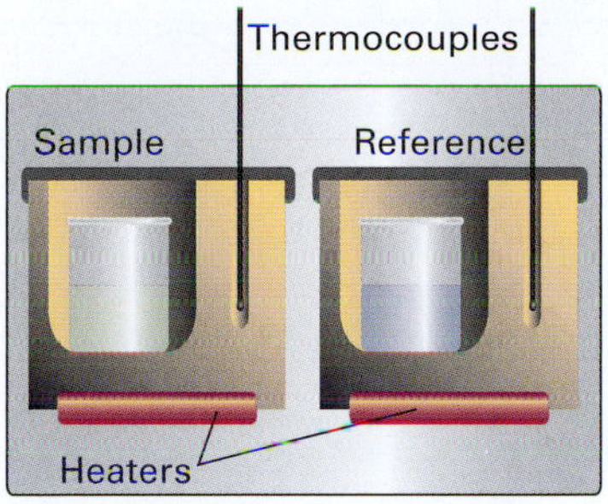

Figure 2C.3 A differential scanning calorimeter. The sample and a reference material are heated in separate but identical metal heat sinks. The output is the difference in power needed to maintain the heat sinks at equal temperatures as the temperature rises.

If no physical or chemical change occurs in the sample at temperature T, we write the heat transferred to the sample as $q_p=C_p\Delta T$, where $\Delta T=T-T_0$ and we have assumed that C_p is independent of temperature. Because $T=T_0+\alpha t$, $\Delta T=\alpha t$. The chemical or physical process requires the transfer of $q_p+q_{p,ex}$, where $q_{p,ex}$ is the excess energy transferred as heat needed to attain the same change in temperature of the sample as the control. The quantity $q_{p,ex}$ is interpreted in terms of an apparent change in the heat capacity at constant pressure of the sample, C_p, during the temperature scan:

$$C_{p,ex}=\frac{q_{p,ex}}{\Delta T}=\frac{q_{p,ex}}{\alpha t}=\frac{P_{ex}}{\alpha} \tag{2C.8}$$

where $P_{ex}=q_{p,ex}/t$ is the excess electrical power necessary to equalize the temperature of the sample and reference compartments. A DSC trace, also called a **thermogram**, consists of a plot of $C_{p,ex}$ against T (Fig. 2C.4). The enthalpy change associated with the process is

$$\Delta H=\int_{T_1}^{T_2} C_{p,ex}\,dT \tag{2C.9}$$

where T_1 and T_2 are, respectively, the temperatures at which the process begins and ends. This relation shows that the enthalpy change is equal to the area under the plot of $C_{p,ex}$ against T.

The technique is used, for instance, to assess the stability of proteins, nucleic acids, and membranes. The thermogram shown in Fig. 2C.4 indicates that the protein ubiquitin undergoes an endothermic conformational change in which a large number of non-covalent interactions (such as hydrogen bonds) are broken simultaneously and result in denaturation, the loss of the protein's three-dimensional structure. The area under the curve represents the heat absorbed in this process and

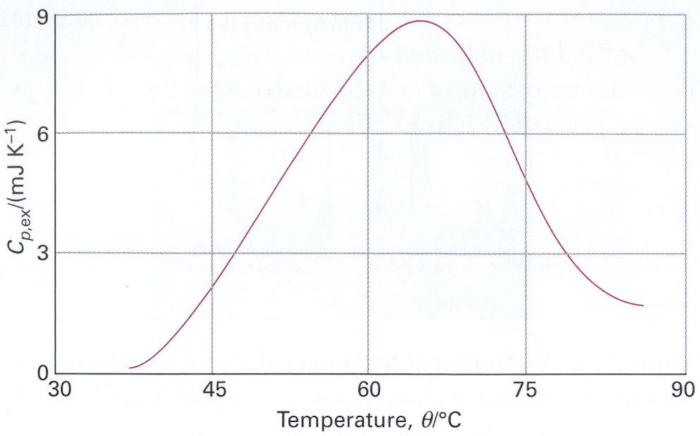

Figure 2C.4 A thermogram for the protein ubiquitin at pH = 2.45. The protein retains its native structure up to about 45 °C and then undergoes an endothermic conformational change. (Adapted from B. Chowdhry and S. LeHarne, *J. Chem. Educ.* **74**, 236 (1997).)

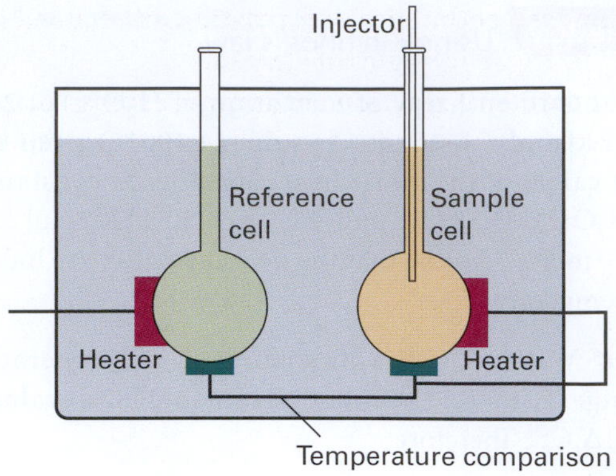

Figure 2C.5 A schematic diagram of the apparatus used for isothermal titration calorimetry.

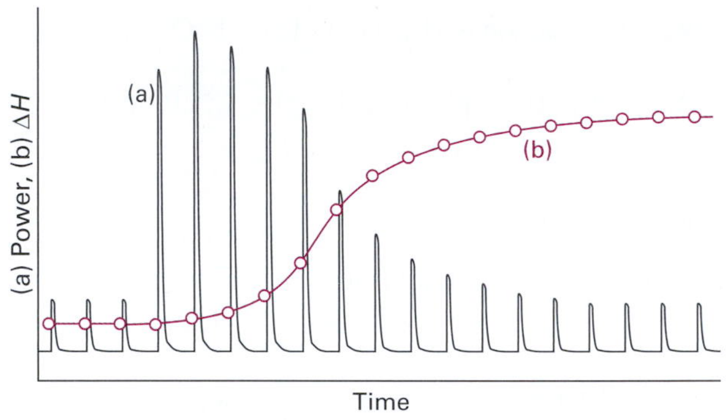

Figure 2C.6 (a) The record of the power applied as each injection is made, and (b) the sum of successive enthalpy changes in the course of the titration.

can be identified with the enthalpy change. The thermogram also reveals the formation of new intermolecular interactions in the denatured form. The increase in heat capacity accompanying the native → denatured transition reflects the change from a more compact native conformation to one in which the more exposed amino acid side chains in the denatured form have more extensive interactions with the surrounding water molecules.

(b) Isothermal titration calorimetry

Isothermal titration calorimetry (ITC) is also a 'differential' technique in which the thermal behaviour of a sample is compared with that of a reference. The apparatus is shown in Fig. 2C.5. One of the thermally conducting vessels, which have a volume of a few millilitres (10^{-6} m^3), contains the reference (water for instance) and a heater rated at a few milliwatts. The second vessel contains one of the reagents, such as a solution of a macromolecule with binding sites; it also contains a heater. At the start of the experiment, both heaters are activated, and then precisely determined amounts (of volume of about a microlitre, 10^{-9} m^3) of the second reagent are added to the reaction cell. The power required to maintain the same temperature differential with the reference cell is monitored. If the reaction is exothermic, less power is needed; if it is endothermic, then more power must be supplied.

A typical result is shown in Fig. 2C.6, which shows the power needed to maintain the temperature differential: from the power and the length of time, δt, for which it is supplied, the heat supplied, δq_i, for the injection i can be calculated from $\delta q_i = P_i \delta t$. If the volume of solution is V and the molar concentration of unreacted reagent A is c_i at the time of the ith injection, then the change in its concentration at that injection is δc_i and the heat generated (or absorbed) by the reaction is $V\Delta_r H \delta c_i = \delta q_i$. The sum of all such quantities, given that the sum of δc_i is the known initial concentration of the reactant, can then be interpreted as the value of $\Delta_r H$ for the reaction.

Checklist of concepts

☐ 1. The **standard enthalpy of transition** is equal to the energy transferred as heat at constant pressure in the transition.

☐ 2. A **thermochemical equation** is a chemical equation and its associated change in enthalpy.

☐ 3. **Hess's law** states that the standard enthalpy of an overall reaction is the sum of the standard enthalpies of the individual reactions into which a reaction may be divided.

- ☐ 4. **Standard enthalpies of formation** are defined in terms of the reference states of elements.
- ☐ 5. The **standard reaction enthalpy** is expressed as the difference of the standard enthalpies of formation of products and reactants.
- ☐ 6. Computer modelling is used to estimate standard enthalpies of formation.
- ☐ 7. The temperature dependence of a reaction enthalpy is expressed by **Kirchhoff's law**.

Checklist of equations

Property	Equation	Comment	Equation number
The standard reaction enthalpy	$\Delta_r H^\ominus = \sum_{\text{Products}} \nu\Delta_f H^\ominus - \sum_{\text{Reactants}} \nu\Delta_f H^\ominus$	ν: stoichiometric coefficients; ν_J: (signed) stoichiometric numbers	2C.5
	$\Delta_r H^\ominus = \sum_J \nu_J \Delta_f H^\ominus(J)$		
Kirchhoff's law	$\Delta_r H^\ominus(T_2) = \Delta_r H^\ominus(T_2) + \int_{T_1}^{T_2} \Delta_r C_p^\ominus \mathrm{d}T$		2C.7a
	$\Delta_r C_{p,m}^\ominus = \sum_J \nu_J C_{p,m}^\ominus(J)$		2C.7c
	$\Delta_r H^\ominus(T_2) = \Delta_r H^\ominus(T_1) + (T_2 - T_1)\Delta_r C_p^\ominus$	If $\Delta_r C_p^\ominus$ independent of temperature	

2D State functions and exact differentials

Contents

➤ Why do you need to know this material?

Thermodynamics gives us the power to derive relations between a variety of properties: this Topic is a first introduction to the manipulation of equations involving state functions. In the process, we obtain such important relations as that between heat capacities. An important technological consequence is the Joule–Thomson effect for cooling gases, which is derived here.

➤ What is the key idea?

The fact that internal energy and enthalpy are state functions leads to relations between thermodynamic properties.

➤ What do you need to know already?

You need to be aware that internal energy and enthalpy are state functions (Topics 2B and 2C) and be familiar with heat capacity. You need to be able to make use of several simple relations involving partial derivatives (*Mathematical background* 2).

A **state function** is a property that depends only on the current state of a system and is independent of its history. The internal energy and enthalpy are two examples of state functions. Physical quantities that do depend on the path between two states are called **path functions**. Examples of path functions are the work and the heating that are done when preparing a state. We do not speak of a system in a particular state as possessing work or heat. In each case, the energy transferred as work or heat relates to the path being taken between states, not the current state itself.

A part of the richness of thermodynamics is that it uses the mathematical properties of state functions to draw far-reaching conclusions about the relations between physical properties and thereby establish connections that may be completely unexpected. The practical importance of this ability is that we can combine measurements of different properties to obtain the value of a property we require.

2D.1 Exact and inexact differentials

Consider a system undergoing the changes depicted in Fig. 2D.1. The initial state of the system is i and in this state the internal energy is U_i. Work is done by the system as it expands adiabatically to a state f. In this state the system has an internal energy U_f and the work done on the system as it changes along Path 1 from i to f is w. Notice our use of language: U is a property of the state; w is a property of the path. Now consider another process, Path 2, in which the initial and final states are the same as those in Path 1 but in which the expansion is not adiabatic. The internal energy of both the initial and the final states are the same as before (because U is a state function). However, in the second path an energy q' enters the system as heat and the work w' is not the same as w. The work and the heat are path functions.

If a system is taken along a path (for example, by heating it), U changes from U_i to U_f, and the overall change is the sum (integral) of all the infinitesimal changes along the path:

$$\Delta U = \int_i^f \mathrm{d}U \qquad (2D.1)$$

The value of ΔU depends on the initial and final states of the system but is independent of the path between them. This path-independence of the integral is expressed by saying that dU

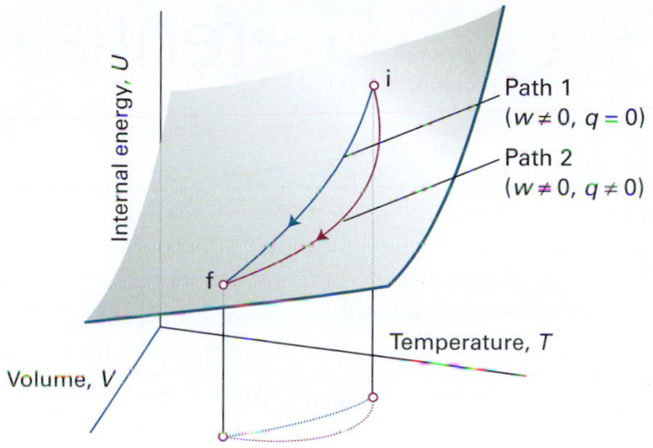

Figure 2D.1 As the volume and temperature of a system are changed, the internal energy changes. An adiabatic and a non-adiabatic path are shown as Path 1 and Path 2, respectively: they correspond to different values of q and w but to the same value of ΔU.

is an 'exact differential'. In general, an **exact differential** is an infinitesimal quantity that, when integrated, gives a result that is independent of the path between the initial and final states.

When a system is heated, the total energy transferred as heat is the sum of all individual contributions at each point of the path:

$$q=\int_{\mathrm{i,path}}^{\mathrm{f}} \mathrm{d}q \qquad (2\mathrm{D}.2)$$

Notice the differences between this equation and eqn 2D.1. First, we do not write Δq, because q is not a state function and the energy supplied as heat cannot be expressed as q_f-q_i. Secondly, we must specify the path of integration because q depends on the path selected (for example, an adiabatic path has $q=0$, whereas a non-adiabatic path between the same two states would have $q\neq 0$). This path-dependence is expressed by saying that $\mathrm{d}q$ is an 'inexact differential'. In general, an **inexact differential** is an infinitesimal quantity that, when integrated, gives a result that depends on the path between the initial and final states. Often $\mathrm{d}q$ is written $đq$ to emphasize that it is inexact and requires the specification of a path.

The work done on a system to change it from one state to another depends on the path taken between the two specified states; for example, in general the work is different if the change takes place adiabatically and non-adiabatically. It follows that $\mathrm{d}w$ is an inexact differential. It is often written $đw$.

Example 2D.1 Calculating work, heat, and change in internal energy

Consider a perfect gas inside a cylinder fitted with a piston. Let the initial state be T,V_i and the final state be T,V_f. The change of state can be brought about in many ways, of which the two simplest are the following: Path 1, in which there is free expansion against zero external pressure; Path 2, in which there is reversible, isothermal expansion. Calculate w, q, and ΔU for each process.

Method To find a starting point for a calculation in thermodynamics, it is often a good idea to go back to first principles and to look for a way of expressing the quantity we are asked to calculate in terms of other quantities that are easier to calculate. It is argued in Topic 2B that the internal energy of a perfect gas depends only on the temperature and is independent of the volume those molecules occupy, so for any isothermal change, $\Delta U=0$. We also know that in general $\Delta U=q+w$. The question depends on being able to combine the two expressions. Topic 2A presents a number of expressions for the work done in a variety of processes, and here we need to select the appropriate ones.

Answer Because $\Delta U=0$ for both paths and $\Delta U=q+w$, in each case $q=-w$. The work of free expansion is zero (eqn 2A.7 of Topic 2A, $w=0$); so in Path 1, $w=0$ and therefore $q=0$ too. For Path 2, the work is given by eqn 2A.9 of Topic 2A ($w=-nRT\ln(V_f/V_i)$) and consequently $q=nRT\ln(V_f/V_i)$.

Self-test 2D.1 Calculate the values of q, w, and ΔU for an irreversible isothermal expansion of a perfect gas against a constant nonzero external pressure.

Answer: $q=p_{ex}\Delta V$, $w=-p_{ex}\Delta V$, $\Delta U=0$

2D.2 Changes in internal energy

We begin to unfold the consequences of $\mathrm{d}U$ being an exact differential by exploring a closed system of constant composition (the only type of system considered in the rest of this Topic). The internal energy U can be regarded as a function of V, T, and p, but, because there is an equation of state (Topic 1A), stating the values of two of the variables fixes the value of the third. Therefore, it is possible to write U in terms of just two independent variables: V and T, p and T, or p and V. Expressing U as a function of volume and temperature fits the purpose of our discussion.

(a) General considerations

Because the internal energy is a function of the volume and the temperature, when these two quantities change, the internal energy changes by

$$\mathrm{d}U=\left(\frac{\partial U}{\partial V}\right)_T \mathrm{d}V+\left(\frac{\partial U}{\partial T}\right)_V \mathrm{d}T \qquad \text{General expression for a change in } U \text{ with } T \text{ and } V \qquad (2\mathrm{D}.3)$$

The interpretation of this equation is that, in a closed system of constant composition, any infinitesimal change in the internal

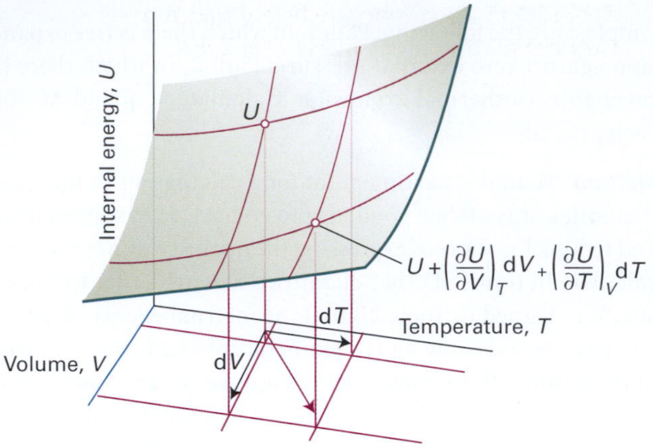

Figure 2D.2 An overall change in U, which is denoted dU, arises when both V and T are allowed to change. If second-order infinitesimals are ignored, the overall change is the sum of changes for each variable separately.

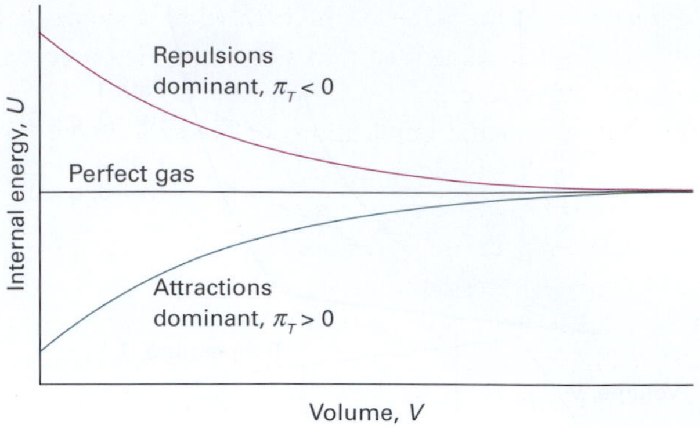

Figure 2D.4 For a perfect gas, the internal energy is independent of the volume (at constant temperature). If attractions are dominant in a real gas, the internal energy increases with volume because the molecules become farther apart on average. If repulsions are dominant, the internal energy decreases as the gas expands.

energy is proportional to the infinitesimal changes of volume and temperature, the coefficients of proportionality being the two partial derivatives (Fig. 2D.2).

In many cases partial derivatives have a straightforward physical interpretation, and thermodynamics gets shapeless and difficult only when that interpretation is not kept in sight. In the present case, we have already met $(\partial U/\partial T)_V$ in Topic 2A, where we saw that it is the constant-volume heat capacity, C_V. The other coefficient, $(\partial U/\partial V)_T$, plays a major role in thermodynamics because it is a measure of the variation of the internal energy of a substance as its volume is changed at constant temperature (Fig. 2D.3). We shall denote it π_T and, because it has the same dimensions as pressure but arises from the interactions between the molecules within the sample, call it the **internal pressure**:

$$\pi_T=\left(\frac{\partial U}{\partial V}\right)_T \qquad \textit{Definition} \quad \text{Internal pressure} \qquad (2D.4)$$

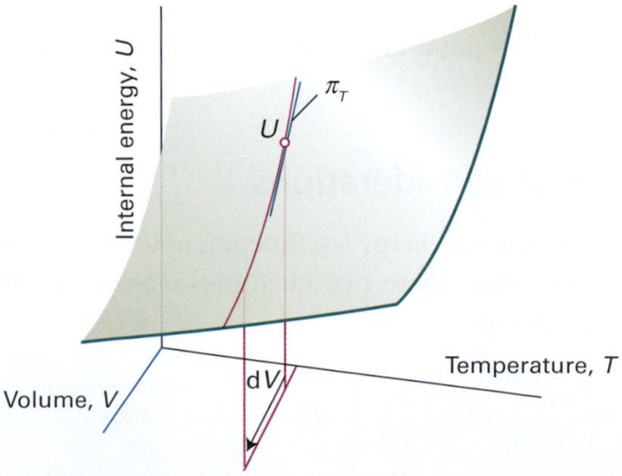

Figure 2D.3 The internal pressure, π_T, is the slope of U with respect to V with the temperature T held constant.

In terms of the notation C_V and π_T, eqn 2D.3 can now be written

$$dU=\pi_T dV+C_V dT \qquad (2D.5)$$

When there are no interactions between the molecules, the internal energy is independent of their separation and hence independent of the volume of the sample. Therefore, for a perfect gas we can write $\pi_T=0$. If the gas is described by the van der Waals equation with a, the parameter corresponding to attractive interactions, dominant, then an increase in volume increases the average separation of the molecules and therefore raises the internal energy. In this case, we expect $\pi_T>0$ (Fig. 2D.4).

The statement $\pi_T=0$ (that is, the internal energy is independent of the volume occupied by the sample) can be taken to be the definition of a perfect gas, for in Topic 3D we see that it implies the equation of state $pV \propto T$.

James Joule thought that he could measure π_T by observing the change in temperature of a gas when it is allowed to expand into a vacuum. He used two metal vessels immersed in a water bath (Fig. 2D.5). One was filled with air at about 22 atm and the other was evacuated. He then tried to measure the change

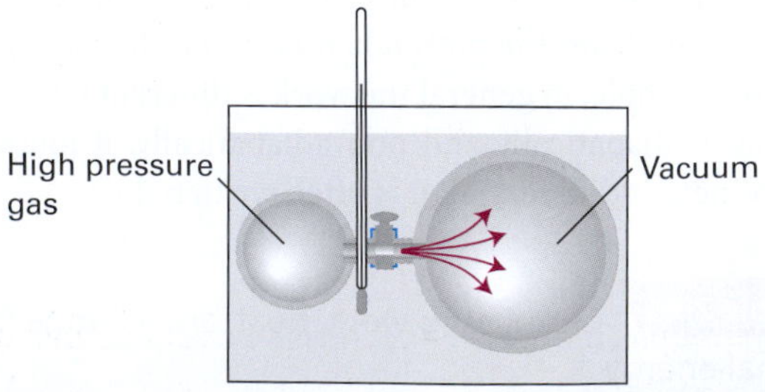

Figure 2D.5 A schematic diagram of the apparatus used by Joule in an attempt to measure the change in internal energy when a gas expands isothermally. The heat absorbed by the gas is proportional to the change in temperature of the bath.

in temperature of the water of the bath when a stopcock was opened and the air expanded into a vacuum. He observed no change in temperature.

The thermodynamic implications of the experiment are as follows. No work was done in the expansion into a vacuum, so $w=0$. No energy entered or left the system (the gas) as heat because the temperature of the bath did not change, so $q=0$. Consequently, within the accuracy of the experiment, $\Delta U=0$. Joule concluded that U does not change when a gas expands isothermally and therefore that $\pi_T=0$. His experiment, however, was crude. In particular, the heat capacity of the apparatus was so large that the temperature change that gases do in fact cause was too small to measure. Nevertheless, from his experiment Joule had extracted an essential limiting property of a gas, a property of a perfect gas, without detecting the small deviations characteristic of real gases.

(b) Changes in internal energy at constant pressure

Partial derivatives have many useful properties and some that we shall draw on frequently are reviewed in *Mathematical background* 2. Skilful use of them can often turn some unfamiliar quantity into a quantity that can be recognized, interpreted, or measured.

As an example, suppose we want to find out how the internal energy varies with temperature when the pressure rather than the volume of the system is kept constant. If we divide both sides of eqn 2D.5 by dT and impose the condition of constant pressure on the resulting differentials, so that dU/dT on the left becomes $(\partial U/\partial T)_p$, we obtain

$$\left(\frac{\partial U}{\partial T}\right)_p = \pi_T \left(\frac{\partial V}{\partial T}\right)_p + C_V$$

It is usually sensible in thermodynamics to inspect the output of a manipulation like this to see if it contains any recognizable physical quantity. The partial derivative on the right in this expression is the slope of the plot of volume against temperature (at constant pressure). This property is normally tabulated as the **expansion coefficient**, α, of a substance, which is defined as

$$\alpha = \frac{1}{V}\left(\frac{\partial V}{\partial T}\right)_p \qquad \textit{Definition} \quad \text{Expansion coefficient} \qquad (2D.6)$$

and physically is the fractional change in volume that accompanies a rise in temperature. A large value of α means that the volume of the sample responds strongly to changes in temperature. Table 2D.1 lists some experimental values of α. For future reference, it also lists the **isothermal compressibility**, κ_T (kappa), which is defined as

Table 2D.1* Expansion coefficients (α) and isothermal compressibilities (κ_T) at 298 K

	$\alpha/(10^{-4}\ \mathrm{K^{-1}})$	$\kappa_T/(10^{-6}\ \mathrm{bar^{-1}})$
Benzene	12.4	90.9
Diamond	0.030	0.185
Lead	0.861	2.18
Water	2.1	49.0

* More values are given in the *Resource section*.

$$\kappa_T = -\frac{1}{V}\left(\frac{\partial V}{\partial p}\right)_T \qquad \textit{Definition} \quad \text{Isothermal compressibility} \qquad (2D.7)$$

The isothermal compressibility is a measure of the fractional change in volume when the pressure is increased by a small amount; the negative sign in the definition ensures that the compressibility is a positive quantity, because an increase of pressure, implying a positive dp, brings about a reduction of volume, a negative dV.

Example 2D.2 Calculating the expansion coefficient of a gas

Derive an expression for the expansion coefficient of a perfect gas.

Method The expansion coefficient is defined in eqn 2D.6. To use this expression, substitute the expression for V in terms of T obtained from the equation of state for the gas. As implied by the subscript in eqn 2D.6, the pressure, p, is treated as a constant.

Answer Because $pV=nRT$, we can write

$$\alpha = \frac{1}{V}\left(\frac{\partial(nRT/p)}{\partial T}\right)_p = \frac{1}{V}\times\frac{nR}{p}\frac{dT}{dT} = \frac{nR}{pV} = \frac{1}{T}$$

The higher the temperature, the less responsive is the volume of a perfect gas to a change in temperature.

Self-test 2D.2 Derive an expression for the isothermal compressibility of a perfect gas.

Answer: $\kappa_T = 1/p$

When we introduce the definition of α into the equation for $(\partial U/\partial T)_p$, we obtain

$$\left(\frac{\partial U}{\partial T}\right)_p = \alpha\pi_T V + C_V \qquad (2D.8)$$

This equation is entirely general (provided the system is closed and its composition is constant). It expresses the dependence of the internal energy on the temperature at constant pressure in terms of C_V, which can be measured in one experiment, in terms of α, which can be measured in another, and in terms of the quantity π_T. For a perfect gas, $\pi_T = 0$, so then

$$\left(\frac{\partial U}{\partial T}\right)_p = C_V \qquad (2D.9)$$

That is, although the constant-volume heat capacity of a perfect gas is defined as the slope of a plot of internal energy against temperature at constant volume, for a perfect gas C_V is also the slope at constant pressure.

Equation 2D.9 provides an easy way to derive the relation between C_p and C_V for a perfect gas. Thus, we can use it to express both heat capacities in terms of derivatives at constant pressure:

$$C_p - C_V = \overbrace{\left(\frac{\partial H}{\partial T}\right)_p}^{\text{Definition of } C_p} - \overbrace{\left(\frac{\partial U}{\partial T}\right)_p}^{\text{eqn 2D.9}}$$

Then we introduce $H = U + pV = U + nRT$ into the first term, which results in

$$C_p - C_V = \left(\frac{\partial (U+nRT)}{\partial T}\right)_p - \left(\frac{\partial U}{\partial T}\right)_p = nR \qquad (2D.10)$$

We show in the following *Justification* that in general

$$C_p - C_V = \frac{\alpha^2 TV}{\kappa_T} \qquad (2D.11)$$

Equation 2D.11 applies to any substance (that is, it is 'universally true'). It reduces to eqn 2D.10 for a perfect gas when we set $\alpha = 1/T$ and $\kappa_T = 1/p$. Because expansion coefficients α of liquids and solids are small, it is tempting to deduce from eqn 2D.11 that for them $C_p \approx C_V$. But this is not always so, because the compressibility κ_T might also be small, so α^2/κ_T might be large. That is, although only a little work need be done to push back the atmosphere, a great deal of work may have to be done to pull atoms apart from one another as the solid expands.

Brief illustration 2D.1 The relation between heat capacities

The expansion coefficient and isothermal compressibility of water at 25 °C are given in Table 2D.1 as $2.1 \times 10^{-4}\,\text{K}^{-1}$ and $4.96 \times 10^{-5}\,\text{atm}^{-1}$ ($4.90 \times 10^{-10}\,\text{Pa}^{-1}$), respectively. The molar volume of water at that temperature, $V_\text{m} = M/\rho$ (where ρ is the mass density) is $18.1\,\text{cm}^3\,\text{mol}^{-1}$ ($1.81 \times 10^{-5}\,\text{m}^3\,\text{mol}^{-1}$). Therefore, from eqn 2D.11, the difference in molar heat capacities (which is given by V_m in place of V) is

$$C_{p,\text{m}} - C_{V,\text{m}} = \frac{(2.1\times10^{-4}\,\text{K}^{-1})^2 \times (298\,\text{K}) \times (1.81\times10^{-5}\,\text{m}^3\,\text{mol}^{-1})}{4.90\times10^{-10}\,\text{Pa}^{-1}}$$
$$= 0.485\,\text{Pa}\,\text{m}^3\,\text{mol}^{-1} = 0.485\,\text{J}\,\text{mol}^{-1}$$

For water, $C_{p,\text{m}} = 75.3\,\text{J}\,\text{K}^{-1}\,\text{mol}^{-1}$, so $C_{V,\text{m}} = 74.8\,\text{J}\,\text{K}^{-1}\,\text{mol}^{-1}$. In some cases, the two heat capacities differ by as much as 30 per cent.

Self-test 2D.3 Evaluate the difference in molar heat capacities for benzene; use data from the *Resource section*.

Answer: $45\,\text{J}\,\text{K}^{-1}\,\text{mol}^{-1}$

Justification 2D.1 The relation between heat capacities

A useful rule when doing a problem in thermodynamics is to go back to first principles. In the present problem we do this twice, first by expressing C_p and C_V in terms of their definitions and then by inserting the definition $H = U + pV$:

$$C_p - C_V = \left(\frac{\partial H}{\partial T}\right)_p - C_V$$
$$= \left(\frac{\partial U}{\partial T}\right)_p + \left(\frac{\partial (pV)}{\partial T}\right)_p - C_V$$

Equation 2D.8, $(\partial U/\partial T)_p = \alpha\pi_T V + C_V$, lets us write the difference of the first and third terms as $\alpha\pi_T V$. We can simplify the remaining term by noting that, because p is constant,

$$\left(\frac{\partial (pV)}{\partial T}\right)_p = p\left(\frac{\partial V}{\partial T}\right)_p = \alpha pV$$

Collecting the two contributions gives

$$C_p - C_V = \alpha(p + \pi_T)V$$

The first term on the right, αpV, is a measure of the work needed to push back the atmosphere; the second term on the right, $\alpha\pi_T V$, is the work required to separate the molecules composing the system.

At this point we can go further by using the (Second Law) result proved in Topic 3D that

$$\pi_T = T\left(\frac{\partial p}{\partial T}\right)_V - p$$

When this expression is inserted in the last equation we obtain

$$C_p - C_V = \alpha TV\left(\frac{\partial p}{\partial T}\right)_V$$

We now transform the remaining partial derivative. With V regarded as a function of p and T, when these two quantities change the resulting change in V is

$$\mathrm{d}V=\left(\frac{\partial V}{\partial T}\right)_p \mathrm{d}T+\left(\frac{\partial V}{\partial p}\right)_T \mathrm{d}p$$

For the volume to be constant, $\mathrm{d}V=0$ implies that

$$\left(\frac{\partial V}{\partial T}\right)_p \mathrm{d}T=-\left(\frac{\partial V}{\partial p}\right)_T \mathrm{d}p \text{ at constant volume}$$

On division by $\mathrm{d}T$, this relation becomes

$$\left(\frac{\partial V}{\partial T}\right)_p=-\left(\frac{\partial V}{\partial p}\right)_T\left(\frac{\partial p}{\partial T}\right)_V$$

and therefore

$$\left(\frac{\partial p}{\partial T}\right)_V=-\frac{(\partial V/\partial T)_p}{(\partial V/\partial p)_T}=\frac{\alpha}{\kappa_T}$$

Insertion of this relation into the expression above for C_p-C_V produces eqn 2D.11.

2D.3 The Joule–Thomson effect

We can carry out a similar set of operations on the enthalpy, $H=U+pV$. The quantities U, p, and V are all state functions; therefore H is also a state function and $\mathrm{d}H$ is an exact differential. It turns out that H is a useful thermodynamic function when the pressure is under our control: we saw a sign of that in the relation $\Delta H=q_p$ (this is eqn 2B.2b of Topic 2B). We shall therefore regard H as a function of p and T, and adapt the argument in Section 2D.2 for the variation of U to find an expression for the variation of H with temperature at constant volume. As explained in the following *Justification*, we find that for a closed system of constant composition,

$$\mathrm{d}H=-\mu C_p\mathrm{d}p+C_p\mathrm{d}T \qquad (2D.12)$$

where the **Joule–Thomson coefficient**, μ (mu), is defined as

$$\mu=\left(\frac{\partial T}{\partial p}\right)_H \qquad \textit{Definition} \quad \text{Joule–Thomson coefficient} \qquad (2D.13)$$

This relation will prove useful for relating the heat capacities at constant pressure and volume and for a discussion of the liquefaction of gases.

Justification 2D.2 The variation of enthalpy with pressure and temperature

Because H is a function of p and T we can write, when these two quantities change by an infinitesimal amount, the enthalpy changes by

$$\mathrm{d}H=\left(\frac{\partial H}{\partial p}\right)_T \mathrm{d}p+\left(\frac{\partial H}{\partial T}\right)_p \mathrm{d}T$$

The second partial derivative is C_p; our task here is to express $(\partial H/\partial p)_T$ in terms of recognizable quantities. If the enthalpy is constant, $\mathrm{d}H=0$ and this expression then requires that

$$\left(\frac{\partial H}{\partial p}\right)_T \mathrm{d}p=-C_p\mathrm{d}T \text{ at constant } H$$

Division of both sides by $\mathrm{d}p$ then gives

$$\left(\frac{\partial H}{\partial p}\right)_T=-C_p\left(\frac{\partial T}{\partial p}\right)_H=-C_p\mu$$

Equation 2D.13 now follows directly.

(a) Observation of the Joule–Thomson effect

The analysis of the Joule–Thomson coefficient is central to the technological problems associated with the liquefaction of gases. We need to be able to interpret it physically and to measure it. As shown in the following *Justification*, the cunning required to impose the constraint of constant enthalpy, so that the process is **isenthalpic**, was supplied by Joule and William Thomson (later Lord Kelvin). They let a gas expand through a porous barrier from one constant pressure to another and monitored the difference of temperature that arose from the expansion (Fig. 2D.6). The whole apparatus was insulated so that the process was adiabatic. They observed a lower temperature on the low pressure side, the difference in temperature being proportional to the pressure difference they maintained.

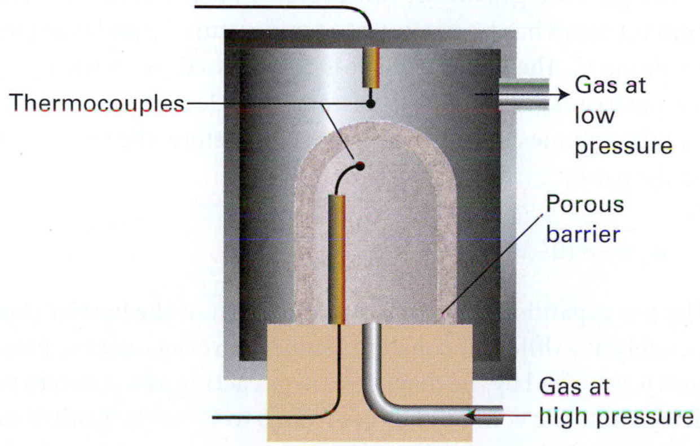

Figure 2D.6 The apparatus used for measuring the Joule–Thomson effect. The gas expands through the porous barrier, which acts as a throttle, and the whole apparatus is thermally insulated. As explained in the text, this arrangement corresponds to an isenthalpic expansion (expansion at constant enthalpy). Whether the expansion results in a heating or a cooling of the gas depends on the conditions.

This cooling by isenthalpic expansion is now called the **Joule–Thomson effect**.

Justification 2D.3 The Joule–Thomson effect

Here we show that the experimental arrangement results in expansion at constant enthalpy. Because all changes to the gas occur adiabatically, $q=0$ implies that $\Delta U=w$. Next, consider the work done as the gas passes through the barrier. We focus on the passage of a fixed amount of gas from the high pressure side, where the pressure is p_i, the temperature T_i, and the gas occupies a volume V_i (Fig. 2D.7).

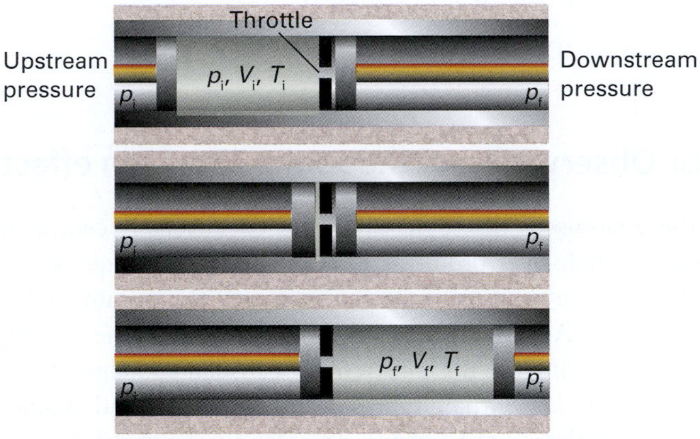

Figure 2D.7 The thermodynamic basis of Joule–Thomson expansion. The pistons represent the upstream and downstream gases, which maintain constant pressures either side of the throttle. The transition from the top diagram to the bottom diagram, which represents the passage of a given amount of gas through the throttle, occurs without change of enthalpy.

The gas emerges on the low pressure side, where the same amount of gas has a pressure p_f, a temperature T_f, and occupies a volume V_f. The gas on the left is compressed isothermally by the upstream gas acting as a piston. The relevant pressure is p_i and the volume changes from V_i to 0; therefore, the work done on the gas is

$$w_1 = -p_i(0-V_i) = p_iV_i$$

The gas expands isothermally on the right of the barrier (but possibly at a different constant temperature) against the pressure p_f provided by the downstream gas acting as a piston to be driven out. The volume changes from 0 to V_f, so the work done on the gas in this stage is

$$w_2 = -p_f(V_f-0) = -p_fV_f$$

The total work done on the gas is the sum of these two quantities, or

$$w_1+w_2 = p_iV_i - p_fV_f$$

It follows that the change of internal energy of the gas as it moves adiabatically from one side of the barrier to the other is

$$U_f - U_i = w = p_iV_i - p_fV_f$$

Reorganization of this expression gives

$$U_f + p_fV_f = U_i + p_iV_i \quad \text{or} \quad H_f = H_i$$

Therefore, the expansion occurs without change of enthalpy.

The property measured in the experiment is the ratio of the temperature change to the change of pressure, $\Delta T/\Delta p$. Adding the constraint of constant enthalpy and taking the limit of small Δp implies that the thermodynamic quantity measured is $(\partial T/\partial p)_H$, which is the Joule–Thomson coefficient, μ. In other words, the physical interpretation of μ is that it is the ratio of the change in temperature to the change in pressure when a gas expands under conditions that ensure there is no change in enthalpy.

The modern method of measuring μ is indirect, and involves measuring the **isothermal Joule–Thomson coefficient**, the quantity

$$\mu_T = \left(\frac{\partial H}{\partial p}\right)_T \qquad \textit{Definition} \quad \text{Isothermal Joule–Thomson coefficient} \qquad (2D.14)$$

which is the slope of a plot of enthalpy against pressure at constant temperature (Fig. 2D.8). Comparing eqns 2D.13 and 2D.14, we see (from the last line of *Justification* 2D.2) that the two coefficients are related by:

$$\mu_T = -C_p\mu \qquad (2D.15)$$

To measure μ_T, the gas is pumped continuously at a steady pressure through a heat exchanger, which brings it to the required temperature, and then through a porous plug inside a thermally insulated container. The steep pressure drop is measured and the cooling effect is exactly offset by an electric heater placed immediately after the plug (Fig. 2D.9). The energy provided

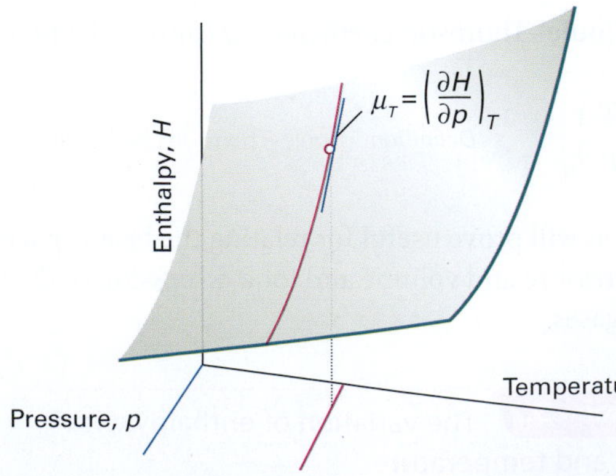

Figure 2D.8 The isothermal Joule–Thomson coefficient is the slope of the enthalpy with respect to changing pressure, the temperature being held constant.

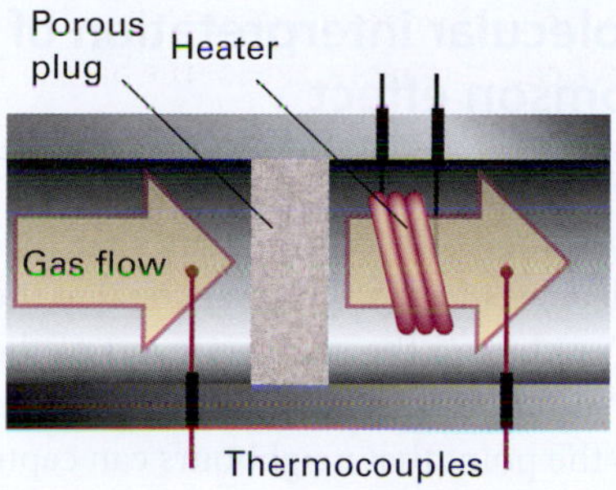

Figure 2D.9 A schematic diagram of the apparatus used for measuring the isothermal Joule–Thomson coefficient. The electrical heating required to offset the cooling arising from expansion is interpreted as ΔH and used to calculate $(\partial H/\partial p)_T$, which is then converted to μ as explained in the text.

Table 2D.2* Inversion temperatures (T_I), normal freezing (T_f) and boiling (T_b) points, and Joule–Thomson coefficients (μ) at 1 atm and 298 K

	T_I/K	T_f/K	T_b/K	μ/(K bar^{-1})
Ar	723	83.8	87.3	
CO_2	1500	194.7	+1.10	
He	40	4.2	−0.060	
N_2	621	63.3	77.4	+0.25

* More values are given in the *Resource section*.

by the heater is monitored. Because $\Delta H = q_p$, the energy transferred as heat can be identified with the value of ΔH. The pressure change Δp is known, so we can find μ_T from the limiting value of $\Delta H/\Delta p$ as $\Delta p \rightarrow 0$ and then convert it to μ. Table 2D.2 lists some values obtained in this way.

Real gases have nonzero Joule–Thomson coefficients. Depending on the identity of the gas, the pressure, the relative magnitudes of the attractive and repulsive intermolecular forces, and the temperature, the sign of the coefficient may be either positive or negative (Fig. 2D.10). A positive sign implies that dT is negative when dp is negative, in which case the gas cools on expansion. Gases that show a heating effect ($\mu<0$) at one temperature show a cooling effect ($\mu>0$) when the temperature is below their upper **inversion temperature**, T_I (Table 2D.2, Fig. 2D.11). As indicated in Fig. 2D.11, a gas typically has two inversion temperatures.

Brief illustration 2D.2 The Joule–Thomson effect

The Joule–Thomson coefficient for nitrogen at 298 K and 1 atm (Table 2D.2) is +0.25 K bar^{-1}. It follows that the change in temperature the gas undergoes when its pressure changes by −10 bar under isenthalpic conditions is

$$\Delta T \approx \mu\Delta p = +(0.25\,\text{K bar}^{-1})\times(-10\,\text{bar}) = -2.5\,\text{K}$$

At the same initial temperature and pressure, the molar isothermal Joule coefficient is

$$\mu_{T,\text{m}} = -C_{p,\text{m}}\,\mu = -(29.1\,\text{J K}^{-1}\,\text{mol}^{-1})\times(+0.25\,\text{K bar}^{-1}) = -7.3\,\text{J bar}^{-1}\,\text{mol}^{-1}$$

Note that μ is an intensive property but μ_T is extensive (but $\mu_{T,\text{m}}$, like all molar quantities, is intensive).

Self-test 2D.4 Evaluate the change in temperature when the pressure of carbon dioxide changes isenthalpically by −10 bar at 300 K and evaluate its molar isothermal Joule coefficient.

Answer: −11 K, 41.2 J bar^{-1} mol^{-1}

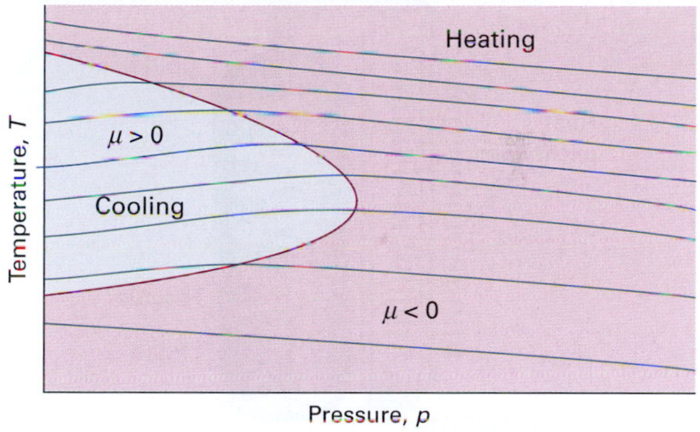

Figure 2D.10 The sign of the Joule–Thomson coefficient, μ, depends on the conditions. Inside the boundary, the blue area, it is positive and outside it is negative. The temperature corresponding to the boundary at a given pressure is the 'inversion temperature' of the gas at that pressure. For a given pressure, the temperature must be below a certain value if cooling is required but, if it becomes too low, the boundary is crossed again and heating occurs. Reduction of pressure under adiabatic conditions moves the system along one of the isenthalps, or curves of constant enthalpy. The inversion temperature curve runs through the points of the isenthalps where their slope changes from negative to positive.

The 'Linde refrigerator' makes use of Joule–Thomson expansion to liquefy gases (Fig. 2D.12). The gas at high pressure is allowed to expand through a throttle; it cools and is circulated past the incoming gas. That gas is cooled, and its subsequent expansion cools it still further. There comes a stage when the circulating gas becomes so cold that it condenses to a liquid.

For a perfect gas, $\mu=0$; hence, the temperature of a perfect gas is unchanged by Joule–Thomson expansion. (Simple adiabatic expansion does cool a perfect gas, because the gas does work, Topic 2E.) This characteristic points clearly to the involvement of intermolecular forces in determining the size of the effect. However, the Joule–Thomson coefficient of a real gas

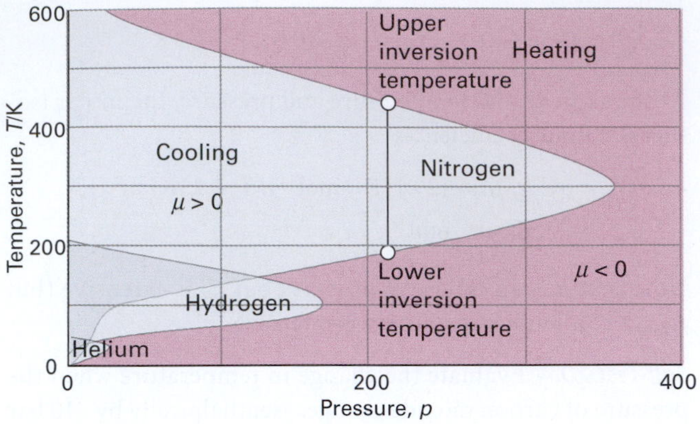

Figure 2D.11 The inversion temperatures for three real gases, nitrogen, hydrogen, and helium.

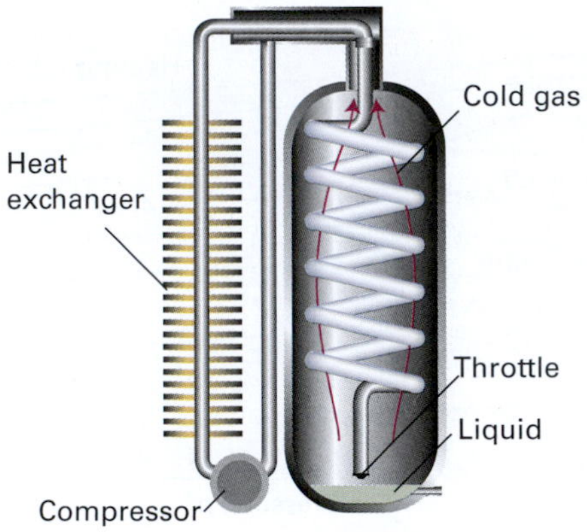

Figure 2D.12 The principle of the Linde refrigerator is shown in this diagram. The gas is recirculated, and so long as it is beneath its inversion temperature it cools on expansion through the throttle. The cooled gas cools the high-pressure gas, which cools still further as it expands. Eventually liquefied gas drips from the throttle.

does not necessarily approach zero as the pressure is reduced even though the equation of state of the gas approaches that of a perfect gas. The coefficient behaves like the properties discussed in Topic 1C in the sense that it depends on derivatives and not on p, V, and T themselves.

(b) The molecular interpretation of the Joule–Thomson effect

The kinetic model of gases (Topic 1B) and the equipartition theorem (*Foundations* B) jointly imply that the mean kinetic energy of molecules in a gas is proportional to the temperature. It follows that reducing the average speed of the molecules is equivalent to cooling the gas. If the speed of the molecules can be reduced to the point that neighbours can capture each other by their intermolecular attractions, then the cooled gas will condense to a liquid.

To slow the gas molecules, we make use of an effect similar to that seen when a ball is thrown into the air: as it rises it slows in response to the gravitational attraction of the Earth and its kinetic energy is converted into potential energy. We saw in Topic 1C that molecules in a real gas attract each other (the attraction is not gravitational, but the effect is the same). It follows that, if we can cause the molecules to move apart from each other, like a ball rising from a planet, then they should slow. It is very easy to move molecules apart from each other: we simply allow the gas to expand, which increases the average separation of the molecules. To cool a gas, therefore, we allow it to expand without allowing any energy to enter from outside as heat. As the gas expands, the molecules move apart to fill the available volume, struggling as they do so against the attraction of their neighbours. Because some kinetic energy must be converted into potential energy to reach greater separations, the molecules travel more slowly as their separation increases. This sequence of molecular events explains the Joule–Thomson effect: the cooling of a real gas by adiabatic expansion. The cooling effect, which corresponds to $\mu > 0$, is observed under conditions when attractive interactions are dominant ($Z < 1$, where Z is the compression factor defined in eqn 1C.1, $Z = V_m/V_m^\circ$), because the molecules have to climb apart against the attractive force in order for them to travel more slowly. For molecules under conditions when repulsions are dominant ($Z > 1$), the Joule–Thomson effect results in the gas becoming warmer, or $\mu < 0$.

Checklist of concepts

- ☐ 1. The quantity $\mathrm{d}U$ is an exact differential; $\mathrm{d}w$ and $\mathrm{d}q$ are not.
- ☐ 2. The change in internal energy may be expressed in terms of changes in temperature and pressure.
- ☐ 3. The **internal pressure** is the variation of internal energy with volume at constant temperature.
- ☐ 4. **Joule's experiment** showed that the internal pressure of a perfect gas is zero.
- ☐ 5. The change in internal energy with pressure and temperature is expressed in terms of the internal pressure and the heat capacity and leads to a general expression for the relation between heat capacities.
- ☐ 6. The **Joule–Thomson effect** is the change in temperature of a gas when it undergoes isenthalpic expansion.

Checklist of equations

Property	Equation	Comment	Equation number
Change in $U(V,T)$	$\mathrm{d}U=(\partial U/\partial V)_T\mathrm{d}V+(\partial U/\partial T)_V\mathrm{d}T$	Constant composition	2D.3
Internal pressure	$\pi_T=(\partial U/\partial V)_T$	Definition; for a perfect gas, $\pi_T=0$	2D.4
Change in $U(V,T)$	$\mathrm{d}U=\pi_T\mathrm{d}V+C_V\mathrm{d}T$	Constant composition	2D.5
Expansion coefficient	$\alpha=(1/V)(\partial V/\partial T)_p$	Definition	2D.6
Isothermal compressibility	$\kappa_T=-(1/V)(\partial V/\partial p)_T$	Definition	2D.7
Relation between heat capacities	$C_p-C_V=nR$	Perfect gas	2D.10
	$C_p-C_V=\alpha^2TV/\kappa_T$		2D.11
Change in $H(p,T)$	$\mathrm{d}H=-\mu C_p\mathrm{d}p+C_p\mathrm{d}T$	Constant composition	2D.12
Joule–Thomson coefficient	$\mu=(\partial T/\partial p)_H$	For a perfect gas, $\mu=0$	2D.13
Isothermal Joule–Thomson coefficient	$\mu_T=(\partial H/\partial p)_T$	For a perfect gas, $\mu_T=0$	2D.14
Relation between coefficients	$\mu_T=-C_p\mu$		2D.15

2E Adiabatic changes

Contents

➤ Why do you need to know this material?

Adiabatic processes complement isothermal processes, and are used in the discussion of the Second Law of thermodynamics.

➤ What is the key idea?

The temperature of a perfect gas falls when it does work in an adiabatic expansion.

➤ What do you need to know already?

This Topic makes use of the discussion of the properties of gases (Topic 1A), particularly the perfect gas law. It also uses the definitions of heat capacity at constant volume (Topic 1B) and constant pressure (Topic 2B), and the relation between them (Topic 2D).

The temperature falls when a gas expands adiabatically (in a thermally insulated container). Work is done, but as no heat enters the system, the internal energy falls, and therefore the temperature of the working gas also falls. In molecular terms, the kinetic energy of the molecules falls as work is done, so their average speed decreases, and hence the temperature falls too.

2E.1 The change in temperature

To calculate the change in temperature that results from a process we focus first on the change in internal energy. The change in internal energy of a perfect gas when the temperature is changed from T_i to T_f and the volume is changed from V_i to V_f can be expressed as the sum of two steps (Fig. 2E.1). In the first step, only the volume changes and the temperature is held constant at its initial value. However, because the internal

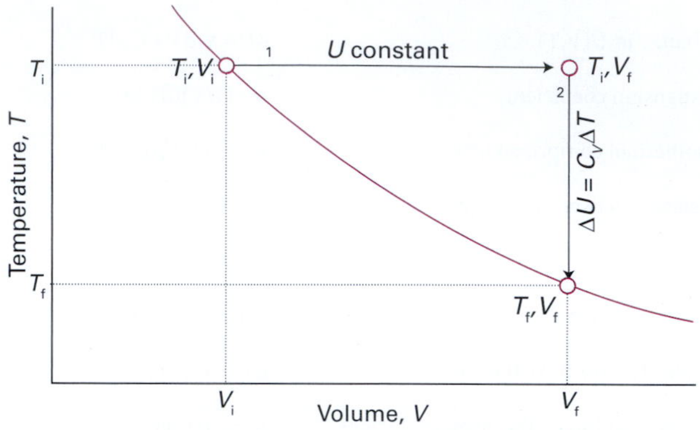

Figure 2E.1 To achieve a change of state from one temperature and volume to another temperature and volume, we may consider the overall change as composed of two steps. In the first step, the system expands at constant temperature; there is no change in internal energy if the system consists of a perfect gas. In the second step, the temperature of the system is reduced at constant volume. The overall change in internal energy is the sum of the changes for the two steps.

energy of a perfect gas is independent of the volume the molecules occupy (Topic 2A), the overall change in internal energy arises solely from the second step, the change in temperature at constant volume. Provided the heat capacity is independent of temperature, this change is

$$\Delta U = (T_f - T_i)C_V = C_V\Delta T$$

Because the expansion is adiabatic, we know that $q=0$; then because $\Delta U = q + w$, it follows that $\Delta U = w_{ad}$. The subscript 'ad' denotes an adiabatic process. Therefore, by equating the two expressions for ΔU, we obtain

$$w_{ad} = C_V\Delta T \qquad \textit{Perfect gas} \qquad \text{Work of adiabatic change} \qquad (2E.1)$$

That is, the work done during an adiabatic expansion of a perfect gas is proportional to the temperature difference between the initial and final states. That is exactly what we expect on molecular grounds, because the mean kinetic energy is proportional to T, so a change in internal energy arising from temperature alone is also expected to be proportional to ΔT.

In the following *Justification* we show that, based on this result, the initial and final temperatures of a perfect gas that undergoes reversible adiabatic expansion (reversible expansion in a thermally insulated container) can be calculated from

$$T_f = T_i\left(\frac{V_i}{V_f}\right)^{1/c} \qquad c = C_{V,m}/R \qquad \textit{Adiabatic, reversible, perfect gas} \qquad \text{Final temperature} \qquad (2E.2a)$$

By raising each side of this expression to the power c, an equivalent expression is

$$V_iT_i^c = V_fT_f^c \qquad c = C_{V,m}/R$$ *Adiabatic, reversible, perfect gas* Final temperature (2E.2b)

This result is often summarized in the form VT^c = constant.

Brief illustration 2E.1 The change in temperature

Consider the adiabatic, reversible expansion of 0.020 mol Ar, initially at 25 °C, from 0.50 dm³ to 1.00 dm³. The molar heat capacity of argon at constant volume is 12.47 J K⁻¹ mol⁻¹, so c = 1.501. Therefore, from eqn 2B.2a,

$$T_f = (298\,\text{K})\left(\frac{0.50\,\text{dm}^3}{1.00\,\text{dm}^3}\right)^{1/1.501} = 188\,\text{K}$$

It follows that $\Delta T = -110$ K, and therefore, from eqn 2E.1, that

$$w = \{(0.020\,\text{mol})\times(12.48\,\text{J K}^{-1}\,\text{mol}^{-1})\}\times(-110\,\text{K}) = -27\,\text{J}$$

Note that temperature change is independent of the amount of gas but the work is not.

Self-test 2E.1 Calculate the final temperature, the work done, and the change of internal energy when ammonia is used in a reversible adiabatic expansion from 0.50 dm³ to 2.00 dm³, the other initial conditions being the same.

Answer: 194 K, −56 J, −56 J

Justification 2E.1 Changes in temperature

Consider a stage in a reversible adiabatic expansion when the pressure inside and out is p. The work done when the gas expands by dV is $dw = -pdV$; however, for a perfect gas, $dU = C_V dT$. Therefore, because for an adiabatic change ($dq = 0$) $dU = dw + dq = dw$, we can equate these two expressions for dU and write

$$C_V dT = -pdV$$

We are dealing with a perfect gas, so we can replace p by nRT/V and obtain

$$\frac{C_V dT}{T} = -\frac{nRdV}{V}$$

To integrate this expression we note that T is equal to T_i when V is equal to V_i, and is equal to T_f when V is equal to V_f at the end of the expansion. Therefore,

$$C_V\int_{T_i}^{T_f}\frac{dT}{T} = -nR\int_{V_i}^{V_f}\frac{dV}{V}$$

(We are taking C_V to be independent of temperature.) Then, because $\int dx/x = \ln x + \text{constant}$, we obtain

$$C_V \ln\frac{T_f}{T_i} = -nR\ln\frac{V_f}{V_i}$$

Because $\ln(x/y) = -\ln(y/x)$, this expression rearranges to

$$\frac{C_V}{nR}\ln\frac{T_f}{T_i} = \ln\frac{V_i}{V_f}$$

With $c = C_V/nR$ we obtain (because $\ln x^a = a\ln x$)

$$\ln\left(\frac{T_f}{T_i}\right)^c = \ln\frac{V_i}{V_f}$$

which implies that $(T_f/T_i)^c = (V_i/V_f)$ and, upon rearrangement, eqn 2E.2.

2E.2 The change in pressure

We show in the following *Justification* that the pressure of a perfect gas that undergoes reversible adiabatic expansion from a volume V_i to a volume V_f is related to its initial pressure by

$$p_fV_f^\gamma = p_iV_i^\gamma$$ *Perfect gas* Reversible adiabatic expansion (2E.3)

where $\gamma = C_{p,m}/C_{V,m}$. This result is commonly summarized in the form pV^γ = constant.

Justification 2E.2 The relation between pressure and volume

The initial and final states of a perfect gas satisfy the perfect gas law regardless of how the change of state takes place, so we can use $pV = nRT$ to write

$$\frac{p_iV_i}{p_fV_f} = \frac{T_i}{T_f}$$

However, from eqn 2E.2 we know that $T_i/T_f = (V_f/V_i)^{1/c}$. Therefore,

$$\frac{p_iV_i}{p_fV_f} = \left(\frac{V_f}{V_i}\right)^{1/c}, \text{ so } \frac{p_i}{p_f}\left(\frac{V_i}{V_f}\right)^{1/c+1} = 1$$

We now use the result from Topic 2B that $C_{p,m} - C_{V,m} = R$ to note that

$$\frac{1}{c} + 1 = \frac{1+c}{c} = \frac{R + C_{V,m}}{C_{V,m}} = \frac{C_{p,m}}{C_{V,m}} = \gamma$$

It follows that

$$\frac{p_i}{p_f}\left(\frac{V_i}{V_f}\right)^{\gamma}=1$$

which rearranges to eqn 2E.3.

For a monatomic perfect gas, $C_{V,m}=\frac{3}{2}R$ (Topic 2A) and $C_{p,m}=\frac{5}{2}R$ (from $C_{p,m}-C_{V,m}=R$), so $\gamma=\frac{5}{3}$. For a gas of nonlinear polyatomic molecules (which can rotate as well as translate; vibrations make little contribution at normal temperatures), $C_{V,m}=3R$ and $C_{p,m}=4R$, so $\gamma=\frac{4}{3}$. The curves of pressure versus volume for adiabatic change are known as **adiabats**, and one for a reversible path is illustrated in Fig. 2E.2. Because $\gamma>1$, an adiabat falls more steeply ($p\propto 1/V^{\gamma}$) than the corresponding isotherm ($p\propto 1/V$). The physical reason for the difference is that, in an isothermal expansion, energy flows into the system as heat and maintains the temperature; as a result, the pressure does not fall as much as in an adiabatic expansion.

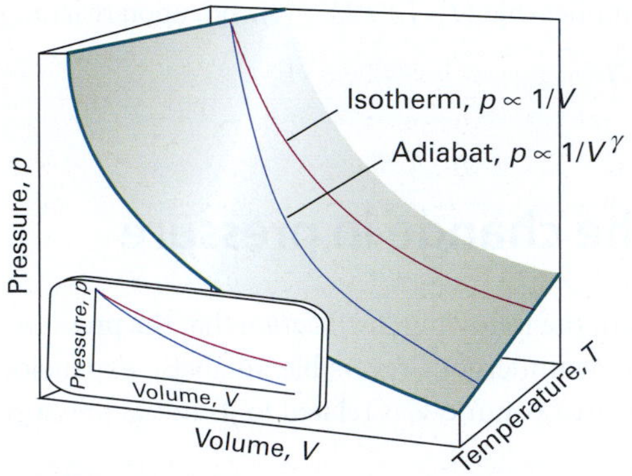

Figure 2E.2 An adiabat depicts the variation of pressure with volume when a gas expands adiabatically. Note that the pressure declines more steeply for an adiabat than it does for an isotherm because the temperature decreases in the former.

Brief illustration 2E.2 Adiabatic expansion

When a sample of argon (for which $\gamma=\frac{5}{3}$) at 100 kPa expands reversibly and adiabatically to twice its initial volume the final pressure will be

$$p_f=\left(\frac{V_i}{V_f}\right)^{\gamma}p_i=\left(\frac{1}{2}\right)^{5/3}\times(100\,\text{kPa})=32\,\text{kPa}$$

For an isothermal doubling of volume, the final pressure would be 50 kPa.

Self-test 2E.2 What is the final pressure when a sample of carbon dioxide at 100 kPa expands reversibly and adiabatically to five times its initial volume?

Answer: 13 kPa

Checklist of concepts

☐ 1. The temperature of a gas falls when it undergoes adiabatic expansion (and does work).

☐ 2. An **adiabat** is a curve showing how pressure varies with volume in an adiabatic process.

Checklist of equations

Property	Equation	Comment	Equation number
Work of adiabatic expansion	$w_{ad}=C_V\Delta T$	Perfect gas	2E.1
Final temperature	$T_f=T_i(V_i/V_f)^{1/c}$ $c=C_{V,m}/R$	Perfect gas, reversible expansion	2E.2a
	$V_iT_i^c=V_fT_f^c$		2E.2b
Adiabats	$p_fV_f^{\gamma}=p_iV_i^{\gamma}$, $\gamma=C_{p,m}/C_{V,m}$		2E.3

CHAPTER 2 The First Law

Assume all gases are perfect unless stated otherwise. Unless otherwise stated, thermochemical data are for 298.15 K.

TOPIC 2A Internal energy

Discussion questions

2A.1 Describe and distinguish the various uses of the words 'system' and 'state' in physical chemistry.

2A.2 Describe the distinction between heat and work in thermodynamic and molecular terms, the latter in terms of populations and energy levels.

2A.3 Identify varieties of additional work.

Exercises

2A.1(a) Use the equipartition theorem to estimate the molar internal energy relative to $U(0)$ of (i) I_2, (ii) CH_4, (iii) C_6H_6 in the gas phase at 25 °C.
2A.1(b) Use the equipartition theorem to estimate the molar internal energy relative to $U(0)$ of (i) O_3, (ii) C_2H_6, (iii) SO_2 in the gas phase at 25 °C.

2A.2(a) Which of (i) pressure, (ii) temperature, (iii) work, (iv) enthalpy are state functions?
2A.2(b) Which of (i) volume, (ii) heat, (iii) internal energy, (iv) density are state functions?

2A.3(a) A chemical reaction takes place in a container of cross-sectional area 50 cm². As a result of the reaction, a piston is pushed out through 15 cm against an external pressure of 1.0 atm. Calculate the work done by the system.
2A.3(b) A chemical reaction takes place in a container of cross-sectional area 75.0 cm². As a result of the reaction, a piston is pushed out through 25.0 cm against an external pressure of 150 kPa. Calculate the work done by the system.

2A.4(a) A sample consisting of 1.00 mol Ar is expanded isothermally at 20 °C from 10.0 dm³ to 30.0 dm³ (i) reversibly, (ii) against a constant external pressure equal to the final pressure of the gas, and (iii) freely (against zero external pressure). For the three processes calculate q, w, and ΔU.
2A.4(b) A sample consisting of 2.00 mol He is expanded isothermally at 0 °C from 5.0 dm³ to 20.0 dm³ (i) reversibly, (ii) against a constant external pressure equal to the final pressure of the gas, and (iii) freely (against zero external pressure). For the three processes calculate q, w, and ΔU.

2A.5(a) A sample consisting of 1.00 mol of perfect gas atoms, for which $C_{V,\mathrm{m}} = \frac{3}{2}R$, initially at $p_1 = 1.00$ atm and $T_1 = 300$ K, is heated reversibly to 400 K at constant volume. Calculate the final pressure, ΔU, q, and w.
2A.5(b) A sample consisting of 2.00 mol of perfect gas molecules, for which $C_{V,\mathrm{m}} = \frac{5}{2}R$, initially at $p_1 = 111$ kPa and $T_1 = 277$ K, is heated reversibly to 356 K at constant volume. Calculate the final pressure, ΔU, q, and w.

2A.6(a) A sample of 4.50 g of methane occupies 12.7 dm³ at 310 K. (i) Calculate the work done when the gas expands isothermally against a constant external pressure of 200 Torr until its volume has increased by 3.3 dm³. (ii) Calculate the work that would be done if the same expansion occurred reversibly.
2A.6(b) A sample of argon of mass 6.56 g occupies 18.5 dm³ at 305 K. (i) Calculate the work done when the gas expands isothermally against a constant external pressure of 7.7 kPa until its volume has increased by 2.5 dm³. (ii) Calculate the work that would be done if the same expansion occurred reversibly.

Problems

2A.1 Calculate the work done during the isothermal reversible expansion of a van der Waals gas (Topic 1C). Plot on the same graph the indicator diagrams (graphs of pressure against volume) for the isothermal reversible expansion of (a) a perfect gas, (b) a van der Waals gas in which $a = 0$ and $b = 5.11 \times 10^{-2}$ dm³ mol⁻¹, and (c) $a = 4.2$ dm⁶ atm mol⁻² and $b = 0$. The values selected exaggerate the imperfections but give rise to significant effects on the indicator diagrams. Take $V_i = 1.0$ dm³, $n = 1.0$ mol, and $T = 298$ K.

2A.2 A sample consisting of 1.0 mol $CaCO_3$(s) was heated to 800 °C, when it decomposed. The heating was carried out in a container fitted with a piston that was initially resting on the solid. Calculate the work done during complete decomposition at 1.0 atm. What work would be done if instead of having a piston the container was open to the atmosphere?

2A.3 Calculate the work done during the isothermal reversible expansion of a gas that satisfies the virial equation of state, eqn 1C.3. Evaluate (a) the work for 1.0 mol Ar at 273 K (for data, see Table 1C.1) and (b) the same amount of a perfect gas. Let the expansion be from 500 cm³ to 1000 cm³ in each case.

2A.4 Express the work of isothermal reversible expansion of a van der Waals gas in reduced variables (Topic 1C) and find a definition of reduced work that makes the overall expression independent of the identity of the gas. Calculate the work of isothermal reversible expansion along the critical isotherm from V_c to xV_c.

2A.5 Suppose that a DNA molecule resists being extended from an equilibrium, more compact conformation with a restoring force $F = -k_f x$, where x is the difference in the end-to-end distance of the chain from an equilibrium value and k_f is the force constant. Use this model to write an expression for the work that must be done to extend a DNA molecule by a distance x. Draw a graph of your conclusion.

2A.6 A better model of a DNA molecule is the 'one-dimensional freely jointed chain', in which a rigid unit of length l can only make an angle of 0° or 180° with an adjacent unit. In this case, the restoring force of a chain extended by $x=nl$ is given by

$$F=\frac{kT}{2l}\ln\left(\frac{1+\nu}{1-\nu}\right) \qquad \nu=\frac{n}{N}$$

where k is Boltzmann's constant. (a) What is the magnitude of the force that must be applied to extend a DNA molecule with $N=200$ by 90 nm? (b) Plot the restoring force against ν, noting that ν can be either positive or negative. How is the variation of the restoring force with end-to-end distance different from that predicted by Hooke's law? (c) Keep in mind that the difference in end-to-end distance from an equilibrium value is $x=nl$ and, consequently, $dx=ldn=Nld\nu$, and write an expression for the work of extending a DNA molecule. (d) Calculate the work of extending a DNA molecule from $\nu=0$ to $\nu=1.0$. *Hint*: You must integrate the expression for w. The task can be accomplished easily with mathematical software.

2A.7 As a continuation of Problem 2A.6, (a) show that for small extensions of the chain, when $\nu \ll 1$, the restoring force is given by

$$F\approx\frac{\nu kT}{l}=\frac{nkT}{Nl}$$

(b) Is the variation of the restoring force with extension of the chain given in part (a) different from that predicted by Hooke's law? Explain your answer.

TOPIC 2B Enthalpy

Discussion questions

2B.1 Explain the difference between the change in internal energy and the change in enthalpy accompanying a process.

2B.2 Why is the heat capacity at constant pressure of a substance normally greater than its heat capacity at constant volume?

Exercises

2B.1(a) When 229 J of energy is supplied as heat to 3.0 mol Ar(g), the temperature of the sample increases by 2.55 K. Calculate the molar heat capacities at constant volume and constant pressure of the gas.
2B.1(b) When 178 J of energy is supplied as heat to 1.9 mol of gas molecules, the temperature of the sample increases by 1.78 K. Calculate the molar heat capacities at constant volume and constant pressure of the gas.

2B.2(a) The constant-pressure heat capacity of a sample of a perfect gas was found to vary with temperature according to the expression $C_p/(\mathrm{J\,K^{-1}})=20.17+0.3665(T/\mathrm{K})$. Calculate q, w, and ΔH when the temperature is raised from 25 °C to 100 °C (i) at constant pressure, (ii) at constant volume.

2B.2(b) The constant-pressure heat capacity of a sample of a perfect gas was found to vary with temperature according to the expression $C_p/(\mathrm{J\,K^{-1}})=20.17+0.4001(T/\mathrm{K})$. Calculate q, w, and ΔH when the temperature is raised from 25 °C to 100 °C (i) at constant pressure, (ii) at constant volume.

2B.3(a) When 3.0 mol O_2 is heated at a constant pressure of 3.25 atm, its temperature increases from 260 K to 285 K. Given that the molar heat capacity of O_2 at constant pressure is 29.4 J K^{-1} mol^{-1}, calculate q, ΔH, and ΔU.
2B.3(b) When 2.0 mol CO_2 is heated at a constant pressure of 1.25 atm, its temperature increases from 250 K to 277 K. Given that the molar heat capacity of CO_2 at constant pressure is 37.11 J K^{-1} mol^{-1}, calculate q, ΔH, and ΔU.

Problems

2B.1 The following data show how the standard molar constant-pressure heat capacity of sulfur dioxide varies with temperature. By how much does the standard molar enthalpy of SO_2(g) increase when the temperature is raised from 298.15 K to 1500 K?

T/K	300	500	700	900	1100	1300	1500
$C^{\ominus}_{p,\mathrm{m}}/(\mathrm{J\,K^{-1}mol^{-1}})$	39.909	46.490	50.829	53.407	54.993	56.033	56.759

2B.2 The following data show how the standard molar constant-pressure heat capacity of ammonia depends on the temperature. Use mathematical software to fit an expression of the form of eqn 2B.8 to the data and determine the values of a, b, and c. Explore whether it would be better to express the data as $C_{p,\mathrm{m}}=\alpha+\beta T+\gamma T^2$, and determine the values of these coefficients.

T/K	300	400	500	600	700	800	900	1000
$C^{\ominus}_{p,\mathrm{m}}/(\mathrm{J\,K^{-1}mol^{-1}})$	35.678	38.674	41.994	45.229	48.269	51.112	53.769	56.244

2B.3 A sample consisting of 2.0 mol CO_2 occupies a fixed volume of 15.0 dm^3 at 300 K. When it is supplied with 2.35 kJ of energy as heat its temperature increases to 341 K. Assume that CO_2 is described by the van der Waals equation of state (Topic 1C) and calculate w, ΔU, and ΔH.

2B.4 (a) Express $(\partial C_V/\partial V)_T$ as a second derivative of U and find its relation to $(\partial U/\partial V)_T$ and $(\partial C_p/\partial p)_T$ as a second derivative of H and find its relation to $(\partial H/\partial p)_T$. (b) From these relations show that $(\partial C_V/\partial V)_T=0$ and $(\partial C_p/\partial p)_T=0$ for a perfect gas.

TOPIC 2C Thermochemistry

Discussion questions

2C.1 Describe two calorimetric methods for the determination of enthalpy changes that accompany chemical processes.

2C.2 Distinguish between 'standard state' and 'reference state', and indicate their applications.

Exercises

2C.1(a) For tetrachloromethane, $\Delta_{vap}H^{\ominus}=30.0\,kJ\,mol^{-1}$. Calculate q, w, ΔH, and ΔU when 0.75 mol $CCl_4(l)$ is vaporized at 250 K and 750 Torr.
2C.1(b) For ethanol, $\Delta_{vap}H^{\ominus}=43.5\,kJ\,mol^{-1}$. Calculate q, w, ΔH, and ΔU when 1.75 mol $C_2H_5OH(l)$ is vaporized at 260 K and 765 Torr.

2C.2(a) The standard enthalpy of formation of ethylbenzene is −12.5 kJ mol⁻¹. Calculate its standard enthalpy of combustion.
2C.2(b) The standard enthalpy of formation of phenol is −165.0 kJ mol⁻¹. Calculate its standard enthalpy of combustion.

2C.3(a) The standard enthalpy of combustion of cyclopropane is −2091 kJ mol⁻¹ at 25 °C. From this information and enthalpy of formation data for $CO_2(g)$ and $H_2O(g)$, calculate the enthalpy of formation of cyclopropane. The enthalpy of formation of propene is +20.42 kJ mol⁻¹. Calculate the enthalpy of isomerization of cyclopropane to propene.
2C.3(b) From the following data, determine $\Delta_f H^{\ominus}$ for diborane, $B_2H_6(g)$, at 298 K:

(1) $B_2H_6(g)+3\,O_2(g)\rightarrow B_2O_3(s)+3\,H_2O(g)$ $\quad\Delta_r H^{\ominus}=-1941\,kJ\,mol^{-1}$

(2) $2\,B(s)+\frac{3}{2}\,O_2(g)\rightarrow B_2O_3(s)$ $\quad\Delta_r H^{\ominus}=-2368\,kJ\,mol^{-1}$

(3) $H_2(g)+\frac{1}{2}\,O_2(g)\rightarrow H_2O(g)$ $\quad\Delta_r H^{\ominus}=-241.8\,kJ\,mol^{-1}$

2C.4(a) Given that the standard enthalpy of formation of HCl(aq) is −167 kJ mol⁻¹, what is the value of $\Delta_f H^{\ominus}(Cl^-, aq)$?
2C.4(b) Given that the standard enthalpy of formation of HI(aq) is −55 kJ mol⁻¹, what is the value of $\Delta_f H^{\ominus}(I^-, aq)$?

2C.5(a) When 120 mg of naphthalene, $C_{10}H_8(s)$, was burned in a bomb calorimeter the temperature rose by 3.05 K. Calculate the calorimeter constant. By how much will the temperature rise when 150 mg of phenol, $C_6H_5OH(s)$, is burned in the calorimeter under the same conditions?
2C.5(b) When 225 mg of anthracene, $C_{14}H_{10}(s)$, was burned in a bomb calorimeter the temperature rose by 1.75 K. Calculate the calorimeter constant. By how much will the temperature rise when 125 mg of phenol, $C_6H_5OH(s)$, is burned in the calorimeter under the same conditions? ($\Delta_c H^{\ominus}(C_{14}H_{10},s)=-7061\,kJ\,mol^{-1}$.)

2C.6(a) Given the reactions (1) and (2) below, determine (i) $\Delta_r H^{\ominus}$ and $\Delta_r U^{\ominus}$ for reaction (3), (ii) $\Delta_f H^{\ominus}$ for both HCl(g) and $H_2O(g)$ all at 298 K.

(1) $H_2(g)+Cl_2(g)\rightarrow 2\,HCl(g)$ $\quad\Delta_r H^{\ominus}=-184.62\,kJ\,mol^{-1}$

(2) $H_2(g)+O_2(g)\rightarrow 2\,H_2O(g)$ $\quad\Delta_r H^{\ominus}=-483.64\,kJ\,mol^{-1}$

(3) $4\,HCl(g)+O_2(g)\rightarrow 2\,Cl_2(g)+2\,H_2O(g)$

2C.6(b) Given the reactions (1) and (2) below, determine (i) $\Delta_r H^{\ominus}$ and $\Delta_r U^{\ominus}$ for reaction (3), (ii) $\Delta_f H^{\ominus}$ for both HI(g) and $H_2O(g)$ all at 298 K.

(1) $H_2(g)+I_2(s)\rightarrow 2\,HI(g)$ $\quad\Delta_r H^{\ominus}=+52.96\,kJ\,mol^{-1}$

(2) $2\,H_2(g)+O_2(g)\rightarrow 2\,H_2O(g)$ $\quad\Delta_r H^{\ominus}=-483.64\,kJ\,mol^{-1}$

(3) $4\,HI(g)+O_2(g)\rightarrow 2\,I_2(s)+2\,H_2O(g)$

2C.7(a) For the reaction $C_2H_5OH(l)+3\,O_2(g)\rightarrow 2\,CO_2(g)+3\,H_2O(g)$, $\Delta_r U^{\ominus}=-1373\,kJ\,mol^{-1}$ at 298 K. Calculate $\Delta_r H^{\ominus}$.
2C.7(b) For the reaction $2\,C_6H_5COOH(s)+15\,O_2(g)\rightarrow 14\,CO_2(g)+6\,H_2O(g)$, $\Delta_r U^{\ominus}=-772.7\,kJ\,mol^{-1}$ at 298 K. Calculate $\Delta_r H^{\ominus}$.

2C.8(a) From the data in Tables 2C.2 and 2C.3, calculate $\Delta_r H^{\ominus}$ and $\Delta_r U^{\ominus}$ at (i) 298 K, (ii) 478 K for the reaction $C(graphite)+H_2O(g)\rightarrow CO(g)+H_2(g)$. Assume all heat capacities to be constant over the temperature range of interest.
2C.8(b) Calculate $\Delta_r H^{\ominus}$ and $\Delta_r U^{\ominus}$ at 298 K and $\Delta_r H^{\ominus}$ at 427 K for the hydrogenation of ethyne (acetylene) to ethene (ethylene) from the enthalpy of combustion and heat capacity data in Tables 2C.5 and 2C.6. Assume the heat capacities to be constant over the temperature range involved.

2C.9(a) Estimate $\Delta_r H^{\ominus}$ (500 K) for the combustion of methane, $CH_4(g)+2\,O_2(g)\rightarrow CO_2(g)+2\,H_2O(g)$ by using the data on the temperature dependence of heat capacities in Table 2B.1.
2C.9(b) Estimate $\Delta_r H^{\ominus}$ (478 K) for the combustion of naphthalene, $C_{10}H_8(l)+12\,O_2(g)\rightarrow 10\,CO_2(g)+4\,H_2O(g)$ by using the data on the temperature dependence of heat capacities in Table 2B.1.

2C.10(a) Set up a thermodynamic cycle for determining the enthalpy of hydration of Mg^{2+} ions using the following data: enthalpy of sublimation of Mg(s), +167.2 kJ mol⁻¹; first and second ionization enthalpies of Mg(g), 7.646 eV and 15.035 eV; dissociation enthalpy of $Cl_2(g)$, +241.6 kJ mol⁻¹; electron gain enthalpy of Cl(g), −3.78 eV; enthalpy of solution of $MgCl_2(s)$, −150.5 kJ mol⁻¹; enthalpy of hydration of $Cl^-(g)$, −383.7 kJ mol⁻¹.
2C.10(b) Set up a thermodynamic cycle for determining the enthalpy of hydration of Ca^{2+} ions using the following data: enthalpy of sublimation of Ca(s), +178.2 kJ mol⁻¹; first and second ionization enthalpies of Ca(g), 589.7 kJ mol⁻¹ and 1145 kJ mol⁻¹; enthalpy of vaporization of bromine, +30.91 kJ mol⁻¹; dissociation enthalpy of $Br_2(g)$, +192.9 kJ mol⁻¹; electron gain enthalpy of Br(g), −331.0 kJ mol⁻¹; enthalpy of solution of $CaBr_2(s)$, −103.1 kJ mol⁻¹; enthalpy of hydration of $Br^-(g)$, −289 kJ mol⁻¹.

Problems

2C.1 A sample of the sugar D-ribose ($C_5H_{10}O_5$) of mass 0.727 g was placed in a constant-volume bomb calorimeter and then ignited in the presence of excess oxygen. The temperature rose by 0.910 K. In a separate experiment in the same calorimeter, the combustion of 0.825 g of benzoic acid, for which the internal energy of combustion is −3251 kJ mol⁻¹, gave a temperature rise of 1.940 K. Calculate the enthalpy of formation of D-ribose.

2C.2 The standard enthalpy of formation of bis(benzene)chromium was measured in a calorimeter. It was found for the reaction $Cr(C_6H_6)_2(s)\rightarrow Cr(s)+2\,C_6H_6(g)$ that $\Delta_r U^{\ominus}(583\,K)=+8.0\,kJ\,mol^{-1}$. Find the corresponding reaction enthalpy and estimate the standard enthalpy of formation of the compound at 583 K. The constant-pressure molar heat capacity of benzene is 136.1 J K⁻¹ mol⁻¹ in its liquid range and 81.67 J K⁻¹ mol⁻¹ as a gas.

2C.3‡ From the enthalpy of combustion data in Table 2C.1 for the alkanes methane through octane, test the extent to which the relation $\Delta_c H^\ominus = k\{(M/(\text{g mol}^{-1})\}^n$ holds and find the numerical values for k and n. Predict $\Delta_c H^\ominus$ for decane and compare to the known value.

2C.4‡ Kolesov et al. reported the standard enthalpy of combustion and of formation of crystalline C_{60} based on calorimetric measurements (V.P. Kolesov et al., *J. Chem. Thermodynamics* **28**, 1121 (1996)). In one of their runs, they found the standard specific internal energy of combustion to be $-36.0334\,\text{kJ g}^{-1}$ at 298.15 K. Compute $\Delta_c H^\ominus$ and $\Delta_f H^\ominus$ of C_{60}.

2C.5‡ A thermodynamic study of $DyCl_3$ (E.H.P. Cordfunke et al., *J. Chem. Thermodynamics* **28**, 1387 (1996)) determined its standard enthalpy of formation from the following information

(1) $DyCl_3(s) \rightarrow DyCl_3(aq, \text{in } 4.0\,\text{M HCl})$ $\quad \Delta_r H^\ominus = -180.06\,\text{kJ mol}^{-1}$

(2) $Dy(s) + 3\,HCl(aq, 4.0\,\text{M}) \rightarrow DyCl_3(aq, \text{in } 4.0\,\text{M}\,HCl(aq)) + \frac{3}{2}H_2(g)$ $\quad \Delta_r H^\ominus = -699.43\,\text{kJ mol}^{-1}$

(3) $\frac{1}{2}H_2(g) + \frac{1}{2}Cl_2(g) \rightarrow HCl(aq, 4.0\,\text{M})$ $\quad \Delta_r H^\ominus = -158.31\,\text{kJ mol}^{-1}$

Determine $\Delta_f H^\ominus(DyCl_3, s)$ from these data.

2C.6‡ Silylene (SiH_2) is a key intermediate in the thermal decomposition of silicon hydrides such as silane (SiH_4) and disilane (Si_2H_6). H.K. Moffat et al. (*J. Phys. Chem.* **95**, 145 (1991)) report $\Delta_f H^\ominus(SiH_2) = +274\,\text{kJ mol}^{-1}$. If $\Delta_f H^\ominus(SiH_4) = +34.3\,\text{kJ mol}^{-1}$ and $\Delta_f H^\ominus(Si_2H_6) = +80.3\,\text{kJ mol}^{-1}$, compute the standard enthalpies of the following reactions:

(a) $SiH_4(g) \rightarrow SiH_2(g) + H_2(g)$
(b) $Si_2H_6(g) \rightarrow SiH_2(g) + SiH_4(g)$

2C.7 As remarked in Problem 2B.2, it is sometimes appropriate to express the temperature dependence of the heat capacity by the empirical expression $C_{p,m} = \alpha + \beta T + \gamma T^2$. Use this expression to estimate the standard enthalpy of combustion of methane at 350 K. Use the following data:

	$\alpha/(\text{J K}^{-1}\,\text{mol}^{-1})$	$\beta/(\text{mJ K}^{-2}\,\text{mol}^{-1})$	$\gamma/(\mu\text{J K}^{-3}\,\text{mol}^{-1})$
$CH_4(g)$	14.16	75.5	−17.99
$CO_2(g)$	26.86	6.97	−0.82
$O_2(g)$	25.72	12.98	−3.862
$H_2O(g)$	30.36	9.61	1.184

2C.8 Figure 2.1 shows the experimental DSC scan of hen white lysozyme (G. Privalov et al., *Anal. Biochem.* **79**, 232 (1995)) converted to joules (from calories). Determine the enthalpy of unfolding of this protein by integration of the curve and the change in heat capacity accompanying the transition.

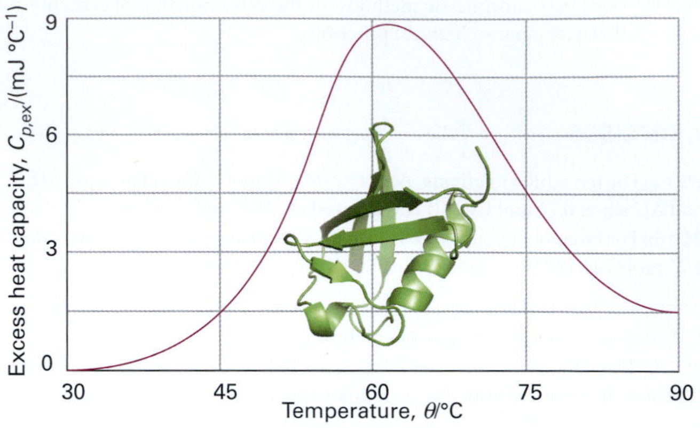

Figure 2.1 The experimental DSC scan of hen white lysozyme.

2C.9 An average human produces about 10 MJ of heat each day through metabolic activity. If a human body were an isolated system of mass 65 kg with the heat capacity of water, what temperature rise would the body experience? Human bodies are actually open systems, and the main mechanism of heat loss is through the evaporation of water. What mass of water should be evaporated each day to maintain constant temperature?

2C.10 In biological cells that have a plentiful supply of oxygen, glucose is oxidized completely to CO_2 and H_2O by a process called *aerobic oxidation*. Muscle cells may be deprived of O_2 during vigorous exercise and, in that case, one molecule of glucose is converted to two molecules of lactic acid ($CH_3CH(OH)COOH$) by a process called *anaerobic glycolysis*. (a) When 0.3212 g of glucose was burned in a bomb calorimeter of calorimeter constant $641\,\text{J K}^{-1}$ the temperature rose by 7.793 K. Calculate (i) the standard molar enthalpy of combustion, (ii) the standard internal energy of combustion, and (iii) the standard enthalpy of formation of glucose. (b) What is the biological advantage (in kilojoules per mole of energy released as heat) of complete aerobic oxidation compared with anaerobic glycolysis to lactic acid?

TOPIC 2D State functions and exact differentials

Discussion questions

2D.1 Suggest (with explanation) how the internal energy of a van der Waals gas should vary with volume at constant temperature.

2D.2 Explain why a perfect gas does not have an inversion temperature.

Exercises

2D.1(a) Estimate the internal pressure, π_T, of water vapour at 1.00 bar and 400 K, treating it as a van der Waals gas. *Hint*: Simplify the approach by estimating the molar volume by treating the gas as perfect.

2D.1(b) Estimate the internal pressure, π_T, of sulfur dioxide at 1.00 bar and 298 K, treating it as a van der Waals gas. *Hint*: Simplify the approach by estimating the molar volume by treating the gas as perfect.

2D.2(a) For a van der Waals gas, $\pi_T = a/V_m^2$. Calculate ΔU_m for the isothermal expansion of nitrogen gas from an initial volume of $1.00\,\text{dm}^3$ to $20.00\,\text{dm}^3$ at 298 K. What are the values of q and w?

2D.2(b) Repeat Exercise 2D.2(a) for argon, from an initial volume of $1.00\,\text{dm}^3$ to $30.00\,\text{dm}^3$ at 298 K.

‡ These problems were provided by Charles Trapp and Carmen Giunta.

2D.3(a) The volume of a certain liquid varies with temperature as

$$V = V^{\ominus}\{0.75 + 3.9 \times 10^{-4}(T/\text{K}) + 1.48 \times 10^{-6}(T/\text{K})^2\}$$

where $V^{\ominus}$ is its volume at 300 K. Calculate its expansion coefficient, α, at 320 K.

2D.3(b) The volume of a certain liquid varies with temperature as

$$V = V^{\ominus}\{0.77 + 3.7 \times 10^{-4}(T/\text{K}) + 1.52 \times 10^{-6}(T/\text{K})^2\}$$

where $V^{\ominus}$ is its volume at 298 K. Calculate its expansion coefficient, α, at 310 K.

2D.4(a) The isothermal compressibility of water at 293 K is 4.96×10^{-5} atm^{-1}. Calculate the pressure that must be applied in order to increase its density by 0.10 per cent.

2D.4(b) The isothermal compressibility of lead at 293 K is 2.21×10^{-6} atm^{-1}. Calculate the pressure that must be applied in order to increase its density by 0.10 per cent.

2D.5(a) Given that $\mu = 0.25$ K atm^{-1} for nitrogen, calculate the value of its isothermal Joule–Thomson coefficient. Calculate the energy that must be supplied as heat to maintain constant temperature when 10.0 mol N_2 flows through a throttle in an isothermal Joule–Thomson experiment and the pressure drop is 85 atm.

2D.5(b) Given that $\mu = 1.11$ K atm^{-1} for carbon dioxide, calculate the value of its isothermal Joule–Thomson coefficient. Calculate the energy that must be supplied as heat to maintain constant temperature when 10.0 mol CO_2 flows through a throttle in an isothermal Joule–Thomson experiment and the pressure drop is 75 atm.

Problems

2D.1‡ In 2006, the Intergovernmental Panel on Climate Change (IPCC) considered a global average temperature rise of 1.0–3.5 °C likely by the year 2100, with 2.0 °C its best estimate. Predict the average rise in sea level due to thermal expansion of sea water based on temperature rises of 1.0 °C, 2.0 °C, and 3.5 °C given that the volume of the Earth's oceans is 1.37×10^9 km^3 and their surface area is 361×10^6 km^2, and state the approximations which go into the estimates.

2D.2 The heat capacity ratio of a gas determines the speed of sound in it through the formula $c_s = (\gamma RT/M)^{1/2}$, where $\gamma = C_p/C_V$ and M is the molar mass of the gas. Deduce an expression for the speed of sound in a perfect gas of (a) diatomic, (b) linear triatomic, (c) nonlinear triatomic molecules at high temperatures (with translation and rotation active). Estimate the speed of sound in air at 25 °C.

2D.3 Starting from the expression $C_p - C_V = T(\partial p/\partial T)_V(\partial V/\partial T)_p$, use the appropriate relations between partial derivatives to show that

$$C_p - C_V = \frac{T(\partial V/\partial T)_p^2}{(\partial V/\partial p)_T}$$

Evaluate $C_p - C_V$ for a perfect gas.

2D.4 (a) Write expressions for dV and dp given that V is a function of p and T and p is a function of V and T. (b) Deduce expressions for d ln V and d ln p in terms of the expansion coefficient and the isothermal compressibility.

2D.5 Rearrange the van der Waals equation of state, $p = nRT/(V - nb) - n^2a/V^2$, to give an expression for T as a function of p and V (with n constant). Calculate $(\partial T/\partial p)_V$ and confirm that $(\partial T/\partial p)_V = 1/(\partial p/\partial T)_V$. Go on to confirm Euler's chain relation (*Mathematical background* 2).

2D.6 Calculate the isothermal compressibility and the expansion coefficient of a van der Waals gas (see Problem 2D.5). Show, using Euler's chain relation (*Mathematical background* 2), that $\kappa_T R = \alpha(V_m - b)$.

2D.7 The speed of sound, c_s, in a gas of molar mass M is related to the ratio of heat capacities γ by $c_s = (\gamma RT/M)^{1/2}$. Show that $c_s = (\gamma p/\rho)^{1/2}$, where ρ is the mass density of the gas. Calculate the speed of sound in argon at 25 °C.

2D.8‡ A gas obeying the equation of state $p(V - nb) = nRT$ is subjected to a Joule–Thomson expansion. Will the temperature increase, decrease, or remain the same?

2D.9 Use the fact that $(\partial U/\partial V)_T = a/V_m^2$ for a van der Waals gas (Topic 1C) to show that $\mu C_{p,m} \approx (2a/RT) - b$ by using the definition of μ and appropriate relations between partial derivatives. *Hint*: Use the approximation $pV_m \approx RT$ when it is justifiable to do so.

2D.10‡ Concerns over the harmful effects of chlorofluorocarbons on stratospheric ozone have motivated a search for new refrigerants. One such alternative is 2,2-dichloro-1,1,1-trifluoroethane (refrigerant 123). Younglove and McLinden published a compendium of thermophysical properties of this substance (B.A. Younglove and M. McLinden, *J. Phys. Chem. Ref. Data* **23**, 7 (1994)), from which properties such as the Joule–Thomson coefficient μ can be computed. (a) Compute μ at 1.00 bar and 50 °C given that $(\partial H/\partial p)_T = -3.29 \times 10^3$ J MPa^{-1} mol^{-1} and $C_{p,m} = 110.0$ J K^{-1} mol^{-1}. (b) Compute the temperature change which would accompany adiabatic expansion of 2.0 mol of this refrigerant from 1.5 bar to 0.5 bar at 50 °C.

2D.11‡ Another alternative refrigerant (see preceding problem) is 1,1,1,2-tetrafluoroethane (refrigerant HFC-134a). A compendium of thermophysical properties of this substance has been published (R. Tillner-Roth and H.D. Baehr, *J. Phys. Chem. Ref. Data* **23**, 657 (1994)) from which properties such as the Joule–Thomson coefficient μ can be computed. (a) Compute μ at 0.100 MPa and 300 K from the following data (all referring to 300 K):

p/MPa	0.080	0.100	0.12
Specific enthalpy/(kJ kg^{-1})	426.48	426.12	425.76

(The specific constant-pressure heat capacity is 0.7649 kJ K^{-1} kg^{-1}.) (b) Compute μ at 1.00 MPa and 350 K from the following data (all referring to 350 K):

p/MPa	0.80	1.00	1.2
Specific enthalpy/(kJ kg^{-1})	461.93	459.12	42B.15

(The specific constant-pressure heat capacity is 1.0392 kJ K^{-1} kg^{-1}.)

TOPIC 2E Adiabatic changes

Discussion questions

2E.1 Why are adiabats steeper than isotherms?

2E.2 Why do heat capacities play a role in the expressions for adiabatic expansion?

Exercises

2E.1(a) Use the equipartition principle to estimate the values of $\gamma = C_p/C_V$ for gaseous ammonia and methane. Do this calculation with and without the vibrational contribution to the energy. Which is closer to the expected experimental value at 25 °C?

2E.1(b) Use the equipartition principle to estimate the value of $\gamma = C_p/C_V$ for carbon dioxide. Do this calculation with and without the vibrational contribution to the energy. Which is closer to the expected experimental value at 25 °C?

2E.2(a) Calculate the final temperature of a sample of argon of mass 12.0 g that is expanded reversibly and adiabatically from 1.0 dm^3 at 273.15 K to 3.0 dm^3.

2E.2(b) Calculate the final temperature of a sample of carbon dioxide of mass 16.0 g that is expanded reversibly and adiabatically from 500 cm^3 at 298.15 K to 2.00 dm^3.

2E.3(a) A sample consisting of 1.0 mol of perfect gas molecules with $C_V = 20.8\ \mathrm{J\ K^{-1}}$ is initially at 4.25 atm and 300 K. It undergoes reversible adiabatic expansion until its pressure reaches 2.50 atm. Calculate the final volume and temperature and the work done.

2E.3(b) A sample consisting of 2.5 mol of perfect gas molecules with $C_{p,\mathrm{m}} = 20.8\ \mathrm{J\ K^{-1}\ mol^{-1}}$ is initially at 240 kPa and 325 K. It undergoes reversible adiabatic expansion until its pressure reaches 150 kPa. Calculate the final volume and temperature and the work done.

2E.4(a) A sample of carbon dioxide of mass 2.45 g at 27.0 °C is allowed to expand reversibly and adiabatically from 500 cm^3 to 3.00 dm^3. What is the work done by the gas?

2E.4(b) A sample of nitrogen of mass 3.12 g at 23.0 °C is allowed to expand reversibly and adiabatically from 400 cm^3 to 2.00 dm^3. What is the work done by the gas?

2E.5(a) Calculate the final pressure of a sample of carbon dioxide that expands reversibly and adiabatically from 67.4 kPa and 0.50 dm^3 to a final volume of 2.00 dm^3. Take $\gamma = 1.4$.

2E.5(b) Calculate the final pressure of a sample of water vapour that expands reversibly and adiabatically from 97.3 Torr and 400 cm^3 to a final volume of 5.0 dm^3. Take $\gamma = 1.3$.

Problem

2E.1 The constant-volume heat capacity of a gas can be measured by observing the decrease in temperature when it expands adiabatically and reversibly. The value of $\gamma = C_p/C_V$ can be inferred if the decrease in pressure is also measured and the constant-pressure heat capacity deduced by combining the two values. A fluorocarbon gas was allowed to expand reversibly and adiabatically to twice its volume; as a result, the temperature fell from 298.15 K to 248.44 K and its pressure fell from 202.94 kPa to 81.840 kPa. Evaluate C_p.

Integrated activities

2.1 Give examples of state functions and discuss why they play a critical role in thermodynamics.

2.2 The thermochemical properties of hydrocarbons are commonly investigated by using molecular modelling methods. (a) Use software to predict $\Delta_c H^\ominus$ values for the alkanes methane through pentane. To calculate $\Delta_c H^\ominus$ values, estimate the standard enthalpy of formation of $C_nH_{2n+2}(g)$ by performing semi-empirical calculations (for example, AM1 or PM3 methods) and use experimental standard enthalpy of formation values for $CO_2(g)$ and $H_2O(l)$. (b) Compare your estimated values with the experimental values of $\Delta_c H^\ominus$ (Table 2C.4) and comment on the reliability of the molecular modelling method. (c) Test the extent to which the relation $\Delta_c H^\ominus = \text{constant} \times \{(M/(\mathrm{g\ mol^{-1}})\}^n$ holds and determine the numerical values of the constant and n.

2.3 Use mathematical software, a spreadsheet, or the *Living graphs* on the web site for this book to:

(a) Calculate the work of isothermal reversible expansion of 1.0 mol $CO_2(g)$ at 298 K from 1.0 m^3 to 3.0 m^3 on the basis that it obeys the van der Waals equation of state.

(b) Explore how the parameter γ affects the dependence of the pressure on the volume. Does the pressure–volume dependence become stronger or weaker with increasing volume?

Mathematical background 2 Multivariate calculus

A thermodynamic property of a system typically depends on a number of variables, such as the internal energy depending on the amount, volume, and temperature. To understand how these properties vary with the conditions we need to understand how to manipulate their derivatives. This is the field of **multivariate calculus**, the calculus of several variables.

MB2.1 Partial derivatives

A **partial derivative** of a function of more than one variable, such as $f(x,y)$, is the slope of the function with respect to one of the variables, all the other variables being held constant (Fig. MB2.1). Although a partial derivative shows how a function changes when one variable changes, it may be used to determine how the function changes when more than one variable changes by an infinitesimal amount. Thus, if f is a function of x and y, then when x and y change by dx and dy, respectively, f changes by

$$\mathrm{d}f = \left(\frac{\partial f}{\partial x}\right)_y \mathrm{d}x + \left(\frac{\partial f}{\partial y}\right)_x \mathrm{d}y \qquad \text{(MB2.1)}$$

where the symbol ∂ ('curly d') is used (instead of d) to denote a partial derivative and the subscript on the parentheses indicates which variable is being held constant. The quantity df is also called the **differential** of f. Successive partial derivatives may be taken in any order:

$$\left(\frac{\partial}{\partial y}\left(\frac{\partial f}{\partial x}\right)_y\right)_x = \left(\frac{\partial}{\partial x}\left(\frac{\partial f}{\partial y}\right)_x\right)_y \qquad \text{(MB2.2)}$$

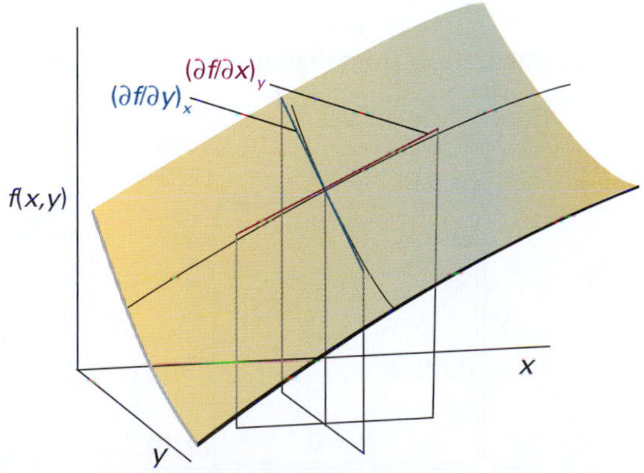

Figure MB2.1 A function of two variables, $f(x,y)$, as depicted by the coloured surface and the two partial derivatives, $(\partial f/\partial x)_y$ and $(\partial f/\partial y)_x$, the slope of the function parallel to the x- and y-axes, respectively. The function plotted here is $f(x,y) = ax^3y + by^2$ with $a = 1$ and $b = -2$.

Brief illustration MB2.1 Partial derivatives

Suppose that $f(x,y) = ax^3y + by^2$ (the function plotted in Fig. MB2.1) then

$$\left(\frac{\partial f}{\partial x}\right)_y = 3ax^2y \qquad \left(\frac{\partial f}{\partial y}\right)_x = ax^3 + 2by$$

Then, when x and y undergo infinitesimal changes, f changes by

$$\mathrm{d}f = 3ax^2y\,\mathrm{d}x + (ax^3 + 2by)\mathrm{d}y$$

To verify that the order of taking the second partial derivative is irrelevant, we form

$$\left(\frac{\partial}{\partial y}\left(\frac{\partial f}{\partial x}\right)_y\right)_x = \left(\frac{\partial(3ax^2y)}{\partial y}\right)_x = 3ax^2$$

$$\left(\frac{\partial}{\partial x}\left(\frac{\partial f}{\partial y}\right)_x\right)_y = \left(\frac{\partial(ax^3+2by)}{\partial x}\right)_y = 3ax^2$$

In the following, z is a variable on which x and y depend (for example, x, y, and z might correspond to p, V, and T).

Relation 1. When x is changed at constant z:

$$\left(\frac{\partial f}{\partial x}\right)_z = \left(\frac{\partial f}{\partial x}\right)_y + \left(\frac{\partial f}{\partial y}\right)_x\left(\frac{\partial y}{\partial x}\right)_z \qquad \text{(MB2.3a)}$$

Relation 2

$$\left(\frac{\partial y}{\partial x}\right)_z = \frac{1}{(\partial x/\partial y)_z} \qquad \text{(MB2.3b)}$$

Relation 3

$$\left(\frac{\partial x}{\partial y}\right)_z = -\left(\frac{\partial x}{\partial z}\right)_y\left(\frac{\partial z}{\partial y}\right)_x \qquad \text{(MB2.3c)}$$

By combining Relations 2 and 3 we obtain the **Euler chain relation**:

$$\left(\frac{\partial y}{\partial x}\right)_z\left(\frac{\partial x}{\partial z}\right)_y\left(\frac{\partial z}{\partial y}\right)_x = -1 \qquad \text{Euler chain relation} \quad \text{(MB2.4)}$$

MB2.2 Exact differentials

The relation in eqn MB2.2 is the basis of a test for an **exact differential**; that is, the test of whether

$$\mathrm{d}f = g(x,y)\mathrm{d}x + h(x,y)\mathrm{d}y \qquad \text{(MB2.5)}$$

has the form in eqn MB2.1. If it has that form, then g can be identified with $(\partial f/\partial x)_y$ and h can be identified with $(\partial f/\partial y)_x$. Then eqn MB2.2 becomes

$$\left(\frac{\partial g}{\partial y}\right)_x = \left(\frac{\partial h}{\partial x}\right)_y \qquad \text{Test for exact differential} \qquad \text{(MB2.6)}$$

Brief illustration MB2.2 Exact differentials

Suppose, instead of the form $df = 3ax^2y dx + (ax^3 + 2by)dy$ in the previous *Brief illustration*, we were presented with the expression

$$df = \overbrace{3ax^2y}^{g(x,y)} dx + \overbrace{(ax^2 + 2by)}^{h(x,y)} dy$$

with ax^2 in place of ax^3 inside the second parentheses. To test whether this is an exact differential, we form

$$\left(\frac{\partial g}{\partial y}\right)_x = \left(\frac{\partial(3ax^2y)}{\partial y}\right)_x = 3ax^2$$

$$\left(\frac{\partial h}{\partial x}\right)_y = \left(\frac{\partial(ax^2+2by)}{\partial x}\right)_y = 2ax$$

These two expressions are not equal, so this form of df is not an exact differential and there is not a corresponding integrated function of the form $f(x,y)$.

If df is exact, then we can do two things:

- From a knowledge of the functions g and h we can reconstruct the function f.
- Be confident that the integral of df between specified limits is independent of the path between those limits.

The first conclusion is best demonstrated with a specific example.

Brief illustration MB2.3 The reconstruction of an equation

We consider the differential $df = 3ax^2y dx + (ax^3 + 2by)dy$, which we know to be exact. Because $(\partial f/\partial x)_y = 3ax^2y$, we can integrate with respect to x with y held constant, to obtain

$$f = \int df = \int 3ax^2y\,dx = 3ay\int x^2\,dx = ax^3y + k$$

where the 'constant' of integration k may depend on y (which has been treated as a constant in the integration), but not on x. To find $k(y)$, we note that $(\partial f/\partial y)_x = ax^3 + 2by$, and therefore

$$\left(\frac{\partial f}{\partial y}\right)_x = \left(\frac{\partial(ax^3y+k)}{\partial y}\right)_x = ax^3 + \frac{dk}{dy} = ax^3 + 2by$$

Therefore

$$\frac{dk}{dy} = 2by$$

from which it follows that $k = by^2 + \text{constant}$. We have found, therefore, that

$$f(x,y) = ax^3y + by^2 + \text{constant}$$

which, apart from the constant, is the original function in the *Brief illustration* MB2.1. The value of the constant is pinned down by stating the boundary conditions; thus, if it is known that $f(0,0) = 0$, then the constant is zero.

To demonstrate that the integral of df is independent of the path is now straight forward. Because df is a differential, its integral between the limits a and b is

$$\int_a^b df = f(b) - f(a)$$

The value of the integral depends only on the values at the end points and is independent of the path between them. If df is not an exact differential, the function f does not exist, and this argument no longer holds. In such cases, the integral of df does depend on the path.

Brief illustration MB2.4 Path-dependent integration

Consider the inexact differential (the expression with ax^2 in place of ax^3 inside the second parentheses):

$$df = 3ax^2y\,dx + (ax^2 + 2by)dy$$

Suppose we integrate df from (0,0) to (2,2) along the two paths shown in Fig. MB2.2. Along Path 1,

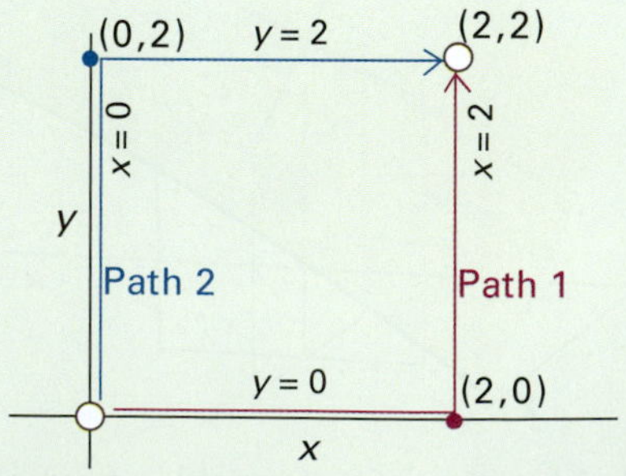

Figure MB2.2 The two integration paths referred to in *Brief illustration* MB2.4.

$$\int_{\text{Path 1}} df = \int_{0,0}^{2,0} 3ax^2 y\,dx + \int_{2,0}^{2,2} (ax^2+2by)\,dy$$

$$= 0 + 4a\int_0^2 dy + 2b\int_0^2 y\,dy = 8a + 4b$$

whereas along Path 2,

$$\int_{\text{Path 2}} df = \int_{0,2}^{2,2} 3ax^2 y\,dx + \int_{0,0}^{0,2} (ax^2+2by)\,dy$$

$$= 6a\int_0^2 x^2\,dx + 0 + 2b\int_0^2 y\,dy = 16a + 4b$$

The two integrals are not the same.

An inexact differential may sometimes be converted into an exact differential by multiplication by a factor known as an *integrating factor*. A physical example is the integrating factor $1/T$ that converts the inexact differential dq_{rev} into the exact differential dS in thermodynamics (Topic 3A).

Brief illustration MB2.5 An integrating factor

We have seen that the differential $df = 3ax^2y\,dx + (ax^2 + 2by)\,dy$ is inexact; the same is true when we set $b = 0$ and consider $df = 3ax^2y\,dx + ax^2\,dy$ instead. Suppose we multiply this df by $x^m y^n$ and write $x^m y^n df = df'$, then we obtain

$$df' = \overbrace{3ax^{m+2}y^{n+1}}^{g(x,y)}\,dx + \overbrace{ax^{m+2}y^n}^{h(x,y)}\,dy$$

We evaluate the following two partial derivatives:

$$\left(\frac{\partial g}{\partial y}\right)_x = \left(\frac{\partial(3ax^{m+2}y^{n+1})}{\partial y}\right)_x = 3a(n+1)x^{m+2}y^n$$

$$\left(\frac{\partial h}{\partial x}\right)_y = \left(\frac{\partial(ax^{m+2}y^n)}{\partial x}\right)_y = a(m+2)x^{m+1}y^n$$

For the new differential to be exact, these two partial derivatives must be equal, so we write

$$3a(n+1)x^{m+2}y^n = a(m+2)x^{m+1}y^n$$

which simplifies to

$$3(n+1)x = m+2$$

The only solution that is independent of x is $n = -1$ and $m = -2$. It follows that

$$df' = 3a\,dx + (a/y)\,dy$$

is an exact differential. By the procedure already illustrated, its integrated form is $f'(x,y) = 3ax + a\ln y + \text{constant}$.

CHAPTER 3

The Second and Third Laws

Some things happen naturally, some things don't. Some aspect of the world determines the **spontaneous** direction of change, the direction of change that does not require work to bring it about. An important point, though, is that throughout this text 'spontaneous' must be interpreted as a natural *tendency* that may or may not be realized in practice. Thermodynamics is silent on the rate at which a spontaneous change in fact occurs, and some spontaneous processes (such as the conversion of diamond to graphite) may be so slow that the tendency is never realized in practice whereas others (such as the expansion of a gas into a vacuum) are almost instantaneous.

3A Entropy

The direction of change is related to the *distribution of energy and matter*, and spontaneous changes are always accompanied by a dispersal of energy or matter. To quantify this concept we introduce the property called 'entropy', which is central to the formulation of the 'Second Law of thermodynamics'. That law governs all spontaneous change.

3B The measurement of entropy

To make the Second Law quantitative, it is necessary to measure the entropy of a substance. We see that measurement, perhaps with calorimetric methods, of the energy transferred as heat during a physical process or chemical reaction leads to determination of the entropy change and, consequently, the direction of spontaneous change. The discussion in this Topic also leads to the 'Third Law of thermodynamics', which helps us to understand the properties of matter at very low temperatures and to set up an absolute measure of the entropy of a substance.

3C Concentrating on the system

One problem with dealing with the entropy is that it requires separate calculations of the changes taking place in the system and the surroundings. Providing we are willing to impose certain restrictions on the system, that problem can be overcome by introducing the 'Gibbs energy'. Indeed, most thermodynamic calculations in chemistry focus on the change in Gibbs energy, not the direct measurement of the entropy change.

3D Combining the First and Second Laws

Finally, we bring the First and Second Laws together and begin to see the considerable power of thermodynamics for accounting for the properties of matter.

What is the impact of this material?

The Second Law is at the heart of the operation of engines of all types, including devices resembling engines that are used to cool objects. See *Impact* I3.1 for an application to the technology of refrigeration. Entropy considerations are also important in modern electronic materials for it permits a quantitative discussion of the concentration of impurities. See *Impact* I3.2 for a note about how measurement of the entropy at low temperatures gives insight into the purity of materials used as superconductors.

To read more about the impact of this material, scan the QR code, or go to bcs.whfreeman.com/webpub/chemistry/pchem10e/impact/pchem-3-1.html

3A Entropy

Contents

➤ Why do you need to know this material?

Entropy is the concept on which almost all applications of thermodynamics in chemistry are based: it explains why some reactions take place and others do not.

➤ What is the key idea?

The change in entropy of a system can be calculated from the heat transferred to it reversibly.

➤ What do you need to know already?

You need to be familiar with the First-Law concepts of work, heat, and internal energy (Topic 2A). The Topic draws on the expression for work of expansion of a perfect gas (Topic 2A) and on the changes in volume and temperature that accompany the reversible adiabatic expansion of a perfect gas (Topic 2D).

What determines the direction of spontaneous change? It is not the total energy of the isolated system. The First Law of thermodynamics states that energy is conserved in any process, and we cannot disregard that law now and say that everything tends towards a state of lower energy. When a change occurs, the total energy of an isolated system remains constant but it is parcelled out in different ways. Can it be, therefore, that the direction of change is related to the *distribution* of energy? We shall see that this idea is the key, and that spontaneous changes are always accompanied by a dispersal of energy or matter.

3A.1 The Second Law

We can begin to understand the role of the dispersal of energy and matter by thinking about a ball (the system) bouncing on a floor (the surroundings). The ball does not rise as high after each bounce because there are inelastic losses in the materials of the ball and floor. The kinetic energy of the ball's overall motion is spread out into the energy of thermal motion of its particles and those of the floor that it hits. The direction of spontaneous change is towards a state in which the ball is at rest with all its energy dispersed into disorderly thermal motion of molecules in the air and of the atoms of the virtually infinite floor (Fig. 3A.1).

A ball resting on a warm floor has never been observed to start bouncing. For bouncing to begin, something rather special would need to happen. In the first place, some of the thermal motion of the atoms in the floor would have to accumulate in a single, small object, the ball. This accumulation requires a spontaneous localization of energy from the myriad vibrations of the atoms of the floor into the much smaller number of atoms that constitute the ball (Fig. 3A.2). Furthermore, whereas the thermal motion is random, for the ball to move upwards its

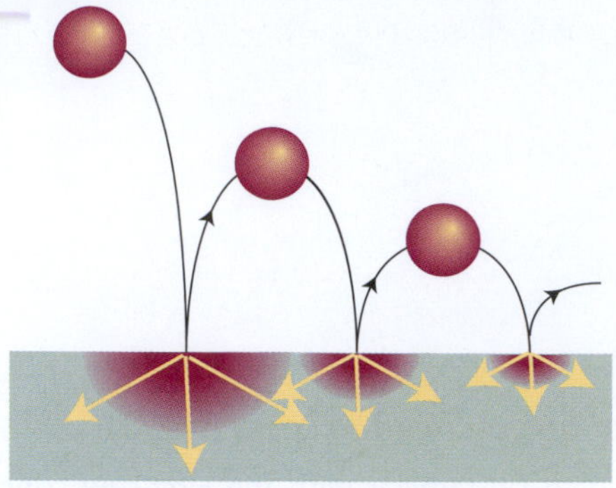

Figure 3A.1 The direction of spontaneous change for a ball bouncing on a floor. On each bounce some of its energy is degraded into the thermal motion of the atoms of the floor, and that energy disperses. The reverse has never been observed to take place on a macroscopic scale.

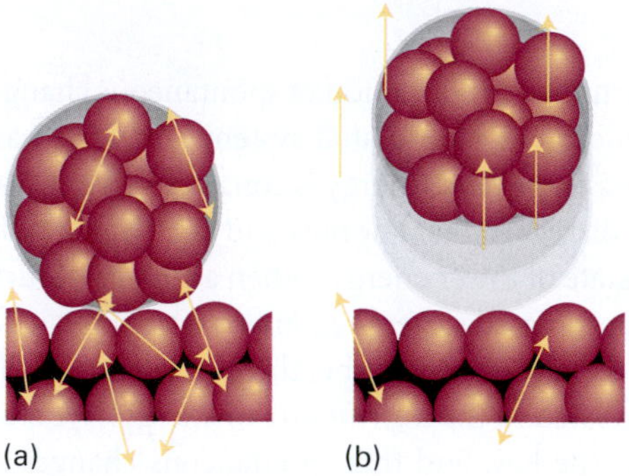

Figure 3A.2 The molecular interpretation of the irreversibility expressed by the Second Law. (a) A ball resting on a warm surface; the atoms are undergoing thermal motion (vibration, in this instance), as indicated by the arrows. (b) For the ball to fly upwards, some of the random vibrational motion would have to change into coordinated, directed motion. Such a conversion is highly improbable.

atoms must all move in the same direction. The localization of random, disorderly motion as concerted, ordered motion is so unlikely that we can dismiss it as virtually impossible.[1]

We appear to have found the signpost of spontaneous change: *we look for the direction of change that leads to dispersal of the total energy of the isolated system.* This principle accounts for the direction of change of the bouncing ball, because its energy is spread out as thermal motion of the atoms of the floor. The reverse process is not spontaneous because it is highly improbable that energy will become localized, leading to uniform motion of the ball's atoms.

Matter also has a tendency to disperse in disorder. A gas does not contract spontaneously because to do so the random motion of its molecules, which spreads out the distribution of molecules throughout the container, would have to take them all into the same region of the container. The opposite change, spontaneous expansion, is a natural consequence of matter becoming more dispersed as the gas molecules occupy a larger volume.

The recognition of two classes of process, spontaneous and non-spontaneous, is summarized by the **Second Law of thermodynamics**. This law may be expressed in a variety of equivalent ways. One statement was formulated by Kelvin:

> No process is possible in which the sole result is the absorption of heat from a reservoir and its complete conversion into work.

For example, it has proved impossible to construct an engine like that shown in Fig. 3A.3, in which heat is drawn from a hot reservoir and completely converted into work. All real heat engines have both a hot source and a cold sink; some energy is always discarded into the cold sink as heat and not converted into work. The Kelvin statement is a generalization of the everyday observation that we have already discussed, that a ball at rest on a surface has never been observed to leap spontaneously upwards. An upward leap of the ball would be equivalent to the conversion of heat from the surface into work. Another statement of the Second Law is due to Rudolf Clausius (Fig. 3A.4):

> Heat does not flow spontaneously from a cool body to a hotter body.

To achieve the transfer of heat to a hotter body, it is necessary to do work on the system, as in a refrigerator.

These two empirical observations turn out to be aspects of a single statement in which the Second Law is expressed in terms of a new state function, the **entropy**, S. We shall see that the entropy (which we shall define shortly, but is a measure of the energy and matter dispersed in a process) lets us assess whether one state is accessible from another by a spontaneous change:

> The entropy of an isolated system increases in the course of a spontaneous change: $\Delta S_{tot} > 0$

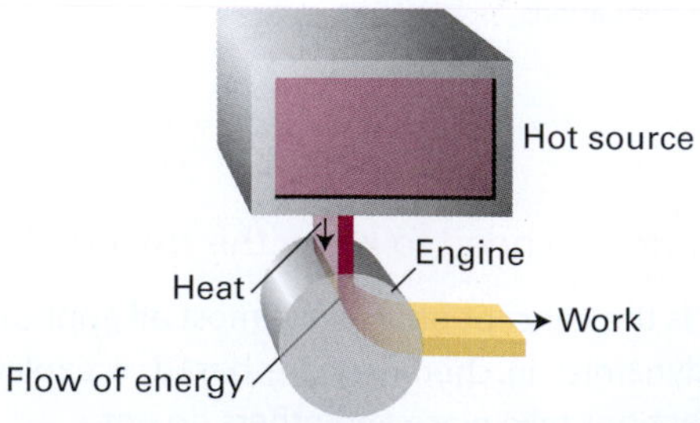

Figure 3A.3 The Kelvin statement of the Second Law denies the possibility of the process illustrated here, in which heat is changed completely into work, there being no other change. The process is not in conflict with the First Law because energy is conserved.

[1] Concerted motion, but on a much smaller scale, is observed as *Brownian motion,* the jittering motion of small particles suspended in a liquid or gas.

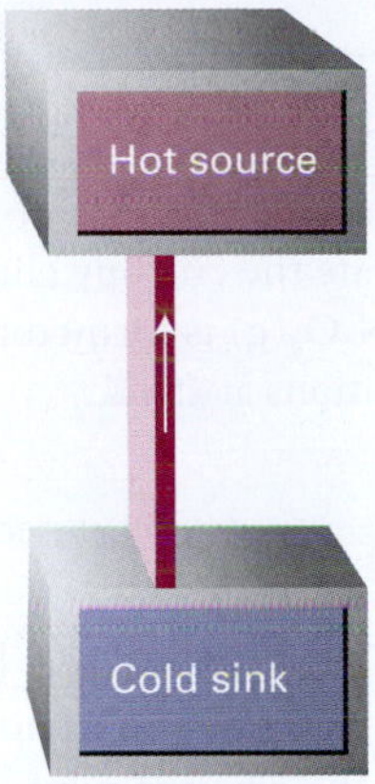

Figure 3A.4 The Clausius statement of the Second Law denies the possibility of the process illustrated here, in which energy as heat migrates from a cool source to a hot sink, there being no other change. The process is not in conflict with the First Law because energy is conserved.

where S_{tot} is the total entropy of the system and its surroundings. Thermodynamically irreversible processes (like cooling to the temperature of the surroundings and the free expansion of gases) are spontaneous processes, and hence must be accompanied by an increase in total entropy.

In summary, the First Law uses the internal energy to identify *permissible* changes; the Second Law uses the entropy to identify the *spontaneous changes* among those permissible changes.

3A.2 The definition of entropy

To make progress, and to turn the Second Law into a quantitatively useful expression, we need to define and then calculate the entropy change accompanying various processes. There are two approaches, one classical and one molecular. They turn out to be equivalent, but each one enriches the other.

(a) The thermodynamic definition of entropy

The thermodynamic definition of entropy concentrates on the change in entropy, dS, that occurs as a result of a physical or chemical change (in general, as a result of a 'process'). The definition is motivated by the idea that a change in the extent to which energy is dispersed depends on how much energy is transferred as heat. As explained in Topic 2A, heat stimulates random motion in the surroundings. On the other hand, work stimulates uniform motion of atoms in the surroundings and so does not change their entropy.

The thermodynamic definition of entropy is based on the expression

$$dS = \frac{dq_{rev}}{T} \quad \textit{Definition} \quad \text{Entropy change} \quad (3A.1)$$

For a measurable change between two states i and f,

$$\Delta S = \int_i^f \frac{dq_{rev}}{T} \quad (3A.2)$$

That is, to calculate the difference in entropy between any two states of a system, we find a *reversible* path between them, and integrate the energy supplied as heat at each stage of the path divided by the temperature at which heating occurs.

A note on good practice According to eqn 3A.1, when the energy transferred as heat is expressed in joules and the temperature is in kelvins, the units of entropy are joules per kelvin ($J K^{-1}$). Entropy is an extensive property. Molar entropy, the entropy divided by the amount of substance, $S_m = S/n$, is expressed in joules per kelvin per mole ($J K^{-1} mol^{-1}$). The units of entropy are the same as those of the gas constant, R, and molar heat capacities. Molar entropy is an intensive property.

Example 3A.1 Calculating the entropy change for the isothermal expansion of a perfect gas

Calculate the entropy change of a sample of perfect gas when it expands isothermally from a volume V_i to a volume V_f.

Method The definition of entropy instructs us to find the energy supplied as heat for a reversible path between the stated initial and final states regardless of the actual manner in which the process takes place. A simplification is that the expansion is isothermal, so the temperature is a constant and may be taken outside the integral in eqn 3A.2. The energy absorbed as heat during a reversible isothermal expansion of a perfect gas can be calculated from $\Delta U = q + w$ and $\Delta U = 0$, which implies that $q = -w$ in general and therefore that $q_{rev} = -w_{rev}$ for a reversible change. The work of reversible isothermal expansion is calculated in Topic 2A. The change in molar entropy is calculated from $\Delta S_m = \Delta S/n$.

Answer Because the temperature is constant, eqn 3A.2 becomes

$$\Delta S = \frac{1}{T}\int_i^f dq_{rev} = \frac{q_{rev}}{T}$$

From Topic 2A we know that

$$q_{rev} = -w_{rev} = nRT \ln \frac{V_f}{V_i}$$

It follows that

$$\Delta S = nR \ln \frac{V_f}{V_i} \quad \text{and} \quad \Delta S_m = R \ln \frac{V_f}{V_i}$$

Self-test 3A.1 Calculate the change in entropy when the pressure of a fixed amount of perfect gas is changed isothermally from p_i to p_f. What is this change due to?

Answer: $\Delta S = nR \ln(p_i/p_f)$; the change in volume when the gas is compressed or expands

The definition in eqn 3A.1 is used to formulate an expression for the change in entropy of the surroundings, ΔS_{sur}. Consider an infinitesimal transfer of heat dq_{sur} to the surroundings. The surroundings consist of a reservoir of constant volume, so the energy supplied to them by heating can be identified with the change in the internal energy of the surroundings, dU_{sur}.[2] The internal energy is a state function, and dU_{sur} is an exact differential. These properties imply that dU_{sur} is independent of how the change is brought about and in particular is independent of whether the process is reversible or irreversible. The same remarks therefore apply to dq_{sur}, to which dU_{sur} is equal. Therefore, we can adapt the definition in eqn 3A.1, delete the constraint 'reversible', and write

$$dS=\frac{dq_{rev,sur}}{T_{sur}}=\frac{dq_{sur}}{T_{sur}} \quad \text{Entropy change of the surroundings} \quad (3A.3a)$$

Furthermore, because the temperature of the surroundings is constant whatever the change, for a measurable change

$$\Delta S_{sur}=\frac{q_{sur}}{T_{sur}} \quad (3A.3b)$$

That is, regardless of how the change is brought about in the system, reversibly or irreversibly, we can calculate the change of entropy of the surroundings by dividing the heat transferred by the temperature at which the transfer takes place.

Equation 3A.3 makes it very simple to calculate the changes in entropy of the surroundings that accompany any process. For instance, for any adiabatic change, $q_{sur}=0$, so

$$\Delta S_{sur}=0 \quad \text{Adiabatic change} \quad (3A.4)$$

This expression is true however the change takes place, reversibly or irreversibly, provided no local hot spots are formed in the surroundings. That is, it is true so long as the surroundings remain in internal equilibrium. If hot spots do form, then the localized energy may subsequently disperse spontaneously and hence generate more entropy.

Brief illustration 3A.1 The entropy change of the surroundings

To calculate the entropy change in the surroundings when 1.00 mol $H_2O(l)$ is formed from its elements under standard conditions at 298 K, we use $\Delta H^{\ominus}=-286\,kJ$ from Table 2C.2. The energy released as heat is supplied to the surroundings, now regarded as being at constant pressure, so $q_{sur}=+286\,kJ$. Therefore,

$$\Delta S_{sur}=\frac{2.86\times10^5\,J\,mol^{-1}}{298\,K}=+960\,J\,K^{-1}$$

This strongly exothermic reaction results in an increase in the entropy of the surroundings as energy is released as heat into them.

Self-test 3A.2 Calculate the entropy change in the surroundings when 1.00 mol $N_2O_4(g)$ is formed from 2.00 mol $NO_2(g)$ under standard conditions at 298 K.

Answer: $-192\,J\,K^{-1}$

We are now in a position to see how the definition of entropy is consistent with Kelvin's and Clausius's statements of the Second Law. In the arrangement shown in Fig. 3A.3, the entropy of the hot source is reduced as energy leaves it as heat, but no other change in entropy occurs (the transfer of energy as work does not result in the production of entropy); consequently the arrangement does not produce work. In Clausius version, the entropy of the cold source in Fig 3A.4 decreases when a certain quantity of energy leaves it as heat, but when that heat enters the hot sink the rise in entropy is not as great. Therefore, overall there is a decrease in entropy: the process is not spontaneous.

(b) The statistical definition of entropy

The entry point into the molecular interpretation of the Second Law of thermodynamics is Boltzmann's insight, first mentioned in *Foundations* B, that an atom or molecule can possess only certain values of the energy, called its 'energy levels'. The continuous thermal agitation that molecules experience at $T>0$ ensures that they are distributed over the available energy levels. Boltzmann also made the link between the distribution of molecules over energy levels and the entropy. He proposed that the entropy of a system is given by

$$S=k\ln \mathcal{W} \quad \text{Boltzmann formula for the entropy} \quad (3A.5)$$

where $k=1.381\times10^{-23}\,J\,K^{-1}$ and $\mathcal{W}$ is the number of **microstates**, the number of ways in which the molecules of a system can be arranged while keeping the total energy constant. Each microstate lasts only for an instant and corresponds to a certain distribution of molecules over the available energy levels. When we measure the properties of a system, we are measuring an average taken over the many microstates the system can occupy under the conditions of the experiment. The concept of the number of microstates makes quantitative the ill-defined qualitative concepts of 'disorder' and 'the dispersal of matter and energy' that are used widely to introduce the concept of entropy: a more disorderly distribution of matter and a greater dispersal of energy corresponds to a greater number of microstates associated with the same total energy. This point is discussed in much greater detail in Topic 15E.

Equation 3A.5 is known as the **Boltzmann formula** and the entropy calculated from it is sometimes called the **statistical**

[2] Alternatively, the surroundings can be regarded as being at constant pressure, in which case we could equate dq_{sur} to dH_{sur}.

entropy. We see that if $\mathcal{W}=1$, which corresponds to one microstate (only one way of achieving a given energy, all molecules in exactly the same state), then $S=0$ because $\ln 1=0$. However, if the system can exist in more than one microstate, then $\mathcal{W}>1$ and $S>0$. If the molecules in the system have access to a greater number of energy levels, then there may be more ways of achieving a given total energy; that is, there are more microstates for a given total energy, $\mathcal{W}$ is greater, and the entropy is greater than when fewer states are accessible. Therefore, the statistical view of entropy summarized by the Boltzmann formula is consistent with our previous statement that the entropy is related to the dispersal of energy and matter. In particular, for a gas of particles in a container, the energy levels become closer together as the container expands (Fig. 3A.5; this is a conclusion from quantum theory that is verified in Topic 8A). As a result, more microstates become possible, $\mathcal{W}$ increases, and the entropy increases, exactly as we inferred from the thermodynamic definition of entropy.

Brief illustration 3A.2 The Boltzmann formula

Suppose that each diatomic molecule in a solid sample can be arranged in either of two orientations and that there are $N=6.022\times10^{23}$ molecules in the sample (that is, 1 mol of molecules). Then $\mathcal{W}=2^N$ and the entropy of the sample is

$$S=k\ln 2^N=Nk\ln 2=(6.022\times10^{23})\times(1.381\times10^{-23}\,\mathrm{J\,K^{-1}})\ln 2$$
$$=5.76\,\mathrm{J\,K^{-1}}$$

Self-test 3A.3 What is the molar entropy of a similar system in which each molecule can be arranged in four different orientations?

Answer: $11.5\,\mathrm{J\,K^{-1}\,mol^{-1}}$

The molecular interpretation of entropy advanced by Boltzmann also suggests the thermodynamic definition given by eqn 3A.1. To appreciate this point, consider that molecules in a system at high temperature can occupy a large number of the available energy levels, so a small additional transfer of energy as heat will lead to a relatively small change in the number of accessible energy levels. Consequently, the number of microstates does not increase appreciably and neither does the entropy of the system. In contrast, the molecules in a system at low temperature have access to far fewer energy levels (at $T=0$, only the lowest level is accessible), and the transfer of the same quantity of energy by heating will increase the number of accessible energy levels and the number of microstates significantly. Hence, the change in entropy upon heating will be greater when the energy is transferred to a cold body than when it is transferred to a hot body. This argument suggests that the change in entropy for a given transfer of energy as heat should be greater at low temperatures than at high, as in eqn 3A.1.

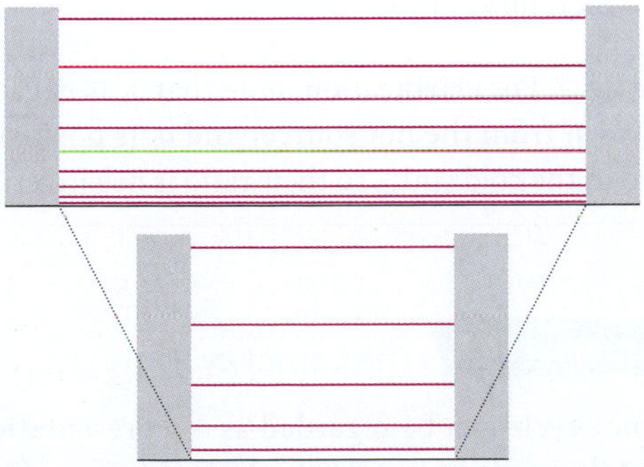

Figure 3A.5 When a box expands, the energy levels move closer together and more become accessible to the molecules. As a result the number of ways of achieving the same energy (the value of $\mathcal{W}$) increases, and so therefore does the entropy.

3A.3 The entropy as a state function

Entropy is a state function. To prove this assertion, we need to show that the integral of dS is independent of path. To do so, it is sufficient to prove that the integral of eqn 3A.1 around an arbitrary cycle is zero, for that guarantees that the entropy is the same at the initial and final states of the system regardless of the path taken between them (Fig. 3A.6). That is, we need to show that

$$\oint \mathrm{d}S=\oint\frac{\mathrm{d}q_{\mathrm{rev}}}{T}=0 \qquad (3A.6)$$

where the symbol $\oint$ denotes integration around a closed path. There are three steps in the argument:

1. First, to show that eqn 3A.6 is true for a special cycle (a 'Carnot cycle') involving a perfect gas.

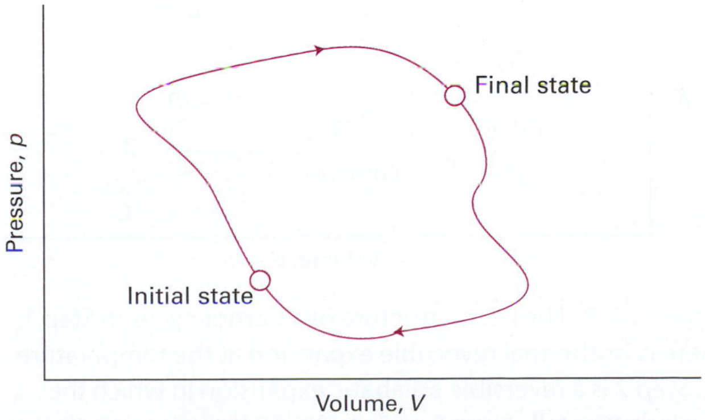

Figure 3A.6 In a thermodynamic cycle, the overall change in a state function (from the initial state to the final state and then back to the initial state again) is zero.

2. Then to show that the result is true whatever the working substance.
3. Finally, to show that the result is true for any cycle.

(a) The Carnot cycle

A **Carnot cycle**, which is named after the French engineer Sadi Carnot, consists of four reversible stages (Fig. 3A.7):

1. Reversible isothermal expansion from A to B at T_h; the entropy change is q_h/T_h, where q_h is the energy supplied to the system as heat from the hot source.
2. Reversible adiabatic expansion from B to C. No energy leaves the system as heat, so the change in entropy is zero. In the course of this expansion, the temperature falls from T_h to T_c, the temperature of the cold sink.
3. Reversible isothermal compression from C to D at T_c. Energy is released as heat to the cold sink; the change in entropy of the system is q_c/T_c; in this expression q_c is negative.
4. Reversible adiabatic compression from D to A. No energy enters the system as heat, so the change in entropy is zero. The temperature rises from T_c to T_h.

The total change in entropy around the cycle is the sum of the changes in each of these four steps:

$$\oint dS = \frac{q_h}{T_h} + \frac{q_c}{T_c}$$

However, we show in the following *Justification* that for a perfect gas

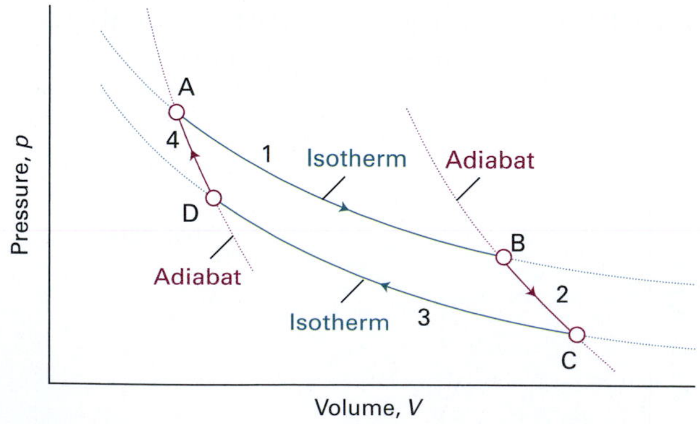

Figure 3A.7 The basic structure of a Carnot cycle. In Step 1, there is isothermal reversible expansion at the temperature T_h. Step 2 is a reversible adiabatic expansion in which the temperature falls from T_h to T_c. In Step 3 there is an isothermal reversible compression at T_c, and that isothermal step is followed by an adiabatic reversible compression, which restores the system to its initial state.

$$\frac{q_h}{q_c} = -\frac{T_h}{T_c} \tag{3A.7}$$

Substitution of this relation into the preceding equation gives zero on the right, which is what we wanted to prove.

Justification 3A.1 Heating accompanying reversible adiabatic expansion

This *Justification* is based on two features of the cycle. One feature is that the two temperatures T_h and T_c in eqn 3A.7 lie on the same adiabat in Fig. 3A.7. The second feature is that the energy transferred as heat during the two isothermal stages are

$$q_h = nRT_h \ln\frac{V_B}{V_A} \qquad q_c = nRT_c \ln\frac{V_D}{V_C}$$

We now show that the two volume ratios are related in a very simple way. From the relation between temperature and volume for reversible adiabatic processes (VT^c=constant, Topic 2D):

$$V_A T_h^c = V_D T_c^c \qquad V_C T_c^c = V_B T_h^c$$

Multiplication of the first of these expressions by the second gives

$$V_A V_C T_h^c T_c^c = V_D V_B T_h^c T_c^c$$

which, on cancellation of the temperatures, simplifies to

$$\frac{V_D}{V_C} = \frac{V_A}{V_B}$$

With this relation established, we can write

$$q_c = nRT_c \ln\frac{V_D}{V_C} = nRT_c \ln\frac{V_A}{V_B} = -nRT_c \ln\frac{V_B}{V_A}$$

and therefore

$$\frac{q_h}{q_c} = \frac{nRT_h \ln(V_B/V_A)}{-nRT_c \ln(V_B/V_A)} = -\frac{T_h}{T_c}$$

as in eqn 3A.7. For clarification, note that q_h is negative (heat is withdrawn from the hot source) and q_c is positive (heat is deposited in the cold sink), so their ratio is negative.

Brief illustration 3A.3 The Carnot cycle

The Carnot cycle can be regarded as a representation of the changes taking place in an actual idealized engine, where heat is converted into work. (However, other cycles are closer approximations to real engines.) In an engine running in accord with the Carnot cycle, 100 J of energy is withdrawn from the hot source ($q_h = -100$ J) at 500 K and some is used to do work, with the remainder deposited in the cold sink at 300 K. According to eqn 3A.7, the amount of heat deposited is

$$q_c = -q_h \times \frac{T_c}{T_h} = -(-100\ \text{J}) \times \frac{300\ \text{K}}{500\ \text{K}} = +60\ \text{J}$$

That means that 40 J was used to do work.

Self-test 3A.4 How much work can be extracted when the temperature of the hot source is increased to 800 K?

Answer: 62 J

In the second step we need to show that eqn 3A.6 applies to any material, not just a perfect gas (which is why, in anticipation, we have not labelled it in blue). We begin this step of the argument by introducing the **efficiency**, η (eta), of a heat engine:

$$\eta = \frac{\text{work performed}}{\text{heat absorbed from hot source}} = \frac{|w|}{|q_h|} \qquad \text{Definition of efficiency} \qquad (3A.8)$$

We are using modulus signs to avoid complications with signs: all efficiencies are positive numbers. The definition implies that the greater the work output for a given supply of heat from the hot reservoir, the greater is the efficiency of the engine. We can express the definition in terms of the heat transactions alone, because (as shown in Fig. 3A.8), the energy supplied as work by the engine is the difference between the energy supplied as heat by the hot reservoir and returned to the cold reservoir:

$$\eta = \frac{|q_h| - |q_c|}{|q_h|} = 1 - \frac{|q_c|}{|q_h|} \qquad (3A.9)$$

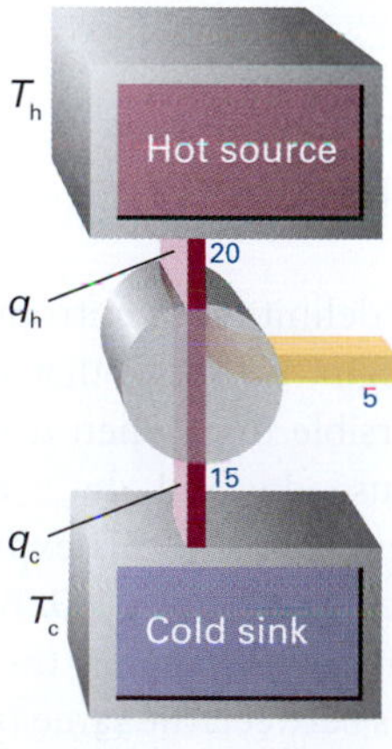

Figure 3A.8 Suppose an energy q_h (for example, 20 kJ) is supplied to the engine and q_c is lost from the engine (for example, $q_c = -15$ kJ) and discarded into the cold reservoir. The work done by the engine is equal to $q_h + q_c$ (for example, 20 kJ + (−15 kJ) = 5 kJ). The efficiency is the work done divided by the energy supplied as heat from the hot source.

It then follows from eqn 3A.7 written as $|q_c|/|q_h| = T_c/T_h$ (see the concluding remark in *Justification* 3A.1) that

$$\eta = 1 - \frac{T_c}{T_h} \qquad \text{Carnot efficiency} \qquad (3A.10)$$

Brief illustration 3A.4 Thermal efficiency

A certain power station operates with superheated steam at 300 °C ($T_h = 573$ K) and discharges the waste heat into the environment at 20 °C ($T_c = 293$ K). The theoretical efficiency is therefore

$$\eta = 1 - \frac{293\ \text{K}}{573\ \text{K}} = 0.489, \text{ or } 48.9 \text{ per cent}$$

In practice, there are other losses due to mechanical friction and the fact that the turbines do not operate reversibly.

Self-test 3A.5 At what temperature of the hot source would the theoretical efficiency reach 80 per cent?

Answer: 1465 K

Now we are ready to generalize this conclusion. The Second Law of thermodynamics implies that *all reversible engines have the same efficiency regardless of their construction*. To see the truth of this statement, suppose two reversible engines are coupled together and run between the same two reservoirs (Fig. 3A.9). The working substances and details of construction of the two engines are entirely arbitrary. Initially, suppose that engine A is more efficient than engine B, and that we choose a setting of the controls that causes engine B to acquire energy as heat q_c from the cold reservoir and to release a certain

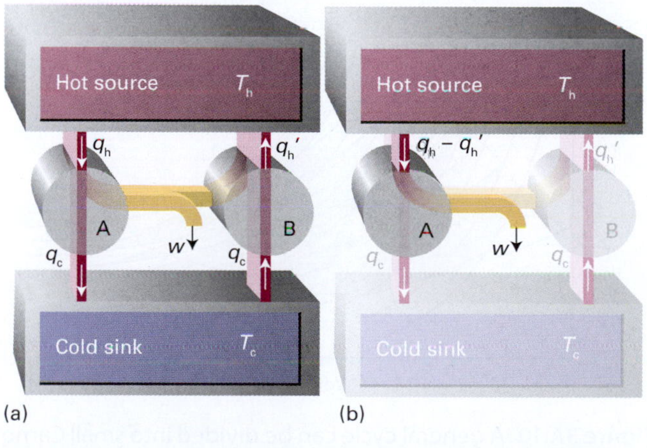

Figure 3A.9 (a) The demonstration of the equivalence of the efficiencies of all reversible engines working between the same thermal reservoirs is based on the flow of energy represented in this diagram. (b) The net effect of the processes is the conversion of heat into work without there being a need for a cold sink: this is contrary to the Kelvin statement of the Second Law.

quantity of energy as heat into the hot reservoir. However, because engine A is more efficient than engine B, not all the work that A produces is needed for this process, and the difference can be used to do work. The net result is that the cold reservoir is unchanged, work has been done, and the hot reservoir has lost a certain amount of energy. This outcome is contrary to the Kelvin statement of the Second Law, because some heat has been converted directly into work. In molecular terms, the random thermal motion of the hot reservoir has been converted into ordered motion characteristic of work. Because the conclusion is contrary to experience, the initial assumption that engines A and B can have different efficiencies must be false. It follows that the relation between the heat transfers and the temperatures must also be independent of the working material, and therefore that eqn 3A.10 is always true for any substance involved in a Carnot cycle.

For the final step in the argument, we note that any reversible cycle can be approximated as a collection of Carnot cycles and the integral around an arbitrary path is the sum of the integrals around each of the Carnot cycles (Fig. 3A.10). This approximation becomes exact as the individual cycles are allowed to become infinitesimal. The entropy change around each individual cycle is zero (as demonstrated above), so the sum of entropy changes for all the cycles is zero. However, in the sum, the entropy change along any individual path is cancelled by the entropy change along the path it shares with the neighbouring cycle. Therefore, all the entropy changes cancel except for those along the perimeter of the overall cycle. That is,

$$\sum_{\text{all}} \frac{q_{\text{rev}}}{T} = \sum_{\text{perimeter}} \frac{q_{\text{rev}}}{T} = 0$$

In the limit of infinitesimal cycles, the non-cancelling edges of the Carnot cycles match the overall cycle exactly, and the sum becomes an integral. Equation 3A.6 then follows immediately. This result implies that $\mathrm{d}S$ is an exact differential and therefore that S is a state function.

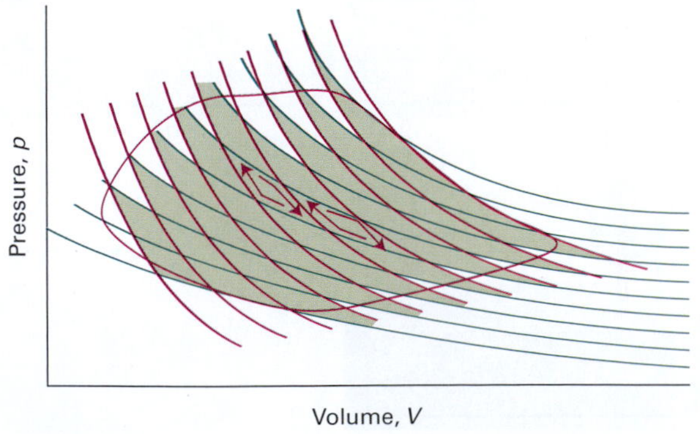

Figure 3A.10 A general cycle can be divided into small Carnot cycles. The match is exact in the limit of infinitesimally small cycles. Paths cancel in the interior of the collection, and only the perimeter, an increasingly good approximation to the true cycle as the number of cycles increases, survives. Because the entropy change around every individual cycle is zero, the integral of the entropy around the perimeter is zero too.

(b) The thermodynamic temperature

Suppose we have an engine that is working reversibly between a hot source at a temperature T_h and a cold sink at a temperature T, then we know from eqn 3A.10 that

$$T = (1-\eta)T_\text{h} \tag{3A.11}$$

This expression enabled Kelvin to define the **thermodynamic temperature scale** in terms of the efficiency of a heat engine: we construct an engine in which the hot source is at a known temperature and the cold sink is the object of interest. The temperature of the latter can then be inferred from the measured efficiency of the engine. The **Kelvin scale** (which is a special case of the thermodynamic temperature scale) is currently defined by using water at its triple point as the notional hot source and defining that temperature as 273.16 K exactly.[3]

Brief illustration 3A.5 The thermodynamic temperature

A heat engine was constructed that used a hot source at the triple point temperature of water and used as a cold source a cooled liquid. The efficiency of the engine was measured as 0.400. The temperature of the liquid is therefore

$$T = (1-0.400)\times(273.16\,\text{K}) = 164\,\text{K}$$

Self-test 3A.6 What temperature would be reported for the hot source if a thermodynamic efficiency of 0.500 was measured when the cold sink was at 273.16 K?

Answer: 546 K

(c) The Clausius inequality

We now show that the definition of entropy is consistent with the Second Law. To begin, we recall that more work is done when a change is reversible than when it is irreversible. That is, $|\mathrm{d}w_\text{rev}| \geq |\mathrm{d}w|$. Because $\mathrm{d}w$ and $\mathrm{d}w_\text{rev}$ are negative when energy leaves the system as work, this expression is the same as $-\mathrm{d}w_\text{rev} \geq -\mathrm{d}w$, and hence $\mathrm{d}w - \mathrm{d}w_\text{rev} \geq 0$. Because the internal energy is a state function, its change is the same for irreversible and reversible paths between the same two states, so we can also write:

$$\mathrm{d}U = \mathrm{d}q + \mathrm{d}w = \mathrm{d}q_\text{rev} + \mathrm{d}w_\text{rev}$$

[3] Discussions are in progress to replace this definition by another that is independent of the specification of a particular substance.

It follows that $dq_{rev} - dq = dw - dw_{rev} \geq 0$, or $dq_{rev} \geq dq$, and therefore that $dq_{rev}/T \geq dq/T$. Now we use the thermodynamic definition of the entropy (eqn 3A.1; $dS = dq_{rev}/T$) to write

$$dS \geq \frac{dq}{T} \qquad \text{Clausius inequality} \qquad (3A.12)$$

This expression is the **Clausius inequality**. It proves to be of great importance for the discussion of the spontaneity of chemical reactions, as is shown in Topic 3C.

Brief illustration 3A.6 The Clausius inequality

Consider the transfer of energy as heat from one system—the hot source—at a temperature T_h to another system—the cold sink—at a temperature T_c (Fig. 3A.11).

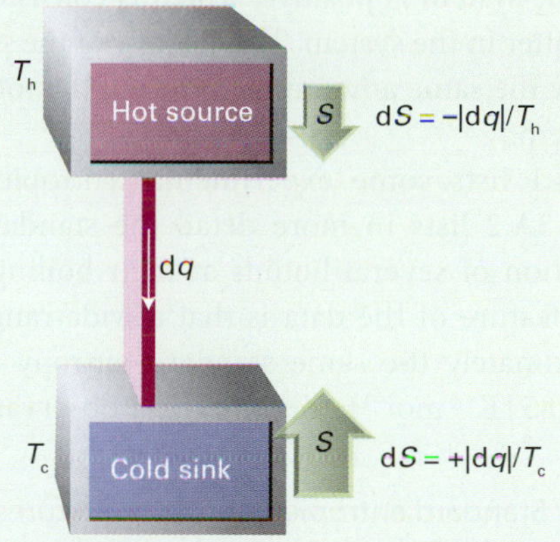

Figure 3A.11 When energy leaves a hot reservoir as heat, the entropy of the reservoir decreases. When the same quantity of energy enters a cooler reservoir, the entropy increases by a larger amount. Hence, overall there is an increase in entropy and the process is spontaneous. Relative changes in entropy are indicated by the sizes of the arrows.

When $|dq|$ leaves the hot source (so $dq_h < 0$), the Clausius inequality implies that $dS \geq dq_h/T_h$. When $|dq|$ enters the cold sink the Clausius inequality implies that $dS \geq dq_c/T_c$ (with $dq_c > 0$). Overall, therefore,

$$dS \geq \frac{dq_h}{T_h} + \frac{dq_c}{T_c}$$

However, $dq_h = -dq_c$, so

$$dS \geq -\frac{dq_c}{T_h} + \frac{dq_c}{T_c} = \left(\frac{1}{T_c} - \frac{1}{T_h}\right) dq_c$$

which is positive (because $dq_c > 0$ and $T_h \geq T_c$). Hence, cooling (the transfer of heat from hot to cold) is spontaneous, as we know from experience.

Self-test 3A.7 What is the change in entropy when 1.0 J of energy as heat transfers from a large block of iron at 30 °C to another large block at 20 °C?

Answer: $+0.1$ mJ K^{-1}

We now suppose that the system is isolated from its surroundings, so that $dq = 0$. The Clausius inequality implies that

$$dS \geq 0 \qquad (3A.13)$$

and we conclude that *in an isolated system the entropy cannot decrease when a spontaneous change occurs*. This statement captures the content of the Second Law.

3A.4 Entropy changes accompanying specific processes

We now see how to calculate the entropy changes that accompany a variety of basic processes.

(a) Expansion

We established in *Example* 3A.1 that the change in entropy of a perfect gas that expands isothermally from V_i to V_f is

$$\Delta S = nR \ln \frac{V_f}{V_i} \qquad \text{Entropy change for the isothermal expansion of a perfect gas} \qquad (3A.14)$$

Because S is a state function, the value of ΔS *of the system* is independent of the path between the initial and final states, so this expression applies whether the change of state occurs reversibly or irreversibly. The logarithmic dependence of entropy on volume is illustrated in Fig. 3A.12.

The *total* change in entropy, however, does depend on how the expansion takes place. For any process the energy lost as heat from the system is acquired by the surroundings, so $dq_{sur} = -dq$. For a reversible change we use the expression in *Example* 3A.1 ($q_{rev} = nRT \ln(V_f/V_i)$); consequently, from eqn 3A.3b

$$\Delta S_{sur} = \frac{q_{sur}}{T} = -\frac{q_{rev}}{T} = -nR \ln \frac{V_f}{V_i} \qquad (3A.15)$$

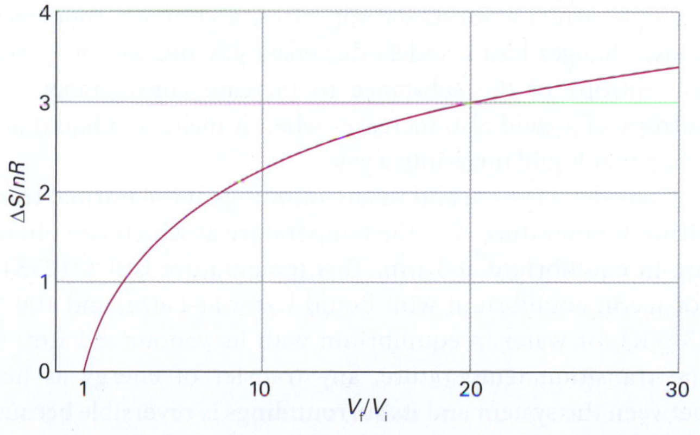

Figure 3A.12 The logarithmic increase in entropy of a perfect gas as it expands isothermally.

This change is the negative of the change in the system, so we can conclude that $\Delta S_{\rm tot}=0$, which is what we should expect for a reversible process. If, on the other hand, the isothermal expansion occurs freely ($w=0$), then $q=0$ (because $\Delta U=0$). Consequently, $\Delta S_{\rm sur}=0$, and the total entropy change is given by eqn 3A.17 itself:

$$\Delta S_{\rm tot}=nR\ln\frac{V_{\rm f}}{V_{\rm i}} \qquad (3A.16)$$

In this case, $\Delta S_{\rm tot}>0$, as we expect for an irreversible process.

Brief illustration 3A.7 Entropy of expansion

When the volume of any perfect gas is doubled at any constant temperature, $V_{\rm f}/V_{\rm i}=2$ and the change in molar entropy of the system is

$$\Delta S_{\rm m}=(8.3145\,{\rm J\,K^{-1}\,mol^{-1}})\times\ln 2=+5.76\,{\rm J\,K^{-1}\,mol^{-1}}$$

If the change is carried out reversibly, the change in entropy of the surroundings is $-5.76\,{\rm J\,K^{-1}\,mol^{-1}}$ (the 'per mole' meaning per mole of gas molecules in the sample). The total change in entropy is 0. If the expansion is free, the change in molar entropy of the gas is still $+5.76\,{\rm J\,K^{-1}\,mol^{-1}}$, but that of the surroundings is 0, and the total change is $+5.76\,{\rm J\,K^{-1}\,mol^{-1}}$.

Self-test 3A.8 Calculate the change in entropy when a perfect gas expands isothermally to 10 times its initial volume (a) reversibly, (b) irreversibly against zero pressure.

Answer: (a) $\Delta S_{\rm m}=+19\,{\rm J\,K^{-1}\,mol^{-1}}$, $\Delta S_{\rm surr}=-19\,{\rm J\,K^{-1}\,mol^{-1}}$, $\Delta S_{\rm tot}=0$; (b) $\Delta S_{\rm m}=+19\,{\rm J\,K^{-1}\,mol^{-1}}$, $\Delta S_{\rm surr}=0$, $\Delta S_{\rm tot}=+19\,{\rm J\,K^{-1}\,mol^{-1}}$

(b) Phase transitions

The degree of dispersal of matter and energy changes when a substance freezes or boils as a result of changes in the order with which the molecules pack together and the extent to which the energy is localized or dispersed. Therefore, we should expect the transition to be accompanied by a change in entropy. For example, when a substance vaporizes, a compact condensed phase changes into a widely dispersed gas and we can expect the entropy of the substance to increase considerably. The entropy of a solid also increases when it melts to a liquid and when that liquid turns into a gas.

Consider a system and its surroundings at the **normal transition temperature**, $T_{\rm trs}$, the temperature at which two phases are in equilibrium at 1 atm. This temperature is 0 °C (273 K) for ice in equilibrium with liquid water at 1 atm, and 100 °C (373 K) for water in equilibrium with its vapour at 1 atm. At the transition temperature, any transfer of energy as heat between the system and its surroundings is reversible because the two phases in the system are in equilibrium. Because at constant pressure $q=\Delta_{\rm trs}H$, the change in molar entropy *of the system* is[4]

$$\Delta_{\rm trs}S=\frac{\Delta_{\rm trs}H}{T_{\rm trs}} \qquad \textit{At the transition temperature} \qquad \text{Entropy of phase transition} \qquad (3A.17)$$

If the phase transition is exothermic ($\Delta_{\rm trs}H<0$, as in freezing or condensing), then the entropy change of the system is negative. This decrease in entropy is consistent with the increased order of a solid compared with a liquid and with the increased order of a liquid compared with a gas. The change in entropy of the surroundings, however, is positive because energy is released as heat into them, and at the transition temperature the total change in entropy is zero. If the transition is endothermic ($\Delta_{\rm trs}H>0$, as in melting and vaporization), then the entropy change of the system is positive, which is consistent with dispersal of matter in the system. The entropy of the surroundings decreases by the same amount, and overall the total change in entropy is zero.

Table 3A.1 lists some experimental entropies of transition. Table 3A.2 lists in more detail the standard entropies of vaporization of several liquids at their boiling points. An interesting feature of the data is that a wide range of liquids give approximately the same standard entropy of vaporization (about $85\,{\rm J\,K^{-1}\,mol^{-1}}$): this empirical observation is called

Table 3A.1* Standard entropies (and temperatures) of phase transitions, $\Delta_{\rm trs}S^{\ominus}/({\rm J\,K^{-1}\,mol^{-1}})$

	Fusion (at $T_{\rm f}$)	Vaporization (at $T_{\rm b}$)
Argon, Ar	14.17 (at 83.8 K)	74.53 (at 87.3 K)
Benzene, C_6H_6	38.00 (at 279 K)	87.19 (at 353 K)
Water, H_2O	22.00 (at 273.15 K)	109.0 (at 373.15 K)
Helium, He	4.8 (at 8 K and 30 bar)	19.9 (at 4.22 K)

* More values are given in the *Resource section*.

Table 3A.2* The standard enthalpies and entropies of vaporization of liquids at their normal boiling points

	$\Delta_{\rm vap}H^{\ominus}/({\rm kJ\,mol^{-1}})$	$\theta_{\rm b}/°{\rm C}$	$\Delta_{\rm vap}S^{\ominus}/({\rm J\,K^{-1}\,mol^{-1}})$
Benzene	30.8	80.1	87.2
Carbon tetrachloride	30	76.7	85.8
Cyclohexane	30.1	80.7	85.1
Hydrogen sulfide	18.7	−60.4	87.9
Methane	8.18	−161.5	73.2
Water	40.7	100.0	109.1

* More values are given in the *Resource section*.

[4] According to Topic 2C, $\Delta_{\rm trs}H$ is an enthalpy change per mole of substance; so $\Delta_{\rm trs}S$ is also a molar quantity.

Trouton's rule. The explanation of Trouton's rule is that a comparable change in volume occurs when any liquid evaporates and becomes a gas. Hence, all liquids can be expected to have similar standard entropies of vaporization. Liquids that show significant deviations from Trouton's rule do so on account of strong molecular interactions that result in a partial ordering of their molecules. As a result, there is a greater change in disorder when the liquid turns into a vapour than for a fully disordered liquid. An example is water, where the large entropy of vaporization reflects the presence of structure arising from hydrogen-bonding in the liquid. Hydrogen bonds tend to organize the molecules in the liquid so that they are less random than, for example, the molecules in liquid hydrogen sulfide (in which there is no hydrogen bonding). Methane has an unusually low entropy of vaporization. A part of the reason is that the entropy of the gas itself is slightly low (186 J K^{-1} mol^{-1} at 298 K); the entropy of N_2 under the same conditions is 192 J K^{-1} mol^{-1}. As explained in Topic 12B, fewer rotational states are accessible at room temperature for molecules with low moments of inertia (like CH_4) than for molecules with relatively high moments of inertia (like N_2), so their molar entropy is slightly lower.

Brief illustration 3A.8 Trouton's rule

There is no hydrogen bonding in liquid bromine and Br_2 is a heavy molecule that is unlikely to display unusual behaviour in the gas phase, so it is safe to use Trouton's rule. To predict the standard molar enthalpy of vaporization of bromine given that it boils at 59.2 °C, we use the rule in the form

$$\Delta_{\text{vap}}H^{\ominus} = T_b \times (85\,\text{J K}^{-1}\text{mol}^{-1})$$

Substitution of the data then gives

$$\Delta_{\text{vap}}H^{\ominus} = (332.4\,\text{K}) \times (85\,\text{J K}^{-1}\text{mol}^{-1}) = +2.8\times10^3\,\text{J mol}^{-1} = +28\,\text{kJ mol}^{-1}$$

The experimental value is +29.45 kJ mol^{-1}.

Self-test 3A.9 Predict the enthalpy of vaporization of ethane from its boiling point, −88.6 °C.

Answer: 16 kJ mol^{-1}

(c) Heating

Equation 3A.2 can be used to calculate the entropy of a system at a temperature T_f from a knowledge of its entropy at another temperature T_i and the heat supplied to change its temperature from one value to the other:

$$S(T_f) = S(T_i) + \int_{T_i}^{T_f} \frac{\mathrm{d}q_{\text{rev}}}{T} \qquad (3\text{A}.18)$$

We shall be particularly interested in the entropy change when the system is subjected to constant pressure (such as from the atmosphere) during the heating. Then, from the definition of constant-pressure heat capacity (eqn 2B.5, $C_p = (\partial H/\partial T)_p$, written as $\mathrm{d}q_{\text{rev}} = C_p\mathrm{d}T$):

$$S(T_f) = S(T_i) + \int_{T_i}^{T_f} \frac{C_p\mathrm{d}T}{T} \qquad \textit{Constant pressure} \quad \text{Entropy variation with temperature} \qquad (3\text{A}.19)$$

The same expression applies at constant volume, but with C_p replaced by C_V. When C_p is independent of temperature in the temperature range of interest, it can be taken outside the integral and we obtain

$$S(T_f) = S(T_i) + C_p\int_{T_i}^{T_f} \frac{\mathrm{d}T}{T} = S(T_i) + C_p \ln\frac{T_f}{T_i} \qquad (3\text{A}.20)$$

with a similar expression for heating at constant volume. The logarithmic dependence of entropy on temperature is illustrated in Fig. 3A.13.

Brief illustration 3A.9 Entropy change on heating

The molar constant-volume heat capacity of water at 298 K is 75.3 J K^{-1} mol^{-1}. The change in molar entropy when it is heated from 20 °C (293 K) to 50 °C (323 K), supposing the heat capacity to be constant in that range, is therefore

$$\Delta S_m = S_m(323\,\text{K}) - S_m(293\,\text{K}) = (75.3\,\text{J K}^{-1}\,\text{mol}^{-1}) \times \ln\frac{323\,\text{K}}{293\,\text{K}} = +7.34\,\text{J K}^{-1}\,\text{mol}^{-1}$$

Self-test 3A.10 What is the change when further heating takes the temperature from 50 °C to 80 °C?

Answer: +5.99 J K^{-1} mol^{-1}

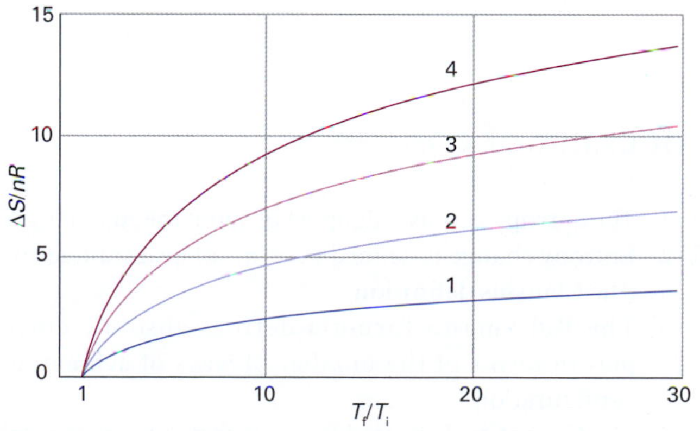

Figure 3A.13 The logarithmic increase in entropy of a substance as it is heated at constant volume. Different curves correspond to different values of the heat capacity (which is assumed constant over the temperature range) expressed as C_m/R.

(d) Composite processes

In many cases, more than one parameter changes. For instance, it might be the case that both the volume and the temperature of a gas are different in the initial and final states. Because S is a state function, we are free to choose the most convenient path from the initial state to the final state, such as reversible isothermal expansion to the final volume, followed by reversible heating at constant volume to the final temperature. Then the total entropy change is the sum of the two contributions.

Example 3A.2 Calculating the entropy change for a composite process

Calculate the entropy change when argon at 25 °C and 1.00 bar in a container of volume 0.500 dm³ is allowed to expand to 1.000 dm³ and is simultaneously heated to 100 °C.

Method As remarked in the text, use reversible isothermal expansion to the final volume, followed by reversible heating at constant volume to the final temperature. The entropy change in the first step is given by eqn 3A.16 and that of the second step, provided C_V is independent of temperature, by eqn 3A.20 (with C_V in place of C_p). In each case we need to know n, the amount of gas molecules, and can calculate it from the perfect gas equation and the data for the initial state from $n=p_iV_i/RT_i$. The molar heat capacity at constant volume is given by the equipartition theorem as $\frac{3}{2}R$. (The equipartition theorem is reliable for monatomic gases: for others and in general use experimental data like that in Tables 2C.1 and 2C.2 of the *Resource section*, converting to the value at constant volume by using the relation $C_{p,\mathrm{m}}-C_{V,\mathrm{m}}=R$.)

Answer From eqn 3A.16 the entropy change in the isothermal expansion from V_i to V_f is

$$\Delta S(\text{Step 1})=nR\ln\frac{V_f}{V_i}$$

From eqn 3A.20, the entropy change in the second step, from T_i to T_f at constant volume, is

$$\Delta S(\text{Step 2})=nC_{V,\mathrm{m}}\ln\frac{T_f}{T_i}=\tfrac{3}{2}nR\ln\frac{T_f}{T_i}=nR\ln\left(\frac{T_f}{T_i}\right)^{3/2}$$

The overall entropy change of the system, the sum of these two changes, is

$$\Delta S=nR\ln\frac{V_f}{V_i}+nR\ln\left(\frac{T_f}{T_i}\right)^{3/2}=nR\ln\frac{V_f}{V_i}\left(\frac{T_f}{T_i}\right)^{3/2}$$

(We have used $\ln x+\ln y=\ln xy$.) Now we substitute $n=p_iV_i/RT_i$ and obtain

$$\Delta S=\frac{p_iV_i}{T_i}\ln\frac{V_f}{V_i}\left(\frac{T_f}{T_i}\right)^{3/2}$$

At this point we substitute the data:

$$\Delta S=\frac{(1.00\times10^5\,\mathrm{Pa})\times(0.500\times10^{-3}\,\mathrm{m^3})}{298\,\mathrm{K}}\times\ln\frac{1.000}{0.500}\left(\frac{373}{298}\right)^{3/2}$$
$$=+0.173\,\mathrm{J\,K^{-1}}$$

A note on good practice It is sensible to proceed as generally as possible before inserting numerical data so that, if required, the formula can be used for other data and to avoid rounding errors.

Self-test 3A.11 Calculate the entropy change when the same initial sample is compressed to 0.0500 dm³ and cooled to −25 °C.

Answer: $-0.44\,\mathrm{J\,K^{-1}}$

Checklist of concepts

- ☐ 1. The **entropy** acts as a signpost of spontaneous change.
- ☐ 2. Entropy change is defined in terms of heat transactions (the **Clausius definition**).
- ☐ 3. The **Boltzmann formula** defines absolute entropies in terms of the number of ways of achieving a configuration.
- ☐ 4. The **Carnot cycle** is used to prove that entropy is a state function.
- ☐ 5. The **efficiency** of a heat engine is the basis of the definition of the thermodynamic temperature scale and one realization, the Kelvin scale.
- ☐ 6. The **Clausius inequality** is used to show that the entropy increases in a spontaneous change and therefore that the Clausius definition is consistent with the Second Law.
- ☐ 7. The entropy of a perfect gas increases when it expands isothermally.
- ☐ 8. The change in entropy of a substance accompanying a change of state at its transition temperature is calculated from its enthalpy of transition.
- ☐ 9. The increase in entropy when a substance is heated is expressed in terms of its heat capacity.

Checklist of equations

Property	Equation	Comment	Equation number
Thermodynamic entropy	$dS = dq_{rev}/T$	Definition	3A.1
Entropy change of surroundings	$\Delta S_{sur} = q_{sur}/T_{sur}$		3A.3b
Boltzmann formula	$S = k \ln W$	Definition	3A.5
Carnot efficiency	$\eta = 1 - T_c/T_h$	Reversible processes	3A.10
Thermodynamic temperature	$T = (1-\eta)T_h$		3A.11
Clausius inequality	$dS \geq dq/T$		3A.12
Entropy of isothermal expansion	$\Delta S = nR \ln(V_f/V_i)$	Perfect gas	3A.14
Entropy of transition	$\Delta_{trs}S = \Delta_{trs}H/T_{trs}$	At the transition temperature	3A.17
Variation of the entropy with temperature	$S(T_f) = S(T_i) + C \ln(T_f/T_i)$	The heat capacity, C, is independent of temperature and no phase transitions occur	3A.20

3B The measurement of entropy

Contents

➤ **Why do you need to know this material?**

For entropy to be a quantitatively useful concept it is important to be able to measure it: the calorimetric procedure is described here. The discussion also introduces the Third Law of thermodynamics, which has important implications for the measurement of entropies and (as shown in later Topics) the attainment of absolute zero.

➤ **What is the key idea?**

The entropy of a perfectly crystalline solid is zero at $T=0$.

➤ **What do you need to know already?**

You need to be familiar with the expression for the temperature dependence of entropy and how entropies of transition are calculated (Topic 3A). The discussion of residual entropy draws on the Boltzmann formula for the entropy (Topic 3A).

The entropy of a substance can be determined in two ways. One, which is the subject of this Topic, is to make calorimetric measurements of the heat required to raise the temperature of a sample from $T=0$ to the temperature of interest. The other, which is described in Topic 15E, is to use calculated parameters or spectroscopic data and to calculate the entropy by using Boltzmann's statistical definition.

3B.1 The calorimetric measurement of entropy

It is established in Topic 3A that the entropy of a system at a temperature T is related to its entropy at $T=0$ by measuring its heat capacity C_p at different temperatures and evaluating the integral in eqn 3A.19 ($S(T_f)=S(T_i)+\int_{T_i}^{T_f} C_p\mathrm{d}T/T$). The entropy of transition ($\Delta_{trs}H/T_{trs}$) for each phase transition between $T=0$ and the temperature of interest must then be included in the overall sum. For example, if a substance melts at T_f and boils at T_b, then its molar entropy above its boiling temperature is given by

$$S_m(T)=S_m(0)+\overbrace{\int_0^{T_f}\frac{C_{p,m}(s,T)}{T}\mathrm{d}T}^{\text{Heat solid to its melting point}}+\overbrace{\frac{\Delta_{fus}H}{T_f}}^{\text{Entropy of fusion}}$$
$$+\overbrace{\int_{T_f}^{T_b}\frac{C_{p,m}(l,T)}{T}\mathrm{d}T}^{\text{Heat liquid to its boiling point}}+\overbrace{\frac{\Delta_{vap}H}{T_b}}^{\text{Entropy of vaporization}} \qquad (3B.1)$$
$$+\overbrace{\int_{T_b}^{T}\frac{C_{p,m}(g,T)}{T}\mathrm{d}T}^{\text{Heat vapour to the final temperature}}$$

All the properties required, except $S_m(0)$, can be measured calorimetrically, and the integrals can be evaluated either graphically or, as is now more usual, by fitting a polynomial to the data and integrating the polynomial analytically. The former procedure is illustrated in Fig. 3B.1: the area under the curve of $C_{p,m}/T$ against T is the integral required. Provided all measurements are made at 1 bar on a pure material, the final value is the **standard entropy**, $S^{\ominus}(T)$ and, on division by the amount of substance n, its **standard molar entropy**, $S_m^{\ominus}(T)=S^{\ominus}(T)/n$. Because $\mathrm{d}T/T=\mathrm{d}\ln T$, an alternative procedure is to evaluate the area under a plot of $C_{p,m}$ against $\ln T$.

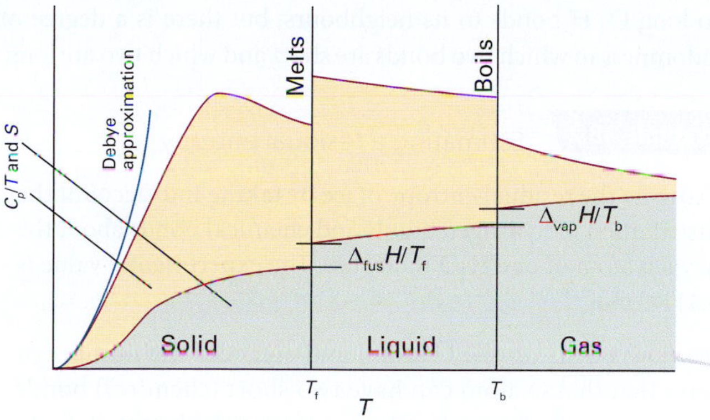

Figure 3B.1 The variation of C_p/T with the temperature for a sample is used to evaluate the entropy, which is equal to the area beneath the upper curve up to the corresponding temperature, plus the entropy of each phase transition passed.

Brief illustration 3B.1 The standard molar entropy

The standard molar entropy of nitrogen gas at 25 °C has been calculated from the following data:

	$S_m^{\ominus}/(\text{J K}^{-1}\,\text{mol}^{-1})$
Debye extrapolation*	1.92
Integration, from 10 K to 35.61 K	25.25
Phase transition at 35.61 K	6.43
Integration, from 35.61 K to 63.14 K	23.38
Fusion at 63.14 K	11.42
Integration, from 63.14 K to 77.32 K	11.41
Vaporization at 77.32 K	72.13
Integration, from 77.32 K to 298.15 K	39.20
Correction for gas imperfection	0.92
Total	192.06

Therefore, $S_m^{\ominus}(298.15\,\text{K}) = S_m(0) + 192.1\,\text{J K}^{-1}\,\text{mol}^{-1}$.

*This extrapolation is explained immediately following.

One problem with the determination of entropy is the difficulty of measuring heat capacities near $T=0$. There are good theoretical grounds for assuming that the heat capacity of a non-metallic solid is proportional to T^3 when T is low (see Topic 7A), and this dependence is the basis of the **Debye extrapolation**. In this method, C_p is measured down to as low a temperature as possible and a curve of the form aT^3 is fitted to the data. That fit determines the value of a, and the expression $C_{p,m}=aT^3$ is assumed valid down to $T=0$.

Example 3B.1 Calculating the entropy at low temperatures

The molar constant-pressure heat capacity of a certain solid at 4.2 K is $0.43\,\text{J K}^{-1}\,\text{mol}^{-1}$. What is its molar entropy at that temperature?

Method Because the temperature is so low, we can assume that the heat capacity varies with temperature as aT^3, in which case we can use eqn 3A.19 (quoted in the opening paragraph of 3B.1) to calculate the entropy at a temperature T in terms of the entropy at $T=0$ and the constant a. When the integration is carried out, it turns out that the result can be expressed in terms of the heat capacity at the temperature T, so the data can be used directly to calculate the entropy.

Answer The integration required is

$$S_m(T) = S_m(0) + \int_0^T \frac{aT^3}{T}\,dT = S_m(0) + a\int_0^T T^2\,dT$$
$$= S_m(0) + \tfrac{1}{3}aT^3 = S_m(0) + \tfrac{1}{3}C_{p,m}(T)$$

from which it follows that

$$S_m(4.2\,\text{K}) = S_m(0) + 0.14\,\text{J K}^{-1}\,\text{mol}^{-1}$$

Self-test 3B.1 For metals, there is also a contribution to the heat capacity from the electrons which is linearly proportional to T when the temperature is low. Find its contribution to the entropy at low temperatures.

Answer: $S(T)=S(0)+C_p(T)$

3B.2 The Third Law

We now address the problem of the value of $S(0)$. At $T=0$, all energy of thermal motion has been quenched, and in a perfect crystal all the atoms or ions are in a regular, uniform array. The localization of matter and the absence of thermal motion suggest that such materials also have zero entropy. This conclusion is consistent with the molecular interpretation of entropy, because $S=0$ if there is only one way of arranging the molecules and only one microstate is accessible (all molecules occupy the ground state, $\mathcal{W}=1$).

(a) The Nernst heat theorem

The experimental observation that turns out to be consistent with the view that the entropy of a regular array of molecules is zero at $T=0$ is summarized by the **Nernst heat theorem**:

> The entropy change accompanying any physical or chemical transformation approaches zero as the temperature approaches zero: $\Delta S \rightarrow 0$ as $T \rightarrow 0$ provided all the substances involved are perfectly ordered.

Nernst heat theorem

Brief illustration 3B.2 The Nernst heat theorem

Consider the entropy of the transition between orthorhombic sulfur, α, and monoclinic sulfur, β, which can be calculated from the transition enthalpy (−402 J mol⁻¹) at the transition temperature (369 K):

$$\Delta_{\rm trs}S = S_{\rm m}(\beta) - S_{\rm m}(\alpha) = \frac{-402\,{\rm J\,mol^{-1}}}{369\,{\rm K}}$$
$$= -1.09\,{\rm J\,K^{-1}\,mol^{-1}}$$

The two individual entropies can also be determined by measuring the heat capacities from $T=0$ up to $T=369$ K. It is found that $S_{\rm m}(\alpha) = S_{\rm m}(\alpha,0) + 37\,{\rm J\,K^{-1}\,mol^{-1}}$ and $S_{\rm m}(\beta) = S_{\rm m}(\beta,0) + 38\,{\rm J\,K^{-1}\,mol^{-1}}$. These two values imply that at the transition temperature

$$\Delta_{\rm trs}S = S_{\rm m}(\alpha,0) - S_{\rm m}(\beta,0) = -1\,{\rm J\,K^{-1}\,mol^{-1}}$$

On comparing this value with the one above, we conclude that $S_{\rm m}(\alpha,0) - S_{\rm m}(\beta,0) \approx 0$, in accord with the theorem.

Self-test 3B.2 Two forms of a metallic solid (see *Self-test* 3B.1) undergo a phase transition at $T_{\rm trs}$, which is close to $T=0$. What is the enthalpy of transition at $T_{\rm trs}$ in terms of the heat capacities of the two polymorphs?

Answer: $\Delta_{\rm trs}H(T_{\rm trs}) = T_{\rm trs}\Delta C_p(T_{\rm trs})$

It follows from the Nernst theorem, that if we arbitrarily ascribe the value zero to the entropies of elements in their perfect crystalline form at $T=0$, then all perfect crystalline compounds also have zero entropy at $T=0$ (because the change in entropy that accompanies the formation of the compounds, like the entropy of all transformations at that temperature, is zero). This conclusion is summarized by the **Third Law of thermodynamics**:

> The entropy of all perfect crystalline substances is zero at $T=0$. — Third Law of thermodynamics

As far as thermodynamics is concerned, choosing this common value as zero is a matter of convenience. The molecular interpretation of entropy, however, justifies the value $S=0$ at $T=0$ because then, as we have remarked, $\mathcal{W}=1$.

In certain cases $\mathcal{W}>1$ at $T=0$ and therefore $S(0)>0$. This is the case if there is no energy advantage in adopting a particular orientation even at absolute zero. For instance, for a diatomic molecule AB there may be almost no energy difference between the arrangements ...AB AB AB... and ...BA AB BA..., so $\mathcal{W}>1$ even at $T=0$. If $S(0)>0$ we say that the substance has a **residual entropy**. Ice has a residual entropy of 3.4 J K⁻¹ mol⁻¹. It stems from the arrangement of the hydrogen bonds between neighbouring water molecules: a given O atom has two short O—H bonds and two long O···H bonds to its neighbours, but there is a degree of randomness in which two bonds are short and which two are long.

Example 3B.2 Estimating a residual entropy

Estimate the residual entropy of ice by taking into account the distribution of hydrogen bonds and chemical bonds about the oxygen atom of one H_2O molecule. The experimental value is 3.4 J K⁻¹ mol⁻¹.

Method Focus on the O atom, and consider the number of ways that that O atom can have two short (chemical) bonds and two long hydrogen bonds to its four neighbours. Refer to Fig. 3B.2.

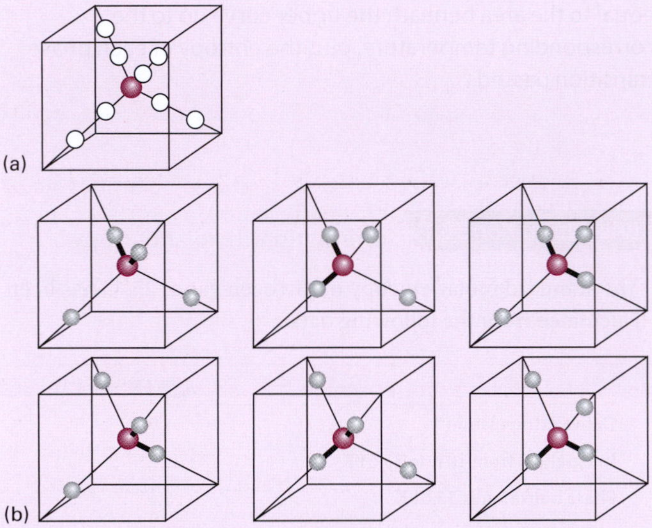

Figure 3B.2 The model of ice showing (a) the local structure of an oxygen atom and (b) the array of chemical and hydrogen bonds used to calculate the residual entropy of ice.

Answer Suppose each H atom can lie either close to or far from its 'parent' O atom, as depicted in Fig. 3B.2. The total number of these conceivable arrangements in a sample that contains N H_2O molecules and therefore $2N$ H atoms is 2^{2N}. Now consider a single central O atom. The total number of possible arrangements of locations of H atoms around the central O atom of one H_2O molecule is $2^4=16$. Of these 16 possibilities, only 6 correspond to two short and two long bonds. That is, only $\frac{6}{16}=\frac{3}{8}$ of all possible arrangements are possible, and for N such molecules only $(3/8)^N$ of all possible arrangements are possible. Therefore, the total number of allowed arrangements in the crystal is $2^{2N}(3/8)^N = 4^N(3/8)^N = (3/2)^N$. If we suppose that all these arrangements are energetically identical, the residual entropy is

$$S(0) = k\ln\left(\tfrac{3}{2}\right)^N = Nk\ln\tfrac{3}{2} = nN_{\rm A}k\ln\tfrac{3}{2} = nR\ln\tfrac{3}{2}$$

and the residual molar entropy would be

$$S_m(0) = R\ln\tfrac{3}{2} = 3.4\,\mathrm{J\,K^{-1}\,mol^{-1}}$$

in accord with the experimental value.

Self-test 3B.3 What would be the residual molar entropy of HCF_3 on the assumption that each molecule could take up one of four tetrahedral orientations in a crystal?

Answer: $11.5\,\mathrm{J\,K^{-1}\,mol^{-1}}$

(b) Third-Law entropies

Entropies reported on the basis that $S(0)=0$ are called **Third-Law entropies** (and commonly just 'entropies'). When the substance is in its standard state at the temperature T, the **standard (Third-Law) entropy** is denoted $S^{\ominus}(T)$. A list of values at 298 K is given in Table 3B.1.

The **standard reaction entropy**, $\Delta_r S^{\ominus}$, is defined, like the standard reaction enthalpy in Topic 2C, as the difference between the molar entropies of the pure, separated products and the pure, separated reactants, all substances being in their standard states at the specified temperature:

$$\Delta_r S^{\ominus} = \sum_{\text{Products}} \nu S_m^{\ominus} - \sum_{\text{Reactants}} \nu S_m^{\ominus} \quad \textit{Definition} \quad \text{Standard reaction entropy} \quad (3B.2a)$$

In this expression, each term is weighted by the appropriate stoichiometric coefficient. A more sophisticated approach is to adopt the notation introduced in Topic 2C and to write

$$\Delta_r S^{\ominus} = \sum_J \nu_J S_m^{\ominus}(J) \quad (3B.2b)$$

where the ν_J are signed (+ for products, − for reactants) stoichiometric numbers. Standard reaction entropies are likely to be positive if there is a net formation of gas in a reaction, and are likely to be negative if there is a net consumption of gas.

Just as in the discussion of enthalpies in Topic 2C, where it is acknowledged that solutions of cations cannot be prepared in the absence of anions, the standard molar entropies of ions in solution are reported on a scale in which the standard entropy of the H^+ ions in water is taken as zero at all temperatures:

$$S^{\ominus}(H^+, aq) = 0 \quad \textit{Convention} \quad \text{Ions in solution} \quad (3B.3)$$

The values based on this choice are listed in Table 2C.5 in the *Resource section*.[1] Because the entropies of ions in water are values relative to the hydrogen ion in water, they may be either positive or negative. A positive entropy means that an ion has a higher molar entropy than H^+ in water and a negative entropy means that the ion has a lower molar entropy than H^+ in water. Ion entropies vary as expected on the basis that they are related to the degree to which the ions order the water molecules around them in the solution. Small, highly charged ions induce local structure in the surrounding water, and the disorder of

Table 3B.1* Standard Third-Law entropies at 298 K, $S_m^{\ominus}/(\mathrm{J\,K^{-1}\,mol^{-1}})$

	$S_m^{\ominus}/(\mathrm{J\,K^{-1}\,mol^{-1}})$
Solids	
Graphite, C(s)	5.7
Diamond, C(s)	2.4
Sucrose, $C_{12}H_{22}O_{11}$(s)	360.2
Iodine, I_2(s)	116.1
Liquids	
Benzene, C_6H_6(l)	173.3
Water, H_2O(l)	69.9
Mercury, Hg(l)	76.0
Gases	
Methane, CH_4(g)	186.3
Carbon dioxide, CO_2(g)	213.7
Hydrogen, H_2(g)	130.7
Helium, He(g)	126.2
Ammonia, NH_3(g)	192.4

* More values are given in the *Resource section*.

Brief illustration 3B.3 The standard reaction entropy

To calculate the standard reaction entropy of $H_2(g)+\tfrac{1}{2}O_2(g)\rightarrow H_2O(l)$ at 298 K, we use the data in Table 2C.5 of the *Resource section* to write

$$\begin{aligned}\Delta_r S^{\ominus} &= S_m^{\ominus}(H_2O,l) - \{S_m^{\ominus}(H_2,g) + \tfrac{1}{2}S_m^{\ominus}(O_2,g)\}\\ &= 69.9\,\mathrm{J\,K^{-1}\,mol^{-1}} - \{130.7 + \tfrac{1}{2}(205.1)\}\,\mathrm{J\,K^{-1}\,mol^{-1}}\\ &= -163.4\,\mathrm{J\,K^{-1}\,mol^{-1}}\end{aligned}$$

The negative value is consistent with the conversion of two gases to a compact liquid.

A note on good practice Do not make the mistake of setting the standard molar entropies of elements equal to zero: they have nonzero values (provided $T>0$), as we have already discussed.

Self-test 3B.4 Calculate the standard reaction entropy for the combustion of methane to carbon dioxide and liquid water at 298 K.

Answer: $-243\,\mathrm{J\,K^{-1}\,mol^{-1}}$

[1] In terms of the language introduced in Topic 5A, the entropies of ions in solution are actually *partial molar entropies*, for their values include the consequences of their presence on the organization of the solvent molecules around them.

the solution is decreased more than in the case of large, singly charged ions. The absolute, Third-Law standard molar entropy of the proton in water can be estimated by proposing a model of the structure it induces, and there is some agreement on the value $-21\,\mathrm{J\,K^{-1}\,mol^{-1}}$. The negative value indicates that the proton induces order in the solvent.

Brief illustration 3B.4 Absolute and relative ion entropies

The standard molar entropy of $Cl^-(aq)$ is $+57\,\mathrm{J\,K^{-1}\,mol^{-1}}$ and that of $Mg^{2+}(aq)$ is $-128\,\mathrm{J\,K^{-1}\,mol^{-1}}$. That is, the partial molar entropy of $Cl^-(aq)$ is $57\,\mathrm{J\,K^{-1}\,mol^{-1}}$ higher than that of the proton in water (presumably because it induces less local structure in the surrounding water), whereas that of $Mg^{2+}(aq)$ is $128\,\mathrm{J\,K^{-1}\,mol^{-1}}$ lower (presumably because its higher charge induces more local structure in the surrounding water).

Self-test 3B.5 Estimate the absolute values of the partial molar entropies of these ions.

Answer: $+36\,\mathrm{J\,K^{-1}\,mol^{-1}}$, $-149\,\mathrm{J\,K^{-1}\,mol^{-1}}$

Checklist of concepts

- ☐ 1. Entropies are determined calorimetrically by measuring the heat capacity of a substance from low temperatures up to the temperature of interest.
- ☐ 2. The **Debye-T^3 law** is used to estimate heat capacities of non-metallic solids close to $T=0$.
- ☐ 3. The **Nernst heat theorem** states that the entropy change accompanying any physical or chemical transformation approaches zero as the temperature approaches zero: $\Delta S \to 0$ as $T \to 0$ provided all the substances involved are perfectly ordered.
- ☐ 4. The **Third Law of thermodynamics** states that the entropy of all perfect crystalline substances is zero at $T=0$.
- ☐ 5. The **residual entropy** of a solid is the entropy arising from disorder that persists at $T=0$.
- ☐ 6. **Third-Law entropies** are entropies based on $S(0)=0$.
- ☐ 7. The **standard entropies of ions in solution** are based on setting $S^\ominus(H^+,aq)=0$ at all temperatures.
- ☐ 8. The **standard reaction entropy**, $\Delta_r S^\ominus$, is the difference between the molar entropies of the pure, separated products and the pure, separated reactants, all substances being in their standard states.

Checklist of equations

Property	Equation	Comment	Equation number
Standard molar entropy from calorimetry	See eqn 3B.1	Sum of contributions from $T=0$ to temperature of interest	3B.1
Standard reaction entropy	$\Delta_r S^\ominus = \sum_{\text{Products}} \nu S_m^\ominus - \sum_{\text{Reactants}} \nu S_m^\ominus$ $\Delta_r S^\ominus = \sum_J \nu_J S_m^\ominus(J)$	ν: (positive) stoichiometric coefficients; ν_J: (signed) stoichiometric numbers	3B.2

3C Concentrating on the system

Contents

➤ Why do you need to know this material?

Most processes of interest in chemistry occur at constant temperature and pressure. Under these conditions, thermodynamic processes are discussed in terms of the Gibbs energy, which is introduced in this Topic. The Gibbs energy is the foundation of the discussion of phase equilibria, chemical equilibrium, and bioenergetics.

➤ What is the key idea?

The Gibbs energy is a signpost of spontaneous change at constant temperature and pressure, and is equal to the maximum non-expansion work that a system can do.

➤ What do you need to know already?

This Topic develops the Clausius inequality (Topic 3A) and draws on information about standard states and reaction enthalpy introduced in Topic 2C. The derivation of the Born equation uses information about the energy of one electric charge in the field of another (*Foundations* B).

Entropy is the basic concept for discussing the direction of natural change, but to use it we have to analyse changes in both the system and its surroundings. In Topic 3A it is shown that it is always very simple to calculate the entropy change in the surroundings (from $\Delta S_{\text{sur}}=q_{\text{sur}}/T_{\text{sur}}$); here we see that it is possible to devise a simple method for taking that contribution into account automatically. This approach focuses our attention on the system and simplifies discussions. Moreover, it is the foundation of all the applications of chemical thermodynamics that follow.

3C.1 The Helmholtz and Gibbs energies

Consider a system in thermal equilibrium with its surroundings at a temperature T. When a change in the system occurs and there is a transfer of energy as heat between the system and the surroundings, the Clausius inequality (eqn 3A.12, $\mathrm{d}S \geq \mathrm{d}q/T$) reads

$$\mathrm{d}S - \frac{\mathrm{d}q}{T} \geq 0 \qquad (3C.1)$$

We can develop this inequality in two ways according to the conditions (of constant volume or constant pressure) under which the process occurs.

(a) Criteria of spontaneity

First, consider heating at constant volume. Then, in the absence of additional (non-expansion) work, we can write $\mathrm{d}q_V = \mathrm{d}U$; consequently

$$\mathrm{d}S - \frac{\mathrm{d}U}{T} \geq 0 \qquad (3C.2)$$

The importance of the inequality in this form is that it expresses the criterion for spontaneous change solely in terms of the state functions of the system. The inequality is easily rearranged into

$$T\mathrm{d}S \geq \mathrm{d}U \quad \text{(constant } V\text{, no additional work)} \qquad (3C.3)$$

Because $T>0$, at either constant internal energy ($\mathrm{d}U=0$) or constant entropy ($\mathrm{d}S=0$) this expression becomes, respectively,

$$\mathrm{d}S_{U,V} \geq 0 \qquad \mathrm{d}U_{S,V} \leq 0 \qquad (3C.4)$$

where the subscripts indicate the constant conditions.

Equation 3C.4 expresses the criteria for spontaneous change in terms of properties relating to the system. The first inequality states that, in a system at constant volume and constant internal energy (such as an isolated system), the entropy increases in a spontaneous change. That statement is essentially the content of the Second Law. The second inequality is less obvious, for it says that if the entropy and volume of the system are constant, then the internal energy must decrease in a spontaneous change. Do not interpret this criterion as a tendency of the system to sink to lower energy. It is a disguised statement about entropy and should be interpreted as implying that if the entropy of the system is unchanged, then there must be an increase in entropy of the surroundings, which can be achieved only if the energy of the system decreases as energy flows out as heat.

When energy is transferred as heat at constant pressure, and there is no work other than expansion work, we can write $\mathrm{d}q_p=\mathrm{d}H$ and obtain

$$T\mathrm{d}S \geq \mathrm{d}H \quad \text{(constant } p\text{, no additional work)} \qquad (3C.5)$$

At either constant enthalpy or constant entropy this inequality becomes, respectively,

$$\mathrm{d}S_{H,p} \geq 0 \qquad \mathrm{d}H_{S,p} \leq 0 \qquad (3C.6)$$

The interpretations of these inequalities are similar to those of eqn 3C.4. The entropy of the system at constant pressure must increase if its enthalpy remains constant (for there can then be no change in entropy of the surroundings). Alternatively, the enthalpy must decrease if the entropy of the system is constant, for then it is essential to have an increase in entropy of the surroundings.

Brief illustration 3C.1 Spontaneous changes at constant volume

A concrete example of the criterion $\mathrm{d}S_{U,V}\geq 0$ is the diffusion of a solute B through a solvent A that form an ideal solution (in the sense of Topic 5B, in which AA, BB, and AB interactions are identical). There is no change in internal energy or volume of the system or the surroundings as B spreads into A, but the process is spontaneous.

Self-test 3C.1 Invent an example of the criterion $\mathrm{d}U_{S,V}\leq 0$.

Answer: A phase change in which one perfectly ordered phase changes into another of lower energy and equal density at $T=0$

Because eqns 3C.4 and 3C.6 have the forms $\mathrm{d}U-T\mathrm{d}S\leq 0$ and $\mathrm{d}H-T\mathrm{d}S\leq 0$, respectively, they can be expressed more simply by introducing two more thermodynamic quantities. One is the **Helmholtz energy**, A, which is defined as

$$A=U-TS \qquad \textit{Definition} \quad \text{Helmholtz energy} \qquad (3C.7)$$

The other is the **Gibbs energy**, G:

$$G=H-TS \qquad \textit{Definition} \quad \text{Gibbs energy} \qquad (3C.8)$$

All the symbols in these two definitions refer to the system.

When the state of the system changes at constant temperature, the two properties change as follows:

$$\text{(a) } \mathrm{d}A=\mathrm{d}U-T\mathrm{d}S \qquad \text{(b) } \mathrm{d}G=\mathrm{d}H-T\mathrm{d}S \qquad (3C.9)$$

When we introduce eqns 3C.4 and 3C.6, respectively, we obtain the criteria of spontaneous change as

$$\text{(a) } \mathrm{d}A_{T,V} \leq 0 \qquad \text{(b) } \mathrm{d}G_{T,p} \leq 0 \qquad \text{Criteria of spontaneous change} \qquad (3C.10)$$

These inequalities, especially the second, are the most important conclusions from thermodynamics for chemistry. They are developed in subsequent sections, Topics, and chapters.

Brief illustration 3C.2 The spontaneity of endothermic reactions

The existence of spontaneous endothermic reactions provides an illustration of the role of G. In such reactions, H increases, the system rises spontaneously to states of higher enthalpy, and $\mathrm{d}H>0$. Because the reaction is spontaneous we know that $\mathrm{d}G<0$ despite $\mathrm{d}H>0$; it follows that the entropy of the system increases so much that $T\mathrm{d}S$ outweighs $\mathrm{d}H$ in $\mathrm{d}G=\mathrm{d}H-T\mathrm{d}S$. Endothermic reactions are therefore driven by the increase of entropy of the system, and this entropy change overcomes the reduction of entropy brought about in the surroundings by the inflow of heat into the system ($\mathrm{d}S_{\text{sur}}=-\mathrm{d}H/T$ at constant pressure).

Self-test 3C.2 Why are so many exothermic reactions spontaneous?

Answer: With $\mathrm{d}H<0$, it is common for $\mathrm{d}G<0$ unless $T\mathrm{d}S$ is strongly negative.

(b) Some remarks on the Helmholtz energy

A change in a system at constant temperature and volume is spontaneous if $dA_{T,V} \leq 0$. That is, a change under these conditions is spontaneous if it corresponds to a decrease in the Helmholtz energy. Such systems move spontaneously towards states of lower A if a path is available. The criterion of equilibrium, when neither the forward nor reverse process has a tendency to occur, is

$$dA_{T,V} = 0 \qquad (3C.11)$$

The expressions $dA = dU - TdS$ and $dA < 0$ are sometimes interpreted as follows. A negative value of dA is favoured by a negative value of dU and a positive value of TdS. This observation suggests that the tendency of a system to move to lower A is due to its tendency to move towards states of lower internal energy and higher entropy. However, this interpretation is false because the tendency to lower A is solely a tendency towards states of greater overall entropy. *Systems change spontaneously if in doing so the total entropy of the system and its surroundings increases, not because they tend to lower internal energy.* The form of dA may give the impression that systems favour lower energy, but that is misleading: dS is the entropy change of the system, $-dU/T$ is the entropy change of the surroundings (when the volume of the system is constant), and their total tends to a maximum.

Brief illustration 3C.3 Spontaneous change at constant volume

A bouncing ball comes to rest. The spontaneous direction of change is one in which the energy of the ball (potential at the top of its bounce, kinetic when it strikes the floor) spreads out into the surroundings on each bounce. When the ball is still, the energy of the universe is the same as initially, but the energy of the ball is dispersed over the surroundings.

Self-test 3C.3 What other spontaneous similar mechanical processes have a similar explanation?

Answer: One example: A pendulum coming to rest through friction.

(c) Maximum work

It turns out, as we show in the following *Justification*, that A carries a greater significance than being simply a signpost of spontaneous change: *the change in the Helmholtz function is equal to the maximum work accompanying a process at constant temperature*:

$$dw_{max} = dA \quad \text{Constant temperature} \quad \text{Maximum work} \qquad (3C.12)$$

As a result, A is sometimes called the 'maximum work function', or the 'work function'.[1]

Justification 3C.1 Maximum work

To demonstrate that maximum work can be expressed in terms of the changes in Helmholtz energy, we combine the Clausius inequality $dS \geq dq/T$ in the form $TdS \geq dq$ with the First Law, $dU = dq + dw$, and obtain

$$dU \leq TdS + dw$$

dU is smaller than the term of the right because dq has been replaced by TdS, which in general is larger than dq. This expression rearranges to

$$dw \geq dU - TdS$$

It follows that the most negative value of dw, and therefore the maximum energy that can be obtained from the system as work, is given by

$$dw_{max} = dU - TdS$$

and that this work is done only when the path is traversed reversibly (because then the equality applies). Because at constant temperature $dA = dU - TdS$, we conclude that $dw_{max} = dA$.

When a macroscopic isothermal change takes place in the system, eqn 3C.12 becomes

$$w_{max} = \Delta A \quad \text{Constant temperature} \quad \text{Maximum work} \qquad (3C.13)$$

with

$$\Delta A = \Delta U - T\Delta S \quad \text{Constant temperature} \qquad (3C.14)$$

This expression shows that, depending on the sign of $T\Delta S$, not all the change in internal energy may be available for doing work. If the change occurs with a decrease in entropy (of the system), in which case $T\Delta S < 0$, then the right-hand side of this equation is not as negative as ΔU itself, and consequently the maximum work is less than ΔU. For the change to be spontaneous, some of the energy must escape as heat in order to generate enough entropy in the surroundings to overcome the reduction in entropy in the system (Fig. 3C.1). In this case, Nature is demanding a tax on the internal energy as it is converted into work. This is the origin of the alternative name 'Helmholtz free energy' for A, because ΔA is that part of the change in internal energy that we are free to use to do work.

Further insight into the relation between the work that a system can do and the Helmholtz energy is to recall that work is

[1] *Arbeit* is the German word for work; hence the symbol A.

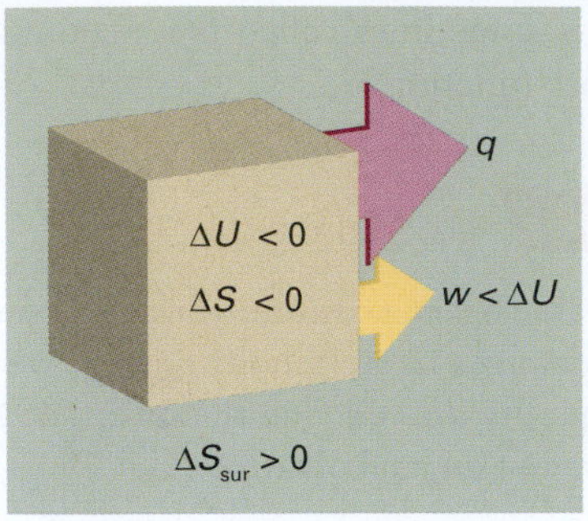

Figure 3C.1 In a system not isolated from its surroundings, the work done may be different from the change in internal energy. Moreover, the process is spontaneous if overall the entropy of the global, isolated system increases. In the process depicted here, the entropy of the system decreases, so that of the surroundings must increase in order for the process to be spontaneous, which means that energy must pass from the system to the surroundings as heat. Therefore, less work than ΔU can be obtained.

energy transferred to the surroundings as the uniform motion of atoms. We can interpret the expression $A=U-TS$ as showing that A is the total internal energy of the system, U, less a contribution that is stored as energy of thermal motion (the quantity TS). Because energy stored in random thermal motion cannot be used to achieve uniform motion in the surroundings, only the part of U that is not stored in that way, the quantity $U-TS$, is available for conversion into work.

If the change occurs with an increase of entropy of the system (in which case $T\Delta S>0$), the right-hand side of the equation is more negative than ΔU. In this case, the maximum work that can be obtained from the system is greater than ΔU. The explanation of this apparent paradox is that the system is not isolated and energy may flow in as heat as work is done. Because the entropy of the system increases, we can afford a reduction of the entropy of the surroundings yet still have, overall, a spontaneous process. Therefore, some energy (no more than the value of $T\Delta S$) may leave the surroundings as heat and contribute to the work the change is generating (Fig. 3C.2). Nature is now providing a tax refund.

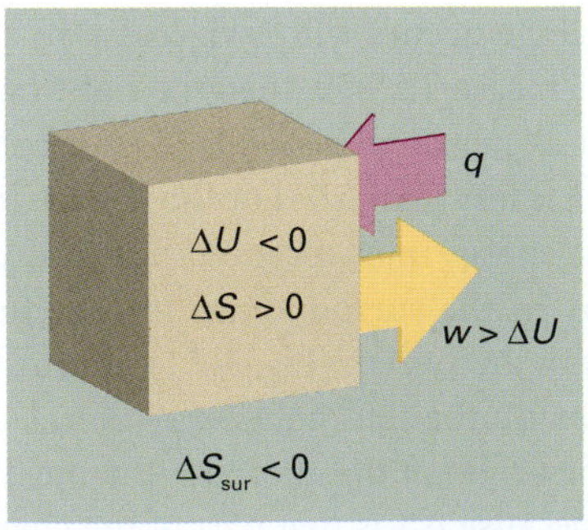

Figure 3C.2 In this process, the entropy of the system increases; hence we can afford to lose some entropy of the surroundings. That is, some of their energy may be lost as heat to the system. This energy can be returned to them as work. Hence the work done can exceed ΔU.

Example 3C.1 Calculating the maximum available work

When 1.000 mol $C_6H_{12}O_6$ (glucose) is oxidized to carbon dioxide and water at 25 °C according to the equation $C_6H_{12}O_6(s) + 6\ O_2(g) \rightarrow 6\ CO_2(g) + 6\ H_2O(l)$ calorimetric measurements give $\Delta_r U^{\ominus} = -2808\ \text{kJ mol}^{-1}$ and $\Delta_r S^{\ominus} = +182.4\ \text{J K}^{-1}\ \text{mol}^{-1}$ at 25 °C. How much of this energy change can be extracted as (a) heat at constant pressure, (b) work?

Method We know that the heat released at constant pressure is equal to the value of ΔH, so we need to relate $\Delta_r H^{\ominus}$ to $\Delta_r U^{\ominus}$, which is given. To do so, we suppose that all the gases involved are perfect, and use eqn 2B.4 ($\Delta H = \Delta U + \Delta n_g RT$) in the form $\Delta_r H = \Delta_r U + \Delta\nu_g RT$. For the maximum work available from the process we use eqn 3C.13.

Answer (a) Because $\Delta\nu_g = 0$, we know that $\Delta_r H^{\ominus} = \Delta_r U^{\ominus} = -2808\ \text{kJ mol}^{-1}$. Therefore, at constant pressure, the energy available as heat is $2808\ \text{kJ mol}^{-1}$. (b) Because $T = 298$ K, the value of $\Delta_r A^{\ominus}$ is

$$\Delta_r A^{\ominus} = \Delta_r U^{\ominus} - T\Delta_r S^{\ominus} = -2862\ \text{kJ mol}^{-1}$$

Therefore, the combustion of 1.000 mol $C_6H_{12}O_6$ can be used to produce up to 2862 kJ of work. The maximum work available is greater than the change in internal energy on account of the positive entropy of reaction (which is partly due to the generation of a large number of small molecules from one big one). The system can therefore draw in energy from the surroundings (so reducing their entropy) and make it available for doing work.

Self-test 3C.4 Repeat the calculation for the combustion of 1.000 mol $CH_4(g)$ under the same conditions, using data from Table 2C.4.

Answer: $|q_p| = 890$ kJ, $|w_{max}| = 813$ kJ

(d) Some remarks on the Gibbs energy

The Gibbs energy (the 'free energy') is more common in chemistry than the Helmholtz energy because, at least in laboratory chemistry, we are usually more interested in changes occurring at constant pressure than at constant volume. The criterion $dG_{T,p} \le 0$ carries over into chemistry as the observation that, *at constant temperature and pressure, chemical reactions are spontaneous in the direction of decreasing Gibbs energy*. Therefore, if we want to know whether a reaction is spontaneous, the pressure and temperature being constant, we assess the change in the Gibbs energy. If G decreases as the reaction proceeds, then the reaction has a spontaneous tendency to convert the reactants into products. If G increases, then the reverse reaction is spontaneous. The criterion for equilibrium, when neither the forward nor reverse process is spontaneous, under conditions of constant temperature and pressure is

$$dG_{T,p}=0 \tag{3C.15}$$

The existence of spontaneous endothermic reactions provides an illustration of the role of G. In such reactions, H increases, the system rises spontaneously to states of higher enthalpy, and $dH>0$. Because the reaction is spontaneous we know that $dG<0$ despite $dH>0$; it follows that the entropy of the system increases so much that TdS outweighs dH in $dG=dH-TdS$. Endothermic reactions are therefore driven by the increase of entropy of the system, and this entropy change overcomes the reduction of entropy brought about in the surroundings by the inflow of heat into the system ($dS_{sur}=-dH/T$ at constant pressure).

(e) Maximum non-expansion work

The analogue of the maximum work interpretation of ΔA, and the origin of the name 'free energy', can be found for ΔG. In the following *Justification*, we show that at constant temperature and pressure, the maximum additional (non-expansion) work, $w_{add,max}$, is given by the change in Gibbs energy:

$$dw_{add,max}=dG \quad \textit{Constant temperature and pressure} \quad \text{Maximum non-expansion work} \tag{3C.16a}$$

The corresponding expression for a measurable change is

$$w_{add,max}=\Delta G \quad \textit{Constant temperature and pressure} \quad \text{Maximum non-expansion work} \tag{3C.16b}$$

This expression is particularly useful for assessing the electrical work that may be produced by fuel cells and electrochemical cells, and we shall see many applications of it.

Justification 3C.2 Maximum non-expansion work

Because $H=U+pV$, the change in enthalpy for a general change in conditions is

$$dH=dq+dw+d(pV)$$

The corresponding change in Gibbs energy ($G=H-TS$) is

$$dG=dH-TdS-SdT=dq+dw+d(pV)-TdS-SdT$$

When the change is isothermal we can set $dT=0$; then

$$dG=dq+dw+d(pV)-TdS$$

When the change is reversible, $dw=dw_{rev}$ and $dq=dq_{rev}=TdS$, so for a reversible, isothermal process

$$dG=TdS+dw_{rev}+d(pV)-TdS=dw_{rev}+d(pV)$$

The work consists of expansion work, which for a reversible change is given by $-pdV$, and possibly some other kind of work (for instance, the electrical work of pushing electrons through a circuit or of raising a column of liquid); this additional work we denote dw_{add}. Therefore, with $d(pV)=pdV+Vdp$,

$$dG=(-pdV+dw_{add,rev})+pdV+Vdp=dw_{add,rev}+Vdp$$

If the change occurs at constant pressure (as well as constant temperature), we can set $dp=0$ and obtain $dG=dw_{add,rev}$. Therefore, at constant temperature and pressure, $dw_{add,rev}=dG$. However, because the process is reversible, the work done must now have its maximum value, so eqn 3C.16 follows.

Example 3C.2 Calculating the maximum non-expansion work of a reaction

How much energy is available for sustaining muscular and nervous activity from the combustion of 1.00 mol of glucose molecules under standard conditions at 37 °C (blood temperature)? The standard entropy of reaction is $+182.4\,J\,K^{-1}\,mol^{-1}$.

Method The non-expansion work available from the reaction is equal to the change in standard Gibbs energy for the reaction ($\Delta_r G^\ominus$, a quantity defined more fully below). To calculate this quantity, it is legitimate to ignore the temperature-dependence of the reaction enthalpy, to obtain $\Delta_r H^\ominus$ from Table 2C.5, and to substitute the data into $\Delta_r G^\ominus=\Delta_r H^\ominus-T\Delta_r S^\ominus$.

Answer Because the standard reaction enthalpy is $-2808\,kJ\,mol^{-1}$, it follows that the standard reaction Gibbs energy is

$$\begin{aligned}\Delta_r G^\ominus &= -2808\,kJ\,mol^{-1}-(310\,K)\times(182.4\,J\,K^{-1}\,mol^{-1})\\ &= -2865\,kJ\,mol^{-1}\end{aligned}$$

Therefore, $w_{add,max}=-2865\,kJ$ for the combustion of 1 mol glucose molecules, and the reaction can be used to do up to 2865 kJ of non-expansion work. To place this result in perspective, consider that a person of mass 70 kg needs to do 2.1 kJ of work to climb vertically through 3.0 m; therefore, at least 0.13 g of glucose is needed to complete the task (and in practice significantly more).

Self-test 3C.5 How much non-expansion work can be obtained from the combustion of 1.00 mol $CH_4(g)$ under standard conditions at 298 K? Use $\Delta_r S^\ominus=-243\,J\,K^{-1}\,mol^{-1}$.

Answer: 818 kJ

3C.2 Standard molar Gibbs energies

Standard entropies and enthalpies of reaction can be combined to obtain the **standard Gibbs energy of reaction** (or 'standard reaction Gibbs energy'), $\Delta_r G^\ominus$:

$$\Delta_r G^\ominus = \Delta_r H^\ominus - T\Delta_r S^\ominus \quad \textit{Definition} \quad \text{Standard Gibbs energy of reaction} \quad (3C.17)$$

The standard Gibbs energy of reaction is the difference in standard molar Gibbs energies of the products and reactants in their standard states at the temperature specified for the reaction as written.

(a) Gibbs energies of formation

As in the case of standard reaction enthalpies, it is convenient to define the **standard Gibbs energies of formation**, $\Delta_f G^\ominus$, the standard reaction Gibbs energy for the formation of a compound from its elements in their reference states.[2] Standard Gibbs energies of formation of the elements in their reference states are zero, because their formation is a 'null' reaction. A selection of values for compounds is given in Table 3C.1. From the values there, it is a simple matter to obtain the standard Gibbs energy of reaction by taking the appropriate combination:

$$\Delta_r G^\ominus = \sum_{\text{Products}} \nu\Delta_f G_m^\ominus - \sum_{\text{Reactants}} \nu\Delta_f G_m^\ominus \quad \textit{Practical implementation} \quad \text{Standard Gibbs energy of reaction} \quad (3C.18a)$$

In the notation introduced in Topic 2C,

$$\Delta_r G^\ominus = \sum_J \nu_J \Delta_f G_m^\ominus(J) \quad (3C.18b)$$

Table 3C.1* Standard Gibbs energies of formation at 298 K, $\Delta_f G^\ominus/(\text{kJ mol}^{-1})$

	$\Delta_f G^\ominus/(\text{kJ mol}^{-1})$
Diamond, C(s)	+2.9
Benzene, C_6H_6(l)	+124.3
Methane, CH_4(g)	−50.7
Carbon dioxide, CO_2(g)	−394.4
Water, H_2O(l)	−237.1
Ammonia, NH_3(g)	−16.5
Sodium chloride, NaCl(s)	−384.1

* More values are given in the *Resource section*.

[2] The reference state of an element is defined in Topic 2C.

Brief illustration 3C.4 The standard reaction Gibbs energy

To calculate the standard Gibbs energy of the reaction $CO(g)+\frac{1}{2}O_2(g)\rightarrow CO_2(g)$ at 25 °C, we write

$$\begin{aligned}\Delta_r G^\ominus &= \Delta_f G^\ominus(CO_2,g) - \{\Delta_f G^\ominus(CO,g) + \tfrac{1}{2}\Delta_f G^\ominus(O_2,g)\}\\ &= -394.4\,\text{kJ mol}^{-1} - \{(-137.2) + \tfrac{1}{2}(0)\}\,\text{kJ mol}^{-1}\\ &= -257.2\,\text{kJ mol}^{-1}\end{aligned}$$

Self-test 3C.6 Calculate the standard reaction Gibbs energy for the combustion of CH_4(g) at 298 K.

Answer: −818 kJ mol^{-1}

Just as is done in Topics 2C and 3B, where it is acknowledged that solutions of cations cannot be prepared without their accompanying anions, we define one ion, conventionally the hydrogen ion, to have zero standard Gibbs energy of formation at all temperatures:

$$\Delta_f G^\ominus(H^+,aq) = 0 \quad \textit{Convention} \quad \text{Ions in solution} \quad (3C.19)$$

In essence, this definition adjusts the actual values of the Gibbs energies of formation of ions by a fixed amount, which is chosen so that the standard value for one of them, H^+(aq), has the value zero.

Brief illustration 3C.5 Gibbs energies of formation of ions

For the reaction

$$\tfrac{1}{2}H_2(g) + \tfrac{1}{2}Cl_2(g) \rightarrow H^+(aq) + Cl^-(aq) \quad \Delta_r G^\ominus = -131.23\,\text{kJ mol}^{-1}$$

we can write

$$\Delta_r G^\ominus = \Delta_f G^\ominus(H^+,aq) + \Delta_f G^\ominus(Cl^-,aq) = \Delta_f G^\ominus(Cl^-,aq)$$

and hence identify $\Delta_f G^\ominus(Cl^-,aq)$ as −131.23 kJ mol^{-1}.

Self-test 3C.7 Evaluate $\Delta_f G^\ominus(Ag^+,aq)$ from $Ag(s)+\frac{1}{2}Cl_2(g)\rightarrow Ag^+(aq)+Cl^-(aq)$, $\Delta_r G^\ominus = -54.12\,\text{kJ mol}^{-1}$

Answer: +77.11 kJ mol^{-1}

The factors responsible for the magnitude of the Gibbs energy of formation of an ion in solution can be identified by analysing it in terms of a thermodynamic cycle. As an illustration, we consider the standard Gibbs energy of formation of Cl^- in water, which is −131 kJ mol^{-1}. We do so by treating the formation reaction

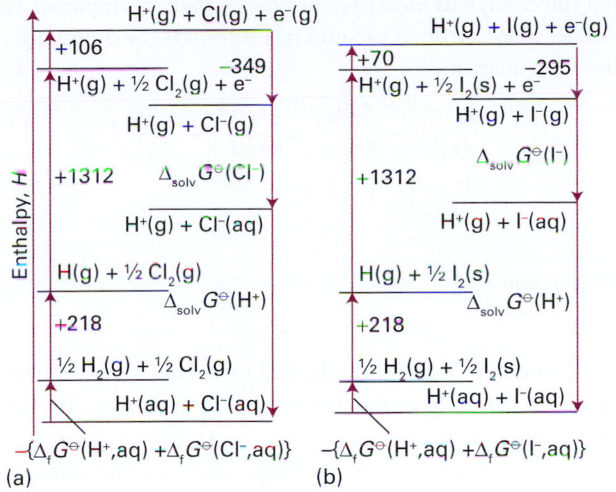

Figure 3C.3 The thermodynamic cycles for the discussion of the Gibbs energies of solvation (hydration) and formation of (a) chloride ions, (b) iodide ions in aqueous solution. The changes in Gibbs energies around the cycle sum to zero because *G* is a state function.

$$\tfrac{1}{2}H_2(g)+\tfrac{1}{2}X_2(g)\rightarrow H^+(aq)+X^-(aq)$$

as the outcome of the sequence of steps shown in Fig. 3C.3 (with values taken from the *Resource section*). The sum of the Gibbs energies for all the steps around a closed cycle is zero, so

$$\Delta_f G^\ominus(Cl^-,aq)=1287\ \text{kJ mol}^{-1}+\Delta_{solv}G^\ominus(H^+)+\Delta_{solv}G^\ominus(Cl^-)$$

The standard Gibbs energies of formation of the gas-phase ions are unknown. We have therefore used ionization energies and electron affinities and have assumed that any differences from the Gibbs energies arising from conversion to enthalpy and the inclusion of entropies to obtain Gibbs energies in the formation of H^+ are cancelled by the corresponding terms in the electron gain of X. The conclusions from the cycles are therefore only approximate. An important point to note is that the value of $\Delta_f G^\ominus$ of an ion X is not determined by the properties of X alone but includes contributions from the dissociation, ionization, and hydration of hydrogen.

(b) The Born equation

Gibbs energies of solvation of individual ions may be estimated from an equation derived by Max Born, who identified $\Delta_{solv}G^\ominus$ with the electrical work of transferring an ion from a vacuum into the solvent treated as a continuous dielectric of relative permittivity ε_r. The resulting **Born equation**, which is derived in the following *Justification*, is

$$\Delta_{solv}G^\ominus=-\frac{z_i^2 e^2 N_A}{8\pi\varepsilon_0 r_i}\left(1-\frac{1}{\varepsilon_r}\right) \quad \text{Born equation} \quad (3C.20a)$$

where z_i is the charge number of the ion and r_i its radius (N_A is Avogadro's constant). Note that $\Delta_{solv}G^\ominus<0$, and that $\Delta_{solv}G^\ominus$ is strongly negative for small, highly charged ions in media of high relative permittivity. For water, for which $\varepsilon_r=78.54$ at 25 °C,

$$\Delta_{solv}G^\ominus=-\frac{z_i^2}{r_i/\text{pm}}\times 6.86\times 10^4\ \text{kJ mol}^{-1} \quad (3C.20b)$$

Brief illustration 3C.6 The Born equation

To see how closely the Born equation reproduces the experimental data, we calculate the difference in the values of $\Delta_f G^\ominus$ for Cl^- and I^- in water at 25 °C, given their radii as 181 pm and 220 pm, respectively, is

$$\Delta_{solv}G^\ominus(Cl^-)-\Delta_{solv}G^\ominus(I^-)=-\left(\frac{1}{181}-\frac{1}{220}\right)\times 6.86\times 10^4\ \text{kJ mol}^{-1}$$
$$=-67\ \text{kJ mol}^{-1}$$

This estimated difference is in good agreement with the experimental difference, which is −61 kJ mol⁻¹.

Self-test 3C.8 Estimate the value of $\Delta_{solv}G^\ominus(Cl^-)-\Delta_{solv}G^\ominus(Br^-)$ in water from experimental data and from the Born equation.

Answer: −26 kJ mol⁻¹ experimental; −29 kJ mol⁻¹ calculated

Justification 3C.3 The Born equation

The strategy of the calculation is to identify the Gibbs energy of solvation with the work of transferring an ion from a vacuum into the solvent. That work is calculated by taking the difference of the work of charging an ion when it is in the solution and the work of charging the same ion when it is in a vacuum.

The Coulomb interaction between two charges Q_1 and Q_2 separated by a distance r is described by the Coulombic potential energy:

$$V(r)=\frac{Q_1 Q_2}{4\pi\varepsilon r}$$

where ε is the medium's permittivity. The relative permittivity (formerly called the 'dielectric constant') of a substance is defined as $\varepsilon_r=\varepsilon/\varepsilon_0$. Ions do not interact as strongly in a solvent of high relative permittivity (such as water, with $\varepsilon_r=80$ at 293 K) as they do in a solvent of lower relative permittivity (such as ethanol, with $\varepsilon_r=25$ at 293 K). See Topic 16A for more details. The potential energy of a charge Q_1 in the presence of a charge Q_2 can be expressed in terms of the Coulomb potential, ϕ:

$$V(r)=Q_1\phi(r) \qquad \phi(r)=\frac{Q_2}{4\pi\varepsilon r}$$

We model an ion as a sphere of radius r_i immersed in a medium of permittivity ε. The charge of the sphere is Q, the electric potential, ϕ, at its surface is the same as the potential due to a point charge at its centre, so we can use the last expression and write

$$\phi(r_i)=\frac{Q}{4\pi\varepsilon r_i}$$

The work of bringing up a charge dQ to the sphere is $\phi(r_i)\mathrm{d}Q$. Therefore, the total work of charging the sphere from 0 to z_ie is

$$w=\int_0^{z_ie}\phi(r_i)\mathrm{d}Q=\frac{1}{4\pi\varepsilon r_i}\int_0^{z_ie}Q\mathrm{d}Q=\frac{z_i^2e^2}{8\pi\varepsilon r_i}$$

This electrical work of charging, when multiplied by Avogadro's constant, is the molar Gibbs energy for charging the ions.

The work of charging an ion in a vacuum is obtained by setting $\varepsilon=\varepsilon_0$, the vacuum permittivity. The corresponding value for charging the ion in a medium is obtained by setting $\varepsilon=\varepsilon_r\varepsilon_0$, where ε_r is the relative permittivity of the medium. It follows that the change in molar Gibbs energy that accompanies the transfer of ions from a vacuum to a solvent is the difference of these two quantities:

$$\Delta_{\mathrm{solv}}G^{\ominus}=\frac{z_i^2e^2N_A}{8\pi\varepsilon r_i}-\frac{z_i^2e^2N_A}{8\pi\varepsilon_0 r_i}=\frac{z_i^2e^2N_A}{8\pi\varepsilon_r\varepsilon_0 r_i}-\frac{z_i^2e^2N_A}{8\pi\varepsilon_0 r_i}$$
$$=-\frac{z_i^2e^2N_A}{8\pi\varepsilon_0 r_i}\left(1-\frac{1}{\varepsilon_r}\right)$$

which is eqn 3B.20.

Calorimetry (for ΔH directly, and for S via heat capacities) is only one of the ways of determining Gibbs energies. They may also be obtained from equilibrium constants (Topic 6A) and electrochemical measurements (Topic 6D), and for gases they may be calculated using data from spectroscopic observations (Topic 15F).

Checklist of concepts

- ☐ 1. The **Clausius inequality** implies a number of criteria for spontaneous change under a variety of conditions that may be expressed in terms of the properties of the system alone; they are summarized by introducing the Helmholtz and Gibbs energies.
- ☐ 2. A **spontaneous process** at constant temperature and volume is accompanied by a decrease in the Helmholtz energy.
- ☐ 3. The change in the Helmholtz function is equal to the **maximum work** accompanying a process at constant temperature.
- ☐ 4. A spontaneous process at constant temperature and pressure is accompanied by a decrease in the Gibbs energy.
- ☐ 5. The change in the Gibbs function is equal to the **maximum non-expansion work** accompanying a process at constant temperature and pressure.
- ☐ 6. **Standard Gibbs energies of formation** are used to calculate the standard Gibbs energies of reactions.
- ☐ 7. The standard Gibbs energies of formation of ions may be estimated from a thermodynamic cycle and the **Born equation**.

Checklist of equations

Property	Equation	Comment	Equation number
Criteria of spontaneity	(a) $\mathrm{d}S_{U,V}\geq 0$ (b) $\mathrm{d}U_{S,V}\leq 0$	Constant volume (etc.)*	3C.4
	(a) $\mathrm{d}S_{H,p}\geq 0$ (b) $\mathrm{d}H_{S,p}\leq 0$	Constant pressure (etc.)	3C.6
Helmholtz energy	$A=U-TS$	Definition	3C.7
Gibbs energy	$G=H-TS$	Definition	3C.8
	(a) $\mathrm{d}A_{T,V}\leq 0$ (b) $\mathrm{d}G_{T,p}\leq 0$	Constant temperature (etc.)	3C.10
Equilibrium	$\mathrm{d}A_{T,V}=0$	Constant volume (etc.)	3C.11
Maximum work	$\mathrm{d}w_{\max}=\mathrm{d}A$, $w_{\max}=\Delta A$	Constant temperature	3C.12

Property	Equation	Comment	Equation number
Equilibrium	$dG_{T,p}=0$	Constant pressure (etc.)	3C.15
Maximum non-expansion work	$dw_{\text{add,max}}=dG$, $w_{\text{add,max}}=\Delta G$	Constant temperature and pressure	3C.16
Standard Gibbs energy of reaction	$\Delta_r G^{\ominus}=\Delta_r H^{\ominus}-T\Delta_r S^{\ominus}$	Definition	3C.17
	$\Delta_r G^{\ominus}=\sum_J \nu_J \Delta_f G_m^{\ominus}(J)$	Practical implementation	3C.18
Ions in solution	$\Delta_f G^{\ominus}(H^+,aq)=0$	Convention	3C.19
Born equation	$\Delta_{\text{solv}} G^{\ominus}=-(z_i^2 e^2 N_A/8\pi\varepsilon_0 r_i)(1-1/\varepsilon_r)$	Solvent a continuum	3C.20

* 'etc.' indicates that the conditions are as expressed by the subscripts.

3D Combining the First and Second Laws

Contents

➤ Why do you need to know this material?

The First and Second Laws of thermodynamics are both relevant to the behaviour of bulk matter, and we can bring the whole force of thermodynamics to bear on a problem by setting up a formulation that combines them.

➤ What is the key idea?

The fact that infinitesimal changes in thermodynamic functions are exact differentials leads to relations between a variety of properties.

➤ What do you need to know already?

You need to be aware of the definitions of the state functions U (Topic 2A), H (Topic 2B), S (Topic 3A), and A and G (Topic 3C). The mathematical derivations in this Topic draw frequently on the properties of partial derivatives and exact differentials, which are described in *Mathematical background* 2.

The First Law of thermodynamics may be written $dU = dq + dw$. For a reversible change in a closed system of constant composition, and in the absence of any additional (non-expansion) work, we may set $dw_{rev} = -pdV$ and (from the definition of entropy) $dq_{rev} = TdS$, where p is the pressure of the system and T its temperature. Therefore, for a reversible change in a closed system,

$$dU = TdS - pdV \qquad \text{The fundamental equation} \qquad (3D.1)$$

However, because dU is an exact differential, its value is independent of path. Therefore, the same value of dU is obtained whether the change is brought about irreversibly or reversibly. Consequently, *eqn 3D.1 applies to any change—reversible or irreversible—of a closed system that does no additional (non-expansion) work*. We shall call this combination of the First and Second Laws the **fundamental equation**.

The fact that the fundamental equation applies to both reversible and irreversible changes may be puzzling at first sight. The reason is that only in the case of a reversible change may TdS be identified with dq and $-pdV$ with dw. When the change is irreversible, $TdS > dq$ (the Clausius inequality) and $-pdV > dw$. The sum of dw and dq remains equal to the sum of TdS and $-pdV$, provided the composition is constant.

3D.1 Properties of the internal energy

Equation 3D.1 shows that the internal energy of a closed system changes in a simple way when either S or V is changed ($dU \propto dS$ and $dU \propto dV$). These simple proportionalities suggest that U is best regarded as a function of S and V. We could regard U as a function of other variables, such as S and p or T and V, because they are all interrelated; but the simplicity of the fundamental equation suggests that $U(S,V)$ is the best choice.

The *mathematical* consequence of U being a function of S and V is that we can express an infinitesimal change dU in terms of changes dS and dV by

$$dU = \left(\frac{\partial U}{\partial S}\right)_V dS + \left(\frac{\partial U}{\partial V}\right)_S dV \qquad (3D.2)$$

The two partial derivatives are the slopes of the plots of U against S and V, respectively. When this expression is compared term-by-term to the *thermodynamic* relation, eqn 3D.1, we see that for systems of constant composition,

$$\left(\frac{\partial U}{\partial S}\right)_V = T \quad \left(\frac{\partial U}{\partial V}\right)_S = -p \qquad (3D.3)$$

The first of these two equations is a purely thermodynamic definition of temperature as the ratio of the changes in the internal energy (a First-Law concept) and entropy (a Second-Law concept) of a constant-volume, closed, constant-composition system. We are beginning to generate relations between the properties of a system and to discover the power of thermodynamics for establishing unexpected relations.

(a) The Maxwell relations

An infinitesimal change in a function $f(x,y)$ can be written $df=gdx+hdy$, where g and h may be functions of x and y. The mathematical criterion for df being an exact differential (in the sense that its integral is independent of path) is that

$$\left(\frac{\partial g}{\partial y}\right)_x = \left(\frac{\partial h}{\partial x}\right)_y \qquad (3D.4)$$

This criterion is discussed in *Mathematical background* 2. Because the fundamental equation, eqn 3D.1, is an expression for an exact differential, the functions multiplying dS and dV (namely T and $-p$) must pass this test. Therefore, it must be the case that

$$\left(\frac{\partial T}{\partial V}\right)_S = -\left(\frac{\partial p}{\partial S}\right)_V \qquad \text{A Maxwell relation} \qquad (3D.5)$$

We have generated a relation between quantities which, at first sight, would not seem to be related.

Equation 3D.5 is an example of a **Maxwell relation**. However, apart from being unexpected, it does not look particularly interesting. Nevertheless, it does suggest that there might be other similar relations that are more useful. Indeed, we can use the fact that H, G, and A are all state functions to derive three more Maxwell relations. The argument to obtain them runs in the same way in each case: because H, G, and A are state functions, the expressions for dH, dG, and dA satisfy relations like eqn 3D.4. All four relations are listed in Table 3D.1.

Example 3D.1 Using the Maxwell relations

Use the Maxwell relations in Table 3D.1 to show that the entropy of a perfect gas is proportional to ln V.

Method The natural place to start, given that you are invited to use the Maxwell relations, is by considering the relation for $(\partial S/\partial V)_T$, as that differential coefficient shows how the entropy varies with volume at constant temperature. Be alert for an opportunity to use the perfect gas equation of state.

Answer From Table 3D.1,

$$\left(\frac{\partial S}{\partial V}\right)_T = \left(\frac{\partial p}{\partial T}\right)_V$$

Now use the perfect gas equation of state to write

$$\left(\frac{\partial p}{\partial T}\right)_V = \left(\frac{\partial (nRT/V)}{\partial T}\right)_V = \frac{nR}{V}$$

At this point, we can write

$$\left(\frac{\partial S}{\partial V}\right)_T = \frac{nR}{V}$$

and therefore, at constant temperature

$$\int dS = nR\int \frac{dV}{V} = nR\ln V + \text{constant}$$

The integral of the left is S+constant, which completes the demonstration.

Self-test 3D.1 How does the entropy depend on the volume of a van der Waals gas? Discuss.

Answer: S varies as $nR\ln(V-nb)$; molecules in a smaller available volume

Table 3D.1 The Maxwell relations

State function	Exact differential	Maxwell relation
U	$dU=TdS-pdV$	$\left(\frac{\partial T}{\partial V}\right)_S=-\left(\frac{\partial p}{\partial S}\right)_V$
H	$dH=TdS+Vdp$	$\left(\frac{\partial T}{\partial p}\right)_S=\left(\frac{\partial V}{\partial S}\right)_p$
A	$dA=-pdV-SdT$	$\left(\frac{\partial p}{\partial T}\right)_V=\left(\frac{\partial S}{\partial V}\right)_T$
G	$dG=Vdp-SdT$	$\left(\frac{\partial V}{\partial T}\right)_p=-\left(\frac{\partial S}{\partial p}\right)_T$

(b) The variation of internal energy with volume

The quantity $\pi_T=(\partial U/\partial V)_T$, which represents how the internal energy changes as the volume of a system is changed isothermally, plays a central role in the manipulation of the First Law, and in Topic 2D we use the relation

$$\pi_T = T\left(\frac{\partial p}{\partial T}\right)_V - p \qquad \text{A thermodynamic equation of state} \qquad (3D.6)$$

This relation is called a **thermodynamic equation of state** because it is an expression for pressure in terms of a variety of thermodynamic properties of the system. We are now ready to derive it by using a Maxwell relation.

Justification 3D.1 The thermodynamic equation of state

We obtain an expression for the coefficient π_T by dividing both sides of eqn 3D.1 by dV, imposing the constraint of constant temperature, which gives

$$\overbrace{\left(\frac{\partial U}{\partial V}\right)_T}^{\pi_T}=\overbrace{\left(\frac{\partial U}{\partial S}\right)_V}^{T}\left(\frac{\partial S}{\partial V}\right)_T+\overbrace{\left(\frac{\partial U}{\partial V}\right)_S}^{p}$$

Next, we introduce the two relations in eqn 3D.3 (as indicated by the annotations) and the definition of π_T to obtain

$$\pi_T=T\left(\frac{\partial S}{\partial V}\right)_T-p$$

The third Maxwell relation in Table 3D.1 turns $(\partial S/\partial V)_T$ into $(\partial p/\partial T)_V$, which completes the derivation of eqn 3D.6.

Example 3D.2 Deriving a thermodynamic relation

Show thermodynamically that $\pi_T=0$ for a perfect gas, and compute its value for a van der Waals gas.

Method Proving a result 'thermodynamically' means basing it entirely on general thermodynamic relations and equations of state, without drawing on molecular arguments (such as the existence of intermolecular forces). We know that for a perfect gas, $p=nRT/V$, so this relation should be used in eqn 3D.6. Similarly, the van der Waals equation is given in Table 1C.3, and for the second part of the question it should be used in eqn 3D.6.

Answer For a perfect gas we write

$$\left(\frac{\partial p}{\partial T}\right)_V=\left(\frac{\partial nRT/V}{\partial T}\right)_V=\frac{nR}{V}$$

Then, eqn 3D.6 becomes

$$\pi_T=\frac{nRT}{V}-p=0$$

The equation of state of a van der Waals gas is

$$p=\frac{nRT}{V-nb}-a\frac{n^2}{V^2}$$

Because a and b are independent of temperature,

$$\left(\frac{\partial p}{\partial T}\right)_V=\left(\frac{\partial nRT/(V-nb)}{\partial T}\right)_V=\frac{nR}{V-nb}$$

Therefore, from eqn 3D.6,

$$\pi_T=\frac{nRT}{V-nb}-p=\frac{nRT}{V-nb}-\left(\frac{nRT}{V-nb}-a\frac{n^2}{V^2}\right)=a\frac{n^2}{V^2}$$

This result for π_T implies that the internal energy of a van der Waals gas increases when it expands isothermally (that is, $(\partial U/\partial V)_T>0$), and that the increase is related to the parameter a, which models the attractive interactions between the particles. A larger molar volume, corresponding to a greater average separation between molecules, implies weaker mean intermolecular attractions, so the total energy is greater.

Self-test 3D.2 Calculate π_T for a gas that obeys the virial equation of state (Table 1C.3).

Answer: $\pi_T=RT^2(\partial B/\partial T)_V/V_m^2+\cdots$

3D.2 Properties of the Gibbs energy

The same arguments that we have used for U can be used for the Gibbs energy $G=H-TS$. They lead to expressions showing how G varies with pressure and temperature that are important for discussing phase transitions and chemical reactions.

(a) General considerations

When the system undergoes a change of state, G may change because H, T, and S all change and

$$\mathrm{d}G=\mathrm{d}H-\mathrm{d}(TS)=\mathrm{d}H-T\mathrm{d}S-S\mathrm{d}T$$

Because $H=U+pV$, we know that

$$\mathrm{d}H=\mathrm{d}U+\mathrm{d}(pV)=\mathrm{d}U+p\mathrm{d}V+V\mathrm{d}p$$

and therefore

$$\mathrm{d}G=\mathrm{d}U+p\mathrm{d}V+V\mathrm{d}p-T\mathrm{d}S-S\mathrm{d}T$$

For a closed system doing no non-expansion work, we can replace dU by the fundamental equation $\mathrm{d}U=T\mathrm{d}S-p\mathrm{d}V$ and obtain

$$\mathrm{d}G=T\mathrm{d}S-p\mathrm{d}V+p\mathrm{d}V+V\mathrm{d}p-T\mathrm{d}S-S\mathrm{d}T$$

Four terms now cancel on the right, and we conclude that for a closed system in the absence of non-expansion work and at constant composition

$$\mathrm{d}G=V\mathrm{d}p-S\mathrm{d}T \qquad \text{The fundamental equation of chemical thermodynamics} \qquad (3D.7)$$

This expression, which shows that a change in G is proportional to a change in p or T, suggests that G may be best regarded as a function of p and T. It may be regarded as the **fundamental equation of chemical thermodynamics** as it is so central to the application of thermodynamics to chemistry: it suggests that G is an important quantity in chemistry because the pressure and temperature are usually the variables under our control. In other words, G carries around the combined consequences of the First and Second Laws in a way that makes it particularly suitable for chemical applications.

The same argument that led to eqn 3D.3, when applied to the exact differential $dG = Vdp - SdT$, now gives

$$\left(\frac{\partial G}{\partial T}\right)_p = -S \qquad \left(\frac{\partial G}{\partial p}\right)_T = V \qquad \text{The variation of } G \text{ with } T \text{ and } p \qquad (3D.8)$$

These relations show how the Gibbs energy varies with temperature and pressure (Fig. 3D.1). The first implies that:

- Because $S>0$ for all substances, G always *decreases* when the temperature is raised (at constant pressure and composition).
- Because $(\partial G/\partial T)_p$ becomes more negative as S increases, G decreases most sharply with increasing temperature when the entropy of the system is large.

Physical interpretation

Therefore, the Gibbs energy of the gaseous phase of a substance, which has a high molar entropy, is more sensitive to temperature than its liquid and solid phases (Fig. 3D.2). Similarly, the second relation implies that:

- Because $V>0$ for all substances, G always *increases* when the pressure of the system is increased (at constant temperature and composition).

Physical interpretation

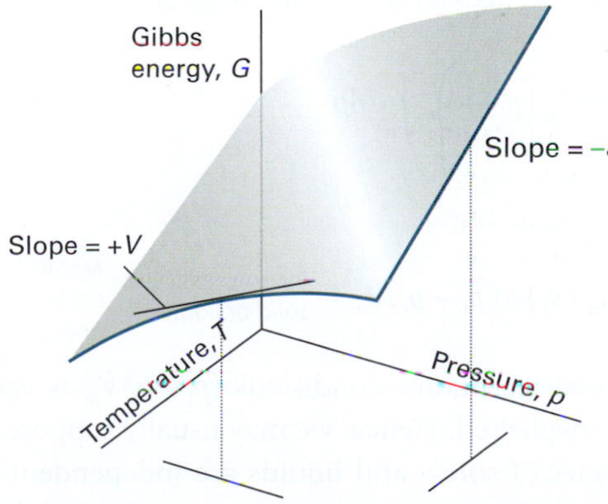

Figure 3D.1 The variation of the Gibbs energy of a system with (a) temperature at constant pressure and (b) pressure at constant temperature. The slope of the former is equal to the negative of the entropy of the system and that of the latter is equal to the volume.

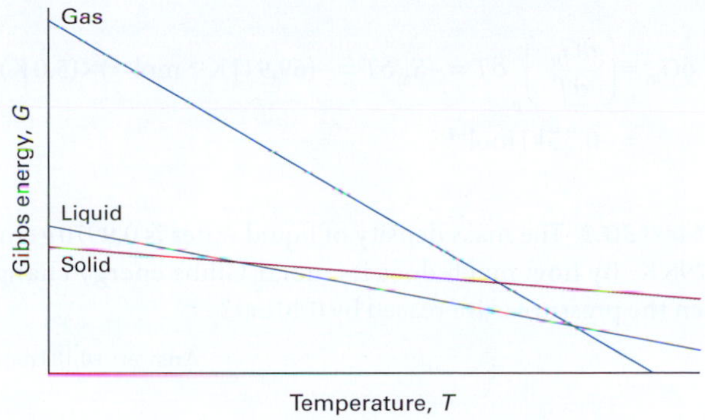

Figure 3D.2 The variation of the Gibbs energy with the temperature is determined by the entropy. Because the entropy of the gaseous phase of a substance is greater than that of the liquid phase, and the entropy of the solid phase is smallest, the Gibbs energy changes most steeply for the gas phase, followed by the liquid phase, and then the solid phase of the substance.

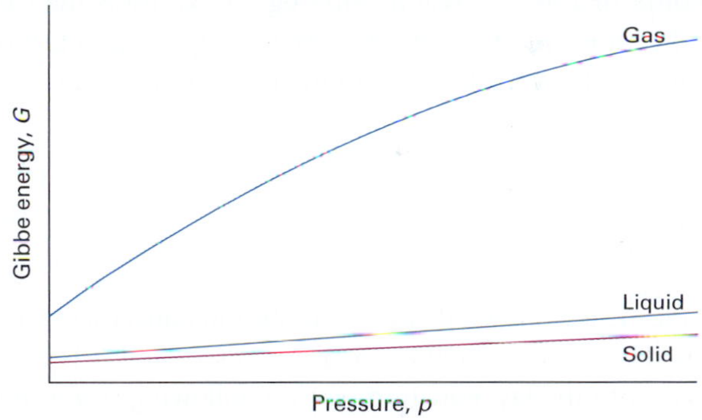

Figure 3D.3 The variation of the Gibbs energy with the pressure is determined by the volume of the sample. Because the volume of the gaseous phase of a substance is greater than that of the same amount of liquid phase, and the entropy of the solid phase is smallest (for most substances), the Gibbs energy changes most steeply for the gas phase, followed by the liquid phase, and then the solid phase of the substance. Because the volumes of the solid and liquid phases of a substance are similar, their molar Gibbs energies vary by similar amounts as the pressure is changed.

- Because $(\partial G/\partial p)_T$ increases with V, G is more sensitive to pressure when the volume of the system is large.

Because the molar volume of the gaseous phase of a substance is greater than that of its condensed phases, the molar Gibbs energy of a gas is more sensitive to pressure than its liquid and solid phases (Fig. 3D.3).

Brief illustration 3D.1 The variation of molar Gibbs energy

The standard molar entropy of liquid water at 298 K is 69.91 J K^{-1} mol^{-1}. It follows that when the temperature is increased by 5.0 K, the molar Gibbs energy changes by

$$\delta G_m \approx \left(\frac{\partial G_m}{\partial T}\right)_p \delta T = -S_m \delta T = -(69.91\,\mathrm{J\,K^{-1}\,mol^{-1}}) \times (5.0\,\mathrm{K})$$
$$= -0.35\,\mathrm{kJ\,mol^{-1}}$$

Self-test 3D.3 The mass density of liquid water is 0.9970 g cm^{-3} at 298 K. By how much does its molar Gibbs energy change when the pressure is increased by 0.10 bar?

Answer: +0.18 J mol^{-1}

(b) The variation of the Gibbs energy with temperature

Because the equilibrium composition of a system depends on the Gibbs energy, to discuss the response of the composition to temperature we need to know how G varies with temperature.

The first relation in eqn 3D.8, $(\partial G/\partial T)_p = -S$, is our starting point for this discussion. Although it expresses the variation of G in terms of the entropy, we can express it in terms of the enthalpy by using the definition of G to write $S = (H - G)/T$. Then

$$\left(\frac{\partial G}{\partial T}\right)_p = \frac{G-H}{T} \tag{3D.9}$$

In Topic 6A it is shown that the equilibrium constant of a reaction is related to G/T rather than to G itself, and it is easy to deduce from the last equation (see the following *Justification*) that

$$\left(\frac{\partial G/T}{\partial T}\right)_p = -\frac{H}{T^2} \qquad \text{Gibbs–Helmholtz equation} \tag{3D.10}$$

This expression is called the **Gibbs–Helmholtz equation**. It shows that if we know the enthalpy of the system, then we know how G/T varies with temperature.

Justification 3D.2 The Gibbs–Helmholtz equation

First, we note that

$$\left(\frac{\partial G/T}{\partial T}\right)_p = \frac{1}{T}\left(\frac{\partial G}{\partial T}\right)_p + G\frac{\mathrm{d}(1/T)}{\mathrm{d}T} = \frac{1}{T}\left(\frac{\partial G}{\partial T}\right)_p - \frac{G}{T^2}$$
$$= \frac{1}{T}\left\{\left(\frac{\partial G}{\partial T}\right)_p - \frac{G}{T}\right\}$$

Then we use eqn 3D.9 to write

$$\left(\frac{\partial G}{\partial T}\right)_p - \frac{G}{T} = \frac{G-H}{T} - \frac{G}{T} = -\frac{H}{T}$$

When this expression is substituted in the preceding one, we obtain eqn 3D.10.

The Gibbs–Helmholtz equation is most useful when it is applied to changes, including changes of physical state and chemical reactions at constant pressure. Then, because $\Delta G = G_f - G_i$ for the change of Gibbs energy between the final and initial states and because the equation applies to both G_f and G_i, we can write

$$\left(\frac{\partial \Delta G/T}{\partial T}\right)_p = -\frac{\Delta H}{T^2} \tag{3D.11}$$

This equation shows that if we know the change in enthalpy of a system that is undergoing some kind of transformation (such as vaporization or reaction), then we know how the corresponding change in Gibbs energy varies with temperature. As we shall see, this is a crucial piece of information in chemistry.

(c) The variation of the Gibbs energy with pressure

To find the Gibbs energy at one pressure in terms of its value at another pressure, the temperature being constant, we set $\mathrm{d}T = 0$ in eqn 3D.7, which gives $\mathrm{d}G = V\mathrm{d}p$, and integrate:

$$G(p_f) = G(p_i) + \int_{p_i}^{p_f} V\mathrm{d}p \tag{3D.12a}$$

For molar quantities,

$$G_m(p_f) = G_m(p_i) + \int_{p_i}^{p_f} V_m\mathrm{d}p \tag{3D.12b}$$

This expression is applicable to any phase of matter, but to evaluate it we need to know how the molar volume, V_m, depends on the pressure.

The molar volume of a condensed phase changes only slightly as the pressure changes (Fig. 3D.4), so we can treat V_m as a constant and take it outside the integral:

$$G_m(p_f) = G_m(p_i) + V_m\int_{p_i}^{p_f} \mathrm{d}p$$

That is,

$$G_m(p_f) = G_m(p_i) + (p_f - p_i)V_m \qquad \textit{Incompressible solid or liquid} \quad \text{Molar Gibbs energy} \tag{3D.13}$$

Under normal laboratory conditions $(p_f - p_i)V_m$ is very small and may be neglected. Hence, we may usually suppose that the Gibbs energies of solids and liquids are independent of pressure. However, if we are interested in geophysical problems, then because pressures in the Earth's interior are huge, their effect on the Gibbs energy cannot be ignored. If the pressures are so great that there are substantial volume changes over the range of integration, then we must use the complete expression, eqn 3D.12.

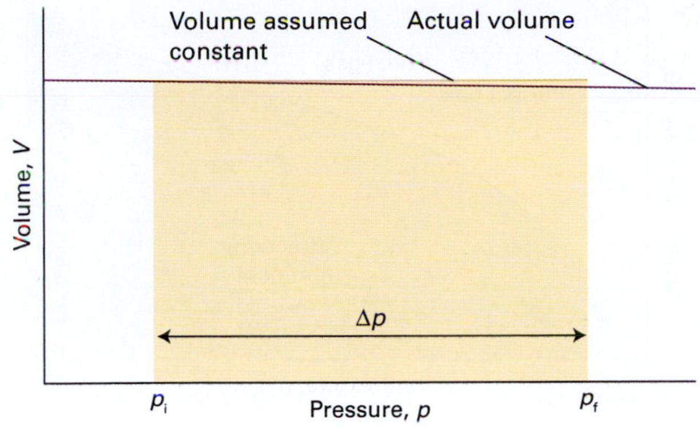

Figure 3D.4 The difference in Gibbs energy of a solid or liquid at two pressures is equal to the rectangular area shown. We have assumed that the variation of volume with pressure is negligible.

Example 3D.3 Evaluating the pressure dependence of a Gibbs energy of transition

Suppose that for a certain phase transition of a solid $\Delta_{trs}V = +1.0\,cm^3\,mol^{-1}$ is independent of pressure. By how much does that Gibbs energy of transition change when the pressure is increased from 1.0 bar (1.0×10^5 Pa) to 3.0 Mbar (3.0×10^{11} Pa)?

Method Start with eqn 3D.12b to obtain expressions for the Gibbs energy of each of the phases 1 and 2 of the solid

$$G_{m,1}(p_f)=G_{m,1}(p_i)+\int_{p_i}^{p_f} V_{m,1}dp$$
$$G_{m,2}(p_f)=G_{m,2}(p_i)+\int_{p_i}^{p_f} V_{m,2}dp$$

Now subtract the second expression from the first, noting that $G_{m,2}-G_{m,1}=\Delta_{trs}G$ and $V_{m,2}-V_{m,1}=\Delta_{trs}V$:

$$\Delta_{trs}G_m(p_f)=\Delta_{trs}G_m(p_i)+\int_{p_i}^{p_f}\Delta_{trs}V_m dp$$

Use the data to complete the calculation.

Answer Because $\Delta_{trs}V$ is independent of pressure, the expression above simplifies to

$$\begin{aligned}\Delta_{trs}G_m(p_f)&=\Delta_{trs}G_m(p_i)+\Delta_{trs}V_m\int_{p_i}^{p_f}dp\\&=\Delta_{trs}G_m(p_i)+\Delta_{trs}V_m(p_f-p_i)\end{aligned}$$

Inserting the data gives

$$\begin{aligned}\Delta_{trs}G(3\,\text{Mbar})&=\Delta_{trs}G(1\,\text{bar})+(1.0\times10^{-6}\,\text{m}^3\,\text{mol}^{-1})\times\\&\quad(3.0\times10^{11}\,\text{Pa}-1.0\times10^{5}\,\text{Pa})\\&=\Delta_{trs}G(1\,\text{bar})+3.0\times10^{2}\,\text{kJ}\,\text{mol}^{-1}\end{aligned}$$

where we have used $1\,Pa\,m^3 = 1\,J$.

Self-test 3D.4 Calculate the change in G_m for ice at −10 °C, with density 917 kg m^{-3}, when the pressure is increased from 1.0 bar to 2.0 bar.

Answer: +2.0 J mol^{-1}

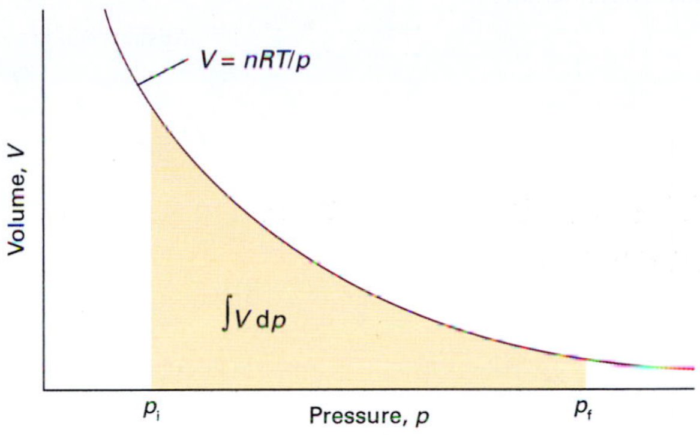

Figure 3D.5 The difference in Gibbs energy for a perfect gas at two pressures is equal to the area shown below the perfect-gas isotherm.

The molar volumes of gases are large, so the Gibbs energy of a gas depends strongly on the pressure. Furthermore, because the volume also varies markedly with the pressure, we cannot treat it as a constant in the integral in eqn 3D.12b (Fig. 3D.5). For a perfect gas we substitute $V_m = RT/p$ into the integral, treat RT as a constant, and find

$$G_m(p_f)=G_m(p_i)+RT\int_{p_i}^{p_f}\frac{1}{p}dp=G_m(p_i)+RT\ln\frac{p_f}{p_i} \qquad (3D.14)$$

This expression shows that when the pressure is increased tenfold at room temperature, the molar Gibbs energy increases by $RT \ln 10 \approx 6\,kJ\,mol^{-1}$. It also follows from this equation that if we set $p_i = p^\ominus$ (the standard pressure of 1 bar), then the molar Gibbs energy of a perfect gas at a pressure p (set $p_f = p$) is related to its standard value by

$$G_m(p)=G_m^\ominus+RT\ln\frac{p}{p^\ominus} \qquad \textit{Perfect gas} \quad \text{Molar Gibbs energy} \qquad (3D.15)$$

Brief illustration 3D.2 The pressure dependence of the Gibbs energy of a gas

Suppose we are interested in the molar Gibbs energy of water vapour (treated as a perfect gas) when the pressure is increased isothermally from 1.0 bar to 2.0 bar at 298 K. According to eqn 3D.14

$$\begin{aligned}G_m(2.0\,\text{bar})&=G_m^\ominus(1.0\,\text{bar})+(8.3145\,\text{J}\,\text{K}^{-1}\,\text{mol}^{-1})\times(298\,\text{K})\times\\&\quad\ln\left(\frac{2.0\,\text{bar}}{1.0\,\text{bar}}\right)=G_m^\ominus(1.0\,\text{bar})+1.7\,\text{kJ}\,\text{mol}^{-1}\end{aligned}$$

Note that whereas the change in molar Gibbs energy for a condensed phase is a few joules per mole, for a gas the change is of the order of kilojoules per mole

Self-test 3D.5 By how much does the molar Gibbs energy of a perfect gas differ from its standard value at 298 K when its pressure is 0.10 bar?

Answer: $-5.7\,\text{kJ mol}^{-1}$

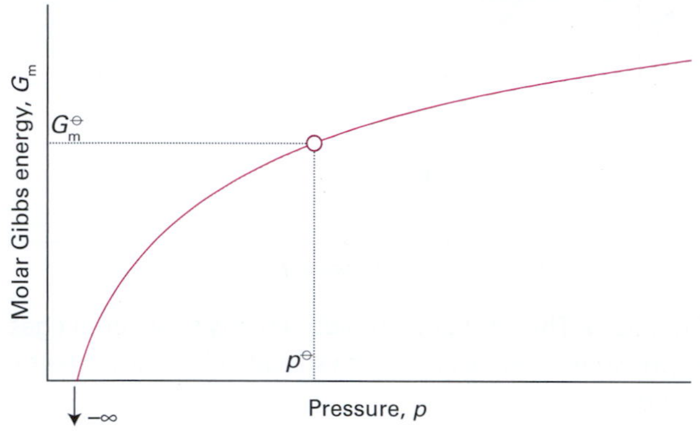

Figure 3D.6 The molar Gibbs energy of a perfect gas varies as ln p, and the standard state is reached at $p^\ominus$. Note that, as $p \to 0$, the molar Gibbs energy becomes negatively infinite.

The logarithmic dependence of the molar Gibbs energy on the pressure predicted by eqn 3D.15 is illustrated in Fig. 3D.6. This very important expression, the consequences of which we unfold in the following chapters, applies to perfect gases (which is usually a good enough approximation). The following section shows how to accommodate imperfections.

(d) The fugacity

At various stages in the development of physical chemistry it is necessary to switch from a consideration of idealized systems to real systems. In many cases it is desirable to preserve the form of the expressions that have been derived for an idealized system. Then deviations from the idealized behaviour can be expressed most simply. For instance, the pressure-dependence of the molar Gibbs energy of a real gas might resemble that shown in Fig. 3D.7. To adapt eqn 3D.14 to this case, we replace the true pressure, p, by an effective pressure, called the **fugacity**,[1] f, and write

$$G_m = G_m^\ominus + RT\ln(f/p^\ominus) \qquad \textit{Definition} \quad \text{Fugacity} \qquad (3D.16)$$

The fugacity, a function of the pressure and temperature, is defined so that this relation is exactly true. A very similar approach is taken in the discussion of real solutions (Topic 5E),

[1] The name 'fugacity' comes from the Latin for 'fleetness' in the sense of 'escaping tendency'; fugacity has the same dimensions as pressure.

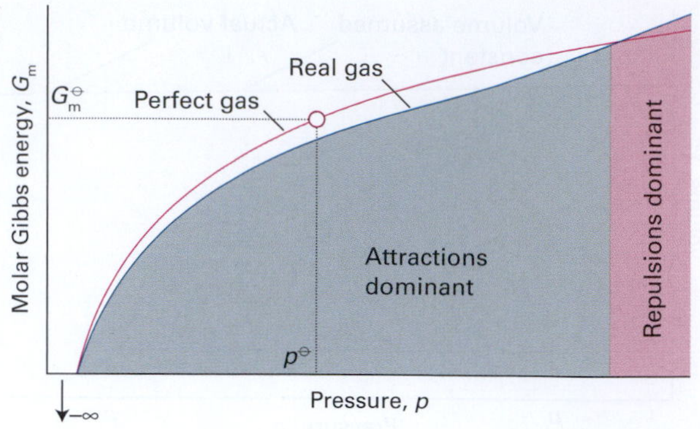

Figure 3D.7 The molar Gibbs energy of a real gas. As $p \to 0$, the molar Gibbs energy coincides with the value for a perfect gas (shown by the purple line). When attractive forces are dominant (at intermediate pressures), the molar Gibbs energy is less than that of a perfect gas and the molecules have a lower 'escaping tendency'. At high pressures, when repulsive forces are dominant, the molar Gibbs energy of a real gas is greater than that of a perfect gas. Then the 'escaping tendency' is increased.

where 'activities' are effective concentrations. Indeed, $f/p^\ominus$ may be regarded as a gas-phase activity.

Although thermodynamic expressions in terms of fugacities derived from this expression are exact, they are useful only if we know how to interpret fugacities in terms of actual pressures. To develop this relation we write the fugacity as

$$f = \phi p \qquad \textit{Definition} \quad \text{Fugacity coefficient} \qquad (3D.17)$$

where ϕ is the dimensionless **fugacity coefficient**, which in general depends on the temperature, the pressure, and the identity of the gas. We show in the following *Justification* that the fugacity coefficient is related to the compression factor, Z, of a gas (Topic 1C) by

$$\ln\phi = \int_0^p \frac{Z-1}{p}\,dp \qquad (3D.18)$$

Provided we know how Z varies with pressure up to the pressure of interest, this expression enable us to determine the fugacity coefficient and hence, through eqn 3D.17, to relate the fugacity to the pressure of the gas.

Justification 3D.3 The fugacity coefficient

Equation 3D.12a is true for all gases whether real or perfect. Expressing it in terms of the fugacity by using eqn 3D.16 turns it into

$$\int_{p'}^{p} V_m\,dp = G_m(p) - G_m(p') = \left\{G_m^\ominus + RT\ln\frac{f}{p^\ominus}\right\} - \left\{G_m^\ominus + RT\ln\frac{f'}{p^\ominus}\right\} = RT\ln\frac{f}{f'}$$

In this expression, f is the fugacity when the pressure is p and f' is the fugacity when the pressure is p'. If the gas were perfect, we would write

$$\int_{p'}^{p} \overbrace{V_{\text{perfect,m}}}^{RT/p} \mathrm{d}p = RT\int_{p'}^{p} \frac{1}{p}\mathrm{d}p = RT \ln\frac{p}{p'}$$

The difference between the two equations is

$$\int_{p'}^{p}(V_{\text{m}} - V_{\text{perfect,m}})\mathrm{d}p = RT\left(\ln\frac{f}{f'} - \ln\frac{p}{p'}\right) = RT\ln\frac{f/f'}{p/p'} = RT\ln\frac{f/p}{f'/p'}$$

which can be rearranged into

$$\ln\frac{f/p}{f'/p'} = \frac{1}{RT}\int_{p'}^{p}(V_{\text{m}} - V_{\text{perfect,m}})\mathrm{d}p$$

When $p' \to 0$, the gas behaves perfectly and f' becomes equal to the pressure, p'. Therefore, $f'/p' \to 1$ as $p' \to 0$. If we take this limit, which means setting $f'/p' = 1$ on the left and $p' = 0$ on the right, the last equation becomes

$$\ln\frac{f}{p} = \frac{1}{RT}\int_{0}^{p}(V_{\text{m}} - V_{\text{perfect,m}})\mathrm{d}p$$

Then, with $\phi = f/p$,

$$\ln\phi = \frac{1}{RT}\int_{0}^{p}(V_{\text{m}} - V_{\text{perfect,m}})\mathrm{d}p$$

For a perfect gas, $V_{\text{perfect,m}} = RT/p$. For a real gas, $V_{\text{m}} = RTZ/p$, where Z is the compression factor of the gas (Topic 1C). With these two substitutions, we obtain eqn 3D.18.

For a perfect gas, $\phi = 1$ at all pressures and temperatures. We know from Fig. 1C.9 that for most gases $Z < 1$ up to moderate pressures, but that $Z > 1$ at higher pressures. If $Z < 1$ throughout the range of integration, then the integrand in eqn 3D.18 is negative and $\phi < 1$. This value implies that $f < p$ (the molecules tend to stick together) and that the molar Gibbs energy of the gas is less than that of a perfect gas. At higher pressures, the range over which $Z > 1$ may dominate the range over which $Z < 1$. The integral is then positive, $\phi > 1$, and $f > p$ (the repulsive interactions are dominant and tend to drive the particles apart). Now the molar Gibbs energy of the gas is greater than that of the perfect gas at the same pressure.

Figure 3D.8, which has been calculated using the full van der Waals equation of state, shows how the fugacity coefficient depends on the pressure in terms of the reduced variables (Topic 1C). Because critical constants are available in Table 1C.2, the graphs can be used for quick estimates of the fugacities of a wide range of gases. Table 3D.2 gives some explicit values for nitrogen.

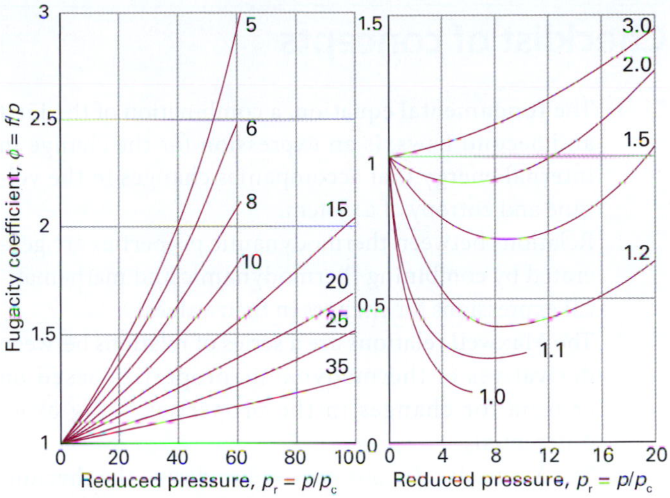

Figure 3D.8 The fugacity coefficient of a van der Waals gas plotted using the reduced variables of the gas. The curves are labelled with the reduced temperature $T_r = T/T_c$.

Table 3D.2* The fugacity of nitrogen at 273 K, f/atm

p/atm	f/atm
1	0.999 55
10	9.9560
100	97.03
1000	1839

* More values are given in the *Resource section*.

Brief illustration 3D.3 The fugacity of a real gas

To use Fig. 3D.8 to estimate the fugacity of carbon dioxide at 400 K and 400 atm, we note from Table 1C.2 that its critical constants are $p_c = 72.85$ atm and $T_c = 304.2$ K. In terms of reduced variables, the gas has $p_r = (400\text{ atm})/(72.85\text{ atm}) = 5.5$ and $T_r = (400\text{ K})/(304.2\text{ K}) = 1.31$. From Fig. 3D.8 (interpolating by eye), these conditions correspond to $\phi \approx 0.4$ and therefore to $f \approx 160$ atm.

Self-test 3D.6 At what temperature would carbon dioxide have a fugacity of 400 atm when its pressure is 400 atm?

Answer: At about $T_r = 2.5$, corresponding to $T = 760$ K

Checklist of concepts

- ☐ 1. The **fundamental equation**, a combination of the First and Second Laws, is an expression for the change in internal energy that accompanies changes in the volume and entropy of a system.
- ☐ 2. Relations between thermodynamic properties are generated by combining thermodynamic and mathematical expressions for changes in their values.
- ☐ 3. The **Maxwell relations** are a series of relations between derivatives of thermodynamic properties based on criteria for changes in the properties being exact differentials.
- ☐ 4. The Maxwell relations are used to derive the **thermodynamic equation of state** and to determine how the internal energy of a substance varies with volume.
- ☐ 5. The variation of the Gibbs energy of a system suggests that it is best regarded as a function of pressure and temperature.
- ☐ 6. The Gibbs energy of a substance decreases with temperature and increases with pressure.
- ☐ 7. The variation of Gibbs energy with temperature is related to the enthalpy by the **Gibbs–Helmholtz equation**.
- ☐ 8. The Gibbs energies of solids and liquids are almost independent of pressure; those of gases vary linearly with the logarithm of the pressure.
- ☐ 9. The **fugacity** is a kind of effective pressure of a real gas.

Checklist of equations

Property	Equation	Comment	Equation number
Fundamental equation	$dU = TdS - pdV$	No additional work	3D.1
Fundamental equation of chemical thermodynamics	$dG = Vdp - SdT$	No additional work	3D.7
Variation of G	$(\partial G/\partial p)_T = V$ and $(\partial G/\partial T)_p = -S$	Composition constant	3D.8
Gibbs–Helmholtz equation	$(\partial(G/T)/\partial T)_p = -H/T^2$	Composition constant	3D.10
Pressure dependence of G	$G_m(p_f) = G_m(p_i) + (p_f - p_i)V_m$	Incompressible substance	3D.13
	$G_m(p_f) = G_m(p_i) + RT\ln(p_f/p_i)$	Perfect gas	3D.14
	$G_m = G_m^{\ominus} + RT\ln(p/p^{\ominus})$	Perfect gas	3D.15
Fugacity	$G_m = G_m^{\ominus} + RT\ln(f/p^{\ominus})$	Definition	3D.16
Fugacity coefficient	$f = \phi p$	Definition	3D.17
	$\ln\phi = \int_0^p \{(Z-1)/p\}dp$	Determination	3D.18

CHAPTER 3 The Second and Third Laws

Assume that all gases are perfect and that data refer to 298.15 K unless otherwise stated.

TOPIC 3A Entropy

Discussion questions

3A.1 The evolution of life requires the organization of a very large number of molecules into biological cells. Does the formation of living organisms violate the Second Law of thermodynamics? State your conclusion clearly and present detailed arguments to support it.

3A.2 Discuss the significance of the terms 'dispersal' and 'disorder' in the context of the Second Law.

3A.3 Discuss the relationships between the various formulations of the Second Law of thermodynamics.

3A.4 Account for deviations from Trouton's rule for liquids such as water and ethanol. Is their entropy of vaporization larger or smaller than $85\,J\,K^{-1}\,mol^{-1}$? Why?

Exercises

3A.1(a) During a hypothetical process, the entropy of a system increases by $125\,J\,K^{-1}$ while the entropy of the surroundings decreases by $125\,J\,K^{-1}$. Is the process spontaneous?

3A.1(b) During a hypothetical process, the entropy of a system increases by $105\,J\,K^{-1}$ while the entropy of the surroundings decreases by $95\,J\,K^{-1}$. Is the process spontaneous?

3A.2(a) A certain ideal heat engine uses water at the triple point as the hot source and an organic liquid as the cold sink. It withdraws 10.00 kJ of heat from the hot source and generates 3.00 kJ of work. What is the temperature of the organic liquid?

3A.2(b) A certain ideal heat engine uses water at the triple point as the hot source and an organic liquid as the cold sink. It withdraws 2.71 kJ of heat from the hot source and generates 0.71 kJ of work. What is the temperature of the organic liquid?

3A.3(a) Calculate the change in entropy when 100 kJ of energy is transferred reversibly and isothermally as heat to a large block of copper at (i) 0 °C, (ii) 50 °C.

3A.3(b) Calculate the change in entropy when 250 kJ of energy is transferred reversibly and isothermally as heat to a large block of lead at (i) 20 °C, (ii) 100 °C.

3A.4(a) Which of $F_2(g)$ and $I_2(g)$ is likely to have the higher standard molar entropy at 298 K?

3A.4(b) Which of $H_2O(g)$ and $CO_2(g)$ is likely to have the higher standard molar entropy at 298 K?

3A.5(a) Calculate the change in entropy when 15 g of carbon dioxide gas is allowed to expand from $1.0\,dm^3$ to $3.0\,dm^3$ at 300 K.

3A.5(b) Calculate the change in entropy when 4.00 g of nitrogen is allowed to expand from $500\,cm^3$ to $750\,cm^3$ at 300 K.

3A.6(a) Predict the enthalpy of vaporization of benzene from its normal boiling point, 80.1 °C.

3A.6(b) Predict the enthalpy of vaporization of cyclohexane from its normal boiling point, 80.7 °C.

3A.7(a) Calculate the molar entropy of a constant-volume sample of neon at 500 K given that it is $146.22\,J\,K^{-1}\,mol^{-1}$ at 298 K.

3A.7(b) Calculate the molar entropy of a constant-volume sample of argon at 250 K given that it is $154.84\,J\,K^{-1}\,mol^{-1}$ at 298 K.

3A.8(a) Calculate ΔS (for the system) when the state of 3.00 mol of perfect gas atoms, for which $C_{p,m} = \frac{5}{2}R$, is changed from 25 °C and 1.00 atm to 125 °C and 5.00 atm. How do you rationalize the sign of ΔS?

3A.8(b) Calculate ΔS (for the system) when the state of 2.00 mol diatomic perfect gas molecules, for which $C_{p,m} = \frac{5}{2}R$, is changed from 25 °C and 1.50 atm to 135 °C and 7.00 atm. How do you rationalize the sign of ΔS?

3A.9(a) Calculate ΔS_{tot} when two copper blocks, each of mass 1.00 kg, one at 50 °C and the other at 0 °C are placed in contact in an isolated container. The specific heat capacity of copper is $0.385\,J\,K^{-1}\,g^{-1}$ and may be assumed constant over the temperature range involved.

3A.9(b) Calculate ΔS_{tot} when two iron blocks, each of mass 10.0 kg , one at 100 °C and the other at 25 °C, are placed in contact in an isolated container. The specific heat capacity of iron is $0.449\,J\,K^{-1}\,g^{-1}$ and may be assumed constant over the temperature range involved.

3A.10(a) Calculate the change in the entropies of the system and the surroundings, and the total change in entropy, when a sample of nitrogen gas of mass 14 g at 298 K and 1.00 bar doubles its volume in (i) an isothermal reversible expansion, (ii) an isothermal irreversible expansion against $p_{ex}=0$, and (iii) an adiabatic reversible expansion.

3A.10(b) Calculate the change in the entropies of the system and the surroundings, and the total change in entropy, when the volume of a sample of argon gas of mass 21 g at 298 K and 1.50 bar increases from $1.20\,dm^3$ to $4.60\,dm^3$ in (i) an isothermal reversible expansion, (ii) an isothermal irreversible expansion against $p_{ex}=0$, and (iii) an adiabatic reversible expansion.

3A.11(a) The enthalpy of vaporization of chloroform ($CHCl_3$) is $29.4\,kJ\,mol^{-1}$ at its normal boiling point of 334.88 K. Calculate (i) the entropy of vaporization of chloroform at this temperature and (ii) the entropy change of the surroundings.

3A.11(b) The enthalpy of vaporization of methanol is $35.27\,kJ\,mol^{-1}$ at its normal boiling point of 64.1 °C. Calculate (i) the entropy of vaporization of methanol at this temperature and (ii) the entropy change of the surroundings.

3A.12(a) Calculate the change in entropy of the system when 10.0 g of ice at −10.0 °C is converted into water vapour at 115.0 °C and at a constant pressure of 1 bar. The constant-pressure molar heat capacity of $H_2O(s)$ and $H_2O(l)$ is $75.291\,J\,K^{-1}\,mol^{-1}$ and that of $H_2O(g)$ is $33.58\,J\,K^{-1}\,mol^{-1}$.

3A.12(b) Calculate the change in entropy of the system when 15.0 g of ice at −12.0 °C is converted to water vapour at 105.0 °C at a constant pressure of 1 bar. For data, see the preceding exercise.

Problems

3A.1 Represent the Carnot cycle on a temperature–entropy diagram and show that the area enclosed by the cycle is equal to the work done.

3A.2 The cycle involved in the operation of an internal combustion engine is called the *Otto cycle*. Air can be considered to be the working substance and can be assumed to be a perfect gas. The cycle consists of the following steps: (1) Reversible adiabatic compression from A to B, (2) reversible constant-volume pressure increase from B to C due to the combustion of a small amount of fuel, (3) reversible adiabatic expansion from C to D, and (4) reversible and constant-volume pressure decrease back to state A. Determine the change in entropy (of the system and of the surroundings) for each step of the cycle and determine an expression for the efficiency of the cycle, assuming that the heat is supplied in Step 2. Evaluate the efficiency for a compression ratio of 10:1. Assume that in state A, $V=4.00\,\mathrm{dm^3}$, $p=1.00\,\mathrm{atm}$, and $T=300\,\mathrm{K}$, that $V_A=10V_B$, $p_C/p_B=5$, and that $C_{p,m}=\frac{7}{2}R$.

3A.3 Prove that two reversible adiabatic paths can never cross. Assume that the energy of the system under consideration is a function of temperature only. (*Hint*: Suppose that two such paths can intersect, and complete a cycle with the two paths plus one isothermal path. Consider the changes accompanying each stage of the cycle and show that they conflict with the Kelvin statement of the Second Law.)

3A.4 To calculate the work required to lower the temperature of an object, we need to consider how the coefficient of performance c (see *Impact* I3.1) changes with the temperature of the object. (a) Find an expression for the work of cooling an object from T_i to T_f when the refrigerator is in a room at a temperature T_h. *Hint*: Write $\mathrm{d}w=\mathrm{d}q/c(T)$, relate $\mathrm{d}q$ to $\mathrm{d}T$ through the heat capacity C_p, and integrate the resulting expression. Assume that the heat capacity is independent of temperature in the range of interest. (b) Use the result in part (a) to calculate the work needed to freeze 250 g of water in a refrigerator at 293 K. How long will it take when the refrigerator operates at 100 W?

3A.5 The expressions that apply to the treatment of refrigerators (Problem 3A.4) also describe the behaviour of heat pumps, where warmth is obtained from the back of a refrigerator while its front is being used to cool the outside world. Heat pumps are popular home heating devices because they are very efficient. Compare heating of a room at 295 K by each of two methods: (a) direct conversion of 1.00 kJ of electrical energy in an electrical heater, and (b) use of 1.00 kJ of electrical energy to run a reversible heat pump with the outside at 260 K. Discuss the origin of the difference in the energy delivered to the interior of the house by the two methods.

3A.6 Calculate the difference in molar entropy (a) between liquid water and ice at −5 °C, (b) between liquid water and its vapour at 95 °C and 1.00 atm. The differences in heat capacities on melting and on vaporization are $37.3\,\mathrm{J\,K^{-1}\,mol^{-1}}$ and $-41.9\,\mathrm{J\,K^{-1}\,mol^{-1}}$, respectively. Distinguish between the entropy changes of the sample, the surroundings, and the total system, and discuss the spontaneity of the transitions at the two temperatures.

3A.7 The molar heat capacity of chloroform (trichloromethane, $CHCl_3$) in the range 240 K to 330 K is given by $C_{p,m}/(\mathrm{J\,K^{-1}\,mol^{-1}})=91.47+7.5\times10^{-2}\,(T/\mathrm{K})$. In a particular experiment, 1.00 mol $CHCl_3$ is heated from 273 K to 300 K. Calculate the change in molar entropy of the sample.

3A.8 A block of copper of mass 2.00 kg ($C_{p,m}=24.44\,\mathrm{J\,K^{-1}\,mol^{-1}}$) and temperature 0 °C is introduced into an insulated container in which there is 1.00 mol $H_2O(g)$ at 100 °C and 1.00 atm. (a) Assuming all the steam is condensed to water, what will be the final temperature of the system, the heat transferred from water to copper, and the entropy change of the water, copper, and the total system? (b) In fact, some water vapour is present at equilibrium. From the vapour pressure of water at the temperature calculated in (a), and assuming that the heat capacities of both gaseous and liquid water are constant and given by their values at that temperature, obtain an improved value of the final temperature, the heat transferred, and the various entropies. (*Hint*: You will need to make plausible approximations.)

3A.9 A sample consisting of 1.00 mol of perfect gas molecules at 27 °C is expanded isothermally from an initial pressure of 3.00 atm to a final pressure of 1.00 atm in two ways: (a) reversibly, and (b) against a constant external pressure of 1.00 atm. Determine the values of q, w, ΔU, ΔH, ΔS, ΔS_{surr}, and ΔS_{tot} for each path.

3A.10 A block of copper of mass 500 g and initially at 293 K is in thermal contact with an electric heater of resistance 1.00 kΩ and negligible mass. A current of 1.00 A is passed for 15.0 s. Calculate the change in entropy of the copper, taking $C_{p,m}=24.4\,\mathrm{J\,K^{-1}\,mol^{-1}}$. The experiment is then repeated with the copper immersed in a stream of water that maintains its temperature at 293 K. Calculate the change in entropy of the copper and the water in this case.

3A.11 Find an expression for the change in entropy when two blocks of the same substance and of equal mass, one at the temperature T_h and the other at T_c, are brought into thermal contact and allowed to reach equilibrium. Evaluate the change for two blocks of copper, each of mass 500 g, with $C_{p,m}=24.4\,\mathrm{J\,K^{-1}\,mol^{-1}}$, taking $T_h=500\,\mathrm{K}$ and $T_c=250\,\mathrm{K}$.

3A.12 According to Newton's law of cooling, the rate of change of temperature is proportional to the temperature difference between the system and its surroundings. Given that $S(T)-S(T_i)=C\ln(T/T_i)$, where T_i is the initial temperature and C the heat capacity, deduce an expression for the rate of change of entropy of the system as it cools.

3A.13 The protein lysozyme unfolds at a transition temperature of 75.5 °C and the standard enthalpy of transition is $509\,\mathrm{kJ\,mol^{-1}}$. Calculate the entropy of unfolding of lysozyme at 25.0 °C, given that the difference in the constant-pressure heat capacities upon unfolding is $6.28\,\mathrm{kJ\,K^{-1}\,mol^{-1}}$ and can be assumed to be independent of temperature. *Hint*: Imagine that the transition at 25.0 °C occurs in three steps: (i) heating of the folded protein from 25.0 °C to the transition temperature, (ii) unfolding at the transition temperature, and (iii) cooling of the unfolded protein to 25.0 °C. Because the entropy is a state function, the entropy change at 25.0 °C is equal to the sum of the entropy changes of the steps.

TOPIC 3B The measurement of entropy

Discussion question

3B.1 Discuss why the standard entropies of ions in solution may be positive, negative, or zero.

Exercises

3B.1(a) Calculate the residual molar entropy of a solid in which the molecules can adopt (i) three, (ii) five, (iii) six orientations of equal energy at $T=0$.

3B.1(b) Suppose that the hexagonal molecule $C_6H_nF_{6-n}$ has a residual entropy on account of the similarity of the H and F atoms. Calculate the residual for each value of n.

3B.2(a) Calculate the standard reaction entropy at 298 K of

(i) $2CH_3CHO(g)+O_2(g)\rightarrow 2CH_3COOH(l)$

(ii) $2AgCl(s)+Br_2(l)\rightarrow 2AgBr(s)+Cl_2(g)$

(iii) $Hg(l)+Cl_2(g)\rightarrow HgCl_2(s)$

3B.2(b) Calculate the standard reaction entropy at 298 K of

(i) $Zn(s)+Cu^{2+}(aq)\rightarrow Zn^{2+}(aq)+Cu(s)$

(ii) $C_{12}H_{22}O_{11}(s)+12O_2(g)\rightarrow 12CO_2(g)+11H_2O(l)$

Problems

3B.1 The standard molar entropy of $NH_3(g)$ is 192.45 J K^{-1} mol^{-1} at 298 K, and its heat capacity is given by eqn 2B.8 with the coefficients given in Table 2B.1. Calculate the standard molar entropy at (a) 100 °C and (b) 500 °C.

3B.2 The molar heat capacity of lead varies with temperature as follows:

T/K	10	15	20	25	30	50
$C_{p,m}$/(J K^{-1} mol^{-1})	2.8	7.0	10.8	14.1	16.5	21.4
T/K	70	100	150	200	250	298
$C_{p,m}$/(J K^{-1} mol^{-1})	23.3	24.5	25.3	25.8	26.2	26.6

Calculate the standard Third-Law entropy of lead at (a) 0 °C and (b) 25 °C.

3B.3 From standard enthalpies of formation, standard entropies, and standard heat capacities available from tables in the *Resource section*, calculate: (a) the standard enthalpies and entropies at 298 K and 398 K for the reaction $CO_2(g)+H_2(g)\rightarrow CO(g)+H_2O(g)$. Assume that the heat capacities are constant over the temperature range involved.

3B.4 The molar heat capacity of anhydrous potassium hexacyanoferrate(II) varies with temperature as follows:

T/K	10	20	30	40	50	60
$C_{p,m}$/(J K^{-1} mol^{-1})	2.09	14.43	36.44	62.55	87.03	111.0
T/K	70	80	90	100	110	150
$C_{p,m}$/(J K^{-1} mol^{-1})	131.4	149.4	165.3	179.6	192.8	237.6
T/K	160	170	180	190	200	
$C_{p,m}$/(J K^{-1} mol^{-1})	247.3	256.5	265.1	273.0	280.3	

Calculate the molar enthalpy relative to its value at $T=0$ and the Third-Law entropy at each of these temperatures.

3B.5 The compound 1,3,5-trichloro-2,4,6-trifluorobenzene is an intermediate in the conversion of hexachlorobenzene to hexafluorobenzene, and its thermodynamic properties have been examined by measuring its heat capacity over a wide temperature range (R.L. Andon and J.F. Martin, *J. Chem. Soc. Faraday Trans. I*, 871 (1973)). Some of the data are as follows:

T/K	14.14	16.33	20.03	31.15	44.08	64.81
$C_{p,m}$/(J K^{-1} mol^{-1})	9.492	12.70	18.18	32.54	46.86	66.36
T/K	100.90	140.86	183.59	225.10	262.99	298.06
$C_{p,m}$/(J K^{-1} mol^{-1})	95.05	121.3	144.4	163.7	180.2	196.4

Calculate the molar enthalpy relative to its value at $T=0$ and the Third-Law molar entropy of the compound at these temperatures.

3B.6‡ Given that $S_m^{\ominus}=29.79$ J K^{-1} mol^{-1} for bismuth at 100 K and the following tabulated heat capacity data (D.G. Archer, *J. Chem. Eng. Data* **40**, 1015 (1995)), compute the standard molar entropy of bismuth at 200 K.

T/K	100	120	140	150	160	180	200
$C_{p,m}$/(J K^{-1} mol^{-1})	23.00	23.74	24.25	24.44	24.61	24.89	25.11

Compare the value to the value that would be obtained by taking the heat capacity to be constant at 24.44 J K^{-1} mol^{-1} over this range.

3B.7 Derive an expression for the molar entropy of a monatomic solid on the basis of the Einstein and Debye models and plot the molar entropy against the temperature (use T/θ in each case, with θ the Einstein or Debye temperature). Use the following expressions for the temperature-dependence of the heat capacities:

$$\text{Einstein: } C_{V,m}(T)=3Rf^{E}(T) \quad f^{E}(T)=\left(\frac{\theta^{E}}{T}\right)^2\left(\frac{e^{\theta^{E}/2T}}{e^{\theta^{E}/T}-1}\right)^2$$

$$\text{Debye: } C_{V,m}(T)=3Rf^{D}(T) \quad f^{D}(T)=3\left(\frac{T}{\theta^{D}}\right)^2\int_0^{\theta^{D}/T}\frac{x^4e^x}{(e^x-1)^2}dx$$

Use mathematical software to evaluate the appropriate expressions.

3B.8 An average human DNA molecule has 5×10^8 binucleotides (rungs on the DNA ladder) of four different kinds. If each rung were a random choice of one of these four possibilities, what would be the residual entropy associated with this typical DNA molecule?

TOPIC 3C Concentrating on the system

Discussion questions

3C.1 The following expressions have been used to establish criteria for spontaneous change: $dA_{T,V}<0$ and $dG_{T,p}<0$. Discuss the origin, significance, and applicability of each criterion.

3C.2 Under what circumstances, and why, can the spontaneity of a process be discussed in terms of the properties of the system alone?

‡ These problems were provided by Charles Trapp and Carmen Giunta.

Exercises

3C.1(a) Combine the reaction entropies calculated in Exercise 3B.2(a) with the reaction enthalpies, and calculate the standard reaction Gibbs energies at 298 K.

3C.1(b) Combine the reaction entropies calculated in Exercise 3B.2(b) with the reaction enthalpies, and calculate the standard reaction Gibbs energies at 298 K.

3C.2(a) Calculate the standard Gibbs energy of the reaction $4\,HI(g)+O_2(g)\rightarrow 2\,I_2(s)+2\,H_2O(l)$ at 298 K, from the standard entropies and enthalpies of formation given in the *Resource section*.

3C.2(b) Calculate the standard Gibbs energy of the reaction $CO(g)+CH_3CH_2OH(l)\rightarrow CH_3CH_2COOH(l)$ at 298 K, from the standard entropies and enthalpies of formation given in the *Resource section*.

3C.3(a) Calculate the maximum non-expansion work per mole that may be obtained from a fuel cell in which the chemical reaction is the combustion of methane at 298 K.

3C.3(b) Calculate the maximum non-expansion work per mole that may be obtained from a fuel cell in which the chemical reaction is the combustion of propane at 298 K.

3C.4(a) Use standard Gibbs energies of formation to calculate the standard reaction Gibbs energies at 298 K of the reactions

(i) $2\,CH_3CHO(g)+O_2(g)\rightarrow 2\,CH_3COOH(l)$
(ii) $2\,AgCl(s)+Br_2(l)\rightarrow 2\,AgBr(s)+Cl_2(g)$
(iii) $Hg(l)+Cl_2(g)\rightarrow HgCl_2(s)$

3C.4(b) Use standard Gibbs energies of formation to calculate the standard reaction Gibbs energies at 298 K of the reactions

(i) $Zn(s)+Cu^{2+}(aq)\rightarrow Zn^{2+}(aq)+Cu(s)$
(ii) $C_{12}H_{22}O_{11}(s)+12\,O_2(g)\rightarrow 12\,CO_2(g)+11\,H_2O(l)$

3C.5(a) The standard enthalpy of combustion of ethyl acetate ($CH_3COOC_2H_5$) is $-2231\,\text{kJ mol}^{-1}$ at 298 K and its standard molar entropy is $259.4\,\text{J K}^{-1}\,\text{mol}^{-1}$. Calculate the standard Gibbs energy of formation of the compound at 298 K.

3C.5(b) The standard enthalpy of combustion of the amino acid glycine (NH_2CH_2COOH) is $-969\,\text{kJ mol}^{-1}$ at 298 K and its standard molar entropy is $103.5\,\text{J K}^{-1}\,\text{mol}^{-1}$. Calculate the standard Gibbs energy of formation of glycine at 298 K.

Problems

3C.1 Consider a perfect gas contained in a cylinder and separated by a frictionless adiabatic piston into two sections A and B. All changes in B are isothermal; that is, a thermostat surrounds B to keep its temperature constant. There is 2.00 mol of the gas molecules in each section. Initially $T_A=T_B=300\,\text{K}$, $V_A=V_B=2.00\,\text{dm}^3$. Energy is supplied as heat to Section A and the piston moves to the right reversibly until the final volume of Section B is $1.00\,\text{dm}^3$. Calculate (a) ΔS_A and ΔS_B, (b) ΔA_A and ΔA_B, (c) ΔG_A and ΔG_B, (d) ΔS of the total system and its surroundings. If numerical values cannot be obtained, indicate whether the values should be positive, negative, or zero or are indeterminate from the information given. (Assume $C_{V,m}=20\,\text{J K}^{-1}\,\text{mol}^{-1}$.)

3C.2 Calculate the molar internal energy, molar entropy, and molar Helmholtz energy of a collection of harmonic oscillators and plot your expressions as a function of T/θ^V, where $\theta^V=h\nu/k$.

3C.3 In biological cells, the energy released by the oxidation of foods is stored in adenosine triphosphate (ATP or ATP^{4-}). The essence of ATP's action is its ability to lose its terminal phosphate group by hydrolysis and to form adenosine diphosphate (ADP or ADP^{3-}):

$$ATP^{4-}(aq)+H_2O(l)\rightarrow ADP^{3-}(aq)+HPO_4^{2-}(aq)+H_3O^+(aq)$$

At pH = 7.0 and 37 °C (310 K, blood temperature) the enthalpy and Gibbs energy of hydrolysis are $\Delta_r H=-20\,\text{kJ mol}^{-1}$ and $\Delta_r G=-31\,\text{kJ mol}^{-1}$, respectively. Under these conditions, the hydrolysis of 1 mol $ATP^{4-}(aq)$ results in the extraction of up to 31 kJ of energy that can be used to do non-expansion work, such as the synthesis of proteins from amino acids, muscular contraction, and the activation of neuronal circuits in our brains. (a) Calculate and account for the sign of the entropy of hydrolysis of ATP at pH = 7.0 and 310 K. (b) Suppose that the radius of a typical biological cell is 10 μm and that inside it 1×10^6 ATP molecules are hydrolysed each second. What is the power density of the cell in watts per cubic metre ($1\,\text{W}=1\,\text{J s}^{-1}$)? A computer battery delivers about 15 W and has a volume of $100\,\text{cm}^3$. Which has the greater power density, the cell or the battery? (c) The formation of glutamine from glutamate and ammonium ions requires $14.2\,\text{kJ mol}^{-1}$ of energy input. It is driven by the hydrolysis of ATP to ADP mediated by the enzyme glutamine synthetase. How many moles of ATP must be hydrolysed to form 1 mol glutamine?

TOPIC 3D Combining the First and Second Laws

Discussion questions

3D.1 Suggest a physical interpretation of the dependence of the Gibbs energy on the temperature.

3D.2 Suggest a physical interpretation of the dependence of the Gibbs energy on the pressure.

Exercises

3D.1(a) Suppose that 2.5 mmol $N_2(g)$ occupies $42\,\text{cm}^3$ at 300 K and expands isothermally to $600\,\text{cm}^3$. Calculate ΔG for the process.

3D.1(b) Suppose that 6.0 mmol Ar(g) occupies $52\,\text{cm}^3$ at 298 K and expands isothermally to $122\,\text{cm}^3$. Calculate ΔG for the process.

3D.2(a) The change in the Gibbs energy of a certain constant-pressure process was found to fit the expression $\Delta G/\text{J}=-85.40+36.5(T/\text{K})$. Calculate the value of ΔS for the process.

3D.2(b) The change in the Gibbs energy of a certain constant-pressure process was found to fit the expression $\Delta G/\mathrm{J} = -73.1 + 42.8(T/\mathrm{K})$. Calculate the value of ΔS for the process.

3D.3(a) Estimate the change in the Gibbs energy and molar Gibbs energy of 1.0 dm^3 of octane when the pressure acting on it is increased from 1.0 atm to 100 atm. The mass density of octane is 0.703 g cm^{-3}.

3D.3(b) Estimate the change in the Gibbs energy and molar Gibbs energy of 100 cm^3 of water when the pressure acting on it is increased from 100 kPa to 500 kPa. The mass density of water is 0.997 g cm^{-3}.

3D.4(a) Calculate the change in the molar Gibbs energy of hydrogen gas when its pressure is increased isothermally from 1.0 atm to 100.0 atm at 298 K.

3D.4(b) Calculate the change in the molar Gibbs energy of oxygen when its pressure is increased isothermally from 50.0 kPa to 100.0 kPa at 500 K.

Problems

3D.1 Calculate $\Delta_r G^\ominus$ (375 K) for the reaction $2\,CO(g) + O_2(g) \rightarrow 2\,CO_2(g)$ from the value of $\Delta_r G^\ominus$ (298 K), $\Delta_r H^\ominus$ (298 K), and the Gibbs–Helmholtz equation.

3D.2 Estimate the standard reaction Gibbs energy of $N_2(g) + 3\,H_2(g) \rightarrow 2\,NH_3(g)$ at (a) 500 K, (b) 1000 K from their values at 298 K.

3D.3 At 298 K the standard enthalpy of combustion of sucrose is −5797 kJ mol^{-1} and the standard Gibbs energy of the reaction is −6333 kJ mol^{-1}. Estimate the additional non-expansion work that may be obtained by raising the temperature to blood temperature, 37 °C.

3D.4 Two empirical equations of state of a real gas are as follows:

$$\text{van der Waals}: p = \frac{RT}{V_\mathrm{m} - b} - \frac{a}{V_\mathrm{m}^2}$$

$$\text{Dieterici}: p = \frac{RT\mathrm{e}^{-a/RTV_\mathrm{m}}}{V_\mathrm{m} - b}$$

Evaluate $(\partial S/\partial V)_T$ for each gas. For an isothermal expansion, for which kind of gas (also consider a perfect gas) will ΔS be greatest? Explain your conclusion.

3D.5 Two of the four Maxwell relations were derived in the text, but two were not. Complete their derivation by showing that $(\partial S/\partial V)_T = (\partial p/\partial T)_V$ and $(\partial T/\partial p)_S = (\partial V/\partial S)_p$.

3D.6 (a) Use the Maxwell relations to express the derivatives $(\partial S/\partial V)_T$, $(\partial V/\partial S)_p$, $(\partial p/\partial S)_V$, and $(\partial V/\partial S)_p$ in terms of the heat capacities, the expansion coefficient $\alpha = (1/V)(\partial V/\partial T)_p$, and the isothermal compressibility, $\kappa_T = -(1/V)(\partial V/\partial p)_T$. (b) The Joule coefficient, μ_J, is defined as $\mu_\mathrm{J} = (\partial T/\partial V)_U$. Show that $\mu_\mathrm{J} C_V = p - \alpha T/\kappa_T$.

3D.7 Suppose that S is regarded as a function of p and T. Show that $T\mathrm{d}S = C_p\mathrm{d}T - \alpha TV\mathrm{d}p$. Hence, show that the energy transferred as heat when the pressure on an incompressible liquid or solid is increased by Δp is equal to $-\alpha TV\Delta p$, where $\alpha = (1/V)(\partial V/\partial T)_p$. Evaluate q when the pressure acting on 100 cm^3 of mercury at 0 °C is increased by 1.0 kbar. ($\alpha = 1.82 \times 10^{-4}$ K^{-1}.)

3D.8 Equation 3D.6 ($\pi_T = T(\partial p/\partial T)_V - p$) expresses the internal pressure π_T in terms of the pressure and its derivative with respect to temperature. Express π_T in terms of the molecular partition function.

3D.9 Explore the consequences of replacing the equation of state of a perfect gas by the van der Waals equation of state for the pressure-dependence of the molar Gibbs energy. Proceed in three steps. First, consider the case when $a = 0$ and only repulsions are significant. Then consider the case when $b = 0$ and only attractions are significant. For the latter, you should consider making the approximation that the attractions are weak. Finally, explore the full expression by using mathematical software. In each case plot your results graphically and account physically for the deviations from the perfect gas expression.

3D.10‡ Nitric acid hydrates have received much attention as possible catalysts for heterogeneous reactions which bring about the Antarctic ozone hole. Worsnop et al. (*Science* **259**, 71 (1993)) investigated the thermodynamic stability of these hydrates under conditions typical of the polar winter stratosphere. They report thermodynamic data for the sublimation of mono-, di-, and trihydrates to nitric acid and water vapour, $HNO_3 \cdot nH_2O(s) \rightarrow HNO_3(g) + nH_2O(g)$, for $n = 1$, 2, and 3. Given $\Delta_r G^\ominus$ and $\Delta_r H^\ominus$ for these reactions at 220 K, use the Gibbs–Helmholtz equation to compute $\Delta_r G^\ominus$ at 190 K.

n	1	2	3
$\Delta_r G^\ominus$/(kJ mol^{-1})	46.2	69.4	93.2
$\Delta_r H^\ominus$/(kJ mol^{-1})	127	188	237

Integrated activities

3.1 A gaseous sample consisting of 1.00 mol molecules is described by the equation of state $pV_\mathrm{m} = RT(1 + Bp)$. Initially at 373 K, it undergoes Joule–Thomson expansion from 100 atm to 1.00 atm. Given that $C_{p,\mathrm{m}} = \frac{5}{2}R$, $\mu = 0.21$ K atm^{-1}, $B = -0.525(\mathrm{K}/T)$ atm^{-1} and that these are constant over the temperature range involved, calculate ΔT and ΔS for the gas.

3.2 Discuss the relationship between the thermodynamic and statistical definitions of entropy.

3.3 Use mathematical software, a spreadsheet, or the *Living graphs* on the web site for this book to:

(a) Evaluate the change in entropy of 1.00 mol CO_2(g) on expansion from 0.001 m^3 to 0.010 m^3 at 298 K, treated as a van der Waals gas.
(b) Allow for the temperature dependence of the heat capacity by writing $C = a + bT + c/T^2$, and plot the change in entropy for different values of the three coefficients (including negative values of c).
(c) Show how the first derivative of G, $(\partial G/\partial p)_T$, varies with pressure, and plot the resulting expression over a pressure range. What is the physical significance of $(\partial G/\partial p)_T$?
(d) Evaluate the fugacity coefficient as a function of the reduced volume of a van der Waals gas and plot the outcome for a selection of reduced temperatures over the range $0.8 \le V_\mathrm{r} \le 3$.

CHAPTER 4

Physical transformations of pure substances

Vaporization, melting (fusion), and the conversion of graphite to diamond are all examples of changes of phase without change of chemical composition. The discussion of the phase transitions of pure substances is among the simplest applications of thermodynamics to chemistry, and is guided by the principle that the tendency of systems at constant temperature and pressure is to minimize their Gibbs energy.

4A Phase diagrams of pure substances

First, we see that one type of phase diagram is a map of the pressures and temperatures at which each phase of a substance is the most stable. The thermodynamic criterion of phase stability enables us to deduce a very general result, the 'phase rule', which summarizes the constraints on the equilibria between phases. In preparation for later chapters, we express the rule in a general way that can be applied to systems of more than one component. Then, we describe the interpretation of empirically determined phase diagrams for a selection of substances.

4B Thermodynamic aspects of phase transitions

Here we consider the factors that determine the positions and shapes of the boundaries between the regions on a phase diagram. The practical importance of the expressions we derive is that they show how the vapour pressure of a substance varies with temperature and how the melting point varies with pressure. Transitions between phases are classified by noting how various thermodynamic functions change when the transition occurs. This chapter also introduces the 'chemical potential', a property that will be at the centre of our discussions of mixtures and chemical reactions.

What is the impact of this material?

The properties of carbon dioxide in its supercritical fluid phase can form the basis for novel and useful chemical separation methods, and have considerable promise for 'green' chemistry synthetic procedures. Its properties and applications are discussed in *Impact* I4.1.

To read more about the impact of this material, scan the QR code, or go to bcs.whfreeman.com/webpub/chemistry/pchem10e/impact/pchem-4-1.html

4A Phase diagrams of pure substances

Contents

➤ Why do you need to know this material?

Phase diagrams summarize the behaviour of substances under different conditions. In metallurgy, the ability to control the microstructure resulting from phase equilibria makes it possible to tailor the mechanical properties of the materials to a particular application.

➤ What is the key idea?

A pure substance tends to adopt the phase with the lowest chemical potential.

➤ What do you need to know already?

This Topic builds on the fact that the Gibbs energy is a signpost of spontaneous change under conditions of constant temperature and pressure (Topic 3C).

One of the most succinct ways of presenting the physical changes of state that a substance can undergo is in terms of its 'phase diagram'. This material is also the basis of the discussion of mixtures in Chapter 5.

4A.1 The stabilities of phases

Thermodynamics provides a powerful language for describing and understanding the stabilities and transformations of phases, but to apply it we need to employ definitions carefully.

(a) The number of phases

A **phase** is a form of matter that is uniform throughout in chemical composition and physical state. Thus, we speak of solid, liquid, and gas phases of a substance, and of its various solid phases, such as the white and black allotropes of phosphorus or the aragonite and calcite polymorphs of calcium carbonate.

A note on good practice An *allotrope* is a particular molecular form of an element (such as O_2 and O_3) and may be solid, liquid, or gas. A *polymorph* is one of a number of solid phases of an element or compound.

The number of phases in a system is denoted P. A gas, or a gaseous mixture, is a single phase ($P=1$), a crystal of a substance is a single phase, and two fully miscible liquids form a single phase.

Brief illustration 4A.1 The number of phases

A solution of sodium chloride in water is a single phase ($P=1$). Ice is a single phase even though it might be chipped into small fragments. A slurry of ice and water is a two-phase system ($P=2$) even though it is difficult to map the physical boundaries between the phases. A system in which calcium carbonate undergoes the thermal decomposition $CaCO_3(s) \rightarrow CaO(s) + CO_2(g)$ consists of two solid phases (one consisting of calcium carbonate and the other of calcium oxide) and one gaseous phase (consisting of carbon dioxide), so $P=3$.

Self-test 4A.1 How many phases are present in a sealed, half-full vessel containing water?

Answer: 2

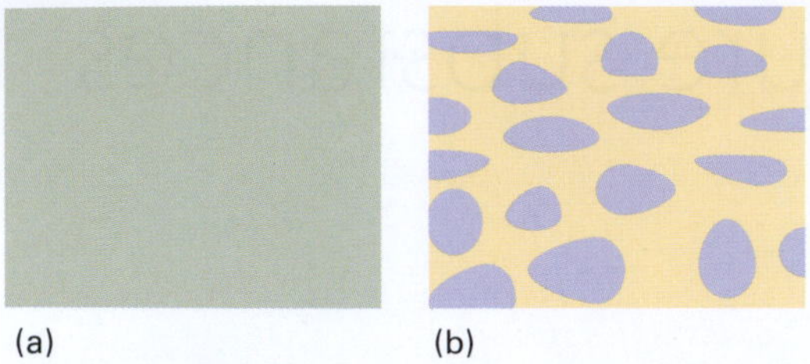

Figure 4A.1 The difference between (a) a single-phase solution, in which the composition is uniform on a microscopic scale, and (b) a dispersion, in which regions of one component are embedded in a matrix of a second component.

Two metals form a two-phase system ($P=2$) if they are immiscible, but a single-phase system ($P=1$), an alloy, if they are miscible. This example shows that it is not always easy to decide whether a system consists of one phase or of two. A solution of solid B in solid A—a homogeneous mixture of the two substances—is uniform on a molecular scale. In a solution, atoms of A are surrounded by atoms of A and B, and any sample cut from the sample, even microscopically small, is representative of the composition of the whole.

A dispersion is uniform on a macroscopic scale but not on a microscopic scale, for it consists of grains or droplets of one substance in a matrix of the other. A small sample could come entirely from one of the minute grains of pure A and would not be representative of the whole (Fig. 4A.1). Dispersions are important because, in many advanced materials (including steels), heat treatment cycles are used to achieve the precipitation of a fine dispersion of particles of one phase (such as a carbide phase) within a matrix formed by a saturated solid solution phase.

(b) Phase transitions

A **phase transition**, the spontaneous conversion of one phase into another phase, occurs at a characteristic temperature for a given pressure. The **transition temperature**, T_{trs}, is the temperature at which the two phases are in equilibrium and the Gibbs energy of the system is minimized at the prevailing pressure.

Brief illustration 4A.2 Phase transitions

At 1 atm, ice is the stable phase of water below 0 °C, but above 0 °C liquid water is more stable. This difference indicates that below 0 °C the Gibbs energy decreases as liquid water changes into ice and that above 0 °C the Gibbs energy decreases as ice changes into liquid water. The numerical values of the Gibbs energies are considered in the next *Brief illustration*.

Self-test 4A.2 Which has the higher standard molar Gibbs energy at 105 °C, liquid water or its vapour?

Answer: Liquid water

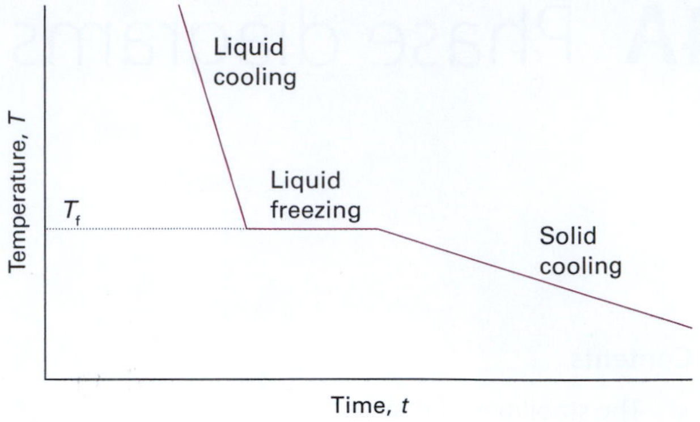

Figure 4A.2 A cooling curve at constant pressure. The flat section corresponds to the pause in the fall of temperature while the first-order exothermic transition (freezing) occurs. This pause enables T_f to be located even if the transition cannot be observed visually.

Detecting a phase transition is not always as simple as seeing water boil in a kettle, so special techniques have been developed. One technique is **thermal analysis**, which takes advantage of the heat that is evolved or absorbed during any transition. The transition is detected by noting that the temperature does not change even though heat is being supplied or removed from the sample (Fig. 4A.2). Differential scanning calorimetry (Topic 2C) is also used. Thermal techniques are useful for solid–solid transitions, where simple visual inspection of the sample may be inadequate. X-ray diffraction (Topic 18A) also reveals the occurrence of a phase transition in a solid, for different structures are found on either side of the transition temperature.

As always, it is important to distinguish between the thermodynamic description of a process and the rate at which the process occurs. A phase transition that is predicted from thermodynamics to be spontaneous may occur too slowly to be significant in practice. For instance, at normal temperatures and pressures the molar Gibbs energy of graphite is lower than that of diamond, so there is a thermodynamic tendency for diamond to change into graphite. However, for this transition to take place, the C atoms must change their locations, which is an immeasurably slow process in a solid except at high temperatures. The discussion of the rate of attainment of equilibrium is a kinetic problem and is outside the range of thermodynamics. In gases and liquids the mobilities of the molecules allow phase transitions to occur rapidly, but in solids thermodynamic instability may be frozen in. Thermodynamically unstable phases that persist because the transition is kinetically hindered are called **metastable phases**. Diamond is a metastable but persistent phase of carbon under normal conditions.

(c) Thermodynamic criteria of phase stability

All our considerations will be based on the Gibbs energy of a substance, and in particular on its molar Gibbs energy, G_m. In

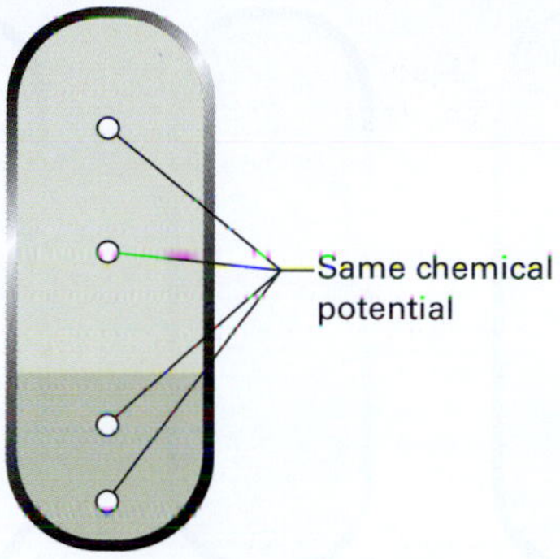

Figure 4A.3 When two or more phases are in equilibrium, the chemical potential of a substance (and, in a mixture, a component) is the same in each phase and is the same at all points in each phase.

fact, this quantity plays such an important role in this chapter and the rest of the text that we give it a special name and symbol, the **chemical potential**, μ (mu). For a one-component system, 'molar Gibbs energy' and 'chemical potential' are synonyms, so $\mu = G_m$, but in Topic 5A we see that chemical potential has a broader significance and a more general definition. The name 'chemical potential' is also instructive: as we develop the concept, we shall see that μ is a measure of the potential that a substance has for undergoing change in a system. In this chapter and Chapter 5, it reflects the potential of a substance to undergo physical change. In Chapter 6, we see that μ is the potential of a substance to undergo chemical change.

We base the entire discussion on the following consequence of the Second Law (Fig. 4A.3):

> At equilibrium, the chemical potential of a substance is the same throughout a sample, regardless of how many phases are present.

Criterion of phase equilibrium

To see the validity of this remark, consider a system in which the chemical potential of a substance is μ_1 at one location and μ_2 at another location. The locations may be in the same or in different phases. When an infinitesimal amount dn of the substance is transferred from one location to the other, the Gibbs energy of the system changes by $-\mu_1 dn$ when material is removed from location 1, and it changes by $+\mu_2 dn$ when that material is added to location 2. The overall change is therefore $dG = (\mu_2 - \mu_1)dn$. If the chemical potential at location 1 is higher than that at location 2, the transfer is accompanied by a decrease in G, and so has a spontaneous tendency to occur. Only if $\mu_1 = \mu_2$ is there no change in G, and only then is the system at equilibrium.

Brief illustration 4A.3 Gibbs energy and phase transition

The standard molar Gibbs energy of formation of water vapour at 298 K (25 °C) is −229 kJ mol^{-1} and that of liquid water at the same temperature is −237 kJ mol^{-1}. It follows that there is a decrease in Gibbs energy when water vapour condenses to the liquid at 298 K, so condensation is spontaneous at that temperature (and 1 bar).

Self-test 4A.3 The standard Gibbs energies of formation of HN_3 at 298 K are +327 kJ mol^{-1} and +328 kJ mol^{-1} for the liquid and gas phases, respectively. Which phase of hydrogen azide is the more stable at that temperature and 1 bar?

Answer: Liquid

4A.2 Phase boundaries

The **phase diagram** of a pure substance shows the regions of pressure and temperature at which its various phases are thermodynamically stable (Fig. 4A.4). In fact, any two intensive variables may be used (such as temperature and magnetic field; in Topic 5A mole fraction is another variable), but in this Topic we concentrate on pressure and temperature. The lines separating the regions, which are called **phase boundaries** (or *coexistence curves*), show the values of p and T at which two phases coexist in equilibrium and their chemical potentials are equal.

(a) Characteristic properties related to phase transitions

Consider a liquid sample of a pure substance in a closed vessel. The pressure of a vapour in equilibrium with the liquid is called the **vapour pressure** of the substance (Fig. 4A.5). Therefore, the liquid–vapour phase boundary in a phase diagram shows how the vapour pressure of the liquid varies with temperature. Similarly, the solid–vapour phase boundary shows the temperature variation of the **sublimation vapour pressure**, the vapour pressure of the solid phase. The vapour pressure of a substance increases with temperature because at higher temperatures

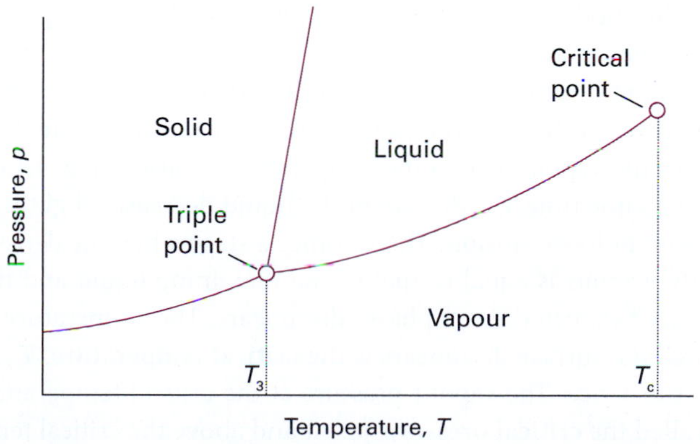

Figure 4A.4 The general regions of pressure and temperature where solid, liquid, or gas is stable (that is, has minimum molar Gibbs energy) are shown on this phase diagram. For example, the solid phase is the most stable phase at low temperatures and high pressures. In the following paragraphs we locate the precise boundaries between the regions.

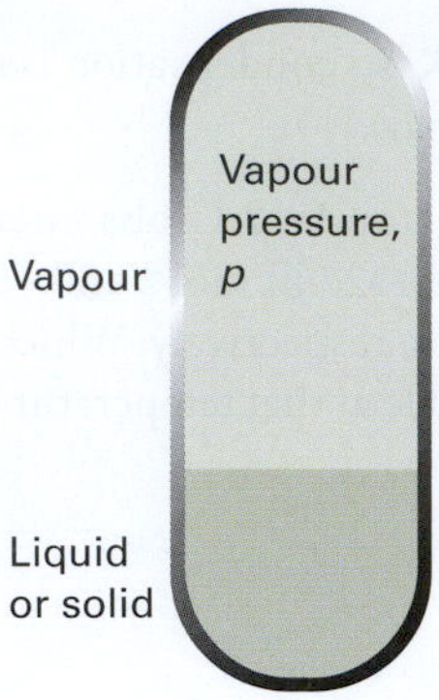

Figure 4A.5 The vapour pressure of a liquid or solid is the pressure exerted by the vapour in equilibrium with the condensed phase.

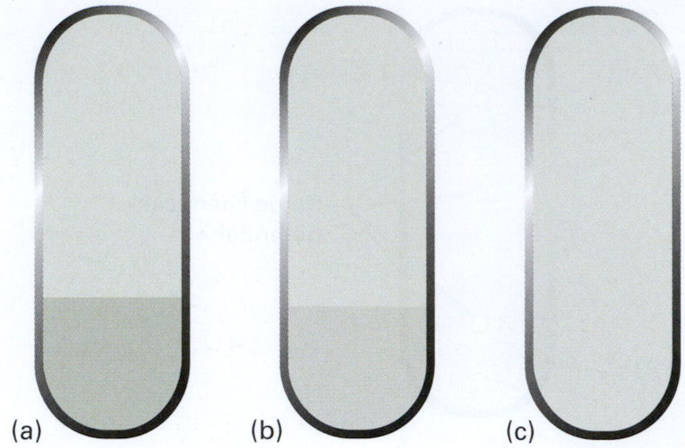

Figure 4A.6 (a) A liquid in equilibrium with its vapour. (b) When a liquid is heated in a sealed container, the density of the vapour phase increases and that of the liquid decreases slightly. There comes a stage (c) at which the two densities are equal and the interface between the fluids disappears. This disappearance occurs at the critical temperature. The container needs to be strong: the critical temperature of water is 374 °C and the vapour pressure is then 218 atm.

more molecules have sufficient energy to escape from their neighbours.

When a liquid is heated in an open vessel, the liquid vaporizes from its surface. When the vapour pressure is equal to the external pressure, vaporization can occur throughout the bulk of the liquid and the vapour can expand freely into the surroundings. The condition of free vaporization throughout the liquid is called **boiling**. The temperature at which the vapour pressure of a liquid is equal to the external pressure is called the **boiling temperature** at that pressure. For the special case of an external pressure of 1 atm, the boiling temperature is called the **normal boiling point**, T_b. With the replacement of 1 atm by 1 bar as standard pressure, there is some advantage in using the **standard boiling point** instead: this is the temperature at which the vapour pressure reaches 1 bar. Because 1 bar is slightly less than 1 atm (1.00 bar = 0.987 atm), the standard boiling point of a liquid is slightly lower than its normal boiling point. The normal boiling point of water is 100.0 °C; its standard boiling point is 99.6 °C. We need to distinguish normal and standard properties only for precise work in thermodynamics because any thermodynamic properties that we intend to add together must refer to the same conditions.

Boiling does not occur when a liquid is heated in a rigid, closed vessel. Instead, the vapour pressure, and hence the density of the vapour, rise as the temperature is raised (Fig. 4A.6). At the same time, the density of the liquid decreases slightly as a result of its expansion. There comes a stage when the density of the vapour is equal to that of the remaining liquid and the surface between the two phases disappears. The temperature at which the surface disappears is the **critical temperature**, T_c, of the substance. The vapour pressure at the critical temperature is called the **critical pressure**, p_c. At and above the critical temperature, a single uniform phase called a **supercritical fluid** fills the container and an interface no longer exists. That is, above the critical temperature, the liquid phase of the substance does not exist.

The temperature at which, under a specified pressure, the liquid and solid phases of a substance coexist in equilibrium is called the **melting temperature**. Because a substance melts at exactly the same temperature as it freezes, the melting temperature of a substance is the same as its **freezing temperature**. The freezing temperature when the pressure is 1 atm is called the **normal freezing point**, T_f, and its freezing point when the pressure is 1 bar is called the **standard freezing point**. The normal and standard freezing points are negligibly different for most purposes. The normal freezing point is also called the **normal melting point**.

There is a set of conditions under which three different phases of a substance (typically solid, liquid, and vapour) all simultaneously coexist in equilibrium. These conditions are represented by the **triple point**, a point at which the three phase boundaries meet. The temperature at the triple point is denoted T_3. The triple point of a pure substance is outside our control: it occurs at a single definite pressure and temperature characteristic of the substance.

As we can see from Fig. 4A.4, the triple point marks the lowest pressure at which a liquid phase of a substance can exist. If (as is common) the slope of the solid–liquid phase boundary is as shown in the diagram, then the triple point also marks the lowest temperature at which the liquid can exist; the critical temperature is the upper limit.

Brief illustration 4A.4 The triple point

The triple point of water lies at 273.16 K and 611 Pa (6.11 mbar, 4.58 Torr), and the three phases of water (ice, liquid water, and water vapour) coexist in equilibrium at no other combination of pressure and temperature. This invariance of the triple point was the basis of its use in the about-to-be superseded definition of the Kelvin scale of temperature (Topic 3A).

Self-test 4A.4 How many triple points are present (as far as it is known) in the full phase diagram for water shown later in this Topic in Fig. 4A.9?

Answer: 6

(b) The phase rule

In one of the most elegant arguments of the whole of chemical thermodynamics, which is presented in the following *Justification*, J.W. Gibbs deduced the **phase rule**, which gives the number of parameters that can be varied independently (at least to a small extent) while the number of phases in equilibrium is preserved. The phase rule is a general relation between the variance, F, the number of components, C, and the number of phases at equilibrium, P, for a system of any composition:

$$F = C - P + 2 \quad \text{The phase rule} \quad (4A.1)$$

A **component** is a *chemically independent* constituent of a system. The number of components, C, in a system is the minimum number of types of independent species (ions or molecules) necessary to define the composition of all the phases present in the system. In this chapter we deal only with one-component systems ($C=1$), so for this chapter

$$F = 3 - P \quad \text{A one-component system} \quad \text{The phase rule} \quad (4A.2)$$

By a **constituent** of a system we mean a chemical species that is present. The **variance** (or *number of degrees of freedom*), F, of a system is the number of intensive variables that can be changed independently without disturbing the number of phases in equilibrium.

Brief illustration 4A.5 The number of components

A mixture of ethanol and water has two constituents. A solution of sodium chloride has three constituents: water, Na^+ ions, and Cl^- ions but only two components because the numbers of Na^+ and Cl^- ions are constrained to be equal by the requirement of charge neutrality.

Self-test 4A.5 How many components are present in an aqueous solution of acetic acid, allowing for its partial deprotonation and the autoprotolysis of water?

Answer: 2

In a single-component, single-phase system ($C=1$, $P=1$), the pressure and temperature may be changed independently without changing the number of phases, so $F=2$. We say that such a system is **bivariant**, or that it has two **degrees of freedom**. On the other hand, if two phases are in equilibrium (a liquid and its vapour, for instance) in a single-component system ($C=1$, $P=2$), the temperature (or the pressure) can be changed at will, but the change in temperature (or pressure) demands an accompanying change in pressure (or temperature) to preserve the number of phases in equilibrium. That is, the variance of the system has fallen to 1.

Justification 4A.1 The phase rule

Consider first the special case of a one-component system for which the phase rule is $F=3-P$. For two phases α and β in equilibrium ($P=2$, $F=1$) at a given pressure and temperature, we can write

$$\mu(\alpha; p,T) = \mu(\beta; p,T)$$

(For instance, when ice and water are in equilibrium, we have $\mu(s; p,T)=\mu(l; p,T)$ for H_2O.) This is an equation relating p and T, so only one of these variables is independent (just as the equation $x+y=xy$ is a relation for y in terms of x: $y=x/(x-1)$). That conclusion is consistent with $F=1$. For three phases of a one-component system in mutual equilibrium ($P=3$, $F=0$),

$$\mu(\alpha; p,T) = \mu(\beta; p,T) = \mu(\gamma; p,T)$$

This relation is actually two equations for two unknowns, $\mu(\alpha; p,T)=\mu(\beta; p,T)$ and $\mu(\beta; p,T)=\mu(\gamma; p,T)$, and therefore has a solution only for a single value of p and T (just as the pair of equations $x+y=xy$ and $3x-y=xy$ has the single solution $x=2$ and $y=2$). That conclusion is consistent with $F=0$. Four phases cannot be in mutual equilibrium in a one-component system because the three equalities

$$\mu(\alpha; p,T) = \mu(\beta; p,T)$$
$$\mu(\beta; p,T) = \mu(\gamma; p,T)$$
$$\mu(\gamma; p,T) = \mu(\delta; p,T)$$

are three equations for two unknowns (p and T) and are not consistent (just as $x+y=xy$, $3x-y=xy$, and $x+y=2xy^2$ have no solution).

Now consider the general case. We begin by counting the total number of intensive variables. The pressure, p, and temperature, T, count as 2. We can specify the composition of a phase by giving the mole fractions of $C-1$ components. We need specify only $C-1$ and not all C mole fractions because $x_1+x_2+\cdots+x_C=1$, and all mole fractions are known if all except one are specified. Because there are P phases, the total number of composition variables is $P(C-1)$. At this stage, the total number of intensive variables is $P(C-1)+2$.

At equilibrium, the chemical potential of a component J must be the same in every phase:

$$\mu(\alpha; p,T) = \mu(\beta; p,T) = \cdots \text{ for } P \text{ phases}$$

That is, there are $P-1$ equations of this kind to be satisfied for each component J. As there are C components, the total number of equations is $C(P-1)$. Each equation reduces our freedom to vary one of the $P(C-1)+2$ intensive variables. It follows that the total number of degrees of freedom is

$$F=P(C-1)+2-C(P-1)=C-P+2$$

which is eqn 4A.1.

4A.3 Three representative phase diagrams

For a one-component system, such as pure water, $F=3-P$. When only one phase is present, $F=2$ and both p and T can be varied independently (at least over a small range) without changing the number of phases. In other words, a single phase is represented by an *area* on a phase diagram. When two phases are in equilibrium $F=1$, which implies that pressure is not freely variable if the temperature is set; indeed, at a given temperature, a liquid has a characteristic vapour pressure. It follows that the equilibrium of two phases is represented by a *line* in the phase diagram. Instead of selecting the temperature, we could select the pressure, but having done so the two phases would be in equilibrium at a single definite temperature. Therefore, freezing (or any other phase transition) occurs at a definite temperature at a given pressure.

When three phases are in equilibrium, $F=0$ and the system is invariant. This special condition can be established only at a definite temperature and pressure that is characteristic of the substance and outside our control. The equilibrium of three phases is therefore represented by a *point*, the triple point, on a phase diagram. Four phases cannot be in equilibrium in a one-component system because F cannot be negative.

Brief illustration 4A.6 Characteristics of phase diagrams

Figure 4A.7 shows a reasonably typical phase diagram of a single pure substance, with one forbidden feature, the 'quadruple point' where phases β, γ, δ, and ε are said to be in equilibrium. Two triple points are shown (for the equilibria $\alpha \rightleftharpoons \beta \rightleftharpoons \gamma$ and $\alpha \rightleftharpoons \beta \rightleftharpoons \delta$, respectively), corresponding to $P=3$ and $F=0$. The lines represent various equilibria, including $\alpha \rightleftharpoons \beta$, $\alpha \rightleftharpoons \delta$, and $\gamma \rightleftharpoons \varepsilon$.

Self-test 4A.6 What is the minimum number of components necessary before five phases can be in mutual equilibrium in a system?

Answer: 3

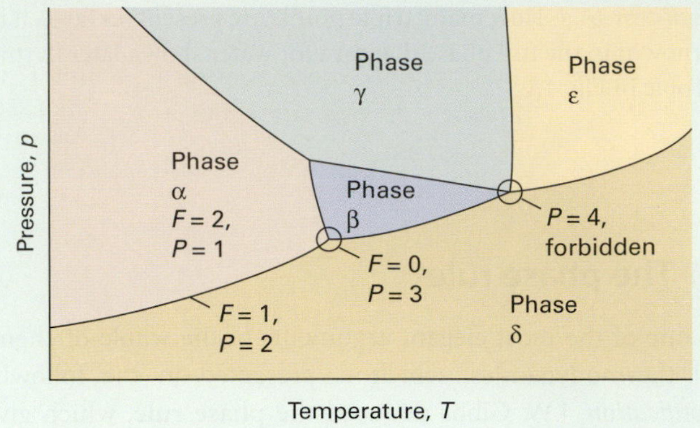

Figure 4A.7 The typical regions of a one-component phase diagram. The lines represent conditions under which the two adjoining phases are in equilibrium. A point represents the unique set of conditions under which three phases coexist in equilibrium. Four phases cannot mutually coexist in equilibrium.

(a) Carbon dioxide

The phase diagram for carbon dioxide is shown in Fig. 4A.8. The features to notice include the positive slope (up from left to right) of the solid–liquid boundary; the direction of this line is characteristic of most substances. This slope indicates that the melting temperature of solid carbon dioxide rises as the pressure is increased. Notice also that, as the triple point lies above 1 atm, the liquid cannot exist at normal atmospheric pressures whatever the temperature. As a result, the solid sublimes when left in the open (hence the name 'dry ice'). To obtain the liquid, it is necessary to exert a pressure of at least 5.11 atm. Cylinders of carbon dioxide generally contain the liquid or compressed gas; at 25 °C that implies a vapour pressure of 67 atm if both

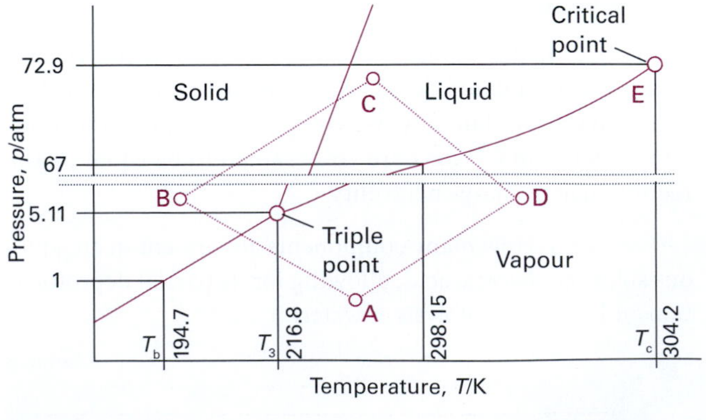

Figure 4A.8 The experimental phase diagram for carbon dioxide. Note that, as the triple point lies at pressures well above atmospheric, liquid carbon dioxide does not exist under normal conditions (a pressure of at least 5.11 atm must be applied). The path ABCD is discussed in *Brief illustration* 4A.7

gas and liquid are present in equilibrium. When the gas squirts through the throttle it cools by the Joule–Thomson effect, so when it emerges into a region where the pressure is only 1 atm, it condenses into a finely divided snow-like solid. That carbon dioxide gas cannot be liquefied except by applying high pressure reflects the weakness of the intermolecular forces between the nonpolar carbon dioxide molecules (Topic 16B).

Brief illustration 4A.7 A phase diagram 1

Consider the path ABCD in Fig. 4A.8. At A the carbon dioxide is a gas. When the temperature and pressure are adjusted to B, the vapour condenses directly to a solid. Increasing the pressure and temperature to C results in the formation of the liquid phase, which evaporates to the vapour when the conditions are changed to D.

Self-test 4A.7 Describe what happens on circulating around the critical point, Path E.

Answer: Liquid → $scCO_2$ → vapour → liquid

(b) Water

Figure 4A.9 shows the phase diagram for water. The liquid–vapour boundary in the phase diagram summarizes how the vapour pressure of liquid water varies with temperature. It also summarizes how the boiling temperature varies with pressure: we simply read off the temperature at which the vapour pressure is equal to the prevailing atmospheric pressure. The solid–liquid boundary shows how the melting temperature varies with the pressure; its very steep slope indicates that enormous pressures are needed to bring about significant changes. Notice that the line has a negative slope (down from left to right) up to 2 kbar, which means that the melting temperature falls as the

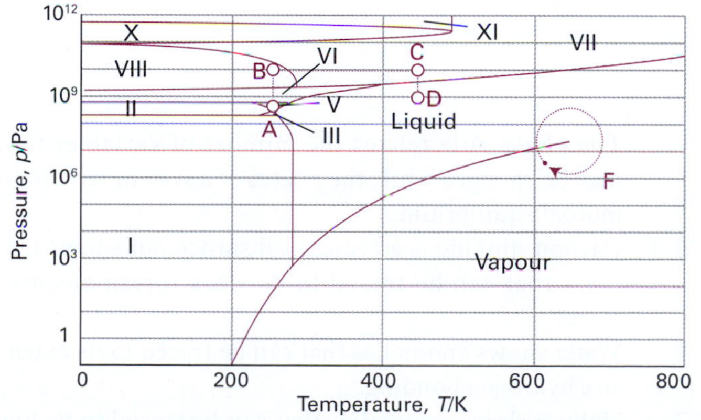

Figure 4A.9 The experimental phase diagram for water showing the different solid phases. The path ABCD is discussed in *Brief illustration* 4A.8.

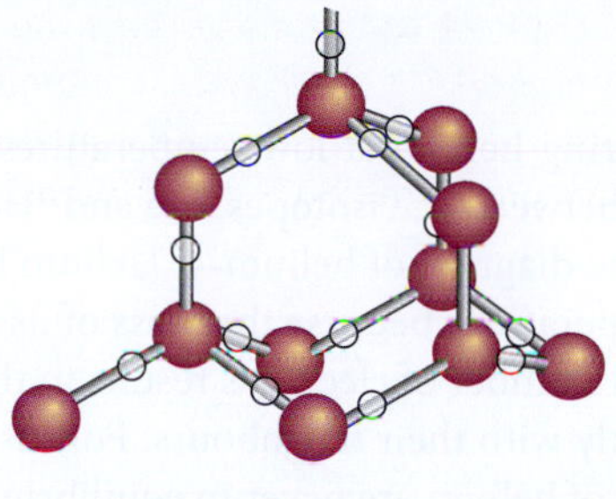

Figure 4A.10 A fragment of the structure of ice (ice-I). Each O atom is linked by two covalent bonds to H atoms and by two hydrogen bonds to a neighbouring O atom, in a tetrahedral array.

pressure is raised. The reason for this almost unique behaviour can be traced to the decrease in volume that occurs on melting: it is more favourable for the solid to transform into the liquid as the pressure is raised. The decrease in volume is a result of the very open structure of ice: as shown in Fig. 4A.10, the water molecules are held apart, as well as together, by the hydrogen bonds between them but the hydrogen-bonded structure partially collapses on melting and the liquid is denser than the solid. Other consequences of its extensive hydrogen bonding are the anomalously high boiling point of water for a molecule of its molar mass and its high critical temperature and pressure.

Figure 4A.9 shows that water has one liquid phase but many different solid phases other than ordinary ice ('ice I'). Some of these phases melt at high temperatures. Ice VII, for instance, melts at 100 °C but exists only above 25 kbar. Two further phases, Ice XIII and XIV, were identified in 2006 at −160 °C but have not yet been allocated regions in the phase diagram. Note that five more triple points occur in the diagram other than the one where vapour, liquid, and ice I coexist. Each one occurs at a definite pressure and temperature that cannot be changed. The solid phases of ice differ in the arrangement of the water molecules: under the influence of very high pressures, hydrogen bonds buckle and the H_2O molecules adopt different arrangements. These polymorphs of ice may contribute to the advance of glaciers, for ice at the bottom of glaciers experiences very high pressures where it rests on jagged rocks.

Brief illustration 4A.8 A phase diagram 2

Consider the path ABCD in Fig. 4A.9. At A, water is present as ice V. Increasing the pressure to B at the same temperature results in the formation of a polymorph, ice VIII. Heating to C leads to the formation of ice VII, and reduction in pressure to D results in the solid melting to liquid.

Self-test 4A.8 Describe what happens on circulating around the critical point, Path F.

Answer: Vapour → liquid → scH_2O → vapour

(c) Helium

When considering helium at low temperatures it is necessary to distinguish between the isotopes ^{3}He and ^{4}He. Figure 4A.11 shows the phase diagram of helium-4. Helium behaves unusually at low temperatures because the mass of its atoms is so low and their small number of electrons results in them interacting only very weakly with their neighbours. For instance, the solid and gas phases of helium are never in equilibrium however low the temperature: the atoms are so light that they vibrate with a large-amplitude motion even at very low temperatures and the solid simply shakes itself apart. Solid helium can be obtained, but only by holding the atoms together by applying pressure. The isotopes of helium behave differently for quantum mechanical reasons that are explained in Part 2. (The difference stems from the different nuclear spins of the isotopes and the role of the Pauli exclusion principle: helium-4 has $I=0$ and is a boson; helium-3 has $I=\frac{1}{2}$ and is a fermion.)

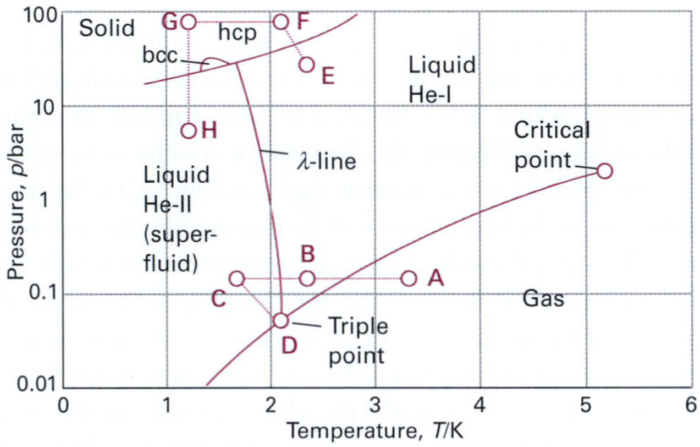

Figure 4A.11 The phase diagram for helium (^{4}He). The λ-line marks the conditions under which the two liquid phases are in equilibrium. Helium-II is the superfluid phase. Note that a pressure of over 20 bar must be exerted before solid helium can be obtained. The labels hcp and bcc denote different solid phases in which the atoms pack together differently: hcp denotes hexagonal closed packing and bcc denotes body-centred cubic (see Topic 18B for a description of these structures). The path ABCD is discussed in *Brief illustration* 4A.9.

Pure helium-4 has two liquid phases. The phase marked He-I in the diagram behaves like a normal liquid; the other phase, He-II, is a **superfluid**; it is so called because it flows without viscosity.[1] Provided we discount the liquid crystalline substances discussed in *Impact* I5.1 on line, helium is the only known substance with a liquid–liquid boundary, shown as the **λ-line** (lambda line) in Fig. 4A.11.

The phase diagram of helium-3 differs from the phase diagram of helium-4, but it also possesses a superfluid phase. Helium-3 is unusual in that melting is exothermic ($\Delta_{fus}H<0$) and therefore (from $\Delta_{fus}S=\Delta_{fus}H/T_f$) at the melting point the entropy of the liquid is lower than that of the solid.

Brief illustration 4A.9 A phase diagram 3

Consider the path ABCD in Fig. 4A.11. At A, helium is present as a vapour. On cooling to B it condenses to helium-I, and further cooling to C results in the formation of helium-II. Adjustment of the pressure and temperature to D results in a system in which three phases, helium-I, helium-II, and vapour, are in mutual equilibrium.

Self-test 4A.9 Describe what happens on the path EFGH.

Answer: He-I $\rightarrow$ solid $\rightarrow$ solid $\rightarrow$ He-II

[1] Water might also have a superfluid liquid phase.

Checklist of concepts

- ☐ 1. A **phase** is a form of matter that is uniform throughout in chemical composition and physical state.
- ☐ 2. A **phase transition** is the spontaneous conversion of one phase into another and may be studied by techniques that include thermal analysis.
- ☐ 3. The thermodynamic analysis of phases is based on the fact that at equilibrium, the chemical potential of a substance is the same throughout a sample.
- ☐ 4. A substance is characterized by a variety of parameters that can be identified on its **phase diagram**.
- ☐ 5. The **phase rule** relates the number of variables that may be changed while the phases of a system remain in mutual equilibrium.
- ☐ 6. Carbon dioxide is a typical substance but shows features that can be traced to its weak intermolecular forces.
- ☐ 7. Water shows anomalies that can be traced to its extensive hydrogen bonding.
- ☐ 8. Helium shows anomalies that can be traced to its low mass and weak interactions.

Checklist of equations

Property	Equation	Comment	Equation number
Chemical potential	$\mu = G_m$	For a pure substance	
Phase rule	$F = C - P + 2$		4A.1

4B Thermodynamic aspects of phase transitions

Contents

➤ Why do you need to know this material?

This Topic illustrates how thermodynamics is used to discuss the equilibria of the phases of one-component systems and shows how to make predictions about the effect of pressure on freezing and boiling points.

➤ What is the key idea?

The effect of temperature and pressure on the chemical potentials of phases in equilibrium is determined by the molar entropy and molar volume, respectively, of the phases.

➤ What do you need to know already?

You need to be aware that phases are in equilibrium when their chemical potentials are equal (Topic 4A) and that the variation of the molar Gibbs energy of a substance depends on its molar volume and entropy (Topic 3D). We draw on expressions for the entropy of transition (Topic 3B) and the perfect gas law (Topic 1A).

As explained in Topic 4A, the thermodynamic criterion of phase equilibrium is the equality of the chemical potentials of each phase. For a one-component system, the chemical potential is the same as the molar Gibbs energy of the phase. As Topic 3D explains how the Gibbs energy varies with temperature and pressure, by combining these two aspects, we can expect to be able to deduce how phase equilibria vary as the conditions are changed.

4B.1 The dependence of stability on the conditions

At very low temperatures and provided the pressure is not too low, the solid phase of a substance has the lowest chemical potential and is therefore the most stable phase. However, the chemical potentials of different phases change with temperature in different ways, and above a certain temperature the chemical potential of another phase (perhaps another solid phase, a liquid, or a gas) may turn out to be the lowest. When that happens, a transition to the second phase is spontaneous and occurs if it is kinetically feasible to do so.

(a) The temperature dependence of phase stability

The temperature dependence of the Gibbs energy is expressed in terms of the entropy of the system by eqn 3D.8 ($(\partial G/\partial T)_p = -S$). Because the chemical potential of a pure substance is just another name for its molar Gibbs energy, it follows that

$$\left(\frac{\partial \mu}{\partial T}\right)_p = -S_m \qquad \text{Variation of chemical potential with } T \qquad (4B.1)$$

This relation shows that, as the temperature is raised, the chemical potential of a pure substance decreases: $S_m > 0$ for all substances, so the slope of a plot of μ against T is negative.

Equation 4B.1 implies that because $S_m(g) > S_m(l)$ the slope of a plot of μ against temperature is steeper for gases than for liquids. Because $S_m(l) > S_m(s)$ almost always, the slope is also steeper for a liquid than the corresponding solid. These features are illustrated in Fig. 4B.1. The steep negative slope of $\mu(l)$ results in it falling below $\mu(s)$ when the temperature is high enough, and then the liquid becomes the stable phase: the solid melts. The chemical potential of the gas phase plunges steeply downwards as the temperature is raised (because the molar entropy of the vapour is so high), and there comes a temperature at which it lies lowest. Then the gas is the stable phase and vaporization is spontaneous.

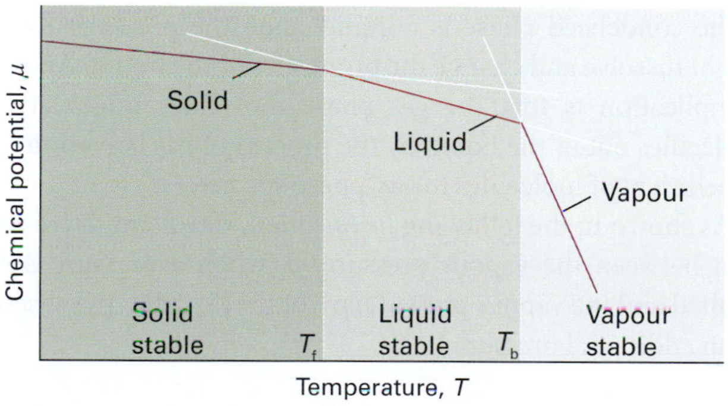

Figure 4B.1 The schematic temperature dependence of the chemical potential of the solid, liquid, and gas phases of a substance (in practice, the lines are curved). The phase with the lowest chemical potential at a specified temperature is the most stable one at that temperature. The transition temperatures, the melting and boiling temperatures (T_f and T_b, respectively), are the temperatures at which the chemical potentials of the two phases are equal.

Brief illustration 4B.1 The temperature variation of μ

The standard molar entropy of liquid water at 100 °C is 86.8 J K^{-1} mol^{-1} and that of water vapour at the same temperature is 195.98 J K^{-1} mol^{-1}. It follows that when the temperature is raised by 1.0 K the changes in chemical potential are

$$\delta\mu(l) \approx S_m(l)\delta T = 87\,\text{J mol}^{-1} \qquad \delta\mu(g) \approx S_m(g)\delta T = 196\,\text{J mol}^{-1}$$

At 100 °C the two phases are in equilibrium with equal chemical potentials, so at 1.0 K higher the chemical potential of the vapour is lower (by 109 J mol^{-1}) than that of the liquid and vaporization is spontaneous.

Self-test 4B.1 The standard molar entropy of liquid water at 0 °C is 65 J K^{-1} mol^{-1} and that of ice at the same temperature is 43 J K^{-1} mol^{-1}. What is the effect of increasing the temperature by 1.0 K?

Answer: $\delta\mu(l) \approx -65$ J mol^{-1}, $\delta\mu(s) \approx -43$ J mol^{-1}; ice melts

(b) The response of melting to applied pressure

Most substances melt at a higher temperature when subjected to pressure. It is as though the pressure is preventing the formation of the less dense liquid phase. Exceptions to this behaviour include water, for which the liquid is denser than the solid. Application of pressure to water encourages the formation of the liquid phase. That is, water freezes and ice melts at a lower temperature when it is under pressure.

We can rationalize the response of melting temperatures to pressure as follows. The variation of the chemical potential with pressure is expressed (from the second of eqns 3D.8, $(\partial G/\partial p)_T = V$) by

$$\left(\frac{\partial \mu}{\partial p}\right)_T = V_m \qquad \text{Variation of chemical potential with } p \qquad (4B.2)$$

This equation shows that the slope of a plot of chemical potential against pressure is equal to the molar volume of the substance. An increase in pressure raises the chemical potential of any pure substance (because $V_m > 0$). In most cases, $V_m(l) > V_m(s)$ and the equation predicts that an increase in pressure increases the chemical potential of the liquid more than that of the solid. As shown in Fig. 4B.2a, the effect of pressure in such a case is to raise the melting temperature slightly. For water, however, $V_m(l) < V_m(s)$, and an increase in pressure increases the chemical potential of the solid more than that of the liquid. In this case, the melting temperature is lowered slightly (Fig. 4B.2b).

Example 4B.1 Assessing the effect of pressure on the chemical potential

Calculate the effect on the chemical potentials of ice and water of increasing the pressure from 1.00 bar to 2.00 bar at 0 °C. The density of ice is 0.917 g cm^{-3} and that of liquid water is 0.999 g cm^{-3} under these conditions.

Method From eqn 4B.2 in the form $d\mu = V_m dp$, we know that the change in chemical potential of an incompressible substance when the pressure is changed by Δp is $\Delta\mu = V_m\Delta p$. Therefore, to answer the question, we need to know the molar volumes of the two phases of water. These values are obtained from the mass density, ρ, and the molar mass, M, by using $V_m = M/\rho$. We therefore use the expression $\Delta\mu = M\Delta p/\rho$.

Answer The molar mass of water is 18.02 g mol⁻¹ (1.802×10^{-2} kg mol⁻¹); therefore,

$$\Delta\mu(\text{ice}) = \frac{(1.802\times10^{-2}\,\text{kg mol}^{-1})\times(1.00\times10^{5}\,\text{Pa})}{917\,\text{kg m}^{-3}} = +1.97\,\text{J mol}^{-1}$$

$$\Delta\mu(\text{water}) = \frac{(1.802\times10^{-2}\,\text{kg mol}^{-1})\times(1.00\times10^{5}\,\text{Pa})}{999\,\text{kg m}^{-3}} = +1.80\,\text{J mol}^{-1}$$

We interpret the numerical results as follows: the chemical potential of ice rises more sharply than that of water, so if they are initially in equilibrium at 1 bar, then there will be a tendency for the ice to melt at 2 bar.

Self-test 4B.2 Calculate the effect of an increase in pressure of 1.00 bar on the liquid and solid phases of carbon dioxide (molar mass 44.0 g mol⁻¹) in equilibrium with densities 2.35 g cm⁻³ and 2.50 g cm⁻³, respectively.

Answer: $\Delta\mu(\text{l}) = +1.87$ J mol⁻¹, $\Delta\mu(\text{s}) = +1.76$ J mol⁻¹; solid forms

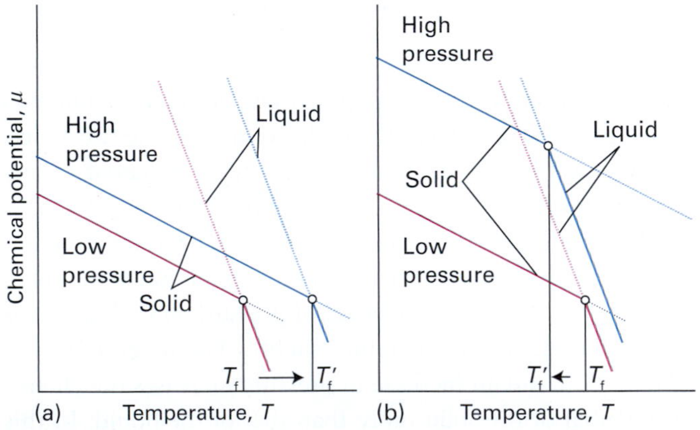

Figure 4B.2 The pressure dependence of the chemical potential of a substance depends on the molar volume of the phase. The lines show schematically the effect of increasing pressure on the chemical potential of the solid and liquid phases (in practice, the lines are curved), and the corresponding effects on the freezing temperatures. (a) In this case the molar volume of the solid is smaller than that of the liquid and μ(s) increases less than μ(l). As a result, the freezing temperature rises. (b) Here the molar volume is greater for the solid than the liquid (as for water), μ(s) increases more strongly than μ(l), and the freezing temperature is lowered.

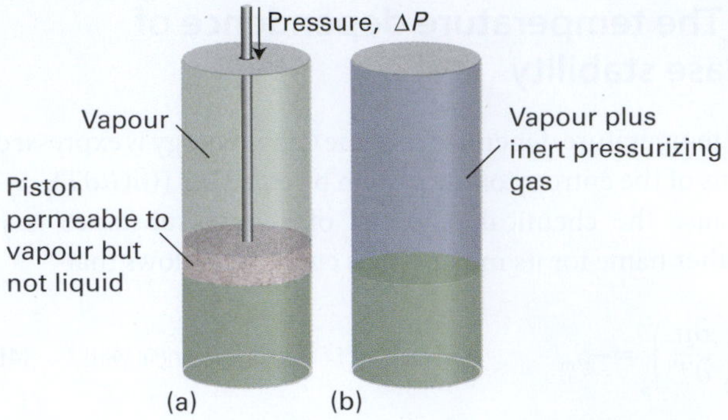

Figure 4B.3 Pressure may be applied to a condensed phase either (a) by compressing the condensed phase or (b) by subjecting it to an inert pressurizing gas. When pressure is applied, the vapour pressure of the condensed phase increases.

(c) The vapour pressure of a liquid subjected to pressure

When pressure is applied to a condensed phase, its vapour pressure rises: in effect, molecules are squeezed out of the phase and escape as a gas. Pressure can be exerted on the condensed phase mechanically or by subjecting it to the applied pressure of an inert gas (Fig. 4B.3). In the latter case, the vapour pressure is the partial pressure of the vapour in equilibrium with the condensed phase. We then speak of the **partial vapour pressure** of the substance. One complication (which we ignore here) is that, if the condensed phase is a liquid, then the pressurizing gas might dissolve and change the properties of the liquid. Another complication is that the gas phase molecules might attract molecules out of the liquid by the process of **gas solvation**, the attachment of molecules to gas-phase species.

As shown in the following *Justification*, the quantitative relation between the vapour pressure, p, when a pressure ΔP is applied and the vapour pressure, p^*, of the liquid in the absence of an additional pressure is

$$p = p^* e^{V_m(\text{l})\Delta P/RT} \qquad \text{Effect of applied pressure } \Delta P \text{ on vapour pressure } p \qquad (4B.3)$$

This equation shows how the vapour pressure increases when the pressure acting on the condensed phase is increased.

Justification 4B.1 The vapour pressure of a pressurized liquid

We calculate the vapour pressure of a pressurized liquid by using the fact that at equilibrium the chemical potentials of the liquid and its vapour are equal: $\mu(\text{l}) = \mu(\text{g})$. It follows that, for any change that preserves equilibrium, the resulting change in μ(l) must be equal to the change in μ(g); therefore, we can write $d\mu(\text{g}) = d\mu(\text{l})$. When the pressure P on the liquid is increased by dP, the chemical potential of the liquid changes by $d\mu(\text{l}) = V_m(\text{l})dP$. The chemical potential of the vapour changes by $d\mu(g)=V_m(g)dp$ where dp is the change in the vapour pressure we are trying to find. If we treat the vapour as a perfect gas, the molar volume can be replaced by $V_m(g)=RT/p$, and we obtain $d\mu(g)=RTdp/p$. Next, we equate the changes in chemical potentials of the vapour and the liquid:

$$\frac{RTdp}{p}=V_m(l)dP$$

We can integrate this expression once we know the limits of integration.

When there is no additional pressure acting on the liquid, P (the pressure experienced by the liquid) is equal to the normal vapour pressure p^*, so when $P=p^*$, $p=p^*$ too. When there is an additional pressure ΔP on the liquid, with the result that $P=p+\Delta P$, the vapour pressure is p (the value we want to find). Provided the effect of pressure on the vapour pressure is small (as will turn out to be the case) a good approximation is to replace the p in $p+\Delta P$ by p^* itself, and to set the upper limit of the integral to $p^*+\Delta P$. The integrations required are therefore as follows:

$$RT\int_{p^*}^{p}\frac{dp}{p}=\int_{p^*}^{p^*+\Delta P}V_m(l)\,dP$$

We now divide both sides by RT and assume that the molar volume of the liquid is the same throughout the small range of pressures involved:

$$\int_{p^*}^{p}\frac{dp}{p}=\frac{1}{RT}\int_{p^*}^{p^*+\Delta P}V_m(l)\,dP=\frac{V_m(l)}{RT}\int_{p^*}^{p^*+\Delta P}dP$$

Then both integrations are straightforward, and lead to

$$\ln\frac{p}{p^*}=\frac{V_m(l)}{RT}\Delta P$$

which rearranges to eqn 4B.3 because $e^{\ln x}=x$.

Brief illustration 4B.2 The effect of pressurization

For water, which has density 0.997 g cm^{-3} at 25 °C and therefore molar volume 18.1 cm^3 mol^{-1}, when the pressure is increased by 10 bar (that is, $\Delta P=1.0\times10^6$ Pa)

$$\frac{V_m(l)\Delta P}{RT}=\frac{(1.81\times10^{-5}\,\text{m}^3\,\text{mol}^{-1})\times(1.0\times10^6\,\text{Pa})}{(8.3145\,\text{J K}^{-1}\,\text{mol}^{-1})\times(298\,\text{K})}$$

$$=\frac{1.81\times1.0\times10^1}{8.3145\times298}=0.0073\ldots$$

where we have used 1 J = 1 Pa m^3. It follows that $p=1.0073p^*$, an increase of 0.73 per cent.

Self-test 4B.3 Calculate the effect of an increase in pressure of 100 bar on the vapour pressure of benzene at 25 °C, which has density 0.879 g cm^{-3}.

Answer: 43 per cent increase

4B.2 The location of phase boundaries

The precise locations of the phase boundaries—the pressures and temperatures at which two phases can coexist—can be found by making use of the fact that, when two phases are in equilibrium, their chemical potentials must be equal. Therefore, where the phases α and β are in equilibrium,

$$\mu(\alpha;p,T)=\mu(\beta;p,T) \qquad (4B.4)$$

By solving this equation for p in terms of T, we get an equation for the phase boundary.

(a) The slopes of the phase boundaries

It turns out to be simplest to discuss the phase boundaries in terms of their slopes, dp/dT. Let p and T be changed infinitesimally, but in such a way that the two phases α and β remain in equilibrium. The chemical potentials of the phases are initially equal (the two phases are in equilibrium). They remain equal when the conditions are changed to another point on the phase boundary, where the two phases continue to be in equilibrium (Fig. 4B.4). Therefore, the changes in the chemical potentials of the two phases must be equal and we can write $d\mu(\alpha)=d\mu(\beta)$. Because, from eqn 3D.7 ($dG=Vdp-SdT$), we know that $d\mu=-S_mdT+V_mdp$ for each phase, it follows that

$$-S_m(\alpha)dT+V_m(\alpha)dp=-S_m(\beta)dT+V_m(\beta)dp$$

where $S_m(\alpha)$ and $S_m(\beta)$ are the molar entropies of the phases and $V_m(\alpha)$ and $V_m(\beta)$ are their molar volumes. Hence

$$\{S_m(\beta)-S_m(\alpha)\}dT=\{V_m(\beta)-V_m(\alpha)\}dp$$

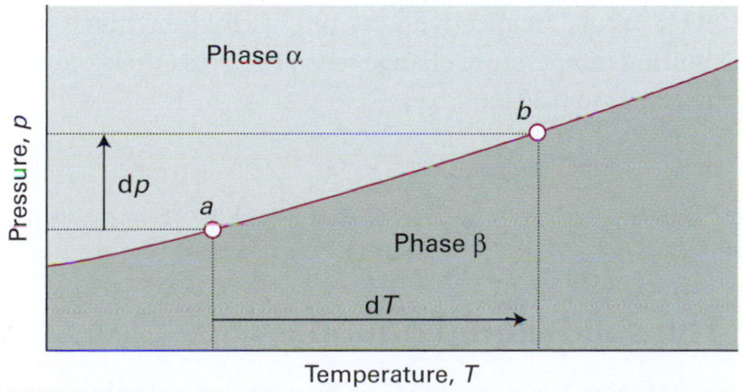

Figure 4B.4 When pressure is applied to a system in which two phases are in equilibrium (at *a*), the equilibrium is disturbed. It can be restored by changing the temperature, so moving the state of the system to *b*. It follows that there is a relation between d*p* and d*T* that ensures that the system remains in equilibrium as either variable is changed.

Then, with $\Delta_{trs}S=S_m(\beta)-S_m(\alpha)$ and $\Delta_{trs}V=V_m(\beta)-V_m(\alpha)$, which are the (molar) entropy and volume of transition, respectively,

$$\Delta_{trs}S\mathrm{d}T=\Delta_{trs}V\mathrm{d}p$$

This relation rearranges into the **Clapeyron equation**:

$$\frac{\mathrm{d}p}{\mathrm{d}T}=\frac{\Delta_{trs}S}{\Delta_{trs}V} \quad \text{Clapeyron equation} \quad (4B.5a)$$

The Clapeyron equation is an exact expression for the slope of the tangent to the boundary at any point and applies to any phase equilibrium of any pure substance. It implies that we can use thermodynamic data to predict the appearance of phase diagrams and to understand their form. A more practical application is to the prediction of the response of freezing and boiling points to the application of pressure, when it can be used in the form obtained by inverting both sides:

$$\frac{\mathrm{d}T}{\mathrm{d}p}=\frac{\Delta_{trs}V}{\Delta_{trs}S} \quad (4B.5b)$$

Brief illustration 4B.3 The Clapeyron equation

The standard volume and entropy of transition of water from ice to liquid are $-1.6\,\mathrm{cm^3\,mol^{-1}}$ and $+22\,\mathrm{J\,K^{-1}\,mol^{-1}}$, respectively, at 0 °C. The slope of the solid–liquid phase boundary at that temperature is therefore

$$\frac{\mathrm{d}T}{\mathrm{d}p}=\frac{-1.6\times10^{-6}\,\mathrm{m^3\,mol^{-1}}}{22\,\mathrm{J^{-1}\,mol^{-1}}}=-7.3\times10^{-8}\frac{\mathrm{K}}{\mathrm{J\,m^{-3}}}=-7.3\times10^{-8}\,\mathrm{K\,Pa^{-1}}$$

which corresponds to $-7.3\,\mathrm{mK\,bar^{-1}}$. An increase of 100 bar therefore results in a lowering of the freezing point of water by 0.73 K.

Self-test 4B.4 The standard volume and entropy of transition of water from liquid to vapour are $+30\,\mathrm{dm^3\,mol^{-1}}$ and $+109\,\mathrm{J\,K^{-1}\,mol^{-1}}$, respectively, at 100 °C. By how much does the boiling temperature change when the pressure is reduced from 1.0 bar to 0.80 bar?

Answer: −5.5 K

(b) The solid–liquid boundary

Melting (fusion) is accompanied by a molar enthalpy change $\Delta_{fus}H$ and occurs at a temperature T. The molar entropy of melting at T is therefore $\Delta_{fus}H/T$ (Topic 3B), and the Clapeyron equation becomes

$$\frac{\mathrm{d}p}{\mathrm{d}T}=\frac{\Delta_{fus}H}{T\Delta_{fus}V} \quad \text{Slope of solid–liquid boundary} \quad (4B.6)$$

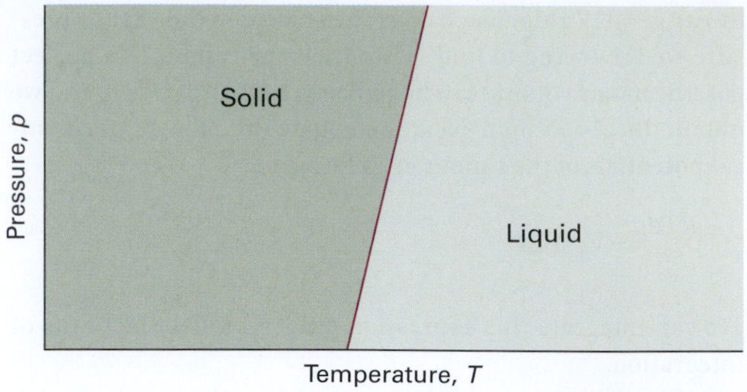

Figure 4B.5 A typical solid–liquid phase boundary slopes steeply upwards. This slope implies that, as the pressure is raised, the melting temperature rises. Most substances behave in this way.

where $\Delta_{fus}V$ is the change in molar volume that occurs on melting. The enthalpy of melting is positive (the only exception is helium-3) and the volume change is usually positive and always small. Consequently, the slope $\mathrm{d}p/\mathrm{d}T$ is steep and usually positive (Fig. 4B.5).

We can obtain the formula for the phase boundary by integrating $\mathrm{d}p/\mathrm{d}T$, assuming that $\Delta_{fus}H$ and $\Delta_{fus}V$ change so little with temperature and pressure that they can be treated as constant. If the melting temperature is T^* when the pressure is p^*, and T when the pressure is p, the integration required is

$$\int_{p^*}^{p}\mathrm{d}p=\frac{\Delta_{fus}H}{\Delta_{fus}V}\int_{T^*}^{T}\frac{\mathrm{d}T}{T}$$

Therefore, the approximate equation of the solid–liquid boundary is

$$p=p^*+\frac{\Delta_{fus}H}{\Delta_{fus}V}\ln\frac{T}{T^*} \quad (4B.7)$$

This equation was originally obtained by yet another Thomson—James, the brother of William, Lord Kelvin. When T is close to T^*, the logarithm can be approximated by using

$$\ln\frac{T}{T^*}=\ln\left(1+\frac{T-T^*}{T^*}\right)\approx\frac{T-T^*}{T^*}$$

where we have used the expansion $\ln(1+x)=x-\frac{1}{2}x^2+\cdots$ (*Mathematical background* 1) and neglected all but the leading term; therefore

$$p\approx p^*+\frac{\Delta_{fus}H}{T^*\Delta_{fus}V}(T-T^*) \quad (4B.8)$$

This expression is the equation of a steep straight line when p is plotted against T (as in Fig. 4B.5).

Brief illustration 4B.4 The solid–liquid boundary

The enthalpy of fusion of ice at 0 °C and 1 bar (273 K) is 6.008 kJ mol^{-1} and the volume of fusion is $-1.6\,\text{cm}^3\,\text{mol}^{-1}$. It follows that the solid–liquid phase boundary is given by the equation

$$p \approx 1\,\text{bar} + \frac{6.008\times10^3\,\text{J mol}^{-1}}{(273\,\text{K})\times(-1.6\times10^{-6}\,\text{m}^3\,\text{mol}^{-1})}(T-T^*)$$
$$\approx 1\,\text{bar} - 1.4\times10^7\,\text{Pa K}^{-1}(T-T^*)$$

That is,

$$p/\text{bar} = 1 - 140(T-T^*)/\text{K}$$

with $T^* = 273\,\text{K}$. This expression is plotted in Fig. 4B.6.

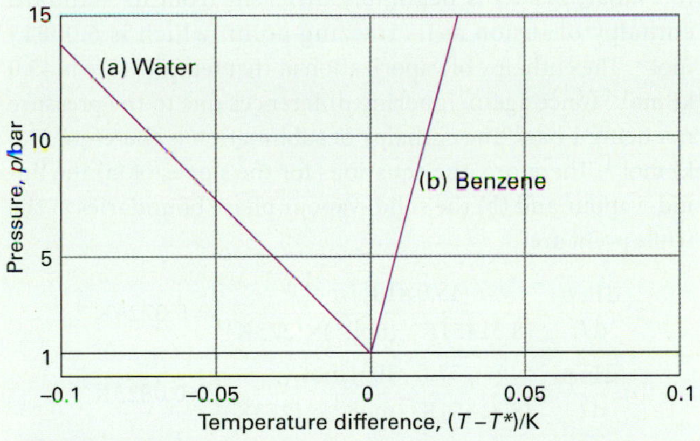

Figure 4B.6 The solid–liquid phase boundaries (the melting point curves) for water and benzene, as calculated in *Brief illustration* 4B.4.

Self-test 4B.5 The enthalpy of fusion of benzene is 10.59 kJ mol^{-1} at its melting point of 279 K and its volume of fusion is close to $+0.50\,\text{cm}^3\,\text{mol}^{-1}$ (an estimated value). What is the equation of its solid–liquid phase boundary?

Answer: $p/\text{bar} = 1 + 760(T-T^*)$, as in Fig. 4B.6

(c) The liquid–vapour boundary

The entropy of vaporization at a temperature T is equal to $\Delta_{\text{vap}}H/T$; the Clapeyron equation for the liquid–vapour boundary is therefore

$$\frac{\text{d}p}{\text{d}T} = \frac{\Delta_{\text{vap}}H}{T\Delta_{\text{vap}}V} \qquad \text{Slope of liquid–vapour boundary} \qquad (4B.9)$$

The enthalpy of vaporization is positive; $\Delta_{\text{vap}}V$ is large and positive. Therefore, $\text{d}p/\text{d}T$ is positive, but it is much smaller than for the solid–liquid boundary. It follows that $\text{d}T/\text{d}p$ is large, and hence that the boiling temperature is more responsive to pressure than the freezing temperature.

Example 4B.2 Estimating the effect of pressure on the boiling temperature

Estimate the typical size of the effect of increasing pressure on the boiling point of a liquid.

Method To use eqn 4B.9 we need to estimate the right-hand side. At the boiling point, the term $\Delta_{\text{vap}}H/T$ is Trouton's constant (Topic 3B). Because the molar volume of a gas is so much greater than the molar volume of a liquid, we can write $\Delta_{\text{vap}}V = V_{\text{m}}(\text{g}) - V_{\text{m}}(\text{l}) \approx V_{\text{m}}(\text{g})$ and take for $V_{\text{m}}(\text{g})$ the molar volume of a perfect gas (at low pressures, at least).

Answer Trouton's constant has the value $85\,\text{J K}^{-1}\,\text{mol}^{-1}$. The molar volume of a perfect gas is about $25\,\text{dm}^3\,\text{mol}^{-1}$ at 1 atm and near but above room temperature. Therefore,

$$\frac{\text{d}p}{\text{d}T} \approx \frac{85\,\text{J K}^{-1}\,\text{mol}^{-1}}{2.5\times10^{-2}\,\text{m}^3\,\text{mol}^{-1}} = 3.4\times10^3\,\text{Pa K}^{-1}$$

We have used $1\,\text{J} = 1\,\text{Pa m}^3$. This value corresponds to 0.034 atm K^{-1} and hence to $\text{d}T/\text{d}p = 29\,\text{K atm}^{-1}$. Therefore, a change of pressure of +0.1 atm can be expected to change a boiling temperature by about +3 K.

Self-test 4B.6 Estimate $\text{d}T/\text{d}p$ for water at its normal boiling point using the information in Table 3A.2 and $V_{\text{m}}(\text{g}) = RT/p$.

Answer: $28\,\text{K atm}^{-1}$

Because the molar volume of a gas is so much greater than the molar volume of a liquid, we can write $\Delta_{\text{vap}}V \approx V_{\text{m}}(\text{g})$ (as in *Example* 4B.2). Moreover, if the gas behaves perfectly, $V_{\text{m}}(\text{g}) = RT/p$. These two approximations turn the exact Clapeyron equation into

$$\frac{\text{d}p}{\text{d}T} = \frac{\Delta_{\text{vap}}H}{T(RT/p)} = \frac{p\Delta_{\text{vap}}H}{RT^2}$$

which, by using $\text{d}x/x = \text{d}\ln x$, rearranges into the **Clausius–Clapeyron equation** for the variation of vapour pressure with temperature:

$$\frac{\text{d}\ln p}{\text{d}T} = \frac{\Delta_{\text{vap}}H}{RT^2} \qquad \textit{Vapour is a perfect gas} \qquad \text{Clausius–Clapeyron equation} \qquad (4B.10)$$

Like the Clapeyron equation (which is exact), the Clausius–Clapeyron equation (which is an approximation) is important for understanding the appearance of phase diagrams, particularly the location and shape of the liquid–vapour and solid–vapour phase boundaries. It lets us predict how the vapour pressure varies with temperature and how the boiling temperature varies with pressure. For instance, if we also assume that the enthalpy of vaporization is independent of temperature, eqn 4B.10 can be integrated as follows:

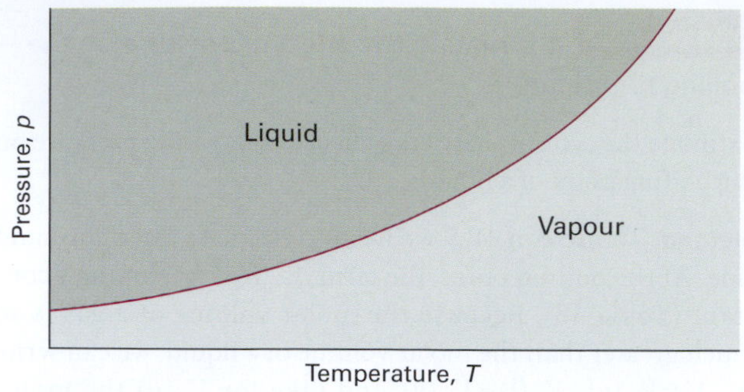

Figure 4B.7 A typical liquid–vapour phase boundary. The boundary can be regarded as a plot of the vapour pressure against the temperature. Note that, in some depictions of phase diagrams in which a logarithmic pressure scale is used, the phase boundary has the opposite curvature (see Fig. 4B.8). This phase boundary terminates at the critical point (not shown).

$$\int_{\ln p^*}^{\ln p} \mathrm{d}\ln p = \frac{\Delta_{\text{vap}}H}{R}\int_{T^*}^{T}\frac{\mathrm{d}T}{T^2} = -\frac{\Delta_{\text{vap}}H}{R}\left(\frac{1}{T}-\frac{1}{T^*}\right)$$

where p^* is the vapour pressure when the temperature is T^* and p the vapour pressure when the temperature is T. Therefore, because the integral on the left evaluates to $\ln(p/p^*)$, the two vapour pressures are related by

$$p = p^*\mathrm{e}^{-\chi} \qquad \chi = \frac{\Delta_{\text{vap}}H}{R}\left(\frac{1}{T}-\frac{1}{T^*}\right) \tag{4B.11}$$

Equation 4B.11 is plotted as the liquid–vapour boundary in Fig. 4B.7. The line does not extend beyond the critical temperature T_c, because above this temperature the liquid does not exist.

Brief illustration 4B.5 The Clausius–Clapeyron equation

Equation 4B.11 can be used to estimate the vapour pressure of a liquid at any temperature from its normal boiling point, the temperature at which the vapour pressure is 1.00 atm (101 kPa). The normal boiling point of benzene is 80 °C (353 K) and (from Table 3A.2), $\Delta_{\text{vap}}H^{\ominus} = 30.8\ \text{kJ mol}^{-1}$. Therefore, to calculate the vapour pressure at 20 °C (293 K), we write

$$\chi = \frac{3.08\times10^4\ \text{J mol}^{-1}}{8.3145\ \text{J K}^{-1}\text{mol}^{-1}}\left(\frac{1}{293\ \text{K}}-\frac{1}{353\ \text{K}}\right) = 2.14\ldots$$

and substitute this value into eqn 4B.11 with $p^* = 101$ kPa. The result is 12 kPa. The experimental value is 10 kPa.

A note on good practice Because exponential functions are so sensitive, it is good practice to carry out numerical calculations like this without evaluating the intermediate steps and using rounded values.

(d) The solid–vapour boundary

The only difference between this case and the last is the replacement of the enthalpy of vaporization by the enthalpy of sublimation, $\Delta_{\text{sub}}H$. Because the enthalpy of sublimation is greater than the enthalpy of vaporization (recall that $\Delta_{\text{sub}}H = \Delta_{\text{fus}}H + \Delta_{\text{vap}}H$), the equation predicts a steeper slope for the sublimation curve than for the vaporization curve at similar temperatures, which is near where they meet at the triple point (Fig. 4B.8).

Brief illustration 4B.6 The solid–vapour boundary

The enthalpy of fusion of ice at the triple point of water (6.1 mbar, 273 K) is negligibly different from its standard enthalpy of fusion at its freezing point, which is 6.008 kJ mol^{-1}. The enthalpy of vaporization at that temperature is 45.0 kJ mol^{-1} (once again, ignoring differences due to the pressure not being 1 bar). The enthalpy of sublimation is therefore 51.0 kJ mol^{-1}. Therefore, the equations for the slopes of (a) the liquid–vapour and (b) the solid–vapour phase boundaries at the triple point are

$$\text{(a)}\ \frac{\mathrm{d}\ln p}{\mathrm{d}T} = \frac{45.0\times10^3\ \text{J mol}^{-1}}{(8.3145\ \text{J K}^{-1}\text{mol}^{-1})\times(273\ \text{K})^2} = 0.0726\ \text{K}^{-1}$$

$$\text{(b)}\ \frac{\mathrm{d}\ln p}{\mathrm{d}T} = \frac{51.0\times10^3\ \text{J mol}^{-1}}{(8.3145\ \text{J K}^{-1}\text{mol}^{-1})\times(273\ \text{K})^2} = 0.0823\ \text{K}^{-1}$$

We see that the slope of $\ln p$ plotted against T is greater for the solid–vapour boundary than for the liquid–vapour boundary at the triple point.

Self-test 4B.7 Confirm that the same may be said for the plot of p against T at the triple point.

Answer: $\mathrm{d}p/\mathrm{d}T = p\,\mathrm{d}\ln p/\mathrm{d}T$, $p = p_3 = 6.1$ mbar

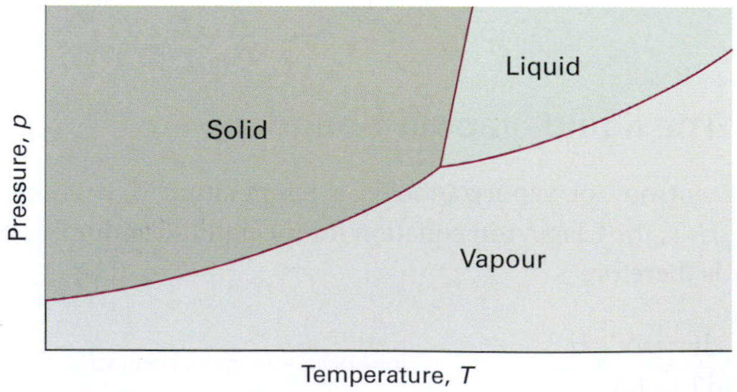

Figure 4B.8 Near the point where they coincide (at the triple point), the solid–vapour boundary has a steeper slope than the liquid–vapour boundary because the enthalpy of sublimation is greater than the enthalpy of vaporization and the temperatures that occur in the Clausius–Clapeyron equation for the slope have similar values.

4B.3 The Ehrenfest classification of phase transitions

There are many different types of phase transition, including the familiar examples of fusion and vaporization and the less familiar examples of solid–solid, conducting–superconducting, and fluid–superfluid transitions. We shall now see that it is possible to use thermodynamic properties of substances, and in particular the behaviour of the chemical potential, to classify phase transitions into different types. Classification is commonly a first step towards a molecular interpretation and the identification of common features. The classification scheme was originally proposed by Paul Ehrenfest, and is known as the **Ehrenfest classification**.

(a) The thermodynamic basis

Many familiar phase transitions, like fusion and vaporization, are accompanied by changes of enthalpy and volume. These changes have implications for the slopes of the chemical potentials of the phases at either side of the phase transition. Thus, at the transition from a phase α to another phase β,

$$\left(\frac{\partial\mu(\beta)}{\partial p}\right)_T-\left(\frac{\partial\mu(\alpha)}{\partial p}\right)_T=V_{\rm m}(\beta)-V_{\rm m}(\alpha)=\Delta_{\rm trs}V$$
$$\left(\frac{\partial\mu(\beta)}{\partial T}\right)_p-\left(\frac{\partial\mu(\alpha)}{\partial T}\right)_p=-S_{\rm m}(\beta)+S_{\rm m}(\alpha)=-\Delta_{\rm trs}S=-\frac{\Delta_{\rm trs}H}{T_{\rm trs}} \qquad (4B.12)$$

Because $\Delta_{\rm trs}V$ and $\Delta_{\rm trs}H$ are non-zero for melting and vaporization, it follows that for such transitions the slopes of the chemical potential plotted against either pressure or temperature are different on either side of the transition (Fig. 4B.9a). In other words, the first derivatives of the chemical potentials with respect to pressure and temperature are discontinuous at the transition.

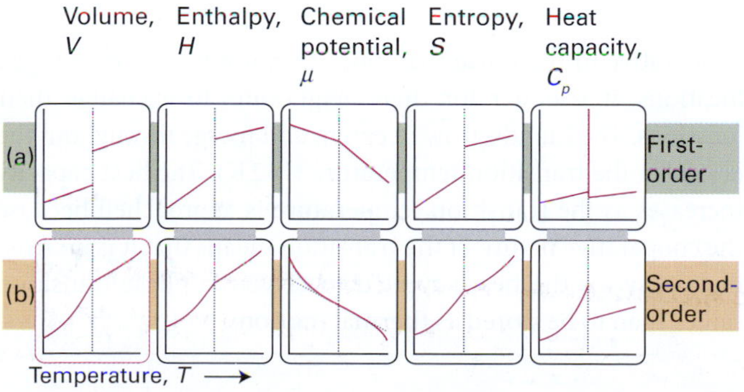

Figure 4B.9 The changes in thermodynamic properties accompanying (a) first-order and (b) second-order phase transitions.

Brief illustration 4B.7 Discontinuities of the transitions

The melting of water at its normal melting point of 0 °C has $\Delta_{\rm trs}V=-1.6\,{\rm cm^3\,mol^{-1}}$ and $\Delta_{\rm trs}H=6.008\,{\rm kJ\,mol^{-1}}$, so

$$\left(\frac{\partial\mu({\rm l})}{\partial p}\right)_T-\left(\frac{\partial\mu({\rm s})}{\partial p}\right)_T=\Delta_{\rm fus}V=-1.6\,{\rm cm^3\,mol^{-1}}$$
$$\left(\frac{\partial\mu({\rm l})}{\partial T}\right)_p-\left(\frac{\partial\mu({\rm s})}{\partial T}\right)_p=-\frac{\Delta_{\rm fus}H}{T_{\rm fus}}=-\frac{6.008\times10^3\,{\rm J\,mol^{-1}}}{273\,{\rm K}}=-22.0\,{\rm J\,mol^{-1}}$$

and both slopes are discontinuous.

Self-test 4B.8 Evaluate the difference in slopes at the normal boiling point.

Answer: $+31\,{\rm dm^3\,mol^{-1}}$, $-109\,{\rm J\,mol^{-1}}$

A transition for which the first derivative of the chemical potential with respect to temperature is discontinuous is classified as a **first-order phase transition**. The constant-pressure heat capacity, C_p, of a substance is the slope of a plot of the enthalpy with respect to temperature. At a first-order phase transition, H changes by a finite amount for an infinitesimal change of temperature. Therefore, at the transition the heat capacity is infinite. The physical reason is that heating drives the transition rather than raising the temperature. For example, boiling water stays at the same temperature even though heat is being supplied.

A **second-order phase transition** in the Ehrenfest sense is one in which the first derivative of μ with respect to temperature is continuous but its second derivative is discontinuous. A continuous slope of μ (a graph with the same slope on either side of the transition) implies that the volume and entropy (and hence the enthalpy) do not change at the transition (Fig. 4B.9b). The heat capacity is discontinuous at the transition but does not become infinite there. An example of a second-order transition is the conducting–superconducting transition in metals at low temperatures.[1]

The term **λ-transition** is applied to a phase transition that is not first-order yet the heat capacity becomes infinite at the transition temperature. Typically, the heat capacity of a system that shows such a transition begins to increase well before the transition (Fig. 4B.10), and the shape of the heat capacity curve resembles the Greek letter lambda. Examples of λ-transitions include order–disorder transitions in alloys, the onset of ferromagnetism, and the fluid–superfluid transition of liquid helium.

[1] A metallic conductor is a substance with an electrical conductivity that decreases as the temperature increases. A superconductor is a solid that conducts electricity without resistance. See Topic 18C for more details.

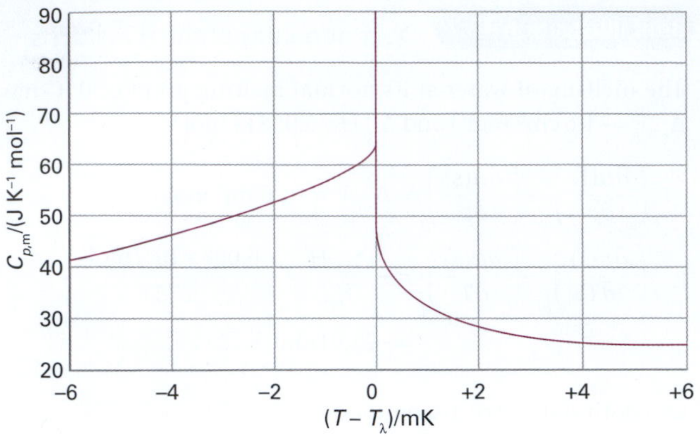

Figure 4B.10 The λ-curve for helium, where the heat capacity rises to infinity. The shape of this curve is the origin of the name λ-transition.

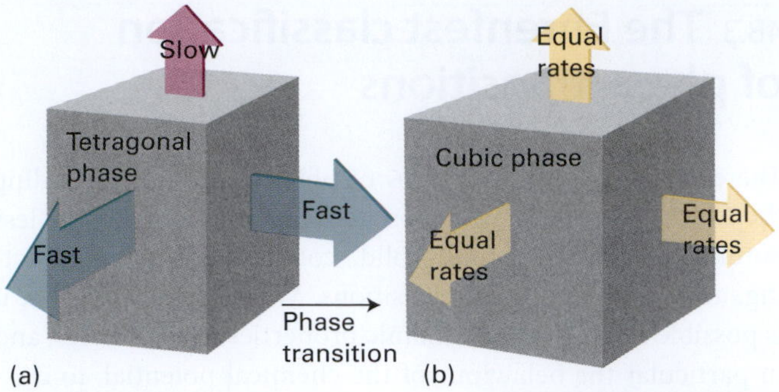

Figure 4B.11 One version of a second-order phase transition in which (a) a tetragonal phase expands more rapidly in two directions than a third, and hence becomes a cubic phase, which (b) expands uniformly in three directions as the temperature is raised. There is no rearrangement of atoms at the transition temperature, and hence no enthalpy of transition.

(b) Molecular interpretation

First-order transitions typically involve the relocation of atoms, molecules, or ions with a consequent change in the energies of their interactions. Thus, vaporization eliminates the attractions between molecules and a first-order phase transition from one ionic polymorph to another (as in the conversion of calcite to aragonite) involves the adjustment of the relative positions of ions.

One type of second-order transition is associated with a change in symmetry of the crystal structure of a solid. Thus, suppose the arrangement of atoms in a solid is like that represented in Fig. 4B.11a, with one dimension (technically, of the unit cell) longer than the other two, which are equal. Such a crystal structure is classified as tetragonal (see Topic 18A). Moreover, suppose the two shorter dimensions increase more than the long dimension when the temperature is raised. There may come a stage when the three dimensions become equal. At that point the crystal has cubic symmetry (Fig. 4B.11b), and at higher temperatures it will expand equally in all three directions (because there is no longer any distinction between them). The tetragonal→cubic phase transition has occurred, but as it has not involved a discontinuity in the interaction energy between the atoms or the volume they occupy, the transition is not first-order.

The order–disorder transition in β-brass (CuZn) is an example of a λ-transition. The low-temperature phase is an orderly array of alternating Cu and Zn atoms. The high-temperature phase is a random array of the atoms (Fig. 4B.12). At $T=0$ the order is perfect, but islands of disorder appear as the temperature is raised. The islands form because the transition is cooperative in the sense that, once two atoms have exchanged locations, it is easier for their neighbours to exchange their locations. The islands grow in extent and merge throughout the crystal at the transition temperature (742 K). The heat capacity increases as the transition temperature is approached because the cooperative nature of the transition means that it is increasingly easy for the heat supplied to drive the phase transition rather than to be stored as thermal motion.

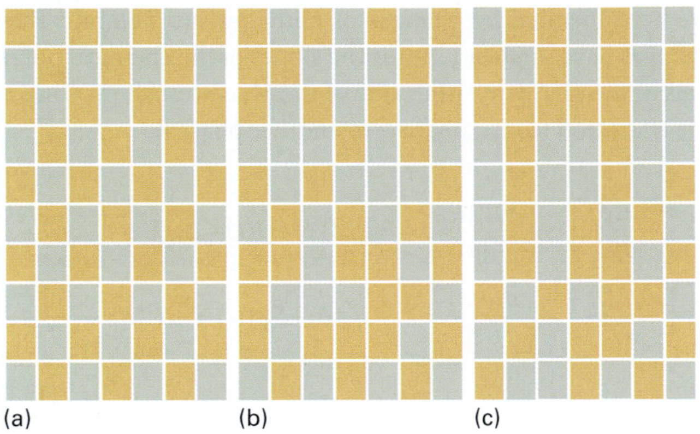

Figure 4B.12 An order–disorder transition. (a) At $T=0$, there is perfect order, with different kinds of atoms occupying alternate sites. (b) As the temperature is increased, atoms exchange locations and islands of each kind of atom form in regions of the solid. Some of the original order survives. (c) At and above the transition temperature, the islands occur at random throughout the sample.

Checklist of concepts

- ☐ 1. The chemical potential of a substance decreases with increasing temperature at a rate determined by its molar entropy.
- ☐ 2. The chemical potential of a substance increases with increasing pressure at a rate determined by its molar volume.
- ☐ 3. When pressure is applied to a condensed phase, its vapour pressure rises.
- ☐ 4. The **Clapeyron equation** is an expression for the slope of a phase boundary.
- ☐ 5. The **Clausius–Clapeyron equation** is an approximation that relates the slope of the liquid–vapour boundary to the enthalpy of vaporization.
- ☐ 6. According to the **Ehrenfest classification**, different types of phase transition are identified by the behaviour of thermodynamic properties at the transition temperature.
- ☐ 7. The classification reveals the type of molecular process occurring at the phase transition.

Checklist of equations

Property	Equation	Comment	Equation number
Variation of μ with temperature	$(\partial\mu/\partial T)_p = -S_m$		4B.1
Variation of μ with pressure	$(\partial\mu/\partial p)_T = V_m$		4B.2
Vapour pressure in the presence of applied pressure	$p = p^* e^{V_m(l)\Delta P/RT}$	$\Delta P = P_{applied} - p^*$	4B.3
Clapeyron equation	$dp/dT = \Delta_{trs}S/\Delta_{trs}V$		4B.5a
Clausius–Clapeyron equation	$d\ln p/dT = \Delta_{vap}H/RT^2$	Assumes $V_m(g) \gg V_m(l)$ and vapour is a perfect gas	4B.10

CHAPTER 4 Physical transformations of pure substances

TOPIC 4A Phase diagrams of pure substances

Discussion questions

4A.1 Describe how the concept of chemical potential unifies the discussion of phase equilibria.

4A.2 Why does the chemical potential change with pressure even if the system is incompressible (that is, remains at the same volume when pressure is applied)?

4A.3 Explain why four phases cannot be in equilibrium in a one-component system.

4A.4 Discuss what would be observed as a sample of water is taken along a path that encircles and is close to its critical point.

Exercises

4A.1(a) How many phases are present at each of the points marked in Fig. 4.1a?
4A.1(b) How many phases are present at each of the points marked in Fig. 4.1b?

4A.2(a) The difference in chemical potential between two regions of a system is +7.1 kJ mol^{-1}. By how much does the Gibbs energy change when 0.10 mmol of a substance is transferred from one region to the other?
4A.2(b) The difference in chemical potential between two regions of a system is −8.3 kJ mol^{-1}. By how much does the Gibbs energy change when 0.15 mmol of a substance is transferred from one region to the other?

4A.3(a) What is the maximum number of phases that can be in mutual equilibrium in a two-component system?
4A.3(b) What is the maximum number of phases that can be in mutual equilibrium in a four-component system?

For problems relating to one-component phase diagrams, see the Integrated activities section of this chapter.

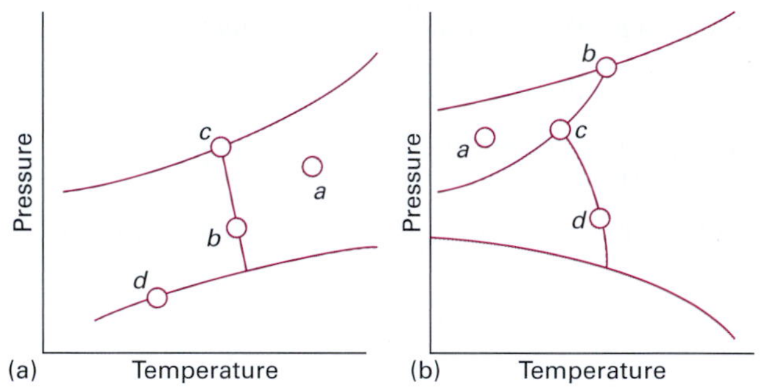

Figure 4.1 The phase diagrams referred to in (a) Exercise 4A.1(a) and (b) Exercise 4A.1(b).

TOPIC 4B Thermodynamic aspects of phase transitions

Discussion questions

4B.1 What is the physical reason for the fact that the chemical potential of a pure substance decreases as the temperatures is raised?

4B.2 What is the physical reason for the fact that the chemical potential of a pure substance increases as the pressure is raised?

4B.3 How may differential scanning calorimetry (DSC) be used to identify phase transitions?

4B.4 Distinguish between a first-order phase transition, a second-order phase transition, and a λ-transition at both molecular and macroscopic levels.

Exercises

4B.1(a) Estimate the difference between the normal and standard melting points of ice.
4B.1(b) Estimate the difference between the normal and standard boiling points of water.

4B.2(a) Water is heated from 25 °C to 100 °C. By how much does its chemical potential change?
4B.2(b) Iron is heated from 100 °C to 1000 °C. By how much does its chemical potential change? Take $S_m^{\ominus} = 53\,\mathrm{J\,K^{-1}\,mol^{-1}}$ for the entire range.

4B.3(a) By how much does the chemical potential of copper change when the pressure exerted on a sample is increased from 100 kPa to 10 MPa?
4B.3(b) By how much does the chemical potential of benzene change when the pressure exerted on a sample is increased from 100 kPa to 10 MPa?

4B.4(a) Pressure was exerted with a piston on water at 20 °C. The vapour pressure of water under 1.0 bar is 2.34 kPa. What is its vapour pressure when the pressure on the liquid is 20 MPa?
4B.4(b) Pressure was exerted with a piston on molten naphthalene at 95 °C. The vapour pressure of naphthalene under 1.0 bar is 2.0 kPa. What is its vapour pressure when the pressure on the liquid is 15 MPa?

4B.5(a) The molar volume of a certain solid is 161.0 $cm^3 mol^{-1}$ at 1.00 atm and 350.75 K, its melting temperature. The molar volume of the liquid at this temperature and pressure is 163.3 $cm^3 mol^{-1}$. At 100 atm the melting temperature changes to 351.26 K. Calculate the enthalpy and entropy of fusion of the solid.
4B.5(b) The molar volume of a certain solid is 142.0 $cm^3 mol^{-1}$ at 1.00 atm and 427.15 K, its melting temperature. The molar volume of the liquid at this temperature and pressure is 152.6 $cm^3 mol^{-1}$. At 1.2 MPa the melting temperature changes to 429.26 K. Calculate the enthalpy and entropy of fusion of the solid.

4B.6(a) The vapour pressure of dichloromethane at 24.1 °C is 53.3 kPa and its enthalpy of vaporization is 28.7 kJ mol^{-1}. Estimate the temperature at which its vapour pressure is 70.0 kPa.
4B.6(b) The vapour pressure of a substance at 20.0 °C is 58.0 kPa and its enthalpy of vaporization is 32.7 kJ mol^{-1}. Estimate the temperature at which its vapour pressure is 66.0 kPa.

4B.7(a) The vapour pressure of a liquid in the temperature range 200 K to 260 K was found to fit the expression $\ln(p/\text{Torr}) = 16.255 - 2501.8/(T/\text{K})$. What is the enthalpy of vaporization of the liquid?
4B.7(b) The vapour pressure of a liquid in the temperature range 200 K to 260 K was found to fit the expression $\ln(p/\text{Torr}) = 18.361 - 3036.8/(T/\text{K})$. What is the enthalpy of vaporization of the liquid?

4B.8(a) The vapour pressure of benzene between 10 °C and 30 °C fits the expression $\log(p/\text{Torr}) = 7.960 - 1780/(T/\text{K})$. Calculate (i) the enthalpy of vaporization and (ii) the normal boiling point of benzene.
4B.8(b) The vapour pressure of a liquid between 15 °C and 35 °C fits the expression $\log(p/\text{Torr}) = 8.750 - 1625/(T/\text{K})$. Calculate (i) the enthalpy of vaporization and (ii) the normal boiling point of the liquid.

4B.9(a) When benzene freezes at 5.5 °C its density changes from 0.879 g cm^{-3} to 0.891 g cm^{-3}. Its enthalpy of fusion is 10.59 kJ mol^{-1}. Estimate the freezing point of benzene at 1000 atm.
4B.9(b) When a certain liquid freezes at −3.65 °C its density changes from 0.789 g cm^{-3} to 0.801 g cm^{-3}. Its enthalpy of fusion is 8.68 kJ mol^{-1}. Estimate the freezing point of the liquid at 100 MPa.

4B.10(a) In July in Los Angeles, the incident sunlight at ground level has a power density of 1.2 kW m^{-2} at noon. A swimming pool of area 50 m^2 is directly exposed to the sun. What is the maximum rate of loss of water? Assume that all the radiant energy is absorbed.
4B.10(b) Suppose the incident sunlight at ground level has a power density of 0.87 kW m^{-2} at noon. What is the maximum rate of loss of water from a lake of area 1.0 ha? (1 ha = 10^4 m^2.) Assume that all the radiant energy is absorbed.

4B.11(a) An open vessel containing (i) water, (ii) benzene, (iii) mercury stands in a laboratory measuring 5.0 m × 5.0 m × 3.0 m at 25 °C. What mass of each substance will be found in the air if there is no ventilation? (The vapour pressures are (i) 3.2 kPa, (ii) 13.1 kPa, (iii) 0.23 Pa.)
4B.11(b) On a cold, dry morning after a frost, the temperature was −5 °C and the partial pressure of water in the atmosphere fell to 0.30 kPa. Will the frost sublime? What partial pressure of water would ensure that the frost remained?

4B.12(a) Naphthalene, $C_{10}H_8$, melts at 80.2 °C. If the vapour pressure of the liquid is 1.3 kPa at 85.8 °C and 5.3 kPa at 119.3 °C, use the Clausius–Clapeyron equation to calculate (i) the enthalpy of vaporization, (ii) the normal boiling point, and (iii) the enthalpy of vaporization at the boiling point.
4B.12(b) The normal boiling point of hexane is 69.0 °C. Estimate (i) its enthalpy of vaporization and (ii) its vapour pressure at 25 °C and 60 °C.

4B.13(a) Calculate the melting point of ice under a pressure of 50 bar. Assume that the density of ice under these conditions is approximately 0.92 g cm^{-3} and that of liquid water is 1.00 g cm^{-3}.
4B.13(b) Calculate the melting point of ice under a pressure of 10 MPa. Assume that the density of ice under these conditions is approximately 0.915 g cm^{-3} and that of liquid water is 0.998 g cm^{-3}.

4B.14(a) What fraction of the enthalpy of vaporization of water is spent on expanding the water vapour?
4B.14(b) What fraction of the enthalpy of vaporization of ethanol is spent on expanding its vapour?

Problems

4B.1 The temperature dependence of the vapour pressure of solid sulfur dioxide can be approximately represented by the relation $\log(p/\text{Torr}) = 10.5916 - 1871.2/(T/\text{K})$ and that of liquid sulfur dioxide by $\log(p/\text{Torr}) = 8.3186 - 1425.7/(T/\text{K})$. Estimate the temperature and pressure of the triple point of sulfur dioxide.

4B.2 Prior to the discovery that freon-12 (CF_2Cl_2) was harmful to the Earth's ozone layer, it was frequently used as the dispersing agent in spray cans for hair spray, etc. Its enthalpy of vaporization at its normal boiling point of −29.2 °C is 20.25 kJ mol^{-1}. Estimate the pressure that a can of hair spray using freon-12 had to withstand at 40 °C, the temperature of a can that has been standing in sunlight. Assume that $\Delta_{vap}H$ is a constant over the temperature range involved and equal to its value at −29.2 °C.

4B.3 The enthalpy of vaporization of a certain liquid is found to be 14.4 kJ mol^{-1} at 180 K, its normal boiling point. The molar volumes of the liquid and the vapour at the boiling point are 115 $cm^3 mol^{-1}$ and 14.5 $dm^3 mol^{-1}$, respectively. (a) Estimate dp/dT from the Clapeyron equation and (b) the percentage error in its value if the Clausius–Clapeyron equation is used instead.

4B.4 Calculate the difference in slope of the chemical potential against temperature on either side of (a) the normal freezing point of water and (b) the normal boiling point of water. (c) By how much does the chemical potential of water supercooled to −5.0 °C exceed that of ice at that temperature?

4B.5 Calculate the difference in slope of the chemical potential against pressure on either side of (a) the normal freezing point of water and (b) the normal boiling point of water. The densities of ice and water at 0 °C are 0.917 g cm^{-3} and 1.000 g cm^{-3}, and those of water and water vapour at 100 °C are 0.958 g cm^{-3} and 0.598 g dm^{-3}, respectively. By how much does the chemical potential of water vapour exceed that of liquid water at 1.2 atm and 100 °C?

4B.6 The enthalpy of fusion of mercury is 2.292 kJ mol^{-1}, and its normal freezing point is 234.3 K with a change in molar volume of +0.517 $cm^{-3} mol^{-1}$ on melting. At what temperature will the bottom of a column of mercury (density 13.6 g cm^{-3}) of height 10.0 m be expected to freeze?

4B.7 50.0 dm^3 of dry air was slowly bubbled through a thermally insulated beaker containing 250 g of water initially at 25 °C. Calculate the final temperature. (The vapour pressure of water is approximately constant at 3.17 kPa throughout, and its heat capacity is 75.5 J $K^{-1} mol^{-1}$. Assume that the air is not heated or cooled and that water vapour is a perfect gas.)

4B.8 The vapour pressure, p, of nitric acid varies with temperature as follows:

θ/°C	0	20	40	50	70	80	90	100
p/kPa	1.92	6.38	17.7	27.7	62.3	89.3	124.9	170.9

What are (a) the normal boiling point and (b) the enthalpy of vaporization of nitric acid?

4B.9 The vapour pressure of the ketone carvone ($M = 150.2\,\mathrm{g\,mol^{-1}}$), a component of oil of spearmint, is as follows:

θ/°C	57.4	100.4	133.0	157.3	203.5	227.5
p/Torr	1.00	10.0	40.0	100	400	760

What are (a) the normal boiling point and (b) the enthalpy of vaporization of carvone?

4B.10‡ In a study of the vapour pressure of chloromethane, A. Bah and N. Dupont-Pavlovsky (*J. Chem. Eng. Data* **40**, 869 (1995)) presented data for the vapour pressure over solid chloromethane at low temperatures. Some of that data is as follows:

T/K	145.94	147.96	149.93	151.94	153.97	154.94
p/Pa	13.07	18.49	25.99	36.76	50.86	59.56

Estimate the standard enthalpy of sublimation of chloromethane at 150 K. (Take the molar volume of the vapour to be that of a perfect gas, and that of the solid to be negligible.)

4B.11 Show that, for a transition between two incompressible solid phases, ΔG is independent of the pressure.

4B.12 The change in enthalpy is given by $\mathrm{d}H = C_p\mathrm{d}T + V\mathrm{d}p$. The Clapeyron equation relates $\mathrm{d}p$ and $\mathrm{d}T$ at equilibrium, and so in combination the two equations can be used to find how the enthalpy changes along a phase boundary as the temperature changes and the two phases remain in equilibrium. Show that $\mathrm{d}(\Delta H/T) = \Delta C_p\,\mathrm{d}\ln T$.

4B.13 In the 'gas saturation method' for the measurement of vapour pressure, a volume V of gas (as measured at a temperature T and a pressure p) is bubbled slowly through the liquid that is maintained at the temperature T, and a mass loss m is measured. Show that the vapour pressure, p, of the liquid is related to its molar mass, M, by $p = AmP/(1 + Am)$, where $A = RT/MPV$. The vapour pressure of geraniol ($M = 154.2\,\mathrm{g\,mol^{-1}}$), which is a component of oil of roses, was measured at 110 °C. It was found that, when 5.00 dm^3 of nitrogen at 760 Torr was passed slowly through the heated liquid, the loss of mass was 0.32 g. Calculate the vapour pressure of geraniol.

4B.14 The vapour pressure of a liquid in a gravitational field varies with the depth below the surface on account of the hydrostatic pressure exerted by the overlying liquid. Adapt eqn. 4B.3 to predict how the vapour pressure of a liquid of molar mass M varies with depth. Estimate the effect on the vapour pressure of water at 25 °C in a column 10 m high.

4B.15 Combine the 'barometric formula', $p = p_0\mathrm{e}^{-a/H}$, where $H = 8$ km, for the dependence of the pressure on altitude, a, with the Clausius–Clapeyron equation, and predict how the boiling temperature of a liquid depends on the altitude and the ambient temperature. Take the mean ambient temperature as 20 °C and predict the boiling temperature of water at 3000 m.

4B.16 Figure 4B.1 gives a schematic representation of how the chemical potentials of the solid, liquid, and gaseous phases of a substance vary with temperature. All have a negative slope, but it is unlikely that they are truly straight lines as indicated in the illustration. Derive an expression for the curvatures (specifically, the second derivatives with respect to temperature) of these lines. Is there a restriction on the curvature of these lines? Which state of matter shows the greatest curvature?

4B.17 The Clapeyron equation does not apply to second-order phase transitions, but there are two analogous equations, the *Ehrenfest equations*, that do. They are:

$$\text{(a)}\ \frac{\mathrm{d}p}{\mathrm{d}T} = \frac{\alpha_2 - \alpha_1}{\kappa_{T,2} - \kappa_{T,1}} \qquad \text{(b)}\ \frac{\mathrm{d}p}{\mathrm{d}T} = \frac{C_{p,\mathrm{m};2} - C_{p,\mathrm{m};1}}{TV_\mathrm{m}(\alpha_2 - \alpha_1)}$$

where α is the expansion coefficient, κ_T the isothermal compressibility, and the subscripts 1 and 2 refer to two different phases. Derive these two equations. Why does the Clapeyron equation not apply to second-order transitions?

4B.18 For a first-order phase transition, to which the Clapeyron equation does apply, prove the relation

$$C_S = C_p - \frac{\alpha V \Delta_{\mathrm{trs}}H}{\Delta_{\mathrm{trs}}V}$$

where $C_S = (\partial q/\partial T)_S$ is the heat capacity along the coexistence curve of two phases.

Integrated activities

4.1 Construct the phase diagram for benzene near its triple point at 36 Torr and 5.50 °C using the following data: $\Delta_{\mathrm{fus}}H = 10.6\,\mathrm{kJ\,mol^{-1}}$, $\Delta_{\mathrm{vap}}H = 30.8\,\mathrm{kJ\,mol^{-1}}$, $\rho(\mathrm{s}) = 0.891\,\mathrm{g\,cm^{-3}}$, $\rho(\mathrm{l}) = 0.879\,\mathrm{g\,cm^{-3}}$.

4.2‡ In an investigation of thermophysical properties of toluene, R.D. Goodwin (*J. Phys. Chem. Ref. Data* **18**, 1565 (1989)) presented expressions for two phase boundaries. The solid–liquid boundary is given by

$$p/\mathrm{bar} = p_3/\mathrm{bar} + 1000(5.60 + 11.727x)x$$

where $x = T/T_3 - 1$ and the triple point pressure and temperature are $p_3 = 0.4362\,\mu\mathrm{bar}$ and $T_3 = 178.15$ K. The liquid–vapour curve is given by:

$$\ln(p/\mathrm{bar}) = -10.418/y + 21.157 - 15.996y + 14.015y^2 - 5.0120y^3 + 4.7334(1-y)^{1.70}$$

where $y = T/T_\mathrm{c} = T/(593.95\,\mathrm{K})$. (a) Plot the solid–liquid and liquid–vapour phase boundaries. (b) Estimate the standard melting point of toluene. (c) Estimate the standard boiling point of toluene. (d) Compute the standard enthalpy of vaporization of toluene, given that the molar volumes of the liquid and vapour at the normal boiling point are 0.12 dm^3 mol^{-1} and 30.3 dm^3 mol^{-1}, respectively.

4.3 Proteins are polymers of amino acids that can exist in ordered structures stabilized by a variety of molecular interactions. However, when certain conditions are changed, the compact structure of a polypeptide chain may collapse into a random coil. This structural change may be regarded as a phase transition occurring at a characteristic transition temperature, the *melting temperature*, T_m, which increases with the strength and number

‡ These problems were supplied by Charles Trapp and Carmen Giunta.

of intermolecular interactions in the chain. A thermodynamic treatment allows predictions to be made of the temperature T_m for the unfolding of a helical polypeptide held together by hydrogen bonds into a random coil. If a polypeptide has N amino acids, $N-4$ hydrogen bonds are formed to form an α-helix, the most common type of helix in naturally occurring proteins (see Topic 17A). Because the first and last residues in the chain are free to move, $N-2$ residues form the compact helix and have restricted motion. Based on these ideas, the molar Gibbs energy of unfolding of a polypeptide with $N \geq 5$ may be written as

$$\Delta_{\text{unfold}}G = (N-4)\Delta_{\text{hb}}H - (N-2)T\Delta_{\text{hb}}S$$

where $\Delta_{hb}H$ and $\Delta_{hb}S$ are, respectively, the molar enthalpy and entropy of dissociation of hydrogen bonds in the polypeptide. (a) Justify the form of the equation for the Gibbs energy of unfolding. That is, why are the enthalpy and entropy terms written as $(N-4)\Delta_{hb}H$ and $(N-2)\Delta_{hb}S$, respectively? (b) Show that T_m may be written as

$$T_{\text{m}} = \frac{(N-4)\Delta_{\text{hb}}H}{(N-2)\Delta_{\text{hb}}S}$$

(c) Plot $T_m/(\Delta_{hb}H_m/\Delta_{hb}S_m)$ for $5 \leq N \leq 20$. At what value of N does T_m change by less than 1 per cent when N increases by 1?

4.4‡ A substance as well-known as methane still receives research attention because it is an important component of natural gas, a commonly used fossil fuel. Friend et al. have published a review of thermophysical properties of methane (D.G. Friend, J.F. Ely, and H. Ingham, *J. Phys. Chem. Ref. Data* **18**, 583 (1989)), which included the following data describing the liquid–vapour phase boundary.

T/K	100	108	110	112	114	120	130	140	150	160	170	190
p/MPa	0.034	0.074	0.088	0.104	0.122	0.192	0.368	0.642	1.041	1.593	2.329	4.521

(a) Plot the liquid–vapour phase boundary. (b) Estimate the standard boiling point of methane. (c) Compute the standard enthalpy of vaporization of methane, given that the molar volumes of the liquid and vapour at the standard boiling point are 3.80×10^{-2} and $8.89\,\text{dm}^3\,\text{mol}^{-1}$, respectively.

4.5‡ Diamond is the hardest substance and the best conductor of heat yet characterized. For these reasons, it is used widely in industrial applications that require a strong abrasive. Unfortunately, it is difficult to synthesize diamond from the more readily available allotropes of carbon, such as graphite. To illustrate this point, calculate the pressure required to convert graphite into diamond at 25 °C. The following data apply to 25 °C and 100 kPa. Assume the specific volume, V_s, and κ_T are constant with respect to pressure changes.

	Graphite	Diamond
$\Delta_f G^{\ominus}/(\text{kJ mol}^{-1})$	0	+2.8678
$V_s/(\text{cm}^3\,\text{g}^{-1})$	0.444	0.284
κ_T/kPa	3.04×10^{-8}	0.187×10^{-8}

CHAPTER 5

Simple mixtures

Mixtures are an essential part of chemistry, either in their own right or as starting materials for chemical reactions. This group of Topics deals with the rich physical properties of mixtures and shows how to express them in terms of thermodynamic quantities.

5A The thermodynamic description of mixtures

The first Topic in this chapter develops the concept of chemical potential as an example of a partial molar quantity and explores how to use the chemical potential of a substance to describe the physical properties of mixtures. The underlying principle to keep in mind is that at equilibrium the chemical potential of a species is the same in every phase. We see, by making use of the experimental observations known as Raoult's and Henry's laws, how to express the chemical potential of a substance in terms of its mole fraction in a mixture.

5B The properties of solutions

In this Topic, the concept of chemical potential is applied to the discussion of the effect of a solute on certain thermodynamic properties of a solution. These properties include the lowering of vapour pressure of the solvent, the elevation of its boiling point, the depression of its freezing point, and the origin of osmotic pressure. We see that it is possible to construct a model of a certain class of real solutions called 'regular solutions', and see how they have properties that diverge from those of ideal solutions.

5C Phase diagrams of binary systems

One widely used device used to summarize the equilibrium properties of mixtures is the phase diagram. We see how to construct and interpret these diagrams. The Topic introduces systems of gradually increasing complexity. In each case we shall see how the phase diagram for the system summarizes empirical observations on the conditions under which the various phases of the system are stable.

5D Phase diagrams of ternary systems

Many modern materials (and ancient ones too) have more than two components. In this Topic we show how phase diagrams are extended to the description of systems of three components and how to interpret triangular phase diagrams.

5E Activities

The extension of the concept of chemical potential to real solutions involves introducing an effective concentration called an 'activity'. We see how the activity may be defined and measured. We shall also see how, in certain cases, the activity may be interpreted in terms of intermolecular interactions.

5F The activities of ions

One of the most important types of mixtures encountered in chemistry is an electrolyte solution. Such solutions often deviate considerably from ideal behaviour on account of the strong, long-range interactions between ions. In this Topic we show how a model can be used to estimate the deviations from ideal behaviour when the solution is very dilute, and how to extend the resulting expressions to more concentrated solutions.

What is the impact of this material?

We consider just two applications of this material, one from biology and the other from materials science, from among the huge number that could be chosen for this centrally important field. In *Impact* I5.1, we see how the phenomenon of osmosis contributes to the ability of biological cells to maintain their shapes. In *Impact* I5.2, we see how phase diagrams are used to describe the properties of the technologically important liquid crystals.

To read more about the impact of this material, scan the QR code, or go to bcs.whfreeman.com/webpub/chemistry/pchem10e/impact/pchem-5-1.html

5A The thermodynamic description of mixtures

Contents

➤ Why do you need to know this material?

Chemistry deals with a wide variety of mixtures, including mixtures of substances that can react together. Therefore, it is important to generalize the concepts introduced in Chapter 4 to deal with substances that are mingled together. This Topic also introduces the fundamental equation of chemical thermodynamics on which many of the applications of thermodynamics to chemistry are based.

➤ What is the key idea?

The chemical potential of a substance in a mixture is a logarithmic function of its concentration.

➤ What do you need to know already?

This Topic extends the concept of chemical potential to substances in mixtures by building on the concept introduced in the context of pure substances (Topic 4A). It makes use of the relation between entropy and the temperature dependence of the Gibbs energy (Topic 3D) and the concept of partial pressure (Topic 1A). It uses the notation of partial derivatives (*Mathematical background* 2) but does not draw on their advanced properties.

As a first step towards dealing with chemical reactions (which are treated in Topic 6A), here we consider mixtures of substances that do not react together. At this stage we deal mainly with **binary mixtures**, which are mixtures of two components, A and B. We shall therefore often be able to simplify equations by making use of the relation $x_A + x_B = 1$. In Topic 1A it is established that the partial pressure, which is the contribution of one component to the total pressure, is used to discuss the properties of mixtures of gases. For a more general description of the thermodynamics of mixtures we need to introduce other analogous 'partial' properties.

One preliminary remark is in order. Throughout this and related Topics we need to refer to various measures of concentration of a solute in a solution. The **molar concentration** (colloquially, the 'molarity', [J] or c_J) is the amount of solute divided by the volume of the solution and is usually expressed in moles per cubic decimetre (mol dm^{-3}; more informally, mol L^{-1}). We write $c^{\ominus} = 1\,\text{mol dm}^{-3}$. The term **molality**, b, is the amount of solute divided by the mass of solvent and is usually expressed in moles per kilogram of solvent (mol kg^{-1}). We write $b^{\ominus} = 1\,\text{mol kg}^{-1}$.

5A.1 Partial molar quantities

The easiest partial molar property to visualize is the 'partial molar volume', the contribution that a component of a mixture makes to the total volume of a sample.

(a) Partial molar volume

Imagine a huge volume of pure water at 25 °C. When a further 1 mol H_2O is added, the volume increases by 18 cm³ and we can report that 18 cm³ mol⁻¹ is the molar volume of pure water. However, when we add 1 mol H_2O to a huge volume of pure ethanol, the volume increases by only 14 cm³. The reason for the different increase in volume is that the volume occupied by a given number of water molecules depends on the identity of the molecules that surround them. In the latter case there is so much ethanol present that each H_2O molecule is surrounded by ethanol molecules. The network of hydrogen bonds that normally hold H_2O molecules at certain distances from each other in pure water does not form. The packing of the molecules in the mixture results in the H_2O molecules increasing the volume by only 14 cm³. The quantity 14 cm³ mol⁻¹ is the partial molar volume of water in pure ethanol. In general, the **partial molar volume** of a substance A in a mixture is the change in volume per mole of A added to a large volume of the mixture.

The partial molar volumes of the components of a mixture vary with composition because the environment of each type of molecule changes as the composition changes from pure A to pure B. It is this changing molecular environment, and the consequential modification of the forces acting between molecules, that results in the variation of the thermodynamic properties of a mixture as its composition is changed. The partial molar volumes of water and ethanol across the full composition range at 25 °C are shown in Fig. 5A.1.

The partial molar volume, V_J, of a substance J at some general composition is defined formally as follows:

$$V_J = \left(\frac{\partial V}{\partial n_J}\right)_{p,T,n'} \quad \text{Definition} \quad \text{Partial molar volume} \quad (5A.1)$$

where the subscript n' signifies that the amounts of all other substances present are constant. The partial molar volume is

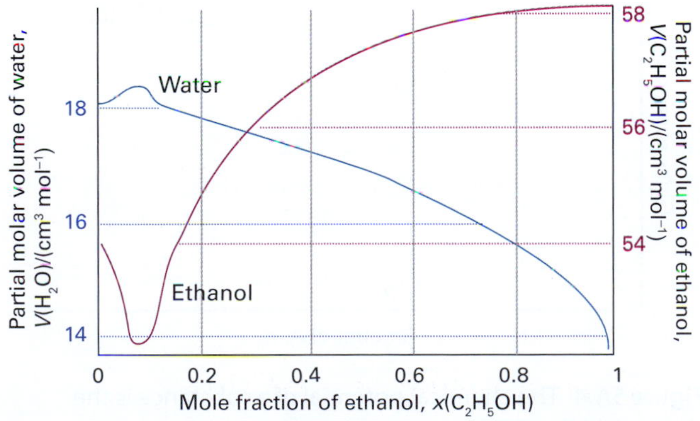

Figure 5A.1 The partial molar volumes of water and ethanol at 25 °C. Note the different scales (water on the left, ethanol on the right).

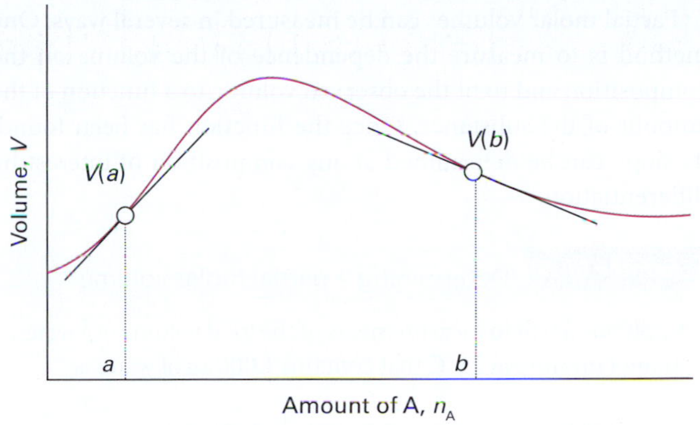

Figure 5A.2 The partial molar volume of a substance is the slope of the variation of the total volume of the sample plotted against the composition. In general, partial molar quantities vary with the composition, as shown by the different slopes at the compositions *a* and *b*. Note that the partial molar volume at *b* is negative: the overall volume of the sample decreases as A is added.

the slope of the plot of the total volume as the amount of J is changed, the pressure, temperature, and amount of the other components being constant (Fig. 5A.2). Its value depends on the composition, as we saw for water and ethanol.

A note on good practice The IUPAC recommendation is to denote a partial molar quantity by $\overline{X}$, but only when there is the possibility of confusion with the quantity X. For instance, to avoid confusion, the partial molar volume of NaCl in water could be written $\overline{V}$ (NaCl, aq) to distinguish it from the total volume of the solution, V.

The definition in eqn 5A.1 implies that when the composition of the mixture is changed by the addition of dn_A of A and dn_B of B, then the total volume of the mixture changes by

$$dV = \left(\frac{\partial V}{\partial n_A}\right)_{p,T,n_B} dn_A + \left(\frac{\partial V}{\partial n_B}\right)_{p,T,n_A} dn_B = V_A dn_A + V_B dn_B \quad (5A.2)$$

Provided the relative composition is held constant as the amounts of A and B are increased, the partial molar volumes are both constant. In that case we can obtain the final volume by integration, treating V_A and V_B as constants:

$$V = \int_0^{n_A} V_A dn_A + \int_0^{n_B} V_B dn_B = V_A \int_0^{n_A} dn_A + V_B \int_0^{n_B} dn_B = V_A n_A + V_B n_B \quad (5A.3)$$

Although we have envisaged the two integrations as being linked (in order to preserve constant relative composition), because V is a state function the final result in eqn 5A.3 is valid however the solution is in fact prepared.

Partial molar volumes can be measured in several ways. One method is to measure the dependence of the volume on the composition and to fit the observed volume to a function of the amount of the substance. Once the function has been found, its slope can be determined at any composition of interest by differentiation.

Example 5A.1 Determining a partial molar volume

A polynomial fit to measurements of the total volume of a water/ethanol mixture at 25 °C that contains 1.000 kg of water is

$$v = 1002.93 + 54.6664x - 0.363\,94x^2 + 0.028\,256x^3$$

where $v = V/\text{cm}^3$, $x = n_E/\text{mol}$, and n_E is the amount of CH_3CH_2OH present. Determine the partial molar volume of ethanol.

Method Apply the definition in eqn 5A.1 taking care to convert the derivative with respect to n to a derivative with respect to x and keeping the units intact.

Answer The partial molar volume of ethanol, V_E, is

$$V_E = \left(\frac{\partial V}{\partial n_E}\right)_{p,T,n_W} = \left(\frac{\partial(V/\text{cm}^3)}{\partial(n_E/\text{mol})}\right)_{p,T,n_W} \frac{\text{cm}^3}{\text{mol}}$$

$$= \left(\frac{\partial v}{\partial x}\right)_{p,T,n_W} \text{cm}^3\,\text{mol}^{-1}$$

Then, because

$$\frac{dv}{dx} = 54.6664 - 2(0.363\,94)x + 3(0.028\,256)x^2$$

we can conclude that

$$V_E/(\text{cm}^3\text{mol}^{-1}) = 54.6664 - 0.727\,88x + 0.084\,768x^2$$

Figure 5A.3 shows a graph of this function.

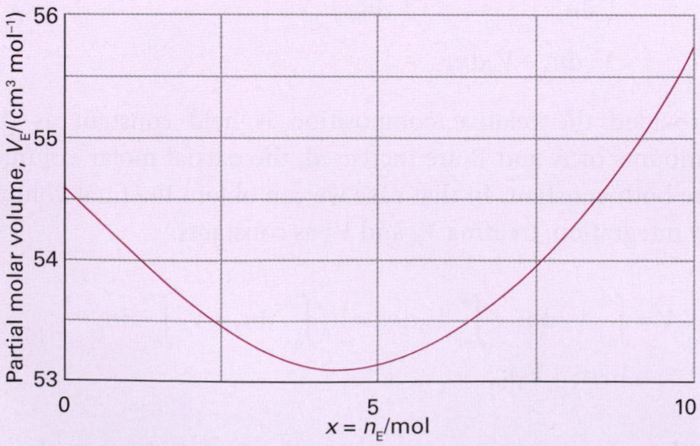

Figure 5A.3 The partial molar volume of ethanol, as expressed by the polynomial in *Example* 5A.1.

Self-test 5A.1 At 25 °C, the density of a 50 per cent by mass ethanol/water solution is 0.914 g cm⁻³. Given that the partial molar volume of water in the solution is 17.4 cm³ mol⁻¹, what is the partial molar volume of the ethanol?

Answer: 56.4 cm³ mol⁻¹; 54.6 cm³ mol⁻¹ by the formula above

Molar volumes are always positive, but partial molar quantities need not be. For example, the limiting partial molar volume of $MgSO_4$ in water (its partial molar volume in the limit of zero concentration) is −1.4 cm³ mol⁻¹, which means that the addition of 1 mol $MgSO_4$ to a large volume of water results in a decrease in volume of 1.4 cm³. The mixture contracts because the salt breaks up the open structure of water as the Mg^{2+} and SO_4^{2-} ions become hydrated, and it collapses slightly.

(b) Partial molar Gibbs energies

The concept of a partial molar quantity can be extended to any extensive state function. For a substance in a mixture, the chemical potential is *defined* as the partial molar Gibbs energy:

$$\mu_J = \left(\frac{\partial G}{\partial n_J}\right)_{p,T,n'} \qquad \textit{Definition} \quad \text{Chemical potential} \quad (5A.4)$$

That is, the chemical potential is the slope of a plot of Gibbs energy against the amount of the component J, with the pressure and temperature (and the amounts of the other substances) held constant (Fig. 5A.4). For a pure substance we can write $G = n_J G_{J,m}$, and from eqn 5A.4 obtain $\mu_J = G_{J,m}$: in this case, the chemical potential is simply the molar Gibbs energy of the substance, as is used in Topic 4B.

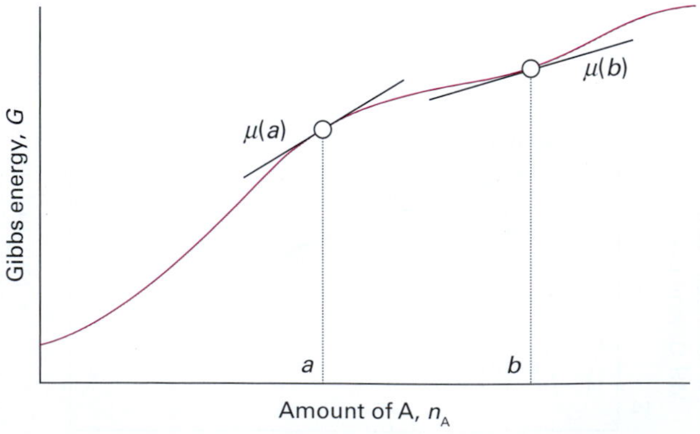

Figure 5A.4 The chemical potential of a substance is the slope of the total Gibbs energy of a mixture with respect to the amount of substance of interest. In general, the chemical potential varies with composition, as shown for the two values at *a* and *b*. In this case, both chemical potentials are positive.

By the same argument that led to eqn 5A.2, it follows that the total Gibbs energy of a binary mixture is

$$G = n_A\mu_A + n_B\mu_B \quad (5A.5)$$

where μ_A and μ_B are the chemical potentials at the composition of the mixture. That is, the chemical potential of a substance in a mixture is the contribution of that substance to the total Gibbs energy of the mixture. Because the chemical potentials depend on composition (and the pressure and temperature), the Gibbs energy of a mixture may change when these variables change, and for a system of components A, B, etc., the equation $dG = Vdp - SdT$ becomes

$$dG = Vdp - SdT + \mu_A dn_A + \mu_B dn_B + \cdots$$

Fundamental equation of chemical thermodynamics (5A.6)

This expression is the **fundamental equation of chemical thermodynamics**. Its implications and consequences are explored and developed in this and the next two chapters.

At constant pressure and temperature, eqn 5A.6 simplifies to

$$dG = \mu_A dn_A + \mu_B dn_B + \cdots \quad (5A.7)$$

We saw in Topic 3C that under the same conditions $dG = dw_{add,max}$. Therefore, at constant temperature and pressure,

$$dw_{add,max} = \mu_A dn_A + \mu_B dn_B + \cdots \quad (5A.8)$$

That is, additional (non-expansion) work can arise from the changing composition of a system. For instance, in an electrochemical cell, the chemical reaction is arranged to take place in two distinct sites (at the two electrodes). The electrical work the cell performs can be traced to its changing composition as products are formed from reactants.

(c) The wider significance of the chemical potential

The chemical potential does more than show how G varies with composition. Because $G = U + pV - TS$, and therefore $U = -pV + TS + G$, we can write a general infinitesimal change in U for a system of variable composition as

$$\begin{aligned} dU &= -pdV - Vdp + SdT + TdS + dG \\ &= -pdV - Vdp + SdT + TdS + \\ &\quad (Vdp - SdT + \mu_A dn_A + \mu_B dn_B + \cdots) \\ &= -pdV + TdS + \mu_A dn_A + \mu_B dn_B + \cdots \end{aligned}$$

This expression is the generalization of eqn 3D.1 (that $dU = TdS - pdV$) to systems in which the composition may change. It follows that at constant volume and entropy,

$$dU = \mu_A dn_A + \mu_B dn_B + \cdots \quad (5A.9)$$

and hence that

$$\mu_J = \left(\frac{\partial U}{\partial n_J}\right)_{S,V,n'} \quad (5A.10)$$

Therefore, not only does the chemical potential show how G changes when the composition changes, it also shows how the internal energy changes too (but under a different set of conditions). In the same way it is possible to deduce that

$$\text{(a)}\ \mu_J = \left(\frac{\partial H}{\partial n_J}\right)_{S,p,n'} \qquad \text{(b)}\ \mu_J = \left(\frac{\partial A}{\partial n_J}\right)_{T,V,n'} \quad (5A.11)$$

Thus we see that the μ_J shows how all the extensive thermodynamic properties U, H, A, and G depend on the composition. This is why the chemical potential is so central to chemistry.

(d) The Gibbs–Duhem equation

Because the total Gibbs energy of a binary mixture is given by eqn 5A.5 and the chemical potentials depend on the composition, when the compositions are changed infinitesimally we might expect G of a binary system to change by

$$dG = \mu_A dn_A + \mu_B dn_B + n_A d\mu_A + n_B d\mu_B$$

However, we have seen that at constant pressure and temperature a change in Gibbs energy is given by eqn 5A.7. Because G is a state function, these two equations must be equal, which implies that at constant temperature and pressure

$$n_A d\mu_A + n_B d\mu_B = 0 \quad (5A.12a)$$

This equation is a special case of the **Gibbs–Duhem equation**:

$$\sum_J n_J d\mu_J = 0$$

Gibbs–Duhem equation (5A.12b)

The significance of the Gibbs–Duhem equation is that the chemical potential of one component of a mixture cannot change independently of the chemical potentials of the other components. In a binary mixture, if one partial molar quantity increases, then the other must decrease, with the two changes related by

$$d\mu_B = -\frac{n_A}{n_B} d\mu_A \quad (5A.13)$$

Brief illustration 5A.1 The Gibbs–Duhem equation

If the composition of a mixture is such that $n_A=2n_B$, and a small change in composition results in μ_A changing by $\delta\mu_A=+1\,\mathrm{J\,mol^{-1}}$, μ_B will change by

$$\delta\mu_B=-2\times(1\,\mathrm{J\,mol^{-1}})=-2\,\mathrm{J\,mol^{-1}}$$

Self-test 5A.2 Suppose that $n_A=0.3n_B$ and a small change in composition results in μ_A changing by $\delta\mu_A=-10\,\mathrm{J\,mol^{-1}}$, by how much will μ_B change?

Answer: $+3\,\mathrm{J\,mol^{-1}}$

The same line of reasoning applies to all partial molar quantities. We can see in Fig. 5A.1, for example, that where the partial molar volume of water increases, that of ethanol decreases. Moreover, as eqn 5A.13 shows, and as we can see from Fig 5A.1, a small change in the partial molar volume of A corresponds to a large change in the partial molar volume of B if n_A/n_B is large, but the opposite is true when this ratio is small. In practice, the Gibbs–Duhem equation is used to determine the partial molar volume of one component of a binary mixture from measurements of the partial molar volume of the second component.

Example 5A.2 Using the Gibbs–Duhem equation

The experimental values of the partial molar volume of $K_2SO_4(aq)$ at 298 K are found to fit the expression

$$v_B=32.280+18.216x^{1/2}$$

where $v_B=V_{K_2SO_4}/(\mathrm{cm^3\,mol^{-1}})$ and x is the numerical value of the molality of K_2SO_4 ($x=b/b^\ominus$; see the remark in the introduction to this chapter). Use the Gibbs–Duhem equation to derive an equation for the molar volume of water in the solution. The molar volume of pure water at 298 K is $18.079\,\mathrm{cm^3\,mol^{-1}}$.

Method Let A denote H_2O, the solvent, and B denote K_2SO_4, the solute. The Gibbs–Duhem equation for the partial molar volumes of two components is $n_A dV_A+n_B dV_B=0$. This relation implies that $dv_A=-(n_B/n_A)dv_B$, and therefore that v_A can be found by integration:

$$v_A=v_A^\star-\int_0^{v_B}\frac{n_B}{n_A}dv_B$$

where $v_A^\star=V_A/(\mathrm{cm^3\,mol^{-1}})$ is the numerical value of the molar volume of pure A. The first step is to change the variable v_B to $x=b/b^\ominus$ and then to integrate the right-hand side between $x=0$ (pure B) and the molality of interest.

Answer It follows from the information in the question that, with $B=K_2SO_4$, $dv_B/dx=9.108x^{-1/2}$. Therefore, the integration required is

$$v_A=v_A^\star-9.108\int_0^{b/b^\ominus}\frac{n_B}{n_A}x^{-1/2}dx$$

However, the ratio of amounts of A (H_2O) and B (K_2SO_4) is related to the molality of B, $b=n_B/(1\,\mathrm{kg\ water})$ and $n_A=(1\,\mathrm{kg\ water})/M_A$ where M_A is the molar mass of water, by

$$\frac{n_B}{n_A}=\frac{n_B}{(1\,\mathrm{kg})/M_A}=\frac{n_BM_A}{1\,\mathrm{kg}}=bM_A=xb^\ominus M_A$$

and hence

$$v_A=v_A^\star-9.108M_Ab^\ominus\int_0^{b/b^\ominus}x^{1/2}dx$$
$$=v_A^\star-\frac{2}{3}(9.108M_Ab^\ominus)(b/b^\ominus)^{3/2}$$

It then follows, by substituting the data (including $M_A=1.802\times10^{-2}\,\mathrm{kg\,mol^{-1}}$, the molar mass of water), that

$$V_A/(\mathrm{cm^3mol^{-1}})=18.079-0.1094(b/b^\ominus)^{3/2}$$

The partial molar volumes are plotted in Fig. 5A.5.

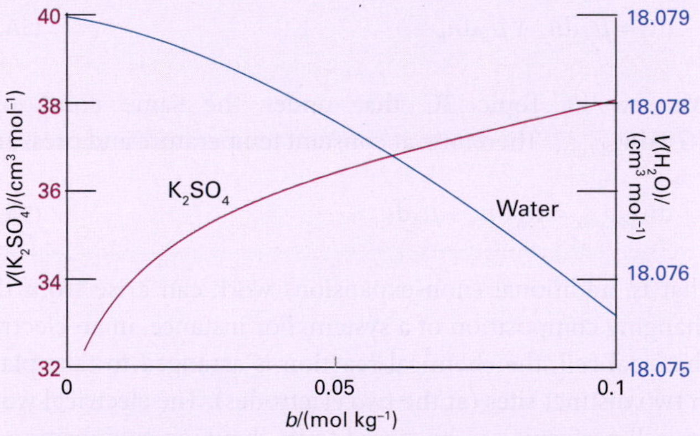

Figure 5A.5 The partial molar volumes of the components of an aqueous solution of potassium sulfate.

Self-test 5A.3 Repeat the calculation for a salt B for which $V_B/(\mathrm{cm^3\,mol^{-1}})=6.218+5.146b-7.147b^2$.

Answer: $V_A/(\mathrm{cm^3\,mol^{-1}})=18.079-0.0464b^2+0.0859b^3$

5A.2 The thermodynamics of mixing

The dependence of the Gibbs energy of a mixture on its composition is given by eqn 5A.5, and we know that at constant temperature and pressure systems tend towards lower Gibbs energy. This is the link we need in order to apply thermodynamics to the discussion of spontaneous changes of composition, as in the mixing of two substances. One simple example of a spontaneous mixing process is that of two gases introduced into the same container. The mixing is spontaneous, so it must correspond to a decrease in G. We shall now see how to express this idea quantitatively.

(a) The Gibbs energy of mixing of perfect gases

Let the amounts of two perfect gases in the two containers be n_A and n_B; both are at a temperature T and a pressure p (Fig. 5A.6). At this stage, the chemical potentials of the two gases have their 'pure' values, which are obtained by applying the definition $\mu = G_m$ to eqn 3D.15 ($G_m(p) = G_m^{\ominus} + RT\ln(p/p^{\ominus})$):

$$\mu = \mu^{\ominus} + RT\ln\frac{p}{p^{\ominus}}$$

Perfect gas Variation of chemical potential with pressure (5A.14a)

where $\mu^{\ominus}$ is the **standard chemical potential**, the chemical potential of the pure gas at 1 bar. It will be much simpler notationally if we agree to let p denote the pressure relative to $p^{\ominus}$; that is, to replace $p/p^{\ominus}$ by p, for then we can write

$$\mu = \mu^{\ominus} + RT\ln p \qquad (5A.14b)$$

To use the equations, we have to remember to replace p by $p/p^{\ominus}$ again. In practice, that simply means using the numerical value of p in bars. The Gibbs energy of the total system is then given by eqn 5A.5 as

$$G_i = n_A\mu_A + n_B\mu_B = n_A(\mu_A^{\ominus} + RT\ln p) + n_B(\mu_B^{\ominus} + RT\ln p) \qquad (5A.15a)$$

After mixing, the partial pressures of the gases are p_A and p_B, with $p_A + p_B = p$. The total Gibbs energy changes to

$$G_f = n_A(\mu_A^{\ominus} + RT\ln p_A) + n_B(\mu_B^{\ominus} + RT\ln p_B) \qquad (5A.15b)$$

The difference $G_f - G_i$, the **Gibbs energy of mixing**, $\Delta_{mix}G$, is therefore

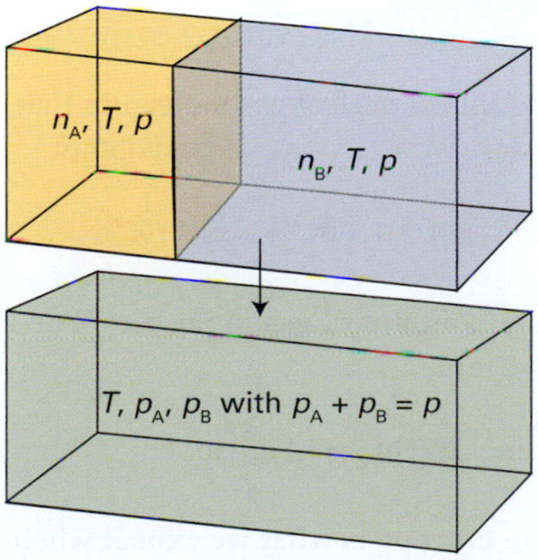

Figure 5A.6 The arrangement for calculating the thermodynamic functions of mixing of two perfect gases.

$$\Delta_{mix}G = n_ART\ln\frac{p_A}{p} + n_BRT\ln\frac{p_B}{p} \qquad (5A.15c)$$

At this point we may replace n_J by x_Jn, where n is the total amount of A and B, and use the relation between partial pressure and mole fraction (Topic 1A, $p_J = x_Jp$) to write $p_J/p = x_J$ for each component, which gives

$$\Delta_{mix}G = nRT(x_A\ln x_A + x_B\ln x_B)$$

Perfect gases Gibbs energy of mixing (5A.16)

Because mole fractions are never greater than 1, the logarithms in this equation are negative, and $\Delta_{mix}G < 0$ (Fig. 5A.7). The conclusion that $\Delta_{mix}G$ is negative for all compositions confirms that perfect gases mix spontaneously in all proportions. However, the equation extends common sense by allowing us to discuss the process quantitatively.

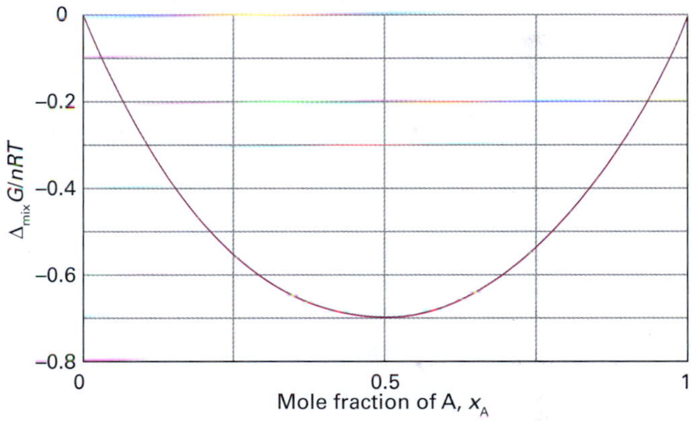

Figure 5A.7 The Gibbs energy of mixing of two perfect gases and (as discussed later) of two liquids that form an ideal solution. The Gibbs energy of mixing is negative for all compositions and temperatures, so perfect gases mix spontaneously in all proportions.

Example 5A.3 Calculating a Gibbs energy of mixing

A container is divided into two equal compartments (Fig. 5A.8). One contains 3.0 mol H_2(g) at 25 °C; the other contains 1.0 mol N_2(g) at 25 °C. Calculate the Gibbs energy of mixing when the partition is removed. Assume perfect behaviour.

Method Equation 5A.16 cannot be used directly because the two gases are initially at different pressures. We proceed by calculating the initial Gibbs energy from the chemical potentials. To do so, we need the pressure of each gas. Write the pressure of nitrogen as p; then the pressure of hydrogen as a multiple of p can be found from the gas laws. Next, calculate the Gibbs energy for the system when the partition is removed. The volume occupied by each gas doubles, so its initial partial pressure is halved.

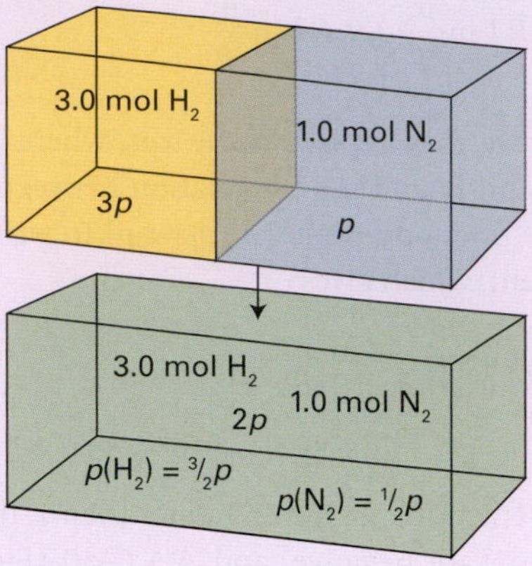

Figure 5A.8 The initial and final states considered in the calculation of the Gibbs energy of mixing of gases at different initial pressures.

Answer Given that the pressure of nitrogen is p, the pressure of hydrogen is $3p$; therefore, the initial Gibbs energy is

$$G_i = (3.0\ \text{mol})\{\mu^\ominus(H_2) + RT\ln 3p\} + (1.0\ \text{mol})\{\mu^\ominus(N_2) + RT\ln p\}$$

When the partition is removed and each gas occupies twice the original volume, the partial pressure of nitrogen falls to $\frac{1}{2}p$ and that of hydrogen falls to $\frac{3}{2}p$. Therefore, the Gibbs energy changes to

$$G_f = (3.0\ \text{mol})\{\mu^\ominus(H_2) + RT\ln \tfrac{3}{2}p\} + (1.0\ \text{mol})\{\mu^\ominus(N_2) + RT\ln \tfrac{1}{2}p\}$$

The Gibbs energy of mixing is the difference of these two quantities:

$$\begin{aligned}\Delta_{\text{mix}}G &= (3.0\,\text{mol})RT\ln\frac{\frac{3}{2}p}{3p} + (1.0\,\text{mol})RT\ln\frac{\frac{1}{2}p}{p}\\ &= -(3.0\,\text{mol})RT\ln 2 - (1.0\,\text{mol})RT\ln 2\\ &= -(4.0\,\text{mol})RT\ln 2 = -6.9\,\text{kJ}\end{aligned}$$

In this example, the value of $\Delta_{\text{mix}}G$ is the sum of two contributions: the mixing itself, and the changes in pressure of the two gases to their final total pressure, $2p$. When 3.0 mol H_2 mixes with 1.0 mol N_2 at the same pressure, with the volumes of the vessels adjusted accordingly, the change of Gibbs energy is -5.6 kJ. However, do not be misled into interpreting this negative change in Gibbs energy as a sign of spontaneity: in this case, the pressure changes, and $\Delta G < 0$ is a signpost of spontaneous change only at constant temperature and pressure.

Self-test 5A.4 Suppose that 2.0 mol H_2 at 2.0 atm and 25 °C and 4.0 mol N_2 at 3.0 atm and 25 °C are mixed by removing the partition between them. Calculate $\Delta_{\text{mix}}G$.

Answer: -9.7 kJ

(b) Other thermodynamic mixing functions

In Topic 3D it is shown that $(\partial G/\partial T)_{p,n} = -S$. It follows immediately from eqn 5A.16 that, for a mixture of perfect gases initially at the same pressure, the entropy of mixing, $\Delta_{\text{mix}}S$, is

$$\Delta_{\text{mix}}S = -\left(\frac{\partial \Delta_{\text{mix}}G}{\partial T}\right)_{p,n_A,n_B} = -nR(x_A\ln x_A + x_B\ln x_B)$$

Perfect gases Entropy of mixing (5A.17)

Because $\ln x < 0$, it follows that $\Delta_{\text{mix}}S > 0$ for all compositions (Fig. 5A.9).

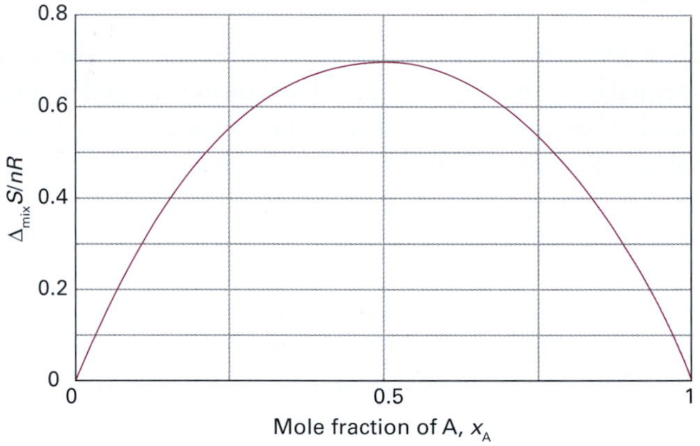

Figure 5A.9 The entropy of mixing of two perfect gases and (as discussed later) of two liquids that form an ideal solution. The entropy increases for all compositions and temperatures, so perfect gases mix spontaneously in all proportions. Because there is no transfer of heat to the surroundings when perfect gases mix, the entropy of the surroundings is unchanged. Hence, the graph also shows the total entropy of the system plus the surroundings when perfect gases mix.

Brief illustration 5A.2 The entropy of mixing

For equal amounts of perfect gas molecules that are mixed at the same pressure we set $x_A = x_B = \frac{1}{2}$ and obtain

$$\Delta_{\text{mix}}S = -nR\{\tfrac{1}{2}\ln\tfrac{1}{2} + \tfrac{1}{2}\ln\tfrac{1}{2}\} = nR\ln 2$$

with n the total amount of gas molecules. For 1 mol of each species, so $n = 2$ mol,

$$\Delta_{\text{mix}}S = (2\ \text{mol})\times R\ln 2 = +11.5\,\text{J mol}^{-1}$$

An increase in entropy is what we expect when one gas disperses into the other and the disorder increases.

Self-test 5A.5 Calculate the change in entropy for the arrangement in *Example* 5A.3.

Answer: +23 J mol⁻¹

We can calculate the isothermal, isobaric (constant pressure) **enthalpy of mixing**, $\Delta_{mix}H$, the enthalpy change accompanying mixing, of two perfect gases from $\Delta G = \Delta H - T\Delta S$. It follows from eqns 5A.16 and 5A.17 that

$$\Delta_{mix}H = 0 \quad \text{Perfect gases} \quad \text{Enthalpy of mixing} \quad (5A.18)$$

The enthalpy of mixing is zero, as we should expect for a system in which there are no interactions between the molecules forming the gaseous mixture. It follows that the whole of the driving force for mixing comes from the increase in entropy of the system because the entropy of the surroundings is unchanged.

5A.3 The chemical potentials of liquids

To discuss the equilibrium properties of liquid mixtures we need to know how the Gibbs energy of a liquid varies with composition. To calculate its value, we use the fact that, as established in Topic 4A, at equilibrium the chemical potential of a substance present as a vapour must be equal to its chemical potential in the liquid.

(a) Ideal solutions

We shall denote quantities relating to pure substances by a superscript *, so the chemical potential of pure A is written μ_A^* and as $\mu_A^*(l)$ when we need to emphasize that A is a liquid. Because the vapour pressure of the pure liquid is p_A^* it follows from eqn 5A.14 that the chemical potential of A in the vapour (treated as a perfect gas) is $\mu_A^{\ominus} = + RT\ln p_A$ (with p_A to be interpreted as the relative pressure, $p_A/p^{\ominus}$). These two chemical potentials are equal at equilibrium (Fig. 5A.10), so we can write

$$\mu_A^* = \mu_A^{\ominus} + RT\ln p_A^* \quad (5A.19a)$$

If another substance, a solute, is also present in the liquid, the chemical potential of A in the liquid is changed to μ_A and its vapour pressure is changed to p_A. The vapour and solvent are still in equilibrium, so we can write

$$\mu_A = \mu_A^{\ominus} + RT\ln p_A \quad (5A.19b)$$

Next, we combine these two equations to eliminate the standard chemical potential of the gas. To do so, we write eqn 5A.19a as $\mu_A^{\ominus} = \mu_A^* - RT\ln p_A^*$ and substitute this expression into eqn 5A.19b to obtain

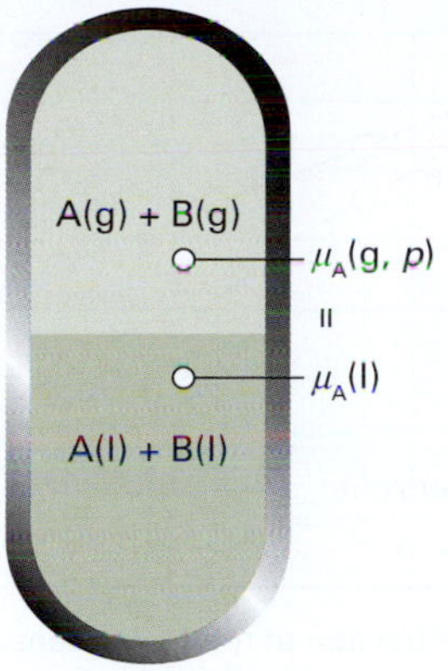

Figure 5A.10 At equilibrium, the chemical potential of the gaseous form of a substance A is equal to the chemical potential of its condensed phase. The equality is preserved if a solute is also present. Because the chemical potential of A in the vapour depends on its partial vapour pressure, it follows that the chemical potential of liquid A can be related to its partial vapour pressure.

$$\mu_A = \mu_A^* - RT\ln p_A^* + RT\ln p_A = \mu_A^* + RT\ln\frac{p_A}{p_A^*} \quad (5A.20)$$

In the final step we draw on additional experimental information about the relation between the ratio of vapour pressures and the composition of the liquid. In a series of experiments on mixtures of closely related liquids (such as benzene and methylbenzene), the French chemist François Raoult found that the ratio of the partial vapour pressure of each component to its vapour pressure as a pure liquid, p_A/p_A^*, is approximately equal to the mole fraction of A in the liquid mixture. That is, he established what we now call **Raoult's law**:

$$p_A = x_A p_A^* \quad \text{Ideal solution} \quad \text{Raoult's law} \quad (5A.21)$$

This law is illustrated in Fig. 5A.11. Some mixtures obey Raoult's law very well, especially when the components are

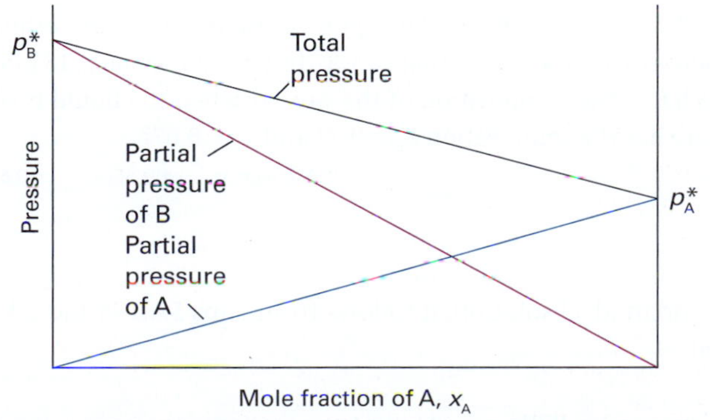

Figure 5A.11 The total vapour pressure and the two partial vapour pressures of an ideal binary mixture are proportional to the mole fractions of the components.

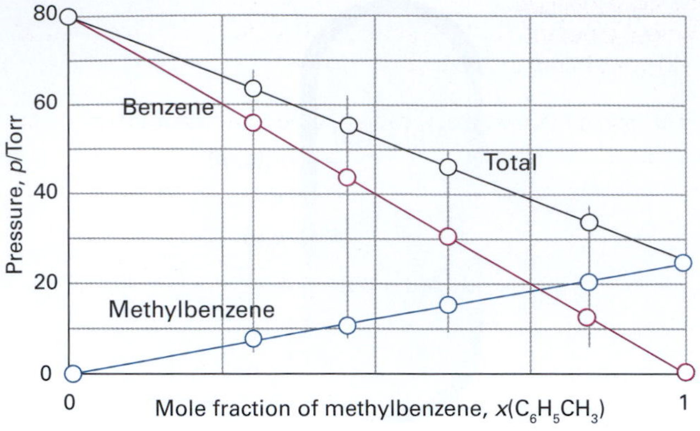

Figure 5A.12 Two similar liquids, in this case benzene and methylbenzene (toluene), behave almost ideally, and the variation of their vapour pressures with composition resembles that for an ideal solution.

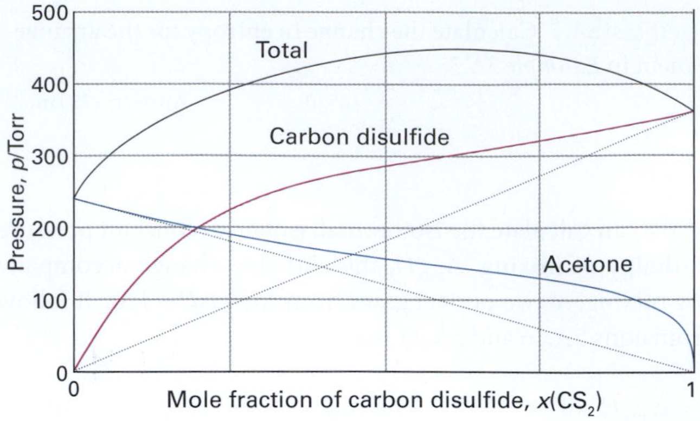

Figure 5A.13 Strong deviations from ideality are shown by dissimilar liquids, in this case carbon disulfide and acetone (propanone).

structurally similar (Fig. 5A.12). Mixtures that obey the law throughout the composition range from pure A to pure B are called **ideal solutions**.

Brief illustration 5A.3 Raoult's law

The vapour pressure of benzene at 20 °C is 75 Torr and that of methylbenzene is 21 Torr at the same temperature. In an equimolar mixture, $x_{\text{benzene}} = x_{\text{methylbenzene}} = \frac{1}{2}$ so the vapour pressure of each one in the mixture is

$$p_{\text{benzene}} = \tfrac{1}{2} \times 75\ \text{Torr} = 38\ \text{Torr}$$
$$p_{\text{methylbenzene}} = \tfrac{1}{2} \times 21\ \text{Torr} = 11\ \text{Torr}$$

The total vapour pressure of the mixture is 49 Torr. Given the two partial vapour pressures, it follows from the definition of partial pressure (Topic 1A) that the mole fractions in the vapour are $x_{\text{vap,benzene}} = (38\ \text{Torr})/(49\ \text{Torr}) = 0.78$ and $x_{\text{vap,methylbenzene}} = (11\ \text{Torr})/(49\ \text{Torr}) = 0.22$. The vapour is richer in the more volatile component (benzene).

Self-test 5A.6 At 90 °C the vapour pressure of 1,2-dimethylbenzene is 20 kPa and that of 1,3-dimethylbenzene is 18 kPa. What is the composition of the vapour when the liquid mixture has the composition $x_{12} = 0.33$ and $x_{13} = 0.67$?

Answer: $x_{\text{vap},12} = 0.35$, $x_{\text{vap},13} = 0.65$

For an ideal solution, it follows from eqns 5A.19a and 5A.21 that

$$\mu_{\text{A}} = \mu_{\text{A}}^{*} + RT \ln x_{\text{A}} \quad \textit{Ideal solution} \quad \text{Chemical potential} \quad (5A.22)$$

This important equation can be used as the *definition* of an ideal solution (so that it implies Raoult's law rather than stemming from it). It is in fact a better definition than eqn 5A.21 because it does not assume that the vapour is a perfect gas.

The molecular origin of Raoult's law is the effect of the solute on the entropy of the solution. In the pure solvent, the molecules have a certain disorder and a corresponding entropy; the vapour pressure then represents the tendency of the system and its surroundings to reach a higher entropy. When a solute is present, the solution has a greater disorder than the pure solvent because we cannot be sure that a molecule chosen at random will be a solvent molecule. Because the entropy of the solution is higher than that of the pure solvent, the solution has a lower tendency to acquire an even higher entropy by the solvent vaporizing. In other words, the vapour pressure of the solvent in the solution is lower than that of the pure solvent.

Some solutions depart significantly from Raoult's law (Fig. 5A.13). Nevertheless, even in these cases the law is obeyed increasingly closely for the component in excess (the solvent) as it approaches purity. The law is another example of a limiting law (in this case, achieving reliability as $x_{\text{A}} \rightarrow 1$) and is a good approximation for the properties of the solvent if the solution is dilute.

(b) Ideal–dilute solutions

In ideal solutions the solute, as well as the solvent, obeys Raoult's law. However, the English chemist William Henry found experimentally that, for real solutions at low concentrations, although the vapour pressure of the solute is proportional to its mole fraction, the constant of proportionality is not the vapour pressure of the pure substance (Fig. 5A.14). **Henry's law** is:

$$p_{\text{B}} = x_{\text{B}} K_{\text{B}} \quad \textit{Ideal–dilute solution} \quad \text{Henry's law} \quad (5A.23)$$

In this expression, x_{B} is the mole fraction of the solute and K_{B} is an empirical constant (with the dimensions of pressure)

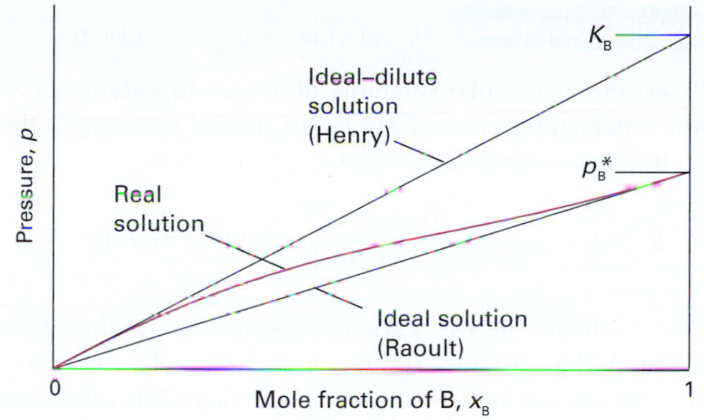

Figure 5A.14 When a component (the solvent) is nearly pure, it has a vapour pressure that is proportional to mole fraction with a slope p_B^* (Raoult's law). When it is the minor component (the solute) its vapour pressure is still proportional to the mole fraction, but the constant of proportionality is now K_B (Henry's law).

chosen so that the plot of the vapour pressure of B against its mole fraction is tangent to the experimental curve at $x_B=0$. Henry's law is therefore also a limiting law, achieving reliability as $x_B \to 0$.

Mixtures for which the solute B obeys Henry's law and the solvent A obeys Raoult's law are called **ideal–dilute solutions**. The difference in behaviour of the solute and solvent at low concentrations (as expressed by Henry's and Raoult's laws, respectively) arises from the fact that in a dilute solution the solvent molecules are in an environment very much like the one they have in the pure liquid (Fig. 5A.15). In contrast, the solute molecules are surrounded by solvent molecules, which is entirely different from their environment when pure. Thus, the solvent behaves like a slightly modified pure liquid, but the solute behaves entirely differently from its pure state unless the solvent and solute molecules happen to be very similar. In the latter case, the solute also obeys Raoult's law.

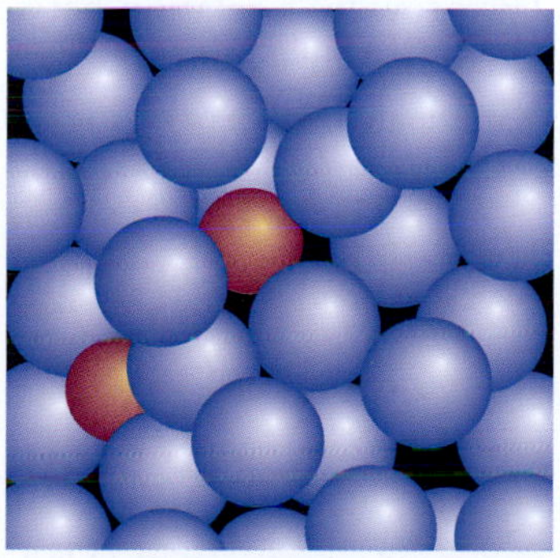

Figure 5A.15 In a dilute solution, the solvent molecules (the blue spheres) are in an environment that differs only slightly from that of the pure solvent. The solute particles (the purple spheres), however, are in an environment totally unlike that of the pure solute.

Example 5A.4 Investigating the validity of Raoult's and Henry's laws

The vapour pressures of each component in a mixture of propanone (acetone, A) and trichloromethane (chloroform, C) were measured at 35 °C with the following results:

x_C	0	0.20	0.40	0.60	0.80	1
p_C/kPa	0	4.7	11	18.9	26.7	36.4
p_A/kPa	46.3	33.3	23.3	12.3	4.9	0

Confirm that the mixture conforms to Raoult's law for the component in large excess and to Henry's law for the minor component. Find the Henry's law constants.

Method Both Raoult's and Henry's laws are statements about the form of the graph of partial vapour pressure against mole fraction. Therefore, plot the partial vapour pressures against mole fraction. Raoult's law is tested by comparing the data with the straight line $p_J = x_J p_J^*$ for each component in the region in which it is in excess (and acting as the solvent). Henry's law is tested by finding a straight line $p_J = x_J K_J^*$ that is tangent to each partial vapour pressure at low x, where the component can be treated as the solute.

Answer The data are plotted in Fig. 5A.16 together with the Raoult's law lines. Henry's law requires $K_A = 16.9$ kPa for propanone and $K_C = 20.4$ kPa for trichloromethane. Notice how the system deviates from both Raoult's and Henry's laws even for quite small departures from $x=1$ and $x=0$, respectively. We deal with these deviations in Topic 5E.

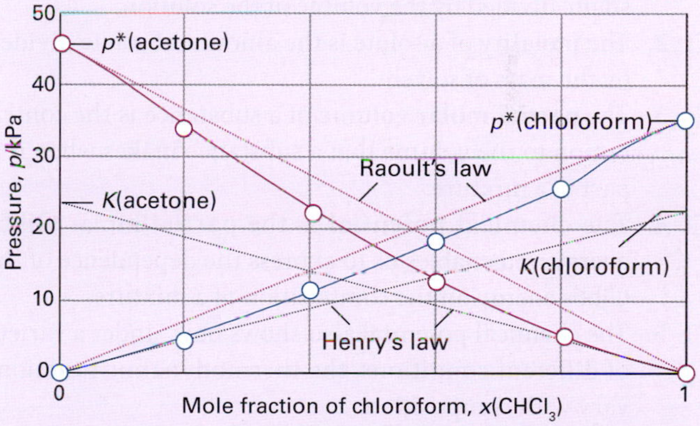

Figure 5A.16 The experimental partial vapour pressures of a mixture of chloroform (trichloromethane) and acetone (propanone) based on the data in *Example* 5A.3. The values of K are obtained by extrapolating the dilute solution vapour pressures, as explained in the *Example*.

Self-test 5A.7 The vapour pressure of chloromethane at various mole fractions in a mixture at 25 °C was found to be as follows:

x	0.005	0.009	0.019	0.024
p/kPa	27.3	48.4	101	126

Estimate Henry's law constant.

Answer: 5 MPa

For practical applications, Henry's law is expressed in terms of the molality, b, of the solute, $p_B = b_B K_B$. Some Henry's law data for this convention are listed in Table 5A.1. As well as providing a link between the mole fraction of solute and its partial pressure, the data in the table may also be used to calculate gas solubilities. A knowledge of Henry's law constants for gases in blood and fats is important for the discussion of respiration, especially when the partial pressure of oxygen is abnormal, as in diving and mountaineering, and for the discussion of the action of gaseous anaesthetics.

Table 5A.1* Henry's law constants for gases in water at 298 K, $K/(\text{kPa kg mol}^{-1})$

	$K/(\text{kPa kg mol}^{-1})$
CO_2	3.01×10^3
H_2	1.28×10^5
N_2	1.56×10^5
O_2	7.92×10^4

* More values are given in the *Resource section*.

Brief illustration 5A.4 Henry's law and gas solubility

To estimate the molar solubility of oxygen in water at 25 °C and a partial pressure of 21 kPa, its partial pressure in the atmosphere at sea level, we write

$$b_{O_2} = \frac{p_{O_2}}{K_{O_2}} = \frac{21\,\text{kPa}}{7.9\times10^4\,\text{kPa kg mol}^{-1}} = 2.9\times10^{-4}\,\text{mol kg}^{-1}$$

The molality of the saturated solution is therefore $0.29\,\text{mmol kg}^{-1}$. To convert this quantity to a molar concentration, we assume that the mass density of this dilute solution is essentially that of pure water at 25 °C, or $\rho = 0.997\,\text{kg dm}^{-3}$. It follows that the molar concentration of oxygen is

$$[O_2] = b_{O_2}\rho = (2.9\times10^{-4}\,\text{mol kg}^{-1})\times(0.997\,\text{kg dm}^{-3})$$
$$= 0.29\,\text{mmol dm}^{-3}$$

Self-test 5A.8 Calculate the molar solubility of nitrogen in water exposed to air at 25 °C; partial pressures were calculated in *Example* 1A.3 of Topic 1A.

Answer: $0.51\,\text{mmol dm}^{-3}$

Checklist of concepts

- ☐ 1. The **molar concentration** of a solute is the amount of solute divided by the volume of the solution.
- ☐ 2. The **molality** of a solute is the amount of solute divided by the mass of solvent.
- ☐ 3. The **partial molar volume** of a substance is the contribution to the volume that a substance makes when it is part of a mixture.
- ☐ 4. The **chemical potential** is the partial molar Gibbs energy and enables us to express the dependence of the Gibbs energy on the composition of a mixture.
- ☐ 5. The chemical potential also shows how, under a variety of different conditions, the thermodynamic functions vary with composition.
- ☐ 6. The **Gibbs–Duhem equation** shows how the changes in chemical potential of the components of a mixture are related.
- ☐ 7. The **Gibbs energy of mixing** is calculated by forming the difference of the Gibbs energies before and after mixing: the quantity is negative for perfect gases at the same pressure.
- ☐ 8. The **entropy of mixing** of perfect gases initially at the same pressure is positive and the enthalpy of mixing is zero.
- ☐ 9. **Raoult's law** provides a relation between the vapour pressure of a substance and its mole fraction in a mixture; it is the basis of the definition of an ideal solution.
- ☐ 10. **Henry's law** provides a relation between the vapour pressure of a solute and its mole fraction in a mixture; it is the basis of the definition of an ideal–dilute solution.

Checklist of equations

Property	Equation	Comment	Equation number
Partial molar volume	$V_J = (\partial V/\partial n_J)_{p,T,n'}$	Definition	5A.1
Chemical potential	$\mu_J = (\partial G/\partial n_J)_{p,T,n'}$	Definition	5A.4
Total Gibbs energy	$G = n_A\mu_A + n_B\mu_B$		5A.5

Property	Equation	Comment	Equation number
Fundamental equation of chemical thermodynamics	$dG = Vdp - SdT + \mu_A dn_A + \mu_B dn_B + \dots$		5A.6
Gibbs–Duhem equation	$\sum_J n_J d\mu_J = 0$		5A.12b
Chemical potential of a gas	$\mu = \mu^\ominus + RT \ln(p/p^\ominus)$	Perfect gas	5A.14a
Gibbs energy of mixing	$\Delta_{mix}G = nRT(x_A \ln x_A + x_B \ln x_B)$	Perfect gases and ideal solutions	5A.16
Entropy of mixing	$\Delta_{mix}S = -nR(x_A \ln x_A + x_B \ln x_B)$	Perfect gases and ideal solutions	5A.17
Enthalpy of mixing	$\Delta_{mix}H = 0$	Perfect gases and ideal solutions	5A.18
Raoult's law	$p_A = x_A p_A^*$	True for ideal solutions; limiting law as $x_A \to 1$	5A.21
Chemical potential of component	$\mu_A = \mu_A^* + RT \ln x_A$	Ideal solution	5A.22
Henry's law	$p_B = x_B K_B$	True for ideal-dilute solutions; limiting law as $x_B \to 0$	5A.23

5B The properties of solutions

Contents

➤ Why do you need to know this material?

Mixtures and solutions play a central role in chemistry, and it is important to understand how their compositions affect their thermodynamic properties, such as their boiling and freezing points. One very important property of a solution is its osmotic pressure, which is used, among other things, to determine the molar masses of macromolecules.

➤ What is the key idea?

The chemical potential of a substance in a mixture is the same in each phase in which it occurs.

➤ What do you need to know already?

This Topic is based on the expression derived from Raoult's law (Topic 5A) in which chemical potential is related to mole fraction. The derivations make use of the Gibbs–Helmholtz equation (Topic 3D) and the effect of pressure on chemical potential (Topic 3D). Some of the derivations are the same as those used in the discussion of the mixing of perfect gases (Topic 5A).

First, we consider the simple case of mixtures of liquids that mix to form an ideal solution. In this way, we identify the thermodynamic consequences of molecules of one species mingling randomly with molecules of the second species. The calculation provides a background for discussing the deviations from ideal behaviour exhibited by real solutions. Then we consider the effect of a solute on the properties of ideal and real solutions.

5B.1 Liquid mixtures

Thermodynamics can provide insight into the properties of liquid mixtures, and a few simple ideas can bring the whole field of study together. The development here is based on the relation derived in Topic 5A between the chemical potential of a component (which here we call J for reasons that will become clear) in an ideal mixture or solution, μ_J, its value when pure, μ_J^*, and its mole fraction in the mixture, x_J:

$$\mu_J = \mu_J^* + RT\ln x_J \qquad \textit{Ideal solution} \quad \text{Chemical potential} \qquad (5B.1)$$

(a) Ideal solutions

The Gibbs energy of mixing of two liquids to form an ideal solution is calculated in exactly the same way as for two gases (Topic 5A). The total Gibbs energy before liquids are mixed is

$$G_i = n_A\mu_A^* + n_B\mu_B^* \qquad (5B.2a)$$

where the * denotes the pure liquid. When they are mixed, the individual chemical potentials are given by eqn 5B.1 and the total Gibbs energy is

$$G_f = n_A(\mu_A^* + RT\ln x_A) + n_B(\mu_B^* + RT\ln x_B) \qquad (5B.2b)$$

Consequently, the Gibbs energy of mixing, the difference of these two quantities, is

$$\Delta_{mix}G = nRT(x_A\ln x_A + x_B\ln x_B) \qquad \textit{Ideal solution} \quad \text{Gibbs energy of mixing} \qquad (5B.3)$$

where $n = n_A + n_B$. As for gases, it follows that the ideal entropy of mixing of two liquids is

$$\Delta_{mix}S = -nR(x_A\ln x_A + x_B\ln x_B) \qquad \textit{Ideal solution} \quad \text{Entropy of mixing} \qquad (5B.4)$$

Because $\Delta_{mix}H = \Delta_{mix}G + T\Delta_{mix}S = 0$, the ideal enthalpy of mixing is zero, $\Delta_{mix}H = 0$. The ideal volume of mixing, the change in volume on mixing, is also zero because it follows from eqn 3D.8 ($(\partial G/\partial p)_T = V$) that $\Delta_{mix}V = (\partial \Delta_{mix}G/\partial p)_T$, but $\Delta_{mix}G$ in eqn 5B.3 is independent of pressure, so the derivative with respect to pressure is zero.

Equations 5B.3 and 5B.4 are the same as those for the mixing of two perfect gases and all the conclusions drawn there are valid here: the driving force for mixing is the increasing entropy of the system as the molecules mingle and the enthalpy of mixing is zero. It should be noted, however, that solution ideality means something different from gas perfection. In a perfect gas there are no forces acting between molecules. In ideal solutions there are interactions, but the average energy of A–B interactions in the mixture is the same as the average energy of A–A and B–B interactions in the pure liquids. The variation of the Gibbs energy and entropy of mixing with composition is the same as that for gases (Figs. 5A.7 and 5A.9); both graphs are repeated here (as Figs. 5B.1 and 5B.2).

A note on good practice It is on the basis of this distinction that the term 'perfect gas' is preferable to the more common 'ideal gas'. In an ideal solution there are interactions, but they are effectively the same between the various species. In a perfect gas, not only are the interactions the same, but they are also zero. Few people, however, trouble to make this valuable distinction.

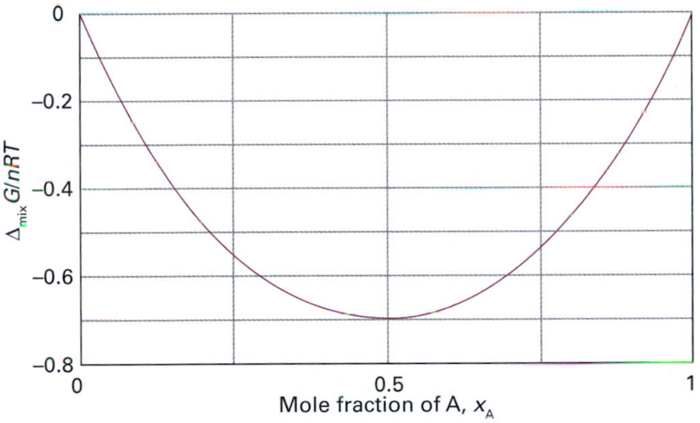

Figure 5B.1 The Gibbs energy of mixing of two liquids that form an ideal solution.

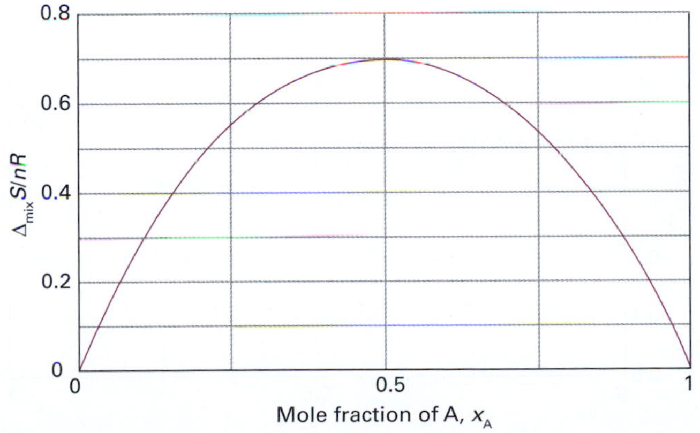

Figure 5B.2 The entropy of mixing of two liquids that form an ideal solution.

Brief illustration 5B.1 Ideal solutions

Consider a mixture of benzene and methylbenzene, which form an approximately ideal solution, and suppose 1.0 mol $C_6H_6(l)$ is mixed with 2.0 mol $C_6H_5CH_3(l)$. For the mixture, $x_{benzene} = 0.33$ and $x_{methylbenzene} = 0.67$. The Gibbs energy and entropy of mixing at 25 °C, when $RT = 2.48\,kJ\,mol^{-1}$, are

$$\Delta_{mix}G/n = (2.48\,kJ\,mol^{-1}) \times (0.33 \ln 0.33 + 0.67 \ln 0.67) = -1.6\,kJ\,mol^{-1}$$

$$\Delta_{mix}S/n = -(8.3145\,J\,K^{-1}\,mol^{-1}) \times (0.33 \ln 0.33 + 0.67 \ln 0.67) = +5.3\,J\,K^{-1}\,mol^{-1}$$

The enthalpy of mixing is zero (presuming that the solution is ideal).

Self-test 5B.1 Calculate the Gibbs energy and entropy of mixing when the proportions are reversed.

Answer: same: $-1.6\,kJ\,mol^{-1}$, $+5.3\,J\,K^{-1}\,mol^{-1}$

Real solutions are composed of particles for which A–A, A–B, and B–B interactions are all different. Not only may there be enthalpy and volume changes when liquids mix, but there may also be an additional contribution to the entropy arising from the way in which the molecules of one type might cluster together instead of mingling freely with the others. If the enthalpy change is large and positive or if the entropy change is adverse (because of a reorganization of the molecules that results in an orderly mixture), then the Gibbs energy might be positive for mixing. In that case, separation is spontaneous and the liquids may be immiscible. Alternatively, the liquids might be **partially miscible**, which means that they are miscible only over a certain range of compositions.

(b) Excess functions and regular solutions

The thermodynamic properties of real solutions are expressed in terms of the **excess functions**, X^E, the difference between the observed thermodynamic function of mixing and the function for an ideal solution:

$$X^E = \Delta_{mix}X - \Delta_{mix}X^{ideal} \qquad \textit{Definition} \quad \text{Excess function} \qquad (5B.5)$$

The **excess entropy**, S^E, for example, is calculated using the value of $\Delta_{mix}S^{ideal}$ given by eqn 5B.4. The excess enthalpy and volume are both equal to the observed enthalpy and volume of mixing, because the ideal values are zero in each case.

Brief illustration 5B.2 Excess functions

Figure 5B.3 shows two examples of the composition dependence of molar excess functions. In Fig 5B.3a, the positive values of H^E, which implies that $\Delta_{mix}H>0$, indicate that the A–B interactions in the mixture are less attractive than the A–A and B–B interactions in the pure liquids (which are benzene and pure cyclohexane). The symmetrical shape of the curve reflects the similar strengths of the A–A and B–B interactions. Figure 5B.3b shows the composition dependence of the excess volume, V^E, of a mixture of tetrachloroethene and cyclopentane. At high mole fractions of cyclopentane, the solution contracts as tetrachloroethene is added because the ring structure of cyclopentane results in inefficient packing of the molecules but as tetrachloroethene is added, the molecules in the mixture pack together more tightly. Similarly, at high mole fractions of tetrachloroethene, the solution expands as cyclopentane is added because tetrachloroethene molecules are nearly flat and pack efficiently in the pure liquid but become disrupted as bulky ring cyclopentane is added.

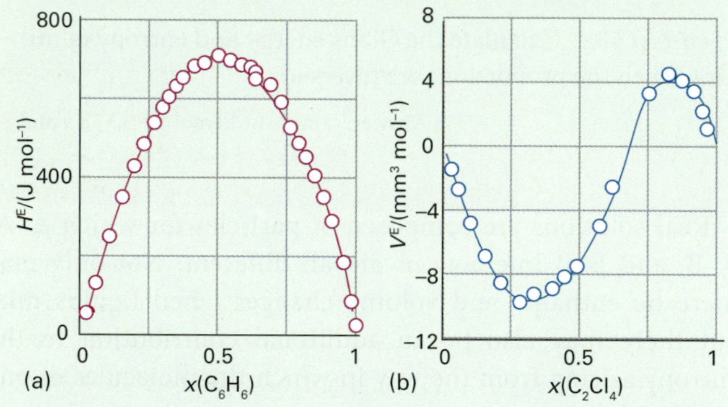

Figure 5B.3 Experimental excess functions at 25 °C. (a) H^E for benzene/cyclohexane; this graph shows that the mixing is endothermic (because $\Delta_{mix}H=0$ for an ideal solution). (b) The excess volume, V^E, for tetrachloroethene/cyclopentane; this graph shows that there is a contraction at low tetrachloroethene mole fractions, but an expansion at high mole fractions (because $\Delta_{mix}V=0$ for an ideal mixture).

Self-test 5B.2 Would you expect the excess volume of mixing of oranges and melons to be positive or negative?

Answer: Positive; close-packing disrupted

Deviations of the excess energies from zero indicate the extent to which the solutions are non-ideal. In this connection a useful model system is the **regular solution**, a solution for which $H^E \neq 0$ but $S^E=0$. We can think of a regular solution as one in which the two kinds of molecules are distributed randomly (as in an ideal solution) but have different energies of interactions with each other. To express this concept more quantitatively we can suppose that the excess enthalpy depends on composition as

$$H^E = n\xi RTx_A x_B \tag{5B.6}$$

where ξ (xi) is a dimensionless parameter that is a measure of the energy of AB interactions relative to that of the AA and BB interactions. (For H^E expressed as a molar quantity, discard the n.) The function given by eqn 5B.6 is plotted in Fig. 5B.4, and we see it resembles the experimental curve in Fig. 5B.3a. If $\xi<0$, mixing is exothermic and the solute–solvent interactions are more favourable than the solvent–solvent and solute–solute interactions. If $\xi>0$, then the mixing is endothermic. Because the entropy of mixing has its ideal value for a regular solution, the excess Gibbs energy is equal to the excess enthalpy, and the Gibbs energy of mixing is

$$\Delta_{mix}G=nRT(x_A \ln x_A + x_B \ln x_B + \xi x_A x_B) \tag{5B.7}$$

Figure 5B.5 shows how $\Delta_{mix}G$ varies with composition for different values of ξ. The important feature is that for $\xi>2$ the graph shows two minima separated by a maximum. The implication of this observation is that, provided $\xi>2$, then the

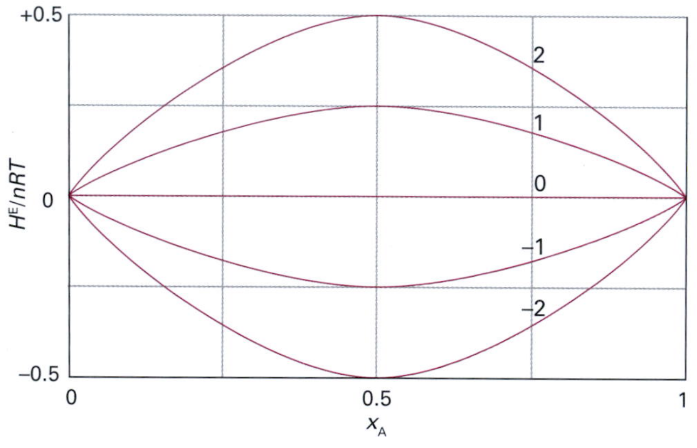

Figure 5B.4 The excess enthalpy according to a model in which it is proportional to $\xi x_A x_B$, for different values of the parameter ξ.

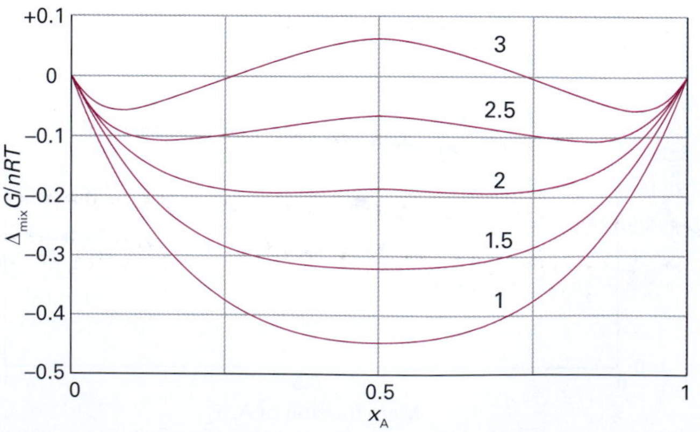

Figure 5B.5 The Gibbs energy of mixing for different values of the parameter ξ.

system will separate spontaneously into two phases with compositions corresponding to the two minima, for that separation corresponds to a reduction in Gibbs energy. We develop this point in Topic 5C.

Example 5B.1 Identifying the parameter for a regular solution

Identify the value of the parameter ξ that would be appropriate to model a mixture of benzene and cyclohexane at 25 °C and estimate the Gibbs energy of mixing to produce an equimolar mixture.

Method Refer to Fig. 5B.3a and identify the value at the curve maximum, and then relate it to eqn 5B.6 written as a molar quantity ($H^{\rm E}=\xi RTx_{\rm A}x_{\rm B}$). For the second part, assume that the solution is regular and that the Gibbs energy of mixing is given by eqn 5B.7.

Answer The experimental value occurs close to $x_{\rm A}=x_{\rm B}=\frac{1}{2}$ and its value is close to 710 J mol⁻¹. It follows that

$$\xi=\frac{H^{\rm E}}{RTx_{\rm A}x_{\rm B}}=\frac{701\,{\rm J\,mol^{-1}}}{(8.3145\,{\rm J\,K^{-1}\,mol^{-1}})\times(298\,{\rm K})\times\frac{1}{2}\times\frac{1}{2}}=1.13$$

The total Gibbs energy of mixing to achieve the stated composition (provided the solution is regular) is therefore

$$\Delta_{\rm mix}G/n=-RT\ln 2+701\,{\rm J\,mol^{-1}}$$
$$=-1.72\,{\rm kJ\,mol^{-1}}+0.701\,{\rm kJ\,mol^{-1}}=-1.02\,{\rm kJ\,mol^{-1}}$$

Self-test 5B.3 Fit the entire data set, as best as can be inferred from the graph in Fig. 5B.3a, to an expression of the form in eqn 5B.6 by a curve-fitting procedure.

Answer: The best fit of the form $Ax(1-x)$ to the data pairs

X	0.1	0.2	0.3	0.4	0.5	0.6	0.7	0.8	0.9
$H^{\rm E}/({\rm J\,mol^{-1}})$	150	350	550	680	700	690	600	500	280

is $A=690\,{\rm J\,mol^{-1}}$

5B.2 Colligative properties

The properties we consider are the lowering of vapour pressure, the elevation of boiling point, the depression of freezing point, and the osmotic pressure arising from the presence of a solute. In dilute solutions these properties depend only on the number of solute particles present, not their identity. For this reason, they are called **colligative properties** (denoting 'depending on the collection'). In this development, we denote the solvent by A and the solute by B.

We assume throughout the following that the solute is not volatile, so it does not contribute to the vapour. We also assume that the solute does not dissolve in the solid solvent: that is, the pure solid solvent separates when the solution is frozen. The latter assumption is quite drastic, although it is true of many mixtures; it can be avoided at the expense of more algebra, but that introduces no new principles.

(a) The common features of colligative properties

All the colligative properties stem from the reduction of the chemical potential of the liquid solvent as a result of the presence of solute. For an ideal solution (one that obeys Raoult's law, Topic 5A; $p_{\rm A}=x_{\rm A}p_{\rm A}^*$), the reduction is from $\mu_{\rm A}^*$ for the pure solvent to $\mu_{\rm A}=\mu_{\rm A}^*+RT\ln x_{\rm A}$ when a solute is present ($\ln x_{\rm A}$ is negative because $x_{\rm A}<1$). There is no direct influence of the solute on the chemical potential of the solvent vapour and the solid solvent because the solute appears in neither the vapour nor the solid. As can be seen from Fig. 5B.6, the reduction in chemical potential of the solvent implies that the liquid–vapour equilibrium occurs at a higher temperature (the boiling point is raised) and the solid–liquid equilibrium occurs at a lower temperature (the freezing point is lowered).

The molecular origin of the lowering of the chemical potential is not the energy of interaction of the solute and solvent particles, because the lowering occurs even in an ideal solution (for which the enthalpy of mixing is zero). If it is not an enthalpy effect, it must be an entropy effect. The vapour pressure of the pure liquid reflects the tendency of the solution towards greater entropy, which can be achieved if the liquid vaporizes to form a gas. When a solute is present, there is an additional contribution to the entropy of the liquid, even in an ideal solution. Because the entropy of the liquid is already higher than that of the pure liquid, there is a weaker tendency to form the gas (Fig. 5B.7). The effect of the solute appears as a lowered vapour pressure, and hence a higher boiling point. Similarly, the enhanced molecular randomness of the solution opposes the tendency to freeze. Consequently, a lower temperature must be reached

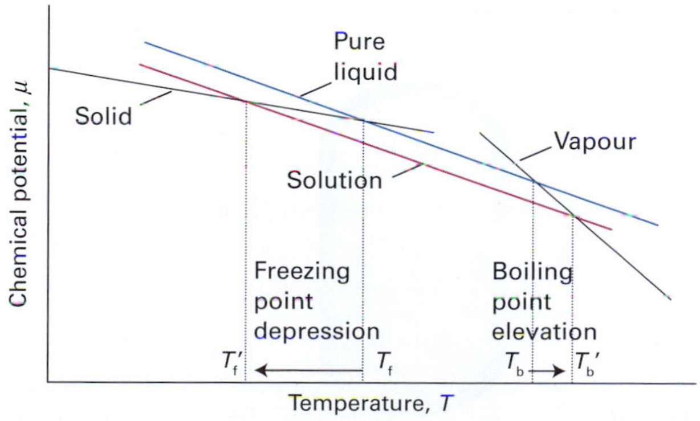

Figure 5B.6 The chemical potential of a solvent in the presence of a solute. The lowering of the liquid's chemical potential has a greater effect on the freezing point than on the boiling point because of the angles at which the lines intersect.

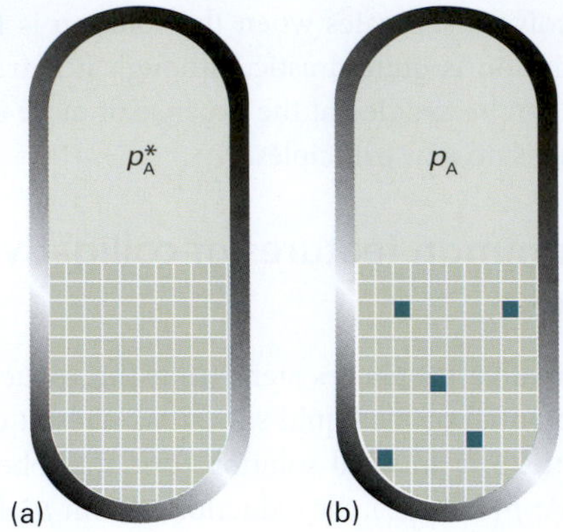

Figure 5B.7 The vapour pressure of a pure liquid represents a balance between the increase in disorder arising from vaporization and the decrease in disorder of the surroundings. (a) Here the structure of the liquid is represented highly schematically by the grid of squares. (b) When solute (the dark squares) is present, the disorder of the condensed phase is higher than that of the pure liquid, and there is a decreased tendency to acquire the disorder characteristic of the vapour.

before equilibrium between solid and solution is achieved. Hence, the freezing point is lowered.

The strategy for the quantitative discussion of the elevation of boiling point and the depression of freezing point is to look for the temperature at which, at 1 atm, one phase (the pure solvent vapour or the pure solid solvent) has the same chemical potential as the solvent in the solution. This is the new equilibrium temperature for the phase transition at 1 atm, and hence corresponds to the new boiling point or the new freezing point of the solvent.

(b) The elevation of boiling point

The heterogeneous equilibrium of interest when considering boiling is between the solvent vapour and the solvent in solution at 1 atm (Fig. 5B.8). The equilibrium is established at a temperature for which

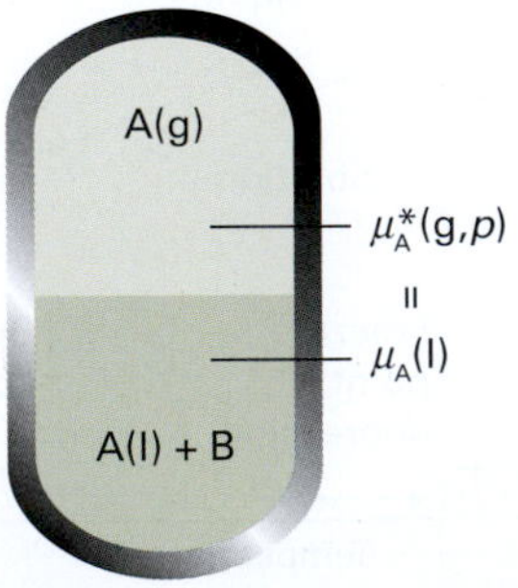

Figure 5B.8 The heterogeneous equilibrium involved in the calculation of the elevation of boiling point is between A in the pure vapour and A in the mixture, A being the solvent and B a non-volatile solute.

$$\mu_A^*(g) = \mu_A^*(l) + RT\ln x_A \quad (5B.8)$$

(The pressure of 1 atm is the same throughout, and will not be written explicitly.) We show in the following *Justification* that this equation implies that the presence of a solute at a mole fraction x_B causes an increase in normal boiling point from T^* to $T^* + \Delta T_b$, where

$$\Delta T_b = Kx_B \quad K = \frac{RT^{*2}}{\Delta_{vap}H} \quad \textit{Ideal solution} \quad \text{Elevation of boiling point} \quad (5B.9)$$

Justification 5B.1 The elevation of the boiling point of a solvent

Equation 5B.8 can be rearranged into

$$\ln x_A = \frac{\mu_A^*(g) - \mu_A^*(l)}{RT} = \frac{\Delta_{vap}G}{RT}$$

where $\Delta_{vap}G$ is the Gibbs energy of vaporization of the pure solvent (A). First, to find the relation between a change in composition and the resulting change in boiling temperature, we differentiate both sides with respect to temperature and use the Gibbs–Helmholtz equation (Topic 3D, $(\partial(G/T)/\partial T)_p = -H/T^2$) to express the term on the right:

$$\frac{d\ln x_A}{dT} = \frac{1}{R}\frac{d(\Delta_{vap}G/T)}{dT} = -\frac{\Delta_{vap}H}{RT^2}$$

Now multiply both sides by dT and integrate from $x_A = 1$, corresponding to $\ln x_A = 0$ (and when $T = T^*$, the boiling point of pure A) to x_A (when the boiling point is T):

$$\int_0^{\ln x_A} d\ln x_A = -\frac{1}{R}\int_{T^*}^{T} \frac{\Delta_{vap}H}{T^2} dT$$

The left-hand side integrates to ln x_A, which is equal to $\ln(1 - x_B)$. The right-hand side can be integrated if we assume that the enthalpy of vaporization is a constant over the small range of temperatures involved and can be taken outside the integral. Thus, we obtain

$$\ln(1 - x_B) = -\frac{\Delta_{vap}H}{R}\int_{T^*}^{T} \frac{1}{T^2} dT$$

and therefore

$$\ln(1 - x_B) = \frac{\Delta_{vap}H}{R}\left(\frac{1}{T} - \frac{1}{T^*}\right)$$

We now suppose that the amount of solute present is so small that $x_B \ll 1$, and use the expansion $\ln(1-x) = -x - \frac{1}{2}x^2 + \cdots \approx -x$ (*Mathematical background* 1) and hence obtain

$$x_B = \frac{\Delta_{vap}H}{R}\left(\frac{1}{T^*} - \frac{1}{T}\right)$$

Finally, because $T \approx T^*$, it also follows that

$$\frac{1}{T^*}-\frac{1}{T}=\frac{T-T^*}{TT^*}\approx\frac{T-T^*}{T^{*2}}=\frac{\Delta T_b}{T^{*2}}$$

with $\Delta T_b = T - T^*$. The previous equation then rearranges into eqn 5B.9.

Because eqn 5B.9 makes no reference to the identity of the solute, only to its mole fraction, we conclude that the elevation of boiling point is a colligative property. The value of ΔT does depend on the properties of the solvent, and the biggest changes occur for solvents with high boiling points. By Trouton's rule (Topic 3B), $\Delta_{vap}H/T^*$ is a constant; therefore eqn 5B.9 has the form $\Delta T \propto T^*$ and is independent of $\Delta_{vap}H$ itself. For practical applications of eqn 5B.9, we note that the mole fraction of B is proportional to its molality, b, in the solution, and write

$$\Delta T_b = K_b b \quad \textit{Empirical relation} \quad \text{Boiling point elevation} \quad (5B.10)$$

where K_b is the empirical **boiling-point constant** of the solvent (Table 5B.1).

Brief illustration 5B.3 Elevation of boiling point

The boiling-point constant of water is 0.51 K kg mol^{-1}, so a solute present at a molality of 0.10 mol kg^{-1} would result in an elevation of boiling point of only 0.051 K. The boiling-point constant of benzene is significantly larger, at 2.53 K kg mol^{-1}, so the elevation would be 0.25 K.

Self-test 5B.4 Identify the feature that accounts for the difference in boiling-point constants of water and benzene.

Answer: High enthalpy of vaporization of water; given molality corresponds to a smaller mole fraction

(c) The depression of freezing point

The heterogeneous equilibrium now of interest is between pure solid solvent A and the solution with solute present at a mole fraction x_B (Fig. 5B.9). At the freezing point, the chemical potentials of A in the two phases are equal:

Table 5B.1* Freezing-point (K_f) and boiling-point (K_b) constants

	K_f/(K kg mol^{-1})	K_b/(K kg mol^{-1})
Benzene	5.12	2.53
Camphor	40	
Phenol	7.27	3.04
Water	1.86	0.51

* More values are given in the *Resource section*.

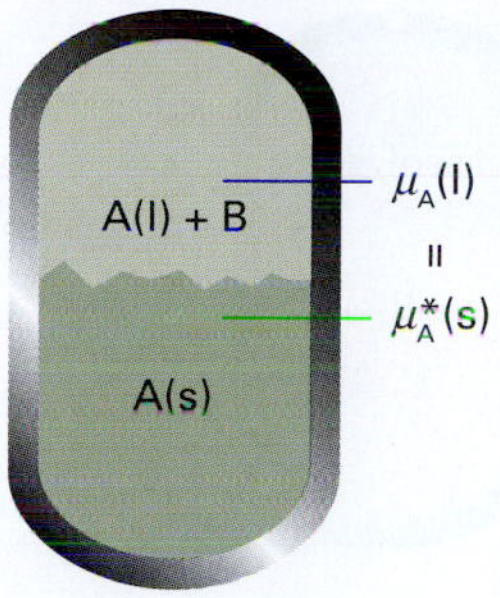

Figure 5B.9 The heterogeneous equilibrium involved in the calculation of the lowering of freezing point is between A in the pure solid and A in the mixture, A being the solvent and B a solute that is insoluble in solid A.

$$\mu_A^*(s) = \mu_A^*(l) + RT \ln x_A \quad (5B.11)$$

The only difference between this calculation and the last is the appearance of the solid's chemical potential in place of the vapour's. Therefore we can write the result directly from eqn 5B.9:

$$\Delta T_f = K' x_B \quad K' = \frac{RT^{*2}}{\Delta_{fus}H} \quad \text{Freezing point depression} \quad (5B.12)$$

where ΔT_f is the freezing point depression, $T^* - T$, and $\Delta_{fus}H$ is the enthalpy of fusion of the solvent. Larger depressions are observed in solvents with low enthalpies of fusion and high melting points. When the solution is dilute, the mole fraction is proportional to the molality of the solute, b, and it is common to write the last equation as

$$\Delta T_f = K_f b \quad \textit{Empirical relation} \quad \text{Freezing point depression} \quad (5B.13)$$

where K_f is the empirical **freezing-point constant** (Table 5B.1). Once the freezing-point constant of a solvent is known, the depression of freezing point may be used to measure the molar mass of a solute in the method known as **cryoscopy**; however, the technique is of little more than historical interest.

Brief illustration 5B.4 Depression of freezing point

The freezing-point constant of water is 1.86 K kg mol^{-1}, so a solute present at a molality of 0.10 mol kg^{-1} would result in a depression of freezing point of only 0.19 K. The freezing-point constant of camphor is significantly larger, at 40 K kg mol^{-1}, so the depression would be 4.0 K. Camphor was once widely used for estimates of molar mass by cryoscopy.

Self-test 5B.5 Why are freezing-point constants typically larger than the corresponding boiling-point constants of a solvent?

Answer: Enthalpy of fusion is smaller than the enthalpy of vaporization of a substance

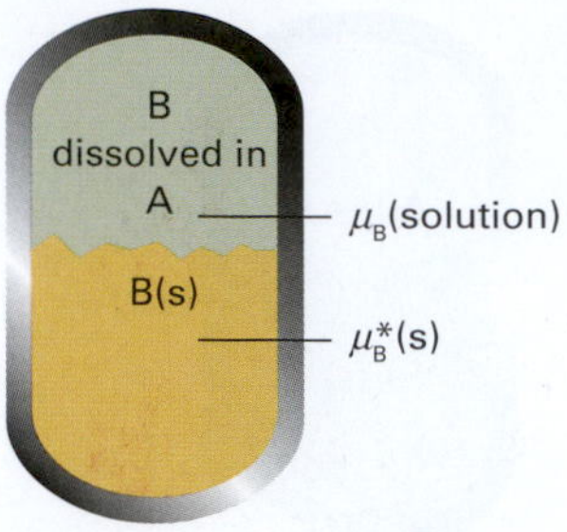

Figure 5B.10 The heterogeneous equilibrium involved in the calculation of the solubility is between pure solid B and B in the mixture.

(d) Solubility

Although solubility is not a colligative property (because solubility varies with the identity of the solute), it may be estimated by the same techniques as we have been using. When a solid solute is left in contact with a solvent, it dissolves until the solution is saturated. Saturation is a state of equilibrium, with the undissolved solute in equilibrium with the dissolved solute. Therefore, in a saturated solution the chemical potential of the pure solid solute, $\mu_B^*(s)$, and the chemical potential of B in solution, μ_B, are equal (Fig. 5B.10). Because the latter is related to the mole fraction in the solution by $\mu_B = \mu_B^*(l) + RT\ln x_B$, we can write

$$\mu_B^*(s) = \mu_B^*(l) + RT\ln x_B \qquad (5B.14)$$

This expression is the same as the starting equation of the last section, except that the quantities refer to the solute B, not the solvent A. We now show in the following *Justification* that

$$\ln x_B = \frac{\Delta_{fus}H}{R}\left(\frac{1}{T_f} - \frac{1}{T}\right) \qquad \text{Ideal solubility} \qquad (5B.15)$$

where $\Delta_{fus}H$ is the enthalpy of fusion of the solute and T_f is its melting point.

Justification 5B.2 The solubility of an ideal solute

The starting point is the same as in *Justification* 5B.1 but the aim is different. In the present case, we want to find the mole fraction of B in solution at equilibrium when the temperature is T. Therefore, we start by rearranging eqn 5B.14 into

$$\ln x_B = \frac{\mu_B^*(s) - \mu_B^*(l)}{RT} = -\frac{\Delta_{fus}G}{RT}$$

As in *Justification* 5B.1, we relate the change in composition $d\ln x_B$ to the change in temperature by differentiation and use of the Gibbs–Helmholtz equation. Then we integrate from the melting temperature of B (when $x_B = 1$ and $\ln x_B = 0$) to the *lower* temperature of interest (when x_B has a value between 0 and 1):

$$\int_0^{\ln x_B} d\ln x_B = \frac{1}{R}\int_{T_f}^{T} \frac{\Delta_{fus}H}{T^2}\,dT$$

If we suppose that the enthalpy of fusion of B is constant over the range of temperatures of interest, it can be taken outside the integral, and we obtain eqn 5B.15.

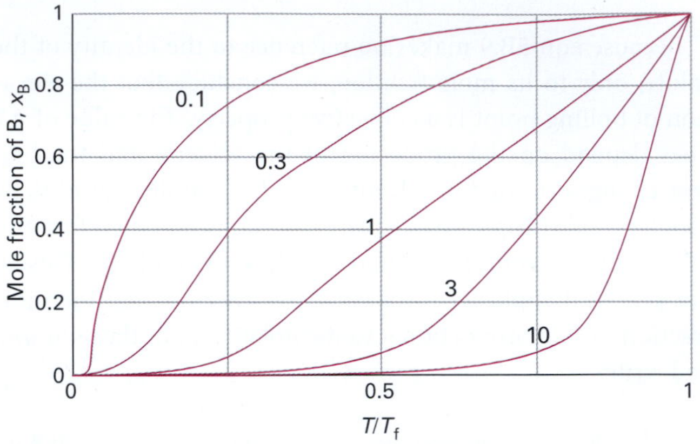

Figure 5B.11 The variation of solubility (the mole fraction of solute in a saturated solution) with temperature (T_f is the freezing temperature of the solute). Individual curves are labelled with the value of $\Delta_{fus}H/RT_f$.

Equation 5B.15 is plotted in Fig. 5B.11. It shows that the solubility of B decreases exponentially as the temperature is lowered from its melting point. The illustration also shows that solutes with high melting points and large enthalpies of melting have low solubilities at normal temperatures. However, the detailed content of eqn 5B.15 should not be treated too seriously because it is based on highly questionable approximations, such as the ideality of the solution. One aspect of its approximate character is that it fails to predict that solutes will have different solubilities in different solvents, for no solvent properties appear in the expression.

Brief illustration 5B.5 Ideal solubility

The ideal solubility of naphthalene in benzene is calculated from eqn 5B.15 by noting that the enthalpy of fusion of naphthalene is 18.80 kJ mol⁻¹ and its melting point is 354 K. Then, at 20 °C,

$$\ln x_{naphthalene} = \frac{1.880\times10^4\,\text{J mol}^{-1}}{8.3145\,\text{J K}^{-1}\,\text{mol}^{-1}}\left(\frac{1}{354\,\text{K}} - \frac{1}{293\,\text{K}}\right) = -1.32\ldots$$

and therefore $x_{naphthalene} = 0.26$. This mole fraction corresponds to a molality of 4.5 mol kg^{-1} (580 g of naphthalene in 1 kg of benzene).

Self-test 5B.6 Plot the solubility of naphthalene as a function of temperature against mole fraction: in Topic 5C we see that such diagrams are 'temperature–composition phase diagrams'.

Answer: See Fig. 5B.12.

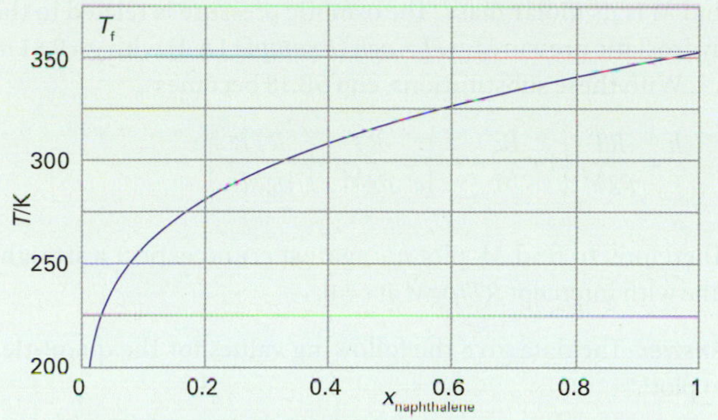

Figure 5B.12 The theoretical solubility of naphthalene in benzene, as calculated in *Self-test* 5B.6.

(e) Osmosis

The phenomenon of **osmosis** (from the Greek word for 'push') is the spontaneous passage of a pure solvent into a solution separated from it by a **semipermeable membrane**, a membrane permeable to the solvent but not to the solute (Fig. 5B.13). The **osmotic pressure**, Π, is the pressure that must be applied to the solution to stop the influx of solvent. Important examples of osmosis include transport of fluids through cell membranes, dialysis and **osmometry**, the determination of molar mass by the measurement of osmotic pressure. Osmometry is widely used to determine the molar masses of macromolecules.

In the simple arrangement shown in Fig. 5B.14, the opposing pressure arises from the head of solution that the osmosis itself produces. Equilibrium is reached when the hydrostatic pressure of the column of solution matches the osmotic pressure.

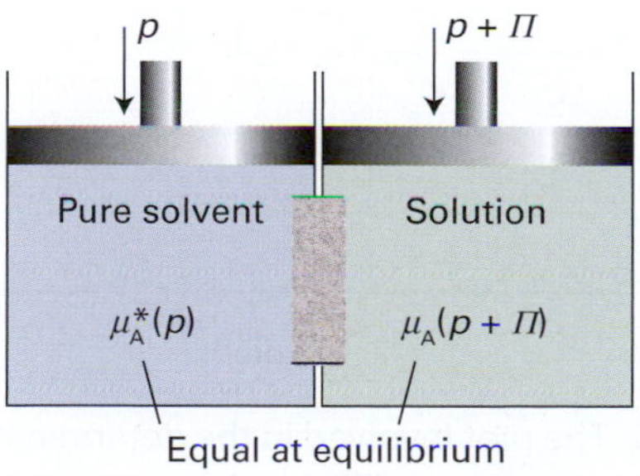

Figure 5B.13 The equilibrium involved in the calculation of osmotic pressure, Π, is between pure solvent A at a pressure p on one side of the semipermeable membrane and A as a component of the mixture on the other side of the membrane, where the pressure is $p+\Pi$.

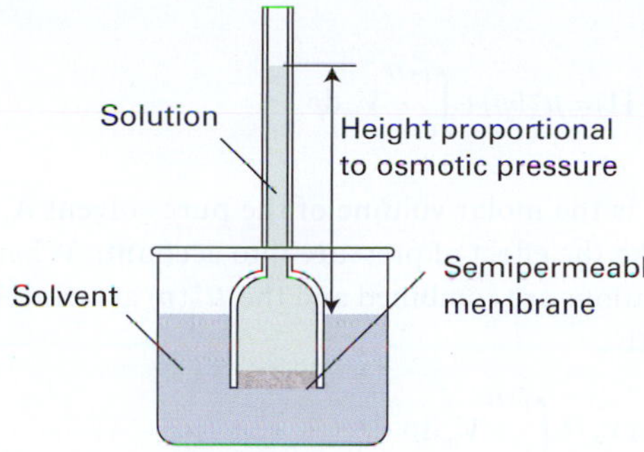

Figure 5B.14 In a simple version of the osmotic pressure experiment, A is at equilibrium on each side of the membrane when enough has passed into the solution to cause a hydrostatic pressure difference.

The complicating feature of this arrangement is that the entry of solvent into the solution results in its dilution, and so it is more difficult to treat than the arrangement in Fig. 5B.13, in which there is no flow and the concentrations remain unchanged.

The thermodynamic treatment of osmosis depends on noting that, at equilibrium, the chemical potential of the solvent must be the same on each side of the membrane. The chemical potential of the solvent is lowered by the solute, but is restored to its 'pure' value by the application of pressure. As shown in the following *Justification*, this equality implies that for dilute solutions the osmotic pressure is given by the **van 't Hoff equation**:

$$\Pi = [B]RT \quad \text{van 't Hoff equation} \quad (5B.16)$$

where $[B]=n_B/V$ is the molar concentration of the solute.

Justification 5B.3 The van 't Hoff equation

On the pure solvent side the chemical potential of the solvent, which is at a pressure p, is $\mu_A^*(p)$. On the solution side, the chemical potential is lowered by the presence of the solute, which reduces the mole fraction of the solvent from 1 to x_A. However, the chemical potential of A is raised on account of the greater pressure, $p+\Pi$, that the solution experiences. At equilibrium the chemical potential of A is the same in both compartments, and we can write

$$\mu_A^*(p)=\mu_A(x_A, p+\Pi)$$

The presence of solute is taken into account in the normal way by using eqn 5B.1:

$$\mu_A(x_A, p+\Pi)=\mu_A^*(p+\Pi)+RT\ln x_A$$

Equation 3D.12b,

$$G_m(p_f)=G_m(p_i)+\int_{p_i}^{p_f} V_m \mathrm{d}p$$

written as

$$\mu_A^*(p+\Pi)=\mu_A^*(p)+\int_p^{p+\Pi} V_m dp$$

where V_m is the molar volume of the pure solvent A, shows how to take the effect of pressure into account:. When these three equations are combined and the $\mu_A^*(p)$ are cancelled we are left with

$$-RT\ln x_A=\int_p^{p+\Pi} V_m dp \qquad (5B.17)$$

This expression enables us to calculate the additional pressure Π that must be applied to the solution to restore the chemical potential of the solvent to its 'pure' value and thus to restore equilibrium across the semipermeable membrane. For dilute solutions, ln x_A may be replaced by $\ln(1-x_B)\approx -x_B$. We may also assume that the pressure range in the integration is so small that the molar volume of the solvent is a constant. That being so, V_m may be taken outside the integral, giving

$$RTx_B=\Pi V_m$$

When the solution is dilute, $x_B\approx n_B/n_A$. Moreover, because $n_A V_m=V$, the total volume of the solvent, the equation simplifies to eqn 5B.16.

Because the effect of osmotic pressure is so readily measurable and large, one of the most common applications of osmometry is to the measurement of molar masses of macromolecules, such as proteins and synthetic polymers. As these huge molecules dissolve to produce solutions that are far from ideal, it is assumed that the van 't Hoff equation is only the first term of a virial-like expansion, much like the extension of the perfect gas equation to real gases (in Topic 1C) to take into account molecular interactions:

$$\Pi=[J]RT\{1+B[J]+\ldots\} \qquad \text{Osmotic virial expansion} \qquad (5B.18)$$

(We have denoted the solute J to avoid too many different Bs in this expression.) The additional terms take the non-ideality into account; the empirical constant B is called the **osmotic virial coefficient**.

Example 5B.2 Using osmometry to determine the molar mass of a macromolecule

The osmotic pressures of solutions of poly(vinyl chloride), PVC, in cyclohexanone at 298 K are given below. The pressures are expressed in terms of the heights of solution (of mass density $\rho=0.980\,\text{g cm}^{-3}$) in balance with the osmotic pressure. Determine the molar mass of the polymer.

$c/(\text{g dm}^{-3})$	1.00	2.00	4.00	7.00	9.00
h/cm	0.28	0.71	2.01	5.10	8.00

Method The osmotic pressure is measured at a series of mass concentrations, c, and a plot of Π/c against c is used to determine the molar mass of the polymer. We use eqn 5B.18 with $[J]=c/M$ where c is the mass concentration of the polymer and M is its molar mass. The osmotic pressure is related to the hydrostatic pressure by $\Pi=\rho gh$ (*Example* 1A.1) with $g=9.81\,\text{m s}^{-2}$. With these substitutions, eqn 5B.18 becomes

$$\frac{h}{c}=\frac{RT}{\rho gM}\left\{1+\frac{Bc}{M}+\cdots\right\}=\frac{RT}{\rho gM}+\left(\frac{RTB}{\rho gM^2}\right)c+\cdots$$

Therefore, to find M, plot h/c against c, and expect a straight line with intercept $RT/\rho gM$ at $c=0$.

Answer The data give the following values for the quantities to plot:

$c/(\text{g dm}^{-3})$	1.00	2.00	4.00	7.00	9.00
$(h/c)/(\text{cm g}^{-1}\,\text{dm}^3)$	0.28	0.36	0.503	0.729	0.889

The points are plotted in Fig. 5B.15. The intercept is at 0.20. Therefore,

$$M=\frac{RT}{\rho g}\times\frac{1}{0.20\,\text{cm g}^{-1}\,\text{dm}^3}$$
$$=\frac{(8.3145\,\text{J K}^{-1}\,\text{mol}^{-1})\times(298\,\text{K})}{(980\,\text{kg m}^{-3})\times(9.81\,\text{m s}^{-2})}\times\frac{1}{2.0\times10^{-3}\,\text{m}^4\,\text{kg}^{-1}}$$
$$=1.3\times10^2\,\text{kg mol}^{-1}$$

where we have used $1\,\text{kg m}^2\,\text{s}^{-2}=1\,\text{J}$. Modern osmometers give readings of osmotic pressure in pascals, so the analysis of the data is more straightforward and eqn 5B.18 can be used directly. As explained in Topic 17D, the value obtained from osmometry is the 'number average molar mass'.

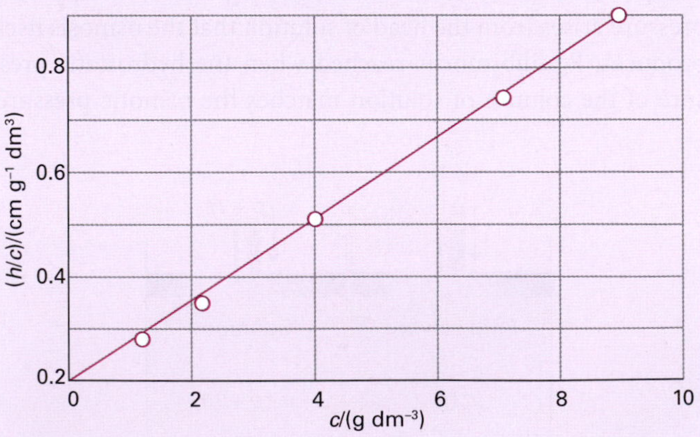

Figure 5B.15 The plot involved in the determination of molar mass by osmometry. The molar mass is calculated from the intercept at $c=0$.

Self-test 5B.7 Estimate the depression of freezing point of the most concentrated of these solutions, taking K_f as about $10\,\text{K/(mol kg}^{-1})$.

Answer: 0.8 mK

Checklist of concepts

- ☐ 1. The **Gibbs energy of mixing** of two liquids to form an ideal solution is calculated in the same way as for two perfect gases.
- ☐ 2. The **enthalpy of mixing** is zero and the Gibbs energy is due entirely to the entropy of mixing.
- ☐ 3. A **regular solution** is one in which the entropy of mixing is the same as for an ideal solution but the enthalpy of mixing is non-zero.
- ☐ 4. A **colligative property** depends only on the number of solute particles present, not their identity.
- ☐ 5. All the colligative properties stem from the reduction of the chemical potential of the liquid solvent as a result of the presence of solute.
- ☐ 6. The **elevation of boiling point** is proportional to the molality of the solute.
- ☐ 7. The **depression of freezing point** is also proportional to the molality of the solute.
- ☐ 8. Solutes with high melting points and large enthalpies of melting have low solubilities at normal temperatures.
- ☐ 9. The **osmotic pressure** is the pressure that when applied to a solution prevents the influx of solvent through a semipermeable membrane.
- ☐ 10. The relation of the osmotic pressure to the molar concentration of the solute is given by the **van 't Hoff equation** and is a sensitive way of determining molar mass.

Checklist of equations

Property	Equation	Comment	Equation number
Gibbs energy of mixing	$\Delta_{mix}G=nRT(x_A \ln x_A+x_B \ln x_B)$	Ideal solutions	5B.3
Entropy of mixing	$\Delta_{mix}S=-nR(x_A \ln x_A+x_B \ln x_B)$	Ideal solutions	5B.4
Enthalpy of mixing	$\Delta_{mix}H=0$	Ideal solutions	
Excess function	$X^E=\Delta_{mix}X-\Delta_{mix}X^{ideal}$	Definition	5B.5
Regular solution ($S^E=0$)	$H^E=n\xi RTx_Ax_B$	Model	5B.6
Elevation of boiling point	$\Delta T_b=K_b b$	Empirical, non-volatile solute	5B.10
Depression of freezing point	$\Delta T_f=K_f b$	Empirical, solute insoluble in solid solvent	5B.13
Ideal solubility	$\ln x_B=(\Delta_{fus}H/R)(1/T_f-1/T)$	Ideal solution	58.15
van 't Hoff equation	$\Pi=[B]RT$	Valid as $[B]\rightarrow 0$	5B.16
Osmotic virial expansion	$\Pi=[J]RT\{1+B[J]+\ldots\}$	Empirical	5B.18

5C Phase diagrams of binary systems

Contents

➤ Why do you need to know this material?

Phase diagrams are used widely in materials science, metallurgy, geology, and the chemical industry to summarize the composition of mixtures and it is important to be able to interpret them.

➤ What is the key idea?

A phase diagram is a map showing the conditions under which each phase of a system is the most stable.

➤ What do you need to know already?

It would be helpful to review the interpretation of one-component phase diagrams and the phase rule (Topic 4A). The early part of this Topic draws on Raoult's law (Topic 4B) and the concept of partial pressure (Topic 1A).

One-component phase diagrams are described in Topic 4A. The phase equilibria of binary systems are more complex because composition is an additional variable. However, they provide very useful summaries of phase equilibria for both ideal and empirically established real systems.

5C.1 Vapour pressure diagrams

The partial vapour pressures of the components of an ideal solution of two volatile liquids are related to the composition of the liquid mixture by Raoult's law (Topic 5A)

$$p_A = x_A p_A^* \qquad p_B = x_B p_B^* \qquad (5C.1)$$

where p_A^* is the vapour pressure of pure A and p_B^* that of pure B. The total vapour pressure p of the mixture is therefore

$$p = p_A + p_B = x_A p_A^* + x_B p_B^* = p_B^* + (p_A^* - p_B^*)x_A \qquad (5C.2)$$

Total vapour pressure

This expression shows that the total vapour pressure (at some fixed temperature) changes linearly with the composition from p_B^* to p_A^* as x_A changes from 0 to 1 (Fig. 5C.1).

(a) The composition of the vapour

The compositions of the liquid and vapour that are in mutual equilibrium are not necessarily the same. Common sense suggests that the vapour should be richer in the more volatile component. This expectation can be confirmed as follows. The partial pressures of the components are given by eqn 1A.8 of Topic 1A ($p_J = x_J p$). It follows from that definition that the mole fractions in the gas, y_A and y_B, are

$$y_A = \frac{p_A}{p} \qquad y_B = \frac{p_B}{p} \qquad (5C.3)$$

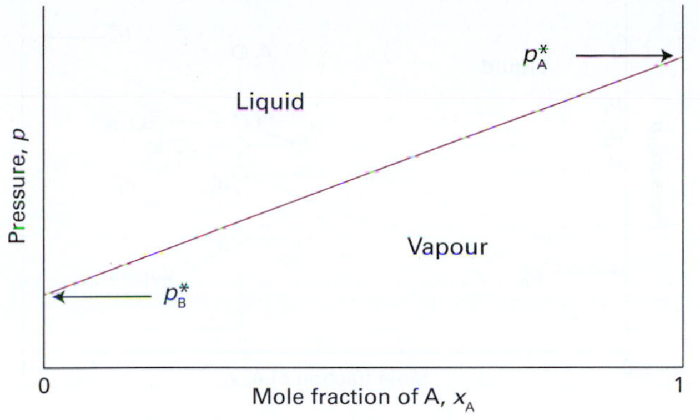

Figure 5C.1 The variation of the total vapour pressure of a binary mixture with the mole fraction of A in the liquid when Raoult's law is obeyed.

Provided the mixture is ideal, the partial pressures and the total pressure may be expressed in terms of the mole fractions in the liquid by using eqn 5C.1 for p_J and eqn 5C.2 for the total vapour pressure p, which gives

$$y_A = \frac{x_A p_A^*}{p_B^* + (p_A^* - p_B^*)x_A} \quad y_B = 1 - y_A \qquad \text{Composition of vapour} \qquad (5C.4)$$

Figure 5C.2 shows the composition of the vapour plotted against the composition of the liquid for various values of $p_A^*/p_B^* > 1$. We see that in all cases $y_A > x_A$, that is, the vapour is richer than the liquid in the more volatile component. Note that if B is non-volatile, so that $p_B^* = 0$ at the temperature of interest, then it makes no contribution to the vapour ($y_B = 0$).

Equation 5C.3 shows how the total vapour pressure of the mixture varies with the composition of the liquid. Because we can relate the composition of the liquid to the composition of

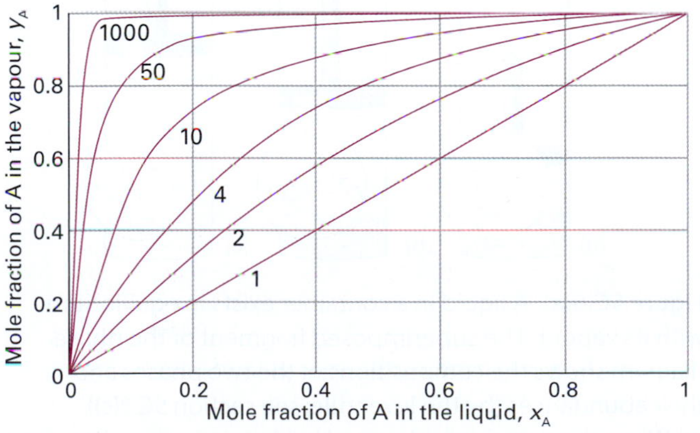

Figure 5C.2 The mole fraction of A in the vapour of a binary ideal solution expressed in terms of its mole fraction in the liquid, calculated using eqn 5C.4 for various values of p_A^*/p_B^* (the label on each curve) with A more volatile than B. In all cases the vapour is richer than the liquid in A.

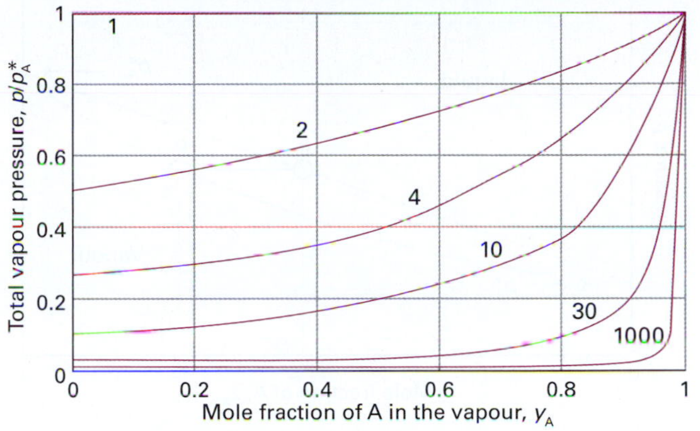

Figure 5C.3 The dependence of the vapour pressure of the same system as in Fig. 5C.2, but expressed in terms of the mole fraction of A in the vapour by using eqn 5C.5. Individual curves are labelled with the value of p_A^*/p_B^*.

the vapour through eqn 5C.3, we can now also relate the total vapour pressure to the composition of the vapour:

$$p = \frac{p_A^* p_B^*}{p_A^* + (p_B^* - p_A^*)y_A} \qquad \text{Total vapour pressure} \qquad (5C.5)$$

This expression is plotted in Fig. 5C.3.

Brief illustration 5C.1 The composition of the vapour

The vapour pressures of benzene and methylbenzene at 20 °C are 75 Torr and 21 Torr, respectively. The composition of the vapour in equilibrium with an equimolar liquid mixture ($x_{\text{benzene}} = x_{\text{methylbenzene}} = \frac{1}{2}$) is

$$y_{\text{benzene}} = \frac{\frac{1}{2}\times(75\,\text{Torr})}{21\,\text{Torr} + (75 - 21\,\text{Torr})\times\frac{1}{2}} = 0.78$$

$$y_{\text{methylbenzene}} = 1 - 0.78 = 0.22$$

The vapour pressure of each component is

$$p_{\text{benzene}} = \tfrac{1}{2}(75\,\text{Torr}) = 38\,\text{Torr}$$

$$p_{\text{methylbenzene}} = \tfrac{1}{2}(21\,\text{Torr}) = 10\,\text{Torr}$$

for a total vapour pressure of 48 Torr.

Self-test 5C.1 What is the composition of the vapour in equilibrium with a mixture in which the mole fraction of benzene is 0.75?

Answer: 0.91, 0.09

(b) The interpretation of the diagrams

If we are interested in distillation, both the vapour and the liquid compositions are of equal interest. It is therefore sensible

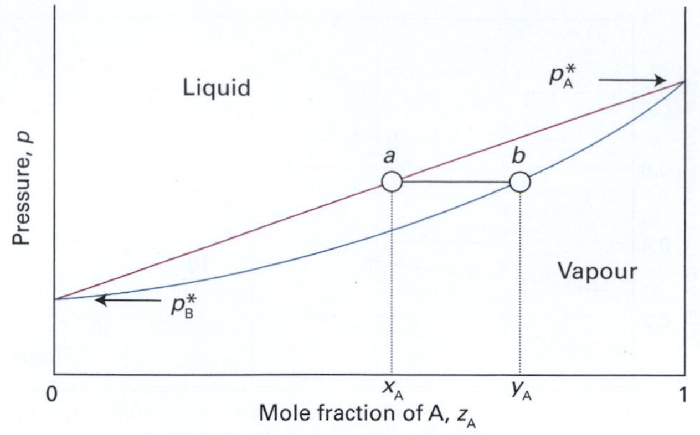

Figure 5C.4 The dependence of the total vapour pressure of an ideal solution on the mole fraction of A in the entire system. A point between the two lines corresponds to both liquid and vapour being present; outside that region there is only one phase present. The mole fraction of A is denoted z_A, as explained in the text.

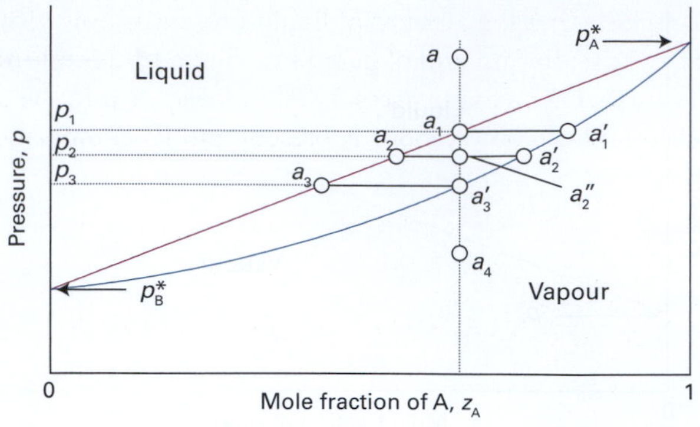

Figure 5C.5 The points of the pressure–composition diagram discussed in the text. The vertical line through a is an isopleth, a line of constant composition of the entire system.

to combine Figs. 5C.2 and 5C.3 into one (Fig. 5C.4). The point a indicates the vapour pressure of a mixture of composition x_A, and the point b indicates the composition of the vapour that is in equilibrium with the liquid at that pressure. A richer interpretation of the phase diagram is obtained, however, if we interpret the horizontal axis as showing the *overall* composition, z_A, of the system (essentially, the mole fraction showing how the mixture was prepared). If the horizontal axis of the vapour pressure diagram is labelled with z_A, then all the points down to the solid diagonal line in the graph correspond to a system that is under such high pressure that it contains only a liquid phase (the applied pressure is higher than the vapour pressure), so $z_A = x_A$, the composition of the liquid. On the other hand, all points below the lower curve correspond to a system that is under such low pressure that it contains only a vapour phase (the applied pressure is lower than the vapour pressure), so $z_A = y_A$.

Points that lie between the two lines correspond to a system in which there are two phases present, one a liquid and the other a vapour. To see this interpretation, consider the effect of lowering the pressure on a liquid mixture of overall composition a in Fig. 5C.5. The lowering of pressure can be achieved by drawing out a piston (Fig. 5C.6). The changes to the system do not affect the overall composition, so the state of the system moves down the vertical line that passes through a. This vertical line is called an **isopleth**, from the Greek words for 'equal abundance'. Until the point a_1 is reached (when the pressure has been reduced to p_1), the sample consists of a single liquid phase. At a_1 the liquid can exist in equilibrium with its vapour. As we have seen, the composition of the vapour phase is given by point a_1'. A line joining two points representing phases in equilibrium is called a **tie line**. The composition of the liquid is the same as initially (a_1 lies on the isopleth through a), so we have to conclude that at this pressure there is virtually no vapour present; however, the tiny amount of vapour that is present has the composition a_1'

Now consider the effect of lowering the pressure to p_2, so taking the system to a pressure and overall composition represented by the point a_2'. This new pressure is below the vapour pressure of the original liquid, so it vaporizes until the vapour pressure of the remaining liquid falls to p_2. Now we know that the composition of such a liquid must be a_2. Moreover, the composition of the vapour in equilibrium with that liquid must be given by the point a_2' at the other end of the tie line. If the pressure is reduced to p_3, a similar readjustment in composition takes place, and now the compositions of the liquid and vapour are represented by the points a_3 and a_3', respectively. The latter point corresponds to a system in which the composition of the vapour is the same as the overall composition, so we have

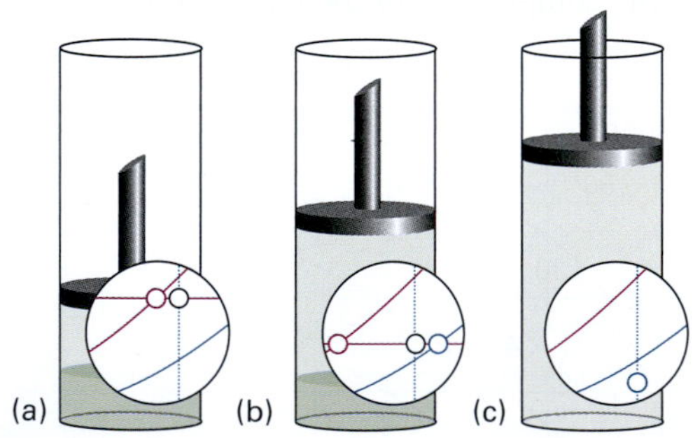

Figure 5C.6(a) A liquid in a container exists in equilibrium with its vapour. The superimposed fragment of the phase diagram shows the compositions of the two phases and their abundances (by the lever rule; see section 5C.1(c)). (b) When the pressure is changed by drawing out a piston, the compositions of the phases adjust as shown by the tie line in the phase diagram. (c) When the piston is pulled so far out that all the liquid has vaporized and only the vapour is present, the pressure falls as the piston is withdrawn and the point on the phase diagram moves into the one-phase region.

to conclude that the amount of liquid present is now virtually zero, but the tiny amount of liquid present has the composition a_3. A further decrease in pressure takes the system to the point a_4; at this stage, only vapour is present and its composition is the same as the initial overall composition of the system (the composition of the original liquid).

Example 5C.1 Constructing a vapour pressure diagram

The following temperature/composition data were obtained for a mixture of octane (O) and methylbenzene (M) at 1.00 atm, where x is the mole fraction in the liquid and y the mole fraction in the vapour at equilibrium.

θ/°C	110.9	112.0	114.0	115.8	117.3	119.0	121.1	123.0
x_M	0.908	0.795	0.615	0.527	0.408	0.300	0.203	0.097
y_M	0.923	0.836	0.698	0.624	0.527	0.410	0.297	0.164

The boiling points are 110.6 °C and 125.6 °C for M and O, respectively. Plot the temperature–composition diagram for the mixture. What is the composition of the vapour in equilibrium with the liquid of composition (a) $x_M=0.250$ and (b) $x_O=0.250$?

Method Plot the composition of each phase (on the horizontal axis) against the temperature (on the vertical axis). The two boiling points give two further points corresponding to $x_M=1$ and $x_M=0$, respectively. Use a curve-fitting program to draw the phase boundaries. For the interpretation, draw the appropriate tie-lines.

Answer The points are plotted in Fig. 5C.7. The two sets of points are fitted to the polynomials $a+bx+cx^2+dx^3$ with

$$\text{For the liquid line: } \theta/°\text{C}=125.422-22.9494x+6.64602x^2 +1.32623x^3+\ldots$$

$$\text{For the vapour line: } \theta/°\text{C}=125.485-11.9387x-12.5626x^2 +9.36542x^3+\ldots$$

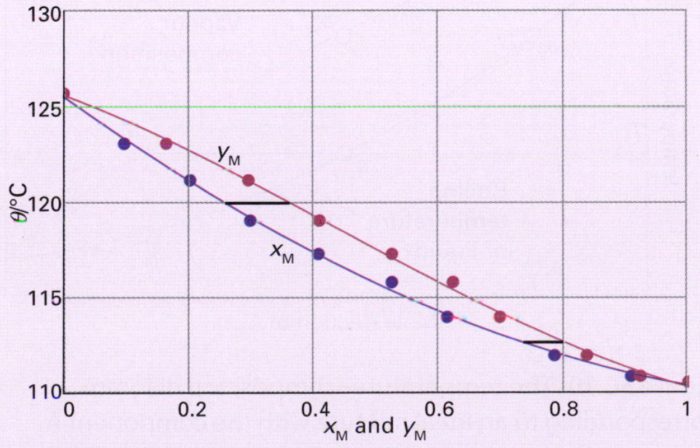

Figure 5C.7 The plot of data and the fitted curves for a mixture of octane and methylbenzene (M) in *Example* 5C.1.

The tie lines at $x_M=0.250$ and $x_O=0.250$ (corresponding to $x_M=0.750$) have been drawn on the graph starting at the lower (liquid curve). They intersect the upper (vapour curve) at $y_M=0.36$ and 0.80, respectively.

Self-test 5C.2 Repeat the analysis for the following data on hexane and heptane at 70 °C:

θ/°C	65	66	70	77	85	100
x_{hexane}	0	0.20	0.40	0.60	0.80	1
y_{hexane}	0	0.02	0.08	0.20	0.48	1

Answer: See Fig. 5C.8.

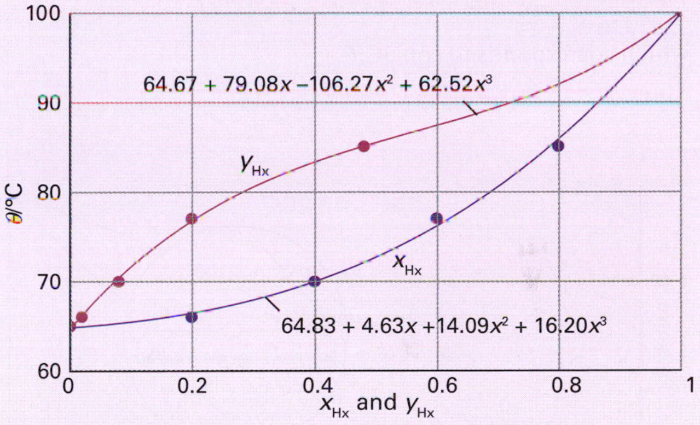

Figure 5C.8 The plot of data and the fitted curves for a mixture of hexane (Hx) and heptane in *Self-test* 5C.2.

(c) The lever rule

A point in the two-phase region of a phase diagram indicates not only qualitatively that both liquid and vapour are present, but represents quantitatively the relative amounts of each. To find the relative amounts of two phases α and β that are in equilibrium, we measure the distances l_α and l_β along the horizontal tie line, and then use the **lever rule** (Fig. 5C.9):

$$n_\alpha l_\alpha = n_\beta l_\beta \qquad \text{Lever rule} \qquad (5C.6)$$

Here n_α is the amount of phase α and n_β the amount of phase β. In the case illustrated in Fig. 5C.9, because $l_\beta \approx 2l_\alpha$, the amount of phase α is about twice the amount of phase β.

Justification 5C.1 The lever rule

To prove the lever rule we write the total amount of A and B molecules as $n=n_\alpha+n_\beta$, where n_α is the amount of molecules in phase α and n_β the amount in phase β. The mole fraction of A in phase α is $x_{A,\alpha}$, so the amount of A in that phase is

$n_\alpha x_{A,\alpha}$. Similarly, the amount of A in phase β is $n_\beta x_{A,\beta}$. The total amount of A is therefore

$$n_A = n_\alpha x_{A,\alpha} + n_\beta x_{A,\beta}$$

Let the composition of the entire mixture be expressed by the mole fraction z_A (this is the label on the horizontal axis, and reflects how the sample is prepared). The total amount of A molecules is therefore

$$n_A = nz_A = n_\alpha z_A + n_\beta z_A$$

By equating these two expressions it follows that

$$n_\alpha(x_{A,\alpha} - z_A) = n_\beta(z_A - x_{A,\beta})$$

which corresponds to eqn 5C.6.

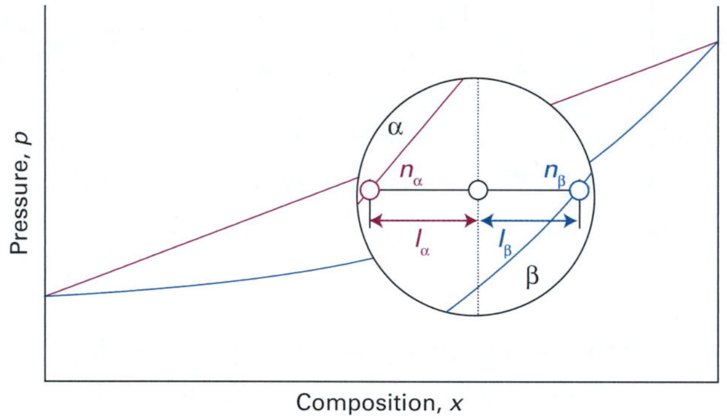

Figure 5C.9 The lever rule. The distances l_α and l_β are used to find the proportions of the amounts of phases α (such as vapour) and β (for example, liquid) present at equilibrium. The lever rule is so called because a similar rule relates the masses at two ends of a lever to their distances from a pivot ($m_\alpha l_\alpha = m_\beta l_\beta$ for balance).

Brief illustration 5C.2 The lever rule

At p_1 in Fig. 5C.5, the ratio l_{vap}/l_{liq} is almost infinite for this tie line, so n_{liq}/n_{vap} is also almost infinite, and there is only a trace of vapour present. When the pressure is reduced to p_2, the value of l_{vap}/l_{liq} is about 0.3, so $n_{liq}/n_{vap} \approx 0.3$ and the amount of liquid is about 0.3 times the amount of vapour. When the pressure has been reduced to p_3, the sample is almost completely gaseous and because $l_{vap}/l_{liq} \approx 0$ we conclude that there is only a trace of liquid present.

Self-test 5C.3 Suppose that in a phase diagram, when the sample was prepared with the mole fraction of component A equal to 0.40 it was found that the compositions of the two phases in equilibrium corresponded to the mole fractions $x_{A,\alpha} = 0.60$ and $x_{A,\beta} = 0.20$. What is the ratio of amounts of the two phases?

Answer: $n_\alpha/n_\beta = 1.0$

5C.2 Temperature–composition diagrams

To discuss distillation we need a **temperature–composition diagram**, a phase diagram in which the boundaries show the composition of the phases that are in equilibrium at various temperatures (and a given pressure, typically 1 atm). An example is shown in Fig. 5C.10. Note that the liquid phase now lies in the lower part of the diagram.

(a) The distillation of mixtures

Consider what happens when a liquid of composition a_1 in Fig. 5C.10 is heated. It boils when the temperature reaches T_2. Then the liquid has composition a_2 (the same as a_1) and the vapour (which is present only as a trace) has composition a_2'. The vapour is richer in the more volatile component A (the component with the lower boiling point). From the location of a_2, we can state the vapour's composition at the boiling point, and from the location of the tie line joining a_2 and a_2' we can read off the boiling temperature (T_2) of the original liquid mixture.

In a **simple distillation**, the vapour is withdrawn and condensed. This technique is used to separate a volatile liquid from a non-volatile solute or solid. In **fractional distillation**, the boiling and condensation cycle is repeated successively. This technique is used to separate volatile liquids. We can follow the changes that occur by seeing what happens when the first condensate of composition a_3 is reheated. The phase diagram shows that this mixture boils at T_3 and yields a vapour of composition a_3', which is even richer in the more volatile component. That vapour is drawn off, and the first drop condenses to a liquid of composition a_4. The cycle can then be repeated until

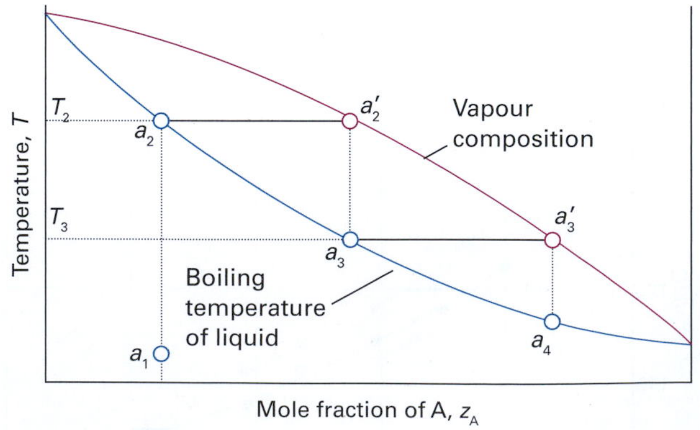

Figure 5C.10 The temperature–composition diagram corresponding to an ideal mixture with the component A more volatile than component B. Successive boilings and condensations of a liquid originally of composition a_1 lead to a condensate that is pure A. The separation technique is called fractional distillation.

in due course almost pure A is obtained in the vapour and pure B remains in the liquid.

The efficiency of a fractionating column is expressed in terms of the number of **theoretical plates**, the number of effective vaporization and condensation steps that are required to achieve a condensate of given composition from a given distillate.

Brief illustration 5C.3 Theoretical plates

To achieve the degree of separation shown in Fig. 5C.11a, the fractionating column must correspond to three theoretical plates. To achieve the same separation for the system shown in Fig. 5C.11b, in which the components have more similar partial pressures, the fractionating column must be designed to correspond to five theoretical plates.

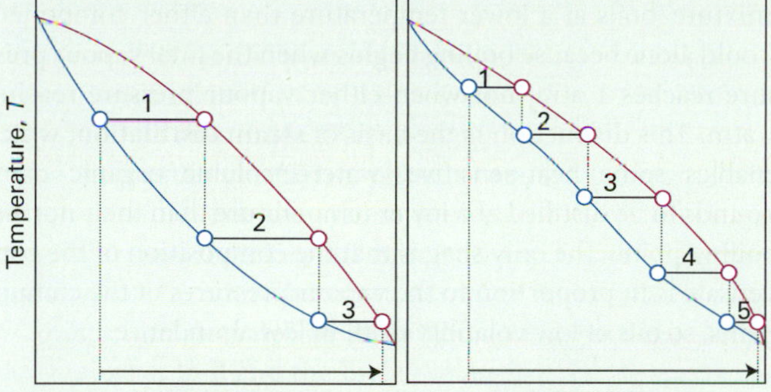

Figure 5C.11 The number of theoretical plates is the number of steps needed to bring about a specified degree of separation of two components in a mixture. The two systems shown correspond to (a) 3, (b) 5 theoretical plates.

Self-test 5C.4 Refer to Fig. 5C.11b: suppose the composition of the mixture corresponds to $z_A = 0.1$; how many theoretical plates would be required to achieve a composition $z_A = 0.9$?

Answer: 5

(b) Azeotropes

Although many liquids have temperature–composition phase diagrams resembling the ideal version in Fig. 5C.10, in a number of important cases there are marked deviations. A maximum in the phase diagram (Fig. 5C.12) may occur when the favourable interactions between A and B molecules reduce the vapour pressure of the mixture below the ideal value and so raise its boiling temperature: in effect, the A–B interactions stabilize the liquid. In such cases the excess Gibbs energy, G^E (Topic 5B), is negative (more favourable to mixing than ideal).

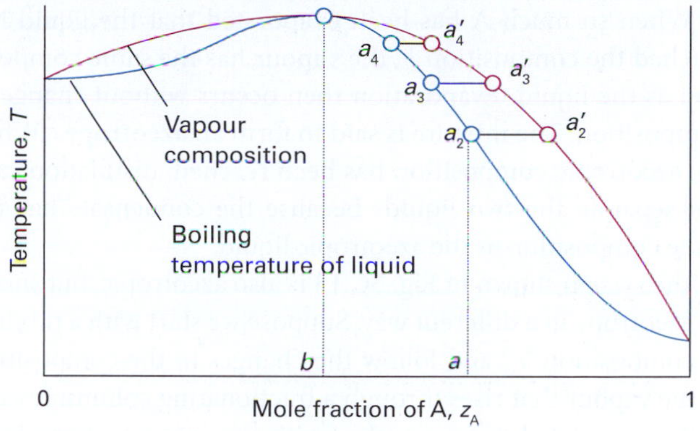

Figure 5C.12 A high-boiling azeotrope. When the liquid of composition a is distilled, the composition of the remaining liquid changes towards b but no further.

Phase diagrams showing a minimum (Fig. 5C.13) indicate that the mixture is destabilized relative to the ideal solution, the A–B interactions then being unfavourable; in this case, the boiling temperature is lowered. For such mixtures G^E is positive (less favourable to mixing than ideal), and there may be contributions from both enthalpy and entropy effects.

Deviations from ideality are not always so strong as to lead to a maximum or minimum in the phase diagram, but when they do there are important consequences for distillation. Consider a liquid of composition a on the right of the maximum in Fig. 5C.12. The vapour (at a_2') of the boiling mixture (at a_2) is richer in A. If that vapour is removed (and condensed elsewhere), then the remaining liquid will move to a composition that is richer in B, such as that represented by a_3, and the vapour in equilibrium with this mixture will have composition a_3'. As that vapour is removed, the composition of the boiling liquid shifts to a point such as a_4, and the composition of the vapour shifts to a_4'. Hence, as evaporation proceeds, the composition of the remaining liquid shifts towards B as A is drawn off. The boiling point of the liquid rises, and the vapour becomes richer in

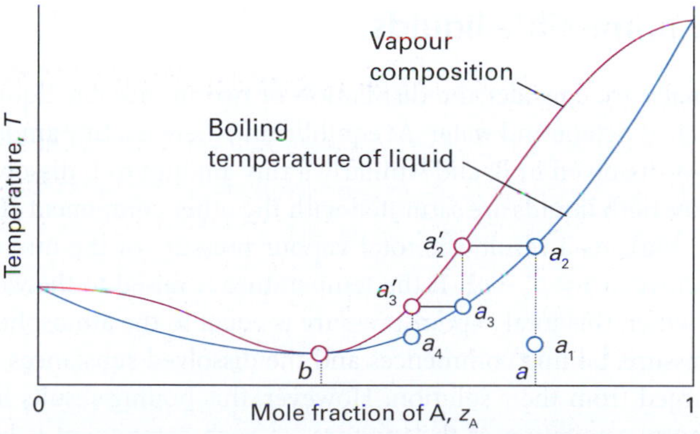

Figure 5C.13 A low-boiling azeotrope. When the mixture at a is fractionally distilled, the vapour in equilibrium in the fractionating column moves towards b and then remains unchanged.

B. When so much A has been evaporated that the liquid has reached the composition b, the vapour has the same composition as the liquid. Evaporation then occurs without change of composition. The mixture is said to form an **azeotrope**.[1] When the azeotropic composition has been reached, distillation cannot separate the two liquids because the condensate has the same composition as the azeotropic liquid.

The system shown in Fig. 5C.13 is also azeotropic, but shows its azeotropy in a different way. Suppose we start with a mixture of composition a_1, and follow the changes in the composition of the vapour that rises through a fractionating column (essentially a vertical glass tube packed with glass rings to give a large surface area). The mixture boils at a_2 to give a vapour of composition a_2'. This vapour condenses in the column to a liquid of the same composition (now marked a_3). That liquid reaches equilibrium with its vapour at a_3', which condenses higher up the tube to give a liquid of the same composition, which we now call a_4. The fractionation therefore shifts the vapour towards the azeotropic composition at b, but not beyond, and the azeotropic vapour emerges from the top of the column.

Brief illustration 5C.4 Azeotropes

Examples of the behaviour of the type shown in Fig. 5C.12 include (a) trichloromethane/propanone and (b) nitric acid/water mixtures. Hydrochloric acid/water is azeotropic at 80 per cent by mass of water and boils unchanged at 108.6 °C. Examples of the behaviour of the type shown in Fig. 5C.13 include (c) dioxane/water and (d) ethanol/water mixtures. Ethanol/water boils unchanged when the water content is 4 per cent by mass and the temperature is 78 °C.

Self-test 5C.5 Suggest a molecular interpretation of the two types of behaviour.

Answer: (a,b) favourable A–B interactions; (c,d) unfavourable A–B interactions

(c) Immiscible liquids

Finally we consider the distillation of two immiscible liquids, such as octane and water. At equilibrium, there is a tiny amount of A dissolved in B, and similarly a tiny amount of B dissolved in A: both liquids are saturated with the other component (Fig. 5C.14a). As a result, the total vapour pressure of the mixture is close to $p = p_A^* + p_B^*$. If the temperature is raised to the value at which this total vapour pressure is equal to the atmospheric pressure, boiling commences and the dissolved substances are purged from their solution. However, this boiling results in a vigorous agitation of the mixture, so each component is kept saturated in the other component, and the purging continues as the very dilute solutions are replenished. This intimate contact is essential: two immiscible liquids heated in a container like that shown in Fig. 5C.14b would not boil at the same temperature. The presence of the saturated solutions means that the 'mixture' boils at a lower temperature than either component would alone because boiling begins when the total vapour pressure reaches 1 atm, not when either vapour pressure reaches 1 atm. This distinction is the basis of **steam distillation**, which enables some heat-sensitive, water-insoluble organic compounds to be distilled at a lower temperature than their normal boiling point. The only snag is that the composition of the condensate is in proportion to the vapour pressures of the components, so oils of low volatility distil in low abundance.

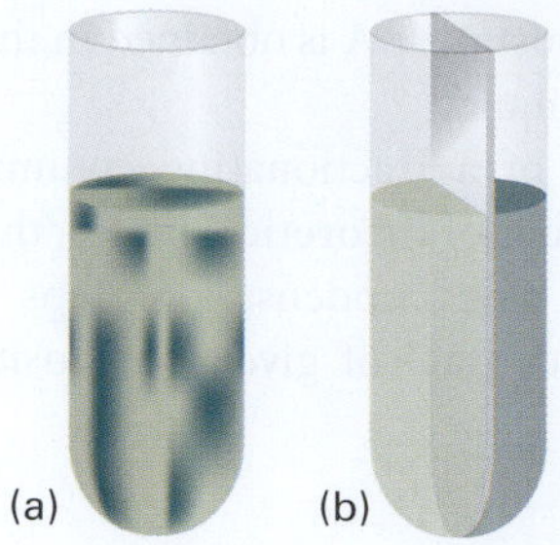

Figure 5C.14 The distillation of (a) two immiscible liquids can be regarded as (b) the joint distillation of the separated components, and boiling occurs when the sum of the partial pressures equals the external pressure.

[1] The name comes from the Greek words for 'boiling without changing'.

5C.3 Liquid–liquid phase diagrams

Now we consider temperature–composition diagrams for systems that consist of pairs of **partially miscible** liquids, which are liquids that do not mix in all proportions at all temperatures. An example is hexane and nitrobenzene. The same principles of interpretation apply as to liquid–vapour diagrams.

(a) Phase separation

Suppose a small amount of a liquid B is added to a sample of another liquid A at a temperature T'. Liquid B dissolves completely, and the binary system remains a single phase. As more B is added, a stage comes at which no more dissolves. The sample now consists of two phases in equilibrium with each other, the most abundant one consisting of A saturated with B, the minor one a trace of B saturated with A. In the temperature–composition diagram drawn in Fig. 5C.15, the composition of the former is represented by the point a' and that of the latter by the point a''. The relative abundances of the two phases are given by the lever rule. When more B is added, A dissolves in it slightly. The compositions of the two phases in equilibrium remain a' and a''. A stage is reached when so much B is present

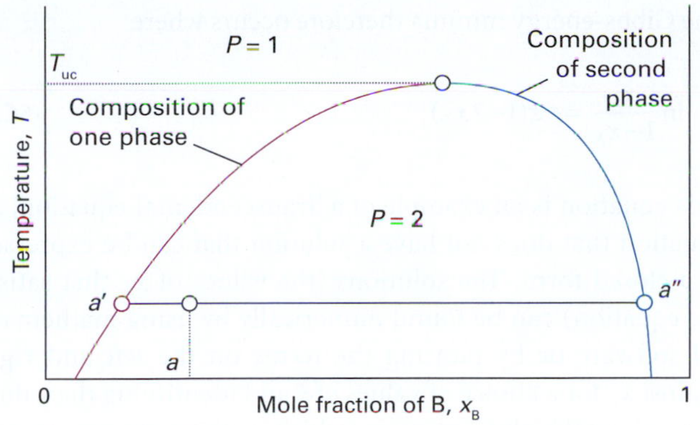

Figure 5C.15 The temperature–composition diagram for a mixture of A and B. The region below the curve corresponds to the compositions and temperatures at which the liquids are partially miscible. The upper critical temperature, T_{uc}, is the temperature above which the two liquids are miscible in all proportions.

that it can dissolve all the A, and the system reverts to a single phase. The addition of more B now simply dilutes the solution, and from then on a single phase remains.

The composition of the two phases at equilibrium varies with the temperature. For the system shown in Fig. 5C.15, raising the temperature increases the miscibility of A and B. The two-phase region therefore becomes narrower because each phase in equilibrium is richer in its minor component: the A-rich phase is richer in B and the B-rich phase is richer in A. We can construct the entire phase diagram by repeating the observations at different temperatures and drawing the envelope of the two-phase region.

Example 5C.2 Interpreting a liquid–liquid phase diagram

The phase diagram for the system nitrobenzene/hexane at 1 atm is shown in Fig. 5C.16. A mixture of 50 g of hexane

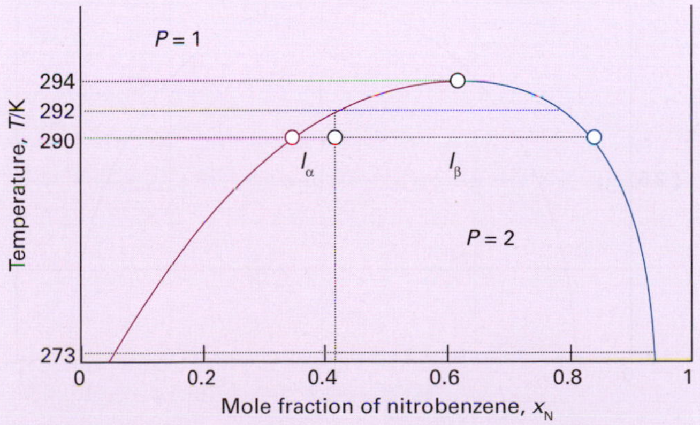

Figure 5C.16 The temperature–composition diagram for hexane and nitrobenzene at 1 atm, with the points and lengths discussed in the text.

(0.59 mol C_6H_{14}) and 50 g of nitrobenzene (0.41 mol $C_6H_5NO_2$) was prepared at 290 K. What are the compositions of the phases, and in what proportions do they occur? To what temperature must the sample be heated in order to obtain a single phase?

Method The compositions of phases in equilibrium are given by the points where the tie-line representing the temperature intersects the phase boundary. Their proportions are given by the lever rule (eqn 5C.6). The temperature at which the components are completely miscible is found by following the isopleth upwards and noting the temperature at which it enters the one-phase region of the phase diagram.

Answer We denote hexane by H and nitrobenzene by N; refer to Fig. 5C.16. The point $x_N = 0.41$, $T = 290$ K occurs in the two-phase region of the phase diagram. The horizontal tie line cuts the phase boundary at $x_N = 0.35$ and $x_N = 0.83$, so those are the compositions of the two phases. According to the lever rule, the ratio of amounts of each phase is equal to the ratio of the distances l_α and l_β:

$$\frac{n_\alpha}{n_\beta} = \frac{l_\beta}{l_\alpha} = \frac{0.83-0.41}{0.41-0.35} = \frac{0.42}{0.06} = 7$$

That is, there is about 7 times more hexane-rich phase than nitrobenzene-rich phase. Heating the sample to 292 K takes it into the single-phase region. Because the phase diagram has been constructed experimentally, these conclusions are not based on any assumptions about ideality. They would be modified if the system were subjected to a different pressure.

Self-test 5C.6 Repeat the problem for 50 g of hexane and 100 g of nitrobenzene at 273 K.

Answer: $x_N = 0.09$ and 0.95 in ratio 1:1.3; 294 K

(b) Critical solution temperatures

The **upper critical solution temperature**, T_{uc} (or *upper consolute temperature*), is the highest temperature at which phase separation occurs. Above the upper critical temperature the two components are fully miscible. This temperature exists because the greater thermal motion overcomes any potential energy advantage in molecules of one type being close together. One example is the nitrobenzene/hexane system shown in Fig. 5C.16. An example of a solid solution is the palladium/hydrogen system, which shows two phases, one a solid solution of hydrogen in palladium and the other a palladium hydride, up to 300 °C but forms a single phase at higher temperatures (Fig. 5C.17).

The thermodynamic interpretation of the upper critical solution temperature focuses on the Gibbs energy of mixing and its variation with temperature. The simple model of a real solution (specifically, of a regular solution) discussed in Topic 5B results in a Gibbs energy of mixing that behaves as shown in

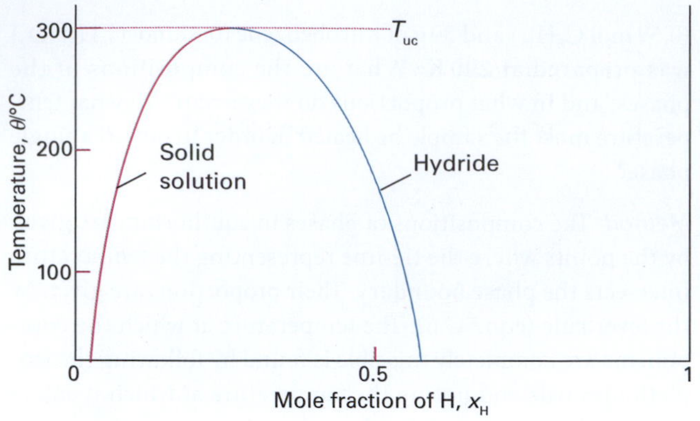

Figure 5C.17 The phase diagram for palladium and palladium hydride, which has an upper critical temperature at 300 °C.

Fig. 5C.18. Provided the parameter ξ introduced in eqn 5B.6 ($H^E = \xi RT x_A x_B$) is greater than 2, the Gibbs energy of mixing has a double minimum. As a result, for $\xi > 2$ we can expect phase separation to occur. The same model shows that the compositions corresponding to the minima are obtained by looking for the conditions at which $\partial\Delta_{mix}G/\partial x = 0$, and a simple manipulation of eqn 5B.7 ($\Delta_{mix}G = nRT(x_A \ln x_A + x_B \ln x_B + \xi x_A x_B)$, with $x_B = 1 - x_A$) shows that we have to solve

$$\left(\frac{\partial\Delta_{mix}G}{\partial x_A}\right)_{T,p}$$
$$= nRT\left(\frac{\partial\{x_A \ln x_A + (1-x_A)\ln(1-x_A) + \xi x_A(1-x_A)\}}{\partial x_A}\right)_{T,p}$$
$$= nRT\{\ln x_A + 1 - \ln(1-x_A) - 1 + \xi(1-2x_A)\}$$
$$= nRT\left\{\ln\frac{x_A}{1-x_A} + \xi(1-2x_A)\right\}$$

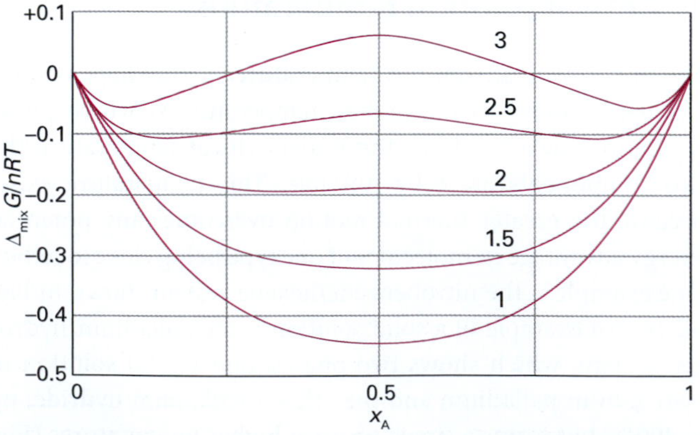

Figure 5C.18 The temperature variation of the Gibbs energy of mixing of a system that is partially miscible at low temperatures. A system of composition in the region $P=2$ forms two phases with compositions corresponding to the two local minima of the curve. This illustration is a duplicate of Fig. 5B.5.

The Gibbs-energy minima therefore occurs where

$$\ln\frac{x_A}{1-x_A} = -\xi(1-2x_A) \qquad (5C.7)$$

This equation is an example of a 'transcendental equation', an equation that does not have a solution that can be expressed in a closed form. The solutions (the values of x_A that satisfy the equation) can be found numerically by using mathematical software or by plotting the terms on the left and right against x_A for a choice of values of ξ and identifying the values of x_A where the plots intersect (which is where the two expressions are equal) (Fig. 5C.19). The solutions found in this way are plotted in Fig. 5C.20. We see that, as ξ decreases, which can be interpreted as an increase in temperature provided the intermolecular forces remain constant (so that H^E remains constant), then the two minima move together and merge when $\xi = 2$.

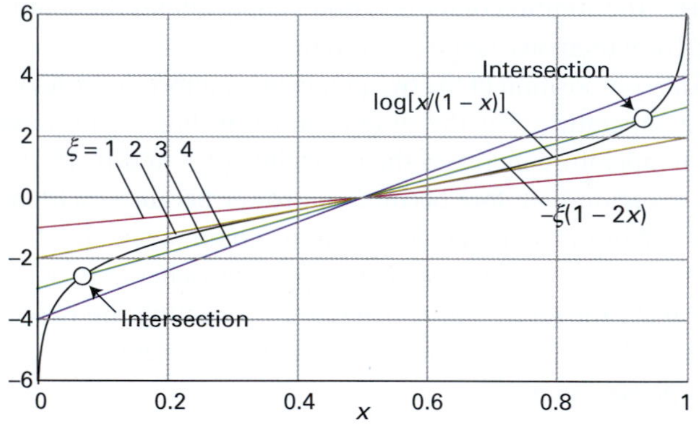

Figure 5C.19 The graphical procedure for solving eqn 5C.7. When $\xi < 2$, the only intersection occurs at $x = 0.5$. When $\xi \geq 2$, there are three solutions (those for $\xi = 3$ are marked).

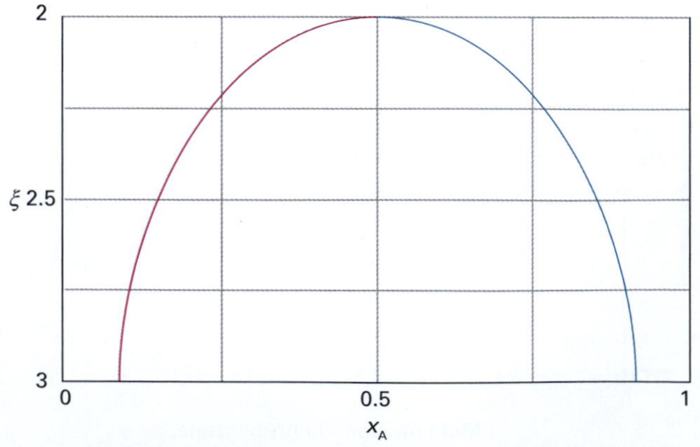

Figure 5C.20 The location of the phase boundary as computed on the basis of the ξ-parameter model introduced in Topic 5B.

Brief illustration 5C.5 Phase separation

In the system composed of benzene and cyclohexane treated in *Example* 5B.1 it is established that $\xi = 1.13$, so we do not expect a two-phase system; that is, the two components are completely miscible at the temperature of the experiment. The single solution of the equation

$$\ln \frac{x_A}{1-x_A} + 1.13(1-2x_A) = 0$$

is $x_A = \frac{1}{2}$, corresponding to a single minimum of the Gibbs energy of mixing, and there is no phase separation.

Self-test 5C.7 Would phase separation be expected if the excess enthalpy were modelled by the expression $H^E = \xi RTx_A^2 x_B^2$ (Fig. 5C.21a)?

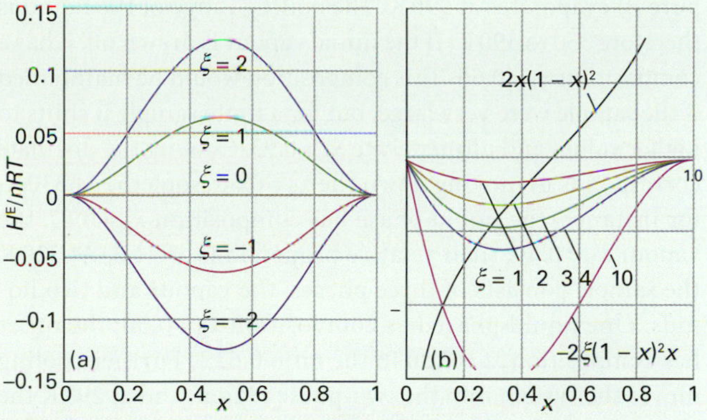

Figure 5C.21(a) The excess enthalpy and (b) the graphical solution of the resulting equation for the minima of the Gibbs energy of mixing in *Self-test* 5C.7.

Answer: No, see Fig. 5C.21b

Some systems show a **lower critical solution temperature**, T_{lc} (or *lower consolute temperature*), below which they mix in all proportions and above which they form two phases. An example is water and triethylamine (Fig. 5C.22). In this case, at low temperatures the two components are more miscible because they form a weak complex; at higher temperatures the complexes break up and the two components are less miscible.

Some systems have both upper and lower critical solution temperatures. They occur because, after the weak complexes have been disrupted, leading to partial miscibility, the thermal motion at higher temperatures homogenizes the mixture again, just as in the case of ordinary partially miscible liquids. The most famous example is nicotine and water, which are partially miscible between 61 °C and 210 °C (Fig. 5C.23).

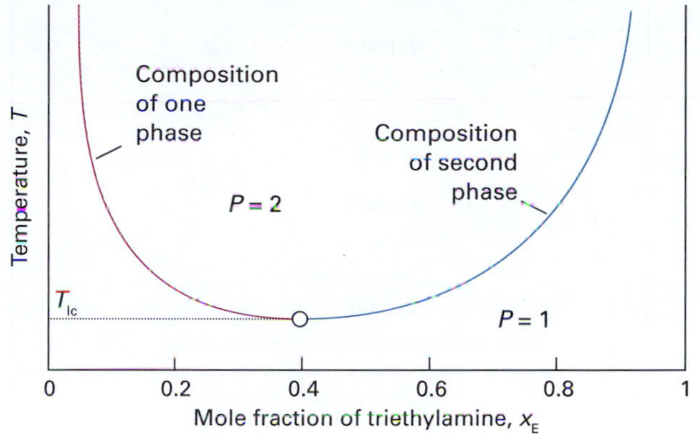

Figure 5C.22 The temperature–composition diagram for water and triethylamine. This system shows a lower critical temperature at 292 K. The labels indicate the interpretation of the boundaries.

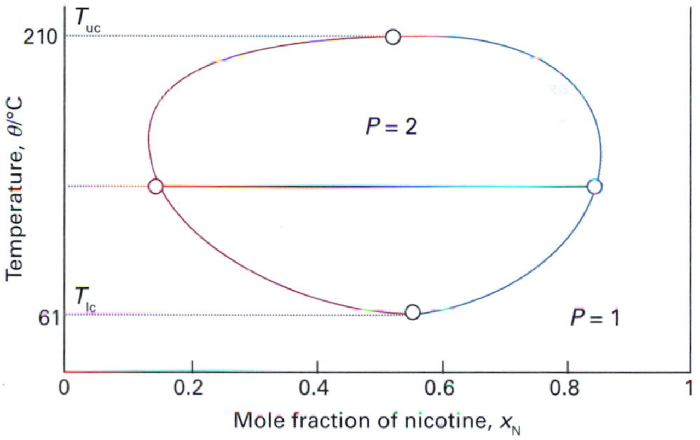

Figure 5C.23 The temperature–composition diagram for water and nicotine, which has both upper and lower critical temperatures. Note the high temperatures for the liquid (especially the water): the diagram corresponds to a sample under pressure.

(c) The distillation of partially miscible liquids

Consider a pair of liquids that are partially miscible and form a low-boiling azeotrope. This combination is quite common because both properties reflect the tendency of the two kinds of molecule to avoid each other. There are two possibilities: one in which the liquids become fully miscible before they boil; the other in which boiling occurs before mixing is complete.

Figure 5C.24 shows the phase diagram for two components that become fully miscible before they boil. Distillation of a mixture of composition a_1 leads to a vapour of composition b_1, which condenses to the completely miscible single-phase solution at b_2. Phase separation occurs only when this distillate is cooled to a point in the two-phase liquid region, such as b_3. This description applies only to the first drop of distillate. If distillation continues, the composition of the remaining liquid changes. In the end, when the whole sample has evaporated and condensed, the composition is back to a_1.

Figure 5C.25 shows the second possibility, in which there is no upper critical solution temperature. The distillate obtained from a liquid initially of composition a_1 has composition b_3 and is a two-phase mixture. One phase has composition b_3' and the other has composition b_3''.

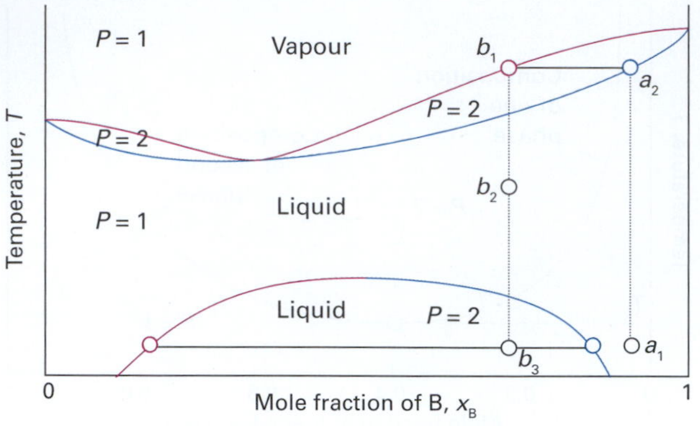

Figure 5C.24 The temperature–composition diagram for a binary system in which the upper critical temperature is less than the boiling point at all compositions. The mixture forms a low-boiling azeotrope.

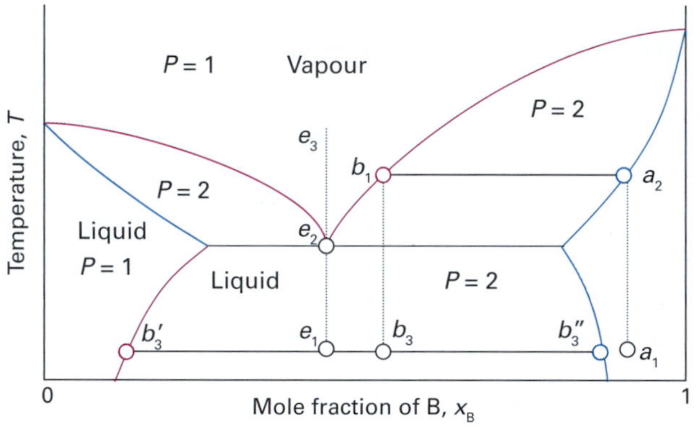

Figure 5C.25 The temperature–composition diagram for a binary system in which boiling occurs before the two liquids are fully miscible.

The behaviour of a system of composition represented by the isopleth *e* in Fig. 5C.25 is interesting. A system at e_1 forms two phases, which persist (but with changing proportions) up to the boiling point at e_2. The vapour of this mixture has the same composition as the liquid (the liquid is an azeotrope). Similarly,

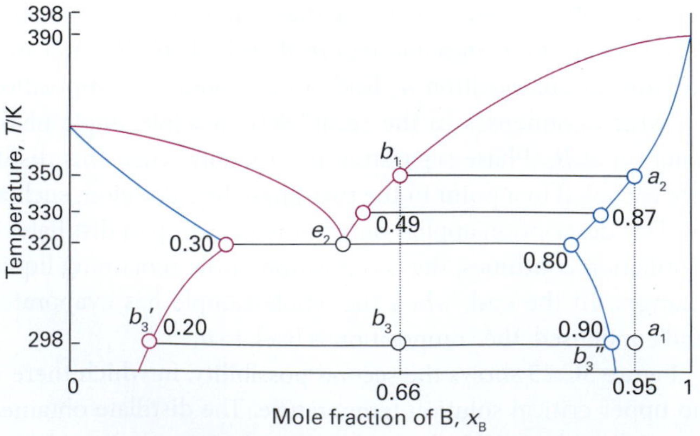

Figure 5C.26 The points of the phase diagram in Fig. 5C.25 that are discussed in *Example* 5C.3.

condensing a vapour of composition e_3 gives a two-phase liquid of the same overall composition. At a fixed temperature, the mixture vaporizes and condenses like a single substance.

Example 5C.3 Interpreting a phase diagram

State the changes that occur when a mixture of composition $x_B=0.95$ (a_1) in Fig. 5C.26 is boiled and the vapour condensed.

Method The area in which the point lies gives the number of phases; the compositions of the phases are given by the points at the intersections of the horizontal tie line with the phase boundaries; the relative abundances are given by the lever rule.

Answer The initial point is in the one-phase region. When heated it boils at 350 K (a_2) giving a vapour of composition $x_B=0.66$ (b_1). The liquid gets richer in B, and the last drop (of pure B) evaporates at 390 K. The boiling range of the liquid is therefore 350 to 390 K. If the initial vapour is drawn off, it has a composition $x_B=0.66$. This composition would be maintained if the sample were very large, but for a finite sample it shifts to higher values and ultimately to $x_B=0.95$. Cooling the distillate corresponds to moving down the $x_B=0.66$ isopleth. At 330 K, for instance, the liquid phase has composition $x_B=0.87$, the vapour $x_B=0.49$; their relative proportions are 1:3. At 320 K the sample consists of three phases: the vapour and two liquids. One liquid phase has composition $x_B=0.30$; the other has composition $x_B=0.80$ in the ratio 0.62:1. Further cooling moves the system into the two-phase region, and at 298 K the compositions are 0.20 and 0.90 in the ratio 0.82:1. As further distillate boils over, the overall composition of the distillate becomes richer in B. When the last drop has been condensed the phase composition is the same as at the beginning.

Self-test 5C.8 Repeat the discussion, beginning at the point $x_B=0.4$, $T=298$ K.

5C.4 Liquid–solid phase diagrams

Knowledge of the temperature–composition diagrams for solid mixtures guides the design of important industrial processes, such as the manufacture of liquid crystal displays and semiconductors. In this section, we shall consider systems where solid and liquid phases may both be present at temperatures below the boiling point.

(a) Eutectics

Consider the two-component liquid of composition a_1 in Fig. 5C.27. The changes that occur as the system is cooled may be expressed as follows:

- $a_1 \rightarrow a_2$. The system enters the two-phase region labelled 'Liquid+B'. Pure solid B begins to come out of solution and the remaining liquid becomes richer in A

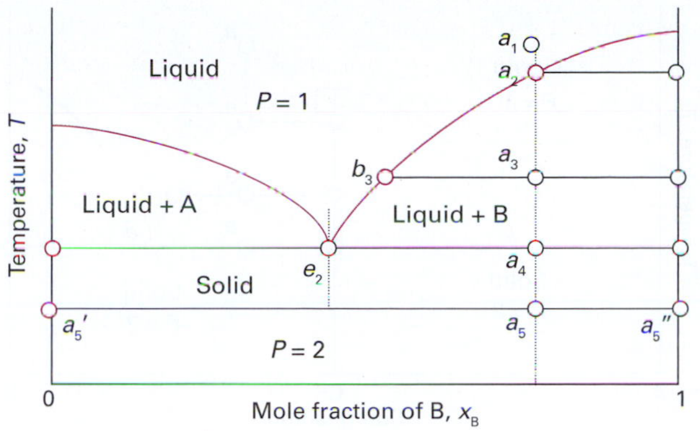

Figure 5C.27 The temperature–composition phase diagram for two almost immiscible solids and their completely miscible liquids. Note the similarity to Fig. 5C.25. The isopleth through e_2 corresponds to the eutectic composition, the mixture with lowest melting point.

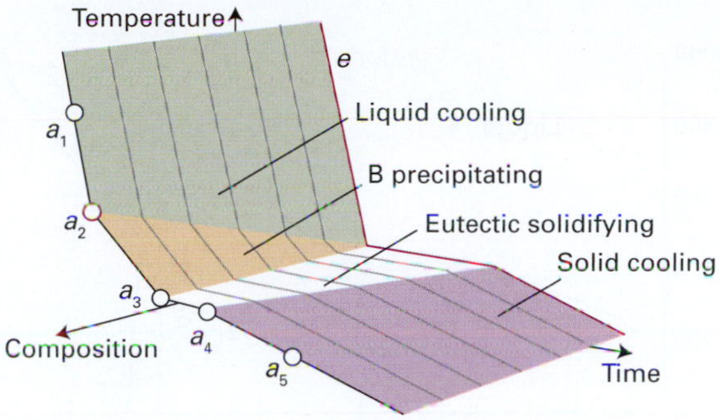

Figure 5C.28 The cooling curves for the system shown in Fig. 5C.27. For isopleth a, the rate of cooling slows at a_2 because solid B deposits from solution. There is a complete halt at a_4 while the eutectic solidifies. This halt is longest for the eutectic isopleth, e. The eutectic halt shortens again for compositions beyond e (richer in A). Cooling curves are used to construct the phase diagram.

- $a_2 \rightarrow a_3$. More of the solid B forms, and the relative amounts of the solid and liquid (which are in equilibrium) are given by the lever rule. At this stage there are roughly equal amounts of each. The liquid phase is richer in A than before (its composition is given by b_3) because some B has been deposited.
- $a_3 \rightarrow a_4$. At the end of this step, there is less liquid than at a_3, and its composition is given by e_2. This liquid now freezes to give a two-phase system of pure B and pure A.

Physical interpretation

The isopleth at e_2 in Fig. 5C.27 corresponds to the **eutectic** composition, the mixture with the lowest melting point.[2] A liquid with the eutectic composition freezes at a single temperature, without previously depositing solid A or B. A solid with the eutectic composition melts, without change of composition, at the lowest temperature of any mixture. Solutions of composition to the right of e_2 deposit B as they cool, and solutions to the left deposit A: only the eutectic mixture (apart from pure A or pure B) solidifies at a single definite temperature without gradually unloading one or other of the components from the liquid.

One technologically important eutectic is solder, which in one form has mass composition of about 67 per cent tin and 33 per cent lead and melts at 183 °C. The eutectic formed by 23 per cent NaCl and 77 per cent H_2O by mass melts at −21.1 °C. When salt is added to ice under isothermal conditions (for example, when spread on an icy road) the mixture melts if the temperature is above −21.1 °C (and the eutectic composition has been achieved). When salt is added to ice under adiabatic conditions (for example, when added to ice in a vacuum flask) the ice melts, but in doing so it absorbs heat from the rest of the mixture. The temperature of the system falls and, if enough salt is added, cooling continues down to the eutectic temperature. Eutectic formation occurs in the great majority of binary alloy systems, and is of great importance for the microstructure of solid materials. Although a eutectic solid is a two-phase system, it crystallizes out in a nearly homogeneous mixture of microcrystals. The two microcrystalline phases can be distinguished by microscopy and structural techniques such as X-ray diffraction (Topic 18A).

Thermal analysis is a very useful practical way of detecting eutectics. We can see how it is used by considering the rate of cooling down the isopleth through a_1 in Fig. 5C.27. The liquid cools steadily until it reaches a_2, when B begins to be deposited (Fig. 5C.28). Cooling is now slower because the solidification of B is exothermic and retards the cooling. When the remaining liquid reaches the eutectic composition, the temperature remains constant until the whole sample has solidified: this region of constant temperature is the eutectic halt. If the liquid has the eutectic composition e initially, the liquid cools steadily down to the freezing temperature of the eutectic, when there is a long **eutectic halt** as the entire sample solidifies (like the freezing of a pure liquid).

Brief illustration 5C.6 Interpreting a binary phase diagram

Figure 5C.29 is the phase diagram for the binary system silver/tin. The regions have been labelled to show which each one represents. When a liquid of composition a is cooled, solid silver with dissolved tin begins to precipitate at a_1 and the sample solidifies completely at a_2.

Self-test 5C.9 Describe what happens when the sample of composition b is cooled.

[2] The name comes from the Greek words for 'easily melted'.

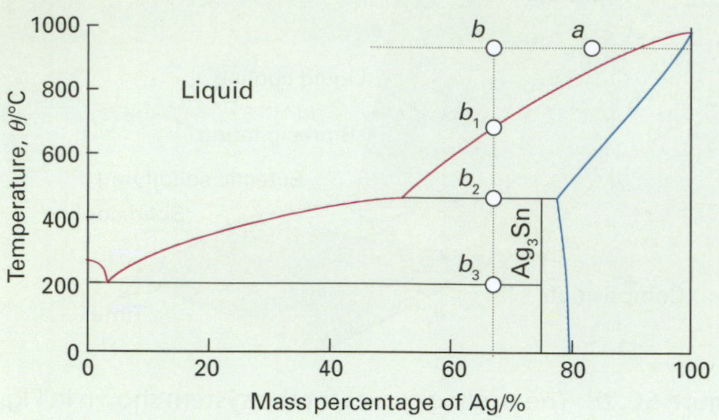

Figure 5C.29 The phase diagram for silver/tin discussed in *Brief illustration* 5C.6.

Answer: Solid Ag with dissolved Sn begins to precipitate at b_1, and the liquid becomes richer in Sn as the temperature falls further. At b_2 solid Ag_3Sn begins to precipitate, and the liquid becomes richer in Sn. At b_3 the system has its eutectic composition (a solid solution of Sn and Ag_3Sn) and it freezes without further change in composition.

Monitoring the cooling curves at different overall compositions gives a clear indication of the structure of the phase diagram. The solid–liquid boundary is given by the points at which the rate of cooling changes. The longest eutectic halt gives the location of the eutectic composition and its melting temperature.

(b) Reacting systems

Many binary mixtures react to produce compounds, and technologically important examples of this behaviour include the Group 13/15 (III/V) semiconductors, such as the gallium arsenide system, which forms the compound GaAs. Although three constituents are present, there are only two components because GaAs is formed from the reaction Ga + As → GaAs. We shall illustrate some of the principles involved with a system that forms a compound C that also forms eutectic mixtures with the species A and B (Fig. 5C.30).

A system prepared by mixing an excess of B with A consists of C and unreacted B. This is a binary B, C system, which we suppose forms a eutectic. The principal change from the eutectic phase diagram in Fig. 5C.27 is that the whole of the phase diagram is squeezed into the range of compositions lying between equal amounts of A and B ($x_B = 0.5$, marked C in Fig. 5C.30) and pure B. The interpretation of the information in the diagram is obtained in the same way as for Fig. 5C.27. The solid deposited on cooling along the isopleth a is the compound C. At temperatures below a_4 there are two solid phases, one consisting of C and the other of B. The pure compound C melts **congruently**, that is, the composition of the liquid it forms is the same as that of the solid compound.

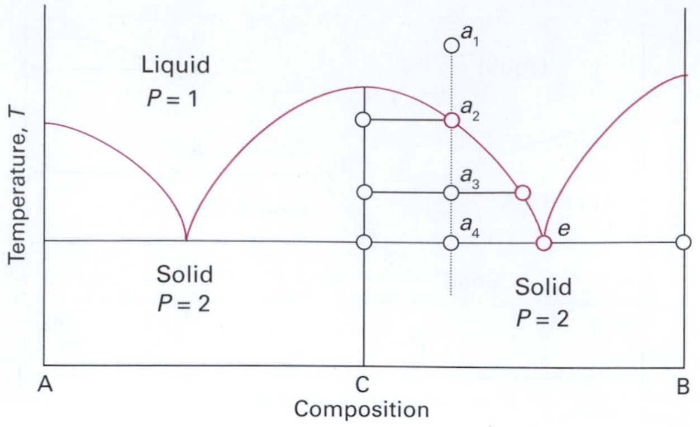

Figure 5C.30 The phase diagram for a system in which A and B react to form a compound C = AB. This resembles two versions of Fig. 5C.27 in each half of the diagram. The constituent C is a true compound, not just an equimolar mixture.

(c) Incongruent melting

In some cases the compound C is not stable as a liquid. An example is the alloy Na_2K, which survives only as a solid (Fig. 5C.31). Consider what happens as a liquid at a_1 is cooled:

- $a_1 \to a_2$. A solid solution rich in Na is deposited, and the remaining liquid is richer in K.
- $a_2 \to$ just below a_3. The sample is now entirely solid and consists of a solid solution rich in Na and solid Na_2K.

Now consider the isopleth through b_1:

- $b_1 \to b_2$. No obvious change occurs until the phase boundary is reached at b_2 when a solid solution rich in Na begins to deposit.

Physical interpretation

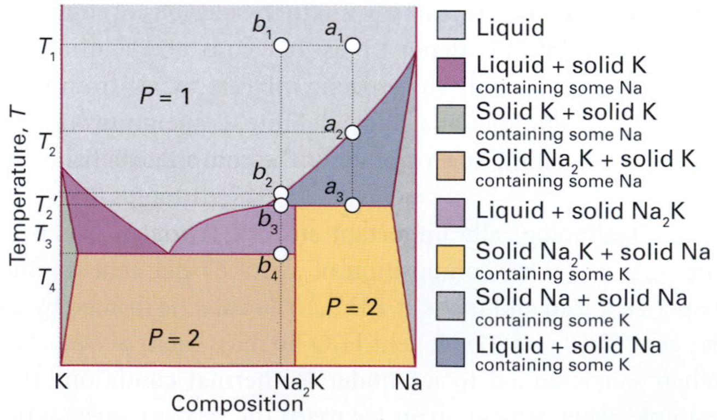

Figure 5C.31 The phase diagram for an actual system (sodium and potassium) like that shown in Fig. 5C.30, but with two differences. One is that the compound is Na_2K, corresponding to A_2B and not AB as in that illustration. The second is that the compound exists only as the solid, not as the liquid. The transformation of the compound at its melting point is an example of incongruent melting.

- $b_2 \rightarrow b_3$. A solid solution rich in Na deposits, but at b_3 a reaction occurs to form Na_2K: this compound is formed by the K atoms diffusing into the solid Na.
- b_3. At b_3, three phases are in mutual equilibrium: the liquid, the compound Na_2K, and a solid solution rich in Na. The horizontal line representing this three-phase equilibrium is called a **peritectic line**. At this stage the liquid Na/K mixture is in equilibrium with a little solid Na_2K, but there is still no liquid compound
- $b_3 \rightarrow b_4$. As cooling continues, the amount of solid compound increases until at b_4 the liquid reaches its eutectic composition. It then solidifies to give a two-phase solid consisting of a solid solution rich in K and solid Na_2K.

Physical interpretation

If the solid is reheated, the sequence of events is reversed. No liquid Na_2K forms at any stage because it is too unstable to exist as a liquid. This behaviour is an example of **incongruent melting**, in which a compound melts into its components and does not itself form a liquid phase.

Checklist of concepts

- ☐ 1. Raoult's law is used to calculate the total vapour pressure of a binary system of two volatile liquids.
- ☐ 2. The composition of the vapour in equilibrium with a binary mixture is calculated by using Dalton's law.
- ☐ 3. The composition of the vapour and the liquid phase in equilibrium are located at each end of a tie line.
- ☐ 4. The **lever rule** is used to deduce the relative abundances of each phase in equilibrium.
- ☐ 5. A phase diagram can be used to discuss the process of **fractional distillation**.
- ☐ 6. Depending on the relative strengths of the intermolecular forces, high- or low-boiling **azeotropes** may be formed.
- ☐ 7. The vapour pressure of a system composed of immiscible liquids is the sum of the vapour pressures of the pure liquids.
- ☐ 8. A phase diagram may be used to discuss the distillation of partially miscible liquids.
- ☐ 9. Phase separation of partially miscible liquids may occur when the temperature is below the upper critical solution temperature or above the lower critical solution temperature; the process may be discussed in terms of the model of a regular solution.
- ☐ 10. A phase diagram summarizes the temperature–composition properties of a binary system with solid and liquid phases; at the **eutectic composition** the liquid phase solidifies without change of composition.
- ☐ 11. The phase equilibria of binary systems in which the components react may also be summarized by a phase diagram.
- ☐ 12. In some cases, a solid compound does not survive melting.

Checklist of equations

Property	Equation	Comment	Equation number
Composition of vapour	$y_A = x_A p_A^* / (p_B^* + (p_A^* - p_B^*)x_A) y_B = 1 - y_A$	Ideal solution	5C.4
Total vapour pressure	$p = p_A^* p_B^* / (p_A^* + (p_B^* - p_A^*) y_A)$	Ideal solution	5C.5
Lever rule	$n_\alpha l_\alpha = n_\beta l_\beta$		5C.6

5D Phase diagrams of ternary systems

Contents

➤ Why do you need to know this material?

Ternary phase diagrams have become important in materials science as more complex materials are investigated, such as the ceramics found to have superconducting properties.

➤ What is the key idea?

A phase diagram is a map showing the conditions under which each phase of a system is the most stable.

➤ What do you need to know already?

It would be helpful to review the interpretation of two-component phase diagrams (Topic 5C) and the phase rule (Topic 5A). The interpretation of the phase diagrams presented here uses the lever rule (Topic 5C).

This short Topic is a brief introduction to the depiction of phases of systems of three components. In terms of the phase rule (Topic 5A), $C=3$, so $F=5-P$. If we restrict systems to constant temperature and pressure, two degrees of freedom are discarded and we are left with $F''=3-P$. An area on a ternary phase diagram therefore represents a region where a single phase is present, a line represents the equilibrium between two phases of varying composition, and a point corresponds to a composition at which three phases are present in equilibrium.

5D.1 Triangular phase diagrams

The mole fractions of the three components of a ternary system satisfy $x_A+x_B+x_C=1$. A phase diagram drawn as an equilateral triangle ensures that this property is satisfied automatically because the sum of the distances to a point inside an equilateral triangle of side 1 and measured parallel to the edges is equal to 1 (Fig. 5D.1).

Figure 5D.1 shows how this approach works in practice. The edge AB corresponds to $x_C=0$, and likewise for the other two edges. Hence, each of the three edges corresponds to one of the three binary systems (A,B), (B,C), and (C,A). An interior point corresponds to a system in which all three components are present. The point P, for instance, represents $x_A=0.50$, $x_B=0.10$, $x_C=0.40$.

Any point on a straight line joining an apex to a point on the opposite edge (the broken line in Fig. 5D.1) represents a composition that is progressively richer in A the closer the point is to the A apex but for which the concentration ratio B:C remains constant. Therefore, if we wish to represent the changing composition of a system as A is added, we draw a line from the A apex to the point on BC representing the initial binary system. Any ternary system formed by adding A then lies at some point on this line.

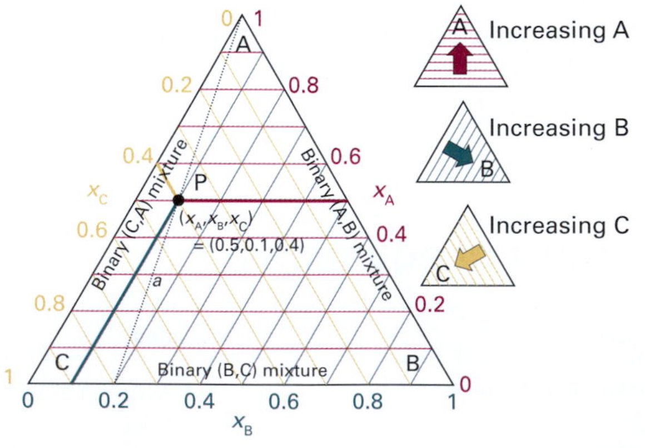

Figure 5D.1 The triangular coordinates used for the discussion of three-component systems. Each edge corresponds to a binary system. All points along the dotted line *a* correspond to mole fractions of C and B in the same ratio.

Brief illustration 5D.1 The representation of composition

The following points are represented on Fig. 5D.2:

Point	x_A	x_B	x_C
a	0.20	0.80	0
b	0.42	0.26	0.32
c	0.80	0.10	0.10
d	0.10	0.20	0.70
e	0.20	0.40	0.40
f	0.30	0.60	0.10

Note that the points d, e, f have $x_A/x_B = 0.50$ and lie on a straight line.

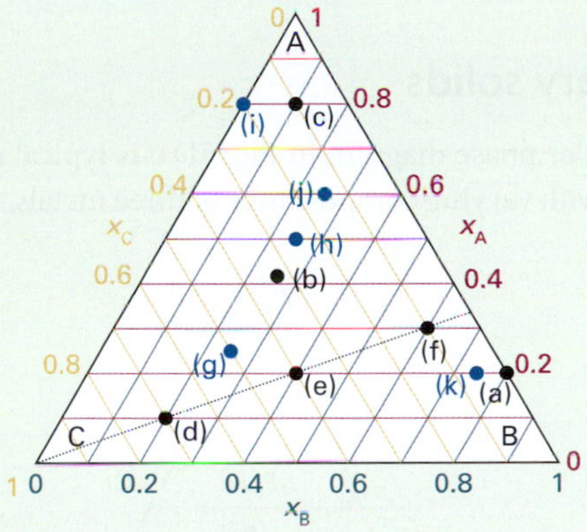

Figure 5D.2 The points referred to in *Brief illustration* 5D.1 (black) and *Self-test* 5D.1 (blue).

Self-test 5D.1 Mark the following points on the triangle.

Point	x_A	x_B	x_C
g	0.25	0.25	0.50
h	0.50	0.25	0.25
i	0.80	0	0.20
j	0.60	0.25	0.15
k	0.20	0.75	0.0.05

Answer: See Fig. 5D.2.

A single triangle represents the equilibria when one of the discarded degrees of freedom (the temperature, for instance) has a certain value. Different temperatures give rise to different equilibrium behaviour and therefore different triangular phase diagrams. Each one may therefore be regarded as a horizontal slice through a three-dimensional triangular prism, such as that shown in Fig. 5D.3.

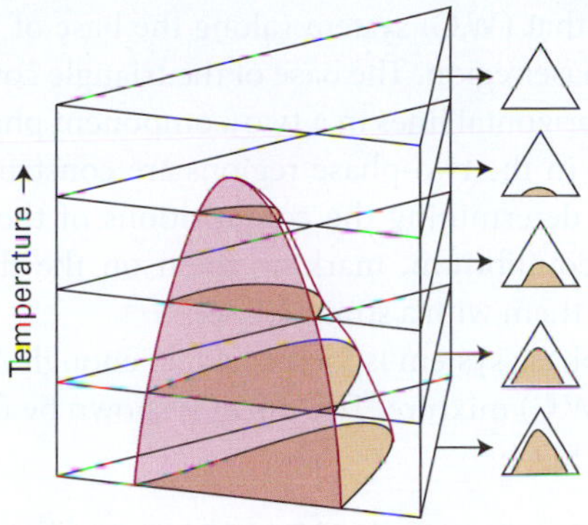

Figure 5D.3 When temperature is included as a variable, the phase diagram becomes a triangular prism. Horizontal sections through the prism correspond to the triangular phase diagrams being discussed.

5D.2 Ternary systems

Ternary phase diagrams are widely used in metallurgy and materials science in general. Although they can become quite complex, they can be interpreted in much the same way as binary diagrams. Here we give two examples.

(a) Partially miscible liquids

The phase diagram for a ternary system in which W (in due course: water) and A (in due course: acetic acid) are fully miscible, A and C (in due course: chloroform) are fully miscible, but W and C are only partially miscible is shown in Fig. 5D.4. This illustration is for the system water/acetic acid/chloroform at room temperature, which behaves in this way. It shows that the two fully miscible pairs, (A,W) and (A,C), form single-phase

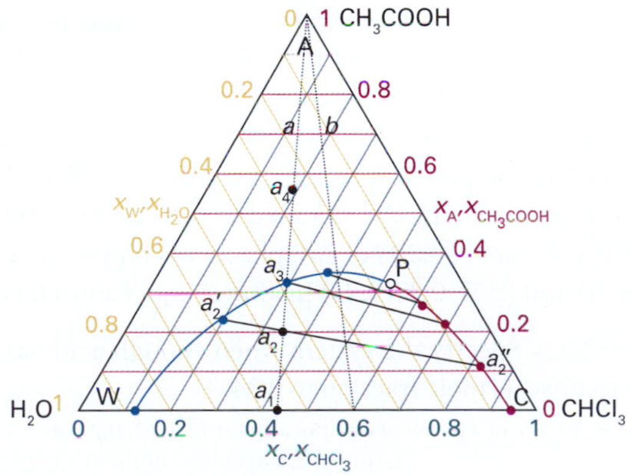

Figure 5D.4 The phase diagram, at fixed temperature and pressure, of the three-component system acetic acid, chloroform, and water. Only some of the tie lines have been drawn in the two-phase region. All points along the line *a* correspond to chloroform and water present in the same ratio.

regions and that (W,C) system (along the base of the triangle) has a two-phase region. The base of the triangle corresponds to one of the horizontal lines in a two-component phase diagram. The tie lines in the two-phase regions are constructed experimentally by determining the compositions of the two phases that are in equilibrium, marking them on the diagram, and then joining them with a straight line.

A single-phase system is formed when enough A is added to the binary (W,C) mixture. This effect is shown by following the line *a* in Fig. 5D.4:

- a_1. The system consists of two phases and the relative amounts of the two phases can be read off by using the lever rule.
- $a_1 \rightarrow a_2$. The addition of A takes the system along the line joining a_1 to the A apex. At a_2 the solution still has two phases, but there is slightly more W in the largely C phase (represented by the point a_2'') and more C in the largely W phase a_2' because the presence of A helps both to dissolve. The phase diagram shows that there is more A in the W-rich phase than in the C-rich phase (a_2' is closer than (a_2'') to the A apex).
- $a_2 \rightarrow a_3$. At a_3 two phases are present, but the C-rich layer is present only as a trace (lever rule).
- $a_3 \rightarrow a_4$. Further addition of A takes the system towards and beyond a_4, and only a single phase is present.

Physical interpretation

Brief illustration 5D.2 Partially miscible liquids

Suppose we have a mixture of water (W in Fig. 5D.4) and chloroform (C) with $x_W = 0.40$ and $x_C = 0.60$, and acetic acid (A) is added to it. The relative proportions of A and C remain constant, so the point representing the overall composition moves along the straight line *b* from $x_C = 0.60$ on the base to the acetic acid apex. The initial composition is in a two-phase region: one phase has the composition $(x_W, x_C, x_A) = (0.95, 0.05, 0)$ and the other has composition $(x_W, x_C, x_A) = (0.12, 0.88, 0)$. When sufficient acetic acid has been added to raise its mole fraction to 0.18 the system consists of two phases of composition (0.07, 0.82, 0.11) and (0.57, 0.20, 0.23) in almost equal abundance.

Self-test 5D.2 Specify the system when enough acid has been added to raise its mole fraction to 0.34.

Answer: A trace of a phase of composition (0.12, 0.61, 0.27) and a dominating phase of composition (0.28, 0.37, 0.35)

The point marked P in Fig. 5D.4 is called the **plait point**. It is yet another example of a critical point. At the plait point, the compositions of the two phases in equilibrium become identical. For convenience, the general interpretation of a triangular phase diagram is summarized in Fig. 5D.5.

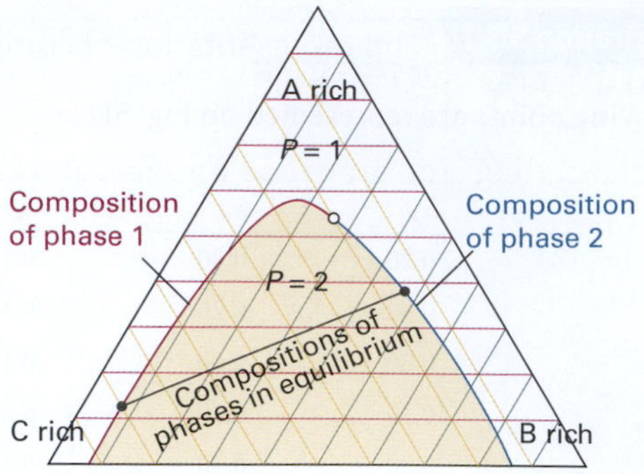

Figure 5D.5 The interpretation of a triangular phase diagram. The region inside the curved line consists of two phases, and the compositions of the two phases in equilibrium are given by the points at the ends of the tie lines (the tie lines are determined experimentally).

(b) Ternary solids

The triangular phase diagram in Fig. 5D.6 is typical of that for a solid alloy with varying compositions of three metals, A, B, and C.

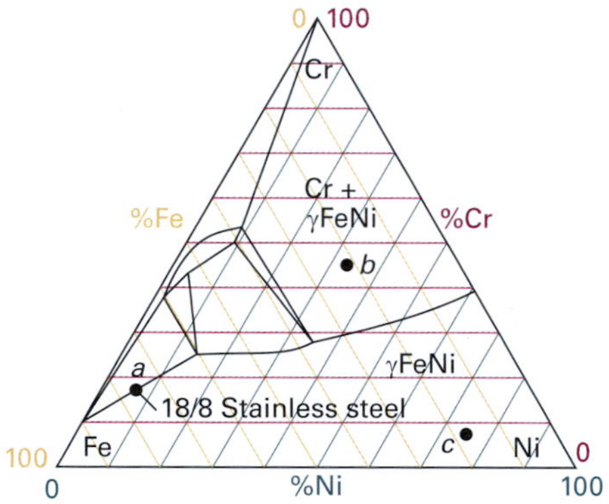

Figure 5D.6 A simplified triangular phase diagram of the ternary system represented by a stainless steel composed of iron, chromium, and nickel.

Brief illustration 5D.3 Stainless steel

Figure 5D.6 is a simplified version of the phase diagram for a stainless steel consisting of iron, chromium, and nickel. The axes are labelled with the mass percentage compositions instead of the mole fractions, but as the three percentages add up to 100 per cent, the interpretation of points in the triangle is essentially the same as for mole fractions. The point *a* corresponds to the composition 74 per cent Fe, 18 per cent Cr, and 8 per cent Ni. It corresponds to the most common form of stainless steel, '18-8 stainless steel'. The composition corresponding to point *b* lies in the two-phase region, one phase consisting of Cr and the other of the alloy γ-FeNi.

Self-test 5D.3 Identify the composition represented by point *c*.

Answer: Three phases, Fe, Ni, and γ-FeNi

Checklist of concepts

- ☐ 1. A phase diagram drawn as an equilateral triangle ensures that the property $x_A + x_B + x_C = 1$ is satisfied automatically.
- ☐ 2. At the **plait point**, the compositions of the two phases in equilibrium are identical.

5E Activities

Contents

➤ Why do you need to know this material?

Ideal solutions are a good starting point for the discussion of mixtures, but to understand real solutions it is important to be able to describe deviations from ideal behaviour and to express them in terms of molecular interactions.

➤ What is the key idea?

The activity of a species, i.e. its effective concentration, helps to preserve the form of the expressions derived on the basis of ideal behaviour but extends their reach to real mixtures.

➤ What do you need to know already?

This Topic is based on the expression for chemical potential of a species derived from Raoult's and Henry's laws (Topic 5A). It also uses the formulation of a model of a regular solution introduced in Topic 5B. You need to be aware of the expression for the Gibbs energy of mixing of an ideal solution (Topic 5B).

In this Topic we see how to adjust the expressions developed in Topics 5A and 5B to take into account deviations from ideal behaviour. In Topic 3D it is remarked that a quantity called 'fugacity' takes into account the effects of gas imperfections in a manner that resulted in the least upset of the form of equations. Here we see how the expressions encountered in the treatment of ideal solutions can also be preserved almost intact by introducing the concept of 'activity'. As in other Topics collected in this chapter, we denote the solvent by A, the solute by B, and a general component by J.

5E.1 The solvent activity

The general form of the chemical potential of a real or ideal solvent is given by a straightforward modification of eqn 5A.14 ($\mu_A = \mu_A^* + RT\ln(p_A/p_A^*)$, where p_A^* is the vapour pressure of pure A and p_A is the vapour pressure of A when it is a component of a solution). The solvent in an ideal solution obeys Raoult's law (Topic 5A, $p_A = x_A p_A^*$) at all concentrations and we can express the chemical potential as eqn 5A.22 (that is, as $\mu_A = \mu_A^* + RT\ln x_A$). The form of the this relation can be preserved when the solution does not obey Raoult's law by writing

$$\mu_A = \mu_A^* + RT\ln a_A \qquad \textit{Definition} \quad \text{Activity of solvent} \qquad (5E.1)$$

The quantity a_A is the **activity** of A, a kind of 'effective' mole fraction, just as the fugacity is an effective pressure.

Because eqn 5E.1 is true for both real and ideal solutions (the only approximation being the use of pressures rather than fugacities), we can conclude by comparing it with $\mu_A = \mu_A^* + RT\ln(p_A/p_A^*)$ that

$$a_A = \frac{p_A}{p_A^*} \qquad \textit{Measurement} \quad \text{Activity of solvent} \qquad (5E.2)$$

We see that there is nothing mysterious about the activity of a solvent: it can be determined experimentally simply by measuring the vapour pressure and then using eqn 5E.2.

Brief illustration 5E.1 Solvent activity

The vapour pressure of 0.500 M KNO_3(aq) at 100 °C is 99.95 kPa, so the activity of water in the solution at this temperature is

$$a_A = \frac{99.95\ \text{kPa}}{101.325\ \text{kPa}} = 0.9864$$

Self-test 5E.1 The vapour pressure of water in a saturated solution of calcium nitrate at 20 °C is 1.381 kPa; the vapour pressure of pure water at that temperature is 2.3393 kPa. What is the activity of water in this solution?

Answer: 0.5903

Because all solvents obey Raoult's law more closely as the concentration of solute approaches zero, the activity of the solvent approaches the mole fraction as $x_A \to 1$:

$$a_A \to x_A \text{ as } x_A \to 1 \qquad (5E.3)$$

A convenient way of expressing this convergence is to introduce the **activity coefficient**, γ (gamma), by the definition

$$a_A = \gamma_A x_A \quad \gamma_A \to 1 \text{ as } x_A \to 1$$

Definition Activity coefficient of solvent (5E.4)

at all temperatures and pressures. The chemical potential of the solvent is then

$$\mu_A = \mu_A^* + RT \ln x_A + RT \ln \gamma_A$$

Chemical potential of solvent (5E.5)

The standard state of the solvent is established when $x_A = 1$ and the pressure is 1 bar.

5E.2 The solute activity

The problem with defining activity coefficients and standard states for solutes is that they approach ideal-dilute (Henry's law) behaviour as $x_B \to 0$, not as $x_B \to 1$ (corresponding to pure solute). We shall show how to set up the definitions for a solute that obeys Henry's law exactly, and then show how to allow for deviations.

(a) Ideal-dilute solutions

A solute B that satisfies Henry's law (Topic 5A) has a vapour pressure given by $p_B = K_B x_B$, where K_B is an empirical constant. In this case, the chemical potential of B is

$$\mu_B = \mu_B^* + RT \ln \frac{p_B}{p_B^*} = \overbrace{\mu_B^* + RT \ln \frac{K_B}{p_B^*}}^{\mu_B^{\ominus}} + RT \ln x_B \qquad (5E.6)$$

Both K_B and p_B^* are characteristics of the solute, so the second term may be combined with the first to give a new standard chemical potential, $\mu_B^{\ominus}$.

$$\mu_B^{\ominus} = \mu_B^* + RT \ln \frac{K_B}{p_B^*} \qquad (5E.7)$$

It then follows that the chemical potential of a solute in an ideal-dilute solution is related to its mole fraction by

$$\mu_B = \mu_B^{\ominus} + RT \ln x_B \qquad (5E.8)$$

If the solution is ideal, $K_B = p_B^*$ and eqn 5E.7 reduces to $\mu_B^{\ominus} = \mu_B^*$, as we should expect.

Brief illustration 5E.2 The solute activity

In *Example* 5A.4 it is established that in a mixture of propanone (acetone, A) and trichloromethane (chloroform, C) at 298 K $K_{\text{propanone}} = 23.3$ kPa whereas $p^*_{\text{propanone}} = 4.63$ kP. It follows from eqn 5E.7 that

$$\begin{aligned}\mu^{\ominus}_{\text{propanone}} &= \mu^*_{\text{propanone}} + RT \ln \frac{23.3\ \text{kPa}}{4.63\ \text{kPa}} \\ &= \mu^*_{\text{propanone}} + (8.3145\ \text{J K}^{-1}\ \text{mol}^{-1}) \times (298\ \text{K}) \times \ln \frac{23.3}{4.63} \\ &= \mu^*_{\text{propanone}} + 4.00\ \text{kJ mol}^{-1}\end{aligned}$$

and the standard value differs from the value for the pure liquid by 4.00 kJ mol^{-1}.

Self-test 5E.2 In the same mixture, with trichloromethane treated as the solute, $K_{\text{trichloromethane}} = 22.0$ kPa, whereas $p^*_{\text{trichloromethane}} = 36.4$ kPa. What is the relation between the standard chemical potential and that of the pure liquid?

Answer: $\mu^{\ominus}_{\text{trichloromethane}} = \mu^*_{\text{trichloromethane}} - 1.25\ \text{kJ mol}^{-1}$

(b) Real solutes

We now permit deviations from ideal-dilute, Henry's law behaviour. For the solute, we introduce a_B in place of x_B in eqn 5E.8, and obtain

$$\mu_B = \mu_B^{\ominus} + RT \ln a_B$$

Definition Activity of solute (5E.9)

The standard state remains unchanged in this last stage, and all the deviations from ideality are captured in the activity a_B. The value of the activity at any concentration can be obtained in the same way as for the solvent, but in place of eqn 5E.2 we use

$$a_B = \frac{p_B}{K_B}$$

Measurement Activity of solute (5E.10)

As we did for the solvent, it is sensible to introduce an activity coefficient through

$$a_B = \gamma_B x_B$$

Definition Activity coefficient of solute (5E.11)

Now all the deviations from ideality are captured in the activity coefficient γ_B. Because the solute obeys Henry's law as its concentration goes to zero, it follows that

$$a_B \to x_B \text{ and } \gamma_B \to 1 \text{ as } x_B \to 0 \qquad (5E.12)$$

at all temperatures and pressures. Deviations of the solute from ideality disappear as zero concentration is approached.

Example 5E.1 Measuring activity

Use the following information to calculate the activity and activity coefficient of trichloromethane (chloroform, C) in propanone (acetone, A) at 25 °C, treating it first as a solvent and then as a solute.

x_C	0	0.20	0.40	0.60	0.80	1
p_C/kPa	0	4.7	11	18.9	26.7	36.4
p_A/kPa	46.3	33.3	23.3	12.3	4.9	0

Method For the activity of chloroform as a solvent (the Raoult's law activity), form $a_C = p_C/p_C^*$ and $\gamma_C = a_C/x_C$. For its activity as a solute (the Henry's law activity), form $a_C = p_C/K_C$ and $\gamma_C = a_C/x_C$ with the new activity.

Answer Because $p_C^* = 36.4\,\text{kPa}$ and $K_C = 22.0\,\text{kPa}$, we can construct the following tables. For instance, at $x_C = 0.20$, in the Raoult's law case we find $a_C = (4.7\,\text{kPa})/(36.4\,\text{kPa}) = 0.13$ and $\gamma_C = 0.13/0.20 = 0.65$; likewise, in the Henry's law case, $a_C = (4.7\,\text{kPa})/(22.0\,\text{kPa}) = 0.21$ and $\gamma_C = 0.21/0.20 = 1.05$.

From Raoult's law (chloroform regarded as the solvent):

a_C	0	0.13	0.30	0.52	0.73	1.00
γ_C		0.65	0.75	0.87	0.91	1.00

From Henry's law (chloroform regarded as the solute):

a_C	0	0.21	0.50	0.86	1.21	1.65
γ_C	1	1.05	1.25	1.43	1.51	1.65

These values are plotted in Fig. 5E.1. Notice that $\gamma_C \to 1$ as $x_C \to 1$ in the Raoult's law case, but that $\gamma_C \to 1$ as $x_C \to 0$ in the Henry's law case.

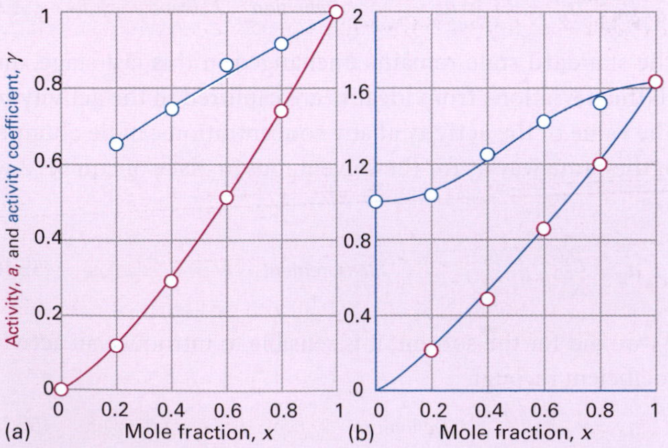

Figure 5E.1 The variation of activity and activity coefficient for a chloroform/acetone (trichloromethane/propanone) solution with composition according to (a) Raoult's law, (b) Henry's law.

Self-test 5E.3 Calculate the activities and activity coefficients for acetone according to the two conventions.

Answer: At $x_A = 0.60$, for instance $a_R = 0.50$; $\gamma_R = 0.83$; $a_H = 1.00$, $\gamma_H = 1.67$

(c) Activities in terms of molalities

The selection of a standard state is entirely arbitrary, so we are free to choose one that best suits our purpose and the description of the composition of the system. Because compositions are often expressed as molalities, b, in place of mole fractions it is convenient to write

$$\mu_B = \mu_B^{\ominus} + RT \ln b_B \tag{5E.13}$$

where $\mu_B^{\ominus}$ has a different value from the standard values introduced earlier. According to this definition, the chemical potential of the solute has its standard value $\mu_B^{\ominus}$ when the molality of B is equal to $b^{\ominus}$ (that is, at $1\,\text{mol kg}^{-1}$). Note that as $b_B \to 0$, $\mu_B \to \infty$; that is, as the solution becomes diluted, so the solute becomes increasingly stabilized. The practical consequence of this result is that it is very difficult to remove the last traces of a solute from a solution.

Now, as before, we incorporate deviations from ideality by introducing a dimensionless activity a_B, a dimensionless activity coefficient γ_B, and writing

$$a_B = \gamma_B \frac{b_B}{b^{\ominus}}, \text{ where } \gamma_B \to 1 \text{ as } b_B \to 0 \tag{5E.14}$$

at all temperatures and pressures. The standard state remains unchanged in this last stage and, as before, all the deviations from ideality are captured in the activity coefficient γ_B. We then arrive at the following succinct expression for the chemical potential of a real solute at any molality:

$$\mu_B = \mu_B^{\ominus} + RT \ln a_B \tag{5E.15}$$

(d) The biological standard state

One important illustration of the ability to choose a standard state to suit the circumstances arises in biological applications. The conventional standard state of hydrogen ions (unit activity, corresponding to pH = 0)[1] is not appropriate to normal biological conditions. Therefore, in biochemistry it is common to adopt the **biological standard state**, in which pH = 7 (an activity of 10^{-7}, neutral solution) and to label the corresponding standard thermodynamic functions as $G^{\oplus}$, $H^{\oplus}$, $\mu^{\oplus}$, and $S^{\oplus}$ (some texts use $X^{\circ\prime}$).

[1] Recall from introductory chemistry courses that pH = −log $a(H_3O^+)$.

To find the relation between the thermodynamic and biological standard values of the chemical potential of hydrogen ions we need to note from eqn 5E.15 that

$$\mu_{H^+} = \mu_{H^+}^{\ominus} + RT \ln a_{H^+} = \mu_{H^+}^{\ominus} - (RT \ln 10)\text{pH}$$

It follows that

$$\mu_{H^+}^{\oplus} = \mu_{H^+}^{\ominus} - 7RT \ln 10 \qquad \text{Relation between standard state and biological standard state} \qquad (5E.16)$$

Brief illustration 5E.3 The biological standard state

At 298 K, $7RT\ln 10 = 39.96$ kJ mol^{-1}, so the two standard values differ by about 40 kJ mol^{-1} and specifically $\mu_{H^+}^{\oplus} = \mu_{H^+}^{\ominus} - 39.96$ kJ mol^{-1}. Thus, in a reaction of the form $A + 2\,H^+(aq) \rightarrow B$, the standard and biological standard Gibbs energies are related as follows:

$$\begin{aligned}\Delta_r G^{\oplus} &= \mu_B^{\ominus} - \{\mu_A^{\ominus} + 2\mu_{H^+}^{\oplus}\} = \mu_B^{\ominus} - \{\mu_A^{\ominus} + 2\mu_{H^+}^{\ominus} - 14RT\ln 10\} \\ &= \mu_B^{\ominus} - \{\mu_A^{\ominus} + 2\mu_{H^+}^{\ominus}\} + 14RT\ln 10 = \Delta_r G^{\ominus} + 14RT\ln 10 \\ &= \Delta_r G^{\ominus} + 79.92\ \text{kJ mol}^{-1}\end{aligned}$$

Self-test 5E.4 Find the relation between the standard and biological standard Gibbs energies of a reaction of the form $A \rightarrow B + 3\,H^+(aq)$.

Answer: $\Delta_r G^{\oplus} = \Delta_r G^{\ominus} - 119.88$ kJ mol^{-1}

5E.3 The activities of regular solutions

The material on regular solutions in Topic 5B gives further insight into the origin of deviations from Raoult's law and its relation to activity coefficients. The starting point is the model expression for the excess (molar) enthalpy (eqn 5B.6, $H^E = \xi RTx_A x_B$) and its implication for the Gibbs energy of mixing for a regular solution. We show in the following *Justification* that for this model the activity coefficients are given by

$$\ln\gamma_A = \xi x_B^2 \qquad \ln\gamma_B = \xi x_A^2 \qquad \text{Margules equations} \qquad (5E.17)$$

These relations are called the **Margules equations**.

Justification 5E.1 The Margules equations

The Gibbs energy of mixing to form an ideal solution is

$$\Delta_{mix}G = nRT\{x_A \ln x_A + x_B \ln x_B\}$$

(This is eqn 5B.16 of Topic 5B.) The corresponding expression for a non-ideal solution is

$$\Delta_{mix}G = nRT\{x_A \ln a_A + x_B \ln a_B\}$$

This relation follows in the same way as for an ideal mixture but with activities in place of mole fractions. If each activity is replaced by γx, this expression becomes

$$\begin{aligned}\Delta_{mix}G &= nRT\{x_A \ln x_A\gamma_A + x_B \ln x_B\gamma_B\} \\ &= nRT\{x_A \ln x_A + x_B \ln x_B + x_A \ln \gamma_A + x_B \ln\gamma_B\}\end{aligned}$$

Now we introduce the two expressions in eqn 5E.17, and use $x_A + x_B = 1$, which gives

$$\begin{aligned}\Delta_{mix}G &= nRT\{x_A \ln x_A + x_B \ln x_B + \xi x_A x_B^2 + \xi x_B x_A^2\} \\ &= nRT\{x_A \ln x_A + x_B \ln x_B + \xi x_A x_B(x_A + x_B)\} \\ &= nRT\{x_A \ln x_A + x_B \ln x_B + \xi x_A x_B\}\end{aligned}$$

Note that the activity coefficients behave correctly for dilute solutions: $\gamma_A \rightarrow 1$ as $x_B \rightarrow 0$ and $\gamma_B \rightarrow 1$ as $x_A \rightarrow 0$.

At this point we can use the Margules equations to write the activity of A as

$$a_A = \gamma_A x_A = x_A e^{\xi x_B^2} = x_A e^{\xi(1-x_A)^2} \qquad (5E.18)$$

with a similar expression for a_B. The activity of A, though, is just the ratio of the vapour pressure of A in the solution to the vapour pressure of pure A (eqn 5E.2, $a_A = p_A/p_A^*$), so we can write

$$p_A = p_A^* x_A e^{\xi(1-x_A)^2} \qquad (5E.19)$$

This function is plotted in Fig. 5E.2. We see that

- When $\xi = 0$, corresponding to an ideal solution, $p_A = p_A^* x_A$, in accord with Raoult's law.
- Positive values of ξ (endothermic mixing, unfavourable solute–solvent interactions) give vapour pressures higher than ideal.
- Negative values of ξ (exothermic mixing, favourable solute–solvent interactions) give a lower vapour pressure.

Physical interpretation

All the plots of eqn 5E.19 approach linearity and coincide with the Raoult's law line as $x_A \rightarrow 1$ and the exponential function in eqn 5E.19 approaches 1. When $x_A \ll 1$, eqn 5E.19 approaches

$$p_A = p_A^* x_A e^{\xi} \qquad (5E.20)$$

This expression has the form of Henry's law once we identify K with $e^{\xi}p_A^*$, which is different for each solute–solvent system.

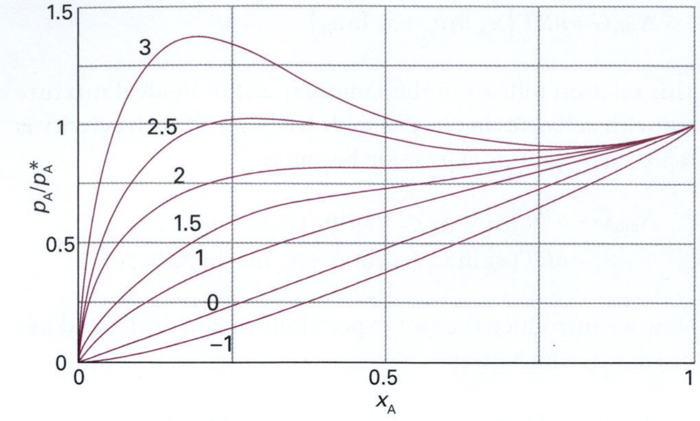

Figure 5E.2 The vapour pressure of a mixture based on a model in which the excess enthalpy is proportional to ξRTx_Ax_B. An ideal solution corresponds to $\xi=0$ and gives a straight line, in accord with Raoult's law. Positive values of ξ give vapour pressures higher than ideal. Negative values of ξ give a lower vapour pressure.

Brief illustration 5E.4 The Margules equations

In *Example* 5B.1 of Topic 5B it is established that $\xi=1.13$ for a mixture of benzene and cyclohexane at 25 °C. Because $\xi>0$ we can expect the vapour pressure of the mixture to be greater than its ideal value. The total vapour pressure of the mixture is therefore

$$p=p^*_{\text{benzene}}x_{\text{benzene}}e^{1.13(1-x_{\text{benzene}})^2}+p^*_{\text{cyclohexane}}x_{\text{cyclohexane}}e^{1.13(1-x_{\text{cyclohexane}})^2}$$

This expression is plotted in Fig. 5E.3a using $p_{\text{benzene}}=10.0\,\text{kPa}$ and $p^*_{\text{cyclohexane}}=10.4\,\text{kPa}$.

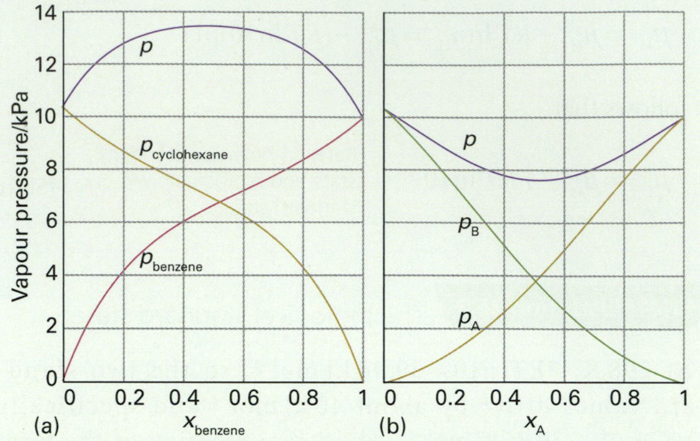

Figure 5E.3 The computed vapour pressure curves for a mixture of benzene and cyclohexane at 25 °C (a) as derived in *Brief illustration* 5E.4 and (b) *Self-test* 5E.5.

Self-test 5E.5 Suppose it is found that for a hypothetical mixture $\xi=-1.13$, but with other properties the same. Draw the vapour pressure diagram.

Answer: See Fig. 5E.3b

Checklist of concepts

☐ 1. The **activity** is an effective concentration that preserves the form of the expression for the chemical potential. See Table 5E.1.

☐ 2. The chemical potential of a solute in an ideal–dilute solution is defined on the basis of Henry's law.

☐ 3. The activity of a solute takes into account departures from Henry's law behaviour.

☐ 4. An alternative approach to the definition of the solute activity is based on the molality of the solute.

☐ 5. The **biological standard state** of a species in solution is defined as pH=7 (and 1 bar).

☐ 6. The **Margules equations** relate the activities of the components of a model regular solution to its composition. They lead to expressions for the vapour pressures of the components of a regular solution.

Table 5E.1 Activities and standard states: a summary*

Component	Basis	Standard state	Activity	Limits
Solid or liquid		Pure, 1 bar	$a=1$	
Solvent	Raoult	Pure solvent, 1 bar	$a=p/p^*$, $a=\gamma x$	$\gamma\to1$ as $x\to1$ (pure solvent)
Solute	Henry	(1) A hypothetical state of the pure solute	$a=p/K$, $a=\gamma x$	$\gamma\to1$ as $x\to0$
		(2) A hypothetical state of the solute at molality $b^{\ominus}$	$a=\gamma b/b^{\ominus}$	$\gamma\to1$ as $b\to0$
Gas	Fugacity†	Pure, a hypothetical state of 1 bar and behaving as a perfect gas	$f=\gamma p$	$\gamma\to1$ as $p\to0$

* In each case, $\mu=\mu^{\ominus}+RT\ln a$.

† Fugacity is discussed in Topic 3D.

Checklist of equations

Property	Equation	Comment	Equation number
Chemical potential of solvent	$\mu_A = \mu_A^* + RT \ln a_A$	Definition	5E.1
Activity of solvent	$a_A = p_A/p_A^*$	$a_A \to x_A$ as $x_A \to 1$	5E.2
Activity coefficient of solvent	$a_A = \gamma_A x_A$	$\gamma_A \to 1$ as $x_A \to 1$	5E.4
Chemical potential of solute	$\mu_B = \mu_B^{\ominus} + RT \ln a_B$	Definition	5E.9
Activity of solute	$a_B = p_B/K_B$	$a_B \to x_B$ as $x_B \to 0$	5E.10
Activity coefficient of solute	$a_B = \gamma_B x_B$	$\gamma_B \to 1$ as $x_B \to 0$	5E.11
Conversion to biological standard state	$\mu_{H^+}^{\oplus} = \mu_{H^+}^{\ominus} - 7RT \ln 10$		5E.16
Margules equations	$\ln \gamma_A = \xi x_B^2$, $\ln \gamma_B = \xi x_A^2$	Regular solution	5E.17
Vapour pressure	$p_A = p_A^* x_A e^{\xi(1-x_A)^2}$	Regular solution	5E.19

5F The activities of ions

Contents

➤ Why do you need to know this material?

Interactions between ions are so strong that the approximation of replacing activities by molalities is valid only in very dilute solutions (less than 1 mmol kg^{-1} in total ion concentration) and in precise work activities themselves must be used. We need, therefore, to pay special attention to the activities of ions in solution, especially in preparation for the discussion of electrochemical phenomena.

➤ What is the key idea?

The chemical potential of an ion is lowered as a result of its electrostatic interaction with its ionic atmosphere.

➤ What do you need to know already?

This Topic builds on the relation between chemical potential and mole fraction (Topic 5A) and on the relation between Gibbs free energy and non-expansion work (Topic 3D). If you intend to work through the derivation of the Debye–Hückel theory, you need to be familiar with some concepts from electrostatics, including the Coulomb potential and its relation to charge density through Poisson's equation (which is explained in the Topic); this Topic also draws on the Boltzmann distribution (*Foundations* B and, in much more detail, Topic 15A).

If the chemical potential of the cation M^+ is denoted μ_+ and that of the anion X^- is denoted μ_-, the total molar Gibbs energy of the ions in the electrically neutral solution is the sum of these partial molar quantities. The molar Gibbs energy of an *ideal* solution of such ions is

$$G_m^{ideal} = \mu_+^{ideal} + \mu_-^{ideal} \tag{5F.1}$$

with $\mu_J^{ideal} = \mu_J^{\ominus} + RT \ln x_J$. However, for a *real* solution of M^+ and X^- of the same molality we write $\mu_J = \mu_J^{\ominus} + RT \ln a_J$ with $a_J = \gamma_J x_J$, which implies that $\mu_J = \mu_J^{ideal} + RT \ln \gamma_J$. It then follows that

$$\begin{aligned} G_m &= \mu_+ + \mu_- = \mu_+^{ideal} + \mu_-^{ideal} + RT\ln\gamma_+ + RT\ln\gamma_- \\ &= G_m^{ideal} + RT\ln\gamma_+\gamma_- \end{aligned} \tag{5F.2}$$

All the deviations from ideality are contained in the last term.

5F.1 Mean activity coefficients

There is no experimental way of separating the product $\gamma_+\gamma_-$ into contributions from the cations and the anions. The best we can do experimentally is to assign responsibility for the non-ideality equally to both kinds of ion. Therefore, for a 1,1-electrolyte, we introduce the 'mean activity coefficient' as the geometric mean of the individual coefficients, where the geometric mean of x^p and y^q is $(x^p y^q)^{1/(p+q)}$. Thus:

$$\gamma_\pm = (\gamma_+\gamma_-)^{1/2} \tag{5F.3}$$

and express the individual chemical potentials of the ions as

$$\mu_+ = \mu_+^{ideal} + RT\ln\gamma_\pm \quad \mu_- = \mu_-^{ideal} + RT\ln\gamma_\pm \tag{5F.4}$$

The sum of these two chemical potentials is the same as before, eqn 5F.2, but now the non-ideality is shared equally.

We can generalize this approach to the case of a compound M_pX_q that dissolves to give a solution of p cations and q anions from each formula unit. The molar Gibbs energy of the ions is the sum of their partial molar Gibbs energies:

$$G_m = p\mu_+ + q\mu_- = G_m^{ideal} + pRT\ln\gamma_+ + qRT\ln\gamma_- \tag{5F.5}$$

If we introduce the **mean activity coefficient** now defined in a more general way as

$$\gamma_{\pm}=(\gamma_{+}^{p}\gamma_{-}^{q})^{1/s} \quad s=p+q \qquad \textit{Definition} \quad \text{Mean activity coefficient} \tag{5F.6}$$

and write the chemical potential of each ion as

$$\mu_i=\mu_i^{\text{ideal}}+RT\ln\gamma_{\pm} \tag{5F.7}$$

we get the same expression as in eqn 5F.2 for G_m when we write $G_m=p\mu_+ + q\mu_-$. However, both types of ion now share equal responsibility for the non-ideality.

(a) The Debye–Hückel limiting law

The long range and strength of the Coulombic interaction between ions means that it is likely to be primarily responsible for the departures from ideality in ionic solutions and to dominate all the other contributions to non-ideality. This domination is the basis of the **Debye–Hückel theory** of ionic solutions, which was devised by Peter Debye and Erich Hückel in 1923. We give here a qualitative account of the theory and its principal conclusions. The calculation itself, which is a profound example of how a seemingly intractable problem can be formulated and then resolved by drawing on physical insight, is described the following section.

Oppositely charged ions attract one another. As a result, anions are more likely to be found near cations in solution, and vice versa (Fig. 5F.1). Overall, the solution is electrically neutral, but near any given ion there is an excess of counter ions (ions of opposite charge). Averaged over time, counter ions are more likely to be found near any given ion. This time-averaged, spherical haze around the central ion, in which counter ions outnumber ions of the same charge as the central ion, has a net charge equal in magnitude but opposite in sign to that on the central ion, and is called its **ionic atmosphere**. The energy, and therefore the chemical potential, of any given central ion is lowered as a result of its electrostatic interaction with its ionic atmosphere. This lowering of energy appears as the difference between the molar Gibbs energy G_m and the ideal value G_m^{ideal} of the solute, and hence can be identified with $RT \ln \gamma_{\pm}$. The stabilization of ions by their interaction with their ionic atmospheres is part of the explanation why chemists commonly use dilute solutions, in which the stabilization is less important, to achieve precipitation of ions from electrolyte solutions.

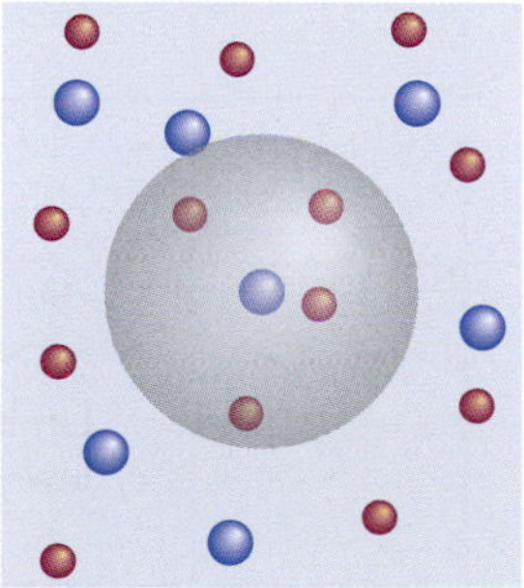

Figure 5F.1 The picture underlying the Debye–Hückel theory is of a tendency for anions to be found around cations, and of cations to be found around anions (one such local clustering region is shown by the grey sphere). The ions are in ceaseless motion, and the diagram represents a snapshot of their motion. The solutions to which the theory applies are far less concentrated than shown here.

The model leads to the result that at very low concentrations the activity coefficient can be calculated from the **Debye–Hückel limiting law**

$$\log\gamma_{\pm}=-A|z_+z_-|I^{1/2} \qquad \text{Debye–Hückel limiting law} \tag{5F.8}$$

where $A=0.509$ for an aqueous solution at 25 °C and I is the dimensionless **ionic strength** of the solution:

$$I=\frac{1}{2}\sum_i z_i^2(b_i/b^{\ominus}) \qquad \textit{Definition} \quad \text{Ionic strength} \tag{5F.9}$$

In this expression, z_i is the charge number of an ion i (positive for cations and negative for anions) and b_i is its molality. The ionic strength occurs widely wherever ionic solutions are discussed, as we shall see. The sum extends over all the ions present in the solution. For solutions consisting of two types of ion at molalities b_+ and b_-,

$$I=\tfrac{1}{2}(b_+z_+^2+b_-z_-^2)/b^{\ominus} \tag{5F.10}$$

The ionic strength emphasizes the charges of the ions because the charge numbers occur as their squares. Table 5F.1 summarizes the relation of ionic strength and molality in an easily usable form.

Brief illustration 5F.1 The limiting law

The mean activity coefficient of 5.0 mmol kg^{-1} KCl(aq) at 25 °C is calculated by writing $I=\tfrac{1}{2}(b_++b_-)/b^{\ominus}=b/b^{\ominus}$, where b is the molality of the solution (and $b_+=b_-=b$). Then, from eqn 5F.8,

$$\log\gamma_{\pm}=-0.509\times(5.0\times10^{-3})^{1/2}=-0.03\ldots$$

Hence, $\gamma_{\pm}=0.92$. The experimental value is 0.927.

Self-test 5F.1 Calculate the ionic strength and the mean activity coefficient of 1.00 mmol kg^{-1} $CaCl_2$(aq) at 25 °C.

Answer: 3.00 mmol kg^{-1}, 0.880

The name 'limiting law' is applied to eqn 5F.8 because ionic solutions of moderate molalities may have activity

Table 5F.1 Ionic strength and molality, $I = kb/b^{\ominus}$

k	X^-	X^{2-}	X^{3-}	X^{4-}
M^+	1	3	6	10
M^{2+}	3	4	15	12
M^{3+}	6	15	9	42
M^{4+}	10	12	42	16

For example, the ionic strength of an M_2X_3 solution of molality b, which is understood to give M^{3+} and X^{2-} ions in solution is $15b/b^{\ominus}$.

Table 5F.2* Mean activity coefficients in water at 298 K

$b/b^{\ominus}$	KCl	$CaCl_2$
0.001	0.966	0.888
0.01	0.902	0.732
0.1	0.770	0.524
1.0	0.607	0.725

* More values are given in the *Resource section*.

coefficients that differ from the values given by this expression, yet all solutions are expected to conform as $b \to 0$. Table 5F.2 lists some experimental values of activity coefficients for salts of various valence types. Figure 5F.2 shows some of these values plotted against $I^{1/2}$, and compares them with the theoretical straight lines calculated from eqn 5F.8. The agreement at very low molalities (less than about 1 mmol kg^{-1}, depending on charge type) is impressive and convincing evidence in support of the model. Nevertheless, the departures from the theoretical curves above these molalities are large, and show that the approximations are valid only at very low concentrations.

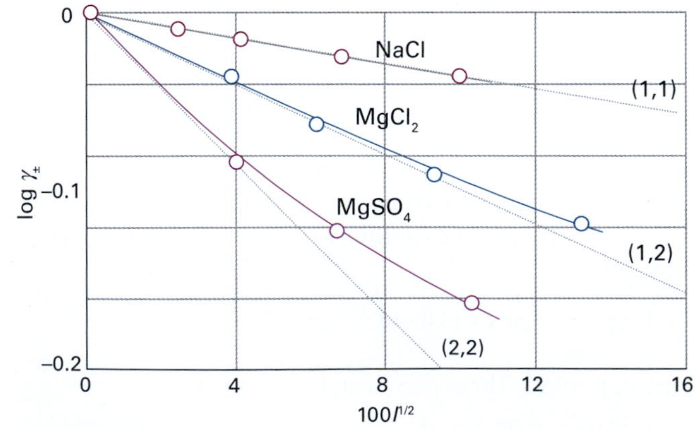

Figure 5F.2 An experimental test of the Debye–Hückel limiting law. Although there are marked deviations for moderate ionic strengths, the limiting slopes as $I \to 0$ are in good agreement with the theory, so the law can be used for extrapolating data to very low molalities.

(b) Extensions of the limiting law

When the ionic strength of the solution is too high for the limiting law to be valid, the activity coefficient may be estimated from the **extended Debye–Hückel law** (sometimes called the *Truesdell–Jones equation*):

$$\log \gamma_\pm = -\frac{A|z_+z_-|I^{1/2}}{1+BI^{1/2}} \quad \text{Extended Debye–Hückel law} \quad (5F.11a)$$

where B is a dimensionless constant. A more flexible extension is the **Davies equation** proposed by C.W. Davies in 1938:

$$\log \gamma_\pm = -\frac{A|z_+z_-|I^{1/2}}{1+BI^{1/2}} + CI \quad \text{Davies equation} \quad (5F.11b)$$

where C is another dimensionless constant. Although B can be interpreted as a measure of the closest approach of the ions, it (like C) is best regarded as an adjustable empirical parameter. A curve drawn on the basis of the Davies equation is shown in Fig. 5F.3. It is clear that eqn 5F.11 accounts for some activity coefficients over a moderate range of dilute solutions (up to about 0.1 mol kg^{-1}); nevertheless it remains very poor near 1 mol kg^{-1}.

Current theories of activity coefficients for ionic solutes take an indirect route. They set up a theory for the dependence of the activity coefficient of the solvent on the concentration of the solute, and then use the Gibbs–Duhem equation (eqn 5A.12a, $n_A \mathrm{d}\mu_A + n_B \mathrm{d}\mu_B = 0$) to estimate the activity coefficient of the solute. The results are reasonably reliable for solutions with molalities greater than about 0.1 mol kg^{-1} and are valuable for the discussion of mixed salt solutions, such as sea water.

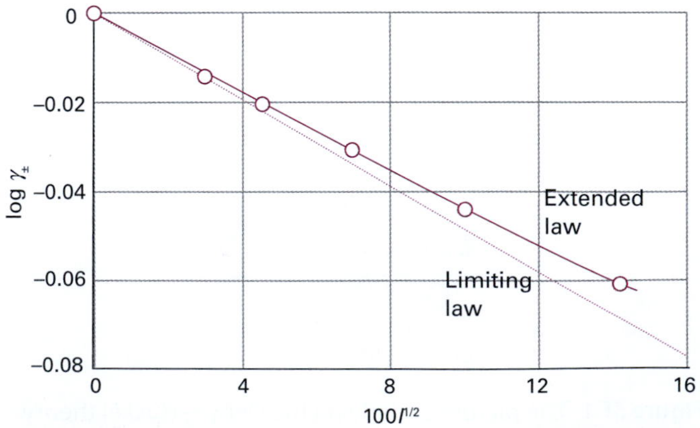

Figure 5F.3 The Davies equation gives agreement with experiment over a wider range of molalities than the limiting law (as shown here for a 1,1-electrolyte), but it fails at higher molalities.

5F.2 The Debye–Hückel theory

The strategy for the calculation is to establish the relation between the work needed to charge an ion and its chemical potential, and then to relate that work to the ion's interaction with the atmosphere of counter ions that has assembled around it as a result of the competition of the attraction between oppositely charged ions, the repulsion of like-charged ions, and the distributing effect of thermal motion.

(a) The work of charging

Imagine a solution in which all the ions have their actual positions, but in which their Coulombic interactions have been turned off and so are behaving 'ideally'. The difference in molar Gibbs energy between the ideal and real solutions is equal to w_e, the electrical work of charging the system in this arrangement. For a salt M_pX_q, we write

$$w_e = \overbrace{(p\mu_+ + q\mu_-)}^{G_m,\ \text{charged}} - \overbrace{(p\mu_+^{\text{ideal}} + q\mu_-^{\text{ideal}})}^{G_m^{\text{ideal}},\ \text{uncharged}}$$
$$= p(\mu_+ - \mu_+^{\text{ideal}}) + q(\mu_- - \mu_-^{\text{ideal}}) \qquad (5F.12)$$

From eqn 5F.7 we write

$$\mu_+ - \mu_+^{\text{ideal}} = \mu_- - \mu_-^{\text{ideal}} = RT\ln\gamma_\pm$$

So it follows that

$$\ln\gamma_\pm = \frac{w_e}{sRT} \qquad s = p+q \qquad (5F.13)$$

This equation tells us that we must first find the final distribution of the ions and then the work of charging them in that distribution.

Brief illustration 5F.2 The work of charging

The measured mean activity coefficient of 5.0 mmol kg^{-1} KCl(aq) at 25 °C is 0.927. It follows that the average work involved in charging the ions in their environment in the solution is given by eqn 5F.13 in the form

$$w_e = sRT\ln\gamma_\pm = 2\times(8.3145\ \text{J K}^{-1}\,\text{mol}^{-1})\times(298\ \text{K})\times\ln 0.927$$
$$= -0.38\ \text{kJ mol}^{-1}$$

Self-test 5F.2 The measured mean activity coefficient of 0.1 mol kg^{-1} Na_2SO_4(aq) at 25 °C is 0.445. What is the work of charging the ions in the solution?

Answer: −6.02 kJ mol^{-1}

(b) The potential due to the charge distribution

As explained in *Foundations* B, the Coulomb potential at a distance r from an isolated ion of charge z_ie in a medium of permittivity ε is

$$\phi_i = \frac{Z_i}{r} \qquad Z_i = \frac{z_ie}{4\pi\varepsilon} \qquad (5F.14)$$

The ionic atmosphere causes the potential to decay with distance more sharply than this expression implies. Such shielding is a familiar problem in electrostatics, and its effect is taken into account by replacing the Coulomb potential by the **shielded Coulomb potential**, an expression of the form

$$\phi_i = \frac{Z_i}{r}e^{-r/r_D} \qquad \text{Shielded Coulomb potential} \qquad (5F.15)$$

where r_D is called the **Debye length**. When r_D is large, the shielded potential is virtually the same as the unshielded potential. When it is small, the shielded potential is much smaller than the unshielded potential, even for short distances (Fig. 5F.4). We establish in the following *Justification* that

$$r_D = \left(\frac{\varepsilon RT}{2\rho F^2 Ib^{\ominus}}\right)^{1/2} \qquad \text{Debye length} \qquad (5F.16)$$

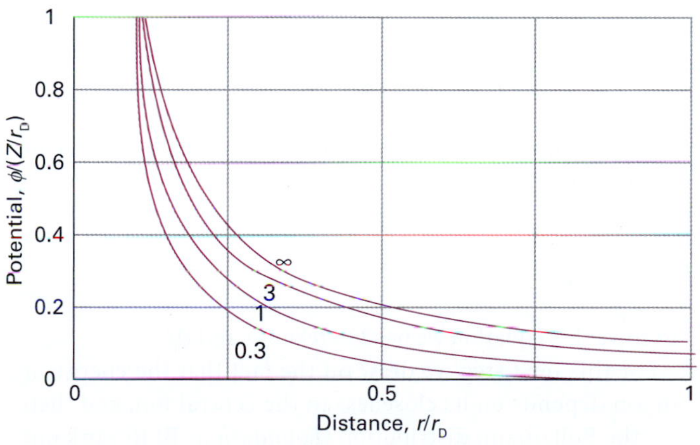

Figure 5F.4 The variation of the shielded Coulomb potential with distance for different values of the Debye length, r_D/a. The smaller the Debye length, the more sharply the potential decays to zero. In each case, a is an arbitrary unit of length.

Brief illustration 5F.3 The Debye length

To estimate the Debye length in an aqueous solution of ionic strength 0.100 and density 1.000 g cm^{-3} at 25 °C we write

$$r_D = \left(\frac{\overbrace{80.10 \times (8.854\times10^{-12}\,\mathrm{J^{-1}\,C^2\,m^{-1}})}^{\varepsilon} \times (8.3145\,\mathrm{J\,K^{-1}\,mol^{-1}}) \times (298\,\mathrm{K})}{2 \times \underbrace{(1000\,\mathrm{kg\,m^{-3}})}_{1.000\,\mathrm{g\,cm^{-3}}} \times (9.649\times10^4\,\mathrm{C\,mol^{-1}})^2 \times (0.100) \times \underbrace{(1\,\mathrm{mol\,kg^{-1}})}_{b^\ominus}} \right)^{1/2}$$

$$= 9.72\times10^{-10}\,\mathrm{m}, \text{ or } 0.972\,\mathrm{nm}$$

Self-test 5F.3 Estimate the Debye length in an ethanol solution of ionic strength 0.100 and density 0.789 g cm^{-3} at 25 °C. Use $\varepsilon_r = 25.3$.

Answer: 0.615 nm

Justification 5F.1 The Debye length

To calculate r_D, we need to know how the **charge density**, ρ_i, of the ionic atmosphere, the charge in a small region divided by the volume of the region, varies with distance from the ion. This step draws on another standard result of electrostatics, in which charge density and potential are related by **Poisson's equation**:

$$\nabla^2\phi = -\frac{\rho}{\varepsilon} \qquad \text{Poisson's equation}$$

where $\nabla^2 = \partial^2/\partial x^2 + \partial^2/\partial y^2 + \partial^2/\partial z^2$. Because we are considering only a spherical ionic atmosphere, we can use a simplified form of this equation in which the charge density varies only with distance from the central ion:

$$\frac{1}{r^2}\frac{\mathrm{d}}{\mathrm{d}r}\left(r^2\frac{\mathrm{d}\phi_i}{\mathrm{d}r}\right) = -\frac{\rho_i}{\varepsilon}$$

Substitution of the expression for the shielded potential, eqn 5F.15, results in

$$r_D^2 = -\frac{\varepsilon\phi_i}{\rho_i}$$

To solve this equation we need to relate ρ_i and ϕ_i.

For this next step we draw on the fact that the energy of an ion depends on its closeness to the central ion, and then use the Boltzmann distribution (*Foundations* B) to work out the probability that an ion will be found at each distance. The energy of an ion j of charge z_je at a distance where it experiences the potential ϕ_i of the central ion i relative to its energy when it is far away in the bulk solution is its charge times the potential $z_je\phi_i$. Therefore, according to the Boltzmann distribution, the ratio of the molar concentration, c_j, of ions at a distance r and the molar concentration in the bulk, c_j°, where the energy is zero, is

$$\frac{c_j}{c_j^\circ} = \mathrm{e}^{-z_je\phi_i/kT}$$

The charge density, ρ_i, at a distance r from the ion i is the molar concentration of each type of ion multiplied by the charge per mole of ions, z_ieN_A. The quantity $eN_A = F$, the magnitude of the charge per mole of electrons, is Faraday's constant. It follows that

$$\rho_i = c_+z_+F + c_-z_-F = c_+^\circ z_+F\mathrm{e}^{-z_+e\phi_i/kT} + c_-^\circ z_-F\mathrm{e}^{-z_-e\phi_i/kT}$$

At this stage we need to simplify the expression to avoid the awkward exponential terms. Because the average electrostatic interaction energy is small compared with kT we may use the expansion $\mathrm{e}^x = 1 + x + \cdots$ and write the charge density as

$$\rho_i = c_+^\circ z_+F\left(1 - \frac{z_+e\phi_i}{kT} + \cdots\right) + c_-^\circ z_-F\left(1 - \frac{z_-e\phi_i}{kT} + \cdots\right)$$

$$= \overbrace{(c_+^\circ z_+ + c_-^\circ z_-)}^{0}F - (c_+^\circ z_+^2 + c_-^\circ z_-^2)\frac{Fe\phi_i}{kT} + \cdots$$

The first term in the expansion is zero because it is the charge density in the bulk, uniform solution, and the solution is electrically neutral. Replacing e by F/N_A and N_Ak by R results in the following expression:

$$\rho_i = -(c_+^\circ z_+^2 + c_-^\circ z_-^2)\frac{F^2\phi_i}{RT}$$

The unwritten terms are assumed to be too small to be significant. This one remaining term, in blue, can be expressed in terms of the ionic strength, eqn 5F.9, by noting that in the dilute aqueous solutions we are considering there is little difference between molality and molar concentration, and $c \approx b\rho$, where ρ is the mass density of the solvent

$$c_+^\circ z_+^2 + c_-^\circ z_-^2 \approx \overbrace{(b_+^\circ z_+^2 + b_-^\circ z_-^2)}^{2Ib^\ominus}\rho = 2Ib^\ominus\rho$$

With these approximations, the last equation becomes

$$\rho_i = -\frac{2Ib^\ominus\rho F^2\phi_i}{RT}$$

We can now substitute this expression into $r_D^2 = -\varepsilon\phi_i/\rho_i$, when the ϕ_i cancel and we obtain eqn 5F.16.

(c) The activity coefficient

To calculate the activity coefficient we need to find the electrical work of charging the central ion when it is surrounded by its atmosphere. This calculation is carried through in the following *Justification*, which leads to the conclusion that the work of charging an ion i when it is surrounded by the atmosphere it has assembled is

$$w_{e,i} = -\frac{z_i^2F^2}{8\pi N_A\varepsilon r_D} \qquad \text{Work of charging} \qquad (5F.17)$$

Justification 5F.2 The work of charging

To calculate the work of charging the central ion we need to know the potential at the ion due to its atmosphere, ϕ_{atmos}. This potential is the difference between the total potential, given by eqn 5F.15, and the potential due to the central ion itself:

$$\phi_{\text{atmosphere}} = \phi - \phi_{\text{central ion}} = Z_i\left(\frac{e^{-r/r_D}}{r} - \frac{1}{r}\right), \quad Z_i = \frac{z_i e}{4\pi\varepsilon}$$

The potential at the central ion (at $r=0$) is obtained by taking the limit of this expression as $r \to 0$ and is

$$\phi_{\text{atmosphere}}(0) = Z_i \lim_{r\to 0}\left(\frac{1 - r/r_D + \cdots}{r} - \frac{1}{r}\right) = -\frac{Z_i}{r_D}$$

This expression shows us that the potential due to the ionic atmosphere is equivalent to the potential arising from a single charge of equal magnitude but opposite sign to that of the central ion and located at a distance r_D from the ion. If the charge of the central ion were Q and not $z_i e$, then the potential due to its atmosphere would be

$$\phi_{\text{atmosphere}}(0) = -\frac{Q}{4\pi\varepsilon r_D}$$

The work of adding a charge dQ to a region where the electrical potential is $\phi_{\text{atmosphere}}(0)$, Table 2A.1 (from $dw = \phi dQ$), is

$$dw_e = \phi_{\text{atmosphere}}(0)dQ$$

Therefore, the total molar work of fully charging the ion i in the presence of its atmosphere is

$$w_{e,i} = N_A \int_0^{z_i e} \phi_{\text{atmosphere}}(0)dQ = -\frac{N_A}{4\pi\varepsilon r_D}\int_0^{z_i e} QdQ$$

$$= -\frac{N_A z_i^2 e^2}{8\pi\varepsilon r_D} = -\frac{z_i^2 F^2}{8\pi N_A \varepsilon r_D}$$

where in the last step we have used $F = N_A e$. This expression is eqn. 5F.17.

We can now collect the various pieces of this calculation and arrive at an expression for the mean activity coefficient. It follows from eqn 5F.13 with $w_e = pw_{e,+} + qw_{e,-}$, the total work of charging p cations and q anions in the presence of their atmospheres, that the mean activity coefficient of the ions is

$$\ln\gamma_\pm = \frac{pw_{e,+} + qw_{e,-}}{sRT} = -\frac{(pz_+^2 + qz_-^2)F^2}{8\pi N_A sRT\varepsilon r_D}$$

However, for neutrality $pz_+ + qz_- = 0$; therefore

$$pz_+^2 + qz_-^2 = \overbrace{pz_+}^{-qz_-} z_+ + \overbrace{qz_-}^{-pz_+} z_- = -\overbrace{(p+q)}^{s}\overbrace{z_+z_-}^{-|z_+z_-|} = s|z_+z_-|$$

It then follows that

$$\ln\gamma_\pm = -\frac{|z_+z_-|F^2}{8\pi N_A RT\varepsilon r_D} \tag{5F.18}$$

The replacement of r_D with the expression in eqn 5F.16 gives

$$\ln\gamma_\pm = -\frac{|z_+z_-|F^2}{8\pi N_A RT\varepsilon}\left(\frac{2\rho F^2 I b^\ominus}{\varepsilon RT}\right)^{1/2} \tag{5F.19a}$$

$$= -|z_+z_-|\left\{\frac{F^3}{4\pi N_A}\left(\frac{\rho b^\ominus}{2\varepsilon^3 R^3 T^3}\right)^{1/2}\right\} I^{1/2}$$

where we have grouped terms in such a way as to show that this expression is beginning to take the form of eqn 5F.8 ($\log\gamma_\pm = -|z_+z_-|AI^{1/2}$). Indeed, conversion to common logarithms (by using $\ln x = \ln 10 \times \log x$) gives

$$\log\gamma_\pm = -|z_+z_-|\overbrace{\left\{\frac{F^3}{4\pi N_A \ln 10}\left(\frac{\rho b^\ominus}{2\varepsilon^3 R^3 T^3}\right)^{1/2}\right\}}^{A} I^{1/2} \tag{5F.19b}$$

which is eqn 5F.8 with

$$A = \frac{F^3}{4\pi N_A \ln 10}\left(\frac{\rho b^\ominus}{2\varepsilon^3 R^3 T^3}\right)^{1/2} \tag{5F.20}$$

Brief illustration 5F.4 The Debye–Hückel constant

To evaluate the constant A for water at 25.00 °C, we use $\rho = 0.9971\ \text{g cm}^{-3}$ and $\varepsilon = 78.54\varepsilon_0$ to find

$$A = \frac{(9.649\times10^4\ \text{C mol}^{-1})^3}{4\pi\times(6.022\times10^{23}\ \text{mol}^{-1})\ln 10}$$

$$\times\left(\frac{(997.1\ \text{kg m}^{-3})\times(1\ \text{mol kg}^{-1})}{2\times(78.54\times8.854\times10^{-12}\ \text{J}^{-1}\text{C}^2\text{m}^{-1})^3\times(8.3145\ \text{J K}^{-1}\text{mol}^{-1})^3\times(298.15\ \text{K})^3}\right)^{1/2}$$

$$= 0.5086$$

Self-test 5F.4 Evaluate the constant A for ethanol at 25 °C, when $\varepsilon_r = 25.3$ and $\rho = 0.789\ \text{g cm}^{-3}$.

Answer: 2.47

Checklist of concepts

- ☐ 1. **Mean activity coefficients** apportion deviations from ideality equally to the cations and anions in an ionic solution.
- ☐ 2. The **Debye–Hückel theory** ascribes deviations from ideality to the Coulombic interaction of an ion with the ionic atmosphere that assembles around it.
- ☐ 3. The **Debye–Hückel limiting law** is extended by including two further empirical constants.

Checklist of equations

Property	Equation	Comment	Equation number
Mean activity coefficient	$\gamma_\pm=(\gamma_+^p\gamma_-^q)^{1/(p+q)}$	Definition	5F.6
Debye–Hückel limiting law	$\log\gamma_\pm=-A\lvert z_+z_-\rvert I^{1/2}$	Valid as $I\to 0$	5F.8
Ionic strength	$I=\frac{1}{2}\sum_i z_i^2(b_i/b^{\ominus})$	Definition	5F.9
Davies equation	$\log\gamma_\pm=-A\lvert z_+z_-\rvert I^{1/2}/(1+BI^{1/2})+CI$	A, B, C empirical constants	5F.11

CHAPTER 5 Simple mixtures

TOPIC 5A The thermodynamic description of mixtures

Discussion questions

5A.1 Explain the concept of partial molar quantity, and justify the remark that the partial molar property of a solute depends on the properties of the solvent too.

5A.2 Explain how thermodynamics relates non-expansion work to a change in composition of a system.

5A.3 Are there any circumstances under which two (real) gases will not mix spontaneously?

5A.4 Explain how Raoult's law and Henry's law are used to specify the chemical potential of a component of a mixture.

5A.5 Explain the molecular origin of Raoult's law and Henry's law.

Exercises

5A.1(a) A polynomial fit to measurements of the total volume of a binary mixture of A and B is

$$v = 987.93 + 35.6774x - 0.45923x^2 + 0.017325x$$

where $v = V/\text{cm}^3$, $x = n_B/\text{mol}$, and n_B is the amount of B present. Determine the partial molar volumes of A and B.

5A.1(b) A polynomial fit to measurements of the total volume of a binary mixture of A and B is

$$v = 778.55 - 22.5749x + 0.56892x^2 + 0.01023x^3 + 0.00234x$$

where $v = V/\text{cm}^3$, $x = n_B/\text{mol}$, and n_B is the amount of B present. Determine the partial molar volumes of A and B.

5A.2(a) The volume of an aqueous solution of NaCl at 25 °C was measured at a series of molalities b, and it was found that the volume fitted the expression $v = 1003 + 16.62x + 1.77x^{3/2} + 0.12x^2$ where $v = V/\text{cm}^3$, V is the volume of a solution formed from 1.000 kg of water and $x = b/b^{\ominus}$. Calculate the partial molar volume of the components in a solution of molality 0.100 mol kg^{-1}.

5A.2(b) At 18 °C the total volume V of a solution formed from $MgSO_4$ and 1.000 kg of water fits the expression $v = 1001.21 + 34.69(x - 0.070)^2$, where $v = V/\text{cm}^3$ and $x = b/bx = b/b^{\ominus}$. Calculate the partial molar volumes of the salt and the solvent when in a solution of molality 0.050 mol kg^{-1}.

5A.3(a) Suppose that $n_A = 0.10n_B$ and a small change in composition results in μ_A changing by $\delta\mu_A = +12\,\text{J mol}^{-1}$, by how much will μ_B change?

5A.3(b) Suppose that $n_A = 0.22n_B$ and a small change in composition results in μ_A changing by $\delta\mu_A = -15\,\text{J mol}^{-1}$, by how much will μ_B change?

5A.4(a) Consider a container of volume 5.0 dm^3 that is divided into two compartments of equal size. In the left compartment there is nitrogen at 1.0 atm and 25 °C; in the right compartment there is hydrogen at the same temperature and pressure. Calculate the entropy and Gibbs energy of mixing when the partition is removed. Assume that the gases are perfect.

5A.4(b) Consider a container of volume 250 cm^3 that is divided into two compartments of equal size. In the left compartment there is argon at 100 kPa and 0 °C; in the right compartment there is neon at the same temperature and pressure. Calculate the entropy and Gibbs energy of mixing when the partition is removed. Assume that the gases are perfect.

5A.5(a) Air is a mixture with mass percentage composition 75.5 (N_2), 23.2 (O_2), 1.3 (Ar). Calculate the entropy of mixing when it is prepared from the pure (and perfect) gases.

5A.5(b) When carbon dioxide is taken into account, the mass percentage composition of air is 75.52 (N_2), 23.15 (O_2), 1.28 (Ar), and 0.046 (CO_2). What is the change in entropy from the value in the preceding exercise?

5A.6(a) The vapour pressure of benzene at 20 °C is 10 kPa and that of methylbenzene is 2.8 kPa at the same temperature. What is the vapour pressure of a mixture of equal masses of each component?

5A.6(b) At 90 °C the vapour pressure of 1,2-dimethylbenzene is 20 kPa and that of 1,3-dimethylbenzene is 18 kPa. What is the composition of the vapour of an equimolar mixture of the two components?

5A.7(a) The partial molar volumes of acetone (propanone) and chloroform (trichloromethane) in a mixture in which the mole fraction of $CHCl_3$ is 0.4693 are 74.166 cm^3 mol^{-1} and 80.235 cm^3 mol^{-1}, respectively. What is the volume of a solution of mass 1.000 kg?

5A.7(b) The partial molar volumes of two liquids A and B in a mixture in which the mole fraction of A is 0.3713 are 188.2 cm^3 mol^{-1} and 176.14 cm^3 mol^{-1}, respectively. The molar masses of the A and B are 241.1 g mol^{-1} and 198.2 g mol^{-1}. What is the volume of a solution of mass 1.000 kg?

5A.8(a) At 25 °C, the density of a 50 per cent by mass ethanol–water solution is 0.914 g cm^{-3}. Given that the partial molar volume of water in the solution is 17.4 cm^3 mol^{-1}, calculate the partial molar volume of the ethanol.

5A.8(b) At 20 °C, the density of a 20 per cent by mass ethanol/water solution is 968.7 kg m^{-3}. Given that the partial molar volume of ethanol in the solution is 52.2 cm^3 mol^{-1}, calculate the partial molar volume of the water.

5A.9(a) At 300 K, the partial vapour pressure of HCl (that is, the partial pressure of the HCl vapour) in liquid $GeCl_4$ is as follows:

x_{HCl}	0.005	0.012	0.019
p_{HCl}/kPa	32.0	76.9	121.8

Show that the solution obeys Henry's law in this range of mole fractions, and calculate Henry's law constant at 300 K.

5A.9(b) At 310 K, the partial vapour pressure of a substance B dissolved in a liquid A is as follows:

x_B	0.010	0.015	0.020
p_B/kPa	82.0	122.0	166.1

Show that the solution obeys Henry's law in this range of mole fractions, and calculate Henry's law constant at 310 K.

5A.10(a) Calculate the molar solubility of nitrogen in benzene exposed to air at 25 °C; partial pressures were calculated in *Example* 1A.3 of Topic 1A.
5A.10(b) Calculate the molar solubility of methane at 1.0 bar in benzene at 25 °C.

5A.11(a) Use Henry's law and the data in Table 5A.1 to calculate the solubility (as a molality) of CO_2 in water at 25 °C when its partial pressure is (i) 0.10 atm, (ii) 1.00 atm.
5A.11(b) The mole fractions of N_2 and O_2 in air at sea level are approximately 0.78 and 0.21. Calculate the molalities of the solution formed in an open flask of water at 25 °C.

5A.12(a) A water carbonating plant is available for use in the home and operates by providing carbon dioxide at 5.0 atm. Estimate the molar concentration of the soda water it produces.
5A.12(b) After some weeks of use, the pressure in the water carbonating plant mentioned in the previous exercise has fallen to 2.0 atm. Estimate the molar concentration of the soda water it produces at this stage.

Problems

5A.1 The experimental values of the partial molar volume of a salt in water are found to fit the expression $v_B = 5.117 + 19.121x^{1/2}$, where $v_B = V_B/(\text{cm}^3\,\text{mol}^{-1})$ and x is the numerical value of the molality of B ($x = b/b^\ominus$). Use the Gibbs–Duhem equation to derive an equation for the molar volume of water in the solution. The molar volume of pure water at the same temperature is 18.079 cm³ mol⁻¹.

5A.2 The compound *p*-azoxyanisole forms a liquid crystal. 5.0 g of the solid was placed in a tube, which was then evacuated and sealed. Use the phase rule to prove that the solid will melt at a definite temperature and that the liquid crystal phase will make a transition to a normal liquid phase at a definite temperature.

5A.3 The following table gives the mole fraction of methylbenzene (A) in liquid and gaseous mixtures (x_A and y_A, respectively) with butanone at equilibrium at 303.15 K and the total pressure p. Take the vapour to be perfect and calculate the partial pressures of the two components. Plot them against their respective mole fractions in the liquid mixture and find the Henry's law constants for the two components.

x_A	0	0.0898	0.2476	0.3577	0.5194	0.6036	0.7188	0.8019	0.9105	1
y_A	0	0.0410	0.1154	0.1762	0.2772	0.3393	0.4450	0.5435	0.7284	1
p/kPa	36.066	34.121	30.900	28.626	25.239	23.402	20.6984	18.592	15.496	12.295

5A.4 The densities of aqueous solutions of copper(II) sulfate at 20 °C were measured as set out below. Determine and plot the partial molar volume of $CuSO_4$ in the range of the measurements.

$m(CuSO_4)$/g	5	10	15	20
ρ/(g cm⁻³)	1.051	1.107	1.167	1.230

where $m(CuSO_4)$ is the mass of $CuSO_4$ dissolved in 100 g of solution.

5A.5 Haemoglobin, the red blood protein responsible for oxygen transport, binds about 1.34 cm³ of oxygen per gram. Normal blood has a haemoglobin concentration of 150 g dm⁻³. Haemoglobin in the lungs is about 97 per cent saturated with oxygen, but in the capillary is only about 75 per cent saturated. What volume of oxygen is given up by 100 cm³ of blood flowing from the lungs in the capillary?

5A.6 Use the data from *Example* 5A.1 to determine the value of b at which V_E has a minimum value.

TOPIC 5B The properties of solutions

Discussion questions

5B.1 Explain what is meant by a regular solution; what additional features distinguish a real solution from a regular solution?

5B.2 Explain the physical origin of colligative properties.

5B.3 Colligative properties are independent of the identity of the solute. Why, then, can osmometry be used to determine the molar mass of a solute?

Exercises

5B.1(a) Predict the partial vapour pressure of HCl above its solution in liquid germanium tetrachloride of molality 0.10 mol kg⁻¹. For data, see Exercise 5A.10(a).
5B.1(b) Predict the partial vapour pressure of the component B above its solution in A in Exercise 5A.10(b) when the molality of B is 0.25 mol kg⁻¹. The molar mass of A is 74.1 g mol⁻¹.

5B.2(a) The vapour pressure of benzene is 53.3 kPa at 60.6 °C, but it fell to 51.5 kPa when 19.0 g of a non-volatile organic compound was dissolved in 500 g of benzene. Calculate the molar mass of the compound.
5B.2(b) The vapour pressure of 2-propanol is 50.00 kPa at 338.8 K, but it fell to 49.62 kPa when 8.69 g of a non-volatile organic compound was dissolved in 250 g of 2-propanol. Calculate the molar mass of the compound.

5B.3(a) The addition of 100 g of a compound to 750 g of CCl_4 lowered the freezing point of the solvent by 10.5 K. Calculate the molar mass of the compound.
5B.3(b) The addition of 5.00 g of a compound to 250 g of naphthalene lowered the freezing point of the solvent by 0.780 K. Calculate the molar mass of the compound.

5B.4(a) The osmotic pressure of an aqueous solution at 300 K is 120 kPa. Calculate the freezing point of the solution.
5B.4(b) The osmotic pressure of an aqueous solution at 288 K is 99.0 kPa. Calculate the freezing point of the solution.

5B.5(a) Calculate the Gibbs energy, entropy, and enthalpy of mixing when 0.50 mol C_6H_{14} (hexane) is mixed with 2.00 mol C_7H_{16} (heptane) at 298 K; treat the solution as ideal.
5B.5(b) Calculate the Gibbs energy, entropy, and enthalpy of mixing when 1.00 mol C_6H_{14} (hexane) is mixed with 1.00 mol C_7H_{16} (heptane) at 298 K; treat the solution as ideal.

5B.6(a) What proportions of hexane and heptane should be mixed (i) by mole fraction, (ii) by mass in order to achieve the greatest entropy of mixing?
5B.6(b) What proportions of benzene and ethylbenzene should be mixed (i) by mole fraction, (ii) by mass in order to achieve the greatest entropy of mixing?

5B.7(a) The enthalpy of fusion of anthracene is 28.8 kJ mol^{-1} and its melting point is 217 °C. Calculate its ideal solubility in benzene at 25 °C.
5B.7(b) Predict the ideal solubility of lead in bismuth at 280 °C given that its melting point is 327 °C and its enthalpy of fusion is 5.2 kJ mol^{-1}.

5B.8(a) The osmotic pressure of solutions of polystyrene in toluene were measured at 25 °C and the pressure was expressed in terms of the height of the solvent of density 1.004 g cm^{-3}:

$c/(\mathrm{g\,dm^{-3}})$	2.042	6.613	9.521	12.602
h/cm	0.592	1.910	2.750	3.600

Calculate the molar mass of the polymer.

5B.8(b) The molar mass of an enzyme was determined by dissolving it in water, measuring the osmotic pressure at 20 °C, and extrapolating the data to zero concentration. The following data were obtained:

$c/(\mathrm{mg\,cm^{-3}})$	3.221	4.618	5.112	6.722
h/cm	5.746	8.238	9.119	11.990

Calculate the molar mass of the enzyme.

5B.9(a) A dilute solution of bromine in carbon tetrachloride behaves as an ideal dilute solution. The vapour pressure of pure CCl_4 is 33.85 Torr at 298 K. The Henry's law constant when the concentration of Br_2 is expressed as a mole fraction is 122.36 Torr. Calculate the vapour pressure of each component, the total pressure, and the composition of the vapour phase when the mole fraction of Br_2 is 0.050, on the assumption that the conditions of the ideal dilute solution are satisfied at this concentration.

5B.9(b) The vapour pressure of a pure liquid A is 23 kPa at 20 °C and its Henry's law constant in liquid B is 73 kPa. Calculate the vapour pressure of each component, the total pressure, and the composition of the vapour phase when the mole fraction of A is 0.066 on the assumption that the conditions of the ideal dilute solution are satisfied at this concentration.

5B.10(a) At 90 °C, the vapour pressure of methylbenzene is 53.3 kPa and that of 1,2-dimethylbenzene is 20.0 kPa. What is the composition of a liquid mixture that boils at 90 °C when the pressure is 0.50 atm? What is the composition of the vapour produced?
5B.10(b) At 90 °C, the vapour pressure of 1,2-dimethylbenzene is 20 kPa and that of 1,3-dimethylbenzene is 18 kPa What is the composition of a liquid mixture that boils at 90 °C when the pressure is 19 kPa? What is the composition of the vapour produced?

5B.11(a) The vapour pressure of pure liquid A at 300 K is 76.7 kPa and that of pure liquid B is 52.0 kPa. These two compounds form ideal liquid and gaseous mixtures. Consider the equilibrium composition of a mixture in which the mole fraction of A in the vapour is 0.350. Calculate the total pressure of the vapour and the composition of the liquid mixture.
5B.11(b) The vapour pressure of pure liquid A at 293 K is 68.8 kPa and that of pure liquid B is 82.1 kPa. These two compounds form ideal liquid and gaseous mixtures. Consider the equilibrium composition of a mixture in which the mole fraction of A in the vapour is 0.612. Calculate the total pressure of the vapour and the composition of the liquid mixture.

5B.12(a) It is found that the boiling point of a binary solution of A and B with $x_A = 0.6589$ is 88 °C. At this temperature the vapour pressures of pure A and B are 127.6 kPa and 50.60 kPa, respectively. (i) Is this solution ideal? (ii) What is the initial composition of the vapour above the solution?
5B.12(b) It is found that the boiling point of a binary solution of A and B with $x_A = 0.4217$ is 96 °C. At this temperature the vapour pressures of pure A and B are 110.1 kPa and 76.5 kPa, respectively. (i) Is this solution ideal? (ii) What is the initial composition of the vapour above the solution?

5B.13(a) Dibromoethene (DE, $p^*_{DE} = 22.9$ kPa at 358 K) and dibromopropene (DP, $p^*_{DP} = 17.1$ kPa at 358 K) form a nearly ideal solution. If $x_{DE} = 0.60$, what is (i) p_{total} when the system is all liquid, (ii) the composition of the vapour when the system is still almost all liquid.
5B.13(b) Benzene and toluene form nearly ideal solutions. Consider an equimolar solution of benzene and toluene. At 20 °C the vapour pressures of pure benzene and toluene are 9.9 kPa and 2.9 kPa, respectively. The solution is boiled by reducing the external pressure below the vapour pressure. Calculate (i) the pressure when boiling begins, (ii) the composition of each component in the vapour, and (iii) the vapour pressure when only a few drops of liquid remain. Assume that the rate of vaporization is low enough for the temperature to remain constant at 20 °C.

Problems

5B.1 Potassium fluoride is very soluble in glacial acetic acid and the solutions have a number of unusual properties. In an attempt to understand them, freezing point depression data were obtained by taking a solution of known molality and then diluting it several times (J. Emsley, *J. Chem. Soc. A*, 2702 (1971)). The following data were obtained:

$b/(\mathrm{mol\,kg^{-1}})$	0.015	0.037	0.077	0.295	0.602
$\Delta T/\mathrm{K}$	0.115	0.295	0.470	1.381	2.67

Calculate the apparent molar mass of the solute and suggest an interpretation. Use $\Delta_{fus}H = 11.4$ kJ mol^{-1} and $T_f^* = 290$ K.

5B.2 In a study of the properties of an aqueous solution of $Th(NO_3)_4$ (by A. Apelblat, D. Azoulay, and A. Sahar, *J. Chem. Soc. Faraday Trans., I*, 1618 (1973)), a freezing point depression of 0.0703 K was observed for an aqueous solution of molality 9.6 mmol kg^{-1}. What is the apparent number of ions per formula unit?

5B.3‡ Aminabhavi et al. examined mixtures of cyclohexane with various long-chain alkanes (T.M. Aminabhavi et al., *J. Chem. Eng. Data* **41**, 526 (1996)). Among their data are the following measurements of the density of a mixture of cyclohexane and pentadecane as a function of mole fraction of cyclohexane (x_c) at 298.15 K:

x_c	0.6965	0.7988	0.9004
$\rho/(\mathrm{g\,cm^{-3}})$	0.7661	0.7674	0.7697

Compute the partial molar volume for each component in a mixture which has a mole fraction of cyclohexane of 0.7988.

5B.4‡ Comelli and Francesconi examined mixtures of propionic acid with various other organic liquids at 313.15 K (F. Comelli and R. Francesconi, *J. Chem. Eng. Data* **41**,101 (1996)). They report the excess volume of mixing

‡ These problems were provided by Charles Trapp and Carmen Giunta.

propionic acid with oxane as $V^E = x_1x_2\{a_0 + a_1(x_1 - x_2)\}$, where x_1 is the mole fraction of propionic acid, x_2 that of oxane, $a_0 = -2.4697\,\mathrm{cm^3\,mol^{-1}}$ and $a_1 = 0.0608\,\mathrm{cm^3\,mol^{-1}}$. The density of propionic acid at this temperature is $0.97174\,\mathrm{g\,cm^{-3}}$; that of oxane is $0.86398\,\mathrm{g\,cm^{-3}}$. (a) Derive an expression for the partial molar volume of each component at this temperature. (b) Compute the partial molar volume for each component in an equimolar mixture.

5B.5‡ Equation 5B.15 indicates, after it has been converted into an expression for x_B, that solubility is an exponential function of temperature. The data in the table below gives the solubility, S, of calcium acetate in water as a function of temperature.

θ/°C	0	20	40	60	80
S/(g (100 g solvent)$^{-1}$)	36.4	34.9	33.7	32.7	31.7

Determine the extent to which the data fit the exponential $S = S_0 e^{\tau/T}$ and obtain values for S_0 and τ. Express these constants in terms of properties of the solute.

5B.6 The excess Gibbs energy of solutions of methylcyclohexane (MCH) and tetrahydrofuran (THF) at 303.15 K were found to fit the expression

$$G^E = RTx(1-x)\left\{0.4857 - 0.1077(2x-1) + 0.0191(2x-1)^2\right\}$$

where x is the mole fraction of the methylcyclohexane. Calculate the Gibbs energy of mixing when a mixture of 1.00 mol of MCH and 3.00 mol of THF is prepared.

5B.7‡ Figure 5.1 shows $\Delta_{\mathrm{mix}}G(x_{\mathrm{Pb}}, T)$ for a mixture of copper and lead. (a) What does the graph reveal about the miscibility of copper and lead and the spontaneity of solution formation? What is the variance (F) at (i) 1500 K, (ii) 1100 K? (b) Suppose that at 1500 K a mixture of composition (i) $x_{\mathrm{Pb}} = 0.1$, (ii) $x_{\mathrm{Pb}} = 0.7$, is slowly cooled to 1100 K. What is the equilibrium composition of the final mixture? Include an estimate of the relative amounts of each phase. (c) What is the solubility of (i) lead in copper, (ii) copper in lead at 1100 K?

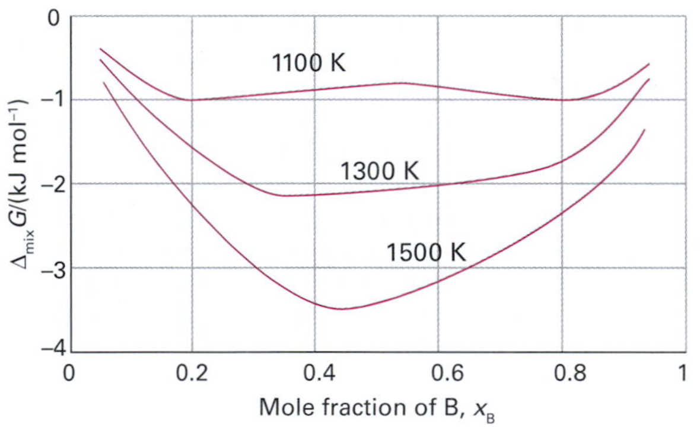

Figure 5.1 The Gibbs energy of mixing of copper and lead.

5B.8 The excess Gibbs energy of a certain binary mixture is equal to $gRTx(1-x)$ where g is a constant and x is the mole fraction of a solute A. Find an expression for the chemical potential of A in the mixture and sketch its dependence on the composition.

5B.9 Use the Gibbs–Helmholtz equation to find an expression for $\mathrm{d}\ln x_A$ in terms of $\mathrm{d}T$. Integrate $\mathrm{d}\ln x_A$ from $x_A = 0$ to the value of interest, and integrate the right-hand side from the transition temperature for the pure liquid A to the value in the solution. Show that, if the enthalpy of transition is constant, then eqns 5B.9 and 5B.12 are obtained.

5B.10‡ Polymer scientists often report their data in a variety of units. For example, in the determination of molar masses of polymers in solution by osmometry, osmotic pressures are often reported in grams per square centimetre ($\mathrm{g\,cm^{-2}}$) and concentrations in grams per cubic centimetre ($\mathrm{g\,cm^{-3}}$). (a) With these choices of units, what would be the units of R in the van 't Hoff equation? (b) The data in the table below on the concentration dependence of the osmotic pressure of polyisobutene in chlorobenzene at 25 °C have been adapted from J. Leonard and H. Daoust (*J. Polymer Sci.* **57**, 53 (1962)). From these data, determine the molar mass of polyisobutene by plotting Π/c against c. (c) 'Theta solvents' are solvents for which the second osmotic coefficient is zero; for 'poor' solvents the plot is linear and for good solvents the plot is nonlinear. From your plot, how would you classify chlorobenzene as a solvent for polyisobutene? Rationalize the result in terms of the molecular structure of the polymer and solvent. (d) Determine the second and third osmotic virial coefficients by fitting the curve to the virial form of the osmotic pressure equation. (e) Experimentally, it is often found that the virial expansion can be represented as

$$\Pi/c = RT/M(1 + B_c' + gB'^2c^2 + \ldots)$$

and in good solvents, the parameter g is often about 0.25. With terms beyond the second power ignored, obtain an equation for $(\Pi/c)^{1/2}$ and plot this quantity against c. Determine the second and third virial coefficients from the plot and compare to the values from the first plot. Does this plot confirm the assumed value of g?

$10^{-2}(\Pi/c)/(\mathrm{g\,cm^{-2}/g\,cm^{-3}})$	2.6	2.9	3.6	4.3	6.0	12.0
$c/(\mathrm{g\,cm^{-3}})$	0.0050	0.010	0.020	0.033	0.057	0.10

$10^{-2}(\Pi/c/(\mathrm{g\,cm^{-2}//g\,cm^{-3}})$	19.0	31.0	38.0	52	63
$c/(\mathrm{g\,cm^{-3}})$	0.145	0.195	0.245	0.27	0.29

5B.11‡ K. Sato, F.R. Eirich, and J.E. Mark (*J. Polymer Sci., Polym. Phys.* **14**, 619 (1976)) have reported the data in the table below for the osmotic pressures of polychloroprene ($\rho = 1.25\,\mathrm{g\,cm^{-3}}$) in toluene ($\rho = 0.858\,\mathrm{g\,cm^{-3}}$) at 30 °C. Determine the molar mass of polychloroprene and its second osmotic virial coefficient.

$c/(\mathrm{mg\,cm^{-3}})$	1.33	2.10	4.52	7.18	9.87
$\Pi/(\mathrm{N\,m^{-2}})$	30	51	132	246	390

5B.12 Use mathematical software, a spreadsheet, or the *Living graphs* on the web site for this book to draw graphs of $\Delta_{\mathrm{mix}}G$ against x_A at different temperatures in the range 298 K to 500 K. For what value of x_A does $\Delta_{\mathrm{mix}}G$ depend on temperature most strongly?

5B.13 Using the graph in Fig. 5B.4, fix ξ and vary the temperature. For what value of x_A does the excess enthalpy depend on temperature most strongly?

5B.14 Derive an expression for the temperature coefficient of the solubility, $\mathrm{d}x_B/\mathrm{d}T$, and plot it as a function of temperature for several values of the enthalpy of fusion.

5B.15 Calculate the osmotic virial coefficient B from the data in *Example* 5B.2.

TOPIC 5C Phase diagrams of binary systems

Discussion questions

5C.1 Draw phase diagrams for the following types of systems. Label the regions of the diagrams, stating what materials (possibly compounds or azeotropes) are present and whether they are solid liquid or gas:

(a) two-component, temperature–composition, solid–liquid diagram, one compound AB formed that melts congruently, negligible solid–solid solubility;

(b) two-component, temperature–composition, solid–liquid diagram, one compound of formula AB_2 that melts incongruently, negligible solid–solid solubility;
(c) two-component, constant temperature–composition, liquid–vapour diagram, formation of an azeotrope at $x_B=0.333$, complete miscibility.

5C.2 What molecular features determine whether a mixture of two liquids will show high- and low-boiling azeotropic behaviour?

5C.3 What factors determine the number of theoretical plates required to achieve a desired degree of separation in fractional distillation?

Exercises

5C.1(a) The following temperature–composition data were obtained for a mixture of octane (O) and methylbenzene (M) at 1.00 atm, where x is the mole fraction in the liquid and y the mole fraction in the vapour at equilibrium.

θ/°C	110.9	112.0	114.0	115.8	117.3	119.0	121.1	123.0
x_M	0.908	0.795	0.615	0.527	0.408	0.300	0.203	0.097
y_M	0.923	0.836	0.698	0.624	0.527	0.410	0.297	0.164

The boiling points are 110.6 °C and 125.6 °C for M and O, respectively. Plot the temperature–composition diagram for the mixture. What is the composition of the vapour in equilibrium with the liquid of composition (i) $x_M=0.250$ and (ii) $x_O=0.250$?

5C.1(b) The following temperature–composition data were obtained for a mixture of two liquids A and B at 1.00 atm, where x is the mole fraction in the liquid and y the mole fraction in the vapour at equilibrium.

θ/°C	125	130	135	140	145	150
x_A	0.91	0.65	0.45	0.30	0.18	0.098
y_A	0.99	0.91	0.77	0.61	0.45	0.25

The boiling points are 124 °C for A and 155 °C for B. Plot the temperature/composition diagram for the mixture. What is the composition of the vapour in equilibrium with the liquid of composition (i) $x_A=0.50$ and (ii) $x_B=0.33$?

5C.2(a) Methylethyl ether (A) and diborane, B_2H_6 (B), form a compound which melts congruently at 133 K. The system exhibits two eutectics, one at 25 mol per cent B and 123 K and a second at 90 mol per cent B and 104 K. The melting points of pure A and B are 131 K and 110 K, respectively. Sketch the phase diagram for this system. Assume negligible solid–solid solubility.

5C.2(b) Sketch the phase diagram of the system NH_3/N_2H_4 given that the two substances do not form a compound with each other, that NH_3 freezes at −78 °C and N_2H_4 freezes at +2 °C, and that a eutectic is formed when the mole fraction of N_2H_4 is 0.07 and that the eutectic melts at −80 °C.

5C.3(a) Figure 5.2 shows the phase diagram for two partially miscible liquids, which can be taken to be that for water (A) and 2-methyl-1-propanol (B). Describe what will be observed when a mixture of composition $x_B=0.8$ is heated, at each stage giving the number, composition, and relative amounts of the phases present.

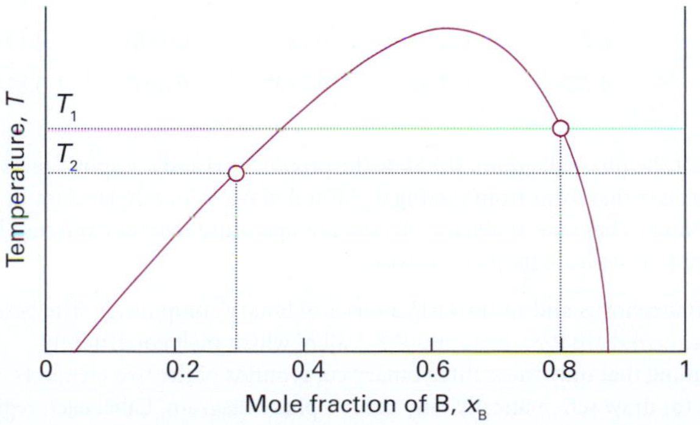

Figure 5.2 The phase diagram for two partially miscible liquids.

5C.3(b) Refer to Fig. 5.2 again. Describe what will be observed when a mixture of composition $x_B=0.3$ is heated, at each stage giving the number, composition, and relative amounts of the phases present.

5C.4(a) Indicate on the phase diagram in Fig. 5.3 the feature that denotes incongruent melting. What is the composition of the eutectic mixture and at what temperature does it melt?

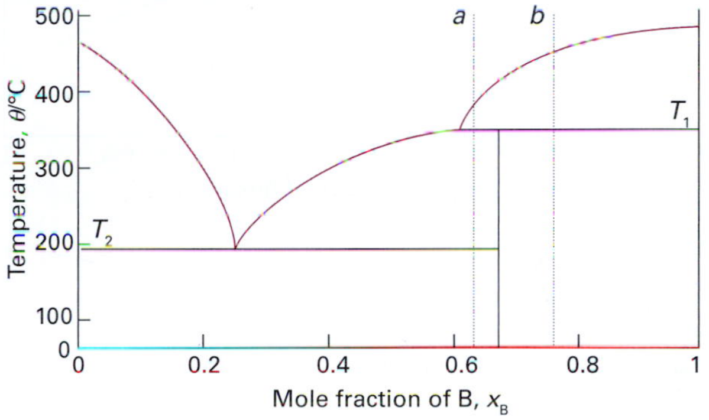

Figure 5.3 The phase diagram referred to in Exercise 5C.4(a).

5C.4(b) Indicate on the phase diagram in Fig. 5.4 the feature that denotes incongruent melting. What are the compositions of any eutectic mixtures and at what temperatures do they melt?

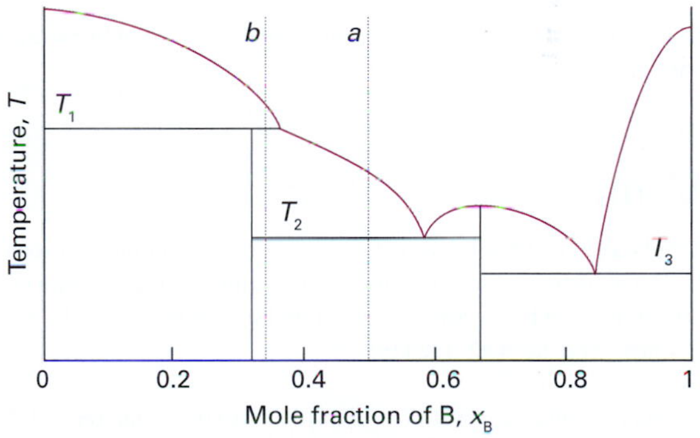

Figure 5.4 The phase diagram referred to in Exercise 5C.4(b).

5C.5(a) Sketch the cooling curves for the isopleths a and b in Fig. 5.3.

5C.5(b) Sketch the cooling curves for the isopleths a and b in Fig. 5.4.

5C.6(a) Use the phase diagram in Fig. 5.3 to state (i) the solubility of Ag in Sn at 800 °C and (ii) the solubility of Ag_3Sn in Ag at 460 °C, (iii) the solubility of Ag_3Sn in Ag at 300 °C.

5C.6(b) Use the phase diagram in Fig. 5.3 to state (i) the solubility of B in A at 500 °C and (ii) the solubility of B in A at 390 °C, (iii) the solubility of AB_2 in B at 300 °C.

5C.7(a) Figure 5.5 shows the experimentally determined phase diagrams for the nearly ideal solution of hexane and heptane. (i) Label the regions of the diagrams to which phases are present. (ii) For a solution containing 1 mol each of hexane and heptane molecules, estimate the vapour pressure at 70 °C when vaporization on reduction of the external pressure just begins. (iii) What is the vapour pressure of the solution at 70 °C when just one drop of liquid remains. (iv) Estimate from the figures the mole fraction of hexane in the liquid and vapour phases for the conditions of part b. (v) What are the mole fractions for the conditions of part c? (vi) At 85 °C and 760 Torr, what are the amounts substance in the liquid and vapour phases when $z_{heptane} = 0.40$?

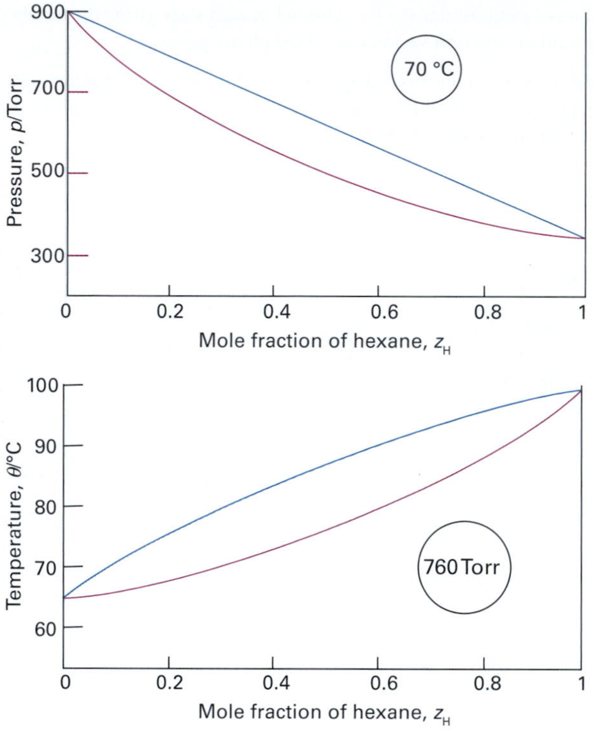

Figure 5.5 Phase diagrams for the nearly ideal solution of hexane and heptane.

5C.7(b) Uranium tetrafluoride and zirconium tetrafluoride melt at 1035 °C and 912 °C respectively. They form a continuous series of solid solutions with a minimum melting temperature of 765 °C and composition $x(ZrF_4) = 0.77$. At 900 °C, the liquid solution of composition $x(ZrF_4) = 0.28$ is in equilibrium with a solid solution of composition $x(ZrF_4) = 0.14$. At 850 °C the two compositions are 0.87 and 0.90, respectively. Sketch the phase diagram for this system and state what is observed when a liquid of composition $x(ZrF_4) = 0.40$ is cooled slowly from 900 °C to 500 °C.

5C.8(a) Methane (melting point 91 K) and tetrafluoromethane (melting point 89 K) do not form solid solutions with each other, and as liquids they are only partially miscible. The upper critical temperature of the liquid mixture is 94 K at $x(CF_4) = 0.43$ and the eutectic temperature is 84 K at $x(CF_4) = 0.88$. At 86 K, the phase in equilibrium with the tetrafluoromethane-rich solution changes from solid methane to a methane-rich liquid. At that temperature, the two liquid solutions that are in mutual equilibrium have the compositions $x(CF_4) = 0.10$ and $x(CF_4) = 0.80$. Sketch the phase diagram.

5C.8(b) Describe the phase changes that take place when a liquid mixture of 4.0 mol B_2H_6 (melting point 131 K) and 1.0 mol CH_3OCH_3 (melting point 135 K) is cooled from 140 K to 90 K. These substances form a compound $(CH_3)_2OB_2H_6$ that melts congruently at 133 K. The system exhibits one eutectic at $x(B_2H_6) = 0.25$ and 123 K and another at $x(B_2H_6) = 0.90$ and 104 K.

5C.9(a) Refer to the information in Exercise 5C.8(a) and sketch the cooling curves for liquid mixtures in which $x(CF_4)$ is (i) 0.10, (ii) 0.30, (iii) 0.50, (iv) 0.80, and (v) 0.95.

5C.9(b) Refer to the information in Exercise 5C.8(b) and sketch the cooling curves for liquid mixtures in which $x(B_2H_6)$ is (i) 0.10, (ii) 0.30, (iii) 0.50, (iv) 0.80, and (v) 0.95.

5C.10(a) Hexane and perfluorohexane show partial miscibility below 22.70 °C. The critical concentration at the upper critical temperature is $x = 0.355$, where x is the mole fraction of C_6F_{14}. At 22.0 °C the two solutions in equilibrium have $x = 0.24$ and $x = 0.48$, respectively, and at 21.5 °C the mole fractions are 0.22 and 0.51. Sketch the phase diagram. Describe the phase changes that occur when perfluorohexane is added to a fixed amount of hexane at (i) 23 °C, (ii) 22 °C.

5C.10(b) Two liquids, A and B, show partial miscibility below 52.4 °C. The critical concentration at the upper critical temperature is $x = 0.459$, where x is the mole fraction of A. At 40.0 °C the two solutions in equilibrium have $x = 0.22$ and $x = 0.60$, respectively, and at 42.5 °C the mole fractions are 0.24 and 0.48. Sketch the phase diagram. Describe the phase changes that occur when B is added to a fixed amount of A at (i) 48 °C, (ii) 52.4 °C.

Problems

5C.1‡ 1-Butanol and chlorobenzene form a minimum-boiling azeotropic system. The mole fraction of 1-butanol in the liquid (x) and vapour (y) phases at 1.000 atm is given below for a variety of boiling temperatures (H. Artigas et al., *J. Chem. Eng. Data* **42**, 132 (1997)).

T/K	396.57	393.94	391.60	390.15	389.03	388.66	388.57
x	0.1065	0.1700	0.2646	0.3687	0.5017	0.6091	0.7171
y	0.2859	0.3691	0.4505	0.5138	0.5840	0.6409	0.7070

Pure chlorobenzene boils at 404.86 K. (a) Construct the chlorobenzene-rich portion of the phase diagram from the data. (b) Estimate the temperature at which a solution for which the mole fraction of 1-butanol is 0.300 begins to boil. (c) State the compositions and relative proportions of the two phases present after a solution initially 0.300 1-butanol is heated to 393.94 K.

5C.2‡ An, Zhao, Jiang, and Shen investigated the liquid–liquid coexistence curve of *N,N*-dimethylacetamide and heptane (X. An et al., *J. Chem. Thermodynamics* **28**, 1221 (1996)). Mole fractions of *N,N*-dimethylacetamide in the upper (x_1) and lower (x_2) phases of a two-phase region are given opposite as a function of temperature:

T/K	309.820	309.422	309.031	308.006	306.686
x_1	0.473	0.400	0.371	0.326	0.239
x_2	0.529	0.601	0.625	0.657	0.690

T/K	304.553	301.803	299.097	296.000	294.534
x_1	0.255	0.218	0.193	0.168	0.157
x_2	0.724	0.758	0.783	0.804	0.814

(a) Plot the phase diagram. (b) State the proportions and compositions of the two phases that form from mixing 0.750 mol of *N,N*-dimethylacetamide with 0.250 mol of heptane at 296.0 K. To what temperature must the mixture be heated to form a single-phase mixture?

5C.3 Phosphorus and sulfur form a series of binary compounds. The best characterized are P_4S_3, P_4S_7, and P_4S_{10}, all of which melt congruently. Assuming that only these three binary compounds of the two elements exist, (a) draw schematically only the P/S phase diagram. Label each region of the diagram with the substance that exists in that region and indicate its phase. Label the horizontal axis as x_S and give the numerical values of x_S that correspond to the compounds. The melting point of pure phosphorus is 44 °C and that of pure sulfur is 119 °C. (b) Draw, schematically, the cooling curve for a mixture of composition $x_S = 0.28$. Assume that a eutectic occurs at $x_S = 0.2$ and negligible solid–solid solubility.

5C.4 The following table gives the break and halt temperatures found in the cooling curves of two metals A and B. Construct a phase diagram consistent with the data of these curves. Label the regions of the diagram, stating what phases and substances are present. Give the probable formulas of any compounds that form.

$100x_B$	θ_{break}/°C	$\theta_{halt,1}$/°C	$\theta_{halt,2}$/°C
0		1100	
10.0	1060	700	
20.0	1000	700	
30.0	940	700	400
40.0	850	700	400
50.0	750	700	400
60.0	670	400	
70.0	550	400	
80.0		400	
90.0	450	400	
100.0		500	

5C.5 Consider the phase diagram in Fig. 5.6, which represents a solid–liquid equilibrium. Label all regions of the diagram according to the chemical species that exist in that region and their phases. Indicate the number of species and phases present at the points labelled *b, d, e, f, g,* and *k*. Sketch cooling curves for compositions $x_B = 0.16$, 0.23, 0.57, 0.67, and 0.84.

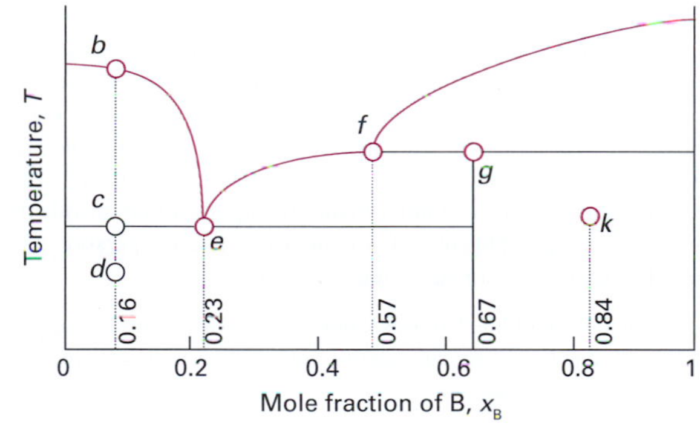

Figure 5.6 The phase diagram referred to in Problem 5C.5.

5C.6 Sketch the phase diagram for the Mg/Cu system using the following information: θ_f(Mg) = 648 °C, θ_f(Cu) = 1085 °C; two intermetallic compounds are formed with $\theta_f(MgCu_2)$ = 800 °C and $\theta_f(Mg_2Cu)$ = 580 °C; eutectics of mass percentage Mg composition and melting points 10 per cent (690 °C), 33 per cent (560 °C), and 65 per cent (380 °C). A sample of Mg/Cu alloy containing 25 per cent Mg by mass was prepared in a crucible heated to 800 °C in an inert atmosphere. Describe what will be observed if the melt is cooled slowly to room temperature. Specify the composition and relative abundances of the phases and sketch the cooling curve.

5C.7‡ The temperature–composition diagram for the Ca/Si binary system is shown in Fig. 5.7. (a) Identify eutectics, congruent melting compounds, and incongruent melting compounds. (b) If a 20 per cent by atom composition melt of silicon at 1500 °C is cooled to 1000 °C, what phases (and phase composition) would be at equilibrium? Estimate the relative amounts of each phase. (c) Describe the equilibrium phases observed when an 80 per cent by atom composition Si melt is cooled to 1030 °C. What phases, and relative amounts, would be at equilibrium at a temperature (i) slightly higher than 1030 °C, (ii) slightly lower than 1030 °C? Draw a graph of the mole percentages of both Si(s) and $CaSi_2$(s) as a function of mole percentage of melt that is freezing at 1030 °C.

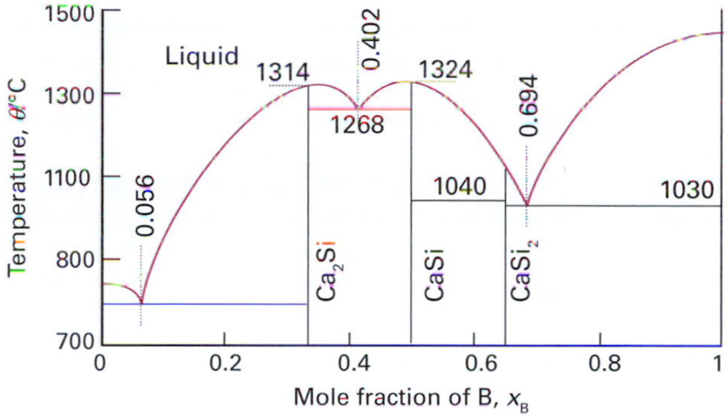

Figure 5.7 The temperature–composition diagram for the Ca/Si binary system.

5C.8 Iron(II) chloride (melting point 677 °C) and potassium chloride (melting point 776 °C) form the compounds $KFeCl_3$ and K_2FeCl_4 at elevated temperatures. $KFeCl_3$ melts congruently at 380 °C and K_2FeCl_4 melts incongruently at 399 °C. Eutectics are formed with compositions $x = 0.38$ (melting point 351 °C) and $x = 0.54$ (melting point 393 °C), where x is the mole fraction of $FeCl_2$. The KCl solubility curve intersects the K_2FeCl_4 curve at $x = 0.34$. Sketch the phase diagram. State the phases that are in equilibrium when a mixture of composition $x = 0.36$ is cooled from 400 °C to 300 °C.

5C.9 To reproduce the results of Fig. 5C.2, first rearrange eqn 5C.4 so that y_A is expressed as a function of x_A and the ratio p_A^*/p_B^*. Then plot y_A against x_A for several values of ratio $p_A^*/p_B^* > 1$.

5C.10 To reproduce the results of Fig. 5C.3, first rearrange eqn 5C.5 so that the ratio p_A^*/p_B^* is expressed as a function of y_A and the ratio p_A^*/p_B^*. Then plot p_A/p_A^* against y_A for several values of $p_A^*/p_B^* > 1$.

5C.11 Working from eqn 5B.7, write an expression for T_{min}, the temperature at which $\Delta_{mix}G$ has a minimum, as a function of ξ and x_A. Then, plot T_{min} against x_A for several values of ξ. Provide a physical interpretation for any maxima or minima that you observe in these plots.

5C.12 Use eqn 5C.7 to generate plots of ξ against x_A by one of two methods: (a) solve the transcendental equation $\ln\{(x/(1-x)\} + \xi(1-2x) = 0$ numerically, or (b) plot the first term of the transcendental equation against the second and identify the points of intersection as ξ is changed.

TOPIC 5D Phase diagrams of ternary systems

Discussion questions

5D.1 What is the maximum number of phases that can be in equilibrium in a ternary system?

5D.2 Does the lever rule apply to a ternary system?

5D.3 Could a regular tetrahedron be used to depict the properties of a four-component system?

Exercises

5D.1(a) Mark the following features on triangular coordinates: (i) the point (0.2, 0.2, 0.6), (ii) the point (0, 0.2, 0.8), (iii) the point at which all three mole fractions are the same.

5D.1(b) Mark the following features on triangular coordinates: (i) the point (0.6, 0.2, 0.2), (ii) the point (0.8, 0.2, 0), (iii) the point (0.25, 0.25, 0.50).

5D.2(a) Mark the following points on a ternary phase diagram for the system $NaCl/Na_2SO_4{\cdot}10H_2O/H_2O$: (i) 25 per cent by mass NaCl, 25 per cent $Na_2SO_4{\cdot}10H_2O$, and the rest H_2O; (ii) the line denoting the same relative composition of the two salts but with changing amounts of water.

5D.2(b) Mark the following points on a ternary phase diagram for the system $NaCl/Na_2SO_4{\cdot}10H_2O/H_2O$: (i) 33 per cent by mass NaCl, 33 per cent $Na_2SO_4{\cdot}10H_2O$, and the rest H_2O; (ii) the line denoting the same relative composition of the two salts but with changing amounts of water.

5D.3(a) Refer to the ternary phase diagram in Fig. 5D.4. How many phases are present, and what are their compositions and relative abundances, in a mixture that contains 2.3 g of water, 9.2 g of chloroform, and 3.1 g of acetic acid? Describe what happens when (i) water, (iii) acetic acid is added to the mixture.

5D.3(b) Refer to the ternary phase diagram in Fig. 5D.4. How many phases are present, and what are their compositions and relative abundances, in a mixture that contains 55.0 g of water, 8.8 g of chloroform, and 3.7 g of acetic acid? Describe what happens when (i) water, (ii) acetic acid is added to the mixture.

5D.4(a) Figure 5.8 shows the phase diagram for the ternary system $NH_4Cl/(NH_4)_2SO_4/H_2O$ at 25 °C. Identify the number of phases present for mixtures of compositions (i) (0.2, 0.4, 0.4), (ii) (0.4, 0.4, 0.2), (iii) (0.2, 0.1, 0.7), (iv) (0.4, 0.16, 0.44). The numbers are mole fractions of the three components in the order (NH_4Cl,$(NH_4)_2SO_4$,H_2O).

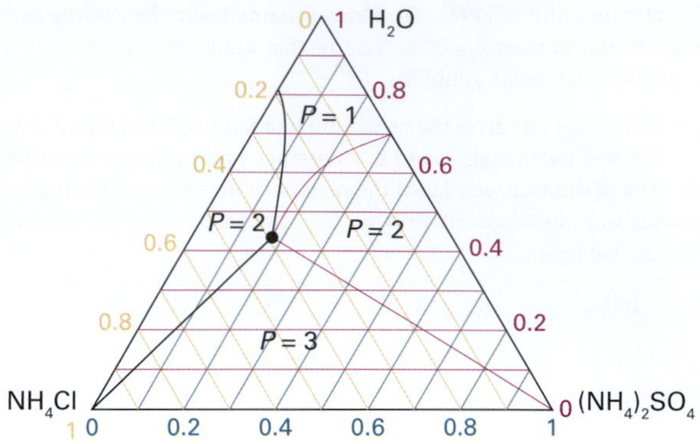

Figure 5.8 The phase diagram for the ternary system $NH_4Cl/(NH_4)_2SO_4/H_2O$ at 25 °C.

5D.4(b) Refer to Fig. 5.8 and identify the number of phases present for mixtures of compositions (i) (0.4, 0.1, 0.5), (ii) (0.8, 0.1, 0.1), (iii) (0, 0.3,0.7), (iv) (0.33, 0.33, 0.34). The numbers are mole fractions of the three components in the order (NH_4Cl,$(NH_4)_2SO_4$,H_2O).

5D.5(a) Referring to Fig. 5.8, deduce the molar solubility of (i) NH_4Cl, (ii) $(NH_4)_2SO_4$ in water at 25 °C.

5D.5(b) Describe what happens when (i) $(NH_4)_2SO_4$ is added to a saturated solution of NH_4Cl in water in the presence of excess NH_4Cl, (ii) water is added to a mixture of 25 g of NH_4Cl and 75 g of $(NH_4)_2SO_4$.

Problems

5D.1 At a certain temperature, the solubility of I_2 in liquid CO_2 is $x(I_2) = 0.03$. At the same temperature its solubility in nitrobenzene is 0.04. Liquid carbon dioxide and nitrobenzene are miscible in all proportions, and the solubility of I_2 in the mixture varies linearly with the proportion of nitrobenzene. Sketch a phase diagram for the ternary system.

5D.2 The binary system nitroethane/decahydronaphthalene (DEC) shows partial miscibility, with the two-phase region lying between $x = 0.08$ and $x = 0.84$, where x is the mole fraction of nitroethane. The binary system liquid carbon dioxide/DEC is also partially miscible, with its two-phase region lying between $y = 0.36$ and $y = 0.80$, where y is the mole fraction of DEC. Nitroethane and liquid carbon dioxide are miscible in all proportions. The addition of liquid carbon dioxide to mixtures of nitroethane and DEC increases the range of miscibility, and the plait point is reached when z, the mole fraction of CO_2, is 0.18 and $x = 0.53$. The addition of nitroethane to mixtures of carbon dioxide and DEC also results in another plait point at $x = 0.08$ and $y = 0.52$. (a) Sketch the phase diagram for the ternary system, (b) For some binary mixtures of nitroethane and liquid carbon dioxide the addition of arbitrary amounts of DEC will not cause phase separation. Find the range of concentration for such binary mixtures.

5D.3 Prove that a straight line from the apex A of a ternary phase diagram to the opposite edge BC represents mixtures of constant ratio of B and C, however much A is present.

TOPIC 5E Activities

Discussion questions

5E.1 What are the contributions that account for the difference between activity and concentration?

5E.2 How is Raoult's law modified so as to describe the vapour pressure of real solutions?

5E.3 Summarize the ways in which activities may be measured.

Exercises

5E.1(a) Substances A and B are both volatile liquids with $p_A^* = 300$ Torr, $p_B^* = 250$ Torr, and $K_B = 200$ Torr (concentration expressed in mole fraction). When $x_A = 0.9$, $b_B = 2.22\ \mathrm{mol\,kg^{-1}}$, $p_A = 250$ Torr, and $p_B = 25$ Torr. Calculate the activities and activity coefficients of A and B. Use the mole fraction, Raoult's law basis system for A and the Henry's law basis system (both mole fractions and molalities) for B.

5E.1(b) Given that $p^*(H_2O)=0.02308$ atm and $p(H_2O)=0.02239$ atm in a solution in which 0.122 kg of a involatile solute ($M=241\ g\ mol^{-1}$) is dissolved in 0.920 kg water at 293 K, calculate the activity and activity coefficient of water in the solution.

5E.2(a) By measuring the equilibrium between liquid and vapour phases of an acetone(A)/methanol(M) solution at 57.2 °C at 1.00 atm, it was found that $x_A=0.400$ when $y_A=0.516$. Calculate the activities and activity coefficients of both components in this solution on the Raoult's law basis. The vapour pressures of the pure components at this temperature are: $p_A^*=105$ kPa and $p_M^*=73.5$ kPa. (x_A is the mole fraction in the liquid and y_A the mole fraction in the vapour.)

5E.2(b) By measuring the equilibrium between liquid and vapour phases of a solution at 30 °C at 1.00 atm, it was found that $x_A=0.220$ when $y_A=0.314$. Calculate the activities and activity coefficients of both components in this solution on the Raoult's law basis. The vapour pressures of the pure components at this temperature are: $p_A^*=73.0$ kPa and $p_B^*=92.1$ kPa. (x_A is the mole fraction in the liquid and y_A the mole fraction in the vapour.)

5E.3(a) Find the relation between the standard and biological standard Gibbs energies of a reaction of the form $A \rightarrow 2B + 2H^+(aq)$.

5E.3(b) Find the relation between the standard and biological standard Gibbs energies of a reaction of the form $2A \rightarrow B + 4H^+(aq)$.

5E.4(a) Suppose it is found that for a hypothetical regular solution that $\xi=1.40$, $p_A^*=15.0$ kPa and $p_B^*=11.6$ kPa. Draw the vapour-pressure diagram.

5E.4(b) Suppose it is found that for a hypothetical regular solution that $\xi=-1.40$, $p_A^*=15.0$ kPa and $p_B^*=11.6$ kPa. Draw the vapour-pressure diagram.

Problems

5E.1‡ Francesconi, Lunelli, and Comelli studied the liquid–vapour equilibria of trichloromethane and 1,2-epoxybutane at several temperatures (Francesconi et al., *J. Chem. Eng. Data* **41**, 310 (1996)). Among their data are the following measurements of the mole fractions of trichloromethane in the liquid phase (x_T) and the vapour phase (y_T) at 298.15 K as a function of pressure.

p/kPa	23.40	21.75	20.25	18.75	18.15	20.25	22.50	26.30
x	0	0.129	0.228	0.353	0.511	0.700	0.810	1
y	0	0.065	0.145	0.285	0.535	0.805	0.915	1

Compute the activity coefficients of both components on the basis of Raoult's law.

5E.2 The *osmotic coefficient* ϕ is defined as $\phi=-(x_A/x_B)\ln a_A$. By writing $r=x_B/x_A$, and using the Gibbs–Duhem equation, show that we can calculate the activity of B from the activities of A over a composition range by using the formula

$$\ln\frac{a_B}{r}=\phi-\phi(0)+\int_0^r\frac{\phi-1}{r}\,dr$$

5E.3 Show that the osmotic pressure of a real solution is given by $\Pi V=-RT\ln a_A$. Go on to show that, provided the concentration of the solution is low, this expression takes the form $\Pi V=\phi RT[B]$ and hence that the osmotic coefficient ϕ (which is defined in Problem 5E.2) may be determined from osmometry.

5E.4 Use mathematical software, a spreadsheet, or the *Living graphs* on the web site for this book to plot p_A/p_A^* against x_A with $\xi=2.5$ by using eqn 5E.19 and then eqn 5E.20. Above what value of x_A do the values of p_A/p_A^* given by these equations differ by more than 10 per cent?

TOPIC 5F The activities of ions

Discussion questions

5F.1 Why do the activity coefficients of ions in solution differ from 1? Why are they less than 1 in dilute solutions?

5F.2 Describe the general features of the Debye–Hückel theory of electrolyte solutions.

5F.3 Suggest an interpretation of the additional terms in extended versions of the Debye–Hückel limiting law.

Exercises

5F.1(a) Calculate the ionic strength of a solution that is 0.10 mol kg^{-1} in KCl(aq) and 0.20 mol kg^{-1} in $CuSO_4$(aq).

5F.1(b) Calculate the ionic strength of a solution that is 0.040 mol kg^{-1} in $K_3[Fe(CN)_6]$(aq), 0.030 mol kg^{-1} in KCl(aq), and 0.050 mol kg^{-1} in NaBr(aq).

5F.2(a) Calculate the masses of (i) $Ca(NO_3)_2$ and, separately, (ii) NaCl to add to a 0.150 mol kg^{-1} solution of KNO_3(aq) containing 500 g of solvent to raise its ionic strength to 0.250.

5F.2(b) Calculate the masses of (i) KNO_3 and, separately, (ii) $Ba(NO_3)_2$ to add to a 0.110 mol kg^{-1} solution of KNO_3(aq) containing 500 g of solvent to raise its ionic strength to 1.00.

5F.3(a) Estimate the mean ionic activity coefficient and activity of a solution at 25 °C that is 0.010 mol kg^{-1} $CaCl_2$(aq) and 0.030 mol kg^{-1} NaF(aq).

5F.3(b) Estimate the mean ionic activity coefficient and activity of a solution at 25 °C that is 0.020 mol kg^{-1} NaCl(aq) and 0.035 mol kg^{-1} $Ca(NO_3)_2$(aq).

5F.4(a) The mean activity coefficients of HBr in three dilute aqueous solutions at 25 °C are 0.930 (at 5.0 mmol kg^{-1}), 0.907 (at 10.0 mmol kg^{-1}), and 0.879 (at 20.0 mmol kg^{-1}). Estimate the value of B in eqn 5F.11a.

5F.4(b) The mean activity coefficients of KCl in three dilute aqueous solutions at 25 °C are 0.927 (at 5.0 mmol kg^{-1}), 0.902 (at 10.0 mmol kg^{-1}), and 0.816 (at 50.0 mmol kg^{-1}). Estimate the value of B in eqn 5F.11a.

Problems

5F.1 The mean activity coefficients for aqueous solutions of NaCl at 25 °C are given opposite. Confirm that they support the Debye–Hückel limiting law and that an improved fit is obtained with the Davies equation.

b/(mmol kg^{-1})	1.0	2.0	5.0	10.0	20.0
$\gamma_\pm$	0.9649	0.9519	0.9275	0.9024	0.8712

5F.2 Consider the plot of log $\gamma_\pm$ against $I^{1/2}$ with $B=1.50$ and $C=0$ in the Davies equation as a representation of experimental data for a certain MX electrolyte. Over what range of ionic strengths does the application of the limiting law lead to an error in the value of the activity coefficient of less than 10 per cent of the value predicted by the extended law?

Integrated activities

5.1 The table below lists the vapour pressures of mixtures of iodoethane (I) and ethyl acetate (A) at 50 °C. Find the activity coefficients of both components on (a) the Raoult's law basis, (b) the Henry's law basis with iodoethane as solute.

x_I	0	0.0579	0.1095	0.1918	0.2353	0.3718
p_I/kPa	0	3.73	7.03	11.7	14.05	20.72
p_A/kPa	37.38	35.48	33.64	30.85	29.44	25.05

x_I	0.5478	0.6349	0.8253	0.9093	1.0000
p_I/kPa	28.44	31.88	39.58	43.00	47.12
p_A/kPa	19.23	16.39	8.88	5.09	0

5.2 Plot the vapour pressure data for a mixture of benzene (B) and acetic acid (A) given below and plot the vapour pressure/composition curve for the mixture at 50 °C. Then confirm that Raoult's and Henry's laws are obeyed in the appropriate regions. Deduce the activities and activity coefficients of the components on the Raoult's law basis and then, taking B as the solute, its activity and activity coefficient on a Henry's law basis. Finally, evaluate the excess Gibbs energy of the mixture over the composition range spanned by the data.

x_A	0.0160	0.0439	0.0835	0.1138	0.1714
p_A/kPa	0.484	0.967	1.535	1.89	2.45
p_B/kPa	35.05	34.29	33.28	32.64	30.90

x_A	0.2973	0.3696	0.5834	0.6604	0.8437	0.9931
p_A/kPa	3.31	3.83	4.84	5.36	6.76	7.29
p_B/kPa	28.16	26.08	20.42	18.01	10.0	0.47

5.3‡ Chen and Lee studied the liquid–vapour equilibria of cyclohexanol with several gases at elevated pressures (J.-T. Chen and M.-J. Lee, *J. Chem. Eng. Data* **41**, 339 (1996)). Among their data are the following measurements of the mole fractions of cyclohexanol in the vapour phase (y) and the liquid phase (x) at 393.15 K as a function of pressure.

p/bar	10.0	20.0	30.0	40.0	60.0	80.0
y_{cyc}	0.0267	0.0149	0.0112	0.009 47	0.008 35	0.009 21
x_{cyc}	0.9741	0.9464	0.9204	0.892	0.836	0.773

Determine the Henry's law constant of CO_2 in cyclohexanol, and compute the activity coefficient of CO_2.

5.4‡ The following data have been obtained for the liquid–vapour equilibrium compositions of mixtures of nitrogen and oxygen at 100 kPa.

T/K	77.3	78	80	82	84	86	88	90.2
$x(O_2)$	0	10	34	54	70	82	92	100
$y(O_2)$	0	2	11	22	35	52	73	100
$p^*(O_2)$/Torr	154	171	225	294	377	479	601	760

Plot the data on a temperature–composition diagram and determine the extent to which it fits the predictions for an ideal solution by calculating the activity coefficients of O_2 at each composition.

5.5 Use the Gibbs–Duhem equation to derive the *Gibbs–Duhem–Margules equation*

$$\left(\frac{\partial \ln f_A}{\partial \ln x_A}\right)_{p,T} = \left(\frac{\partial \ln f_B}{\partial \ln x_B}\right)_{p,T}$$

where f is the fugacity. Use the relation to show that when the fugacities are replaced by pressures, that if Raoult's law applies to one component in a mixture it must also apply to the other.

5.6 Use the Gibbs–Duhem equation to show that the partial molar volume (or any partial molar property) of a component B can be obtained if the partial molar volume (or other property) of A is known for all compositions up to the one of interest. Do this by proving that

$$V_B = V_B^* - \int_{V_A^*}^{V_A} \frac{x_A}{1-x_A} \mathrm{d}V_A$$

where the x_A are functions of the V_A. Use the following data (which are for 298 K) to evaluate the integral graphically to find the partial molar volume of acetone at $x = 0.500$.

$x(CHCl_3)$	0	0.194	0.385	0.559	0.788	0.889	1.000
$V_m/(cm^3\,mol^{-1})$	73.99	75.29	76.50	77.55	79.08	79.82	80.67

5.7 Show that the freezing-point depression of a real solution in which the solvent of molar mass M has activity a_A obeys

$$\frac{\mathrm{d}\ln a_A}{\mathrm{d}(\Delta T)} = -\frac{M}{K_f}$$

and use the Gibbs–Duhem equation to show that

$$\frac{\mathrm{d}\ln a_B}{\mathrm{d}(\Delta T)} = -\frac{1}{b_B K_f}$$

where a_B is the solute activity and b_B is its molality. Use the Debye–Hückel limiting law to show that the osmotic coefficient (ϕ, Problem 5E.2) is given by $\phi = 1 - \frac{1}{3}A'I$ with $A' = 2.303A$ and $I = b/b^\ominus$.

5.8 For the calculation of the solubility c of a gas in a solvent, it is often convenient to use the expression $c = Kp$, where K is the Henry's law constant. Breathing air at high pressures, such as in scuba diving, results in an increased concentration of dissolved nitrogen. The Henry's law constant for the solubility of nitrogen is 0.18 μg/(g H_2O atm). What mass of nitrogen is dissolved in 100 g of water saturated with air at 4.0 atm and 20 °C? Compare your answer to that for 100 g of water saturated with air at 1.0 atm. (Air is 78.08 mole per cent N_2.) If nitrogen is four times as soluble in fatty tissues as in water, what is the increase in nitrogen concentration in fatty tissue in going from 1 atm to 4 atm?

5.9 Dialysis may be used to study the binding of small molecules to macromolecules, such as an inhibitor to an enzyme, an antibiotic to DNA, and any other instance of cooperation or inhibition by small molecules attaching to large ones. To see how this is possible, suppose inside the dialysis bag the molar concentration of the macromolecule M is [M] and the total concentration of small molecule A is $[A]_{in}$. This total concentration is the sum of the concentrations of free A and bound A, which we write $[A]_{free}$ and $[A]_{bound}$, respectively. At equilibrium, $\mu_{A,free} = \mu_{A,out}$, which implies that $[A]_{free} = [A]_{out}$, provided the activity coefficient of A is the same in both solutions. Therefore, by measuring the concentration of A in the solution outside the bag, we can find the concentration of unbound A in the macromolecule solution and, from the difference $[A]_{in} - [A]_{free} = [A]_{in} - [A]_{out}$, the concentration of bound A. Now we explore the quantitative consequences of the experimental arrangement just described.

(a) The average number of A molecules bound to M molecules, ν, is

$$\nu = \frac{[A]_{bound}}{[M]} = \frac{[A]_{in} - [A]_{out}}{[M]}$$

The bound and unbound A molecules are in equilibrium, $M + A \rightleftharpoons MA$. Recall from introductory chemistry that we may write the equilibrium constant for binding, *K*, as

$$K = \frac{[MA]}{[M]_{free}[A]_{free}}$$

Now show that

$$K = \frac{\nu}{(1-\nu)[A]_{out}}$$

(b) If there are *N identical* and *independent* binding sites on each macromolecule, each macromolecule behaves like *N* separate smaller macromolecules, with the same value of *K* for each site. It follows that the average number of A molecules per site is ν/N. Show that, in this case, we may write the *Scatchard equation*:

$$\frac{\nu}{[A]_{out}} = KN - K\nu$$

(c) To apply the Scatchard equation, consider the binding of ethidium bromide (E^-) to a short piece of DNA by a process called *intercalation*, in which the aromatic ethidium cation fits between two adjacent DNA base pairs. An equilibrium dialysis experiment was used to study the binding of ethidium bromide (EB) to a short piece of DNA. A 1.00 μmol dm^{-3} aqueous solution of the DNA sample was dialysed against an excess of EB. The following data were obtained for the total concentration of EB:

[EB]/(μmol dm^{-3})					
Side without DNA	0.042	0.092	0.204	0.526	1.150
Side with DNA	0.292	0.590	1.204	2.531	4.150

From these data, make a Scatchard plot and evaluate the intrinsic equilibrium constant, *K*, and total number of sites per DNA molecule. Is the identical and independent sites model for binding applicable?

5.10 The form of the Scatchard equation given Problem 5.9 applies only when the macromolecule has identical and independent binding sites. For non-identical independent binding sites, the Scatchard equation is

$$\frac{\nu}{[A]_{out}} = \sum_i \frac{N_i K_i}{1 + K_i[A]_{out}}$$

Plot $\nu/[A]$ for the following cases. (a) There are four independent sites on an enzyme molecule and the intrinsic binding constant is $K = 1.0 \times 10^7$. (b) There are a total of six sites per polymer. Four of the sites are identical and have an intrinsic binding constant of 1×10^5. The binding constants for the other two sites are 2×10^6.

5.11 The addition of a small amount of a salt, such as $(NH_4)_2SO_4$, to a solution containing a charged protein increases the solubility of the protein in water. This observation is called the *salting-in effect*. However, the addition of large amounts of salt can decrease the solubility of the protein to such an extent that the protein precipitates from solution. This observation is called the *salting-out effect* and is used widely by biochemists to isolate and purify proteins. Consider the equilibrium $PX_\nu(s) \rightleftharpoons P^{\nu+}(aq) + \nu\, X^-(aq)$, where $P^{\nu+}$ is a polycationic protein of charge $\nu+$ and X^- is its counter ion. Use Le Chatelier's principle and the physical principles behind the Debye–Hückel theory to provide a molecular interpretation for the salting-in and salting-out effects.

5.12 Some polymers can form liquid crystal mesophases with unusual physical properties. For example, liquid crystalline Kevlar (1) is strong enough to be the material of choice for bulletproof vests and is stable at temperatures up to 600 K. What molecular interactions contribute to the formation, thermal stability, and mechanical strength of liquid crystal mesophases in Kevlar?

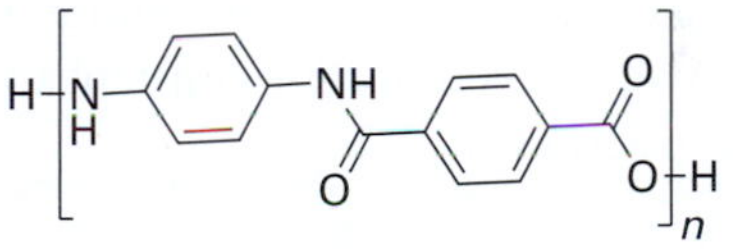

1 Kevlar

CHAPTER 6

Chemical equilibrium

Chemical reactions tend to move towards a dynamic equilibrium in which both reactants and products are present but have no further tendency to undergo net change. In some cases, the concentration of products in the equilibrium mixture is so much greater than that of the unchanged reactants that for all practical purposes the reaction is 'complete'. However, in many important cases the equilibrium mixture has significant concentrations of both reactants and products.

6A The equilibrium constant

This Topic develops the concept of chemical potential and shows how it is used to account for the equilibrium composition of chemical reactions. The equilibrium composition corresponds to a minimum in the Gibbs energy plotted against the extent of reaction. By locating this minimum we establish the relation between the equilibrium constant and the standard Gibbs energy of reaction.

6B The response of equilibria to the conditions

The thermodynamic formulation of equilibrium enables us to establish the quantitative effects of changes in the conditions. One very important aspect of equilibrium is the control that can be exercised by varying the conditions, such as the pressure or temperature.

6C Electrochemical cells

Because many reactions involve the transfer of electrons, they can be studied (and utilized) by allowing them to take place in a cell equipped with electrodes, with the spontaneous reaction forcing electrons through an external circuit. We shall see that the electric potential of the cell is related to the reaction Gibbs energy, so providing an electrical procedure for the determination of thermodynamic quantities.

6D Electrode potentials

Electrochemistry is in part a major application of thermodynamic concepts to chemical equilibria as well as being of great technological importance. As elsewhere in thermodynamics, we see how to report electrochemical data in a compact form and apply it to problems of real chemical significance, especially to the prediction of the spontaneous direction of reactions and the calculation of equilibrium constants.

What is the impact of this material?

The thermodynamic description of spontaneous reactions has numerous practical and theoretical applications. We highlight two applications. One is to the discussion of biochemical processes, where one reaction drives another (*Impact* I6.1). That, ultimately, is why we have to eat, for we see that the reaction that takes place when one substance is oxidized can drive non-spontaneous reactions, such as protein synthesis, forward. Another makes use of the great sensitivity of electrochemical processes to the concentration of electroactive materials, and we see how specially designed electrodes are used in analysis (*Impact* I6.2).

To read more about the impact of this material, scan the QR code, or go to bcs.whfreeman.com/webpub/chemistry/pchem10e/impact/pchem-6-1.html

6A The equilibrium constant

Contents

➤ **Why do you need to know this material?**

Equilibrium constants lie at the heart of chemistry and are a key point of contact between thermodynamics and laboratory chemistry. The material in this Topic shows how they arise and explains the thermodynamic properties that determine their values.

➤ **What is the key idea?**

The composition of a reaction mixture tends to change until the Gibbs energy is a minimum.

➤ **What do you need to know already?**

Underlying the whole discussion is the expression of the direction of spontaneous change in terms of the Gibbs energy of a system (Topic 3C).This material draws on the concept of chemical potential and its dependence on the concentration or pressure of the substance (Topic 5A). You need to know how to express the total Gibbs energy of a mixture in terms of the chemical potentials of its components (Topic 5A).

As explained in Topic 3C, the direction of spontaneous change at constant temperature and pressure is towards lower values of the Gibbs energy, *G*. The idea is entirely general, and in this Topic we apply it to the discussion of chemical reactions. There is a tendency of a mixture of reactants to undergo reaction until the Gibbs energy of the mixture has reached a minimum: that state corresponds to a state of chemical equilibrium. The equilibrium is dynamic in the sense that the forward and reverse reactions continue, but at matching rates. As always in the application of thermodynamics, spontaneity is a *tendency*: there might be kinetic reasons why that tendency is not realized.

6A.1 The Gibbs energy minimum

We locate the equilibrium composition of a reaction mixture by calculating the Gibbs energy of the reaction mixture and identifying the composition that corresponds to minimum *G*. Here we proceed in two steps: first, we consider a very simple equilibrium, and then we generalize it.

(a) The reaction Gibbs energy

Consider the equilibrium $A \rightleftharpoons B$. Even though this reaction looks trivial, there are many examples of it, such as the isomerization of pentane to 2-methylbutane and the conversion of L-alanine to D-alanine.

Suppose an infinitesimal amount $d\xi$ of A turns into B, then the change in the amount of A present is $dn_A = -d\xi$ and the change in the amount of B present is $dn_B = +d\xi$. The quantity ξ (xi) is called the **extent of reaction;** it has the dimensions of amount of substance and is reported in moles. When the extent of reaction changes by a measurable amount $\Delta\xi$, the amount of A present changes from $n_{A,0}$ to $n_{A,0} - \Delta\xi$ and the amount of B changes from $n_{B,0}$ to $n_{B,0} + \Delta\xi$. In general, the amount of a component J changes by $\nu_J\Delta\xi$, where ν_J is the stoichiometric number of the species J (positive for products, negative for reactants).

Brief illustration 6A.1 The extent of reaction

If initially 2.0 mol A is present and we wait until $\Delta\xi = +1.5$ mol, then the amount of A remaining will be 0.5 mol. The amount of B formed will be 1.5 mol.

Self-test 6A.1 Suppose the reaction is 3 A→2 B and that initially 2.5 mol A is present. What is the composition when $\Delta\xi = +0.5$ mol?

Answer: 1.0 mol A, 1.0 mol B

The **reaction Gibbs energy**, $\Delta_r G$, is defined as the slope of the graph of the Gibbs energy plotted against the extent of reaction:

$$\Delta_r G = \left(\frac{\partial G}{\partial \xi}\right)_{p,T} \quad \textit{Definition} \quad \text{Reaction Gibbs energy} \quad (6A.1)$$

Although Δ normally signifies a *difference* in values, here it signifies a *derivative*, the slope of G with respect to ξ. However, to see that there is a close relationship with the normal usage, suppose the reaction advances by $d\xi$. The corresponding change in Gibbs energy is

$$dG = \mu_A dn_A + \mu_B dn_B = -\mu_A d\xi + \mu_B d\xi = (\mu_B - \mu_A)d\xi$$

This equation can be reorganized into

$$\left(\frac{\partial G}{\partial \xi}\right)_{p,T} = \mu_B - \mu_A$$

That is,

$$\Delta_r G = \mu_B - \mu_A \quad (6A.2)$$

We see that $\Delta_r G$ can also be interpreted as the difference between the chemical potentials (the partial molar Gibbs energies) of the reactants and products *at the current composition of the reaction mixture.*

Because chemical potentials vary with composition, the slope of the plot of Gibbs energy against extent of reaction, and therefore the reaction Gibbs energy, changes as the reaction proceeds. The spontaneous direction of reaction lies in the direction of decreasing G (that is, down the slope of G plotted against ξ). Thus we see from eqn 6A.2 that the reaction A→B is spontaneous when $\mu_A > \mu_B$, whereas the reverse reaction is spontaneous when $\mu_B > \mu_A$. The slope is zero, and the reaction is at equilibrium and spontaneous in neither direction, when

$$\Delta_r G = 0 \quad \text{Condition of equilibrium} \quad (6A.3)$$

This condition occurs when $\mu_B = \mu_A$ (Fig. 6A.1). It follows that, if we can find the composition of the reaction mixture that ensures $\mu_B = \mu_A$, then we can identify the composition of the reaction mixture at equilibrium. Note that the chemical potential is now fulfilling the role its name suggests: it represents the potential for chemical change, and equilibrium is attained when these potentials are in balance.

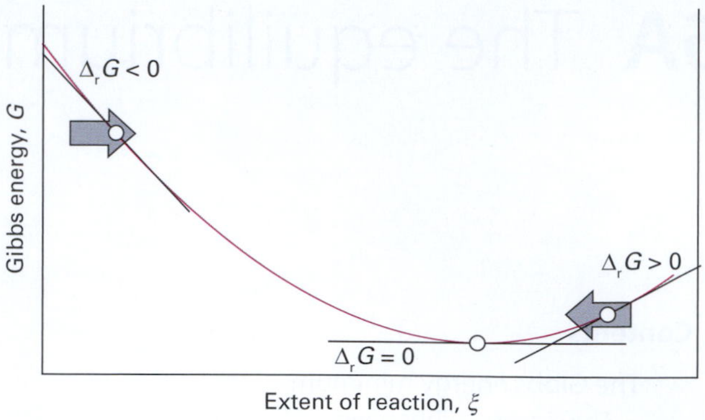

Figure 6A.1 As the reaction advances (represented by motion from left to right along the horizontal axis) the slope of the Gibbs energy changes. Equilibrium corresponds to zero slope at the foot of the valley.

(b) Exergonic and endergonic reactions

The spontaneity of a reaction at constant temperature and pressure can be expressed in terms of the reaction Gibbs energy:

- If $\Delta_r G < 0$, the forward reaction is spontaneous.
- If $\Delta_r G > 0$, the reverse reaction is spontaneous.
- If $\Delta_r G = 0$, the reaction is at equilibrium.

A reaction for which $\Delta_r G < 0$ is called **exergonic** (from the Greek words for work producing). The name signifies that, because the process is spontaneous, it can be used to drive another process, such as another reaction, or used to do non-expansion work. A simple mechanical analogy is a pair of weights joined by a string (Fig. 6A.2): the lighter of the pair of weights will be pulled up as the heavier weight falls down. Although the lighter weight has a natural tendency to move downward, its coupling to the heavier weight results in it being raised. In biological cells, the oxidation of carbohydrates act as

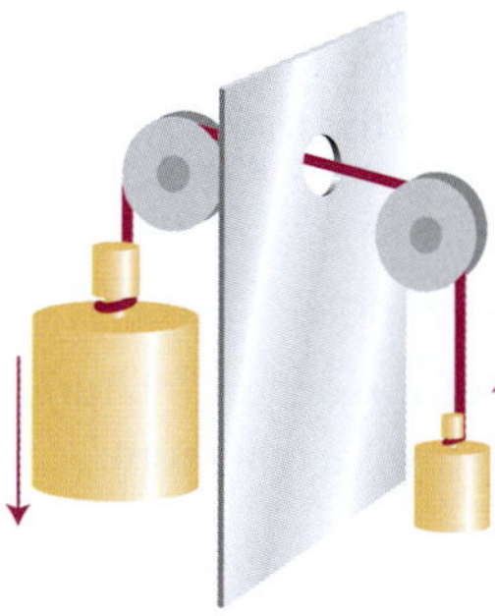

Figure 6A.2 If two weights are coupled as shown here, then the heavier weight will move the lighter weight in its non-spontaneous direction: overall, the process is still spontaneous. The weights are the analogues of two chemical reactions: a reaction with a large negative ΔG can force another reaction with a smaller ΔG to run in its non-spontaneous direction.

the heavy weight that drives other reactions forward and results in the formation of proteins from amino acids, muscle contraction, and brain activity. A reaction for which $\Delta_r G > 0$ is called **endergonic** (signifying work consuming). The reaction can be made to occur only by doing work on it, such as electrolysing water to reverse its spontaneous formation reaction.

Brief illustration 6A.2 Exergonic and endergonic reactions

The standard Gibbs energy of the reaction $H_2(g) + \frac{1}{2}O_2(g) \rightarrow H_2O(l)$ at 298 K is $-237\,\text{kJ mol}^{-1}$, so the reaction is exergonic and in a suitable device (a fuel cell, for instance) operating at constant temperature and pressure could produce 237 kJ of electrical work for each mole of H_2 molecules that react. The reverse reaction, for which $\Delta_r G^\ominus = +237\,\text{kJ mol}^{-1}$ is endergonic and at least 237 kJ of work must be done to achieve it.

Self-test 6A.2 Classify the formation of methane from its elements as exergonic or endergonic under standard conditions at 298 K.

Answer: Endergonic

6A.2 The description of equilibrium

With the background established, we are now ready to see how to apply thermodynamics to the description of chemical equilibrium.

(a) Perfect gas equilibria

When A and B are perfect gases we can use eqn 5A.14b ($\mu = \mu^\ominus + RT\ln p$, with p interpreted as $p/p^\ominus$) to write

$$\begin{aligned}\Delta_r G &= \mu_B - \mu_A = (\mu_B^\ominus + RT\ln p_B) - (\mu_A^\ominus + RT\ln p_A)\\ &= \Delta_r G^\ominus + RT\ln\frac{p_B}{p_A}\end{aligned} \qquad (6A.4)$$

If we denote the ratio of partial pressures by Q, we obtain

$$\Delta_r G = \Delta_r G^\ominus + RT\ln Q \qquad Q = \frac{p_B}{p_A} \qquad (6A.5)$$

The ratio Q is an example of a 'reaction quotient', a quantity we define more formally shortly. It ranges from 0 when $p_B = 0$ (corresponding to pure A) to infinity when $p_A = 0$ (corresponding to pure B). The standard reaction Gibbs energy, $\Delta_r G^\ominus$ (Topic 3C), is the difference in the standard molar Gibbs energies of the reactants and products, so for our reaction

$$\Delta_r G^\ominus = G_m^\ominus(B) - G_m^\ominus(A) = \mu_B^\ominus - \mu_A^\ominus \qquad (6A.6)$$

Note that in the definition of $\Delta_r G^\ominus$, the Δ_r has its normal meaning as the difference 'products − reactants'. In Topic 3C we saw that the difference in standard molar Gibbs energies of the products and reactants is equal to the difference in their standard Gibbs energies of formation, so in practice we calculate $\Delta_r G^\ominus$ from

$$\Delta_r G^\ominus = \Delta_f G^\ominus(B) - \Delta_f G^\ominus(A) \qquad (6A.7)$$

At equilibrium, $\Delta_r G = 0$. The ratio of partial pressures at equilibrium is denoted K, and eqn 6A.5 becomes

$$0 = \Delta_r G^\ominus + RT\ln K$$

which rearranges to

$$RT\ln K = -\Delta_r G^\ominus \qquad K = \left(\frac{p_B}{p_A}\right)_{\text{equilibrium}} \qquad (6A.8)$$

This relation is a special case of one of the most important equations in chemical thermodynamics: it is the link between tables of thermodynamic data, such as those in the *Resource section*, and the chemically important 'equilibrium constant', K (again, a quantity we define formally shortly).

Brief illustration 6A.3 The equilibrium constant

The standard Gibbs energy of the isomerization of pentane to 2-methylbutane at 298 K, the reaction $CH_3(CH_2)_3CH_3(g) \rightarrow (CH_3)_2CHCH_2CH_3(g)$, is close to $-6.7\,\text{kJ mol}^{-1}$ (this is an estimate based on enthalpies of formation; its actual value is not listed). Therefore, the equilibrium constant for the reaction is

$$K = e^{-(-6.7\times10^3\,\text{J mol}^{-1})/(8.3145\,\text{J K}^{-1}\,\text{mol}^{-1})\times(298\,\text{K})} = e^{2.7\ldots} = 15$$

Self-test 6A.3 Suppose it is found that at equilibrium the partial pressures of A and B in the gas-phase reaction $A \rightleftharpoons B$ are equal. What is the value of $\Delta_r G^\ominus$?

Answer: 0

In molecular terms, the minimum in the Gibbs energy, which corresponds to $\Delta_r G = 0$, stems from the Gibbs energy of mixing of the two gases. To see the role of mixing, consider the reaction $A \rightarrow B$. If only the enthalpy were important, then H and therefore G would change linearly from its value for pure reactants to its value for pure products. The slope of this straight line is a constant and equal to $\Delta_r G^\ominus$ at all stages of the reaction and there is no intermediate minimum in the graph (Fig. 6A.3). However, when the entropy is taken into account, there is an additional contribution to the Gibbs energy that is given by eqn 5A.16 ($\Delta_{mix}G = nRT(x_A \ln x_A + x_B \ln x_B)$). This expression makes a U-shaped contribution to the total change in Gibbs energy.

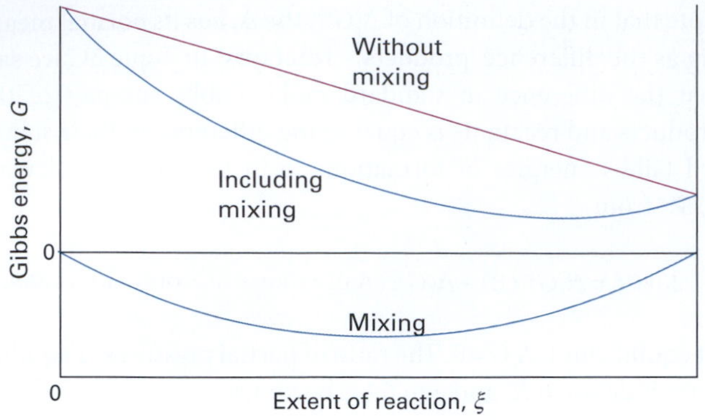

Figure 6A.3 If the mixing of reactants and products is ignored, then the Gibbs energy changes linearly from its initial value (pure reactants) to its final value (pure products) and the slope of the line is $\Delta_r G^\ominus$. However, as products are produced, there is a further contribution to the Gibbs energy arising from their mixing (lowest curve). The sum of the two contributions has a minimum. That minimum corresponds to the equilibrium composition of the system.

As can be seen from Fig. 6A.3, when it is included there is an intermediate minimum in the total Gibbs energy, and its position corresponds to the equilibrium composition of the reaction mixture.

We see from eqn 6A.8 that, when $\Delta_r G^\ominus > 0$, $K < 1$. Therefore, at equilibrium the partial pressure of A exceeds that of B, which means that the reactant A is favoured in the equilibrium. When $\Delta_r G^\ominus < 0$, $K > 1$, so at equilibrium the partial pressure of B exceeds that of A. Now the product B is favoured in the equilibrium.

A note on good practice A common remark is that 'a reaction is spontaneous if $\Delta_r G^\ominus < 0$'. However, whether or not a reaction is spontaneous at a particular composition depends on the value of $\Delta_r G$ at that composition, not $\Delta_r G^\ominus$. It is far better to interpret the sign of $\Delta_r G^\ominus$ as indicating whether K is greater or smaller than 1. The forward reaction is spontaneous ($\Delta_r G < 0$) when $Q < K$ and the reverse reaction is spontaneous when $Q > K$.

(b) The general case of a reaction

We can now extend the argument that led to eqn 6A.8 to a general reaction. First, we note that a chemical reaction may be expressed symbolically in terms of (signed) stoichiometric numbers as

$$0 = \sum_J \nu_J J$$ *Symbolic form* Chemical equation (6A.9)

where J denotes the substances and the ν_J are the corresponding stoichiometric numbers in the chemical equation. In the reaction $2\,A + B \rightarrow 3\,C + D$, for instance, these numbers have the values $\nu_A = -2$, $\nu_B = -1$, $\nu_C = +3$, and $\nu_D = +1$. A stoichiometric number is positive for products and negative for reactants. Then we define the extent of reaction ξ so that, if it changes by $\Delta\xi$, then the change in the amount of any species J is $\nu_J \Delta\xi$.

With these points in mind and with the reaction Gibbs energy, $\Delta_r G$, defined in the same way as before (eqn 6A.1) we show in the following *Justification* that the Gibbs energy of reaction can always be written

$$\Delta_r G = \Delta_r G^\ominus + RT \ln Q$$ Reaction Gibbs energy at an arbitrary stage (6A.10)

with the standard reaction Gibbs energy calculated from

$$\Delta_r G^\ominus = \sum_{\text{Products}} \nu \Delta_f G^\ominus - \sum_{\text{Reactants}} \nu \Delta_f G^\ominus$$ *Practical implementation* Reaction Gibbs energy (6A.11a)

where the ν are the (positive) stoichiometric coefficients. More formally,

$$\Delta_r G^\ominus = \sum_J \nu_J \Delta_f G^\ominus(J)$$ *Formal expression* Reaction Gibbs energy (6A.11b)

where the ν_J are the (signed) stoichiometric numbers. The reaction quotient, Q, has the form

$$Q = \frac{\text{activities of products}}{\text{activities of reactants}}$$ *General form* Reaction quotient (6A.12a)

with each species raised to the power given by its stoichiometric coefficient. More formally, to write the general expression for Q we introduce the symbol Π to denote the product of what follows it (just as Σ denotes the sum), and define Q as

$$Q = \prod_J a_J^{\nu_J}$$ *Definition* Reaction quotient (6A.12b)

Because reactants have negative stoichiometric numbers, they automatically appear as the denominator when the product is written out explicitly. Recall from Table 5E.1 that, for pure solids and liquids, the activity is 1, so such substances make no contribution to Q even though they may appear in the chemical equation.

Brief illustration 6A.4 The reaction quotient

Consider the reaction $2\,A + 3\,B \rightarrow C + 2\,D$, in which case $\nu_A = -2$, $\nu_B = -3$, $\nu_C = +1$, and $\nu_D = +2$. The reaction quotient is then

$$Q = a_A^{-2} a_B^{-3} a_C a_D^2 = \frac{a_C a_D^2}{a_A^2 a_B^3}$$

Self-test 6A.4 Write the reaction quotient for $A + 2\,B \rightarrow 3\,C$.

Answer: $Q = a_C^3 / a_A a_B^2$

Justification 6A.1 The dependence of the reaction Gibbs energy on the reaction quotient

Consider a reaction with stoichiometric numbers ν_J. When the reaction advances by $d\xi$, the amounts of reactants and products change by $dn_J = \nu_J d\xi$. The resulting infinitesimal change in the Gibbs energy at constant temperature and pressure is

$$dG = \sum_J \mu_J dn_J = \sum_J \mu_J \nu_J d\xi = \left(\sum_J \mu_J \nu_J\right) d\xi$$

It follows that

$$\Delta_r G = \left(\frac{\partial G}{\partial \xi}\right)_{p,T} = \sum_J \nu_J \mu_J$$

To make progress, we note that the chemical potential of a species J is related to its activity by eqn 5E.9 ($\mu_J = \mu_J^{\ominus} + RT\ln a_J$). When this expression is substituted into eqn 6A.11 we obtain

$$\Delta_r G = \overbrace{\sum_J \nu_J \mu_J^{\ominus}}^{\Delta_r G^{\ominus}} + RT\sum_J \nu_J \ln a_J$$

$$= \Delta_r G^{\ominus} + RT\sum_J \nu_J \ln a_J = \Delta_r G^{\ominus} + RT\ln \overbrace{\prod_J a_J^{\nu_J}}^{Q}$$

$$= \Delta_r G^{\ominus} + RT\ln Q$$

In the second line we use first $a \ln x = \ln x^a$ and then $\ln x + \ln y + \ldots = \ln xy\ldots$, so

$$\sum_i \ln x_i = \ln\left(\prod_i x_i\right)$$

Now we conclude the argument, starting from eqn 6A.10. At equilibrium, the slope of G is zero: $\Delta_r G = 0$. The activities then have their equilibrium values and we can write

$$K = \left(\prod_J a_J^{\nu_J}\right)_{\text{equilibrium}} \quad \textit{Definition} \quad \text{Equilibrium constant} \quad (6A.13)$$

This expression has the same form as Q but is evaluated using equilibrium activities. From now on, we shall not write the 'equilibrium' subscript explicitly, and will rely on the context to make it clear that for K we use equilibrium values and for Q we use the values at the specified stage of the reaction. An equilibrium constant K expressed in terms of activities (or fugacities) is called a **thermodynamic equilibrium constant**. Note that, because activities are dimensionless numbers, the thermodynamic equilibrium constant is also dimensionless. In elementary applications, the activities that occur in eqn 6A.13 are often replaced as follows:

State	Measure	Approximation for a_J	Definition
Solute	molality	$b_J/b_J^{\ominus}$	$b^{\ominus} = 1\ \text{mol kg}^{-1}$
	molar concentration	$[J]/c^{\ominus}$	$c^{\ominus} = 1\ \text{mol dm}^{-3}$
Gas phase	partial pressure	$p_J/p^{\ominus}$	$p^{\ominus} = 1\ \text{bar}$

In such cases, the resulting expressions are only approximations. The approximation is particularly severe for electrolyte solutions, for in them activity coefficients differ from 1 even in very dilute solutions (Topic 5F).

Brief illustration 6A.5 The equilibrium constant

The equilibrium constant for the heterogeneous equilibrium $CaCO_3(s) \rightleftharpoons CaO(s) + CO_2(g)$ is

$$K = a_{CaCO_3(s)}^{-1} a_{CaO(s)} a_{CO_2(g)} = \frac{\overbrace{a_{CaO(s)}}^{1} a_{CO_2(g)}}{\underbrace{a_{CaCO_3(s)}}_{1}} = a_{CO_2(g)}$$

Provided the carbon dioxide can be treated as a perfect gas, we can go on to write

$$K = p_{CO_2}/p^{\ominus}$$

and conclude that in this case the equilibrium constant is the numerical value of the decomposition vapour pressure of calcium carbonate.

Self-test 6A.5 Write the equilibrium constant for the reaction $N_2(g) + 3\,H_2(g) \rightleftharpoons 2\,NH_3(g)$, with the gases treated as perfect.

Answer: $K = a_{NH_3}^2/a_{N_2}a_{H_2}^3 = p_{NH_3}^2 p^{\ominus 2}/p_{N_2}p_{H_2}^3$

At this point we set $\Delta_r G = 0$ in eqn 6A.10 and replace Q by K. We immediately obtain

$$\Delta_r G^{\ominus} = -RT\ln K \quad \text{Thermodynamic equilibrium constant} \quad (6A.14)$$

This is an exact and highly important thermodynamic relation, for it enables us to calculate the equilibrium constant of any reaction from tables of thermodynamic data, and hence to predict the equilibrium composition of the reaction mixture. In Topic 15F we see that the right-hand side of eqn 6A.14 may be expressed in terms of spectroscopic data for gas-phase species; so this expression also provides a link between spectroscopy and equilibrium composition.

Example 6A.1 Calculating an equilibrium constant

Calculate the equilibrium constant for the ammonia synthesis reaction, $N_2(g) + 3\,H_2(g) \rightleftharpoons 2\,NH_3(g)$, at 298 K and show how K is related to the partial pressures of the species at equilibrium when the overall pressure is low enough for the gases to be treated as perfect.

Method Calculate the standard reaction Gibbs energy from eqn 6A.10 and convert it to the value of the equilibrium constant by using eqn 6A.14. The expression for the equilibrium constant is obtained from eqn 6A.13, and because the gases are taken to be perfect, we replace each activity by the ratio $p_J/p^{\ominus}$, where p_J is the partial pressure of species J.

Answer The standard Gibbs energy of the reaction is

$$\begin{aligned}\Delta_r G^{\ominus} &= 2\Delta_f G^{\ominus}(NH_3,g) - \{\Delta_f G^{\ominus}(N_2,g) + 3\Delta_f G^{\ominus}(H_2,g)\}\\ &= 2\Delta_f G^{\ominus}(NH_3,g) = 2\times(-16.45\,kJ\,mol^{-1})\end{aligned}$$

Then,

$$\ln K = -\frac{2\times(-1.645\times10^4\,J\,mol^{-1})}{(8.3145\,J\,K^{-1}\,mol^{-1})\times(298\,K)} = \frac{2\times1.645\times10^4}{8.3145\times298} = 13.2\ldots$$

Hence, $K = 5.8\times10^5$. This result is thermodynamically exact. The thermodynamic equilibrium constant for the reaction is

$$K = \frac{a_{NH_3}^2}{a_{N_2}a_{H_2}^3}$$

and this ratio has the value we have just calculated. At low overall pressures, the activities can be replaced by the ratios $p_J/p^{\ominus}$ and an approximate form of the equilibrium constant is

$$K = \frac{(p_{NH_3}/p^{\ominus})^2}{(p_{N_2}/p^{\ominus})(p_{H_2}/p^{\ominus})^3} = \frac{p_{NH_3}^2 p^{\ominus 2}}{p_{N_2}p_{H_2}^3}$$

Self-test 6A.6 Evaluate the equilibrium constant for $N_2O_4(g) \rightleftharpoons 2\,NO_2(g)$ at 298 K.

Answer: $K = 0.15$

Example 6A.2 Estimating the degree of dissociation at equilibrium

The *degree of dissociation* (or *extent of dissociation*, α) is defined as the fraction of reactant that has decomposed; if the initial amount of reactant is n and the amount at equilibrium is n_{eq}, then $\alpha = (n - n_{eq})/n$. The standard reaction Gibbs energy for the decomposition $H_2O(g) \rightarrow H_2(g) + \frac{1}{2}O_2(g)$ is +118.08 kJ mol^{-1} at 2300 K. What is the degree of dissociation of H_2O at 2300 K and 1.00 bar?

Method The equilibrium constant is obtained from the standard Gibbs energy of reaction by using eqn 6A.11, so the task is to relate the degree of dissociation, α, to K and then to find its numerical value. Proceed by expressing the equilibrium compositions in terms of α, and solve for α in terms of K. Because the standard reaction Gibbs energy is large and positive, we can anticipate that K will be small, and hence that $\alpha \ll 1$, which opens the way to making approximations to obtain its numerical value.

Answer The equilibrium constant is obtained from eqn 6A.14 in the form

$$\begin{aligned}\ln K &= -\frac{\Delta_r G^{\ominus}}{RT} = -\frac{1.1808\times10^5\,J\,mol^{-1}}{(8.3145\,J\,K^{-1}\,mol^{-1})\times(2300\,K)}\\ &= -\frac{1.1808\times10^5}{8.3145\times2300} = -6.17\ldots\end{aligned}$$

It follows that $K = 2.08\times10^{-3}$. The equilibrium composition can be expressed in terms of α by drawing up the following table:

	H_2O	H_2	$+\frac{1}{2}O_2$	
Initial amount	n	0	0	
Change to reach equilibrium	$-\alpha n$	$+\alpha n$	$+\frac{1}{2}\alpha n$	
Amount at equilibrium	$(1-\alpha)n$	αn	$\frac{1}{2}\alpha n$	Total: $(1+\frac{1}{2}\alpha)n$
Mole fraction, x_J	$\frac{1-\alpha}{1+\frac{1}{2}\alpha}$	$\frac{\alpha}{1+\frac{1}{2}}$	$\frac{\frac{1}{2}\alpha}{1+\frac{1}{2}\alpha}$	
Partial pressure, p_J	$\frac{(1-\alpha)p}{1+\frac{1}{2}\alpha}$	$\frac{\alpha p}{1+\frac{1}{2}}$	$\frac{\frac{1}{2}\alpha p}{1+\frac{1}{2}\alpha}$	

where, for the entries in the last row, we have used $p_J = x_J p$ (eqn 1A.8). The equilibrium constant is therefore

$$K = \frac{p_{H_2}p_{O_2}^{1/2}}{p_{H_2O}} = \frac{\alpha^{3/2}p^{1/2}}{(1-\alpha)(2+\alpha)^{1/2}}$$

In this expression, we have written p in place of $p/p^{\ominus}$, to simplify its appearance. Now make the approximation that $\alpha \ll 1$, and hence obtain

$$K \approx \frac{\alpha^{3/2}p^{1/2}}{2^{1/2}}$$

Under the stated condition, $p = 1.00$ bar (that is, $p/p^{\ominus} = 1.00$), so $\alpha \approx (2^{1/2}K)^{2/3} = 0.0205$. That is, about 2 per cent of the water has decomposed.

A note on good practice Always check that the approximation is consistent with the final answer. In this case $\alpha \ll 1$, in accord with the original assumption.

Self-test 6A.7 Given that the standard Gibbs energy of reaction at 2000 K is +135.2 kJ mol^{-1} for the same reaction, suppose that steam at 200 kPa is passed through a furnace tube at that temperature. Calculate the mole fraction of O_2 present in the output gas stream.

Answer: 0.00221

(c) The relation between equilibrium constants

Equilibrium constants in terms of activities are exact, but it is often necessary to relate them to concentrations. Formally, we need to know the activity coefficients, and then to use $a_J = \gamma_J x_J$, $a_J = \gamma_J b_J / b^{\ominus}$, or $a_J = [J]/c^{\ominus}$, where x_J is a mole fraction, b_J is a molality, and [J] is a molar concentration. For example, if we were interested in the composition in terms of molality for an equilibrium of the form $A + B \rightleftharpoons C + D$, where all four species are solutes, we would write

$$K = \frac{a_C a_D}{a_A a_B} = \frac{\gamma_C \gamma_D}{\gamma_A \gamma_B} \times \frac{b_C b_D}{b_A b_B} = K_\gamma K_b \quad (6A.15)$$

The activity coefficients must be evaluated at the equilibrium composition of the mixture (for instance, by using one of the Debye–Hückel expressions, Topic 5F), which may involve a complicated calculation, because the activity coefficients are known only if the equilibrium composition is already known. In elementary applications, and to begin the iterative calculation of the concentrations in a real example, the assumption is often made that the activity coefficients are all so close to unity that $K_\gamma = 1$. Then we obtain the result widely used in elementary chemistry that $K \approx K_b$, and equilibria are discussed in terms of molalities (or molar concentrations) themselves.

A special case arises when we need to express the equilibrium constant of a gas-phase reaction in terms of molar concentrations instead of the partial pressures that appear in the thermodynamic equilibrium constant. Provided we can treat the gases as perfect, the p_J that appear in K can be replaced by $[J]RT$, and

$$K = \prod_J a_J^{\nu_J} = \prod_J \left(\frac{p_J}{p^{\ominus}}\right)^{\nu_J} = \prod_J [J]^{\nu_J} \left(\frac{RT}{p^{\ominus}}\right)^{\nu_J}$$
$$= \prod_J [J]^{\nu_J} \times \prod_J \left(\frac{RT}{p^{\ominus}}\right)^{\nu_J}$$

(Products can always be factorized like that: *abcdef* is the same as *abc*×*def*.) The (dimensionless) equilibrium constant K_c is defined as

$$K_c = \prod_J \left(\frac{[J]}{c^{\ominus}}\right)^{\nu_J} \quad \textit{Definition} \quad K_c \text{ for gas-phase reactions} \quad (6A.16)$$

It follows that

$$K = K_c \times \prod_J \left(\frac{c^{\ominus} RT}{p^{\ominus}}\right)^{\nu_J} \quad (6A.17a)$$

If now we write $\Delta\nu = \sum_J \nu_J$, which is easier to think of as ν(products) − ν(reactants), then the relation between K and K_c for a gas-phase reaction is

$$K = K_c \times \left(\frac{c^{\ominus} RT}{p^{\ominus}}\right)^{\Delta\nu} \quad \text{Relation between } K \text{ and } K_c \text{ for gas-phase reactions} \quad (6A.17b)$$

The term in parentheses works out as $T/(12.03\,\text{K})$.

Brief illustration 6A.6 The relation between equilibrium constants

For the reaction $N_2(g) + 3\,H_2(g) \rightarrow 2\,NH_3(g)$, $\Delta\nu = 2 - 4 = -2$, so

$$K = K_c \times \left(\frac{T}{12.03\,\text{K}}\right)^{-2} = K_c \times \left(\frac{12.03\,\text{K}}{T}\right)^2$$

At 298.15 K the relation is

$$K = K_c \times \left(\frac{12.03\,\text{K}}{298.15\,\text{K}}\right)^2 = \frac{K_c}{614.2}$$

so $K_c = 614.2K$. Note that both K and K_c are dimensionless.

Self-test 6A.8 Find the relation between K and K_c for the equilibrium $H_2(g) + \frac{1}{2}O_2(g) \rightarrow H_2O(l)$ at 298 K.

Answer: $K_c = 123K$

(d) Molecular interpretation of the equilibrium constant

Deeper insight into the origin and significance of the equilibrium constant can be obtained by considering the Boltzmann distribution of molecules over the available states of a system composed of reactants and products (*Foundations* B). When atoms can exchange partners, as in a reaction, the available states of the system include arrangements in which the atoms are present in the form of reactants and in the form of products: these arrangements have their characteristic sets of energy levels, but the Boltzmann distribution does not distinguish between their identities, only their energies. The atoms distribute themselves over both sets of energy levels in accord with the Boltzmann distribution (Fig. 6A.4). At a given temperature, there will be a specific distribution of populations, and hence a specific composition of the reaction mixture.

It can be appreciated from the illustration that, if the reactants and products both have similar arrays of molecular energy levels, then the dominant species in a reaction mixture at equilibrium will be the species with the lower set of energy levels. However, the fact that the Gibbs energy occurs in the expression is a signal that entropy plays a role as well as energy. Its role can be appreciated by referring to Fig. 6A.4. In Fig. 6A.4b we see that, although the B energy levels lie higher than the A energy levels, in this instance they are much more closely spaced. As a result, their total population may be considerable and B could even dominate in the reaction mixture at equilibrium. Closely spaced energy levels correlate with a high

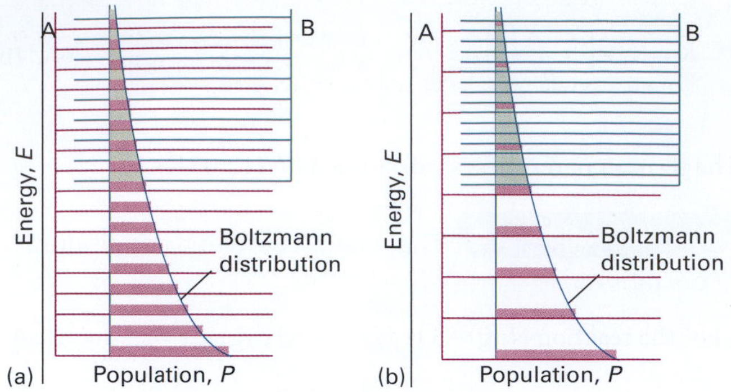

Figure 6A.4 The Boltzmann distribution of populations over the energy levels of two species A and B with similar densities of energy levels. The reaction A→B is endothermic in this example. (a) The bulk of the population is associated with the species A, so that species is dominant at equilibrium. (b) Even though the reaction A→B is endothermic, the density of energy levels in B is so much greater than that in A that the population associated with B is greater than that associated with A, so B is dominant at equilibrium.

entropy (Topic 15E), so in this case we see that entropy effects dominate adverse energy effects. This competition is mirrored in eqn 6A.14, as can be seen most clearly by using $\Delta_r G^\ominus = \Delta_r H^\ominus - T\Delta_r S^\ominus$ and writing it in the form

$$K = e^{-\Delta_r H^\ominus/RT} e^{\Delta_r S^\ominus/R} \qquad (6A.18)$$

Note that a positive reaction enthalpy results in a lowering of the equilibrium constant (that is, an endothermic reaction can be expected to have an equilibrium composition that favours the reactants). However, if there is positive reaction entropy, then the equilibrium composition may favour products, despite the endothermic character of the reaction.

Brief illustration 6A.7 Contributions to K

In *Example* 6A.1 it is established that $\Delta_r G^\ominus = -33.0\,\text{kJ mol}^{-1}$ for the reaction $N_2(g) + 3\,H_2(g) \rightleftharpoons 2\,NH_3(g)$ at 298 K. From the tables of data in the *Resource section*, we can find that $\Delta_r H^\ominus = -92.2\,\text{kJ mol}^{-1}$ and $\Delta_r S^\ominus = -198.8\,\text{J K}^{-1}\,\text{mol}^{-1}$. The contributions to K are therefore

$$\begin{aligned} K &= e^{-(-9.22\times10^4\,\text{J mol}^{-1})/(8.3145\,\text{J K}^{-1}\,\text{mol}^{-1})\times(298\,\text{K})} \\ &\quad \times e^{(-198.8\,\text{J K}^{-1}\,\text{mol}^{-1})/(8.3145\,\text{J K}^{-1}\,\text{mol}^{-1})} \\ &= e^{37.2\ldots} \times e^{-23.9\ldots} \end{aligned}$$

We see that the exothermic character of the reaction encourages the formation of products (it results in a large increase in entropy of the surroundings) but the decrease in entropy of the system as H atoms are pinned to N atoms opposes their formation.

Self-test 6A.9 Analyse the equilibrium $N_2O_4(g) \rightleftharpoons 2\,NO_2(g)$ similarly.

Answer: $K = e^{-26.7\ldots} \times e^{21.1\ldots}$; enthalpy opposes, entropy encourages

Checklist of concepts

- ☐ 1. The **reaction Gibbs energy** is the slope of the plot of Gibbs energy against extent of reaction.
- ☐ 2. Reactions are either **exergonic** or **endergonic**.
- ☐ 3. The reaction Gibbs energy depends logarithmically on the **reaction quotient**.
- ☐ 4. When the reaction Gibbs energy is zero the reaction quotient has a value called the **equilibrium constant**.
- ☐ 5. Under ideal conditions, the thermodynamic equilibrium constant may be approximated by expressing it in terms of concentrations and partial pressures.

Checklist of equations

Property	Equation	Comment	Equation number
Reaction Gibbs energy	$\Delta_r G = (\partial G/\partial \xi)_{p,T}$	Definition	6A.1
Reaction Gibbs energy	$\Delta_r G = \Delta_r G^\ominus + RT \ln Q$		6A.10
Standard reaction Gibbs energy	$\Delta_r G^\ominus = \sum_{\text{Products}} \nu \Delta_f G^\ominus - \sum_{\text{Reactants}} \nu \Delta_f G^\ominus = \sum_J \nu_J \Delta_f G^\ominus(J)$	ν are positive; ν_J are signed	6A.11

Property	Equation	Comment	Equation number
Reaction quotient	$Q=\prod_J a_J^{\nu_J}$	Definition; evaluated at arbitrary stage of reaction	6A.12
Thermodynamic equilibrium constant	$K=\left(\prod_J a_J^{\nu_J}\right)_{\text{equilibrium}}$	Definition	6A.13
Equilibrium constant	$\Delta_r G^\ominus=-RT\ln K$		6A.14
Relation between K and K_c	$K=K_c(c^\ominus RT/p^\ominus)^{\Delta\nu}$	Gas-phase reactions; perfect gases	6A.17b

6B The response of equilibria to the conditions

Contents

➤ **Why do you need to know this material?**

Chemists, and chemical engineers designing a chemical plant, need to know how an equilibrium will respond to changes in the conditions, such as a change in pressure or temperature. The variation with temperature also provides a way to determine the enthalpy and entropy of a reaction.

➤ **What is the key idea?**

A system at equilibrium, when subjected to a disturbance, responds in a way that tends to minimize the effect of the disturbance.

➤ **What do you need to know already?**

This Topic builds on the relation between the equilibrium constant and the standard Gibbs energy of reaction (Topic 6A). To express the temperature dependence of K it draws on the Gibbs–Helmholtz equation (Topic 3D).

The equilibrium constant for a reaction is not affected by the presence of a catalyst or an enzyme (a biological catalyst). As explained in detail in Topics 20H and 22C, catalysts increase the rate at which equilibrium is attained but do not affect its position. However, it is important to note that in industry reactions rarely reach equilibrium, partly on account of the rates at which reactants mix. The equilibrium constant is also independent of pressure, but as we shall see, that does not necessarily mean that the composition at equilibrium is independent of pressure. The equilibrium constant does depend on the temperature in a manner that can be predicted from the standard reaction enthalpy.

6B.1 The response to pressure

The equilibrium constant depends on the value of $\Delta_r G^\ominus$, which is defined at a single, standard pressure. The value of $\Delta_r G^\ominus$, and hence of K, is therefore independent of the pressure at which the equilibrium is actually established. In other words, at a given temperature K is a constant.

The conclusion that K is independent of pressure does not necessarily mean that the equilibrium composition is independent of the pressure, and the effect depends on how the pressure is applied.

The pressure within a reaction vessel can be increased by injecting an inert gas into it. However, so long as the gases are perfect, this addition of gas leaves all the partial pressures of the reacting gases unchanged: the partial pressures of a perfect gas is the pressure it would exert if it were alone in the container, so the presence of another gas has no effect. It follows that pressurization by the addition of an inert gas has no effect on the equilibrium composition of the system (provided the gases are perfect).

Alternatively, the pressure of the system may be increased by confining the gases to a smaller volume (that is, by compression). Now the individual partial pressures are changed but their ratio (as it appears in the equilibrium constant) remains the same. Consider, for instance, the perfect gas equilibrium $A \rightleftharpoons 2\,B$, for which the equilibrium constant is

$$K = \frac{p_B^2}{p_A p^\ominus}$$

The right-hand side of this expression remains constant only if an increase in p_A cancels an increase in the *square* of p_B. This relatively steep increase of p_A compared to p_B will occur if the equilibrium composition shifts in favour of A at the expense of B. Then the number of A molecules will increase as the volume of the container is decreased and its partial pressure will rise

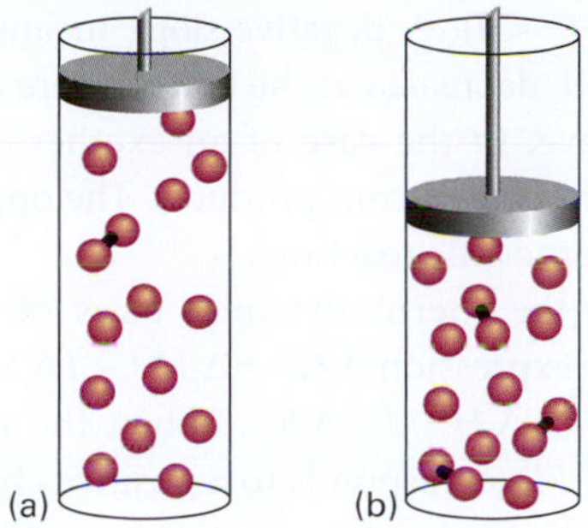

Figure 6B.1 When a reaction at equilibrium is compressed (from *a* to *b*), the reaction responds by reducing the number of molecules in the gas phase (in this case by producing the dimers represented by the linked spheres).

more rapidly than can be ascribed to a simple change in volume alone (Fig. 6B.1).

The increase in the number of A molecules and the corresponding decrease in the number of B molecules in the equilibrium $A \rightleftharpoons 2\,B$ is a special case of a principle proposed by the French chemist Henri Le Chatelier, which states that:

> A system at equilibrium, when subjected to a disturbance, responds in a way that tends to minimize the effect of the disturbance.

Le Chatelier's principle

The principle implies that, if a system at equilibrium is compressed, then the reaction will adjust so as to minimize the increase in pressure. This it can do by reducing the number of particles in the gas phase, which implies a shift $A \leftarrow 2\,B$.

To treat the effect of compression quantitatively, we suppose that there is an amount n of A present initially (and no B). At equilibrium the amount of A is $(1-\alpha)n$ and the amount of B is $2\alpha n$, where α is the degree of dissociation of A into 2B. It follows that the mole fractions present at equilibrium are

$$x_A = \frac{(1-\alpha)n}{(1-\alpha)n+2\alpha n} = \frac{1-\alpha}{1+\alpha} \qquad x_B = \frac{2\alpha}{1+\alpha}$$

The equilibrium constant for the reaction is

$$K = \frac{p_B^2}{p_A p^{\ominus}} = \frac{x_B^2 p^2}{x_A p p^{\ominus}} = \frac{4\alpha^2 (p/p^{\ominus})}{1-\alpha^2}$$

which rearranges to

$$\alpha = \left(\frac{1}{1+4p/Kp^{\ominus}}\right)^{1/2} \qquad (6B.1)$$

This formula shows that, even though K is independent of pressure, the amounts of A and B do depend on pressure (Fig. 6B.2). It also shows that as p is increased, α decreases, in accord with Le Chatelier's principle.

Brief illustration 6B.1 Le Chatelier's principle

To predict the effect of an increase in pressure on the composition of the ammonia synthesis at equilibrium, *Example* 6A.1, we note that the number of gas molecules decreases (from 4 to 2). So, Le Chatelier's principle predicts that an increase in pressure will favour the product. The equilibrium constant is

$$K = \frac{p_{NH_3}^2 p^{\ominus}}{p_{N_2} p_{H_2}^3} = \frac{x_{NH_3}^2 p^2 p^{\ominus 2}}{x_{N_2} x_{H_2}^3 p^4} = \frac{x_{NH_3}^2 p^{\ominus 2}}{x_{N_2} x_{H_2}^3 p^2} = K_x \times \frac{p^{\ominus 2}}{p^2}$$

where K_x is the part of the equilibrium constant expression that contains the equilibrium mole fractions of reactants and products (note that, unlike K itself, K_x is not an equilibrium constant). Therefore, doubling the pressure must increase K_x by a factor of 4 to preserve the value of K.

Self-test 6B.1 Predict the effect of a tenfold pressure increase on the equilibrium composition of the reaction $3\,N_2(g)+H_2(g) \rightleftharpoons 2\,N_3H(g)$.

Answer: 100-fold increase in K_x

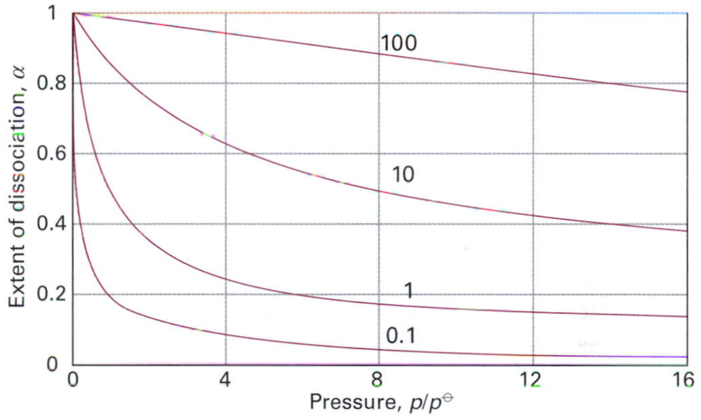

Figure 6B.2 The pressure dependence of the degree of dissociation, α, at equilibrium for an $A(g) \rightleftharpoons 2\,B(g)$ reaction for different values of the equilibrium constant K. The value $\alpha=0$ corresponds to pure A; $\alpha=1$ corresponds to pure B

6B.2 The response to temperature

Le Chatelier's principle predicts that a system at equilibrium will tend to shift in the endothermic direction if the temperature is raised, for then energy is absorbed as heat and the rise in temperature is opposed. Conversely, an equilibrium can be expected to shift in the exothermic direction if the temperature is lowered, for then energy is released and the reduction in temperature is opposed. These conclusions can be summarized as follows:

> Exothermic reactions: increased temperature favours the reactants.

Endothermic reactions: increased temperature favours the products.

We shall now justify these remarks thermodynamically and see how to express the changes quantitatively.

(a) The van 't Hoff equation

The **van 't Hoff equation**, which is derived in the following *Justification*, is an expression for the slope of a plot of the equilibrium constant (specifically, ln *K*) as a function of temperature. It may be expressed in either of two ways:

$$\frac{\mathrm{d}\ln K}{\mathrm{d}T}=\frac{\Delta_r H^{\ominus}}{RT^2}$$ van 't Hoff equation (6B.2a)

$$\frac{\mathrm{d}\ln K}{\mathrm{d}(1/T)}=-\frac{\Delta_r H^{\ominus}}{R}$$ Alternative version van 't Hoff equation (6B.2b)

Justification 6B.1 The van 't Hoff equation

From eqn 6A.14, we know that

$$\ln K=-\frac{\Delta_r G^{\ominus}}{RT}$$

Differentiation of ln *K* with respect to temperature then gives

$$\frac{\mathrm{d}\ln K}{\mathrm{d}T}=-\frac{1}{R}\frac{\mathrm{d}(\Delta_r G^{\ominus}/T)}{\mathrm{d}T}$$

The differentials are complete (that is, they are not partial derivatives) because *K* and $\Delta_r G^{\ominus}$ depend only on temperature, not on pressure. To develop this equation we use the Gibbs–Helmholtz equation (eqn 3D.10, $\mathrm{d}(\Delta G/T)=-\Delta H/T^2$) in the form

$$\frac{\mathrm{d}(\Delta_r G^{\ominus}/T)}{\mathrm{d}T}=-\frac{\Delta_r H^{\ominus}}{R}$$

where $\Delta_r H^{\ominus}$ is the standard reaction enthalpy at the temperature *T*. Combining the two equations gives the van 't Hoff equation, eqn 6B.2a. The second form of the equation is obtained by noting that

$$\frac{\mathrm{d}(1/T)}{\mathrm{d}T}=-\frac{1}{T^2},\quad \text{so } \mathrm{d}T=-T^2\mathrm{d}(1/T)$$

It follows that eqn 6B.2a can be rewritten as

$$-\frac{\mathrm{d}\ln K}{T^2\mathrm{d}(1/T)}=\frac{\Delta_r H^{\ominus}}{RT^2}$$

which simplifies into eqn 6B.2b.

Equation 6B.2a shows that $\mathrm{d}\ln K/\mathrm{d}T<0$ (and therefore that $\mathrm{d}K/\mathrm{d}T<0$) for a reaction that is exothermic under standard conditions ($\Delta_r H^{\ominus}<0$). A negative slope means that ln *K*, and therefore *K* itself, decreases as the temperature rises. Therefore, as asserted above, in the case of an exothermic reaction the equilibrium shifts away from products. The opposite occurs in the case of endothermic reactions.

Insight into the thermodynamic basis of this behaviour comes from the expression $\Delta_r G^{\ominus}=\Delta_r H^{\ominus}-T\Delta_r S^{\ominus}$ written in the form $-\Delta_r G^{\ominus}/T=-\Delta_r H^{\ominus}/T+\Delta_r S^{\ominus}$. When the reaction is exothermic, $-\Delta_r H^{\ominus}/T$ corresponds to a positive change of entropy of the surroundings and favours the formation of products. When the temperature is raised, $-\Delta_r H^{\ominus}/T$ decreases and the increasing entropy of the surroundings has a less important role. As a result, the equilibrium lies less to the right. When the reaction is endothermic, the principal factor is the increasing entropy of the reaction system. The importance of the unfavourable change of entropy of the surroundings is reduced if the temperature is raised (because then $\Delta_r H^{\ominus}/T$ is smaller), and the reaction is able to shift towards products.

These remarks have a molecular basis that stems from the Boltzmann distribution of molecules over the available energy levels (*Foundations* B, and in more detail in Topic 15F). The typical arrangement of energy levels for an endothermic reaction is shown in Fig. 6B.3a. When the temperature is increased, the Boltzmann distribution adjusts and the populations change as shown. The change corresponds to an increased population of the higher energy states at the expense of the population of the lower energy states. We see that the states that arise from the B molecules become more populated at the expense of the A molecules. Therefore, the total population of B states increases, and B becomes more abundant in the equilibrium mixture. Conversely, if the reaction is exothermic (Fig. 6B.3b),

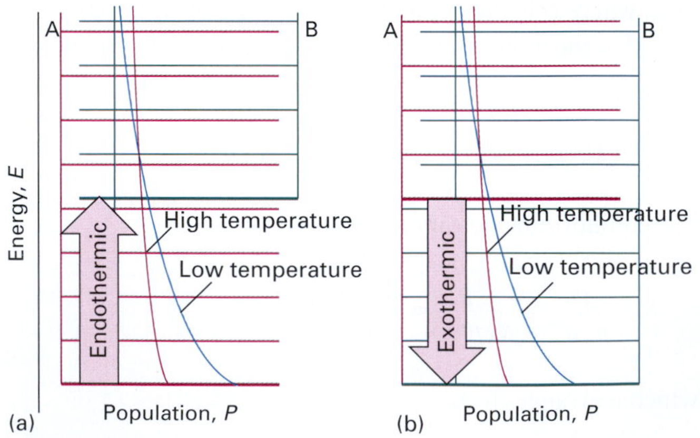

Figure 6B.3 The effect of temperature on a chemical equilibrium can be interpreted in terms of the change in the Boltzmann distribution with temperature and the effect of that change in the population of the species. (a) In an endothermic reaction, the population of B increases at the expense of A as the temperature is raised. (b) In an exothermic reaction, the opposite happens.

then an increase in temperature increases the population of the A states (which start at higher energy) at the expense of the B states, so the reactants become more abundant.

The temperature dependence of the equilibrium constant provides a non-calorimetric method of determining $\Delta_r H^{\ominus}$. A drawback is that the reaction enthalpy is actually temperature-dependent, so the plot is not expected to be perfectly linear. However, the temperature dependence is weak in many cases, so the plot is reasonably straight. In practice, the method is not very accurate, but it is often the only method available.

Example 6B.1 Measuring a reaction enthalpy

The data below show the temperature variation of the equilibrium constant of the reaction $Ag_2CO_3(s) \rightleftharpoons Ag_2O(s) + CO_2(g)$. Calculate the standard reaction enthalpy of the decomposition.

T/K	350	400	450	500
K	3.98×10^{-4}	1.41×10^{-2}	1.86×10^{-1}	1.48

Method It follows from eqn 6B.2b that, provided the reaction enthalpy can be assumed to be independent of temperature, a plot of $-\ln K$ against $1/T$ should be a straight line of slope $\Delta_r H^{\ominus}/R$.

Answer We draw up the following table:

T/K	350	400	450	500
$(10^3\,\text{K})/T$	2.86	2.50	2.22	2.00
$-\ln K$	6.83	4.26	1.68	−0.39

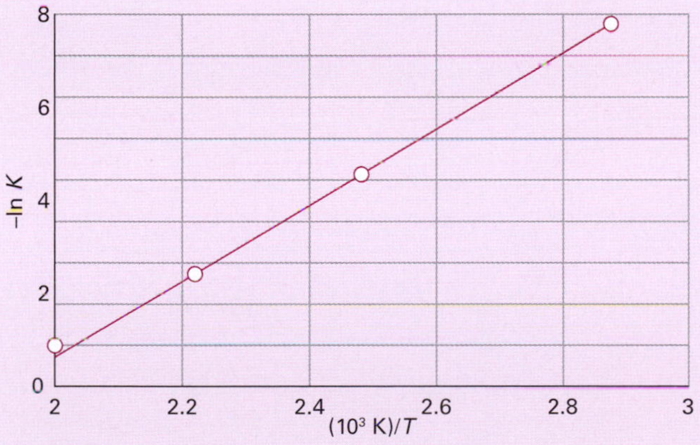

Figure 6B.4 When $-\ln K$ is plotted against $1/T$, a straight line is expected with slope equal to $\Delta_r H^{\ominus}/R$ if the standard reaction enthalpy does not vary appreciably with temperature. This is a non-calorimetric method for the measurement of reaction enthalpies.

These points are plotted in Fig. 6B.4. The slope of the graph is $+9.6\times10^3$, so

$$\Delta_r H^{\ominus} = (+9.6\times10^3\,\text{K})\times R = +80\,\text{kJ mol}^{-1}$$

Self-test 6B.2 The equilibrium constant of the reaction $2\,SO_2(g) + O_2(g) \rightleftharpoons 2\,SO_3(g)$ is 4.0×10^{24} at 300 K, 2.5×10^{10} at 500 K, and 3.0×10^{4} at 700 K. Estimate the reaction enthalpy at 500 K.

Answer: $-200\,\text{kJ mol}^{-1}$

(b) The value of K at different temperatures

To find the value of the equilibrium constant at a temperature T_2 in terms of its value K_1 at another temperature T_1, we integrate eqn 6B.2b between these two temperatures:

$$\ln K_2 - \ln K_1 = -\frac{1}{R}\int_{1/T_1}^{1/T_2} \Delta_r H^{\ominus}\,\text{d}(1/T) \qquad (6\text{B}.4)$$

If we suppose that $\Delta_r H^{\ominus}$ varies only slightly with temperature over the temperature range of interest, then we may take it outside the integral. It follows that

$$\ln K_2 - \ln K_1 = -\frac{\Delta_r H^{\ominus}}{R}\left(\frac{1}{T_2} - \frac{1}{T_1}\right) \qquad \text{Temperature dependence of } K \qquad (6\text{B}.5)$$

Brief illustration 6B.2 The temperature dependence of K

To estimate the equilibrium constant or the synthesis of ammonia at 500 K from its value at 298 K (6.1×10^5 for the reaction written as $N_2(g) + 3\,H_2(g) \rightleftharpoons 2\,NH_3(g)$) we use the standard reaction enthalpy, which can be obtained from Table 2C.2 in the *Resource section* by using $\Delta_r H^{\ominus} = 2\Delta_f H^{\ominus}(NH_3,g)$ and assume that its value is constant over the range of temperatures. Then, with $\Delta_r H^{\ominus} = -92.2\,\text{kJ mol}^{-1}$, from eqn 6B.3 we find

$$\ln K_2 = \ln(6.1\times10^5) - \left(\frac{-9.22\times10^4\,\text{J mol}^{-1}}{8.3145\,\text{J K}^{-1}\,\text{mol}^{-1}}\right)\times\left(\frac{1}{500\,\text{K}} - \frac{1}{298\,\text{K}}\right)$$
$$= -1.7\ldots$$

It follows that $K_2 = 0.18$, a lower value than at 298 K, as expected for this exothermic reaction.

Self-test 6B.3 The equilibrium constant for $N_2O_4(g) \rightleftharpoons 2\,NO_2(g)$ was calculated in *Self-test* 6A.6. Estimate its value at 100 °C.

Answer: 15

Checklist of concepts

- ☐ 1. The thermodynamic equilibrium constant is independent of pressure.
- ☐ 2. The response of composition to changes in the conditions is summarized by **Le Chatelier's principle**.
- ☐ 3. The dependence of the equilibrium constant on the temperature is expressed by the **van 't Hoff equation** and can be explained in terms of the distribution of molecules over the available states.

Checklist of equations

Property	Equation	Comment	Equation number
van 't Hoff equation	$\mathrm{d}\ln K/\mathrm{d}T = \Delta_r H^\ominus/RT^2$		6B.2a
	$\mathrm{d}\ln K/\mathrm{d}(1/T) = -\Delta_r H^\ominus/R$	Alternative version	6B.2b
Temperature dependence of equilibrium constant	$\ln K_2 - \ln K_1 = -(\Delta_r H^\ominus/R)(1/T_2 - 1/T_1)$	$\Delta_r H^\ominus$ assumed constant	6B.5

6C Electrochemical cells

Contents

➤ Why do you need to know this material?

One very special case of the material treated in Topic 6B that has enormous fundamental, technological, and economic significance concerns reactions that take place in electrochemical cells. Moreover, the ability to make very precise measurements of potential differences ('voltages') means that electrochemical methods can be used to determine thermodynamic properties of reactions that may be inaccessible by other methods.

➤ What is the key idea?

The electrical work that a reaction can perform at constant pressure and temperature is equal to the reaction Gibbs energy.

➤ What do you need to know already?

This Topic develops the relation between the Gibbs energy and non-expansion work (Topic 3C). You need to be aware of how to calculate the work of moving a charge through an electrical potential difference (Topic 2A). The equations make use of the definition of the reaction quotient Q and the equilibrium constant K (Topic 6A).

An **electrochemical cell** consists of two **electrodes**, or metallic conductors, in contact with an **electrolyte**, an ionic conductor (which may be a solution, a liquid, or a solid). An electrode and its electrolyte comprise an **electrode compartment**. The two electrodes may share the same compartment. The various kinds of electrode are summarized in Table 6C.1. Any 'inert metal' shown as part of the specification is present to act as a source or sink of electrons, but takes no other part in the reaction other than acting as a catalyst for it. If the electrolytes are different, the two compartments may be joined by a **salt bridge**, which is a tube containing a concentrated electrolyte solution (for instance, potassium chloride in agar jelly) that completes the electrical circuit and enables the cell to function. A **galvanic cell** is an electrochemical cell that produces electricity as a result of the spontaneous reaction occurring inside it. An **electrolytic cell** is an electrochemical cell in which a non-spontaneous reaction is driven by an external source of current.

6C.1 Half-reactions and electrodes

It will be familiar from introductory chemistry courses that **oxidation** is the removal of electrons from a species, a **reduction** is the addition of electrons to a species, and a **redox reaction** is a

Table 6C.1 Varieties of electrode

Electrode type	Designation	Redox couple	Half-reaction
Metal/metal ion	$M(s)\|M^+(aq)$	M^+/M	$M^+(aq)+e^- \rightarrow M(s)$
Gas	$Pt(s)\|X_2(g)\|X^+(aq)$	X^+/X_2	$X^+(aq)+e^- \rightarrow \frac{1}{2}X_2(g)$
	$Pt(s)\|X_2(g)\|X^-(aq)$	X_2/X^-	$\frac{1}{2}X_2(g)+e^- \rightarrow X^-(aq)$
Metal/insoluble salt	$M(s)\|MX(s)\|X^-(aq)$	$MX/M,X^-$	$MX(s)+e^- \rightarrow M(s)+X^-(aq)$
Redox	$Pt(s)\|M^+(aq),M^{2+}(aq)$	M^{2+}/M^+	$M^{2+}(aq)+e^- \rightarrow M^+(aq)$

reaction in which there is a transfer of electrons from one species to another. The electron transfer may be accompanied by other events, such as atom or ion transfer, but the net effect is electron transfer and hence a change in oxidation number of an element. The **reducing agent** (or *reductant*) is the electron donor; the **oxidizing agent** (or *oxidant*) is the electron acceptor. It should also be familiar that any redox reaction may be expressed as the difference of two reduction **half-reactions**, which are conceptual reactions showing the gain of electrons. Even reactions that are not redox reactions may often be expressed as the difference of two reduction half-reactions. The reduced and oxidized species in a half-reaction form a **redox couple**. In general we write a couple as Ox/Red and the corresponding reduction half-reaction as

$$\mathrm{Ox} + \nu e^- \rightarrow \mathrm{Red} \qquad (6C.1)$$

Brief illustration 6C.1 Redox couples

The dissolution of silver chloride in water $AgCl(s) \rightarrow Ag^+(aq) + Cl^-(aq)$, which is not a redox reaction, can be expressed as the difference of the following two reduction half-reactions:

$$AgCl(s) + e^- \rightarrow Ag(s) + Cl^-(aq)$$
$$Ag^+(aq) + e^- \rightarrow Ag(s)$$

The redox couples are $AgCl/Ag,Cl^-$ and Ag^+/Ag, respectively.

Self-test 6C.1 Express the formation of H_2O from H_2 and O_2 in acidic solution (a redox reaction) as the difference of two reduction half-reactions.

Answer: $4\,H^+(aq) + 4\,e^- \rightarrow 2\,H_2(g)$, $O_2(g) + 4\,H^+(aq) + 4\,e^- \rightarrow 2\,H_2O(l)$

We shall often find it useful to express the composition of an electrode compartment in terms of the reaction quotient, Q, for the half-reaction. This quotient is defined like the reaction quotient for the overall reaction (Topic 6A, $Q = \prod_J a_J^{\nu_J}$), but the electrons are ignored because they are stateless.

Brief illustration 6C.2 The reaction quotient

The reaction quotient for the reduction of O_2 to H_2O in acid solution, $O_2(g) + 4\,H^+(aq) + 4\,e^- \rightarrow 2\,H_2O(l)$, is

$$Q = \frac{a_{H_2O}^2}{a_{H^+}^4 a_{O_2}} \approx \frac{p^\ominus}{a_{H^+}^4 p_{O_2}}$$

The approximations used in the second step are that the activity of water is 1 (because the solution is dilute) and the oxygen behaves as a perfect gas, so $a_{O_2} \approx p_{O_2}/p^\ominus$.

Self-test 6C.2 Write the half-reaction and the reaction quotient for a chlorine gas electrode.

Answer: $Cl_2(g) + 2\,e^- \rightarrow 2\,Cl^-(aq)$, $Q \approx a_{Cl^-}^2 p^\ominus / p_{Cl_2}$

The reduction and oxidation processes responsible for the overall reaction in a cell are separated in space: oxidation takes place at one electrode and reduction takes place at the other. As the reaction proceeds, the electrons released in the oxidation $\mathrm{Red}_1 \rightarrow \mathrm{Ox}_1 + \nu e^-$ at one electrode travel through the external circuit and re-enter the cell through the other electrode. There they bring about reduction $\mathrm{Ox}_2 + \nu e^- \rightarrow \mathrm{Red}_2$. The electrode at which oxidation occurs is called the **anode**; the electrode at which reduction occurs is called the **cathode**. In a galvanic cell, the cathode has a higher potential than the anode: the species undergoing reduction, Ox_2, withdraws electrons from its electrode (the cathode, Fig. 6C.1), so leaving a relative positive charge on it (corresponding to a high potential). At the anode, oxidation results in the transfer of electrons to the electrode, so giving it a relative negative charge (corresponding to a low potential).

6C.2 Varieties of cells

The simplest type of cell has a single electrolyte common to both electrodes (as in Fig. 6C.1). In some cases it is necessary to immerse the electrodes in different electrolytes, as in the 'Daniell cell' in which the redox couple at one electrode is Cu^{2+}/Cu and at the other is Zn^{2+}/Zn (Fig. 6C.2). In an **electrolyte concentration cell**, the electrode compartments are identical except for the concentrations of the electrolytes. In an **electrode concentration cell** the electrodes themselves have different concentrations, either because they are gas electrodes operating at different pressures or because they are amalgams (solutions in mercury) with different concentrations.

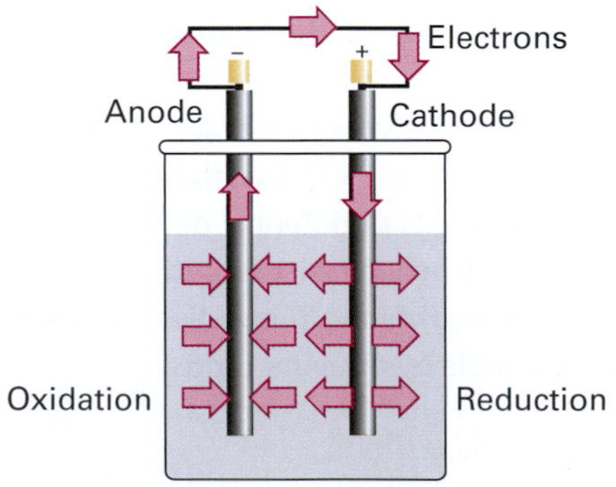

Figure 6C.1 When a spontaneous reaction takes place in a galvanic cell, electrons are deposited in one electrode (the site of oxidation, the anode) and collected from another (the site of reduction, the cathode), and so there is a net flow of current which can be used to do work. Note that the + sign of the cathode can be interpreted as indicating the electrode at which electrons enter the cell, and the − sign of the anode is where the electrons leave the cell.

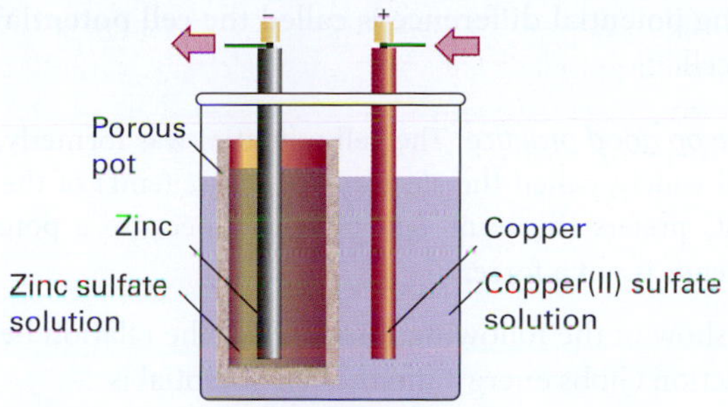

Figure 6C.2 One version of the Daniell cell. The copper electrode is the cathode and the zinc electrode is the anode. Electrons leave the cell from the zinc electrode and enter it again through the copper electrode.

(a) Liquid junction potentials

In a cell with two different electrolyte solutions in contact, as in the Daniell cell, there is an additional source of potential difference across the interface of the two electrolytes. This potential is called the **liquid junction potential**, E_{lj}. Another example of a junction potential is that between different concentrations of hydrochloric acid. At the junction, the mobile H^+ ions diffuse into the more dilute solution. The bulkier Cl^- ions follow, but initially do so more slowly, which results in a potential difference at the junction. The potential then settles down to a value such that, after that brief initial period, the ions diffuse at the same rates. Electrolyte concentration cells always have a liquid junction; electrode concentration cells do not.

The contribution of the liquid junction to the potential can be reduced (to about 1 to 2 mV) by joining the electrolyte compartments through a salt bridge (Fig. 6C.3). The reason for the success of the salt bridge is that provided the ions dissolved in the jelly have similar mobilities, then the liquid junction potentials at either end are largely independent of the concentrations of the two dilute solutions, and so nearly cancel.

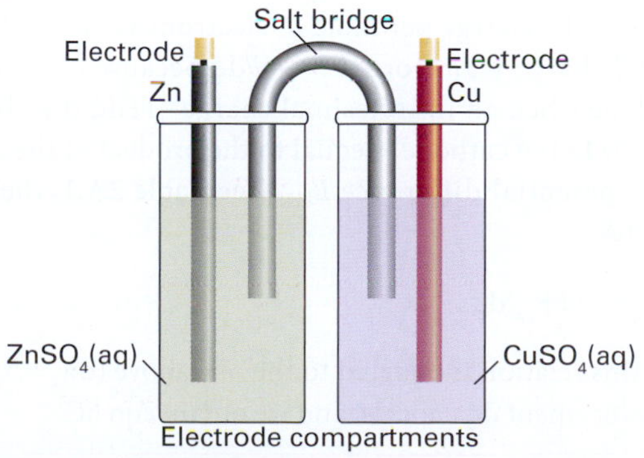

Figure 6C.3 The salt bridge, essentially an inverted U-tube full of concentrated salt solution in a jelly, has two opposing liquid junction potentials that almost cancel.

(b) Notation

We use the following notation for cells:

|	A phase boundary
⋮	A liquid junction
||	An interface for which it is assumed that the junction potential has been eliminated

Brief illustration 6C.3 Cell notation

A cell in which two electrodes share the same electrolyte is

$$Pt(s)|H_2(g)|HCl(aq)|AgCl|Ag(s)$$

The cell in Fig. 6C.2 is denoted

$$Zn(s)|ZnSO_4(aq)⋮CuSO_4(aq)|Cu(s)$$

The cell in Fig. 6C.3 is denoted

$$Zn(s)|ZnSO_4(aq)||CuSO_4(aq)|Cu(s)$$

An example of an electrolyte concentration cell in which the liquid junction potential is assumed to be eliminated is

$$Pt(s)|H_2(g)|HCl(aq,b_1)||HCl(aq,b_2)|H_2(g)|Pt(s)$$

Self-test 6C.3 Write the symbolism for a cell in which the half-reactions are $4\,H^+(aq)+4\,e^- \rightarrow 2\,H_2(g)$ and $O_2(g)+4\,H^+(aq)+4\,e^- \rightarrow 2\,H_2O(l)$, (a) with a common electrolyte, (b) with separate compartments joined by a salt bridge.

Answer: (a) $Pt(s)|H_2(g)|HCl(aq,b)|O_2(g)|Pt(s)$; (b) $Pt(s)|H_2(g)|HCl(aq,b_1)||HCl(aq,b_2)|O_2(g)|Pt(s)$

6C.3 The cell potential

The current produced by a galvanic cell arises from the spontaneous chemical reaction taking place inside it. The **cell reaction** is the reaction in the cell written on the assumption that the right-hand electrode is the cathode, and hence that the spontaneous reaction is one in which reduction is taking place in the right-hand compartment. Later we see how to predict if the right-hand electrode is in fact the cathode; if it is, then the cell reaction is spontaneous as written. If the left-hand electrode turns out to be the cathode, then the reverse of the corresponding cell reaction is spontaneous.

To write the cell reaction corresponding to a cell diagram, we first write the right-hand half-reaction as a reduction (because we have assumed that to be spontaneous). Then we subtract from it the left-hand reduction half-reaction (for, by implication, that electrode is the site of oxidation).

Brief illustration 6C.4 The cell reaction

For the cell $Zn(s)|ZnSO_4(aq)||CuSO_4(aq)|Cu(s)$ the two electrodes and their reduction half-reactions are

$$\text{Right-hand electrode: } Cu^{2+}(aq)+2e^- \rightarrow Cu(s)$$
$$\text{Left-hand electrode: } Zn^{2+}(aq)+2e^- \rightarrow Zn(s)$$

Hence, the overall cell reaction is the difference Right − Left:

$$Cu^{2+}(aq)+2e^- - Zn^{2+}(aq)-2e^- \rightarrow Cu(s)-Zn(s)$$

which, after cancellation of the $2e^-$, rearranges to

$$Cu^{2+}(aq)+Zn(s)\rightarrow Cu(s)+Zn^{2+}(aq)$$

Self-test 6C.4 Construct the overall cell reaction for the cells:

(a) $Pt(s)|H_2(g)|HCl(aq,b)|O_2(g)|Pt(s)$;

(b) $Pt(s)|H_2(g)|HCl(aq,b_L)||HCl(aq,b_R)|O_2(g)|Pt(s)$.

Answer: (a) $2\,H_2(g)+O_2(g)\rightarrow 2\,H_2O(l)$;
(b) $2\,H_2(g)+O_2(g)+4\,H^+(b_R)\rightarrow 2\,H_2O(l)+4\,H^+(b_L)$

(a) The Nernst equation

A cell in which the overall cell reaction has not reached chemical equilibrium can do electrical work as the reaction drives electrons through an external circuit. The work that a given transfer of electrons can accomplish depends on the potential difference between the two electrodes. When the potential difference is large, a given number of electrons travelling between the electrodes can do a large amount of electrical work. When the potential difference is small, the same number of electrons can do only a small amount of work. A cell in which the overall reaction is at equilibrium can do no work, and then the potential difference is zero.

According to the discussion in Topic 3C, we know that the maximum non-expansion work a system can do at constant temperature and pressure is given by eqn 3C.16b ($w_{e,max}=\Delta G$). In electrochemistry, the non-expansion work is identified with electrical work, the system is the cell, and ΔG is the Gibbs energy of the cell reaction, $\Delta_r G$. Maximum work is produced when a change occurs reversibly. It follows that, to draw thermodynamic conclusions from measurements of the work that a cell can do, we must ensure that the cell is operating reversibly. Moreover, it is established in Topic 6A that the reaction Gibbs energy is actually a property related, through $RT \ln Q$, to a specified composition of the reaction mixture. Therefore, to make use of $\Delta_r G$ we must ensure that the cell is operating reversibly at a specific, constant composition. Both these conditions are achieved by measuring the cell potential when it is balanced by an exactly opposing source of potential so that the cell reaction occurs reversibly, the composition is constant, and no current flows: in effect, the cell reaction is poised for change, but not actually changing. The resulting potential difference is called the **cell potential**, E_{cell}, of the cell.

A note on good practice The cell potential was formerly, and is still widely, called the *electromotive force* (emf) of the cell. IUPAC prefers the term 'cell potential' because a potential difference is not a force.

As we show in the following *Justification*, the relation between the reaction Gibbs energy and the cell potential is

$$-\nu FE_{cell}=\Delta_r G \qquad \text{The cell potential} \quad (6C.2)$$

where F is Faraday's constant, $F=eN_A$, and ν is the stoichiometric coefficient of the electrons in the half-reactions into which the cell reaction can be divided. This equation is the key connection between electrical measurements on the one hand and thermodynamic properties on the other. It will be the basis of all that follows.

Justification 6C.1 The relation between the cell potential and the reaction Gibbs energy

We consider the change in G when the cell reaction advances by an infinitesimal amount $d\xi$ at some composition. From *Justification* 6A.1, specifically the equation $\Delta_r G=(\partial G/\partial\xi)_{T,p}$, we can write (at constant temperature and pressure)

$$dG=\Delta_r G d\xi$$

The maximum non-expansion (electrical) work that the reaction can do as it advances by $d\xi$ at constant temperature and pressure is therefore

$$dw_e=\Delta_r G d\xi$$

This work is infinitesimal, and the composition of the system is virtually constant when it occurs.

Suppose that the reaction advances by $d\xi$, then $\nu d\xi$ electrons must travel from the anode to the cathode. The total charge transported between the electrodes when this change occurs is $-\nu eN_A d\xi$ (because $\nu d\xi$ is the amount of electrons in moles and the charge per mole of electrons is $-eN_A$). Hence, the total charge transported is $-\nu Fd\xi$ because $eN_A=F$. The work done when an infinitesimal charge $-\nu Fd\xi$ travels from the anode to the cathode is equal to the product of the charge and the potential difference E_{cell} (see Table 2A.1, the entry $dw=Qd\phi$):

$$dw_e=-\nu FE_{cell}d\xi$$

When this relation is equated to the one above ($dw_e=\Delta_r G d\xi$), the advancement $d\xi$ cancels, and we obtain eqn 6C.2.

It follows from eqn 6C.2 that, by knowing the reaction Gibbs energy at a specified composition, we can state the cell

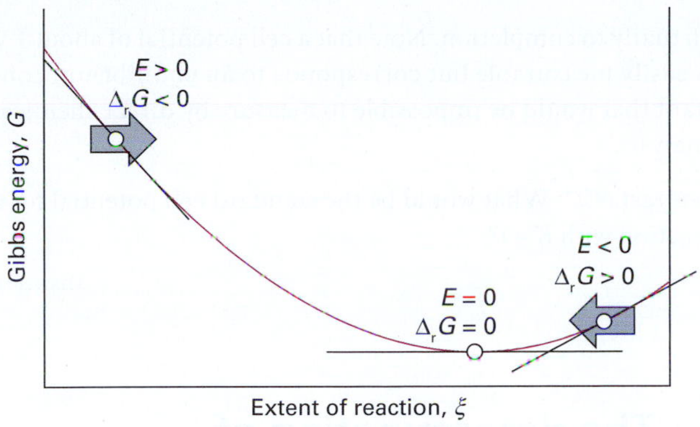

Figure 6C.4 A spontaneous reaction occurs in the direction of decreasing Gibbs energy. When expressed in terms of a cell potential, the spontaneous direction of change can be expressed in terms of the cell potential, E_{cell}. The reaction is spontaneous as written (from left to right on the illustration) when $E_{cell} > 0$. The reverse reaction is spontaneous when $E_{cell} < 0$. When the cell reaction is at equilibrium, the cell potential is zero.

potential at that composition. Note that a negative reaction Gibbs energy, corresponding to a spontaneous cell reaction, corresponds to a positive cell potential. Another way of looking at the content of eqn 6C.2 is that it shows that the driving power of a cell (that is, its potential) is proportional to the slope of the Gibbs energy with respect to the extent of reaction. It is plausible that a reaction that is far from equilibrium (when the slope is steep) has a strong tendency to drive electrons through an external circuit (Fig. 6C.4). When the slope is close to zero (when the cell reaction is close to equilibrium), the cell potential is small.

Brief illustration 6C.5 The reaction Gibbs energy

Equation 6C.2 provides an electrical method for measuring a reaction Gibbs energy at any composition of the reaction mixture: we simply measure the cell potential and convert it to $\Delta_r G$. Conversely, if we know the value of $\Delta_r G$ at a particular composition, then we can predict the cell potential. For example, if $\Delta_r G = -1\times10^2\,\text{kJ mol}^{-1}$ and $\nu = 1$, then

$$E_{cell} = -\frac{\Delta_r G}{\nu F} = -\frac{(-1\times10^5\,\text{J mol}^{-1})}{1\times(9.6485\times10^4\,\text{C mol}^{-1})} = 1\,\text{V}$$

where we have used 1 J = 1 C V.

Self-test 6C.5 Estimate the potential of a fuel cell in which the reaction is $H_2(g) + \frac{1}{2}O_2(g) \rightarrow H_2O(l)$.

Answer: 1.2 V

We can go on to relate the cell potential to the activities of the participants in the cell reaction. We know that the reaction Gibbs energy is related to the composition of the reaction mixture by eqn 6A.10 ($\Delta_r G = \Delta_r G^{\ominus} + RT\ln Q$); it follows, on division of both sides by $-\nu F$ and recognizing that $\Delta_r G/(-\nu F) = E_{cell}$, that

$$E_{cell} = -\frac{\Delta_r G^{\ominus}}{\nu F} - \frac{RT}{\nu F}\ln Q$$

The first term on the right is written

$$E^{\ominus}_{cell} = -\frac{\Delta_r G^{\ominus}}{\nu F} \qquad \textit{Definition} \quad \text{Standard cell potential} \qquad (6C.3)$$

and called the **standard cell potential**. That is, the standard cell potential is the standard reaction Gibbs energy expressed as a potential difference (in volts). It follows that

$$E_{cell} = E^{\ominus}_{cell} - \frac{RT}{\nu F}\ln Q \qquad \text{Nernst equation} \qquad (6C.4)$$

This equation for the cell potential in terms of the composition is called the **Nernst equation**; the dependence that it predicts is summarized in Fig. 6C.5. One important application of the Nernst equation is to the determination of the pH of a solution and, with a suitable choice of electrodes, of the concentration of other ions (Topic 6D).

We see from eqn 6C.4 that the standard cell potential can be interpreted as the cell potential when all the reactants and products in the cell reaction are in their standard states, for then all activities are 1, so $Q = 1$ and $\ln Q = 0$. However, the fact that the standard cell potential is merely a disguised form of the standard reaction Gibbs energy (eqn 6C.3) should always be kept in mind and underlies all its applications.

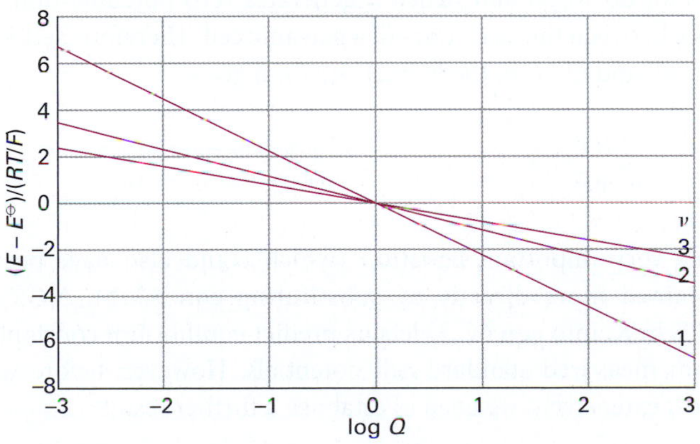

Figure 6C.5 The variation of cell potential with the value of the reaction quotient for the cell reaction for different values of ν (the number of electrons transferred). At 298 K, $RT/F = 25.69$ mV, so the vertical scale refers to multiples of this value.

Brief illustration 6C.6 The Nernst equation

Because $RT/F = 25.7$ mV at 25 °C, a practical form of the Nernst equation is

$$E_{\text{cell}} = E^{\ominus}_{\text{cell}} - \frac{25.7\,\text{mV}}{\nu} \ln Q$$

It then follows that, for a reaction in which $\nu = 1$, if Q is increased by a factor of 10, then the cell potential decreases by 59.2 mV.

Self-test 6C.6 By how much does the cell potential change when Q is decreased by a factor of 10 for a reaction in which $\nu = 2$?

Answer: −29.6 V

An important feature of a standard cell potential is that it is unchanged if the chemical equation for the cell reaction is multiplied by a numerical factor. A numerical factor increases the value of the standard Gibbs energy for the reaction. However, it also increases the number of electrons transferred by the same factor, and by eqn 6D.2 the value of $E^{\ominus}_{\text{cell}}$ remains unchanged. A practical consequence is that a cell potential is independent of the physical size of the cell. In other words, the cell potential is an intensive property.

(b) Cells at equilibrium

A special case of the Nernst equation has great importance in electrochemistry and provides a link to the discussion of equilibrium in Topic 6A. Suppose the reaction has reached equilibrium; then $Q = K$, where K is the equilibrium constant of the cell reaction. However, a chemical reaction at equilibrium cannot do work, and hence it generates zero potential difference between the electrodes of a galvanic cell. Therefore, setting $E_{\text{cell}} = 0$ and $Q = K$ in the Nernst equation gives

$$E^{\ominus}_{\text{cell}} = \frac{RT}{\nu F} \ln K \qquad \text{Equilibrium constant and standard cell potential} \qquad (6C.5)$$

This very important equation (which could also have been obtained more directly by substituting eqn 6A.14, $\Delta_r G^{\ominus} = -RT \ln K$, into eqn 6C.3) lets us predict equilibrium constants from measured standard cell potentials. However, before we use it extensively, we need to establish a further result.

Brief illustration 6C.7 Equilibrium constants

Because the standard potential of the Daniell cell is +1.10 V, the equilibrium constant for the cell reaction $Cu^{2+}(aq) + Zn(s) \rightarrow Cu(s) + Zn^{2+}(aq)$, for which $\nu = 2$, is $K = 1.5 \times 10^{37}$ at 298 K. We conclude that the displacement of copper by zinc goes virtually to completion. Note that a cell potential of about 1 V is easily measurable but corresponds to an equilibrium constant that would be impossible to measure by direct chemical analysis.

Self-test 6C.7 What would be the standard cell potential for a reaction with $K = 1$?

Answer: 0

6C.4 The determination of thermodynamic functions

The standard potential of a cell is related to the standard reaction Gibbs energy through eqn 6C.3 (written as $-\nu F E^{\ominus}_{\text{cell}} = \Delta_r G^{\ominus}$). Therefore, by measuring $E^{\ominus}_{\text{cell}}$ we can obtain this important thermodynamic quantity. Its value can then be used to calculate the Gibbs energy of formation of ions by using the convention explained in Topic 3C, that $\Delta_f G^{\ominus}(H^+,aq) = 0$.

Brief illustration 6C.8 The reaction Gibbs energy

The reaction taking place in the cell

$$Pt(s)|H_2(g)|H^+(aq)||Ag^+(aq)|Ag(s) \quad E^{\ominus}_{\text{cell}} = +0.7996\,\text{V}$$

is

$$Ag^+(aq) + \tfrac{1}{2}H_2(g) \rightarrow H^+(aq) + Ag(s) \quad \Delta_r G^{\ominus} = -\Delta_f G^{\ominus}(Ag^+,aq)$$

Therefore, with $\nu = 1$, we find

$$\begin{aligned}\Delta_f G^{\ominus}(Ag^+,aq) &= -(-FE^{\ominus}) \\ &= (9.6485\times10^4\,\text{C mol}^{-1})\times(0.7996\,\text{V}) \\ &= +77.15\,\text{kJ mol}^{-1}\end{aligned}$$

which is in close agreement with the value in Table 2C.2 of the *Resource section*.

Self-test 6C.8 Derive the value of $\Delta_f G^{\ominus}(H_2O, l)$ at 298 K from the standard potential of the cell $Pt(s)|H_2(g)|HCl(aq)|O_2(g)|Pt$, $E^{\ominus}_{\text{cell}} = +1.23$ V.

Answer: −237 kJ mol⁻¹

The temperature coefficient of the standard cell potential, $dE^{\ominus}_{\text{cell}}/dT$, gives the standard entropy of the cell reaction. This conclusion follows from the thermodynamic relation $(\partial G/\partial T)_p = -S$ derived in Topic 3D and eqn 6C.3, which combine to give

$$\frac{dE^{\ominus}_{\text{cell}}}{dT} = \frac{\Delta_r S^{\ominus}}{\nu F} \qquad \text{Temperature coefficient of standard cell potential} \qquad (6C.6)$$

The derivative is complete (not partial) because $E^{\ominus}_{\text{cell}}$ like $\Delta_r G^{\ominus}$, is independent of the pressure. Hence we have an electrochemical technique for obtaining standard reaction entropies and through them the entropies of ions in solution.

Finally, we can combine the results obtained so far and use them to obtain the standard reaction enthalpy:

$$\Delta_r H^{\ominus} = \Delta_r G^{\ominus} + T\Delta_r S^{\ominus} = -\nu F\left(E^{\ominus}_{\text{cell}} - T\frac{dE^{\ominus}_{\text{cell}}}{dT}\right) \qquad (6C.7)$$

This expression provides a non-calorimetric method for measuring $\Delta_r H^{\ominus}$ and, through the convention $\Delta_f H^{\ominus}(H^+, aq)=0$ the standard enthalpies of formation of ions in solution (Topic 2C).

Example 6C.1 Using the temperature coefficient of the cell potential

The standard potential of the cell Pt(s)|H_2(g)|HBr(aq)|AgBr(s)|Ag(s) was measured over a range of temperatures, and the data were found to fit the following polynomial:

$$E^{\ominus}_{\text{cell}}/\text{V} = 0.07131 - 4.99\times10^{-4}(T/\text{K}-298) - 3.45\times10^{-6}(T/\text{K}-298)^2$$

The cell reaction is $\text{AgBr(s)} + \tfrac{1}{2}\text{H}_2\text{(g)} \rightarrow \text{Ag(s)} + \text{HBr(aq)}$. Evaluate the standard reaction Gibbs energy, enthalpy, and entropy at 298 K.

Method The standard Gibbs energy of reaction is obtained by using eqn 6C.2 after evaluating $E^{\ominus}_{\text{cell}}$ at 298 K and by using 1 V C = 1 J. The standard entropy of reaction is obtained by using eqn 6C.6, which involves differentiating the polynomial with respect to T and then setting T = 298 K. The reaction enthalpy is obtained by combining the values of the standard Gibbs energy and entropy.

Answer At T = 298 K, $E^{\ominus}_{\text{cell}}/\text{V} = 0.07131\,\text{V}$, so

$$\Delta_r G^{\ominus} = -\nu F E^{\ominus}_{\text{cell}} = -(1)\times(9.6485\times10^4\,\text{C mol}^{-1})\times(0.07131\,\text{V})$$
$$= -6.880\times10^3\,\text{C V mol}^{-1} = -6.880\,\text{kJ mol}^{-1}$$

The temperature coefficient of the cell potential is

$$\frac{dE^{\ominus}_{\text{cell}}}{dT} = -4.99\times10^{-4}\,\text{V K}^{-1} - 2(3.45\times10^{-6})(T/\text{K}-298)\,\text{V K}^{-1}$$

At T = 298 K this expression evaluates to

$$\frac{dE^{\ominus}_{\text{cell}}}{dT} = -4.99\times10^{-4}\,\text{V K}^{-1}$$

So, from eqn 6C.6 the reaction entropy is

$$\Delta_r S^{\ominus} = \nu F\frac{dE^{\ominus}_{\text{cell}}}{dT} = (1)\times(9.6485\times10^4\,\text{C mol}^{-1})\times(-4.99\times10^{-4}\,\text{V})$$
$$= -48.2\,\text{J K}^{-1}\,\text{mol}^{-1}$$

The negative value stems in part from the elimination of gas in the cell reaction. It then follows that

$$\Delta_r H^{\ominus} = \Delta_r G^{\ominus} + T\Delta_r S^{\ominus} = -6.880\,\text{kJ mol}^{-1} + (298\,\text{K})\times(-0.0482\,\text{kJ K}^{-1}\,\text{mol}^{-1})$$
$$= -21.2\,\text{kJ mol}^{-1}$$

One difficulty with this procedure lies in the accurate measurement of small temperature coefficients of cell potential. Nevertheless, it is another example of the striking ability of thermodynamics to relate the apparently unrelated, in this case to relate electrical measurements to thermal properties.

Self-test 6C.9 Predict the standard potential of the Harned cell at 303 K from tables of thermodynamic data.

Answer: +0.2222 V

Checklist of concepts

- ☐ 1. A **redox reaction** is expressed as the difference of two reduction **half-reactions**; each one defines a redox couple.
- ☐ 2. **Galvanic cells** are classified as **electrolyte concentration** and **electrode concentration cells**.
- ☐ 3. A **liquid junction potential** arises at the junction of two electrolyte solutions.
- ☐ 4. The cell notation specifies the structure of a cell.
- ☐ 5. The **Nernst equation** relates the cell potential to the composition of the reaction mixture.
- ☐ 6. The **standard cell potential** may be used to calculate the equilibrium constant of the cell reaction.
- ☐ 7. The temperature coefficient of the cell potential is used to measure thermodynamic properties of electroactive species.

Checklist of equations

Property	Equation	Comment	Equation number
Cell potential and reaction Gibbs energy	$-\nu FE_{cell}=\Delta_r G$	Constant temperature and pressure	6C.2
Standard cell potential	$E^{\ominus}_{cell}=-\Delta_r G^{\ominus}/\nu F$	Definition	6C.3
Nernst equation	$E_{cell}=E^{\ominus}_{cell}-(RT/\nu F)\ln Q$		6C.4
Equilibrium constant of cell reaction	$E^{\ominus}_{cell}=(RT/\nu F)\ln K$		6C.5
Temperature coefficient of cell potential	$dE^{\ominus}_{cell}/dT=\Delta_r S^{\ominus}/\nu F$		6C.6

6D Electrode potentials

Contents

➤ Why do you need to know this material?

A very powerful, compact, and widely used way to report standard cell potentials is to ascribe a potential to each electrode. Electrode potentials are widely used in chemistry to assess the oxidizing and reducing power of redox couples and to infer thermodynamic properties, including equilibrium constants.

➤ What is the key idea?

Each electrode of a cell can be supposed to make a characteristic contribution to the cell potential; redox couples with low electrode potentials tend to reduce those with higher potentials.

➤ What do you need to know already?

This Topic develops the concepts in Topic 6D, so you need to understand the concept of cell potential and standard cell potential (Topic 6D); it makes use of the Nernst equation (Topic 6D). The measurement of standard potentials makes use of the Debye–Hückel limiting law (Topic 5F).

As explained in Topic 6C, a galvanic cell is a combination of two electrodes each of which can be considered to make a characteristic contribution to the overall cell potential. Although it is not possible to measure the contribution of a single electrode, we can define the potential of one of the electrodes as zero and then assign values to others on that basis.

6D.1 Standard potentials

The specially selected electrode is the **standard hydrogen electrode** (SHE):

$$\mathrm{Pt(s)|H_2(g)|H^+(aq)} \quad E^{\ominus}=0 \qquad \textit{Convention} \quad \text{Standard potentials} \qquad (6D.1)$$

at all temperatures. To achieve the standard conditions, the activity of the hydrogen ions must be 1 (that is, pH=0) and the pressure (more precisely, the fugacity) of the hydrogen gas must be 1 bar. The **standard potential**, $E^{\ominus}(\mathrm{X})$, of another couple X is then assigned by constructing a cell in which it is the right-hand electrode and the standard hydrogen electrode is the left-hand electrode:

$$\mathrm{Pt(s)|H_2(g)|H^+(aq)||X} \quad E^{\ominus}(\mathrm{X})=E^{\ominus}_{\mathrm{cell}} \qquad \textit{Convention} \quad \text{Standard potentials} \qquad (6D.2)$$

The standard potential of a cell of the form L||R, where L is the left-hand electrode of the cell as written (not as arranged on the bench) and R is the right-hand electrode, is then given by the difference of the two standard potentials:

$$\mathrm{L||R} \quad E^{\ominus}_{\mathrm{cell}}=E^{\ominus}(\mathrm{R})-E^{\ominus}(\mathrm{L}) \qquad \text{Standard cell potential} \qquad (6D.3)$$

A list of standard potentials at 298 K is given in Table 6D.1, and longer lists in numerical and alphabetical order are in the *Resource section.*

Table 6D.1* Standard potentials at 298 K, $E^{\ominus}$/V

Couple	$E^{\ominus}$/V
$\mathrm{Ce^{4+}(aq)+e^- \rightarrow Ce^{3+}(aq)}$	+1.61
$\mathrm{Cu^{2+}(aq)+2\,e^- \rightarrow Cu(s)}$	+0.34
$\mathrm{H^+(aq)+e^- \rightarrow \frac{1}{2}H_2(g)}$	0
$\mathrm{AgCl(s)+e^- \rightarrow Ag(s)+Cl^-(aq)}$	+0.22
$\mathrm{Zn^{2+}(aq)+2\,e^- \rightarrow Zn(s)}$	−0.76
$\mathrm{Na^+(aq)+e^- \rightarrow Na(s)}$	−2.71

* More values are given in the *Resource section.*

Brief illustration 6D.1 Standard electrode potentials

The cell $Ag(s)|AgCl(s)|HCl(aq)|O_2(g)|Pt(s)$ can be regarded as formed from the following two electrodes, with their standard potentials taken from the *Resource section*:

Electrode	Half-reaction	Standard potential
R: $Pt(s)|O_2(g)|H^+(aq)$	$O_2(g)+4\,H^+(aq)+4\,e^- \to 2\,H_2O(l)$	+1.23 V
L: $Ag(s)|AgCl(s)|Cl^-(aq)$	$AgCl(s)+e^- \to Ag(s)+Cl^-(aq)$	+0.22 V
	$E^\ominus_{cell}$ =	+1.01 V

Self-test 6D.1 What is the standard potential of the cell $Pt(s)|Fe^{2+}(aq),Fe^{3+}(aq)||Ce^{4+}(aq),Ce^{3+}(aq)|Pt(s)$?

Answer: +0.84 V

(a) The measurement procedure

The procedure for measuring a standard potential can be illustrated by considering a specific case, the silver chloride electrode. The measurement is made on the 'Harned cell':

$$Pt(s)|H_2(g)|HCl(aq,b)|AgCl(s)|Ag(s)$$
$$\tfrac{1}{2}H_2(g)+AgCl(s)\to HCl(aq)+Ag(s)$$
$$E^\ominus_{cell}=E^\ominus(AgCl/Ag,Cl^-)-E^\ominus(SHE)=E^\ominus(AgCl/Ag,Cl^-),\nu=1$$

for which the Nernst equation is

$$E_{cell}=E^\ominus(AgCl/Ag,Cl^-)-\frac{RT}{F}\ln\frac{a_{H^+}a_{Cl^-}}{a_{H_2}^{1/2}}$$

We shall set $a_{H_2}=1$ from now on, and for simplicity write the standard potential of the $AgCl/Ag,Cl^-$ electrode as $E^\ominus$; then

$$E_{cell}=E^\ominus-\frac{RT}{F}\ln a_{H^+}a_{Cl^-}$$

We show in the following *Justification* that as the molality $b\to 0$,

$$\overbrace{E_{cell}+\frac{2RT}{F}\ln b}^{y}=\overbrace{E^\ominus}^{\text{intercept}}+\overbrace{C\times b^{1/2}}^{\text{slope}\times x} \qquad (6D.4)$$

where C is a constant. To use this equation, which has the form y=intercept+slope$\times x$ with $x=b^{1/2}$, the expression on the left is evaluated at a range of molalities, plotted against $b^{1/2}$, and extrapolated to $b=0$. The intercept at $b^{1/2}=0$ is the value of $E^\ominus$ for the silver/silver chloride electrode. In precise work, the $b^{1/2}$ term is brought to the left, and a higher-order correction term from extended versions of the Debye–Hückel law is used on the right.

Justification 6D.1 The Harned cell potential

The activities in the expression for E_{cell} can be expressed in terms of the molality b of HCl(aq) through $a_{H^+}=\gamma_\pm b/b^\ominus$ and $a_{Cl^-}=\gamma_\pm b/b^\ominus$ as established in Topic 5E, so, with $b/b^\ominus$ replaced by b

$$E_{cell}=E^\ominus-\frac{RT}{F}\ln b^2-\frac{RT}{F}\ln\gamma_\pm^2$$
$$=E^\ominus-\frac{2RT}{F}\ln b-\frac{2RT}{F}\ln\gamma_\pm$$

and therefore

$$E_{cell}+\frac{2RT}{F}\ln b=E^\ominus-\frac{2RT}{F}\ln\gamma_\pm$$

From the Debye–Hückel limiting law for a 1,1-electrolyte (eqn 5F.8, $\log\gamma_\pm=-A|z_+z_-|I^{1/2}$), as $b\to 0$

$$\log\gamma_\pm=-A|z_+z_-|I^{1/2}=-A(b/b^\ominus)^{1/2}$$

Therefore, because $\ln x=\ln 10\log x$,

$$\ln\gamma_\pm=\ln 10\log\gamma_\pm=-A\ln 10(b/b^\ominus)^{1/2}$$

and the equation for E_{cell} becomes

$$E_{cell}+\frac{2RT}{F}\ln b=E^\ominus+\frac{2ART\ln 10}{F(b^\ominus)^{1/2}}b^{1/2}\quad \text{as } b\to 0$$

With the term in blue denoted C, this equation becomes eqn 6D.4.

Example 6D.1 Evaluating a standard electrode potential

The potential of the Harned cell at 25 °C has the following values:

$b/(10^{-3}b^\ominus)$	3.215	5.619	9.138	25.63
E_{cell}/V	0.520 53	0.492 57	0.468 60	0.418 24

Determine the standard potential of the silver–silver chloride electrode.

Method As explained in the text, evaluate $y=E_{cell}+(2RT/F)\ln b$ and plot it against $b^{1/2}$; then extrapolate to $b=0$. Use $2RT/F=0.051\,39$ V.

Answer To determine the standard potential of the cell we draw up the following table:

$b/(10^{-3}b^\ominus)$	3.215	5.619	9.138	25.63
$\{b/(10^{-3}b^\ominus)\}^{1/2}$	1.793	2.370	3.023	5.063
E_{cell}/V	0.520 53	0.492 57	0.468 60	0.418 24
y/V	0.2256	0.2263	0.2273	0.2299

The data are plotted in Fig. 6D.1; as can be seen, they extrapolate to $E^\ominus=+0.2232$ V (the value obtained, to preserve the precision of the data, by linear regression).

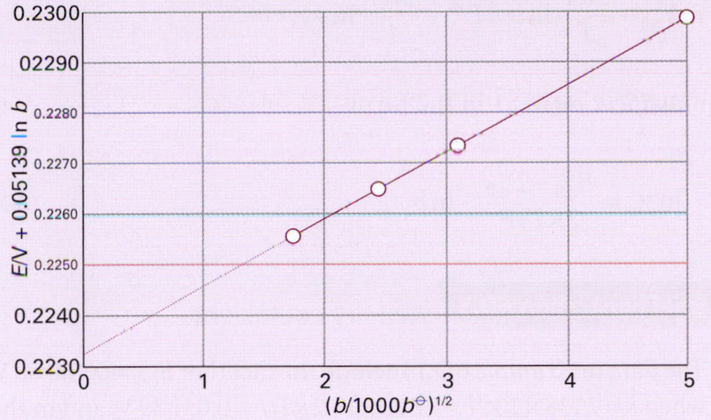

Figure 6D.1 The plot and the extrapolation used for the experimental measurement of a standard cell potential. The intercept at $b^{1/2}=0$ is $E^{\ominus}_{\text{cell}}$.

Self-test 6D.2 The data below are for the cell $\text{Pt(s)}|\text{H}_2\text{(g)}|\text{HBr(aq},b)|\text{AgBr(s)}|\text{Ag(s)}$ at 25 °C. Determine the standard cell potential.

$b/(10^{-4}b^{\ominus})$	4.042	8.444	37.19
E_{cell}/V	0.47381	0.43636	0.36173

Answer: +0.071 V

(b) Combining measured values

The standard potentials in Table 6D.1 may be combined to give values for couples that are not listed there. However, to do so, we must take into account the fact that different couples may correspond to the transfer of different numbers of electrons. The procedure is illustrated in the following *Example*.

Example 6D.2 Evaluating a standard potential from two others

Given that the standard potentials of the Cu^{2+}/Cu and Cu^{+}/Cu couples are +0.340 V and +0.522 V, respectively, evaluate $E^{\ominus}(Cu^{2+}, Cu^{+})$.

Method First, we note that reaction Gibbs energies may be added (as in a Hess's law analysis of reaction enthalpies). Therefore, we should convert the $E^{\ominus}$ values to $\Delta_r G^{\ominus}$ values by using eqn 6C.2 ($-\nu FE^{\ominus}=\Delta_r G^{\ominus}$), add them appropriately, and then convert the overall $\Delta_r G^{\ominus}$ to the required $E^{\ominus}$ by using eqn 6D.2 again. This roundabout procedure is necessary because, as we shall see, although the factor F cancels, the factor ν in general does not.

Answer The electrode half-reactions are as follows:

(a) $\text{Cu}^{2+}\text{(aq)}+2\,\text{e}^- \rightarrow \text{Cu(s)} \quad E^{\ominus}(\text{a})=+0.340\,\text{V},$

$\text{so}\,\Delta_r G^{\ominus}(\text{a})=-2(0.340\,\text{V})F$

(b) $\text{Cu}^{+}\text{(aq)}+\text{e}^- \rightarrow \text{Cu(s)} \quad E^{\ominus}(\text{b})=+0.522\,\text{V},$

$\text{so}\,\Delta_r G^{\ominus}(\text{b})=-(0.522\,\text{V})F$

The required reaction is

$$\text{(c)}\quad \text{Cu}^{2+}\text{(aq)}+\text{e}^- \rightarrow \text{Cu}^+\text{(aq)} \quad E^{\ominus}(\text{c})=-\Delta_r G^{\ominus}(\text{c})/F$$

Because (c) = (a) − (b), the standard Gibbs energy of reaction (c) is

$$-\Delta_r G^{\ominus}(\text{c})=\Delta_r G^{\ominus}(\text{a})-\Delta_r G^{\ominus}(\text{b})=-(0.680\,\text{V})F-(-0.522\,\text{V})F$$
$$=(+0.158\,\text{V})F$$

Therefore, $E^{\ominus}(\text{c})=+0.158$ V.

Self-test 6D.3 Evaluate $E^{\ominus}(Fe^{3+}, Fe^{2+})$ from $E^{\ominus}(Fe^{3+}, Fe)$ and $E^{\ominus}(Fe^{2+}, Fe)$.

Answer: +0.76 V

The generalization of the calculation in *Example* 6D.2 is

$$\nu_c E^{\ominus}(\text{c})=\nu_a E^{\ominus}(\text{a})-\nu_b E^{\ominus}(\text{b}) \qquad \text{Combination of standard potentials} \qquad (6\text{D}.5)$$

with the ν_r the stoichiometric coefficients of the electrons in each half-reaction

6D.2 Applications of standard potentials

Cell potentials are a convenient source of data on equilibrium constants and the Gibbs energies, enthalpies, and entropies of reactions. In practice the standard values of these quantities are the ones normally determined.

(a) The electrochemical series

We have seen that for two redox couples, Ox_L/Red_L and Ox_R/Red_R, and the cell

$$\text{L}\|\text{R}=\text{Ox}_L/\text{Red}\|\text{Ox}_R/\text{Red}_R$$
$$\text{Ox}_R+\nu\text{e}^- \rightarrow \text{Red}_R \quad \text{Ox}_L+\nu\text{e}^- \rightarrow \text{Red}_L \qquad \text{Cell convention} \qquad (6\text{D}.6\text{a})$$
$$E^{\ominus}_{\text{cell}}=E^{\ominus}(\text{R})-E^{\ominus}(\text{L})$$

that the cell reaction

$$\text{R}-\text{L}:\ \text{Red}_L+\text{Ox}_R \rightarrow \text{Ox}_L+\text{Red}_R \qquad (6\text{D}.6\text{b})$$

has $K>1$ as written if $E^{\ominus}_{\text{cell}}>0$, and therefore if $E^{\ominus}(\text{L})<E^{\ominus}(\text{R})$. Because in the cell reaction Red_L reduces Ox_R, we can conclude that:

> Red_L has a thermodynamic tendency (in the sense $K>1$) to reduce Ox_R if $E^{\ominus}(\text{L})<E^{\ominus}(\text{R})$

More briefly: low reduces high.

Table 6D.2* The electrochemical series of the metals

Least strongly reducing
Gold
Platinum
Silver
Mercury
Copper
(Hydrogen)
Lead
Tin
Nickel
Iron
Zinc
Chromium
Aluminium
Magnesium
Sodium
Calcium
Potassium
Most strongly reducing

* The complete series can be inferred from Table 6D.1 in the *Resource section*.

Table 6D.2 shows a part of the **electrochemical series**, the metallic elements (and hydrogen) arranged in the order of their reducing power as measured by their standard potentials in aqueous solution. A metal low in the series (with a lower standard potential) can reduce the ions of metals with higher standard potentials. This conclusion is qualitative. The quantitative value of K is obtained by doing the calculations we have described previously and review below.

Brief illustration 6D.2 The electrochemical series

Zinc lies above magnesium in the electrochemical series, so zinc cannot reduce magnesium ions in aqueous solution. Zinc can reduce hydrogen ions, because hydrogen lies higher in the series. However, even for reactions that are thermodynamically favourable, there may be kinetic factors that result in very slow rates of reaction.

Self-test 6D.4 Can nickel reduce hydrogen ions to hydrogen gas?

Answer: Yes

(b) The determination of activity coefficients

Once the standard potential of an electrode in a cell is known, we can use it to determine mean activity coefficients by measuring the cell potential with the ions at the concentration of interest. For example, the mean activity coefficient of the ions in hydrochloric acid of molality b is obtained from the relation

$$E_{\text{cell}} + \frac{2RT}{F}\ln b = E^{\ominus} - \frac{2RT}{F}\ln \gamma_{\pm}$$

in *Justification* 6D.1 in the form

$$\ln \gamma_{\pm} = \frac{E^{\ominus}_{\text{cell}} - E_{\text{cell}}}{2RT/F} - \ln b \qquad (6D.7)$$

Brief illustration 6D.3 Activity coefficients

The data in *Example* 6D.1 include the fact that $E_{\text{cell}} = 0.468\,60$ V when $b = 9.138\times10^{-3}b^{\ominus}$. Because $2RT/F = 0.051\,39$ V, and in the *Example* it is established that $E^{\ominus}_{\text{cell}} = 0.2232$ V, the mean activity coefficient at this molality is

$$\ln \gamma_{\pm} = \frac{0.2232\,\text{V} - 0.468\,60\,\text{V}}{0.05139\,\text{V}} - \ln(9.138\times10^{-3}) = -0.0788\ldots$$

Therefore, $\gamma_{\pm} = 0.9242$.

Self-test 6D.5 Evaluate the mean activity coefficient when $b = 5.619\times10^{-3}b^{\ominus}$.

Answer: 0.9417

(c) The determination of equilibrium constants

The principal use for standard potentials is to calculate the standard potential of a cell formed from any two electrodes. To do so, we construct $E^{\ominus}_{\text{cell}} = E^{\ominus}(\text{R}) - E^{\ominus}(\text{L})$ and use eqn 6C.5 of Topic 6C ($E^{\ominus}_{\text{cell}} = (RT/\nu F)\ln K$, arranged into $\ln K = \nu F E^{\ominus}_{\text{cell}}/RT$).

Brief illustration 6D.4 Equilibrium constants

A disproportionation reaction is a reaction in which a species is both oxidized and reduced. To study the disproportionation $2\,Cu^{+}(aq) \rightarrow Cu(s) + Cu^{2+}(aq)$ at 298 K we combine the following electrodes:

R: $Cu(s)|Cu^{+}(aq)$	$Cu^{+}(aq) + e^{-} \rightarrow Cu(s)$	$E^{\ominus}(\text{R}) = +0.52$ V
L: $Pt(s)|Cu^{2+}(aq),Cu^{+}(aq)$	$Cu^{2+}(aq) + e^{-} \rightarrow Cu^{+}(s)$	$E^{\ominus}(\text{R}) = +0.16$ V

The standard potential of the cell is therefore

$$E^{\ominus}_{\text{cell}} = 0.52\,\text{V} - 0.16\,\text{V} = +0.36\,\text{V}$$

We can now calculate the equilibrium constant of the cell reaction. Because $\nu = 1$, from eqn 6C.5 with $RT/F = 0.025\,693$ V,

$$\ln K = \frac{0.36\,\text{V}}{0.025693\,\text{V}} = 14.0\ldots$$

Hence, $K = 1.2\times10^{6}$.

Self-test 6D.6 Evaluate the equilibrium constant for the reaction $Sn(s) + Sn^{4+}(aq) \rightleftharpoons 2\,Sn^{2+}(aq)$.

Answer: 6.5×10^{9}

Checklist of concepts

- ☐ 1. The **standard potential** of a couple is the cell potential in which it forms the right-hand electrode and the left-hand electrode is a **standard hydrogen electrode**.
- ☐ 2. The **electrochemical series** lists the metallic elements in the order of their reducing power as measured by their standard potentials in aqueous solution: low reduces high.
- ☐ 3. The cell potential is used to measure the activity coefficient of electroactive ions.
- ☐ 4. The standard cell potential is used to infer the equilibrium constant of the cell reaction.

Checklist of equations

Property	Equation	Comment	Equation number
Standard cell potential	$E^{\ominus}_{\text{cell}} = E^{\ominus}(\text{R}) - E^{\ominus}(\text{L})$	Cell: L‖R	6D.3
Combined potentials	$\nu_c E^{\ominus}(\text{c}) = \nu_a E^{\ominus}(\text{a}) - \nu_b E^{\ominus}(\text{b})$		6D.5

CHAPTER 6 Chemical equilibrium

TOPIC 6A The equilibrium constant

Discussion questions

6A.1 Explain how the mixing of reactants and products affects the position of chemical equilibrium.

6A.2 What is the justification for not including a pure liquid or solid in the expression for an equilibrium constant?

Exercises

6A.1(a) Consider the reaction A→2 B. Initially, 1.50 mol A is present and no B. What are the amounts of A and B when the extent of reaction is 0.60 mol?
6A.1(b) Consider the reaction 2 A→B. Initially, 1.75 mol A and 0.12 mol B are present. What are the amounts of A and B when the extent of reaction is 0.30 mol?

6A.2(a) When the reaction A→2 B advances by 0.10 mol (that is, $\Delta\xi = +0.10$ mol) the Gibbs energy of the system changes by −6.4 kJ mol⁻¹. What is the Gibbs energy of reaction at this stage of the reaction?
6A.2(b) When the reaction 2 A→B advances by 0.051 mol (that is, $\Delta\xi = +0.051$ mol) the Gibbs energy of the system changes by −2.41 kJ mol⁻¹. What is the Gibbs energy of reaction at this stage of the reaction?

6A.3(a) The standard Gibbs energy of the reaction $N_2(g) + 3\,H_2(g) \rightarrow 2\,NH_3(g)$ is −32.9 kJ mol⁻¹ at 298 K. What is the value of $\Delta_r G$ when Q = (i) 0.010, (ii) 1.0, (iii) 10.0, (iv) 100 000, (v) 1 000 000? Estimate (by interpolation) the value of K from the values you calculate. What is the actual value of K?
6A.3(b) The standard Gibbs energy of the reaction $2\,NO_2(g) \rightarrow N_2O_4(g)$ is −4.73 kJ mol⁻¹ at 298 K. What is the value of $\Delta_r G$ when Q = (i) 0.10, (ii) 1.0, (iii) 10, (iv) 100? Estimate (by interpolation) the value of K from the values you calculate. What is the actual value of K?

6A.4(a) At 2257 K and 1.00 bar total pressure, water is 1.77 per cent dissociated at equilibrium by way of the reaction $2\,H_2O(g) \rightleftharpoons 2\,H_2(g) + O_2(g)$. Calculate K.
6A.4(b) For the equilibrium, $N_2O_4(g) \rightleftharpoons 2\,NO_2(g)$, the degree of dissociation, α, at 298 K is 0.201 at 1.00 bar total pressure. Calculate K.

6A.5(a) Dinitrogen tetroxide is 18.46 per cent dissociated at 25 °C and 1.00 bar in the equilibrium $N_2O_4(g) \rightleftharpoons 2\,NO_2(g)$. Calculate K at (i) 25 °C, (ii) 100 °C given that $\Delta_r H^\ominus = +56.2$ kJ mol⁻¹ over the temperature range.
6A.5(b) Molecular bromine is 24 per cent dissociated at 1600 K and 1.00 bar in the equilibrium $Br_2(g) \rightleftharpoons 2\,Br(g)$. Calculate K at (i) 1600 °C, (ii) 2000 °C given that $\Delta_r H^\ominus = +112$ kJ mol⁻¹ over the temperature range.

6A.6(a) From information in the *Resource section*, calculate the standard Gibbs energy and the equilibrium constant at (i) 298 K and (ii) 400 K for the reaction $PbO(s) + CO(g) \rightleftharpoons Pb(s) + CO_2(g)$. Assume that the reaction enthalpy is independent of temperature.
6A.6(b) From information in the *Resource section*, calculate the standard Gibbs energy and the equilibrium constant at (i) 25 °C and (ii) 50 °C for the reaction $CH_4(g) + 3\,Cl_2(g) \rightleftharpoons CHCl_3(l) + 3\,HCl(g)$. Assume that the reaction enthalpy is independent of temperature.

6A.7(a) Establish the relation between K and K_c for the reaction $H_2CO(g) \rightleftharpoons CO(g) + H_2(g)$.
6A.7(b) Establish the relation between K and K_c for the reaction $3\,N_2(g) + H_2(g) \rightleftharpoons 2\,HN_3(g)$.

6A.8(a) In the gas-phase reaction $2\,A + B \rightleftharpoons 3\,C + 2\,D$, it was found that, when 1.00 mol A, 2.00 mol B, and 1.00 mol D were mixed and allowed to come to equilibrium at 25 °C, the resulting mixture contained 0.90 mol C at a total pressure of 1.00 bar. Calculate (i) the mole fractions of each species at equilibrium, (ii) K_x, (iii) K, and (iv) $\Delta_r G^\ominus$.
6A.8(b) In the gas-phase reaction $A + B \rightleftharpoons C + 2\,D$, it was found that, when 2.00 mol A, 1.00 mol B, and 3.00 mol D were mixed and allowed to come to equilibrium at 25 °C, the resulting mixture contained 0.79 mol C at a total pressure of 1.00 bar. Calculate (i) the mole fractions of each species at equilibrium, (ii) K_x, (iii) K, and (iv) $\Delta_r G^\ominus$.

6A.9(a) The standard reaction Gibbs energy of the isomerization of borneol ($C_{10}H_{17}OH$) to isoborneol in the gas phase at 503 K is +9.4 kJ mol⁻¹. Calculate the reaction Gibbs energy in a mixture consisting of 0.15 mol of borneol and 0.30 mol of isoborneol when the total pressure is 600 Torr.
6A.9(b) The equilibrium pressure of H_2 over solid uranium and uranium hydride, UH_3, at 500 K is 139 Pa. Calculate the standard Gibbs energy of formation of $UH_3(s)$ at 500 K.

6A.10(a) The standard Gibbs energy of formation of $NH_3(g)$ is −16.5 kJ mol⁻¹ at 298 K. What is the reaction Gibbs energy when the partial pressures of the N_2, H_2, and NH_3 (treated as perfect gases) are 3.0 bar, 1.0 bar, and 4.0 bar, respectively? What is the spontaneous direction of the reaction in this case?
6A.10(b) The standard Gibbs energy of formation of $PH_3(g)$ is +13.4 kJ mol⁻¹ at 298 K. What is the reaction Gibbs energy when the partial pressures of the H_2 and PH_3 (treated as perfect gases) are 1.0 bar and 0.60 bar, respectively? What is the spontaneous direction of the reaction in this case?

6A.11(a) For $CaF_2(s) \rightleftharpoons Ca^{2+}(aq) + 2\,F^-(aq)$, $K = 3.9 \times 10^{-11}$ at 25 °C and the standard Gibbs energy of formation of $CaF_2(s)$ is −1167 kJ mol⁻¹. Calculate the standard Gibbs energy of formation of $CaF_2(aq)$.
6A.11(b) For $PbI_2(s) \rightleftharpoons Pb^{2+}(aq) + 2\,I^-(aq)$, $K = 1.4 \times 10^{-8}$ at 25 °C and the standard Gibbs energy of formation of $PbI_2(s)$ is −173.64 kJ mol⁻¹. Calculate the standard Gibbs energy of formation of $PbI_2(aq)$.

Problems

6A.1 The equilibrium constant for the reaction $I_2(s) + Br_2(g) \rightleftharpoons 2\,IBr(g)$ is 0.164 at 25 °C. (a) Calculate $\Delta_r G^\ominus$ for this reaction. (b) Bromine gas is introduced into a container with excess solid iodine. The pressure and temperature are held at 0.164 atm and 25 °C, respectively. Find the partial pressure of IBr(g) at equilibrium. Assume that all the bromine is in the liquid form and that the vapour pressure of iodine is negligible. (c) In fact, solid iodine has a measurable vapour pressure at 25 °C. In this case, how would the calculation have to be modified?

6A.2 Calculate the equilibrium constant of the reaction $CO(g) + H_2(g) \rightleftharpoons H_2CO(g)$ given that, for the production of liquid formaldehyde, $\Delta_r G^\ominus = +28.95$ kJ mol⁻¹ at 298 K and that the vapour pressure of formaldehyde is 1500 Torr at that temperature.

6A.3 A sealed container was filled with 0.300 mol $H_2(g)$, 0.400 mol $I_2(g)$, and 0.200 mol HI(g) at 870 K and total pressure 1.00 bar. Calculate the amounts of the components in the mixture at equilibrium given that $K=870$ for the reaction $H_2(g)+I_2(g) \rightleftharpoons 2\,HI(g)$.

6A.4‡ Nitric acid hydrates have received much attention as possible catalysts for heterogeneous reactions that bring about the Antarctic ozone hole. Standard reaction Gibbs energies are as follows:

(i) $H_2O(g) \rightarrow H_2O(s)$ $\quad \Delta_r G^\ominus$ −23.6 kJ mol^{-1}

(ii) $H_2O(g)+HNO_3(g) \rightarrow HNO_3{\cdot}H_2O(s)$ $\quad \Delta_r G^\ominus$ −57.2 kJ mol^{-1}

(iii) $2\,H_2O(g)+HNO_3(g) \rightarrow HNO_3{\cdot}2H_2O(s)$ $\quad \Delta_r G^\ominus$ −85.6 kJ mol^{-1}

(iv) $3\,H_2O(g)+HNO_3(g) \rightarrow HNO_3{\cdot}3H_2O(s)$ $\quad \Delta_r G^\ominus$ −112.8 kJ mol^{-1}

Which solid is thermodynamically most stable at 190 K if $p_{H_2}=0.13\,\mu$ bar and $p_{HNO_3}=0.41$ nbar *Hint:* Try computing $\Delta_r G$ for each reaction under the prevailing conditions; if more than one solid form spontaneously, examine $\Delta_r G$ for the conversion of one solid to another.

6A.5 Express the equilibrium constant of a gas-phase reaction $A+3\,B \rightleftharpoons 2\,C$ in terms of the equilibrium value of the extent of reaction, ξ, given that initially A and B were present in stoichiometric proportions. Find an expression for ξ as a function of the total pressure, p, of the reaction mixture and sketch a graph of the expression obtained.

TOPIC 6B The response to equilibria to the conditions

Discussion questions

6B.1 Suggest how the thermodynamic equilibrium constant may respond differently to changes in pressure and temperature from the equilibrium constant expressed in terms of partial pressures.

6B.2 Account for Le Chatelier's principle in terms of thermodynamic quantities.

6B.3 Explain the molecular basis of the van 't Hoff equation for the temperature dependence of K.

Exercises

6B.1(a) The standard reaction enthalpy of $Zn(s)+H_2O(g) \rightarrow ZnO(s)+H_2(g)$ is approximately constant at +224 kJ mol^{-1} from 920 K up to 1280 K. The standard reaction Gibbs energy is +33 kJ mol^{-1} at 1280 K. Estimate the temperature at which the equilibrium constant becomes greater than 1.

6B.1(b) The standard enthalpy of a certain reaction is approximately constant at +125 kJ mol^{-1} from 800 K up to 1500 K. The standard reaction Gibbs energy is +22 kJ mol^{-1}at 1120 K. Estimate the temperature at which the equilibrium constant becomes greater than 1.

6B.2(a) The equilibrium constant of the reaction $2\,C_3H_6(g) \rightleftharpoons C_2H_4(g)+C_4H_8(g)$ is found to fit the expression $\ln K=A+B/T+C/T^2$ between 300 K and 600 K, with $A=-1.04$, $B=-1088$ K, and $C=1.51\times10^5$ K^2. Calculate the standard reaction enthalpy and standard reaction entropy at 400 K.

6B.2(b) The equilibrium constant of a reaction is found to fit the expression $\ln K=A+B/T+C/T^3$ between 400 K and 500 K with $A=-2.04$, $B=-1176$ K, and $C=2.1\times10^7$ K^3. Calculate the standard reaction enthalpy and standard reaction entropy at 450 K.

6B.3(a) Calculate the percentage change in K_x for the reaction $H_2CO(g) \rightleftharpoons CO(g)+H_2(g)$ when the total pressure is increased from 1.0 bar to 2.0 bar at constant temperature.

6B.3(b) Calculate the percentage change in K_x for the reaction $CH_3OH(g)+NOCl(g) \rightleftharpoons HCl(g)+CH_3NO_2(g)$ when the total pressure is increased from 1.0 bar to 2.0 bar at constant temperature.

6B.4(a) The equilibrium constant for the gas-phase isomerization of borneol ($C_{10}H_{17}OH$) to isoborneol at 503 K is 0.106. A mixture consisting of 7.50 g of borneol and 14.0 g of isoborneol in a container of volume 5.0 dm^3 is heated to 503 K and allowed to come to equilibrium. Calculate the mole fractions of the two substances at equilibrium.

6B.4(b) The equilibrium constant for the reaction $N_2(g)+O_2(g) \rightleftharpoons 2\,NO(g)$ is 1.69×10^{-3} at 2300 K. A mixture consisting of 5.0 g of nitrogen and 2.0 g of oxygen in a container of volume 1.0 dm^3 is heated to 2300 K and allowed to come to equilibrium. Calculate the mole fraction of NO at equilibrium.

6B.5(a) What is the standard enthalpy of a reaction for which the equilibrium constant is (i) doubled, (ii) halved when the temperature is increased by 10 K at 298 K?

6B.5(b) What is the standard enthalpy of a reaction for which the equilibrium constant is (i) doubled, (ii) halved when the temperature is increased by 15 K at 310 K?

6B.6(a) Estimate the temperature at which $CaCO_3$(calcite) decomposes.

6B.6(b) Estimate the temperature at which $CuSO_4{\cdot}5H_2O$ undergoes dehydration.

6B.7(a) The dissociation vapour pressure of a salt $A_2B(s) \rightleftharpoons A_2(g)+B(g)$ at 367 °C is 208 kPa but at 477 °C it has risen to 547 kPa. Calculate (i) the equilibrium constant, (ii) the standard reaction Gibbs energy, (iii) the standard enthalpy, (iv) the standard entropy of dissociation, all at 422 °C. Assume that the vapour behaves as a perfect gas and that $\Delta H^\ominus$ and $\Delta S^\ominus$ are independent of temperature in the range given.

6B.7(b) The dissociation vapour pressure of NH_4Cl at 427 °C is 608 kPa but at 459 °C it has risen to 1115 kPa. Calculate (i) the equilibrium constant, (ii) the standard reaction Gibbs energy, (iii) the standard enthalpy, (iv) the standard entropy of dissociation, all at 427 °C. Assume that the vapour behaves as a perfect gas and that $\Delta H^\ominus$ and $\Delta S^\ominus$ are independent of temperature in the range given.

Problems

6B.1 Consider the dissociation of methane, $CH_4(g)$, into the elements $H_2(g)$ and C(s, graphite). (a) Given that $\Delta_f H^\ominus(CH_4,g)=-74.85$ kJ mol^{-1} and that $\Delta_r S^\ominus=-80.67$ J K^{-1} mol^{-1} at 298 K, calculate the value of the equilibrium constant at 298 K. (b) Assuming that $\Delta_r H^\ominus$ is independent of temperature, calculate K at 50 °C. (c) Calculate the degree of dissociation, α, of methane

‡ These problems were supplied by Charles Trapp and Carmen Giunta.

at 25 °C and a total pressure of 0.010 bar. (d) Without doing any numerical calculations, explain how the degree of dissociation for this reaction will change as the pressure and temperature are varied.

6B.2 The equilibrium pressure of H_2 over U(s) and UH_3(s) between 450 K and 715 K fits the expression $\ln(p/\text{Pa})=A+B/T+C\ln(T/\text{K})$, with $A=69.32$, $B=-1.464\times10^4$ K, and $C=-5.65$. Find an expression for the standard enthalpy of formation of UH_3(s) and from it calculate $\Delta_f C_p^{\ominus}$.

6B.3 The degree of dissociation, α, of CO_2(g) into CO(g) and O_2(g) at high temperatures was found to vary with temperature as follows:

T/K	1395	1443	1498
$\alpha/10^{-4}$	1.44	2.50	4.71

Assuming $\Delta_r H^{\ominus}$ to be constant over this temperature range, calculate K, $\Delta_r G^{\ominus}$, $\Delta_r H^{\ominus}$, and $\Delta_r S^{\ominus}$. Make any justifiable approximations.

6B.4 The standard reaction enthalpy for the decomposition of $CaCl_2{\cdot}NH_3$(s) into $CaCl_2$(s) and NH_3(g) is nearly constant at +78 kJ mol^{-1} between 350 K and 470 K. The equilibrium pressure of NH_3 in the presence of $CaCl_2{\cdot}NH_3$ is 1.71 kPa at 400 K. Find an expression for the temperature dependence of $\Delta_r G^{\ominus}$ in the same range.

6B.5 Acetic acid was evaporated in a container of volume 21.45 cm^3 at 437 K and at an external pressure of 101.9 kPa, and the container was then sealed. The mass of acid present in the sealed container was 0.0519 g. The experiment was repeated with the same container but at 471 K, and it was found that 0.0380 g of acetic acid was present. Calculate the equilibrium constant for the dimerization of the acid in the vapour and the enthalpy of vaporization.

6B.6 The dissociation of I_2 can be monitored by measuring the total pressure, and three sets of results are as follows:

T/K	973	1073	1173
$100p$/atm	6.244	6.500	9.181
$10^4 n_I$	2.4709	2.4555	2.4366

where n_I is the amount of I atoms per mole of I_2 molecules in the mixture, which occupied 342.68 cm^3. Calculate the equilibrium constants of the dissociation and the standard enthalpy of dissociation at the mean temperature.

6B.7‡ The 1980s saw reports of $\Delta_f G^{\ominus}(SiH_2)$ ranging from 243 to 289 kJ mol^{-1}. If the standard enthalpy of formation is uncertain by this amount, by what factor is the equilibrium constant for the formation of SiH_2 from its elements uncertain at (a) 298 K, (b) 700 K?

6B.8 Find an expression for the standard reaction Gibbs energy at a temperature T' in terms of its value at another temperature T and the coefficients a, b, and c in the expression for the molar heat capacity listed in Table 2B.1. Evaluate the standard Gibbs energy of formation of H_2O(l) at 372 K from its value at 298 K.

6B.9 Derive an expression for the temperature dependence of K_c for a gas-phase reaction.

TOPIC 6C Electrochemical cells

Discussion questions

6C.1 Explain why reactions that are not redox reactions may be used to generate an electric current.

6C.2 Explain the role of a salt bridge.

6C.3 Why is it necessary to measure the cell potential under zero-current conditions?

6C.4 Can you identify other contributions to the cell potential when a current is being drawn from the cell?

Exercises

6C.1(a) Write the cell reaction and electrode half-reactions and calculate the standard potential of each of the following cells:

(i) Zn|$ZnSO_4$(aq)||$AgNO_3$(aq)|Ag
(ii) Cd|$CdCl_2$(aq)||HNO_3(aq)|H_2(g)|Pt
(iii) Pt|$K_3[Fe(CN)_6]$(aq),$K_4[Fe(CN)_6]$(aq)||$CrCl_3$(aq)|Cr

6C.1(b) Write the cell reaction and electrode half-reactions and calculate the standard potential of each the following cells:

(i) Pt|Cl_2(g)|HCl(aq)||K_2CrO_4(aq)|Ag_2CrO_4(s)|Ag
(ii) Pt|Fe^{3+}(aq),Fe^{2+}(aq)||Sn^{4+}(aq),Sn^{2+}(aq)|Pt
(iii) Cu|Cu^{2+}(aq)||Mn^{2+}(aq),H^+(aq)|MnO_2(s)|Pt

6C.2(a) Devise cells in which the following are the reactions and calculate the standard cell potential in each case:

(i) Zn(s) + $CuSO_4$(aq) → $ZnSO_4$(aq) + Cu(s)
(ii) 2 AgCl(s) + H_2(g) → 2 HCl(aq) + 2 Ag(s)
(iii) 2 H_2(g) + O_2(g) → 2 H_2O(l)

6C.2(b) Devise cells in which the following are the reactions and calculate the standard cell potential in each case:

(i) 2 Na(s) + 2 H_2O(l) → 2 NaOH(aq) + H_2(g)
(ii) H_2(g) + I_2(g) → 2 HI(aq)
(iii) H_3O^+(aq) + OH^-(aq) → 2 H_2O(l)

6C.3(a) Use the Debye–Hückel limiting law and the Nernst equation to estimate the potential of the cell Ag|AgBr(s)|KBr(aq, 0.050 mol kg^{-1})||Cd(NO_3)$_2$(aq, 0.010 mol kg^{-1})|Cd at 25 °C.

6C.3(b) Consider the cell Pt|H_2(g,$p^{\ominus}$)| HCl(aq)|AgCl(s)|Ag, for which the cell reaction is 2 AgCl(s) + H_2(g) → 2 Ag(s) + 2 HCl(aq). At 25 °C and a molality of HCl of 0.010 mol kg^{-1}, E_{cell} = +0.4658 V. (i) Write the Nernst equation for the cell reaction. (ii) Calculate $\Delta_r G$ for the cell reaction. (iii) Assuming that the Debye–Hückel limiting law holds at this concentration, calculate $E^{\ominus}$(Cl$^-$, AgCl, Ag).

Problems

6C.1 A fuel cell develops an electric potential from the chemical reaction between reagents supplied from an outside source. What is the cell potential of a cell fuelled by (a) hydrogen and oxygen, (b) the combustion of butane at 1.0 bar and 298 K?

6C.2 Although the hydrogen electrode may be conceptually the simplest electrode and is the basis for our reference state of electrical potential in electrochemical systems, it is cumbersome to use. Therefore, several substitutes for it have been devised. One of these alternatives is the quinhydrone electrode (quinhydrone, Q·QH$_2$, is a complex of quinone, $C_6H_4O_2$=Q, and hydroquinone, $C_6H_4O_2H_2$=QH$_2$). The electrode half-reaction is $Q(aq)+2\,H^+(aq)+2\,e^- \rightarrow QH_2(aq)$, $E^\ominus=+0.6994$ V. If the cell Hg|Hg$_2$Cl$_2$(s)|HCl(aq)|Q·QH$_2$|Au is prepared, and the measured cell potential is +0.190 V, what is the pH of the HCl solution? Assume that the Debye–Hückel limiting law is applicable.

6C.3 Fuel cells provide electrical power for spacecraft (as in the NASA space shuttles) and also show promise as power sources for automobiles. Hydrogen and carbon monoxide have been investigated for use in fuel cells, so their solubilities in molten salts are of interest. Their solubilities in a molten $NaNO_3/KNO_3$ mixture were found to fit the following expressions:

$$\log s_{H_2} = -5.39 - \frac{980}{T/K} \qquad \log s_{CO} = -5.98 - \frac{980}{T/K}$$

where s is the solubility in mol cm^{-3} bar^{-1}. Calculate the standard molar enthalpies of solution of the two gases at 570 K.

TOPIC 6D Electrode potentials

Discussion questions

6D.1 Describe a method for the determination of the standard potential of a redox couple.

6D.2 Devise a method for the determination of the pH of an aqueous solution.

Exercises

6D.1(a) Calculate the equilibrium constants of the following reactions at 25 °C from standard potential data:

(i) $Sn(s)+Sn^{4+}(aq) \rightleftharpoons 2\,Sn^{2+}(aq)$
(ii) $Sn(s)+2\,AgCl(s) \rightleftharpoons SnCl_2(aq)+2\,Ag(s)$

6D.1(b) Calculate the equilibrium constants of the following reactions at 25 °C from standard potential data:

(i) $Sn(s)+CuSO_4(aq) \rightleftharpoons Cu(s)+SnSO_4(aq)$
(ii) $Cu^{2+}(aq)+Cu(s) \rightleftharpoons 2\,Cu^+(aq)$

6D.2(a) The potential of the cell Ag|AgI(s)|AgI(aq)|Ag is +0.9509 V at 25 °C. Calculate (i) the solubility product of AgI and (ii) its solubility.

6D.2(b) The potential of the cell Bi|Bi$_2$S$_3$(s)|Bi$_2$S$_3$(aq)|Bi is −0.96 V at 25 °C. Calculate (i) the solubility product of Bi_2S_3 and (ii) its solubility.

Problems

6D.1 The potential of the cell Pt|H$_2$(g,$p^\ominus$)|HCl(aq,b)|Hg$_2$Cl$_2$(s)|Hg(l) has been measured with high precision with the following results at 25 °C:

b/(mmol kg^{-1})	1.6077	3.0769	5.0403	7.6938	10.9474
E/V	0.60080	0.56825	0.54366	0.52267	0.50532

Determine the standard cell potential and the mean activity coefficient of HCl at these molalities. (Make a least-squares fit of the data to the best straight line.)

6D.2 The standard potential of the AgCl/Ag,Cl$^-$ couple fits the expression

$$E^\ominus/\text{V} = 0.236\,59 - 4.8564\times10^{-4}(\theta/°\text{C}) - 3.4205\times10^{-6}(\theta/°\text{C})^2 + 5.869\times10^{-9}(\theta/°\text{C})^3$$

Calculate the standard Gibbs energy and enthalpy of formation of Cl$^-$(aq) and its entropy at 298 K.

Integrated activities

6.1‡ Thorn et al. (*J. Phys. Chem.* **100**, 14178 (1996)) carried out a study of Cl_2O(g) by photoelectron ionization. From their measurements, they report $\Delta_f H^\ominus(Cl_2O)=+77.2$ kJ mol^{-1}. They combined this measurement with literature data on the reaction $Cl_2O(g)+H_2O(g) \rightarrow 2\,HOCl(g)$, for which $K=8.2\times10^{-2}$ and $\Delta_r S^\ominus=+16.38$ J K^{-1} mol^{-1}, and with readily available thermodynamic data on water vapour to report a value for $\Delta_f H^\ominus$(HOCl). Calculate that value. All quantities refer to 298 K.

6.2 Given that $\Delta_r G^\ominus=-212.7$ kJ mol^{-1} for the reaction in the Daniell cell at 25 °C, and $b(CuSO_4)=1.0\times10^{-3}$ mol kg^{-1} and $b(ZnSO_4)=3.0\times10^{-3}$ mol kg^{-1}, calculate (a) the ionic strengths of the solutions, (b) the mean ionic activity coefficients in the compartments, (c) the reaction quotient, (d) the standard cell potential, and (e) the cell potential. (Take $\gamma_+=\gamma_-=\gamma_\pm$ in the respective compartments.)

6.3 Consider the cell, $Zn(s)|ZnCl_2(0.0050\,mol\,kg^{-1})|Hg_2Cl_2(s)|Hg(l)$, for which the cell reaction is $Hg_2Cl_2(s)+Zn(s)\rightarrow 2\,Hg(l)+2\,Cl^-(aq)+Zn^{2+}(aq)$. Given that $E^\ominus(Zn^{2+},Zn)=-0.7628\,V$, $E^\ominus(Hg_2Cl_2,Hg)=+0.2676\,V$, and that the cell potential is +1.2272 V, (a) write the Nernst equation for the cell. Determine (b) the standard cell potential, (c) $\Delta_r G$, $\Delta_r G^\ominus$, and K for the cell reaction, (d) the mean ionic activity and activity coefficient of $ZnCl_2$ from the measured cell potential, and (e) the mean ionic activity coefficient of $ZnCl_2$ from the Debye–Hückel limiting law. (f) Given that $(\partial E_{cell}/\partial T)_p=-4.52\times10^{-4}\,V\,K^{-1}$, calculate $\Delta_r S$ and $\Delta_r H$.

6.4 Careful measurements of the potential of the cell $Pt|H_2(g,p^\ominus)|NaOH(aq, 0.0100\,mol\,kg^{-1}),Nacl(aq, 0.011\,25\,mol\,kg^{-1})|AgCl(s)|Ag(s)$ have been reported. Among the data is the following information:

θ/°C	20.0	25.0	30.0
E_{cell}/V	1.04774	1.04864	1.04942

Calculate pK_w at these temperatures and the standard enthalpy and entropy of the autoprotolysis of water at 25.0 °C.

6.5 Measurements of the potential of cells of the type $Ag|AgX(s)|MX(b_1)|M_xHg|MX(b_2)|AgX(s)|Ag$, where M_xHg denotes an amalgam and the electrolyte is an LiCl in ethylene glycol, are given below. Estimate the activity coefficient at the concentration marked * and then use this value to calculate activity coefficients from the measured cell potential at the other concentrations. Base your answer on the Davies equation (eqn 5F.11) with $A=1.461$, $B=1.70$, $C=0.20$, and $I=b/b^\ominus$. For $b_2=0.09141\,mol\,kg^{-1}$:

b_1/(mol kg^{-1})	0.0555	0.09141	0.1652	0.2171	1.040	1.350*
E/V	−0.0220	0.0000	0.0263	0.0379	0.1156	0.1336

6.6‡ The table below summarizes the potential of the cell $Pd|H_2(g, 1\,bar)|BH(aq, b), B(aq, b)|AgCl(s)|Ag$. Each measurement is made at equimolar concentrations of 2-aminopyridinium chloride (BH) and 2-aminopyridine (B). The data are for 25 °C and it is found that $E^\ominus=0.222\,51\,V$. Use the data to determine pK_a for the acid at 25 °C and the mean activity coefficient ($\gamma_\pm$) of BH as a function of molality (b) and ionic strength (I). Use the Davies equation (eqn 5F.11) with $A=0.5091$ and B and C are parameters that depend upon the ions. Draw a graph of the mean activity coefficient with $b=0.04\,mol\,kg^{-1}$ and $0\le I\le 0.1$.

b/(mol kg^{-1})	0.01	0.02	0.03	0.04	0.05
E_{cell}(25 °C)/V	0.74452	0.72853	0.71928	0.71314	0.70809
b/(mol kg^{-1})	0.06	0.07	0.08	0.09	0.10
E_{cell}(25 °C)/V	0.70380	0.70059	0.69790	0.69571	0.69338

Hint: Use mathematical software or a spreadsheet.

6.7 Here we investigate the molecular basis for the observation that the hydrolysis of ATP is exergonic at pH = 7.0 and 310 K. (a) It is thought that the exergonicity of ATP hydrolysis is due in part to the fact that the standard entropies of hydrolysis of polyphosphates are positive. Why would an increase in entropy accompany the hydrolysis of a triphosphate group into a diphosphate and a phosphate group? (b) Under identical conditions, the Gibbs energies of hydrolysis of H_4ATP and $MgATP^{2-}$, a complex between the Mg^{2+} ion and ATP^{4-}, are less negative than the Gibbs energy of hydrolysis of ATP^{4-}. This observation has been used to support the hypothesis that electrostatic repulsion between adjacent phosphate groups is a factor that controls the exergonicity of ATP hydrolysis. Provide a rationale for the hypothesis and discuss how the experimental evidence supports it. Do these electrostatic effects contribute to the $\Delta_r H$ or $\Delta_r S$ terms that determine the exergonicity of the reaction? *Hint*. In the $MgATP^{2-}$ complex, the Mg^{2+} ion and ATP^{4-} anion form two bonds: one that involves a negatively charged oxygen belonging to the terminal phosphate group of ATP^{4-} and another that involves a negatively charged oxygen belonging to the phosphate group adjacent to the terminal phosphate group of ATP^{4-}.

6.8 To get a sense of the effect of cellular conditions on the ability of ATP to drive biochemical processes, compare the standard Gibbs energy of hydrolysis of ATP to ADP with the reaction Gibbs energy in an environment at 37 °C in which pH = 7.0 and the ATP, ADP, and P_i^- concentrations are all 1.0 mmol dm^{-3}.

6.9 Under biochemical standard conditions, aerobic respiration produces approximately 38 molecules of ATP per molecule of glucose that is completely oxidized. (a) What is the percentage efficiency of aerobic respiration under biochemical standard conditions? (b) The following conditions are more likely to be observed in a living cell: $p_{CO_2}=5.3\times10^{-2}$ atm, $p_{O_2}=0.132$ atm, [glucose] = 5.6 pmol dm^{-3}, [ATP] = [ADP] = [P$_i$] = 0.10 mmol dm^{-3}, pH = 7.4, $T=310$ K. Assuming that activities can be replaced by the numerical values of molar concentrations, calculate the efficiency of aerobic respiration under these physiological conditions. (c) A typical diesel engine operates between $T_c=873$ K and $T_h=1923$ K with an efficiency that is approximately 75 per cent of the theoretical limit of $1-T_c/T_h$ (see Topic 3A). Compare the efficiency of a typical diesel engine with that of aerobic respiration under typical physiological conditions (see part b). Why is biological energy conversion more or less efficient than energy conversion in a diesel engine?

6.10 In anaerobic bacteria, the source of carbon may be a molecule other than glucose and the final electron acceptor is some molecule other than O_2. Could a bacterium evolve to use the ethanol/nitrate pair instead of the glucose/O_2 pair as a source of metabolic energy?

6.11 The standard potentials of proteins are not commonly measured by the methods described in this chapter because proteins often lose their native structure and function when they react on the surfaces of electrodes. In an alternative method, the oxidized protein is allowed to react with an appropriate electron donor in solution. The standard potential of the protein is then determined from the Nernst equation, the equilibrium concentrations of all species in solution, and the known standard potential of the electron donor. We illustrate this method with the protein cytochrome *c*. The one-electron reaction between cytochrome *c*, cyt, and 2,6-dichloroindophenol, D, can be followed spectrophotometrically because each of the four species in solution has a distinct absorption spectrum. We write the reaction as $cyt_{ox}+D_{red}\rightleftharpoons cyt_{red}+D_{ox}$, where the subscripts 'ox' and 'red' refer to oxidized and reduced states, respectively. (a) Consider $E^\ominus_{cyt}$ and $E^\ominus_D$ to be the standard potentials of cytochrome *c* and D, respectively. Show that, at equilibrium, a plot of $\ln([D_{ox}]_{eq}/[D_{red}]_{eq})$ versus $\ln([cyt_{ox}]_{eq}/[cyt_{red}]_{eq})$ is linear with slope of 1 and *y*-intercept $F(E^\ominus_{cyt}-E^\ominus_D)/RT$, where equilibrium activities are replaced by the numerical values of equilibrium molar concentrations. (b) The following data were obtained for the reaction between oxidized cytochrome *c* and reduced D in a pH 6.5 buffer at 298 K. The ratios $[D_{ox}]_{eq}/[D_{red}]_{eq}$ and $[cyt_{ox}]_{eq}/[cyt_{red}]_{eq}$ were adjusted by titrating a solution containing oxidized cytochrome *c* and reduced D with a solution of sodium ascorbate, which is a strong reductant. From the data and the standard potential of D of 0.237 V, determine the standard potential cytochrome *c* at pH 6.5 and 298 K.

$[D_{ox}]_{eq}/[D_{red}]_{eq}$	0.00279	0.00843	0.0257	0.0497	0.0748	0.238	0.534
$[cyt_{ox}]_{eq}/[cyt_{red}]_{eq}$	0.0106	0.0230	0.0894	0.197	0.335	0.809	1.39

6.12‡ The dimerization of ClO in the Antarctic winter stratosphere is believed to play an important part in that region's severe seasonal depletion of ozone. The following equilibrium constants are based on measurements on the reaction $2\,ClO(g)\rightarrow(ClO)_2(g)$.

T/K	233	248	258	268	273	280
K	4.13×10^8	5.00×10^7	1.45×10^7	5.37×10^6	3.20×10^6	9.62×10^5
T/K	288	295	303			
K	4.28×10^5	1.67×10^5	6.02×10^4			

(a) Derive the values of $\Delta_r H^\ominus$ and $\Delta_r S^\ominus$ for this reaction. (b) Compute the standard enthalpy of formation and the standard molar entropy of $(ClO)_2$ given $\Delta_f H^\ominus(ClO,g)=+101.8\,kJ\,mol^{-1}$ and $S^\ominus_m(ClO,g)=266.6\,J\,K^{-1}\,mol^{-1}$.

6.13‡ Suppose that an iron catalyst at a particular manufacturing plant produces ammonia in the most cost-effective manner at 450 °C when the pressure is such that $\Delta_r G$ for the reaction $\frac{1}{2}N_2(g)+\frac{3}{2}H_2(g)\rightarrow NH_3(g)$ is equal to $-500\,J\,mol^{-1}$. (a) What pressure is needed? (b) Now suppose that a new catalyst is developed that is most cost-effective at 400 °C when the pressure gives the same value of $\Delta_r G$. What pressure is needed when the new catalyst is used? What are the advantages of the new catalyst? Assume that (i) all gases are perfect gases or that (ii) all gases are van der Waals gases. Isotherms of $\Delta_r G(T, p)$ in the pressure range $100\,atm \le p \le 400\,atm$ are needed to derive the answer. (c) Do the isotherms you plotted confirm Le Chatelier's principle concerning the response of equilibrium changes in temperature and pressure?

PART TWO

Structure

In Part 1, we examined the properties of bulk matter from the viewpoint of thermodynamics. In Part 2 we examine the structures and properties of individual atoms and molecules from the viewpoint of quantum mechanics and explain how their structures are determined spectroscopically. The two viewpoints, the macroscopic and the microscopic, merge in Chapter 15, where we show how structural data are used to predict and explain the bulk thermodynamic properties encountered in Part 1. The final three chapters of this Part focus on the way that intermolecular forces lead to the aggregation of molecules, how molecular properties influence the properties of the resulting liquids and solids, and how the structures of these condensed phases are determined.

CHAPTER 7

Introduction to quantum theory

It was once thought that the motion of atoms and subatomic particles could be expressed using 'classical mechanics', the laws of motion introduced in the seventeenth century by Isaac Newton, for these laws were very successful at explaining the motion of everyday objects and planets. However, a proper description of electrons, atoms, and molecules requires a different kind of mechanics, 'quantum mechanics', which we introduce in this chapter and then apply throughout the remainder of the text.

7A The origins of quantum mechanics

Experimental evidence accumulated towards the end of the nineteenth century showed that classical mechanics failed when it was applied to particles as small as electrons. More specifically, careful measurements led to the conclusion that particles may not have an arbitrary energy and that the classical concepts of particle and wave blend together. In this Topic we see how these observations set the stage for the development of the concepts and equations of quantum mechanics through the early twentieth century.

7B Dynamics of microscopic systems

In quantum mechanics, all the properties of a system are expressed in terms of a wavefunction which is obtained by solving the 'Schrödinger equation'. In this Topic we see how to interpret wavefunctions.

7C The principles of quantum theory

This Topic introduces some of the mathematical techniques of quantum mechanics in terms of operators. We also see that quantum theory introduces the 'uncertainty principle', one of the most profound departures from classical mechanics.

What is the impact of this material?

In *Impact* I7.1 we highlight an application of quantum mechanics that still requires much research before it becomes a useful technology. It is based on the speculation that through 'quantum computing' calculations can be carried out on many states of a system simultaneously, leading to a new generation of very fast computers.

To read more about the impact of this material, scan the QR code, or go to bcs.whfreeman.com/webpub/chemistry/pchem10e/impact/pchem-7-1.html

7A The origins of quantum mechanics

Contents

➤ **Why do you need to know this material?**

You should know how experimental results motivated the development of quantum theory, which underlies all descriptions of the structure of atoms and molecules and pervades the whole of spectroscopy and chemistry in general.

➤ **What is the key idea?**

Experimental evidence led to the conclusions that energy cannot be continuously varied and that the classical concepts of a 'particle' and a 'wave' blend together when applied to light, atoms, and molecules.

➤ **What do you need to know already?**

You should be familiar with the basic principles of classical mechanics, which are reviewed in *Foundations* B. The discussion of heat capacities of solids formally makes use of material in Topic 2A but is introduced independently here.

The basic principles of classical mechanics are reviewed in *Foundations* B. In brief, they show that classical physics (1) predicts a precise trajectory for particles, with precisely specified locations and momenta at each instant, and (2) allows the translational, rotational, and vibrational modes of motion to be excited to any energy simply by controlling the forces that are applied. These conclusions agree with everyday experience. Everyday experience, however, does not extend to individual atoms, and careful experiments have shown that classical mechanics fails when applied to the transfers of very small energies and to objects of very small mass.

We also investigate the properties of light. The classical view, discussed in *Foundations* C, is of light as an oscillating electromagnetic field that spreads as a wave through empty space with a wavelength, λ (lambda), a frequency, ν (nu), and a constant speed, c (Fig. C.1). Again, a number of experimental results are not consistent with this interpretation.

This Topic describes the experiments that revealed limitations of classical physics. The remaining Topics of the Chapter show how a new picture of light and matter led to the formulation of an entirely new and hugely successful theory called **quantum mechanics**.

7A.1 Energy quantization

Here we outline three experiments conducted near the end of the nineteenth century and which drove scientists to the view that energy can be transferred only in discrete amounts.

(a) Black-body radiation

A hot object emits electromagnetic radiation. At high temperatures, an appreciable proportion of the radiation is in the visible region of the spectrum and a higher proportion of short-wavelength blue light is generated as the temperature is raised. This behaviour is seen when a heated metal bar glowing red hot becomes white hot when heated further. The dependence is illustrated in Fig. 7A.1, which shows how the energy output varies with wavelength at several temperatures. The curves are those of an ideal emitter called a **black body**, which is an object capable of emitting and absorbing all wavelengths of radiation uniformly. A good approximation to a black body is a pinhole in an empty container maintained at a constant temperature: any radiation leaking out of the hole has been absorbed and re-emitted inside so many times as it reflected around inside

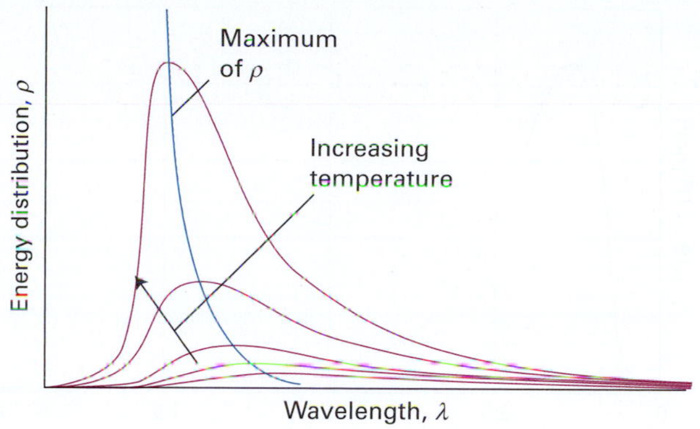

Figure 7A.1 The energy distribution in a black-body cavity at several temperatures. Note how the spectral density of states increases in the region of shorter wavelength as the temperature is raised, and how the peak shifts to shorter wavelengths.

the container that it has come to thermal equilibrium with the walls (Fig. 7A.2).

The approach adopted by nineteenth-century scientists to explain black-body radiation was to calculate the **energy density**, d$\mathcal{E}$, the total energy in a region of the electromagnetic field divided by the volume of the region (units: joules per metre-cubed, J m^{-3}), due to all the oscillators corresponding to wavelengths between λ and $\lambda+d\lambda$. This energy density is proportional to the width, dλ, of this range, and is written

$$d\mathcal{E}=\rho(\lambda,T)d\lambda \qquad (7A.1)$$

where ρ (rho), the constant of proportionality between d$\mathcal{E}$ and dλ, is called the **density of states** (units: joules per metre^4, J m^{-4}). A high density of states at the wavelength λ and temperature T simply means that there is a lot of energy associated with wavelengths lying between λ and $\lambda+d\lambda$ at that temperature. The total energy density in a region is the integral over all wavelengths:

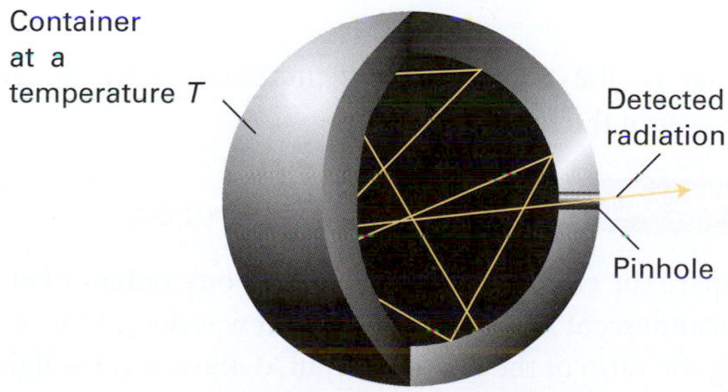

Figure 7A.2 An experimental representation of a black body is a pinhole in an otherwise closed container. The radiation is reflected many times within the container and comes to thermal equilibrium with the walls. Radiation leaking out through the pinhole is characteristic of the radiation within the container.

$$\mathcal{E}(T)=\int_0^\infty \rho(\lambda,T)d\lambda \qquad (7A.2)$$

It depends on the temperature: the higher the temperature, the greater the energy density. Just as the mass of an object is its mass density multiplied by its volume, the total energy within a region of volume V is this energy density multiplied by the volume:

$$E(T)=V\mathcal{E}(T) \qquad (7A.3)$$

The physicist Lord Rayleigh thought of the electromagnetic field as a collection of oscillators of all possible frequencies. He regarded the presence of radiation of frequency ν (and therefore of wavelength $\lambda=c/\nu$, eqn C.3) as signifying that the electromagnetic oscillator of that frequency had been excited (Fig. 7A.3). Rayleigh knew that according to the classical equipartition principle (*Foundations* B), the average energy of each oscillator, regardless of its frequency, is kT. On that basis, with minor help from James Jeans, he arrived at the **Rayleigh–Jeans law** for the density of states:

$$\rho(\lambda,T)=\frac{8\pi kT}{\lambda^4} \qquad \text{Rayleigh–Jeans law} \qquad (7A.4)$$

where k is Boltzmann's constant ($k=1.381\times10^{-23}$ J K^{-1}).

Although the Rayleigh–Jeans law is quite successful at long wavelengths (low frequencies), it fails badly at short wavelengths (high frequencies). Thus, as λ decreases, ρ increases without going through a maximum (Fig. 7A.4). The equation therefore predicts that oscillators of very short wavelength (corresponding to ultraviolet radiation, X-rays, and even γ-rays) are strongly excited even at room temperature. The total energy density in a region, the integral in eqn 7A.2, is also predicted to be infinite at all temperatures above zero. This absurd result, which implies that a large amount of energy is radiated in the high-frequency region of the electromagnetic spectrum, is called the **ultraviolet catastrophe**. According to classical

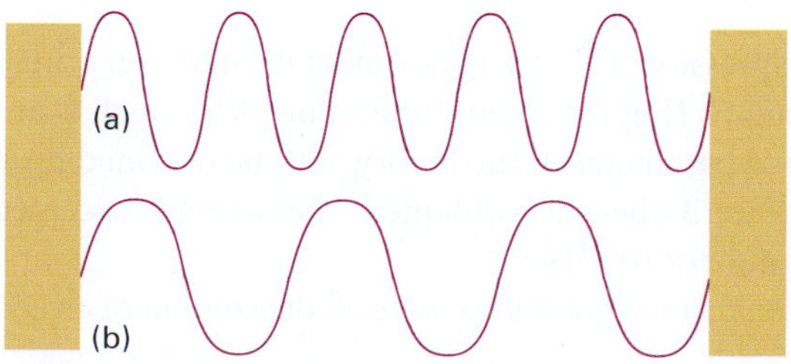

Figure 7A.3 The electromagnetic vacuum can be regarded as able to support oscillations of the electromagnetic field. When a high-frequency, short-wavelength oscillator (a) is excited, that frequency of radiation is present. The presence of low-frequency, long-wavelength radiation (b) signifies that an oscillator of the corresponding frequency has been excited.

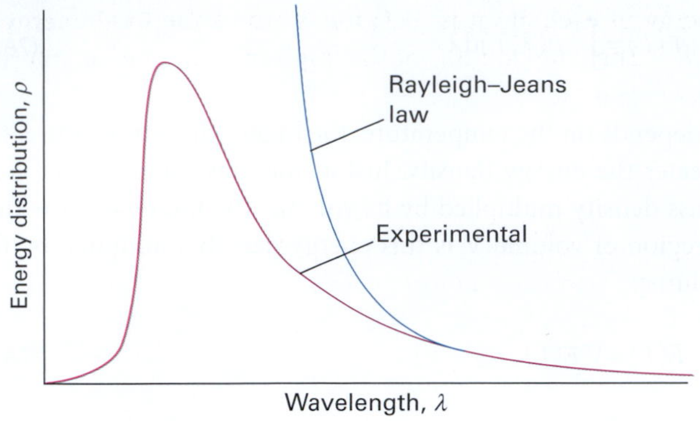

Figure 7A.4 The Rayleigh–Jeans law (eqn 7A.4) predicts an infinite spectral density of states at short wavelengths. This approach to infinity is called the ultraviolet catastrophe.

physics, even cool objects should radiate in the visible and ultraviolet regions, so objects should glow in the dark; there should in fact be no darkness.

In 1900, the German physicist Max Planck found that he could account for the experimental observations by proposing that the energy of each electromagnetic oscillator is limited to discrete values and cannot be varied arbitrarily. This proposal is contrary to the viewpoint of classical physics in which all possible energies are allowed and every oscillator has a mean energy kT. The limitation of energies to discrete values is called the **quantization of energy**. In particular, Planck found that he could account for the observed distribution of energy if he supposed that the permitted energies of an electromagnetic oscillator of frequency ν are integer multiples of $h\nu$:

$$E = nh\nu \qquad n = 0,1,2,\ldots \tag{7A.5}$$

where h is a fundamental constant now known as **Planck's constant**. On the basis of this assumption, Planck was able to derive what is now called the **Planck distribution**:

$$\rho(\lambda,T) = \frac{8\pi hc}{\lambda^5(e^{hc/\lambda kT}-1)} \qquad \text{Planck distribution} \tag{7A.6}$$

This expression fits the experimental curve very well at all wavelengths (Fig. 7A.5), and the value of h, which is an undetermined parameter in the theory, may be obtained by varying its value until a best fit is obtained. The currently accepted value for h is 6.626×10^{-34} J s.

As usual, it is a good idea to 'read' the content of an equation:

- The Planck distribution resembles the Rayleigh–Jeans law (eqn 7A.4) apart from the all-important exponential factor in the denominator. For short wavelengths, $hc/\nu kT \gg 1$ and $e^{hc/\lambda kT}\to\infty$ faster than $\lambda^5\to0$; therefore $\rho\to0$ as $\lambda\to0$ or $\nu\to\infty$. Hence, the energy density approaches zero at high frequencies, in agreement with observation.

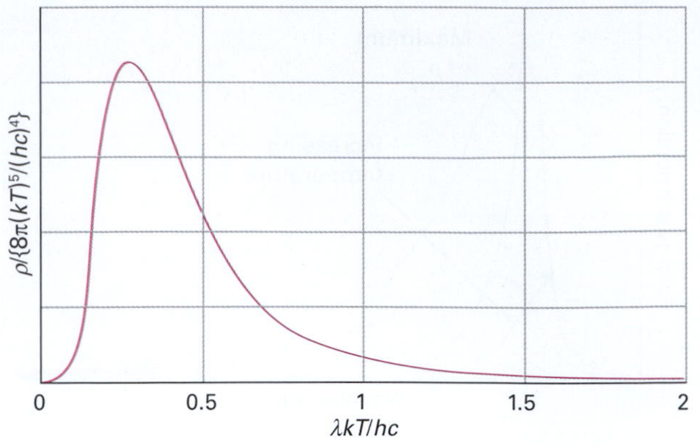

Figure 7A.5 The Planck distribution (eqn 7A.6) accounts very well for the experimentally determined distribution of black-body radiation. Planck's quantization hypothesis essentially quenches the contributions of high frequency, short wavelength oscillators. The distribution coincides with the Rayleigh–Jeans distribution at long wavelengths.

- For long wavelengths, $hc/\lambda kT \ll 1$, and the denominator in the Planck distribution can be replaced by (see *Mathematical background* 1)

$$e^{hc/\lambda kT}-1 = \left(1+\frac{hc}{\lambda kT}+\cdots\right)-1 \approx \frac{hc}{\lambda kT}$$

When this approximation is substituted into eqn 7A.6, we find that the Planck distribution reduces to the Rayleigh–Jeans law.

- As we should infer from the graph in Fig. 7A.5, the total energy density (the integral in eqn 7A.2 and therefore the area under the curve) is no longer infinite, and in fact

$$\mathcal{E}(T) = \int_0^\infty \frac{8\pi hc}{\lambda^5(e^{hc/\lambda kT}-1)}\,d\lambda = aT^4 \quad \text{with} \quad a = \frac{8\pi^5k^4}{15(hc)^3} \tag{7A.7}$$

That is, the energy density increases as the fourth power of the temperature.

Physical interpretation

Example 7A.1 Using the Planck distribution

Compare the energy output of a black-body radiator (such as an incandescent lamp) at two different wavelengths by calculating the ratio of the energy output at 450 nm (blue light) to that at 700 nm (red light) at 298 K.

Method Use eqn 7A.6. At a temperature T, the ratio of the spectral density of states at a wavelength λ_1 to that at λ_2 is

$$\frac{\rho(\lambda_1,T)}{\rho(\lambda_2,T)} = \left(\frac{\lambda_2}{\lambda_1}\right)^5 \times \frac{(e^{hc/\lambda_2 kT}-1)}{(e^{hc/\lambda_1 kT}-1)}$$

Insert the data to evaluate this ratio.

Answer With $\lambda_1 = 450\,\text{nm}$ and $\lambda_2 = 700\,\text{nm}$:

$$\frac{hc}{\lambda_1 kT} = \frac{(6.626\times10^{-34}\,\text{J s})\times(2.998\times10^{8}\,\text{m s}^{-1})}{(450\times10^{-9}\,\text{m})\times(1.381\times10^{-23}\,\text{J K}^{-1})\times(298\,\text{K})} = 107.2\ldots$$

$$\frac{hc}{\lambda_2 kT} = \frac{(6.626\times10^{-34}\,\text{J s})\times(2.998\times10^{8}\,\text{m s}^{-1})}{(700\times10^{-9}\,\text{m})\times(1.381\times10^{-23}\,\text{J K}^{-1})\times(298\,\text{K})} = 68.9\ldots$$

and therefore

$$\frac{\rho(450\,\text{nm},298\,\text{K})}{\rho(700\,\text{nm},298\,\text{K})} = \left(\frac{700\times10^{-9}\,\text{m}}{450\times10^{-9}\,\text{m}}\right)^5 \times \frac{(\text{e}^{68.9\ldots}-1)}{(\text{e}^{107.2\ldots}-1)}$$
$$= 9.11\times(2.30\times10^{-17}) = 2.10\times10^{-16}$$

At room temperature, the proportion of short wavelength radiation is insignificant.

Self-test 7A.1 Repeat the calculation for a temperature of 13.6 MK, which is close to the temperature at the core of the Sun.

Answer: 5.85

It is easy to see why Planck's approach was successful whereas Rayleigh's was not. The thermal motion of the atoms in the walls of the black body excites the oscillators of the electromagnetic field. According to classical mechanics, all the oscillators of the field share equally in the energy supplied by the walls, so even the highest frequencies are excited. The excitation of very high frequency oscillators results in the ultraviolet catastrophe. According to Planck's hypothesis, however, oscillators are excited only if they can acquire an energy of at least $h\nu$. This energy is too large for the walls to supply in the case of the very high frequency oscillators, so the latter remain unexcited. The effect of quantization is to reduce the contribution from the high frequency oscillators, for they cannot be significantly excited with the energy available.

(b) Heat capacities

In the early nineteenth century, the French scientists Pierre-Louis Dulong and Alexis-Thérèse Petit determined the heat capacities, $C_V = (\partial U/\partial T)_V$ (Topic 2A), of a number of monatomic solids. On the basis of some somewhat slender experimental evidence, they proposed that the molar heat capacities of all monatomic solids are the same and (in modern units) close to $25\,\text{J K}^{-1}\,\text{mol}^{-1}$.

Dulong and Petit's law is easy to justify in terms of classical physics in much the same way as Rayleigh attempted to explain black-body radiation. If classical physics were valid, the equipartition principle could be used to infer that the mean energy of an atom as it oscillates about its mean position in a solid is kT for each direction of displacement. As each atom can oscillate in three dimensions, the average energy of each atom is $3kT$; for N atoms the total energy is $3NkT$. The contribution of this motion to the molar internal energy is therefore

$$U_\text{m} = 3N_\text{A}kT = 3RT \qquad \text{(7A.8a)}$$

because $N_\text{A}k = R$, the gas constant. The molar constant volume heat capacity is then predicted to be

$$C_{V,\text{m}} = \left(\frac{\partial U_\text{m}}{\partial T}\right)_V = 3R \qquad \text{(7A.8b)}$$

This result, with $3R = 24.9\,\text{J K}^{-1}\,\text{mol}^{-1}$, is in striking accord with Dulong and Petit's value.

Unfortunately (for Dulong and Petit), significant deviations from their law were observed when advances in refrigeration techniques made it possible to measure heat capacities at low temperatures. It was found that the molar heat capacities of all monatomic solids are lower than $3R$ at low temperatures, and that the values approach zero as $T\to0$. To account for these observations, Einstein (in 1905) assumed that each atom oscillated about its equilibrium position with a single frequency ν. He then invoked Planck's hypothesis to assert that the energy of oscillation is confined to discrete values, and specifically to $nh\nu$, where n is an integer. Einstein discarded the equipartition result, calculated the vibrational contribution of the atoms to the total molar internal energy of the solid (by a method described in Topic 15E), and obtained the expression now known as the **Einstein formula**:

$$C_{V,\text{m}}(T) = 3Rf_\text{E}(T) \quad f_\text{E}(T) = \left(\frac{\theta_\text{E}}{T}\right)^2\left(\frac{\text{e}^{\theta_\text{E}/2T}}{\text{e}^{\theta_\text{E}/T}-1}\right)^2 \qquad \text{Einstein formula} \qquad \text{(7A.9)}$$

The **Einstein temperature**, $\theta_\text{E} = h\nu/k$, is a way of expressing the frequency of oscillation of the atoms as a temperature and allows us to be quantitative about what we mean by 'high temperature' ($T \gg \theta_\text{E}$) and 'low temperature' ($T \ll \theta_\text{E}$) in this context. Note that a high vibrational frequency corresponds to a high Einstein temperature.

As before, we now 'read' this expression:

Physical interpretation

- At high temperatures (when $T \gg \theta_\text{E}$) the exponentials in f_E can be expanded as $1+\theta_\text{E}/T+\ldots$ and higher terms ignored. The result is

$$f_\text{E}(T) = \left(\frac{\theta_\text{E}}{T}\right)^2\left\{\frac{1+\theta_\text{E}/2T+\cdots}{(1+\theta_\text{E}/T+\cdots)-1}\right\}^2 \approx 1 \qquad \text{(7A.10a)}$$

  Consequently, the classical result ($C_{V,\text{m}} = 3R$) is obtained at high temperatures.

- At low temperatures (when $T \ll \theta_\text{E}$) and $\text{e}^{\theta_\text{E}/T} \gg 1$,

$$f_\text{E}(T) \approx \left(\frac{\theta_\text{E}}{T}\right)^2\left(\frac{\text{e}^{\theta_\text{E}/2T}}{\text{e}^{\theta_\text{E}/T}}\right)^2 = \left(\frac{\theta_\text{E}}{T}\right)^2\text{e}^{-\theta_\text{E}/T} \qquad \text{(7A.10b)}$$

The strongly decaying exponential function goes to zero more rapidly than $1/T$ goes to infinity; so $f_E \to 0$ as $T \to 0$, and the heat capacity therefore approaches zero too.

We see that Einstein's formula accounts for the decrease of heat capacity at low temperatures. The physical reason for this success is that at low temperatures only a few oscillators possess enough energy to oscillate significantly so the solid behaves as though it contains far fewer atoms than is actually the case. At higher temperatures, there is enough energy available for all the oscillators to become active: all $3N$ oscillators contribute, many of their energy levels are accessible, and the heat capacity approaches its classical value.

Figure 7A.6 shows the temperature dependence of the heat capacity predicted by the Einstein formula. The general shape of the curve is satisfactory, but the numerical agreement is in fact quite poor. The poor fit arises from Einstein's assumption that all the atoms oscillate with the same frequency, whereas in fact they oscillate over a range of frequencies from zero up to a maximum value, ν_D. This complication is taken into account by averaging over all the frequencies present, the final result being the **Debye formula**:

$$C_{V,\mathrm{m}}(T) = 3Rf_D(T) \quad f_D(T) = 3\left(\frac{T}{\theta_D}\right)^3 \int_0^{\theta_D/T} \frac{x^4 e^x}{(e^x - 1)^2} dx$$

Debye formula (7A.11)

where $\theta_D = h\nu_D/k$ is the **Debye temperature**. The integral in eqn 7A.11 has to be evaluated numerically, but that is simple with mathematical software. The details of this modification, which, as Fig. 7A.7 shows, gives improved agreement with experiment, need not distract us at this stage from the main conclusion, which is that quantization must be introduced in order to explain the thermal properties of solids.

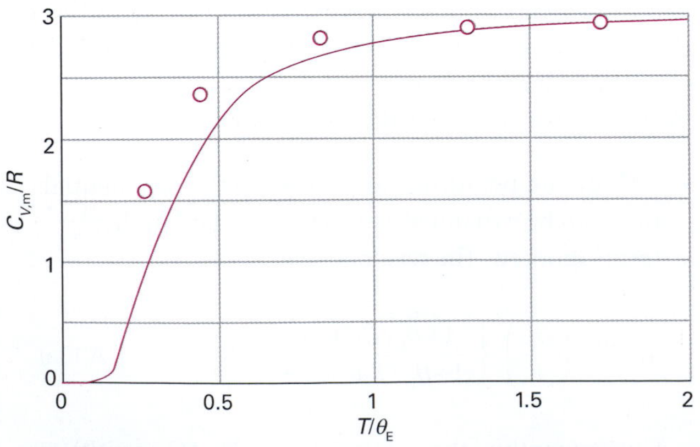

Figure 7A.6 Experimental low-temperature molar heat capacities and the temperature dependence predicted on the basis of Einstein's theory. His equation (eqn 7A.10) accounts for the dependence fairly well, but is everywhere too low.

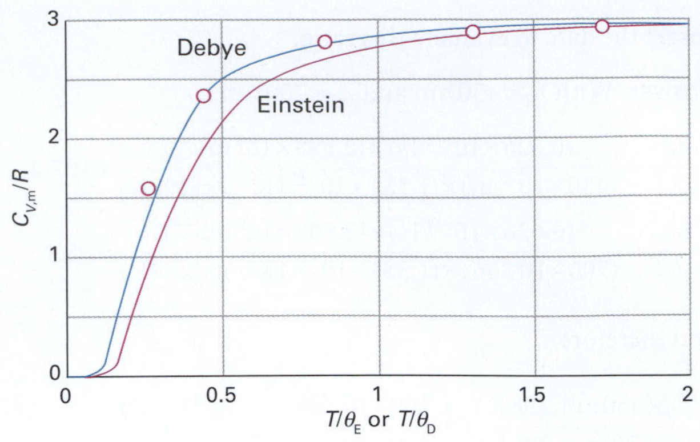

Figure 7A.7 Debye's modification of Einstein's calculation (eqn 7A.11) gives very good agreement with experiment. For copper, $T/\theta_D = 2$ corresponds to about 170 K, so the detection of deviations from Dulong and Petit's law had to await advances in low-temperature physics.

Brief illustration 7A.1 The Debye formula

The Debye temperature for lead is 105 K, corresponding to a vibrational frequency of 2.2×10^{12} Hz. As we see from Fig. 7A.7, $f_D \approx 1$ for $T > \theta_D$ and the heat capacity is almost classical. For lead at 25 °C, corresponding to $T/\theta_D = 2.8$, $f_D = 0.99$ and the heat capacity has almost its classical value.

Self-test 7A.2 Evaluate the Debye temperature for diamond ($\nu_D = 4.6 \times 10^{13}$ Hz). What fraction of the classical value of the heat capacity does diamond reach at 25 °C?

Answer: 2230 K; 15 per cent

(c) Atomic and molecular spectra

The most compelling and direct evidence for the quantization of energy comes from **spectroscopy**, the detection and analysis of the electromagnetic radiation absorbed, emitted, or scattered by a substance. The record of the intensity of light intensity transmitted or scattered by a molecule as a function of frequency (ν), wavelength (λ), or wavenumber ($\tilde{\nu} = \nu/c$) is called its **spectrum** (from the Latin word for appearance).

A typical atomic spectrum is shown in Fig. 7A.8, and a typical molecular spectrum is shown in Fig. 7A.9. The obvious feature of both is that radiation is emitted or absorbed at a series of discrete frequencies. This observation can be understood if the energy of the atoms or molecules is also confined to discrete values, for then energy can be discarded or absorbed only in discrete amounts (Fig. 7A.10). Then, if the energy of an atom decreases by ΔE, the energy is carried away as radiation of frequency ν, and an emission 'line', a sharply defined peak, appears in the spectrum. We say that a molecule undergoes a **spectroscopic transition**, a change of state, when the **Bohr frequency condition**

$$\Delta E = h\nu$$

Bohr frequency condition (7A.12)

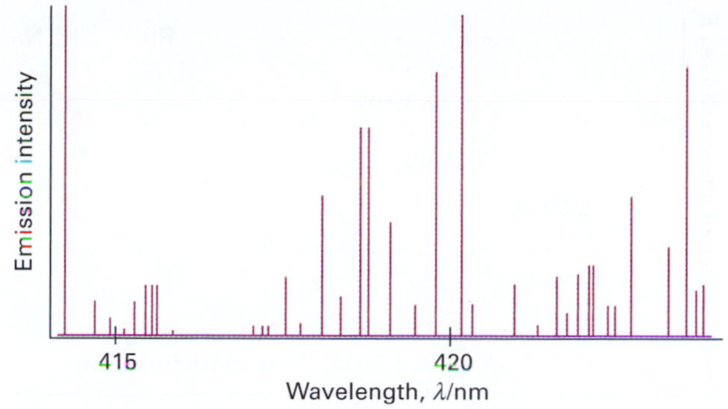

Figure 7A.8 A region of the spectrum of radiation emitted by excited iron atoms consists of radiation at a series of discrete wavelengths (or frequencies).

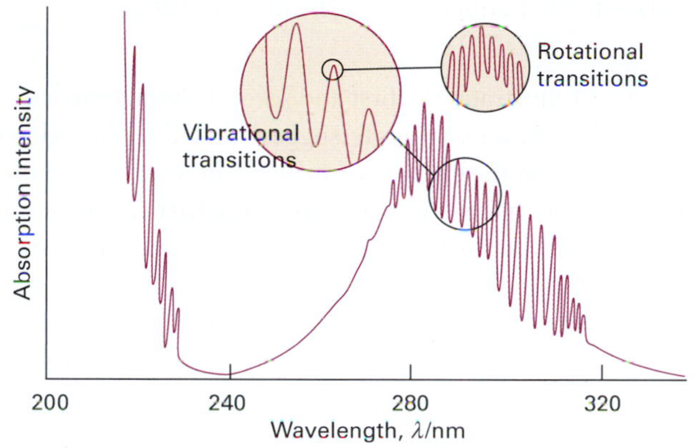

Figure 7A.9 When a molecule changes its state, it does so by absorbing radiation at definite frequencies. This spectrum is part of that due to the electronic, vibrational, and rotational excitation of sulfur dioxide (SO_2) molecules. This observation suggests that molecules can possess only discrete energies, not an arbitrary energy.

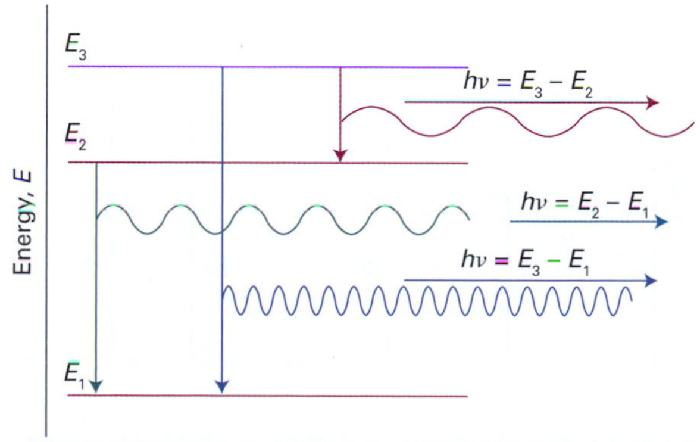

Figure 7A.10 Spectroscopic transitions, such as those shown above, can be accounted for if we assume that a molecule emits electromagnetic radiation as it changes between discrete energy levels. Note that high-frequency radiation is emitted when the energy change is large.

is fulfilled. We develop the principles and applications of atomic spectroscopy in Topics 9A–9C and of molecular spectroscopy in Topics 12A–14D.

Brief illustration 7A.2 The Bohr frequency condition

Atomic sodium produces a yellow glow (as in some street lamps) resulting from the emission of radiation of 590 nm. The spectroscopic transition responsible for the emission involves electronic energy levels that have a separation given by eqn 7A.12:

$$\Delta E = h\nu = \frac{hc}{\lambda} = \frac{(6.626\times10^{-34}\,\mathrm{J\,s})\times(2.998\times10^{8}\,\mathrm{m\,s^{-1}})}{590\times10^{-9}\,\mathrm{m}}$$
$$= 3.37\times10^{-19}\,\mathrm{J}$$

This energy difference can be expressed in a variety of ways. For instance, multiplication by Avogadro's constant results in an energy separation per mole of atoms, of 203 kJ mol^{-1}, comparable to the energy of a weak chemical bond. The calculated value of ΔE also corresponds to 2.10 eV (*Foundations* B).

Self-test 7A.3 Neon lamps emit red radiation of wavelength 736 nm. What is the energy separation of the levels in joules, kilojoules per mole, and electronvolts responsible for the emission?

Answer: 2.70×10^{-19} J, 163 kJ mol^{-1}, 1.69 eV

7A.2 Wave–particle duality

At this stage we have established that the energies of the electromagnetic field and of oscillating atoms are quantized. In this section we see the experimental evidence that led to the revision of two other basic concepts concerning natural phenomena. One experiment shows that electromagnetic radiation—which classical physics treats as wave-like—actually also displays the characteristics of particles. Another experiment shows that electrons—which classical physics treats as particles—also display the characteristics of waves.

(a) The particle character of electromagnetic radiation

The observation that electromagnetic radiation of frequency ν can possess only the energies 0, $h\nu$, $2h\nu$, ... suggests (and at this stage it is only a suggestion) that it can be thought of as consisting of 0, 1, 2, ... particles, each particle having an energy $h\nu$. Then, if one of these particles is present, the energy is $h\nu$, if two are present the energy is $2h\nu$, and so on. These particles of electromagnetic radiation are now called **photons**. The observation of discrete spectra from atoms and molecules can be pictured as the atom or molecule generating a photon of energy $h\nu$ when it discards an energy of magnitude ΔE, with $\Delta E = h\nu$.

Example 7A.2 Calculating the number of photons

Calculate the number of photons emitted by a 100 W yellow lamp in 1.0 s. Take the wavelength of yellow light as 560 nm and assume 100 per cent efficiency.

Method Each photon has an energy $h\nu$, so the total number of photons needed to produce an energy E is $E/h\nu$. To use this equation, we need to know the frequency of the radiation (from $\nu = c/\lambda$) and the total energy emitted by the lamp. The latter is given by the product of the power (P, in watts) and the time interval for which the lamp is turned on ($E = P\Delta t$).

Answer The number of photons is

$$N = \frac{E}{h\nu} = \frac{P\Delta t}{h(c/\lambda)} = \frac{\lambda P\Delta t}{hc}$$

Substitution of the data gives

$$N = \frac{(5.60\times10^{-7}\,\text{m})\times(100\,\text{J s}^{-1})\times(1.0\,\text{s})}{(6.626\times10^{-34}\,\text{J s})\times(2.998\times10^{8}\,\text{m s}^{-1})} = 2.8\times10^{20}$$

Note that it would take the lamp nearly 40 min to produce 1 mol of these photons.

A note on good practice To avoid rounding and other numerical errors, it is best to carry out algebraic calculations first, and to substitute numerical values into a single, final formula. Moreover, an analytical result may be used for other data without having to repeat the entire calculation.

Self-test 7A.4 How many photons does a monochromatic (single frequency) infrared rangefinder of power 1 mW and wavelength 1000 nm emit in 0.1 s?

Answer: 5×10^{14}

So far, the existence of photons is only a suggestion. Experimental evidence for their existence comes from the measurement of the energies of electrons produced in the **photoelectric effect**. This effect is the ejection of electrons from metals when they are exposed to ultraviolet radiation. The experimental characteristics of the photoelectric effect are as follows:

- No electrons are ejected, regardless of the intensity of the radiation, unless its frequency exceeds a threshold value characteristic of the metal.
- The kinetic energy of the ejected electrons increases linearly with the frequency of the incident radiation but is independent of the intensity of the radiation.
- Even at low light intensities, electrons are ejected immediately if the frequency is above the threshold.

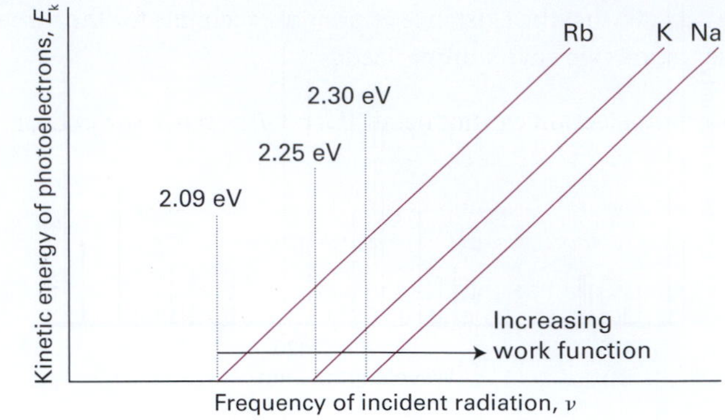

Figure 7A.11 In the photoelectric effect, it is found that no electrons are ejected when the incident radiation has a frequency below a value characteristic of the metal, and, above that value, the kinetic energy of the photoelectrons varies linearly with the frequency of the incident radiation.

Figure 7A.11 illustrates the first and second characteristics.

These observations strongly suggest that the photoelectric effect depends on the ejection of an electron when it is involved in a collision with a particle-like projectile that carries enough energy to eject the electron from the metal. If we suppose that the projectile is a photon of energy $h\nu$, where ν is the frequency of the radiation, then the conservation of energy requires that the kinetic energy of the ejected electron ($E_k = \frac{1}{2}m_e v^2$) should obey

$$E_k = \tfrac{1}{2}m_e v^2 = h\nu - \Phi \qquad \text{Photoelectric effect} \qquad (7A.13)$$

In this expression, Φ (uppercase phi) is a characteristic of the metal called its **work function**, the energy required to remove an electron from the metal to infinity (Fig. 7A.12), the analogue of the ionization energy of an individual atom or molecule. We

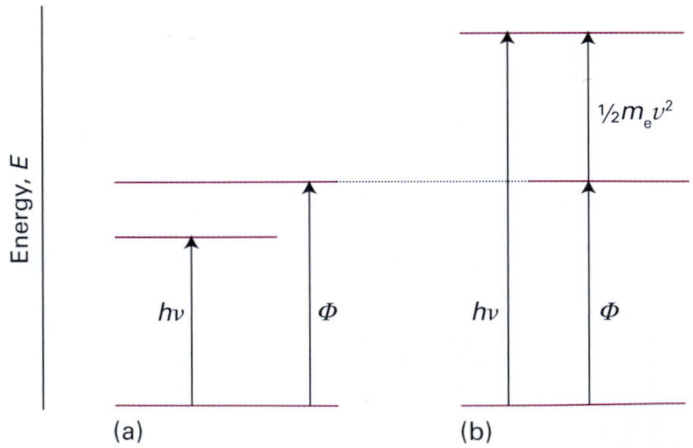

Figure 7A.12 The photoelectric effect can be explained if it is supposed that the incident radiation is composed of photons that have energy proportional to the frequency of the radiation. (a) The energy of the photon is insufficient to drive an electron out of the metal. (b) The energy of the photon is more than enough to eject an electron, and the excess energy is carried away as the kinetic energy of the photoelectron.

can now see that the existence of photons accounts for the three observations we have summarized:

- Photoejection cannot occur if $h\nu < \Phi$ because the photon brings insufficient energy.
- Equation 7A.13 predicts that the kinetic energy of an ejected electron should increase linearly with frequency.
- When a photon collides with an electron, it gives up all its energy, so we should expect electrons to appear as soon as the collisions begin, provided the photons have sufficient energy.

A practical application of eqn 7A.13 is that it provides a technique for the determination of Planck's constant, for the slopes of the lines in Fig. 7A.11 are all equal to h.

Example 7A.3 Calculating the maximum wavelength capable of photoejection

A photon of radiation of wavelength 305 nm ejects an electron from a metal with a kinetic energy of 1.77 eV. Calculate the maximum wavelength of radiation capable of ejecting an electron from the metal.

Method Use eqn 7A.13 rearranged into $\Phi = h\nu - E_k$ with $\nu = c/\lambda$ to calculate the work function of the metal from the data. The threshold for photoejection, the frequency able to remove the electron but not give it any excess energy, then corresponds to radiation of frequency $\nu_{min} = \Phi/h$. Use this value of the frequency to calculate the maximum wavelength capable of photoejection.

Answer From the expression for the work function $\Phi = h\nu - E_k$ the minimum frequency for photoejection is

$$\nu_{min} = \frac{\Phi}{h} = \frac{h\nu - E_k}{h} \overset{\nu=c/\lambda}{=} \frac{c}{\lambda} - \frac{E_k}{h}$$

The maximum wavelength is therefore

$$\lambda_{max} = \frac{c}{\nu_{min}} = \frac{c}{c/\lambda - E_k/h} = \frac{1}{1/\lambda - E_k/hc}$$

Now we substitute the data. The kinetic energy of the electron is

$$E_k = 1.77\,\text{eV} \times (1.602 \times 10^{-19}\,\text{J eV}^{-1}) = 2.83\ldots \times 10^{-19}\,\text{J}$$

$$\frac{E_k}{hc} = \frac{2.83\ldots \times 10^{-19}\,\text{J}}{(6.626 \times 10^{-34}\,\text{J s}) \times (2.998 \times 10^{8}\,\text{m s}^{-1})} = 1.42\ldots \times 10^{6}\,\text{m}^{-1}$$

Therefore, with $1/\lambda = 1/305\,\text{nm} = 3.27\ldots \times 10^{6}\,\text{m}^{-1}$,

$$\lambda_{max} = \frac{1}{(3.27\ldots \times 10^{6}\,\text{m}^{-1}) - (1.42\ldots \times 10^{6}\,\text{m}^{-1})} = 5.40 \times 10^{-7}\,\text{m}$$

or 540 nm.

Self-test 7A.5 When ultraviolet radiation of wavelength 165 nm strikes a certain metal surface, electrons are ejected with a speed of $1.24\,\text{Mm s}^{-1}$. Calculate the speed of electrons ejected by radiation of wavelength 265 nm.

Answer: $735\,\text{km s}^{-1}$

(b) The wave character of particles

Although contrary to the long-established wave theory of light, the view that light consists of particles had been held before, but discarded. No significant scientist, however, had taken the view that matter is wave-like. Nevertheless, experiments carried out in 1925 forced people to consider that possibility. The crucial experiment was performed by the American physicists Clinton Davisson and Lester Germer, who observed the diffraction of electrons by a crystal (Fig. 7A.13). **Diffraction** is the interference caused by an object in the path of waves. Depending on whether the interference is constructive or destructive, the result is a region of enhanced or diminished intensity of the wave. Davisson and Germer's success was a lucky accident, because a chance rise of temperature caused their polycrystalline sample to anneal, and the ordered planes of atoms then acted as a diffraction grating. At almost the same time, G.P. Thomson, working in Scotland, showed that a beam of electrons was diffracted when passed through a thin gold foil.

The Davisson–Germer experiment, which has since been repeated with other particles (including α particles and molecular hydrogen), shows clearly that particles have wave-like properties, and the diffraction of neutrons is a well-established technique for investigating the structures and dynamics of condensed phases (Topic 18A). We have also seen that waves of electromagnetic radiation have particle-like properties. Thus we are brought to the heart of modern

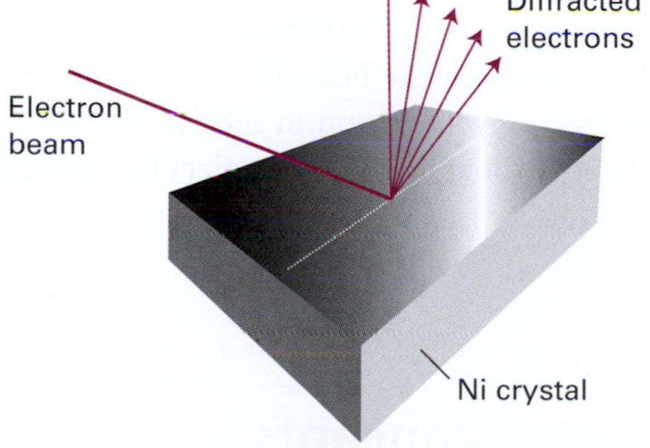

Figure 7A.13 The Davisson–Germer experiment. The scattering of an electron beam from a nickel crystal shows a variation of intensity characteristic of a diffraction experiment in which waves interfere constructively and destructively in different directions.

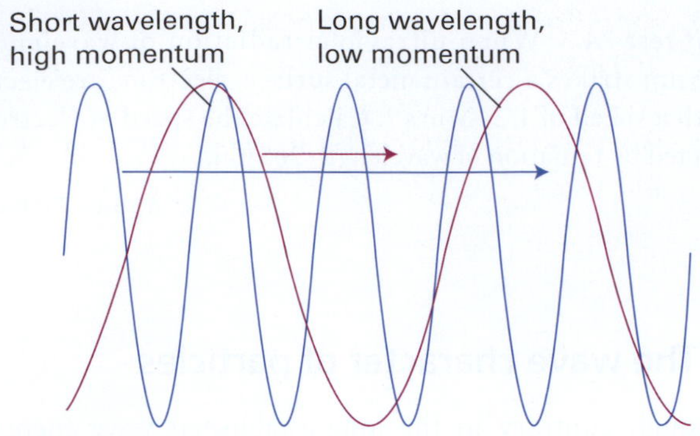

Figure 7A.14 An illustration of the de Broglie relation between momentum and wavelength. The wave is associated with a particle. A particle with high momentum corresponds to a wave with a short wavelength, and vice versa.

physics. When examined on an atomic scale, the classical concepts of particle and wave melt together, particles taking on the characteristics of waves, and waves the characteristics of particles.

Some progress towards coordinating these properties had already been made by the French physicist Louis de Broglie when, in 1924, he suggested that any particle, not only photons, travelling with a linear momentum $p=mv$ (with m the mass and v the speed of the particle) should have in some sense a wavelength given by what is now called the **de Broglie relation**:

$$\lambda=\frac{h}{p} \qquad \text{de Broglie relation} \qquad (7A.14)$$

That is, a particle with a high linear momentum has a short wavelength (Fig. 7A.14). Macroscopic bodies have such high momenta even when they are moving slowly (because their mass is so great), that their wavelengths are undetectably small, and the wavelike properties cannot be observed. This undetectability is why, in spite of its deficiencies, classical mechanics can be used to explain the behaviour of macroscopic bodies. It is necessary to invoke quantum mechanics only for microscopic systems, such as atoms and molecules, in which masses are small.

Example 7A.4 Estimating the de Broglie wavelength

Estimate the wavelength of electrons that have been accelerated from rest through a potential difference of 40 kV.

Method To use the de Broglie relation, we need to know the linear momentum, p, of the electrons. To calculate the linear momentum, we note that the energy acquired by an electron accelerated through a potential difference $\Delta\phi$ is $e\Delta\phi$, where e is the magnitude of its charge. At the end of the period of acceleration, all the acquired energy is in the form of kinetic energy, $E_k = \frac{1}{2}m_ev^2 = p^2/2m_e$, so we can determine p by setting $p^2/2m_e$ equal to $e\Delta\phi$. As before, carry through the calculation algebraically before substituting the data.

Answer The expression $p^2/2m_e=e\Delta\phi$ solves to $p=(2m_ee\Delta\phi)^{1/2}$; then, from the de Broglie relation $\lambda=h/p$,

$$\lambda=\frac{h}{(2m_ee\Delta\phi)^{1/2}}$$

Substitution of the data and the fundamental constants (from inside the front cover) gives

$$\lambda=\frac{6.626\times10^{-34}\,\text{J s}}{\{2\times(9.109\times10^{-31}\,\text{kg})\times(1.609\times10^{-19}\,\text{C})\times(4.0\times10^{4}\,\text{V})\}^{1/2}}$$
$$=6.1\times10^{-12}\,\text{m}$$

For the manipulation of units we have used 1 V C = 1 J and 1 J = 1 kg m^2 s^{-2}. The wavelength of 6.1 pm is shorter than typical bond lengths in molecules (about 100 pm). Electrons accelerated in this way are used in the technique of electron diffraction for the visualization of biological systems (*Impact* I7.1) and the determination of the structures of solid surfaces (Topic 22A).

Self-test 7A.6 Calculate the wavelength of (a) a neutron with a translational kinetic energy equal to kT at 300 K, (b) a tennis ball of mass 57 g travelling at 80 km h^{-1}.

Answer: (a) 178 pm, (b) 5.2×10^{-34} m

We now have to conclude that not only has electromagnetic radiation the character classically ascribed to particles, but electrons (and all other particles) have the characteristics classically ascribed to waves. This joint particle and wave character of matter and radiation is called **wave–particle duality**.

Checklist of concepts

☐ 1. A **black body** is an object capable of emitting and absorbing all wavelengths of radiation uniformly.

☐ 2. The vibrations of atoms can take up energy only in discrete amounts.

☐ 3. Atomic and molecular spectra show that atoms and molecules can take up energy only in discrete amounts.

☐ 4. The **photoelectric effect** establishes the view that electromagnetic radiation, regarded in classical physics as wavelike, consists of particles (photons).

☐ 5. The diffraction of electrons establishes the view that electrons, regarded in classical physics as particles, are wavelike with a wavelength given by the **de Broglie relation**.

☐ 6. **Wave–particle duality** is the recognition that the concepts of particle and wave blend together.

Checklist of equations

Property	Equation	Comment	Equation number
Planck distribution	$\rho(\lambda,T)=8\pi hc/\{\lambda^5(e^{hc/\lambda kT}-1)\}$		7A.6
Heat capacity	$C_{V,m}(T)=3Rf(T)$	$f=f_E$ or f_D	
Einstein formula	$f_E(T)=(\theta_E/T)^2\{e^{\theta_E/2T}/(e^{\theta_E/T}-1)\}^2$	Einstein temperature: $\theta_E=h\nu/k$	7A.9
Debye formula	$f_D(T)=3(T/\theta_D)^3\int_0^{\theta_D/T} x^4e^x/(e^x-1)^2\,dx$	Debye temperature: $\theta_D=h\nu_D/k$	7A.11
Bohr frequency condition	$\Delta E=h\nu$	Conservation of energy	7A.12
Photoelectric effect	$E_k=\frac{1}{2}m_ev^2=h\nu-\Phi$	Φ is the work function	7A.13
de Broglie relation	$\lambda=h/p$	λ is the wavelength of a particle of linear momentum p	7A.14

7B Dynamics of microscopic systems

Contents

➤ Why do you need to know this material?

Quantum theory provides the essential foundation for understanding of the properties of electrons in atoms and molecules.

➤ What is the key idea?

All the dynamical properties of a system are contained in the wavefunction, which is obtained by solving the Schrödinger equation.

➤ What do you need to know already?

You need to be aware of the shortcomings of classical physics that drove the development of quantum theory (Topic 7A).

Wave–particle duality (Topic 7A) strikes at the heart of classical physics, where particles and waves are treated as entirely distinct entities. Experiments have also shown that the energies of electromagnetic radiation and of matter cannot be varied continuously, and that for small objects the discreteness of energy is highly significant. In classical mechanics, in contrast, energies can be varied continuously. Such total failure of classical physics for small objects implied that its basic concepts are false. A new mechanics—quantum mechanics—had to be devised to take its place.

A new mechanics can be constructed from the ashes of classical physics by supposing that, rather than travelling along a definite path, a particle is distributed through space like a wave. This remark may seem mysterious: it will be interpreted more fully shortly. The mathematical representation of the wave that in quantum mechanics replaces the classical concept of trajectory is called a **wavefunction**, ψ (psi), a function that contains all the dynamical information about a system, such as its location and momentum.

7B.1 The Schrödinger equation

In 1926, the Austrian physicist Erwin Schrödinger proposed an equation for finding the wavefunction of any system. The **time-independent Schrödinger equation** for a particle of mass m moving in one dimension with energy E in a system that does not change with time (for instance, its volume remains constant) is

$$-\frac{\hbar^2}{2m}\frac{\mathrm{d}^2\psi}{\mathrm{d}x^2}+V(x)\psi=E\psi \qquad \text{Time-independent Schrödinger equation} \qquad (7B.1)$$

The factor $V(x)$ is the potential energy of the particle at the point x; because the total energy E is the sum of potential and kinetic energies, the first term must be related (in a manner we explore later) to the kinetic energy of the particle; $\hbar=h/2\pi$ (which is read h-cross or h-bar) is a convenient modification of Planck's constant with the value 1.055×10^{-34} J s. Three simple but important general forms of the potential energy are (the explicit forms are found in the corresponding Topics):

- For a particle moving freely in one dimension the potential energy is constant, so $V(x)=V$. It is often convenient to write $V=0$ (Topic 8A).
- For a particle free to oscillate to-and-fro near a point x_0, $V(x)\propto(x-x_0)^2$ (Topic 8B).
- For two electric charges Q_1 and Q_2 separated by a distance x, $V(x)\propto Q_1Q_2/x$ (*Foundations* B).

The following *Justification* shows that the Schrödinger equation is plausible and the discussions later in the chapter will help to overcome its apparent arbitrariness. For the present, we

Table 7B.1 The Schrödinger equation

Expression	Equation	Comment
Time-independent Schrödinger equation	$\hat{H}\psi = E\psi$	General case
	$-\dfrac{\hbar^2}{2m}\dfrac{d^2\psi}{dx^2}+V(x)\psi(x)=E\psi(x)$	One dimension
	$-\dfrac{\hbar^2}{2m}\left(\dfrac{\partial^2\psi}{\partial x^2}+\dfrac{\partial^2\psi}{\partial y^2}\right)+V(x,y)\psi(x,y)$ $=E\psi(x,y)$	Two dimensions
	$-\dfrac{\hbar^2}{2m}\nabla^2\psi+V\psi=E\psi$	Three dimensions
Laplacian operator	$\nabla^2=\dfrac{\partial^2}{\partial x^2}+\dfrac{\partial^2}{\partial y^2}+\dfrac{\partial^2}{\partial z^2}$	
	$\nabla^2=\dfrac{1}{r}\dfrac{\partial^2}{\partial r^2}r+\dfrac{1}{r^2}\Lambda^2$ $=\dfrac{\partial^2}{\partial r^2}+\dfrac{2}{r}\dfrac{\partial}{\partial r}+\dfrac{1}{r^2}\Lambda^2$ $=\dfrac{1}{r^2}\dfrac{\partial}{\partial r}r^2\dfrac{\partial}{\partial r}+\dfrac{1}{r^2}\Lambda^2$	Alternative forms
Legendrian operator	$\Lambda^2=\dfrac{1}{\sin^2\theta}\dfrac{\partial^2}{\partial\phi^2}+\dfrac{1}{\sin\theta}\dfrac{\partial}{\partial\theta}\sin\theta\dfrac{\partial}{\partial\theta}$	
Time-dependent Schrödinger equation	$\hat{H}\Psi=i\hbar\dfrac{\partial\Psi}{\partial t}$	

shall treat the equation simply as a quantum-mechanical postulate that replaces Newton's postulate of his apparently equally arbitrary equation of motion (that force = mass × acceleration). Various ways of expressing the Schrödinger equation, of incorporating the time-dependence of the wavefunction, and of extending it to more dimensions, are collected in Table 7B.1. In the Topics of Chapter 8 we solve the equation for a number of important cases; in this chapter we are mainly concerned with its significance, the interpretation of its solutions, and seeing how it implies that energy is quantized.

Justification 7B.1 The plausibility of the Schrödinger equation

The Schrödinger equation can be seen to be plausible by noting that it implies the de Broglie relation (eqn 7A.14, $p=h/\lambda$) for a freely moving particle. After writing $V(x)=V$, we can rearrange eqn 7B.1 into

$$\frac{d^2\psi}{dx^2}=\frac{2m}{\hbar^2}(V-E)\psi$$

General strategies for solving differential equations of this and other types that occur frequently in physical chemistry are treated in *Mathematical background* 4 at the end of Chapter 8; we need only the simplest procedures in this Topic. In this case a solution is

$$\psi=\cos kx \qquad k=\left\{\frac{2m(E-V)}{\hbar^2}\right\}^{1/2}$$

We now recognize that cos kx is a wave of wavelength $\lambda=2\pi/k$, as can be seen by comparing cos kx with the standard form of a harmonic wave, $\cos(2\pi x/\lambda)$ (*Foundations* C). The quantity $E-V$ is equal to the kinetic energy of the particle, E_k, so $k=(2mE_k/\hbar^2)^{1/2}$, which implies that $E_k=k^2\hbar^2/2m$. Because $E_k=p^2/2m$ (*Foundations* B), it follows that $p=k\hbar$. Therefore, the linear momentum is related to the wavelength of the wavefunction by

$$p=\frac{2\pi}{\lambda}\times\frac{h}{2\pi}=\frac{h}{\lambda}$$

which is the de Broglie relation.

7B.2 The Born interpretation of the wavefunction

A central principle of quantum mechanics is that *the wavefunction contains all the dynamical information about the system it describes*. Here we concentrate on the information it carries about the location of the particle.

The interpretation of the wavefunction in terms of the location of the particle is based on a suggestion made by Max Born. He made use of an analogy with the wave theory of light, in which the square of the amplitude of an electromagnetic wave in a region is interpreted as its intensity and therefore (in quantum terms) as a measure of the probability of finding a photon present in the region. The **Born interpretation** of the wavefunction focuses on the square of the wavefunction (or the square modulus, $|\psi|^2=\psi^*\psi$, if ψ is complex; see *Mathematical background* 3). For a one-dimensional system (Fig. 7B.1):

> If the wavefunction of a particle has the value ψ at some point x, then the probability of finding the particle between x and $x+dx$ is proportional to $|\psi|^2dx$.

Born interpretation

Thus, $|\psi|^2$ is the **probability density**, and to obtain the probability it must be multiplied by the length of the infinitesimal region dx. The wavefunction ψ itself is called the **probability amplitude**. For a particle free to move in three dimensions (for example, an electron near a nucleus in an atom), the

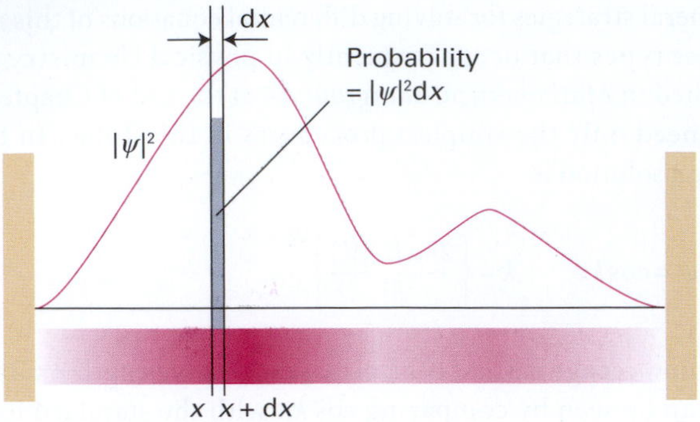

Figure 7B.1 The wavefunction ψ is a probability amplitude in the sense that its square modulus ($\psi^*\psi$ or $|\psi|^2$) is a probability density. The probability of finding a particle in the region dx located at x is proportional to $|\psi|^2$dx. We represent the probability density by the density of shading in the superimposed band.

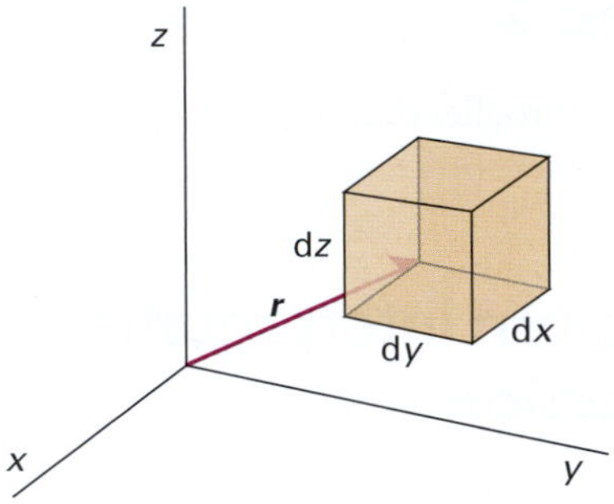

Figure 7B.2 The Born interpretation of the wavefunction in three-dimensional space implies that the probability of finding the particle in the volume element dτ=dxdydz at some location r is proportional to the product of dτ and the value of $|\psi|^2$ at that location.

wavefunction depends on the point r with coordinates x, y, and z, and the interpretation of $\psi(r)$ is as follows (Fig. 7B.2):

> If the wavefunction of a particle has the value ψ at some point r, then the probability of finding the particle in an infinitesimal volume dτ=dxdydz at that point is proportional to $|\psi|^2$dτ.

The Born interpretation does away with any worry about the significance of a negative (and, in general, complex) value of ψ because $|\psi|^2$ is real and never negative. There is no *direct* significance in the negative (or complex) value of a wavefunction: only the square modulus, a positive quantity, is directly physically significant, and both negative and positive regions of a wavefunction may correspond to a high probability of finding a particle in a region (Fig. 7B.3). However, later we shall see that the presence of positive and negative regions of a wavefunction is of great *indirect* significance, because it gives rise to the possibility of constructive and destructive interference between different wavefunctions.

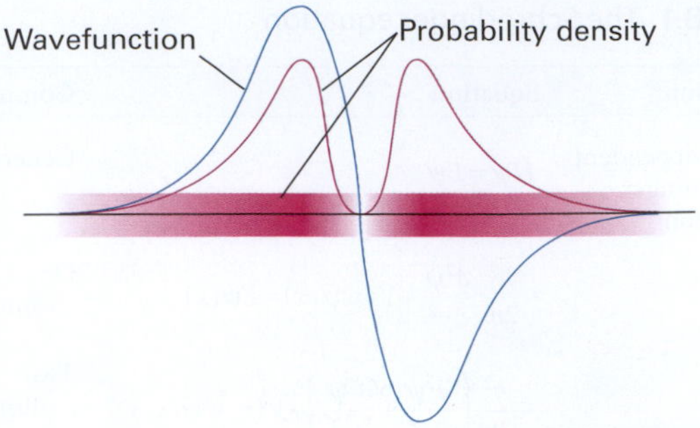

Figure 7B.3 The sign of a wavefunction has no direct physical significance: the positive and negative regions of this wavefunction both correspond to the same probability distribution (as given by the square modulus of ψ and depicted by the density of the shading).

Example 7B.1 Interpreting a wavefunction

In Topic 9A it is shown that the wavefunction of an electron in the lowest energy state of a hydrogen atom is proportional to e^{-r/a_0}, with a_0 a constant and r the distance from the nucleus. Calculate the relative probabilities of finding the electron inside a region of volume δV=1.0 pm^3, which is small even on the scale of the atom, located at (a) the nucleus, (b) a distance a_0 from the nucleus.

Method The region of interest is so small on the scale of the atom that we can ignore the variation of ψ within it and write the probability, P, as proportional to the probability density (ψ^2; note that ψ is real) evaluated at the point of interest multiplied by the volume of interest, δV. That is, $P \propto \psi^2\delta V$, with $\psi^2 \propto e^{-2r/a_0}$.

Answer In each case δV=1.0 pm^3. (a) At the nucleus, r=0, so

$$P \propto e^0 \times (1.0\,\text{pm}^3) = (1.0)\times(1.0\,\text{pm}^3)$$

(b) At a distance $r=a_0$ in an arbitrary direction,

$$P \propto e^{-2} \times (1.0\,\text{pm}^3) = (0.14)\times(1.0\,\text{pm}^3)$$

Therefore, the ratio of probabilities is 1.0/0.14=7.1. Note that it is more probable (by a factor of 7) that the electron will be found at the nucleus than in a volume element of the same size located at a distance a_0 from the nucleus. The negatively charged electron is attracted to the positively charged nucleus, and is likely to be found close to it.

A note on good practice The square of a wavefunction is a probability density, and (in three dimensions) has the dimensions of 1/length^3. It becomes a (unitless) probability when multiplied by a volume. In general, we have to take into account the variation of the amplitude of the wavefunction over the volume of interest, but here we are supposing that the volume is so small that the variation of ψ in the region can be ignored.

Self-test 7B.1 The wavefunction for the electron in its lowest energy state in the ion He^+ is proportional to e^{-2r/a_0}. Repeat the calculation for this ion. Any comment?

Answer: 55; more compact wavefunction

(a) Normalization

A mathematical feature of the Schrödinger equation is that if ψ is a solution, then so is $N\psi$, where N is any constant. This feature is confirmed by noting that ψ occurs in every term in eqn 7B.1, so any constant factor can be cancelled. This freedom to vary the wavefunction by a constant factor means that it is always possible to find a **normalization constant**, N, such that the proportionality of the Born interpretation becomes an equality.

We find the normalization constant by noting that, for a normalized wavefunction $N\psi$, the probability that a particle is in the region dx is equal to $(N\psi^*)(N\psi)\mathrm{d}x$ (we are taking N to be real). Furthermore, the sum over all space of these individual probabilities must be 1 (the probability of the particle being somewhere is 1). Expressed mathematically, the latter requirement is

$$N^2\int_{-\infty}^{\infty}\psi^*\psi\mathrm{d}x=1 \tag{7B.2}$$

Wavefunctions for which the integral in eqn 7B.2 exists (in the sense of having a finite value) are said to be 'square-integrable'. It follows that

$$N=\frac{1}{\left(\int_{-\infty}^{\infty}\psi^*\psi\mathrm{d}x\right)^{1/2}} \tag{7B.3}$$

Therefore, by evaluating the integral, we can find the value of N and hence 'normalize' the wavefunction. From now on, unless we state otherwise, we always use wavefunctions that have been normalized to 1; that is, from now on we assume that ψ already includes a factor which ensures that (in one dimension)

$$\int_{-\infty}^{\infty}\psi^*\psi\mathrm{d}x=1 \tag{7B.4a}$$

In three dimensions, the wavefunction is normalized if

$$\int_{-\infty}^{\infty}\int_{-\infty}^{\infty}\int_{-\infty}^{\infty}\psi^*\psi\,\mathrm{d}x\mathrm{d}y\mathrm{d}z=1 \tag{7B.4b}$$

or, more succinctly, if

$$\int\psi^*\psi\,\mathrm{d}\tau=1 \quad \text{Normalization integral} \tag{7B.4c}$$

where $\mathrm{d}\tau=\mathrm{d}x\mathrm{d}y\mathrm{d}z$ and the limits of this definite integral are not written explicitly: in all such integrals, the integration is over all the space accessible to the particle. For systems with spherical symmetry it is best to work in spherical polar coordinates (*The chemist's toolkit* 7B.1), so the explicit form of eqn 7B.4c is

$$\int_0^{\infty}\int_0^{\pi}\int_0^{2\pi}\psi^*\psi\,r^2\mathrm{d}r\sin\theta\,\mathrm{d}\theta\,\mathrm{d}\phi=1 \tag{7B.4d}$$

The chemist's toolkit 7B.1 Spherical polar coordinates

For systems with spherical symmetry it is best to work in **spherical polar coordinates** r, θ, and ϕ (Sketch 1)

$$x=r\sin\theta\cos\phi,\ y=r\sin\theta\sin\phi,\ z=r\cos\theta \quad \text{Spherical polar coordinates}$$

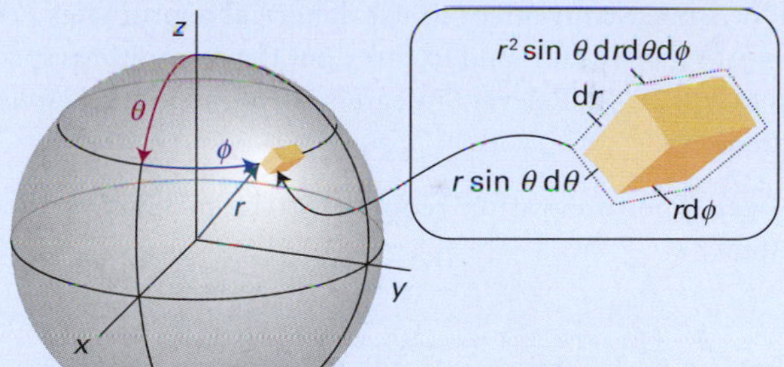

Sketch 1 The spherical polar coordinates used for discussing systems with spherical symmetry.

where:

- r, the radius, ranges from 0 to ∞
- θ, the colatitude, ranges from 0 to π
- ϕ, the azimuth, ranges from 0 to 2π

That these ranges cover space is illustrated in Sketch 2. Standard manipulations then yield

$$\mathrm{d}\tau=r^2\sin\theta\,\mathrm{d}r\,\mathrm{d}\theta\,\mathrm{d}\phi$$

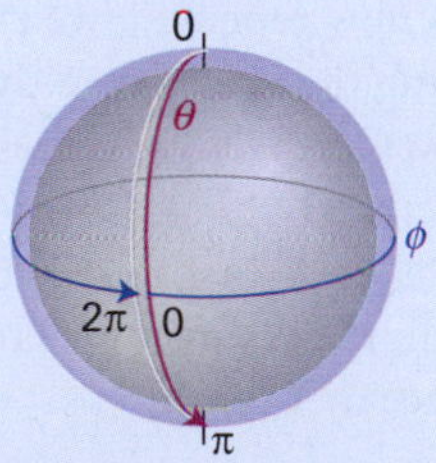

Sketch 2 The surface of a sphere is covered by allowing θ to range from 0 to π, and then sweeping that arc around a complete circle by allowing ϕ to range from 0 to 2π.

In these coordinates, the integral of a function $f(r,\theta,\phi)$ over all space takes the form

$$\int_0^{\infty}\int_0^{\pi}\int_0^{2\pi} f(r,\theta,\phi)r^2\,\mathrm{d}r\sin\theta\,\mathrm{d}\theta\,\mathrm{d}\phi$$

where the limits on the first integral sign refer to r, those on the second to θ, and those on the third to ϕ.

Example 7B.2 Normalizing a wavefunction

Normalize the wavefunction used for the hydrogen atom in *Example* 7B.1.

Method We need to find the factor N that guarantees that the integral in eqn 7B.4c is equal to 1. Because the system is spherical, it is most convenient to use spherical coordinates (*The chemist's toolkit* 7B.1) and to carry out the integrations specified in eqn 7B.4d. Relevant integrals are found in the *Resource section*.

Answer The integration required is the product of three factors:

$$\int\psi^*\psi\,\mathrm{d}\tau = N^2\overbrace{\int_0^{\infty} r^2\mathrm{e}^{-2r/a_0}\,\mathrm{d}r}^{\frac{1}{4}a_0^3}\overbrace{\int_0^{\pi}\sin\theta\,\mathrm{d}\theta}^{2}\overbrace{\int_0^{2\pi}\mathrm{d}\phi}^{2\pi} = \pi a_0^3 N^2$$

Therefore, for this integral to equal 1, we must set

$$N = \left(\frac{1}{\pi a_0^3}\right)^{1/2}$$

and the normalized wavefunction is

$$\psi = \left(\frac{1}{\pi a_0^3}\right)^{1/2}\mathrm{e}^{-r/a_0}$$

Note that because a_0 is a length, the dimensions of ψ are $1/\text{length}^{3/2}$ and therefore those of ψ^2 are $1/\text{length}^3$ (for instance, $1/\text{m}^3$) as is appropriate for a probability density.

If *Example* 7B.1 is now repeated, we can obtain the actual probabilities of finding the electron in the volume element at each location, not just their relative values. Given (from inside the front cover) that $a_0 = 52.9$ pm, the results are (a) 2.2×10^{-6}, corresponding to 1 chance in about 500 000 inspections of finding the electron in the test volume, and (b) 2.9×10^{-7}, corresponding to 1 chance in 3.4 million.

Self-test 7B.2 Normalize the wavefunction given in *Self-test* 7B.1.

Answer: $N = (8/\pi a_0^3)^{1/2}$

(b) Constraints on the wavefunction

The Born interpretation puts severe restrictions on the acceptability of wavefunctions. The principal constraint is that ψ must not be infinite over a finite region. If it were, it would not be square-integrable, and the normalization constant would be zero. The normalized function would then be zero everywhere, except where it is infinite, which would be unacceptable (the particle must be somewhere). Note that infinitely sharp spikes are acceptable provided they have zero width.

The requirement that ψ is finite everywhere rules out many possible solutions of the Schrödinger equation, because many mathematically acceptable solutions rise to infinity and are therefore physically unacceptable. We could imagine a solution of the Schrödinger equation that gives rise to more than one value of $|\psi|^2$ at a single point. The Born interpretation implies that such solutions are unacceptable, because it would be absurd to have more than one probability that a particle is at the same point. This restriction is expressed by saying that the wavefunction must be *single-valued*; that is, have only one value at each point of space.

The Schrödinger equation itself also implies some mathematical restrictions on the type of functions that can occur. Because it is a second-order differential equation, the second derivative of ψ must be well-defined if the equation is to be applicable everywhere. We can take the second derivative of a function only if it is continuous (so there are no sharp steps in it, Fig. 7B.4) and if its first derivative, its slope, is continuous (so there are no kinks in the wavefunction).

There are cases, and we shall meet them, where acceptable wavefunctions have kinks. These cases arise when the potential energy has peculiar properties, such as rising abruptly to infinity. When the potential energy is smoothly well-behaved and finite, the slope of the wavefunction must be continuous; if the potential energy becomes infinite, then the slope of the wavefunction need not be continuous. There are only two cases

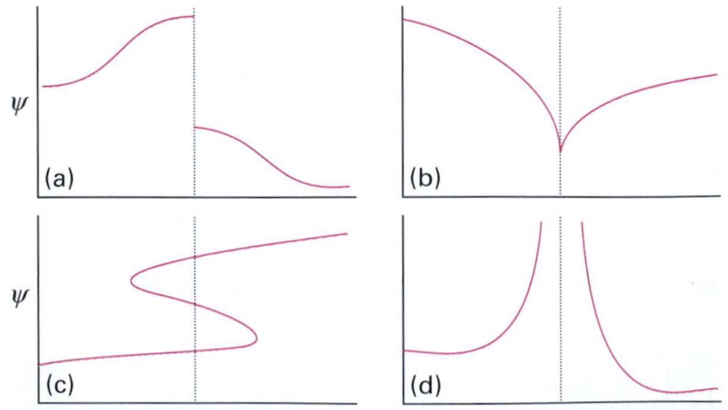

Figure 7B.4 The wavefunction must satisfy stringent conditions for it to be acceptable: (a) unacceptable because it is not continuous; (b) unacceptable because its slope is discontinuous; (c) unacceptable because it is not single-valued; (d) unacceptable because it is infinite over a finite region.

of this behaviour in elementary quantum mechanics, and the peculiarity will be mentioned when we meet them.

At this stage we see that ψ:

- must not be infinite over a non-infinitesimal region
- must be single-valued
- must be continuous
- must have a continuous slope.

Constraints on the wavefunction

(c) Quantization

The restrictions just noted are so severe that acceptable solutions of the Schrödinger equation do not in general exist for arbitrary values of the energy E. In other words, a particle may possess only certain energies, for otherwise its wavefunction would be physically unacceptable. That is, *as a consequence of the restrictions on its wavefunction, the energy of a particle is quantized*. We can find the acceptable energies by solving the Schrödinger equation for motion of various kinds, and selecting the solutions that conform to the restrictions listed above. That task is taken forward in Chapter 8.

7B.3 The probability density

Once we have obtained the normalized wavefunction, we can then proceed to determine the probability density. As an example, consider a particle of mass m free to move parallel to the x-axis with zero potential energy. The Schrödinger equation is obtained from eqn 7B.1 by setting $V=0$, and is

$$-\frac{\hbar^2}{2m}\frac{d^2\psi(x)}{dx^2}=E\psi(x) \quad (7B.5)$$

As shown in the following *Justification*, the solutions of this equation have the form

$$\psi(x)=Ae^{ikx}+Be^{-ikx} \qquad E=\frac{k^2\hbar^2}{2m} \quad (7B.6)$$

where A and B are constants. (See *Mathematical background* 3 at the end of this chapter for more on complex numbers.)

Justification 7B.2 The wavefunction of a free particle in one dimension

To verify that $\psi(x)$ in eqn 7B.6 is a solution of eqn 7B.5, we simply substitute it into the left-hand side of the equation and show that $E=k^2\hbar^2/2m$. To begin, we write

$$-\frac{\hbar^2}{2m}\frac{d^2\psi(x)}{dx^2}=-\frac{\hbar^2}{2m}\frac{d^2}{dx^2}(Ae^{ikx}+Be^{-ikx})$$

Because $de^{\pm ax}/dx=\pm ae^{\pm ax}$ and $i^2=-1$, the second derivatives evaluate to

$$-\frac{\hbar^2}{2m}\{A(ik)^2e^{ikx}+B(-ik)^2e^{-ikx}\}=\overbrace{\frac{k^2\hbar^2}{2m}}^{E}\overbrace{(Ae^{ikx}+Be^{-ikx})}^{\psi(x)}=E\psi(x)$$

We see in Topic 8A what determines the values of A and B; here we can treat them as arbitrary constants that we can vary at will. Suppose that $B=0$ in eqn 7B.6, then the wavefunction is simply

$$\psi(x)=Ae^{ikx} \quad (7B.7)$$

Where is the particle? To find out, we calculate the probability density:

$$|\psi(x)|^2=(Ae^{ikx})^*(Ae^{ikx})=(A^*e^{-ikx})(Ae^{ikx})=|A|^2 \quad (7B.8)$$

This probability density is independent of x; so, wherever we look in a region of fixed length located anywhere along the x-axis, there is an equal probability of finding the particle (Fig. 7B.5a). In other words, if the wavefunction of the particle is given by eqn 7B.7, then we cannot predict where we will find it. The same would be true if the wavefunction in eqn 7B.6 had $A=0$; then the probability density would be $|B|^2$, a constant.

Now suppose that in the wavefunction $A=B$. Then, because $\cos kx=\frac{1}{2}(e^{ikx}+e^{-ikx})$ (*Mathematical background* 3), eqn 7B.6 becomes

$$\psi(x)=A(e^{ikx}+e^{-ikx})=2A\cos kx \quad (7B.9)$$

The probability density now has the form

$$|\psi(x)|^2=(2A\cos kx)^*(2A\cos kx)=4|A|^2\cos^2 kx \quad (7B.10)$$

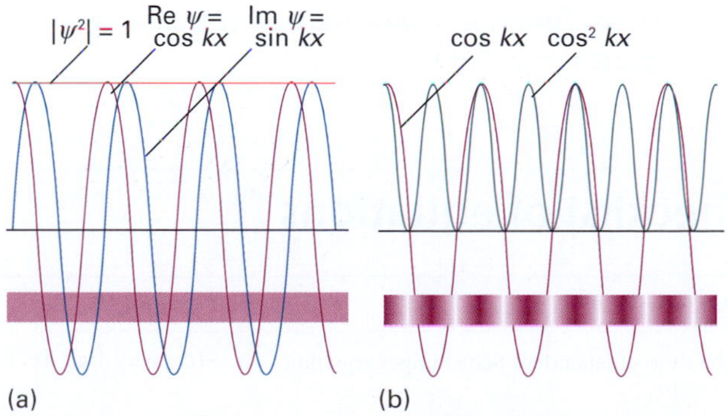

Figure 7B.5 (a) The square modulus of a wavefunction corresponding to the wavefunction in eqn 7B.7 is a constant; so it corresponds to a uniform probability of finding the particle anywhere. (b) The probability distribution corresponding to the wavefunction in eqn 7B.7.

This function is illustrated in Fig. 7B.5b. As we see, the probability density periodically varies between 0 and $4|A|^2$. The locations where the probability density is zero correspond to *nodes* in the wavefunction. Specifically, a **node** is a point where a wavefunction passes *through* zero. The location where a wavefunction approaches zero without actually passing through zero is not a node.

To calculate the probability of finding the system in a region of space that is not infinitesimal we sum (that is, we integrate) the probability density over the region of space of interest. For example, for a one-dimensional wavefunction, the probability P of finding the particle between x_1 and x_2 is given by

$$P=\int_{x_1}^{x_2}|\psi(x)|^2\,\mathrm{d}x \quad \textit{One-dimensional region} \quad \text{Probability} \qquad (7B.11)$$

Example 7B.3 Determining a probability

The lowest-energy electrons of a carbon nanotube can described by the normalized wavefunction $(2/L)^{1/2}\sin(\pi x/L)$, where L is the length of the nanotube. What is the probability of finding the electron between $x=L/4$ and $x=L/2$?

Method Use eqn 7B.11 and the normalized wavefunction to write an expression for the probability of finding the electron in the region of interest. Relevant integrals are given in the *Resource section*.

Answer From eqn 7B.11 and the wavefunction provided, the expression for the probability is

$$P=\int_{L/4}^{L/2}\left(\frac{2}{L}\right)\sin^2(\pi x/L)\,\mathrm{d}x$$

It follows that

$$P \overset{\text{Integral T.2}}{=} \left(\frac{2}{L}\right)\left(\frac{x}{2}-\frac{\sin(2\pi x/L)}{4\pi/L}\right)\Bigg|_{L/4}^{L/2}=\left(\frac{2}{L}\right)\left(\frac{L}{4}-\frac{L}{8}-0+\frac{L}{4\pi}\right)$$
$$=0.409$$

There is a chance of about 41 per cent that the electron will be found between $x=L/4$ and $x=L/2$ along the nanotube.

Self-test 7B.3 The next higher energy wavefunction of the electron in the nanotube is described by the normalized wavefunction $(2/L)^{1/2}\sin(2\pi x/L)$. What is the probability of finding the electron between $x=L/4$ and $x=L/2$?

Answer: 0.25

Checklist of concepts

- ☐ 1. A **wavefunction** is a mathematical function that contains all the dynamical information about a system.
- ☐ 2. The **Schrödinger equation** is a second-order differential equation used to calculate the wavefunction of a system.
- ☐ 3. According to the **Born interpretation**, the probability density is proportional to the square of the wavefunction.
- ☐ 4. A wavefunction is **normalized** if the integral of its square is equal to 1.
- ☐ 5. A wavefunction must be single-valued, continuous, not infinite over a non-infinitesimal region of space, and have a continuous slope.
- ☐ 6. The quantization of energy stems from the constraints that an acceptable wavefunction must satisfy.
- ☐ 7. A **node** is a point where a wavefunction passes through zero.

Checklist of equations

Property	Equation	Comment	Equation number
The time-independent Schrödinger equation	$-(\hbar^2/2m)(\mathrm{d}^2\psi/\mathrm{d}x^2)+V(x)\psi=E\psi$, or $\hat{H}\psi=E\psi$	One-dimensional system	7B.1
Normalization integral	$\int\psi^*\psi\,\mathrm{d}\tau=1$	Integration over all space	7B.4c
Probability of locating a particle	$P=\int_{x_1}^{x_2}\|\psi(x)\|^2\,\mathrm{d}x$	One-dimensional region	7B.11

7C The principles of quantum theory

Contents

➤ Why do you need to know this material?

The wavefunction is the central feature in quantum mechanics, so you need to know how to extract dynamical information from it. The procedures described here allow you to predict the results of measurements of observables.

➤ What is the key idea?

The wavefunction is obtained by solving the Schrödinger equation, and the dynamical information it contains is extracted by determining the eigenvalues of hermitian operators.

➤ What do you need to know already?

You need to know that the state of a system is fully described by a wavefunction (Topic 7B). You also need to be familiar with elementary integration (*Mathematical background* 1) and manipulation of complex functions (*Mathematical background* 3).

A wavefunction contains all the information it is possible to obtain about the dynamical properties of the particle (for example, its location and momentum). The Born interpretation (Topic 7B) tells us as much as we can know about location, but how do we extract any additional dynamical information?

7C.1 Operators

To formulate a systematic way of extracting information from the wavefunction, we first note that any Schrödinger equation may be written in the succinct form

$$\hat{H}\psi = E\psi \quad \text{Operator form of Schrödinger equation} \quad (7C.1a)$$

with (in one dimension)

$$\hat{H} = -\frac{\hbar^2}{2m}\frac{d^2}{dx^2} + V(x) \quad \text{Hamiltonian operator} \quad (7C.1b)$$

The quantity $\hat{H}$ (commonly read aitch-hat) is an **operator**, something that carries out a mathematical operation on the function ψ. In this case, the operation is to take the second derivative of ψ and (after multiplication by $-\hbar^2/2m$) to add the result to the outcome of multiplying ψ by $V(x)$.

The operator $\hat{H}$ plays a special role in quantum mechanics, and is called the **hamiltonian operator** after the nineteenth-century mathematician William Hamilton, who developed a form of classical mechanics which, it subsequently turned out, is well suited to the formulation of quantum mechanics. The hamiltonian operator is the operator corresponding to the total energy of the system, the sum of the kinetic and potential energies. Consequently, we can infer that the first term in eqn 7C.1b (the term proportional to the second derivative) must be the operator for the kinetic energy.

(a) Eigenvalue equations

When the Schrödinger equation is written as in eqn 7C.1a, it is seen to be an **eigenvalue equation**, an equation of the form

$$(\text{Operator})(\text{function}) = (\text{constant factor}) \times (\text{same function}) \quad (7C.2a)$$

If we denote a general operator by $\hat{\Omega}$ (where Ω is uppercase omega) and a constant factor by ω (lowercase omega), then an eigenvalue equation has the form

$$\hat{\Omega}\psi = \omega\psi \quad \text{Eigenvalue equation} \quad (7C.2b)$$

The factor ω is called the **eigenvalue** of the operator. The eigenvalue in eqn 7C.1a is the energy. The function ψ in an equation of this kind is called an **eigenfunction** of the operator $\hat{\Omega}$ and is different for each eigenvalue. So, in this technical language, we would write eqn 7C.2a as

$$\text{(Operator)(eigenfunction)} = \text{(eigenvalue)} \times \text{(eigenfunction)} \qquad (7C.2c)$$

The eigenfunction in eqn 7C.1a is the wavefunction corresponding to the energy E. It follows that another way of saying 'solve the Schrödinger equation' is to say 'find the eigenvalues and eigenfunctions of the hamiltonian operator for the system'.

Example 7C.1 Identifying an eigenfunction

Show that e^{ax} is an eigenfunction of the operator d/dx, and find the corresponding eigenvalue. Show that e^{ax^2} is not an eigenfunction of d/dx.

Method We need to operate on the function with the operator and check whether the result is a constant factor times the original function.

Answer For $\hat{\Omega} = d/dx$ (the operation 'differentiate with respect to x') and $\psi = e^{ax}$:

$$\hat{\Omega}\psi = \frac{d}{dx}e^{ax} = ae^{ax} = a\psi$$

Therefore e^{ax} is indeed an eigenfunction of d/dx, and its eigenvalue is a. For $\psi = e^{ax^2}$,

$$\hat{\Omega}\psi = \frac{d}{dx}e^{ax^2} = 2axe^{ax^2} = 2ax \times \psi$$

which is not an eigenvalue equation of $\hat{\Omega}$. Even though the same function ψ occurs on the right, ψ is now multiplied by a variable factor ($2ax$), not a constant factor. Alternatively, if the right hand side is written $2a(xe^{ax^2})$, we see that it is a constant ($2a$) times a *different* function.

Self-test 7C.1 Is the function $\cos ax$ an eigenfunction of (a) d/dx, (b) d^2/dx^2?

Answer: (a) No, (b) yes

(b) The construction of operators

The importance of eigenvalue equations is that the pattern

$$\text{(Energy operator)}\,\psi = \text{(energy)} \times \psi$$

exemplified by the Schrödinger equation is repeated for other **observables**, or measurable properties of a system, such as the momentum or the electric dipole moment. Thus, it is often the case that we can write

$$\text{(Operator corresponding to an observable)}\psi = \text{(value of observable)} \times \psi$$

The symbol $\hat{\Omega}$ in eqn 7C.2b is then interpreted as an operator (for example, the hamiltonian operator) corresponding to an observable (for example, the energy), and the eigenvalue ω is the value of that observable (for example, the value of the energy, E). Therefore, if we know both the wavefunction ψ and the operator $\hat{\Omega}$ corresponding to the observable Ω of interest, and the wavefunction is an eigenfunction of the operator $\hat{\Omega}$, then we can predict the outcome of an observation of the property Ω (for example, an atom's energy) by picking out the factor ω in the eigenvalue equation, eqn 7C.2b.

A basic postulate of quantum mechanics tells us how to set up the operator corresponding to a given observable:

Observables, Ω, are represented by operators, $\hat{\Omega}$, built from the following position and momentum operators:

$$\hat{x} = x\times \qquad \hat{p}_x = \frac{\hbar}{i}\frac{d}{dx} \qquad \text{Specification of operators} \quad (7C.3)$$

That is, the operator for location along the x-axis is multiplication (of the wavefunction) by x and the operator for linear momentum parallel to the x-axis is proportional to taking the derivative (of the wavefunction) with respect to x.

Example 7C.2 Determining the value of an observable

What is the linear momentum of a free particle described by the wavefunction $\psi(x) = Ae^{ikx} + Be^{-ikx}$ (eqn 7B.6) with (a) $B = 0$, (b) $A = 0$?

Method We operate on ψ with the operator corresponding to linear momentum (eqn 7C.3), and inspect the result. If the outcome is the original wavefunction multiplied by a constant (that is, we generate an eigenvalue equation), then the constant is identified with the value of the observable.

Answer (a) With $B = 0$,

$$\hat{p}_x\psi = \frac{\hbar}{i}\frac{d\psi}{dx} = \frac{\hbar}{i}A\frac{de^{ikx}}{dx} = \frac{\hbar}{i}A \times ike^{ikx} = \overbrace{k\hbar}^{\text{Eigenvalue}}\,\psi$$

This is an eigenvalue equation, and by comparing it with eqn 7C.2b we find that $p_x = +k\hbar$.

(b) For the wavefunction with $A = 0$,

$$\hat{p}_x\psi = \frac{\hbar}{i}\frac{d\psi}{dx} = \frac{\hbar}{i}B\frac{de^{-ikx}}{dx} = \frac{\hbar}{i}A \times (-ik)e^{ikx} = \overbrace{-k\hbar}^{\text{Eigenvalue}}\,\psi$$

The magnitude of the linear momentum is the same in each case ($k\hbar$), but the signs are different: in (a) the particle is travelling to the right (positive x) but in (b) it is travelling to the left (negative x).

Self-test 7C.2 The operator for the angular momentum of a particle travelling in a circle in the xy-plane is $\hat{l}_z = (\hbar/\mathrm{i})\mathrm{d}/\mathrm{d}\phi$, where ϕ is its angular position. What is the angular momentum of a particle described by the wavefunction $\mathrm{e}^{-2\mathrm{i}\phi}$?

Answer: $l_z = -2\hbar$

We use the definitions in eqn 7C.3 to construct operators for other spatial observables. For example, suppose we wanted the operator for a potential energy of the form $V(x) = \frac{1}{2}k_f x^2$, with k_f a constant (later, we shall see that this potential energy describes the vibrations of atoms in molecules). Then it follows from eqn 7C.3 that the operator corresponding to $V(x)$ is multiplication by x^2:

$$\hat{V}(x) = \tfrac{1}{2}k_f x^2 \times \qquad (7C.4)$$

In normal practice, the multiplication sign is omitted. To construct the operator for kinetic energy, we make use of the classical relation between kinetic energy and linear momentum, which in one dimension is $E_k = p_x^2/2m$ (*Foundations* B). Then, by using the operator for p_x in eqn 7C.3 we find:

$$\hat{E}_k = \frac{1}{2m}\left(\frac{\hbar}{\mathrm{i}}\frac{\mathrm{d}}{\mathrm{d}x}\right)\left(\frac{\hbar}{\mathrm{i}}\frac{\mathrm{d}}{\mathrm{d}x}\right) = -\frac{\hbar^2}{2m}\frac{\mathrm{d}^2}{\mathrm{d}x^2} \qquad (7C.5)$$

It follows that the operator for the total energy, the hamiltonian operator, is

$$\hat{H} = \hat{E}_k + \hat{V} = -\frac{\hbar^2}{2m}\frac{\mathrm{d}^2}{\mathrm{d}x^2} + \hat{V}(x) \qquad \text{Hamiltonian operator} \qquad (7C.6)$$

with $\hat{V}(x)$ the multiplicative operator in eqn 7C.4 (or some other appropriate expression for the potential energy).

The expression for the kinetic energy operator, eqn 7C.5, enables us to develop an important point about the Schrödinger equation. In mathematics, the second derivative of a function is a measure of its curvature: a large second derivative indicates a sharply curved function (Fig. 7C.1). It follows that a sharply curved wavefunction is associated with a high kinetic energy, and one with a low curvature is associated with a low kinetic energy. This interpretation is consistent with the de Broglie relation, which predicts a short wavelength (a sharply curved wavefunction) when the linear momentum (and hence the kinetic energy) is high. However, it extends the interpretation to wavefunctions that do not spread through space and resemble those shown in Fig. 7C.1. The curvature of a wavefunction in general varies from place to place. Wherever a wavefunction is sharply curved, its contribution to the total kinetic energy is large (Fig. 7C.2). Wherever the wavefunction is not sharply curved, its contribution to the overall kinetic energy is low. As we shall shortly see, the observed kinetic energy of the particle is an integral of all the contributions of the kinetic energy from each region. Hence, we can expect a particle to have a high kinetic energy if the average curvature of its wavefunction is high. Locally there can be both positive and negative contributions to the kinetic energy (because the curvature can be either positive, ∪, or negative, ∩), but the average is always positive (see Problem 7C.12).

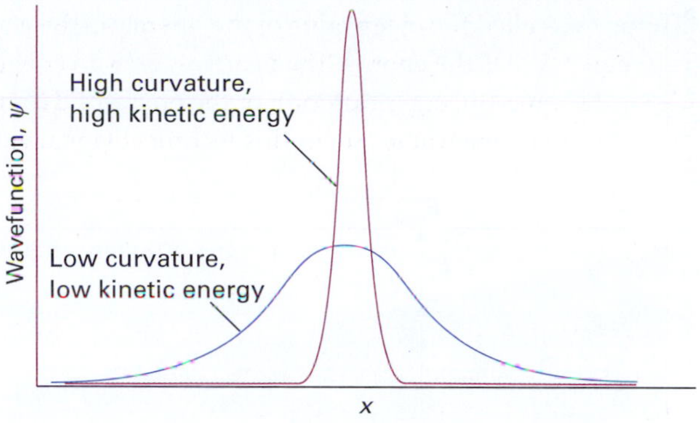

Figure 7C.1 Even if the wavefunction does not have the form of a periodic wave, it is still possible to infer from it the average kinetic energy of a particle by noting its average curvature. This figure shows two wavefunctions: the sharply curved function corresponds to a higher kinetic energy than the less sharply curved function.

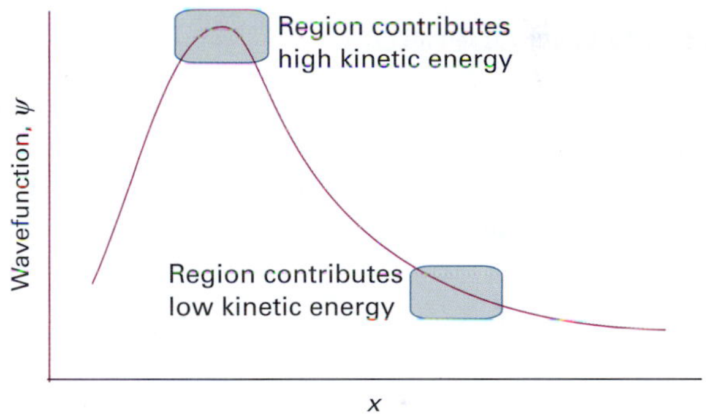

Figure 7C.2 The observed kinetic energy of a particle is an average of contributions from the entire space covered by the wavefunction. Sharply curved regions contribute a high kinetic energy to the average; slightly curved regions contribute only a small kinetic energy.

The association of high curvature with high kinetic energy will turn out to be a valuable guide to the interpretation of wavefunctions and the prediction of their shapes. For example, suppose we need to know the wavefunction of a particle with a given total energy and a potential energy that decreases with increasing x (Fig. 7C.3). Because the difference $E - V = E_k$ increases from left to right, the wavefunction must become more sharply curved as x increases: its wavelength decreases as the local contributions to its kinetic energy increase. We can therefore guess that the wavefunction will look like the function sketched in the illustration, and more detailed calculation confirms this to be so.

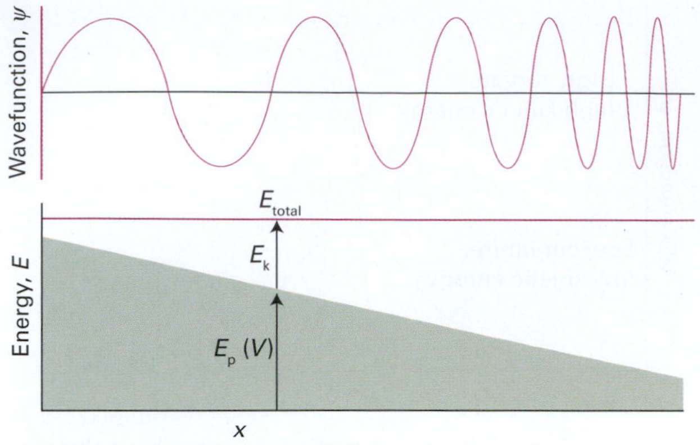

Figure 7C.3 The wavefunction of a particle in a potential decreasing towards the right and hence subjected to a constant force to the right. Only the real part of the wavefunction is shown, the imaginary part is similar, but displaced to the right.

(c) Hermitian operators

All the quantum mechanical operators that correspond to observables have a very special mathematical property: they are 'hermitian'. An **hermitian operator** is one for which the following relation is true:

$$\int \psi_i^* \hat{\Omega} \psi_j \, \mathrm{d}\tau = \left\{ \int \psi_j^* \hat{\Omega} \psi_i \, \mathrm{d}\tau \right\}^* \qquad \textit{Definition} \quad \text{Hermiticity} \quad (7\text{C}.7)$$

That is, the same result is obtained by letting the operator act on ψ_j and then integrating or by letting it act on ψ_i instead, integrating, and then taking the complex conjugate of the result. One trivial consequence of hermiticity is that it reduces the number of integrals we need to evaluate. However, as we shall see, hermiticity has much more profound implications.

It is easy to confirm that the position operator ($x \times$) is hermitian because we are free to change the order of the factors in the integrand:

$$\int \psi_i^* x \psi_j \, \mathrm{d}\tau = \int \psi_j x \psi_i^* \, \mathrm{d}\tau = \left\{ \int \psi_j^* x \psi_i \, \mathrm{d}\tau \right\}^*$$

The demonstration that the linear momentum operator is hermitian is more involved because we cannot just alter the order of functions we differentiate; but it is hermitian, as we show in the following *Justification*.

Justification 7C.1 The hermiticity of the linear momentum operator

Our task is to show that

$$\int \psi_i^* \hat{p}_x \psi_j \, \mathrm{d}\tau = \left\{ \int \psi_j^* \hat{p}_x \psi_i \, \mathrm{d}\tau \right\}^*$$

with $\hat{p}_x$ given in eqn 7C.3. To do so, we use 'integration by parts' (see *Mathematical background* 1), the relation

$$\int f \frac{\mathrm{d}g}{\mathrm{d}x} \mathrm{d}x = fg - \int g \frac{\mathrm{d}f}{\mathrm{d}x} \mathrm{d}x$$

In the present case we write

$$\int \psi_i^* \hat{p}_x \psi_j \, \mathrm{d}\tau = \frac{\hbar}{\mathrm{i}} \int_{-\infty}^{\infty} \overbrace{\psi_i^*}^{f} \overbrace{\frac{\mathrm{d}\psi_j}{\mathrm{d}x}}^{\mathrm{d}g/\mathrm{d}x} \mathrm{d}x$$

$$= \overbrace{\frac{\hbar}{\mathrm{i}} \overbrace{\psi_i^* \psi_j}^{fg} \Big|_{-\infty}^{\infty}}^{0} - \frac{\hbar}{\mathrm{i}} \int_{-\infty}^{\infty} \overbrace{\psi_j}^{g} \overbrace{\frac{\mathrm{d}\psi_i^*}{\mathrm{d}x}}^{\mathrm{d}f/\mathrm{d}x} \mathrm{d}x$$

The first term on the right of the second equality is zero, because all wavefunctions are either zero or converge to the same value at infinity in either direction, so we are left with

$$\int \psi_i^* \hat{p}_x \psi_j \, \mathrm{d}\tau = -\frac{\hbar}{\mathrm{i}} \int_{-\infty}^{\infty} \psi_j \frac{\mathrm{d}\psi_i^*}{\mathrm{d}x} \mathrm{d}x = \left\{ \frac{\hbar}{\mathrm{i}} \int_{-\infty}^{\infty} \psi_j^* \frac{\mathrm{d}\psi_i}{\mathrm{d}x} \mathrm{d}x \right\}^*$$

$$= \left\{ \int \psi_j^* \hat{p}_x \psi_i \, \mathrm{d}\tau \right\}^*$$

as we set out to prove. In the final line we have used $(\psi^*)^* = \psi$.

Hermitian operators are enormously important by virtue of two properties:

- The eigenvalues of hermitian operators are real: $\omega^* = \omega$ (as we prove in the following *Justification*).
- The eigenfunctions of hermitian operators are 'orthogonal' in the sense defined below.

All observables have real values (in the mathematical sense, such as $x = 2\,\mathrm{m}$ and $E = 10\,\mathrm{J}$), so all observables are represented by hermitian operators.

Justification 7C.2 The reality of eigenvalues

For a wavefunction ψ that is normalized and is an eigenfunction of an hermitian operator $\hat{\Omega}$ with eigenvalue ω, we can write

$$\int \psi^* \hat{\Omega} \psi \, \mathrm{d}\tau = \int \psi^* \omega \psi \, \mathrm{d}\tau = \omega \int \psi^* \psi \, \mathrm{d}\tau = \omega$$

However, by taking the complex conjugate we can write

$$\omega^* = \left\{ \int \psi^* \hat{\Omega} \psi \, \mathrm{d}\tau \right\}^* \overset{\text{hermiticity}}{=} \int \psi^* \hat{\Omega} \psi \, \mathrm{d}\tau = \omega$$

The conclusion that $\omega^* = \omega$ confirms that ω is real.

(d) Orthogonality

To say that two different functions ψ_i and ψ_j are **orthogonal** means that the integral (over all space) of their product is zero:

$$\int \psi_i^* \psi_j \, d\tau = 0 \quad \text{for } i \neq j \qquad \textit{Definition} \quad \text{Orthogonality} \qquad (7C.8)$$

A general feature of quantum mechanics which we prove in the following *Justification* is that *wavefunctions corresponding to different eigenvalues of a hermitian operator are orthogonal.* For example, the hamiltonian operator is hermitian (it corresponds to an observable, the energy). Therefore, if ψ_1 corresponds to one energy, and ψ_2 corresponds to a different energy, then we know at once that the two functions are orthogonal and that the integral (over all space) of their product is zero.

Justification 7C.3 The orthogonality of wavefunctions

Suppose we have two eigenfunctions of $\hat{\Omega}$, with unequal eigenvalues:

$$\hat{\Omega}\psi_i = \omega_i \psi_i \quad \text{and} \quad \hat{\Omega}\psi_j = \omega_j \psi_j$$

with ω_i not equal to ω_j. Multiply the first of these eigenvalue equations on both sides by ψ_j^* and the second by ψ_i^*, and integrate over all space:

$$\int \psi_j^* \hat{\Omega} \psi_i \, d\tau = \omega_i \int \psi_j^* \psi_i \, d\tau$$

$$\int \psi_i^* \hat{\Omega} \psi_j \, d\tau = \omega_j \int \psi_i^* \psi_j \, d\tau$$

Now take the complex conjugate of the first of these two expressions (noting that, by the hermiticity of $\hat{\Omega}$, the eigenvalues are real):

$$\left\{\int \psi_j^* \hat{\Omega} \psi_i \, d\tau\right\}^* = \omega_i \int \psi_j \psi_i^* \, d\tau = \omega_i \int \psi_i^* \psi_j \, d\tau$$

However, by hermiticity, the first term on the left is

$$\left\{\int \psi_j^* \hat{\Omega} \psi_i \, d\tau\right\}^* = \int \psi_i^* \hat{\Omega} \psi_j \, d\tau = \omega_j \int \psi_i^* \psi_j \, d\tau$$

Subtraction of this line from the preceding line then gives

$$0 = (\omega_i - \omega_j) \int \psi_i^* \psi_j \, d\tau$$

But we know that the two eigenvalues are not equal, so the integral must be zero, as we set out to prove.

The property of orthogonality is of great importance in quantum mechanics because it enables us to eliminate a large number of integrals from calculations. Orthogonality plays a central role in the theory of chemical bonding (Chapter 10) and spectroscopy (Chapters 12–14). Sets of functions that are normalized and mutually orthogonal are called **orthonormal**.

Example 7C.3 Verifying orthogonality

It is shown in Topic 8A that two possible wavefunctions for an electron confined to a one-dimensional quantum dot (a collection of atoms with dimensions in the range of nanometres and of great interest in nanotechnology) are of the form sin x and sin $2x$. These two wavefunctions are eigenfunctions of the kinetic energy operator, which is hermitian, and correspond to different eigenvalues:

$$\hat{E}_k \sin x = -\frac{\hbar^2}{2m_e}\frac{d^2 \sin x}{dx^2} = \frac{\hbar^2}{2m_e}\sin x$$

$$\hat{E}_k \sin 2x = -\frac{\hbar^2}{2m_e}\frac{d^2 \sin 2x}{dx^2} = \frac{2\hbar^2}{m_e}\sin 2x$$

Verify that the two wavefunctions are mutually orthogonal.

Method To verify the orthogonality of two functions, we integrate their product, sin $2x$ sin x, over all space, which we may take to span from $x=0$ to $x=2\pi$, because both functions repeat themselves outside that range. Hence proving that the integral of their product is zero within that range implies that the integral over the whole of space is also zero (Fig. 7C.4). Relevant integrals are given in the *Resource section.*

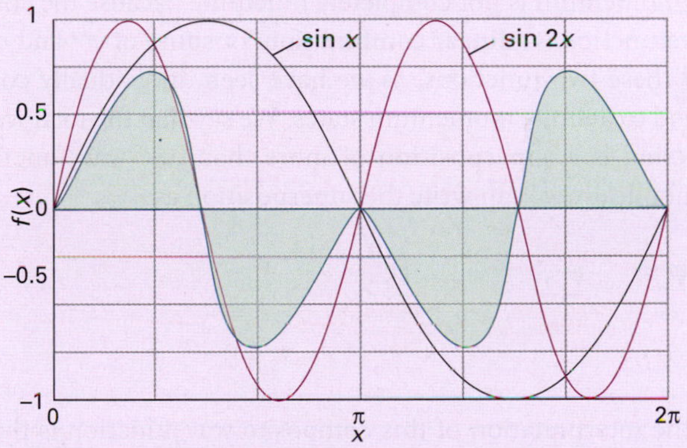

Figure 7C.4 The integral of the function $f(x) = \sin 2x \sin x$ is equal to the area (tinted) below the green curve, and is zero, as can be inferred by symmetry. The function, and the value of the integral, repeats itself for all replications of the section between 0 and 2π, so the integral from −∞ to +∞ is zero.

Answer It follows that, for $a=2$ and $b=1$, and given the fact that sin 0=0, sin 2π=0, and sin 6π=0,

$$\int_0^{2\pi} \sin 2x \sin x \, dx \overset{\text{Integral T.5}}{=} \left.\frac{\sin x}{2}\right|_0^{2\pi} - \left.\frac{\sin 3x}{6}\right|_0^{2\pi} = 0$$

and the two functions are mutually orthogonal.

Self-test 7C.3 When the electron is excited to higher energies, its wavefunction may become sin $3x$. Confirm that the functions sin x and sin $3x$ are mutually orthogonal.

Answer: $\int_0^{2\pi} \sin 3x \sin x \, dx = 0$

7C.2 Superpositions and expectation values

Suppose that the wavefunction of a free particle is $\psi(x)=2A\cos kx$ (this is one of the possibilities treated in Topic 7B, eqn 7B.9). What is the linear momentum of the particle it describes? We quickly run into trouble if we use the operator technique. When we operate with $\hat{p}_x$, we find

$$\hat{p}_x\psi=\frac{\hbar}{i}\frac{d\psi}{dx}=\frac{2\hbar}{i}A\frac{d\cos kx}{dx}=-\frac{2k\hbar}{i}A\sin kx \qquad (7C.9)$$

This expression is not an eigenvalue equation, because the function on the right (sin kx) is different from that on the left (cos kx).

When the wavefunction of a particle is not an eigenfunction of an operator, the property to which the operator corresponds does not have a definite value. However, in the current example the momentum is not completely indefinite because the cosine wavefunction is a **linear combination**, or sum,[1] of e^{ikx} and e^{-ikx}, and these two functions, as we have seen, individually correspond to definite momentum states. We say that the total wavefunction is a **superposition** of more than one wavefunction. Symbolically we can write the superposition as

$$\psi=\underbrace{\psi_{\rightarrow}}_{\substack{\text{Particle with}\\ \text{linear momentum}\\ +k\hbar}}+\underbrace{\psi_{\leftarrow}}_{\substack{\text{Particle with}\\ \text{linear momentum}\\ -k\hbar}}$$

The interpretation of this composite wavefunction is that if the momentum of the particle is repeatedly measured in a long series of observations, then its magnitude will found to be $k\hbar$ in all the measurements (because that is the value for each component of the wavefunction). However, because the two component wavefunctions occur equally in the superposition, half the measurements will show that the particle is moving to the right ($p_x=+k\hbar$), and half the measurements will show that it is moving to the left ($p_x=-k\hbar$). According to quantum mechanics, we cannot predict in which direction the particle will in fact be found to be travelling; all we can say is that, in a long series of observations, if the particle is described by this wavefunction, then there are equal probabilities of finding the particle travelling to the right and to the left.

The same interpretation applies to any wavefunction written as a linear combination of eigenfunctions of an operator. Thus, suppose the wavefunction is known to be a superposition of many different linear momentum eigenfunctions and written as the linear combination

$$\psi=c_1\psi_1+c_2\psi_2+\cdots=\sum_k c_k\psi_k \qquad \text{Linear combination of basis functions} \qquad (7C.10)$$

where the c_k are numerical (possibly complex) coefficients and the ψ_k correspond to different momentum states. The functions ψ_k are said to form a **complete set** in the sense that any arbitrary function can be expressed as a linear combination of them. Then according to quantum mechanics:

- When the momentum is measured, in a single observation one of the eigenvalues corresponding to the ψ_k that contribute to the superposition will be found.
- The probability of measuring a particular eigenvalue in a series of observations is proportional to the square modulus ($|c_k|^2$) of the corresponding coefficient in the linear combination.
- The average value of a large number of observations is given by the expectation value, $\langle\Omega\rangle$, of the operator corresponding to the observable of interest.

Physical interpretation

The **expectation value** of an operator $\hat{\Omega}$ is defined as

$$\langle\Omega\rangle=\int\psi^*\hat{\Omega}\psi\,d\tau \qquad \textit{Definition} \quad \text{Expectation value} \qquad (7C.11)$$

This formula is valid only for normalized wavefunctions. As we see in the following *Justification*, an expectation value is the weighted average of a large number of observations of a property.

Justification 7C.4 The expectation value of an operator

If ψ is an eigenfunction of $\hat{\Omega}$ with eigenvalue ω, the expectation value of $\hat{\Omega}$ is

$$\langle\Omega\rangle=\int\psi^*\overbrace{\hat{\Omega}\psi}^{\omega\psi}d\tau=\int\psi^*\omega\psi\,d\tau=\omega\int\psi^*\psi\,d\tau=\omega$$

because ω is a constant and may be taken outside the integral, and the resulting integral is equal to 1 for a normalized wavefunction. The interpretation of this expression is that, because every observation of the property Ω results in the value ω (because the wavefunction is an eigenfunction of $\hat{\Omega}$), the mean value of all the observations is also ω.

A wavefunction that is not an eigenfunction of the operator of interest can be written as a linear combination of

[1] A linear combination is more general than a sum, for it includes weighted sums of the form $ax+by+\ldots$ where a, b, … are constants. A sum is a linear combination with $a=b=\cdots=1$.

eigenfunctions. For simplicity, suppose the wavefunction is the sum of two eigenfunctions (the general case, eqn 7C.10, can be developed analogously). Then

$$\begin{aligned}\langle\Omega\rangle &= \int (c_1\psi_1+c_2\psi_2)^*\,\hat{\Omega}(c_1\psi_1+c_2\psi_2)\mathrm{d}\tau \\ &= \int (c_1\psi_1+c_2\psi_2)^*(c_1\hat{\Omega}\psi_1+c_2\hat{\Omega}\psi_2)\,\mathrm{d}\tau \\ &= \int (c_1\psi_1+c_2\psi_2)^*(c_1\omega_1\psi_1+c_2\omega_2\psi_2)\mathrm{d}\tau \\ &= c_1^*c_1\omega_1\overbrace{\int\psi_1^*\psi_1\,\mathrm{d}\tau}^{1}+c_2^*c_2\omega_2\overbrace{\int\psi_2^*\psi_2\,\mathrm{d}\tau}^{1} \\ &\quad +c_1^*c_2\omega_2\overbrace{\int\psi_1^*\psi_2\,\mathrm{d}\tau}^{0}+c_2^*c_1\omega_1\overbrace{\int\psi_2^*\psi_1\,\mathrm{d}\tau}^{0}\end{aligned}$$

The first two integrals on the right are both equal to 1 because the wavefunctions are individually normalized. Because ψ_1 and ψ_2 correspond to different eigenvalues of an hermitian operator, they are orthogonal, so the third and fourth integrals on the right are zero. We can conclude that

$$\langle\Omega\rangle=|c_1|^2\omega_1+|c_2|^2\omega_2$$

This expression shows that the expectation value is the sum of the two eigenvalues weighted by the probabilities that each one will be found in a series of measurements. Hence, the expectation value is the weighted mean of a series of observations.

Example 7C.4 Calculating an expectation value

Calculate the average value of the distance of an electron from the nucleus in the hydrogen atom in its state of lowest energy.

Method The average radius is the expectation value of the operator corresponding to the distance from the nucleus, which is multiplication by r. To evaluate $\langle r\rangle$, we need to know the normalized wavefunction (from *Example* 7B.2) and then evaluate the integral in eqn 7C.11.

Answer The average value is given by the expectation value

$$\langle r\rangle=\int\psi^* r\psi\,\mathrm{d}\tau=\int r|\psi|^2\,\mathrm{d}\tau$$

which we evaluate by using spherical polar coordinates and the appropriate expression for the volume element, $\mathrm{d}\tau = r^2\mathrm{d}r\,\sin\theta\,\mathrm{d}\theta\mathrm{d}\phi$ (*The chemist's toolkit* 7B.1). Using the normalized function in *Example* 7B.2 and a standard integral from the *Resource section*, gives

$$\langle r\rangle=\frac{1}{4\pi a_0^3}\overbrace{\int_0^\infty r^3\mathrm{e}^{-2r/a_0}\,\mathrm{d}r}^{\text{Use Integral E1, } 3!a_0^4/2^4}\overbrace{\int_0^\pi \sin\theta\,\mathrm{d}\theta}^{2}\overbrace{\int_0^{2\pi}\mathrm{d}\phi}^{2\pi}=\tfrac{3}{2}a_0$$

Because a_0=52.9 pm (see inside the front cover), $\langle r\rangle$=79.4 pm. This result means that if a very large number of measurements of the distance of the electron from the nucleus are made, then their mean value will be 79.4 pm. However, each different observation will give a different and unpredictable individual result because the wavefunction is not an eigenfunction of the operator corresponding to r.

Self-test 7C.4 Evaluate the root mean square distance, $\langle r^2\rangle^{1/2}$, of the electron from the nucleus in the hydrogen atom.

Answer: $3^{1/2}a_0$=91.6 pm

The mean kinetic energy of a particle in one dimension is the expectation value of the operator given in eqn 7C.5. Therefore, we can write

$$\langle E_\mathrm{k}\rangle=\int\psi^*\hat{E}_\mathrm{k}\psi\,\mathrm{d}x=-\frac{\hbar^2}{2m}\int\psi^*\frac{\mathrm{d}^2\psi}{\mathrm{d}x^2}\mathrm{d}x \qquad (7\text{C}.12)$$

This conclusion confirms the previous assertion that the kinetic energy is a kind of average over the curvature of the wavefunction: we get a large contribution to the observed value from regions where the wavefunction is sharply curved (so $\mathrm{d}^2\psi/\mathrm{d}x^2$ is large) and the wavefunction itself is large (so that ψ^* is large too).

7C.3 The uncertainty principle

We have seen that if the wavefunction is Ae^{ikx}, then the particle it describes has a definite state of linear momentum, namely travelling to the right with momentum $p_x=+k\hbar$. However, we have also seen that the position of the particle described by this wavefunction is completely unpredictable. In other words, if the momentum is specified precisely, it is impossible to predict the location of the particle. This statement is one half of a special case of the **Heisenberg uncertainty principle**, one of the most celebrated results of quantum mechanics:

> It is impossible to specify simultaneously, with arbitrary precision, both the momentum and the position of a particle.

Heisenberg uncertainty principle

Before discussing the principle further, we must establish its other half: that if we know the position of a particle exactly, then we can say nothing about its momentum. The argument draws on the idea of regarding a wavefunction as a superposition of eigenfunctions, and runs as follows.

If we know that the particle is at a definite location, its wavefunction must be large there and zero everywhere else (Fig. 7C.5). Such a wavefunction can be created by superimposing a large number of harmonic (sine and cosine) functions, or,

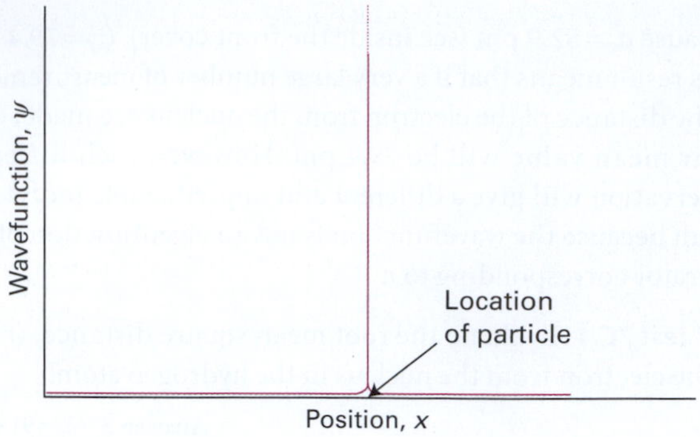

Figure 7C.5 The wavefunction for a particle at a well-defined location is a sharply spiked function which has zero amplitude everywhere except at the particle's position.

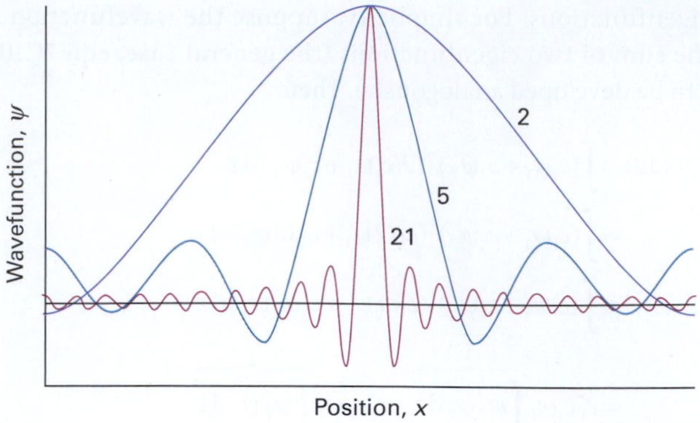

Figure 7C.6 The wavefunction for a particle with an ill-defined location can be regarded as the linear combination of several wavefunctions of definite wavelength that interfere constructively in one place but destructively elsewhere. As more waves are used in the superposition (as given by the numbers attached to the curves), the location becomes more precise at the expense of uncertainty in the particle's momentum. An infinite number of waves are needed in the superposition to construct the wavefunction of the perfectly localized particle.

equivalently, a number of e^{ikx} functions. In other words, we can create a sharply localized wavefunction, called a **wavepacket**, by forming a linear combination of wavefunctions that correspond to many different linear momenta. The superposition of a few harmonic functions gives a wavefunction that spreads over a range of locations (Fig. 7C.6). However, as the number of wavefunctions in the superposition increases, the wavepacket becomes sharper on account of the more complete interference between the positive and negative regions of the individual waves. When an infinite number of components are used, the wavepacket is a sharp, infinitely narrow spike, which corresponds to perfect localization of the particle. Now the particle is perfectly localized. However, we have lost all information about its momentum because, as we saw above, a measurement of the momentum will give a result corresponding to any one of the infinite number of waves in the superposition, and which one it will give is unpredictable. Hence, if we know the location of the particle precisely (implying that its wavefunction is a superposition of an infinite number of momentum eigenfunctions), then its momentum is completely unpredictable.

A quantitative version of this result is

$$\Delta p \Delta q \geq \tfrac{1}{2}\hbar \qquad \text{Heisenberg uncertainty principle} \qquad (7C.13a)$$

In this expression Δp is the 'uncertainty' in the linear momentum parallel to the axis q, and Δq is the uncertainty in position along that axis. These 'uncertainties' are precisely defined, for they are the root mean square deviations of the properties from their mean values:

$$\Delta p = \{\langle p^2\rangle - \langle p\rangle^2\}^{1/2} \qquad \Delta q = \{\langle q^2\rangle - \langle q\rangle^2\}^{1/2} \qquad (7C.13b)$$

If there is complete certainty about the position of the particle ($\Delta q = 0$), then the only way that eqn 7C.13a can be satisfied is for $\Delta p = \infty$, which implies complete uncertainty about the momentum. Conversely, if the momentum parallel to an axis is known exactly ($\Delta p = 0$), then the position along that axis must be completely uncertain ($\Delta q = \infty$).

The p and q that appear in eqn 7C.13 refer to the same direction in space. Therefore, whereas simultaneous specification of the position on the x-axis and momentum parallel to the x-axis are restricted by the uncertainty relation, simultaneous location of position on x and motion parallel to y or z are not restricted. The restrictions that the uncertainty principle implies are summarized in Table 7C.1.

Example 7C.5 Using the uncertainty principle

Suppose the speed of a projectile of mass 1.0 g is known to within $1\,\mu\text{m s}^{-1}$. Calculate the minimum uncertainty in its position.

Method Estimate Δp from $m\Delta v$, where Δv is the uncertainty in the speed; then use eqn 7C.13a to estimate the minimum uncertainty in position, Δq.

Answer The minimum uncertainty in position is

$$\Delta q = \frac{\hbar}{2m\Delta v} = \frac{1.055\times10^{-34}\,\text{J s}}{2\times(1.0\times10^{-3}\,\text{kg})\times(1\times10^{-6}\,\text{m s}^{-1})} = 5\times10^{-26}\,\text{m}$$

where we have used $1\,\text{J} = 1\,\text{kg m}^2\,\text{s}^{-2}$. The uncertainty is completely negligible for all practical purposes concerning macroscopic objects. However, if the mass is that of an electron, then the same uncertainty in speed implies an uncertainty in position far larger than the diameter of an atom (the analogous calculation gives $\Delta q = 60$ m); so the concept of a trajectory, the simultaneous possession of a precise position and momentum, is untenable.

Self-test 7C.5 Estimate the minimum uncertainty in the speed of an electron in a one-dimensional region of length $2a_0$.

Answer: $500\,\text{km s}^{-1}$

Table 7C.1 Constraints of the uncertainty principle*

	Variable 1					
Variable 2	x	y	z	p_x	p_y	p_z
x				■		
y					■	
z						■
p_x	■					
p_y		■				
p_z			■			

* Pairs of observables that cannot be determined simultaneously with arbitrary precision are marked with a blue rectangle; all others are unrestricted.

The Heisenberg uncertainty principle is more general than eqn 7C.13 suggests. It applies to any pair of observables called **complementary observables**, which are defined in terms of the properties of their operators. Specifically, two observables Ω_1 and Ω_2 are complementary if

$$\hat{\Omega}_1\hat{\Omega}_2\psi \neq \hat{\Omega}_2\hat{\Omega}_1\psi \quad \text{Complementarity of observables} \quad (7C.14)$$

where the term on the left implies that $\hat{\Omega}_2$ acts first, then $\hat{\Omega}_1$ acts on the result, and the term on the right implies that the operations are performed in the opposite order. When the effect of two operators applied in succession depends on their order (as this equation implies), we say that they do not **commute**. The different outcomes of the effect of applying $\hat{\Omega}_1$ and $\hat{\Omega}_2$ in a different order are expressed by introducing the **commutator** of the two operators, which is defined as

$$[\hat{\Omega}_1,\hat{\Omega}_2] = \hat{\Omega}_1\hat{\Omega}_2 - \hat{\Omega}_2\hat{\Omega}_1 \quad \textit{Definition} \quad \text{Commutator} \quad (7C.15)$$

We show in the following *Justification* that the commutator of the operators for position and linear momentum is

$$[\hat{x},\hat{p}_x] = \mathrm{i}\hbar \quad (7C.16)$$

Justification 7C.5 The commutator of position and momentum

To show that the operators for position and momentum do not commute (and hence are complementary observables) we consider the effect of $\hat{x}\hat{p}_x$ (that is, the effect of $\hat{p}_x$ followed by the effect on the outcome of multiplication by x) on a wavefunction ψ:

$$\hat{x}\hat{p}_x\psi = x\times\frac{\hbar}{\mathrm{i}}\frac{\mathrm{d}\psi}{\mathrm{d}x}$$

Next, we consider the effect of $\hat{p}_x\hat{x}$ on the same function (that is, the effect of multiplication by x followed by the effect of $\hat{p}_x$ on the outcome):

$$\hat{p}_x\hat{x}\psi = \frac{\hbar}{\mathrm{i}}\frac{\mathrm{d}(x\psi)}{\mathrm{d}x} = \frac{\hbar}{\mathrm{i}}\left(\psi + x\frac{\mathrm{d}\psi}{\mathrm{d}x}\right)$$

For this step we have used the standard rule about differentiating a product of functions ($\mathrm{d}(fg)/\mathrm{d}x = f\mathrm{d}g/\mathrm{d}x + g\mathrm{d}f/\mathrm{d}x$). The second expression is clearly different from the first, so the two operators do not commute. Their commutator can be inferred from the difference of the two expressions:

$$\hat{x}\hat{p}_x\psi - \hat{p}_x\hat{x}\psi = -\frac{\hbar}{\mathrm{i}}\psi = \mathrm{i}\hbar\psi$$

This relation is true for any wavefunction ψ, so the operator relation in eqn 7C.16 follows immediately.

The commutator in eqn 7C.16 is of such vital significance in quantum mechanics that it is taken as a fundamental distinction between classical mechanics and quantum mechanics. In fact, this commutator may be taken as a postulate of quantum mechanics, and is used to justify the choice of the operators for position and linear momentum given in eqn 7C.3.

With the concept of commutator established, the Heisenberg uncertainty principle can be given its most general form. For *any* two pairs of observables, Ω_1 and Ω_2, the uncertainties (to be precise, the root mean square deviations of their values from the mean) in simultaneous determinations are related by

$$\Delta\Omega_1\Delta\Omega_2 \geq \tfrac{1}{2}\left|\langle[\hat{\Omega}_1,\hat{\Omega}_2]\rangle\right| \quad (7C.17)$$

We obtain the special case of eqn 7C.13a when we identify the observables with x and p_x and use eqn 7C.16 for their commutator. (See *Mathematical background* 3 for the meaning of the $|\ldots|$ notation.)

Complementary observables are observables with non-commuting operators. With the discovery that some pairs of observables are complementary (we meet more examples in Topic 8C), we are at the heart of the difference between classical and quantum mechanics. Classical mechanics supposed, falsely as we now know, that the position and momentum of a particle could be specified simultaneously with arbitrary precision. However, quantum mechanics shows that position and momentum are complementary, and that we have to make a choice: we can specify position at the expense of momentum, or momentum at the expense of position.

The realization that some observables are complementary allows us to make considerable progress with the calculation of atomic and molecular properties; but it does away with some of the most cherished concepts of classical physics.

7C.4 The postulates of quantum mechanics

Here and in Topic 7B we have developed the principles of quantum theory. They can be summarized as a series of postulates, which will form the basis for chemical applications of quantum mechanics through the text.

The wavefunction: All dynamical information is contained in the wavefunction ψ for the system, which is a mathematical function found by solving the Schrödinger equation for the system. In one dimension:

$$-\frac{\hbar^2}{2m}\frac{d^2\psi}{dx^2}+V(x)\psi=E\psi$$

The Born interpretation: If the wavefunction of a particle has the value ψ at some point $\boldsymbol{r}$, then the probability of finding the particle in an infinitesimal volume $d\tau = dxdydz$ at that point is proportional to $|\psi|^2 d\tau$.

Acceptable wavefunctions: An acceptable wavefunction must be single-valued, continuous, not infinite over a finite region of space, and have a continuous slope.

Observables: Observables, Ω, are represented by operators, $\hat{\Omega}$, built from the following position and momentum operators:

$$\hat{x}=x\times \qquad \hat{p}_x=\frac{\hbar}{i}\frac{d}{dx}$$

or, more generally, from operators that satisfy the commutation relation

$$[\hat{x},\hat{p}_x]=i\hbar$$

The Heisenberg uncertainty relation: It is impossible to specify simultaneously, with arbitrary precision, both the momentum and the position of a particle and, more generally, any pair of observables with operators that do not commute.

Checklist of concepts

- ☐ 1. The Schrödinger equation is an **eigenvalue equation.**
- ☐ 2. An **operator** carries out a mathematical operation on a function.
- ☐ 3. The **hamiltonian operator** is the operator corresponding to the total energy of the system, the sum of the kinetic and potential energies.
- ☐ 4. The wavefunction corresponding to a specific energy is an **eigenfunction** of the hamiltonian operator.
- ☐ 5. The value of an observable is an **eigenvalue** of the corresponding operator constructed from the operators for position and linear momentum.
- ☐ 6. Two different functions are **orthogonal** if the integral (over all space) of their product is zero.
- ☐ 7. **Hermitian operators** have real eigenvalues and orthogonal eigenfunctions.
- ☐ 8. **Observables** are represented by hermitian operators.
- ☐ 9. Sets of functions that are normalized and mutually orthogonal are called **orthonormal**.
- ☐ 10. When the system is not described by a single eigenfunction of an operator, it may be expressed as a **superposition** of such eigenfunctions.
- ☐ 11. The mean value of a series of observations is given by the **expectation value** of the corresponding operator.
- ☐ 12. The **uncertainty principle** restricts the precision with which complementary observables may be specified and measured simultaneously.
- ☐ 13. **Complementary observables** are observables for which the corresponding operators do not commute.

Checklist of equations

Property	Equation	Comment	Equation number
Hermiticity	$\int \psi_i^* \hat{\Omega}\psi_j\, d\tau = \left\{\int \psi_j^* \hat{\Omega}\psi_i\, d\tau\right\}^*$	Real eigenvalues, orthogonal eigenfunctions	7C.7
Orthogonality	$\int \psi_i^* \psi_j\, d\tau = 0$ for $i \neq j$	Integration over all space	7C.8

Property	Equation	Comment	Equation number
Expectation value	$\langle \Omega \rangle = \int \psi^* \hat{\Omega} \psi \, d\tau$	Definition	7C.11
Commutator of two operators	$[\hat{\Omega}_1, \hat{\Omega}_2] = \hat{\Omega}_1\hat{\Omega}_2 - \hat{\Omega}_2\hat{\Omega}_1$ Special case: $[\hat{x}, \hat{p}_x] = i\hbar$	The observables are complementary if $[\hat{\Omega}_1, \hat{\Omega}_2] \neq 0$	7C.15
Heisenberg uncertainty principle	$\Delta\Omega_1 \Delta\Omega_2 \geq \frac{1}{2} \lvert \langle [\hat{\Omega}_1, \hat{\Omega}_2] \rangle \rvert$ Special case: $\Delta p \Delta q \geq \frac{1}{2}\hbar$		7C.17

CHAPTER 7 Introduction to quantum theory

TOPIC 7A The origins of quantum mechanics

Discussion questions

7A.1 Summarize the evidence that led to the introduction of quantum mechanics.

7A.2 Explain how Planck's introduction of quantization accounted for the properties of black-body radiation.

7A.3 Explain how Einstein's introduction of quantization accounted for the properties of heat capacities at low temperatures.

7A.4 Explain the meaning and consequences of wave–particle duality.

Exercises

7A.1(a) Calculate the size of the quantum involved in the excitation of (i) an electronic oscillation of period 1.0 fs, (ii) a molecular vibration of period 10 fs, (iii) a pendulum of period 1.0 s. Express the results in joules and kilojoules per mole.

7A.1(b) Calculate the size of the quantum involved in the excitation of (i) an electronic oscillation of period 2.50 fs, (ii) a molecular vibration of period 2.21 fs, (iii) a balance wheel of period 1.0 ms. Express the results in joules and kilojoules per mole.

7A.2(a) Calculate the energy per photon and the energy per mole of photons for radiation of wavelength (i) 600 nm (red), (ii) 550 nm (yellow), (iii) 400 nm (blue).

7A.2(b) Calculate the energy per photon and the energy per mole of photons for radiation of wavelength (i) 200 nm (ultraviolet), (ii) 150 pm (X-ray), (iii) 1.00 cm (microwave).

7A.3(a) Calculate the speed to which a stationary H atom would be accelerated if it absorbed each of the photons used in Exercise 7A.2(a).

7A.3(b) Calculate the speed to which a stationary ^{4}He atom (mass $4.0026m_u$) would be accelerated if it absorbed each of the photons used in Exercise 7A.2(b).

7A.4(a) A glow-worm of mass 5.0 g emits red light (650 nm) with a power of 0.10 W entirely in the backward direction. To what speed will it have accelerated after 10 y if released into free space and assumed to live?

7A.4(b) A photon-powered spacecraft of mass 10.0 kg emits radiation of wavelength 225 nm with a power of 1.50 kW entirely in the backward direction. To what speed will it have accelerated after 10.0 y if released into free space?

7A.5(a) A sodium lamp emits yellow light (550 nm). How many photons does it emit each second if its power is (i) 1.0 W, (ii) 100 W?

7A.5(b) A laser used to read CDs emits red light of wavelength 700 nm. How many photons does it emit each second if its power is (i) 0.10 W, (ii) 1.0 W?

7A.6(a) The work function for metallic caesium is 2.14 eV. Calculate the kinetic energy and the speed of the electrons ejected by light of wavelength (i) 700 nm, (ii) 300 nm.

7A.6(b) The work function for metallic rubidium is 2.09 eV. Calculate the kinetic energy and the speed of the electrons ejected by light of wavelength (i) 650 nm, (ii) 195 nm.

7A.7(a) In an X-ray photoelectron experiment, a photon of wavelength 150 pm ejects an electron from the inner shell of an atom and it emerges with a speed of 21.4 Mm s^{-1}. Calculate the binding energy of the electron.

7A.7(b) In an X-ray photoelectron experiment, a photon of wavelength 121 pm ejects an electron from the inner shell of an atom and it emerges with a speed of 56.9 Mm s^{-1}. Calculate the binding energy of the electron.

7A.8(a) To what speed must an electron be accelerated for it to have a wavelength of 100 pm? What accelerating potential difference is needed?

7A.8(b) To what speed must a proton be accelerated for it to have a wavelength of 100 pm? What accelerating potential difference is needed?

7A.9(a) To what speed must an electron be accelerated for it to have a wavelength of 3.0 cm?

7A.9(b) To what speed must a proton be accelerated for it to have a wavelength of 3.0 cm?

7A.10(a) The fine-structure constant, α, plays a special role in the structure of matter; its approximate value is 1/137. What is the wavelength of an electron travelling at a speed αc, where c is the speed of light?

7A.10(b) Calculate the linear momentum of photons of wavelength 350 nm. What speed does a hydrogen molecule need to travel to have the same linear momentum?

7A.11(a) Calculate the de Broglie wavelength of (i) a mass of 1.0 g travelling at 1.0 cm s^{-1}, (ii) the same, travelling at 100 km s^{-1}, (iii) an He atom travelling at 1000 m s^{-1} (a typical speed at room temperature).

7A.11(b) Calculate the de Broglie wavelength of an electron accelerated from rest through a potential difference of (i) 100 V, (ii) 1.0 kV, (iii) 100 kV.

Problems

7A.1 The Planck distribution gives the energy in the wavelength range dλ at the wavelength λ. Calculate the energy density in the range 650 nm to 655 nm inside a cavity of volume 100 cm^3 when its temperature is (a) 25 °C, (b) 3000 °C.

7A.2 Demonstrate that the Planck distribution reduces to the Rayleigh–Jeans law at long wavelengths.

7A.3 Derive *Wien's law*, that $\lambda_{max}T$ is a constant, where λ_{max} is the wavelength corresponding to maximum in the Planck distribution at the temperature T, and deduce an expression for the constant as a multiple of the second radiation constant, $c_2 = hc/k$.

7A.4 For a black body, the temperature and the wavelength of emission maximum, λ_{max}, are related by Wien's law, $\lambda_{max}T = \frac{1}{5}c_2$, where $c_2 = hc/k$ (see Problem 7A.3). Values of λ_{max} from a small pinhole in an electrically heated container were determined at a series of temperatures, and the results are given in the following table. Deduce a value for Planck's constant.

$\theta/°C$	1000	1500	2000	2500	3000	3500
λ_{max}/nm	2181	1600	1240	1035	878	763

7A.5‡ Solar energy strikes the top of the Earth's atmosphere at a rate of 343 W m^{-2}. About 30 per cent of this energy is reflected directly back into space by the Earth or the atmosphere. The Earth–atmosphere system absorbs the remaining energy and re-radiates it into space as black-body radiation. What is the average black-body temperature of the Earth? What is the wavelength of the most plentiful of the Earth's black-body radiation? *Hint*: Use Wien's law, Problem 7A.3.

7A.6 Use the Planck distribution to deduce the *Stefan–Boltzmann law* that the total energy density of black-body radiation is proportional to T^4, and find the constant of proportionality.

7A.7‡ Prior to Planck's derivation of the distribution law for black-body radiation, Wien found empirically a closely related distribution function which is very nearly but not exactly in agreement with the experimental results, namely $\rho=(a/\lambda^5)e^{-b/\lambda kT}$. This formula shows small deviations from Planck's at long wavelengths. (i) By fitting Wien's empirical formula to Planck's at short wavelengths determine the constants a and b. (ii) Demonstrate that Wien's formula is consistent with Wien's law (Problem 7A.3) and with the Stefan–Boltzmann law (Problem 7A.6).

7A.8‡ The temperature of the Sun's surface is approximately 5800 K. On the assumption that the human eye evolved to be most sensitive at the wavelength of light corresponding to the maximum in the Sun's radiant energy distribution, determine the colour of light to which the eye is the most sensitive.

7A.9 The Einstein frequency is often expressed in terms of an equivalent temperature θ_E, where $\theta_E=h\nu/k$. Confirm that θ_E has the dimensions of temperature, and express the criterion for the validity of the high-temperature form of the Einstein equation in terms of it. Evaluate θ_E for (a) diamond, for which ν=46.5 THz and (b) for copper, for which ν=7.15 THz. What fraction of the Dulong and Petit value of the heat capacity does each substance reach at 25 °C?

TOPIC 7B Dynamics of microscopic systems

Discussion questions

7B.1 Describe how a wavefunction summarizes the dynamical properties of a system and how those properties may be predicted.

7B.2 Discuss the relation between probability amplitude, probability density, and probability.

7B.3 Describe the constraints that the Born interpretation puts on acceptable wavefunctions.

7B.4 What are the advantages of working with normalized wavefunctions?

Exercises

7B.1(a) Consider a time-independent wavefunction of a particle moving in three-dimensional space. Identify the variables upon which the wavefunction depends.
7B.1(b) Consider a time-dependent wavefunction of a particle moving in two-dimensional space. Identify the variables upon which the wavefunction depends.

7B.2(a) Consider a time-independent wavefunction of a hydrogen atom. Identify the variables upon which the wavefunction depends. Use spherical polar coordinates.
7B.2(b) Consider a time-dependent wavefunction of a helium atom. Identify the variables upon which the wavefunction depends. Use spherical polar coordinates.

7B.3(a) An unnormalized wavefunction for a light atom rotating around a heavy atom to which it is bonded is $\psi(\phi)=e^{i\phi}$ with $0\le\phi\le2\pi$. Normalize this wavefunction.
7B.3(b) An unnormalized wavefunction for an electron in a carbon nanotube of length L is $\sin(2\pi x/L)$. Normalize this wavefunction.

7B.4(a) For the system described in Exercise 7B.3(a), what is the probability of finding the light atom in the volume element $d\phi$ at $\phi=\pi$?
7B.4(b) For the system described in Exercise 7B.3(b), what is the probability of finding the electron in the range dx at $x=L/2$?

7B.5(a) For the system described in Exercise 7B.3(a), what is the probability of finding the light atom between $\phi=\pi/2$ and $\phi=3\pi/2$?
7B.5(b) For the system described in Exercise 7B.3(b), what is the probability of finding the electron between $x=L/4$ and $x=L/2$?

Problems

7B.1 Normalize the following wavefunctions: (i) $\sin(n\pi x/L)$ in the range $0\le x\le L$, where $n=1, 2, 3, \ldots$ (this wavefunction can be used to describe delocalized electrons in a linear polyene), (ii) a constant in the range $-L\le x\le L$, (iii) $e^{-r/a}$ in three-dimensional space (this wavefunction can be used to describe the electron in the ion He^+), (iv) $xe^{-r/2a}$ in three-dimensional space. *Hint*: The volume element in three dimensions is $d\tau=r^2dr\sin\theta\, d\theta\, d\phi$, with $0\le r<\infty$, $0\le\theta\le\pi$, $0\le\phi\le2\pi$.

7B.2 Two (unnormalized) excited state wavefunctions of the H atom are

$$\text{(i) } \psi(r)=\left(2-\frac{r}{a_0}\right)e^{-r/2a_0} \qquad \text{(ii) } \psi(r,\theta,\phi)=r\sin\theta\cos\phi\, e^{-r/2a_0}$$

(a) Normalize both functions to 1. (b) Confirm that these two functions are mutually orthogonal.

7B.3 A particle free to move along one dimension x (with $0\le x<\infty$) is described by the unnormalized wavefunction $\psi(x)=e^{-ax}$ with $a=2\,m^{-1}$. What is the probability of finding the particle at a distance $x\ge1$ m?

7B.4 The ground-state wavefunction for a particle confined to a one-dimensional box of length L is $\psi=(2/L)^{1/2}\sin(\pi x/L)$. Suppose the box is

‡ These problems were supplied by Charles Trapp and Carmen Giunta.

10.0 nm long. Calculate the probability that the particle is (a) between $x=4.95$ nm and 5.05 nm, (b) between $x=1.95$ nm and 2.05 nm, (c) between $x=9.90$ nm and 10.00 nm, (d) in the right half of the box, (e) in the central third of the box.

7B.5 The ground-state wavefunction of a hydrogen atom is $\psi=(1/\pi a_0^3)^{1/2}\mathrm{e}^{-r/a_0}$ where $a_0=53$ pm (the Bohr radius). (a) Calculate the probability that the electron will be found somewhere within a small sphere of radius 1.0 pm centred on the nucleus. (b) Now suppose that the same sphere is located at $r=a_0$. What is the probability that the electron is inside it?

7B.6 Atoms in a chemical bond vibrate around the equilibrium bond length. An atom undergoing vibrational motion is described by the wavefunction $\psi(x)=N\mathrm{e}^{-x^2/2a^2}$, where a is a constant and $-\infty<x<\infty$. (a) Normalize this function. (b) Calculate the probability of finding the particle in the range $-a\le x\le a$. *Hint*: The integral encountered in part (ii) is the error function. It is provided in most mathematical software packages.

7B.7 Suppose that the state of the vibrating atom in Problem 7B.6 is described by the wavefunction $\psi(x)=Nx\mathrm{e}^{-x^2/2a^2}$. Where is the most probable location of the particle?

TOPIC 7C The principles of quantum theory

Discussion questions

7C.1 Suggest how the general shape of a wavefunction can be predicted without solving the Schrödinger equation explicitly.

7C.2 Describe the relationship between operators and observables in quantum mechanics.

7C.3 Account for the uncertainty relation between position and linear momentum in terms of the shape of the wavefunction.

7C.4 Describe the properties of wavepackets in terms of the Heisenberg uncertainty principle.

Exercises

7C.1(a) Construct the potential energy operator of a particle subjected to a harmonic oscillator potential (see Topic 8B).
7C.1(b) Construct the potential energy operator of a particle subjected to a Coulomb potential.

7C.2(a) Confirm that the kinetic energy operator, $-(\hbar^2/2m)\mathrm{d}^2/\mathrm{d}x^2$, is hermitian.
7C.2(b) The operator corresponding to the angular momentum of a particle is $(\hbar/\mathrm{i})\mathrm{d}/\mathrm{d}\phi$, where ϕ is an angle. Is this operator hermitian?

7C.3(a) Functions of the form $\sin(n\pi x/L)$ can be used to model the wavefunctions of electrons in a carbon nanotube of length L. Show that the wavefunctions $\sin(n\pi x/L)$ and $\sin(m\pi x/L)$, where $n\neq m$, are orthogonal for a particle confined to the region $0\le x\le L$.
7C.3(b) Functions of the form $\cos(n\pi x/L)$ can be used to model the wavefunctions of electrons in metals. Show that the wavefunctions $\cos(n\pi x/L)$ and $\cos(m\pi x/L)$, where $n\neq m$, are orthogonal for a particle confined to the region $0\le x\le L$.

7C.4(a) A light atom rotating around a heavy atom to which it is bonded is described by a wavefunction of the form $\psi(\phi)=\mathrm{e}^{\mathrm{i}m\phi}$ with $0\le\phi\le 2\pi$ and m an integer. Show that the $m=+1$ and $m=+2$ wavefunctions are orthogonal.
7C.4(b) Repeat Exercise 7C.4(a) for the $m=+1$ and $m=-1$ wavefunctions.

7C.5(a) An electron in a carbon nanotube of length L is described by the wavefunction $\psi(x)=\sin(2\pi x/L)$. Compute the expectation value of the position of the electron.
7C.5(b) An electron in a carbon nanotube of length L is described by the wavefunction $\psi(x)=(2/L)^{1/2}\sin(\pi x/L)$. Compute the expectation value of the kinetic energy of the electron.

7C.6(a) An electron in a one-dimensional metal of length L is described by the wavefunction $\psi(x)=\sin(\pi x/L)$. Compute the expectation value of the momentum of the electron.
7C.6(b) A light atom rotating around a heavy atom to which it is bonded is described by a wavefunction of the form $\psi(\phi)=\mathrm{e}^{\mathrm{i}\phi}$ with $0\le\phi\le 2\pi$. If the operator corresponding to angular momentum is given by $(\hbar/\mathrm{i})\mathrm{d}/\mathrm{d}\phi$, compute the expectation value of the angular momentum of the light atom.

7C.7(a) Calculate the minimum uncertainty in the speed of a ball of mass 500 g that is known to be within 1.0 μm of a certain point on a bat. What is the minimum uncertainty in the position of a bullet of mass 5.0 g that is known to have a speed somewhere between 350.000 01 m s^{-1} and 350.000 00 m s^{-1}?
7C.7(b) An electron is confined to a linear region with a length of the same order as the diameter of an atom (about 100 pm). Calculate the minimum uncertainties in its position and speed.

7C.8(a) The speed of a certain proton is 0.45 Mm s^{-1}. If the uncertainty in its momentum is to be reduced to 0.0100 per cent, what uncertainty in its location must be tolerated?
7C.8(b) The speed of a certain electron is 995 km s^{-1}. If the uncertainty in its momentum is to be reduced to 0.0010 per cent, what uncertainty in its location must be tolerated?

7C.9(a) Determine the commutators of the operators (i) $\mathrm{d}/\mathrm{d}x$ and $1/x$, (ii) $\mathrm{d}/\mathrm{d}x$ and x^2.
7C.9(b) Determine the commutators of the operators a and a^+, where $a=(x+\mathrm{i}p)/2^{1/2}$ and $a^+=(x-\mathrm{i}p)/2^{1/2}$.

Problems

7C.1 Write the time-independent Schrödinger equations for (a) an electron moving in one dimension about a stationary proton and subjected to a Coulomb potential, (b) a free particle, (c) a particle subjected to a constant, uniform force.

7C.2 Construct quantum mechanical operators for the following observables: (a) kinetic energy in one and in three dimensions, (b) the inverse separation, $1/x$, (c) electric dipole moment in one dimension, (d) the mean square deviations of the position and momentum of a particle (in one dimension) from the mean values.

7C.3 Identify which of the following functions are eigenfunctions of the operator $\mathrm{d}/\mathrm{d}x$: (a) $\mathrm{e}^{\mathrm{i}kx}$, (b) k, (c) kx, (d) e^{-ax^2}. Give the corresponding eigenvalue where appropriate.

7C.4 Determine which of the following functions are eigenfunctions of the inversion operator $\hat{\imath}$ which has the effect of making the replacement $x \to -x$): (a) $x^3 - kx$, (b) $\cos kx$, (c) x^2+3x-1. State the eigenvalue of $\hat{\imath}$ when relevant.

7C.5 Which of the functions in Problem 7C.3 are (a) also eigenfunctions of d^2/dx^2 and (b) only eigenfunctions of d^2/dx^2? Give the eigenvalues where appropriate.

7C.6 Show that the product of a hermitian operator with itself is also a hermitian operator.

7C.7 Calculate the average linear momentum of a particle described by the following wavefunctions: (a) e^{ikx}, (b) $\cos kx$, (c) e^{-ax^2}, where in each one x ranges from $-\infty$ to $+\infty$.

7C.8 The normalized wavefunctions for a particle confined to move on a circle are $\psi(\phi)=(1/2\pi)^{1/2}\,e^{-im\phi}$, where $m=0, \pm1, \pm2, \pm3, \ldots$ and $0\le\phi\le 2\pi$. Determine $\langle\phi\rangle$.

7C.9 A particle freely moving in one dimension x with $0\le x<\infty$ is in a state described by the wavefunction $\psi(x)=a^{1/2}e^{-ax/2}$, where a is a constant. Determine the expectation value of the position operator.

7C.10 The wavefunction of an electron in a linear accelerator is $\psi=(\cos\chi)e^{ikx}+(\sin\chi)e^{-ikx}$, where χ (chi) is a parameter. (i) What is the probability that the electron will be found with a linear momentum (a) $+k\hbar$, (b) $-k\hbar$? (c) What form would the wavefunction have if it were 90 per cent certain that the electron had linear momentum $+k\hbar$? (c) Evaluate the kinetic energy of the electron.

7C.11 Two (unnormalized) excited state wavefunctions of the H atom are (i) $\psi=(2-r/a_0)e^{-r/2a_0}$ and (ii) $\psi=r\sin\theta\cos\phi\, e^{-r/2a_0}$. (a) Normalize both functions to 1. (b) Confirm that these two functions are mutually orthogonal. (c) Evaluate the expectation values of r and r^2 for the atom.

7C.12 The ground-state wavefunction of a hydrogen atom is $\psi=(1/\pi a_0^3)^{1/2}e^{-r/a_0}$. Calculate (a) the mean potential energy and (b) the mean kinetic energy of an electron in the ground state of a hydrogenic atom.

7C.13 Show that the expectation value of an operator that can be written as the square of an hermitian operator is positive.

7C.14 A particle is in a state described by the wavefunction $\psi(x)=(2a/\pi)^{1/4}e^{-ax^2}$, where a is a constant and $-\infty\le x\le\infty$. Verify that the value of the product $\Delta p\Delta x$ is consistent with the predictions from the uncertainty principle.

7C.15 A particle is in a state described by the wavefunction $\psi(x)=(2a)^{1/2}e^{-ax}$, where a is a constant and $0\le x\le\infty$. Determine the expectation value of the commutator of the position and momentum operators.

7C.16 Evaluate the commutators (a) $[\hat{H},\hat{p}_x]$ and (b) $[\hat{H},\hat{x}]$ where $\hat{H}=\hat{p}_x^2/2m+\hat{V}(x)$. Choose (i) $V(x)=V$, a constant, (ii) $V(x)=\frac{1}{2}k_f x^2$.

7C.17 (a) Given that any operators used to represent observables must satisfy the commutation relation in eqn 7C.16, what would be the operator for position if the choice had been made to represent linear momentum parallel to the x-axis by multiplication by the linear momentum. These different choices are all valid 'representations' of quantum mechanics. (b) With the identification of $\hat{x}$ in this representation, what would be the operator for $1/x$? *Hint*: Think of $1/x$ as x^{-1}.

Integrated activities

7.1‡ A star too small and cold to shine has been found by S. Kulkarni et al. (*Science* **270**, 1478 (1995)). The spectrum of the object shows the presence of methane which, according to the authors, would not exist at temperatures much above 1000 K. The mass of the star, as determined from its gravitational effect on a companion star, is roughly 20 times the mass of Jupiter. The star is considered to be a brown dwarf, the coolest ever found. (a) From available thermodynamic data, test the stability of methane at temperatures above 1000 K. (b) What is λ_{max} for this star? (c) What is the energy density of the star relative to that of the Sun (6000 K)? (d) To determine whether the star will shine, estimate the fraction of the energy density of the star in the visible region of the spectrum.

7.2 Suppose that the wavefunction of an electron in a carbon nanotube is a linear combination of $\cos(nx)$ functions. (a) Use mathematical software, a spreadsheet, or the *Living graphs* on the web site of this book to construct superpositions of cosine functions as

$$\psi(x)=\frac{1}{N}\sum_{k=1}^{N}\cos(k\pi x)$$

where the constant $1/N$ is introduced to keep the superpositions with the same overall magnitude. Set $x=0$ at the centre of the screen and build the superposition there. (b) Explore how the probability density $\psi^2(x)$ changes with the value of N. (c) Evaluate the root mean square location of the packet, $\langle x^2\rangle^{1/2}$. (d) Determine the probability that a given momentum will be observed.

Mathematical background 3 Complex numbers

We describe here general properties of complex numbers and functions, which are mathematical constructs frequently encountered in quantum mechanics.

MB3.1 Definitions

Complex numbers have the general form

$$z = x + iy \qquad \text{General form of a complex number} \qquad \text{(MB3.1)}$$

where $i = (-1)^{1/2}$. The real numbers x and y are, respectively, the real and imaginary parts of z, denoted $\text{Re}(z)$ and $\text{Im}(z)$. When $y = 0$, $z = x$ is a real number; when $x = 0$, $z = iy$ is a pure imaginary number. Two complex numbers $z_1 = x_1 + iy_1$ and $z_2 = x_2 + iy_2$ are equal when $x_1 = x_2$ and $y_1 = y_2$. Although the general form of the imaginary part of a complex number is written iy, a specific numerical value is typically written in the reverse order; for instance, as $3i$.

The **complex conjugate** of z, denoted z^*, is formed by replacing i by –i:

$$z^* = x - iy \qquad \text{Complex conjugate} \qquad \text{(MB3.2)}$$

The product of z^* and z is denoted $|z|^2$ and is called the **square modulus** of z. From eqns MB3.1 and MB3.2,

$$|z|^2 = (x+iy)(x-iy) = x^2 + y^2 \qquad \text{Square modulus} \qquad \text{(MB3.3)}$$

since $i^2 = -1$. The square modulus is a real number. The **absolute value** or **modulus** is itself denoted $|z|$ and is given by:

$$|z| = (z^*z)^{1/2} = (x^2+y^2)^{1/2} \qquad \text{Absolute value or modulus} \qquad \text{(MB3.4)}$$

Since $zz^* = |z|^2$ it follows that $z \times (z^*/|z|^2) = 1$, from which we can identify the (multiplicative) **inverse** of z (which exists for all nonzero complex numbers):

$$z^{-1} = \frac{z^*}{|z|^2} \qquad \text{Inverse of a complex number} \qquad \text{(MB3.5)}$$

Brief illustration MB3.1 Inverse

Consider the complex number $z = 8 - 3i$. Its square modulus is

$$|z|^2 = z^*z = (8-3i)^*(8-3i) = (8+3i)(8-3i) = 64 + 9 = 73$$

The modulus is therefore $|z| = 73^{1/2}$. From eqn MB3.5, the inverse of z is

$$z^{-1} = \frac{8+3i}{73} = \frac{8}{73} + \frac{3}{73}i$$

MB3.2 Polar representation

The complex number $z = x + iy$ can be represented as a point in a plane, the **complex plane**, with $\text{Re}(z)$ along the x-axis and $\text{Im}(z)$ along the y-axis (Fig. MB3.1). If, as shown in the figure, r and ϕ denote the polar coordinates of the point, then since $x = r\cos\phi$ and $y = r\sin\phi$, we can express the complex number in **polar form** as

$$z = r(\cos\phi + i\sin\phi) \qquad \text{Polar form of a complex number} \qquad \text{(MB3.6)}$$

The angle ϕ, called the **argument** of z, is the angle that z makes with the x-axis. Because $y/x = \tan\phi$, it follows that the polar form can be constructed from

$$r = (x^2+y^2)^{1/2} = |z| \qquad \phi = \arctan\frac{y}{x} \qquad \text{(MB3.7a)}$$

To convert from polar to Cartesian form, use

$$x = r\cos\phi \text{ and } y = r\sin\phi \text{ to form } z = x + iy \qquad \text{(MB3.7b)}$$

One of the most useful relations involving complex numbers is **Euler's formula**:

$$e^{i\phi} = \cos\phi + i\sin\phi \qquad \text{Euler's formula} \qquad \text{(MB3.8a)}$$

The simplest proof of this relation is to expand the exponential function as a power series and to collect real and imaginary terms. It follows that

$$\cos\phi = \tfrac{1}{2}(e^{i\phi} + e^{-i\phi}) \qquad \sin\phi = -\tfrac{1}{2}i(e^{i\phi} - e^{-i\phi}) \qquad \text{(MB3.8b)}$$

The polar form in eqn MB3.6 then becomes

$$z = re^{i\phi} \qquad \text{(MB3.9)}$$

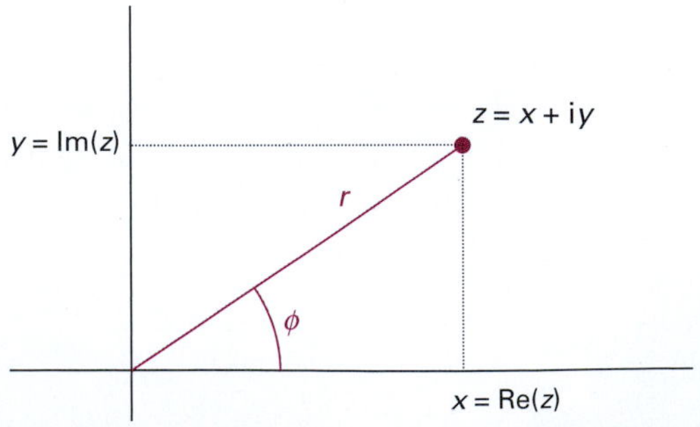

Figure MB3.1 The representation of a complex number z as a point in the complex plane using Cartesian coordinates (x,y) or polar coordinates (r,ϕ).

Brief illustration MB3.2 Polar representation

Consider the complex number $z=8-3\mathrm{i}$. From *Brief illustration* MB3.1, $r=|z|=73^{1/2}$. The argument of z is

$$\theta=\arctan\left(\frac{-3}{8}\right)=-0.359\,\mathrm{rad}, \mathrm{or}\,-20.6°$$

The polar form of the number is therefore

$$z=73^{1/2}\mathrm{e}^{-0.359\mathrm{i}}$$

MB3.3 Operations

The following rules apply for arithmetic operations for the complex numbers $z_1=x_1+\mathrm{i}y_1$ and $z_2=x_2+\mathrm{i}y_2$.

1. Addition: $z_1+z_2=(x_1+x_2)+\mathrm{i}(y_1+y_2)$ (MB3.10a)
2. Subtraction: $z_1-z_2=(x_1-x_2)+\mathrm{i}(y_1-y_2)$ (MB3.10b)
3. Multiplication:

$$\begin{aligned}z_1z_2&=(x_1+\mathrm{i}y_1)(x_2+\mathrm{i}y_2)\\&=(x_1x_2-y_1y_2)+\mathrm{i}(x_1y_2+y_1x_2)\end{aligned} \quad \text{(MB3.10c)}$$

4. Division: We interpret z_1/z_2 as $z_1z_2^{-1}$ and use eqn MB3.5 for the inverse:

$$\frac{z_1}{z_2}=z_1z_2^{-1}=\frac{z_1z_2^*}{|z_2|^2} \quad \text{(MB3.10d)}$$

Brief illustration MB3.3 Operations with numbers

Consider the complex numbers $z_1=6+2\mathrm{i}$ and $z_2=-4-3\mathrm{i}$. Then

$$z_1+z_2=(6-4)+(2-3)\mathrm{i}=2-\mathrm{i}$$
$$z_1-z_2=10+5\mathrm{i}$$
$$z_1z_2=\{6(-4)-2(-3)\}+\{6(-3)+2(-4)\}\mathrm{i}=-18-26\mathrm{i}$$
$$\frac{z_1}{z_2}=(6+2\mathrm{i})\left(\frac{-4+3\mathrm{i}}{25}\right)=-\frac{6}{5}+\frac{2}{5}\mathrm{i}$$

The polar form of a complex number is commonly used to perform arithmetical operations. For instance the product of two complex numbers in polar form is

$$z_1z_2=(r_1\mathrm{e}^{\mathrm{i}\phi_1})(r_2\mathrm{e}^{\mathrm{i}\phi_2})=r_1r_2\mathrm{e}^{\mathrm{i}(\phi_1+\phi_2)} \quad \text{(MB3.11)}$$

This multiplication can be depicted in the complex plane, as shown in Fig. MB3.2.

The nth power and the nth root of a complex number are

$$z^n=(r\mathrm{e}^{\mathrm{i}\phi})^n=r^n\mathrm{e}^{\mathrm{i}n\phi} \qquad z^{1/n}=(r\mathrm{e}^{\mathrm{i}\phi})^{1/n}=r^{1/n}\mathrm{e}^{\mathrm{i}\phi/n} \quad \text{(MB3.12)}$$

The depictions in the complex plane are shown in Fig. MB3.3.

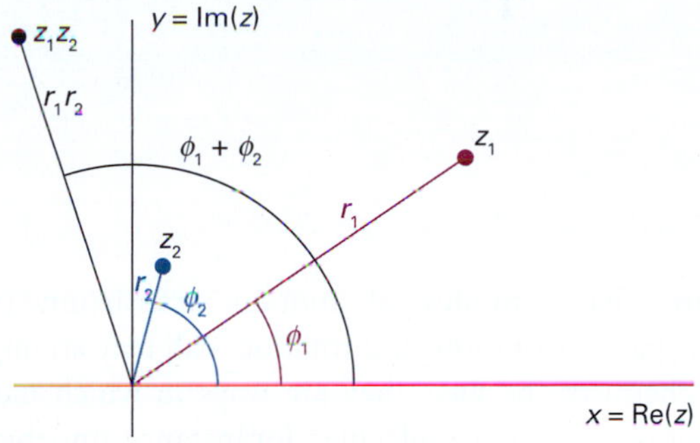

Figure MB3.2 The multiplication of two complex numbers depicted in the complex plane.

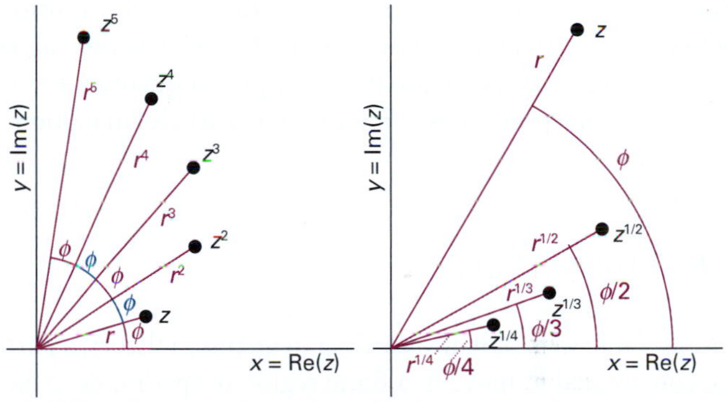

Figure MB3.3 The nth powers ($n=1, 2, 3, 4, 5$) and the nth roots ($n=1, 2, 3, 4$) of a complex number depicted in the complex plane.

Brief illustration MB3.4 Roots

To determine the 5th root of $z=8-3\mathrm{i}$, we note that from *Brief illustration* MB3.2 its polar form is

$$z=73^{1/2}\mathrm{e}^{-0.359\mathrm{i}}=8.544\mathrm{e}^{-0.359\mathrm{i}}$$

The 5th root is therefore

$$z^{1/5}=(8.544\mathrm{e}^{-0.359\mathrm{i}})^{1/5}=8.544^{1/5}\mathrm{e}^{-0.359\mathrm{i}/5}=1.536\mathrm{e}^{-0.0718\mathrm{i}}$$

It follows that $x=1.536\cos(-0.0718)=1.532$ and $y=1.536\sin(-0.0718)=-0.110$ (note that we work in radians), so

$$(8-3\mathrm{i})^{1/5}=1.532-0.110\mathrm{i}$$

CHAPTER 8

The quantum theory of motion

The three basic modes of motion—translation (motion through space), vibration, and rotation—all play an important role in chemistry because they are ways in which molecules store energy. Gas-phase molecules, for instance, undergo translational motion and their kinetic energy is a contribution to the total internal energy of a sample. Molecules can also store energy as rotational kinetic energy and transitions between their rotational energy states can be observed spectroscopically. Energy is also stored as molecular vibration, and transitions between vibrational states also give rise to spectroscopic signatures. In this Chapter we use the principles of quantum theory to calculate the properties of microscopic particles in motion.

8A Translation

In this Topic we see that, according to quantum theory, a particle constrained to move in a finite region of space is described by only certain wavefunctions and their corresponding energies. Hence, quantization emerges as a natural consequence of solving the Schrödinger equation and the conditions imposed on it. The solutions also bring to light a number of non-classical features of particles, especially their ability to tunnel into and through regions where classical physics would forbid them to be found.

8B Vibrational motion

This Topic introduces the 'harmonic oscillator', a simple but very important model for the description of molecular vibrations. We see that the energies of oscillator are quantized. The acceptable wavefunctions also show that the oscillator may be found at extensions and compressions that are forbidden by classical physics.

8C Rotational motion

The energy of a rotating particle is quantized, but in this Topic we see that its angular momentum is also restricted to certain values. The quantization of angular momentum is a very important aspect of the quantum theory of electrons in atoms and of rotating molecules.

What is the impact of this material?

'Nanoscience' is the study of atomic and molecular assemblies with dimensions ranging from 1 nm to about 100 nm and 'nanotechnology' is concerned with the incorporation of such assemblies into devices. We encounter several concepts of nanoscience throughout the text. In *Impact* I8.1 we explore quantum mechanical effects that render the properties of a nanometre-sized assembly dependent on its size.

To read more about the impact of this material, scan the QR code, or go to bcs.whfreeman.com/webpub/chemistry/pchem10e/impact/pchem-8-1.html

8A Translation

Contents

➤ Why do you need to know this material?

The application of quantum theory to translation reveals the origin of quantization and other non-classical features of physical and chemical phenomena. This material is important for the discussion of atoms and molecules that are free to move within a restricted volume, such as a gas in a container.

➤ What is the key idea?

The translational energy levels of a particle confined to a finite region of space are quantized, and under certain conditions particles can pass into and through classically forbidden regions.

➤ What do you need to know already?

You should know that the wavefunction is the solution of the Schrödinger equation (Topic 7B), and be familiar with the techniques of deriving dynamical properties from the wavefunction by using operators corresponding to the observables (Topic 7C).

In this Topic we present the essential features of the solutions of the Schrödinger equation for translation, one of the basic types of motion. We see that quantization emerges as a natural consequence of the Schrödinger equation and conditions imposed on it. The solutions also bring to light a number of non-classical features of particles, especially their ability to tunnel into and through regions where classical physics would forbid them to be found.

8A.1 Free motion in one dimension

The Schrödinger equation for a particle of mass m moving freely in one dimension is (Topic 7B)

$$-\frac{\hbar^2}{2m}\frac{\mathrm{d}^2\psi(x)}{\mathrm{d}x^2}=E\psi(x)$$ *Free motion in one dimension* | Schrödinger equation | (8A.1)

and the solutions are (as in eqn 7B.6)

$$\psi_k = A\mathrm{e}^{ikx} + B\mathrm{e}^{-ikx} \qquad E_k=\frac{k^2\hbar^2}{2m}$$ *Free motion in one dimension* | Wavefunctions and energies | (8A.2)

with A and B constants. Note that we are now labelling both the wavefunctions and the energies with the index k. The wavefunctions in eqn 8A.2 are continuous, have continuous slope everywhere, are single-valued, and do not go to infinity, and so—in the absence of any other information—are acceptable for all values of k. Because the energy of the particle is proportional to k^2, all non-negative values, including zero, of the energy are permitted. *It follows that the translational energy of a free particle is not quantized.*

The values of the constants A and B depend on how the state of motion of the particle is prepared:

- If it is shot towards positive x, then its linear momentum is $+k\hbar$ (Topic 7C), and its wavefunction is proportional to e^{ikx}. In this case $B=0$ and A is a normalization factor.
- If the particle is shot in the opposite direction, towards negative x, then its linear momentum is $-k\hbar$ and its wavefunction is proportional to e^{-ikx}. In this case, $A=0$ and B is the normalization factor.

Physical interpretation

Brief illustration 8A.1 The wavefunction of a freely-moving particle

An electron at rest that is shot out of an accelerator towards positive x through a potential difference of 1.0 V acquires a kinetic energy of 1.0 eV or 0.16 aJ (1.6×10^{-19} J). The wavefunction for such a particle is given by eqn 8A.3 with $B=0$ and k given by rearranging the expression for the energy in eqn 8A.2 into

$$k=\left(\frac{2m_eE_k}{\hbar^2}\right)^{1/2}=\left(\frac{2\times(9.109\times10^{-31}\,\text{kg})\times(1.6\times10^{-19}\,\text{J})}{(1.055\times10^{-34}\,\text{J s})^2}\right)^{1/2}$$
$$=5.1\times10^{9}\,\text{m}^{-1}$$

or 5.1 nm^{-1} (with 1 nm$=10^{-9}$ m). Therefore the wavefunction is $\psi(x)=Ae^{5.1ix/\text{nm}}$.

Self-test 8A.1 Write the wavefunction for an electron travelling to the left (negative x) after being accelerated through a potential difference of 10 kV.

Answer: $\psi(x)=Be^{-510ix/\text{nm}}$

The probability density $|\psi|^2$ is uniform if the particle is in either of the pure momentum states e^{ikx} or e^{-ikx}. According to the Born interpretation (Topic 7B), nothing further can be said about the location of the particle. That conclusion is consistent with the uncertainty principle, because if the momentum is certain, then the position cannot be specified (the operators corresponding to x and p do not commute and thus correspond to complementary observables, Topic 8C).

8A.2 Confined motion in one dimension

Consider a **particle in a box** in which a particle of mass m is confined to a finite region of space between two impenetrable walls. The potential energy is zero inside the box but rises abruptly to infinity at the walls at $x=0$ and $x=L$ (Fig. 8A.1).

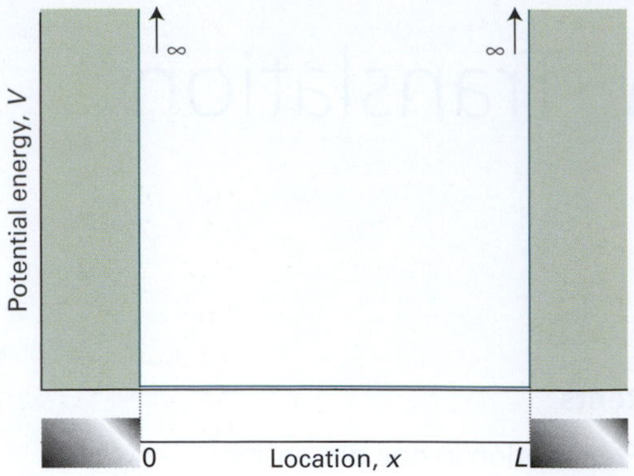

Figure 8A.1 A particle in a one-dimensional region with impenetrable walls. Its potential energy is zero between $x=0$ and $x=L$, and rises abruptly to infinity as soon as it touches the walls.

When the particle is between the walls, the Schrödinger equation is the same as for a free particle (eqn 8A.1), so the general solutions given in eqn 8A.2 are also the same. However, it will prove convenient to use $e^{\pm ikx}=\cos kx\pm i\sin kx$ (*Mathematical background* 3) to write

$$\psi_k(x)=Ae^{ikx}+Be^{-ikx}=A(\cos kx+i\sin kx)+B(\cos kx-i\sin kx)$$
$$=(A+B)\cos kx+(A-B)i\sin kx$$

If we write $C=(A-B)i$ and $D=A+B$ the general solutions take the form

$$\psi_k(x)=C\sin kx+D\cos kx \qquad \text{General solution for } 0\le x\le L \qquad (8A.3)$$

Outside the box the wavefunctions must be zero as the particle will not be found in a region where its potential energy is infinite:

$$\text{For } x<0 \text{ and } x>L,\quad \psi_k(x)=0 \qquad (8A.4)$$

At this point, there are no restrictions on the value of k and all solutions appear to be acceptable.

(a) The acceptable solutions

The requirement of the continuity of the wavefunction (Topic 7B) implies that $\psi_k(x)$ as given by eqn 8A.3 must be zero at the walls, for it must match the wavefunction inside the material of the walls where the functions meet. That is, the wavefunction must satisfy the following two **boundary conditions**, or constraints on the function at certain locations:

$$\psi_k(0)=0 \text{ and } \psi_k(L)=0 \qquad \textit{Particle in a one-dimensional box} \quad \text{Boundary conditions} \qquad (8A.5)$$

As we show in the following *Justification*, the requirement that the wavefunction satisfy these boundary conditions implies that only certain wavefunctions are acceptable and that the only permitted wavefunctions and energies of the particle are

$$\psi_n(x) = C\sin\left(\frac{n\pi x}{L}\right) \qquad n=1,2,\cdots \tag{8A.6a}$$

$$E_n = \frac{n^2h^2}{8mL^2} \qquad n=1,2,\cdots \tag{8A.6b}$$

where C is an as yet undetermined constant. Note that the wavefunctions and energy are now labelled with the dimensionless integer n instead of the quantity k.

Justification 8A.1 The energy levels and wavefunctions of a particle in a one-dimensional box

From the boundary condition $\psi_k(0)=0$ and the fact that, from eqn 8A.3, $\psi_k(0)=D$ (because $\sin 0=0$ and $\cos 0=1$), we can conclude that $D=0$. It follows that the wavefunction must be of the form $\psi_k(x)=C\sin kx$. From the second boundary condition, $\psi_k(L)=0$, we know that $\psi_k(L)=C\sin kL=0$. We could take $C=0$, but doing so would give $\psi_k(x)=0$ for all x, which would conflict with the Born interpretation (the particle must be somewhere). The alternative is to require that kL be chosen so that $\sin kL=0$. This condition is satisfied if

$$kL = n\pi \qquad n=1,2,\ldots$$

The value $n=0$ is ruled out, because it implies $k=0$ and $\psi_k(x)=0$ everywhere (because $\sin 0=0$), which is unacceptable. Negative values of n merely change the sign of $\sin kL$ (because $\sin(-x)=-\sin x$) and do not result in new solutions. The wavefunctions are therefore

$$\psi_n(x) = C\sin(n\pi x/L) \qquad n=1,2,\ldots$$

as in eqn 8A.6a. At this stage we have begun to label the solutions with the index n instead of k. Because k and E_k are related by eqn 8A.2, and k and n are related by $kL=n\pi$, it follows that the energy of the particle is limited to $E_n=n^2h^2/8mL^2$, as in eqn 8A.6b.

We conclude that the energy of the particle in a one-dimensional box is quantized and that this quantization arises from the boundary conditions that ψ must satisfy. This is a general conclusion: *the need to satisfy boundary conditions implies that only certain wavefunctions are acceptable, and hence restricts observables to discrete values.* So far, only energy has been quantized; shortly we shall see that other physical observables may also be quantized.

We need to determine the constant C in eqn 8A.6a. To do so, we normalize the wavefunction to 1 by using a standard integral from the *Resource section*. Because the wavefunction is zero outside the range $0\le x\le L$, we use

$$\int_0^L \psi^2 dx = C^2\int_0^L \sin^2\left(\frac{n\pi x}{L}\right)dx \overset{\text{Integral T.2}}{=} C^2\times\frac{L}{2}=1, \quad \text{so } C=\left(\frac{2}{L}\right)^{1/2}$$

for all n. Therefore, the complete solution for the particle in a box is

$$\psi_n(x) = \left(\frac{2}{L}\right)^{1/2}\sin\left(\frac{n\pi x}{L}\right) \quad \text{for } 0\le x\le L$$
$$\psi_n(x)=0 \quad \text{for } x<0 \text{ and } x>L \tag{8A.7a}$$

One-dimensional box — Wavefunctions

$$E_n = \frac{n^2h^2}{8mL^2} \qquad n=1,2,\ldots \tag{8A.7b}$$

One-dimensional box — Energy levels

where the energies and wavefunctions are labelled with the quantum number n. A **quantum number** is an integer (in some cases, as we see in Topic 9B, a half-integer) that labels the state of the system. For a particle in a one-dimensional box there is an infinite number of acceptable solutions, and the quantum number n specifies the one of interest (Fig. 8A.2). As well as acting as a label, a quantum number can often be used to calculate the energy corresponding to the state and to write down the wavefunction explicitly (in the present example, by using the relations in eqn 8A.7).

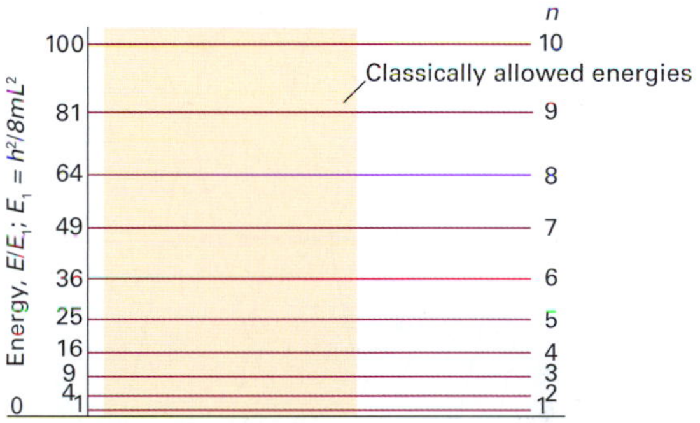

Figure 8A.2 The allowed energy levels for a particle in a box. Note that the energy levels increase as n^2, and that their separation increases as the quantum number increases. Classically, the particle is allowed to have any value of the energy in the continuum shown as a shaded area.

Brief illustration 8A.2 The energy of a particle in a box

A long carbon nanotube can be modelled as a one-dimensional structure and its electrons described by particle-in-a-box wavefunctions. The lowest energy of an electron in a carbon nanotube of length 100 nm is given by eqn 8A.7b with $n=1$:

$$E_1 = \frac{(1)^2\times(6.626\times10^{-34}\ \overset{\text{kg m}^2\text{ s}^{-2}}{\widehat{\text{J}}}\ \text{s})^2}{8\times(9.109\times10^{-31}\text{ kg})\times(100\times10^{-9}\text{ m})^2} = 6.02\times10^{-24}\text{ J}$$

or 0.00602 zJ and its wavefunction is $\psi_1(x)=(2/L)^{1/2}\sin(\pi x/L)$.

Self-test 8A.2 What are the energy and wavefunction for the next higher energy electron of the system described in this *Brief illustration*?

Answer: $E_2=0.0241$ zJ, $\psi_2(x)=(2/L)^{1/2}\sin(2\pi x/L)$

(b) The properties of the wavefunctions

Figure 8A.3 shows some of the wavefunctions of a particle in a one-dimensional box. We see that:

- The wavefunctions are all sine functions with the same amplitude but different wavelengths.
- Shortening the wavelength results in a sharper average curvature of the wavefunction and therefore an increase in the kinetic energy of the particle (its only source of energy because $V=0$ inside the box).
- The number of nodes also increases as n increase; the wavefunction ψ_n has $n-1$ nodes.
- Increasing the number of nodes between walls of a given separation increases the average curvature of the wavefunction and hence the kinetic energy of the particle.
- The probability density for a particle in a one-dimensional box is

Physical interpretation

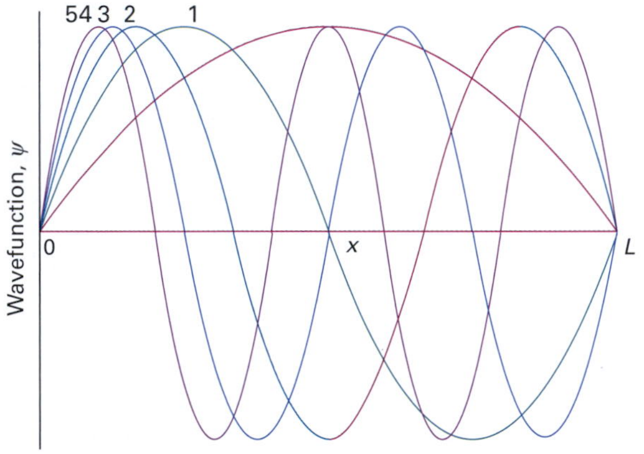

Figure 8A.3 The first five normalized wavefunctions of a particle in a box. Each wavefunction is a standing wave; successive functions possess one more half wave and a correspondingly shorter wavelength.

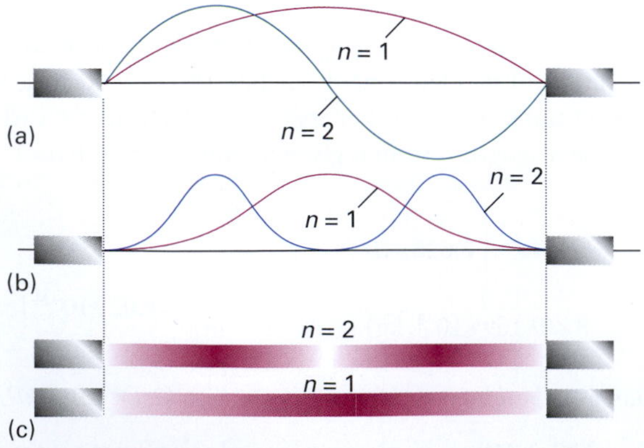

Figure 8A.4 (a) The first two wavefunctions, (b) the corresponding probability densities, and (c) a representation of the probability density in terms of the darkness of shading.

$$\psi_n^2(x)=\frac{2}{L}\sin^2\left(\frac{n\pi x}{L}\right) \qquad (8A.8)$$

- and varies with position. The non-uniformity in the probability density is pronounced when n is small (Fig. 8A.4). The most probable locations of the particle correspond to the maxima in the probability density.

Physical interpretation

Example 8A.1 Determining the probability of finding the particle in a finite region

The wavefunctions of an electron in a conjugated polyene can be approximated by particle-in-a-box wavefunctions. What is the probability, P, of locating the electron between $x=0$ (the left-hand end of a molecule) and $x=0.2$ nm in its lowest energy state in a conjugated molecule of length 1.0 nm?

Method According to the Born interpretation, $\psi(x)^2\mathrm{d}x$ is the probability of finding the particle in the small region dx located at x; therefore, the total probability of finding the electron in the specified region is the integral of $\psi(x)^2\mathrm{d}x$ over that region, as given in eqn 7B.11. The wavefunction of the electron is given in eqn 8A.7a with $n=1$. The integral you need is in the *Resource section*:

Answer The probability of finding the particle in a region between $x=0$ and $x=l$ is

$$P=\int_0^l \psi_n^2\mathrm{d}x=\frac{2}{L}\int_0^l \sin^2\left(\frac{n\pi x}{L}\right)\mathrm{d}x=\frac{l}{L}-\frac{1}{2n\pi}\sin\left(\frac{2\pi nl}{L}\right)$$

Now set $n=1$, $L=1.0$ nm, and $l=0.2$ nm, which gives $P=0.05$. The result corresponds to a chance of 1 in 20 of finding the electron in the region. As n becomes infinite, the sine term, which is multiplied by $1/n$, makes no contribution to P and the classical result for a uniformly distributed particle, $P=l/L$, is obtained.

Self-test 8A.3 Calculate the probability that an electron in the state with $n=1$ will be found between $x=0.25L$ and $x=0.75L$ in a conjugated molecule of length L (with $x=0$ at the left-hand end of the molecule).

Answer: $P=0.82$

The probability density $\psi_n^2(x)$ becomes more uniform as n increases provided we ignore the fine detail of the increasingly rapid oscillations (Fig. 8A.5). The probability density at high quantum numbers reflects the classical result that a particle bouncing between the walls spends, on the average, equal times at all points. That the quantum result corresponds to the classical prediction at high quantum numbers is an illustration of the **correspondence principle**, which states that classical mechanics emerges from quantum mechanics as high quantum numbers are reached.

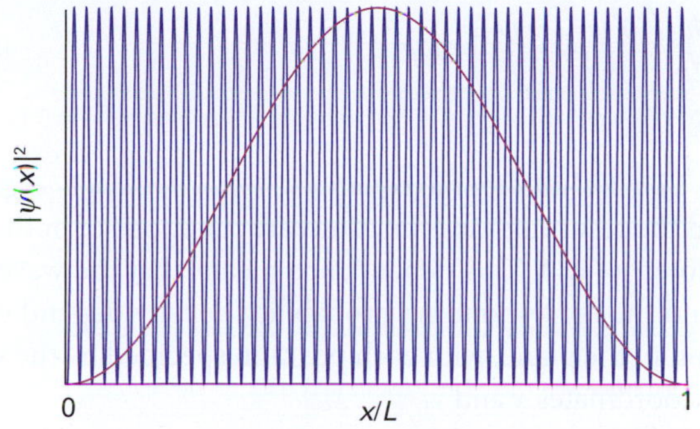

Figure 8A.5 The probability density $\psi^2(x)$ for large quantum number (here $n=50$, blue, compared with $n=1$, red). Notice that for high n the probability density is nearly uniform, provided we ignore the fine detail of the increasingly rapid oscillations.

(c) The properties of observables

The linear momentum of a particle in a box is not well defined because the wavefunction sin kx is not an eigenfunction of the linear momentum operator. However, each wavefunction is a linear combination of the linear momentum eigenfunctions e^{ikx} and e^{-ikx}. Then, because $\sin x=(e^{ix}-e^{-ix})/2i$, we can write

$$\psi_n(x)=\left(\frac{2}{L}\right)^{1/2}\sin\left(\frac{n\pi x}{L}\right)=\frac{1}{2i}\left(\frac{2}{L}\right)^{1/2}(e^{ikx}-e^{-ikx})\qquad k=\frac{n\pi}{L}\qquad \text{(8A.9)}$$

It follows from the discussion in Topic 7C that half the measurements of the linear momentum will give the value $+k\hbar$ and $-k\hbar$ for the other half. This detection of opposite directions of travel with equal probability is the quantum mechanical version of the classical picture that a particle in a one-dimensional box rattles from wall to wall and in any given period spends half its time travelling to the left and half travelling to the right.

Because n cannot be zero, the lowest energy that the particle may possess is not zero (as would be allowed by classical mechanics, corresponding to a stationary particle) but

$$E_1=\frac{h^2}{8mL^2}\qquad \textit{Particle in a box}\quad \text{Zero-point energy}\qquad \text{(8A.10)}$$

This lowest, irremovable energy is called the **zero-point energy**. The physical origin of the zero-point energy can be explained in two ways:

Physical interpretation

- The Heisenberg uncertainty principle requires a particle to possess kinetic energy if it is confined to a finite region: the location of the particle is not completely indefinite ($\Delta x\neq\infty$), so the uncertainty in its momentum cannot be precisely zero ($\Delta p\neq 0$). Because $\Delta p=(\langle p^2\rangle-\langle p\rangle^2)^{1/2}=\langle p^2\rangle^{1/2}$ in this case, $\Delta p\neq 0$ implies that $\langle p^2\rangle\neq 0$, which implies that the particle must always have nonzero kinetic energy.
- If the wavefunction is to be zero at the walls, but smooth, continuous, and not zero everywhere, then it must be curved, and curvature in a wavefunction implies the possession of kinetic energy.

The separation between adjacent energy levels with quantum numbers n and $n+1$ is

$$E_{n+1}-E_n=\frac{(n+1)^2h^2}{8mL^2}-\frac{n^2h^2}{8mL^2}=(2n+1)\frac{h^2}{8mL^2}\qquad \text{(8A.11)}$$

This separation decreases as the length of the container increases, and is very small when the container has macroscopic dimensions. The separation of adjacent levels becomes zero when the walls are infinitely far apart. Atoms and molecules free to move in normal laboratory-sized vessels may therefore be treated as though their translational energy is not quantized.

Example 8A.2 Estimating an absorption wavelength

β-Carotene (**1**) is a linear polyene in which 10 single and 11 double bonds alternate along a chain of 22 carbon atoms. If we take each C—C bond length to be about 140 pm, then the length L of the molecular box in β-carotene is $L=2.94$ nm. Estimate the wavelength of the light absorbed by this molecule from its ground state to the next higher excited state.

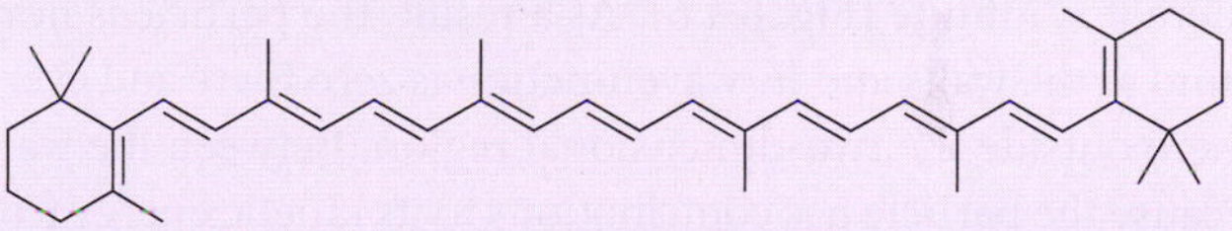

1 β-Carotene

Method For reasons that will be familiar from introductory chemistry, each C atom contributes one p electron to the π-orbitals. Use eqn 8A.11 to calculate the energy separation between the highest occupied and the lowest unoccupied levels, and convert that energy to a wavelength by using the Bohr frequency relation (eqn 7A.12).

Answer There are 22 C atoms in the conjugated chain; each contributes one p electron to the levels, so each level up to $n=11$ is occupied by two electrons. The separation in energy between the ground state and the state in which one electron is promoted from $n=11$ to $n=12$ is

$$\begin{aligned}\Delta E&=E_{12}-E_{11}\\&=(2\times11+1)\frac{(6.626\times10^{-34}\,\text{J s})^2}{8\times(9.109\times10^{-31}\,\text{kg})\times(2.94\times10^{-9}\,\text{m})^2}\\&=1.60\times10^{-19}\,\text{J}\end{aligned}$$

or 0.160 aJ. It follows from the Bohr frequency condition ($\Delta E = h\nu$) that the frequency of radiation required to cause this transition is

$$\nu = \frac{\Delta E}{h} = \frac{1.60\times10^{-19}\,\mathrm{J}}{6.626\times10^{-34}\,\mathrm{J\,s}} = 2.41\times10^{14}\,\mathrm{s^{-1}}$$

or 241 THz (1 THz $= 10^{12}$ Hz), corresponding to a wavelength $\lambda = 1240$ nm. The experimental value is 603 THz ($\lambda = 497$ nm), corresponding to radiation in the visible range of the electromagnetic spectrum. Considering the crudeness of the model we have adopted here, we should be encouraged that the computed and observed frequencies agree to within a factor of 2.5.

Self-test 8A.4 Estimate a typical nuclear excitation energy in electronvolts (1 eV $= 1.602\times10^{-19}$ J; 1 GeV $= 10^9$ eV) by calculating the first excitation energy of a proton confined to a one-dimensional box with a length equal to the diameter of a nucleus (approximately 1×10^{-15} m, or 1 fm).

Answer: 0.6 GeV

8A.3 Confined motion in two or more dimensions

Now consider a rectangular two-dimensional region of a surface with length L_1 in the x-direction and L_2 in the y-direction; the potential energy is zero everywhere except at the walls, where it is infinite (Fig. 8A.6). As a result, the particle is never found at the walls and its wavefunction is zero there and everywhere outside the two-dimensional region. Between the walls, because the particle has contributions to its kinetic energy from its motion in both the x and y directions, the Schrödinger equation has two kinetic energy terms, one for each axis. For a particle of mass m the equation is

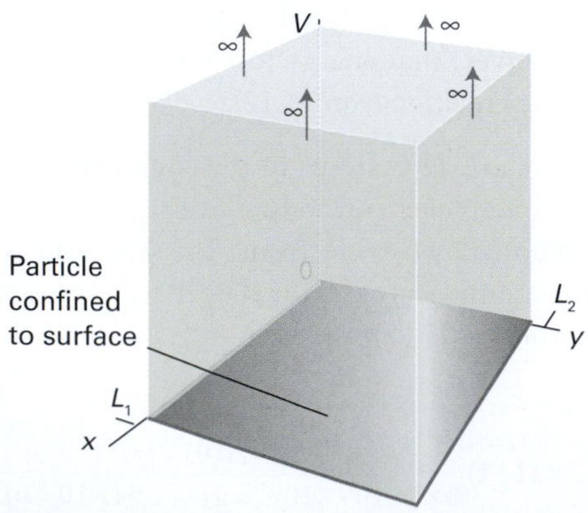

Figure 8A.6 A two-dimensional square well. The particle is confined to the plane bounded by impenetrable walls. As soon as it touches the walls, its potential energy rises to infinity.

$$-\frac{\hbar^2}{2m}\left(\frac{\partial^2\psi}{\partial x^2}+\frac{\partial^2\psi}{\partial y^2}\right) = E\psi \qquad (8A.12)$$

This is a *partial* differential equation (*Mathematical background* 4), and the resulting wavefunctions are functions of both x and y, denoted $\psi(x,y)$. This dependence means that the wavefunction and the corresponding probability density depend on the location in the plane, with each position specified by the values of the coordinates x and y.

(a) Separation of variables

A partial differential equation of the form of eqn 8A.12 can be simplified by the **separation of variables technique** (*Mathematical background* 4), which divides the equation into two or more ordinary differential equations, one for each variable. We show in the *Justification* below using this technique that the wavefunction can be written as a product of functions, one depending only on x and the other only on y:

$$\psi(x,y) = X(x)Y(y) \qquad (8A.13a)$$

and that the total energy is given by

$$E = E_X + E_Y \qquad (8A.13b)$$

where E_X is the energy associated with the motion of the particle parallel to the x-axis, and likewise for E_Y and motion parallel to the y-axis.

Justification 8A.2 The separation of variables technique applied to the particle in a two-dimensional box

We follow the procedure in *Mathematical background* 4 and apply it to eqn 8A.12. The first step to confirm that the Schrödinger equation can be separated and the wavefunction can be factored into the product of two functions X and Y is to note that, because X is independent of y and Y is independent of x, we can write

$$\frac{\partial^2\psi}{\partial x^2} = \frac{\partial^2 XY}{\partial x^2} = Y\frac{\mathrm{d}^2X}{\mathrm{d}x^2} \qquad \frac{\partial^2\psi}{\partial y^2} = \frac{\partial^2 XY}{\partial y^2} = X\frac{\mathrm{d}^2Y}{\mathrm{d}y^2}$$

Note the replacement of the partial derivatives by ordinary derivatives in each case. Then eqn 8A.12 becomes

$$-\frac{\hbar^2}{2m}\left(Y\frac{\mathrm{d}^2X}{\mathrm{d}x^2} + X\frac{\mathrm{d}^2Y}{\mathrm{d}y^2}\right) = EXY$$

Next, we divide both sides by XY, and rearrange the resulting equation into

$$\frac{1}{X}\frac{\mathrm{d}^2X}{\mathrm{d}x^2} + \frac{1}{Y}\frac{\mathrm{d}^2Y}{\mathrm{d}y^2} = -\frac{2mE}{\hbar^2}$$

The first term on the left, $(1/X)(\mathrm{d}^2X/\mathrm{d}x^2)$, is independent of y, so if y is varied only the second term on the left, $(1/Y)(\mathrm{d}^2Y/\mathrm{d}y^2)$, can change. But the sum of these two terms is a constant, $2mE/\hbar^2$, given by the right-hand side of the equation. Therefore, if the second term did change, then the right-hand side could not be constant. Consequently, even the second term cannot change when y is changed. In other words, the second term, $(1/Y)(\mathrm{d}^2Y/\mathrm{d}y^2)$, is a constant, which we write $-2mE_Y/\hbar^2$. By a similar argument, the first term, $(1/X)(\mathrm{d}^2X/\mathrm{d}x^2)$, is a constant when x changes, and we write it $-2mE_X/\hbar^2$, with $E=E_X+E_Y$. Therefore, we can write

$$\frac{1}{X}\frac{\mathrm{d}^2X}{\mathrm{d}x^2}=-\frac{2mE_X}{\hbar^2} \qquad \frac{1}{Y}\frac{\mathrm{d}^2Y}{\mathrm{d}y^2}=-\frac{2mE_Y}{\hbar^2}$$

These expressions rearrange into the two ordinary (that is, single-variable) differential equations

$$-\frac{\hbar^2}{2m}\frac{\mathrm{d}^2X}{\mathrm{d}x^2}=E_XX \qquad -\frac{\hbar^2}{2m}\frac{\mathrm{d}^2Y}{\mathrm{d}y^2}=E_YY \qquad (8A.14)$$

Each of the two ordinary differential equations in eqn 8A.14 is the same as the one-dimensional particle-in-a-box Schrödinger equation (Section 8A.2). The boundary conditions are also the same, apart from the detail of requiring $X(x)$ to be zero at $x=0$ and L_1, and $Y(y)$ to be zero at $y=0$ and L_2. We can therefore adapt eqn 8A.7a without further calculation:

$$X_{n_1}(x)=\left(\frac{2}{L_1}\right)^{1/2}\sin\left(\frac{n_1\pi x}{L_1}\right) \quad \text{for } 0\le x\le L_1$$

$$Y_{n_2}(y)=\left(\frac{2}{L_2}\right)^{1/2}\sin\left(\frac{n_2\pi y}{L_2}\right) \quad \text{for } 0\le y\le L_2$$

Then, because $\psi=XY$,

$$\psi_{n_1,n_2}(x,y)=\frac{2}{(L_1L_2)^{1/2}}\times \sin\left(\frac{n_1\pi x}{L_1}\right)\sin\left(\frac{n_2\pi y}{L_2}\right) \quad \text{for } 0\le x\le L_1, 0\le y\le L_2$$

$$\psi_{n_1,n_2}(x,y)=0 \quad \text{outside box}$$

Two-dimensional box | Wavefunctions (8A.15a)

Similarly, because $E=E_X+E_Y$, the energy of the particle is limited to the values

$$E_{n_1,n_2}=\left(\frac{n_1^2}{L_1^2}+\frac{n_2^2}{L_2^2}\right)\frac{h^2}{8m}$$

Two-dimensional box | Energy levels (8A.15b)

with the two quantum numbers taking the values $n_1=1, 2, \ldots$ and $n_2=1, 2, \ldots$ independently. The state of lowest energy is $(n_1=1, n_2=1)$ and $E_{1,1}$ is the zero-point energy.

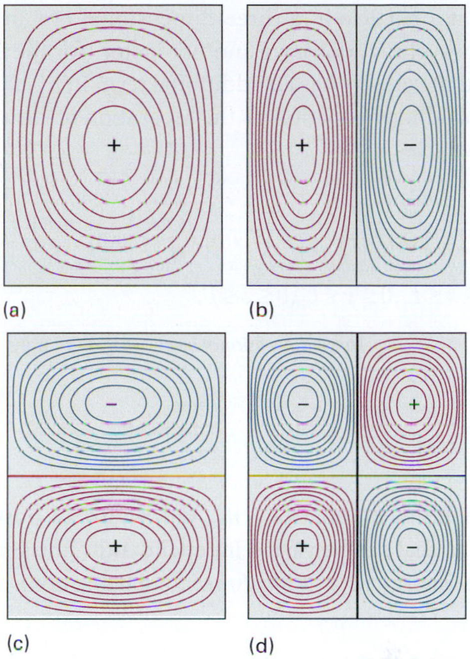

Figure 8A.7 The wavefunctions for a particle confined to a rectangular surface depicted as contours of equal amplitude. (a) $n_1=1, n_2=1$, the state of lowest energy; (b) $n_1=1, n_2=2$; (c) $n_1=2, n_2=1$; (d) $n_1=2, n_2=2$.

Some of the wavefunctions are plotted as contours in Fig. 8A.7. They are the two-dimensional versions of the wavefunctions shown in Fig. 8A.3. Whereas in one dimension the wavefunctions resemble states of a vibrating string with ends fixed, in two dimensions the wavefunctions correspond to vibrations of a rectangular plate with fixed edges.

Brief illustration 8A.3 The distribution of a particle in a two-dimensional box

Consider an electron confined to a square cavity of side L, and in the state with quantum numbers $n_1=1, n_2=2$. Because the probability density is

$$\psi_{1,2}^2(x,y)=\frac{4}{L^2}\sin^2\left(\frac{\pi x}{L}\right)\sin^2\left(\frac{2\pi y}{L}\right)$$

the most probable locations correspond to $\sin^2(\pi x/L)=1$ and $\sin^2(2\pi x/L)=1$, or $(x,y)=(L/2, L/4)$ and $(L/2, 3L/4)$. The least probable locations (the nodes, where the wavefunction passes through zero) correspond to zeroes in the probability density within the box, which occur along the line $y=L/2$.

Self-test 8A.5 Determine the most probable locations of an electron in a square cavity of side L when it is in the state with quantum numbers $n_1=2, n_2=3$.

Answer: points ($x=L/4$ and $3L/4$; $y=L/6$, $L/2$, and $5L/6$)

We treat a particle in a three-dimensional box in the same way. The wavefunctions have another factor (for the z-dependence), and the energy has an additional term in n_3^2/L_3^2. Solution of the Schrödinger equation by the separation of variables technique then gives

$$\psi_{n_1,n_2,n_3}(x,y,z)=\left(\frac{8}{L_1L_2L_3}\right)^{1/2}\sin\left(\frac{n_1\pi x}{L_1}\right)\sin\left(\frac{n_2\pi y}{L_2}\right)\sin\left(\frac{n_3\pi z}{L_3}\right)$$
$$\text{for } 0\le x\le L_1, 0\le y\le L_2, 0\le z\le L_3$$

Three-dimensional box Wavefunctions (8A.16a)

$$E_{n_1,n_2,n_3}=\left(\frac{n_1^2}{L_1^2}+\frac{n_2^2}{L_2^2}+\frac{n_3^2}{L_3^2}\right)\frac{h^2}{8m}$$

Three-dimensional box Energy levels (8A.16b)

The quantum numbers n_1, n_2, and n_3 are all positive integers 1, 2, … that can be varied independently. The system has a zero-point energy ($E_{1,1,1}=3h^2/8mL^2$ for a cubic box).

(b) Degeneracy

A special feature of the solutions arises when a two-dimensional box is not merely rectangular but square, with $L_1=L_2=L$. Then the wavefunctions and their energies are

$$\psi_{n_1,n_2}(x,y)=\frac{2}{L}\sin\left(\frac{n_1\pi x}{L}\right)\sin\left(\frac{n_2\pi y}{L}\right)$$
$$\text{for } 0\le x\le L, 0\le y\le L$$
$$\psi_{n_1,n_2}(x,y)=0 \quad \text{outside box}$$

Square box Wavefunctions (8.17a)

$$E_{n_1,n_2}=\left(n_1^2+n_2^2\right)\frac{h^2}{8mL^2}$$

Square box Energy levels (8.17b)

Consider the cases $n_1=1$, $n_2=2$ and $n_1=2$, $n_2=1$:

$$\psi_{1,2}=\frac{2}{L}\sin\left(\frac{\pi x}{L}\right)\sin\left(\frac{2\pi y}{L}\right) \qquad E_{1,2}=\frac{5h^2}{8mL^2}$$

$$\psi_{2,1}=\frac{2}{L}\sin\left(\frac{2\pi x}{L}\right)\sin\left(\frac{\pi y}{L}\right) \qquad E_{2,1}=\frac{5h^2}{8mL^2}$$

Although the wavefunctions are different, they have the same energy. The technical term for different wavefunctions corresponding to the same energy is **degeneracy**, and in this case we say that the state with energy $5h^2/8mL^2$ is 'doubly degenerate'. In general, if N wavefunctions correspond to the same energy, then we say that the state is 'N-fold degenerate'.

The occurrence of degeneracy is related to the symmetry of the system. Figure 8A.8 shows contour diagrams of the two degenerate functions $\psi_{1,2}$ and $\psi_{2,1}$. Because the box is square, one wavefunction can be converted into the other simply by rotating the plane by 90°. Interconversion by rotation through 90° is not possible when the plane is not square, and $\psi_{1,2}$ and $\psi_{2,1}$ are then not degenerate. Similar arguments account for the degeneracy of states in a cubic box. Other examples of degeneracy occur in quantum mechanical systems (for instance, in the hydrogen atom, Topic 9A), and all of them can be traced to the symmetry properties of the system.

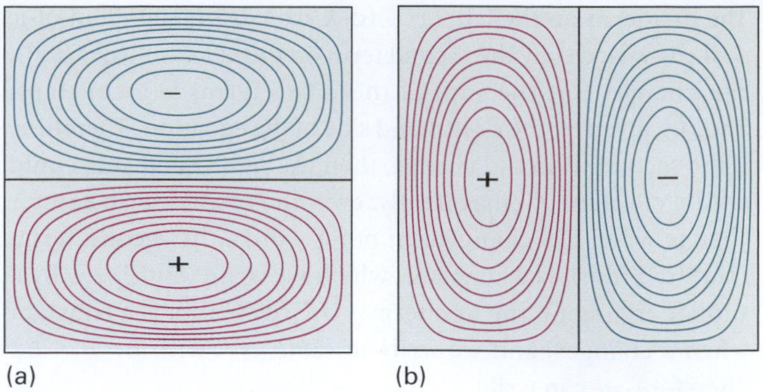

Figure 8A.8 The wavefunctions for a particle confined to a square well. Note that one wavefunction can be converted into the other by rotation of the box by 90°. The two functions correspond to the same energy. True degeneracy is a consequence of symmetry.

Brief illustration 8A.4 Degeneracies in a two-dimensional box

The energy of a particle in a two-dimensional square box of side L in the state with $n_1=1$, $n_2=7$ is

$$E_{1,7}=(1^2+7^2)\frac{h^2}{8mL^2}=\frac{50h^2}{8mL^2}$$

This state is degenerate with the state with $n_1=7$ and $n_2=1$. Thus, at first sight the energy level $50h^2/8mL^2$ is doubly degenerate. However, in certain systems there may be states that are not apparently related by symmetry but are 'accidentally' degenerate. Such is the case here, for the state with $n_1=5$ and $n_2=5$ also has energy $50h^2/8mL^2$. Accidental degeneracy is also encountered in the hydrogen atom (Topic 9A).

Self-test 8A.6 Find a state (n_1, n_2) for a particle in a rectangular box with sides of length $L_1=L$ and $L_2=2L$ that is accidentally degenerate with the state (4,4).

Answer: ($n_1=2$, $n_2=8$)

8A.4 Tunnelling

If the potential energy of a particle does not rise to infinity when it is in the wall of the container, and $E<V$, the wavefunction does not decay abruptly to zero. If the walls are thin (so that the potential energy falls to zero again after a finite distance), then the wavefunction oscillates inside the box, varies

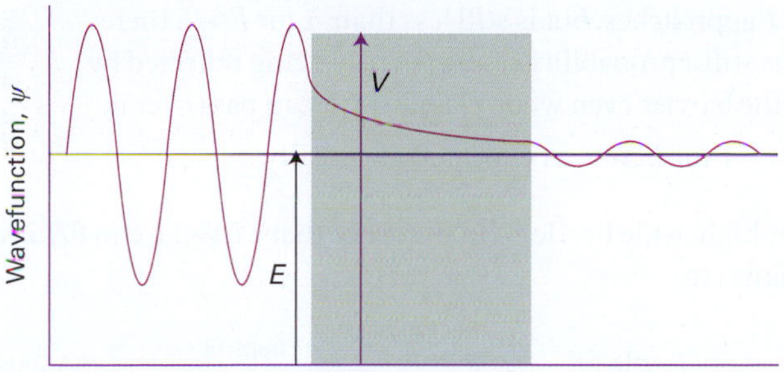

Figure 8A.9 A particle incident on a barrier from the left has an oscillating wave function, but inside the barrier there are no oscillations (for $E<V$). If the barrier is not too thick, the wavefunction is nonzero at its opposite face, and so oscillations begin again there. (Only the real component of the wavefunction is shown.)

smoothly inside the region representing the wall, and oscillates again on the other side of the wall outside the box (Fig. 8A.9). Hence the particle might be found on the outside of a container even though according to classical mechanics it has insufficient energy to escape. Such leakage by penetration through a classically forbidden region is called **tunnelling**.

The Schrödinger equation is used to calculate the probability of tunnelling of a particle of mass m incident from the left on a rectangular potential energy barrier that extends from $x=0$ to $x=L$. On the left of the barrier ($x<0$) the wavefunctions are those of a particle with $V=0$, so from eqn 8A.2 we can write

$$\psi = A\mathrm{e}^{\mathrm{i}kx} + B\mathrm{e}^{-\mathrm{i}kx} \qquad k\hbar = (2mE)^{1/2} \tag{8A.18}$$

Particle in a rectangular barrier — Wavefunction left of barrier

The Schrödinger equation for the region representing the barrier ($0 \le x \le L$), where the potential energy is the constant V, is

$$-\frac{\hbar^2}{2m}\frac{\mathrm{d}^2\psi(x)}{\mathrm{d}x^2} + V\psi(x) = E\psi(x) \tag{8A.19}$$

We shall consider particles that have $E<V$ (so, according to classical physics, the particle has insufficient energy to pass through the barrier), and therefore for which $V-E>0$. The general solutions of this equation are

$$\psi = C\mathrm{e}^{\kappa x} + D\mathrm{e}^{-\kappa x} \qquad \kappa\hbar = \{2m(V-E)\}^{1/2} \tag{8A.20}$$

Particle in a rectangular barrier — Wavefunction inside barrier

as can be verified by differentiating Ψ twice with respect to x. The important feature to note is that the two exponentials in eqn 8A.20 are now real functions, as distinct from the complex, oscillating functions for the region where $V=0$. To the right of the barrier ($x>L$), where $V=0$ again, the wavefunctions are

$$\psi = A'\mathrm{e}^{\mathrm{i}kx} \qquad k\hbar = (2mE)^{1/2} \tag{8A.21}$$

Particle in a rectangular barrier — Wavefunction right of barrier

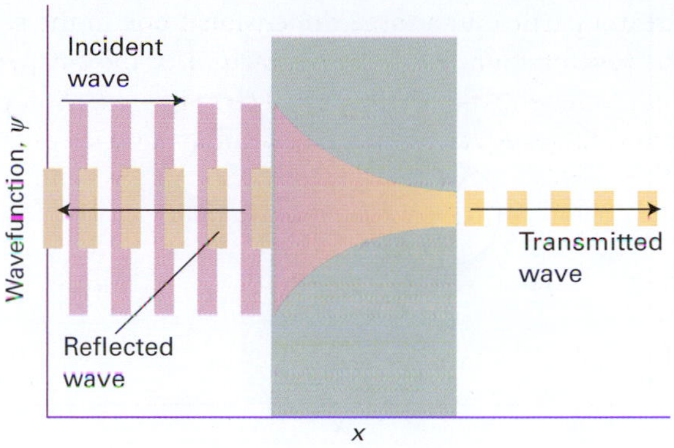

Figure 8A.10 When a particle is incident on a barrier from the left, the wavefunction consists of a wave representing linear momentum to the right, a reflected component representing momentum to the left, a varying but not oscillating component inside the barrier, and a (weak) wave representing motion to the right on the far side of the barrier.

Note that to the right of the barrier, the particle can only be moving to the right and therefore terms of the form $\mathrm{e}^{-\mathrm{i}kx}$ do not contribute to the wavefunction in eqn 8A.21.

The complete wavefunction for a particle incident from the left consists of (Fig. 8A.10):

Physical interpretation

- an incident wave ($A\mathrm{e}^{\mathrm{i}kx}$ corresponds to positive momentum);
- a wave reflected from the barrier ($B\mathrm{e}^{-\mathrm{i}kx}$ corresponds to negative momentum, motion to the left);
- the exponentially changing amplitudes inside the barrier (eqn 8A.20);
- an oscillating wave (eqn 8A.21) representing the propagation of the particle to the right after tunnelling through the barrier successfully.

The probability that a particle is travelling towards positive x (to the right) on the left of the barrier ($x<0$) is proportional to $|A|^2$, and the probability that it is travelling to the right on the right of the barrier ($x>L$) is $|A'|^2$. The ratio of these two probabilities, $|A'|^2/|A|^2$, which reflects the probability of the particle tunnelling through the barrier, is called the **transmission probability**, T.

To determine the relationship between $|A'|^2$ and $|A|^2$, we need to investigate the relationships between the coefficients A, B, C, D, and A'. Since the acceptable wavefunctions must be continuous at the edges of the barrier (at $x=0$ and $x=L$, remembering that $\mathrm{e}^0=1$):

$$A+B=C+D \qquad C\mathrm{e}^{\kappa L} + D\mathrm{e}^{-\kappa L} = A'\mathrm{e}^{\mathrm{i}kL} \tag{8A.22a}$$

Their slopes (their first derivatives) must also be continuous there (Fig. 8A.11):

$$\mathrm{i}kA - \mathrm{i}kB = \kappa C - \kappa D \qquad \kappa C\mathrm{e}^{\kappa L} - \kappa D\mathrm{e}^{-\kappa L} = \mathrm{i}kA'\mathrm{e}^{\mathrm{i}kL} \tag{8A.22b}$$

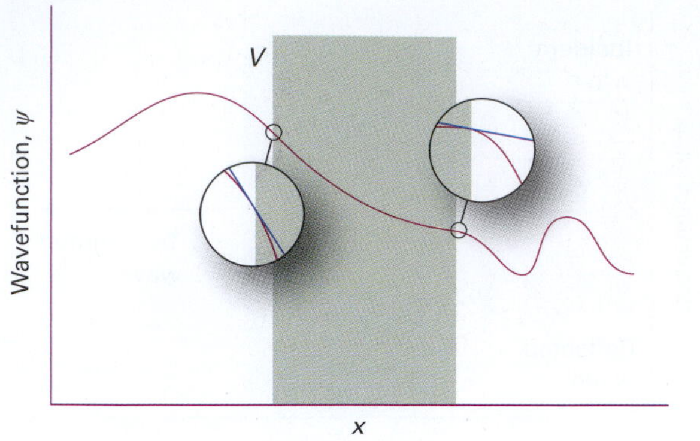

Figure 8A.11 The wavefunction and its slope must be continuous at the edges of the barrier. The conditions for continuity enable us to connect the wavefunctions in the three zones and hence to obtain relations between the coefficients that appear in the solutions of the Schrödinger equation.

After straightforward but lengthy algebraic manipulations of the above set of equations 8A.22 (see Problem 8A.6), we find

$$T=\left\{1+\frac{(e^{\kappa L}-e^{-\kappa L})^2}{16\varepsilon(1-\varepsilon)}\right\}^{-1}$$ *Rectangular potential barrier* Transmission probability (8A.23a)

where $\varepsilon = E/V$. This function is plotted in Fig. 8A.12. The transmission probability for $E > V$ is shown there too. The transmission probability has the following properties:

Physical interpretation

- $T \approx 0$ for $E \ll V$;
- T increases as E approaches V: the probability of tunnelling increases;

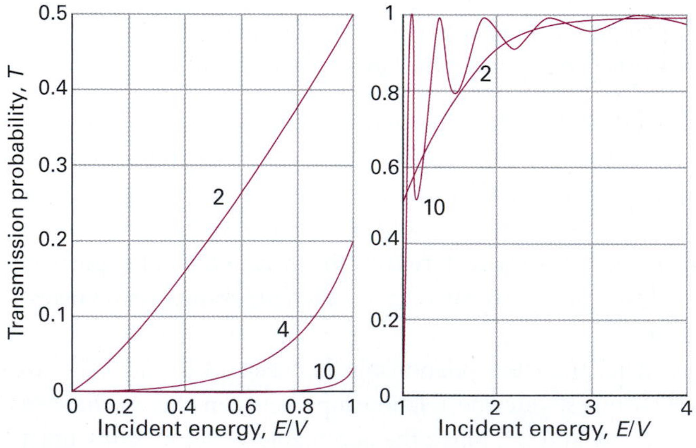

Figure 8A.12 The transmission probabilities for passage through a rectangular potential barrier. The horizontal axis is the energy of the incident particle expressed as a multiple of the barrier height. The curves are labelled with the value of $L(2mV)^{1/2}/\hbar$. The graph on the left is for $E < V$ and that on the right for $E > V$. Note that $T > 0$ for $E < V$ whereas classically T would be zero. However, $T < 1$ for $E > V$, whereas classically T would be 1.

- T approaches, but is still less than, 1 for $E > V$: there is still a probability of the particle being reflected by the barrier even when classically it can pass over it;
- $T \approx 1$ for $E \gg V$, as expected classically.

Physical interpretation

For high, wide barriers (in the sense that $\kappa L \gg 1$), eqn 8A.23a simplifies to

$$T \approx 16\varepsilon(1-\varepsilon)e^{-2\kappa L}$$ *Rectangular potential barrier; $\kappa L \gg 1$* Transmission probability (8A.23b)

The transmission probability decreases exponentially with the thickness of the barrier and with $m^{1/2}$. It follows that particles of low mass are more able to tunnel through barriers than heavy ones (Fig. 8A.13). Tunnelling is very important for electrons and muons ($m_\mu \approx 207\,m_e$), and moderately important for protons ($m_p \approx 1840 m_e$); for heavier particles it is less important.

A number of effects in chemistry (for example, the isotope-dependence of some reaction rates) depend on the ability of the proton to tunnel more readily than the deuteron. The very rapid equilibration of proton transfer reactions is also a manifestation of the ability of protons to tunnel through barriers and transfer quickly from an acid to a base. Tunnelling of protons between acidic and basic groups is also an important feature of the mechanism of some enzyme-catalysed reactions.

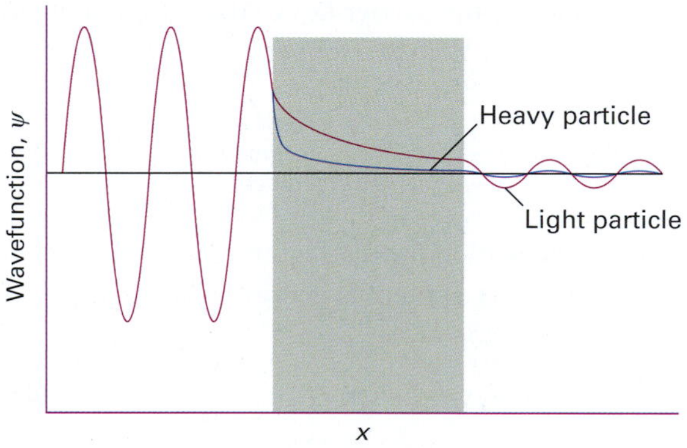

Figure 8A.13 The wavefunction of a heavy particle decays more rapidly inside a barrier than that of a light particle. Consequently, a light particle has a greater probability of tunnelling through the barrier.

Brief illustration 8A.5 Transmission probabilities for a rectangular barrier

Suppose that a proton of an acidic hydrogen atom is confined to an acid that can be represented by a barrier of height 2.000 eV and length 100 pm. The probability that a proton with energy 1.995 eV (corresponding to 0.3195 aJ) can escape from the acid is computed using 8A.23a, with $\varepsilon = E/V = 1.995\ \text{eV}/2.000\ \text{eV} = 0.9975$ and $V - E = 0.005$ eV (corresponding to 8.0×10^{-22} J).

$$\kappa = \frac{\{2\times(1.67\times10^{-27}\,\text{kg})\times(8.0\times10^{-22}\,\text{J})\}^{1/2}}{1.055\times10^{-34}\,\text{J s}} = 1.55\ldots\times10^{10}\,\text{m}^{-1}$$

We have used $1\,\text{J}=1\,\text{kg m}^2\,\text{s}^{-2}$. It follows that

$$\kappa L = (1.55\ldots\times10^{10}\,\text{m}^{-1})\times(100\times10^{-12}\,\text{m}) = 1.55\ldots$$

Equation 8A.23a then yields

$$T = \left\{1 + \frac{(e^{1.55\ldots} - e^{-1.55\ldots})^2}{16\times0.9975\times(1-0.9975)}\right\}^{-1} = 1.96\times10^{-3}$$

The larger the value of $L(2mV)^{1/2}/\hbar$ (here, 31) the smaller is the value of T for energies close to, but below, the barrier height.

Self-test 8A.7 Suppose that the junction between two semiconductors can be represented by a barrier of height 2.00 eV and length 100 pm. Calculate the probability that an electron of energy 1.95 eV can tunnel through the barrier.

Answer: $T=0.881$

A problem related to tunnelling is that of a particle in a square-well potential of finite depth (Fig. 8A.14). In this kind of potential, the wavefunction penetrates into the walls, where it decays exponentially towards zero, and oscillates within the well. The wavefunctions are found by ensuring, as in the discussion of tunnelling, that they and their slopes are continuous at the edges of the potential. Some of the lowest energy solutions are shown in Fig. 8A.15. A further difference from the solutions for an infinitely deep well is that there is only a finite number of bound states. Regardless of the depth and length of the well, however, there is always at least one bound state. Detailed consideration of the Schrödinger equation for the problem shows that in general the number of levels is equal to N, with

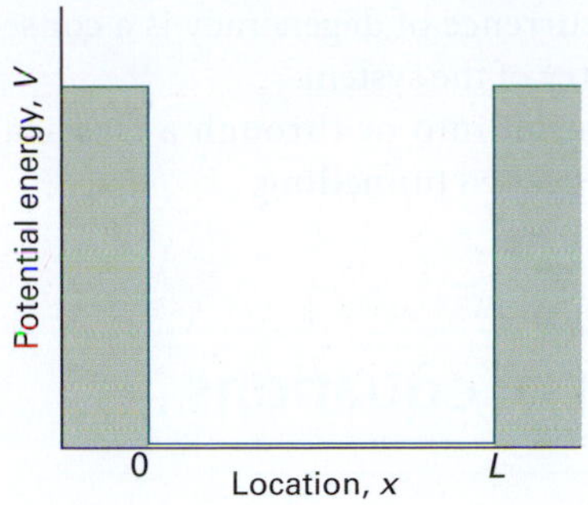

Figure 8A.14 A potential well with a finite depth.

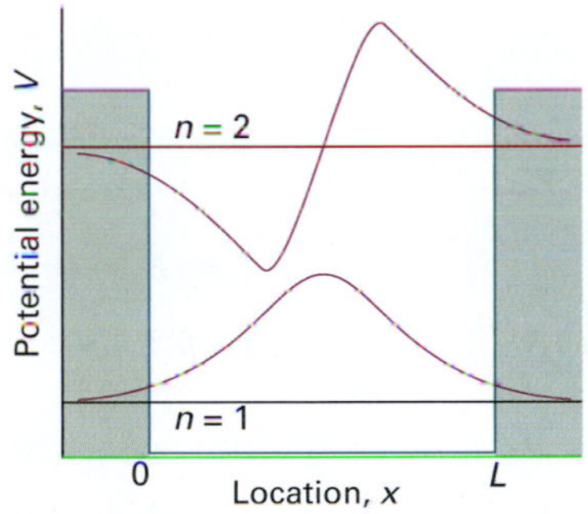

Figure 8A.15 The lowest two bound-state wavefunctions for a particle in the well shown in Fig. 8A.14.

$$N-1 < \frac{(8mVL)^{1/2}}{h} < N \qquad (8A.24)$$

where V is the depth of the well and L is its length. We see that the deeper and wider the well, the greater the number of bound states. As the depth becomes infinite, so the number of bound states also becomes infinite, as we have already seen.

Checklist of concepts

- ☐ 1. The translational energy of a free particle is not quantized.
- ☐ 2. The need to satisfy **boundary conditions** implies that only certain wavefunctions are acceptable and therefore restricts observables to discrete values.
- ☐ 3. A **quantum number** is an integer (in certain cases, a half-integer) that labels the state of the system.
- ☐ 4. A particle in a box possesses a **zero-point energy**, an irremovable minimum energy.
- ☐ 5. The **correspondence principle** states that classical mechanics emerges from quantum mechanics as high quantum numbers are reached.
- ☐ 6. The wavefunction for a particle in a two- or three-dimensional box is the product of wavefunctions for the particle in a one-dimensional box.
- ☐ 7. The energy of a particle in a two- or three-dimensional box is the sum of energies for the particle in two or three one-dimensional boxes.
- ☐ 8. The zero-point energy for a particle in a two-dimensional box corresponds to the state with quantum numbers ($n_1=1$, $n_2=1$); for three dimensions, ($n_1=1$, $n_2=1$, $n_3=1$).
- ☐ 9. **Degeneracy** occurs when different wavefunctions correspond to the same energy.

☐ 10. The occurrence of degeneracy is a consequence of the symmetry of the system.

☐ 11. Penetration into or through a classically forbidden region is called **tunnelling**.

☐ 12. The probability of tunnelling decreases with an increase in the height and width of the potential barrier.

☐ 13. Light particles are more able to tunnel through barriers than heavy ones.

Checklist of equations

Property	Equation	Comment	Equation number
Free-particle wavefunctions and energies	$\psi_k = Ae^{ikx} + Be^{-ikx}$ $\quad E_k = k^2\hbar^2/2m$	All values of k allowed	8A.2
Particle in a box			
One dimension: Wavefunctions	$\psi_n(x) = (2/L)^{1/2}\sin(n\pi x/L), \quad 0 \le x \le L$ $\psi_n(x) = 0, \quad x < 0$ and $x > L$	$n = 1, 2, \ldots$	8A.7a
Energies	$E_n = n^2h^2/8mL^2$		8A.7b
Zero-point energy	$E_1 = h^2/8mL^2$		8A.10
Two dimensions: Wavefunctions	$\psi_{n_1,n_2}(x,y) = (2/(L_1L_2)^{1/2})\sin(n_1\pi x/L_1)\sin(n_2\pi y/L_2)$ $0 \le x \le L_1, 0 \le y \le L_2$ $\psi_{n_1,n_2}(x,y) = 0$ outside box	$n_1, n_2 = 1, 2, \ldots$	8A.15a
Energies	$E_{n_1,n_2} = (n_1^2/L_1^2 + n_2^2/L_2^2)h^2/8m$		8A.15b
Three dimensions: Wavefunctions	$\psi_{n_1n_2n_3}(x,y,z) = 8/(L_1L_2L_3)^{1/2} \times$ $\sin(n_1\pi x/L_1)\sin(n_2\pi y/L_2)\sin(n_3\pi z/L_3)$, $0 \le x \le L_1, 0 \le y \le L_2, 0 \le z \le L_3$ $\psi_{n_1,n_2,n_3}(x,y,z) = 0$ outside box	$n_1, n_2, n_3 = 1, 2, \ldots$	8A.16a
Energies	$E_{n_1,n_2,n_3} = (n_1^2/L_1^2 + n_2^2/L_2^2 + n_3^2/L_3^2)h^2/8m$		8A.16b
Transmission probability	$T = \{1 + (e^{\kappa L} - e^{-\kappa L})^2/16\varepsilon(1-\varepsilon)\}^{-1}$	Rectangular potential barrier	8A.23a
	$T = 16\varepsilon(1-\varepsilon)e^{-2\kappa L}$	High, wide rectangular barrier	8A.23b

8B Vibrational motion

Contents

➤ Why do you need to know this material?

The detection and interpretation of vibrational frequencies is the basis of infrared spectroscopy (Topics 12D and 12E). Molecular vibration plays a role in the interpretation of thermodynamic properties, such as heat capacities (Topics 5E and 15F), and of the rates of chemical reactions (Topic 21C).

➤ What is the key idea?

The quantum mechanical treatment of the simplest model of vibrational motion, the harmonic oscillator, reveals that the energy is quantized and the wavefunctions are products of a polynomial and a Gaussian (bell-shaped) function.

➤ What do you need to know already?

You should know how to formulate the Schrödinger equation given a potential energy function. You should also be familiar with the concepts of tunnelling (Topic 8A) and the expectation value of an observable (Topic 7B).

Atoms in molecules and solids vibrate around their mean positions as bonds stretch, compress, and bend. Here we consider one particular type of vibrational motion, that of 'harmonic motion' in one dimension.

8B.1 The harmonic oscillator

A particle undergoes **harmonic motion**, and is said to be a **harmonic oscillator**, if it experiences a restoring force proportional to its displacement:

$$F = -k_f x \qquad \textit{Harmonic motion} \quad \text{Restoring force} \quad (8B.1)$$

where k_f is the **force constant**: the stiffer the 'spring', the greater the value of k_f. Because force is related to potential energy by $F = -dV/dx$ (see *Foundations* B), the force in eqn 8B.1 corresponds to the particle having a potential energy

$$V(x) = \tfrac{1}{2}k_f x^2 \qquad \text{Parabolic potential energy} \quad (8B.2)$$

when it is displaced through a distance x from its equilibrium position. This expression, which is the equation of a parabola (Fig. 8B.1), is the origin of the term 'parabolic potential energy' for the potential energy characteristic of a harmonic oscillator. The Schrödinger equation for the particle of mass m is therefore

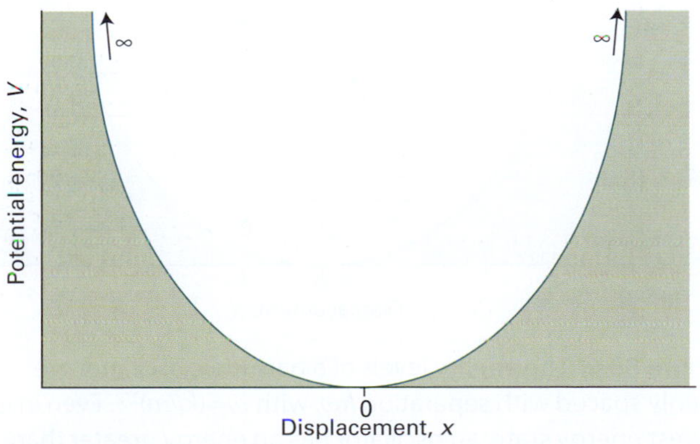

Figure 8B.1 The parabolic potential energy $V = \frac{1}{2}k_f x^2$ of a harmonic oscillator, where x is the displacement from equilibrium. The narrowness of the curve depends on the force constant k_f: the larger the value of k_f, the narrower the well.

$$-\frac{\hbar^2}{2m}\frac{d^2\psi(x)}{dx^2}+\tfrac{1}{2}k_f x^2\psi(x)=E\psi(x)$$ *Harmonic oscillator* Schrödinger equation (8B.3)

We can anticipate that the energy of an oscillator will be quantized because the wavefunction has to satisfy boundary conditions (as in Topic 8A for a particle in a box): it will not be found with very large extensions because its potential energy rises to infinity there. That is, when we impose the boundary conditions $\psi=0$ at $x=\pm\infty$, we can expect to find that only certain wavefunctions and their corresponding energies are possible.

(a) The energy levels

Equation 8B.3 is a standard equation in the theory of differential equations and its solutions are well known to mathematicians.[1] The permitted energy levels are

$$E_v=\left(v+\tfrac{1}{2}\right)\hbar\omega \qquad \omega=(k_f/m)^{1/2} \qquad v=0,1,2,\ldots$$ *Harmonic oscillator* Energy levels (8B.4)

where ω (omega) is the frequency of oscillation of a classical harmonic oscillator of the same mass and force constant. Note that ω is large when the force constant is large and the mass small. It follows that the separation between adjacent levels is

$$E_{v+1}-E_v=\hbar\omega \tag{8B.5}$$

which is the same for all v. Therefore, the energy levels form a uniform ladder of spacing $\hbar\omega$ (Fig. 8B.2). The energy separation $\hbar\omega$ is negligibly small for macroscopic objects (with large mass) for which classical mechanics is adequate for describing vibrational motion; however, the energy separation is of great importance for objects with mass similar to that of atoms.

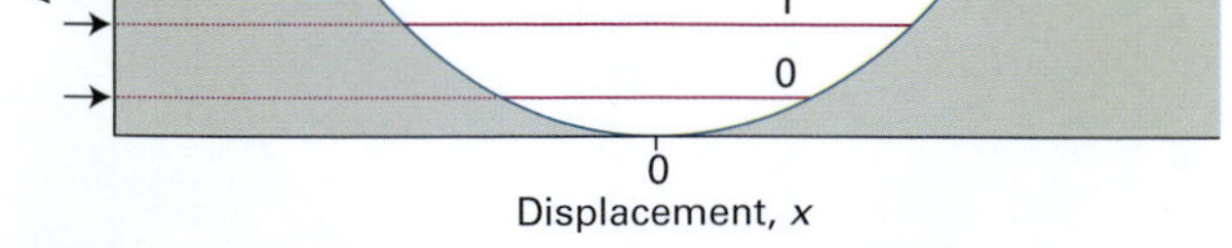

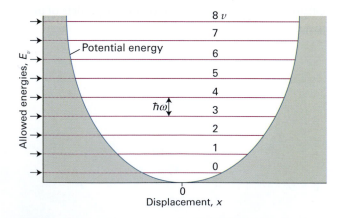

Figure 8B.2 The energy levels of a harmonic oscillator are evenly spaced with separation $\hbar\omega$, with $\omega=(k_f/m)^{1/2}$. Even in its lowest energy state, an oscillator has an energy greater than zero.

Because the smallest permitted value of v is 0, it follows from eqn 8B.4 that a harmonic oscillator has a zero-point energy

$$E_0=\tfrac{1}{2}\hbar\omega$$ *Harmonic oscillator* Zero-point energy (8B.6)

The mathematical reason for the zero-point energy is that v cannot take negative values, for if it did the wavefunction would not obey the boundary conditions. The physical reason is the same as for the particle in a box (Topic 8A): the particle is confined, its position is not completely uncertain, and therefore its momentum, and hence its kinetic energy, cannot be exactly zero. We can picture this zero-point state as one in which the particle fluctuates incessantly around its equilibrium position; classical mechanics would allow the particle to be perfectly still.

Atoms vibrate relative to one another in molecules with the bond acting like a spring. The question then arises as to what mass to use to predict the frequency of the vibration. In general, the relevant mass is a complicated combination of the masses of all the atoms that move, with each contribution weighted by the amplitude of the atom's motion. That amplitude depends on the mode of motion, such as whether the vibration is a bending motion or a stretching motion, so each mode of vibration has a characteristic 'effective mass'. For a diatomic molecule AB, however, for which there is only one mode of vibration, corresponding to the stretching and compression of the bond, the **effective mass**, μ, has a very simple form:

$$\mu=\frac{m_A m_B}{m_A+m_B}$$ *Diatomic molecule* Effective mass (8B.7)

When A is much heavier than B, m_B can be neglected in the denominator and the effective mass is $\mu\approx m_B$, the mass of the lighter atom. This result is plausible, for in the limit of the heavy atom being like a brick wall, only the lighter atom moves and hence determines the vibrational frequency.

Brief illustration 8B.1 The vibration of a diatomic molecule

The effective mass of $^1H^{35}Cl$ is

$$\mu=\frac{m_H m_{Cl}}{m_H+m_{Cl}}=\frac{(1.0078m_u)\times(34.9688m_u)}{(1.0078m_u)+(34.9688m_u)}=0.9796m_u$$

which is close to the mass of the proton. The force constant of the bond is $k_f=516.3\,\mathrm{N\,m^{-1}}$. It follows from eqn 8B.4, with μ in place of m, that

[1] For details, see our *Molecular quantum mechanics*, Oxford University Press, Oxford (2011).

$$\omega = \left(\frac{k_f}{\mu}\right)^{1/2} = \left(\frac{516.3\,\mathrm{N\,m^{-1}}}{0.9796\times(1.66054\times10^{-27}\,\mathrm{kg})}\right)^{1/2} = 5.634\times10^{14}\,\mathrm{s}$$

or 563.4 THz. (We have used 1 N = 1 kg m s^{-2}.) Therefore the separation of adjacent levels is (eqn 8B.5)

$$E_{v+1} - E_v = (1.054\,57\times10^{-34}\,\mathrm{J\,s})\times(5.634\times10^{14}\,\mathrm{s}) = 5.941\times10^{-20}\,\mathrm{J}$$

or 59.41 zJ, about 0.37 eV. This energy separation corresponds to 36 kJ mol^{-1}, which is chemically significant. The zero-point energy, eqn 8B.6, of this molecular oscillator is 29.71 zJ, which corresponds to 0.19 eV, or 18 kJ mol^{-1}.

Self-test 8B.1 Suppose a hydrogen atom is adsorbed on the surface of a gold nanoparticle by a bond of force constant 855 N m^{-1}. Calculate its zero-point vibrational energy.

Answer: 37.7 zJ, 22.7 kJ mol^{-1}, 0.24 eV

The result in *Brief illustration* 8B.1 implies that excitation requires radiation of frequency $\nu = \Delta E/h = 90$ THz and wavelength $\lambda = c/\nu = 3.3$ μm. It follows that transitions between adjacent vibrational energy levels of molecules are stimulated by or emit infrared radiation (Topics 12D and 12E).

(b) The wavefunctions

Like the particle in a box (Topic 8A), a particle undergoing harmonic motion is trapped in a symmetrical well in which the potential energy rises to large values (and ultimately to infinity) for sufficiently large displacements (compare Figs. 8A.1 and 8B.1). However, there are two important differences:

- Because the potential energy climbs towards infinity only as x^2 and not abruptly, the wavefunction approaches zero more slowly at large displacements than for the particle in a box.
- As the kinetic energy of the oscillator depends on the displacement in a more complex way (on account of the variation of the potential energy), the curvature of the wavefunction also varies in a more complex way.

Physical interpretation

The detailed solution of eqn 8B.3 confirms these points and shows that the wavefunctions for a harmonic oscillator have the form

$$\psi(x) = N\times(\text{polynomial in } x)\times(\text{bell-shaped Gaussian function})$$

where N is a normalization constant. A Gaussian function is a bell-shaped function of the form e^{-x^2} (Fig. 8B.3). The precise form of the wavefunctions is

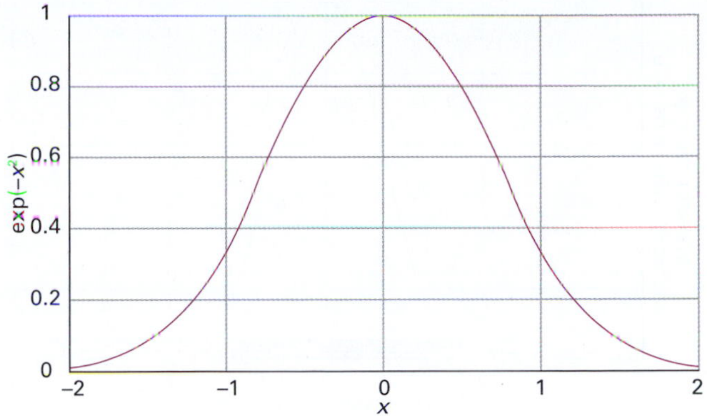

Figure 8B.3 The graph of the Gaussian function, $f(x) = e^{-x^2}$.

$$\psi_v(x) = N_v H_v(y) e^{-y^2/2} \qquad y = \frac{x}{\alpha} \qquad \alpha = \left(\frac{\hbar^2}{mk_f}\right)^{1/4} \qquad \textit{Harmonic oscillator} \quad \text{Wavefunctions} \qquad (8B.8)$$

The factor $H_v(y)$ is a **Hermite polynomial;** the form of these polynomials and some of their properties are listed in Table 8B.1. Hermite polynomials, which are members of a class of functions called 'orthogonal polynomials', have a wide range of important properties which allow a number of quantum mechanical calculations to be done with relative ease. Note that the first few Hermite polynomials are very simple: for instance, $H_0(y) = 1$ and $H_1(y) = 2y$.

Because $H_0(y) = 1$, the wavefunction for the ground state (the lowest energy state, with $v = 0$) of the harmonic oscillator is

$$\psi_0(x) = N_0 e^{-y^2/2} = N_0 e^{-x^2/2\alpha^2} \qquad \textit{Harmonic oscillator} \quad \text{Ground-state wavefunction} \qquad (8B.9a)$$

Table 8B.1 The Hermite polynomials, $H_v(y)$*

v	$H_v(y)$
0	1
1	$2y$
2	$4y^2-2$
3	$8y^3-12y$
4	$16y^4-48y^2+12$
5	$32y^5-160y^3+120y$
6	$64y^6-480y^4+720y^2-120$

* The Hermite polynomials are solutions of the differential equation

$$H_v'' - 2yH_v' + 2vH_v = 0$$

where primes denote differentiation. They satisfy the recursion relation

$$H_{v+1} - 2yH_v + 2vH_{v-1} = 0$$

An important integral is

$$\int_{-\infty}^{\infty} H_{v'}H_v e^{-y^2}\,dy = \begin{cases} 0 & \text{if } v' \neq v \\ \pi^{1/2}2^v v! & \text{if } v' = v \end{cases}$$

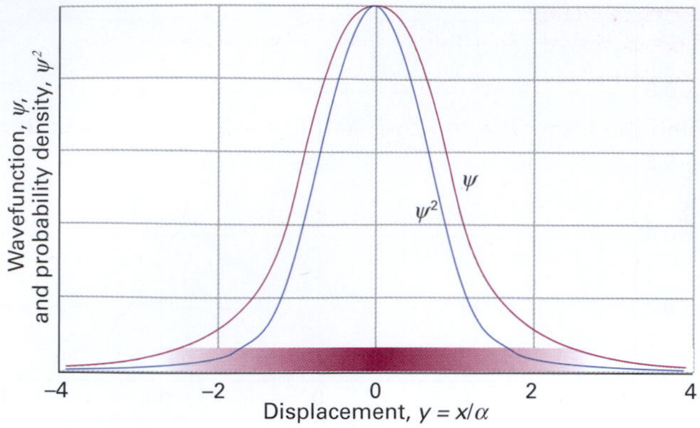

Figure 8B.4 The normalized wavefunction and probability density (shown also by shading) for the lowest energy state of a harmonic oscillator.

and the corresponding probability density is

$$\psi_0^2(x)=N_0^2\mathrm{e}^{-y^2}=N_0^2\mathrm{e}^{-x^2/\alpha^2} \quad \textit{Harmonic oscillator} \quad \text{Ground-state probability density} \quad (8B.9b)$$

The wavefunction and the probability density are shown in Fig. 8B.4. Both curves have their largest values at zero displacement (at $x=0$), so they capture the classical picture of the zero-point energy as arising from the ceaseless fluctuation of the particle about its equilibrium position.

The wavefunction for the first excited state of the oscillator, the state with $v=1$, is

$$\psi_1(x)=N_1(2y)\mathrm{e}^{-y^2/2} =N_1\left(\frac{2}{\alpha}\right)x\mathrm{e}^{-x^2/2\alpha^2} \quad \textit{Harmonic oscillator} \quad \text{First excited-state wavefunction} \quad (8B.10)$$

This function has a node at zero displacement ($x=0$), and the probability density has maxima at $x=\pm\alpha$ (Fig. 8B.5).

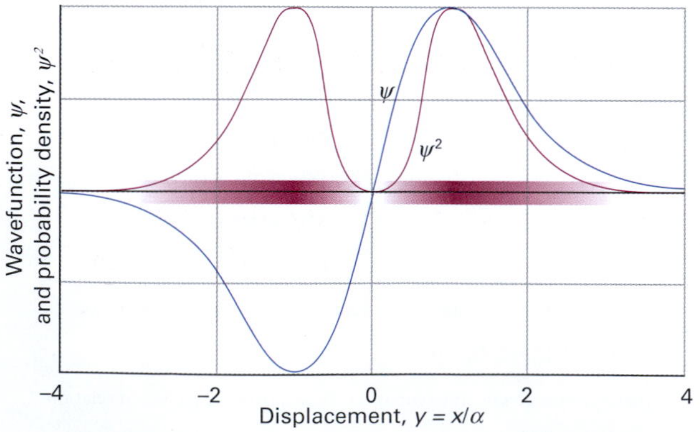

Figure 8B.5 The normalized wavefunction and probability density (shown also by shading) for the first excited state of a harmonic oscillator.

Example 8B.1 Confirming that a wavefunction is a solution of the Schrödinger equation

Confirm that the ground-state wavefunction (eqn 8B.9a) is a solution of the Schrödinger equation, eqn 8B.3.

Method Substitute the wavefunction given in eqn 8B.9a into eqn 8B.3. Use the definition of α given in eqn 8B.8 to determine the energy on the right-hand side of eqn 8B.3 and confirm that it matches the zero-point energy given in eqn 8B.6.

Answer We need to evaluate the second derivative of the ground-state wavefunction:

$$\frac{\mathrm{d}}{\mathrm{d}x}N_0\mathrm{e}^{-x^2/2\alpha^2}=-N_0\left(\frac{x}{\alpha^2}\right)\mathrm{e}^{-x^2/2\alpha^2}$$

$$\frac{\mathrm{d}^2}{\mathrm{d}x^2}N_0\mathrm{e}^{-x^2/2\alpha^2}=\frac{\mathrm{d}}{\mathrm{d}x}\left\{-N_0\left(\frac{x}{\alpha^2}\right)\mathrm{e}^{-x^2/2\alpha^2}\right\}$$

$$=-\frac{N_0}{\alpha^2}\mathrm{e}^{-x^2/2\alpha^2}+N_0\left(\frac{x}{\alpha^2}\right)^2\mathrm{e}^{-x^2/2\alpha^2}$$

$$=-(1/\alpha^2)\psi_0+(x^2/\alpha^4)\psi_0$$

Substituting ψ_0 into eqn 8B.3 and using the definition of α (eqn 8B.8), we obtain

$$\frac{\hbar^2}{2m}\left(\frac{mk_\mathrm{f}}{\hbar^2}\right)^{1/2}\psi_0-\frac{\hbar^2}{2m}\left(\frac{mk_\mathrm{f}}{\hbar^2}\right)x^2\psi_0+\tfrac{1}{2}k_\mathrm{f}x^2\psi_0=E\psi_0$$

and therefore

$$\frac{\hbar}{2}\left(\frac{k_\mathrm{f}}{m}\right)^{1/2}\psi_0-\tfrac{1}{2}k_\mathrm{f}x^2\psi_0+\tfrac{1}{2}k_\mathrm{f}x^2\psi_0=E\psi_0$$

The second and third terms on the left-hand side (in blue) cancel and we obtain $E=\frac{1}{2}\hbar(k_\mathrm{f}/m)^{1/2}$ in accord with eqn 8B.6 for the zero-point energy.

Self-test 8B.2 Confirm that the wavefunction in eqn 8B.10 is a solution of eqn 8B.3 and determine its energy.

Answer: yes, with $E=\frac{3}{2}\hbar\omega$

The shapes of several wavefunctions are shown in Fig. 8B.6 and the corresponding probability densities are shown in Fig. 8B.7. At high quantum numbers, harmonic oscillator wavefunctions have their largest amplitudes near the turning points of the classical motion (the locations at which $V=E$, so the kinetic energy is zero). We see classical properties emerging in the correspondence principle limit of high quantum numbers (Topic 8A), for a classical particle is most likely to be found at the turning points (where it travels most slowly) and is least likely to be found at zero displacement (where it travels most rapidly).

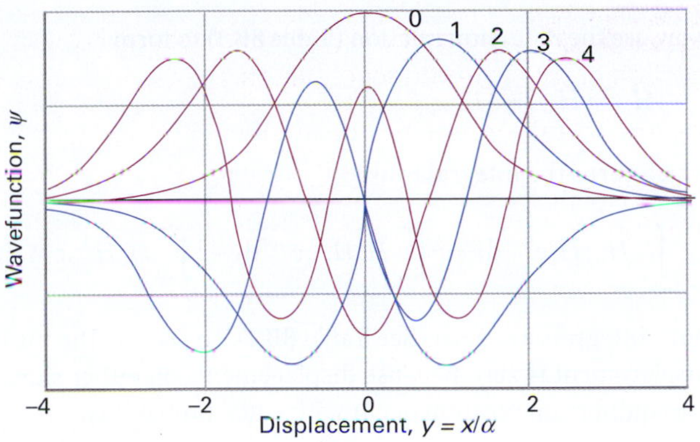

Figure 8B.6 The normalized wavefunctions for the first five states of a harmonic oscillator. Note that the number of nodes is equal to v and that alternate wavefunctions are symmetrical or asymmetrical about $y=0$ (zero displacement).

Note the following features of the wavefunctions:

Physical interpretation

- The Gaussian function goes very strongly to zero as the displacement increases (in either direction, extension or compression), so all the wavefunctions approach zero at large displacements.
- The exponent y^2 is proportional to $x^2 \times (mk_f)^{1/2}$, so the wavefunctions decay more rapidly for large masses and stiff springs.
- As v increases, the Hermite polynomials become larger at large displacements (as x^v), so the wavefunctions grow large before the Gaussian function damps them down to zero: as a result, the wavefunctions spread over a wider range as v increases (Fig. 8B.7).

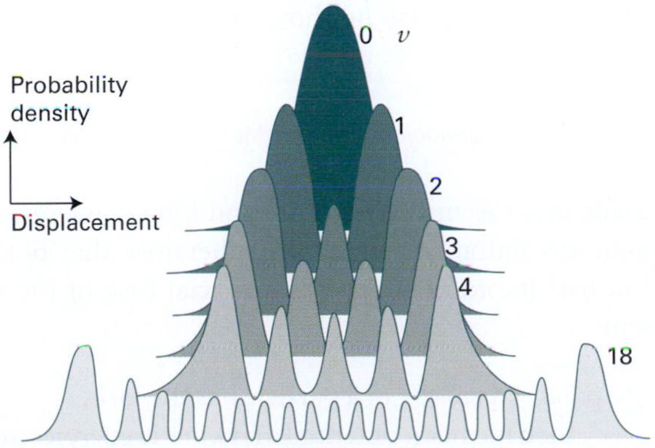

Figure 8B.7 The probability densities for the first five states of a harmonic oscillator and the state with $v=18$. Note how the regions of highest probability density move towards the turning points of the classical motion as v increases.

Example 8B.2 Normalizing a harmonic oscillator wavefunction

Find the normalization constant for the harmonic oscillator wavefunctions.

Method Normalization is carried out by evaluating the integral of $|\psi|^2$ over all space and then finding the normalization factor from eqn 7B.3 ($N=1/(\int\psi^*\psi \mathrm{d}\tau)^{1/2}$). The normalized wavefunction is then equal to $N\psi$. In this one-dimensional problem, the volume element is $\mathrm{d}x$ and the integration is from $-\infty$ to $+\infty$. The wavefunctions are expressed in terms of the dimensionless variable $y=x/\alpha$, so begin by expressing the integral in terms of y by using $\mathrm{d}x=\alpha\mathrm{d}y$. The integrals required are given in Table 8B.1.

Answer The unnormalized wavefunction is

$$\psi_v(x)=H_v(y)\mathrm{e}^{-y^2/2}$$

It follows from the integrals given in Table 8B.1 that

$$\int_{-\infty}^{\infty}\psi_v^*\psi_v\mathrm{d}x=\alpha\int_{-\infty}^{\infty}\psi_v^*\psi_v\mathrm{d}y=\alpha\int_{-\infty}^{\infty}H_v^2(y)\mathrm{e}^{-y^2}\mathrm{d}y=\alpha\pi^{1/2}2^v v!$$

where $v!=v(v-1)(v-2)\ldots1$. Therefore,

$$N_v=\left(\frac{1}{\alpha\pi^{1/2}2^v v!}\right)^{1/2}$$

Note that, unlike the normalization constant for a particle in a box, for a harmonic oscillator N_v is different for each value of v.

Self-test 8B.3 Confirm, by explicit evaluation of the integral, that ψ_0 and ψ_1 are orthogonal.

Answer: Show that $\int_{-\infty}^{\infty}\psi_0^*\psi_1\mathrm{d}x=0$ by using the information in Table 8B.1

8B.2 The properties of oscillators

The average value of a property is calculated by evaluating the expectation value of the corresponding operator (eqn 7C.11, $\langle\Omega\rangle=\int\psi^*\hat{\Omega}\psi\mathrm{d}\tau$). Now that we know the wavefunctions of the harmonic oscillator, we can start to explore its properties by evaluating integrals of the type

$$\langle\Omega\rangle=\int_{-\infty}^{\infty}\psi_v^*\hat{\Omega}\psi_v\mathrm{d}x \qquad (8B.11)$$

(Here and henceforth, the wavefunctions are all taken to be normalized to 1.) When the explicit wavefunctions are substituted, the integrals look fearsome, but the Hermite polynomials have many simplifying features.

(a) Mean values

We show in the following example that the mean displacement, $\langle x\rangle$, and the mean square displacement, $\langle x^2\rangle$, of the oscillator when it is in the state with quantum number v are

$$\langle x\rangle = 0 \qquad \text{Harmonic oscillator} \quad \text{Mean displacement} \qquad (8B.12a)$$

$$\langle x^2\rangle = \left(v+\tfrac{1}{2}\right)\frac{\hbar}{(mk_f)^{1/2}} \qquad \text{Harmonic oscillator} \quad \text{Mean square displacement} \qquad (8B.12b)$$

The result for $\langle x\rangle$ shows that the oscillator is equally likely to be found on either side of $x=0$ (like a classical oscillator). The result for $\langle x^2\rangle$ shows that the mean square displacement increases with v. This increase is apparent from the probability densities in Fig. 8B.7, and corresponds to the classical amplitude of swing increasing as the oscillator becomes more highly excited.

Example 8B.3 Calculating properties of a harmonic oscillator

Consider the harmonic oscillator motion of the H—Cl molecule in *Brief illustration* 8B.1. Calculate the mean displacement of the oscillator when it is in a state with quantum number v.

Method Normalized wavefunctions must be used to calculate the expectation value. The operator for position along x is multiplication by the value of x (Topic 7C). The resulting integral can be evaluated either

- by inspection (the integrand is the product of an odd and an even function), or
- by explicit evaluation using the formulas in Table 8B.1.

The former procedure makes use of the definitions that an even function is one for which $f(-x)=f(x)$ and an odd function is one for which $f(-x)=-f(x)$. Therefore, the product of an odd and even function is itself odd, and the integral of an odd function over a symmetrical range about $x=0$ is zero. The latter procedure using explicit integration is illustrated here to give practice in the calculation of expectation values. We shall need the relation $x=\alpha y$, which implies that $dx=\alpha dy$.

Answer The integral we require is

$$\begin{aligned}\langle x\rangle &= \int_{-\infty}^{\infty}\psi_v^* x\psi_v \mathrm{d}x = N_v^2\int_{-\infty}^{\infty}(H_v \mathrm{e}^{-y^2/2})x(H_v \mathrm{e}^{-y^2/2})\mathrm{d}x \\ &= \alpha^2 N_v^2\int_{-\infty}^{\infty}(H_v \mathrm{e}^{-y^2/2})y(H_v \mathrm{e}^{-y^2/2})\mathrm{d}y \\ &= \alpha^2 N_v^2\int_{-\infty}^{\infty}H_v y H_v \mathrm{e}^{-y^2}\mathrm{d}y\end{aligned}$$

Now use the recursion relation (Table 8B.1) to form

$$yH_v = vH_{v-1}+\tfrac{1}{2}H_{v+1}$$

which turns the integral into

$$\int_{-\infty}^{\infty}H_v y H_v \mathrm{e}^{-y^2}\mathrm{d}y = v\int_{-\infty}^{\infty}H_v H_{v-1}\mathrm{e}^{-y^2}\mathrm{d}y + \tfrac{1}{2}\int_{-\infty}^{\infty}H_v H_{v+1}\mathrm{e}^{-y^2}\mathrm{d}y$$

Both integrals are zero (See Table 8B.1), so $\langle x\rangle=0$. The mean displacement is zero because displacements on either side of the equilibrium position occur with equal probability.

Self-test 8B.4 Calculate the mean square displacement, $\langle x^2\rangle$, of the H—Cl bond distance from its equilibrium position by using the recursion relation in Table 8B.1 twice.

Answer: $(v+\tfrac{1}{2})\times 115\,\text{pm}^2$; eqn 9.12b, with μ in place of m

The mean potential energy of an oscillator, the expectation value of $V=\tfrac{1}{2}k_f x^2$, can now be calculated very easily:

$$\langle V\rangle = \tfrac{1}{2}k_f\langle x^2\rangle = \tfrac{1}{2}\left(v+\tfrac{1}{2}\right)\hbar\left(\frac{k_f}{m}\right)^{1/2}$$

or

$$\langle V\rangle = \tfrac{1}{2}\left(v+\tfrac{1}{2}\right)\hbar\omega \qquad \text{Harmonic oscillator} \quad \text{Mean potential energy} \qquad (8B.13a)$$

Because the total energy in the state with quantum number v is $(v+\tfrac{1}{2})\hbar\omega$, it follows that

$$\langle V\rangle = \tfrac{1}{2}E_v \qquad \text{Harmonic oscillator} \quad \text{Mean potential energy} \qquad (8B.13b)$$

The total energy is the sum of the potential and kinetic energies, so it follows at once that the mean kinetic energy of the oscillator is (as could also be shown using the kinetic energy operator)

$$\langle E_k\rangle = \tfrac{1}{2}E_v \qquad \text{Harmonic oscillator} \quad \text{Mean kinetic energy} \qquad (8B.13c)$$

The result that the mean potential and kinetic energies of a harmonic oscillator are equal (and therefore that both are equal to half the total energy) is a special case of the **virial theorem**:

> If the potential energy of a particle has the form $V=ax^b$, then its mean potential and kinetic energies are related by

$$2\langle E_k\rangle = b\langle V\rangle \qquad \text{Virial theorem} \qquad (8B.14)$$

For a harmonic oscillator $b=2$, so $\langle E_k\rangle=\langle V\rangle$, as we have found. The virial theorem is a short cut to the establishment of a number of useful results, and we use it elsewhere (for example, in Topic 9A).

(b) Tunnelling

An oscillator may be found at extensions with $V>E$, which are forbidden by classical physics, for they correspond to negative kinetic energy; this is an example of the phenomenon of tunnelling (Topic 8A). As shown in *Example* 8B.4, for the lowest energy state of the harmonic oscillator, there is about an 8 per cent chance of finding the oscillator stretched beyond its classical limit and an 8 per cent chance of finding it with a classically forbidden compression. These tunnelling probabilities are independent of the force constant and mass of the oscillator.

Example 8B.4 Calculating the tunnelling probability for the harmonic oscillator

Calculate the probability that the ground-state harmonic oscillator will be found in a classically forbidden region.

Method Find the expression for the classical turning point, x_{tp}, where the kinetic energy vanishes, by equating the potential energy to the total energy E of the harmonic oscillator. The probability of finding the oscillator stretched beyond a displacement x_{tp} is the sum of the probabilities $\psi^2 dx$ of finding it in any of the intervals dx lying between x_{tp} and infinity, so evaluate the integral

$$P=\int_{x_{tp}}^{\infty}\psi_v^2 dx$$

The variable of integration is best expressed in terms of $y=x/\alpha$ and the integral to be evaluated is a special case of the *error function*, erf z, defined as

$$\text{erf}\,(z)=1-\frac{2}{\pi^{1/2}}\int_z^{\infty}e^{-y^2}dy$$

and evaluated for some values of z in Table 8B.2 (this function is commonly available in mathematical software packages). By symmetry, the probability of being found stretched into a classically forbidden region is the same as that of being found compressed into a classically forbidden region.

Answer According to classical mechanics, the turning point, x_{tp}, of an oscillator occurs when its kinetic energy is zero, which is when its potential energy $\frac{1}{2}k_f x^2$ is equal to its total energy E. This equality occurs when

$$x_{tp}^2=\frac{2E}{k_f}\quad\text{or}\quad x_{tp}=\pm\left(\frac{2E}{k_f}\right)^{1/2}$$

with E given by eqn 8B.4. The variable of integration in the integral P is best expressed in terms of $y=x/\alpha$ with $\alpha=(\hbar^2/mk_f)^{1/4}$, and then the right-hand turning point lies at

$$y_{tp}=\frac{x_{tp}}{\alpha}=\left\{\frac{2(v+\frac{1}{2})\hbar\omega}{\alpha^2 k_f}\right\}^{1/2}\overset{\omega=(k_f/m)^{1/2}}{=}(2v+1)^{1/2}$$

For the state of lowest energy ($v=0$), $y_{tp}=1$ and the probability of being beyond that point is

$$P=\int_{x_{tp}}^{\infty}\psi_0^2 dx=\alpha N_0^2\int_1^{\infty}e^{-y^2}dy$$

The normalization constant N_0 is calculated from the expression for N_v in *Example* 8B.2 ($N_v=1/(\alpha\pi^{1/2}2^v v!)^{1/2}$):

$$N_0=\left(\frac{1}{\alpha\pi^{1/2}2^0 0!}\right)^{1/2}=\left(\frac{1}{\alpha\pi^{1/2}}\right)^{1/2}$$

The integral in the expression for P is written in terms of the error function erf(1) as

$$\text{erf}(1)=1-\frac{2}{\pi^{1/2}}\int_1^{\infty}e^{-y^2}dy\quad\text{so}\quad\int_1^{\infty}e^{-y^2}dy=\frac{1}{2}\pi^{1/2}(1-\text{erf}(1))$$

It follows that

$$P=\alpha\times\overbrace{\frac{1}{\alpha\pi^{1/2}}}^{N_0^2}\times\overbrace{\tfrac{1}{2}\pi^{1/2}(1-\text{erf}\,(1))}^{\int_1^{\infty}e^{-y^2}dy}=\tfrac{1}{2}\left(1-\overbrace{\text{erf}(1)}^{0.843}\right)=0.079$$

In 7.9 per cent of a large number of observations, any oscillator in the state with quantum number $v=0$ will be found stretched to a classically forbidden extent. There is the same probability of finding the oscillator with a classically forbidden compression. The total probability of finding the oscillator tunnelled into a classically forbidden region (stretched or compressed) is about 16 per cent.

Self-test 8B.5 Calculate the probability that a harmonic oscillator in the state with quantum number $v=1$ will be found at a classically forbidden extension. (Follow the argument given in *Example* 8B.4 and use the method of integration by parts (*Mathematical background* 1) to obtain an integral that can be expressed in terms of the error function.)

Answer: $P=0.056$

Table 8B.2 The error function, erf(z)*

z	erf(z)
0	0
0.01	0.0113
0.05	0.0564
0.10	0.1125
0.50	0.5205
1.00	0.8427
1.50	0.9661
2.00	0.9953

*More values are available in mathematical software packages.

The probability of finding the oscillator in classically forbidden regions decreases quickly with increasing v, and vanishes entirely as v approaches infinity, as we would expect from the correspondence principle. Macroscopic oscillators (such as pendulums) are in states with very high quantum numbers, so the tunnelling probability is wholly negligible and classical mechanics is reliable. Molecules, however, are normally in their vibrational ground states, and for them the probability is very significant and classical mechanics is misleading.

Checklist of concepts

- ☐ 1. A particle undergoing **harmonic motion** is called a **harmonic oscillator** and experiences a restoring force proportional to its displacement;
- ☐ 2. The potential energy of a harmonic oscillator is a parabolic function of the displacement from equilibrium.
- ☐ 3. The energy levels of a harmonic oscillator form an evenly spaced ladder.
- ☐ 4. The wavefunctions of a harmonic oscillator are products of a **Hermite polynomial** and a Gaussian (bell-shaped) function.
- ☐ 5. There is a **zero-point energy**, an irremovable minimum energy, which is consistent with, and can be interpreted in terms of, the uncertainty principle.
- ☐ 6. The probability of finding the harmonic oscillator in classically forbidden regions is significant for the ground vibrational state ($v=0$) but decreases quickly with increasing v.

Checklist of equations

Property	Equation	Comment	Equation number
Energy levels of harmonic oscillator	$E_v=(v+\frac{1}{2})\hbar\omega, \quad \omega=(k_f/m)^{1/2}$	$v=0, 1, 2,\ldots$	8B.4
Zero-point energy of harmonic oscillator	$E_0=\frac{1}{2}\hbar\omega$		8B.6
Wavefunction of harmonic oscillator	$\psi_v(x)=N_vH_v(y)e^{-y^2/2}$ $y=x/\alpha, \quad \alpha=(\hbar^2/mk_f)^{1/4}$ $N_v=(1/\alpha\pi^{1/2}2^v v!)^{1/2}$	$v=0, 1, 2,\ldots$	8B.8
Mean displacement of harmonic oscillator	$\langle x\rangle=0$		8B.12a
Mean square displacement of harmonic oscillator	$\langle x^2\rangle=(v+\frac{1}{2})\hbar/(mk_f)^{1/2}$		8B.12b
Virial theorem	$2\langle E_k\rangle=b\langle V\rangle$	$V=ax^b$	8B.14

8C Rotational motion

Contents

➤ Why do you need to know this material?

The investigation of rotational motion introduces the concept of angular momentum, which is central to the description of the electronic structure of atoms and molecules and the interpretation of details observed in molecular spectra.

➤ What is the main idea?

The energy and the angular momentum of a rotating object are quantized.

➤ What do you need to know already?

You should know the postulates of quantum mechanics (Topic 7C), and be familiar with the concept of angular momentum in classical physics (*Foundations* B).

This Topic provides a quantum mechanical description of rotation in two and three dimensions. The concepts developed here form the basis for discussion of atomic structure (Topics 9A and 9B) and molecular rotation (Topic 12B).

8C.1 Rotation in two dimensions

We consider a particle of mass m constrained to move in a circular path (a 'ring') of radius r in the xy-plane with constant potential energy, which may be taken to be zero (Fig. 8C.1). The total energy is equal to the kinetic energy, because $V=0$ everywhere. We can therefore write $E=p^2/2m$. According to classical mechanics (*Foundations* B), the **angular momentum**, J_z, around the z-axis (which lies perpendicular to the xy-plane) is $J_z=\pm pr$, so the energy can be expressed as $J_z^2/2mr^2$. Because mr^2 is the **moment of inertia**, I, of the mass on its path, it follows that

$$E=\frac{J_z^2}{2I} \qquad I=mr^2 \qquad \text{Particle on a ring, classical expression} \qquad \text{Energy} \qquad (8C.1)$$

We shall now see that not all the values of the angular momentum are permitted in quantum mechanics, and therefore that both angular momentum and rotational energy are quantized.

(a) The qualitative origin of quantized rotation

Because $J_z=\pm pr$, and since the de Broglie relation gives $p=h/\lambda$ (Topic 7A) the angular momentum about the z-axis is

$$J_z=\pm\frac{hr}{\lambda} \qquad (8C.2)$$

Opposite signs correspond to opposite directions of travel. This equation shows that the shorter the wavelength of the particle on a circular path of given radius, the greater the angular momentum of the particle. It follows that if we can see why the wavelength is restricted to discrete values, then we shall understand why the angular momentum is quantized.

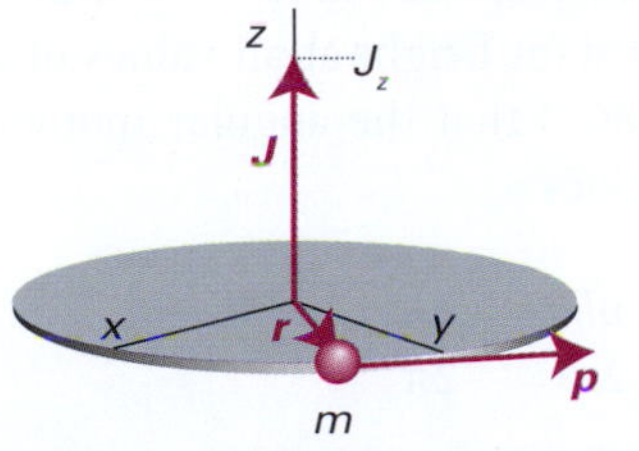

Figure 8C.1 The angular momentum of a particle of mass m on a circular path of radius r in the xy-plane is represented by a vector $\boldsymbol{J}$ with the single non-zero component J_z of magnitude pr perpendicular to the plane.

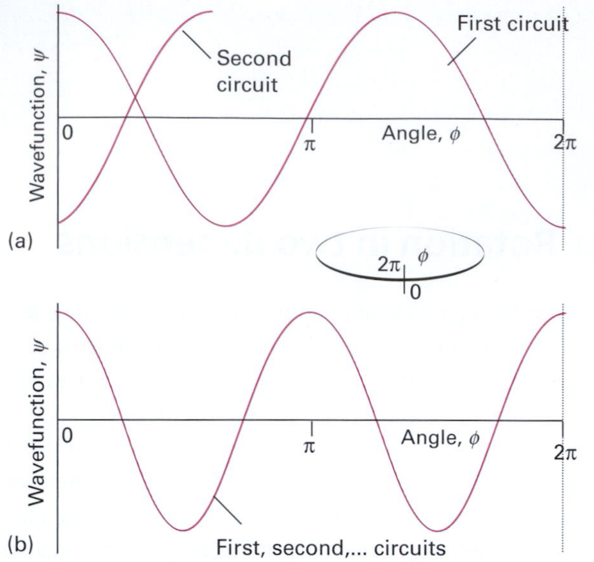

Figure 8C.2 Two solutions of the Schrödinger equation for a particle on a ring. The circumference has been opened out into a straight line; the points at $\phi=0$ and 2π are identical. The solution in (a) is unacceptable because it is not single-valued. Moreover, on successive circuits it interferes destructively with itself, and does not survive. The solution in (b) is acceptable: it is single-valued, and on successive circuits it reproduces itself.

Suppose for the moment that λ can take an arbitrary value. In that case, the wavefunction depends on the azimuthal angle ϕ as shown in Fig. 8C.2a. When ϕ increases beyond 2π, the wavefunction continues to change, but for an arbitrary wavelength it gives rise to a different value at a given point after each circuit, which is unacceptable because a wavefunction must be single-valued. An acceptable solution is obtained only if the wavefunction reproduces itself on successive circuits, as in Fig. 8C.2b. Because only some wavefunctions have this property, it follows that only some angular momenta are acceptable and therefore that only certain rotational energies are allowed. That is, the energy of the particle is quantized. Specifically, an integer number of wavelengths must fit the circumference of the ring (which is $2\pi r$):

$$n\lambda=2\pi r \quad n=0,1,2,\ldots \tag{8C.3}$$

The value $n=0$ corresponds to $\lambda=\infty$; a 'wave' of infinite wavelength has a constant height at all values of ϕ. It follows from eqns 8C.2 and 8C.3 that the angular momentum is therefore limited to the values

$$J_z=\pm\frac{hr}{\lambda}=\pm\frac{nhr}{2\pi r}=\pm\frac{nh}{2\pi} \quad n=0,1,2,\ldots$$

The sign of J_z (which indicated the sense of the rotation) can be absorbed into the quantum number by replacing n by $m_l=0, \pm1, \pm2, \ldots$ where we have allowed m_l (the conventional notation for this quantum number) to have positive and negative integer values. At the same time we recognize the presence of $h/2\pi=\hbar$ and obtain

$$J_z=m_l\hbar \qquad m_l=0,\pm1,\pm2,\ldots \tag{8C.4}$$

Particle on a ring — Angular momenta

Positive values of m_l correspond to rotation in a clockwise sense around the z-axis (as viewed in the direction of increasing z, Fig. 8C.3) and negative values of m_l correspond to counter-clockwise rotation around z. It then follows from eqns 8C.1 and 8C.4 that the energy is limited to the values

$$E_{m_l}=\frac{m_l^2\hbar^2}{2I} \qquad m_l=0,\pm1,\pm2,\ldots \tag{8C.5}$$

Particle on a ring — Energy levels

We explore this result further by noting that:

- The energies, labelled by m_l, are quantized because m_l must be an integer.
- The occurrence of m_l as its square means that the energy of rotation is independent of the sense of rotation (the sign of m_l), as we expect physically. That is, states with a given nonzero value of $|m_l|$ are doubly degenerate.
- The state described by $m_l=0$ is non-degenerate, consistent with the interpretation that, when m_l is zero, the particle has an infinite wavelength and is 'stationary'; the question of the direction of rotation does not arise.
- There is no zero-point energy in this system: the lowest possible energy is $E_0=0$.

Physical interpretation

(b) The solutions of the Schrödinger equation

To obtain the wavefunctions for the particle on a ring and to confirm that the energies from eqn 8C.5 are correct, we need

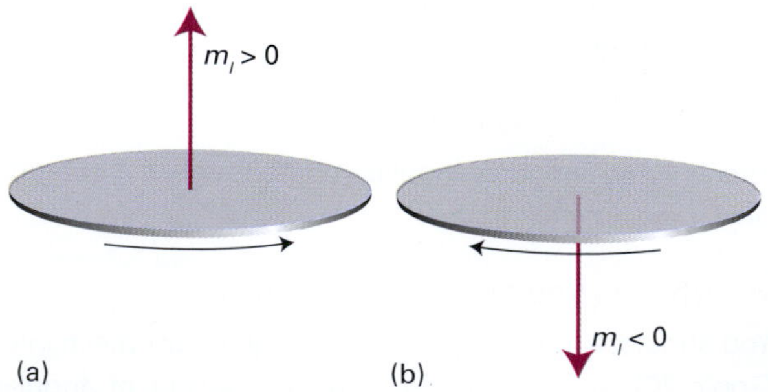

Figure 8C.3 The angular momentum of a particle confined to a plane can be represented by a vector of length $|m_l|$ units along the z-axis and with an orientation that indicates the direction of motion of the particle. The direction is given by the right-hand screw rule, so (a) corresponds to $m_l>0$, clockwise as seen from below and (b) corresponds to $m_l<0$, anticlockwise as seen from below.

to solve the Schrödinger equation explicitly. We show in the following *Justification* that the normalized wavefunctions and corresponding energies are

$$\psi_{m_l}(\phi)=\frac{e^{im_l\phi}}{(2\pi)^{1/2}} \qquad m_l=0,\pm1,\pm2,\ldots$$ *Particle on a ring* Wavefunctions (8C.6a)

$$E_{m_l}=\frac{m_l^2\hbar^2}{2I}$$ *Particle on a ring* Energy levels (8C.6b)

The wavefunction with $m_l=0$ is $\psi_0(\phi)=1/(2\pi)^{1/2}$, corresponding to uniform amplitude around the ring, and its energy is $E_0=0$.

Justification 8C.1 The solutions of the Schrödinger equation for a particle on a ring

The hamiltonian for a particle of mass m travelling on a circle in the xy-plane (with $V=0$) is the same as that for free motion in a plane (eqn 8A.1 of Topic 8A),

$$\hat{H}=-\frac{\hbar^2}{2m}\left(\frac{\partial^2}{\partial x^2}+\frac{\partial^2}{\partial y^2}\right) \tag{8C.7}$$

but with the constraint to a path of constant radius r. It is always a good idea to use coordinates that reflect the full symmetry of the system, so we introduce the coordinates r and ϕ (*The chemist's toolkit* 8C.1). By standard manipulations we can write

$$\frac{\partial^2}{\partial x^2}+\frac{\partial^2}{\partial y^2}=\frac{\partial^2}{\partial r^2}+\frac{1}{r}\frac{\partial}{\partial r}+\frac{1}{r^2}\frac{\partial^2}{\partial\phi^2} \tag{8C.8}$$

However, because the radius of the path is fixed, the (blue) derivatives with respect to r can be discarded. Only the last term in eqn 8C.8 then survives and the hamiltonian becomes simply

$$\hat{H}=-\frac{\hbar^2}{2mr^2}\frac{d^2}{d\phi^2}$$ *Particle on a ring* Hamiltonian (8C.9a)

The partial derivative has been replaced by a complete derivative because ϕ is now the only variable. The moment of inertia, $I=mr^2$ has appeared automatically so $\hat{H}$ may be written

$$\hat{H}=-\frac{\hbar^2}{2I}\frac{d^2}{d\phi^2}$$ *Particle on a ring* Hamiltonian (8C.9b)

and the Schrödinger equation is

$$-\frac{\hbar^2}{2I}\frac{d^2\psi}{d\phi^2}=E\psi$$ *Particle on a ring* Schrödinger equation (8C.10a)

We rewrite this equation as

$$\frac{d^2\psi}{d\phi^2}=-\frac{2IE}{\hbar^2}\psi$$

For a given energy, $2IE/\hbar^2$ is a constant, which for convenience (and an eye on the future) we write as m_l^2. At this stage m_l is just a dimensionless number with no restrictions. Then the equation becomes

$$\frac{d^2\psi}{d\phi^2}=-m_l^2\psi \tag{8C.10b}$$

The (unnormalized) general solutions of this equation are

$$\psi_{m_l}(\phi)=e^{im_l\phi} \tag{8C.11}$$

as can be verified by substitution.

We now select the acceptable solutions from among these general solutions by imposing the condition that the wavefunction should be single-valued. That is, the wavefunction ψ must satisfy a **cyclic boundary condition**, and match at points separated by a complete revolution: $\psi(\phi+2\pi)=\psi(\phi)$. On substituting the general wavefunction into this condition, we find

$$\psi_{m_l}(\phi+2\pi)=e^{im_l(\phi+2\pi)}=e^{im_l\phi}e^{2\pi im_l}=\psi_{m_l}(\phi)e^{2\pi im_l}$$
$$=\psi_{m_l}(\phi)(e^{\pi i})^{2m_l}$$

As $e^{i\pi}=-1$ (Euler's formula, *Mathematical background* 3), this relation is equivalent to

$$\psi_{m_l}(\phi+2\pi)=(-1)^{2m_l}\psi_{m_l}(\phi)$$

Because cyclic boundary conditions require $(-1)^{2m_l}=1$, $2m_l$ must be a positive or a negative even integer (including 0), and therefore m_l must be an integer: $m_l=0,\pm1,\pm2,\ldots$.

We now normalize the wavefunction by finding the normalization constant N given by eqn 7B.3 $(N=(\int_{-\infty}^{\infty}\psi^*\psi\,dx)^{-1/2})$, which in this case becomes:

$$N=\frac{1}{\left(\int_0^{2\pi}\psi^*\psi\,d\phi\right)^{1/2}}=\frac{1}{\left(\int_0^{2\pi}\overbrace{e^{-im_l\phi}e^{im_l\phi}}^{1}\,d\phi\right)^{1/2}}=\frac{1}{(2\pi)^{1/2}} \tag{8C.12}$$

and the normalized wavefunctions for a particle on a ring are those given by eqn 8C.6a. The expression for the energies of the states (eqn 8C.6b) is obtained by rearranging the relation $m_l^2=2IE/\hbar^2$ into $E=m_l^2\hbar^2/2I$.

The chemist's toolkit 8C.1 Cylindrical coordinates

For systems with cylindrical symmetry it is best to work in **cylindrical coordinates**, r, ϕ, and z (Sketch 1), with

$$x=r\cos\theta \qquad y=r\sin\phi$$

and where

r ranges from 0 to ∞	ϕ ranges from 0 to 2π	z ranges from $-\infty$ to ∞

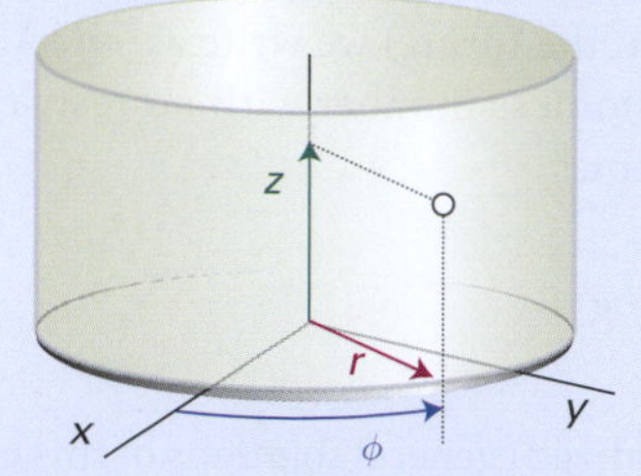

Sketch 1 Cylindrical coordinates

The volume element is

$$d\tau = r\,dr\,d\phi\,dz$$

For motion in a plane we set $z=0$ and for the volume element use

$$d\tau = r\,dr\,d\phi$$

Self-test 8C.1 Use the particle on a ring model to calculate the minimum energy required for the excitation of a π electron in coronene, $C_{24}H_{12}$ (**1**). Assume that the radius of the ring is three times the carbon–carbon bond length in benzene and that the electrons are confined to the periphery of the molecule.

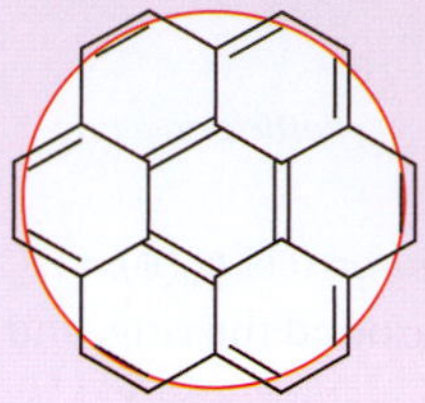

1 Coronene
(model ring in red)

Answer: For transition from $m_l=+3$ to $m_l=+4$: $\Delta E=0.0147$ zJ or 8.83 J mol^{-1}

Example 8C.1 Using the particle-on-a-ring model

The particle-on-a-ring model is a crude but illustrative model of cyclic, conjugated molecular systems. Treat the π electrons in benzene as particles freely moving over a circular ring of carbon atoms and calculate the minimum energy required for the excitation of a π electron. The carbon–carbon bond length in benzene is 140 pm.

Method Because each carbon atom contributes one π electron, six electrons in the conjugated system move along the perimeter of the ring. Each state is occupied by two electrons, so only the $m_l=0$, $+1$, and -1 states are occupied (with the last two being degenerate). The minimum energy required for excitation corresponds to a transition of an electron from the $m_l=+1$ (or -1) state to the $m_l=+2$ (or -2) state. Use eqn 8C.6b, and the mass of the electron, to calculate the energies of the states.

Answer From eqn 8C.6b, the energy separation between the $m_l=+1$ and the $m_l=+2$ states is

$$\Delta E = E_{+2} - E_{-1} = (4-1)\times\frac{(1.055\times10^{-34}\,\mathrm{J\,s})^2}{2\times(9.109\times10^{-31}\,\mathrm{kg})\times(1.40\times10^{-10}\,\mathrm{m})^2}$$
$$= 9.35\times10^{-19}\,\mathrm{J}$$

Therefore the minimum energy required to excite an electron is 0.935 aJ or 563 kJ mol^{-1}. This energy separation corresponds to an absorption frequency of 1.41 PHz (1 PHz $=10^{15}$ Hz) and a wavelength of 213 nm; the experimental value for a transition of this kind is 260 nm. That such a primitive model gives relatively good agreement is encouraging. In addition, even though the model is primitive, it gives insight into the origin of the quantized π electron energy levels in cyclic conjugated systems (Topic 10D).

A note on good practice Note that, when quoting the value of m_l, it is good practice always to give the sign, even if m_l is positive. Thus, we write $m_l=+1$, not $m_l=1$.

(c) Quantization of angular momentum

We have seen that the angular momentum around the z-axis is quantized and confined to the values given in eqn 8C.4 ($J_z=m_l\hbar$). The wavefunction for the particle on a ring is given by eqn 8C.6a:

$$\psi_{m_l}(\phi)=\frac{e^{im_l\phi}}{(2\pi)^{1/2}}=\frac{1}{(2\pi)^{1/2}}(\cos m_l\phi+i\sin m_l\phi)$$

Therefore, as $|m_l|$ increases, the increasing angular momentum is associated with:

- an increase in the number of nodes in the real ($\cos m_l\phi$) and imaginary ($\sin m_l\phi$) parts of the wavefunction (the complex function does not have nodes but each of its real and imaginary components does);
- a decrease in the wavelength and, by the de Broglie relation, an increase in the linear momentum with which the particle travels round the ring (Fig. 8C.4).

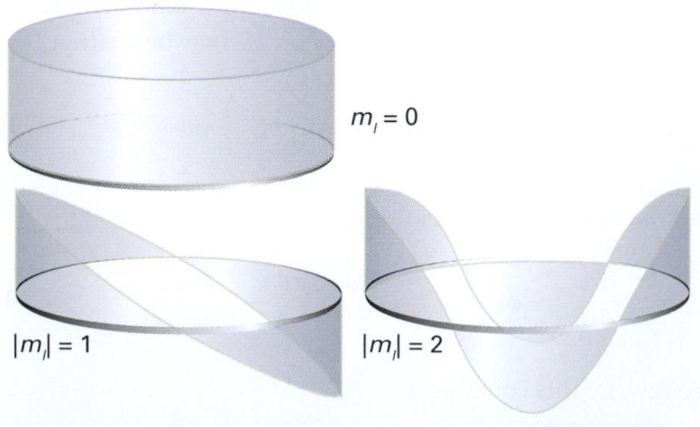

Figure 8C.4 The real parts of the wavefunctions of a particle on a ring. As shorter wavelengths are achieved, the magnitude of the angular momentum around the z-axis grows in steps of $\hbar$.

Brief illustration 8C.1 Nodes in the wavefunction

Whereas the $m_l=0$ ground-state wavefunction has no nodes, the $m_l=+1$ wavefunction

$$\psi_{+1}(\phi)=\frac{e^{i\phi}}{(2\pi)^{1/2}}=\frac{1}{(2\pi)^{1/2}}(\cos\phi+i\sin\phi)$$

has nodes at $\phi=\pi/2$ and $3\pi/2$ in its real part and at $\phi=0$ and π in its imaginary part. An increase in the number of nodes results in greater curvature of the (real and imaginary parts of the) wavefunction, consistent with an increase in kinetic and, in this case, total energy. Note that the sense of rotation (clockwise seen along the z-axis) is reflected in the fact that the imaginary component precedes the real component in phase: the real chases the imaginary.

Self-test 8C.2 Determine the number of nodes in the real and imaginary parts of the wavefunction for a state of general m_l.

Answer: $2m_l$ nodes each in real and imaginary part

As we show in the following *Justification*, we can come to the same conclusion about quantization of the z-component of angular momentum more formally by using the argument about the relation between eigenvalues and the values of observables established in Topic 7C.

Justification 8C.2 The quantization of angular momentum

In classical mechanics, the angular momentum $\boldsymbol{l}$ of a particle with position $\boldsymbol{r}$ and linear momentum $\boldsymbol{p}$ is given by the vector product $\boldsymbol{l}=\boldsymbol{r}\times\boldsymbol{p}$ (see *Mathematical background* 5 following Chapter 9 for a reminder about vector products). For motion restricted to two dimensions, with $\boldsymbol{i}$ and $\boldsymbol{j}$ denoting unit vectors (vectors of length 1) pointing along the positive directions on the x- and y-axes, respectively,

$$\boldsymbol{r}=x\boldsymbol{i}+y\boldsymbol{j} \qquad \boldsymbol{p}=p_x\boldsymbol{i}+p_y\boldsymbol{j}$$

where p_x is the component of linear momentum parallel to the x-axis and p_y is the component parallel to the y-axis. Therefore,

$$\boldsymbol{l}=\boldsymbol{r}\times\boldsymbol{p}=(x\boldsymbol{i}+y\boldsymbol{j})\times(p_x\boldsymbol{i}+p_y\boldsymbol{j})=(xp_y-yp_x)\boldsymbol{k}$$

where $\boldsymbol{k}$ is the unit vector pointing along the positive z-axis. For a particle rotating in the xy-plane, the angular momentum vector lies entirely along the z-axis with a magnitude given by $|xp_y-yp_x|$ (Fig. 8C.5).

The operators for the linear momentum components p_x and p_y are given in Topic 7C, so the operator for angular momentum about the z-axis is

$$\hat{l}_z=\frac{\hbar}{i}\left(x\frac{\partial}{\partial y}-y\frac{\partial}{\partial x}\right) \qquad \text{z-Component of the angular momentum operator} \qquad (8C.13a)$$

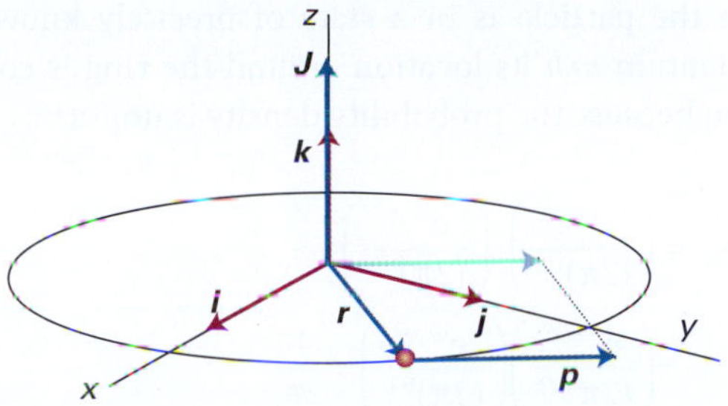

Figure 8C.5 The classical angular momentum $\boldsymbol{l}$ of a particle with position $\boldsymbol{r}$ and linear momentum $\boldsymbol{p}$ is given by the vector product $\boldsymbol{l}=\boldsymbol{r}\times\boldsymbol{p}$. For the motion restricted to the xy-plane as depicted here, $\boldsymbol{r}=x\boldsymbol{i}+y\boldsymbol{j}$, $\boldsymbol{p}=p_x\boldsymbol{i}+p_y\boldsymbol{j}$, and $\boldsymbol{l}=(xp_y-yp_x)\boldsymbol{k}$, with $\boldsymbol{i}$, $\boldsymbol{j}$, and $\boldsymbol{k}$ denoting unit vectors along the positive x-, y-, and z-axes.

When expressed in terms of the cylindrical coordinates r and ϕ (*The chemist's toolkit* 8C.1), this equation becomes

$$\hat{l}_z=\frac{\hbar}{i}\frac{\partial}{\partial\phi} \qquad (8C.13b)$$

With the angular momentum operator available, we can test if the wavefunction in eqn 8B.6a is an eigenfunction. Because the wavefunction depends on only the coordinate ϕ, the partial derivative in eqn 8C.13b can be replaced by a complete derivative and we find

$$\hat{l}_z\psi_{m_l}=\frac{\hbar}{i}\frac{d}{d\phi}\psi_{m_l}=\frac{\hbar}{i}\frac{d}{d\phi}\frac{e^{im_l\phi}}{(2\pi)^{1/2}}=im_l\frac{\hbar}{i}\frac{e^{im_l\phi}}{(2\pi)^{1/2}}=m_l\hbar\psi_{m_l} \qquad (8C.14)$$

That is, ψ_{m_l} is an eigenfunction of $\hat{l}_z$, and corresponds to an angular momentum $m_l\hbar$, in accord with eqn 8C.4. When m_l is positive, the angular momentum is positive (clockwise rotation when seen from below); when m_l is negative, the angular momentum is negative (anticlockwise when seen from below). These features are the origin of the vector representation of angular momentum, in which the magnitude is represented by the length of a vector and the direction of motion by its orientation (Fig. 8C.6)

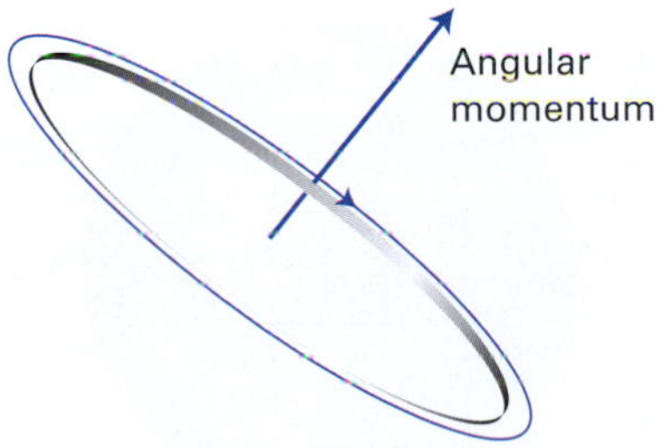

Figure 8C.6 The basic ideas of the vector representation of angular momentum: the magnitude of the angular momentum is represented by the length of the vector, and the orientation of the motion in space is represented by the orientation of the vector (using the right-hand screw rule).

When the particle is in a state of precisely known angular momentum $m_l\hbar$ its location around the ring is completely unknown because the probability density is uniform:

$$\psi_{m_l}^{*}\psi_{m_l}=\left(\frac{e^{im_l\phi}}{(2\pi)^{1/2}}\right)^{*}\left(\frac{e^{im_l\phi}}{(2\pi)^{1/2}}\right)=\left(\frac{e^{-im_l\phi}}{(2\pi)^{1/2}}\right)\left(\frac{e^{im_l\phi}}{(2\pi)^{1/2}}\right)=\frac{1}{2\pi}$$

Angular momentum and angular position are a pair of complementary observables (in the sense defined in Topic 7C), and the inability to specify them simultaneously with arbitrary precision is another example of the uncertainty principle.

8C.2 Rotation in three dimensions

We now consider a particle of mass m that is free to move anywhere on the surface of a sphere of radius r. The sphere can be thought of as a three-dimensional stack of rings with the additional freedom for the particle to migrate from one ring to another. The cyclic boundary condition for the particle on each ring leads to the quantum number m_l that is encountered for motion on an individual ring. The requirement that the wavefunction must match as a path is traced over the poles as well as round the equator of the sphere surrounding the central point introduces a second cyclic boundary condition and therefore a second quantum number (Fig. 8C.7).

(a) The wavefunctions

The hamiltonian operator for motion in three dimensions (Table 7B.1) is

$$\hat{H}=-\frac{\hbar^2}{2m}\nabla^2+V \qquad \textit{Three dimensions} \quad \text{Hamiltonian operator} \qquad (8C.15a)$$

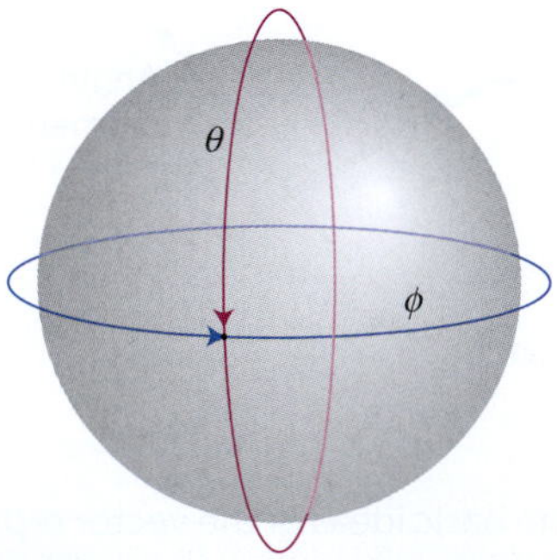

Figure 8C.7 The wavefunction of a particle on the surface of a sphere must satisfy two cyclic boundary conditions. This requirement leads to two quantum numbers for its state of angular momentum.

$$\nabla^2=\frac{\partial^2}{\partial x^2}+\frac{\partial^2}{\partial y^2}+\frac{\partial^2}{\partial z^2} \qquad \textit{Three dimensions} \quad \text{Laplacian} \qquad (8C.15b)$$

The laplacian, ∇^2 (read 'del squared'), is a convenient abbreviation for the sum of the three second derivatives. For the particle confined to a spherical surface, $V=0$ wherever it is free to travel and r is a constant. To take advantage of the symmetry of the problem and the fact that r is constant for a particle on a sphere, we use spherical polar coordinates (*The chemist's toolkit* 7B.1). The wavefunction is therefore a function of the colatitude θ, and the azimuth ϕ, and we write it $\psi(\theta,\phi)$. The Schrödinger equation is therefore

$$-\frac{\hbar^2}{2m}\nabla^2\psi=E\psi \qquad \textit{Particle on a sphere} \quad \text{Schrödinger equation} \qquad (8C.16)$$

This Schrödinger equation is solved by using the technique of separation of variables (*Mathematical background* 4), which confirms that, as shown in the following *Justification*, the wavefunction can be written as a product of functions

$$\psi(\theta,\phi)=\Theta(\theta)\Phi(\phi) \qquad (8C.17)$$

where Θ is a function only of θ and Φ is a function only of ϕ. As confirmed in the *Justification*, the Φ are the solutions for a particle on a ring (Section 8C.1) and the overall solutions are specified by the **orbital angular momentum quantum number** l and the **magnetic quantum number** m_l. These quantum numbers are restricted to the values

$$l=0,1,2,\ldots \qquad m_l=l,l-1,\ldots,-l$$

The quantum number l is a non-negative integer and, for a given value of l, there are $2l+1$ permitted values of m_l.

Justification 8C.3 The solutions of the Schrödinger equation for a particle on a sphere

Because r is constant, we can discard the part of the laplacian that involves differentiation with respect to r, and so write the Schrödinger equation as

$$-\frac{\hbar^2}{2mr^2}\Lambda^2\psi=E\psi \qquad \textit{Particle on a sphere} \quad \text{Schrödinger equation} \qquad (8C.18)$$

The moment of inertia, $I=mr^2$, has appeared. This expression can be rearranged into

$$\Lambda^2\psi=-\varepsilon\psi \qquad \varepsilon=\frac{2IE}{\hbar^2}$$

To verify that this expression is separable, we try the substitution $\psi=\Theta\Phi$ and use the form of the legendrian in Table 7B.1:

$$\Lambda^2\Theta\Phi=\frac{1}{\sin^2\theta}\frac{\partial^2(\Theta\Phi)}{\partial\phi^2}+\frac{1}{\sin\theta}\frac{\partial}{\partial\theta}\sin\theta\frac{\partial(\Theta\Phi)}{\partial\theta}=-\varepsilon\Theta\Phi$$

We now use the fact that Θ and Φ are each functions of one variable, so the partial derivatives become complete derivatives:

$$\frac{\Theta}{\sin^2\theta}\frac{\mathrm{d}^2\Phi}{\mathrm{d}\phi^2}+\frac{\Phi}{\sin\theta}\frac{\mathrm{d}}{\mathrm{d}\theta}\sin\theta\frac{\mathrm{d}\Theta}{\mathrm{d}\theta}=-\varepsilon\Theta\Phi$$

Division through by $\Theta\Phi$ and multiplication by $\sin^2\theta$ gives

$$\frac{1}{\Phi}\frac{\mathrm{d}^2\Phi}{\mathrm{d}\phi^2}+\frac{\sin\theta}{\Theta}\frac{\mathrm{d}}{\mathrm{d}\theta}\sin\theta\frac{\mathrm{d}\Theta}{\mathrm{d}\theta}=-\varepsilon\sin^2\theta$$

and, after minor rearrangement,

$$\frac{1}{\Phi}\frac{\mathrm{d}^2\Phi}{\mathrm{d}\phi^2}+\frac{\sin\theta}{\Theta}\frac{\mathrm{d}}{\mathrm{d}\theta}\sin\theta\frac{\mathrm{d}\Theta}{\mathrm{d}\theta}+\varepsilon\sin^2\theta=0$$

The first term on the left depends only on ϕ and the remaining two terms depend only on θ. By the argument presented in *Mathematical background* 4, each term is equal to a constant. Thus, if we set the first term equal to the constant $-m_l^2$ (using a notation chosen with an eye to the future), the separated equations are

$$\frac{1}{\Phi}\frac{\mathrm{d}^2\Phi}{\mathrm{d}\phi^2}=-m_l^2 \qquad \frac{\sin\theta}{\Theta}\frac{\mathrm{d}}{\mathrm{d}\theta}\sin\theta\frac{\mathrm{d}\Theta}{\mathrm{d}\theta}+\varepsilon\sin^2\theta=m_l^2$$

The first of these two equations is the same as that encountered for the particle on a ring and has the same solutions:

$$\Phi=\frac{1}{(2\pi)^{1/2}}\mathrm{e}^{\mathrm{i}m_l\phi} \qquad m_l=0,\pm1,\pm2,\ldots$$

(Shortly we shall see that m_l is in fact bounded for a three-dimensional system.) The second equation is new, but its solutions are well known to mathematicians as 'associated Legendre functions'. The cyclic boundary condition for the matching of the wavefunction at $\phi=0$ and 2π restricts m_l to positive and negative integer values (including 0), as for a particle on a ring. The additional requirement that the wavefunction also match on a journey over the poles (Fig. 8C.7) results in the introduction of the second quantum number, l, with non-negative integer values. However, the presence of the quantum number m_l in the second equation implies that the ranges of the two quantum numbers are linked, and it turns out that for a given value of l, m_l ranges in integer steps from $-l$ to $+l$, as quoted in the text.

The normalized wavefunctions $\psi(\theta,\phi)$ for a given l and m_l are usually denoted $Y_{l,m_l}(\theta,\phi)$ and are called the **spherical harmonics** (Table 8C.1). They are as fundamental to the description of waves on spherical surfaces as the harmonic (sine and cosine) functions are to the description of waves on lines and planes. These important functions satisfy the equation[1]

$$\Lambda^2 Y_{l,m_l}(\theta,\phi)=-l(l+1)Y_{l,m_l}(\theta,\phi) \qquad \text{(8C.19)}$$

[1] For a full account of the solution, see our *Molecular quantum mechanics* (2011).

Table 8C.1 The spherical harmonics, $Y_{l,m_l}(\theta,\phi)$

l	m_l	$Y_{l,m_l}(\theta,\phi)$
0	0	$\left(\frac{1}{4\pi}\right)^{1/2}$
1	0	$\left(\frac{3}{4\pi}\right)^{1/2}\cos\theta$
	±1	$\mp\left(\frac{3}{8\pi}\right)^{1/2}\sin\theta\,\mathrm{e}^{\pm\mathrm{i}\phi}$
2	0	$\left(\frac{5}{16\pi}\right)^{1/2}(3\cos^2\theta-1)$
	±1	$\mp\left(\frac{15}{8\pi}\right)^{1/2}\cos\theta\sin\theta\,\mathrm{e}^{\pm\mathrm{i}\phi}$
	±2	$\left(\frac{15}{32\pi}\right)^{1/2}\sin^2\theta\,\mathrm{e}^{\pm2\mathrm{i}\phi}$
3	0	$\left(\frac{7}{16\pi}\right)^{1/2}(5\cos^3\theta-3\cos\theta)$
	±1	$\mp\left(\frac{21}{64\pi}\right)^{1/2}(5\cos^2\theta-1)\sin\theta\,\mathrm{e}^{\pm\mathrm{i}\phi}$
	±2	$\left(\frac{105}{32\pi}\right)^{1/2}\sin^2\theta\cos\theta\,\mathrm{e}^{\pm2\mathrm{i}\phi}$
	±3	$\mp\left(\frac{35}{64\pi}\right)^{1/2}\sin^3\theta\,\mathrm{e}^{\pm3\mathrm{i}\phi}$

Figure 8C.8 is a representation of the spherical harmonics for $l=0$ to 4 and $m_l=0$; the use of different tints of shading, which correspond to different signs of the wavefunction, emphasizes

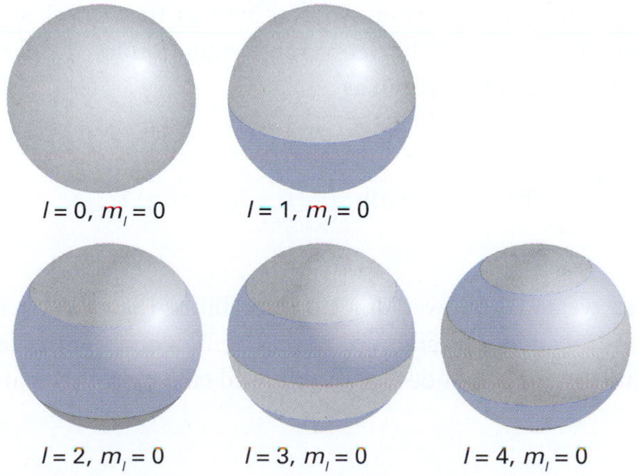

Figure 8C.8 A representation of the wavefunctions of a particle on the surface of a sphere that emphasizes the location of angular nodes: different colours of shading correspond to different signs of the wavefunction. Note that the number of nodes increases as the value of l increases. All these wavefunctions correspond to $m_l=0$; a path round the vertical z-axis of the sphere does not cut through any nodes.

the location of the angular nodes (the positions at which the wavefunction passes through zero). Note that:

- There are no angular nodes around the z-axis for functions with $m_l=0$. The spherical harmonic with $l=0$, $m_l=0$ has no nodes at all: it is a 'wave' of constant height at all positions of the surface.
- The spherical harmonic with $l=1$, $m_l=0$ has a single angular node at $\theta=\pi/2$; therefore, the equatorial plane is a nodal plane.
- The spherical harmonic with $l=2$, $m_l=0$ has two angular nodes.

Physical interpretation

Brief illustration 8C.2 The angular nodes of the spherical harmonics

For the spherical harmonic with $l=2$, $m_l=0$, the angular nodes correspond to angles where (see Table 8C.1) $3\cos^2\theta-1=0$, or $\cos\theta=\pm1/3^{1/2}$. The angular nodes are therefore at 54.7° and 125.3°.

Self-test 8C.3 Find the angular nodes for the spherical harmonic $l=3$, $m_l=0$.

Answer: $\theta=39.2°, 90°, 140.8°$

(b) The energies

In general, the number of angular nodes is equal to l. As the number of nodes increases, the wavefunctions become more buckled, and with this increasing curvature we can anticipate that the kinetic energy of the particle (and therefore its total energy because the potential energy is zero) increases.

With the wavefunctions ψ identified as the spherical harmonics Y, eqn 8C.18, the Schrödinger equation for the particle on a sphere, can be written as

$$-\frac{\hbar^2}{2\underbrace{mr^2}_{I}}\overbrace{\Lambda^2 Y_{l,m_l}}^{-l(l+1)Y_{l,m_l}}=E\psi,\text{ or } l(l+1)\frac{\hbar^2}{2I}Y_{l,m_l}=E_{l,m_l}Y_{l,m_l}$$

where we have allowed for the possibility that the energies depend on the two quantum numbers. By equating the terms in blue we can conclude that the allowed energies of the particle are

$$E_{l,m_l}=l(l+1)\frac{\hbar^2}{2I}$$ *Particle on a sphere* Energy levels (8C.20)

According to this equation:

- The energies are quantized because $l=0, 1, 2, \ldots$.
- The energies are independent of the value of m_l, and henceforth we shall denote them simply as E_l.
- Because there are $2l+1$ different wavefunctions (one for each value of m_l) that correspond to the same energy, it follows that a level with quantum number l is $(2l+1)$-fold degenerate.
- There is no zero-point energy, and $E_0=0$.

Physical interpretation

Example 8C.2 Using the rotational energy levels

Under certain circumstances, the particle on a sphere is a reasonable model for the description of the rotation of diatomic molecules. Consider, for example, the rotation of a $^1H^{127}I$ molecule: because of the large difference in atomic masses, it is appropriate to picture the ^1H atom as orbiting a stationary ^{127}I atom at a distance $r=160$ pm, the equilibrium bond distance. Determine the energies and degeneracies of the lowest four energy levels of an $^1H^{127}I$ molecule freely rotating in three dimensions. What is the frequency of the transition between the lowest two rotational levels?

Method Because in this model the ^{127}I atom is stationary, the moment of inertia is $I=m_{^1H}r^2$, with $r=160$ pm. The rotational energies are given in eqn 8C.20; but for reasons that are developed in Topic 12B, the angular momentum quantum number of rotating molecules is denoted J in place of l, and we use that symbol here. The degeneracy of a level with quantum number J is $2J+1$, the analogue of $2l+1$. A transition between two rotational levels can be brought about by the emission or absorption of a photon with a frequency given by the Bohr frequency condition (Topic 7A, $h\nu=\Delta E$).

Answer The moment of inertia is

$$I=\overbrace{(1.675\times10^{-27}\text{ kg})}^{m_{^1H}}\times\overbrace{(1.60\times10^{-12}\text{ m})^2}^{r^2}=4.29\times10^{-47}\text{ kg m}^2$$

It follows that

$$\frac{\hbar^2}{2I}=\frac{(1.055\times10^{-34}\text{ J s})^2}{2\times(4.29\times10^{-47}\text{ kg m}^2)}=1.30\times10^{-22}\text{ J}$$

or 0.130 zJ. We now draw up the following table, where the molar energies are obtained by multiplying the individual energies by Avogadro's constant:

J	E/zJ	E/(J mol^{-1})	Degeneracy
0	0	0	1
1	0.260	156	3
2	0.780	470	5
3	1.56	939	7

The energy separation between the two lowest rotational energy levels ($J=0$ and 1) is 2.60×10^{-22} J, which corresponds to a photon frequency of

$$\nu=\frac{\Delta E}{h}=\frac{2.60\times10^{-22}\text{ J}}{6.626\times10^{-34}\text{ J s}}=3.92\times10^{11}\overbrace{\text{s}^{-1}}^{\text{Hz}}=392\text{ GHz}$$

Radiation of this frequency belongs to the microwave region of the electromagnetic spectrum, so microwave spectroscopy is used to study molecular rotations (Topic 12C). Because the transition frequencies depend on the moment of inertia and frequencies can be measured with great precision, microwave spectroscopy is a very precise technique for the determination of bond lengths.

Self-test 8C.4 What is the frequency of the transition between the lowest two rotational levels in $^2H^{127}I$? (Same bond length as $^1H^{127}I$.)

Answer: 196 GHz

(c) Angular momentum

The energy of a rotating particle is related classically to its angular momentum J by $E=J^2/2I$ (eqn 8C.1). Therefore, by comparing this equation with eqn 8C.20, we can deduce that because the energy is quantized, then so too is the magnitude of the angular momentum, and confined to the values

$$\{l(l+1)\}^{1/2}\hbar \qquad l=0,1,2,\ldots \qquad \textit{Particle on a sphere} \qquad \text{Magnitude of angular momentum} \qquad (8C.21a)$$

We show in the following *Justification* that the angular momentum about the z-axis is also quantized, with the values

$$m_l\hbar \qquad m_l=l,l-1,\ldots,-l \qquad \textit{Particle on a sphere} \qquad z\text{-Component of angular momentum} \qquad (8C.21b)$$

Justification 8C.4 The z-component of angular momentum for a particle on a sphere

The operator for the z-component of the angular momentum in polar coordinates is

$$\hat{l}_z=\frac{\hbar}{\mathrm{i}}\frac{\partial}{\partial\phi}$$

With this operator available, we can test if the wavefunction in eqn 8C.17 is an eigenfunction:

$$\hat{l}_z\psi=\hat{l}_z\Theta\Phi=\frac{\hbar}{\mathrm{i}}\frac{\partial}{\partial\phi}\Theta\Phi=\Theta\overbrace{\frac{\hbar}{\mathrm{i}}\frac{\mathrm{d}}{\mathrm{d}\phi}\Phi}^{\mathrm{i}m_l\Phi}=\Theta\times m_l\hbar\Phi=m_l\hbar\psi$$

The partial derivative has been replaced above by a full derivative because Θ is independent of ϕ and we have used the result, as given in *Justification* 8C.1, that $\Phi\propto e^{\mathrm{i}m_l\phi}$. Therefore, the wavefunctions are eigenfunctions of $\hat{l}_z$, and correspond to an angular momentum around the z-axis of $m_l\hbar$, in accord with eqn 8C.21b.

Brief illustration 8C.3 The magnitude of the angular momentum

The lowest four rotational energy levels of the $^1H^{127}I$ molecule of *Example* 8C.2 correspond to $J=0, 1, 2, 3$. Using equations 8C.21a and 8C.21b, we can draw up the following table:

J	Magnitude of angular momentum/$\hbar$	Degeneracy	z-Component of angular momentum/$\hbar$
0	0	1	0
1	$2^{1/2}$	3	+1, 0, −1
2	$6^{1/2}$	5	+2, +1, 0, −1, −2
3	$12^{1/2}$	7	+3, +2, +1, 0, −1, −2, −3

Self-test 8C.5 What is the degeneracy and magnitude of the angular momentum for $J=5$?

Answer: 11, $30^{1/2}\hbar$

(d) Space quantization

The result that m_l is confined to the values l, $l-1$, …, $-l$ for a given value of l means that the component of angular momentum about the z-axis—the contribution to the total angular momentum of rotation around that axis—may take only $2l+1$ values. If we represent the angular momentum by a vector of length $\{l(l+1)\}^{1/2}$, then it follows that this vector must be oriented so that its projection on the z-axis is m_l and that it can have only $2l+1$ orientations rather than the continuous range of orientations of a rotating classical body (Fig. 8C.9). The remarkable implication is that *the orientation of a rotating body is quantized.*

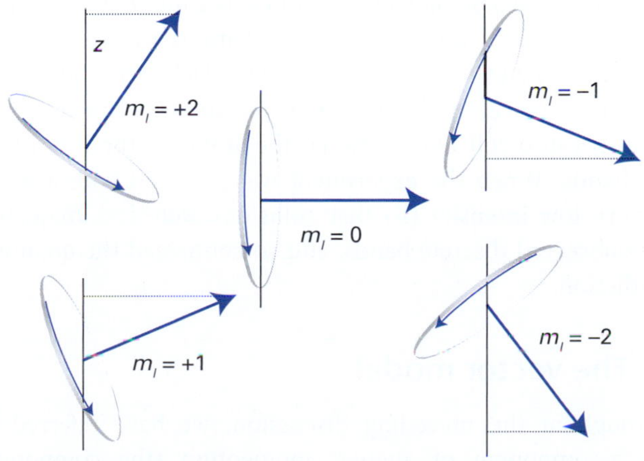

Figure 8C.9 The permitted orientations of angular momentum when $l=2$. We shall see soon that this representation is too specific because the azimuthal orientation of the vector (its angle around z) is indeterminate.

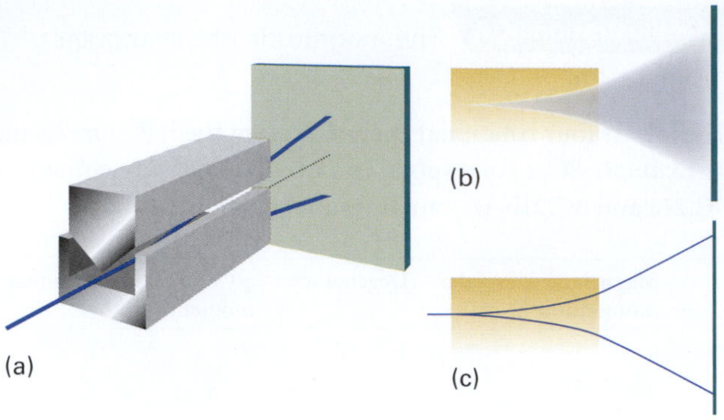

Figure 8C.10 (a) The experimental arrangement for the Stern–Gerlach experiment: the magnet provides an inhomogeneous field. (b) The classically expected result. (c) The observed outcome using silver atoms.

The quantum mechanical result that a rotating body may not take up an arbitrary orientation with respect to some specified axis (for example, an axis defined by the direction of an externally applied electric or magnetic field) is called **space quantization**. It was confirmed by an experiment first performed by Otto Stern and Walther Gerlach in 1921, who shot a beam of silver atoms through an inhomogeneous magnetic field (Fig. 8C.10). The idea behind the experiment was that a silver atom behaves like a magnet and interacts with the applied field (a point explored in more detail in the discussion of 'spin' in Topic 9B.). According to classical mechanics, because the orientation of the angular momentum can take any value, the associated magnet can take any orientation. Because the direction in which the magnet is driven by the inhomogeneous field depends on the magnet's orientation, it follows that a broad band of atoms is expected to emerge from the region where the magnetic field acts. According to quantum mechanics, however, because the orientation of the angular momentum is quantized, the associated magnet lies in a number of discrete orientations, so several sharp bands of atoms are expected.

In their first experiment, Stern and Gerlach appeared to confirm the classical prediction. However, the experiment is difficult because collisions between the atoms in the beam blur the bands. When the experiment was repeated with a beam of very low intensity (so that collisions were less frequent), they observed discrete bands, and so confirmed the quantum prediction.

(e) The vector model

Throughout the preceding discussion, we have referred to the z-component of angular momentum (the component about an arbitrary axis, which is conventionally denoted z), and have made no reference to the x- and y-components (the components about the two axes perpendicular to z). The reason for this omission is found by examining the operators for the three components, each one being given by a term like that in eqn 8C.13a:

$$\hat{l}_x = \frac{\hbar}{\mathrm{i}}\left(y\frac{\partial}{\partial z} - z\frac{\partial}{\partial y}\right)$$
$$\hat{l}_y = \frac{\hbar}{\mathrm{i}}\left(z\frac{\partial}{\partial x} - x\frac{\partial}{\partial z}\right) \qquad \text{Angular momentum operators} \qquad (8C.22)$$
$$\hat{l}_z = \frac{\hbar}{\mathrm{i}}\left(x\frac{\partial}{\partial y} - y\frac{\partial}{\partial x}\right) = \frac{\hbar}{\mathrm{i}}\frac{\partial}{\partial \phi}$$

The commutation relations among the three operators, which you are invited to derive in Problem 8C.11, are

$$[\hat{l}_x,\hat{l}_y] = \mathrm{i}\hbar\hat{l}_z \quad [\hat{l}_y,\hat{l}_z] = \mathrm{i}\hbar\hat{l}_x \quad [\hat{l}_z,\hat{l}_x] = \mathrm{i}\hbar l_y \qquad \text{Angular momentum commutators} \qquad (8C.23)$$

Therefore, we cannot specify more than one component (unless $l=0$). In other words, l_x, l_y, and l_z are complementary observables. On the other hand, the operator for the square of the magnitude of the angular momentum is

$$\hat{l}^2 = \hat{l}_x^2 + \hat{l}_y^2 + \hat{l}_z^2 \qquad \text{Operator for square of magnitude of angular momentum} \qquad (8C.24)$$

This operator commutes with all three components (see Problem 8C.11):

$$[\hat{l}^2,\hat{l}_q] = 0 \qquad q = x, y, \text{and } z \qquad \text{Commutators of angular momentum operators} \qquad (8C.25)$$

Therefore, although we may specify the magnitude of the angular momentum and any of its components, if l_z is known, then it is impossible to ascribe values to the other two components. It follows that the illustration in Fig. 8C.9, which is summarized in Fig. 8C.11a, gives a false impression of the state of the system, because it suggests definite values for the x- and y-components. A better picture must reflect the impossibility of specifying l_x and l_y if l_z is known.

The **vector model** of angular momentum uses pictures like that in Fig. 8C.11b. The cones are drawn with side $\{l(l+1)\}^{1/2}$ units, and represent the magnitude of the angular momentum. Each cone has a definite projection (of m_l units) on the z-axis, representing the system's precise value of l_z. The l_x and l_y projections, however, are indefinite. The vector representing the state of angular momentum can be thought of as lying with its tip on any point on the mouth of the cone. At this stage it should not be thought of as sweeping round the cone; that aspect of the model will be added when we allow the picture to convey more information (Topics 9B and 9C).

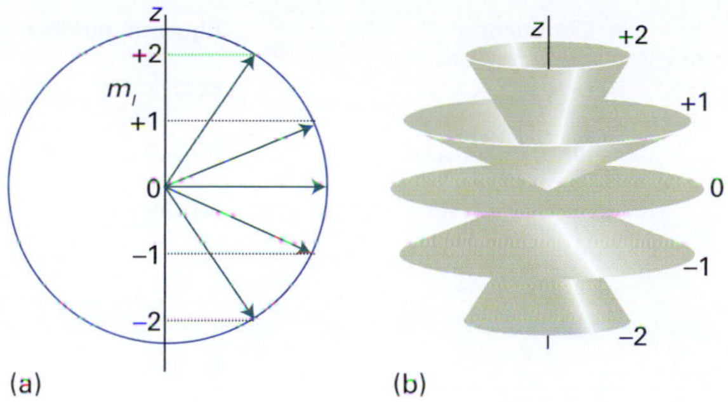

Figure 8C.11 (a) A summary of Fig. 8C.9. However, because the azimuthal angle of the vector around the *z*-axis is indeterminate, a better representation is as in (b), where each vector lies at an unspecified azimuthal angle on its cone.

Brief illustration 8C.4 The vector model of the angular momentum

If the wavefunction of a rotating molecule is given by the spherical harmonic $Y_{3,+2}$ then the angular momentum can be represented by a cone

- with a side of length $12^{1/2}$ (representing the magnitude of $12^{1/2}\hbar$); and
- with a projection of +2 on the *z*-axis (representing the *z*-component of $+2\hbar$).

Self-test 8C.6 Analyse the vector model of angular momentum if the wavefunction is given by the spherical harmonic $Y_{3,-1}$.

Answer: length is $12^{1/2}$, projection is −1

Checklist of concepts

☐ 1. The energy and angular momentum for a particle rotating in two- or three-dimensions are quantized; quantization results from the requirement that the wavefunction satisfies a **cyclic boundary condition**.

☐ 2. All energy levels of a particle rotating in two dimensions are doubly-degenerate except for the lowest level ($m_l=0$).

☐ 3. There is no zero-point energy for a particle rotating in a plane or on a sphere.

☐ 4. It is impossible to specify the angular momentum and location of the particle rotating in two dimensions simultaneously with arbitrary precision.

☐ 5. For a particle rotating in three dimensions, the cyclic boundary conditions imply that the magnitude and *z*-component of the angular momentum are quantized.

☐ 6. **Space quantization** refers to the quantum mechanical result that a rotating body may not take up an arbitrary orientation with respect to some specified axis.

☐ 7. Angular momentum and orientation are complementary observables.

☐ 8. Because the components of angular momentum do not commute, only the magnitude of the angular momentum and one of its components can be specified simultaneously.

☐ 9. In the **vector model** of angular momentum, the angular momentum is represented by a cone with a side of length $\{l(l+1)\}^{1/2}$ and a projection of m_l on the *z*-axis. The vector can be thought of as lying with its tip on an indeterminate point on the mouth of the cone.

Checklist of equations

Property	Equation	Comment	Equation number
Wavefunction of particle on ring	$\psi_{m_l}(\phi)=e^{im_l\phi}/(2\pi)^{1/2}$	$m_l=0, \pm1, \pm2, \ldots$	8C.6a
Energy levels of particle on ring	$E_{m_l}=m_l^2\hbar^2/2I$	$m_l=0, \pm1, \pm2, \ldots$; $I=mr^2$	8C.6b
z-Component of angular momentum of particle on ring	$l_z=m_l\hbar$	$m_l=0, \pm1, \pm2, \ldots$	8C.14
Wavefunction of particle on sphere	$\psi(\theta,\phi)=Y_{l,m_l}(\theta,\phi)$	Y is a spherical harmonic	
Energy levels of particle on sphere	$E_{l,m_l}=l(l+1)\hbar^2/2I$	$l=0, 1, 2, \ldots$	8C.20

Property	Equation	Comment	Equation number
Magnitude of angular momentum	$\{l(l+1)\}^{1/2}\hbar$	$l=0, 1, 2, \ldots$	8C.21a
z-Component of angular momentum	$m_l\hbar$	$m_l=l, l-1, \ldots, -l$	8C.21b
Angular momentum commutation relations	$[l_x,l_y]=\mathrm{i}\hbar l_z$		8C.23
	$[l_y,l_z]=\mathrm{i}\hbar l_x$		
	$[l_z,l_x]=\mathrm{i}\hbar l_y$		
	$[l^2,l_q]=0, \quad q=x, y, \text{and } z$		8C.25

CHAPTER 8 The quantum theory of motion

TOPIC 8A Translation

Discussion questions

8A.1 Discuss the physical origin of quantization energy for a particle confined to moving inside a one-dimensional box.

8A.2 Describe the features of the solution of the particle in a one-dimensional box that appear in the solutions of the particle in two- and three-dimensional boxes. What concept applies to the latter but not to a one-dimensional box?

8A.3 Discuss the physical origins of quantum mechanical tunnelling. Why is tunnelling more likely to contribute to the mechanisms of electron transfer and proton transfer processes than to mechanisms of group transfer reactions, such as AB + C → A + BC (where A, B, and C are large molecular groups)?

Exercises

8A.1(a) Determine the linear momentum and kinetic energy of a free electron described by the wavefunction e^{ikx} with $k = 3\,\text{nm}^{-1}$.
8A.1(b) Determine the linear momentum and kinetic energy of a free proton described by the wavefunction e^{-ikx} with $k = 5\,\text{nm}^{-1}$.

8A.2(a) Write the wavefunction for a particle of mass 2.0 g travelling to the left with a kinetic energy of 20 J.
8A.2(b) Write the wavefunction for a particle of mass 1.0 g travelling to the right at $10\,\text{m s}^{-1}$.

8A.3(a) Calculate the energy separations in joules, kilojoules per mole, electronvolts, and reciprocal centimetres between the levels (i) $n = 2$ and $n = 1$, (ii) $n = 6$ and $n = 5$ of an electron in a box of length 1.0 nm.
8A.3(b) Calculate the energy separations in joules, kilojoules per mole, electronvolts, and reciprocal centimetres between the levels (i) $n = 3$ and $n = 1$, (ii) $n = 7$ and $n = 6$ of an electron in a box of length 1.50 nm.

8A.4(a) Calculate the probability that a particle will be found between $0.49L$ and $0.51L$ in a box of length L when it has (i) $n = 1$, (ii) $n = 2$. Take the wavefunction to be a constant in this range.
8A.4(b) Calculate the probability that a particle will be found between $0.65L$ and $0.67L$ in a box of length L when it has (i) $n = 1$, (ii) $n = 2$. Take the wavefunction to be a constant in this range.

8A.5(a) Calculate the expectation values of $\hat{p}$ and $\hat{p}^2$ for a particle in the state $n = 1$ in a one-dimensional square-well potential.
8A.5(b) Calculate the expectation values of $\hat{p}$ and $\hat{p}^2$ for a particle in the state $n = 2$ in a one-dimensional square-well potential.

8A.6(a) Calculate the expectation values of $\hat{x}$ and $\hat{x}^2$ for a particle in the state $n = 1$ in a one-dimensional square-well potential.
8A.6(b) Calculate the expectation values of of $\hat{x}$ and $\hat{x}^2$ for a particle in the state $n = 2$ in a one-dimensional square-well potential.

8A.7(a) An electron is confined to a square well of length L. What would be the length of the box such that the zero-point energy of the electron is equal to its rest mass energy, m_ec^2? Express your answer in terms of the parameter $\lambda_C = h/m_ec$, the 'Compton wavelength' of the electron.
8A.7(b) Repeat Exercise 8A.7(a) for a general particle of mass m in a cubic box.

8A.8(a) What are the most likely locations of a particle in a box of length L in the state $n = 3$?
8A.8(b) What are the most likely locations of a particle in a box of length L in the state $n = 5$?

8A.9(a) Calculate the percentage change in a given energy level of a particle in a one-dimensional box when the length of the box is increased by 10 per cent.
8A.9(b) Calculate the percentage change in a given energy level of a particle in a cubic box when the length of the edge of the cube is decreased by 10 per cent in each direction.

8A.10(a) What is the value of n of a particle in a one-dimensional box such that the separation between neighbouring levels is equal to the mean energy of thermal motion ($\frac{1}{2}kT$).
8A.10(b) A nitrogen molecule is confined in a cubic box of volume $1.00\,\text{m}^3$. (i) Assuming that the molecule has an energy equal to $\frac{3}{2}kT$ at $T = 300\,\text{K}$, what is the value of $n = (n_x^2 + n_y^2 + n_z^2)^{1/2}$ for this molecule? (ii) What is the energy separation between the levels n and $n + 1$? (iii) What is its de Broglie wavelength?

8A.11(a) For a particle in a rectangular box with sides of length $L_1 = L$ and $L_2 = 2L$, find a state that is degenerate with the state $n_1 = n_2 = 2$. Degeneracy is normally associated with symmetry; why, then, are these two states degenerate?
8A.11(b) For a particle in a rectangular box with sides of length $L_1 = L$ and $L_2 = 2L$, find a state that is degenerate with the state $n_1 = 2$, $n_2 = 8$. Degeneracy is normally associated with symmetry; why, then, are these two states degenerate?

8A.12(a) Consider a particle in a cubic box. What is the degeneracy of the level that has an energy three times that of the lowest level?
8A.12(b) Consider a particle in a cubic box. What is the degeneracy of the level that has an energy $\frac{14}{3}$ times that of the lowest level?

8A.13(a) Suppose that the junction between two semiconductors can be represented by a barrier of height 2.0 eV and length 100 pm. Calculate the transmission probability of an electron with energy 1.5 eV.
8A.13(b) Suppose that a proton of an acidic hydrogen atom is confined to an acid that can be represented by a barrier of height 2.0 eV and length 100 pm. Calculate the probability that a proton with energy 1.5 eV can escape from the acid.

Problems

8A.1 Calculate the separation between the two lowest levels for an O_2 molecule in a one-dimensional container of length 5.0 cm. At what value of n does the energy of the molecule reach $\frac{1}{2}kT$ at 300 K, and what is the separation of this level from the one immediately below?

8A.2 When β-carotene (**1**) is oxidized *in vivo*, it breaks in half and forms two molecules of retinal (vitamin A), which is a precursor to the pigment in the retina responsible for vision.

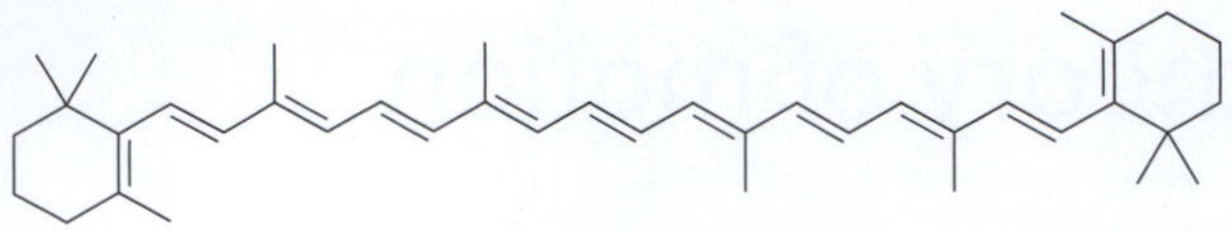

1 β-Carotene

The conjugated system of retinal consists of 11 C atoms and one O atom. In the ground state of retinal, each level up to $n=6$ is occupied by two electrons. Assuming an average internuclear distance of 140 pm, calculate (a) the separation in energy between the ground state and the first excited state in which one electron occupies the state with $n=7$, and (b) the frequency of the radiation required to produce a transition between these two states. (c) Using your results, choose among the words in parentheses to generate a rule for the prediction of frequency shifts in the absorption spectra of linear polyenes:

> The absorption spectrum of a linear polyene shifts to (higher/lower) frequency as the number of conjugated atoms (increases/decreases).

8A.3‡ A particle is confined to move in a one-dimensional box of length L. (a) If the particle is classical, show that the average value of x is $\frac{1}{2}L$ and that the root-mean square value is $L/3^{1/2}$. (b) Show that for large values of n, a quantum particle approaches the classical values. This result is an example of the correspondence principle, which states that, for very large values of the quantum numbers, the predictions of quantum mechanics approach those of classical mechanics.

8A.4 Here we explore further the idea introduced in *Impact* I8.1 that quantum mechanical effects need to be invoked in the description of the electronic properties of metallic nanocrystals, here modelled as three-dimensional boxes. (a) Set up the Schrödinger equation for a particle of mass m in a three-dimensional rectangular box with sides L_1, L_2, and L_3. Show that the Schrödinger equation is separable. (b) Show that the wavefunction and the energy are defined by three quantum numbers. (c) Specialize the result from part (b) to an electron moving in a cubic box of side $L=5$ nm and draw an energy diagram resembling Fig. 8A.2 and showing the first 15 energy levels. Note that each energy level may consist of degenerate energy states. (d) Compare the energy level diagram from part (c) with the energy level diagram for an electron in a one-dimensional box of length $L=5$ nm. Are the energy levels more or less sparsely distributed in the cubic box than in the one-dimensional box?

8A.5 Many biological electron transfer reactions, such as those associated with biological energy conversion, may be visualized as arising from electron tunnelling between protein-bound cofactors, such as cytochromes, quinones, flavins, and chlorophylls. This tunnelling occurs over distances that are often greater than 1.0 nm, with sections of protein separating electron donor from acceptor. For a specific combination of donor and acceptor, the rate of electron tunnelling is proportional to the transmission probability, with $\kappa \approx 7\,\text{nm}^{-1}$ (eqn 8A.23). By what factor does the rate of electron tunnelling between two cofactors increase as the distance between them changes from 2.0 nm to 1.0 nm?

8A.6 Derive eqn 8A.23a, the expression for the transmission probability and show that then $\kappa L \gg 1$ it reduces to eqn 8A.23b.

8A.7‡ Consider the one-dimensional space in which a particle has one of three potential energies depending upon its position. They are: $V=0$ for $-\infty < x \le 0$, $V=V_2$ for $0 \le x \le L$, and $V=V_3$ for $L \le x < \infty$. The particle wavefunction has both a component e^{ik_1x} that is incident upon the barrier V_2 and a reflected component e^{-ik_1x} in Zone 1 ($-\infty < x \le 0$). In Zone 2 ($0 \le x \le L$) the wavefunction has components e^{k_2x} and e^{-k_2x}. In Zone 3 the wavefunction has only a forward component, e^{ik_3x}, which represents a particle that has traversed the barrier. The energy of the particle, E, is somewhere in the range $V_2 > E > V_3$. The transmission probability, T, is the ratio of the square modulus of Zone 3 amplitude to the square modulus of the incident amplitude. (a) Base your calculation on the continuity of the amplitude and slope of the wavefunction at the locations of the zone boundaries and derive a general equation for T. (b) Show that the general equation for T reduces to eqn 8A.23b in the high, wide barrier limit when $V_1 = V_3 = 0$. (c) Draw a graph of the probability of proton tunnelling when $V_3=0$, $L=50$ pm, and $E=10\,\text{kJ mol}^{-1}$ in the barrier range $E < V_2 < 2E$.

8A.8 The wavefunction inside a long barrier of height V is $\psi = Ne^{-\kappa x}$. Calculate (a) the probability that the particle is inside the barrier and (b) the average penetration depth of the particle into the barrier.

TOPIC 8B Vibrational motion

Discussion questions

8B.1 Describe the variation of the separation of the vibrational energy levels with the mass and force constant of the harmonic oscillator.

8B.2 In what ways does the quantum mechanical description of a harmonic oscillator merge with its classical description at high quantum numbers?

8B.3 What is the physical reason for the existence of a zero-point vibrational energy?

Exercises

8B.1(a) Calculate the zero-point energy of a harmonic oscillator consisting of a particle of mass 2.33×10^{-26} kg and force constant $155\,\text{N m}^{-1}$.
8B.1(b) Calculate the zero-point energy of a harmonic oscillator consisting of a particle of mass 5.16×10^{-26} kg and force constant $285\,\text{N m}^{-1}$.

8B.2(a) For a certain harmonic oscillator of effective mass 1.33×10^{-25} kg, the difference in adjacent energy levels is 4.82 zJ. Calculate the force constant of the oscillator.

8B.2(b) For a certain harmonic oscillator of effective mass 2.88×10^{-25} kg, the difference in adjacent energy levels is 3.17 zJ. Calculate the force constant of the oscillator.

8B.3(a) Calculate the wavelength of a photon needed to excite a transition between neighbouring energy levels of a harmonic oscillator of effective mass equal to that of a proton ($1.0078m_u$) and force constant $855\,\text{N m}^{-1}$.
8B.3(b) Calculate the wavelength of a photon needed to excite a transition between neighbouring energy levels of a harmonic oscillator of effective mass equal to that of an oxygen atom ($15.9949m_u$) and force constant $544\,\text{N m}^{-1}$.

‡ These problems were supplied by Charles Trapp and Carmen Giunta.

8B.4(a) The vibrational frequency of H_2 is 131.9 THz. What is the vibrational frequency of D_2 ($D = {}^2H$)?
8B.4(b) The vibrational frequency of H_2 is 131.9 THz. What is the vibrational frequency of T_2 ($T = {}^3H$)?

8B.5(a) Calculate the minimum excitation energies of (i) a pendulum of length 1.0 m on the surface of the Earth, (ii) the balance-wheel of a clockwork watch ($\nu = 5$ Hz).
8B.5(b) Calculate the minimum excitation energies of (i) the 33 kHz quartz crystal of a watch, (ii) the bond between two O atoms in O_2, for which $k_f = 1177\,\mathrm{N\,m^{-1}}$.

8B.6(a) Assuming that the vibrations of a $^{35}Cl_2$ molecule are equivalent to those of a harmonic oscillator with a force constant $k_f = 329\,\mathrm{N\,m^{-1}}$, what is the zero-point energy of vibration of this molecule? The effective mass of a homonuclear diatomic molecule is half its total mass, and $m(^{35}\mathrm{Cl}) = 34.9688m_u$.
8B.6(b) Assuming that the vibrations of a $^{14}N_2$ molecule are equivalent to those of a harmonic oscillator with a force constant $k_f = 2293.8\,\mathrm{N\,m^{-1}}$, what is the zero-point energy of vibration of this molecule? The effective mass of a homonuclear diatomic molecule is half its total mass, and $m(^{14}\mathrm{N}) = 14.0031m_u$.

8B.7(a) Locate the nodes of the harmonic oscillator wavefunction with $v = 4$.
8B.7(b) Locate the nodes of the harmonic oscillator wavefunction with $v = 5$.

8B.8(a) What are the most probable displacements of a harmonic oscillator with $v = 1$?
8B.8(b) What are the most probable displacements of a harmonic oscillator with $v = 3$?

8B.9(a) Calculate the probability that an O—H bond treated as an harmonic oscillator will be found at a classically forbidden extension when $v = 1$.
8B.9(b) Calculate the probability that an O—H bond treated as an harmonic oscillator will be found at a classically forbidden extension when $v = 2$.

Problems

8B.1 The mass to use in the expression for the vibrational frequency of a diatomic molecule is the effective mass $\mu = m_A m_B/(m_A + m_B)$, where m_A and m_B are the masses of the individual atoms. The following data on the infrared absorption wavenumbers (wavenumbers in $\mathrm{cm^{-1}}$) of molecules are taken from *Spectra of diatomic molecules*, G. Herzberg, van Nostrand (1950):

$H^{35}Cl$	$H^{81}Br$	HI	CO	NO
2990	2650	2310	2170	1904

Calculate the force constants of the bonds and arrange them in order of increasing stiffness.

8B.2 Carbon monoxide binds strongly to the Fe^{2+} ion of the haem group of the protein myoglobin. Estimate the vibrational frequency of CO bound to myoglobin by using the data in Problem 8B.1 and by making the following assumptions: the atom that binds to the haem group is immobilized, the protein is infinitely more massive than either the C or O atom, the C atom binds to the Fe^{2+} ion, and binding of CO to the protein does not alter the force constant of the C=O bond.

8B.3 Of the four assumptions made in Problem 8B.2, the last two are questionable. Suppose that the first two assumptions are still reasonable and that you have at your disposal a supply of myoglobin, a suitable buffer in which to suspend the protein, $^{12}C^{16}O$, $^{13}C^{16}O$, $^{12}C^{18}O$, $^{13}C^{18}O$, and an infrared spectrometer. Assuming that isotopic substitution does not affect the force constant of the C=O bond, describe a set of experiments that: (a) proves which atom, C or O, binds to the haem group of myoglobin, and (b) allows for the determination of the force constant of the C=O bond for myoglobin-bound carbon monoxide.

8B.4 Confirm that a function of the form e^{-gx^2} is a solution of the Schrödinger equation for the ground state of a harmonic oscillator and find an expression for g in terms of the mass and force constant of the oscillator.

8B.5 Calculate the mean kinetic energy of a harmonic oscillator by using the relations in Table 8B.1.

8B.6 Calculate the values of $\langle x^3\rangle$ and $\langle x^4\rangle$ for a harmonic oscillator by using the relations in Table 8B.1.

8B.7 Extend the calculation in *Example* 8B.4 by using mathematical software to calculate the probability that a harmonic oscillator will be found outside the classically allowed displacements for general v and plot the probability as a function of v.

8B.8 The intensities of spectroscopic transitions between the vibrational states of a molecule are proportional to the square of the integral $\int \psi_{v'} x \psi_v \mathrm{d}x$ over all space. Use the relations between Hermite polynomials given in Table 8B.1 to show that the only permitted transitions are those for which $v' = v \pm 1$ and evaluate the integral in these cases.

8B.9 Use mathematical software to construct a harmonic oscillator wavepacket of the form

$$\Psi(x,t) = \sum_{v=0}^{N} c_v \psi_v(x) e^{-iE_v t/\hbar}$$

where the wavefunctions and energies are those of a harmonic oscillator and with coefficients of your choice (for example, all equal). Explore how the wavepacket oscillates to and fro with time.

8B.10 Show that, whatever superposition of harmonic oscillator states is used to construct a wavepacket (as in Problem 8B.9), it is localized at the same place at the times 0, T, $2T$, …, where T is the classical period of the oscillator.

8B.11 The potential energy of the rotation of one CH_3 group relative to its neighbour in ethane can be expressed as $V(\phi) = V_0 \cos 3\phi$. Show that for small displacements the motion of the group is quantized and calculate the energy of excitation from $v = 0$ to $v = 1$. What do you expect to happen to the energy levels and wavefunctions as the excitation increases to high quantum numbers?

8B.12 Use the virial theorem to obtain an expression for the relation between the mean kinetic and potential energies of an electron in a hydrogen atom.

TOPIC 8C Rotational motion

Discussion questions

8C.1 Discuss the physical origin of quantization of energy for a particle confined to motion around a ring.

8C.2 Describe the features of the solution of the particle on a ring that appear in the solution of the particle on a sphere. What concept applies to the latter but not to the former?

8C.3 Describe the vector model of angular momentum in quantum mechanics. What features does it capture? What is its status as a model?

Exercises

8C.1(a) The rotation of a molecule can be represented by the motion of a point mass moving over the surface of a sphere. Calculate the magnitude of its angular momentum when $l=1$ and the possible components of the angular momentum on an arbitrary axis. Express your results as multiples of $\hbar$.

8C.1(b) The rotation of a molecule can be represented by the motion of a point mass moving over the surface of a sphere with angular momentum quantum number $l=2$. Calculate the magnitude of its angular momentum and the possible components of the angular momentum on an arbitrary axis. Express your results as multiples of $\hbar$.

8C.2(a) The wavefunction, $\psi(\phi)$, for the motion of a particle in a ring is of the form $\psi=Ne^{im\phi}$. Determine the normalization constant, N.

8C.2(b) Confirm that wavefunctions for a particle in a ring with different values of the quantum number m_l are mutually orthogonal.

8C.3(a) Calculate the minimum excitation energy of a proton constrained to rotate in a circle of radius 100 pm around a fixed point.

8C.3(b) Calculate the value of $|m_l|$ for the system described in the preceding exercise corresponding to a rotational energy equal to the classical average energy at 25 °C (which is equal to $\frac{1}{2}kT$).

8C.4(a) The moment of inertia of a CH_4 molecule is 5.27×10^{-47} kg m^2. What is the minimum energy needed to start it rotating?

8C.4(b) The moment of inertia of an SF_6 molecule is 3.07×10^{-45} kg m^2. What is the minimum energy needed to start it rotating?

8C.5(a) Use the data in Exercise 8C.4(a) to calculate the energy needed to excite a CH_4 molecule from a state with $l=1$ to a state with $l=2$.

8C.5(b) Use the data in Exercise 8C.4(b) to calculate the energy needed to excite an SF_6 molecule from a state with $l=2$ to a state with $l=3$.

8C.6(a) What is the magnitude of the angular momentum of a CH_4 molecule when it is rotating with its minimum energy?

8C.6(b) What is the magnitude of the angular momentum of an SF_6 molecule when it is rotating with its minimum energy?

8C.7(a) Draw scale vector diagrams to represent the states (i) $l=1$, $m_l=+1$, (ii) $l=2$, $m_l=0$.

8C.7(b) Draw the vector diagram for all the permitted states of a particle with $l=6$.

8C.8(a) The number of states corresponding to a given energy plays a crucial role in atomic structure and thermodynamic properties. Determine the degeneracy of a body rotating with $l=3$.

8C.8(b) The number of states corresponding to a given energy plays a crucial role in atomic structure and thermodynamic properties. Determine the degeneracy of a body rotating with $l=4$.

Problems

8C.1 The particle on a ring is a useful model for the motion of electrons around the porphine ring (2), the conjugated macrocycle that forms the structural basis of the haem group and the chlorophylls.

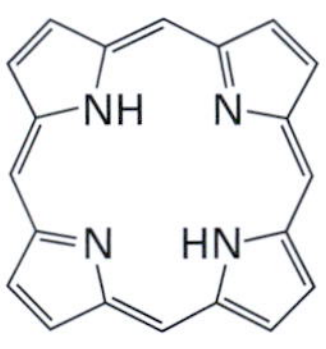

2 Porphine (porphin) ring

We may treat the group as a circular ring of radius 440 pm, with 22 electrons in the conjugated system moving along the perimeter of the ring. In the ground state of the molecule each state is occupied by two electrons. (a) Calculate the energy and angular momentum of an electron in the highest occupied level. (b) Calculate the frequency of radiation that can induce a transition between the highest occupied and lowest unoccupied levels.

8C.2 Use mathematical software to construct a wavepacket for a particle moving on a circular ring of the form

$$\Psi(\phi,t)=\sum_{m_l=0}^{m_{l,\max}} c_{m_l}e^{i(m_l\phi-E_{m_l}t/\hbar)} \qquad E_{m_l}=m_l^2\hbar^2/2I$$

with coefficients c of your choice (for example, all equal). Explore how the wavepacket migrates on the ring but spreads with time.

8C.3 Evaluate the z-component of the angular momentum and the kinetic energy of a particle on a ring that is described by the (unnormalized) wavefunctions (a) $e^{i\phi}$, (b) $e^{-2i\phi}$, (c) $\cos\phi$, and (d) $(\cos\chi)e^{i\phi}+(\sin\chi)e^{-i\phi}$.

8C.4 Is the Schrödinger equation for a particle on an elliptical ring of semi-major axes a and b separable? *Hint:* Although r varies with angle ϕ, the two are related by $r^2=a^2\sin^2\phi+b^2\cos^2\phi$.

8C.5 Calculate the energies of the first four rotational levels of $^1H^{127}I$ free to rotate in three dimensions, using for its moment of inertia $I=\mu R^2$, with $\mu=m_Hm_I/(m_H+m_I)$ and $R=160$ pm.

8C.6 Confirm that the spherical harmonics (a) $Y_{0,0}$, (b) $Y_{2,-1}$, and (c) $Y_{3,+3}$ satisfy the Schrödinger equation for a particle free to rotate in three dimensions, and find its energy and angular momentum in each case.

8C.7 Confirm that $Y_{3,+3}$ is normalized to 1. (The integration required is over the surface of a sphere.)

8C.8 Show that the function $f=\cos ax\cos by\cos cz$ is an eigenfunction of ∇^2, and determine its eigenvalue.

8C.9 Develop an expression (in Cartesian coordinates) for the quantum mechanical operators for the three components of angular momentum starting from the classical definition of angular momentum, $\boldsymbol{l}=\boldsymbol{r}\times\boldsymbol{p}$. Show that any two of the components do not mutually commute, and find their commutator.

8C.10 Starting from the operator $\hat{l}_z=\hat{x}\hat{p}_y-\hat{y}\hat{p}_x$, prove that in spherical polar coordinates $\hat{l}_z=-i\hbar\partial/\partial\phi$.

8C.11 Show that $[l^2,l_z]=0$, and then, without further calculation, justify the remark that $\hat{l}^2,\hat{l}_q=0$ for all $q=x$, y, and z.

8C.12 A particle confined to within a spherical cavity is a reasonable starting point for the discussion of the electronic properties of spherical metal nanoparticles (*Impact* I8.1). Here, you are invited to show in a series of steps that the $l=0$ energy levels of an electron in a spherical cavity of radius R are quantized and given by

$$E_n=\frac{n^2h^2}{8m_eR^2}$$

(a) The hamiltonian for a particle free to move inside a sphere of radius a is

$$\hat{H}=-\frac{\hbar^2}{2m}\nabla^2$$

Show that the Schrödinger equation is separable into radial and angular components. That is, begin by writing $\psi(r,\theta,\phi)=R(r)Y(\theta,\phi)$, where $R(r)$ depends only on the distance of the particle from the centre of the sphere, and $Y(\theta,\phi)$ is a spherical harmonic. Then show that the Schrödinger equation can be separated into two equations, one for $R(r)$, the radial equation, and the other for $Y(\theta,\phi)$, the angular equation. (b) Consider the case $l=0$. Show by differentiation that the solution of the radial equation has the form

$$R(r)=(2\pi a)^{-1/2}\frac{\sin(n\pi r/a)}{r}$$

(c) Now go on to show (by acknowledging the appropriate boundary conditions) that the allowed energies are given by $E_n=n^2h^2/8ma^2$. With substitution of m_e for m and of R for a, this is the equation given above for the energy.

Integrated activities

8.1 Describe the features that stem from nanometre-scale dimensions that are not found in macroscopic objects.

8.2 Explain why the particle in a box and the harmonic oscillator are useful models for quantum mechanical systems: what chemically significant systems can they be used to represent?

8.3 Suppose that 1.0 mol perfect gas molecules all occupy the lowest energy level of a cubic box. (a) How much work must be done to change the volume of the box by ΔV? (b) Would the work be different if the molecules all occupied a state $n\neq 1$? (c) What is the relevance of this discussion to the expression for the expansion work discussed in Topic 2A? (d) Can you identify a distinction between adiabatic and isothermal expansion?

8.4 Determine the values of $\Delta x=(\langle x^2\rangle-\langle x\rangle^2)^{1/2}$ and $\Delta p=(\langle p^2\rangle-\langle p\rangle^2)^{1/2}$ for the ground state of (a) a particle in a box of length L and (b) an harmonic oscillator. Discuss these quantities with reference to the uncertainty principle.

8.5 Repeat Problem 8.4 for (a) a particle in a box and (b) a harmonic oscillator in a general quantum state (n and v, respectively).

8.6 Use mathematical software, a spreadsheet, or the *Living graphs* on the web site of this book for the following exercises:
(a) Plot the probability density for a particle in a box with $n=1, 2, \ldots, 5$ and $n=50$. How do your plots illustrate the correspondence principle?
(b) Plot the transmission probability T against E/V for passage by (i) a hydrogen molecule, (ii) a proton, and (iii) an electron through a barrier of height V.
(c) To gain some insight into the origins of the nodes in the harmonic oscillator wavefunctions, plot the Hermite polynomials $H_v(y)$ for $v=0$ through 5.
(d) Use mathematical software to generate three-dimensional plots of the wavefunctions for a particle confined to a rectangular surface with (i) $n_1=1$, $n_2=1$, the state of lowest energy, (ii) $n_1=1$, $n_2=2$, (iii) $n_1=2$, $n_2=1$, and (iv) $n_1=2$, $n_2=2$. Deduce a rule for the number of nodal lines in a wavefunction as a function of the values of n_1 and n_2.

Mathematical background 4 Differential equations

A **differential equation** is a relation between a function and its derivatives, as in

$$a\frac{d^2f}{dx^2}+b\frac{df}{dx}+cf=0 \qquad \text{(MB4.1)}$$

where f is a function of the variable x and the factors a, b, c may be either constants or functions of x. If the unknown function depends on only one variable, as in this example, the equation is called an **ordinary differential equation**; if it depends on more than one variable, as in

$$a\frac{\partial^2 f}{\partial x^2}+b\frac{\partial^2 f}{\partial y^2}+cf=0 \qquad \text{(MB4.2)}$$

it is called a **partial differential equation**. Here, f is a function of x and y, and the factors a, b, c may be either constants or functions of both variables. Note the change in symbol from d to ∂ to signify a *partial derivative* (see *Mathematical background* 2).

MB4.1 The structure of differential equations

The **order** of the differential equation is the order of the highest derivative that occurs in it: both examples above are second-order equations. Only rarely in science is a differential equation of order higher than two encountered.

A **linear differential equation** is one for which if f is a solution then so is constant$\times f$. Both examples above are linear. If the 0 on the right were replaced by a different number or a function other than f, then they would cease to be linear.

Solving a differential equation means something different from solving an algebraic equation. In the latter case, the solution is a value of the variable x (as in the solution $x=2$ of the quadratic equation $x^2-4=0$). The solution of a differential equation is the entire function that satisfies the equation, as in

$$\frac{d^2f}{dx^2}+f=0, \quad f(x)=A\sin x+B\cos x \qquad \text{(MB4.3)}$$

with A and B constants. The process of finding a solution of a differential equation is called **integrating** the equation. The solution in eqn MB4.3 is an example of a **general solution** of a differential equation; that is, it is the most general solution of the equation and is expressed in terms of a number of constants (A and B in this case). When the constants are chosen to accord with certain specified **initial conditions** (if one variable is the time) or certain **boundary conditions** (to fulfil certain spatial restrictions on the solutions), we obtain the **particular solution** of the equation. The particular solution of a first-order differential equation requires one such condition; a second-order differential equation requires two.

Brief illustration MB4.1 Particular solutions

If we are informed that $f(0)=0$, then because from eqn MB4.3 it follows that $f(0)=B$, we can conclude that $B=0$. That still leaves A undetermined. If we are also told that $df/dx=2$ at $x=0$ (that is, $f'(0)=2$, where the prime denotes a first derivative), then because the general solution (but with $B=0$) implies that $f'(x)=A\cos x$, we know that $f'(0)=A$, and therefore $A=2$. The particular solution is therefore $f(x)=2\sin x$. Figure MB4.1 shows a series of particular solutions corresponding to different boundary conditions.

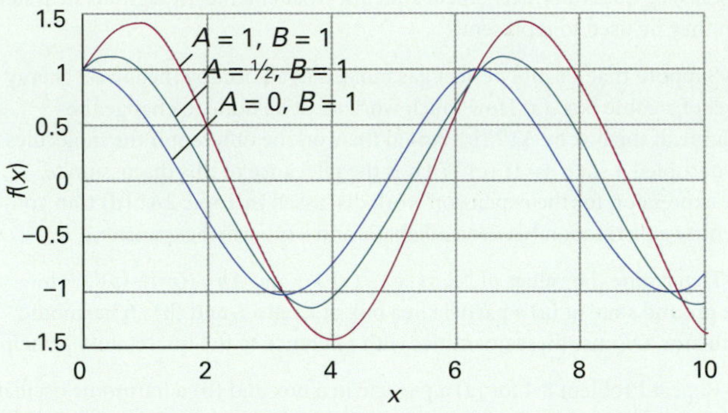

Figure MB4.1 The solution of the differential equation in *Brief illustration* MB4.1 with three different boundary conditions (as indicated by the resulting values of the constants A and B).

MB4.2 The solution of ordinary differential equations

The first-order linear differential equation

$$\frac{df}{dx}+af=0 \qquad \text{(MB4.4a)}$$

with a a function of x or a constant can be solved by direct integration. To proceed, we use the fact that the quantities df and dx (called *differentials*) can be treated algebraically like any quantity and rearrange the equation into

$$\frac{df}{f}=-a\,dx \qquad \text{(MB4.4b)}$$

and integrate both sides. For the left-hand side, we use the familiar result $\int dy/y=\ln y+\text{constant}$. After pooling all the constants into a single constant C, we obtain:

$$\ln f(x)=-\int a\,dx+C \qquad \text{(MB4.4c)}$$

Brief illustration MB4.2 The solution of a first-order equation

Suppose that in eqn MB4.4a the factor $a=2x$; then the general solution, eqn MB4.4c, is

$$\ln f(x)=-2\int x\,\mathrm{d}x+C=-x^2+C$$

(We have absorbed the constant of integration into the constant C.) Therefore

$$f(x)=N\mathrm{e}^{-x^2},\quad N=\mathrm{e}^C$$

If we are told that $f(0)=1$, then we can infer that $N=1$ and therefore that $f(x)=e^{-x^2}$.

Even the solutions of first-order differential equations quickly become more complicated. A nonlinear first-order equation of the form

$$\frac{\mathrm{d}f}{\mathrm{d}x}+af=b \qquad \text{(MB4.5a)}$$

with a and b functions of x (or constants) has a solution of the form

$$f(x)\mathrm{e}^{\int a\,\mathrm{d}x}=\int \mathrm{e}^{\int a\,\mathrm{d}x}b\,\mathrm{d}x+C \qquad \text{(MB4.5b)}$$

as may be verified by differentiation. Mathematical software packages can often perform the required integrations.

Second-order differential equations are in general much more difficult to solve than first-order equations. One powerful approach commonly used to lay siege to second-order differential equations is to express the solution as a power series:

$$f(x)=\sum_{n=0}^{\infty}c_n x^n \qquad \text{(MB4.6)}$$

and then to use the differential equation to find a relation between the coefficients. This approach results, for instance, in the Hermite polynomials that form part of the solution of the Schrödinger equation for the harmonic oscillator (Topic 8B). Many of the second-order differential equations that occur in this text are tabulated in compilations of solutions or can be solved with mathematical software, and the specialized techniques that are needed to establish the form of the solutions may be found in mathematical texts.

MB4.3 The solution of partial differential equations

The only partial differential equations that we need to solve are those that can be separated into two or more ordinary differential equations by the technique known as **separation of variables**. To discover if the differential equation in eqn MB4.2 can be solved by this method we suppose that the full solution can be factored into functions that depend only on x or only on y, and write $f(x,y)=X(x)Y(y)$. At this stage there is no guarantee that the solution can be written in this way. Substituting this trial solution into the equation and recognizing that

$$\frac{\partial^2 XY}{\partial x^2}=Y\frac{\mathrm{d}^2X}{\mathrm{d}x^2}\quad \frac{\partial^2 XY}{\partial y^2}=X\frac{\mathrm{d}^2Y}{\mathrm{d}y^2}$$

we obtain

$$aY\frac{\mathrm{d}^2X}{\mathrm{d}x^2}+bX\frac{\mathrm{d}^2Y}{\mathrm{d}y^2}+cXY=0$$

We are using d instead of ∂ at this stage to denote differentials because each of the functions X and Y depends on one variable, x and y, respectively. Division through by XY turns this equation into

$$\frac{a}{X}\frac{\mathrm{d}^2X}{\mathrm{d}x^2}+\frac{b}{Y}\frac{\mathrm{d}^2Y}{\mathrm{d}y^2}+c=0$$

Now suppose that a is a function only of x, b a function of y, and c a constant. (There are various other possibilities that permit the argument to continue.) Then the first term depends only on x and the second only on y. If x is varied, only the first term can change. But as the other two terms do not change and the sum of the three terms is a constant (0), even that first term must be a constant. The same is true of the second term. Therefore because each term is equal to a constant, we can write

$$\frac{a}{X}\frac{\mathrm{d}^2X}{\mathrm{d}x^2}=c_1\quad \frac{b}{Y}\frac{\mathrm{d}^2Y}{\mathrm{d}y^2}=c_2\quad \text{with}\; c_1+c_2=-c$$

We now have two ordinary differential equations to solve by the techniques described in Section MB4.2. An example of this procedure is given in Topic 8A, for a particle in a two-dimensional region.

CHAPTER 9

Atomic structure and spectra

In this chapter we see how to use quantum mechanics to describe and investigate the electronic structure of an atom, the arrangement of electrons around a nucleus. The concepts we meet are of central importance for understanding the structures and reactions of atoms and molecules, and hence have extensive chemical applications.

9A Hydrogenic atoms

In this Topic we use the principles of quantum mechanics introduced in Chapters 7 and 8 to describe the internal structures of atoms. We start with the simplest type of atom. A 'hydrogenic atom' is a one-electron atom or ion of general atomic number Z; examples are H, He^{+}, Li^{2+}, O^{7+}, and even U^{91+}. Hydrogenic atoms are important because their Schrödinger equations can be solved exactly. They also provide a set of concepts that are used to describe the structures of many-electron atoms and, as we see in the Topics of Chapter 10, the structures of molecules too. We see what experimental information is available from a study of the spectrum of atomic hydrogen. Then we set up the Schrödinger equation for an electron in an atom and separate it into angular and radial parts. The wavefunctions obtained are the hugely important 'atomic orbitals' of hydrogenic atoms.

9B Many-electron atoms

A 'many-electron atom' (or *polyelectronic atom*) is an atom or ion with more than one electron; examples include all neutral atoms other than H. So even He, with only two electrons, is a many-electron atom. In this Topic we use hydrogenic atomic orbitals to describe the structures of many-electron atoms. Then, in conjunction with the concept of spin and the Pauli exclusion principle, we account for the periodicity of atomic properties and the structure of the periodic table.

9C Atomic spectra

The spectra of many-electron atoms are more complicated than those of hydrogen, but the same principles apply. In this Topic we see how such spectra are described by using term symbols, and the origin of their finer details.

What is the impact of this material?

In *Impact* I9.1, we focus on the use of atomic spectroscopy to examine stars. By analysing their spectra we see that it is possible to determine the composition of their outer layers and the surrounding gases and to determine features of their physical state.

To read more about the impact of this material, scan the QR code, or go to bcs.whfreeman.com/webpub/chemistry/pchem10e/impact/pchem-9-1.html

9A Hydrogenic atoms

Contents

➤ Why do you need to know this material?

An understanding of the structure of the hydrogen atom is central to the understanding of all other atoms, the periodic table, and bonding. All accounts of the structures of molecules are based on the language and concepts it introduces.

➤ What is the key idea?

Atomic orbitals are labelled by three quantum numbers that specify the energy and angular momentum of an electron in a hydrogenic atom.

➤ What do you need to know already?

You need to be aware of the concept of wavefunction (Topic 7B) and its interpretation. You need to know how to set up a Schrödinger equation and how boundary conditions limit its solutions (Topic 8A).

When an electric discharge is passed through gaseous hydrogen, the H_2 molecules are dissociated and the energetically excited H atoms that are produced emit light of discrete frequencies, producing a spectrum of a series of 'lines' (Fig. 9A.1). The Swedish spectroscopist Johannes Rydberg noted (in 1890) that all the lines are described by the expression

$$\tilde{\nu} = \tilde{R}_H \left(\frac{1}{n_1^2} - \frac{1}{n_2^2} \right) \qquad \text{Spectral lines of a hydrogen atom} \qquad (9A.1)$$

with $n_1=1$ (the *Lyman series*), 2 (the *Balmer series*), and 3 (the *Paschen series*), and that in each case $n_2=n_1+1, n_1+2, \ldots$. The constant $\tilde{R}_H$ is now called the **Rydberg constant** for the hydrogen atom and is found empirically to have the value $109\,677\ \text{cm}^{-1}$.

As eqn 9A.1 suggests, each spectral line can be written as the difference of two **terms**, each of the form

$$T_n = \frac{\tilde{R}_H}{n^2} \qquad (9A.2)$$

The **Ritz combination principle** states that *the wavenumber of any spectral line (of any atom, not just hydrogenic atoms) is the difference between two terms.* We say that two terms T_1 and T_2 'combine' to produce a spectral line of wavenumber

$$\tilde{\nu} = T_1 - T_2 \qquad \text{Ritz combination principle} \qquad (9A.3)$$

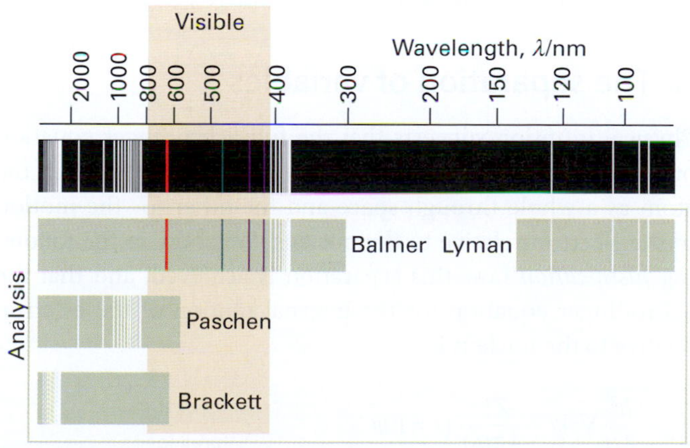

Figure 9A.1 The spectrum of atomic hydrogen. Both the observed spectrum and its resolution into overlapping series are shown. Note that the Balmer series lies in the visible region.

Thus, if each spectroscopic term represents an energy hcT, the difference in energy when the atom undergoes a transition between two terms is $\Delta E = hcT_1 - hcT_2$ and, according to the Bohr frequency condition ($\Delta E = h\nu$, Topic 7A), the frequency of the radiation emitted is given by $\nu = cT_1 - cT_2$. This expression rearranges into the Ritz formula when expressed in terms of wavenumbers (on division by c; $\tilde{\nu} = \nu/c$). The Ritz combination principle applies to all types of atoms and molecules, but only for hydrogenic atoms do the terms have the simple form (constant)/n^2.

Because spectroscopic observations show that electromagnetic radiation is absorbed and emitted by atoms only at certain wavenumbers, it follows that only certain energy states of atoms are permitted. Our tasks in this Topic are to determine the origin of this energy quantization, to find the permitted energy levels, and to account for the value of $\tilde{R}_H$. The spectra of more complex atoms are treated in Topic 9C.

9A.1 The structure of hydrogenic atoms

The Coulomb potential energy of an electron in a hydrogenic atom of atomic number Z and therefore nuclear charge Ze is

$$V(r) = -\frac{Ze^2}{4\pi\varepsilon_0 r} \qquad (9A.4)$$

where r is the distance of the electron from the nucleus and ε_0 is the vacuum permittivity. The hamiltonian for the electron and a nucleus of mass m_N is therefore

$$\hat{H} = \hat{E}_{k,\text{electron}} + \hat{E}_{k,\text{nucleus}} + \hat{V}(r) = -\frac{\hbar^2}{2m_e}\nabla_e^2 - \frac{\hbar^2}{2m_N}\nabla_N^2 - \frac{Ze^2}{4\pi\varepsilon_0 r} \qquad (9A.5)$$

Hamiltonian for a hydrogenic atom

The subscripts e and N on ∇^2 indicate differentiation with respect to the electron or nuclear coordinates, respectively.

(a) The separation of variables

Physical intuition suggests that the full Schrödinger equation ought to separate into two equations, one for the motion of the atom as a whole through space and the other for the motion of the electron relative to the nucleus. We show in the following *Justification* how this separation is achieved, and that the Schrödinger equation for the internal motion of the electron relative to the nucleus is

$$-\frac{\hbar^2}{2\mu}\nabla^2\psi - \frac{Ze^2}{4\pi\varepsilon_0 r}\psi = E\psi \qquad \frac{1}{\mu} = \frac{1}{m_e} + \frac{1}{m_N} \qquad (9A.6)$$

Schrödinger equation for a hydrogenic atom

where differentiation is now with respect to the coordinates of the electron relative to the nucleus. The quantity μ is called the **reduced mass**. The reduced mass is very similar to the electron mass because m_N, the mass of the nucleus, is much larger than the mass of an electron, so $1/\mu \approx 1/m_e$ and therefore $\mu \approx m_e$. In all except the most precise work, the reduced mass can be replaced by m_e.

Justification 9A.1 The separation of internal and external motion

Consider a one-dimensional system in which the potential energy depends only on the separation of the two particles. The total energy is

$$E = \frac{p_1^2}{2m_1} + \frac{p_2^2}{2m_2} + V(x_1 - x_2)$$

where $p_1 = m_1(\mathrm{d}x_1/\mathrm{d}t)$ and $p_2 = m_2(\mathrm{d}x_2/\mathrm{d}t)$. The centre of mass (Fig. 9A.2) is located at

$$X = \frac{m_1}{m}x_1 + \frac{m_2}{m}x_2 \qquad m = m_1 + m_2$$

and the separation of the particles is $x = x_1 - x_2$. It follows that

$$x_1 = X + \frac{m_2}{m}x \qquad x_2 = X - \frac{m_1}{m}x$$

The linear momenta of the particles can now be expressed in terms of the rates of change of x and X:

$$p_1 = m_1\frac{\mathrm{d}x_1}{\mathrm{d}t} = m_1\frac{\mathrm{d}X}{\mathrm{d}t} + \frac{m_1 m_2}{m}\frac{\mathrm{d}x}{\mathrm{d}t}$$

$$p_2 = m_2\frac{\mathrm{d}x_2}{\mathrm{d}t} = m_2\frac{\mathrm{d}X}{\mathrm{d}t} - \frac{m_1 m_2}{m}\frac{\mathrm{d}x}{\mathrm{d}t}$$

Then it follows that

$$\frac{p_1^2}{2m_1} + \frac{p_2^2}{2m_2} = \tfrac{1}{2}m\left(\frac{\mathrm{d}X}{\mathrm{d}t}\right)^2 + \tfrac{1}{2}\mu\left(\frac{\mathrm{d}x}{\mathrm{d}t}\right)^2$$

where μ is given in eqn 9A.6. By writing $P = m(\mathrm{d}X/\mathrm{d}t)$ for the linear momentum of the system as a whole and $p = \mu(\mathrm{d}x/\mathrm{d}t)$, we find

$$E = \frac{P^2}{2m} + \frac{p^2}{2\mu} + V(x)$$

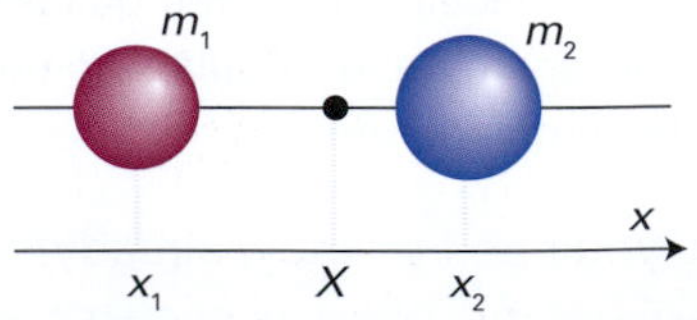

Figure 9A.2 The coordinates used for discussing the separation of the relative motion of two particles from the motion of the centre of mass.

The corresponding hamiltonian (generalized to three dimensions) is therefore

$$\hat{H}=-\frac{\hbar^2}{2m}\nabla^2_{\text{c.m.}}-\frac{\hbar^2}{2\mu}\nabla^2$$

where the first term differentiates with respect to the centre of mass coordinates and the second with respect to the relative coordinates.

Now we write the overall wavefunction as the product $\psi_{\text{total}}(X,x)=\psi_{\text{c.m.}}(X)\psi(x)$, where the first factor is a function of only the centre of mass coordinates and the second is a function of only the relative coordinates. The overall Schrödinger equation, $\hat{H}\psi_{\text{total}}=E_{\text{total}}\psi_{\text{total}}$, then separates by the argument that we have used in Topics 8A and 8C, with $E_{\text{total}}=E_{\text{c.m.}}+E$.

Because the potential energy is centrosymmetric (independent of angle), we can suspect that the equation for the wavefunction is separable into radial and angular components. Therefore, we write

$$\psi(r,\theta,\phi)=R(r)Y(\theta,\phi) \quad (9A.7)$$

and examine whether the Schrödinger equation can be separated into two equations, one for the **radial wavefunction** $R(r)$ and the other for the **angular wavefunction** $Y(\theta,\phi)$. As shown in the following *Justification*, the equation does separate, and the equations we have to solve are

$$\Lambda^2 Y=-l(l+1)Y \quad (9A.8a)$$

$$-\frac{\hbar^2}{2\mu}\frac{\mathrm{d}^2u}{\mathrm{d}r^2}+V_{\text{eff}}u=Eu \quad (9A.8b)$$

where $u(r)=rR(r)$ and

$$V_{\text{eff}}(r)=-\frac{Ze^2}{4\pi\varepsilon_0 r}+\frac{l(l+1)\hbar^2}{2\mu r^2} \quad (9A.8c)$$

Justification 9A.2 The separation of angular and radial motion

The laplacian in three dimensions is given in Table 7B.1. It follows that the Schrödinger equation in eqn 9A.6 is

$$-\frac{\hbar^2}{2\mu}\nabla^2 RY+VRY=-\frac{\hbar^2}{2\mu}\left(\frac{\partial^2}{\partial r^2}+\frac{2}{r}\frac{\partial}{\partial r}+\frac{1}{r^2}\Lambda^2\right)RY+VRY=ERY$$

Because R depends only on r and Y depends only on the angular coordinates, this equation becomes

$$-\frac{\hbar^2}{2\mu}\left(Y\frac{\mathrm{d}^2R}{\mathrm{d}r^2}+\frac{2Y}{r}\frac{\mathrm{d}R}{\mathrm{d}r}+\frac{R}{r^2}\Lambda^2 Y\right)+VRY=ERY$$

where the partial derivatives with respect to r have been replaced by complete derivatives because R depends only on r. If we multiply through by r^2/RY, we obtain

$$-\frac{\hbar^2}{2\mu R}\left(r^2\frac{\mathrm{d}^2R}{\mathrm{d}r^2}+2r\frac{\mathrm{d}R}{\mathrm{d}r}\right)+Vr^2-\overbrace{\frac{\hbar^2}{2\mu Y}\Lambda^2 Y}^{\text{Depends on }\theta,\phi}=Er^2$$

At this point we employ the usual argument. The term in Y is the only one that depends on the angular variables, so it must be a constant. When we write this constant as $\hbar^2 l(l+1)/2\mu$, eqn 9A.8c follows immediately.

Equation 9A.8a is the same as the Schrödinger equation for a particle free to move round a central point, and is considered in Topic 8C. The solutions are the spherical harmonics (Table 8C.1), and are specified by the quantum numbers l and m_l. We consider them in more detail shortly. Equation 9A.8b is called the **radial wave equation**. The radial wave equation is the description of the motion of a particle of mass μ in a one-dimensional region $0\le r<\infty$ where the potential energy is $V_{\text{eff}}(r)$.

(b) The radial solutions

We can anticipate some features of the shapes of the radial wavefunctions by analysing the form of V_{eff}. The first term in eqn 9A.8c is the Coulomb potential energy of the electron in the field of the nucleus. The second term stems from what in classical physics would be called the centrifugal force that arises from the angular momentum of the electron around the nucleus. When $l=0$, the electron has no angular momentum, and the effective potential energy is purely Coulombic and attractive at all radii (Fig. 9A.3). When $l\neq 0$, the centrifugal term gives a positive (repulsive) contribution to the effective potential energy. When the electron is close to the nucleus ($r\approx 0$), this repulsive term, which is proportional to $1/r^2$, dominates the attractive Coulombic component, which is proportional to $1/r$, and the net result is an effective repulsion of the electron from the nucleus. The two effective potential energies, the one for $l=0$ and the one for $l\neq 0$, are therefore qualitatively very different close to the nucleus. However, they are similar at large distances because the centrifugal contribution tends to zero more rapidly (as $1/r^2$) than the Coulombic contribution (as $1/r$). Therefore, we can expect the solutions with $l=0$ and $l\neq 0$ to be quite different near the nucleus but similar far away from it. There are two important features of the radial wavefunction:

- Close to the nucleus the radial wavefunction is proportional to r^l, and the higher the orbital angular momentum, the less likely it is that the electron will be found there (Fig. 9A.4).

Physical interpretation

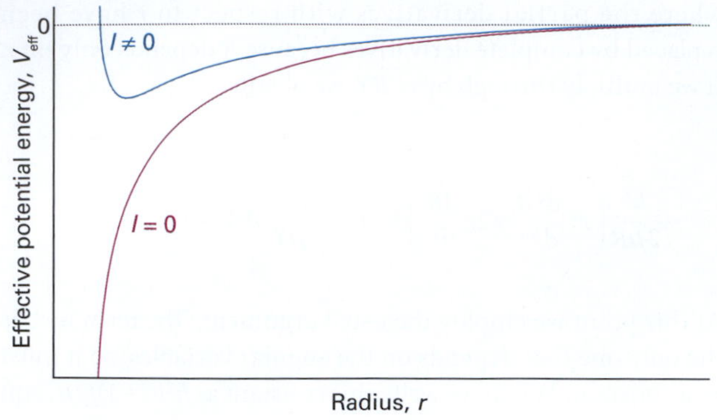

Figure 9A.3 The effective potential energy of an electron in the hydrogen atom. When the electron has zero orbital angular momentum, the effective potential energy is the Coulombic potential energy. When the electron has nonzero orbital angular momentum, the centrifugal effect gives rise to a positive contribution which is very large close to the nucleus. The $l=0$ and $l\neq 0$ wavefunctions are therefore very different near the nucleus.

- Far from the nucleus all radial wavefunctions approach zero exponentially.

We shall not go through the technical steps of solving the radial equation for the full range of radii and seeing how the form r^l close to the nucleus blends into the exponentially decaying form at great distances. For our purposes it is sufficient to know that the two limits can be bridged only for integral values of a quantum number n, and that the allowed energies corresponding to the allowed solutions are

$$E_n = -\frac{\mu e^4}{32\pi^2\varepsilon_0^2\hbar^2}\times\frac{Z^2}{n^2} \quad \text{Bound state energies} \quad (9A.9)$$

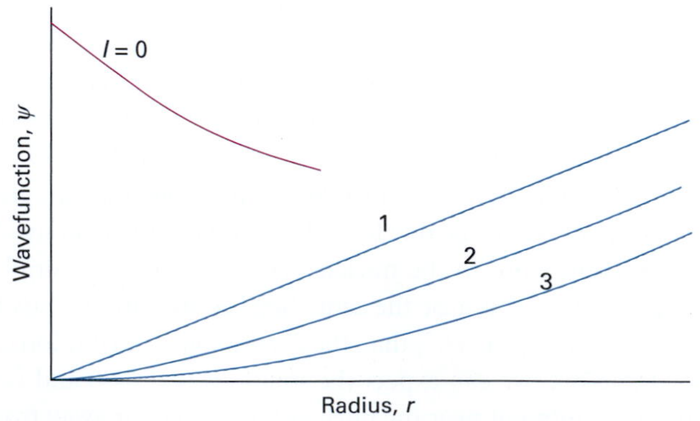

Figure 9A.4 Close to the nucleus, orbitals with $l=1$ are proportional to r, orbitals with $l=2$ are proportional to r^2, and orbitals with $l=3$ are proportional to r^3. Electrons are progressively excluded from the neighbourhood of the nucleus as l increases. An orbital with $l=0$ has a finite, nonzero value at the nucleus.

with $n=1, 2, \ldots$. Likewise, the radial wavefunctions depend on the values of both n and l (but not on m_l because only l appears in the radial wave equation), and all of them have the form

$$R(r) = \overbrace{r^l}^{\text{Dominant close to the nucleus}} \times \overbrace{(\text{polynomial in } r)}^{\text{Bridges the two ends of the function}} \times \overbrace{(\text{decaying exponential in } r)}^{\text{Dominant far from the nucleus}} \quad (9A.10)$$

and therefore look like

$$R(r) = r^l L(r)\mathrm{e}^{-r}$$

with various constants and where $L(r)$ is the bridging polynomial. The specific forms of the functions are most simply written in terms of the dimensionless quantity ρ (rho), where

$$\rho = \frac{2Zr}{na} \qquad a = \frac{m_\mathrm{e}}{\mu}a_0 = \frac{m_\mathrm{e}+m_\mathrm{N}}{m_\mathrm{N}}a_0 \qquad a_0 = \frac{4\pi\varepsilon_0\hbar^2}{m_\mathrm{e}e^2} \quad (9A.11)$$

The **Bohr radius**, a_0, has the value 52.9 pm; it is so called because the same quantity appeared in Bohr's early model of the hydrogen atom as the radius of the electron orbit of lowest energy. In practice, because $m_\mathrm{e} \ll m_\mathrm{N}$ there is so little difference between a and a_0 that it is safe to use a_0 in the definition of ρ for all atoms (even for ^{1}H, $a = 1.0005a_0$). Specifically, the radial wavefunctions for an electron with quantum numbers n and l are the (real) functions

$$R_{n,l}(r) = N_{n,l}\rho^l L_{n-l-1}^{2l+1}(\rho)\mathrm{e}^{-\rho/2} \quad \text{Radial wavefunctions} \quad (9A.12)$$

where $L(\rho)$ is an *associated Laguerre polynomial*. The notation might look fearsome, but the polynomials have quite simple forms, such as 1, ρ, and $2-\rho$ (they can be picked out in Table 9A.1). The factor N ensures that the radial wavefunction is normalized to 1 in the sense that

$$\int_0^\infty R_{n,l}(r)^2 r^2 \mathrm{d}r = 1 \quad (9A.13)$$

The r^2 comes from the volume element in spherical polar coordinates (*The chemist's toolkit* 7B.1). Specifically, we can interpret the components of eqn 9A.12 as follows:

Physical interpretation

- The exponential factor ensures that the wavefunction approaches zero far from the nucleus.
- The factor ρ^l ensures that (provided $l>0$) the wavefunction vanishes at the nucleus. The zero at $r=0$ is not a radial node because the radial wavefunction does not pass through zero at that point (because r cannot be negative). Nodes passing through the nucleus are all angular nodes.
- The associated Laguerre polynomial is a function that in general oscillates from positive to negative values and accounts for the presence of radial nodes.

Table 9A.1 Hydrogenic radial wavefunctions, $R_{n,l}(r)$

n	l	$R_{n,l}(r)$
1	0	$2\left(\frac{Z}{a}\right)^{3/2} e^{-\rho/2}$
2	0	$\frac{1}{8^{1/2}}\left(\frac{Z}{a}\right)^{3/2}(2-\rho)e^{-\rho/2}$
2	1	$\frac{1}{24^{1/2}}\left(\frac{Z}{a}\right)^{3/2}\rho e^{-\rho/2}$
3	0	$\frac{1}{243^{1/2}}\left(\frac{Z}{a}\right)^{3/2}(6-6\rho+\rho^2)e^{-\rho/2}$
3	1	$\frac{1}{486^{1/2}}\left(\frac{Z}{a}\right)^{3/2}(4-\rho)\rho e^{-\rho/2}$
3	2	$\frac{1}{2430^{1/2}}\left(\frac{Z}{a}\right)^{3/2}\rho^2 e^{-\rho/2}$

$\rho=(2Z/na)r$ with $a=4\pi\varepsilon_0\hbar^2/\mu e^2$. For an infinitely heavy nucleus (or one that may be assumed to be), $\mu=m_e$ and $a=a_0$, the Bohr radius.

Expressions for some radial wavefunctions are given in Table 9A.1 and illustrated in Fig. 9A.5.

Brief illustration 9A.1 Probability densities

To calculate the probability density at the nucleus for an electron with $n=1$, $l=0$, and $m_l=0$, we evaluate ψ at $r=0$:

$$\psi_{1,0,0}(0,\theta,\phi)=R_{1,0}(0)Y_{0,0}(\theta,\phi)=2\left(\frac{Z}{a_0}\right)^{3/2}\left(\frac{1}{4\pi}\right)^{1/2}$$

The probability density is therefore

$$\psi_{1,0,0}(0,\theta,\phi)^2=\frac{Z^3}{\pi a_0^3}$$

which evaluates to 2.15×10^{-6} pm^{-3} when $Z=1$.

Self-test 9A.1 Evaluate the probability density at the nucleus of the electron for an electron with $n=2$, $l=0$, $m_l=0$.

Answer: $(Z/a_0)^3/8\pi$

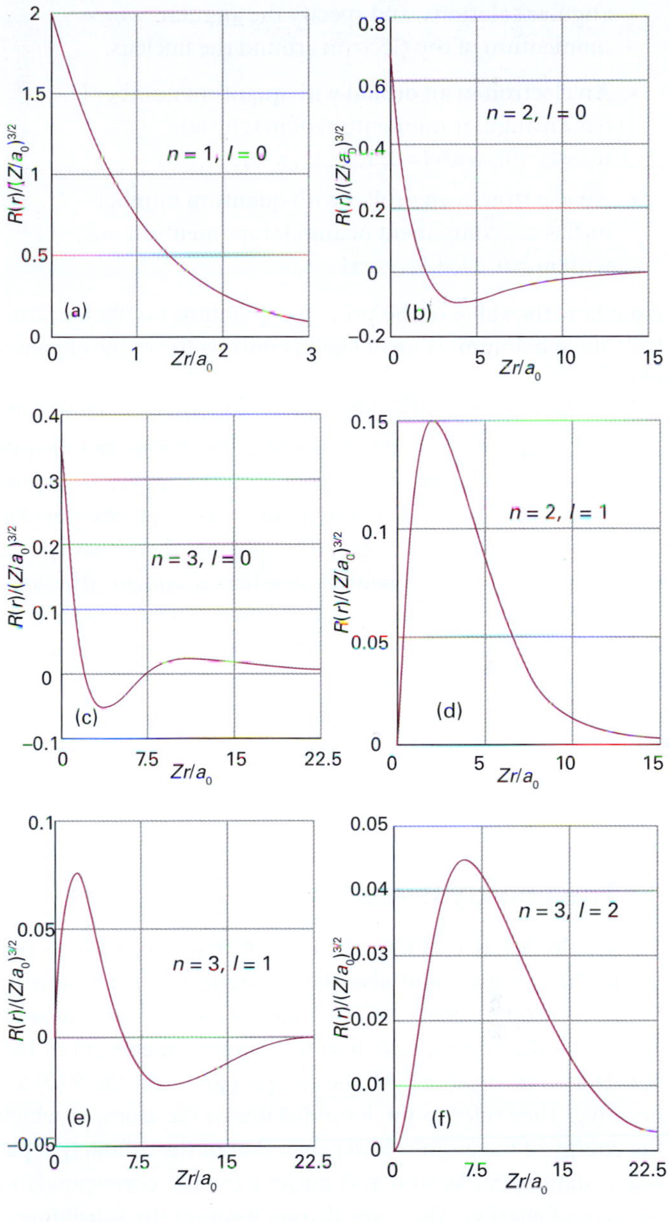

Figure 9A.5 The radial wavefunctions of the first few states of hydrogenic atoms of atomic number Z. Note that the orbitals with $l=0$ have a nonzero and finite value at the nucleus. The horizontal scales are different in each case: orbitals with high principal quantum numbers are relatively distant from the nucleus.

9A.2 Atomic orbitals and their energies

An **atomic orbital** is a one-electron wavefunction for an electron in an atom. Each hydrogenic atomic orbital is defined by three quantum numbers, designated n, l, and m_l. When an electron is described by one of these wavefunctions, we say that it 'occupies' that orbital. We could go on to say that the electron is in the state $|n,l,m_l\rangle$. For instance, an electron described by the wavefunction $\psi_{1,0,0}$ and in the state $|1,0,0\rangle$ is said to 'occupy' the orbital with $n=0$, $l=0$, and $m_l=0$.

(a) The specification of orbitals

The quantum number n is called the **principal quantum number**; it can take the value $n=1, 2, 3, \ldots$ and determines the energy of the electron:

- An electron in an orbital with quantum number n has an energy given by eqn 9A.9. The two other quantum numbers, l and m_l, come from the

Physical interpretation

angular solutions, and specify the angular momentum of the electron around the nucleus.

- An electron in an orbital with quantum number l has an angular momentum of magnitude $\{l(l+1)\}^{1/2}\hbar$, with $l=0, 1, 2, \ldots, n-1$.
- An electron in an orbital with quantum number m_l has a z-component of angular momentum $m_l\hbar$, with $m_l=0, \pm1, \pm2, \ldots, \pm l$.

Physical interpretation

Note how the value of the principal quantum number, n, controls the maximum value of l and l controls the range of values of m_l.

To define the state of an electron in a hydrogenic atom fully we need to specify not only the orbital it occupies but also its spin state. In Topic 8C it is mentioned that an electron possesses an intrinsic angular momentum, its 'spin'. We develop this property further in Topic 9B and show there that spin is described by the two quantum numbers s and m_s (the analogues of l and m_l). The value of s is fixed at $\frac{1}{2}$ for an electron, so we do not need to consider it further at this stage. However, m_s may be either $+\frac{1}{2}$ or $-\frac{1}{2}$, and to specify the state of an electron in a hydrogenic atom we need to specify which of these values describes it. It follows that, to specify the state of an electron in a hydrogenic atom, we need to give the values of four quantum numbers, namely n, l, m_l, and m_s.

(b) The energy levels

The energy levels predicted by eqn 9A.9 are depicted in Fig. 9A.6. The energies, and also the separation of neighbouring levels, are proportional to Z^2, so the levels are four times as wide apart (and the ground state four times lower in energy) in He^+ ($Z=2$) than in H ($Z=1$). All the energies given by eqn 9A.9 are negative. They refer to the **bound states** of the atom, in which the energy of the atom is lower than that of the infinitely separated, stationary electron and nucleus (which corresponds to the zero of energy). There are also solutions of the Schrödinger equation with positive energies. These solutions correspond to **unbound states** of the electron, the states to which an electron is raised when it is ejected from the atom by a high-energy collision or photon. The energies of the unbound electron are not quantized and form the continuum states of the atom.

Equation 9A.9, which we can write as

$$E_n = -\frac{hcZ^2\tilde{R}_N}{n^2} \qquad \tilde{R}_N = \frac{\mu e^4}{32\pi^2\varepsilon_0^2\hbar^2} \quad \text{Bound state energies} \quad (9A.14)$$

is consistent with the spectroscopic result summarized by eqn 9A.1, and we can identify the Rydberg constant for the atom as

$$\tilde{R}_N = \frac{\mu}{m_e}\times\tilde{R}_\infty \qquad \tilde{R}_\infty = \frac{m_e e^4}{8\varepsilon_0^2 h^3 c} \quad \text{Rydberg constant} \quad (9A.15)$$

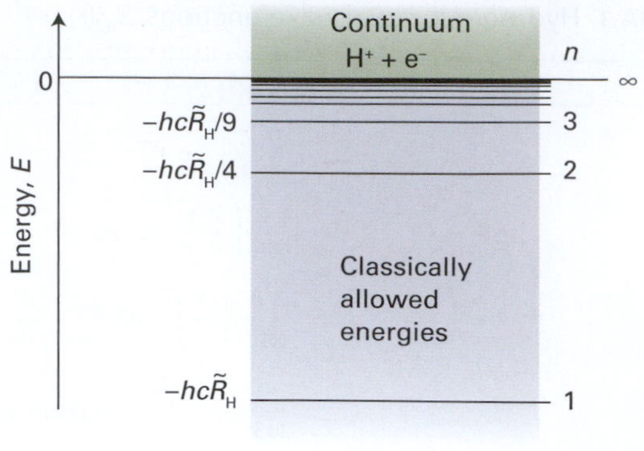

Figure 9A.6 The energy levels of a hydrogen atom. The values are relative to an infinitely separated, stationary electron and a proton.

where μ is the reduced mass of the atom and $\tilde{R}_\infty$ is the **Rydberg constant**. Insertion of the values of the fundamental constants into the expression for $\tilde{R}_H$ gives very close agreement with the experimental value for hydrogen. The only discrepancies arise from the neglect of relativistic corrections (in simple terms, the increase of mass with speed), which the non-relativistic Schrödinger equation ignores.

Brief illustration 9A.2 The energy levels

The value of $\tilde{R}_\infty$ is given inside the front cover and is 109 737 cm^{-1}. The reduced mass of a hydrogen atom with $m_p=1.672\,62\times10^{-27}$ kg and $m_e=9.109\,38\times10^{-31}$ kg is

$$\mu=\frac{m_e m_p}{m_e+m_p}=\frac{(9.109\,38\times10^{-31}\,\text{kg})\times(1.672\,62\times10^{-27}\,\text{kg})}{(9.109\,38\times10^{-31}\,\text{kg})+(1.672\,62\times10^{-27}\,\text{kg})}$$
$$=9.104\,42\times10^{-31}\,\text{kg}$$

It then follows that

$$\tilde{R}_H=\frac{9.104\,42\times10^{-31}\,\text{kg}}{9.109\,38\times10^{-31}\,\text{kg}}\times109\,737\,\text{cm}^{-1}=109\,677\,\text{cm}^{-1}$$

and that the ground state of the electron ($n=1$) lies at

$$E=-hc\tilde{R}_H=-(6.626\,08\times10^{-34}\,\text{J s})\times(2.997\,945\times10^{10}\,\text{cm s}^{-1})$$
$$\times(109\,677\,\text{cm}^{-1})=-2.178\,69\times10^{-18}\,\text{J}\ \ (-2.178\,69\,\text{aJ})$$

This energy corresponds to −13.598 eV.

Self-test 9A.2 What is the corresponding value for a deuterium atom? Take $m_D=2.013\,55m_u$.

Answer: −13.602 eV

(c) Ionization energies

The **ionization energy**, I, of an element is the minimum energy required to remove an electron from the ground state, the state of lowest energy, of one of its atoms in the gas phase. Because the ground state of hydrogen is the state with $n=1$, with energy $E_1 = -hc\tilde{R}_H$ and the atom is ionized when the electron has been excited to the level corresponding to $n=\infty$ (see Fig. 9A.6), the energy that must be supplied is

$$I = hc\tilde{R}_H \qquad (9A.16)$$

The value of I is 2.179 aJ (1 aJ $=10^{-18}$ J), which corresponds to 13.60 eV.

A note on good practice Ionization energies are sometimes referred to as *ionization potentials*. That is incorrect, but not uncommon. If the term is used at all, it should denote the potential difference through which an electron must be moved for its potential energy to change by an amount equal to the ionization energy, and reported in volts.

Example 9A.1 Measuring an ionization energy spectroscopically

The emission spectrum of atomic hydrogen shows lines at 82 259, 97 492, 102 824, 105 292, 106 632, and 107 440 cm^{-1}, which correspond to transitions to the same lower state. Determine (a) the ionization energy of the lower state, (b) the value of the Rydberg constant for hydrogen.

Method The spectroscopic determination of ionization energies depends on the determination of the series limit, the wavenumber at which the series terminates and becomes a continuum. If the upper state lies at an energy $-hc\tilde{R}_H/n^2$, then, when the atom makes a transition to $E_{lower} = -I$ a photon of wavenumber

$$\tilde{\nu} = -\frac{\tilde{R}_H}{n^2} - \frac{E_{lower}}{hc} = -\frac{\tilde{R}_H}{n^2} + \frac{I}{hc}$$

A plot of the wavenumbers against $1/n^2$ should give a straight line of slope $-\tilde{R}_H$ and intercept I/hc. Use a computer to make a least-squares fit of the data in order to obtain a result that reflects the precision of the data.

Answer The wavenumbers are plotted against $1/n^2$ in Fig. 9A.7. (a) The (least-squares) intercept lies at 109 679 cm^{-1}, so (b) the ionization energy is

$$I = hc\tilde{R}_H = (6.626\,08\times10^{-34}\,\text{J s})\times(2.997\,945\times10^{10}\,\text{cm s}^{-1})$$
$$\times 109\,679\,\text{cm}^{-1} = 2.1787\times10^{-18}\,\text{J}$$

or 2.1787 aJ, corresponding to 1312.1 kJ mol^{-1} (the negative of the value of E calculated in *Brief illustration* 9A.2).

Self-test 9A.3 The emission spectrum of atomic deuterium shows lines at 15 238, 20 571, 23 039, and 24 380 cm^{-1}, which correspond to transitions to the same lower state. Determine (a) the ionization energy of the lower state, (b) the ionization energy of the ground state, (c) the mass of the deuteron (by expressing the Rydberg constant in terms of the reduced mass of the electron and the deuteron, and solving for the mass of the deuteron).

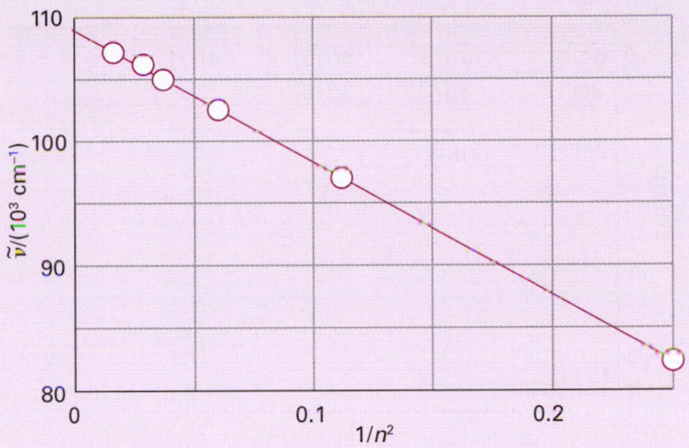

Figure 9A.7 The plot of the data in *Example* 9A.1 used to determine the ionization energy of an atom (in this case, of H).

Answer: (a) 328.1 kJ mol^{-1}, (b) 1312.4 kJ mol^{-1}, (c) 2.8×10^{-27} kg, a result very sensitive to $\tilde{R}_D$

(d) Shells and subshells

All the orbitals of a given value of n are said to form a single **shell** of the atom. In a hydrogenic atom (and only in a hydrogenic atom), all orbitals of given n, and therefore belonging to the same shell, have the same energy. It is common to refer to successive shells by letters:

$n=$	1	2	3	4…
	K	L	M	N…

Specification of shells

Thus, all the orbitals of the shell with $n=2$ form the L shell of the atom, and so on.

The orbitals with the same value of n but different values of l are said to form a **subshell** of a given shell. These subshells are generally referred to by letters:

$l=$	0	1	2	3	4	5	6…
	s	p	d	f	g	h	i…

Specification of subshells

All orbitals of the same subshell have the same energy in both hydrogenic and many-electron atoms. The letters then run alphabetically (j is not used because in some languages i and j are not distinguished). Figure 9A.8 is a version of Fig. 9A.6 which shows the subshells explicitly. Because l can range from 0 to $n-1$, giving n values in all, it follows that there are n subshells of a shell with principal quantum number n. The organization of orbitals in the shells is summarized in Fig. 9A.9. In general, the number of orbitals in a shell of principal quantum number n is n^2, so in a hydrogenic atom each energy level is n^2-fold degenerate.

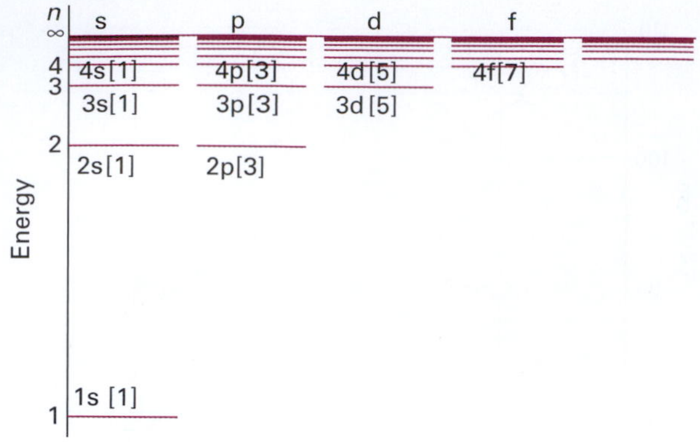

Figure 9A.8 The energy levels of a hydrogenic atom showing the subshells and (in square brackets) the numbers of orbitals in each subshell. All orbitals of a given shell have the same energy.

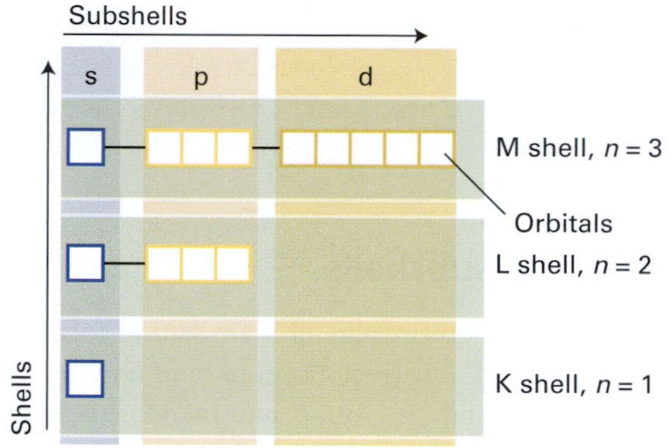

Figure 9A.9 The organization of orbitals (white squares) into subshells (characterized by l) and shells (characterized by n).

Brief illustration 9A.3 Shells, subshells, and orbitals

When $n=1$ there is only one subshell, that with $l=0$, and that subshell contains only one orbital, with $m_l=0$ (the only value of m_l permitted). When $n=2$, there are four orbitals, one in the s subshell with $l=0$ and $m_l=0$, and three in the $l=1$ subshell with $m_l=+1, 0, -1$. When $n=3$ there are nine orbitals (one with $l=0$, three with $l=1$, and five with $l=2$).

Self-test 9A.4 What subshells and orbitals are available in the N shell?

Answer: s (1), p (3), d (5), f (7)

(e) s Orbitals

The orbital occupied in the ground state is the one with $n=1$ (and therefore with $l=0$ and $m_l=0$, the only possible values of these quantum numbers when $n=1$). From Table 9A.1 and $Y_{0,0}=1/2\pi^{1/2}$ we can write (for $Z=1$):

$$\psi = \frac{1}{(\pi a_0^3)^{1/2}} e^{-r/a_0} \quad (9A.17)$$

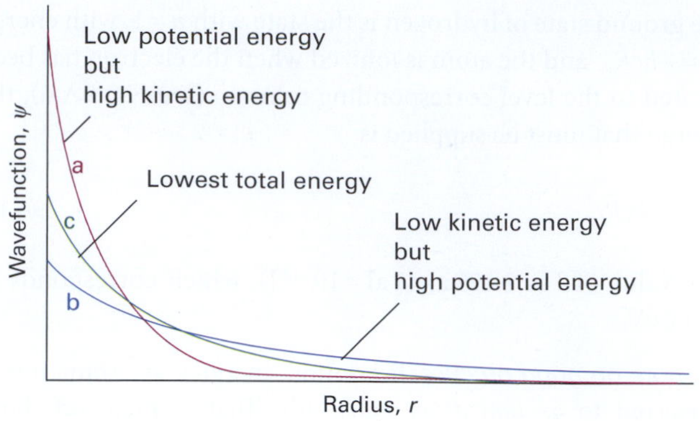

Figure 9A.10 The balance of kinetic and potential energies that accounts for the structure of the ground state of hydrogenic atoms. (a) The sharply curved but localized orbital has high mean kinetic energy, but low mean potential energy; (b) the mean kinetic energy is low, but the potential energy is not very favourable; (c) the compromise of moderate kinetic energy and moderately favourable potential energy.

This wavefunction is independent of angle and has the same value at all points of constant radius; that is, the 1s orbital is 'spherically symmetrical'. The wavefunction decays exponentially from a maximum value of $1/(\pi a_0^3)^{1/2}$ at the nucleus (at $r=0$). It follows that the probability density of the electron is greatest at the nucleus itself, where it has the value $1/\pi a_0^3 = 2.15\times10^{-6}\,\text{pm}^{-3}$.

We can understand the general form of the ground-state wavefunction by considering the contributions of the potential and kinetic energies to the total energy of the atom. The closer the electron is to the nucleus on average, the lower its average potential energy. This dependence suggests that the lowest potential energy should be obtained with a sharply peaked wavefunction that has a large amplitude at the nucleus and is zero everywhere else (Fig. 9A.10). However, this shape implies a high kinetic energy, because such a wavefunction has a very high average curvature. The electron would have very low kinetic energy if its wavefunction had only a very low average curvature. However, such a wavefunction spreads to great distances from the nucleus and the average potential energy of the electron is correspondingly high. The actual ground-state wavefunction is a compromise between these two extremes: the wavefunction spreads away from the nucleus (so the expectation value of the potential energy is not as low as in the first example, but nor is it very high) and has a reasonably low average curvature (so the expectation of the kinetic energy is not very low, but nor is it as high as in the first example).

One way of depicting the probability density of the electron is to represent $|\psi|^2$ by the density of shading (Fig. 9A.11). A simpler procedure is to show only the **boundary surface**, the surface that captures a high proportion (typically about 90 per cent) of the electron probability. For the 1s orbital, the boundary surface is a sphere centred on the nucleus (Fig. 9A.12).

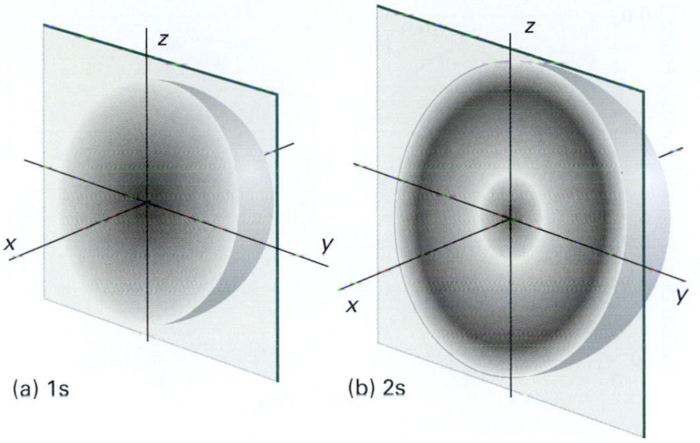

Figure 9A.11 Representations of cross-sections through the (a) 1s and (b) 2s hydrogenic atomic orbitals in terms of their electron probability densities (as represented by the density of shading).

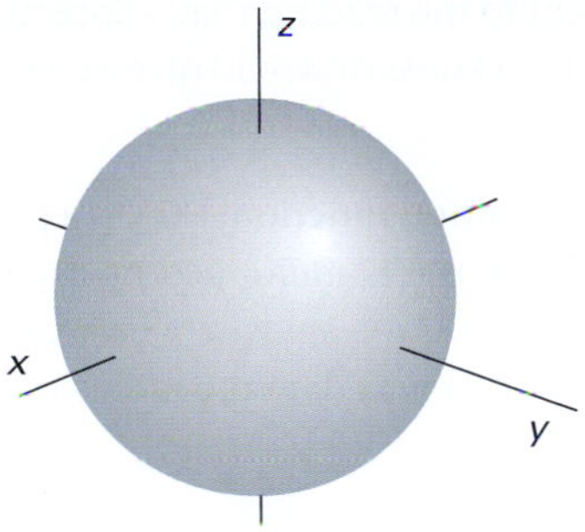

Figure 9A.12 The boundary surface of a 1s orbital, within which there is a 90 per cent probability of finding the electron. All s orbitals have spherical boundary surfaces.

Example 9A.2 Calculating the mean radius of an orbital

Use hydrogenic orbitals to calculate the mean radius of a 1s orbital.

Method The mean radius is the expectation value

$$\langle r\rangle=\int\psi^{*}r\psi\,\mathrm{d}\tau=\int r|\psi|^{2}\,\mathrm{d}\tau$$

We therefore need to evaluate the integral using the wavefunctions given in Table 9A.1 and $\mathrm{d}\tau=r^2\mathrm{d}r\sin\theta\,\mathrm{d}\theta\,\mathrm{d}\phi$. The angular parts of the wavefunction (Table 8C.1) are normalized in the sense that

$$\int_0^{\pi}\int_0^{2\pi}\left|Y_{l,m_l}\right|^2\sin\theta\,\mathrm{d}\theta\,\mathrm{d}\phi=1$$

The integral over r required is given in the *Resource section*.

Answer With the wavefunction written in the form $\psi=RY$, the integration (with the integral over the angular variables, which is equal to 1, in blue) is

$$\langle r\rangle=\int_0^{\infty}\int_0^{\pi}\int_0^{2\pi}rR_{n,l}^2\left|Y_{l,m_l}\right|^2r^2\mathrm{d}r\,\sin\theta\,\mathrm{d}\theta\,\mathrm{d}\phi=\int_0^{\infty}r^3R_{n,l}^2\mathrm{d}r$$

For a 1s orbital

$$R_{1,0}=2\left(\frac{Z}{a_0}\right)^{3/2}\mathrm{e}^{-Zr/a_0}$$

Hence

$$\langle r\rangle=\frac{4Z^3}{a_0^3}\int_0^{\infty}r^3\mathrm{e}^{-2Zr/a_0}\mathrm{d}r\overset{\text{Integral E.1}}{=}\frac{4Z^3}{a_0^3}\times\frac{3!}{(2Z/a_0)^4}=\frac{3a_0}{2Z}$$

Self-test 9A.5 Evaluate the mean radius of a 3s orbital by integration.

Answer: $27a_0/2Z$

All s orbitals are spherically symmetric, but differ in the number of radial nodes. For example, the 1s, 2s, and 3s orbitals have 0, 1, and 2 radial nodes, respectively. In general, an ns orbital has $n-1$ radial nodes. As n increases, the radius of the spherical boundary surface that captures a given fraction of the probability also increases.

Brief illustration 9A.4 The location of radial nodes

The radial nodes of a 2s orbital lie at the locations where the Legendre polynomial factor (Table 9A.1) is equal to zero. In this case the factor is simply $\rho-2$ so there is a node at $\rho=2$. For a 2s orbital, $\rho=Zr/a_0$, so the radial node occurs at $r=2a_0/Z$ (see Fig. 9A.5).

Self-test 9A.6 Locate the two nodes of a 3s orbital.

Answer: $1.90a_0/Z$ and $7.10a_0/Z$

(f) Radial distribution functions

The wavefunction tells us, through the value of $|\psi|^2$, the probability of finding an electron in any region. As we have stressed, $|\psi|^2$ is a probability *density* (dimensions: 1/volume) and can be interpreted as a (dimensionless) probability when multiplied by the (infinitesimal) volume of interest. Thus, we can imagine a probe with a fixed volume $\mathrm{d}\tau$ and sensitive to electrons, and which we can move around near the nucleus of a hydrogen atom. Because the probability density in the ground state of the atom is proportional to e^{-2Zr/a_0}, the reading from the detector decreases exponentially as the probe is moved out along any radius but is constant if the probe is moved on a circle of constant radius (Fig. 9A.13).

Now consider the total probability of finding the electron *anywhere* between the two walls of a spherical shell of thickness

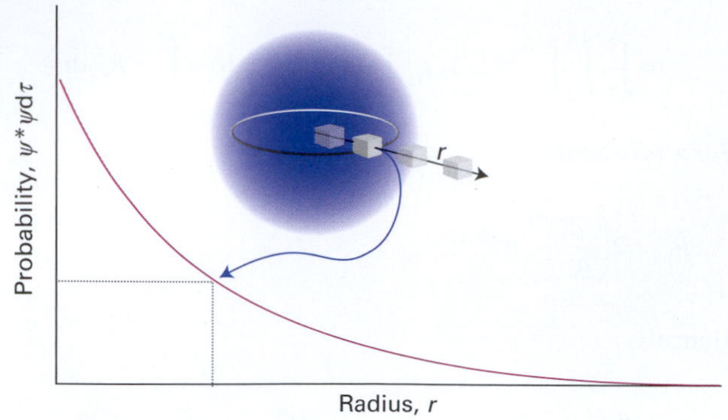

Figure 9A.13 A constant-volume electron-sensitive detector (the small cube) gives its greatest reading at the nucleus, and a smaller reading elsewhere. The same reading is obtained anywhere on a circle of given radius: the s orbital is spherically symmetrical.

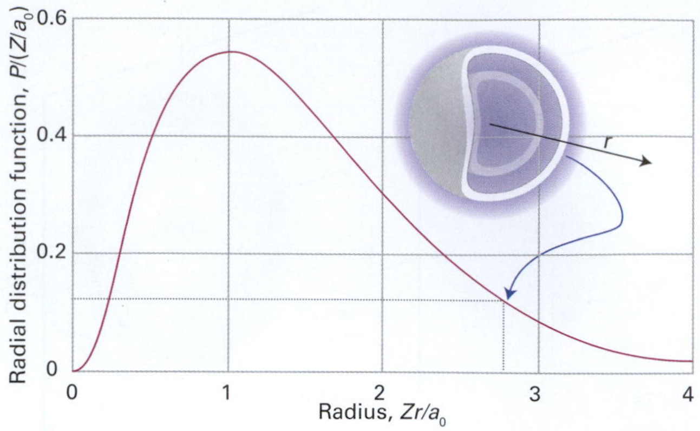

Figure 9A.14 The radial distribution function $P(r)$ is the probability density that the electron will be found anywhere in a shell of radius r; the probability itself is $P(r)\mathrm{d}r$, where $\mathrm{d}r$ is the thickness of the shell. For a 1s electron in hydrogen, $P(r)$ is a maximum when r is equal to the Bohr radius a_0. The value of $P(r)\mathrm{d}r$ is equivalent to the reading that a detector shaped like a spherical shell of thickness $\mathrm{d}r$ would give as its radius is varied.

$\mathrm{d}r$ at a radius r. The sensitive volume of the probe is now the volume of the shell (Fig. 9A.14), which is $4\pi r^2\mathrm{d}r$ (the product of its surface area, $4\pi r^2$, and its thickness, $\mathrm{d}r$). Note that the volume probed increases with distance from the nucleus and is zero at the nucleus itself, when $r=0$. The probability that the electron will be found between the inner and outer surfaces of this shell is the probability density at the radius r multiplied by the volume of the probe, or $|\psi|^2\times 4\pi r^2\mathrm{d}r$. This expression has the form $P(r)\mathrm{d}r$, where

$$P(r)=4\pi r^2|\psi|^2 \qquad (9A.18a)$$

The more general expression, which also applies to orbitals that are not spherically symmetrical is derived in the following *Justification*, and is

$$P(r)=r^2R(r)^2 \qquad \text{Radial distribution function} \qquad (9A.18b)$$

where $R(r)$ is the radial wavefunction for the orbital in question.

Justification 9A.3 The general form of the radial distribution function

The probability of finding an electron in a volume element $\mathrm{d}\tau$ when its wavefunction is $\psi=RY$ is $|RY|^2\mathrm{d}\tau$ with $\mathrm{d}\tau=r^2\mathrm{d}r\sin\theta\,\mathrm{d}\theta\,\mathrm{d}\phi$. The total probability of finding the electron at any angle at a constant radius is the integral of this probability over the surface of a sphere of radius r, and is written $P(r)\mathrm{d}r$; so

$$P(r)\mathrm{d}r=\int_0^{\pi}\int_0^{2\pi}R(r)^2\left|Y_{l,m_l}\right|^2 r^2\mathrm{d}r\sin\theta\,\mathrm{d}\theta\mathrm{d}\phi=r^2R(r)^2$$

The last equality follows from the fact that the spherical harmonics are normalized to 1 (the blue integration, as in *Example* 9A.1).

The **radial distribution function**, $P(r)$, is a probability density in the sense that, when it is multiplied by $\mathrm{d}r$, it gives the probability of finding the electron anywhere between the two walls of a spherical shell of thickness $\mathrm{d}r$ at the radius r. For a 1 s orbital,

$$P(r)=\frac{4Z^3}{a_0^3}r^2\mathrm{e}^{-2Zr/a_0} \qquad (9A.19)$$

Let's interpret this expression:

Physical interpretation

- Because $r^2=0$ at the nucleus, $P(0)=0$. The volume of the shell of inspection is zero when $r=0$.
- As $r\to\infty$, $P(r)\to 0$ on account of the exponential term. The wavefunction has fallen to zero at great distances from the nucleus.
- The increase in r^2 and the decrease in the exponential factor means that P passes through a maximum at an intermediate radius (see Fig. 9A.14).

The maximum of $P(r)$, which can be found by differentiation, marks the most probable radius at which the electron will be found, and for a 1s orbital in hydrogen occurs at $r=a_0$, the Bohr radius. When we carry through the same calculation for the radial distribution function of the 2s orbital in hydrogen, we find that the most probable radius is $5.2a_0=275$ pm. This larger value reflects the expansion of the atom as its energy increases.

Example 9A.3 Calculating the most probable radius

Calculate the most probable radius, r^*, at which an electron will be found when it occupies a 1s orbital of a hydrogenic atom of atomic number Z, and tabulate the values for the one-electron species from H to Ne^{9+}.

Method We find the radius at which the radial distribution function of the hydrogenic 1s orbital has a maximum value by solving $dP/dr=0$. If there are several maxima, then we choose the one corresponding to the greatest amplitude.

Answer The radial distribution function is given in eqn 9A.19A. It follows that

$$\frac{dP}{dr}=\frac{4Z^3}{a_0^3}\left(2r-\frac{2Zr^2}{a_0}\right)e^{-2Zr/a_0}$$

This function is zero where the term in parentheses is zero, which (other than at $r=0$) is at

$$r^*=\frac{a_0}{Z}$$

Then, with $a_0=52.9$ pm, the most probable radius is

	H	He^+	Li^{2+}	Be^{3+}	B^{4+}	C^{5+}	N^{6+}	O^{7+}	F^{8+}	Ne^{9+}
r^*/pm	52.9	26.5	17.6	13.2	10.6	8.82	7.56	6.61	5.88	5.29

Notice how the 1s orbital is drawn towards the nucleus as the nuclear charge increases. At uranium the most probable radius is only 0.58 pm, almost 100 times closer than for hydrogen. (On a scale where $r^*=10$ cm for H, $r^*=1$ mm for U.) We need to be cautious, though, in extending this result to very heavy atoms because relativistic effects are then important and complicate the calculation.

Self-test 9A.7 Find the most probable distance of a 2s electron from the nucleus in a hydrogenic atom.

Answer: $(3+5^{1/2})a_0/Z=5.24a_0/Z$

(g) p Orbitals

The three 2p orbitals are distinguished by the three different values that m_l can take when $l=1$. Because the quantum number m_l tells us the orbital angular momentum around an axis, these different values of m_l denote orbitals in which the electron has different orbital angular momenta around an arbitrary z-axis but the same magnitude of that momentum (because l is the same for all three). The orbital with $m_l=0$, for instance, has zero angular momentum around the z-axis. Its angular variation is given by the spherical harmonic $Y_{1,0}$, which is proportional to cos θ (see Table 8C.1). Therefore, the probability density, which is proportional to $\cos^2\theta$, has its maximum value on either side of the nucleus along the z-axis (at $\theta=0$ and 180°). Specifically, the wavefunction of a 2p orbital with $m_l=0$ is

$$\psi_{2,1,0}=R_{2,1}(r)Y_{1,0}(\theta,\phi)=\frac{1}{4(2\pi)^{1/2}}\left(\frac{Z}{a_0}\right)^{5/2}r\cos\theta\,e^{-Zr/2a_0}$$
$$=r\cos\theta\,f(r) \qquad (9A.20a)$$

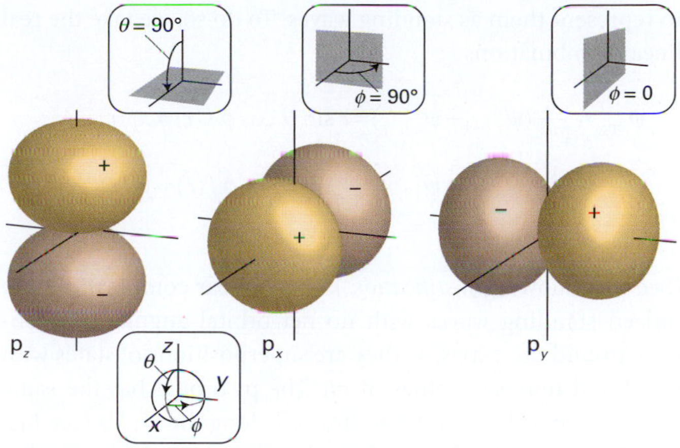

Figure 9A.15 The boundary surfaces of 2p orbitals. A nodal plane passes through the nucleus and separates the two lobes of each orbital. The dark and light areas denote regions of opposite sign of the wavefunction. The angles of the spherical polar coordinate system are also shown. All p orbitals have boundary surfaces like those shown here.

where $f(r)$ is a function only of r. Because in spherical polar coordinates $z=r\cos\theta$, this wavefunction may also be written

$$\psi_{2,1,0}=zf(r) \qquad (9A.20b)$$

All p orbitals with $m_l=0$ have wavefunctions of this form, but $f(r)$ depends on the value of n. This way of writing the orbital is the origin of the name 'p_z orbital': its boundary surface is shown in Fig. 9A.15. The wavefunction is zero everywhere in the xy-plane, where $z=0$, so the xy-plane is a **nodal plane** of the orbital: the wavefunction changes sign on going from one side of the plane to the other.

The wavefunctions of 2p orbitals with $m_l=\pm1$ have the following form:

$$\psi_{2,1,\pm1}=R_{2,1}(r)Y_{1,\pm1}(\theta,\phi)$$
$$=\mp\frac{1}{8\pi^{1/2}}\left(\frac{Z}{a_0}\right)^{5/2}r\sin\theta\,e^{\pm i\phi}e^{-Zr/2a_0} \qquad (9A.21)$$
$$=\mp\frac{1}{2^{1/2}}r\sin\theta\,e^{\pm i\phi}f(r)$$

In Topic 8A it is shown that a particle that has net motion is described by a complex wavefunction. In the present case, the functions correspond to non-zero angular momentum about the z-axis: $e^{+i\phi}$ corresponds to clockwise rotation when viewed from below, and $e^{-i\phi}$ corresponds to anticlockwise rotation (from the same viewpoint). They have zero amplitude where $\theta=0$ and 180° (along the z-axis) and maximum amplitude at 90°, which is in the xy-plane. To draw the functions it is usual to represent them as standing waves. To do so, we take the real linear combinations

$$\psi_{2p_x}=\frac{1}{2^{1/2}}(\psi_{2,1,+1}-\psi_{2,1,-1})=r\sin\theta\cos\phi\,f(r)=xf(r)$$
$$\psi_{2p_y}=\frac{\mathrm{i}}{2^{1/2}}(\psi_{2,1,+1}+\psi_{2,1,-1})=r\sin\theta\sin\phi\,f(r)=yf(r) \quad (9A.22)$$

(See the following *Justification.*) These linear combinations are indeed standing waves with no net orbital angular momentum around the z-axis, as they are superpositions of states with equal and opposite values of m_l. The p_x orbital has the same shape as a p_z orbital, but it is directed along the x-axis (see Fig. 9A.15); the p_y orbital is similarly directed along the y-axis. The wavefunction of any p orbital of a given shell can be written as a product of x, y, or z and the same function f (which depends on the value of n).

Justification 9A.4 The linear combination of degenerate wavefunctions

We justify here the step of taking linear combinations of degenerate orbitals when we want to indicate a particular point. The freedom to do so rests on the fact, as we show below, that whenever two or more wavefunctions correspond to the same energy, then any linear combination of them is an equally valid solution of the Schrödinger equation.

Suppose ψ_1 and ψ_2 are both solutions of the Schrödinger equation with energy E; then we know that $\hat{H}\psi_1=E\psi_1$ and $\hat{H}\psi_2=E\psi_2$. Now consider the linear combination $\psi=c_1\psi_1+c_2\psi_2$ where c_1 and c_2 are arbitrary coefficients. Then it follows that

$$\hat{H}\psi=\hat{H}(c_1\psi_1+c_2\psi_2)=c_1\hat{H}\psi_1+c_2\hat{H}\psi_2=c_1E\psi_1+c_2E\psi_2=E\psi$$

Hence, the linear combination is also a solution corresponding to the same energy E.

(h) d Orbitals

When $n=3$, l can be 0, 1, or 2. As a result, this shell consists of one 3s orbital, three 3p orbitals, and five 3d orbitals. Each value of the quantum number $m_l=+2, +1, 0, -1, -2$ corresponds to a different value for the component of the angular momentum about the z-axis. As for the p orbitals, d orbitals with opposite values of m_l (and hence opposite senses of motion around the z-axis) may be combined in pairs to give real standing waves, and the boundary surfaces of the resulting shapes are shown in Fig. 9A.16. The real linear combinations have the following forms, with the function f depending on the value of n:

$$\psi_{d_{xy}}=xyf(r)\quad \psi_{d_{yz}}=yzf(r)\quad \psi_{d_{zx}}=zxf(r)$$
$$\psi_{d_{x^2-y^2}}=\tfrac{1}{2}(x^2-y^2)f(r)\quad \psi_{d_{z^2}}=\frac{3^{1/2}}{2}(3z^2-r^2)f(r) \quad (9A.23)$$

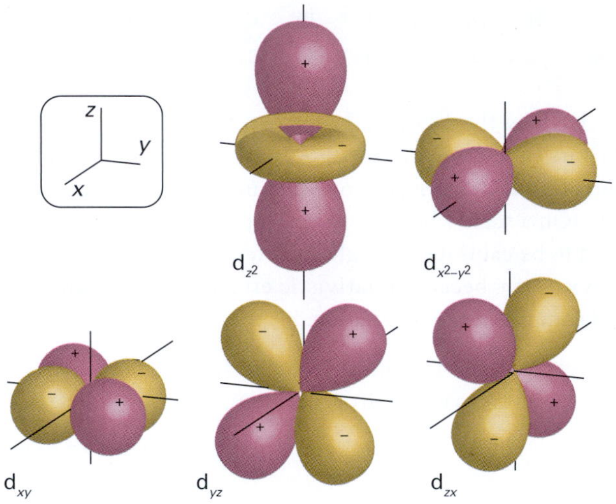

Figure 9A.16 The boundary surfaces of 3d orbitals. Two nodal planes in each orbital intersect at the nucleus and separate the lobes of each orbital. The dark and light areas denote regions of opposite sign of the wavefunction. All d orbitals have boundary surfaces like those shown here.

Checklist of concepts

- ☐ 1. The **Ritz combination principle** states that the wavenumber of any spectral line is the difference between two terms.
- ☐ 2. The Schrödinger equation for hydrogenic atoms separates into two equations: the solutions of one give the angular variation of the wavefunction and the solution of the other gives its radial dependence.
- ☐ 3. Close to the nucleus the radial wavefunction is proportional to r^l; far from the nucleus all wavefunctions approach zero exponentially.
- ☐ 4. An **atomic orbital** is a one-electron wavefunction for an electron in an atom.
- ☐ 5. Atomic orbitals are specified by the **quantum numbers** n, l, and m_l.
- ☐ 6. The energies of the bound states of hydrogenic atoms are proportional to Z^2/n^2.
- ☐ 7. The **ionization energy** of an element is the minimum energy required to remove an electron from the ground state of one of its atoms.

☐ 8. Orbitals of a given value of n form a **shell** of an atom, and within that shell orbitals of the same value of l form **subshells**.

☐ 9. Orbitals of the same shell all have the same energy in hydrogenic atoms; orbitals of the same subshell of a shell are degenerate in all types of atoms.

☐ 10. **s Orbitals** are spherically symmetrical and have nonzero probability density at the nucleus.

☐ 11. A **radial distribution function** is the probability density for the distribution of the electron as a function of distance from the nucleus.

☐ 12. There are three **p orbitals** in a given subshell; each one has an angular node.

☐ 13. There are five **d orbitals** in a given subshell; each one has two angular nodes.

Checklist of equations

Property	Equation	Comment	Equation number
Wavenumbers of the spectral lines of a hydrogen atom	$\tilde{\nu}=\tilde{R}_H(1/n_1^2-1/n_2^2)$	$\tilde{R}_H$ is the Rydberg constant for hydrogen (expressed as a wavenumber)	9A.1
Wavefunctions of hydrogenic atoms	$\psi(r,\theta,\phi)=R(r)Y(\theta,\phi)$	Y are spherical harmonics	9A.7
Bohr radius	$a_0=4\pi\varepsilon_0\hbar^2/m_e e^2$	$a_0=52.9$ pm; the most probable radius for a 1s electron in hydrogen	9A.11
Rydberg constant for an atom N	$\tilde{R}_N=\mu e^4/32\pi^2\varepsilon_0^2\hbar^2$	$\tilde{R}_N\approx\tilde{R}_\infty$, the Rydberg constant; $\mu=m_e m_N/(m_e+m_N)$	9A.14
Energies of hydrogenic atoms	$E_n=-hcZ^2\tilde{R}_N/n^2$	$\tilde{R}_N$ is the for the atom N	9A.14
Radial distribution function	$P(r)=r^2R(r)^2$	$P(r)=4\pi r^2\psi^2$ for s orbitals	9A.18b

9B Many-electron atoms

Contents

➤ Why do you need to know this material?

Many-electron atoms are the building blocks of all compounds, and to understand their properties, including their ability to participate in chemical bonding, it is essential to understand their electronic structure. Moreover, a knowledge of that structure explains the structure of the periodic table and all that it summarizes.

➤ What is the key idea?

Electrons occupy the lowest energy available orbital subject to the requirements of the Pauli exclusion principle.

➤ What do you need to know already?

This Topic builds on the account of the structure of hydrogenic atoms (Topic 9A), especially their shell structure. In the discussion of ionization energies and electron affinities it makes use of the properties of standard reaction enthalpy (Topic 2C).

The Schrödinger equation for a many-electron atom is highly complicated because all the electrons interact with one another. One very important consequence of these interactions is that orbitals of the same value of n but different values of l are no longer degenerate in a many-electron atom. Moreover, even for a helium atom, with its two electrons, no analytical expression for the orbitals and energies can be given, and we are forced to make approximations. We adopt a simple approach based on the structure of hydrogenic atoms (Topic 9A). In the final section we see the kind of numerical computations that are currently used to obtain accurate wavefunctions and energies.

9B.1 The orbital approximation

The wavefunction of a many-electron atom is a very complicated function of the coordinates of all the electrons, and we should write it $\Psi(r_1,r_2,\ldots)$, where r_i is the vector from the nucleus to electron i (uppercase psi, Ψ, is commonly used to denote a many-electron wavefunction). However, in the **orbital approximation** we suppose that a reasonable first approximation to this exact wavefunction is obtained by thinking of each electron as occupying its 'own' orbital, and write

$$\Psi(r_1,r_2,\ldots)=\psi(r_1)\psi(r_2)\ldots \quad \text{Orbital approximation} \quad (9B.1)$$

We can think of the individual orbitals as resembling the hydrogenic orbitals, but corresponding to nuclear charges modified by the presence of all the other electrons in the atom. This description is only approximate, as the following *Justification* reveals, but it is a useful model for discussing the chemical properties of atoms, and is the starting point for more sophisticated descriptions of atomic structure.

Justification 9B.1 The orbital approximation

The orbital approximation would be exact if there were no interactions between electrons. To demonstrate the validity of this remark, we need to consider a system in which the hamiltonian for the energy is the sum of two contributions, one for electron 1 and the other for electron 2: $\hat{H}=\hat{H}_1+\hat{H}_2$. In an actual atom (such as helium atom), there is an additional term (proportional to $1/r_{12}$) corresponding to the interaction of the two electrons:

$$\hat{H}=\overbrace{-\frac{\hbar^2}{2m_e}\nabla_1^2-\frac{2e^2}{4\pi\varepsilon_0 r_1}}^{\hat{H}_1}\overbrace{-\frac{\hbar^2}{2m_e}\nabla_2^2-\frac{2e^2}{4\pi\varepsilon_0 r_2}}^{\hat{H}_2}+\frac{e^2}{4\pi\varepsilon_0 r_{12}}$$

but we are ignoring that term. We shall now show that if $\psi(r_1)$ is an eigenfunction of $\hat{H}_1$ with energy E_1, and $\psi(r_2)$ is an eigenfunction of $\hat{H}_2$ with energy E_2, then the product $\Psi(r_1,r_2)=\psi(r_1)\psi(r_2)$ is an eigenfunction of the combined hamiltonian $\hat{H}$. To do so we write

$$\begin{aligned}\hat{H}\Psi(r_1,r_2)&=(\hat{H}_1+\hat{H}_2)\psi(r_1)\psi(r_2)=\hat{H}_1\psi(r_1)\psi(r_2)+\psi(r_1)\hat{H}_2\psi(r_2)\\&=E_1\psi(r_1)\psi(r_2)+\psi(r_1)E_2\psi(r_2)=(E_1+E_2)\psi(r_1)\psi(r_2)\\&=E\Psi(r_1,r_2)\end{aligned}$$

where $E=E_1+E_2$. This is the result we need to prove. However, if the electrons interact (as they do in fact), then the proof fails.

(a) The helium atom

The orbital approximation allows us to express the electronic structure of an atom by reporting its **configuration**, a statement of its occupied orbitals (usually, but not necessarily, in its ground state). Thus, as the ground state of a hydrogenic atom consists of the single electron in a 1s orbital, we report its configuration as $1s^1$ (read 'one-ess-one').

A He atom has two electrons. We can imagine forming the atom by adding the electrons in succession to the orbitals of the bare nucleus (of charge $2e$). The first electron occupies a 1s hydrogenic orbital, but because $Z=2$ that orbital is more compact than in H itself. The second electron joins the first in the 1s orbital, so the electron configuration of the ground state of He is $1s^2$.

Brief illustration 9B.1 Helium wavefunctions

According to the orbital approximation, each electron occupies a hydrogenic 1s orbital of the kind given in Topic 9A. If we anticipate (see below) that the electrons experience an effective nuclear charge $Z_{\text{eff}}e$ rather than its actual charge Ze (specifically, as we shall see, 1.69e rather than 2e), then the two-electron wavefunction of the atom is

$$\begin{aligned}\Psi(r_1,r_2)&=\overbrace{\frac{Z_{\text{eff}}^{3/2}}{(\pi a_0^3)^{1/2}}e^{-Z_{\text{eff}}r_1/a_0}}^{\psi_{1s}(r_1)}\times\overbrace{\frac{Z_{\text{eff}}^{3/2}}{(\pi a_0^3)^{1/2}}e^{-Z_{\text{eff}}r_2/a_0}}^{\psi_{1s}(r_2)}\\&=\frac{Z_{\text{eff}}^3}{\pi a_0^3}e^{-Z_{\text{eff}}(r_1+r_2)/a_0}\end{aligned}$$

As can be seen, there is nothing particularly mysterious about a two-electron wavefunction: in this case it is a simple exponential function of the distances of the two electrons from the nucleus.

Self-test 9B.1 Construct the wavefunction for an excited state of the He atom with configuration $1s^12s^1$. Use $Z_{\text{eff}}=2$ for the 1s electron and $Z_{\text{eff}}=1$ for the 2s electron. Why those values should become clear shortly.

Answer: $\Psi(r_1,r_2)=(1/2\pi a_0^3)(2-r_2/a_0)e^{-(2r_1+r_2/2)/a_0}$

It is tempting to suppose that the electronic configurations of the atoms of successive elements with atomic numbers $Z=3$, 4, …, and therefore with Z electrons, are simply $1s^Z$. That, however, is not the case. The reason lies in two aspects of nature: that electrons possess 'spin' and must obey the very fundamental 'Pauli principle'.

(b) Spin

The quantum mechanical property of electron **spin**, the possession of an intrinsic angular momentum, was identified by the experiment performed by Otto Stern and Walther Gerlach in 1921, who shot a beam of silver atoms through an inhomogeneous magnetic field, as explained in Topic 8C. Stern and Gerlach observed *two* bands of Ag atoms in their experiment. This observation seems to conflict with one of the predictions of quantum mechanics, because an angular momentum l gives rise to $2l+1$ orientations, which is equal to 2 only if $l=\frac{1}{2}$, contrary to the conclusion that l must be an integer. The conflict was resolved by the suggestion that the angular momentum they were observing was not due to orbital angular momentum (the motion of an electron around the atomic nucleus) but arose instead from the motion of the electron about its own axis. This intrinsic angular momentum of the electron, or 'spin', also emerged when Dirac combined quantum mechanics with special relativity and established the theory of relativistic quantum mechanics.

The spin of an electron about its own axis does not have to satisfy the same boundary conditions as those for a particle circulating around a central point, so the quantum number for spin angular momentum is subject to different restrictions. To distinguish this spin angular momentum from orbital angular momentum we use the **spin quantum number** s (in place of the l in Topic 9A; like l, s is a non-negative number) and m_s, the **spin magnetic quantum number**, for the projection on the z-axis. The magnitude of the spin angular momentum is $\{s(s+1)\}^{1/2}\hbar$ and the component $m_s\hbar$ is restricted to the $2s+1$ values $m_s=s, s-1, \cdots, -s$. To account for Stern and Gerlach's observation, $s=\frac{1}{2}$ and $m_s=\pm\frac{1}{2}$.

A note on good practice You will sometimes see the quantum number s used in place of m_s, and written $s=\pm\frac{1}{2}$. That is wrong: like l, s is never negative and denotes the magnitude of the spin angular momentum. For the z-component, use m_s.

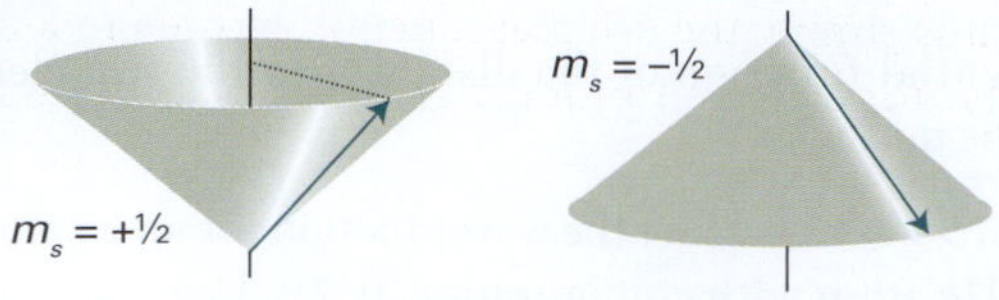

Figure 9B.1 The vector representation of the spin of an electron. The length of the side of the cone is $3^{1/2}/2$ units and the projections are $\pm\frac{1}{2}$ units.

The detailed analysis of the spin of a particle is sophisticated and shows that the property should not be taken to be an actual spinning motion. It is better to regard 'spin' as an intrinsic property like mass and charge: every electron has exactly the same value and the magnitude of the spin angular momentum of an electron cannot be changed. However, the picture of an actual spinning motion can be very useful when used with care. On the vector model of angular momentum (Topic 8C), the spin may lie in two different orientations (Fig. 9B.1). One orientation corresponds to $m_s = +\frac{1}{2}$ (this state is often denoted α or ↑); the other orientation corresponds to $m_s = -\frac{1}{2}$ (this state is denoted β or ↓).

Other elementary particles have characteristic spin. For example, protons and neutrons are spin-$\frac{1}{2}$ particles (that is, $s = \frac{1}{2}$) and invariably spin with the same angular momentum. Because the masses of a proton and a neutron are so much greater than the mass of an electron, yet they all have the same spin angular momentum, the classical picture would be of these two particles spinning much more slowly than an electron. Some mesons, another variety of fundamental particle, are spin-1 particles (that is, $s = 1$), as are some atomic nuclei, but for our purposes the most important spin-1 particle is the photon. The importance of photon spin in spectroscopy is explained in Topic 12A; proton spin is the basis of Topic 14A (magnetic resonance).

Brief illustration 9B.2 Spin

The magnitude of the spin angular momentum, like any angular momentum, is $\{s(s+1)\}^{1/2}\hbar$. For any spin-$\frac{1}{2}$ particle, not only electrons, this angular momentum is $(\frac{3}{4})^{1/2}\hbar = 0.866\hbar$, or 9.13×10^{-35} J s. The component on the z-axis is $m_s\hbar$, which for a spin-$\frac{1}{2}$ particle is $\pm\frac{1}{2}\hbar$, or $\pm 5.27 \times 10^{-35}$ J s.

Self-test 9B.2 Evaluate the spin angular momentum of a photon.

Answer: $2^{1/2}\hbar = 1.49 \times 10^{-34}$ J s

Particles with half-integral spin are called **fermions** and those with integral spin (including 0) are called **bosons**. Thus, electrons and protons are fermions and photons are bosons. It is a very deep feature of nature that all the elementary particles that constitute matter are fermions whereas the elementary particles that transmit the forces that bind fermions together are all bosons. Photons, for example, transmit the electromagnetic force that binds together electrically charged particles. Matter, therefore, is an assembly of fermions held together by forces conveyed by bosons.

(c) The Pauli principle

With the concept of spin established, we can resume our discussion of the electronic structures of atoms. Lithium, with $Z = 3$, has three electrons. The first two occupy a 1s orbital drawn even more closely than in He around the more highly charged nucleus. The third electron, however, does not join the first two in the 1s orbital because that configuration is forbidden by the **Pauli exclusion principle**:

> No more than two electrons may occupy any given orbital, and if two do occupy one orbital, then their spins must be paired.

Pauli exclusion principle

Electrons with paired spins, denoted ↑↓, have zero net spin angular momentum because the spin of one electron is cancelled by the spin of the other. Specifically, one electron has $m_s = +\frac{1}{2}$ the other has $m_s = -\frac{1}{2}$ and in the vector model they are orientated on their respective cones so that the resultant spin is zero (Fig. 9B.2). The exclusion principle is the key to the structure of complex atoms, to chemical periodicity, and to molecular structure. It was proposed by Wolfgang Pauli in 1924 when he was trying to account for the absence of some lines in the spectrum of helium. Later he was able to derive a very general form of the principle from theoretical considerations.

The Pauli exclusion principle in fact applies to any pair of identical fermions. Thus it applies to protons, neutrons, and ^{13}C nuclei (all of which have $s = \frac{1}{2}$) and to ^{35}Cl nuclei (which have $s = \frac{3}{2}$). It does not apply to identical bosons, which include photons ($s = 1$) and ^{12}C nuclei ($s = 0$). Any number of identical bosons may occupy the same state (that is, be described by the same wavefunction).

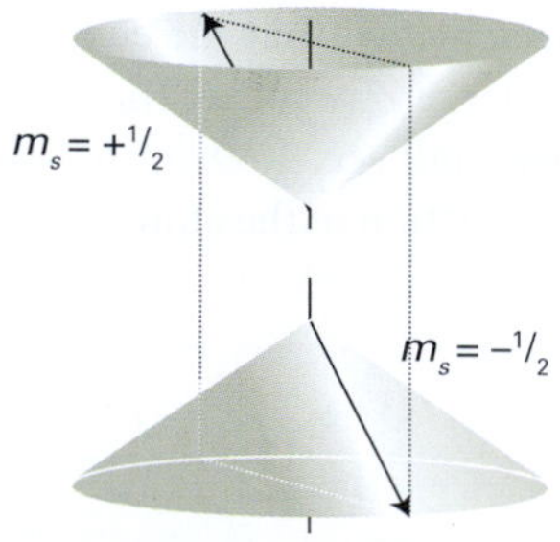

Figure 9B.2 Electrons with paired spins have zero resultant spin angular momentum. They can be represented by two vectors that lie at an indeterminate position on the cones shown here, but wherever one lies on its cone, the other points in the opposite direction; their resultant is zero.

The Pauli *exclusion* principle is a special case of a general statement called the **Pauli principle**:

> When the labels of any two identical fermions are exchanged, the total wavefunction changes sign; when the labels of any two identical bosons are exchanged, the sign of the total wavefunction remains the same.

Pauli principle

By 'total wavefunction' is meant the entire wavefunction, including the spin of the particles.

To see that the Pauli principle implies the Pauli exclusion principle, we consider the wavefunction for two electrons $\Psi(1,2)$. The Pauli principle implies that it is a fact of nature (which has its roots in the theory of relativity) that the wavefunction must change sign if we interchange the labels 1 and 2 wherever they occur in the function:

$$\Psi(1,2)=-\Psi(2,1) \tag{9B.2}$$

Suppose the two electrons in an atom occupy an orbital ψ, then in the orbital approximation the overall wavefunction is $\psi(1)\psi(2)$. To apply the Pauli principle, we must deal with the *total* wavefunction, the wavefunction including spin. There are several possibilities for two spins: both α, denoted $\alpha(1)\alpha(2)$, both β, denoted $\beta(1)\beta(2)$, and one α the other β, denoted either $\alpha(1)\beta(2)$ or $\alpha(2)\beta(1)$. Because we cannot tell which electron is α and which is β, in the last case it is appropriate to express the spin states as the (normalized) linear combinations

$$\begin{aligned}\sigma_+(1,2)&=(1/2^{1/2})\{\alpha(1)\beta(2)+\beta(1)\alpha(2)\}\\ \sigma_-(1,2)&=(1/2^{1/2})\{\alpha(1)\beta(2)-\beta(1)\alpha(2)\}\end{aligned} \tag{9B.3}$$

(A stronger justification for taking these linear combinations is that they correspond to eigenfunctions of the total spin operators S^2 and S_z, with $M_S=0$ and, respectively, $S=1$ and 0.) These combinations allow one spin to be α and the other β with equal probability. The total wavefunction of the system is therefore the product of the orbital part and one of the four spin states:

$$\begin{aligned}&\psi(1)\psi(2)\alpha(1)\alpha(2) \quad \psi(1)\psi(2)\beta(1)\beta(2)\\ &\psi(1)\psi(2)\sigma_+(1,2) \quad \psi(1)\psi(2)\sigma_-(1,2)\end{aligned} \tag{9B.4}$$

The Pauli principle says that for a wavefunction to be acceptable (for electrons), it must change sign when the electrons are exchanged. In each case, exchanging the labels 1 and 2 converts the factor $\psi(1)\psi(2)$ into $\psi(2)\psi(1)$, which is the same, because the order of multiplying the functions does not change the value of the product. The same is true of $\alpha(1)\alpha(2)$ and $\beta(1)\beta(2)$. Therefore, the first two overall products are not allowed, because they do not change sign. The combination $\sigma_+(1,2)$ changes to

$$\sigma_+(2,1)=(1/2^{1/2})\{\alpha(2)\beta(1)+\beta(2)\alpha(1)\}=\sigma_+(1,2)$$

because it is simply the original function written in a different order. The third overall product is therefore also disallowed. Finally, consider $\sigma_-(1,2)$:

$$\begin{aligned}\sigma_-(2,1)&=(1/2^{1/2})\{\alpha(2)\beta(1)-\beta(2)\alpha(1)\}\\ &=-(1/2^{1/2})\{\alpha(1)\beta(2)-\beta(1)\alpha(2)\}=-\sigma_-(1,2)\end{aligned}$$

This combination does change sign (it is 'antisymmetric'). The product $\psi(1)\psi(2)\sigma_-(1,2)$ also changes sign under particle exchange, and therefore it is acceptable.

Now we see that only one of the four possible states is allowed by the Pauli principle, and the one that survives has paired α and β spins. This is the content of the Pauli exclusion principle. The exclusion principle (but not the more general Pauli principle) is irrelevant when the orbitals occupied by the electrons are different, and both electrons may then have, but need not have, the same spin state. In each case the overall wavefunction must still be antisymmetric overall and must satisfy the Pauli principle itself.

A final point in this connection is that the acceptable product wavefunction $\psi(1)\psi(2)\sigma_-(1,2)$ can be expressed as a determinant (see *The chemist's toolkit* 9B.1):

$$\begin{aligned}&\frac{1}{2^{1/2}}\begin{vmatrix}\psi(1)\alpha(1) & \psi(2)\alpha(2)\\ \psi(1)\beta(1) & \psi(2)\beta(2)\end{vmatrix}\\ &=\frac{1}{2^{1/2}}\{\psi(1)\alpha(1)\psi(2)\beta(2)-\psi(2)\alpha(2)\psi(1)\beta(1)\}\\ &=\psi(1)\psi(2)\sigma_-(1,2)\end{aligned}$$

Any acceptable wavefunction for a closed-shell species can be expressed as a **Slater determinant**, as such determinants are known. In general, for N electrons in orbitals ψ_a, ψ_b, …

$$\Psi(1,2,\ldots,N) =\frac{1}{(N!)^{1/2}}\begin{vmatrix}\psi_a(1)\alpha(1) & \psi_a(2)\alpha(2) & \psi_a(3)\alpha(3)\ldots\psi_a(N)\alpha(N)\\ \psi_a(1)\beta(1) & \psi_a(2)\beta(2) & \psi_a(3)\beta(3)\ldots\psi_a(N)\beta(N)\\ \psi_b(1)\alpha(1) & \psi_b(2)\alpha(2) & \psi_b(3)\alpha(3)\ldots\psi_b(N)\alpha(N)\\ \vdots & \vdots & \vdots \qquad \ldots\\ \psi_z(1)\beta(1) & \psi_z(2)\beta(2) & \psi_z(3)\beta(3)\ldots\psi_z(N)\beta(N)\end{vmatrix} \tag{9B.5a}$$

Writing a many-electron wavefunction in this way ensures that it is antisymmetric under the interchange of any pair of electrons (see Problem 9B.2). Because a Slater determinant takes up a lot of space, it is normally reported by writing only its diagonal elements, as in

$$\Psi(1,2,\ldots N)=(1/N!)^{1/2}\det|\psi_a^\alpha(1)\psi_a^\beta(2)\psi_b^\alpha(3)\ldots\psi_z^\beta(N)| \tag{9B.5b}$$

Notation for a Slater determinant

The chemist's toolkit 9B.1 Determinants

A 2×2 determinant is the quantity

$$\begin{vmatrix} a & b \\ c & d \end{vmatrix} = ad - bc \quad \text{2×2 Determinant}$$

A 3×3 determinant is evaluated by expanding it as a sum of 2×2 determinants:

$$\begin{vmatrix} a & b & c \\ d & e & f \\ g & h & i \end{vmatrix} = a\begin{vmatrix} e & f \\ h & i \end{vmatrix} - b\begin{vmatrix} d & f \\ g & i \end{vmatrix} + c\begin{vmatrix} d & e \\ g & h \end{vmatrix} \quad \text{3×3 Determinant}$$

$$= a(ei - fh) - b(di - fg) + c(dh - eg)$$

Note the sign change in alternate columns (b occurs with a negative sign in the expansion). An important property of a determinant is that if any two rows or any two columns are interchanged, then the determinant changes sign:

$$\text{Exchange columns:} \begin{vmatrix} b & a \\ d & c \end{vmatrix} = bc - ad = -(ad - bc) = -\begin{vmatrix} a & b \\ c & d \end{vmatrix}$$

$$\text{Exchange rows:} \begin{vmatrix} c & d \\ a & b \end{vmatrix} = cd - da = -(ad - bc) = -\begin{vmatrix} a & b \\ c & d \end{vmatrix}$$

Now we can return to lithium. In Li ($Z=3$), the third electron cannot enter the 1s orbital because that orbital is already full: we say the K shell (the orbital with $n=1$, Topic 9A) is *complete* and that the two electrons form a **closed shell**. Because a similar closed shell is characteristic of the He atom, we denote it [He]. The third electron is excluded from the K shell and must occupy the next available orbital, which is one with $n=2$ and hence belonging to the L shell (which consists of the four orbitals with $n=2$). However, we now have to decide whether the next available orbital is the 2s orbital or a 2p orbital, and therefore whether the lowest energy configuration of the atom is $[He]2s^1$ or $[He]2p^1$.

(d) Penetration and shielding

Unlike in hydrogenic atoms, the 2s and 2p orbitals (and, in general, all subshells of a given shell) are not degenerate in many-electron atoms. An electron in a many-electron atom experiences a Coulombic repulsion from all the other electrons present. If it is at a distance r from the nucleus, it experiences an average repulsion that can be represented by a point negative charge located at the nucleus and equal in magnitude to the total charge of the electrons within a sphere of radius r (Fig. 9B.3). The effect of this point negative charge, when averaged over all the locations of the electron, is to reduce the full charge of the nucleus from Ze to $Z_{\text{eff}}e$, the **effective nuclear charge**. In everyday parlance, Z_{eff} itself is commonly referred to as the 'effective nuclear charge'. We say that the electron experiences a **shielded** nuclear charge, and the difference between Z and Z_{eff} is called the **shielding constant**, σ:

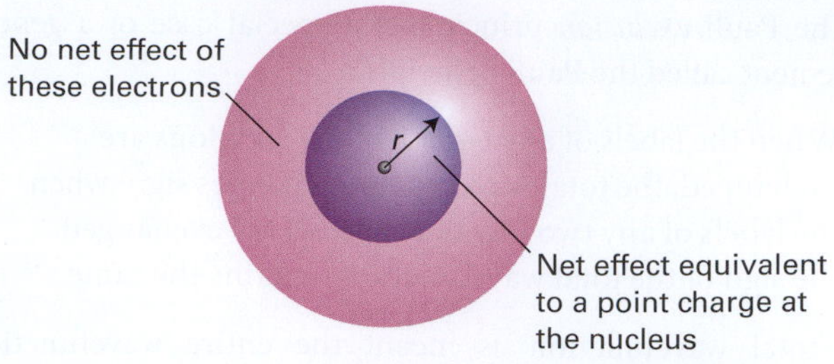

Figure 9B.3 An electron at a distance r from the nucleus experiences a Coulombic repulsion from all the electrons within a sphere of radius r and which is equivalent to a point negative charge located on the nucleus. The negative charge reduces the effective nuclear charge of the nucleus from Ze to $Z_{\text{eff}}e$.

$$Z_{\text{eff}} = Z - \sigma \quad \text{Effective nuclear charge} \quad (9B.6)$$

The electrons do not actually 'block' the full Coulombic attraction of the nucleus: the shielding constant is simply a way of expressing the net outcome of the nuclear attraction and the electronic repulsions in terms of a single equivalent charge at the centre of the atom.

The shielding constant is different for s and p electrons because they have different radial distributions (Fig. 9B.4). An

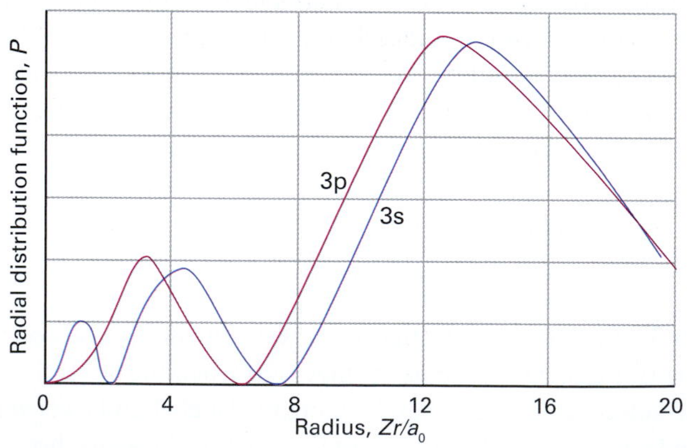

Figure 9B.4 An electron in an s orbital (here a 3s orbital) is more likely to be found close to the nucleus than an electron in a p orbital of the same shell (note the closeness of the innermost peak of the 3s orbital to the nucleus at $r=0$). Hence an s electron experiences less shielding and is more tightly bound than a p electron.

s electron has a greater **penetration** through inner shells than a p electron, in the sense that it is more likely to be found close to the nucleus than a p electron of the same shell (the wavefunction of a p orbital, remember, is zero at the nucleus). Because only electrons inside the sphere defined by the location of the electron contribute to shielding, an s electron experiences less shielding than a p electron. Consequently, by the combined effects of penetration and shielding, an s electron is more tightly bound than a p electron of the same shell. Similarly, a d electron penetrates less than a p electron of the same shell (recall that the wavefunction of a d orbital varies as r^2 close to the nucleus, whereas a p orbital varies as r), and therefore experiences more shielding.

Shielding constants for different types of electrons in atoms have been calculated from their wavefunctions obtained by numerical solution of the Schrödinger equation for the atom (Table 9B.1). We see that, in general, valence-shell s electrons do experience higher effective nuclear charges than p electrons, although there are some discrepancies. We return to this point shortly.

Table 9B.1* Effective nuclear charge, $Z_{\text{eff}}=Z-\sigma$

Element	Z	Orbital	Z_{eff}
He	2	1s	1.6875
C	6	1s	5.6727
		2s	3.2166
		2p	3.1358

* More values are given in the *Resource section*.

Brief illustration 9B.3 Penetration and shielding

The effective nuclear charge for 1s, 2s, and 2p electrons in a carbon atom are 5.6727, 3.2166, and 3.1358, respectively. The radial distribution functions for these orbitals (Topic 9A) are generated by forming $P(r)=r^2R(r)^2$, where $R(r)$ is the radial wavefunction, which are given in Table 9A.1. The three radial distribution functions are plotted in Fig. 9B.5. As can be seen, the s orbital has greater penetration than the p orbital. The average radii of the 2s and 2p orbitals are 99 pm and 84 pm, respectively, which shows that the average distance of a 2s electron from the nucleus is greater than that of a 2p orbital. To account for the lower energy of the 2s orbital we see that the extent of penetration is more important than the average distance.

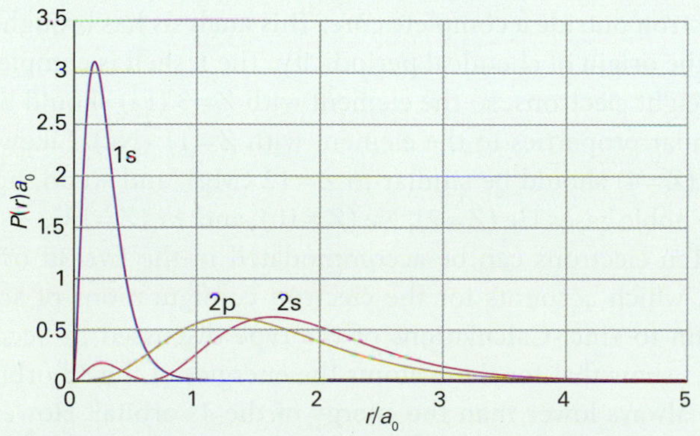

Figure 9B.5 The radial distribution functions for electrons in a carbon atom, as calculated in *Brief illustration* 9B.3.

Self-test 9B.3 Confirm the values for the average radii. Instead of carrying out the integrations, you might prefer to use the general formula $\langle r\rangle_{n,l}=(n^2a_0/Z)\{1+\frac{1}{2}[1-l(l+1)/n^2]\}$.

Answer: 2s: $1.865a_0$; 2p: $1.595a_0$

The consequence of penetration and shielding is that the energies of subshells of a shell in a many-electron atom (those with the same values of n but different values of l) in general lie in the order s<p<d<f. The individual orbitals of a given subshell (those with the same value of l but different values of m_l) remain degenerate because they all have the same radial characteristics and so experience the same effective nuclear charge.

We can now complete the Li story. Because the shell with $n=2$ consists of two non-degenerate subshells, with the 2s orbital lower in energy than the three 2p orbitals, the third electron occupies the 2s orbital. This occupation results in the ground-state configuration $1s^22s^1$, with the central nucleus surrounded by a complete helium-like shell of two 1s electrons, and around that a more diffuse 2s electron. The electrons in the outermost shell of an atom in its ground state are called the **valence electrons** because they are largely responsible for the chemical bonds that the atom forms. Thus, the valence electron in Li is a 2s electron and its other two electrons belong to its core.

9B.2 The building-up principle

The extension of the argument used to account for the structures of H, He, and Li is called the **building-up principle**, or the *Aufbau principle*, from the German word for building up, which will be familiar from introductory courses. In brief, we imagine the bare nucleus of atomic number Z, and then feed into the orbitals Z electrons in succession. The order of occupation is

1s 2s 2p 3s 3p 4s 3d 4p 5s 4d 5p 6s

Each orbital may accommodate up to two electrons.

Brief illustration 9B.4 The building-up principle

Consider the carbon atom, for which $Z=6$ and there are six electrons to accommodate. Two electrons enter and fill the 1s orbital, two enter and fill the 2s orbital, leaving two electrons to occupy the orbitals of the 2p subshell. Hence the ground-state configuration of C is $1s^2 2s^2 2p^2$, or more succinctly $[He]2s^2 2p^2$, with [He] the helium-like $1s^2$ core.

Self-test 9B.4 What is the ground-state configuration of a Mg atom?

Answer: $[Ne]3s^2$

(a) Hund's rules

We can be more precise about the configuration of a carbon atom than in *Brief illustration* 9B.4: we can expect the last two electrons to occupy different 2p orbitals because they will then be further apart on average and repel each other less than if they were in the same orbital. Thus, one electron can be thought of as occupying the $2p_x$ orbital and the other the $2p_y$ orbital (the *x, y, z* designation is arbitrary, and it would be equally valid to use the complex forms of these orbitals), and the lowest energy configuration of the atom is $[He]2s^2 2p_x^1 2p_y^1$. The same rule applies whenever degenerate orbitals of a subshell are available for occupation. Thus, another rule of the building-up principle is:

> Electrons occupy different orbitals of a given subshell before doubly occupying any one of them.

For instance, nitrogen ($Z=7$) has the ground-state configuration $[He]2s^2 2p_x^1 2p_y^1 2p_z^1$, and only when we get to oxygen ($Z=8$) is a 2p orbital doubly occupied, giving $[He]2s^2 2p_x^2 2p_y^1 2p_z^1$.

When electrons occupy orbitals singly we invoke **Hund's maximum multiplicity rule**:

> An atom in its ground state adopts a configuration with the greatest number of unpaired electrons.

Hund's maximum multiplicity rule

The explanation of Hund's rule is subtle, but it reflects the quantum mechanical property of **spin correlation**, that, as we demonstrate in the following *Justification*, electrons with parallel spins behave as if they have a tendency to stay well apart, and hence repel each other less. In essence, the effect of spin correlation is to allow the atom to shrink slightly, so the electron–nucleus interaction is improved when the spins are parallel. We can now conclude that, in the ground state of the carbon atom, the two 2p electrons have the same spin, that all three 2p electrons in the N atoms have the same spin (that is, they are parallel), and that the two 2p electrons in different orbitals in the O atom have the same spin (the two in the $2p_x$ orbital are necessarily paired).

Justification 9B.2 Spin correlation

Suppose electron 1 is described by a wavefunction $\psi_a(r_1)$ and electron 2 is described by a wavefunction $\psi_b(r_2)$; then, in the orbital approximation, the joint wavefunction of the electrons is the product $\Psi = \psi_a(r_1)\psi_b(r_2)$. However, this wavefunction is not acceptable, because it suggests that we know which electron is in which orbital, whereas we cannot keep track of electrons. According to quantum mechanics, the correct description is either of the two following wavefunctions:

$$\Psi_{\pm} = (1/2^{1/2})\{\psi_a(\boldsymbol{r}_1)\psi_b(\boldsymbol{r}_2) \pm \psi_b(\boldsymbol{r}_1)\psi_1(\boldsymbol{r}_2)\}$$

According to the Pauli principle, because Ψ_+ is symmetrical under particle interchange, it must be multiplied by an antisymmetric spin function (the one denoted σ_-). That combination corresponds to a spin-paired state. Conversely, Ψ_- is antisymmetric, so it must be multiplied by one of the three symmetric spin states. These three symmetric states correspond to electrons with parallel spins (see Section 9C.2 for an explanation of this point).

Now consider the values of the two combinations when one electron approaches another, and $r_1 = r_2$. We see that Ψ_- vanishes, which means that there is zero probability of finding the two electrons at the same point in space when they have parallel spins. The other combination does not vanish when the two electrons are at the same point in space. Because the two electrons have different relative spatial distributions depending on whether their spins are parallel or not, it follows that their Coulombic interaction is different, and hence that the two states have different energies.

Neon, with $Z=10$, has the configuration $[He]2s^2 2p^6$, which completes the L shell. This closed-shell configuration is denoted [Ne], and acts as a core for subsequent elements. The next electron must enter the 3s orbital and begin a new shell, so an Na atom, with $Z=11$, has the configuration $[Ne]3s^1$. Like lithium with the configuration $[He]2s^1$, sodium has a single s electron outside a complete core. This analysis has brought us to the origin of chemical periodicity. The L shell is completed by eight electrons, so the element with $Z=3$ (Li) should have similar properties to the element with $Z=11$ (Na). Likewise, Be ($Z=4$) should be similar to $Z=12$ (Mg), and so on, up to the noble gases He ($Z=2$), Ne ($Z=10$), and Ar ($Z=18$).

Ten electrons can be accommodated in the five 3d orbitals, which accounts for the electron configurations of scandium to zinc. Calculations of the type discussed in Section 9B.3 show that for these atoms the energies of the 3d orbitals are always lower than the energy of the 4s orbital. However, spectroscopic results show that Sc has the configuration $[Ar]3d^1 4s^2$, instead of $[Ar]3d^3$ or $[Ar]3d^2 4s^1$. To understand this observation, we have to consider the nature of electron–electron repulsions in 3d and 4s orbitals. The most probable distance of a 3d electron from the nucleus is less than that for a 4s electron, so two 3d electrons repel each other more strongly than two 4s electrons. As a result, Sc has the configuration $[Ar]3d^14s^2$ rather than the two alternatives, for then the strong electron–electron repulsions in the 3d orbitals are minimized. The total energy of the atom is least despite the cost of allowing electrons to populate the high energy 4s orbital (Fig. 9B.6). The effect just described is generally true for scandium through zinc, so their electron configurations are of the form $[Ar]3d^n4s^2$, where $n=1$ for scandium and $n=10$ for zinc. Two notable exceptions, which are observed experimentally, are Cr, with electron configuration $[Ar]3d^54s^1$, and Cu, with electron configuration $[Ar]3d^{10}4s^1$.

At gallium, the building-up principle is used in the same way as in preceding periods. Now the 4s and 4p subshells constitute the valence shell, and the period terminates with krypton. Because 18 electrons have intervened since argon, this row is the first 'long period' of the periodic table. The existence of the d-block elements (the 'transition metals') reflects the stepwise occupation of the 3d orbitals, and the subtle shades of energy differences and effects of electron–electron repulsion along this series gives rise to the rich complexity of inorganic d-metal chemistry. A similar intrusion of the f orbitals in Periods 6 and 7 accounts for the existence of the f block of the periodic table (the lanthanoids and actinoids).

We derive the configurations of cations of elements in the s, p, and d blocks of the periodic table by removing electrons from the ground-state configuration of the neutral atom in a specific order. First, we remove valence p electrons, then valence s electrons, and then as many d electrons as are necessary to achieve the specified charge. The configurations of anions of the p-block elements are derived by continuing the building-up procedure and adding electrons to the neutral atom until the configuration of the next noble gas has been reached.

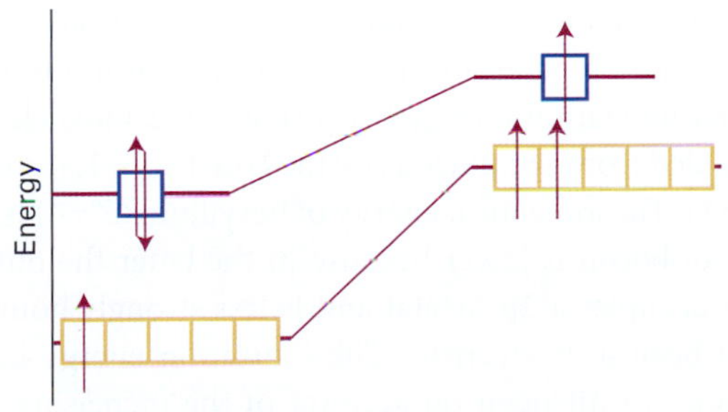

Figure 9B.6 Strong electron–electron repulsions in the 3d orbitals are minimized in the ground state of Sc if the atom has the configuration $[Ar]3d^14s^2$ (shown on the left) instead of $[Ar]3d^24s^1$ (shown on the right). The total energy of the atom is lower when it has the $[Ar]3d^14s^2$ configuration despite the cost of populating the high energy 4s orbital.

Brief illustration 9B.5 Ion configurations

Because the configuration of vanadium is $[Ar]3d^34s^2$, the V^{2+} cation has the configuration $[Ar]3d^3$. It is reasonable that we remove the more energetic 4s electrons in order to form the cation, but it is not obvious why the $[Ar]3d^3$ configuration is preferred in V^{2+} over the $[Ar]3d^14s^2$ configuration, which is found in the isoelectronic Sc atom. Calculations show that the energy difference between $[Ar]3d^3$ and $[Ar]3d^14s^2$ depends on Z_{eff}. As Z_{eff} increases, transfer of a 4s electron to a 3d orbital becomes more favourable because the electron–electron repulsions are compensated by attractive interactions between the nucleus and the electrons in the spatially compact 3d orbital. Indeed, calculations reveal that, for a sufficiently large Z_{eff}, $[Ar]3d^3$ is lower in energy than $[Ar]3d^14s^2$. This conclusion explains why V^{2+} has a $[Ar]3d^3$ configuration and also accounts for the observed $[Ar]4s^03d^n$ configurations of the M^{2+} cations of Sc through Zn.

Self-test 9B.5 Write the ground state configuration of the O^{2-} ion.

Answer: $[He]2s^22p^6$

(b) Ionization energies and electron affinities

The minimum energy necessary to remove an electron from a many-electron atom in the gas phase is the **first ionization energy**, I_1, of the element. The **second ionization energy**, I_2, is the minimum energy needed to remove a second electron (from the singly charged cation). The variation of the first ionization energy through the periodic table is shown in Fig. 9B.7 and some numerical values are given in Table 9B.2. In thermodynamic calculations we often need the **standard enthalpy of ionization**, $\Delta_{ion}H^{\ominus}$. As shown in the following *Justification*, the two are related by

$$\Delta_{ion}H^{\ominus}(T)=I+\tfrac{5}{2}RT \qquad \text{Enthalpy of ionization} \qquad (9B.7a)$$

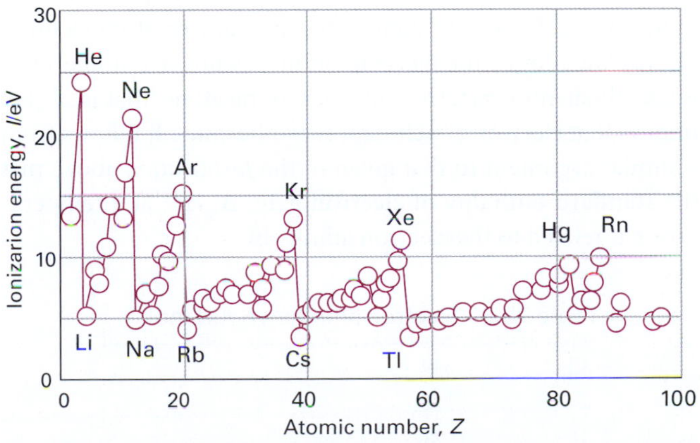

Figure 9B.7 The first ionization energies of the elements plotted against atomic number.

Table 9B.2* First and second ionization energies, I/(kJ mol^{-1})

Element	I_1/(kJ mol^{-1})	I_2/(kJ mol^{-1})
H	1312	
He	2372	5251
Mg	738	1451
Na	496	4562

* More values are given in the *Resource section*.

At 298 K, the difference between the ionization enthalpy and the corresponding ionization energy is 6.20 kJ mol^{-1}. The same expression applies to each successive ionization step, so the overall ionization enthalpy for the formation of M^{2+} is

$$\Delta_{\text{ion}}H^{\ominus}(T)=I_1+I_2+5RT \quad (9B.7b)$$

Justification 9B.3 The ionization enthalpy and the ionization energy

It follows from Kirchhoff's law (Topic 2C, eqn 2C.7a) that the reaction enthalpy, the enthalpy of ionization, for

$$M(g) \rightarrow M^+(g)+e^-(g)$$

at a temperature T is related to the value at $T=0$ by

$$\Delta_{\text{ion}}H^{\ominus}(T)=\overbrace{\Delta_{\text{ion}}H^{\ominus}(0)}^{I}+\int_0^T \Delta_{\text{ion}}C_p^{\ominus}\,dT$$

The molar constant pressure heat capacity of each species in the reaction is $\frac{5}{2}R$, so $\Delta_{\text{ion}}C_p^{\ominus}=+\frac{5}{2}R$. The integral in this expression therefore evaluates to $\frac{5}{2}RT$. The reaction enthalpy at $T=0$ is the same as the (molar) ionization energy, I. Equation 9B.7a then follows.

The **electron affinity**, E_{ea}, is the energy released when an electron attaches to a gas-phase atom (Table 9B.3). In a common, logical (given its name), but not universal convention (which we adopt), the electron affinity is positive if energy is released when the electron attaches to the atom (that is, $E_{\text{ea}}>0$ implies that electron attachment is exothermic). It follows from a similar argument to that given in the *Justification* above that the **standard enthalpy of electron gain**, $\Delta_{\text{eg}}H^{\ominus}$, at a temperature T is related to the electron affinity by

Table 9B.3* Electron affinities, E_a/(kJ mol^{-1})

Cl	349		
F	322		
H	73		
O	141	O^-	−844

* More values are given in the *Resource section*.

$$\Delta_{\text{eg}}H^{\ominus}(T)=-E_{\text{ea}}-\tfrac{5}{2}RT \quad \text{Enthalpy of electron gain} \quad (9B.8)$$

Note the change of sign. In typical thermodynamic cycles the $\frac{5}{2}RT$ that appears in eqn 9B.7 cancels that in eqn 9B.8, so ionization energies and electron affinities can be used directly. A final preliminary point is that the electron-gain enthalpy of a species X is the negative of the ionization enthalpy of its negative ion:

$$\Delta_{\text{eg}}H^{\ominus}(X)=-\Delta_{\text{ion}}H^{\ominus}(X^-) \quad (9B.9)$$

As ionization energy is often easier to measure than electron affinity, this relation can be used to determine numerical values of the latter.

Brief illustration 9B.6 Ionization energy and electron affinity

Tables of thermodynamic data give the standard enthalpies of formation of Na(g) and Na^+(g) at 298.15 K as +107.32 kJ mol^{-1} and +609.358 kJ mol^{-1}, respectively. Therefore, the standard enthalpy of ionization is the difference, +502.04 kJ mol^{-1}. The ionization energy is therefore

$$\begin{aligned} I&=\Delta_{\text{ion}}H^{\ominus}(298.15\,\text{K})-\tfrac{5}{2}R\times(298.15\,\text{K})\\ &=502.04\,\text{kJ mol}^{-1}-6.197\,\text{kJ mol}^{-1}\\ &=495.84\,\text{kJ mol}^{-1}\ (\text{or } 5.139\ \text{eV}) \end{aligned}$$

as in Table 9B.2.

Self-test 9B.6 The standard enthalpies of formation of Cl(g) and Cl^-(g) at 298.15 K are +121.679 kJ mol^{-1} and −233.13 kJ mol^{-1}, respectively. What is the electron affinity of chlorine atoms?

Answer: +348.61 kJ mol^{-1}, +3.613 eV

As will be familiar from introductory chemistry, ionization energies and electron affinities show periodicities. The former is more regular and we concentrate on it. Lithium has a low first ionization energy because its outermost electron is well shielded from the nucleus by the core ($Z_{\text{eff}}=1.3$, compared with $Z=3$). The ionization energy of beryllium ($Z=4$) is greater but that of boron is lower because in the latter the outermost electron occupies a 2p orbital and is less strongly bound than if it had been a 2s electron. The ionization energy increases from boron to nitrogen on account of the increasing nuclear charge. However, the ionization energy of oxygen is less than would be expected by simple extrapolation. The explanation is that at oxygen a 2p orbital must become doubly occupied, and the electron–electron repulsions are increased above what would be expected by simple extrapolation along the row. In addition, the loss of a 2p electron results in a configuration with a half-filled subshell (like that of N), which is an arrangement of low energy, so the energy of $O^+ + e^-$ is lower than might be expected, and the ionization energy is correspondingly low too. (The kink is less pronounced in the next row, between phosphorus and sulfur because their orbitals are more diffuse.) The values for oxygen, fluorine, and neon fall roughly on the same line, the increase of their ionization energies reflecting the increasing attraction of the more highly charged nuclei for the outermost electrons.

The outermost electron in sodium ($Z=11$) is 3s. It is far from the nucleus, and the latter's charge is shielded by the compact, complete neon-like core, with the result that $Z_{\text{eff}} \approx 2.5$. As a result, the ionization energy of sodium is substantially lower than that of neon ($Z=10$, $Z_{\text{eff}} \approx 5.8$). The periodic cycle starts again along this row, and the variation of the ionization energy can be traced to similar reasons.

Electron affinities are greatest close to fluorine, for the incoming electron enters a vacancy in a compact valence shell and can interact strongly with the nucleus. The attachment of an electron to an anion (as in the formation of O^{2-} from O^-) is invariably endothermic, so E_{ea} is negative. The incoming electron is repelled by the charge already present. Electron affinities are also small, and may be negative, when an electron enters an orbital that is far from the nucleus (as in the heavier alkali metal atoms) or is forced by the Pauli principle to occupy a new shell (as in the noble gas atoms).

9B.3 Self-consistent field orbitals

The central difficulty of the Schrödinger equation is the presence of the electron–electron interaction terms. The potential energy of all the electrons in an N-electron atom is

$$V = -\overbrace{\sum_{i=1}^{N} \frac{Ze^2}{4\pi\varepsilon_0 r_i}}^{\text{Attraction to the nucleus}} + \tfrac{1}{2}\overbrace{\sum_{i,j=1}^{N}{}' \frac{e^2}{4\pi\varepsilon_0 r_{ij}}}^{\text{Repulsion between electrons}} \qquad (9B.10)$$

The first term on the right is the total attractive interaction between the electrons and the nucleus. The second term is the total repulsive interaction between the electrons; r_{ij} is the distance between electrons i and j. The prime on the second sum indicates that contributions with $i=j$ are excluded, and the factor of one-half prevents double-counting of electron pair repulsions (1 interacting with 2 is the same as 2 interacting with 1). It is hopeless to expect to find analytical solutions of a Schrödinger equation with such a complicated potential energy term, but computational techniques are available that give very detailed and reliable numerical solutions for the wavefunctions and energies. The techniques were originally introduced by D.R. Hartree (before computers were available) and then modified by V. Fock to take into account the Pauli principle. In broad outline, the **Hartree–Fock self-consistent field** (HF-SCF) procedure is as follows.

Imagine that we have a rough idea of the structure of the atom. In the Ne atom, for instance, the orbital approximation suggests the configuration $1s^22s^22p^6$ with the orbitals approximated by hydrogenic atomic orbitals. Now consider one of the 2p electrons. A Schrödinger equation can be written for this electron by ascribing to it a potential energy due to the nuclear attraction and the repulsion from the other electrons. This equation has the form

$$\begin{aligned}&\hat{H}(1)\psi_{2p}(1) + V(\text{other electrons})\psi_{2p}(1)\\ &\quad - V(\text{exchange correction})\psi_{2p}(1)\\ &\quad = E_{2p}\psi_{2p}(1)\end{aligned} \qquad (9B.11)$$

Although the equation is for an electron in the 2p orbital, it depends on the wavefunctions of all the other occupied orbitals in the atom, and similar equations can be written for them too. The various terms are as follows:

Physical interpretation

- The first term on the left is the contribution of the kinetic energy and the attraction of the electron to the nucleus, just as in a hydrogenic atom
- The second term takes into account the potential energy of the electron of interest due to the electrons in the other occupied orbitals.
- The third term is an *exchange correction* that takes into account the spin correlation effects discussed earlier.

There is no hope of solving eqn 9B.11 analytically. However, it can be solved numerically if we guess an approximate form of

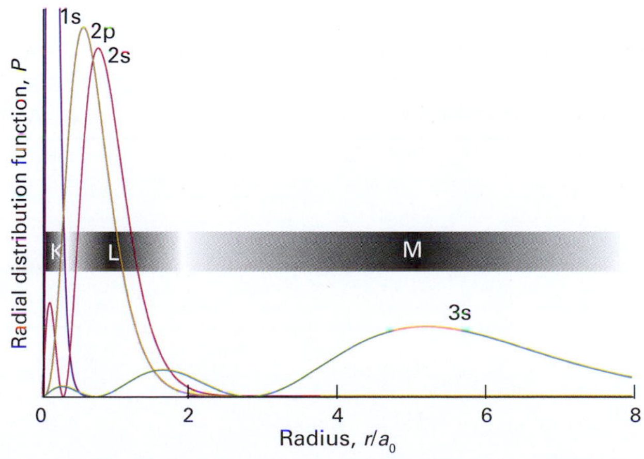

Figure 9B.8 The radial distribution functions for the orbitals of Na based on SCF calculations. Note the shell-like structure, with the 3s orbital outside the inner K and L shells.

the wavefunctions of all the orbitals except 2p. The procedure is then repeated for the other orbitals in the atom, the 1s and 2s orbitals. This sequence of calculations gives the form of the 2p, 2s, and 1s orbitals, and in general they will differ from the set used initially to start the calculation. These improved orbitals can be used in another cycle of calculation, and a second improved set of orbitals is obtained. The recycling continues until the orbitals and energies obtained are insignificantly different from those used at the start of the current cycle. The solutions are then 'self-consistent' and accepted as solutions of the problem.

Figure 9B.8 shows plots of some of the HF-SCF radial distribution functions for sodium. They show the grouping of electron density into shells, as was anticipated by the early chemists, and the differences of penetration as discussed above. These SCF calculations therefore support the qualitative discussions that are used to explain chemical periodicity. They also considerably extend that discussion by providing detailed wavefunctions and precise energies.

Checklist of concepts

- ☐ 1. In the **orbital approximation**, each electron is regarded as occupying its own orbital.
- ☐ 2. A **configuration** is a statement of the occupied orbitals.
- ☐ 3. The **Pauli exclusion principle**, a special case of the Pauli principle, limits to two the number of electrons that can occupy a given orbital.
- ☐ 4. In many-electron atoms, s orbitals lie at a lower energy than p orbitals of the same shell due to the combined effects of **penetration** and **shielding**.
- ☐ 5. The **building-up principle** is a procedure for predicting the ground state electron configuration of an atom.
- ☐ 6. Electrons occupy different orbitals of a given subshell before doubly occupying any one of them.
- ☐ 7. An atom in its ground state adopts a configuration with the greatest number of unpaired electrons.
- ☐ 8. The **ionization energy** and **electron affinity** vary periodically through the periodic table.
- ☐ 9. The Schrödinger equation for many-electron atoms is solved numerically and iteratively until the solutions are self-consistent.

Checklist of equations

Property	Equation	Comment	Equation number
Orbital approximation	$\Psi(r_1,r_2,\ldots)=\psi(r_1)\psi(r_2)\ldots$		9B.1
Effective nuclear charge	$Z_{\text{eff}}=Z-\sigma$	The charge is this number times e	9B.6
Relation between enthalpy of ionization and ionization energy	$\Delta_{\text{ion}}H^{\ominus}(T)=I+\frac{5}{2}RT$		9B.7a
Relation between electron-gain enthalpy and electron affinity	$\Delta_{\text{eg}}H^{\ominus}(T)=-E_{\text{ea}}-\frac{5}{2}RT$		9B.8
Relation between enthalpies	$\Delta_{\text{eg}}H^{\ominus}(\text{X})=-\Delta_{\text{ion}}H^{\ominus}(\text{X}^-)$		9B.9

9C Atomic spectra

Contents

➤ **Why do you need to know this material?**

A knowledge of the energies of electrons in atoms is essential for understanding many chemical properties and concepts, such as chemical bonding and the structure of the periodic table.

➤ **What is the key idea?**

The frequency and wavenumber of radiation emitted when transitions take place provide information on the electronic energy states of atoms.

➤ **What do you need to know already?**

This Topic draws on knowledge of the energy levels of hydrogenic atoms (Topic 9A) and the configurations of many-electron atoms (Topic 9B). In places, it uses the properties of angular momentum (Topic 8C).

The general idea behind atomic spectroscopy is straightforward: lines in the spectrum (in either emission or absorption) occur when the electron distribution in an atom undergoes a transition with a change of energy $|\Delta E|$, and emits or absorbs a photon of frequency $\nu = |\Delta E|/h$ and wavenumber $\tilde{\nu} = |\Delta E|/hc$. Hence, we can expect the spectrum to give information about the energies of electrons in atoms.

9C.1 The spectra of hydrogenic atoms

The energies of the hydrogenic atoms are given in Topic 9A ($E_n = -hcZ^2\tilde{R}_N/n^2$). When the electron undergoes a **transition**, a change of state, from an orbital with quantum numbers n_1, l_1, m_{l1} to another (lower energy) orbital with quantum numbers n_2, l_2, m_{l2}, it undergoes a change of energy ΔE and discards the excess energy as a photon of electromagnetic radiation with a frequency ν given by the Bohr frequency condition (Topic 7A, eqn 7A.12; $\Delta E = h\nu$).

Not all transitions are observed. A photon has an intrinsic spin angular momentum corresponding to $s=1$ (Topic 9B). Because total angular momentum is conserved, the change in angular momentum of the electron must compensate for the angular momentum carried away by the photon. Thus, an electron in a d orbital ($l=2$) cannot make a transition into an s orbital ($l=0$) because the photon cannot carry away enough angular momentum. Similarly, an s electron cannot make a transition to another s orbital, because there would then be no change in the angular momentum of the electron to make up for the angular momentum carried away by the photon. It follows that some spectroscopic transitions are **allowed**, meaning that they can occur, whereas others are **forbidden**, meaning that they cannot occur.

A **selection rule** is a statement about which transitions are allowed. They are derived (for atoms) by identifying the transitions that conserve angular momentum when a photon is emitted or absorbed. We show in the following *Justification* that the selection rules for hydrogenic atoms are

$$\Delta l = \pm 1 \quad \Delta m_l = 0, \pm 1 \qquad \text{Selection rules for hydrogenic atoms} \qquad (9C.1)$$

The principal quantum number n can change by any amount consistent with the Δl for the transition, because it does not relate directly to the angular momentum.

Brief illustration 9C.1 Selection rules

To identify the orbitals to which a 4d electron may make radiative transitions, we first identify the value of l and then apply the selection rule for this quantum number. Because $l=2$, the final orbital must have $l=1$ or 3. Thus, an electron may make a transition from a 4d orbital to any np orbital (subject to $\Delta m_l=0, \pm1$) and to any nf orbital (subject to the same rule). However, it cannot undergo a transition to any other orbital, so a transition to any ns orbital or to another nd orbital is forbidden.

Self-test 9C.1 To what orbitals may a 4s electron make electric-dipole allowed radiative transitions?

Answer: to np orbitals only

Justification 9C.1 The identification of selection rules

The underlying classical idea behind a spectroscopic transition is that, for an atom or molecule to be able to interact with the electromagnetic field and absorb or create a photon of frequency ν, it must possess, at least transiently, a dipole oscillating at that frequency. This transient dipole is expressed quantum mechanically in terms of the **transition dipole moment,** $\boldsymbol{\mu}_{\rm fi}$, between the initial and final states, where[1]

$$\boldsymbol{\mu}_{\rm fi}=\int \psi_{\rm f}^{*}\hat{\boldsymbol{\mu}}\psi_{\rm i}\,{\rm d}\tau \qquad (9C.2)$$

and $\hat{\boldsymbol{\mu}}$ is the electric dipole moment operator. For a one-electron atom $\hat{\boldsymbol{\mu}}$ is multiplication by $-e\boldsymbol{r}$ with components $\mu_x=-ex$, $\mu_y=-ey$, and $\mu_z=-ez$. If the transition dipole moment is zero, then the transition is forbidden; the transition is allowed if the transition moment is non-zero.

To evaluate a transition dipole moment, we consider each component in turn. For example, for the z-component,

$$\mu_{z,\rm fi}=-e\int \psi_{\rm f}^{*}\, z\psi_{\rm i}\,{\rm d}\tau$$

To evaluate the integral, we note from Table 8C.1 that $z=r\cos\theta=(4\pi/3)^{1/2}rY_{1,0}$, so

$$\int \psi_{\rm f}^{*} z\psi_{\rm i}\,{\rm d}\tau=$$

$$\int_0^\infty\int_0^{2\pi}\int_0^\pi \overbrace{R_{n_{\rm f},l_{\rm f}}Y_{l_{\rm f},m_{l,\rm f}}^{*}}^{\psi_{\rm f}^{*}}\overbrace{\left(\frac{4\pi}{3}\right)^{1/2}rY_{1,0}}^{z}\overbrace{R_{n_{\rm i},l_{\rm i}}Y_{l_{\rm i},m_{l,\rm i}}}^{\psi_{\rm i}}\overbrace{r^2{\rm d}r\sin\theta\,{\rm d}\theta\,{\rm d}\phi}^{{\rm d}\tau}$$

This multiple integral is the product of three factors, an integral over r and two integrals (in blue) over the angles, so the factors on the right can be grouped as follows:

$$\int \psi_{\rm f}^{*} z\psi_{\rm i}\,{\rm d}\tau=$$

$$\left(\frac{4\pi}{3}\right)^{1/2}\int_0^\infty R_{n_{\rm f},l_{\rm f}}r^3R_{n_{\rm i},l_{\rm i}}{\rm d}r\int_0^{2\pi}\int_0^\pi Y_{l_{\rm f},m_{l,\rm f}}^{*}Y_{1,0}Y_{l_{\rm i},m_{l,\rm i}}\sin\theta\,{\rm d}\theta\,{\rm d}\phi$$

It follows from the properties of the spherical harmonics that the integral

$$I=\int_0^{2\pi}\int_0^\pi Y_{l_{\rm f},m_{l,\rm f}}^{*}Y_{l,m}Y_{l_{\rm i},m_{l,\rm i}}\sin\theta\,{\rm d}\theta\,{\rm d}\phi$$

is zero unless $l_{\rm f}=l_{\rm i}\pm1$ and $m_{l,\rm f}=m_{l,\rm i}+m$. Because $m=0$ in the present case, the angular integral, and hence the z-component of the transition dipole moment, is zero unless $\Delta l=\pm1$ and $\Delta m_l=0$, which is a part of the set of selection rules. The same procedure, but considering the x- and y-components, results in the complete set of rules.

[1] See our *Physical chemistry: Quanta, matter, and change* (2014) for a detailed development of the form of eqn 9C.2.

The selection rules and the atomic energy levels jointly account for the structure of a **Grotrian diagram** (Fig. 9C.1), which summarizes the energies of the states and the transitions between them. The thicknesses of the transition lines in the diagram denote their relative intensities in the spectrum.

9C.2 The spectra of complex atoms

The spectra of atoms rapidly become very complicated as the number of electrons increases, but there are some important and moderately simple features that make atomic spectroscopy useful in the study of the composition of samples as large and as complex as stars. However, the actual energy levels are not given solely by the energies of the orbitals, because the electrons interact with one another in various ways.

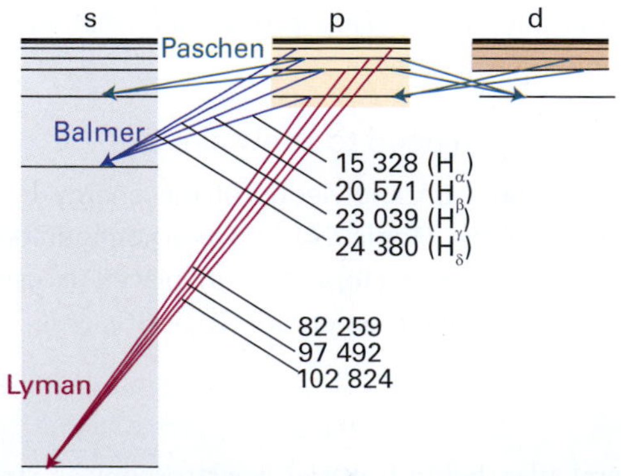

Figure 9C.1 A Grotrian diagram that summarizes the appearance and analysis of the spectrum of atomic hydrogen. The wavenumbers of the transitions (in cm^{-1}) are indicated.

(a) Singlet and triplet states

Suppose we were interested in the energy levels of a He atom, with its two electrons. We know that the ground-state configuration is $1s^2$, and can anticipate that an excited configuration will be one in which one of the electrons has been promoted into a 2s orbital, giving the configuration $1s^12s^1$. The two electrons need not be paired because they occupy different orbitals. According to Hund's maximum multiplicity rule (Topic 9B), the state of the atom with the spins parallel lies lower in energy than the state in which they are paired. Both states are permissible, and can contribute to the spectrum of the atom.

Parallel and antiparallel (paired) spins differ in their overall spin angular momentum. In the paired case, the two spin momenta cancel each other, and there is zero net spin (as depicted in Fig. 9B.2). The paired-spin arrangement is called a **singlet**. Its spin state is the one denoted σ_- in the discussion of the Pauli principle:

$$\sigma_-(2,1)=(1/2^{1/2})\{\alpha(2)\beta(1)-\beta(2)\alpha(1)\} \quad \text{Singlet spin function} \quad (9C.3a)$$

The angular momenta of two parallel spins add together to give a nonzero total spin, and the resulting state is called a **triplet**. As illustrated in Fig. 9C.2, there are three ways of achieving a nonzero total spin, but only one way to achieve zero spin. The three spin states are the symmetric combinations introduced in Topic 9B:

$$\begin{aligned}&\alpha(1)\alpha(2)\\ &\sigma_+(1,2)=(1/2^{1/2})\{\alpha(1)\beta(2)+\beta(1)\alpha(2)\} \quad \text{Triplet spin functions} \quad (9C.3b)\\ &\beta(1)\beta(2)\end{aligned}$$

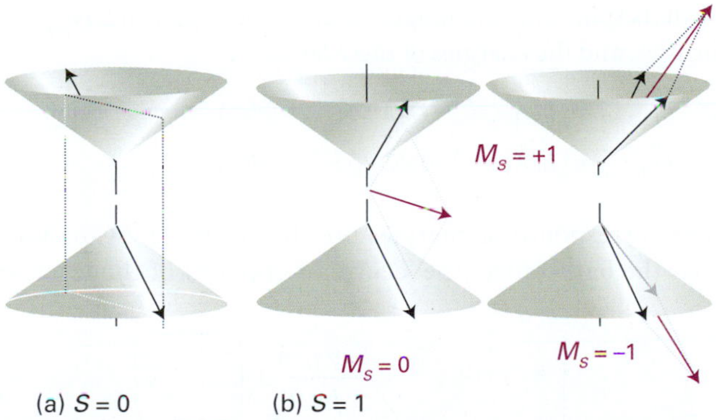

Figure 9C.2 (a) Electrons with paired spins have zero resultant spin angular momentum ($S=0$). They can be represented by two vectors that lie at an indeterminate position on the cones shown here, but wherever one lies on its cone, the other points in the opposite direction; their resultant is zero. (b) When two electrons have parallel spins, they have a nonzero total spin angular momentum ($S=1$). There are three ways of achieving this resultant, which are shown by these vector representations. Note that, whereas two paired spins are precisely antiparallel, two 'parallel' spins are not strictly parallel.

The fact that the parallel arrangement of spins in the $1s^12s^1$ configuration of the He atom lies lower in energy than the antiparallel arrangement can now be expressed by saying that the triplet state of the $1s^12s^1$ configuration of He lies lower in energy than the singlet state. This is a general conclusion that applies to other atoms (and molecules) and, *for states arising from the same configuration, the triplet state generally lies lower than the singlet state*. The origin of the energy difference lies in the effect of spin correlation on the Coulombic interactions between electrons, as in the case of Hund's maximum multiplicity rule for ground-state configurations (Topic 9B). Because the Coulombic interaction between electrons in an atom is strong, the difference in energies between singlet and triplet states of the same configuration can be large. The two states of $1s^12s^1$ He, for instance, differ by 6421 cm^{-1} (corresponding to 0.80 eV).

The spectrum of atomic helium is more complicated than that of atomic hydrogen, but there are two simplifying features. One is that the only excited configurations it is necessary to consider are of the form $1s^1nl^1$; that is, only one electron is excited. Excitation of two electrons requires an energy greater than the ionization energy of the atom, so the He^+ ion is formed instead of the doubly excited atom. Second, no radiative transitions take place between singlet and triplet states because the relative orientation of the two electron spins cannot change during a transition. Thus, there is a spectrum arising from transitions between singlet states (including the ground state) and between triplet states, but not between the two. Spectroscopically, helium behaves like two distinct species, and the early spectroscopists actually thought of helium as consisting of 'parahelium' and 'orthohelium'. The Grotrian diagram for helium in Fig. 9C.3 shows the two sets of transitions.

(b) Spin–orbit coupling

An electron has a magnetic moment that arises from its spin. Similarly, an electron with orbital angular momentum (that is, an electron in an orbital with $l>0$) is in effect a circulating current, and possesses a magnetic moment that arises from its orbital momentum. The interaction of the spin magnetic

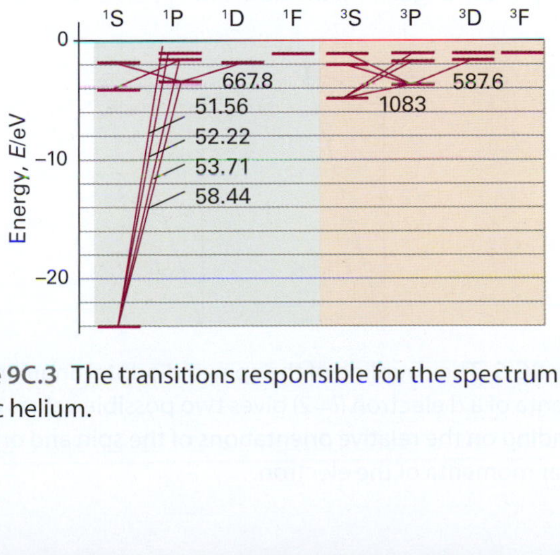

Figure 9C.3 The transitions responsible for the spectrum of atomic helium.

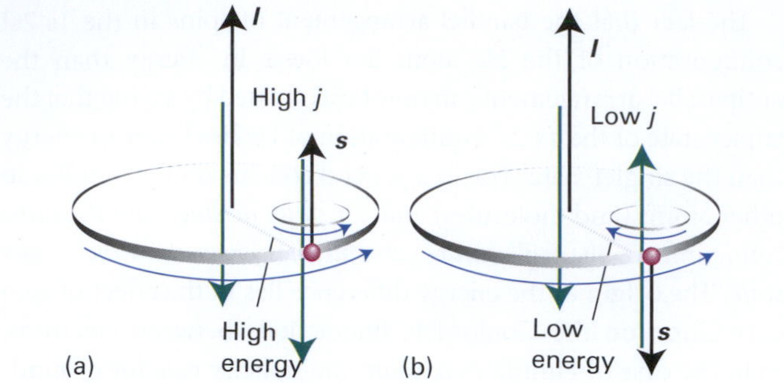

Figure 9C.4 Spin–orbit coupling is a magnetic interaction between spin and orbital magnetic moments. When the angular momenta are parallel, as in (a), the magnetic moments are aligned unfavourably; when they are opposed, as in (b), the interaction is favourable. This magnetic coupling is the cause of the splitting of a configuration into levels.

moment with the magnetic field arising from the orbital angular momentum is called **spin–orbit coupling**. The strength of the coupling, and its effect on the energy levels of the atom, depend on the relative orientations of the spin and orbital magnetic moments, and therefore on the relative orientations of the two angular momenta (Fig. 9C.4).

One way of expressing the dependence of the spin–orbit interaction on the relative orientation of the spin and orbital momenta is to say that it depends on the total angular momentum of the electron, the vector sum of its spin and orbital momenta. Thus, when the spin and orbital angular momenta are nearly parallel, the total angular momentum is high; when the two angular momenta are opposed, the total angular momentum is low.

The total angular momentum of an electron is described by the quantum numbers j and m_j, with $j=l+\frac{1}{2}$ (when the two angular momenta are in the same direction) or $j=l-\frac{1}{2}$ (when they are opposed, as in Fig. 9C.5). The different values of j that can arise for a given value of l label **levels** of a term. For $l=0$, the only permitted value is $j=\frac{1}{2}$ (the total angular momentum is the same as the spin angular momentum because there is no other source of angular momentum in the atom). When $l=1$, j may be either $\frac{3}{2}$ (the spin and orbital angular momenta are in the same sense) or $\frac{1}{2}$ (the spin and angular momenta are in opposite senses).

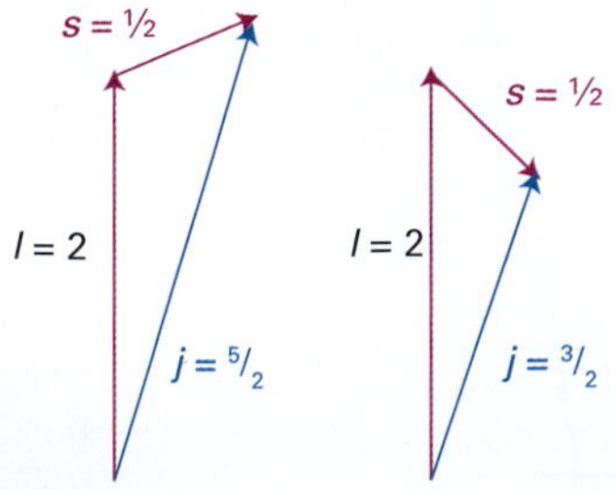

Figure 9C.5 The coupling of the spin and orbital angular momenta of a d electron ($l=2$) gives two possible values of j depending on the relative orientations of the spin and orbital angular momenta of the electron.

Brief illustration 9C.2 The levels of a configuration

To identify the levels that may arise from the configurations (a) d^1, (b) s^1 we need to identify the value of l and then the possible values of j. (a) For a d electron, $l=2$ and there are two levels in the configuration, one with $j=2+\frac{1}{2}=\frac{5}{2}$ and the other with $j=2-\frac{1}{2}=\frac{3}{2}$. (b) For an s electron $l=0$, so only one level is possible, and $j=\frac{1}{2}$.

Self-test 9C.2 Identify the levels of the configurations (a) p^1 and (b) f^1.

Answer: (a) $\frac{3}{2},\frac{1}{2}$; (b) $\frac{7}{2},\frac{5}{2}$

The dependence of the spin–orbit interaction on the value of j is expressed in terms of the **spin–orbit coupling constant**, $\tilde{A}$ (which is typically expressed as a wavenumber). The calculation in the following *Justification* leads to the result that the energies of the levels with quantum numbers s, l, and j are given by

$$E_{l,s,j}=\tfrac{1}{2}hc\tilde{A}\{j(j+1)-l(l+1)-s(s+1)\} \qquad (9C.4)$$

Brief illustration 9C.3 Spin–orbit coupling

The unpaired electron in the ground state of an alkali metal atom has $l=0$, so $j=\frac{1}{2}$. Because the orbital angular momentum is zero in this state, the spin–orbit coupling energy is zero (as is confirmed by setting $j=s$ and $l=0$ in eqn 9C.4). When the electron is excited to an orbital with $l=1$, it has orbital angular momentum and can give rise to a magnetic field that interacts with its spin. In this configuration the electron can have $j=\frac{3}{2}$ or $j=\frac{1}{2}$, and the energies of these levels are

$$E_{1,1/2,3/2}=\tfrac{1}{2}hc\tilde{A}\{\tfrac{3}{2}\times\tfrac{5}{2}-1\times2-\tfrac{1}{2}\times\tfrac{3}{2}\}=\tfrac{1}{2}hc\tilde{A}$$
$$E_{1,1/2,1/2}=\tfrac{1}{2}hc\tilde{A}\{\tfrac{1}{2}\times\tfrac{3}{2}-1\times2-\tfrac{1}{2}\times\tfrac{3}{2}\}=-hc\tilde{A}$$

The corresponding energies are shown in Fig. 9C.6. Note that the barycentre (the 'centre of gravity') of the levels is

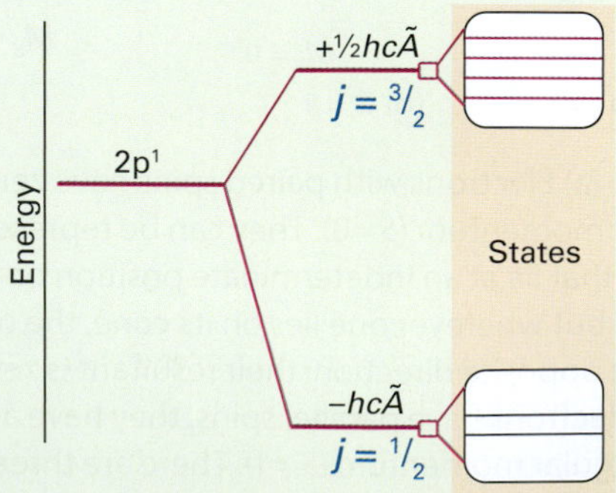

Figure 9C.6 The levels of a ^{2}P term arising from spin–orbit coupling. Note that the low-j level lies below the high-j level in energy.

unchanged, because there are four states of energy $\frac{1}{2}hc\tilde{A}$ and two of energy $-hc\tilde{A}$.

Self-test 9C.3 What are the energies of the two terms that can arise from a d^1 configuration?

Answer: $E_{2,1/2,5/2}=2hc\tilde{A}$, $E_{2,1/2,3/2}=-3hc\tilde{A}$

Justification 9C.2 The energy of spin–orbit interaction

The energy of a magnetic moment $\boldsymbol{\mu}$ in a magnetic field $\boldsymbol{\mathcal{B}}$ is equal to their scalar product $-\boldsymbol{\mu}\cdot\boldsymbol{\mathcal{B}}$. If the magnetic field arises from the orbital angular momentum of the electron, it is proportional to $\boldsymbol{l}$; if the magnetic moment $\boldsymbol{\mu}$ is that of the electron spin, then it is proportional to $\boldsymbol{s}$. It then follows that the energy of interaction is proportional to the scalar product $\boldsymbol{s}\cdot\boldsymbol{l}$:

$$\text{Energy of interaction} = -\boldsymbol{\mu}\cdot\boldsymbol{\mathcal{B}} \propto \boldsymbol{s}\cdot\boldsymbol{l}$$

(For the various vector manipulations used in this section, see *Mathematical background* 5.) Next, we note that the total angular momentum is the vector sum of the spin and orbital momenta: $\boldsymbol{j}=\boldsymbol{l}+\boldsymbol{s}$. The magnitude of the vector $\boldsymbol{j}$ is calculated by evaluating

$$\boldsymbol{j}\cdot\boldsymbol{j}=(\boldsymbol{l}+\boldsymbol{s})\cdot(\boldsymbol{l}+\boldsymbol{s})=\boldsymbol{l}\cdot\boldsymbol{l}+\boldsymbol{s}\cdot\boldsymbol{s}+2\boldsymbol{s}\cdot\boldsymbol{l}$$

so that

$$j^2=l^2+s^2+2\boldsymbol{s}\cdot\boldsymbol{l}$$

That is,

$$\boldsymbol{s}\cdot\boldsymbol{l}=\tfrac{1}{2}\{j^2-l^2-s^2\}$$

This equation is a classical result. To make the transition to quantum mechanics, we replace all the quantities on the right with their quantum-mechanical values (Topic 8C):

$$\boldsymbol{s}\cdot\boldsymbol{l}=\tfrac{1}{2}\{j(j+1)-l(l+1)-s(s+1)\}\hbar^2$$

Then, by inserting this expression into the formula for the energy of interaction ($E\propto\boldsymbol{s}\cdot\boldsymbol{l}$) and writing the constant of proportionality as $hc\tilde{A}/\hbar^2$, we obtain eqn 9C.4.

The strength of the spin–orbit coupling depends on the nuclear charge. To understand why this is so, imagine riding on the orbiting electron and seeing a charged nucleus apparently orbiting around us (like the Sun rising and setting). As a result, we find ourselves at the centre of a ring of current. The greater the nuclear charge, the greater this current, and therefore the stronger the magnetic field we detect. Because the spin magnetic moment of the electron interacts with this orbital magnetic field, it follows that the greater the nuclear charge, the stronger the spin–orbit interaction. The coupling increases sharply with atomic number (as Z^4). Whereas it is only small in H (giving rise to shifts of energy levels of no more than about 0.4 cm^{-1}), in heavy atoms like Pb it is very large (giving shifts of the order of thousands of reciprocal centimetres).

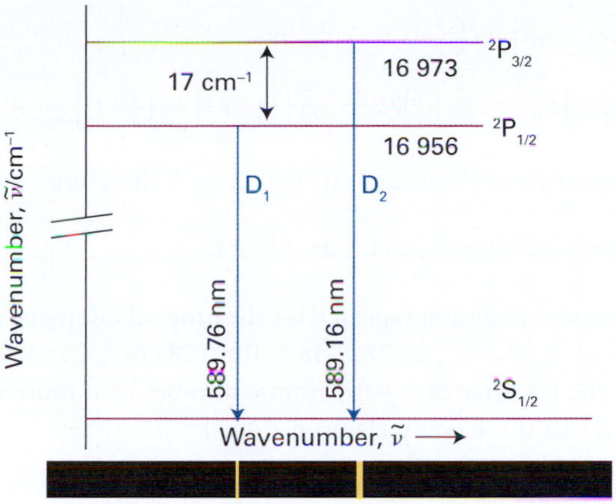

Figure 9C.7 The energy-level diagram for the formation of the sodium D lines. The splitting of the spectral lines (by 17 cm^{-1}) reflects the splitting of the levels of the ^{2}P term.

Two spectral lines are observed when the p electron of an electronically excited alkali metal atom undergoes a transition and falls into a lower *s* orbital. One line is due to a transition starting in a $j=\frac{3}{2}$ level and the other line is due to a transition starting in the $j=\frac{1}{2}$ level of the same configuration. The two lines are jointly an example of the **fine structure** of a spectrum, the structure in a spectrum due to spin–orbit coupling. Fine structure can be clearly seen in the emission spectrum from sodium vapour excited by an electric discharge (for example, in one kind of street lighting). The yellow line at 589 nm (close to 17 000 cm^{-1}) is actually a doublet composed of one line at 589.76 nm (16 956.2 cm^{-1}) and another at 589.16 nm (16 973.4 cm^{-1}); the components of this doublet are the 'D lines' of the spectrum (Fig. 9C.7). Therefore, in Na, the spin–orbit coupling affects the energies by about 17 cm^{-1}.

Example 9C.1 Analysing a spectrum for the spin–orbit coupling constant

The origin of the D lines in the spectrum of atomic sodium is shown in Fig. 9C.7. Calculate the spin–orbit coupling constant for the upper configuration of the Na atom.

Method We see from Fig. 9C.7 that the splitting of the lines is equal to the energy separation of the $j=\frac{3}{2}$ and $\frac{1}{2}$ levels of the excited configuration. This separation can be expressed in terms of $\tilde{A}$ by using eqn 9C.4. Therefore, set the observed splitting equal to the energy separation calculated from eqn 9C.4 and solve the equation for $\tilde{A}$.

Answer The two levels are split by

$$\Delta\tilde{\nu}=\left(E_{0,\frac{1}{2},\frac{3}{2}}-E_{0,\frac{1}{2},\frac{1}{2}}\right)/hc=\tfrac{1}{2}\tilde{A}\left\{\tfrac{3}{2}\left(\tfrac{3}{2}+1\right)-\tfrac{1}{2}\left(\tfrac{1}{2}+1\right)\right\}=\tfrac{3}{2}\tilde{A}$$

The experimental value of $\Delta\tilde{\nu}$ is 17.2 cm^{-1}; therefore

$$\tilde{A}=\tfrac{2}{3}\times(17.2\,\text{cm}^{-1})=11.5\,\text{cm}^{-1}$$

The same calculation repeated for the other alkali metal atoms gives Li: 0.23 cm^{-1}, K: 38.5 cm^{-1}, Rb: 158 cm^{-1}, Cs: 370 cm^{-1}. Note the increase of A with atomic number (but more slowly than Z^4 for these many-electron atoms).

Self-test 9C.4 The configuration ...4p^{6}5d^1 of rubidium has two levels at 25 700.56 cm^{-1} and 25 703.52 cm^{-1} above the ground state. What is the spin–orbit coupling constant in this excited state?

Answer: 1.18 cm^{-1}

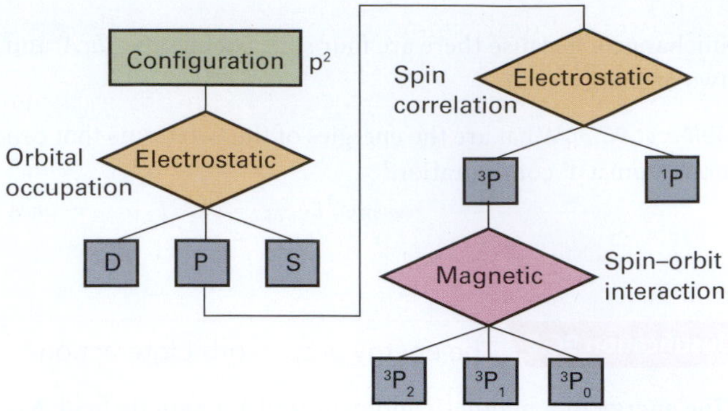

Figure 9C.8 A summary of the types of interaction that are responsible for the various kinds of splitting of energy levels in atoms. For light atoms, magnetic interactions are small, but in heavy atoms they may dominate the electrostatic (charge–charge) interactions.

(c) Term symbols

We have used expressions such as 'the $j=\frac{3}{2}$ level of a doublet term with $L=1$'. A **term symbol**, which is a symbol looking like $^2P_{3/2}$ or $^3D_2$, conveys this information, specifically the total spin, total orbital angular momentum, and total overall angular momentum, very succinctly.

A term symbol gives three pieces of information:

- The letter (P or D in the examples) indicates the total orbital angular momentum quantum number, L.
- The left superscript in the term symbol (the 2 in $^2P_{3/2}$) gives the multiplicity of the term.
- The right subscript on the term symbol (the $\frac{3}{2}$ in $^2P_{3/2}$) is the value of the total angular momentum quantum number, J.

We shall now say what each of these statements means; the contributions to the energies which we are about to discuss are summarized in Fig. 9C.8.

When several electrons are present, it is necessary to judge how their individual orbital angular momenta add together to augment or oppose each other. The **total orbital angular momentum quantum number**, L, tells us the magnitude of the angular momentum through $\{L(L+1)\}^{1/2}\hbar$. It has $2L+1$ orientations distinguished by the quantum number M_L, which can take the values $L, L-1, \ldots, -L$. Similar remarks apply to the **total spin quantum number**, S, and the quantum number M_S, and the **total angular momentum quantum number**, J, and the quantum number M_J.

The value of L (a non-negative integer) is obtained by coupling the individual orbital angular momenta by using the **Clebsch–Gordan series**:

$$L=l_1+l_2, l_1+l_2-1, \ldots, |l_1-l_2| \qquad \text{Clebsch–Gordan series} \qquad (9C.5)$$

The modulus signs are attached to l_1-l_2 because L is non-negative. The maximum value, $L=l_1+l_2$, is obtained when the two orbital angular momenta are in the same direction; the lowest value, $|l_1-l_2|$, is obtained when they are in opposite directions. The intermediate values represent possible intermediate relative orientations of the two momenta (Fig. 9C.9). For two p electrons (for which $l_1=l_2=1$), $L=2, 1, 0$. The code for converting the value of L into a letter is the same as for the s, p, d, f, ... designation of orbitals, but uses uppercase Roman letters (the convention of using lowercase letters to label orbitals and uppercase letters to label overall states applies throughout spectroscopy, not just to atoms):

L:	0	1	2	3	4	5	6...
	S	P	D	F	G	H	I...

Thus, a p^2 configuration for which $L=2, 1, 0$ can give rise to D, P, and S terms. The terms differ in energy on account of the

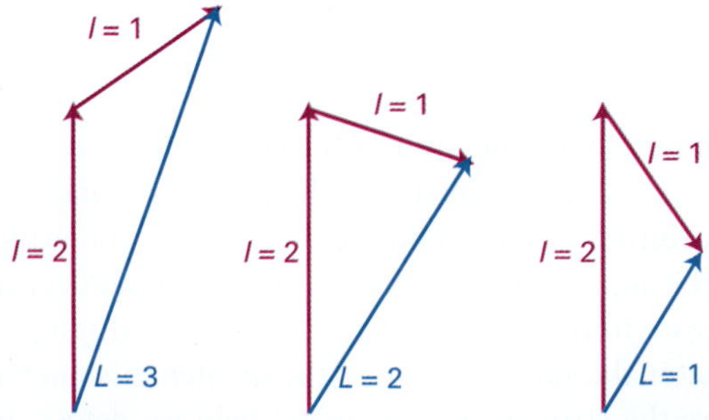

Figure 9C.9 The total angular orbital momenta of a p electron and a d electron correspond to $L=3$, 2, and 1 and reflect the different relative orientations of the two momenta.

different spatial distribution of the electrons and the consequent differences in repulsion between them.

A closed shell has zero orbital angular momentum because all the individual orbital angular momenta sum to zero. Therefore, when working out term symbols, we need consider only the electrons of the unfilled shell. In the case of a single electron outside a closed shell, the value of L is the same as the value of l; so the configuration [Ne]$3s^1$ has only an S term.

Example 9C.2 Deriving the total orbital angular momentum of a configuration

Find the terms that can arise from the configurations (a) d^2, (b) p^3.

Method Use the Clebsch–Gordan series and begin by finding the minimum value of L (so that we know where the series terminates). When there are more than two electrons to couple together, use two series in succession: first couple two electrons, and then couple the third to each combined state, and so on.

Answer (a) The minimum value is $|l_1 - l_2| = |2-2| = 0$. Therefore,

$$L = 2+2, 2+2-1, \ldots, 0 = 4, 3, 2, 1, 0$$

corresponding to G, F, D, P, S terms, respectively. (b) Coupling two electrons gives a minimum value of $|1-1|=0$. Therefore,

$$L' = 1+1, 1+1-1, \ldots, 0 = 2, 1, 0$$

Now couple l_3 with $L'=2$, to give $L=3, 2, 1$; with $L'=1$, to give $L=2, 1, 0$; and with $L'=0$, to give $L=1$. The overall result is

$$L = 3, 2, 2, 1, 1, 1, 0$$

giving one F, two D, three P, and one S term.

Self-test 9C.5 Repeat the question for the configurations (a) f^1d^1 and (b) d^3.

Answer: (a) H, G, F, D, P; (b) I, 2H, 3G, 4F, 5D, 3P, S

A note on good practice Throughout our discussion of atomic spectroscopy, distinguish italic S, the total spin quantum number, from Roman S, the term label.

When there are several electrons to be taken into account, we must assess their total spin angular momentum quantum number, S (a non-negative integer or half integer). Once again, we use the Clebsch–Gordan series in the form

$$S = s_1 + s_2, s_1 + s_2 - 1, \ldots, |s_1 - s_2| \quad (9C.6)$$

to decide on the value of S, noting that each electron has $s=\frac{1}{2}$, which gives $S=1, 0$ for two electrons (Fig. 9C.10). If there are three electrons, the total spin angular momentum is obtained by coupling the third spin to each of the values of S for the first two spins, which results in $S=\frac{3}{2}$ and $\frac{1}{2}$.

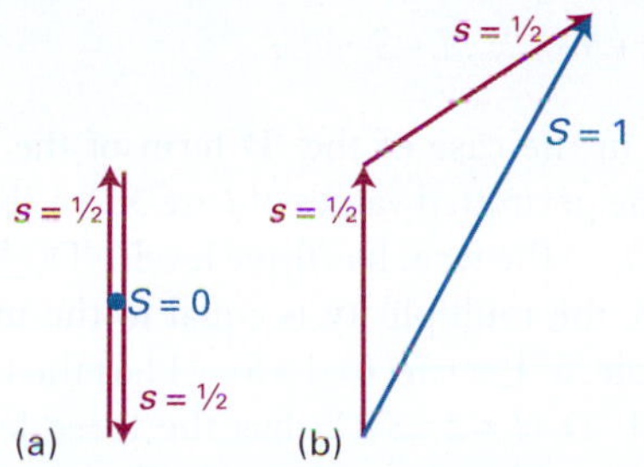

Figure 9C.10 For two electrons (each of which has $s=\frac{1}{2}$, only two total spin states are permitted ($S=0, 1$). (a) The state with $S=0$ can have only one value of M_S ($M_S=0$) and is a singlet; (b) the state with $S=1$ can have any of three values of M_S (+1, 0, −1) and is a triplet. The vector representations of the singlet and triplet states are shown in Figs. 9C.2.

The **multiplicity** of a term is the value of $2S+1$. When $S=0$ (as for a closed shell, like $1s^2$) the electrons are all paired and there is no net spin: this arrangement gives a singlet term, ^{1}S. A single electron has $S=s=\frac{1}{2}$, so a configuration such as [Ne]$3s^1$ can give rise to a doublet term, ^{2}S. Likewise, the configuration [Ne]$3p^1$ is a doublet, ^{2}P. When there are two unpaired electrons $S=1$, so $2S+1=3$, giving a triplet term, such as ^{3}D. We discussed the relative energies of singlets and triplets earlier in the Topic and saw that their energies differ on account of the different effects of spin correlation.

As we have seen, the quantum number j tells us the relative orientation of the spin and orbital angular momenta of a single electron. The **total angular momentum quantum number**, J (a non-negative integer or half integer), does the same for several electrons. If there is a single electron outside a closed shell, $J=j$, with j either $l=\frac{1}{2}$ or $|l-\frac{1}{2}|$. The [Ne]$3s^1$ configuration has $j=\frac{1}{2}$ (because $l=0$ and $s=\frac{1}{2}$), so the ^{2}S term has a single level, which we denote $^2S_{1/2}$. The [Ne]$3p^1$ configuration has $l=1$; therefore $j=\frac{3}{2}$ and $\frac{1}{2}$; the ^{2}P term therefore has two levels, $^2P_{3/2}$ and $^2P_{1/2}$. These levels lie at different energies on account of the magnetic spin–orbit interaction.

If there are several electrons outside a closed shell we have to consider the coupling of all the spins and all the orbital angular momenta. This complicated problem can be simplified when the spin–orbit coupling is weak (for atoms of low atomic number), for then we can use the **Russell–Saunders coupling** scheme. This scheme is based on the view that, if spin–orbit coupling is weak, then it is effective only when all the orbital momenta are operating cooperatively. We therefore imagine that all the orbital angular momenta of the electrons couple to give a total L, and that all the spins are similarly coupled to give a total S. Only at this stage do we imagine the two kinds of momenta coupling through the spin–orbit interaction to give a total J. The permitted values of J are given by the Clebsch–Gordan series

$$J = L+S, L+S-1, \ldots, |L-S| \tag{9C.7}$$

For example, in the case of the ^{3}D term of the configuration [Ne]$2p^13p^1$, the permitted values of J are 3, 2, 1 (because ^{3}D has $L=2$ and $S=1$), so the term has three levels, $^3D_3$, $^3D_2$, and $^3D_1$.

When $L \geq S$, the multiplicity is equal to the number of levels. For example, a ^{2}P term ($L=1>S=\frac{1}{2}$) has the two levels $^2P_{3/2}$ and $^2P_{1/2}$, and ^{3}D ($L=2>S=1$) has the three levels $^3D_3$, $^3D_2$, and $^3D_1$. However, this is not the case when $L<S$: the term ^{2}S ($L=0<S=\frac{1}{2}$), for example, has only the one level $^2S_{1/2}$.

Example 9C.3 Deriving term symbols

Write the term symbols arising from the ground-state configurations of (a) Na and (b) F, and (c) the excited configuration $1s^22s^22p^13p^1$ of C.

Method Begin by writing the configurations, but ignore inner closed shells. Then couple the orbital momenta to find L and the spins to find S. Next, couple L and S to find J. Finally, express the term as $^{2S+1}\{L\}_J$, where $\{L\}$ is the appropriate letter. For F, for which the valence configuration is $2p^5$, treat the single gap in the closed-shell $2p^6$ configuration as a single spin-$\frac{1}{2}$ particle.

Answer (a) For Na, the configuration is [Ne]$3s^1$, and we consider the single 3s electron. Because $L=l=0$ and $S=s=\frac{1}{2}$, it is possible for $J=j=s=\frac{1}{2}$ only. Hence the term symbol is $^2S_{1/2}$. (b) For F, the configuration is [He]$2s^22p^5$, which we can treat as [Ne]$2p^{-1}$ (where the notation $2p^{-1}$ signifies the absence of a 2p electron). Hence $L=1$, and $S=s=\frac{1}{2}$. Two values of $J=j$ are allowed: $J=\frac{3}{2},\frac{1}{2}$. Hence, the term symbols for the two levels are $^2P_{3/2}$, $^2P_{1/2}$. (c) We are treating an excited configuration of carbon because, in the ground configuration, $2p^2$, the Pauli principle forbids some terms, and deciding which survive (^{1}D, ^{3}P, ^{1}S, in fact) is quite complicated. That is, there is a distinction between 'equivalent electrons', which are electrons that occupy the same orbitals, and 'inequivalent electrons', which are electrons that occupy different orbitals. The excited configuration of C under consideration is effectively $2p^13p^1$. This is a two-electron problem, and $l_1=l_2=1$, $s_1=s_2=\frac{1}{2}$. It follows that $L=2, 1, 0$ and $S=1, 0$. The terms are therefore ^{3}D and ^{1}D, ^{3}P and ^{1}P, and ^{3}S and ^{1}S. For ^{3}D, $L=2$ and $S=1$; hence $J=3, 2, 1$ and the levels are $^3D_3$, $^3D_2$, and $^3D_1$. For ^{1}D, $L=2$ and $S=0$, so the single level is $^1D_2$. The triplet of levels of ^{3}P is $^3P_2$, $^3P_1$, and $^3P_0$, and the singlet is $^1P_1$. For the ^{3}S term there is only one level, $^3S_1$ (because $J=1$ only), and the singlet term is $^1S_0$.

Self-test 9C.6 Write down the terms arising from the configurations (a) $2s^12p^1$, (b) $2p^13d^1$.

Answer: (a) $^3P_2$, $^3P_1$, $^3P_0$, $^1P_1$; (b) $^3F_4$, $^3F_3$, $^3F_2$, $^1F_3$, $^3D_3$, $^3D_2$, $^3D_1$, $^1D_2$, $^3P_1$, $^3P_0$, $^1P_1$

Russell–Saunders coupling fails when the spin–orbit coupling is large (in heavy atoms, those with high Z). In that case, the individual spin and orbital momenta of the electrons are coupled into individual j values; then these momenta are combined into a grand total, J. This scheme is called ***jj*-coupling**. For example, in a p^2 configuration, the individual values of j are $\frac{3}{2}$ and $\frac{1}{2}$ for each electron. If the spin and the orbital angular momentum of each electron are coupled together strongly, it is best to consider each electron as a particle with angular momentum $j=\frac{3}{2}$ or $\frac{1}{2}$. These individual total momenta then couple as follows:

j_1	j_2	J
$\frac{3}{2}$	$\frac{3}{2}$	3, 2, 1, 0
$\frac{3}{2}$	$\frac{1}{2}$	2, 1
$\frac{1}{2}$	$\frac{3}{2}$	2, 1
$\frac{3}{2}$	$\frac{3}{2}$	1, 0

For heavy atoms, in which jj-coupling is appropriate, it is best to discuss their energies using these quantum numbers.

Although jj-coupling should be used for assessing the energies of heavy atoms, the term symbols derived from Russell–Saunders coupling can still be used as labels. To see why this procedure is valid, we need to examine how the energies of the atomic states change as the spin–orbit coupling increases in strength. Such a **correlation diagram** is shown in Fig. 9C.11. It shows that there is a correspondence between the low spin–orbit coupling (Russell–Saunders coupling) and high spin–orbit coupling (jj-coupling) schemes, so the labels derived by using the Russell–Saunders scheme can be used to label the states of the jj-coupling scheme.

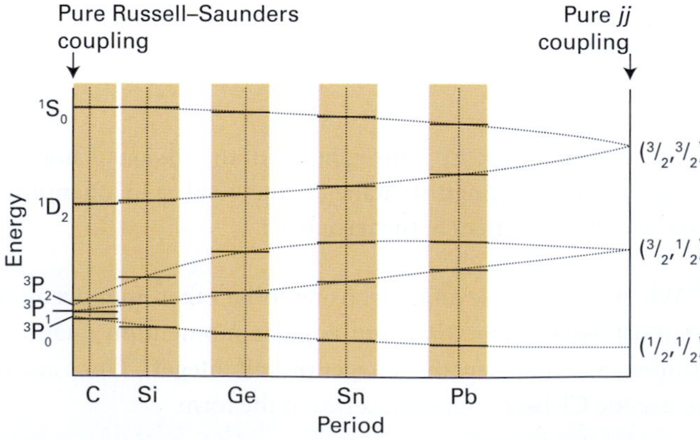

Figure 9C.11 The correlation diagram for some of the states of a two-electron system. All atoms lie between the two extremes, but the heavier the atom, the closer it lies to the pure jj-coupling case.

(d) Hund's rules

As we have remarked, the terms arising from a given configuration differ in energy because they represent different relative orientations of the angular momenta of the electrons and therefore different spatial distributions. The terms arising from the ground-state configuration of an atom (and less reliably from other configurations) can be put into the order of increasing energy by using **Hund's rules**, which summarize the preceding discussion:

1. For a given configuration, the term of greatest multiplicity lies lowest in energy.

As discussed in Topic 9B, this rule is a consequence of electron spin correlation, the quantum-mechanical tendency of electrons of the same spin orientation to stay apart from one another.

2. For a given multiplicity, the term with the highest value of L lies lowest in energy.

This rule can be explained classically by noting that two electrons have a high orbital angular momentum if they circulate in the same direction, in which case they can stay apart. If they circulate in opposite directions, they meet. Thus, a D term is expected to lie lower in energy than an S term of the same multiplicity.

3. For atoms with less than half-filled shells, the level with the lowest value of J lies lowest in energy; for more than half-filled shells, the highest value of J.

This rule stems from considerations of spin–orbit coupling. Thus, for a state of low J, the orbital and spin angular momenta lie in opposite directions, and so too do the corresponding magnetic moments. In classical terms the magnetic moments are then antiparallel, with the N pole of one close to the S pole of the other, which is a low-energy arrangement.

(e) Selection rules

Any state of the atom, and any spectral transition, can be specified by using term symbols. For example, the transitions giving rise to the yellow sodium doublet (which were shown in Fig. 9C.7) are

$$3p^1\,{}^2P_{3/2} \rightarrow 3s^1\,{}^2S_{1/2} \quad 3p^1\,{}^2P_{1/2} \rightarrow 3s^1\,{}^2S_{1/2}$$

By convention, the upper term precedes the lower. The corresponding absorptions are therefore denoted ${}^2P_{3/2} \leftarrow {}^2S_{1/2}$ and ${}^2P_{1/2} \leftarrow {}^2S_{1/2}$. (The configurations have been omitted.)

We have seen (in Section 9C.1) that selection rules arise from the conservation of angular momentum during a transition and from the fact that a photon has a spin of 1. They can therefore be expressed in terms of the term symbols, because the latter carry information about angular momentum. A detailed analysis leads to the following rules:

$$\Delta S = 0 \quad \Delta L = 0, \pm 1 \quad \Delta J = 0, \pm 1 \quad \text{but } J = 0 \leftarrow\!|\!\rightarrow J = 0$$

Selection rules for atoms (9C.8)

where the symbol $\leftarrow\!|\!\rightarrow$ denotes a forbidden transition. The rule about ΔS (no change of overall spin) stems from the fact that the light does not affect the spin directly. The rules about ΔL and Δl express the fact that the orbital angular momentum of an individual electron must change (so $\Delta l = \pm 1$), but whether or not this results in an overall change of orbital momentum depends on the coupling.

The selection rules given above apply when Russell–Saunders coupling is valid (in light atoms, those of low Z). If we insist on labelling the terms of heavy atoms with symbols like ${}^3D$, then we shall find that the selection rules progressively fail as the atomic number increases because the quantum numbers S and L become ill defined as jj-coupling becomes more appropriate. As explained above, Russell–Saunders term symbols are only a convenient way of labelling the terms of heavy atoms: they do not bear any direct relation to the actual angular momenta of the electrons in a heavy atom. For this reason, transitions between singlet and triplet states (for which $\Delta S = \pm 1$), while forbidden in light atoms, are allowed in heavy atoms.

Checklist of concepts

- ☐ 1. Two electrons with paired spins form a singlet state; if their spins are parallel, they form a **triplet state**.
- ☐ 2. The orbital and spin angular momenta interact magnetically.
- ☐ 3. **Spin–orbit coupling** results in the levels of a term having different energies.
- ☐ 4. Fine structure in a spectrum is due to transitions to different levels of a term.
- ☐ 5. A **term symbol** specifies the angular momentum states of an atom.
- ☐ 6. Angular momenta are combined into a resultant by using the **Clebsch–Gordan series**.
- ☐ 7. The **multiplicity** of a term is the value of $2S+1$.
- ☐ 8. The total angular momentum in light atoms is obtained on the basis of **Russell–Saunders coupling**; in heavy atoms, ***jj*-coupling** is used.

☐ 9. Selection rules for light atoms include the fact that changes of total spin do not occur.

☐ 10. **Hund's rules** can be expressed as:

- The term with the maximum multiplicity lies lowest in energy.
- For a given multiplicity, the term with the highest value of L lies lowest in energy.
- For atoms with less than half-filled shells, the level with the lowest value of J lies lowest in energy; for more than half-filled shells, the highest value of J.

Checklist of equations

Property	Equation	Comment	Equation number
Spin–orbit coupling energies	$E_{l,s,j}=\frac{1}{2}hc\tilde{A}\{j(j+1)-l(l+1)-s(s+1)\}$		9C.4
Clebsch–Gordan series	$J=j_1+j_2, j_1+j_2-1, \ldots, \lvert j_1-j_2\rvert$	J, j denote any kind of angular momentum	9C.5
Selection rules	$\Delta S=0$, $\Delta L=0, \pm1$, $\Delta l=\pm1$, $\Delta J=0, \pm1$, but $J=0 \leftarrow\mid\rightarrow J=0$	Light atoms	9C.8

CHAPTER 9 Atomic structure and atomic spectra

TOPIC 9A Hydrogenic atoms

Discussion questions

9A.1 Describe the separation of variables procedure as it is applied to simplify the description of a hydrogenic atom free to move through space.

9A.2 List and describe the significance of the quantum numbers needed to specify the internal state of a hydrogenic atom.

9A.3 Explain the significance of (a) a boundary surface and (b) the radial distribution function for hydrogenic orbitals.

Exercises

9A.1(a) State the orbital degeneracy of the levels in a hydrogen atom that have energy (i) $-hc\tilde{R}_H$; (ii) $-\frac{1}{9}hc\tilde{R}_H$; (iii) $-\frac{1}{25}hc\tilde{R}_H$.
9A.1(b) State the orbital degeneracy of the levels in a hydrogenic atom (Z in parentheses) that have energy (i) $-4hc\tilde{R}_N$, (2); (ii) $-\frac{1}{4}hc\tilde{R}_N$ (4), and (iii) $-hc\tilde{R}_N$ (5).

9A.2(a) The wavefunction for the ground state of a hydrogen atom is Ne^{-r/a_0}. Determine the normalization constant N.
9A.2(b) The wavefunction for the 2s orbital of a hydrogen atom is $N(2-r/a_0)e^{-r/2a_0}$. Determine the normalization constant N.

9A.3(a) By differentiation of the 2s radial wavefunction, show that it has two extrema in its amplitude, and locate them.
9A.3(b) By differentiation of the 3s radial wavefunction, show that it has three extrema in its amplitude, and locate them.

9A.4(a) At what radius does the probability of finding an electron at a point in the H atom fall to 50 per cent of its maximum value?
9A.4(b) At what radius in the H atom does the radial distribution function of the ground state have (i) 50 per cent, (ii) 75 per cent of its maximum value.

9A.5(a) Locate the radial nodes in the 3s orbital of an H atom.
9A.5(b) Locate the radial nodes in the 4p orbital of an H atom. A 4p orbital is proportional to $(20-10\rho+\rho^2)\rho e^{-\rho/2}$.

9A.6(a) Calculate the average kinetic and potential energies of an electron in the ground state of a hydrogen atom.
9A.6(b) Calculate the average kinetic and potential energies of a 2s electron in a hydrogenic atom of atomic number Z.

9A.7(a) Write down the expression for the radial distribution function of a 2s electron in a hydrogenic atom of atomic number Z and determine the radius at which the electron is most likely to be found.
9A.7(b) Write down the expression for the radial distribution function of a 3s electron in a hydrogenic atom of atomic number Z and determine the radius at which the electron is most likely to be found.

9A.8(a) Write down the expression for the radial distribution function of a 2p electron in a hydrogenic atom of atomic number Z and determine the radius at which the electron is most likely to be found.
9A.8(b) Write down the expression for the radial distribution function of a 3p electron in a hydrogenic atom of atomic number Z and determine the radius at which the electron is most likely to be found.

9A.9(a) What is the orbital angular momentum of an electron in the orbitals (i) 1s, (ii) 3s, (iii) 3d? Give the numbers of angular and radial nodes in each case.
9A.9(b) What is the orbital angular momentum of an electron in the orbitals (i) 4d, (ii) 2p, (iii) 3p? Give the numbers of angular and radial nodes in each case.

9A.10(a) Locate the radial nodes and nodal planes of each of the 3p orbitals of a hydrogenic atom of atomic number Z. To locate the angular nodes, give the angle that the plane makes with the z-axis.
9A.10(b) Locate the radial nodes and nodal planes of each of the 4d orbitals of a hydrogenic atom of atomic number Z. To locate the angular nodes, give the angle that the plane or cone makes with the z-axis. A 4d wavefunction is proportional to $(6-\rho)^{2e^{-\rho/2}}$.

Problems

9A.1 What is the most probable point (not radius) that a 2p electron will be found in the hydrogen atom?

9A.2 Show by explicit integration that (a) hydrogenic 1s and 2s orbitals, (b) $2p_x$ and $2p_y$ orbitals are mutually orthogonal.

9A.3 Explicit expressions for hydrogenic orbitals are given in Tables 8C.1 (for the angular component) and 9A.1 (for the radial component). (a) Verify both that the $3p_x$ orbital is normalized (to 1) and that $3p_x$ and $3d_{xy}$ are mutually orthogonal. (b) Determine the positions of both the radial nodes and nodal planes of the 3s, $3p_x$ and $3d_{xy}$ orbitals. (c) Determine the mean radius of the 3s orbital. (d) Draw a graph of the radial distribution function for the three orbitals (of part (b)) and discuss the significance of the graphs for interpreting the properties of many-electron atoms. (e) Create both xy-plane polar plots and boundary surface plots for these orbitals. Construct the boundary plots so that the distance from the origin to the surface is the absolute value of the angular part of the wavefunction. Compare the s, p, and d boundary surface plots with that of an f orbital; e.g., $\psi_f \propto x(5z^2-r^2) \propto \sin\theta\,(5\cos^2\theta-1)\cos\phi$.

9A.4 Determine whether the p_x and p_y orbitals are eigenfunctions of l_z. If not, does a linear combination exist that is an eigenfunction of l_z?

9A.5 Show that l_z and l^2 both commute with the hamiltonian for a hydrogen atom. What is the significance of this result?

9A.6 The 'size' of an atom is sometimes considered to be measured by the radius of a sphere that contains 90 per cent of the charge density of the electrons in the outermost occupied orbital. Calculate the 'size' of a hydrogen atom in its ground state according to this definition. Go on to explore how the 'size' varies as the definition is changed to other percentages, and plot your conclusion.

9A.7 Some atomic properties depend on the average value of $1/r$ rather than the average value of r itself. Evaluate the expectation value of $1/r$ for (a) a hydrogen 1s orbital, (b) a hydrogenic 2s orbital, (c) a hydrogenic 2p orbital. (d) Does $\langle 1/r\rangle = 1/\langle r\rangle$?

9A.8 One of the most famous of the obsolete theories of the hydrogen atom was proposed by Bohr. It has been replaced by quantum mechanics, but by a remarkable coincidence (not the only one where the Coulomb potential is concerned), the energies it predicts agree exactly with those obtained from the Schrödinger equation. In the Bohr atom, an electron travels in a circle around the nucleus. The Coulombic force of attraction ($Ze^2/4\pi\varepsilon_0 r^2$) is balanced by the centrifugal effect of the orbital motion. Bohr proposed that the angular momentum is limited to integral values of $\hbar$. When the two forces are balanced, the atom remains in a stationary state until it makes a spectral transition. Calculate the energies of a hydrogenic atom using the Bohr model.

9A.9 The Bohr model of the atom is specified in Problem 9A.8. (a) What features of it are untenable according to quantum mechanics? (b) How does the Bohr ground state differ from the actual ground state? (c) Is there an experimental distinction between the Bohr and quantum mechanical models of the ground state?

9A.10 Atomic units of length and energy may be based on the properties of a particular atom. The usual choice is that of a hydrogen atom, with the unit of length being the Bohr radius, a_0, and the unit of energy being the 'hartree', E_h, which is equal to twice the (negative of the) energy of the 1s orbital (specifically, and more precisely, $E_h = 2hc\tilde{R}_\infty$). If the positronium atom (e^+,e^-) were used instead, with analogous definitions of units of length and energy, what would be the relation between these two sets of atomic units?

9A.11 The distribution of isotopes of an element may yield clues about the nuclear reactions that occur in the interior of a star. Show that it is possible to use spectroscopy to confirm the presence of both $^4He^+$ and $^3He^+$ in a star by calculating the wavenumbers of the $n=3 \rightarrow n=2$ and of the $n=2 \rightarrow n=1$ transitions for each isotope.

TOPIC 9B Many-electron atoms

Discussion questions

9B.1 Outline the electron configurations of many-electron atoms in terms of their location in the periodic table.

9B.2 Describe and account for the variation of first ionization energies along Period 2 of the periodic table. Would you expect the same variation in Period 3?

9B.3 Describe the orbital approximation for the wavefunction of a many-electron atom. What are the limitations of the approximation?

Exercises

9B.1(a) Write the ground-state electron configurations of the d-metals from scandium to zinc.
9B.1(b) Write the ground-state electron configurations of the d-metals from yttrium to cadmium.

9B.2(a) (i) Write the electronic configuration of the Ni^{2+} ion. (ii) What are the possible values of the total spin quantum numbers S and M_S for this ion?
9B.2(b) (i) Write the electronic configuration of the V^{2+} ion. (ii) What are the possible values of the total spin quantum numbers S and M_S for this ion?

Problems

9B.1 In 1976 it was mistakenly believed that the first of the 'superheavy' elements had been discovered in a sample of mica. Its atomic number was believed to be 126. Without taking relativistic effects into account, calculate the most probable distance of the innermost electrons from the nucleus of an atom of this element? Does your result suggest that relativistic effects should be included in the calculation?

9B.2 The wavefunction of a many-electron closed-shell atom can expressed as a Slater determinant. A useful property of determinants is that interchanging any two rows or columns changes their sign and therefore that if any two rows or columns are identical, then the determinant vanishes. Use this property to show that (a) the wavefunction is antisymmetric under particle exchange, (b) no two electrons can occupy the same orbital with the same spin.

9B.3 The d-metals iron, copper, and manganese form cations with different oxidation states. For this reason, they are found in many oxidoreductases and in several proteins of oxidative phosphorylation and photosynthesis. Explain why many d-metals form cations with different oxidation states.

9B.4 Thallium, a neurotoxin, is the heaviest member of Group 13 of the periodic table and is found most usually in the +1 oxidation state. Aluminium, which causes anaemia and dementia, is also a member of the group but its chemical properties are dominated by the +3 oxidation state. Examine this issue by plotting the first, second, and third ionization energies for the Group 13 elements against atomic number. Explain the trends you observe. *Hints*: The third ionization energy, I_3, is the minimum energy needed to remove an electron from the doubly charged cation): $E^{2+}(g) \rightarrow E^{3+}(g) + e^-(g)$, $I_3 = E(E^{3+}) - E(E^{2+})$. For data, see the links to databases of atomic properties provided in the text's web site.

TOPIC 9C Atomic spectra

Discussion questions

9C.1 Discuss the origin of the series of lines in the emission spectra of hydrogen. What region of the electromagnetic spectrum is associated with each of the series shown in Fig. 9C.1?

9C.2 Specify and account for the selection rules for transitions in hydrogenic atoms.

9C.3 Explain the origin of spin–orbit coupling and how it affects the appearance of a spectrum.

Exercises

9C.1(a) Identify the shortest and longest wavelength lines in the Lyman series.

9C.1(b) The Pfund series has $n_1=5$. Identify the shortest and longest wavelength lines in the Pfund series.

9C.2(a) Compute the wavelength, frequency, and wavenumber of the $n=2 \rightarrow n=1$ transition in He^+.

9C.2(b) Compute the wavelength, frequency, and wavenumber of the $n=5 \rightarrow n=4$ transition in Li^{+2}.

9C.3(a) When ultraviolet radiation of wavelength 58.4 nm from a helium lamp is directed on to a sample of krypton, electrons are ejected with a speed of 1.59 Mm s^{-1}. Calculate the ionization energy of krypton.

9C.3(b) When ultraviolet radiation of wavelength 58.4 nm from a helium lamp is directed on to a sample of xenon, electrons are ejected with a speed of 1.79 Mm s^{-1}. Calculate the ionization energy of xenon.

9C.4(a) Which of the following transitions are allowed in the normal electronic emission spectrum of an atom: (i) 2s$\rightarrow$1s, (ii) 2p$\rightarrow$1s, (iii) 3d$\rightarrow$2p?

9C.4(b) Which of the following transitions are allowed in the normal electronic emission spectrum of an atom: (i) 5d$\rightarrow$2s, (ii) 5p$\rightarrow$3s, (iii) 6p$\rightarrow$4f?

9C.5(a) Calculate the permitted values of j for (i) a d electron, (ii) an f electron.

9C.5(b) Calculate the permitted values of j for (i) a p electron, (ii) an h electron.

9C.6(a) An electron in two different states of an atom is known to have $j=\frac{3}{2}$ and $\frac{1}{2}$. What is its orbital angular momentum quantum number in each case?

9C.6(b) What are the allowed total angular momentum quantum numbers of a composite system in which $j_1=5$ and $j_2=3$?

9C.7(a) What information does the term symbol $^1D_2$ provide about the angular momentum of an atom?

9C.7(b) What information does the term symbol $^3F_4$ provide about the angular momentum of an atom?

9C.8(a) Suppose that an atom has (i) 2, (ii) 3 electrons in different orbitals. What are the possible values of the total spin quantum number S? What is the multiplicity in each case?

9C.8(b) Suppose that an atom has (i) 4, (ii) 5, electrons in different orbitals. What are the possible values of the total spin quantum number S? What is the multiplicity in each case?

9C.9(a) What atomic terms are possible for the electron configuration ns^1nd^1? Which term is likely to lie lowest in energy?

9C.9(b) What atomic terms are possible for the electron configuration np^1nd^1? Which term is likely to lie lowest in energy?

9C.10(a) What values of J may occur in the terms (i) ^1S, (ii) ^2P, (iii) ^3P? How many states (distinguished by the quantum number M_J) belong to each level?

9C.10(b) What values of J may occur in the terms (i) ^3D, (ii) ^4D, (iii) ^2G? How many states (distinguished by the quantum number M_J) belong to each level?

9C.11(a) Give the possible term symbols for (i) Li [He]2s^1, (ii) Na [Ne]3p^1.

9C.11(b) Give the possible term symbols for (i) Sc [Ar]3d^{10}4s^2, (ii) Br [Ar]3d^{10}4s^{2}4p^5.

9C.12(a) Which of the following transitions between terms are allowed in the normal electronic emission spectrum of a many-electron atom: (i) $^3D_2 \rightarrow {}^3P_1$, (ii) $^3P_2 \rightarrow {}^1S_0$, (iii) $^3F_4 \rightarrow {}^3D_3$?

9C.12(b) Which of the following transitions between terms are allowed in the normal electronic emission spectrum of a many-electron atom: (i) $^2P_{3/2} \rightarrow {}^2S_{1/2}$, (ii) $^3P_0 \rightarrow {}^3S_1$, (iii) $^3D_3 \rightarrow {}^1P_1$?

Problems

9C.1 The *Humphreys series* is a group of lines in the spectrum of atomic hydrogen. It begins at 12 368 nm and has been traced to 3281.4 nm. What are the transitions involved? What are the wavelengths of the intermediate transitions?

9C.2 A series of lines in the spectrum of atomic hydrogen lies at 656.46 nm, 486.27 nm, 434.17 nm, and 410.29 nm. What is the wavelength of the next line in the series? What is the ionization energy of the atom when it is in the lower state of the transitions?

9C.3 The Li^{2+} ion is hydrogenic and has a Lyman series at 740 747 cm^{-1}, 877 924 cm^{-1}, 925 933 cm^{-1}, and beyond. Show that the energy levels are of the form $-hc\tilde{R}_{Li}/n^2$ and find the value of $\tilde{R}_{Li}$ for this ion. Go on to predict the wavenumbers of the two longest-wavelength transitions of the Balmer series of the ion and find the ionization energy of the ion.

9C.4 A series of lines in the spectrum of neutral Li atoms rise from combinations of 1s^{2}2p^1 ^{2}P with 1s^2nd^1 ^{2}D and occur at 610.36 nm, 460.29 nm, and 413.23 nm. The d orbitals are hydrogenic. It is known that the ^{2}P term lies at 670.78 nm above the ground state, which is 1s^{2}2s^1 ^{2}S. Calculate the ionization energy of the ground-state atom.

9C.5‡ W.P. Wijesundera et al. (*Phys. Rev.* A **51**, 278 (1995)) attempted to determine the electron configuration of the ground state of lawrencium, element 103. The two contending configurations are [Rn]5f^{14}7s^{2}7p^1 and [Rn]5f^{14}6d^{1}7s^2. Write down the term symbols for each of these configurations, and identify the lowest level within each configuration. Which level would be lowest according to a simple estimate of spin–orbit coupling?

‡ These problems were supplied by Charles Trapp and Carmen Giunta.

9C.6 An emission line from K atoms is found to have two closely spaced components, one at 766.70 nm and the other at 770.11 nm. Account for this observation, and deduce what information you can.

9C.7 Calculate the mass of the deuteron given that the first line in the Lyman series of H lies at 82 259.098 cm^{-1} whereas that of D lies at 82 281.476 cm^{-1}. Calculate the ratio of the ionization energies of H and D.

9C.8 Positronium consists of an electron and a positron (same mass, opposite charge) orbiting round their common centre of mass. The broad features of the spectrum are therefore expected to be hydrogen-like, the differences arising largely from the mass differences. Predict the wavenumbers of the first three lines of the Balmer series of positronium. What is the binding energy of the ground state of positronium?

9C.9 The *Zeeman effect* is the modification of an atomic spectrum by the application of a strong magnetic field. It arises from the interaction between applied magnetic fields and the magnetic moments due to orbital and spin angular momenta (recall the evidence provided for electron spin by the Stern–Gerlach experiment, Topic 8C). To gain some appreciation for the so-called *normal Zeeman effect*, which is observed in transitions involving singlet states, consider a p electron, with $l=1$ and $m_l=0, \pm1$. In the absence of a magnetic field, these three states are degenerate. When a field of magnitude $\mathcal{B}$ is present, the degeneracy is removed and it is observed that the state with $m_l=+1$ moves up in energy by $\mu_B\mathcal{B}$, the state with $m_l=0$ is unchanged, and the state with $m_l=-1$ moves down in energy by $\mu_B\mathcal{B}$, where $\mu_B=e\hbar/2m_e=9.274\times10^{-24}$ J T^{-1} is the 'Bohr magneton'. Therefore, a transition between a $^1S_0$ term and a $^1P_1$ term consists of three spectral lines in the presence of a magnetic field where, in the absence of the magnetic field, there is only one. (a) Calculate the splitting in reciprocal centimetres between the three spectral lines of a transition between a $^1S_0$ term and a $^1P_1$ term in the presence of a magnetic field of 2 T (where

$1\,T=1\,kg\,s^{-2}\,A^{-1}$). (b) Compare the value you calculated in (a) with typical optical transition wavenumbers, such as those for the Balmer series of the H atom. Is the line splitting caused by the normal Zeeman effect relatively small or relatively large?

9C.10 Some of the selection rules for hydrogenic atoms were derived in *Justification* 9C.1. Complete the derivation by considering the *x*- and *y*-components of the electric dipole moment operator.

9C.11‡ Stern–Gerlach splittings of atomic beams are small and require either large magnetic field gradients or long magnets for their observation. For a beam of atoms with zero orbital angular momentum, such as H or Ag, the deflection is given by $x=\pm(\mu_B L^2/4E_k)\mathrm{d}\mathcal{B}/\mathrm{d}z$, where μ_B is the Bohr magneton (Problem 9C.9), L is the length of the magnet, E_k is the average kinetic energy of the atoms in the beam, and $\mathrm{d}\mathcal{B}/\mathrm{d}z$ is the magnetic field gradient across the beam. (a) Use the Maxwell–Boltzmann velocity distribution (eqn 1B.4 of Topic 1B) to show that the average translational kinetic energy of the atoms emerging as a beam from a pinhole in an oven at temperature T is $2kT$. (b) Calculate the magnetic field gradient required to produce a splitting of 1.00 mm in a beam of Ag atoms from an oven at 1000 K with a magnet of length 50 cm.

9C.12 Hydrogen is the most abundant element in all stars. However, neither absorption nor emission lines due to neutral hydrogen are found in the spectra of stars with effective temperatures higher than 25 000 K. Account for this observation.

Integrated activities

9.1 An electron in the ground-state He^+ ion undergoes a transition to a state described by the wavefunction $R_{4,1}(r)Y_{1,1}(\theta,\phi)$. (a) Describe the transition using term symbols. (b) Compute the wavelength, frequency, and wavenumber of the transition. (c) By how much does the mean radius of the electron change due to the transition?

9.2‡ Highly excited atoms have electrons with large principal quantum numbers. Such *Rydberg atoms* have unique properties and are of interest to astrophysicists. (a) For hydrogen atoms with large n, derive a relation for the separation of energy levels. (b) Calculate this separation for $n=100$; also calculate the average radius, the geometric cross section, and the ionization energy. (c) Could a thermal collision with another hydrogen atom ionize this Rydberg atom? (d) What minimum velocity of the second atom is required? (e) Could a normal sized neutral H atom simply pass through the Rydberg atom leaving it undisturbed? (f) What might the radial wavefunction for a 100s orbital be like?

Mathematical background 5 Vectors

A **scalar physical property** (such as temperature) in general varies through space and is represented by a single value at each point of space. A **vector physical property** (such as the velocity or the electric field strength) can also vary through space, but in general has a different direction as well as a different magnitude at each point.

MB5.1 Definitions

A **vector** $\boldsymbol{v}$ has the general form (in three dimensions):

$$\boldsymbol{v}=v_x\boldsymbol{i}+v_y\boldsymbol{j}+v_z\boldsymbol{k} \qquad \text{(MB5.1)}$$

where $\boldsymbol{i}$, $\boldsymbol{j}$, and $\boldsymbol{k}$ are **unit vectors**, vectors of magnitude 1, pointing along the positive directions on the x, y, and z axes and v_x, v_y, and v_z are the **components** of the vector on each axis (Fig. MB5.1). The **magnitude** of the vector is denoted v or $|v|$ and is given by

$$v=(v_x^2+v_y^2+v_z^2)^{1/2} \qquad \text{Magnitude} \qquad \text{(MB5.2)}$$

The vector makes an angle θ to the z-axis and an angle ϕ to the x-axis in the xy-plane. It follows that

$$v_x=v\sin\theta\cos\phi \quad v_y=v\sin\theta\sin\phi \quad v_z=v\cos\theta \qquad \text{Orientation} \qquad \text{(MB5.3a)}$$

and therefore that

$$\theta=\arccos(v_z/v) \quad \phi=\arctan(v_y/v_x) \qquad \text{(MB5.3b)}$$

Brief illustration MB5.1 Vector orientation

The vector $\boldsymbol{v}=2\boldsymbol{i}+3\boldsymbol{j}-\boldsymbol{k}$ has magnitude

$$v=\{2^2+3^2+(-1)^2\}^{1/2}=14^{1/2}=3.74$$

Its direction is given by

$$\theta=\arccos\left(-1/14^{1/2}\right)=105.5° \quad \phi=\arctan(3/2)=56.3°$$

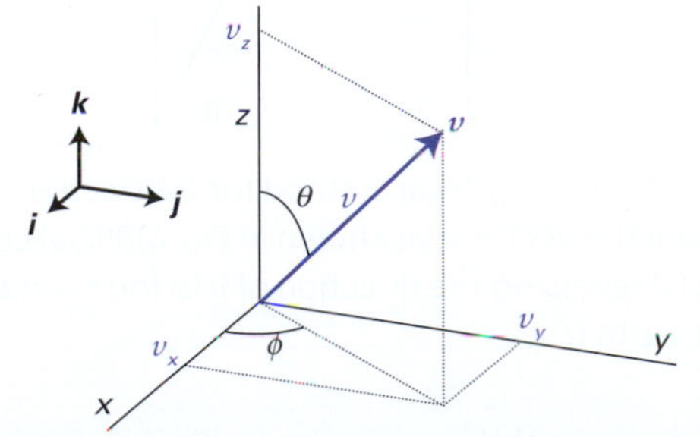

Figure MB5.1 The vector $\boldsymbol{v}$ has components v_x, v_y, and v_z on the x, y, and z axes, respectively. It has a magnitude v and makes an angle θ to the z-axis and an angle θ to the x-axis in the xy-plane.

MB5.2 Operations

Consider the two vectors

$$\boldsymbol{u}=u_x\boldsymbol{i}+u_y\boldsymbol{j}+u_z\boldsymbol{k} \qquad \boldsymbol{v}=v_x\boldsymbol{i}+v_y\boldsymbol{j}+v_z\boldsymbol{k}$$

The operations of addition, subtraction, and multiplication are as follows:

1. *Addition*:

$$\boldsymbol{v}+\boldsymbol{u}=(v_x+u_x)\boldsymbol{i}+(v_y+u_y)\boldsymbol{j}+(v_z+u_z)\boldsymbol{k} \qquad \text{(MB5.4a)}$$

2. *Subtraction*:

$$\boldsymbol{v}-\boldsymbol{u}=(v_x-u_x)\boldsymbol{i}+(v_y-u_y)\boldsymbol{j}+(v_z-u_z)\boldsymbol{k} \qquad \text{(MB5.4b)}$$

Brief illustration MB5.2 Addition and subtraction

Consider the vectors $\boldsymbol{u}=\boldsymbol{i}-4\boldsymbol{j}+\boldsymbol{k}$ (of magnitude 4.24) and $\boldsymbol{v}=-4\boldsymbol{i}+2\boldsymbol{j}+3\boldsymbol{k}$ (of magnitude 5.39). Their sum is

$$\boldsymbol{u}+\boldsymbol{v}=(1-4)\boldsymbol{i}+(-4+2)\boldsymbol{j}+(1+3)\boldsymbol{k}=-3\boldsymbol{i}-2\boldsymbol{j}+4\boldsymbol{k}$$

The magnitude of the resultant vector is $29^{1/2}=5.39$. The difference of the two vectors is

$$\boldsymbol{u}-\boldsymbol{v}=(1+4)\boldsymbol{i}+(-4-2)\boldsymbol{j}+(1-3)\boldsymbol{k}=5\boldsymbol{i}-6\boldsymbol{j}-2\boldsymbol{k}$$

The magnitude of this resultant is 8.06. Note that in this case the difference is longer than either individual vector.

3. *Multiplication*:

(a) The **scalar product**, or *dot product*, of the two vectors $\boldsymbol{u}$ and $\boldsymbol{v}$ is

$$\boldsymbol{u}\cdot\boldsymbol{v}=u_xv_x+u_yv_y+u_zv_z \qquad \text{Scalar product} \qquad \text{(MB5.4c)}$$

and is itself a scalar quantity. We can always choose a new coordinate system—we shall write it X, Y, Z—in which the Z-axis lies parallel to $\boldsymbol{u}$, so $\boldsymbol{u}=u\boldsymbol{K}$, where $\boldsymbol{K}$ is the unit vector parallel to $\boldsymbol{u}$. It then follows from eqn MB5.4c that $\boldsymbol{u}\cdot\boldsymbol{v}=uv_Z$. Then, with $v_Z=v\cos\theta$, where θ is the angle between $\boldsymbol{u}$ and $\boldsymbol{v}$, we find

$$\boldsymbol{u}\cdot\boldsymbol{v}=uv\cos\theta \qquad \text{Scalar product} \qquad \text{(MB5.4d)}$$

(b) The **vector product**, or *cross product*, of two vectors is

$$\boldsymbol{u}\times\boldsymbol{v}=\begin{vmatrix}\boldsymbol{i} & \boldsymbol{j} & \boldsymbol{k}\\ u_x & u_y & u_z\\ v_x & v_y & v_z\end{vmatrix}$$

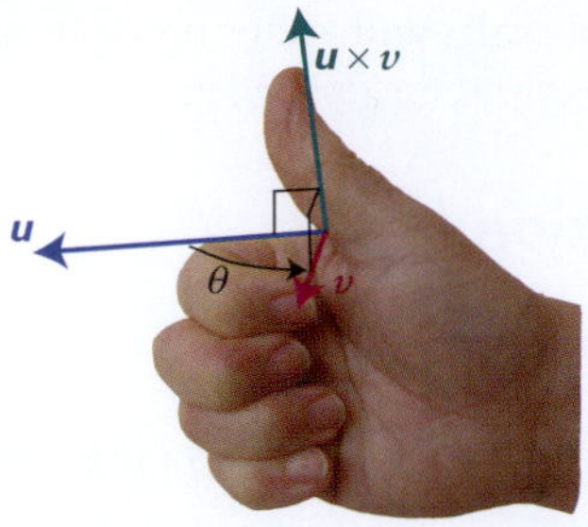

Figure MB5.2 A depiction of the 'right-hand rule'. When the fingers of the right hand rotate $\boldsymbol{u}$ into $\boldsymbol{v}$, the thumb points in the direction of $\boldsymbol{u}\times\boldsymbol{v}$.

$$=(u_yv_z-u_zv_y)\boldsymbol{i}-(u_xv_z-u_zv_x)\boldsymbol{j}+(u_xv_y-u_yv_x)\boldsymbol{k}$$

Vector product (MB5.4e)

(Determinants are discussed in *The chemist's toolkit* 9B.1.) Once again, choosing the coordinate system so that $\boldsymbol{u}=u\boldsymbol{K}$, leads to the simple expression:

$$\boldsymbol{u}\times\boldsymbol{v}=(uv\sin\theta)\boldsymbol{l}$$

Vector product (MB5.4f)

where θ is the angle between the two vectors and $\boldsymbol{l}$ is a unit vector perpendicular to both $\boldsymbol{u}$ and $\boldsymbol{v}$, with a direction determined by the 'right-hand rule' as in Fig. MB5.2. A special case is when each vector is a unit vector, for then

$$\boldsymbol{i}\times\boldsymbol{j}=\boldsymbol{k}\quad \boldsymbol{j}\times\boldsymbol{k}=\boldsymbol{i}\quad \boldsymbol{k}\times\boldsymbol{i}=\boldsymbol{j}$$

(MB5.5)

It is important to note that the order of vector multiplication is important and that $\boldsymbol{u}\times\boldsymbol{v}=-\boldsymbol{v}\times\boldsymbol{u}$.

Brief illustration MB5.3 Scalar and vector products

The scalar and vector products of the two vectors in *Brief illustration* MB5.2, $\boldsymbol{u}=\boldsymbol{i}-4\boldsymbol{j}+\boldsymbol{k}$ (of magnitude 4.24) and $\boldsymbol{v}=-4\boldsymbol{i}+2\boldsymbol{j}+3\boldsymbol{k}$ (of magnitude 5.39) are

$$\boldsymbol{u}\cdot\boldsymbol{v}=\{1\times(-4)\}+\{(-4)\times 2\}+\{1\times 3\}=-9$$

$$\boldsymbol{u}\times\boldsymbol{v}=\begin{vmatrix}\boldsymbol{i} & \boldsymbol{j} & \boldsymbol{k}\\ 1 & -4 & 1\\ -4 & 2 & 3\end{vmatrix}$$
$$=\{(-4)(3)-(1)(2)\}\boldsymbol{i}-\{(1)(3)-(1)(-4)\}\boldsymbol{j}+\{(1)(2)-(-4)(-4)\}\boldsymbol{k}$$
$$=-14\boldsymbol{i}-7\boldsymbol{j}-14\boldsymbol{k}$$

The vector product is a vector of magnitude 21.00 pointing in a direction perpendicular to the plane defined by the two individual vectors.

MB5.3 The graphical representation of vector operations

Consider two vectors $\boldsymbol{v}$ and $\boldsymbol{u}$ making an angle θ (Fig. MB5.3). The first step in the addition of $\boldsymbol{u}$ to $\boldsymbol{v}$ consists of joining the tip (the 'head') of $\boldsymbol{u}$ to the starting point (the 'tail') of $\boldsymbol{v}$. In the second step, we draw a vector $\boldsymbol{v}_{res}$, the **resultant vector**, originating from the tail of $\boldsymbol{u}$ to the head of $\boldsymbol{v}$. Reversing the order of addition leads to the same result; that is, we obtain the same $\boldsymbol{v}_{res}$ whether we add $\boldsymbol{u}$ to $\boldsymbol{v}$ or $\boldsymbol{v}$ to $\boldsymbol{u}$. To calculate the magnitude of v_{res}, we note that

$$v_{res}^2=(\boldsymbol{u}+\boldsymbol{v})\cdot(\boldsymbol{u}+\boldsymbol{v})=\boldsymbol{u}\cdot\boldsymbol{u}+\boldsymbol{v}\cdot\boldsymbol{v}+2\boldsymbol{u}\cdot\boldsymbol{v}=u^2+v^2+2uv\cos\theta$$

where θ is the angle between $\boldsymbol{u}$ and $\boldsymbol{v}$. In terms of the angle $\theta'=\pi-\theta$ shown in the figure, and $\cos(\pi-\theta)=-\cos\theta$, we obtain the **law of cosines**:

$$v_{res}^2=u^2+v^2-2uv\cos\theta$$

Law of cosines (MB5.6)

for the relation between the lengths of the sides of a triangle.

Subtraction of $\boldsymbol{v}$ from $\boldsymbol{u}$ amounts to addition of $-\boldsymbol{v}$ to $\boldsymbol{u}$. It follows that in the first step of subtraction we draw $-\boldsymbol{v}$ by reversing

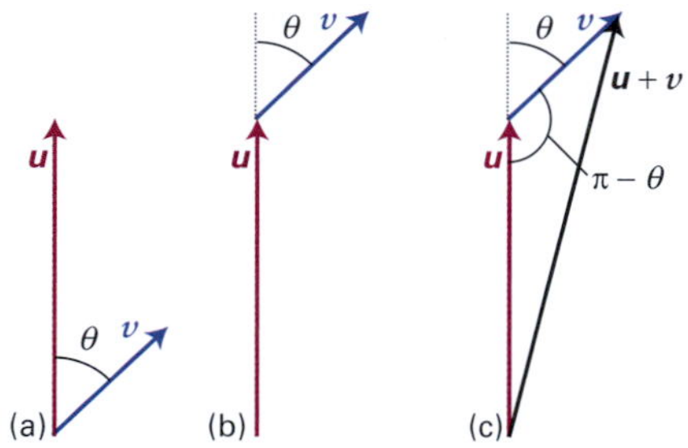

Figure MB5.3 (a) The vectors $\boldsymbol{v}$ and $\boldsymbol{u}$ make an angle θ. (b) To add $\boldsymbol{u}$ to $\boldsymbol{v}$, we first join the head of $\boldsymbol{u}$ to the tail of $\boldsymbol{v}$, making sure that the angle θ between the vectors remains unchanged. (c) To finish the process, we draw the resultant vector by joining the tail of $\boldsymbol{u}$ to the head of $\boldsymbol{v}$.

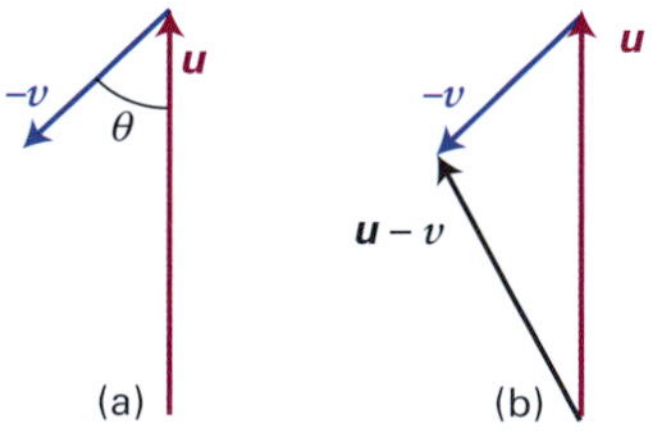

Figure MB5.4 The graphical method for subtraction of the vector $\boldsymbol{v}$ from the vector $\boldsymbol{u}$ (as shown in Fig. MB5.3a) consists of two steps: (a) reversing the direction of $\boldsymbol{v}$ to form $-\boldsymbol{v}$, and (b) adding $-\boldsymbol{v}$ to $\boldsymbol{u}$.

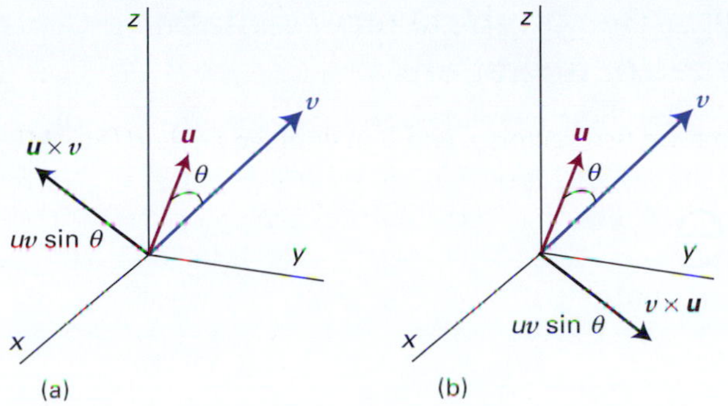

Figure MB5.5 The direction of the cross products of two vectors $\boldsymbol{u}$ and $\boldsymbol{v}$ with an angle θ between them: (a) $\boldsymbol{u}\times\boldsymbol{v}$ and (b) $\boldsymbol{v}\times\boldsymbol{u}$. Note that the cross product, and the unit vector $\boldsymbol{l}$ of eqn MB5.4f, are perpendicular to both $\boldsymbol{u}$ and $\boldsymbol{v}$ but the direction depends on the order in which the product is taken. The magnitude of the cross product, in either case, is $uv \sin\theta$.

the direction of $\boldsymbol{v}$ (Fig. MB5.4). Then, the second step consists of adding $-\boldsymbol{v}$ to $\boldsymbol{u}$ by using the strategy shown in the figure; we draw a resultant vector $\boldsymbol{v}_{\text{res}}$ by joining the tail of $-\boldsymbol{v}$ to the head of $\boldsymbol{u}$.

Vector multiplication is represented graphically by drawing a vector (using the right-hand rule) perpendicular to the plane defined by the vectors $\boldsymbol{u}$ and $\boldsymbol{v}$, as shown in Fig. MB5.5. Its length is equal to $uv \sin\theta$, where θ is the angle between $\boldsymbol{u}$ and $\boldsymbol{v}$.

MB5.4 Vector differentiation

The derivative $\mathrm{d}\boldsymbol{v}/\mathrm{d}t$, where the components v_x, v_y, and v_z are themselves functions of t, is

$$\frac{\mathrm{d}\boldsymbol{v}}{\mathrm{d}t}=\left(\frac{\mathrm{d}v_x}{\mathrm{d}t}\right)\boldsymbol{i}+\left(\frac{\mathrm{d}v_y}{\mathrm{d}t}\right)\boldsymbol{j}+\left(\frac{\mathrm{d}v_z}{\mathrm{d}t}\right)\boldsymbol{k} \quad \text{Derivative} \quad \text{(MB5.7)}$$

The derivatives of scalar and vector products are obtained using the rules of differentiating a product:

$$\frac{\mathrm{d}\boldsymbol{u}\cdot\boldsymbol{v}}{\mathrm{d}t}=\left(\frac{\mathrm{d}\boldsymbol{u}}{\mathrm{d}t}\right)\cdot\boldsymbol{v}+\boldsymbol{u}\cdot\left(\frac{\mathrm{d}\boldsymbol{v}}{\mathrm{d}t}\right) \quad \text{(MB5.8a)}$$

$$\frac{\mathrm{d}\boldsymbol{u}\times\boldsymbol{v}}{\mathrm{d}t}=\left(\frac{\mathrm{d}\boldsymbol{u}}{\mathrm{d}t}\right)\times\boldsymbol{v}+\boldsymbol{u}\times\left(\frac{\mathrm{d}\boldsymbol{v}}{\mathrm{d}t}\right) \quad \text{(MB5.8b)}$$

In the latter, note the importance of preserving the order of vectors.

The **gradient** of a scalar function $f(x,y,z)$, denoted grad f or ∇f, is

$$\nabla f=\left(\frac{\partial f}{\partial x}\right)\boldsymbol{i}+\left(\frac{\partial f}{\partial y}\right)\boldsymbol{j}+\left(\frac{\partial f}{\partial z}\right)\boldsymbol{k} \quad \text{Gradient} \quad \text{(MB5.9)}$$

where partial derivatives are described in *Mathematical background* 2. Note that the gradient of a scalar function is a vector. We can treat ∇ as a vector operator (in the sense that it operates on a function and results in a vector), and write

$$\nabla=\boldsymbol{i}\frac{\partial}{\partial x}+\boldsymbol{j}\frac{\partial}{\partial y}+\boldsymbol{k}\frac{\partial}{\partial z} \quad \text{(MB5.10)}$$

The scalar product of ∇ and ∇f, using eqns MB5.9 and MB5.10, is

$$\nabla\cdot\nabla f=\left(\boldsymbol{i}\frac{\partial}{\partial x}+\boldsymbol{j}\frac{\partial}{\partial y}+\boldsymbol{k}\frac{\partial}{\partial z}\right)\cdot\left(\boldsymbol{i}\frac{\partial}{\partial x}+\boldsymbol{j}\frac{\partial}{\partial y}+\boldsymbol{k}\frac{\partial}{\partial z}\right)f$$
$$=\frac{\partial^2 f}{\partial x^2}+\frac{\partial^2 f}{\partial y^2}+\frac{\partial^2 f}{\partial z^2} \quad \text{Laplacian} \quad \text{(MB5.11)}$$

Equation MB5.11 defines the **Laplacian** ($\nabla^2=\nabla\cdot\nabla$) of a function.

CHAPTER 10

Molecular structure

The concepts developed in Chapter 9, particularly those of orbitals, can be extended to a description of the electronic structures of molecules. There are two principal quantum mechanical theories of molecular electronic structure. In 'valence-bond theory', the starting point is the concept of the shared electron pair.

10A Valence-bond theory

In this Topic we see how to write the wavefunction for a shared electron pair, and how it may be extended to account for the structures of a wide variety of molecules. The theory introduces the concepts of σ and π bonds, promotion, and hybridization that are used widely in chemistry.

10B Principles of molecular orbital theory

Almost all modern computational work makes use of molecular orbital theory (MO theory), and we concentrate on that theory in this chapter. In MO theory, the concept of atomic orbital is extended to that of 'molecular orbital', which is a wavefunction that spreads over all the atoms in a molecule. The Topic begins with an account of the hydrogen molecule, which sets the scene for the application of MO theory to more complicated molecules.

10C Homonuclear diatomic molecules

The principles established for the hydrogen molecule are readily extended to other homonuclear diatomic molecules, the principal difference being that more types of atomic orbital must be included to give a more varied collection of molecular orbitals. The building-up principle for atoms is extended to the occupation of molecular orbitals and used to predict the electronic structure of molecules.

10D Heteronuclear diatomic molecules

The MO theory of heteronuclear diatomic molecules introduces the possibility that the atomic orbitals on the two atoms contribute unequally to the molecular orbital. As a result, the molecule is polar. The polarity can be expressed in terms of the concept of electronegativity.

10E Polyatomic molecules

Most molecules are polyatomic, so it is important to be able to account for their electronic structure. An early approach to the electronic structure of planar conjugated polyenes is the 'Hückel method'. This procedure introduces severe approximations, but sets the scene for more sophisticated procedures. These more sophisticated procedures have given rise to what is essentially a huge and vibrant theoretical chemistry industry in which elaborate computations are used to predict molecular properties. In this Topic we see a little of how those calculations are formulated.

What is the impact of this material?

The concepts introduced in this chapter pervade the whole of chemistry and are encountered throughout the text. We focus on two biochemical aspects here. In *Impact* I10.1 we see how simple concepts account for the reactivity of small molecules that occur in organisms. In *Impact* I10.2 we see a little of the contribution of computational chemistry to the explanation of the thermodynamic and spectroscopic properties of several biologically significant molecules.

To read more about the impact of this material, scan the QR code, or go to bcs.whfreeman.com/webpub/chemistry/pchem10e/impact/pchem-10-1.html

10A Valence-bond theory

Contents

➤ Why do you need to know this material?

Valence-bond theory was the first quantum mechanical theory of bonding to be developed. The language it introduced, which includes concepts such as spin pairing, σ and π bonds, and hybridization, is widely used throughout chemistry, especially in the description of the properties and reactions of organic compounds.

➤ What is the key idea?

A bond forms when an electron in an atomic orbital on one atom pairs its spin with that of an electron in an atomic orbital on another atom.

➤ What do you need to know already?

You need to know about atomic orbitals (Topic 9A) and the concepts of normalization and orthogonality (Topic 7C). This Topic also makes use of the Pauli principle (Topic 9B).

Here we summarize essential topics of valence-bond theory (VB theory) that should be familiar from introductory chemistry and set the stage for the development of molecular orbital theory (MO theory). However, there is an important preliminary point. All theories of molecular structure make the same simplification at the outset. Whereas the Schrödinger equation for a hydrogen atom can be solved exactly, an exact solution is not possible for any molecule because even the simplest molecule consists of three particles (two nuclei and one electron). We therefore adopt the **Born–Oppenheimer approximation** in which it is supposed that the nuclei, being so much heavier than an electron, move relatively slowly and may be treated as stationary while the electrons move in their field. That is, we think of the nuclei as fixed at arbitrary locations, and then solve the Schrödinger equation for the wavefunction of the electrons alone.

The Born–Oppenheimer approximation allows us to select an internuclear separation in a diatomic molecule and then to solve the Schrödinger equation for the electrons at that nuclear separation. Then we choose a different separation and repeat the calculation, and so on. In this way we can explore how the energy of the molecule varies with bond length and obtain a **molecular potential energy curve** (Fig. 10A.1). It is called a *potential* energy curve because the kinetic energy of the stationary nuclei is zero. Once the curve has been calculated or determined experimentally (by using the spectroscopic techniques described in Topics 12C–12E and 13A), we can identify the **equilibrium bond length**, R_e, the internuclear separation at the minimum of the curve, and the **bond dissociation energy**, D_0, which is closely related to the depth, D_e, of the minimum below the energy of the infinitely widely separated and stationary atoms. When more than one molecular parameter is changed in a polyatomic molecule, such as its various bond lengths and angles, we obtain a potential energy *surface*; the overall equilibrium shape of the molecule corresponds to the global minimum of the surface.

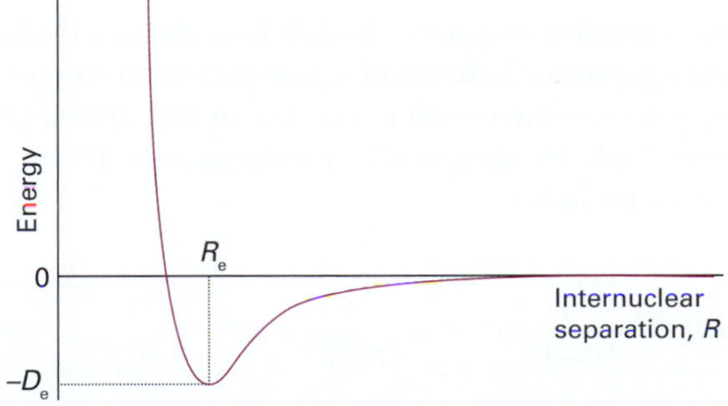

Figure 10A.1 A molecular potential energy curve. The equilibrium bond length corresponds to the energy minimum.

10A.1 Diatomic molecules

We begin the account of VB theory by considering the simplest possible chemical bond, the one in molecular hydrogen, H_2.

(a) The basic formulation

The spatial wavefunction for an electron on each of two widely separated H atoms is

$$\Psi(1,2)=\chi_{\mathrm{H1s_A}}(r_1)\chi_{\mathrm{H1s_B}}(r_2) \qquad (10\mathrm{A}.1)$$

if electron 1 is on atom A and electron 2 is on atom B; in this chapter, and as is common in the chemical literature, we use χ (chi) to denote atomic orbitals. For simplicity, we shall write this wavefunction as $\Psi(1,2)=A(1)B(2)$. When the atoms are close, it is not possible to know whether it is electron 1 or electron 2 that is on A. An equally valid description is therefore $\Psi(1,2)=A(2)B(1)$, in which electron 2 is on A and electron 1 is on B. When two outcomes are equally probable, quantum mechanics instructs us to describe the true state of the system as a superposition of the wavefunctions for each possibility (Topic 7C), so a better description of the molecule than either wavefunction alone is one of the (unnormalized) linear combinations $\Psi(1,2)=A(1)B(2)\pm A(2)B(1)$. The combination with lower energy is the one with a + sign, so the valence-bond wavefunction of the electrons in an H_2 molecule is

$$\Psi(1,2)=A(1)B(2)+A(2)B(1) \qquad \text{A valence-bond wavefunction} \qquad (10\mathrm{A}.2)$$

The reason why this linear combination has a lower energy than either the separate atoms or the linear combination with a negative sign can be traced to the constructive interference between the wave patterns represented by the terms $A(1)B(2)$ and $A(2)B(1)$, and the resulting enhancement of the probability density of the electrons in the internuclear region (Fig. 10A.2).

Brief illustration 10.A1 A valence-bond wavefunction

The wavefunction in eqn 10A.2 might look abstract, but in fact it can be expressed in terms of simple exponential functions. Thus, if we use the wavefunction for an H1s orbital ($Z=1$) given in Topic 9A, then, with the radii measured from their respective nuclei (**1**),

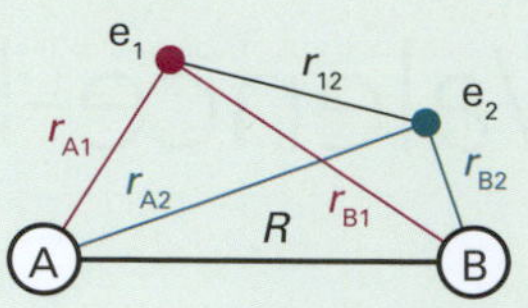

1

$$\Psi(1,2)=\overbrace{\frac{1}{(\pi a_0^3)^{1/2}}\mathrm{e}^{-r_{\mathrm{A1}}/a_0}}^{A(1)}\times\overbrace{\frac{1}{(\pi a_0^3)^{1/2}}\mathrm{e}^{-r_{\mathrm{B2}}/a_0}}^{B(2)}+\overbrace{\frac{1}{(\pi a_0^3)^{1/2}}\mathrm{e}^{-r_{\mathrm{A2}}/a_0}}^{A(2)}$$
$$\times\overbrace{\frac{1}{(\pi a_0^3)^{1/2}}\mathrm{e}^{-r_{\mathrm{B1}}/a_0}}^{B(1)}=\frac{1}{\pi a_0^3}\{\mathrm{e}^{-(r_{\mathrm{A1}}+r_{\mathrm{B2}})/a_0}+\mathrm{e}^{-(r_{\mathrm{A2}}+r_{\mathrm{B1}})/a_0}\}$$

Self-test 10A.1 Express this wavefunction in terms of the Cartesian coordinates of each electron given that the internuclear separation (along the z-axis) is R.

Answer: $r_{\mathrm{A}i}=(x_i^2+y_i^2+z_i^2)^{1/2}$, $r_{\mathrm{B}i}=(x_i^2+y_i^2+(z_i-R)^2)^{1/2}$

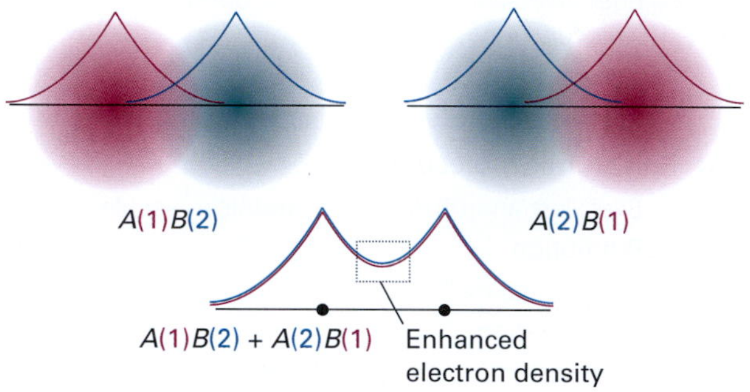

Figure 10A.2 It is very difficult to represent valence-bond wavefunctions because they refer to two electrons simultaneously. However, this illustration is an attempt. The atomic orbital for electron 1 is represented by the purple shading, and that of electron 2 is represented by the green shading. The left illustration represents $A(1)B(2)$, and the right illustration represents the contribution $A(2)B(1)$. When the two contributions are superimposed, there is interference between the purple contributions and between the green contributions, resulting in an enhanced (two-electron) density in the internuclear region.

The electron distribution described by the wavefunction in eqn 10A.2 is called a **σ bond**. A σ bond has cylindrical symmetry around the internuclear axis, and is so called because, when viewed along the internuclear axis, it resembles a pair of electrons in an s orbital (and σ is the Greek equivalent of s).

A chemist's picture of a covalent bond is one in which the spins of two electrons pair as the atomic orbitals overlap. The origin of the role of spin, as we show in the following *Justification*, is that the wavefunction in eqn 10A.2 can be formed only by a pair of spin-paired electrons. Spin pairing is not an end in itself: it is a means of achieving a wavefunction and the probability distribution it implies that corresponds to a low energy.

Justification 10.A1 Electron pairing in VB theory

The Pauli principle requires the overall wavefunction of two electrons, the wavefunction including spin, to change sign when the labels of the electrons are interchanged (Topic 9B). The overall VB wavefunction for two electrons is

$$\Psi(1,2)=\{A(1)B(2)+A(2)B(1)\}\sigma(1,2)$$

where σ represents the spin component of the wavefunction. When the labels 1 and 2 are interchanged, this wavefunction becomes

$$\Psi(2,1)=\{A(2)B(1)+A(1)B(2)\}\sigma(2,1)$$
$$=\{A(1)B(2)+A(2)B(1)\}\sigma(2,1)$$

The Pauli principle requires that $\Psi(2,1)=-\Psi(1,2)$, which is satisfied only if $\sigma(2,1)=-\sigma(1,2)$. The combination of two spins that has this property is

$$\sigma_-(1,2)=(1/2^{1/2})\{\alpha(1)\beta(2)-\beta(1)\alpha(2)\}$$

which corresponds to paired electron spins (Topic 9C). Therefore, we conclude that the state of lower energy (and hence the formation of a chemical bond) is achieved if the electron spins are paired.

The VB description of H_2 can be applied to other homonuclear diatomic molecules. For N_2, for instance, we consider the valence electron configuration of each atom, which is $2s^2 2p_x^1 2p_y^1 2p_z^1$. It is conventional to take the *z*-axis to be the internuclear axis, so we can imagine each atom as having a $2p_z$ orbital pointing towards a $2p_z$ orbital on the other atom (Fig. 10A.3), with the $2p_x$ and $2p_y$ orbitals perpendicular to the axis. A σ bond is then formed by spin pairing between the two electrons in the two $2p_z$ orbitals. Its spatial wavefunction is given by eqn 10A.2, but now *A* and *B* stand for the two $2p_z$ orbitals.

The remaining N2p orbitals cannot merge to give σ bonds as they do not have cylindrical symmetry around the internuclear axis. Instead, they merge to form two π bonds. A **π bond** arises from the spin pairing of electrons in two p orbitals that approach side-by-side (Fig. 10A.4). It is so called because, viewed along the inter-nuclear axis, a π bond resembles a pair of electrons in a p orbital (and π is the Greek equivalent of p).

There are two π bonds in N_2, one formed by spin pairing in two neighbouring $2p_x$ orbitals and the other by spin pairing in two neighbouring $2p_y$ orbitals. The overall bonding pattern in N_2 is therefore a σ bond plus two π bonds (Fig. 10A.5), which is consistent with the Lewis structure :N≡N: for nitrogen.

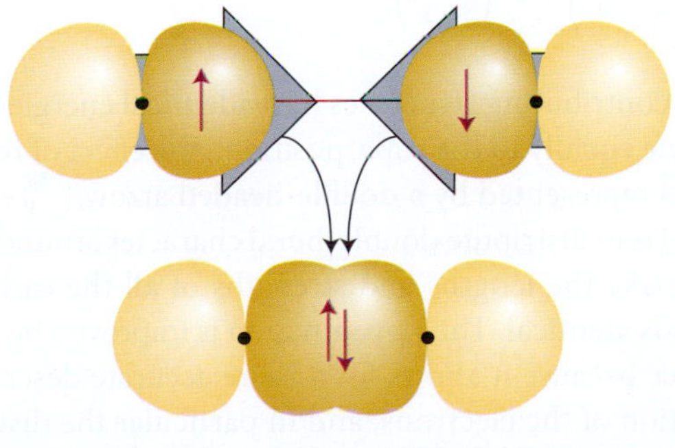
Figure 10A.3 The orbital overlap and spin pairing between electrons in two collinear p orbitals that results in the formation of a σ bond.

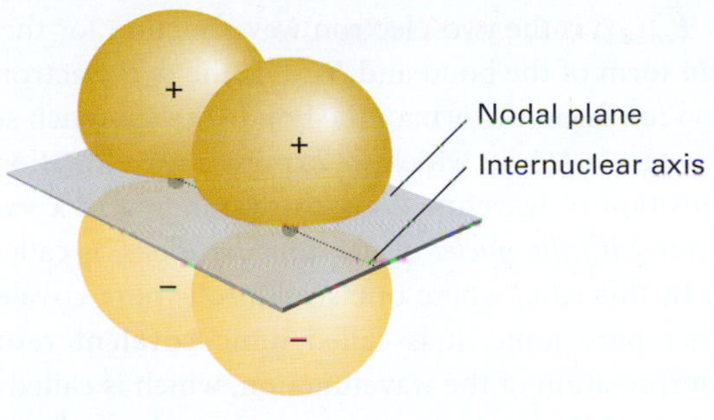

Figure 10A.4 A π bond results from orbital overlap and spin pairing between electrons in p orbitals with their axes perpendicular to the internuclear axis. The bond has two lobes of electron density separated by a nodal plane.

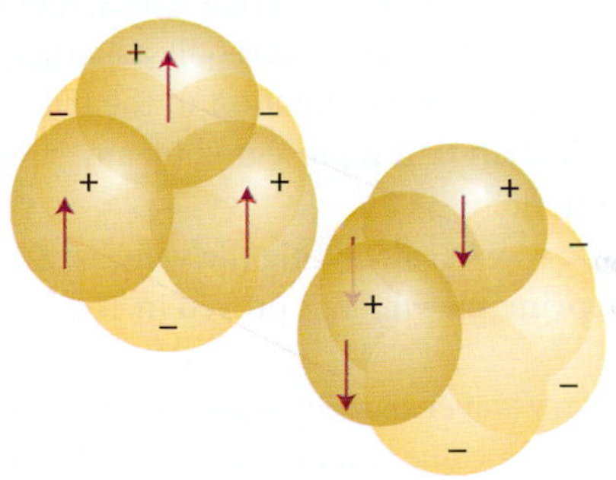
Figure 10A.5 The structure of bonds in a nitrogen molecule, with one σ bond and two π bonds. The overall electron density has cylindrical symmetry around the internuclear axis.

(b) Resonance

Another term introduced into chemistry by VB theory is **resonance**, the superposition of the wavefunctions representing different electron distributions in the same nuclear framework. To understand what this means, consider the VB description of a purely covalently bonded HCl molecule, which could be written as $\Psi=A(1)B(2)+A(2)B(1)$, with *A* now a H1s orbital and *B* a Cl2p orbital. However, this description is physically unlikely: it allows electron 1 to be on the H atom when electron 2 is on the Cl atom, and vice versa, but it does not allow for the possibility that both electrons are on the Cl atom ($\Psi=B(1)B(2)$, representing H^+Cl^-) or even on the H atom ($\Psi=A(1)A(2)$, representing the much less likely H^-Cl^+). A better description of the wavefunction for the molecule is as a superposition of the covalent and ionic descriptions, and we write (with a slightly simplified notation, and ignoring the less likely H^-Cl^+ possibility) $\Psi_{HCl}=\Psi_{H-Cl}+\lambda\Psi_{H^+Cl^-}$ with λ (lambda) some numerical coefficient. In general, we write

$$\Psi=\Psi_{\text{covalent}}+\lambda\Psi_{\text{ionic}} \qquad (10A.3)$$

where $\Psi_{covalent}$ is the two-electron wavefunction for the purely covalent form of the bond and Ψ_{ionic} is the two-electron wavefunction for the ionic form of the bond. The approach summarized by eqn 10A.3, in which we express a wavefunction as the superposition of wavefunctions corresponding to a variety of structures *with the nuclei in the same locations*, is called **resonance**. In this case, where one structure is pure covalent and the other pure ionic, it is called **ionic–covalent resonance**. The interpretation of the wavefunction, which is called a **resonance hybrid**, is that if we were to inspect the molecule, then the probability that it would be found with an ionic structure is proportional to λ^2. If λ^2 is very small, the covalent description is dominant. If λ^2 is very large, the ionic description is dominant. Resonance is not a flickering between the contributing states: it is a blending of their characteristics, much as a mule is a blend of a horse and a donkey. It is only a mathematical device for achieving a closer approximation to the true wavefunction of the molecule than that represented by any single contributing structure alone.

A systematic way of calculating the value of λ is provided by the **variation principle** which is proved in Topic 10C:

> If an arbitrary wavefunction is used to calculate the energy, then the value calculated is never less than the true energy.

Variation principle

The arbitrary wavefunction is called the **trial wavefunction**. The principle implies that, if we vary the parameter λ in the trial wavefunction until the lowest energy is achieved (by evaluating the expectation value of the hamiltonian for the wavefunction), then that value of λ will be the best and through λ^2 represents the appropriate contribution of the ionic wavefunction to the resonance hybrid.

Brief illustration 10A.2 / Resonance hybrids

Consider a bond described by eqn 10A.3. We might find that the lowest energy is reached when $\lambda = 0.1$, so the best description of the bond in the molecule is a resonance structure described by the wavefunction $\Psi = \Psi_{covalent} + 0.1\Psi_{ionic}$. This wavefunction implies that the probabilities of finding the molecule in its covalent and ionic forms are in the ratio 100:1 (because $0.1^2 = 0.01$).

10A.2 Polyatomic molecules

Each σ bond in a polyatomic molecule is formed by the spin pairing of electrons in atomic orbitals with cylindrical symmetry around the relevant internuclear axis. Likewise, π bonds are formed by pairing electrons that occupy atomic orbitals of the appropriate symmetry.

Brief illustration 10A.3 A polyatomic molecule

The VB description of H_2O will make this approach clear. The valence-electron configuration of an O atom is $2s^2 2p_x^2 2p_y^1 2p_z^1$. The two unpaired electrons in the O2p orbitals can each pair with an electron in an H1s orbital, and each combination results in the formation of a σ bond (each bond has cylindrical symmetry about the respective O—H internuclear axis). Because the $2p_x$ and $2p_y$ orbitals lie at 90° to each other, the two σ bonds also lie at 90° to each other (Fig. 10A.6). We predict, therefore, that H_2O should be an angular molecule, which it is. However, the theory predicts a bond angle of 90°, whereas the actual bond angle is 104.5°.

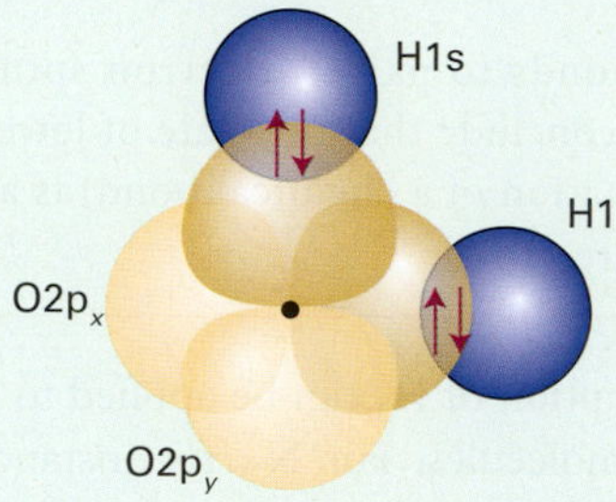

Figure 10A.6 In a primitive view of the structure of an H_2O molecule, each bond is formed by the overlap and spin pairing of an H1s electron and an O2p electron.

Self-test 10A.2 Use VB theory to suggest a shape for the ammonia molecule, NH_3.

Answer: Trigonal pyramidal with HNH bond angle 90°; experimental: 107°

Resonance plays an important role in the valence-bond description of polyatomic molecules. One of the most famous examples of resonance is in the VB description of benzene, where the wavefunction of the molecule is written as a superposition of the many-electron wavefunctions of the two covalent Kekulé structures:

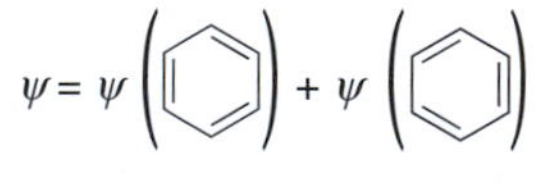

$$\psi = \psi\left(\bigcirc\right) + \psi\left(\bigcirc\right) \tag{10A.4}$$

The two contributing structures have identical energies, so they contribute equally to the superposition. The effect of resonance (which is represented by a double-headed arrow, $\bigcirc \leftrightarrow \bigcirc$, in this case) is to distribute double-bond character around the ring and to make the lengths and strengths of all the carbon–carbon bonds identical. The wavefunction is improved by allowing resonance because it allows for a more accurate description of the location of the electrons, and in particular the distribution can adjust into a state of lower energy. This lowering is called the **resonance stabilization** of the molecule and, in the context of VB theory, is largely responsible for the unusual stability of

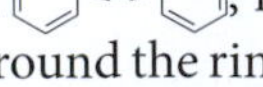

aromatic rings. Resonance always lowers the energy, and the lowering is greatest when the contributing structures have similar energies. The wavefunction of benzene is improved still further, and the calculated energy of the molecule is lowered further still, if we allow ionic–covalent resonance too, by allowing a small admixture of structures such as 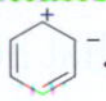.

(a) Promotion

As pointed out in *Brief illustration* 10A.3, simple VB theory predicts a bond angle of 90°, whereas the actual bond angle is 104.5°. Another deficiency of this initial formulation of VB theory is its inability to account for carbon's tetravalence (its ability to form four bonds). The ground-state configuration of C is $2s^2 2p_x^1 2p_y^1$, which suggests that a carbon atom should be capable of forming only two bonds, not four.

This deficiency is overcome by allowing for **promotion**, the excitation of an electron to an orbital of higher energy. In carbon, for example, the promotion of a 2s electron to a 2p orbital can be thought of as leading to the configuration $2s^1 2p_x^1 2p_y^1 2p_z^1$, with four unpaired electrons in separate orbitals. These electrons may pair with four electrons in orbitals provided by four other atoms (such as four H1s orbitals if the molecule is CH_4), and hence form four σ bonds. Although energy was required to promote the electron, it is more than recovered by the promoted atom's ability to form four bonds in place of the two bonds of the unpromoted atom.

Promotion, and the formation of four bonds, is a characteristic feature of carbon because the promotion energy is quite small: the promoted electron leaves a doubly occupied 2s orbital and enters a vacant 2p orbital, hence significantly relieving the electron–electron repulsion it experiences in the former. However, it is important to remember that promotion is not a 'real' process in which an atom somehow becomes excited and then forms bonds: it is a notional contribution to the overall energy change that occurs when bonds form.

Brief illustration 10A.4 Promotion

Sulfur can form six bonds (an 'expanded octet'), as in the molecule SF_6. Because the ground-state electron configuration of sulfur is $[Ne]3s^2 3p^4$, this bonding pattern requires the promotion of a 3s electron and a 3p electron to two different 3d orbitals, which are nearby in energy, to produce the notional configuration $[Ne]3s^1 3p^3 3d^2$ with all six of the valence electrons in different orbitals and capable of bond formation with six electrons provided by six F atoms.

Self-test 10A.3 Account for the ability of phosphorus to form five bonds, as in PF_5.

Answer: Promotion of a 3s electron from $[Ne]3s^2 3p^3$ to $[Ne]3s^1 3p^3 3d^1$

(b) Hybridization

The description of the bonding in CH_4 (and other alkanes) is still incomplete because it implies the presence of three σ bonds of one type (formed from H1s and C2p orbitals) and a fourth σ bond of a distinctly different character (formed from H1s and C2s). This problem is overcome by realizing that the electron density distribution in the promoted atom is equivalent to the electron density in which each electron occupies a **hybrid orbital** formed by interference between the C2s and C2p orbitals of the same atom. The origin of the hybridization can be appreciated by thinking of the four atomic orbitals centred on a nucleus as waves that interfere destructively and constructively in different regions, and give rise to four new shapes.

As we show in the following *Justification*, the specific linear combinations that give rise to four equivalent hybrid orbitals are

$$h_1 = s + p_x + p_y + p_z \qquad h_2 = s - p_x - p_y + p_z$$
$$h_3 = s - p_x + p_y - p_z \qquad h_4 = s + p_x - p_y - p_z \qquad \text{sp}^3 \text{ hybrid orbitals} \qquad (10A.5)$$

As a result of the interference between the component orbitals, each hybrid orbital consists of a large lobe pointing in the direction of one corner of a regular tetrahedron (Fig. 10A.7). The angle between the axes of the hybrid orbitals is the tetrahedral angle, $\arccos(-1/3) = 109.47°$. Because each hybrid is built from one s orbital and three p orbitals, it is called an **sp³ hybrid orbital**.

Justification 10A.2 Determining the form of tetrahedral hybrid orbitals

We begin by supposing that each hybrid can be written in the form $h = as + b_x p_x + b_y p_y + b_z p_z$. The hybrid h_1 that points to the (1,1,1) corner of a cube must have equal contributions from all three p orbitals, so we can set the three b coefficients equal to each other and write $h_1 = as + b(p_x + p_y + p_z)$. The other three hybrids have the same composition (they are equivalent, apart from their direction in space), but are orthogonal to h_1. This orthogonality is achieved by choosing different signs for the p-orbitals but the same overall composition. For instance, we might choose $h_2 = as + b(-p_x - p_y + p_z)$, in which case the orthogonality condition is

$$\int h_1 h_2 \mathrm{d}\tau = \int (as + b(p_x + p_y + p_z))(as + b(-p_x - p_y + p_z))\mathrm{d}\tau$$
$$= a^2 \overbrace{\int s^2 \mathrm{d}\tau}^{1} - b^2 \overbrace{\int p_x^2 \mathrm{d}\tau}^{1} - \cdots - ab \overbrace{\int s p_x \mathrm{d}\tau}^{0} - \cdots$$
$$- b^2 \overbrace{\int p_x p_y \mathrm{d}\tau}^{0} + \cdots = a^2 - b^2 - b^2 + b^2 = a^2 - b^2 = 0$$

We conclude that a solution is $a = b$ (the alternative solution, $a = -b$, simply corresponds to choosing different absolute phases for the p orbitals) and the two hybrid orbitals are the h_1 and h_2 in eqn 10A.3. A similar argument but with $h_3 = as + b(-p_x + p_y - p_z)$ or $h_4 = as + b(p_x - p_y - p_z)$ leads to the other two hybrids in eqn 10A.3.

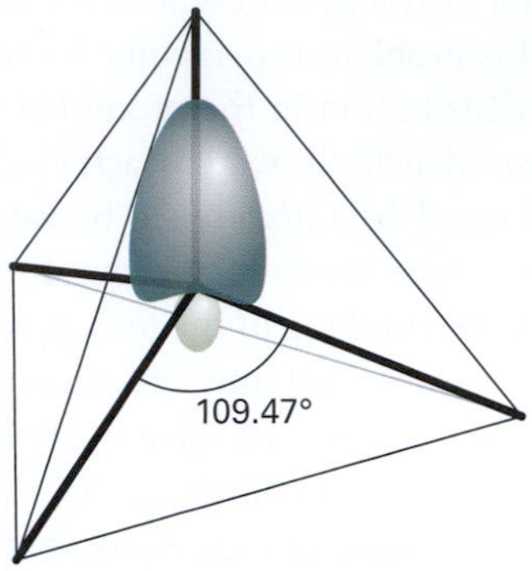

Figure 10A.7 An sp^3 hybrid orbital formed from the superposition of s and p orbitals on the same atom. There are four such hybrids: each one points towards the corner of a regular tetrahedron. The overall electron density remains spherically symmetrical.

It is now easy to see how the valence-bond description of the CH_4 molecule leads to a tetrahedral molecule containing four equivalent C—H bonds. Each hybrid orbital of the promoted C atom contains a single unpaired electron; an H1s electron can pair with each one, giving rise to a σ bond pointing in a tetrahedral direction. For example, the (un-normalized) two-electron wavefunction for the bond formed by the hybrid orbital h_1 and the $1s_A$ orbital (with wavefunction that we shall denote A) is

$$\Psi(1,2) = h_1(1)A(2) + h_1(2)A(1) \qquad (10A.6)$$

As for H_2, to achieve this wavefunction, the two electrons it describes must be paired. Because each sp^3 hybrid orbital has the same composition, all four σ bonds are identical apart from their orientation in space (Fig. 10A.8).

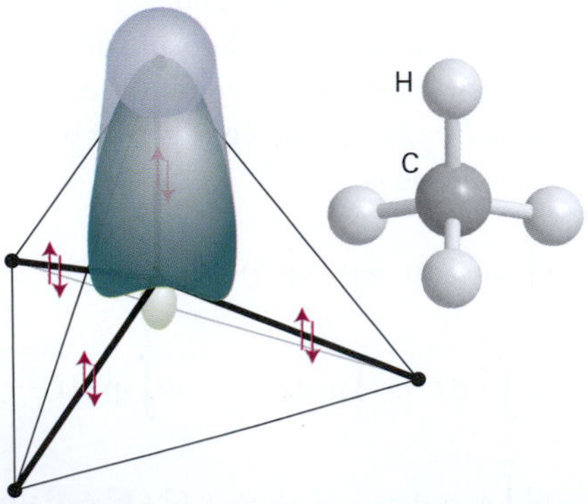

Figure 10A.8 Each sp^3 hybrid orbital forms a σ bond by overlap with an H1s orbital located at the corner of the tetrahedron. This model accounts for the equivalence of the four bonds in CH_4.

A hybrid orbital has enhanced amplitude in the internuclear region, which arises from the constructive interference between the s orbital and the positive lobes of the p orbitals. As a result, the bond strength is greater than for a bond formed from an s or p orbital alone. This increased bond strength is another factor that helps to repay the promotion energy.

Hybridization is used to describe the structure of an ethene molecule, $H_2C{=}CH_2$, and the torsional rigidity of double bonds. An ethene molecule is planar, with HCH and HCC bond angles close to 120°. To reproduce the σ bonding structure, each C atom is regarded as promoted to a $2s^12p^3$ configuration. However, instead of using all four orbitals to form hybrids, we form **sp^2 hybrid orbitals**:

$$\begin{aligned} h_1 &= s + 2^{1/2}p_y \\ h_2 &= s + \left(\tfrac{3}{2}\right)^{1/2}p_x - \left(\tfrac{1}{2}\right)^{1/2}p_y \\ h_3 &= s - \left(\tfrac{3}{2}\right)^{1/2}p_x - \left(\tfrac{1}{2}\right)^{1/2}p_y \end{aligned} \qquad sp^2 \text{ hybrid orbitals} \qquad (10A.7)$$

These hybrids lie in a plane and point towards the corners of an equilateral triangle at 120° to each other (Fig. 10A.9 and Problem 10A.3). The third 2p orbital ($2p_z$) is not included in the hybridization; its axis is perpendicular to the plane in which the hybrids lie. The different signs of the coefficients, as well as ensuring that the hybrids are mutually orthogonal, also ensure that constructive interference takes place in different regions of space, so giving the patterns in the illustration. The sp^2-hybridized C atoms each form three σ bonds by spin pairing with either the h_1 hybrid of the other C atom or with H1s orbitals. The σ framework therefore consists of C—H and C—C σ bonds at 120° to each other. When the two CH_2 groups lie in the same plane, the two electrons in the unhybridized p orbitals can pair and form a π bond (Fig. 10A.10). The formation of this π bond locks the framework into the planar arrangement, for any rotation of one CH_2 group relative to the other leads to a weakening of the π bond (and consequently an increase in energy of the molecule).

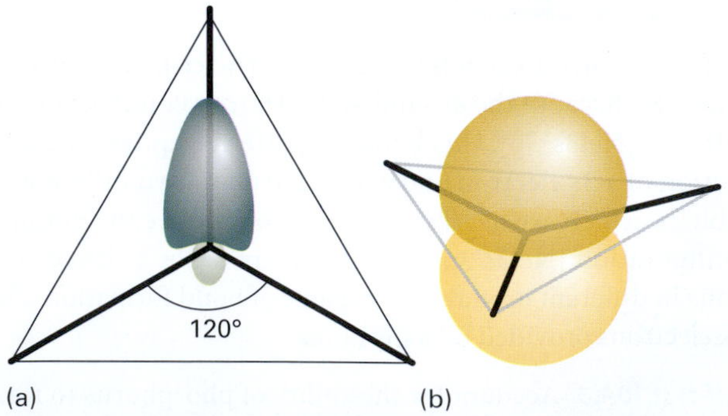

Figure 10A.9 (a) An s orbital and two p orbitals can be hybridized to form three equivalent orbitals that point towards the corners of an equilateral triangle. (b) The remaining, unhybridized p orbital is perpendicular to the plane.

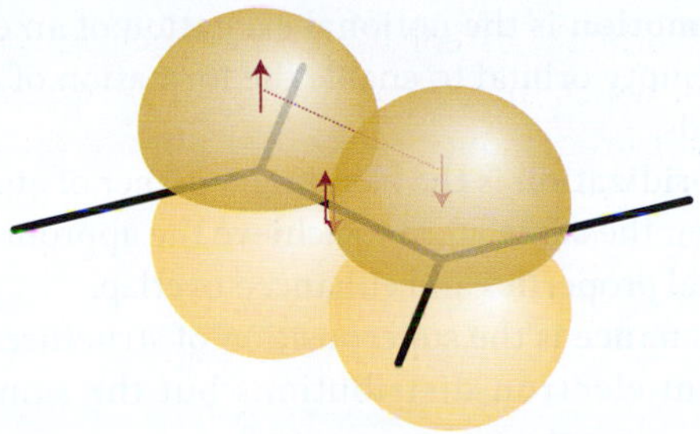

Figure 10A.10 A representation of the structure of a double bond in ethene; only the π bond is shown explicitly.

A similar description applies to ethyne, HC≡CH, a linear molecule. Now the C atoms are **sp hybridized**, and the σ bonds are formed using hybrid atomic orbitals of the form

$$h_1 = \mathrm{s}+\mathrm{p}_z \quad h_2 = \mathrm{s}-\mathrm{p}_z \qquad \text{sp hybrid orbitals} \qquad (10\text{A}.8)$$

These two hybrids lie along the internuclear axis. The electrons in them pair either with an electron in the corresponding hybrid orbital on the other C atom or with an electron in one of the H1s orbitals. Electrons in the two remaining p orbitals on each atom, which are perpendicular to the molecular axis, pair to form two perpendicular π bonds (Fig. 10A.11).

Other hybridization schemes, particularly those involving d orbitals, are often invoked in elementary descriptions of molecular structure to be consistent with other molecular geometries (Table 10A.1). The hybridization of *N* atomic orbitals always results in the formation of *N* hybrid orbitals, which may either form bonds or may contain lone pairs of electrons.

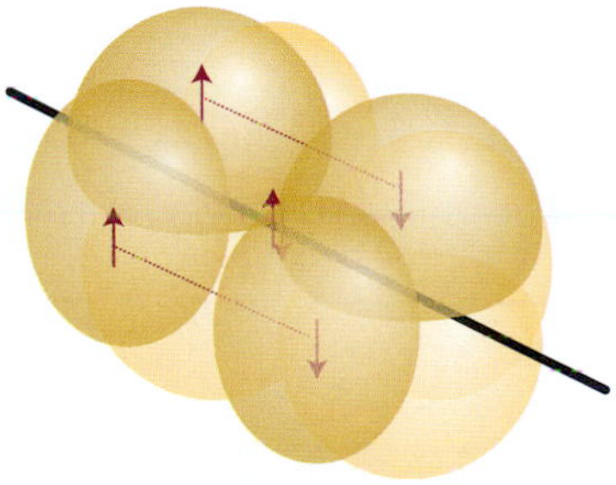

Figure 10A.11 A representation of the structure of a triple bond in ethyne; only the π bonds are shown explicitly. The overall electron density has cylindrical symmetry around the axis of the molecule.

Table 10A.1 Some hybridization schemes

Coordination number	Arrangement	Composition
2	Linear	sp, pd, sd
	Angular	sd
3	Trigonal planar	sp^2, p^2d
	Unsymmetrical planar	spd
	Trigonal pyramidal	pd^2
4	Tetrahedral	sp^3, sd^3
	Irregular tetrahedral	spd^2, p^3d, dp^3
	Square planar	p^2d^2, sp^2d
5	Trigonal bipyramidal	sp^3d, spd^3
	Tetragonal pyramidal	sp^2d^2, sd^4, pd^4, p^3d^2
	Pentagonal planar	p^2d^3
6	Octahedral	sp^3d^2
	Trigonal prismatic	spd^4, pd^5
	Trigonal antiprismatic	p^3d^3

Brief illustration 10A.5 Hybrid structures

For example, sp^3d^2 hybridization results in six equivalent hybrid orbitals pointing towards the corners of a regular octahedron; it is sometimes invoked to account for the structure of octahedral molecules, such as SF_6 (recall the promotion of sulfur's electrons in *Brief illustration* 10A.4). Hybrid orbitals do not always form bonds: they may also contain lone pairs of electrons. For example, in the hydrogen peroxide molecule, H_2O_2, each O atom can be regarded as sp^3 hybridized. Two of the hybrid orbitals form bonds, one O—O bond and one O—H bond at approximately 109° (the experimental value is much less, at 94.8°). The remaining two hybrids on each atom accommodate lone pairs of electrons. Rotation around the O—O bond is possible, so the molecule is conformationally mobile.

Self-test 10A.4 Account for the structure of methylamine, CH_3NH_2.

Answer: C, N both sp^3 hybridized; a lone pair on N

Checklist of concepts

☐ 1. The **Born–Oppenheimer approximation** treats the nuclei as stationary while the electrons move in their field.

☐ 2. A **molecular potential energy curve** depicts the variation of the energy of the molecule as a function of bond length.

☐ 3. The **equilibrium bond length** is the internuclear separation at the minimum of the curve.

☐ 4. The **bond dissociation energy** is the minimum energy need to separate the two atoms of a molecule.

☐ 5. A bond forms when an electron in an atomic orbital on one atom pairs its spin with that of an electron in an atomic orbital on another atom.

☐ 6. A **σ bond** has cylindrical symmetry around the internuclear axis.

☐ 7. A **π bond** has symmetry like that of a p orbital perpendicular to the internuclear axis.

☐ 8. **Promotion** is the notional excitation of an electron to an empty orbital to enable the formation of additional bonds.

☐ 9. **Hybridization** is the blending together of atomic orbitals on the same atom to achieve the appropriate directional properties and enhanced overlap.

☐ 10. **Resonance** is the superposition of structures with different electron distributions but the same nuclear arrangement.

Checklist of equations

Property	Equation	Comment	Equation number
Valence-bond wavefunction	$\Psi = A(1)B(2) + A(2)B(1)$		10A.2
Resonance	$\Psi = \Psi_{\text{covalent}} + \lambda\Psi_{\text{ionic}}$	Ionic–covalent resonance	10A.3
Hybridization	$h = \sum_i c_i \chi_i$	All atomic orbitals on the same atom; specific forms in the text	10A.5 10A.6

10B Principles of molecular orbital theory

Contents

➤ Why do you need to know this material?

Molecular orbital theory is the basis of almost all descriptions of chemical bonding, including that of individual molecules and of solids. It is the basis of almost all computational techniques for the prediction and analysis of the properties of molecules.

➤ What is the key idea?

Molecular orbitals are wavefunctions that spread over all the atom in a molecule and each one can accommodate up to two electrons.

➤ What do you need to know already?

You need to be familiar with the shapes of atomic orbitals (Topic 9B) and how an energy is calculated from a wavefunction (Topic 7C). The entire discussion is within the framework of the Born–Oppenheimer approximation (Topic 10A).

In **molecular orbital theory** (MO theory), electrons do not belong to particular bonds but spread throughout the entire molecule. This theory has been more fully developed than valence-bond theory (Topic 10A) and provides the language that is widely used in modern discussions of bonding. To introduce it, we follow the same strategy as in Topic 9B, where the one-electron H atom was taken as the fundamental species for discussing atomic structure and then developed into a description of many-electron atoms. In this chapter we use the simplest molecular species of all, the hydrogen molecule-ion, H_2^+, to introduce the essential features of bonding and then use it to describe the structures of more complex systems.

10B.1 Linear combinations of atomic orbitals

The hamiltonian for the single electron in H_2^+ is

$$\hat{H}=-\frac{\hbar^2}{2m_e}\nabla_1^2+V \quad V=-\frac{e^2}{4\pi\varepsilon_0}\left(\frac{1}{r_{A1}}+\frac{1}{r_{B1}}-\frac{1}{R}\right) \tag{10B.1}$$

where r_{A1} and r_{B1} are the distances of the electron from the two nuclei A and B (**1**) and R is the distance between the two nuclei. In the expression for V, the first two terms in parentheses are the attractive contribution from the interaction between the electron and the nuclei; the remaining term is the repulsive interaction between the nuclei. The collection of fundamental constant $e^2/4\pi\varepsilon_0$ occurs widely throughout this chapter, and we shall denote it j_0.

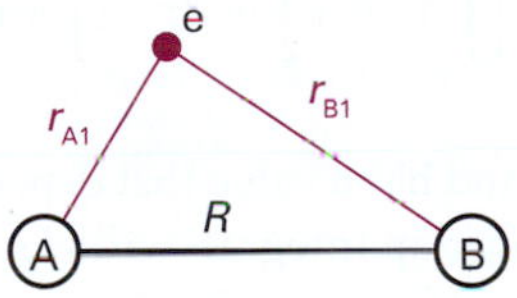

1

The one-electron wavefunctions obtained by solving the Schrödinger equation $H\psi=E\psi$ are called **molecular orbitals** (MOs). A molecular orbital ψ gives, through the value of $|\psi|^2$, the distribution of the electron in the molecule. A molecular orbital is like an atomic orbital, but spreads throughout the molecule.

(a) The construction of linear combinations

The Schrödinger equation can be solved analytically for H_2^+ (within the Born–Oppenheimer approximation), but the wavefunctions are very complicated functions; moreover, the solution cannot be extended to polyatomic systems. Therefore, we adopt a simpler procedure that, while more approximate, can be extended readily to other molecules.

If an electron can be found in an atomic orbital belonging to atom A and also in an atomic orbital belonging to atom B, then the overall wavefunction is a superposition of the two atomic orbitals:

$$\psi_{\pm} = N(A \pm B) \qquad \text{Linear combination of atomic orbitals} \qquad (10B.2)$$

where, for H_2^+, A denotes a 1s atomic orbital on atom A, which we denote (as in Topic 10A) χ_{H1s_A}, B likewise denotes χ_{H1s_B}, and N is a normalization factor. The technical term for the superposition in eqn 10B.2 is a **linear combination of atomic orbitals** (LCAO). An approximate molecular orbital formed from a linear combination of atomic orbitals is called an **LCAO-MO**. A molecular orbital that has cylindrical symmetry around the internuclear axis, such as the one we are discussing, is called a **σ orbital** because it resembles an s orbital when viewed along the axis and, more precisely, because it has zero orbital angular momentum around the internuclear axis.

Example 10.B1 Normalizing a molecular orbital

Normalize the molecular orbital ψ_+ in eqn 10B.2.

Method We need to find the factor N such that $\int \psi^* \psi \, d\tau = 1$, where the integration is over the whole of space. To proceed, substitute the LCAO into this integral, and make use of the fact that the atomic orbitals are individually normalized.

Answer Substitution of the wavefunction gives

$$\int \psi^* \psi \, d\tau = N^2 \left\{ \overbrace{\int A^2 \, d\tau}^{1} + \overbrace{\int B^2 \, d\tau}^{1} + 2\overbrace{\int AB \, d\tau}^{S} \right\} = 2(1+S)N^2$$

where $S = \int AB \, d\tau$ and has a value that depends on the nuclear separation (this 'overlap integral' will play a significant role later). For the integral to be equal to 1, we require

$$N = \frac{1}{\{2(1+S)\}^{1/2}}$$

In H_2^+, $S \approx 0.59$, so $N = 0.56$.

Self-test 10B.1 Normalize the orbital ψ_- in eqn 10B.2.

Answer: $N = 1/\{2(1-S)\}^{1/2}$, so $N = 1.10$

Figure 10B.1 shows the contours of constant amplitude for the molecular orbital ψ_+ in eqn 10B.2. Plots like these are readily obtained using commercially available software. The calculation is quite straightforward, because all we need do is feed in the mathematical forms of the two atomic orbitals and then let the program do the rest.

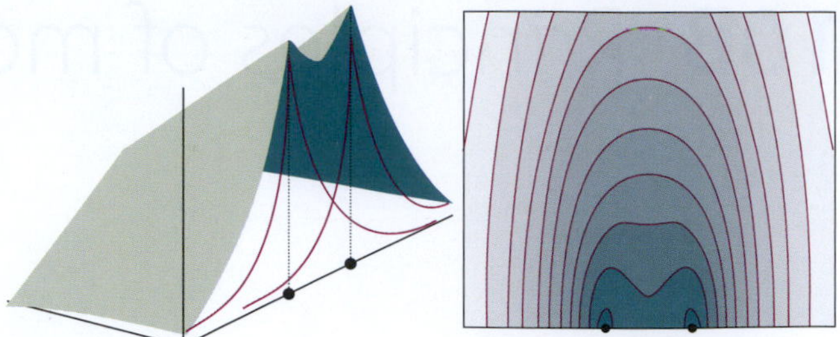

(a) (b)

Figure 10B.1 (a) The amplitude of the bonding molecular orbital in a hydrogen molecule-ion in a plane containing the two nuclei and (b) a contour representation of the amplitude.

Brief illustration 10B.1 A molecular orbital

We can use the same two H1s orbitals as in Topic 10A, namely

$$A = \frac{1}{(\pi a_0^3)^{1/2}} e^{-r_{A1}/a_0} \qquad B = \frac{1}{(\pi a_0^3)^{1/2}} e^{-r_{B1}/a_0}$$

and note that r_A and r_B are not independent (**1**), but when expressed in Cartesian coordinates based on atom A (**2**) are related by $r_{A1} = \{x^2 + y^2 + z^2\}^{1/2}$ and $r_{B1} = \{x^2 + y^2 + (z-R)^2\}^{1/2}$, where R is the bond length. The resulting surfaces of constant amplitude are shown in Fig. 10B.2.

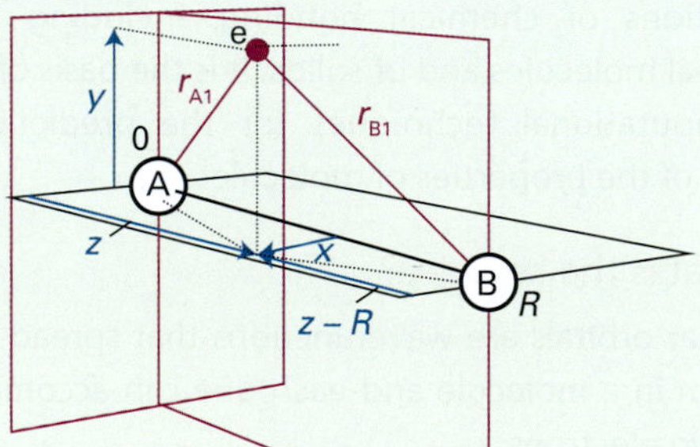

2

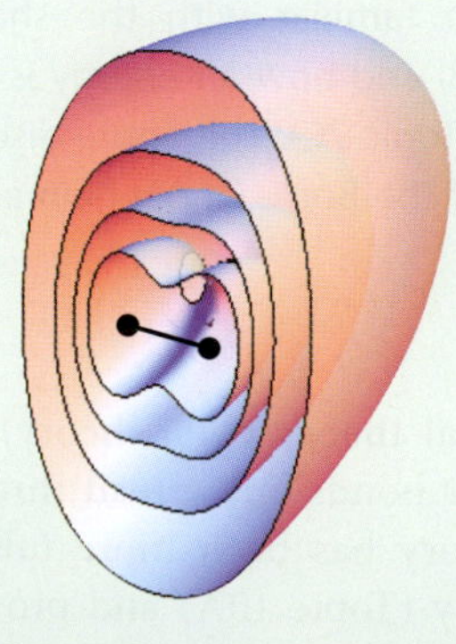

Figure 10B.2 Surfaces of constant amplitude of the wavefunction ψ_+ of the hydrogen molecule-ion.

Self-test 10B.2 Repeat the analysis for ψ_-.

Answer: See Fig. 10B.3.

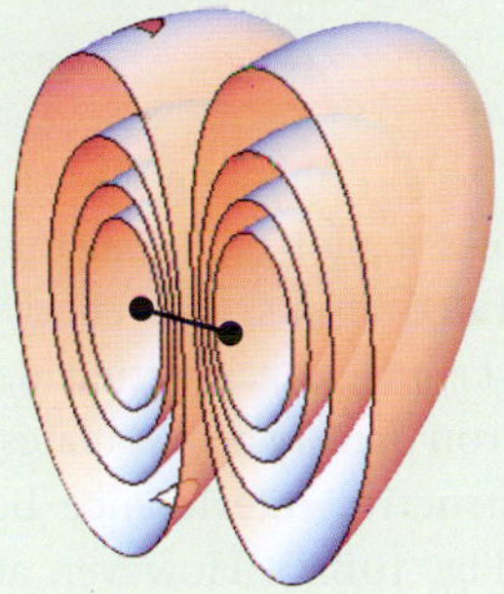

Figure 10B.3 Surfaces of constant amplitude of the wavefunction ψ_- of the hydrogen molecule-ion.

(b) Bonding orbitals

According to the Born interpretation, the probability density of the electron at each point in H_2^+ is proportional to the square modulus of its wavefunction at that point. The probability density corresponding to the (real) wavefunction ψ_+ in eqn 10B.2 is

$$\psi_+^2 = N^2(A^2 + B^2 + 2AB) \quad \text{Bonding probability density} \quad (10B.3)$$

This probability density is plotted in Fig. 10B.4. An important feature becomes apparent when we examine the internuclear region, where both atomic orbitals have similar amplitudes. According to eqn 10B.3, the total probability density is proportional to the sum of:

- A^2, the probability density if the electron were confined to the atomic orbital A.
- B^2, the probability density if the electron were confined to the atomic orbital B.
- $2AB$, an extra contribution to the density from both atomic orbitals.

Physical interpretation

The last contribution, the **overlap density**, is crucial, because it represents an enhancement of the probability of finding the electron in the internuclear region. The enhancement can be traced to the constructive interference of the two atomic orbitals: each has a positive amplitude in the internuclear region, so the total amplitude is greater there than if the electron were confined to a single atomic orbital.

We shall frequently make use of the observation *bonds form due to a build-up of electron density where atomic orbitals overlap and interfere constructively*. The conventional explanation of this observation is based on the notion that accumulation of electron density between the nuclei puts the electron in a position where it interacts strongly with both nuclei. Hence,

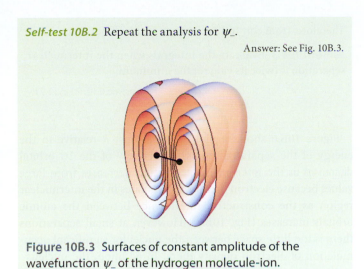

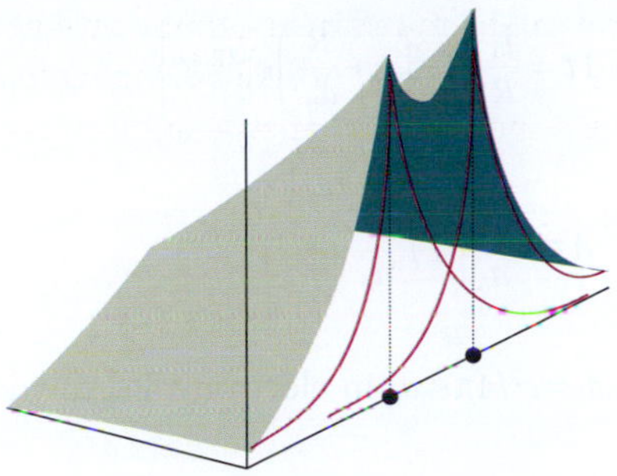

Figure 10B.4 The electron density calculated by forming the square of the wavefunction used to construct Fig. 10B.3. Note the accumulation of electron density in the internuclear region.

the energy of the molecule is lower than that of the separate atoms, where each electron can interact strongly with only one nucleus. This conventional explanation, however, has been called into question, because shifting an electron away from a nucleus into the internuclear region *raises* its potential energy. The modern (and still controversial) explanation does not emerge from the simple LCAO treatment given here. It seems that, at the same time as the electron shifts into the internuclear region, the atomic orbitals shrink. This orbital shrinkage improves the electron–nucleus attraction more than it is decreased by the migration to the internuclear region, so there is a net lowering of potential energy. The kinetic energy of the electron is also modified because the curvature of the wavefunction is changed, but the change in kinetic energy is dominated by the change in potential energy. Throughout the following discussion we ascribe the strength of chemical bonds to the accumulation of electron density in the internuclear region. We leave open the question whether in molecules more complicated than H_2^+ the true source of energy lowering is that accumulation itself or some indirect but related effect.

The σ orbital we have described is an example of a **bonding orbital**, an orbital which, if occupied, helps to bind two atoms together. Specifically, we label it 1σ as it is the σ orbital of lowest energy. An electron that occupies a σ orbital is called a **σ electron**, and if that is the only electron present in the molecule (as in the ground state of H_2^+), then we report the configuration of the molecule as $1\sigma^1$.

The energy $E_{1\sigma}$ of the 1σ orbital is (see Problem 10B.3):

$$E_{1\sigma} = E_{\text{H1s}} + \frac{j_0}{R} - \frac{j+k}{1+S} \quad \text{Energy of bonding orbital} \quad (10B.4)$$

where E_{H1s} is the energy of a H1s orbital, j_0/R is the potential energy of repulsion between the two nuclei (remember that j_0 is shorthand for $e^2/4\pi\varepsilon_0$), and

$$S = \int AB\,\mathrm{d}\tau = \left\{1 + \frac{R}{a_0} + \frac{1}{3}\left(\frac{R}{a_0}\right)^2\right\}\mathrm{e}^{-R/a_0} \quad (10B.5a)$$

$$j = j_0 \int \frac{A^2}{r_B} d\tau = \frac{j_0}{R}\left\{1 - \left(1 + \frac{R}{a_0}\right)e^{-2R/a_0}\right\} \quad (10B.5b)$$

$$k = j_0 \int \frac{AB}{r_B} d\tau = \frac{j_0}{a_0}\left(1 + \frac{R}{a_0}\right)e^{-R/a_0} \quad (10B.5c)$$

To express $j_0/a_0 = e^2/4\pi\varepsilon_0 a_0$ in electronvolts, divide it by e, and then find

$$\frac{j_0}{ea_0} = \frac{e}{4\pi\varepsilon_0 a_0} = \frac{e}{4\pi\varepsilon_0} \times \frac{\pi m_e e^2}{\varepsilon_0 h^2} = \frac{m_e e^3}{4\varepsilon_0^2 h^2} = 27.211\ldots \text{V} \quad (10B.5d)$$

This value should be recognized as $2hc\tilde{R}_\infty$. The integrals are plotted in Fig. 10B.5. We can interpret them as follows:

Physical interpretation

- All three integrals are positive and decline towards zero at large internuclear separations (S and k on account of the exponential term, j on account of the factor $1/R$). The integral S is discussed in more detail in Topic 10B.4c.
- The integral j is a measure of the interaction between a nucleus and electron density centred on the other nucleus.
- The integral k is a measure of the interaction between a nucleus and the excess electron density in the internuclear region arising from overlap.

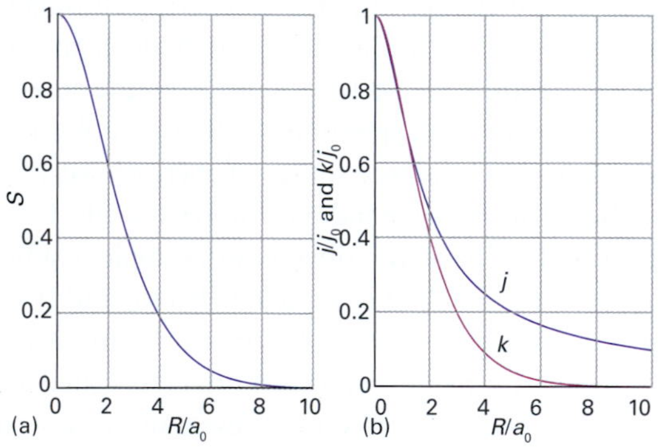

Figure 10B.5 The integrals (a) S, (b) j and k calculated for H_2^+ as a function of internuclear distance.

Brief illustration 10B.2 Molecular integrals

It turns out (see next paragraph of text) that the minimum value of $E_{1\sigma}$ occurs at $R = 2.45a_0$. At this separation

$$S = \left\{1 + 2.45 + \frac{2.45^2}{3}\right\}e^{-2.45} = 0.47$$

$$j = \frac{j_0/a_0}{2.45}\{1 - 3.45e^{-4.90}\} = 0.40\, j_0/a_0$$

$$k = \frac{j_0}{a_0}(1 + 2.45)e^{-2.45} = 0.30\, j_0/a_0$$

Therefore, from eqn 10B.5d, $j = 11$ eV and $k = 8.2$ eV.

Self-test 10B.3 Evaluate the integrals when the internuclear separation is twice its value at the minimum.

Answer: 0.10, 5.5 eV, 1.2 Ev

Figure 10B.6 shows a plot of $E_{1\sigma}$ against R relative to the energy of the separated atoms. The energy of the 1σ orbital decreases as the internuclear separation decreases from large values because electron density accumulates in the internuclear region as the constructive interference between the atomic orbitals increases (Fig. 10B.7). However, at small separations there is too little space between the nuclei for significant accumulation of electron density there. In addition, the nucleus–nucleus repulsion (which is proportional to $1/R$) becomes large. As a result, the energy of the molecule rises at short distances, and there is a minimum in the potential energy curve. Calculations on H_2^+ give $R_e = 2.45a_0 = 130$ pm and $D_e = 1.76$ eV (171 kJ mol^{-1}); the experimental values are 106 pm and 2.6 eV,

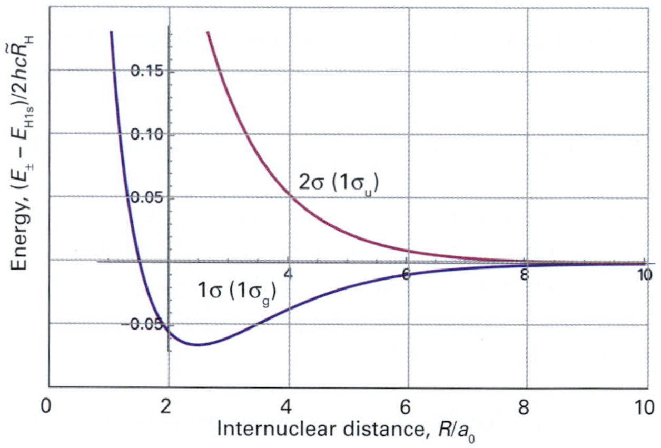

Figure 10B.6 The calculated molecular potential energy curves for a hydrogen molecule-ion showing the variation of the energies of the bonding and antibonding orbitals as the bond length is changed. The alternative notation of the orbitals is explained later.

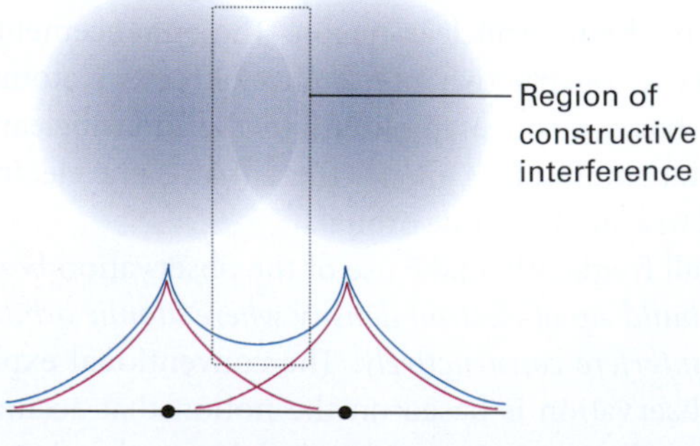

Figure 10B.7 A representation of the constructive interference that occurs when two H1s orbitals overlap and form a bonding σ orbital.

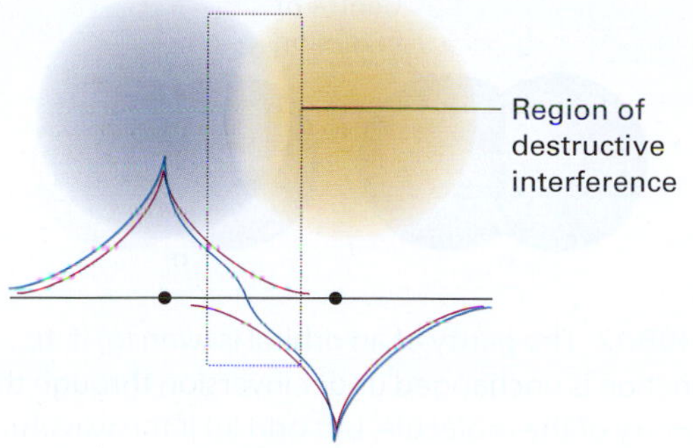

Figure 10B.8 A representation of the destructive interference that occurs when two H1s orbitals overlap and form an antibonding 2σ orbital.

so this simple LCAO-MO description of the molecule, while inaccurate, is not absurdly wrong.

(c) Antibonding orbitals

The linear combination ψ_- in eqn 10B.2 corresponds to an energy higher than that of ψ_+. Because it is also a σ orbital we label it 2σ. This orbital has an internuclear nodal plane where A and B cancel exactly (Figs. 10B.8 and 10B.9). The probability density is

$$\psi_-^2 = N^2(A^2 + B^2 - 2AB) \qquad \text{Antibonding probability density} \qquad (10B.6)$$

There is a reduction in probability density between the nuclei due to the $-2AB$ term (Fig. 10B.10); in physical terms, there is destructive interference where the two atomic orbitals overlap. The 2σ orbital is an example of an **antibonding orbital**, an orbital that, if occupied, contributes to a reduction in the cohesion between two atoms and helps to raise the energy of the molecule relative to the separated atoms.

The energy $E_{2\sigma}$ of the 2σ antibonding orbital is given by (see Problem 10B.3)

$$E_{2\sigma} = E_{\mathrm{H1s}} + \frac{j_0}{R} - \frac{j-k}{1-S} \qquad (10B.7)$$

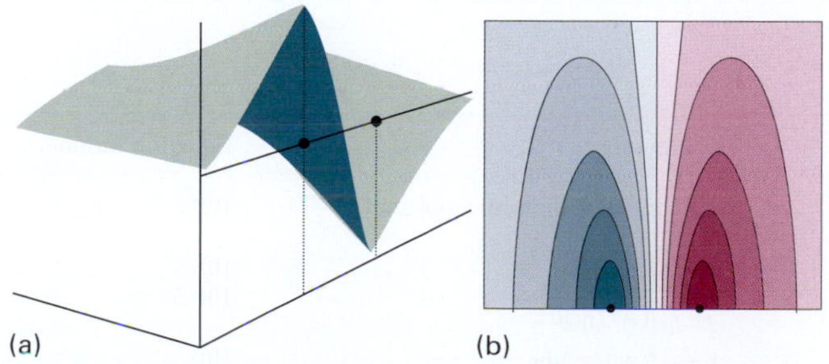

Figure 10B.9 (a) The amplitude of the antibonding molecular orbital in a hydrogen molecule-ion in a plane containing the two nuclei and (b) a contour representation of the amplitude. Note the internuclear node.

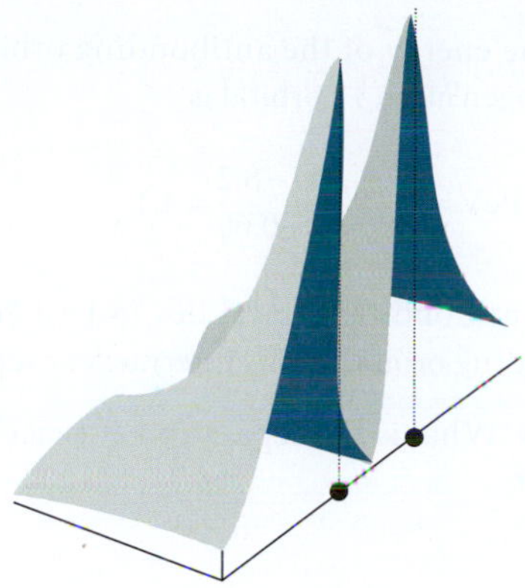

Figure 10B.10 The electron density calculated by forming the square of the wavefunction used to construct Fig.10B.9. Note the reduction of electron density in the internuclear region.

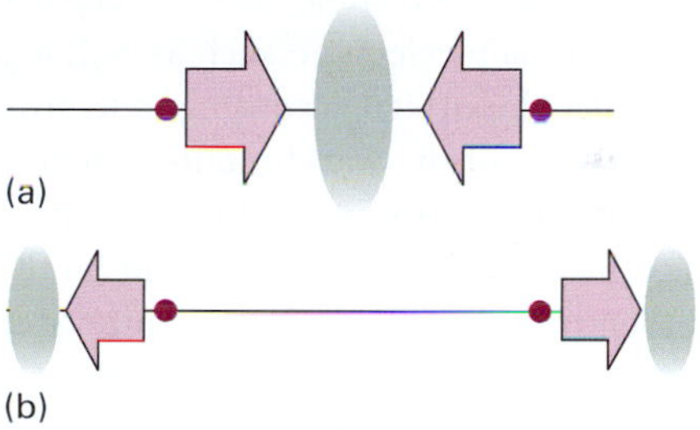

Figure 10B.11 A partial explanation of the origin of bonding and antibonding effects. (a) In a bonding orbital, the nuclei are attracted to the accumulation of electron density in the internuclear region. (b) In an antibonding orbital, the nuclei are attracted to an accumulation of electron density outside the internuclear region.

where the integrals S, j, and k are the same as before (eqn 10B.5). The variation of $E_{2\sigma}$ with R is shown in Fig. 10B.6, where we see the destabilizing effect of an antibonding electron. The effect is partly due to the fact that an antibonding electron is excluded from the internuclear region and hence is distributed largely outside the bonding region. In effect, whereas a bonding electron pulls two nuclei together, an antibonding electron pulls the nuclei apart (Fig. 10B.11). The illustration also shows another feature that we draw on later: $|E_- - E_{\mathrm{H1s}}| > |E_+ - E_{\mathrm{H1s}}|$, which indicates that *the antibonding orbital is more antibonding than the bonding orbital is bonding*. This important conclusion stems in part from the presence of the nucleus–nucleus repulsion (j_0/R): this contribution raises the energy of both molecular orbitals. Antibonding orbitals are often labelled with an asterisk (*), so the 2σ orbital could also be denoted 2σ* (and read '2 sigma star').

Brief illustration 10B.3 Antibonding energies

At the minimum of the bonding orbital energy we have seen that $R = 2.45$, and from *Brief illustration* 10B.2 we know that $S = 0.60$, $j = 11$ eV, and $k = 8.2$ eV. It follows that at that separation, the energy of the antibonding orbital relative to that of a hydrogen atom 1s orbital is

$$(E_{2\sigma}-E_{\mathrm{H1s}})/\mathrm{eV}=\frac{27.2}{2.45}-\frac{11-8.2}{1-0.60}=4.1$$

That is, the antibonding orbital lies (4.1 + 1.76) eV = 5.9 eV above the bonding orbital at this internuclear separation.

Self-test 10B.4 What is the separation at twice that internuclear distance?

Answer: 1.4 Ev

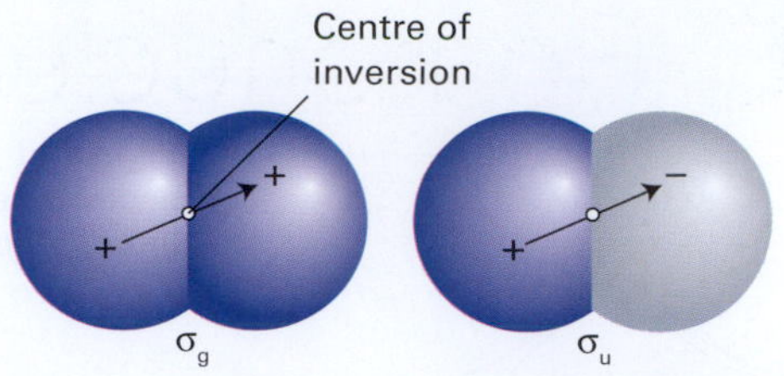

Figure 10B.12 The parity of an orbital is even (g) if its wavefunction is unchanged under inversion through the centre of symmetry of the molecule, but odd (u) if the wavefunction changes sign. Heteronuclear diatomic molecules do not have a centre of inversion, so for them the g, u classification is irrelevant.

10B.2 Orbital notation

For homonuclear diatomic molecules (molecules consisting of two atoms of the same element, such as N_2), it proves helpful to label a molecular orbital according to its **inversion symmetry**, the behaviour of the wavefunction when it is inverted through the centre (more formally, the centre of inversion) of the molecule. Thus, if we consider any point on the bonding σ orbital, and then project it through the centre of the molecule and out an equal distance on the other side, then we arrive at an identical value of the wavefunction (Fig. 10B.12). This so-called **gerade symmetry** (from the German word for 'even') is denoted by a subscript g, as in σ_g. The same procedure applied to the antibonding 2σ orbital results in the same amplitude but opposite sign of the wavefunction. This **ungerade symmetry** ('odd symmetry') is denoted by a subscript u, as in σ_u.

When using the g,u notation, each set of orbitals of the same inversion symmetry is labelled separately so, whereas 1σ becomes $1\sigma_g$, its antibonding partner, which so far we have called 2σ, is the first orbital of a different symmetry, and is denoted $1\sigma_u$. The general rule is that *each set of orbitals of the same symmetry designation is labelled separately*. This point is developed in Topic 10C. The inversion symmetry classification is not applicable to heteronuclear diatomic molecules (diatomic molecules formed by atoms from two different elements, such as CO) because these molecules do not have a centre of inversion.

Checklist of concepts

- ☐ 1. A **molecular orbital** is constructed as a linear combination of atomic orbitals.
- ☐ 2. A **bonding orbital** arises from the constructive overlap of neighbouring atomic orbitals.
- ☐ 3. An **antibonding orbital** arises from the destructive overlap of neighbouring atomic orbitals.
- ☐ 4. **σ Orbitals** have cylindrical symmetry and zero orbital angular momentum around the internuclear axis.
- ☐ 5. A molecular orbital in a homonuclear diatomic molecule is labelled 'gerade' (g) or 'ungerade' (u) according to its behaviour under **inversion symmetry**.

Checklist of equations

Property	Equation	Comment	Equation number
Linear combination of atomic orbitals	$\psi_{\pm}=N(A\pm B)$	Homonuclear diatomic molecule	10B.2
Energies of σ orbitals	$E_{1\sigma}=E_{\mathrm{H1s}}+j_0/R-(j+k)/(1+S)$	$S=\int AB\,\mathrm{d}\tau$, $j=j_0\int(A^2/r_B)\mathrm{d}\tau$	10B.4 10B.5
	$E_{2\sigma}=E_{\mathrm{H1s}}+j_0/R-(j-k)/(1-S)$	$k=j_0\int(AB/r_B)\mathrm{d}\tau$	10B.7

10C Homonuclear diatomic molecules

Contents

➤ Why do you need to know this material?

Although the hydrogen molecule-ion establishes the basic approach to the construction of molecular orbitals, almost all chemically significant molecules have more than one electron, and we need to see how to construct their electron configurations. Homonuclear diatomic molecules are a good starting point, not only because they are simple to describe but because they include such important species as H_2, N_2, O_2, and the dihalogens.

➤ What is the key idea?

Each molecular orbital can accommodate up to two electrons.

➤ What do you need to know already?

You need to be familiar with the discussion of the bonding and antibonding linear combinations of atomic orbitals in Topic 10B and the building-up principle for atoms (Topic 9B).

In Topic 9C the hydrogenic atomic orbitals and the building-up principle are used as a basis for the discussion and prediction of the ground electronic configurations of many-electron atoms. We now do the same for many-electron diatomic molecules by using the H_2^+ molecular orbitals developed in Topic 10B as a basis for their discussion.

10C.1 Electron configurations

The starting point of the building-up principle for diatomic molecules is the construction of molecular orbitals by combining the available atomic orbitals. Once they are available, we adopt the following procedure, which is essentially the same as the building-up principle for atoms (Topic 9B):

Building-up principle for molecules

- The electrons supplied by the atoms are accommodated in the orbitals so as to achieve the lowest overall energy subject to the constraint of the Pauli exclusion principle, that no more than two electrons may occupy a single orbital (and then must be paired).
- If several degenerate molecular orbitals are available, electrons are added singly to each individual orbital before doubly occupying any one orbital (because that minimizes electron–electron repulsions).
- According to Hund's maximum multiplicity rule (Topic 9B), if two electrons do occupy different degenerate orbitals, then a lower energy is obtained if they do so with parallel spins.

(a) σ Orbitals and π orbitals

Consider H_2, the simplest many-electron diatomic molecule. Each H atom contributes a 1s orbital (as in H_2^+), so we can form the $1\sigma_g$ and $1\sigma_u$ orbitals from them, as explained in Topic 10B. At the experimental internuclear separation these orbitals will have the energies shown in Fig. 10C.1, which is called a **molecular orbital energy level diagram**. Note that from two atomic orbitals we can build two molecular orbitals. In general, from N atomic orbitals we can build N molecular orbitals.

There are two electrons to accommodate, and both can enter $1\sigma_g$ by pairing their spins, as required by the Pauli principle (just as for atoms, Topic 9B). The ground-state configuration is therefore $1\sigma_g^2$ and the atoms are joined by a bond consisting of an electron pair in a bonding σ orbital. This approach shows that an electron pair, which was the focus of Lewis's account of chemical bonding, represents the maximum number of electrons that can enter a bonding molecular orbital.

The same argument explains why He does not form diatomic molecules. Each He atom contributes a 1s orbital, so $1\sigma_g$ and $1\sigma_u$ molecular orbitals can be constructed. Although these orbitals differ in detail from those in H_2, their general shapes

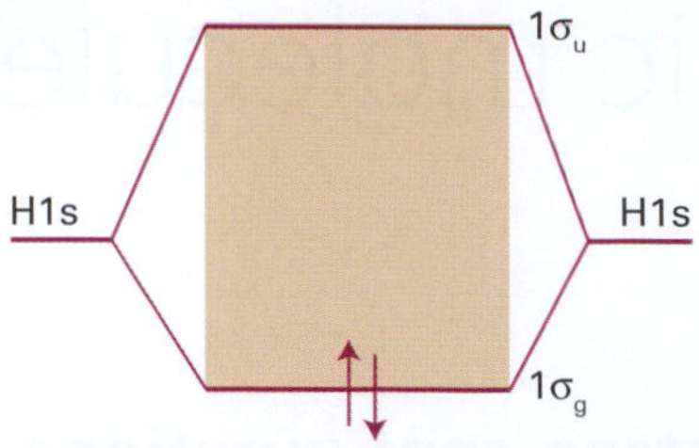

Figure 10C.1 A molecular orbital energy level diagram for orbitals constructed from the overlap of H1s orbitals; the separation of the levels corresponds to that found at the equilibrium bond length. The ground electronic configuration of H_2 is obtained by accommodating the two electrons in the lowest available orbital (the bonding orbital).

are the same and we can use the same qualitative energy level diagram in the discussion. There are four electrons to accommodate. Two can enter the $1\sigma_g$ orbital, but then it is full, and the next two must enter the $1\sigma_u$ orbital (Fig. 10C.2). The ground electronic configuration of He_2 is therefore $1\sigma_g^2 1\sigma_u^2$. We see that there is one bond and one antibond. Because $1\sigma_u$ is raised in energy relative to the separate atoms more than $1\sigma_g$ is lowered, an He_2 molecule has a higher energy than the separated atoms, so it is unstable relative to them.

We shall now see how the concepts we have introduced apply to homonuclear diatomic molecules in general. In elementary treatments, only the orbitals of the valence shell are used to form molecular orbitals so, for molecules formed with atoms from Period 2 elements, only the 2s and 2p atomic orbitals are considered. We shall make that approximation here too.

A general principle of molecular orbital theory is that *all orbitals of the appropriate symmetry* contribute to a molecular orbital. Thus, to build σ orbitals, we form linear combinations of all atomic orbitals that have cylindrical symmetry about the internuclear axis. These orbitals include the 2s orbitals on each atom and the $2p_z$ orbitals on the two atoms (Fig. 10C.3). The general form of the σ orbitals that may be formed is therefore

$$\psi = c_{A2s}\chi_{A2s} + c_{B2s}\chi_{B2s} + c_{A2p_z}\chi_{A2p_z} + c_{B2p_z}\chi_{B2p_z} \qquad (10C.1)$$

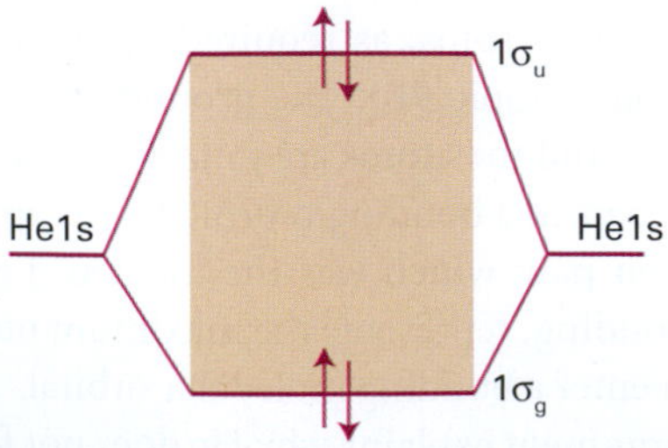

Figure 10C.2 The ground electronic configuration of the hypothetical four-electron molecule He_2 has two bonding electrons and two antibonding electrons. It has a higher energy than the separated atoms, and so is unstable.

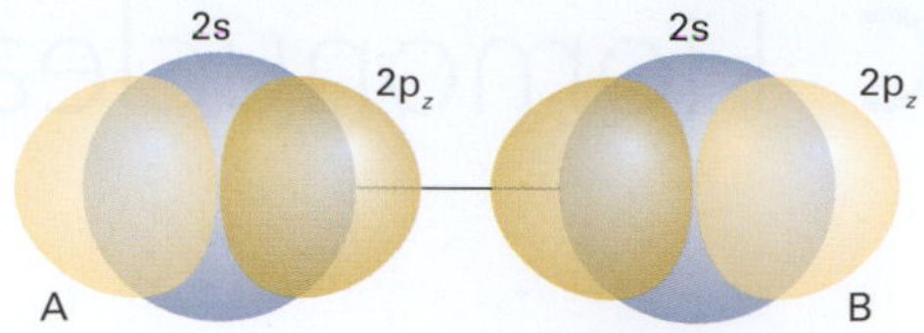

Figure 10C.3 According to molecular orbital theory, σ orbitals are built from all orbitals that have the appropriate symmetry. In homonuclear diatomic molecules of Period 2, that means that two 2s and two $2p_z$ orbitals should be used. From these four orbitals, four molecular orbitals can be built.

From these four atomic orbitals we can form four molecular orbitals of σ symmetry by an appropriate choice of the coefficients *c*.

The procedure for calculating the coefficients is described in Topic 10D and more fully in Topic 10E. Here we adopt a simpler route, and suppose that, because the 2s and $2p_z$ orbitals have distinctly different energies, they may be treated separately. That is, the four σ orbitals fall approximately into two sets, one consisting of two molecular orbitals of the form

$$\psi = c_{A2s}\chi_{A2s} + c_{B2s}\chi_{B2s} \qquad (10C.2a)$$

and another consisting of two orbitals of the form

$$\psi = c_{A2p_z}\chi_{A2p_z} + c_{B2p_z}\chi_{B2p_z} \qquad (10C.2b)$$

Because atoms A and B are identical, the energies of their 2s orbitals are the same, so the coefficients are equal (apart from a possible difference in sign); the same is true of the $2p_z$ orbitals. Therefore, the two sets of orbitals have the form $\chi_{A2s} \pm \chi_{B2s}$ and $\chi_{A2p_z} \pm \chi_{B2p_z}$.

The 2s orbitals on the two atoms overlap to give a bonding and an antibonding σ orbital ($1\sigma_g$ and $1\sigma_u$, respectively) in exactly the same way as we have already seen for 1s orbitals. The two $2p_z$ orbitals directed along the internuclear axis overlap strongly. They may interfere either constructively or destructively, and give a bonding or antibonding σ orbital (Fig. 10C.4). These two σ orbitals are labelled $2\sigma_g$ and $2\sigma_u$, respectively. In general, note how the numbering follows the order of increasing energy. We number only the molecular orbitals formed from atomic orbitals in the valence shell and ignore any combinations of core atomic orbitals.

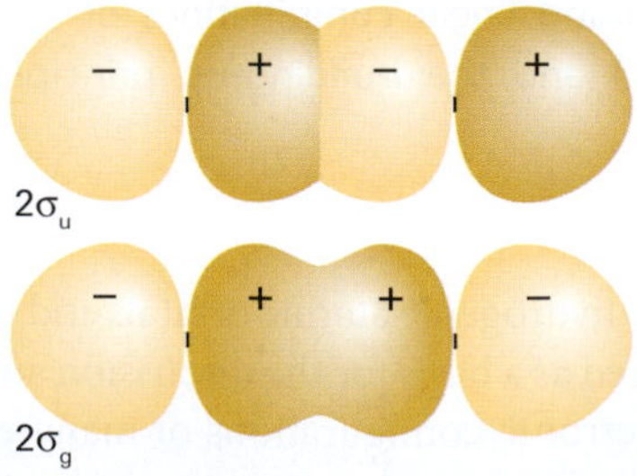

Figure 10C.4 A representation of the composition of bonding and antibonding σ orbitals built from the overlap of p orbitals. These illustrations are schematic.

Brief illustration 10C.1 Ground-state configurations

The valence configuration of a sodium atom is [Ne]$3s^1$, so 3s and 3p orbitals are used to construct molecular orbitals. At this level of approximation, we consider (3s,3s)- and (3p,3p)-overlap separately. In fact, because there are only two electrons to accommodate (one from each 3s orbital), we need consider only the former. That overlap results in $1\sigma_g$ and $1\sigma_u$ molecular orbitals. The only two valence electrons occupy the former, so the ground-state configuration of Na_2 is $1\sigma_g^2$.

Self-test 10C.1 Identify the ground-state configuration of Be_2.

Answer: $1\sigma_g^2 1\sigma_u^2$ built from Be2s orbitals

Now consider the $2p_x$ and $2p_y$ orbitals of each atom. These orbitals are perpendicular to the internuclear axis and may overlap broadside-on. This overlap may be constructive or destructive and results in a bonding or an antibonding **π orbital** (Fig. 10C.5). The notation π is the analogue of p in atoms, for when viewed along the axis of the molecule, a π orbital looks like a p orbital and has one unit of orbital angular momentum around the internuclear axis. The two neighbouring $2p_x$ orbitals overlap to give a bonding and antibonding π_x orbital, and the two $2p_y$ orbitals overlap to give two π_y orbitals. The π_x and π_y bonding orbitals are degenerate; so too are their antibonding partners. We also see from Fig. 10C.5 that a bonding π orbital has odd parity (Topic 10B) and is denoted π_u and an antibonding π orbital has even parity, denoted π_g.

(b) The overlap integral

The extent to which two atomic orbitals on different atoms overlap is measured by the **overlap integral**, *S*:

$$S=\int \chi_A^* \chi_B \,d\tau \quad \text{Definition} \quad \text{Overlap integral} \quad (10C.3)$$

This integral also occurs in Topic 10B (in *Example* 10B.1 and eqn 10B.5a). If the atomic orbital χ_A on A is small wherever the orbital χ_B on B is large, or vice versa, then the product of their amplitudes is everywhere small and the integral—the sum of these products—is small (Fig. 10C.6). If χ_A and χ_B are both large in some region of space, then *S* may be large. If the two normalized atomic orbitals are identical (for instance, 1s orbitals on the same nucleus), then $S=1$. In some cases, simple formulas can be given for overlap integrals. For instance, the variation of *S* with internuclear separation for hydrogenic 1s orbitals on atoms of atomic number *Z* is given by

$$S(1s,1s)=\left\{1+\frac{ZR}{a_0}+\frac{1}{3}\left(\frac{ZR}{a_0}\right)^2\right\}e^{-ZR/a_0} \quad \text{(1s,1s)-overlap integral} \quad (10C.4)$$

and is plotted in Fig. 10C.7 (eqn 10C.4 is a generalization of eqn 10B.5a, which is for H1s orbitals).

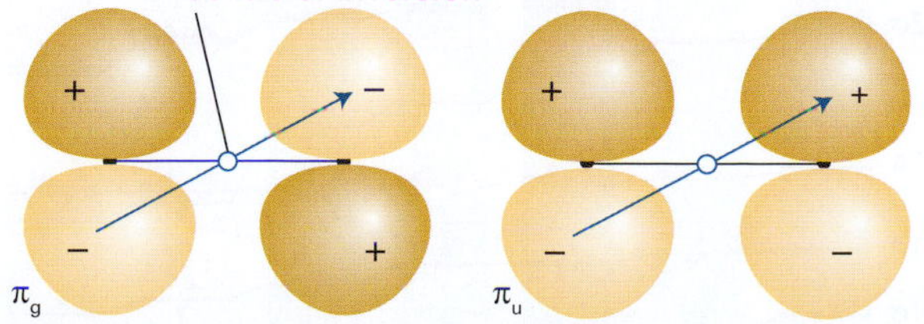

Figure 10C.5 A schematic representation of the structure of π bonding and antibonding molecular orbitals. The figure also shows that the bonding π orbital has odd parity, whereas the antibonding π orbital has even parity.

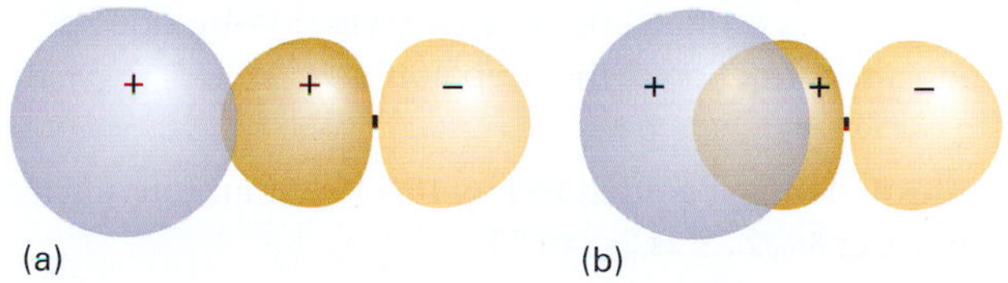

Figure 10C.6 (a) When two orbitals are on atoms that are far apart, the wavefunctions are small where they overlap, so *S* is small. (b) When the atoms are closer, both orbitals have significant amplitudes where they overlap, and *S* may approach 1. Note that *S* will decrease again as the two atoms approach more closely than shown here because the region of negative amplitude of the p orbital starts to overlap the positive amplitude of the s orbital. When the centres of the atoms coincide, $S=0$.

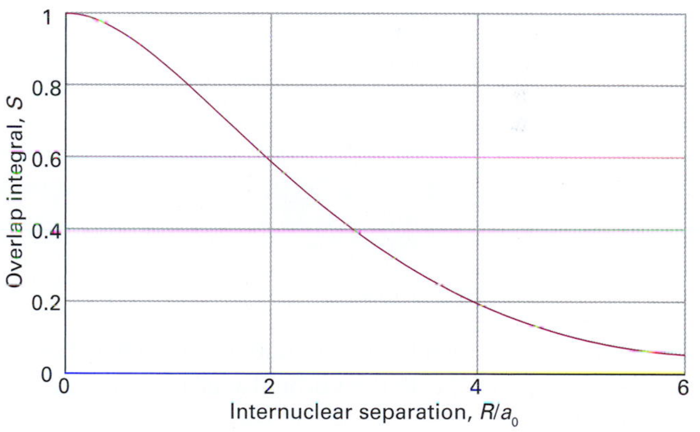

Figure 10C.7 The overlap integral, *S*, between two H1s orbitals as a function of their separation *R*.

Brief illustration 10C.2 Overlap integrals

Familiarity with the magnitudes of overlap integrals is useful when considering bonding abilities of atoms, and hydrogenic orbitals give an indication of their values. The overlap integral between two hydrogenic 2s orbitals is

$$S(2s,2s)=\left\{1+\frac{ZR}{2a_0}+\frac{1}{12}\left(\frac{ZR}{a_0}\right)^2+\frac{1}{240}\left(\frac{ZR}{a_0}\right)^4\right\}e^{-ZR/2a_0}$$

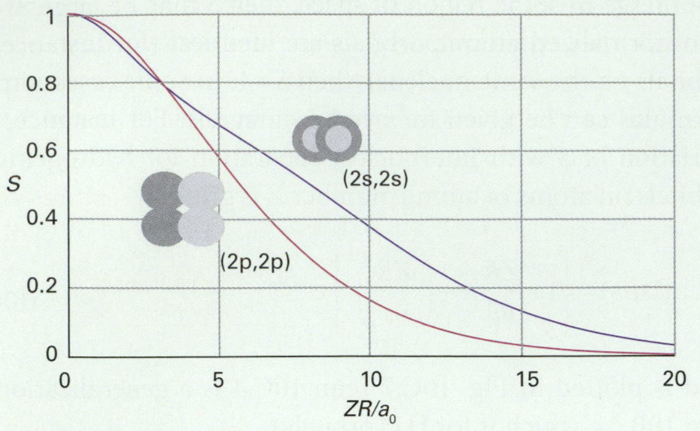

Figure 10C.8 The overlap integral, S, between two hydrogenic 2s orbitals and between two side-by-side 2p orbitals as a function of their separation R.

This expression is plotted in Fig. 10C.8. For an internuclear distance of $8a_0/Z$, $S(2s,2s)=0.50$.

Self-test 10C.2 The side-by-side overlap of two 2p orbitals of atoms of atomic number Z is

$$S(2p,2p)=\left\{1+\frac{ZR}{2a_0}+\frac{1}{10}\left(\frac{ZR}{a_0}\right)^2+\frac{1}{120}\left(\frac{ZR}{a_0}\right)^3\right\}e^{-ZR/2a_0}$$

Evaluate this overlap integral for $R=8a_0/Z$.

Answer: See Fig. 10C.8, 0.29

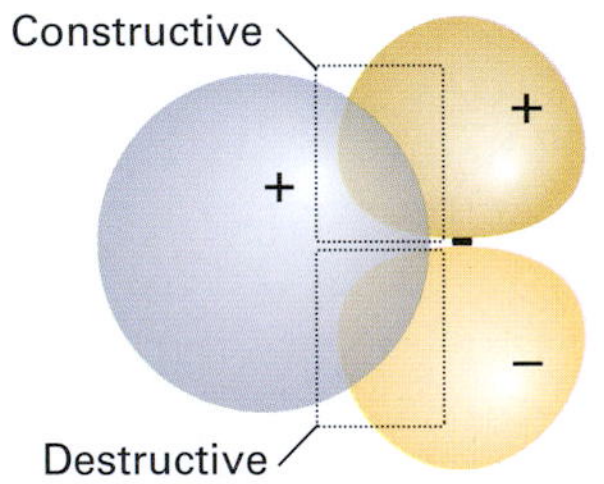

Figure 10C.9 A p orbital in the orientation shown here has zero net overlap ($S=0$) with the s orbital at all internuclear separations.

Now consider the arrangement in which an s orbital is superimposed on a p_x orbital of a different atom (Fig. 10C.9). The integral over the region where the product of orbitals is positive exactly cancels the integral over the region where the product of orbitals is negative, so overall $S=0$ exactly. Therefore, there is no net overlap between the s and p orbitals in this arrangement.

(c) Period 2 diatomic molecules

To construct the molecular orbital energy level diagram for Period 2 homonuclear diatomic molecules, we form eight molecular orbitals from the eight valence shell orbitals (four from each atom). In some cases, π orbitals are less strongly bonding than σ orbitals because their maximum overlap occurs off-axis. This relative weakness suggests that the molecular orbital energy level diagram ought to be as shown in Fig. 10C.10. However, we must remember that we have assumed that 2s and $2p_z$ orbitals contribute to different sets of molecular orbitals whereas in fact all four atomic orbitals have the same symmetry around the internuclear axis and contribute jointly to the four σ orbitals. Hence, there is no guarantee that this order of energies should prevail, and it is found experimentally (by spectroscopy) and by detailed calculation that the order varies along Period 2 (Fig. 10C.11). The order shown in Fig. 10C.12 is appropriate as far as N_2, and Fig. 10C.10 is appropriate for O_2 and F_2. The relative order is controlled by the separation of the 2s and 2p orbitals in the atoms, which increases across the group. The consequent switch in order occurs at about N_2.

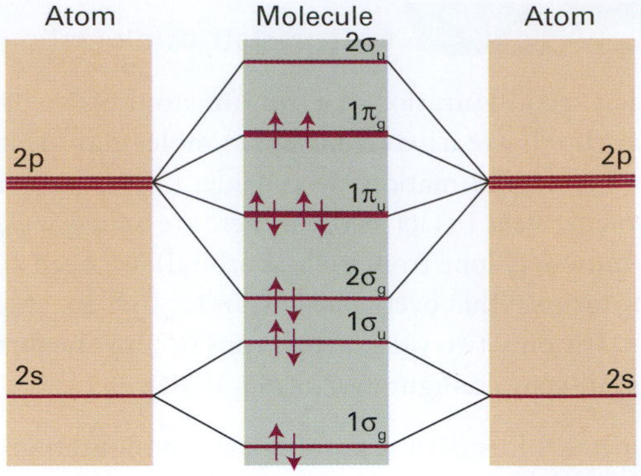

Figure 10C.10 The molecular orbital energy level diagram for homonuclear diatomic molecules. The lines in the middle are an indication of the energies of the molecular orbitals that can be formed by overlap of atomic orbitals. As remarked in the text, this diagram should be used for O_2 (the configuration shown) and F_2.

With the molecular orbital energy level diagram established, we can deduce the probable ground configurations of the molecules by adding the appropriate number of electrons to the orbitals and following the building-up rules. Anionic species (such as the peroxide ion, O_2^{2-}) need more electrons than the

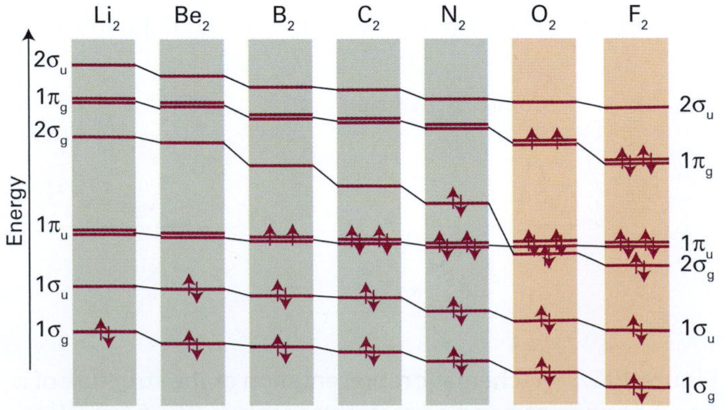

Figure 10C.11 The variation of the orbital energies of Period 2 homonuclear diatomics

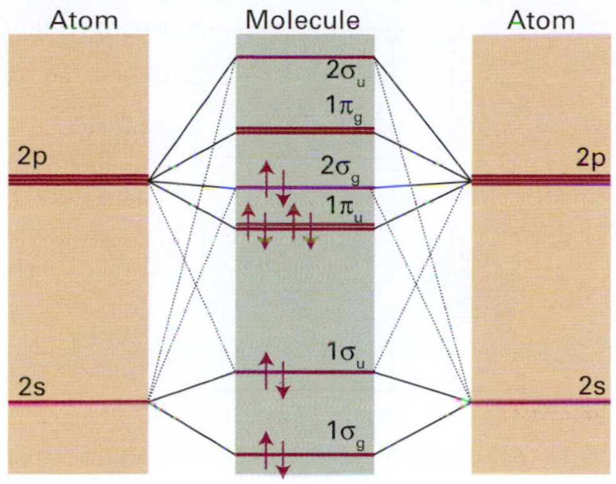

Figure 10C.12 An alternative molecular orbital energy level diagram for homonuclear diatomic molecules. As remarked in the text, this diagram should be used for diatomics up to and including N_2 (the configuration shown).

parent neutral molecules; cationic species (such as O_2^+) need fewer.

Consider N_2, which has 10 valence electrons. Two electrons pair, occupy, and fill the $1\sigma_g$ orbital; the next two occupy and fill the $1\sigma_u$ orbital. Six electrons remain. There are two $1\pi_u$ orbitals, so four electrons can be accommodated in them. The last two enter the $2\sigma_g$ orbital. Therefore, the ground-state configuration of N_2 is $1\sigma_g^2 1\sigma_u^2 1\pi_u^4 2\sigma_g^2$. It is sometimes helpful to include an asterisk to denote an antibonding orbital, in which case this configuration would be denoted $1\sigma_g^2 1\sigma_u^{*2} 1\pi_u^4 2\sigma_g^2$.

A measure of the net bonding in a diatomic molecule is its **bond order**, b:

$$b = \frac{1}{2}(N - N^*) \qquad \text{Definition} \quad \text{Bond order} \quad (10C.5)$$

where N is the number of electrons in bonding orbitals and N^* is the number of electrons in antibonding orbitals.

Brief illustration 10C.3 Bond order

Each electron pair in a bonding orbital increases the bond order by 1 and each pair in an antibonding orbital decreases b by 1. For H_2, $b=1$, corresponding to a single bond, H–H, between the two atoms. In He_2, $b=0$, and there is no bond. In N_2, $b=\frac{1}{2}(8-2)=3$. This bond order accords with the Lewis structure of the molecule (:N≡N:).

Self-test 10C.3 Evaluate the bond orders of O_2, O_2^+, and O_2^-.

Answer: 2, $\frac{5}{2}$, 1

The ground-state electron configuration of O_2, with 12 valence electrons, is based on Fig. 10C.10, and is $1\sigma_g^2 1\sigma_u^2 2\sigma_g^2 1\pi_u^4 1\pi_g^2$ (or $1\sigma_g^2 1\sigma_u^{*2} 2\sigma_g^2 1\pi_u^4 1\pi_g^{*2}$). Its bond order is 2. According to the building-up principle, however, the two $1\pi_g$ electrons occupy different orbitals: one will enter $1\pi_{g,x}$ and the other will enter $1\pi_{g,y}$. Because the electrons are in different orbitals, they will have parallel spins. Therefore, we can predict that an O_2 molecule will have a net spin angular momentum $S=1$ and, in the language introduced in Topic 9C, be in a triplet state. As electron spin is the source of a magnetic moment, we can go on to predict that oxygen should be paramagnetic, a substance that tends to move into a magnetic field (see Topic 18C). This prediction, which VB theory does not make, is confirmed by experiment.

An F_2 molecule has two more electrons than an O_2 molecule. Its configuration is therefore $1\sigma_g^2 1\sigma_u^{*2} 2\sigma_g^2 1\pi_u^4 1\pi_g^{*4}$ and $b=1$. We conclude that F_2 is a singly-bonded molecule, in agreement with its Lewis structure. The hypothetical molecule dineon, Ne_2, has two additional electrons: its configuration is $1\sigma_g^2 1\sigma_u^{*2} 2\sigma_g^2 1\pi_u^4 1\pi_g^{*4} 2\sigma_u^{*2}$ and $b=0$. The zero bond order is consistent with the monatomic nature of Ne.

The bond order is a useful parameter for discussing the characteristics of bonds, because it correlates with bond length and bond strength. For bonds between atoms of a given pair of elements:

- The greater the bond order, the shorter the bond.
- The greater the bond order, the greater the bond strength.

Physical interpretation

Table 10C.1 lists some typical bond lengths in diatomic and polyatomic molecules. The strength of a bond is measured by its bond dissociation energy, D_0, the energy required to separate the atoms to infinity or by the well depth D_e, with $D_0 = D_e - \frac{1}{2}\hbar\omega$. Table 10C.2 lists some experimental values of D_0.

Example 10C.1 Judging the relative bond strengths of molecules and ions

Predict whether N_2^+ is likely to have a larger or smaller dissociation energy than N_2.

Method Because the molecule with the higher bond order is likely to have the higher dissociation energy, compare their electronic configurations and assess their bond orders.

Answer From Fig. 10C.12, the electron configurations and bond orders are

$$N_2 \quad 1\sigma_g^2 1\sigma_u^{*2} 1\pi_u^4 2\sigma_g^2 \quad b=3$$
$$N_2^+ \quad 1\sigma_g^2 1\sigma_u^{*2} 1\pi_u^4 2\sigma_g^1 \quad b=2\tfrac{1}{2}$$

Because the cation has the smaller bond order, we expect it to have the smaller dissociation energy. The experimental dissociation energies are 945 kJ mol^{-1} for N_2 and 842 kJ mol^{-1} for N_2^+.

Self-test 10C.4 Which can be expected to have the higher dissociation energy, F_2 or F_2^+?

Answer: F_2^+

Table 10C.1* Bond lengths, R_e/pm

Bond	Order	R_e/pm
HH	1	74.14
NN	3	109.76
HCl	1	127.45
CH	1	*114*
CC	1	*154*
CC	2	*134*
CC	3	*120*

* More values will be found in the *Resource section*. Numbers in italics are mean values for polyatomic molecules.

Table 10C.2* Bond dissociation energies, D_0/(kJ mol^{-1})

Bond	Order	D_0/(kJ mol^{-1})
HH	1	432.1
NN	3	941.7
HCl	1	427.7
CH	1	*435*
CC	1	*368*
CC	2	*720*
CC	3	*962*

* More values will be found in the *Resource section*. Numbers in italics are mean values for polyatomic molecules.

10C.2 Photoelectron spectroscopy

So far we have treated molecular orbitals as purely theoretical constructs, but is there experimental evidence for their existence? **Photoelectron spectroscopy** (PES) measures the ionization energies of molecules when electrons are ejected from different orbitals by absorption of a photon of known energy, and uses the information to infer the energies of molecular orbitals. The technique is also used to study solids, and in Topic 22A we see the important information that it gives about species at or on surfaces.

Because energy is conserved when a photon ionizes a sample, the sum of the ionization energy, I, of the sample and the kinetic energy of the **photoelectron**, the ejected electron, must be equal to the energy of the incident photon $h\nu$ (Fig. 10C.13):

$$h\nu = \frac{1}{2}m_e v^2 + I \qquad (10C.6)$$

This equation (which is like the one used for the photoelectric effect, eqn 7A.13 of Topic 7A, $E_k = \frac{1}{2}m_e v^2 = h\nu - \Phi$, written as $h\nu = \frac{1}{2}m_e v^2 + \Phi$) can be refined in two ways. First, photoelectrons may originate from one of a number of different orbitals, and each one has a different ionization energy. Hence, a series of different kinetic energies of the photoelectrons will be obtained, each one satisfying $h\nu = \frac{1}{2}m_e v^2 + I_i$, where I_i is the ionization energy for ejection of an electron from an orbital i. Therefore, by measuring the kinetic energies of the

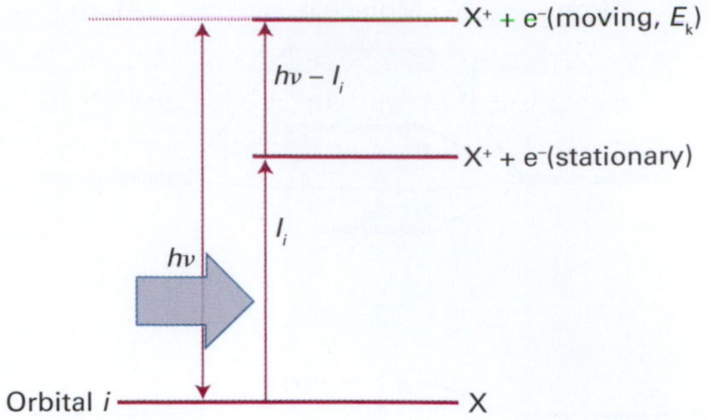

Figure 10C.13 An incoming photon carries an energy $h\nu$; an energy I_i is needed to remove an electron from an orbital i, and the difference appears as the kinetic energy of the electron.

photoelectrons, and knowing the frequency ν, these ionization energies can be determined. Photoelectron spectra are interpreted in terms of an approximation called **Koopmans' theorem**, which states that the ionization energy I_i is equal to the orbital energy of the ejected electron (formally: $I_i = -\varepsilon_i$). That is, we can identify the ionization energy with the energy of the orbital from which it is ejected. The theorem is only an approximation because it ignores the fact that the remaining electrons adjust their distributions when ionization occurs.

The ionization energies of molecules are several electronvolts even for valence electrons, so it is essential to work in at least the ultraviolet region of the spectrum and with wavelengths of less than about 200 nm. Much work has been done with radiation generated by a discharge through helium: the He(I) line ($1s^12p^1 \rightarrow 1s^2$) lies at 58.43 nm, corresponding to a photon energy of 21.22 eV. Its use gives rise to the technique of **ultraviolet photoelectron spectroscopy** (UPS). When core electrons are being studied, photons of even higher energy are needed to expel them: X-rays are used, and the technique is denoted XPS.

The kinetic energies of the photoelectrons are measured using an electrostatic deflector that produces different deflections in the paths of the photoelectrons as they pass between charged

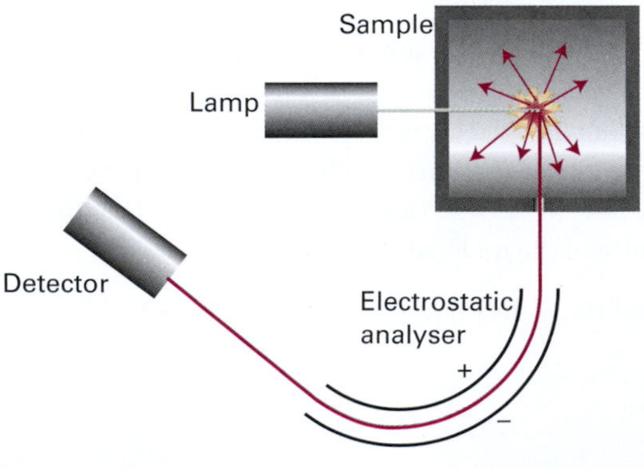

Figure 10C.14 A photoelectron spectrometer consists of a source of ionizing radiation (such as a helium discharge lamp for UPS and an X-ray source for XPS), an electrostatic analyser, and an electron detector. The deflection of the electron path caused by the analyser depends on the speed of the electrons.

plates (Fig. 10C.14). As the field strength is increased, electrons of different speeds, and therefore kinetic energies, reach the detector. The electron flux can be recorded and plotted against kinetic energy to obtain the photoelectron spectrum.

Brief illustration 10C.4 A photoelectron spectrum

Photoelectrons ejected from N_2 with He(I) radiation have kinetic energies of 5.63 eV (1 eV = 8065.5 cm^{-1}, Fig. 10C.15). Helium(I) radiation of wavelength 58.43 nm has wavenumber 1.711×10^5 cm^{-1} and therefore corresponds to an energy of 21.22 eV. Then, from eqn 10C.6 with I_i in place of I, 21.22 eV = 5.63 eV + I_i, so I_i = 15.59 eV. This ionization energy is the energy needed to remove an electron from the occupied molecular orbital with the highest energy of the N_2 molecule, the $2\sigma_g$ bonding orbital.

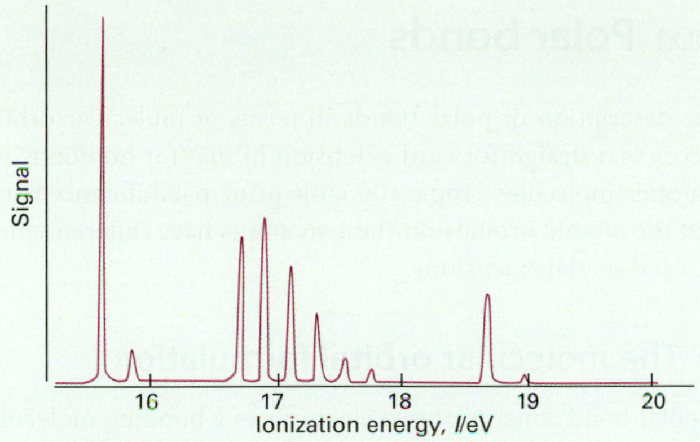

Figure 10C.15 The photoelectron spectrum of N_2.

Self-test 10C.5 Under the same circumstances, photoelectrons are also detected at 4.53 eV. To what ionization energy does that correspond? Suggest an origin.

Answer: 16.7 eV, $1\pi_u$

It is often observed that photoejection results in cations that are excited vibrationally. Because different energies are needed to excite different vibrational states of the ion, the photoelectrons appear with different kinetic energies. The result is **vibrational fine structure**, a progression of lines with a frequency spacing that corresponds to the vibrational frequency of the molecule. Figure 10C.16 shows an example of vibrational fine structure in the photoelectron spectrum of HBr.

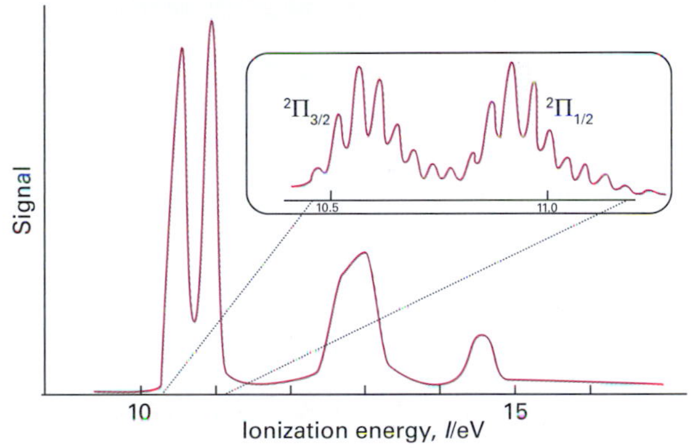

Figure 10C.16 The photoelectron spectrum of HBr.

Checklist of concepts

- ☐ 1. Electrons are added to available molecular orbitals in a manner that achieves the lowest total energy.
- ☐ 2. As a first approximation, σ orbitals are constructed separately from valence s and p orbitals.
- ☐ 3. An **overlap integral** is a measure of the extent of orbital overlap.
- ☐ 4. The greater the **bond order** of a molecule, the shorter and stronger is the bond.
- ☐ 5. **Photoelectron spectroscopy** is a technique for determining the energies of electrons in molecular orbitals.

Checklist of equations

Property	Equation	Comment	Equation number
Overlap integral	$S=\int \chi_A^* \chi_B \, d\tau$	Integration over all space	10C.3
Bond order	$b=\frac{1}{2}(N-N^*)$	N and N^* are the numbers of electrons in bonding and antibonding orbitals, respectively	10C.5
Photoelectron spectroscopy	$h\nu=\frac{1}{2}m_e v^2+I$	Interpret I as I_i, the ionization energy from orbital i.	10C.6

10D Heteronuclear diatomic molecules

Contents

➤ Why do you need to know this material?

Most molecules are heteronuclear, so you need to appreciate the differences in their electronic structure from homonuclear species, and how to treat those differences quantitatively.

➤ What is the key idea?

The bonding molecular orbital of a heteronuclear diatomic molecule is composed mostly of the atomic orbital of the more electronegative atom; the opposite is true of the antibonding orbital.

➤ What do you need to know already?

You need to know about the molecular orbitals of homonuclear diatomic molecules (Topic 10C) and the concepts of normalization and orthogonality (Topic 7C). This Topic makes use of determinants (*The chemist's toolkit* 9B.1) and the rules of differentiation (*Mathematical background* 1).

The electron distribution in a covalent bond in a heteronuclear diatomic molecule is not shared equally by the atoms because it is energetically favourable for the electron pair to be found closer to one atom than to the other. This imbalance results in a **polar bond**, a covalent bond in which the electron pair is shared unequally by the two atoms. The bond in HF, for instance, is polar, with the electron pair closer to the F atom. The accumulation of the electron pair near the F atom results in that atom having a net negative charge, which is called a **partial negative charge** and denoted δ–. There is a matching **partial positive charge**, δ+, on the H atom (Fig. 10D.1).

10D.1 Polar bonds

The description of polar bonds in terms of molecular orbital theory is a straightforward extension of that for homonuclear diatomic molecules (Topic 10C), the principal difference being that the atomic orbitals on the two atoms have different energies and spatial extensions.

(a) The molecular orbital formulation

A polar bond consists of two electrons in a bonding molecular orbital of the form

$$\psi = c_A A + c_B B \qquad \text{Wavefunction of a polar bond} \qquad (10D.1)$$

with unequal coefficients. The proportion of the atomic orbital A in the bond is $|c_A|^2$ and that of B is $|c_B|^2$. A nonpolar bond has $|c_A|^2=|c_B|^2$ and a pure ionic bond has one coefficient zero (so the species A^+B^- would have $c_A=0$ and $c_B=1$). The atomic orbital with the lower energy makes the larger contribution

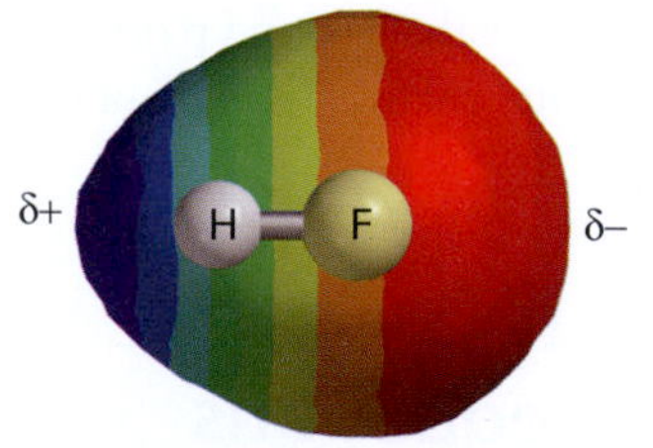

Figure 10D.1 The electron density of the molecule HF, computed with one of the methods described in Topic 10E. Different colours show the distribution of electrostatic potential and hence net charge, with blue representing the region with largest partial positive charge, and red the region with largest partial negative charge.

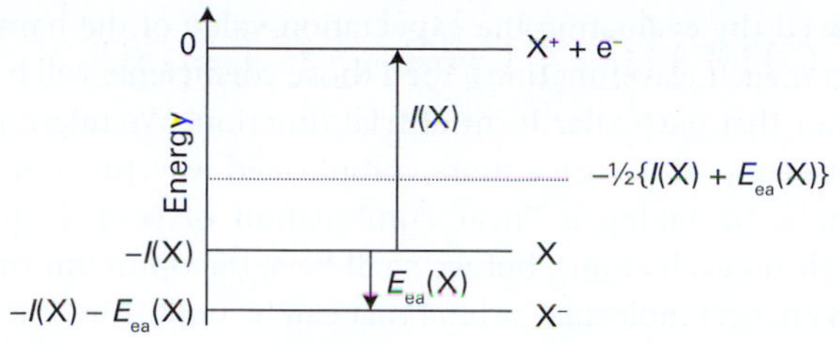

Figure 10D.2 The procedure for estimating the energy of an atomic orbital in a molecule.

to the bonding molecular orbital. The opposite is true of the antibonding orbital, for which the dominant component comes from the atomic orbital with higher energy.

Deciding what values to use for the energies of the atomic orbitals in eqn 10D.1 presents a dilemma because they are known only after a complicated calculation of the kind described in Topic 10E has been performed. An alternative, one that gives some insight into the origin of the energies, is to estimate them from ionization energies and electron affinities. Thus, the extreme cases of an atom X in a molecule are X^+ if it has lost control of the electron it supplied, X if it is sharing the electron pair equally with its bonded partner, and X^- if it has gained control of both electrons in the bond. If X^+ is taken as defining the energy 0, then X lies at $-I(X)$ and X^- lies at $-\{I(X)+E_{ea}(X)\}$, where I is the ionization energy and E_{ea} the electron affinity (Fig. 10D.2). The actual energy of the orbital lies at an intermediate value, and in the absence of further information, we shall estimate it as half-way down to the lowest of these values, namely $-\frac{1}{2}\{I(X)+E_{ea}(X)\}$. Then, to establish the MO composition and energies, we form linear combinations of atomic orbitals with these values of the energy and anticipate that the atom with the more negative value of $-\frac{1}{2}\{I(X)+E_{ea}(X)\}$ contributes the greater amount to the bonding orbital. As we shall see shortly, the quantity $\frac{1}{2}\{I(X)+E_{ea}(X)\}$ also has a further significance.

Brief illustration 10D.1 Heteronuclear diatomic molecules 1

These points can be illustrated by considering HF. The general form of the molecular orbital is $\psi = c_H\chi_H + c_F\chi_F$, where χ_H is an H1s orbital and χ_F is an F2p$_z$ orbital (with z along the internuclear axis, the convention for linear molecules). The relevant data are as follows:

	I/eV	E_{ea}/eV	$\frac{1}{2}\{I+E_{ea}\}$/eV
H	13.6	0.75	7.2
F	17.4	3.34	10.4

We see that the electron distribution in HF is likely to be predominantly on the F atom. We take the calculation further below (in *Brief illustrations* 10D.3 and 10D.4).

Self-test 10D.1 Which atomic orbital, H1s or N2p$_z$, makes the dominant contribution to the bonding σ orbital in the HN molecular radical? For data, see Tables 9B.2 and 9B.3.

Answer: N2p$_z$

(b) Electronegativity

The charge distribution in bonds is commonly discussed in terms of the **electronegativity**, χ (chi), of the elements involved (there should be little danger of confusing this use of χ with its use to denote an atomic orbital, which is another common convention). The electronegativity is a parameter introduced by Linus Pauling as a measure of the power of an atom to attract electrons to itself when it is part of a compound. Pauling used valence-bond arguments to suggest that an appropriate numerical scale of electronegativities could be defined in terms of bond dissociation energies, D_0, and proposed that the difference in electronegativities could be expressed as

$$|\chi_A - \chi_B| = \{D_0(AB) - \tfrac{1}{2}[D_0(AA) + D_0(BB)]\}^{1/2}$$

Definition Pauling electronegativity (10D.2)

where D_0(AA) and D_0(BB) are the dissociation energies of A—A and B—B bonds and D_0(AB) is the dissociation energy of an A—B bond, all in electronvolts. (In later work Pauling used the geometrical mean of dissociation energies in place of the arithmetic mean.) This expression gives differences of electronegativities; to establish an absolute scale Pauling chose individual values that gave the best match to the values obtained from eqn 10D.2. Electronegativities based on this definition are called **Pauling electronegativities** (Table 10D.1). The most electronegative elements are those close to F (excluding the noble gases); the least are those close to Cs. It is found that the greater the difference in electronegativities, the greater the polar character of the bond. The difference for HF, for instance, is 1.78; a C—H bond, which is commonly regarded as almost nonpolar, has an electronegativity difference of 0.51.

Table 10D.1* Pauling electronegativities

Element	χ_P
H	2.2
C	2.6
N	3.0
O	3.4
F	4.0
Cl	3.2
Cs	0.79

* More values will be found in the *Resource section*.

Brief illustration 10D.2 Electronegativity

The bond dissociation energies of hydrogen, chlorine, and hydrogen chloride are 4.52 eV, 2.51 eV, and 4.47 eV, respectively. From eqn 10D.2 we find

$$|\chi_{\text{Pauling}}(\text{H})-\chi_{\text{Pauling}}(\text{Cl})|=\{4.47-\tfrac{1}{2}(4.52+2.51)\}^{1/2}=0.98\approx 1.0$$

Self-test 10D.2 Repeat the analysis for HBr. Use data from Table 10D.1.

Answer: $|\chi_{\text{Pauling}}(\text{H})-\chi_{\text{Pauling}}(\text{Br})|=0.73$

The spectroscopist Robert Mulliken proposed an alternative definition of electronegativity. He argued that an element is likely to be highly electronegative if it has a high ionization energy (so it will not release electrons readily) and a high electron affinity (so it is energetically favourable to acquire electrons). The **Mulliken electronegativity scale** is therefore based on the definition

$$\chi=\tfrac{1}{2}(I+E_{\text{ea}}) \qquad \textit{Definition} \quad \text{Mulliken electronegativity} \qquad (10\text{D}.3)$$

where I is the ionization energy of the element and E_{ea} is its electron affinity (both in electronvolts). It will be recognized that this combination of energies is precisely the one we have used to estimate the energy of an atomic orbital in a molecule, and can therefore see that the greater the value of the Mulliken electronegativity the greater is the contribution of that atom to the electron distribution in the bond. There is one word of caution: the values of I and E_{ea} in eqn 10D.3 are strictly those for a special 'valence state' of the atom, not a true spectroscopic state. We ignore that complication here. The Mulliken and Pauling scales are approximately in line with each other. A reasonably reliable conversion relation between the two is

$$\chi_{\text{Pauling}}=1.35\chi_{\text{Mulliken}}^{1/2}-1.37 \qquad (10\text{D}.4)$$

10D.2 The variation principle

A more systematic way of discussing bond polarity and finding the coefficients in the linear combinations used to build molecular orbitals is provided by the **variation principle** which is proved in the following *Justification*:

> If an arbitrary wavefunction is used to calculate the energy, the value calculated is never less than the true energy. *Variation principle*

This principle is the basis of all modern molecular structure calculations. The arbitrary wavefunction is called the **trial wavefunction**. The principle implies that, if we vary the coefficients in the trial wavefunction until the lowest energy is achieved (by evaluating the expectation value of the hamiltonian for each wavefunction), then those coefficients will be the best for that particular form of trial function. We might get a lower energy if we use a more complicated wavefunction (for example, by taking a linear combination of several atomic orbitals on each atom), but we shall have the optimum (minimum energy) molecular orbital that can be built from the chosen **basis set**, the given set of atomic orbitals.

Justification 10D.1 The variation principle

To justify the variation principle, consider a trial (normalized) wavefunction written as a linear combination $\psi_{\text{trial}}=\sum_n c_n\psi_n$ of the true (but unknown), normalized, and orthogonal eigenfunctions of the hamiltonian, $\hat{H}$. The energy associated with this trial function is the expectation value

$$E=\int\psi_{\text{trial}}^*\hat{H}\psi_{\text{trial}}\,\mathrm{d}\tau$$

The true lowest energy of the system is E_0, the eigenvalue corresponding to ψ_0. Consider the following difference:

$$\begin{aligned}E-E_0&=\int\psi_{\text{trial}}^*\hat{H}\psi_{\text{trial}}\,\mathrm{d}\tau-E_0\overbrace{\int\psi_{\text{trial}}^*\psi_{\text{trial}}\,\mathrm{d}\tau}^{1}\\&=\int\psi_{\text{trial}}^*\hat{H}\psi_{\text{trial}}\,\mathrm{d}\tau-\int\psi_{\text{trial}}^*E_0\psi_{\text{trial}}\,\mathrm{d}\tau\\&=\int\psi_{\text{trial}}^*(\hat{H}-E_0)\psi_{\text{trial}}\,\mathrm{d}\tau\\&=\int\left(\sum_n c_n^*\psi_n^*\right)(\hat{H}-E_0)\left(\sum_{n'}c_{n'}\psi_{n'}\right)\mathrm{d}\tau\\&=\sum_{n,n'}c_n^*c_{n'}\int\psi_n^*(\hat{H}-E_0)\psi_{n'}\,\mathrm{d}\tau\end{aligned}$$

Because $\int\psi_n^*\hat{H}\psi_{n'}\,\mathrm{d}\tau=E_{n'}\int\psi_n^*\psi_{n'}\,\mathrm{d}\tau$ and $\int\psi_n^*E_0\psi_{n'}\,\mathrm{d}\tau=E_0\int\psi_n^*\psi_{n'}\,\mathrm{d}\tau$, we write

$$\int\psi_n^*(\hat{H}-E_0)\psi_{n'}\,\mathrm{d}\tau=(E_{n'}-E_0)\int\psi_n^*\psi_{n'}\,\mathrm{d}\tau$$

and

$$E-E_0=\sum_{n,n'}c_n^*c_{n'}(E_{n'}-E_0)\overbrace{\int\psi_n^*\psi_{n'}\,\mathrm{d}\tau}^{0\text{ unless }n'=n}$$

The eigenfunctions are orthogonal, so only $n'=n$ contributes to this sum, and as each eigenfunction is normalized, each surviving integral is 1. Consequently

$$E-E_0=\sum_n\overbrace{c_n^*c_n}^{\geq 0}\overbrace{(E_n-E_0)}^{\geq 0}\geq 0$$

That is, $E\geq E_0$, as we set out to prove.

(a) The procedure

The method can be illustrated by the trial wavefunction in eqn 10D.1. We show in the following *Justification* that the coefficients are given by the solutions of the two **secular equations**[1]

$$(\alpha_A - E)c_A + (\beta - ES)c_B = 0 \qquad (10D.5a)$$

$$(\beta - ES)c_A + (\alpha_A - E)c_B = 0 \qquad (10D.5b)$$

where

$$\alpha_A = \int A\hat{H}A\,d\tau \quad \alpha_B = \int B\hat{H}B\,d\tau \qquad \text{Coulomb integrals} \qquad (10D.5c)$$

$$\beta = \int A\hat{H}B\,d\tau = \int B\hat{H}A\,d\tau \qquad \text{Resonance integral} \qquad (10D.5d)$$

The parameter α is called a **Coulomb integral**. It is negative and can be interpreted as the energy of the electron when it occupies A (for α_A) or B (for α_B). In a homonuclear diatomic molecule, $\alpha_A = \alpha_B$. The parameter β is called a **resonance integral** (for classical reasons). It vanishes when the orbitals do not overlap, and at equilibrium bond lengths it is normally negative.

Justification 10D.2 The variation principle applied to a heteronuclear diatomic molecule

The trial wavefunction in eqn 10D.1 is real but not normalized because at this stage the coefficients can take arbitrary values. Therefore, we can write $\psi^* = \psi$ but do not assume that $\int \psi^2 d\tau = 1$. When a wavefunction is not normalized, we write the expression for the energy as

$$E = \frac{\int \psi^* \hat{H}\psi\,d\tau}{\int \psi^*\psi\,d\tau} \xrightarrow{\psi\text{ real}} \frac{\int \psi \hat{H}\psi\,d\tau}{\int \psi^2\,d\tau} \qquad \text{Energy} \qquad (10D.6)$$

We now search for values of the coefficients in the trial function that minimize the value of E. This is a standard problem in calculus, and is solved by finding the coefficients for which

$$\frac{\partial E}{\partial c_A} = 0 \qquad \frac{\partial E}{\partial c_B} = 0$$

The first step is to express the two integrals in eqn 10D.6 in terms of the coefficients. The denominator is

$$\int \psi^2 d\tau = \int (c_A A + c_B B)^2 d\tau$$

$$= c_A^2 \overbrace{\int A^2 d\tau}^{1} + c_B^2 \overbrace{\int B^2 d\tau}^{1} + 2c_A c_B \overbrace{\int AB\,d\tau}^{S} = c_A^2 + c_B^2 + 2c_A c_B S$$

because the individual atomic orbitals are normalized and the third integral is the overlap integral S (eqn 10C.3, $S = \int \chi_A \chi_B\,d\tau$). The numerator is

$$\int \psi \hat{H}\psi\,d\tau = \int (c_A A + c_B B)\hat{H}(c_A A + c_B B)d\tau$$

$$= c_A^2 \overbrace{\int A\hat{H}A\,d\tau}^{\alpha_A} + c_B^2 \overbrace{\int B\hat{H}B\,d\tau}^{\alpha_B} + c_A c_B \overbrace{\int A\hat{H}B\,d\tau}^{\beta} + c_A c_B \overbrace{\int B\hat{H}A\,d\tau}^{\beta}$$

With the integrals written as shown (the two β integrals are equal by hermiticity, Topic 7C), the numerator is

$$\int \psi \hat{H}\psi\,d\tau = c_A^2\alpha_A + c_B^2\alpha_B + 2c_A c_B\beta$$

At this point we can write the complete expression for E as

$$E = \frac{c_A^2\alpha_A + c_B^2\alpha_B + 2c_A c_B\beta}{c_A^2 + c_B^2 + 2c_A c_B S}$$

Its minimum is found by differentiation with respect to the two coefficients and setting the results equal to 0. After some straightforward work we obtain

$$\frac{\partial E}{\partial c_A} = \frac{2\{(\alpha_A - E)c_A + (\beta - SE)c_B\}}{c_A^2 + c_B^2 + 2c_A c_B S}$$

$$\frac{\partial E}{\partial c_B} = \frac{2\{(\alpha_B - E)c_B + (\beta - SE)c_A\}}{c_A^2 + c_B^2 + 2c_A c_B S}$$

For the derivatives to be equal to 0, the numerators, and specifically the terms in blue, of these expressions must vanish. That is, we must find values of c_A and c_B that satisfy the conditions

$$(\alpha_A - E)c_A + (\beta - SE)c_B = 0$$

$$(\alpha_B - E)c_B + (\beta - SE)c_A = 0$$

which are the secular equations (eqn 10D.5).

To solve the secular equations for the coefficients we need to know the energy E of the orbital. As for any set of simultaneous equations, the secular equations have a solution if the **secular determinant**, the determinant of the coefficients, is zero; that is, if

$$\begin{vmatrix} \alpha_A - E & \beta - SE \\ \beta - SE & \alpha_B - E \end{vmatrix} = (\alpha_A - E)(\alpha_B - E) - (\beta - SE)^2$$

$$= (1 - S^2)E^2 + \{2\beta S - (\alpha_A + \alpha_B)\}E + (\alpha_A\alpha_B - \beta^2)$$

$$= 0 \qquad (10D.7)$$

Because a quadratic equation of the form $ax^2 + bx + c = 0$ has the solutions

$$x = \frac{-b \pm (b^2 - 4ac)^{1/2}}{2a}$$

[1] The name 'secular' is derived from the Latin word for age or generation. The term comes from astronomy, where the same equations appear in connection with slowly accumulating modifications of planetary orbits.

this quadratic equation for E with $a=1-S^2$, $b=2\beta S-(\alpha_A+\alpha_B)$, and $c=\alpha_A\alpha_B-\beta^2$ has the solutions

$$E_{\pm}=\frac{\alpha_A+\alpha_B-2\beta S\pm\{(\alpha_A+\alpha_B-2\beta S)^2-4(1-S^2)(\alpha_A\alpha_B-\beta^2)\}^{1/2}}{2(1-S^2)} \quad (10D.8a)$$

which are the energies of the bonding and antibonding molecular orbitals formed from the two atomic orbitals.

Equation 10D.8a becomes easier to understand in two cases. For a *homonuclear diatomic molecule* we can set $\alpha_A=\alpha_B=\alpha$ and obtain

$$E_{\pm}=\frac{2\alpha-2\beta S\pm\{\overbrace{(2\alpha-2\beta S)^2-4(1-S^2)(\alpha^2-\beta^2)}^{(2\beta-2\alpha S)^2}\}^{1/2}}{2\underbrace{(1-S^2)}_{(1+S)(1-S)}}$$

$$=\frac{\alpha-\beta S\pm(\beta-\alpha S)}{(1+S)(1-S)}=\frac{(\alpha\pm\beta)(1\mp S)}{(1+S)(1-S)}$$

and therefore

$$E_+=\frac{\alpha_A+\beta}{1+S} \qquad E_-=\frac{\alpha_A-\beta}{1-S} \quad (10D.8b)$$

Homonuclear diatomic molecules Molecular orbital energies

For $\beta<0$, E_+ is the lower energy solution. For *heteronuclear diatomic molecules* we can make the approximation that $S=0$ (simply to get a more transparent expression), and find

$$E_{\pm}=\tfrac{1}{2}(\alpha_A+\alpha_B)\pm\tfrac{1}{2}(\alpha_A-\alpha_B)\left\{1+\left(\frac{2\beta}{\alpha_A-\alpha_B}\right)^2\right\}^{1/2} \quad (10D.8c)$$

Zero overlap approximation

Brief illustration 10D.3 Heteronuclear diatomic molecules 2

In *Brief illustration* 10D.1 we estimated the H1s and F2p orbital energies in HF as −7.2 eV and −10.4 eV, respectively. Therefore we set $\alpha_H=-7.2$ eV and $\alpha_F=-10.4$ eV. We take $\beta=-1.0$ eV as a typical value and $S=0$. Substitution of these values into eqn 10D.8c gives

$$E_{\pm}/\text{eV}=\tfrac{1}{2}(-7.2-10.4)\pm\tfrac{1}{2}(-7.2+10.4)\left\{1+\left(\frac{-2.0}{-7.2+10.4}\right)^2\right\}^{1/2}$$
$$=-8.8\pm1.9=-10.7 \text{ and } -6.9$$

These values, representing a bonding orbital at −10.7 eV and an antibonding orbital at −6.9 eV, are shown in Fig. 10D.3.

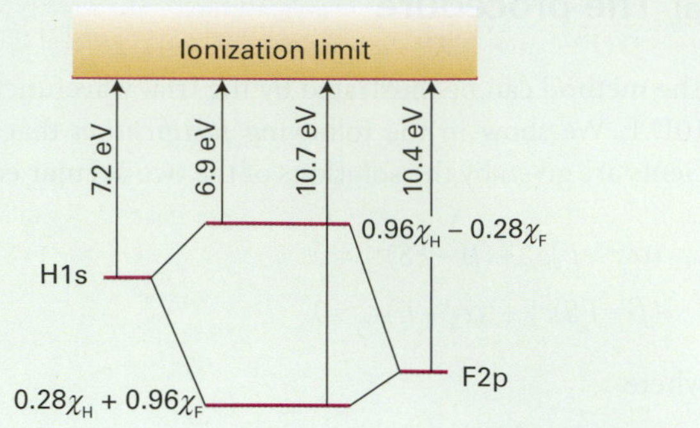

Figure 10D.3 The estimated energies of the atomic orbitals in HF and the molecular orbitals they form.

Self-test 10D.3 Does the neglect of overlap make much difference? Use $S=0.20$ (a typical value), to find the two energies.

Answer: $E_+=-10.8$ eV, $E_-=-7.1$ eV

(b) The features of the solutions

An important feature of eqn 10D.8c is that as the energy difference $|\alpha_A-\alpha_B|$ between the interacting atomic orbitals increases, the bonding and antibonding effects decrease (Fig. 10D.4). Thus, when $|\alpha_B-\alpha_A|\gg 2|\beta|$ we can make the approximation $(1+x)^{1/2}\approx 1+\tfrac{1}{2}x$ and obtain

$$E_+\approx\alpha_A+\frac{\beta^2}{\alpha_A-\alpha_B} \qquad E_-\approx\alpha_B-\frac{\beta^2}{\alpha_A-\alpha_B} \quad (10D.9)$$

As these expressions show, and as can be seen from the graph, when the energy difference is very large, the energies of the resulting molecular orbitals differ only slightly from those of the atomic orbitals, which implies in turn that the bonding and antibonding effects are small. That is:

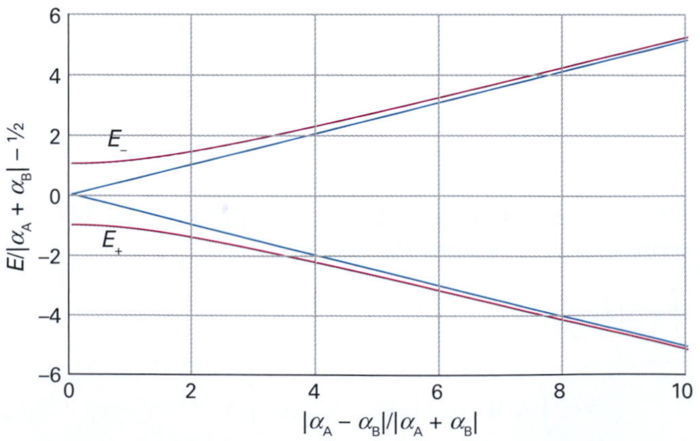

Figure 10D.4 The variation of the energies of the molecular orbitals as the energy difference of the contributing atomic orbitals is changed. The plots are for $\beta=-1$; the blue lines are for the energies in the absence of mixing (that is, $\beta=0$).

The strongest bonding and antibonding effects are obtained when the two contributing orbitals have similar energies. *Orbital contribution criterion*

The large difference in energy between core and valence orbitals is the justification for neglecting the contribution of core orbitals to molecular orbitals constructed from valence atomic orbitals. Although the core orbitals of one atom have a similar energy to the core orbitals of the other atom, so might be expected to combine strongly, core–core interaction is largely negligible because the overlap between them (and hence the value of β) is so small.

The values of the coefficients in the linear combination in eqn 10D.1 are obtained by solving the secular equations using the two energies obtained from the secular determinant. The lower energy, E_+, gives the coefficients for the bonding molecular orbital, the upper energy, E_-, the coefficients for the antibonding molecular orbital. The secular equations give expressions for the ratio of the coefficients. Thus, the first of the two secular equations in eqn 10D.5a, $(\alpha_A-E)c_A+(\beta-ES)c_B=0$, gives

$$c_B=-\left(\frac{\alpha_A-E}{\beta-ES}\right)c_A \tag{10D.10}$$

The wavefunction should also be normalized. This condition means that the term $c_A^2+c_B^2+2c_Ac_BS$ established in the preceding *Justification* must satisfy

$$c_A^2+c_B^2+2c_Ac_BS=1 \tag{10D.11}$$

When the preceding relation is substituted into this expression, we find

$$c_A=\frac{1}{\left\{1+\left(\frac{\alpha_A-E}{\beta-ES}\right)^2-2S\left(\frac{\alpha_A-E}{\beta-ES}\right)\right\}^{1/2}} \tag{10D.12}$$

which, together with eqn 10D.10, gives explicit expressions for the coefficients once we substitute the appropriate values of $E=E_\pm$ given in eqn 10D.8a.

As before, this expression becomes more transparent in two cases. First, for a homonuclear diatomic molecule, with $\alpha_A=\alpha_B=\alpha$ and $E_\pm$ given in eqn 10D.8b we find

$$E_+=\frac{\alpha+\beta}{1+S},\quad c_A=\frac{1}{\{2(1+S)\}^{1/2}},\quad c_B=c_A \tag{10D.13a}$$

$$E_-=\frac{\alpha-\beta}{1-S},\quad c_A=\frac{1}{\{2(1-S)\}^{1/2}},\quad c_B=-c_A \tag{10D.13b}$$

Homonuclear diatomic molecules

For a heteronuclear diatomic molecule with $S=0$, the coefficients are given by

$$c_A=\frac{1}{\left\{1+\left(\frac{\alpha_A-E}{\beta}\right)^2\right\}^{1/2}},\quad c_B=\frac{1}{\left\{1+\left(\frac{\beta}{\alpha_A-E}\right)^2\right\}^{1/2}} \tag{10D.14}$$

Zero overlap approximation

with the appropriate values of $E=E_\pm$ taken from eqn 10D.8c.

Brief illustration 10D.4 Heteronuclear diatomic molecules 3

Here we continue the previous *Brief illustration* using HF. With $\alpha_H=-7.2$ eV, $\alpha_F=-10.4$ eV, $\beta=-1.0$ eV, and $S=0$ the two orbital energies were found to be $E_+=-10.7$ eV and $E_-=-6.9$ eV. When these values are substituted into eqn 10D.14 we find the following coefficients:

$E_+=-10.7$ eV $\quad \psi_+=0.28\chi_H+0.96\chi_F$

$E_-=-6.9$ eV $\quad \psi_-=0.96\chi_H-0.28\chi_F$

Notice how the lower energy orbital (the one with energy -10.7 eV) has a composition that is more F2p orbital than H1s, and that the opposite is true of the higher energy, antibonding orbital.

Self-test 10D.4 Find the energies and forms of the σ orbitals in the HCl molecule using $\beta=-1.0$ eV and $S=0$. Use data from Tables 9C.2 and 9C.3.

Answer: $E_+=-8.9$ eV, $E_-=-6.6$ eV; $\psi_-=0.86\chi_H-0.51\chi_{Cl}$; $\psi_+=0.51\chi_H+0.86\chi_{Cl}$

Checklist of concepts

☐ 1. A **polar bond** can be regarded as arising from a molecular orbital that is concentrated more on one atom than its partner.

☐ 2. The **electronegativity** of an element is a measure of the power of an atom to attract electrons to itself when it is part of a compound.

- ☐ 3. The **variation principle** provides a criterion of acceptability of an approximate wavefunction.
- ☐ 4. A **basis set** refers to the given set of atomic orbitals from which the molecular orbitals are constructed.
- ☐ 5. The bonding and antibonding effects are strongest when contributing atomic orbitals have similar energies.

Checklist of equations

Property	Equation	Comment	Equation number
Molecular orbital	$\psi = c_A A + c_B B$		10D.1
Pauling electronegativity	$\lvert \chi_A - \chi_B \rvert = \{D_0(AB) - \frac{1}{2}[D_0(AA) + D_0(BB)]\}^{1/2}$	All D_0 in electronvolts	10D.2
Mulliken electronegativity	$\chi = \frac{1}{2}(I + E_{ea})$	I and E_{ea} in electronvolts	10D.3
Coulomb integral	$\alpha_A = \int A\hat{H}A\,d\tau$	Definition	10D.5c
Resonance integral	$\beta = \int A\hat{H}B\,d\tau = \int B\hat{H}A\,d\tau$	Definition	10D.5d
Energy	$E = \int \psi\hat{H}\psi\,d\tau / \int \psi^2\,d\tau$	Unnormalized real wavefunction	10D.6
Variation principle	$\partial E/\partial c = 0$	Minimization of energy	

10E Polyatomic molecules

Contents

➤ Why do you need to know this material?

Most molecules of interest in chemistry are polyatomic, so it is important to be able to discuss their electronic structure. Although computational procedures are now widely available, to understand them it is helpful to see how they emerged from the more primitive approach described here.

➤ What is the key idea?

Molecular orbitals can be expressed as linear combinations of all the atomic orbitals of the appropriate symmetry.

➤ What do you need to know already?

This Topic extends the approach used for heteronuclear diatomic molecules in Topic 10D, particularly the concepts of secular determinants and secular equations. The principal mathematical technique used is matrix algebra (*Mathematical background* 6); you should be, or become, familiar with the use of mathematical software to manipulate matrices numerically.

The molecular orbitals of polyatomic molecules are built in the same way as in diatomic molecules (Topic 10D), the only difference being that more atomic orbitals are used to construct them. As for diatomic molecules, polyatomic molecular orbitals spread over the entire molecule. A molecular orbital has the general form

$$\psi = \sum_o c_o \chi_o \qquad \text{General form of LCAO} \qquad (10\text{E}.1)$$

where χ_o is an atomic orbital and the sum extends over all the valence orbitals of all the atoms in the molecule. To find the coefficients, we set up the secular equations and the secular determinant, just as for diatomic molecules, solve the latter for the energies, and then use these energies in the secular equations to find the coefficients of the atomic orbitals for each molecular orbital.

The principal difference between diatomic and polyatomic molecules lies in the greater range of shapes that are possible: a diatomic molecule is necessarily linear, but a triatomic molecule, for instance, may be either linear or angular (bent) with a characteristic bond angle. The shape of a polyatomic molecule—the specification of its bond lengths and its bond angles—can be predicted by calculating the total energy of the molecule for a variety of nuclear positions, and then identifying the conformation that corresponds to the lowest energy. Such calculations are best done using the latest software, but a more primitive approach gives useful insight for conjugated polyenes, in which there is an alternation of single and double bonds along a chain of carbon atoms. We focus on them in the first two sections of this Topic, to set the scene for the more sophisticated approaches mentioned in Section 10E.3.

The planarity of conjugated polyenes is an aspect of their symmetry, and considerations of molecular symmetry play a vital role in setting up and labelling molecular orbitals (see Topic 11B). In the present case, planarity provides a distinction between the σ and π orbitals of the molecule, and in elementary approaches such molecules are commonly discussed in terms of the characteristics of their π orbitals with the σ bonds providing a rigid framework that determines the general shape of the molecule.

10E.1 The Hückel approximation

The π molecular orbital energy level diagrams of conjugated molecules can be constructed using a set of approximations suggested by Erich Hückel in 1931. All the C atoms are treated identically, so all the Coulomb integrals α for the atomic orbitals that contribute to the π orbitals are set equal. For example, in ethene, which we use to introduce the method, we take the σ bonds as fixed, and concentrate on finding the energies of the single π bond and its companion antibond.

(a) An introduction to the method

We express the π orbitals as linear combinations of the C2p orbitals that lie perpendicular to the molecular plane. In ethene, for instance, we would write

$$\psi = c_A A + c_B B \qquad (10E.2)$$

where the A is a C2p orbital on atom A, and likewise for B. Next, the optimum coefficients and energies are found by the variation principle as explained in Topic 10D. That is, we solve the secular determinant, which in the case of ethene is eqn 10D.7 with $\alpha_A = \alpha_B = \alpha$:

$$\begin{vmatrix} \alpha - E & \beta - ES \\ \beta - ES & \alpha - E \end{vmatrix} = 0 \qquad (10E.3)$$

In a modern computation all the resonance integrals and overlap integrals would be included, but an indication of the molecular orbital energy level diagram can be obtained very readily if we make the following additional **Hückel approximations**:

Hückel approximation

- All overlap integrals are set equal to zero.
- All resonance integrals between non-neighbours are set equal to zero.
- All remaining resonance integrals are set equal (to β).

These approximations are obviously very severe, but they let us calculate at least a general picture of the molecular orbital energy levels with very little work. The approximations result in the following structure of the secular determinant:

- All diagonal elements: $\alpha - E$.
- Off-diagonal elements between neighbouring atoms: β.
- All other elements: 0.

These approximations convert eqn 10E.3 to

$$\begin{vmatrix} \alpha - E & \beta \\ \beta & \alpha - E \end{vmatrix} = (\alpha - E)^2 - \beta^2 = (\alpha - E + \beta)(\alpha - E - \beta) = 0 \qquad (10E.4)$$

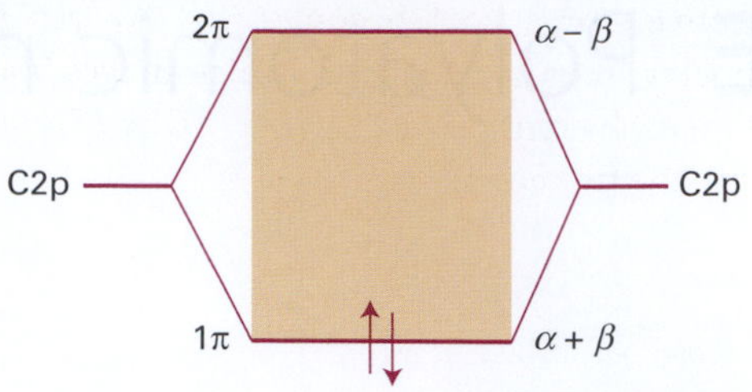

Figure 10E.1 The Hückel molecular orbital energy levels of ethene. Two electrons occupy the lower π orbital.

The roots of the equation are $E_{\pm} = \alpha \pm \beta$. The + sign corresponds to the bonding combination (β is negative) and the − sign corresponds to the antibonding combination (Fig. 10E.1).

The building-up principle leads to the configuration $1\pi^2$, because each carbon atom supplies one electron to the π system and both electrons can occupy the bonding orbital. The **highest occupied molecular orbital** in ethene, its HOMO, is the 1π orbital; the **lowest unoccupied molecular orbital**, its LUMO, is the 2π orbital (or, as it is sometimes denoted, the $2\pi^*$ orbital). These two orbitals jointly form the **frontier orbitals** of the molecule. The frontier orbitals are important because they are largely responsible for many of the chemical and spectroscopic properties of this and analogous molecules.

Brief illustration 10E.1 Ethene

We can estimate that the $\pi^* \leftarrow \pi$ excitation energy of ethene is $2|\beta|$, the energy required to excite an electron from the 1π to the 2π orbital. This transition occurs at close to 40 000 cm^{-1}, corresponding to 4.8 eV. It follows that a plausible value of β is about −2.4 eV (−230 kJ mol^{-1}).

Self-test 10E.1 The ionization energy of ethene is 10.5 eV. Estimate α.

Answer: −8.1 eV

(b) The matrix formulation of the method

In preparation for making Hückel theory more sophisticated and readily applicable to bigger molecules, we need to reformulate it in terms of matrices (see *Mathematical background* 6 following this chapter). Our starting point is the pair of secular equations developed for a heteronuclear diatomic molecule in Topic 10D:

$$(\alpha_A - E)c_A + (\beta - ES)c_B = 0$$
$$(\beta - ES)c_A + (\alpha_B - E)c_B = 0$$

To prepare to generalize this expression we shall write $\alpha_J = H_{JJ}$ (with J = A or B), $\beta = H_{AB}$, and label the overlap integrals with their respective atoms, so S becomes S_{AB}. We can introduce more symmetry into the equations (which makes it simpler to generalize them) by replacing the E in $\alpha_J - E$ by ES_{JJ}, with $S_{JJ}=1$. At this point, the two equations are

$$(H_{AA}-ES_{AA})c_A+(H_{AB}-ES_{AB})c_B=0$$
$$(H_{BA}-ES_{BA})c_A+(H_{BB}-ES_{BB})c_B=0$$

There is one further notational change. The coefficients c_J depend on the value of E, so we need to distinguish the two sets corresponding to the two energies, which we denote E_i with $i=1$ and 2. We therefore write the coefficients as $c_{i,J}$, with $i=1$ (the coefficients c_{1A} and c_{1B} for energy E_1) or 2 (the coefficients c_{2A} and c_{2B} for energy E_2). With this notational change, the two equations become

$$(H_{AA}-E_iS_{AA})c_{i,A}+(H_{AB}-E_iS_{AB})c_{i,B}=0 \quad (10E.5a)$$

$$(H_{BA}-E_iS_{BA})c_{i,A}+(H_{BB}-E_iS_{BB})c_{i,B}=0 \quad (10E.5b)$$

with $i=1$ and 2, giving four equations in all. Each pair of equations can be written in matrix form as

$$\begin{pmatrix} H_{AA}-E_iS_{AA} & H_{AB}-E_iS_{AB} \\ H_{BA}-E_iS_{BA} & H_{BB}-E_iS_{BB} \end{pmatrix}\begin{pmatrix} c_{i,A} \\ c_{i,B} \end{pmatrix}=0 \quad (10E.5c)$$

because multiplying out the matrices gives the two expressions in eqns 10E.5a and 10.5b. If we introduce the following matrices

$$\boldsymbol{H}=\begin{pmatrix} H_{AA} & H_{AB} \\ H_{BA} & H_{BB} \end{pmatrix} \quad \boldsymbol{S}=\begin{pmatrix} S_{AA} & S_{AB} \\ S_{BA} & S_{BB} \end{pmatrix} \quad \boldsymbol{c}_i=\begin{pmatrix} c_{i,A} \\ c_{i,B} \end{pmatrix} \quad (10E.6)$$

so that

$$\boldsymbol{H}-E_i\boldsymbol{S}=\begin{pmatrix} H_{AA}-E_iS_{AA} & H_{AB}-E_iS_{AB} \\ H_{BA}-E_iS_{BA} & H_{BB}-E_iS_{BB} \end{pmatrix}$$

then eqn 10E.5c may be written more succinctly as

$$(\boldsymbol{H}-E_i\boldsymbol{S})\boldsymbol{c}_i=0 \quad \text{or} \quad \boldsymbol{H}\boldsymbol{c}_i=\boldsymbol{S}\boldsymbol{c}_iE_i \quad (10E.7)$$

As shown in the following *Justification*, these two sets of equations (with $i=1$ and 2) can be combined into a single matrix equation by introducing the matrices

$$\boldsymbol{c}=(\boldsymbol{c}_1 \quad \boldsymbol{c}_2)=\begin{pmatrix} c_{1,A} & c_{2,A} \\ c_{1,B} & c_{2,B} \end{pmatrix} \quad \boldsymbol{E}=\begin{pmatrix} E_1 & 0 \\ 0 & E_2 \end{pmatrix} \quad (10E.8)$$

for then all four equations in eqn 10E.7 are summarized by the single expression

$$\boldsymbol{Hc}=\boldsymbol{ScE} \quad (10E.9)$$

Justification 10E.1 The matrix formulation

Substitution of the matrices defined in eqn 10E.8 into eqn 10E.9 gives

$$\overbrace{\begin{pmatrix} H_{AA} & H_{AB} \\ H_{BA} & H_{BB} \end{pmatrix}}^{\boldsymbol{H}}\overbrace{\begin{pmatrix} c_{1,A} & c_{2,A} \\ c_{1,B} & c_{2,B} \end{pmatrix}}^{\boldsymbol{c}}=\overbrace{\begin{pmatrix} S_{AA} & S_{AB} \\ S_{BA} & S_{BB} \end{pmatrix}}^{\boldsymbol{S}}\overbrace{\begin{pmatrix} c_{1,A} & c_{2,A} \\ c_{1,B} & c_{2,B} \end{pmatrix}}^{\boldsymbol{c}}\overbrace{\begin{pmatrix} E_1 & 0 \\ 0 & E_2 \end{pmatrix}}^{\boldsymbol{E}}$$

The product on the left is

$$\begin{pmatrix} H_{AA} & H_{AB} \\ H_{BA} & H_{BB} \end{pmatrix}\begin{pmatrix} c_{1,A} & c_{2,A} \\ c_{1,B} & c_{2,B} \end{pmatrix}=\begin{pmatrix} H_{AA}c_{1,A}+H_{AB}c_{1,B} & H_{AA}c_{2,A}+H_{AB}c_{2,B} \\ H_{BA}c_{1,A}+H_{BB}c_{1,B} & H_{BA}c_{2,A}+H_{BB}c_{2,B} \end{pmatrix}$$

The product on the right is

$$\begin{pmatrix} S_{AA} & S_{AB} \\ S_{BA} & S_{BB} \end{pmatrix}\begin{pmatrix} c_{1,A} & c_{2,A} \\ c_{1,B} & c_{2,B} \end{pmatrix}\begin{pmatrix} E_1 & 0 \\ 0 & E_2 \end{pmatrix}=\begin{pmatrix} S_{AA} & S_{AB} \\ S_{BA} & S_{BB} \end{pmatrix}\begin{pmatrix} c_{1,A}E_1 & c_{2,A}E_2 \\ c_{1,B}E_1 & c_{2,B}E_2 \end{pmatrix}$$
$$=\begin{pmatrix} E_1S_{AA}c_{1,A}+E_1S_{AB}c_{1,B} & E_2S_{AA}c_{2,A}+E_2S_{AB}c_{2,B} \\ E_1S_{BA}c_{1,A}+E_1S_{BB}c_{1,B} & E_2S_{BA}c_{2,A}+E_2S_{BB}c_{2,B} \end{pmatrix}$$

Comparison of matching terms (like those in blue) recreates the four secular equations (two for each value of i).

In the Hückel approximation, $H_{AA}=H_{BB}=\alpha$, $H_{AB}=H_{BA}=\beta$, and we neglect overlap, setting $\boldsymbol{S}=\boldsymbol{1}$, the unit matrix (with 1 on the diagonal and 0 elsewhere). Then

$$\boldsymbol{Hc}=\boldsymbol{cE}$$

At this point, we multiply from the left by the inverse matrix $\boldsymbol{c}^{-1}$, use $\boldsymbol{c}^{-1}\boldsymbol{c}=\boldsymbol{1}$, and find

$$\boldsymbol{c}^{-1}\boldsymbol{Hc}=\boldsymbol{E} \quad (10E.10)$$

In other words, to find the eigenvalues E_i, we have to find a transformation of $\boldsymbol{H}$ that makes it diagonal. This procedure is called **matrix diagonalization**. The diagonal elements then correspond to the eigenvalues E_i and the columns of the matrix $\boldsymbol{c}$ that brings about this diagonalization are the coefficients of the members of the **basis set**, the set of atomic orbitals used in the calculation, and hence give us the composition of the molecular orbitals.

Example 10E.1 Finding molecular orbitals by matrix diagonalization

Set up and solve the matrix equations within the Hückel approximation for the π orbitals of butadiene (**1**).

1 Butadiene

Method The matrices will be four-dimensional for this four-atom system. Ignore overlap, and construct the matrix $\boldsymbol{H}$ by using the Hückel approximation and the parameters α and β. Find the matrix $\boldsymbol{c}$ that diagonalizes $\boldsymbol{H}$: for this step, use mathematical software. Full details are given in *Mathematical background* 6, but note that if $\boldsymbol{H}=\alpha\mathbf{1}+\boldsymbol{M}$, where $\boldsymbol{M}$ is a non-diagonal matrix, then because $\alpha\boldsymbol{c}^{-1}\mathbf{1}\boldsymbol{c}=\alpha\boldsymbol{c}^{-1}\boldsymbol{c}\mathbf{1}=\alpha\mathbf{1}$, whatever matrix $\boldsymbol{c}$ diagonalizes $\boldsymbol{M}$ leaves $\alpha\mathbf{1}$ unchanged, so to achieve the overall diagonalization of $\boldsymbol{H}$ we need to diagonalize only $\boldsymbol{M}$.

Answer The hamiltonian matrix $\boldsymbol{H}$ is

$$\boldsymbol{H}=\begin{pmatrix} \overbrace{H_{11}}^{\alpha} & \overbrace{H_{12}}^{\beta} & \overbrace{H_{13}}^{0} & \overbrace{H_{14}}^{0} \\ H_{21} & H_{22} & H_{23} & H_{24} \\ H_{31} & H_{32} & H_{33} & H_{34} \\ H_{41} & H_{42} & H_{43} & H_{44} \end{pmatrix} \xrightarrow{\text{Hückel approximation}} \begin{pmatrix} \alpha & \beta & 0 & 0 \\ \beta & \alpha & \beta & 0 \\ 0 & \beta & \alpha & \beta \\ 0 & 0 & \beta & \alpha \end{pmatrix}$$

which we write as

$$\boldsymbol{H}=\alpha\mathbf{1}+\beta\overbrace{\begin{pmatrix} 0 & 1 & 0 & 0 \\ 1 & 0 & 1 & 0 \\ 0 & 1 & 0 & 1 \\ 0 & 0 & 1 & 0 \end{pmatrix}}^{\boldsymbol{M}}$$

because most mathematical software can deal only with numerical matrices. The diagonalized form of the matrix $\boldsymbol{M}$ is

$$\begin{pmatrix} +1.62 & 0 & 0 & 0 \\ 0 & -0.62 & 0 & 0 \\ 0 & 0 & -0.62 & 0 \\ 0 & 0 & 0 & -0.62 \end{pmatrix}$$

so we can infer that the diagonalized Hamiltonian matrix is

$$\boldsymbol{E}=\begin{pmatrix} \alpha+1.62\beta & 0 & 0 & 0 \\ 0 & \alpha+0.62\beta & 0 & 0 \\ 0 & 0 & \alpha-0.62\beta & 0 \\ 0 & 0 & 0 & \alpha-1.62\beta \end{pmatrix}$$

The matrix that achieves the diagonalization is

$$\boldsymbol{c}=\begin{pmatrix} 0.372 & 0.602 & 0.602 & -0.372 \\ 0.602 & 0.372 & -0.372 & 0.602 \\ 0.602 & -0.372 & -0.372 & -0.602 \\ 0.372 & -0.602 & 0.602 & -0.372 \end{pmatrix}$$

with each column giving the coefficients of the atomic orbitals for the corresponding molecular orbital. We can conclude that the energies and molecular orbitals are

$$E_1=\alpha+1.62\beta \quad \psi_1=0.372\chi_A+0.602\chi_B+0.602\chi_C+0.372\chi_D$$
$$E_2=\alpha+0.62\beta \quad \psi_2=0.602\chi_A+0.372\chi_B-0.372\chi_C-0.602\chi_D$$
$$E_3=\alpha-0.62\beta \quad \psi_3=0.602\chi_A-0.372\chi_B-0.372\chi_C+0.602\chi_D$$
$$E_4=\alpha-1.62\beta \quad \psi_4=-0.372\chi_A+0.602\chi_B-0.602\chi_C+0.372\chi_D$$

where the C2p atomic orbitals are denoted by $\chi_A, \ldots, \chi_D$. Note that the molecular orbitals are mutually orthogonal and, with overlap neglected, normalized.

Self-test 10E.2 Repeat the exercise for the allyl radical, $\cdot CH_2-CH=CH_2$.

Answer: $E=\alpha+1.41\beta, \alpha, \alpha, \alpha-1.41\beta$; $\psi_1=0.500\chi_A+0.707\chi_B+0.500\chi_C$, $\psi_2=0.707\chi_A-0.707\chi_C$, $\psi_3=0.500\chi_A-0.707\chi_B+0.500\chi_C$

10E.2 Applications

Although the Hückel method is very primitive, it can be used to account for some of the properties of conjugated polyenes.

(a) Butadiene and π-electron binding energy

As we saw in *Example* 10E.1, the energies of the four LCAO-MOs for butadiene are

$$E=\alpha\pm1.62\beta, \quad \alpha\pm0.62\beta \qquad (10E.11)$$

These orbitals and their energies are drawn in Fig. 10E.2. Note that the greater the number of internuclear nodes, the higher the energy of the orbital. There are four electrons to accommodate, so the ground-state configuration is $1\pi^2 2\pi^2$. The frontier orbitals of butadiene are the 2π orbital (the HOMO, which is largely bonding) and the 3π orbital (the LUMO, which is largely antibonding). 'Largely' bonding means that an orbital has both bonding and antibonding interactions between various neighbours, but the bonding effects dominate. 'Largely antibonding' indicates that the antibonding effects dominate.

An important point emerges when we calculate the total **π-electron binding energy**, E_π, the sum of the energies of each π electron, and compare it with what we find in ethene. In ethene the total energy is

$$E_\pi=2(\alpha+\beta)=2\alpha+2\beta$$

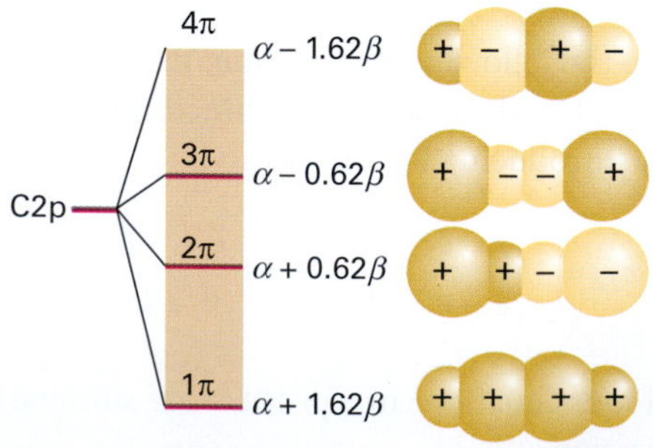

Figure 10E.2 The Hückel molecular orbital energy levels of butadiene and the top view of the corresponding π orbitals. The four p electrons (one supplied by each C) occupy the two lower π orbitals. Note that the orbitals are delocalized.

In butadiene it is

$$E_\pi = 2(\alpha+1.62\beta)+2(\alpha+0.62\beta)=4\alpha+4.48\beta$$

Therefore, the energy of the butadiene molecule lies lower by 0.48β (about 110 kJ mol^{-1}) than the sum of two individual π bonds. This extra stabilization of a conjugated system compared with a set of localized π bonds is called the **delocalization energy** of the molecule.

A closely related quantity is the **π-bond formation energy**, $E_{\rm bf}$, the energy released when a π bond is formed. Because the contribution of α is the same in the molecule as in the atoms, we can find the π-bond formation energy from the π-electron binding energy by writing

$$E_{\rm bf} = E_\pi - N_{\rm C}\alpha \qquad \textit{Definition} \qquad \text{π-Bond formation energy} \qquad (10E.12)$$

where $N_{\rm C}$ is the number of carbon atoms in the molecule. The π-bond formation energy in butadiene, for instance, is 4.48β.

Example 10E.2 Estimating the delocalization energy

Use the Hückel approximation to find the energies of the π orbitals of cyclobutadiene, and estimate the delocalization energy.

Method Set up the secular determinant using the same basis as for butadiene, but note that atoms A and D are also now neighbours. Then solve for the roots of the secular equation and assess the total π-bond energy. For the delocalization energy, subtract from the total π-bond energy the energy of two π-bonds.

Answer The hamiltonian matrix is

$$H=\begin{pmatrix}\alpha & \beta & 0 & \beta\\ \beta & \alpha & \beta & 0\\ 0 & \beta & \alpha & \beta\\ \beta & 0 & \beta & \alpha\end{pmatrix}$$

$$=\alpha 1+\beta\begin{pmatrix}0&1&0&1\\1&0&1&0\\0&1&0&1\\1&0&1&0\end{pmatrix} \xrightarrow{\text{Diagonalize}} \begin{pmatrix}2&0&0&0\\0&0&0&0\\0&0&0&0\\0&0&0&-2\end{pmatrix}$$

Diagonalization gives the energies of the orbitals as

$$E=\alpha+2\beta,\alpha,\alpha,\alpha-2\beta$$

Four electrons must be accommodated. Two occupy the lowest orbital (of energy $\alpha+2\beta$), and two occupy the doubly degenerate orbitals (of energy α). The total energy is therefore $4\alpha+4\beta$. Two isolated π bonds would have an energy $4\alpha+4\beta$; therefore, in this case, the delocalization energy is zero.

Self-test 10E.3 Repeat the calculation for benzene (use software!).

Answer: See next subsection

(b) Benzene and aromatic stability

The most notable example of delocalization conferring extra stability is benzene and the aromatic molecules based on its structure. In elementary accounts, benzene, and other aromatic compounds, is often expressed in a mixture of valence-bond and molecular orbital terms, with typically valence-bond language used for its σ framework and molecular orbital language used to describe its π electrons.

First, the valence-bond component. The six C atoms are regarded as sp^2 hybridized, with a single unhybridized perpendicular 2p orbital. One H atom is bonded by (Csp^2,H1s) overlap to each C carbon, and the remaining hybrids overlap to give a regular hexagon of atoms (Fig. 10E.3). The internal angle of a regular hexagon is 120°, so sp^2 hybridization is ideally suited for forming σ bonds. We see that the hexagonal shape of benzene permits strain-free σ bonding.

Now consider the molecular orbital component of the description. The six C2p orbitals overlap to give six π orbitals that spread all round the ring. Their energies are calculated within the Hückel approximation by diagonalizing the hamiltonian matrix

$$H=\begin{pmatrix}\alpha&\beta&0&0&0&\beta\\ \beta&\alpha&\beta&0&0&0\\ 0&\beta&\alpha&\beta&0&0\\ 0&0&\beta&\alpha&\beta&0\\ 0&0&0&\beta&\alpha&\beta\\ \beta&0&0&0&\beta&\alpha\end{pmatrix}$$

$$=\alpha 1+\beta\begin{pmatrix}0&1&0&0&0&1\\1&0&1&0&0&0\\0&1&0&1&0&0\\0&0&1&0&1&0\\0&0&0&1&0&1\\1&0&0&0&1&0\end{pmatrix} \xrightarrow{\text{Diagonalize}} \begin{pmatrix}2&0&0&0&0&0\\0&1&0&0&0&0\\0&0&1&0&0&0\\0&0&0&-1&0&0\\0&0&0&0&-1&0\\0&0&0&0&0&-2\end{pmatrix}$$

The MO energies, the eigenvalues of this matrix, are simply

$$E=\alpha\pm2\beta,\alpha\pm\beta,\alpha\pm\beta \qquad (10E.13)$$

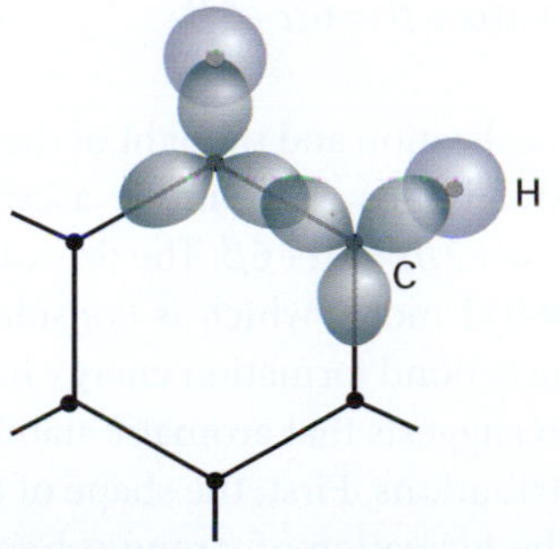

Figure 10E.3 The σ framework of benzene is formed by the overlap of Csp^2 hybrids, which fit without strain into a hexagonal arrangement.

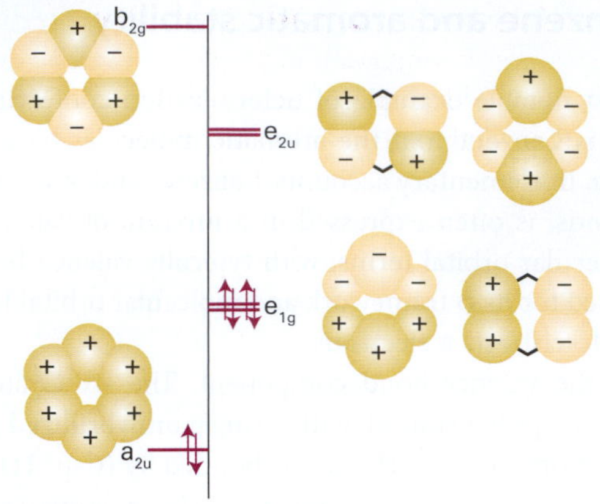

Figure 10E.4 The Hückel orbitals of benzene and the corresponding energy levels. The symmetry labels are explained in Topic 11B. The bonding and antibonding character of the delocalized orbitals reflects the numbers of nodes between the atoms. In the ground state, only the bonding orbitals are occupied.

as shown in Fig. 10E.4. The orbitals there have been given symmetry labels that are explained in Topic 11B. Note that the lowest energy orbital is bonding between all neighbouring atoms, the highest energy orbital is antibonding between each pair of neighbours, and the intermediate orbitals are a mixture of bonding, nonbonding, and antibonding character between adjacent atoms.

The simple form of the eigenvalues in eqn 10E.13 suggests that there is a more direct way of determining them than by using mathematical software. That is in fact the case, for symmetry arguments of the kind described in Topic 11B show that the 6×6 matrix can be factorized into two 1×1 matrices and two 2×2 matrices, which are very easy to deal with.

We now apply the building-up principle to the π system. There are six electrons to accommodate (one from each C atom), so the three lowest orbitals (a_{2u} and the doubly-degenerate pair e_{1g}) are fully occupied, giving the ground-state configuration $a_{2u}^2e_{1g}^4$. A significant point is that the only molecular orbitals occupied are those with net bonding character (the analogy with the very stable N_2 molecule, Topic 10B, should be noted).

The π-electron energy of benzene is

$$E=2(\alpha+2\beta)+4(\alpha+\beta)=6\alpha+8\beta$$

If we ignored delocalization and thought of the molecule as having three isolated π bonds, it would be ascribed a π-electron energy of only $3(2\alpha+2\beta)=6\alpha+6\beta$. The delocalization energy is therefore $2\beta\approx-460\,\text{kJ mol}^{-1}$, which is considerably more than for butadiene. The π-bond formation energy in benzene is 8β.

This discussion suggests that aromatic stability can be traced to two main contributions. First, the shape of the regular hexagon is ideal for the formation of strong σ bonds: the σ framework is relaxed and without strain. Second, the π orbitals are such as to be able to accommodate all the electrons in bonding orbitals, and the delocalization energy is large.

Example 10E.3 Judging the aromatic character of a molecule

Decide whether the molecules C_4H_4 and the molecular ion $C_4H_4^{2+}$ are aromatic when planar.

Method Follow the procedure for benzene. Set up and solve the secular equations within the Hückel approximation, assuming a planar σ framework, and then decide whether the ion has nonzero delocalization energy. Use mathematical software to diagonalize the hamiltonian (in Topic 11B it is shown how to use symmetry to arrive at the eigenvalues more simply.)

Answer The hamiltonian matrix is

$$H=\begin{pmatrix}\alpha & \beta & 0 & \beta\\ \beta & \alpha & \beta & 0\\ 0 & \beta & \alpha & \beta\\ \beta & 0 & \beta & \alpha\end{pmatrix}=\alpha 1+\beta\begin{pmatrix}0&1&0&1\\1&0&1&0\\0&1&0&1\\1&0&1&0\end{pmatrix}$$

The matrix multiplying β diagonalizes as follows:

$$\begin{pmatrix}0&1&0&1\\1&0&1&0\\0&1&0&1\\1&0&1&0\end{pmatrix}\xrightarrow{\text{Diagonalize}}\begin{pmatrix}-2&0&0&0\\0&0&0&0\\0&0&0&0\\0&0&0&2\end{pmatrix}$$

It follows that the energy levels of the two species are $E=\alpha\pm 2\beta, \alpha, \alpha$. There are four π electrons to accommodate in C_4H_4, so the total π-bonding energy is $2(\alpha+2\beta)+2\alpha=4(\alpha+\beta)$. The energy of two localized π-bonds is $4(\alpha+\beta)$. Therefore, the delocalization energy is zero, so the molecule is not aromatic. There are only two π electrons to accommodate in $C_4H_4^{2+}$, so the total π-bonding energy is $2(\alpha+2\beta)=2\alpha+4\beta$. The energy of a single localized π-bond is $2(\alpha+\beta)$, so the delocalization energy is 2β and the molecular-ion is aromatic.

Self-test 10E.4 What is the total π-bonding energy of $C_3H_3^-$?

Answer: $4\alpha+2\beta$

10E.3 Computational chemistry

The severe assumptions of the Hückel method are now easy to avoid by using a variety of software packages that can be used not only to calculate the shapes and energies of molecular orbitals but also to predict with reasonable accuracy the structure and reactivity of molecules. The full treatment of molecular electronic structure has received an enormous amount of attention by chemists and has become a keystone of modern chemical research. However, the calculations are very complex, and all this section seeks to do is to provide a brief introduction.[1] In every case, the procedures focus on the calculation or estimation of integrals like H_{JJ} and H_{IJ} rather than setting them equal to the constants α or β, or ignoring them entirely.

In all cases the Schrödinger equation is solved iteratively and self-consistently, just as for the self-consistent field (SCF) approach to atoms (Topic 9B). First, the molecular orbitals for the electrons present in the molecule are formulated as LCAOs. One molecular orbital is then selected and all the others are used to set up an expression for the potential energy of an electron in the chosen orbital. The resulting Schrödinger equation is then solved numerically to obtain a better version of the chosen molecular orbital and its energy. The procedure is repeated for all the molecular orbitals and used to calculate the total energy of the molecule. The process is repeated until the computed orbitals and energy are constant to within some tolerance.

(a) Semi-empirical and *ab initio* methods

In a **semi-empirical method**, many of the integrals are estimated by appealing to spectroscopic data or physical properties such as ionization energies, and using a series of rules to set certain integrals equal to zero. A primitive form of this procedure is used in *Brief illustration* 10D.1 of Topic 10D where we identify the integral α with a combination of the ionization energy and electron affinity of an atom. In an ***ab initio* method** an attempt is made to calculate all the integrals, including overlap integrals. Both procedures employ a great deal of computational effort and, along with cryptanalysts and meteorologists, theoretical chemists are among the heaviest users of the fastest computers. The integrals that are required involve atomic orbitals that in general may be centred on different nuclei. It can be appreciated that, if there are several dozen atomic orbitals used to build the molecular orbitals, then there will be tens of thousands of integrals of this form to evaluate (the number of integrals increases as the fourth power of the number of atomic orbitals in the basis). Some kind of approximation scheme is necessary.

One severe semi-empirical approximation used in the early days of computational chemistry was called **complete neglect of differential overlap** (CNDO), in which all molecular integrals are set to zero unless A and B are the same orbitals centred on the same nucleus, and likewise for C and D. The surviving integrals are then adjusted until the energy levels are in good agreement with experiment or the computed enthalpy of formation of the compound is in agreement with experiment. More recent semi-empirical methods make less draconian decisions about which integrals are to be ignored, but they are all descendants of the early CNDO technique.

Commercial packages are also available for *ab initio* calculations. Here the problem is to evaluate as efficiently as possible thousands of integrals that arise from the Coulombic interaction between two electrons and have the form

$$(AB|CD)=j_0\int\int A(1)B(1)\frac{1}{r_{12}}C(2)D(2)\mathrm{d}\tau_1\mathrm{d}\tau_2$$

Notation Molecular integral (10E.14a)

with $j_0=e^2/4\pi\varepsilon_0$ and the possibility that each of the atomic orbitals A, B, C, D is centred on a different atom, a so-called 'four-centre integral'. This task is greatly facilitated by expressing the atomic orbitals used in the LCAOs as linear combinations of Gaussian orbitals. A **Gaussian type orbital** (GTO) is a function of the form e^{-r^2}. The advantage of GTOs over the correct orbitals (which for hydrogenic systems are proportional to exponential functions of the form e^{-r}) is that the product of two Gaussian functions is itself a Gaussian function that lies between the centres of the two contributing functions (Fig. 10E.5). In this way, the four-centre integrals like these become two-centre integrals of the form

$$(AB|CD)=j_0\int\int X(1)\frac{1}{r_{12}}Y(2)\mathrm{d}\tau_1\mathrm{d}\tau_2 \qquad (10E.14b)$$

where X is the Gaussian corresponding to the product AB and Y is the corresponding Gaussian from CD. Integrals of this form are much easier and faster to evaluate numerically than the original four-centre integrals. Although more GTOs have to be used to simulate the atomic orbitals, there is an overall increase in speed of computation.

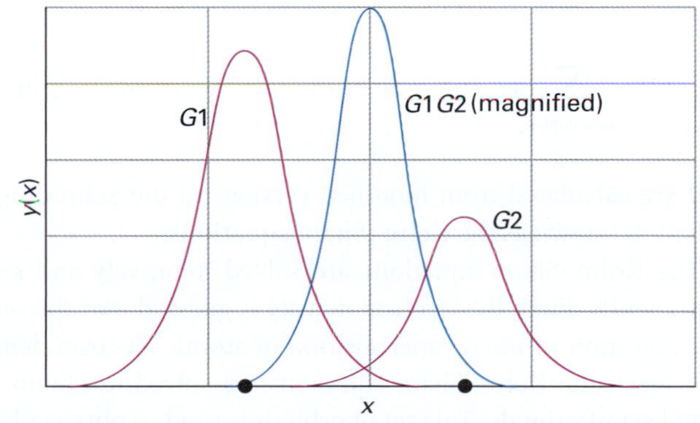

Figure 10E.5 The product of two Gaussian functions on different centres is itself a Gaussian function located at a point between the two contributing Gaussians. The scale of the product has been increased relative to that of its two components.

[1] A more complete account with detailed examples will be found in our companion volume, *Physical chemistry: Quanta, matter, and change* (2014).

Brief illustration 10E.2 Gaussian type orbitals

Suppose we consider a one-dimensional 'homonuclear' system, with GTOs of the form e^{-ax^2} located at 0 and R. Then one of the integrals that would have to be evaluated would include the term

$$\chi_A(1)\chi_B(1)=e^{-ax^2}e^{-a(x-R)^2}=e^{-2ax^2+2axR-aR^2}$$

Next we note that $-2a(x-\frac{1}{2}R)^2=-2ax^2+2axR-\frac{1}{2}aR^2$, so we can write

$$\chi_A(1)\chi_B(1)=e^{-2a(x-\frac{1}{2}R)^2-\frac{1}{2}aR^2}=e^{-2a(x-\frac{1}{2}R)^2}e^{-\frac{1}{2}aR^2}$$

which is proportional to a single Gaussian (the term in blue) centred on the mid-point of the internuclear separation, at $x=\frac{1}{2}R$.

Self-test 10E.5 Repeat the analysis for a heteronuclear species with GTOs of the form e^{-ax^2} and e^{-bx^2}.

Answer: $\chi_A(1)\chi_B(1)=e^{-(cx-bR/c)^2-a^2R^2/c^2}$, $c=(a+b)^{1/2}$

(b) Density functional theory

A technique that has gained considerable ground in recent years to become one of the most widely used techniques for the calculation of molecular structure is **density functional theory** (DFT). Its advantages include less demanding computational effort, less computer time, and—in some cases (particularly d-metal complexes)—better agreement with experimental values than is obtained from other procedures.

The central focus of DFT is the electron density, ρ, rather than the wavefunction, ψ. The 'functional' part of the name comes from the fact that the energy of the molecule is a function of the electron density, written $E[\rho]$, the electron density is itself a function of position, $\rho(r)$, and in mathematics a function of a function is called a 'functional'. The occupied orbitals are used to construct the electron density from

$$\rho(r)=\sum_{m,\text{occupied}}|\psi_m(r)|^2 \qquad \text{Electron probability density} \qquad (10E.15)$$

and are calculated from modified versions of the Schrödinger equation known as the **Kohn–Sham equations**.

The Kohn–Sham equations are solved iteratively and self-consistently. First, the electron density is guessed. For this step it is common to use a superposition of atomic electron densities. Next, the Kohn–Sham equations are solved to obtain an initial set of orbitals. This set of orbitals is used to obtain a better approximation to the electron density and the process is repeated until the density and the computed energy are constant to within some tolerance.

(c) Graphical representations

One of the most significant developments in computational chemistry has been the introduction of graphical representations of molecular orbitals and electron densities. The raw output of a molecular structure calculation is a list of the coefficients of the atomic orbitals in each molecular orbital and the energies of these orbitals. The graphical representation of a molecular orbital uses stylized shapes to represent the basis set, and then scales their size to indicate the coefficient in the linear combination. Different signs of the wavefunctions are represented by different colours.

Once the coefficients are known, it is possible to construct a representation of the electron density in the molecule by noting which orbitals are occupied and then forming the squares of those orbitals. The total electron density at any point is then the sum of the squares of the wavefunctions evaluated at that point (as in eqn 10E.15). The outcome is commonly represented by an **isodensity surface**, a surface of constant total electron density (Fig. 10E.6). As shown in the illustration, there are several styles of representing an isodensity surface, as a solid form, as a transparent form with a ball-and-stick representation of the molecule within, or as a mesh. A related representation is a **solvent-accessible surface** in which the shape represents the shape of the molecule by imagining a sphere representing a solvent molecule rolling across the surface and plotting the locations of the centre of that sphere.

One of the most important aspects of a molecule other than its geometrical shape is the distribution of charge over its surface, which is commonly depicted as an **electrostatic potential surface** (an 'elpot surface'). The potential energy, E_p, of an imaginary positive charge Q at a point is calculated by taking into account its interaction with the nuclei and the electron density throughout the molecule. Then, because $E_p=Q\phi$, where ϕ is the electric potential, the potential energy can be interpreted as a potential and depicted as an appropriate colour (Fig. 10E.7). Electron-rich regions usually have negative potentials and electron-poor regions usually have positive potentials.

Figure 10E.6 Various representations of an isodensity surface of ethanol: (a) solid surface, (b) transparent surface, and (c) mesh surface.

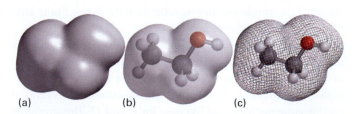

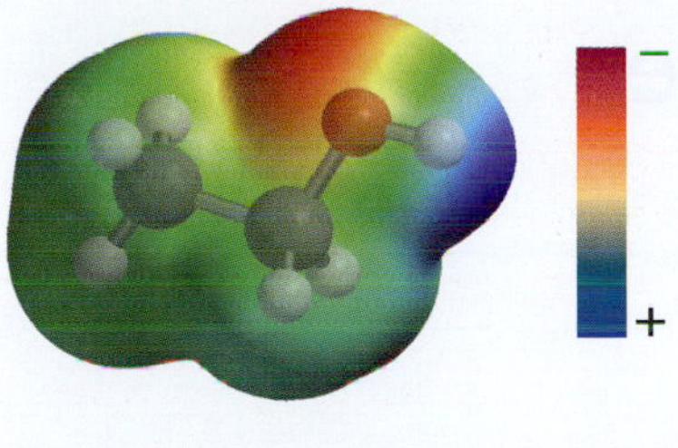

Figure 10E.7 An elpot diagram of ethanol; the molecule has the same orientation as in Fig. 10E.6. Red denotes regions of negative electrostatic potential and blue regions of positive potential (as in $^{\delta-}$O–H$^{\delta+}$).

Representations such as those we have illustrated are of critical importance in a number of fields. For instance, they may be used to identify an electron-poor region of a molecule that is susceptible to association with or chemical attack by an electron-rich region of another molecule. Such considerations are important for assessing the pharmacological activity of potential drugs.

Checklist of concepts

☐ 1. The **Hückel method** neglects overlap and interactions between atoms that are not neighbours.
☐ 2. The Hückel method may be expressed in a compact manner by introducing matrices.
☐ 3. The **π-bond formation energy** is the energy released when a π bond is formed.
☐ 4. The **π-electron binding energy** is the sum of the energies of each π electron.
☐ 5. The **delocalization energy** is the difference between the π-electron energy and the energy of the same molecule with localized π bonds.
☐ 6. The highest occupied molecular orbital (HOMO) and the lowest unoccupied molecular orbital (LUMO) form the **frontier orbitals** of a molecule.
☐ 7. The stability of benzene arises from the geometry of the ring and the high delocalization energy.
☐ 8. **Semi-empirical calculations** approximate integrals by estimating integrals using empirical data; ***ab initio* methods** evaluate all integrals numerically.
☐ 9. **Density functional theories** develop equations based on the electron density rather than the wavefunction itself.
☐ 10. Graphical techniques are used to plot a variety of surfaces based on electronic structure calculations.

Checklist of equations

Property	Equation	Comment	Equation number
LCAO	$\psi=\sum_o c_o\chi_o$	χ_0 are atomic orbitals	10E.1
Hückel equations	$\boldsymbol{Hc}=\boldsymbol{ScE}$	Hückel approximation: $S=0$ except between neighbours	10E.9
Diagonalization	$\boldsymbol{c}^{-1}\boldsymbol{Hc}=\boldsymbol{E}$		10E.10
π-Bond formation energy	$E_{\text{bf}}=E_\pi-N_{\text{C}}\alpha$	Definition; N_{C} is the number of carbon atoms	10E.12
Molecular integrals	$(AB\|CD)=j_0\int\int A(1)B(1)(1/r_{12})C(2)D(2)\,\text{d}\tau_1\text{d}\tau_2$	A, B, C, D are atomic orbitals	10E.14a
Electron probability density	$\rho(\boldsymbol{r})=\sum_{m,\text{occ}}\|\psi_m(\boldsymbol{r})\|^2$	Sum over occupied molecular orbitals m	10E.15

CHAPTER 10 Molecular structure

TOPIC 10A Valence-bond theory

Discussion questions

10A.1 Discuss the role of the Born–Oppenheimer approximation in the calculation of a molecular potential energy curve or surface.

10A.2 Why are promotion and hybridization invoked in valence-bond theory?

10A.3 Describe the various types of hybrid orbitals and how they are used to describe the bonding in alkanes, alkenes, and alkynes. How does hybridization explain that in allene, $CH_2{=}C{=}CH_2$, the two CH_2 groups lie in perpendicular planes?

10A.4 Why is spin-pairing so common a features of bond formation (in the context of valence-bond theory)?

10A.5 What are the consequences of resonance?

Exercises

10A.1(a) Write the valence-bond wavefunction for the single bond in HF.
10A.1(b) Write the valence-bond wavefunction for the triple bond in N_2.

10A.2(a) Write the valence-bond wavefunction for the resonance hybrid $HF \leftrightarrow H^+F^- \leftrightarrow H^-F^+$ (allow for different contributions of each structure).
10A.2(b) Write the valence-bond wavefunction for the resonance hybrid $N_2 \leftrightarrow N^+N^- \leftrightarrow N^{2-}N^{2+} \leftrightarrow$ structures of similar energy.

10A.3(a) Describe the structure of a P_2 molecule in valence-bond terms. Why is P_4 a more stable form of molecular phosphorus than P_2?
10A.3(b) Describe the structures of SO_2 and SO_3 in terms of valence bond theory.

10A.4(a) Describe the bonding in 1,3-butadiene using hybrid orbitals.
10A.4(b) Describe the bonding in 1,3-pentadiene using hybrid orbitals.

10A.5(a) Show that the linear combinations $h_1 = s + p_x + p_y + p_z$ and $h_2 = s - p_x - p_y + p_z$ are mutually orthogonal.
10A.5(b) Show that the linear combinations $h_1 = (\sin\zeta)s + (\cos\zeta)p$ and $h_2 = (\cos\zeta)s - (\sin\zeta)p$ are mutually orthogonal for all values of the angle ζ (zeta).

10A.6(a) Normalize the sp^2 hybrid orbital $h = s + 2^{1/2}p$ given that the s and p orbitals are each normalized to 1.
10A.6(b) Normalize the linear combinations in Exercise 10A.5b given that the s and p orbitals are each normalized to 1.

Problems

10A.1 An sp^2 hybrid orbital that lies in the xy plane and makes an angle of 120° to the x-axis has the form

$$\psi = \frac{1}{3^{1/2}}\left(s - \frac{1}{2^{1/2}}p_x + \frac{3^{1/2}}{2^{1/2}}p_y\right)$$

Use hydrogenic atomic orbitals to write the explicit form of the hybrid orbital. Show that it has its maximum amplitude in the direction specified.

10A.2 Confirm that the hybrid orbitals in eqn 10A.5 make angles of 120° to each other.

10A.3 Show that two equivalent hybrid orbitals of the form sp^λ make an angle θ to each other, then $\lambda = -1/\cos\theta$. Plot a graph of λ against θ and confirm that $\theta = 180°$ when no s orbital is included and $\theta = 120°$ when $\lambda = 2$.

TOPIC 10B Principles of molecular orbital theory

Discussion questions

10B.1 What feature of molecular orbital theory is responsible for bond formation?

10B.2 Why is spin-pairing so common a features of bond formation (in the context of molecular orbital theory)?

Exercises

10B.1(a) Normalize the molecular orbital $\psi = \psi_A + \lambda\psi_B$ in terms of the parameter λ and the overlap integral S.
10B.1(b) A better description of the molecule in Exercise 10B.1(a) might be obtained by including more orbitals on each atom in the linear combination. Normalize the molecular orbital $\psi = \psi_A + \lambda\psi_B + \lambda'\psi_B'$ in terms of the parameters λ and λ' and the appropriate overlap integrals S, where ψ_B and ψ_B' are mutually orthogonal orbitals on atom B.

10B.2(a) Suppose that a molecular orbital has the (unnormalized) form $0.145A + 0.844B$. Find a linear combination of the orbitals A and B that is orthogonal to this combination and determine the normalization constants of both combinations using $S = 0.250$.

10B.2(b) Suppose that a molecular orbital has the (unnormalized) form $0.727A+0.144B$. Find a linear combination of the orbitals A and B that is orthogonal to this combination and determine the normalization constants of both combinations using $S=0.117$.

10B.3(a) The energy of H_2^+ with internuclear separation R is given by eqn 10B.4. The values of the contributions are given below. Plot the molecular potential energy curve and find the bond dissociation energy (in electronvolts) and the equilibrium bond length.

R/a_0	0	1	2	3	4
j/E_h	1.000	0.729	0.472	0.330	0.250
k/E_h	1.000	0.736	0.406	0.199	0.092
S	1.000	0.858	0.587	0.349	0.189

where $E_h=27.2\,\text{eV}$, $a_0=52.9\,\text{pm}$, and $E_H=-\frac{1}{2}E_h$.

10B.3(b) The same data as in Exercise 10B.3(a) may be used to calculate the molecular potential energy curve for the antibonding orbital, which is given by eqn 10B.7. Plot the curve.

10B.4(a) Identify the g or u character of bonding and antibonding π orbitals formed by side-by-side overlap of p atomic orbitals.

10B.4(b) Identify the g or u character of bonding and antibonding δ orbitals formed by face-to-face overlap of d atomic orbitals.

Problems

10B.1 Calculate the (molar) energy of electrostatic repulsion between two hydrogen nuclei at the separation in H_2 (74.1 pm). The result is the energy that must be overcome by the attraction from the electrons that form the bond. Does the gravitational attraction between them play any significant role? *Hint*: The gravitational potential energy of two masses is equal to $-Gm_1m_2/r$; G is listed inside the front cover.

10B.2 Imagine a small electron-sensitive probe of volume $1.00\,\text{pm}^3$ inserted into an H_2^+ molecule-ion in its ground state. Calculate the probability that it will register the presence of an electron at the following positions: (a) at nucleus A, (b) at nucleus B, (c) half way between A and B, (c) at a point 20 pm along the bond from A and 10 pm perpendicularly. Do the same for the molecule-ion the instant after the electron has been excited into the antibonding LCAO-MO.

10B.3 Derive eqns 10B.4 and 10B.7 by working with the normalized LCAO-MOs for the H_2^+ molecule-ion. Proceed by evaluating the expectation value of the hamiltonian for the ion. Make use of the fact that A and B each individually satisfy the Schrödinger equation for an isolated H atom.

10B.4 Examine whether occupation of the bonding orbital with one electron (as calculated in the preceding problem) has a greater or lesser bonding effect than occupation of the antibonding orbital with one electron. Is that true at all internuclear separations?

10B.5‡ The LCAO-MO approach described in the text can be used to introduce numerical methods needed in quantum chemistry. In this problem we evaluate the overlap, Coulomb, and resonance integrals numerically and compare the results with the analytical equations (eqns 10B.5). (a) Use the LCAO-MO wavefunction and the H_2^+ hamiltonian to derive equations for the relevant integrals and use mathematical software or an electronic spreadsheet to evaluate the overlap, Coulomb, and resonance integrals numerically, and the total energy for the $1s\sigma_g$ MO in the range $a_0<R<4a_0$. Compare the results obtained by numerical integration with results obtained analytically. (b) Use the results of the numerical integrations to draw a graph of the total energy, $E(R)$, and determine the minimum of total energy, the equilibrium internuclear distance, and the dissociation energy (D_0).

10B.6 (a) Calculate the total amplitude of the normalized bonding and antibonding LCAO-MOs that may be formed from two H1s orbitals at a separation of $2a_0=106\,\text{pm}$. Plot the two amplitudes for positions along the molecular axis both inside and outside the internuclear region. (b) Plot the probability densities of the two orbitals. Then form the *difference density*, the difference between ψ^2 and $\frac{1}{2}(\psi_A^2+\psi_B^2)$.

TOPIC 10C Homonuclear diatomic molecules

Discussion questions

10C.1 Draw diagrams to show the various orientations in which a p orbital and a d orbital on adjacent atoms may form bonding and antibonding molecular orbitals.

10C.2 Outline the rules of the building-up principle for homonuclear diatomic molecules.

10C.3 What is the role of the Born–Oppenheimer approximation in molecular orbital theory?

10C.4 What is the justification for treating s and p atomic orbital contributions to molecular orbitals separately?

10C.5 To what extent can orbital overlap be related to bond strength?

Exercises

10C.1(a) Give the ground-state electron configurations and bond orders of (i) Li_2, (ii) Be_2, and (iii) C_2.

10C.1(b) Give the ground-state electron configurations of (i) F_2^-, (ii) N_2, and (iii) O_2^{2-}.

10C.2(a) From the ground-state electron configurations of B_2 and C_2, predict which molecule should have the greater dissociation energy.

10C.2(b) From the ground-state electron configurations of Li_2 and Be_2, predict which molecule should have the greater dissociation energy.

10C.3(a) Which has the higher dissociation energy, F_2 or F_2^+?

‡ These problems were supplied by Charles Trapp and Carmen Giunta.

10C.3(b) Arrange the species $O_2^+, O_2, O_2^-, O_2^{2-}$ in order of increasing bond length.

10C.4(a) Evaluate the bond order of each Period 2 homonuclear diatomic molecule.

10C.4(b) Evaluate the bond order of each Period 2 homonuclear diatomic cation, X_2^+, and anion, X_2^-.

10C.5(a) For each of the species in Exercise 10C.4(b), specify which molecular orbital is the HOMO.

10C.5(b) For each of the species in Exercise 10C.4(b), specify which molecular orbital is the LUMO.

10C.6(a) What is the speed of a photoelectron ejected from an orbital of ionization energy 12.0 eV by a photon of radiation of wavelength 100 nm?

10C.6(b) What is the speed of a photoelectron ejected from a molecule with radiation of energy 21 eV and known to come from an orbital of ionization energy 12 eV?

10C.7(a) The overlap integral between two hydrogenic 1s orbitals on nuclei separated by a distance R is given by eqn 10C.4. At what separation is $S=0.20$ for (i) H_2, (ii) He_2?

10C.7(b) The overlap integral between two hydrogenic 2s orbitals on nuclei separated by a distance R is given by the expression in *Brief illustration* 10C.2. At what separation is $S=0.20$ for (i) H_2, (ii) He_2?

Problems

10C.1 Before doing the calculation below, sketch how the overlap between a 1s orbital and a 2p orbital directed towards it can be expected to depend on their separation. The overlap integral between an H1s orbital and an H2p orbital directed towards it on nuclei separated by a distance R is $S=(R/a_0)\{1+(R/a_0)+\frac{1}{3}(R/a_0)^2\}e^{-R/a_0}$. Plot this function, and find the separation for which the overlap is a maximum.

10C.2‡ Use the $2p_x$ and $2p_z$ hydrogenic atomic orbitals to construct simple LCAO descriptions of $2p\sigma$ and $2p\pi$ molecular orbitals. (a) Make a probability density plot, and both surface and contour plots of the xz-plane amplitudes of the $2p_z\sigma$ and $2p_z\sigma^*$ molecular orbitals. (b) Make surface and contour plots of the xz-plane amplitudes of the $2p_x\pi$ and $2p_x\pi^*$ molecular orbitals. Include plots for both an internuclear distance, R, of $10a_0$ and $3a_0$, where $a_0=52.9$ pm. Interpret the graphs, and explain why this graphical information is useful.

10C.3 Show, if overlap is ignored, (a) that any molecular orbital expressed as a linear combination of two atomic orbitals may be written in the form $\psi=\psi_A\cos\theta+\psi_B\sin\theta$, where θ is a parameter that varies between 0 and π, and (b) that if ψ_A and ψ_B are orthogonal and normalized to 1, then ψ is also normalized to 1. (c) To what values of θ do the bonding and antibonding orbitals in a homonuclear diatomic molecule correspond?

10C.4 In a particular photoelectron spectrum using 21.21 eV photons, electrons were ejected with kinetic energies of 11.01 eV, 8.23 eV, and 5.22 eV. Sketch the molecular orbital energy level diagram for the species, showing the ionization energies of the three identifiable orbitals.

TOPIC 10D Heteronuclear diatomic molecules

Discussion questions

10D.1 Describe the Pauling and Mulliken electronegativity scales. Why should they be approximately in step?

10D.2 Why do both ionization energy and electron affinity play a role in estimating the energy of an atomic orbital to use in a molecular structure calculation?

10D.3 Discuss the steps involved in the calculation of the energy of a system by using the variation principle. Are any assumptions involved?

10D.4 What is the physical significance of the Coulomb and resonance integrals?

10D.5 Discuss how the properties of carbon explain the bonding features that make it an ideal biological building block.

Exercises

10D.1(a) Give the ground-state electron configurations of (i) CO, (ii) NO, and (iii) CN^-.

10D.1(b) Give the ground-state electron configurations of (i) XeF, (ii) PN, and (iii) SO^-.

10D.2(a) Sketch the molecular orbital energy level diagram for XeF and deduce its ground-state electron configuration. Is XeF likely to have a shorter bond length than XeF^+?

10D.2(b) Sketch the molecular orbital energy level diagram for IF and deduce its ground-state electron configuration. Is IF likely to have a shorter bond length than IF^- or IF^+?

10D.3(a) Use the electron configurations of NO^- and NO^+ to predict which is likely to have the shorter bond length.

10D.3(b) Use the electron configurations of SO^- and SO^+ to predict which is likely to have the shorter bond length.

10D.4(a) A reasonably reliable conversion between the Mulliken and Pauling electronegativity scales is given by eqn 10D.4. Use Table 10D.1 in the *Resource section* to assess how good the conversion formula is for Period 2 elements.

10D.4(b) A reasonably reliable conversion between the Mulliken and Pauling electronegativity scales is given by eqn 10D.4. Use Table 10D.1 in the *Resource section* to assess how good the conversion formula is for Period 3 elements.

10D.5(a) Estimate the orbital energies to use in a calculation of the molecular orbitals of HCl. For data, see Tables 9B.2 and 9B.3.

10D.5(b) Estimate the orbital energies to use in a calculation of the molecular orbitals of HBr. For data, see Tables 9B.2 and 9B.3.

10D.6(a) Use the values derived in Exercise 10D.5(a) to estimate the molecular orbital energies in HCl; Use $S=0$.

10D.6(b) Use the values derived in Exercise 10D.5(b) to estimate the molecular orbital energies in HBr; Use $S=0$.

10D.7(a) Now repeat Exercise 10D.6(a), but with $S=0.20$.
10D.7(b) Now repeat Exercise 10D.6(b), but with $S=0.20$.

Problems

10D.1 Equation 10D.9 follows from eqn 10D.8a by making the approximation $|\alpha_B - \alpha_A| \gg 2|\beta|$ and setting $S=0$. Explore the consequences of not setting $S=0$.

10D.2 Suppose that a molecular orbital of a heteronuclear diatomic molecule is built from the orbital basis *A*, *B*, and *C*, where *B* and *C* are both on one atom (they can be envisaged as F2s and F2p in HF, for instance). Set up the secular equations for the optimum values of the coefficients and the corresponding secular determinant.

10D.3 Continue the preceding problem by setting $\alpha_A=-7.2\,\text{eV}$, $\alpha_B=-10.4\,\text{eV}$, $\alpha_B=-8.4\,\text{eV}$, $\beta_{AB}=-1.0\,\text{eV}$, $\beta_{AC}=-0.8\,\text{eV}$, and calculate the orbital energies and coefficients with (i) both $S=0$, (ii) both $S=0.2$.

10D.4 As a variation of the preceding problem explore the consequences of increasing the energy separation of the *B* and *C* orbitals (use $S=0$ for this stage of the calculation). Are you justified in ignoring orbital *C* at any stage?

TOPIC 10E Polyatomic molecules

Discussion questions

10E.1 Discuss the scope, consequences, and limitations of the approximations on which the Hückel method is based.

10E.2 Distinguish between delocalization energy, π-electron binding energy, and π-bond formation energy. Explain how each concept is employed.

10E.3 Outline the computational steps used in the self-consistent field approach to electronic structure calculations.

10E.4 Explain why the use of Gaussian-type orbitals is generally preferred over the use of hydrogenic (exponential) orbitals in basis sets.

10E.5 Distinguish between semi-empirical, *ab initio*, and density functional theory methods of electronic structure determination.

Exercises

10E.1(a) Write down the secular determinants for (i) linear H_3, (ii) cyclic H_3 within the Hückel approximation.
10E.1(b) Write down the secular determinants for (i) linear H_4, (ii) cyclic H_4 within the Hückel approximation.

10E.2(a) Predict the electron configurations of (i) the benzene anion, (ii) the benzene cation. Estimate the π-electron binding energy in each case.
10E.2(b) Predict the electron configurations of (i) the allyl radical, (ii) the cyclobutadiene cation. Estimate the π-electron binding energy in each case.

10E.3(a) Compute the delocalization energy and π-bond formation energy of (i) the benzene anion, (ii) the benzene cation.
10E.3(b) Compute the delocalization energy and π-bond formation energy of (i) the allyl radical, (ii) the cyclobutadiene cation.

10E.4(a) Write down the secular determinants for (i) anthracene (**1**), (ii) phenanthrene (**2**) within the Hückel approximation and using the C2p orbitals as the basis set.

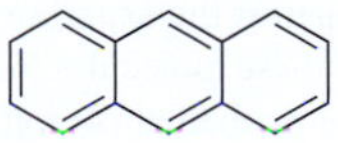

1 Anthracene

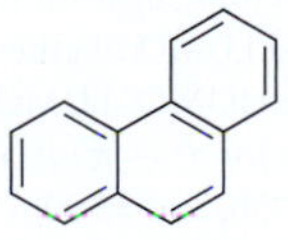

2 Phenanthrene

10E.4(b) Write down the secular determinants for (i) azulene (**3**), (ii) acenaphthalene (**4**) within the Hückel approximation and using the C2p orbitals as the basis set.

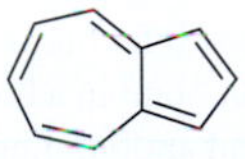

3 Azulene

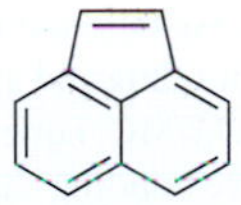

4 Acenaphthalene

10E.5(a) Use mathematical software to estimate the π-electron binding energy of (i) anthracene (**1**), (ii) phenanthrene (**2**) within the Hückel approximation.
10E.5(b) Use mathematical software to estimate the π-electron binding energy of (i) azulene (**3**), (ii) acenaphthalene (**4**) within the Hückel approximation.

10E.6(a) Write the electronic hamiltonian for HeH^+.
10E.6(b) Write the electronic hamiltonian for LiH^{2+}.

Problems

10E.1 Set up and solve the Hückel secular equations for the π electrons of CO_3^{2-}. Express the energies in terms of the Coulomb integrals α_O and α_C and the resonance integral β. Determine the delocalization energy of the ion.

10E.2 For monocyclic conjugated polyenes (such as cyclobutadiene and benzene) with each of N carbon atoms contributing an electron in a 2p orbital, simple Hückel theory gives the following expression for the energies E_k of the resulting π molecular orbitals:

$$E_k = \alpha + 2\beta\cos\frac{2k\pi}{N} \quad k=0,\pm1,\ldots\pm N/2 \text{ for } N \text{ even}$$
$$k=0,\pm1,\ldots\pm(N-1)/2 \text{ for } N \text{ odd}$$

(a) Calculate the energies of the π molecular orbitals of benzene and cyclooctatetraene (**5**). Comment on the presence or absence of degenerate energy levels. (b) Calculate and compare the delocalization energies of benzene (using the expression above) and hexatriene. What do you conclude from your results? (c) Calculate and compare the delocalization energies of cyclooctatetraene and octatetraene. Are your conclusions for this pair of molecules the same as for the pair of molecules investigated in part (b)?

5 Cyclooctatetraene

10E.3 Set up the secular determinants for the homologous series consisting of ethene, butadiene, hexatriene, and octatetraene and diagonalize them by using mathematical software. Use your results to show that the π molecular orbitals of linear polyenes obey the following rules:

- The π molecular orbital with lowest energy is delocalized over all carbon atoms in the chain.
- The number of nodal planes between C2p orbitals increases with the energy of the π molecular orbital.

10E.4 Set up the secular determinants for cyclobutadiene, benzene, and cyclooctatetraene and diagonalize them by using mathematical software. Use your results to show that the π molecular orbitals of monocyclic polyenes with an even number of carbon atoms follow a pattern in which:

- The π molecular orbitals of lowest and highest energy are non-degenerate.
- The remaining π molecular orbitals exist as degenerate pairs.

10E.5 Electronic excitation of a molecule may weaken or strengthen some bonds because bonding and antibonding characteristics differ between the HOMO and the LUMO. For example, a carbon–carbon bond in a linear polyene may have bonding character in the HOMO and antibonding character in the LUMO. Therefore, promotion of an electron from the HOMO to the LUMO weakens this carbon–carbon bond in the excited electronic state, relative to the ground electronic state. Consult Figs. 10E.2 and 10E.4 and discuss in detail any changes in bond order that accompany the $\pi^* \leftarrow \pi$ ultraviolet absorptions in butadiene and benzene.

10E.6‡ Prove that for an open chain of N conjugated carbons the characteristic polynomial of the secular determinant (the polynomial obtained by expanding the determinant), $P_N(x)$, where $x=(\alpha-\beta)/\beta$, obeys the recurrence relation $P_N = xP_{N-1} - P_{N-2}$, with $P_1 = x$ and $P_0 = 1$.

10E.7 The standard potential of a redox couple is a measure of the thermodynamic tendency of an atom, ion, or molecule to accept an electron (Topic 6D). Studies indicate that there is a correlation between the LUMO energy and the standard potential of aromatic hydrocarbons. Do you expect the standard potential to increase or decrease as the LUMO energy decreases? Explain your answer.

10E.8‡ In Exercise 10E.1(a) you are invited to set up the Hückel secular determinant for linear and cyclic H_3. The same secular determinant applies to the molecular ions H_3^+ and D_3^+. The molecular ion H_3^+ was discovered as long ago as 1912 by J.J. Thomson but the equilateral triangular structure was confirmed by M.J. Gaillard et al. much more recently (*Phys. Rev. A* **17**, 1797 (1978)). The molecular ion H_3^+ is the simplest polyatomic species with a confirmed existence and plays an important role in chemical reactions occurring in interstellar clouds that may lead to the formation of water, carbon monoxide, and ethyl alcohol. The H_3^+ ion has also been found in the atmospheres of Jupiter, Saturn, and Uranus. (a) Solve the Hückel secular equations for the energies of the H_3 system in terms of the parameters α and β, draw an energy level diagram for the orbitals, and determine the binding energies of H_3^+, H_3, and H_3^-. (b) Accurate quantum mechanical calculations by G.D. Carney and R.N. Porter (*J. Chem. Phys.* **65**, 3547 (1976)) give the dissociation energy for the process $H_3^+ \rightarrow H + H + H^+$ as 849 kJ mol^{-1}. From this information and data in Table 10C.2, calculate the enthalpy of the reaction $H^+(g) + H_2(g) \rightarrow H_3^+(g)$. (c) From your equations and the information given, calculate a value for the resonance integral β in H_3^+. Then go on to calculate the binding energies of the other H_3 species in (a).

10E.9‡ There is some indication that other hydrogen ring compounds and ions in addition to H_3 and D_3 species may play a role in interstellar chemistry. According to J.S. Wright and G.A. DiLabio (*J. Phys. Chem.* **96**, 10793 (1992)), H_5^-, H_6, and H_7^+ are particularly stable whereas H_4 and H_5^+ are not. Confirm these statements by Hückel calculations.

10E.10 Use appropriate electronic structure software and basis sets of your or your instructor's choosing, perform self-consistent field calculations for the ground electronic states of H_2 and F_2. Determine ground-state energies and equilibrium geometries. Compare computed equilibrium bond lengths to experimental values.

10E.11 Use an appropriate semi-empirical method to compute the equilibrium bond lengths and standard enthalpies of formation of (a) ethanol, C_2H_5OH, (b) 1,4-dichlorobenzene, $C_6H_4Cl_2$. Compare to experimental values and suggest reasons for any discrepancies.

10E.12 Molecular electronic structure methods may be used to estimate the standard enthalpy of formation of molecules in the gas phase. (a) Use a semi-empirical method of your choice or your instructor's suggestion to calculate the standard enthalpy of formation of ethene, butadiene, hexatriene, and octatetraene in the gas phase. (b) Consult a database of thermochemical data, and, for each molecule in part (a), calculate the difference between the calculated and experimental values of the standard enthalpy of formation. (c) A good thermochemical database will also report the uncertainty in the experimental value of the standard enthalpy of formation. Compare experimental uncertainties with the relative errors calculated in part (b) and discuss the reliability of your chosen semi-empirical method for the estimation of thermochemical properties of linear polyenes.

10E.13 Molecular orbital calculations based on semi-empirical, *ab initio*, and DFT methods describe the spectroscopic properties of conjugated molecules better than simple Hückel theory. (a) Use the computational method of your choice (semi-empirical, *ab initio*, or density functional methods) or your instructor's suggestion to calculate the energy separation between the HOMO and LUMO of ethene, butadiene, hexatriene, and octatetraene. (b) Plot the HOMO–LUMO energy separations against the experimental frequencies for $\pi^* \leftarrow \pi$ ultraviolet absorptions for these molecules (61 500, 46 080, 39 750, and 32 900 cm^{-1}, respectively). Use mathematical software to find the polynomial equation that best fits the data. (c) Use your polynomial fit from part (b) to estimate the wavenumber and wavelength of the $\pi^* \leftarrow \pi$ ultraviolet absorption of decapentaene from the calculated HOMO–LUMO energy separation. (d) Discuss why the calibration procedure of part (b) is necessary.

Integrated activities

10.1 The languages of valence-bond theory and molecular orbital theory are commonly combined when discussing unsaturated organic compounds. Construct the molecular orbital energy level diagrams of ethene on the basis that the molecule is formed from the appropriately hybridized CH_2 or CH fragments.

10.2 Here we develop a molecular orbital theory treatment of the peptide group (**6**), which links amino acids in proteins, and establish the features that stabilize its planar conformation. (a) It will be familiar from introductory chemistry that valence bond theory explains the planar conformation by invoking delocalization of the π bond over the oxygen, carbon, and nitrogen atoms by resonance:

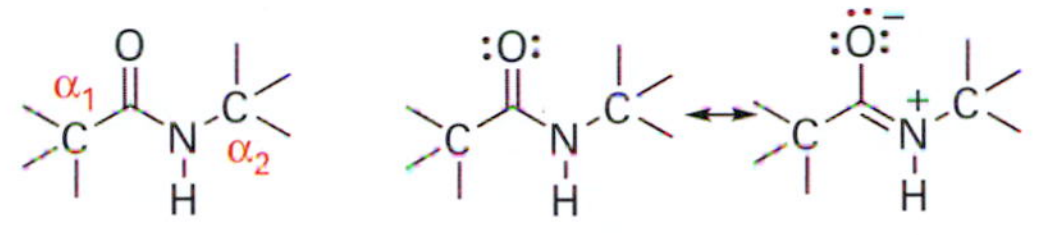

6 Peptide group

It follows that we can model the peptide group using molecular orbital theory by making LCAO-MOs from 2p orbitals perpendicular to the plane defined by the O, C, and N atoms. The three combinations have the form:

$$\psi_1 = a\psi_O + b\psi_C + c\psi_N \quad \psi_2 = d\psi_O - e\psi_N \quad \psi_3 = f\psi_O - g\psi_C + h\psi_N$$

where the coefficients *a* to *h* are all positive. Sketch the orbitals ψ_1, ψ_2, and ψ_3 and characterize them as bonding, non-bonding, or antibonding molecular orbitals. In a non-bonding molecular orbital, a pair of electrons resides in an orbital confined largely to one atom and not appreciably involved in bond formation. (b) Show that this treatment is consistent only with a planar conformation of the peptide link. (c) Draw a diagram showing the relative energies of these molecular orbitals and determine the occupancy of the orbitals. *Hint*: Convince yourself that there are four electrons to be distributed among the molecular orbitals. (d) Now consider a non-planar conformation of the peptide link, in which the O2p and C2p orbitals are perpendicular to the plane defined by the O, C, and N atoms, but the N2p orbital lies on that plane. The LCAO-MOs are given by

$$\psi_4 = a\psi_O + b\psi_C \quad \psi_5 = e\psi_N \quad \psi_6 = f\psi_O - g\psi_C$$

Just as before, sketch these molecular orbitals and characterize them as bonding, non-bonding, or antibonding. Also, draw an energy level diagram and determine the occupancy of the orbitals. (e) Why is this arrangement of atomic orbitals consistent with a non-planar conformation for the peptide link? (f) Does the bonding MO associated with the planar conformation have the same energy as the bonding MO associated with the non-planar conformation? If not, which bonding MO is lower in energy? Repeat the analysis for the non-bonding and antibonding molecular orbitals. (g) Use your results from parts (a)–(f) to construct arguments that support the planar model for the peptide link.

10.3 Molecular orbital calculations may be used to predict trends in the standard potentials of conjugated molecules, such as the quinones and flavins, that are involved in biological electron transfer reactions. It is commonly assumed that decreasing the energy of the LUMO enhances the ability of a molecule to accept an electron into the LUMO, with an accompanying increase in the value of the molecule's standard potential. Furthermore, a number of studies indicate that there is a linear correlation between the LUMO energy and the reduction potential of aromatic hydrocarbons. (a) The standard potentials at pH=7 for the one-electron reduction of methyl-substituted 1,4-benzoquinones (**7**) to their respective semiquinone radical anions are:

R_2	R_3	R_5	R_6	$E^{\ominus}$/V
H	H	H	H	0.078
CH_3	H	H	H	0.023
CH_3	H	CH_3	H	−0.067
CH_3	CH_3	CH_3	H	−0.165
CH_3	CH_3	CH_3	CH_3	−0.260

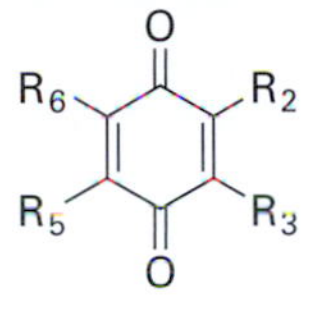

7

Using the computational method of your choice (semi-empirical, *ab initio*, or density functional theory methods), calculate E_{LUMO}, the energy of the LUMO of each substituted 1,4-benzoquinone, and plot E_{LUMO} against $E^{\ominus}$. Do your calculations support a linear relation between E_{LUMO} and $E^{\ominus}$? (b) The 1,4-benzoquinone for which $R_2 = R_3 = CH_3$ and $R_5 = R_6 = OCH_3$ is a suitable model of ubiquinone, a component of the respiratory electron transport chain (*Impact* I17.3). Determine E_{LUMO} of this quinone and then use your results from part (a) to estimate its standard potential. (c) The 1,4-benzoquinone for which $R_2 = R_3 = R_5 = CH_3$ and $R_6 = H$ is a suitable model of plastoquinone, an electron carrier in photosynthesis. Determine E_{LUMO} of this quinone and then use your results from part (a) to estimate its standard potential. Is plastoquinone expected to be a better or worse oxidizing agent than ubiquinone?

10.4 The variation principle can be used to formulate the wavefunctions of electrons in atoms as well as molecules. Suppose that the function $\psi_{trial} = N(\alpha)e^{-\alpha r^2}$ with $N(\alpha)$ the normalization constant and α an adjustable parameter, is used as a trial wavefunction for the 1s orbital of the hydrogen atom. Show that

$$E(\alpha) = \frac{3\alpha\hbar^2}{2\mu} - 2e^2\left(\frac{2\alpha}{\pi}\right)^{1/2}$$

where *e* is the fundamental charge and μ is the reduced mass for the H atom. What is the minimum energy associated with this trial wavefunction?

10.5 The particle-in-a-box wavefunctions can be used as a crude approximation to the molecular orbitals of conjugated polyenes, when it is known as the *free-electron molecular orbital* (FEMO) method. (a) For a linear conjugated polyene with each of *N* carbon atoms contributing an electron in a 2p orbital, the energies E_k of the resulting π molecular orbitals are given by:

$$E_k = \alpha + 2\beta\cos\frac{k\pi}{N+1} \qquad k = 1, 2, \ldots, N$$

Use this expression to determine a reasonable empirical estimate of the resonance integral β for the homologous series consisting of ethene, butadiene, hexatriene, and octatetraene given that $\pi^* \leftarrow \pi$ ultraviolet absorptions from the HOMO to the LUMO occur at 61 500, 46 080, 39 750, and 32 900 cm^{-1}, respectively. (b) Calculate the π-electron delocalization energy, $E_{deloc} = E_\pi - n(\alpha + \beta)$, of octatetraene, where E_π is the total π-electron binding energy and *n* is the total number of π-electrons. (c) In the context of this Hückel model, the π molecular orbitals are written as linear combinations of the carbon 2p orbitals. The coefficient of the *j*th atomic orbital in the *k*th molecular orbital is given by:

$$c_{kj} = \left(\frac{2}{N+1}\right)^{1/2} \sin\frac{jk\pi}{N+1} \quad j=1,2,\ldots,N$$

Determine the values of the coefficients of each of the six 2p orbitals in each of the six π molecular orbitals of hexatriene. Match each set of coefficients (that is, each molecular orbital) with a value of the energy calculated with the expression given in part (a) of the molecular orbital. Comment on trends that relate the energy of a molecular orbital with its 'shape', which can be inferred from the magnitudes and signs of the coefficients in the linear combination that describes the molecular orbital.

10.6 Use mathematical software, a spreadsheet, or the *Living graphs* on the web site for this book to:

(a) Plot the 1σ orbital (eqn 10B.2, with the atomic orbitals given in *Brief illustration* 10B.1) for different values of the internuclear distance. Point to the features of the 1σ orbital that lead to bonding.
(b) Plot the 2σ orbital (eqn 10B.2, with the atomic orbitals given in *Brief illustration* 10B.1) for different values of the internuclear distance. Point to the features of the 2σ orbital that lead to antibonding.

Mathematical background 6 Matrices

A **matrix** is an array of numbers. We shall consider only square matrices, which have the numbers arranged in the same number of rows and columns. By using matrices, we can manipulate large numbers of ordinary numbers simultaneously. A **determinant** is a particular combination of the numbers that appear in a matrix and is used to manipulate the matrix.

Matrices may be combined together by addition or multiplication according to generalizations of the rules for ordinary numbers. Although we describe below the key algebraic procedures involving matrices, it is important to note that most numerical matrix manipulations are now carried out with mathematical software. You are encouraged to use such software, if it is available to you.

MB6.1 Definitions

Consider a square matrix M of n^2 numbers arranged in n columns and n rows. These n^2 numbers are the **elements** of the matrix, and may be specified by stating the row, r, and column, c, at which they occur. Each element is therefore denoted M_{rc}. A **diagonal matrix** is a matrix in which the only nonzero elements lie on the major diagonal (the diagonal from M_{11} to M_{nn}). Thus, the matrix

$$D=\begin{pmatrix}1 & 0 & 0\\ 0 & 2 & 0\\ 0 & 0 & 1\end{pmatrix}$$

is a 3×3 diagonal square matrix. The condition may be written

$$M_{rc}=m_r\delta_{rc} \qquad \text{(MB6.1)}$$

where δ_{rc} is the **Kronecker delta**, which is equal to 1 for $r=c$ and to 0 for $r\neq c$. In the above example, $m_1=1$, $m_2=2$, and $m_3=1$. The **unit matrix**, **1** (and occasionally **I**), is a special case of a diagonal matrix in which all on the major diagonal are 1.

The **transpose** of a matrix M is denoted M^{T} and is defined by

$$M_{mn}^{\mathrm{T}}=M_{nm} \qquad \text{Transpose} \quad \text{(MB6.2)}$$

That is, the element in row n, column m of the original matrix becomes the element in row m, column n of the transpose (in effect, the elements are reflected across the diagonal). The **determinant**, $|M|$, of the matrix M is a real number arising from a specific procedure for taking sums and differences of products of matrix elements, as described in *The chemist's toolkit* 9B.1. For convenience, that discussion is repeated here. Thus, a 2×2 determinant is evaluated as

$$\begin{vmatrix}a & b\\ c & d\end{vmatrix}=ad-bc \qquad \text{2×2 Determinant} \quad \text{(MB6.3a)}$$

and a 3×3 determinant is evaluated by expanding it as a sum of 2×2 determinants:

$$\begin{vmatrix}a & b & c\\ d & e & f\\ g & h & i\end{vmatrix}=a\begin{vmatrix}e & f\\ h & i\end{vmatrix}-b\begin{vmatrix}d & f\\ g & i\end{vmatrix}+c\begin{vmatrix}d & e\\ g & h\end{vmatrix}$$
$$=a(ei-fh)-b(di-fg)+c(dh-eg) \qquad \text{3×3 Determinant} \quad \text{(MB6.3b)}$$

Note the sign change in alternate columns (b occurs with a negative sign in the expansion). An important property of a determinant is that if any two rows or any two columns are interchanged, then the determinant changes sign.

Brief illustration MB6.1 Matrix manipulations

The following grid illustrates the features so far:

Matrix	Transpose	Determinant
M	M^{T}	$\lvert M\rvert$
$\begin{pmatrix}1 & 2\\ 3 & 4\end{pmatrix}$	$\begin{pmatrix}1 & 3\\ 2 & 4\end{pmatrix}$	$\begin{vmatrix}1 & 2\\ 3 & 4\end{vmatrix}=1\times4-2\times3=-2$

MB6.2 Matrix addition and multiplication

Two matrices M and N may be added to give the sum $S=M+N$, according to the rule

$$S_{rc}=M_{rc}+N_{rc} \qquad \text{Matrix addition} \quad \text{(MB6.4)}$$

That is, corresponding elements are added. Two matrices may also be multiplied to give the product $P=MN$ according to the rule

$$P_{rc}=\sum_n M_{rn}N_{nc} \qquad \text{Matrix multiplication} \quad \text{(MB6.5)}$$

These procedures are illustrated in Fig. MB6.1. It should be noticed that in general $MN\neq NM$, and matrix multiplication is in general non-commutative (that is, the result depends on the order of multiplication).

Brief illustration MB6.2 Matrix addition and multiplication

Consider the matrices

$$M=\begin{pmatrix}1 & 2\\ 3 & 4\end{pmatrix} \quad \text{and} \quad N=\begin{pmatrix}5 & 6\\ 7 & 8\end{pmatrix}$$

Their sum is

$$S=\begin{pmatrix}1 & 2\\ 3 & 4\end{pmatrix}+\begin{pmatrix}5 & 6\\ 7 & 8\end{pmatrix}=\begin{pmatrix}6 & 8\\ 10 & 12\end{pmatrix}$$

and their product is

$$P=\begin{pmatrix}1 & 2\\ 3 & 4\end{pmatrix}\begin{pmatrix}5 & 6\\ 7 & 8\end{pmatrix}=\begin{pmatrix}1\times5+2\times7 & 1\times6+2\times8\\ 3\times5+4\times7 & 3\times6+4\times8\end{pmatrix}=\begin{pmatrix}19 & 22\\ 43 & 50\end{pmatrix}$$

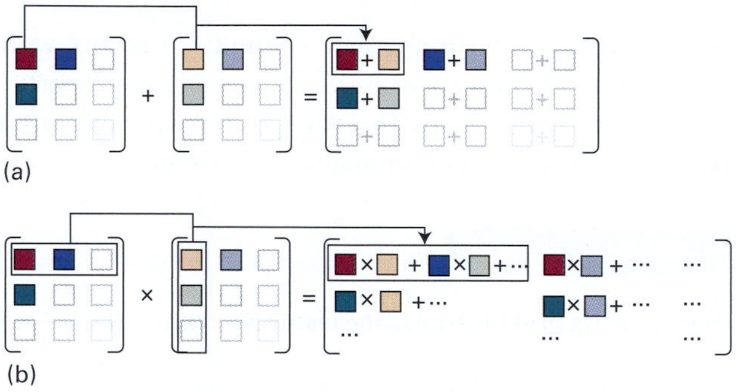

Figure MB6.1 A diagrammatic representation of (a) matrix addition, (b) matrix multiplication.

The **inverse** of a matrix M is denoted M^{-1}, and is defined so that

$$MM^{-1}=M^{-1}M=1 \qquad \text{Inverse} \qquad \text{(MB6.6)}$$

The inverse of a matrix is best constructed by using mathematical software and the tedious analytical approach is rarely necessary.

Brief illustration MB6.3 Inversion

Mathematical software gives the following inversion of M:

Matrix	Inverse
M	M^{-1}
$\begin{pmatrix}1 & 2\\ 3 & 4\end{pmatrix}$	$\begin{pmatrix}-2 & 1\\ \frac{3}{2} & -\frac{1}{2}\end{pmatrix}$

MB6.3 Eigenvalue equations

An **eigenvalue equation** is an equation of the form

$$Mx=\lambda x \qquad \text{Eigenvalue equation} \qquad \text{(MB6.7a)}$$

where M is a square matrix with n rows and n columns, λ is a constant, the **eigenvalue**, and x is the **eigenvector**, an $n\times1$ (column) matrix that satisfies the conditions of the eigenvalue equation and has the form:

$$x=\begin{pmatrix}x_1\\ x_2\\ \vdots\\ x_n\end{pmatrix}$$

In general, there are n eigenvalues $\lambda^{(i)}$, $i=1, 2, \ldots, n$, and n corresponding eigenvectors $x^{(i)}$. We write eqn MB6.8a as (noting that $1x=x$)

$$(M-\lambda 1)x=0 \qquad \text{(MB6.7b)}$$

Equation MB6.7b has a solution only if the determinant $|M-\lambda 1|$ of the coefficients of the matrix $M-\lambda 1$ is zero. It follows that the n eigenvalues may be found from the solution of the **secular equation:**

$$|M-\lambda 1|=0 \qquad \text{(MB6.8)}$$

If the inverse of the matrix $M-\lambda 1$ exists, then, from eqn MB6.7b, $(M-\lambda 1)^{-1}(M-\lambda 1)x=x=0$, a trivial solution. For a nontrivial solution, $(M-\lambda 1)^{-1}$ must not exist, which is the case if eqn MB6.9 holds.

Brief illustration MB6.4 Simultaneous equations

Once again we use the matrix M in *Brief illustration* MB6.1, and write eqn MB6.7a as

$$\begin{pmatrix}1 & 2\\ 3 & 4\end{pmatrix}\begin{pmatrix}x_1\\ x_2\end{pmatrix}=\lambda\begin{pmatrix}x_1\\ x_2\end{pmatrix} \text{ rearranged into } \begin{pmatrix}1-\lambda & 2\\ 3 & 4-\lambda\end{pmatrix}\begin{pmatrix}x_1\\ x_2\end{pmatrix}=0$$

From the rules of matrix multiplication, the latter form expands into

$$\begin{pmatrix}(1-\lambda)x_1+2x_2\\ 3x_1+(4-\lambda)x_2\end{pmatrix}=0$$

which is simply a statement of the two simultaneous equations

$$(1-\lambda)x_1+2x_2=0 \text{ and } 3x_1+(4-\lambda)x_2=0$$

The condition for these two equations to have solutions is

$$|M-\lambda 1|=\begin{vmatrix}1-\lambda & 2\\ 3 & 4-\lambda\end{vmatrix}=(1-\lambda)(4-\lambda)-6=0$$

This condition corresponds to the quadratic equation

$$\lambda^2-5\lambda-2=0$$

with solutions $\lambda=+5.372$ and $\lambda=-0.372$, the two eigenvalues of the original equation.

The n eigenvalues found by solving the secular equations are used to find the corresponding eigenvectors. To do so, we begin by considering an $n\times n$ matrix X which will be formed from the eigenvectors corresponding to all the eigenvalues. Thus, if the eigenvalues are λ_1, λ_2, ..., and the corresponding eigenvectors are

$$x^{(1)}=\begin{pmatrix} x_1^{(1)} \\ x_2^{(1)} \\ \vdots \\ x_n^{(1)} \end{pmatrix} \quad x^{(2)}=\begin{pmatrix} x_1^{(2)} \\ x_2^{(2)} \\ \vdots \\ x_n^{(2)} \end{pmatrix} \quad \cdots \quad x^{(n)}=\begin{pmatrix} x_1^{(n)} \\ x_2^{(n)} \\ \vdots \\ x_n^{(n)} \end{pmatrix} \qquad \text{(MB6.9a)}$$

the matrix X is

$$X=(x^{(1)}x^{(2)}\cdots x^{(n)})=\begin{pmatrix} x_1^{(1)} & x_1^{(2)} & \cdots & x_1^{(n)} \\ x_2^{(1)} & x_2^{(2)} & \cdots & x_2^{(n)} \\ \vdots & \vdots & & \vdots \\ x_n^{(1)} & x_n^{(2)} & \cdots & x_n^{(n)} \end{pmatrix} \qquad \text{(MB6.9b)}$$

Similarly, we form an $n\times n$ matrix Λ with the eigenvalues λ along the diagonal and zeroes elsewhere:

$$\Lambda=\begin{pmatrix} \lambda_1 & 0 & \cdots & 0 \\ 0 & \lambda_2 & \cdots & 0 \\ \vdots & \vdots & \vdots & \\ 0 & 0 & \cdots & \lambda_n \end{pmatrix} \qquad \text{(MB6.10)}$$

Now all the eigenvalue equations $Mx^{(i)}=\lambda_i x^{(i)}$ may be combined into the single matrix equation

$$MX=X\Lambda \qquad \text{(MB6.11)}$$

Brief illustration MB6.5 Eigenvalue equations

In *Brief illustration* MB6.4 we established that if $M=\begin{pmatrix} 1 & 2 \\ 3 & 4 \end{pmatrix}$ then $\lambda_1=+5.372$ and $\lambda_2=-0.372$, with eigenvectors $x^{(1)}=\begin{pmatrix} x_1^{(1)} \\ x_2^{(1)} \end{pmatrix}$ and $x^{(2)}=\begin{pmatrix} x_1^{(2)} \\ x_2^{(2)} \end{pmatrix}$. We form

$$X=\begin{pmatrix} x_1^{(1)} & x_1^{(2)} \\ x_2^{(1)} & x_2^{(2)} \end{pmatrix} \quad \Lambda=\begin{pmatrix} 5.372 & 0 \\ 0 & -0.372 \end{pmatrix}$$

The expression $MX=X\Lambda$ becomes

$$\begin{pmatrix} 1 & 2 \\ 3 & 4 \end{pmatrix}\begin{pmatrix} x_1^{(1)} & x_1^{(2)} \\ x_2^{(1)} & x_2^{(2)} \end{pmatrix}=\begin{pmatrix} x_1^{(1)} & x_1^{(2)} \\ x_2^{(1)} & x_2^{(2)} \end{pmatrix}\begin{pmatrix} 5.372 & 0 \\ 0 & -0.372 \end{pmatrix}$$

which expands to

$$\begin{pmatrix} x_1^{(1)}+2x_2^{(1)} & x_1^{(2)}+2x_2^{(2)} \\ 3x_1^{(1)}+4x_2^{(1)} & 3x_1^{(2)}+4x_2^{(2)} \end{pmatrix}=\begin{pmatrix} 5.372x_1^{(1)} & -0.372x_1^{(2)} \\ 5.372x_2^{(1)} & -0.372x_2^{(2)} \end{pmatrix}$$

This is a compact way of writing the four equations

$$x_1^{(1)}+2x_2^{(1)}=5.372x_1^{(1)} \quad x_1^{(2)}+2x_2^{(2)}=-0.372x_1^{(2)}$$
$$3x_1^{(1)}+4x_2^{(1)}=5.372x_2^{(1)} \quad 3x_1^{(2)}+4x_2^{(2)}=-0.372x_2^{(2)}$$

corresponding to the two original simultaneous equations and their two roots.

Finally, we form X^{-1} from X and multiply eqn MB6.13 by it from the left:

$$X^{-1}MX=X^{-1}X\Lambda=\Lambda \qquad \text{(MB6.12)}$$

A structure of the form $X^{-1}MX$ is called a **similarity transformation**. In this case the similarity transformation $X^{-1}MX$ makes M diagonal (because Λ is diagonal). It follows that if the matrix X that causes $X^{-1}MX$ to be diagonal is known, then the problem is solved: the diagonal matrix so produced has the eigenvalues as its only nonzero elements, and the matrix X used to bring about the transformation has the corresponding eigenvectors as its columns. As will be appreciated once again, the solutions of eigenvalue equations are best found by using mathematical software.

Brief illustration MB6.6 Similarity transformation

To apply the similarity transformation, eqn MB6.12, to the matrix $\begin{pmatrix} 1 & 2 \\ 3 & 4 \end{pmatrix}$ from *Brief illustration* MB6.1 it is best to use mathematical software to find the form of X. The result is

$$X=\begin{pmatrix} 0.416 & 0.825 \\ 0.909 & -0.566 \end{pmatrix} \quad X^{-1}=\begin{pmatrix} 0.574 & 0.837 \\ 0.922 & -0.422 \end{pmatrix}$$

This result can be verified by carrying out the multiplication

$$X^{-1}MX=\begin{pmatrix} 0.574 & 0.837 \\ 0.922 & -0.422 \end{pmatrix}\begin{pmatrix} 1 & 2 \\ 3 & 4 \end{pmatrix}\begin{pmatrix} 0.416 & 0.825 \\ 0.909 & -0.566 \end{pmatrix}$$
$$=\begin{pmatrix} 5.372 & 0 \\ 0 & -0.372 \end{pmatrix}$$

The result is indeed the diagonal matrix Λ calculated in *Brief illustration* MB6.4. It follows that the eigenvectors $x^{(1)}$ and $x^{(2)}$ are

$$x^{(1)}=\begin{pmatrix} 0.416 \\ 0.909 \end{pmatrix} \quad x^{(2)}=\begin{pmatrix} 0.825 \\ -0.566 \end{pmatrix}$$

CHAPTER 11

Molecular symmetry

In this chapter we sharpen the concept of 'shape' into a precise definition of 'symmetry', and show that symmetry may be discussed systematically.

11A Symmetry elements

We see how to classify any molecule according to its symmetry and how to use this classification to discuss the polarity and chirality of molecules.

11B Group theory

The systematic treatment of symmetry is 'group theory'. We show that it is possible to represent the outcome of symmetry operations (such as rotations and reflections) by matrices. That step allows us to express symmetry operations numerically and therefore to perform numerical manipulations. One important outcome is the ability to classify various combinations of atomic orbitals according to their symmetries. It also introduces the hugely important concept of a 'character table', which is the concept most widely employed in chemical applications of group theory.

11C Applications of symmetry

The symmetry analysis described in the preceding two Topics is now put to use. We see that it provides simple criteria for deciding whether certain integrals necessarily vanish. One important integral is the overlap integral between two orbitals. By knowing which atomic orbitals may have nonzero overlap, we can decide which ones can contribute to molecular orbitals. We also see how to select linear combinations of atomic orbitals that match the symmetry of the nuclear framework. Finally, by considering the symmetry properties of integrals, we see that it is possible to derive the selection rules that govern spectroscopic transitions.

11A Symmetry elements

Contents

➤ Why do you need to know this material?

Symmetry arguments can be used to make immediate assessments of the properties of molecules, and when expressed quantitatively (Topic 11B) can be used to save a great deal of calculation.

➤ What is the key idea?

Molecules can be classified into groups according to their symmetry elements.

➤ What do you need to know already?

This Topic does not draw on others directly, but it will be useful to be aware of the shapes of a variety of simple molecules and ions encountered in introductory chemistry courses.

Some objects are 'more symmetrical' than others. A sphere is more symmetrical than a cube because it looks the same after it has been rotated through any angle about any diameter. A cube looks the same only if it is rotated through certain angles about specific axes, such as 90°, 180°, or 270° about an axis passing through the centres of any of its opposite faces (Fig. 11A.1), or by 120° or 240° about an axis passing through any of its opposite corners. Similarly, an NH_3 molecule is 'more symmetrical' than an H_2O molecule because NH_3 looks the same after rotations of 120° or 240° about the axis shown in Fig. 11A.2, whereas H_2O looks the same only after a rotation of 180°.

This Topic puts these intuitive notions on a more formal foundation. In it, we see that molecules can be grouped together according to their symmetry, with the tetrahedral species CH_4 and SO_4^{2-} in one group and the pyramidal species NH_3 and SO_3^{2-} in another. It turns out that molecules in the same group share certain physical properties, so powerful predictions can be made about whole series of molecules once we know the group to which they belong.

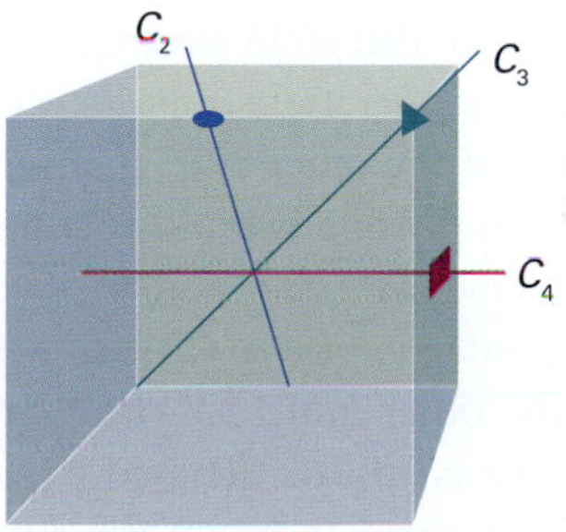

Figure 11A.1 Some of the symmetry elements of a cube. The twofold, threefold, and fourfold axes are labelled with the conventional symbols.

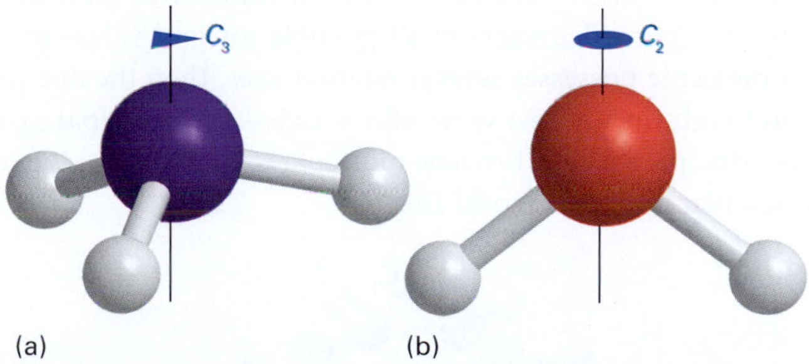

Figure 11A.2 (a) An NH_3 molecule has a threefold (C_3) axis and (b) an H_2O molecule has a twofold (C_2) axis. Both have other symmetry elements too.

We have slipped in the term 'group' in its conventional sense. In fact, a group in mathematics has a precise formal significance and considerable power and gives rise to the name 'group theory' for the quantitative study of symmetry. This power is revealed in Topic 11B.

11A.1 Symmetry operations and symmetry elements

An action that leaves an object looking the same after it has been carried out is called a **symmetry operation**. Typical symmetry operations include rotations, reflections, and inversions. There is a corresponding **symmetry element** for each symmetry operation, which is the point, line, or plane with respect to which the symmetry operation is performed. For instance, a rotation (a symmetry operation) is carried out around an axis (the corresponding symmetry element). We shall see that we can classify molecules by identifying all their symmetry elements, and grouping together molecules that possess the same set of symmetry elements. This procedure, for example, puts the trigonal pyramidal species NH_3 and SO_3^{2-} into one group and the angular species H_2O and SO_2 into another group.

An ***n*-fold rotation** (the operation) about an ***n*-fold axis of symmetry**, C_n (the corresponding element), is a rotation through 360°/n. An H_2O molecule has one twofold axis, C_2. An NH_3 molecule has one threefold axis, C_3, with which is associated two symmetry operations, one being 120° rotation in a clockwise sense and the other 120° rotation in an anticlockwise sense. There is only one twofold rotation associated with a C_2 axis because clockwise and anticlockwise 180° rotations are identical. A pentagon has a C_5 axis, with two rotations (one clockwise, the other anticlockwise) through 72° associated with it. It also has an axis denoted C_5^2, corresponding to two successive C_5 rotations; there are two such operations, one through 144° in a clockwise sense and the other through 144° in a anticlockwise sense. A cube has three C_4 axes, four C_3 axes, and six C_2 axes. However, even this high symmetry is exceeded by a sphere, which possesses an infinite number of symmetry axes (along any diameter) of all possible integral values of n. If a molecule possesses several rotation axes, then the one (or more) with the greatest value of n is called the **principal axis**. The principal axis of a benzene molecule is the sixfold axis perpendicular to the hexagonal ring (**1**).

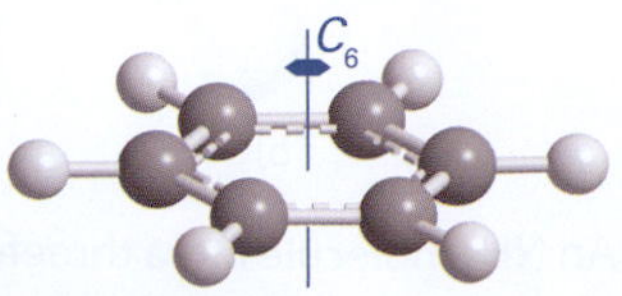

1 Benzene, C_6H_6

A **reflection** (the operation) in a **mirror plane**, σ (the element), may contain the principal axis of a molecule or be perpendicular to it. If the plane contains the principal axis, it is called 'vertical' and denoted σ_v. An H_2O molecule has two vertical planes of symmetry (Fig. 11A.3) and an NH_3 molecule has three. A vertical mirror plane that bisects the angle between two C_2 axes is called a 'dihedral plane' and is denoted σ_d (Fig. 11A.4). When the plane of symmetry is perpendicular to the principal axis it is called 'horizontal' and denoted σ_h. A C_6H_6 molecule has a C_6 principal axis and a horizontal mirror plane (as well as several other symmetry elements).

In an **inversion** (the operation) through a **centre of symmetry**, i (the element), we imagine taking each point in a molecule, moving it to the centre of the molecule, and then moving it out the same distance on the other side; that is, the point (x, y, z) is taken into the point $(-x, -y, -z)$. Neither an H_2O molecule nor an NH_3 molecule has a centre of inversion, but a sphere and a cube do have one. A C_6H_6 molecule does have a centre of inversion, so does a regular octahedron (Fig. 11A.5); a regular tetrahedron and a CH_4 molecule do not.

An ***n*-fold improper rotation** (the operation) about an ***n*-fold axis of improper rotation** or an ***n*-fold improper rotation axis**, S_n, (the symmetry element) is composed of two successive transformations. The first component is a rotation through 360°/n, and the second is a reflection through a plane perpendicular to the axis of that rotation; neither operation alone needs to be a symmetry operation. A CH_4 molecule has three S_4 axes (Fig. 11A.6).

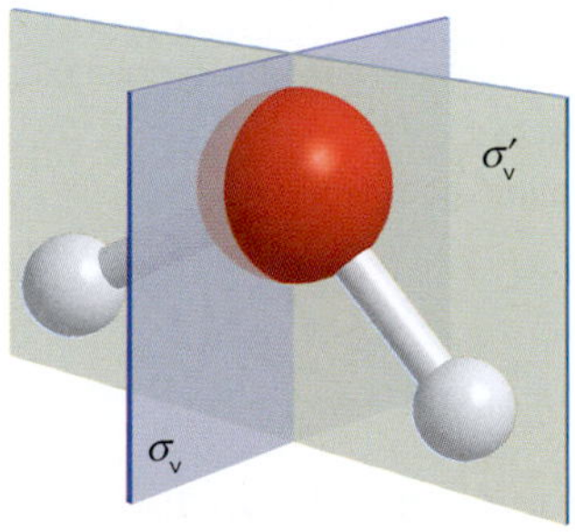

Figure 11A.3 An H_2O molecule has two mirror planes. They are both vertical (that is, contain the principal axis), so are denoted σ_v and σ_v'.

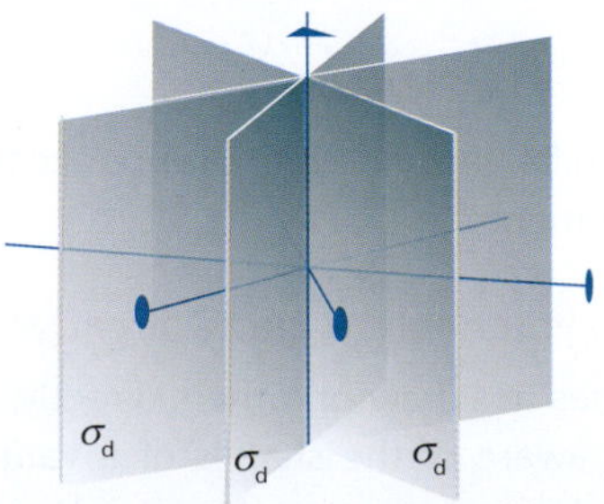

Figure 11A.4 Dihedral mirror planes (σ_d) bisect the C_2 axes perpendicular to the principal axis.

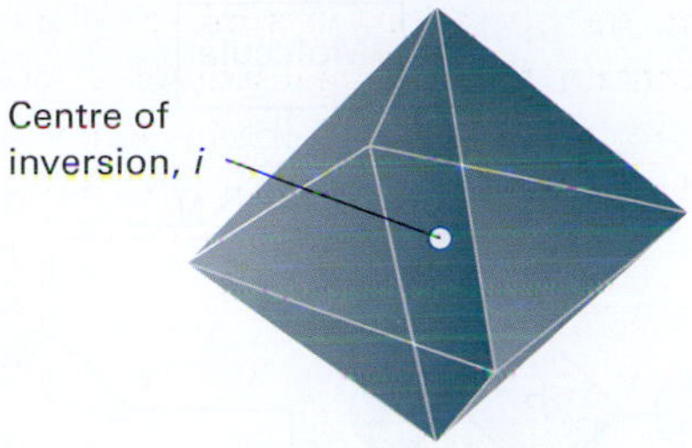

Figure 11A.5 A regular octahedron has a centre of inversion (i).

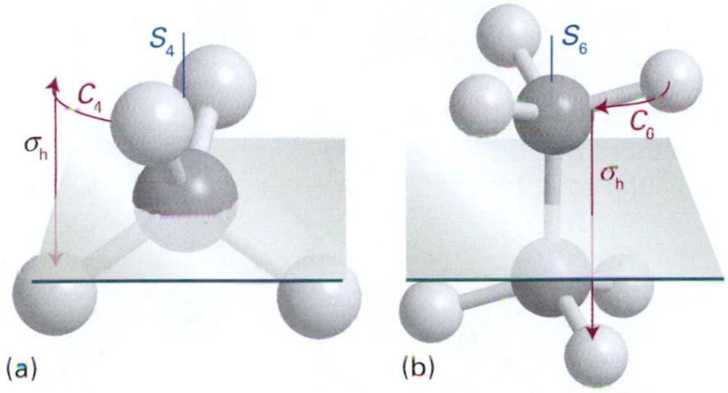

Figure 11A.6 (a) A CH_4 molecule has a fourfold improper rotation axis (S_4): the molecule is indistinguishable after a 90° rotation followed by a reflection across the horizontal plane, but neither operation alone is a symmetry operation. (b) The staggered form of ethane has an S_6 axis composed of a 60° rotation followed by a reflection.

The **identity**, E, consists of doing nothing; the corresponding symmetry element is the entire object. Because every molecule is indistinguishable from itself if nothing is done to it, every object possesses at least the identity element. One reason for including the identity is that some molecules have only this symmetry element (**2**).

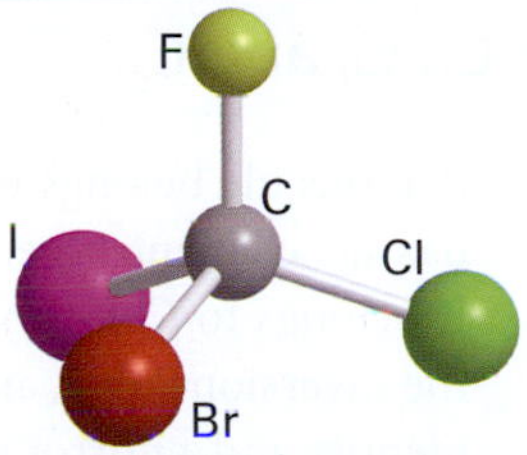

2 CBrClFI

Brief illustration 11A.1 Symmetry elements

To identify the symmetry elements of a naphthalene molecule (**3**) we first note that, like all molecules, it has the identity element, E. There is one twofold axis of rotation, C_2, perpendicular to the plane and two others, C_2', lying in the plane. There is a mirror plane in the plane of the molecule, σ_h, and two perpendicular planes, σ_v, containing the C_2 rotation axis. There is also a centre of inversion, i, through the mid-point of the molecule. Note that some of these elements are implied by others: the centre of inversion, for instance, is implied by a σ_v plane and a C_2', axis.

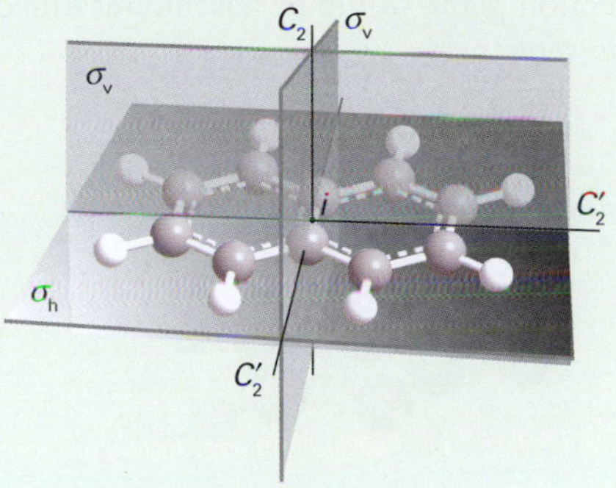

3 Naphthalene, $C_{10}H_8$

Self-test 11A.1 Identify the symmetry elements of an SF_6 molecule.

Answer: E, $3S_4$, $3C_4$, $6C_2$, $4S_6$, $4C_3$, $3\sigma_h$, $6\sigma_d$, i

11A.2 The symmetry classification of molecules

The classification of objects according to symmetry elements corresponding to operations that leave at least one common point unchanged gives rise to the **point groups**. There are five kinds of symmetry operation (and five kinds of symmetry element) of this kind. When we consider crystals (Topic 18A), we meet symmetries arising from translation through space. These more extensive groups are called **space groups**.

To classify molecules according to their symmetries, we list their symmetry elements and collect together molecules with the same list of elements. The name of the group to which a molecule belongs is determined by the symmetry elements it possesses. There are two systems of notation (Table 11A.1). The **Schoenflies system** (in which a name looks like C_{4v}) is more common for the discussion of individual molecules, and the **Hermann–Mauguin system**, or **International system** (in which a name looks like $4mm$), is used almost exclusively in the discussion of crystal symmetry. The identification of a molecule's point group according to the Schoenflies system is simplified by referring to the flow diagram in Fig. 11A.7 and the shapes shown in Fig. 11A.8.

Brief illustration 11A.2 Symmetry classification

To identify the point group to which a ruthenocene molecule (**4**) belongs we use the flow diagram in Fig. 11A.7. The path to trace is shown by a blue line; it ends at D_{nh}. Because the molecule has a fivefold axis, it belongs to the group D_{5h}. If the rings were staggered, as they are in an excited state of ferrocene that lies 4 kJ mol^{-1} above the ground state (5), the horizontal reflection plane would be absent, but dihedral planes would be present.

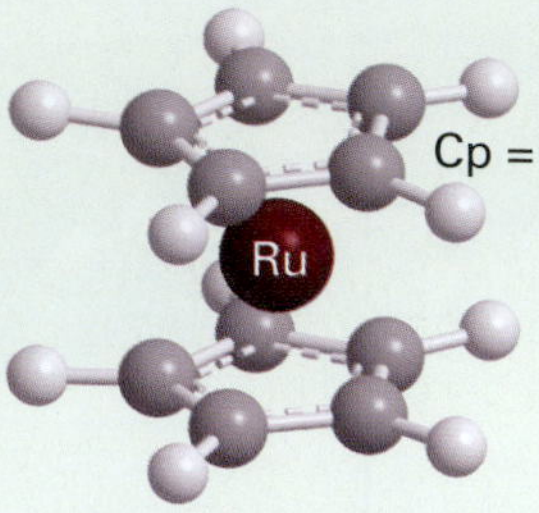

4 Ruthenocene, Ru(Cp)$_2$

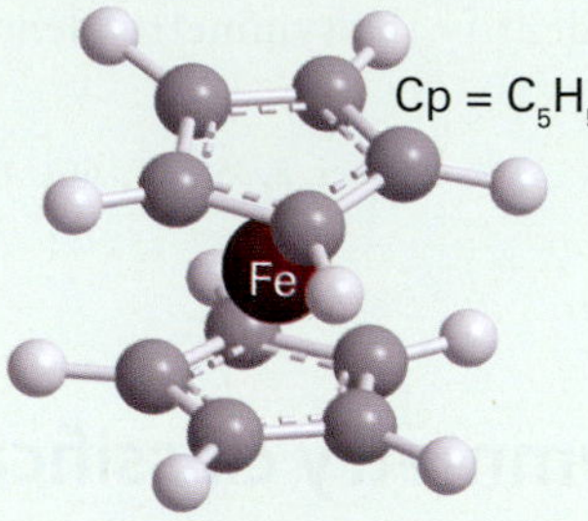

5 Ferrocene, Fe(Cp)$_2$
(excited state)

Self-test 11A.2 Classify the pentagonal antiprismatic excited state of ferrocene (5).

Answer: D_{5d}

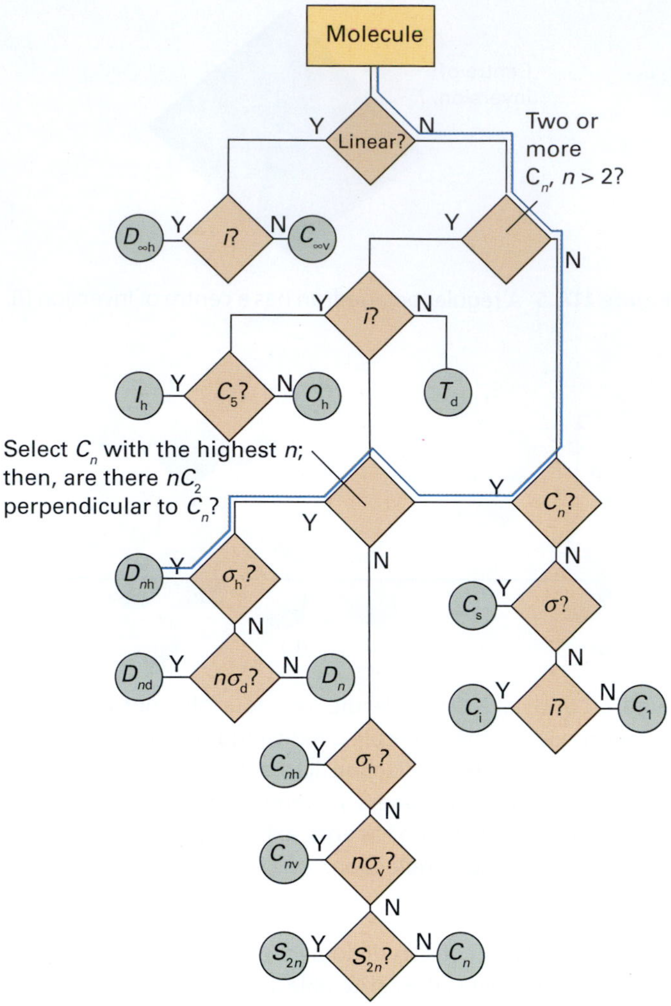

Figure 11A.7 A flow diagram for determining the point group of a molecule. Start at the top and answer the question posed in each diamond (Y=yes, N=no).

Table 11A.1 The notations for point groups*

C_i	$\bar{1}$								
C_s	m								
C_1	1	C_2	2	C_3	3	C_4	4	C_6	6
		C_{2v}	$2mm$	C_{3v}	$3m$	C_{4v}	$4mm$	C_{6v}	$6mm$
		C_{2h}	$2/m$	C_{3h}	$\bar{6}$	C_{4h}	$4/m$	C_{6h}	$6/m$
		D_2	222	D_3	32	D_4	422	D_6	622
		D_{2h}	mmm	D_{3h}	$\bar{6}2m$	D_{4h}	$4/mmm$	D_{6h}	$6/mmm$
		D_{2d}	$\bar{4}2m$	D_{3d}	$\bar{3}m$	S_4	$\bar{4}/m$	S_6	$\bar{3}$
T	23	T_d	$\bar{4}3m$	T_h	$m3$				
O	432	O_h	$m3m$						

* Shoenflies notation in black, Hermann–Mauguin (International system) in blue. In the Hermann–Mauguin system, a number n denotes the presence of an n-fold axis and m denotes a mirror plane. A slash (/) indicates that the mirror plane is perpendicular to the symmetry axis. It is important to distinguish symmetry elements of the same type but of different classes, as in $4/mmm$, in which there are three classes of mirror plane. A bar over a number indicates that the element is combined with an inversion. The only groups listed here are the so-called 'crystallographic point groups'.

(a) The groups C_1, C_i, and C_s

Name	Elements
C_1	E
C_i	E, i
C_s	E, σ

A molecule belongs to the group C_1 if it has no element other than the identity. It belongs to C_i if it has the identity and the inversion alone, and to C_s if it has the identity and a mirror plane alone.

Brief illustration 11A.3 C_1, C_i, and C_s

The CBrClFI molecule (2) has only the identity element, and so belongs to the group C_1. Meso-tartaric acid (6) has the identity and inversion elements, and so belongs to the group C_i. Quinoline (7) has the elements (E,σ), and so belongs to the group C_s.

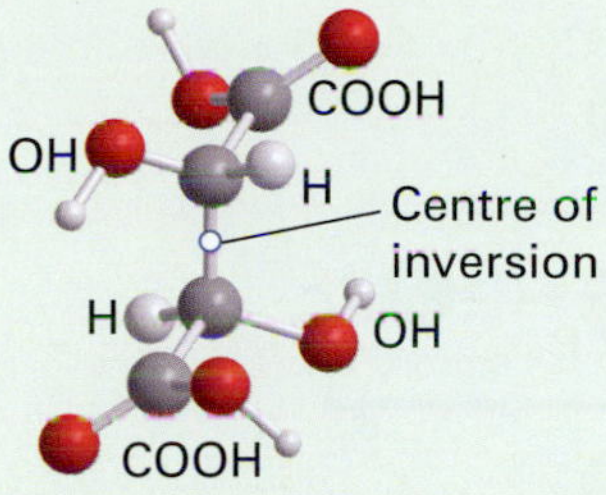

6 Meso-tartaric acid, HOOCCH(OH)CH(OH)COOH

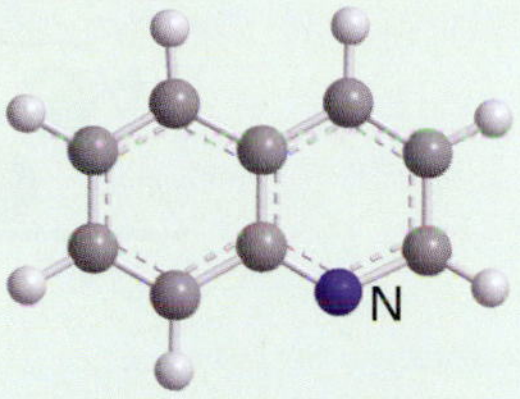

7 Quinoline, C_9H_7N

Self-test 11A.3 Identify the group to which the molecule (**8**) belongs.

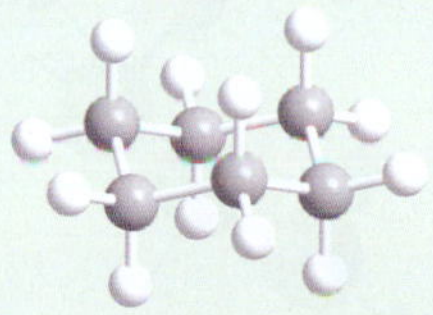

8

Answer: C_{2v}

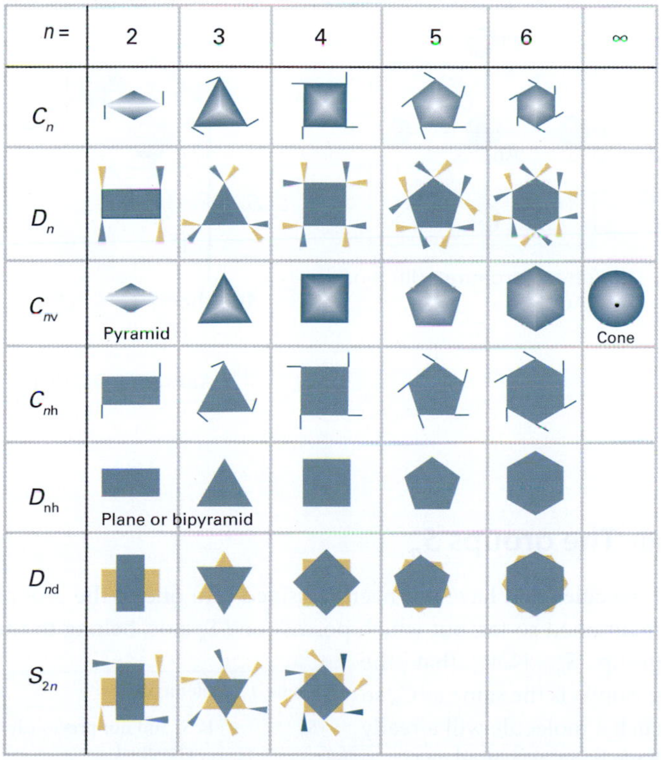

Figure 11A.8 A summary of the shapes corresponding to different point groups. The group to which a molecule belongs can often be identified from this diagram without going through the formal procedure in Fig. 11A.7.

(b) The groups C_n, C_{nv}, and C_{nh}

A molecule belongs to the group C_n if it possesses an n-fold axis. Note that symbol C_n is now playing a triple role: as the label of a symmetry element, a symmetry operation, and a group. If in addition to the identity and a C_n axis a molecule has n vertical mirror planes σ_v, then it belongs to the group C_{nv}. Objects that in addition to the identity and an n-fold principal axis also have a horizontal mirror plane σ_h belong to the groups C_{nh}. The presence of certain symmetry elements may be implied by the presence of others: thus, in C_{2h} the elements C_2 and σ_h jointly imply the presence of a centre of inversion (Fig. 11A.9). Note also that the tables specify the *elements*, not the *operations*: for instance, there are two operations associated with a single C_3 axis (rotations by +120° and −120°).

Name	Elements
C_n	E, C_n
C_{nv}	E, C_n, $n\sigma_v$
C_{nh}	E, C_n, σ_h

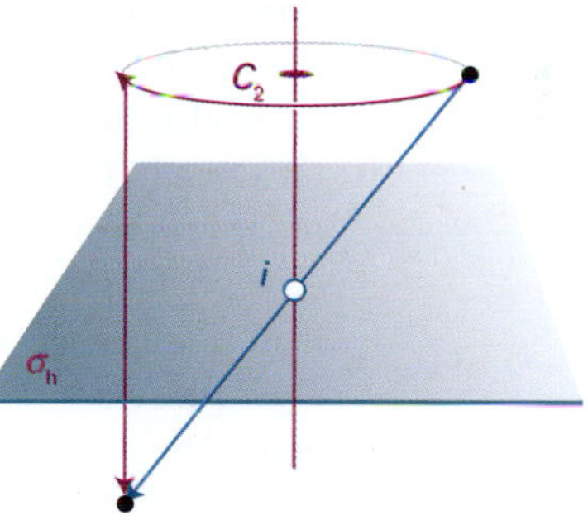

Figure 11A.9 The presence of a twofold axis and a horizontal mirror plane jointly imply the presence of a centre of inversion in the molecule.

Brief illustration 11A.4 C_n, C_{nv}, and C_{nh}

An H_2O_2 molecule (**9**) has the symmetry elements E and C_2, so belongs to the group C_2. An H_2O molecule has the symmetry elements E, C_2, and $2\sigma_v$, so it belongs to the group C_{2v}. An NH_3 molecule has the elements E, C_3, and $3\sigma_v$, so it belongs to the group C_{3v}. A heteronuclear diatomic molecule such as HCl belongs to the group $C_{\infty v}$ because rotations around the axis by any angle and reflections in all the infinite number of planes that contain the axis are symmetry operations. Other members of the group $C_{\infty v}$ include the linear OCS molecule and a cone. The molecule *trans*-CHCl=CHCl (**10**) has the elements E, C_2, and σ_h, so belongs to the group C_{2h}.

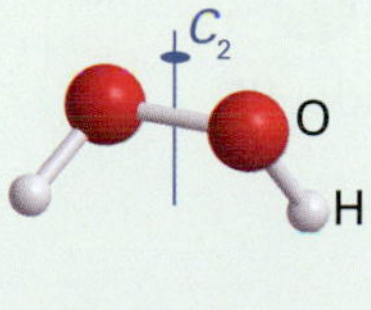

9 Hydrogen peroxide, H_2O_2

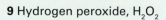

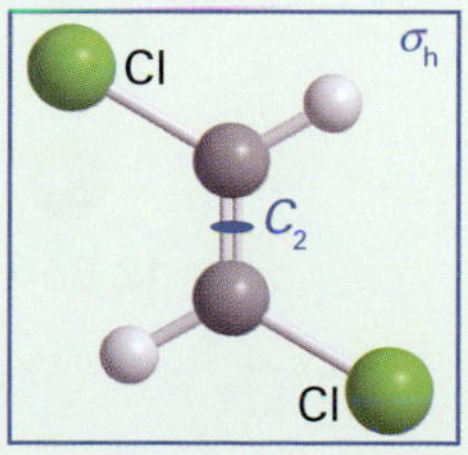

10 *trans*-CHCl=CHCl

Self-test 11A.4 Identify the group to which the molecule $B(OH)_3$ in the conformation shown in (**11**) belongs.

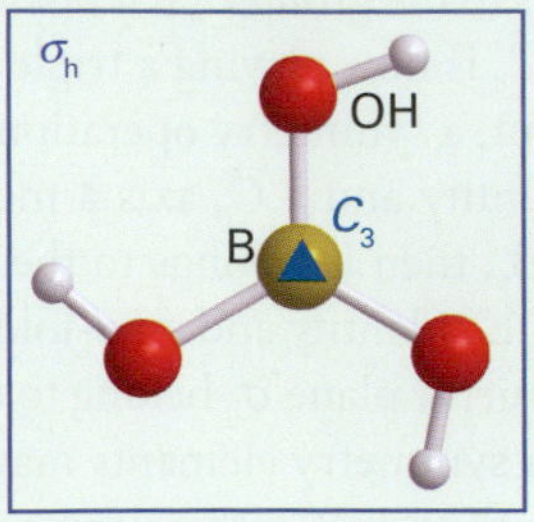

11 $B(OH)_3$

Answer: C_{3h}

(c) The groups D_n, D_{nh}, and D_{nd}

We see from Fig. 11A.7 that a molecule that has an n-fold principal axis and n twofold axes perpendicular to C_n belongs to the group D_n. A molecule belongs to D_{nh} if it also possesses a horizontal mirror plane. $D_{\infty h}$ is also the group of the linear OCO and HCCH molecules, and of a uniform cylinder. A molecule belongs to the group D_{nd} if in addition to the elements of D_n it possesses n dihedral mirror planes σ_d.

Name	Elements
D_n	E, C_n, nC_2'
D_{nh}	E, C_n, nC_2', σ_h
D_{nd}	$E, C_n, nC_2', n\sigma_d$

Brief illustration 11A.5 D_n, D_{nh}, and D_{nd}

The planar trigonal BF_3 molecule has the elements E, C_3, $3C_2$, and σ_h (with one C_2 axis along each B—F bond), so belongs to D_{3h} (**12**). The C_6H_6 molecule has the elements E, C_6, $3C_2$, $3C_2'$, and σ_h together with some others that these elements imply, so it belongs to D_{6h}. Three of the C_2 axes bisect C—C bonds and the other three pass through vertices of the hexagon formed by the carbon framework of the molecule. The prime on $3C_2'$ indicates that the three C_2 axes are different from the other three C_2 axes. All homonuclear diatomic molecules, such as N_2, belong to the group $D_{\infty h}$ because all rotations around the axis are symmetry operations, as are end-to-end rotation and end-to-end reflection. Another example of a D_{nh} species is (**13**). The twisted, 90° allene (**14**) belongs to D_{2d}.

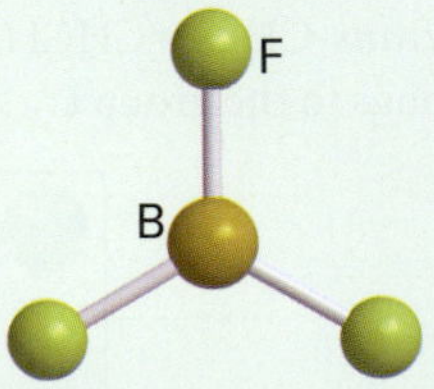

12 Boron trifluoride, BF_3

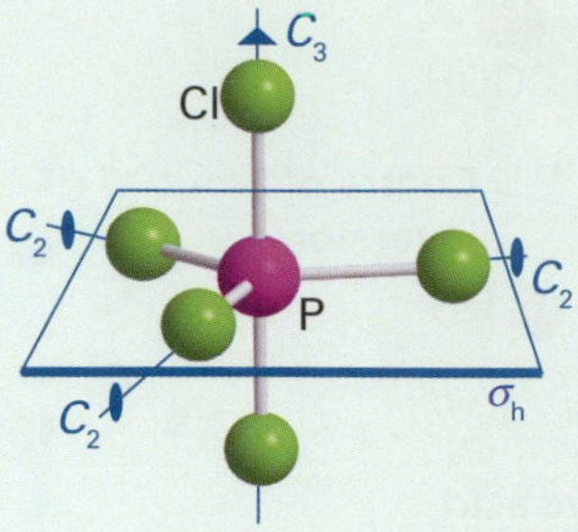

13 Phosphorus pentachloride, PCl_5 (D_{3h})

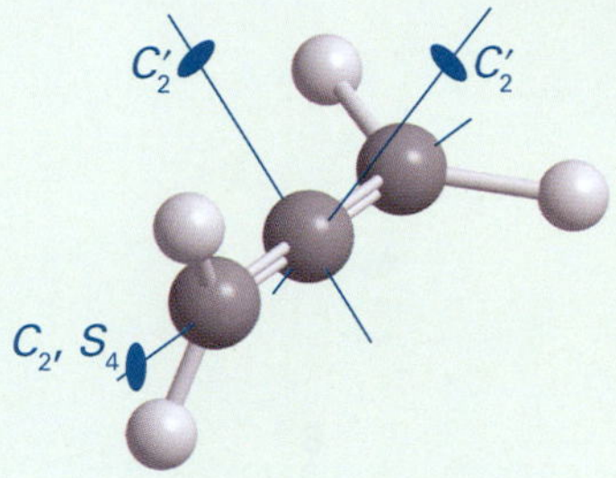

14 Allene, C_3H_4 (D_{2d})

Self-test 11A.5 Identify the groups to which (a) the tetrachloroaurate(III) ion (**15**) and (b) the staggered conformation of ethane (**16**) belong.

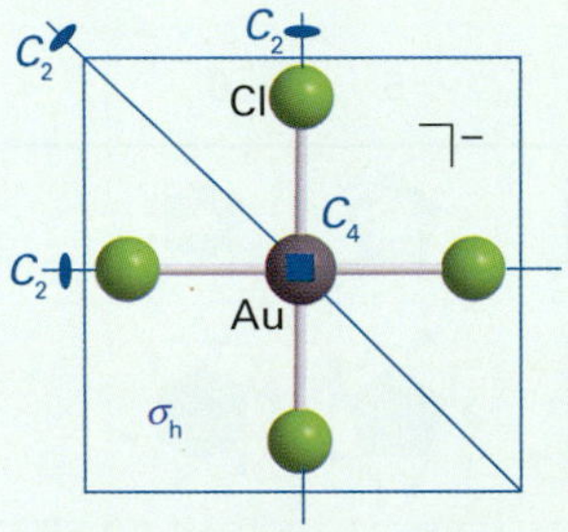

15 Tetrachloroaurate(III) ion, $[AuCl_4]^-$

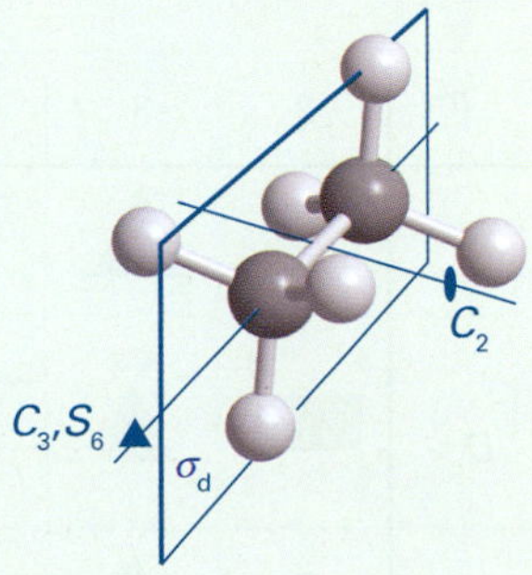

16 Ethane, C_2H_6 (D_{3d})

Answer: (a) D_{4h}, (b) D_{3d}

(d) The groups S_n

Molecules that have not been classified into one of the groups mentioned so far, but which possess one S_n axis, belong to the group S_n. Note that the group S_2 is the same as C_i, so such a molecule will already have been classified as C_i.

Name	Elements
S_n	E, S_n and not previously classified

Brief illustration 11A.6 S_n

Tetraphenylmethane (**17**) belongs to the point group S_4. Molecules belonging to S_n with $n>4$ are rare.

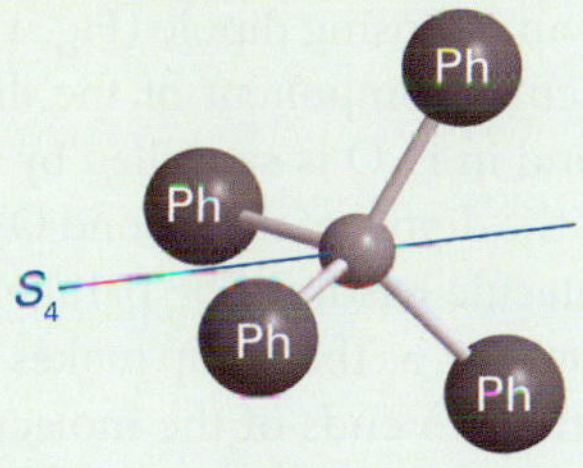

17 Tetraphenylmethane, $C(C_6H_5)_4$ (S_4)

Self-test 11A.6 Identify the group to which the ion in (**18**) belongs.

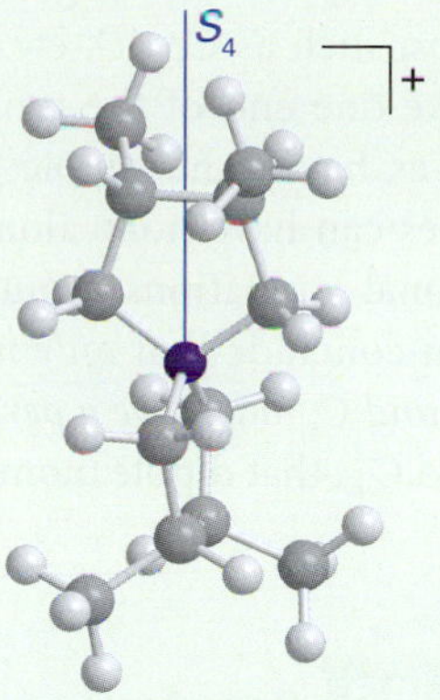

18 $N(CH_2CH(CH_3)CH(CH_3)CH_2)_2^+$

Answer: S_4

(e) The cubic groups

A number of very important molecules possess more than one principal axis. Most belong to the **cubic groups**, and in particular to the **tetrahedral groups** T, T_d, and T_h (Fig. 11A.10a) or to the **octahedral groups** O and O_h (Fig. 11A.10b). A few icosahedral (20-faced) molecules belonging to the **icosahedral group**, I (Fig. 11A.10c), are also known. The groups T_d and O_h are the groups of the regular tetrahedron and the regular octahedron, respectively. If the object possesses the rotational symmetry of the tetrahedron or the octahedron, but none of their planes of reflection, then it belongs to the simpler groups T or O (Fig. 11A.11). The group T_h is based on T but also contains a centre of inversion (Fig. 11A.12).

Name	Elements
T	$E, 4C_3, 3C_2$
T_d	$E, 3C_2, 4C_3, 3S_4, 6\sigma_d$
T_h	$E, 3C_2, 4C_3, i, 4S_6, 3\sigma_h$
O	$E, 3C_4, 4C_3, 6C_2$
O_h	$E, 3S_4, 3C_4, 6C_2, 4S_6, 4C_3, 3\sigma_h, 6\sigma_d, i$
I	$E, 6C_5, 10C_3, 15C_2$
I_h	$E, 6S_{10}, 10S_6, 6C_5, 10C_3, 15C_2, 15\sigma, i$

Brief illustration 11A.7 The cubic groups

The molecules CH_4 and SF_6 belong, respectively, to the groups T_d and O_h. Molecules belonging to the icosahedral group I include some of the boranes and buckminsterfullerene, C_{60} (**19**). The molecules shown in Fig. 11A.11 belong to the groups T and O, respectively.

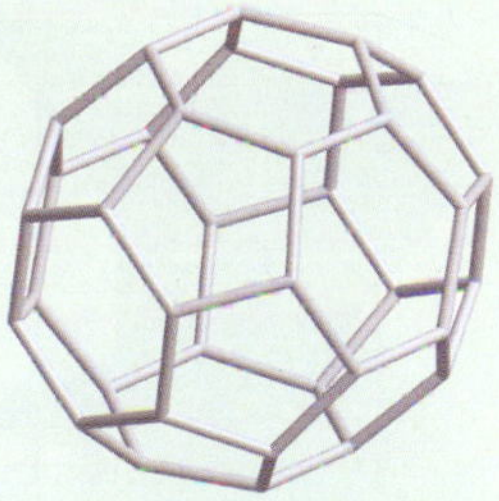

19 Buckminsterfullerene, C_{60} (I)

Self-test 11A.7 Identify the group to which the object shown in **20** belongs.

20

Answer: T_h

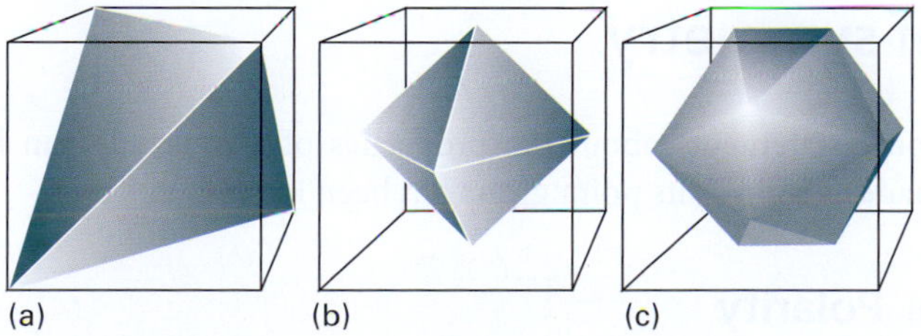

Figure 11A.10 (a) Tetrahedral, (b) octahedral, and (c) icosahedral molecules are drawn in a way that shows their relation to a cube: they belong to the cubic groups T_d, O_h, and I_h, respectively.

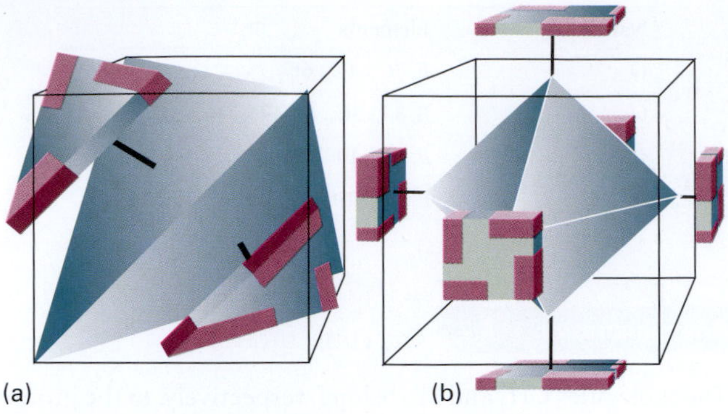

Figure 11A.11 Shapes corresponding to the point groups (a) *T* and (b) *O*. the presence of the decorated slabs reduces the symmetry of the object from T_d and O_h, respectively.

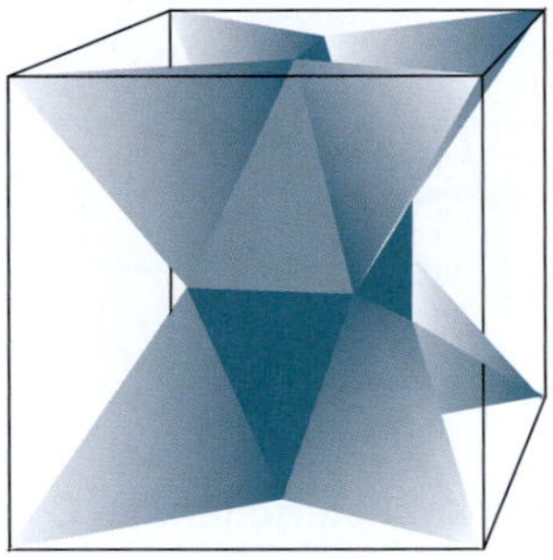

Figure 11A.12 The shape of an object belonging to the group T_h.

(f) The full rotation group

The **full rotation group**, R_3 (the 3 refers to rotation in three dimensions), consists of an infinite number of rotation axes with all possible values of n. A sphere and an atom belong to R_3, but no molecule does. Exploring the consequences of R_3 is a very important way of applying symmetry arguments to atoms, and is an alternative approach to the theory of orbital angular momentum.

Name	Elements
R_3	$E, \infty C_2, \infty C_3, \ldots$

11A.3 Some immediate consequences of symmetry

Some statements about the properties of a molecule can be made as soon as its point group has been identified.

(a) Polarity

A **polar molecule** is one with a permanent electric dipole moment (HCl, O_3, and NH_3 are examples). If the molecule belongs to the group C_n with $n>1$, it cannot possess a charge distribution with a dipole moment perpendicular to the symmetry axis because the symmetry of the molecule implies that any dipole that exists in one direction perpendicular to the axis is cancelled by an opposing dipole (Fig. 11A.13a). For example, the perpendicular component of the dipole associated with one O—H bond in H_2O is cancelled by an equal but opposite component of the dipole of the second O—H bond, so any dipole that the molecule has must be parallel to the twofold symmetry axis. However, as the group makes no reference to operations relating the two ends of the molecule, a charge distribution may exist that results in a dipole along the axis (Fig. 11A.13b), and H_2O has a dipole moment parallel to its twofold symmetry axis.

The same remarks apply generally to the group C_{nv}, so molecules belonging to any of the C_{nv} groups may be polar. In all the other groups, such as C_{3h}, D, etc., there are symmetry operations that take one end of the molecule into the other. Therefore, as well as having no dipole perpendicular to the axis, such molecules can have none along the axis, for otherwise these additional operations would not be symmetry operations. We can conclude that *only molecules belonging to the groups C_n, C_{nv}, and C_s may have a permanent electric dipole moment.* For C_n and C_{nv}, that dipole moment must lie along the symmetry axis.

Brief illustration 11A.8 Polar molecules

Ozone, O_3, which is angular and belongs to the group C_{2v}, may be polar (and is), but carbon dioxide, CO_2, which is linear and belongs to the group $D_{\infty h}$, is not.

Self-test 11A.8 Is tetraphenylmethane polar?

Answer: No (S_4)

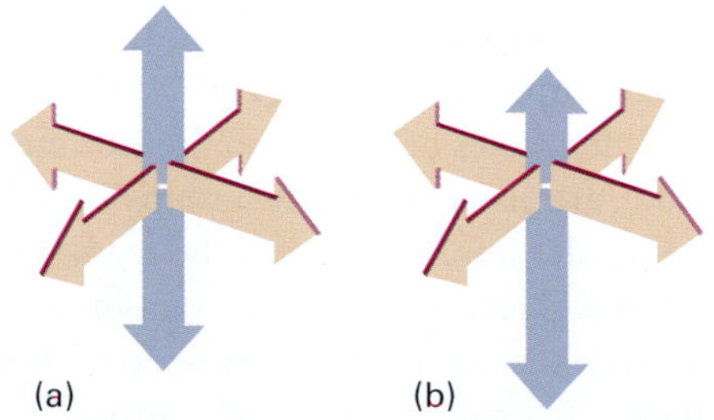

Figure 11A.13 (a) A molecule with a C_n axis cannot have a dipole perpendicular to the axis, but (b) it may have one parallel to the axis. The arrows represent local contributions to the overall electric dipole, such as may arise from bonds between pairs of neighbouring atoms with different electronegativities.

(b) Chirality

A **chiral molecule** (from the Greek word for 'hand') is a molecule that cannot be superimposed on its mirror image. An **achiral molecule** is a molecule that can be superimposed on its mirror image. Chiral molecules are **optically active** in the sense that they rotate the plane of polarized light. A chiral molecule and its mirror-image partner constitute an **enantiomeric pair** of isomers and rotate the plane of polarization in equal but opposite directions.

A molecule may be chiral, and therefore optically active, only if it does not possess an axis of improper rotation, S_n. We need to be aware that an S_n improper rotation axis may be present under a different name, and be implied by other symmetry elements that are present. For example, molecules belonging to the groups C_{nh} possess an S_n axis implicitly because they possess both C_n and σ_h, which are the two components of an improper rotation axis. Any molecule containing a centre of inversion, i, also possesses an S_2 axis, because i is equivalent to C_2 in conjunction with σ_h, and that combination of elements is S_2 (Fig. 11A.14). It follows that all molecules with centres of inversion are achiral and hence optically inactive. Similarly, because $S_1 = \sigma$, it follows that any molecule with a mirror plane is achiral.

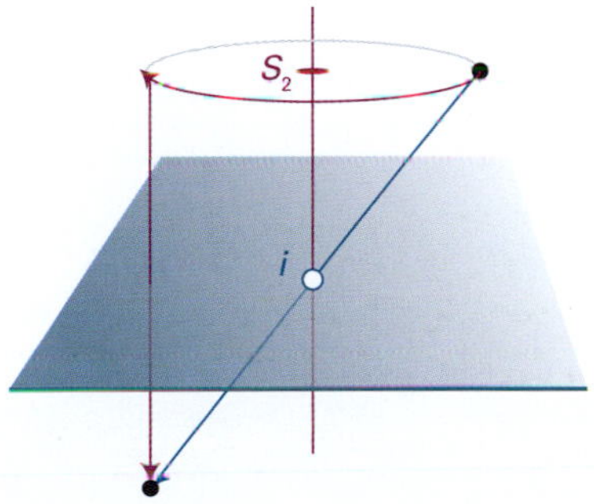

Figure 11A.14 Some symmetry elements are implied by the other symmetry elements in a group. Any molecule containing an inversion also possesses at least an S_2 element because i and S_2 are equivalent.

Brief illustration 11A.9 Chiral molecules

A molecule may be chiral if it does not have a centre of inversion or a mirror plane, which is the case with the amino acid alanine (**21**), but not with glycine (**22**). However, a molecule may be achiral even though it does not have a centre of inversion. For example, the S_4 species (**18**) is achiral and optically inactive: though it lacks i (that is, S_2) it does have an S_4 axis.

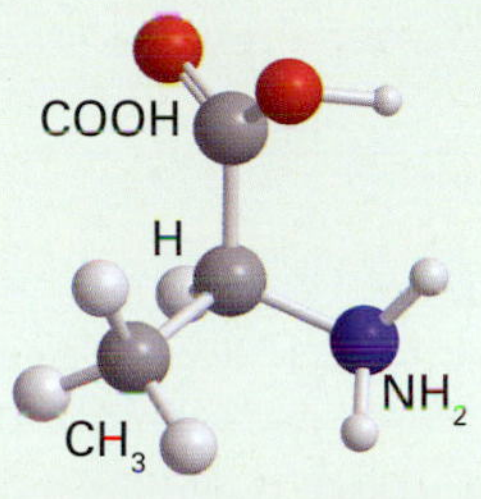

21 L-Alanine, $NH_2CH(CH_3)COOH$

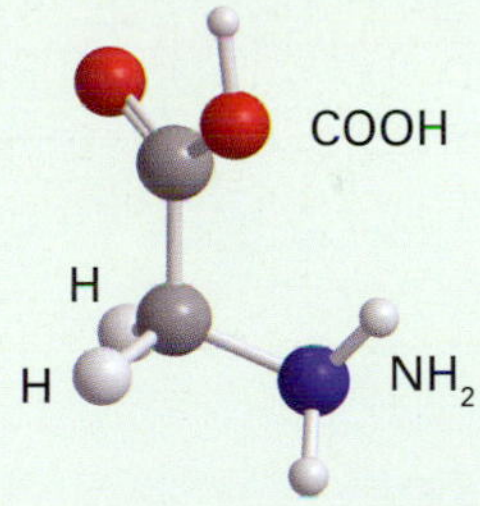

22 Glycine, NH_2CH_2COOH

Self-test 11A.9 Is tetraphenylmethane chiral?

Answer: No (S_4)

Checklist of concepts

- ☐ 1. A **symmetry operation** is an action that leaves an object looking the same after it has been carried out.
- ☐ 2. A **symmetry element** is a point, line, or plane with respect to which a symmetry operation is performed.
- ☐ 3. The **notation** for point groups commonly used for molecules and solids is summarized in Table 11A1.
- ☐ 4. To be **polar** a molecule must belong to C_n, C_{nv}, or C_s (and have no higher symmetry).
- ☐ 5. A molecule may be **chiral** only if it does not possess an axis of improper rotation, S_n.

Checklist of operations and elements

Symmetry operation	Symbol	Symmetry element
n-Fold rotation	C_n	n-fold axis of rotation
Reflection	σ	mirror plane
Inversion	i	centre of symmetry
n-Fold improper rotation	S_n	n-fold improper axis of rotation
Identity	E	entire object

11B Group theory

Contents

➤ Why do you need to know this material?

Group theory puts qualitative ideas about symmetry on to a systematic basis that can be applied to a wide variety of calculations; it is used to draw conclusions that might not be immediately obvious and as a result can greatly simplify calculations. It is also the basis of the labelling of atomic and molecular orbitals that is used throughout chemistry.

➤ What is the key idea?

Symmetry operations may be represented by the effect of matrices on a basis.

➤ What do you need to know already?

You need to know about the types of symmetry operation and element introduced in Topic 11A. This discussion draws heavily on matrix algebra, especially matrix multiplication, as set out in *Mathematical background* 6.

The systematic discussion of symmetry is called **group theory**. Much of group theory is a summary of common sense about the symmetries of objects. However, because group theory is systematic, its rules can be applied in a straightforward, mechanical way. In most cases the theory gives a simple, direct method for arriving at useful conclusions with the minimum of calculation, and this is the aspect we stress here. In some cases, though, they lead to unexpected results.

11B.1 The elements of group theory

A **group** in mathematics is a collection of transformations that satisfy four criteria. Thus, if we write the transformations as R, R', … (which we can think of as reflections, rotations, and so on, of the kind introduced in Topic 11A), then they form a group if:

1. One of the transformations is the identity (that is: 'do nothing').
2. For every transformation R, the inverse transformation R^{-1} is included in the collection so that the combination RR^{-1} (the transformation R^{-1} followed by R) is equivalent to the identity.
3. The combination RR' (the transformation R' followed by R) is equivalent to a single member of the collection of transformations.
4. The combination $R(R'R'')$, the transformation $(R'R'')$ followed by R, is equivalent to $(RR')R''$, the transformation R'' followed by (RR').

Example 11B.1 Showing that symmetry operations form a group

Show that $C_{2v}=\{E,C_2,2\sigma_v\}$ (specified by its elements) and consisting of the operations $\{E,C_2,\sigma_v,\sigma_v'\}$ is indeed a group in the mathematical sense.

Method We need to show that combinations of the operations match the criteria set out above. The operations are specified in Topic 11A.

Answer Criterion 1 is fulfilled because the collection of symmetry operations includes the identity E. Criterion 2 is fulfilled because in each case the inverse of an operation is the operation itself. Thus, two successive twofold rotations is equivalent to the identity: $C_2C_2=E$ and likewise for the two reflections and the identity itself. Criterion 3 is fulfilled, because in each case one operation followed by another is the same as one of the four symmetry operations. For instance, a twofold rotation C_2 followed by the reflection σ_v' is the same as the single reflection σ_v (Fig. 11B.1). Thus: $\sigma_v'C_2=\sigma_v$. Criterion 4 is fulfilled, as it is immaterial how the operations are grouped together. The following group multiplication table for the point group can be constructed similarly, where the entries are the product symmetry operations RR':

$R\downarrow R'\rightarrow$	E	C_2	σ_v	σ_v'
E	E	C_2	σ_v	σ_v'
C_2	C_2	E	σ_v'	σ_v
σ_v	σ_v	σ_v'	E	C_2
σ_v'	σ_v'	σ_v	C_2	E

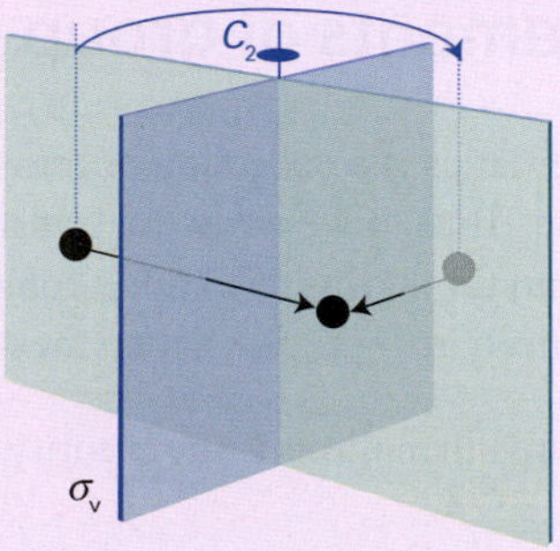

Figure 11B.1 A twofold rotation C_2 followed by the reflection σ_v' is the same as the single reflection σ_v.

Self-test 11B.1 Confirm that $C_{3v}=\{E,C_3,3\sigma_v\}$ and consisting of the operations $\{E,2C_3,3\sigma_v\}$ is a group.

Answer: Criteria are fulfilled

There is one potentially very confusing point that needs to be clarified at the outset. The entities that make up a group are its 'elements'. In chemistry, these elements are almost always symmetry operations. However, as explained in Topic 11A, we distinguish 'symmetry operations' from 'symmetry elements', the axes, planes, and so on with respect to which the operation is carried out. Finally, there is a third use of the word 'element', to denote the number lying in a particular location in a matrix. Be very careful to distinguish *element* (of a group), *symmetry element*, and *matrix element*.

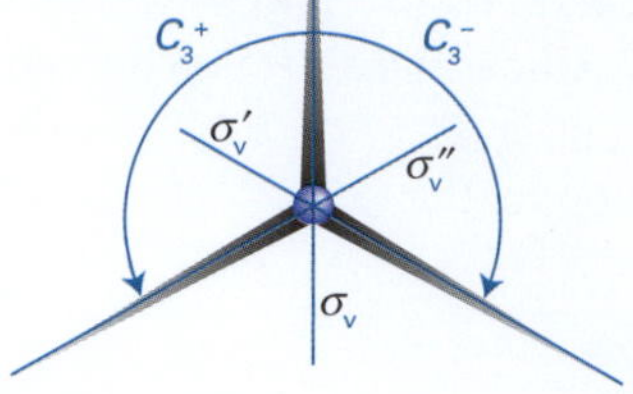

Figure 11B.2 Symmetry operations in the same class are related to one another by the symmetry operations of the group. Thus, the three mirror planes shown here are related by threefold rotations, and the two rotations shown here are related by reflection in σ_v.

Symmetry operations fall into the same **class** if they are of the same type (for example, rotations) and can be transformed into one another by a symmetry operation of the group. The two threefold rotations in C_{3v} belong to the same class because one can be converted into the other by a reflection (Fig. 11B.2); the three reflections all belong to the same class because each can be rotated into another by a threefold rotation. The formal definition of a class is that two operations R and R' belong to the same class if there is a member S of the group such that

$$R'=S^{-1}RS \qquad \text{Membership of a class} \qquad (11B.1)$$

where S^{-1} is the inverse of S.

Brief illustration 11B.1 Classes

To show that C_3^+ and C_3^- belong to the same class in C_{3v} (which intuitively we know to be the case as they are both rotations around the same axis), take $S=\sigma_v$. The reciprocal of a reflection is the reflection itself, so $\sigma_v^{-1}=\sigma_v$. It follows by using the relations derived to confirm the result of *Self-test* 11B.1 that

$$\overbrace{\sigma_v^{-1}}^{\sigma_v}C_3^+\sigma_v=\sigma_v\overbrace{C_3^+\sigma_v}^{\sigma_v'}=\overbrace{\sigma_v\sigma_v'}^{C_3^-}=C_3^-$$

Therefore, C_3^+ and C_3^- are related by an equation of the form of eqn 11B.1 and hence belong to the same class.

Self-test 11B.2 Show that the two reflections of the group C_{2v} fall into different classes.

Answer: No operation of the group takes $\sigma_v\rightarrow\sigma_v'$

11B.2 Matrix representations

Group theory takes on great power when the notional ideas presented so far are expressed in terms of collections of numbers in the form of matrices.

(a) Representatives of operations

Consider the set of three p orbitals shown on the C_{2v} SO_2 molecule in Fig. 11B.3. Under the reflection operation σ_v, the change $(p_S, p_B, p_A) \leftarrow (p_S, p_A, p_B)$ takes place. We can express this transformation by using matrix multiplication (*Mathematics background* 6):

$$(p_S, p_B, p_A) = (p_S, p_A, p_B)\overbrace{\begin{pmatrix} 1 & 0 & 0 \\ 0 & 0 & 1 \\ 0 & 1 & 0 \end{pmatrix}}^{D(\sigma_v)} = (p_S, p_A, p_B)D(\sigma_v) \qquad (11B.2a)$$

The matrix $D(\sigma_v)$ is called a **representative** of the operation σ_v. Representatives take different forms according to the **basis**, the set of orbitals that has been adopted. In this case, the basis is (p_S, p_A, p_B).

Brief illustration 11B.2 Representatives

We use the same technique to find matrices that reproduce the other symmetry operations. For instance, C_2 has the effect $(-p_S, -p_B, -p_A) \leftarrow (p_S, p_A, p_B)$, and its representative is

$$D(C_2) = \begin{pmatrix} -1 & 0 & 0 \\ 0 & 0 & -1 \\ 0 & -1 & 0 \end{pmatrix} \qquad (11B.2b)$$

The effect of σ'_v is $(-p_S, -p_A, -p_B) \leftarrow (p_S, p_A, p_B)$, and its representative is

$$D(\sigma'_v) = \begin{pmatrix} -1 & 0 & 0 \\ 0 & -1 & 0 \\ 0 & 0 & -1 \end{pmatrix} \qquad (11B.2c)$$

The identity operation has no effect on the basis, so its representative is the 3×3 unit matrix:

$$D(E) = \begin{pmatrix} 1 & 0 & 0 \\ 0 & 1 & 0 \\ 0 & 0 & 1 \end{pmatrix} \qquad (11B.2d)$$

Self-test 11B.3 Find the representative of the one remaining operation of the group, the reflection σ_v.

Answer: $D(\sigma_v) = \begin{pmatrix} 1 & 0 & 0 \\ 0 & 0 & 1 \\ 0 & 1 & 0 \end{pmatrix}$

(b) The representation of a group

The set of matrices that represents *all* the operations of the group is called a **matrix representation**, Γ (uppercase gamma),

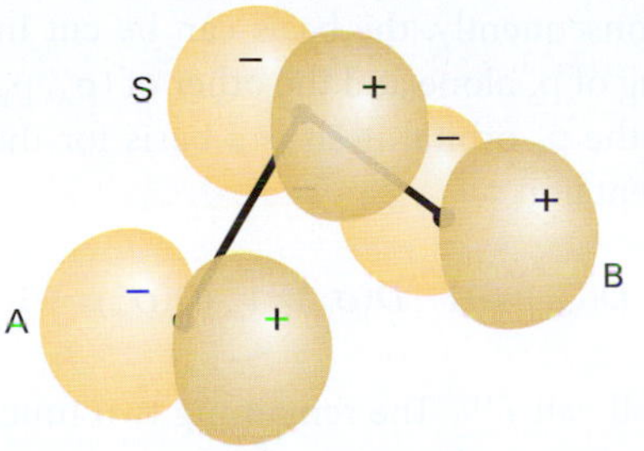

Figure 11B.3 The three p_x orbitals that are used to illustrate the construction of a matrix representation in a C_{2v} molecule (SO_2).

of the group for the basis that has been chosen. We denote this 'three-dimensional' representation (a representation consisting of 3×3 matrices) by $\Gamma^{(3)}$. The matrices of a representation multiply together in the same way as the operations they represent. Thus, if for any two operations R and R' we know that $RR' = R''$, then $D(R)D(R') = D(R'')$ for a given basis.

Brief illustration 11B.3 Matrix representations

In the group C_{2v}, a twofold rotation followed by a reflection in a mirror plane is equivalent to a reflection in the second mirror plane: specifically, $\sigma'_v C_2 = \sigma_v$. When we use the representatives specified above, we find

$$D(\sigma'_v)D(C_2) = \begin{pmatrix} -1 & 0 & 0 \\ 0 & -1 & 0 \\ 0 & 0 & -1 \end{pmatrix}\begin{pmatrix} -1 & 0 & 0 \\ 0 & 0 & -1 \\ 0 & -1 & 0 \end{pmatrix} = \begin{pmatrix} 1 & 0 & 0 \\ 0 & 0 & 1 \\ 0 & 1 & 0 \end{pmatrix}$$
$$= D(\sigma_v)$$

This multiplication reproduces the group multiplication. The same is true of all pairs of representative multiplications, so the four matrices form a representation of the group.

Self-test 11B.4 Confirm the result that $\sigma_v \sigma'_v = C_2$ by using the matrix representations developed here.

The discovery of a matrix representation of the group means that we have found a link between symbolic manipulations of operations and algebraic manipulations of numbers.

(c) Irreducible representations

Inspection of the representatives shows that they are all of **block-diagonal form**:

$$D = \begin{pmatrix} \blacksquare & 0 & 0 \\ 0 & \blacksquare & \blacksquare \\ 0 & \blacksquare & \blacksquare \end{pmatrix} \qquad \text{Block-diagonal form} \qquad (11B.3)$$

The block-diagonal form of the representatives shows us that the symmetry operations of C_{2v} never mix p_S with the other two functions. Consequently, the basis can be cut into two parts, one consisting of p_S alone and the other of (p_A, p_B). It is readily verified that the p_S orbital itself is a basis for the one-dimensional representaion

$$D(E)=1 \quad D(C_2)=-1 \quad D(\sigma_v)=1 \quad D(\sigma_v')=-1$$

which we shall call $\Gamma^{(1)}$. The remaining two functions (p_A, p_B) are a basis for the two-dimensional representation $\Gamma^{(2)}$:

$$D(E)=\begin{pmatrix}1 & 0\\ 0 & 1\end{pmatrix} \quad D(C_2)=\begin{pmatrix}0 & -1\\ -1 & 0\end{pmatrix}$$
$$D(\sigma_v)=\begin{pmatrix}0 & 1\\ 1 & 0\end{pmatrix} \quad D(\sigma_v')=\begin{pmatrix}-1 & 0\\ 0 & -1\end{pmatrix}$$

These matrices are the same as those of the original three-dimensional representation, except for the loss of the first row and column. We say that the original three-dimensional representation has been **reduced** to the 'direct sum' of a one-dimensional representation 'spanned' by p_S, and a two-dimensional representation spanned by (p_A, p_B). This reduction is consistent with the common sense view that the central orbital plays a role different from the other two. We denote the reduction symbolically by writing[1]

$$\Gamma^{(3)} = \Gamma^{(1)} + \Gamma^{(2)} \qquad \text{Direct sum} \qquad (11B.4)$$

The one-dimensional representation $\Gamma^{(1)}$ cannot be reduced any further, and is called an **irreducible representation** of the group (an 'irrep'). We can demonstrate that the two-dimensional representation $\Gamma^{(2)}$ is reducible (for this basis in this group) by switching attention to the linear combinations $p_1=p_A+p_B$ and $p_2=p_A-p_B$. These combinations are sketched in Fig. 11B.4. The representatives in the new basis can be constructed from the old by noting, for example, that under σ_v, $(p_B, p_A) \leftarrow (p_A, p_B)$. In this way we find the following representation in the new basis:

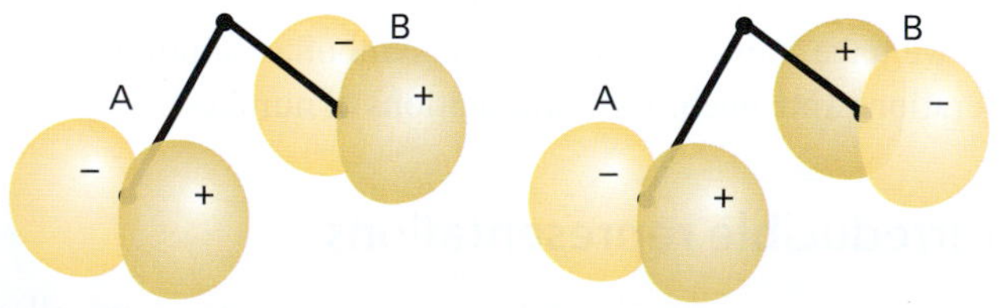

Figure 11B.4 Two symmetry-adapted linear combinations of the basis orbitals shown in Fig. 11B.3. The two combinations each span a one-dimensional irreducible representation, and their symmetry species are different.

[1] The symbol ⊕ is sometimes used to denote a direct sum to distinguish it from an ordinary sum, in which case eqn 11B.4 would be written $\Gamma^{(3)}=\Gamma^{(1)}\oplus\Gamma^{(2)}$.

$$D(E)=\begin{pmatrix}1 & 0\\ 0 & 1\end{pmatrix} \quad D(C_2)=\begin{pmatrix}-1 & 0\\ 0 & 1\end{pmatrix}$$
$$D(\sigma_v)=\begin{pmatrix}1 & 0\\ 0 & -1\end{pmatrix} \quad D(\sigma_v')=\begin{pmatrix}-1 & 0\\ 0 & -1\end{pmatrix}$$

The new representatives are all in block-diagonal form, in this case $\begin{pmatrix}\blacksquare & 0\\ 0 & \blacksquare\end{pmatrix}$, and the two combinations are not mixed with each other by any operation of the group. We have therefore achieved the reduction of $\Gamma^{(2)}$ to the sum of two one-dimensional representations. Thus, p_1 spans

$$D(E)=1 \quad D(C_2)=-1 \quad D(\sigma_v)=1 \quad D(\sigma_v')=-1$$

which is the same one-dimensional representation as that spanned by p_S, and p_2 spans

$$D(E)=1 \quad D(C_2)=1 \quad D(\sigma_v)=-1 \quad D(\sigma_v')=-1$$

which is a different one-dimensional representation; we shall denote these two representations $\Gamma^{(1)\prime}$ and $\Gamma^{(1)\prime\prime}$, respectively. At this stage we have reduced the original representation as follows:

$$\Gamma^{(3)} = \Gamma^{(1)} + \Gamma^{(1)\prime} + \Gamma^{(1)\prime\prime}$$

(d) Characters and symmetry species

The **character**, χ (chi), of an operation in a particular matrix representation is the sum of the diagonal elements of the representative of that operation. Thus, in the original basis we are using, the characters of the representatives are

R	E	C_2	σ_v	σ_v'
$D(R)$	$\begin{pmatrix}1&0&0\\0&1&0\\0&0&1\end{pmatrix}$	$\begin{pmatrix}-1&0&0\\0&0&-1\\0&-1&0\end{pmatrix}$	$\begin{pmatrix}1&0&0\\0&0&1\\0&1&0\end{pmatrix}$	$\begin{pmatrix}-1&0&0\\0&-1&0\\0&0&-1\end{pmatrix}$
$\chi(R)$	3	−1	1	−3

The characters of one-dimensional representatives are just the representatives themselves. The sum of the characters of the reduced representation is unchanged by the reduction:

R	E	C_2	σ_v	σ_v'
$\chi(R)$ for $\Gamma^{(1)}$	1	−1	1	−1
$\chi(R)$ for $\Gamma^{(1)\prime}$	1	−1	1	−1
$\chi(R)$ for $\Gamma^{(1)\prime\prime}$	1	1	−1	−1
Sum:	3	−1	1	−3

Although the notation $\Gamma^{(n)}$ can be used for general representations, it is common in chemical applications of group theory to use the labels A, B, E, and T to denote the **symmetry species** of the representation:

A: one-dimensional representation, character +1 under the principal rotation

B: one-dimensional representation, character −1 under the principal rotation

E: two-dimensional irreducible representation

T: three-dimensional irreducible representation

Subscripts are used to distinguish the irreducible representations if there is more than one of the same type: A_1 is reserved for the representation with character +1 for all operations; A_2 has +1 for the principal rotation but −1 for reflections. All the irreducible representations of C_{2v} are one-dimensional, and the table above is labelled as follows:

Symmetry species	E	C_2	σ_v	σ'_v
B_2	1	−1	1	−1
B_1	1	−1	1	−1
A_2	1	1	−1	−1

At this point we have found three irreducible representations of the group C_{2v}. Are these the only irreducible representations of the group C_{2v}? There is in fact only one more species of irreducible representations of this group, for a surprising theorem of group theory states that

$$\text{Number of symmetry species} = \text{number of classes}$$

Number of species (11B.5)

In C_{2v}, for instance, there are four classes of operation (four columns in the character table), so there are only four species of irreducible representation. The character table therefore shows the characters of all the irreducible representations of this group. Another powerful result relates the sum of the dimensions, d_i, of all the symmetry species $\Gamma^{(i)}$ to the **order** of the group, the total number of symmetry operations, h:

$$\sum_{\text{Species } i} d_i^2 = h$$

Dimensionality and order (11B.6)

Brief illustration 11B.4 Symmetry species

There are three classes of operation in the group C_{3v} with operations $\{E,2C_3,3\sigma_v\}$, so there are three symmetry species (they turn out to be A_1, A_2, and E). The order of the group is 6, so if we already knew that two of the symmetry species are one dimensional, we could infer that the remaining irreducible representation is two-dimensional (E) from $1^2+1^2+d^2=6$.

Self-test 11B.5 How many symmetry species are there for the group T_d, with operations $\{E,8C_3,3C_2,6\sigma_d,6S_4\}$? Can you infer their dimensionalities?

Answer: 5 species; 2A+E+2T for $h=24$

11B.3 Character tables

The tables we have been constructing are called **character tables** and from now on move to the centre of the discussion. The columns of a character table are labelled with the symmetry operations of the group. For instance, for the group C_{3v} the columns are headed E, $2C_3$, and $3\sigma_v$ (Table 11B.1). The numbers multiplying each operation are the numbers of members of each class. The rows under the labels for the operations summarize the symmetry properties of the orbitals. They are labelled with the symmetry species.

(a) Character tables and orbital degeneracy

The character of the identity operation E tells us the degeneracy of the orbitals. Thus, in a C_{3v} molecule, any orbital with a symmetry label A_1 or A_2 is non-degenerate. Any doubly degenerate pair of orbitals in C_{3v} must be labelled E because, in this group, only E symmetry species have characters greater than 1. (Take care to distinguish the identity operation E (italic, a column heading) from the symmetry label E (roman, a row label).)

Because there are no characters greater than 2 in the column headed E in C_{3v}, we know that there can be no triply degenerate orbitals in a C_{3v} molecule. This last point is a powerful result of group theory, for it means that with a glance at the character table of a molecule, we can state the maximum possible degeneracy of its orbitals.

Table 11B.1* The C_{3v} character table

C_{3v}, $3m$	E	$2C_3$	$3\sigma_v$	$h=6$	
A_1	1	1	1	z	z^2, x^2+y^2
A_2	1	1	−1		
E	2	−1	0	(x, y)	(xy, x^2-y^2), (yz, zx)

* More character tables are given in the *Resource section*.

Example 11B.2 Using a character table to judge degeneracy

Can a trigonal planar molecule such as BF_3 have triply degenerate orbitals? What is the minimum number of atoms from which a molecule can be built that does display triple degeneracy?

Method First identify the point group, and then refer to the corresponding character table in the *Resource section*. The maximum number in the column headed by the identity E is the maximum orbital degeneracy possible in a molecule of that point group. For the second part, consider the shapes that can be built from two, three, etc. atoms, and decide which number can be used to form a molecule that can have orbitals of symmetry species T.

Answer Trigonal planar molecules belong to the point group D_{3h}. Reference to the character table for this group shows that the maximum degeneracy is 2, as no character exceeds 2 in the column headed E. Therefore, the orbitals cannot be triply degenerate. A tetrahedral molecule (symmetry group T) has an irreducible representation with a T symmetry species. The minimum number of atoms needed to build such a molecule is four (as in P_4, for instance).

Self-test 11B.6 A buckminsterfullerene molecule, C_{60}, belongs to the icosahedral point group. What is the maximum possible degree of degeneracy of its orbitals?

Answer: 5

(b) The symmetry species of atomic orbitals

The characters in the rows labelled A and B and in the columns headed by symmetry operations other than the identity E indicate the behaviour of an orbital under the corresponding operations: a +1 indicates that an orbital is unchanged, and a −1 indicates that it changes sign. It follows that we can identify the symmetry label of the orbital by comparing the changes that occur to an orbital under each operation, and then comparing the resulting +1 or −1 with the entries in a row of the character table for the point group concerned. By convention, irreducible representations are labelled with upper case Roman letters (such as A_1 and E) and the orbitals to which they apply are labelled with the lower case equivalents (so an orbital of symmetry species A_1 is called an a_1 orbital). Examples of each type of orbital are shown in Fig. 11B.5.

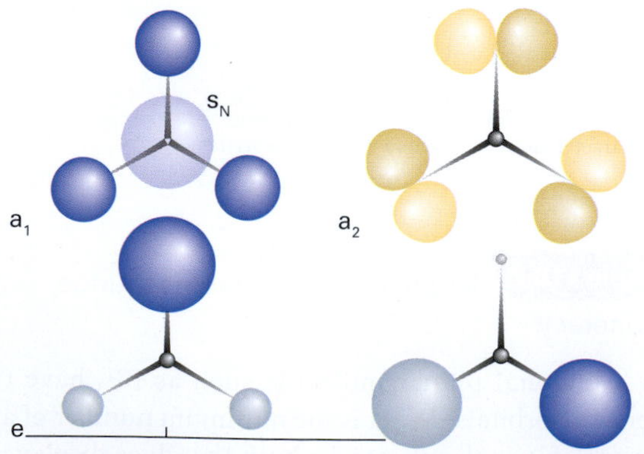

Figure 11B.5 Typical symmetry-adapted linear combinations of orbitals in a C_{3v} molecule.

Brief illustration 11B.5 Symmetry species of atomic orbitals

Consider the O2p$_x$ orbital in H_2O (the x-axis is perpendicular to the molecular plane; the y-axis is parallel to the H–H direction; the z-axis bisects the HOH angle). Because H_2O belongs to the point group C_{2v}, we know by referring to the C_{2v} character table (Table 11B.2) that the labels available for the orbitals are a_1, a_2, b_1, and b_2. We can decide the appropriate label for O2p$_x$ by noting that under a 180° rotation (C_2) the orbital changes sign (Fig. 11B.6), so it must be either B_1 or B_2, as only these two symmetry types have character −1 under C_2. The O2p$_x$ orbital also changes sign under the reflection σ'_v, which identifies it as B_1. As we shall see, any molecular orbital built from this atomic orbital will also be a b_1 orbital. Similarly, O2p$_y$ changes sign under C_2 but not under σ'_v; therefore, it can contribute to b_2 orbitals.

Table 11B.2* The C_{2v} character table

C_{2v}, $2mm$	E	C_2	σ_v	σ'_v	$h=4$	
A_1	1	1	1	1	z	z^2, y^2, x^2
A_2	1	1	−1	−1		xy
B_1	1	−1	1	−1	x	zx
B_2	1	−1	−1	1	y	yz

* More character tables are given in the *Resource section*.

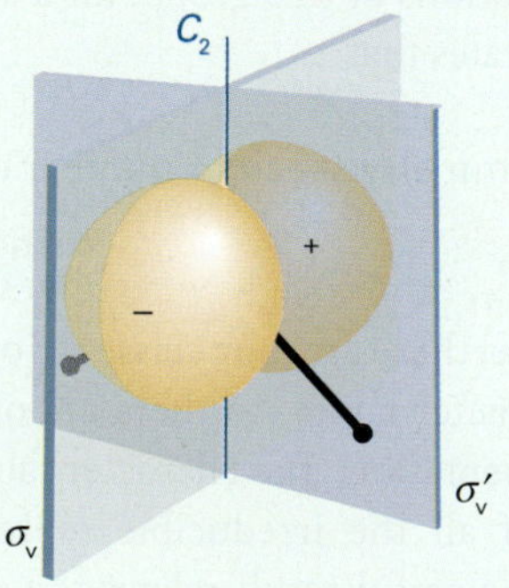

Figure 11B.6 A p$_x$ orbital on the central atom of a C_{2v} molecule and the symmetry elements of the group.

Self-test 11B.7 Identify the symmetry species of d orbitals on the central atom of a square-planar (D_{4h}) complex.

Answer: $A_{1g}+B_{1g}+B_{2g}+E_g$

For the rows labelled E or T (which refer to the behaviour of sets of doubly and triply degenerate orbitals, respectively), the characters in a row of the table are the sums of the characters summarizing the behaviour of the individual orbitals in the basis. Thus, if one member of a doubly degenerate pair remains unchanged under a symmetry operation but the other changes sign (Fig. 11B.7), then the entry is reported as $\chi=1-1=0$. Care must be exercised with these characters because the

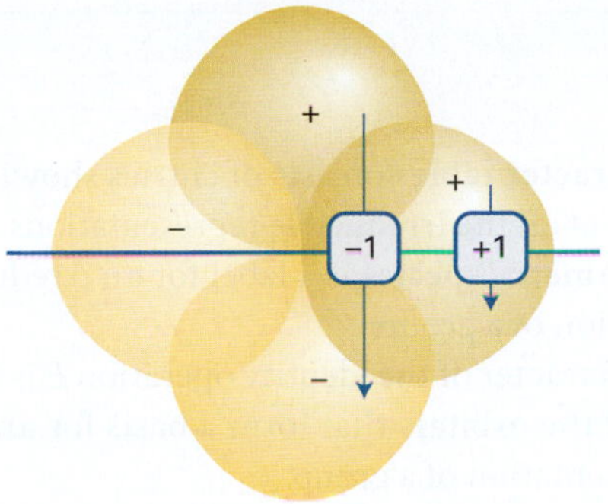

Figure 11B.7 The two orbitals shown here have different properties under reflection through the mirror plane: one changes sign (character −1), the other does not (character +1).

transformations of orbitals can be quite complicated; nevertheless, the sums of the individual characters are integers.

The behaviour of s, p, and d orbitals on a central atom under the symmetry operations of the molecule is so important that the symmetry species of these orbitals are generally indicated in a character table. To make these allocations, we look at the symmetry species of x, y, and z, which appear on the right hand side of the character table. Thus, the position of z in Table 11B.1 shows that p_z (which is proportional to $zf(r)$), has symmetry species A_1 in C_{3v}, whereas p_x and p_y (which are proportional to $xf(r)$ and $yf(r)$, respectively) are jointly of E symmetry. In technical terms, we say that p_x and p_y jointly **span** an irreducible representation of symmetry species E. An s orbital on the central atom always spans the fully symmetrical irreducible representation (typically labelled A_1 but sometimes A_1') of a group as it is unchanged under all symmetry operations.

The five d orbitals of a shell are represented by xy for d_{xy}, etc., and are also listed on the right of the character table. We can see at a glance that in C_{3v}, d_{xy} and $d_{x^2-y^2}$ on a central atom jointly belong to E and hence form a doubly degenerate pair.

(c) The symmetry species of linear combinations of orbitals

So far, we have dealt with the symmetry classification of individual orbitals. The same technique may be applied to linear combinations of orbitals on atoms that are related by symmetry transformations of the molecule, such as the combination $\psi_1 = \psi_A + \psi_B + \psi_C$ of the three H1s orbitals in the C_{3v} molecule NH_3 (Fig. 11B.8). This combination remains unchanged under a C_3 rotation and under any of the three vertical reflections of the group, so its characters are

$$\chi(E)=1 \quad \chi(C_3)=1 \quad \chi(\sigma_v)=-1$$

Comparison with the C_{3v} character table shows that ψ_1 is of symmetry species A_1, and therefore that it contributes to a_1 molecular orbitals in NH_3.

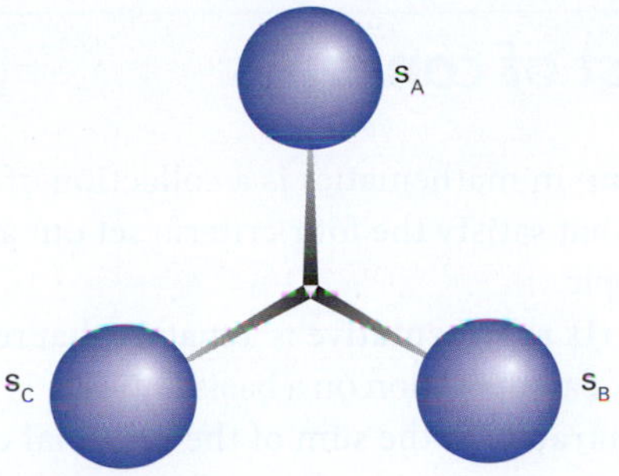

Figure 11B.8 The three H1s orbitals used to construct symmetry-adapted linear combinations in a C_{3v} molecule such as NH_3.

Example 11B.3 Identifying the symmetry species of orbitals

Identify the symmetry species of the orbital $\psi = \psi_A - \psi_B$ in a C_{2v} NO_2 molecule, where ψ_A is an $O2p_x$ orbital on one O atom and ψ_B that on the other O atom.

Method The negative sign in ψ indicates that the sign of ψ_B is opposite to that of ψ_A. We need to consider how the combination changes under each operation of the group, and then write the character as +1, −1, or 0 as specified above. Then we compare the resulting characters with each row in the character table for the point group, and hence identify the symmetry species.

Answer The combination is shown in Fig. 11B.9. Under C_2, ψ changes into itself, implying a character of +1. Under the reflection σ_v, both orbitals change sign, so $\psi \rightarrow -\psi$, implying a character of −1. Under σ_v', $\psi \rightarrow -\psi$, so the character for this operation is also −1. The characters are therefore

$$\chi(E)=1 \quad \chi(C_2)=1 \quad \chi(\sigma_v)=-1 \quad \chi(\sigma_v')=-1$$

These values match the characters of the A_2 symmetry species, so ψ can contribute to an a_2 orbital.

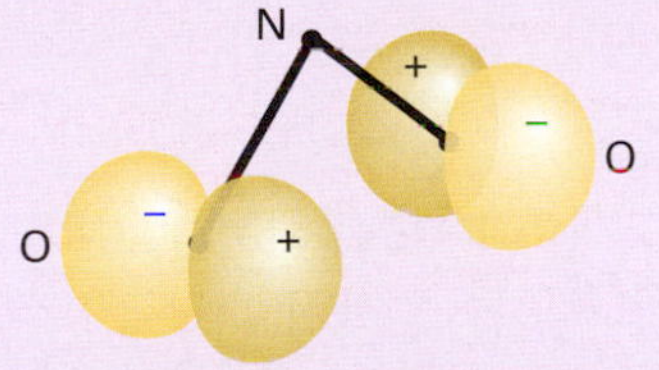

Figure 11B.9 One symmetry-adapted linear combination of $O2p_x$ orbitals in the C_{2v} NO_2 molecule.

Self-test 11B.8 Consider $PtCl_4^-$, in which the Cl ligands form a square planar array of point group D_{4h} (**1**). Identify the symmetry type of the combination $\psi_A - \psi_B + \psi_C - \psi_D$.

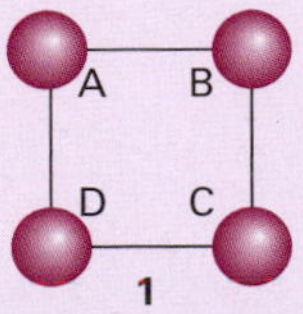

1

Answer: B_{2g}

Checklist of concepts

- ☐ 1. A **group** in mathematics is a collection of transformations that satisfy the four criteria set out at the start of the Topic.
- ☐ 2. A **matrix representative** is a matrix that represents the effect of an operation on a basis.
- ☐ 3. The **character** is the sum of the diagonal elements of a matrix representative of an operation.
- ☐ 4. A **matrix representation** is the collection of matrix representatives for the operations in the group.
- ☐ 5. A **character table** consists of entries showing the characters of all the irreducible representations of a group.
- ☐ 6. A **symmetry species** is a label for an irreducible representation of a group.
- ☐ 7. The character of the identity operation E is the degeneracy of the orbitals that form a basis for an irreducible representation of a group.

Checklist of equations

Property	Equation	Comment	Equation number
Class membership	$R' = S^{-1}RS$	All elements are members of the group	11B.1
Number of species rule	Number of symmetry species = number of classes		11B.5
Character and order	$\sum_{\text{Species } i} d_i^2 = h$	h is the order of the group	11B.6

11C Applications of symmetry

Contents

➤ **Why do you need to know this material?**

This Topic explains how the concepts introduced in Topics 11A and 11B are put to use. The arguments here are essential for understanding how molecular orbitals are constructed and underlie the whole of spectroscopy.

➤ **What is the key idea?**

An integral is invariant under symmetry transformations of a molecule.

➤ **What do you need to know already?**

This Topic develops the material that began in Topic 11A, where the symmetry classification of molecules is introduced on the basis of their symmetry elements, and draws heavily on the properties of characters and character tables described in Topic 11B. You need to be aware that many quantum-mechanical properties, including transition probabilities (Topic 9C), depend on integrals over pairs of wavefunctions (Topic 7C).

Group theory shows its power when brought to bear on a variety of problems in chemistry, among them the construction of molecular orbitals and the formulation of spectroscopic selection rules. This Topic describes these two applications after establishing a general result relating to integrals. In Topic 7C it is explained how integrals ('matrix elements') are central to the formulation of quantum mechanics, and knowing with very little calculation that various integrals are necessarily zero can save a great deal of calculational effort as well as adding to insight about the origin of properties.

11C.1 Vanishing integrals

An integral, which we shall denote I, in one dimension is equal to the area beneath the curve. In higher dimensions, it is equal to volume and various generalizations of volume. The key point is that the value of the area, volume, etc. is independent of the orientation of the axes used to express the function being integrated, the 'integrand' (Fig. 11C.1). In group theory we express this point by saying that *I is invariant under any symmetry operation*, and that each symmetry operation brings about the trivial transformation $I \rightarrow I$.

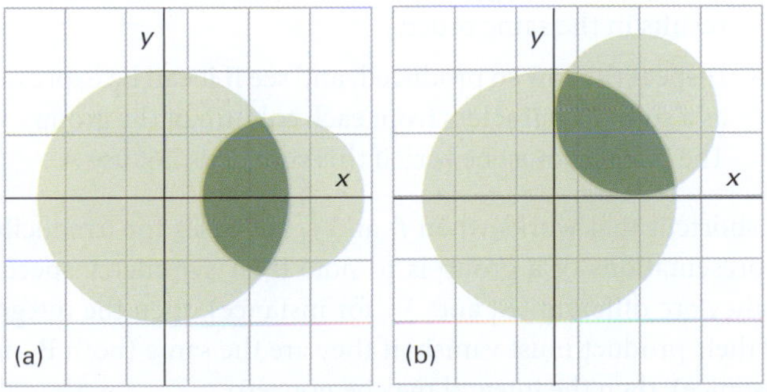

Figure 11C.1 The value of an integral I (for example, an area) is independent of the coordinate system used to evaluate it. That is, I is a basis of a representation of symmetry species A_1 (or its equivalent).

(a) Integrals over the product of two functions

Suppose we had to evaluate the integral

$$I = \int f_1 f_2 \mathrm{d}\tau \tag{11C.1}$$

where f_1 and f_2 are functions and the integration is over all space. For example, f_1 might be an atomic orbital A on one atom and f_2 an atomic orbital B on another atom, in which case I would be their overlap integral. If we knew that the integral is zero, we could say at once that a molecular orbital does not result from (A,B) overlap in that molecule. We shall now see that the character tables introduced in Topic 11B provide a quick way of judging whether an integral is necessarily zero.

The volume element $\mathrm{d}\tau$ is invariant under any symmetry operation. It follows that the integral is nonzero only if the integrand itself, the product $f_1 f_2$, is unchanged by any symmetry operation of the molecular point group. If the integrand changed sign under a symmetry operation, the integral would be the sum of equal and opposite contributions, and hence would be zero. It follows that the only contribution to a nonzero integral comes from functions for which under any symmetry operation of the molecular point group $f_1 f_2 \rightarrow f_1 f_2$, and hence for which the characters of the operations are all equal to +1. Therefore, for I not to be zero, *the integrand $f_1 f_2$ must have symmetry species A_1* (or its equivalent in the specific molecular point group).

The following procedure is used to deduce the symmetry species spanned by the product $f_1 f_2$ and hence to see whether it does indeed span A_1.

- Identify the symmetry species of the individual functions f_1 and f_2 by reference to the character table for the molecular point group in question and write their characters in two rows in the same order as in the table.
- Multiply the two numbers in each column, writing the results in the same order.
- Inspect the row so produced, and see if it can be expressed as a sum of characters from each column of the group. The integral must be zero if this sum does not use A_1.

A shortcut that works when f_1 and f_2 are bases for irreducible representations of a group is to note their symmetry species; if they are different (B_1 and A_2, for instance), then the integral of their product must vanish; if they are the same (both B_1, for instance), then the integral may be nonzero.

It is important to note that group theory is specific about when an integral must be zero, but integrals that it allows to be nonzero may be zero for reasons unrelated to symmetry. For example, the N–H distance in ammonia may be so great that the (s_1,s_N) overlap integral is zero simply because the orbitals are so far apart.

Example 11C.1 Deciding if an integral must be zero 1

May the integral of the function $f=xy$ be nonzero when evaluated over a region the shape of an equilateral triangle centred on the origin (Fig. 11C.2)?

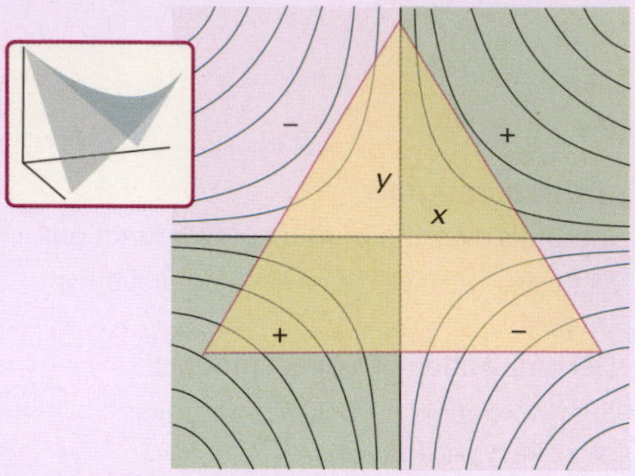

Figure 11C.2 The integral of the function $f=xy$ over the tinted region is zero. In this case, the result is obvious by inspection, but group theory can be used to establish similar results in less obvious cases. The insert shows the shape of the function in three dimensions.

Method First, note that an integral over a single function f is included in the previous discussion if we take $f_1=f$ and $f_2=1$ in eqn 11C.1. Therefore, we need to judge whether f alone belongs to the symmetry species A_1 (or its equivalent) in the point group of the system. To decide that, we identify the point group and then examine the character table to see whether f belongs to A_1 (or its equivalent).

Answer An equilateral triangle has the point-group symmetry D_{3h}. If we refer to the character table of the group, we see that xy is a member of a basis that spans the irreducible representation E′. Therefore, its integral must be zero, because the integrand has no component that spans A_1'.

Self-test 11C.1 Can the function x^2+y^2 have a nonzero integral when integrated over a regular pentagon centred on the origin?

Answer: Yes, see Fig. 11C.3.

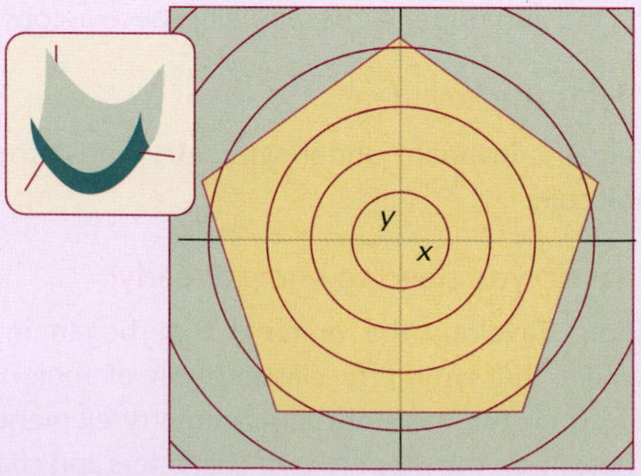

Figure 11C.3 The integration of a function over a pentagonal region. The insert shows the shape of the function in three dimensions.

(b) Decomposition of a direct product

In many cases, the product of functions f_1 and f_2 spans a sum of irreducible representations. For instance, in C_{2v} we may find the characters 2,0,0,−2 when we multiply the characters of f_1 and f_2 together. In this case, we note that these characters are the sum of the characters for A_2 and B_1:

	E	C_{2v}	σ_v	σ'_v
A_2	1	1	−1	−1
B_1	1	−1	1	−1
A_2+B_1	2	0	0	−2

To summarize this result we write the symbolic expression $A_2\times B_1=A_2+B_1$, which is called the **decomposition of a direct product**. This expression is symbolic. The × and + signs in this expression are not ordinary multiplication and addition signs: formally, they denote technical procedures with matrices called a 'direct product' and a 'direct sum'.[1] Because the sum on the right does not include a component that is a basis for an irreducible representation of symmetry species A_1, we can conclude that the integral of f_1f_2 over all space is zero in a C_{2v} molecule.

Whereas the decomposition of the characters 2,0,0,−2 can be done by inspection in this simple case, in other cases and more complex groups the decomposition is often far from obvious. For example, if we found the characters 8,−2,−6,4, it might not be obvious that the sum contains A_1. Group theory, however, provides a systematic way of using the characters of the representation spanned by a product to find the symmetry species of the irreducible representations. The formal recipe is

$$n(\Gamma)=\frac{1}{h}\sum_R \chi^{(\Gamma)}(R)\chi(R) \quad \text{Decomposition of direct product} \tag{11C.2}$$

We implement this expression as follows:

1. Write down a table with columns headed by the symmetry operations, R, of the group. Include a column for every operation, not just the classes.
2. In the first row write down the characters of the representation we want to analyse; these are the $\chi(R)$.
3. In the second row, write down the characters of the irreducible representation Γ we are interested in; these are the $\chi^{(\Gamma)}(R)$.
4. Multiply the two rows together, add the products together, and divide by the order of the group, h.

The resulting number, $n(\Gamma)$, is the number of times Γ occurs in the decomposition.

[1] As mentioned in Topic 11B, for this reason a direct sum is sometimes denoted ⊕; likewise, a direct product is sometimes denoted ⊗.

Brief illustration 11C.1 Decomposition of a direct product

To find whether A_1 does indeed occur in the product with characters 8,−2,−6,4 in C_{2v}, we draw up the following table:

	E	C_{2v}	σ_v	$4\sigma'_v$	$h=4$ (the order of the group)
f_1f_2	8	−2	−6	4	(the characters of the product)
A_1	1	1	1	1	(the symmetry species we are interested in)
	8	−2	−6	4	(the product of the two sets of characters)

The sum of the numbers in the last line is 4; when that number is divided by the order of the group, we get 1, so A_1 occurs once in the decomposition. When the procedure is repeated for all four symmetry species, we find that f_1f_2 spans $A_1+2A_2+5B_2$.

Self-test 11C.2 Does A_2 occur among the symmetry species of the irreducible representations spanned by a product with characters 7,−3,−1,5 in the group C_{2v}?

Answer: No

(c) Integrals over products of three functions

Integrals of the form

$$I=\int f_1f_2f_3\,d\tau \tag{11C.3}$$

are also common in quantum mechanics for they include matrix elements of operators (Topic 7C), and it is important to know when they are necessarily zero. As for integrals over two functions, for I to be nonzero, *the product $f_1f_2f_3$ must span A_1 (or its equivalent) or contain a component that spans A_1*. To test whether this is so, the characters of all three functions are multiplied together in the same way as in the rules set out earlier.

Example 11C.2 Deciding if an integral must be zero 2

Does the integral $\int(d_{z^2})x(d_{xy})d\tau$ vanish in a C_{2v} molecule?

Method We must refer to the C_{2v} character table (Table 11B.2) and the characters of the irreducible representations spanned by $3z^2-r^2$ (the form of the d_{z^2} orbital), x, and xy; then we can use the procedure set out above (with one more row of multiplication).

Answer We draw up the following table:

	E	C_2	σ_v	σ'_v	
$f_3=d_{xy}$	1	1	−1	−1	A_2
$f_2=x$	1	−1	1	−1	B_1
$f_1=d_{z^2}$	1	1	1	1	A_1
$f_1f_2f_3$	1	−1	−1	1	

The characters are those of B_2. Therefore, the integral is necessarily zero.

Self-test 11C.3 Does the integral $\int(p_x)y(p_z)d\tau$ necessarily vanish in an octahedral environment?

Answer: No

11C.2 Applications to orbitals

The rules we have outlined let us decide which atomic orbitals may have nonzero overlap in a molecule. It is also very useful to have a set of procedures to construct linear combinations of atomic orbitals (LCAOs) to have a certain symmetry, and thus to know in advance whether or not they will have nonzero overlap with other orbitals.

(a) Orbital overlap

An overlap integral, S, between two sets of atomic orbitals ψ_1 and ψ_2 is

$$S=\int\psi_2^*\psi_1 d\tau \qquad \text{Overlap integral} \qquad (11C.4)$$

and clearly has the same form as eqn 11C.1. It follows from that discussion that *only orbitals of the same symmetry species may have nonzero overlap* ($S\neq 0$), so only orbitals of the same symmetry species form bonding and antibonding combinations. It is explained in Topics 10B–10D that the selection of atomic orbitals that had mutual nonzero overlap is the central and initial step in the construction of molecular orbitals by the LCAO procedure. We are therefore at the point of contact between group theory and the material introduced in those Topics.

Example 11C.3 Determining which orbitals can contribute to bonding

The four H1s orbitals of methane span A_1+T_2. With which of the C atom orbitals can they overlap? What bonding pattern would be possible if the C atom had d orbitals available?

Method Refer to the T_d character table (in the *Resource section*) and look for s, p, and d orbitals spanning A_1 or T_2.

Answer An s orbital spans A_1 in the group T_d, so it may have nonzero overlap with the A_1 combination of H1s orbitals. The C2p orbitals span T_2, so they may have nonzero overlap with the T_2 combination. The d_{xy}, d_{yz}, and d_{zx} orbitals span T_2, so they may overlap the same combination. Neither of the other two d orbitals spans A_1 (they span E), so they remain nonbonding orbitals. It follows that in methane there are (C2s,H1s)-overlap a_1 orbitals and (C2p,H1s)-overlap t_2 orbitals. The C3d orbitals might contribute to the latter. The lowest energy configuration is probably $a_1^2t_2^6$, with all bonding orbitals occupied.

Self-test 11C.4 Consider the octahedral SF_6 molecule, with the bonding arising from overlap of S orbitals and a 2p orbital on each fluorine directed towards the central sulfur atom. The latter span $A_{1g}+E_g+T_{1u}$. What S orbitals have nonzero overlap? Suggest what the ground-state configuration is likely to be.

Answer: $3s(A_{1g})$, $3p(T_{1u})$, $3d(E_g)$; $a_{1g}^2t_{1u}^6e_g^4$

(b) Symmetry-adapted linear combinations

In the discussion of the molecular orbitals of NH_3 (Topic 10C) we encounter molecular orbitals of the form $\psi=c_1s_N+c_2(s_1+s_2+s_3)$, where s_N is an N2s atomic orbital and s_1, s_2, and s_3 are H1s orbitals. The s_N orbital has nonzero overlap with the combination of H1s orbitals as the latter has matching symmetry. The combination of H1s orbitals is an example of a **symmetry-adapted linear combination** (SALC), which are orbitals constructed from equivalent atoms and having a specified symmetry. Group theory also provides machinery that takes an arbitrary **basis**, or set of atomic orbitals (s_A, etc.), as input and generates combinations of the specified symmetry. As illustrated by the example of NH_3, SALCs are the building blocks of LCAO molecular orbitals and their construction is the first step in any molecular orbital treatment of molecules.

The technique for building SALCs is derived by using the full power of group theory and involves the use of a **projection operator**, $P^{(\Gamma)}$, an operator that takes one of the basis orbitals and generates from it—projects from it—an SALC of the symmetry species Γ:

$$P^{(\Gamma)}=\frac{1}{h}\sum_R\chi^{(\Gamma)}(R)R \quad \text{for} \quad \psi_m^{(\Gamma)}=P^{(\Gamma)}\chi_o \qquad \text{Projection operator} \qquad (11C.5)$$

To implement this rule, do the following:

1. Write each basis orbital at the head of a column and in successive rows show the effect of each operation R on each orbital. Treat each operation individually.
2. Multiply each member of the column by the character, $\chi^{(\Gamma)}(R)$, of the corresponding operation.
3. Add together all the orbitals in each column with the factors as determined in (2).
4. Divide the sum by the order of the group, h.

Example 11C.4 Constructing symmetry-adapted orbitals

Construct the A_1 symmetry-adapted linear combination of H1s orbitals for NH_3.

Method Identify the point group of the molecule and have available its character table. Then apply the projection operator technique.

Answer From the (s_N,s_A,s_B,s_C) basis in NH_3 we form the following table with each row showing the effect of the operation shown on the left.

	s_N	s_A	s_B	s_C
E	s_N	s_A	s_B	s_C
C_3^+	s_N	s_B	s_C	s_A
C_3^-	s_N	s_C	s_A	s_B
σ_v	s_N	s_A	s_C	s_B
σ_v'	s_N	s_B	s_A	s_C
σ_v''	s_N	s_C	s_B	s_A

To generate the A_1 combination, we take the characters for A_1 (1,1,1,1,1,1); then rules 2 and 3 lead to $\psi \propto s_N + s_N + \ldots = 6s_N$. The order of the group (the number of elements) is 6, so the combination of A_1 symmetry that can be generated from s_N is s_N itself. Applying the same technique to the column under s_A gives

$$\psi = \tfrac{1}{6}(s_A + s_B + s_C + s_A + s_B + s_C) = \tfrac{1}{3}(s_A + s_B + s_C)$$

The same combination is built from the other two columns, so they give no further information. The combination we have just formed is the s_1 combination used in Topic 10D (apart from the numerical factor).

Self-test 11C.5 Construct the A_1 symmetry-adapted linear combinations of H1s orbitals for CH_4.

Answer: $\tfrac{1}{4}(s_A + s_B + s_C + s_D)$

We now form the overall molecular orbital by forming a linear combination of all the SALCs of the specified symmetry species. In this case, therefore, the a_1 molecular orbital is $\psi = c_N s_N + c_1 s_1$, as specified above. This is as far as group theory can take us. The coefficients are found by solving the Schrödinger equation; they do not come directly from the symmetry of the system.

We run into a problem when we try to generate an SALC of symmetry species E, because, for representations of dimension 2 or more, the rules generate sums of SALCs. This problem can be illustrated as follows. In C_{3v}, the E characters are 2,–1,–1,0,0,0, so the column under s_N gives

$$\psi = \tfrac{1}{6}(2s_N - s_N - s_N + 0 + 0 + 0) = 0$$

The other columns give

$$\tfrac{1}{6}(2s_A - s_B - s_C) \qquad \tfrac{1}{6}(2s_B - s_A - s_C) \qquad \tfrac{1}{6}(2s_C - s_B - s_A)$$

However, any one of these three expressions can be expressed as a sum of the other two (they are not 'linearly independent'). The difference of the second and third gives $\tfrac{1}{2}(s_B - s_C)$, and this combination and the first, $\tfrac{1}{6}(2s_A - s_B - s_C)$ are the two (now linearly independent) SALCs we have used in the discussion of e orbitals.

11C.3 Selection rules

It is explained in Topic 9C and developed further in Topics 12A, 12C–12E, and 13A that the intensity of a spectral line arising from a molecular transition between some initial state with wavefunction ψ_i and a final state with wavefunction ψ_f depends on the (electric) transition dipole moment, $\boldsymbol{\mu}_{fi}$. The z-component of this vector is defined through

$$\mu_{z,fi} = -e\int \psi_f^* z \psi_i \mathrm{d}\tau \qquad \text{Transition dipole moment} \qquad (11C.6)$$

where $-e$ is the charge of the electron. The transition moment has the form of the integral in eqn 11C.3; so, once we know the symmetry species of the states, we can use group theory to formulate the selection rules for the transitions.

Example 11C.5 Deducing a selection rule

Is $p_x \rightarrow p_y$ an allowed transition in a tetrahedral environment?

Method We must decide whether the product $p_y q p_x$, with $q = x$, y, or z, spans A_1 by using the T_d character table.

Answer The procedure works out as follows:

	E	$8C_3$	$3C_2$	$6\sigma_d$	$6S_4$	
$f_3(p_y)$	3	0	−1	1	−1	T_2
$f_2(q)$	3	0	−1	1	−1	T_2
$f_1(p_x)$	3	0	−1	1	−1	T_2
$f_1f_2f_3$	27	0	−1	1	−1	

We now use the decomposition procedure described to deduce that A_1 occurs (once) in this set of characters, so $p_x \rightarrow p_y$ is allowed. A more detailed analysis (using the matrix representatives rather than the characters) shows that only $q = z$ gives an integral that may be nonzero, so the transition is z-polarized. That is, the electromagnetic radiation involved in the transition has a component of its electric vector in the z-direction.

Self-test 11C.6 What are the allowed transitions, and their polarizations, of an electron in a b_1 orbital in a C_{4v} molecule?

Answer: $b_1 \rightarrow b_1(z)$; $b_1 \rightarrow e(x,y)$

Checklist of concepts

- ☐ 1. Character tables are used to decide whether an integral is necessarily zero.
- ☐ 2. To be nonzero, an integrand must include a component that is a basis for the totally symmetric representation.
- ☐ 3. Only orbitals of the same symmetry species may have nonzero overlap.
- ☐ 4. A **symmetry-adapted linear combination** (SALC) is a linear combination of atomic orbitals constructed from equivalent atoms and having a specified symmetry.

Checklist of equations

Property	Equation	Comment	Equation number
Decomposition of direct product	$n(\Gamma)=(1/h)\sum_R \chi^{(\Gamma)}(R)\chi(R)$	Real characters*	11C.2
Overlap integral	$S=\int \psi_2^*\psi_1 d\tau$	Definition	11C.4
Projection operator	$P^{(\Gamma)}=(1/h)\sum_R \chi^{(\Gamma)}(R)R$	To generate $\psi_m^{(\Gamma)}=P^{(\Gamma)}\chi_o$	11C.5
Transition dipole moment	$\mu_{z,\mathrm{fi}}=-e\int \psi_\mathrm{f}^* z\psi_\mathrm{i}\mathrm{d}\tau$	z-Component	11C.6

* In general, characters may have complex values; throughout this text we encounter only real values.

CHAPTER 11 Molecular symmetry

TOPIC 11A Symmetry elements

Discussion questions

11A.1 Explain how a molecule is assigned to a point group.

11A.2 List the symmetry operations and the corresponding symmetry elements of the point groups.

11A.3 State and explain the symmetry criteria that allow a molecule to be polar.

11A.4 State the symmetry criteria that allow a molecule to be optically active.

Exercises

11A.1(a) The CH_3Cl molecule belongs to the point group C_{3v}. List the symmetry elements of the group and locate them in a drawing of the molecule.
11A.1(b) The CCl_4 molecule belongs to the point group T_d. List the symmetry elements of the group and locate them in a drawing of the molecule.

11A.2(a) Identify the group to which the naphthalene molecule belongs and locate the symmetry elements in a drawing of the molecule.
11A.2(b) Identify the group to which the anthracene molecule belongs and locate the symmetry elements in a drawing of the molecule.

11A.3(a) Identify the point groups to which the following objects belong: (i) a sphere, (ii) an isosceles triangle, (iii) an equilateral triangle, (iv) an unsharpened cylindrical pencil.
11A.3(b) Identify the point groups to which the following objects belong: (i) a sharpened cylindrical pencil, (ii) a three-bladed propeller, (iii) a four-legged table, (iv) yourself (approximately).

11A.4(a) List the symmetry elements of the following molecules and name the point groups to which they belong: (i) NO_2, (ii) N_2O, (iii) $CHCl_3$, (iv) $CH_2{=}CH_2$.
11A.4(b) List the symmetry elements of the following molecules and name the point groups to which they belong: (i) furan (**1**), (ii) γ-pyran (**2**), (iii) 1,2,5-trichlorobenzene.

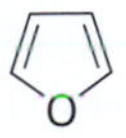

1 Furan

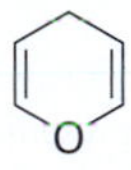

2 γ-Pyran

11A.5(a) Assign (i) *cis*-dichloroethene and (ii) *trans*-dichloroethene to point groups.
11A.5(b) Assign the following molecules to point groups: (i) HF, (ii) IF_7 (pentagonal bipyramid), (iii) XeO_2F_2 (see-saw), (iv) $Fe_2(CO)_9$ (**3**), (v) cubane, C_8H_8, (vi) tetrafluorocubane, $C_8H_4F_4$ (**4**).

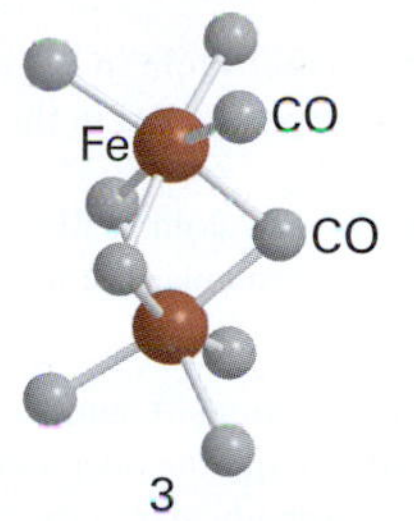

3

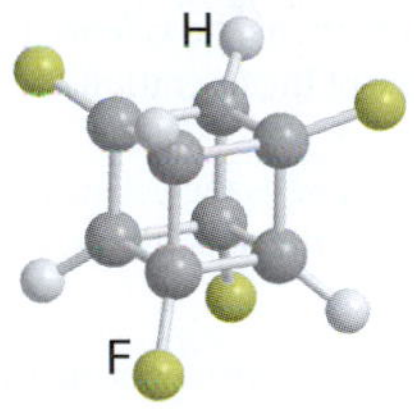

4

11A.6(a) Which of the following molecules may be polar? (i) pyridine, (ii) nitroethane, (iii) gas-phase $HgBr_2$, (iv) $B_3N_3H_6$.
11A.6(b) Which of the following molecules may be polar? (i) CH_3Cl, (ii) $HW_2(CO)_{10}$, (iii) $SnCl_4$.

11A.7(a) Identify the point groups to which all isomers of dichloronaphthalene belong.
11A.7(b) Identify the point groups to which all isomers of dichloroanthracene belong.

11A.8(a) Can molecules belonging to the point groups D_{2h} or C_{3h} be chiral? Explain your answer.
11A.8(b) Can molecules belonging to the point groups T_h or T_d be chiral? Explain your answer.

Problems

11A.1 List the symmetry elements of the following molecules and name the point groups to which they belong: (a) staggered CH_3CH_3, (b) chair and boat cyclohexane, (c) B_2H_6, (d) $[Co(en)_3]^{3+}$, where en is ethylenediamine (1,2-diaminoethane; ignore its detailed structure), (e) crown-shaped S_8. Which of these molecules can be (i) polar, (ii) chiral?

11A.2‡ In the square-planar complex anion $[trans\text{-}Ag(CF_3)_2(CN)_2]^-$, the Ag–CN groups are collinear. (a) Assume free rotation of the CF_3 groups (that is, disregarding the AgCF and AgCH angles) and name the point group of this complex ion. (b) Now suppose the CF_3 groups cannot rotate freely (because the ion was in a solid, for example). Structure (**5**) shows a plane which bisects the NC–Ag–CN axis and is perpendicular to it. Name the point group of the complex if each CF_3 group has a CF bond in that plane (so the CF_3 groups do not point to either CN group preferentially) and the CF_3 groups are (i) staggered, (ii) eclipsed.

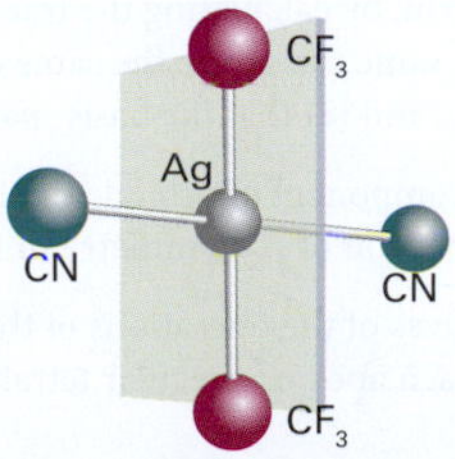

5

11A.3‡ B.A. Bovenzi and G.A. Pearse, Jr. (*J. Chem. Soc. Dalton Trans.*, 2763 (1997)) synthesized coordination compounds of the tridentate ligand pyridine-2,6-diamidoxime ($C_7H_9N_5O_2$, **6**). Reaction with $NiSO_4$ produced a complex in which two of the essentially planar ligands are bonded at right angles to a single Ni atom. Name the point group and the symmetry operations of the resulting $[Ni(C_7H_9N_5O_2)_2]^{2+}$ complex cation.

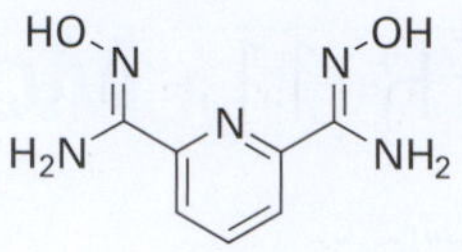

6

TOPIC 11B Group theory

Discussion questions

11B.1 Explain what is meant by a 'group'.

11B.2 Explain what is meant by (a) a representative and (b) a representation in the context of group theory.

11B.3 Explain the construction and content of a character table.

11B.4 Explain what is meant by the reduction of a representation to a direct sum of representations.

11B.5 Discuss the significance of the letters and subscripts used to denote the symmetry species of a representation.

Exercises

11B.1(a) Use as a basis the valence p_z orbitals on each atom in BF_3 to find the representative of the operation σ_h. Take z as perpendicular to the molecular plane.
11B.1(b) Use as a basis the valence p_z orbitals on each atom in BF_3 to find the representative of the operation C_3. Take z as perpendicular to the molecular plane.

11B.2(a) Use the matrix representatives of the operations σ_h and C_3 in a basis of valence p_z orbitals on each atom in BF_3 to find the operation and its representative resulting from $\sigma_h C_3$. Take z as perpendicular to the molecular plane.
11B.2(b) Use the matrix representatives of the operations σ_h and C_3 in a basis of valence p_z orbitals on each atom in BF_3 to find the operation and its representative resulting from $C_3\sigma_h$. Take z as perpendicular to the molecular plane.

11B.3(a) Show that all three C_2 operations in the group D_{3h} belong to the same class.
11B.3(b) Show that all three σ_v operations in the group D_{3h} belong to the same class.

11B.4(a) What is the maximum degeneracy of a particle confined to the interior of an octahedral hole in a crystal?
11B.4(b) What is the maximum degeneracy of a particle confined to the interior of an icosahedral nanoparticle?

11B.5(a) What is the maximum possible degree of degeneracy of the orbitals in benzene?
11B.5(b) What is the maximum possible degree of degeneracy of the orbitals in 1,4-dichlorobenzene?

Problems

11B.1 The group C_{2h} consists of the elements E, C_2, σ_h, i. Construct the group multiplication table and find an example of a molecule that belongs to the group.

11B.2 The group D_{2h} has a C_2 axis perpendicular to the principal axis and a horizontal mirror plane. Show that the group must therefore have a centre of inversion.

11B.3 Consider the H_2O molecule, which belongs to the group C_{2v}. Take as a basis the two H1s orbitals and the four valence orbitals of the O atom and set up the 6×6 matrices that represent the group in this basis. Confirm by explicit matrix multiplication that the group multiplications (a) $C_2\sigma_v = \sigma_v'$ and (b) $\sigma_v\sigma_v' = C_2$. Confirm, by calculating the traces of the matrices, (a) that symmetry elements in the same class have the same character, (b) that the representation is reducible, and (c) that the basis spans $3A_1 + B_1 + 2B_2$.

11B.4 Confirm that the z-component of orbital angular momentum is a basis for an irreducible representation of A_2 symmetry in C_{3v}.

11B.5 Find the representatives of the operations of the group T_d in a basis of four H1s orbitals, one at each apex of a regular tetrahedron (as in CH_4).

‡ These problems were provided by Charles Trapp and Carmen Giunta.

11B.6 Confirm that the representatives constructed in Problem 11B.5 reproduce the group multiplications $C_3^+C_3^- = E$, $S_4C_3 = S_4'$, and $S_4C_3 = \sigma_d$.

11B.7 The (one-dimensional) matrices $D(C_3) = 1$ and $D(C_2) = 1$, and $D(C_3) = 1$ and $D(C_2) = -1$ both represent the group multiplication $C_3C_2 = C_6$ in the group C_{6v} with $D(C_6) = +1$ and -1, respectively. Use the character tale to confirm these remarks. What are the representatives of σ_v and σ_d in each case?

11B.8 Construct the multiplication table of the Pauli spin matrices, $\boldsymbol{\sigma}$, and the 2×2 unit matrix:

$$\sigma_x = \begin{pmatrix} 0 & 1 \\ 1 & 0 \end{pmatrix} \quad \sigma_y = \begin{pmatrix} 0 & -i \\ i & 0 \end{pmatrix} \quad \sigma_z = \begin{pmatrix} 1 & 0 \\ 0 & -1 \end{pmatrix} \quad \sigma_0 = \begin{pmatrix} 1 & 0 \\ 0 & 1 \end{pmatrix}$$

Do the four matrices from a group under multiplication?

11B.9 The algebraic forms of the f orbitals are a radial function multiplied by one of the factors (a) $z(5z^2 - 3r^2)$, (b) $y(5y^2 - 3r^2)$, (c) $x(5x^2 - 3r^2)$, (d) $z(x^2 - y^2)$, (e) $y(x^2 - z^2)$, (f) $x(z^2 - y^2)$, (g) xyz. Identify the irreducible representations spanned by these orbitals in (a) C_{2v}, (b) C_{3v}, (c) T_d, (d) O_h. Consider a lanthanoid ion at the centre of (a) a tetrahedral complex, (b) an octahedral complex. What sets of orbitals do the seven f orbitals split into?

TOPIC 11C Applications of symmetry

Discussion questions

11C.1 Identify and list four applications of character tables.

11C.2 Explain how symmetry arguments are used to construct molecular orbitals.

Exercises

11C.1(a) Use symmetry properties to determine whether or not the integral $\int p_x z p_z \mathrm{d}\tau$ is necessarily zero in a molecule with symmetry C_{2v}.
11C.1(b) Use symmetry properties to determine whether or not the integral $\int p_x z p_z \mathrm{d}\tau$ is necessarily zero in a molecule with symmetry D_{3h}.

11C.2(a) Is the transition $A_1 \rightarrow A_2$ forbidden for electric dipole transitions in a C_{3v} molecule?
11C.2(b) Is the transition $A_{1g} \rightarrow E_{2u}$ forbidden for electric dipole transitions in a D_{6h} molecule?

11C.3(a) Show that the function xy has symmetry species B_2 in the group C_{4v}.
11C.3(b) Show that the function xyz has symmetry species A_1 in the group D_2.

11C.4(a) Consider the C_{2v} molecule NO_2. The combination $p_x(A)-p_x(B)$ of the two O atoms (with x perpendicular to the plane) spans A_2. Is there any orbital of the central N atom that can have a nonzero overlap with that combination of O orbitals? What would be the case in SO_2, where 3d orbitals might be available?
11C.4(b) Consider the C_{3v} ion NO_3^-. Is there any orbital of the central N atom that can have a nonzero overlap with the combination $2p_z(A)-p_z(B)-p_z(C)$ of the three O atoms (with z perpendicular to the plane). What would be the case in SO_3, where 3d orbitals might be available?

11C.5(a) The ground state of NO_2 is A_1 in the group C_{2v}. To what excited states may it be excited by electric dipole transitions, and what polarization of light is it necessary to use?
11C.5(b) The ClO_2 molecule (which belongs to the group C_{2v}) was trapped in a solid. Its ground state is known to be B_1. Light polarized parallel to the y-axis (parallel to the OO separation) excited the molecule to an upper state. What is the symmetry species of that state?

11C.6(a) A set of basis functions is found to span a reducible representation of the group C_{4v} with characters 4,1,1,3,1 (in the order of operations in the character table in the *Resource section*). What irreducible representations does it span?
11C.6(b) A set of basis functions is found to span a reducible representation of the group D_2 with characters 6,–2,0,0 (in the order of operations in the character table in the *Resource section*). What irreducible representations does it span?

11C.7(a) What states of (i) benzene, (ii) naphthalene may be reached by electric dipole transitions from their (totally symmetrical) ground states?
11C.7(b) What states of (i) anthracene, (ii) coronene (**7**) may be reached by electric dipole transitions from their (totally symmetrical) ground states?

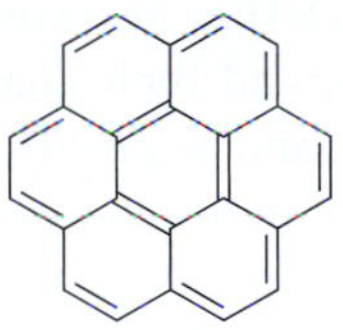

7 Coronene

11C.8(a) Write $f_1 = \sin\theta$ and $f_2 = \cos\theta$, and show by symmetry arguments using the group C_s that the integral of their product over a symmetrical range around $\theta = 0$ is zero.
11C.8(b) Write $f_1 = x$ and $f_2 = 3x^2 - 1$, and show by symmetry arguments using the group C_s that the integral of their product over a symmetrical range around $x = 0$ is zero.

Problems

11C.1 What irreducible representations do the four H1s orbitals of CH_4 span? Are there s and p orbitals of the central C atom that may form molecular orbitals with them? Could d orbitals, even if they were present on the C atom, play a role in orbital formation in CH_4?

11C.2 Suppose that a methane molecule became distorted to (a) C_{3v} symmetry by the lengthening of one bond, (b) C_{2v} symmetry, by a kind of scissors action in which one bond angle opened and another closed slightly. Would more d orbitals become available for bonding?

11C.3 Does the product $3x^2 - 1$ necessarily vanish when integrated over (a) a cube, (b) a tetrahedron, (c) a hexagonal prism, each centred on the origin?

11C.4‡ In a spectroscopic study of C_{60}, Negri et al. (*J. Phys. Chem.* **100**, 10849 (1996)) assigned peaks in the fluorescence spectrum. The molecule has icosahedral symmetry (I_h). The ground electronic state is A_{1g}, and the lowest-lying excited states are T_{1g} and G_g. (a) Are photon-induced transitions allowed from the ground state to either of these excited states? Explain your answer. (b) What if the molecule is distorted slightly so as to remove its centre of inversion?

11C.5 In the square planar XeF_4 molecule, consider the symmetry-adapted linear combination $p_1 = p_A - p_B + p_C - p_D$ where p_A, p_B, p_C, and p_D are $2p_z$ atomic orbitals on the fluorine atoms (clockwise labelling of the F atoms). Using the reduced point group D_4 rather than the full symmetry point group of the molecule, determine which of the various s, p, and d atomic orbitals on the central Xe atom can form molecular orbitals with p_1.

11C.6 The chlorophylls that participate in photosynthesis and the haem groups of cytochromes are derived from the porphine dianion group (**8**), which belongs to the D_{4h} point group. The ground electronic state is A_{1g} and the lowest-lying excited state is E_u. Is a photon-induced transition allowed from the ground state to the excited state? Explain your answer.

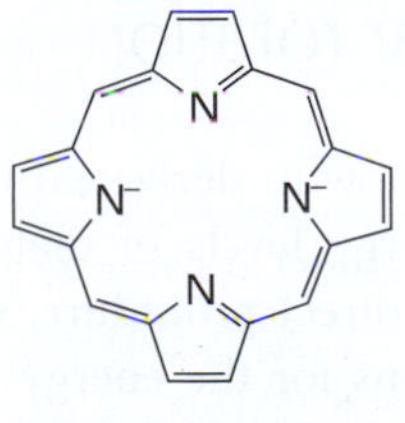

8

CHAPTER 12

Rotational and vibrational spectra

The origin of spectral lines in molecular spectroscopy is the absorption, emission, or scattering of a photon when the energy of a molecule changes. The difference from atomic spectroscopy (Topic 9C) is that the energy of a molecule can change not only as a result of electronic transitions but also because it can undergo changes of rotational and vibrational state. Molecular spectra are therefore more complex than atomic spectra. However, they also contain information relating to more properties, and their analysis leads to values of bond strengths, lengths, and angles. They also provide a way of determining a variety of molecular properties, such as dipole moments.

The general strategy we adopt in this chapter is to set up expressions for the energy levels of molecules and then infer the form of rotational and vibrational spectra. Electronic spectra are considered in Chapter 13.

12A General features of molecular spectroscopy

This Topic begins with a discussion of the theory of absorption and emission of radiation, leading to the factors that determine the intensities and widths of spectral lines. Then we describe features of instrumentation used to monitor the absorption, emission, and scattering of radiation spanning a wide range of frequencies.

12B Molecular rotation

In this Topic we see how to derive expressions for the values of the rotational energy levels of diatomic and polyatomic molecules. The most direct procedure, which we adopt, is to identify the expressions for the energy and angular momentum obtained in classical physics, and then to transform these expressions into their quantum mechanical counterparts.

12C Rotational spectroscopy

This Topic focuses on the interpretation of pure rotational and rotational Raman spectra, in which only the rotational state of a molecule changes. We explain in terms of nuclear spin and the Pauli principle the observation that not all molecules can occupy all rotational states.

12D Vibrational spectroscopy of diatomic molecules

In this Topic we consider the vibrational energy levels of diatomic molecules and see that we can use the properties of harmonic oscillators developed in Topic 8B, but must also take into account deviations from harmonic oscillation. We also see that vibrational spectra of gaseous samples show features that arise from the rotational transitions that accompany the excitation of vibrations.

12E Vibrational spectroscopy of polyatomic molecules

The vibrational spectra of polyatomic molecules may be discussed as though they consisted of a set of independent harmonic oscillators, so the same approach as employed for diatomic molecules may be used. We also see that the symmetry properties of the atomic displacements of polyatomic molecules are helpful for deciding which modes of vibration can be studied spectroscopically.

What is the impact of this material?

Molecular spectroscopy is also useful to astrophysicists and environmental scientists. In *Impact* I12.1 we see how the identities of molecules found in interstellar space can be inferred from their rotational and vibrational spectra. In *Impact* I12.2 we turn our attention back towards the Earth and see how the vibrational properties of its atmospheric constituents can affect its climate.

To read more about the impact of this material, scan the QR code, or go to bcs.whfreeman.com/webpub/chemistry/pchem10e/impact/pchem-12-1.html

12A General features of molecular spectroscopy

Contents

➤ Why do you need to know this material?

To interpret data from the wide range of varieties of molecular spectroscopy you need to understand the experimental and theoretical features that all types of spectra share.

➤ What is the key idea?

A transition from a low energy state to one of higher energy can be stimulated by absorption of electromagnetic radiation; a transition from a higher to a lower state may be either spontaneous (resulting in emission of radiation) or stimulated by radiation.

➤ What do you need to know already?

You need to be familiar with quantization of energy in molecules (Topics 8A–8C), and the concept of selection rules in spectroscopy (Topic 9C).

In **emission spectroscopy**, a molecule undergoes a transition from a state of high energy E_1 to a state of lower energy E_2 and emits the excess energy as a photon. In **absorption spectroscopy**, the net absorption of incident radiation is monitored as its frequency is varied. We say *net* absorption, because, when a sample is irradiated, both absorption and emission at a given frequency are stimulated, and the detector measures the difference, the net absorption. In **Raman spectroscopy**, changes in molecular state are explored by examining the frequencies present in the radiation scattered by molecules.

The energy, $h\nu$, of the photon emitted or absorbed, and therefore the frequency ν of the radiation emitted or absorbed, is given by the Bohr frequency condition (eqn 7A.12 of Topic 7A, $h\nu=|E_1-E_2|$). Emission and absorption spectroscopy give the same information about electronic, vibrational, or rotational energy level separations, but practical considerations generally determine which technique is employed.

In Raman spectroscopy the difference between the frequencies of the scattered and incident radiation is determined by the transitions that take place within the molecule as a result of the impact of the incoming photon; this technique is used to study molecular vibrations and rotations. About 1 in 10^7 of the incident photons that collide with the molecules, give up some of their energy, and emerge with a lower energy. These scattered photons constitute the lower-frequency **Stokes radiation** from the sample (Fig. 12A.1). Other incident photons may collect energy from the molecules (if they are already excited), and emerge as higher-frequency **anti-Stokes radiation**. The component scattered without change of frequency is called **Rayleigh radiation**.

Atomic spectroscopy is discussed in Topic 9C. Here we set the stage for detailed discussion of rotational (Topics 12B

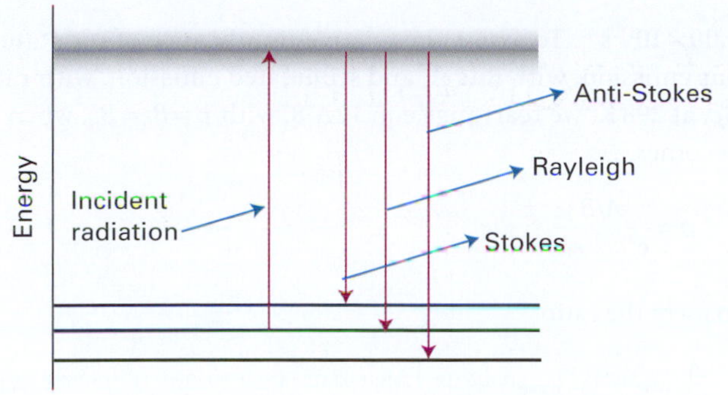

Figure 12A.1 In Raman spectroscopy, an incident photon is scattered from a molecule with either an increase in frequency (if the radiation collects energy from the molecule) or with a lower frequency (if it loses energy to the molecule) to give the anti-Stokes and Stokes lines, respectively. Scattering without change of frequency results in the Rayleigh lines. The process can be regarded as taking place by an excitation of the molecule to a wide range of states (represented by the shaded band), and the subsequent return of the molecule to a lower state; the net energy change is then carried away by the photon.

and 12C), vibrational (Topics 12D and 12E), and electronic (the several Topics of Chapter 13) transitions in molecules. Techniques that probe transitions between spin states of electrons and nuclei are also useful. They rely on special experimental approaches and theoretical considerations described in Chapter 14.

12A.1 The absorption and emission of radiation

As mentioned in *Foundations* B, the separation of rotational energy levels (in small molecules, $\Delta E \approx 0.01$ zJ, corresponding to about 0.01 kJ mol^{-1}) is smaller than that of vibrational energy levels ($\Delta E \approx 10$ zJ, corresponding to 10 kJ mol^{-1}), which itself is smaller than that of electronic energy levels ($\Delta E \approx 0.1$–1 aJ, corresponding to about 10^2–10^3 kJ mol^{-1}). From $\nu = \Delta E/h$, it follows that rotational, vibrational, and electronic transitions result from the absorption of emission of microwave, infrared, and ultraviolet/visible/far infrared radiation, respectively (see also Chapter 8). Here we turn our attention to the origins of spectroscopic transitions, focusing on concepts that apply generally to all varieties of spectroscopy.

(a) Stimulated and spontaneous radiative processes

Albert Einstein identified three contributions to the transitions between states. First, he recognized the transition from a low energy state to one of higher energy that is driven by the electromagnetic field oscillating at the transition frequency. This process is called **stimulated absorption**. The rate of this type of transition is proportional to the intensity of the incident radiation: the more intense the incident radiation, the greater is the rate of the transition and the stronger is the absorption by the sample. Einstein wrote this transition rate as

$$w_{f\leftarrow i} = B_{fi}\rho \qquad \textit{Stimulated absorption} \quad \text{Transition rate} \qquad (12A.1)$$

The constant B_{fi} is the **Einstein coefficient of stimulated absorption** and $\rho d\nu$ is the energy density of radiation in the frequency range from ν to $\nu + d\nu$, where ν is the frequency of the transition. For instance, when the atom or molecule is exposed to black-body radiation from a source of temperature T, ρ is given by the Planck distribution (eqn 7A.6 of Topic 7A):

$$\rho = \frac{8\pi h\nu^3/c^3}{e^{h\nu/kT}-1} \qquad \text{Planck distribution} \qquad (12A.2)$$

At this stage B_{fi} is an empirical parameter that characterizes the transition: if it is large, then a given intensity of incident radiation will induce transitions strongly and the sample will be strongly absorbing. The **total rate of absorption**, $W_{f\leftarrow i}$, is the transition rate of a single molecule multiplied by the number of molecules N_i in the lower state:

$$W_{f\leftarrow i} = N_i w_{f\leftarrow i} = N_i B_{fi}\rho \qquad \text{Total absorption rate} \qquad (12A.3)$$

Einstein considered that the radiation was also able to induce the molecule in the upper state to undergo a transition to the lower state, and hence to generate a photon of frequency ν. Thus, he wrote the rate of this **stimulated emission** as

$$w_{f\rightarrow i} = B_{if}\rho \qquad \textit{Stimulated emission} \quad \text{Transition rate} \qquad (12A.4)$$

where B_{if} is the **Einstein coefficient of stimulated emission**. This coefficient is in fact equal to the coefficient of stimulated absorption as we shall see below. Moreover, only radiation of the same frequency as the transition can stimulate an excited state to fall to a lower state. At this point, it is tempting to suppose that the total rate of emission is this individual rate multiplied by the number of molecules in the upper state, N_f, and therefore to write $W_{f\rightarrow i} = N_f B_{if}\rho$. But here we encounter a problem: at equilibrium (as in a black-body container), the rate of emission is equal to the rate of absorption, so $N_i B_{fi}\rho = N_f B_{if}\rho$ and therefore, since $B_{if} = B_{fi}$, $N_i = N_f$. The conclusion that the populations must be equal at equilibrium is in conflict with another very fundamental conclusion, that the ratio of populations is given by the Boltzmann distribution (*Foundations* B and Topic 15A) which implies that $N_i \neq N_f$.

Einstein realized that to bring the analysis of transition rates into alignment with the Boltzmann distribution there must be another route for the upper state to decay into the lower state, and wrote

$$w_{f\to i} = A + B_{if}\rho \qquad \text{Emission rate} \qquad (12A.5)$$

The constant A is the **Einstein coefficient of spontaneous emission**. The **total rate of emission,** $W_{f\to i}$, is therefore

$$W_{f\to i} = N_f w_{f\to i} = N_f(A + B_{if}\rho) \qquad \text{Total emission rate} \qquad (12A.6)$$

At thermal equilibrium, N_i and N_f do not change over time. This condition is reached when the total rates of emission and absorption are equal:

$$N_i B_{fi}\rho = N_f(A + B_{if}\rho) \qquad \text{Thermal equilibrium} \qquad (12A.7)$$

and therefore

$$\rho = \frac{N_f A}{N_i B_{fi} - N_f B_{if}} \overset{\text{divide by } N_f B_{fi}}{=} \frac{A/B_{fi}}{N_i/N_f - B_{if}/B_{fi}} = \frac{A/B_{fi}}{e^{h\nu/kT} - B_{if}/B_{fi}} \qquad (12A.8)$$

We have used the Boltzmann expression (*Foundations* B and Topic 15A) for the ratio of populations of the upper state (of energy E_f) and lower state (of energy E_i):

$$\frac{N_f}{N_i} = e^{-\overbrace{(E_f - E_i)}^{h\nu}/kT}$$

This result has the same form as the Planck distribution (eqn 12A.2), which describes the radiation density at thermal equilibrium. Indeed when we compare eqns 12A.2 and 12A.8, we can conclude that $B_{if} = B_{fi}$ (as we promised to show) and that

$$A = \left(\frac{8\pi h\nu^3}{c^3}\right)B \qquad (12A.9)$$

The important point about eqn 12A.9 is that it shows that the relative importance of spontaneous emission increases as the cube of the transition frequency and therefore that it is therefore potentially of great importance at very high frequencies. Conversely, spontaneous emission can be ignored at low transition frequencies, in which case intensities of those transitions can be discussed in terms of stimulated emission and absorption alone.

Brief illustration 12A.1 The Einstein coefficients

For a transition in the microwave region of the electromagnetic spectrum (corresponding to an excitation of a molecular rotation), a typical frequency is 600 GHz (1 GHz = 10^9 Hz), or 6.00×10^{11} s^{-1}. To assess the relative significance of spontaneous emission, with rate A, and stimulated emission, with rate $B\rho$, at 298 K, we rearrange eqn 12A.8, with $B = B_{fi} = B_{if}$, when it becomes

$$\rho = \frac{A/B}{e^{h\nu/kT} - 1}$$

to form the ratio

$$\frac{A}{B\rho} = e^{h\nu/kT} - 1 = e^{(6.626\times10^{-34}\,\text{J s})\times(6.00\times10^{11}\,\text{s}^{-1})/(1.381\times10^{-23}\,\text{J K}^{-1})\times(298\,\text{K})} - 1$$
$$= 0.101$$

and both spontaneous and stimulated emission are significant at this wavelength.

Self-test 12A.1 Calculate the ratio $A/B\rho$ at 298 K for a transition in the infrared region of the electromagnetic spectrum, corresponding to excitation of a molecular vibration, with wavenumber 2000 cm^{-1}. What conclusion can you draw?

Answer: $A/B\rho = 1.6\times10^4$; for vibrational transitions spontaneous emission is more significant than stimulated emission

(b) Selection rules and transition moments

We first met the concept of a 'selection rule' in Topic 9C as a statement about whether a transition is forbidden or allowed. Selection rules also apply to molecular spectra, and the form they take depends on the type of transition. The underlying classical idea is that, for the molecule to be able to interact with the electromagnetic field and absorb or create a photon of frequency ν, it must possess, at least transiently, a dipole oscillating at that frequency. In Topic 9C it is shown that this transient dipole is expressed quantum mechanically in terms of the transition dipole moment, $\boldsymbol{\mu}_{fi}$, between states ψ_i and ψ_f:

$$\boldsymbol{\mu}_{fi} = \int \psi_f^* \hat{\boldsymbol{\mu}} \psi_i \, d\tau \qquad \textit{Definition} \qquad \text{Transition dipole moment} \qquad (12A.10)$$

where $\hat{\boldsymbol{\mu}}$ is the electric dipole moment operator. The size of the transition dipole can be regarded as a measure of the charge redistribution that accompanies a transition: a transition is active (and generates or absorbs photons) only if the accompanying charge redistribution is dipolar (Fig. 12A.2). Only if the transition dipole moment is nonzero does the transition contribute to the spectrum. It follows that, to identify the selection rules, we must establish the conditions for which $\boldsymbol{\mu}_{fi} \neq 0$.

A **gross selection rule** specifies the general features that a molecule must have if it is to have a spectrum of a given kind. For instance, in Topic 12C it is shown that a molecule gives a rotational spectrum only if it has a permanent electric dipole moment. This rule and others like it for other types of transition

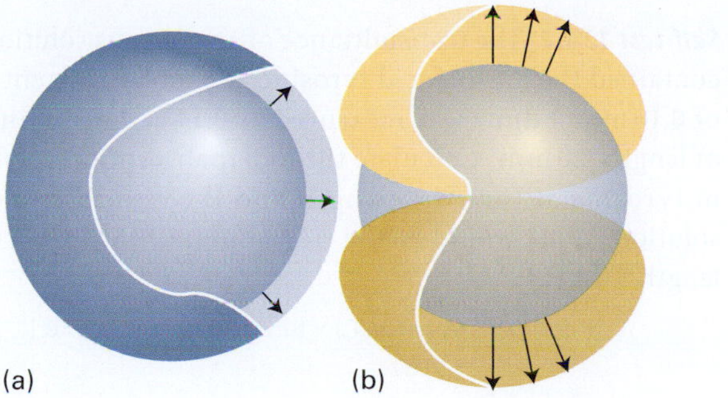

Figure 12A.2 (a) When a 1s electron becomes a 2s electron, there is a spherical migration of charge. There is no dipole moment associated with this migration of charge, so this transition is electric-dipole forbidden. (b) In contrast, when a 1s electron becomes a 2p electron, there is a dipole associated with the charge migration; this transition is allowed.

are explained in relevant Topics. A detailed study of the transition moment leads to the **specific selection rules** that express the allowed transitions in terms of the changes in quantum numbers.

(c) The Beer–Lambert law

Consider the absorption of radiation by a sample. It is found empirically that the transmitted intensity I varies with the length, L, of the sample and the molar concentration, [J], of the absorbing species J in accord with the **Beer–Lambert law**:

$$I=I_0 10^{-\varepsilon[\mathrm{J}]L} \quad \text{Beer–Lambert law} \quad (12\text{A}.11)$$

where I_0 is the incident intensity. The quantity ε (epsilon) is called the **molar absorption coefficient** (formerly, and still widely, the 'extinction coefficient'). The molar absorption coefficient depends on the frequency of the incident radiation and is greatest where the absorption is most intense. Its dimensions are 1/(concentration×length), and it is normally convenient to express it in cubic decimetres per mole per centimetre ($\mathrm{dm^3\ mol^{-1}\ cm^{-1}}$); in SI base units it is expressed in metres-squared per mole ($\mathrm{m^2\ mol^{-1}}$). The latter units imply that ε may be regarded as a (molar) cross-section for absorption and that the greater the cross-sectional area of the molecule for absorption, the greater is its ability to block the passage of the incident radiation at a given frequency. The Beer–Lambert law is an empirical result. However, it is simple to account for its form as we show in the following *Justification*.

Justification 12A.1 The Beer–Lambert law

We think of the sample as consisting of a stack of infinitesimal slices, like sliced bread (Fig. 12A.3). The thickness of each layer is dx. The change in intensity, dI, that occurs when electromagnetic radiation passes through one particular slice is

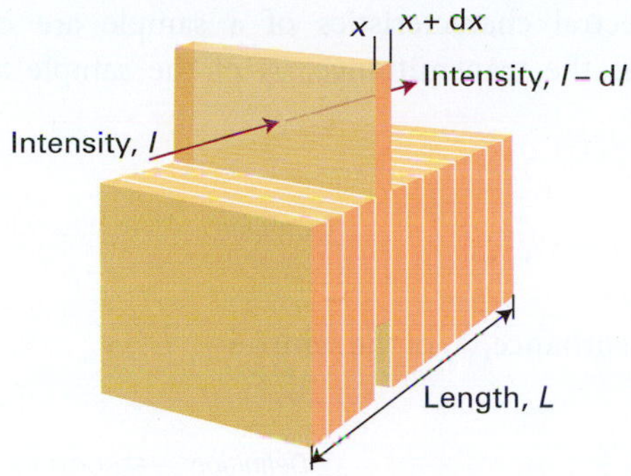

Figure 12A.3 To establish the Beer–Lambert law, the sample is supposed to be sliced into a large number of planes. The reduction in intensity caused by one plane is proportional to the intensity incident on it (after passing through the preceding planes), the thickness of the plane, and the concentration of absorbing species.

proportional to the thickness of the slice, the concentration of the absorber J, and the intensity of the incident radiation at that slice of the sample, so $\mathrm{d}I \propto [\mathrm{J}]I\mathrm{d}x$. Because d$I$ is negative (the intensity is reduced by absorption), we can write

$$\mathrm{d}I=-\kappa[\mathrm{J}]I\mathrm{d}x$$

where κ (kappa) is the proportionality coefficient. Division of both sides by I gives

$$\frac{\mathrm{d}I}{I}=-\kappa[\mathrm{J}]\mathrm{d}x$$

This expression applies to each successive slice.

To obtain the intensity that emerges from a sample of thickness L when the intensity incident on one face of the sample is I_0, we sum all the successive changes. Because a sum over infinitesimally small increments is an integral, we write:

$$\overbrace{\int_{I_0}^{I}\frac{\mathrm{d}I}{I}}^{\ln(I/I_0)\ \text{Integral A.2}}=-\kappa\int_0^L[\mathrm{J}]\mathrm{d}x \overset{[\mathrm{J}]\ \text{uniform}}{=} -\kappa[\mathrm{J}]\overbrace{\int_0^L\mathrm{d}x}^{L\ \text{Integral A.1}}$$

in the second step we have supposed that the concentration is uniform, so [J] is independent of x and can be taken outside the integral. Therefore

$$\ln\frac{I}{I_0}=-\kappa[\mathrm{J}]L$$

Because $\ln x=(\ln 10)\log x$, we can write $\varepsilon=\kappa/\ln 10$ and obtain

$$\log\frac{I}{I_0}=-\varepsilon[\mathrm{J}]L$$

which, on taking (common) antilogarithms, is the Beer–Lambert law (eqn 12A.11).

The spectral characteristics of a sample are commonly reported as the **transmittance**, T, of the sample at a given frequency:

$$T=\frac{I}{I_0} \qquad \textit{Definition} \quad \text{Transmittance} \quad (12A.12)$$

and the **absorbance**, A, of the sample:

$$A=\log\frac{I_0}{I} \qquad \textit{Definition} \quad \text{Absorbance} \quad (12A.13)$$

The two quantities are related by $A=-\log T$ (note the common logarithm) and the Beer–Lambert law becomes

$$A=\varepsilon[\mathrm{J}]L \qquad (12A.14)$$

The product $\varepsilon[\mathrm{J}]L$ was known formerly as the *optical density* of the sample.

Example 12A.1 Determining a molar absorption coefficient

Radiation of wavelength 280 nm passed through 1.0 mm of a solution that contained an aqueous solution of the amino acid tryptophan at a concentration of 0.50 mmol dm^{-3}. The light intensity is reduced to 54 per cent of its initial value (so $T=0.54$). Calculate the absorbance and the molar absorption coefficient of tryptophan at 280 nm. What would be the transmittance through a cell of thickness 2.0 mm?

Method From $A=-\log T=\varepsilon[\mathrm{J}]L$, it follows that $\varepsilon = -(\log T)/[\mathrm{J}]L$. For the transmittance through the thicker cell, we use $T=10^{-A}$ and the value of ε calculated here.

Solution The molar absorption coefficient is

$$\varepsilon=\frac{-\log 0.54}{(5.0\times10^{-4}\,\mathrm{mol\,dm^{-3}})\times(1.0\,\mathrm{mm})}=5.4\times10^{2}\,\mathrm{dm^3\,mol^{-1}\,mm^{-1}}$$

These units are convenient for the rest of the calculation (but the outcome could be reported as 5.4×10^3 dm^3 mol^{-1} cm^{-1} if desired). The absorbance is

$$A=-\log 0.54=0.27$$

The absorbance of a sample of length 2.0 mm is

$$A=(5.4\times10^{2}\,\mathrm{dm^3\,mol^{-1}\,mm^{-1}})\times(5.0\times10^{-4}\,\mathrm{mol\,dm^{-3}})\times(2.0\,\mathrm{mm})=0.54$$

It follows that the transmittance is now

$$T=10^{-A}=10^{-0.54}=0.29$$

That is, the emergent light is reduced to 29 per cent of its incident intensity.

Self-test 12A.2 The transmittance of an aqueous solution that contained the amino acid tyrosine at a molar concentration of 0.10 mmol dm^{-3} was measured as 0.14 at 240 nm in a cell of length 5.0 mm. Calculate the molar absorption coefficient of tyrosine at that wavelength and the absorbance of the solution. What would be the transmittance through a cell of length 1.0 mm?

Answer: 1.1×10^4 dm^3 mol^{-1} cm^{-1}, $A=0.17$, $T=0.68$

The maximum value of the molar absorption coefficient, $\varepsilon_{\max}$, is an indication of the intensity of a transition. However, as absorption bands generally spread over a range of wavenumbers, quoting the absorption coefficient at a single wavenumber might not give a true indication of the intensity of a transition. The **integrated absorption coefficient**, $\mathcal{A}$, is the sum of the absorption coefficients over the entire band (Fig. 12A.4), and corresponds to the area under the plot of the molar absorption coefficient against wavenumber:

$$\mathcal{A}=\int_{\text{band}}\varepsilon(\tilde{\nu})\mathrm{d}\tilde{\nu} \qquad \textit{Definition} \quad \text{Integrated absorption coefficient} \quad (12A.15)$$

For lines of similar widths, the integrated absorption coefficients are proportional to the heights of the lines.

12A.2 Spectral linewidths

A number of effects contribute to the widths of spectroscopic lines. Some contributions to linewidths can be modified by changing the conditions, and to achieve high resolutions we need to know how to minimize these contributions. Other contributions cannot be changed, and represent an inherent limitation on resolution.

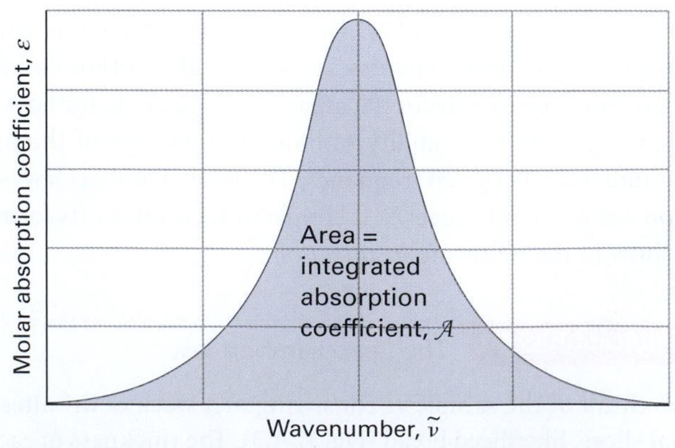

Figure 12A.4 The integrated absorption coefficient of a transition is the area under a plot of the molar absorption coefficient against the wavenumber of the incident radiation.

(a) Doppler broadening

One important broadening process in gaseous samples is the **Doppler effect**, in which radiation is shifted in frequency when the source is moving towards or away from the observer. When a source emitting electromagnetic radiation of frequency ν moves with a speed s relative to an observer, the observer detects radiation of frequency

$$\nu_{\text{receding}} = \left(\frac{1-s/c}{1+s/c}\right)^{1/2}\nu \qquad \nu_{\text{approaching}} = \left(\frac{1+s/c}{1-s/c}\right)^{1/2}\nu$$

Doppler shifts (12A.16a)

where c is the speed of light. For nonrelativistic speeds ($s \ll c$), these expressions simplify to

$$\nu_{\text{receding}} \approx \frac{\nu}{1+s/c} \qquad \nu_{\text{approaching}} \approx \frac{\nu}{1-s/c} \qquad (12\text{A}.16\text{b})$$

Atoms and molecules reach high speeds in all directions in a gas, and a stationary observer detects the corresponding Doppler-shifted range of frequencies. Some molecules approach the observer, some move away; some move quickly, others slowly. The detected spectral 'line' is the absorption or emission profile arising from all the resulting Doppler shifts. As shown in the following *Justification*, the profile reflects the distribution of velocities parallel to the line of sight, which is a bell-shaped Gaussian curve. The Doppler line shape is therefore also a Gaussian (Fig. 12A.5), and we show in the *Justification* that, when the temperature is T and the mass of the atom or molecule is m, then the observed width of the line at half-height (in terms of frequency or wavelength) is

$$\delta\nu_{\text{obs}} = \frac{2\nu}{c}\left(\frac{2kT\ln 2}{m}\right)^{1/2} \qquad \delta\lambda_{\text{obs}} = \frac{2\lambda}{c}\left(\frac{2kT\ln 2}{m}\right)^{1/2}$$

Doppler broadening (12A.17)

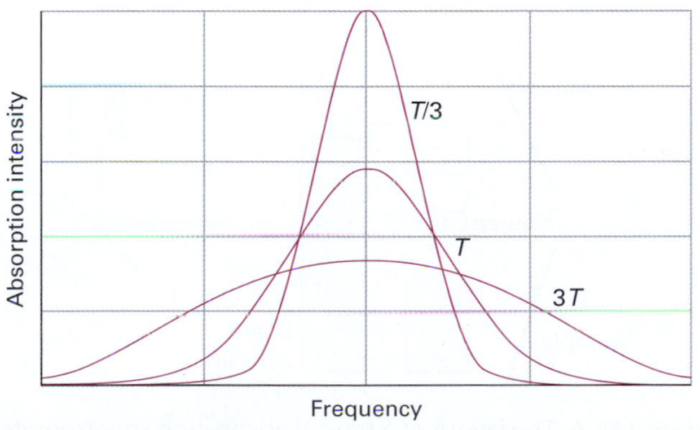

Figure 12A.5 The Gaussian shape of a Doppler-broadened spectral line reflects the Maxwell distribution of speeds in the sample at the temperature of the experiment. Notice that the line broadens as the temperature is increased.

Doppler broadening increases with temperature because the molecules acquire a wider range of speeds. Therefore, to obtain spectra of maximum sharpness, it is best to work with cool samples.

Brief illustration 12A.2 Doppler broadening

For a molecule like N_2 at $T = 300\,\text{K}$,

$$\frac{\delta\nu_{\text{obs}}}{\nu} = \frac{2}{c}\left(\frac{2kT\ln 2}{m_{N_2}}\right)^{1/2} = \frac{2}{2.998\times10^{8}\,\text{m s}^{-1}} \times \left(\frac{2\times\left(1.380\times10^{-23}\,\overbrace{\text{J}}^{\text{kg m}^2\,\text{s}^{-2}}\,\text{K}^{-1}\right)\times(300\,\text{K})\times\ln 2}{4.653\times10^{-26}\,\text{kg}}\right)^{1/2}$$

$$= 2.34\times10^{-6}$$

For a transition wavenumber of $2331\,\text{cm}^{-1}$ (from the Raman spectrum of N_2), corresponding to a frequency of 69.9 THz ($1\,\text{THz} = 10^{12}\,\text{Hz}$), the linewidth is 164 MHz.

Self-test 12A.3 What is the Doppler-broadened linewidth of the transition at 821 nm in atomic hydrogen at 300 K?

Answer: 4.38 GHz

Justification 12A.2 Doppler broadening

It follows from the Boltzmann distribution (*Foundations* B and Topic 15A) that the probability that an atom or molecule of mass m and speed s in a gas phase sample at a temperature T has kinetic energy $E_k = \frac{1}{2}ms^2$ is proportional to $e^{-ms^2/2kT}$. The observed frequencies, ν_{obs}, emitted or absorbed by the molecule are related to its speed by eqn 12A.16b. When $s \ll c$, the Doppler shift in the frequency is

$$\nu_{\text{obs}} - \nu \approx \pm \nu s/c$$

More specifically, the intensity I of a transition at ν_{obs} is proportional to the probability of there being an atom that emits or absorbs at ν_{obs}, so it follows from the Boltzmann distribution and the expression for the Doppler shift in the form $s = (\nu_{\text{obs}} - \nu)c/\nu$ that

$$I(\nu_{\text{obs}}) \propto e^{-mc^2(\nu_{\text{obs}}-\nu)^2/2\nu^2 kT} \qquad (12\text{A}.18)$$

which has the form of a Gaussian function. Because the width at half-height of a Gaussian function $ae^{-(x-b)^2/2\sigma^2}$ (where a, b, and σ are constants) is $\delta x = 2\sigma(2\ln 2)^{1/2}$, $\delta\nu_{\text{obs}}$ can be inferred directly from the exponent of eqn 12A.18 to give eqn 12A.17.

(b) Lifetime broadening

It is found that spectroscopic lines from gas-phase samples are not infinitely sharp even when Doppler broadening has been largely eliminated by working at low temperatures. This residual broadening is due to quantum mechanical effects. Specifically, when the Schrödinger equation is solved for a system that is changing with time, it is found that it is impossible to specify the energy levels exactly. If on average a system survives in a state for a time τ, the lifetime of the state, then its energy levels are blurred to an extent of order $\delta E \approx \hbar/\tau$. With the energy spread expressed as a wavenumber through $\delta E = hc\delta\tilde{\nu}$, and the values of the fundamental constants introduced, this relation becomes

$$\delta\tilde{\nu} \approx \frac{5.3\ \text{cm}^{-1}}{\tau/\text{ps}} \quad \text{Lifetime broadening} \quad (12\text{A}.19)$$

and given an indication of **lifetime broadening** of spectral lines. No excited state has an infinite lifetime; therefore, all states are subject to some lifetime broadening and the shorter the lifetimes of the states involved in a transition the broader are the corresponding spectral lines.

Brief illustration 12A.3 Lifetime broadening

A typical electronic excited state natural lifetime is about $\tau = 10^{-8}$ s $= 1.0 \times 10^4$ ps, corresponding to a linewidth of

$$\delta\tilde{\nu} \approx \frac{5.3\ \text{cm}^{-1}}{1.0\times 10^4} = 5.3\times 10^{-4}\ \text{cm}^{-1}$$

which corresponds to 16 MHz.

Self-test 12A.4 Consider a molecular rotation with a lifetime of about 10^3 s. What is the linewidth of the spectral line?

Answer: $5\times10^{-15}\ \text{cm}^{-1}$ (of the order of 10^{-4} Hz)

Two processes are responsible for the finite lifetimes of excited states. The dominant one for low frequency transitions is **collisional deactivation**, which arises from collisions between atoms or with the walls of the container. If the **collisional lifetime**, the mean time between collisions, is τ_{col}, the resulting collisional linewidth is $\delta E_{\text{col}} \approx \hbar/\tau_{\text{col}}$. Because $\tau_{\text{col}} = 1/z$, where z is the collision frequency, and from the kinetic model of gases (Topic 1B), which implies that z is proportional to the pressure, we conclude that the collisional linewidth is proportional to the pressure. The collisional linewidth can therefore be minimized by working at low pressures.

The rate of spontaneous emission cannot be changed. It is a natural limit to the lifetime of an excited state which cannot be changed by modifying the conditions, and the resulting lifetime broadening is the **natural linewidth** of the transition. Because the rate of spontaneous emission increases as ν^3, the lifetime of the excited state decreases as ν^3, and the natural linewidth increases with the transition frequency. Thus, rotational (microwave) transitions occur at much lower frequencies than vibrational (infrared) transitions and consequently have much longer lifetimes and hence much smaller natural linewidths: at low pressures rotational linewidths are due principally to Doppler broadening.

12A.3 Experimental techniques

We now turn to practical aspects of molecular spectroscopy. Common to all spectroscopic techniques is a **spectrometer**, an instrument that detects the characteristics of radiation scattered, emitted, or absorbed by atoms and molecules. As an example, Fig. 12A.6 shows the general layout of an absorption spectrometer. Radiation from an appropriate source is directed toward a sample and the radiation transmitted strikes a device that separates it into different frequencies. The intensity of radiation at each frequency is then analysed by a suitable detector.

(a) Sources of radiation

Sources of radiation are either *monochromatic*, those spanning a very narrow range of frequencies around a central value, or *polychromatic*, those spanning a wide range of frequencies. Monochromatic sources that can be tuned over a range of frequencies include the *klystron* and the *Gunn diode*, which operate in the microwave range, and lasers (Topic 13C).

Polychromatic sources that take advantage of black-body radiation from hot materials (Topic 7A) can be used from the infrared to the ultraviolet regions of the electromagnetic spectrum. Examples include mercury arcs inside a quartz envelope ($35\ \text{cm}^{-1} < \tilde{\nu} < 200\ \text{cm}^{-1}$), *Nernst filaments* and *globars* ($200\ \text{cm}^{-1} < \tilde{\nu} < 4000\ \text{cm}^{-1}$), and *quartz–tungsten–halogen lamps* ($320\ \text{nm} < \lambda < 2500\ \text{nm}$).

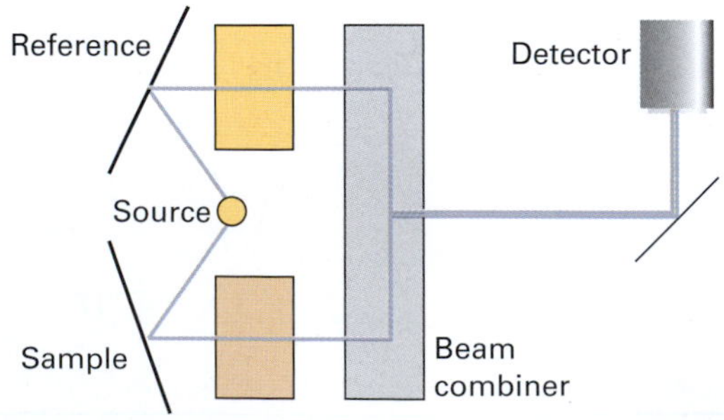

Figure 12A.6 The layout of a typical absorption spectrometer, in which the exciting beams of radiation pass alternately through a sample and a reference cell, and the detector is synchronized with them so that the relative absorption can be determined.

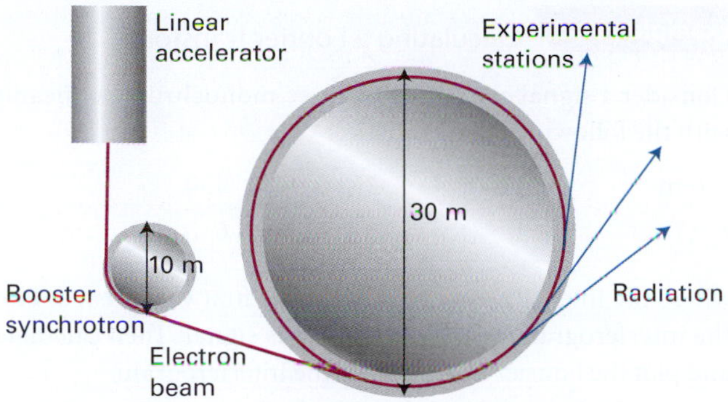

Figure 12A.7 A simple synchrotron storage ring. The electrons injected into the ring from the linear accelerator and booster synchrotron are accelerated to high speed in the main ring. An electron in a curved path is subject to constant acceleration, and an accelerated charge radiates electromagnetic energy. Different versions of synchrotrons use different strategies for generating radiation across a wide spectral range, so that experiments at different frequencies can be conducted simultaneously.

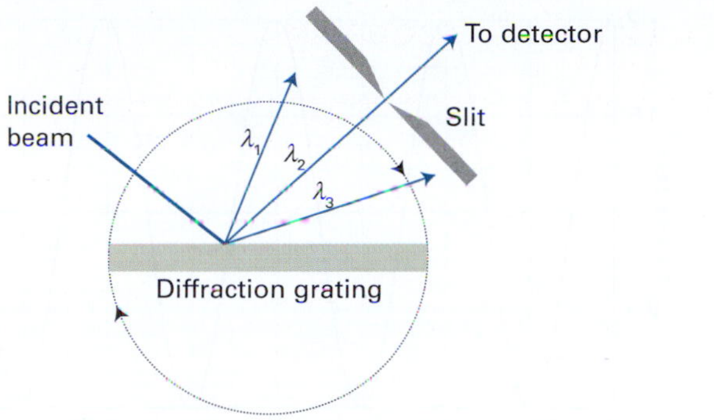

Figure 12A.8 A polychromatic beam is dispersed by a diffraction grating into three component wavelengths λ_1, λ_2, and λ_3. In the configuration shown, only radiation with λ_2 passes through a narrow slit and reaches the detector. Rotating the diffraction grating (as shown by the arrows on the dotted circle) allows λ_1 or λ_3 to reach the detector.

A *gas discharge lamp* is a common source of ultraviolet and visible radiation. In a *xenon discharge lamp*, an electrical discharge excites xenon atoms to excited states, which then emit ultraviolet radiation. In a *deuterium lamp*, excited D_2 molecules dissociate into electronically excited D atoms, which emit intense radiation in the range 200–400 nm.

For certain applications, radiation is generated in a *synchrotron storage ring*, which consists of an electron beam travelling in a circular path with circumferences of up to several hundred metres. As electrons travelling in a circle are constantly accelerated by the forces that constrain them to their path, they generate radiation (Fig. 12A.7). This **synchrotron radiation** spans a wide range of frequencies, including the infrared and X-rays. Except in the microwave region, synchrotron radiation is much more intense than can be obtained by most conventional sources.

(b) Spectral analysis

A common device for the analysis of the wavelengths (or wavenumbers) in a beam of radiation is a *diffraction grating*, which consists of a glass or ceramic plate into which fine grooves have been cut and covered with a reflective aluminium coating. For work in the visible region of the spectrum, the grooves are cut about 1000 nm apart (a spacing comparable to the wavelength of visible light). The grating causes interference between waves reflected from its surface, and constructive interference occurs at specific angles that depend on the frequency of the radiation being used. Thus, each wavelength of light is directed into a specific direction (Fig. 12A.8). In a *monochromator*, a narrow exit slit allows only a narrow range of wavelengths to reach the detector. Turning the grating around an axis perpendicular to the incident and diffracted beams allows different wavelengths to be analysed; in this way, the absorption spectrum is built up one narrow wavelength range at a time. In a *polychromator*, there is no slit and a broad range of wavelengths can be analysed simultaneously by *array detectors*, such as those discussed below.

Many spectrometers, particularly those operating in the infrared and near-infrared, now almost always use **Fourier transform techniques** of spectral detection and analysis. The heart of a Fourier transform spectrometer is a *Michelson interferometer*, a device for analysing the frequencies present in a composite signal. The total signal from a sample is like a chord played on a piano, and the Fourier transform of the signal is equivalent to the separation of the chord into its individual notes, its spectrum.

The Michelson interferometer works by splitting the beam from the sample into two and introducing a varying path difference, *p*, into one of them (Fig. 12A.9). When the two components recombine, there is a phase difference between them, and

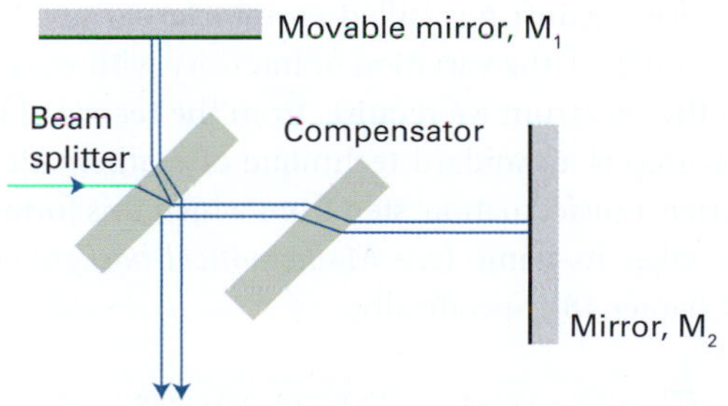

Figure 12A.9 A Michelson interferometer. The beam-splitting element divides the incident beam into two beams with a path difference that depends on the location of the mirror M_1. The compensator ensures that both beams pass through the same thickness of material.

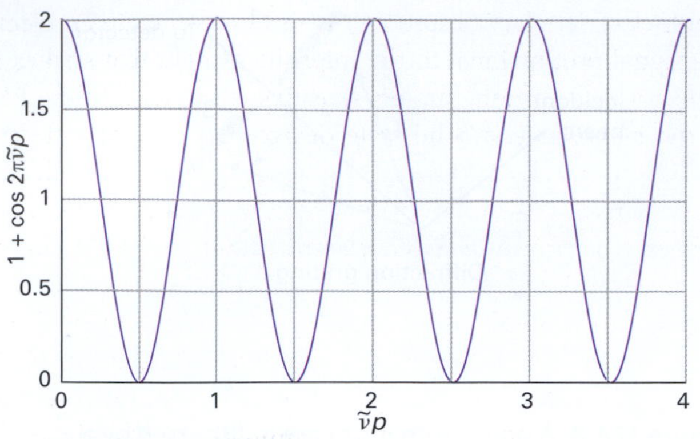

Figure 12A.10 An interferogram produced as the path length p is changed in the interferometer shown in Fig. 12A.9. Only a single frequency component is present in the signal, so the graph is a plot of the function $I(p)=I_0(1+\cos 2\pi\tilde{\nu}p)$, where I_0 is the intensity of the radiation.

they interfere either constructively or destructively depending on the difference in path lengths. The detected signal oscillates as the two components alternately come into and out of phase as the path difference is changed (Fig. 12A.10). If the radiation has wavenumber $\tilde{\nu}$, the intensity of the detected signal due to radiation in the range of wavenumbers $\tilde{\nu}$ to $\tilde{\nu}+d\tilde{\nu}$, which we denote $I(p,\tilde{\nu})d\tilde{\nu}$, varies with p as

$$I(p,\tilde{\nu})d\tilde{\nu}=I(\tilde{\nu})(1+\cos 2\pi\tilde{\nu}p)d\tilde{\nu} \qquad (12A.20)$$

Hence, the interferometer converts the presence of a particular wavenumber component in the signal into a variation in intensity of the radiation reaching the detector. An actual signal consists of radiation spanning a large number of wavenumbers, and the total intensity at the detector, which we write $I(p)$, is the sum of contributions from all the wavenumbers present in the signal:

$$I(p)=\int_0^\infty I(p,\tilde{\nu})d\tilde{\nu}=\int_0^\infty I(\tilde{\nu})(1+\cos 2\pi\tilde{\nu}p)d\tilde{\nu} \qquad (12A.21)$$

A plot of I(p) against p is called an **interferogram**. The problem is to find $I(\tilde{\nu})$, the variation of intensity with wavenumber, which is the spectrum we require, from the record of values of $I(p)$. This step is a standard technique of mathematics, and is the 'Fourier transformation' step from which this form of spectroscopy takes its name (see *Mathematical background* 7 following Chapter 18). Specifically:

$$I(\tilde{\nu})=4\int_0^\infty \{I(p)-\tfrac{1}{2}I(0)\}\cos 2\pi\tilde{\nu}p\, dp \qquad \text{Fourier transformation} \qquad (12A.22)$$

where $I(0)$ is given by eqn 12A.21 with $p=0$. This integration is carried out numerically in a computer connected to the spectrometer, and the output, $I(\tilde{\nu})$, is the transmission spectrum of the sample.

Example 12A.2 Calculating a Fourier transform

Consider a signal consisting of three monochromatic beams with the following characteristics:

$\tilde{\nu}_i/\text{cm}^{-1}$	150	250	450
$I(\tilde{\nu}_i)$	1	3	6

where the intensities are relative to the first value listed. Plot the interferogram associated with this signal. Then calculate and plot the Fourier transform of the interferogram.

Method For a signal consisting of only a few monochromatic beams, the integral in eqn 12A.21 can be replaced by a sum over the finite number of wavenumbers. It follows that the interferogram is

$$I(p)=\sum_i I(\tilde{\nu}_i)(1+\cos 2\pi\tilde{\nu}_i p)$$

In practice, the path difference p does not vary continuously, so the integral over p in eqn 12A.22 must be replaced by a sum over discrete path lengths p_j, in which case the equation to use to generate the Fourier transform of $I(p)$ is

$$I(\tilde{\nu})=4\sum_j \{I(p_j)-\tfrac{1}{2}I(0)\}\cos 2\pi\tilde{\nu}_i p_j$$

It is best to sum over a large number N of data points spanning a relatively large overall path difference P, with $p_j=jP/N$ (see Problem 12A.13). For example, let j range from 0 to 1000 with $P=1.0$ cm, so that the path length difference increases in steps of (1.0/1000) cm = 10 μm.

Answer From the data, the interferogram is

$$\begin{aligned}I(p)&=(1+\cos 2\pi\tilde{\nu}_1 p)+3\times(1+\cos 2\pi\tilde{\nu}_2 p)+6\times(1+\cos 2\pi\tilde{\nu}_3 p)\\&=10+\cos 2\pi\tilde{\nu}_1 p+3\cos 2\pi\tilde{\nu}_2 p+6\cos 2\pi\tilde{\nu}_3 p\end{aligned}$$

This function is plotted in Fig. 12A.11. The calculation of the Fourier transform $I(\tilde{\nu})$ is made easier by the use of mathematical software. The result is shown in Fig. 12A.12.

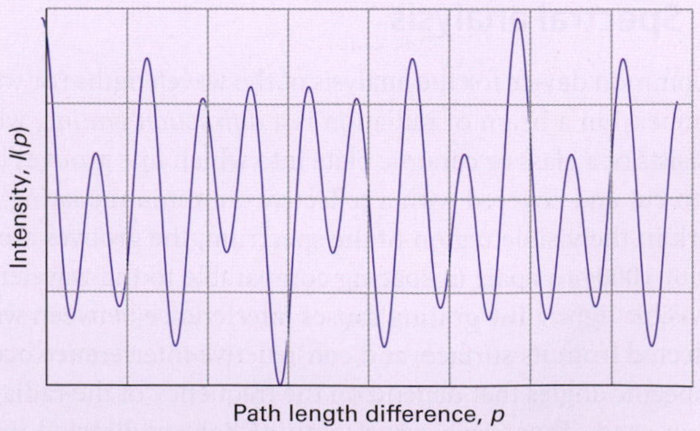

Figure 12A.11 The interferogram calculated from data in *Example* 12A.2.

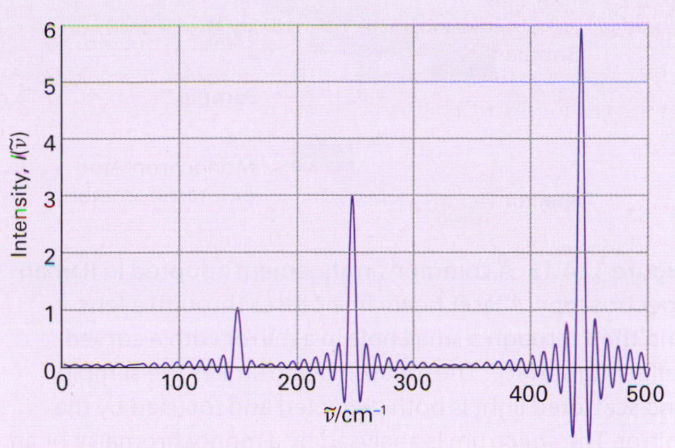

Figure 12A.12 The Fourier transform of the interferogram shown in Fig. 12A.11.

Self-test 12A.5 Explore the effect of varying the wavenumbers of the three components of the radiation on the shape of the interferogram by changing the value of $\tilde{\nu}_3$ to 550 cm^{-1}.

Answer: See Fig. 12A.13.

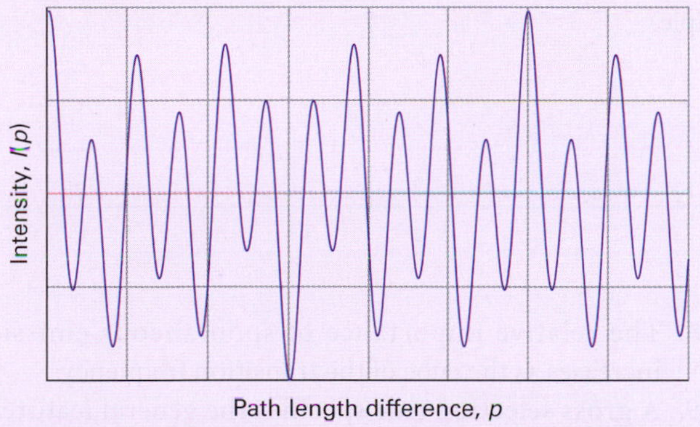

Figure 12A.13 The interferogram calculated from data in *Self-test* 12A.5.

(c) Detectors

A **detector** is a device that converts radiation into an electric signal for appropriate processing and display. Detectors may consist of a single radiation sensing element or of several small elements arranged in one- or two-dimensional arrays.

A microwave detector is typically a *crystal diode* consisting of a tungsten tip in contact with a semiconductor. The most common detectors found in commercial infrared spectrometers are sensitive in the mid-infrared region. In a *photovoltaic device* the potential difference changes upon exposure to infrared radiation. In a *pyroelectric device* the capacitance is sensitive to temperature and hence the presence of infrared radiation.

A common detector for work in the ultraviolet and visible ranges is the *photomultiplier tube* (PMT), in which the photoelectric effect (Topic 7A) is used to generate an electrical signal proportional to the intensity of light that strikes the detector. A common, but less sensitive, alternative to the PMT is the *photodiode*, a solid-state device that conducts electricity when struck by photons because light-induced electron transfer reactions in the detector material create mobile charge carriers (negatively charged electrons and positively charged 'holes').

The *charge-coupled device* (CCD) is a two-dimensional array of several million small photodiode detectors. With a CCD, a wide range of wavelengths that emerge from a polychromator are detected simultaneously, thus eliminating the need to measure light intensity one narrow wavelength range at a time. CCD detectors are the imaging devices in digital cameras, but are also used widely in spectroscopy to measure absorption, emission, and Raman scattering.

(d) Examples of spectrometers

With a proper choice of spectrometer, absorption spectroscopy can probe electronic, vibrational, and rotational transitions in molecules. It is often necessary to modify the general design of Fig. 12A.6 in order to detect weak signals. For example, to detect rotational transitions with a microwave spectrometer it is useful to modulate the transmitted intensity by varying the energy levels with an oscillating electric field. In this **Stark modulation**, an electric field of about 10^5 V m^{-1} (1 kV cm^{-1}) and a frequency of between 10 and 100 kHz is applied to the sample.

Virtually every commercial absorption spectrometer operating in the infrared region and designed for the study of vibrational transitions uses Fourier transform techniques. Their major advantage is that all the radiation emitted by the source is monitored continuously, in contrast to a spectrometer in which a monochromator discards most of the generated radiation. As a result, Fourier transform spectrometers have a higher sensitivity than conventional spectrometers.

Rotational, vibrational, and electronic transitions can be explored by monitoring the spectrum of radiation emitted by a sample. Emission by electronic excited states of molecules has two forms: **fluorescence**, which ceases within a few nanoseconds of the exciting radiation being extinguished, and **phosphorescence**, which may persist for long periods (Topic 13B). In a conventional fluorescence experiment, the source is tuned, often with the use of a monochromator, to a wavelength that causes electronic excitation of the molecule. Typically, the emitted radiation is detected perpendicular to the direction of the exciting beam of radiation, and analysed with a second monochromator (Fig. 12A.14).

In a typical Raman spectroscopy experiment, a monochromatic incident laser beam is passed through the sample and the radiation scattered from the front face of sample is monitored (Fig. 12A.15). Lasers are used as the source of the incident

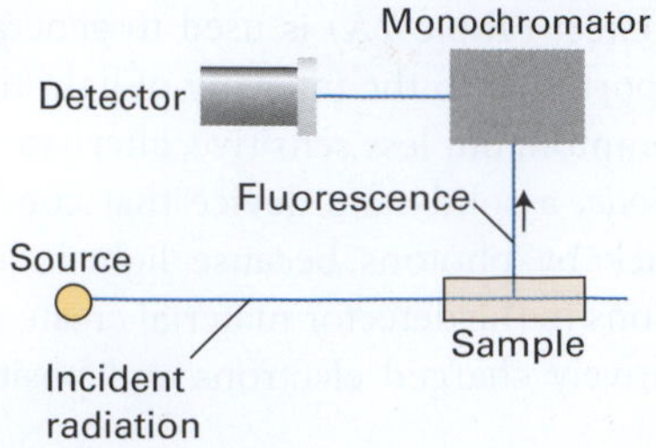

Figure 12A.14 A simple emission spectrometer for monitoring fluorescence, where light emitted by the sample is detected at right angles to the direction of propagation of an incident beam of radiation.

radiation because an intense beam increases the intensity of scattered radiation. The monochromaticity of laser radiation makes possible the observation of frequencies of scattered light that differs only slightly from that of the incident radiation. Such high resolution is particularly useful for observing rotational transitions by Raman spectroscopy. The monochromaticity of laser radiation also allows observations to be made very close to absorption frequencies. Fourier transform instruments are common, as are spectrometers using polychromators connected to CCD detectors.

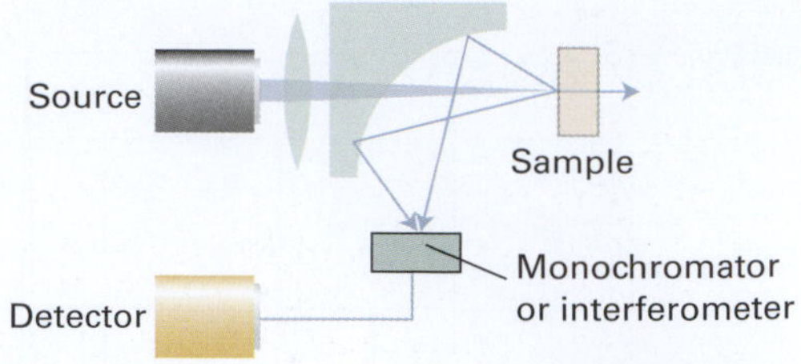

Figure 12A.15 A common arrangement adopted in Raman spectroscopy. A laser beam first passes through a lens and then through a small hole in a mirror with a curved reflecting surface. The focused beam strikes the sample and scattered light is both deflected and focused by the mirror. The spectrum is analysed by a monochromator or an interferometer.

Raman spectroscopy can be used to study rotational and vibrational transitions in molecules. Most commercial instruments are designed for vibrational studies, which lead to applications in biochemistry, art restoration, and monitoring of industrial processes. Raman spectrometers can also be coupled to microscopes, resulting in spectra of very small regions of a sample.

Checklist of concepts

- ☐ 1. In **Raman spectroscopy**, changes in molecular states are explored by examining the frequencies present in the radiation scattered by molecules.
- ☐ 2. **Stokes radiation** is the result of Raman scattering of photons that give up some of their energy during (and emerge with lower frequency after) collisions with molecules.
- ☐ 3. **Anti-Stokes radiation** is the result of Raman scattering of photons that collect some energy during (and emerge with higher frequency after) collisions with molecules.
- ☐ 4. The component of radiation scattered without change of frequency is called **Rayleigh radiation.**
- ☐ 5. A transition from a low energy state to one of higher energy that is driven by an oscillating electromagnetic field is called **stimulated absorption.**
- ☐ 6. A transition driven from high energy to low energy is called **stimulated emission.**
- ☐ 7. A transition from a high energy state to a low energy state occurs by the process of **spontaneous emission** at a rate independent of any radiation also present.
- ☐ 8. The relative importance of spontaneous emission increases as the cube of the transition frequency.
- ☐ 9. A **gross selection rule** specifies the general features a molecule must have if it is to have a spectrum of a given kind.
- ☐ 10. A **specific selection rule** expresses the allowed transitions in terms of the changes in quantum numbers.
- ☐ 11. **Doppler broadening** of a spectral line is caused by the distribution of molecular and atomic speeds in a sample.
- ☐ 12. **Lifetime broadening** arises from the finite lifetime of an excited state and a consequent blurring of energy levels.
- ☐ 13. Collisions between atoms can affect excited state lifetimes and spectral linewidths.
- ☐ 14. The **natural linewidth** of a transition is an intrinsic property that depends on the rate of spontaneous emission at the transition frequency.
- ☐ 15. **A spectrometer** is an instrument that detects the characteristics of radiation scattered, emitted, or absorbed by atoms and molecules.

Checklist of equations

Property	Equation	Comment	Equation number
Ratio of Einstein coefficients of spontaneous and stimulated emission	$A/B=8\pi h\nu^3/c^3$	$B_{fi}=B_{if}$ (= B)	12A.9
Transition dipole moment	$\mu_{fi}=\int \psi_f^* \hat{\mu}\psi_i\,d\tau$	Electric dipole transitions	12A.10
Beer–Lambert law	$I=I_0 10^{-\varepsilon[J]L}$	Uniform sample	12A.11
Absorbance	$A=\log(I_0/I)=-\log T$	Definition	12A.13
Integrated absorption coefficient	$\mathcal{A}=\int_{band}\varepsilon(\tilde{\nu})d\tilde{\nu}$	Definition	12A.15
Doppler broadening	$\delta\nu_{obs}=(2\nu/c)(2kT\ln 2/m)^{1/2}$ $\delta\lambda_{obs}=(2\lambda/c)(2kT\ln 2/m)^{1/2}$		12A.17
Lifetime broadening	$\delta\tilde{\nu}\approx 5.3\ \text{cm}^{-1}/(\tau/\text{ps})$		12A.19
Fourier transformation	$I(\tilde{\nu})=4\int_0^{\infty}\{I(p)-\frac{1}{2}I(0)\}\cos 2\pi\tilde{\nu}p\,dp$	Spectral data collected with a Michelson interferometer	12A.22

12B Molecular rotation

Contents

➤ Why do you need to know this material?

To understand the origin of rotational spectra and to derive structural-information, such as bond lengths, about molecules from them, you need to understand the quantum mechanical treatment of rotation of polyatomic molecules.

➤ What is the key idea?

The energy levels of a molecule modelled as a rigid rotor may be expressed in terms of quantum numbers and parameters related to its moments of inertia.

➤ What do you need to know already?

You need to be familiar with the classical description of rotational motion (*Foundations* B). You also need to be familiar with the particle on a ring and particle on a sphere as quantum mechanical models of rotational motion (Topic 8C).

Topic 8C explores the rotational states of diatomic molecules by using the particle on a ring and particle on a sphere, respectively, as models. Here we use a related but more sophisticated model that can be applied to the rotation of polyatomic molecules.

12B.1 Moments of inertia

The key molecular parameter we need for the description of molecular rotation is the **moment of inertia**, I, of the molecule. The moment of inertia of a molecule is defined as the mass of each atom multiplied by the square of its distance from the rotational axis passing through the centre of mass of the molecule (Fig. 12B.1):

$$I=\sum_i m_i x_i^2 \qquad \textit{Definition} \quad \text{Moment of inertia} \qquad (12B.1)$$

where x_i is the perpendicular distance of the atom i from the axis of rotation. The moment of inertia depends on the masses of the atoms present and the molecular geometry, so we can suspect (and see explicitly in Topic 12C) that microwave spectroscopy will give information about bond lengths and bond angles.

In general, the rotational properties of any molecule can be expressed in terms of the moments of inertia about three perpendicular axes set in the molecule (Fig. 12B.2). The convention is to label the moments of inertia I_a, I_b, and I_c, with the axes chosen so that $I_c \geq I_b \geq I_a$. For linear molecules, the moment of inertia around the internuclear axis is zero (because $x_i=0$ for all the atoms) and the two remaining moments of inertia, which are equal, are denoted simply I. The explicit expressions for the moments of inertia of some symmetrical molecules are given in Table 12B.1.

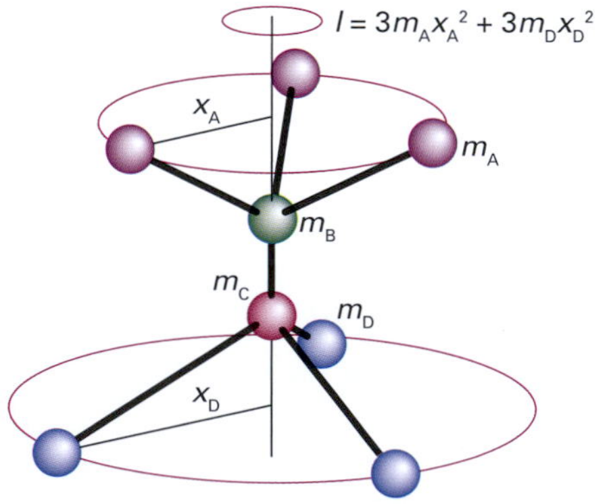

Figure 12B.1 The definition of moment of inertia. In this molecule there are three identical atoms attached to the B atom and three different but mutually identical atoms attached to the C atom. In this example, the centre of mass lies on an axis passing through the B and C atoms, and the perpendicular distances are measured from this axis.

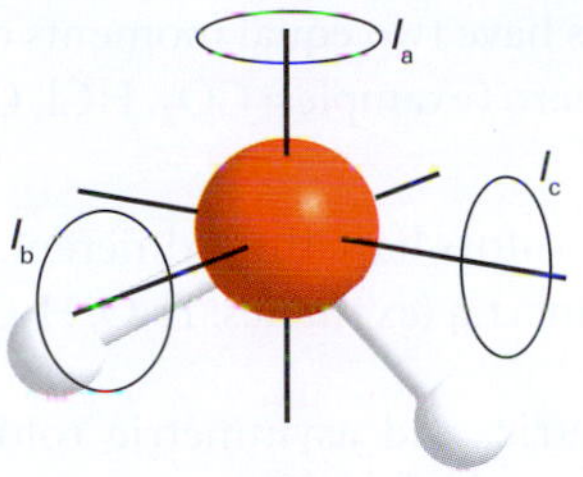

Figure 12B.2 An asymmetric rotor has three different moments of inertia; all three rotation axes coincide at the centre of mass of the molecule.

Table 12B.1 Moments of inertia*

1. *Diatomic molecules*	
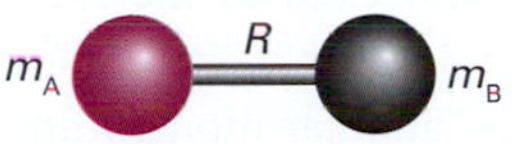	$I=\mu R^2 \qquad \mu=\frac{m_A m_B}{m}$
2. *Triatomic linear rotors*	
	$I=m_A R^2+m_C R'^2-\frac{(m_A R-m_C R')^2}{m}$
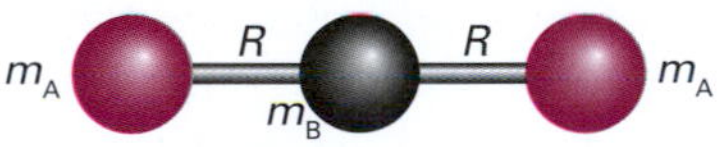	$I=2m_A R^2$
3. *Symmetric rotors*	
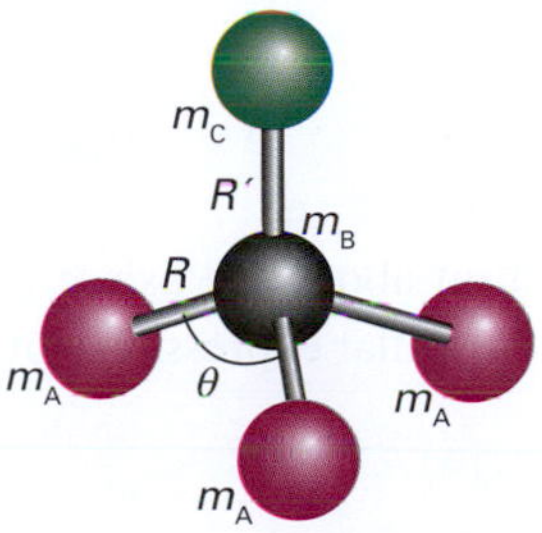	$I_{\parallel}=2m_A(1-\cos\theta)R^2$ $I_{\perp}=m_A(1-\cos\theta)R^2+\frac{m_A}{m}\times(m_B+m_A)(1+2\cos\theta)R^2+\frac{m_C}{m}\{(3m_A+m_B)R'+6m_A R\left[\frac{1}{3}(1+2\cos\theta)\right]^{1/2}\}R'$
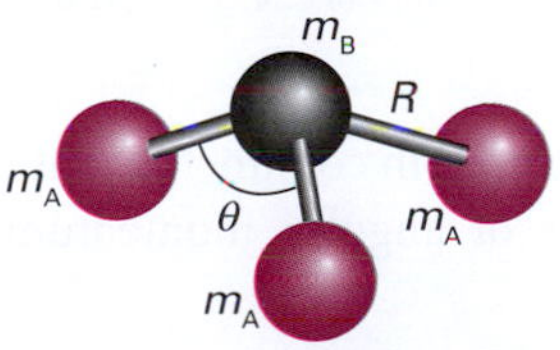	$I_{\parallel}=2m_A(1-\cos\theta)R^2$ $I_{\perp}=m_A(1-\cos\theta)R^2+\frac{m_A m_B}{m}(1+2\cos\theta)R^2$
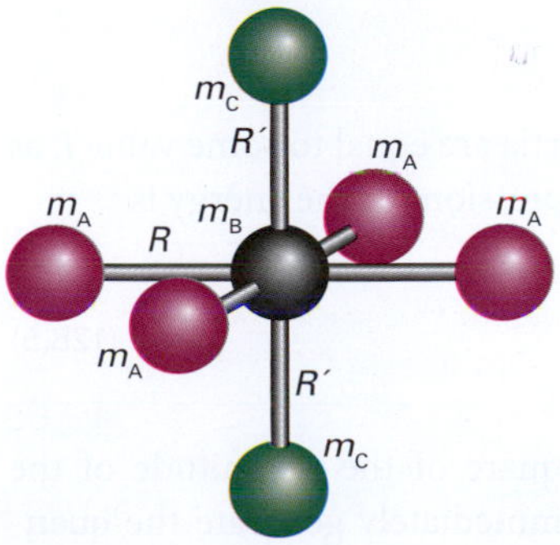	$I_{\parallel}=4m_A R^2$ $I_{\perp}=2m_A R^2+2m_C R'^2$
4. *Spherical rotors*	
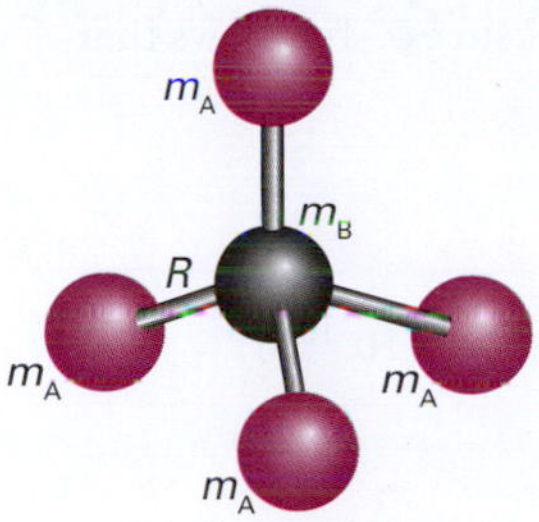	$I=\frac{8}{3}m_A R^2$
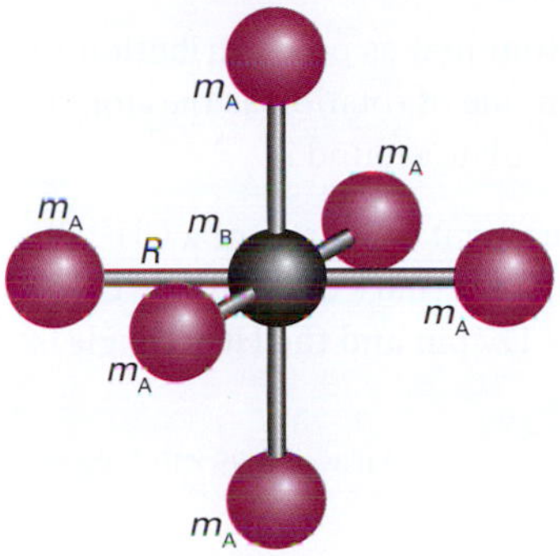	$I=4m_A R^2$

* In each case, m is the total mass of the molecule.

Example 12B.1 Calculating the moment of inertia of a molecule

Calculate the moment of inertia of an H_2O molecule around the axis defined by the bisector of the HOH angle (**1**). The HOH bond angle is 104.5° and the bond length is 95.7 pm. Use $m(^1H)=1.0078m_u$.

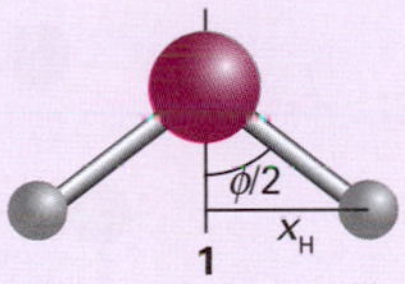

1

Method According to eqn 12B.1, the moment of inertia is the sum of the masses multiplied by the squares of their distances from the axis of rotation. The latter can be expressed by using trigonometry and the bond angle and bond length.

A note on good practice The mass to use in the calculation of the moment of inertia is the actual atomic mass, not the element's molar mass; don't forget to convert from relative masses to actual masses by using the atomic mass constant m_u.

Answer From eqn 12B.1,

$$I=\sum_i m_i x_i^2=m_H x_H^2+0+m_H x_H^2=2m_H x_H^2$$

If the bond angle of the molecule is denoted ϕ and the bond length is R, trigonometry gives $x_H = R\sin\frac{1}{2}\phi$. It follows that

$$I = 2m_H R^2 \sin^2\tfrac{1}{2}\phi$$

Substitution of the data gives

$$\begin{aligned}I &= 2\times(1.0078\times1.6605\times10^{-27}\,\text{kg})\times(9.57\times10^{-11}\,\text{m})^2\\ &\quad\times\sin^2(\tfrac{1}{2}\times104.5°)\\ &= 1.92\times10^{-47}\,\text{kg m}^2\end{aligned}$$

Note that the mass of the O atom makes no contribution to the moment of inertia for this mode of rotation as the atom is immobile while the H atoms circulate around it.

Self-test 12B.1 Calculate the moment of inertia of a $CH^{35}Cl_3$ molecule around a rotational axis that contains the C—H bond. The C—Cl bond length is 177 pm and the HCCl angle is 107°; $m(^{35}Cl) = 34.97m_u$.

Answer: $4.99\times10^{-45}\,\text{kg m}^2$

We shall suppose initially that molecules are **rigid rotors**, bodies that do not distort under the stress of rotation. Rigid rotors can be classified into four types (Fig. 12B.3):

Spherical rotors have three equal moments of inertia (examples: CH_4, SiH_4, and SF_6).

Symmetric rotors have two equal moments of inertia and a third that is non-zero (examples: NH_3, CH_3Cl, and CH_3CN).

Linear rotors have two equal moments of inertia and a third that is zero (examples: CO_2, HCl, OCS, and HC≡CH).

Asymmetric rotors have three different and non-zero moments of inertia (examples: H_2O, H_2CO, and CH_3OH).

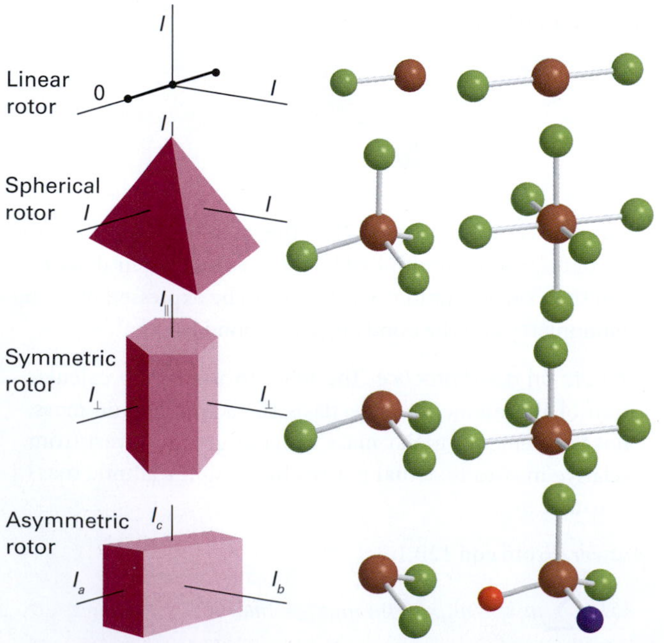

Figure 12B.3 A schematic illustration of the classification of rigid rotors.

Spherical, symmetric, and asymmetric rotors are also called *spherical tops*.

12B.2 The rotational energy levels

The rotational energy levels of a rigid rotor may be obtained by solving the appropriate Schrödinger equation. Fortunately, however, there is a much less onerous short cut to the exact expressions: we note the classical expression for the energy of a rotating body, express it in terms of the angular momentum, and then import the quantum mechanical properties of angular momentum into the equations.

The classical expression for the energy of a body rotating about an axis a is

$$E_a = \tfrac{1}{2}I_a\omega_a^2 \tag{12B.2}$$

where ω_a is the angular velocity about that axis and I_a is the corresponding moment of inertia. A body free to rotate about three axes has an energy

$$E = \tfrac{1}{2}I_a\omega_a^2 + \tfrac{1}{2}I_b\omega_b^2 + \tfrac{1}{2}I_c\omega_c^2 \tag{12B.3}$$

Because the classical angular momentum about the axis a is $J_a = I_a\omega_a$ (eqn B.3 of *Foundations* B) with similar expressions for the other axes, it follows that

$$E = \frac{J_a^2}{2I_a} + \frac{J_b^2}{2I_b} + \frac{J_c^2}{2I_c} \quad \textit{Classical expression} \quad \text{Rotational energy} \tag{12B.4}$$

This is the key equation, which can be used in conjunction with the quantum mechanical properties of angular momentum developed in Topic 8C.

(a) Spherical rotors

When all three moments of inertia are equal to some value I, as in CH_4 and SF_6, the classical expression for the energy is

$$E = \frac{J_a^2 + J_b^2 + J_c^2}{2I} = \frac{\mathcal{J}^2}{2I} \tag{12B.5}$$

where $\mathcal{J}^2 = J_a^2 + J_b^2 + J_c^2$ is the square of the magnitude of the angular momentum. We can immediately generate the quantum expression by making the replacement

$$\mathcal{J}^2 \to J(J+1)\hbar^2 \quad J=0,1,2,\ldots$$

where J is the angular momentum quantum number. Therefore, the energy of a spherical rotor is confined to the values

$$E_J = J(J+1)\frac{\hbar^2}{2I} \qquad J=0,1,2,\ldots \qquad \textit{Spherical rotor} \quad \text{Rotational energy levels} \qquad (12B.6)$$

The resulting ladder of energy levels is illustrated in Fig. 12B.4. The energy is normally expressed in terms of the **rotational constant**, $\tilde{B}$, of the molecule, where

$$hc\tilde{B}=\frac{\hbar^2}{2I} \quad \text{so} \quad \tilde{B}=\frac{\hbar}{4\pi cI} \qquad \textit{Spherical rotor} \quad \text{Rotational constant} \qquad (12B.7)$$

It follows that $\tilde{B}$ is a wavenumber. The expression for the energy is then

$$E_J = hc\tilde{B}J(J+1) \quad J=0,1,2,\ldots \qquad \textit{Spherical rotor} \quad \text{Energy levels} \qquad (12B.8)$$

It is also common to express the rotational constant as a frequency and to denote it simply B. Then $B=\hbar/4\pi I$ and the energy is $E=hBJ(J+1)$. The two quantities are related by $B=c\tilde{B}$.

The energy of a rotational state is normally reported as the **rotational term**, $\tilde{F}(J)$, a wavenumber, by division of both sides of eqn 12B.8 by hc:

$$\tilde{F}(J)=\tilde{B}J(J+1) \qquad \textit{Spherical rotor} \quad \text{Rotational terms} \qquad (12B.9)$$

To express the rotational term as a frequency, use $F=c\tilde{F}$. The separation of adjacent levels is

$$\tilde{F}(J+1)-\tilde{F}(J)=\tilde{B}(J+1)(J+2)-\tilde{B}J(J+1)=2\tilde{B}(J+1) \qquad (12B.10)$$

Because the rotational constant is inversely proportional to I, large molecules have closely spaced rotational energy levels.

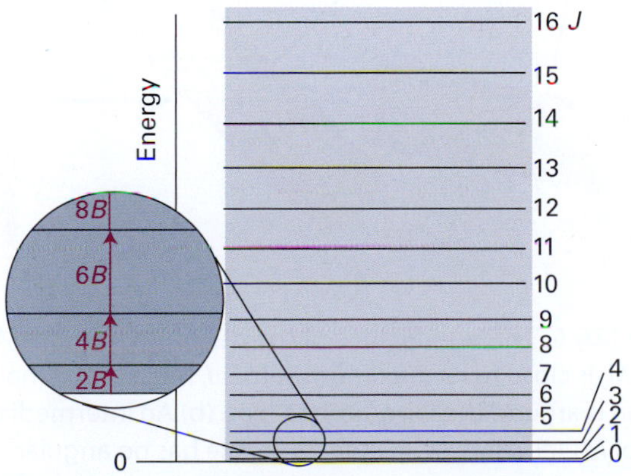

Figure 12B.4 The rotational energy levels of a linear or spherical rotor. Note that the energy separation between neighbouring levels increases as J increases.

Brief illustration 12B.1 Spherical rotors

Consider $^{12}C^{35}Cl_4$: from Table 12B.1, the C−Cl bond length ($R_{C-Cl}=177$ pm) and the mass of the ^{35}Cl nuclide ($m(^{35}Cl)=34.97m_u$), we find

$$I=\tfrac{8}{3}m(^{35}\text{Cl})R_{\text{C-Cl}}^2$$

$$=\tfrac{8}{3}\times\overbrace{(5.807\times10^{-26}\text{ kg})}^{34.97\times(1.66054\times10^{-27}\text{ kg})}\times(1.77\times10^{-10}\text{ m})^2$$

$$=4.85\times10^{-45}\text{ kg m}^2$$

and, from eqn 12B.7,

$$\tilde{B}=\frac{1.05457\times10^{-34}\ \overset{\text{kg m}^2\text{s}^{-2}}{\widehat{\text{J}}}\ \text{s}}{4\pi\times(2.998\times10^8\text{ m s}^{-1})\times(4.85\times10^{-45}\text{ kg m}^2)}$$

$$=5.77\text{ m}^{-1}=0.0577\text{ cm}^{-1}$$

It follows from eqn 12B.10 that the energy separation between the $J=0$ and $J=1$ levels is $\tilde{F}(1)-\tilde{F}(0)=2\tilde{B}=0.1154\text{ cm}^{-1}$.

Self-test 12B.2 Calculate $\tilde{F}(2)-\tilde{F}(0)$ for $^{12}C^{35}Cl_4$.

Answer: $6\tilde{B}=0.3462\text{ cm}^{-1}$

(b) Symmetric rotors

In symmetric rotors, all three moments of inertia are non-zero but two are the same and different from the third (as in CH_3Cl, NH_3, and C_6H_6); the unique axis of the molecule is its **principal axis** (or *figure axis*). We shall write the unique moment of inertia (that about the principal axis) as $I_{\parallel}$ and the other two as $I_{\perp}$. If $I_{\parallel}>I_{\perp}$, the rotor is classified as **oblate** (like a pancake, and C_6H_6); if $I_{\parallel}<I_{\perp}$ it is classified as **prolate** (like a cigar, and CH_3Cl). The classical expression for the energy, eqn 12B.5, becomes

$$E=\frac{J_b^2+J_c^2}{2I_{\perp}}+\frac{J_a^2}{2I_{\parallel}} \qquad (12B.11)$$

Again, this expression can be written in terms of $\mathcal{J}^2=J_a^2+J_b^2+J_c^2$:

$$E=\frac{\mathcal{J}^2-J_a^2}{2I_{\perp}}+\frac{J_a^2}{2I_{\parallel}}=\frac{\mathcal{J}^2}{2I_{\perp}}+\left(\frac{1}{2I_{\parallel}}-\frac{1}{2I_{\perp}}\right)J_a^2 \qquad (12B.12)$$

Now we generate the quantum expression by first replacing $\mathcal{J}^2$ by $J(J+1)\hbar^2$. Then, using the quantum theory of angular momentum (Topic 8C), we note that the component of angular momentum about any axis is restricted to the values $K\hbar$, with $K=0, \pm1, \ldots, \pm J$. (K is the quantum number used to signify a component on the principal axis; M_J is reserved for a component on an externally defined axis.) Therefore, we also replace J_a^2 by $K^2\hbar^2$. It follows that the rotational terms are

$$\tilde{F}(J,K)=\tilde{B}J(J+1)+(\tilde{A}-\tilde{B})K^2 \qquad J=0,1,2,\ldots \quad K=0,\pm1,\ldots,\pm J$$ *Symmetric rotor* Rotational terms (12B.13)

with

$$\tilde{A}=\frac{\hbar}{4\pi c I_{\parallel}} \qquad \tilde{B}=\frac{\hbar}{4\pi c I_{\perp}} \tag{12B.14}$$

As eqn 12B.13 matches what we should expect for the dependence of the energy levels on the two distinct moments of inertia of the molecule:

Physical interpretation

- When $K=0$, there is no component of angular momentum about the principal axis, and the energy levels depend only on $I_{\perp}$ (Fig. 12B.5).
- When $K=\pm J$, almost all the angular momentum arises from rotation around the principal axis, and the energy levels are determined largely by $I_{\parallel}$.
- The sign of K does not affect the energy because opposite values of K correspond to opposite senses of rotation, and the energy does not depend on the sense of rotation.

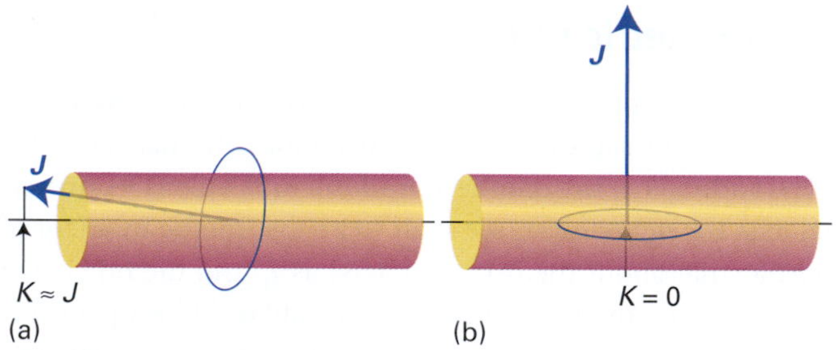

Figure 12B.5 The significance of the quantum number K. (a) When $|K|$ is close to its maximum value, J, most of the molecular rotation is around the figure axis. (b) When $K=0$ the molecule has no angular momentum about its principal axis: it is undergoing end-over-end rotation.

Example 12B.2 Calculating the rotational energy levels of a symmetric rotor

A $^{14}NH_3$ molecule is a symmetric rotor with bond length 101.2 pm and HNH bond angle 106.7°. Calculate its rotational terms.

A note on good practice To calculate moments of inertia precisely, it is necessary to specify the nuclide.

Method Begin by calculating the rotational constants $\tilde{A}$ and $\tilde{B}$ by using the expressions for moments of inertia given in Table 12B.1 and eqn 12B.14. Then use eqn 12B.13 to find the rotational terms.

Answer Substitution of $m_A=1.0078m_u$, $m_B=14.0031\,m_u$, $R=101.2$ pm, and $\theta=106.7°$ into the second of the symmetric rotor expressions in Table 12B.1 gives $I_{\parallel}=4.4128\times10^{-47}$ kg m^2 and $I_{\perp}=2.8059\times10^{-47}$ kg m^2. Hence, by the same kind of calculations as in *Brief illustration* 12B.1, $\tilde{A}=6.344$ cm^{-1} and $\tilde{B}=9.977$ cm^{-1}. It follows from eqn 12B.13 that

$$\tilde{F}(J,K)/\text{cm}^{-1}=9.977\times J(J+1)-3.933K^2$$

Multiplication by c converts $\tilde{F}(J,K)$ to a frequency, denoted $F(J,K)$:

$$F(J,K)/\text{GHz}=299.1\times J(J+1)-108.9K^2$$

For $J=1$, the energy needed for the molecule to rotate mainly about its figure axis ($K=\pm J$) is equivalent to 16.32 cm^{-1} (489.3 GHz), but end-over-end rotation ($K=0$) corresponds to 19.95 cm^{-1} (598.1 GHz).

Self-test 12B.3 A $CH_3{}^{35}Cl$ molecule has a C—Cl bond length of 178 pm, a C—H bond length of 111 pm, and an HCH angle of 110.5°. Calculate its rotational energy terms.

Answer: $\tilde{F}(J,K)/\text{cm}^{-1}=0.444J(J+1)+4.58K^2$; also $F(J,K)/\text{GHz}=13.3J(J+1)+137K^2$

The energy of a symmetric rotor depends on J and K, and each level except those with $K=0$ is doubly degenerate: the states with K and $-K$ have the same energy. However, we must not forget that the angular momentum of the molecule has a component on an external, laboratory-fixed axis. This component is quantized, and its permitted values are $M_J\hbar$, with $M_J=0$, ±1, …, $\pm J$, giving $2J+1$ values in all (Fig. 12B.6). The quantum number M_J does not appear in the expression for the energy, but it is necessary for a complete specification of the state of

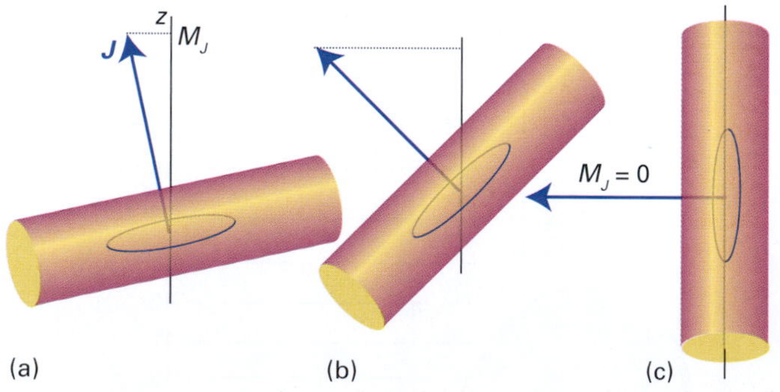

Figure 12B.6 The significance of the quantum number M_J. (a) When M_J is close to its maximum value, J, most of the molecular rotation is around the laboratory z-axis. (b) An intermediate value of M_J. (c) When $M_J=0$ the molecule has no angular momentum about the z-axis. All three diagrams correspond to a state with $K=0$; there are corresponding diagrams for different values of K, in which the angular momentum makes a different angle to the molecule's principal axis.

the rotor. Consequently, all $2J+1$ orientations of the rotating molecule have the same energy. It follows that a symmetric rotor level is $2(2J+1)$-fold degenerate for $K\neq 0$ and $(2J+1)$-fold degenerate for $K=0$.

A spherical rotor can be regarded as a version of a symmetric rotor in which $\tilde{A}=\tilde{B}$. The quantum number K may still take any one of $2J+1$ values, but the energy is independent of which value it takes. Therefore, as well as having a $(2J+1)$-fold degeneracy arising from its orientation in space, the rotor also has a $(2J+1)$-fold degeneracy arising from its orientation with respect to an arbitrary axis in the molecule. The overall degeneracy of a symmetric rotor with quantum number J is therefore $(2J+1)^2$. This degeneracy increases very rapidly: when $J=10$, for instance, there are 441 states of the same energy.

(c) Linear rotors

For a linear rotor (such as CO_2, HCl, and C_2H_2), in which the nuclei are regarded as mass points, the rotation occurs only about an axis perpendicular to the line of atoms and there is zero angular momentum around the line. Therefore, the component of angular momentum around the figure axis of a linear rotor is identically zero, and $K\equiv 0$ in eqn 12B.13. The rotational terms of a linear molecule are therefore

$$\tilde{F}(J)=\tilde{B}J(J+1) \qquad J=0,1,2,\ldots \qquad \textit{Linear rotor} \quad \text{Rotational terms} \qquad (12B.15)$$

This expression is the same as eqn 12B.9 but we have arrived at it in a significantly different way: here $K\equiv 0$ but for a spherical rotor $\tilde{A}=\tilde{B}$. A linear rotor has $2J+1$ components on the laboratory axis, so its degeneracy is $2J+1$.

Brief illustration 12B.2 Linear rotors

Equation 12B.10 for the energy separation of adjacent levels of a spherical rotor also applies to linear rotors. For $^{1}H^{35}Cl$, $\tilde{F}(3)-\tilde{F}(2)=63.56\ cm^{-1}$, and it follows that $6\tilde{B}=63.56\ cm^{-1}$ and $\tilde{B}=10.59\ cm^{-1}$.

Self-test 12B.4 For $^{1}H^{81}Br$, $\tilde{F}(1)-\tilde{F}(0)=16.93\ cm^{-1}$. Determine the value of $\tilde{B}$.

Answer: $8.465\ cm^{-1}$

(d) Centrifugal distortion

We have treated molecules as rigid rotors. However, the atoms of rotating molecules are subject to centrifugal forces that tend to distort the molecular geometry and change the moments of inertia (Fig. 12B.7). The effect of centrifugal distortion on a diatomic molecule is to stretch the bond and hence to increase the moment of inertia. As a result, centrifugal distortion reduces the rotational constant and consequently the energy levels are slightly closer than the rigid-rotor expressions predict. The effect is usually taken into account largely empirically by subtracting a term from the energy and writing

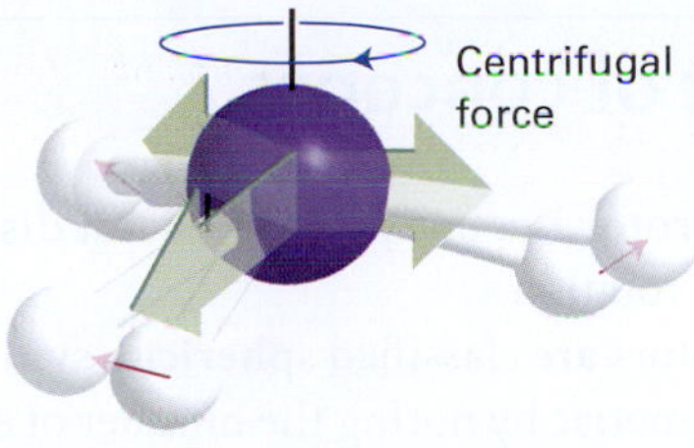

Figure 12B.7 The effect of rotation on a molecule. The centrifugal force arising from rotation distorts the molecule, opening out bond angles and stretching bonds slightly. The effect is to increase the moment of inertia of the molecule and hence to decrease its rotational constant.

$$\tilde{F}(J)=\tilde{B}J(J+1)-\tilde{D}_J J^2(J+1)^2 \qquad \text{Rotational terms affected by centrifugal distortion} \qquad (12B.16)$$

The parameter $\tilde{D}_J$ is the **centrifugal distortion constant**. It is large when the bond is easily stretched. The centrifugal distortion constant of a diatomic molecule is related to the vibrational wavenumber of the bond, $\tilde{\nu}$ (which, as we see in Topic 12D, is a measure of its stiffness), through the approximate relation (see Problem 12.2)

$$\tilde{D}_J=\frac{4\tilde{B}^3}{\tilde{\nu}^2} \qquad \text{Centrifugal distortion constant} \qquad (12B.17)$$

Hence the observation of the convergence of the rotational levels as J increases can be interpreted in terms of the rigidity of the bond.

Brief illustration 12B.3 The effect of centrifugal distortion

For $^{12}C^{16}O$, $\tilde{B}=1.931\ cm^{-1}$ and $\tilde{\nu}=2170\ cm^{-1}$. It follows that

$$\tilde{D}_J=\frac{4\times(1.931\ cm^{-1})^3}{(2170\ cm^{-1})^2}=6.116\times 10^{-6}\ cm^{-1}$$

and that, because $\tilde{D}_J \ll \tilde{B}$, centrifugal distortion has a very small effect on the energy levels.

Self-test 12B.5 Does centrifugal distortion increase or decrease the separation between adjacent energy levels?

Answer: decrease

Checklist of concepts

- ☐ 1. A **rigid rotor** is a body that does not distort under the stress of rotation.
- ☐ 2. Rigid rotors are classified **spherical, symmetric, linear,** or **asymmetric** by noting the number of equal principal moments of inertia.
- ☐ 3. **Symmetric rotors** are classified as prolate or oblate.
- ☐ 4. A linear rotor rotates only about an axis perpendicular to the line of atoms.
- ☐ 5. The **degeneracies** of spherical, symmetric ($K \neq 0$), and linear rotors are $(2J+1)^2$, $2(2J+1)$, and $2J+1$, respectively.
- ☐ 6. **Centrifugal distortion** arises from forces that change the geometry of a molecule.

Checklist of equations

Property	Equation	Comment	Equation number
Moment of inertia	$I=\sum_i m_i x_i^2$	x_i is perpendicular distance of atom i from the axis of rotation	12B.1
Rotational terms of a spherical or linear rotor	$\tilde{F}(J)=\tilde{B}J(J+1)$	$J=0, 1, 2, \ldots$ $\tilde{B}=\hbar/4\pi cI$	12B.9, 12B.15
Rotational terms of a symmetric rotor	$\tilde{F}(J,K)=\tilde{B}J(J+1)+(\tilde{A}-\tilde{B})K^2$	$J=0, 1, 2, \ldots$ $K=0, \pm1, \ldots, \pm J$ $\tilde{A}=\hbar/4\pi cI_{\parallel}$ $\tilde{B}=\hbar/4\pi cI_{\perp}$	12B.13
Rotational terms of a spherical or linear rotor affected by centrifugal distortion	$\tilde{F}(J)=\tilde{B}J(J+1)-\tilde{D}_J J^2(J+1)^2$		12B.16
Centrifugal distortion constant	$\tilde{D}_J=4\tilde{B}^3/\tilde{\nu}^2$		12B.17

12C Rotational spectroscopy

Contents

➤ Why do you need to know this material?

Rotational spectroscopy provides very precise details of bond lengths and bond angles of molecules in the gas phase. Transitions between rotational levels also contribute to vibrational and electronic spectra and are used in the investigation of gas-phase reactions such as those taking place in the atmosphere.

➤ What is the key idea?

Analysis of rotational spectra yields the bond lengths and dipole moments of molecules in the gas phase.

➤ What do you need to know already?

You should be familiar with the quantum mechanical treatment of molecular rotation (Topic 12B), the general principles of molecular spectroscopy (Topic 12A), and the Pauli principle (Topic 9B).

Pure rotational spectra, in which only the rotational state of a molecule changes, can be observed only in the gas phase. In spite of this limitation, rotational spectroscopy can provide a wealth of information about molecules, including precise bond lengths and dipole moments.

Our approach to the description of rotational spectra consists of developing the gross and specific selection rules for rotational transitions, examining the appearance of rotational spectra, and exploring the information that can be obtained from the spectra. This material is also used in the discussion of the fine details observed in vibrational spectra (Topic 12D) and electronic spectra (Topic 13A). Our discussion of the appearance of rotational spectra is based principally on the expression for the rotational terms of a linear rotor developed in Topic 12B:

$$\text{Eqn. 12B.9:}\quad \tilde{F}(J)=\tilde{B}J(J+1) \qquad \tilde{B}=\hbar/4\pi cI$$

where I is the moment of inertia of the molecule (the energies themselves are $E_J=hc\tilde{F}(J)$). The same expression applies to spherical rotors; the expression for symmetric rotors is slightly more elaborate:

$$\text{Eqn.12B.13:}\quad \tilde{F}(J,K)=\tilde{B}J(J+1)+(\tilde{A}-\tilde{B})K^2$$
$$\tilde{A}=\hbar/4\pi cI_{\parallel} \quad \tilde{B}=\hbar/4\pi cI_{\perp}$$

The values allowed to J, K, and M_J (which does not affect the energy but is needed to define the state fully) are set out in Topic 12B.

12C.1 Microwave spectroscopy

Typical values of the rotational constant $\tilde{B}$ for small molecules are in the region of 0.1–10 cm^{-1} (Topic 12B); two examples are 0.356 cm^{-1} for NF_3 and 10.59 cm^{-1} for HCl. It follows that rotational transitions can be studied with **microwave spectroscopy,** a technique that monitors the absorption or emission of radiation in the microwave region of the spectrum.

(a) Selection rules

We show in the following *Justification* that the gross selection rule for the observation of a pure rotational transition in a microwave spectrum is that a molecule must have a permanent electric dipole moment. That is, *to absorb or emit microwave radiation and undergo a pure rotational transition, a molecule must be polar.* The classical basis of this rule is that a polar molecule appears to possess a fluctuating dipole when rotating but a nonpolar molecule does not (Fig. 12C.1). The permanent

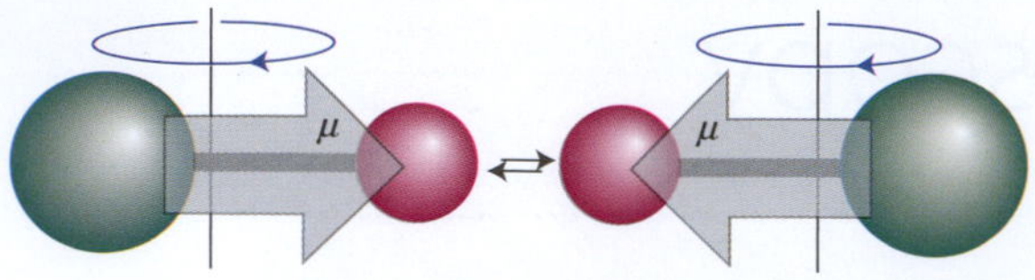

Figure 12C.1 To a stationary observer, a rotating polar molecule looks like an oscillating dipole which can stir the electromagnetic field into oscillation (and vice versa for absorption). This picture is the classical origin of the gross selection rule for rotational transitions.

dipole can be regarded as a handle with which the molecule stirs the electromagnetic field into oscillation (and vice versa for absorption).

Brief illustration 12C.1 Gross selection rules for microwave spectroscopy

Homonuclear diatomic molecules and nonpolar polyatomic molecules, such as CO_2, $CH_2{=}CH_2$, and C_6H_6, are rotationally inactive. On the other hand, OCS and H_2O are polar, and have microwave spectra. Spherical rotors cannot have electric dipole moments unless they become distorted by rotation, so they are rotationally inactive except in special cases. An example of a spherical rotor that does become sufficiently distorted for it to acquire a dipole moment is SiH_4, which has a dipole moment of about 8.3 μD by virtue of its rotation when $J \approx 10$ (for comparison, HCl has a permanent dipole moment of 1.1 D; molecular dipole moments and their units are discussed in Topic 16A).

Self-test 12C.1 Which of the molecules H_2, NO, N_2O, CH_4 can have a pure rotational spectrum?

Answer: NO, N_2O

The specific rotational selection rules are found by evaluating the transition dipole moment (Topic 12A) between rotational states. We show in the following *Justification* that, for a linear molecule, the transition moment vanishes unless the following conditions are fulfilled:

$$\Delta J = \pm 1 \quad \Delta M_J = 0, \pm 1 \quad \textit{Linear rotors} \quad \text{Rotational selection rules} \quad (12C.1)$$

The transition $\Delta J = +1$ corresponds to absorption and the transition $\Delta J = -1$ corresponds to emission.

- The allowed change in J arises from the conservation of angular momentum when a photon, a spin-1 particle, is emitted or absorbed (Fig. 12C.2).
- The allowed change in M_J also arises from the conservation of angular momentum when a photon is emitted into or absorbed from a specific direction.

Physical interpretation

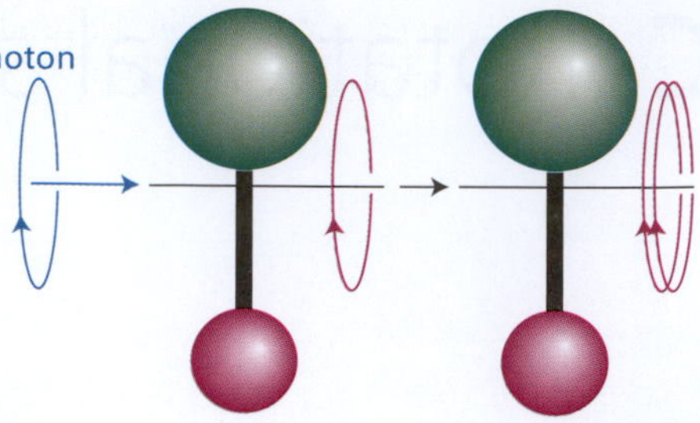

Figure 12C.2 When a photon is absorbed by a molecule, the angular momentum of the combined system is conserved. If the molecule is rotating in the same sense as the spin of the incoming photon, then J increases by 1.

Justification 12C.1 Selection rules for microwave spectra

The starting point for any discussion about selection rules is the transition dipole moment (Topic 12A) and the total wavefunction for a molecule, which can be written as $\psi_{\text{total}} = \psi_{\text{c.m.}}\psi$, where $\psi_{\text{c.m.}}$ describes the motion of the centre of mass and ψ describes the internal motion of the molecule. The Born–Oppenheimer approximation (Topic 10A) allows us to write ψ as the product of an electronic part, ψ_ε, a vibrational part, ψ_ν, and a rotational part, which for a diatomic molecule can be represented by the spherical harmonics $Y_{J,M_J}(\theta,\phi)$ (Topic 8C). The transition dipole moment for the spectroscopic transition i→f can now be written as

$$\boldsymbol{\mu}_{\text{fi}} = \int \psi^*_{\varepsilon_\text{f}} \psi^*_{\nu_\text{f}} Y^*_{J_\text{f},M_{J,\text{f}}} \hat{\boldsymbol{\mu}} \psi_{\varepsilon_\text{i}} \psi_{\nu_\text{i}} Y_{J_\text{i},M_{J,\text{i}}} \text{d}\tau \quad (12C.2)$$

and our task is to explore conditions for which this integral does not vanish.

For a pure rotational transition the initial and final electronic and vibrational states are the same, and we identify $\boldsymbol{\mu}_\text{i} = \int \psi^*_{\varepsilon_\text{i}} \psi^*_{\nu_\text{i}} \hat{\boldsymbol{\mu}} \psi_{\varepsilon_\text{i}} \psi_{\nu_\text{i}} \text{d}\tau$ with the *permanent* electric dipole moment of the molecule in the state i. Equation 12C.2 then becomes

$$\boldsymbol{\mu}_{\text{fi}} = \int Y^*_{J_\text{f},M_{J,\text{f}}} \boldsymbol{\mu}_\text{i} Y_{J_\text{i},M_{J,\text{i}}} \text{d}\tau \quad (12C.3)$$

The remaining integration is over the angles representing the orientation of the molecule. We see immediately that the molecule must have a permanent dipole moment in order to have a microwave spectrum. This is the gross selection rule for microwave spectroscopy.

From this point on, the deduction of the specific selection rules proceeds as in the case of atomic transitions (Topic 9C), and makes use of the fact that the three components of the dipole moment (Fig. 12C.3) are

$$\mu_{\text{i},x} = \mu_0 \sin\theta\cos\phi \quad \mu_{\text{i},y} = \mu_0 \sin\theta\sin\phi \quad \mu_{\text{i},z} = \mu_0 \cos\theta \quad (12C.4)$$

and can be expressed in terms of the spherical harmonics $Y_{j,m}$, with $j = 1$ and $m = 0, \pm 1$ (see *Justification* 9C.1). The condition for the non-vanishing of the integral over the product of three spherical harmonics, which is described in Topic 9C, then implies that

$$\int Y^*_{J_f,M_{J,f}} Y_{j,m} Y_{J_i,M_{J,i}} \mathrm{d}\tau_{\text{angles}} = 0 \qquad (12\text{C}.5)$$

unless $M_{J,f} = M_{J,i} + m$ and lines of length J_f, J_i, and j can form a triangle (such as 1, 2, and 3, or 1, 1, and 1, but not 1, 2, and 4). By exactly the same argument as in *Justification* 9C.1, we conclude that $J_f - J_i = \pm 1$ and $M_{J,f} - M_{J,i} = 0$ or ± 1.

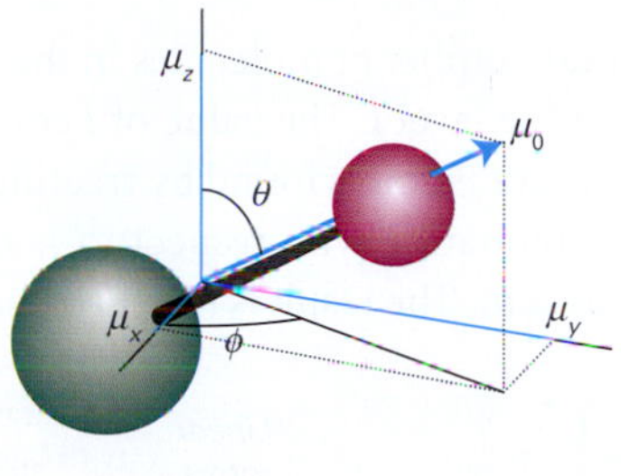

Figure 12C.3 The axis system used in the calculation of the transition dipole moment.

When the transition moment is evaluated for all possible relative orientations of the molecule to the line of flight of the photon, it is found that the total $J+1 \leftrightarrow J$ transition intensity is proportional to

$$|\mu_{J+1,J}|^2 = \left(\frac{J+1}{2J+1}\right)\mu_0^2 \qquad (12\text{C}.6)$$

where μ_0 is the permanent electric dipole moment of the molecule. The intensity is proportional to the square of μ_0, so strongly polar molecules give rise to much more intense rotational lines than less polar molecules.

For symmetric rotors, an additional selection rule states that $\Delta K = 0$. To understand this rule, consider the symmetric rotor NH_3, where the electric dipole moment lies parallel to the figure axis. Such a molecule cannot be accelerated into different states of rotation around this axis by the absorption of radiation, so $\Delta K = 0$. Therefore, for symmetric rotors the selection rules are:

$$\Delta J = \pm 1 \quad \Delta M_J = 0, \pm 1 \quad \Delta K = 0$$ *Symmetric rotors* — Rotational selection rules (12C.7)

The degeneracy associated with the quantum number M_J (the orientation of the rotation in space) is partly removed when an electric field is applied to a polar molecule (for example, HCl or NH_3), as illustrated in Fig. 12C.4. The splitting of states by an electric field is called the **Stark effect**. The energy shift depends on the permanent electric dipole moment, μ_0, so the observation of the Stark effect can be used to measure the magnitudes of electric dipole moments with a rotational spectrum.

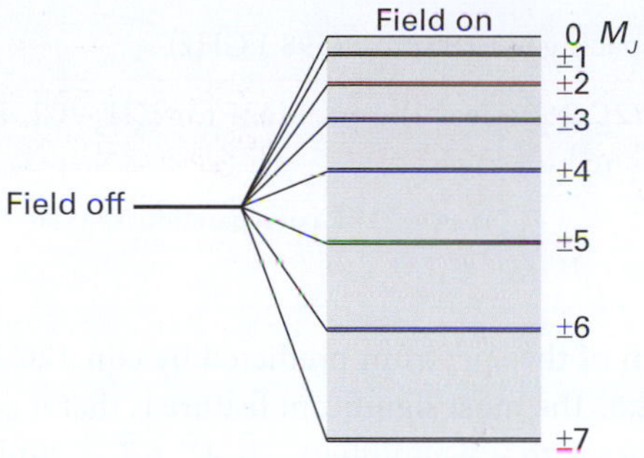

Figure 12C.4 The effect of an electric field on the energy levels of a polar linear rotor. All levels are doubly degenerate except that with $M_J = 0$.

(b) The appearance of microwave spectra

When the selection rules are applied to the expressions for the energy levels of a rigid spherical or linear rotor, it follows that the wavenumbers of the allowed $J+1 \leftarrow J$ absorptions are

$$\tilde{\nu}(J+1 \leftarrow J) = \tilde{F}(J+1) - \tilde{F}(J) = 2\tilde{B}(J+1) \qquad J = 0,1,2,\ldots$$ *Linear and spherical rotors* — Wavenumbers of rotational transitions (12C.8a)

When centrifugal distortion (Topic 12B) is taken into account, the corresponding expression obtained from eqn 12B.16 is

$$\tilde{\nu}(J+1 \leftarrow J) = 2\tilde{B}(J+1) - 4\tilde{D}_J(J+1)^3 \qquad (12\text{C}.8\text{b})$$

However, because the second term is typically very small compared with the first (see *Brief illustration* 12B.3), the appearance of the spectrum closely resembles that predicted from eqn 12C.8a.

Example 12C.1 Predicting the appearance of a rotational spectrum

Predict the form of the rotational spectrum of $^{14}NH_3$, for which $\tilde{B} = 9.977\ \text{cm}^{-1}$.

Method The $^{14}NH_3$ molecule is a polar symmetric rotor, so the rotational terms are given by $\tilde{F}(J,K) = \tilde{B}J(J+1) + (\tilde{A} - \tilde{B})K^2$. Because $\Delta J = \pm 1$ and $\Delta K = 0$, the expression for the wavenumbers of the rotational transitions is identical to eqn 12C.8a and depends only on $\tilde{B}$. For absorption, $\Delta J = +1$.

Answer We can draw up the following table for the $J+1 \leftarrow J$ transitions.

J	0	1	2	3	…
$\tilde{\nu}/\text{cm}^{-1}$	19.95	39.91	59.86	79.82	…
ν/GHz	598.1	1197	1795	2393	…

The line spacing is 19.95 cm^{-1} (598.1 GHz).

Self-test 12C.2 Repeat the problem for $CH_3{}^{35}Cl$, for which $\tilde{B} = 0.444\ cm^{-1}$.

Answer: Lines of separation 0.888 cm^{-1} (26.6 GHz)

The form of the spectrum predicted by eqn 12C.8 is shown in Fig. 12C.5. The most significant feature is that it consists of a series of lines with wavenumbers $2\tilde{B}, 4\tilde{B}, 6\tilde{B}, \ldots$ and of separation $2\tilde{B}$. The measurement of the line spacing therefore gives $\tilde{B}$, and hence the moment of inertia perpendicular to the figure axis of the molecule. Because the masses of the atoms are known, it is a simple matter to deduce the bond length of a diatomic molecule. However, in the case of a polyatomic molecule such as OCS or NH_3, the analysis gives only a single quantity, $I_\perp$, and it is not possible to infer both bond lengths (in OCS) or the bond length and bond angle (in NH_3). This difficulty can be overcome by using isotopologues, isotopically substituted molecules, such as ABC and A′BC; then, by assuming that $R(\text{A–B}) = R(\text{A}'\text{–B})$, both A–B and B–C bond lengths can be extracted from the two moments of inertia. A famous example of this procedure is the study of OCS; the actual calculation is worked through in Problem 12C.5. The assumption that bond lengths are unchanged in isotopologues is only an approximation, but it is a good approximation in most cases. Nuclear spin (Topic 14A), which differs from one isotope to another, also affects the appearance of high-resolution rotational spectra because spin is a source of angular momentum and can couple with the rotation of the molecule itself and hence affect the rotational energy levels.

The intensities of spectral lines increase with increasing J and pass through a maximum before tailing off as J becomes large. The most important reason for the maximum in intensity is the existence of a maximum in the population of rotational levels. The Boltzmann distribution (*Foundations* B and Topic 15A) implies that the population of each state decreases exponentially with increasing J, but the degeneracy of the levels increases. These two opposite trends result in the population of the energy levels (as distinct from the individual states) passing through a maximum. Specifically, the population of a rotational energy level J is given by the Boltzmann expression

$$N_J \propto N g_J e^{-E_J/kT}$$

where N is the total number of molecules in the sample and g_J is the degeneracy of the level J. The value of J corresponding to a maximum of this expression is found by treating J as a continuous variable, differentiating with respect to J, and then setting the result equal to zero. The result is (see Problem 12C.9)

$$J_{max} \approx \left(\frac{kT}{2hc\tilde{B}}\right)^{1/2} - \frac{1}{2} \quad \text{Linear rotors} \quad \text{Rotational state with largest population} \qquad (12C.9)$$

For a typical molecule (for example, OCS, with $\tilde{B} = 0.2\ cm^{-1}$) at room temperature, $kT \approx 1000hc\tilde{B}$, so $J_{max} \approx 30$. However, it must be recalled that the intensity of each transition also depends on the value of J (eqn 12C.6) and on the population difference between the two states involved in the transition. Hence the value of J corresponding to the most intense line is not quite the same as the value of J for the most highly populated level.

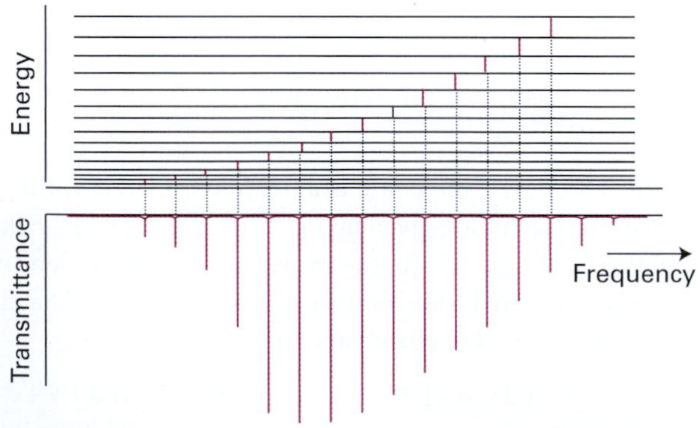

Figure 12C.5 The rotational energy levels of a linear rotor, the transitions allowed by the selection rule $\Delta J = +1$, and a typical pure rotational absorption spectrum (displayed here in terms of the radiation transmitted through the sample). The intensities reflect the populations of the initial level in each case and the strengths of the transition dipole moments.

12C.2 Rotational Raman spectroscopy

Raman scattering (Topic 12A) can also lead to rotational transitions. The gross selection rule for rotational Raman transitions is *that the molecule must be anisotropically polarizable*. To understand this criterion we need to know that the distortion of a molecule in an electric field is determined by its polarizability, α (Topic 16A). More precisely, if the strength of the field is $\mathcal{E}$, then the molecule acquires an induced dipole moment of magnitude

$$\mu = \alpha \mathcal{E} \qquad (12C.10)$$

in addition to any permanent dipole moment it might have. An atom is isotropically polarizable. That is, the same distortion is induced whatever the direction of the applied field. The polarizability of a spherical rotor is also isotropic. However, non-spherical rotors have polarizabilities that do depend on the direction of the field relative to the molecule, so these molecules are anisotropically polarizable (Fig. 12C.6). The electron distribution in H_2, for example, is more distorted when the field is applied parallel to the bond than when it is applied perpendicular to it, and we write $\alpha_\parallel > \alpha_\perp$.

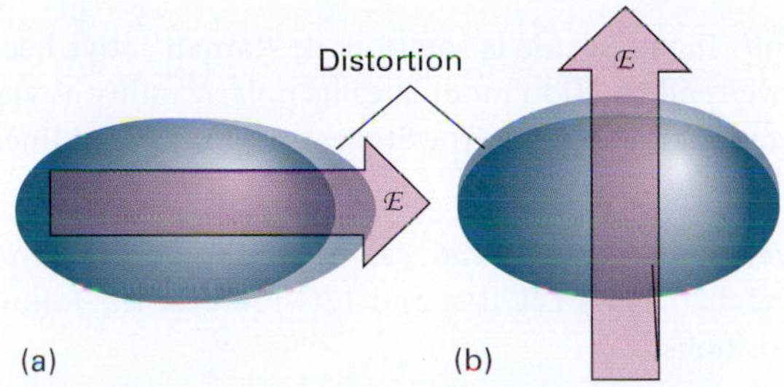

Figure 12C.6 An electric field applied to a molecule results in its distortion, and the distorted molecule acquires a contribution to its dipole moment (even if it is nonpolar initially). The polarizability may be different when the field is applied (a) parallel or (b) perpendicular to the molecular axis (or, in general, in different directions relative to the molecule); if that is so, then the molecule has an anisotropic polarizability.

All linear and diatomic molecules (whether homonuclear or heteronuclear) have anisotropic polarizabilities, and so are rotationally Raman active. This activity is one reason for the importance of rotational Raman spectroscopy, for the technique can be used to study many of the molecules that are inaccessible to microwave spectroscopy. Spherical rotors such as CH_4 and SF_6, however, are rotationally Raman inactive as well as microwave inactive. This inactivity does not mean that such molecules are never found in rotationally excited states. Molecular collisions do not have to obey such restrictive selection rules, and hence collisions between molecules can result in the population of any rotational state.

The specific rotational Raman selection rules are:

$$\begin{aligned}&\text{Linear rotors:} && \Delta J = 0, \pm 2\\ &\text{Symmetric rotors:} && \Delta J = 0, \pm 1, \pm 2\\ & && \Delta K = 0\end{aligned}$$

Rotational Raman selection rules (12C.11)

The $\Delta J = 0$ transitions do not lead to a shift in frequency of the scattered photon in pure rotational Raman spectroscopy, and contribute to the unshifted radiation (the Rayleigh radiation, Topic 12A). The specific selection rule for linear rotors is explored in the following *Justification*.

Justification 12C.2 Selection rules for rotational Raman spectra

The origin of the gross and specific selection rules for rotational Raman spectroscopy can be illustrated by using a diatomic molecule as an example. The incident electric field, $\mathcal{E}$, of a wave of electromagnetic radiation of frequency ω_i induces a molecular dipole moment that is given by

$$\mu_{\text{ind}} = \alpha\mathcal{E}(t) = \alpha\mathcal{E}\cos\omega_i t \qquad (12C.12)$$

If the molecule is rotating at an angular frequency ω_R, to an external observer its polarizability is also time dependent (if it is anisotropic), and we can write

$$\alpha = \alpha_0 + \Delta\alpha\cos 2\omega_R t \qquad (12C.13)$$

where $\Delta\alpha = \alpha_{\parallel} - \alpha_{\perp}$ and α ranges from $\alpha_0 + \Delta\alpha$ to $\alpha_0 - \Delta\alpha$ as the molecule rotates. The $2\omega_R$ appears because the polarizability returns to its initial value twice each revolution (Fig. 12C.7). Substituting this expression into the expression for the induced dipole moment gives

$$\begin{aligned}\mu_{\text{ind}} &= (\alpha_0 + \Delta\alpha\cos 2\omega_R t)\times(\mathcal{E}\cos\omega_i t)\\ &= \alpha_0\mathcal{E}\cos\omega_i t + \mathcal{E}\Delta\alpha\cos 2\omega_R t\cos\omega_i t\\ &\overset{\cos x\cos y = \frac{1}{2}\{\cos(x+y)+\cos(x-y)\}}{=} \alpha_0\mathcal{E}\cos\omega_i t + \tfrac{1}{2}\mathcal{E}\Delta\alpha\{\cos(\omega_i + 2\omega_R)t\\ &\qquad + \cos(\omega_i - 2\omega_R)t\}\end{aligned} \qquad (12C.14)$$

This calculation shows that the induced dipole has a component oscillating at the incident frequency (which generates Rayleigh radiation), and that it also has two components at $\omega_i \pm 2\omega_R$, which give rise to the shifted Raman lines. These lines appear only if $\Delta\alpha \neq 0$; hence the polarizability must be anisotropic for there to be Raman lines. This is the gross selection rule for rotational Raman spectroscopy.

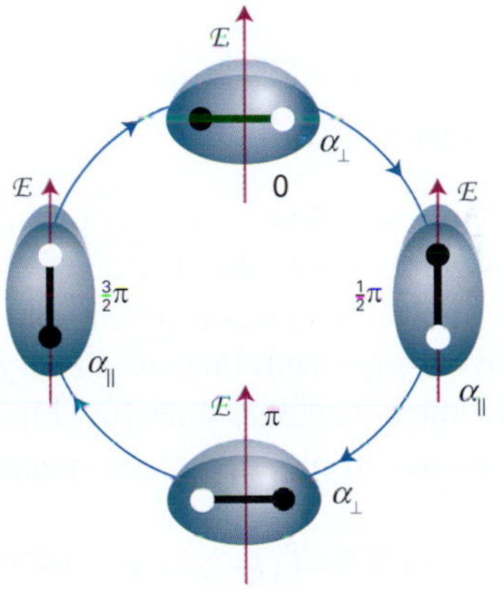

Figure 12C.7 The distortion induced in a molecule by an applied electric field returns the polarizability to its initial value after a rotation of only 180° (that is, twice a revolution). This is the origin of the $\Delta J = \pm 2$ selection rule in rotational Raman spectroscopy.

We also see that the distortion induced in the molecule by the incident electric field returns to its initial value after a rotation of 180° (that is, twice a revolution). This is the classical origin of the specific selection rule $\Delta J = \pm 2$.[1]

[1] See our *Molecular Quantum Mechanics* (2011) for the quantum mechanical calculation of the selection rules for rotational Raman spectroscopy.

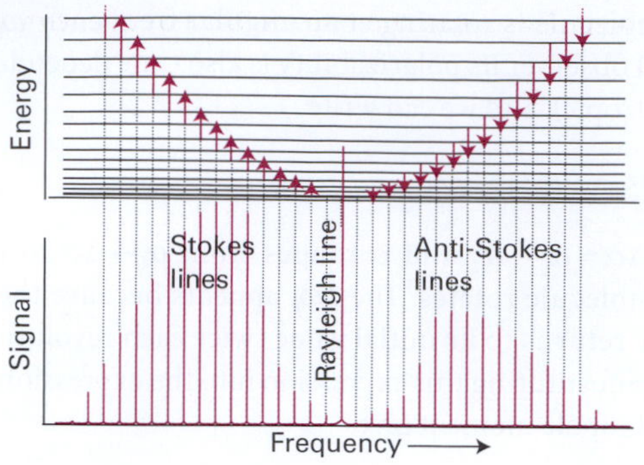

Figure 12C.8 The rotational energy levels of a linear rotor and the transitions allowed by the $\Delta J=\pm 2$ Raman selection rules. The form of a typical rotational Raman spectrum is also shown. The Rayleigh line is much stronger than depicted in the figure; it is shown as a weaker line to improve visualization of the Raman lines.

To predict the form of the Raman spectrum of a linear rotor we apply the selection rule $\Delta J=\pm 2$ to the rotational energy levels (Fig. 12C.8). When the molecule makes a transition with $\Delta J=+2$, the scattered radiation leaves the molecule in a higher rotational state, so the wavenumber of the incident radiation, initially $\tilde{\nu}_i$, is decreased. These transitions account for the Stokes lines (the lines at lower than the incident frequency, Topic 12A) in the spectrum:

$$\tilde{\nu}(J+2\leftarrow J)=\tilde{\nu}_i-\{\tilde{F}(J+2)-\tilde{F}(J)\} =\tilde{\nu}_i-2\tilde{B}(2J+3)$$ *Linear rotors* Wavenumbers of Stokes lines (12C.15a)

The Stokes lines appear to low frequency of the incident radiation and at displacements $6\tilde{B}, 10\tilde{B}, 14\tilde{B},\ldots$ from $\tilde{\nu}_i$: for $J=0$, 1, 2, …. When the molecule makes a transition with $\Delta J=-2$, the scattered photon emerges with increased energy. These transitions account for the anti-Stokes lines (the lines at higher than the incident frequency, Topic 12A) of the spectrum:

$$\tilde{\nu}(J-2\leftarrow J)=\tilde{\nu}_i+\{\tilde{F}(J)-\tilde{F}(J-2)\}=\tilde{\nu}_i+2\tilde{B}(2J-1)$$ *Linear rotors* Wavenumbers of anti-Stokes lines (12C.15b)

The anti-Stokes lines occur at displacements of $6\tilde{B}, 10\tilde{B}, 14\tilde{B},\ldots$ (for $J=2$, 3, 4, …; $J=2$ is the lowest state that can contribute under the selection rule $\Delta J=-2$) to high frequency of the incident radiation. The separation of adjacent lines in both the Stokes and the anti-Stokes regions is $4\tilde{B}$, so from its measurement $I_\perp$ can be determined and then used to find the bond lengths exactly as in the case of microwave spectroscopy.

Example 12C.2 Predicting the form of a Raman spectrum

Predict the form of the rotational Raman spectrum of $^{14}N_2$, for which $\tilde{B}=1.99\ cm^{-1}$, when it is exposed to 336.732 nm laser radiation.

Method The molecule is rotationally Raman active because end-over-end rotation modulates its polarizability as viewed by a stationary observer. The Stokes and anti-Stokes lines are given by eqn 12C.15.

Answer Because $\lambda_i=336.732$ nm corresponds to $\tilde{\nu}_i=29697.2\ cm^{-1}$, eqns 12C.15a and 12C.15b give the following line positions:

J	0	1	2	3
Stokes lines				
$\tilde{\nu}/cm^{-1}$	29 685.3	29 677.3	29 669.3	29 661.4
λ/nm	336.867	336.958	337.048	337.139
Anti-Stokes lines				
$\tilde{\nu}/cm^{-1}$			29 709.1	29 717.1
λ/nm			336.597	336.507

There will be a strong central line at 336.732 nm accompanied on either side by lines of increasing and then decreasing intensity (as a result of transition moment and population effects). The spread of the entire spectrum is very small, so the incident light must be highly monochromatic.

Self-test 12C.3 Repeat the calculation for the rotational Raman spectrum of NH_3 ($\tilde{B}=9.977\ cm^{-1}$).

Answer: Stokes lines at 29 637.3, 29 597.4, 29 557.5, 29 517.6 cm^{-1}; anti-Stokes lines at 29 757.1, 29 797.0 cm^{-1}

12C.3 Nuclear statistics and rotational states

If eqn 12C.15 is used in conjunction with the rotational Raman spectrum of CO_2, the rotational constant is inconsistent with other measurements of C—O bond lengths. The results are consistent only if it is supposed that the molecule can exist in states with even values of J, so the Stokes lines are $2\leftarrow 0, 4\leftarrow 2, \ldots$ and not $2\leftarrow 0, 3\leftarrow 1, 4\leftarrow 2, 5\leftarrow 3, \ldots$.

The explanation of the missing lines is the Pauli principle (Topic 9B) and the fact that ^{16}O nuclei are spin-0 bosons: just as the Pauli principle excludes certain electronic states, so too does it exclude certain molecular rotational states. The form of the Pauli principle given in Topic 9B states that, when two identical bosons are exchanged, the overall wavefunction must remain unchanged in every respect, including sign. When a CO_2 molecule rotates through 180°, two identical O nuclei are interchanged, so the overall wavefunction of the molecule must remain unchanged. However, inspection of the form of the rotational wavefunctions (which have the same form as the s, p, etc. orbitals of atoms) shows that they change sign by $(-1)^J$ under such a rotation (Fig. 12C.9). Therefore, only even values of J are permissible for CO_2, and hence the Raman spectrum shows only alternate lines.

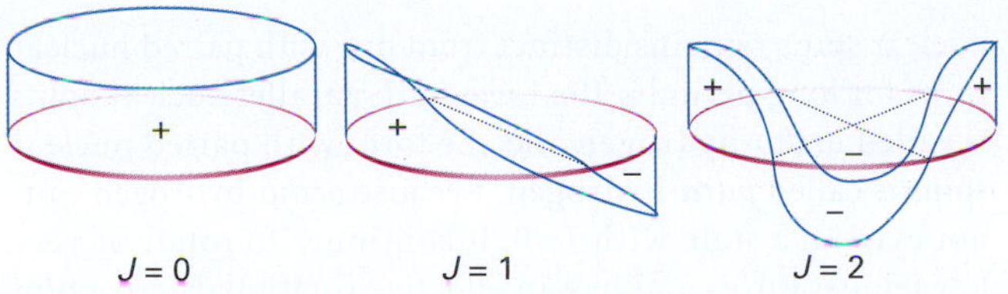

Figure 12C.9 The symmetries of rotational wavefunctions (shown here, for simplicity as a two-dimensional rotor) under a rotation through 180°. Wavefunctions with J even do not change sign; those with J odd do change sign.

The selective occupation of rotational states that stems from the Pauli principle is termed **nuclear statistics**. Nuclear statistics must be taken into account whenever a rotation interchanges equivalent nuclei. However, the consequences are not always as simple as for CO_2 because there are complicating features when the nuclei have nonzero spin: there may be several different relative nuclear spin orientations consistent with even values of J and a different number of spin orientations consistent with odd values of J. For molecular hydrogen and fluorine, for instance, with their two identical spin-$\frac{1}{2}$ nuclei, we show in the following *Justification* that there are three times as many ways of achieving a state with odd J than with even J, and there is a corresponding 3:1 alternation in intensity in their rotational Raman spectra (Fig. 12C.10). In general, for a homonuclear diatomic molecule with nuclei of spin I, the numbers of ways of achieving states of odd and even J are in the ratio

$$\frac{\text{Number of ways of achieving odd } J}{\text{Number of ways of achieving even } J} = \begin{cases} (I+1)/I \text{ for half-integral spin nuclei} \\ I/(I+1) \text{ for integral spin nuclei} \end{cases}$$

Homonuclear diatomic molecules Nuclear statistics (12C.16)

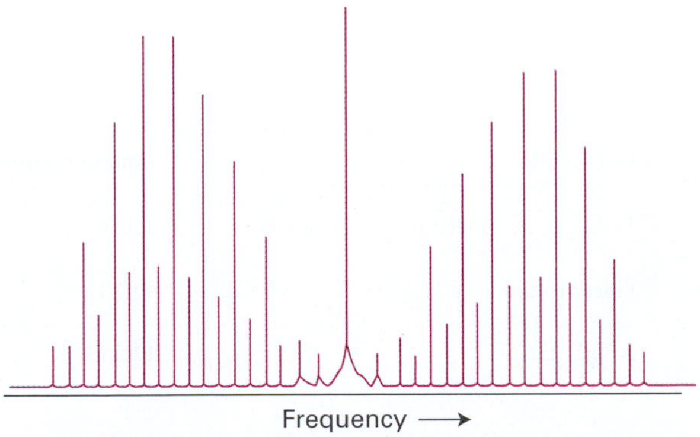

Figure 12C.10 The rotational Raman spectrum of a diatomic molecule with two identical spin-$\frac{1}{2}$ nuclei shows an alternation in intensity as a result of nuclear statistics. The Rayleigh line is much stronger than depicted in the figure; it is shown as a weaker line to improve visualization of the Raman lines.

For hydrogen, $I=\frac{1}{2}$, and the ratio is 3:1. For N_2, with $I=1$, the ratio is 1:2.

Justification 12C.3 The effect of nuclear statistics on rotational spectra

Hydrogen nuclei are fermions, so the Pauli principle requires the overall wavefunction to change sign under particle interchange. However, the rotation of an H_2 molecule through 180° has a more complicated effect than merely relabelling the nuclei, because it interchanges their spin states too if the nuclear spins are paired ($\uparrow\downarrow$; $I_{total}=0$) but not if they are parallel ($\uparrow\uparrow$, $I_{total}=1$).

First, consider the case when the spins are parallel and their state is $\alpha(A)\alpha(B)$, $\alpha(A)\beta(B)+\alpha(B)\beta(A)$, or $\beta(A)\beta(B)$. The $\alpha(A)\alpha(B)$ and $\beta(A)\beta(B)$ combinations are unchanged when the molecule rotates through 180° so the rotational wavefunction must change sign to achieve an overall change of sign. Hence, only odd values of J are allowed. Although at first sight the spins must be interchanged in the combination $\alpha(A)\beta(B)+\alpha(B)\beta(A)$ so as to achieve a simple $A \leftrightarrow B$ interchange of labels (Fig. 12C.11), $\beta(A)\alpha(B)+\beta(B)\alpha(A)$ is the same as $\alpha(A)\beta(B)+\alpha(B)\beta(A)$ apart from the order of terms, so only odd values of J are allowed for it too. In contrast, if the nuclear spins are paired, their wavefunction is $\alpha(A)\beta(B)-\alpha(B)\beta(A)$. This combination changes sign when α and β are exchanged (in order to achieve a simple $A \leftrightarrow B$ interchange overall). Therefore, for the overall wavefunction to change sign in this case requires the rotational wavefunction *not* to change sign. Hence, only even values of J are allowed if the nuclear spins are paired. In accord with the prediction of eqn 12C.16, there are three ways of achieving odd J but only one of achieving even J.

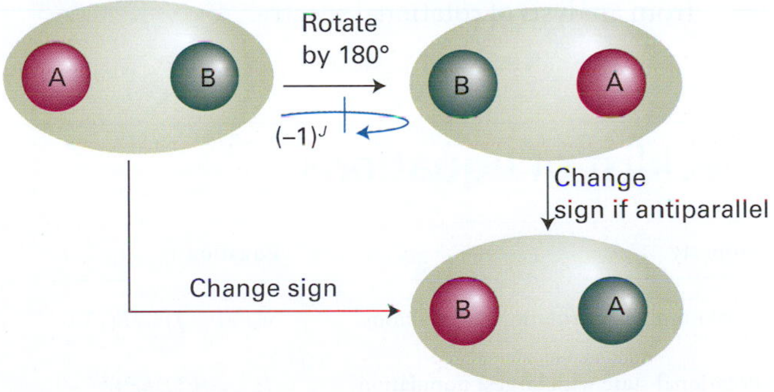

Figure 12C.11 The interchange of two identical fermion nuclei results in the change in sign of the overall wavefunction. The relabelling can be thought of as occurring in two steps: the first is a rotation of the molecule; the second is the interchange of unlike spins (represented by the different colours of the nuclei). The wavefunction changes sign in the second step if the nuclei have antiparallel spins.

Brief illustration 12C.2 *Ortho*- and *para*-hydrogen

Different relative nuclear spin orientations change into one another only very slowly, so an H_2 molecule with parallel nuclear spins remains distinct from one with paired nuclear spins for long periods. The form with parallel nuclear spins is called ***ortho*-hydrogen** and the form with paired nuclear spins is called ***para*-hydrogen**. Because *ortho*-hydrogen cannot exist in a state with $J=0$, it continues to rotate at very low temperatures and has an effective rotational zero-point energy (Fig. 12C.12).

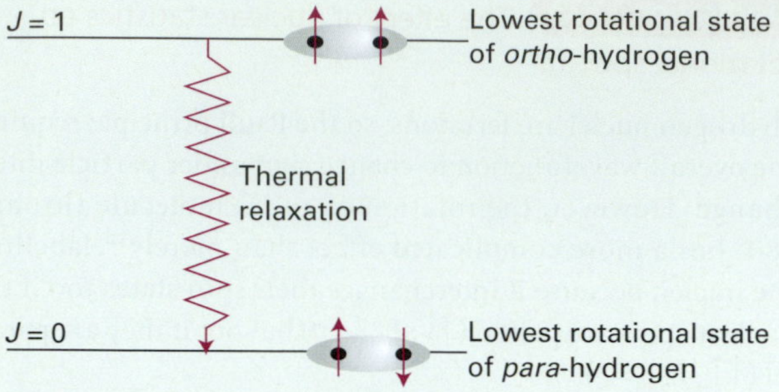

Figure 12C.12 When hydrogen is cooled, the molecules with parallel nuclear spins accumulate in their lowest available rotational state, the one with $J=1$. They can enter the lowest rotational state ($J=0$) only if the spins change their relative orientation and become antiparallel. This is a slow process under normal circumstances, so energy is slowly released.

Self-test 12C.4 Does BeF_2 exist in *ortho* and *para* forms? *Hints*: (a) Determine the geometry of BeF_2, then (b) decide whether fluorine nuclei are fermions or bosons.

Answer: Yes

Checklist of concepts

- ☐ 1. Pure rotational transitions can be studied with **microwave spectroscopy** and **rotational Raman spectroscopy**.
- ☐ 2. For a molecule to give a pure rotational spectrum, it must be polar.
- ☐ 3. The **specific selection rules** for microwave spectroscopy are $\Delta J=\pm1$, $\Delta M_J=0,\pm1$, $\Delta K=0$.
- ☐ 4. Bond lengths and dipole moments may be obtained from analysis of rotational spectra.
- ☐ 5. A molecule must be **anisotropically polarizable** for it to be rotationally Raman active.
- ☐ 6. The **specific selection rules** for rotational Raman spectroscopy are: (i) linear rotors, $\Delta J=0,\pm2$; (ii) symmetric rotors, $\Delta J=0,\pm1,\pm2$; $\Delta K=0$.
- ☐ 7. The appearance of rotational spectra is affected by **nuclear statistics**, the selective occupation of rotational states that stems from the Pauli principle.

Checklist of equations

Property	Equation	Comment	Equation number
Wavenumbers of rotational transitions	$\tilde{\nu}(J+1\leftarrow J)=2\tilde{B}(J+1)$	$J=0, 1, 2, \ldots$ spherical and linear rotors	12C.8a
Rotational state with largest population	$J_{\max}\approx(kT/2hc\tilde{B})^{1/2}-\frac{1}{2}$	Linear rotors	12C.9
Wavenumbers of (i) Stokes and (ii) anti-Stokes lines in the rotational Raman spectrum of linear rotors	(i) $\tilde{\nu}(J+2\leftarrow J)=\tilde{\nu}_i-2\tilde{B}(2J+3)$ (ii) $\tilde{\nu}(J-2\leftarrow J)=\tilde{\nu}_i+2\tilde{B}(2J-1)$	$J=0, 1, 2, \ldots$	12C.15
Nuclear statistics	$\dfrac{\text{Number of ways of achieving odd } J}{\text{Number of ways of achieving even } J}=\begin{cases}(I+1)/I \text{ for half-integral spin nuclei}\\ I/(I+1) \text{ for integral spin nuclei}\end{cases}$	Homonuclear diatomic molecules	12C.16

12D Vibrational spectroscopy of diatomic molecules

Contents

➤ Why do you need to know this material?

The observation of the frequencies of transitions between the vibrational states of a molecule gives information about the identity of the molecule and provides quantitative information about the flexibility of its bonds. Infrared spectroscopy is a valuable analytical tool and is widely used in chemical laboratories.

➤ What is the key idea?

The vibrational spectrum of a diatomic molecule can be interpreted by using the harmonic oscillator model, with modifications that account for bond dissociation and the coupling of rotational and vibrational motion.

➤ What do you need to know already?

You need to be familiar with the harmonic oscillator (Topic 8B) and rigid rotor (Topic 12B) models of molecular motion, the general principles of spectroscopy (Topic 12A), and the interpretation of rotational spectra (Topic 12C).

Here we explore the vibrational energy levels of diatomic molecules and establish the selection rules for spectroscopic transitions between these levels. We also see how the simultaneous excitation of rotation modifies the appearance of a vibrational spectrum. This material sets the stage for the discussion of vibrations of polyatomic molecules in Topic 12E.

12D.1 Vibrational motion

We base our discussion on Fig. 12D.1, which shows a typical potential energy curve of a diatomic molecule (it is essentially a reproduction of Fig. 8B.1 of Topic 8B). In regions close to R_e (at the minimum of the curve) the potential energy can be approximated by a parabola, so we can write

$$V = \tfrac{1}{2}k_f x^2 \qquad x = R - R_e \qquad \text{Parabolic potential energy} \qquad (12D.1)$$

where k_f is the **force constant** of the bond. The steeper the walls of the potential (the stiffer the bond), the greater the force constant.

To see the connection between the shape of the molecular potential energy curve and the value of k_f, note that we can expand the potential energy around its minimum by using a Taylor series (*Mathematical background* 1 following Chapter 1), which is a common way of expressing how a function varies near a selected point (in this case, the minimum of the curve at $x=0$):

$$V(x) = V(0) + \left(\frac{dV}{dx}\right)_0 x + \tfrac{1}{2}\left(\frac{d^2V}{dx^2}\right)_0 x^2 + \cdots \qquad (12D.2)$$

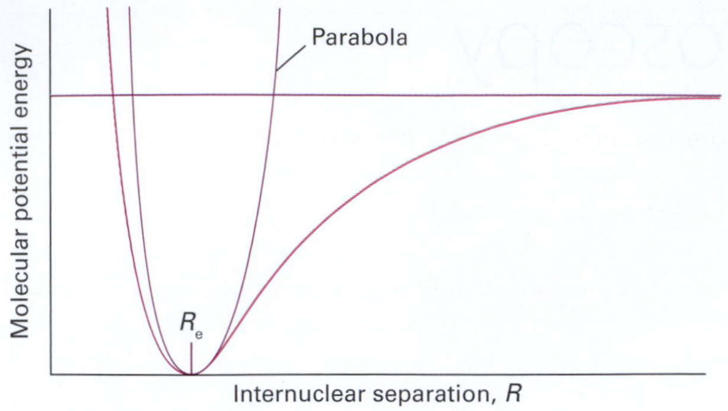

Figure 12D.1 A molecular potential energy curve can be approximated by a parabola near the bottom of the well. The parabolic potential energy results in harmonic oscillations. At high excitation energies the parabolic approximation is poor (the true potential energy is less confining), and is totally wrong near the dissociation limit.

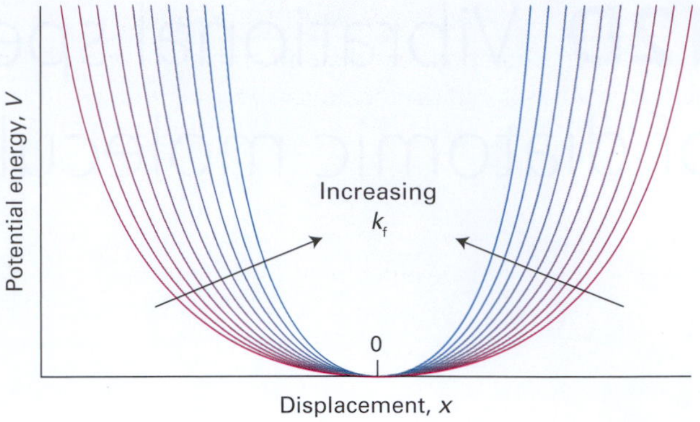

Figure 12D.2 The force constant is a measure of the curvature of the potential energy close to the equilibrium extension of the bond. A strongly confining well (one with steep sides, a stiff bond) corresponds to high values of k_f.

The notation $(\ldots)_0$ means that the derivatives are first evaluated and then x is set equal to 0. The term $V(0)$ can be set arbitrarily to zero. The first derivative of V is zero at the minimum. Therefore, the first surviving term is proportional to the square of the displacement. For small displacements we can ignore all the higher terms, and so write

$$V(x) \approx \tfrac{1}{2}\left(\frac{d^2V}{dx^2}\right)_0 x^2 \qquad (12D.3)$$

Therefore, the first approximation to a molecular potential energy curve is a parabolic potential, and we can identify the force constant as

$$k_f = \left(\frac{d^2V}{dx^2}\right)_0 \qquad \textit{Formal definition} \quad \text{Force constant} \qquad (12D.4)$$

We see that if the potential energy curve is sharply curved close to its minimum, then k_f will be large and the bond stiff. Conversely, if the potential energy curve is wide and shallow, then k_f will be small and the bond easily stretched or compressed (Fig. 12D.2).

The Schrödinger equation for the relative motion of two atoms of masses m_1 and m_2 with a parabolic potential energy is

$$-\frac{\hbar^2}{2m_{\text{eff}}}\frac{d^2\psi}{dx^2} + \tfrac{1}{2}k_f x^2 \psi = E\psi \qquad (12D.5)$$

where m_{eff} is the **effective mass**:

$$m_{\text{eff}} = \frac{m_1 m_2}{m_1 + m_2} \qquad \textit{Definition} \quad \text{Effective mass} \qquad (12D.6)$$

These equations are derived in the same way as in Topic 8B, but here the separation of variables procedure is used to separate the relative motion of the atoms from the motion of the molecule as a whole.

A note on good practice Distinguish *effective mass* from *reduced mass*. The former is a measure of the mass that is moved during a vibration. The latter is the quantity that emerges from the separation of relative internal and overall translational motion. For a diatomic molecule the two are the same, but that is not true in general for vibrations of polyatomic molecules. Many, however, do not make this distinction and refer to both quantities as the 'reduced mass'.

The Schrödinger equation in eqn 12D.5 is the same as eqn 8B.3 for a particle of mass m undergoing harmonic motion. Therefore, we can use the results of Topic 8B to write down the permitted vibrational energy levels:

$$E_v = \left(v + \tfrac{1}{2}\right)\hbar\omega \qquad \omega = \left(\frac{k_f}{m_{\text{eff}}}\right)^{1/2} \qquad v = 0, 1, 2, \ldots$$

$$\textit{Diatomic molecule} \quad \text{Vibrational energy levels} \qquad (12D.7)$$

The **vibrational terms** of a molecule, the energies of its vibrational states expressed as wavenumbers, are denoted $\tilde{G}(v)$, with $E_v = hc\tilde{G}(v)$, so

$$\tilde{G}(v) = \left(v + \tfrac{1}{2}\right)\tilde{\nu} \qquad \tilde{\nu} = \frac{1}{2\pi c}\left(\frac{k_f}{m_{\text{eff}}}\right)^{1/2}$$

$$\textit{Diatomic molecule} \quad \text{Vibrational terms} \qquad (12D.8)$$

The vibrational wavefunctions are the same as those discussed in Topic 8B for a harmonic oscillator.

It is important to note that the vibrational terms depend on the *effective* mass of the molecule, not directly on its total mass. This dependence is physically reasonable, for if atom 1 were as heavy as a brick wall, then we would find $m_{\text{eff}} \approx m_2$, the mass of the lighter atom. The vibration would then be that of a light atom relative to that of a stationary wall (this is approximately the case in HI, for example, where the I atom barely moves and $m_{\text{eff}} \approx m_{\text{H}}$). For a homonuclear diatomic molecule $m_1 = m_2$, and the effective mass is half the total mass: $m_{\text{eff}} = \frac{1}{2}m$.

Brief illustration 12D.1 The vibrational frequency of a diatomic molecule

The force constant of the bond in HCl is 516 N m^{-1}, a reasonably typical value for a single bond. The effective mass of ^{1}H^{35}Cl is 1.63×10^{-27} kg (note that this mass is very close to the mass of the hydrogen atom, 1.67×10^{-27} kg, so the Cl atom is acting like a brick wall). These values imply that

$$\omega = \left(\frac{516 \overbrace{\text{N}}^{\text{kg m s}^{-2}} \text{m}^{-1}}{1.63 \times 10^{-27}\ \text{kg}}\right)^{1/2} = 5.63 \times 10^{14}\ \text{s}^{-1}$$

or $\nu = \omega/2\pi = 89.5$ THz (1 THz $= 10^{12}$ Hz).

Self-test 12D.1 The vibrational frequency ν of ^{35}Cl$_2$ is 16.94 THz. Calculate the force constant of the bond.

Answer: 327.8 N m^{-1}

12D.2 Infrared spectroscopy

The gross selection rule for a change in vibrational state brought about by absorption or emission of radiation is that *the electric dipole moment of the molecule must change when the atoms are displaced relative to one another*. Such vibrations are said to be **infrared active**. The classical basis of this rule is that the molecule can shake the electromagnetic field into oscillation if its dipole changes as it vibrates, and vice versa (Fig. 12D.3); its formal basis is given in the following *Justification*. Note that the molecule need not have a permanent dipole: the rule requires only a *change* in dipole moment, possibly from zero. Some vibrations do not affect the molecule's dipole moment (for instance, the stretching motion of a homonuclear diatomic molecule), so they neither absorb nor generate radiation: such vibrations are said to be **infrared inactive**.

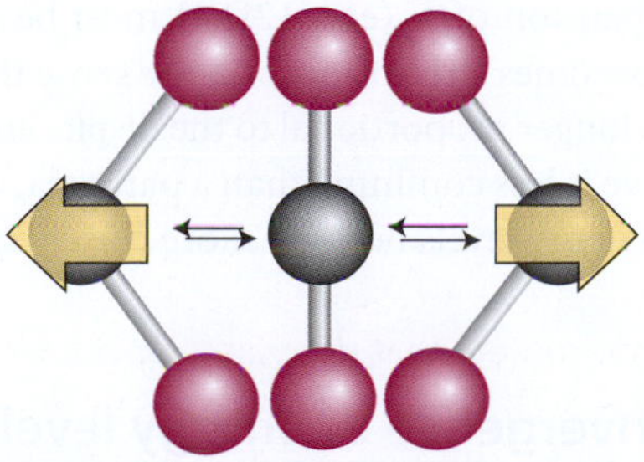

Figure 12D.3 The oscillation of a molecule, even if it is nonpolar, may result in an oscillating electric dipole moment that can interact with the electromagnetic field.

Brief illustration 12D.2 The gross selection rule for infrared spectroscopy

Homonuclear diatomic molecules are infrared inactive because their dipole moments remain zero however long the bond; heteronuclear diatomic molecules are infrared active. Weak infrared transitions can be observed from homonuclear diatomic molecules trapped within various nanomaterials. For instance, when incorporated into solid C$_{60}$, H$_2$ molecules interact through van der Waals forces with the surrounding C$_{60}$ molecules and acquire dipole moments, with the result that they have observable infrared spectra.

Self-test 12D.2 Identify the infrared active molecules in the group: N$_2$, NO, and CO.

Answer: NO and CO

The specific selection rule, which is obtained from an analysis of the expression for the transition moment and the properties of integrals over harmonic oscillator wavefunctions (as shown in the following *Justification*), is

$$\Delta v = \pm 1 \quad \textit{Infrared spectroscopy} \quad \text{Specific selection rule} \qquad (12\text{D}.9)$$

Justification 12D.1 Gross and specific selection rules for infrared spectra

The gross selection rule for infrared spectroscopy is based on an analysis of the transition dipole moment $\mu_{\text{fi}} = \int \psi_{v_f}^* \hat{\mu} \psi_{v_i} \text{d}\tau$, (Topic 12B) which arises from eqn 12B.2 of that Topic ($\mu_{\text{fi}} = \int \psi_{\varepsilon_f}^* \psi_{v_f}^* Y_{J_f,M_{J,f}}^* \hat{\mu} \psi_{\varepsilon_i} \psi_{v_i} Y_{J_i,M_{J,i}} \text{d}\tau$) when the molecule does not change electronic or rotational states. For simplicity, we consider a one-dimensional oscillator (like a diatomic molecule, which can stretch and compress only along the direction parallel to its bond). The electric dipole moment operator depends on the location of all the electrons and all the nuclei in the molecule, so it varies as the internuclear separation changes (Fig. 12D.4). We can write its variation with displacement from the equilibrium separation, x, as

$$\hat{\mu} = \hat{\mu}_0 + \left(\frac{\text{d}\hat{\mu}}{\text{d}x}\right)_0 x + \cdots$$

where $\hat{\mu}_0$ is the electric dipole moment operator when the nuclei have their equilibrium separation. It then follows that, with f≠i and keeping only the term linear in the small displacement x,

$$\mu_{\text{fi}} = \int \psi_{v_f}^* \hat{\mu} \psi_{v_i} \text{d}x = \mu_0 \overbrace{\int \psi_{v_f}^* \psi_{v_i} \text{d}x}^{0} + \left(\frac{\text{d}\mu}{\text{d}x}\right)_0 \int \psi_{v_f}^* x \psi_{v_i} \text{d}x$$

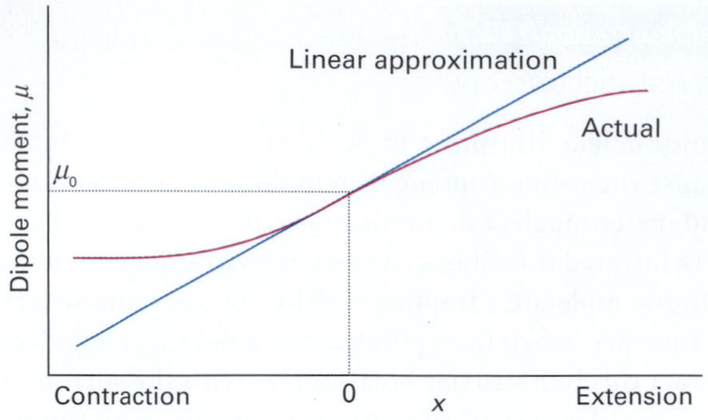

Figure 12D.4 The electric dipole moment of a heteronuclear diatomic molecule varies as shown by the purple curve. For small displacements the change in dipole moment is proportional to the displacement.

The term multiplying μ_0 is zero because the states with different values of v are orthogonal (Topic 8B). It follows that the transition dipole moment is

$$\mu_{fi} = \left(\frac{d\mu}{dx}\right)_0 \int \psi^*_{v_f} x \psi_{v_i} dx$$

We see that the right-hand side is zero unless the dipole moment varies with displacement. This is the gross selection rule for infrared spectroscopy.

The specific selection rule is determined by considering the value of $\int \psi^*_{v_f} x \psi_{v_i} dx$. We need to write out the wavefunctions in terms of the Hermite polynomials given in Topic 8B and then to use their properties. We note that $x = \alpha y$ with $\alpha = (\hbar^2/m_{eff}k_f)^{1/4}$ (this is eqn 8B.8 of Topic 8B). Then we write

$$\int \psi^*_{v_f} x \psi_{v_i} dx = N_{v_f} N_{v_i} \int_{-\infty}^{\infty} H_{v_f} x H_{v_i} e^{-y^2} dx$$
$$= \alpha^2 N_{v_f} N_{v_i} \int_{-\infty}^{\infty} H_{v_f} y H_{v_i} e^{-y^2} dy$$

To evaluate the integral we use the 'recursion' relation

$$yH_v = vH_{v-1} + \tfrac{1}{2}H_{v+1}$$

which leads to

$$\int \psi^*_{v_f} x \psi_{v_i} dx$$
$$= \alpha^2 N_{v_f} N_{v_i} \left\{ v_i \int_{-\infty}^{\infty} H_{v_f} H_{v_i-1} e^{-y^2} dy + \tfrac{1}{2} \int_{-\infty}^{\infty} H_{v_f} H_{v_i+1} e^{-y^2} dy \right\}$$

The first integral is zero unless $v_f = v_i - 1$ and the second is zero unless $v_f = v_i + 1$ (Table 8B.1). It follows that the transition dipole moment is zero unless $\Delta v = \pm 1$.

Transitions for which $\Delta v = +1$ correspond to absorption and those with $\Delta v = -1$ correspond to emission. It follows that the wavenumbers of allowed vibrational transitions, which are denoted $\Delta\tilde{G}_{v+\frac{1}{2}}$ for the transition $v+1 \leftarrow v$, are

$$\Delta\tilde{G}_{v+\frac{1}{2}} = \tilde{G}(v+1) - \tilde{G}(v) = \tilde{\nu} \qquad (12D.10)$$

The wavenumbers of vibrational transitions correspond to those of radiation in the infrared region of the electromagnetic spectrum, so vibrational transitions absorb and generate infrared radiation.

At room temperature $kT/hc \approx 200\,cm^{-1}$, and most vibrational wavenumbers are significantly greater than $200\,cm^{-1}$. It follows from the Boltzmann distribution (*Foundations* B and Topic 15A) that at room temperature almost all the molecules are in their vibrational ground states. Hence, the dominant spectral transition will be the **fundamental transition**, $1 \leftarrow 0$. As a result, the spectrum is expected to consist of a single absorption line. If the molecules are formed in a vibrationally excited state, such as when vibrationally excited HF molecules are formed in the reaction $H_2 + F_2 \rightarrow 2\ HF^*$, where the star indicates a vibrationally 'hot' molecule, the transitions $5 \rightarrow 4$, $4 \rightarrow 3$, ... may also appear (in emission). In the harmonic approximation, all these lines lie at the same frequency, and the spectrum is also a single line. However, as we shall now show, the breakdown of the harmonic approximation causes the transitions to lie at slightly different frequencies, so several lines are observed.

12D.3 Anharmonicity

The vibrational terms in eqn 12D.8 are only approximate because they are based on a parabolic approximation to the actual potential energy curve. A parabola cannot be correct at all extensions because it does not allow a bond to dissociate. At high vibrational excitations the swing of the atoms (more precisely, the spread of the vibrational wavefunction) allows the molecule to explore regions of the potential energy curve where the parabolic approximation is poor and additional terms in the Taylor expansion of V (eqn 12D.2) must be retained. The motion then becomes **anharmonic**, in the sense that the restoring force is no longer proportional to the displacement. Because the actual curve is less confining than a parabola, we can anticipate that the energy levels become more closely spaced at high excitations.

(a) The convergence of energy levels

One approach to the calculation of the energy levels in the presence of anharmonicity is to use a function that resembles the true potential energy more closely. The **Morse potential energy** is

$$V = hc\tilde{D}_e\{1-e^{-a(R-R_e)}\}^2 \qquad a=\left(\frac{m_{\text{eff}}\omega^2}{2hc\tilde{D}_e}\right)^{1/2}$$

Morse potential energy (12D.11)

where $\tilde{D}_e$ is the depth of the potential minimum (Fig. 12D.5). Near the well minimum the variation of V with displacement resembles a parabola (as can be checked by expanding the exponential as far as the first term) but, unlike a parabola, eqn 12D.11 allows for dissociation at large displacements. The Schrödinger equation can be solved for the Morse potential and the permitted energy levels are

$$\tilde{G}(v)=\left(v+\tfrac{1}{2}\right)\tilde{\nu}-\left(v+\tfrac{1}{2}\right)^2 x_e\tilde{\nu} \qquad x_e=\frac{a^2\hbar}{2m_{\text{eff}}\omega}=\frac{\tilde{\nu}}{4\tilde{D}_e}$$

Morse potential energy Vibrational terms (12D.12)

The dimensionless parameter x_e is called the **anharmonicity constant**. The number of vibrational levels of a Morse oscillator is finite, and $v=0, 1, 2, \ldots, v_{\max}$, as shown in Fig. 12D.6 (see also Problem 12D.7). The second term in the expression for $\tilde{G}$ subtracts from the first with increasing effect as v increases, and hence gives rise to the convergence of the levels at high quantum numbers.

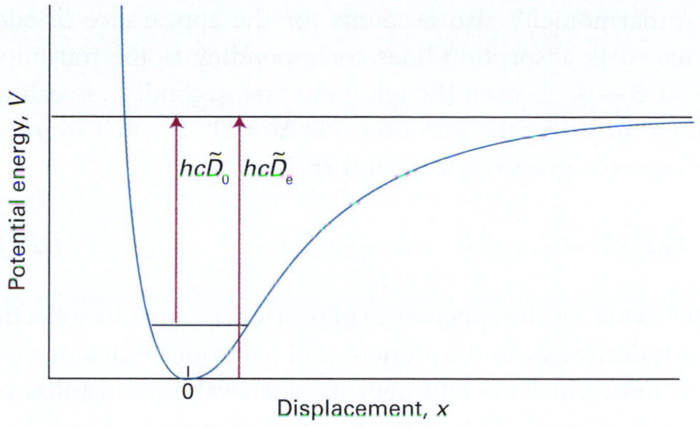

Figure 12D.5 The dissociation energy of a molecule, $hc\tilde{D}_0$, differs from the depth of the potential well, $hc\tilde{D}_e$, on account of the zero-point energy of the vibrations of the bond.

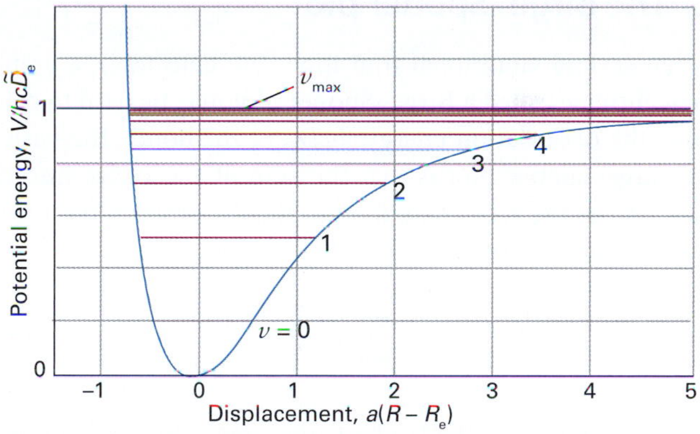

Figure 12D.6 The Morse potential energy curve reproduces the general shape of a molecular potential energy curve. The corresponding Schrödinger equation can be solved, and the values of the energies obtained. The number of bound levels is finite.

Example 12D.1 Estimating an anharmonicity constant

Estimate the anharmonicity constant x_e for $^1H^{19}F$ from the data in Table 12D.1 of the *Resource section*.

Method The anharmonicity constant is evaluated from $\tilde{\nu}$, $\tilde{D}_e$, and eqn 12D.12. However, note that Table 12D.1 lists values of $\tilde{D}_0=\tilde{D}_e-\tfrac{1}{2}\tilde{\nu}$ (Fig. 12D.5), so calculate $\tilde{D}_e$ first before using eqn 12D.12. A useful conversion factor is $1\,\text{kJ mol}^{-1}=83.593\,\text{cm}^{-1}$.

Answer The depth of the potential minimum is

$$\tilde{D}_e=\tilde{D}_0+\tfrac{1}{2}\tilde{\nu}=\overbrace{(4.718\times10^4\,\text{cm}^{-1})}^{564.4\,\text{kJmol}^{-1}\times\frac{83.593\,\text{cm}^{-1}}{1\,\text{kJmol}^{-1}}}+\tfrac{1}{2}\times(4138.32\,\text{cm}^{-1})$$
$$=(4.718\times10^4+\tfrac{1}{2}\times4138.32)\,\text{cm}^{-1}$$

It follows from eqn 12D.12 that the anharmonicity constant is

$$x_e=\frac{4138.32\,\text{cm}^{-1}}{(4.718\times10^4+\tfrac{1}{2}\times4138.32)\,\text{cm}^{-1}}=2.101\times10^{-2}$$

Self-test 12D.3 Estimate the anharmonicity constant for $^1H^{81}Br$.

Answer: 2.093×10^{-2}

Although the Morse oscillator is quite useful theoretically, in practice the more general expression

$$\tilde{G}(v)=\left(v+\tfrac{1}{2}\right)\tilde{\nu}-\left(v+\tfrac{1}{2}\right)^2 x_e\tilde{\nu}+\left(v+\tfrac{1}{2}\right)^3 y_e\tilde{\nu}+\cdots \qquad (12D.13)$$

where x_e, y_e, … are empirical dimensionless constants characteristic of the molecule, is used to fit the experimental data and to find the dissociation energy of the molecule. When anharmonicities are present, the wavenumbers of transitions with $\Delta v=+1$ are

$$\Delta\tilde{G}_{v+\frac{1}{2}}=\tilde{G}(v+1)-\tilde{G}(v)=\tilde{\nu}-2(v+1)x_e\tilde{\nu}+\cdots \qquad (12D.14)$$

Equation 12D.14 shows that, when $x_e>0$, the transitions move to lower wavenumbers as v increases.

Anharmonicity also accounts for the appearance of additional weak absorption lines corresponding to the transitions $2\leftarrow0$, $3\leftarrow0$, …, even though these first, second, … **overtones** are forbidden by the selection rule $\Delta v=\pm1$. The first overtone, for example, gives rise to an absorption at

$$\tilde{G}(v+2)-\tilde{G}(v)=2\tilde{\nu}-2(2v+3)x_e\tilde{\nu}+\cdots \qquad (12D.15)$$

The reason for the appearance of overtones is that the selection rule is derived from the properties of harmonic oscillator wavefunctions, which are only approximately valid when anharmonicity is present. Therefore, the selection rule is also only an approximation. For an anharmonic oscillator, all values of Δv are allowed, but transitions with $\Delta v>1$ are allowed only weakly if the anharmonicity is slight.

(b) The Birge–Sponer plot

When several vibrational transitions are detectable, a graphical technique called a **Birge–Sponer plot** may be used to determine the dissociation energy, $hc\tilde{D}_0$, of the bond. The basis of the Birge–Sponer plot is that the sum of successive intervals $\Delta\tilde{G}_{v+\frac{1}{2}}$ from the zero-point level to the dissociation limit is the dissociation energy:

$$\tilde{D}_0=\Delta\tilde{G}_{1/2}+\Delta\tilde{G}_{3/2}+\cdots=\sum_v \Delta\tilde{G}_{v+1/2} \qquad (12D.16)$$

just as the height of the ladder is the sum of the separations of its rungs (Fig. 12D.7). The construction in Fig. 12D.8 shows that the area under the plot of $\Delta\tilde{G}_{v+\frac{1}{2}}$ against $v+\frac{1}{2}$ is equal to the sum, and therefore to $\tilde{D}_0$. The successive terms decrease linearly when only the x_e anharmonicity constant is taken into account and the inaccessible part of the spectrum can be estimated by linear extrapolation. Most actual plots differ from the linear plot as shown in Fig. 12D.8, so the value of $\tilde{D}_0$ obtained in this way is usually an overestimate of the true value.

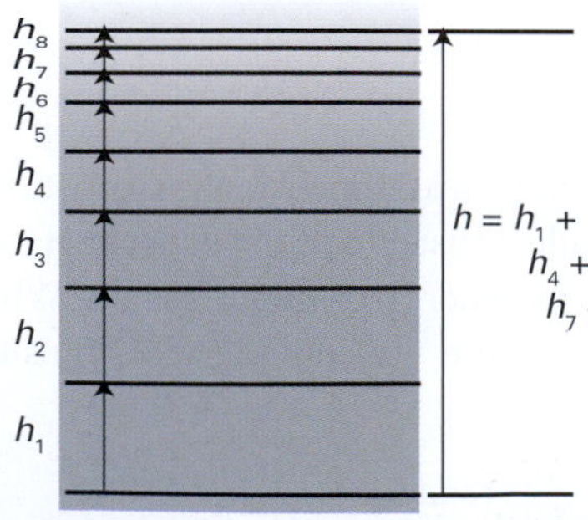

Figure 12D.7 The dissociation energy is the sum of the separations h_i of the vibrational energy levels up to the dissociation limit just as the length of a ladder is the sum of the separations of its rungs.

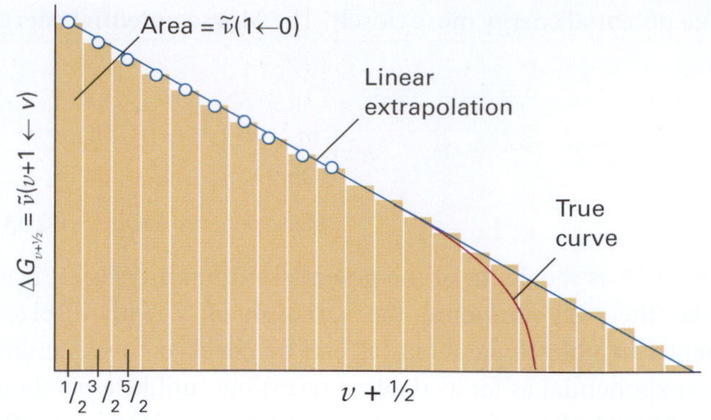

Figure 12D.8 The area under a plot of transition wavenumber against vibrational quantum number is equal to the dissociation energy of the molecule. The assumption that the differences approach zero linearly is the basis of the Birge–Sponer extrapolation.

Example 12D.2 Using a Birge–Sponer plot

The observed vibrational intervals of H_2^+ lie at the following values for $1\leftarrow0$, $2\leftarrow1$, …, respectively (in cm^{-1}): 2191, 2064, 1941, 1821, 1705, 1591, 1479, 1368, 1257, 1145, 1033, 918, 800, 677, 548, 411. Determine the dissociation energy of the molecule.

Method Plot the separations against $v+\frac{1}{2}$, extrapolate linearly to the point cutting the horizontal axis, and then measure the area under the curve.

Answer The points are plotted in Fig. 12D.9, and a linear extrapolation is shown as a dotted line. The area under the curve (use the formula for the area of a triangle or count the squares) is 214. Each square corresponds to $100\,cm^{-1}$ (refer to the scale of the vertical axis); hence the dissociation energy is $21\,400\,cm^{-1}$ (corresponding to $256\,kJ\,mol^{-1}$).

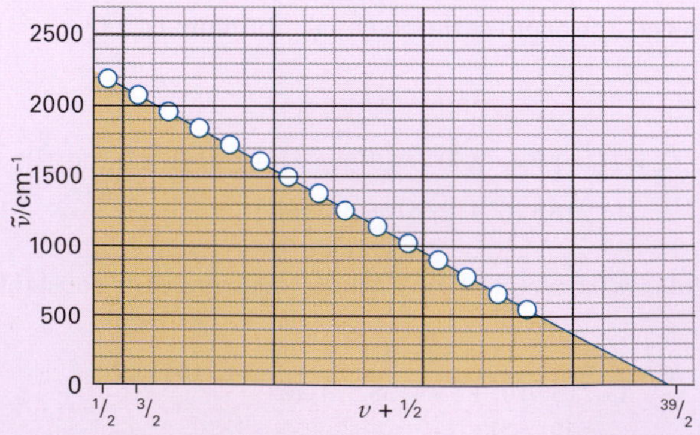

Figure 12D.9 The Birge–Sponer plot used in *Example* 12D.2. The area is obtained simply by counting the squares beneath the line or using the formula for the area of a right triangle (area $=\frac{1}{2}\times$base$\times$height).

Self-test 12D.4 The vibrational levels of HgH converge rapidly, and successive intervals are 1203.7 (which corresponds to the $1 \leftarrow 0$ transition), 965.6, 632.4, and $172\,\mathrm{cm^{-1}}$. Estimate the dissociation energy.

Answer: $35.6\,\mathrm{kJ\,mol^{-1}}$

12D.4 Vibration–rotation spectra

Each line of the high resolution vibrational spectrum of a gas-phase heteronuclear diatomic molecule is found to consist of a large number of closely spaced components (Fig. 12D.10). Hence, molecular spectra are often called **band spectra**. The separation between the components is less than $10\,\mathrm{cm^{-1}}$, which suggests that the structure is due to rotational transitions accompanying the vibrational transition. A rotational change should be expected because classically we can think of the vibrational transition as leading to a sudden increase or decrease in the instantaneous bond length. Just as ice-skaters rotate more rapidly when they bring their arms in, and more slowly when they throw them out, so the molecular rotation is either accelerated or retarded by a vibrational transition.

(a) Spectral branches

A detailed analysis of the quantum mechanics of simultaneous vibrational and rotational changes shows that the rotational quantum number J changes by ± 1 during the vibrational transition of a diatomic molecule. If the molecule also possesses angular momentum about its axis, as in the case of the electronic orbital angular momentum of the paramagnetic molecule NO with its configuration $\ldots\pi^1$, then the selection rules also allow $\Delta J = 0$.

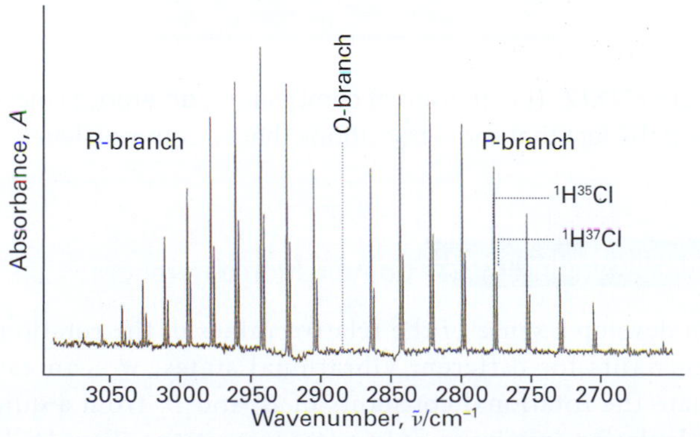

Figure 12D.10 A high-resolution vibration–rotation spectrum of HCl. The lines appear in pairs because $H^{35}Cl$ and $H^{37}Cl$ both contribute (their abundance ratio is 3:1). There is no Q branch, because $\Delta J = 0$ is forbidden for this molecule.

The appearance of the vibration–rotation spectrum of a diatomic molecule can be discussed in terms of the combined vibration–rotation terms, $\tilde{S}$:

$$\tilde{S}(v,J) = \tilde{G}(v) + \tilde{F}(J) \qquad (12D.17)$$

If we ignore anharmonicity and centrifugal distortion we can use eqn 12D.8 for the first term on the right and eqn 12B.9 ($\tilde{F}(J) = \tilde{B}J(J+1)$) for the second, and obtain

$$\tilde{S}(v,J) = \left(v + \tfrac{1}{2}\right)\tilde{\nu} + \tilde{B}J(J+1) \qquad (12D.18)$$

In a more detailed treatment, $\tilde{B}$ is allowed to depend on the vibrational state because, as v increases, the molecule swells slightly and the moment of inertia changes. We shall continue with the simple expression initially.

When the vibrational transition $v+1 \leftarrow v$ occurs, J changes by ± 1 and in some cases by 0 (when $\Delta J = 0$ is allowed). The absorptions then fall into three groups called **branches** of the spectrum. The **P branch** consists of all transitions with $\Delta J = -1$:

$$\tilde{\nu}_P(J) = \tilde{S}(v+1, J-1) - \tilde{S}(v,J) = \tilde{\nu} - 2\tilde{B}J \qquad \text{P branch transitions} \qquad (12D.19a)$$

This branch consists of lines at $\tilde{\nu} - 2\tilde{B}, \tilde{\nu} - 4\tilde{B}, \ldots$ with an intensity distribution reflecting both the populations of the rotational levels and the magnitude of the $J-1 \leftarrow J$ transition moment (Fig. 12D.11). The **Q branch** consists of all lines with $\Delta J = 0$, and its wavenumbers are all

$$\tilde{\nu}_Q(J) = \tilde{S}(v+1, J) - \tilde{S}(v,J) = \tilde{\nu} \qquad \text{Q branch transitions} \qquad (12D.19b)$$

for all values of J. This branch, when it is allowed (as in NO), appears at the vibrational transition wavenumber. In Fig. 12D.11 there is a gap at the expected location of the Q branch

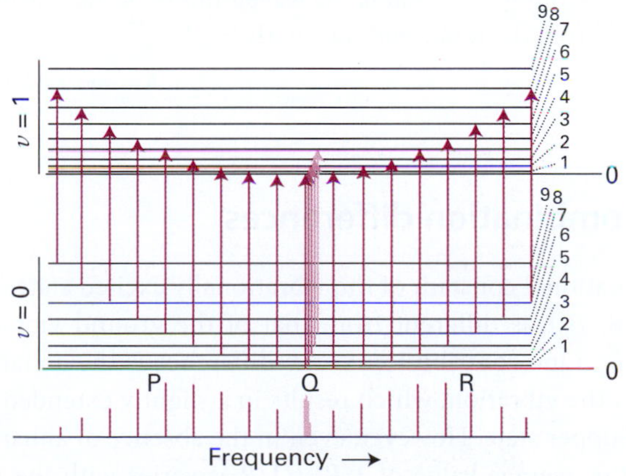

Figure 12D.11 The formation of P, Q, and R branches in a vibration–rotation spectrum. The intensities reflect the populations of the initial rotational levels and magnitudes of the transition moments.

Table 12D.1* Properties of diatomic molecules

	$\tilde{\nu}$/cm^{-1}	R_e/pm	$\tilde{B}$/cm^{-1}	k_f/(N m^{-1})	$hc\tilde{D}_0$/(kJ mol^{-1})
$^1H_2$	4400	74	60.86	575	432
$^1H^{35}Cl$	2991	127	10.59	516	428
$^1H^{127}I$	2308	161	6.51	314	295
$^{35}Cl_2$	560	199	0.244	323	239

* More values are given in the *Resource section*.

because it is forbidden in HCl. The **R branch** consists of lines with $\Delta J=+1$:

$$\tilde{\nu}_R(J)=\tilde{S}(v+1,J+1)-\tilde{S}(v,J)=\tilde{\nu}+2\tilde{B}(J+1) \quad \text{R branch transitions} \quad (12D.19c)$$

This branch consists of lines displaced from $\tilde{\nu}$ to high wavenumber by $2\tilde{B}, 4\tilde{B}, \ldots$.

The separation between the lines in the P and R branches of a vibrational transition gives the value of $\tilde{B}$. Therefore, the bond length can be deduced without needing to take a pure rotational microwave spectrum. However, the latter is more precise because microwave frequencies can be measured with greater precision than infrared frequencies.

Brief illustration 12D.3 The wavenumber of an R branch transition

Infrared absorption by $^1H^{81}Br$ gives rise to an R branch from $v=0$. It follows from eqn 12D.19c and the data in Table 12D.1 (in the *Resource section*) that the wavenumber of the line originating from the rotational state with $J=2$ is

$$\tilde{\nu}_R(2)=\tilde{\nu}+6\tilde{B}=(2648.98\ \text{cm}^{-1})+6\times(8.465\ \text{cm}^{-1}) = 2699.77\ \text{cm}^{-1}$$

Self-test 12D.5 Infrared absorption by $^1H^{127}I$ gives rise to an R branch from $v=0$. What is the wavenumber of the line originating from the rotational state with $J=2$?

Answer: 2347.16 cm^{-1}

(b) Combination differences

The rotational constant of the vibrationally excited state, $\tilde{B}_1$ (in general, $\tilde{B}_v$), is different from that of the ground vibrational state, $\tilde{B}_0$. One contribution to the difference is the anharmonicity of the vibration, which results in a slightly extended bond in the upper state. However, even in the absence of anharmonicity, the average value of $1/R^2$ ($\langle 1/R^2\rangle$) varies with the vibrational state (see Problems 12D.12 and 12D.13). As a result, the Q branch (if it exists) consists of a series of closely spaced lines. The lines of the R branch converge slightly as J increases; and those of the P branch diverge:

$$\begin{aligned}\tilde{\nu}_P(J)&=\tilde{\nu}-(\tilde{B}_1+\tilde{B}_0)J+(\tilde{B}_1-\tilde{B}_0)J^2\\ \tilde{\nu}_Q(J)&=\tilde{\nu}+(\tilde{B}_1-\tilde{B}_0)J(J+1)\\ \tilde{\nu}_R(J)&=\tilde{\nu}+(\tilde{B}_1+\tilde{B}_0)(J+1)+(\tilde{B}_1-\tilde{B}_0)(J+1)^2\end{aligned} \quad (12D.20)$$

To determine the two rotational constants individually, we use the method of **combination differences**. This procedure is used widely in spectroscopy to extract information about a particular state. It involves setting up expressions for the difference in the wavenumbers of transitions to a common state; the resulting expression then depends solely on properties of the other state.

As can be seen from Fig. 12D.12, the transitions $\tilde{\nu}_R(J-1)$ and $\tilde{\nu}_P(J+1)$ have a common upper state, and hence can be anticipated to depend on $\tilde{B}_0$. Indeed, it is easy to show from eqn 12D.20 that

$$\tilde{\nu}_R(J-1)-\tilde{\nu}_P(J+1)=4\tilde{B}_0(J+\tfrac{1}{2}) \quad (12D.21a)$$

Therefore, a plot of the combination difference against $J+\frac{1}{2}$ should be a straight line of slope $4\tilde{B}_0$, so the rotational constant of the molecule in the state $v=0$ can be determined. (Any deviation from a straight line is a consequence of centrifugal distortion, so that effect can be investigated too.) Similarly, $\tilde{\nu}_R(J)$ and $\tilde{\nu}_P(J)$ have a common lower state, and hence their combination difference gives information about the upper state:

$$\tilde{\nu}_R(J)-\tilde{\nu}_P(J)=4\tilde{B}_1(J+\tfrac{1}{2}) \quad (12D.21b)$$

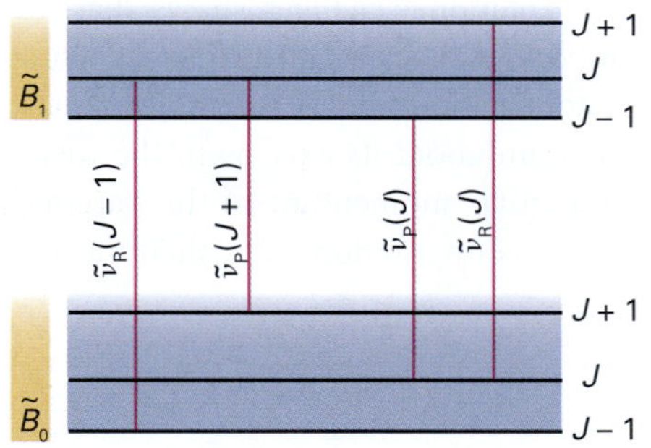

Figure 12D.12 The method of combination differences makes use of the fact that some transitions share a common level.

Brief illustration 12D.4 Combination differences

To develop a sense of the relative values of the rotational constants for different vibrational states, we can estimate the rotational constants of $\tilde{B}_0$ and $\tilde{B}_1$ from a quick calculation involving only a few transitions. For $^1H^{35}Cl$, $\tilde{\nu}_R(0)-\tilde{\nu}_P(2)=62.6\ \text{cm}^{-1}$, and it follows from eqn 12D.21a, with $J=1$, that $\tilde{B}_0=62.6/\{4\times(1+\frac{1}{2})\}\ \text{cm}^{-1}=10.4\ \text{cm}^{-1}$. Similarly, $\tilde{\nu}_R(1)-\tilde{\nu}_P(1)=60.8\ \text{cm}^{-1}$, and it follows from eqn 12D.21b, again with $J=1$ that $\tilde{B}_1=60.8/\{4\times(1+\frac{1}{2})\}$ cm^{-1}=10.1 cm^{-1}. The linear least-squares procedure applied to a richer data set gives $\tilde{B}_0=10.440$ cm^{-1} and $\tilde{B}_1=10.136$ cm^{-1}. We see that the two rotational constants do not differ by much.

Self-test 12D.6 For $^{12}C^{16}O$, $\tilde{\nu}_R(0)=2147.084$ cm^{-1}, $\tilde{\nu}_R(1)=2150.858$ cm^{-1}, $\tilde{\nu}_P(1)=2139.427$ cm^{-1}, and $\tilde{\nu}_P(2)=2135.548$ cm^{-1}. Estimate the values of $\tilde{B}_0$ and $\tilde{B}_1$.

Answer: $\tilde{B}_0=1.923$ cm^{-1}, $\tilde{B}_1=1.905$ cm^{-1}

12D.5 Vibrational Raman spectra

The gross selection rule for vibrational Raman transitions (see the following *Justification*) is that *the polarizability should change as the molecule vibrates*. The polarizability plays a role in vibrational Raman spectroscopy because the molecule must be squeezed and stretched by the incident radiation in order that a vibrational excitation may occur during the photon–molecule collision.

Brief illustration 12D.5 The gross selection rule for vibrational Raman spectra

Both homonuclear and heteronuclear diatomic molecules swell and contract during a vibration, the control of the nuclei over the electrons varies, and hence the molecular polarizability changes. Both types of diatomic molecule are therefore vibrationally Raman active.

Self-test 12D.7 Can a linear, nonpolar molecule like CO_2 have a Raman spectrum?

Answer: Yes

The specific selection rule for vibrational Raman transitions in the harmonic approximation is $\Delta v=\pm1$. The formal basis for the gross and specific selection rules is given in the following *Justification*.

Justification 12D.2 Gross and specific selection rules for vibrational Raman spectra

For simplicity, we consider a one-dimensional harmonic oscillator (like a diatomic molecule). First, we note that the oscillating electric field, $\mathcal{E}(t)$, of the incident electromagnetic radiation can induce a dipole moment that is proportional to the strength of the field. We write $\hat{\mu}=\alpha(x)\mathcal{E}(t)$, where $\alpha(x)$ is the polarizability of the molecule (Topic 12B). The transition dipole moment is then

$$\mu_{fi}=\int \psi_{v_f}^*\hat{\mu}\psi_{v_i}\,d\tau=\int \psi_{v_f}^*\alpha(x)\mathcal{E}(t)\psi_{v_i}\,dx=\mathcal{E}(t)\int \psi_{v_f}^*\alpha(x)\psi_{v_i}\,dx$$

The polarizability varies with the length of the bond because the control of the nuclei over the electrons varies as their position changes, so $\alpha(x)=\alpha_0+(d\alpha/dx)_0x+\ldots$. Now the calculation proceeds as in *Justification* 12D.1, but $(d\mu/dx)_0$ is replaced by $\mathcal{E}(t)(d\alpha/dx)_0$ in the expression for μ_{fi}. For f≠i,

$$\mu_{fi}=\mathcal{E}(t)\left(\frac{d\alpha}{dx}\right)_0\int \psi_{v_f}^* x\psi_{v_i}\,dx$$

Therefore, the vibration is Raman active only if $(d\alpha/dx)_0\neq0$; that is, only if the polarizability varies with displacement, and, as we saw in *Justification* 12D.1, if $v_f-v_i=\pm1$.

The lines to high frequency of the incident radiation, in the language introduced in Topic 12A, the 'anti-Stokes lines', are those for which $\Delta v=-1$. The lines to low frequency, the 'Stokes lines', correspond to $\Delta v=+1$. The intensities of the anti-Stokes and Stokes lines are governed largely by the Boltzmann populations of the vibrational states involved in the transition. It follows that anti-Stokes lines are usually weak because very few molecules are in an excited vibrational state initially.

In gas-phase spectra, the Stokes and anti-Stokes lines have a branch structure arising from the simultaneous rotational transitions that accompany the vibrational excitation (Fig. 12D.13). The selection rules are $\Delta J=0, \pm2$ (as in pure rotational Raman spectroscopy), and give rise to the **O branch** ($\Delta J=-2$), the **Q branch** ($\Delta J=0$), and the **S branch** ($\Delta J=+2$):

$$\begin{aligned}\tilde{\nu}_O(J)&=\tilde{\nu}_i-\tilde{\nu}-2\tilde{B}+4\tilde{B}J && \text{O branch transitions}\\ \tilde{\nu}_Q(J)&=\tilde{\nu}_i-\tilde{\nu} && \text{Q branch transitions}\\ \tilde{\nu}_S(J)&=\tilde{\nu}_i-\tilde{\nu}-6\tilde{B}-4\tilde{B}J && \text{S branch transitions}\end{aligned} \qquad (12D.22)$$

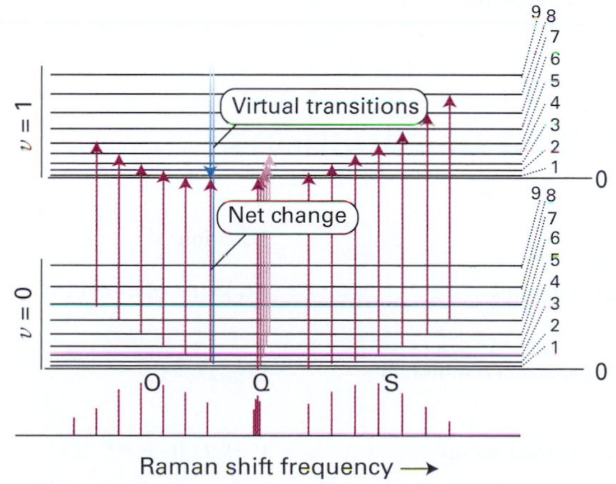

Figure 12D.13 The formation of O, Q, and S branches in a vibration–rotation Raman spectrum of a linear rotor. Note that the frequency scale runs in the opposite direction to that in Fig. 12D.11, because the higher energy transitions (on the right) extract more energy from the incident beam and leave it at lower frequency.

where $\tilde{\nu}_i$ is the wavenumber of the incident radiation. Note that, unlike in infrared spectroscopy, a Q branch is obtained for all linear molecules. The spectrum of CO, for instance, is shown in Fig. 12D.14: the structure of the Q branch arises from the differences in rotational constants of the upper and lower vibrational states.

The information available from vibrational Raman spectra adds to that from infrared spectroscopy because homonuclear diatomics can also be studied. The spectra can be interpreted in terms of the force constants, dissociation energies, and bond lengths, and some of the information obtained is included in Table 12D.1.

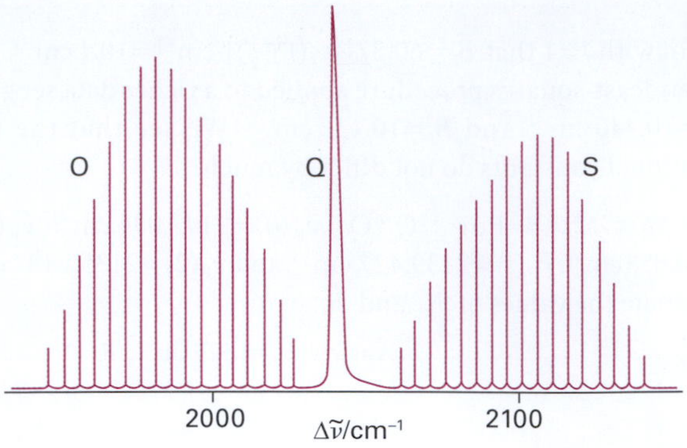

Figure 12D.14 The structure of a vibrational line in the vibrational Raman spectrum of carbon monoxide, showing the O, Q, and S branches. The horizontal axis represents the wavenumber difference between the incident and scattered radiation.

Checklist of concepts

☐ 1. The vibrational energy levels of a diatomic molecule modelled as a harmonic oscillator depend on a **force constant** k_f (a measure of the bond's stiffness) and the **effective mass** of the vibration.

☐ 2. The **gross selection rule** for infrared spectra is that the electric dipole moment of the molecule must change when the atoms are displaced relative to one another.

☐ 3. The **specific selection rule** for infrared spectra (within the harmonic approximation) is $\Delta v = \pm 1$.

☐ 4. The **Morse potential energy function** can be used to model anharmonic motion.

☐ 5. The strongest infrared transitions are the **fundamental transitions** ($v=1 \leftarrow v=0$).

☐ 6. Anharmonicity gives rise to weaker **overtone transitions** ($v=2 \leftarrow v=0$, $v=3 \leftarrow v=0$, etc.).

☐ 7. A **Birge–Sponer** plot may be used to determine the dissociation energy of the bond in a diatomic molecule.

☐ 8. In the gas phase vibrational transitions have a **P, Q, R branch structure** due to simultaneous rotational transitions.

☐ 9. For a vibration to be **Raman active**, the polarizability must change as the molecule vibrates.

☐ 10. The **specific selection rule** for vibrational Raman spectra (within the harmonic approximation) is $\Delta v = \pm 1$.

☐ 11. In gas-phase spectra, the Stokes and anti-Stokes lines in a Raman spectrum have an **O, Q, S branch structure**.

Checklist of equations

Property	Equation	Comment	Equation number
Vibrational terms	$\tilde{G}(v)=(v+\frac{1}{2})\tilde{\nu}, \tilde{\nu}=(1/2\pi c)(k_f/m_{eff})^{1/2}$	Diatomic molecules; simple harmonic oscillator	12D.8
Infrared spectra (vibrational)	$\Delta\tilde{G}_{v+\frac{1}{2}}=\tilde{\nu}$	Diatomic molecules; simple harmonic oscillator	12D.10
Morse potential energy	$V=hc\tilde{D}_e\{1-e^{-a(R-R_e)}\}^2$, $a=(m_{eff}\omega^2/2hc\tilde{D}_e)^{1/2}$, $m_{eff}=m_1m_2/(m_1+m_2)$		12D.11
Vibrational terms (diatomic molecules)	$\tilde{G}(v)=(v+\frac{1}{2})\tilde{\nu}-(v+\frac{1}{2})^2x_e\tilde{\nu}$, $x_e=\tilde{\nu}/4\tilde{D}_e$	Morse potential energy	12D.12

Property	Equation	Comment	Equation number
Infrared spectra (vibrational)	$\Delta\tilde{G}_{v+\frac{1}{2}}=\tilde{\nu}-2(v+1)x_e\tilde{\nu}+\cdots$	Anharmonic oscillator	12D.14
	$\tilde{G}(v+2)-\tilde{G}(v)=2\tilde{\nu}-2(2v+3)x_e\tilde{\nu}+\cdots$	Overtones	12D.15
Dissociation energy	$\tilde{D}_0=\Delta\tilde{G}_{1/2}+\Delta\tilde{G}_{3/2}+\cdots=\sum_v \Delta\tilde{G}_{v+\frac{1}{2}}$	Birge–Sponer plot	12D.16
Vibration–rotation terms (diatomic molecules)	$\tilde{S}(v,J)=(v+\frac{1}{2})\tilde{\nu}+\tilde{B}J(J+1)$	Rotation coupled to vibration	12D.18
Infrared spectra (vibration–rotation)	$\tilde{\nu}_P(J)=\tilde{S}(v+1,J-1)-\tilde{S}(v,J)=\tilde{\nu}-2\tilde{B}J$	P branch ($\Delta J=-1$)	12D.19a
	$\tilde{\nu}_Q(J)=\tilde{S}(v+1,J)-\tilde{S}(v,J)=\tilde{\nu}$	Q branch ($\Delta J=0$)	12D.19b
	$\tilde{\nu}_R(J)=\tilde{S}(v+1,J+1)-\tilde{S}(v,J)=\tilde{\nu}+2\tilde{B}(J+1)$	R branch ($\Delta J=+1$)	12D.19c
	$\tilde{\nu}_R(J-1)-\tilde{\nu}_P(J+1)=4\tilde{B}_0(J+\frac{1}{2})$ $\tilde{\nu}_R(J)-\tilde{\nu}_P(J)=4\tilde{B}_1(J+\frac{1}{2})$	Combination differences	12D.21
Raman spectra (vibration-rotation)	$\tilde{\nu}_O(J)=\tilde{\nu}_i-\tilde{\nu}-2\tilde{B}+4\tilde{B}J$	O branch ($\Delta J=-2$)	12D.22
	$\tilde{\nu}_Q(J)=\tilde{\nu}_i-\tilde{\nu}$	Q branch ($\Delta J=0$)	
	$\tilde{\nu}_S(J)=\tilde{\nu}_i-\tilde{\nu}-6\tilde{B}-4\tilde{B}J$	S branch ($\Delta J=+2$)	

12E Vibrational spectroscopy of polyatomic molecules

Contents

➤ Why do you need to know this material?

The analysis of vibrational spectra provides information about the identity and conformation of polyatomic molecules in the gas and condensed phases. Even complex systems, such as synthetic materials and biological cells, can be studied.

➤ What is the key idea?

The vibrational spectrum of a polyatomic molecule can be interpreted in terms of the coordinated, collective harmonic motion of groups of atoms.

➤ What do you need to know already?

You need to be familiar with the harmonic oscillator (Topic 8B), the general principles of spectroscopy (Topic 12A), and the selection rules for infrared and Raman spectroscopy (Topic 12D). The treatment of the symmetry aspects of infrared and Raman active vibrations requires concepts from Chapter 11.

There is only one mode of vibration for a diatomic molecule: bond stretch. In polyatomic molecules there are several, sometimes hundreds, of modes of vibration because all the bond lengths and angles may change. Consequently, the vibrational spectra are very complex. Nonetheless, infrared and Raman spectroscopy can be used to obtain information about the structure of systems as large as animal and plant tissues. Raman spectroscopy is particularly useful for characterizing nanomaterials, especially carbon nanotubes.

12E.1 Normal modes

We begin by calculating the total number of vibrational modes of a polyatomic molecule. We then see that we can choose combinations of these atomic displacements that give the simplest description of the vibrations.

As we show in the following *Justification*, the number of independent modes of motion of an N-atom molecule depends on whether it is linear or nonlinear:

Linear molecule:	$3N-5$
Nonlinear molecule:	$3N-6$

Brief illustration 12E.1 The number of normal modes

Water, H_2O, is a nonlinear triatomic molecule, $N=3$, and has $3N-6=3$ modes of vibration (and three modes of rotation); CO_2 is a linear triatomic molecule, and has $3N-5=4$ modes of vibration (and only two modes of rotation). A biological macromolecule with $N \approx 500$ atoms can vibrate in nearly 1500 different independent ways.

Self-test 12E.1 How many normal modes does naphthalene ($C_{10}H_8$) have?

Answer: 48

Justification 12E.1 The number of vibrational modes

The total number of coordinates needed to specify the locations of N atoms is $3N$. Each atom may change its location by varying one of its three coordinates (x, y, and z), so the total number of displacements available is $3N$. These displacements can be grouped together in a physically sensible way. For example, three coordinates are needed to specify the location of the centre of mass of the molecule, so three of the $3N$ displacements correspond to the translational motion of the molecule as a whole. The remaining $3N - 3$ displacements are non-translational 'internal' modes of the molecule.

Two angles are needed to specify the orientation of a linear molecule in space: in effect, we need to give only the latitude and longitude of the direction in which the molecular axis is pointing (Fig. 12E.1a). However, three angles are needed for a nonlinear molecule because we also need to specify the orientation of the molecule around the direction defined by the latitude and longitude (Fig. 12E.1b). Therefore, two (linear) or three (nonlinear) of the $3N-3$ internal displacements are rotational. This leaves $3N-5$ (linear) or $3N-6$ (nonlinear) displacements of the atoms relative to one another: these are the vibrational modes. It follows that the number of modes of vibration N_{vib} is $3N-5$ for linear molecules and $3N-6$ for nonlinear molecules.

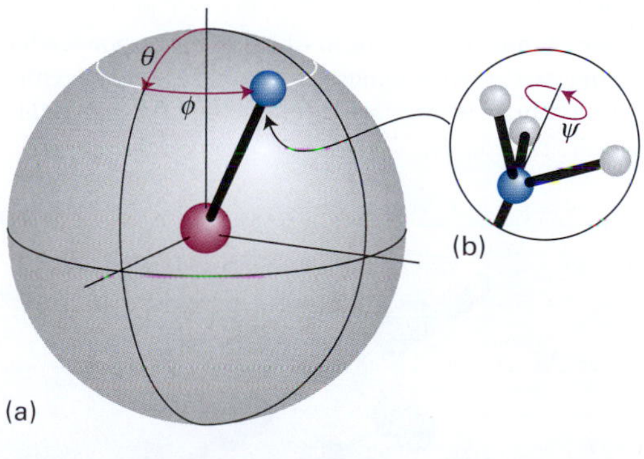

Figure 12E.1 (a) The orientation of a linear molecule requires the specification of two angles. (b) The orientation of a nonlinear molecule requires the specification of three angles.

The next step is to find the best description of the modes. One choice for the four modes of CO_2, for example, might be the ones in Fig. 12E.2a. This illustration shows the stretching of one bond (the mode ν_L), the stretching of the other (ν_R), and the two perpendicular bending modes (ν_2). The description, while permissible, has a disadvantage: when one CO bond vibration is excited, the motion of the C atom sets the other CO bond in motion, so energy flows backwards and forwards between ν_L and ν_R. Moreover, the position of the centre of mass of the molecule varies in the course of either vibration.

The description of the vibrational motion is much simpler if linear combinations of ν_L and ν_R are taken. For example, one combination is ν_1 in Fig. 12E.2b: this mode is the **symmetric stretch**. In this mode, the C atom is buffeted simultaneously from each side and the motion continues indefinitely. Another mode is ν_2, the **antisymmetric stretch**, in which the two O atoms always move in the same direction as each other and opposite to that of the C atom. Both modes are independent in the sense that, if one is excited, then it does not excite the other. They are two of the 'normal modes' of the molecule, its independent, collective vibrational displacements. The two other normal modes are the two bending modes ν_3. In general, a **normal mode** is an independent, synchronous motion of atoms or groups of atoms that may be excited without leading to the excitation of any other normal mode and without involving translation or rotation of the molecule as a whole.

The four normal modes of CO_2, and the N_{vib} normal modes of polyatomics in general, are the key to the description of molecular vibrations. Each normal mode, q, behaves like an independent harmonic oscillator (if anharmonicities are neglected), so each has a series of terms

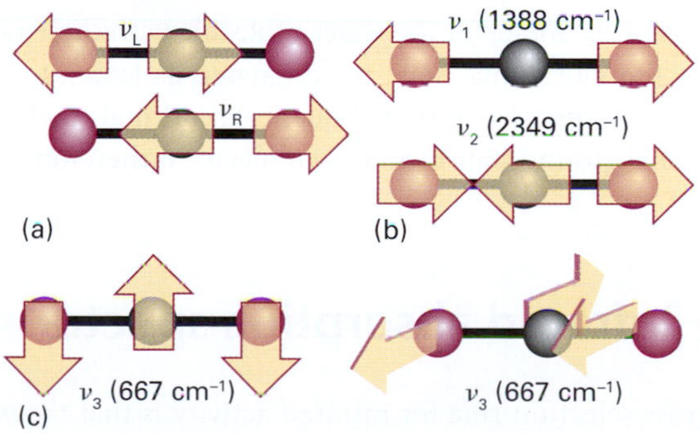

Figure 12E.2 Alternative descriptions of the vibrations of CO_2. (a) The stretching modes are not independent, and if one C—O group is excited the other begins to vibrate. They are not normal modes of vibration of the molecule. (b) The symmetric and antisymmetric stretches are independent, and one can be excited without affecting the other: they are normal modes. (c) The two perpendicular bending motions are also normal modes.

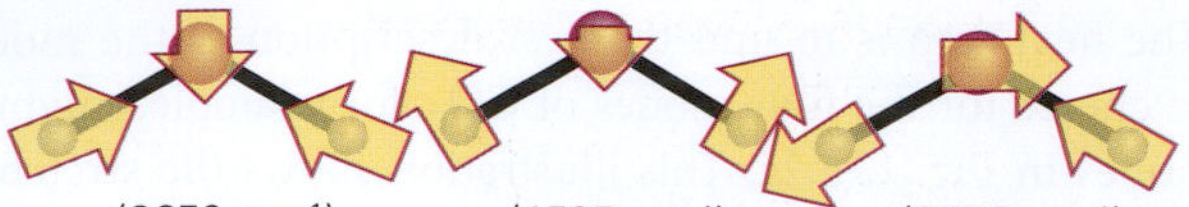

Figure 12E.3 The three normal modes of H_2O. The mode ν_2 is predominantly bending, and occurs at lower wavenumber than the other two.

$$\tilde{G}_q(v)=(v+\tfrac{1}{2})\tilde{\nu}_q \qquad \tilde{\nu}_q=\frac{1}{2\pi c}\left(\frac{k_{f,q}}{m_q}\right)^{1/2} \qquad \text{Vibrational terms of normal modes} \qquad (12E.1)$$

where $\tilde{\nu}_q$ is the wavenumber of mode q and depends on the force constant $k_{f,q}$ for the mode and on the effective mass m_q of the mode. The effective mass of the mode is a measure of the mass that is swung about by the vibration and in general is a combination of the masses of the atoms. For example, in the symmetric stretch of CO_2, the C atom is stationary, and the effective mass depends on the masses of only the O atoms. In the antisymmetric stretch and in the bends, all three atoms move, so all contribute to the effective mass. The three normal modes of H_2O are shown in Fig. 12E.3: note that the predominantly bending mode (ν_2) has a lower frequency than the others, which are predominantly stretching modes. It is generally the case that the frequencies of bending motions are lower than those of stretching modes. One point that must be appreciated is that only in special cases (such as the CO_2 molecule) are the normal modes purely stretches or purely bends. In general, a normal mode is a composite motion of simultaneous stretching and bending of bonds. Another point in this connection is that heavy atoms generally move less than light atoms in normal modes.

The vibrational state of a polyatomic molecule is specified by the vibrational quantum number v for each of the normal modes. For example, for the water molecule with three normal modes, the vibrational state is designated as (v_1,v_2,v_3) where v_i is the number of vibrational quanta in normal mode i. The vibrational ground state of an H_2O molecule is therefore (0,0,0).

12E.2 Infrared absorption spectra

The gross selection rule for infrared activity is that *the motion corresponding to a normal mode should be accompanied by a change of dipole moment.* Simple inspection of atomic motions is sometimes all that is needed in order to assess whether a normal mode is infrared active. For example, the symmetric stretch of CO_2 leaves the dipole moment unchanged (at zero, see Fig. 12E.2), so this mode is infrared inactive. The antisymmetric stretch, however, changes the dipole moment because the molecule becomes unsymmetrical as it vibrates, so this mode is infrared active. Because the dipole moment change is parallel to the principal axis, the transitions arising from this mode are classified as **parallel bands** in the spectrum. Both bending modes are infrared active: they are accompanied by a changing dipole perpendicular to the principal axis, so transitions involving them lead to a **perpendicular band** in the spectrum.

Example 12E.1 Using the gross selection rule for infrared spectroscopy

State which of the following molecules are infrared active: N_2O, OCS, H_2O, $CH_2{=}CH_2$.

Method Molecules that are infrared active have dipole moments that change during the course of a vibration. Therefore, judge whether a distortion of the molecule can change its dipole moment (including changing it from zero).

Answer All the molecules possess at least one normal mode that results in a change of dipole moment, so all are infrared active. Note that not all the modes of complicated molecules are infrared active. For example, a vibration of $CH_2{=}CH_2$ in which the C=C bond stretches and contracts (while the C—H bonds either do not vibrate or stretch and contract synchronously) is inactive because it leaves the dipole moment unchanged (at zero, Fig. 12E.4).

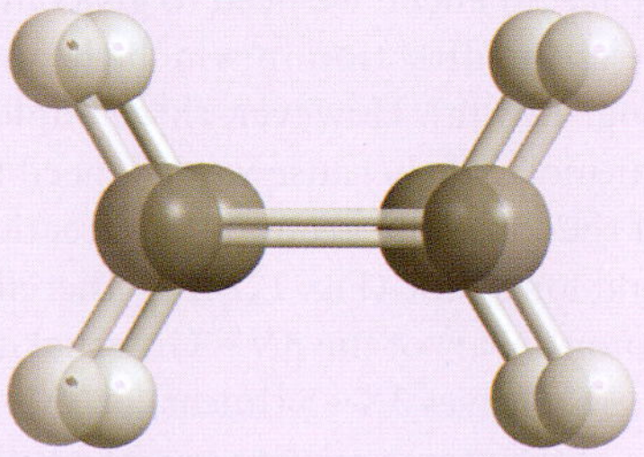

Figure 12E.4 A normal mode of $CH_2{=}CH_2$ (ethene) that is not infrared active.

Self-test 12E.2 Identify a normal mode of C_6H_6 that is not infrared active

Answer: A 'breathing' mode in which all the carbon–carbon bonds contract and stretch synchronously, while the C—H bonds either do not vibrate or stretch and contract synchronously (Fig. 12E.5)

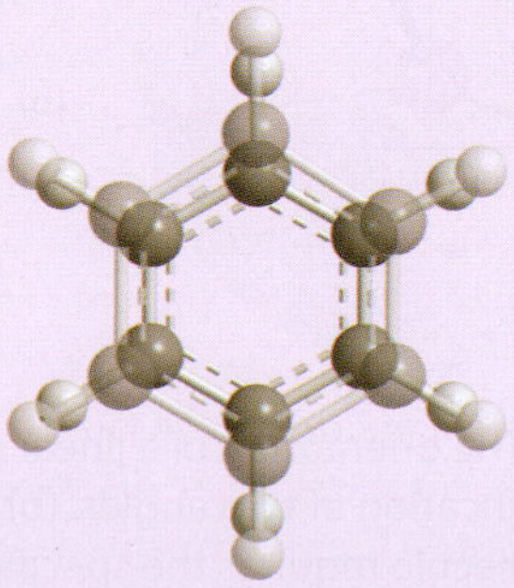

Figure 12E.5 A normal mode of C_6H_6 (benzene) that is not infrared active.

The active modes are subject to the specific selection rule $\Delta v_q = \pm 1$ in the harmonic approximation, so the wavenumber of the fundamental transition (the 'first harmonic') of each active mode is $\tilde{\nu}_q$. A polyatomic molecule has several fundamental transitions. For example, the spectrum of a molecule with three infrared active normal modes features three fundamental transitions: $(1,0,0) \leftarrow (0,0,0)$, $(0,1,0) \leftarrow (0,0,0)$, and $(0,0,1) \leftarrow (0,0,0)$. Also possible are **combination bands** corresponding to the excitation of more than one normal mode in the transition, as in $(1,1,0) \leftarrow (0,0,0)$. Moreover, overtone transitions, such as $(2,0,0) \leftarrow (0,0,0)$, can appear in the spectrum when anharmonicity is important (Topic 12D).

From the analysis of the spectrum, a picture may be constructed of the stiffness of various parts of the molecule, that is, we can establish its **force field**, the set of force constants corresponding to all the displacements of the atoms. The force field may also be estimated by using the computational techniques described in Topic 10E. Superimposed on the simple force-field scheme are the complications arising from anharmonicities and the effects of molecular rotation. In the gas phase, rotational transitions affect the spectrum in a way similar to their effect on diatomic molecules (Topic 12D), but as polyatomic molecules are typically asymmetric rotors, the resulting band structure is very complex.

Molecules are unable to rotate freely in a liquid or a solid. In a liquid, for example, a molecule may be able to rotate through only a few degrees before it is struck by another, so it changes its rotational state frequently. This random changing of orientation is called **tumbling**. As a result of this intermolecular buffeting, the lifetimes of rotational states in liquids are very short, so in most cases the rotational energies are ill-defined. Collisions occur at a rate of about 10^{13} s^{-1} and, even allowing for only a 10 per cent success rate in knocking the molecule into another rotational state, a lifetime broadening (eqn 12A.19, in the form $\delta\tilde{\nu} \approx 1/2\pi c\tau$) of more than 1 cm^{-1} can easily result. The rotational structure of the vibrational spectrum is blurred by this effect, so the infrared spectra of molecules in condensed phases usually consist of broad lines spanning the entire range of the resolved gas-phase spectrum, and showing no branch structure.

One very important application of infrared spectroscopy to condensed phase samples, and for which the blurring of the rotational structure by random collisions is a welcome simplification, is to chemical analysis. The vibrational spectra of different groups in a molecule give rise to absorptions at characteristic frequencies because a normal mode of even a very large molecule is often dominated by the motion of a small group of atoms. The intensities of the vibrational bands that can be identified with the motions of small groups are also transferable between molecules. Consequently, the molecules in a sample can often be identified by examining its infrared spectrum and referring to a table of characteristic frequencies and intensities (Table 12E.1).

Table 12E.1* Typical vibrational wavenumbers, $\tilde{\nu}$/cm^{-1}

Vibration type	$\tilde{\nu}$/cm^{-1}
C—H stretch	2850–2960
C—H bend	1340–1465
C—C stretch, bend	700–1250
C=C stretch	1620–1680

* More values are given in the *Resource section*.

Example 12E.2 Interpreting an infrared spectrum

The infrared spectrum of an organic compound is shown in Fig. 12E.6. Suggest an identification.

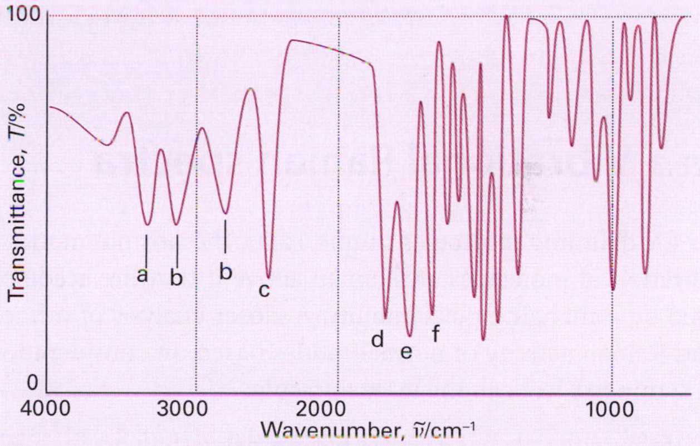

Figure 12E.6 A typical infrared absorption spectrum taken by forming a sample into a disk with potassium bromide. As explained *Example* 12E.2, the substance can be identified as $O_2NC_6H_4$—C≡C—COOH.

Method Some of the features at wavenumbers above 1500 cm^{-1} can be identified by comparison with the data in Table 12E.1.

Answer The features of the spectrum include: (a) C—H stretch of a benzene ring, indicating a substituted benzene; (b) carboxylic acid O—H stretch, indicating a carboxylic acid; (c) the strong absorption of a conjugated C≡C group, indicating a substituted alkyne; (d) this strong absorption is also characteristic of a carboxylic acid that is conjugated to a carbon–carbon multiple bond; (e) a characteristic vibration of a benzene ring, confirming the deduction drawn from (a); (f) a characteristic absorption of a nitro group ($-NO_2$) connected to a multiply bonded carbon–carbon system, suggesting a nitro-substituted benzene. The molecule contains as components a benzene ring, an aromatic carbon–carbon bond, a -COOH group, and a $-NO_2$ group. The molecule is in fact $O_2N-C_6H_4-$C≡C—COOH. A more detailed analysis shows it to be the 1,4-isomer.

Self-test 12E.3 Suggest an identification of the organic compound responsible for the spectrum shown in Fig. 12E.7. (*Hint*: The molecular formula of the compound is C_3H_5ClO.)

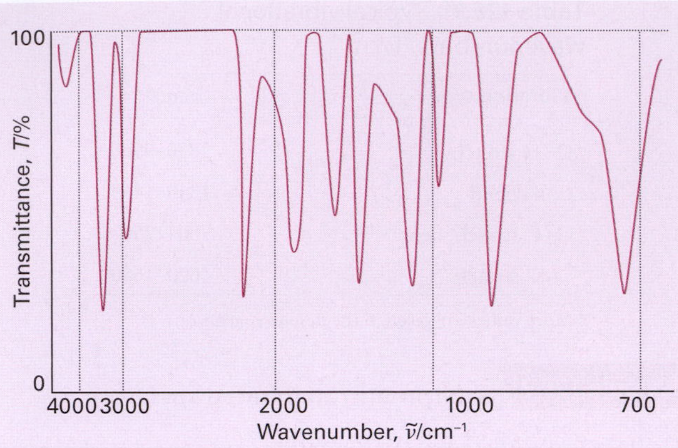

Figure 12E.7 The spectrum considered in *Self-test* 12E.3.

Answer: $CH_2{=}CClCH_2OH$

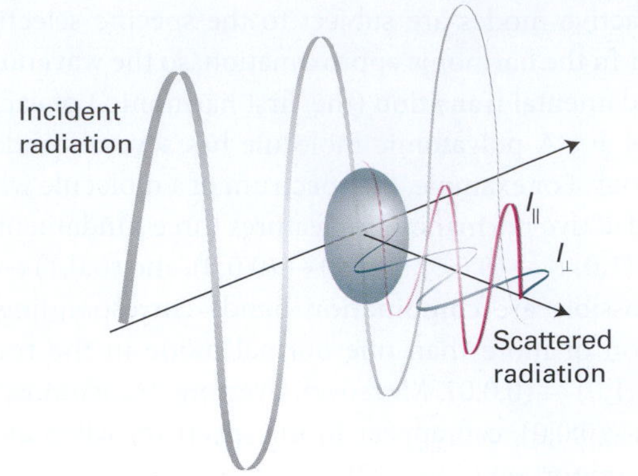

Figure 12E.8 The definition of the planes used for the specification of the depolarization ratio, ρ, in Raman scattering.

12E.3 Vibrational Raman spectra

As for diatomic molecules (Topic 12D), the normal modes of vibration of molecules are Raman active if they are accompanied by a changing polarizability. A closer analysis of infrared and Raman activity of normal modes based on considerations of symmetry leads to the **exclusion rule**:

> If the molecule has a centre of symmetry then no modes can be both infrared and Raman active. — Exclusion rule

(A mode may be inactive in both.) Because it is often possible to judge intuitively if a mode changes the molecular dipole moment, we can use this rule to identify modes that are not Raman active.

Brief illustration 12E.2 Raman active modes of polyatomic molecules

The symmetric stretch of CO_2 alternately swells and contracts the molecule: this motion changes the size and hence the polarizability of the molecule, so the mode is Raman active. The other modes of CO_2 leave the polarizability unchanged, so they are Raman inactive. Furthermore, the exclusion rule applies to CO_2 because it has a centre of symmetry.

Self-test 12E.4 Does the exclusion rule apply to H_2O or CH_4?

Answer: No; neither molecule has centre of symmetry

(a) Depolarization

The assignment of Raman lines to particular vibrational modes is aided by noting the state of polarization of the scattered light. The **depolarization ratio**, ρ, of a line is the ratio of the intensities, I, of the scattered light with polarizations perpendicular and parallel to the plane of polarization of the incident radiation:

$$\rho = \frac{I_{\perp}}{I_{\parallel}} \qquad \textit{Definition} \quad \text{Depolarization ratio} \qquad (12E.2)$$

To measure ρ, the intensity of a Raman line is measured with a polarizing filter (a 'half-wave plate') first parallel and then perpendicular to the polarization of the incident beam. If the emergent light is not polarized, then both intensities are the same and ρ is close to 1; if the light retains its initial polarization, then $I_{\perp}=0$, so $\rho=0$ (Fig. 12E.8). A line is classified as **depolarized** if it has ρ close to or greater than 0.75 and as **polarized** if $\rho<0.75$. Only totally symmetrical vibrations give rise to polarized lines in which the incident polarization is largely preserved. Vibrations that are not totally symmetrical give rise to depolarized lines because the incident radiation can give rise to radiation in the perpendicular direction too.

(b) Resonance Raman spectra

A modification of the basic Raman effect involves using incident radiation that nearly coincides with the frequency of an electronic transition of the sample (Fig. 12E.9). The technique is then called **resonance Raman spectroscopy**. It is characterized by a much greater intensity in the scattered radiation. Furthermore, because it is often the case that only a few vibrational modes contribute to the more intense scattering, the spectrum is greatly simplified.

Resonance Raman spectroscopy is used to study biological molecules that absorb strongly in the ultraviolet and visible regions of the spectrum. Examples include the pigments β-carotene and chlorophyll, which capture solar energy during plant photosynthesis. The resonance Raman spectra of Fig. 12E.10 show vibrational transitions from only the few pigment molecules that are bound to very large proteins dissolved

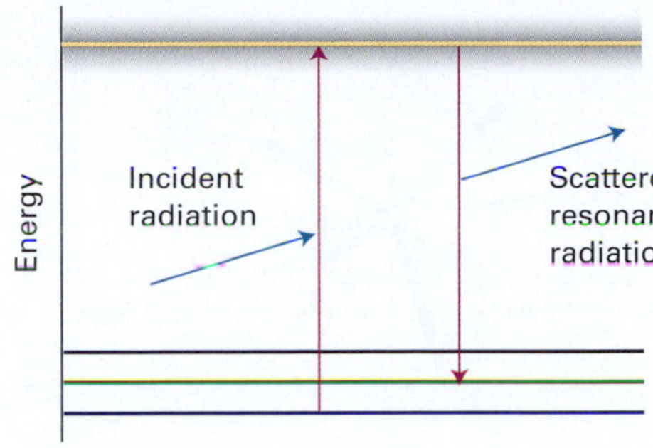

Figure 12E.9 In the resonance Raman effect the incident radiation has a frequency close to an actual electronic excitation of the molecule. A photon is emitted when the excited state returns to a state close to the ground state.

in an aqueous buffer solution. This selectivity arises from the fact that water (the solvent), amino acid residues, and the peptide group do not have electronic transitions at the laser wavelengths used in the experiment, so their conventional Raman spectra are weak compared to the enhanced spectra of the pigments. Comparison of the spectra in Figs. 12E.10a and 12E.10b also shows that, with proper choice of excitation wavelength, it is possible to examine individual classes of pigments bound to the same protein: excitation at 488 nm, where β-carotene absorbs strongly, shows vibrational bands from β-carotene only, whereas excitation at 407 nm, where chlorophyll *a* and β-carotene absorb, reveals features from both types of pigments.

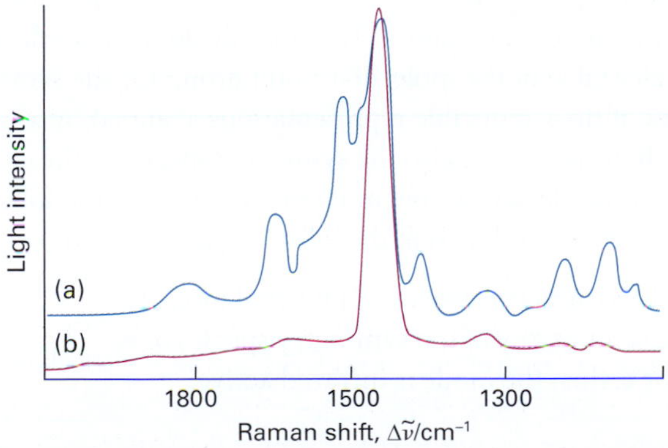

Figure 12E.10 The resonance Raman spectra of a protein complex that is responsible for some of the initial electron transfer events in plant photosynthesis. (a) Laser excitation of the sample at 407 nm shows Raman bands due to both chlorophyll *a* and β-carotene bound to the protein because both pigments absorb light at this wavelength. (b) Laser excitation at 488 nm shows Raman bands from β-carotene only because chlorophyll *a* does not absorb light very strongly at this wavelength. (Adapted from D.F. Ghanotakis et al., *Biochim. Biophys. Acta* **974**, 44 (1989).)

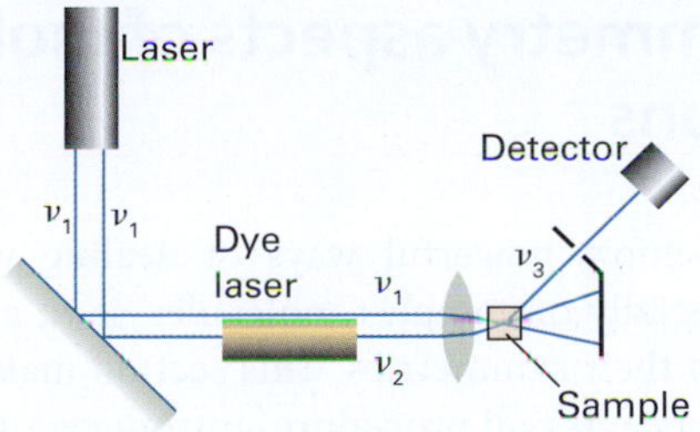

Figure 12E.11 The experimental arrangement for the CARS experiment.

(c) Coherent anti-Stokes Raman spectroscopy

The intensity of Raman transitions may be enhanced by **coherent anti-Stokes Raman spectroscopy** (CARS, Fig. 12E.11). The technique relies on the fact that, if two laser beams of frequencies ν_1 and ν_2 pass through a sample, then they may mix together and give rise to coherent radiation of several different frequencies, one of which is

$$\nu' = 2\nu_1 - \nu_2 \qquad (12E.3a)$$

Suppose that ν_2 is varied until it matches any Stokes line from the sample, such as the one with frequency $\nu_1 - \Delta\nu$; then the coherent emission will have frequency

$$\nu' = 2\nu_1 - (\nu_1 - \Delta\nu) = \nu_1 + \Delta\nu \qquad (12E.3b)$$

which is the frequency of the corresponding anti-Stokes line. This coherent radiation forms a narrow beam of high intensity.

An advantage of CARS is that it can be used to study Raman transitions in the presence of competing incoherent background radiation, and so can be used to observe the Raman spectra of species in flames. One example is the vibration–rotation CARS spectrum of N_2 gas in a methane–air flame shown in Fig 12E.12.

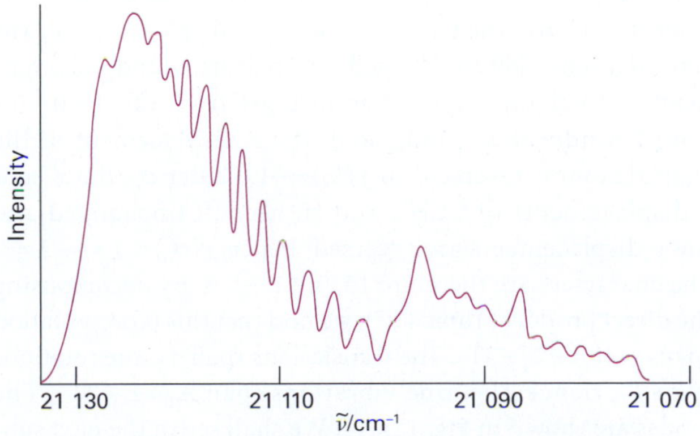

Figure 12E.12 CARS spectrum of a methane–air flame at 2104 K. The peaks correspond to the Q branch of the vibration–rotation spectrum of N_2 gas. (Adapted from J.F. Verdieck et al., *J. Chem. Ed.* **59**, 495 (1982).)

12E.4 Symmetry aspects of molecular vibrations

One of the most powerful ways of dealing with normal modes, especially of complex molecules, is to classify them according to their symmetries. This section makes extensive use of the concepts and procedures introduced in Topic 11C, which is an essential background to this discussion. In particular, each normal mode must belong to one of the symmetry species of the molecular point group, as discussed in that Topic.

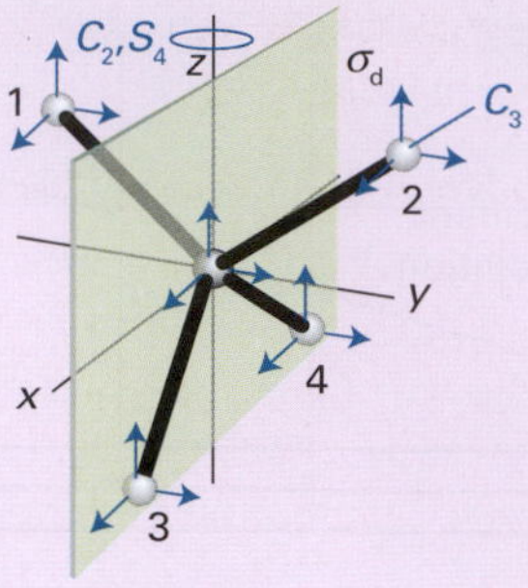

Figure 12E.13 The atomic displacements of CH_4 and the symmetry elements used to calculate the characters.

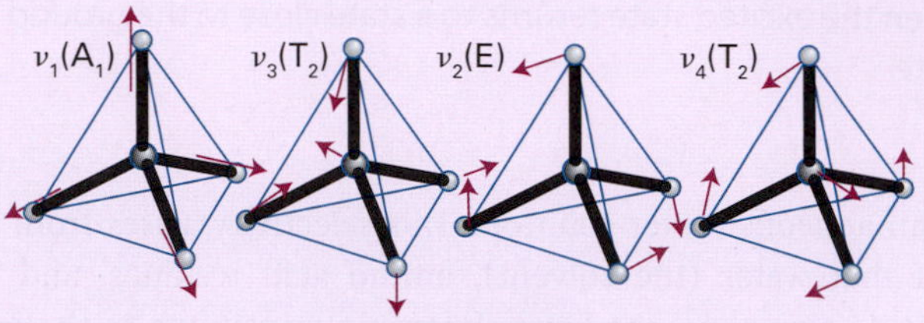

Figure 12E.14 Typical normal modes of vibration of a tetrahedral molecule. There are in fact two modes of symmetry species E and three modes of each T_2 symmetry species.

Example 12E.3 Identifying the symmetry species of a normal mode

Establish the symmetry species of the normal mode vibrations of CH_4, which belongs to the group T_d.

Method The first step in the procedure is to identify the symmetry species of the irreducible representations spanned by all the $3N$ displacements of the atoms, using the characters of the molecular point group. Find these characters by counting 1 if the displacement is unchanged under a symmetry operation, −1 if it changes sign, and 0 if it is changed into some other displacement. Next, subtract the symmetry species of the translations. Translational displacements span the same symmetry species as x, y, and z, so they can be obtained from the right-most column of the character table. Finally, subtract the symmetry species of the rotations, which are also given in the character table (and denoted there by R_x, R_y, and R_z).

Answer There are $3\times5=15$ degrees of freedom, of which $(3\times5)-6=9$ are vibrations. Refer to Fig. 12E.13. Under E, no displacement coordinates are changed, so the character is 15. Under C_3, no displacements are left unchanged, so the character is 0. Under the C_2 indicated, the z-displacement of the central atom is left unchanged, whereas its x- and y-components both change sign. Therefore $\chi(C_2)=1-1-1+0+0+\ldots=-1$. Under the S_4 indicated, the z-displacement of the central atom is reversed, so $\chi(S_4)=-1$. Under σ_d, the x- and z-displacements of C, H_3, and H_4 are left unchanged and the y-displacements are reversed; hence $\chi(\sigma_d)=3+3-3=3$. The characters are therefore 15, 0, −1, −1, 3. By decomposing the direct product (Topic 11C), we find that this representation spans $A_1+E+T_1+3T_2$. The translations span T_2; the rotations span T_1. Hence, the nine vibrations span A_1+E+2T_2. The modes are shown in Fig. 12E.14. We shall see in the next subsection that symmetry analysis gives a quick way of deciding which modes are active.

Self-test 12E.5 Establish the symmetry species of the normal modes of H_2O.

Answer: $2A_1+B_2$

(a) Infrared activity of normal modes

It is best to use group theory to judge the activities of more complex modes of vibration. This is easily done by checking the character table of the molecular point group for the symmetry species of the irreducible representations spanned by x, y, and z, for their species are also the symmetry species of the components of the electric dipole moment. Then apply the following rule, which is developed in the following *Justification*:

If the symmetry species of a normal mode is the same as any of the symmetry species of x, y, or z, then the mode is infrared active. Symmetry test for IR activity

Brief illustration 12E.3 The infrared activity of normal modes

To decide which normal modes of CH_4 are IR active, we note that we found in *Example* 12E.3 that their symmetry species are A_1+E+2T_2. Therefore, because x, y, and z span T_2 in the group T_d, only the T_2 modes are infrared active. The distortions accompanying these modes lead to a changing dipole moment. The A_1 mode, which is inactive, is the symmetrical 'breathing' mode of the molecule.

Self-test 12E.6 Which of the normal modes of H_2O are infrared active?

Answer: All three

Justification 12E.2 Using symmetry to identify infrared active normal modes

The rule hinges on the form of the transition dipole moment (Topic 12A): $\mu_{fi,x} \propto \int \psi_{v_f}^* x \psi_{v_i} dx$ in the x-direction, with similar expressions for the two other components of the transition moment. Consider a harmonic oscillator in the x-direction undergoing a transition from the ground vibrational state ($v_i = 0$) to the first excited state ($v_f = 1$). Because $\psi_0 \propto e^{-x^2}$ and $\psi_1 \propto xe^{-x^2}$ (Topic 8B), the components of the transition dipole moment take the following forms:

- $\int_{-\infty}^{+\infty} \overbrace{xe^{-x^2}}^{\text{Form of } \psi_1} \overbrace{x}^{\text{Form of } \mu_x} \overbrace{e^{-x^2}}^{\text{Form of } \psi_0} dx = \int_{-\infty}^{+\infty} x^2 e^{-2x^2} dx$ in the x-direction. As can be verified by direct calculation, this integral does not vanish.
- $\int_{-\infty}^{+\infty} xye^{-2x^2} dx$ and $\int_{-\infty}^{+\infty} xze^{-2x^2} dx$ in the y- and z-directions, respectively. A direct calculation shows that both integrals vanish.

Consequently, the excited state wavefunction must have the same symmetry as the displacement x.

(b) Raman activity of normal modes

Group theory provides an explicit recipe for judging the Raman activity of a normal mode. First, we need to know that the polarizability transforms in the same way as the quadratic forms (x^2, xy, etc.) listed in character tables. Then we use the following rule:

> If the symmetry species of a normal mode is the same as the symmetry species of a quadratic form, then the mode is Raman active.

Symmetry test for Raman activity

Brief illustration 12E.4 The Raman activity of normal modes

To decide which of the normal modes of CH_4 are Raman active, refer to the T_d character table. It was established in *Example* 12E.3 that the symmetry species of the normal modes are A_1+E+2T_2. Because the quadratic forms span A_1+E+T_2, all the normal modes are Raman active. By combining this information with that in *Brief illustration* 12E.3, we see how the infrared and Raman spectra of CH_4 are assigned. The assignment of spectral features to the T_2 modes is straightforward because these are the only modes that are both infrared and Raman active. This leaves the A_1 and E modes to be assigned in the Raman spectrum. Measurement of the depolarization ratio distinguishes between these modes because the A_1 mode, being totally symmetric, is polarized and the E mode is depolarized.

Self-test 12E.7 Which of the vibrational modes of H_2O are Raman active?

Answer: All three

Checklist of concepts

- ☐ 1. A **normal mode** is an independent, synchronous motion of atoms or groups of atoms that may be excited without leading to the excitation of any other normal mode.
- ☐ 2. The **number of normal modes** is $3N-6$ (for nonlinear molecules) or $3N-5$ (linear molecules).
- ☐ 3. A normal mode is **infrared active** if it is accompanied by a change of dipole moment; the specific selection rule is $\Delta v_q = \pm 1$.
- ☐ 4. The **exclusion rule** states that, if the molecule has a centre of symmetry, then no modes can be both infrared and Raman active.
- ☐ 5. Totally symmetrical vibrations give rise to **polarized lines**.
- ☐ 6. A normal mode is infrared active if its symmetry species is the same as any of the symmetry species of x, y, or z.
- ☐ 7. A normal mode is Raman active if its symmetry species is the same as the symmetry species of a quadratic form.

Checklist of equations

Property	Equation	Comment	Equation number
Vibrational terms of normal modes	$\tilde{G}_q(v)=(v+\frac{1}{2})\tilde{\nu}_q$, $\tilde{\nu}_q=(1/2\pi c)(k_{f,q}/m_q)^{1/2}$		12E.1
Depolarization ratio	$\rho=I_{\perp}/I_{\parallel}$	Depolarized lines: ρ close to or greater than 0.75. Polarized lines: $\rho<0.75$	12E.2

CHAPTER 12 Rotational and vibrational spectra

Note: The masses of nuclides are listed in Table 0.2 of the *Resource section*.

TOPIC 12A General features of molecular spectroscopy

Discussion questions

12A.1 What is the physical interpretation of a selection rule?

12A.2 Describe the physical origins of linewidths in absorption and emission spectra. Do you expect the same contributions for species in condensed and gas phases?

12A.3 Describe the basic experimental arrangements commonly used for absorption, emission, and Raman spectroscopy.

Exercises

12A.1(a) Calculate the ratio A/B for transitions with the following characteristics: (i) 70.8 pm X-rays, (ii) 500 nm visible light, (iii) 3000 cm^{-1} infrared radiation.

12A.1(b) Calculate the ratio A/B for transitions with the following characteristics: (i) 500 MHz radiofrequency radiation, (ii) 3.0 cm microwave radiation.

12A.2(a) The molar absorption coefficient of a substance dissolved in hexane is known to be 723 dm^3 mol^{-1} cm^{-1} at 260 nm. Calculate the percentage reduction in intensity when light of that wavelength passes through 2.50 mm of a solution of concentration 4.25 mmol dm^{-3}.

12A.2(b) The molar absorption coefficient of a substance dissolved in hexane is known to be 227 dm^3 mol^{-1} cm^{-1} at 290 nm. Calculate the percentage reduction in intensity when light of that wavelength passes through 2.00 mm of a solution of concentration 2.52 mmol dm^{-3}.

12A.3(a) A solution of an unknown component of a biological sample when placed in an absorption cell of path length 1.00 cm transmits 18.1 per cent of light of 320 nm incident upon it. If the concentration of the component is 0.139 mmol dm^{-3}, what is the molar absorption coefficient?

12A.3(b) When light of wavelength 400 nm passes through 2.50 mm of a solution of an absorbing substance at a concentration 0.717 mmol dm^{-3}, the transmission is 61.5 per cent. Calculate the molar absorption coefficient of the solute at this wavelength and express the answer in cm^2 mol^{-1}.

12A.4(a) The molar absorption coefficient of a solute at 540 nm is 386 dm^3 mol^{-1} cm^{-1}. When light of that wavelength passes through a 5.00 mm cell containing a solution of the solute, 38.5 per cent of the light is absorbed. What is the molar concentration of the solute?

12A.4(b) The molar absorption coefficient of a solute at 440 nm is 423 dm^3 mol^{-1} cm^{-1}. When light of that wavelength passes through a 6.50 mm cell containing a solution of the solute, 48.3 per cent of the light is absorbed. What is the molar concentration of the solute?

12A.5(a) The following data were obtained for the absorption by Br_2 in carbon tetrachloride using a 2.0 mm cell. Calculate the molar absorption coefficient of bromine at the wavelength employed:

$[Br_2]$/(mol dm^{-3})	0.0010	0.0050	0.0100	0.0500
T/(per cent)	81.4	35.6	12.7	3.0×10^{-3}

12A.5(b) The following data were obtained for the absorption by a dye dissolved in methylbenzene using a 2.50 mm cell. Calculate the molar absorption coefficient of the dye at the wavelength employed:

[dye]/(mol dm^{-3})	0.0010	0.0050	0.0100	0.0500
T/(per cent)	68	18	3.7	1.03×10^{-5}

12A.6(a) A 2.0 mm cell was filled with a solution of benzene in a non-absorbing solvent. The concentration of the benzene was 0.010 mol dm^{-3} and the wavelength of the radiation was 256 nm (where there is a maximum in the absorption). Calculate the molar absorption coefficient of benzene at this wavelength given that the transmission was 48 per cent. What will the transmittance be in a 4.0 mm cell at the same wavelength?

12A.6(b) A 5.00 mm cell was filled with a solution of a dye. The concentration of the dye was 18.5 mmol dm^{-3}. Calculate the molar absorption coefficient of the dye at this wavelength given that the transmission was 29 per cent. What will the transmittance be in a 2.50 mm cell at the same wavelength?

12A.7(a) A swimmer enters a gloomier world (in one sense) on diving to greater depths. Given that the mean molar absorption coefficient of sea water in the visible region is 6.2×10^{-3} dm^3 mol^{-1} cm^{-1}, calculate the depth at which a diver will experience (i) half the surface intensity of light, (ii) one tenth the surface intensity.

12A.7(b) Given that the maximum molar absorption coefficient of a molecule containing a carbonyl group is 30 dm^3 mol^{-1} cm^{-1} near 280 nm, calculate the thickness of a sample that will result in (i) half the initial intensity of radiation, (ii) one tenth the initial intensity.

12A.8(a) The absorption associated with a particular transition begins at 220 nm, peaks sharply at 270 nm, and ends at 300 nm. The maximum value of the molar absorption coefficient is 2.21×10^4 dm^3 mol^{-1} cm^{-1}. Estimate the integrated absorption coefficient of the transition assuming a symmetrical triangular lineshape.

12A.8(b) The absorption associated with a certain transition begins at 156 nm, peaks sharply at 210 nm, and ends at 275 nm. The maximum value of the molar absorption coefficient is 3.35×10^4 dm^3 mol^{-1} cm^{-1}. Estimate the integrated absorption coefficient of the transition assuming an inverted parabolic lineshape (Fig. 12.1).

12A.9(a) The electronic absorption bands of many molecules in solution have half-widths at half-height of about 5000 cm^{-1}. Estimate the integrated absorption coefficients of bands for which (i) $\varepsilon_{max}\approx1\times10^4$ dm^3 mol^{-1} cm^{-1}, (ii) $\varepsilon_{max}\approx5\times10^2$ dm^3 mol^{-1} cm^{-1}.

12A.9(b) The electronic absorption band of a compound in solution had a Gaussian lineshape and a half-width at half-height of 4233 cm^{-1} and $\varepsilon_{max}=1.54\times10^4$ dm^3 mol^{-1} cm^{-1}. Estimate the integrated absorption coefficient.

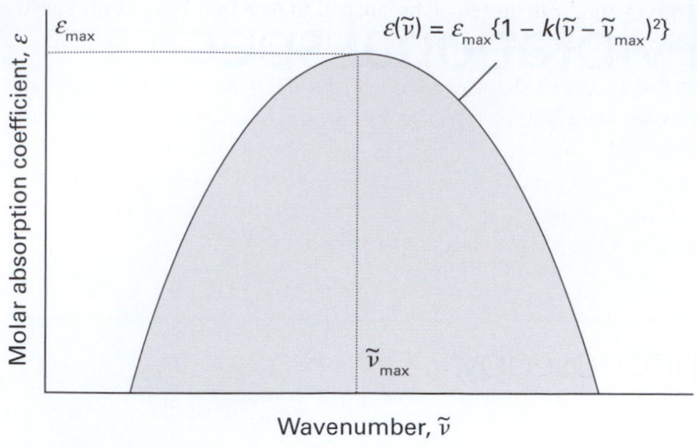

Figure 12.1 A model parabolic absorption lineshape.

12A.10(a) What is the Doppler-shifted wavelength of a red (680 nm) traffic light approached at 60 km h^{-1}?

12A.10(b) At what speed of approach would a red (680 nm) traffic light appear green (530 nm)?

12A.11(a) Estimate the lifetime of a state that gives rise to a line of width (i) 0.20 cm^{-1}, (ii) 2.0 cm^{-1}.

12A.11(b) Estimate the lifetime of a state that gives rise to a line of width (i) 200 MHz, (ii) 2.45 cm^{-1}.

12A.12(a) A molecule in a liquid undergoes about 1.0×10^{13} collisions in each second. Suppose that (i) every collision is effective in deactivating the molecule vibrationally and (ii) that one collision in 100 is effective. Calculate the width (in cm^{-1}) of vibrational transitions in the molecule.

12A.12(b) A molecule in a gas undergoes about 1.0×10^{9} collisions in each second. Suppose that (i) every collision is effective in deactivating the molecule rotationally and (ii) that one collision in 10 is effective. Calculate the width (in hertz) of rotational transitions in the molecule.

Problems

12A.1 The flux of visible photons reaching Earth from the North Star is about 4×10^3 mm^{-2} s^{-1}. Of these photons, 30 per cent are absorbed or scattered by the atmosphere and 25 per cent of the surviving photons are scattered by the surface of the cornea of the eye. A further 9 per cent are absorbed inside the cornea. The area of the pupil at night is about 40 mm^2 and the response time of the eye is about 0.1 s. Of the photons passing through the pupil, about 43 per cent are absorbed in the ocular medium. How many photons from the North Star are focused onto the retina in 0.1 s? For a continuation of this story, see R.W. Rodieck, *The first steps in seeing*, Sinauer, Sunderland (1998).

12A.2 A *Dubosq colorimeter* consists of a cell of fixed path length and a cell of variable path length. By adjusting the length of the latter until the transmission through the two cells is the same, the concentration of the second solution can be inferred from that of the former. Suppose that a plant dye of concentration 25 μg dm^{-3} is added to the fixed cell, the length of which is 1.55 cm. Then a solution of the same dye, but of unknown concentration, is added to the second cell. It is found that the same transmittance is obtained when the length of the second cell is adjusted to 1.18 cm. What is the concentration of the second solution?

12A.3 The Beer–Lambert law is derived on the basis that the concentration of absorbing species is uniform. Suppose, instead, that the concentration falls exponentially as $[J]=[J]_0e^{-x/\lambda}$. Develop an expression for the variation of I with sample length; suppose that $L\gg\lambda$.

12A.4 It is common to make measurements of absorbance at two wavelengths and use them to find the individual concentrations of two components A and B in a mixture. Show that the molar concentrations of A and B are

$$[A]=\frac{\varepsilon_{B2}A_1-\varepsilon_{B1}A_2}{(\varepsilon_{A1}\varepsilon_{B2}-\varepsilon_{A2}\varepsilon_{B2})L} \qquad [B]=\frac{\varepsilon_{A1}A_2-\varepsilon_{A2}A_1}{(\varepsilon_{A1}\varepsilon_{B2}-\varepsilon_{A2}\varepsilon_{B1})L}$$

where A_1 and A_2 are absorbances of the mixture at wavelengths λ_1 and λ_2, and the molar extinction coefficients of A (and B) at these wavelengths are ε_{A1} and ε_{A2} (and ε_{B1} and ε_{B2}).

12A.5 When pyridine is added to a solution of iodine in carbon tetrachloride the 520 nm band of absorption shifts toward 450 nm. However, the absorbance of the solution at 490 nm remains constant: this feature is called an *isosbestic point*. Show that an isosbestic point should occur when two absorbing species are in equilibrium.

12A.6‡ Ozone absorbs ultraviolet radiation in a part of the electromagnetic spectrum energetic enough to disrupt DNA in biological organisms and that is absorbed by no other abundant atmospheric constituent. This spectral range, denoted UV-B, spans the wavelengths of about 290 nm to 320 nm. The molar extinction coefficient of ozone over this range is given in the table below (DeMore et al., *Chemical kinetics and photochemical data for use in stratospheric modeling: Evaluation Number 11*, JPL Publication 94–26 (1994)).

λ/nm	292.0	296.3	300.8	305.4	310.1	315.0	320.0
ε/(dm^3 mol^{-1} cm^{-1})	1512	865	477	257	135.9	69.5	34.5

Compute the integrated absorption coefficient of ozone over the wavelength range 290–320 nm. (*Hint:* $\varepsilon(\tilde{\nu})$ can be fitted to an exponential function quite well.)

12A.7 In many cases it is possible to assume that an absorption band has a Gaussian lineshape (one proportional to e^{-x^2}) centred on the band maximum. Assume such a line shape, and show that $\mathcal{A}=\int\varepsilon(\tilde{\nu})d\tilde{\nu}\approx1.0645\,\varepsilon_{max}\Delta\tilde{\nu}_{1/2}$, where $\Delta\tilde{\nu}_{1/2}$ is the width at half-height. The absorption spectrum of azoethane ($CH_3CH_2N_2$) between 24 000 cm^{-1} and 34 000 cm^{-1} is shown in Fig. 12.2. First, estimate $\mathcal{A}$ for the band by assuming that it is Gaussian. Then use mathematical software to fit a polynomial to the absorption band (or a Gaussian), and integrate the result analytically.

12A.8‡ Wachewsky et al. (*J. Phys. Chem.* **100**, 11559 (1996)) examined the UV absorption spectrum of CH_3I, a species of interest in connection with stratospheric ozone chemistry. They found the integrated absorption coefficient to be dependent on temperature and pressure to an extent inconsistent with internal structural changes in isolated CH_3I molecules; they explained the changes as due to dimerization of a substantial fraction

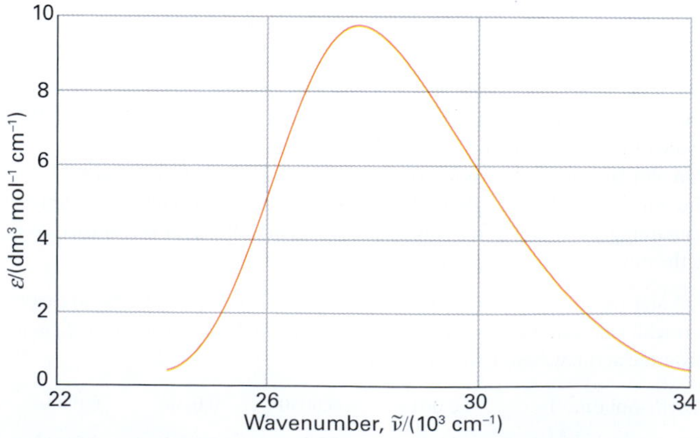

Figure 12.2 The absorption spectrum of azoethane.

‡ These problems were supplied by Charles Trapp and Carmen Giunta.

of the CH_3I, a process which would naturally be pressure and temperature dependent. (a) Compute the integrated absorption coefficient over a triangular lineshape in the range 31 250 to 34 483 cm^{-1} and a maximal molar absorption coefficient of 150 dm^3 mol^{-1} cm^{-1} at 31 250 cm^{-1}. (b) Suppose 1.0 per cent of the CH_3I units in a sample at 2.4 Torr and 373 K exists as dimers. Compute the absorbance expected at 31 250 cm^{-1} in a sample cell of length 12.0 cm. (c) Suppose 18 per cent of the CH_3I units in a sample at 100 Torr and 373 K exists as dimers. Compute the absorbance expected at 31 250 cm^{-1} in a sample cell of length 12.0 cm; compute the molar absorption coefficient which would be inferred from this absorbance if dimerization was not considered.

12A.9 The spectrum of a star is used to measure its *radial velocity* with respect to the Sun, the component of the star's velocity vector that is parallel to a vector connecting the star's centre to the centre of the Sun. The measurement relies on the Doppler effect. When a star emitting electromagnetic radiation of frequency ν moves with a speed s relative to an observer, the observer detects radiation of frequency $\nu_{\text{receding}} = \nu f$ or $\nu_{\text{approaching}} = \nu/f$, where $f = \{(1 - s/c)/(1 + s/c)\}^{1/2}$ and c is the speed of light. (a) Three Fe I lines of the star HDE 271 182, which belongs to the Large Magellanic Cloud, occur at 438.882 nm, 441.000 nm, and 442.020 nm. The same lines occur at 438.392 nm, 440.510 nm, and 441.510 nm in the spectrum of an Earth-bound iron arc. Determine whether HDE 271 182 is receding from or approaching the Earth and estimate the star's radial speed with respect to the Earth. (b) What additional information would you need to calculate the radial velocity of HDE 271 182 with respect to the Sun?

12A.10 In Problem 12A.9, we saw that Doppler shifts of atomic spectral lines are used to estimate the speed of recession or approach of a star. A spectral line of $^{48}Ti^{8+}$ (of mass $47.95m_u$) in a distant star was found to be shifted from 654.2 nm to 706.5 nm and to be broadened to 61.8 pm. What is the speed of recession and the surface temperature of the star?

12A.11 The Gaussian shape of a Doppler-broadened spectral line reflects the Maxwell distribution of speeds in the sample at the temperature of the experiment. In a spectrometer that makes use of *phase-sensitive detection* the output signal is proportional to the first derivative of the signal intensity, $dI/d\nu$. Plot the resulting line shape for various temperatures. How is the separation of the peaks related to the temperature?

12A.12 The collision frequency z of a molecule of mass m in a gas at a pressure p is $z = 4\sigma(kT/\pi m)^{1/2}p/kT$, where σ is the collision cross-section. Find an expression for the collision-limited lifetime of an excited state assuming that every collision is effective. Estimate the width of rotational transition in HCl ($\sigma = 0.30$ nm^2) at 25 °C and 1.0 atm. To what value must the pressure of the gas be reduced in order to ensure that collision broadening is less important than Doppler broadening?

12A.13 Refer to Fig. 12A.9, which depicts a Michelson interferometer. The mirror M_1 moves in finite distance increments, so the path difference p is also incremented in finite steps. Explore the effect of increasing the step size on the shape of the interferogram for a monochromatic beam of wavenumber $\tilde{\nu}$ and intensity I_0. That is, draw plots of $I(p)/I_0$ against $\tilde{\nu}p$, each with a different number of data points spanning the same total distance path taken by the movable mirror M_1.

12A.14 Using mathematical software, elaborate on the results of *Example* 12A.2 by: (a) exploring the effect of varying the wavenumbers and intensities of the three components of the radiation on the shape of the interferogram; and (b) calculating the Fourier transforms of the functions you generated in part (a).

TOPIC 12B Molecular rotation

Discussion questions

12B.1 Account for the rotational degeneracy of the various types of rigid rotor. Would their lack of rigidity affect your conclusions?

12B.2 Describe the differences between an oblate and a prolate symmetric rotor and give several examples of each.

Exercises

12B.1(a) Calculate the moment of inertia around the C_2 axis (the bisector of the OOO angle) and the corresponding rotational constant of an $^{16}O_3$ molecule (bond angle 117°; OO bond length 128 pm).
12B.1(b) Calculate the moment of inertia around the C_3 axis (the threefold symmetry axis) and the corresponding rotational constant of a $^{31}P^1H_3$ molecule (bond angle 93.5°; PH bond length 142 pm).

12B.2(a) Plot the expressions for the two moments of inertia of a C_{3v} symmetric top version of an AB_4 molecule (Table 12B.1) with equal bond lengths but with the angle θ increasing from 90° to the tetrahedral angle.
12B.2(b) Plot the expressions for the two moments of inertia of a C_{3v} symmetric top version of an AB_4 molecule (Table 12B.1) with θ equal to the tetrahedral angle but with one A–B bond varying. *Hint*: Write $\rho = R'_{AB}/R_{AB}$, and allow ρ to vary from 2 to 1.

12B.3(a) Classify the following rotors: (i) O_3, (ii) CH_3CH_3, (iii) XeO_4, (iv) $FeCp_2$ (Cp denotes the cyclopentadienyl group, C_5H_5).
12B.3(b) Classify the following rotors: (i) $CH_2{=}CH_2$, (ii) SO_3, (iii) ClF_3, (iv) N_2O.

12B.4(a) Determine the HC and CN bond lengths in HCN from the rotational constants $B(^1H^{12}C^{14}N) = 44.316$ GHz and $B(^2H^{12}C^{14}N) = 36.208$ GHz.
12B.4(b) Determine the CO and CS bond lengths in OCS from the rotational constants $B(^{16}O^{12}C^{32}S) = 6081.5$ MHz, $B(^{16}O^{12}C^{34}S) = 5932.8$ MHz.

12B.5(a) Estimate the centrifugal distortion constant for $^1H^{127}I$, for which $\tilde{B} = 6.511$ cm^{-1} and $\tilde{\nu} = 2308$ cm^{-1}. By what factor would the constant change when ^2H is substituted for ^1H?
12B.5(b) Estimate the centrifugal distortion constant for $^{79}Br^{81}Br$, for which $\tilde{B} = 0.0809$ cm^{-1} and $\tilde{\nu} = 323.2$ cm^{-1}. By what factor would the constant change when the ^{79}Br is replaced by ^{81}Br?

Problems

12B.1 Show that the moment of inertia of a diatomic molecule composed of atoms of masses m_A and m_B and bond length R is equal to $m_{\text{eff}}R^2$, where $m_{\text{eff}} = m_Am_B/(m_A + m_B)$.

12B.2 Confirm the expression given in Table 12B.1 for the moment of inertia of a linear ABC molecule. *Hint*: Begin by locating the centre of mass.

TOPIC 12C Rotational spectroscopy

Discussion questions

12C.1 Describe the physical origins of the gross selection rules for microwave spectroscopy.

12C.2 Describe the physical origins of the gross selection rules for rotational Raman spectroscopy.

12C.3 Describe the role of nuclear statistics in the occupation of energy levels in $^1H^{12}C{\equiv}^{12}C^1H$, $^1H^{13}C{\equiv}^{13}C^1H$, and $^2H^{12}C{\equiv}^{12}C^2H$. For nuclear spin data, see Table 14A.2.

12C.4 Account for the existence of a rotational zero-point energy in molecular hydrogen.

Exercises

12C.1(a) Which of the following molecules may show a pure rotational microwave absorption spectrum: (i) H_2, (ii) HCl, (iii) CH_4, (iv) CH_3Cl, (v) CH_2Cl_2?
12C.1(b) Which of the following molecules may show a pure rotational microwave absorption spectrum: (i) H_2O, (ii) H_2O_2, (iii) NH_3, (iv) N_2O?

12C.2(a) Calculate the frequency and wavenumber of the $J=3\leftarrow 2$ transition in the pure rotational spectrum of $^{14}N^{16}O$. The equilibrium bond length is 115 pm. Does the frequency increase or decrease if centrifugal distortion is considered?
12C.2(b) Calculate the frequency and wavenumber of the $J=2\leftarrow 1$ transition in the pure rotational spectrum of $^{12}C^{16}O$. The equilibrium bond length is 112.81 pm. Does the frequency increase or decrease if centrifugal distortion is considered?

12C.3(a) The wavenumber of the $J=3\leftarrow 2$ rotational transition of $^1H^{35}Cl$ considered as a rigid rotor is 63.56 cm^{-1}; what is the H–Cl bond length?
12C.3(b) The wavenumber of the $J=1\leftarrow 0$ rotational transition of $^1H^{81}Br$ considered as a rigid rotor is 16.93 cm^{-1}; what is the H–Br bond length?

12C.4(a) The spacing of lines in the microwave spectrum of $^{27}Al^1H$ is 12.604 cm^{-1}; calculate the moment of inertia and bond length of the molecule.
12C.4(b) The spacing of lines in the microwave spectrum of $^{35}Cl^{19}F$ is 1.033 cm^{-1}; calculate the moment of inertia and bond length of the molecule.

12C.5(a) What is the most highly populated rotational level of Cl_2 at (i) 25 °C, (ii) 100 °C? Take $\tilde{B}=0.244$ cm^{-1}.
12C.5(b) What is the most highly populated rotational level of Br_2 at (i) 25 °C, (ii) 100 °C? Take $\tilde{B}=0.0809$ cm^{-1}.

12C.6(a) Which of the following molecules may show a pure rotational Raman spectrum: (i) H_2, (ii) HCl, (iii) CH_4, (iv) CH_3Cl?
12C.6(b) Which of the following molecules may show a pure rotational Raman spectrum: (i) CH_2Cl_2, (ii) CH_3CH_3, (iii) SF_6, (iv) N_2O?

12C.7(a) The wavenumber of the incident radiation in a Raman spectrometer is 20 487 cm^{-1}. What is the wavenumber of the scattered Stokes radiation for the $J=2\leftarrow 0$ transition of $^{14}N_2$?
12C.7(b) The wavenumber of the incident radiation in a Raman spectrometer is 20 623 cm^{-1}. What is the wavenumber of the scattered Stokes radiation for the $J=4\leftarrow 2$ transition of $^{16}O_2$?

12C.8(a) The rotational Raman spectrum of $^{35}Cl_2$ shows a series of Stokes lines separated by 0.9752 cm^{-1} and a similar series of anti-Stokes lines. Calculate the bond length of the molecule.
12C.8(b) The rotational Raman spectrum of $^{19}F_2$ shows a series of Stokes lines separated by 3.5312 cm^{-1} and a similar series of anti-Stokes lines. Calculate the bond length of the molecule.

12C.9(a) What is the ratio of weights of populations due to the effects of nuclear statistics for $^{35}Cl_2$?
12C.9(b) What is the ratio of weights of populations due to the effects of nuclear statistics for $^{12}C^{32}S_2$? What effect would be observed when ^{12}C is replaced by ^{13}C? For nuclear spin data, see Table 14A.2.

Problems

12C.1 The rotational constant of NH_3 is 298 GHz. Compute the separation of the pure rotational spectrum lines as a frequency (in GHz), a wavenumber (in cm^{-1}), and a wavelength (in mm), and show that the value of B is consistent with an N–H bond length of 101.4 pm and a bond angle of 106.78°.

12C.2 Rotational absorption lines from $^1H^{35}Cl$ gas were found at the following wavenumbers (R.L. Hausler and R.A. Oetjen, *J. Chem. Phys.* **21**, 1340 (1953)): 83.32, 104.13, 124.73, 145.37, 165.89, 186.23, 206.60, 226.86 cm^{-1}. Calculate the moment of inertia and the bond length of the molecule. Predict the positions of the corresponding lines in $^2H^{35}Cl$.

12C.3 Is the bond length in HCl the same as that in DCl? The wavenumbers of the $J=1\leftarrow 0$ rotational transitions for $H^{35}Cl$ and $^2H^{35}Cl$ are 20.8784 and 10.7840 cm^{-1}, respectively. Accurate atomic masses are 1.007 825m_u and 2.0140m_u for ^1H and ^2H, respectively. The mass of ^{35}Cl is 34.96885m_u. Based on this information alone, can you conclude that the bond lengths are the same or different in the two molecules?

12C.4 Thermodynamic considerations suggest that the copper monohalides CuX should exist mainly as polymers in the gas phase, and indeed it proved difficult to obtain the monomers in sufficient abundance to detect spectroscopically. This problem was overcome by flowing the halogen gas over copper heated to 1100 K (Manson et al. (*J. Chem. Phys.* **63**, 2724 (1975))). For CuBr the $J=13$–14, 14–15, and 15–16 transitions occurred at 84 421.34, 90 449.25, and 96 476.72 MHz, respectively. Calculate the rotational constant and bond length of CuBr.

12C.5 The microwave spectrum of $^{16}O^{12}CS$ gave absorption lines (in GHz) as follows:

J	1	2	3	4
^{32}S	24.325 92	36.48882	48.651 64	60.814 08
^{34}S	23.732 33		47.462 40	

Using the expressions for moments of inertia in Table 12B.1 and assuming that the bond lengths are unchanged by substitution, calculate the CO and CS bond lengths in OCS.

12C.6 Equation 12C.8b may be rearranged into

$$\tilde{\nu}(J+1\leftarrow J)/\{2(J+1)\}=\tilde{B}-2\tilde{D}_J(J+1)^2$$

which is the equation of a straight line when the left-hand side is plotted against $(J+1)^2$. The following wavenumbers of transitions (in cm^{-1}) were observed for $^{12}C^{16}O$:

J:	0	1	2	3	4
	3.845 033	7.689 919	11.534 510	15.378 662	19.222 223

Determine $\tilde{B}$, $\tilde{D}_J$, and the equilibrium bond length of CO.

12C.7‡ In a study of the rotational spectrum of the linear FeCO radical, Tanaka et al. (*J. Chem. Phys.* **106**, 6820 (1997)) report the following $J+1 \leftarrow J$ transitions:

J	24	25	26	27	28	29
MHz	214 777.7	223 379.0	231 981.2	240 584.4	249 188.5	257 793.5

Evaluate the rotational constant of the molecule. Also, estimate the value of J for the most highly populated rotational energy level at 298 K and at 100 K.

12C.8 The rotational terms of a symmetric top, allowing for centrifugal distortion, are commonly written

$$\tilde{F}(J,K) = \tilde{B}J(J+1) + (\tilde{A}-\tilde{B})K^2 - \tilde{D}_J J^2(J+1)^2 - \tilde{D}_{JK}J(J+1)K^2 - \tilde{D}_K K^4$$

(a) Develop an expression for the wavenumbers of the allowed rotational transitions. (b) The following transition frequencies (in gigahertz, GHz) were observed for CH_3F:

51.0718	102.1426	102.1408	153.2103	153.2076

Determine the values of as many constants in the expression for the rotational terms as these values permit.

12C.9 Develop an expression for the value of J corresponding to the most highly populated rotational energy level of a diatomic rotor at a temperature T remembering that the degeneracy of each level is $2J+1$. Evaluate the expression for ICl (for which $\tilde{B}=0.1142\ \text{cm}^{-1}$) at 25 °C. Repeat the problem for the most highly populated level of a spherical rotor, taking note of the fact that each level is $(2J+1)^2$-fold degenerate. Evaluate the expression for CH_4 (for which $\tilde{B}=5.24\ \text{cm}^{-1}$) at 25 °C.

12C.10 A. Dalgarno, in *Chemistry in the interstellar medium, Frontiers of Astrophysics*, ed. E.H. Avrett, Harvard University Press, Cambridge (1976), notes that although both CH and CN spectra show up strongly in the interstellar medium in the constellation Ophiuchus, the CN spectrum has become the standard for the determination of the temperature of the cosmic microwave background radiation. Demonstrate through a calculation why CH would not be as useful for this purpose as CN. The rotational constant $\tilde{B}_0$ for CH is 14.190 cm^{-1}.

12C.11 The space immediately surrounding stars, the *circumstellar space*, is significantly warmer because stars are very intense black-body emitters with temperatures of several thousand kelvin. Discuss how such factors as cloud temperature, particle density, and particle velocity may affect the rotational spectrum of CO in an interstellar cloud. What new features in the spectrum of CO can be observed in gas ejected from and still near a star with temperatures of about 1000 K, relative to gas in a cloud with temperature of about 10 K? Explain how these features may be used to distinguish between circumstellar and interstellar material on the basis of the rotational spectrum of CO.

12C.12 Pure rotational Raman spectra of gaseous C_6H_6 and C_6D_6 yield the following rotational constants: $\tilde{B}(C_6H_6)=0.18960\ \text{cm}^{-1}$, $\tilde{B}(C_6D_6)=0.15681\ \text{cm}^{-1}$. The moments of inertia of the molecules about any axis perpendicular to the C_6 axis were calculated from these data as $I(C_6H_6)=1.4759\times10^{-45}\ \text{kg m}^2$, $I(C_6D_6)=1.7845\times10^{-45}\ \text{kg m}^2$. Calculate the CC, CH, and CD bond lengths.

TOPIC 12D Vibrational spectroscopy of diatomic molecules

Discussion questions

12D.1 Discuss the strengths and limitations of the parabolic and Morse functions as descriptors of the potential energy curve of a diatomic molecule.

12D.2 Describe the effect of vibrational excitation on the rotational constant of a diatomic molecule.

12D.3 How is the method of combination differences used in rotation–vibration spectroscopy to determine rotational constants?

12D.4 In what ways may the rotational and vibrational spectra of molecules change as a result of isotopic substitution?

Exercises

12D.1(a) An object of mass 100 g suspended from the end of a rubber band has a vibrational frequency of 2.0 Hz. Calculate the force constant of the rubber band.
12D.1(b) An object of mass 1.0 g suspended from the end of a spring has a vibrational frequency of 10.0 Hz. Calculate the force constant of the spring.

12D.2(a) Calculate the percentage difference in the fundamental vibrational wavenumbers of $^{23}Na^{35}Cl$ and $^{23}Na^{37}Cl$ on the assumption that their force constants are the same.
12D.2(b) Calculate the percentage difference in the fundamental vibrational wavenumbers of $^{1}H^{35}Cl$ and $^{2}H^{37}Cl$ on the assumption that their force constants are the same.

12D.3(a) The wavenumber of the fundamental vibrational transition of $^{35}Cl_2$ is 564.9 cm^{-1}. Calculate the force constant of the bond.
12D.3(b) The wavenumber of the fundamental vibrational transition of $^{79}Br^{81}Br$ is 323.2 cm^{-1}. Calculate the force constant of the bond.

12D.4(a) The hydrogen halides have the following fundamental vibrational wavenumbers: 4141.3 cm^{-1} (HF); 2988.9 cm^{-1} ($H^{35}Cl$); 2649.7 cm^{-1} ($H^{81}Br$); 2309.5 cm^{-1} ($H^{127}I$). Calculate the force constants of the hydrogen–halogen bonds.
12D.4(b) From the data in Exercise 12D.4(a), predict the fundamental vibrational wavenumbers of the deuterium halides.

12D.5(a) Calculate the relative numbers of Cl_2 molecules ($\tilde{\nu}=559.7\ \text{cm}^{-1}$) in the ground and first excited vibrational states at (i) 298 K, (ii) 500 K.
12.5(b) Calculate the relative numbers of Br_2 molecules ($\tilde{\nu}=321\ \text{cm}^{-1}$) in the second and first excited vibrational states at (i) 298 K, (ii) 800 K.

12D.6(a) For $^{16}O_2$, $\Delta\tilde{G}$ values for the transitions $v=1\leftarrow0$, $2\leftarrow0$, and $3\leftarrow0$ are, respectively, 1556.22, 3088.28, and 4596.21 cm^{-1}. Calculate $\tilde{\nu}$ and x_e. Assume y_e to be zero.
12D.6(b) For $^{14}N_2$, $\Delta\tilde{G}$ values for the transitions $v=1\leftarrow0$, $2\leftarrow0$, and $3\leftarrow0$ are, respectively, 2329.91, 4631.20, and 6903.69 cm^{-1}. Calculate $\tilde{\nu}$ and x_e. Assume y_e to be zero.

12D.7(a) The first five vibrational energy levels of HCl are at 1481.86, 4367.50, 7149.04, 9826.48, and 12 399.8 cm^{-1}. Calculate the dissociation energy of the molecule in reciprocal centimetres and electronvolts.

12D.7(b) The first five vibrational energy levels of HI are at 1144.83, 3374.90, 5525.51, 7596.66, and 9588.35 cm^{-1}. Calculate the dissociation energy of the molecule in reciprocal centimetres and electronvolts.

Problems

12D.1 Derive an expression for the force constant of an oscillator that can be modelled by a Morse potential (eqn 12D.14).

12D.2 Suppose a particle confined to a cavity in a microporous material has a potential energy of the form $V(x)=V_0(e^{-a^2/x^2}-1)$. Sketch $V(x)$. What is the value of the force constant corresponding to this potential energy? Would the particle undergo simple harmonic motion? Sketch the likely form of the first two vibrational wavefunctions.

12D.3 The vibrational levels of NaI lie at the wavenumbers 142.81, 427.31, 710.31, and 991.81 cm^{-1}. Show that they fit the expression $(v+\frac{1}{2})\tilde{\nu}-(v+\frac{1}{2})^2 x_e\tilde{\nu}$, and deduce the force constant, zero-point energy, and dissociation energy of the molecule.

12D.4 The HCl molecule is quite well described by the Morse potential with $hc\tilde{D}_e=5.33$ eV, $\tilde{\nu}=2989.7$ cm^{-1}, and $x_e\tilde{\nu}=52.05$ cm^{-1}. Assuming that the potential is unchanged on deuteration, predict the dissociation energies ($hc\tilde{D}_0$, in electronvolts) of (a) HCl, (b) DCl.

12D.5 The Morse potential (eqn 12D.14) is very useful as a simple representation of the actual molecular potential energy. When RbH was studied, it was found that $\tilde{\nu}=936.8$ cm^{-1} and $x_e\tilde{\nu}=14.15$ cm^{-1}. Plot the potential energy curve from 50 pm to 800 pm around $R_e=236.7$ pm. Then go on to explore how the rotation of a molecule may weaken its bond by allowing for the kinetic energy of rotation of a molecule and plotting $V^*=V+hc\tilde{B}J(J+1)$ with $\tilde{B}=\hbar/4\pi c\mu R^2$. Plot these curves on the same diagram for $J=40$, 80, and 100, and observe how the dissociation energy is affected by the rotation. (Taking $\tilde{B}=3.020$ cm^{-1} at the equilibrium bond length will greatly simplify the calculation.)

12D.6‡ Luo et al. (*J. Chem. Phys.* **98**, 3564 (1993)) reported experimental observation of the He_2 complex, a species which had escaped detection for a long time. The fact that the observation required temperatures in the neighbourhood of 1 mK is consistent with computational studies which suggest that $hc\tilde{D}_e$ for He_2 is about 1.51×10^{-23} J, $hc\tilde{D}_0\approx2\times10^{-26}$ J, and R_e about 297 pm. (a) Estimate the fundamental vibrational wavenumber, force constant, moment of inertia, and rotational constant based on the harmonic oscillator and rigid-rotor approximations. (b) Such a weakly bound complex is hardly likely to be rigid. Estimate the vibrational wavenumber and anharmonicity constant based on the Morse potential energy.

12D.7 Confirm that a Morse oscillator has a finite number of bound states. Determine the value of v_{max} for the highest bound state.

12D.8 Provided higher order terms are neglected, eqn 12D.17, for the vibrational wavenumbers of an anharmonic oscillator, $\Delta\tilde{G}_{v+\frac{1}{2}}=\tilde{\nu}-2(v+1)x_e\tilde{\nu}+\cdots$, is the equation of a straight line when the left-hand side is plotted against $v+1$. Use the following data on CO to determine the values of $\tilde{\nu}$ and $x_e\tilde{\nu}$ for CO:

v	0	1	2	3	4
$\Delta\tilde{G}_{v+\frac{1}{2}}/cm^{-1}$	2143.1	2116.1	2088.9	2061.3	2033.5

12D.9 The rotational constant for CO is 1.9314 cm^{-1} and 1.6116 cm^{-1} in the ground and first excited vibrational states, respectively. By how much does the internuclear distance change as a result of this transition?

12D.10 The average spacing between the rotational lines of the P and R branches of $^{12}C_2{}^1H_2$ and $^{12}C_2{}^2H_2$ is 2.352 cm^{-1} and 1.696 cm^{-1}, respectively. Estimate the CC and CH bond lengths.

12D.11 Absorptions in the $v=1\leftarrow0$ vibration–rotation spectrum of $^1H^{35}Cl$ were observed at the following wavenumbers (in cm^{-1}):

2998.05	2981.05	2963.35	2944.99	2925.92
2906.25	2865.14	2843.63	2821.59	2799.00

Assign the rotational quantum numbers and use the method of combination differences to determine the rotational constants of the two vibrational levels.

12D.12 Suppose that the internuclear distance may be written $R=R_e+x$ where R_e is the equilibrium bond length. Also suppose that the potential well is symmetrical and confines the oscillator to small displacements. Deduce expressions for $1/\langle R\rangle^2$, $1/\langle R^2\rangle$, and $\langle 1/R^2\rangle$ to the lowest non-zero power of $\langle x^2\rangle/R_e^2$ and confirm that values are not the same.

12D.13 Continue the development of Problem 12D.12 by using the virial expression to relate $\langle x^2\rangle$ to the vibrational quantum number. Does your result imply that the rotational constant increases or decreases as the oscillator becomes excited to higher quantum states. What would be the effect of anharmonicity?

12D.14 The rotational constant for a diatomic molecule in the vibrational state with quantum number v typically fits the expression $\tilde{B}_v=\tilde{B}_e-a(v+\frac{1}{2})$. For the interhalogen molecule IF it is found that $\tilde{B}_e=0.27971$ cm^{-1} and $a=0.187$ m^{-1} (note the change of units). Calculate $\tilde{B}_0$ and $\tilde{B}_1$ and use these values to calculate the wavenumbers of the $J'\rightarrow3$ transitions of the P and R branches. You will need the following additional information: $\tilde{\nu}=610.258$ cm^{-1} and $x_e\tilde{\nu}=3.141$ cm^{-1}. Estimate the dissociation energy of the IF molecule.

12D.15 At low resolution, the strongest absorption band in the infrared absorption spectrum of $^{12}C^{16}O$ is centred at 2150 cm^{-1}. Upon closer examination at higher resolution, this band is observed to be split into two sets of closely spaced peaks, one on each side of the centre of the spectrum at 2143.26 cm^{-1}. The separation between the peaks immediately to the right and left of the centre is 7.655 cm^{-1}. Make the harmonic oscillator and rigid rotor approximations and calculate from these data: (a) the vibrational wavenumber of a CO molecule, (b) its molar zero-point vibrational energy, (c) the force constant of the CO bond, (d) the rotational constant $\tilde{B}$, and (e) the bond length of CO.

12D.16 The analysis of combination differences summarized in the text considered the R and P branches. Extend the analysis to the O and S branches of a Raman spectrum.

TOPIC 12E Vibrational spectroscopy of polyatomic molecules

Discussion questions

12E.1 Describe the physical origins of the gross selection rules for infrared spectroscopy.

12E.2 Describe the physical origins of the gross selection rules for vibrational Raman spectroscopy.

12E.3 Suppose that you wish to characterize the normal modes of benzene in the gas phase. Why is it important to obtain both infrared absorption and Raman spectra of your sample?

Exercises

12E.1(a) Which of the following molecules may show infrared absorption spectra: (i) H_2, (ii) HCl, (iii) CO_2, (iv) H_2O?
12E.1(b) Which of the following molecules may show infrared absorption spectra: (i) CH_3CH_3, (ii) CH_4, (iii) CH_3Cl, (iv) N_2?

12E.2(a) How many normal modes of vibration are there for the following molecules: (i) H_2O, (ii) H_2O_2, (iii) C_2H_4?
12E.2(b) How many normal modes of vibration are there for the following molecules: (i) C_6H_6, (ii) $C_6H_5H_3$, (iii) HC≡C–C≡C–H?

12E.3(a) How many vibrational modes are there for the molecule NC–(C≡C–C≡C–)$_{10}$CN detected in an interstellar cloud?
12E.3(b) How many vibrational modes are there for the molecule NC–(C≡C–C≡C–)$_8$CN detected in an interstellar cloud?

12E.4(a) Write an expression for the vibrational term for the ground vibrational state of H_2O in terms of the wavenumbers of the normal modes. Neglect anharmonicities, as in eqn 12E.1.
12E.4(b) Write an expression for the vibrational term for the ground vibrational state of SO_2 in terms of the wavenumbers of the normal modes. Neglect anharmonicities, as in eqn 12E.1.

12E.5(a) Which of the three vibrations of an AB_2 molecule are infrared or Raman active when it is (i) angular, (ii) linear?
12E.5(b) Which of the vibrations of an AB_3 molecule are infrared or Raman active when it is (i) trigonal planar, (ii) trigonal pyramidal?

12E.6(a) Consider the vibrational mode that corresponds to the uniform expansion of the benzene ring. Is it (i) Raman, (ii) infrared active?
12E.6(b) Consider the vibrational mode that corresponds to the boat-like bending of a benzene ring. Is it (i) Raman, (ii) infrared active?

12E.7(a) The molecule CH_2Cl_2 belongs to the point group C_{2v}. The displacements of the atoms span $5A_1+2A_2+4B_1+4B_2$. What are the symmetry species of the normal modes of vibration?
12E.7(b) A carbon disulfide molecule belongs to the point group $D_{\infty h}$. The nine displacements of the three atoms span $A_{1g}+A_{1u}+A_{2g}+2E_{1u}+E_{1g}$. What are the symmetry species of the normal modes of vibration?

12E.8(a) Which of the normal modes of CH_2Cl_2 (Exercise 12E.7(a)) are infrared active? Which are Raman active?
12E.8(b) Which of the normal modes of carbon disulfide (Exercise 12E.7(b)) are infrared active? Which are Raman active?

Problems

12E.1 Suppose that the out-of-plane distortion of a planar molecule are described by a potential energy $V=V_0(1-e^{-bh^4})$, where h is the distance by which the central atom is displaced. Sketch this potential energy as a function of h (allow h to be both negative and positive). What could be said about (a) the force constant, (b) the vibrations? Sketch the form of the ground-state wavefunction.

12E.2 Predict the shape of the nitronium ion, NO_2^+, from its Lewis structure and the VSEPR model. It has one Raman active vibrational mode at 1400 cm^{-1}, two strong IR active modes at 2360 and 540 cm^{-1}, and one weak IR mode at 3735 cm^{-1}. Are these data consistent with the predicted shape of the molecule? Assign the vibrational wavenumbers to the modes from which they arise.

12E.3 Consider the molecule CH_3Cl. (a) To what point group does the molecule belong? (b) How many normal modes of vibration does the molecule have? (c) What are the symmetry species of the normal modes of vibration for this molecule? (d) Which of the vibrational modes of this molecule are infrared active? (e) Which of the vibrational modes of this molecule are Raman active?

12E.4 Suppose that three conformations are proposed for the nonlinear molecule H_2O_2 (**1**, **2**, and **3**). The infrared absorption spectrum of gaseous H_2O_2 has bands at 870, 1370, 2869, and 3417 cm^{-1}. The Raman spectrum of the same sample has bands at 877, 1408, 1435, and 3407 cm^{-1}. All bands correspond to fundamental vibrational wavenumbers and you may assume that: (a) the 870 and 877 cm^{-1} bands arise from the same normal mode, and (b) the 3417 and 3407 cm^{-1} bands arise from the same normal mode. (i) If H_2O_2 were linear, how many normal modes of vibration would it have? (ii) Give the symmetry point group of each of the three proposed conformations of nonlinear H_2O_2. (iii) Determine which of the proposed conformations is inconsistent with the spectroscopic data. Explain your reasoning.

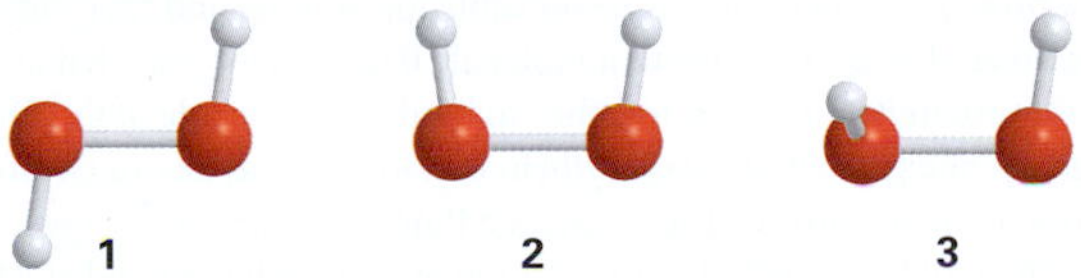

1 **2** **3**

Integrated activities

12.1 In the group theoretical language developed in Topics 11A–11C, a spherical rotor is a molecule that belongs to a cubic or icosahedral point group, a symmetric rotor is a molecule with at least a threefold axis of symmetry, and an asymmetric rotor is a molecule without a threefold (or higher) axis. Linear molecules are linear rotors. Classify each of the following molecules as a spherical, symmetric, linear, or asymmetric rotor and justify your answers with group theoretical arguments: (a) CH_4, (b) CH_3CN, (c) CO_2, (d) CH_3OH, (e) benzene, (f) pyridine.

12.2 Derive eqn 12B.17 ($\tilde{D}_J=4\tilde{B}^3/\tilde{\nu}^2$) for the centrifugal distortion constant D_J of a diatomic molecule of effective mass m_{eff}. Treat the bond as an elastic spring with force constant k_f and equilibrium length r_e that is subjected to a centrifugal distortion to a new length r_c. Begin the derivation by letting the particles experience a restoring force of magnitude $k_f(r_c-r_e)$ that is countered perfectly by a centrifugal force $m_{\text{eff}}\omega^2 r_c$, where ω is the angular velocity of the rotating molecule. Then introduce quantum mechanical effects by writing the angular momentum as $\{J(J+1)\}^{1/2}\hbar$. Finally, write an expression for the energy of the rotating molecule, compare it with eqn 12B.16, and infer an expression for $\tilde{D}_J$.

12.3‡ The H_3^+ ion has recently been found in the interstellar medium and in the atmospheres of Jupiter, Saturn, and Uranus. The rotational energy levels of H_3^+, an oblate symmetric rotor, are given by eqn 12B.13, with $\tilde{C}$ replacing $\tilde{A}$, when centrifugal distortion and other complications are ignored. Experimental values for vibrational–rotational constants are $\tilde{\nu}(E')=2521.6\,cm^{-1}$, $\tilde{B}=43.55\,cm^{-1}$, and $\tilde{C}=20.71\,cm^{-1}$. (a) Show that for a nonlinear planar molecule (such as H_3^+) that $I_C=2I_B$. The rather large discrepancy with the experimental values is due to factors ignored in eqn 12B.13. (b) Calculate an approximate value of the H–H bond length in H_3^+. (c) The value of R_e obtained from the best quantum mechanical calculations by J.B. Anderson (*J. Chem. Phys.* **96**, 3702 (1991)) is 87.32 pm. Use this result to calculate the values of the rotational constants $\tilde{B}$ and $\tilde{C}$. (d) Assuming that the geometry and force constants are the same in D_3^+ and H_3^+, calculate the spectroscopic constants of D_3^+. The molecular ion D_3^+ was first produced by Shy et al. (*Phys. Rev. Lett* **45**, 535 (1980)) who observed the $\nu_2(E')$ band in the infrared.

12.4 Use molecular modelling software and the computational method of your choice to construct molecular potential energy curves like the one shown in Fig. 12D.1. Consider the hydrogen halides (HF, HCl, HBr, and HI): (a) plot the calculated energy of each molecule against the bond length, and (b) identify the order of force constants of the H–Hal bonds.

12.5 The computational methods discussed in Topic 10E can be used to simulate the vibrational spectrum of a molecule, and it is then possible to determine the correspondence between a vibrational frequency and the atomic displacements that give rise to a normal mode. (a) Using molecular modelling software and the computational method of your choice, calculate the fundamental vibrational wavenumbers and depict the vibrational normal modes of SO_2 in the gas phase graphically. (b) The experimental values of the fundamental vibrational wavenumbers of SO_2 in the gas phase are $525\,cm^{-1}$, $1151\,cm^{-1}$, and $1336\,cm^{-1}$. Compare the calculated and experimental values. Even if agreement is poor, is it possible to establish a correlation between an experimental value of the vibrational wavenumber with a specific vibrational normal mode?

12.6 Use appropriate electronic structure software to perform calculations on H_2O and CO_2 with basis sets of your or your instructor's choosing. (a) Compute ground-state energies, equilibrium geometries and vibrational frequencies for each molecule. (b) Compute the magnitude of the dipole moment of H_2O; the experimental value is 1.854 D. (c) Compare computed values to experiment and suggest reasons for any discrepancies.

12.7 The protein haemerythrin is responsible for binding and carrying O_2 in some invertebrates. Each protein molecule has two Fe^{2+} ions that are in very close proximity and work together to bind one molecule of O_2. The Fe_2O_2 group of oxygenated haemerythrin is coloured and has an electronic absorption band at 500 nm. The resonance Raman spectrum of oxygenated haemerythrin obtained with laser excitation at 500 nm has a band at $844\,cm^{-1}$ that has been attributed to the O–O stretching mode of bound $^{16}O_2$. (a) Why is resonance Raman spectroscopy and not infrared spectroscopy the method of choice for the study of the binding of O_2 to haemerythrin? (b) Proof that the $844\,cm^{-1}$ band arises from a bound O_2 species may be obtained by conducting experiments on samples of haemerythrin that have been mixed with $^{18}O_2$, instead of $^{16}O_2$. Predict the fundamental vibrational wavenumber of the ^{18}O–^{18}O stretching mode in a sample of haemerythrin that has been treated with $^{18}O_2$. (c) The fundamental vibrational wavenumbers for the O–O stretching modes of O_2, O_2^- (superoxide anion), and O_2^{2-} (peroxide anion) are 1555, 1107, and $878\,cm^{-1}$, respectively. Explain this trend in terms of the electronic structures of O_2, O_2^-, and O_2^{2-}. *Hint*: Review Topic 10C. What are the bond orders of O_2, O_2^-, and O_2^{2-}? (d) Based on the data given above, which of the following species best describes the Fe_2O_2 group of haemerythrin: $Fe_2^{2+}O_2$, $Fe^{2+}Fe^{3+}O_2^-$, or $Fe_2^{3+}O_2^{2-}$? Explain your reasoning. (e) The resonance Raman spectrum of haemerythrin mixed with $^{16}O^{18}O$ has two bands that can be attributed to the O–O stretching mode of bound oxygen. Discuss how this observation may be used to exclude one or more of the four proposed schemes (4–7) for binding of O_2 to the Fe_2 site of haemerythrin.

4

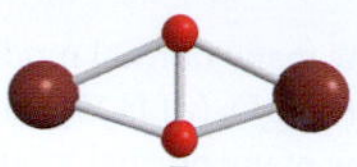
5

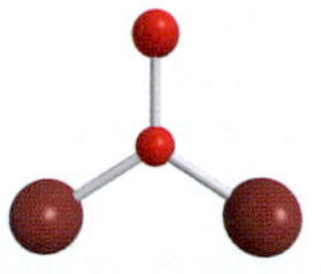
6

7

12.8 The moments of inertia of the linear mercury(II) halides are very large, so the O and S branches of their vibrational Raman spectra show little rotational structure. Nevertheless, the peaks of both branches can be identified and have been used to measure the rotational constants of the molecules (R.J.H. Clark and D.M. Rippon, *J. Chem. Soc. Faraday Soc. II*, **69**, 1496 (1973)). Show, from a knowledge of the value of J corresponding to the intensity maximum, that the separation of the peaks of the O and S branches is given by the Placzek–Teller relation $\delta=(32\tilde{B}kT/hc)^{1/2}$. The following widths were obtained at the temperatures stated:

	$HgCl_2$	$HgBr_2$	HgI_2
θ/°C	282	292	292
δ/cm^{-1}	23.8	15.2	11.4

Calculate the bond lengths in the three molecules.

12.9‡ A mixture of carbon dioxide (2.1 per cent) and helium, at 1.00 bar and 298 K in a gas cell of length 10 cm has an infrared absorption band centred at $2349\,cm^{-1}$ with absorbances, $A(\tilde{\nu})$, described by:

$$A(\tilde{\nu})=\frac{a_1}{1+a_2(\tilde{\nu}-a_3)^2}+\frac{a_4}{1+a_5(\tilde{\nu}-a_6)^2}$$

where the coefficients are $a_1=0.932$, $a_2=0.005050\,cm^2$, $a_3=2333\,cm^{-1}$, $a_4=1.504$, $a_5=0.01521\,cm^2$, $a_6=2362\,cm^{-1}$. (a) Draw graphs of $A(\tilde{\nu})$ and $\varepsilon(\tilde{\nu})$. What is the origin of both the band and the band width? What are the allowed and forbidden transitions of this band? (b) Calculate the transition wavenumbers and absorbances of the band with a simple harmonic oscillator–rigid rotor model and compare the result with the experimental spectra. The CO bond length is 116.2 pm. (c) Within what height, h, is basically all the infrared emission from the Earth in this band absorbed by atmospheric carbon dioxide? The mole fraction of CO_2 in the atmosphere is 3.3×10^{-4} and $T/K=288-0.0065(h/m)$ below 10 km. Draw a surface plot of the atmospheric transmittance of the band as a function of both height and wavenumber.

CHAPTER 13

Electronic transitions

Unlike for rotational and vibrational modes, simple analytical expressions for the electronic energy levels of molecules cannot be given. Therefore, this chapter concentrates on the qualitative features of electronic transitions.

13A Electronic spectra

A common theme throughout the chapter is that electronic transitions occur within a stationary nuclear framework. This Topic begins with a discussion of the electronic spectra of diatomic molecules, and we see that in the gas phase it is possible to observe simultaneous vibrational and rotational transitions that accompany the electronic transition. Then we describe features of the electronic spectra of polyatomic molecules.

13B Decay of excited states

We begin this Topic with an account of spontaneous emission by molecules, including the phenomena of 'fluorescence' and 'phosphorescence'. Then we see how non-radiative decay of excited states can result in transfer of energy as heat to the surroundings or can result in molecular dissociation.

13C Lasers

A specially important example of stimulated radiative decay is that responsible for the action of lasers, and in this Topic we see how this stimulated emission may be achieved and employed.

What is the impact of this material?

Absorption and emission spectroscopy is also useful to biochemists. In *Impact* I13.1 we describe how the absorption of visible radiation by special molecules in the eye initiates the process of vision. In *Impact* I13.2 we see how fluorescence techniques can be used to make very small samples visible, ranging from specialized compartments inside biological cells to single molecules.

To read more about the impact of this material, scan the QR code, or go to bcs.whfreeman.com/webpub/chemistry/pchem10e/impact/pchem-13-1.html

13A Electronic spectra

Contents

➤ Why do you need to know this material?

Many of the colours of the objects in the world around us stem from transitions in which an electron is promoted from one orbital of a molecule or ion into another. In some cases the relocation of an electron may be so extensive that it results in the breaking of a bond and the initiation of a chemical reaction. To understand these physical and chemical phenomena, you need to explore the origins of electronic transitions in molecules.

➤ What is the key idea?

Electronic transitions occur within a stationary nuclear framework.

➤ What do you need to know already?

You need to be familiar with the general features of spectroscopy (Topic 12A), the quantum mechanical origins of selection rules (Topics 9C, 12C, and 12D), and vibration–rotation spectra (Topic 12D); it would be helpful to be aware of atomic term symbols (Topic 9C). One example uses the method of combination differences described in Topic 12D.

Consider a molecule in the lowest vibrational state of its ground electronic state. The nuclei are (in a classical sense) at their equilibrium locations and experience no net force from the electrons and other nuclei in the molecule. The electron distribution is changed when an electronic transition occurs and the nuclei become subjected to different forces. In response, they start to vibrate around their new equilibrium locations. The resulting vibrational transitions that accompany the electronic transition give rise to the **vibrational structure** of the electronic transition. This structure can be resolved for gaseous samples, but in a liquid or solid the lines usually merge together and result in a broad, almost featureless band (Fig. 13A.1).

The energies needed to change the electron distributions of molecules are of the order of several electronvolts (1 eV is equivalent to about 8000 cm^{-1} or 100 kJ mol^{-1}). Consequently, the photons emitted or absorbed when such changes occur

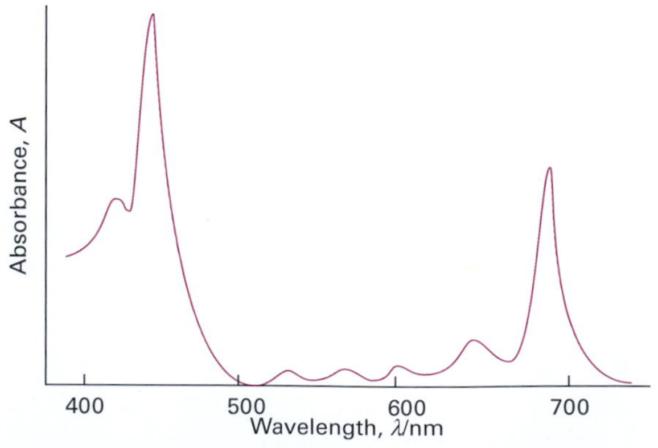

Figure 13A.1 The absorption spectrum of chlorophyll in the visible region. Note that it absorbs in the red and blue regions, and that green light is not absorbed.

Table 13A.1* Colour, wavelength, frequency, and energy of light

Colour	λ/nm	$\nu/(10^{14}$ Hz)	E/(kJ mol^{-1})
Infrared	>1000	<3.0	<120
Red	700	4.3	170
Yellow	580	5.2	210
Blue	470	6.4	250
Ultraviolet	<400	>7.5	>300

* More values are given in the *Resource section*.

lie in the visible and ultraviolet regions of the spectrum (Table 13A.1). What follows is a discussion of absorption processes. Emission processes are discussed in Topic 13B.

13A.1 Diatomic molecules

Topic 9C explains how the states of atoms are expressed by using term symbols and how the selection rules for electronic transitions can be expressed in terms of these term symbols. Much the same is true of diatomic molecules, one principal difference being the replacement of full spherical symmetry of atoms by the cylindrical symmetry defined by the axis of the molecule. The second principal difference is the fact that a diatomic molecule can vibrate and rotate.

(a) Term symbols

The term symbols of linear molecules (the analogues of the symbols ^{2}P, etc. for atoms) are constructed in a similar way to those for atoms, with the Roman uppercase letter (the P in this instance for atoms) representing the total orbital angular momentum of the electrons around the nucleus. In a linear molecule, and specifically a diatomic molecule, a Greek uppercase letter represents the total orbital angular momentum of the electrons around the internuclear axis. If this component of orbital angular momentum is $\Lambda\hbar$ with $\Lambda=0, \pm1, \pm2 \ldots$, we use the following designation:

$\lvert\Lambda\rvert$	0	1	2	…
	Σ	Π	Δ	…

These labels are the analogues of S, P, D, … for atoms for states with $L=0, 1, 2, \ldots$. To decide on the value of L for atoms we had to use the Clebsch–Gordan series (Topic 9C) to couple the individual angular momenta. The procedure to determine Λ is much simpler in a diatomic molecule because we simply add the values of the individual components of each electron, $\lambda\hbar$:

$$\Lambda=\lambda_1+\lambda_2+\cdots \tag{13A.1}$$

We note the following:

- A single electron in a σ orbital has $\lambda=0$.

The orbital is cylindrically symmetrical and has no angular nodes when viewed along the internuclear axis. Therefore, if that is the only type of electron present, $\Lambda=0$. The term symbol for the ground state of H_2^+ with electron configuration $1\sigma_g^2$ is therefore Σ.

- A π electron in a diatomic molecule has one unit of orbital angular momentum about the internuclear axis ($\lambda=\pm1$).

If it is the only electron outside a closed shell, it gives rise to a Π term. If there are two π electrons (as in the ground state of O_2, with configuration $\ldots1\pi_g^2$), there are two possible outcomes. If the electrons are travelling in opposite directions, then $\lambda_1=+1$ and $\lambda_2=-1$ (or vice versa) and $\Lambda=0$, corresponding to a Σ term. Alternatively, the electrons might occupy the same π orbital and $\lambda_1=\lambda_2=+1$ (or -1), and $\Lambda=\pm2$, corresponding to a Δ term. In O_2 it is energetically favourable for the electrons to occupy different orbitals, so the ground term is Σ.

As in atoms, we use a left superscript with the value of $2S+1$ to denote the multiplicity of the term, where S is the total spin quantum number of the electrons.

Brief illustration 13A.1 The multiplicity of a term

It follows from the procedure for assigning multiplicity of terms that for $S=s=\frac{1}{2}$ because there is only one electron, and the term symbol is $^2\Sigma$, a doublet term. In O_2, because in the ground state the two π electrons occupy different orbitals (as we saw above), they may have either parallel or antiparallel spins; the lower energy is obtained (as in atoms) if the spins are parallel, so $S=1$ and the ground state is $^3\Sigma$.

Self-test 13A.1 What is the value of S and the term symbol for the ground-state of H_2?

Answer: $S=0$, $^1\Sigma$

The overall parity of the state (its symmetry under inversion through the centre of the molecule, if it has one) is added as a right subscript to the term symbol. For H_2^+ in its ground state, the parity of the only occupied orbital ($1\sigma_g$) is g, so the term itself is also g, and in full dress is $^2\Sigma_g$. If there are several electrons, the overall parity is calculated by noting the parity of each occupied orbital and using

$$\mathrm{g}\times\mathrm{g}=\mathrm{g} \quad \mathrm{u}\times\mathrm{u}=\mathrm{g} \quad \mathrm{u}\times\mathrm{g}=\mathrm{u} \tag{13A.2}$$

These rules are generated by interpreting g as +1 and u as −1. As a consequence:

- The term symbol for the ground state of any closed-shell homonuclear diatomic molecule is $^1\Sigma_g$ because the spin is zero (a singlet term in which all electrons paired), there is no orbital angular momentum from a closed shell, and the overall parity is g.

- If the molecule is heteronuclear, parity is irrelevant and the ground state of a closed-shell species, such as CO, is $^1\Sigma$.

Brief illustration 13A.2 Term symbol of O_2 1

The parity of the ground state of O_2 is g×g=g, so it is denoted $^3\Sigma_g$. An excited configuration of O_2 is $\ldots 1\pi_g^2$, with both π electrons in the same orbital. As we have seen, $|\Lambda|=2$, represented by Δ. The two electrons must be paired if they occupy the same orbital, so $S=0$. The overall parity is g×g=g. Therefore, the term symbol is $^1\Delta_g$.

Self-test 13A.2 The term symbol for one of the lowest excited states of H_2 is $^3\Pi_u$. To which excited-state configuration does this term symbol correspond?

Answer: $1\sigma_g^1 1\pi_u^1$

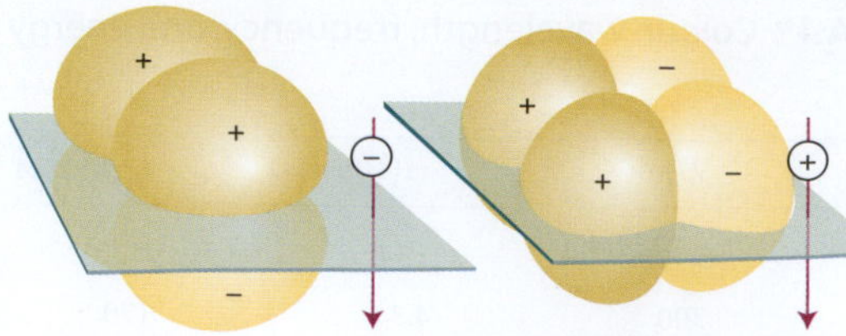

Figure 13A.2 The + or – on a term symbol refers to the overall symmetry of an electronic wavefunction under reflection in a plane containing the two nuclei.

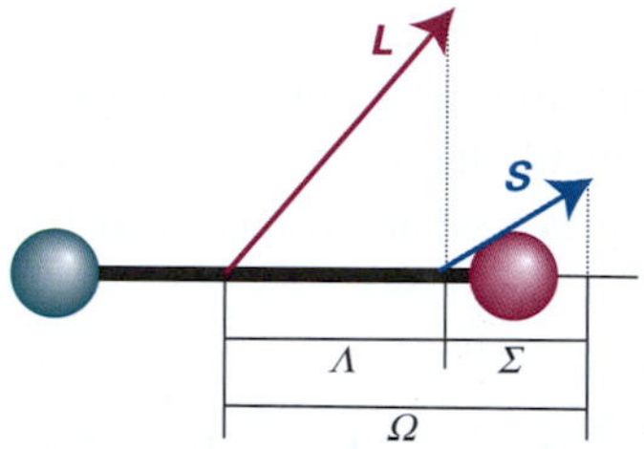

Figure 13A.3 The coupling of spin and orbital angular momenta in a linear molecule: only the components along the internuclear axis are conserved.

There is an additional symmetry operation that distinguishes different types of Σ term: reflection in a plane containing the internuclear axis. A + right superscript on Σ is used to denote a wavefunction that does not change sign under this reflection and a – sign is used if the wavefunction changes sign (Fig. 13A.2).

Brief illustration 13A.3 Term symbol of O_2 2

If we think of O_2 in its ground state as having one electron in $1\pi_{g,x}$, which changes sign under reflection in the *yz*-plane, and the other electron in $1\pi_{g,y}$, which does not change sign under reflection in the same plane, then the overall reflection symmetry is (closed shell)×(+)×(–)=(–), and the full term symbol of the ground electronic state of O_2 is $^3\Sigma_g^-$. Alternatively, if we consider the configuration to be $1\pi_+^1 1\pi_-^1$, with $\pi_\pm \propto \pi_{g,x} \pm i\pi_{g,y}$ being two states of definite but opposite orbital angular momentum around the axis, then for the triplet state we must take the linear combination $\Psi(1,2) \propto \pi_+(1)\pi_-(2) - \pi_+(2)\pi_-(1)$. Because under reflection in the *yz*-plane $\pi_+ \rightarrow -\pi_-$ and $\pi_- \rightarrow -\pi_+$, $\Psi(1,2) \rightarrow \pi_-(1)\pi_+(2) - \pi_-(2)\pi_+(1) = -\Psi(1,2)$, and the state is also (–).

Self-test 13A.3 What is the full term symbol of the ground electronic state of Li_2^+?

Answer: $^2\Sigma_g^+$

As for atoms, sometimes it is necessary to specify the total electronic angular momentum. In atoms we use the quantum number *J*, which appears as a right subscript in the term symbol, as in $^2P_{1/2}$, with different values of *J* corresponding to different *levels* of a term. In a linear molecule, only the electronic angular momentum about the internuclear axis is well defined, and has the value $\Omega\hbar$. For light molecules, where the spin–orbit coupling is weak, Ω is obtained by adding together the components of orbital angular momentum around the axis (the value of Λ) and the component of the electron spin on that axis (Fig. 13A.3). The latter is denoted Σ, where $\Sigma = S, S-1, S-2, \ldots, -S$. (It is important to distinguish between the upright term symbol Σ and the sloping quantum number Σ.) Then

$$\Omega = \Lambda + \Sigma \qquad (13A.3)$$

The value of $|\Omega|$ may then be attached to the term symbol as a right subscript (just like *J* is used in atoms) to denote the different levels. These levels differ in energy, as in atoms, as a result of spin–orbit coupling.

Brief illustration 13A.4 The term symbol of NO

The ground-state configuration of NO is $\ldots\pi_g^1$, so it is a $^2\Pi$ term with $\Lambda = \pm 1$ and $\Sigma = \pm\frac{1}{2}$. Therefore, there are two levels of the term, one with $\Omega = \pm\frac{1}{2}$ and the other with $\pm\frac{3}{2}$, denoted $^2\Pi_{1/2}$ and $^2\Pi_{3/2}$, respectively. Each level is doubly degenerate (corresponding to the opposite signs of Ω). In NO, $^2\Pi_{1/2}$ lies slightly lower than $^2\Pi_{3/2}$.

Self-test 13A.4 What are the levels of the term for the ground electronic state of O_2^-?

Answer: $^2\Pi_{1/2}$, $^2\Pi_{3/2}$

(b) Selection rules

A number of selection rules govern which transitions can be observed in the electronic spectrum of a molecule. The selection rules concerned with changes in angular momentum are

$$\Delta\Lambda=0,\pm1 \quad \Delta S=0 \qquad \Delta\Sigma=0 \quad \Delta\Omega=0,\pm1 \qquad \textit{Linear molecules} \quad \text{Selection rules for electronic spectra} \qquad (13A.4)$$

As in atoms (Topic 9C), the origins of these rules are conservation of angular momentum during a transition and the fact that a photon has a spin of 1.

There are two selection rules concerned with changes in symmetry. First, as we show in the following *Justification*,

For Σ terms, only $\Sigma^+\leftrightarrow\Sigma^+$ and $\Sigma^-\leftrightarrow\Sigma^-$ are allowed.

Second, the **Laporte selection rule** for centrosymmetric molecules (those with a centre of inversion) states that *the only allowed transitions are transitions that are accompanied by a change of parity*. That is,

For centrosymmetric molecules, only $u\rightarrow g$ and $g\rightarrow u$ are allowed — Laporte selection rule

Justification 13A.1 Symmetry-based selection rules

The last two selection rules result from the fact that the electric-dipole transition moment introduced in Topic 9C, $\mu_{fi}=\int\psi_f^*\hat{\mu}\psi_i d\tau$ vanishes unless the integrand is invariant under all symmetry operations of the molecule.

The z-component of the electric dipole moment operator is the component of μ responsible for $\Sigma\leftrightarrow\Sigma$ transitions (the other components have Π symmetry and cannot make a contribution). The z-component of μ has (+) symmetry with respect to reflection in a plane containing the internuclear axis. Therefore, for a $(+)\leftrightarrow(-)$ transition, the overall symmetry of the transition dipole moment is $(+)\times(+)\times(-)=(-)$, so it must be zero and hence $\Sigma^+\leftrightarrow\Sigma^-$ transitions are not allowed. The integrals for $\Sigma^+\leftrightarrow\Sigma^+$ and $\Sigma^-\leftrightarrow\Sigma^-$ transform as $(+)\times(+)\times(+)=(+)$ and $(-)\times(+)\times(-)=(+)$, respectively, and so both transitions are allowed.

The three components of the dipole moment operator transform like x, y, and z, and in a centrosymmetric molecule are all u. Therefore, for a $g\rightarrow g$ transition, the overall parity of the transition dipole moment is $g\times u\times g=u$, so it must be zero. Likewise, for a $u\rightarrow u$ transition, the overall parity is $u\times u\times u=u$, so the transition dipole moment must also vanish. Hence, transitions without a change of parity are forbidden. For a $g\leftrightarrow u$ transition the integral transforms as $g\times u\times u=g$, and is allowed.

A forbidden $g\rightarrow g$ transition can become allowed if the centre of symmetry is eliminated by an asymmetrical vibration, such as the one shown in Fig. 13A.4. When the centre of symmetry is lost, $g\rightarrow g$ and $u\rightarrow u$ transitions are no longer parity-forbidden and become weakly allowed. A transition that derives its intensity from an asymmetrical vibration of a molecule is called a **vibronic transition**.

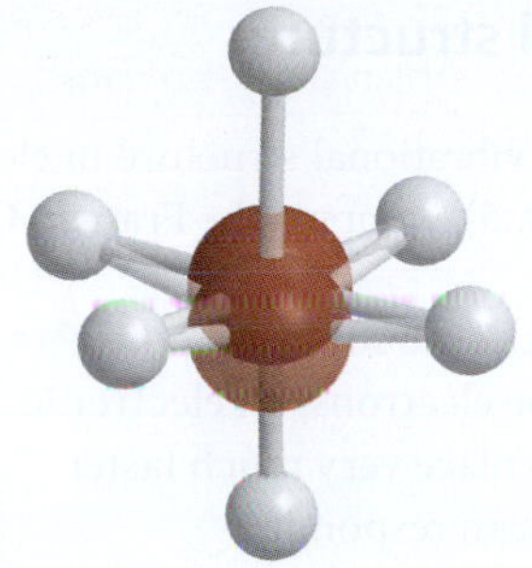

Figure 13A.4 A d–d transition is parity-forbidden because it corresponds to a g–g transition. However, a vibration of the molecule can destroy the inversion symmetry of the molecule and the g,u classification no longer applies. The removal of the centre of symmetry gives rise to a vibronically allowed transition.

Brief illustration 13A.5 Allowed transitions of O_2

If we were presented with the following possible transitions in the electronic spectrum of O_2, namely $^3\Sigma_g^-\leftarrow{}^3\Sigma_u^-$, $^3\Sigma_g^-\leftarrow{}^1\Delta_g$, $^3\Sigma_g^-\leftarrow{}^3\Sigma_u^+$, we could decide which are allowed by constructing the following table and referring to the rules. Forbidden values are in red.

	ΔS	$\Delta\Lambda$	$\Sigma^\pm\leftarrow\Sigma^\pm$	Change of parity	
$^3\Sigma_g^-\leftarrow{}^3\Sigma_u^-$	0	0	$\Sigma^-\leftarrow\Sigma^-$	$g\leftarrow u$	Allowed
$^3\Sigma_g^-\leftarrow{}^1\Delta_g$	+1	−2	Not applicable	$g\leftarrow g$	Forbidden
$^3\Sigma_g^-\leftarrow{}^3\Sigma_u^+$	0	0	$\Sigma^-\leftarrow\Sigma^+$	$g\leftarrow u$	Forbidden

Self-test 13A.5 Which of the following electronic transitions are allowed in O_2: $^3\Sigma_g^-\leftrightarrow{}^1\Sigma_g^+$ and $^3\Sigma_g^-\leftrightarrow{}^3\Delta_u$?

Answer: None

The large number of photons in an incident beam generated by a laser gives rise to a qualitatively different branch of spectroscopy, for the photon density is so great that more than one photon may be absorbed by a single molecule and give rise to **multiphoton processes**. One application of multiphoton processes is that states inaccessible by conventional one-photon spectroscopy become observable because the overall transition occurs with no change of parity. For example, in one-photon spectroscopy, only $g\leftrightarrow u$ transitions are observable; in two-photon spectroscopy, however, the overall outcome of absorbing two photons is a $g\leftarrow g$ or a $u\leftarrow u$ transition.

(c) Vibrational structure

To account for the vibrational structure in electronic spectra of molecules (Fig. 13A.5), we apply the **Franck–Condon principle**:

> Because the nuclei are so much more massive than the electrons, an electronic transition takes place very much faster than the nuclei can respond.

Franck–Condon principle

As a result of the transition, electron density is rapidly built up in new regions of the molecule and removed from others. In classical terms, the initially stationary nuclei suddenly experience a new force field, to which they respond by beginning to vibrate and (in classical terms) swing backwards and forwards from their original separation (which was maintained during the rapid electronic excitation). The stationary equilibrium separation of the nuclei in the initial electronic state therefore becomes a stationary turning point in the final electronic state (Fig. 13A.6). We can imagine the transition as taking place up the vertical line in Fig. 13A.6. This interpretation is the origin of the expression **vertical transition**, which denotes an electronic transition that occurs without change of nuclear geometry and in classical terms, the nuclei remain stationary.

The vibrational structure of the spectrum depends on the relative horizontal position of the two potential energy curves, and a long **vibrational progression**, a lot of vibrational structure, is stimulated if the upper potential energy curve is appreciably displaced horizontally from the lower. The upper curve is usually displaced to greater equilibrium bond lengths because electronically excited states usually have more antibonding character than electronic ground states. The separation of the vibrational lines depends on the vibrational energies of the *upper* electronic state.

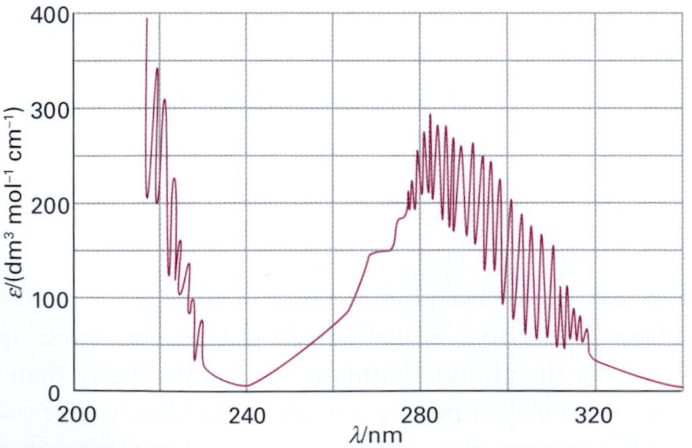

Figure 13A.5 The electronic spectra of some molecules show significant vibrational structure. Shown here is the ultraviolet spectrum of gaseous SO_2 at 298 K. As explained in the text, the sharp lines in this spectrum are due to transitions from a lower electronic state to different vibrational levels of a higher electronic state. Vibrational structure due to transitions to two different excited electronic states is apparent.

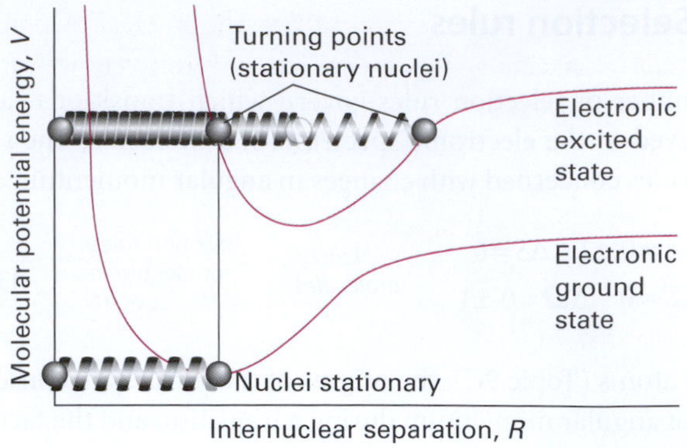

Figure 13A.6 According to the Franck–Condon principle, the most intense vibronic transition is from the ground vibrational state to the vibrational state lying vertically above it. As a result of the vertical transition, the nuclei suddenly experience a new force field, to which they respond through their vibrational motion. The equilibrium separation of the nuclei in the initial electronic state therefore becomes a turning point in the final electronic state. Transitions to other vibrational levels also occur, but with lower intensity.

The quantum mechanical version of the Franck–Condon principle refines this picture. Instead of saying that the nuclei stay at the same locations and are stationary during the transition, we say that *they retain their initial dynamic state*. In quantum mechanics, the dynamical state is expressed by the wavefunction, so an equivalent statement is that the nuclear wavefunction does not change during the electronic transition. Initially the molecule is in the lowest vibrational state of its ground electronic state with a bell-shaped wavefunction centred on the equilibrium bond length (Fig. 13A.7). To find the nuclear state to which the transition takes place, we look for the vibrational wavefunction that most closely resembles this initial

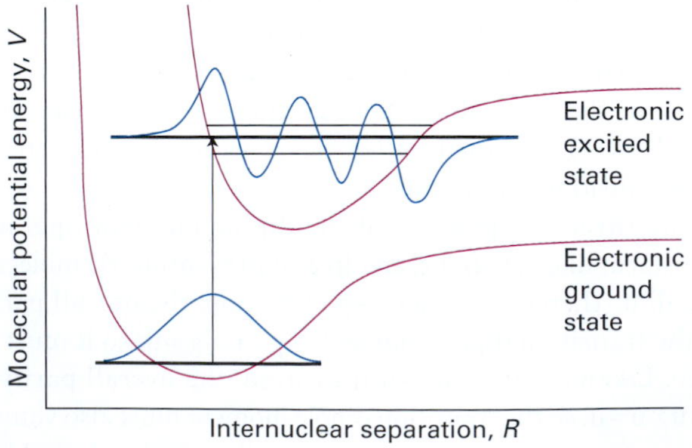

Figure 13A.7 In the quantum mechanical version of the Franck–Condon principle, the molecule undergoes a transition to the upper vibrational state that most closely resembles the vibrational wavefunction of the vibrational ground state of the lower electronic state. The two wavefunctions shown here have the greatest overlap integral of all the vibrational states of the upper electronic state and hence are most closely similar.

wavefunction, for that corresponds to the nuclear dynamical state that is least changed in the transition. Intuitively, we can see that the final wavefunction is the one with a large peak close to the position of the initial bell-shaped function. As explained in Topic 8B, provided the vibrational quantum number is not zero, the biggest peaks of vibrational wavefunctions occur close to the edges of the confining potential, so we can expect the transition to occur to those vibrational states, in accord with the classical description. However, several vibrational states have their major peaks in similar positions, so we should expect transitions to occur to a range of vibrational states, as is observed.

The quantitative form of the Franck–Condon principle and the justification of the preceding description is derived from the expression for the transition dipole moment (as in *Justification* 13A.1). The electric dipole moment operator is a sum over all nuclei and electrons in the molecule:

$$\hat{\mu} = -e\sum_i r_i + e\sum_I Z_I R_I \qquad (13A.5)$$

where the vectors are the distances from the centre of charge of the molecule. The intensity of the transition is proportional to the square modulus, $|\mu_{fi}|^2$, of the magnitude of the transition dipole moment, and we show in the following *Justification* that this intensity is proportional to the square modulus of the overlap integral, $S(v_f,v_i)$, between the vibrational states of the initial and final electronic states. This overlap integral is a measure of the match between the vibrational wavefunctions in the upper and lower electronic states: $S=1$ for a perfect match and $S=0$ when there is no similarity.

Justification 13A.2 The Franck–Condon approximation

The overall state of the molecule consists of an electronic part, labelled with ε, and a vibrational part, labelled with v. Therefore, within the Born–Oppenheimer approximation, the transition dipole moment factorizes as follows:

$$\begin{aligned}\mu_{fi} &= \int \psi^*_{\varepsilon,f}\psi^*_{v,f}\left\{-e\sum_i r_i + e\sum_I Z_I R_I\right\}\psi_{\varepsilon,i}\psi_{v,i}\,d\tau \\ &= -e\sum_i \int \psi^*_{\varepsilon,f} r_i \psi_{\varepsilon,i}\,d\tau_e \int \psi^*_{v,f}\psi_{v,i}\,d\tau_n \\ &\quad + e\sum_I Z_I \overbrace{\int \psi^*_{\varepsilon,f}\psi_{\varepsilon,i}\,d\tau_e}^{0} \int \psi^*_{v,f} R_I \psi_{v,i}\,d\tau_n\end{aligned}$$

The second term on the right of the second row (including the term in blue) is zero, because two different electronic states are orthogonal. Therefore,

$$\mu_{fi} = -e\sum_i \overbrace{\int \psi^*_{\varepsilon,f} r_i \psi_{\varepsilon,i}\,d\tau_e}^{\mu_{\varepsilon,fi}} \overbrace{\int \psi^*_{v,f}\psi_{v,i}\,d\tau_n}^{S(v_f,v_i)} = \mu_{\varepsilon,fi} S(v_f,v_i)$$

The quantity $\mu_{\varepsilon,fi}$ is the electric-dipole transition moment arising from the redistribution of electrons (and a measure of the 'kick' this redistribution gives to the electromagnetic field, and vice versa for absorption). The factor $S(v_f,v_i)$, is the overlap integral between the vibrational state with quantum number v_i in the initial electronic state of the molecule, and the vibrational state with quantum number v_f in the final electronic state of the molecule.

Because the transition intensity is proportional to the square of the magnitude of the transition dipole moment, the intensity of an absorption is proportional to $|S(v_f,v_i)|^2$, which is known as the **Franck–Condon factor** for the transition:

$$|S(v_f,v_i)|^2 = \left(\int \psi^*_{v,f}\psi_{v,i}\,d\tau_n\right)^2 \qquad \text{Franck–Condon factor} \qquad (13A.6)$$

It follows that, the greater the overlap of the vibrational state wavefunction in the upper electronic state with the vibrational wavefunction in the lower electronic state, the greater the absorption intensity of that particular simultaneous electronic and vibrational transition.

Example 13A.1 Calculating a Franck–Condon factor

Consider the transition from one electronic state to another, their bond lengths being R_e and R_e' and their force constants equal. Calculate the Franck–Condon factor for the 0–0 transition and show that the transition is most intense when the bond lengths are equal.

Method We need to calculate $S(0,0)$, the overlap integral of the two ground-state vibrational wavefunctions, and then take its square. The difference between harmonic and anharmonic vibrational wavefunctions is negligible for $v=0$, so harmonic oscillator wavefunctions can be used (Table 8B.1).

Answer We use the (real) wavefunctions

$$\psi_0 = \left(\frac{1}{\alpha\pi^{1/2}}\right)^{1/2} e^{-x^2/2\alpha^2} \qquad \psi_0' = \left(\frac{1}{\alpha\pi^{1/2}}\right)^{1/2} e^{-x'^2/2\alpha^2}$$

where $x=R-R_e$ and $x'=R-R_e'$, with $\alpha=(\hbar^2/mk_f)^{1/4}$ (Topic 8B). The overlap integral is

$$S(0,0) = \int_{-\infty}^{\infty} \psi_0'\psi_0\,dR = \frac{1}{\alpha\pi^{1/2}}\int_{-\infty}^{\infty} e^{-(x^2+x'^2)/2\alpha^2}\,dx$$

We now write $\alpha z = R-\frac{1}{2}(R_e+R_e')$ and manipulate this expression into

$$S(0,0) = \frac{1}{\pi^{1/2}} e^{-(R_e-R_e')^2/4\alpha^2} \overbrace{\int_{-\infty}^{\infty} e^{-z^2}\,dz}^{\pi^{1/2}} = e^{-(R_e-R_e')^2/4\alpha^2}$$

and the Franck–Condon factor is

$$S(0,0)^2 = e^{-(R_e - R_e')^2/2\alpha^2}$$

This factor is equal to 1 when $R_e' = R_e$ and decreases as the equilibrium bond lengths diverge from each other (Fig. 13A.8).

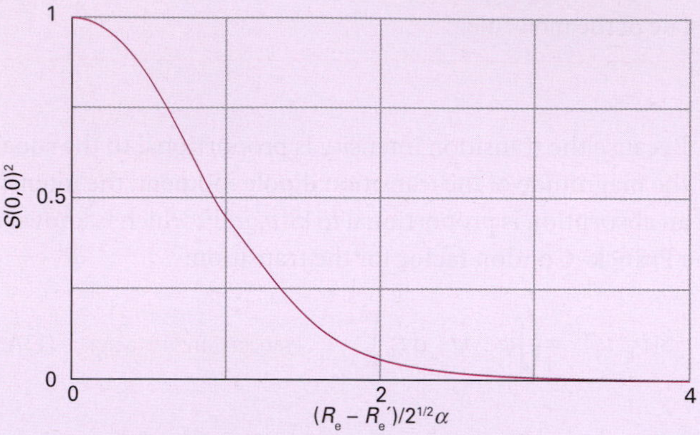

Figure 13A.8 The Franck–Condon factor for the arrangement discussed in *Example* 13A.1.

For Br_2, $R_e = 228$ pm and there is an upper state with $R_e' = 266$ pm. Taking the vibrational wavenumber as 250 cm^{-1} gives $S(0,0)^2 = 5.1 \times 10^{-10}$, so the intensity of the 0–0 transition is only 5.1×10^{-10} what it would have been if the potential curves had been directly above each other.

Self-test 13A.6 Suppose the vibrational wavefunctions can be approximated by rectangular functions of width W and W', centred on the equilibrium bond lengths (Fig. 13A.9). Find the corresponding Franck–Condon factors when the centres are coincident and $W' < W$.

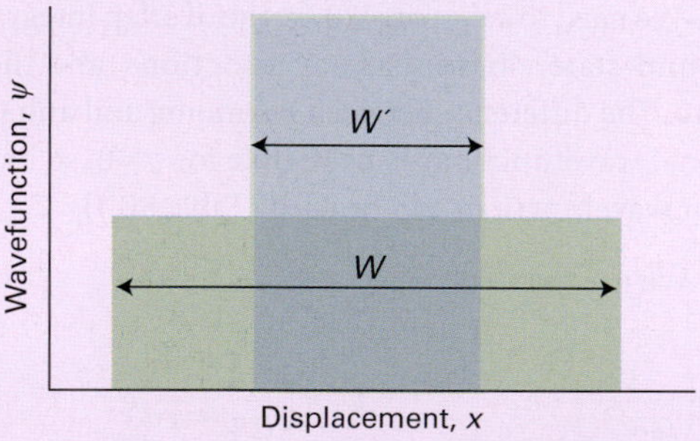

Figure 13A.9 The model wavefunctions used in *Self-test* 13A.6.

Answer: $S^2 = W'/W$

(d) Rotational structure

Just as in vibrational spectroscopy, where a vibrational transition is accompanied by rotational excitation, so rotational transitions accompany the excitation of the vibrational excitation that accompanies electronic excitation. We therefore see P, Q, and R branches for each vibrational transition, and the electronic transition has a very rich structure. However, the principal difference is that electronic excitation can result in much larger changes in bond length than vibrational excitation causes alone, and the rotational branches have a more complex structure than in vibration–rotation spectra.

We suppose that the rotational constants of the electronic ground and excited states are $\tilde{B}$ and $\tilde{B}'$, respectively. The rotational energy levels of the initial and final states are

$$E(J) = hc\tilde{B}J(J+1) \qquad E(J') = hc\tilde{B}'J'(J'+1) \qquad (13A.7)$$

When a transition occurs with $\Delta J = -1$ the wavenumber of the vibrational component of the electronic transition is shifted from $\tilde{\nu}$ to

$$\tilde{\nu} + \tilde{B}'(J-1)J - \tilde{B}J(J+1) = \tilde{\nu} - (\tilde{B}' + \tilde{B})J + (\tilde{B}' - \tilde{B})J^2$$

This transition is a contribution to the P branch (just as in Topic 12D). There are corresponding transitions to the Q and R branches with wavenumbers that may be calculated in a similar way. All three branches are:

$$\text{P branch } (\Delta J = -1): \quad \tilde{\nu}_P(J) = \tilde{\nu} - (\tilde{B}' + \tilde{B})J + (\tilde{B}' - \tilde{B})J^2$$

Branch structure (13A.8a)

$$\text{Q branch } (\Delta J = 0): \quad \tilde{\nu}_Q(J) = \tilde{\nu} + (\tilde{B}' - \tilde{B})J(J+1) \qquad (13A.8b)$$

$$\text{R branch } (\Delta J = +1): \quad \tilde{\nu}_R(J) = \tilde{\nu} + (\tilde{B}' + \tilde{B})(J+1) + (\tilde{B}' - \tilde{B})(J+1)^2 \qquad (13A.8c)$$

These expressions are the analogues of eqn 12D.19.

Example 13A.2 Estimating rotational constants from electronic spectra

The following rotational transitions were observed in the 0–0 band of the $^1\Sigma^+ \leftarrow {}^1\Sigma^+$ electronic transition of $^{63}Cu^2H$: $\tilde{\nu}_R(3) = 23\,347.69\ \text{cm}^{-1}$, $\tilde{\nu}_P(3) = 23\,298.85\ \text{cm}^{-1}$, and $\tilde{\nu}_P(5) = 23\,275.77\ \text{cm}^{-1}$. Estimate the values of $\tilde{B}'$ and $\tilde{B}$.

Method Use the method of combination differences introduced in Topic 12D: form the differences $\tilde{\nu}_R(J) - \tilde{\nu}_P(J)$ and $\tilde{\nu}_R(J-1) - \tilde{\nu}_P(J+1)$ from eqns 13A.8a and 13A.8b, then use the resulting expressions to calculate the rotational constants $\tilde{B}'$ and $\tilde{B}$ from the wavenumbers provided.

Answer From eqns 13A.8a and 13A.8b it follows that

$$\begin{aligned}\tilde{\nu}_R(J) - \tilde{\nu}_P(J) &= (\tilde{B}' + \tilde{B})(J+1) + (\tilde{B}' - \tilde{B})(J+1)^2 \\ &\quad - \{-(\tilde{B}' + \tilde{B})J + (\tilde{B}' - \tilde{B})J^2\} = 4\tilde{B}'(J + \tfrac{1}{2})\end{aligned}$$

$$\begin{aligned}\tilde{\nu}_R(J-1) - \tilde{\nu}_P(J+1) &= (\tilde{B}' + \tilde{B})J + (\tilde{B}' - \tilde{B})J^2 \\ &\quad - \{-(\tilde{B}' + \tilde{B})(J+1) + (\tilde{B}' - \tilde{B})(J+1)^2\} = 4\tilde{B}(J + \tfrac{1}{2})\end{aligned}$$

(These equations are analogous to eqns 12D.21a and 12D.21b.) After using the data provided, we obtain:

$$\text{For } J=3: \quad \tilde{\nu}_R(3)-\tilde{\nu}_P(3)=\overbrace{48.84}^{23\,347.69-23\,298.85}\ \text{cm}^{-1}=14\tilde{B}'$$

$$\text{For } J=4: \quad \tilde{\nu}_R(3)-\tilde{\nu}_P(5)=\overbrace{71.92}^{23\,347.69-23\,275.77}\ \text{cm}^{-1}=18\tilde{B}$$

and calculate $\tilde{B}'=3.489\,\text{cm}^{-1}$ and $\tilde{B}=3.996\,\text{cm}^{-1}$.

Self-test 13A.7 The following rotational transitions were observed in the $^1\Sigma^+ \leftarrow {}^1\Sigma^+$ electronic transition of RhN: $\tilde{\nu}_R(5)=22\,387.06\,\text{cm}^{-1}$, $\tilde{\nu}_P(5)=22\,376.87\,\text{cm}^{-1}$ and $\tilde{\nu}_P(7)=22\,373.95\,\text{cm}^{-1}$. Estimate the values of $\tilde{B}'$ and $\tilde{B}$.

Answer: $\tilde{B}'=0.4632\,\text{cm}^{-1}$, $\tilde{B}=0.5042\,\text{cm}^{-1}$

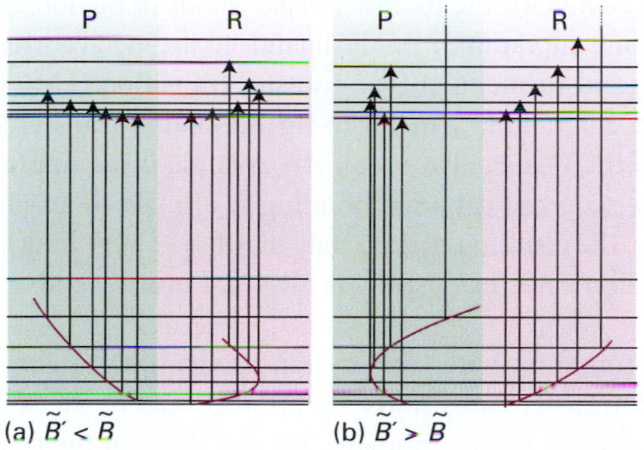

Figure 13A.10 When the rotational constants of a diatomic molecule differ significantly in the initial and final states of an electronic transition, the P and R branches show a head. (a) The formation of a head in the R branch when $\tilde{B}'<\tilde{B}$; (b) the formation of a head in the P branch when $\tilde{B}'>\tilde{B}$.

Suppose that the bond length in the electronically excited state is greater than that in the ground state; then $\tilde{B}'<\tilde{B}$ and $\tilde{B}'-\tilde{B}$ is negative. In this case the lines of the R branch converge with increasing J and when J is such that $|\tilde{B}'-\tilde{B}|(J+1)>\tilde{B}'+\tilde{B}$ the lines start to appear at successively decreasing wavenumbers. That is, the R branch has a **band head** (Fig. 13A.10a). When the bond is shorter in the excited state than in the ground state, $\tilde{B}'>\tilde{B}$ and $\tilde{B}'-\tilde{B}$ is positive. In this case, the lines of the P branch begin to converge and go through a head when J is such that $|\tilde{B}'-\tilde{B}|J>\tilde{B}'+\tilde{B}$ (Fig. 13A.10b).

13A.2 Polyatomic molecules

The absorption of a photon can often be traced to the excitation of specific types of electrons or to electrons that belong to a small group of atoms in a polyatomic molecule. For example, when a carbonyl group (>C=O) is present, an absorption at about 290 nm is normally observed, although its precise location depends on the nature of the rest of the molecule. Groups with characteristic optical absorptions are called **chromophores** (from the Greek for 'colour bringer'), and their presence often accounts for the colours of substances (Table 13A.2).

Table 13A.2* Absorption characteristics of some groups and molecules

Group	$\tilde{\nu}/\text{cm}^{-1}$	λ_{max}/nm	$\varepsilon_{max}/(\text{dm}^3\ \text{mol}^{-1}\ \text{cm}^{-1})$
C=C ($\pi^* \leftarrow \pi$)	61 000	163	15 000
C=O ($\pi^* \leftarrow$ n)	35 000–37 000	270–290	10–20
H_2O ($\pi^* \leftarrow$ n)	60 000	167	7 000

* More values are given in the *Resource section*.

(a) d-Metal complexes

In a free atom, all five d orbitals of a given shell are degenerate. In a d-metal complex, where the immediate environment of the atom is no longer spherical, the d orbitals are not all degenerate, and electrons can absorb energy by making transitions between them.

To see the origin of this splitting in an octahedral complex such as $[Ti(OH_2)_6]^{3+}$ (**1**), we regard the six ligands as point negative charges that repel the d electrons of the central ion (Fig. 13A.11). As a result, the orbitals fall into two groups, with $d_{x^2-y^2}$ and d_{z^2} pointing directly towards the ligand positions, and d_{xy}, d_{yz}, and d_{zx} pointing between them. An electron occupying an orbital of the former group has a less favourable potential energy than when it occupies any of the three orbitals of the other group, and so the d orbitals split into the two sets shown in (**2**) with an energy difference Δ_O: a triply degenerate set comprising the d_{xy}, d_{yz}, and d_{zx} orbitals and labelled t_{2g}, and a doubly degenerate set comprising the with $d_{x^2-y^2}$ and d_{z^2} orbitals and labelled e_g. The three t_{2g} orbitals lie below the two e_g orbitals in energy; the difference in energy Δ_O is called the **ligand-field splitting parameter** (the O denoting octahedral symmetry). The ligand field splitting is typically about 10 per cent of the overall energy of interaction between the ligands and the central metal atom, which is largely responsible for the existence of the complex. The d orbitals also divide into two sets in a tetrahedral complex, but in this case the e orbitals lie below the t_2 orbitals (the g,u classification is no longer relevant as a tetrahedral complex has no centre of inversion) and their separation is written Δ_T.

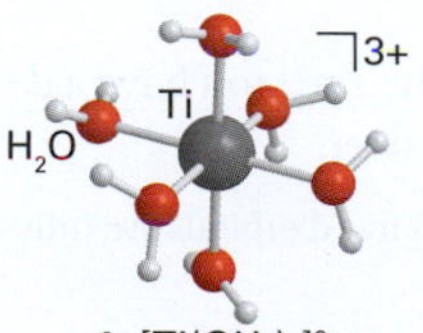

1 $[Ti(OH_2)_6]^{3+}$

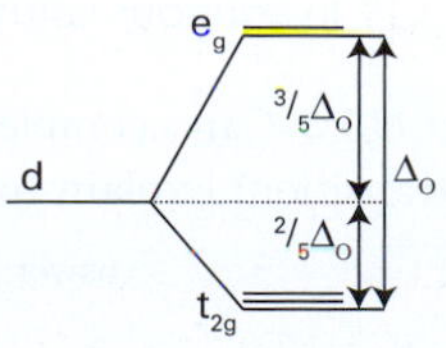

2

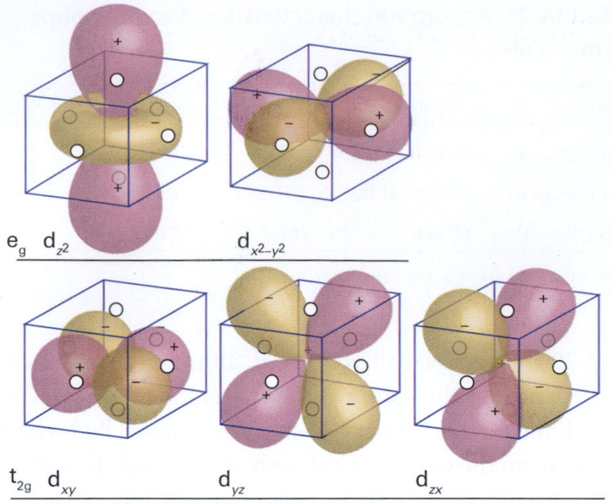

Figure 13A.11 The classification of d orbitals in an octahedral environment. The open circles represent the positions of the six (point-charge) ligands.

Neither Δ_O nor Δ_T is large, so transitions between the two sets of orbitals typically occur in the visible region of the spectrum. The transitions are responsible for many of the colours that are so characteristic of d-metal complexes.

Brief illustration 13A.6 The electronic spectrum of a d-metal complex

The spectrum of $[Ti(OH_2)_6]^{3+}$ (**1**) near 20 000 cm^{-1} (500 nm) is shown in Fig. 13A.12, and can be ascribed to the promotion of its single d electron from a t_{2g} orbital to an e_g orbital. The wavenumber of the absorption maximum suggests that $\Delta_O \approx$ 20 000 cm^{-1} for this complex, which corresponds to about 2.5 eV.

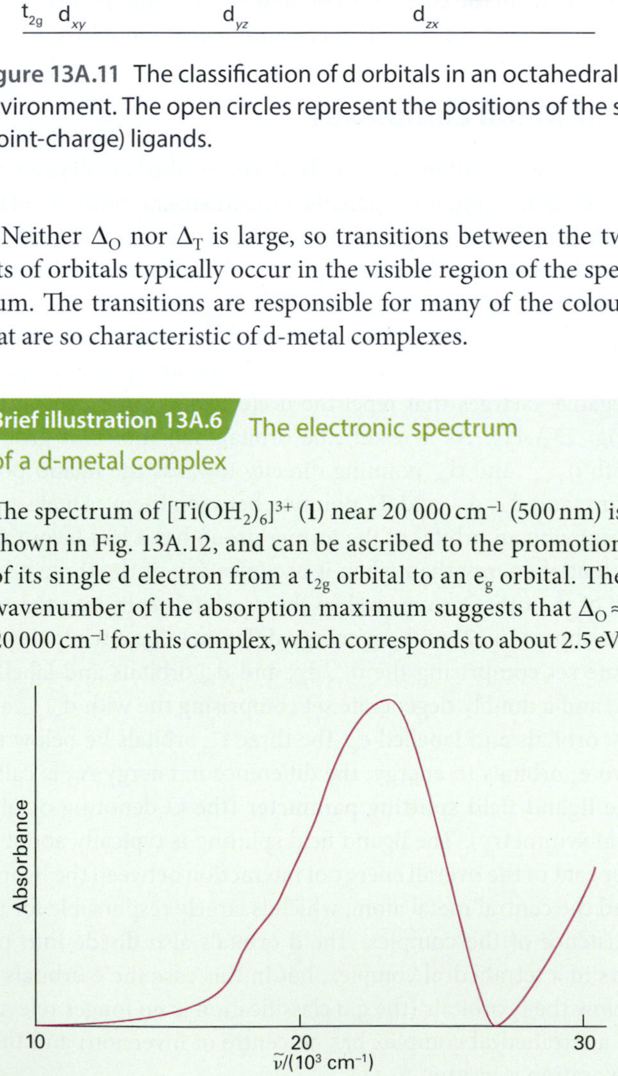

Figure 13A.12 The electronic absorption spectrum of $[Ti(OH_2)_6]^{3+}$ in aqueous solution.

Self-test 13A.8 Can a complex of the Zn^{2+} ion have a d−d electronic transition? Explain your answer.

Answer: No; all five d orbitals are fully occupied

According to the Laporte rule (Section 13A.1b), d−d transitions are parity-forbidden in octahedral complexes because they are g→g transitions (more specifically $e_g \leftarrow t_{2g}$ transitions). However, d−d transitions become weakly allowed as **vibronic transitions**, joint vibrational and electronic transitions, as a result of coupling to asymmetrical vibrations such as that shown in Fig. 13A.4.

A d-metal complex may also absorb radiation as a result of the transfer of an electron from the ligands into the d orbitals of the central atom, or vice versa. In such **charge-transfer transitions** the electron moves through a considerable distance, which means that the transition dipole moment may be large and the absorption correspondingly intense. In the permanganate ion, MnO_4^-, the charge redistribution that accompanies the migration of an electron from the O atoms to the central Mn atom results in strong transition in the range 420–700 nm that accounts for the intense purple colour of the ion. Such an electronic migration from the ligands to the metal corresponds to a **ligand-to-metal charge-transfer transition** (LMCT). The reverse migration, a **metal-to-ligand charge-transfer transition** (MLCT), can also occur. An example is the migration of a d electron onto the antibonding π orbitals of an aromatic ligand. The resulting excited state may have a very long lifetime if the electron is extensively delocalized over several aromatic rings.

In common with other transitions, the intensities of charge-transfer transitions are proportional to the square of the transition dipole moment. We can think of the transition moment as a measure of the distance moved by the electron as it migrates from metal to ligand or vice versa, with a large distance of migration corresponding to a large transition dipole moment and therefore a high intensity of absorption. However, because the integrand in the transition dipole is proportional to the product of the initial and final wavefunctions, it is zero unless the two wavefunctions have nonzero values in the same region of space. Therefore, although large distances of migration favour high intensities, the diminished overlap of the initial and final wavefunctions for large separations of metal and ligands favours low intensities (see Problem 13A.9).

(b) $\pi^* \leftarrow \pi$ and $\pi^* \leftarrow n$ transitions

Absorption by a C=C double bond results in the excitation of a π electron into an antibonding π* orbital (Fig. 13A.13). The chromophore activity is therefore due to a $\boldsymbol{\pi^* \leftarrow \pi}$ **transition** (which is normally read 'π to π-star transition'). Its energy is about 7 eV for an unconjugated double bond, which corresponds to an absorption at 180 nm (in the ultraviolet). When the double bond is part of a conjugated chain, the energies of the molecular orbitals lie closer together and the $\pi^* \leftarrow \pi$ transition moves to longer wavelength; it may even lie in the visible region if the conjugated system is long enough.

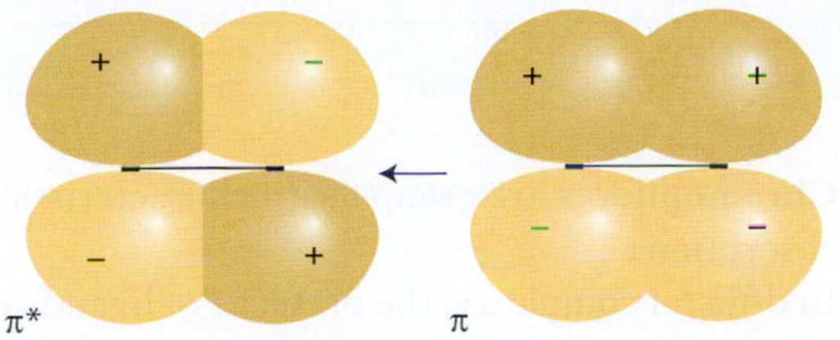

Figure 13A.13 A C=C double bond acts as a chromophore. One of its important transitions is the $\pi^* \leftarrow \pi$ transition illustrated here, in which an electron is promoted from a π orbital to the corresponding antibonding orbital.

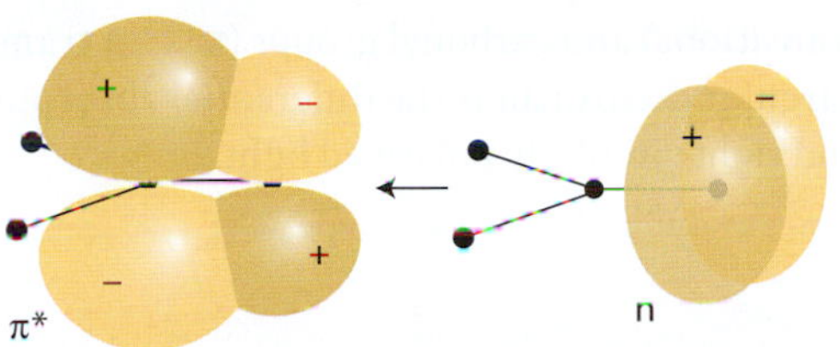

Figure 13A.14 A carbonyl group (C=O) acts as a chromophore partly on account of the excitation of a nonbonding O lone-pair electron to an antibonding CO π orbital.

One of the transitions responsible for absorption in carbonyl compounds can be traced to the lone pairs of electrons on the O atom. The Lewis concept of a 'lone pair' of electrons is represented in molecular orbital theory by a pair of electrons in an orbital confined largely to one atom and not appreciably involved in bond formation. One of these electrons may be excited into an empty π* orbital of the carbonyl group (Fig. 13A.14), which gives rise to an **$\pi^* \leftarrow$ n transition** (an 'n to π-star transition'). Typical absorption energies are about 4eV (290 nm). Because $\pi^* \leftarrow$ n transitions in carbonyls are symmetry forbidden, the absorptions are weak. By contrast, the $\pi^* \leftarrow \pi$ transition in a carbonyl, which corresponds to excitation of a π electron of the C=O double bond, is allowed by symmetry and results in relatively strong absorption.

Brief illustration 13A.7 $\pi^* \leftarrow \pi$ and $\pi^* \leftarrow$ n transitions

The compound $CH_3CH{=}CHCHO$ has a strong absorption in the ultraviolet at 46 950 cm^{-1} (213 nm) and a weak absorption at 30 000 cm^{-1} (330 nm). The former is a $\pi^* \leftarrow \pi$ transition associated with the delocalized π system C=C—C=O. Delocalization extends the range of the C=O $\pi^* \leftarrow \pi$ transition to lower wavenumbers (longer wavelengths). The latter is an $\pi^* \leftarrow$ n transition associated with the carbonyl chromophore.

Self-test 13A.9 Account for the observation that propanone (acetone, $(CH_3)_2CO$) has a strong absorption at 189 nm and a weaker absorption at 280 nm.

Answer: Both transitions are associated with the C=O chromophore, with the weaker being an $\pi^* \leftarrow$ n transition and the stronger a $\pi^* \leftarrow \pi$ transition.

(c) Circular dichroism

Electronic spectra can reveal additional details of molecular structure when experiments are conducted with **polarized light**, electromagnetic radiation with electric and magnetic fields that oscillate only in certain directions. A mode of polarization is **circular polarization**, in which the electric and magnetic fields rotate around the direction of propagation in either a clockwise or a counter-clockwise sense but remain perpendicular to it and each other (Fig. 13A.15). Chiral molecules exhibit **circular dichroism**, meaning that they absorb left and right circularly polarized light to different extents. For example, the circular dichroism (CD) spectra of the enantiomeric pairs of chiral d-metal complexes are distinctly different, whereas there is little difference between their absorption spectra (Fig. 13A.16).

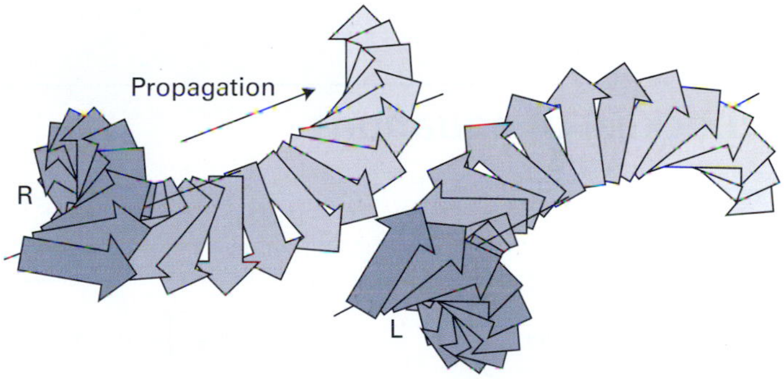

Figure 13A.15 In circularly polarized light, the electric field rotates at different angles around the direction of propagation. The arrays of arrows in these illustrations show the view of the electric field (a) right-circularly polarized, (b) left-circularly polarized light.

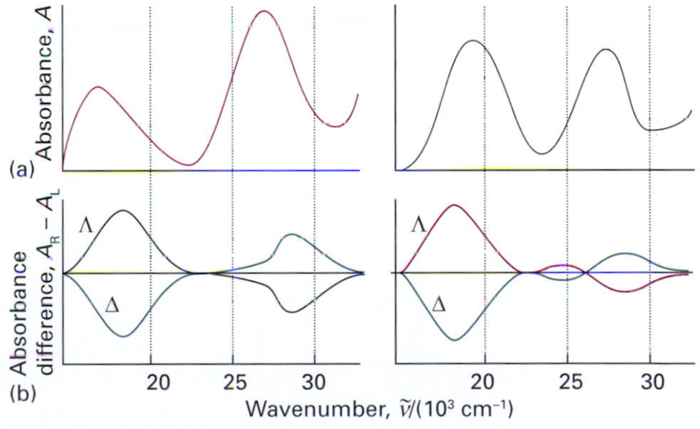

Figure 13A.16 (a) The absorption spectra of two isomers, denoted mer and fac, of $[Co(ala)_3]$, where ala is the conjugate base of alanine, and (b) the corresponding CD spectra. The left- and right-handed forms of these isomers give identical absorption spectra. However, the CD spectra are distinctly different, and the absolute configurations (denoted Λ and Δ) have been assigned by comparison with the CD spectra of a complex of known absolute configuration.

Checklist of concepts

- ☐ 1. The term symbols of diatomic molecules express the components of electronic angular momentum around the internuclear axis.
- ☐ 2. Selection rules for electronic transitions are based on considerations of angular momentum and symmetry.
- ☐ 3. The **Laporte selection rule** states that, for centrosymmetric molecules, only u→g and g→u transitions are allowed.
- ☐ 4. The **Franck–Condon principle** provides a basis for explaining the vibrational structure of electronic transitions.
- ☐ 5. In gas phase samples, rotational structure is present too and can give rise to **band heads**.
- ☐ 6. **Chromophores** are groups with characteristic optical absorptions.
- ☐ 7. In d-metal complexes, the presence of ligands removes the degeneracy of d orbitals and vibrationally allowed **d–d transitions** can occur between them.
- ☐ 8. **Charge-transfer transitions** typically involve the migration of electrons between the ligands and the central metal atom.
- ☐ 9. Other chromophores include double bonds (**$\pi^* \leftarrow \pi$ transitions**) and carbonyl groups (**$\pi^* \leftarrow$ n transitions**).
- ☐ 10. **Circular dichroism** is the differential absorption of left and right circularly polarized light.

Checklist of equations

Property	Equation	Comment	Equation number
Selection rules (angular momentum)	$\Delta\Lambda=0, \pm1; \Delta S=0; \Delta\Sigma=0; \Delta\Omega=0, \pm1$	Linear molecules	13A.4
Franck–Condon factor	$\lvert S(v_f, v_i)\rvert^2 = \left(\int \psi^*_{v,f}\psi_{v,i}\mathrm{d}\tau_n\right)^2$	Assumes Franck–Condon principle applies	13A.6
Rotational structure of electronic spectra (diatomic molecules)	$\tilde{\nu}_P(J)=\tilde{\nu}-(\tilde{B}'+\tilde{B})J+(\tilde{B}'-\tilde{B})J^2$	P branch ($\Delta J=-1$)	13A.8a
	$\tilde{\nu}_Q(J)=\tilde{\nu}+(\tilde{B}'-\tilde{B})J(J+1)$	Q branch ($\Delta J=0$)	13A.8b
	$\tilde{\nu}_R(J)=\tilde{\nu}+(\tilde{B}'+\tilde{B})(J+1)+(\tilde{B}'-\tilde{B})(J+1)^2$	R branch ($\Delta J=+1$)	13A.8c

13B Decay of excited states

Contents

➤ Why do you need to know this material?

Considerable information about the electronic structure of a molecule can be obtained from the photons emitted when excited electronic states decay radiatively back to the ground state.

➤ What is the key idea?

Molecules in excited electronic states discard their excess energy by emission of electromagnetic radiation, transfer as heat to the surroundings, or fragmentation.

➤ What do you need to know already?

You need to be familiar with electronic transitions in molecules (Topic 13A), the difference between spontaneous and stimulated emission of radiation (Topic 12A), and the general features of spectroscopy (Topic 12A). You need to be aware of the difference between singlet and triplet states (Topic 9C) and of the Franck–Condon principle (Topic 13A).

A **radiative decay process** is a process in which a molecule discards its excitation energy as a photon (Topic 12A). In this Topic we pay particular attention to spontaneous radiative decay processes, which include fluorescence and phosphorescence. A more common fate of an electronically excited molecule is **non-radiative decay**, in which the excess energy is transferred into the vibration, rotation, and translation of the surrounding molecules. This thermal degradation converts the excitation energy into thermal motion of the environment (that is, to 'heat'). An excited molecule may also dissociate or take part in a chemical reaction (Topic 20G).

13B.1 Fluorescence and phosphorescence

In **fluorescence**, spontaneous emission of radiation occurs while the sample is being irradiated and ceases within nanoseconds to milliseconds of the exciting radiation being extinguished (Fig. 13B.1). In **phosphorescence**, the spontaneous emission may persist for long periods (even hours, but characteristically seconds or fractions of seconds). The difference suggests that fluorescence is a fast conversion of absorbed radiation into re-emitted energy, and that phosphorescence involves the storage of energy in a reservoir from which it slowly leaks.

Figure 13B.2 shows the sequence of steps involved in fluorescence of chromophores in solution. The initial stimulated absorption takes the molecule to an excited electronic state, and if the absorption spectrum were monitored it would look like the one shown in Fig. 13B.3a. The excited molecule is subjected to collisions with the surrounding molecules, and as it gives up energy nonradiatively it steps down (typically within picoseconds) the ladder of vibrational levels to the lowest vibrational level of the electronically excited molecular state. The surrounding molecules, however, might now be unable to accept the larger energy difference needed to lower the molecule to the ground electronic state. It might therefore survive long enough to undergo spontaneous emission and emit the remaining excess energy as radiation. The downward electronic transition is vertical, in accord with the Franck–Condon principle (Topic

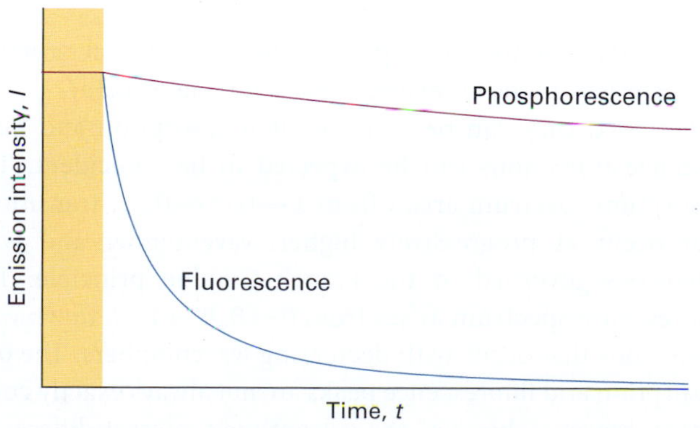

Figure 13B.1 The empirical (observation-based) distinction between fluorescence and phosphorescence is that the former is extinguished very quickly after the exciting source is removed, whereas the latter continues with relatively slowly diminishing intensity.

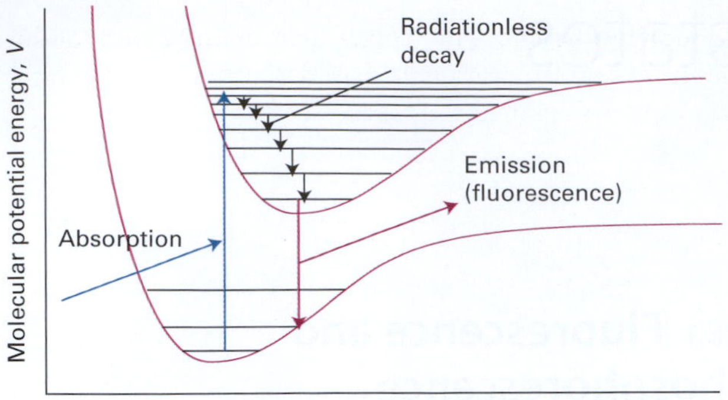

Figure 13B.2 The sequence of steps leading to fluorescence by chromophores in solution. After the initial absorption, the upper vibrational states undergo radiationless decay by giving up energy to the surrounding molecules. A radiative transition then occurs from the vibrational ground state of the upper electronic state.

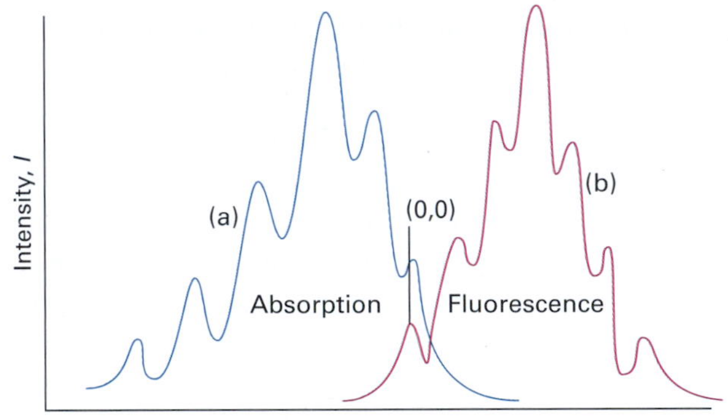

Figure 13B.3 An absorption spectrum (a) shows a vibrational structure characteristic of the upper state. A fluorescence spectrum (b) shows a structure characteristic of the lower state; it is also displaced to lower frequencies (but the 0–0 transitions are coincident) and resembles a mirror image of the absorption.

13A), and the fluorescence spectrum has a vibrational structure characteristic of the *lower* electronic state (Fig. 13B.3b).

Provided they can be seen, the 0–0 absorption and fluorescence transitions can be expected to be coincident. The absorption spectrum arises from $1 \leftarrow 0$, $2 \leftarrow 0$, … transitions that occur at progressively higher wavenumber and with intensities governed by the Franck–Condon principle. The fluorescence spectrum arises from $0 \rightarrow 0$, $0 \rightarrow 1$, … *downward* transitions that occur with decreasing wavenumbers. The 0–0 absorption and fluorescence peaks are not always exactly coincident, however, because the solvent may interact differently with the solute in the ground and excited states (for instance, the hydrogen bonding pattern might differ). Because the solvent molecules do not have time to rearrange during the transition, the absorption occurs in an environment characteristic

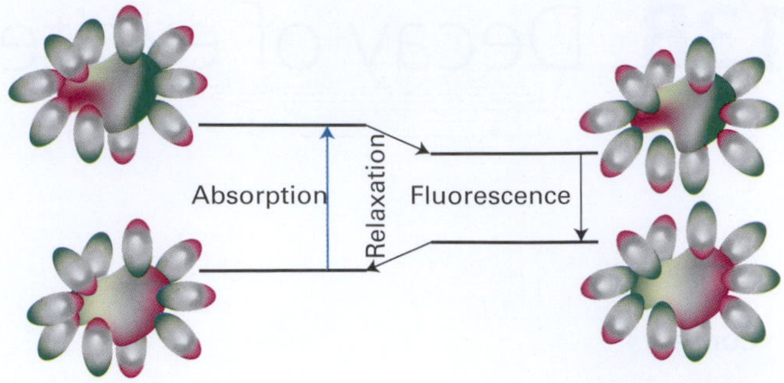

Figure 13B.4 The solvent can shift the fluorescence spectrum relative to the absorption spectrum. On the left we see that the absorption occurs with the solvent (depicted by the ellipses) in the arrangement characteristic of the ground electronic state of the molecule (the sphere). However, before fluorescence occurs, the solvent molecules relax into a new arrangement, and that arrangement is preserved during the subsequent radiative transition.

of the solvated ground state; however, the fluorescence occurs in an environment characteristic of the solvated excited state (Fig. 13B.4).

Fluorescence occurs at lower frequencies (longer wavelengths) than the incident radiation because the emissive transition occurs after some vibrational energy has been discarded into the surroundings. The vivid oranges and greens of fluorescent dyes are an everyday manifestation of this effect: they absorb in the ultraviolet and blue, and fluoresce in the visible. The mechanism also suggests that the intensity of the fluorescence ought to depend on the ability of the solvent molecules to accept the electronic and vibrational quanta. It is indeed found that a solvent composed of molecules with widely spaced vibrational levels (such as water) can in some cases accept the large quantum of electronic energy and so extinguish, or 'quench', the fluorescence. The rate at which fluorescence is quenched by other molecules also gives valuable kinetic information (Topic 20G).

Figure 13B.5 shows the sequence of events leading to phosphorescence for a molecule with a singlet ground state. The first steps are the same as in fluorescence, but the presence of a triplet excited state at an energy close to that of the singlet excited state plays a decisive role. The singlet and triplet excited states share a common geometry at the point where their potential energy curves intersect. Hence, if there is a mechanism for unpairing two electron spins (and achieving the conversion of $\uparrow\downarrow$ to $\uparrow\uparrow$), the molecule may undergo **intersystem crossing**, a non-radiative transition between states of different multiplicity, and become a triplet state. As in the discussion of atomic spectra (Topic 9C), singlet–triplet transitions may occur in the presence of spin–orbit coupling. Intersystem crossing is expected to be important when a molecule contains a moderately heavy atom (such as sulfur), because then the spin–orbit coupling is large.

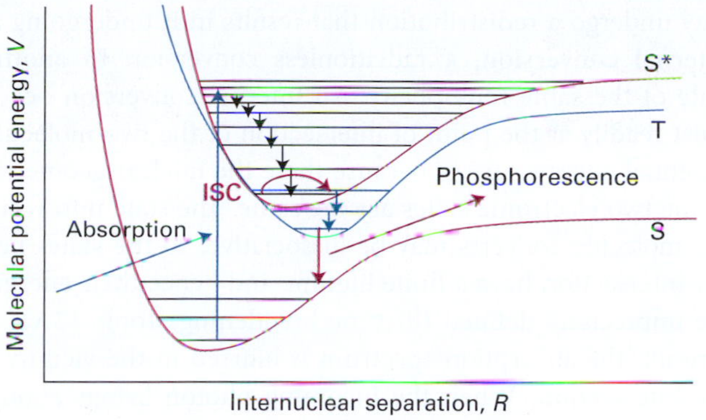

Figure 13B.5 The sequence of steps leading to phosphorescence. The important step is the intersystem crossing (ISC), the switch from a singlet state to a triplet state brought about by spin–orbit coupling. The triplet state acts as a slowly radiating reservoir because the return to the ground state is spin-forbidden.

If an excited molecule crosses into a triplet state, it continues to discard energy into the surroundings. However, it is now stepping down the triplet's vibrational ladder, and at the lowest energy level it is trapped because the triplet state is at a lower energy than the corresponding singlet (Hund's rule, Topic 9B). The solvent cannot absorb the final, large quantum of electronic excitation energy, and the molecule cannot radiate its energy because return to the ground state is spin-forbidden. The radiative transition, however, is not totally forbidden because the spin–orbit coupling that was responsible for the intersystem crossing also breaks the selection rule. The molecules are therefore able to emit weakly, and the emission may continue long after the original excited state was formed.

The mechanism accounts for the observation that the excitation energy seems to get trapped in a slowly leaking reservoir. It also suggests (as is confirmed experimentally) that phosphorescence should be most intense from solid samples: energy transfer is then less efficient and intersystem crossing has time to occur as the singlet excited state steps slowly past the intersection point. The mechanism also suggests that the phosphorescence efficiency should depend on the presence of a moderately heavy atom (with strong spin–orbit coupling), which is in fact the case.

The various types of non-radiative and radiative transitions that can occur in molecules are often represented on a schematic **Jablonski diagram** of the type shown in Fig. 13B.6.

Brief illustration 13B.1 Fluorescence and phosphorescence of organic molecules

Fluorescence efficiency decreases, and the phosphorescence efficiency increases, in the series of compounds: naphthalene, 1-chloronaphthalene, 1-bromonaphthalene, 1-iodonaphthalene. The replacement of an H atom by successively heavier atoms enhances both intersystem crossing from the first excited singlet state to the first excited triplet state (thereby decreasing the efficiency of fluorescence) and the radiative transition from the first excited triplet state to the ground singlet state (thereby increasing the efficiency of phosphorescence).

Self-test 13B.1 Consider an aqueous solution of a chromophore that fluoresces strongly. Is the addition of iodide ion to the solution likely to increase or decrease the efficiency of phosphorescence the chromophore?

Answer: increase

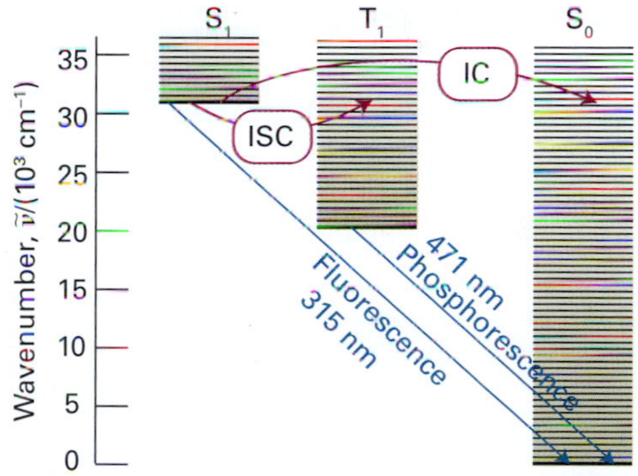

Figure 13B.6 A Jablonski diagram (here, for naphthalene) is a simplified portrayal of the relative positions of the electronic energy levels of a molecule. Vibrational levels of states of a given electronic state lie above each other, but the relative horizontal locations of the columns bear no relation to the nuclear separations in the states. The ground vibrational states of each electronic state are correctly located vertically but the other vibrational states are shown only schematically. (IC: internal conversion; ISC: intersystem crossing.)

13B.2 Dissociation and predissociation

Another fate for an electronically excited molecule is **dissociation**, the breaking of bonds (Fig. 13B.7). The onset of dissociation can be detected in an absorption spectrum by seeing that the vibrational structure of a band terminates at a certain energy. Absorption occurs in a continuous band above this **dissociation limit** because the final state is an unquantized translational motion of the fragments. Locating the dissociation limit is a valuable way of determining the bond dissociation energy.

In some cases, the vibrational structure disappears but resumes at higher photon energies. This effect provides evidence of **predissociation**, which can be interpreted in terms of the molecular potential energy curves shown in Fig. 13B.8. When a molecule is excited to a vibrational level, its electrons

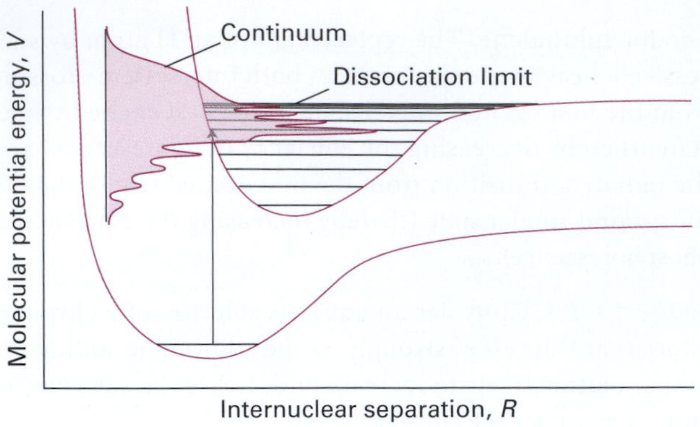

Figure 13B.7 When absorption occurs to unbound states of the upper electronic state, the molecule dissociates and the absorption is a continuum. Below the dissociation limit the electronic spectrum shows a normal vibrational structure.

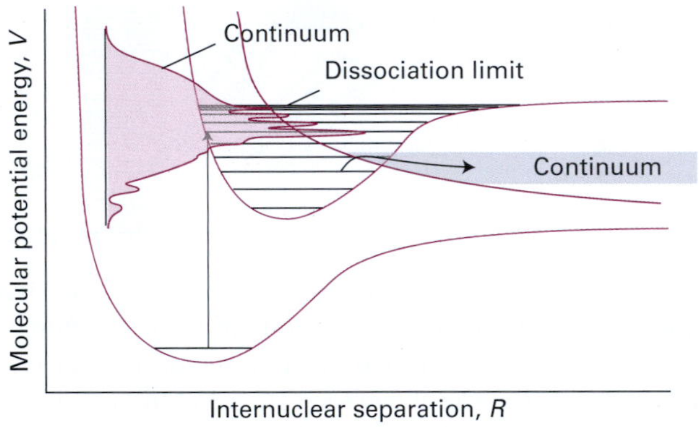

Figure 13B.8 When a dissociative state crosses a bound state, molecules excited to levels near the crossing may dissociate. This process is called predissociation, and is detected in the spectrum as a loss of vibrational structure that resumes at higher frequencies.

may undergo a redistribution that results in it undergoing an **internal conversion**, a radiationless conversion to another state of the same multiplicity. An internal conversion occurs most readily at the point of intersection of the two molecular potential energy curves, because there the nuclear geometries of the two electronic states are the same. The state into which the molecule converts may be dissociative, so the states near the intersection have a finite lifetime and hence their energies are imprecisely defined (lifetime broadening, Topic 12A). As a result, the absorption spectrum is blurred in the vicinity of the intersection. When the incoming photon brings enough energy to excite the molecule to a vibrational level high above the intersection, the internal conversion does not occur (the nuclei are unlikely to have the same geometry). Consequently, the levels resume their well-defined, vibrational character with correspondingly well-defined energies, and the line structure resumes on the high-frequency side of the blurred region.

Brief illustration 13B.2 The effect of predissociation on an electronic spectrum

The O_2 molecule absorbs ultraviolet radiation in a transition from its ${}^3\Sigma_g^-$ ground electronic state to a ${}^3\Sigma_u^-$ excited state that is energetically close to a dissociative ${}^3\Pi_u$ state. In this case, the effect of predissociation is more subtle than the abrupt loss of vibrational–rotational structure in the spectrum; instead, the vibrational structure simply broadens rather than being lost completely. As before, the broadening is explained by short lifetimes of the excited vibrational states near the intersection of the curves describing the bound and dissociative excited electronic states.

Self-test 13B.2 What can be estimated from the wavenumber of onset of predissociation?

Answer: See Fig. 13B.8; an upper limit on the dissociation energy of the ground electronic state

Checklist of concepts

- ☐ 1. **Fluorescence** is radiative decay between states of the same multiplicity; it ceases as soon as the exciting source is removed.
- ☐ 2. **Phosphorescence** is radiative decay between states of different multiplicity; it persists after the exciting radiation is removed.
- ☐ 3. **Intersystem crossing** is the non-radiative conversion to a state of different multiplicity.
- ☐ 4. A **Jablonski diagram** is a schematic diagram of the types of non-radiative and radiative transitions that can occur in molecules.
- ☐ 5. An additional fate of an electronically excited species is **dissociation**.
- ☐ 6. **Internal conversion** is a non-radiative conversion to a state of the same multiplicity.
- ☐ 7. **Predissociation** is the observation of the effects of dissociation before the dissociation limit is reached.

13C Lasers

Contents

➤ Why do you need to know this material?

Radiative decay has great technological importance: lasers have brought unprecedented precision to spectroscopy and are used in medicine, telecommunications, and many aspects of everyday life.

➤ What is the key idea?

Laser action is the stimulated emission of coherent radiation taking place between states related by a population inversion.

➤ What do you need to know already?

You need to be familiar with electronic transitions in molecules (Topic 13A), the difference between spontaneous and stimulated emission of radiation (Topic 12A), and the general features of spectroscopy (Topics 12A and 13B).

The word laser is an acronym formed from light amplification by stimulated emission of radiation. In stimulated emission (Topic 12A), an excited state is stimulated to emit a photon by radiation of the same frequency: the more photons that are present, the greater the probability of the emission. The essential feature of laser action is positive-feedback: the greater the number of photons present of the appropriate frequency, the greater the rate at which even more photons of that frequency will be stimulated to form.

Laser radiation has a number of striking characteristics (Table 13C.1). Each of them (sometimes in combination with the others) opens up interesting opportunities in physical chemistry. Raman spectroscopy has flourished on account of the high intensity monochromatic radiation available from lasers (Topic 12A), and the ultra-short pulses that lasers can generate make possible the study of light-initiated reactions on timescales of femtoseconds and even attoseconds.

13C.1 Population inversion

One requirement of laser action is the existence of a **metastable excited state**, an excited state with a long enough lifetime for it to participate in stimulated emission. Another requirement is the existence of a greater population in the metastable state than in the lower state where the transition terminates, for

Table 13C.1 Characteristics of laser radiation and their chemical applications

Characteristic	Advantage	Application
High power	Multiphoton process	Spectroscopy
	Low detector noise	Improved sensitivity
	High scattering intensity	Raman spectroscopy (Topics 12C–12 E)
Monochromatic	High resolution	Spectroscopy
	State selection	Photochemical studies (Topic 20G)
		Reaction dynamics (Topic 21D)
Collimated beam	Long path lengths	Improved sensitivity
	Forward-scattering observable	Raman spectroscopy (Topics 12C–12E)
Coherent	Interference between separate beams	CARS (Topic 12E)
Pulsed	Precise timing of excitation	Fast reactions (Topics 13C, 20G, and 21C)
		Relaxation (Topic 20C)
		Energy transfer (Topic 20C)

then there will be a net emission of radiation. Because at thermal equilibrium the opposite is true, it is necessary to achieve a **population inversion** in which there are more molecules in the upper state than in the lower.

One way of achieving population inversion is illustrated in Fig. 13C.1. The molecule is excited to an intermediate state I, which then gives up some of its energy non-radiatively and changes into a lower state A; the laser transition is the return of A to the ground state X. Because three energy levels are involved overall, this arrangement leads to a **three-level laser**. In practice, I consists of many states, all of which can convert to the upper of the two laser states A. The I←X transition is stimulated with an intense flash of light in the process called **pumping**. The pumping is often achieved with an electric discharge through xenon or with the light of another laser. The conversion of I to A should be rapid, and the laser transitions from A to X should be relatively slow.

The disadvantage of the three-level arrangement is that it is difficult to achieve population inversion, because so many ground-state molecules must be converted to the excited state by the pumping action. The arrangement adopted in a **four-level laser** simplifies this task by having the laser transition terminate in a state A′ other than the ground state (Fig. 13C.2).

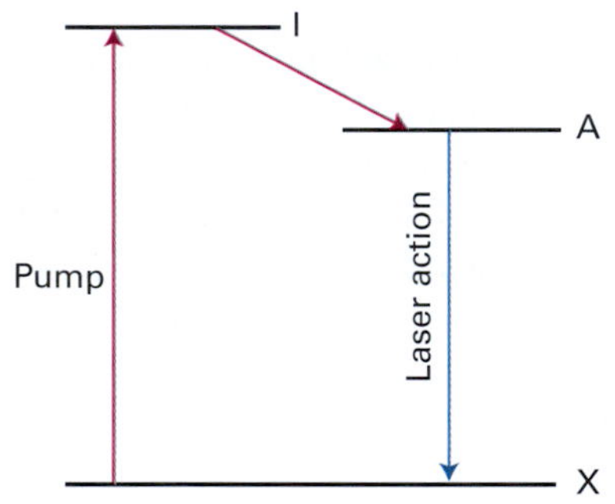

Figure 13C.1 The transitions involved in one kind of three-level laser. The pumping pulse populates the intermediate state I, which in turn populates the metastable state A. The laser transition is the stimulated emission A→X.

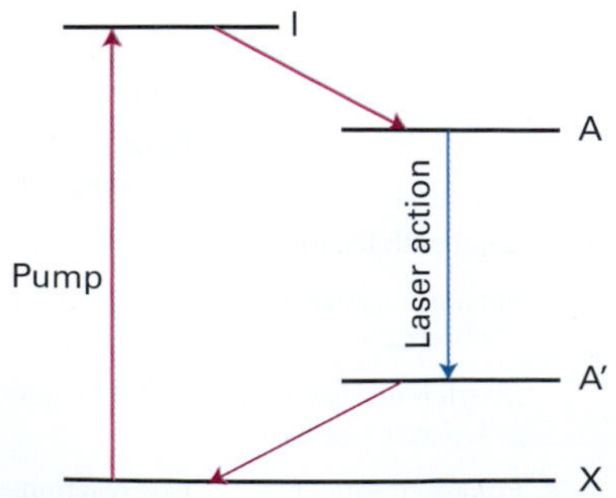

Figure 13C.2 The transitions involved in a four-level laser. Because the laser transition terminates in an excited state (A′), the population inversion between A and A′ is much easier to achieve.

Because A′ is unpopulated initially, any population in A corresponds to a population inversion and we can expect laser action if A is sufficiently metastable. Moreover, this population inversion can be maintained if the A′→X transitions are rapid, for these transitions will deplete any population in A′ that stems from the laser transition, and keep the state A′ relatively empty.

Brief illustration 13C.1 Simple lasers

The ruby laser is an example of a three-level laser (Fig. 13C.3). Ruby is Al_2O_3 containing a small proportion of Cr^{3+} ions. The lower level of the laser transition is the $^4A_2$ ground state of the Cr^{3+} ion. The process of pumping a majority of the Cr^{3+} ions into the $^4T_2$ and $^4T_1$ excited states is followed by a radiationless transition to the ^2E excited state. The laser transition is $^2E \rightarrow {}^4A_2$, and gives rise to red 694 nm radiation.

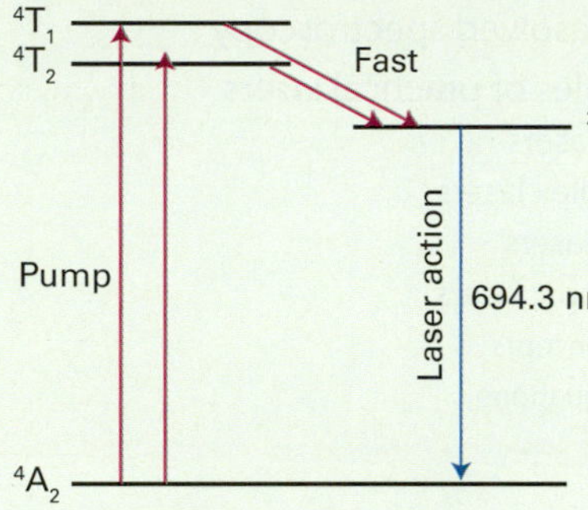

Figure 13C.3 The transitions involved in a ruby laser.

The neodymium laser is an example of a four-level laser (Fig 13C.4). In one form it consists of Nd^{3+} ions at low concentration in yttrium aluminium garnet (YAG, specifically $Y_3Al_5O_{12}$), and is then known as a Nd:YAG laser. A neodymium laser operates at a number of wavelengths in the infrared, the band at 1064 nm being most common.

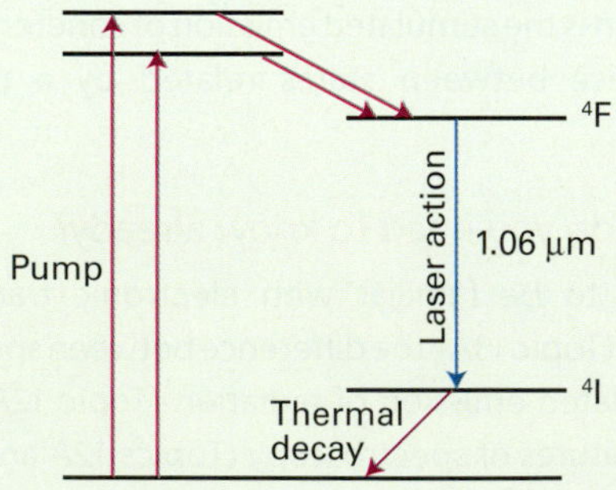

Figure 13C.4 The transitions involved in a neodymium laser.

Self-test 13C.1 In the arrangement discussed here, does a ruby laser generate pulses of light or a continuous beam of light?

Answer: Pulses

13C.2 Cavity and mode characteristics

The laser medium is confined to a cavity that ensures that only certain photons of a particular frequency, direction of travel, and state of polarization are generated abundantly. The cavity is essentially a region between two mirrors, which reflect the light back and forth. This arrangement can be regarded as a version of the particle in a box, with the particle now being a photon. As in the treatment of a particle in a box (Topic 8A), the only wavelengths that can be sustained satisfy

$$n\times\tfrac{1}{2}\lambda = L \qquad \text{Resonant modes} \qquad (13C.1)$$

where n is an integer and L is the length of the cavity. That is, only an integral number of half-wavelengths fit into the cavity; all other waves undergo destructive interference with themselves. In addition, not all wavelengths that can be sustained by the cavity are amplified by the laser medium (many fall outside the range of frequencies of the laser transitions), so only a few contribute to the laser radiation. These wavelengths are the **resonant modes** of the laser.

Brief illustration 13C.2 Resonant modes

It follows from eqn 13C.1 that the frequencies of the resonant modes are $\nu = c/\lambda = (c/2L)\times n$. For a laser cavity of length 30.0 cm, the allowed frequencies are

$$\nu = \frac{\overbrace{2.998\times10^{8}\,\mathrm{m\,s^{-1}}}^{c}}{2\times\underbrace{(0.300\,\mathrm{m})}_{L}}\times n = (5.00\times10^{8}\,\mathrm{s^{-1}})\times n = (500\,\mathrm{MHz})\times n$$

with $n=1, 2, \ldots$, and therefore $\nu = 500$ MHz, 1000 MHz, ….

Self-test 13C.2 Consider a laser cavity of length 1.0 m. What is the frequency difference between successive resonant modes?

Answer: 150 MHz

Photons with the correct wavelength for the resonant modes of the cavity and the correct frequency to stimulate the laser transition are highly amplified. One photon might be generated spontaneously and travel through the medium. It stimulates the emission of another photon, which in turn stimulates more (Fig. 13C.5). The cascade of energy builds up rapidly, and soon the cavity is an intense reservoir of radiation at all the resonant modes it can sustain. Some of this radiation can be withdrawn if one of the mirrors is partially transmitting.

The resonant modes of the cavity have various natural characteristics, and to some extent may be selected. Only photons that are travelling strictly parallel to the axis of the cavity undergo more than a couple of reflections, so only they are amplified, all others simply vanishing into the surroundings. Hence, laser light generally forms a beam with very low divergence. It may also be polarized, with its electric vector in a particular plane (or in some other state of polarization), by including a polarizing filter into the cavity or by making use of polarized transitions in a solid medium.

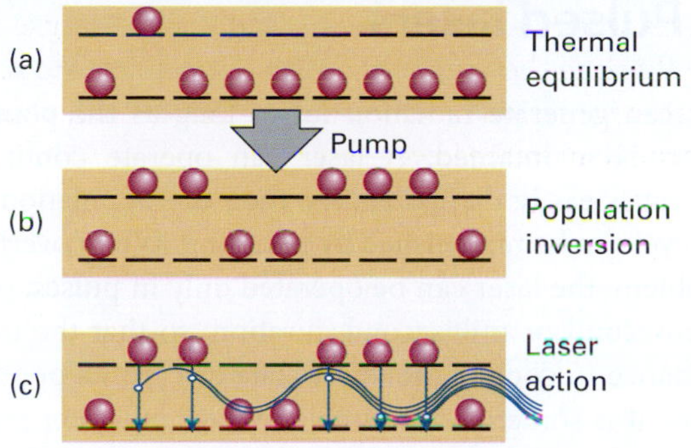

Figure 13C.5 A schematic illustration of the steps leading to laser action. (a) The Boltzmann population of states, with more atoms in the ground state. (b) When the initial state absorbs, the populations are inverted (the atoms are pumped to the excited state). (c) A cascade of radiation then occurs, as one emitted photon stimulates another atom to emit, and so on. The radiation is coherent (phases in step).

Laser radiation is **coherent** in the sense that the electromagnetic waves are all in step. In **spatial coherence** the waves are in step across the cross-section of the beam emerging from the cavity. In **temporal coherence** the waves remain in step along the beam. The former is normally expressed in terms of a **coherence length**, l_C, the distance across the beam over which the waves remain coherent, and is related to the range of wavelengths, $\Delta\lambda$, present in the beam:

$$l_C = \frac{\lambda^2}{2\Delta\lambda} \qquad \text{Coherence length} \qquad (13C.2)$$

When many wavelengths are present, and $\Delta\lambda$ is large, the waves get out of step in a short distance and the coherence length is small.

Brief illustration 13C.3 Coherence length

A typical light bulb gives out light with a coherence length of only about 400 nm. By contrast, a He–Ne laser with $\lambda = 633$ nm and $\Delta\lambda = 2.0$ pm has a coherence length of

$$l_C = \frac{\overbrace{(633\,\mathrm{nm})^2}^{\lambda^2}}{2\times\underbrace{(0.0020\,\mathrm{nm})}_{\Delta\lambda}} = 1.0\times10^{8}\,\mathrm{nm} = 0.10\,\mathrm{m} = 10\,\mathrm{cm}$$

Self-test 13C.3 What is the condition that would lead to an infinite coherence length?

Answer: A perfectly monochromatic beam, or $\Delta\lambda = 0$

13C.3 Pulsed lasers

A laser can generate radiation for as long as the population inversion is maintained. A laser can operate continuously when heat is easily dissipated, for then the population of the upper level can be replenished by pumping. When overheating is a problem, the laser can be operated only in pulses, perhaps of microsecond or millisecond duration, so that the medium has a chance to cool or the lower state discard its population. However, it is sometimes desirable to have pulses of radiation rather than a continuous output, with a lot of power concentrated into a brief pulse. One way of achieving pulses is by **Q-switching**, the modification of the resonance characteristics of the laser cavity. The name comes from the 'Q-factor' used as a measure of the quality of a resonance cavity in microwave engineering.

Example 13C.1 Relating the power and energy of a laser

A certain laser can generate radiation in 3.0 ns pulses, each of which delivers an energy of 0.10 J, at a pulse repetition frequency of 10 Hz. Assuming that the pulses are rectangular, calculate the peak power and the average power of this laser.

Method Power is the energy released in an interval divided by the duration of the interval, and is expressed in watts ($1\,\mathrm{W}=1\,\mathrm{J\,s^{-1}}$). The peak power, P_{peak}, of a rectangular pulse is defined as the energy delivered in a pulse divided by its duration. The average power, P_{average}, is the total energy delivered by a large number of pulses divided by the duration of the time interval over which that total energy is measured. If each pulse delivers an energy E_{pulse} and in an interval Δt there are N pulses, the total energy delivered is NE_{pulse} and the average power is $P_{\text{average}}=NE_{\text{pulse}}/\Delta t$. However, $\Delta t/N$ is the interval between pulses and therefore the inverse of the pulse repetition frequency, $\nu_{\text{repetition}}$. It follows that $P_{\text{average}}=E_{\text{pulse}}\nu_{\text{repetition}}$.

Answer From the data,

$$P_{\text{peak}}=\frac{0.10\,\mathrm{J}}{3.0\times10^{-9}\,\mathrm{s}}=3.3\times10^{7}\,\mathrm{J\,s^{-1}}=33\,\mathrm{MJ\,s^{-1}}=33\,\mathrm{MW}$$

The pulse repetition frequency rate is 10 Hz. It follows that the average power is

$$P_{\text{average}}=0.10\,\mathrm{J}\times10\,\mathrm{s^{-1}}=1.0\,\mathrm{J\,s^{-1}}=1.0\,\mathrm{W}$$

The peak power is much higher than the average power because this laser emits light for only 30 ns during each second of operation.

Self-test 13C.4 Calculate the peak power and average power of a laser with a pulse energy of 2.0 mJ, a pulse duration of 30 ps, and a pulse repetition rate of 38 MHz.

Answer: $P_{\text{peak}}=67\,\mathrm{MW}$, $P_{\text{average}}=76\,\mathrm{kW}$

The aim of Q-switching is to achieve a healthy population inversion in the absence of the resonant cavity, then to plunge the population-inverted medium into a cavity and hence to obtain a sudden pulse of radiation. The switching may be achieved by impairing the resonance characteristics of the cavity in some way while the pumping pulse is active and then suddenly to improve them (Fig. 13C.6). One technique is to use the ability of some crystals to change their optical properties when an electrical potential difference is applied. For example, a crystal of potassium dihydrogenphosphate (KH_2PO_4) rotates the plane of polarization of light to different extents when a potential difference is switched on and off. In this way energy can be stored or released in a laser cavity, resulting in an intense pulse of stimulated emission.

The technique of **mode locking** can produce pulses of picosecond duration and less. A laser radiates at a number of different frequencies, depending on the precise details of the resonance characteristics of the cavity and in particular on the number of half-wavelengths of radiation that can be trapped between the mirrors (the cavity modes). The resonant modes differ in frequency by multiples of $c/2L$ (*Brief illustration* 13C.4). Normally, these modes have random phases relative to each other. However, it is possible to lock their phases together. As we show in the following *Justification*, interference then occurs to give a series of sharp peaks, and the energy of the laser is obtained in short bursts (Fig. 13C.7). More specifically, the intensity, I, of the radiation varies with time as

$$I(t)\propto\mathcal{E}_0^2\,\frac{\sin^2(N\pi ct/2L)}{\sin^2(\pi ct/2L)}\qquad\text{Mode-locked laser output}\qquad(13C.3)$$

where $\mathcal{E}_0$ is the amplitude of the electromagnetic wave describing the laser beam and N is the number of locked modes. This function is shown in Fig. 13C.8. We see that it is a series of peaks with maxima separated by $t=2L/c$, the round-trip transit time of the light in the cavity, and that the peaks become sharper as N is increased. In a laser with a cavity of length 30 cm, the peaks are separated by 2 ns. If 1000 modes contribute, the width of the pulses is 4 ps.

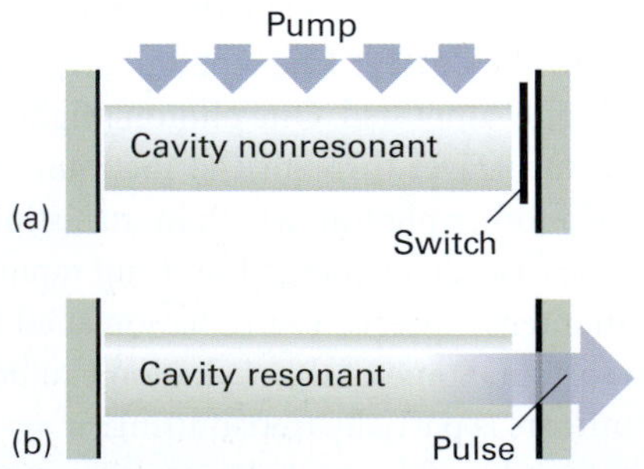

Figure 13C.6 The principle of Q-switching. (a) The excited state is populated while the cavity is non-resonant. (b) Then the resonance characteristics are suddenly restored, and the stimulated emission emerges in a giant pulse.

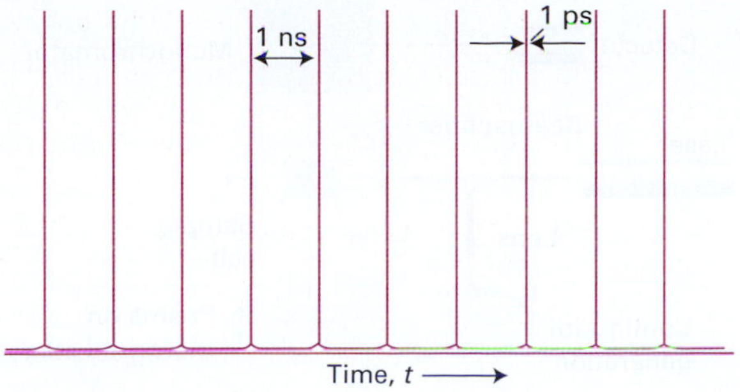

Figure 13C.7 The output of a mode-locked laser consists of a stream of very narrow pulses (here 1 ps in duration) separated by an interval equal to the time it takes for light to make a round trip inside the cavity (here 1 ns).

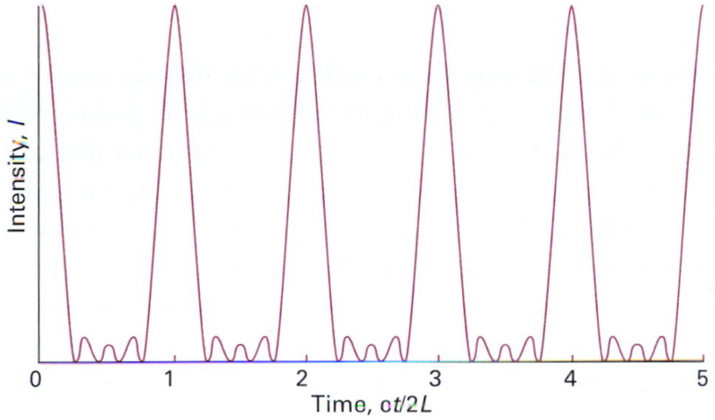

Figure 13C.8 The structure of the pulses generated by a mode-locked laser.

Justification 13C.1 The origin of mode locking

The general expression for a (complex) wave of amplitude $\mathcal{E}_0$ and frequency ω is $\mathcal{E}_0 e^{i\omega t}$. Therefore, each wave that can be supported by a cavity of length L has the form

$$\mathcal{E}_n(t)=\mathcal{E}_0 e^{2\pi i(\nu+nc/2L)t}$$

where ν is the lowest frequency. A wave formed by superimposing N modes with $n=0, 1, \ldots, N-1$ has the form

$$\mathcal{E}(t)=\sum_{n=0}^{N-1}\mathcal{E}_n(t)=\mathcal{E}_0 e^{2\pi i\nu t}\overbrace{\sum_{n=0}^{N-1}e^{i\pi nct/L}}^{S(N)}=\mathcal{E}_0 e^{2\pi i\nu t}S(N)$$

The sum simplifies to:

$$S(N)=1+e^{i\pi ct/L}+e^{2i\pi ct/L}+\cdots+e^{(N-1)i\pi ct/L}$$

The sum of a geometric series is

$$1+e^{x}+e^{2x}+\cdots+e^{(N-1)x}=\frac{e^{Nx}-1}{e^{x}-1}$$

so, with $x=i\pi ct/L$,

$$S(N)=\frac{e^{Ni\pi ct/L}-1}{e^{i\pi ct/L}-1}$$

On multiplication of both the numerator and denominator by $e^{-i\pi ct/2L}$ and a little rearrangement this expression becomes

$$S(N)=\frac{e^{Ni\pi ct/2L}-e^{-Ni\pi ct/2L}}{e^{i\pi ct/2L}-e^{-i\pi ct/2L}}\times e^{(N-1)i\pi ct/2L}$$

At this point we use $\sin x=(1/2i)(e^{ix}-e^{-ix})$, and obtain

$$S(N)=\frac{\sin(N\pi ct/2L)}{\sin(\pi ct/2L)}\times e^{(N-1)i\pi ct/2L}$$

The intensity, $I(t)$, of the radiation is proportional to the square modulus of the total amplitude, so

$$I(t)\propto \mathcal{E}^{*}\mathcal{E}=\mathcal{E}_0^2\frac{\sin^2(N\pi ct/2L)}{\sin^2(\pi ct/2L)}$$

which is eqn 13C.3.

Mode locking is achieved by varying the Q-factor of the cavity periodically at the frequency $c/2L$. The modulation can be pictured as the opening of a shutter in synchrony with the round-trip travel time of the photons in the cavity, so only photons making the journey in that time are amplified. The modulation can be achieved by linking a prism in the cavity to a transducer driven by a radiofrequency source at a frequency $c/2L$. The transducer sets up standing-wave vibrations in the prism and modulates the loss it introduces into the cavity.

Another mechanism for mode-locking lasers is based on the **optical Kerr effect**, which arises from a change in refractive index of a well-chosen medium, the **Kerr medium**, when it is exposed to intense laser pulses. Because a beam of light changes direction when it passes from a region of one refractive index to a region with a different refractive index, changes in refractive index result in the self-focussing of an intense laser pulse as it travels through the Kerr medium (Fig. 13C.9).

To bring about mode-locking, a Kerr medium is included in the laser cavity and next to it is a small aperture. The procedure makes use of the fact that the **gain**, the growth in intensity, of a frequency component of the radiation in the cavity is very sensitive to amplification, and once a particular frequency begins to grow, it can quickly dominate. When the power inside the cavity is low, a portion of the photons will be blocked by the aperture, creating a significant loss. A spontaneous fluctuation in intensity—a bunching of photons—may begin to turn on the optical Kerr effect and the changes in the refractive index of the Kerr medium will result in a **Kerr lens,** which is the self-focusing of the laser beam. The bunch of photons can pass through and travel to the far end of the cavity, amplifying as it goes.

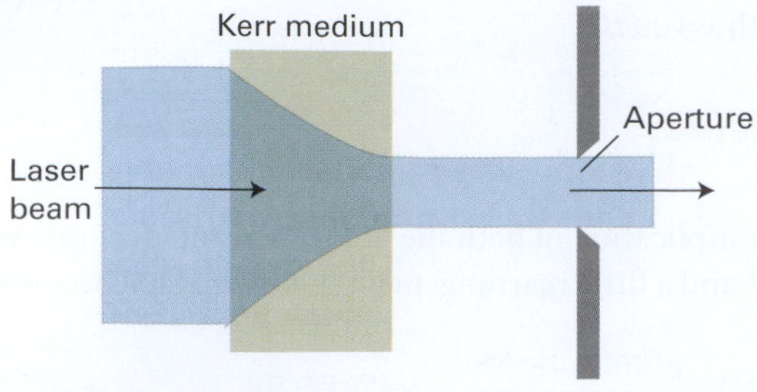

Figure 13C.9 An illustration of the Kerr effect. An intense laser beam is focused inside a Kerr medium and passes through a small aperture in the laser cavity. This effect may be used to mode-lock a laser, as explained in the text.

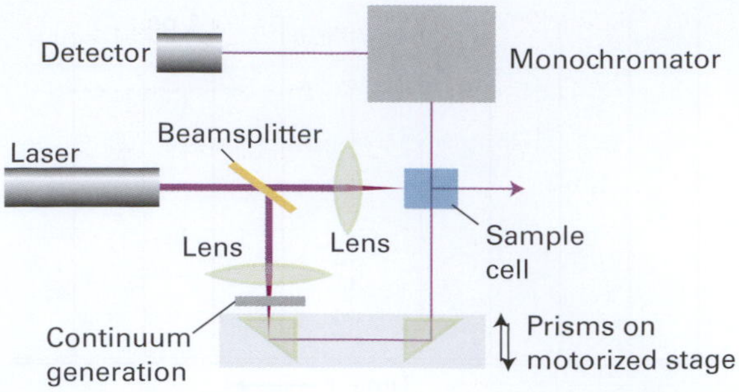

Figure 13C.10 A configuration used for time-resolved absorption spectroscopy, in which the same pulsed laser is used to generate a monochromatic pump pulse and, after continuum generation in a suitable liquid, a 'white' light probe pulse. The time delay between the pump and probe pulses may be varied.

The Kerr lens immediately disappears (if the medium is well chosen), but is re-created when the intense pulse returns from the mirror at the far end. In this way, that particular bunch of photons may grow to considerable intensity because it alone is stimulating emission in the cavity.

13C.4 Time-resolved spectroscopy

The ability of lasers to produce pulses of very short duration is particularly useful in chemistry when we want to monitor processes in time. In **time-resolved spectroscopy,** laser pulses are used to obtain the absorption, emission, or Raman spectrum of reactants, intermediates, products, and even transition states of reactions. It is also possible to study energy transfer, molecular rotations, vibrations, and conversion from one mode of motion to another.

The arrangement shown in Fig. 13C.10 is often used to study ultrafast chemical reactions that can be initiated by light (Topic 20G). A strong and short laser pulse, the *pump*, promotes a molecule A to an excited electronic state A* that can either emit a photon (as fluorescence or phosphorescence) or react with another species B to yield a product C:

$$A + h\nu \rightarrow A^* \qquad \text{(absorption)}$$
$$A^* \rightarrow A \qquad \text{(emission)}$$
$$A^* + B \rightarrow [AB] \rightarrow C \qquad \text{(reaction)}$$

Here [AB] denotes either an intermediate or an activated complex. The rates of appearance and disappearance of the various species are determined by observing time-dependent changes in the absorption spectrum of the sample during the course of the reaction. This monitoring is done by passing a weak pulse of white light, the *probe*, through the sample at different times after the laser pulse. Pulsed 'white' light can be generated directly from the laser pulse by the phenomenon of **continuum generation**, in which focusing a short laser pulse on a vessel containing water, carbon tetrachloride, or sapphire results in an outgoing beam with a wide distribution of frequencies. A time delay between the strong laser pulse and the 'white' light pulse can be introduced by allowing one of the beams to travel a longer distance before reaching the sample. For example, a difference in travel distance of $\Delta d = 3\,\text{mm}$ corresponds to a time delay $\Delta t = \Delta d/c \approx 10\,\text{ps}$ between two beams, where c is the speed of light. The relative distances travelled by the two beams in Fig 13C.10 are controlled by directing the 'white' light beam to a motorized stage carrying a pair of mirrors.

Variations of the arrangement in Fig 13C.10 can be used for the observation of the decay of an excited state and of time-resolved Raman spectra during the course of the reaction. The lifetime of A* can be determined by exciting A as before and measuring the decay of the fluorescence intensity after the pulse with a fast photodetector system. In this case, continuum generation is not necessary. Time-resolved resonance Raman spectra of A, A*, B, [AB], or C can be obtained by initiating the reaction with a strong laser pulse of a certain wavelength and then, sometime later, irradiating the sample with another laser pulse that can excite the resonance Raman spectrum of the desired species. Also in this case continuum generation is not necessary.

13C.5 Examples of practical lasers

Figure 13C.11 summarizes the requirements for an efficient laser. In practice, the requirements can be satisfied by using a variety of different systems. We have already considered the ruby and neodymium lasers, and here we review other arrangements that are commonly available. We also include some lasers that operate by using other than electronic transitions. Noticeably absent from this discussion are the ubiquitous diode lasers, which we discuss in Topic 18D.

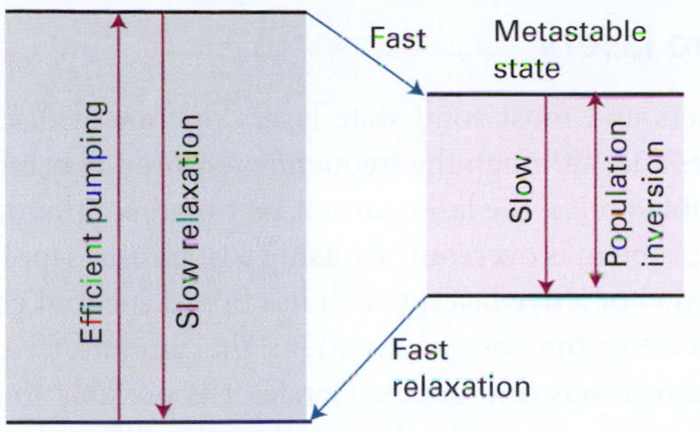

Figure 13C.11 A summary of the features needed for efficient laser action.

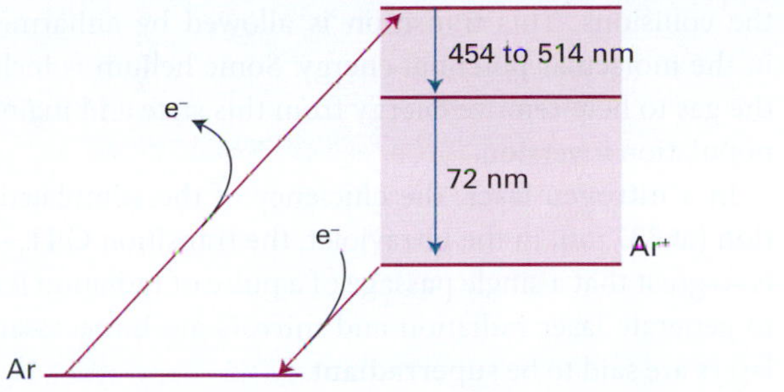

Figure 13C.13 The transitions involved in an argon-ion laser.

(a) Gas lasers

Because gas lasers can be cooled by a rapid flow of the gas through the cavity, they can be used to generate high powers. The pumping is normally achieved using a gas that is different from the gas responsible for the laser emission itself.

In the **helium–neon laser** the active medium is a mixture of helium and neon in a mole ratio of about 5:1 (Fig. 13C.12). The initial step is the excitation of an He atom to the metastable $1s^12s^1$ configuration by using an electric discharge (the collisions of electrons and ions cause transitions that are not restricted by electric-dipole selection rules). The excitation energy of this transition happens to match an excitation energy of neon, and during an He–Ne collision efficient transfer of energy may occur, leading to the production of highly excited, metastable Ne atoms with unpopulated intermediate states. Laser action generating 633 nm radiation (among about 100 other lines) then occurs.

The **argon-ion laser** (Fig. 13C.13), one of a number of 'ion lasers', consists of argon at about 1 Torr, through which is passed an electric discharge. The discharge results in the formation of Ar^+ and Ar^{2+} ions in excited states, which undergo a laser transition to a lower state. These ions then revert to their ground states by emitting hard (short wavelength) ultraviolet radiation (at 72 nm), and are then neutralized by a series of electrodes in the laser cavity. One of the design problems is to find materials that can withstand this damaging residual radiation. There are many lines in the laser transition because the excited ions may make transitions to many lower states, but two strong emissions from Ar^+ are at 488 nm (blue) and 514 nm (green); other transitions occur elsewhere in the visible region, in the infrared, and in the ultraviolet. The **krypton-ion laser** works similarly. It is less efficient, but gives a wider range of wavelengths, the most intense being at 647 nm (red), but it can also generate yellow, green, and violet light.

The **carbon dioxide laser** works on a slightly different principle (Fig. 13C.14), for its radiation (between 9.2 μm and 10.8 μm, with the strongest emission at 10.6 μm, in the infrared) arises from vibrational transitions. Most of the working gas is nitrogen, which becomes vibrationally excited by electronic and ionic collisions in an electric discharge. The vibrational levels happen to coincide with the ladder of antisymmetric stretch (ν_3, see Fig. 12E.2) energy levels of CO_2, which pick up the energy during a collision. Laser action then occurs from the lowest excited level of ν_3 to the lowest excited level of the symmetric stretch (ν_1), which has remained unpopulated during

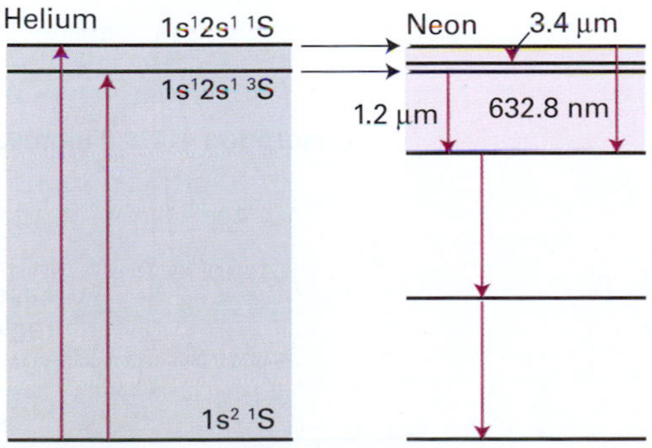

Figure 13C.12 The transitions involved in a helium–neon laser. The pumping (of the neon) depends on a coincidental matching of the helium and neon energy separations, so excited He atoms can transfer their excess energy to Ne atoms during a collision.

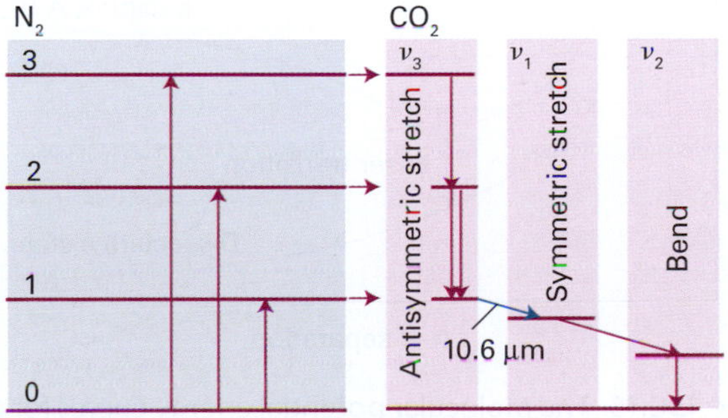

Figure 13C.14 The transitions involved in a carbon dioxide laser. The pumping also depends on the coincidental matching of energy separations; in this case the vibrationally excited N_2 molecules have excess energies that correspond to a vibrational excitation of the antisymmetric stretch of CO_2. The laser transition is from $v=1$ of mode ν_3 to $v=1$ of mode ν_1.

the collisions. This transition is allowed by anharmonicities in the molecular potential energy. Some helium is included in the gas to help remove energy from this state and maintain the population inversion.

In a **nitrogen laser**, the efficiency of the stimulated transition (at 337 nm, in the ultraviolet, the transition $C^3\Pi_u \rightarrow B^3\Pi_g$) is so great that a single passage of a pulse of radiation is enough to generate laser radiation and mirrors are unnecessary: such lasers are said to be **superradiant**.

(b) Exciplex lasers

The population inversion needed for laser action is achieved in an underhand way in **exciplex lasers**, for in these (as we shall see) the lower state does not effectively exist. This odd situation is achieved by forming an **exciplex**, a combination of two atoms that survives only in an excited state and which dissociates as soon as the excitation energy has been discarded. An exciplex can be formed in a mixture of xenon, chlorine, and neon (which acts as a buffer gas). An electric discharge through the mixture produces excited Cl atoms, which attach to the Xe atoms to give the exciplex XeCl*. The exciplex survives for about 10 ns, which is time for it to participate in laser action at 308 nm (in the ultraviolet). As soon as XeCl* has discarded a photon, the atoms separate because the molecular potential energy curve of the ground state is dissociative, and the ground state of the exciplex cannot become populated (Fig. 13C.15). The KrF* exciplex laser is another example: it produces radiation at 249 nm.

The term 'excimer laser' is also widely encountered and used loosely when 'exciplex laser' is more appropriate. An exciplex has the form AB* whereas an excimer, an excited dimer, is AA*.

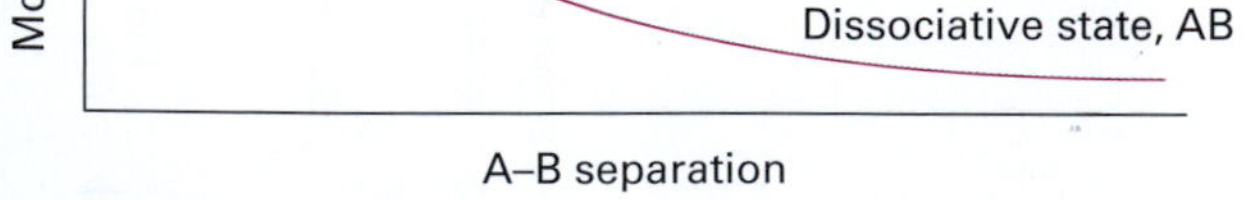

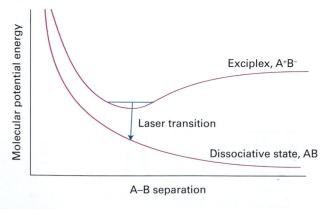

Figure 13C.15 The molecular potential energy curves for an exciplex. The species can survive only as an excited state (in this case a charge-transfer complex, A^+B^-), because on discarding its energy it enters the lower, dissociative state. Because only the upper state can exist, there is never any population in the lower state.

(c) Dye lasers

Gas lasers and most solid state lasers operate at discrete frequencies and, although the frequency required may be selected by suitable optics, the laser cannot be tuned continuously. The tuning problem is overcome by using a titanium–sapphire laser (see below) or a **dye laser**, which has broad spectral characteristics because the solvent broadens the vibrational structure of the transitions into bands. Hence, it is possible to scan the wavelength continuously (by rotating the diffraction grating in the cavity) and achieve laser action at any chosen wavelength. A commonly used dye is rhodamine 6G in methanol (Fig. 13C.16). As the gain is very high, only a short length of the optical path need be through the dye. The excited states of the active medium, the dye, are sustained by another laser or a flash lamp, and the dye solution is flowed through the laser cavity to avoid thermal degradation.

(d) Vibronic lasers

The **titanium–sapphire laser** ('Ti:sapphire laser'), which consists of Ti^{3+} ions at low concentration in an alumina (Al_2O_3) crystal. The electronic absorption spectrum of Ti^{3+} ion in sapphire is very similar to that shown in Fig. 13A.12, with a broad absorption band centred at around 500 nm that arises from vibronically allowed d–d transitions of the Ti^{3+} ion in an octahedral environment provided by oxygen atoms of the host lattice. As a result, the emission spectrum of Ti^{3+} in sapphire is also broad and laser action occurs over a wide range of wavelengths (Fig. 13C.17). Therefore, the titanium sapphire laser is an example of a **vibronic laser**, in which the laser transitions originate from vibronic transitions in the laser medium. The titanium sapphire laser is usually pumped by another laser, such as a Nd:YAG laser or an argon-ion laser, and can be operated in either a continuous or pulsed fashion.

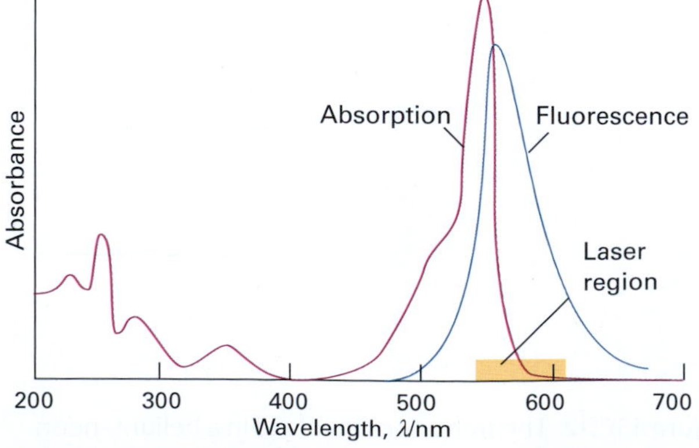

Figure 13C.16 The optical absorption spectrum of the dye rhodamine 6G and the region used for laser action.

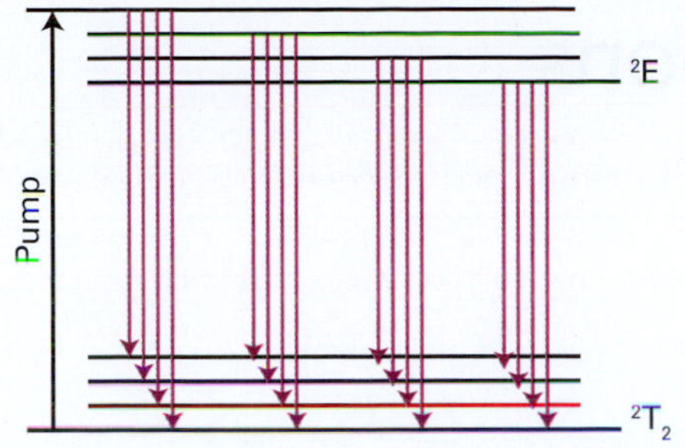

Figure 13C.17 The transitions involved in a Ti:sapphire laser. Monochromatic light from a pump laser induces a $^2E \leftarrow {}^2T_2$ transition in a Ti^{3+} ion that resides in a site with octahedral symmetry. After radiationless vibrational excitation in the ^2E state, laser emission occurs from a very large number of closely spaced vibronic states of the medium. As a result, the laser emits radiation over a broad spectrum that spans from about 700 nm to about 1000 nm.

Sapphire is an example of a Kerr medium that facilitates the mode locking of titanium sapphire lasers, resulting in very short (20–100 fs, 1 fs $= 10^{-15}$ s) pulses. When considered together with broad wavelength tunability (700–1000 nm), these features of the titanium sapphire laser justify its wide use in modern spectroscopy and photochemistry.

Checklist of concepts

☐ 1. **Laser action** is the stimulated emission of coherent radiation between states related by a population inversion.

☐ 2. A **population inversion** is a condition in which the population of an upper state is greater than that of a relevant lower state.

☐ 3. The **resonant modes** of a laser are the wavelengths of radiation sustained inside a laser cavity.

☐ 4. Laser pulses are generated by the techniques of ***Q*-switching** and **mode locking**.

☐ 5. In **time-resolved spectroscopy**, laser pulses are used to obtain the absorption, emission, or Raman spectrum of reactants, intermediates, products, and even transition states of reactions.

☐ 6. Practical lasers include **gas**, **dye**, **exciplex**, and **vibronic lasers**.

Checklist of equations

Property	Equation	Comment	Equation number
Resonant modes	$n \times \frac{1}{2}\lambda = L$	Laser cavity of length L	13C.1
Coherence length	$l_C = \lambda^2/2\Delta\lambda$		13C.2
Mode-locked laser output	$I(t) \propto \mathcal{E}_0^2\{\sin^2(N\pi ct/2L)/\sin^2(\pi ct/2L)\}$	N locked modes	13C.3

CHAPTER 13 Electronic transitions

TOPIC 13A Electronic spectra

Discussion questions

13A.1 Explain the origin of the term symbol $^3\Sigma_g^-$ for the ground state of dioxygen.

13A.2 Explain the basis of the Franck–Condon principle and how it leads to the formation of a vibrational progression.

13A.3 How do the band heads in P and R branches arise? Could the Q branch show a head?

13A.4 Explain how colour can arise from molecules.

13A.5 Suppose that you are a colour chemist and had been asked to intensify the colour of a dye without changing the type of compound, and that the dye in question was a conjugated polyene. (a) Would you choose to lengthen or to shorten the chain? (b) Would the modification to the length shift the apparent colour of the dye towards the red or the blue?

Exercises

13A.1(a) One of the excited states of the C_2 molecule has the valence electron configuration $1\sigma_g^2 1\sigma_u^2 1\pi_u^3 1\pi_g^1$. Give the multiplicity and parity of the term.
13A.1(b) Another of the excited states of the C_2 molecule has the valence electron configuration $1\sigma_g^2 1\sigma_u^2 1\pi_u^2 1\pi_g^2$. Give the multiplicity and parity of the term.

13A.2(a) Which of the following transitions are electric-dipole allowed? (i) $^2\Pi \leftrightarrow {}^2\Pi$, (ii) $^1\Sigma \leftrightarrow {}^1\Sigma$, (iii) $\Sigma \leftrightarrow \Delta$, (iv) $\Sigma^+ \leftrightarrow \Sigma^-$, (v) $\Sigma^+ \leftrightarrow \Sigma^+$.
13A.2(b) Which of the following transitions are electric-dipole allowed? (i) $^1\Sigma_g^+ \leftrightarrow {}^1\Sigma_u^+$, (ii) $^3\Sigma_g^+ \leftrightarrow {}^3\Sigma_u^+$, (iii) $\pi^* \leftrightarrow n$.

13A.3(a) The ground-state wavefunction of a certain molecule is described by the vibrational wavefunction $\psi_0 = N_0 e^{-ax^2}$. Calculate the Franck–Condon factor for a transition to a vibrational state described by the wavefunction $\psi_v = N_v e^{-a(x-x_0)^2/2}$.
13A.3(b) The ground-state wavefunction of a certain molecule is described by the vibrational wavefunction $\psi_0 = N_0 e^{-ax^2}$. Calculate the Franck–Condon factor for a transition to a vibrational state described by the wavefunction $\psi_v = N_v x e^{-a(x-x_0)^2/2}$.

13A.4(a) Suppose that the ground vibrational state of a molecule is modelled by using the particle-in-a-box wavefunction $\psi_0 = (2/L)^{1/2}\sin(\pi x/L)$ for $0 \le x \le L$ and 0 elsewhere. Calculate the Franck–Condon factor for a transition to a vibrational state described by the wavefunction $\psi' = (2/L)^{1/2}\sin\{\pi(x-L/4)/L\}$ for $L/4 \le x \le 5L/4$ and 0 elsewhere.
13A.4(b) Suppose that the ground vibrational state of a molecule is modelled by using the particle-in-a-box wavefunction $\psi_0 = (2/L)^{1/2}\sin(\pi x/L)$ for $0 \le x \le L$ and 0 elsewhere. Calculate the Franck–Condon factor for a transition to a vibrational state described by the wavefunction $\psi' = (2/L)^{1/2}\sin\{\pi(x-L/2)/L\}$ for $L/2 \le x \le 3L/2$ and 0 elsewhere.

13A.5(a) Use eqn 13A.8a to infer the value of J corresponding to the location of the band head of the P branch of a transition.
13A.5(b) Use eqn 13A.8c to infer the value of J corresponding to the location of the band head of the R branch of a transition.

13A.6(a) The following parameters describe the electronic ground state and an excited electronic state of SnO: $\tilde{B} = 0.3540\,\text{cm}^{-1}$, $\tilde{B}' = 0.3101\,\text{cm}^{-1}$. Which branch of the transition between them shows a head? At what value of J will it occur?
13A.6(b) The following parameters describe the electronic ground state and an excited electronic state of BeH: $\tilde{B} = 10.308\,\text{cm}^{-1}$, $\tilde{B}' = 10.470\,\text{cm}^{-1}$. Which branch of the transition between them shows a head? At what value of J will it occur?

13A.7(a) The R-branch of the $^1\Pi_u \leftarrow {}^1\Sigma_g^+$ transition of H_2 shows a band head at the very low value of $J = 1$. The rotational constant of the ground state is $60.80\,\text{cm}^{-1}$. What is the rotational constant of the upper state? Has the bond length increased or decreased in the transition?
13A.7(b) The P-branch of the $^2\Pi \leftarrow {}^2\Sigma^+$ transition of CdH shows a band head at $J = 25$. The rotational constant of the ground state is $5.437\,\text{cm}^{-1}$. What is the rotational constant of the upper state? Has the bond length increased or decreased in the transition?

13A.8(a) The complex ion $[Fe(OH_2)_6]^{3+}$ has an electronic absorption spectrum with a maximum at 700 nm. Estimate a value of Δ_O for the complex.
13A.8(b) The complex ion $[Fe(CN)_6]^{3-}$ has an electronic absorption spectrum with a maximum at 305 nm. Estimate a value of Δ_O for the complex.

13A.9(a) Suppose that we can model a charge-transfer transition in a one-dimensional system as a process in which a rectangular wavefunction that is nonzero in the range $0 \le x \le a$ makes a transition to another rectangular wavefunction that is nonzero in the range $\frac{1}{2}a \le x \le b$. Evaluate the transition moment $\int \psi_f x \psi_i \,dx$. (Assume $a < b$.)
13A.9(b) Suppose that we can model a charge-transfer transition in a one-dimensional system as a process in which an electron described by a rectangular wavefunction that is nonzero in the range $0 \le x \le a$ makes a transition to another rectangular wavefunction that is nonzero in the range $ca \le x \le a$ where $0 \le c \le 1$. Evaluate the transition moment $\int \psi_f x \psi_i \,dx$ and explore its dependence on c.

13A.10(a) Suppose that we can model a charge-transfer transition in a one-dimensional system as a process in which a Gaussian wavefunction centred on $x = 0$ and width a makes a transition to another Gaussian wavefunction of the same width centred on $x = \frac{1}{2}a$. Evaluate the transition moment $\int \psi_f x \psi_i \,dx$.
13A.10(b) Suppose that we can model a charge-transfer transition in a one-dimensional system as a process in which an electron described by a Gaussian wavefunction centred on $x = 0$ and width a makes a transition to another Gaussian wavefunction of width $a/2$ and centred on $x = 0$. Evaluate the transition moment $\int \psi_f x \psi_i \,dx$.

13A.11(a) The two compounds 2,3-dimethyl-2-butene (**1**) and 2,5-dimethyl-2,4-hexadiene (**2**) are to be distinguished by their ultraviolet absorption spectra. The maximum absorption in one compound occurs at 192 nm and in the other at 243 nm. Match the maxima to the compounds and justify the assignment.

1 2,3-Dimethyl-2-butene

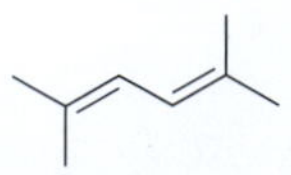

2 2,5-Dimethyl-2,4-hexadiene

13A.11(b) 3-Buten-2-one (**3**) has a strong absorption at 213 nm and a weaker absorption at 320 nm. Justify these features and assign the ultraviolet absorption transitions.

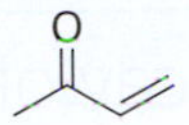

3 3-Buten-2-one

Problems

13A.1‡ J.G. Dojahn et al. (*J. Phys. Chem.* **100**, 9649 (1996)) characterized the potential energy curves of the ground and electronic states of homonuclear diatomic halogen anions. These anions have a $^2\Sigma_u^+$ ground state and $^2\Pi_g$, $^2\Pi_u$, and $^2\Sigma_g^+$ excited states. To which of the excited states are electric-dipole transitions allowed? Explain your conclusion.

13A.2 The vibrational wavenumber of the oxygen molecule in its electronic ground state is 1580 cm^{-1}, whereas that in the excited state ($B^3\Sigma_u^-$), to which there is an allowed electronic transition, is 700 cm^{-1}. Given that the separation in energy between the minima in their respective potential energy curves of these two electronic states is 6.175 eV, what is the wavenumber of the lowest energy transition in the band of transitions originating from the $v=0$ vibrational state of the electronic ground state to this excited state? Ignore any rotational structure or anharmonicity.

13A.3 A transition of particular importance in O_2 gives rise to the Schumann–Runge band in the ultraviolet region. The wavenumbers (in cm^{-1}) of transitions from the ground state to the vibrational levels of the first excited state ($^3\Sigma_u^-$) are 50 062.6, 50 725.4, 51 369.0, 51 988.6, 52 579.0, 53 143.4, 53 679.6, 54 177.0, 54 641.8, 55 078.2, 55 460.0, 55 803.1, 56 107.3, 56 360.3, 56 570.6. What is the dissociation energy of the upper electronic state? (Use a Birge–Sponer plot, Topic 12D.) The same excited state is known to dissociate into one ground state O atom and one excited state atom with an energy 190 kJ mol^{-1} above the ground state. (This excited atom is responsible for a great deal of photochemical mischief in the atmosphere.) Ground state O_2 dissociates into two ground state atoms. Use this information to calculate the dissociation energy of ground-state O_2 from the Schumann–Runge data.

13A.4 You are now ready to understand more deeply the features of photoelectron spectra (Topic 10B). Figure 13.1 shows the photoelectron spectrum of HBr. Disregarding for now the fine structure, the HBr lines fall into two main groups. The least tightly bound electrons (with the lowest ionization energies and hence highest kinetic energies when ejected) are those in the lone pairs of the Br atom. The next ionization energy lies at 15.2 eV, and corresponds to the removal of an electron from the HBr σ bond. (a) The spectrum shows that ejection of a σ electron is accompanied by a considerable amount of vibrational excitation. Use the Franck–Condon principle to account for this observation. (b) Go on to explain why the lack of much vibrational structure in the other band is consistent with the nonbonding role of the Br4p_x and Br4p_y lone-pair electrons.

13A.5 The highest kinetic energy electrons in the photoelectron spectrum of H_2O using 21.22 eV radiation are at about 9 eV and show a large vibrational spacing of 0.41 eV. The symmetric stretching mode of the neutral H_2O molecule lies at 3652 cm^{-1}. (a) What conclusions can be drawn from the nature of the orbital from which the electron is ejected? (b) In the same spectrum of H_2O, the band near 7.0 eV shows a long vibrational series with spacing 0.125 eV. The bending mode of H_2O lies at 1596 cm^{-1}. What conclusions can you draw about the characteristics of the orbital occupied by the photoelectron?

13A.6 A lot of information about the energy levels and wavefunctions of small inorganic molecules can be obtained from their ultraviolet spectra. An example of a spectrum with considerable vibrational structure, that of gaseous SO_2 at 25 °C, is shown in Fig. 13A.5. Estimate the integrated absorption coefficient for the transition. What electronic states are accessible from the A_1 ground state of this C_{2v} molecule by electric dipole transitions?

‡ These problems were supplied by Charles Trapp and Carmen Giunta.

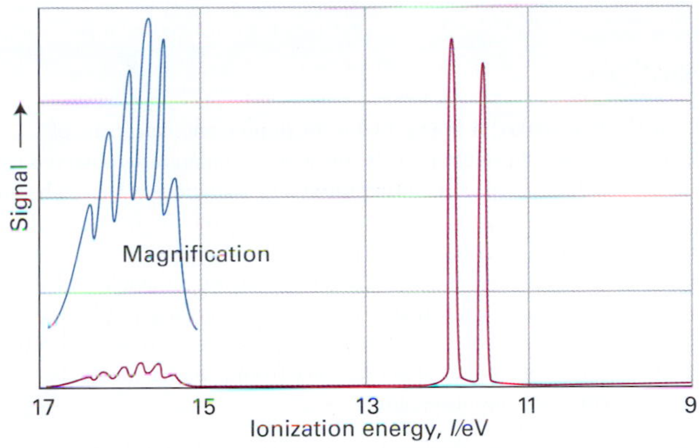

Figure 13.1 The photoelectron spectrum of HBr.

13A.7 Assume that the electronic states of the π electrons of a conjugated molecule can be approximated by the wavefunctions of a particle in a one-dimensional box, and that the magnitude of the dipole moment can be related to the displacement along this length by $\mu=-ex$. Show that the transition probability for the transition $n=1\rightarrow n=2$ is nonzero, whereas that for $n=1\rightarrow n=3$ is zero. *Hints*: The following relation will be useful: $\sin x \sin y=\frac{1}{2}\cos(x-y)-\frac{1}{2}\cos(x+y)$. Relevant integrals are found in the *Resource section*.

13A.8 1,3,5-Hexatriene (a kind of 'linear' benzene) was converted into benzene itself. On the basis of a free-electron molecular orbital model (in which hexatriene is treated as a linear box and benzene as a ring), would you expect the lowest energy absorption to rise or fall in energy?

13A.9 Estimate the magnitude of the transition dipole moment of a charge-transfer transition modelled as the migration of an electron from a H1s orbital on one atom to another H1s orbital on an atom a distance R away. Approximate the transition moment by $-eRS$ where S is the overlap integral of the two orbitals. Sketch the transition moment as a function of R using the curve for S given in Fig. 10C.7. Why does the intensity of a charge-transfer transition fall to zero as R approaches 0 and infinity?

13A.10 Figure 13.2 shows the UV-visible absorption spectra of a selection of amino acids. Suggest reasons for their different appearances in terms of the structures of the molecules.

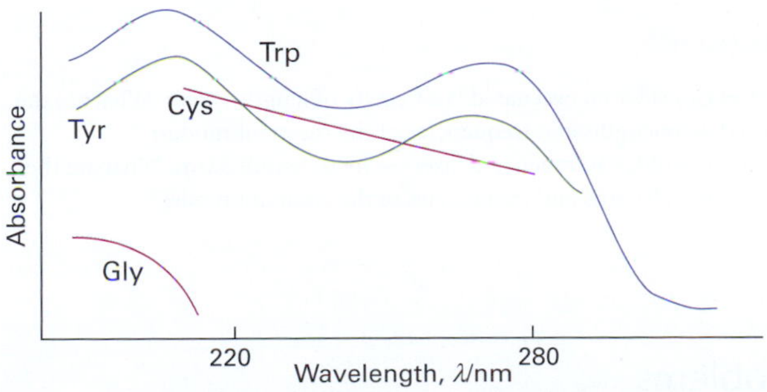

Figure 13.2 Electronic absorption spectra of selected amino acids.

TOPIC 13B Decay of excited states

Discussion questions

13B.1 Describe the mechanism of fluorescence. In what respects is a fluorescence spectrum not the exact mirror image of the corresponding absorption spectrum?

13B.2 What is the evidence for the correctness of the mechanism of fluorescence?

Exercises

13B.1(a) The line marked A in Fig. 13.3 is the fluorescence spectrum of benzophenone in solid solution in ethanol at low temperatures observed when the sample is illuminated with 360 nm ultraviolet radiation. What can be said about the vibrational energy levels of the carbonyl group in (i) its ground electronic state and (ii) its excited electronic state?

13B.1(b) When naphthalene is illuminated with 360 nm ultraviolet radiation it does not absorb, but the line marked B in Fig. 13.3 is the phosphorescence spectrum of a solid solution of a mixture of naphthalene and benzophenone in ethanol. Now a component of fluorescence from naphthalene can be detected. Account for this observation.

13B.2(a) The oxygen molecule absorbs ultraviolet radiation in a transition from its ${}^3\Sigma_g^-$ ground electronic state to an excited state that is energetically close to a dissociative ${}^5\Pi_u$ state. The absorption band has a relatively large experimental linewidth. Account for this observation.

13B.2(b) The hydrogen molecule absorbs ultraviolet radiation in a transition from its ${}^1\Sigma_g^+$ ground electronic state to an excited state that is energetically close to a dissociative ${}^1\Sigma_u^+$ state. The absorption band has a relatively large experimental linewidth. Account for this observation.

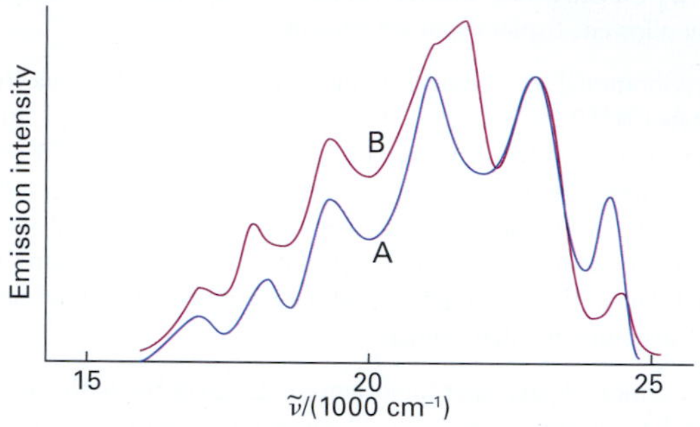

Figure 13.3 The fluorescence and phosphorescence spectra of two solutions.

Problem

13B.1 The fluorescence spectrum of anthracene vapour shows a series of peaks of increasing intensity with individual maxima at 440 nm, 410 nm, 390 nm, and 370 nm followed by a sharp cut-off at shorter wavelengths. The absorption spectrum rises sharply from zero to a maximum at 360 nm with a trail of peaks of lessening intensity at 345 nm, 330 nm, and 305 nm. Account for these observations.

TOPIC 13C Lasers

Discussion questions

13C.1 Describe the principles of (a) continuous-wave and (b) pulsed laser action.

13C.2 How might you use a *Q*-switched or mode-locked laser in the study of a very fast chemical reaction that can be initiated by absorption of light?

Exercises

13C.1(a) Consider an evacuated laser cavity of length 1.0 cm. What are the allowed wavelengths and frequencies of the resonant modes?

13C.1(b) Consider an evacuated laser cavity of length 3.0 m. What are the allowed wavelengths and frequencies of the resonant modes?

13C.2(a) A certain laser can generate radiation in pulses, each of which delivers an energy of 0.10 mJ, with peak power of 5.0 MW and average power of 7.0 kW. What are the pulse duration and repetition frequency?

13C.2(b) A certain laser can generate radiation in pulses, each of which delivers an energy of 20.0 μJ, with peak power of 100 kW and average power of 0.40 mW. What are the pulse duration and repetition frequency?

Problems

13C.1 Light-induced degradation of molecules, also called *photobleaching*, is a serious problem in fluorescence microscopy. A molecule of a fluorescent dye commonly used to label biopolymers can withstand about 10^6 excitations by photons before light-induced reactions destroy its π system and the molecule no longer fluoresces. For how long will a single dye molecule fluoresce while being excited by 1.0 mW of 488 nm radiation from a continuous-wave argon ion laser? You may assume that the dye has an absorption spectrum that peaks at 488 nm and that every photon delivered by the laser is absorbed by the molecule.

13C.2 Use mathematical software or an electronic spreadsheet to simulate the output of a mode-locked laser (that is, plots such as that shown in Fig. 13C.8) for $L = 30$ cm and $N = 100$ and 1000.

Integrated activities

13.1‡ One of the principal methods for obtaining the electronic spectra of unstable radicals is to study the spectra of comets, which are almost entirely due to radicals. Many radical spectra have been detected in comets, including that due to CN. These radicals are produced in comets by the absorption of far-ultraviolet solar radiation by their parent compounds. Subsequently, their fluorescence is excited by sunlight of longer wavelength. The spectra of comet Hale–Bopp (C/1995 O1) have been the subject of many recent studies. One such study is that of the fluorescence spectrum of CN in the comet at large heliocentric distances by R.M. Wagner and D.G. Schleicher (*Science* **275**, 1918 (1997)), in which the authors determine the spatial distribution and rate of production of CN in the coma (the cloud constituting the major part of the head of the comet). The (0–0) vibrational band is centred on 387.6 nm and the weaker (1–1) band with relative intensity 0.1 is centred on 386.4 nm. The band heads for (0–0) and (0–1) are known to be 388.3 and 421.6 nm, respectively. From these data, calculate the energy of the excited S_1 state relative to the ground S_0 state, the vibrational wavenumbers and the difference in the vibrational wavenumbers of the two states, and the relative populations of the $v = 0$ and $v = 1$ vibrational levels of the S_1 state. Also estimate the effective temperature of the molecule in the excited S_1 state. Only eight rotational levels of the S_1 state are thought to be populated. Is that observation consistent with the effective temperature of the S_1 state?

13.2 Use a group theoretical argument to decide which of the following transitions are electric-dipole allowed: (a) the $\pi^* \leftarrow \pi$ transition in ethene, (b) the $\pi^* \leftarrow n$ transition in a carbonyl group in a C_{2v} environment.

13.3 Use molecule (4) as a model of the *trans* conformation of the chromophore found in rhodopsin. In this model, the methyl group bound to the nitrogen atom of the protonated Schiff's base replaces the protein. (a) Using molecular modelling software and the computational method of your instructor's choice, calculate the energy separation between the HOMO and LUMO of (4). (b) Repeat the calculation for the 11-*cis* form of (4). (c) Based on your results from parts (a) and (b), do you expect the experimental frequency for the $\pi^* \leftarrow \pi$ visible absorption of the *trans* form of (4) to be higher or lower than that for the 11-*cis* form of (4)?

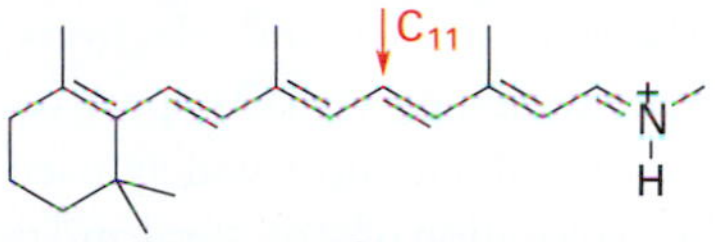

4

13.4 Aromatic hydrocarbons and I_2 form complexes from which charge-transfer electronic transitions are observed. The hydrocarbon acts as an electron donor and I_2 as an electron acceptor. The energies $h\nu_{max}$ of the charge-transfer transitions for a number of hydrocarbon–I_2 complexes are given below:

Hydrocarbon	benzene	biphenyl	naphthalene	phenanthrene	pyrene	anthracene
$h\nu_{max}$/eV	4.184	3.654	3.452	3.288	2.989	2.890

Investigate the hypothesis that there is a correlation between the energy of the HOMO of the hydrocarbon (from which the electron comes in the charge-transfer transition) and $h\nu_{max}$. Use one of the computational methods discussed in Topic 10E to determine the energy of the HOMO of each hydrocarbon in the data set.

13.5 Spin angular momentum is conserved when a molecule dissociates into atoms. What atom multiplicities are permitted when (a) an O_2 molecule, (b) an N_2 molecule dissociates into atoms?

CHAPTER 14

Magnetic resonance

The techniques of 'magnetic resonance' probe transitions between spin states of nuclei and electrons in molecules. 'Nuclear magnetic resonance' (NMR) spectroscopy, the focus of this chapter, is one of the most widely used procedures in chemistry for the exploration of structural and dynamical properties of molecules of all sizes, up to as large as biopolymers.

14A General principles

The chapter begins with an account of the principles that govern spectroscopic transitions between spin states of nuclei and electrons in molecules. It also describes simple experimental arrangements for the detection of these transitions. The concepts developed in this Topic prepare the ground for a discussion of the chemical applications of NMR and 'electron paramagnetic resonance' (EPR).

14B Features of NMR spectra

This Topic contains a discussion of conventional NMR, showing how the properties of a magnetic nucleus are affected by its electronic environment and the presence of magnetic nuclei in its vicinity. These concepts lead to understanding of how molecular structure governs the appearance of NMR spectra.

14C Pulse techniques in NMR

In this Topic we turn to the modern versions of NMR, which are based on the use of pulses of electromagnetic radiation and the processing of the resulting signal by 'Fourier transform' techniques. It is through the application of these pulse techniques that NMR spectroscopy can probe a vast array of small and large molecules in a variety of environments.

14D Electron paramagnetic resonance

The experimental techniques for EPR resemble those used in the early days of NMR. The information obtained is used to investigate species with unpaired electrons. This Topic includes a brief survey of the applications of EPR to the study of organic radicals and d-metal complexes.

What is the impact of this material?

Magnetic resonance techniques are ubiquitous in chemistry, as they are an enormously powerful analytical and structural technique, especially in organic chemistry and biochemistry. One of the most striking applications of nuclear magnetic resonance is in medicine. 'Magnetic resonance imaging' (MRI) is a portrayal of the concentrations of protons in a solid object (*Impact* I14.1). The technique is particularly useful for diagnosing disease. In *Impact* I14.2 we highlight an application of electron paramagnetic resonance in materials science and biochemistry: the use of a 'spin probe', a radical that interacts with biopolymer or a nanostructure and has an EPR spectrum that reveals its structural and dynamical properties.

To read more about the impact of this material, scan the QR code, or go to bcs.whfreeman.com/webpub/chemistry/pchem10e/impact/pchem-14-1.html

14A General principles

Contents

➤ Why do you need to know this material?

Nuclear magnetic resonance spectroscopy is used widely in chemistry and medicine. To understand the power of magnetic resonance, you need to understand the principles that govern spectroscopic transitions between spin states of electrons and nuclei in molecules.

➤ What is the key idea?

Resonant absorption occurs when the separation of the energy levels of spins in a magnetic field matches the energy of incident photons.

➤ What do you need to know already?

You need to be familiar with the quantum mechanical concept of spin (Topic 9B), the Boltzmann distribution (*Foundations* B and Topic 15A), and the general features of spectroscopy (Topic 12A).

When two pendulums share a slightly flexible support and one is set in motion, the other is forced into oscillation by the motion of the common axle. As a result, energy flows between the two pendulums. The energy transfer occurs most efficiently when the frequencies of the two pendulums are identical. The condition of strong effective coupling when the frequencies of two oscillators are identical is called **resonance**. Resonance is the basis of a number of everyday phenomena, including the response of radios to the weak oscillations of the electromagnetic field generated by a distant transmitter. Historically, spectroscopic techniques that measure transitions between nuclear and electron spin states have carried the term 'resonance' in their names because they have depended on matching a set of energy levels to a source of monochromatic radiation and observing the strong absorption that occurs at resonance. In fact, all spectroscopy is a form of resonant coupling between the electromagnetic field and the molecules; what distinguishes **magnetic resonance** is that the energy levels themselves are modified by the application of a magnetic field.

The Stern–Gerlach experiment (Topic 9B) provided evidence for electron spin. It turns out that many nuclei also possess spin angular momentum. Orbital and spin angular momenta give rise to magnetic moments, and to say that electrons and nuclei have magnetic moments means that, to some extent, they behave like small bar magnets with energies that depend on their orientation in an applied magnetic field. Here we establish how the energies of electrons and nuclei depend on the applied field. This material sets the stage for the exploration of the structure and dynamics of many kinds of molecules by magnetic resonance spectroscopy (Topics 14B–14D).

14A.1 Nuclear magnetic resonance

The application of resonance that we describe here depends on the fact that many nuclei possess spin angular momentum characterized by a **nuclear spin quantum number** I (the analogue of s for electrons). To understand the **nuclear magnetic resonance** (NMR) experiment we need to describe the behaviour of nuclei in magnetic fields and then the basic techniques for detecting spectroscopic transitions.

(a) The energies of nuclei in magnetic fields

The nuclear spin quantum number, I, is a fixed characteristic property of a nucleus in its ground state (the only state we consider) and, depending on the nuclide, is either an integer or a half-integer (Table 14A.1). A nucleus with spin quantum number I has the following properties:

Table 14A.1 Nuclear constitution and the nuclear spin quantum number*

Number of protons	Number of neutrons	I
Even	Even	0
Odd	Odd	Integer (1, 2, 3, …)
Even	Odd	Half-integer ($\frac{1}{2}, \frac{3}{2}, \frac{5}{2}, \ldots$)
Odd	Even	Half-integer ($\frac{1}{2}, \frac{3}{2}, \frac{5}{2}, \ldots$)

* The spin of a nucleus may be different if it is in an excited state; throughout this chapter we deal only with the ground state of nuclei.

- An angular momentum of magnitude $\{I(I+1)\}^{1/2}\hbar$.
- A component of angular momentum $m_I\hbar$ on a specified axis ('the z-axis'), where $m_I = I, I-1, \ldots, -I$.
- If $I>0$, a magnetic moment with a constant magnitude and an orientation that is determined by the value of m_I.

Physical interpretation

According to the second property, the spin, and hence the magnetic moment, of the nucleus may lie in $2I+1$ different orientations relative to an axis. A proton has $I=\frac{1}{2}$ and its spin may adopt either of two orientations; a ^{14}N nucleus has $I=1$ and its spin may adopt any of three orientations; both ^{12}C and ^{16}O have $I=0$ and hence zero magnetic moment.

Classically, the energy of a magnetic moment μ in a magnetic field $\mathcal{B}$ is equal to the scalar product (*Mathematical background* 5 following Chapter 9)

$$E = -\boldsymbol{\mu}\cdot\boldsymbol{\mathcal{B}} \quad (14A.1)$$

More formally, $\mathcal{B}$ is the magnetic induction and is measured in tesla, T; $1\,\text{T} = 1\,\text{kg}\,\text{s}^{-2}\,\text{A}^{-1}$. The (non-SI) unit gauss, G, is also occasionally used: $1\,\text{T} = 10^4\,\text{G}$. Quantum mechanically, we write the hamiltonian as

$$\hat{H} = -\hat{\boldsymbol{\mu}}\cdot\boldsymbol{\mathcal{B}} \quad (14A.2)$$

To write an expression for $\hat{\boldsymbol{\mu}}$, we use the fact that, just as for electrons (Topic 9B), the magnetic moment of a nucleus is proportional to its angular momentum. The operators in eqn 14A.2 are then:

$$\hat{\boldsymbol{\mu}} = \gamma_N \hat{\boldsymbol{I}} \quad \text{and} \quad \hat{H} = -\gamma_N \boldsymbol{\mathcal{B}}\cdot\hat{\boldsymbol{I}} \quad (14A.3a)$$

where γ_N is the **nuclear magnetogyric ratio** of the specified nucleus, an empirically determined characteristic arising from its internal structure (Table 14A.2). For a magnetic field of magnitude $\mathcal{B}_0$ along the z-direction, the hamiltonian in eqn 14A.3a becomes

$$\hat{H} = -\gamma_N \mathcal{B}_0 \hat{I}_z \quad (14A.3b)$$

Because the eigenvalues of the operator $\hat{I}_z$ are $m_I\hbar$, the eigenvalues of this hamiltonian are

$$E_{m_I} = -\gamma_N \hbar \mathcal{B}_0 m_I \quad \text{Energies of a nuclear spin in a magnetic field} \quad (14A.4a)$$

When written in terms of the **nuclear magneton**, μ_N,

$$\mu_N = \frac{e\hbar}{2m_p} = 5.051\times10^{-27}\,\text{J}\,\text{T}^{-1} \quad \text{Nuclear magneton} \quad (14A.4b)$$

(where m_p is the mass of the proton) and an empirical constant called the **nuclear *g*-factor**, g_I, the energy in eqn 14A.4a becomes

$$E_{m_I} = -g_I \mu_N \mathcal{B}_0 m_I \quad g_I = \frac{\gamma_N \hbar}{\mu_N} \quad \text{Energies of a nuclear spin in a magnetic field} \quad (14A.4c)$$

Nuclear g-factors are experimentally determined dimensionless quantities with values typically between −6 and +6 (Table 14A.2). Positive values of g_I and γ_N denote a magnetic moment that lies in the same direction as the spin angular momentum vector; negative values indicate that the magnetic moment and spin lie in opposite directions. A nuclear magnet is about 2000 times weaker than the magnet associated with electron spin.

For the remainder of our discussion of nuclear magnetic resonance we assume that γ_N is positive, as is the case for the majority of nuclei. In such cases, it follows from eqn 14A.4c

Table 14A.2* Nuclear spin properties

Nuclide	Natural abundance/%	Spin I	g-factor, g_I	Magnetogyric ratio, $\gamma_N/(10^7\,\text{T}^{-1}\,\text{s}^{-1})$	NMR frequency at 1 T, ν/MHz
^{1}n		$\frac{1}{2}$	−3.826	−18.32	29.164
^{1}H	99.98	$\frac{1}{2}$	5.586	26.75	42.576
^{2}H	0.02	1	0.857	4.11	6.536
^{13}C	1.11	$\frac{1}{2}$	1.405	6.73	10.708
^{14}N	99.64	1	0.404	1.93	3.078

* More values are given in the *Resource section*.

that states with $m_I > 0$ lie below states with $m_I < 0$. It follows that the energy separation between the lower $m_I = +\frac{1}{2}(\alpha)$ and upper $m_I = -\frac{1}{2}(\beta)$ states of a **spin-$\frac{1}{2}$ nucleus**, a nucleus with $I = \frac{1}{2}$, is

$$\Delta E = E_{-1/2} - E_{+1/2} = \tfrac{1}{2}\gamma_N \hbar \mathcal{B}_0 - (-\tfrac{1}{2}\gamma_N \hbar \mathcal{B}_0) = \gamma_N \hbar \mathcal{B}_0 \quad (14A.5)$$

and resonant absorption occurs when the resonance condition

$$h\nu = \gamma_N \hbar \mathcal{B}_0 \quad \text{or} \quad \nu = \frac{\gamma_N \mathcal{B}_0}{2\pi} \quad \textit{Spin-}\tfrac{1}{2}\textit{ nuclei} \quad \text{Resonance condition} \quad (14A.6)$$

is fulfilled (Fig. 14A.1). At resonance there is strong coupling between the spins and the radiation, and absorption occurs as the spins flip from the lower energy state to the upper state.

It is sometimes useful to compare the quantum mechanical and classical pictures of magnetic nuclei pictured as tiny bar magnets. A bar magnet in an externally applied magnetic field undergoes the motion called **precession** as it twists round the direction of the field (Fig. 14A.2). The rate of precession ν_L is called the **Larmor precession frequency**:

$$\nu_L = \frac{\gamma_N \mathcal{B}_0}{2\pi} \quad \textit{Definition} \quad \text{Larmor frequency of a nucleus} \quad (14A.7)$$

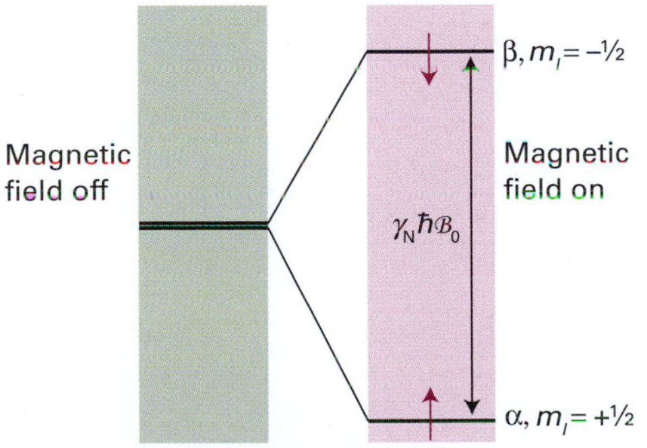

Figure 14A.1 The nuclear spin energy levels of a spin-$\frac{1}{2}$ nucleus with positive magnetogyric ratio (for example, ^{1}H or ^{13}C) in a magnetic field. Resonance occurs when the energy separation of the levels matches the energy of the photons in the electromagnetic field.

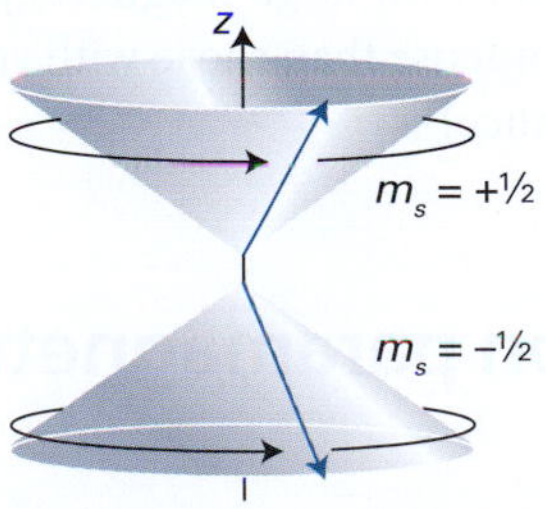

Figure 14A.2 The interactions between the m_I states of a spin-$\frac{1}{2}$ nucleus and an external magnetic field may be visualized as the precession of the vectors representing the angular momentum.

It follows by comparing this expression with eqn 14A.6 that resonance absorption by spin-$\frac{1}{2}$ nuclei occurs when the Larmor precession frequency ν_L is the same as the frequency of the applied electromagnetic field, ν.

Brief illustration 14A.1 The resonance condition in NMR

To calculate the frequency at which radiation comes into resonance with proton ($I = \frac{1}{2}$) spins in a 12.0 T magnetic field we use eqn 14A.6 as follows:

$$\nu = \frac{\overbrace{(2.6752 \times 10^8\ \mathrm{T^{-1}\,s^{-1}})}^{\gamma_N} \times \overbrace{(12.0\ \mathrm{T})}^{\mathcal{B}_0}}{2\pi} = 5.11 \times 10^8\ \mathrm{s^{-1}}$$
$$= 511\ \mathrm{MHz}$$

Self-test 14A.1 Determine the resonance frequency for ^{31}P nuclei, for which $\gamma_N = 1.0841 \times 10^8\ \mathrm{T^{-1}\,s^{-1}}$, under the same conditions.

Answer: 207 MHz

(b) The NMR spectrometer

In its simplest form, NMR is the study of the properties of molecules containing magnetic nuclei by applying a magnetic field and observing the frequency of the resonant electromagnetic field. Larmor frequencies of nuclei at the fields normally employed (about 12 T) typically lie in the radiofrequency region of the electromagnetic spectrum (close to 500 MHz), so NMR is a radiofrequency technique. For much of our discussion we consider spin-$\frac{1}{2}$ nuclei, but NMR is applicable to nuclei with any non-zero spin. As well as protons, which are the most common nuclei studied by NMR, spin-$\frac{1}{2}$ nuclei include ^{13}C, ^{19}F, and ^{31}P.

An NMR spectrometer consists of the appropriate sources of radiofrequency radiation and a magnet that can produce a uniform, intense field. Most modern instruments use a superconducting magnet capable of producing fields of the order of 10 T and more (Fig. 14A.3). The sample is rotated rapidly to average out magnetic inhomogeneities; however, although sample spinning is essential for the investigation of small molecules, for large molecules it can lead to irreproducible results and is often avoided. Although a superconducting magnet (Topic 18C) operates at the temperature of liquid helium (4 K), the sample itself is normally at room temperature or held in a variable temperature enclosure between, typically, −150 and +100 °C.

Modern NMR spectroscopy uses pulses of radiofrequency radiation. These techniques of Fourier-transform (FT) NMR make possible the determination of structures of very large molecules in solution and in solids. They are discussed in Topic 14C.

The intensity of an NMR transition depends on a number of factors. We show in the following *Justification* that

$$\text{Intensity} \propto (N_\alpha - N_\beta)\mathcal{B}_0 \quad (14A.8a)$$

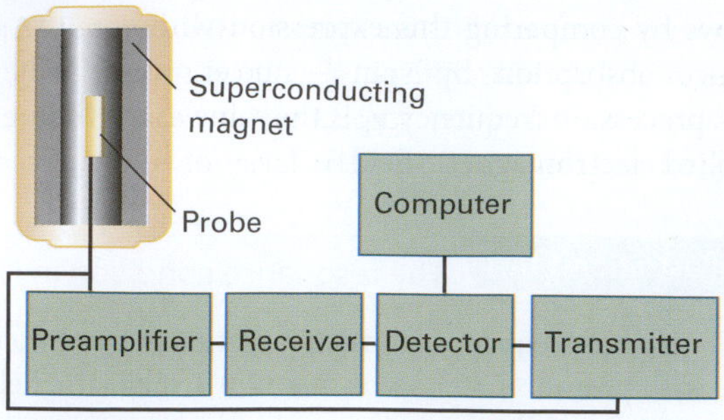

Figure 14A.3 The layout of a typical NMR spectrometer. The link from the transmitter to the detector indicates that the high frequency of the transmitter is subtracted from the high frequency received signal to give a low frequency signal for processing.

where

$$N_\alpha - N_\beta \approx \frac{N\gamma_N \hbar \mathcal{B}_0}{2kT} \quad \textit{Nuclei} \quad \text{Population difference} \quad (14A.8b)$$

with N the total number of spins ($N=N_\alpha+N_\beta$). It follows that decreasing the temperature increases the intensity by increasing the population difference.

Brief illustration 14A.2 Nuclear spin populations

For protons $\gamma_N = 2.675\times10^8\ \mathrm{T^{-1}\,s^{-1}}$. Therefore, for 1 000 000 protons in a field of 10 T at 20 °C

$$N_\alpha - N_\beta \approx \frac{\overbrace{1000000}^{N}\times\overbrace{(2.675\times10^8\ \mathrm{T^{-1}\,s^{-1}})}^{\gamma_N}\times\overbrace{(1.055\times10^{-34}\ \mathrm{J\,s})}^{\hbar}\times\overbrace{10\,\mathrm{T}}^{\mathcal{B}_0}}{2\times\underbrace{(1.381\times10^{-23}\ \mathrm{J\,K^{-1}})}_{k}\times\underbrace{(293\,\mathrm{K})}_{T}}$$

$$\approx 35$$

Even in such a strong field there is only a tiny imbalance of population of about 35 in a million.

Self-test 14A.2 For ^{13}C nuclei, $\gamma_N = 6.7283\times10^7\ \mathrm{T^{-1}\,s^{-1}}$. Determine the magnetic field that would need to be achieved in order to induce the same imbalance in the distribution of ^{13}C spins at 20 °C.

Answer: 40 T

Justification 14A.1 Intensities in NMR spectra

From the general considerations of transition intensities in Topic 12A, we know that the rate of absorption of electromagnetic radiation is proportional to the population of the lower energy state (N_α in the case of a proton NMR transition) and the rate of stimulated emission is proportional to the population of the upper state (N_β). At the low frequencies typical of magnetic resonance, spontaneous emission can be neglected as it is very slow. Therefore, the net rate of absorption is proportional to the difference in populations, and we can write

$$\text{Rate of absorption} \propto N_\alpha - N_\beta$$

The intensity of absorption, the rate at which energy is absorbed, is proportional to the product of the rate of absorption (the rate at which photons are absorbed) and the energy of each photon. The latter is proportional to the frequency ν of the incident radiation (through $E=h\nu$). At resonance, this frequency is proportional to the applied magnetic field (through $\nu=\gamma_N\mathcal{B}_0/2\pi$), so we can write

$$\text{Rate of absorption} \propto (N_\alpha - N_\beta)\mathcal{B}_0$$

as in eqn 14A.8a. To write an expression for the population difference, we use the Boltzmann distribution (*Foundations* B and Topic 15A) to write the ratio of populations as

$$\frac{N_\beta}{N_\alpha} = e^{-\overbrace{\gamma_N\hbar\mathcal{B}_0}^{\Delta E}/kT} \overset{e^{-x}=1-x+\cdots}{\approx} 1-\frac{\gamma_N\hbar\mathcal{B}_0}{kT}$$

The expansion of the exponential term is appropriate for $\Delta E = \gamma_N\hbar\mathcal{B}_0 \ll kT$, a condition usually met for nuclear spins. It follows that

$$\frac{N_\alpha - N_\beta}{\underbrace{N_\alpha + N_\beta}_{N}} = \frac{N_\alpha(1-N_\beta/N_\alpha)}{N_\alpha(1+N_\beta/N_\alpha)} = \frac{1-\overbrace{N_\beta/N_\alpha}^{1-\gamma_N\hbar\mathcal{B}_0/kT}}{1+\underbrace{N_\beta/N_\alpha}_{1-\gamma_N\hbar\mathcal{B}_0/kT}}$$

$$\approx \frac{1-(1-\gamma_N\hbar\mathcal{B}_0/kT)}{\underbrace{1+(1-\gamma_N\hbar\mathcal{B}_0/kT)}_{\approx 1}} = \frac{\gamma_N\hbar\mathcal{B}_0/kT}{2}$$

which is eqn 14A.8b.

By combining eqns 14A.8a and 14A.8b we see that the intensity is proportional to $\mathcal{B}_0^2$, so NMR transitions can be enhanced significantly by increasing the strength of the applied magnetic field. The use of high magnetic fields also simplifies the appearance of spectra (a point explained in Topic 14B) and so allows them to be interpreted more readily. We can also conclude that absorptions of nuclei with large magnetogyric ratios (^{1}H, for instance) are more intense than those with small magnetogyric ratios (^{13}C, for instance).

14A.2 Electron paramagnetic resonance

Electron paramagnetic resonance (EPR), or **electron spin resonance** (ESR), is the study of molecules and ions containing unpaired electrons by observing the magnetic field at which they come into resonance with radiation of known frequency.

As we have done for NMR, we write expressions for the resonance condition in EPR and then describe the general features of EPR spectrometers.

(a) The energies of electrons in magnetic fields

The spin magnetic moment of an electron, which has a spin quantum number $s=\frac{1}{2}$ (Topic 9B), is proportional to its spin angular momentum. The spin magnetic moment and hamiltonian operators are, respectively,

$$\hat{\mu}=\gamma_e\hat{s} \quad \text{and} \quad \hat{H}=-\gamma_e\mathcal{B}\cdot\hat{s} \qquad (14A.9a)$$

where $\hat{s}$ is the spin angular momentum operator, and γ_e is the **magnetogyric ratio of the electron**:

$$\gamma_e=-\frac{g_e e}{2m_e} \qquad \textit{Electrons} \quad \text{Magnetogyric ratio} \qquad (14A.9b)$$

with g_e=2.002 319… as the **g-value of the electron**. (Note that the current convention is to include the g-value in the definition of the magnetogyric ratio.) Dirac's relativistic theory, his modification of the Schrödinger equation to make it consistent with Einstein's special relativity, gives g_e=2; the additional 0.002 319… arises from interactions of the electron with the electromagnetic fluctuations of the vacuum that surrounds the electron. The negative sign of γ_e (arising from the sign of the electron's charge) shows that the magnetic moment is opposite in direction to the angular momentum vector.

For a magnetic field of magnitude $\mathcal{B}_0$ in the z-direction,

$$\hat{H}=-\gamma_e\mathcal{B}_0\hat{s}_z \qquad (14A.10)$$

Because the eigenvalues of the operator $\hat{s}_z$ are $m_s\hbar$ with $m_s=+\frac{1}{2}(\alpha)$ and $m_s=-\frac{1}{2}(\beta)$, it follows that the energies of an electron spin in a magnetic field are

$$E_{m_s}=-\gamma_e\hbar\mathcal{B}_0 m_s \qquad \text{Energies of an electron spin in a magnetic field} \qquad (14A.11a)$$

They can also be expressed in terms of the **Bohr magneton**, μ_B, as

$$E_{m_s}=g_e\mu_B\mathcal{B}_0 m_s \qquad \text{Energies of an electron spin in a magnetic field} \qquad (14A.11b)$$

where

$$\mu_B=\frac{e\hbar}{2m_e}=9.274\times10^{-24}\,\text{J}\,\text{T}^{-1} \qquad \text{Bohr magneton} \qquad (14A.11c)$$

The Bohr magneton, a positive quantity, is often regarded as the fundamental quantum of magnetic moment.

In the absence of a magnetic field, the states with different values of m_s are degenerate. When a field is present, the degeneracy is removed: the state with $m_s=+\frac{1}{2}$ moves up in energy by $\frac{1}{2}g_e\mu_B\mathcal{B}_0$ and the state with $m_s=-\frac{1}{2}$ moves down by $\frac{1}{2}g_e\mu_B\mathcal{B}_0$. From eqn 14A.11b, the separation between the (upper) $m_s=+\frac{1}{2}(\alpha)$ and (lower) $m_s=-\frac{1}{2}(\beta)$ levels of an electron spin in a magnetic field of magnitude $\mathcal{B}_0$ in the z-direction is

$$\Delta E=E_{+1/2}-E_{-1/2}=\tfrac{1}{2}g_e\mu_B\mathcal{B}_0-(-\tfrac{1}{2}g_e\mu_B\mathcal{B}_0)=g_e\mu_B\mathcal{B}_0 \qquad (14A.12a)$$

The energy separations come into resonance with the electromagnetic radiation of frequency ν when

$$h\nu=g_e\mu_B\mathcal{B}_0 \qquad \textit{Electrons} \quad \text{Resonance condition} \qquad (14A.12b)$$

This is the resonance condition for EPR (Fig. 14A.4). At resonance there is strong coupling between the electron spins and the radiation, and strong absorption occurs as the spins make the transition $\alpha\leftarrow\beta$.

Brief illustration 14A.3 The resonance condition in EPR

Magnetic fields of about 0.30 T (the value used in most commercial EPR spectrometers) correspond to resonance at

$$\nu=\frac{\overbrace{(2.0023)}^{g_e}\times\overbrace{(9.274\times10^{-24}\,\text{J}\,\text{T}^{-1})}^{\mu_B}\times\overbrace{(0.30\,\text{T})}^{\mathcal{B}_0}}{\underbrace{6.626\times10^{-34}\,\text{J}\,\text{s}}_{h}}=8.4\times10^{9}\,\text{s}^{-1}=8.4\,\text{GHz}$$

which corresponds to a wavelength of 3.6 cm.

Self-test 14A.3 Determine the magnetic field for EPR transitions in a spectrometer that uses radiation of wavelength 0.88 cm.

Answer: 1.2 T

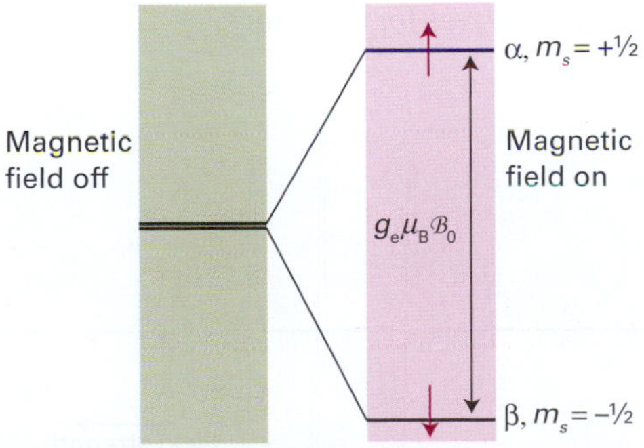

Figure 14A.4 Electron spin levels in a magnetic field. Note that the β state is lower in energy than the α state (because the magnetogyric ratio of an electron is negative). Resonance is achieved when the frequency of the incident radiation matches the frequency corresponding to the energy separation.

(b) The EPR spectrometer

It follows from *Brief illustration* 14A.3 that most commercial EPR spectrometers operate at wavelengths of approximately 3 cm. Because 3 cm radiation falls in the microwave region of the electromagnetic spectrum, EPR is a microwave technique.

Both Fourier-transform (FT) and continuous wave (CW) EPR spectrometers are available. The FT-EPR instrument is based on the concepts developed in Topic 14C for NMR spectroscopy, except that pulses of microwaves are used to excite electron spins in the sample. The layout of the more common CW-EPR spectrometer is shown in Fig. 14A.5. It consists of a microwave source (a klystron or a Gunn oscillator), a cavity in which the sample is inserted in a glass or quartz container, a microwave detector, and an electromagnet with a field that can be varied in the region of 0.3 T. The EPR spectrum is obtained by monitoring the microwave absorption as the field is changed, and a typical spectrum (of the benzene radical anion, $C_6H_6^-$) is shown in Fig. 14A.6. The peculiar appearance of the spectrum, which is in fact displayed as the first-derivative of the absorption, arises from the detection technique, which is sensitive to the slope of the absorption curve (Fig. 14A.7).

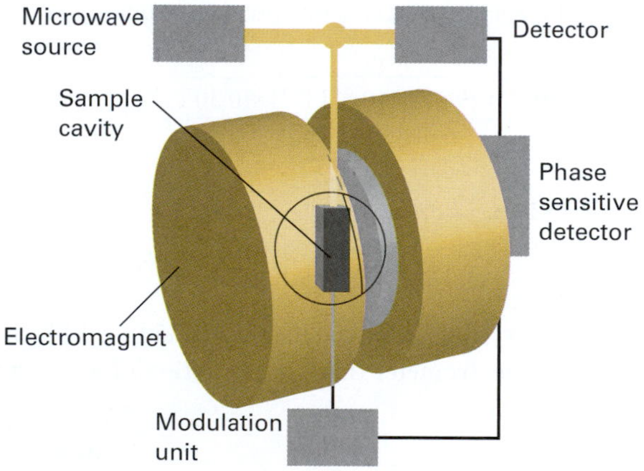

Figure 14A.5 The layout of a continuous-wave EPR spectrometer. A typical magnetic field is 0.3 T, which requires 9 GHz (3 cm) microwaves for resonance.

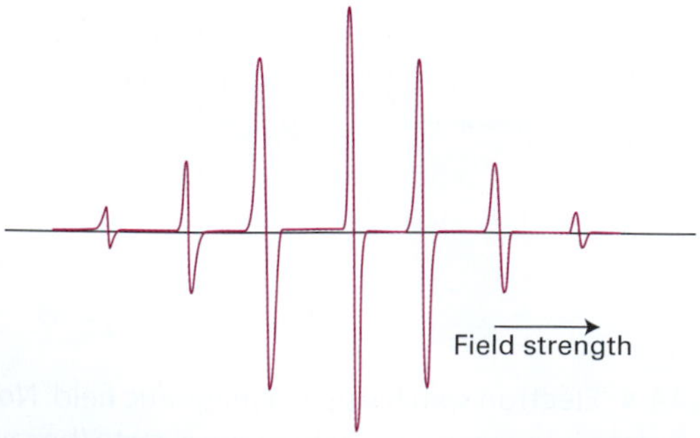

Figure 14A.6 The EPR spectrum of the benzene radical anion, $C_6H_6^-$, in fluid solution.

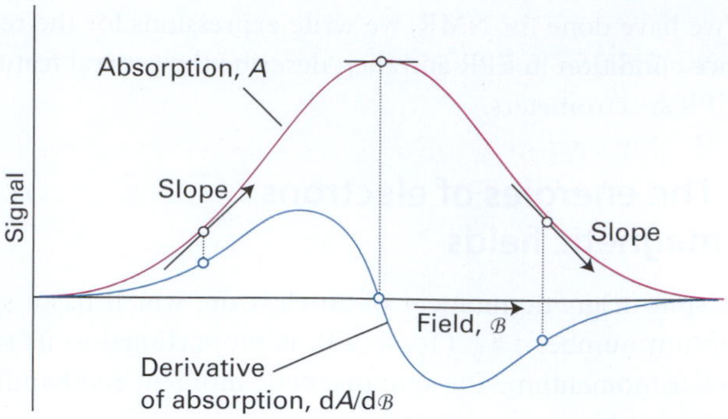

Figure 14A.7 When phase-sensitive detection is used, the signal is the first derivative of the absorption intensity. Note that the peak of the absorption corresponds to the point where the derivative passes through zero.

As usual, the intensities of spectral lines in EPR depend on the difference in populations between the ground and excited states. For an electron, the β state lies below the α state in energy and, by a similar argument to that for nuclei,

$$N_\beta - N_\alpha \approx \frac{N g_e \mu_B \mathcal{B}_0}{2kT} \quad \textit{Electrons} \quad \text{Population difference} \qquad (14A.13)$$

where N is the total number of spins.

Brief illustration 14A.4 Electron spin populations

When 1000 electron spins are exposed to a 1.0 T magnetic field at 20 °C (293 K),

$$N_\beta - N_\alpha \approx \frac{\overbrace{1000}^{N}\times\overbrace{2.0023}^{g_e}\times\overbrace{(9.274\times10^{-24}\,\mathrm{J\,T^{-1}})}^{\mu_B}\times\overbrace{(1.0\,\mathrm{T})}^{\mathcal{B}_0}}{2\times\underbrace{(1.381\times10^{-23}\,\mathrm{J\,K^{-1}})}_{k}\times\underbrace{(293\,\mathrm{K})}_{T}}$$

$$\approx 2.3$$

There is an imbalance of populations of only about 2 electrons in a thousand. However, the imbalance is much larger for electron spins than for nuclear spins (*Brief illustration* 14A.2) because the energy separation between the spin states of electrons is larger than that for nuclear spins even at the lower magnetic field strengths normally employed.

Self-test 14A.4 It is common to conduct EPR experiments at very low temperatures. At what temperature would the imbalance in spin populations be 5 electrons in 100, with $\mathcal{B}_0 = 0.30$ T?

Answer: 4 K

Checklist of concepts

- ☐ 1. The **nuclear spin quantum number**, I, of a nucleus is either a non-negative integer or half-integer.
- ☐ 2. Nuclei with different values of m_I have different energies in the presence of a magnetic field.
- ☐ 3. **Nuclear magnetic resonance** (NMR) is the observation of resonant absorption of radiofrequency electromagnetic radiation by nuclei in a magnetic field.
- ☐ 4. NMR spectrometers consist of a source of radiofrequency radiation and a magnet that provides a strong, uniform field.
- ☐ 5. The resonance absorption intensity increases with the strength of the applied magnetic field (as $\mathcal{B}_0^2$).
- ☐ 6. Electrons with different values of m_s have different energies in the presence of a magnetic field.
- ☐ 7. **Electron paramagnetic resonance** (EPR) is the observation of resonant absorption of microwave electromagnetic radiation by unpaired electrons in a magnetic field.
- ☐ 8. EPR spectrometers consist of a microwave source, a cavity in which the sample is inserted, a microwave detector, and an electromagnet.

Checklist of equations

Property	Equation	Comment	Equation number
Nuclear magneton	$\mu_N = e\hbar/2m_p$	$\mu_N = 5.051\times10^{-27}\,\mathrm{J\,T^{-1}}$	14A.4b
Energies of a nuclear spin in a magnetic field	$E_{m_I} = -\gamma_N\hbar\mathcal{B}_0 m_I$ $= -g_I\mu_N\mathcal{B}_0 m_I$		14A.4c
Resonance condition (spin-$\frac{1}{2}$ nuclei)	$h\nu = \gamma_N\hbar\mathcal{B}_0$	$\gamma_N > 0$	14A.6
Larmor frequency	$\nu_L = \gamma_N\mathcal{B}_0/2\pi$	$\gamma_N > 0$	14A.7
Population difference (nuclei)	$N_\alpha - N_\beta \approx N\gamma_N\hbar\mathcal{B}_0/2kT$		14A.8b
Magnetogyric ratio (electron)	$\gamma_e = -g_e e/2m_e$	$g_e = 2.002\,319$	14A.9b
Energies of an electron spin in a magnetic field	$E_{m_s} = -\gamma_e\hbar\mathcal{B}_0 m_s$ $= g_e\mu_B\mathcal{B}_0 m_s$		14A.11b
Bohr magneton	$\mu_B = e\hbar/2m_e$	$\mu_B = 9.274\times10^{-24}\,\mathrm{J\,T^{-1}}$	14A.11c
Resonance condition (electrons)	$h\nu = g_e\mu_B\mathcal{B}_0$		14A.12b
Population difference (electrons)	$N_\beta - N_\alpha \approx Ng_e\mu_B\mathcal{B}_0/2kT$		14A.13

14B Features of NMR spectra

Contents

➤ Why do you need to know this material?

To make progress with the analysis of NMR spectra and extract the wealth of information they contain you need to understand how the appearance of a spectrum correlates with molecular structure.

➤ What is the key idea?

The resonance frequency of a magnetic nucleus is affected by its electronic environment and the presence of magnetic nuclei in its vicinity.

➤ What do you need to know already?

You need to be familiar with the general principles of magnetic resonance (Topic 14A) and specifically that resonance occurs when the frequency of the radiofrequency field matches the Larmor frequency.

Nuclear magnetic moments interact with the *local* magnetic field. The local field may differ from the applied field because the latter induces electronic orbital angular momentum (that is, the circulation of electronic currents) which gives rise to a small additional magnetic field $\delta\mathcal{B}$ at the nuclei. This additional field is proportional to the applied field, and it is conventional to write

$$\delta\mathcal{B} = -\sigma\mathcal{B}_0 \qquad \textit{Definition} \quad \text{Shielding constant} \qquad (14\text{B}.1)$$

where the dimensionless quantity σ is called the **shielding constant** of the nucleus (σ is usually positive but may be negative). The ability of the applied field to induce an electronic current in the molecule, and hence affect the strength of the resulting local magnetic field experienced by the nucleus, depends on the details of the electronic structure near the magnetic nucleus of interest, so nuclei in different chemical groups have different shielding constants. The calculation of reliable values of the shielding constant is very difficult, but trends in it are quite well understood and we concentrate on them.

14B.1 The chemical shift

Because the total local field $\mathcal{B}_{\text{loc}}$ is

$$\mathcal{B}_{\text{loc}} = \mathcal{B}_0 + \delta\mathcal{B} = (1-\sigma)\mathcal{B}_0 \qquad (14\text{B}.2)$$

the nuclear Larmor frequency (eqn 14A.7 of Topic 14A, $\nu_{\text{L}} = \gamma_{\text{N}}\mathcal{B}/2\pi$) becomes

$$\nu_L = \frac{\gamma_N \mathcal{B}_{loc}}{2\pi} = \frac{\gamma_N \mathcal{B}_0}{2\pi}(1-\sigma) \qquad (14B.3)$$

This frequency is different for nuclei in different environments. Hence, different nuclei, even of the same element, come into resonance at different frequencies if they are in different molecular environments.

The **chemical shift** of a nucleus is the difference between its resonance frequency and that of a reference standard. The standard for protons is the proton resonance in tetramethylsilane, $Si(CH_3)_4$, commonly referred to as TMS, which bristles with protons and dissolves without reaction in many solutions. For ^{13}C, the reference frequency is the ^{13}C resonance in TMS, and for ^{31}P it is the ^{31}P resonance in 85 per cent $H_3PO_4(aq)$. Other references are used for other nuclei. The separation of the resonance of a particular group of nuclei from the standard increases with the strength of the applied magnetic field because the induced field is proportional to the applied field; the stronger the latter, the greater the shift.

Chemical shifts are reported on the **δ scale**, which is defined as

$$\delta = \frac{\nu - \nu^\circ}{\nu^\circ} \times 10^6 \qquad \textit{Definition} \quad \delta\text{ Scale} \qquad (14B.4)$$

where ν° is the resonance frequency of the standard. The advantage of the δ scale is that shifts reported on it are independent of the applied field (because both numerator and denominator are proportional to the applied field). The resonance frequencies themselves, however, do depend on the applied field through

$$\nu = \nu^\circ + (\nu^\circ/10^6)\delta \qquad (14B.5)$$

Brief illustration 14B.1 The δ scale

A nucleus with $\delta = 1.00$ in a spectrometer where $\nu^\circ = 500$ MHz (a '500 MHz NMR spectrometer'), will have a shift relative to the reference equal to

$$\nu - \nu^\circ = (500\,\text{MHz}/10^6) \times 1.00 = (500\,\text{Hz}) \times 1.00 = 500\,\text{Hz}$$

because $1\,\text{MHz} = 10^6\,\text{Hz}$. In a spectrometer operating at $\nu^\circ = 100$ MHz, the shift relative to the reference would be only 100 Hz.

A note on good practice In much of the literature, chemical shifts are reported in parts per million, ppm, in recognition of the factor of 10^6 in the definition; this is unnecessary. If you see '$\delta = 10$ ppm', interpret it, and use it in eqn 14B.5, as $\delta = 10$.

Self-test 14B.1 What is the shift of the resonance from TMS of a group of nuclei with $\delta = 3.50$ and an operating frequency of 350 MHz?

Answer: 1.23 kHz

The relation between δ and σ is obtained by substituting eqn 14B.3 into eqn 14B.4:

$$\delta = \frac{(1-\sigma)\mathcal{B}_0 - (1-\sigma^\circ)\mathcal{B}_0}{(1-\sigma^\circ)\mathcal{B}_0} \times 10^6 = \frac{\sigma^\circ - \sigma}{1-\sigma^\circ} \times 10^6 \approx (\sigma^\circ - \sigma) \times 10^6 \qquad \text{Relation between } \delta \text{ and } \sigma \quad (14B.6)$$

The last line follows from $\sigma^\circ \ll 1$. As the shielding constant σ, gets smaller, δ increases. Therefore, we speak of nuclei with large chemical shifts as being strongly **deshielded.** Some typical chemical shifts are given in Fig. 14B.1. As can be seen from the illustration, the nuclei of different elements have very different ranges of chemical shifts. The ranges exhibit the variety of electronic environments of the nuclei in molecules: the higher the atomic number of the element, the greater the number of electrons around the nucleus and hence the greater the range of the extent of shielding. By convention, NMR spectra are plotted with δ increasing from right to left.

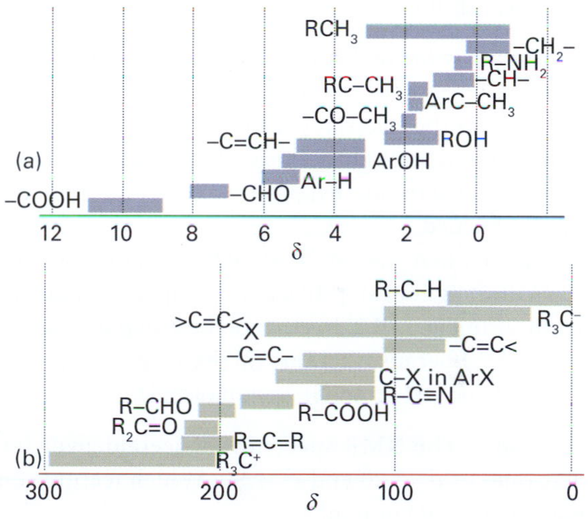

Figure 14B.1 The range of typical chemical shifts for (a) ^1H resonances and (b) ^{13}C resonances.

Example 14B.1 Interpreting the NMR spectrum of ethanol

Figure 14B.2 shows the NMR spectrum of ethanol. Account for the observed chemical shifts.

Method Consider the effect of an electron-withdrawing atom: it deshields strongly those protons to which it is bound, and has a smaller effect on distant protons.

Answer The spectrum is consistent with the following assignments:

- The CH_3 protons form one group of nuclei with $\delta = 1$.

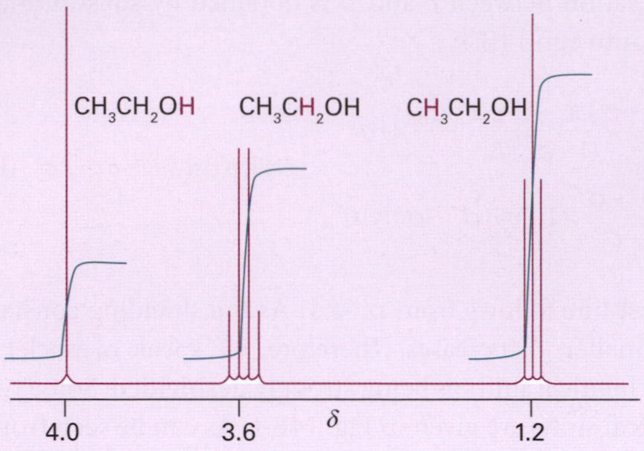

Figure 14B.2 The ^{1}H-NMR spectrum of ethanol. The bold letters denote the protons giving rise to the resonance peak, and the step-like curve is the integrated signal.

- The two CH_2 protons are in a different part of the molecule, experience a different local magnetic field, and resonate at $\delta=3$.
- The OH proton is in another environment, and has a chemical shift of $\delta=4$.

The increasing value of δ (that is, the decrease in shielding) is consistent with the electron-withdrawing power of the O atom: it reduces the electron density of the OH proton most, and that proton is strongly deshielded. It reduces the electron density of the distant methyl protons least, and those nuclei are least deshielded.

The relative intensities of the signals are commonly represented as the height of step-like curves superimposed on the spectrum, as in Fig. 14B.2. In ethanol the group intensities are in the ratio 3:2:1 because there are three CH_3 protons, two CH_2 protons, and one OH proton in each molecule.

Self-test 14B.2 The NMR spectrum of acetaldehyde (ethanal) has lines at $\delta=2.20$ and $\delta=9.80$. Which feature can be assigned to the CHO proton?

Answer: $\delta=9.80$

14B.2 The origin of shielding constants

The calculation of shielding constants is difficult, even for small molecules, for it requires detailed information (using the techniques outlined in Topic 10E) about the distribution of electron density in the ground and excited states and the excitation energies of the molecule. Nevertheless, considerable success has been achieved with small molecules such as H_2O and CH_4 and even large molecules, such as proteins, are within the scope of some types of calculation. However, it is easier to understand the different contributions to chemical shifts by studying the large body of empirical information now available.

The empirical approach supposes that the observed shielding constant is the sum of three contributions:

$$\sigma=\sigma(\text{local})+\sigma(\text{neighbour})+\sigma(\text{solvent}) \quad (14B.7)$$

The **local contribution**, σ(local), is essentially the contribution of the electrons of the atom that contains the nucleus in question. The **neighbouring group contribution**, σ(neighbour), is the contribution from the groups of atoms that form the rest of the molecule. The **solvent contribution**, σ(solvent), is the contribution from the solvent molecules.

(a) The local contribution

It is convenient to regard the local contribution to the shielding constant as the sum of a **diamagnetic contribution**, σ_d, and a **paramagnetic contribution**, σ_p:

$$\sigma(\text{local})=\sigma_d+\sigma_p \quad \text{Local contribution to the shielding constant} \quad (14B.8)$$

A diamagnetic contribution to σ (local) opposes the applied magnetic field and shields the nucleus in question. A paramagnetic contribution to σ (local) reinforces the applied magnetic field and deshields the nucleus in question. Therefore, $\sigma_d>0$ and $\sigma_p<0$. The total local contribution is positive if the diamagnetic contribution dominates, and is negative if the paramagnetic contribution dominates.

The diamagnetic contribution arises from the ability of the applied field to generate a circulation of charge in the ground-state electron distribution of the atom. The circulation generates a magnetic field that opposes the applied field and hence shields the nucleus. The magnitude of σ_d depends on the electron density close to the nucleus and can be calculated from the **Lamb formula**:[1]

$$\sigma_d=\frac{e^2\mu_0}{12\pi m_e}\left\langle\frac{1}{r}\right\rangle \quad \text{Lamb formula} \quad (14B.9)$$

where μ_0 is the vacuum permeability (a fundamental constant, see inside the front cover) and r is the electron–nucleus distance.

Example 14B.2 Using the Lamb formula

Calculate the shielding constant for the proton in a free H atom.

Method To calculate σ_d from the Lamb formula, calculate the expectation value of $1/r$ for a hydrogen 1s orbital. Wavefunctions are given in Table 9A.1.

Answer The wavefunction for a hydrogen 1s orbital is

[1] For a derivation, see our *Molecular quantum mechanics* (2011).

$$\psi = \left(\frac{1}{\pi a_0^3}\right)^{1/2} e^{-r/a_0}$$

so, because $d\tau = r^2 dr \sin\theta\, d\theta d\phi$, the expectation value of $1/r$ is

$$\left\langle \frac{1}{r} \right\rangle = \int \frac{\psi^* \psi}{r} d\tau = \frac{1}{\pi a_0^3} \overbrace{\int_0^{2\pi} d\phi}^{2\pi} \overbrace{\int_0^{\pi} \sin\theta\, d\theta}^{2} \overbrace{\int_0^{\infty} r e^{-2r/a_0} dr}^{\text{Integral E.1}}$$

$$= \frac{1}{\pi a_0^3} \times 4\pi \times \frac{a_0^2}{4} = \frac{1}{a_0}$$

Therefore,

$$\sigma_d = \frac{e^2 \mu_0}{12\pi m_e a_0} = \frac{(1.602\times10^{-19}\,\text{C})^2 \times \left(4\pi\times10^{-7}\,\overbrace{\text{J}}^{\text{kg m}^2\,\text{s}^{-2}}\,\text{s}^2\,\text{C}^{-2}\,\text{m}^{-1}\right)}{12\pi\times(9.109\times10^{-31}\,\text{kg})\times(5.292\times10^{-11}\,\text{m})}$$

$$= 1.775\times10^{-5}$$

Self-test 14B.3 Derive a general expression for σ_d that applies to all hydrogenic atoms.

Answer: $Ze^2\mu_0/12\pi m_e a_0$

The diamagnetic contribution is the only contribution in closed-shell free atoms. It is also the only contribution to the local shielding for electron distributions that have spherical or cylindrical symmetry. Thus, it is the only contribution to the local shielding from inner cores of atoms, for cores remain nearly spherical even though the atom may be a component of a molecule and its valence electron distribution is highly distorted. The diamagnetic contribution is broadly proportional to the electron density of the atom containing the nucleus of interest. It follows that the shielding is decreased if the electron density on the atom is reduced by the influence of an electronegative atom nearby. That reduction in shielding as the electronegativity of a neighbouring atom increases translates into an increase in the chemical shift δ (Fig. 14B.3).

The local paramagnetic contribution, σ_p, arises from the ability of the applied field to force electrons to circulate through the molecule by making use of orbitals that are unoccupied in the ground state. It is zero in free atoms and around the axes of linear molecules (such as ethyne, HC≡CH) where the electrons can circulate freely and a field applied along the internuclear axis is unable to force them into other orbitals. We can expect large paramagnetic contributions from small atoms (because the induced currents are then close to the nucleus) in molecules with low lying excited states (because an applied field can then induce significant currents). In fact, the paramagnetic contribution is the dominant local contribution for atoms other than hydrogen.

(b) Neighbouring group contributions

The neighbouring group contribution arises from the currents induced in nearby groups of atoms. Consider the influence

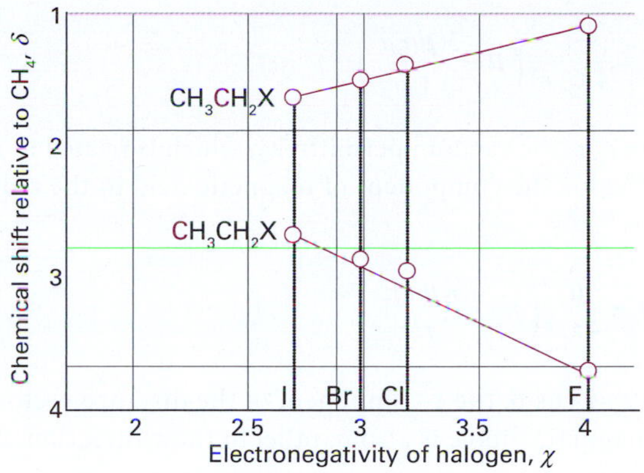

Figure 14B.3 The variation of chemical shielding with electronegativity. The shifts for the methyl protons agree with the trend expected with increasing electronegativity. However, to emphasize that chemical shifts are subtle phenomena, notice that the trend for the methylene protons is opposite to that expected. For these protons another contribution (the magnetic anisotropy of C–H and C–X bonds) is dominant.

of the neighbouring group X on the proton H in a molecule such as H–X. The applied field generates currents in the electron distribution of X and gives rise to an induced magnetic moment proportional to the applied field; the constant of proportionality is the magnetic susceptibility, χ (chi), of the group X: $\mu_{\text{induced}} = \chi \mathcal{B}_0$. The susceptibility is negative for a diamagnetic group because the induced moment is opposite to the direction of the applied field. The induced moment gives rise to a magnetic field with a component parallel to the applied field and at a distance r and angle θ (1) that has the form (*The chemist's toolkit* 14B.1):

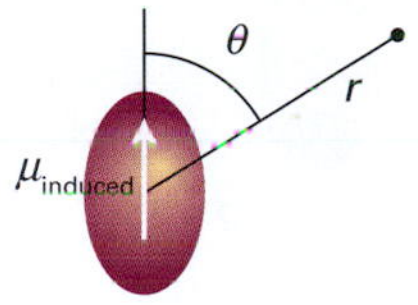

1

$$\mathcal{B}_{\text{local}} \propto \frac{\mu_{\text{induced}}}{r^3}(1 - 3\cos^2\theta) \qquad \text{Local dipolar field} \qquad (14\text{B}.10\text{a})$$

The chemist's toolkit 14B.1 Dipolar fields

Standard electromagnetic theory gives the magnetic field at a point $\boldsymbol{r}$ from a point magnetic dipole $\boldsymbol{\mu}$ as

$$\mathcal{B} = \frac{\mu_0}{4\pi r^3}\left(\boldsymbol{\mu} - \frac{3(\boldsymbol{\mu}\cdot\boldsymbol{r})\boldsymbol{r}}{r^2}\right)$$

where μ_0 is the vacuum permeability (a fundamental constant with the defined value $4\pi\times10^{-7}\,\text{T}^2\,\text{J}^{-1}\,\text{m}^3$). The electric field due to a point electric dipole is given by a similar expression:

$$\mathcal{E}=\frac{1}{4\pi\varepsilon_0 r^3}\left(\boldsymbol{\mu}-\frac{3(\boldsymbol{\mu}\cdot\boldsymbol{r})\boldsymbol{r}}{r^2}\right)$$

where ε_0 is the vacuum permittivity, which is related to μ_0 by $\varepsilon_0=1/\mu_0 c^2$. The component of magnetic field in the z-direction is

$$\mathcal{B}_z=\frac{\mu_0}{4\pi r^3}\left(\mu_z-\frac{3(\boldsymbol{\mu}\cdot\boldsymbol{r})z}{r^2}\right)$$

with $z=r\cos\theta$, the z-component of the distance vector $\boldsymbol{r}$. If the magnetic dipole is also parallel to the z-direction, it follows that

$$\mathcal{B}_z=\frac{\mu_0}{4\pi r^3}\left(\overbrace{\mu}^{\mu_z}-\frac{3\overbrace{(\mu r\cos\theta)}^{\boldsymbol{\mu}\cdot\boldsymbol{r}}\overbrace{(r\cos\theta)}^{z}}{r^2}\right)=\frac{\mu\mu_0}{4\pi r^3}(1-3\cos^2\theta)$$

We see that the strength of the additional magnetic field experienced by the proton is inversely proportional to the cube of the distance r between H and X. If the magnetic susceptibility is independent of the orientation of the molecule (is 'isotropic'), the local field averages to zero because $1-3\cos^2\theta$ is zero when averaged over a sphere (see Problem 14B.7). To a good approximation, the shielding constant σ(neighbour) depends on the distance r

$$\sigma(\text{neighbour})\propto(\chi_{\parallel}-\chi_{\perp})\left(\frac{1-3\cos^2\theta}{r^3}\right) \quad \text{Neighbouring group contribution} \quad (14B.10b)$$

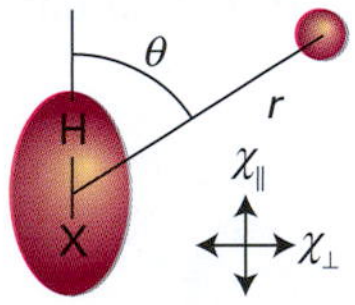

2

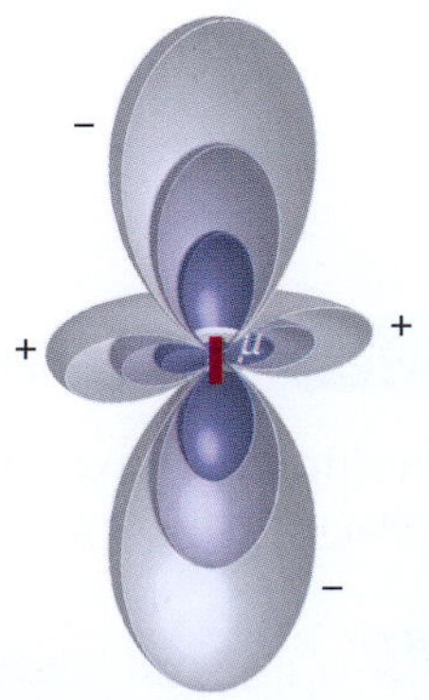

Figure 14B.4 A depiction of the field arising from a point magnetic dipole. The three shades of colour represent the strength of field declining with distance (as $1/r^3$), and each surface shows the angle dependence of the z-component of the field for each distance.

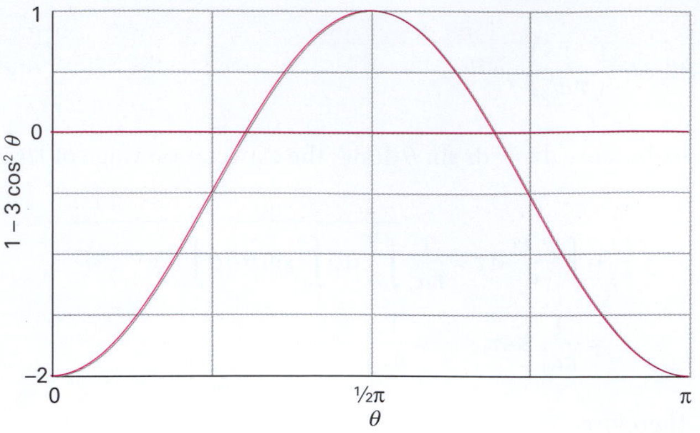

Figure 14B.5 The variation of the function $1-3\cos^2\theta$ with the angle θ.

where $\chi_{\parallel}$ and $\chi_{\perp}$ are, respectively, the parallel and perpendicular components of the magnetic susceptibility, and θ is the angle between the X—H axis and the symmetry axis of the neighbouring group (**2**). Equation 14B.10b shows that the neighbouring group contribution may be positive or negative according to the relative magnitudes of the two magnetic susceptibilities and the relative orientation of the nucleus with respect to X. If $54.7° < \theta < 125.3°$, then $1-3\cos^2\theta$ is positive, but it is negative otherwise (Figs. 14B.4 and 14B.5).

Brief illustration 14B.2 Ring currents

A special case of a neighbouring group effect is found in aromatic compounds. The strong anisotropy of the magnetic susceptibility of the benzene ring is ascribed to the ability of the field to induce a *ring current*, a circulation of electrons around the ring, when it is applied perpendicular to the molecular plane. Protons in the plane are deshielded (Fig. 14B.6), but any that happen to lie above or below the plane (as members of substituents of the ring) are shielded.

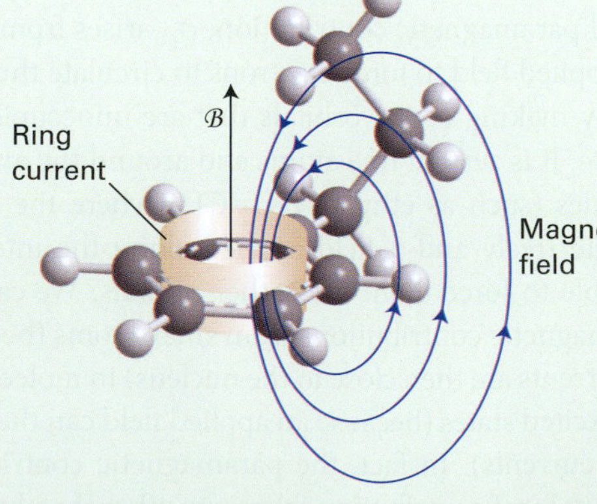

Figure 14B.6 The shielding and deshielding effects of the ring current induced in the benzene ring by the applied field. Protons attached to the ring are deshielded but a proton attached to a substituent that projects above the ring is shielded.

Self-test 14B.4 Consider ethyne, HC≡CH. Are its protons shielded or deshielded by currents induced by the triple bond?

Answer: Shielded

(c) The solvent contribution

A solvent can influence the local magnetic field experienced by a nucleus in a variety of ways. Some of these effects arise from specific interactions between the solute and the solvent (such as hydrogen-bond formation and other forms of Lewis acid–base complex formation). The anisotropy of the magnetic susceptibility of the solvent molecules, especially if they are aromatic, can also be the source of a local magnetic field. Moreover, if there are steric interactions that result in a loose but specific interaction between a solute molecule and a solvent molecule, then protons in the solute molecule may experience shielding or deshielding effects according to their location relative to the solvent molecule.

Brief illustration 14B.3 The effect of aromatic solvents

An aromatic solvent like benzene can give rise to local currents that shield or deshield a proton in a solute molecule. The arrangement shown in Fig. 14B.7 leads to shielding of a proton on the solute molecule.

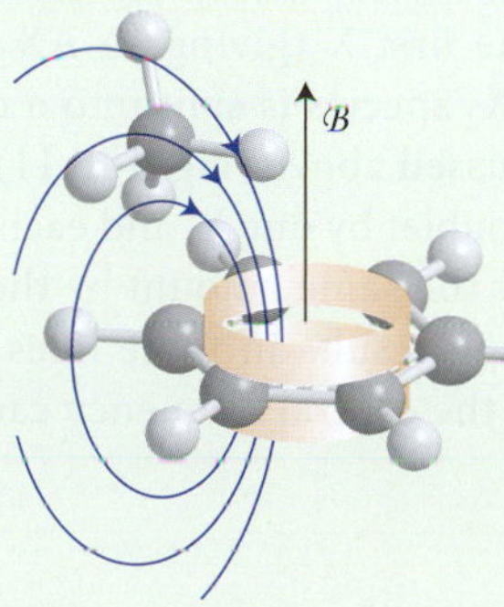

Figure 14B.7 An aromatic solvent (benzene here) can give rise to local currents that shield or deshield a proton in a solute molecule. In this relative orientation of the solvent and solute, the proton on the solute molecule is shielded.

Self-test 14B.5 Refer to Fig. 14B.7 and suggest an arrangement that leads to deshielding of a proton on the solute molecule.

Answer: Proton on the solute molecule coplanar with the benzene ring

14B.3 The fine structure

The splitting of resonances into individual lines by spin–spin coupling shown in Fig. 14B.2 is called the **fine structure** of the spectrum. It arises because each magnetic nucleus may contribute to the local field experienced by the other nuclei and so modify their resonance frequencies. The strength of the interaction is expressed in terms of the **scalar coupling constant**, J. The scalar coupling constant is so called because the energy of interaction it describes is proportional to the scalar product of the two interacting spins: $E \propto \boldsymbol{I}_1 \cdot \boldsymbol{I}_2$. As explained in *Mathematical background* 5, a scalar product depends on the angle between the two vectors, so writing the energy in this way is simply a way of saying that the energy of interaction between two spins depends on their relative orientation. The constant of proportionality in this expression is written $hJ/\hbar^2$ (so $E = (hJ/\hbar^2)\boldsymbol{I}_1 \cdot \boldsymbol{I}_2$): because each spin angular momentum is proportional to $\hbar$, E is then proportional to hJ and J is a frequency (with units hertz, Hz). For nuclei that are constrained to align with the applied field in the z-direction, the only contribution to $\boldsymbol{I}_1 \cdot \boldsymbol{I}_2$ is $I_{1z}I_{2z}$, with eigenvalues $m_1 m_2 \hbar^2$, so in that case the energy due to spin–spin coupling is

$$E_{m_1 m_2} = hJm_1 m_2 \quad \text{Spin–spin coupling energy} \quad (14B.11)$$

(a) The appearance of the spectrum

In NMR, letters far apart in the alphabet (typically A and X) are used to indicate nuclei with very different chemical shifts; letters close together (such as A and B) are used for nuclei with similar chemical shifts. We shall consider first an AX system, a molecule that contains two spin-$\frac{1}{2}$ nuclei A and X with very different chemical shifts in the sense that the difference in chemical shift corresponds to a frequency that is large compared to J.

For a spin-$\frac{1}{2}$ AX system there are four spin states: $\alpha_A\alpha_X$, $\alpha_A\beta_X$, $\beta_A\alpha_X$, $\beta_A\beta_X$. The energy depends on the orientation of the spins in the external magnetic field, and if spin–spin coupling is neglected

$$\begin{aligned} E_{m_A m_X} &= -\gamma_N \hbar(1-\sigma_A)\mathcal{B}_0 m_A - \gamma_N \hbar(1-\sigma_X)\mathcal{B}_0 m_X \\ &= -h\nu_A m_A - h\nu_X m_X \end{aligned} \quad (14B.12a)$$

where ν_A and ν_X are the Larmor frequencies of A and X and m_A and m_X are their quantum numbers ($m_A = \pm\frac{1}{2}, m_X = \pm\frac{1}{2}$). This expression gives the four lines on the left of Fig. 14B.8. When spin–spin coupling is included (by using eqn 14B.11), the energy levels are

$$E_{m_A m_X} = -h\nu_A m_A - h\nu_X m_X + hJm_A m_X \quad (14B.12b)$$

If $J>0$, a lower energy is obtained when $m_A m_X < 0$, which is the case if one spin is α and the other is β. A higher energy is obtained if both spins are α or both spins are β. The opposite is true if $J<0$. The resulting energy level diagram (for $J>0$) is shown on the right of Fig. 14B.8. We see that the αα and ββ

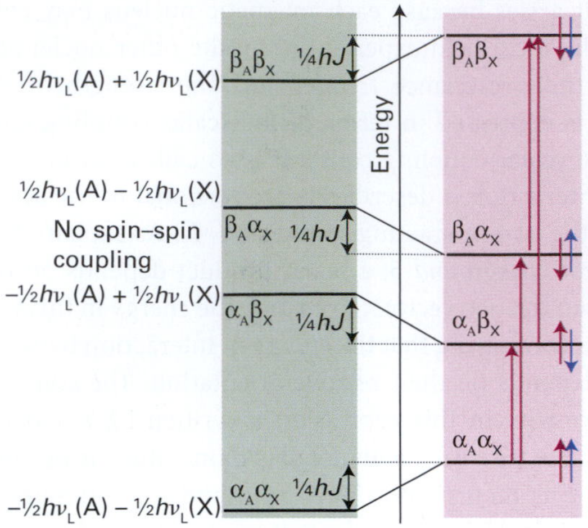

Figure 14B.8 The energy levels of an AX system. The four levels on the left are those of the two spins in the absence of spin–spin coupling. The four levels on the right show how a positive spin–spin coupling constant affects the energies. The transitions shown are for $\beta \leftarrow \alpha$ of A or X, the other nucleus (X or A, respectively) remaining unchanged. We have exaggerated the effect for clarity. In practice, the splitting caused by spin–spin coupling is much smaller than that caused by the applied field.

states are both raised by $\frac{1}{4}hJ$ and that the $\alpha\beta$ and $\beta\alpha$ states are both lowered by $\frac{1}{4}hJ$.

When a transition of nucleus A occurs, nucleus X remains unchanged. Therefore, the A resonance is a transition for which $\Delta m_A = +1$ and $\Delta m_X = 0$. There are two such transitions, one in which $\beta_A \leftarrow \alpha_A$ occurs when the X nucleus is α, and the other in which $\beta_A \leftarrow \alpha_A$ occurs when the X nucleus is β. They are shown in Fig. 14B.8 and in a slightly different form in Fig. 14B.9. The energies of the transitions are

$$\Delta E = h\nu_A \pm \tfrac{1}{2}hJ \qquad (14B.13a)$$

Therefore, the A resonance consists of a doublet of separation J centred on the chemical shift of A (Fig. 14B.10). Similar remarks apply to the X resonance, which consists of two

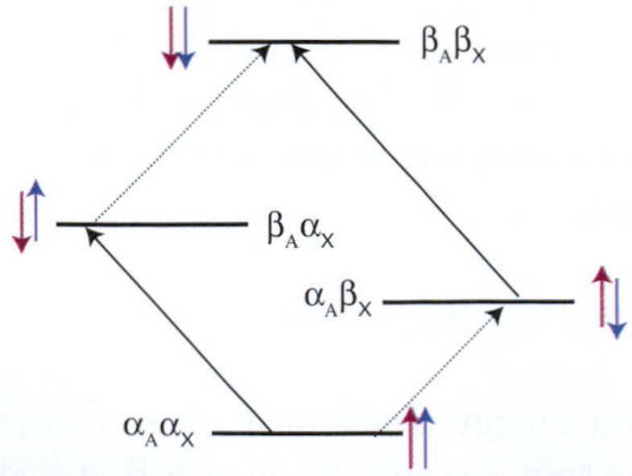

Figure 14B.9 An alternative depiction of the energy levels and transitions shown in Fig. 14B.8. Once again, we have exaggerated the effect of spin–spin coupling.

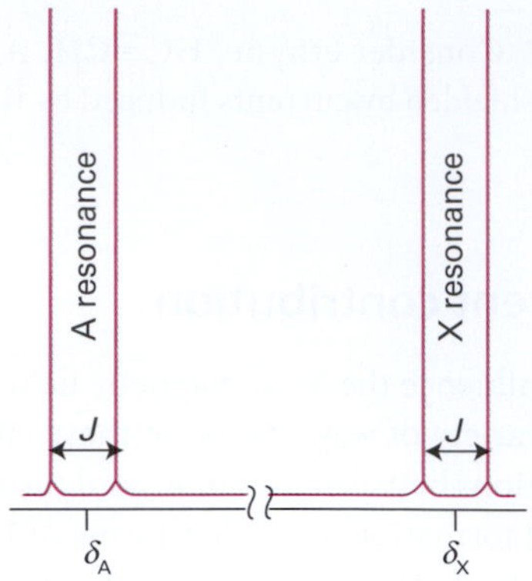

Figure 14B.10 The effect of spin–spin coupling on an AX spectrum. Each resonance is split into two lines separated by J. The pairs of resonances are centred on the chemical shifts of the protons in the absence of spin–spin coupling.

transitions according to whether the A nucleus is α or β (as shown in Fig. 14B.9). The transition energies are

$$\Delta E = h\nu_X \pm \tfrac{1}{2}hJ \qquad (14B.13b)$$

It follows that the X resonance also consists of two lines of the same separation J, but they are centred on the chemical shift of X (as shown in Fig. 14B.10).

If there is another X nucleus in the molecule with the same chemical shift as the first X (giving an AX_2 species), the X resonance of the AX_2 species is split into a doublet by A, as in the AX case discussed above (Fig. 14B.11). The resonance of A is split into a doublet by one X, and each line of the doublet is split again by the same amount by the second X (Fig. 14B.12). This splitting results in three lines in the intensity ratio 1:2:1 (because the central frequency can be obtained in two ways).

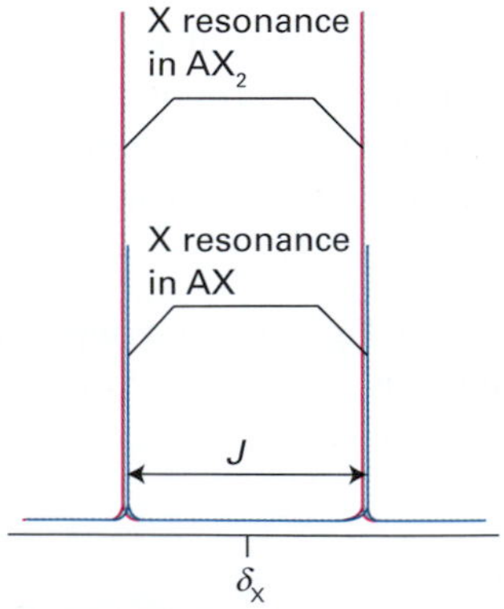

Figure 14B.11 The X resonance of an AX_2 species is also a doublet, because the two equivalent X nuclei behave like a single nucleus; however, the overall absorption is twice as intense as that of an AX species.

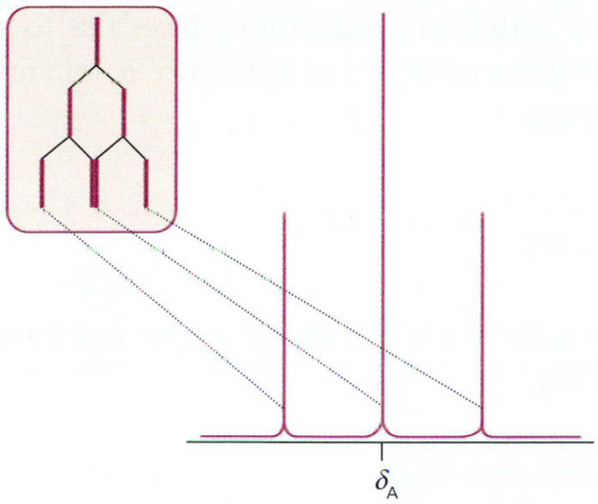

Figure 14B.12 The origin of the 1:2:1 triplet in the A resonance of an AX_2 species. The resonance of A is split into two by coupling with one X nucleus (as shown in the inset), and then each of those two lines is split into two by coupling to the second X nucleus. Because each X nucleus causes the same splitting, the two central transitions are coincident and give rise to an absorption line of double the intensity of the outer lines.

Three equivalent X nuclei (an AX_3 species) split the resonance of A into four lines of intensity ratio 1:3:3:1 (Fig. 14B.13). The X resonance remains a doublet as a result of the splitting caused by A. In general, N equivalent spin-$\frac{1}{2}$ nuclei split the resonance of a nearby spin or group of equivalent spins into $N+1$ lines with an intensity distribution given by Pascal's triangle (**3**). Successive rows of this triangle are formed by adding together the two adjacent numbers in the line above.

1
1 1
1 2 1
1 3 3 1
1 4 6 4 1
1 5 10 10 5 1

3

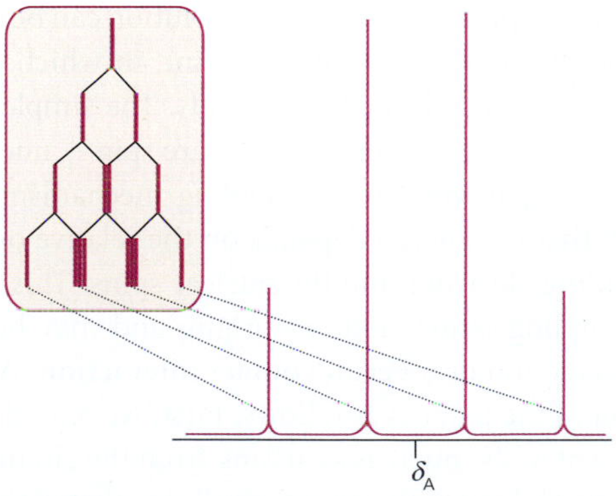

Figure 14B.13 The origin of the 1:3:3:1 quartet in the A resonance of an AX_3 species. The third X nucleus splits each of the lines shown in Fig. 14B.11 for an AX_2 species into a doublet, and the intensity distribution reflects the number of transitions that have the same energy.

Example 14B.3 Accounting for the fine structure in a spectrum

Account for the fine structure in the NMR spectrum of the C–H protons of ethanol.

Method Consider how each group of equivalent protons (for instance, three methyl protons) split the resonances of the other groups of protons. There is no splitting within groups of equivalent protons. Each splitting pattern can be decided by referring to Pascal's triangle.

Answer The three protons of the CH_3 group split the resonance of the CH_2 protons into a 1:3:3:1 quartet with a splitting J. Likewise, the two protons of the CH_2 group split the resonance of the CH_3 protons into a 1:2:1 triplet with the same splitting J. The OH resonance is not split because the OH protons migrate rapidly from molecule to molecule (including molecules of impurities in the sample) and their effect averages to zero. In gaseous ethanol, where this migration does not occur, the OH resonance appears as a triplet, showing that the CH_2 protons interact with the OH proton.

Self-test 14B.6 What fine structure can be expected for the protons in $^{14}NH_4^+$? The spin quantum number of nitrogen-14 is 1.

Answer: 1:1:1 triplet from N

(b) The magnitudes of coupling constants

The scalar coupling constant of two nuclei separated by N bonds is denoted ^NJ, with subscripts for the types of nuclei involved. Thus, $^1J_{CH}$ is the coupling constant for a proton joined directly to a ^{13}C atom, and $^2J_{CH}$ is the coupling constant when the same two nuclei are separated by two bonds (as in $^{13}C{-}C{-}H$). A typical value of $^1J_{CH}$ is in the range 120 to 250 Hz; $^2J_{CH}$ is between 10 and 20 Hz. Both ^3J and ^4J can give detectable effects in a spectrum, but couplings over larger numbers of bonds can generally be ignored. One of the longest range couplings that has been detected is $^9J_{HH}=0.4$ Hz between the CH_3 and CH_2 protons in $CH_3C{\equiv}C{-}C{\equiv}C{-}C{\equiv}C{-}CH_2OH$.

As remarked (in the discussion following eqn 14B.12b), the sign of J_{XY} indicates whether the energy of two spins is lower when they are parallel ($J<0$) or when they are antiparallel ($J>0$). It is found that $^1J_{CH}$ is often positive, $^2J_{HH}$ is often negative, $^3J_{HH}$ is often positive, and so on. An additional point is that J varies with the angle between the bonds (Fig. 14B.14). Thus, a $^3J_{HH}$ coupling constant is often found to depend on the dihedral angle ϕ (**4**) according to the **Karplus equation**:

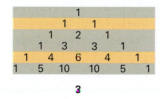

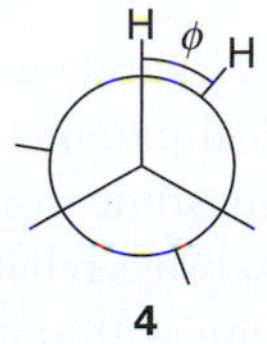

4

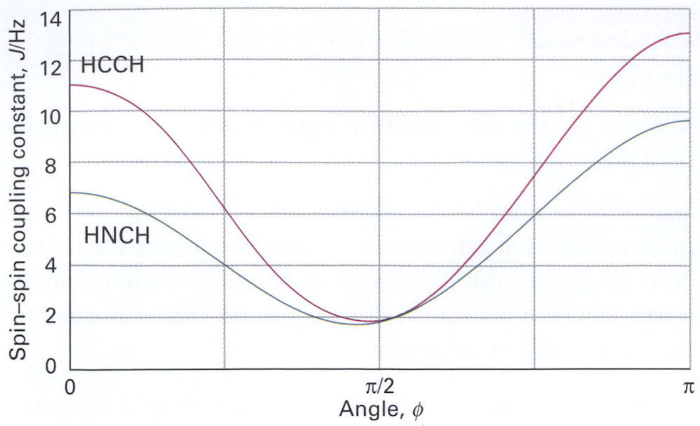

Figure 14B.14 The variation of the spin–spin coupling constant with angle predicted by the Karplus equation for an HCCH group and an HNCH group.

$$^3J_{HH} = A + B\cos\phi + C\cos 2\phi \quad \text{Karplus equation} \quad (14B.14)$$

with *A*, *B*, and *C* empirical constants with values close to +7 Hz, −1 Hz, and +5 Hz, respectively, for an HCCH fragment. It follows that the measurement of $^3J_{HH}$ in a series of related compounds can be used to determine their conformations. The coupling constant $^1J_{CH}$ also depends on the hybridization of the C atom, as the following values indicate:

	sp	sp^2	sp^3
$^1J_{CH}$/Hz	250	160	125

Brief illustration 14B.4 The Karplus equation

The investigation of H—N—C—H couplings in polypeptides can help reveal their conformation. For $^3J_{HH}$ coupling in such a group, $A = +5.1$ Hz, $B = -1.4$ Hz, and $C = +3.2$ Hz. For a helical polymer, ϕ is close to 120°, which would give $^3J_{HH} \approx 4$ Hz. For the sheet-like conformation, ϕ is close to 180°, which would give $^3J_{HH} \approx 10$ Hz.

Self-test 14B.7 NMR experiments reveal that for H—C—C—H coupling in polypeptides, $A = +3.5$ Hz, $B = -1.6$ Hz, and $C = +4.3$ Hz. In an investigation of the polypeptide flavodoxin, the $^3J_{HH}$ coupling constant for such a grouping was determined to be 2.1 Hz. Is this value consistent with a helical or sheet conformation?

Answer: Helical conformation

(c) The origin of spin–spin coupling

Spin–spin coupling is a very subtle phenomenon and it is better to treat *J* as an empirical parameter than to use calculated values. However, we can get some insight into its origins, if not its precise magnitude—or always reliably its sign—by considering the magnetic interactions within molecules.

A nucleus with spin projection m_I gives rise to a magnetic field with *z*-component $\mathcal{B}_{nuc}$ at a distance *R*, where, to a good approximation,

$$\mathcal{B}_{nuc} = -\frac{\gamma_N \hbar \mu_0}{4\pi R^3}(1 - 3\cos^2\theta)m_I \quad (14B.15)$$

The angle θ is defined in (**1**); we saw a version of this expression in eqn 14B.10a.

Brief illustration 14B.5 Magnetic fields from nuclei

The *z*-component of the magnetic field arising from a proton ($m_I = \frac{1}{2}$) at $R = 0.30$ nm, with its magnetic moment parallel to the *z*-axis ($\theta = 0$) is

$$\mathcal{B}_{nuc} = -\frac{\overbrace{(2.821\times10^{-26}\,\mathrm{J\,T^{-1}})}^{\gamma_N\hbar}\times\overbrace{4\pi\times10^{-7}\,\mathrm{T^2\,J^{-1}\,m^3}}^{\mu_0}}{4\pi\times\underbrace{(3.0\times10^{-10}\,\mathrm{m})}_{R}{}^3}\times\overbrace{(-1)}^{(1-3\cos^2\theta)m_I}$$

$$= 1.0\times10^{-4}\ \mathrm{T} = 0.10\,\mathrm{mT}$$

A field of this magnitude can give rise to the splitting of resonance signals in solid samples. In a liquid, the angle θ sweeps over all values as the molecule tumbles, and the factor $1 - 3\cos^2\theta$ averages to zero. Hence the direct dipolar interaction between spins cannot account for the fine structure of the spectra of rapidly tumbling molecules.

Self-test 14B.8 In gypsum, $CaSO_4{\cdot}2H_2O$, the splitting in the H_2O resonance can be interpreted in terms of a magnetic field of 0.715 mT generated by one proton and experienced by the other. With $\theta = 0$, what is the separation of the protons in the H_2O molecule?

Answer: 158 pm

Spin–spin coupling in molecules in solution can be explained in terms of the **polarization mechanism**, in which the interaction is transmitted through the bonds. The simplest case to consider is that of $^1J_{XY}$, where X and Y are spin-$\frac{1}{2}$ nuclei joined by an electron-pair bond. The coupling mechanism depends on the fact that the energy depends on the relative orientation of the bonding electrons and the nuclear spins. This electron–nucleus coupling is magnetic in origin, and may be either a dipolar interaction or a **Fermi contact interaction**. A pictorial description of the latter is as follows. First, we regard the magnetic moment of the nucleus as arising from the circulation of a current in a tiny loop with a radius similar to that of the nucleus (Fig. 14B.15). Far from the nucleus the field generated by this loop is indistinguishable from the field generated by a point magnetic dipole. Close to the loop, however, the field differs from that of a point dipole. The magnetic interaction between this non-dipolar field and the electron's magnetic moment is

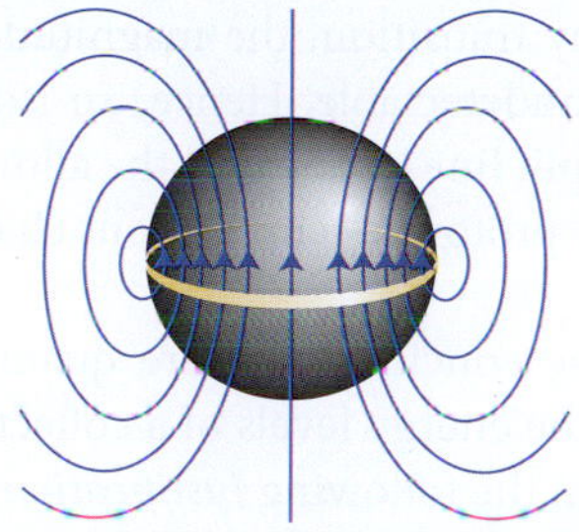

Figure 14B.15 The origin of the Fermi contact interaction. From far away, the magnetic field pattern arising from a ring of current (representing the rotating charge of the nucleus, the pale grey sphere) is that of a point dipole. However, if an electron can sample the field close to the region indicated by the sphere, the field distribution differs significantly from that of a point dipole. For example, if the electron can penetrate the sphere, then the spherical average of the field it experiences is not zero.

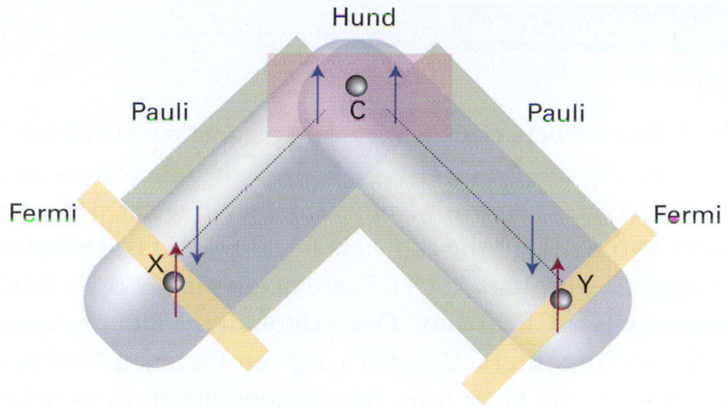

Figure 14B.17 The polarization mechanism for $^2J_{HH}$ spin–spin coupling. The spin information is transmitted from one bond to the next by a version of the mechanism that accounts for the lower energy of electrons with parallel spins in different atomic orbitals (Hund's rule of maximum multiplicity). In this case, $J < 0$, corresponding to a lower energy when the nuclear spins are parallel.

the contact interaction. The contact interaction—essentially the failure of the point-dipole approximation—depends on the very close approach of an electron to the nucleus and hence can occur only if the electron occupies an s orbital (which is the reason why $^1J_{CH}$ depends on the hybridization ratio). We shall suppose that it is energetically favourable for an electron spin and a nuclear spin to be antiparallel (as is the case for a proton and an electron in a hydrogen atom).

If the X nucleus is α, a β electron of the bonding pair will tend to be found nearby, because that is an energetically favourable arrangement (Fig. 14B.16). The second electron in the bond, which must have α spin if the other is β (by the Pauli principle; Topic 9B), will be found mainly at the far end of the bond because electrons tend to stay apart to reduce their mutual repulsion. Because it is energetically favourable for the spin of Y to be antiparallel to an electron spin, a Y nucleus with β spin has a lower energy than when it has α spin. The opposite is true when X is β, for now the α spin of Y has the lower energy. In other words, the antiparallel arrangement of nuclear spins lies lower in energy than the parallel arrangement as a

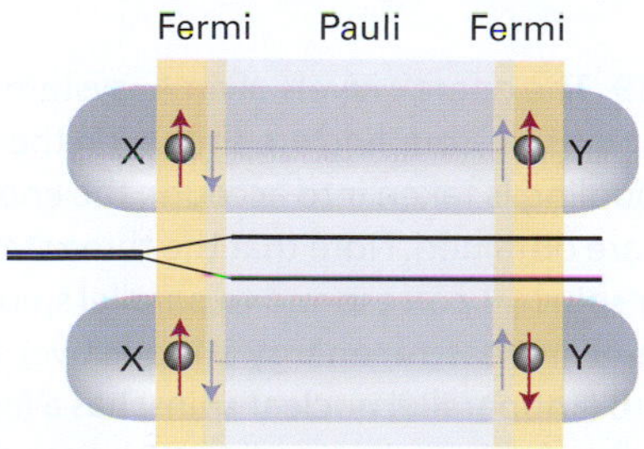

Figure 14B.16 The polarization mechanism for spin–spin coupling ($^1J_{HH}$). The two arrangements have slightly different energies. In this case, J is positive, corresponding to a lower energy when the nuclear spins are antiparallel.

result of their magnetic coupling with the bond electrons. That is, $^1J_{CH}$ is positive.

To account for the value of $^2J_{XY}$, as for $^2J_{HH}$ in H—C—H, we need a mechanism that can transmit the spin alignments through the central C atom (which may be ^{12}C, with no nuclear spin of its own). In this case (Fig. 14B.17), an X nucleus with α spin polarizes the electrons in its bond, and the α electron is likely to be found closer to the C nucleus. The more favourable arrangement of two electrons on the same atom is with their spins parallel (Hund's rule, Topic 9B), so the more favourable arrangement is for the α electron of the neighbouring bond to be close to the C nucleus. Consequently, the β electron of that bond is more likely to be found close to the Y nucleus, and therefore that nucleus will have a lower energy if it is α. Hence, according to this mechanism, the lower energy will be obtained if the Y spin is parallel to that of X. That is, $^2J_{HH}$ is negative.

The coupling of nuclear spin to electron spin by the Fermi contact interaction is most important for proton spins, but it is not necessarily the most important mechanism for other nuclei. These nuclei may also interact by a dipolar mechanism with the electron magnetic moments and with their orbital motion, and there is no simple way of specifying whether J will be positive or negative.

(d) Equivalent nuclei

A group of nuclei are **chemically equivalent** if they are related by a symmetry operation of the molecule and have the same chemical shifts. Chemically equivalent nuclei are nuclei that would be regarded as 'equivalent' according to ordinary chemical criteria. Nuclei are **magnetically equivalent** if, as well as being chemically equivalent, they also have identical spin–spin interactions with any other magnetic nuclei in the molecule.

Brief illustration 14B.6 Chemical and magnetic equivalence

The difference between chemical and magnetic equivalence is illustrated by CH_2F_2 and $H_2C{=}CF_2$. In each of these molecules the protons are chemically equivalent: they are related by symmetry and undergo the same chemical reactions. However, although the protons in CH_2F_2 are magnetically equivalent, those in $CH_2{=}CF_2$ are not. One proton in the latter has a *cis* spin-coupling interaction with a given F nucleus whereas the other proton has a *trans* interaction with it. In contrast, in CH_2F_2 both protons are connected to a given F nucleus by identical bonds, so there is no distinction between them.

Self-test 14B.9 Are the CH_3 protons in ethanol magnetically inequivalent?

Answer: Yes, on account of their different interactions with the CH_2 protons in the next group

Strictly speaking, CH_3 protons are magnetically inequivalent. However, they are in practice made magnetically equivalent by the rapid rotation of the CH_3 group, which averages out any differences. Magnetically inequivalent species can give very complicated spectra (for instance, the proton and ^{19}F spectra of $H_2C{=}CF_2$ each consist of 12 lines), and we shall not consider them further.

An important feature of chemically equivalent magnetic nuclei is that, although they do couple together, the coupling has no effect on the appearance of the spectrum. The qualitative reason for the invisibility of the coupling is that all allowed nuclear spin transitions are *collective* reorientations of groups of equivalent nuclear spins that do not change the relative orientations of the spins within the group (Fig. 14B.18). Then, because the relative orientations of nuclear spins are not changed in any transition, the magnitude of the coupling between them is undetectable. Hence, an isolated CH_3 group gives a single, unsplit line because all the allowed transitions of the group of three protons occur without change of their relative orientations.

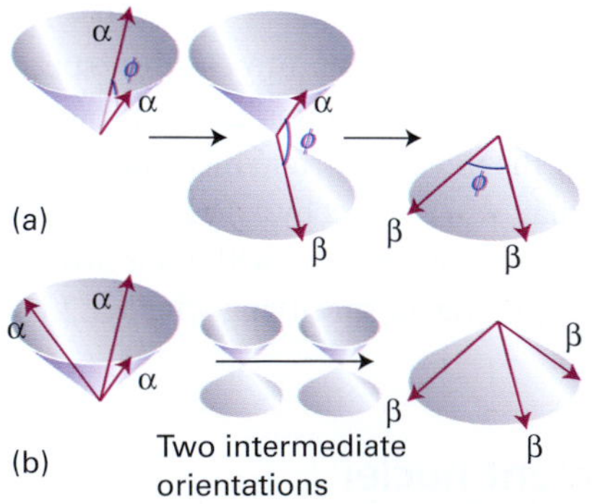

Figure 14B.18 (a) A group of two equivalent nuclei realigns as a group, without change of angle between the spins, when a resonant absorption occurs. Hence it behaves like a single nucleus and the spin–spin coupling between the individual spins of the group is undetectable. (b) Three equivalent nuclei also realign as a group without change of their relative orientations.

To express these conclusions more quantitatively, we first need to establish the energy levels of a collection of equivalent nuclei. As shown in the following *Justification* for an A_2 system, they have the values depicted on the right of Fig. 14B.19.

Justification 14B.1 The energy levels of an A_2 system

Consider an A_2 system of two spin-$\frac{1}{2}$ nuclei. First, consider the energy levels in the absence of spin–spin coupling. There are four spin states that (just as for two electrons) can be classified according to their total spin I (the analogue of S for two electrons) and their total projection M_I on the z-axis. The states are analogous to those for two electrons in singlet and triplet states (Topic 9C):

Spins parallel, $I=1$:	$M_I=+1$	$\alpha\alpha$
	$M_I=0$	$(1/2^{1/2})\{\alpha\beta+\beta\alpha\}$
	$M_I=-1$	$\beta\beta$
Spins paired, $I=0$:	$M_I=0$	$(1/2^{1/2})\{\alpha\beta-\beta\alpha\}$

The sign in $\alpha\beta+\beta\alpha$ signifies an in-phase alignment of spins and $I=1$; the $-$ sign in $\alpha\beta-\beta\alpha$ signifies an alignment out of phase by π, and hence $I=0$. The effect of a magnetic field on these four states is shown in Fig. 14B.19: the energies of the

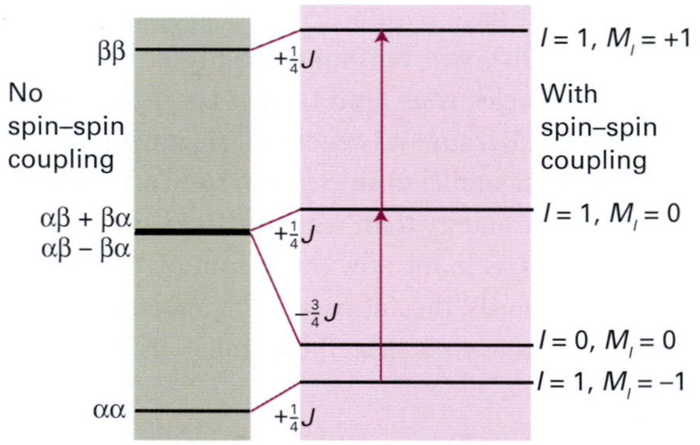

Figure 14B.19 The energy levels of an A_2 system in the absence of spin–spin coupling are shown on the left. When spin–spin coupling is taken into account, the energy levels on the right are obtained. Note that the three states with total nuclear spin $I=1$ correspond to parallel spins and give rise to the same increase in energy (J is positive); the one state with $I=0$ (antiparallel nuclear spins) has a lower energy in the presence of spin–spin coupling. The only allowed transitions are those that preserve the angle between the spins, and so take place between the three states with $I=1$. They occur at the same resonance frequency as they would have in the absence of spin–spin coupling.

two states with $M_I=0$ are unchanged by the field because they are composed of equal proportions of α and β spins.

The spin–spin coupling energy is proportional to the scalar product of the vectors representing the spins, $E=(hJ/\hbar^2)\boldsymbol{I}_1\cdot\boldsymbol{I}_2$. The scalar product can be expressed in terms of the total nuclear spin $\boldsymbol{I}=\boldsymbol{I}_1+\boldsymbol{I}_2$ by noting that

$$I^2=(\boldsymbol{I}_1+\boldsymbol{I}_2)\cdot(\boldsymbol{I}_1+\boldsymbol{I}_2)=I_1^2+I_2^2+2\boldsymbol{I}_1\cdot\boldsymbol{I}_2$$

rearranging this expression to

$$\boldsymbol{I}_1\cdot\boldsymbol{I}_2=\tfrac{1}{2}\{I^2-I_1^2-I_2^2\}$$

and replacing the magnitudes by their quantum mechanical values:

$$\boldsymbol{I}_1\cdot\boldsymbol{I}_2=\tfrac{1}{2}\{I(I+1)-I_1(I_1+1)-I_2(I_2+1)\}\hbar^2$$

Then, because $I_1=I_2=\frac{1}{2}$, it follows that

$$E=\tfrac{1}{2}hJ\{I(I+1)-\tfrac{3}{2}\}$$

For parallel spins, $I=1$ and $E=+\frac{1}{4}hJ$; for antiparallel spins $I=0$ and $E=-\frac{3}{4}hJ$, as in Fig. 14B.19. We see that three of the states move in energy in one direction and the fourth (the one with antiparallel spins) moves three times as much in the opposite direction.

We now consider the allowed transitions between the states of an A_2 system shown in Fig. 14B.18. The radiofrequency field affects the two equivalent protons equally, so it cannot change the orientation of one proton relative to the other; therefore, the transitions take place within the set of states that correspond to parallel spin (those labelled $I=1$), and no spin-parallel state can change to a spin-antiparallel state (the state with $I=0$). Put another way, the allowed transitions are subject to the selection rule $\Delta I=0$. This selection rule is in addition to the rule $\Delta M_I=\pm1$ that arises from the conservation of angular momentum and the unit spin of the photon. The allowed transitions are shown in Fig. 14B.19: we see that there are only two transitions, and that they occur at the same resonance frequency that the nuclei would have in the absence of spin–spin coupling. Hence, the spin–spin coupling interaction does not affect the appearance of the spectrum.

(e) Strongly coupled nuclei

NMR spectra are usually much more complex than the foregoing simple analysis suggests. We have described the extreme case in which the differences in chemical shifts are much greater than the spin–spin coupling constants. In such cases it is simple to identify groups of magnetically equivalent nuclei and to think of the groups of nuclear spins as reorienting relative to each other. The spectra that result are called **first-order spectra**.

Transitions cannot be allocated to definite groups when the differences in their chemical shifts are comparable to their spin–spin coupling interactions. The complicated spectra that are then obtained are called **strongly coupled spectra** (or 'second-order spectra') and are much more difficult to analyse.

Brief illustration 14B.7 Strongly coupled spectra

Figure 14B.20 shows NMR spectra of an A_2 system (top) and an AX system (bottom). Both are simple 'first-order' spectra. At intermediate relative values of the chemical shift difference and the spin–spin coupling, complex 'strongly coupled' spectra are obtained. Note how the inner two lines of the bottom spectrum move together, grow in intensity, and form the single central line of the top spectrum. The two outer lines diminish in intensity and are absent in the top spectrum.

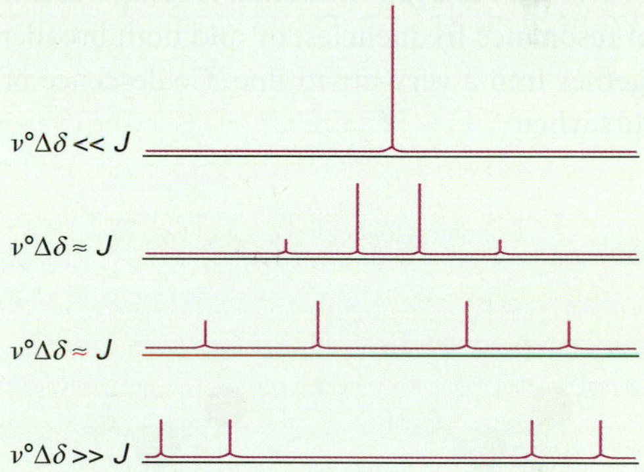

Figure 14B.20 The NMR spectra of an A_2 system (top) and an AX system (bottom) are simple 'first-order' spectra.

Self-test 14B.10 Explain why, in some cases, a second-order spectrum may become simpler (and first-order) at high fields.

Answer: The difference in resonance frequencies increases with field, but spin–spin coupling constants are independent of it

A clue to the type of analysis that is appropriate is given by the notation for the types of spins involved. Thus, an AX spin system (which consists of two nuclei with a large chemical shift difference) has a first-order spectrum. An AB system, on the other hand (with two nuclei of similar chemical shifts), gives a spectrum typical of a strongly coupled system. An AX system may have widely different Larmor frequencies because A and X are nuclei of different elements (such as ^{13}C and ^{1}H), in which case they form a **heteronuclear spin system**. AX may also denote a **homonuclear spin system** in which the nuclei are of the same element but in markedly different environments.

14B.4 Conformational conversion and exchange processes

The appearance of an NMR spectrum is changed if magnetic nuclei can jump rapidly between different environments. Consider a molecule, such as *N,N*-dimethylformamide, that can jump between conformations; in its case, the methyl shifts depend on whether they are *cis* or *trans* to the carbonyl group (Fig. 14B.21). When the jumping rate is low, the spectrum shows two sets of lines, one each from molecules in each conformation. When the interconversion is fast, the spectrum shows a single line at the mean of the two chemical shifts. At intermediate inversion rates, the line is very broad. This maximum broadening occurs when the lifetime, τ, of a conformation gives rise to a linewidth that is comparable to the difference of resonance frequencies, $\delta\nu$ and both broadened lines blend together into a very broad line. Coalescence of the two lines occurs when

$$\tau = \frac{2^{1/2}}{\pi\delta\nu} \qquad \text{Condition for coalescence of two NMR lines} \qquad (14B.16)$$

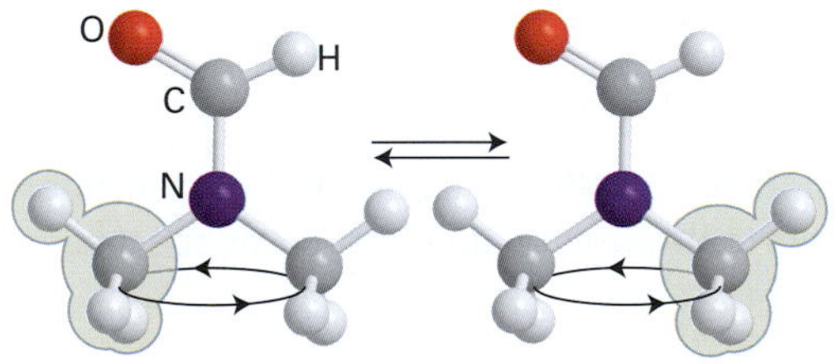

Figure 14B.21 When a molecule changes from one conformation to another, the positions of its protons are interchanged and the protons jump between magnetically distinct environments.

Brief illustration 14B.8 The effect of chemical exchange on NMR spectra

The NO group in *N,N*-dimethylnitrosamine, $(CH_3)_2N{-}NO$ (5), rotates about the N—N bond and, as a result, the magnetic environments of the two CH_3 groups are interchanged. The two CH_3 resonances are separated by 390 Hz in a 600 MHz spectrometer. According to eqn 14B.16,

$$\tau = \frac{2^{1/2}}{\pi \times (390\,\text{s}^{-1})} = 1.2\,\text{ms}$$

It follows that the signal will collapse to a single line when the interconversion rate exceeds about $1/\tau = 830\,\text{s}^{-1}$.

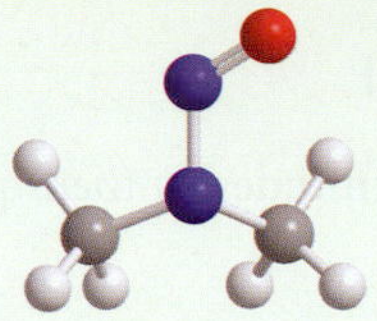

5 *N,N*-Dimethylnitrosamine

Self-test 14B.11 What would you deduce from the observation of a single line from the same molecule in a 300 MHz spectrometer?

Answer: Conformation lifetime less than 2.3 ms

A similar explanation accounts for the loss of fine structure in solvents able to exchange protons with the sample. For example, hydroxyl protons are able to exchange with water protons. When this **chemical exchange** occurs, a molecule ROH with an α-spin proton (we write this ROH_α) rapidly converts to ROH_β and then perhaps to ROH_α again because the protons provided by the solvent molecules in successive exchanges have random spin orientations. Therefore, instead of seeing a spectrum composed of contributions from both ROH_α and ROH_β molecules (that is, a spectrum showing a doublet structure due to the OH proton) we see a spectrum that shows no splitting caused by coupling of the OH proton (as in Fig. 14B.2 and as discussed in Example 14B.3). The effect is observed when the lifetime of a molecule due to this chemical exchange is so short that the lifetime broadening is greater than the doublet splitting. Because this splitting is often very small (a few hertz), a proton must remain attached to the same molecule for longer than about 0.1 s for the splitting to be observable. In water, the exchange rate is much faster than that, so alcohols show no splitting from the OH protons. In dry dimethylsulfoxide (DMSO), the exchange rate may be slow enough for the splitting to be detected.

Checklist of concepts

- ☐ 1. The **chemical shift** of a nucleus is the difference between its resonance frequency and that of a reference standard.
- ☐ 2. The **shielding constant** is the sum of a local contribution, a neighbouring group contribution, and a solvent contribution.
- ☐ 3. The **local contribution** is the sum of a diamagnetic contribution and a paramagnetic contribution.
- ☐ 4. The **neighbouring** group contribution arises from the currents induced in nearby groups of atoms.
- ☐ 5. The **solvent contribution** can arise from specific molecular interactions between the solute and the solvent.
- ☐ 6. **Fine structure** is the splitting of resonances into individual lines by spin–spin coupling.
- ☐ 7. **Spin–spin coupling** is expressed in terms of the **spin–spin coupling constant** J and depends on the relative orientation of two nuclear spins.
- ☐ 8. The coupling constant decreases as the number of bonds separating two nuclei increases.
- ☐ 9. Spin–spin coupling can be explained in terms of the **polarization mechanism** and the **Fermi contact interaction**.
- ☐ 10. Chemically and magnetically equivalent nuclei have the same chemical shifts.
- ☐ 11. In **strongly coupled spectra**, transitions cannot be allocated to definite groups.
- ☐ 12. Coalescence of two NMR lines occurs when a conformational interchange or chemical exchange of nuclei is fast.

Checklist of equations

Property	Equation	Comment	Equation number
δ Scale of chemical shifts	$\delta=\{(\nu-\nu^\circ)/\nu^\circ\}\times10^6$	Definition	14B.4
Relation between chemical shift and shielding constant	$\delta\approx(\sigma^\circ-\sigma)\times10^6$		14B.6
Local contribution to the shielding constant	$\sigma(\text{local})=\sigma_d+\sigma_p$		14B.8
Lamb formula	$\sigma_d=(e^2\mu_0/12\pi m_e)\langle 1/r\rangle$		14B.9
Neighbouring group contribution to the shielding constant	$\sigma(\text{neighbour})\propto(\chi_\parallel-\chi_\perp)\{(1-3\cos^2\theta)/r^3\}$	The angle θ is defined in (1)	14B.10b
Karplus equation	${}^3J_{HH}=A+B\cos\phi+C\cos2\phi$	A, B, and C are empirical constants	14B.14
Condition for coalescence of two NMR lines	$\tau=2^{1/2}/\pi\delta\nu$	Conformational conversions and exchange processes	14B.16

14C Pulse techniques in NMR

Contents

➤ Why do you need to know this material?

To understand how nuclear magnetic resonance spectroscopy is used to study large molecules and even diagnose disease, you need to understand how spectral information is obtained by analysing the response of nuclei to the application of strong pulses of radiofrequency radiation.

➤ What is the key idea?

Fourier-transform NMR spectroscopy is the analysis of the radiation emitted by nuclear spins as they return to equilibrium after stimulation by one or more pulses of radiofrequency radiation.

➤ What do you need to know already?

You need to be familiar with the general principles of magnetic resonance (Topic 14A), the features of NMR spectra (Topics 14B), the vector model of angular momentum (Topic 8B), the magnetic properties of molecules (Topic 18C), and Fourier transforms (Topic 12A and *Mathematical background* 7). The development makes use of the concept of precession at the Larmor frequency (Topic 14A).

The common method of detecting the energy separation between nuclear spin states in NMR spectroscopy is more sophisticated than simply looking for the frequency at which resonance occurs. One of the best analogies that have been suggested to illustrate the preferred way of observing an NMR spectrum is that of detecting the spectrum of vibrations of a bell. We could stimulate the bell with a gentle vibration at a gradually increasing frequency, and note the frequencies at which it resonated with the stimulation. A lot of time would be spent getting zero response when the stimulating frequency was between the bell's vibrational modes. However, if we were simply to hit the bell with a hammer, we would immediately obtain a clang composed of all the frequencies that the bell can produce. The equivalent in NMR is to monitor the radiation nuclear spins emit as they return to equilibrium after the appropriate stimulation. The resulting **Fourier-transform NMR** (FT-NMR) spectroscopy gives greatly increased sensitivity, so opening up much of the periodic table to the technique. Moreover, multiple-pulse FT-NMR gives chemists unparalleled control over the information content and display of spectra.

14C.1 The magnetization vector

Consider a sample composed of many identical spin-$\frac{1}{2}$ nuclei. By analogy with the discussion of angular momenta in Topic 8C, a nuclear spin can be represented by a vector of length $\{I(I+1)\}^{1/2}$ units with a component of length m_I units along the z-axis. As the uncertainty principle does not allow us to specify the x- and y-components of the angular momentum, all we know is that the vector lies somewhere on a cone around the z-axis. For $I=\frac{1}{2}$, the length of the vector is $\frac{1}{2}3^{1/2}$ and when $m_I=+\frac{1}{2}$ it makes an angle of $\arccos(\frac{1}{2}/(\frac{1}{2}3^{1/2}))=55°$ to the z-axis (Fig. 14C.1).

In the absence of a magnetic field, the sample consists of equal numbers of α and β nuclear spins with their vectors lying at random angles on the cones. These angles are unpredictable, and at this stage we picture the spin vectors as stationary.

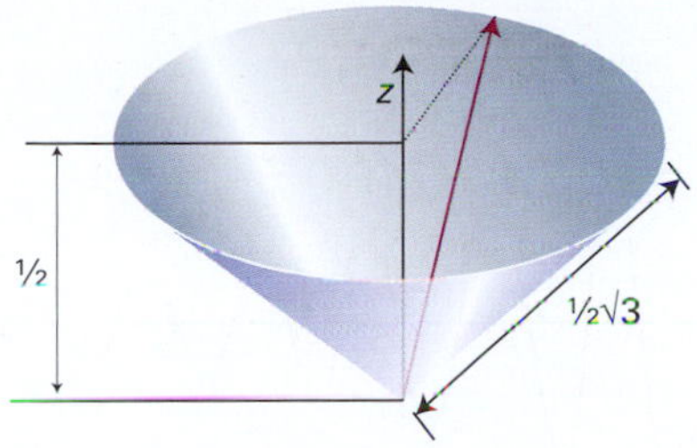

Figure 14C.1 The vector model of angular momentum for a single spin-$\frac{1}{2}$ nucleus. The angle around the z-axis is indeterminate.

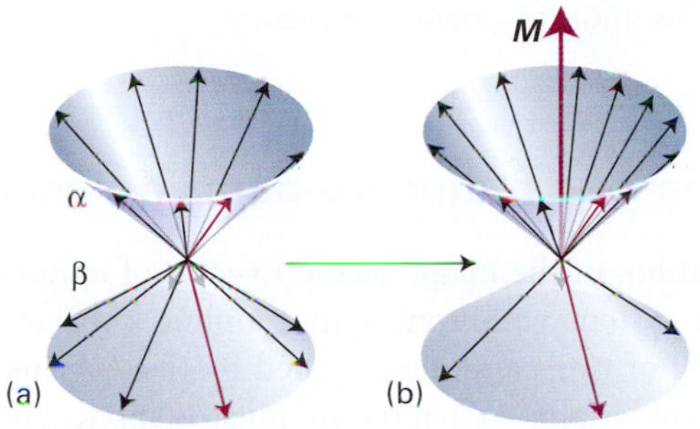

Figure 14C.2 The magnetization of a sample of spin-$\frac{1}{2}$ nuclei is the resultant of all their magnetic moments. (a) In the absence of an externally applied field, there are equal numbers of α and β spins at random angles around the z-axis (the field direction) and the magnetization is zero. (b) In the presence of a field, the spins precess around their cones (that is, there is an energy difference between the α and β states) and there are slightly more α spins than β spins. As a result, there is a net magnetization along the z-axis.

The **magnetization**, $\boldsymbol{M}$, of the sample, its net nuclear magnetic moment, is zero (Fig. 14C.2a).

Two changes occur in the magnetization when a magnetic field of magnitude $\mathcal{B}_0$ is present and aligned in the z-direction:

- The energies of the two orientations change, the α spins moving to low energy and the β spins to high energy (provided $\gamma_N > 0$).

At 10 T, the Larmor frequency for protons is 427 MHz, and in the vector model the individual vectors are pictured as precessing at this rate (Topic 14A). This motion is a pictorial representation of the difference in energy of the spin states (it is not an actual representation of reality but is inspired by the actual motion of a classical bar magnet in a magnetic field). As the field is increased, the Larmor frequency increases and the precession becomes faster.

- The populations of the two spin states (the numbers of α and β spins) at thermal equilibrium change, with slightly more α spins than β spins (see Topic 14A).

Despite its smallness, this imbalance means that there is a net magnetization that we can represent by a vector $\boldsymbol{M}$ pointing in the z-direction and with a length proportional to the population difference (Fig. 14C.2b).

(a) The effect of the radiofrequency field

Now consider the effect of a radiofrequency field circularly polarized in the xy-plane, so that the magnetic component of the electromagnetic field (the only component we need to consider) is rotating around the z-direction in the same sense as the Larmor precession of the nuclei. The strength of the rotating magnetic field is $\mathcal{B}_1$.

To interpret the effects of radiofrequency pulses on the magnetization, it is useful to imagine stepping on to a platform, a so-called **rotating frame**, that rotates around the direction of the applied field. Suppose we choose the frequency of the radiofrequency field to be equal to the Larmor frequency of the spins, $\nu_L = \gamma_N \mathcal{B}_0/2\pi$; this choice is equivalent to selecting the resonance condition in the conventional experiment. The rotating magnetic field is in step with the precessing spins, the nuclei experience a steady $\mathcal{B}_1$ field, and precess about it at a frequency $\gamma_N \mathcal{B}_1/2\pi$ (Fig. 14C.3). Now suppose that the $\mathcal{B}_1$ field is applied in a pulse of duration $\Delta\tau = \frac{1}{4} \times 2\pi/\gamma_N \mathcal{B}_1$, the magnetization tips through an angle of $\frac{1}{4} \times 2\pi = \pi/2$ (90°) away from the vertical z-direction and we say that we have applied a **90° pulse**, or a 'π/2 pulse' (Fig. 14C.4a).

Brief illustration 14C.1 Radiofrequency pulses

The duration of a radiofrequency pulse depends on the strength of the $\mathcal{B}_1$ field. If a 90° pulse requires 10 μs, then for protons

$$\mathcal{B}_1 = \frac{\pi}{2 \times \underbrace{(2.675 \times 10^{-8}\,\mathrm{T^{-1}\,s^{-1}})}_{\gamma_N} \times \underbrace{(1.0 \times 10^{-5}\,\mathrm{s})}_{\Delta\tau}} = 5.9 \times 10^{-4}\,\mathrm{T}$$

or 0.59 mT.

Self-test 14C.1 How long would a 180° pulse require for protons?

Answer: 20 μs

Now imagine stepping out of the rotating frame. To a fixed external observer (the role played by a radiofrequency coil), the magnetization vector is now rotating at the Larmor frequency in the xy-plane (Fig. 14C.4b). The rotating magnetization induces in the coil a signal that oscillates at the Larmor frequency and which can be amplified and processed. In practice, the processing takes place after subtraction of a constant high frequency component (the radiofrequency used for $\mathcal{B}_1$), so that all the signal manipulation takes place at frequencies of a few kilohertz.

As time passes, the individual spins move out of step (partly because they are precessing at slightly different rates, as we

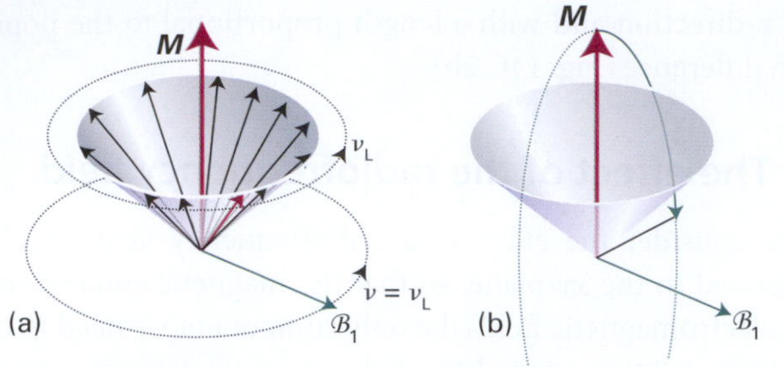

Figure 14C.3 (a) In a resonance experiment, a circularly polarized radiofrequency magnetic field $\mathcal{B}_1$ is applied in the *xy*-plane (the magnetization vector lies along the *z*-axis). (b) If we step into a frame rotating at the radiofrequency, $\mathcal{B}_1$ appears to be stationary, as does the magnetization ***M*** if the Larmor frequency is equal to the radiofrequency. When the two frequencies coincide, the magnetization vector of the sample rotates around the direction of the $\mathcal{B}_1$ field.

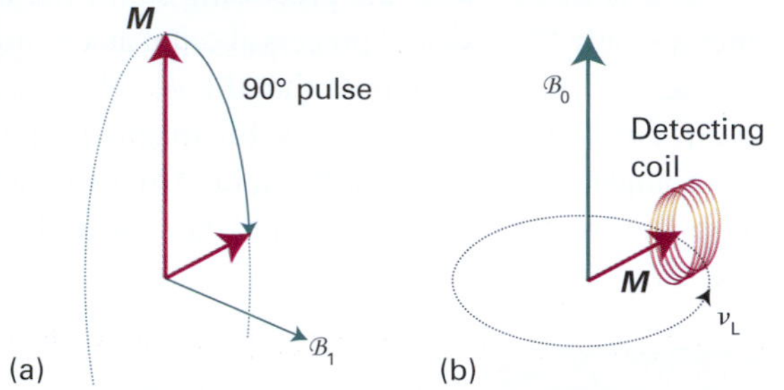

Figure 14C.4 (a) If the radiofrequency field is applied for a certain time, the magnetization vector is rotated into the *xy*-plane. (b) To an external stationary observer (the coil), the magnetization vector is rotating at the Larmor frequency, and can induce a signal in the coil.

explain later), so the magnetization vector shrinks exponentially with a time constant T_2 and induces an ever weaker signal in the detector coil. The form of the signal that we can expect is therefore the oscillating-decaying **free-induction decay** (FID) shown in Fig. 14C.5. The *y*-component of the magnetization varies as

$$M_y(t) = M_0 \cos(2\pi\nu_L t)e^{-t/T_2} \quad \text{Free-induction decay} \quad (14C.1)$$

We have considered the effect of a $\mathcal{B}_1$ pulse applied at exactly the Larmor frequency. However, virtually the same effect is obtained off resonance, provided that the pulse is applied close to ν_L. If the difference in frequency is small compared to the inverse of the duration of the 90° pulse, the magnetization will end up in the *xy*-plane. Note that we do not need to know the Larmor frequency beforehand: the short pulse is the analogue of the hammer blow on the bell, exciting a range of frequencies. The detected signal shows that a particular resonant frequency is present.

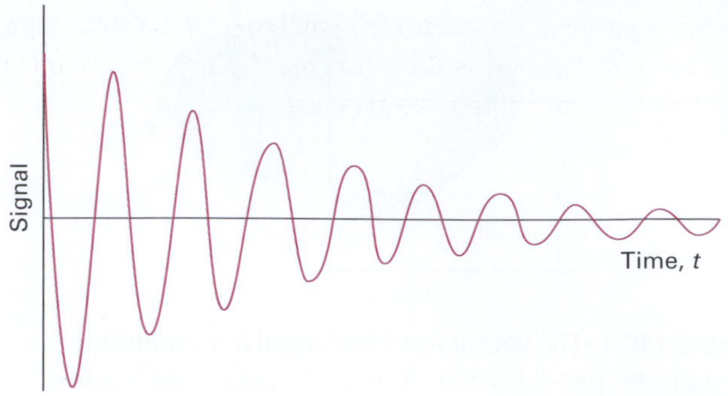

Figure 14C.5 A simple free-induction decay of a sample of spins with a single resonance frequency.

(b) Time- and frequency-domain signals

We can think of the magnetization vector of a homonuclear AX spin system with spin–spin coupling constant $J = 0$ as consisting of two parts, one formed by the A spins and the other by the X spins. When the 90° pulse is applied, both magnetization vectors are rotated into the *xy*-plane. However, because the A and X nuclei precess at different frequencies, they induce two signals in the detector coils, and the overall FID curve may resemble that in Fig. 14C.6a. The composite FID curve is the analogue of the struck bell emitting a rich tone composed of all the frequencies (in this case, just the two resonance frequencies of the uncoupled A and X nuclei) at which it can vibrate.

The problem we must address is how to recover the resonance frequencies present in a free-induction decay. We know that the FID curve is a sum of decaying oscillating functions, so the problem is to analyse it into its components by carrying out a Fourier transformation. The analysis of the FID curve is achieved by the standard mathematical technique of Fourier transformation, which is explained more fully in *Mathematical background* 7 following chapter 18.

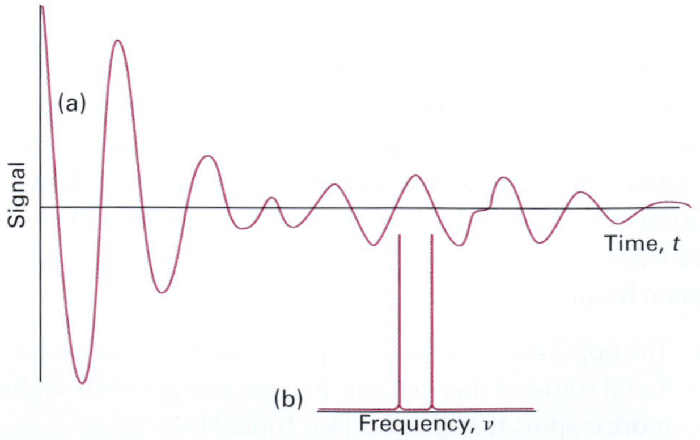

Figure 14C.6 (a) A free-induction decay signal of a sample of AX species and (b) its analysis into its frequency components.

We start by noting that the signal $S(t)$ in the time domain, the total FID curve, is the sum (more precisely, the integral) over all the contributing frequencies

$$S(t)=\int_{-\infty}^{\infty} I(\nu)e^{-2\pi i\nu t}d\nu \qquad (14C.2)$$

Because $e^{2\pi i\nu t}=\cos(2\pi\nu t)+i\sin(2\pi\nu t)$, this expression is a sum over harmonically oscillating functions, with each one weighted by the intensity $I(\nu)$.

We need $I(\nu)$, the spectrum in the frequency domain; it is obtained by evaluating the integral

$$I(\nu)=2\mathrm{Re}\int_{0}^{\infty} S(t)e^{2\pi i\nu t}dt \qquad (14C.3)$$

where Re means take the real part of the following expression. This integral gives a nonzero value if $S(t)$ contains a component that matches the oscillating function $e^{2i\pi\nu t}$. The integration is carried out at a series of frequencies ν on a computer that is built into the spectrometer. When the signal in Fig. 14C.6a is transformed in this way, we get the frequency-domain spectrum shown in Fig. 14C.6b. One line represents the Larmor frequency of the A nuclei and the other that of the X nuclei.

Brief illustration 14C.2 Fourier analysis

Fourier analysis is a common feature of most mathematical software packages, but one simple example is the Fourier transform of the function

$$S(t)=S(0)\cos(2\pi\nu_L t)e^{-t/T_2}$$

which describes the behaviour of the FID signal in eqn 14C.1. The result is (Problem 14C.3)

$$I(\nu)=\frac{S(0)T_2}{1+(\nu_L-\nu)^2(2\pi T_2)^2}$$

which has the so-called 'Lorentzian' shape, with a maximum intensity at $I(\nu_L)=S(0)T_2$.

Self-test 14C.2 What is the width at half-height $\Delta\nu_{1/2}$ of the Lorentzian function above?

Answer: $\Delta\nu_{1/2}=1/\pi T_2$

The FID curve in Fig. 14C.7 is obtained from a sample of ethanol. The frequency-domain spectrum obtained from it by Fourier transformation is the one discussed in Topic 14B (see Fig. 14B.2). We can now see why the FID curve in Fig. 14C.7 is so complex: it arises from the precession of a magnetization vector that is composed of eight components, each with a characteristic frequency.

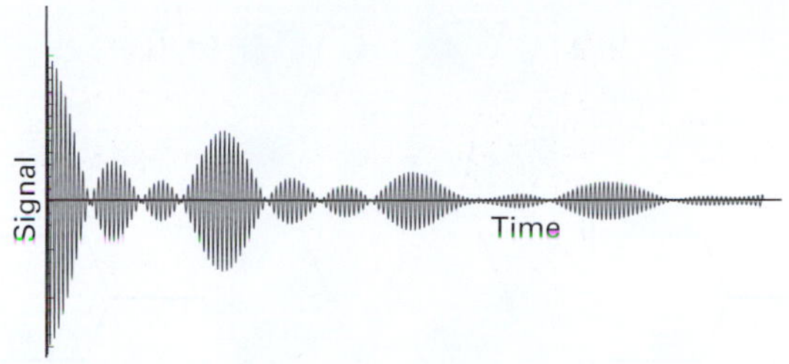

Figure 14C.7 A free-induction decay signal of a sample of ethanol. Its Fourier transform is the frequency-domain spectrum shown in Fig. 14B.2. The total length of the image corresponds to about 1 s.

14C.2 Spin relaxation

There are two reasons why the component of the magnetization vector in the *xy*-plane shrinks. Both reflect the fact that the nuclear spins are not in thermal equilibrium with their surroundings (for then M lies parallel to z). At thermal equilibrium the spins have a Boltzmann distribution, with more α spins than β spins and lie at random orientations on their precessional cones. The return to equilibrium is the process called **spin relaxation**.

(a) Longitudinal and transverse relaxation

Consider the effect of a 180° pulse, which may be visualized in the rotating frame as a flip of the net magnetization vector from one direction along the *z*-axis (with more α spins than β spins) to the opposite direction (with more β spins than α spins). After the pulse, the populations revert to their thermal equilibrium values exponentially. As they do so, the *z*-component of magnetization reverts to its equilibrium value M_0 with a time constant called the **longitudinal relaxation time**, T_1 (Fig. 14C.8):

$$M_z(t)-M_0\propto e^{-t/T_1} \qquad \textit{Definition} \quad \text{Longitudinal relaxation time} \qquad (14C.4)$$

Because this relaxation process involves giving up energy to the surroundings (the 'lattice') as β spins revert to α spins, the time constant T_1 is also called the **spin–lattice relaxation time**. Spin–lattice relaxation is caused by local magnetic fields that fluctuate at a frequency close to the resonance frequency of the β→α transition. Such fields can arise from the tumbling motion of molecules in a fluid sample. If molecular tumbling is too slow or too fast compared to the resonance frequency, it will give rise to a fluctuating magnetic field with a frequency that is either too low or too high to stimulate a spin change from β to α, so T_1 will be long. Only if the molecule tumbles at about the resonance frequency will the fluctuating magnetic

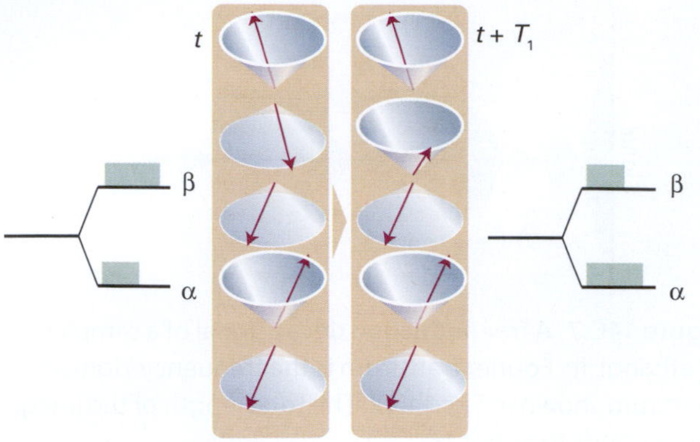

Figure 14C.8 In longitudinal relaxation the spins relax back towards their thermal equilibrium populations. On the left we see the precessional cones representing spin-$\frac{1}{2}$ angular momenta, and they do not have their thermal equilibrium populations (there are more β spins than α spins). On the right, which represents the sample a long time after a time T_1 has elapsed, the populations are those characteristic of a Boltzmann distribution. In actuality, T_1 is the time constant for relaxation to the arrangement on the right and T_1 ln 2 is the half-life of the arrangement on the left.

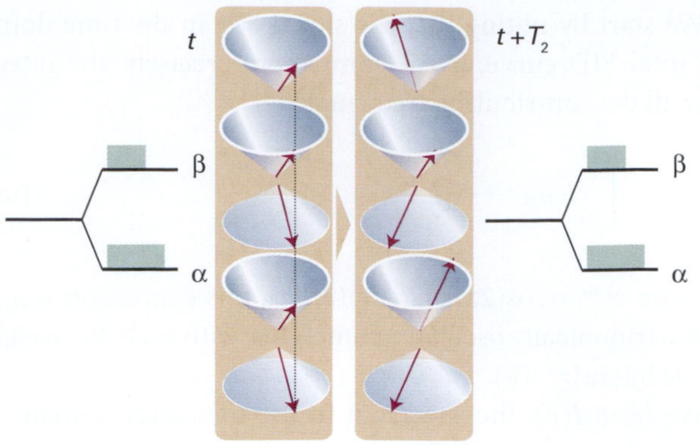

Figure 14C.10 The transverse relaxation time, T_2, is the time constant for the phases of the spins to become randomized (another condition for equilibrium) and to change from the orderly arrangement shown on the left to the disorderly arrangement on the right (long after a time T_2 has elapsed). Note that the populations of the states remain the same; only the relative phase of the spins relaxes. In actuality, T_2 is the time constant for relaxation to the arrangement on the right and T_2 ln 2 is the half-life of the arrangement on the left.

field be able to induce spin changes effectively, and only then will T_1 be short. The rate of molecular tumbling increases with temperature and with reducing viscosity of the solvent, so we can expect a dependence like that shown in Fig. 14C.9. The quantitative treatment of relaxation times depends on setting up models of molecular motion and using, for instance, the diffusion equation (Topic 19C) adapted for rotational motion.

Now consider the events following a 90° pulse. The magnetization vector in the *xy*-plane is large when the spins are bunched together immediately after the pulse. However, this orderly bunching of spins is not at equilibrium and, even if there were no spin–lattice relaxation, we would expect the individual spins to spread out until they were uniformly distributed with all possible angles around the *z*-axis (Fig. 14C.10). At that stage, the component of magnetization vector in the plane would be zero. The randomization of the spin directions occurs exponentially with a time constant called the **transverse relaxation time**, T_2:

$$M_y(t) \propto e^{-t/T_2} \qquad \textit{Definition} \quad \text{Transverse relaxation time} \qquad (14C.5)$$

Because the relaxation involves the relative orientation of the spins around their respective cones, T_2 is also known as the **spin–spin relaxation time**. Any relaxation process that changes the balance between α and β spins will also contribute to this randomization, so the time constant T_2 is almost always less than or equal to T_1.

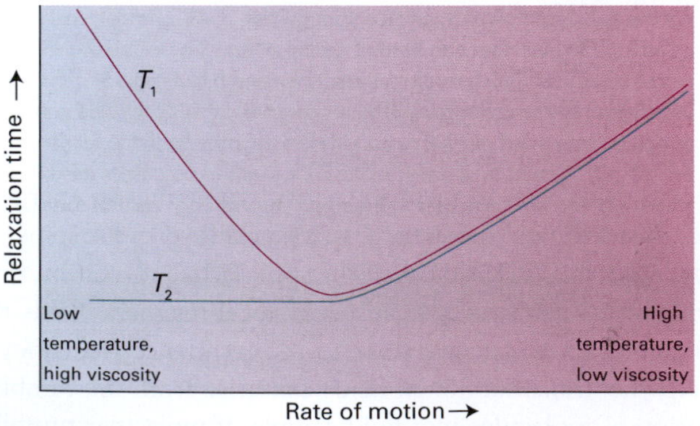

Figure 14C.9 The variation of the two relaxation times with the rate at which the molecules move (either by tumbling or migrating through the solution). The horizontal axis can be interpreted as representing temperature or viscosity. Note that at rapid rates of motion, the two relaxation times coincide.

Local magnetic fields also affect spin–spin relaxation. When the fluctuations are slow, each molecule lingers in its local magnetic environment and the spin orientations randomize quickly around their cones. If the molecules move rapidly from one magnetic environment to another, the effects of differences in local magnetic field average to zero: individual spins do not precess at very different rates, they can remain bunched for longer, and spin–spin relaxation does not take place as quickly. In other words, slow molecular motion corresponds to short T_2 and fast motion corresponds to long T_2 (as shown in Fig. 14C.9). Calculations show that, when the motion is fast, the main orientational randomizing effect arises from $\beta \rightarrow \alpha$ transitions rather than different precession rates on the cones, and then $T_2 \approx T_1$.

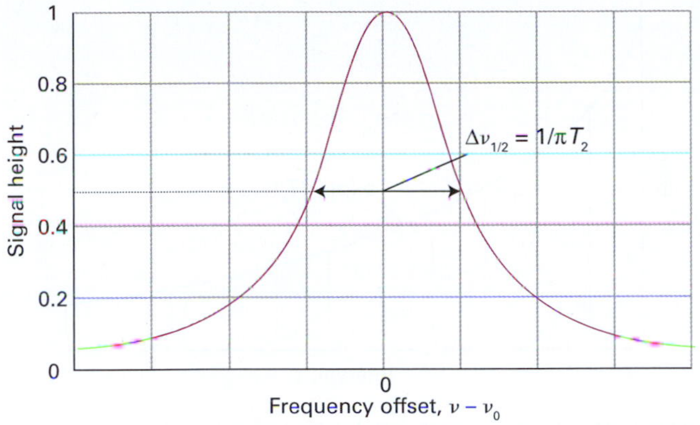

Figure 14C.11 A Lorentzian absorption line. The width at half-height is inversely proportional to the parameter T_2 and the longer the transverse relaxation time, the narrower the line.

If the y-component of magnetization decays with a time constant T_2, the spectral line is broadened (Fig. 14C.11), and its width at half-height becomes (see *Brief illustration* 14C.2)

$$\Delta\nu_{1/2} = \frac{1}{\pi T_2} \qquad \text{Width at half-height of an NMR line} \qquad (14C.6)$$

Typical values of T_2 in proton NMR are of the order of seconds, so linewidths of around 0.1 Hz can be anticipated, in broad agreement with observation.

So far, we have assumed that the equipment, and in particular the magnet, is perfect, and that the differences in Larmor frequencies arise solely from interactions within the sample. In practice, the magnet is not perfect, and the field is different at different locations in the sample. The inhomogeneity broadens the resonance, and in most cases this **inhomogeneous broadening** dominates the broadening we have discussed so far. It is common to express the extent of inhomogeneous broadening in terms of an **effective transverse relaxation time**, T_2^*, by using a relation like eqn 14C.6, but writing

$$T_2^* = \frac{1}{\pi\Delta\nu_{1/2}} \qquad \textit{Definition} \qquad \text{Effective transverse relaxation time} \qquad 14C.7)$$

where $\Delta\nu_{1/2}$ is the observed width at half-height of a line with a Lorentzian shape of the form $I \propto 1/(1+\nu^2)$.

Brief illustration 14C.3 Inhomogeneous broadening

Consider a line in a spectrum with a width of 10 Hz. It follows from eqn 14C.7 that the effective transverse relaxation time is

$$T_2^* = \frac{1}{\pi\times(10\,\mathrm{s}^{-1})} = 32\,\mathrm{ms}$$

Self-test 14C.3 Name two processes that could contribute to further broadening of the NMR line.

Answer: Conformational conversion or chemical exchange (Topic 14B)

(b) The measurement of T_1 and T_2

The longitudinal relaxation time T_1 can be measured by the **inversion recovery technique**. The first step is to apply a 180° pulse to the sample. A 180° pulse is achieved by applying the $\mathcal{B}_1$ field for twice as long as for a 90° pulse, so the magnetization vector precesses through 180° and points in the z-direction (Fig. 14C.12). No signal can be seen at this stage because there is no component of magnetization in the xy-plane (where the coil can detect it). The β spins begin to relax back into α spins, and the magnetization vector first shrinks exponentially, falling through zero to its thermal equilibrium value, M_0. After an interval τ, a 90° pulse is applied that rotates the remaining magnetization into the xy-plane, where it generates an FID signal. The frequency-domain spectrum is then obtained by Fourier transformation.

The intensity of the spectrum obtained in this way depends on the length of the magnetization vector that is rotated into the xy-plane. The length of that vector changes exponentially as the interval between the two pulses is increased, so the intensity of the spectrum also changes exponentially with increasing τ. We can therefore measure T_1 by fitting an exponential curve to the series of spectra obtained with different values of τ.

The measurement of T_2 (as distinct from T_2^*) depends on being able to eliminate the effects of inhomogeneous broadening. The cunning required is at the root of some of the most important advances that have been made in NMR since its introduction.

A **spin echo** is the magnetic analogue of an audible echo: transverse magnetization is created by a radiofrequency pulse, decays away, is reflected by a second pulse, and grows back to form an echo. The sequence of events is shown in Fig. 14C.13. We can consider the overall magnetization as being made up of a number of different magnetizations, each of which arises from a **spin packet** of nuclei with very similar precession frequencies. The spread in these frequencies arises because the applied field $\mathcal{B}_0$ is inhomogeneous, so different parts of the sample experience different fields. The precession frequencies also differ if there is more than one chemical shift present. As will be seen,

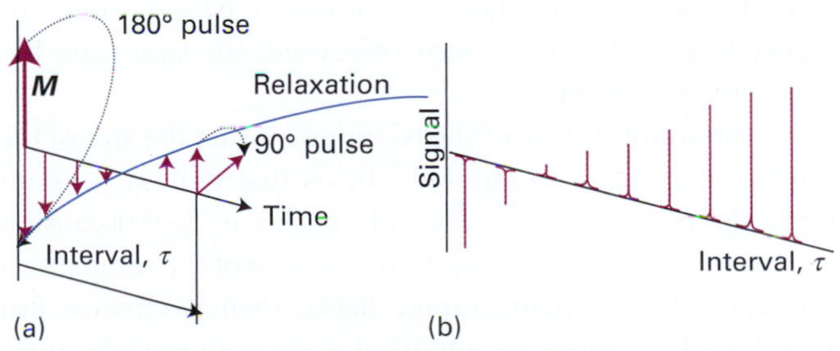

Figure 14C.12 (a) The result of applying a 180° pulse to the magnetization in the rotating frame and the effect of a subsequent 90° pulse. (b) The amplitude of the frequency-domain spectrum varies with the interval between the two pulses because spin–lattice relaxation has time to occur.

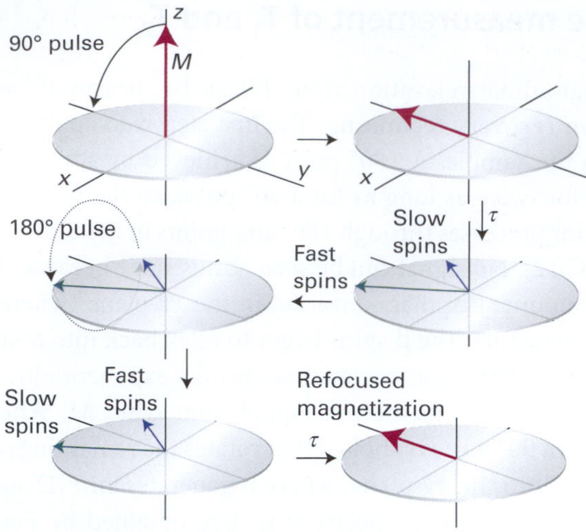

Figure 14C.13 The sequence of pulses leading to the observation of a spin echo.

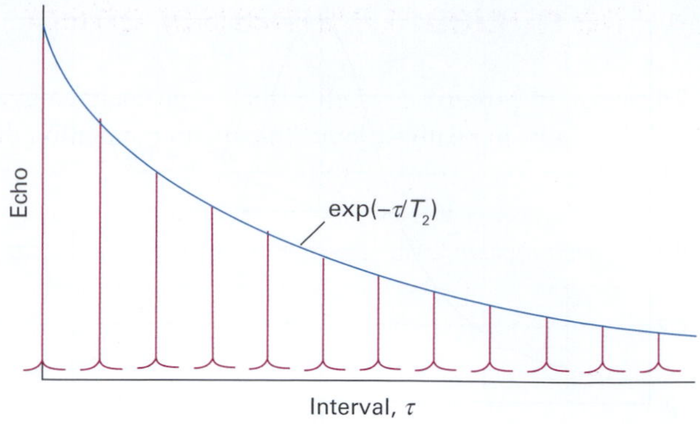

Figure 14C.14 The exponential decay of spin echoes can be used to determine the transverse relaxation time.

the importance of a spin echo is that it can suppress the effects of both field inhomogeneities and chemical shifts.

First, a 90° pulse is applied to the sample. We follow events by using the rotating frame, in which $\mathcal{B}_1$ is stationary along the rotating *x*-axis and causes the magnetization to rotate into the *xy*-plane. The spin packets now begin to fan out because they have different Larmor frequencies, with some above the radiofrequency and some below. The detected signal depends on the resultant of the spin-packet magnetization vectors, and decays with a time-constant T_2^* because of the combined effects of field inhomogeneity and spin–spin relaxation.

After an evolution period τ, a 180° pulse is applied to the sample; this time, about the *y*-axis of the rotating frame (the axis of the pulse is changed from *x* to *y* by a 90° phase shift of the radiofrequency radiation). The pulse rotates the magnetization vectors of the faster spin packets into the positions previously occupied by the slower spin packets, and vice versa. Thus, as the vectors continue to precess, the fast vectors are now behind the slow; the fan begins to close up again, and the resultant signal begins to grow back into an echo. After another interval of length τ, all the vectors will once more be aligned along the *y*-axis, and the fanning out caused by the field inhomogeneity is said to have been **refocused**: the spin echo has reached its maximum.

The important feature of the technique is that the size of the echo is independent of any local fields that remain constant during the two τ intervals. If a spin packet is 'fast' because it happens to be composed of spins in a region of the sample that experiences higher than average fields, then it remains fast throughout both intervals, and what it gains on the first interval it loses on the second interval. Hence, the size of the echo is independent of inhomogeneities in the magnetic field, for these remain constant. The true transverse relaxation arises from fields that vary on a molecular distance scale, and there is no guarantee that an individual 'fast' spin will remain 'fast' in the refocusing phase: the spins within the packets therefore spread with a time constant T_2. Hence, the effects of the true relaxation are not refocused, and the size of the echo decays with the time constant T_2 (Fig. 14C.14).

14C.3 Spin decoupling

Carbon-13 is a **dilute-spin species** in the sense that it is unlikely that more than one ^{13}C nucleus will be found in any given small molecule (provided the sample has not been enriched with that isotope; the natural abundance of ^{13}C is only 1.1 per cent). Even in large molecules, although more than one ^{13}C nucleus may be present, it is unlikely that they will be close enough to give an observable splitting. Hence, it is not normally necessary to take into account ^{13}C–^{13}C spin–spin coupling within a molecule.

Protons are **abundant-spin species** in the sense that a molecule is likely to contain many of them. Protons are abundant-spin species. If we were observing a ^{13}C-NMR spectrum, we would obtain a very complex spectrum on account of the coupling of the one ^{13}C nucleus with many of the protons that are present. To avoid this difficulty, ^{13}C-NMR spectra are normally observed using the technique of **proton decoupling**. Thus, if the CH_3 protons of ethanol are irradiated with a second, strong, resonant radiofrequency pulse, they undergo rapid spin reorientations and the ^{13}C nucleus senses an average orientation. As a result, its resonance is a single line and not a 1:3:3:1 quartet. Proton decoupling has the additional advantage of enhancing sensitivity, because the intensity is concentrated into a single transition frequency instead of being spread over several transition frequencies. If care is taken to ensure that the other parameters on which the strength of the signal depends are kept constant, the intensities of proton-decoupled spectra are proportional to the number of ^{13}C nuclei present. The technique is widely used to characterize synthetic polymers.

14C.4 The nuclear Overhauser effect

One advantage of protons in NMR is their high magnetogyric ratio, which results in relatively large Boltzmann population differences and also strong coupling to the radiofrequency field, and hence greater resonance intensities than for most other nuclei. In the steady-state **nuclear Overhauser effect** (NOE), spin relaxation processes involving internuclear dipole–dipole interactions are used to transfer this population advantage to another nucleus (such as ^{13}C or another proton), so that the latter's resonances are modified. In a dipole–dipole interaction between two nuclei, one nucleus influences the behaviour of another nucleus in much the same way that the orientation of a bar magnet is influenced by the presence of another bar magnet nearby.

To understand the effect, consider the populations of the four levels of a homonuclear (for instance, proton) AX system; these levels are shown in Fig. 14B.9. At thermal equilibrium, the population of the $\alpha_A\alpha_X$ level is the greatest, and that of the $\beta_A\beta_X$ level is the least; the other two levels have the same energy and an intermediate population. The thermal equilibrium absorption intensities reflect these populations, as shown in Fig. 14C.15. Now consider the combined effect of spin relaxation and keeping the X spins saturated. When we saturate the X transition, the populations of the X levels are equalized ($N_{\alpha X}=N_{\beta X}$) and all transitions involving $\alpha_X \leftrightarrow \beta_X$ spin flips are no longer observed. At this stage there is no change in the populations of the A levels. If that were all there were to happen, all we would see would be the loss of the X resonance and no effect on the A resonance.

Now consider the effect of spin relaxation. Relaxation can occur in a variety of ways if there is a dipolar interaction between the A and X spins. One possibility is for the magnetic field acting between the two spins to cause them *both* to flip simultaneously from β to α, so the $\alpha_A\alpha_X$ and $\beta_A\beta_X$ states regain their thermal equilibrium populations. However, the populations of the $\alpha_A\beta_X$ and $\beta_A\alpha_X$ levels remain unchanged at the values characteristic of saturation. As we see from Fig. 14C.16, the population difference between the states joined by

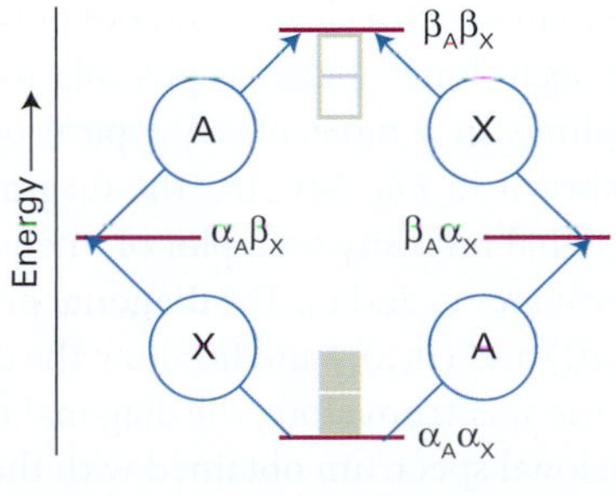

Figure 14C.15 The energy levels of an AX system and an indication of their relative populations. Each green square above the line represents an excess population and each white square below the line represents a population deficit. The transitions of A and X are marked.

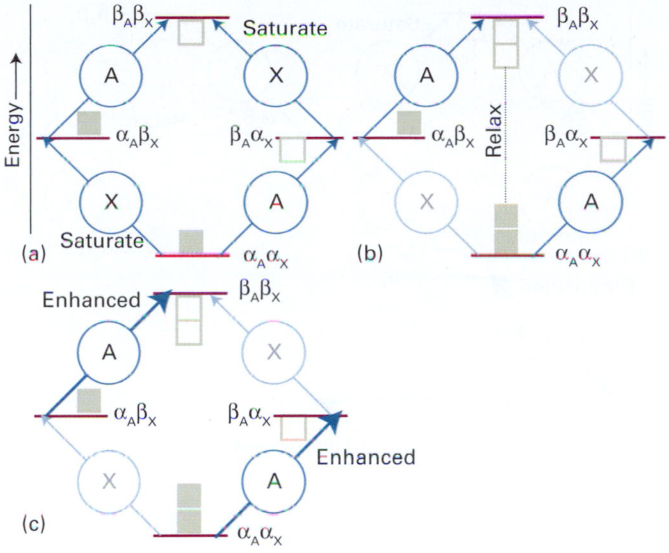

Figure 14C.16 (a) When the X transition is saturated, the populations of its two states are equalized and the population excess and deficit become as shown (using the same symbols as in Fig. 14C.15). (b) Dipole–dipole relaxation relaxes the populations of the highest and lowest states, and they regain their original populations. (c) The A transitions reflect the difference in populations resulting from the preceding changes, and are enhanced compared with those shown in Fig. 14C.15.

transitions of A is now greater than at equilibrium, so the resonance absorption is enhanced. Another possibility is for the dipolar interaction between the two spins to cause α_A to flip to β_A and simultaneously β_X to flip to α_X (or vice versa). This transition equilibrates the populations of $\alpha_A\beta_X$ and $\beta_A\alpha_X$ but leaves the $\alpha_A\alpha_X$ and $\beta_A\beta_X$ populations unchanged. Now we see from the illustration that the population differences in the states involved in the A transitions are decreased, so the resonance absorption is diminished.

Which effect wins? Does the NOE enhance the A absorption or does it diminish it? As in the discussion of relaxation times in Section 14C.2, the efficiency of the intensity-enhancing $\beta_A\beta_X \leftrightarrow \beta_A\alpha_X$ relaxation is high if the dipole field oscillates close to the transition frequency, which in this case is about 2ν; likewise, the efficiency of the intensity-diminishing $\alpha_A\beta_X \leftrightarrow \beta_A\alpha_X$ relaxation is high if the dipole field is stationary (as there is no frequency difference between the initial and final states). A large molecule rotates so slowly that there is very little motion at 2ν, so we expect an intensity decrease (Fig. 14C.17). A small molecule rotating rapidly can be expected to have substantial motion at 2ν, and a consequent enhancement of the signal. In practice, the enhancement lies somewhere between the two extremes and is reported in terms of the parameter η (eta), where

$$\eta = \frac{I_A - I_A^\circ}{I_A^\circ} \qquad \text{NOE enhancement parameter} \qquad (14C.8)$$

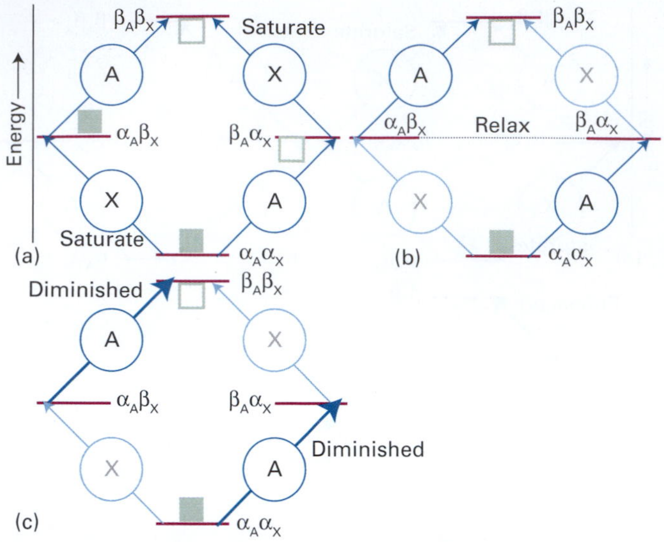

Figure 14C.17 (a) When the X transition is saturated, just as in Fig. 14C.16 the populations of its two states are equalized and the population excess and deficit become as shown. (b) Dipole–dipole relaxation relaxes the populations of the two intermediate states, and they regain their original populations. (c) The A transitions reflect the difference in populations resulting from the preceding changes, and are diminished compared with those shown in Fig. 14C.15.

Here I_A° and I_A are the intensities of the NMR signals due to nucleus A before and after application of the long ($> T_1$) radiofrequency pulse that saturates transitions due to the X nucleus. When A and X are nuclei of the same species, such as protons, η lies between −1 (diminution) and $+\frac{1}{2}$ (enhancement). However, η also depends on the values of the magnetogyric ratios of A and X. In the case of maximal enhancement it is possible to show that

$$\eta=\frac{\gamma_X}{2\gamma_A} \qquad (14C.9)$$

where γ_A and γ_X are the magnetogyric ratios of nuclei A and X, respectively.

Brief illustration 14C.4 NOE enhancement

From eqn 14C.9 and the data in Table 14A.2, the NOE enhancement parameter for ^{13}C close to a saturated proton is

$$\eta=\frac{\overbrace{2.675\times10^{8}\ \mathrm{T^{-1}\,s^{-1}}}^{\gamma_{^1H}}}{\underbrace{2\times(6.73\times10^{7}\ \mathrm{T^{-1}\,s^{-1}})}_{\gamma_{^{13}C}}}=1.99$$

which shows that an enhancement of about a factor of 2 can be achieved.

Self-test 14C.4 Interpret the following features of the NMR spectra of a protein: (a) saturation of a proton resonance assigned to the side chain of a methionine residue changes the intensities of proton resonances assigned to the side chains of a tryptophan and a tyrosine residue; (b) saturation of proton resonances assigned to the tryptophan residue did not affect the spectrum of the tyrosine residue.

Answer: The tryptophan and tyrosine residues are close to the methionine residue, but are far from each other

The NOE is also used to determine inter-proton distances. The Overhauser enhancement of a proton A generated by saturating a spin X depends on the fraction of A's spin–lattice relaxation that is caused by its dipolar interaction with X. Because the dipolar field is proportional to r^{-3}, where r is the internuclear distance, and the relaxation effect is proportional to the square of the field, and therefore to r^{-6}, the NOE may be used to determine the geometries of molecules in solution. The determination of the structure of a small protein in solution involves the use of several hundred NOE measurements, effectively casting a net over the protons present. The enormous importance of this procedure is that we can determine the conformation of biological macromolecules in an aqueous environment and do not need to try to make the single crystals that are essential for an X-ray diffraction investigation (Topic 18A).

14C.5 Two-dimensional NMR

An NMR spectrum contains a great deal of information and, if many protons are present, is very complex when the fine structures of different groups of lines overlap. The complexity would be reduced if we could use two axes to display the data, with resonances belonging to different groups lying at different locations on the second axis. This separation is essentially what is achieved in **two-dimensional NMR**.

Much modern NMR work makes use of **correlation spectroscopy** (COSY) in which a clever choice of pulses and Fourier transformation techniques makes it possible to determine all spin–spin couplings in a molecule. A typical outcome for an AX system is shown in Fig. 14C.18. The diagram shows contours of equal signal intensity on a plot of intensity against the frequency coordinates ν_1 and ν_2. The **diagonal peaks** are signals centred on (δ_A,δ_A) and (δ_X,δ_X) and lie along the diagonal where $\nu_1=\nu_2$. That is, the spectrum along the diagonal is equivalent to the one-dimensional spectrum obtained with the conventional NMR technique (as in Fig. 14B.2). The **cross peaks** (or *off-diagonal peaks*) are signals centred on (δ_A,δ_X) and (δ_X,δ_A) and owe their existence to the coupling between the A and X nuclei.

Although information from two-dimensional NMR spectroscopy is trivial in an AX system, it can be of enormous help

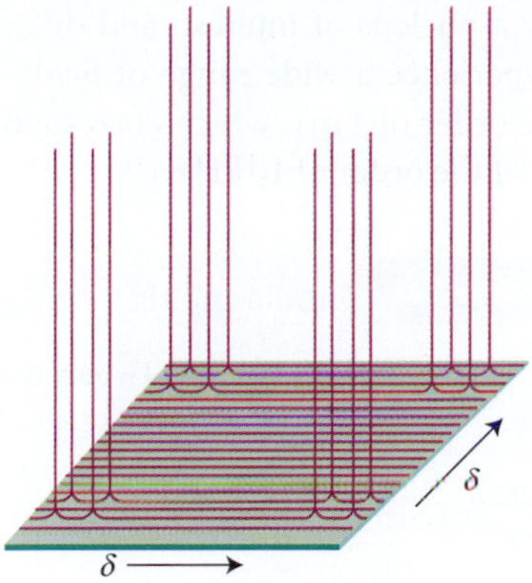

Figure 14C.18 An idealization of the COSY spectrum of an AX spin system.

in the interpretation of more complex spectra, leading to a map of the couplings between spins and to the determination of the bonding network in complex molecules. Indeed, the spectrum of a synthetic or biological polymer that would be impossible to interpret in one-dimensional NMR can often be interpreted reasonably rapidly by two-dimensional NMR.

Example 14C.1 Interpreting a two-dimensional NMR spectrum

Figure 14C.19 is a portion of the COSY spectrum of the amino acid isoleucine (**1**). Assign the resonances to protons bound to the carbon atoms.

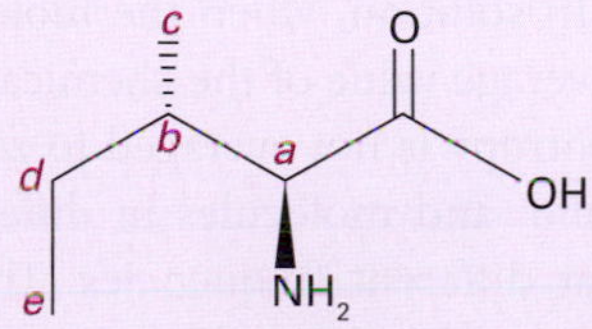

1 Isoleucine

Method Cross peaks in this spectrum arise from the coupling of protons that are separated as in H—C—C—H. That is, the fine structure in the spectrum is determined by the values of $^3J_{HH}$ coupling constants (Topic 14B). Identify expected couplings from the arrangement of bonds in the molecular structure. Then match spectral and molecular features, taking into consideration the effects of chemical and magnetic equivalence (Topic 14B). For example, expect two cross-peaks to arise from coupling of a proton to two inequivalent protons, even if both protons are bound to the same carbon atom.

Answer From the molecular structure, we expect that: (i) the C_a—H proton is coupled only to the C_b—H proton, (ii) the C_b—H protons are coupled to the C_a—H, C_c—H, and C_d—H protons, and (iii) the inequivalent C_d—H protons are coupled

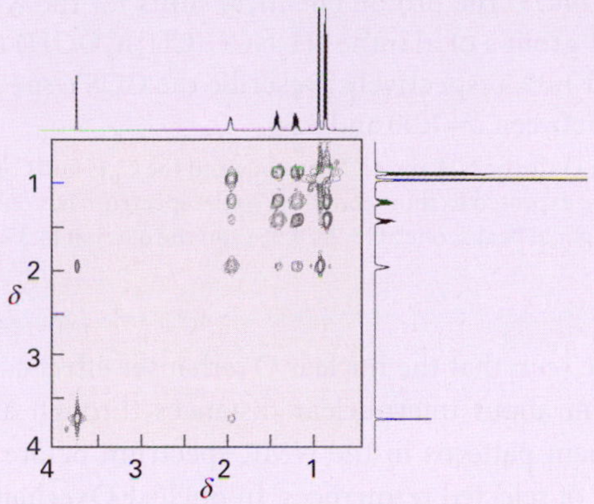

Figure 14C.19 Proton COSY spectrum of isoleucine. (The *Example* and corresponding spectrum are adapted from K.E. van Holde et al., *Principles of physical biochemistry*, Prentice Hall, Upper Saddle River (1998).)

to the C_b—H and C_e—H protons. We proceed with the assignments by noting that:

- The resonance at $\delta = 1.9$ shares cross-peaks with resonances at $\delta = 3.6$, 1.4, 1.2, and 0.9.

Only the C_b—H proton is coupled to protons with four different resonances: the C_a—H proton, the equivalent C_c—H protons, and the two inequivalent C_d—H protons. It follows that the resonance at $\delta = 1.9$ corresponds to the C_b—H proton.

- The resonance at $\delta = 3.6$ shares a cross-peak with only one other resonance at $\delta = 1.9$.

We already know that the resonance at $\delta = 1.9$ corresponds to the C_b—H proton. Only the C_a—H proton is coupled to the C_b—H proton and to no other protons. It follows that the resonance at $\delta = 3.6$ corresponds to the C_a—H proton.

- The proton with resonance at $\delta = 0.8$ is not coupled to the proton with resonance at $\delta = 1.9$, which we have assigned to the C_b—H proton.

Only the equivalent C_e—H protons are not coupled to the C_b—H proton, which we have already assigned to the resonance at $\delta = 1.9$. Hence we assign the resonance at $\delta = 0.8$ to the C_e—H protons.

- The resonances at $\delta = 1.4$ and 1.2 do not share cross-peaks with the resonance at $\delta = 0.9$.

Yet to be assigned are the resonances corresponding to the C_c—H and C_d—H protons. In the light of the expected couplings, the resonances from the inequivalent C_d—H protons do not share cross-peaks with the resonance from the equivalent C_c—H protons. It follows that the resonance at $\delta = 0.9$ can be assigned to the equivalent C_c—H protons and the resonances at $\delta = 1.4$ and 1.2 to the inequivalent C_d—H protons.

Self-test 14C.5 The proton chemical shifts for the NH, $C_\alpha H$, and $C_\beta H$ groups of alanine ($H_2NCH(CH_3)COOH$) are 8.25, 4.35, and 1.39, respectively. Describe the COSY spectrum of alanine between $\delta = 1.00$ and 8.50.

Answer: Only the NH and $C_\alpha H$ protons and the $C_\alpha H$ and $C_\beta H$ protons are expected to show coupling, so the spectrum has only two off-diagonal peaks, one at (8.25, 4.35) and the other at (4.35, and 1.39)

We have seen that the nuclear Overhauser effect can provide information about internuclear distances through analysis of enhancement patterns in the NMR spectrum before and after saturation of selected resonances. In **nuclear Overhauser effect spectroscopy** (NOESY) a map of all possible NOE interactions is obtained by using a proper choice of radiofrequency pulses and Fourier transformation techniques. Like a COSY spectrum, a NOESY spectrum consists of a series of diagonal peaks that correspond to the one-dimensional NMR spectrum of the sample. The off-diagonal peaks indicate which nuclei are close enough to each other to give rise to a nuclear Overhauser effect. NOESY data provide internuclear distances of up to about 0.5 nm.

14C.6 Solid-state NMR

The principal difficulty with the application of NMR to solids is the low resolution characteristic of solid samples. Nevertheless, there are good reasons for seeking to overcome these difficulties. They include the possibility that a compound of interest is unstable in solution or that it is insoluble, so conventional solution NMR cannot be employed. Moreover, many species, such as polymers and nanomaterials, are intrinsically interesting as solids, and it is important to be able to determine their structures and dynamics when X-ray diffraction techniques fail.

There are three principal contributions to the linewidths of solids. One is the direct magnetic dipolar interaction between nuclear spins. As pointed out in the discussion of spin–spin coupling (Topic 14B), a nuclear magnetic moment gives rise to a local magnetic field which points in different directions at different locations around the nucleus. If we are interested only in the component parallel to the direction of the applied magnetic field (because only this component has a significant effect), then, provided certain subtle effects arising from transformation from the static to the rotating frame are neglected, we can use a classical expression in *The chemist's toolkit* 14B.1 to write the magnitude of the local magnetic field as

$$\mathcal{B}_{\text{loc}} = -\frac{\gamma_N \hbar \mu_0 m_I}{4\pi R^3}(1 - 3\cos^2\theta) \qquad (14C.10)$$

Unlike in solution, in a solid this field is not motionally averaged to zero. Many nuclei may contribute to the total local field experienced by a nucleus of interest, and different nuclei in a sample may experience a wide range of fields. Typical dipole fields are of the order of 1 mT, which corresponds to splittings and linewidths of the order of 10 kHz.

Brief illustration 14C.5 Dipolar fields in solids

When the angle θ can vary only between 0 and $\theta_{\max}$, eqn 14C.10 becomes

$$\mathcal{B}_{\text{loc}} = \frac{\gamma_N \hbar \mu_0 m_I}{4\pi R^3}(\cos^2\theta_{\max} + \cos\theta_{\max})$$

When $\theta_{\max} = 30°$ and $R = 160$ pm, the local field generated by a proton is

$$\mathcal{B}_{\text{loc}} = \frac{\overbrace{(3.546\cdots\times10^{-32}\,\mathrm{T\,m^3})}^{\gamma_N\hbar\mu_0}\times\overbrace{(\tfrac{1}{2})}^{m_I}\times\overbrace{1.616}^{\cos^2\theta_{\max}+\cos\theta_{\max}}}{4\pi\times\underbrace{(1.60\times10^{-10}\,\mathrm{m})^3}_{R}}$$
$$= 5.57\times10^{-4}\,\mathrm{T} = 0.557\,\mathrm{mT}$$

Self-test 14C.6 Calculate the distance at which the local field from a ^{13}C nucleus is 0.50 mT, with $\theta_{\max} = 40°$.

Answer: $R = 99$ pm

A second source of linewidth is the anisotropy of the chemical shift. Chemical shifts arise from the ability of the applied field to generate electron currents in molecules. In general, this ability depends on the orientation of the molecule relative to the applied field. In solution, when the molecule is tumbling rapidly, only the average value of the chemical shift is relevant. However, the anisotropy is not averaged to zero for stationary molecules in a solid, and molecules in different orientations have resonances at different frequencies. The chemical shift

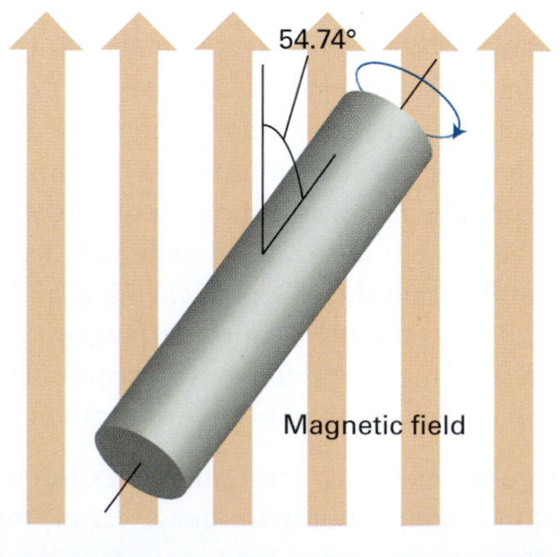

Figure 14C.20 In magic-angle spinning, the sample spins at 54.74° (that is, arccos $1/3^{1/2}$) to the applied magnetic field. Rapid motion at this angle averages dipole–dipole interactions and chemical shift anisotropies to zero.

anisotropy also varies with the angle between the applied field and the principal axis of the molecule as $1-3\cos^2\theta$.

The third contribution is the electric quadrupole interaction. Nuclei with $I>\frac{1}{2}$ have an electric quadrupole moment, a measure of the extent to which the distribution of charge over the nucleus is not uniform (for instance, the positive charge may be concentrated around the equator or at the poles). An electric quadrupole interacts with an electric field gradient, such as may arise from a non-spherical distribution of charge around the nucleus. This interaction also varies as $1-3\cos^2\theta$.

Fortunately, there are techniques available for reducing the linewidths of solid samples. One technique, **magic-angle spinning** (MAS), takes note of the $1-3\cos^2\theta$ dependence of the dipole–dipole interaction, the chemical shift anisotropy, and the electric quadrupole interaction. The 'magic angle' is the angle at which $1-3\cos^2\theta=0$, and corresponds to 54.74°. In the technique, the sample is spun at high speed at the magic angle to the applied field (Fig. 14C.20). All the dipolar interactions and the anisotropies average to the value they would have at the magic angle, but at that angle they are zero. The difficulty with MAS is that the spinning frequency must not be less than the width of the spectrum, which is of the order of kilohertz. However, gas-driven sample spinners that can be rotated at up to 25 kHz are now routinely available, and a considerable body of work has been done.

Pulsed techniques similar to those described in the previous section may also be used to reduce linewidths. Elaborate pulse sequences have also been devised that reduce linewidths by averaging procedures that make use of twisting the magnetization vector through an elaborate series of angles.

Checklist of concepts

- ☐ 1. **Free-induction decay** (FID) is the decay of the magnetization after the application of a radiofrequency pulse.
- ☐ 2. Fourier transformation of the FID curve gives the NMR spectrum.
- ☐ 3. During **longitudinal** (or **spin–lattice**) **relaxation**, β spins revert to α spins.
- ☐ 4. **Transverse** (or **spin–spin**) **relaxation** is the randomization of spin directions around the z-axis.
- ☐ 5. The **longitudinal relaxation time** T_1 can be measured by the **inversion recovery technique**.
- ☐ 6. The **transverse relaxation time** T_2 can be measured by observing **spin echoes**.
- ☐ 7. In **proton decoupling** of ^{13}C-NMR spectra, protons are made to undergo rapid spin reorientations and the ^{13}C nucleus senses an average orientation.
- ☐ 8. The **nuclear Overhauser effect** (NOE) is the modification of the intensity of one resonance by the saturation of another.
- ☐ 9. In **two-dimensional NMR**, spectra are displayed in two axes, with resonances belonging to different groups lying at different locations on the second axis.
- ☐ 10. **Magic-angle spinning** (MAS) is technique in which the NMR linewidths in a solid sample are reduced by spinning at an angle of 54.74° to the applied magnetic field.

Checklist of equations

Property	Equation	Comment	Equation number
Free-induction decay	$M_y(t)=M_0\cos(2\pi\nu_L t)e^{-t/T_2}$	T_2 is the transverse relaxation time	14C.1
Longitudinal relaxation	$M_z(t)-M_0\propto e^{-t/T_1}$	T_1 is the spin–lattice relaxation time	14C.4
Transverse relaxation	$M_y(t)\propto e^{-t/T_2}$		14C.5
Width at half-height of an NMR line	$\Delta\nu_{1/2}=1/\pi T_2$		14C.6
Effective transverse relaxation time	$T_2^*=1/\pi\Delta\nu_{1/2}$	Definition; inhomogeneous broadening	14C.7
NOE enhancement parameter	$\eta=(I_A-I_A^\circ)/I_A^\circ$	Definition	14C.8

14D Electron paramagnetic resonance

Contents

➤ Why do you need to know this material?

Many materials and biological systems contain species bearing unpaired electrons. Furthermore, some chemical reactions generate intermediates that contain unpaired electrons. You need to know how to characterize the structures of such species with special spectroscopic techniques.

➤ What is the key idea?

The electron paramagnetic resonance spectrum of a radical arises from the ability of the applied magnetic field to induce local electron currents and the magnetic interaction between the unpaired electron and nuclei with spin.

➤ What do you need to know already?

You need to be familiar with the concepts of electron spin (Topic 9B) and the general principles of magnetic resonance (Topic 14A). The discussion refers to spin–orbit coupling in atoms (Topic 9C) and the Fermi contact interaction in molecules (Topic 14B).

Electron paramagnetic resonance (EPR), which is also known as electron spin resonance (ESR), is used to study radicals formed during chemical reactions or by radiation, radicals that act as probes of biological structure, many d-metal complexes, and molecules in triplet states (such as those involved in phosphorescence, Topic 13B). The sample may be a gas, a liquid, or a solid, but the free rotation of molecules in the gas phase gives rise to complications.

14D.1 The *g*-value

The resonance frequency for a transition between the $m_s=-\frac{1}{2}$ and the $m_s=+\frac{1}{2}$ levels of an electron is

$$h\nu = g_e\mu_B\mathcal{B}_0 \qquad \textit{Free electron} \qquad \text{Resonance condition} \qquad (14D.1)$$

where $g_e \approx 2.0023$ (Topic 14A). The magnetic moment of an unpaired electron in a radical also interacts with an external field, but the field it experiences differs from the applied field due to the presence of local magnetic fields arising from electron currents induced in the molecular framework. This difference is taken into account by replacing g_e by g and expressing the resonance condition as

$$h\nu = g\mu_B\mathcal{B}_0 \qquad \text{EPR resonance condition} \qquad (14D.2)$$

where g is the **g-value** of the radical.

Brief illustration 14D.1 The *g*-value of a radical

The centre of the EPR spectrum of the methyl radical occurred at 329.40 mT in a spectrometer operating at 9.2330 GHz (radiation belonging to the X band of the microwave region). Its *g*-value is therefore

$$g=\frac{\overbrace{(6.626\,08\times10^{-34}\,\mathrm{J\,s})}^{h}\times\overbrace{(9.2330\times10^{9}\,\mathrm{s^{-1}})}^{\nu}}{\underbrace{(9.2740\times10^{-24}\,\mathrm{J\,T^{-1}})}_{\mu_B}\times\underbrace{(0.329\,40\,\mathrm{T})}_{\mathcal{B}_0}}=2.0027$$

Self-test 14D.1 At what magnetic field would the methyl radical come into resonance in a spectrometer operating at 34.000 GHz (radiation belonging to the Q band of the microwave region)?

Answer: 1.213 T

The *g*-value is related to the ease with which the applied field can stir up currents through the molecular framework and the strength of the magnetic field the currents generate. Therefore,

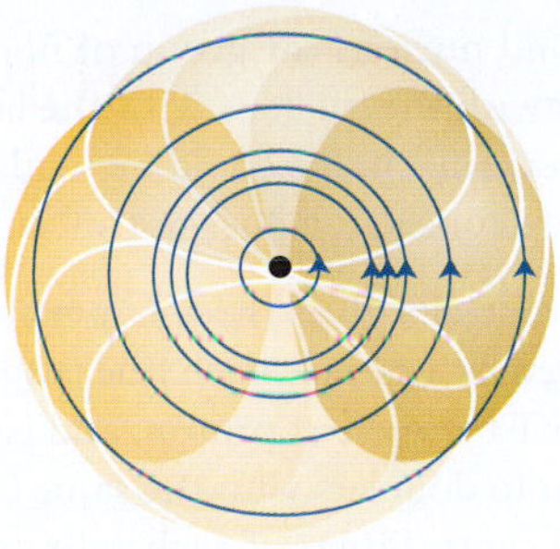

Figure 14D.1 An applied magnetic field can induce circulation of electrons that makes use of excited state orbitals (shown with a white line).

the g-value gives some information about electronic structure and plays a similar role in EPR to that played by shielding constants in NMR.

Two factors are responsible for the difference of the g-value from g_e. Electrons migrate through the molecular framework by making use of excited states (Fig. 14D.1). This circulation gives rise to a local magnetic field that adds to the applied field. The extent to which these currents are induced is inversely proportional to the separation of energy levels, ΔE, in the radical or complex. Secondly, the strength of the field experienced by the electron spin as a result of these electronic currents is proportional to the spin–orbit coupling constant, ξ (Topic 9C). We can conclude that the g-value differs from g_e by an amount that is proportional to $\xi/\Delta E$. This proportionality is widely observed. Many organic radicals, for which ΔE is large and ξ (for carbon) is small, have g-values close to 2.0027, not far removed from g_e itself. Inorganic radicals, which commonly are built from heavier atoms and therefore have larger spin–orbit coupling constants, have g-values typically in the range 1.9 to 2.1. The g-values of paramagnetic d-metal complexes often differ considerably from g_e, varying from 0 to 6, because in them ΔE is small on account of the small splitting of d-orbitals brought about by interactions with ligands (Topic 13A).

The g-value is anisotropic: that is, its magnitude depends on the orientation of the radical with respect to the applied field. The anisotropy arises from the fact that the extent to which an applied field induces currents in the molecule, and therefore the magnitude of the local field, depends on the relative orientation of the molecules and the field. In solution, when the molecule is tumbling rapidly, only the average value of the g-value is observed. Therefore, anisotropy of the g-value is observed only for radicals trapped in solids.

14D.2 Hyperfine structure

The most important feature of an EPR spectrum is its **hyperfine structure**, the splitting of individual resonance lines into components. In general in spectroscopy, the term 'hyperfine structure' means the structure of a spectrum that can be traced to interactions of the electrons with nuclei other than as a result of the latter's point electric charge. The source of the hyperfine structure in EPR is the magnetic interaction between the electron spin and the magnetic dipole moments of the nuclei present in the radical which give rise to local magnetic fields.

(a) The effects of nuclear spin

Consider the effect on the EPR spectrum of a single H nucleus located somewhere in a radical. The proton spin is a source of magnetic field and, depending on the orientation of the nuclear spin, the field it generates adds to or subtracts from the applied field. The total local field is therefore

$$\mathcal{B}_{\text{loc}} = \mathcal{B}_0 + am_I \quad m_I = \pm\tfrac{1}{2} \tag{14D.3}$$

where a is the **hyperfine coupling constant**. Half the radicals in a sample have $m_I = +\frac{1}{2}$, so half resonate when the applied field satisfies the condition

$$h\nu = g\mu_B(\mathcal{B}_0 + \tfrac{1}{2}a), \quad \text{or} \quad \mathcal{B}_0 = \frac{h\nu}{g\mu_B} - \tfrac{1}{2}a \tag{14D.4a}$$

The other half (which have $m_I = -\frac{1}{2}$) resonate when

$$h\nu = g\mu_B(\mathcal{B}_0 - \tfrac{1}{2}a), \quad \text{or} \quad \mathcal{B}_0 = \frac{h\nu}{g\mu_B} + \tfrac{1}{2}a \tag{14D.4b}$$

Therefore, instead of a single line, the spectrum shows two lines of half the original intensity separated by a and centred on the field determined by g (Fig. 14D.2).

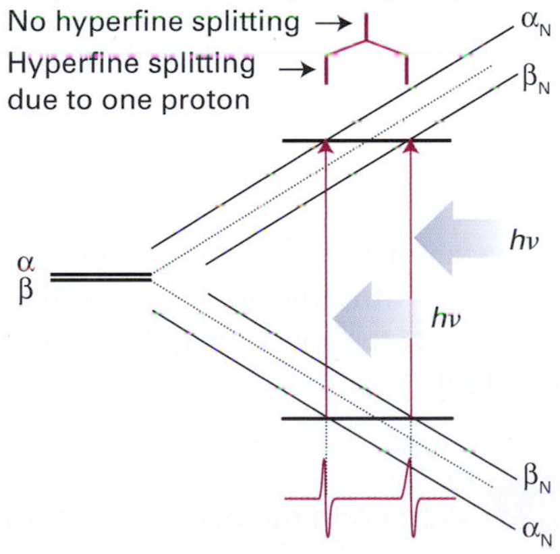

Figure 14D.2 The hyperfine interaction between an electron and a spin-$\frac{1}{2}$ nucleus results in four energy levels in place of the original two. As a result, the spectrum consists of two lines (of equal intensity) instead of one. The intensity distribution can be summarized by a simple stick diagram. The diagonal lines show the energies of the states as the applied field is increased, and resonance occurs when the separation of states matches the fixed energy of the microwave photon.

If the radical contains an ^{14}N atom ($I=1$), its EPR spectrum consists of three lines of equal intensity, because the ^{14}N nucleus has three possible spin orientations, and each spin orientation is possessed by one-third of all the radicals in the sample. In general, a spin-I nucleus splits the spectrum into $2I+1$ hyperfine lines of equal intensity.

When there are several magnetic nuclei present in the radical, each one contributes to the hyperfine structure. In the case of equivalent protons (for example, the two CH_2 protons in the radical CH_3CH_2) some of the hyperfine lines are coincident. If the radical contains N equivalent protons, then there are $N+1$ hyperfine lines with an intensity distribution given by Pascal's triangle (Topic 14B, reproduced here as **1**). The spectrum of the benzene radical anion in Fig. 14D.3, which has seven lines with intensity ratio 1:6:15:20:15:6:1, is consistent with a radical containing six equivalent protons. More generally, if the radical contains N equivalent nuclei with spin quantum number I, then there are $2NI+1$ hyperfine lines with an intensity distribution based on a modified version of Pascal's triangle as shown in the following *Example*.

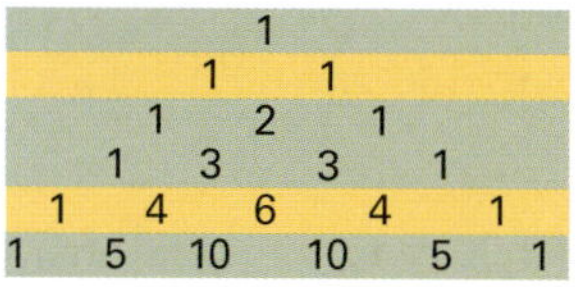

1

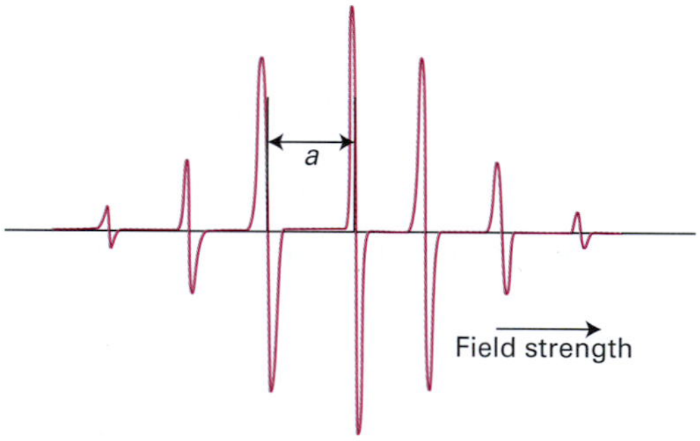

Figure 14D.3 The EPR spectrum of the benzene radical anion, $C_6H_6^-$, in fluid solution, with a the hyperfine splitting of the spectrum. The centre of the spectrum is determined by the g-value of the radical.

Example 14D.1 Predicting the hyperfine structure of an EPR spectrum

A radical contains one ^{14}N nucleus ($I=1$) with hyperfine constant 1.61 mT and two equivalent protons ($I=\frac{1}{2}$) with hyperfine constant 0.35 mT. Predict the form of the EPR spectrum.

Method Consider the hyperfine structure that arises from each type of nucleus or group of equivalent nuclei in succession. So, split a line with one nucleus, then each of those lines is split by a second nucleus (or group of nuclei), and so on. It is best to start with the nucleus with the largest hyperfine splitting; however, any choice could be made, and the order in which nuclei are considered does not affect the conclusion.

Answer The ^{14}N nucleus gives three hyperfine lines of equal intensity separated by 1.61 mT. Each line is split into doublets of spacing 0.35 mT by the first proton, and each line of these doublets is split into doublets with the same 0.35 mT splitting (Fig. 14D.4). The central lines of each split doublet coincide, so the proton splitting gives 1:2:1 triplets of internal splitting 0.35 mT. Therefore, the spectrum consists of three equivalent 1:2:1 triplets.

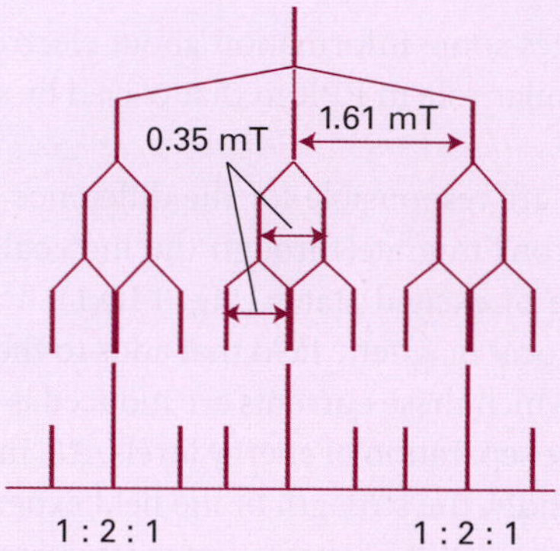

Figure 14D.4 The analysis of the hyperfine structure of radicals containing one ^{14}N nucleus ($I=1$) and two equivalent protons.

Self-test 14D.2 Predict the form of the EPR spectrum of a radical containing three equivalent ^{14}N nuclei.

Answer: See Fig. 14D.5.

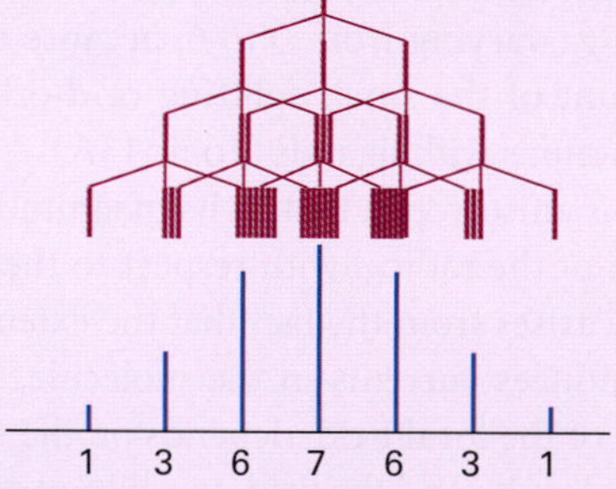

Figure 14D.5 The analysis of the hyperfine structure of radicals containing three equivalent ^{14}N nuclei.

(b) The McConnell equation

The hyperfine structure of an EPR spectrum is a kind of fingerprint that helps to identify the radicals present in a sample. Moreover, because the magnitude of the splitting depends on the distribution of the unpaired electron in the vicinity of the magnetic nuclei, the spectrum can be used to map the molecular orbital occupied by the unpaired electron. For example, because the hyperfine splitting in $C_6H_6^-$ is 0.375 mT, and one proton is close to a C atom that has one-sixth the unpaired electron spin density (because the electron is spread uniformly around the ring), the hyperfine splitting caused by a proton in the electron spin entirely confined to a single adjacent C atom should be $6\times0.375\,\text{mT}=2.25\,\text{mT}$. If in another aromatic radical we find a hyperfine splitting constant a, then the **spin density**, ρ, the probability that an unpaired electron is on the atom, can be calculated from the **McConnell equation**:

$$a=Q\rho \qquad \text{McConnell equation} \qquad (14D.5)$$

with $Q=2.25\,\text{mT}$. In this equation, ρ is the spin density on a C atom and a is the hyperfine splitting observed for the H atom to which it is attached. This expression simply represents the fact that the hyperfine coupling to the H atom is likely to be proportional to the spin density on the C atom to which it is attached.

Brief illustration 14D.2 The McConnell equation

The hyperfine structure of the EPR spectrum of $C_{10}H_8^-$, the naphthalene radical anion, can be interpreted as arising from two groups of four equivalent protons. Those at the 1, 4, 5, and 8 positions in the ring have $a=0.490\,\text{mT}$ and those in the 2, 3, 6, and 7 positions have $a=0.183\,\text{mT}$. The densities obtained by using the McConnell equation are, respectively (2),

$$\rho=\frac{\overbrace{0.490\,\text{mT}}^{a}}{\underbrace{2.25\,\text{mT}}_{Q}}=0.218 \quad \text{and} \quad \rho=\frac{0.183\,\text{mT}}{2.25\,\text{mT}}=0.0813$$

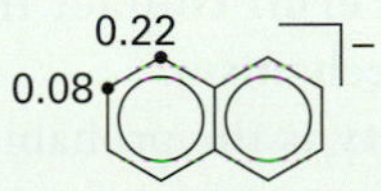

2

Self-test 14D.3 The spin density in $C_{14}H_{10}^-$, the anthracene radical anion, is shown in (3). Predict the form of its EPR spectrum.

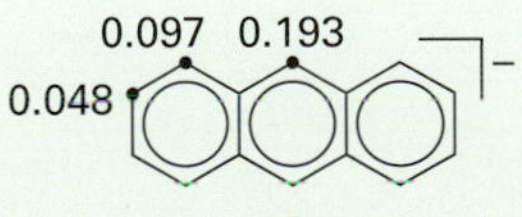

3

Answer: A 1:2:1 triplet of splitting 0.43 mT split into a 1:4:6:4:1 quintet of splitting 0.22 mT, split into a 1:4:6:4:1 quintet of splitting 0.11 mT, $3\times5\times5=75$ lines in all

(c) The origin of the hyperfine interaction

The hyperfine interaction is an interaction between the magnetic moments of the unpaired electron and the nuclei. There are two contributions to the interaction.

An electron in a p orbital centred on a nucleus does not approach the nucleus very closely, so it experiences a field that appears to arise from a point magnetic dipole. The resulting interaction is called the **dipole–dipole interaction**. The contribution of a magnetic nucleus to the local field experienced by the unpaired electron is given by an expression like that in eqn 14B.10a (a dependence proportional to $(1-3\cos^2\theta)/r^3$). A characteristic of this type of interaction is that it is anisotropic and averages to zero when the radical is free to tumble. Therefore, hyperfine structure due to the dipole–dipole interaction is observed only for radicals trapped in solids.

An s electron is spherically distributed around a nucleus and so has zero average dipole–dipole interaction with the nucleus even in a solid sample. However, because it has a nonzero probability of being at the nucleus, it is incorrect to treat the interaction as one between two point dipoles. As explained in Topic 14B, an s electron has a Fermi contact interaction with the nucleus, a magnetic interaction that occurs when the point dipole approximation fails. The contact interaction is isotropic (that is, independent of the radical's orientation), and consequently is shown even by rapidly tumbling molecules in fluids (provided the spin density has some s character).

The dipole–dipole interactions of p electrons and the Fermi contact interaction of s electrons can be quite large. For example, a 2p electron in a nitrogen atom experiences an average field of about 3.4 mT from the ^{14}N nucleus. A 1s electron in a hydrogen atom experiences a field of about 50 mT as a result of its Fermi contact interaction with the central proton. More values are listed in Table 14D.1. The magnitudes of the contact interactions in radicals can be interpreted in terms of the s orbital character of the molecular orbital occupied by the unpaired electron, and the dipole–dipole interaction can be interpreted in terms of the p character. The analysis of hyperfine structure therefore gives information about the composition of the orbital, and especially the hybridization of the atomic orbitals.

Table 14D.1* Hyperfine coupling constants for atoms, a/mT

Nuclide	Isotropic coupling	Anisotropic coupling
^1H	50.8 (1s)	
^2H	7.8 (1s)	
^{14}N	55.2 (2s)	4.8 (2p)
^{19}F	1720 (2s)	108.4 (2p)

*More values are given in the *Resource section*.

Brief illustration 14D.3 The composition of a molecular orbital from analysis of the hyperfine structure

From Table 14D.1, the hyperfine interaction between a 2s electron and the nucleus of a nitrogen atom is 55.2 mT. The EPR spectrum of NO_2 shows an isotropic hyperfine interaction of

5.7 mT. The s character of the molecular orbital occupied by the unpaired electron is the ratio 5.7/55.2 = 0.10. For a continuation of this story, see Problem 14D.6.

Self-test 14D.4 In NO_2 the anisotropic part of the hyperfine coupling is 1.3 mT. What is the p character of the molecular orbital occupied by the unpaired electron?

Answer: 0.38

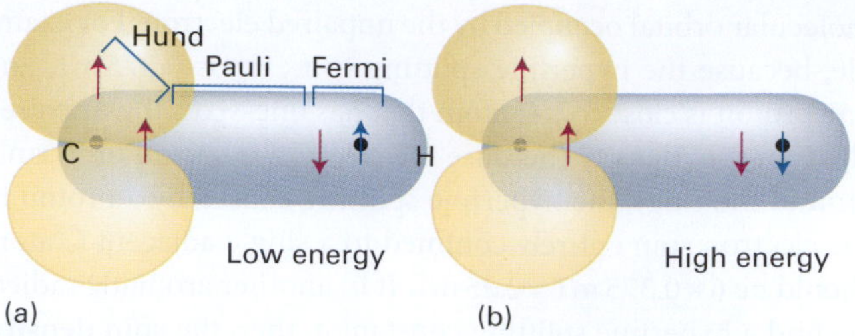

Figure 14D.6 The polarization mechanism for the hyperfine interaction in π-electron radicals. The arrangement in (a) is lower in energy than that in (b), so there is an effective coupling between the unpaired electron and the proton.

We still have the source of the hyperfine structure of the $C_6H_6^-$ anion and other aromatic radical anions to explain. The sample is fluid, and as the radicals are tumbling the hyperfine structure cannot be due to the dipole–dipole interaction. Moreover, the protons lie in the nodal plane of the π orbital occupied by the unpaired electron, so the structure cannot be due to a Fermi contact interaction. The explanation lies in a **polarization mechanism** similar to the one responsible for spin–spin coupling in NMR. There is a magnetic interaction between a proton and the α electrons ($m_s = \pm\frac{1}{2}$) which results in one of the electrons tending to be found with a greater probability nearby (Fig. 14D.6). The electron with opposite spin is therefore more likely to be close to the C atom at the other end of the bond. The unpaired electron on the C atom has a lower energy if it is parallel to that electron (Hund's rule favours parallel electrons on atoms), so the unpaired electron can detect the spin of the proton indirectly. Calculation using this model leads to a hyperfine interaction in agreement with the observed value of 2.25 mT.

Checklist of concepts

- ☐ 1. The EPR resonance condition is written in terms of the ***g*-value** of the radical.
- ☐ 2. The value of g depends on the ability of the applied field to induce local electron currents in the radical.
- ☐ 3. The **hyperfine structure** of an EPR spectrum is its splitting of individual resonance lines into components by the magnetic interaction between the electron and nuclei with spin.
- ☐ 4. If a radical contains N equivalent nuclei with spin quantum number I, then there are $2NI+1$ hyperfine lines with an intensity distribution given by a modified version of Pascal's triangle.
- ☐ 5. Hyperfine structure can be explained by **dipole–dipole interactions**, **Fermi contact interactions**, and the **polarization mechanism**.
- ☐ 6. The **spin density** is the probability that an unpaired electron is on the atom.

Checklist of equations

Property	Equation	Comment	Equation number
EPR resonance condition	$h\nu = g\mu_B\mathcal{B}_0$	No hyperfine interaction	14D.2
	$h\nu = g\mu_B(\mathcal{B}_0 \pm \frac{1}{2}a)$	Hyperfine interaction between an electron and a proton	14D.4
McConnell equation	$a = Q\rho$	Q = 2.25 mT	14D.5

CHAPTER 14 Magnetic resonance

TOPIC 14A General principles

Discussion questions

14A.1 To determine the structures of macromolecules by NMR spectroscopy, chemists use spectrometers that operate at the highest available fields and frequencies. Justify this choice.

14A.2 Compare the effects of magnetic fields on the energies of nuclei and the energies of electrons.

14A.3 What is the Larmor frequency? What is its role in magnetic resonance?

Exercises

14A.1(a) Given that g is a dimensionless number, what are the units of γ_N expressed in tesla and hertz?
14A.1(b) Given that g is a dimensionless number, what are the units of γ_N expressed in SI base units?

14A.2(a) For a proton, what are the magnitude of the spin angular momentum and its allowed components along the z-axis? What are the possible orientations of the angular momentum in terms of the angle it makes with the z-axis?
14A.2(b) For a ^{14}N nucleus, what are the magnitude of the spin angular momentum and its allowed components along the z-axis? What are the possible orientations of the angular momentum in terms of the angle it makes with the z-axis?

14A.3(a) What is the resonance frequency of a proton in a magnetic field of 13.5 T?
14A.3(b) What is the resonance frequency of a ^{19}F nucleus in a magnetic field of 17.1 T?

14A.4(a) ^{33}S has a nuclear spin of $\frac{3}{2}$ and a nuclear g-factor of 0.4289. Calculate the energies of the nuclear spin states in a magnetic field of 6.800 T.
14A.4(b) ^{14}N has a nuclear spin of 1 and a nuclear g-factor of 0.404. Calculate the energies of the nuclear spin states in a magnetic field of 10.50 T.

14A.5(a) Calculate the frequency separation of the nuclear spin levels of a ^{13}C nucleus in a magnetic field of 15.4 T given that the magnetogyric ratio is $6.73\times10^{-7}\,T^{-1}s^{-1}$.
14A.5(b) Calculate the frequency separation of the nuclear spin levels of a ^{14}N nucleus in a magnetic field of 14.4 T given that the magnetogyric ratio is $1.93\times10^{-7}\,T^{-1}s^{-1}$.

14A.6(a) In which of the following systems is the energy level separation larger? (i) A proton in a 600 MHz NMR spectrometer, (ii) a deuteron in the same spectrometer.
14A.6(b) In which of the following systems is the energy level separation larger? (i) A ^{14}N nucleus in (for protons) a 600 MHz NMR spectrometer, (ii) an electron in a radical in a field of 0.300 T.

14A.7(a) Calculate the relative population differences ($\delta N/N$, where δN denotes a small difference $N_\alpha - N_\beta$) for protons in fields of (i) 0.30 T, (ii) 1.5 T, and (iii) 10 T at 25 °C.
14A.7(b) Calculate the relative population differences ($\delta N/N$, where δN denotes a small difference $N_\alpha - N_\beta$) for ^{13}C nuclei in fields of (i) 0.50 T, (ii) 2.5 T, and (iii) 15.5 T at 25 °C.

14A.8(a) The first generally available NMR spectrometers operated at a frequency of 60 MHz; today it is not uncommon to use a spectrometer that operates at 800 MHz. What are the relative population differences of ^{13}C spin states in these two spectrometers at 25 °C?
14A.8(b) What are the relative population differences of ^{19}F spin states in spectrometers operating at 60 MHz and 450 MHz at 25 °C?

14A.9(a) What magnetic field would be required in order to use an EPR X-band spectrometer (9 GHz) to observe ^1H-NMR and a 300 MHz spectrometer to observe EPR?
14A.9(b) Some commercial EPR spectrometers use 8 mm microwave radiation (the 'Q band'). What magnetic field is needed to satisfy the resonance condition?

Problems

14A.1 A scientist investigates the possibility of neutron spin resonance, and has available a commercial NMR spectrometer operating at 300 MHz. What field is required for resonance? What is the relative population difference at room temperature? Which is the lower energy spin state of the neutron?

14A.2‡ The relative sensitivity of NMR lines for equal numbers of different nuclei at constant temperature for a given frequency is $R_\nu \propto (I+1)\mu^3$ whereas for a given field it is $R_\mathcal{B} \propto \{(I+1)/I^2\}\mu^3$. (a) From the data in Table 14A.2, calculate these sensitivities for the deuteron, ^{13}C, ^{14}N, ^{19}F, and ^{31}P relative to the proton. (b) Develop the equation for $R_\mathcal{B}$ from the equation for R_ν.

14A.3 With special techniques, known collectively as magnetic resonance imaging (MRI), it is possible to obtain NMR spectra of entire organisms. A key to MRI is the application of a magnetic field that varies linearly across the specimen. Consider a flask of water held in a field that varies in the z-direction according to $\mathcal{B}_0 + \mathcal{G}_z z$, where $\mathcal{G}_z$ is the field gradient along the z-direction. Then the water protons will be resonant at the frequencies

$$\nu_L(z)=\frac{\gamma_N}{2\pi}(\mathcal{B}_0+\mathcal{G}_z z)$$

(Similar equations may be written for gradients along the x- and y-directions.) Application of a 90° radiofrequency pulse with $\nu=\nu_L(z)$ will result in a signal with an intensity that is proportional to the numbers of protons at the position z. Now suppose a uniform disk-shaped organ is in a linear field gradient, and that the MRI signal is proportional to the number of protons in a slice of width δz at each horizontal distance z from the centre of the disk. Sketch the shape of the absorption intensity for the MRI image of the disk before any computer manipulation has been carried out.

‡ These problems were supplied by Charles Trapp and Carmen Giunta.

TOPIC 14B Features of NMR spectra

Discussion questions

14B.1 Describe the significance of the chemical shift in relation to the terms 'high-field' and 'low-field'.

14B.2 Discuss in detail the origins of the local, neighbouring group, and solvent contributions to the shielding constant.

14B.3 Explain why groups of equivalent protons do not exhibit the spin–spin coupling that exists between them.

14B.4 Explain the difference between magnetically equivalent and chemically equivalent nuclei, and give two examples of each.

14B.5 Discuss how the Fermi contact interaction and the polarization mechanism contribute to spin–spin couplings in NMR.

Exercises

14B.1(a) What are the relative values of the chemical shifts observed for nuclei in the spectrometers mentioned in Exercise 14A.9a in terms of (i) δ values, (ii) frequencies?
14B.1(b) What are the relative values of the chemical shifts observed for nuclei in the spectrometers mentioned in Exercise 14A.9b in terms of (i) δ values, (ii) frequencies?

14B.2(a) The chemical shift of the CH_3 protons in acetaldehyde (ethanal) is $\delta = 2.20$ and that of the CHO proton is 9.80. What is the difference in local magnetic field between the two regions of the molecule when the applied field is (i) 1.5 T, (ii) 15 T?
14B.2(b) The chemical shift of the CH_3 protons in diethyl ether is $\delta = 1.16$ and that of the CH_2 protons is 3.36. What is the difference in local magnetic field between the two regions of the molecule when the applied field is (i) 1.9 T, (ii) 16.5 T?

14B.3(a) Sketch the appearance of the ^{1}H-NMR spectrum of acetaldehyde (ethanal) using $J = 2.90$ Hz and the data in Exercise 14B.2(a) in a spectrometer operating at (i) 250 MHz, (ii) 800 MHz.
14B.3(b) Sketch the appearance of the ^{1}H-NMR spectrum of diethyl ether using $J = 6.97$ Hz and the data in Exercise 14B.2(b) in a spectrometer operating at (i) 400 MHz, (ii) 650 MHz.

14B.4(a) Sketch the form of the ^{19}F-NMR spectra of a natural sample of $^{10}BF_4^-$ and $^{11}BF_4^-$.
14B.4(b) Sketch the form of the ^{31}P-NMR spectra of a sample of $^{31}PF_6^-$.

14B.5(a) From the data in Table 14A.2, predict the frequency needed for ^{19}F-NMR in an NMR spectrometer designed to observe proton resonance at 800 MHz. Sketch the proton and ^{19}F resonances in the NMR spectrum of FH_2^+.
14B.5(b) From the data in Table 14A.2, predict the frequency needed for ^{31}P-NMR in an NMR spectrometer designed to observe proton resonance at 500 MHz. Sketch the proton and ^{31}P resonances in the NMR spectrum of PH_4^+.

14B.6(a) Construct a version of Pascal's triangle to show the fine structure that might arise from spin–spin coupling to a group of four spin-$\frac{3}{2}$ nuclei.
14B.6(b) Construct a version of Pascal's triangle to show the fine structure that might arise from spin–spin coupling to a group of three spin-$\frac{5}{2}$ nuclei.

14B.7(a) Sketch the form of an $A_3M_2X_4$ spectrum, where A, M, and X are protons with distinctly different chemical shifts and $J_{AM} > J_{AX} > J_{MX}$.
14B.7(b) Sketch the form of an $A_2M_2X_5$ spectrum, where A, M, and X are protons with distinctly different chemical shifts and $J_{AM} > J_{AX} > J_{MX}$.

14B.8(a) Which of the following molecules have sets of nuclei that are chemically but not magnetically equivalent? (i) CH_3CH_3, (ii) $CH_2{=}CH_2$.
14B.8(b) Which of the following molecules have sets of nuclei that are chemically but not magnetically equivalent? (i) $CH_2{=}C{=}CF_2$, (ii) *cis*- and *trans*-$[Mo(CO)_4(PH_3)_2]$.

14B.9(a) A proton jumps between two sites with $\delta = 2.7$ and $\delta = 4.8$. At what rate of interconversion will the two signals collapse to a single line in a spectrometer operating at 550 MHz?
14B.9(b) A proton jumps between two sites with $\delta = 4.2$ and $\delta = 5.5$. At what rate of interconversion will the two signals collapse to a single line in a spectrometer operating at 350 MHz?

Problems

14B.1 You are designing an MRI spectrometer (see Problem 14A.3). What field gradient (in microtesla per metre, $\mu T\,m^{-1}$) is required to produce a separation of 100 Hz between two protons separated by the long diameter of a human kidney (taken as 8 cm) given that they are in environments with $\delta = 3.4$? The radiofrequency field of the spectrometer is at 400 MHz and the applied field is 9.4 T.

14B.2 Refer to Fig. 14B.14 and use mathematical software, a spreadsheet, or the *Living graphs* on the web site of this book to draw a family of curves showing the variation of $^3J_{HH}$ with ϕ for which $A = +7.0$ Hz, $B = -1.0$ Hz, and C varies slightly from a typical value of +5.0 Hz. What is the effect of changing the value of the parameter C on the shape of the curve? In a similar fashion, explore the effect of the values of A and B on the shape of the curve.

14B.3‡ Various versions of the Karplus equation (eqn 14B.14) have been used to correlate data on vicinal proton coupling constants in systems of the type $R_1R_2CHCHR_3R_4$. The original version, (M. Karplus, *J. Am. Chem.* Soc. **85**, 2870 (1963)), is $^3J_{HH} = A\cos^2\phi_{HH} + B$. When $R_3 = R_4 = H$, $^3J_{HH} = 7.3$ Hz; when $R_3 = CH_3$ and $R_4 = H$, $^3J_{HH} = 8.0$ Hz; when $R_3 = R_4 = CH_3$, $^3J_{HH} = 11.2$ Hz. Assume that only staggered conformations are important and determine which version of the Karplus equation fits the data better.

14B.4‡ It might be unexpected that the Karplus equation, which was first derived for $^3J_{HH}$ coupling constants, should also apply to vicinal coupling between the nuclei of metals such as tin. T.N. Mitchell and B. Kowall (*Magn. Reson. Chem.* **33**, 325 (1995)) have studied the relation between $^3J_{HH}$ and $^3J_{SnSn}$ in compounds of the type $Me_3SnCH_2CHRSnMe_3$ and find that $^3J_{SnSn} = 78.86\,^3J_{HH} + 27.84$ Hz. (a) Does this result support a Karplus type equation for tin? Explain your reasoning. (b) Obtain the Karplus equation for $^3J_{SnSn}$ and plot it as a function of the dihedral angle. (c) Draw the preferred conformation.

14B.5 Show that the coupling constant as expressed by the Karplus equation passes through a minimum when $\cos\phi = B/4C$.

14B.6 In a liquid, the dipolar magnetic field averages to zero: show this result by evaluating the average of the field given in eqn 14B.15. *Hint*: the surface area element is $\sin\theta\,d\theta d\phi$ in polar coordinates.

TOPIC 14C Pulse techniques in NMR

Discussion questions

14C.1 Discuss in detail the effects of a 90° pulse and of a 180° pulse on a system of spin-$\frac{1}{2}$ nuclei in a static magnetic field.

14C.2 Suggest a reason why the relaxation times of ^{13}C nuclei are typically much longer than those of ^{1}H nuclei.

14C.3 Suggest a reason why the spin–lattice relaxation time of a small molecule (like benzene) in a mobile, deuterated hydrocarbon solvent increases whereas that of a large molecule (like a polymer) decreases.

14C.4 Discuss the origin of the nuclear Overhauser effect and how it can be used to measure distances between protons in a biopolymer.

14C.5 Discuss the origins of diagonal and cross peaks in the COSY spectrum of an AX system.

Exercises

14C.1(a) The duration of a 90° or 180° pulse depends on the strength of the $\mathcal{B}_1$ field. If a 180° pulse requires 12.5 µs, what is the strength of the $\mathcal{B}_1$ field? How long would the corresponding 90° pulse require?
14C.1(b) The duration of a 90° or 180° pulse depends on the strength of the $\mathcal{B}_1$ field. If a 90° pulse requires 5 µs, what is the strength of the $\mathcal{B}_1$ field? How long would the corresponding 180° pulse require?

14C.2(a) What is the effective transverse relaxation time when the width of a resonance line is 1.5 Hz?
14C.2(b) What is the effective transverse relaxation time when the width of a resonance line is 12 Hz?

14C.3(a) Predict the maximum enhancement (as the value of η) that could be obtained in a NOE observation in which ^{31}P is coupled to protons.
14C.3(b) Predict the maximum enhancement (as the value of η) that could be obtained in a NOE observation in which ^{19}F is coupled to protons.

14C.4(a) Figure 14.1 shows the proton COSY spectrum of 1-nitropropane. Account for the appearance of off-diagonal peaks in the spectrum.
14C.4(b) The proton chemical shifts for the NH, C$_\alpha$H, and C$_\beta$H groups of alanine are 8.25 ppm, 4.35 ppm, and 1.39 ppm, respectively. Sketch the COSY spectrum of alanine between 1.00 and 8.50 ppm.

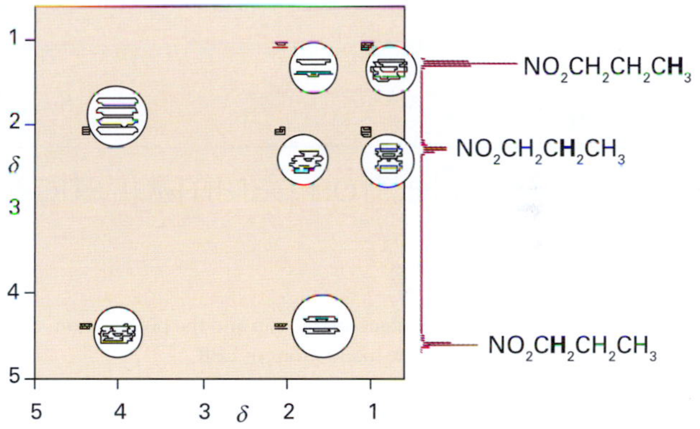

Figure 14.1 The COSY spectrum of 1-nitropropane ($NO_2CH_2CH_2CH_3$). The circles show enhanced views of the spectral features. (Spectrum provided by Prof. G. Morris.)

Problems

14C.1‡ Suppose that the FID in Fig. 14C.5 was recorded in a 400 MHz spectrometer, and that the interval between maxima in the oscillations in the FID is 0.12s. What is the Larmor frequency of the nuclei and the spin–spin relaxation time?

14C.2 Use mathematical software to construct the FID curve for a set of three nuclei with resonances at δ = 3.2, 4.1, and 5.0 in a spectrometer operating at 800 MHz. Suppose that T_1 = 1.0s. Go on to plot FID curves that show how they vary as the magnetic field of the spectrometer is changed.

14C.3 To gain some appreciation for the numerical work done by computers interfaced to NMR spectrometers, perform the following calculations. (a) The total FID $F(t)$ of a signal containing many frequencies, each corresponding to a different nucleus, is given by

$$F(t)=\sum_j S_{0j}\cos(2\pi\nu_{Lj}t)e^{-t/T_{2j}}$$

where, for each nucleus j, S_{0j} is the maximum intensity of the signal, ν_{Lj} is the Larmor frequency, and T_{2j} is the spin–spin relaxation time. Plot the FID for the case

$$S_{01}=1.0 \quad \nu_{L1}=50\,\text{MHz} \quad T_{21}=0.50\,\mu\text{s}$$
$$S_{02}=3.0 \quad \nu_{L2}=10\,\text{MHz} \quad T_{22}=1.0\,\mu\text{s}$$

(b) Explore how the shape of the FID curve changes with changes in the Larmor frequency and the spin–spin relaxation time. (c) Use mathematical software to calculate and plot the Fourier transforms of the FID curves you calculated in parts (a) and (b). How do spectral linewidths vary with the value of T_2? *Hint*: This operation can be performed with the 'fast Fourier transform' routine available in most mathematical software packages. Please consult the package's user manual for details.

14C.4 (a) In many instances it is possible to approximate the NMR lineshape by using a *Lorentzian function* of the form

$$I_{\text{Lorentzian}}(\omega)=\frac{S_0T_2}{1+T_2^2(\omega-\omega_0)^2}$$

where $I(\omega)$ is the intensity as a function of the angular frequency $\omega = 2\pi\nu$, ω_0 is the resonance frequency, S_0 is a constant, and T_2 is the spin–spin relaxation time. Confirm that for this lineshape the width at half-height is $1/\pi T_2$.
(b) Under certain circumstances, NMR lines are Gaussian functions of the frequency, given by

$$I_{\text{Gaussian}}(\omega)=S_0T_2e^{-T_2^2(\omega-\omega_0)^2}$$

Confirm that for the Gaussian lineshape the width at half-height is equal to $2(\ln 2)^{1/2}/T_2$. (c) Compare and contrast the shapes of Lorentzian and Gaussian lines by plotting two lines with the same values of S_0, T_2, and ω_0.

14C.5 The shape of a spectral line, $I(\omega)$, is related to the free induction decay signal $G(t)$ by

$$I(\omega)=a\,\mathrm{Re}\int_0^{\infty} G(t)\mathrm{e}^{\mathrm{i}\omega t}\,\mathrm{d}t$$

where a is a constant and 'Re' means take the real part of what follows. Calculate the lineshape corresponding to an oscillating, decaying function $G(t)=\cos\omega t\,\mathrm{e}^{-t/\tau}$.

14C.6 In the language of Problem 14C.5, show that if $G(t)=(a\cos\omega t+b\cos\omega t)\mathrm{e}^{-t/\tau}$, then the spectrum consists of two lines with intensities proportional to a and b and located at $\omega=\omega_1$ and ω_2, respectively.

14C.7 The z-component of the magnetic field at a distance R from a magnetic moment parallel to the z-axis is given by eqn 14C.10. In a solid, a proton at a distance R from another can experience such a field and the measurement of the splitting it causes in the spectrum can be used to calculate R. In gypsum, for instance, the splitting in the H_2O resonance can be interpreted in terms of a magnetic field of 0.715 mT generated by one proton and experienced by the other. What is the separation of the protons in the H_2O molecule?

14C.8 Interpret the following features of the NMR spectra of hen lysozyme: (a) saturation of a proton resonance assigned to the side chain of methionine-105 changes the intensities of proton resonances assigned to the side chains of tryptophan-28 and tyrosine-23; (b) saturation of proton resonances assigned to tryptophan-28 did not affect the spectrum of tyrosine-23.

14C.9 In a liquid crystal a molecule might not rotate freely in all directions and the dipolar interaction might not average to zero. Suppose a molecule is trapped so that, although the vector separating two protons may rotate freely around the z-axis, the colatitude may vary only between 0 and θ'. Use mathematical software to average the dipolar field over this restricted range of orientation and confirm that the average vanishes when θ' is equal to π (corresponding to free rotation over a sphere). What is the average value of the local dipolar field for the H_2O molecule in Problem 14C.7 if it is dissolved in a liquid crystal that enables it to rotate up to $\theta'=30°$?

TOPIC 14D Electron paramagnetic resonance

Discussion questions

14D.1 Describe how the Fermi contact interaction and the polarization mechanism contribute to hyperfine interactions in EPR.

14D.2 Explain how the EPR spectrum of an organic radical can be used to identify and map the molecular orbital occupied by the unpaired electron.

Exercises

14D.1(a) The centre of the EPR spectrum of atomic hydrogen lies at 329.12 mT in a spectrometer operating at 9.2231 GHz. What is the g value of the electron in the atom?

14D.1(b) The centre of the EPR spectrum of atomic deuterium lies at 330.02 mT in a spectrometer operating at 9.2482 GHz. What is the g value of the electron in the atom?

14D.2(a) A radical containing two equivalent protons shows a three-line spectrum with an intensity distribution 1:2:1. The lines occur at 330.2 mT, 332.5 mT, and 334.8 mT. What is the hyperfine coupling constant for each proton? What is the g value of the radical given that the spectrometer is operating at 9.319 GHz?

14D.2(b) A radical containing three equivalent protons shows a four-line spectrum with an intensity distribution 1:3:3:1. The lines occur at 331.4 mT, 333.6 mT, 335.8 mT, and 338.0 mT. What is the hyperfine coupling constant for each proton? What is the g value of the radical given that the spectrometer is operating at 9.332 GHz?

14D.3(a) A radical containing two inequivalent protons with hyperfine constants 2.0 mT and 2.6 mT gives a spectrum centred on 332.5 mT. At what fields do the hyperfine lines occur and what are their relative intensities?

14D.3(b) A radical containing three inequivalent protons with hyperfine constants 2.11 mT, 2.87 mT, and 2.89 mT gives a spectrum centred on 332.8 mT. At what fields do the hyperfine lines occur and what are their relative intensities?

14D.4(a) Predict the intensity distribution in the hyperfine lines of the EPR spectra of (i) $\cdot CH_3$, (ii) $\cdot CD_3$.

14D.4(b) Predict the intensity distribution in the hyperfine lines of the EPR spectra of (i) $\cdot CH_2CH_3$, (ii) $\cdot CD_2CD_3$.

14D.5(a) The benzene radical anion has $g=2.0025$. At what field should you search for resonance in a spectrometer operating at (i) 9.313 GHz, (ii) 33.80 GHz?

14D.5(b) The naphthalene radical anion has $g=2.0024$. At what field should you search for resonance in a spectrometer operating at (i) 9.501 GHz, (ii) 34.77 GHz?

14D.6(a) The EPR spectrum of a radical with a single magnetic nucleus is split into four lines of equal intensity. What is the nuclear spin of the nucleus?

14D.6(b) The EPR spectrum of a radical with two equivalent nuclei of a particular kind is split into five lines of intensity ratio 1:2:3:2:1. What is the spin of the nuclei?

14D.7(a) Sketch the form of the hyperfine structures of radicals XH_2 and XD_2, where the nucleus X has $I=\frac{5}{2}$.

14D.7(b) Sketch the form of the hyperfine structures of radicals XH_3 and XD_3, where the nucleus X has $I=\frac{3}{2}$.

Problems

14D.1 It is possible to produce very high magnetic fields over small volumes by special techniques. What would be the resonance frequency of an electron spin in an organic radical in a field of 1.0 kT? How does this frequency compare to typical molecular rotational, vibrational, and electronic energy-level separations?

14D.2 The angular NO_2 molecule has a single unpaired electron and can be trapped in a solid matrix or prepared inside a nitrite crystal by radiation damage of NO_2^- ions. When the applied field is parallel to the OO direction the centre of the spectrum lies at 333.64 mT in a spectrometer operating at 9.302 GHz. When the field lies along the bisector of the ONO angle, the resonance lies at 331.94 mT. What are the g values in the two orientations?

14D.3 The hyperfine coupling constant in $\cdot CH_3$ is 2.3 mT. Use the information in Table 14D.1 to predict the splitting between the hyperfine lines of the spectrum of $\cdot CD_3$. What are the overall widths of the hyperfine spectra in each case?

14D.4 The *p*-dinitrobenzene radical anion can be prepared by reduction of *p*-dinitrobenzene. The radical anion has two equivalent N nuclei ($I = 1$) and four equivalent protons. Predict the form of the EPR spectrum using $a(N) = 0.148$ mT and $a(H) = 0.112$ mT.

14D.5 The hyperfine coupling constants observed in the radical anions **1**, **2**, and **3** are shown (in millitesla, mT). Use the value for the benzene radical anion to map the probability of finding the unpaired electron in the π orbital on each C atom.

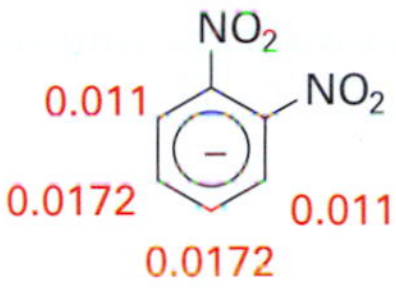

1

2

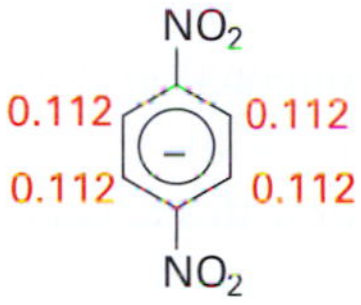

3

14D.6 When an electron occupies a 2s orbital on an N atom it has a hyperfine interaction of 55.2 mT with the nucleus. The spectrum of NO_2 shows an isotropic hyperfine interaction of 5.7 mT. For what proportion of its time is the unpaired electron of NO_2 occupying a 2s orbital? The hyperfine coupling constant for an electron in a 2p orbital of an N atom is 3.4 mT. In NO_2 the anisotropic part of the hyperfine coupling is 1.3 mT. What proportion of its time does the unpaired electron spend in the 2p orbital of the N atom in NO_2? What is the total probability that the electron will be found on (a) the N atoms, (b) the O atoms? What is the hybridization ratio of the N atom? Does the hybridization support the view that NO_2 is angular?

14D.7 Sketch the EPR spectra of the di-*tert*-butyl nitroxide radical (**4**) at 292 K in the limits of very low concentration (at which electron exchange is negligible), moderate concentration (at which electron exchange effects begin to be observed), and high concentration (at which electron exchange effects predominate).

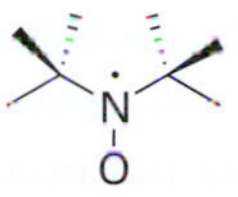

4 di-*tert*-Butyl nitroxide

Integrated activities

14.1 Consider the following series of molecules: benzene, methylbenzene, trifluoromethylbenzene, benzonitrile, and nitrobenzene in which the substituents *para* to the C atom of interest are H, CH_3, CF_3, CN, and NO_2, respectively. (a) Use the computational method of your choice to calculate the net charge at the C atom *para* to these substituents in this series of organic molecules. (b) It is found empirically that the ^{13}C chemical shift of the *para* C atom increases in the order: methylbenzene, benzene, trifluoromethylbenzene, benzonitrile, nitrobenzene. Is there a correlation between the behaviour of the ^{13}C chemical shift and the computed net charge on the ^{13}C atom? (c) The ^{13}C chemical shifts of the *para* C atoms in each of the molecules that you examined computationally are as follows:

Substituent	CH_3	H	CF_3	CN	NO_2
δ	128.4	128.5	128.9	129.1	129.4

Is there a linear correlation between net charge and ^{13}C chemical shift of the *para* C atom in this series of molecules? (d) If you did find a correlation in part (c), explain the physical origins of the correlation.

14.2 The computational techniques described in Topic 10E have shown that the amino acid tyrosine participates in a number of biological electron transfer reactions, including the processes of water oxidation to O_2 in plant photosynthesis and of O_2 reduction to water in oxidative phosphorylation. During the course of these electron transfer reactions, a tyrosine radical forms with spin density delocalized over the side chain of the amino acid. (a) The phenoxy radical shown in **5** is a suitable model of the tyrosine radical. Using molecular modelling software and the computational method of your choice (semi-empirical or *ab initio* methods), calculate the spin densities at the O atom and at all of the C atoms in **5**. (b) Predict the form of the EPR spectrum of **5**.

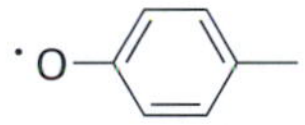

5 Phenoxy radical

14.3 Two groups of protons have $\delta = 4.0$ and $\delta = 5.2$ and are interconverted by a conformational change of a fluxional molecule. In a 60 MHz spectrometer the spectrum collapsed into a single line at 280 K but at 300 MHz the collapse did not occur until the temperature had been raised to 300 K. What is the activation energy of the interconversion?

14.4 NMR spectroscopy may be used to determine the equilibrium constant for dissociation of a complex between a small molecule, such as an enzyme inhibitor I, and a protein, such as an enzyme E:

$$EI \rightleftharpoons E + I \qquad K_I = [E][I]/[EI]$$

In the limit of slow chemical exchange, the NMR spectrum of a proton in I would consist of two resonances: one at ν_I for free I and another at ν_{EI} for bound I. When chemical exchange is fast, the NMR spectrum of the same proton in I consists of a single peak with a resonance frequency ν given by $\nu = f_I\nu_I + f_{EI}\nu_{EI}$, where $f_I = [I]/([I] + [EI])$ and $f_{EI} = [EI]/([I] + [EI])$ are, respectively, the fractions of free I and bound I. For the purposes of analysing the data, it is also useful to define the frequency differences $\delta\nu = \nu - \nu_I$ and $\Delta\nu = \nu_{EI} - \nu_I$. Show that when the initial concentration of I, $[I]_0$, is much greater than the initial concentration of E, $[E]_0$, a plot of $[I]_0$ against $\delta\nu^{-1}$ is a straight line with slope $[E]_0\Delta\nu$ and y-intercept $-K_I$.

CHAPTER 15

Statistical thermodynamics

Statistical thermodynamics provides the link between the microscopic properties of matter and its bulk properties. It provides a means of calculating thermodynamic properties from structural and spectroscopic data and gives insight into the molecular origins of chemical properties.

15A The Boltzmann distribution

The 'Boltzmann distribution', which is used to predict the populations of states in systems at thermal equilibrium, is among the most important equations in chemistry for it summarizes the populations of states; it also provides insight into the nature of 'temperature'. The structure of the Topic separates its key implications from its rather heavy derivation.

15B Partition functions

The Boltzmann distribution introduces the concept of a 'partition function', which is the central mathematical concept of the rest of the chapter. We see how to interpret the partition function and how to calculate it in a number of simple cases.

15C Molecular energies

A partition function is the thermodynamic version of a wavefunction, and contains all the thermodynamic information about a system. As a first step in extracting that information, we see how to use partition functions to calculate the mean values of the basic modes of motion of a collection of independent molecules.

15D The canonical ensemble

Molecules do interact with one another, and statistical thermodynamics would be incomplete without being able to take these interactions into account. This Topic shows how that is done in principle by introducing the 'canonical ensemble', and hints at how this concept can be used.

15E The internal energy and the entropy

The main work of the chapter is to show how molecular partition functions are used to calculate (and give insight into) the two basic thermodynamic functions, the internal energy and the entropy. The latter is based on another central equation introduced by Boltzmann, his definition of 'statistical entropy'.

15F Derived functions

With expressions relating internal energy and entropy to partition functions, we are ready to develop expressions for the derived thermodynamic functions, such as the Helmholtz and Gibbs energies. Then, with the Gibbs energy available, we can make the final step into the calculations of chemically significant expressions by showing how equilibrium constants can be calculated from structural and spectroscopic data.

What is the impact of this material?

There are numerous applications of statistical arguments in biochemistry. We have selected one of the most directly related to partition functions: *Impact* I15.1 describes the helix–coil equilibrium in a polypeptide and the role of cooperative behaviour.

To read more about the impact of this material, scan the QR code, or go to bcs.whfreeman.com/webpub/chemistry/pchem10e/impact/pchem-15-1.html

15A The Boltzmann distribution

Contents

➤ Why do you need to know this material?

The Boltzmann distribution is the key to understanding a great deal of chemistry. All thermodynamic properties can be interpreted in its terms, as can the temperature dependence of equilibrium constants and the rates of chemical reactions. It also illuminates the meaning of 'temperature'. There is, perhaps, no more important unifying concept in chemistry.

➤ What is the key idea?

The most probable distribution of molecules over the available energy levels subject to certain restraints depends on a single parameter, the temperature.

➤ What do you need to know already?

You need to be aware that molecules can exist only in certain discrete energy levels (*Foundations* B and Topic 7A) and that in some cases more than one state has the same energy. The principal mathematical tools used in this Topic are simple probability theory and Lagrange multipliers; the latter is explained in *The chemist's toolkit* 15A.1.

The problem we address in this Topic is the calculation of the populations of states for any type of molecule in any mode of motion at any temperature. The only restriction is that the molecules should be independent, in the sense that the total energy of the system is a sum of their individual energies. We are discounting (at this stage) the possibility that in a real system a contribution to the total energy may arise from interactions between molecules. We also adopt the **principle of equal *a priori* probabilities**, the assumption that all possibilities for the distribution of energy are equally probable. '*A priori*' in this context loosely means 'as far as one knows'. We have no reason to presume otherwise than that for a collection of molecules at thermal equilibrium, a vibrational state of a certain energy, for instance, is as likely to be populated as a rotational state of the same energy.

One very important conclusion that will emerge from the following analysis is that the overwhelmingly most probable populations of the available states depend on a single parameter, the 'temperature'. That is, the work we do here provides a molecular justification for the concept of temperature and some insight into this crucially important quantity.

15A.1 Configurations and weights

Any individual molecule may exist in states with energies ε_0, ε_1, …. For reasons that will become clear, we shall always take the lowest available state as the zero of energy (that is, we set $\varepsilon_0=0$), and measure all other energies relative to that state. To obtain the actual energy of the system we may have to add a constant to the energy calculated on this basis. For example, if we are considering the vibrational contribution to the energy, then we must add the total zero-point energy of any oscillators in the system.

(a) Instantaneous configurations

At any instant there will be N_0 molecules in the state 0 with energy ε_0, N_1 in the state 1 with ε_1, and so on, with $N_0+N_1+\cdots=N$, the total number of molecules in the system. Initially we suppose that all the states have exactly the same energy. The specification of the set of populations N_0, N_1,… in the form $\{N_0,N_1,\ldots\}$ is a statement of the instantaneous **configuration** of the system. The instantaneous configuration fluctuates with time because the populations change, perhaps as a result of collisions. At this stage the energies of all the configurations are identical so there is no restriction on how many of the N molecules are in each state.

We can picture a large number of different instantaneous configurations. One, for example, might be $\{N,0,0,\ldots\}$, corresponding to every molecule being in state 0. Another might be $\{N-2,2,0,0,\ldots\}$, in which two molecules are in state 1. The latter configuration is intrinsically more likely to be found than the former because it can be achieved in more ways: $\{N,0,0,\ldots\}$ can be achieved in only one way, but $\{N-2,2,0,\ldots\}$ can be achieved in $\frac{1}{2}N(N-1)$ different ways (Fig. 15A.1; see the following *Justification*). If, as a result of collisions, the system were to fluctuate between the configurations $\{N,0,0,\ldots\}$ and $\{N-2,2,0,\ldots\}$, it would almost always be found in the second, more likely configuration, especially if N were large. In other words, a system free to switch between the two configurations would show properties characteristic almost exclusively of the second configuration. A general configuration $\{N_0,N_1,\ldots\}$ can be achieved in $\mathcal{W}$ different ways, where $\mathcal{W}$ is called the **weight** of the configuration. The weight of the configuration $\{N_0,N_1,\ldots\}$ is given by the expression

$$\mathcal{W}=\frac{N!}{N_0!N_1!N_2!\cdots} \quad \text{Weight of a configuration} \quad (15A.1)$$

with $x!=x(x-1)\ldots1$ and by definition $0!=1$. Equation 15A.1 is a generalization of the formula $\mathcal{W}=\frac{1}{2}N(N-1)$, and reduces to it for the configuration $\{N-2,2,0,\ldots\}$.

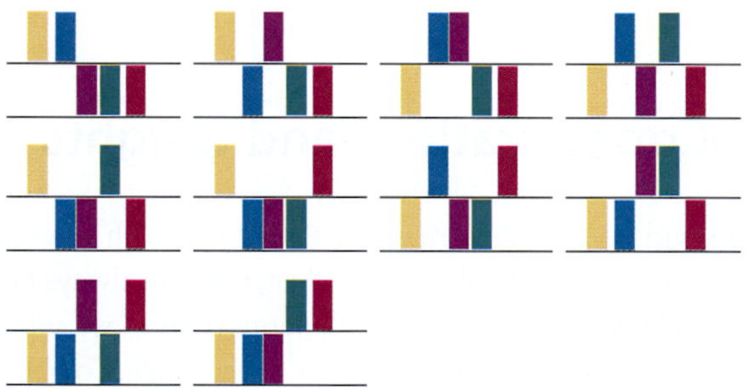

Figure 15A.1 Whereas a configuration $\{5,0,0,\ldots\}$ can be achieved in only one way, a configuration $\{3,2,0,\ldots\}$ can be achieved in the ten different ways shown here, where the tinted blocks represent different molecules.

Brief illustration 15A.1 The weight of a configuration

To calculate the number of ways of distributing 20 identical objects with the arrangement 1, 0, 3, 5, 10, 1, we note that the configuration is $\{1,0,3,5,10,1\}$ with $N=20$; therefore the weight is

$$\mathcal{W}=\frac{20!}{1!0!3!5!10!1!}=9.31\times10^8$$

Self-test 15A.1 Calculate the weight of the configuration in which 20 objects are distributed in the arrangement 0, 1, 5, 0, 8, 0, 3, 2, 0, 1.

Answer: 4.19×10^{10}

Justification 15A.1 The weight of a configuration

First, consider the weight of the configuration $\{N-2,2,0,0,\ldots\}$, which is prepared from the configuration $\{N,0,0,0,\ldots\}$ by the migration of two molecules from state 0 into state 2. One candidate for migration to state 1 can be selected in N ways. There are $N-1$ candidates for the second choice, so the total number of choices is $N(N-1)$. However, we should not distinguish the choice (Jack, Jill) from the choice (Jill, Jack) because they lead to the same configurations. Therefore, only half the choices lead to distinguishable configurations, and the total number of distinguishable choices is $\frac{1}{2}N(N-1)$.

Now we generalize this remark. Consider the number of ways of distributing N balls into bins. The first ball can be selected in N different ways, the next ball in $N-1$ different ways for the balls remaining, and so on. Therefore, there are $N(N-1)\ldots1=N!$ ways of selecting the balls for distribution over the bins. However, if there are N_0 balls in the bin labelled ε_0, there would be $N_0!$ different ways in which the same balls could have been chosen (Fig. 15A.2). Similarly, there are $N_1!$ ways in which the N_1 balls in the bin labelled ε_1 can be chosen, and so on. Therefore, the total number of distinguishable ways

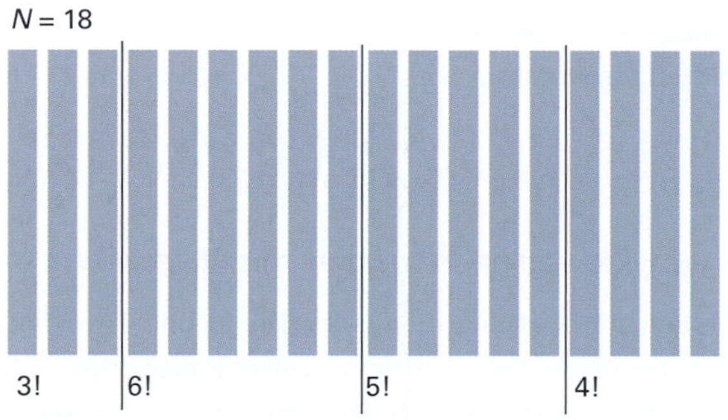

Figure 15A.2 The 18 molecules shown here can be distributed into four receptacles (distinguished by the three vertical lines) in 18! different ways. However, 3! of the selections that put three molecules in the first receptacle are equivalent, 6! that put six molecules into the second receptacle are equivalent, and so on. Hence the number of distinguishable arrangements is 18!/3!6!5!4!, or about 515 million.

of distributing the balls so that there are N_0 in bin ε_0, N_1 in bin ε_1, etc. regardless of the order in which the balls were chosen is $N!/N_0!N_1!\ldots$, which is the content of eqn 15A.1.

It will turn out to be more convenient to deal with the natural logarithm of the weight, ln $\mathcal{W}$, rather than with the weight itself. We shall therefore need the expression

$$\ln\mathcal{W}=\ln\frac{N!}{N_0!N_1!N_2!\cdots}\overset{\ln(x/y)=\ln x-\ln y}{=}\ln N!-\ln N_0!N_1!N_2!\cdots$$

$$\overset{\ln xy=\ln x+\ln y}{=}\ln N!-\ln N_0!-\ln N_1!-\ln N_2!-\cdots=\ln N!-\sum_i \ln N_i!$$

One reason for introducing ln $\mathcal{W}$ is that it is easier to make approximations. In particular, we can simplify the factorials by using *Stirling's approximation*

$$\ln x! \approx \left(x+\tfrac{1}{2}\right)\ln x - x + \tfrac{1}{2}\ln 2\pi \qquad x \gg 1 \qquad \text{Stirling's approximation} \qquad (15A.2a)$$

This approximation is in error by less than 1 per cent when x is greater than about 10. We deal with far larger values of x, and the simplified version

$$\ln x! \approx x\ln x - x \qquad x \gg 1 \qquad \text{Stirling's approximation} \qquad (15A.2b)$$

is adequate. Then the approximate expression for the weight is

$$\begin{aligned}\ln \mathcal{W} &= \{N\ln N - N\} - \sum_i \{N_i \ln N_i - N_i\} \\ &= N\ln N - N - \sum_i N_i \ln N_i + N \quad [\text{because} \sum_i N_i = N] \\ &= N\ln N - \sum_i N_i \ln N_i \end{aligned} \qquad (15A.3)$$

(b) The most probable distribution

We have seen that the configuration $\{N-2,2,0,\ldots\}$ dominates $\{N,0,0,\ldots\}$, and it should be easy to believe that there may be other configurations that have a much greater weight than both. We shall see, in fact, that there is a configuration with so great a weight that it overwhelms all the rest in importance to such an extent that the system will almost always be found in it. The properties of the system will therefore be characteristic of that particular dominating configuration. This dominating configuration can be found by looking for the values of N_i that lead to a maximum value of $\mathcal{W}$. Because $\mathcal{W}$ is a function of all the N_i, we can do this search by varying the N_i and looking for the values that correspond to $d\mathcal{W}=0$ (just as in the search for the maximum of any function), or equivalently a maximum value of ln $\mathcal{W}$. However, there are two difficulties with this procedure.

At this point we allow the states to have different energies. The first difficulty that results from this change is the need to take into account the fact that the only permitted configurations are those corresponding to the specified, constant, total energy of the system. This requirement rules out many configurations; $\{N,0,0,\ldots\}$ and $\{N-2,2,0,\ldots\}$, for instance, have different energies (unless ε_0 and ε_1 happen to have the same energy), so both cannot occur in the same isolated system. It follows that in looking for the configuration with the greatest weight, we must ensure that the configuration also satisfies the condition

$$\sum_i N_i \varepsilon_i = E \qquad \textit{Constant energy} \qquad \text{Energy constraint} \qquad (15A.4)$$

where E is the total energy of the system.

The second constraint is that, because the total number of molecules present is also fixed (at N), we cannot arbitrarily vary all the populations simultaneously. Thus, increasing the population of one state by 1 demands that the population of another state must be reduced by 1. Therefore, the search for the maximum value of $\mathcal{W}$ is also subject to the condition

$$\sum_i N_i = N \qquad \textit{Constant total number of molecules} \qquad \text{Number constraint} \qquad (15A.5)$$

We show in the next section that the populations in the configuration of greatest weight, subject to the two constraints in eqns 15A.4 and 15A.5, depend on the energy of the state according to the **Boltzmann distribution**:

$$\frac{N_i}{N} = \frac{e^{-\beta\varepsilon_i}}{\sum_i e^{-\beta\varepsilon_i}} \qquad \text{Boltzmann distribution} \qquad (15A.6a)$$

The denominator on eqn 15A.6a is denoted q and called the **partition function**:

$$q = \sum_i e^{-\beta\varepsilon_i} \qquad \textit{Definition} \qquad \text{Partition function} \qquad (15A.6b)$$

At this stage the partition function is no more than a convenient abbreviation for the sum; but in Topic 15B we see that it is central to the statistical interpretation of thermodynamic properties.

Equation 15A.6a is the justification of the remark that a single parameter, here denoted β, determines the most probable populations of the states of the system. We confirm in Topic 15D and anticipate throughout this Topic that

$$\beta = \frac{1}{kT} \qquad (15A.7)$$

where T is the thermodynamic temperature and k is Boltzmann's constant. In other words:

> The temperature is the unique parameter that governs the most probable populations of states of a system at thermal equilibrium.

Brief illustration 15A.2 The Boltzmann distribution

Suppose that two conformations of a molecule differ in energy by 5.0 kJ mol^{-1} (corresponding to 8.3 zJ for a single molecule; 1 zJ = 10^{-21} J), so conformation A lies at energy 0 and conformation B lies at ε = 8 zJ. At 20 °C (293 K) the denominator in eqn 15A.6a is

$$\sum_i e^{-\beta\varepsilon_i} = 1 + e^{-\varepsilon/kT} = 1 + e^{-(8.3\times10^{-21}\,\text{J})/(1.381\times10^{-23}\,\text{J K}^{-1})\times(293\,\text{K})} = 1.13$$

The proportion of molecules in conformation B at this temperature is therefore

$$\frac{N_B}{N}=\frac{e^{-(8.3\times10^{-21}\,J)/(1.381\times10^{-23}\,J\,K^{-1})\times(293\,K)}}{1.13}=0.11$$

or 11 per cent of the molecules.

Self-Test 15A.2 Suppose that there is a third conformation a further 0.50 kJ mol^{-1} above B. What proportion of molecules will now be in conformation B?

Answer: 0.10, 10 per cent

(c) The relative population of states

If we are interested only in the relative populations of states, the sum in the denominator of the Boltzmann distribution need not be evaluated, because it cancels when the ratio is taken:

$$\frac{N_i}{N_j}=\frac{e^{-\beta\varepsilon_i}}{e^{-\beta\varepsilon_j}}=e^{-\beta(\varepsilon_i-\varepsilon_j)}$$ *Thermal equilibrium* Boltzmann population ratio (15A.8a)

That $\beta\propto 1/T$ is plausible is demonstrated by noting from eqn 15A.8a that for a given energy separation the ratio of populations N_1/N_0 decreases as β increases, which is what is expected as the temperature decreases. At $T=0$ ($\beta=\infty$) all the population is in the ground state and the ratio is zero. Equation 15A.8a is enormously important for understanding a wide range of chemical phenomena and is the form in which the Boltzmann distribution is commonly employed (for instance, in the discussion of the intensities of spectral transitions, Topics 12A and 14A). It tells us that the relative population of two states falls off exponentially with their difference in energy.

A very important point to note is that the Boltzmann distribution gives the relative populations of *states*, not energy *levels*. Several states might correspond to the same energy, and each state has a population given by eqn 15A.6. If we want to consider the relative populations of energy levels rather than states, then we need to take into account this degeneracy. Thus, if the level of energy ε_i is g_i-fold degenerate (in the sense that there are g_i states with that energy), and the level of energy ε_j is g_j-fold degenerate, then the relative total populations of the levels are given by

$$\frac{N_i}{N_j}=\frac{g_i e^{-\beta\varepsilon_i}}{g_j e^{-\beta\varepsilon_j}}=\frac{g_i}{g_j}e^{-\beta(\varepsilon_i-\varepsilon_j)}$$ *Thermal equilibrium, degeneracies* Boltzmann population ratio (15A.8b)

Example 15A.1 Calculating the relative populations of rotational states

Calculate the relative populations of the $J=1$ and $J=0$ rotational states of HCl at 25 °C.

Method Although the ground state is non-degenerate, the level with $J=1$ is triply degenerate ($M_J=0, \pm1$); see Topic 12B. From Topic 12B, the energy of state with quantum number J is $\varepsilon_J=hc\tilde{B}J(J+1)$. Use $\tilde{B}=10.591\,cm^{-1}$ A useful relation is $kT/hc=207.22\,cm^{-1}$ at 298.15 K.

Answer The energy separation of states with $J=1$ and $J=0$ is

$$\varepsilon_1-\varepsilon_0=2hc\tilde{B}$$

The ratio of the population of a state with $J=1$ and any *one* of its three states M_J to the population of the single state with $J=0$ is therefore

$$\frac{N_{J,M_J}}{N_0}=e^{-2hc\tilde{B}\beta}$$

The relative populations of the *levels*, taking into account the three-fold degeneracy of the upper state, is

$$\frac{N_J}{N_0}=3e^{-2hc\tilde{B}\beta}$$

Insertion of $hc\tilde{B}\beta=hc\tilde{B}/kT=(10.591\,cm^{-1})/(207.22\,cm^{-1})=0.0511\ldots$ then gives

$$\frac{N_J}{N_0}=3e^{-2\times0.0511\ldots}=2.708$$

We see that because the $J=1$ level is triply degenerate, it has a higher population than the level with $J=0$, despite being of higher energy. As the example illustrates, it is very important to take note of whether you are asked for the relative populations of individual states or of a (possibly degenerate) energy level.

Self-test 15A.3 What is the ratio of the populations of the levels with $J=2$ and $J=1$ at the same temperature?

Answer: 1.359

15A.2 The derivation of the Boltzmann distribution

We have remarked that ln $\mathcal{W}$ is easier to handle than $\mathcal{W}$. Therefore, to find the form of the Boltzmann distribution, we look for the condition for ln $\mathcal{W}$ being a maximum rather than dealing directly with $\mathcal{W}$. Because ln $\mathcal{W}$ depends on all the N_i, when a configuration changes and the N_i change to N_i+dN_i, the function ln $\mathcal{W}$ changes to ln $\mathcal{W}$+d ln $\mathcal{W}$, where

$$d\ln\mathcal{W}=\sum_i\left(\frac{\partial\ln\mathcal{W}}{\partial N_i}\right)dN_i$$

All this expression states is that a change in ln $\mathcal{W}$ is the sum of contributions arising from changes in each value of N_i.

(a) The role of constraints

At a maximum, d ln $\mathcal{W}=0$. However, when the N_i change, they do so subject to the two constraints

$$\sum_i \varepsilon_i \mathrm{d}N_i = 0 \qquad \sum_i \mathrm{d}N_i = 0 \qquad \text{Constraints} \quad (15\text{A}.9)$$

The first constraint recognizes that the total energy must not change, and the second recognizes that the total number of molecules must not change. These two constraints prevent us from solving d ln $\mathcal{W}=0$ simply by setting all $(\partial \ln \mathcal{W}/\partial N_i)=0$ because the $\mathrm{d}N_i$ are not all independent.

The way to take constraints into account was devised by the French mathematician Lagrange, and is called the **method of undetermined multipliers** (*The chemist's toolkit* 15A.1). All we need of that method is as follows:

- Each constraint is multiplied by a constant and then added to the main variation equation.
- The variables are then treated as though they were all independent.
- The constants are evaluated at the end of the calculation.

The chemist's toolkit 15A.1 The method of undetermined multipliers

Suppose we need to find the maximum (or minimum) value of some function f that depends on several variables $x_1, x_2, \ldots, x_n$. When the variables undergo a small change from x_i to $x_i+\delta x_i$ the function changes from f to $f+\delta f$, where

$$\delta f = \sum_{i=1}^{n} \left(\frac{\partial f}{\partial x_i}\right)\delta x_i$$

At a minimum or maximum, $\delta f=0$, so then

$$\sum_{i=1}^{n} \left(\frac{\partial f}{\partial x_i}\right)\delta x_i = 0$$

If the x_i were all independent, all the δx_i would be arbitrary, and this equation could be solved by setting each $(\partial f/\partial x_i)=0$ individually. When the x_i are not all independent, the δx_i are not all independent, and the simple solution is no longer valid. We proceed as follows.

Let the constraint connecting the variables be an equation of the form $g=0$. The constraint $g=0$ is always valid, so g remains unchanged when the x_i are varied:

$$\delta g = \sum_{i=1}^{n} \left(\frac{\partial g}{\partial x_i}\right)\delta x_i = 0$$

Because δg is zero, we can multiply it by a parameter, λ, and add it to the preceding equation:

$$\sum_{i=1}^{n} \left\{\left(\frac{\partial f}{\partial x_i}\right) + \lambda\left(\frac{\partial g}{\partial x_i}\right)\right\}\delta x_i = 0$$

This equation can be solved for one of the δx, δx_n for instance, in terms of all the other δx_i. All those other δx_i ($i=1, 2, \ldots, n-1$) are independent, because there is only one constraint on the system. But λ is arbitrary; therefore we can choose it so that

$$\left(\frac{\partial f}{\partial x_n}\right) + \lambda\left(\frac{\partial g}{\partial x_n}\right) = 0 \qquad \text{(A)}$$

Then

$$\sum_{i=1}^{n-1} \left\{\left(\frac{\partial f}{\partial x_i}\right) + \lambda\left(\frac{\partial g}{\partial x_i}\right)\right\}\delta x_i = 0$$

Now the $n-1$ variations δx_i *are* independent, so the solution of this equation is

$$\left(\frac{\partial f}{\partial x_i}\right) + \lambda\left(\frac{\partial g}{\partial x_i}\right) = 0 \quad i=1, 2, \ldots, n-1$$

However, eqn A has exactly the same form as this equation, so the maximum or minimum of f can be found by solving

$$\left(\frac{\partial f}{\partial x_i}\right) + \lambda\left(\frac{\partial g}{\partial x_i}\right) = 0 \quad i=1, 2, \ldots, n$$

If there is more than one constraint, $g_1=0, g_2=0, \ldots$, and this final result generalizes to

$$\left(\frac{\partial f}{\partial x_i}\right) + \lambda_1\left(\frac{\partial g_1}{\partial x_i}\right) + \lambda_2\left(\frac{\partial g_2}{\partial x_i}\right) + \cdots = 0, \quad i=1, 2, \ldots, n$$

with a corresponding multiplier, $\lambda_1, \lambda_2, \ldots$ for each constraint.

Thus, as there are two constraints we introduce the two constants α and $-\beta$ and write

$$\sum_i \left(\frac{\partial \ln \mathcal{W}}{\partial N_i}\right)\mathrm{d}N_i + \alpha\sum_i \mathrm{d}N_i - \beta\sum_i \varepsilon_i \mathrm{d}N_i$$
$$= \sum_i \left\{\left(\frac{\partial \ln \mathcal{W}}{\partial N_i}\right) + \alpha - \beta\varepsilon_i\right\}\mathrm{d}N_i = 0$$

All the $\mathrm{d}N_i$ are now treated as independent. Hence the only way of satisfying d ln $\mathcal{W}=0$ is to require that for each i,

$$\left(\frac{\partial \ln \mathcal{W}}{\partial N_i}\right) + \alpha - \beta\varepsilon_i = 0 \qquad (15\text{A}.10)$$

when the N_i have their most probable values. We show in the following *Justification* that

$$\frac{\partial \ln \mathcal{W}}{\partial N_i} = -\ln \frac{N_i}{N} \qquad (15A.11)$$

It follows from eqn 15A.10 that

$$-\ln \frac{N_i}{N} + \alpha - \beta\varepsilon_i = 0$$

and therefore that

$$\frac{N_i}{N} = e^{\alpha - \beta\varepsilon_i} \qquad (15A.12)$$

which is very close to being the Boltzmann distribution.

Justification 15.A.2 The derivative of the weight

Equation 15A.3 for $\mathcal{W}$ is

$$\ln \mathcal{W} = N \ln N - \sum_i N_i \ln N_i$$

There is a small housekeeping step to take before differentiating $\ln \mathcal{W}$ with respect to N_i: this equation is identical to

$$\ln \mathcal{W} = N \ln N - \sum_j N_j \ln N_j$$

because all we have done is to change the 'name' of the states from i to j. This step makes sure that we do not confuse the i in the differentiation variable (N_i) with the i in the summation. Now differentiation of this expression gives

$$\frac{\partial \ln \mathcal{W}}{\partial N_i} = \frac{\partial (N \ln N)}{\partial N_i} - \sum_j \frac{\partial (N_j \ln N_j)}{\partial N_i}$$

The derivative of the (blue) first term on the right is obtained as follows:

$$\frac{\partial (N \ln N)}{\partial N_i} = \overbrace{\left(\frac{\partial N}{\partial N_i}\right)}^{1} \ln N + N \overbrace{\left(\frac{\partial \ln N}{\partial N_i}\right)}^{(1/N)\partial N/\partial N_i}$$

$$= \ln N + \overbrace{\frac{\partial N}{\partial N_i}}^{1} = \ln N + 1$$

The (blue) $\ln N$ in the first term on the right in the second line arises because $N = N_1 + N_2 + \cdots$ and so the derivative of N with respect to any of the N_i is 1: that is, $\partial N/\partial N_i = 1$. The second term on the right in the second line arises because $\partial(\ln N)/\partial N_i = (1/N)\partial N/\partial N_i$. The final 1 is then obtained in the same way as in the preceding remark, by using $\partial N/\partial N_i = 1$.

For the derivative of the second term we first note that

$$\frac{\partial \ln N_j}{\partial N_i} = \frac{1}{N_j}\left(\frac{\partial N_j}{\partial N_i}\right)$$

If $i \neq j$, N_j is independent of N_i, so $\partial N_j/\partial N_i = 0$. However, if $i = j$, $\partial N_i/\partial N_i = 1$. Therefore,

$$\frac{\partial N_j}{\partial N_i} = \delta_{ij}$$

with δ_{ij} the Kronecker delta ($\delta_{ij} = 1$ if $i = j$, $\delta_{ij} = 0$ otherwise). Then

$$\sum_j \frac{\partial N_j \ln N_j}{\partial N_i} = \sum_j \left\{ \overbrace{\left(\frac{\partial N_j}{\partial N_i}\right)}^{\delta_{ij}} \ln N_j + N_j \overbrace{\left(\frac{\partial \ln N_j}{\partial N_i}\right)}^{(1/N_j)\partial N_j/\partial N_i} \right\}$$

$$= \sum_j \left\{ \delta_{ij} \ln N_j + \overbrace{\left(\frac{\partial N_j}{\partial N_i}\right)}^{\delta_{ij}} \right\} = \sum_j \delta_{ij}(\ln N_j + 1)$$

$$= \ln N_i + 1$$

On bringing the two terms together we can write

$$\frac{\partial \ln \mathcal{W}}{\partial N_i} = \ln N + 1 - (\ln N_i + 1) = -\ln \frac{N_i}{N}$$

as in eqn 15A.11.

(b) The values of the constants

At this stage we note that

$$N = \sum_i N_i = \sum_i N e^{\alpha - \beta\varepsilon_i} = N e^{\alpha} \sum_i e^{-\beta\varepsilon_i}$$

Because the N cancels on each side of this equality, it follows that

$$e^{\alpha} = \frac{1}{\sum_i e^{-\beta\varepsilon_i}} \qquad (15A.13)$$

and therefore

$$\frac{N_i}{N} = e^{\alpha - \beta\varepsilon_i} = e^{\alpha} e^{-\beta\varepsilon_i} = \frac{e^{-\beta\varepsilon_i}}{\sum_i e^{-\beta\varepsilon_i}} \qquad \text{Boltzmann distribution} \qquad (15A.14)$$

which is eqn 15A.6a.

The development of statistical concepts of thermodynamics begins with the Boltzmann distribution, with quantum theory (Chapter 7) providing insight into and ways of calculating the energies ε_i in eqn 15A.14.

Checklist of concepts

- ☐ 1. The **principle of equal *a priori* probabilities** assumes that all possibilities for the distribution of energy are equally probable.
- ☐ 2. The **instantaneous configuration** of a system of N molecules is the specification of the set of populations N_0, N_1, … of the energy levels ε_0, ε_1, ….
- ☐ 3. The **Boltzmann distribution** gives the numbers of molecules in each state of a system at any temperature.
- ☐ 4. The **relative populations** of energy levels, as opposed to states, must take into account the degeneracies of the energy levels.

Checklist of equations

Property	Equation	Comment	Equation number
Boltzmann distribution	$N_i/N = e^{-\beta\varepsilon_i}/q$	$\beta = 1/kT$	15A.6a
Partition function	$q = \sum_i e^{-\beta\varepsilon_i}$	see Topic 15B	15A.6b
Boltzmann population ratio	$N_i/N_j = (g_i/g_j)e^{-\beta(\varepsilon_i-\varepsilon_j)}$	g_i, g_j are degeneracies	15A.8b

15B Molecular partition functions

Contents

➤ Why do you need to know this material?

Statistical thermodynamics provides the link between molecular properties that have been calculated or derived from spectroscopy and thermodynamic properties, including equilibrium concepts. The connection is the partition function. Therefore, this material is an essential foundation for understanding physical and chemical properties of bulk matter in terms of the properties of the constituent molecules.

➤ What is the key idea?

The partition function is calculated by drawing on calculated or spectroscopically derived structural information about molecules.

➤ What do you need to know already?

You need to know that the Boltzmann distribution expresses the most probable distribution of molecules over the available energy levels (Topic 15A). In that Topic we introduce the concept of partition function, which is developed here. You need to be aware of the expressions for the rotational and vibrational levels of molecules (Topics 12B and 12D) and the energy levels of a particle in a box (Topic 8A).

The partition function $q=\sum_i e^{-\beta\varepsilon_i}$ is introduced in Topic 15A simply as a symbol to denote the sum over states that occurs in the denominator of the Boltzmann distribution (eqn 15A.6a, $p_i=e^{-\beta\varepsilon_i}/q$, with $p_i=N_i/N$), but it is far more important than that might suggest. For instance, it contains all the information needed to calculate the bulk properties of a system of independent particles. In this respect q plays a role for bulk matter very similar to that played by the wavefunction in quantum mechanics for individual molecules: q is a kind of thermal wavefunction. This Topic shows how the partition function is calculated in a variety of important cases in preparation for seeing how thermodynamic information is extracted (in Topics 15C and 15E).

15B.1 The significance of the partition function

The **molecular partition function** is

$$q=\sum_{\text{states } i} e^{-\beta\varepsilon_i} \qquad \textit{Definition} \quad \text{Molecular partition function} \qquad (15B.1a)$$

where $\beta=1/kT$. As emphasized in Topic 15A, the sum is over *states*, not energy *levels*. If g_i states have the same energy ε_i (so the level is g_i-fold degenerate), we write

$$q=\sum_{\text{levels } i} g_i e^{-\beta\varepsilon_i} \qquad \textit{Alternative definition} \quad \text{Molecular partition function} \qquad (15B.1b)$$

where the sum is now over energy levels (sets of states with the same energy), not individual states. Also as emphasized in Topic 15A, we always take the lowest available state as the zero of energy and set $\varepsilon_0=0$.

Brief illustration 15B.1 A partition function

Suppose a molecule is confined to the following non-degenerate energy levels: 0, ε, 2ε, … (Fig. 15B.1; later we shall see that this array of levels is used when considering molecular vibration). Then the molecular partition function is

$$q=1+e^{-\beta\varepsilon}+e^{-2\beta\varepsilon}+\cdots=1+e^{-\beta\varepsilon}+(e^{-\beta\varepsilon})^2+\cdots$$

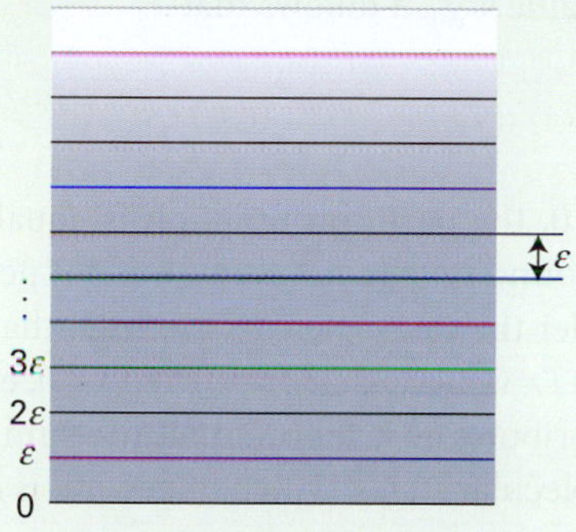

Figure 15B.1 The equally-spaced infinite array of energy levels used in the calculation of the partition function. A harmonic oscillator has the same spectrum of levels.

The sum of the geometrical series $1+x+x^2+\cdots$ is $1/(1-x)$, so in this case

$$q=\frac{1}{1-e^{-\beta\varepsilon}}$$

This function is plotted in Fig. 15B.2.

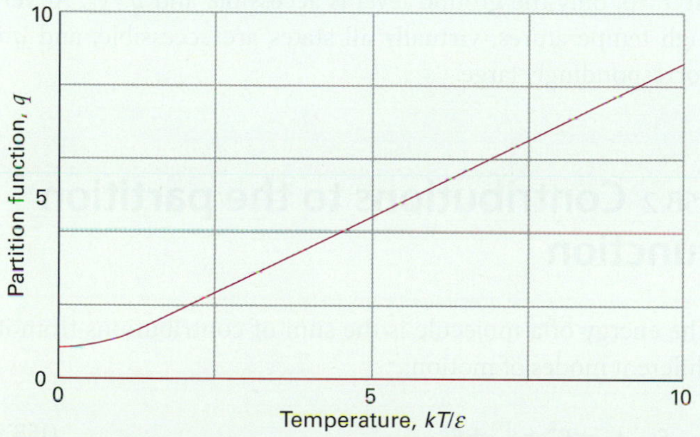

Figure 15B.2 The partition function for the system shown in Fig. 15B.1 (a harmonic oscillator) as a function of temperature.

Self-test 15B.1 Suppose the molecule can exist in only two states, with energies 0 and ε. Derive and plot the expression for the partition function.

Answer: $q=1+e^{-\beta\varepsilon}$, Fig.15 B.3

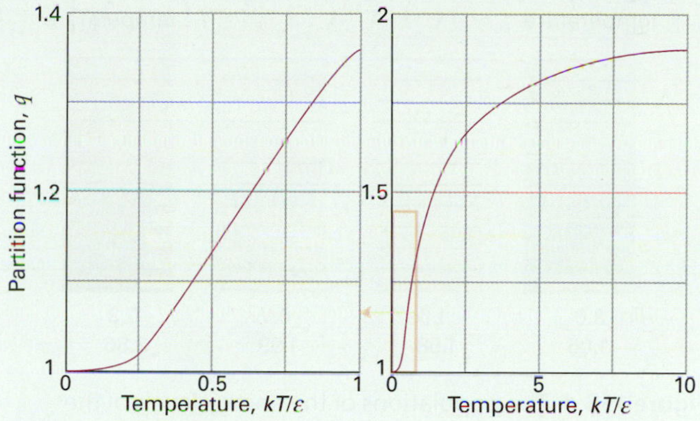

Figure 15B.3 The partition function for a two-level system as a function of temperature. The two graphs differ in the scale of the temperature axis to show the approach to 1 as $T\to 0$ and the slow approach to 2 as $T\to\infty$.

We have derived the following important expression for the partition function for a uniform ladder of states of spacing ε:

$$q=\frac{1}{1-e^{-\beta\varepsilon}} \qquad \textit{Uniform ladder} \quad \text{Partition function} \qquad (15B.2a)$$

We can use this expression to interpret the physical significance of a partition function. To do so, we first note that the Boltzmann distribution for this arrangement of energy levels gives the fraction, $p_i=N_i/N$, of molecules in the state with energy ε_i as

$$p_i=\frac{e^{-\beta\varepsilon_i}}{q}=(1-e^{-\beta\varepsilon})e^{-\beta\varepsilon_i} \qquad \textit{Uniform ladder} \quad \text{Population} \qquad (15B.2b)$$

Figure 15B.4 shows how p_i varies with temperature. At very low temperatures (large β), where q is close to 1, only the lowest state is significantly populated. As the temperature is raised, the population breaks out of the lowest state, and the upper states become progressively more highly populated. At the same time, the partition function rises from 1 towards 2, so we see that its value gives an indication of the range of states populated at any given temperature. The name 'partition function' reflects the sense in which q measures how the total number of molecules is distributed—partitioned—over the available states.

The corresponding expressions for a two-level system derived in *Self-test* 15B.1 are

$$q=1+e^{-\beta\varepsilon} \qquad \textit{Two-level system} \quad \text{Partition function} \qquad (15B.3a)$$

$$p_i=\frac{e^{-\beta\varepsilon_i}}{q}=\frac{e^{-\beta\varepsilon_i}}{1+e^{-\beta\varepsilon}} \qquad \textit{Two-level system} \quad \text{Population} \qquad (15B.3b)$$

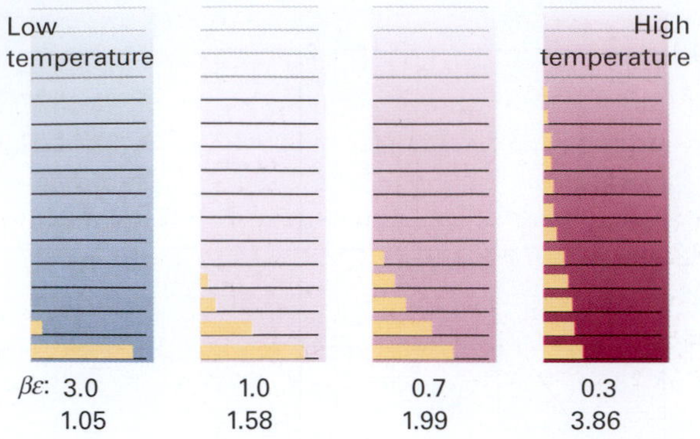

Figure 15B.4 The populations of the energy levels of the system shown in Fig.15B.1 at different temperatures, and the corresponding values of the partition function as calculated from eqn 15B.2b. Note that $\beta = 1/kT$.

In this case, because $\varepsilon_0 = 0$ and $\varepsilon_1 = \varepsilon$,

$$p_0 = \frac{1}{1+e^{-\beta\varepsilon}} \qquad p_1 = \frac{e^{-\beta\varepsilon}}{1+e^{-\beta\varepsilon}} \tag{15B.4}$$

These functions are plotted in Fig. 15B.5. Notice how the populations are $p_0 = 1$ and $p_1 = 0$ and the partition function is $q = 1$ (one state occupied) at $T = 0$. However, the populations tend towards equality ($p_0 = \frac{1}{2}, p_1 = \frac{1}{2}$) and $q = 2$ (two states occupied) as $T \to \infty$.

A note on good practice A common error is to suppose that when $T = \infty$ all the molecules in the system will be found in the upper energy state; however, we see from eqn 15B.4 that as $T \to \infty$, the populations of states become equal. The same conclusion is true of multi-level systems too: as $T \to \infty$, all states become equally populated.

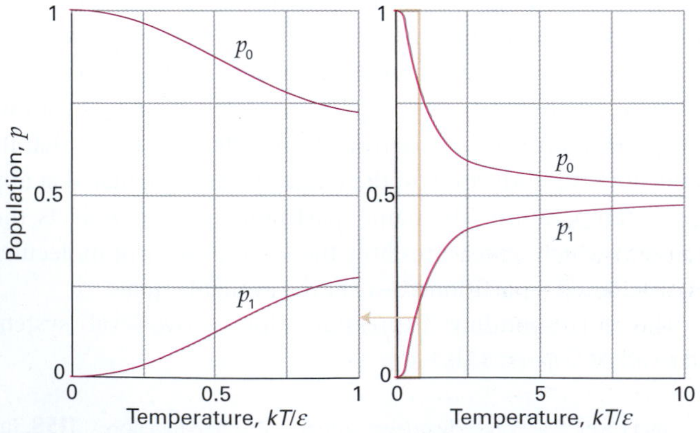

Figure 15B.5 The fraction of populations of the two states of a two-level system as a function of temperature (eqn 15B.4). Note that as the temperature approaches infinity, the populations of the two states become equal (and the fractions both approach 0.5).

We can now generalize the conclusion that the partition function indicates the number of thermally accessible states. When T is close to zero, the parameter $\beta = 1/kT$ is close to infinity. Then every term except one in the sum defining q is zero because each one has the form e^{-x} with $x \to \infty$. The exception is the term with $\varepsilon_0 \equiv 0$ (or the g_0 states at zero energy if they are g_0-fold degenerate), because then $\varepsilon_0/kT \equiv 0$ whatever the temperature, including zero. As there is only one surviving term when $T = 0$, and its value is g_0, it follows that

$$\lim_{T \to 0} q = g_0$$

That is, at $T = 0$, the partition function is equal to the degeneracy of the ground state (commonly, but not necessarily, 1).

Now consider the case when T is so high that for each term in the sum $\varepsilon_j/kT \approx 0$. Because $e^{-x} = 1$ when $x = 0$, each term in the sum now contributes 1. It follows that the sum is equal to the number of molecular states, which in general is infinite:

$$\lim_{T \to \infty} q = \infty$$

In some idealized cases, the molecule may have only a finite number of states; then the upper limit of q is equal to the number of states, as we saw for the two-level system.

In summary, we see that:

The molecular partition function gives an indication of the number of states that are thermally accessible to a molecule at the temperature of the system.

At $T = 0$, only the ground level is accessible and $q = g_0$. At very high temperatures, virtually all states are accessible, and q is correspondingly large.

15B.2 Contributions to the partition function

The energy of a molecule is the sum of contributions from its different modes of motion:

$$\varepsilon_i = \varepsilon_i^{T} + \varepsilon_i^{R} + \varepsilon_i^{V} + \varepsilon_i^{E} \tag{15B.5}$$

where T denotes translation, R rotation, V vibration, and E the electronic contribution. The electronic contribution is not actually a 'mode of motion', but it is convenient to include it here. The separation of terms in eqn 15B.5 is only approximate (except for translation) because the modes are not completely independent, but in most cases it is satisfactory. The separation of the electronic and vibrational motions is justified provided only the ground electronic state is occupied (for otherwise the vibrational characteristics depend on the electronic state) and, for the electronic ground state, that the Born–Oppenheimer approximation is valid (Topic 10A). The separation of the vibrational and rotational modes is justified to the extent that the rotational constant (Topic 12B) is independent of the vibrational state.

Given that the energy is a sum of independent contributions, the partition function factorizes into a product of contributions:

$$q=\sum_i e^{-\beta\varepsilon_i}=\sum_{i\,\text{(all states)}} e^{-\beta\varepsilon_i^{T}-\beta\varepsilon_i^{R}-\beta\varepsilon_i^{V}-\beta\varepsilon_i^{E}}$$

$$=\sum_{i\,\text{(translational)}}\sum_{i\,\text{(rotational)}}\sum_{i\,\text{(vibrational)}}\sum_{i\,\text{(electronic)}} e^{-\beta\varepsilon_i^{T}-\beta\varepsilon_i^{R}-\beta\varepsilon_i^{V}-\beta\varepsilon_i^{E}}$$

$$=\left(\sum_{i\,\text{(translational)}} e^{-\beta\varepsilon_i^{T}}\right)\left(\sum_{i\,\text{(rotational)}} e^{-\beta\varepsilon_i^{R}}\right)\left(\sum_{i\,\text{(vibrational)}} e^{-\beta\varepsilon_i^{V}}\right)$$

$$\times\left(\sum_{i\,\text{(electronic)}} e^{-\beta\varepsilon_i^{E}}\right)$$

That is,

$$q=q^{T}q^{R}q^{V}q^{E} \qquad \text{Factorization of the partition function} \qquad (15B.6)$$

This factorization means that we can investigate each contribution separately. In general, exact analytical expressions for partition functions cannot be obtained. However, approximate expressions can often be found and prove to be very important for understanding chemical phenomena; they are derived in the following sections and collected at the end of the Topic.

(a) The translational contribution

The translational partition function for a particle of mass m free to move in a one-dimensional container of length X can be evaluated by making use of the fact that the separation of energy levels is very small and that large numbers of states are accessible at normal temperatures. As shown in the following *Justification*, in this case

$$q=\frac{X}{\Lambda} \qquad q_X^{T}=\left(\frac{2\pi m}{h^2\beta}\right)^{1/2}X \qquad \textit{One-dimensional container} \quad \text{Translational partition function} \qquad (15B.7a)$$

It will prove convenient to anticipate once again that $\beta=1/kT$ and to write this expression as $q_X^{T}=X/\Lambda$, with

$$\Lambda=\frac{h}{(2\pi mkT)^{1/2}} \qquad \textit{Definition} \quad \text{Thermal wavelength} \qquad (15B.7b)$$

The quantity Λ (uppercase lambda) has the dimensions of length and is called the **thermal wavelength** (sometimes the 'thermal de Broglie wavelength') of the molecule. The thermal wavelength decreases with increasing mass and temperature.

This expression shows that the partition function for translational motion increases with the length of the box and the mass of the particle, for in each case the separation of the energy levels becomes smaller and more levels become thermally accessible. For a given mass and length of the box, the partition function also increases with increasing temperature (decreasing β), because more states become accessible.

Justification 15B.1 The partition function for a particle in a one-dimensional box

The energy levels of a molecule of mass m in a container of length X are given by eqn 8A.6b ($E_n=n^2h^2/8mL^2$) with $L=X$:

$$E_n=\frac{n^2h^2}{8mX^2}$$

The lowest level ($n=1$) has energy $h^2/8mX^2$, so the energies relative to that level are

$$\varepsilon_n=(n^2-1)\varepsilon \qquad \varepsilon=h^2/8mX^2$$

The sum to evaluate is therefore

$$q_X^{T}=\sum_{n=1}^{\infty} e^{-(n^2-1)\beta\varepsilon}$$

The translational energy levels are very close together in a container the size of a typical laboratory vessel; therefore, the sum can be approximated by an integral:

$$q_X^{T}=\int_1^{\infty} e^{-(n^2-1)\beta\varepsilon}\,dn\approx\int_0^{\infty} e^{-n^2\beta\varepsilon}\,dn$$

The extension of the lower limit to $n=0$ and the replacement of n^2-1 by n^2 introduces negligible error but turns the integral into standard form. We make the substitution $x^2=n^2\beta\varepsilon$, implying $dn=dx/(\beta\varepsilon)^{1/2}$, and therefore that

$$q_X^{T}=\left(\frac{1}{\beta\varepsilon}\right)^{1/2}\overbrace{\int_0^{\infty} e^{-x^2}dx}^{\text{Integral G.1}\;\pi^{1/2}/2}=\left(\frac{1}{\beta\varepsilon}\right)^{1/2}\frac{\pi^{1/2}}{2}=\left(\frac{2\pi m}{h^2\beta}\right)^{1/2}X$$

This relation has the form of eqn 15B.7a, $q=X/\Lambda$, provided Λ is identified as

$$\Lambda=\left(\frac{h^2\beta}{2\pi m}\right)^{1/2}\overset{\beta=1/kT}{=}\frac{h}{(2\pi mkT)^{1/2}}$$

as in eqn 15B.7b.

The total energy of a molecule free to move in three dimensions is the sum of its translational energies in all three directions:

$$\varepsilon_{n_1n_2n_3}=\varepsilon_{n_1}^{(X)}+\varepsilon_{n_2}^{(Y)}+\varepsilon_{n_3}^{(Z)} \qquad (15B.8)$$

where n_1, n_2, and n_3 are the quantum numbers for motion in the x-, y-, and z-directions, respectively. Therefore, because $e^{a+b+c}=e^a e^b e^c$, the partition function factorizes as follows:

$$q^{\mathrm{T}}=\sum_{\text{all } n} e^{-\beta\varepsilon_{n_1}^{(X)}-\beta\varepsilon_{n_2}^{(Y)}-\beta\varepsilon_{n_3}^{(Z)}}=\sum_{\text{all } n} e^{-\beta\varepsilon_{n_1}^{(X)}}e^{-\beta\varepsilon_{n_2}^{(Y)}}e^{-\beta\varepsilon_{n_3}^{(Z)}}$$
$$=\left(\sum_{n_1} e^{-\beta\varepsilon_{n_1}^{(X)}}\right)\left(\sum_{n_2} e^{-\beta\varepsilon_{n_2}^{(Y)}}\right)\left(\sum_{n_3} e^{-\beta\varepsilon_{n_3}^{(Z)}}\right)$$

That is,

$$q^{\mathrm{T}}=q_X^{\mathrm{T}}q_Y^{\mathrm{T}}q_Z^{\mathrm{T}} \qquad (15B.9)$$

Equation 15B.7a gives the partition function for translational motion in the x-direction. The only change for the other two directions is to replace the length X by the lengths Y or Z. Hence the partition function for motion in three dimensions is

$$q^{\mathrm{T}}=\left(\frac{2\pi m}{h^2\beta}\right)^{3/2}XYZ=\frac{(2\pi mkT)^{3/2}}{h^3}XYZ \qquad (15B.10a)$$

The product of lengths XYZ is the volume, V, of the container, so we can write

$$q^{\mathrm{T}}=\frac{V}{\Lambda^3} \qquad \textit{Three-dimensional container} \qquad \text{Translational partition function} \qquad (15B.10b)$$

with Λ as defined in eqn 15B.7b. As in the one-dimensional case, the partition function increases with the mass of the particle (as $m^{3/2}$) and the volume of the container (as V); for a given mass and volume, the partition function increases with temperature (as $T^{3/2}$). As in one dimension, $q^{\mathrm{T}}\to\infty$ as $T\to\infty$ because an infinite number of states becomes accessible as the temperature is raised. Even at room temperature, $q^{\mathrm{T}}\approx 2\times10^{28}$ for an O_2 molecule in a vessel of volume 100 cm³.

Brief illustration 15B.2 The translational partition function

To calculate the translational partition function of an H_2 molecule confined to a 100 cm³ vessel at 25 °C we use $m=2.016m_u$; then, from $\Lambda=h/(2\pi mkT)^{1/2}$,

$$\Lambda=\frac{6.626\times10^{-34}\ \overset{1\,\mathrm{J}=1\,\mathrm{kg\,m^2\,s^{-2}}}{\mathrm{J}}\ \mathrm{s}}{\left\{2\pi\times(2.016\times1.6605\times10^{-27}\ \mathrm{kg})\times\left(1.381\times10^{-23}\ \underset{1\,\mathrm{J}=1\,\mathrm{kg\,m^2\,s^{-2}}}{\mathrm{J}}\ \mathrm{K^{-1}}\right)\times(298\ \mathrm{K})\right\}^{1/2}}$$
$$=7.12\times10^{-11}\ \mathrm{m}$$

Therefore,

$$q^{\mathrm{T}}=\frac{1.00\times10^{-4}\ \mathrm{m^3}}{(7.12\times10^{-11}\ \mathrm{m})^3}=2.77\times10^{26}$$

About 10^{26} quantum states are thermally accessible, even at room temperature and for this light molecule. Many states are occupied if the thermal wavelength (which in this case is 71.2 pm) is small compared with the linear dimensions of the container.

Self-test 15B.2 Calculate the translational partition function for a D_2 molecule under the same conditions.

Answer: $q^{\mathrm{T}}=7.8\times10^{26}$, $2^{3/2}$ times larger

The validity of the approximations that led to eqn 15B.10 can be expressed in terms of the average separation, d, of the particles in the container. Because q is the total number of accessible states, the average number of translational states per molecule is q^{T}/N. For this quantity to be large, we require $V/N\Lambda^3\gg1$. However, V/N is the volume occupied by a single particle, and therefore the average separation of the particles is $d=(V/N)^{1/3}$. The condition for there being many states available per molecule is therefore $d^3/\Lambda^3\gg1$, and therefore $d\gg\Lambda$. That is, for eqn 15B.10 to be valid, *the average separation of the particles must be much greater than their thermal wavelength*. For H_2 molecules at 1 bar and 298 K, the average separation is 3 nm, which is significantly larger than their thermal wavelength (71.2 pm).

The validity of eqn 15B.10 can be expressed in a different way by noting that the approximations that led to it are valid if many states are occupied, which requires V/Λ^3 to be large. That will be so if Λ is small compared with the linear dimensions of the container. For H_2 at 25 °C, $\Lambda=71$ pm, which is far smaller than any conventional container is likely to be (but comparable to pores in zeolites or cavities in clathrates). For O_2, a heavier molecule, $\Lambda=18$ pm.

(b) The rotational contribution

The energy levels of a linear rotor are $\varepsilon_J=hc\tilde{B}J(J+1)$, with $J=0, 1, 2, \ldots$ (Topic 12B). The state of lowest energy has zero energy, so no adjustment need be made to the energies given by this expression. Each level consists of $2J+1$ degenerate states. Therefore, the partition function of a non-symmetrical (AB) linear rotor is

$$q^{\mathrm{R}}=\sum_J \overbrace{(2J+1)}^{g_J}e^{-\beta\overbrace{hc\tilde{B}J(J+1)}^{\varepsilon_J}} \qquad (15B.11)$$

The direct method of calculating q^{R} is to substitute the experimental values of the rotational energy levels into this expression and to sum the series numerically.

Example 15B.1 Evaluating the rotational partition function explicitly

Evaluate the rotational partition function of $^1H^{35}Cl$ at 25 °C, given that $\tilde{B}=10.591\,\text{cm}^{-1}$.

Method We need to evaluate eqn 15B.11 term by term. We use $kT/hc=207.224\,\text{cm}^{-1}$ at 298.15 K. The sum is readily evaluated by using mathematical software.

Answer To show how successive terms contribute, we draw up the following table by using $hc\tilde{B}/kT=0.05111$ (Fig. 15B.6):

J	0	1	2	3	4	…	10
$(2J+1)e^{-0.05111J(J+1)}$	1	2.71	3.68	3.79	3.24	…	0.08

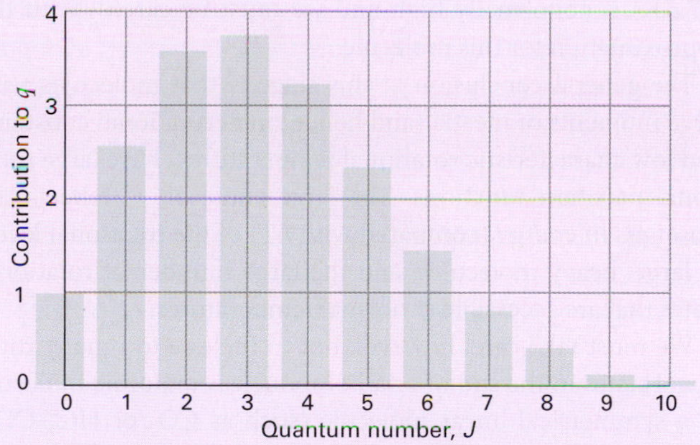

Figure 15B.6 The contributions to the rotational partition function of an HCl molecule at 25 °C. The vertical axis is the value of $(2J+1)e^{-\beta hc\tilde{B}J(J+1)}$. Successive terms (which are proportional to the populations of the levels) pass through a maximum because the population of individual states decreases exponentially, but the degeneracy of the levels increases with J.

The sum required by eqn 15B.11 (the sum of the numbers in the second row of the table) is 19.9, hence $q^R=19.9$ at this temperature. Taking J up to 50 gives $q^R=19.902$. Notice that about ten J-levels are significantly populated but the number of populated *states* is larger on account of the $(2J+1)$-fold degeneracy of each level.

Self-test 15B.3 Evaluate the rotational partition function for $^1H^{35}Cl$ at 0 °C.

Answer: 18.26

At room temperature, $kT/hc\approx 200\,\text{cm}^{-1}$. The rotational constants of many molecules are close to $1\,\text{cm}^{-1}$ (Table 12D.1) and often smaller (though the very light H_2 molecule, for which $\tilde{B}=60.9\,\text{cm}^{-1}$, is one exception). It follows that many rotational levels are populated at normal temperatures. When this is the case, we show in the following two *Justifications* that the partition function may be approximated by

$$q^R=\frac{kT}{hc\tilde{B}} \qquad \text{Linear rotor} \qquad \text{Rotational partition function} \qquad (15B.12a)$$

$$q^R=\left(\frac{kT}{hc}\right)^{3/2}\left(\frac{\pi}{\tilde{A}\tilde{B}\tilde{C}}\right)^{1/2} \qquad \text{Nonlinear rotor} \qquad \text{Rotational partition function} \qquad (15B.12b)$$

where $\tilde{A}, \tilde{B}$, and $\tilde{C}$ are the rotational constants of the molecule expressed as wavenumbers. However, before using these expressions, read on (to eqns 15B.13 and 15B.14).

Justification 15B.2 The rotational contribution for linear molecules

When many rotational states are occupied and kT is much larger than the separation between neighbouring states, the sum in the partition function can be approximated by an integral, much as we did for translational motion:

$$q^R=\int_0^\infty (2J+1)e^{-\beta hc\tilde{B}J(J+1)}\,dJ$$

This integral can be evaluated without much effort by making the substitution $x=\beta hc\tilde{B}J(J+1)$, so that $dx/dJ=\beta hc\tilde{B}(2J+1)$ and therefore $(2J+1)dJ=dx/\beta hc\tilde{B}$. Then

$$q^R=\frac{1}{\beta hc\tilde{B}}\overbrace{\int_0^\infty e^{-x}dx}^{\text{Integral E.1}\;1}=\frac{1}{\beta hc\tilde{B}}$$

which (because $\beta=1/kT$) is eqn 15B.12a.

Brief illustration 15B.3 The rotational contribution

For $^1H^{35}Cl$ at 298.15 K we use $kT/hc=207.224\,\text{cm}^{-1}$ and $\tilde{B}=10.591\,\text{cm}^{-1}$. Then

$$q^R=\frac{kT}{hc\tilde{B}}=\frac{207.224\ \text{cm}^{-1}}{10.591\,\text{cm}^{-1}}=19.59$$

The value is in good agreement with the exact value (19.02) and with much less work.

Self-test 15B.4 Evaluate the rotational contribution to the partition function for $^1H^{35}Cl$ at 0 °C.

Answer: 17.93

Justification 15B.3 The rotational contribution for nonlinear molecules

The energies of a symmetric rotor (Topic 12B) are

$$E_{J,K,M_J}=hc\tilde{B}J(J+1)+hc\left(\tilde{A}-\tilde{B}\right)K^2$$

with $J=0, 1, 2, \ldots, K=J, J-1, \ldots, -J$, and $M_J=J, J-1, \ldots, -J$. Instead of considering these ranges, the same values can be

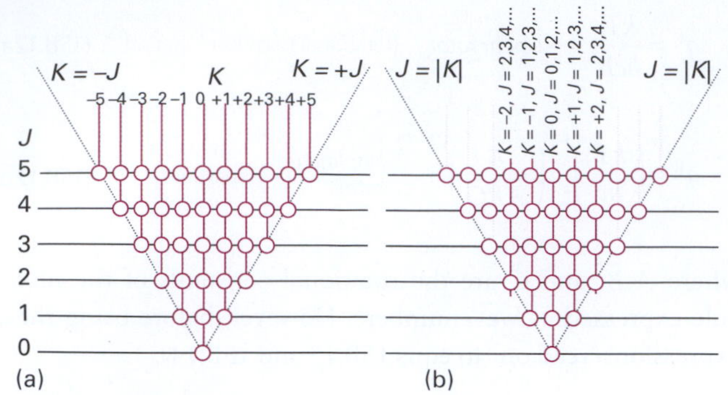

Figure 15B.7 (a) The sum over $J=0, 1, 2, \ldots$ and $K=J, J-1, \ldots, -J$ (depicted by the circles) can be covered (b) by allowing K to range from $-\infty$ to ∞, with J confined to $|K|, |K|+1, \ldots, \infty$ for each value of K.

covered by allowing K to range from $-\infty$ to ∞, with J confined to $|K|, |K|+1, \ldots, \infty$ for each value of K (Fig. 15B.7). Because the energy is independent of M_J, and there are $2J+1$ values of M_J for each value of J, each value of J is $(2J+1)$-fold degenerate. It follows that the partition function

$$q=\sum_{J=0}^{\infty}\sum_{K=-J}^{J}\sum_{M_J=-J}^{J} e^{-\beta E_{J,K,M_J}}$$

can be written equivalently as

$$q=\sum_{J=0}^{\infty}\sum_{K=-J}^{J}(2J+1)e^{-\beta E_{J,K,M_J}}=\sum_{K=-\infty}^{\infty}\sum_{J=|K|}^{\infty}(2J+1)e^{-\beta E_{J,K,M_J}}$$
$$=\sum_{K=-\infty}^{\infty} e^{-hc\beta(\tilde{A}-\tilde{B})K^2}\sum_{J=|K|}^{\infty}(2J+1)e^{-hc\beta\tilde{B}J(J+1)}$$

As in *Justification* 15B.2 we assume that the temperature is so high that numerous states are occupied and that the sums may be approximated by integrals. Then

$$q=\int_{-\infty}^{\infty} e^{-hc\beta(\tilde{A}-\tilde{B})K^2}\int_{|K|}^{\infty}(2J+1)e^{-hc\beta\tilde{B}J(J+1)}\mathrm{d}J\,\mathrm{d}K$$

As before, the integral over J can be recognized as the integral of the derivative of a function, which is the function itself, so, as you should verify,

$$\int_{|K|}^{\infty}(2J+1)e^{-hc\beta\tilde{B}J(J+1)}\,\mathrm{d}J=\frac{1}{hc\beta\tilde{B}}e^{-hc\beta\tilde{B}K^2}$$

We have also supposed that $|K|\gg 1$ for most contributions and replaced $|K|(|K|+1)$ by K^2. Now we can write

$$q=\frac{1}{hc\beta\tilde{B}}\int_{-\infty}^{\infty} e^{-hc\beta(\tilde{A}-\tilde{B})K^2}e^{-hc\beta\tilde{B}K^2}\,\mathrm{d}K=\frac{1}{hc\beta\tilde{B}}\overbrace{\int_{-\infty}^{\infty} e^{-hc\beta\tilde{A}K^2}\mathrm{d}K}^{\text{Integral G.1}}$$
$$=\frac{1}{hc\beta\tilde{B}}\left(\frac{\pi}{hc\beta\tilde{A}}\right)^{1/2}$$

We conclude that

$$q=\frac{1}{(hc\beta)^{3/2}}\left(\frac{\pi}{\tilde{A}\tilde{B}^2}\right)^{1/2}=\left(\frac{kT}{hc}\right)^{3/2}\left(\frac{\pi}{\tilde{A}\tilde{B}^2}\right)^{1/2}$$

For an asymmetric rotor, one of the $\tilde{B}$ is replaced by $\tilde{C}$, to give eqn 15B.12b.

A useful way of expressing the temperature above which the rotational approximation is valid is to introduce the **characteristic rotational temperature**, $\theta^{R}=hc\tilde{B}/k$. Then 'high temperature' means $T\gg\theta^{R}$ and under these conditions the rotational partition function of a linear molecule is simply T/θ^{R}. Some typical values of θ^{R} are given in Table 15B.1. The value for $^1H_2$ (87.6 K) is abnormally high and we must be careful with the approximation for this molecule.

The general conclusion at this stage is that molecules with large moments of inertia (and hence small rotational constants and low characteristic rotational temperatures) have large rotational partition functions. The large value of q^{R} reflects the closeness in energy (compared with kT) of the rotational levels in large, heavy molecules, and the large number of rotational states that are accessible at normal temperatures.

We must take care, however, not to include too many rotational states in the sum. For a homonuclear diatomic molecule or a symmetrical linear molecule (such as CO_2 or HC≡CH), a rotation through 180° results in an indistinguishable state of the molecule. Hence, the number of thermally accessible states is only half the number that can be occupied by a heteronuclear diatomic molecule, where rotation through 180° does result in a distinguishable state. Therefore, for a symmetrical linear molecule,

$$q^{R}=\frac{kT}{2hc\tilde{B}}=\frac{T}{2\theta^{R}} \qquad \textit{Symmetrical linear rotor} \qquad \text{Rotational partition function} \qquad (15B.13a)$$

The equations for symmetrical and non-symmetrical molecules can be combined into a single expression by introducing the **symmetry number**, σ, which is the number of indistinguishable orientations of the molecule. Then

$$q^{R}=\frac{T}{\sigma\theta^{R}} \qquad \textit{Linear rotor} \qquad \text{Rotational partition function} \qquad (15B.13b)$$

Table 15B.1* Rotational temperatures of diatomic molecules

	θ^{R}/K
$^1H_2$	87.6
$^1H^{35}Cl$	15.2
$^{14}N_2$	2.88
$^{35}Cl_2$	0.351

* More values are given in the *Resource section*, Table 12D.1.

For a heteronuclear diatomic molecule $\sigma = 1$; for a homonuclear diatomic molecule or a symmetrical linear molecule, $\sigma = 2$.

Justification 15B.4 The origin of the symmetry number

The quantum mechanical origin of the symmetry number is the Pauli principle, which forbids the occupation of certain states. It is shown in Topic 12C, for example, that H_2 may occupy rotational states with even J only if its nuclear spins are paired (*para*-hydrogen), and odd J states only if its nuclear spins are parallel (*ortho*-hydrogen). There are three states of *ortho*-H_2 to each value of J (because there are three parallel spin states of the two nuclei).

To set up the rotational partition function we note that 'ordinary' molecular hydrogen is a mixture of one part *para*-H_2 (with only its even-J rotational states occupied) and three parts *ortho*-H_2 (with only its odd-J rotational states occupied). Therefore, the average partition function per molecule is

$$q^{\mathrm{R}} = \frac{1}{4}\sum_{\text{even } J}(2J+1)\mathrm{e}^{-\beta hc\tilde{B}J(J+1)} + \frac{3}{4}\sum_{\text{odd } J}(2J+1)\mathrm{e}^{-\beta hc\tilde{B}J(J+1)}$$

The odd-J states are more heavily weighted than the even-J states (Fig. 15B.8). From the illustration we see that we would obtain approximately the same answer for the partition function (the sum of all the populations) if each J term contributed half its normal value to the sum. That is, the last equation can be approximated as

$$q^{\mathrm{R}} = \frac{1}{2}\sum_{J}(2J+1)\mathrm{e}^{-\beta hc\tilde{B}J(J+1)}$$

and this approximation is very good when many terms contribute (at high temperatures, $T \gg 87.6\,\mathrm{K}$).

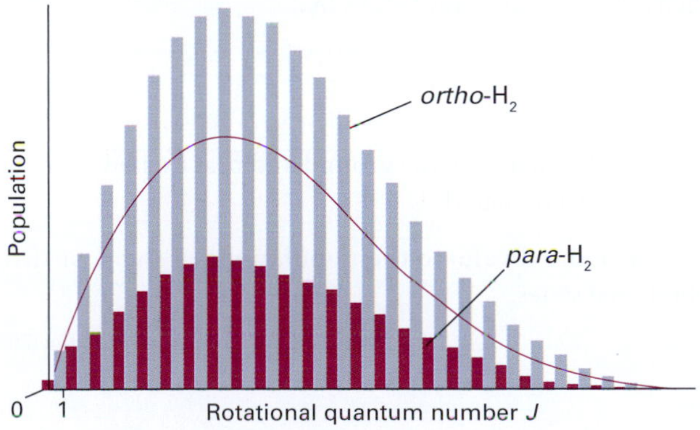

Figure 15B.8 The values of the individual terms $(2J+1)\mathrm{e}^{-\beta hc\tilde{B}J(J+1)}$ contributing to the mean partition function of a 3:1 mixture of *ortho*- and *para*-H_2. The partition function is the sum of all these terms. At high temperatures, the sum is approximately equal to the sum of the terms over all values of J, each with a weight of $\frac{1}{2}$. This is the sum of the contributions indicated by the curve.

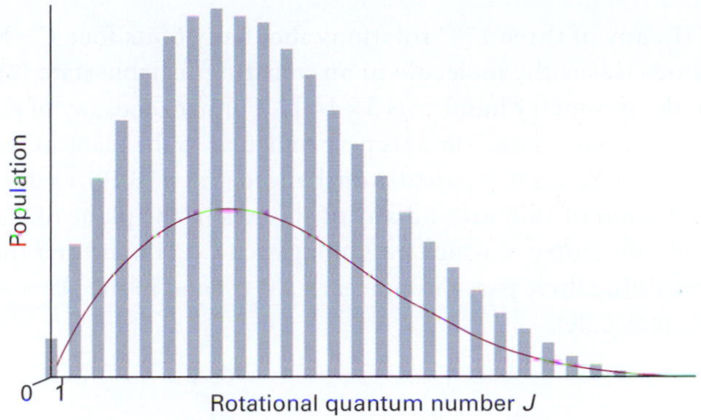

Figure 15B.9 The relative populations of the rotational energy levels of CO_2. Only states with even J values are occupied. The full line shows the smoothed, averaged population of levels.

The same type of argument may be used for linear symmetrical molecules in which identical bosons are interchanged by rotation (such as CO_2). As pointed out in Topic 12C, if the nuclear spin of the bosons is 0, then only even-J states are admissible. Because only half the rotational states are occupied, the rotational partition function is only half the value of the sum obtained by allowing all values of J to contribute (Fig. 15B.9).

The same care must be exercised for other types of symmetrical molecule, and for a nonlinear molecule we write

$$q^{\mathrm{R}} = \frac{1}{\sigma}\left(\frac{kT}{hc}\right)^{3/2}\left(\frac{\pi}{\tilde{A}\tilde{B}\tilde{C}}\right)^{1/2} \quad \textit{Nonlinear rotor} \quad \text{Rotational partition function} \quad (15\text{B}.14)$$

Some typical values of the symmetry numbers are given in Table 15B.2. For the way that group theory is used to identify the value of the symmetry number, see Problem 15B.9.

Table 15B.2* Symmetry numbers of molecules

	σ
$^1H_2$	2
$^1H^2H$	1
NH_3	3
C_6H_6	12

* More values are given in the *Resource section*, Table 12D.1.

Brief illustration 15B.4 The symmetry number

The value $\sigma(H_2O) = 2$ reflects the fact that a 180° rotation about the bisector of the H—O—H angle interchanges two indistinguishable atoms. In NH_3, there are three indistinguishable orientations around the axis, as shown in **1**. For

CH_4, any of three 120° rotations about any of its four C—H bonds leaves the molecule in an indistinguishable state (2), so the symmetry number is $3\times4=12$. For benzene, any of six orientations around the axis perpendicular to the plane of the molecule leaves it apparently unchanged (Fig. 15B.10), as does a rotation of 180° around any of six axes in the plane of the molecule (three of which pass along each C—H bond and the remaining three pass through each C—C bond in the plane of the molecule).

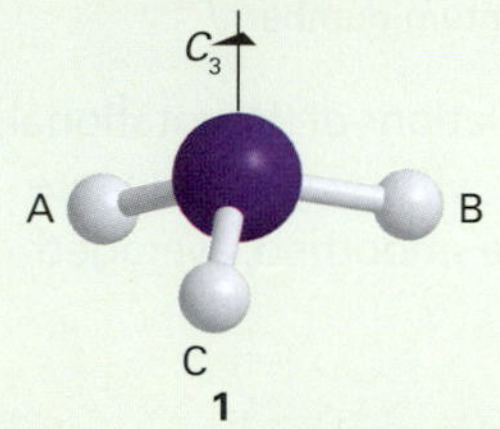

1

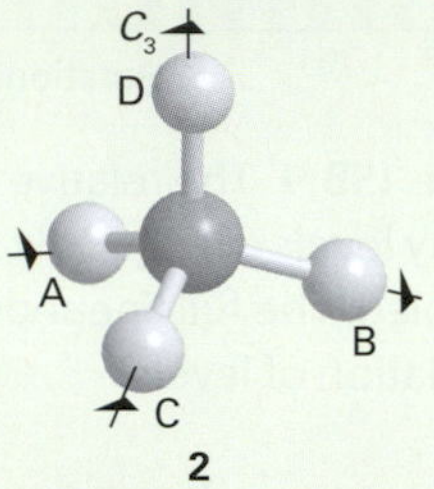

2

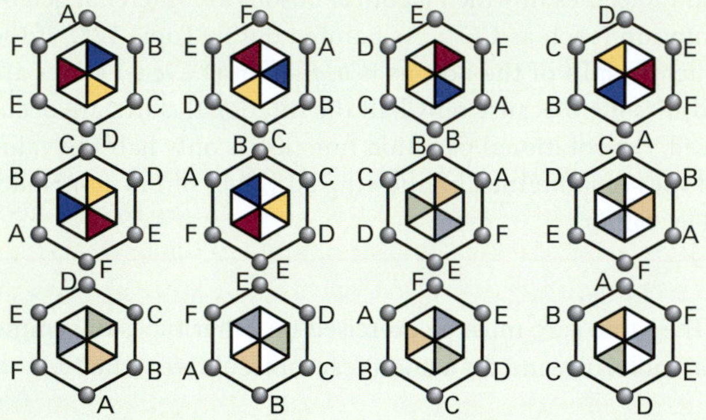

Figure 15B.10 The 12 equivalent orientations of a benzene molecule that can be reached by pure rotations, and give rise to a symmetry number of 12. The six pale colours are the underside of the hexagon after that face has been rotated into view.

Self-test 15B.5 What is the symmetry number for a naphthalene molecule?

Answer: 3

(c) The vibrational contribution

The vibrational partition function of a molecule is calculated by substituting the measured vibrational energy levels into the exponentials appearing in the definition of q^V, and summing them numerically. However, provided it is permissible to assume that the vibrations are harmonic, there is a much simpler way. In that case, the vibrational energy levels form a uniform ladder of separation $hc\tilde{\nu}$ (Topics 8B and 12D), which is exactly the problem treated in *Brief illustration* 15B.1 and summarized in eqn 15B.2a. Therefore we can use that result with $\varepsilon=hc\tilde{\nu}$ and conclude immediately that

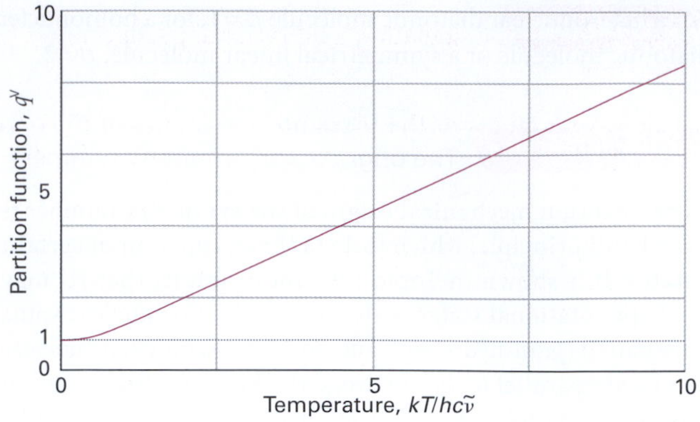

Figure 15B.11 The vibrational partition function of a molecule in the harmonic approximation. Note that the partition function is linearly proportional to the temperature when the temperature is high ($T \gg \theta^V$).

$$q^V=\frac{1}{1-e^{-\beta hc\tilde{\nu}}} \qquad \textit{Harmonic approximation} \qquad \text{Vibrational partition function} \qquad (15B.15)$$

This function is plotted in Fig. 15B.11 (which is essentially the same as Fig. 15B.1). Similarly, the population of each state is given by eqn 15B.2b.

Brief illustration 15B.5 The vibrational partition function

To calculate the partition function of I_2 molecules at 298.15 K we note from Table 12D.1 that their vibrational wavenumber is 214.6 cm^{-1}. Then, because at 298.15 K, $kT/hc=207.224\,\text{cm}^{-1}$, we have

$$\beta\varepsilon=\frac{hc\tilde{\nu}}{kT}=\frac{214.6\,\text{cm}^{-1}}{207.244\,\text{cm}^{-1}}=1.035\ldots$$

Then it follows from eqn 15B.15 that

$$q^V=\frac{1}{1-e^{-1.035\ldots}}=1.55$$

We can infer that only the ground and first excited states are significantly populated.

Self-test 15B.6 Evaluate the populations of the first three vibrational states.

Answer: $p_0=0.645$, $p_1=0.229$, $p_2=0.081$

In a polyatomic molecule, each normal mode (Topic 12E) has its own partition function (provided the anharmonicities are so small that the modes are independent). The overall vibrational partition function is the product of the individual partition functions, and we can write $q^V=q^V(1)q^V(2)\ldots$, where $q^V(K)$ is the partition function for the Kth normal mode and is calculated by direct summation of the observed spectroscopic levels.

Example 15B.2 Calculating a vibrational partition function

The wavenumbers of the three normal modes of H_2O are $3656.7\,\text{cm}^{-1}$, $1594.8\,\text{cm}^{-1}$, and $3755.8\,\text{cm}^{-1}$. Evaluate the vibrational partition function at 1500 K.

Method Use eqn 15B.15 for each mode, and then form the product of the three contributions. At 1500 K, $kT/hc = 1042.6\,\text{cm}^{-1}$.

Answer We draw up the following table displaying the contributions of each mode:

Mode:	1	2	3
$\tilde{\nu}/\text{cm}^{-1}$	3656.7	1594.8	3755.8
$hc\tilde{\nu}/kT$	3.507	1.530	3.602
q^V	1.031	1.276	1.028

The overall vibrational partition function is therefore

$$q^V = 1.031 \times 1.276 \times 1.028 = 1.352$$

The three normal modes of H_2O are at such high wavenumbers that even at 1500 K most of the molecules are in their vibrational ground state. However, there may be so many normal modes in a large molecule that their overall contribution may be significant even though each mode is not appreciably excited. For example, a nonlinear molecule containing 10 atoms has $3N-6=24$ normal modes (Topic 12E). If we assume a value of about 1.1 for the vibrational partition function of one normal mode, the overall vibrational partition function is about $q^V \approx (1.1)^{24} = 9.8$, which indicates significant vibrational excitation relative to a smaller molecule, such as H_2O.

Self-test 15B.7 Repeat the calculation for CO_2, where the vibrational wavenumbers are $1388\,\text{cm}^{-1}$, $667.4\,\text{cm}^{-1}$, and $2349\,\text{cm}^{-1}$, the second being the doubly-degenerate bending mode.

Answer: 6.79

In many molecules the vibrational wavenumbers are so great that $\beta hc\tilde{\nu} > 1$. For example, the lowest vibrational wavenumber of CH_4 is $1306\,\text{cm}^{-1}$, so $\beta hc\tilde{\nu} = 6.3$ at room temperature. Most C—H stretches normally lie in the range 2850 to $2960\,\text{cm}^{-1}$, so for them $\beta hc\tilde{\nu} \approx 14$. In these cases, $e^{-\beta hc\tilde{\nu}}$ in the denominator of q^V is very close to zero (for example, $e^{-6.3} = 0.002$), and the vibrational partition function for a single mode is very close to 1 ($q^V = 1.002$ when $\beta hc\tilde{\nu} = 6.3$), implying that only the zero-point level is significantly occupied.

Now consider the case of bonds with such low vibrational frequencies that $\beta hc\tilde{\nu} \ll 1$. When this condition is satisfied,

Table 15B.3* Vibrational temperatures of diatomic molecules

	θ^V/K
$^1H_2$	6332
$^1H^{35}Cl$	4304
$^{14}N_2$	3393
$^{35}Cl_2$	805

* More values are given in the *Resource section*, Table 12D.1.

the partition function may be approximated by expanding the exponential ($e^x = 1 + x + \cdots$):

$$q^V = \frac{1}{1-e^{-\beta hc\tilde{\nu}}} = \frac{1}{1-(1-\beta hc\tilde{\nu}+\cdots)}$$

That is, for weak bonds at high temperatures,

$$q^V \approx \frac{kT}{hc\tilde{\nu}} \qquad \textit{High-temperature approximation} \qquad \text{Vibrational partition function} \qquad (15B.16)$$

The temperatures for which eqn 15B.16 is valid can be expressed in terms of the **characteristic vibrational temperature**, $\theta^V = hc\tilde{\nu}/k$ (Table 15B.3). The value for H_2 (6332 K) is abnormally high because the atoms are so light and the vibrational frequency is correspondingly high. In terms of the vibrational temperature, 'high temperature' means $T \gg \theta^V$ and when this condition is satisfied, $q^V = T/\theta^V$ (the analogue of the rotational expression).

(d) The electronic contribution

Electronic energy separations from the ground state are usually very large, so for most cases $q^E = 1$ because only the ground state is occupied. An important exception arises in the case of atoms and molecules having electronically degenerate ground states, in which case $q^E = g^E$, where g^E is the degeneracy of the electronic ground state. Alkali metal atoms, for example, have doubly degenerate ground states (corresponding to the two orientations of their electron spin), so $q^E = 2$.

Brief illustration 15B.6 The electronic partition function

Some atoms and molecules have low lying electronically excited states. An example is NO, which has a configuration of the form $\ldots\pi^1$ (Topic 10C). The energy of the two degenerate states in which the orbital and spin momenta are parallel (giving the $^2\Pi_{3/2}$ term, Fig. 15B.12) is slightly greater than that of the two degenerate states in which they are antiparallel (giving the $^2\Pi_{1/2}$ term).

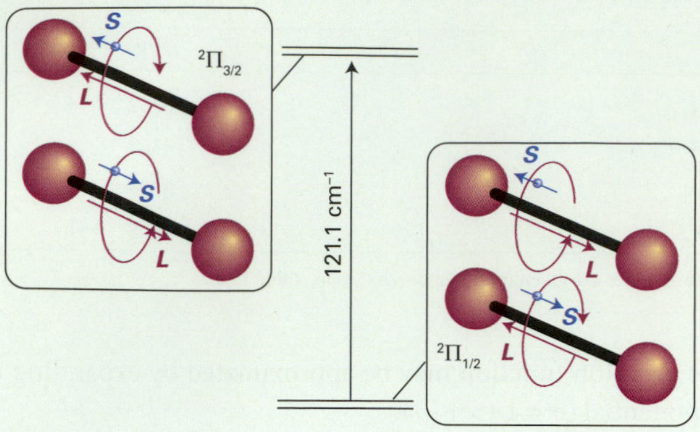

Figure 15B.12 The doubly-degenerate ground electronic level of NO (with the spin and orbital angular momentum around the axis in opposite directions) and the doubly-degenerate first excited level (with the spin and orbital momenta parallel). The upper level is thermally accessible at room temperature.

The separation, which arises from spin–orbit coupling, is only 121 cm^{-1}. If we denote the energies of the two levels as $E_{1/2}=0$ and $E_{3/2}=\varepsilon$, the partition function is

$$q^{\rm E}=\sum_{\text{levels } i} g_i e^{-\beta\varepsilon_i}=2+2e^{-\beta\varepsilon}$$

This function is plotted in Fig. 15B.13. At $T=0$, $q^{\rm E}=2$, because only the doubly degenerate ground state is accessible. At high temperatures, $q^{\rm E}$ approaches 4 because all four states are accessible. At 25 °C, $q^{\rm E}=3.1$.

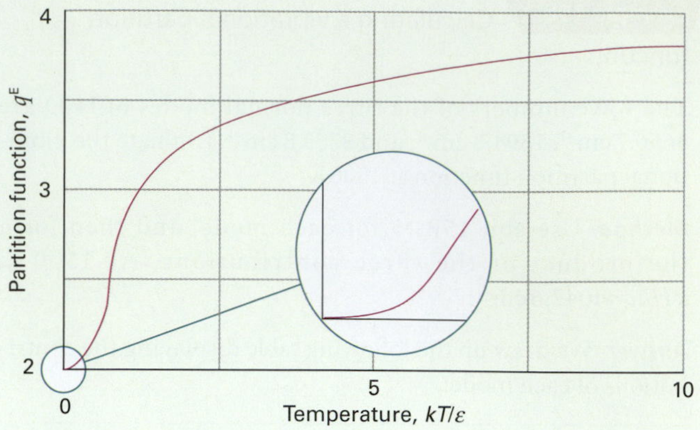

Figure 15B.13 The variation with temperature of the electronic partition function of an NO molecule. Note that the curve resembles that for a two-level system (Fig.15B.3), but rises from 2 (the degeneracy of the lower level) and approaches 4 (the total number of states) at high temperatures.

Self-test 15B.8 A certain atom has a fourfold degenerate ground state and a sixfold degenerate excited state at 400 cm^{-1} above the ground state. Calculate its electronic partition function at 25 °C.

Answer: 4.87

Checklist of concepts

☐ 1. The **molecular partition function** is an indication of the number of thermally accessible states at the temperature of interest.

☐ 2. If the **energy of a molecule** is given by the sum of contributions, then the molecular partition function is a product of contributions from the different modes.

☐ 3. The **symmetry number** takes into account the number of indistinguishable orientations of a symmetrical molecule.

☐ 4. The **vibrational partition function** of a molecule may be approximated by that of an harmonic oscillator.

☐ 5. Because electronic energy separations from the ground state are usually very big, in most cases the **electronic partition function** is equal to the degeneracy of the electronic ground state.

Checklist of equations

Property	Equation	Comment	Equation number
Molecular partition function	$q=\sum_{\text{states } i} e^{-\beta\varepsilon_i}$	Definition, independent molecules	15B.1a
	$q=\sum_{\text{levels } i} g_i e^{-\beta\varepsilon_i}$	Definition, independent molecules	15B.1b

Property	Equation	Comment	Equation number
Uniform ladder	$q=1/(1-\mathrm{e}^{-\beta\varepsilon})$		15B.2a
Two-level system	$q=1+\mathrm{e}^{-\beta\varepsilon}$		15B.3a
Thermal wavelength	$\Lambda=h/(2\pi mkT)^{1/2}$		15B.7b
Translation	$q^{\mathrm{T}}=V/\Lambda^3$		15B.10b
Rotation	$q^{\mathrm{R}}=kT/\sigma hc\tilde{B}$	$T\gg\theta^{\mathrm{R}}$, linear rotor	15B.13
	$q^{\mathrm{R}}=(1/\sigma)(kT/hc)^{3/2}(\pi/\tilde{A}\tilde{B}\tilde{C})^{1/2}$	$T\gg\theta^{\mathrm{R}}$, nonlinear rotor, $\theta^{\mathrm{R}}=hc\tilde{B}/k$	15B.14
Vibration	$q^{\mathrm{V}}=1/(1-\mathrm{e}^{-\beta hc\tilde{\nu}})$	Harmonic approximation, $\theta^{\mathrm{V}}=hc\tilde{\nu}/k$	15B.15

15C Molecular energies

Contents

➤ Why do you need to know this material?

The partition function contains thermodynamic information, but it needs to be extracted. Here we show how to extract one particular property: the average energy of molecules, which plays a central role in thermodynamics.

➤ What is the key idea?

The average energy of a molecule in a collection of independent molecules can be calculated from the molecular partition function alone.

➤ What do you need to know already?

You need know how to calculate the molecular partition function from calculated or spectroscopic data (Topic 15B) and its significance as a measure of the number of accessible states. The Topic also draws on expressions for the rotational and vibrational energies of molecules (Topics 12B and 12D).

This Topic sets up the basic equations that show how to use the molecular partition function to calculate the mean energy of a collection of independent molecules. In Topic 15E we see how those mean energies are used to calculate thermodynamic properties. The equations for collections of interacting molecules are very similar (Topic 15D), but much more difficult to implement.

15C.1 The basic equations

We begin by considering a collection of N molecules that do not interact with one another. Any member of the collection can exist in a state i of energy ε_i measured from the lowest energy state of the molecule. The mean energy of a molecule, $\langle\varepsilon\rangle$, relative to its energy in its ground state is the total energy of the collection, E, divided by the total number of molecules:

$$\langle\varepsilon\rangle=\frac{E}{N}=\frac{1}{N}\sum_i N_i\varepsilon_i \tag{15C.1}$$

In Topic 15A it is shown that the overwhelmingly most probable population of a state in a collection at a temperature T is given by the Boltzmann distribution, eqn 15A.6a ($N_i/N=(1/q)\mathrm{e}^{-\beta\varepsilon_i}$), so we can write

$$\langle\varepsilon\rangle=\frac{1}{q}\sum_i \varepsilon_i \mathrm{e}^{-\beta\varepsilon_i} \tag{15C.2}$$

with $\beta=1/kT$. To manipulate this expression into a form involving only q we note that

$$\varepsilon_i \mathrm{e}^{-\beta\varepsilon_i}=-\frac{\mathrm{d}}{\mathrm{d}\beta}\mathrm{e}^{-\beta\varepsilon_i}$$

It follows that

$$\langle\varepsilon\rangle=-\frac{1}{q}\sum_i \frac{\mathrm{d}}{\mathrm{d}\beta}\mathrm{e}^{-\beta\varepsilon_i}=-\frac{1}{q}\frac{\mathrm{d}}{\mathrm{d}\beta}\sum_i \mathrm{e}^{-\beta\varepsilon_i}=-\frac{1}{q}\frac{\mathrm{d}q}{\mathrm{d}\beta} \tag{15C.3}$$

Several points in relation to eqn 15C.3 need to be made. Because $\varepsilon_0=0$, (we measure all energies from the lowest available level), $\langle\varepsilon\rangle$ should be interpreted as the value of the mean energy relative to its ground-state energy. If the lowest energy of the molecule is in fact $\varepsilon_{\mathrm{gs}}$ rather than 0, then the true mean energy is $\varepsilon_{\mathrm{gs}}+\langle\varepsilon\rangle$. For instance, for an harmonic oscillator, we would set $\varepsilon_{\mathrm{gs}}$ equal to the zero-point energy, $\tfrac{1}{2}hc\tilde{\nu}$. Secondly, because the partition function may depend on variables other than the temperature (for example, the volume), the derivative with respect to β in eqn 15C.3 is actually a *partial* derivative with these other variables held constant. The complete expression relating the molecular partition function to the mean energy of a molecule is therefore

$$\langle\varepsilon\rangle=\varepsilon_{gs}-\frac{1}{q}\left(\frac{\partial q}{\partial\beta}\right)_V \quad \text{Mean molecular energy} \quad (15C.4a)$$

An equivalent form is obtained by noting that $dx/x=d\ln x$:

$$\langle\varepsilon\rangle=\varepsilon_{gs}-\left(\frac{\partial\ln q}{\partial\beta}\right)_V \quad \text{Mean molecular energy} \quad (15C.4b)$$

These two equations confirm that we need know only the partition function (as a function of temperature) to calculate the mean energy.

Brief illustration 15C.1 Mean energy of a two-level system

If a molecule has only two available energy levels, one at 0 and the other at an energy ε, its partition function is

$$q=1+e^{-\beta\varepsilon}$$

Therefore, the mean energy of a collection of these molecules at a temperature T is

$$\langle\varepsilon\rangle=-\frac{1}{1+e^{-\beta\varepsilon}}\frac{d(1+e^{-\beta\varepsilon})}{d\beta}=\frac{\varepsilon e^{-\beta\varepsilon}}{1+e^{-\beta\varepsilon}}=\frac{\varepsilon}{e^{\beta\varepsilon}+1}$$

This function is plotted in Fig. 15C.1. Notice how the mean energy is zero at $T=0$, when only the lower state (at the zero of energy) is occupied, and rises to $\frac{1}{2}\varepsilon$ as $T\to\infty$, when the two levels become equally populated.

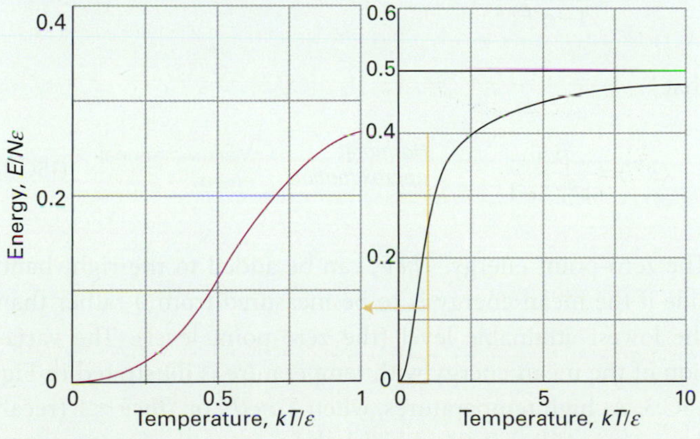

Figure 15C.1 The total energy of a two-level system (expressed as a multiple of $N\varepsilon$) as a function of temperature, on two temperature scales. The graph on the left shows the slow rise away from zero energy at low temperatures; the slope of the graph at $T=0$ is 0. The graph on the right shows the slow rise to 0.5 as $T\to\infty$ as both states become equally populated.

Self-test 15C.1 Deduce an expression for the mean energy when each molecule can exist in states with energies 0, ε, and 2ε.

Answer: $\langle\varepsilon\rangle=\varepsilon(1+2x)x/(1+x+x^2)$, $x=e^{-\beta\varepsilon}$

15C.2 Contributions of the fundamental modes of motion

In the remainder of this Topic we establish expressions for three fundamental types of motion, translation (T), rotation (R), and vibration (V), and then see how to incorporate the electronic states of molecules (E) and the spin of electrons or nuclei (S).

(a) The translational contribution

For a one-dimensional container of length X, for which $q^T=X/\Lambda$ with $\Lambda=h(\beta/2\pi m)^{1/2}$ (Topic 15B), we note that Λ is a constant multiplied by $\beta^{1/2}$, and obtain

$$\langle\varepsilon^T\rangle=-\frac{1}{q^T}\left(\frac{\partial q^T}{\partial\beta}\right)_V=-\frac{\Lambda}{X}\left(\frac{\partial}{\partial\beta}\frac{X}{\Lambda}\right)_V$$
$$=-\frac{\text{constant}\times\beta^{1/2}}{X}\times X\times\frac{d}{d\beta}\left(\frac{1}{\text{constant}\times\beta^{1/2}}\right)$$
$$=-\beta^{1/2}\overbrace{\frac{d}{d\beta}\frac{1}{\beta^{1/2}}}^{-\frac{1}{2}\beta^{-\frac{3}{2}}}=\frac{1}{2\beta}$$

That is,

$$\langle\varepsilon^T\rangle=\tfrac{1}{2}kT \quad \textit{One dimension} \quad \text{Mean translational energy} \quad (15C.5a)$$

For a molecule free to move in three dimensions, the analogous calculation leads to

$$\langle\varepsilon^T\rangle=\tfrac{3}{2}kT \quad \textit{Three dimensions} \quad \text{Mean translational energy} \quad (15C.5b)$$

(b) The rotational contribution

The mean rotational energy of a linear molecule is obtained from the rotational partition function (eqn 15B.11):

$$q^R=\sum_J(2J+1)e^{-\beta hc\tilde{B}J(J+1)}$$

When the temperature is low (in the sense $T<\theta^R=hc\tilde{B}/k$) the series must be summed term by term, which for a heteronuclear diatomic molecule or other non-symmetrical linear molecule gives

$$q^{\mathrm{R}}=1+3\mathrm{e}^{-2\beta hc\tilde{B}}+5\mathrm{e}^{-6\beta hc\tilde{B}}+\cdots$$

Hence, because

$$\frac{\mathrm{d}q^{\mathrm{R}}}{\mathrm{d}\beta}=-hc\tilde{B}(6\mathrm{e}^{-2\beta hc\tilde{B}}+30\mathrm{e}^{-6\beta hc\tilde{B}}+\cdots)$$

(q^{R} is independent of V, so the partial derivative has been replaced by a complete derivative) we find

$$\langle\varepsilon^{\mathrm{R}}\rangle=-\frac{1}{q^{\mathrm{R}}}\frac{\mathrm{d}q^{\mathrm{R}}}{\mathrm{d}\beta}=\frac{hc\tilde{B}\left(6\mathrm{e}^{-2\beta hc\tilde{B}}+30\mathrm{e}^{-6\beta hc\tilde{B}}+\cdots\right)}{1+3\mathrm{e}^{-2\beta hc\tilde{B}}+5\mathrm{e}^{-6\beta hc\tilde{B}}+\cdots}$$

Unsymmetrical linear molecule | Mean rotational energy | (15C.6a)

This ungainly function is plotted in Fig. 15C.2. At high temperatures ($T\gg\theta^{\mathrm{R}}$), q^{R} is given by eqn 15B.13b ($q^{\mathrm{R}}=T/\sigma\theta^{\mathrm{R}}$) in the form $q^{\mathrm{R}}=1/\sigma\beta hc\tilde{B}$, where $\sigma=1$ for a heteronuclear diatomic molecule. It then follows that

$$\langle\varepsilon^{\mathrm{R}}\rangle=-\frac{1}{q^{\mathrm{R}}}\frac{\mathrm{d}q^{\mathrm{R}}}{\mathrm{d}\beta}=-\sigma\beta hc\tilde{B}\frac{\mathrm{d}}{\mathrm{d}\beta}\left(\frac{1}{\sigma\beta hc\tilde{B}}\right)=-\beta\overbrace{\frac{\mathrm{d}}{\mathrm{d}\beta}\frac{1}{\beta}}^{-1/\beta^2}$$

and therefore that

$$\langle\varepsilon^{\mathrm{R}}\rangle=\frac{1}{\beta}=kT$$

Linear molecule, high temperature ($T\gg\theta^{\mathrm{R}}$) | Mean rotational energy | (15C.6b)

The high-temperature result, which is valid when many rotational states are occupied, is also in agreement with the equipartition theorem, because the classical expression for the energy of a linear rotor is $E_{\mathrm{k}}=\frac{1}{2}I_{\perp}\omega_a^2+\frac{1}{2}I_{\perp}\omega_b^2$ and therefore has two quadratic contributions. (There is no rotation around the line of atoms.) It follows from the equipartition theorem (*Foundations* B) that the mean rotational energy is $2\times\frac{1}{2}kT=kT$.

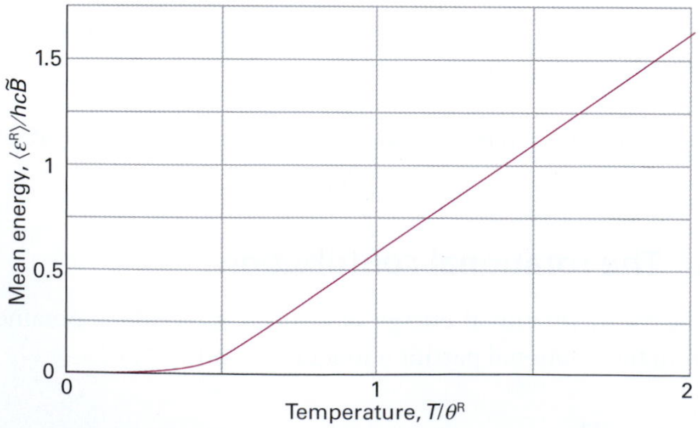

Figure 15C.2 The mean rotational energy of a non-symmetrical linear rotor as a function of temperature. At high temperatures ($T\gg\theta^{\mathrm{R}}$), the energy is linearly proportional to the temperature, in accord with the equipartition theorem.

Brief illustration 15C.2 Mean rotational energy

To estimate the mean energy of a nonlinear molecule we recognize that its rotational kinetic energy (the only contribution to its rotational energy) is $E_{\mathrm{k}}=\frac{1}{2}I_a\omega_a^2+\frac{1}{2}I_b\omega_b^2+\frac{1}{2}I_c\omega_c^2$. As there are three quadratic contributions, its mean rotational energy is $\frac{3}{2}kT$. The molar contribution is $\frac{3}{2}RT$. At 25 °C, this contribution is 3.7 kJ mol^{-1}, the same as the translational contribution, for a total of 7.4 kJ mol^{-1}. A monatomic gas has no rotational contribution.

Self-test 15C.2 How much energy does it take to raise the temperature of 1.0 mol H_2O(g) from 100 °C to 200 °C? Consider only translational and rotational contributions to the heat capacity.

Answer: 2.5 Kj

(c) The vibrational contribution

The vibrational partition function in the harmonic approximation is given in eqn 15B.15 ($q^{\mathrm{V}}=1/(1-\mathrm{e}^{-\beta hc\tilde{\nu}})$). Because q^{V} is independent of the volume, it follows that

$$\frac{\mathrm{d}q^{\mathrm{V}}}{\mathrm{d}\beta}=\frac{\mathrm{d}}{\mathrm{d}\beta}\left(\frac{1}{1-\mathrm{e}^{-\beta hc\tilde{\nu}}}\right)=-\frac{hc\tilde{\nu}\mathrm{e}^{-\beta hc\tilde{\nu}}}{(1-\mathrm{e}^{-\beta hc\tilde{\nu}})^2} \qquad (15C.7)$$

and hence from

$$\langle\varepsilon^{\mathrm{V}}\rangle=-\frac{1}{q^{\mathrm{V}}}\frac{\mathrm{d}q^{\mathrm{V}}}{\mathrm{d}\beta}=\left(1-\mathrm{e}^{-\beta hc\tilde{\nu}}\right)\frac{hc\tilde{\nu}\mathrm{e}^{-\beta hc\tilde{\nu}}}{\left(1-\mathrm{e}^{-\beta hc\tilde{\nu}}\right)^2}=\frac{hc\tilde{\nu}\mathrm{e}^{-\beta hc\tilde{\nu}}}{1-\mathrm{e}^{-\beta hc\tilde{\nu}}}$$

that

$$\langle\varepsilon^{\mathrm{V}}\rangle=\frac{hc\tilde{\nu}}{\mathrm{e}^{\beta hc\tilde{\nu}}-1}$$

Harmonic approximation | Mean vibrational energy | (15C.8)

The zero-point energy, $\frac{1}{2}hc\tilde{\nu}$, can be added to the right-hand side if the mean energy is to be measured from 0 rather than the lowest attainable level (the zero-point level). The variation of the mean energy with temperature is illustrated in Fig. 15C.3. At high temperatures, when $T\gg\theta^{\mathrm{V}}$, or $\beta hc\tilde{\nu}\ll1$ (recall from Topic 15B that $\theta^{\mathrm{V}}=hc\tilde{\nu}/k$), the exponential functions can be expanded ($\mathrm{e}^x=1+x+\cdots$) and all but the leading terms discarded. This approximation leads to

$$\langle\varepsilon^{\mathrm{V}}\rangle=\frac{hc\tilde{\nu}}{(1+\beta hc\tilde{\nu}+\cdots)-1}\approx\frac{1}{\beta}=kT$$

High temperature approximation ($T\gg\theta^{\mathrm{V}}$) | Mean vibrational energy | (15C.9)

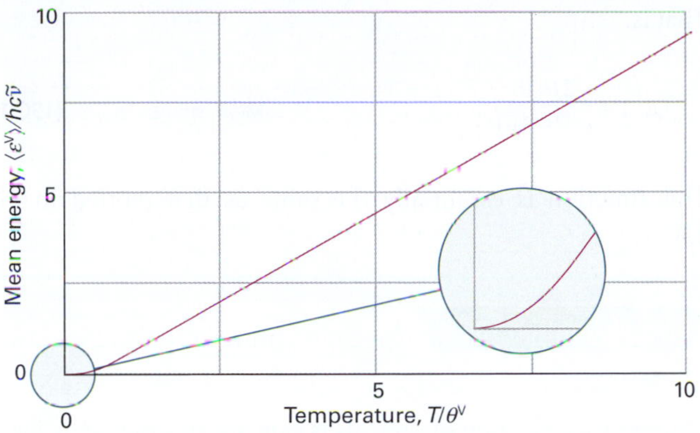

Figure 15C.3 The mean vibrational energy of a molecule in the harmonic approximation as a function of temperature. At high temperatures ($T \gg \theta^V$), the energy is linearly proportional to the temperature, in accord with the equipartition theorem.

This result is in agreement with the value predicted by the classical equipartition theorem, because the energy of a one-dimensional oscillator is $E=\frac{1}{2}mv_x^2+\frac{1}{2}k_f x^2$ and the mean energy of each quadratic term is $\frac{1}{2}kT$. Bear in mind, however, that the condition $T \gg \theta^V$ is rarely satisfied.

Brief illustration 15C.3 The mean vibrational energy

To calculate the mean vibrational energy of I_2 molecules at 298.15 K we note from Table 12D.1 that their vibrational wavenumber is 214.6 cm^{-1}. Then, because at 298.15 K, kT/hc=207.224 cm^{-1}, from eqn 15C.8 with

$$\beta\varepsilon=\frac{hc\tilde{\nu}}{kT}=\frac{214.6\,\mathrm{cm}^{-1}}{207.244\,\mathrm{cm}^{-1}}=1.036$$

it follows from eqn 15C.8 that

$$\langle\varepsilon^V\rangle/hc=\frac{214.6\,\mathrm{cm}^{-1}}{\mathrm{e}^{1.036}-1}=118.0\,\mathrm{cm}^{-1}$$

The addition of the zero-point energy (corresponding to $\frac{1}{2}\times214.6\,\mathrm{cm}^{-1}$) increases this value to 225.3 cm^{-1}. The equipartition result is 207.224 cm^{-1}, the discrepancy reflecting the fact that in this case it is not true that $T \gg \theta^V$ and only the ground and first excited states are significantly populated.

Self-test 15C.3 What must the temperature be before the energy estimated from the equipartition theorem is within 2 per cent of the energy given by eqn 15C.8?

Answer: 625 K; use a spreadsheet

When there are several normal modes that can be treated as harmonic, the overall vibrational partition function is the product of each individual partition function, and the total mean vibrational energy is the sum of the mean energy of each mode.

(d) The electronic contribution

We shall consider two types of electronic contribution: one arising from the electronically excited states of a molecule and one from the spin contribution.

In most cases of interest, the electronic states of atoms and molecules are so widely separated that only the electronic ground state is occupied. As we are adopting the convention that all energies are measured from the ground state of each mode, we can write

$$\langle\varepsilon^E\rangle=0 \qquad \text{Mean electronic energy} \qquad (15C.10)$$

In certain cases, there are thermally accessible states at the temperature of interest. In that case, the partition function and hence the mean electronic energy are best calculated by direct summation over the available states. Care must be taken to take any degeneracies into account, as we illustrate in the following *Example*.

Example 15C.1 Calculating the electronic contribution to the energy

A certain atom has a doubly degenerate electronic ground state and a fourfold degenerate excited state at 600 cm^{-1} above the ground state. What is its mean electronic energy at 25 °C expressed as a wavenumber?

Method Write down the expression for the partition function at a general temperature T (in terms of β) and then derive the mean energy by differentiating with respect to β. Finally, substitute the data. Use $\varepsilon=hc\tilde{\nu}$, $\langle\varepsilon^E\rangle=hc\langle\tilde{\nu}^E\rangle$, and (from inside the front cover), kT/hc=207.226 cm^{-1} at 25 °C.

Answer The partition function is $q^E=2+4\mathrm{e}^{-\beta\varepsilon}$. The mean energy is therefore

$$\langle\varepsilon^E\rangle=-\frac{1}{q^E}\frac{\mathrm{d}q^E}{\mathrm{d}\beta}=-\frac{1}{2+4\mathrm{e}^{-\beta\varepsilon}}\overbrace{\frac{\mathrm{d}}{\mathrm{d}\beta}(2+4\mathrm{e}^{-\beta\varepsilon})}^{-4\varepsilon\mathrm{e}^{-\beta\varepsilon}}$$

$$=\frac{4\varepsilon\mathrm{e}^{-\beta\varepsilon}}{2+4\mathrm{e}^{-\beta\varepsilon}}=\frac{\varepsilon}{\frac{1}{2}\mathrm{e}^{\beta\varepsilon}+1}$$

and expressed as a wavenumber

$$\langle\tilde{\nu}^E\rangle=\frac{\tilde{\nu}}{\frac{1}{2}\mathrm{e}^{hc\tilde{\nu}/kT}+1}$$

From the data,

$$\langle\tilde{\nu}^E\rangle=\frac{600\,\mathrm{cm}^{-1}}{\frac{1}{2}\mathrm{e}^{600/207.226}+1}=59.7\,\mathrm{cm}^{-1}$$

Self-test 15C.4 Repeat the problem for an atom that has a threefold degenerate ground state and a sevenfold degenerate excited state 400 cm^{-1} above.

Answer: 101 cm^{-1}

(e) The spin contribution

An electron spin in a magnetic field $\mathcal{B}$ has two possible energy states that depend on its orientation as expressed by the magnetic quantum number m_s and which are given by

$$E_{m_s} = 2\mu_B \mathcal{B} m_s \qquad \text{Electron spin energies} \qquad (15C.11)$$

where μ_B is the Bohr magneton (see inside the front cover). These energies are discussed in more detail in Topic 14A where we see that the integer 2 needs to be replaced by a number very close to 2. The lower state has $m_s = -\frac{1}{2}$, so the two energy levels available to the electron lie (according to our convention) at $\varepsilon_{-1/2} = 0$ and at $\varepsilon_{+1/2} = 2\mu_B\mathcal{B}$. The spin partition function is therefore

$$q^S = \sum_{m_s} e^{-\beta\varepsilon_{m_s}} = 1 + e^{-2\beta\mu_B\mathcal{B}} \qquad \text{Spin partition function} \qquad (15C.12)$$

The mean energy of the spin is therefore

$$\langle\varepsilon^S\rangle = -\frac{1}{q^S}\frac{dq^S}{d\beta} = -\frac{1}{1+e^{-2\beta\mu_B\mathcal{B}}}\overbrace{\frac{d}{d\beta}\left(1+e^{-2\beta\mu_B\mathcal{B}}\right)}^{-2\mu_B\mathcal{B}e^{-2\beta\mu_B\mathcal{B}}}$$
$$= \frac{2\mu_B\mathcal{B}e^{-2\beta\mu_B\mathcal{B}}}{1+e^{-2\beta\mu_B\mathcal{B}}}$$

That is,

$$\langle\varepsilon^S\rangle = \frac{2\mu_B\mathcal{B}}{e^{2\beta\mu_B\mathcal{B}}+1} \qquad \text{Mean spin energy} \qquad (15C.13)$$

This function is essentially the same as that plotted in Fig. 15C.1.

Brief illustration 15C.4 The spin contribution to the energy

Suppose a collection of radicals is exposed to a magnetic field of 2.5 T (T denotes tesla). With $\mu_B = 9.274\times10^{-24}\,\mathrm{J\,T^{-1}}$ and a temperature of 25 °C,

$$2\mu_B\mathcal{B} = 2\times(9.274\times10^{-24}\,\mathrm{J\,T^{-1}})\times2.5\,\mathrm{T} = 4.6\ldots\times10^{-23}\,\mathrm{J}$$
$$2\beta\mu_B\mathcal{B} = \frac{2\,(9.274\;10^{-24}\,\mathrm{J\,T^{-1}})\times(2.5\,\mathrm{T})}{(1.381\times10^{-23}\,\mathrm{J\,K^{-1}})\times(298\,\mathrm{K})} = 0.011\ldots$$

The mean energy is therefore

$$\langle\varepsilon^S\rangle = \frac{4.6\ldots\times10^{-23}\,\mathrm{J}}{e^{0.011\ldots}+1} = 2.3\;10^{-23}\,\mathrm{J}$$

This energy is equivalent to 14 J mol^{-1} (note joules, not kilojoules).

Self-test 15C.5 Repeat the calculation for a species with $S = 1$ in the same magnetic field.

Answer: 0.0046 zJ, 28 J mol^{-1}

Check list of concepts

☐ 1. The **mean molecular energy** can be calculated from the molecular partition function.

☐ 2. The molecular partition function is calculated from molecular structural parameters obtained from spectroscopy or computation.

Checklist of equations

Property	Equation	Comment	Equation number
Mean molecular energy	$\langle\varepsilon\rangle = \varepsilon_{gs} - (1/q)(\partial q/\partial\beta)_V$		15C.4a
	$\langle\varepsilon\rangle = \varepsilon_{gs} - (\partial \ln q/\partial\beta)_V$	Alternative version	15C.4b
Translation	$\langle\varepsilon^T\rangle = \frac{d}{2}kT$	In d dimensions, $d = 1, 3$	15C.5

Property	Equation	Comment	Equation number
Rotation	$\langle \varepsilon^{R} \rangle = kT$	Linear molecule, $T \gg \theta^{R}$	15C.6b
Vibration	$\langle \varepsilon^{V} \rangle = hc\tilde{\nu}/(e^{\beta hc\tilde{\nu}} - 1)$	Harmonic approximation	15C.8
	$\langle \varepsilon^{V} \rangle = kT$	$T \gg \theta^{V}$	15C.9
Spin	$\langle \varepsilon^{S} \rangle = 2\mu_B \mathcal{B}/(e^{2\beta\mu_B \mathcal{B}} + 1)$	$s = \frac{1}{2}$	15C.13

15D The canonical ensemble

Contents

➤ **Why do you need to know this material?**

Whereas Topics 15B and 15C deal with independent molecules, in practice molecules do interact. Therefore, this material is essential for constructing models of real gases, liquids, and solids and of any system in which intermolecular interactions cannot be neglected.

➤ **What is the main idea?**

A system composed of interacting molecules is described in terms of a canonical partition function, from which its thermodynamic properties may be deduced.

➤ **What do you need to know already?**

This material draws on the calculations in Topic 15A: the calculations here are analogous to those, and are not repeated in detail. This Topic also draws on the calculation of energies from partition functions (Topic 15C); here too the calculations are analogous to those presented there.

Here we consider the formalism appropriate to systems in which the molecules interact with one another, as in real gases and liquids. The crucial concept we need when treating systems of interacting particles is the 'ensemble'. Like so many scientific terms, the term has basically its normal meaning of 'collection', but it has been sharpened and refined into a precise significance.

15D.1 The concept of ensemble

To set up an ensemble, we take a closed system of specified volume, composition, and temperature, and think of it as replicated $\tilde{N}$ times (Fig. 15D.1). All the identical closed systems are regarded as being in thermal contact with one another, so they can exchange energy. The total energy of all the systems is $\tilde{E}$ and, because they are in thermal equilibrium with one another, they all have the same temperature, T. The volume of each member of the ensemble is the same, so the energy levels available to the molecules are the same in each system, and each member contains the same number of molecules, so there is a fixed number of molecules to distribute within each system. This imaginary collection of replications of the actual system with a common temperature is called the **canonical ensemble**. The word 'canon' means 'according to a rule'.

There are two other important ensembles. In the **microcanonical ensemble** the condition of constant temperature is replaced by the requirement that all the systems should have exactly the same energy: each system is individually isolated. In the **grand canonical ensemble** the volume and temperature

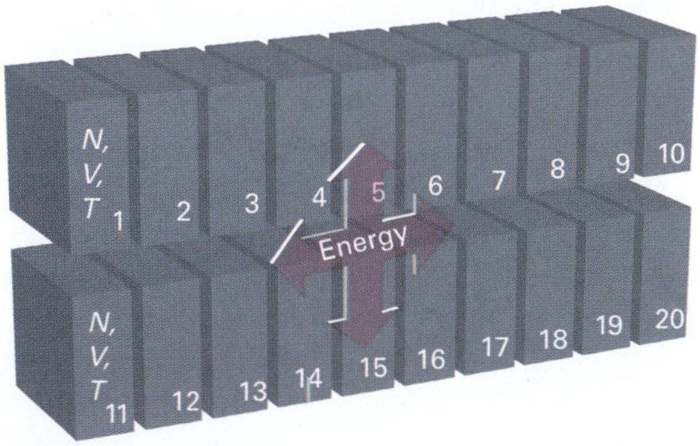

Figure 15D.1 A representation of the canonical ensemble, in this case for $\tilde{N}=20$. The individual replications of the actual system all have the same composition and volume. They are all in mutual thermal contact, and so all have the same temperature. Energy may be transferred between them as heat, and so they do not all have the same energy. The total $\tilde{E}$ of all 20 replications is a constant because the ensemble is isolated overall.

of each system is the same, but they are open, which means that matter can be imagined as able to pass between the systems; the composition of each one may fluctuate, but now the property known as the chemical potential (Topic 5A) is the same in each system. In summary:

Ensemble	Common properties
Microcanonical	V, E, N
Canonical	V, T, N
Grand canonical	V, T, μ

The microcanonical ensemble is the basis of the discussion in Topic 15A; we shall not consider the grand canonical ensemble explicitly.

The important point about an ensemble is that it is a collection of *imaginary* replications of the system, so we are free to let the number of members be as large as we like; when appropriate, we can let $\tilde{N}$ become infinite. The number of members of the ensemble in a state with energy E_i is denoted $\tilde{N}_i$, and we can speak of the configuration of the ensemble (by analogy with the configuration of the system used in Topic 15A) and its weight, $\tilde{\mathcal{W}}$. Note that $\tilde{N}$ is unrelated to N, the number of molecules in the actual system; $\tilde{N}$ is the number of *imaginary* replications of that system.

(a) Dominating configurations

Just as in Topic 15A, some of the configurations of the canonical ensemble will be very much more probable than others. For instance, it is very unlikely that the whole of the total energy, $\tilde{E}$, will accumulate in one system. By analogy with the discussion in Topic 15A, we can anticipate that there will be a dominating configuration, and that we can evaluate the thermodynamic properties by taking the average over the ensemble using that single, most probable, configuration. In the **thermodynamic limit** of $\tilde{N}\rightarrow\infty$, this dominating configuration is overwhelmingly the most probable, and it dominates the properties of the system virtually completely.

The quantitative discussion follows the argument in Topic 15A with the modification that N and N_i are replaced by $\tilde{N}$ and $\tilde{N}_i$. The weight $\tilde{\mathcal{W}}$ of a configuration $\{\tilde{N}_0, \tilde{N}_1, \ldots\}$ is

$$\tilde{\mathcal{W}}=\frac{\tilde{N}!}{\tilde{N}_1!\tilde{N}_2!\cdots} \qquad \text{Weight} \qquad (15\text{D}.1)$$

The configuration of greatest weight, subject to the constraints that the total energy of the ensemble is constant at $\tilde{E}$ and that the total number of members is fixed at $\tilde{N}$, is given by the **canonical distribution**:

$$\frac{\tilde{N}_i}{\tilde{N}}=\frac{e^{-\beta E_i}}{Q} \qquad \text{Canonical distribution} \qquad (15\text{D}.2\text{a})$$

in which

$$Q=\sum_i e^{-\beta E_i} \qquad \textit{Definition} \quad \text{Canonical partition function} \qquad (15\text{D}.2\text{b})$$

where the sum is over all members of the ensemble, each one having an energy E_i. The quantity Q, which is a function of the temperature, is called the **canonical partition function**. Like the molecular partition function, the canonical partition function contains all the thermodynamic information about a system, but in this case, allowing for the possibility of interactions between the constituent molecules.

Brief illustration 15D.1 The canonical distribution

Suppose that we are considering a sample of a monatomic real gas that contains 1.00 mol atoms, then at 298 K its total energy is close to $\frac{3}{2}nRT=\frac{3}{2}(1.00\,\text{mol})\times(8.3145\,\text{J K}^{-1}\,\text{mol}^{-1})\times(298\,\text{K})=3.72\,\text{kJ}$. Suppose that for an instant the molecules are present at separations where the total energy is 3.72 kJ and an instant later are present at separations where the total energy is lower than 3.72 kJ by 0.00 000 001 per cent (that is, by 3.72×10^{-7} J). To predict the ratio of numbers of members of the ensemble with these two energies we use eqn 15D.2a in the form

$$\frac{\tilde{N}(\text{lower})}{\tilde{N}(\text{higher})}=e^{-(-3.70\times10^{-7}\,\text{J})/(1.381\times10^{-23}\,\text{J K}^{-1})\times(298\,\text{K})}$$
$$=e^{3.33\times10^{7}}$$

At first sight, the number of members that have the lower energy vastly outweighs the number with the higher energy. That that is not necessarily the case is explained below.

Self-test 15D.1 Repeat the calculation for members of the same ensemble with energies that differ by 1.0×10^{-20} per cent.

Answer: $\tilde{N}(\text{lower})/\tilde{N}(\text{higher})=e^{90}\approx1\times10^{39}$

(b) Fluctuations from the most probable distribution

The canonical distribution in eqn 15D.2a is only apparently an exponentially decreasing function of the energy of the system. We must appreciate that the equation gives the probability of occurrence of members in a single state i of the entire system of energy E_i. There may in fact be numerous states with almost identical energies. For example, in a gas the identities of the molecules moving slowly or quickly can change without necessarily affecting the total energy. The **density of states**, the number of states in an energy range divided by the width of the range (Fig. 15D.2), is a very sharply increasing function of energy. It follows that the probability of a member of an ensemble having a specified energy (as distinct from being in a specified state) is given by

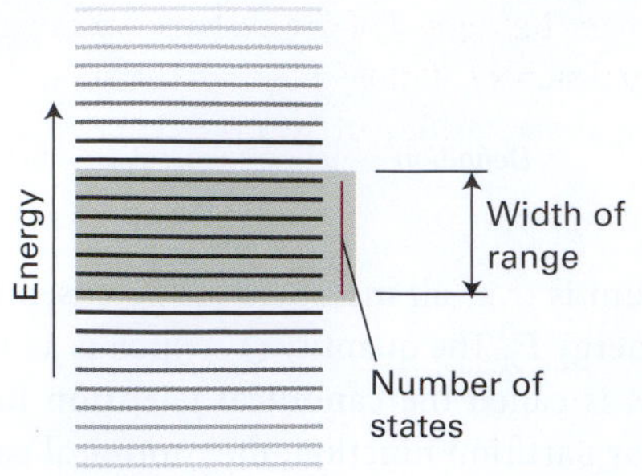

Figure 15D.2 The energy density of states is the number of states in an energy range divided by the width of the range.

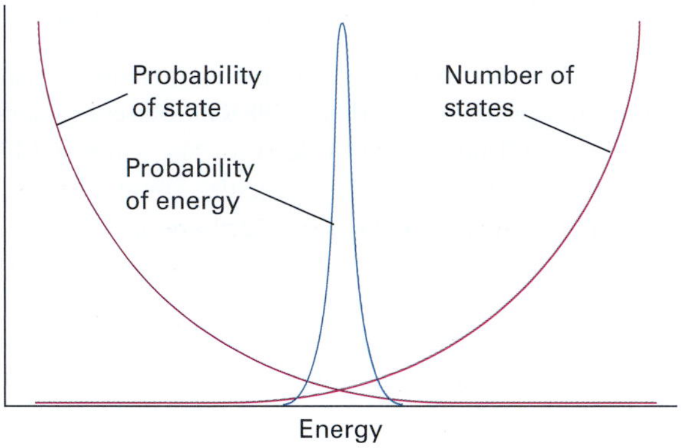

Figure 15D.3 To construct the form of the distribution of members of the canonical ensemble in terms of their energies, we multiply the probability that any one is in a state of given energy, eqn 15D.2a, by the number of states corresponding to that energy (a steeply rising function). The product is a sharply peaked function at the mean energy, which shows that almost all the members of the ensemble have that energy.

eqn 15D.2a, a sharply decreasing function, multiplied by a sharply increasing function (Fig. 15D.3). Therefore, the overall distribution is a sharply peaked function. We conclude that most members of the ensemble have an energy very close to the mean value.

Brief illustration 15D.2 The role of the density of states

A function that increases rapidly is x^N, with N a large value. A function that decreases rapidly is e^{-Nx}, once again, with N a large value. The product of these two functions, normalized so that the maxima for different values of N all coincide,

$$f(x)=e^N x^N e^{-Nx}$$

is plotted for three values of N in Fig. 15D.4. We see that the width of the product does indeed decrease as N increases.

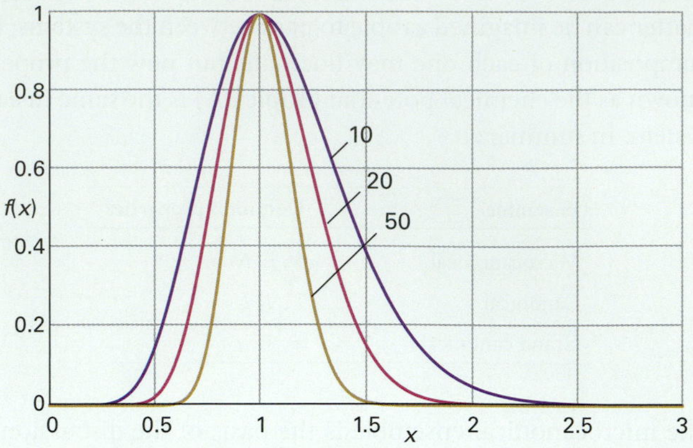

Figure 15D.4 The product of the two functions discussed in *Brief illustration* 15D.2, for three different values of N.

Self-test 15D.2 Show that the product of the functions x^{2N} and e^{-Nx}, suitably normalized, behaves similarly.

Answer: Plot $f(x)=(1/2)^{2N}e^{2N}x^{2N}e^{-Nx}$ for $0\le x\le 4$

15D.2 The mean energy of a system

Just as the molecular partition function can be used to calculate the mean value of a molecular property, so the canonical partition function can be used to calculate the mean energy of an entire system composed of molecules (that might or might not be interacting with one another). Thus, Q is more general than q because it does not assume that the molecules are independent. We can therefore use Q to discuss the properties of condensed phases and real gases where molecular interactions are important.

If the total energy of the ensemble is $\tilde{E}$, and there are $\tilde{N}$ members, the average energy of a member is $\langle E\rangle=\tilde{E}/\tilde{N}$. Because the fraction, $\tilde{p}_i$, of members of the ensemble in a state i with energy E_i is given by the analogue of eqn 15A.6 ($p_i=e^{-\beta\varepsilon_i}/q$ with $p_i=N_i/N$) as

$$\tilde{p}_i=\frac{e^{-\beta E_i}}{Q} \qquad (15D.3)$$

it follows that

$$\langle E\rangle=\sum_i \tilde{p}_i E_i=\frac{1}{Q}\sum_i E_i e^{-\beta E_i} \qquad (15D.4)$$

By the same argument that led to eqn 15C.4 ($\langle\varepsilon\rangle=-(1/q)(\partial q/\partial\beta)_V$, when $\varepsilon_{gs}=0$),

$$E=-\frac{1}{Q}\left(\frac{\partial Q}{\partial\beta}\right)_V=-\left(\frac{\partial \ln Q}{\partial\beta}\right)_V \qquad \text{Mean energy of a system} \qquad (15D.5)$$

As in the case of the mean molecular energy, we must add to this expression the ground-state energy of the entire system if that is not zero.

Brief illustration 15D.3 The expression for the energy

If the canonical partition function is a product of the molecular partition function of each molecule (which we see below is the case when the N molecules of the system are independent), then we can write $Q=q^N$, and infer that the energy of the system is

$$\langle E\rangle=-\frac{1}{q^N}\left(\frac{\partial q^N}{\partial\beta}\right)_V=-\frac{Nq^{N-1}}{q^N}\left(\frac{\partial q}{\partial\beta}\right)_V=-\frac{N}{q}\left(\frac{\partial q}{\partial\beta}\right)_V=N\langle\varepsilon\rangle$$

That is, the mean energy of the system is N times the mean energy of a single molecule.

Self-test 15D.3 Confirm that the same expression is obtained if $Q=q^N/N!$, which is another case described below.

15D.3 Independent molecules revisited

We shall now see how to recover the molecular partition function from the more general canonical partition function when the molecules are independent. We show in the following *Justification* that, when the molecules are independent and distinguishable (in the sense to be described), the relation between Q and q is

$$Q=q^N \tag{15D.6}$$

Justification 15D.1 The relation between Q and q

The total energy of a collection of N independent molecules is the sum of the energies of the molecules. Therefore, we can write the total energy of a state i of the system as

$$E_i=\varepsilon_i(1)+\varepsilon_i(2)+\cdots+\varepsilon_i(N)$$

In this expression, $\varepsilon_i(1)$ is the energy of molecule 1 when the system is in the state i, $\varepsilon_i(2)$ the energy of molecule 2 when the system is in the same state i, and so on. The canonical partition function is then

$$Q=\sum_i \mathrm{e}^{-\beta\varepsilon_i(1)-\beta\varepsilon_i(2)-\cdots-\beta\varepsilon_i(N)}$$

The sum over the states of the system can be reproduced by letting each molecule enter all its own individual states (although we meet an important proviso shortly). Therefore, instead of summing over the states i of the system, we can sum over all the individual states j of molecule 1, all the states j of molecule 2, and so on. This rewriting of the original expression leads to

$$Q=\left(\sum_j \mathrm{e}^{-\beta\varepsilon_j}\right)\left(\sum_j \mathrm{e}^{-\beta\varepsilon_j}\right)\cdots\left(\sum_j \mathrm{e}^{-\beta\varepsilon_j}\right)=q^N$$

If all the molecules are identical and free to move through space, we cannot distinguish them and the relation $Q=q^N$ is not valid. Suppose that molecule 1 is in some state a, molecule 2 is in b, and molecule 3 is in c, then one member of the ensemble has an energy $E=\varepsilon_a+\varepsilon_b+\varepsilon_c$. This member, however, is indistinguishable from one formed by putting molecule 1 in state b, molecule 2 in state c, and molecule 3 in state a, or some other permutation. There are six such permutations in all, and $N!$ in general. In the case of indistinguishable molecules, it follows that we have counted too many states in going from the sum over system states to the sum over molecular states, so writing $Q=q^N$ overestimates the value of Q. The detailed argument is quite involved, but at all except very low temperatures it turns out that the correction factor is $1/N!$. Therefore:

$$\text{For distinguishable independent molecules: } Q=q^N \tag{15D.7a}$$

$$\text{For indistinguishable independent molecules: } Q=q^N/N! \tag{15D.7b}$$

Brief illustration 15D.4 Indistinguishability

For molecules to be indistinguishable, they must be of the same kind: an Ar atom is never indistinguishable from a Ne atom. Their identity, however, is not the only criterion. Each identical molecule in a crystal lattice, for instance, can be 'named' with a set of coordinates. Identical molecules in a lattice can therefore be treated as distinguishable because their sites are distinguishable, and we use eqn 15D.7a. On the other hand, identical molecules in a gas are free to move to different locations, and there is no way of keeping track of the identity of a given molecule; we therefore use eqn 15D.7b.

Self-test 15D.4 Are identical molecules in a liquid indistinguishable?

Answer: Yes

15D.4 The variation of energy with volume

When there are interactions between molecules, the energy of a collection depends on the average distance between them, and therefore on the volume that a fixed number occupy. This dependence on volume is particularly important for the discussion of real gases (Topic 1C).

We need to evaluate $(\partial\langle E\rangle/\partial V)_T$, the variation in energy of a system with volume at constant temperature. (In Topics 2D and 3D, this quantity is identified with the 'internal pressure' of a gas and denoted π_T.) To proceed, we substitute eqn 15D.5 and obtain

$$\left(\frac{\partial E}{\partial V}\right)_T = -\left(\frac{\partial}{\partial V}\left(\frac{\partial \ln Q}{\partial \beta}\right)_V\right)_T \quad (15D.8)$$

We need to consider the translational contribution to Q since translational energy levels depend on volume, but to develop eqn 15D.8, we also need to find a way to build an intermolecular potential energy into the expression for Q.

The total kinetic energy of a gas is the sum of the kinetic energies of the individual molecules. Therefore, even in a real gas the canonical partition function factorizes into a part arising from the kinetic energy, which for the perfect gas is $Q=V^N/\Lambda^{3N}N!$, where Λ is the thermal wavelength, eqn 15B.7b ($\Lambda=h/(2\pi mkT)^{1/2}$), and a factor called the **configuration integral**, $\mathcal{Z}$, which depends on the intermolecular potentials (don't confuse this $\mathcal{Z}$ with the compression factor Z in Topic 1C). We therefore write

$$Q=\frac{\mathcal{Z}}{\Lambda^{3N}} \quad (15D.9)$$

with $\mathcal{Z}$ replacing $V^N/N!$ and expect $\mathcal{Z}$ to equal $V^N/N!$ for a perfect gas (see the next *Brief illustration*). It then follows that

$$\begin{aligned}\left(\frac{\partial E}{\partial V}\right)_T &= -\left(\frac{\partial}{\partial V}\left(\frac{\partial \ln(\mathcal{Z}/\Lambda^{3N})}{\partial \beta}\right)_V\right)_T \\ &= -\left(\frac{\partial}{\partial V}\left(\frac{\partial \ln \mathcal{Z}}{\partial \beta}\right)_V\right)_T - \left(\frac{\partial}{\partial V}\left(\frac{\partial \ln(1/\Lambda^{3N})}{\partial \beta}\right)_V\right)_T \\ &= -\left(\frac{\partial}{\partial V}\left(\frac{\partial \ln \mathcal{Z}}{\partial \beta}\right)_V\right)_T - \left(\frac{\partial}{\partial \beta}\overbrace{\left(\frac{\partial \ln(1/\Lambda^{3N})}{\partial V}\right)_T}^{0}\right)_V \\ &= -\left(\frac{\partial}{\partial V}\left(\frac{\partial \ln \mathcal{Z}}{\partial \beta}\right)_V\right)_T = -\left(\frac{\partial}{\partial V}\frac{1}{\mathcal{Z}}\left(\frac{\partial \mathcal{Z}}{\partial \beta}\right)_V\right)_T\end{aligned} \quad (15D.10)$$

In the third line, to obtain and evaluate the blue term we have used the relation $(\partial^2 f/\partial x\partial y)=(\partial^2 f/\partial y\partial x)$ and then noted that Λ is independent of volume, so its derivative with respect to volume is zero.

For a real gas of atoms (for which the intermolecular interactions are isotropic), $\mathcal{Z}$ is related to the total potential energy E_P of interaction of all the particles, which depends on all their relative locations, by

$$\mathcal{Z}=\frac{1}{N!}\int e^{-\beta E_P}\,d\tau_1 d\tau_2\cdots d\tau_N \quad \text{Configuration integral} \quad (15D.11)$$

where $d\tau_i$ is the volume element for atom i. The physical origin of this term is that the probability of occurrence of each arrangement of molecules possible in the sample is given by a Boltzmann distribution in which the exponent is given by the potential energy corresponding to that arrangement.

Brief illustration 15D.5 A configuration integral

Equation 15D.11 is very difficult to manipulate in practice, even for quite simple intermolecular potentials, except for a perfect gas for which $E_P=0$. In that case, the exponential function becomes 1 and

$$\mathcal{Z}=\frac{1}{N!}\int d\tau_1 d\tau_2\cdots d\tau_N=\frac{1}{N!}\left(\int d\tau\right)^N=\frac{V^N}{N!}$$

just as it should be for a perfect gas.

Self-test 15D.5 Go on to show that for a perfect gas, $(\partial\langle E\rangle/\partial V)_T=0$.

Answer: $\mathcal{Z}$ in this case is independent of temperature

If the potential has the form of a central hard sphere surrounded by a shallow attractive well (Fig. 15D.5), then detailed calculation, which is too involved to reproduce here, leads to

$$\left(\frac{\partial\langle E\rangle}{\partial V}\right)_T=\frac{an^2}{V^2} \quad \text{Attractive potential} \quad (15D.12)$$

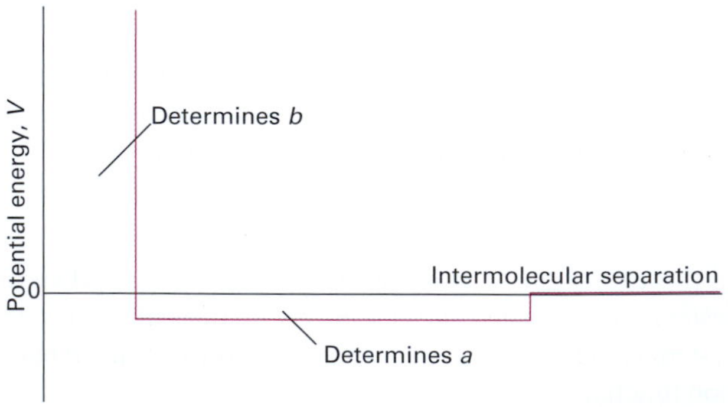

Figure 15D.5 The intermolecular potential energy of molecules in a real gas can be modelled by a central hard sphere of range b surrounded by a shallow attractive well with an area proportional to a. As discussed in the text, calculations based on this model yield results that are consistent with the van der Waals equation of state (Topic 1C).

where n is the amount of molecules present in the volume V and a is a constant that is proportional to the area under the attractive part of the potential. In *Example* 3D.2 of Topic 3D we derive exactly the same expression (in the form $\pi_T = an^2/V^2$) from the van der Waals equation of state. At this point we can conclude that if there are attractive interactions between molecules in a gas, then its energy increases as it expands isothermally (because $(\partial\langle E\rangle/\partial V)_T > 0$, and the slope of $\langle E\rangle$ with respect to V is positive). The energy rises because, at greater average separations, the molecules spend less time in regions where they interact favourably.

Checklist of concepts

- ☐ 1. The **canonical ensemble** is an imaginary collection of replications of the actual system with a common temperature.
- ☐ 2. The **canonical distribution** gives the number of members of the ensemble with a specified total energy.
- ☐ 3. The mean energy of the members of the ensemble can be calculated from the **canonical partition function**.

Checklist of equations

Property	Equation	Comment	Equation number
Canonical distribution	$\tilde{N}_i/\tilde{N} = e^{-\beta E_i}/Q$		15D.2a
Canonical partition function	$Q = \sum_i e^{-\beta E_i}$	Definition	15D.2b
Mean energy	$\langle E\rangle = -(1/Q)(\partial Q/\partial\beta)_V = -(\partial Q/\partial\beta)_V$		15D.5
Configuration integral	$Q = Z/\Lambda^{3N}$ $Z = \frac{1}{N!}\int e^{-\beta E_P}\,d\tau_1 d\tau_2 \ldots d\tau_N$	Isotropic interaction	15D.11
Variation of mean energy with volume	$(\partial\langle E\rangle/\partial V)_T = an^2/V^2$	van der Waals gas	15D.12

15E The internal energy and the entropy

Contents

➤ Why do you need to know this material?

The importance of the molecular partition function is that it contains all the information needed to calculate the thermodynamic properties of a system of independent particles. In this respect, it plays a role in statistical thermodynamics very similar to that played by the wavefunction in quantum mechanics. The importance of this discussion is also the insight that a molecular interpretation provides into thermodynamic properties.

➤ What is the key idea?

The partition function contains all the thermodynamic information about a system and thus provides a bridge between spectroscopy and thermodynamics.

➤ What do you need to know already?

You need to know how to calculate a molecular partition function from structural data (Topic 15B); you should also be familiar with the concepts of internal energy (Topic 2A) and entropy (Topic 3A). This Topic makes use of the calculations of mean molecular energies in Topic 15C.

In this Topic we see how to obtain any thermodynamic function once we know the partition function. The two fundamental properties of thermodynamics are the internal energy, U, and the entropy, S. Once these two properties have been calculated, it is possible to turn to the derived functions, such as the Gibbs energy, G (Topic 15F).

15E.1 The internal energy

We begin to unfold the importance of q by showing how to derive an expression for the internal energy of the system.

(a) The calculation of internal energy

It is established in Topic 15C that the mean energy of a collection of independent molecules is related to the molecular partition function by

$$\langle\varepsilon\rangle=-\frac{1}{q}\left(\frac{\partial q}{\partial\beta}\right)_V \qquad (15E.1)$$

with $\beta=1/kT$. The total energy of a system composed of N molecules is therefore $N\langle\varepsilon\rangle$ and so the internal energy, $U(T)=U(0)+N\langle\varepsilon\rangle$ is related to the molecular partition function by

$$U(T)=U(0)+N\langle\varepsilon\rangle=U(0)-\frac{N}{q}\left(\frac{\partial q}{\partial\beta}\right)_V$$

Independent molecules Internal energy (15E.2a)

In many cases, the expression for $\langle\varepsilon\rangle$ already established for each mode of motion in Topic 15C can be used and it is not necessary to go back to the expression for q except for some formal manipulations. An alternative form of this relation is

$$U(T)=U(0)-N\left(\frac{\partial\ln q}{\partial\beta}\right)_V$$

Independent molecules Internal energy (15E.2b)

A very similar expression is used for a system of interacting molecules. In that case we use the canonical partition function, Q, and write

$$U(T)=U(0)-\left(\frac{\partial \ln Q}{\partial \beta}\right)_V \quad \text{Interacting molecules} \quad \text{Internal energy} \qquad (15E.2c)$$

Brief illustration 15E.1 The internal energy of a collection of oscillators

It is established in Topic 15C (eqn 15C.8) that the mean energy of a collection of harmonic oscillators is $\langle \varepsilon^V\rangle = hc\tilde{\nu}/(e^{\beta hc\tilde{\nu}}-1)$. It follows that the molar internal energy of such a collection is

$$U_m^V(T)=U_m^V(0)+\frac{N_A hc\tilde{\nu}}{e^{\beta hc\tilde{\nu}}-1}$$

For I_2 molecules at 298.15 K we note from Table 12D.1 that their vibrational wavenumber is 214.6 cm^{-1}. Then, because at 298.15 K, $kT/hc=207.224$ cm^{-1}, $hc\tilde{\nu}=4.26$ zJ, and

$$\beta hc\tilde{\nu}=\frac{hc\tilde{\nu}}{kT}=\frac{214.6\,\text{cm}^{-1}}{207.244\,\text{cm}^{-1}}=1.035\ldots$$

Then it follows that the vibrational contribution to the molar internal energy is

$$U_m^V(T)=U_m^V(0)+\frac{(6.022\times10^{23}\,\text{mol}^{-1})\times(4.26\times10^{-21}\,\text{J})}{e^{1.035\ldots}-1}$$
$$=U_m^V(0)+1.41\,\text{kJ mol}^{-1}$$

Self-test 15E.1 What is the molar internal energy of a gas of linear molecules?

Answer: $U_m(T)=U_m(0)+\frac{5}{2}RT$

(b) Heat capacity

The constant-volume heat capacity (Topic 2A) is defined as $C_V=(\partial U/\partial T)_V$. Thus, because the mean vibrational energy of a collection of harmonic oscillators (eqn 15C.8, $\langle \varepsilon^V\rangle = hc\tilde{\nu}/(e^{\beta hc\tilde{\nu}}-1)$) can be written in terms of the vibrational temperature $\theta^V=hc\tilde{\nu}/k$ as

$$\langle \varepsilon^V\rangle=\frac{k\theta^V}{e^{\theta^V/T}-1}$$

it follows that the vibrational contribution to the molar constant-volume heat capacity is

$$C_{V,m}^V=\frac{dN_A\langle\varepsilon^V\rangle}{dT}=R\theta^V\frac{d}{dT}\frac{1}{e^{\theta^V/T}-1}=R\left(\frac{\theta^V}{T}\right)^2\frac{e^{\theta^V/T}}{(e^{\theta^V/T}-1)^2}$$

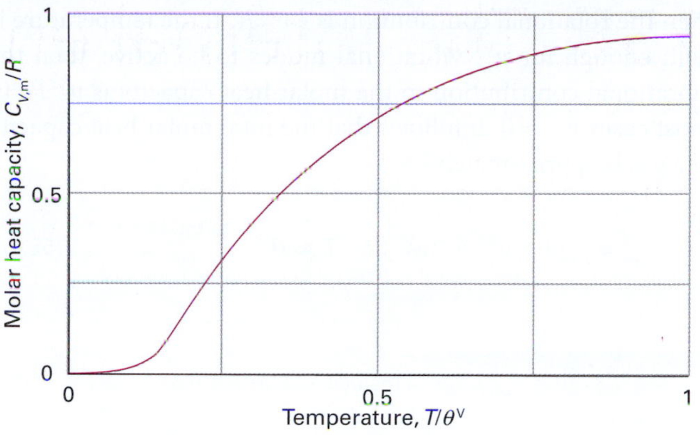

Figure 15E.1 The temperature dependence of the vibrational heat capacity of a molecule in the harmonic approximation calculated by using eqn 15E.3. Note that the heat capacity is within 10 per cent of its classical value for temperatures greater than θ^V.

This expression can be rearranged into

$$C_{V,m}^V=Rf(T) \qquad f(T)=\left(\frac{\theta^V}{T}\right)^2\left(\frac{e^{-\theta^V/2T}}{1-e^{-\theta^V/T}}\right)^2 \quad \text{Vibrational contribution to } C_V \qquad (15E.3)$$

The curve in Fig. 15E.1 shows how the vibrational heat capacity depends on temperature. Note that even when the temperature is only slightly above θ^V the heat capacity is close to its equipartition value. Equation 15E.3 is essentially the same as the Einstein formula for the heat capacity of a solid (eqn 7A.9) with θ^V the Einstein temperature, θ_E. The only difference is that vibrations can take place in three dimensions in a solid.

It is sometimes more convenient to convert the derivative with respect to T into a derivative with respect to β by using

$$\frac{d}{dT}=\frac{d\beta}{dT}\frac{d}{d\beta}=-\frac{1}{kT^2}\frac{d}{d\beta}=-k\beta^2\frac{d}{d\beta} \qquad (15E.4)$$

It follows that

$$C_V=-k\beta^2\left(\frac{\partial U}{\partial \beta}\right)_V=-Nk\beta^2\left(\frac{\partial\langle\varepsilon\rangle}{\partial\beta}\right)_V \overset{\text{eqn 15E.1}}{=} Nk\beta^2\left(\frac{\partial^2 \ln q}{\partial\beta^2}\right)_V \quad \text{Heat capacity} \qquad (15E.5)$$

When equipartition is valid, which is the case when $T \gg \theta^M$, with the characteristic temperature of the mode M ($\theta^V=hc\tilde{\nu}/k$ for vibration, $\theta^R=hc\tilde{B}/k$ for rotation), there is a much simpler route. We can then estimate the heat capacity by counting the numbers of modes that are active. In gases, all three translational modes are always active and contribute $\frac{3}{2}R$ to the molar heat capacity. If we denote the number of active rotational modes by ν^{R*} (so for most molecules at normal temperatures $\nu^{R*}=2$ for linear molecules, and 3 for nonlinear molecules), then the rotational contribution is $\frac{1}{2}\nu^{R*}R$. If the temperature is high enough for ν^{V*} vibrational modes to be active, then the vibrational contribution to the molar heat capacity is $\nu^{V*}R$. In most cases $\nu^{V*} \approx 0$. It follows that the total molar heat capacity of a gas is approximately

$$C_{V,\mathrm{m}} = \tfrac{1}{2}(2 + \nu^{R*} + 2\nu^{V*})R \quad T \gg \theta^{\mathrm{M}}$$ Total heat capacity (15E.6)

Brief illustration 15E.2 The constant-volume heat capacity

The characteristic temperatures (in round numbers) of the vibrations of H_2O are 5300 K, 2300 K, and 5400 K; the vibrations are therefore not excited at 373 K. The three rotational modes of H_2O have characteristic temperatures 40 K, 21 K, and 13 K, so they are fully excited, like the three translational modes. The translational contribution is $\frac{3}{2}R = 12.5\,\mathrm{J\,K^{-1}\,mol^{-1}}$. Fully excited rotations contribute a further $12.5\,\mathrm{J\,K^{-1}\,mol^{-1}}$. Therefore, a value close to $25\,\mathrm{J\,K^{-1}\,mol^{-1}}$ is predicted. The experimental value is $26.1\,\mathrm{J\,K^{-1}\,mol^{-1}}$. The discrepancy is probably due to deviations from perfect gas behaviour.

Self-test 15E.2 Estimate the molar constant-volume heat capacity of gaseous I_2 at 25 °C ($\tilde{B} = 0.037\,\mathrm{cm^{-1}}$; see Table 12D.1 for more data).

Answer: $29\,\mathrm{J\,K^{-1}\,mol^{-1}}$

15E.2 The entropy

One of the most celebrated equations in statistical thermodynamics is the **Boltzmann formula** for the entropy:

$$S = k \ln \mathcal{W}$$ Boltzmann formula for the entropy (15E.7)

In this expression, which is derived in the following *Justification*, $\mathcal{W}$ is the weight of the most probable configuration of the system (as discussed in Topic 15A).

Justification 15E.1 The Boltzmann formula

The interenal energy $U(T) = U(0) + N\langle\varepsilon\rangle$, with $\langle\varepsilon\rangle = (1/N)\sum_i N_i\varepsilon_i$ can be written

$$U(T) = U(0) + \sum_i N_i\varepsilon_i$$

A change in $U(T)$ may arise from either a modification of the energy levels of a system (when ε_i changes to $\varepsilon_i + \mathrm{d}\varepsilon_i$) or from a modification of the populations (when N_i changes to $N_i + \mathrm{d}N_i$). The most general change is therefore

$$\mathrm{d}U(T) = \mathrm{d}U(0) + \sum_i N_i\,\mathrm{d}\varepsilon_i + \sum_i \varepsilon_i\,\mathrm{d}N_i$$

Because the energy levels do not change when a system is heated at constant volume (Fig. 15E.2), in the absence of all changes other than heating, only the third (blue) term on the right survives. We know from thermodynamics, specifically eqn 3D.1 ($\mathrm{d}U = T\mathrm{d}S - p\mathrm{d}V$), that under the same conditions $\mathrm{d}U = T\mathrm{d}S$. Therefore,

$$\mathrm{d}S = \frac{\mathrm{d}U}{T} = \frac{1}{T}\sum_i \varepsilon_i\,\mathrm{d}N_i = k\beta\sum_i \varepsilon_i\,\mathrm{d}N_i$$

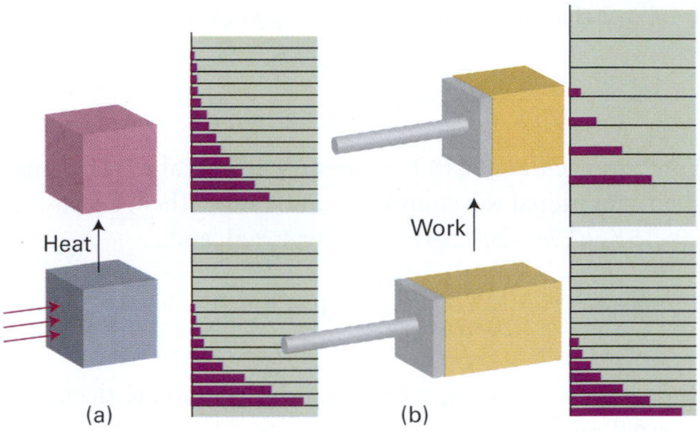

Figure 15E.2 (a) When a system is heated, the energy levels are unchanged but their populations are changed. (b) When work is done on a system, the energy levels themselves are changed. The levels in this case are the one-dimensional particle-in-a-box energy levels of Topic 8A: they depend on the size of the container and move apart as its length is decreased.

For changes in the most probable configuration (the only one we need consider), we know from eqn 15A.10 $(\partial(\ln\mathcal{W})/\partial N_i + \alpha - \beta\varepsilon_i = 0)$ that $\beta\varepsilon_i = \partial(\ln\mathcal{W})/\partial N_i + \alpha$; therefore

$$\mathrm{d}S = k\overbrace{\sum_i\left(\frac{\partial\ln\mathcal{W}}{\partial N_i}\right)\mathrm{d}N_i}^{\mathrm{d}\ln\mathcal{W}} + k\alpha\overbrace{\sum_i \mathrm{d}N_i}^{0} = k(\mathrm{d}\ln\mathcal{W})$$

This relation strongly suggests the definition $S = k\ln$ $S = k\ln\mathcal{W}$, as in eqn 15E.7.

(a) Entropy and the partition function

The statistical entropy behaves in exactly the same way as the thermodynamic entropy. Thus, as the temperature is lowered, the value of $\mathcal{W}$, and hence of S, decreases because fewer configurations are consistent with the total energy. In the limit $T \rightarrow 0$, $\mathcal{W} = 1$, so $\ln\mathcal{W} = 0$, because only one configuration (every molecule in the lowest level) is compatible with $E = 0$. It follows that $S\to 0$ as $T\to 0$, which is compatible with the Third Law of thermodynamics, that the entropies of all perfect crystals approach the same value as $T\to 0$ (Topic 3B).

Now we relate the Boltzmann formula for the entropy to the partition function. As shown in the following *Justification*, the relation for a system of non-interacting *distinguishable* molecules is

$$S(T)=\frac{U(T)-U(0)}{T}+Nk\ln q$$

Independent, distinguishable molecules Entropy (15E.8a)

For indistinguishable molecules (like those in a gas of identical molecules)

$$S(T)=\frac{U(T)-U(0)}{T}+Nk\ln\frac{q}{N}$$

Independent, indistinguishable molecules Entropy (15E.8b)

The corresponding expression for interacting molecules is based on the canonical partition function, and is

$$S(T)=\frac{U(T)-U(0)}{T}+k\ln Q$$

Interacting molecules Entropy (15E.8c)

Justification 15E.2 The statistical entropy

For a system composed of N distinguishable molecules, eqn 15A.3 ($\ln\mathcal{W}=N\ln N-\sum_i N_i\ln N_i$) with $N=\sum_i N_i$ is

$$\ln\mathcal{W}=\sum_i N_i\ln N-\sum_i N_i\ln N_i$$

$$=\sum_i N_i(\ln N-\ln N_i)=-\sum_i N_i\ln\frac{N_i}{N}$$

Equation 15E.7 ($S=k\ln\mathcal{W}$) then becomes

$$S=-k\sum_i N_i\ln\frac{N_i}{N}$$

The value of N_i/N for the most probable distribution is given by the Boltzmann distribution, $N_i/N=\mathrm{e}^{-\beta\varepsilon_i}/q$, and so

$$\ln\frac{N_i}{N}=\ln\mathrm{e}^{-\beta\varepsilon_i}-\ln q=-\beta\varepsilon_i-\ln q$$

Therefore,

$$S=k\beta\overbrace{\sum_i N_i\varepsilon_i}^{N\langle\varepsilon\rangle}+k\sum_i N_i\ln q=Nk\beta\langle\varepsilon\rangle+Nk\ln q$$

Finally, because $N\langle\varepsilon\rangle=U-U(0)$ and $\beta=1/kT$, we obtain eqn 15E.8a.

To treat a system composed of N *in*distinguishable molecules, we need to reduce the weight $\mathcal{W}$ by a factor of $1/N!$ because the $N!$ permutations of the molecules among the states result in the same state of the system. Then, because $\ln(\mathcal{W}/N!)=\ln\mathcal{W}-\ln N!$, the equation in the first line of this *Justification* becomes

$$\ln\mathcal{W}=N\ln N-\sum_i N_i\ln N_i-\overbrace{\ln N!}^{N\ln N-N}$$

$$=\sum_i N_i\ln N-\sum_i N_i\ln N_i-\overbrace{N}^{\sum_i N_i}\ln N+N$$

$$=-\sum_i N_i\ln N_i+N$$

where we have used Stirling's approximation to write $\ln N!=N\ln N-N$. As before, we replace N_i by the Boltzmann value, $N_i=N\mathrm{e}^{-\beta\varepsilon_i}/q$:

$$\sum_i N_i\ln N_i=\sum_i N_i(\ln N-\beta\varepsilon_i-\ln q)$$

$$=N\ln N-N\beta\langle\varepsilon\rangle-N\ln q=-N\beta\langle\varepsilon\rangle-N\ln\frac{q}{N}$$

The entropy in this case is therefore

$$S=Nk\beta\langle\varepsilon\rangle+Nk\ln\frac{q}{N}+Nk$$

Now note that Nk can be written $Nk\ln\mathrm{e}$ and $Nk\ln q/N+Nk\ln\mathrm{e}=Nk\ln q\mathrm{e}/N$, which gives eqn 15E.8b.

Equation 15E.8a expresses the entropy of a collection of independent molecules in terms of the internal energy and the molecular partition function. However, it is shown in Topic 15C that, to a good approximation, the energy of a molecule is a sum of independent contributions, such as translational (T), rotational (R), vibrational (V), and electronic (E), and therefore the partition function factorizes into a product of contributions. As a result, the entropy is also the sum of the individual contributions. For independent, distinguishable particles, each contribution is of the form of eqn 15E.8a, and for a mode M we write

$$S^{\mathrm{M}}=\frac{\{U-U(0)\}^{\mathrm{M}}}{T}+Nk\ln q^{\mathrm{M}}$$

Independent, distinguishable particles, M≠T Entropy due to mode M (15E.9)

This expression applies to M=R, V, and E; the analogous version of eqn 15E.8b should be used for M=T, for the molecules are then indistinguishable.

Brief illustration 15E.3 The entropy of a two-level system

From Topics 15B and 15C, the partition function and mean energy are $q=1+\mathrm{e}^{-\beta\varepsilon}$ and $\langle\varepsilon^{\mathrm{S}}\rangle=\varepsilon/(\mathrm{e}^{\beta\varepsilon}+1)$. The contribution to the molar entropy, with $1/T=k\beta$, is therefore

$$S_{\mathrm{m}}=R\left\{\frac{\beta\varepsilon}{1+\mathrm{e}^{\beta\varepsilon}}+\ln(1+\mathrm{e}^{-\beta\varepsilon})\right\}$$

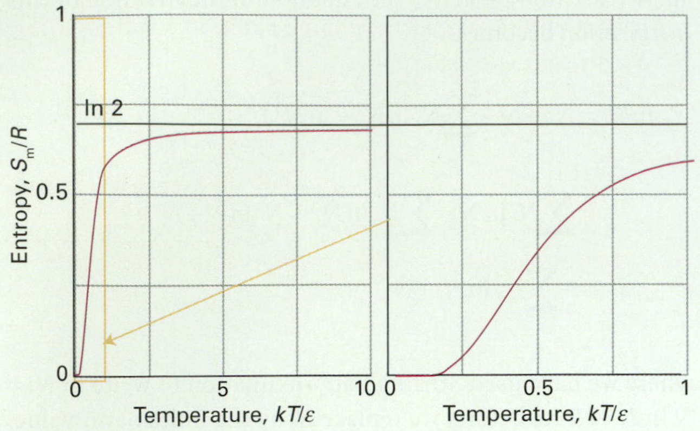

Figure 15E.3 The temperature variation of the molar entropy of a collection of two-level systems expressed as a multiple of $R=Nk$. As $T\rightarrow\infty$ the two states become equally populated and S_m approaches $R\ln 2$.

This awkward function is plotted in Fig. 15E.3. It should be noted that as $T\rightarrow\infty$ (corresponding to $\beta\rightarrow 0$), the molar entropy approaches $R\ln 2$.

Self-test 15E.3 Derive an expression for the molar entropy of an equally spaced three-level system.

Answer: $S_m/R=\beta\varepsilon/(1+e^{-\beta\varepsilon}+e^{-2\beta\varepsilon})+\ln(1+e^{-\beta\varepsilon}+e^{-2\beta\varepsilon})$

(b) The translational contribution

The expressions we have derived for the entropy accord with what we should expect for entropy if it is a measure of the spread of the populations of molecules over the available states. For instance, we show in the following *Justification* that the **Sackur–Tetrode equation** for the molar entropy of a monatomic perfect gas, where the only motion is translation in three dimensions, is

$$S_m=R\ln\left(\frac{V_m e^{5/2}}{N_A \Lambda^3}\right) \qquad \textit{Monatomic perfect gas} \quad \text{Sackur–Tetrode equation} \qquad (15E.10a)$$

where Λ is the thermal wavelength ($\Lambda=h/(2\pi mkT)^{1/2}$). To calculate the standard molar entropy, we note that $V_m=RT/p$, and set $p=p^{\ominus}$:

$$S_m^{\ominus}=R\ln\left(\frac{RTe^{5/2}}{p^{\ominus}N_A\Lambda^3}\right)=R\ln\left(\frac{kTe^{5/2}}{p^{\ominus}\Lambda^3}\right) \qquad (15E.10b)$$

We have used $R/N_A=k$. These expressions are based on the high-temperature approximation of the partition functions, which assumes that many levels are occupied; therefore, they do not apply when T is equal to or very close to zero.

Brief illustration 15E.4 The molar entropy of a gas

To calculate the standard molar entropy of gaseous argon at 25 °C, we use eqn 15E.10b with $\Lambda=h/(2\pi mkT)^{1/2}$. The mass of an Ar atom is $m=39.95m_u$. At 25 °C, its thermal wavelength is 16.0 pm and $kT=4.12\times10^{-21}$ J. Therefore,

$$S_m^{\ominus}=R\ln\left\{\frac{(4.12\times10^{-21}\,\mathrm{J})\times e^{5/2}}{(10^5\,\mathrm{N\,m^{-2}})\times(1.60\times10^{-11}\,\mathrm{m})^3}\right\}$$
$$=18.6R=155\,\mathrm{J\,K^{-1}\,mol^{-1}}$$

We can anticipate, on the basis of the number of accessible states for a lighter molecule, that the standard molar entropy of Ne is likely to be smaller than for Ar; its actual value is 17.60R at 298 K.

Self-test 15E.4 Calculate the translational contribution to the standard molar entropy of H_2 at 25 °C.

Answer: 14.2R

The physical interpretation of these equations is as follows:

Physical interpretation

- Because the molecular mass appears in the numerator (because it appears in the denominator of Λ), the molar entropy of a perfect gas of heavy molecules is greater than that of a perfect gas of light molecules under the same conditions. We can understand this feature in terms of the energy levels of a particle in a box being closer together for heavy particles than for light particles, so more states are thermally accessible.
- Because the molar volume appears in the numerator, the molar entropy increases with the molar volume of the gas. The reason is similar: large containers have more closely spaced energy levels than small containers, so once again more states are thermally accessible.
- Because the temperature appears in the numerator (because, like m, it appears in the denominator of Λ), the molar entropy increases with increasing temperature. The reason for this behaviour is that more energy levels become accessible as the temperature is raised.

Justification 15E.3 The Sackur–Tetrode equation

We start with eqn 15E.8b for a collection of independent, indistinguishable particles and write $N=nN_A$, where N_A is Avogadro's constant. The only mode of motion for a gas of atoms is translation and $U-U(0)=\frac{3}{2}nRT$. The partition function is $q = V/\Lambda^3$ (eqn 15B.7a), where Λ is the thermal wavelength. Therefore,

$$S = \frac{\overbrace{U-U(0)}^{\frac{3}{2}nRT}}{T} + Nk\ln\frac{qe}{N} = \tfrac{3}{2}nR + \overbrace{Nk}^{nR}\ln\frac{Ve}{nN_A\Lambda^3}$$

$$= nR\left\{\overbrace{\tfrac{3}{2}}^{\ln e^{3/2}} + \ln\frac{V_m e}{N_A\Lambda^3}\right\} = nR\ln\frac{V_m e^{5/2}}{N_A\Lambda^3}$$

where $V_m = V/n$ is the molar volume of the gas and we have used $\frac{3}{2} = \ln e^{\frac{3}{2}}$. Division of both sides by n then results in eqn 15E.10a.

The Sackur–Tetrode equation written in the form

$$S = nR\ln\frac{Ve^{5/2}}{nN_A\Lambda^3} = nR\ln aV, \quad a = \frac{e^{5/2}}{nN_A\Lambda^3}$$

implies that when a monatomic perfect gas expands isothermally from V_i to V_f, its entropy changes by

$$\Delta S = nR\ln aV_f - nR\ln aV_i = nR\ln\frac{V_f}{V_i}$$

Perfect gas, isothermal — Change of entropy on expansion (15E.11)

This expression is the same as that obtained starting from the thermodynamic definition of entropy (Topic 3A).

(c) The rotational contribution

The rotational contribution to the molar entropy, S_m^R, can be calculated once we know the molecular partition function. For a linear molecule, the high-temperature limit of q is $kT/\sigma hc\tilde{B}$ (eqn 15B.13b, $q^R = T/\sigma\theta^R$ with $\theta^R = hc\tilde{B}/k$) and the equipartition theorem gives the rotational contribution to the molar internal energy as RT; therefore, from eqn 15E.8a:

$$S_m^R = \frac{\overbrace{U_m - U_m(0)}^{RT}}{T} + R\ln\overbrace{q^R}^{kT/\sigma hc\tilde{B}}$$

and the contribution at high temperatures is

$$S_m^R = R\left\{1 + \ln\frac{kT}{\sigma hc\tilde{B}}\right\}$$

Linear molecule, high temperature ($T \gg \theta^R$) — Rotational contribution (15E.12a)

In terms of this rotational temperature,

$$S_m^R = R\left\{1 + \ln\frac{T}{\sigma\theta^R}\right\}$$

Linear molecule, high temperature ($T \gg \theta^R$) — Rotational contribution (15E.12b)

This function is plotted in Fig. 15E.4. We see that:

- The rotational contribution to the entropy increases with temperature because more rotational states become accessible.
- The rotational contribution is large when $\tilde{B}$ is small, because then the rotational energy levels are close together.

Physical interpretation

Thus, large, heavy molecules have a large rotational contribution to their entropy. As we show in the following *Brief illustration*, the rotational contribution to the molar entropy of $^{35}Cl_2$ is 58.6 J K^{-1} mol^{-1} whereas that for H_2 is only 12.7 J K^{-1} mol^{-1}. We can regard Cl_2 as a more rotationally disordered gas than H_2, in the sense that at a given temperature Cl_2 occupies a greater number of rotational states than H_2 does.

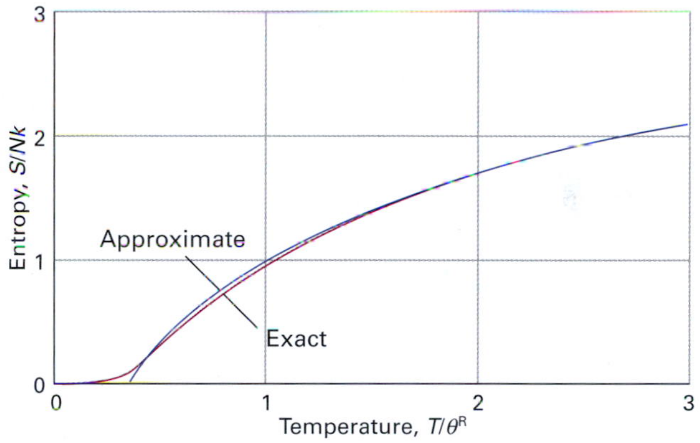

Figure 15E.4 The variation of the rotational contribution to the entropy of a linear molecule ($\sigma = 1$) using the high-temperature approximation and the exact expression (the latter evaluated up to $J = 20$).

Brief illustration 15E.5 The rotational contribution to the entropy

The rotational contribution for $^{35}Cl_2$ at 25 °C, for instance, is calculated by noting that $\sigma = 2$ for this homonuclear diatomic molecule and taking $\tilde{B} = 0.2441\,cm^{-1}$ (corresponding to 24.42 m^{-1}). The rotational temperature of the molecule is

$$\theta^R = \frac{(6.626\times10^{-34}\,J\,s)\times(2.998\times10^8\,m\,s^{-1})\times(24.42\,m^{-1})}{1.381\times10^{-23}\,J\,K^{-1}} = 0.351\,K$$

Therefore,

$$S_m^R = R\left\{1 + \ln\frac{298\,K}{2\times(0.351\,K)}\right\} = 7.05R = 58.6\,J\,K^{-1}\,mol^{-1}$$

Self-test 15E.5 Calculate the rotational contribution to the molar entropy of H_2.

Answer: 12.7 J K^{-1} mol^{-1}

Equation 15E.12 is valid at high temperatures ($T \gg \theta^R$); to track the rotational contribution down to low temperatures it would be necessary to use the full form of the rotational partition function (Topic 15B; see Problem 15E.10); the resulting curve has the form shown in Fig. 15E.4. We see, in fact, that the approximate curve matches the exact curve very well for T/θ^R greater than about 1.

(d) The vibrational contribution

The vibrational contribution to the molar entropy, S_m^V, is obtained by combining the expression for the molecular partition function (eqn 15B.15, $q^V = 1/(1-e^{-\beta\varepsilon})$) with the expression for the mean energy (eqn 15C.8, $\langle\varepsilon^V\rangle = \varepsilon/(e^{\beta\varepsilon}-1)$), to obtain

$$S_m^V = \frac{\overbrace{U_m - U_m(0)}^{N_A\langle\varepsilon^V\rangle}}{\underbrace{T}_{1/k\beta}} + R\ln q^V = \frac{\overbrace{N_A k}^{R}\beta\varepsilon}{e^{\beta\varepsilon}-1} + R\ln\frac{1}{1-e^{-\beta\varepsilon}}$$

$$= R\left\{\frac{\beta\varepsilon}{e^{\beta\varepsilon}-1} - \ln\left(1-e^{-\beta\varepsilon}\right)\right\}$$

Now we recognize that $\varepsilon = hc\tilde{\nu}$ and obtain

$$S_m^V = R\left\{\frac{\beta hc\tilde{\nu}}{e^{\beta hc\tilde{\nu}}-1} - \ln\left(1-e^{-\beta hc\tilde{\nu}}\right)\right\} \quad \text{Vibrational contribution to the entropy} \quad (15E.13a)$$

Once again it is convenient to express this formula in terms of a characteristic temperature, in this case the vibrational temperature $\theta^V = hc\tilde{\nu}/k$:

$$S_m^V = R\left\{\frac{\theta^V/T}{e^{\theta^V/T}-1} - \ln\left(1-e^{-\theta^V/T}\right)\right\} \quad \text{Vibrational contribution to the entropy} \quad (15E.13b)$$

This function is plotted in Fig. 15E.5. As usual, it is helpful to interpret it, with the graph in mind:

Physical interpretation

- Both terms multiplying R become zero as $T \to 0$, so the entropy is zero at $T = 0$.
- The molar entropy rises as the temperature is increased as more vibrational states becoming accessible.
- The molar entropy is higher at a given temperature for molecules with heavy atoms or low force constant than one with light atoms or high force constant. The vibrational energy levels are closer together in the former case than in the latter, so more are thermally accessible

Brief illustration 15E.6 The vibrational contribution to the entropy

The vibrational wavenumber of I_2 is 214.5 cm^{-1}, corresponding to 2.145×10^4 m^{-1}, so its vibrational temperature is 309 K. Therefore, at 25 °C, for instance, $\beta\varepsilon = 1.036$, so

$$S_m^V = R\left\{\frac{309/298}{e^{309/298}-1} - \ln\left(1-e^{-309/298}\right)\right\} = 1.01R = 8.38\,\mathrm{J\,K^{-1}\,mol^{-1}}$$

Self-test 15E.6 Calculate the vibrational contribution to the molar entropy of $^1H_2$ at 25 °C ($\theta^V = 6332$ K).

Answer: 0.11 μJ K^{-1}

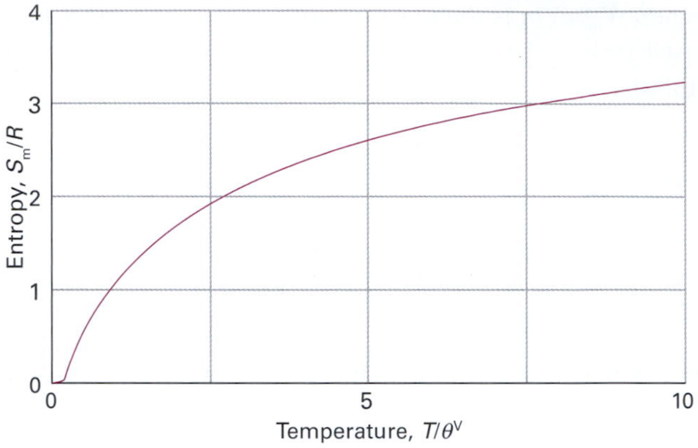

Figure 15E.5 The temperature variation of the molar entropy of a collection of harmonic oscillators expressed as a multiple of $R = Nk$. The molar entropy approaches zero as $T \to 0$, and increases without limit as $T \to \infty$.

(e) Residual entropies

Entropies may be calculated from spectroscopic data; they may also be measured experimentally (Topic 3B). In many cases there is good agreement, but in some the experimental entropy is less than the calculated value. One possibility is that the experimental determination failed to take a phase transition into account and a contribution of the form $\Delta_{trs}H/T_{trs}$ was incorrectly omitted from the sum. Another possibility is that some disorder is present in the solid even at $T = 0$. The entropy at $T = 0$ is then greater than zero and is called the **residual entropy**.

The origin and magnitude of the residual entropy can be explained by considering a crystal composed of AB molecules, where A and B are similar atoms (such as CO, with its very small electric dipole moment). There may be so little energy difference between … AB AB AB AB …, … AB BA BA AB …, and other arrangements that the molecules adopt the orientations AB and BA at random in the solid. We can readily calculate the entropy arising from residual disorder by using the Boltzmann formula $S = k\ln\mathcal{W}$. To do so, we suppose that two orientations are equally probable, and that the sample consists of N molecules. Because the same energy can be achieved in 2^N different ways (because each molecule can take either of two orientations), the total number of ways of achieving the same energy is $\mathcal{W} = 2^N$. It follows that

$$S = k\ln 2^N = Nk\ln 2 = nR\ln 2 \quad (15E.14a)$$

We can therefore expect a residual molar entropy of $R \ln 2 = 5.8\,\mathrm{J\,K^{-1}\,mol^{-1}}$ for solids composed of molecules that can adopt either of two orientations at $T=0$. If s orientations are possible, the residual molar entropy will be

$$S_m(0) = R \ln s \qquad \text{Residual entropy} \qquad (15E.14b)$$

For CO, the measured residual entropy is $5\,\mathrm{J\,K^{-1}\,mol^{-1}}$, which is close to $R \ln 2$, the value expected for a random structure of the form ... CO CO OC CO OC OC ....

Brief illustration 15E.7 Residual entropy

Consider a sample of ice with N H_2O molecules. Each O atom is surrounded tetrahedrally by four H atoms, two of which are attached by short σ bonds, the other two being attached by long hydrogen bonds (Fig. 15E.6). It follows that each of the $2N$ H atoms can be in one of two positions (either close to or far from an O atom as shown in Fig. 15E6), resulting in 2^{2N} possible arrangements.

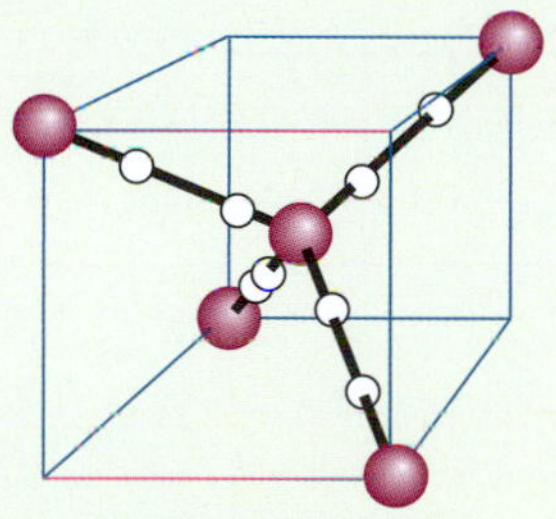

Figure 15E.6 The possible locations of H atoms around a central O atom in an ice crystal are shown by the white spheres. Only one of the locations on each bond may be occupied by an atom, and two H atoms must be close to the O atom and two H atoms must be distant from it.

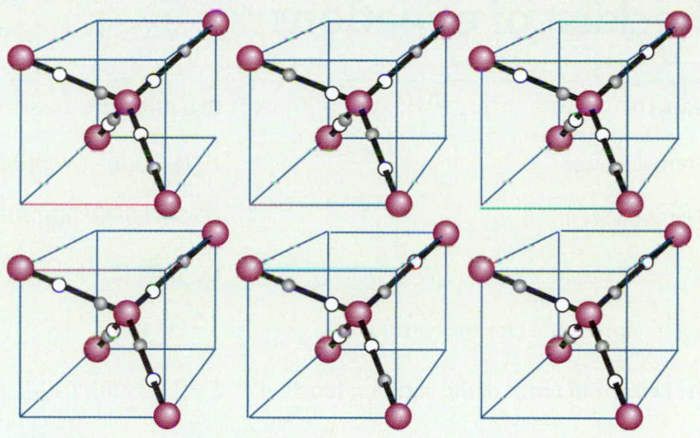

Figure 15E.7 The six possible arrangements of H atoms in the locations identified in Fig.15E.6. Occupied locations are denoted by grey spheres and unoccupied locations by white spheres.

However, not all these arrangements are acceptable. Indeed, of the $2^4 = 16$ ways of arranging four H atoms around one O atom, only 6 have two short and two long OH distances and hence are acceptable (Fig. 15E.7). Therefore, the number of permitted arrangements is $\mathcal{W} = 2^{2N}(6/16)^N = (\frac{3}{2})^N$. It then follows that the residual entropy is $S(0) \approx k \ln(\frac{3}{2})^N = kN \ln \frac{3}{2}$, and its molar value is $S(0) \approx R \ln \frac{3}{2} = 3.4\,\mathrm{J\,K^{-1}\,mol^{-1}}$, which is in good agreement with the experimental value of $3.4\,\mathrm{J\,K^{-1}\,mol^{-1}}$. The model, however, is not exact because it ignores the possibility that next-nearest neighbours and those beyond can influence the local arrangement of bonds.

Self-test 15E.7 Estimate the molar residual entropy of $FClO_3$; each molecule can adopt four orientations with about the same energy.

Answer: $R \ln 4 = 11.5\,\mathrm{J\,K^{-1}\,mol^{-1}}$; experimental: $10.1\,\mathrm{J\,K^{-1}\,mol^{-1}}$

Checklist of concepts

- ☐ 1. The **internal energy** is proportional to the derivative of the partition function with respect to temperature.
- ☐ 2. The **constant-volume heat capacity** can be calculated from the molecular partition function.
- ☐ 3. The **total heat capacity** of a molecular substance is the sum of the contribution of each mode.
- ☐ 4. The **statistical entropy** is defined by the **Boltzmann formula** and may be expressed in terms of the molecular partition function
- ☐ 5. The **residual entropy** is a nonzero entropy at $T=0$ arising from molecular disorder.

Checklist of equations

Property	Equation	Comment	Equation number
Internal energy	$U(T)=U(0)-(N/q)(\partial q/\partial\beta)_V=-N(\partial \ln q/\partial\beta)_V$	Independent molecules	15E.2b
Heat capacity	$C_V=Nk\beta^2(\partial^2 \ln q/\partial\beta^2)_V$	Independent molecules	15E.5
	$C_{V,\mathrm{m}}=\frac{1}{2}(2+\nu^{\mathrm{R}*}+2\nu^{\mathrm{V}*})R$	$T\gg\theta^{\mathrm{M}}$	15E.6
Boltzmann formula for the entropy	$S=k\ln\mathcal{W}$	Definition	15E.7
The entropy in terms of the partition function	$S=\{U-U(0)\}/T+Nk\ln q$	Distinguishable molecules	15E.8a
	$S=\{U-U(0)\}/T+Nk\ln(q/N)$	Indistinguishable molecules	15E.8b
Sackur–Tetrode equation	$S_\mathrm{m}(T)=R\ln(V_\mathrm{m}\mathrm{e}^{5/2}/N_\mathrm{A}\Lambda^3)$	Entropy of a monatomic perfect gas	15E.10a
Residual molar entropy	$S_\mathrm{m}(0)=R\ln s$	s is the number of equivalent sites	15E.14b

15F Derived functions

Contents

➤ Why do you need to know this material?

The power of chemical thermodynamics stems from its deployment of a variety of derived functions, particularly the enthalpy and Gibbs energy. It is therefore important to relate these to structural features through partition functions. One hugely important quantity that is illuminated in this way is the equilibrium constant.

➤ What is the key idea?

The partition function provides a link between spectroscopic and structural data and the derived functions of thermodynamics, particularly the equilibrium constant.

➤ What do you need to know already?

This Topic develops the discussion of internal energy and entropy (Topic 15E). You need to know the relations between those properties and the enthalpy (Topic 2B) and the Helmholtz and Gibbs energies (Topic 3C). The final section makes use of the relation between standard Gibbs energy and the equilibrium constant (Topic 6A).

Classical thermodynamics makes extensive use of various derived functions. Thus, in thermochemistry it focuses on the enthalpy and, provided the pressure and temperature are constant, in discussions of spontaneity it focuses on the Gibbs energy. In this Topic we see how these properties can be related to and understood in terms of partition functions. All these properties are derived from the internal energy and the entropy, which in terms of the canonical partition function are given by

$$U(T)=U(0)-\left(\frac{\partial \ln Q}{\partial \beta}\right)_V \qquad \text{Internal energy} \qquad (15\text{F}.1\text{a})$$

$$S(T)=\frac{U(T)-U(0)}{T}+k\ln Q \qquad \text{Entropy} \qquad (15\text{F}.1\text{b})$$

These two general expressions can be adapted for collections of independent molecules by writing $Q=q^N$ for distinguishable molecules and $Q=q^N/N!$ for indistinguishable molecules (as in a gas).

15F.1 The derivations

The Helmholtz energy, A, is defined as $A=U-TS$. This relation implies that $A(0)=U(0)$, so substitution of the expressions for $U(T)$ and $S(T)$ leads to the very simple expression

$$A(T)=A(0)-kT\ln Q \qquad \text{Helmholtz energy} \qquad (15\text{F}.2)$$

An infinitesimal change in conditions changes the Helmoltz energy by $\mathrm{d}A=-p\mathrm{d}V-S\mathrm{d}T$ (this is the analogue of the expression for dG derived in Topic 3D (eqn 3D.7, $\mathrm{d}G=V\mathrm{d}p-S\mathrm{d}T$). Therefore, it follows that on imposing constant temperature ($\mathrm{d}T=0$), the pressure and the Helmholtz energy are related by $p=-(\partial A/\partial V)_T$. It then follows from eqn 15F.2 that

$$p=kT\left(\frac{\partial \ln Q}{\partial V}\right)_T \qquad \text{Pressure} \qquad (15\text{F}.3)$$

This relation is entirely general, and may be used for any type of substance, including perfect gases, real gases, and liquids. Because Q is in general a function of the volume, temperature, and amount of substance, eqn 15F.3 is an equation of state of the kind discussed in Topic 1C.

Example 15F.1 Deriving an equation of state

Derive an expression for the pressure of a gas of independent particles.

Method We should suspect that the pressure is that given by the perfect gas law. To proceed systematically, substitute the explicit formula for Q for a gas of independent, indistinguishable molecules.

Answer For a gas of independent molecules, $Q=q^N/N!$ with $q=V/\Lambda^3$:

$$p=kT\left(\frac{\partial \ln Q}{\partial V}\right)_T=\frac{kT}{Q}\left(\frac{\partial Q}{\partial V}\right)_T=\frac{N!kT}{q^N}\left(\frac{\partial(q^N/N!)}{\partial V}\right)_T$$
$$=\frac{kT}{q^N}\left(\frac{\partial q^N}{\partial V}\right)_T=\frac{NkT}{q}\left(\frac{\partial q}{\partial V}\right)_T$$
$$=\frac{NkT}{V/\Lambda^3}\left(\frac{\partial(V/\Lambda^3)}{\partial V}\right)_T=\frac{NkT}{V}=\frac{nN_AkT}{V}=\frac{nRT}{V}$$

The calculation shows that the equation of state of a gas of independent particles is indeed the perfect gas law, $pV=nRT$.

Self-test 15F.1 Derive the equation of state of a sample for which $Q=q^Nf/N!$, with $q=V/\Lambda^3$, where f depends on the volume.

Answer: $p=nRT/V+kT(\partial \ln f/\partial V)_T$

At this stage we can use the expressions for U and p in the definition $H=U+pV$ to obtain an expression for the enthalpy, H, of any substance:

$$H(T)=H(0)-\left(\frac{\partial \ln Q}{\partial \beta}\right)_V+kTV\left(\frac{\partial \ln Q}{\partial V}\right)_T \quad \text{Enthalpy} \quad (15F.4)$$

The fact that eqn 15F.4 is rather cumbersome is a sign that the enthalpy is not a fundamental property: as shown in Topic 2B, it is more of an accounting convenience. For a gas of independent particles $U-U(0)=\frac{3}{2}nRT$ and $pV=nRT$. Therefore, for such a gas,

$$H-H(0)=\tfrac{5}{2}nRT \quad (15F.5)$$

One of the most important thermodynamic functions for chemistry is the Gibbs energy, $G=H-TS=A+pV$. We can now express this function in terms of the partition function by combining the expressions for A and p:

$$G(T)=G(0)-kT\ln Q+kTV\left(\frac{\partial \ln Q}{\partial V}\right)_T \quad \text{Gibbs energy} \quad (15F.6)$$

This expression takes a simple form for a gas of independent molecules because pV in the expression $G=A+pV$ can be replaced by nRT:

$$G(T)=G(0)-kT\ln Q+nRT \quad (15F.7)$$

Furthermore, because $Q=q^N/N!$, and therefore $\ln Q=N\ln q-\ln N!$, it follows by using Stirling's approximation ($\ln N!=N\ln N-N$) that we can write

$$G(T)=G(0)-NkT\ln q+kT\ln N!+nRT \quad (15F.8)$$
$$=G(0)-nRT\ln q+kT(N\ln N-N)+nRT$$
$$=G(0)-nRT\ln\frac{q}{N}$$

with $N=nN_A$. Now we see another interpretation of the Gibbs energy: because q is the number of thermally accessible states and N is the number of molecules, the difference $G(T)-G(0)$ is proportional to the logarithm of the average number of thermally accessible states per molecule.

It will turn out to be convenient to define the **molar partition function,** $q_m=q/n$ (with units mol^{-1}), for then

$$G(T)=G(0)-nRT\ln\frac{q_m}{N_A} \quad \textit{Independent molecules} \quad \text{Gibbs energy} \quad (15F.9)$$

To use this expression, $G(0)$ is identified with the energy of the system when all the molecules are in their ground state, E_0. To calculate the standard Gibbs energy, the partition function has its standard value, $q_m^{\ominus}$, which is evaluated by setting the molar volume in the translational contribution equal to the standard molar volume, so $q_m^{\ominus}=(V_m^{\ominus}/\Lambda^3)q^Rq^V$ with $V_m^{\ominus}=RT/p^{\ominus}$.

Example 15F.2 Calculating a standard Gibbs energy of formation from partition functions

Calculate the standard Gibbs energy of formation of $H_2O(g)$ at 25 °C.

Method Write the chemical equation for the formation reaction, and then the expression for the standard Gibbs energy of formation in terms of the Gibbs energy of each molecule; then express those Gibbs energies in terms of the molecular partition function of each species. Ignore molecular vibration as it is unlikely to be excited at 25 °C. Take numerical values from the *Resource section* together with the following rotational constants of H_2O: 27.877, 14.512, and 9.285 cm^{-1}. Take for the atomization energy of H_2O the value −237 kJ mol^{-1}.

Answer The chemical reaction is $H_2(g)+\frac{1}{2}O_2(g)\rightarrow H_2O(g)$. Therefore,

$$\Delta_f G^{\ominus}=G_m^{\ominus}(H_2O,g)-G_m^{\ominus}(H_2,g)-\tfrac{1}{2}G_m^{\ominus}(O_2,g)$$

Now write the standard molar Gibbs energies in terms of the standard molar partition functions of each species J:

$$G_m^{\ominus}(J)=E_{0,m}(J)-RT\ln\frac{q_m^{\ominus}(J)}{N_A}$$
$$q_m^{\ominus}(J)=q_m^{T\ominus}(J)q^R(J)=\frac{V_m^{\ominus}}{\Lambda(J)^3}q^R(J)$$

Therefore

$$\Delta_f G^\ominus = \left\{ E_{0,m}(H_2O) - RT\ln\frac{q_m^\ominus(H_2O)}{N_A} \right\}$$

$$-\left\{ E_{0,m}(H_2) - RT\ln\frac{q_m^\ominus(H_2)}{N_A} \right\}$$

$$-\tfrac{1}{2}\left\{ E_{0,m}(O_2) - RT\ln\frac{q_m^\ominus(O_2)}{N_A} \right\}$$

$$= \Delta E_{0,m} - RT\ln\frac{\overbrace{\{V_m^\ominus/N_A\Lambda(H_2O)^3\}q^R(H_2O)}^{q_m^\ominus(H_2O)/N_A}}{\left[\underbrace{\{V_m^\ominus/N_A\Lambda(H_2)^3\}q^R(H_2)}_{q_m^\ominus(H_2)/N_A}\right]}$$

$$\times\left[\underbrace{\{V_m^\ominus/N_A\Lambda(O_2)^3\}q^R(O_2)}_{q_m^\ominus(O_2)/N_A}\right]^{1/2}$$

$$= \Delta E_m - RT\ln\frac{N_A^{1/2}\{\Lambda(H_2)\Lambda(O_2)^{1/2}/\Lambda(H_2O)\}^3}{V_m^{\ominus 1/2}\{q^R(H_2)q^R(O_2)^{1/2}/q^R(H_2O)\}}$$

where

$$\Delta E_{0,m} = E_{0,m}(H_2O) - E_{0,m}(H_2) - \tfrac{1}{2}E_{0,m}(O_2)$$

At this point we introduce the thermal wavelengths and the rotational partition functions from Topic 15B:

$$\Lambda(J) = \frac{h}{\{2\pi m(J)kT\}^{1/2}}$$

$$\overbrace{q^R = \frac{kT}{2hc\tilde{B}}}^{\text{Linear molecule, } \sigma=2} \quad\text{and}\quad \overbrace{q^R = \frac{1}{2}\left(\frac{kT}{hc}\right)^{3/2}\left(\frac{\pi}{\tilde{A}\tilde{B}\tilde{C}}\right)^{1/2}}^{\text{Nonlinear molecule. } \sigma=2}$$

Now substitute the data, and find

$\Lambda(H_2) = 71.21$ pm $\quad \Lambda(O_2) = 17.87$ pm $\quad \Lambda(H_2O) = 23.82$ pm
$q^R(H_2) = 1.702 \quad q^R(O_2) = 71.60 \quad q^R(H_2O) = 42.13$

It then follows that

$$\Delta_f G^\ominus = \Delta E_{0,m} - RT\ln 0.0291 = \Delta E_{0,m} + 8.77\,\text{kJ mol}^{-1}$$

Now use $\Delta E_{0,m} = -237$ kJ mol^{-1} and obtain $\Delta_f G^\ominus = -228$ kJ mol^{-1}. The value quoted in Table 2C.1 of the *Resource section* is -228.57 kJ mol^{-1}.

Self-test 15E.2 Estimate the standard Gibbs energy of formation of NH_3(g) at 25 °C. Take the atomization energy to be +79 kJ mol^{-1}.

Answer: -16 kJ mol^{-1}

15F.2 Equilibrium constants

The Gibbs energy of a gas of independent molecules is given by eqn 15F.9 in terms of the molar partition function, $q_m = q/n$. The equilibrium constant K of a reaction is related to the standard Gibbs energy of reaction by eqn 6A.14 of Topic 6A ($\Delta_r G^\ominus = -RT \ln K$). To calculate the equilibrium constant, we need to combine these two equations. We shall consider gas phase reactions in which the equilibrium constant is expressed in terms of the partial pressures of the reactants and products.

(a) The relation between *K* and the partition function

To find an expression for the standard reaction Gibbs energy we need expressions for the standard molar Gibbs energies, $G^\ominus/n$, of each species. It then follows that, as shown in the following *Justification*, the equilibrium constant for the reaction aA + bB → cC + dD is given by the expression

$$K = \frac{(q_{C,m}^\ominus/N_A)^c\,(q_{D,m}^\ominus/N_A)^d}{(q_{A,m}^\ominus/N_A)^a\,(q_{B,m}^\ominus/N_A)^b}\,e^{-\Delta_r E_0/RT} \qquad (15F.10a)$$

where $\Delta_r E_0$ is the difference in molar energies of the ground states of the products and reactants (this term is defined more precisely in the *Justification*), and is calculated from the bond dissociation energies of the species (Fig. 15F.1). In terms of the (signed) stoichiometric numbers introduced in Topic 2B, we would write

$$K = \left\{\prod_J \left(\frac{q_{J,m}^\ominus}{N_A}\right)^{\nu_J}\right\} e^{-\Delta_r E_0/RT} \qquad \text{Equilibrium constant} \qquad (15F.10b)$$

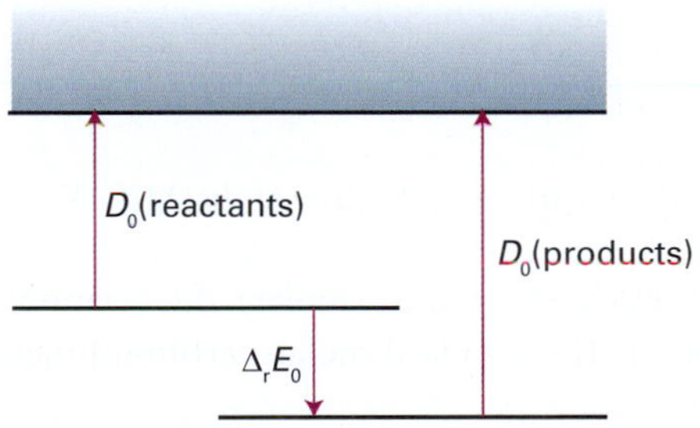

Figure 15F.1 The definition of $\Delta_r E_0$ for the calculation of equilibrium constants.

Justification 15F.1 The equilibrium constant in terms of the partition function 1

The standard molar reaction Gibbs energy for the reaction is

$$\Delta_r G^\ominus = cG_m^\ominus(C) + dG_m^\ominus(D) - \{aG_m^\ominus(A) + bG_m^\ominus(B)\}$$

$$= cG_m^\ominus(C,0) + dG_m^\ominus(D,0) - \{aG_m^\ominus(A,0) + bG_m^\ominus(B,0)\}$$

$$-RT\left\{c\ln\frac{q_{C,m}^\ominus}{N_A} + d\ln\frac{q_{D,m}^\ominus}{N_A} - a\ln\frac{q_{A,m}^\ominus}{N_A} - b\ln\frac{q_{B,m}^\ominus}{N_A}\right\}$$

Because $G(\mathrm{J},0)=E_{0,\mathrm{m}}(\mathrm{J})$, the molar ground state energy of the species J, the first (blue) term on the right is

$$cE_{0,\mathrm{m}}(\mathrm{C},0)+dE_{0,\mathrm{m}}(\mathrm{D},0)-\{aE_{0,\mathrm{m}}(\mathrm{A},0)+bE_{0,\mathrm{m}}(\mathrm{B},0)\}=\Delta_r E_0$$

Then, by using $a\ln x=\ln x^a$ and $\ln x+\ln y=\ln xy$, we can write

$$\Delta_r G^\ominus=\Delta_r E_0-RT\ln\frac{(q^\ominus_{\mathrm{C,m}}/N_A)^c(q^\ominus_{\mathrm{D,m}}/N_A)^d}{(q^\ominus_{\mathrm{A,m}}/N_A)^a(q^\ominus_{\mathrm{B,m}}/N_A)^b}$$

$$=-RT\left\{-\frac{\Delta_r E_0}{RT}+\ln\frac{(q^\ominus_{\mathrm{C,m}}/N_A)^c(q^\ominus_{\mathrm{D,m}}/N_A)^d}{(q^\ominus_{\mathrm{A,m}}/N_A)^a(q^\ominus_{\mathrm{B,m}}/N_A)^b}\right\}$$

At this stage we can pick out an expression for K by comparing this equation with ($\Delta_r G^\ominus=-RT\ln K$, which gives

$$\ln K=-\frac{\Delta_r E_0}{RT}+\ln\frac{(q^\ominus_{\mathrm{C,m}}/N_A)^c(q^\ominus_{\mathrm{D,m}}/N_A)^d}{(q^\ominus_{\mathrm{A,m}}/N_A)^a(q^\ominus_{\mathrm{B,m}}/N_A)^b}$$

This expression is easily rearranged into eqn 15F.10a by forming the exponential of both sides.

(b) A dissociation equilibrium

We illustrate the application of eqn 15F.10 to an equilibrium in which a diatomic molecule X_2 dissociates into its atoms:

$$X_2(g)\rightleftharpoons 2X(g) \qquad K=\frac{p_X^2}{p_{X_2}p^\ominus}$$

According to eqn 15F.10 (with $a=1$, $b=0$, $c=2$, and $d=0$):

$$K=\frac{(q^\ominus_{\mathrm{X,m}}/N_A)^2}{q^\ominus_{\mathrm{X_2,m}}/N_A}e^{-\Delta_r E_0/RT}=\frac{(q^\ominus_{\mathrm{X,m}})^2}{q^\ominus_{\mathrm{X_2,m}}N_A}e^{-\Delta_r E_0/RT} \qquad (15F.11a)$$

with

$$\Delta_r E_0=2E_{0,\mathrm{m}}(\mathrm{X},0)-E_{0,\mathrm{m}}(\mathrm{X_2},0)=N_A hc\tilde{D}_0(\mathrm{X{-}X}) \qquad (15F.11b)$$

where $N_A hc\tilde{D}_0(\mathrm{X{-}X})$ is the (molar) dissociation energy of the X—X bond. The standard molar partition functions of the atoms X are

$$q^\ominus_{\mathrm{X,m}}=\overbrace{g_X}^{q^E}\times\overbrace{\frac{V^\ominus_m}{\Lambda_X^3}}^{q^T}=\frac{g_X RT}{p^\ominus\Lambda_X^3}$$

where g_X is the degeneracy of the electronic ground state of X. The diatomic molecule X_2 also has rotational and vibrational degrees of freedom, so its standard molar partition function is

$$q^\ominus_{\mathrm{X_2,m}}=g_{X_2}\frac{V^\ominus_m}{\Lambda_{X_2}^3}q^R_{X_2}q^V_{X_2}=\frac{g_{X_2}RTq^R_{X_2}q^V_{X_2}}{p^\ominus\Lambda_{X_2}^3}$$

where g_{X_2} is the degeneracy of the electronic ground state of X_2. It follows that

$$K=\frac{(g_X RT/p^\ominus\Lambda_X^3)^2}{g_{X_2}N_A RTq^R_{X_2}q^V_{X_2}/p^\ominus\Lambda_{X_2}^3}e^{-N_A hc\tilde{D}_0/RT}$$
$$=\frac{g_X^2 kT\Lambda_{X_2}^3}{g_{X_2}p^\ominus q^R_{X_2}q^V_{X_2}\Lambda_X^6}e^{-hc\tilde{D}_0/kT} \qquad (15F.12)$$

where we have used $R/N_A=k$. All the quantities in this expression can be calculated from spectroscopic data.

Example 15F.3 Evaluating an equilibrium constant

Evaluate the equilibrium constant for the dissociation $Na_2(g) \rightleftharpoons 2\,Na(g)$ at 1000 K.

Method Assemble the data from the *Resource section*, noting that Na has a double ground state. Use eqn 15F.12 and the expressions for the partition functions assembled in Topic 15B.

Answer Use the following data: $\tilde{B}=0.1547\,\mathrm{cm^{-1}}$, $\tilde{\nu}=159.2\,\mathrm{cm^{-1}}$, $N_A hc\tilde{D}_0=70.4\,\mathrm{kJ\,mol^{-1}}$. For a homonuclear diatomic molecule, $\sigma=2$. The partition functions and other quantities required are as follows:

$\Lambda(Na_2)=8.14\,\mathrm{pm}$	$\Lambda(Na)=11.5\,\mathrm{pm}$
$q^R(Na_2)=2246$	$q^V(Na_2)=4.885$
$g(Na)=2$	$g(Na_2)=1$

Then, from eqn 15F.12,

$$K=\frac{2^2\times(1.381\times10^{-19}\,\mathrm{J\,K^{-1}})\times(1000\,\mathrm{K})\times(8.14\times10^{-12}\,\mathrm{m})}{(10^5\,\mathrm{Pa})\times2246\times4.885\times(1.15\times10^{-11}\,\mathrm{m})}\times e^{-8.47\ldots}$$
$$=2.45$$

where we have used $1\,\mathrm{J}=1\,\mathrm{kg\,m^2\,s^{-2}}$ and $1\,\mathrm{Pa}=1\,\mathrm{kg\,m^{-1}\,s^{-1}}$.

Self-test 15F.3 Evaluate K at 1500 K. Is the answer consistent with the dissociation being endothermic?

Answer: 52; yes

(c) Contributions to the equilibrium constant

We are now in a position to appreciate the physical basis of equilibrium constants. To see what is involved, consider a simple $R\rightleftharpoons P$ gas-phase equilibrium (R for reactants, P for products).

Figure 15F.2 shows two sets of energy levels; one set of states belongs to R, and the other belongs to P. The populations of the states are given by the Boltzmann distribution, and are independent of whether any given state happens to belong to R or to P. We can therefore imagine a single Boltzmann distribution

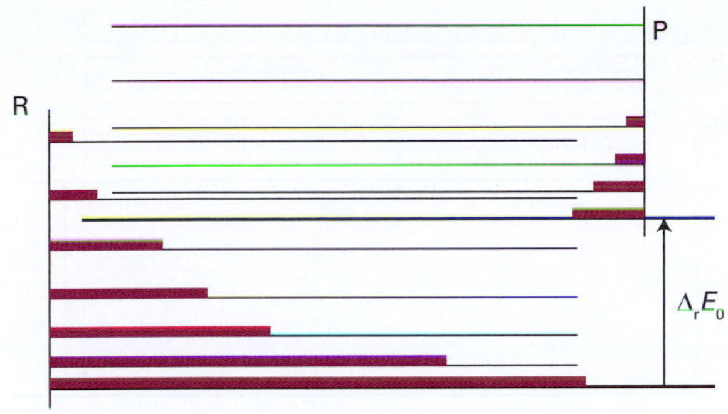

Figure 15F.2 The array of R(eactants) and P(roducts) energy levels. At equilibrium all are accessible (to differing extents, depending on the temperature), and the equilibrium composition of the system reflects the overall Boltzmann distribution of populations. As $\Delta_r E_0$ increases, R becomes dominant.

spreading, without distinction, over the two sets of states. If the spacings of R and P are similar (as in Fig. 15F.2), and P lies above R, the diagram indicates that R will dominate in the equilibrium mixture. However, if P has a high density of states (a large number of states in a given energy range, as in Fig. 15F.3), then, even though its zero-point energy lies above that of R, the species P might still dominate at equilibrium.

It is quite easy to show (see the following *Justification*) that the ratio of numbers of R and P molecules at equilibrium is given by

$$\frac{N_P}{N_R} = \frac{q_P}{q_R} e^{-\Delta_r E_0/RT} \qquad (15F.13a)$$

and therefore that the equilibrium constant for the reaction is

$$K = \frac{q_P}{q_R} e^{-\Delta_r E_0/RT} \qquad (15F.13b)$$

just as would be obtained from eqn 15F.12.

Justification 15F.2 The equilibrium constant in terms of the partition function 2

The population in a state i of the composite (R,P) system is $N_i = Ne^{-\beta\varepsilon_i}/q$, where N is the total number of molecules. The total number of R molecules is the sum of these populations taken over the states belonging to R; these states we label r with energies ε_r. The total number of P molecules is the sum over the states belonging to P; these states we label p with energies ε_p' (the prime is explained in a moment):

$$N_R = \sum_r N_r = \frac{N}{q}\sum_r e^{-\beta\varepsilon_r} \qquad N_P = \sum_p N_p = \frac{N}{q}\sum_p e^{-\beta\varepsilon_p'}$$

The sum over the states of R is its partition function, q_R, so $N_R = Nq_R/q$. The sum over the states of P is also a partition function, but the energies are measured from the ground state of the combined system, which is the ground state of R. However, because $\varepsilon_p' = \varepsilon_p + \Delta\varepsilon_0$, where $\Delta\varepsilon_0$ is the separation of zero-point energies (as in Fig. 15F.3),

$$N_P = \frac{N}{q}\sum_p e^{-\beta(\varepsilon_p+\Delta\varepsilon_0)} = \frac{N}{q}\left(\sum_p e^{-\beta\varepsilon_p}\right)e^{-\beta\Delta\varepsilon_0} = \frac{Nq_P}{q}e^{-\beta\Delta\varepsilon_0}$$
$$= \frac{Nq_P}{q}e^{-\Delta_r E_0/RT}$$

The switch from $\Delta\varepsilon_0/k$ to $\Delta_r E_0/R$ in the last step is the conversion of molecular energies to molar energies.

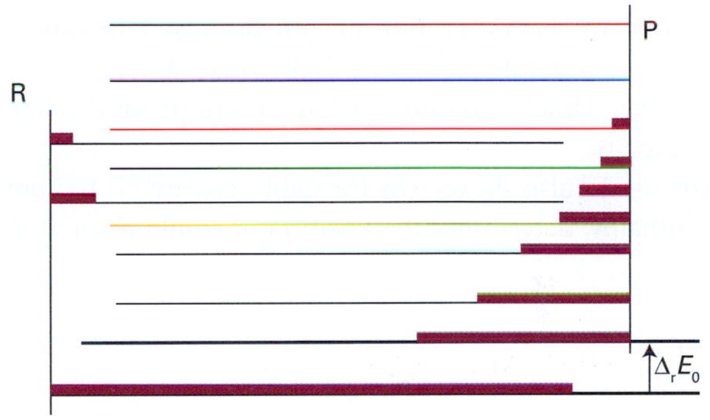

Figure 15F.3 It is important to take into account the densities of states of the molecules. Even though P might lie above R in energy (that is, $\Delta_r E_0$ is positive), P might have so many states that its total population dominates in the mixture. In classical thermodynamic terms, we have to take entropies into account as well as enthalpies when considering equilibria.

The equilibrium constant of the R ⇌ P reaction is proportional to the ratio of the numbers of the two types of molecule. Therefore,

$$K = \frac{N_P}{N_R} = \frac{q_P}{q} e^{-\Delta_r E_0/RT}$$

as in eqn 15F.13b. For an R ⇌ P equilibrium, the V factors in the partition functions cancel, so the appearance of q in place of $q^{\ominus}$ has no effect. In the case of a more general reaction, the conversion from q to $q^{\ominus}$ comes about at the stage of converting the pressures that occur in K to numbers of molecules.

The content of eqn 15F.13 can be seen most clearly by exaggerating the molecular features that contribute to it. We shall suppose that R has only a single accessible level, which implies that $q_R = 1$. We also suppose that P has a large number of evenly, closely spaced levels (Fig. 15F.4). The partition function of P is then $q_P = kT/\varepsilon$. In this model system, the equilibrium constant is

$$K=\frac{kT}{\varepsilon}\mathrm{e}^{-\Delta_r E_0/RT} \qquad (15F.14)$$

When $\Delta_r E_0$ is very large, the exponential term dominates and $K \ll 1$, which implies that very little P is present at equilibrium. When $\Delta_r E_0$ is small but still positive, K can exceed 1 because the factor kT/ε may be large enough to overcome the small size of the exponential term. The size of K then reflects the predominance of P at equilibrium on account of its high density of states. At low temperatures $K \ll 1$ and the system consists entirely of R. At high temperatures the exponential function approaches 1 and the pre-exponential factor is large. Hence P becomes dominant. We see that, in this endothermic reaction (endothermic because P lies above R), a rise in temperature favours P, because its states become accessible. This behaviour is what we saw, from the outside, in Topic 6B.

The model also shows why the Gibbs energy, G, and not just the enthalpy, determines the position of equilibrium. It shows that the density of states (and hence the entropy) of each species as well as their relative energies controls the distribution of populations and hence the value of the equilibrium constant.

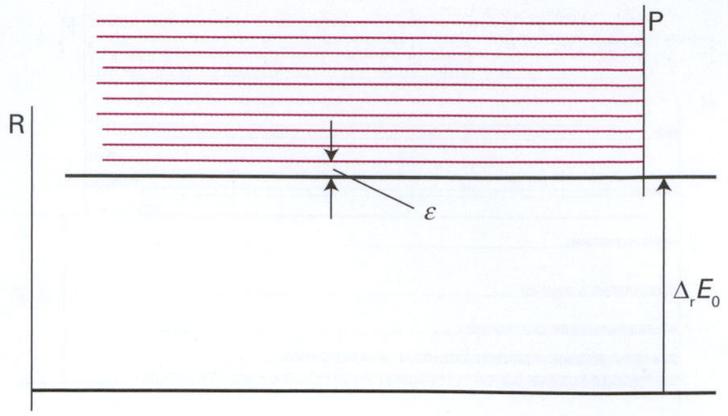

Figure 15F.4 The model used in the text for exploring the effects of energy separations and densities of states on equilibria. The products P can dominate provided $\Delta_r E_0$ is not too large and P has an appreciable density of states.

Checklist of concepts

☐ 1. The **thermodynamic functions** A, p, H, and G can be calculated from the canonical partition function.

☐ 2. For a perfect gas, G depends on the logarithm of the molecular partition function.

☐ 3. The **equilibrium constant** can be written in terms of the partition function.

☐ 4. The equilibrium constant for dissociation of a diatomic molecule in the gas phase may be calculated from spectroscopic data.

☐ 5. The **physical basis of chemical equilibrium** can be understood in terms of a competition between energy separations and densities of states.

Checklist of equations

Property	Equation	Comment	Equation number
Helmholtz energy	$A(T)=A(0)-kT\ln Q$		15F.2
Pressure	$p=kT(\partial \ln Q/\partial V)_T$		15F.3
Enthalpy	$H(T)=H(0)-(\partial \ln Q/\partial \beta)_V+kTV(\partial \ln Q/\partial V)_T$		15F.4
Gibbs energy	$G(T)=G(0)-kT\ln Q+kTV(\partial \ln Q/\partial V)_T$		15F.6
	$G(T)=G(0)-nRT\ln(q_m/N_A)$	Perfect gas	15F.9
Equilibrium constant	$K=\left\{\prod_J (q^{\ominus}_{J,m}/N_A)^{\nu_J}\right\}\mathrm{e}^{-\Delta_r E_0/RT}$	Perfect gas	15F.10b

CHAPTER 15 Statistical thermodynamics

Assume that all gases are perfect and that data refer to 298 K unless otherwise stated.

TOPIC 15A The Boltzmann distribution

Discussion questions

15A.1 Discuss the relationship between 'population', 'configuration', and 'weight'. What is the significance of the most probable configuration?

15A.2 What is the significance and importance of the principle of equal *a priori* probabilities?

15A.3 What is temperature?

15A.4 Summarize the role of the Boltzmann distribution in chemistry.

Exercises

15A.1(a) Calculate the weight of the configuration in which 16 objects are distributed in the arrangement 0, 1, 2, 3, 8, 0, 0, 0, 0, 2.
15A.1(b) Calculate the weight of the configuration in which 21 objects are distributed in the arrangement 6, 0, 5, 0, 4, 0, 3, 0, 2, 0, 0, 1.

15A.2(a) Evaluate 8! by using (i) the exact formula, (ii) Stirling's approximation, eqn 15A.2b; (iii) the more accurate version of Stirling's approximation, eqn 15A.2a.
15A.2(b) Evaluate 10! by using (i) the exact formula, eqn 15A.2b; (ii) Stirling's approximation, (iii) the more accurate version of Stirling's approximation, eqn 15A.2a.

15A.3(a) What are the relative populations of the states of a two-level system when the temperature is infinite?
15A.3(b) What are the relative populations of the states of a two-level system as the temperature approaches zero?

15A.4(a) What is the temperature of a two-level system of energy separation equivalent to 400 cm^{-1} when the population of the upper state is one-third that of the lower state?
15A.4(b) What is the temperature of a two-level system of energy separation equivalent to 300 cm^{-1} when the population of the upper state is one-half that of the lower state?

15A.5(a) Calculate the relative populations of a linear rotor in the levels with $J=0$ and $J=5$, given that $\tilde{B}=2.71\,cm^{-1}$ and a temperature of 298 K.
15A.5(b) Calculate the relative populations of a spherical rotor in the levels with $J=0$ and $J=5$, given that $\tilde{B}=2.71\,cm^{-1}$ and a temperature of 298 K.

15A.6(a) A certain molecule has a non-degenerate excited state lying at 540 cm^{-1} above the non-degenerate ground state. At what temperature will 10 per cent of the molecules be in the upper level?
15A.6(b) A certain molecule has a doubly degenerate excited state lying at 360 cm^{-1} above the non-degenerate ground state. At what temperature will 15 per cent of the molecules be in the upper level?

Problems

15A.1 A sample consisting of five molecules has a total energy 5ε. Each molecule is able to occupy states of energy $j\varepsilon$, with $j=0, 1, 2, \ldots$. (a) Calculate the weight of the configuration in which the molecules are distributed evenly over the available states. (b) Draw up a table with columns headed by the energy of the states and write beneath them all configurations that are consistent with the total energy. Calculate the weights of each configuration and identify the most probable configurations.

15A.2 A sample of nine molecules is numerically tractable but on the verge of being thermodynamically significant. Draw up a table of configurations for $N=9$, total energy 9ε in a system with energy levels $j\varepsilon$ (as in Problem 15A.1). Before evaluating the weights of the configurations, guess (by looking for the most 'exponential' distribution of populations) which of the configurations will turn out to be the most probable. Go on to calculate the weights and identify the most probable configuration.

15A.3 Use mathematical software to evaluate $\mathcal{W}$ for $N=20$ for a series of distributions over a uniform ladder of energy levels, ensuring that the total energy is constant. Identify the configuration of greatest weight and compare it to the distribution predicted by the Boltzmann expression. Explore what happens as the value of the total energy is changed.

15A.4 A certain atom has a doubly degenerate ground state and an upper level of four degenerate states at 450 cm^{-1} above the ground state. In an atomic beam study of the atoms it was observed that 30 per cent of the atoms were in the upper level, and the translational temperature of the beam was 300 K. Are the electronic states of the atoms in thermal equilibrium with the translational states?

15A.5 Explore the consequences of using the full version of Stirling's approximation, $x! \approx (2\pi)^{1/2}x^{x+1/2}e^{-x}$, in the development of the expression for the configuration of greatest weight. Does the more accurate approximation have a significant effect on the form of the Boltzmann distribution?

15A.6 The most probable configuration is characterised by a parameter we know as the 'temperature'. The temperatures of the system specified in Problems 15A.1 and 15A.2 must be such as to give a mean value of ε for the energy of each molecule and a total energy $N\varepsilon$ for the system. (a) Show that the temperature can be obtained by plotting p_j against j, where p_j is the (most probable) fraction of molecules in the state with energy $j\varepsilon$. Apply the procedure to the system in Problem 15A.2. What is the temperature of the system when ε corresponds to 50 cm^{-1}? (b) Choose configurations other than the most probable, and show that the same procedure gives a worse straight line, indicating that a temperature is not well-defined for them.

15A.7‡ The variation of the atmospheric pressure p with altitude h is predicted by the *barometric formula* to be $p=p_0e^{-h/H}$ where p_0 is the pressure at sea level and $H=RT/Mg$ with M the average molar mass of air and T the average temperature. Obtain the barometric formula from the Boltzmann distribution. Recall that the potential energy of a particle at height h above the surface of the Earth is mgh. Convert the barometric formula from pressure to number density, $\mathcal{N}$. Compare the relative number densities, $\mathcal{N}(h)/\mathcal{N}(0)$, for O_2 and H_2O at $h=8.0$ km, a typical cruising altitude for commercial aircraft.

15A.8‡ Planets lose their atmospheres over time unless they are replenished. A complete analysis of the overall process is very complicated and depends upon the radius of the planet, temperature, atmospheric composition, and other factors. Prove that the atmosphere of planets cannot be in an equilibrium state by demonstrating that the Boltzmann distribution leads to a uniform finite number density as $r\rightarrow\infty$. *Hint*: Recall that in a gravitational field the potential energy is $V(r)=-GMm/r$, where G is the gravitational constant, M is the mass of the planet, and m the mass of the particle.

TOPIC 15B Molecular partition functions

Discussion questions

15B.1 Describe the physical significance of the partition function.

15B.2 Describe how the mean energy of a system composed of two levels varies with temperature.

15B.3 What is the difference between a 'state' and an 'energy level'? Why is it important to make this distinction?

15B.4 Why and when is it necessary to include a symmetry number in the calculation of a partition function?

Exercises

15B.1(a) Calculate (i) the thermal wavelength, (ii) the translational partition function at 300 K and 3000 K of a molecule of molar mass 150 g mol^{-1} in a container of volume 1.00 cm^3.
15B.1(b) Calculate (i) the thermal wavelength, (ii) the translational partition function of a Ne atom in a cubic box of side 1.00 cm at 300 K and 3000 K.

15B.2(a) Calculate the ratio of the translational partition functions of H_2 and He at the same temperature and volume.
15B.2(b) Calculate the ratio of the translational partition functions of Ar and Ne at the same temperature and volume.

15B.3(a) The bond length of O_2 is 120.75 pm. Use the high-temperature approximation to calculate the rotational partition function of the molecule at 300 K.
15B.3(b) The bond length of N_2 is 109.75 pm. Use the high-temperature approximation to calculate the rotational partition function of the molecule at 300 K.

15B.4(a) The NOF molecule is an asymmetric rotor with rotational constants 3.1752 cm^{-1}, 0.3951 cm^{-1}, and 0.3505 cm^{-1}. Calculate the rotational partition function of the molecule at (i) 25 °C, (ii) 100 °C.
15B.4(b) The H_2O molecule is an asymmetric rotor with rotational constants 27.877 cm^{-1}, 14.512 cm^{-1}, and 9.285 cm^{-1}. Calculate the rotational partition function of the molecule at (i) 25 °C, (ii) 100 °C.

15B.5(a) The rotational constant of CO is 1.931 cm^{-1}. Evaluate the rotational partition function explicitly (without approximation) and plot its value as a function of temperature. At what temperature is the value within 5 per cent of the value calculated from the approximate formula?
15B.5(b) The rotational constant of HI is 6.511 cm^{-1}. Evaluate the rotational partition function explicitly (without approximation) and plot its value as a function of temperature. At what temperature is the value within 5 per cent of the value calculated from the approximate formula?

15B.6(a) The rotational constant of CH_4 is 5.241 cm^{-1}. Evaluate the rotational partition function explicitly (without approximation but ignoring the role of nuclear statistics) and plot its value as a function of temperature. At what temperature is the value within 5 per cent of the value calculated from the approximate formula?
15B.6(b) The rotational constant of CCl_4 is 0.0572 cm^{-1}. Evaluate the rotational partition function explicitly (without approximation but ignoring the role of nuclear statistics) and plot its value as a function of temperature. At what temperature is the value within 5 per cent of the value calculated from the approximate formula?

15B.7(a) The rotational constants of CH_3Cl are $\tilde{A}=5.097$ cm^{-1} and $\tilde{B}=0.443$ cm^{-1}. Evaluate the rotational partition function explicitly (without approximation but ignoring the role of nuclear statistics) and plot its value as a function of temperature. At what temperature is the value within 5 per cent of the value calculated from the approximate formula?
15B.7(b) The rotational constants of NH_3 are $\tilde{A}=6.196$ cm^{-1} and $\tilde{B}=9.444$ cm^{-1}. Evaluate the rotational partition function explicitly (without approximation but ignoring the role of nuclear statistics) and plot its value as a function of temperature. At what temperature is the value within 5 per cent of the value calculated from the approximate formula?

15B.8(a) Give the symmetry number for each of the following molecules: (i) CO, (ii) O_2, (iii) H_2S, (iv) SiH_4, and (v) $CHCl_3$.
15B.8(b) Give the symmetry number for each of the following molecules: (i) CO_2, (ii) O_3, (iii) SO_3, (iv) SF_6, and (v) Al_2Cl_6.

15B.9(a) Estimate the rotational partition function of ethene at 25 °C given that $\tilde{A}=4.828$ cm^{-1}, $\tilde{B}=1.0012$ cm^{-1}, and $\tilde{C}=0.8282$ cm^{-1}. Take the symmetry number into account.
15B.9(b) Evaluate the rotational partition function of pyridine, C_5H_5N, at 25 °C given that $\tilde{A}=0.2014$ cm^{-1}, $\tilde{B}=0.1936$ cm^{-1}, $\tilde{C}=0.0987$ cm^{-1}. Take the symmetry number into account.

15B.10(a) The vibrational wavenumber of Br_2 is 323.2 cm^{-1}. Evaluate the vibrational partition function explicitly (without approximation) and plot its value as a function of temperature. At what temperature is the value within 5 per cent of the value calculated from the approximate formula?
15B.10(b) The vibrational wavenumber of I_2 is 214.5 cm^{-1}. Evaluate the vibrational partition function explicitly (without approximation) and plot its value as a function of temperature. At what temperature is the value within 5 per cent of the value calculated from the approximate formula?

‡ These problems were supplied by Charles Trapp and Carmen Giunta.

15B.11(a) Calculate the vibrational partition function of CS_2 at 500 K given the wavenumbers 658 cm^{-1} (symmetric stretch), 397 cm^{-1} (bend; two modes), 1535 cm^{-1} (asymmetric stretch).
15B.11(b) Calculate the vibrational partition function of HCN at 900 K given the wavenumbers 3311 cm^{-1} (symmetric stretch), 712 cm^{-1} (bend; two modes), 2097 cm^{-1} (asymmetric stretch).

15B.12(a) Calculate the vibrational partition function of CCl_4 at 500 K given the wavenumbers 459 cm^{-1} (symmetric stretch, A), 217 cm^{-1} (deformation, E), 776 cm^{-1} (deformation, T), 314 cm^{-1} (deformation, T).
15B.12(b) Calculate the vibrational partition function of CI_4 at 500 K given the wavenumbers 178 cm^{-1} (symmetric stretch, A), 90 cm^{-1} (deformation, E), 555 cm^{-1} (deformation, T), 125 cm^{-1} (deformation, T).

15B.13(a) A certain atom has a fourfold degenerate ground level, a non-degenerate electronically excited level at 2500 cm^{-1}, and a twofold degenerate level at 3500 cm^{-1}. Calculate the partition function of these electronic states at 1900 K. What is the relative population of each level at 1900 K?
15B.13(b) A certain atom has a triply degenerate ground level, a non-degenerate electronically excited level at 850 cm^{-1}, and a fivefold degenerate level at 1100 cm^{-1}. Calculate the partition function of these electronic states at 2000 K. What is the relative population of each level at 2000 K?

Problems

15B.1 This problem is best done using mathematical software. Equation 15B.15 is the partition function for a harmonic oscillator. Consider a Morse oscillator (Topic 2D) in which the energy levels are given by eqn 2D.12:

$$E_v = \left(v+\tfrac{1}{2}\right)hc\tilde{\nu} - \left(v+\tfrac{1}{2}\right)^2 hcx_e\tilde{\nu}$$

Evaluate the partition function for this oscillator, remembering to measure energies from the lowest level and to note that there is only a finite number of bound-state levels. Plot the partition function against temperature for a variety of values of x_e, and—on the same graph—compare your results with that for an harmonic oscillator.

15B.2 Explore the conditions under which the 'integral' approximation for the translational partition function is not valid by considering the translational partition function of an H atom in a one-dimensional box of side comparable to that of a typical nanoparticle, 100 nm. Estimate the temperature at which, according to the integral approximation, $q = 10$ and evaluate the exact partition function at that temperature.

15B.3 (a) Calculate the electronic partition function of a tellurium atom at (i) 298 K, (ii) 5000 K by direct summation using the following data:

Term	Degeneracy	Wavenumber/cm^{-1}
Ground	5	0
1	1	4707
2	3	4751
3	5	10 559

(b) What proportion of the Te atoms are in the ground term and in the term labelled 2 at the two temperatures?

15B.4 The four lowest electronic levels of a Ti atom are $^3F_2$, $^3F_3$, $^3F_4$, and $^5F_1$, at 0, 170, 387, and 6557 cm^{-1}, respectively. There are many other electronic states at higher energies. The boiling point of titanium is 3287 °C. What are the relative populations of these levels at the boiling point? *Hint*: the degeneracies of the levels are $2J+1$.

15B.5‡ J. Sugar and A. Musgrove (*J. Phys. Chem. Ref. Data* **22**, 1213 (1993)) have published tables of energy levels for germanium atoms and cations from Ge^{+} to Ge^{+31}. The lowest-lying energy levels in neutral Ge are as follows:

	$^3P_0$	$^3P_1$	$^3P_2$	$^1D_2$	$^1S_0$
(E/hc)/cm^{-1}	0	557.1	1410.0	7125.3	16367.3

Calculate the electronic partition function at 298 K and 1000 K by direct summation. *Hint*: the degeneracy of a level J is $2J+1$.

15B.6 The pure rotational microwave spectrum of HCl has absorption lines at the following wavenumbers (in cm^{-1}): 21.19, 42.37, 63.56, 84.75, 105.93, 127.12, 148.31, 169.49, 190.68, 211.87, 233.06, 254.24, 275.43, 296.62, 317.80, 338.99, 360.18, 381.36, 402.55, 423.74, 444.92, 466.11, 487.30, 508.48. Calculate the rotational partition function at 25 °C by direct summation.

15B.7 Calculate, by explicit summation, the vibrational partition function and the vibrational contribution to the energy of I_2 molecules at (a) 100 K, (b) 298 K given that its vibrational energy levels lie at the following wavenumbers above the zero-point energy level: 0, 213.30, 425.39, 636.27, 845.93 cm^{-1}. What proportion of I_2 molecules are in the ground and first two excited levels at the two temperatures?

15B.8‡ Consider the electronic partition function of a perfect atomic hydrogen gas at a density of 1.99×10^{-4} kg m^{-3} and 5780 K. These are the mean conditions within the Sun's photosphere, the surface layer of the Sun that is about 190 km thick. (a) Show that this partition function, which involves a sum over an infinite number of quantum states that are solutions to the Schrödinger equation for an isolated atomic hydrogen atom, is infinite. (b) Develop a theoretical argument for truncating the sum and estimate the maximum number of quantum states that contribute to the sum. (c) Calculate the equilibrium probability that an atomic hydrogen electron is in each quantum state. Are there any general implications concerning electronic states that will be observed for other atoms and molecules? Is it wise to apply these calculations in the study of the Sun's photosphere?

15B.9 A formal way of arriving at the value of the symmetry number is to note that σ is the order (the number of elements) of the *rotational subgroup* of the molecule, the point group of the molecule with all but the identity and the rotations removed. The rotational subgroup of H_2O is $\{E,C_2\}$, so $\sigma=2$. The rotational subgroup of NH_3 is $\{E,2C_3\}$, so $\sigma=3$. This recipe makes it easy to find the symmetry numbers for more complicated molecules. The rotational subgroup of CH_4 is obtained from the T character table as $\{E,8C_3,3C_2\}$, so $\sigma=12$. For benzene, the rotational subgroup of D_{6h} is $\{E,2C_6,2C_3,C_2,3C_2',3C_2''\}$ so $\sigma=12$. (a) Estimate the rotational partition function of ethene at 25 °C given that $\tilde{A}=4.828$ cm^{-1}, $\tilde{B}=1.0012$ cm^{-1}, and $\tilde{C}=0.8282$ cm^{-1}. (b) Evaluate the rotational partition function of pyridine, C_5H_5N, at room temperature ($\tilde{A}=0.2014$ cm^{-1}, $\tilde{B}=0.1936$ cm^{-1}, $\tilde{C}=0.0987$ cm^{-1}).

TOPIC 15C Molecular energies

Discussion question

15C.1 Identify the conditions under which energies predicted from the equipartition theorem coincide with energies computed by using partition functions.

Exercises

15C.1(a) Compute the mean energy at 298 K of a two-level system of energy separation equivalent to 500 cm^{-1}.

15C.1(b) Compute the mean energy at 400 K of a two-level system of energy separation equivalent to 600 cm^{-1}.

15C.2(a) Evaluate, by explicit summation, the mean rotational energy of CO and plot its value as a function of temperature. At what temperature is the equipartition value within 5 per cent of the accurate value? $\tilde{B}(\text{CO})=1.931\,\text{cm}^{-1}$.

15C.2(b) Evaluate, by explicit summation, the mean rotational energy of HI and plot its value as a function of temperature. At what temperature is the equipartition value within 5 per cent of the accurate value? $\tilde{B}(\text{HI})=6.511\,\text{cm}^{-1}$.

15C.3(a) Evaluate, by explicit summation, the mean rotational energy of CH_4 and plot its value as a function of temperature. At what temperature is the equipartition value within 5 per cent of the accurate value? $\tilde{B}(\text{CH}_4)=5.241\,\text{cm}^{-1}$.

15C.3(b) Evaluate, by explicit summation, the mean rotational energy of CCl_4 and plot its value as a function of temperature. At what temperature is the equipartition value within 5 per cent of the accurate value? $\tilde{B}(\text{CCl}_4)=0.0572\,\text{cm}^{-1}$.

15C.4(a) Evaluate, by explicit summation, the mean rotational energy of CH_3Cl and plot its value as a function of temperature. At what temperature is the equipartition value within 5 per cent of the accurate value? $\tilde{A}=5.097\,\text{cm}^{-1}$ and $\tilde{B}=0.443\,\text{cm}^{-1}$.

15C.4(b) Evaluate, by explicit summation, the mean rotational energy of NH_3 and plot its value as a function of temperature. At what temperature is the equipartition value within 5 per cent of the accurate value? $\tilde{A}=6.196\,\text{cm}^{-1}$ and $\tilde{B}=9.444\,\text{cm}^{-1}$.

15C.5(a) Evaluate, by explicit summation, the mean vibrational energy of Br_2 and plot its value as a function of temperature. At what temperature is the equipartition value within 5 per cent of the accurate value? Use $\tilde{\nu}=323.2\,\text{cm}^{-1}$.

15C.5(b) Evaluate, by explicit summation, the mean vibrational energy of I_2 and plot its value as a function of temperature. At what temperature is the equipartition value within 5 per cent of the accurate value? Use $\tilde{\nu}=214.5\,\text{cm}^{-1}$.

15C.6(a) Evaluate, by explicit summation, the mean vibrational energy of CS_2 and plot its value as a function of temperature. At what temperature is the equipartition value within 5 per cent of the accurate value? Use the wavenumbers 658 cm^{-1} (symmetric stretch), 397 cm^{-1} (bend; two modes), 1535 cm^{-1} (asymmetric stretch). The A modes are non-degenerate, E modes are doubly degenerate, and T modes are triply degenerate.

15C.6(b) Evaluate, by explicit summation, the mean vibrational energy of HCN and plot its value as a function of temperature. At what temperature is the equipartition value within 5 per cent of the accurate value? Use the wavenumbers 3311 cm^{-1} (symmetric stretch), 712 cm^{-1} (bend; two modes), 2097 cm^{-1} (asymmetric stretch). A modes are non-degenerate, E modes are doubly degenerate, and T modes are triply degenerate.

15C.7(a) Evaluate, by explicit summation, the mean vibrational energy of CCl_4 and plot its value as a function of temperature. At what temperature is the equipartition value within 5 per cent of the accurate value? Use the wavenumbers 459 cm^{-1} (symmetric stretch, A), 217 cm^{-1} (deformation, E), 776 cm^{-1} (deformation, T), 314 cm^{-1} (deformation, T).

15C.7(b) Evaluate, by explicit summation, the mean vibrational energy of CI_4 and plot its value as a function of temperature. At what temperature is the equipartition value within 5 per cent of the accurate value? Use the wavenumbers 178 cm^{-1} (symmetric stretch, A), 90 cm^{-1} (deformation, E), 555 cm^{-1} (deformation, T), 125 cm^{-1} (deformation, T).

15C.8(a) Calculate the mean contribution to the electronic energy at 1900 K for a sample composed of the atoms specified in Exercise 15B.13(a).

15C.8(b) Calculate the mean contribution to the electronic energy at 2000 K for a sample composed of the atoms specified in Exercise 15B.13(b).

Problems

15C.1 An electron trapped in an infinitely deep spherical well of radius R, such as may be encountered in the investigation of nanoparticles, has energies given by the expression $E_{nl}=\hbar^2X_{nl}^2/2m_eR^2$, with X_{nl} the value obtained by searching for the zeroes of the spherical Bessel functions. The first six values (with a degeneracy of the corresponding energy level equal to $2l+1$) are as follows:

n	1	1	1	2	1	2
l	0	1	2	0	3	1
X_{nl}	3.142	4.493	5.763	6.283	6.988	7.725

Evaluate the partition function and mean energy of an electron as a function of temperature. Choose the temperature range and radius to be so low that only these six energy levels need be considered. *Hint*: Remember to measure energies from the lowest level.

15C.2 The NO molecule has a doubly degenerate excited electronic level 121.1 cm^{-1} above the doubly degenerate electronic ground term. Calculate and plot the electronic partition function of NO from $T=0$ to 1000 K. Evaluate (a) the term populations and (b) the mean electronic energy at 300 K.

15C.3 Consider a system with energy levels $\varepsilon_j=j\varepsilon$ and N molecules. (a) Show that if the mean energy per molecule is $a\varepsilon$, then the temperature is given by

$$\beta=\frac{1}{\varepsilon}\ln\left(1+\frac{1}{a}\right)$$

Evaluate the temperature for a system in which the mean energy is ε, taking ε equivalent to 50 cm^{-1}. (b) Calculate the molecular partition function q for the system when its mean energy is $a\varepsilon$.

15C.4 Deduce an expression for the root mean square energy, $\langle \varepsilon^2 \rangle^{1/2}$, in terms of the partition function and hence an expression for the root mean square deviation from the mean, $\Delta\varepsilon = (\langle \varepsilon^2 \rangle - \langle \varepsilon \rangle^2)^{1/2}$. Evaluate the resulting expression for a harmonic oscillator.

TOPIC 15D The canonical ensemble

Discussion questions

15D.1 Why is the concept of a canonical ensemble required?

15D.2 Explain what is meant by an ensemble and why it is useful in statistical thermodynamics.

15D.3 Under what circumstances may identical particles be regarded as distinguishable?

15D.4 What is meant by the 'thermodynamic limit'?

Exercises

15D.1(a) Identify the systems for which it is essential to include a factor of $1/N!$ on going from Q to q: (i) a sample of helium gas, (ii) a sample of carbon monoxide gas, (iii) a solid sample of carbon monoxide, (iv) water vapour.

15D.1(b) Identify the systems for which it is essential to include a factor of $1/N!$ on going from Q to q: (i) a sample of carbon dioxide gas, (ii) a sample of graphite, (iii) a sample of diamond, (iv) ice.

Problem

15D.1‡ For a perfect gas, the canonical partition function, Q, is related to the molecular partition function q by $Q = q^N/N!$. In Topic 15F it is established that $p = kT(\partial \ln Q/\partial V)_T$. Use the expression for q to derive the perfect gas law $pV = nRT$.

TOPIC 15E The internal energy and the entropy

Discussion questions

15E.1 Describe the molecular features that determine the magnitudes of the constant-volume molar heat capacity of a molecular substance.

15E.2 Discuss and illustrate the proposition that $1/T$ is a more natural measurement of temperature than T itself.

15E.3 Discuss the relationship between the thermodynamic and statistical definitions of entropy.

15E.4 Justify the differences between the partition-function expression for the entropy for distinguishable particles and the expression for indistinguishable particles.

15E.5 Account for the temperature and volume dependence of the entropy of a perfect gas in terms of the Boltzmann distribution.

15E.6 Explain the origin of residual entropy.

Exercises

15E.1(a) Use the equipartition theorem to estimate the constant-volume molar heat capacity of (i) I_2, (ii) CH_4, (iii) C_6H_6 in the gas phase at 25 °C.
15E.1(b) Use the equipartition theorem to estimate the constant-volume molar heat capacity of (i) O_3, (ii) C_2H_6, (iii) CO_2 in the gas phase at 25 °C.

15E.2(a) Estimate the values of $\gamma = C_p/C_V$ for gaseous ammonia and methane. Do this calculation with and without the vibrational contribution to the energy. Which is closer to the expected experimental value at 25 °C?
15E.2(b) Estimate the value of $\gamma = C_p/C_V$ for carbon dioxide. Do this calculation with and without the vibrational contribution to the energy. Which is closer to the expected experimental value at 25 °C?

15E.3(a) The ground level of Cl is $^2P_{3/2}$ and a $^2P_{1/2}$ level lies 881 cm^{-1} above it. Calculate the electronic contribution to the heat capacity of Cl atoms at (i) 500 K and (ii) 900 K.

15E.3(b) The first electronically excited state of O_2 is $^1\Delta_g$ and lies 7918.1 cm^{-1} above the ground state, which is $^3\Sigma_g^-$. Calculate the electronic contribution to the heat capacity of O_2 at 400 K.

15E.4(a) Plot the molar heat capacity of a collection of harmonic oscillators as a function of T/θ^V, and predict the vibrational heat capacity of ethyne at (i) 298 K, (ii) 500 K. The normal modes (and their degeneracies in parentheses) occur at wavenumbers 612(2), 729(2), 1974, 3287, and 3374 cm^{-1}.
15E.4(b) Plot the molar entropy of a collection of harmonic oscillators as a function of T/θ^V, and predict the standard molar entropy of ethyne at (i) 298 K, (ii) 500 K. For data, see the preceding exercise.

15E.5(a) Calculate the standard molar entropy at 298 K of (i) gaseous helium, (ii) gaseous xenon.

15E.5(b) Calculate the translational contribution to the standard molar entropy at 298 K of (i) $H_2O(g)$, (ii) $CO_2(g)$.

15E.6(a) At what temperature is the standard molar entropy of helium equal to that of xenon at 298 K?

15E.6(b) At what temperature is the translational contribution to the standard molar entropy of $CO_2(g)$ equal to that of $H_2O(g)$ at 298 K?

15E.7(a) Calculate the rotational partition function of H_2O at 298 K from its rotational constants 27.878 cm^{-1}, 14.509 cm^{-1}, and 9.287 cm^{-1} and use your result to calculate the rotational contribution to the molar entropy of gaseous water at 25 °C.

15E.7(b) Calculate the rotational partition function of SO_2 at 298 K from its rotational constants 2.027 36 cm^{-1}, 0.344 17 cm^{-1}, and 0.293 535 cm^{-1} and use your result to calculate the rotational contribution to the molar entropy of sulfur dioxide at 25 °C.

15E.8(a) The ground state of the Co^{2+} ion in $CoSO_4 \cdot 7H_2O$ may be regarded as $^4T_{9/2}$. The entropy of the solid at temperatures below 1 K is derived almost entirely from the electron spin. Estimate the molar entropy of the solid at these temperatures.

15E.8(b) Estimate the contribution of the spin to the molar entropy of a solid sample of a d-metal complex with $S=\frac{5}{2}$.

15E.9(a) Predict the standard molar entropy of methanoic acid (formic acid, HCOOH) at (i) 298 K, (ii) 500 K. The normal modes occur at wavenumbers 3570, 2943, 1770, 1387, 1229, 1105, 625, 1033, 638 cm^{-1}.

15E.9(b) Predict the standard molar entropy of ethyne at (i) 298 K, (ii) 500 K. The normal modes (and their degeneracies in parentheses) occur at wavenumbers 612(2), 729(2), 1974, 3287, and 3374 cm^{-1}.

Problems

15E.1 The NO molecule has a doubly degenerate electronic ground state and a doubly degenerate excited state at 121.1 cm^{-1}. Calculate and plot the electronic contribution to the molar heat capacity of the molecule up to 500 K.

15E.2 Explore whether a magnetic field can influence the heat capacity of a paramagnetic molecule by calculating the electronic contribution to the heat capacity of an NO_2 molecule in a magnetic field. Estimate the total constant-volume heat capacity using equipartition, and calculate the percentage change in heat capacity brought about by a 5.0 T magnetic field at (a) 50 K, (b) 298 K.

15E.3 The energy levels of a CH_3 group attached to a larger fragment are given by the expression for a particle on a ring, provided the group is rotating freely. What is the high-temperature contribution to the heat capacity and entropy of such a freely rotating group at 25 °C? The moment of inertia of CH_3 about its three-fold rotation axis (the axis that passes through the C atom and the centre of the equilateral triangle formed by the H atoms) is 5.341×10^{-47} kg m^2).

15E.4 Calculate the temperature dependence of the heat capacity of *p*-H_2 (in which only rotational states with even values of J are populated) at low temperatures on the basis that its rotational levels $J=0$ and $J=2$ constitute a system that resembles a two-level system except for the degeneracy of the upper level. Use $\tilde{B}=60.864$ cm^{-1} and sketch the heat capacity curve. The experimental heat capacity of *p*-H_2 does in fact show a peak at low temperatures.

15E.5‡ In a spectroscopic study of buckminsterfullerene C_{60}, F. Negri et al. (*J. Phys. Chem.* **100**, 10849 (1996)) reviewed the wavenumbers of all the vibrational modes of the molecule:

Mode	Number	Degeneracy	Wavenumber/cm^{-1}
A_u	1	1	976
T_{1u}	4	3	525, 578, 1180, and 1430
T_{2u}	5	3	354, 715, 1037, 1190, 1540
G_u	6	4	345, 757, 776, 963, 1315, 1410
H_u	7	5	403, 525, 667, 738, 1215, 1342, 1566

How many modes have a vibrational temperature θ^V below 1000 K? Estimate the molar constant-volume heat capacity of C_{60} at 1000 K, counting as active all modes with θ^V below this temperature.

15E.6 Use mathematical software to evaluate the heat capacity of the bound states of a Morse oscillator (see Problem 15B.1). Plot the heat capacity as a function of temperature. Can you devise a way to include the unbound states that lie above the dissociation limit?

15E.7 Although expressions like $\langle\varepsilon\rangle = -\mathrm{d}\ln q/\mathrm{d}\beta$ are useful for formal manipulations in statistical thermodynamics, and for expressing thermodynamic functions in neat formulas, they are sometimes more trouble than they are worth in practical applications. When presented with a table of energy levels, it is often much more convenient to evaluate the following sums directly:

$$q=\sum_j \mathrm{e}^{-\beta\varepsilon_j} \quad \dot{q}=\sum_j \beta\varepsilon_j \mathrm{e}^{-\beta\varepsilon_j} \quad \ddot{q}=\sum_j (\beta\varepsilon_j)^2 \mathrm{e}^{-\beta\varepsilon_j}$$

(a) Derive expressions for the internal energy, heat capacity, and entropy in terms of these three functions. (b) Apply the technique to the calculation of the electronic contribution to the constant-volume molar heat capacity of magnesium vapour at 5000 K using the following data:

Term	^1S	$^3P_0$	$^3P_1$	$^3P_2$	$^1P_1$	$^3S_1$
Degeneracy	1	1	3	5	3	3
$\tilde{\nu}$/cm^{-1}	0	21 850	21 870	21 911	35 051	41 197

15E.8 Show how the heat capacity of a linear rotor is related to the following sum:

$$\xi(\beta)=\frac{1}{q^2}\sum_{J,J'}\{\varepsilon(J)-\varepsilon(J')\}^2 g(J')\mathrm{e}^{-\beta\{\varepsilon(J)+\varepsilon(J')\}}$$

by

$$C=\frac{1}{2}Nk\beta^2\xi(\beta)$$

where the $\varepsilon(J)$ are the rotational energy levels and $g(J)$ their degeneracies. Then go on to show graphically that the total contribution to the heat capacity of a linear rotor can be regarded as a sum of contributions due to transitions $0\to1$, $0\to2$, $1\to2$, $1\to3$, etc. In this way, construct Fig. 15.1 for the rotational heat capacities of a linear molecule.

15E.9 Set up a calculation like that in Problem 15E.8 to analyse the vibrational contribution to the heat capacity in terms of excitations between levels and illustrate your results graphically in terms of a diagram like that in Fig. 15.1.

15E.10 Use the accurate expression for the rotational partition function calculated in Problem 15B.6 for HCl(g) to calculate the rotational contribution to the molar entropy over a range of temperature and plot the contribution as a function of temperature.

15E.11 Calculate the standard molar entropy of $N_2(g)$ at 298 K from its rotational constant $\tilde{B}=1.9987$ cm^{-1} and its vibrational wavenumber $\tilde{\nu}=2358$ cm^{-1}. The thermochemical value is 192.1 J K^{-1} mol^{-1}. What does this suggest about the solid at $T=0$?

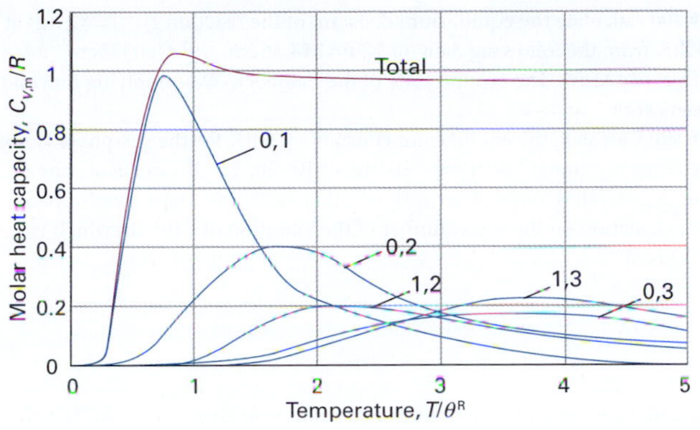

Figure 15.1 Contributions to the rotational heat capacity of a linear molecule.

15E.12‡ J.G. Dojahn et al. (*J. Phys. Chem.* **100**, 9649 (1996)) characterized the potential energy curves of the ground and electronic states of homonuclear diatomic halogen anions. The ground state of F_2^- is $^2\Sigma_u^+$ with a fundamental vibrational wavenumber of 450.0 cm^{-1} and equilibrium internuclear distance of 190.0 pm. The first two excited states are at 1.609 and 1.702 eV above the ground state. Compute the standard molar entropy of F_2^- at 298 K.

15E.13‡ Treat carbon monoxide as a perfect gas and apply equilibrium statistical thermodynamics to the study of its properties, as specified below, in the temperature range 100–1000 K at 1 bar. $\tilde{\nu}=2169.8\,\text{cm}^{-1}$, $\tilde{B}=1.931\,\text{cm}^{-1}$, and $hc\tilde{D}_0=11.09\,\text{eV}$; neglect anharmonicity and centrifugal distortion. (a) Examine the probability distribution of molecules over available rotational and vibrational states. (b) Explore numerically the differences, if any, between the rotational molecular partition function as calculated with the discrete energy distribution with that calculated with the classical, continuous energy distribution. (c) Calculate the individual contributions to $U_m(T)-U_m(100\,\text{K})$, $C_{V,m}(T)$, and $S_m(T)-S_m(100\,\text{K})$ made by the translational, rotational, and vibrational degrees of freedom.

15E.14 The energy levels of a Morse oscillator are given in Problem 15B.1. Set up the expression for the molar entropy of a collection of Morse oscillators and plot it as a function of temperature for a series of anharmonicities. Take into account only the finite number of bound states. On the same graph plot the entropy of an harmonic oscillator and investigate how the two diverge.

15E.15 Explore how the entropy of a collection of two-level systems behaves when the temperature is formally allowed to become negative. You should also construct a graph in which the temperature is replaced by the variable $\beta=1/kT$. Account for the appearance of the graphs physically.

15E.16 Derive the Sackur–Tetrode equation for a monatomic gas confined to a two-dimensional surface, and hence derive an expression for the standard molar entropy of condensation to form a mobile surface film.

15E.17‡ For H_2 at very low temperatures, only translational motion contributes to the heat capacity. At temperatures above $\theta^R=hc\tilde{B}/k$, the rotational contribution to the heat capacity becomes significant. At still higher temperatures, above $\theta^V=h\nu/k$, the vibrations contribute. But at this latter temperature, dissociation of the molecule into the atoms must be considered. (a) Explain the origin of the expressions for θ^R and θ^V, and calculate their values for hydrogen. (b) Obtain an expression for the molar constant-pressure heat capacity of hydrogen at all temperatures taking into account the dissociation of hydrogen. (c) Make a plot of the molar constant-pressure heat capacity as a function of temperature in the high-temperature region where dissociation of the molecule is significant.

15E.18 The heat capacity ratio of a gas determines the speed of sound in it through the formula $c_s=(\gamma RT/M)^{1/2}$, where $\gamma=C_p/C_V$ and M is the molar mass of the gas. Deduce an expression for the speed of sound in a perfect gas of (a) diatomic, (b) linear triatomic, (c) nonlinear triatomic molecules at high temperatures (with translation and rotation active). Estimate the speed of sound in air at 25 °C.

15E.19 An average human DNA molecule has 5×10^8 binucleotides (rungs on the DNA ladder) of four different kinds. If each rung were a random choice of one of these four possibilities, what would be the residual entropy associated with this typical DNA molecule?

15E.20 It is possible to write an approximate expression for the partition function of a protein molecule by including contributions from only two states: the native and denatured forms of the polymer. Proceeding with this crude model gives us insight into the contribution of denaturation to the heat capacity of a protein. According to this model, the total energy of a system of N protein molecules is

$$E=\frac{N\varepsilon e^{-\varepsilon/kT}}{1+e^{-\varepsilon/kT}}$$

where ε is the energy separation between the denatured and native forms. (a) Show that the constant-volume molar heat capacity is

$$C_{V,m}=f(T)R \quad f(T)=\frac{(\varepsilon/kT)^2 e^{-\varepsilon/kT}}{(1+e^{-\varepsilon/kT})^2}$$

(b) Plot the variation of $C_{V,m}$ with temperature. (c) If the function $C_{V,m}(T)$ has a maximum or minimum, derive an expression for the temperature at which it occurs.

TOPIC 15F Derived functions

Discussion questions

15F.1 Suggest a physical interpretation of the relation between pressure and the partition function.

15F.2 Suggest a physical interpretation of the relation between equilibrium constant and the partition functions of the reactants and products in a reaction.

15F.3 How does a statistical analysis of the equilibrium constant account for the latter's temperature dependence?

Exercises

15F.1(a) A CO_2 molecule is linear, and its vibrational wavenumbers are 1388.2 cm^{-1}, 2349.2 cm^{-1}, and 667.4 cm^{-1}, the last being doubly degenerate and the others non-degenerate. The rotational constant of the molecule is 0.3902 cm^{-1}. Calculate the rotational and vibrational contributions to the molar Gibbs energy at 298 K.

15F.1(b) An O_3 molecule is angular, and its vibrational wavenumbers are 1110 cm^{-1}, 705 cm^{-1}, and 1042 cm^{-1}. The rotational constants of the molecule are 3.553 cm^{-1}, 0.4452 cm^{-1}, and 0.3948 cm^{-1}. Calculate the rotational and vibrational contributions to the molar Gibbs energy at 298 K.

15F.2(a) Use the information in Exercise 15E.3(a) to calculate the electronic contribution to the molar Gibbs energy of Cl atoms at (i) 500 K and (ii) 900 K.
15F.2(b) Use the information in Exercise 15E.3(b) to calculate the electronic contribution to the molar Gibbs energy of O_2 at 400 K.

15F.3(a) Calculate the equilibrium constant of the reaction $I_2(g) \rightleftharpoons 2I(g)$ at 1000 K from the following data for I_2, $\tilde{\nu}=214.36$ cm^{-1}, $\tilde{B}=0.0373$ cm^{-1}, $hc\tilde{D}_e=1.5422$ eV. The ground state of the I atoms is $^2P_{3/2}$, implying fourfold degeneracy.
15F.3(b) Calculate the equilibrium constant at 298 K for the gas-phase isotopic exchange reaction $2\,^{79}Br^{81}Br \rightleftharpoons {}^{79}Br^{79}Br + {}^{81}Br^{81}Br$. The Br_2 molecule has a non-degenerate ground state, with no other electronic states nearby. Base the calculation on the wavenumber of the vibration of $^{79}Br^{81}Br$, which is 323.33 cm^{-1}.

Problems

15F.1 Calculate and plot as a function of temperature, in the range 300 K to 1000 K, the equilibrium constant for the reaction $CD_4(g) + HCl(g) \rightleftharpoons CHD_3(g) + DCl(g)$ using the following data (numbers in parentheses are degeneracies):

Molecule	$\tilde{\nu}$/cm^{-1}	$\tilde{B}$/cm^{-1}	$\tilde{A}$/cm^{-1}
CHD_3	2993 (1), 2142 (1), 1003 (3), 1291 (2), 1036 (2)	3.28	2.63
CD_4	2109 (1), 1092 (2), 2259 (3), 996 (3)	2.63	
HCl	2991 (1)	10.59	
DCl	2145 (1)	5.445	

15F.2 The exchange of deuterium between acid and water is an important type of equilibrium, and we can examine it using spectroscopic data on the molecules. Calculate the equilibrium constant at (a) 298 K and (b) 800 K for the gas-phase exchange reaction $H_2O + DCl \rightleftharpoons HDO + HCl$ from the following data:

Molecule	$\tilde{\nu}$/cm^{-1}	$\tilde{A}$/cm^{-1}	$\tilde{B}$/cm^{-1}	$\tilde{C}$/cm^{-1}
H_2O	3656.7, 1594.8, 3755.8	27.88	14.51	9.29
HDO	2726.7, 1402.2, 3707.5	23.38	9.102	6.417
HCl	2991		10.59	
DCl	2145		5.449	

15F.3 Determine whether a magnetic field can influence the value of an equilibrium constant. Consider the equilibrium $I_2(g) \rightleftharpoons 2\,I(g)$ at 1000 K, and calculate the ratio of equilibrium constants $K(\mathcal{B})/K$, where $K(\mathcal{B})$ is the equilibrium constant when a magnetic field $\mathcal{B}$ is present and removes the degeneracy of the four states of the $^2P_{3/2}$ level. Data on the species are given in Exercise 15F.3(a). The electronic g value of the atoms is $\frac{4}{3}$. Calculate the field required to change the equilibrium constant by 1 per cent.

15F.4‡ R. Viswanathan et al. (*J. Phys. Chem.* **100**, 10784 (1996)) studied thermodynamic properties of several boron–silicon gas-phase species experimentally and theoretically. These species can occur in the high-temperature chemical vapour deposition (CVD) of silicon-based semiconductors. Among the computations they reported was computation of the Gibbs energy of BSi(g) at several temperatures based on a $^4\Sigma^-$ ground state with equilibrium internuclear distance of 190.5 pm and fundamental vibrational wavenumber of 772 cm^{-1} and a $^2P_0$ first excited level 8000 cm^{-1} above the ground level. Compute the standard molar Gibbs energy $G_m^{\ominus}(2000\,K) - G_m^{\ominus}(0)$.

15F.5‡ The molecule Cl_2O_2, which is believed to participate in the seasonal depletion of ozone over Antarctica, has been studied by several means. M. Birk et al. (*J. Chem. Phys.* **91**, 6588 (1989)) report its rotational constants (B) as 13 109.4, 2409.8, and 2139.7 MHz. They also report that its rotational spectrum indicates a molecule with a symmetry number of 2. J. Jacobs et al. (*J. Amer. Chem. Soc.* **116**, 1106 (1994)) report its vibrational wavenumbers as 753, 542, 310, 127, 646, and 419 cm^{-1}. Compute $G_m^{\ominus}(200\,K) - G_m^{\ominus}(0)$ of Cl_2O_2.

15F.6‡ J. Hutter et al. (*J. Amer. Chem. Soc.* **116**, 750 (1994)) examined the geometric and vibrational structure of several carbon molecules of formula C_n. Given that the ground state of C_3, a molecule found in interstellar space and in flames, is an angular singlet with moments of inertia 39.340, 39.032, and $0.3082 m_u$ Å^2 (where 1 Å $= 10^{-10}$ m) and with vibrational wavenumbers of 63.4, 1224.5, and 2040 cm^{-1}, compute $G_m^{\ominus}(10.00\,K) - G_m^{\ominus}(0)$ and $G_m^{\ominus}(100.0\,K) - G_m^{\ominus}(0)$ for C_3.

Integrated activity

15.1 Use mathematical software, a spreadsheet, or the *Living graphs* on the web site for this book to: (a) Consider a three-level system with levels 0, ε, and 2ε. Plot the partition function against kT/ε. (b) Plot the function dS/dT for a two-level system, the temperature coefficient of its entropy, against kT/ε. Is there a temperature at which this coefficient passes through a maximum? If you find a maximum, explain its physical origins. (c) Plot the temperature dependence of the vibrational contribution to the molecular partition function for several values of the vibrational wavenumber. Estimate from your plots the temperature above which the harmonic oscillator is in the high-temperature limit.

CHAPTER 16

Molecular interactions

In this chapter we examine molecular interactions and interpret them in terms of electric properties of molecules. We see here, and in more detail in Chapter 17, that molecular interactions govern the structures and functions of molecular assemblies.

16A Electric properties of molecules

The chapter begins with an account of the electric properties of molecules, such as 'electric dipole moments' and 'polarizabilities'. All these properties reflect the degree to which the nuclei of atoms exert control over the electrons in a molecule, either by causing electrons to accumulate in particular regions, or by permitting them to respond more or less strongly to the effects of external electric fields.

16B Interactions between molecules

This Topic describes the basic theory of several important molecular interactions, with a special focus on 'van der Waals interactions' between closed-shell molecules. Also discussed are 'hydrogen bonding' and the 'hydrophobic interaction'. All liquids and solids are bound together by one or more of the cohesive interactions we explore in this Topic. Moreover, these interactions are also important for the structural organization of macromolecules.

16C Liquids

This Topic begins with the basic theory of molecular interactions in liquids, then turns to a description of the properties of liquid surfaces. We see how important effects, such as 'surface tension', 'capillary action', the formation of 'surface films', and condensation, can be explained by thermodynamics arguments.

What is the impact of this material?

Molecular interactions play important roles in biochemistry and biomedicine. In *Impact* I16.1 we focus on the binding of a drug, a small molecule or protein, to a specific receptor site of a target molecule, such as a larger protein or nucleic acid. The chemical result of the formation of this assembly is the inhibition of the progress of disease. We also discuss (in *Impact* I16.2) an example where manipulation of molecular interactions could have significant technological consequences: the design of assemblies that can store and deliver hydrogen gas efficiently, thereby making it a viable fuel for commercial development of a host of devices.

To read more about the impact of this material, scan the QR code, or go to bcs.whfreeman.com/webpub/chemistry/pchem10e/impact/pchem-16-1.html

16A Electric properties of molecules

Contents

➤ Why do you need to know this material?

Because the molecular interactions responsible for the formation of condensed phases and large molecular assemblies arise from the electric properties of molecules, you need to know how the electronic structures of molecules lead to these properties.

➤ What is the key idea?

The nuclei of atoms exert control over the electrons in a molecule, and can cause electrons to accumulate in particular regions, or permit them to respond more or less strongly to external fields.

➤ What do you need to know already?

You need to be familiar with the Coulomb law (*Foundations* B), molecular geometry, and molecular orbital theory, especially the relevance of the energy gap between a HOMO and LUMO (Topic 10E).

The electric properties of molecules are responsible for many of the properties of bulk matter. The small imbalances of charge distributions in molecules allow them to interact with one another and to respond to externally applied fields.

16A.1 Electric dipole moments

An **electric dipole** consists of two electric charges $+Q$ and $-Q$ with a separation R. A **point electric dipole** is an electric dipole in which R is very small compared with its distance from the observer. The **electric dipole moment** is a vector $\boldsymbol{\mu}$ (**1**) that points from the negative charge to the positive charge and has a magnitude given by

$$\mu = QR \quad \textit{Definition} \quad \text{Magnitude of the electric dipole moment} \quad (16A.1)$$

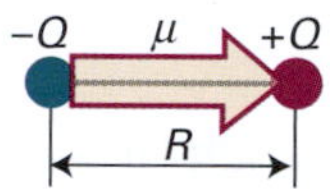

1 Electric dipole

Although the SI unit of dipole moment is coulomb metre (C m), it is still commonly reported in the non-SI unit debye, D, named after Peter Debye, a pioneer in the study of dipole moments of molecules:

$$1\,\text{D} = 3.335\,64 \times 10^{-30}\,\text{C m} \quad (16A.2)$$

The magnitude of the dipole moment formed by a pair of charges $+e$ and $-e$ separated by 100 pm 1.6×10^{-29} C m, corresponding to 4.8 D. The magnitudes of the dipole moments of small molecules are typically about 1 D.[1]

A **polar molecule** is a molecule with a permanent electric dipole moment. A **permanent dipole moment** stems from the partial charges on the atoms in the molecule that arise from differences in electronegativity or, in more sophisticated treatments, variations in electron density through the molecule (Topic 10E). Nonpolar molecules acquire an **induced dipole moment** in an electric field on account of the distortion the field causes in their electronic distributions and nuclear positions. However, this induced moment is only temporary, and disappears as soon as the perturbing field is removed. Polar molecules also have their existing dipole moments temporarily modified by an applied field.

All heteronuclear diatomic molecules are polar, and typical values of μ are 1.08 D for HCl and 0.42 D for HI (Table 16A.1). Molecular symmetry is of the greatest importance in deciding

[1] The conversion factor in eqn 16A.2 stems from the original definition of the debye in terms of c.g.s. units: 1 D is the dipole moment of two equal and opposite charges of magnitude 1 e.s.u. separated by 1 Å.

Table 16A.1* Dipole moments (μ) and polarizability volumes (α')

	μ/D	$\alpha'/(10^{-30}\ \mathrm{m^3})$
CCl_4	0	10.5
H_2	0	0.819
H_2O	1.85	1.48
HCl	1.08	2.63
HI	0.42	5.45

* More values will be found in the *Resource section*.

whether a polyatomic molecule is polar or not (see also Topic 11A). Indeed, molecular symmetry is more important than the question of whether or not the atoms in the molecule belong to the same element. For this reason, and as we see in the following *Brief illustration*, homonuclear polyatomic molecules may be polar if they have low symmetry and the atoms are in inequivalent positions.

Brief illustration 16A.1 Symmetry and the polarity of molecules

The angular molecule ozone (**2**) is homonuclear. However, it is polar because the central O atom is different from the outer two (it is bonded to two atoms, which are bonded only to one). Moreover, the dipole moments associated with each bond make an angle to each other and do not cancel. The heteronuclear linear triatomic molecule CO_2 is nonpolar because, although there are partial charges on all three atoms, the dipole moment associated with the OC bond points in the opposite direction to the dipole moment associated with the CO bond, and the two cancel (**3**).

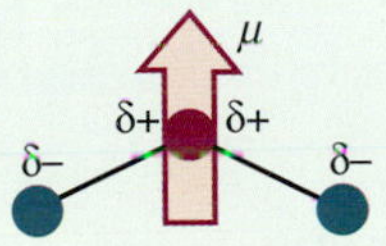

2 Ozone, O_3

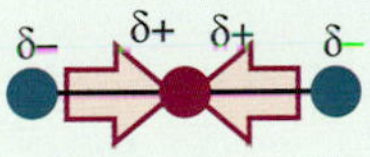

3 Carbon dioxide, CO_2

Self-test 16A.1 Is SO_2 polar?

Answer: Yes

The dipole moment of a polyatomic molecule can be resolved into contributions from various groups of atoms in the molecule and their relative locations (Fig. 16A.1). Thus, 1,4-dichlorobenzene is nonpolar by symmetry on account of the cancellation of two equal but opposing C–Cl moments (exactly as in carbon dioxide). 1,2-Dichlorobenzene, however, has a dipole moment which is approximately the resultant of two chlorobenzene dipole moments arranged at 60° to each other. This technique of 'vector addition' can be applied with fair success to other series of related molecules, and the magnitude of the resultant moment $\boldsymbol{\mu}_{res}$ of $\boldsymbol{\mu}_1$ and $\boldsymbol{\mu}_2$ that make an

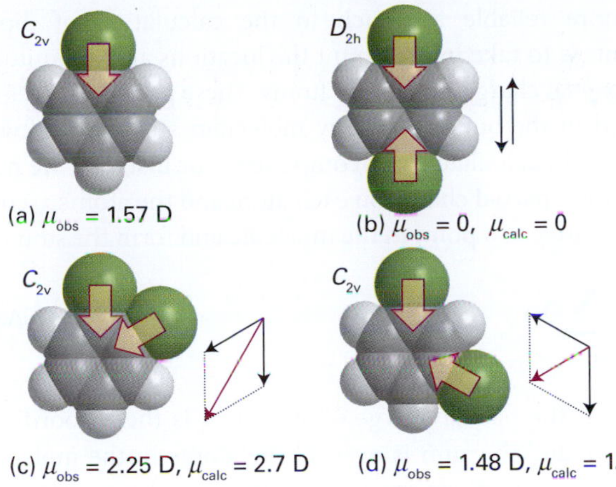

Figure 16A.1 The resultant dipole moments (purple in (c) and (d)) of the dichlorobenzene isomers, (b) to (d), can be obtained approximately by vectorial addition of two chlorobenzene dipole moments (shown in (a), with $\mu_{obs} = 1.57$ D). (The point groups of the molecules are also indicated.)

angle θ to each other (**4**) is approximately (see *Mathematical background* 5 following Chapter 9)

$$\mu_{res} \approx \left(\mu_1^2 + \mu_2^2 + 2\mu_1\mu_2\cos\theta\right)^{1/2} \quad (16A.3a)$$

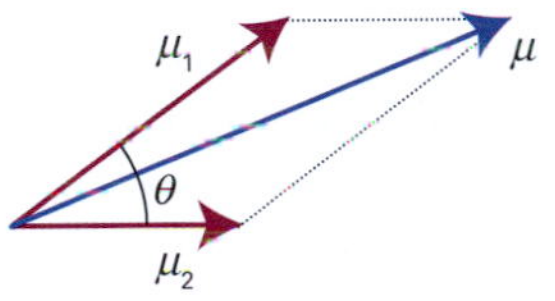

4 Addition of dipole moments

When the two contributing dipole moments have the same magnitude (as in the dichlorobenzenes), this equation simplifies to

$1+\cos\theta = 2\cos^2\tfrac{1}{2}\theta$

$$\mu_{res} \approx \{2\mu_1^2(1+\cos\theta)\}^{1/2} = 2\mu_1\cos\tfrac{1}{2}\theta \quad (16A.3b)$$

Brief illustration 16A.2 Molecular dipole moments

Consider *ortho* (1,2-) and *meta* (1,3-) disubstituted benzenes, for which $\theta_{ortho} = 60°$ and $\theta_{meta} = 120°$. It follows from eqn 16A.3b that the ratio of the magnitudes of the electric dipole moments is:

$$\frac{\mu_{res,\,ortho}}{\mu_{res,meta}} = \frac{\cos\frac{1}{2}\theta_{ortho}}{\cos\frac{1}{2}\theta_{meta}} = \frac{\cos\frac{1}{2}(60°)}{\cos\frac{1}{2}(120°)} = \frac{3^{1/2}/2}{1/2} = 3^{1/2} \approx 1.7$$

Self-test 16A.2 Calculate the resultant of two dipole moments of magnitude 1.5 D and 0.80 D that make an angle of 109.5° to each other.

Answer: 1.4 D

A more reliable approach to the calculation of dipole moments is to take into account the locations and magnitudes of the partial charges on all the atoms. These partial charges are included in the output of many molecular structure software packages. To calculate the x-component, for instance, we need to know the partial charge on each atom and the atom's x-coordinate relative to a point in the molecule and form the sum

$$\mu_x = \sum_J Q_J x_J \qquad (16A.4a)$$

Here Q_J is the partial charge of atom J, x_J is the x-coordinate of atom J, and the sum is over all the atoms in the molecule. Analogous expressions are used for the y- and z-components. For an electrically neutral molecule, the origin of the coordinates is arbitrary, so it is best chosen to simplify the measurements. In common with all vectors, the magnitude of $\boldsymbol{\mu}$ is related to the three components μ_x, μ_y, and μ_z by

$$\mu = \left(\mu_x^2 + \mu_y^2 + \mu_z^2\right)^{1/2} \qquad (16A.4b)$$

Example 16A.1 Calculating a molecular dipole moment

Estimate the magnitude and orientation of the electric dipole moment of the amide group shown in 5 by using the partial charges (as multiples of e) and the locations of the atoms shown, with distances in picometres.

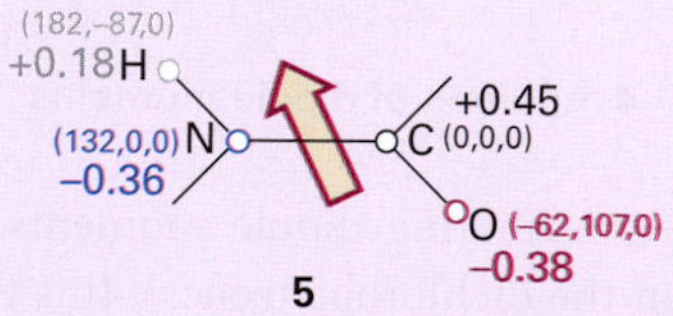

5

Method Use eqn 16A.4a to calculate each of the components of the dipole moment and then eqn 16A.4b to assemble the three components into the magnitude of the dipole moment. Note that the partial charges are multiples of the fundamental charge, $e = 1.609 \times 10^{-19}$ C.

Answer The expression for μ_x is

$$\begin{aligned}\mu_x &= (-0.36e)\times(132\,\text{pm}) + (0.45e)\times(0\,\text{pm}) + (0.18e)\times(182\,\text{pm}) \\ &\quad + (-0.38e)\times(-62.0\,\text{pm}) \\ &= 8.8e\,\text{pm} \\ &= 8.8\times(1.602\times10^{-19}\,\text{C})\times(10^{-12}\,\text{m}) = 1.4\times10^{-30}\,\text{C m}\end{aligned}$$

corresponding to $\mu_x = +0.42$ D. The expression for μ_y is:

$$\begin{aligned}\mu_y &= (-0.36e)\times(0\,\text{pm}) + (0.45e)\times(0\,\text{pm}) + (0.18e)\times(-87\,\text{pm}) \\ &\quad + (-0.38e)\times(107\,\text{pm}) \\ &= -56e\,\text{pm} \\ &= -19.0\times10^{-30}\,\text{C m}\end{aligned}$$

It follows that $\mu_y = -2.7$D. The amide group is planar, so $\mu_z = 0$ and

$$\mu = \{(0.42\,\text{D})^2 + (-2.7\,\text{D})^2\}^{1/2} = 2.7\,\text{D}$$

We can find the orientation of the dipole moment by arranging an arrow of length 2.7 units of length to have x, y, and z components of 0.42, −2.7, and 0 units; the orientation is superimposed on **5**.

Self-test 16A.3 Calculate the magnitude of the electric dipole moment of formaldehyde by using the information in **6**.

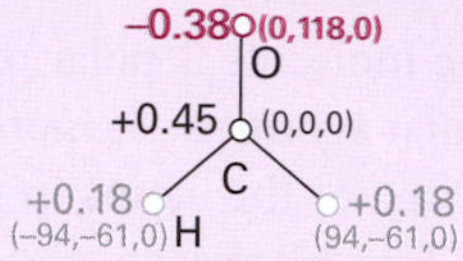

6

Answer: 2.3 D

Molecules may have higher **multipoles**, or arrays of point charges (Fig. 16A.2). Specifically, an ***n*-pole** is an array of point charges with an n-pole moment but no lower moment. Thus, a **monopole** ($n=1$) is a point charge, and the monopole moment is what we normally call the overall charge. A dipole ($n=2$), as we have seen, is an array of charges that has no monopole moment (no net charge). A **quadrupole** ($n=3$) consists of an array of point charges that has neither net charge nor dipole moment (as for CO_2 molecules, **3**). An **octupole** ($n=4$) consists of an array of point charges that sum to zero and which has neither a dipole moment nor a quadrupole moment (as for CH_4 molecules, **7**).

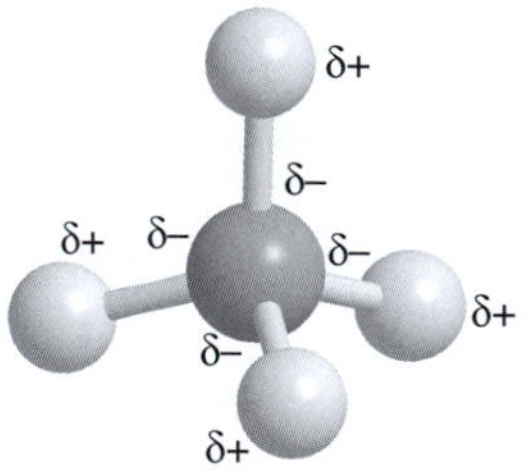

7 Methane, CH_4

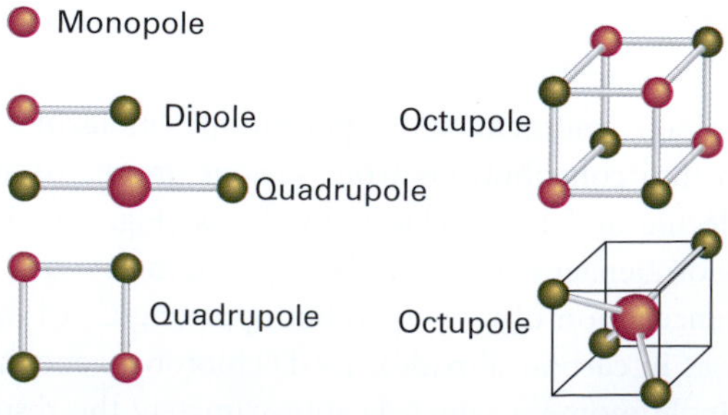

Figure 16A.2 Typical charge arrays corresponding to electric multipoles. The field arising from an arbitrary finite charge distribution can be expressed as the superposition of the fields arising from a superposition of multipoles.

16A.2 Polarizabilities

The failure of nuclear charges to control the surrounding electrons totally means that those electrons can respond to external fields. Therefore, an applied electric field can distort a molecule as well as align its permanent electric dipole moment. The magnitude μ^* of the **induced dipole moment, μ^***, is generally proportional to the field strength, $\mathcal{E}$, and we write

$$\mu^* = \alpha\mathcal{E} \quad \text{Definition} \quad \text{Polarizability} \quad (16A.5a)$$

The constant of proportionality α is the **polarizability** of the molecule. The greater the polarizability, the larger is the induced dipole moment for a given applied field. In a formal treatment, we should use vector quantities and allow for the possibility that the induced dipole moment might not lie parallel to the applied field, in which case the scalar α is replaced by $\boldsymbol{\alpha}$, a 3×3 matrix. We ignore this complication.

When the applied field is very strong (as in tightly focused laser beams), the magnitude of the induced dipole moment is not strictly linear in the strength of the field, and we write

$$\mu^* = \alpha\mathcal{E} + \tfrac{1}{2}\beta\mathcal{E}^2 + \cdots \quad \text{Definition} \quad \text{Hyperpolarizability} \quad (16A.5b)$$

The coefficient β is the (first) **hyperpolarizability** of the molecule.

Polarizability has the units (coulomb metre)² per joule ($C^2\ m^2\ J^{-1}$). That collection of units is awkward, so α is often expressed as a **polarizability volume**, α' by using the relation

$$\alpha' = \frac{\alpha}{4\pi\varepsilon_0} \quad \text{Definition} \quad \text{Polarizability volume} \quad (16A.6)$$

where ε_0 is the vacuum permittivity (*Foundations* B). Because the units of $4\pi\varepsilon_0$ are coulomb-squared per joule per metre ($C^2\ J^{-1}\ m^{-1}$), it follows that α' has the dimensions of volume (hence its name). Polarizability volumes are similar in magnitude to actual molecular volumes (of the order of $10^{-30}\ m^3$, $10^{-3}\ nm^3$, 1 Å³).

Brief illustration 16A.3 The induced dipole moment

The polarizability volume of H_2O is $1.48\times10^{-30}\ m^3$. It follows from eqns 16A.4a and 16A.5 that $\mu^* = 4\pi\varepsilon_0\alpha'\mathcal{E}$ and the magnitude of the dipole moment of the molecule (in addition to the permanent dipole moment) induced by an applied electric field of strength $1.0\times10^5\ V\ m^{-1}$ is

$$\mu^* = 4\pi\times(8.854\times10^{-12}\,J^{-1}\,C^2 m^{-1})\times(1.48\times10^{-30}\,m^3)\times(1.0\times10^5\,J\,C^{-1}m^{-1})$$

$$\overset{1\,V = 1\,J\,C^{-1}}{=} 31.6\times10^{-35}\,C\,m = 4.9\times10^{-6}\,D = 4.9\,\mu D$$

Self-test 16A.4 What strength of electric field is required to induce an electric dipole moment of magnitude 1.0 μD in a molecule of polarizability volume $2.6\times10^{-30}\ m^3$ (like CO_2)?

Answer: $11\ kV\ m^{-1}$

The polarizability volumes of some molecules are given in Table 16A.1. As shown in the following *Justification*, polarizability volumes correlate with the HOMO–LUMO separations in atoms and molecules (Topic 10E). The electron distribution can be distorted readily if the LUMO lies close to the HOMO in energy, so the polarizability is then large. If the LUMO lies high above the HOMO, an applied field cannot perturb the electron distribution significantly, and the polarizability is low. Molecules with small HOMO–LUMO gaps are typically large, and have numerous electrons.

Justification 16A.1 Polarizability and molecular structure

The quantum mechanical expression for the molecular polarizability in the z-direction is[2]

$$\alpha = 2\sum_{n\neq0}\frac{|\mu_{z,0n}|^2}{E_n^{(0)} - E_0^{(0)}}$$

where $\mu_{z,n0} = \int\psi_n^*\hat{\mu}_z\psi_0\,d\tau$ is the z-component of the *transition* electric dipole moment, a measure of the extent to which electric charge is shifted when an electron migrates from the ground state to create an excited state. The sum is over the excited states, with energies E_n. The content of this equation can be appreciated by approximating the excitation energies by a mean value ΔE (an indication of the HOMO–LUMO separation) and supposing that the most important transition dipole moment is approximately equal to the charge of an electron multiplied by the molecular radius R. Then

$$\alpha \approx \frac{2e^2R^2}{\Delta E}$$

This expression shows that α increases with the size of the molecule and with the ease with which it can be excited (the smaller the value of ΔE).

If the excitation energy is approximated by the energy needed to remove an electron to infinity from a distance R from a single positive charge, we can write $\Delta E \approx e^2/4\pi\varepsilon_0 R$. When this expression is substituted into $\alpha \approx 2e^2R^2/\Delta E$ to obtain $\alpha \approx 2(4\pi\varepsilon_0)R^3$, then both sides are divided by $4\pi\varepsilon_0$ and the factor of 2 ignored in this approximation, we obtain $\alpha' \approx R^3$, which is of the same order of magnitude as the molecular volume.

[2] For a derivation of this equation see our *Physical chemistry: Quanta, matter, and change* (2014).

For most molecules, the polarizability is anisotropic, by which is meant that its value depends on the orientation of the molecule relative to the field. The polarizability volume of benzene when the field is applied perpendicular to the ring is 0.0067 nm^3 and it is 0.0123 nm^3 when the field is applied in the plane of the ring. The anisotropy of the polarizability determines whether a molecule is rotationally Raman active (Topic 12C).

16A.3 Polarization

The **polarization**, P, of a sample is the electric dipole moment density, the mean electric dipole moment of the molecules, $\langle\mu\rangle$, multiplied by the number density, $\mathcal{N}$:

$$P=\langle\mu\rangle\mathcal{N} \qquad \textit{Definition} \quad \text{Polarization} \quad (16A.7)$$

In the following pages we refer to the sample as a **dielectric**, by which is meant a polarizable, non-conducting medium.

(a) The frequency dependence of the polarization

The polarization of an isotropic fluid sample is zero in the absence of an applied field because the molecules adopt ceaselessly changing random orientations due to thermal motion, so $\langle\mu\rangle=0$. In the presence of a weak electric field, the orientations of the molecular dipoles fluctuate but we show in the following *Justification* that the mean value of the dipole moment for the sample at a temperature T is

$$\langle\mu_z\rangle=\frac{\mu^2\mathcal{E}}{3kT} \qquad \textit{Weak electric field} \quad \text{Mean value of the dipole moment} \quad (16A.8)$$

where z is the direction of the applied field $\mathcal{E}$. At very high electric fields the orientations of molecular dipole moments fluctuate about the field direction to a lesser extent and the mean dipole moment approaches its maximum value of $\langle\mu_z\rangle=\mu$.

Justification 16A.2 The thermally averaged dipole moment

The probability dp that a dipole has an orientation in the range θ to $\theta+d\theta$ is given by the Boltzmann distribution (Topic 15A), which in this case is

$$dp=\frac{e^{-E(\theta)/kT}\sin\theta\, d\theta}{\int_0^{\pi} e^{-E(\theta)/kT}\sin\theta\, d\theta}$$

where $E(\theta)$ is the energy of the dipole in the field: $E(\theta)=-\mu\mathcal{E}\cos\theta$, with $0\le\theta\le\pi$. The average value of the component of the dipole moment parallel to the applied electric field is therefore

$$\langle\mu_z\rangle=\int\mu\cos\theta\, dp=\mu\int\cos\theta\, dp=\frac{\mu\int_0^{\pi} e^{x\cos\theta}\cos\theta\sin\theta\, d\theta}{\int_0^{\pi} e^{x\cos\theta}\sin\theta\, d\theta}$$

with $x=\mu\mathcal{E}/kT$. The integral takes on a simpler appearance when we write $y=\cos\theta$ and $dy=-\sin\theta\, d\theta$, and change the limits of integration to $y=-1$ (at $\theta=\pi$) and $y=1$ (at $\theta=0$):

$$\langle\mu_z\rangle=\frac{\mu\overbrace{\int_{-1}^{1} ye^{xy}dy}^{\text{Integral E.4}}}{\underbrace{\int_{-1}^{1} e^{xy}dy}_{\text{Integral E.3}}}$$

It is then straightforward algebra to obtain

$$\langle\mu_z\rangle=\mu L(x) \qquad L(x)=\frac{e^x+e^{-x}}{e^x-e^{-x}}-\frac{1}{x} \qquad x=\frac{\mu\mathcal{E}}{kT}$$

$L(x)$ is called the **Langevin function**.

Under most circumstances, x is very small (for example, if $\mu=1$ D and $T=300$ K, then x exceeds 0.01 only if the field strength exceeds 100 kV cm^{-1}, and most measurements are done at much lower strengths). The exponentials in the Langevin function can be expanded as $e^x=1+x+\frac{1}{2}x^2+\frac{1}{6}x^3+\cdots$ when the field is so weak that $x\ll 1$, the largest term that survives is

$$L(x)=\tfrac{1}{3}x+\cdots$$

Therefore, the average molecular dipole moment is given by eqn 16A.8.

When the applied field changes direction slowly, the permanent dipole moment has time to reorient—the whole molecule rotates into a new direction—and follows the field. However, when the frequency of the field is high, a molecule cannot change direction fast enough to follow the change in direction of the applied field and the permanent dipole moment then makes no contribution to the polarization of the sample. We say that the **orientation polarization**, the polarization arising from the permanent dipole moments, is lost at such high frequencies. Because a molecule takes about 1 ps to turn through about 1 radian in a fluid, the loss of the contribution of orientation polarization to the total polarization occurs when measurements are made at frequencies greater than about 10^{11} Hz (in the microwave region).

The next contribution to the polarization to be lost as the frequency is raised is the **distortion polarization**, the polarization that arises from the distortion of the positions of the nuclei by the applied field. The molecule is bent and stretched by the applied field, and the molecular dipole moment changes accordingly. The time taken for a molecule to bend is approximately the inverse of the molecular vibrational frequency, so the distortion polarization disappears when the frequency of the radiation is increased through the infrared.

The disappearance of polarization occurs in stages: as shown in the following *Justification*, each successive stage occurs as the incident frequency rises above the frequency of a particular mode of vibration. At even higher frequencies, in the visible region, only the electrons are mobile enough to respond to the rapidly changing direction of the applied field. The polarization that remains is now due entirely to the distortion of the electron distribution, and the surviving contribution to the molecular polarizability is called the **electronic polarizability**.

Justification 16A.3 The frequency dependence of polarizabilities

The quantum mechanical expression for the polarizability of a molecule in the presence of an electric field that is oscillating at a frequency ω in the z-direction is[3]

$$\alpha(\omega)=\frac{2}{\hbar}\sum_{n\neq 0}\frac{\omega_{n0}\left|\mu_{z,0n}\right|^2}{\omega_{n0}^2-\omega^2}$$

The quantities in this expression (which is valid provided that ω is not close to ω_{n0}) are the same as those in *Justification* 16A.1, with $\hbar\omega_{n0}=E_n-E_0$. As $\omega\rightarrow 0$, the equation reduces to the expression for the static polarizability in *Justification* 16A.1. As ω becomes very high (and much higher than any excitation frequency of the molecule so that the ω_{n0}^2 in the denominator can be ignored), the polarizability becomes

$$\alpha(\omega)=-\frac{2}{\hbar\omega^2}\sum_{n}\omega_{n0}\left|\mu_{0n}\right|^2\rightarrow 0 \qquad \text{as} \quad \omega\rightarrow\infty$$

That is, when the incident frequency is much higher than any excitation frequency, the polarizability becomes zero. The argument applies to each type of excitation, vibrational as well as electronic, and accounts for the successive decreases in polarizability as the frequency is increased.

(b) Molar polarization

When two charges Q_1 and Q_2 are separated by a distance r in a vacuum, the Coulomb potential energy of their interaction is (*Foundations* B)

$$V=\frac{Q_1Q_2}{4\pi\varepsilon_0 r} \tag{16A.9a}$$

[3] For a derivation of this equation see our *Physical chemistry: Quanta, matter, and change* (2014).

When the same two charges are immersed in a medium (such as air or a liquid), their potential energy is reduced to

$$V=\frac{Q_1Q_2}{4\pi\varepsilon r} \tag{16A.9b}$$

where ε is the **permittivity** of the medium. The permittivity is normally expressed in terms of the dimensionless **relative permittivity**, ε_r, (formerly and still widely called the *dielectric constant*) of the medium:

$$\varepsilon_r=\frac{\varepsilon}{\varepsilon_0} \qquad \textit{Definition} \quad \text{Relative permittivity} \tag{16A.10}$$

The relative permittivity can have a very significant effect on the strength of the interactions between ions in solution. For instance, water has a relative permittivity of 78 at 25 °C, so the interionic Coulombic interaction energy is reduced by nearly two orders of magnitude from its vacuum value. Some of the consequences of this reduction for electrolyte solutions are explored in Topic 5F.

The relative permittivity of a substance is large if its molecules are polar or highly polarizable. The quantitative relation between the relative permittivity and the electric properties of the molecules is obtained by considering the polarization of a medium, and is expressed by the **Debye equation**:

$$\frac{\varepsilon_r-1}{\varepsilon_r+2}=\frac{\rho P_m}{M} \qquad \text{Debye equation} \tag{16A.11}$$

where ρ is the mass density of the sample, M is the molar mass of the molecules, and P_m is the **molar polarization**, which is defined as

$$P_m=\frac{N_A}{3\varepsilon_0}\left(\alpha+\frac{\mu^2}{3kT}\right) \qquad \textit{Definition} \quad \text{Molar polarization} \tag{16A.12}$$

(where α is the polarizability, not the polarizability volume α').The term $\mu^2/3kT$ stems from the thermal averaging of the electric dipole moment in the presence of the applied field (eqn 16A.8). The corresponding expression without the contribution from the permanent dipole moment is called the **Clausius–Mossotti equation**:

$$\frac{\varepsilon_r-1}{\varepsilon_r+2}=\frac{\rho N_A\alpha}{3M\varepsilon_0} \qquad \text{Clausius–Mossotti equation} \tag{16A.13}$$

The Clausius–Mossotti equation is used when there is no contribution from permanent electric dipole moments to the polarization, either because the molecules are non-polar or because the frequency of the applied field is so high that the molecules cannot orientate quickly enough to follow the change in direction of the field.

Example 16A.2 Determining dipole moment and polarizability

The relative permittivity of a substance is measured by comparing the capacitance of a capacitor with and without the sample present (C and C_0, respectively) and using $\varepsilon_r = C/C_0$. The relative permittivity of camphor (**8**) was measured at a series of temperatures with the results given below. Determine the dipole moment and the polarizability volume of the molecule.

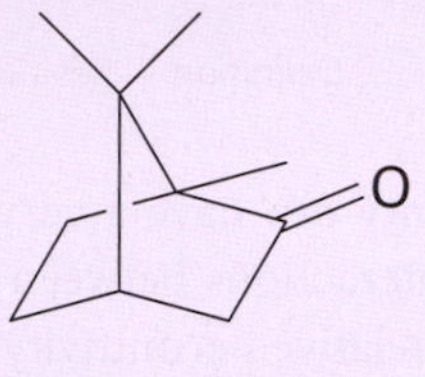

8 Camphor

θ/°C	ρ/(g cm^{-3})	ε_r
0	0.99	12.5
20	0.99	11.4
40	0.99	10.8
60	0.99	10.0
80	0.99	9.50
100	0.99	8.90
120	0.97	8.10
140	0.96	7.60
160	0.95	7.11
200	0.91	6.21

Method The relative permittivity depends on the molar polarization (eqn 16A.11), which in turn depends on the temperature, polarizability, and the magnitude of the permanent dipole moment (eqn 16A.12). It follows that the polarizability and permanent electric dipole moment of the molecules in a sample can be determined by:

- Measuring ε_r at a series of temperatures, calculate $(\varepsilon_r - 1)/(\varepsilon_r + 2)$ at each temperature, and then multiply by M/ρ to form P_m from eqn 16A.11;
- Plotting P_m against $1/T$. Because eqn 16A.16 rearranges to

$$P_m = \overbrace{\frac{N_A\alpha}{3\varepsilon_0}}^{\text{intercept}} + \overbrace{\frac{N_A\mu^2}{9\varepsilon_0 k}}^{\text{slope}} \times \frac{1}{T}$$

The slope of the graph is $N_A\mu^2/9\varepsilon_0 k$ and its intercept at $1/T = 0$ is $N_A\alpha/3\varepsilon_0$.

Answer For camphor, $M = 152.23\ \text{g mol}^{-1}$. We can therefore use the data to draw up the following table:

θ/°C	$(10^3\text{K})/T$	ε_r	$(\varepsilon_r - 1)/(\varepsilon_r + 2)$	P_m/(cm^3 mol^{-1})
0	3.66	12.5	0.793	122
20	3.41	11.4	0.776	119
40	3.19	10.8	0.766	118
60	3.00	10.0	0.750	115
80	2.83	9.50	0.739	114
100	2.68	8.90	0.725	111
120	2.54	8.10	0.703	110
140	2.42	7.60	0.688	109
160	2.31	7.11	0.670	107
200	2.11	6.21	0.634	106

The points are plotted in Fig. 16A.3. The intercept on the vertical axis lies at $P_m/(\text{cm}^3\ \text{mol}^{-1}) = 82.9$, so $N_A\alpha/3\varepsilon_0 = 82.9\ \text{cm}^3\ \text{mol}^{-1} = 8.29\times10^{-4}\ \text{m}^3\ \text{mol}^{-1}$; it then follows that

$$\alpha = \frac{3\times\overbrace{(8.854\times10^{-12}\ \text{J}^{-1}\text{C}^2\text{m}^{-1})}^{\varepsilon_0}}{\underbrace{6.02\times10^{23}\ \text{mol}^{-1}}_{N_A}} \times \overbrace{8.29\times10^{-4}\ \text{m}^3\ \text{mol}^{-1}}^{\text{intercept}}$$

$$= 3.53\times10^{-38}\ \text{C}^2\ \text{m}^2\ \text{J}^{-1}$$

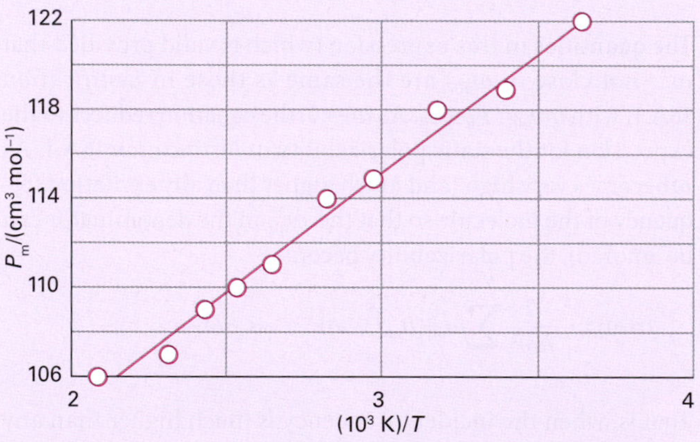

Figure 16A.3 The plot of P_m/(cm^3 mol^{-1}) against (10^3 K)/T used in *Example* 16A.2 for the determination of the polarizability and the magnitude of the dipole moment of camphor.

From eqn 16A.6, it follows that $\alpha' = 3.18\times10^{-28}\ \text{m}^3 = 3.18\times10^{-23}\ \text{cm}^3$. The slope is 10.7, so $N_A\mu^2/9\varepsilon_0 k = 10.7\ \text{cm}^3\ \text{mol}^{-1}\ \text{K} = 1.07\times10^{-4}\ \text{m}^3\ \text{mol}^{-1}\ \text{K}$, so from the expression for P_m in the *Method* it follows that

$$\mu = \left(\frac{9\times\overbrace{(8.854\times10^{-12}\ \text{J}^{-1}\text{C}^2\text{m}^{-1})}^{\varepsilon_0}\times\overbrace{(1.381\times10^{-23}\ \text{J K}^{-1})}^{k}}{\underbrace{6.022\times10^{23}\ \text{mol}^{-1}}_{N_A}}\right)^{1/2} \times \left(\overbrace{1.07\times10^{-4}\ \text{m}^3\ \text{mol}^{-1}\ \text{K}}^{\text{slope}}\right)^{1/2}$$

$$= 4.42\times10^{-31}\ \text{C m} = 0.134\ \text{D}$$

Because the Debye equation describes molecules that are free to rotate, the data show that camphor, which does not melt until 175 °C, is rotating even in the solid. It is an approximately spherical molecule.

Self-test 16A.5 The relative permittivity of chlorobenzene is 5.71 at 20 °C and 5.62 at 25 °C. Assuming a constant density (1.11 g cm^{-3}), estimate its polarizability volume and the magnitude of its dipole moment.

Answer: 1.4×10^{-23} cm^3, 1.2 D

According to Maxwell's theory of electromagnetic radiation, the refractive index at a (visible or ultraviolet) specified wavelength is related to the relative permittivity at that frequency by:

$$n_r = \varepsilon_r^{1/2} \qquad \text{Relation between refractive index and relative permittivity} \qquad (16A.14)$$

where the refractive index, n_r, of the medium is the ratio of the speed of light in a vacuum, c, to its speed c' in the medium: $n_r = c/c'$. A beam of light changes direction ('bends') when it passes from a region of one refractive index to a region with a different refractive index. Therefore, the molar polarization, P_m, and the molecular polarizability, α, can be measured at frequencies typical of visible light (about 10^{15} to 10^{16} Hz) by measuring the refractive index of the sample and using the Clausius–Mossotti equation.

Checklist of concepts

☐ 1. An **electric dipole** consists of two electric charges $+Q$ and $-Q$ separated by a vector $\boldsymbol{R}$.

☐ 2. The **electric dipole moment** $\boldsymbol{\mu}$ is a vector that points from the negative charge to the positive charge of a dipole; its magnitude is μ.

☐ 3. A **polar molecule** is a molecule with a permanent electric dipole moment.

☐ 4. Molecules may have higher electric multipoles: an ***n*-pole** is an array of point charges with an n-pole moment but no lower moment.

☐ 5. The **polarizability** is a measure of the ability of an electric field to induce a dipole moment in a molecule.

☐ 6. **Polarizabilities** (and polarizability volumes) correlate with the HOMO–LUMO separations in atoms and molecules.

☐ 7. For most molecules, the polarizability is anisotropic.

☐ 8. The **polarization** of a medium is the electric dipole moment density.

☐ 9. **Orientation polarization** is the polarization arising from the permanent dipole moments.

☐ 10. **Distortion polarization** is the polarization arising from the distortion of the positions of the nuclei by the applied field.

☐ 11. **Electronic polarizability** is the polarizability due to the distortion of the electron distribution.

Checklist of equations

Property	Equation	Comment	Equation number
Magnitude of the electric dipole moment	$\mu = QR$	Definition	16A.1
Magnitude of the resultant of two dipole moments	$\mu_{res} \approx (\mu_1^2+\mu_2^2+2\mu_1\mu_2\cos\theta)^{1/2}$		16A.3a
Magnitude of the induced dipole moment	$\mu^* = \alpha E$	Linear approximation; α is the polarizability	16A.5a
	$\mu^* = \alpha E + \frac{1}{2}\beta E^2$	Quadratic approximation; β is the hyperpolarizability	16A.5b
Polarizability volume	$\alpha' = \alpha/4\pi\varepsilon_0$	Definition	16A.6
Polarization	$P = \langle\mu\rangle\mathcal{N}$	Definition	16A.7
Potential energy of interaction between two charges in a medium	$V = Q_1Q_2/4\pi\varepsilon r$	The relative permittivity of the medium is $\varepsilon_r = \varepsilon/\varepsilon_0$	16A.9b
Debye equation	$(\varepsilon_r-1)/(\varepsilon_r+2) = \rho P_m/M$		16A.11
Molar polarization	$P_m = (N_A/3\varepsilon_0)(\alpha + \mu^2/3kT)$	Definition	16A.12
Clausius–Mossotti equation	$(\varepsilon_r-1)/(\varepsilon_r+2) = \rho N_A\alpha/3M\varepsilon_0$		16A.13

16B Interactions between molecules

Contents

➤ Why do you need to know this material?

You need to understand the many types of molecular interactions responsible for the formation of condensed phases and large molecular assemblies. The molecular interactions described here are of prime importance for solving one of the great problems of molecular biology: how complex molecules, like proteins and nucleic acids, fold into their three-dimensional structures.

➤ What is the key idea?

Attractive interactions result in cohesion, but repulsive interactions prevent the complete collapse of matter to nuclear densities.

➤ What do you need to know already?

You need to be familiar with elementary aspects of electrostatics, specifically the Coulomb interaction (*Foundations* B), and the relationships between the structure and electric properties of a molecule, specifically its dipole moment and polarizability (Topic 16A).

We begin by examining the interactions between the partial charges of polar molecules. Then we discuss **van der Waals interactions**: attractive interactions between closed-shell molecules that depend on the separation of the molecules as the inverse sixth power ($V \propto 1/r^6$), although this precise criterion is often relaxed to include all non-bonding interactions. Finally, we see that repulsive interactions arise from Coulomb forces and, indirectly, from the Pauli principle (Topic 9B) and the exclusion of electrons from regions of space where the orbitals of neighbouring species overlap.

16B.1 Interactions between partial charges

In general, atoms in molecules have partial charges arising from the spatial variation in electron density in the ground state. If these charges were separated by a medium, they would attract or repel each other in accord with Coulomb's law, and we would write (as in Topic 16A):

$$V = \frac{Q_1 Q_2}{4\pi\varepsilon r} \qquad \text{Coulomb potential energy in a medium} \qquad (16B.1)$$

where Q_1 and Q_2 are the partial charges, r is their separation, and ε is the permittivity of the medium lying between the charges. The following *Brief illustration* examines effect of the permittivity of the medium on the strength of the interaction.

Brief illustration 16B.1 The interaction energy of two partial charges

Different values of the permittivity of the medium take into account the possibility that other parts of the molecule, or other molecules, lie between the charges. For example, the energy of interaction between a partial charge of −0.36 (that is, $Q_1=-0.36e$) on the N atom of an amide group and the partial charge of +0.45 ($Q_2=+0.45e$) on the carbonyl C atom at a distance of 3.0 nm, on the assumption that the medium between them is a vacuum, is

$$V=\frac{(-0.36e)\times(0.45e)}{4\pi\varepsilon_0\times(3.0\,\text{nm})}$$

$$=-\frac{0.36\times0.45\times(1.602\times10^{-19}\,\text{C})^2}{4\pi\times(8.854\times10^{-2}\,\text{J}^{-1}\text{C}^{-2}\text{m}^{-1})\times(3.0\times10^{-9}\,\text{m})}$$

$$=-1.2\times10^{-20}\,\text{J}$$

where ε_0 is the vacuum permittivity. This energy (after multiplication by Avogadro's constant) corresponds to −7.5 kJ mol^{-1}. However, if the medium has a 'typical' relative permittivity $\varepsilon_r=\varepsilon/\varepsilon_0=3.5$ (Topic 16A), then the interaction energy is reduced by that factor to −2.1 kJ mol^{-1}.

Self-test 16B.1 Repeat the calculation for bulk water as the medium.

Answer: −0.96 kJ mol^{-1}

16B.2 The interactions of dipoles

Most of the discussion in this and the following sections is based on the Coulombic potential energy of interaction between two charges (eqn 16B.1). This expression can be adapted to find the potential energy of a point charge and a dipole and extend it to the interaction between two dipoles.

(a) Charge–dipole interactions

A **point dipole** is a dipole in which the separation between the charges is much smaller than the distance at which the dipole is being observed ($l \ll r$). We show in the following *Justification* that the potential energy of interaction between a point dipole with a dipole moment of magnitude $\mu_1=Q_1l$, and the point charge Q_2 in the arrangement shown in **1** is

$$V=-\frac{\mu_1 Q_2}{4\pi\varepsilon_0 r^2} \qquad \text{Energy of interaction between a point dipole and a point charge} \qquad (16B.2)$$

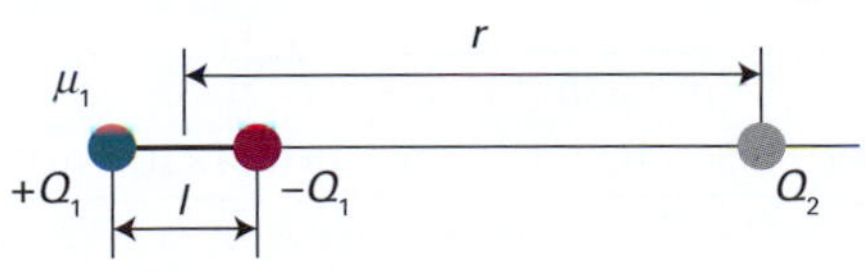

1

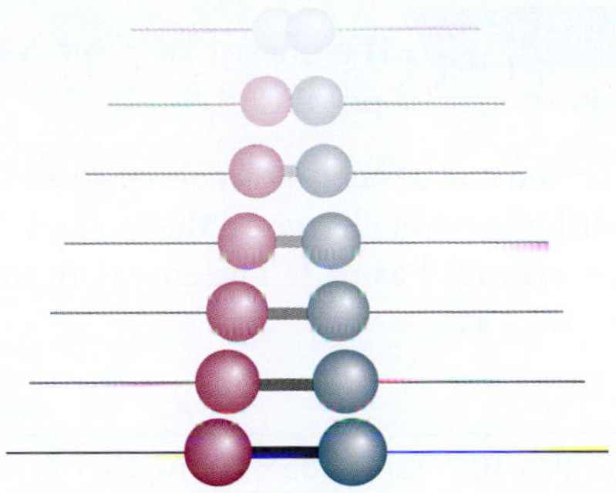

Figure 16B.1 There are two contributions to the diminishing field of an electric dipole with distance (here seen from the side). The potentials of the charges decrease (shown here by a fading intensity) and the two charges appear to merge, so their combined effect approaches zero more rapidly than by the distance effect alone.

With μ_1 in coulomb metres, Q_2 in coulombs, and r in metres, V is obtained in joules (and in the orientation shown in **1** is negative, representing a net attraction). The potential energy rises towards zero (the value at infinite separation of the charge and the dipole) more rapidly (as $1/r^2$) than that between two point charges (which varies as $1/r$) because, from the viewpoint of the point charge, the partial charges of the dipole seem to merge and cancel as the distance r increases (Fig. 16B.1).

Justification 16B.1 The interaction between a point charge and a point dipole

The sum of the potential energies of repulsion between like charges and attraction between opposite charges in the orientation shown in **1** is

$$V=\frac{1}{4\pi\varepsilon_0}\left(-\frac{Q_1Q_2}{r-\frac{1}{2}l}+\frac{Q_1Q_2}{r+\frac{1}{2}l}\right)=\frac{Q_1Q_2}{4\pi\varepsilon_0 r}\left(-\frac{1}{1-x}+\frac{1}{1+x}\right)$$

where $x=l/2r$. Because $l \ll r$ for a point dipole, this expression can be simplified by expanding the terms in x by using (*Mathematical background* 1)

$$\frac{1}{1+x}=1-x+x^2-\cdots \qquad \frac{1}{1-x}=1+x+x^2+\cdots$$

and retaining only the leading surviving term:

$$V=\frac{Q_1Q_2}{4\pi\varepsilon_0 r}\{-(1+x+\cdots)+(1-x+\cdots)\}\approx-\frac{2xQ_1Q_2}{4\pi\varepsilon_0 r}=-\frac{Q_1Q_2 l}{4\pi\varepsilon_0 r^2}$$

With $\mu_1=Q_1l$, this expression becomes eqn 16B.2. The equation should be multiplied by cos θ when the point charge lies at an angle θ to the axis of the dipole.

Brief illustration 16B.2 The energy of interaction of a point charge and a point dipole

Consider a Li^+ ion and a water molecule ($\mu = 1.85\,D$) separated by 1.0 nm, with the point charge on the ion and the dipole of the molecule arranged as in **1**. The energy of interaction is given by eqn 16B.2 as

$$V = -\frac{\overbrace{(1.602\times10^{-19}\,\text{C})}^{Q_{\text{Li}^+}}\times\overbrace{(1.85\times3.336\times10^{-30}\,\text{C m})}^{\mu_{\text{H}_2\text{O}}}}{4\pi\times\underbrace{(8.854\times10^{-12}\,\text{J}^{-1}\text{C}^{-1}\text{m}^{-1})}_{\varepsilon_0}\times\underbrace{(1.0\times10^{-9}\,\text{m})^2}_{r}}$$

$$= -8.9\times10^{-21}\,\text{J}$$

This energy corresponds to $-5.4\,\text{kJ mol}^{-1}$.

Self-test 16B.2 Consider the arrangement in **1** and calculate the molar energy required to reverse the direction of the water molecule when it is at 300 pm from the Li^+ ion.

Answer: $119\,\text{kJ mol}^{-1}$

(b) Dipole–dipole interactions

We show in the following *Justification* that the preceding discussion can be extended to the interaction of two dipoles arranged as in **2**. The result is

$$V = -\frac{\mu_1\mu_2}{2\pi\varepsilon_0 r^3}$$ *Arrangement as in 2* Energy of interaction between two dipoles (16B.3)

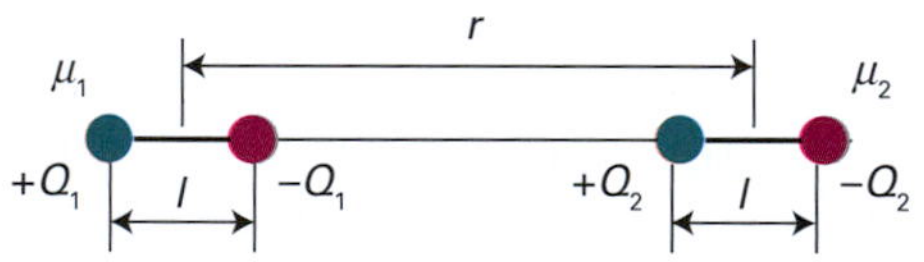

2

This interaction energy approaches zero more rapidly (as $1/r^3$) than for the previous case: now both interacting entities appear neutral to each other at large separations.

Justification 16B.2 The interaction energy of two dipoles

To calculate the potential energy of interaction of two dipoles separated by r in the arrangement shown in **2** we proceed in exactly the same way as in *Justification* 16B.1, but now the total interaction energy is the sum of four pairwise terms, two attractions between opposite charges, which contribute negative terms to the potential energy, and two repulsions between like charges, which contribute positive terms.

The sum of the four contributions is

$$V = \frac{1}{4\pi\varepsilon_0}\left(-\frac{Q_1Q_2}{r+l} + \frac{Q_1Q_2}{r} + \frac{Q_1Q_2}{r} - \frac{Q_1Q_2}{r-l}\right)$$

$$= -\frac{Q_1Q_2}{4\pi\varepsilon_0 r}\left(\frac{1}{1+x} - 2 + \frac{1}{1-x}\right)$$

with $x = l/r$. As before, provided $l \ll r$ we can expand the two terms in x and retain only the first surviving term, which is equal to $2x^2$. This step results in the expression

$$V = -\frac{2x^2 Q_1Q_2}{4\pi\varepsilon_0 r}$$

Therefore, because $\mu_1 = Q_1 l$ and $\mu_2 = Q_2 l$, the potential energy of interaction in the alignment shown in **2** is given by eqn 16B.3.

The preceding *Justification* represents only one possible orientation of two dipoles. More generally, the potential energy of interaction between two polar molecules is a complicated function of their relative orientation. When the two dipoles are parallel and arranged as in **3** , the potential energy is simply

$$V = \frac{\mu_1\mu_2 f(\theta)}{4\pi\varepsilon_0 r^3} \qquad f(\theta) = 1 - 3\cos^2\theta$$ Energy of interaction between two fixed parallel dipoles (16B.4)

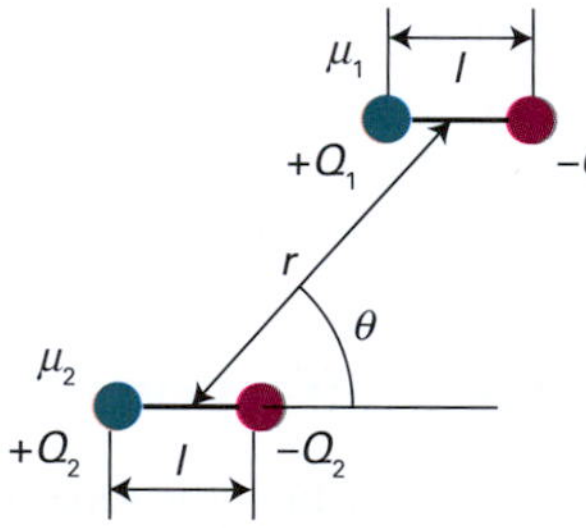

3

Brief illustration 16B.3 The dipolar interaction

We can use eqn 16B.4 to calculate the molar potential energy of the dipolar interaction between two amide groups. Supposing that the groups are separated by 3.0 nm with $\theta = 180°$ (so that $\cos\theta = -1$ and $1 - 3\cos^2\theta = -2$), we take $\mu_1 = \mu_2 = 2.7\,D$, corresponding to 9.1×10^{-30} C m, and find

$$V = \frac{\overbrace{(9.1\times10^{-30}\,\text{C m})^2}^{\mu_1\mu_2}\times\overbrace{(-2)}^{1-3\cos^2\theta}}{4\pi\times\underbrace{(8.854\times10^{-12}\,\text{J}^{-1}\text{C}^2\text{m}^{-1})}_{\varepsilon_0}\times\underbrace{(3.0\times10^{-9}\,\text{m})^3}_{r^3}}$$

$$= \frac{(9.1\times10^{-30})^2\times(-2)}{4\pi\times(8.854\times10^{-12})\times(3.0\times10^{-9})^3}\,\frac{\text{C}^2\text{m}^2}{\text{J}^{-1}\text{C}^2\text{m}^{-1}\text{m}^3}$$

$$= -5.5\times10^{-23}\,\text{J}$$

This value corresponds to −33 J mol⁻¹. Note that this energy is considerably less than that between two partial charges at the same separation (see *Brief illustration* 16B.1).

Self-test 16B.3 Repeat the calculation for an amide group and a water molecule separated by 3.5 nm with $\theta = 90°$, in a medium with relative permittivity of 3.5.

Answer: −2.1 J mol⁻¹

Equation 16B.4 applies to polar molecules in a fixed, parallel, orientation in a solid. In a fluid of freely rotating molecules, the interaction between dipoles averages to zero because $f(\theta)$ changes sign as the orientation changes, and its average value is zero. Physically, the like partial charges of two freely rotating molecules are close together as much as the two opposite partial charges, and the repulsion of the former is cancelled by the attraction of the latter. Mathematically, this result arises from the fact that, as we show in the following *Justification*, the average of the function $1 - 3\cos^2\theta$ is zero.

Justification 16B.3 The dipolar interaction between two freely rotating molecules

Consider the unit sphere shown in Fig. 16B.2. The average value of $f(\theta) = 1 - 3\cos^2\theta$ is the sum of its values in each of the infinitesimal regions on the surface of the sphere (that is, the integral of the function over the surface) divided by the surface area of the sphere (which is equal to 4π).

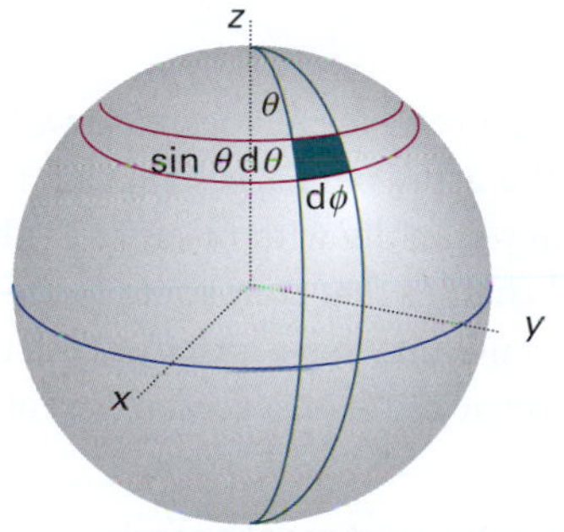

Figure 16B.2 A unit sphere showing the area element $\sin\theta\, d\theta\, d\phi$.

With the area element in spherical polar coordinates as $\sin\theta\, d\theta\, d\phi$, θ ranging from 0 to π, and ϕ ranging from 0 to 2π, the average value $\langle f(\theta)\rangle$ of $f(\theta)$ is

$$\begin{aligned}\langle f(\theta)\rangle &= \frac{1}{4\pi}\int_0^{2\pi}\int_0^{\pi}(1-3\cos^2\theta)\sin\theta\, d\theta\, d\phi \\ &= \frac{1}{4\pi}\int_0^{2\pi} d\phi \int_0^{\pi}(1-3\cos^2\theta)\sin\theta\, d\theta \\ &= \frac{1}{2}\int_0^{\pi}(1-3\cos^2\theta)\sin\theta\, d\theta\end{aligned}$$

The integral is calculated as follows:

$$\begin{aligned}\int_0^{\pi}(1-3\cos^2\theta)\sin\theta\, d\theta &= \int_0^{\pi}\sin\theta\, d\theta - 3\int_0^{\pi}\cos^2\theta\overbrace{\sin\theta\, d\theta}^{-d\cos\theta} \\ &\overset{\text{Integrals T.1 and A.1}}{=} -\cos\theta\Big|_0^{\pi} - 3\left(-\frac{1}{3}\cos^3\theta\Big|_0^{\pi}\right) \\ &= \overbrace{-\cos\theta\Big|_0^{\pi}}^{+2} + \overbrace{\cos^3\theta\Big|_0^{\pi}}^{-2} = 0\end{aligned}$$

It follows that $\langle f(\theta)\rangle = 0$, and, from eqn 16B.6, that the dipolar interaction between two freely rotating molecules vanishes when averaged over a sphere.

The average interaction energy of two *freely* rotating dipoles is zero. However, because their mutual potential energy depends on their relative orientation, the molecules do not in fact rotate completely freely, even in a gas. In fact, the lower energy orientations are marginally favoured, so there is a nonzero average interaction between polar molecules. We show in the following *Justification* that the average potential energy of two rotating molecules that are separated by a distance r is

$$\langle V\rangle = -\frac{C}{r^6} \qquad C = \frac{2\mu_1^2\mu_1^2}{3(4\pi\varepsilon_0)^2 kT} \qquad \text{Average potential energy of two rotating polar molecules} \qquad (16B.5)$$

This expression describes the **Keesom interaction**, and is the first of the contributions to the van der Waals interaction (when that is taken to be a $1/r^6$ interaction).

Justification 16B.4 The Keesom interaction

The detailed calculation of the Keesom interaction energy is quite complicated, but the form of the final answer can be constructed quite simply. First, we note that the average interaction energy of two polar molecules rotating at a fixed separation r is given by

$$\langle V\rangle = \frac{\mu_1\mu_2\langle f(\theta)\rangle}{4\pi\varepsilon_0 r^3}$$

where $\langle f(\theta)\rangle$ now includes a weighting factor in the averaging that is equal to the probability that a particular orientation will be adopted. This probability is given by the Boltzmann distribution, $p \propto e^{-E/kT}$, with E interpreted as the potential energy of interaction of the two dipoles in that orientation. That is,

$$p \propto e^{-V/kT} \qquad V = \frac{\mu_1\mu_2 f(\theta)}{4\pi\varepsilon_0 r^3}$$

When the potential energy of interaction of the two dipoles is very small compared with the energy of thermal motion, we can use $V \ll kT$, expand the exponential function in p, and retain only the first two terms:

$$p \propto 1 - V/kT + \cdots$$

We now write the weighted average of $f(\theta)$ as

$$\langle f(\theta)\rangle = \frac{\int_0^\pi f(\theta)p\mathrm{d}\theta}{\int_0^\pi \mathrm{d}\theta} = \frac{1}{\pi}\int_0^\pi f(\theta)p\mathrm{d}\theta = \frac{1}{\pi}\int_0^\pi f(\theta)(1 - V/kT)\mathrm{d}\theta + \cdots$$

It follows that

$$\begin{aligned}\langle f(\theta)\rangle &= \frac{1}{\pi}\int_0^\pi f(\theta)p\mathrm{d}\theta - \frac{1}{\pi}\int_0^\pi f(\theta)(V/kT)\mathrm{d}\theta + \cdots \\ &= \frac{1}{\pi}\int_0^\pi f(\theta)p\mathrm{d}\theta - \frac{1}{\pi}\int_0^\pi \frac{\mu_1^2\mu_2^2}{(4\pi\varepsilon_0)^2kTr^6} f(\theta)^2\mathrm{d}\theta + \cdots \\ &= \overbrace{\frac{1}{\pi}\int_0^\pi f(\theta)p\mathrm{d}\theta}^{\langle f(\theta)\rangle_0} - \frac{\mu_1^2\mu_2^2}{(4\pi\varepsilon_0)^2kTr^6}\overbrace{\left(\int_0^\pi \frac{1}{\pi} f(\theta)^2\mathrm{d}\theta\right)}^{\langle f(\theta)^2\rangle_0} + \cdots \\ &= \langle f(\theta)_0\rangle - \frac{\mu_1^2\mu_2^2}{(4\pi\varepsilon_0)^2kTr^6}\langle f(\theta)^2\rangle_0 + \cdots\end{aligned}$$

where $\langle\ldots\rangle_0$ denotes an unweighted spherical average. The spherical average of $f(\theta)$ is zero (as in *Justification* 16B.3), so the first term in the expression for $\langle f(\theta)\rangle$ vanishes. However, the average value of $f(\theta)^2$ is nonzero because $f(\theta)^2$ is positive at all orientations, so we can write

$$\langle V\rangle = -\frac{\mu_1^2\mu_2^2\langle f(\theta)^2\rangle_0}{(4\pi\varepsilon_0)^2kTr^6}$$

The average value $\langle f(\theta)^2\rangle_0$ turns out to be $\frac{2}{3}$ when the calculation is carried through in detail. The final result is that quoted in eqn 16B.5.

The important features of eqn 16B.5 are

Physical interpretation

- The negative sign shows that the average interaction is attractive.
- The dependence of the average interaction energy on the inverse sixth power of the separation identifies it as a van der Waals interaction.
- The inverse dependence on the temperature reflects the way that the greater thermal motion overcomes the mutual orientating effects of the dipoles at higher temperatures.
- The inverse sixth power arises from the inverse third power of the interaction potential energy that is weighted by the energy in the Boltzmann term, which is also proportional to the inverse third power of the separation.

Brief illustration 16B.4 The Keesom interaction

Suppose a water molecule ($\mu_1 = 1.85\,\mathrm{D}$) can rotate 1.0 nm from an amide group ($\mu_2 = 2.7\,\mathrm{D}$). The average energy of their interaction at 25 °C (298 K) is

$$\langle V\rangle = -\frac{2\times\overbrace{(1.85\times 3.336\times 10^{-30}\,\mathrm{C\,m})}^{\mu_1}{}^2\times\overbrace{(2.7\times 3.336\times 10^{-30}\,\mathrm{C\,m})}^{\mu_2}{}^2}{3\times\underbrace{(1.710\times 10^{-43}\,\mathrm{J^{-1}\,C^4\,m^{-2}\,K^{-1}})}_{(4\pi\varepsilon_0)^2k}\times\underbrace{(298\,\mathrm{K})}_{T}\times\Big(\underbrace{1.0\times 10^{-9}\,\mathrm{m}}_{r}\Big)^6} = -4.0\times 10^{-23}\,\mathrm{J}$$

This interaction energy corresponds (after multiplication by Avogadro's constant) to $-24\,\mathrm{J\,mol^{-1}}$, and it is much smaller than the energies involved in the making and breaking of chemical bonds.

Self-test 16B.4 Calculate the average interaction energy for pairs of molecules in the gas phase with $\mu = 1\,\mathrm{D}$ when the separation is 0.5 nm at 298 K. Compare this energy with the average molar kinetic energy of the molecules.

Answer: $\langle V\rangle = -0.07\,\mathrm{kJ\,mol^{-1}} \ll \frac{3}{2}RT = 3.7\,\mathrm{kJ\,mol^{-1}}$

Table 16B.1 summarizes the various expressions for the interaction of charges and dipoles. It is quite easy to extend the formulas given there to obtain expressions for the energy of interaction of higher multipoles (electric multipoles are described in Topic 16A). The feature to remember is that the interaction energy falls off more rapidly the higher the order of the multipole. For the interaction of an n-pole with an m-pole, the potential energy varies with distance as

$$V \propto \frac{1}{r^{n+m-1}} \quad \text{Energy of interaction between multipoles} \quad (16B.6)$$

The reason for the even steeper decrease with distance is the same as before: the array of charges appears to blend together into neutrality more rapidly with distance the higher the number of individual charges that contribute to the multipole. Note that a given molecule may have a charge distribution that corresponds to a superposition of several different multipoles, and

Table 16B.1 Interaction potential energies

Interaction type	Distance dependence of potential energy	Typical energy/ (kJ mol^{-1})	Comment
Ion–ion	$1/r$	250	Only between ions
Hydrogen bond		20	Occurs in X–H···Y, where X, Y=N, O, or F
Ion–dipole	$1/r^2$	15	
Dipole–dipole	$1/r^3$	2	Between stationary polar molecules
	$1/r^6$	0.3	Between rotating polar molecules
London (dispersion)	$1/r^6$	2	Between all types of molecules and ions

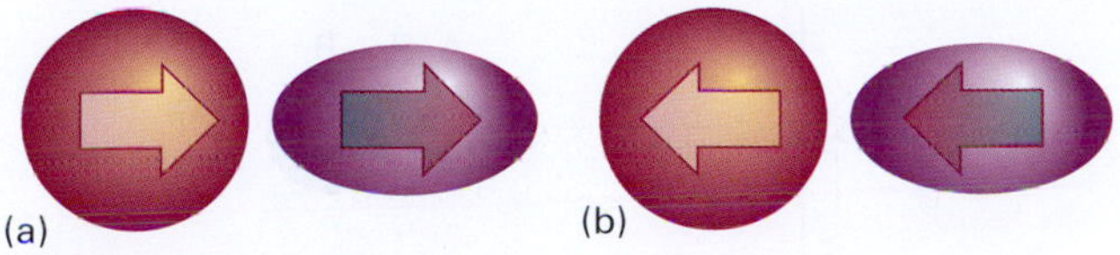
(a) (b)

Figure 16B.3 (a) A polar molecule (full arrow) can induce a dipole (outline arrow) in a nonpolar molecule, and (b) the orientation of the latter follows that of the former, so the interaction does not average to zero.

in such cases the energy of interaction is the sum of terms given by eqn 16B.6.

(c) Dipole–induced dipole interactions

A polar molecule can induce a dipole in a neighbouring polarizable molecule (Fig. 16B.3). The induced dipole interacts with the permanent dipole of the first molecule, and the two are attracted together. The average interaction energy when the separation of the centres of the molecules is r is

$$V=-\frac{C}{r^6} \qquad C=\frac{\mu_1^2\alpha_2'}{4\pi\varepsilon_0} \qquad \text{Potential energy of a polar molecule and a polarizable molecule} \qquad (16B.7)$$

where α_2' is the polarizability volume (Topic 16A) of molecule 2 and μ_1 is the magnitude of the permanent dipole moment of molecule 1. Note that the C in this expression is different from the C in eqn 16B.5 and other expressions below: we are using the same symbol in C/r^6 to emphasize the similarity of form of each expression.

The dipole–induced dipole interaction energy is independent of the temperature because thermal motion has no effect on the averaging process. Moreover, like the dipole–dipole interaction, the potential energy depends on $1/r^6$: this distance dependence stems from the $1/r^3$ dependence of the field (and hence the magnitude of the induced dipole) and the $1/r^3$ dependence of the potential energy of interaction between the permanent and induced dipoles.

Brief illustration 16B.5 The dipole–induced dipole interaction

For a molecule with $\mu=1.0$ D (3.3×10^{-30} C m, such as HCl) separated by 0.30 nm from a molecule of polarizability volume $\alpha'=10\times10^{-30}$ m^3 (such as benzene, Table 16A.1), the average interaction energy is

$$V=-\frac{(3.3\times10^{-30}\,\text{C m})^2\times(10\times10^{-30}\,\text{m}^3)}{4\pi\times(8.854\times10^{-12}\,\text{J}^{-1}\,\text{C}^2\,\text{m}^{-1})\times(3.0\times10^{-10}\,\text{m})^6}$$
$$=-1.4\times10^{-21}\,\text{J}$$

which, upon multiplication by Avogadro's constant, corresponds to -0.83 kJ mol^{-1}.

Self-test 16B.5 Calculate the average interaction energy, in units of joules per mole (J mol^{-1}), between a water molecule and a benzene molecule separated by 1.0 nm.

Answer: -2.1 J mol^{-1}

(d) Induced dipole–induced dipole interactions

Nonpolar molecules (including closed-shell atoms, such as Ar) attract one another even though neither has a permanent dipole moment. The abundant evidence for the existence of interactions between them is the formation of condensed phases of non-polar substances, such as the condensation of hydrogen or argon to a liquid at low temperatures and the fact that benzene is a liquid at normal temperatures.

The interaction between nonpolar molecules arises from the transient dipoles that all molecules possess as a result of fluctuations in the instantaneous positions of electrons. To appreciate the origin of the interaction, suppose that the electrons in one molecule flicker into an arrangement that gives the molecule an instantaneous dipole moment $\boldsymbol{\mu}_1^*$. This dipole generates an electric field that polarizes the other molecule, and induces in that molecule an instantaneous dipole moment $\boldsymbol{\mu}_2$. The two dipoles attract each other and the potential energy of the pair is lowered. Although the first molecule will go on to change the size and direction of its instantaneous dipole, the electron distribution of the second molecule will follow; that is, the two dipoles are correlated in direction (Fig. 16B.4). Because of this correlation, the attraction between the two instantaneous dipoles does not average to zero, and gives rise to an induced dipole–induced dipole interaction. This interaction is called either the **dispersion interaction** or the **London interaction** (for Fritz London, who first described it).

The strength of the dispersion interaction depends on the polarizability of the first molecule because the instantaneous dipole moment of magnitude μ_1^* depends on the looseness of the control that the nuclear charge exercises over the outer electrons. The strength of the interaction also depends on the polarizability of the second molecule, for that polarizability determines how readily a dipole can be induced by another

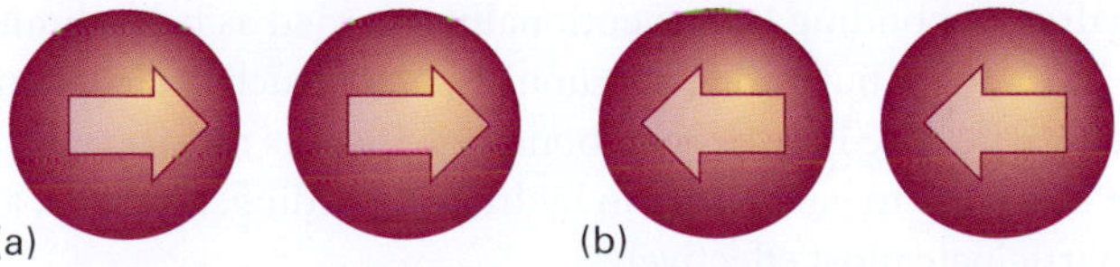
(a) (b)

Figure 16B.4 (a) In the dispersion interaction, an instantaneous dipole on one molecule induces a dipole on another molecule, and the two dipoles then interact to lower the energy. (b) The two instantaneous dipoles are correlated, and although they occur in different orientations at different instants, the interaction does not average to zero.

molecule. The actual calculation of the dispersion interaction is quite involved, but a reasonable approximation to the interaction energy is given by the **London formula**:

$$V=-\frac{C}{r^6} \qquad C=\tfrac{3}{2}\alpha_1'\alpha_2'\frac{I_1I_2}{I_1+I_2} \qquad \text{London formula} \qquad (16B.8)$$

where I_1 and I_2 are the ionization energies of the two molecules (Table 9B.2). This interaction energy is also proportional to the inverse sixth power of the separation of the molecules, which identifies it as a third contribution to the van der Waals interaction. The dispersion interaction generally dominates all the interactions between molecules other than hydrogen bonds.

Brief illustration 16B.6 The London interaction

For two CH_4 molecules separated by 0.30 nm, we can use eqn 16B.9 with $\alpha'=2.6\times10^{-30}\,\text{m}^3$ and $I\approx700\,\text{kJ mol}^{-1}$ and obtain

$$V=-\frac{\tfrac{3}{2}\times(2.6\times10^{-30}\,\text{m}^3)^2}{(0.30\times10^{-9}\,\text{m})^6}\times\frac{(7.00\times10^5\,\text{J mol}^{-1})^2}{2\times(7.00\times10^5\,\text{J mol}^{-1})}$$
$$=-4.9\,\text{kJ mol}^{-1}$$

A very approximate check on this figure is the enthalpy of vaporization of methane, which is $8.2\,\text{kJ mol}^{-1}$. However, this comparison is questionable, partly because the enthalpy of vaporization is a many-body quantity and partly because the long-distance assumption breaks down.

Self-test 16B.6 Estimate the energy of the London interaction for two He atoms separated by 1.0 nm.

Answer: $-0.071\,\text{J mol}^{-1}$

16B.3 Hydrogen bonding

The interactions described so far are universal in the sense that they are possessed by all molecules independent of their specific identity. However, there is a type of interaction possessed by molecules that have a particular constitution. A **hydrogen bond** is an attractive interaction between two species that arises from a link of the form A–H···B, where A and B are highly electronegative elements and B possesses a lone pair of electrons. Hydrogen bonding is conventionally regarded as being limited to N, O, and F but, if B is an anionic species (such as Cl^-), it may also participate in hydrogen bonding. There is no strict cut-off for an ability to participate in hydrogen bonding, but N, O, and F participate most effectively.

The formation of a hydrogen bond can be regarded either as the approach between a partial positive charge of H and a partial negative charge of B or as a particular example of delocalized molecular orbital formation in which A, H, and B each supply one atomic orbital from which three molecular orbitals are constructed (Fig. 16B.5). Experimental evidence and theoretical arguments have been presented in favour of both views and the matter has not yet been resolved. The electrostatic interaction model can be understood readily in terms of the discussion in Section 16B.1. Here we develop the molecular orbital model.

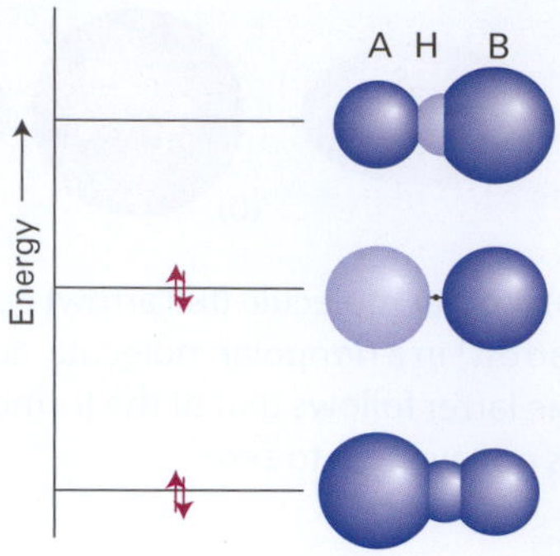

Figure 16B.5 The molecular orbital interpretation of the formation of an A–H···B hydrogen bond. From the three A, H, and B orbitals, three molecular orbitals can be formed (their relative contributions are represented by the sizes of the spheres). Only the two lower energy orbitals are occupied, and there may therefore be a net lowering of energy compared with the separate AH and B species.

Thus, if the A–H bond is regarded as formed from the overlap of an orbital on A, χ_A, and a hydrogen 1s orbital, χ_H, and the lone pair on B occupies an orbital on B, χ_B, then, when the two molecules are close together, we can build three molecular orbitals from the three basis orbitals:

$$\psi=c_1\chi_A+c_2\chi_H+c_3\chi_B$$

One of the molecular orbitals is bonding, one almost nonbonding, and the third anti-bonding. These three orbitals need to accommodate four electrons (two from the original A–H bond and two from the lone pair of B), so two enter the bonding orbital and two enter the nonbonding orbital. Because the anti-bonding orbital remains empty, the net effect—depending on the precise energy of the almost nonbonding orbital—may be a lowering of energy.

In practice, the strength of the bond is found to be about 20 kJ mol^{-1} (there are two hydrogen bonds per molecule in liquid water, and its standard enthalpy of vaporization, from Table 2C.2, is 44 kJ mol^{-1}). Because the bonding depends on orbital overlap, it is virtually a contact-like interaction that is turned on when AH touches B and is zero as soon as the contact is broken. If hydrogen bonding is present, it dominates the other intermolecular interactions. The properties of liquid and solid water, for example, are dominated by the hydrogen bonding between H_2O molecules. The structure of DNA and hence the transmission of genetic information is crucially dependent on the strength of hydrogen bonds between base pairs. The structural evidence for hydrogen bonding comes from noting that the internuclear distance between formally non-bonded atoms is less than expected on the basis of their van der Waals radii, which suggests that a dominating attractive interaction is present. For example, the O–O distance in O–H···O is expected to be 280 pm on the basis of van der Waals radii, but is found to be 270 pm in typical compounds. Moreover, the H···O distance is expected to be 260 pm but is found to be only 170 pm.

Hydrogen bonds may be either symmetric or non-symmetric. In a symmetric hydrogen bond, the H atom lies midway between the two other atoms. This arrangement is rare, but occurs in $F–H\cdots F^-$, where both bond lengths are 120 pm. More common is the non-symmetric arrangement, where the A–H bond is shorter than the H···B bond. Simple electrostatic arguments, treating A–H···B as an array of point charges (partial negative charges on A and B, partial positive on H) suggest that the lowest energy is achieved when the bond is linear, because then the two partial negative charges are furthest apart. The experimental evidence from structural studies supports a linear or near-linear arrangement.

Brief illustration 16B.7 The hydrogen bond

A common hydrogen bond is that formed between O–H groups and O atoms, as in liquid water and ice. In Problem 16B.6, you are invited to use the electrostatic model to calculate the dependence of the potential energy of interaction on the OOH angle, denoted θ in **4**, and the results are plotted in Fig. 16B.6. We see that the strength of bonding is greatest at $\theta=0$ when the OHO atoms lie in a straight line; the molar potential energy is then $-19\,\text{kJ mol}^{-1}$.

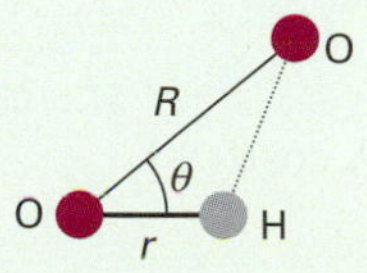

4

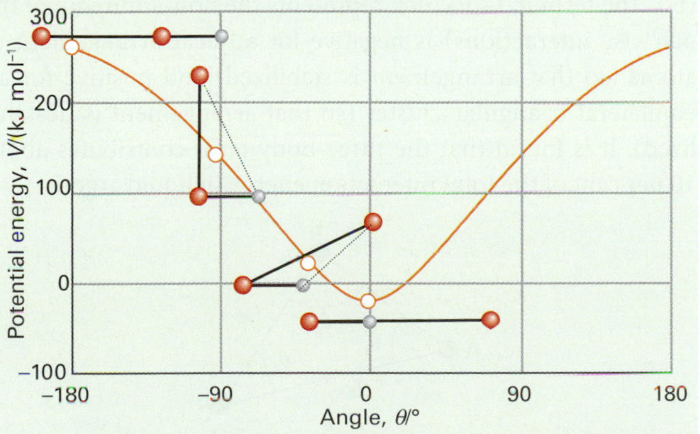

Figure 16B.6 The variation of the energy of interaction (according to the electrostatic model) of a hydrogen bond as the angle between the O–H and :O groups is changed.

Self-test 16B.7 Use Fig. 16B.6 to explore the dependence of the interaction energy on angle: at what angle does the interaction energy become negative?

Answer: Only ±12°, so that the energy is negative (and the interaction is attractive) only when the atoms are close to a linear arrangement

16B.4 The hydrophobic interaction

Nonpolar molecules do dissolve slightly in polar solvents, but strong interactions between solute and solvent are not possible and as a result it is found that each individual solute molecule is surrounded by a solvent cage (Fig. 16B.7). To understand the consequences of this effect, consider the thermodynamics of transfer of a nonpolar hydrocarbon solute from a nonpolar solvent to water, a polar solvent. Experiments indicate that the process is endergonic ($\Delta_{\text{transfer}}G>0$), as expected on the basis of the increase in polarity of the solvent, but exothermic ($\Delta_{\text{transfer}}H<0$). Therefore, it is a large decrease in the entropy of the system ($\Delta_{\text{transfer}}S<0$) that accounts for the positive Gibbs energy of transfer. For example, the process

$$CH_4(\text{in } CCl_4) \rightarrow CH_4(aq)$$

has $\Delta_{\text{transfer}}G=+12\,\text{kJ mol}^{-1}$, $\Delta_{\text{transfer}}H=-10\,\text{kJ mol}^{-1}$, and $\Delta_{\text{transfer}}S=-75\,\text{J K}^{-1}\,\text{mol}^{-1}$ at 298 K. Substances characterized by a positive Gibbs energy of transfer from a nonpolar to a polar solvent are called **hydrophobic**.

It is possible to quantify the hydrophobicity of a small molecular group R by defining the **hydrophobicity constant**, π, as

$$\pi = \log \frac{S}{S_0} \qquad \textit{Definition} \quad \text{Hydrophobicity constant} \qquad (16B.9)$$

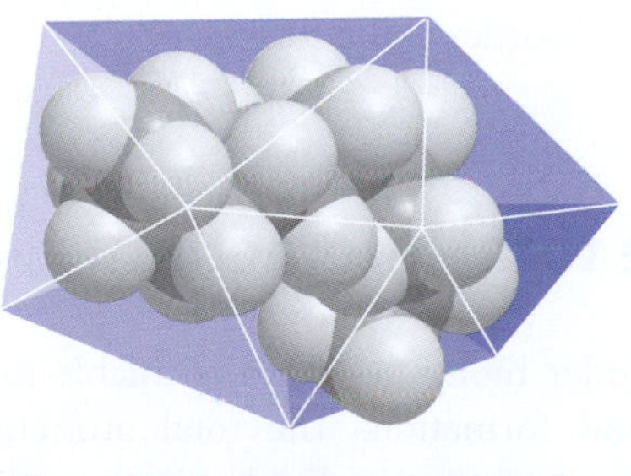

Figure 16B.7 When a hydrocarbon molecule is surrounded by water, the H_2O molecules form a cage. As a result of this acquisition of structure, the entropy of the water decreases, so the dispersal of the hydrocarbon into the water is entropy-opposed; its coalescence is entropy-favoured.

where S is the ratio of the molar solubility of the compound R—A in octanol, a nonpolar solvent, to that in water, and S_0 is the ratio of the molar solubility of the compound H–A in octanol to that in water. Therefore, positive values of π indicate hydrophobicity and negative values of π indicate hydrophilicity, the thermodynamic preference for water as a solvent. It is observed experimentally that the π values of most groups do not depend on the nature of A. However, measurements do suggest group additivity of π values, as the following data show:

R	CH_3	CH_3CH_2	$CH_3(CH_2)_2$	$CH_3(CH_2)_3$	$CH_3(CH_2)_4$
π	0.5	1.0	1.5	2.0	2.5

Thus, acyclic saturated hydrocarbons become more hydrophobic as the carbon chain length increases. This trend can be rationalized by $\Delta_{\text{transfer}}H$ becoming more positive and $\Delta_{\text{transfer}}S$ more negative as the number of carbon atoms in the chain increases.

At the molecular level, formation of a solvent cage around a hydrophobic molecule involves the formation of new hydrogen bonds among solvent molecules. This process is exothermic and accounts for the negative values of $\Delta_{\text{transfer}}H$. On the other hand, the increase in order associated with formation of a very large number of small solvent cages decreases the entropy of the system and accounts for the negative values of $\Delta_{\text{transfer}}S$. However, when many solute molecules cluster together, fewer (albeit larger) cages are required and more solvent molecules are free to move. The net effect of formation of large clusters of hydrophobic molecules is then a decrease in the organization of the solvent and therefore a net *increase* in entropy of the system. This increase in entropy of the solvent is large enough to render spontaneous the association of hydrophobic molecules in a polar solvent.

The increase in entropy that results from fewer structural demands on the solvent placed by the clustering of nonpolar molecules is the origin of the **hydrophobic interaction**, which tends to stabilize aggregation of hydrophobic groups in micelles and biopolymers (Topic 17C). The hydrophobic interaction is an example of an ordering process that is driven by a tendency toward greater disorder of the solvent.

16B.5 The total interaction

Here we consider molecules that are unable to participate in hydrogen bond formation. The total attractive interaction energy between rotating molecules is then the sum of the dipole–dipole, dipole–induced dipole, and dispersion interactions. Only the dispersion interaction contributes if both molecules are nonpolar. In a fluid phase, all three contributions to the potential energy vary as the inverse sixth power of the separation of the molecules, so we may write

$$V = -\frac{C_6}{r^6} \tag{16B.10}$$

where C_6 is a coefficient that depends on the identity of the molecules.

Although attractive interactions between molecules are often expressed as in eqn 16B.10 we must remember that this equation has only limited validity. First, we have taken into account only dipolar interactions of various kinds, for they have the longest range and are dominant if the average separation of the molecules is large. However, in a complete treatment we should also consider quadrupolar and higher-order multipole interactions, particularly if the molecules do not have permanent dipole moments. Secondly, the expressions have been derived by assuming that the molecules can rotate reasonably freely. That is not the case in most solids, and in rigid media the dipole–dipole interaction is proportional to $1/r^3$ (as in eqn 16B.3) because the Boltzmann averaging procedure is irrelevant when the molecules are trapped into a fixed orientation.

A different kind of limitation is that eqn 16B.10 relates to the interactions of pairs of molecules. There is no reason to suppose that the energy of interaction of three (or more) molecules is the sum of the pairwise interaction energies alone. The total dispersion energy of three closed-shell atoms, for instance, is given approximately by the **Axilrod–Teller formula**:

$$V = -\frac{C_6}{r_{AB}^6} - \frac{C_6}{r_{BC}^6} - \frac{C_6}{r_{CA}^6} + \frac{C'}{(r_{AB}r_{BC}r_{CA})^3} \quad \text{Axilrod–Teller formula} \tag{16B.11a}$$

where

$$C' = a(3\cos\theta_A \cos\theta_B \cos\theta_C + 1) \tag{16B.11b}$$

The parameter a is approximately equal to $\frac{3}{4}\alpha' C_6$; the angles θ are the internal angles of the triangle formed by the three atoms (5). The term in C' (which represents the non-additivity of the pairwise interactions) is negative for a linear arrangement of atoms (so that arrangement is stabilized) and positive for an equilateral triangular cluster (so that arrangement is destabilized). It is found that the three-body term contributes about 10 per cent of the total interaction energy in liquid argon.

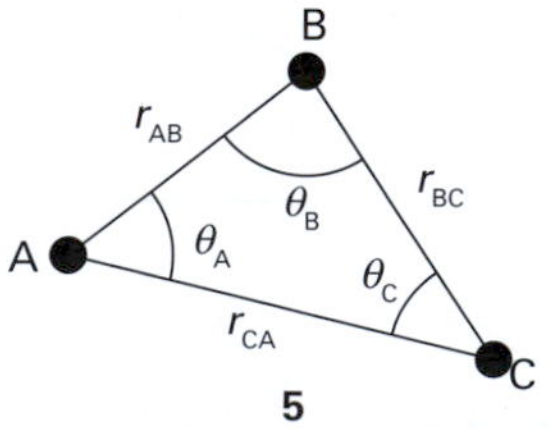

5

When molecules are squeezed together, the nuclear and electronic repulsions begin to dominate the attractive forces. The repulsions increase steeply with decreasing separation in a way that can be deduced only by very extensive, complicated

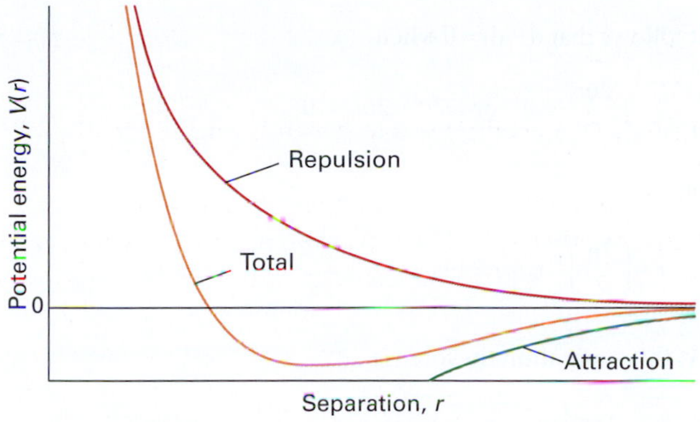

Figure 16B.8 The general form of an intermolecular potential energy curve (the graph of the potential energy of two closed shell species as the distance between them is changed). The attractive (negative) contribution has a long range, but the repulsive (positive) interaction increases more sharply once the molecules come into contact.

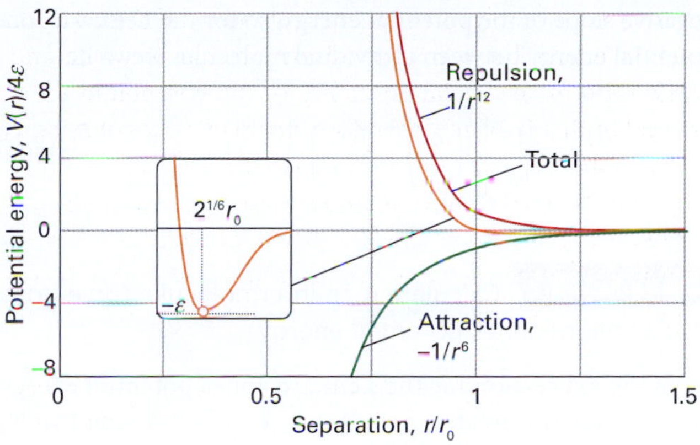

Figure 16B.9 The Lennard-Jones potential energy is another approximation to the true intermolecular potential energy curves. It models the attractive component by a contribution that is proportional to $1/r^6$ and the repulsive component by a contribution that is proportional to $1/r^{12}$. Specifically, these choices result in the Lennard-Jones (12,6) potential energy. Although there are good theoretical reasons for the former, there is plenty of evidence to show that $1/r^{12}$ is only a very poor approximation to the repulsive part of the curve.

molecular structure calculations of the kind described in Topic 10E (Fig. 16B.8).

In many cases, however, progress can be made by using a greatly simplified representation of the potential energy, where the details are ignored and the general features expressed by a few adjustable parameters. One such approximation is the **hard-sphere potential energy**, in which it is assumed that the potential energy rises abruptly to infinity as soon as the particles come within a separation d:

$$V=\infty \text{ for } r\leq d \qquad V=0 \text{ for } r>d \qquad \text{Hard-sphere potential energy} \qquad (16B.12)$$

This very simple expression for the potential energy is surprisingly useful for assessing a number of properties. Another widely used approximation is the **Mie potential energy**:

$$V=\frac{C_n}{r^n}-\frac{C_m}{r^m} \qquad \text{Mie potential energy} \qquad (16B.13)$$

with $n>m$. The first term represents repulsions and the second term attractions. The **Lennard-Jones potential energy** is a special case of the Mie potential energy with $n=12$ and $m=6$ (Fig. 16B.9); it is often written in the form

$$V=4\varepsilon\left\{\left(\frac{r_0}{r}\right)^{12}-\left(\frac{r_0}{r}\right)^{6}\right\} \qquad \text{Lennard-Jones potential energy} \qquad (16B.14)$$

The two parameters are ε, the depth of the well (not to be confused with the symbol of the permittivity of a medium used in Section 16B.1), and r_0, the separation at which $V=0$ (Table 16B.2).

Although the Lennard-Jones potential energy has been used in many calculations, there is plenty of evidence to show that

Table 16B.2* Lennard-Jones parameters for the (12,6) potential

	$\varepsilon/(\text{kJ mol}^{-1})$	r_0/pm
Ar	128	342
Br_2	536	427
C_6H_6	454	527
Cl_2	368	412
H_2	34	297
He	11	258
Xe	236	406

* More values are given in the *Resource section*.

$1/r^{12}$ is a very poor representation of the repulsive potential energy, and that an exponential form, e^{-r/r_0}, is greatly superior. An exponential function is more faithful to the exponential decay of atomic wavefunctions at large distances, and hence to the overlap that is responsible for repulsion. The potential energy with an exponential repulsive term and a $1/r^6$ attractive term is known as an **exp-6 potential energy**. These expressions for the potential energy can be used to calculate the virial coefficients of gases, as explained in Topic 1C, and through them various properties of real gases, such as the Joule–Thomson coefficient (Topic 2D). They are also used to model the structures of condensed fluids.

With the advent of **atomic force microscopy** (AFM), in which the force between a molecular sized probe and a surface is monitored (Topic 22A), it has become possible to measure directly the forces acting between molecules. The force, F, is the negative slope of the potential energy, so for the Lennard-Jones potential energy between individual molecules we write

$$F=-\frac{\mathrm{d}V}{\mathrm{d}r}=\frac{24\varepsilon}{r_0}\left\{2\left(\frac{r_0}{r}\right)^{13}-\left(\frac{r_0}{r}\right)^{7}\right\} \qquad (16B.15)$$

Example 16B.1 Calculating an intermolecular force from the Lennard-Jones potential energy

Use the expression for the Lennard-Jones potential energy to estimate the greatest net attractive force between two N_2 molecules.

Method The magnitude of the force is greatest at the distance r at which $\mathrm{d}F/\mathrm{d}r=0$. Therefore differentiate eqn 16B.15 with respect to r, set the resulting expression to zero, and solve for r. Finally, use the value of r in eqn 16B.15 to calculate the corresponding value of F.

Answer Because $\mathrm{d}x^n/\mathrm{d}x=nx^{n-1}$, the derivative of F with respect to r is

$$\frac{\mathrm{d}F}{\mathrm{d}r}=\frac{24\varepsilon}{r_0}\left\{2\left(-\frac{13r_0^{13}}{r^{14}}\right)-\left(-\frac{7r_0^{7}}{r^{8}}\right)\right\}=24\varepsilon r_0^6\left\{\frac{7}{r^8}-\frac{26r_0^6}{r^{14}}\right\}$$

It follows that $\mathrm{d}F/\mathrm{d}r=0$ when

$$\frac{7}{r^8}-\frac{26r_0^6}{r^{14}}=0 \quad \text{or} \quad 7r^6-26r_0^6=0$$

or

$$r=\left(\frac{26}{7}\right)^{1/6}r_0=1.244r_0$$

At this separation the force is

$$F=\frac{24\varepsilon}{r_0}\left\{2\left(\frac{r_0}{1.244r_0}\right)^{13}-\left(\frac{r_0}{1.244r_0}\right)^{7}\right\}=-\frac{2.396\varepsilon}{r_0}$$

From Table 16B.2, $\varepsilon=1.268\times10^{-21}$ J and $r_0=3.919\times10^{-10}$ m. It follows that

$$F=-\frac{2.396\times(1.268\times10^{-21}\,\mathrm{J})}{3.919\times10^{-10}\,\mathrm{m}}\overset{1\,\mathrm{N}=1\,\mathrm{J\,m^{-1}}}{=}-7.752\times10^{-12}\,\mathrm{N}$$

That is, the magnitude of the force is about 8 pN.

Self-test 16B.8 At what separation r_e does the minimum of the potential energy curve occur for a Lennard-Jones potential?

Answer: $r_e=2^{1/6}r_0$

Checklist of concepts

- ☐ 1. A **van der Waals interaction** between closed-shell molecules is inversely proportional to the sixth power of their separation.
- ☐ 2. The following molecular interactions are important: **charge–charge**, **charge–dipole**, **dipole–dipole**, **dipole–induced dipole**, **dispersion (London)**, **hydrogen bonding**, and the **hydrophobic interaction**.
- ☐ 3. A hydrogen bond is an interaction of the form X–H···Y, where X and Y are typically N, O, or F.
- ☐ 4. The **hydrophobic interaction** fosters clustering of nonpolar molecules in polar solvents.
- ☐ 5. The **Lennard-Jones potential energy** function is a model of the total intermolecular potential energy.

Checklist of equations

Property	Equation	Comment	Equation number
Potential energy of interaction between two point charges in a medium	$V=Q_1Q_2/4\pi\varepsilon r$	The relative permittivity of the medium is $\varepsilon_r=\varepsilon/\varepsilon_0$	16B.1
Energy of interaction between a point dipole and a point charge	$V=-\mu_1Q_2/4\pi\varepsilon_0r^2$		16B.2
Energy of interaction between two fixed dipoles	$V=\mu_1\mu_2f(\theta)/4\pi\varepsilon_0r^3$, $f(\theta)=1-3\cos^2\theta$	Parallel dipoles	16B.4
Energy of interaction between two rotating dipoles	$\langle V\rangle=-2\mu_1^2\mu_2^2/3(4\pi\varepsilon_0)^2kTr^6$		16B.5

Property	Equation	Comment	Equation number
Energy of interaction between a polar molecule and a polarizable molecule	$V=-\mu_1^2\alpha_2'/4\pi\varepsilon_0 r^6$		16B.7
London formula	$V=-\frac{3}{2}\alpha_1'\alpha_2'\{(I_1I_2/(I_1+I_2)\}/r^6$		16B.8
Hydrophobicity constant	$\pi=\log(S/S_0)$	Definition	16B.9
Axilrod–Teller formula	$V=-C_6/r_{AB}^6-C_6/r_{BC}^6-C_6/r_{CA}^6+C'/(r_{AB}r_{BC}r_{CA})^3$	Applies to closed shell atoms	16B.11a
Lennard-Jones potential energy	$V=4\varepsilon\{(r_0/r)^{12}-(r_0/r)^6\}$		16B.14

16C Liquids

Contents

➤ Why do you need to know this material?

Many chemical reactions, including those in biological cells and chemical reactors, occur in liquids, so you need to understand the structure of liquids, how molecular interactions foster condensation of a gas into a liquid, the thermodynamic properties of liquids, and how liquid surfaces are formed.

➤ What is the key idea?

The attractions between molecules are responsible for the condensation of gases into liquids at low temperatures.

➤ What do you need to know already?

You need to understand the nature of molecular interactions (Topic 16B), the concepts of Helmholtz and Gibbs energies (Topic 3C), and the Boltzmann distribution (Topic 15A).

At low enough temperatures the molecules of a gas have insufficient kinetic energy to escape from each other's attraction and they stick together. But although molecules attract each other when they are a few diameters apart, as soon as they come into contact they repel each other. This repulsion is responsible for the fact that liquids and solids have a definite bulk and do not collapse to an infinitesimal point. The molecules are held together by molecular interactions, but their kinetic energies are comparable to their potential energies. As a result, although the molecules of a liquid are not free to escape completely from the bulk, the whole structure is very mobile and we can speak only of the *average* relative locations of molecules. In this Topic we build on those concepts and add thermodynamic arguments to describe the surface of a liquid and the condensation of a gas into a liquid.

16C.1 Molecular interactions in liquids

The starting point for the discussion of solids is the well-ordered structure of a perfect crystal, which is discussed in Topic 18A. The starting point for the discussion of gases is the completely disordered distribution of the molecules of a perfect gas (Topic 1A). Liquids lie between these two extremes. We see that the structural and thermodynamic properties of liquids depend on the nature of intermolecular interactions and that an equation of state can be built in a similar way to that just demonstrated for real gases.

(a) The radial distribution function

The average relative locations of the particles of a liquid are expressed in terms of the **radial distribution function**, $g(r)$. This function is defined so that $\mathcal{N}g(r)r^2\mathrm{d}r$, where $\mathcal{N}$ is the number density of molecules ($\mathcal{N}=N/V$), is the probability that a molecule will be found in the range $\mathrm{d}r$ at a distance r from another molecule. In a perfect crystal, $g(r)$ is a periodic array of sharp spikes, representing the certainty (in the absence of defects and thermal motion) that molecules (or ions) lie at definite locations. This regularity continues out to the edges of the crystal, so we say that crystals have **long-range order**. When the crystal melts, the long-range order is lost and, wherever we look at long distances from a given molecule, there is equal probability of finding a second molecule. Close to the first molecule, though, the nearest neighbours might still adopt approximately their original relative positions and, even if they are displaced by newcomers, the new particles might adopt their vacated positions. It is still possible to detect a sphere of nearest neighbours at a distance r_1, and perhaps beyond them a sphere of next-nearest neighbours at r_2. The existence of this **short-range order** means that the radial distribution function can be expected to oscillate at short distances, with a peak at r_1, a smaller peak at r_2, and perhaps some more structure beyond that.

The radial distribution function of the oxygen atoms in liquid water is shown in Fig. 16C.1. Closer analysis shows that any given H_2O molecule is surrounded by other molecules at the corners of a tetrahedron. The form of $g(r)$ at 100 °C shows that the intermolecular interactions (in this case, principally by hydrogen bonds) are strong enough to affect the local structure right up to the boiling point. Raman spectra indicate that in liquid water most molecules participate in either three or four hydrogen bonds. Infrared spectra show that about 90 per cent of hydrogen bonds are intact at the melting point of ice, falling to about 20 per cent at the boiling point.

The formal expression for the radial distribution function for molecules 1 and 2 in a fluid consisting of N particles is the somewhat fearsome equation

$$g(r)=\frac{N(N-1)\iint\cdots\int e^{-\beta V_N}\,d\tau_3 d\tau_4\cdots d\tau_N}{\mathcal{N}^2\iint\cdots\int e^{-\beta V_N}\,d\tau_1 d\tau_2\cdots d\tau_N}$$

Radial distribution function (16C.1)

where $\beta=1/kT$ and V_N is the N-particle potential energy. Although fearsome, this expression is nothing more than the Boltzmann distribution for the relative locations of two molecules in a field provided by all the other molecules in the system.

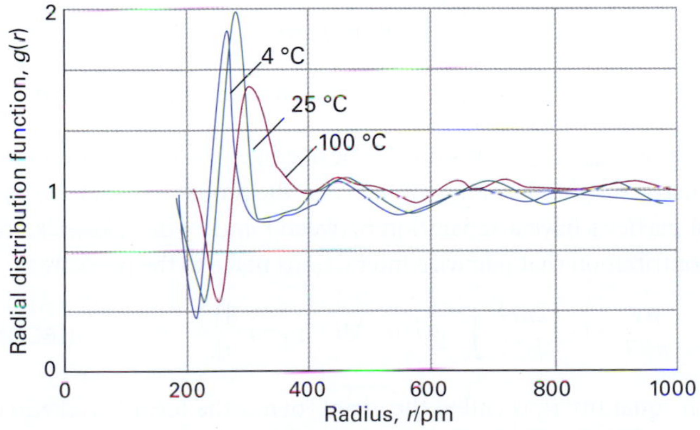

Figure 16C.1 The radial distribution function of the oxygen atoms in liquid water at three temperatures. Note the expansion as the temperature is raised. (Based on A.H. Narten, M.D. Danford, and H.A. Levy, *Discuss. Faraday. Soc.* **43**, 97 (1967).)

(b) The calculation of $g(r)$

Because the radial distribution function can be calculated by making assumptions about the intermolecular interactions, it can be used to test theories of liquid structure. However, even a fluid of hard spheres without attractive interactions (a collection of ball-bearings in a container) gives a function that oscillates near the origin (Fig. 16C.2), and one of the factors influencing, and sometimes dominating, the structure of a liquid is the geometrical problem of stacking together reasonably hard spheres. Indeed, the radial distribution function of a liquid of hard spheres shows more pronounced oscillations at a given temperature than that of any other type of liquid. The attractive part of the potential modifies this basic structure, but sometimes only quite weakly. One of the reasons behind the difficulty of describing liquids theoretically is the similar importance of both the attractive and repulsive (hard core) components of the potential.

There are several ways of building the intermolecular potential into the calculation of $g(r)$. Numerical methods take a box of about 10^3 particles (the number increases as computers grow more powerful), and the rest of the liquid is simulated by surrounding the box with replications of the original box (Fig. 16C.3). Then, whenever a particle leaves the box through one of its faces, its image arrives through the opposite face. When calculating the interactions of a molecule in a box, it interacts with all the molecules in the box and all the periodic replications of those molecules and itself in the other boxes.

In the **Monte Carlo method**, the particles in the box are moved through small but otherwise random distances, and the change in total potential energy of the N particles in the box, ΔV_N, is calculated. Whether or not this new configuration is accepted is then judged from the following rules:

- If the potential energy is not greater than before the change, then the configuration is accepted.

If the potential energy is greater than before the change, then it is necessary to check if the new configuration is reasonable and

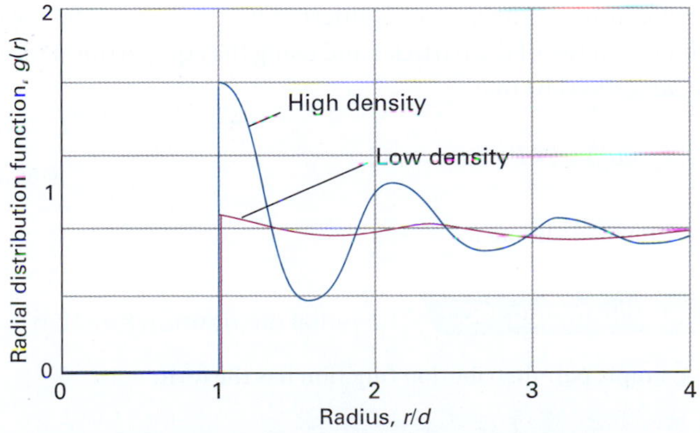

Figure 16C.2 The radial distribution function for a simulation of a liquid using impenetrable hard spheres (ball bearings).

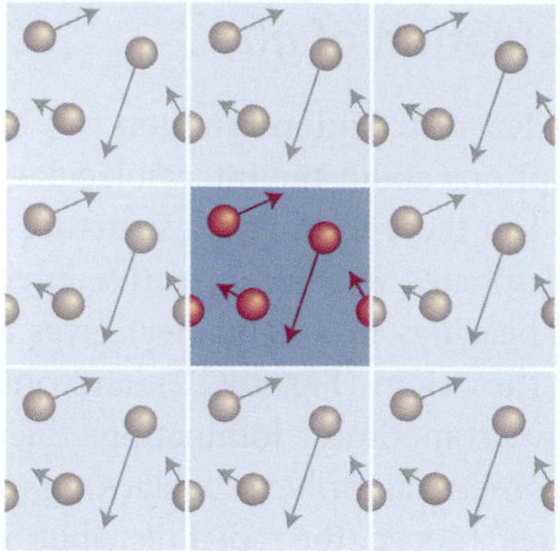

Figure 16C.3 In a two-dimensional simulation of a liquid that uses periodic boundary conditions, when one particle leaves the cell its mirror image enters through the opposite face.

can exist in equilibrium with configurations of lower potential energy at a given temperature. To make progress, we use the result that, at equilibrium, the ratio of populations of two states with energy separation ΔV_N is $e^{-\Delta V_N/kT}$. Because we are testing the viability of a configuration with a higher potential energy than the previous configuration in the calculation, $\Delta V_N>0$ and the exponential factor varies between 0 and 1. In the Monte Carlo method, the second rule, therefore, is:

- The exponential factor is compared with a random number between 0 and 1; if the factor is larger than the random number, then the configuration is accepted; if the factor is not larger, the configuration is rejected.

The configurations generated with Monte Carlo calculations can be used to construct $g(r)$ simply by counting the number of pairs of particles with a separation r and averaging the result over the whole collection of configurations.

In the **molecular dynamics** approach, the history of an initial arrangement is followed by calculating the trajectories of all the particles under the influence of the intermolecular potentials and the forces they exert. The calculation gives a series of snapshots of the liquid, and $g(r)$ can be calculated as before. The temperature of the system is inferred by computing the mean kinetic energy of the particles and using the equipartition result (*Foundations* B) that

$$\langle \tfrac{1}{2}mv_q^2\rangle=\tfrac{1}{2}kT \qquad (16C.2)$$

for each coordinate q.

Brief illustration 16C.1 The radial distribution function

A simple pair distribution function has the form

$$g(r)=1+\cos\left(\frac{4r}{r_0}-4\right)e^{-(r/r_0-1)}$$

for $r\geq r_0$ and $g(r)=0$ for $r<r_0$. Here the parameter r_0 is the separation at which the Lennard-Jones potential energy function (eqn 16B.13; $V=4\varepsilon\{(r_0/r)^{1/2}-(r_0/r)^6\}$ is equal to zero. The function $g(r)$ is plotted in Fig. 16C.4, and we see that it resembles the form shown in Fig. 16C.2.

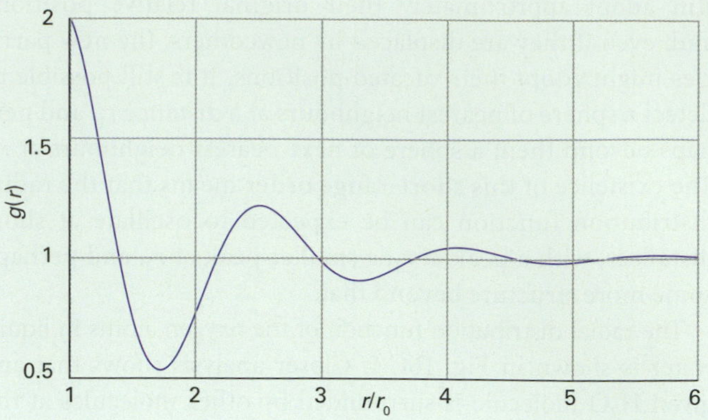

Figure 16C.4 An example of a radial distribution function for two particles with an interaction energy given by the Lennard-Jones potential energy function.

Self-test 16C.1 Use mathematical software to plot the function $v_2(r)=r(dV/dr)$. The significance of this function will become apparent soon.

Answer: See Fig. 16C.5.

(c) The thermodynamic properties of liquids

Once $g(r)$ is known it can be used to calculate the thermodynamic properties of liquids. For example, the contribution of the pairwise additive intermolecular potential, V_2, to the internal energy is given by the integral

$$U_{\text{interaction}}(T)=\frac{2\pi N^2}{V}\int_0^\infty g(r)V_2r^2dr$$

Contribution of pairwise interactions to the internal energy (16C.3)

That is, $U_{\text{interaction}}$ is essentially the average two-particle potential energy weighted by $g(r)r^2dr$, which is the probability that the pair of particles have a separation between r and $r+dr$. Likewise, the contribution that pairwise interactions make to the pressure is

$$\frac{pV}{nRT}=1-\frac{2\pi N}{3kTV}\int_0^\infty g(r)v_2r^2dr \qquad v_2=r\frac{dV_2}{dr} \qquad (16C.4a)$$

The quantity v_2 is called the **virial** (hence the term 'virial equation of state'). The dependence of v_2 on r is shown for a simple form of $g(r)$ in Fig. 16C.5. To understand the physical content of this expression, we rewrite it as

$$p=\frac{nRT}{V}-\frac{2\pi}{3}\left(\frac{N}{V}\right)^2\int_0^\infty g(r)v_2r^2dr$$

Pressure in terms of $g(r)$ (16C.4b)

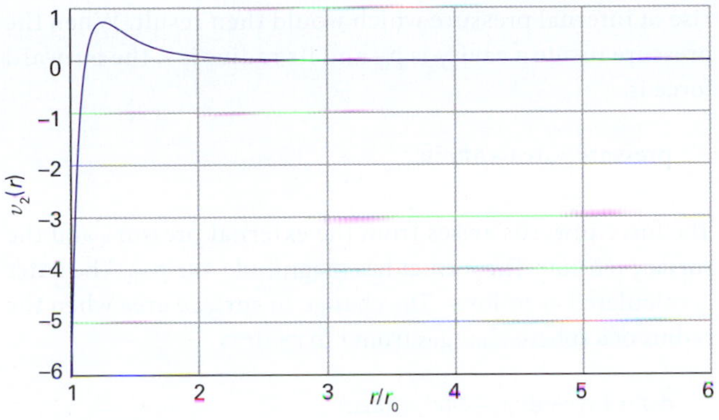

Figure 16C.5 The virial $v_2(r)$ associated with the Lennard-Jones potential energy function.

and then note that

- The first term on the right is the **kinetic pressure**, the contribution to the pressure from the impact of the molecules in free flight.
- The second term is essentially the internal pressure, $\pi_T = (\partial U/\partial V)_T$ (Topic 2D), representing the contribution to the pressure from the intermolecular forces.

Physical interpretation

To see the connection to the internal pressure, we should recognize $-dV_2/dr$ (in v_2) as the force required to move two molecules apart, and therefore $-r(dV_2/dr)$ as the work required to separate the molecules through a distance r. The second term is therefore the average of this work over the range of pairwise separations in the liquid as represented by the probability of finding two molecules at separations between r and $r+dr$, which is $g(r)r^2dr$. In brief, the integral, when multiplied by the square of the number density, is the change in internal energy of the system as it expands, and therefore is equal to the internal pressure.

16C.2 The liquid–vapour interface

We turn our attention to the physical boundary between phases, such as the surface where solid is in contact with liquid or liquid is in contact with its vapour. In this Topic we concentrate on the liquid–vapour interface, which is interesting because it is so mobile. Topic 22A deals with solid surfaces.

(a) Surface tension

Liquids tend to adopt shapes that minimize their surface area, for then the maximum number of molecules is in the bulk and hence surrounded by and interacting with neighbours. Droplets of liquids therefore tend to be spherical, because a sphere is the shape with the smallest surface-to-volume ratio. However, there may be other forces present that compete against the tendency to form this ideal shape, and in particular gravity may flatten spheres into puddles or oceans.

Table 16C.1* Surface tensions of liquids at 293 K, $\gamma/(\text{mN m}^{-1})$

	$\gamma/(\text{mN m}^{-1})$
Benzene	28.88
Mercury	472
Methanol	22.6
Water	72.75

* More values are given in the *Resource section*. Note that $1\,\text{N m}^{-1} = 1\,\text{J m}^{-2}$.

Surface effects may be expressed in the language of Helmholtz and Gibbs energies (Topic 3C). The link between these quantities and the surface area is the work needed to change the area by a given amount, and the fact that dA and dG are equal (under different conditions) to the work done in changing the energy of a system. The work needed to change the surface area, σ, of a sample by an infinitesimal amount $d\sigma$ is proportional to $d\sigma$, and we write

$$dw = \gamma d\sigma \quad \textit{Definition} \quad \text{Surface tension} \quad (16C.5)$$

The constant of proportionality, γ, is called the **surface tension**; its dimensions are energy/area and its units are typically joules per metre squared (J m^{-2}). However, as in Table 16C.1, values of γ are usually reported in newtons per metre (because $1\,\text{J} = 1\,\text{N m}$, it follows that $1\,\text{J m}^{-2} = 1\,\text{N m}^{-1}$). The work of surface formation at constant volume and temperature can be identified with the change in the Helmholtz energy, and we can write

$$dA = \gamma d\sigma \quad (16C.6)$$

Because the Helmholtz energy decreases ($dA<0$) if the surface area decreases ($d\sigma<0$), surfaces have a natural tendency to contract. This is a more formal way of expressing what we have already described.

Example 16C.1 Using the surface tension

Calculate the work needed to raise a wire of length l and to stretch the surface of a liquid through a height h in the arrangement shown in Fig. 16C.6. Disregard gravitational potential energy.

Method According to eqn 16C.5, the work required to create a surface of area σ given that the surface tension does not vary as the surface is formed is $w = \gamma\sigma$. Therefore, all we need do is to calculate the surface area of the two-sided rectangle formed as the frame is withdrawn from the liquid.

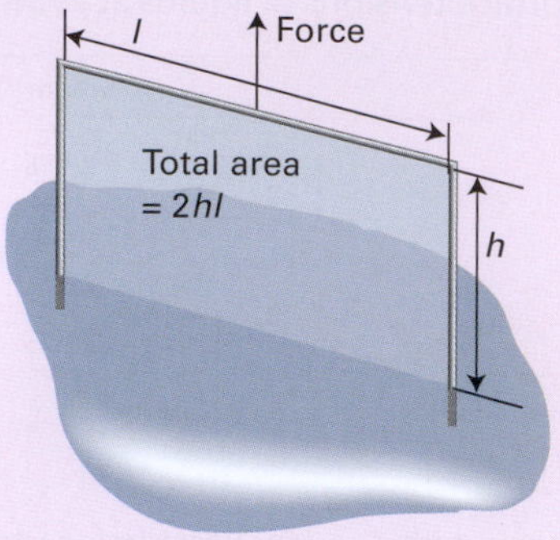

Figure 16C.6 The model used for calculating the work of forming a liquid film when a wire of length *l* is raised and pulls the surface with it through a height *h*.

Answer When the wire of length l is raised through a height h it increases the area of the liquid by twice the area of the rectangle (because there is a surface on each side). The total increase is therefore $2lh$ and the work done is $2\gamma lh$.

The expression $2\gamma lh$ can be expressed as force×distance by writing it as $2\gamma l \times h$, and identifying γl as the opposing force on the wire of length l. This interpretation is why γ is called a tension and why its units are often chosen to be newtons per metre (so γl is a force in newtons).

Self-test 16C.2 Calculate the work of creating a spherical cavity of radius r in a liquid of surface tension γ.

Answer: $4\pi r^2\gamma$

(b) Curved surfaces

The minimization of the surface area of a liquid may result in the formation of a curved surface. A **bubble** is a region in which vapour (and possibly air too) is trapped by a thin film; a **cavity** is a vapour-filled hole in a liquid. What are widely called 'bubbles' in liquids are therefore strictly cavities. True bubbles have two surfaces (one on each side of the film); cavities have only one. The treatments of both are similar, but a factor of 2 is required for bubbles to take into account the doubled surface area. A **droplet** is a small volume of liquid at equilibrium surrounded by its vapour (and possibly also air).

The pressure on the concave side of an interface, p_{in}, is always greater than the pressure on the convex side, p_{out}. This relation is expressed by the **Laplace equation**, which is derived in the following *Justification*:

$$p_{in} = p_{out} + \frac{2\gamma}{r} \qquad \text{Laplace equation} \qquad (16C.7)$$

Justification 16C.1 The Laplace equation

The cavities in a liquid are at equilibrium when the tendency for their surface area to decrease is balanced by the rise of internal pressure which would then result. When the pressure inside a cavity is p_{in} and its radius is r, the outward force is

$$\text{pressure}\times\text{area} = 4\pi r^2 p_{in}$$

The force inwards arises from the external pressure and the surface tension. The former has magnitude $4\pi r^2 p_{out}$. The latter is calculated as follows. The change in surface area when the radius of a sphere changes from r to $r+dr$ is

$$d\sigma = 4\pi(r+dr)^2 - 4\pi r^2 = 8\pi r dr$$

(The second-order infinitesimal, $(dr)^2$, is ignored.) The work done when the surface is stretched by this amount is therefore

$$dw = 8\pi\gamma r dr$$

As force×distance is work, the force opposing stretching through a distance dr when the radius is r is

$$F = 8\pi\gamma r$$

The total inward force is therefore $4\pi r^2 p_{out} + 8\pi\gamma r$. At equilibrium, the outward and inward forces are balanced, so we can write

$$4\pi r^2 p_{in} = 4\pi r^2 p_{out} + 8\pi\gamma r$$

which rearranges into eqn 16C.7.

The Laplace equation shows that the difference in pressure decreases to zero as the radius of curvature becomes infinite (when the surface is flat, Fig. 16C.7). Small cavities have small radii of curvature, so the pressure difference across their surface is quite large.

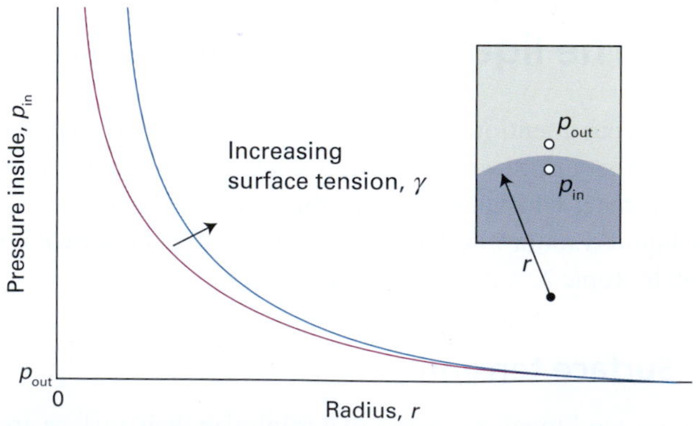

Figure 16C.7 The dependence of the pressure inside a curved surface on the radius of the surface, for two different values of the surface tension.

Brief illustration 16C.2 The Laplace equation

The pressure differential of water across the surface of a spherical droplet of radius 200 nm at 20 °C is

$$p_{in} - p_{out} = \frac{\overbrace{2\times(72.75\times10^{-3}\,\mathrm{N\,m^{-1}})}^{\gamma_{\text{water}}\text{ at }20\,°\mathrm{C}}}{\underbrace{2.00\times10^{-7}\,\mathrm{m}}_{r}}$$

$$=7.28\times10^{5}\,\mathrm{N\,m^{-2}}=728\,\mathrm{kPa}$$

Self-test 16C.3 Calculate the pressure differential of ethanol across the surface of a spherical droplet of radius 220 nm at 20 °C. The surface tension of ethanol at that temperature is 22.39 mN m^{-1}.

Answer: 204 kPa

(c) Capillary action

The tendency of liquids to rise up capillary tubes (tubes of narrow bore; the name comes from the Latin word for 'hair'), which is called **capillary action**, is a consequence of surface tension. Consider what happens when a glass capillary tube is first immersed in water or any liquid that has a tendency to adhere to the walls. The energy is lowest when a thin film covers as much of the glass as possible. As this film creeps up the inside wall it has the effect of curving the surface of the liquid inside the tube. This curvature implies that the pressure just beneath the curving meniscus is less than the atmospheric pressure by approximately $2\gamma/r$, where r is the radius of the tube and we assume a hemispherical surface. The pressure immediately under the flat surface outside the tube is p, the atmospheric pressure; but inside the tube under the curved surface it is only $P-2\gamma/r$. The excess external pressure presses the liquid up the tube until hydrostatic equilibrium (equal pressures at equal depths) has been reached (Fig. 16C.8).

To calculate the height to which the liquid rises, we note that the pressure exerted by a column of liquid of mass density ρ and height h is

$$p=\rho gh \tag{16C.8}$$

This hydrostatic pressure matches the pressure difference $2\gamma/r$ at equilibrium. Therefore, the height of the column at equilibrium is obtained by equating $2\gamma/r$ and ρgh, which gives

$$h=\frac{2\gamma}{\rho gr} \tag{16C.9}$$

This simple expression provides a reasonably accurate way of measuring the surface tension of liquids. Surface tension decreases with increasing temperature (Fig. 16C.9).

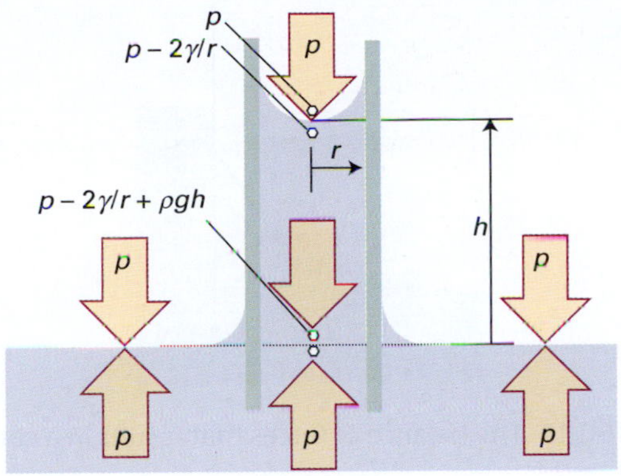

Figure 16C.8 When a capillary tube is first stood in a liquid, the latter climbs up the walls, so curving the surface. The pressure just under the meniscus is less than that arising from the atmosphere by $2\gamma/r$. The pressure is equal at equal heights throughout the liquid provided the hydrostatic pressure (which is equal to ρgh) cancels the pressure difference arising from the curvature.

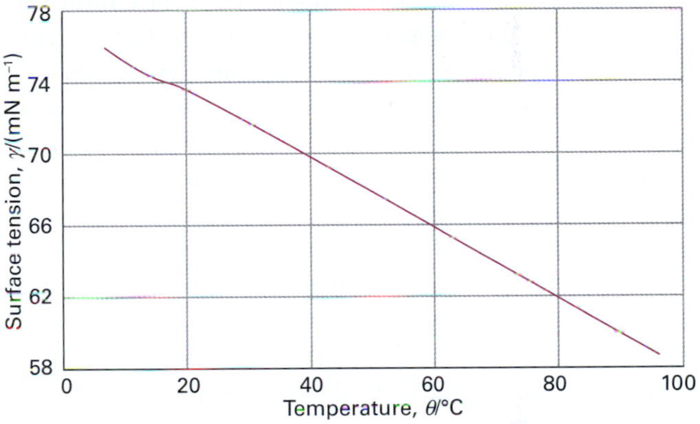

Figure 16C.9 The variation of the surface tension of water with temperature.

Brief illustration 16C.3 Capillary action

If water at 25 °C rises through 7.36 cm in a capillary of radius 0.20 mm, its surface tension at that temperature is

$$\gamma=\tfrac{1}{2}\rho ghr$$
$$=\tfrac{1}{2}\times(997.1\,\mathrm{kg\,m^{-3}})\times(9.81\,\mathrm{m\,s^{-2}})\times(7.36\times10^{-2}\,\mathrm{m})$$
$$\times(2.0\times10^{-4}\,\mathrm{m})$$
$$\overset{1\,\mathrm{kg\,m\,s^{-2}}=1\,\mathrm{N}}{=}\quad 72\,\mathrm{mN\,m^{-1}}$$

Self-test 16C.4 Calculate the surface tension of water at 30 °C given that at that temperature water climbs to a height of 9.11 cm in a clean glass capillary tube of internal radius 0.320 mm. The density of water at 30 °C is 0.9956 g cm^{-3}.

Answer: 142 mN m^{-1}

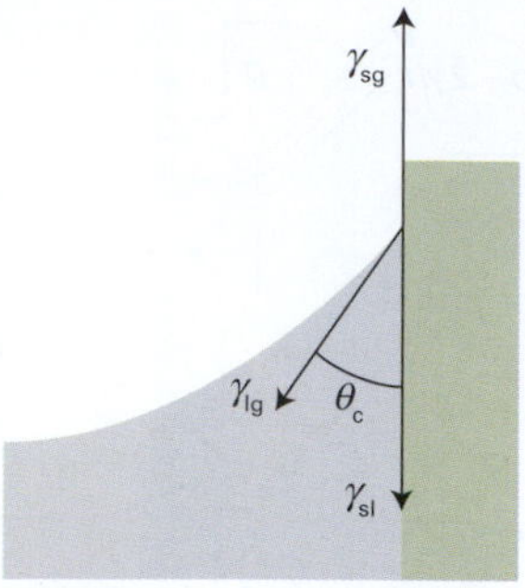

Figure 16C.10 The balance of forces that results in a contact angle, θ_c.

When the adhesive forces between the liquid and the material of the capillary wall are weaker than the cohesive forces within the liquid (as for mercury in glass), the liquid in the tube retracts from the walls. This retraction curves the surface with the concave, high pressure side downwards. To equalize the pressure at the same depth throughout the liquid the surface must fall to compensate for the heightened pressure arising from its curvature. This compensation results in a capillary depression.

In many cases there is a nonzero angle between the edge of the meniscus and the wall. If this contact angle is θ_c, then eqn 16C.9 should be modified by multiplying the right-hand side by cos θ_c. The origin of the contact angle can be traced to the balance of forces at the line of contact between the liquid and the solid (Fig. 16C.10). If the solid–gas, solid–liquid, and liquid–gas surface tensions (essentially the energy needed to create unit area of each of the interfaces) are denoted γ_{sg}, γ_{sl}, and γ_{lg}, respectively, then the vertical forces are in balance if

$$\gamma_{sg} = \gamma_{sl} + \gamma_{lg}\cos\theta_c \qquad (16C.10)$$

This expression solves to

$$\cos\theta_c = \frac{\gamma_{sg} - \gamma_{sl}}{\gamma_{lg}} \qquad (16C.11)$$

If we note that the superficial work of adhesion of the liquid to the solid (the work of adhesion divided by the area of contact) is

$$w_{ad} = \gamma_{sg} + \gamma_{lg} - \gamma_{sl} \qquad (16C.12)$$

eqn 16C.11 can be written

$$\cos\theta_c = \frac{w_{ad}}{\gamma_{lg}} - 1 \qquad \text{Contact angle} \qquad (16C.13)$$

We now see that:

- The liquid 'wets' (spreads over) the surface, corresponding to $0 < \theta_c < 90°$, when $1 < w_{ad}/\gamma_{lg} < 2$ (Fig. 16C.11).
- The liquid does not wet the surface, corresponding to $90° < \theta_c < 180°$, when $0 < w_{ad}/\gamma_{lg} < 1$.

Physical interpretation

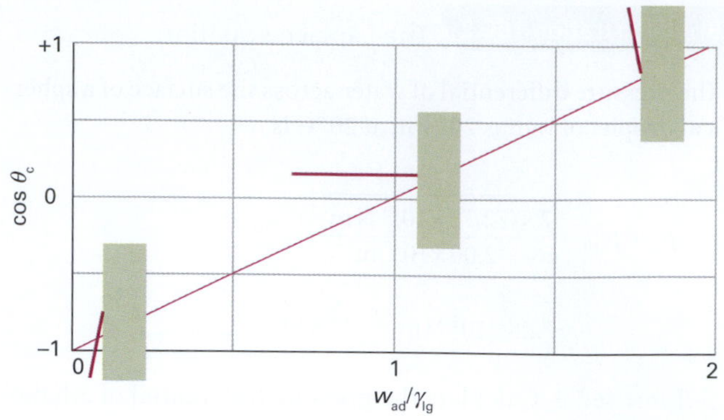

Figure 16C.11 The variation of contact angle (shown by the semaphore-like object) as the ratio w_{ad}/γ_{lg} changes.

For mercury in contact with glass, $\theta_c = 140°$, which corresponds to $w_{ad}/\gamma_{lg} = 0.23$, indicating a relatively low work of adhesion of the mercury to glass on account of the strong cohesive forces within mercury.

16C.3 Surface films

The compositions of surface layers have been investigated by the simple but technically elegant procedure of slicing thin layers off the surfaces of solutions and analysing their compositions. The physical properties of surface films have also been investigated. Surface films one molecule thick are called **monolayers**. When a monolayer has been transferred to a solid support, it is called a **Langmuir–Blodgett film**, after Irving Langmuir and Katherine Blodgett, who developed experimental techniques for studying them.

(a) Surface pressure

The principal apparatus used for the study of surface monolayers is a **surface film balance** (Fig. 16C.12). This device consists of a shallow trough and a barrier that can be moved along the surface of the liquid in the trough, and hence compress any monolayer on the surface. The **surface pressure**, π, the difference between the surface tension of the pure solvent and the solution ($\pi = \gamma^* - \gamma$) is measured by using a torsion wire attached to a strip of mica that rests on the surface and pressing against one edge of the monolayer. The parts of the apparatus that are in touch with liquids are coated in polytetrafluoroethene to eliminate effects arising from the liquid–solid interface. In an actual experiment, a small amount (about 0.01 mg) of the surfactant under investigation is dissolved in a volatile solvent and then poured on to the surface of the water; the compression barrier is then moved across the surface and the surface pressure exerted on the mica bar is monitored.

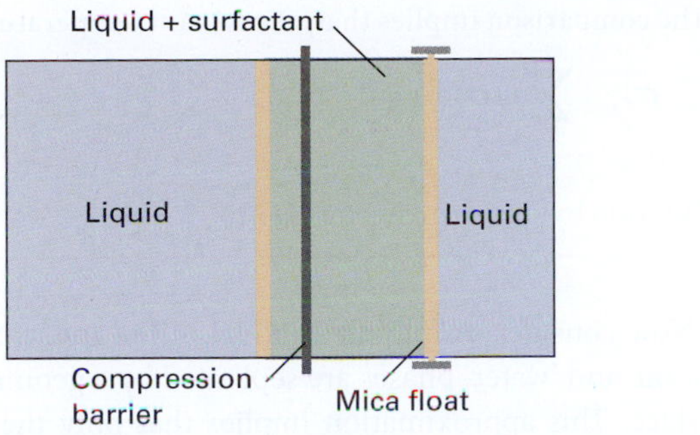

Figure 16C.12 A schematic diagram of the apparatus used to measure the surface pressure and other characteristics of a surface film. The surfactant is spread on the surface of the liquid in the trough, and then compressed horizontally by moving the compression barrier towards the mica float. The latter is connected to a torsion wire, so the difference in force on either side of the float can be monitored.

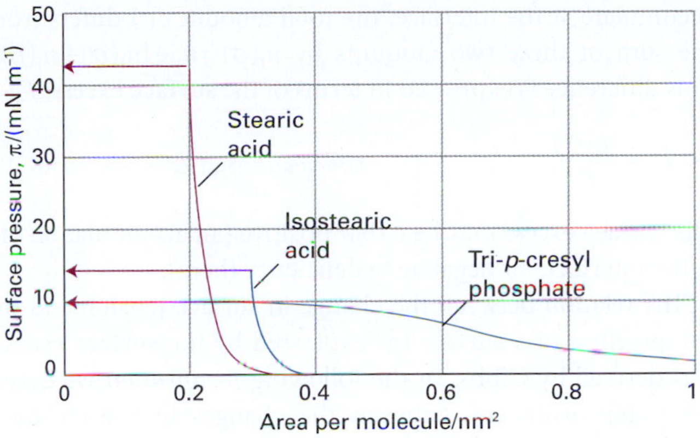

Figure 16C.13 The variation of surface pressure with the area occupied by each surfactant molecule. The collapse pressures are indicated by the horizontal lines.

Some typical results are shown in Fig. 16C.13. One parameter obtained from the isotherms is the area occupied by the molecules when the monolayer is closely packed. This quantity is obtained from the extrapolation of the steepest part of the isotherm to the horizontal axis. As can be seen from the illustration, even though stearic acid (1) and isostearic acid (2) are chemically very similar (they differ only in the location of a methyl group at the end of a long hydrocarbon chain), they occupy significantly different areas in the monolayer. Neither, though, occupies as much area as the tri-*p*-cresyl phosphate molecule (3), which is like a wide bush rather than a lanky tree.

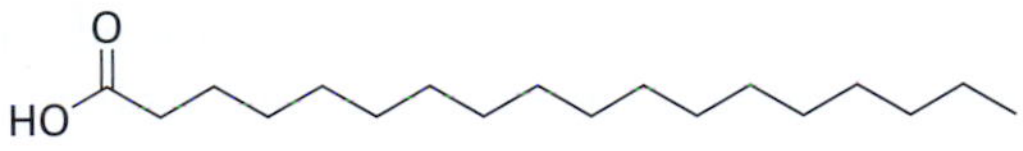

1 Stearic acid, $C_{17}H_{35}COOH$

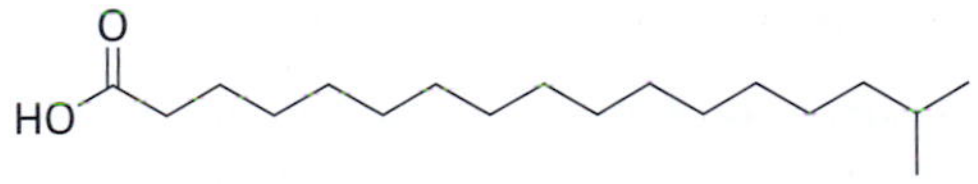

2 Isostearic acid, $C_{17}H_{35}COOH$

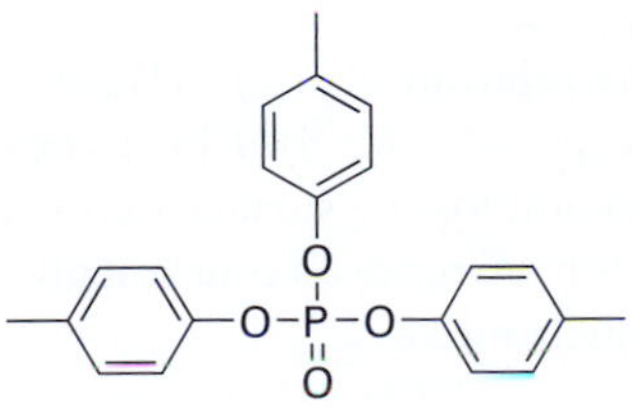

3 Tri-*p*-cresylphosphate

The second feature to note from Fig. 16C.13 is that the tri-*p*-cresyl phosphate isotherm is much less steep than the stearic acid isotherms. This difference indicates that the tri-*p*-cresyl phosphate film is more compressible than the stearic acid films, which is consistent with their different molecular structures.

A third feature of the isotherms is the **collapse pressure**, the highest surface pressure. When the monolayer is compressed beyond the point represented by the collapse pressure, the monolayer buckles and collapses into a film several molecules thick. As can be seen from the isotherms in Fig. 16C.13, stearic acid has a high collapse pressure, but that of tri-*p*-cresyl phosphate is significantly smaller, indicating a much weaker film.

(b) The thermodynamics of surface layers

A **surfactant** is a species that is active at the interface between two phases, such as at the interface between hydrophilic and hydrophobic phases. A surfactant accumulates at the interface, and modifies its surface tension and hence the surface pressure. To establish the relation between the concentration of surfactant at a surface and the change in surface tension it brings about, we consider two phases α and β in contact and suppose that the system consists of several components J, each one present in an overall amount n_J. If the components were distributed uniformly through the two phases right up to the interface, which is taken to be a plane of surface area σ, the total Gibbs energy, G, would be the sum of the Gibbs energies of both phases, $G=G(\alpha)+G(\beta)$. However, the components are not uniformly distributed because one may accumulate at the interface. As a result, the sum of the two Gibbs energies differs from G by an amount called the **surface Gibbs energy**, $G(\sigma)$:

$$G(\sigma)=G-\{G(\alpha)+G(\beta)\} \qquad \textit{Definition} \qquad \text{Surface Gibbs energy} \qquad (16C.14)$$

Similarly, if it is supposed that the concentration of a species J is uniform right up to the interface, then from its volume we would conclude that it contains an amount $n_J(\alpha)$ of J in phase α and an amount $n_J(\beta)$ in phase β. However, because a species may accumulate at the interface, the total amount of J differs from the sum of these two amounts by $n_J(\sigma)=n_J-\{n_J(\alpha)+n_J(\beta)\}$. This difference is expressed in terms of the **surface excess**, Γ_J:

$$\Gamma_J=\frac{n_J(\sigma)}{\sigma} \qquad \textit{Definition} \quad \text{Surface excess} \quad (16C.15)$$

The surface excess may be either positive (an accumulation of J at the interface) or negative (a deficiency there).

The relation between the change in surface tension and the composition of a surface (as expressed by the surface excess) was derived by Gibbs. In the following *Justification* we derive the **Gibbs isotherm**, between the changes in the chemical potentials of the substances present in the interface and the change in surface tension:

$$\mathrm{d}\gamma=-\sum_J \Gamma_J \mathrm{d}\mu_J \qquad \text{Gibbs isotherm} \quad (16C.16)$$

Justification 16C.2 The Gibbs isotherm

A general change in G is brought about by changes in T, p, and the n_J:

$$\mathrm{d}G=-S\mathrm{d}T+V\mathrm{d}p+\gamma\mathrm{d}\sigma+\sum_J \mu_J \mathrm{d}n_J$$

When this relation is applied to G, $G(\alpha)$, and $G(\beta)$ we find

$$\mathrm{d}G(\sigma)=-S(\sigma)\mathrm{d}T+\gamma\mathrm{d}\sigma+\sum_J \mu_J \mathrm{d}n_J(\sigma)$$

because at equilibrium the chemical potential of each component is the same in every phase, $\mu_J(\alpha)=\mu_J(\beta)=\mu_J(\sigma)$. Just as in the discussion of partial molar quantities (Chapter 5), the last equation integrates at constant temperature to

$$G(\sigma)=\gamma\sigma+\sum_J \mu_J n_J(\sigma)$$

We are seeking a connection between the change of surface tension $\mathrm{d}\gamma$ and the change of composition at the interface. Therefore, we use the argument that in Topic 5A led to the Gibbs–Duhem equation (eqn 5A.12b), but this time we compare the expression

$$\mathrm{d}G(\sigma)=\gamma\mathrm{d}\sigma+\sum_J \mu_J \mathrm{d}n_J(\sigma)$$

(which is valid at constant temperature) with the expression for the same quantity but derived from the preceding equation:

$$\mathrm{d}G(\sigma)=\gamma\mathrm{d}\sigma+\sigma\mathrm{d}\gamma+\sum_J \mu_J \mathrm{d}n_J(\sigma)+\sum_J n_J(\sigma)\mathrm{d}\mu_J$$

The comparison implies that, at constant temperature,

$$\sigma\mathrm{d}\gamma+\sum_J n_J(\sigma)\mathrm{d}\mu_J=0$$

Division by σ then gives eqn 16C.16.

Now consider a simplified model of the interface in which the 'oil' and 'water' phases are separated by a geometrically flat surface. This approximation implies that only the surfactant, S, accumulates at the surface, and hence that Γ_{oil} and Γ_{water} are both zero. Then the Gibbs isotherm equation becomes

$$\mathrm{d}\gamma=-\Gamma_S \mathrm{d}\mu_S \qquad (16C.17)$$

For dilute solutions,

$$\mathrm{d}\mu_S=RT\ln c \qquad (16C.18)$$

where c is the molar concentration of the surfactant. It follows that

$$\mathrm{d}\gamma=-RT\Gamma_S\frac{\mathrm{d}c}{c}$$

at constant temperature, or

$$\left(\frac{\partial\gamma}{\partial c}\right)_T=-\frac{RT\Gamma_S}{c} \qquad \text{Dependence of the surface tension on surfactant concentration} \quad (16C.19)$$

If the surfactant accumulates at the interface, its surface excess is positive and eqn 16C.19 implies that $(\partial\gamma/\partial c)_T<0$. That is, the surface tension decreases when a solute accumulates at a surface. Conversely, if the concentration dependence of γ is known, then the surface excess may be predicted and used to infer the area occupied by each surfactant molecule on the surface.

Example 16C.2 Calculating the surface excess

Calculate the surface excess of 1-aminobutanoic acid in a 0.10 mol dm^{-3} aqueous solution at 20 °C given that $\mathrm{d}\gamma/\mathrm{d}(\ln c)=-40\,\mu\mathrm{N\,m^{-1}}$. Convert the answer to the number of molecules per square metre.

Method Use the relation $\mathrm{d}(\ln x)=(1/x)\mathrm{d}x$ to convert eqn 16C.19 into an expression for $\partial\gamma/\partial(\ln c)$, then rearrange it to obtain an expression for the surface excess Γ_S. Multiplying the surface excess by Avogadro's constant gives the number of molecules per square metre.

Answer Because $\mathrm{d}(\ln c)=(1/c)\mathrm{d}c$ and $\mathrm{d}c=c\mathrm{d}(\ln c)$, eqn 16C.19 may be written as

$$\left(\frac{\partial\gamma}{\partial(\ln c)}\right)_T=-RT\Gamma_S$$

It follows that

$$\Gamma_S = \frac{1}{RT}\left(\frac{\partial\gamma}{\partial(\ln c)}\right)_T$$

$$= \frac{1}{(8.314\,\mathrm{J\,K^{-1}\,mol^{-1}})\times(293\,\mathrm{K})}\times(-4.0\times10^{-5}\,\mathrm{N\,m^{-1}})$$

$$= 1.6\times10^{-8}\,\mathrm{mol\,m^{-2}}$$

The number of molecules per square metre, $\mathcal{N}$, is

$$\mathcal{N} = N_A\Gamma_S = (6.02\times10^{23}\,\mathrm{mol^{-1}})\times(1.6\times10^{-8}\,\mathrm{mol\,m^{-2}})$$

$$= 9.6\times10^{15}\,\mathrm{m^{-2}}$$

Self-test 16C.5 Use the result from *Example* 16C.2 to calculate the area occupied by a molecule.

Answer: $1.0\times10^2\,\mathrm{nm^2}$

16C.4 Condensation

We now bring together concepts from this Topic and Topic 4B to explain the condensation of a gas to a liquid. We saw in Topic 4B that the vapour pressure of a liquid depends on the pressure applied to the liquid. Because curving a surface gives rise to a pressure differential of $2\gamma/r$, we can expect the vapour pressure above a curved surface to be different from that above a flat surface. By substituting this value of the pressure difference into eqn 4B.3 ($p = p^* e^{V_m\Delta P/RT}$, where p^* is the vapour pressure when the pressure difference is zero) we obtain the **Kelvin equation** for the vapour pressure of a liquid when it is dispersed as droplets of radius r:

$$p = p^* e^{2\gamma V_m/rT} \quad \text{Kelvin equation} \quad (16C.20)$$

The analogous expression for the vapour pressure inside a cavity can be written at once. The pressure of the liquid outside the cavity is less than the pressure inside, so the only change is in the sign of the exponent in the last expression. For droplets of water of radius 1 µm and 1 nm the ratios p/p^* at 25 °C are about 1.001 and 3, respectively. The second figure, although quite large, is unreliable because at that radius the droplet is less than about 10 molecules in diameter and the basis of the calculation is suspect. The first figure shows that the effect is usually small; nevertheless it may have important consequences.

Consider the formation of a cloud. Warm, moist air rises into the cooler regions higher in the atmosphere. At some altitude the temperature is so low that the vapour becomes thermodynamically unstable with respect to the liquid and we expect it to condense into a cloud of liquid droplets. The initial step can be imagined as a swarm of water molecules congregating into a microscopic droplet. Because the initial droplet is so small, it has an enhanced vapour pressure. Therefore, instead of growing it evaporates. This effect stabilizes the vapour because an initial tendency to condense is overcome by a heightened tendency to evaporate. The vapour phase is then said to be **supersaturated**. It is thermodynamically unstable with respect to the liquid but not unstable with respect to the small droplets that need to form before the bulk liquid phase can appear, so the formation of the latter by a simple, direct mechanism is hindered.

Clouds do form, so there must be a mechanism. Two processes are responsible. The first is that a sufficiently large number of molecules might congregate into a droplet so big that the enhanced evaporative effect is unimportant. The chance of one of these **spontaneous nucleation centres** forming is low, and in rain formation it is not a dominant mechanism. The more important process depends on the presence of minute dust particles or other kinds of foreign matter. These **nucleate** the condensation (that is, provide centres at which it can occur) by providing surfaces to which the water molecules can attach.

Liquids may be **superheated** above their boiling temperatures and **supercooled** below their freezing temperatures. In each case the thermodynamically stable phase is not achieved on account of the kinetic stabilization that occurs in the absence of nucleation centres. For example, superheating occurs because the vapour pressure inside a cavity is artificially low, so any cavity that does form tends to collapse. This instability is encountered when an unstirred beaker of water is heated, for its temperature may be raised above its boiling point. Violent bumping often ensues as spontaneous nucleation leads to bubbles big enough to survive. To ensure smooth boiling at the true boiling temperature, nucleation centres, such as small pieces of sharp-edged glass or bubbles (cavities) of air, should be introduced.

Checklist of concepts

- ☐ 1. The **radial distribution function**, $g(r)$, is a probability density in the sense that $g(r)\mathrm{d}r$ is the probability that a molecule will be found in the range $\mathrm{d}r$ at a distance r from another molecule.
- ☐ 2. The radial distribution function may be calculated with **Monte Carlo** and **molecular dynamics** techniques.
- ☐ 3. The internal energy and pressure of a fluid may be expressed in terms of the radial distribution function.
- ☐ 4. Liquids tend to adopt shapes that minimize their surface area.
- ☐ 5. The minimization of surface area results in the formation of bubbles, cavities, and droplets.

☐ 6. **Capillary action** is the tendency of liquids to rise up narrow tubes.
☐ 7. The **surface pressure** is the difference between the surface tension of the pure solvent and the solution.
☐ 8. The **collapse pressure** is the highest surface pressure that a surface film can sustain.
☐ 9. A **surfactant** modifies the surface tension and surface pressure.
☐ 10. **Nucleation** provides surfaces to which molecules can attach and thereby induce **condensation**.

Checklist of equations

Property	Equation	Comment	Equation number
Radial distribution function	$g(r_{12})=A/B$, $A=N(N-1)\iint\cdots\int e^{-\beta V_N}d\tau_3 d\tau_4\ldots d\tau_N$, $B=\mathcal{N}^2\iint\cdots\int e^{-\beta V_N}d\tau_1 d\tau_2\ldots d\tau_N$		16C.1
Contribution of interactions to the internal energy	$U_{\text{interaction}}(T)=(2\pi N^2/V)\int_0^\infty g(r)V_2 r^2 dr$	V_2 is the intermolecular potential energy	16C.3
Pressure in terms of $g(r)$	$p=nRT/V-(2\pi/3)(N/V)^2\int_0^\infty g(r)v_2 r^2 dr$		16C.4b
Laplace equation	$p_{\text{in}}=p_{\text{out}}+2\gamma/r$	γ is the surface tension	16C.7
Contact angle	$\cos\theta_c=(w_{\text{ad}}/\gamma_{\text{lg}})-1$		16C.13
Surface Gibbs energy	$G(\sigma)=G-\{G(\alpha)+G(\beta)\}$	Definition	16C.14
Excess energy	$\Gamma_J=n_J(\sigma)/\sigma$	Definition	16C.15
Gibbs isotherm	$d\gamma=-\sum_J \Gamma_J d\mu_J$		16C.16
Dependence of the surface tension on surfactant concentration	$(\partial\gamma/\partial c)_T=-RT\Gamma_S/c$		16C.19
Kelvin equation	$p=p^* e^{2\gamma V_m/rRT}$		16C.20

CHAPTER 16 Molecular interactions

TOPIC 16A Electric properties of molecules

Discussion questions

16A.1 Explain how the permanent dipole moment and the polarizability of a molecule arise.

16A.2 Explain why the polarizability of a molecule decreases at high frequencies.

16A.3 Describe the experimental procedures available for determining the electric dipole moment of a molecule.

Exercises

16A.1(a) Which of the following molecules may be polar: ClF_3, O_3, H_2O_2?
16A.1(b) Which of the following molecules may be polar: SO_3, XeF_4, SF_4?

16A.2(a) Calculate the resultant of two dipole moments of magnitude 1.5 D and 0.80 D that make an angle of 109.5° to each other.
16A.2(b) Calculate the resultant of two dipole moments of magnitude 2.5 D and 0.50 D that make an angle of 120° to each other.

16A.3(a) Calculate the magnitude and direction of the dipole moment of the following arrangement of charges in the xy-plane: $3e$ at (0,0), $-e$ at (0.32 nm, 0), and $-2e$ at an angle of 20° from the x-axis and a distance of 0.23 nm from the origin.
16A.3(b) Calculate the magnitude and direction of the dipole moment of the following arrangement of charges in the xy-plane: $4e$ at (0, 0), $-2e$ at (162 pm, 0), and $-2e$ at an angle of 30° from the x-axis and a distance of 143 pm from the origin.

16A.4(a) The molar polarization of fluorobenzene vapour varies linearly with T^{-1}, and is 70.62 $cm^3\,mol^{-1}$ at 351.0 K and 62.47 $cm^3\,mol^{-1}$ at 423.2 K. Calculate the polarizability and dipole moment of the molecule.
16A.4(b) The molar polarization of the vapour of a compound was found to vary linearly with T^{-1}, and is 75.74 $cm^3\,mol^{-1}$ at 320.0 K and 71.43 $cm^3\,mol^{-1}$ at 421.7 K. Calculate the polarizability and dipole moment of the molecule.

16A.5(a) At 0 °C, the molar polarization of liquid chlorine trifluoride is 27.18 $cm^3\,mol^{-1}$ and its density is 1.89 $g\,cm^{-3}$. Calculate the relative permittivity of the liquid.
16A.5(b) At 0 °C, the molar polarization of a liquid is 32.16 $cm^3\,mol^{-1}$ and its density is 1.92 $g\,cm^{-3}$. Calculate the relative permittivity of the liquid. Take M=85.0 $g\,mol^{-1}$.

16A.6(a) The refractive index of CH_2I_2 is 1.732 for 656 nm light. Its density at 20 °C is 3.32 $g\,cm^{-3}$. Calculate the polarizability of the molecule at this wavelength.
16A.6(b) The refractive index of a compound is 1.622 for 643 nm light. Its density at 20 °C is 2.99 $g\,cm^{-3}$. Calculate the polarizability of the molecule at this wavelength. Take M=65.5 $g\,mol^{-1}$.

16A.7(a) The polarizability volume of H_2O at optical frequencies is $1.5\times10^{-24}\,cm^3$: estimate the refractive index of water. The experimental value is 1.33; what may be the origin of the discrepancy?
16A.7(b) The polarizability volume of a liquid of molar mass 72.3 $g\,mol^{-1}$ and density 865 $kg\,m^{-3}$ at optical frequencies is $2.2\times10^{-30}\,m^3$: estimate the refractive index of the liquid.

16A.8(a) The dipole moment of chlorobenzene is 1.57 D and its polarizability volume is $1.23\times10^{-23}\,cm^3$. Estimate its relative permittivity at 25 °C, when its density is 1.173 $g\,cm^{-3}$.
16A.8(b) The dipole moment of bromobenzene is 5.17×10^{-30} C m and its polarizability volume is approximately $1.5\times10^{-19}\,m^3$. Estimate its relative permittivity at 25 °C, when its density is 1491 $kg\,m^{-3}$.

Problems

16A.1 The electric dipole moment of toluene (methylbenzene) is 0.4 D. Estimate the dipole moments of the three xylenes (dimethylbenzene). About which answer can you be sure?

16A.2 Plot the magnitude of the electric dipole moment of hydrogen peroxide as the H—O—O—H (azimuthal) angle ϕ changes from 0 to 2π. Use the dimensions shown in **1**.

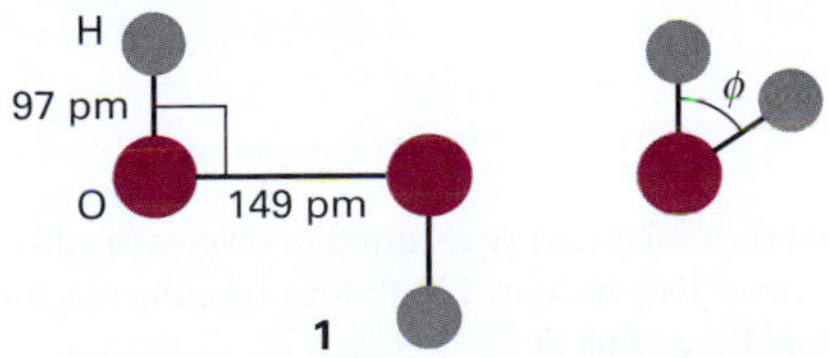

1

16A.3 Acetic acid vapour contains a proportion of planar, hydrogen bonded dimers (**2**). The apparent dipole moment of molecules in pure gaseous acetic acid has a magnitude that increases with increasing temperature. Suggest an interpretation of this observation.

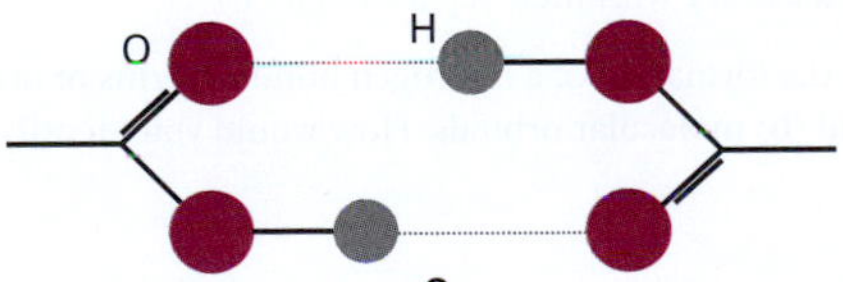

2

16A.4‡ D.D. Nelson et al. (*Science* **238**, 1670 (1987)) examined several weakly bound gas-phase complexes of ammonia in search of examples in which the H atoms in NH_3 formed hydrogen bonds, but found none. For example, they found that the complex of NH_3 and CO_2 has the carbon atom nearest the nitrogen (299 pm away): the CO_2 molecule is at right angles to the C—N 'bond',

‡ These problems were supplied by Charles Trapp and Carmen Giunta

and the H atoms of NH_3 are pointing away from the CO_2. The magnitude of the permanent dipole moment of this complex is reported as 1.77 D. If the N and C atoms are the centres of the negative and positive charge distributions, respectively, what is the magnitude of those partial charges (as multiples of e)?

16A.5 The polarizability volume of NH_3 is $2.22\times10^{-30}\,m^3$; calculate the dipole moment of the molecule (in addition to the permanent dipole moment) induced by an applied electric field of strength $15.0\,kV\,m^{-1}$.

16A.6 The magnitude of the electric field at a distance r from a point charge Q is equal to $Q/4\pi\varepsilon_0 r^2$. How close to a water molecule (of polarizability volume $1.48\times10^{-30}\,m^3$) must a proton approach before the dipole moment it induces has a magnitude equal to that of the permanent dipole moment of the molecule (1.85 D)?

16A.7 The relative permittivity of chloroform was measured over a range of temperatures with the following results:

$\theta/°C$	−80	−70	−60	−40	−20	0	20
ε	3.1	3.1	7.0	6.5	6.0	5.5	5.0
$\rho/(g\,cm^{-3})$	1.65	1.64	1.64	1.61	1.57	1.53	1.50

The freezing point of chloroform is −64 °C. Account for these results and calculate the dipole moment and polarizability volume of the molecule.

16A.8 The relative permittivities of methanol (with a melting point of −95 °C) corrected for density variation are given below. What molecular information can be deduced from these values? Take $\rho=0.791\,g\,cm^{-3}$ at 200 °C.

$\theta/°C$	−185	−170	−150	−140	−110	−80	−50	−20	0	20
ε_r	3.2	3.6	4.0	5.1	67	57	49	43	38	34

16A.9 In his classic book *Polar molecules*, Debye reports some early measurements of the polarizability of ammonia. From the selection below, determine the dipole moment and the polarizability volume of the molecule.

T/K	292.2	309.0	333.0	387.0	413.0	446.0
$P_m/(cm^3\,mol^{-1})$	57.57	55.01	51.22	44.99	42.51	39.59

The refractive index of ammonia at 273 K and 100 kPa is 1.000 379 (for yellow sodium light). Calculate the molar polarizability of the gas at this temperature and at 292.2 K. Combine the value calculated with the static molar polarizability at 292.2 K and deduce from this information alone the molecular dipole moment.

16A.10 Values of the molar polarization of gaseous water at 100 kPa as determined from capacitance measurements are given below as a function of temperature.

T/K	384.3	420.1	444.7	484.1	521.0
$P_m/(cm^3\,mol^{-1})$	57.4	53.5	50.1	46.8	43.1

Calculate the dipole moment of H_2O and its polarizability volume.

16A.11 From data in Table 16A.1 calculate the molar polarization, relative permittivity, and refractive index of methanol at 20 °C. Its density at that temperature is $0.7914\,g\,cm^{-3}$.

16A.12 Show that, in a gas (for which the refractive index is close to 1), the refractive index depends on the pressure as $n_r=1+\text{const}\times p$, and find the constant of proportionality. Go on to show how to deduce the polarizability volume of a molecule from measurements of the refractive index of a gaseous sample.

16A.13 Acetic acid vapour contains a proportion of planar, hydrogen-bonded dimers. The relative permittivity of pure liquid acetic acid is 7.14 at 290 K and increases with increasing temperature. Suggest an interpretation of the latter observation. What effect should isothermal dilution have on the relative permittivity of solutions of acetic acid in benzene?

TOPIC 16B Interactions between molecules

Discussion questions

16B.1 Identify the terms in and limit the generality of the following expressions: (a) $V=-Q_2\mu_1/4\pi\varepsilon_0 r^2$, (b) $V=-Q_2\mu_1\cos\theta/4\pi\varepsilon_0 r^2$, and (c) $V=\mu_2\mu_1(1-3\cos^2\theta)/4\pi\varepsilon_0 r^3$.

16B.2 Draw examples of the arrangements of electrical charges that correspond to monopoles, dipoles, quadrupoles, and octupoles. Suggest a reason for the different distance dependencies of their electric fields.

16B.3 Account for the theoretical conclusion that many attractive interactions between molecules vary with their separation as $1/r^6$.

16B.4 Describe the formation of a hydrogen bond in terms of (a) electrostatic interactions and (b) molecular orbitals. How would you identify the better model?

16B.5 Account for the hydrophobic interaction and discuss its manifestations.

16B.6 Some polymers have unusual properties. For example, Kevlar (3) is strong enough to be the material of choice for bulletproof vests and is stable at temperatures up to 600 K. What molecular interactions contribute to the formation and thermal stability of this polymer?

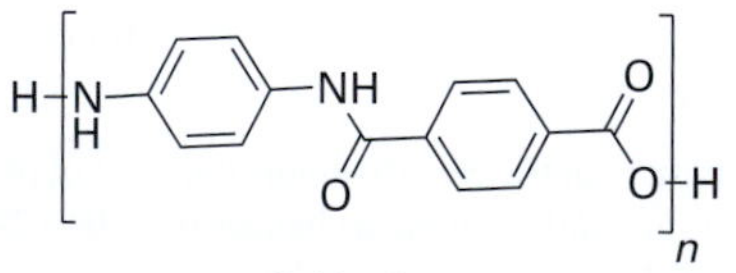

3 Kevlar

Exercises

16B.1(a) Calculate the molar energy required to reverse the direction of an H_2O molecule located 100 pm from a Li^+ ion. Take the magnitude of the dipole moment of water as 1.85 D.

16B.1(b) Calculate the molar energy required to reverse the direction of an HCl molecule located 300 pm from a Mg^{2+} ion. Take the magnitude of the dipole moment of HCl as 1.08 D.

16B.2(a) Calculate the potential energy of the interaction between two linear quadrupoles when they are collinear and their centres are separated by a distance r.

16B.2(b) Calculate the potential energy of the interaction between two linear quadrupoles when they are parallel and separated by a distance r.

16B.3(a) Estimate the energy of the dispersion interaction (use the London formula) for two He atoms separated by 1.0 nm. Relevant data can be found in the *Resource section*.

16B.3(b) Estimate the energy of the dispersion interaction (use the London formula) for two Ar atoms separated by 1.0 nm. Relevant data can be found in the *Resource section*.

16B.4(a) How much energy (in kJ mol^{-1}) is required to break the hydrogen bond in a vacuum ($\varepsilon_r = 1$)? Use the electrostatic model of the hydrogen bond.

16B.4(b) How much energy (in kJ mol^{-1}) is required to break the hydrogen bond in water ($\varepsilon_r \approx 80.0$)? Use the electrostatic model of the hydrogen bond.

Problems

16B.1 An H_2O molecule is aligned by an external electric field of strength 1.0 kV m^{-1} and an Ar atom ($\alpha' = 1.66 \times 10^{-30}$ m^3) is brought up slowly from one side. At what separation is it energetically favourable for the H_2O molecule to flip over and point towards the approaching Ar atom?

16B.2 Suppose an H_2O molecule ($\mu = 1.85$ D) approaches an anion. What is the favourable orientation of the molecule? Calculate the electric field (in volts per metre) experienced by the anion when the water dipole is (a) 1.0 nm, (b) 0.3 nm, (c) 30 nm from the ion.

16B.3 Phenylalanine (Phe, 4) is a naturally occurring amino acid. What is the energy of interaction between its phenyl group and the electric dipole moment of a neighbouring peptide group? Take the distance between the groups as 4.0 nm and treat the phenyl group as a benzene molecule. The dipole moment of the peptide group is $\mu = 2.7$ D and the polarizability volume of benzene is $\alpha' = 1.04 \times 10^{-29}$ m^3.

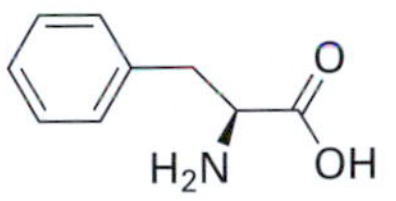

4 Phenylalanine

16B.4 Now consider the London interaction between the phenyl groups of two Phe residues (see Problem 16B.3). (a) Estimate the potential energy of interaction between two such rings (treated as benzene molecules) separated by 4.0 nm. For the ionization energy, use $I = 5.0$ eV. (b) Given that force is the negative slope of the potential, calculate the distance-dependence of the force acting between two non-bonded groups of atoms, such as the phenyl groups of Phe, in a polypeptide chain that can have a London dispersion interaction with each other. What is the separation at which the force between the phenyl groups (treated as benzene molecules) of two Phe residues is zero? (*Hint*: Calculate the slope by considering the potential energy at r and $r + \delta r$, with $\delta r \ll r$, and evaluating $\{V(r+\delta r) - V(r)\}/\delta r$. At the end of the calculation, let δr become vanishingly small.)

16B.5 Given that $F = -\mathrm{d}V/\mathrm{d}r$, calculate the distance dependence of the force acting between two non-bonded groups of atoms in a polymer chain that have a London dispersion interaction with each other.

16B.6 Consider the arrangement shown in 5 for a system consisting of an O—H group and an O atom, and then use the electrostatic model of the hydrogen bond to calculate the dependence of the molar potential energy of interaction on the angle θ.

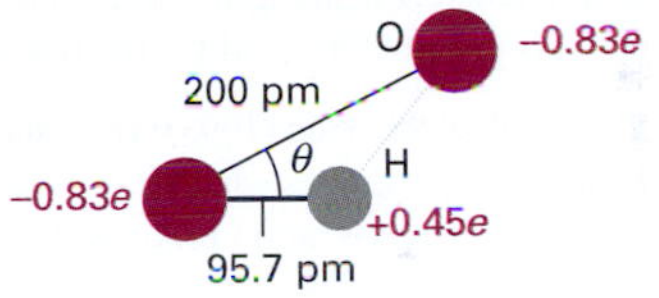

5

16B.7 Suppose you distrusted the Lennard-Jones (12,6) potential for assessing a particular polypeptide conformation, and replaced the repulsive term by an exponential function of the form e^{-r/r_0}. (a) Sketch the form of the potential energy and locate the distance at which it is a minimum. (b) Identify the distance at which the exponential-6 potential is a minimum.

16B.8 The *cohesive energy density*, $\mathcal{U}$, is defined as U/V, where U is the mean potential energy of attraction within the sample and V its volume. Show that $\mathcal{U} = \frac{1}{2}\mathcal{N}\int V(R)\mathrm{d}\tau$ where $\mathcal{N}$ is the number density of the molecules and $V(R)$ is their attractive potential energy and where the integration ranges from d to infinity and over all angles. Go on to show that the cohesive energy density of a uniform distribution of molecules that interact by a van der Waals attraction of the form $-C_6/R^6$ is equal to $(2\pi/3)(N_A^2/d^3M^2)\rho^2 C_6$, where ρ is the mass density of the solid sample and M is the molar mass of the molecules.

TOPIC 16C Liquids

Discussion question

16C.1 Describe the process of condensation.

Exercises

16C.1(a) Calculate the vapour pressure of a spherical droplet of water of radius 10 nm at 20 °C. The vapour pressure of bulk water at that temperature is 2.3 kPa and its density is 0.9982 g cm^{-3}.

16C.1(b) Calculate the vapour pressure of a spherical droplet of water of radius 20.0 nm at 35.0 °C. The vapour pressure of bulk water at that temperature is 5.623 kPa and its density is 994.0 kg m^{-3}.

16C.2(a) The contact angle for water on clean glass is close to zero. Calculate the surface tension of water at 20 °C given that at that temperature water climbs to a height of 4.96 cm in a clean glass capillary tube of internal radius 0.300 mm. The density of water at 20 °C is 998.2 kg m^{-3}.

16C.2(b) The contact angle for water on clean glass is close to zero. Calculate the surface tension of water at 30 °C given that at that temperature water climbs to a height of 9.11 cm in a clean glass capillary tube of internal diameter 0.320 mm. The density of water at 30 °C is 0.9956 g cm^{-3}.

16C.3(a) Calculate the pressure differential of water across the surface of a spherical droplet of radius 200 nm at 20 °C.

16C.3(b) Calculate the pressure differential of ethanol across the surface of a spherical droplet of radius 220 nm at 20 °C. The surface tension of ethanol at that temperature is 22.39 mN m^{-1}.

Problem

16C.1 The surface tensions of a series of aqueous solutions of a surfactant A were measured at 20 °C, with the following results:

[A]/(mol dm^{-3})	0	0.10	0.20	0.30	0.40	0.50
γ/(mN m^{-1})	72.8	70.2	67.7	65.1	62.8	59.8

Calculate the surface excess concentration.

Integrated activities

16.1 Show that the mean interaction energy of N atoms of diameter d interacting with a potential energy of the form C_6/R^6 is given by $U = -2N^2C_6/3Vd^3$, where V is the volume in which the molecules are confined and all effects of clustering are ignored. Hence, find a connection between the van der Waals parameter a and C_6, from $n^2a/V^2 = (\partial U/\partial V)_T$.

16.2‡ F. Luo et al. (*J. Chem. Phys.* **98**, 3564 (1993)) reported experimental observation of the He_2 complex, a species which had escaped detection for a long time. The fact that the observation required temperatures in the neighbourhood of 1 mK is consistent with computational studies which suggest that $hc\tilde{D}_e$ for He_2 is about 1.51×10^{-23} J, $hc\tilde{D}_0$ about 2×10^{-26} J, and R about 297 pm. (a) Determine the Lennard-Jones parameters r_0, and a and plot the Lennard–Jones potential for He—He interactions. (b) Plot the Morse potential given that $a = 5.79\times10^{10}$ m^{-1}.

16.3 Molecular orbital calculations may be used to predict structures of intermolecular complexes. Hydrogen bonds between purine and pyrimidine bases are responsible for the double helix structure of DNA (see Topic 17A). Consider methyladenine (**6**, with R=CH_3) and methylthymine (**7**, with R=CH_3) as models of two bases that can form hydrogen bonds in DNA. (a) Using molecular modelling software and the computational method of your choice, calculate the atomic charges of all atoms in methyladenine and methylthymine. (b) Based on your tabulation of atomic charges, identify the atoms in methyladenine and methylthymine that are likely to participate in hydrogen bonds. (c) Draw all possible adenine–thymine pairs that can be linked by hydrogen bonds, keeping in mind that linear arrangements of the A—H···B fragments are preferred in DNA. For this step, you may want to use your molecular modelling software to align the molecules properly. (d) Consult Topic 17A and determine which of the pairs that you drew in part (c) occur naturally in DNA molecules. (e) Repeat parts (a)–(d) for cytosine and guanine, which also form base pairs in DNA (see Topic 17A for the structures of these bases).

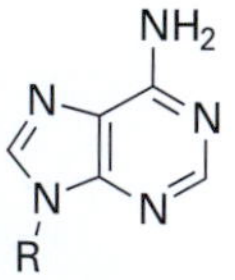

6

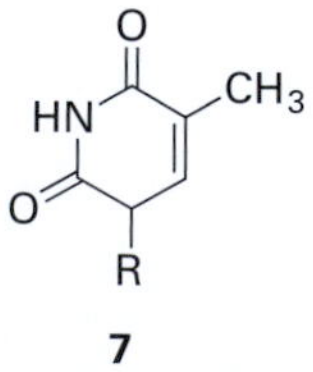

7

16.4 Molecular orbital calculations may be used to predict the dipole moments of molecules. (a) Using molecular modelling software and the computational method of your choice, calculate the dipole moment of the peptide link, modelled as a *trans-N*-methylacetamide (**8**). Plot the energy of interaction between these dipoles against the angle θ for $r = 3.0$ nm (see eqn 16B.4). (b) Compare the maximum value of the dipole–dipole interaction energy from part (a) to 20 kJ mol^{-1}, a typical value for the energy of a hydrogen bonding interaction in biological systems.

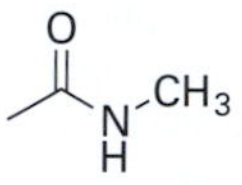

8 trans-*N*-methylacetamide

16.5 This problem gives a simple example of a quantitative structure–activity relation (QSAR). The binding of nonpolar groups of amino acid to hydrophobic sites in the interior of proteins is governed largely by hydrophobic interactions. (a) Consider a family of hydrocarbons R—H. The hydrophobicity constants, π, for R=CH_3, CH_2CH_3, $(CH_2)_2CH_3$, $(CH_2)_3CH_3$, and $(CH_2)_4CH_3$ are, respectively, 0.5, 1.0, 1.5, 2.0, and 2.5. Use these data to predict the π value for $(CH_2)_6CH_3$. (b) The equilibrium constants K_I for the dissociation of inhibitors (**9**) from the enzyme chymotrypsin were measured for different substituents R:

R	CH_3CO	CN	NO_2	CH_3	Cl
π	−0.20	−0.025	0.33	0.5	0.9
log K_I	−1.73	−1.90	−2.43	−2.55	−3.40

Plot log K_I against π. Does the plot suggest a linear relationship? If so, what are the slope and intercept to the log K_I axis of the line that best fits the data? (c) Predict the value of K_I for the case R=H.

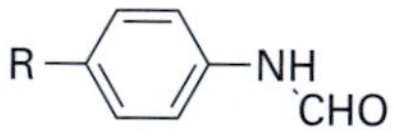

9

16.6 Derivatives of the compound TIBO (**10**) inhibit the enzyme reverse transcriptase, which catalyses the conversion of retroviral RNA to DNA. A QSAR analysis of the activity A of a number of TIBO derivatives suggests the following equation:

$$\log A = b_0 + b_1S + b_2W$$

where S is a parameter related to the drug's solubility in water and W is a parameter related to the width of the first atom in a substituent X shown in **10**. (a) Use the following data to determine the values of b_0, b_1, and b_2. *Hint*: The QSAR equation relates one dependent variable, log A, to two independent variables, S and W. To fit the data, you must use the mathematical procedure of *multiple regression*, which can be performed with mathematical software or an electronic spreadsheet.

X	H	Cl	SCH_3	OCH_3	CN	CHO	Br	CH_3	CCH
log *A*	7.36	8.37	8.3	7.47	7.25	6.73	8.52	7.87	7.53
S	3.53	4.24	4.09	3.45	2.96	2.89	4.39	4.03	3.80
W	1.00	1.80	1.70	1.35	1.60	1.60	1.95	1.60	1.60

(b) What should be the value of W for a drug with $S = 4.84$ and $\log A = 7.60$?

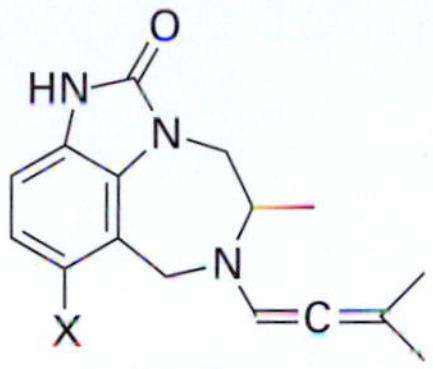

10

CHAPTER 17

Macromolecules and self-assembly

Macromolecules are built from covalently linked components. They are everywhere, inside us and outside us. Some are natural: they include polysaccharides (such as cellulose), polypeptides (such as protein enzymes), and polynucleotides (such as deoxyribonucleic acid, DNA). Others are synthetic (such as nylon and polystyrene). Molecules both large and small may also gather together in a process called 'self-assembly' and give rise to aggregates that to some extent behave like macromolecules. One example is the assembly of the protein actin into filaments in muscle tissue. In this chapter we examine the structures and properties of macromolecules and aggregates.

17A The structures of macromolecules

Macromolecules adopt shapes that are governed by the molecular interactions described in Topic 16B. The overall shape of a protein, for instance, is maintained by van der Waals interactions, hydrogen bonding, and the hydrophobic effect. In this Topic we consider a range of structures, beginning with a structureless 'random coil', partially structured coils, and then the structurally precise proteins and nucleic acids.

17B Properties of macromolecules

Natural macromolecules differ in certain respects from synthetic macromolecules, particularly in their composition and the resulting structure, but the two share a number of common properties. In this Topic we concentrate on mechanical, thermal, and electrical properties.

17C Self-assembly

Atoms, small molecules, and macromolecules can form large aggregates, sometimes by processes involving self-assembly, that are held together by one or more of the molecular interactions described in Topic 16B. In this Topic we explore 'colloids', 'micelles', and biological membranes, which are assemblies with some of the typical properties of molecules but also with their own characteristic features. We also consider examples in which the controlled design of new materials with enhanced properties is informed by understanding of the principles underlying self-assembly.

17D Determination of size and shape

Macromolecules, whether natural or synthetic, and aggregates need to be characterized in terms of their molar mass, their size, and their shape. This Topic considers how these features are determined experimentally.

What is the impact of this material?

The impact of this material is immense as it underlies the discussion of biological phenomena and the properties of many modern materials. However, the applications are embedded in the development of the concepts and are not found on the web site.

17A The structures of macromolecules

Contents

➤ Why do you need to know this material?

Macromolecules give rise to special problems that include the investigation and description of their shapes and the determination of their sizes. You need to know how to describe the structural features of macromolecules in order to understand their physical and chemical properties.

➤ What is the key idea?

The structure of a macromolecule takes on different meanings at the different levels at which the arrangement of the chain or network of its building blocks is considered.

➤ What do you need to know already?

You need to be familiar with statistical arguments (Topic 15A). The discussion of the shapes of biological macromolecules depends on an understanding of the nonbonding interactions that act between molecules (Topic 16B).

Macromolecules are very large molecules assembled from smaller molecules biosynthetically in organisms, by chemists in the laboratory, or in an industrial reactor. Naturally occurring macromolecules include polysaccharides such as cellulose, polypeptides such as protein enzymes, and polynucleotides such as deoxyribonucleic acid (DNA). Synthetic macromolecules include **polymers** such as nylon and polystyrene that are manufactured by stringing together and in some cases cross-linking smaller units known as **monomers** (Fig. 17A.1).

17A.1 The different levels of structure

The concept of the 'structure' of a macromolecule takes on different meanings at the different levels at which we think about the arrangement of the chain or network of monomers. The **primary structure** of a macromolecule is the sequence of small molecular residues making up the polymer. The residues may form either a chain, as in polyethene, or a more complex network in which cross-links connect different chains, as in cross-linked polyacrylamide. In a synthetic polymer, virtually all the residues are identical and it is sufficient to name the monomer used in the synthesis. Thus, the repeating unit of polyethene and its derivatives is $-CHXCH_2-$, and the primary structure of the chain is specified by denoting it as $-(CHXCH_2)_n-$.

The concept of primary structure ceases to be trivial in the case of synthetic copolymers and biological macromolecules, for in general these substances are chains formed from different

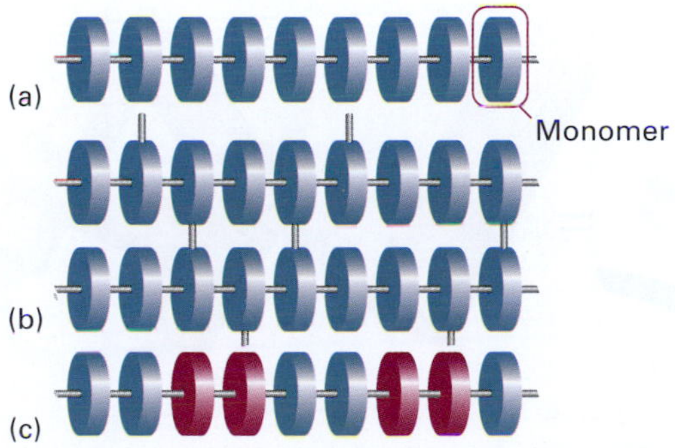

Figure 17A.1 Three varieties of polymer: (a) a simple linear polymer, (b) a cross-linked polymer, and (c) one variety of copolymer.

molecules. For example, proteins are **polypeptides** formed from different amino acids (about twenty occur naturally) strung together by the **peptide link**, —CONH—. The determination of the primary structure is then a highly complex problem of chemical analysis called **sequencing**. The **degradation** of a polymer is a disruption of its primary structure, when the chain breaks into shorter components.

The term **conformation** refers to the spatial arrangement of the different parts of a chain, and one conformation can be changed into another by rotating one part of a chain around a bond. The conformation of a macromolecule is relevant at three levels of structure. The **secondary structure** of a macromolecule is the (often local) spatial arrangement of a chain. The secondary structure of a molecule of polyethene in a good solvent is typically a random coil in the absence of a solvent polyethene forms crystals consisting of stacked sheets with a hairpin-like bend about every 100 monomer units, presumably because for that number of monomers the intermolecular (in this case *intra*molecular) potential energy is sufficient to overcome thermal disordering. The secondary structure of a protein is a highly organized arrangement determined largely by hydrogen bonds, and taking the form of random coils, helices (Fig. 17A.2a), or sheets in various segments of the molecule.

The **tertiary structure** is the overall three-dimensional structure of a macromolecule. For instance, the hypothetical protein shown in Fig. 17A.2b has helical regions connected by short random-coil sections. The helices interact to form a compact tertiary structure. Denaturation may also occur at this level, with tertiary structure lost but secondary structure largely retained.

The **quaternary structure** of a macromolecule is the manner in which large molecules are formed by the aggregation of others. Figure 17A.3 shows how four molecular subunits, each with a specific tertiary structure, aggregate. Quaternary structure can be very important in biology. For example, the oxygen-transport protein haemoglobin consists of four myoglobin-like subunits that work cooperatively to take up and release O_2.

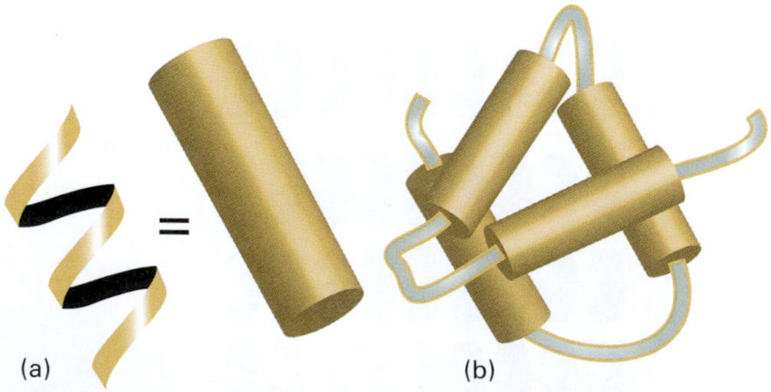

Figure 17A.2 (a) A polymer adopts a highly organized helical conformation, an example of a secondary structure. The helix is represented as a cylinder. (b) Several helical segments connected by short random coils pack together, providing an example of tertiary structure.

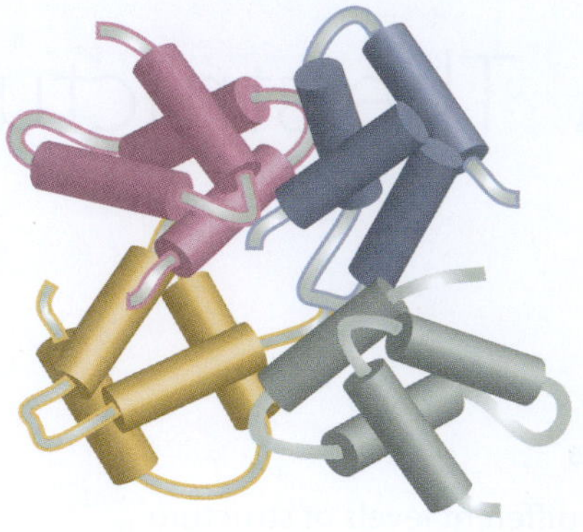

Figure 17A.3 Several subunits with specific tertiary structures pack together, providing an example of quaternary structure.

17A.2 Random coils

The most likely conformation of a chain of identical units not capable of forming hydrogen bonds or any other type of specific bond is a **random coil**. Polyethene is a simple example. The random coil model is a helpful starting point for estimating the orders of magnitude of the hydrodynamic properties of polymers and denatured proteins in solution.

The simplest model of a random coil is a **freely-jointed chain**, in which any bond is free to make any angle with respect to the preceding one (Fig. 17A.4). We assume that the residues occupy zero volume, so different parts of the chain can occupy the same region of space. The model is obviously an oversimplification because a bond is actually constrained to a cone of angles around a direction defined by its neighbour (Fig. 17A.5) and real chains are self-avoiding in the sense that distant parts of the same chain cannot fold back and occupy the same space.

In a hypothetical one-dimensional freely jointed chain all the residues lie in a straight line, and the angle between neighbours is either 0° or 180°. The residues in a three-dimensional freely jointed chain are not restricted to lie in a line or a plane.

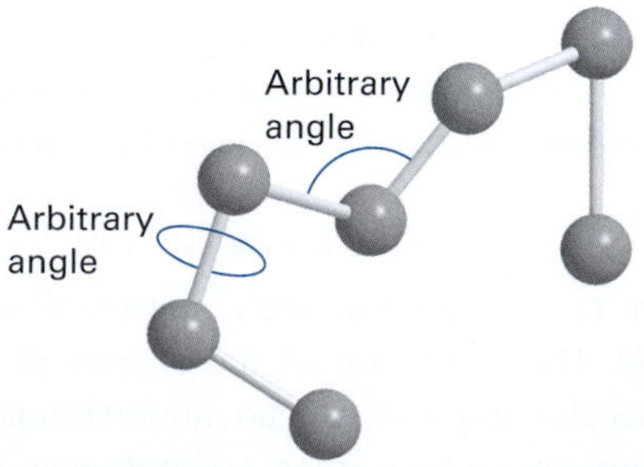

Figure 17A.4 A freely-jointed chain is like a three-dimensional random walk, each step being in an arbitrary direction but of the same length.

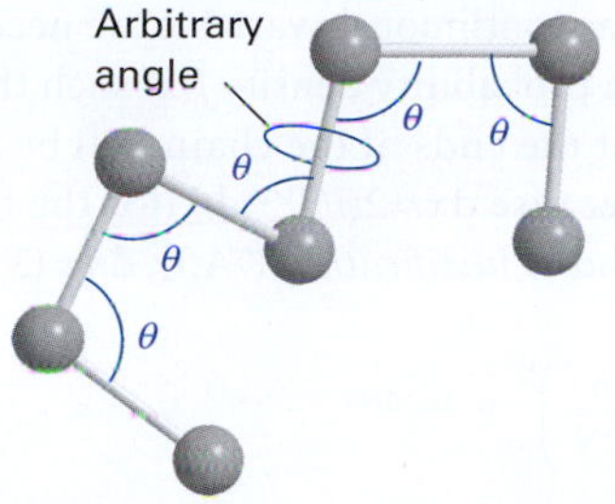

Figure 17A.5 A better description is obtained by fixing the bond angle (for example, at the tetrahedral angle) and allowing free rotation about a bond direction.

(a) Measures of size

As shown in the following *Justification*, the probability, P, that the ends of a long one-dimensional freely jointed chain composed of N units of length l (and therefore of total length Nl) are a distance nl apart is

$$P=\left(\frac{2}{\pi N}\right)^{1/2} e^{-n^2/2N} \quad \text{One-dimensional random coil} \quad \text{Probability distribution} \qquad (17A.1)$$

This function is plotted in Fig. 17A.6.

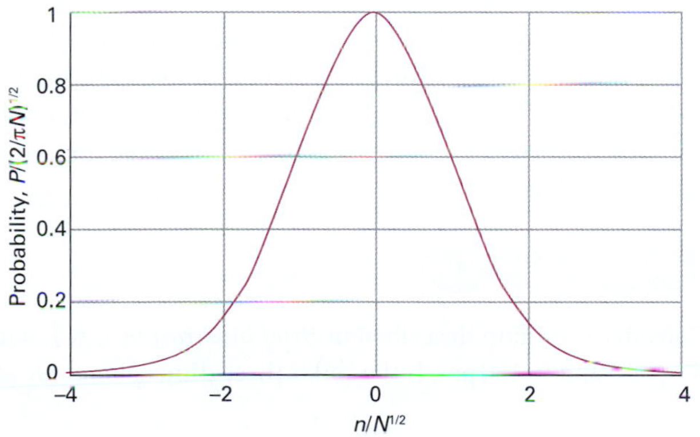

Figure 17A.6 The probability distribution for the separation of the ends of a one-dimensional random coil. The separation of the ends is Nl, where l is the bond length.

Brief illustration 17A.1 One-dimensional random coils

Suppose that $N=1000$ and $l=150$ pm, then the probability that the ends of a one-dimensional random coil are $nl=3.00$ nm apart is given by eqn 17A.1 by setting $n=(3.00\times10^3\ \text{pm})/(150\ \text{pm})=20.0$:

$$P=\left(\frac{2}{\pi\times1000}\right)^{1/2} e^{-20.0^2/(2\times1000)}=0.0207$$

meaning that there is a 1 in 48 chance of being found there.

Self-test 17A.1 What is the probability that the ends of a polyethene chain of molar mass 85 kg mol^{-1} are 10 nm apart when the polymer is treated as a one-dimensional freely jointed chain?

Answer: 9.8×10^{-3}

Justification 17A.1 One-dimensional random coils

Consider a one-dimensional freely jointed polymer. We can specify its conformation by stating the number of bonds pointing to the right (N_R) and the number pointing to the left (N_L). The distance between the ends of the chain is $(N_R - N_L)l$, where l is the length of an individual bond. We write $n=N_R-N_L$ and the total number of bonds as $N=N_R+N_L$. Later in the calculation we use $N_R=\frac{1}{2}(N+n)$ and $N_L=\frac{1}{2}(N-n)$.

The probability, P, that the end-to-end separation of a randomly selected polymer is nl is

$$P=\frac{\text{number of polymers with end-to-end distance } nl}{\text{total number of possible conformations}}$$

Each of the N bonds of the polymer may in principle lie to the left or the right, so the total number of possible conformations is 2^N. The total number of ways, W, of forming a chain of N bonds with the end-to-end distance nl is the number of ways of having N_R right-pointing bonds, the rest being left-pointing bonds. Therefore, to calculate W we need to determine the number of ways of achieving N_R right-pointing bonds given a total of N bonds. This is the same problem as selecting N_R objects from a collection of N objects (see Topic 15A), and is

$$W=\frac{N!}{N_R!(N-N_R)!}=\frac{N!}{N_R!N_L!}=\frac{N!}{\{\frac{1}{2}(N+n)\}!\{\frac{1}{2}(N-n)\}!}$$

It now follows that

$$P=\frac{W}{2^N}=\frac{N!}{\{\frac{1}{2}(N+n)\}!\{\frac{1}{2}(N-n)\}!2^N}$$

When the chain is compact in the sense that $n\ll N$, it is more convenient to evaluate $\ln P$: the factorials are then large and we can use Stirling's approximation (Topic 15A) in the form

$$\ln x!\approx\ln(2\pi)^{1/2}+\left(x+\tfrac{1}{2}\right)\ln x-x$$

The result, after quite a lot of algebra (see Problem 17A.7), is

$$\ln P\approx\ln\left(\frac{2}{\pi N}\right)^{1/2}-\tfrac{1}{2}(N+n+1)\ln(1+\nu)-\tfrac{1}{2}(N-n+1)\ln(1-\nu)$$

where $\nu = n/N$. For a compact coil ($\nu \ll 1$) we use the approximation $\ln(1\pm\nu) \approx \pm\nu - \frac{1}{2}\nu^2$ and so obtain

$$\ln P \approx \ln\left(\frac{2}{\pi N}\right)^{1/2} - \tfrac{1}{2}N\nu^2$$

which rearranges into eqn 17A.1.

To confirm that the total probability of the chain ends being at any separation is 1, we integrate P over all values of n. However, because n can change only in steps of 2, the integration step size is $\frac{1}{2}\mathrm{d}n$, not $\mathrm{d}n$ itself. Then (with N allowed to become infinite),

$$\sum_{n=-N}^{N} P \to \int_{-\infty}^{\infty} P(n)(\tfrac{1}{2}\mathrm{d}n) = \tfrac{1}{2}\left(\frac{2}{\pi N}\right)^{1/2} \int_{-\infty}^{\infty} \mathrm{e}^{-n^2/2N}\mathrm{d}n \overset{\text{Integral G.1}}{=} 1$$

We show in the following *Justification* that eqn 17A.1 can be used to calculate the probability that the ends of a long three-dimensional freely jointed chain lie in the range r to $r+\mathrm{d}r$. We write this probability as $f(r)\mathrm{d}r$, where

$$f(r) = 4\pi\left(\frac{a}{\pi^{1/2}}\right)^3 r^2 \mathrm{e}^{-a^2r^2} \qquad a = \left(\frac{3}{2Nl^2}\right)^{1/2}$$

Three-dimensional random coil — Probability distribution (17A.2)

For a narrow range of distances δr, the probability density can be treated as a constant and the probability calculated from $f(r)\delta r$. An alternative interpretation of this expression is to regard each coil in a sample as ceaselessly writhing from one conformation to another; then $f(r)\mathrm{d}r$ is the probability that at any instant the chain will be found with the separation of its ends between r and $r+\mathrm{d}r$.

Justification 17A.2 Three-dimensional random coils

The formation of a three-dimensional random coil can be regarded as the outcome of a three-dimensional random walk, in which each bond of length l represents a step of length l taken in a random direction. The length of the step can be expressed in terms of its projections on each of three orthogonal axes as $l^2 = l_x^2 + l_y^2 + l_z^2$. The average values of l_x^2, l_y^2, and l_z^2 are all the same in a spherically symmetric environment, so the average length of a step in the x-direction (or any of the other two directions) can be obtained by writing $l^2 = 3\langle l_x^2\rangle$, and is $x = \langle l_x^2\rangle^{1/2} = l/3^{1/2}$. The probability that the random walk will end up at a distance x from the origin is given by eqn 17A.1 with $n = x/(l/3^{1/2}) = 3^{1/2}x/l$:

$$P(x) = \left(\frac{2}{\pi N}\right)^{1/2} \mathrm{e}^{-3x^2/2Nl^2}$$

If x is regarded as continuously variable, we need to replace this probability by a probability density $f(x)$ such that $f(x)\mathrm{d}x$ is the probability that the ends of the chain will be found between x and $x+\mathrm{d}x$. Because $\mathrm{d}x = 2(l/3^{1/2})\mathrm{d}n$ (for the factor 2, see the remark at the end of *Justification* 17A.1), $\mathrm{d}n = (3^{1/2}/2l)\mathrm{d}x$, so

$$f(x) = \frac{1}{2l}\left(\frac{6}{\pi N}\right)^{1/2} \mathrm{e}^{-3x^2/2Nl^2}$$

Because the probabilities of making steps along all three coordinates are independent, the probability of finding the ends of the chain in a region of volume $\mathrm{d}V = \mathrm{d}x\mathrm{d}y\mathrm{d}z$ at a distance r is the product of these densities:

$$f(x,y,z)\mathrm{d}V = f(x)f(y)f(z)\mathrm{d}x\mathrm{d}y\mathrm{d}z = \frac{1}{8l^3}\left(\frac{6}{\pi N}\right)^{3/2} \mathrm{e}^{-3r^2/2Nl^2}\mathrm{d}V$$

The volume of a spherical shell at a distance r is $4\pi r^2$, so the total probability of finding the ends at a separation between r and $r+\mathrm{d}r$, regardless of orientation, is

$$f(r)\mathrm{d}r = \frac{4\pi}{8l^3}\left(\frac{6}{\pi N}\right)^{3/2} r^2\mathrm{e}^{-3r^2/2Nl^2}\mathrm{d}r$$

from which $f(r)$ can be identified (in blue), as in eqn 17A.2.

In some coils, the ends may be far apart whereas in others their separation is small. Here and elsewhere we are ignoring the fact that the chain cannot be longer than Nl. Although eqn 17A.2 gives a nonzero probability for $r > Nl$, the values are so small that the errors in pretending that r can range up to infinity are negligible.

Brief illustration 17A.2 Three-dimensional random coils

Consider the chain described in *Brief illustration* 17A.1, with $N = 1000$ and $l = 150$ pm. If the coil is three dimensional, we set

$$a = \left(\frac{3}{2\times1000\times(150\,\mathrm{pm})^2}\right)^{1/2} = 2.58\ldots\times10^{-4}\,\mathrm{pm}^{-1}$$

Then the probability density at $r = 3.00$ nm is given by eqn 17A.2 as

$$\begin{aligned} f(3.00\,\mathrm{nm}) &= 4\pi\times\left(\frac{2.58\ldots\times10^{-4}\,\mathrm{pm}^{-1}}{\pi^{1/2}}\right)^3 \\ &\quad\times(3.00\times10^3\,\mathrm{pm})^2\times\mathrm{e}^{-(2.58\ldots\times10^{-4}\,\mathrm{pm}^{-1})^2(3.00\times10^3\,\mathrm{pm})^2} \\ &= 1.92\times10^{-4}\,\mathrm{pm}^{-1} \end{aligned}$$

The probability that the ends will be found in a narrow range of width $\delta r = 10.0$ pm at 3.00 nm (regardless of direction) is therefore

$$f(3.00\,\mathrm{nm})\delta r = (1.92\times10^{-4}\,\mathrm{pm}^{-1})\times(10.0\,\mathrm{pm}) = 1.92\times10^{-3}$$

or about 1 in 5200.

Self-test 17A.2 What is the probability that the ends of a polyethene chain of molar mass 85 kg mol^{-1} are between 15.0 nm and 15.1 nm apart when the polymer is treated as a three-dimensional freely jointed chain?

Answer: 5.9×10^{-3}

There are several measures of the geometrical size of a random coil. The **contour length**, R_c, is the length of the macromolecule measured along its backbone from atom to atom. For a polymer of N monomer units each of length l, the contour length is

$$R_c = Nl \quad \textit{Random coil} \quad \text{Contour length} \quad (17A.3)$$

The **root-mean-square separation**, R_{rms}, is a measure of the average separation of the ends of a random coil: it is the square root of the mean value of R^2. To determine its value we note that the vector joining the two ends of the chain is the vector sum of the vectors joining neighbouring monomers: $\boldsymbol{R}=\sum_{i=1}^{N}\boldsymbol{r}_i$ (Fig. 17A.7). The mean square separation of the ends of the chain is therefore

$$\langle R^2\rangle=\langle \boldsymbol{R}\cdot\boldsymbol{R}\rangle=\sum_{i,j}^{N}\langle \boldsymbol{r}_i\cdot\boldsymbol{r}_j\rangle=\sum_{i}^{N}\overbrace{\langle r_i^2\rangle}^{l^2}+\sum_{i\neq j}^{N}\langle \boldsymbol{r}_i\cdot\boldsymbol{r}_j\rangle$$

When N is large (which we assume throughout) the second sum (in blue) vanishes because the individual vectors all lie in random directions. The remaining sum is equal to Nl^2 as all bond lengths are the same (and equal to l); so, after taking square roots to obtain $R_{rms}=\langle R^2\rangle^{1/2}$, we conclude that

$$R_{rms} = N^{1/2}l \quad \textit{Random coil} \quad \text{Root-mean-square separation} \quad (17A.4)$$

We see that, as the number of monomer units increases, the root-mean-square separation of its end increases as $N^{1/2}$ (Fig. 17A.8), and consequently its volume increases as $N^{3/2}$. The result must be multiplied by a factor when the chain is not freely jointed (see next section).

Another convenient measure of size is the **radius of gyration**, R_g, which is the radius of a hollow sphere that has the same moment of inertia (and therefore rotational characteristics) as the actual molecule of the same mass. We show in the following *Justification* that

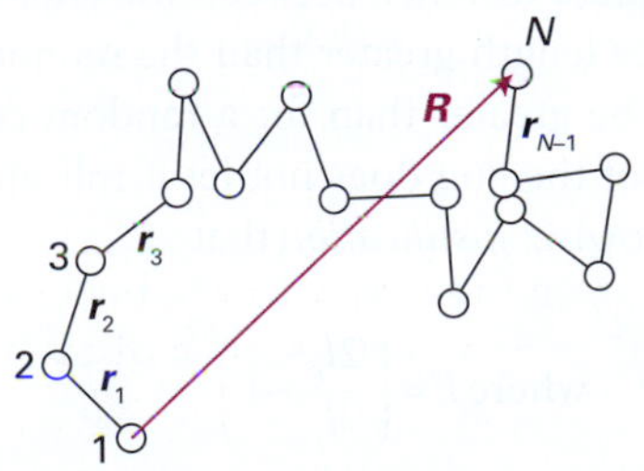

Figure 17A.7 A schematic illustration of the calculation of the root-mean-square separation of the ends of a random coil.

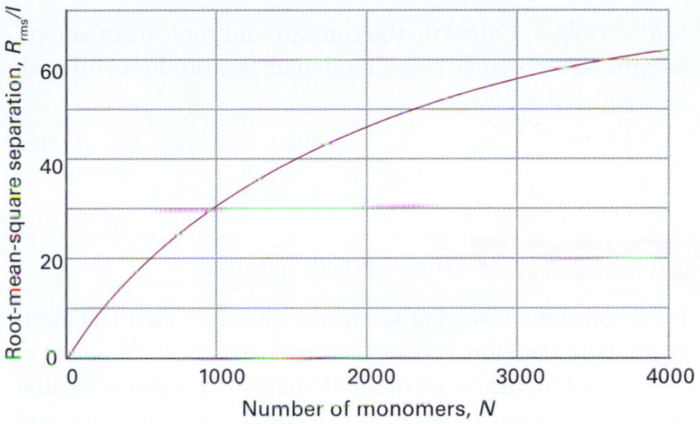

Figure 17A.8 The variation of the root-mean-square separation of the ends of a three-dimensional random coil, R_{rms}, with the number of monomers.

$$R_g = N^{1/2}l \quad \textit{One-dimensional random coil} \quad \text{Radius of gyration} \quad (17A.5)$$

A similar calculation for a three-dimensional random coil gives

$$R_g = \left(\frac{N}{6}\right)^{1/2} l \quad \textit{Three-dimensional random coil} \quad \text{Radius of gyration} \quad (17A.6)$$

The radius of gyration is smaller in this case because the extra dimensions enable the coil to be more compact. The radius of gyration may also be calculated for other geometries. For example, a solid uniform sphere of radius R has $R_g = (\frac{3}{5})^{1/2}R$, and a long thin uniform rod of length l has $R_g = l/(12)^{1/2}$ for rotation about an axis perpendicular to the long axis. A solid sphere with the same radius and mass as a random coil has a greater radius of gyration as it is entirely dense throughout.

Brief illustration 17A.3 Measures of size of a random coil

With a powerful microscope it is possible to see that a long piece of DNA is flexible and writhes as if it were a random coil. However, small segments of the macromolecule resist bending, so it is more appropriate to visualize DNA as a freely jointed chain with N and l as the number and length, respectively, of these rigid units. The length l of a rigid unit is approximately 45 nm. It follows that for a piece of DNA with $N=200$, we estimate (by using 10^3 nm $=1\,\mu$m)

From eqn 17A.3: $R_c = 200\times45\,\text{nm} = 9.0\,\mu\text{m}$

From eqn 17A.4: $R_{rms} = (200)^{1/2}\times45\,\text{nm} = 0.64\,\mu\text{m}$

From eqn 17A.6: $R_g = \left(\frac{200}{6}\right)^{1/2}\times45\,\text{nm} = 0.26\,\mu\text{m}$

Self-test 17A.3 Calculate the contour and root-mean-square lengths of a polymer chain modelled as a random coil with $N=1000$ and $l=150$ pm.

Answer: $R_c=150$ nm, $R_{rms}=4.74$ nm

Justification 17A.3 The radius of gyration

For a one-dimensional random coil with $N+1$ identical monomers (and therefore N bonds) each of mass m, the moment of inertia around the centre of the chain (which is also at the first monomer, because steps occur in equal numbers to left and right) is

$$I=\sum_{i=0}^{N} m_i r_i^2 = m\sum_{i=0}^{N} r_i^2$$

This moment of inertia is set equal to $m_{tot}R_g^2$, where m_{tot} is the total mass of the polymer, $m_{tot}=(N+1)m$. Therefore, after averaging over all conformations,

$$R_g^2=\frac{1}{N+1}\sum_{i=0}^{N}\langle r_i^2\rangle$$

For a linear random chain, $\langle r_i^2\rangle = Nl^2$, and as there are $N+1$ such terms in the sum, we find

$$R_g^2 = Nl^2$$

Equation 17A.5 then follows after taking the square root of each side.

The random coil model ignores the role of the solvent: a poor solvent will tend to cause the coil to tighten so that solute–solvent contacts are minimized; a good solvent does the opposite. Therefore, calculations based on this model are better regarded as lower bounds to the dimensions for a polymer in a good solvent and as an upper bound for a polymer in a poor solvent. The model is most reliable for a polymer in a bulk solid sample, where the coil is likely to have its natural dimensions.

(b) Constrained chains

The freely jointed chain model is improved by removing the freedom of bond angles to take any value. For long chains, we can simply take groups of neighbouring bonds and consider the direction of their resultant. Although each successive individual bond is constrained to a single cone of angle θ relative to its neighbour, the resultant of several bonds lies in a random direction. By concentrating on such groups rather than individuals, it turns out that for long chains the expressions for the root-mean-square separation and the radius of gyration given above should be multiplied by

$$F=\left(\frac{1-\cos\theta}{1+\cos\theta}\right)^{1/2} \qquad (17A.7)$$

For tetrahedral bonds, for which $\cos\theta=\frac{1}{3}$ (that is, $\theta=109.5°$), $F=2^{1/2}$. Therefore:

$$R_{rms}=(2N)^{1/2}l \qquad R_g=\left(\frac{N}{3}\right)^{1/2}l \qquad \text{Dimensions of a tetrahedrally constrained chain} \qquad (17A.8)$$

The model of a randomly coiled molecule is still an approximation, even after the bond angles have been restricted, because it does not take into account the impossibility of two or more atoms occupying the same place. Such self-avoidance tends to swell the coil, so (in the absence of solvent effects) it is better to regard R_{rms} and R_g as lower bounds to the actual values.

(c) Partly rigid coils

An important measure of the flexibility of a chain is the **persistence length**, l_p, a measure of the length over which the direction of the first monomer–monomer direction is sustained. If the chain is a rigid rod, then the persistence length is the same as the contour length. For a freely-jointed random coil, the persistence length is just the length of the monomer–monomer bond. Therefore, the persistence length can be regarded as a measure of the stiffness of the chain. In general, the persistence length of a chain of identical monomers of length l is defined as the average value of the projection of the end-to-end vector on the first bond of the chain (Fig. 17A.9):

$$l_p=\left\langle\frac{\boldsymbol{r}_1}{l}\cdot\boldsymbol{R}\right\rangle=\frac{1}{l}\sum_{i=1}^{N-1}\langle\boldsymbol{r}_1\cdot\boldsymbol{r}_i\rangle \qquad \textit{Definition} \quad \text{Persistence length} \qquad (17A.9)$$

(The sum ends at $N-1$ because the last atom is atom N and the last bond is from atom $N-1$ to atom N.) Experimental values of persistence lengths are as follows:

poly(glycine)	poly(L-alanine)	poly(L-proline)
0.6 nm	2 nm	22 nm

These values suggest that the stiffness of the chain increases from left to right along the series.

The mean square distance between the ends of a chain that has a persistence length greater than the monomer length can be expected to be greater than for a random coil because the partial rigidity of the coil does not let it roll up so tightly. We show in the following *Justification* that

$$R_{rms}=N^{1/2}lF \quad \text{where } F=\left(\frac{2l_p}{l}-1\right)^{1/2} \qquad (17A.10)$$

For a random coil, $l_p=l$, so $R_{rms}=N^{1/2}l$, as we have already found. For $l_p>l$, $F>1$, so the coil has swollen, as we anticipated.

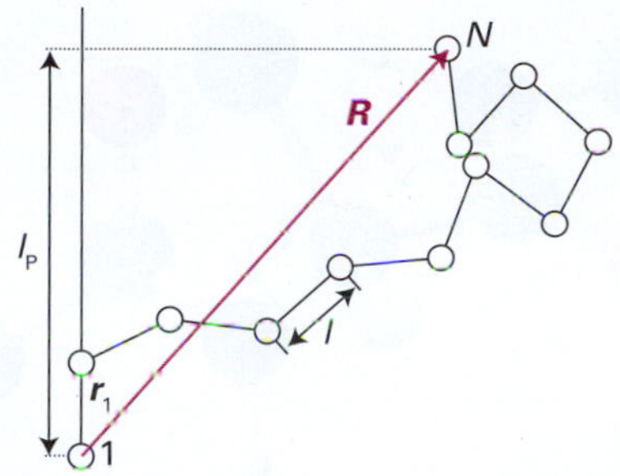

Figure 17A.9 The persistence length is defined as the average value of the projection of the end-to-end vector on the first bond of the chain.

Justification 17A.4 Partly rigid coils

In each of the following steps we use $N\to\infty$ when necessary. We start from

$$\langle R^2\rangle=\sum_i \overbrace{\langle r_i^2\rangle}^{l^2}+\sum_{i\neq j}^{N}\langle r_i\cdot r_j\rangle$$

The first term on the right is Nl^2 regardless of the rigidity of the coil. The second term can be written as follows:

$$\sum_{i\neq j}^{N}\langle r_i\cdot r_j\rangle=2\sum_{i=2}^{N}\langle r_1\cdot r_i\rangle+2\sum_{i=3}^{N}\langle r_2\cdot r_i\rangle+\ldots$$

There are $N-1$ such terms, and provided we allow N to become infinite, all the sums on the right have the same value, so

$$\sum_{i\neq j}^{N}\langle r_i\cdot r_j\rangle=2(N-1)\sum_{i=2}^{N}\langle r_1\cdot r_i\rangle\approx 2N\sum_{i=2}^{N}\langle r_1\cdot r_i\rangle$$

The final (blue) sum on the right is close to being the square of the persistence length. Specifically, from eqn 17A.9,

$$\sum_{j=2}^{N}\langle r_1\cdot r_j\rangle=\sum_{j=1}^{N}\langle r_1\cdot r_j\rangle-\langle r_1^2\rangle=ll_p-l^2$$

Now we bring the three pieces of the calculation together:

$$\langle R^2\rangle=Nl^2+2N(ll_p-l^2)=2Nll_p-Nl^2$$

which, on taking the square root of both sides, is eqn 17A.10.

Example 17A.1 Calculating the root-mean-square separation of a partly rigid coil

By what percentage does the root-mean-square separation of the ends of a polymer chain with $N=1000$ increase or decrease when the persistence length changes from l (the bond length) to 2.5 per cent of the contour length?

Method When $l_p=l$, the chain is a random coil. From eqn 17A.4, write the root-mean-square separation of the ends of the chain in the random coil limit as $R_{\text{rms,random coil}}=N^{1/2}l$. Then it follows from eqn 17A.10 that the root-mean-square separation of the ends of the chain, R_{rms}, of the partly rigid coil with persistence length l_p is

$$R_{\text{rms}}=R_{\text{rms,random coil}}\left(\frac{2l_p}{l}-1\right)^{1/2}$$

From eqn 17A.3, we write $R_c=Nl$ and we are given that $l_p=0.025R_c$, which can therefore be interpreted as $0.025Nl$. From these expressions, calculate the fractional change in the root-mean-square separation, $(R_{\text{rms}}-R_{\text{rms,random coil}})/R_{\text{rms,random coil}}$, and express the result as a percentage.

Answer We write the fractional change in the root-mean-square separation as

$$\begin{aligned}\frac{R_{\text{rms}}-R_{\text{rms,random coil}}}{R_{\text{rms,random coil}}}&=\frac{R_{\text{rms}}}{R_{\text{rms,random coil}}}-1\\&=\left(\frac{2l_p}{l}-1\right)^{1/2}-1\\&=\left(\frac{2\times0.025Nl}{l}-1\right)^{1/2}-1\\&=(0.050N-1)^{1/2}-1\end{aligned}$$

With $N=1000$, the fractional change is 6.00, so the root-mean-square separation increases by 600 per cent.

Self-test 17A.4 Calculate the fractional change in the volume of the same coil.

Answer: 340

17A.3 Biological macromolecules

A protein is a polypeptide composed of linked α-amino acids, $NH_2CHRCOOH$, where R is one of about 20 groups. For a protein to function correctly, it needs to have the correct conformation. For example, an enzyme has its greatest catalytic efficiency only when it is in a specific conformation. The amino acid sequence of a protein contains the necessary information to create the active conformation of the protein as it is formed. However, the prediction of the observed conformation from the primary structure, the so-called *protein folding problem*, is extraordinarily difficult and is still the focus of much research. The other class of biological macromolecules we consider are the nucleic acids, which are key components of the mechanism of storage and transfer of genetic information in biological cells. Deoxyribonucleic acid (DNA) contains the instructions for protein synthesis, which is carried out by different forms of ribonucleic acid (RNA).

(a) Proteins

The origin of the secondary structures of proteins is found in the rules formulated by Linus Pauling and Robert Corey in 1951 that seek to identify the principal contributions to the lowering of energy of the molecule by focusing on the role of hydrogen bonds and the peptide link, −CONH−. The latter can act both as a donor of the H atom (the NH part of the link) and as an acceptor (the CO part). The **Corey–Pauling rules** are as follows (Fig. 17A.10):

1. The four atoms of the peptide link lie in a relatively rigid plane.

The planarity of the link is due to delocalization of π electrons over the O, C, and N atoms and the maintenance of maximum overlap of their p orbitals.

2. The N, H, and O atoms of a hydrogen bond lie in a straight line (with displacements of H tolerated up to not more than 30° from the N−O direction).
3. All NH and CO groups of the peptide links are engaged in hydrogen bonding.

The rules are satisfied by two structures. One, in which hydrogen bonding between peptide links leads to a helical structure, is a *helix*, which can be arranged as either a right-or a left-handed screw. The other, in which hydrogen bonding between peptide links leads to a planar structure, is a *sheet*; this form is the secondary structure of the protein fibroin, the constituent of silk.

Because the planar peptide link is relatively rigid, the geometry of a polypeptide chain can be specified by the two angles that two neighbouring planar peptide links make to each other. Figure 17A.11 shows the two angles ϕ and ψ commonly used to specify this relative orientation. The sign convention is that a positive angle means that the front atom must be rotated clockwise to bring it into an eclipsed position relative to the rear atom. For an all-*trans* form of the chain, all ϕ and ψ are 180°. A helix is obtained when all the ϕ are equal and when all the ψ are equal. For a right-handed helix, an α helix (Fig. 17A.12), all $\phi=57°$ and all $\psi=-47°$. For a left-handed helix, both angles are positive. The torsional contribution to the total potential energy is

$$V_{\text{torsion}} = A(1+\cos 3\phi)+B(1+\cos 3\psi) \qquad (17\text{A}.11)$$

in which A and B are constants of the order of 1 kJ mol^{-1}. Because only two angles are needed to specify the conformation of a helix, and they range from −180° to +180°, the torsional potential energy of the entire molecule can be represented on a **Ramachandran plot**, a contour diagram in which one axis represents ϕ and the other represents ψ.

Figure 17A.13 shows the Ramachandran plots for the helical form of polypeptide chains formed from the non-chiral amino acid glycine (R=H) and the chiral amino acid L-alanine (R]CH$_3$). The glycine map is symmetrical, with minima of equal depth at $\phi=-80°$, $\psi=+90°$ and at $\phi=+80°$, $\psi=-90°$. In contrast, the map for L-alanine is unsymmetrical, and there are three distinct low-energy conformations (marked I, II, III). The minima of regions I and II lie close to the angles typical of

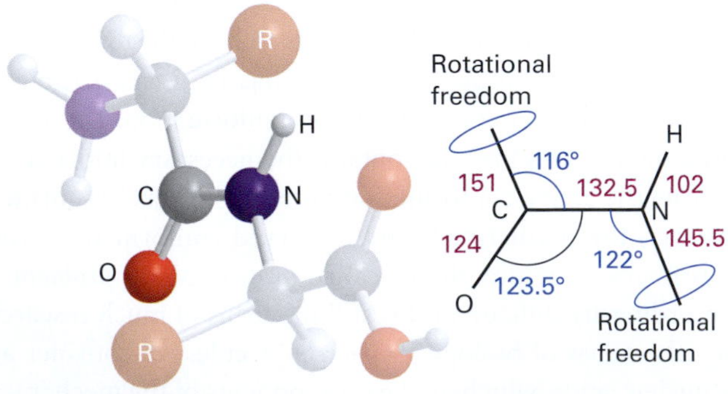

Figure 17A.10 The dimensions that characterize the peptide link. The C−NH−CO−C atoms define a plane (the C−N bond has partial double-bond character), but there is rotational freedom around the C−CO and N−C bonds.

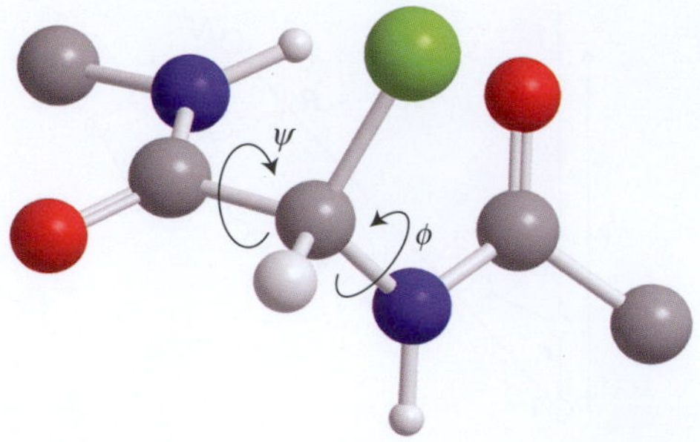

Figure 17A.11 The definition of the torsional angles ψ and ϕ between two peptide units. In this case (an α-L-polypeptide) the chain has been drawn in its all-trans form, with $\psi=\phi=180°$.

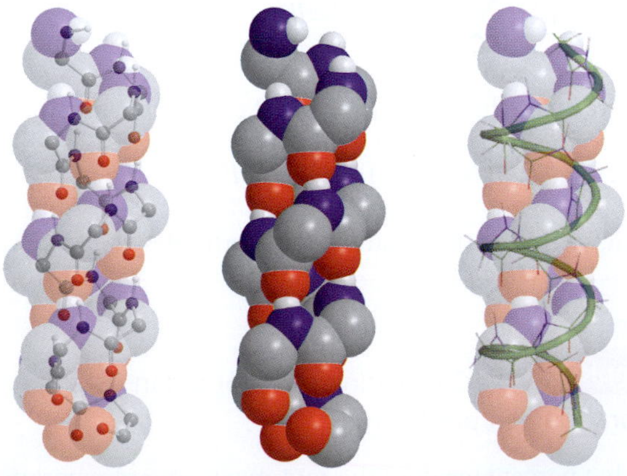

Figure 17A.12 The polypeptide α helix, with poly-L-glycine as an example. There are 3.6 residues per turn, and a translation along the helix of 150 pm per residue, giving a pitch of 540 pm. The diameter (ignoring side chains) is about 600 pm.

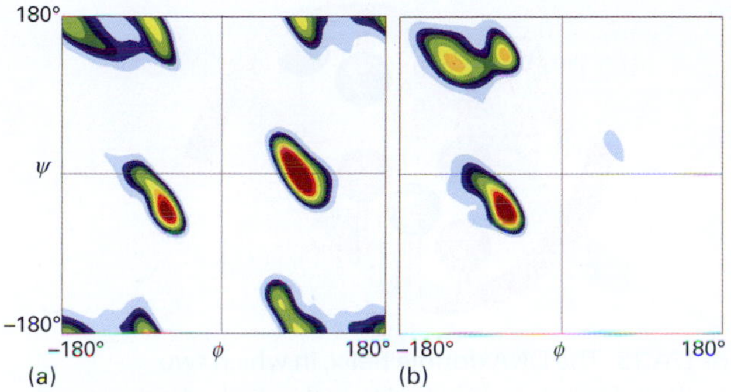

Figure 17A.13 Contour plots of potential energy against the angles ψ and ϕ, also known as a Ramachandran diagram, for (a) a glycyl residue of a polypeptide chain and (b) an alanyl residue. The glycyl diagram is symmetrical, but that for alanyl is unsymmetrical and the global minimum corresponds to an α-helix. (T. Hovmoller et al., *Acta Cryst.* **D58**, 768 (2002).)

right- and left-handed helices, but the former has a lower minimum. This result is consistent with the observation that polypeptides of the naturally occurring L-amino acids tend to form right-handed helices.

A **β-sheet** (also called the **β-pleated sheet**) is formed by hydrogen bonding between two extended polypeptide chains (large absolute values of the torsion angles ϕ and ψ). In an **antiparallel** β-sheet (Fig. 17A.14a), $\phi = 139°$, $\psi = 113°$, and the N—H···O atoms of the hydrogen bonds form a straight line. This arrangement is a consequence of the antiparallel arrangement of the chains: every N—H bond on one chain is aligned with a C—O bond from another chain. Antiparallel β-sheets are very common in proteins. In a **parallel β-sheet** (Fig. 17A.14b), $\phi = 119°$, $\psi = 113°$, and the N—H···O atoms of the hydrogen bonds are not perfectly aligned. This arrangement is a result of the parallel arrangement of the chains: each N—H bond on one chain is aligned with a N—H bond of another chain and, as a result, each C—O bond of one chain is aligned with a C—O bond of another chain. These structures are not common in proteins.

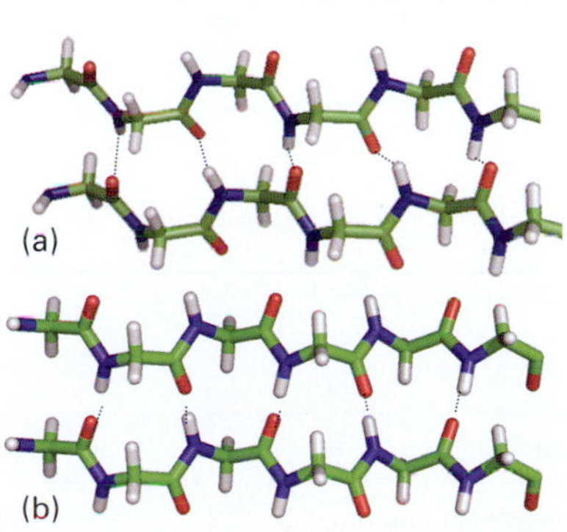

Figure 17A.14 The two types of β-sheets: (a) antiparallel ($\phi = -139°$, $\psi = 113°$), in which the N—H—O atoms of the hydrogen bonds form a straight-line; (b) parallel ($\phi = -119°$, $\psi = 113°$) in which the N—H—O atoms of the hydrogen bonds are not perfectly aligned.

Covalent and non-covalent interactions may cause polypeptide chains with well-defined secondary structures to fold into tertiary structures. Although the rules that govern protein folding are still being elucidated, a few general conclusions may be drawn from X-ray diffraction studies of water-soluble natural proteins and synthetic polypeptides. In an aqueous environment, the chains fold in such a way as to place nonpolar R groups in the interior (which is often not very accessible to solvent) and charged R groups on the surface (in direct contact with the polar solvent). Other factors that promote the folding of proteins include covalent disulfide (—S—S—) links, Coulombic interactions between ions (which depend on the degree of protonation of groups and therefore on the pH), van der Waals interactions, and hydrophobic interactions (Topic 16B). The clustering of nonpolar, hydrophobic, amino acids into the interior of a protein is driven primarily by hydrophobic interactions.

(b) Nucleic acids

Both DNA and RNA are *polynucleotides* (**1**), in which base–sugar–phosphate units are linked by phosphodiester bonds. In RNA the sugar is β-D-ribose and in DNA it is β-D-2-deoxyribose (as shown in **1**). The most common bases are adenine (A, **2**), cytosine (C, **3**), guanine (G, **4**), thymine (T, found in DNA only, **5**), and uracil (U, found in RNA only, **6**). At physiological pH, each phosphate group of the chain carries a negative charge and the bases are deprotonated and neutral. This charge distribution leads to two important properties. One is that the polynucleotide chain is a **polyelectrolyte**, a macromolecule with many different charged sites, with a large and negative overall surface charge. The second is that the bases can interact by hydrogen bonding, as shown for A—T (**7**) and C—G base pairs (**8**). The secondary and tertiary structures of DNA and RNA arise primarily from the pattern of this hydrogen bonding between bases of one or more chains.

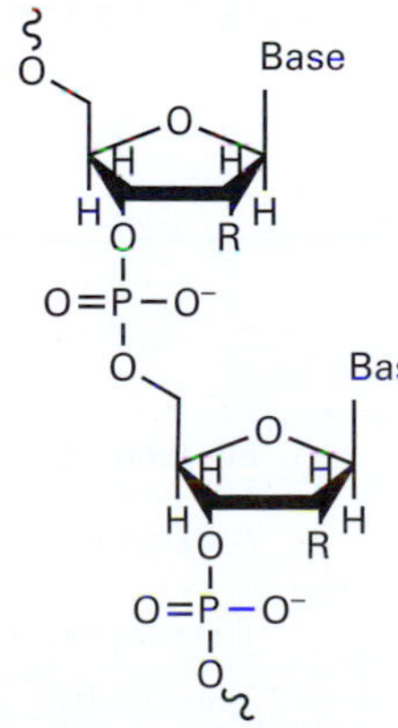

1 D-ribose (R = OH) and 2′-deoxy-D-ribose (R = H)

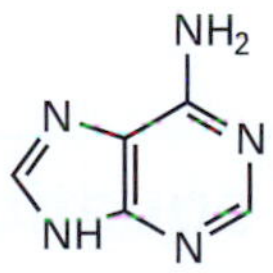

2 Adenine, A

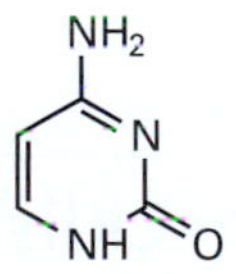

3 Cytosine, C

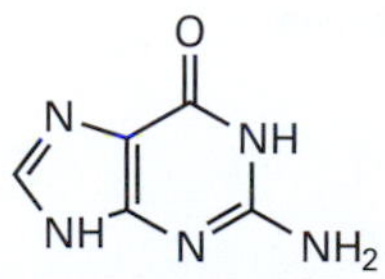

4 Guanine, G

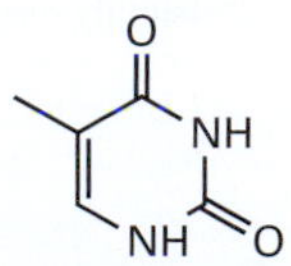

5 Thymine, T

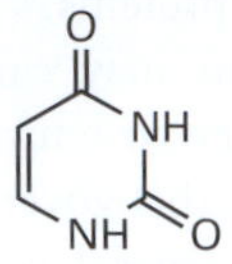

6 Uracil

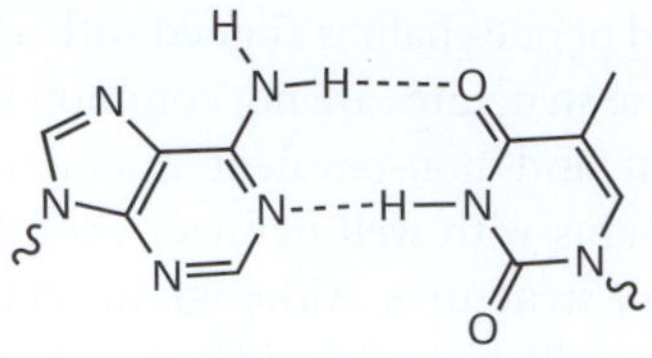

7 A-T base pair

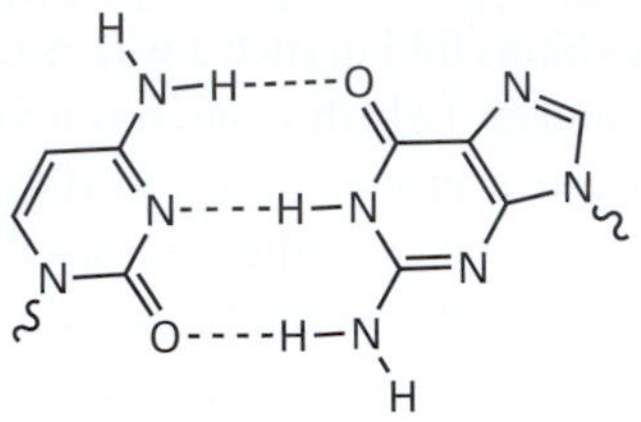

8 C-G base pair

In DNA, two polynucleotide chains wind around each other to form a double helix (Fig. 17A.15). The chains are held together by links involving A—T and C—G base pairs that lie parallel to each other and perpendicular to the axis of the helix. The structure is stabilized further by interactions between the planar π systems of the bases. In B-DNA, the most common form of DNA found in biological cells, the helix is right-handed with a diameter of 2.0 nm and a pitch of 3.4 nm.

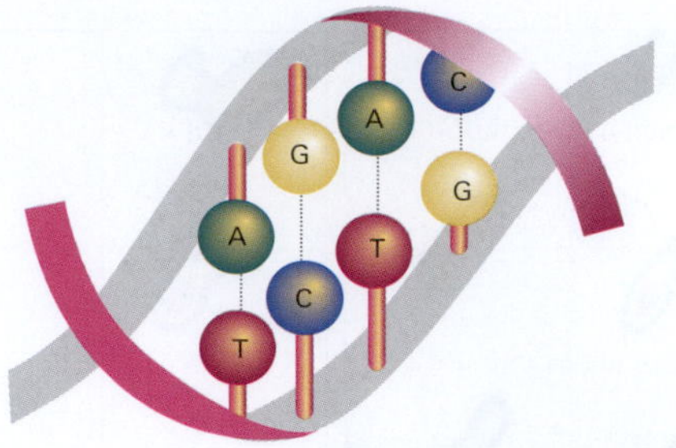

Figure 17A.15 The DNA double helix, in which two polynucleotide chains are linked together by hydrogen bonds between adenine (A) and thymine (T) and between cytosine (C) and guanine (G).

Checklist of concepts

- ☐ 1. The **primary structure** of a macromolecule is the sequence of small molecular residues making up the polymer.
- ☐ 2. The **secondary structure** is the spatial arrangement of a chain of residues.
- ☐ 3. The **tertiary structure** is the overall three-dimensional structure of a macromolecule.
- ☐ 4. The **quaternary structure** is the manner in which large molecules are formed by the aggregation of others.
- ☐ 5. In a **freely jointed chain** any bond in a polymer is free to make any angle with respect to the preceding one.
- ☐ 6. The freely jointed chain model is improved by removing the freedom of bond angles to take any value.
- ☐ 7. The least structured conformation of a macromolecule is a **random coil**, which can be modelled as a freely jointed chain.
- ☐ 8. The secondary structure of a protein is the spatial arrangement of the polypeptide chain and includes **helices** and the **β-sheet**.
- ☐ 9. Helical and sheet-like polypeptide chains are folded into a tertiary structure by bonding influences between the residues of the chain.
- ☐ 10. Some proteins have a quaternary structure as aggregates of two or more polypeptide chains.
- ☐ 11. In DNA, two polynucleotide chains held together by hydrogen bonded base pairs wind around each other to form a double helix.
- ☐ 12. In RNA, single chains fold into complex structures by formation of specific base pairs.

Checklist of equations

Property	Equation	Comment	Equation number
Probability distribution	$P=(2/\pi N)^{1/2}e^{-n^2/2N}$	One-dimensional random coil	17A.1
	$f(r)=4\pi(a/\pi^{1/2})^3 r^2 e^{-a^2r^2}$ $a=(\frac{3}{2}Nl^2)^{1/2}$	Three-dimensional random coil	17A.2
Contour length of a random coil	$R_c=Nl$		17A.3

Property	Equation	Comment	Equation number
Root-mean-square separation of a random coil	$R_{rms}=N^{1/2}l$	Unconstrained chain	17A.4
Radius of gyration of a random coil	$R_g=N^{1/2}l$	Unconstrained one-dimensional chain	17A.5
	$R_g=(N/6)^{1/2}l$	Unconstrained three-dimensional chain	17A.6
Root-mean-square separation of a random coil	$R_{rms}=(2N)^{1/2}l$	Constrained tetrahedral chain	17A.8
Radius of gyration of a random coil	$R_g=(N/3)^{1/2}l$	Constrained tetrahedral chain	17A.8

17B Properties of macromolecules

Contents

➤ Why do you need to know this material?

Macromolecules are important in modern technology. They are also building blocks of biological cells. To understand why this is so, you need to explore the characteristic physical properties of macromolecules.

➤ What is the key idea?

The unique properties of macromolecules are related to their unique structural features.

➤ What do you need to know already?

You need to be familiar with the structural features of macromolecules (Topic 17A), particularly the properties of a random coil. You also need to be familiar with the statistical interpretation of entropy (Topic 15E) and the concept of internal energy (Topic 2A).

Macromolecules have special physical properties that arise from details of their structures. In this Topic we explore the physical mechanical, thermal, and electrical properties of synthetic and biological macromolecules.

17B.1 Mechanical properties

Significant insight into the consequences of stretching and contracting a polymer can be obtained on the basis of the freely jointed chain as a model (Topic 17A).

(a) Conformational entropy

The random coil is the least structured conformation of a polymer chain and corresponds to the state of greatest entropy. Any stretching of the coil introduces order and reduces the entropy. Conversely, the formation of a random coil from a more extended form is a spontaneous process (provided enthalpy contributions do not interfere). As shown in the following *Justification*, we can use the same model to deduce that the change in **conformational entropy**, the statistical entropy arising from the arrangement of bonds, when a one-dimensional chain containing N bonds of length l is stretched or compressed by nl is

$$\Delta S = -\tfrac{1}{2}kN\ln\{(1+\nu)^{1+\nu}(1-\nu)^{1-\nu}\} \qquad \nu = n/N \qquad \textit{Random coil} \quad \text{Conformational entropy} \qquad (17B.1)$$

This function is plotted in Fig. 17B.1, and we see that minimum extension corresponds to maximum entropy.

Brief illustration 17B.1 Conformational entropy

Suppose that $N=1000$ and $l=150\,\text{pm}$. The change in entropy when the (one-dimensional) random coil is stretched through 1.5 nm (corresponding to $n=10$ and $\nu=1/100$) is

$$\Delta S = -\tfrac{1}{2}k\times(1000)\times(\ln\left\{\left(1+\frac{1}{100}\right)^{1+(1/100)}\left(1-\frac{1}{100}\right)^{1-(1/100)}\right\}$$
$$= -0.050k$$

Because $R=N_A k$, the change in molar entropy is $\Delta S_m = -0.050R$, or $-0.42\,\text{J K}^{-1}\,\text{mol}^{-1}$.

Self-test 17B.1 What is the change in conformational entropy when the same random coil is stretched from fully coiled by 10 per cent?

Answer: $-0.042\,\text{J K}^{-1}\,\text{mol}^{-1}$

Justification 17B.1 The conformational entropy of a freely jointed chain

The conformational entropy of the chain is given by the Boltzmann formula, $S=k\ln\mathcal{W}$ (eqn 15E.7). In the present case, we identify $\mathcal{W}$ with the number of ways of achieving a coil with a given extension, the W calculated in eqn 17A.2:

$$W=\frac{N!}{\{\frac{1}{2}(N+n)\}!\{\frac{1}{2}(N-n)\}!}$$

Therefore,

$$S/k=\ln N!-\ln\{\tfrac{1}{2}(N+n)\}!-\ln\{\tfrac{1}{2}(N-n)\}!$$

Because the factorials are large (except for large extensions), we can use Stirling's approximation (Topic 15A, $\ln x!\approx(x+\frac{1}{2})\ln x-x+\frac{1}{2}\ln 2\pi$) to obtain

$$S/k=\ln(2\pi)^{1/2}+(N+1)\ln 2-(N+\tfrac{1}{2})\ln N-\tfrac{1}{2}\ln\{(N+n)^{N+n+1}(N+n)^{N-n+1}\}$$

We have seen that the most probable conformation of a one-dimensional chain is the one with the ends close together ($n=0$). This conformation also corresponds to maximum entropy, as may be confirmed by differentiation. Therefore, the maximum entropy is

$$S/k=\ln(2\pi)^{1/2}+(N+1)\ln 2+\tfrac{1}{2}\ln N$$

The change in entropy when the chain is stretched or compressed by nl is therefore the difference of these two quantities, and the resulting expression, after some algebraic manipulation, is eqn 17B.1.

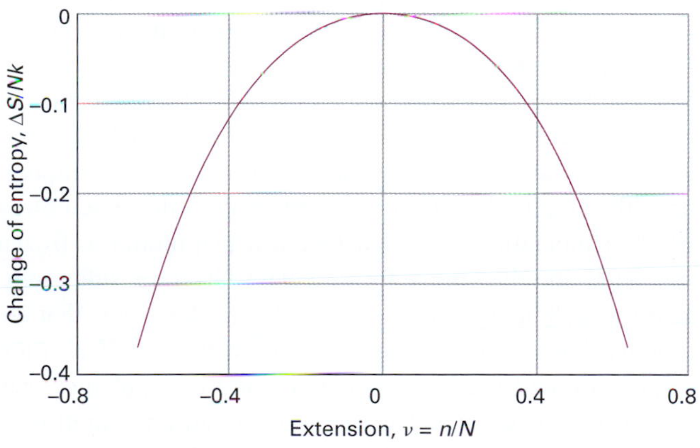

Figure 17B.1 The change in molar entropy of a perfect elastomer as its extension changes; $\nu=1$ corresponds to complete extension; $\nu=0$, the conformation of highest entropy, corresponds to a random coil.

(b) Elastomers

The fundamental concepts for the discussion of the mechanical properties of solids are stress and strain. The **stress** on an object is the applied force divided by the area to which it is applied. The **strain** is the resulting distortion of the sample. The general field of the relations between stress and strain is called **rheology**.

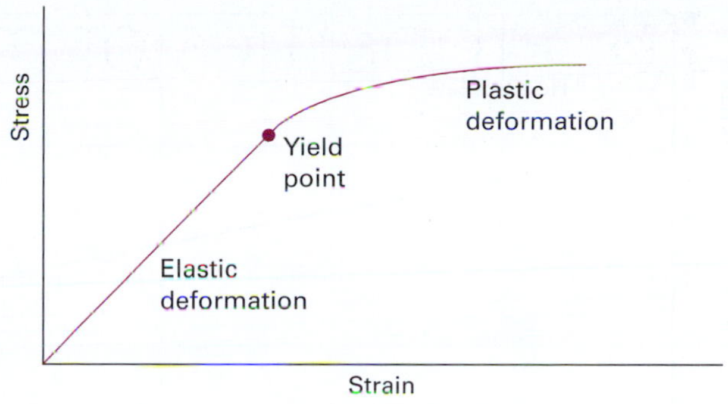

Figure 17B.2 A typical stress–strain curve.

The stress–strain curve in Fig. 17B.2 shows how a material responds to stress. The region of **elastic deformation** is where the strain is proportional to the stress and is reversible: when the stress is removed, the sample returns to its initial shape. As we see in more detail in Topic 18C, the slope of the stress–strain curve in this region is 'Young's modulus', E, for the material. At the **yield point**, the reversible, linear deformation gives way to **plastic deformation**, where the strain is no longer linearly proportional to the stress and the initial shape of the sample is not recovered when the stress is removed. Thermosetting plastics have only a very short elastic range; thermoplastics typically (but not universally) have a long plastic range. An **elastomer** is specifically a polymer with a long elastic range. They typically have numerous cross-links (such as the sulfur links in vulcanized rubber) that pull them back into their original shape when the stress is removed.

Although practical elastomers are typically extensively cross-linked, even a freely jointed chain behaves as an elastomer for small extensions. It is a model of a **perfect elastomer,** a polymer in which the internal energy is independent of the extension. In the following *Justification* we also see that the restoring force, F, of a one-dimensional random coil when the chain is stretched or compressed by nl is

$$F=\frac{kT}{2l}\ln\frac{1+\nu}{1-\nu}\qquad \nu=n/N$$ *One-dimensional random coil* — Restoring force (17B.2a)

where N is the total number of bonds each of length l. This function is plotted in Fig. 17B.3. At low extensions, when $\nu \ll 1$ we can use $\ln(1+x)=x-\frac{1}{2}x^2+\cdots$ and find (retaining only linear terms) that

$$F\approx\frac{\nu kT}{l}=\frac{nkT}{Nl}$$ *One-dimensional random coil, small extensions* — Restoring force (17B.2b)

That is, for small displacements the sample obeys Hooke's law: the restoring force is proportional to the displacement (which is proportional to n). For small displacements, therefore, the whole coil shakes with simple harmonic motion. When this equation is rearranged to

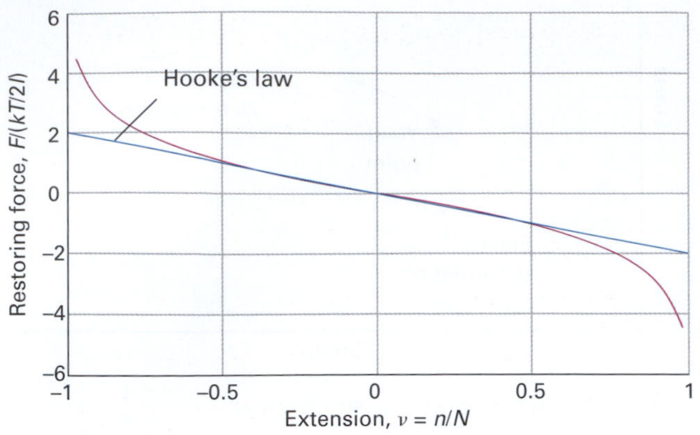

Figure 17B.3 The restoring force, F, of a one-dimensional perfect elastomer. For small strains, F is linearly proportional to the extension, corresponding to Hooke's law.

$$nl=\left(\frac{Nl^2}{kT}\right)F \tag{17B.2c}$$

we see that for small displacements, the strain, as measured by the extension nl, is proportional to the applied force, as is characteristic of the elastic deformation region of an elastomer.

Brief illustration 17B.2 The restoring force

Consider a polymer chain with $N=5000$ and $l=0.15$ nm. If the ends of the chain are moved apart by 1.5 nm, then $n=(1.5\,\text{nm})/(0.15\,\text{nm})=10$ and $\nu=n/N=10/5000=2.0\times10^{-3}$. Because $\nu \ll 1$, the restoring force at 293 K is given by eqn 17B.2b as

$$F=\frac{10\times(1.381\times10^{-23}\overbrace{\text{J}}^{\text{N m}}\text{ K}^{-1})\times(293\,\text{K})}{5000\times(1.5\times10^{-10}\,\text{m})}=5.4\times10^{-14}\,\text{N}$$

or 54 fN.

Self-test 17B.2 Repeat the calculation for $N=6.0\times10^3$ and a displacement of 2.0 nm at 298 K.

Answer: 61 fN

Justification 17B.2 Hooke's law

The work done on an elastomer when it is extended through a distance dx is Fdx, where F is the restoring force. The change in internal energy is therefore

$$\mathrm{d}U=T\mathrm{d}S+F\mathrm{d}x$$

It follows that

$$\left(\frac{\partial U}{\partial x}\right)_T=T\left(\frac{\partial S}{\partial x}\right)_T+F$$

In a perfect elastomer, as in a perfect gas, the internal energy is independent of the dimensions (at constant temperature), so $(\partial U/\partial x)_T=0$. The restoring force is therefore

$$F=-T\left(\frac{\partial S}{\partial x}\right)_T$$

If now we substitute eqn 17B.1 into this expression, we obtain

$$F=-\frac{T}{l}\left(\frac{\partial S}{\partial n}\right)_T=\frac{T}{Nl}\left(\frac{\partial S}{\partial \nu}\right)_T=\frac{kT}{2l}\ln\left(\frac{1+\nu}{1-\nu}\right)$$

as in eqn 17B.2a.

17B.2 Thermal properties

The crystallinity of synthetic polymers can be destroyed by thermal motion at sufficiently high temperatures. This change in crystallinity may be thought of as a kind of intramolecular melting from a crystalline solid to a more fluid random coil. Polymer melting also occurs at a specific **melting temperature**, T_m, which increases with the strength and number of intermolecular interactions in the material. Thus, polyethene, which has chains that interact only weakly in the solid, has $T_m=414$ K and nylon-66 fibre, in which there are strong hydrogen bonds between chains, has $T_m=530$ K. High melting temperatures are desirable in most practical applications involving fibres and plastics.

All synthetic polymers undergo a transition from a state of high to low chain mobility at the **glass transition temperature**, T_g. To visualize the glass transition, we consider what happens to an elastomer as we lower its temperature. There is sufficient energy available at normal temperatures for limited bond rotation to occur and the flexible chains writhe. At lower temperatures, the amplitudes of the writhing motion decrease until a specific temperature, T_g, is reached at which motion is frozen completely and the sample forms a glass. Glass transition temperatures well below 300 K are desirable in elastomers that are to be used at normal temperatures. Both the glass transition temperature and the melting temperature of a polymer may be measured by calorimetric methods. Because the motion of the segments of a polymer chain increase at the glass transition temperature, T_g may also be determined from a plot of the specific volume of a polymer (the reciprocal of its mass density) against temperature (Fig. 17B.4).

Proteins and nucleic acids are relatively unstable towards chemical and thermal **denaturation**, the loss of structure. Thermal denaturation is similar to the melting of synthetic polymers. Denaturation is a **cooperative process** in the sense that the biopolymer becomes increasingly more susceptible to denaturation once the process begins. This cooperativity is observed as a sharp step in a plot of fraction of unfolded polymer against temperature. The melting temperature, T_m, is the temperature at which the fraction of denatured polymer is 0.5

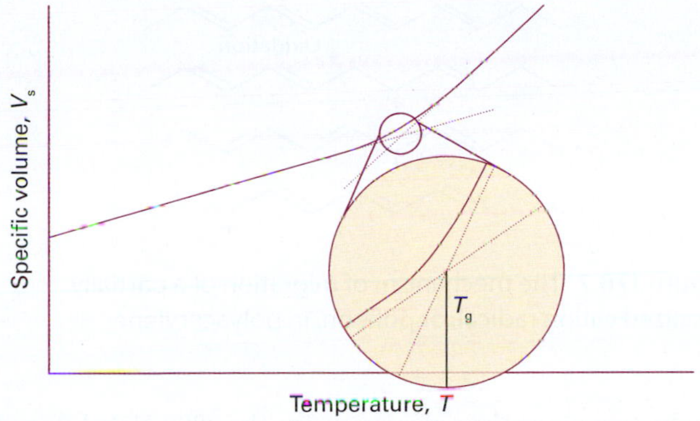

Figure 17B.4 The variation of specific volume with temperature of a synthetic polymer. The glass transition temperature, T_g, is at the point of intersection of extrapolations of the two linear parts of the curve.

(Fig. 17B.5). For example, $T_m = 320\,K$ for ribonuclease T_1 (an enzyme that cleaves RNA in the cell), which is not far above the temperature at which the enzyme must operate (close to body temperature, 310 K). More surprisingly, the Gibbs energy for the denaturation of ribonuclease T_1 at pH = 7.0 and 298 K is only $19.5\,kJ\,mol^{-1}$, which is comparable to the energy required to break a single hydrogen bond (about $20\,kJ\,mol^{-1}$). The stability of a protein does not increase in a simple way with the number of hydrogen bonding interactions. While the reasons for the low stability of proteins are not known, the answer probably lies in a delicate balance of the intra- and intermolecular interactions that allow a protein to fold into its active conformation, and the role of the aqueous environment. By contrast, the melting temperature of DNA can be predicted with reasonable accuracy by examining its structure, as we see in the following *Example*.

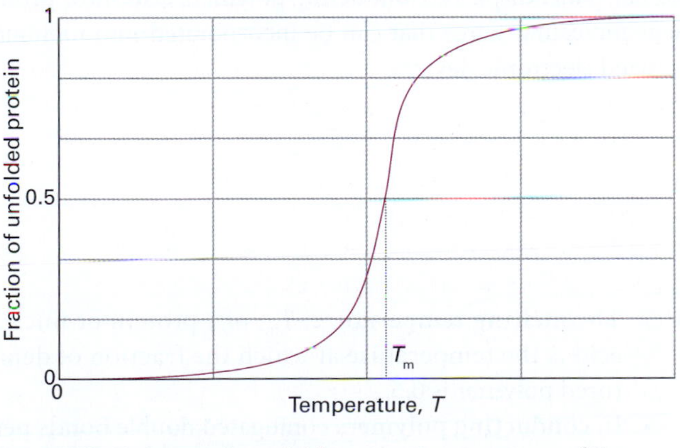

Figure 17B.5 A protein unfolds as the temperature of the sample increases. The sharp step in the plot of fraction of unfolded protein against temperature indicated that the transition is cooperative. The melting temperature, T_m, is the temperature at which the fraction of unfolded polymer is 0.5.

Example 17B.1 Predicting the melting temperature of DNA

The melting temperature of a DNA molecule (in the sense of the temperature at which it undergoes denaturation) can be determined by calorimetric methods. The following data were obtained in 0.010 M $Na_3PO_4(aq)$ for a series of DNA molecules with varying base pair composition, with f the fraction of G—C base pairs:

f	0.375	0.509	0.589	0.688	0.750
T_m/K	339	344	348	351	354

Estimate the melting temperature of a DNA molecule containing 40.0 per cent G—C base pairs.

Method Look for a quantitative relationship between the melting temperature and the composition of DNA. Begin by plotting T_m against fraction of G—C base pairs and examining the shape of the curve. If visual inspection of the plot suggests a linear relationship, then the melting point at any composition can be predicted from the equation of the straight line that fits the data.

Answer Figure 17B.6 shows that T_m varies linearly with the fraction of G—C base pairs, at least in this range of composition. The equation of the line that fits the data is

$$T_m/K = 325 + 39.7f$$

It follows that $T_m = 341\,K$ for 40.0 per cent G—C base pairs (at $f = 0.400$). The thermal stability of DNA increases with the number of C—G base pairs in the sequence because each G—C base pair has three hydrogen bonds, whereas each T—A base pair has only two (Topic 17A). More energy, and therefore a higher temperature, is required to unravel a double helix that

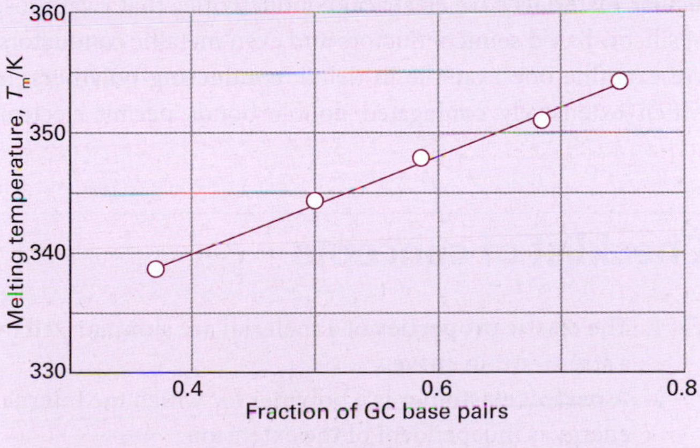

Figure 17B.6 Data for *Example* 17B.1 showing the variation of the melting temperature of DNA molecules with the fraction of G—C base pairs. All the samples also contain $1.0 \times 10^{-2}\,mol\,dm^{-3}\,Na_3PO_4$.

has a higher proportion of hydrogen bonding interactions per base pair.

A note on good practice In this example we do not have a good theory to guide us in the choice of a mathematical model to describe the behaviour of the system over a wide range of conditions. We are limited to finding a purely empirical relation—in this case a simple first-order polynomial equation—that fits the available data. It follows that we should not attempt to predict the property of a system that falls outside the narrow range of the data used to generate the fit because the mathematical model may have to be enhanced (for example, by using higher-order polynomial equations) to describe the system over a wider range of conditions. In the present case, we should not attempt to predict the T_m of DNA molecules outside the range $0.375<f<0.750$.

Self-test 17B.3 The following calorimetric data were obtained in solutions containing 0.15 M NaCl(aq) for the same series of DNA molecules studied in *Example* 17B.1. Estimate the melting temperature of a DNA molecule containing 40.0 per cent G—C base pairs under these conditions.

f	0.375	0.509	0.589	0.688	0.750
T_m/K	359	364	368	371	374

Answer: 360 K

17B.3 Electrical properties

Most of the macromolecules and self-assembled structures considered in this chapter are insulators, or very poor electrical conductors. However, a variety of newly developed macromolecular materials have electrical conductivities that rival those of silicon-based semiconductors and even metallic conductors. We examine one example in detail: **conducting polymers**, in which extensively conjugated double bonds permit electron conduction along the polymer chain. The 2000 Nobel Prize in chemistry was awarded to A.J. Heeger, A.G. MacDiarmid, and H. Shirakawa for their pioneering work in the synthesis and characterization of conducting polymers.

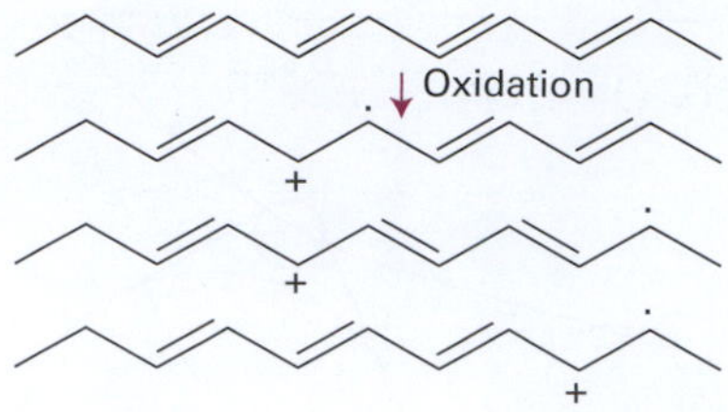

Figure 17B.7 The mechanism of migration of a partially localized cation radical, or polaron, in polyacetylene.

One example of a conducting polymer is polyacetylene (polyethyne, Fig. 17B.7). Whereas the delocalized π bonds do suggest that electrons can move up and down the chain, the electrical conductivity of polyacetylene increases significantly when it is partially oxidized by I_2 and other strong oxidants. The product is a **polaron**, a partially localized cation radical that travels virtually (by exchanging its identity with a neighbour) through the chain, as shown in Fig. 17B.7. Further oxidation of the polymer forms either **bipolarons**, a di-cation that moves virtually (in a similar way) as a unit through the chain, or **solitons**, two separate cation radicals that move independently. Polarons and solitons contribute to the mechanism of charge conduction in polyacetylene.

Conducting polymers are slightly better electrical conductors than silicon semiconductors but are far worse than metallic conductors. They are currently used in a number of devices, such as electrodes in batteries, electrolytic capacitors, and sensors. Recent studies of photon emission by conducting polymers may lead to new technologies for light-emitting diodes and flat-panel displays. Conducting polymers also show promise as molecular wires that can be incorporated into nanometre-sized electronic devices.

Checklist of concepts

- ☐ 1. The **elastic properties** of a material are summarized by a stress–strain curve.
- ☐ 2. A **perfect elastomer** is a polymer for which the internal energy is independent of the extension.
- ☐ 3. The disruption of long-range order in a polymer occurs at a **melting temperature**.
- ☐ 4. Synthetic polymers undergo a transition from a state of high to low chain mobility at the **glass transition temperature**.
- ☐ 5. The **melting temperature**, T_m, of a protein or nucleic acid is the temperature at which the fraction of denatured polymer is 0.5.
- ☐ 6. In **conducting polymers** conjugated double bonds permit electron conduction along the chain.

Checklist of equations

Property	Equation	Comment	Equation number
Conformational entropy of a random coil	$\Delta S = -\frac{1}{2} kN \ln\{(1+\nu)^{1+\nu}(1-\nu)^{1-\nu}\}$	$\nu = n/N$	17B.1
Restoring force of a one-dimensional random coil	$F = (kT/2l)\ln\{(1+\nu)/(1-\nu)\}$		17B.2a
	$F \approx nkT/Nl$	Small extensions	17B.2b

17C Self-assembly

Contents

➤ **Why do you need to know this material?**

Aggregates of small and large molecules form the basis of many technologies (such as detergents and nanotechnology) and are abundant in biological cells. To see why this is the case, you need to understand their structures and properties.

➤ **What is the key idea?**

Colloids, micelles, and biological membranes form spontaneously by self-assembly of molecules or macromolecules and are held together by molecular interactions.

➤ **What do you need to know already?**

You need to be familiar with molecular interactions (Topic 16B), the formation of liquids (Topic 16C), and interactions between ions (Topic 5F).

Self-assembly is the spontaneous formation of complex structures of molecules or macromolecules that are held together by molecular interactions, such as Coulombic, dispersion, hydrogen bonding, or hydrophobic interactions. Examples of self-assembly include the formation of liquid crystals, of protein quaternary structures from two or more polypeptide chains (Topic 17A), and (by implication) of a DNA double helix from two polynucleotide chains (Topic 17A). Here we concentrate on the specific properties of additional self-assembled systems, including some that are becoming important in the development of nanotechnology.

17C.1 Colloids

A **colloid**, or **disperse phase**, is a dispersion of small particles of one material in another that does not settle out under gravity. In this context, 'small' means that one dimension at least is smaller than about 500 nm in diameter (about the wavelength of visible light). Many colloids are suspensions of nanoparticles (particles of diameter up to about 100 nm). In general, colloidal particles are aggregates of numerous atoms or molecules, but are commonly but not universally too small to be seen with an ordinary optical microscope. They pass through most filter papers, but can be detected by light-scattering and sedimentation (Topic 17D).

(a) Classification and preparation

The name given to the colloid depends on the two phases involved:

- A **sol** is a dispersion of a solid in a liquid (such as clusters of gold atoms in water) or of a solid in a solid (such as ruby glass, which is a gold-in-glass sol, and achieves its colour by light scattering).
- An **aerosol** is a dispersion of a liquid in a gas (like fog and many sprays) or a solid in a gas (such as smoke): the particles are often large enough to be seen with a microscope.
- An **emulsion** is a dispersion of a liquid in a liquid (such as milk). A **foam** is a dispersion of a gas in a liquid.

A further classification of colloids is as **lyophilic**, or solvent attracting, and **lyophobic**, solvent repelling. If the solvent is water, the terms **hydrophilic** and **hydrophobic**, respectively, are used instead. Lyophobic colloids include the metal sols. Lyophilic colloids generally have some chemical similarity to the solvent, such as —OH groups able to form hydrogen bonds. A **gel** is a semi-rigid mass of a lyophilic sol.

The preparation of aerosols can be as simple as sneezing (which produces an imperfect aerosol). Laboratory and commercial methods make use of several techniques. Material (for example, quartz) may be ground in the presence of the dispersion medium. Passing a heavy electric current through a cell may lead to the sputtering (crumbling) of an electrode into colloidal particles. Arcing between electrodes immersed in the support medium also produces a colloid. Chemical precipitation sometimes results in a colloid. A precipitate (for example, silver iodide) already formed may be dispersed by the addition of a peptizing agent (for example, potassium iodide). Clays may be peptized by alkalis, the OH^- ion being the active agent.

Emulsions are normally prepared by shaking the two components together vigorously, although some kind of emulsifying agent usually has to be added to stabilize the product. This emulsifying agent may be a soap (the salt of a long-chain carboxylic acid) or other **surfactant** (surface active) species, or a lyophilic sol that forms a protective film around the dispersed phase. In milk, which is an emulsion of fats in water, the emulsifying agent is casein, a protein containing phosphate groups. It is clear from the formation of cream on the surface of milk that casein is not completely successful in stabilizing milk: the dispersed fats coalesce into oily droplets which float to the surface. This coagulation may be prevented by ensuring that the emulsion is dispersed very finely initially: intense agitation with ultrasonics brings this dispersion about, the product being 'homogenized' milk.

One way to form an aerosol is to tear apart a spray of liquid with a jet of gas. The dispersal is aided if a charge is applied to the liquid, for then electrostatic repulsions help to blast it apart into droplets. This procedure may also be used to produce emulsions, for the charged liquid phase may be directed into another liquid.

Colloids are often purified by **dialysis**, the process of squeezing the solution though a membrane. The aim is to remove much (but not all, for reasons explained later) of the ionic material that may have accompanied their formation. A membrane (for example, cellulose) is selected that is permeable to solvent and ions, but not to the colloid particles. Dialysis is very slow, and is normally accelerated by applying an electric field and making use of the charges carried by many colloidal particles; the technique is then called **electrodialysis**.

(b) Structure and stability

Colloids are thermodynamically unstable with respect to the bulk. This instability can be expressed thermodynamically by noting that because the change in Helmholtz energy, dA, when the surface area of the sample changes by $d\sigma$ at constant temperature and pressure is $dA = \gamma d\sigma$, where γ is the interfacial surface tension (Topic 16C), it follows that $dA < 0$ if $d\sigma < 0$. That is, the contraction of the surface ($d\sigma < 0$) is spontaneous ($dA < 0$). The survival of colloids must therefore be a consequence of the kinetics of collapse: colloids are thermodynamically unstable but kinetically non-labile.

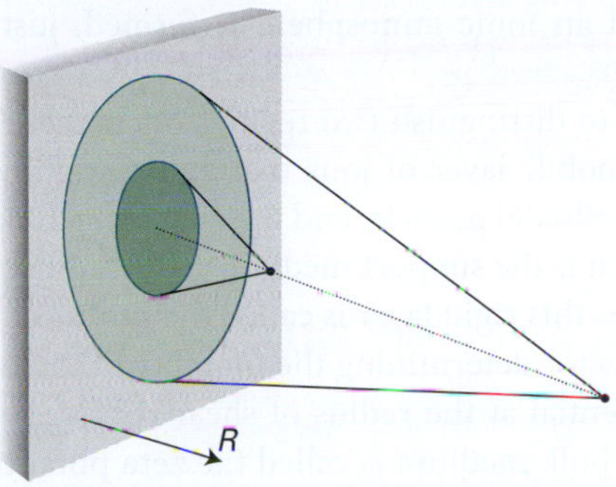

Figure 17C.1 Although the attraction between individual molecules is proportional to $1/R^6$, more molecules are within range at large separations (pale region) than at small separation (dark region), so the total interaction energy declines more slowly and is proportional to a lower power of R.

At first sight, even the kinetic argument seems to fail: colloidal particles attract each other over large distances, so there is a long-range force that tends to condense them into a single blob. The reasoning behind this remark is as follows. The energy of attraction between two individual atoms i and j separated by a distance R_{ij}, one in each colloidal particle, varies with their separation as $1/R_{ij}^6$ (Topic 16B). The sum of all these pairwise interactions, however, decreases only as approximately $1/R^2$ (the precise variation depending on the shape of the particles and their closeness), where R is the separation of the centres of the particles. The change in the power from 6 to 2 stems from the fact that at short distances only a few molecules interact but at large distances many individual molecules are at about the same distance from one another, and contribute equally to the sum (Fig. 17C.1), so the total interaction does not fall off as fast as the single molecule–molecule interaction.

Several factors oppose the long-range dispersion attraction. For example, there may be a protective film at the surface of the colloid particles that stabilizes the interface and cannot be penetrated when two particles touch. Thus the surface atoms of a platinum sol in water react chemically and are turned into $-Pt(OH)_3H_3$, and this layer encases the particle like a shell. A fat can be emulsified by a soap because the long hydrocarbon tails penetrate the oil droplet but the carboxylate head groups (or other hydrophilic groups in synthetic detergents) surround the surface, form hydrogen bonds with water, and give rise to a shell of negative charge that repels a possible approach from another similarly charged particle.

(c) The electrical double layer

A major source of kinetic non-lability of colloids is the existence of an electric charge on the surfaces of the particles. On account of this charge, ions of opposite charge tend to cluster nearby, and an ionic atmosphere is formed, just as for ions (Topic 5F).

We need to distinguish two regions of charge. First, there is a fairly immobile layer of ions that adhere tightly to the surface of the colloidal particle, and which may include water molecules (if that is the support medium). The radius of the sphere that captures this rigid layer is called the **radius of shear** and is the major factor determining the mobility of the particles. The electric potential at the radius of shear relative to its value in the distant, bulk medium is called the **zeta potential**, ζ, or the **electrokinetic potential**. Second, the charged unit attracts an oppositely charged atmosphere of mobile ions. The inner shell of charge and the outer ionic atmosphere is called the **electrical double layer**.

The theory of the stability of lyophobic dispersions was developed by B. Derjaguin and L. Landau and independently by E. Verwey and J.T.G. Overbeek, and is known as the **DLVO theory**.[1] It assumes that there is a balance between the repulsive interaction between the charges of the electric double layers on neighbouring particles and the attractive interactions arising from van der Waals interactions between the molecules in the particles. The potential energy arising from the repulsion of double layers on particles of radius a has the form

$$V_{\text{repulsion}} = +\frac{Aa^2\zeta^2}{R}e^{-s/r_D} \qquad (17C.1)$$

where A is a constant, ζ is the zeta potential, R is the separation of centres, s is the separation of the surfaces of the two particles ($s = R - 2a$ for spherical particles of radius a), and r_D is the thickness of the double layer. This expression is valid for small particles with a thick double layer ($a \ll r_D$). When the double layer is thin ($r_D \ll a$), the expression is replaced by

$$V_{\text{repulsion}} = +\tfrac{1}{2}Aa^2\zeta^2\ln(1+e^{-s/r_D}). \qquad (17C.2)$$

In each case, the thickness of the double layer can be estimated from an expression like that derived for the thickness of the ionic atmosphere in the Debye–Hückel theory (Topic 5F) in which there is a competition between the assembling influences of the attraction between opposite charges and the disruptive effect of thermal motion:

$$r_D = \left(\frac{\varepsilon RT}{2\rho F^2 I b^{\ominus}}\right)^{1/2} \quad \text{Thickness of the electrical double layer} \qquad (17C.3)$$

where I is the ionic strength of the solution, ρ its mass density, and $b^{\ominus} = 1\ \text{mol kg}^{-1}$ (F is Faraday's constant and ε is the permittivity, $\varepsilon = \varepsilon_r\varepsilon_0$). The potential energy arising from the attractive interaction has the form

$$V_{\text{attraction}} = -\frac{B}{s} \qquad (17C.4)$$

where B is another constant. The variation of the total potential energy with separation is shown in Fig. 17C.2.

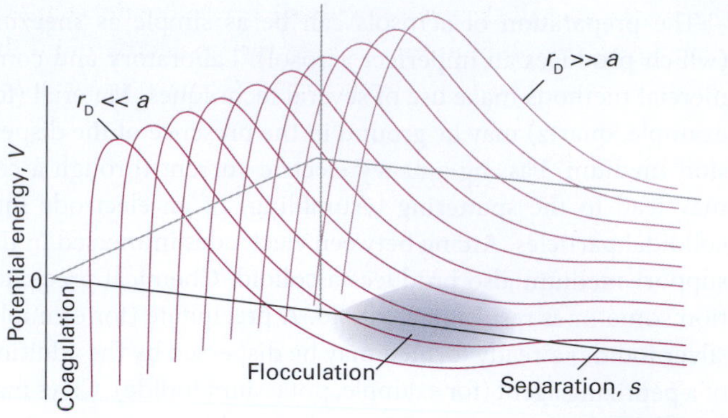

Figure 17C.2 The potential energy of interaction as a function of the separation of the centres of the two particles and its variation with the ratio of the particle size to the thickness a of the electric double layer, r_D. The regions labelled coagulation and flocculation show the dips in the potential energy curves where these processes occur.

At high ionic strengths, the ionic atmosphere is dense and the potential shows a secondary minimum at large separations. Aggregation of the particles arising from the stabilizing effect of this secondary minimum is called **flocculation**. The flocculated material can often be redispersed by agitation because the well is so shallow. **Coagulation**, the irreversible aggregation of distinct particles into large particles, occurs when the separation of the particles is so small that they enter the primary minimum of the potential energy curve and van der Waals forces are dominant.

The ionic strength is increased by the addition of ions, particularly those of high charge type, so such ions act as flocculating agents. This increase is the basis of the empirical **Schulze–Hardy rule**, that hydrophobic colloids are flocculated most efficiently by ions of opposite charge type and high charge number. The Al^{3+} ions in alum are very effective, and are used to induce the congealing of blood. When river water containing colloidal clay flows into the sea, the salt water induces flocculation and coagulation, and is a major cause of silting in estuaries.

Metal oxide sols tend to be positively charged whereas sulfur and the noble metals tend to be negatively charged. Naturally occurring macromolecules also acquire a charge when dispersed in water, and an important feature of proteins and other natural macromolecules is that their overall charge depends on the pH of the medium. For instance, in acidic environments protons attach to basic groups, and the net charge of

[1] The derivation of the expressions quoted is too complicated to include here. For a full description, see Volume 1 of R. J. Hunter, *Foundations of colloid science*, Oxford University Press (1987).

the macromolecule is positive; in basic media the net charge is negative as a result of proton loss. At the **isoelectric point** the pH is such that there is no net charge on the macromolecule.

Example 17C.1 Determining the isoelectric point of a protein

The speed with which bovine serum albumin (BSA) moves through water under the influence of an electric field was monitored at several values of pH, and the data are listed below. What is the isoelectric point of the protein?

pH	4.20	4.56	5.20	5.65	6.30	7.00
Velocity/(μm s^{-1})	0.50	0.18	−0.25	−0.65	−0.90	−1.25

Method If we plot speed against pH, we can use interpolation to find the pH at which the speed is zero, which is the pH at which the molecule has zero net charge.

Answer The data are plotted in Fig. 17C.3. The velocity passes through zero at pH = 4.8; hence pH = 4.8 is the isoelectric point.

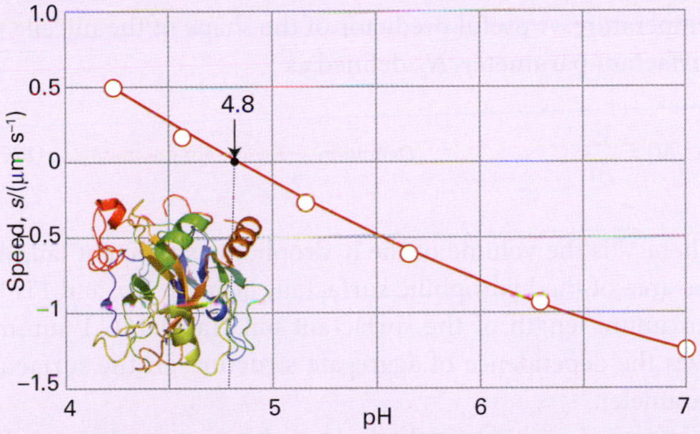

Figure 17C.3 The plot of the speed of a moving macromolecule against pH allows the isoelectric point to be detected as the pH at which the speed is zero. The data are from *Example* 17C.1.

Self-test 17C.1 The following data were obtained for another protein:

pH	3.5	4.5	5.0	5.5	6.0
Velocity/(μm s^{-1})	0.10	−0.10	−0.20	−0.30	−0.40

Estimate the pH of the isoelectric point.

Answer: 4.0

The primary role of the electric double layer is to confer kinetic non-lability. Colliding colloidal particles break through the double layer and coalesce only if the collision is sufficiently energetic to disrupt the layers of ions and solvating molecules, or if thermal motion has stirred away the surface accumulation of charge. This disruption may occur at high temperatures, which is one reason why sols precipitate when they are heated.

17C.2 Micelles and biological membranes

In aqueous solutions surfactant molecules or ions can cluster together as **micelles**, which are colloid-sized clusters of molecules, for their hydrophobic tails tend to congregate (through hydrophobic interactions: see Topic 16B), and their hydrophilic head groups provide protection (Fig. 17C.4).

(a) Micelle formation

Micelles form only above the **critical micelle concentration** (CMC) and above the **Krafft temperature**. The CMC is detected by noting a pronounced change in physical properties of the solution, particularly the molar conductivity (Fig. 17C.5). There is no abrupt change in properties at the CMC; rather, there is a

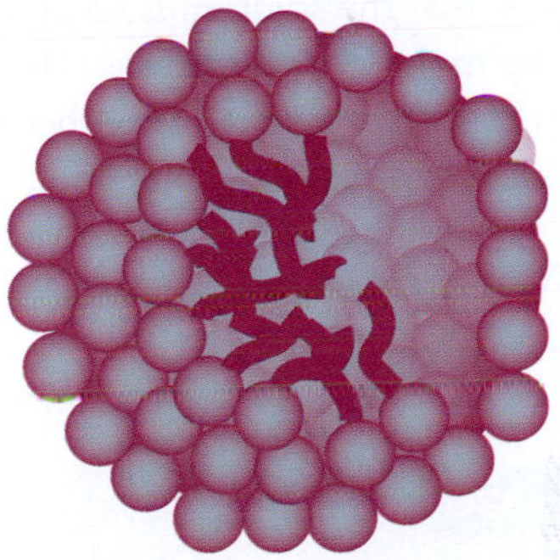

Figure 17C.4 A schematic version of a spherical micelle. The hydrophilic groups are represented by spheres and the hydrophobic hydrocarbon chains are represented by the stalks; these stalks are mobile.

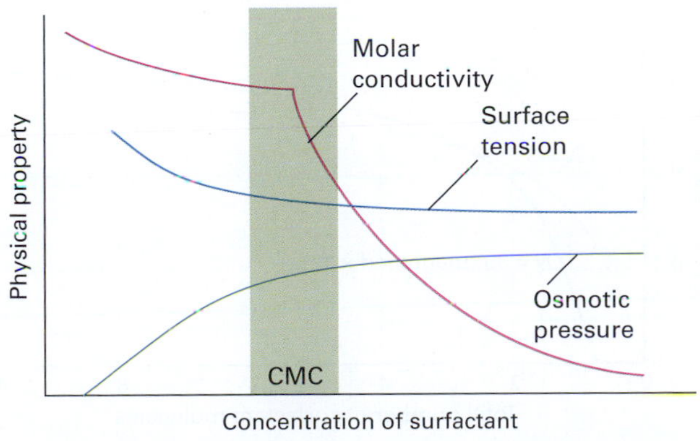

Figure 17C.5 The typical variation of some physical properties of an aqueous solution of sodium dodecylsulfate close to the critical micelle concentration (CMC).

transition region corresponding to a range of concentrations around the CMC where physical properties vary smoothly but nonlinearly with the concentration. The hydrocarbon interior of a micelle is like a droplet of oil. Nuclear magnetic resonance shows that the hydrocarbon tails are mobile, but slightly more restricted than in the bulk. Micelles are important in industry and biology on account of their solubilizing function: matter can be transported by water after it has been dissolved in their hydrocarbon interiors. For this reason, micellar systems are used as detergents, for organic synthesis, froth flotation, and petroleum recovery.

The self-assembly of a micelle has the characteristics of a cooperative process in which the addition of a surfactant molecule to a cluster that is forming becomes more probable the larger the size of the aggregate, so after a slow start there is a cascade of formation of micelles. If we suppose that the dominant micelle M_N consists of N monomers M, then the dominant equilibrium we have to consider is

$$NM \rightleftharpoons M_N \qquad K=\frac{[M_N]}{[M]^N} \tag{17C.5a}$$

We have assumed, probably dangerously on account of the large sizes of monomers, that the solution is ideal and that activities can be replaced by molar concentrations. The total concentration of surfactant is $[M]_{total}=[M]+N[M_N]$ because each micelle consists of N monomer molecules. Therefore,

$$K=\frac{[M_N]}{([M]_{total}-N[M_N])^N} \tag{17C.5b}$$

Brief illustration 17C.1 The fraction of surfactant molecules in micelles

Equation 17C.5b can be solved numerically for the micelle concentration as a function of the total surfactant concentration and some results for $K=1$ are shown in Fig. 17C.6. We see that for large N, there is a reasonably sharp transition in the relative concentrations of surfactant molecules that are present in micelles, which corresponds to the existence of a CMC.

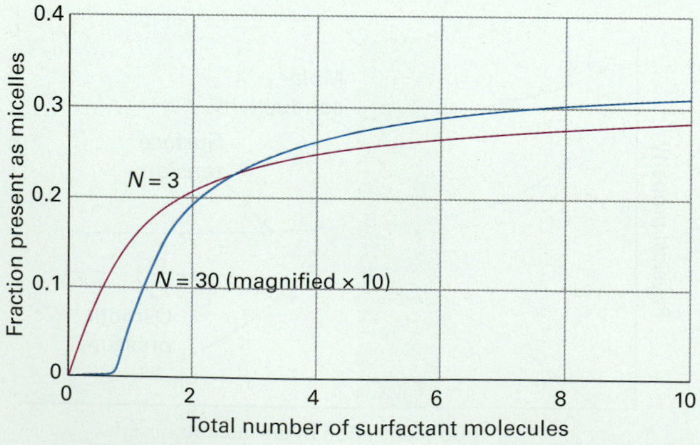

Figure 17C.6 The micelle concentration as a function of the total surfactant concentration for $K=1$.

Self-test 17C.2 Equation 17C.5b is surprisingly tricky to solve, but it is possible to make good progress with simple cases. With $N=2$ and $K=1$, find an expression for $[M_2]$.

Answer: $[M_2]=[M_{total}]-\frac{1}{4}\{(1+8[M_{total}])^{1/2}-1\}$

Non-ionic surfactant molecules may cluster together in clumps of 1000 or more, but ionic species tend to be disrupted by the electrostatic repulsions between head groups and are normally limited to groups of less than about 100. However, the disruptive effect depends more on the effective size of the head group than the charge. For example, ionic surfactants such as sodium dodecyl sulfate (SDS) and cetyltrimethylammonium bromide (CTAB) form rods at moderate concentrations whereas sugar surfactants form small, approximately spherical micelles. The micelle population is often polydisperse, and the shapes of the individual micelles vary with shape of the constituent surfactant molecules, surfactant concentration, and temperature. A useful predictor of the shape of the micelle the **surfactant parameter**, N_s, defined as

$$N_s=\frac{V}{Al} \qquad \textit{Definition} \quad \text{Surfactant parameter} \tag{17C.6}$$

where V is the volume of the hydrophobic surfactant tail, A is the area of the hydrophilic surfactant head group, and l is the maximum length of the surfactant tail. Table 17C.1 summarizes the dependence of aggregate structure on the surfactant parameter.

In aqueous solutions spherical micelles form, as shown in Fig. 17C.4, with the polar head groups of the surfactant molecules on the micellar surface and interacting favourably with solvent and ions in solution. Hydrophobic interactions stabilize the aggregation of the hydrophobic surfactant tails in the micellar core. Under certain experimental conditions, a **liposome** may form, with an inward pointing inner surface of molecules surrounded by an outward pointing outer layer (Fig. 17C.7). Liposomes may be used to carry nonpolar drug molecules in blood.

Table 17C.1 Variation of micelle shape with the surfactant parameter

N_s	Micelle shape
<0.33	Spherical
0.33 to 0.50	Cylindrical rods
0.50 to 1.00	Vesicles
1.00	Planar bilayers
>1.00	Reverse micelles and other shapes

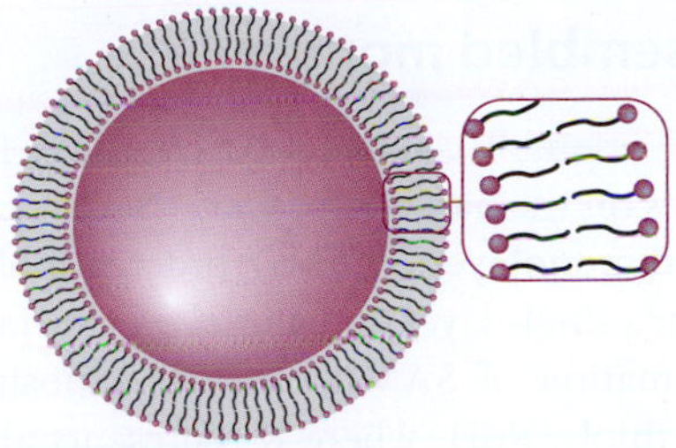

Figure 17C.7 The cross-sectional structure of a spherical liposome.

Increasing the ionic strength of the aqueous solution reduces repulsions between surface head groups, and cylindrical micelles can form. These cylinders may stack together in reasonably close-packed (hexagonal) arrays, forming **lyotropic mesomorphs** and, more colloquially, 'liquid crystalline phases'.

Reverse micelles form in nonpolar solvents, with small polar surfactant head groups in a micellar core and more voluminous hydrophobic surfactant tails extending into the organic bulk phase. These spherical aggregates can solubilize water in organic solvents by creating a pool of trapped water molecules in the micellar core. As aggregates arrange at high surfactant concentrations to yield long-range positional order, many other types of structures are possible including cubic and hexagonal shapes.

The enthalpy of micelle formation reflects the contributions of interactions between micelle chains within the micelles and between the polar head groups and the surrounding medium. Consequently, enthalpies of micelle formation display no readily discernible pattern and may be positive (endothermic) or negative (exothermic). Many non-ionic micelles form endothermically, with ΔH of the order of 10 kJ per mole of surfactant molecules. That such micelles do form above the CMC indicates that the entropy change accompanying their formation must then be positive, and measurements suggest a value of about $+140\,\mathrm{J\,K^{-1}\,mol^{-1}}$ at room temperature. The fact that the entropy change is positive, even though the molecules are clustering together, shows that hydrophobic interactions are important in the formation of micelles (in the sense that water molecules are released to become more disordered in the process and hence make an overwhelming contribution to the entropy change).

(b) Bilayers, vesicles, and membranes

Some micelles at concentrations well above the CMC form extended parallel sheets two molecules thick, called **planar bilayers**. The individual molecules lie perpendicular to the sheets, with hydrophilic groups on the outside in aqueous solution and on the inside in nonpolar media. When segments of planar bilayers fold back on themselves, **unilamellar vesicles** may form where the spherical hydrophobic bilayer shell separates an inner aqueous compartment from the external aqueous environment.

Bilayers show a close resemblance to biological membranes, and are often a useful model on which to base investigations of biological structures. However, actual membranes are highly sophisticated structures. The basic structural element of a membrane is a phospholipid, such as phosphatidyl choline (**1**), which contains long hydrocarbon chains (typically in the range C_{14}–C_{24}) and a variety of polar groups, such as $-CH_2CH_2N(CH_3)_3^+$. The hydrophobic chains stack together to form an extensive layer about 5 nm across. The lipid molecules form layers instead of micelles because the hydrocarbon chains are too bulky to allow packing into nearly spherical clusters.

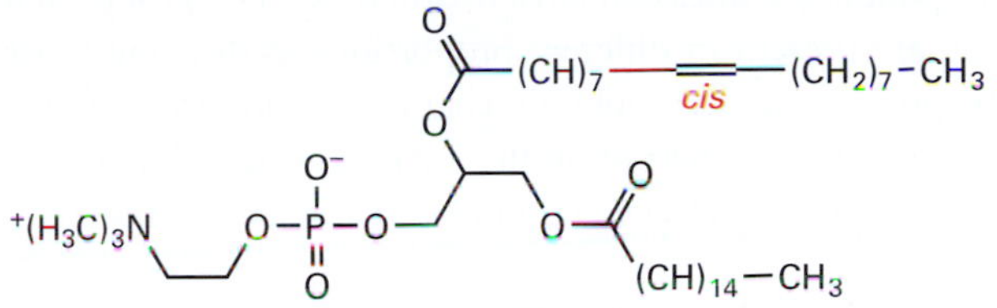

1 Phosphatidyl choline

The bilayer is a highly mobile structure. Not only are the hydrocarbon chains ceaselessly twisting and turning in the region between the polar groups, but the phospholipid and cholesterol molecules migrate over the surface. It is better to think of the membrane as a viscous fluid rather than a permanent structure, with a viscosity about 100 times that of water. Typically, a phospholipid molecule in a membrane migrates through about 1 μm in about 1 min.

All lipid bilayers undergo a transition from a state of high to low chain mobility at a temperature that depends on the structure of the lipid. To visualize the transition, we consider what happens to a membrane as we lower its temperature (Fig. 17C.8). There is sufficient energy available at normal temperatures for limited bond rotation to occur and the flexible chains writhe. However, the membrane is still highly organized in the sense that the bilayer structure does

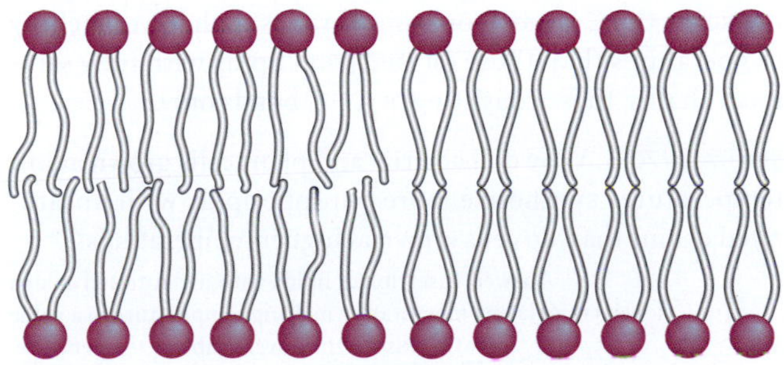

Figure 17C.8 A depiction of the variation with temperature of the flexibility of hydrocarbon chains in a lipid bilayer. (a) At physiological temperature, the bilayer exists as a liquid crystal, in which some order exists but the chains writhe. (b) At a specific temperature, the chains are largely frozen and the bilayer is said to exist as a gel.

not come apart and the system is best described as a liquid crystal. At lower temperatures, the amplitudes of the writhing motion decrease until a specific temperature is reached at which motion is largely frozen. The membrane is then said to exist as a gel. Biological membranes exist as liquid crystals at physiological temperatures.

Phase transitions in membranes are often observed as 'melting' from gel to liquid crystal by calorimetric methods. The data show relations between the structure of the lipid and the melting temperature. Interspersed among the phospholipids of biological membranes are sterols, such as cholesterol (**2**), which is largely hydrophobic but does contain a hydrophilic –OH group. Sterols, which are present in different proportions in different types of cells, prevent the hydrophobic chains of lipids from 'freezing' into a gel and, by disrupting the packing of the chains, spread the melting point of the membrane over a range of temperatures.

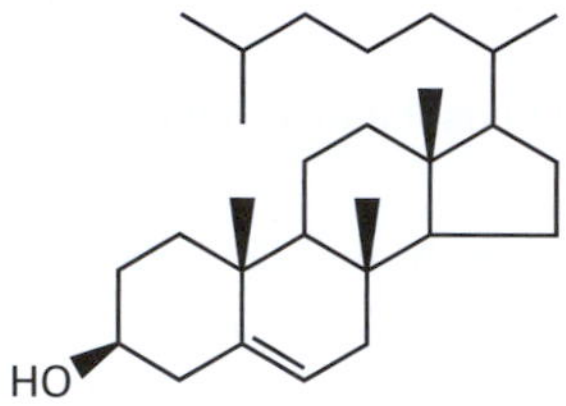

2 Cholesterol

Brief illustration 17C.2 The melting temperatures of membranes

To predict trends in melting temperatures we need to assess the strengths of the interactions between molecules. Longer chains can be expected to be held together more strongly by hydrophobic interactions than shorter chains, so we should expect the melting temperature to increase with the length of the hydrophobic chain of the lipid. On the other hand, any structural elements that prevent alignment of the hydrophobic chains in the gel phase lead to low melting temperatures. Indeed, lipids containing unsaturated chains, those containing some C=C bonds, form membranes with lower melting temperatures than those formed from lipids with fully saturated chains, those consisting of C–C bonds only.

Self-Test 17C.3 Why do bacterial and plant cells grown at low temperatures synthesize more phospholipids with unsaturated chains than do cells grown at higher temperatures?

Answer: Insertion of lipids with unsaturated chains lowers the plasma membrane's melting temperature to a value that is close to the lower ambient temperature.

(c) Self-assembled monolayers

Molecular self-assembly can be used as the basis for manipulation of surfaces on the nanometre scale. Of current interest are **self-assembled monolayers** (SAMs), ordered molecular aggregates that form a single layer of material on a surface. To understand the formation of SAMs, consider exposing molecules such as alkyl thiols, RSH, where R represents an alkyl chain, to an Au(0) surface. The thiols react with the surface, forming $RS^-Au(I)$ adducts:

$$\mathrm{RSH}+\mathrm{Au}(0)_n \rightarrow \mathrm{RS}^-\mathrm{Au(I)}\cdot\mathrm{Au}(0)_{n-1}+\tfrac{1}{2}\mathrm{H}_2$$

If R is a sufficiently long chain, van der Waals interactions between the adsorbed RS units lead to the formation of a highly ordered monolayer on the surface (Fig. 17C.9). The Gibbs energy of formation of SAMs increases with the length of the alkyl chain, with each methylene group contributing 0.4–4 kJ mol^{-1}.

A self-assembled monolayer alters the properties of the surface. For example, a hydrophilic surface may be rendered hydrophobic once covered with a SAM. Furthermore, attaching functional groups to the exposed ends of the alkyl groups may impart specific chemical reactivity or ligand-binding properties to the surface, leading to applications in chemical (or biochemical) sensors and reactors.

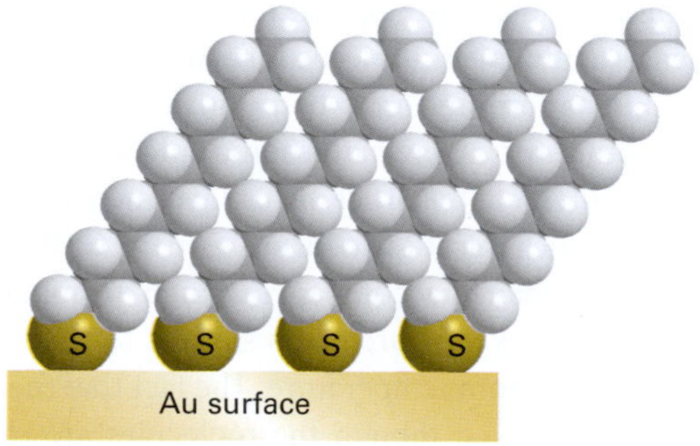

Figure 17C.9 Self-assembled monolayers of alkylthiols formed onto a gold surface by reaction of the thiol groups with the surface and aggregation of the alkyl chains.

Checklist of concepts

☐ 1. A **disperse system** is a dispersion of small particles of one material in another.

☐ 2. **Colloids** are classified as lyophilic and lyophobic.

- ☐ 3. A **surfactant** is a species that accumulates at the interface of two phases or substances.
- ☐ 4. Many colloid particles are thermodynamically unstable but kinetically non-labile.
- ☐ 5. The **radius of shear** is the radius of the sphere that captures the rigid layer of charge attached to a colloid particle.
- ☐ 6. The **zeta potential** is the electric potential at the radius of shear relative to its value in the distant, bulk medium.
- ☐ 7. The inner shell of charge and the outer atmosphere jointly constitute the **electric double layer**.
- ☐ 8. **Flocculation** is the reversible aggregation of colloidal particles.
- ☐ 9. **Coagulation** is the irreversible aggregation of colloidal particles.
- ☐ 10. The **Schultze–Hardy rule** states that hydrophobic colloids are flocculated most efficiently by ions of opposite charge type and high charge number.
- ☐ 11. A **micelle** is a colloid-sized cluster of molecules that forms at and above the **critical micelle concentration** and the **Krafft temperature**.
- ☐ 12. Micelles can assume a number of shapes, depending on temperature, shape and concentration of constituent molecules.
- ☐ 13. **Planar bilayers** are micelles that exist as extended parallel sheets two molecules thick that are extended.
- ☐ 14. **Unilamellar vesicles** are micelles that exist as extended parallel sheets two molecules thick that fold back on themselves.
- ☐ 15. **Self-assembled monolayers** are ordered molecular aggregates that form a single layer of material on a surface spontaneously.

Checklist of equations

Property	Equation	Comment	Equation number
Thickness of the electrical double layer	$r_D = (\varepsilon RT/2\rho F^2 I b^{\ominus})^{1/2}$	Debye–Hückel theory	17C.3
Surfactant parameter	$N_s = V/Al$	Definition	17C.6

17D Determination of size and shape

Contents

➤ Why do you need to know this material?

To appreciate modern work on macromolecules in technology and biochemistry, you need to understand several experimental techniques that are used to determine the molar masses and shapes of synthetic and biological polymers.

➤ What is the key idea?

Mass spectrometry, laser light scattering, ultracentrifugation, and viscosity measurements are useful techniques for the determination of size and shape of macromolecules.

➤ What do you need to know already?

You need to be familiar with structures of macromolecules (Topic 17A) and aggregates (Topic 17C).

X-ray diffraction (Topic 18A) can reveal the position of almost every atom other than hydrogen even in very large molecules. However, there are several reasons why other techniques must also be used. In the first place, the sample might be a mixture of molecules with different chain lengths and extents of cross-linking, in which case sharp X-ray images are not obtained. Even if all the molecules in the sample are identical, it might prove impossible to obtain a single crystal, which is essential for diffraction studies because only then does the electron density (which is responsible for the scattering) have a large-scale periodic variation. Furthermore, although work on proteins and DNA has shown how immensely interesting and motivating the data can be, the information is incomplete. For instance, what can be said about the shape of the molecule in its natural environment, a biological cell? What can be said about the response of its shape to changes in its environment?

17D.1 Mean molar masses

A pure protein is **monodisperse**, meaning that it has a single, definite molar mass. There may be small variations, such as one amino acid replacing another, depending on the source of the sample. A synthetic polymer, however, is **polydisperse**, in the sense that a sample is a mixture of molecules with various chain lengths and molar masses. The various techniques that are used to measure molar mass result in different types of mean values of polydisperse systems.

The mean obtained from the determination of molar mass by osmometry (Topic 5B) is the **number-average molar mass**, $\bar{M}_n$, which is the value obtained by weighting each molar mass by the number of molecules of that mass present in the sample:

$$\bar{M}_{\mathrm{n}} = \frac{1}{N}\sum_i N_i M_i = \langle M \rangle \qquad \textit{Definition} \qquad \text{Number-average molar mass} \qquad (17D.1)$$

where N_i is the number of molecules with molar mass M_i and there are N molecules in all. The notation $\langle X \rangle$ denotes the usual (number average) of a property X, and we shall use it again below. For reasons related to the ways in which macromolecules contribute to physical properties, viscosity measurements give the **viscosity-average molar mass**, $\bar{M}_v$, light-scattering experiments give the **weight-average molar mass**, $\bar{M}_w$, and sedimentation experiments give the ***Z*-average molar mass**, $\bar{M}_Z$. (The name is derived from the *z*-coordinate used to depict data in a procedure for determining the average.) Although such averages are often best left as empirical quantities, some may be interpreted in terms of the composition of the sample. Thus, the weight-average molar mass is the average calculated by weighting the molar masses of the molecules by the mass of each one present in the sample:

$$\bar{M}_{\rm w}=\frac{1}{m}\sum_i m_i M_i \quad \text{Definition} \quad \text{Weight-average molar mass} \quad (17D.2a)$$

In this expression, m_i is the total mass of molecules of molar mass M_i and m is the total mass of the sample. Because $m_i=N_iM_i/N_A$, we can also express this average as

$$\bar{M}_{\rm w}=\frac{\sum_i N_iM_i^2}{\sum_i N_iM_i}=\frac{\langle M^2\rangle}{\langle M\rangle} \quad \text{Interpretation} \quad \text{Weight-average molar mass} \quad (17D.2b)$$

This expression shows that the weight-average molar mass is proportional to the mean square molar mass. Similarly, the Z-average molar mass turns out to be proportional to the mean cubic molar mass:

$$\bar{M}_{\rm Z}=\frac{\sum_i N_iM_i^3}{\sum_i N_iM_i^2}=\frac{\langle M^3\rangle}{\langle M^2\rangle} \quad \text{Interpretation} \quad Z\text{-average molar mass} \quad (17D.2c)$$

Example 17D.1 Calculating number and mass averages

Determine the number-average and the weight-average molar masses of a sample of poly(vinyl chloride) from the following data:

Molar mass interval/(kg mol⁻¹)	Average molar mass within interval/(kg mol⁻¹)	Mass of sample within interval/g
5–9	7.5	9.6
10–14	12.5	8.7
15–19	17.5	8.9
20–24	22.5	5.6
25–29	27.5	3.1
30–35	32.5	1.7

Method The relevant equations are eqns 17D.2a and 17D.2b. Calculate the two averages by weighting the molar mass within each interval by the number and mass, respectively, of the molecules in each interval. Obtain the numbers in each interval by dividing the mass of the sample in each interval by the average molar mass for that interval. Because the number of molecules is proportional to the amount of substance (the number of moles), the number-weighted average can be obtained directly from the amounts in each interval. That is, on dividing the numerator and denominator by Avogadro's constant N_A, and writing $n=N/N_A$, eqn 17D.1 becomes

$$\bar{M}_{\rm n}=\frac{1}{N/N_{\rm A}}\sum_i(N_i/N_{\rm A})M_i=\frac{1}{n}\sum_i n_iM_i$$

Answer The amounts in each interval are as follows:

Interval	5–9	10–14	15–19	20–24	25–29	30–35
Molar mass/(kg mol⁻¹)	7.5	12.5	17.5	22.5	27.5	32.5
Amount/mmol	1.3	0.70	0.51	0.25	0.11	0.052
					Total:	2.92

The number-average molar mass is therefore

$$\bar{M}_{\rm n}/(\text{kg mol}^{-1})=\frac{1}{2.92}(1.3\times7.5+0.70\times12.5+0.51\times17.5+0.25\times22.5+0.11\times27.5+0.052\times32.5)=13$$

The weight-average molar mass is calculated directly from the data after noting that the total mass of the sample is 37.6 g:

$$\bar{M}_{\rm w}/(\text{kg mol}^{-1})=\frac{1}{37.6}(9.6\times7.5+8.7\times12.5+8.9\times17.5+5.6\times22.5+3.1\times27.5+1.7\times32.5)=16$$

Note the different values of the two averages. In this instance, $\bar{M}_{\rm w}/\bar{M}_{\rm n}=1.2$.

Self-test 17D.1 Evaluate the Z-average molar mass of the sample.

Answer: 19 kg mol⁻¹

The ratio $\bar{M}_{\rm w}/\bar{M}_{\rm n}$ is called the (molar-mass) **dispersity** (previously the 'polydispersity index', PDI) and denoted $Đ$ (read 'D-stroke'). It follows from eqns 17D.1 and 17D.2 that

$$Đ=\frac{\bar{M}_{\rm w}}{\bar{M}_{\rm n}} \quad \text{Definition} \quad \text{Dispersity} \quad (17D.3a)$$

It then follows from the interpretation of the weight and number averages that

$$Đ=\frac{\langle M^2\rangle}{\langle M\rangle^2} \quad \text{Interpretation} \quad \text{Dispersity} \quad (17D.3b)$$

That is, the dispersity is proportional to the ratio of the mean square molar mass to the square of the mean molar mass. In the determination of protein molar masses we expect the various averages to be the same because the sample is monodisperse (unless there has been degradation). A synthetic polymer normally spans a range of molar masses and the different averages yield different values. Typical synthetic materials have $Đ\approx4$ but much recent research has been devoted to developing methods that give much lower dispersities. The term 'monodisperse' is conventionally applied to synthetic polymers in which the dispersity is less than 1.1; commercial polyethene samples might be much more heterogeneous, with a dispersity close to 30. One consequence of a narrow molar mass distribution for synthetic polymers is often a higher degree of three-dimensional long-range order in the solid and therefore higher density and melting point. The spread of values is controlled by the choice of catalyst and reaction conditions. In practice, it is found that long-range order is determined more by structural factors (branching, for instance) than by molar mass.

17D.2 The techniques

Average molar masses may be determined by osmotic pressure of polymer solutions. The upper limit for the reliability of membrane osmometry is about 1000 kg mol^{-1}. A major problem for macromolecules of relatively low molar mass (less than about 10 kg mol^{-1}) is their ability to percolate through the membrane. One consequence of this partial permeability is that membrane osmometry tends to overestimate the average molar mass of a polydisperse mixture. Several techniques for the determination of molar mass and dispersity that are not so limited include mass spectrometry, laser light scattering, ultracentrifugation, and viscosity measurements.

A note on good practice The masses of macromolecules are often reported in daltons (Da), where 1 Da$=m_u$ (with $m_u=1.661\times10^{-27}$ kg). Note that 1 Da is a measure of *molecular* mass not of *molar* mass. We might say that the mass (not the molar mass) of a certain macromolecule is 100 kDa (that is, its mass is $100\times10^3\times m_u$); we could also say that its molar mass is 100 kg mol^{-1}; we should not say (even though it is common practice) that its molar mass is 100 kDa.

(a) Mass spectrometry

Mass spectrometry is among the most accurate techniques for the determination of molar masses. The procedure consists of ionizing the sample in the gas phase and then measuring the mass-to-charge number ratio (m/z; more precisely, the dimensionless ratio m/zm_u) of all ions. Macromolecules present a challenge because it is difficult to produce gaseous ions of large species without fragmentation. However, two techniques have emerged that circumvent this problem: **matrix-assisted laser desorption/ionization** (MALDI) and **electrospray ionization**. We shall discuss **MALDI-TOF mass spectrometry**, so called because the MALDI technique is coupled to a time-of-flight (TOF) ion detector.

Figure 17D.1 shows a schematic view of a MALDI-TOF mass spectrometer. The macromolecule is first embedded in a solid matrix which typically consists of an organic material such as *trans*-3-indoleacrylic acid and inorganic salts such as sodium chloride or silver trifluoroacetate. This sample is then irradiated with a pulsed laser. The laser energy ejects electronically excited matrix ions, cations, and neutral macromolecules, thus creating a dense gas plume above the sample surface. The macromolecule is ionized by collisions and complexation with small cations, such as H^+, Na^+, and Ag^+.

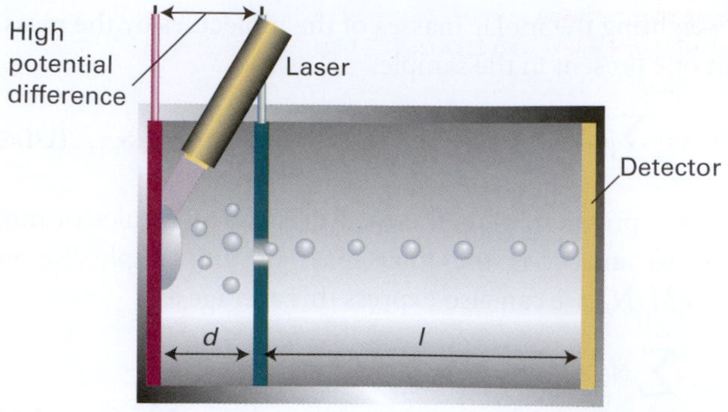

Figure 17D.1 A matrix-assisted laser desorption/ionization time-of-flight (MALDI-TOF) mass spectrometer. A laser beam ejects macromolecules and ions from the solid matrix. The ionized macromolecules are accelerated by an electrical potential difference over a distance *d* and then travel through a drift region of length *l*. Ions with the smallest mass to charge ratio (*m*/*z*) reach the detector first.

Figure 17D.2 shows the MALDI-TOF mass spectrum of a polydisperse sample of poly(butylene adipate) (PBA, **1**). The MALDI technique produces mostly singly charged molecular ions that are not fragmented. Therefore, the multiple peaks in the spectrum arise from polymers of different lengths, with the intensity of each peak being proportional to the abundance of each polymer in the sample. Values of $\bar{M}_n$, $\bar{M}_w$ and the dispersity can be calculated from the data. It is also possible to use the mass spectrum to verify the structure of a polymer, as shown in the following example.

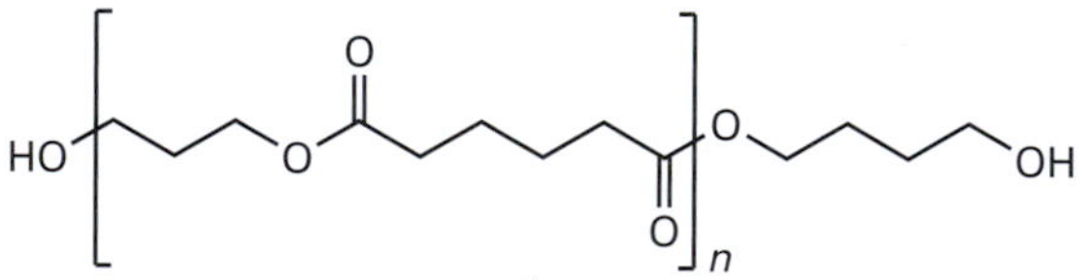

1

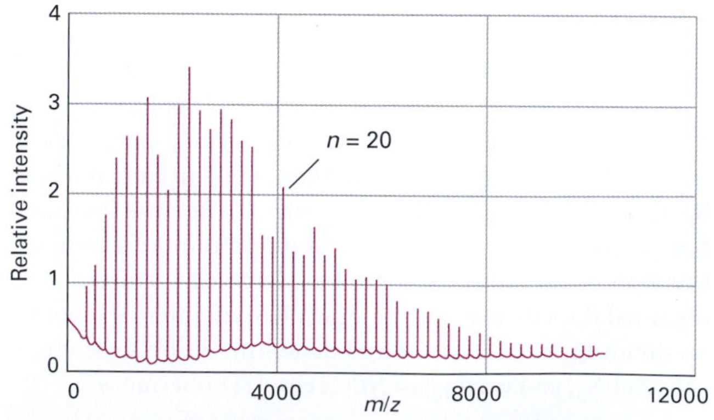

Figure 17D.2 MALDI-TOF spectrum of a sample of poly(butylene adipate) (**1**) with $\bar{M}_n=4525$ g mol^{-1} (Adapted from D.C. Mudiman et al., *J. Chem. Educ.* **74**, 1288 (1997).)

Example 17D.2 Interpreting the mass spectrum of a polymer

The mass spectrum in Fig. 17D.2 consists of peaks spaced by 200 g mol^{-1}. The peak at 4113 g mol^{-1} corresponds to the polymer for which $n=20$. From these data, verify that the sample consists of polymers with the general structure given by **1**.

Method Because each peak corresponds to a different value of n, the molar mass difference, ΔM, between peaks corresponds to the molar mass, M, of the repeating unit (the group inside the brackets in **1**). Furthermore, the molar mass of the terminal groups (the groups outside the brackets in **1**) may be obtained from the molar mass of any peak by using

$$M(\text{terminal groups}) = M(\text{polymer with } n \text{ repeating units}) - n\Delta M - M(\text{cation})$$

where the last term corresponds to the molar mass of the cation that attaches to the macromolecule during ionization.

Answer The value of ΔM is consistent with the molar mass of the repeating unit shown in **1**, which is 200 g mol^{-1}. The molar mass of the terminal group is calculated by recalling that Na^+ is the cation in the matrix:

$$M(\text{terminal group}) = 4113\,\text{g mol}^{-1} - 20(200\,\text{g mol}^{-1}) - 23\,\text{g mol}^{-1} = 90\,\text{g mol}^{-1}$$

The result is consistent with the molar mass of the $—O(CH_2)_4OH$ terminal group (89 g mol^{-1}) plus the molar mass of the —H terminal group (1 g mol^{-1}).

Self-test 17D.2 What would be the molar mass of the $n=20$ polymer if silver trifluoroacetate were used instead of NaCl in the preparation of the matrix?

Answer: 4198 g mol^{-1}

(b) Laser light scattering

The scattering of light by particles with diameters much smaller than the wavelength of the incident radiation is called **Rayleigh scattering**. The intensity of the scattered light is proportional to the molar mass of the particle and to λ^{-4}, so shorter-wavelength radiation is scattered more intensely than longer wavelengths. For example, the blue of the sky arises from the more intense scattering of the blue component of white sunlight by the molecules of the atmosphere.

Consider the experimental arrangement shown in Fig. 17D.3 for the measurement of light scattering from solutions of macromolecules. Typically, the sample is irradiated with monochromatic light from a laser. The intensity of scattered light is then measured as a function of the angle θ that the direction of the laser beam makes with the direction of the detector from the sample at a distance r. Under these conditions, the intensity, $I(\theta)$, of light scattered is written as the **Rayleigh ratio**:

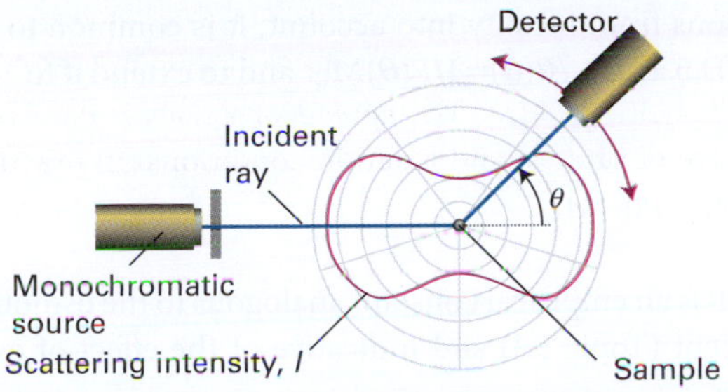

Figure 17D.3 Rayleigh scattering from a sample of point-like particles. The intensity of scattered light depends on the angle θ between the incident and scattered beams.

$$R(\theta) = \frac{I(\theta)}{I_0} \times r^2 \quad \textit{Definition} \quad \text{Rayleigh ratio} \quad (17D.4)$$

where I_0 is the intensity of the incident laser radiation. The factor r^2 occurs in the definition of the Rayleigh ratio because the light wave spreads out over a sphere of radius r and surface area $4\pi r^2$, so any sample of the radiation has its intensity $I(\theta)$ decreased by a factor proportional to r^2.

A detailed examination of the scattering shows that the Rayleigh ratio depends on the mass concentration, c_P (units: kg m^{-3}), of the macromolecule and its weight-average molar mass $\bar{M}_w$ as:

$$R(\theta) = KP(\theta)c_P\bar{M}_w \quad \text{Relation of Rayleigh ratio to molar mass} \quad (17D.5)$$

where the parameter K depends on the refractive index of the solution, the incident wavelength, and the distance between the detector and the sample, which is held constant during the experiment. The parameter $P(\theta)$ is the **structure factor**, which is related to the size of the molecule. When the radius of gyration, R_g, of the molecule (Topic 17A and Table 17D.1) is much smaller than the wavelength of the light,

$$P(\theta) \approx 1 - p(\theta) \quad \text{with } p(\theta) = \frac{16\pi^2 R_g^2 \sin^2 \frac{1}{2}\theta}{3\lambda^2} \quad \textit{Small macromolecules} \quad \text{Structure factor} \quad (17D.6)$$

Equation 17D.5 applies only to ideal solutions. In practice, even relatively dilute solutions of macromolecules can deviate considerably from ideality. Being so large, macromolecules displace a large quantity of solvent instead of replacing individual solvent molecules with negligible disturbance. To take

Table 17D.1* Radius of gyration

	M/(kg mol^{-1})	R_g/nm
Serum albumin	66	2.9
Polystyrene	3.2×10^3	50†
DNA	4×10^3	117

* More values are given in the *Resource section*.
† In a poor solvent.

deviations from ideality into account, it is common to rewrite eqn 17D.5 as $Kc_P/R(\theta)=1/P(\theta)\bar{M}_w$ and to extend it to

$$\frac{Kc_P}{R(\theta)}=\frac{1}{P(\theta)\bar{M}_w}+Bc_P \qquad (17D.7)$$

where B is an empirical constant analogous to the osmotic virial coefficient (Topic 5B) and indicative of the effect of excluded volume.

The preceding discussion shows that structural properties, such as size and the molar mass of a macromolecule, can be obtained from measurements of light scattering by a sample at several angles θ relative to the direction of propagation on an incident beam. In modern instruments, lasers are used as the radiation sources.

Example 17D.3 Determining the size of a polymer by light scattering

The following data for a sample of polystyrene in butanone were obtained at 20 °C with plane-polarized light at $\lambda=546$ nm.

$\theta/°$	26.0	36.9	66.4	90.0	113.6
$R(\theta)/m^2$	19.7	18.8	17.1	16.0	14.4

In separate experiments, it was determined that $K=6.42\times10^{-5}$ mol m^5 kg^{-2}. From this information, calculate R_g and $\bar{M}_w$ for the sample. Assume that B is negligibly small, and that the polymer is small enough that eqn 17D.6 holds.

Method Substituting the result of eqn 17D.6 into eqn 17D.5 gives, after some rearrangement

$$\frac{1}{R(\theta)}=\frac{1}{Kc_P\bar{M}_w}+\left(\frac{16\pi^2R_g^2}{3\lambda^2}\right)\frac{1}{R(\theta)}\sin^2\tfrac{1}{2}\theta$$

Hence, a plot of $1/R(\theta)$ against $\{1/R(\theta)\}\sin^2\frac{1}{2}\theta$ should be a straight line with slope $16\pi^2R_g^2/3\lambda^2$ and y-intercept $1/Kc_p\bar{M}_w$.

Answer We construct a table of values of $1/R(\theta)$ and $(\sin^2\frac{1}{2}\theta)/R(\theta)$ and plot the data (Fig. 17D.4).

$\theta/°$	26.0	36.9	66.4	90.0	113.6
$\{10^2/R(\theta)\}/m^{-2}$	5.06	5.32	5.83	6.25	6.96
$\{10^3\times(\sin^2\frac{1}{2}\theta)/R(\theta)\}/m^{-2}$	2.56	5.33	17.5	31.3	48.7

The best straight line through the data has a slope of 0.391 and a y-intercept of 5.06×10^{-2}. From these values and the value of K, we calculate $R_g=4.71\times10^{-8}$ m $=47.1$ nm and $\bar{M}_w=987$ kg mol^{-1}.

Self-test 17D.3 The following data for an aqueous solution of a protein with $c_P=2.0$ kg m^{-3} were obtained at 20 °C with laser light at $\lambda=532$ nm:

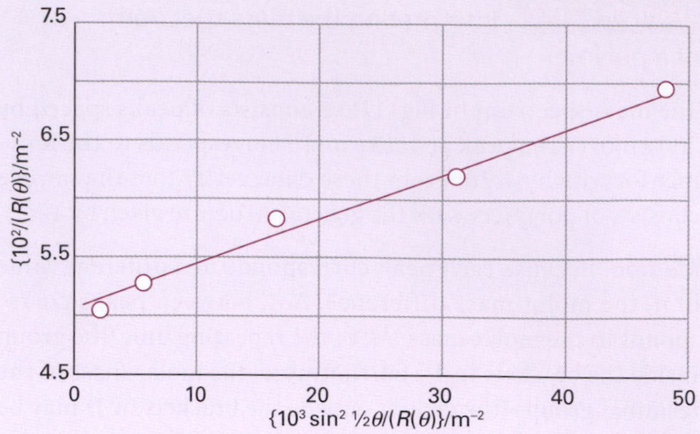

Figure 17D.4 Plot of the data for *Example* 17D.3.

$\theta/°$	15.0	45.0	70.0	85.0	90.0
$R(\theta)/m^2$	23.8	22.9	21.6	20.7	20.4

In a separate experiment, it was determined that $K=2.40\times10^2$ mol m^5 kg^{-2}. From this information, calculate the radius of gyration and the molar mass of the protein. Assume the protein is small enough that eqn 17D.6 holds.

Answer: $R_g=39.8$ nm; $\bar{M}_w=498$ kg mol^{-1}

(c) Sedimentation

In a gravitational field, heavy particles settle towards the foot of a column of solution by the process called **sedimentation**. The rate of sedimentation depends on the strength of the field and on the masses and shapes of the particles. Spherical molecules (and compact molecules in general) sediment faster than rod-like and extended molecules. When the sample is at equilibrium, the particles are dispersed over a range of heights in accord with the Boltzmann distribution (because the gravitational field competes with the stirring effect of thermal motion). The spread of heights depends on the masses of the molecules, so the equilibrium distribution is another way to determine molar mass.

Sedimentation is normally very slow, but it can be accelerated by **ultracentrifugation**, a technique that replaces the gravitational field with a centrifugal field. The effect can be achieved in an ultracentrifuge, which is essentially a cylinder that can be rotated at high speed about its axis with a sample in a cell near its outer edge. Modern ultracentrifuges can produce accelerations equivalent to about 10^5 that of gravity ('10^5 g'). Initially the sample is uniform, but the 'top' (innermost) boundary of the solute moves outwards as sedimentation proceeds.

A solute particle of mass m has an effective mass $m_{eff}=bm$ on account of the buoyancy of the medium, with

$$b=1-\rho v_s \qquad (17D.8)$$

where ρ is the solution density, v_s is the partial specific volume of the solute ($v_s = (\partial V/\partial m_B)_T$, with m_B the total mass of solute), and the dimensionless quantity ρv_s takes into account the mass of solvent displaced by the solute. The solute particles at a distance r from the axis of a rotor spinning at an angular velocity ω experience a centrifugal force of magnitude $m_{eff}r\omega^2$. The acceleration outwards is countered by a frictional force proportional to the speed, $s = dr/dt$, of the particles through the medium. This force is written fs, where f is the **frictional coefficient**. The particles therefore adopt a **drift speed**, a constant speed through the medium, which is found by equating the two forces $m_{eff}r\omega^2$ and fs. The forces are equal when

$$s = \frac{m_{eff}r\omega^2}{f} = \frac{bmr\omega^2}{f} \quad (17D.9)$$

The drift speed depends on the angular velocity and the radius, and it is convenient to define the **sedimentation constant**, S, as

$$S = \frac{s}{r\omega^2} \quad \text{Definition} \quad \text{Sedimentation constant} \quad (17D.10)$$

Then, because the average molecular mass is related to the average molar mass $\bar{M}_n$ through $m = \bar{M}_w/N_A$

$$S = \frac{b\bar{M}_n}{fN_A} \quad (17D.11)$$

For a spherical particle of radius a in a solvent of viscosity η, the frictional coefficient f is given by **Stokes' relation**:

$$f = 6\pi a\eta \quad \text{Stokes' relation} \quad (17D.12)$$

On substituting this expression into eqn 17D.12, we obtain

$$S = \frac{b\bar{M}_n}{6\pi a\eta N_A} \quad \text{Spherical polymer} \quad \text{Relation between } S \text{ and the molar mass} \quad (17D.13)$$

and S may be used to determine either $\bar{M}_n$ or a. Again, if the molecules are not spherical, we use the appropriate value of f given in Table 17D.2. As always when dealing with macromolecules, the measurements must be carried out at a series of concentrations and then extrapolated to zero concentration to avoid the complications that arise from the interference between bulky molecules.

Table 17D.2* Frictional coefficients and molecular geometry†

a/b	Prolate	Oblate
2	1.04	1.04
3	1.18	1.17
6	1.31	1.28
8	1.43	1.37
10	1.54	1.46

* More values and analytical expressions are given in the *Resource section*.

† Entries are the ratio f/f_0, where $f_0 = 6\pi\eta c$, where $c = (ab^2)^{1/3}$ for prolate ellipsoids and $c = (a^2b)^{1/3}$ for oblate ellipsoids; $2a$ is the major axis and $2b$ is the minor axis.

Example 17D.4 Determining a sedimentation constant

The sedimentation of the protein bovine serum albumin (BSA) was monitored at 25 °C. The initial location of the solute surface was at 5.50 cm from the axis of rotation, and during centrifugation at 56 850 r.p.m. it receded as follows:

t/s	0	500	1000	2000	3000	4000	5000
r/cm	5.50	5.55	5.60	5.70	5.80	5.91	6.01

Calculate the sedimentation coefficient.

Method Equation 17D.10 can be interpreted as a differential equation for $s = dr/dt$ in terms of r; so integrate it to obtain a formula for r in terms of t. The integrated expression, an expression for r as a function of t, will suggest how to plot the data and obtain from it the sedimentation constant.

Answer Equation 17D.10 may be written

$$\frac{dr}{dt} = r\omega^2 S \quad \text{and hence} \quad \frac{dr}{r} = \omega^2 S dt$$

If at $t = 0$ the surface is at r_0 and at a later time t it is at r, this equation integrates to

$$\ln\frac{r}{r_0} = \omega^2 St$$

It follows that a plot of $\ln(r/r_0)$ against t should be a straight line of slope $\omega^2 S$. Use $\omega = 2\pi\nu$, where ν is in cycles per second, and draw up the following table:

t/s	0	500	1000	2000	3000	4000	5000
$10^2 \ln(r/r_0)$	0	0.905	1.80	3.57	5.31	7.19	8.87

The straight-line graph (Fig. 17D.5) has a slope of 1.78×10^{-5}; so $\omega^2 S = 1.78 \times 10^{-5}\ s^{-1}$. Because $\omega = 2\pi \times (56\,850/60)\ s^{-1} =$

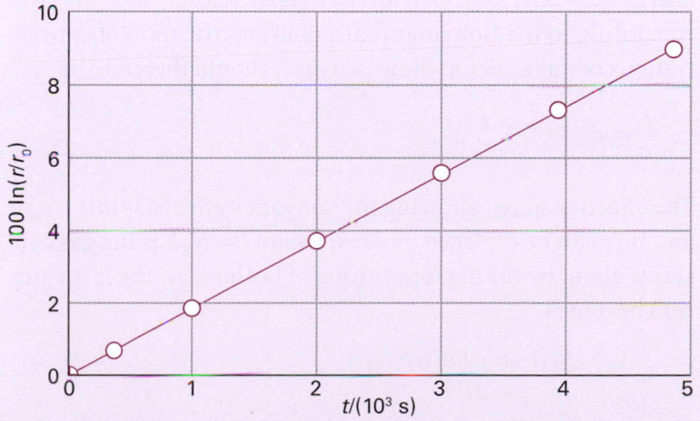

Figure 17D.5 A plot of the data for *Example* 17D.4.

$5.95\times10^3\,\text{s}^{-1}$, it follows that $S=5.02\times10^{-13}\,\text{s}$. The unit $10^{-13}\,\text{s}$ is sometimes called a 'svedberg' and denoted Sv; in this case $S=5.02\,\text{Sv}$.

Self-test 17D.4 Calculate the sedimentation constant given the following data (the other conditions being the same as above):

t/s	0	500	1000	2000	3000	4000	5000
r/cm	5.65	5.68	5.71	5.77	5.84	5.9	5.97

Answer: 3.11 Sv

The difficulty with using sedimentation rates to measure molar masses lies in the inaccuracies inherent in the determination of diffusion coefficients of disperse systems. This problem can be avoided by allowing the system to reach equilibrium, for the transport property D is then no longer relevant. As we show in the following *Justification*, the weight-average molar mass can be obtained from the ratio of concentrations of the macromolecules at two different radii in a centrifuge operating at angular frequency ω:

$$\bar{M}_w=\frac{2RT}{(r_2^2-r_1^2)b\omega^2}\ln\frac{c_2}{c_1} \tag{17D.14}$$

An alternative treatment of the data leads to the Z-average molar mass. The centrifuge is run more slowly in this technique than in the sedimentation rate method to avoid having all the solute pressed in a thin film against the bottom of the cell. At these slower speeds, several days may be needed for equilibrium to be reached.

Justification 17D.1 The weight-average molar mass from sedimentation experiments

The centrifugal force acting on a molecule at a radius r when it is rotating around the axis of the centrifuge at a frequency ω is $m\omega^2 r$. This force corresponds to a difference in potential energy (using $F=-dV/dr$) of $-\frac{1}{2}m\omega^2r^2$. The difference in potential energy between r_1 and r_2 (with $r_2>r_1$) is therefore $\frac{1}{2}m\omega^2(r_1^2-r_2^2)$. According to the Boltzmann distribution, the ratio of concentrations of molecules at these two radii should therefore be

$$\frac{c_2}{c_1}=e^{-\frac{1}{2}m_{\text{eff}}\omega^2(r_1^2-r_2^2)/kT}$$

The effective mass, allowing for buoyancy effects, is $m(1-v_s\rho)$, and m/k can be replaced by M/R, where $R=N_Ak$ is the gas constant. Then, by taking logarithms of both sides, the last equation becomes

$$\ln\frac{c_2}{c_1}=\frac{M(1-v_s\rho)\omega^2(r_2^2-r_1^2)}{2RT}$$

which rearranges into eqn 17D.14.

(d) Viscosity

The formal definition of viscosity is given in Topic 19A; for now, we need to know that highly viscous liquids flow slowly and retard the motion of objects through them. The presence of a macromolecular solute increases the viscosity of a solution. The effect is large even at low concentration, because big molecules affect the fluid flow over an extensive region surrounding them. At low concentrations the viscosity, η, of the solution is related to the viscosity of the pure solvent, η_0, by

$$\eta=\eta_0(1+[\eta]c+[\eta]'c^2+\cdots) \tag{17D.15}$$

The **intrinsic viscosity**, $[\eta]$, is the analogue of a virial coefficient like that encountered in the description of real gases (and has dimensions of 1/concentration). It follows from eqn 17D.15 that

$$[\eta]=\lim_{c\to0}\left(\frac{\eta-\eta_0}{c\eta_0}\right)=\lim_{c\to0}\left(\frac{\eta/\eta_0-1}{c}\right) \quad \textit{Definition} \quad \text{Intrinsic viscosity} \tag{17D.16}$$

Viscosities are measured in several ways. In the **Ostwald viscometer** shown in Fig. 17D.6, the time taken for a solution to flow through the capillary is noted, and compared with a standard sample. The method is well suited to the determination of $[\eta]$ because the ratio of the viscosities of the solution and the pure solvent is proportional to the drainage time t and t_0 after correcting for different densities ρ and ρ_0:

$$\frac{\eta}{\eta_0}=\frac{t}{t_0}\times\frac{\rho}{\rho_0} \tag{17D.17}$$

This ratio can be used directly in eqn 17D.16. Viscometers in the form of rotating concentric cylinders are also used (Fig. 17D.7), and the torque on the inner cylinder is monitored while the outer one is rotated. Such **rotating rheometers** (as in this case, some instruments for the measurement of viscosity are also called *rheometers*, from the Greek word for

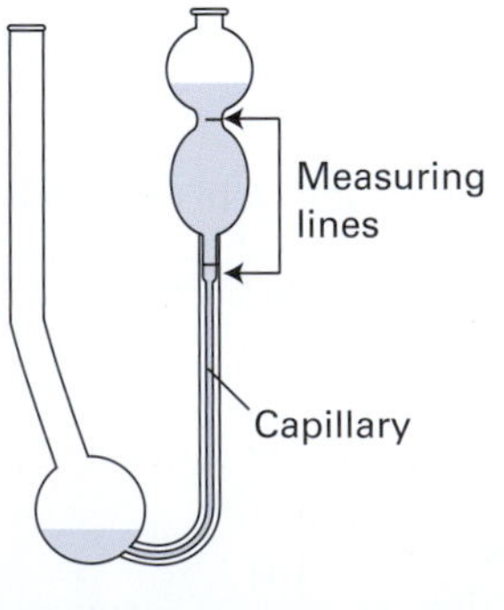

Figure 17D.6 An Ostwald viscometer. The viscosity is measured by noting the time required for the liquid to drain between the two marks.

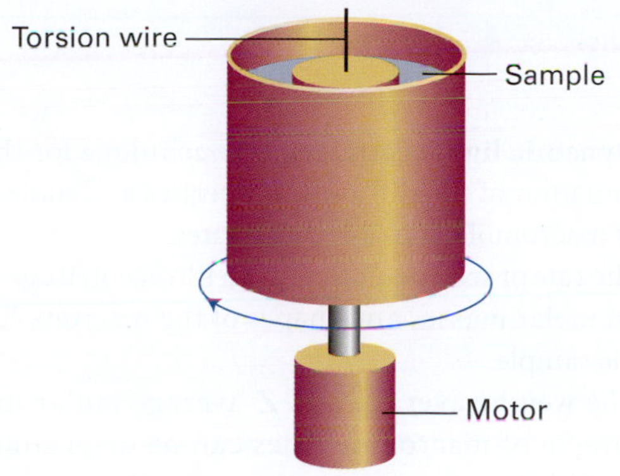

Figure 17D.7 A rotating rheometer. The torque on the inner drum is observed when the outer container is rotated.

'flow') have the advantage over the Ostwald viscometer that the shear gradient between the cylinders is simpler than in the capillary and effects of the kind discussed shortly can be studied more easily.

There are many complications in the interpretation of viscosity measurements. Much of the work is based on empirical observations, and the determination of molar mass is usually based on comparisons with standard, nearly monodisperse samples. Some regularities are observed that help in the determination. For example, it is found that solutions of macromolecules that behave nearly ideally often fit the empirical **Mark–Kuhn–Houwink–Sakurada equation**:

$$[\eta]=K\bar{M}_v^a \qquad \textit{Mark–Kuhn–Houwink–Sakurada equation} \quad (17D.18)$$

where K and a are constants that depend on the solvent and type of macromolecule (Table 17D.3); the viscosity-average molar mass, $\bar{M}_v$ appears in this expression.

Table 17D.3* Intrinsic viscosity

	Solvent	θ/°C	K/(cm^3 g^{-1})	a
Polystyrene	Benzene	25	9.5×10^{-3}	0.74
Polyisobutylene	Benzene	23	8.3×10^{-2}	0.50
Various proteins	Guanidine hydrochloride + $HSCH_2CH_2OH$		7.2×10^{-3}	0.66

* More values are given in the *Resource section*.

Example 17D.5 Using intrinsic viscosity to measure molar mass

The viscosities of a series of solutions of polystyrene in toluene were measured at 25 °C with the following results:

c/(g dm^{-3})	0	2	4	6	8	10
η/(10^{-4} kg m^{-1} s^{-1})	5.58	6.15	6.74	7.35	7.98	8.64

Calculate the intrinsic viscosity and estimate the molar mass of the polymer by using eqn 17D.19 with $K=3.80\times10^5$ dm^3 g^{-1} and $a=0.63$.

Method The intrinsic viscosity is defined in eqn 17D.16; therefore, form this ratio at the series of data points and extrapolate to $c=0$. Interpret $\bar{M}_v$, as $\bar{M}_v$/(g mol^{-1}) in eqn 17D.18.

Answer We draw up the following table:

c/(g dm^{-3})	0	2	4	6	8	10
η/η_0	1	1.102	1.208	1.317	1.43	1.549
$100[(\eta/\eta_0)-1]/(c/$g dm$^{-3})$		5.11	5.2	5.28	5.38	5.49

The points are plotted in Fig. 17D.8. The extrapolated intercept at $c=0$ is 0.0504, so $[\eta]=0.0504$ dm^3 g^{-1}. Therefore,

$$\bar{M}_v=\left(\frac{[\eta]}{K}\right)^{1/a}=9.0\times10^4\ \text{g mol}^{-1}$$

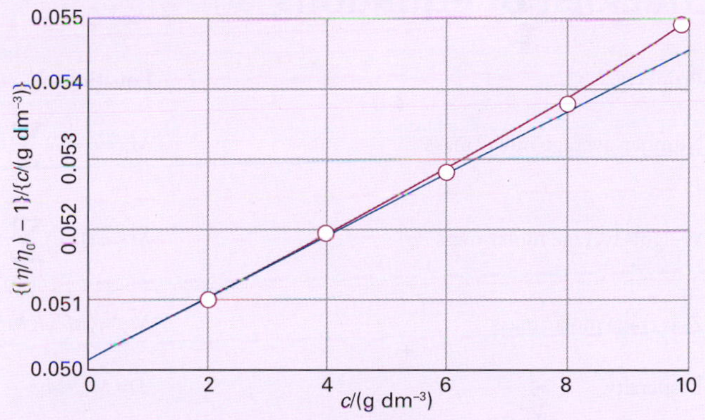

Figure 17D.8 The plot used for the determination of intrinsic viscosity, which is taken from the intercept at $c=0$; see *Example* 17D.5.

Self-test 17D.5 Show that the intrinsic viscosity may also be obtained as $[\eta]=\lim_{c\to0}\ln(\eta/\eta_0)$ and evaluate the viscosity-average molar mass by using this relation.

Answer: 90 kg mol^{-1}

In some cases, the flow is non-Newtonian in the sense that the viscosity of the solution changes as the rate of flow increases. A decrease in viscosity with increasing rate of flow indicates the presence of long rod-like molecules that are orientated by the flow and hence slide past each other more freely. In some somewhat rare cases the stresses set up by the flow are so great that long molecules are broken up, with further consequences for the viscosity.

Checklist of concepts

- ☐ 1. Macromolecules can be **monodisperse**, with a single molar mass, or **polydisperse**, with various molar masses.
- ☐ 2. In the **MALDI-TOF technique**, matrix-assisted laser desorption/ionization is coupled with a time-of-flight mass spectrometer to measure the molar masses of macromolecules.
- ☐ 3. The intensity of **Rayleigh light scattering** by a sample increases with decreasing wavelength of the incident radiation and increasing size of the particles in the sample.
- ☐ 4. Analysis of Rayleigh scattering leads to the determination of the molar mass of a macromolecule or aggregate.
- ☐ 5. **Dynamic light scattering** is a technique for the determination of the diffusion properties and molar masses of macromolecules and aggregates.
- ☐ 6. The rate of sedimentation in an ultracentrifuge depends on molar masses and shapes of the macromolecules in the sample.
- ☐ 7. The **weight-average** and ***Z*-average molar mass** of a sample of macromolecules can be determined from equilibrium measurements of sedimentation in an ultracentrifuge.
- ☐ 8. The **viscosity-average molar mass** can be determined from measurements of the viscosity of solutions of macromolecules.

Checklist of equations

Property	Equation	Comment	Equation number
Number-average molar mass	$\bar{M}_n = (1/N)\sum_i N_i M_i$	Definition	17D.1
Weight-average molar mass	$\bar{M}_w = (1/m)\sum_i m_i M_i$	Definition	17D.2a
Z-average molar mass	$\bar{M}_Z = \langle M^3\rangle/\langle M^2\rangle$	Interpretation	17D.2c
Dispersity	$Đ = \bar{M}_w/\bar{M}_n$	Definition	17D.3a
	$Đ = \langle M^2\rangle/\langle M\rangle^2$	Interpretation	17D.3b
Rayleigh ratio	$R(\theta) = (I(\theta)/I_0)r^2$	Definition	17D.4
	$R(\theta) = KP(\theta)c_P\bar{M}_w$	Experimental implementation; ideal solutions	17D.5
Structure factor	$P(\theta) \approx 1 - p(\theta)$ $p(\theta) = (16\pi^2 R_g^2/3\lambda^2)\sin^2\frac{1}{2}\theta$	Small macromolecules	17D.6
Sedimentation constant	$S = s/r\omega^2$	Definition	17D.10
Stokes relation	$f = 6\pi a\eta$	*f* is the frictional coefficient	17D.12
Relation between the sedimentation constant and the molar mass of a polymer	$S = b\bar{M}_n/6\pi a\eta N_A$	Spherical polymer	17D.13
Intrinsic viscosity	$[\eta] = \lim_{c\to 0}\{(\eta/\eta_0 - 1)/c\}$	Definition	17D.16
Mark–Kuhn–Houwink–Sakurada equation	$[\eta] = K\bar{M}_v^a$	Nearly ideal solutions	17D.18

CHAPTER 17 Macromolecules and self-assembly

TOPIC 17A The structures of macromolecules

Discussion questions

17A.1 Distinguish between the four levels of structure of a macromolecule: primary, secondary, tertiary, and quaternary.

17A.2 What are the consequences of there being partial rigidity in an otherwise random coil?

17A.3 Define the terms in, and identify the limits of the generality of, the following expressions: (a) $R_c = Nl$, (b) $R_{rms} = N^{1/2}l$, (c) $R_{rms} = (2N)^{1/2}l$, (d) $R_{rms} = N^{1/2}lF$, (e) $R_g = N^{1/2}l$, (f) $R_g = (N/6)^{1/2}l$, (g) $R_g = (N/3)^{1/2}l$.

17A.4 Summarize the Core–Pauling rules and explain how they explain the helical and sheet-like structures of polypeptides.

Exercises

17A.1(a) A one-dimensional polymer chain consists of 700 segments, each 0.90 nm long. If the chain were ideally flexible, what would be the r.m.s. separation of the ends of the chain?
17A.1(b) A one-dimensional polymer chain consists of 1200 segments, each 1.125 nm long. If the chain were ideally flexible, what would be the r.m.s. separation of the ends of the chain?

17A.2(a) Calculate the contour length (the length of the extended chain) and the root mean square separation (the end-to-end distance) for polyethylene with a molar mass of 280 kg mol^{-1}, modelled as a one-dimensional chain.
17A.2(b) Calculate the contour length (the length of the extended chain) and the root mean square separation (the end-to-end distance) for polypropylene of molar mass 174 kg mol^{-1}, modelled as a one-dimensional chain.

17A.3(a) The radius of gyration of a long one-dimensional chain molecule is found to be 7.3 nm. The chain consists of C—C links. Assume that the chain is randomly coiled and estimate the number of links in the chain.
17A.3(b) The radius of gyration of a long one-dimensional chain molecule is found to be 18.9 nm. The chain consists of links of length 450 pm. Assume that the chain is randomly coiled and estimate the number of links in the chain.

17A.4(a) What is the probability that the ends of a polyethene chain of molar mass 65 kg mol^{-1} are 10 nm apart when the polymer is treated as a one-dimensional freely jointed chain?
17A.4(b) What is the probability that the ends of a polyethene chain of molar mass 85 kg mol^{-1} are 15 nm apart when the polymer is treated as a one-dimensional freely jointed chain?

17A.5(a) What is the probability that the ends of a polyethene chain of molar mass 65 kg mol^{-1} are between 10.0 nm and 10.1 nm apart when the polymer is treated as a three-dimensional freely jointed chain?
17A.5(b) What is the probability that the ends of a polyethene chain of molar mass 75 kg mol^{-1} are between 14.0 nm and 14.1 nm apart when the polymer is treated as a three-dimensional freely jointed chain?

17A.6(a) By what percentage does the radius of gyration of a one-dimensional polymer chain increase (+) or decrease (−) when the bond angle between units is limited to 109°? What is the percentage change in volume of the coil?
17A.6(b) By what percentage does the root mean square separation of the ends of a one-dimensional polymer chain increase (+) or decrease (−) when the bond angle between units is limited to 120°? What is the percentage change in volume of the coil?

17A.7(a) By what percentage does the radius of gyration of a one-dimensional polymer chain increase (+) or decrease (−) when the persistence length changes from l (the bond length) to 5.0 per cent of the contour length? What is the percentage change in volume of the coil?
17A.7(b) By what percentage does the root mean square separation of the ends of a one-dimensional polymer chain increase (+) or decrease (−) when the persistence length changes from l (the bond length) to 2.5 per cent of the contour length? What is the percentage change in volume of the coil?

17A.8(a) The radius of gyration of a three-dimensional partially rigid polymer of 1000 units each of length 150 pm was measured as 2.1 nm. What is the persistence length of the polymer?
17A.8(b) The radius of gyration of a three-dimensional partially rigid polymer of 1500 units each of length 164 pm was measured as 3.0 nm. What is the persistence length of the polymer?

Problems

17A.1 Evaluate the radius of gyration, R_g, of (a) a solid sphere of radius a, (b) a long straight rod of radius a and length l. Show that in the case of a solid sphere of specific volume v_s, $R_g/\text{nm} \approx 0.056902 \times \{(v_s/\text{cm}^3\,\text{g}^{-1})(M/\text{g mol}^{-1})\}^{1/3}$. Evaluate R_g for a species with $M = 100$ kg mol^{-1}, $v_s = 0.750$ cm^3 g^{-1}, and, in the case of the rod, of radius 0.50 nm.

17A.2 Use eqn 17A.2 to deduce expressions for (a) the root mean square separation of the ends of the chain, (b) the mean separation of the ends, and (c) their most probable separation. Evaluate these three quantities for a fully flexible chain with $N = 4000$ and $l = 154$ pm.

17A.3 Deduce the relation $\langle r_i^2 \rangle = Nl^2$ for the mean square distance of a monomer from the origin in a freely jointed chain of N units each of length l. *Hint*: Use the distribution in eqn 17A.2.

17A.4 Deduce an expression for the radius of gyration of a three-dimensional freely-jointed chain (eqn 17A.6).

17A.5 Derive expressions for the moments of inertia and hence the radii of gyration of (a) a uniform thin disk, (b) a long uniform rod, (c) a uniform sphere.

17A.6 Construct a two-dimensional random walk by using a random number generating routine with mathematical software or electronic spreadsheet. Construct a walk of 50 and 100 steps. If there are many people working on the problem, investigate the mean and most probable separations in the plots by direct measurement. Do they vary as $N^{1/2}$?

17A.7 Confirm that for one-dimensional random coils, $\ln P \approx \ln(2/\pi N)^{1/2} - \frac{1}{2}(N+n+1)\ln(1+\nu) - \frac{1}{2}(N-n+1)\ln(1-\nu)$. *Hint*: See Justification 17A.1.

17A.8 The radius of gyration is defined in *Justification* 17A.3. Show that an equivalent definition is that R_g is the average root mean square distance of the atoms or groups (all assumed to be of the same mass), that is, that $R_g^2 = (1/N)\sum_j R_j^2$, where R_j is the distance of atom j from the centre of mass.

17A.9 Use the following information and the expression for R_g of a solid sphere quoted in the text (following eqn 17A.6), to classify the given species as globular or rod-like.

	$M/(\text{g mol}^{-1})$	$v_s/(\text{cm}^3\,\text{g}^{-1})$	R_g/nm
Serum albumin	66×10^3	0.752	2.98
Bushy stunt virus	10.6×10^6	0.741	12.0
DNA	4×10^6	0.556	117.0

TOPIC 17B Properties of macromolecules

Discussion questions

17B.1 Distinguish between stress and strain.

17B.2 Distinguish between elastic and plastic deformation.

17B.3 Distinguish between the melting temperature and the glass transition temperature of a polymer.

17B.4 Describe the mechanism of electrical conductivity in conducting polymers.

Exercises

17B.1(a) Calculate the change in molar entropy when the ends of a one-dimensional polyethene chain of molar mass 65 kg mol^{-1} are moved apart by 1.0 nm.

17B.1(b) Calculate the change in molar entropy when the ends of a one-dimensional polyethene chain of molar mass 85 kg mol^{-1} are moved apart by 2.0 nm.

17B.2(a) Calculate the restoring force when the ends of a one-dimensional polyethene chain of molar mass 65 kg mol^{-1} are moved apart by 1.0 nm at 20 °C.

17B.2(b) Calculate the restoring force when the ends of a one-dimensional polyethene chain of molar mass 85 kg mol^{-1} are moved apart by 2.0 nm at 25 °C.

Problems

17B.1 Develop an expression for the fundamental vibrational frequency of a one-dimensional random coil that has been slightly stretched and then released. Evaluate this frequency for a sample of polyethene of molar mass 65 kg mol^{-1} at 20 °C. Account physically for the dependence of frequency on temperature and molar mass.

17B.2 On the assumption that the tension, t, required to keep a sample at a constant length is proportional to the temperature ($t = aT$, the analogue of $p \propto T$), show that the tension can be ascribed to the dependence of the entropy on the length of the sample. Account for this result in terms of the molecular nature of the sample.

17B.3 The following table lists the glass transition temperatures, T_g, of several polymers. Discuss the reasons why the structure of the monomer unit has an effect on the value of T_g.

Polymer	Poly (oxymethylene)	Polyethene	Poly(vinyl chloride)	Polystyrene
Structure	$-(\text{OCH}_2)_n-$	$-(\text{CH}_2\text{CH}_2)_n-$	$-(\text{CH}_2-\text{CHCl})_n-$	$-(\text{CH}_2-\text{CH}(\text{C}_6\text{H}_5))_n-$
T_g/K	198	253	354	381

TOPIC 17C Self-assembly

Discussion questions

17C.1 Distinguish between a sol, an emulsion, and a foam. Provide examples of each.

17C.2 It is observed that the critical micelle concentration of sodium dodecyl sulfate in aqueous solution decreases as the concentration of added sodium chloride increases. Explain this effect.

17C.3 What effect is the inclusion of cholesterol likely to have on the transition temperatures of a lipid bilayer?

Exercises

17C.1(a) The velocity v with which a protein moves through water under the influence of an electric field varied with values of pH in the range $3.0 < pH < 7.0$ according to the expression $v/(\mu m\,s^{-1}) = a + b(pH) + c(pH)^2 + d(pH)^3$ with $a = 0.50$, $b = -0.10$, $c = -3.0 \times 10^{-3}$, and $d = 5.0 \times 10^{-4}$. Identify the isoelectric point of the protein.

17C.1(b) The velocity v with which a protein moves through water under the influence of an electric field varied with values of pH in the range $3.0 < pH < 5.0$ according to the expression $v/(\mu m\,s^{-1}) = a + b(pH) + c(pH)^2$ with $a = 0.80$, $b = -4.0 \times 10^{-3}$, and $c = -5.0 \times 10^{-2}$. Identify the isoelectric point of the protein.

Problem

17C.1 Use mathematical software to reproduce the features in Fig. 17C.6.

TOPIC 17D Determination of size and shape

Discussion questions

17D.1 Distinguish between number-average, weight-average, and Z-average molar masses. Identify experimental techniques that can measure each of these properties.

17D.2 Suggest reasons why different techniques produce different molar mass averages.

Exercises

17D.1(a) Calculate the number-average molar mass and the weight-average molar mass of a mixture of equal amounts of two polymers, one having $M = 62\,kg\,mol^{-1}$ and the other $M = 78\,kg\,mol^{-1}$.

17D.1(b) Calculate the number-average molar mass and the weight-average molar mass of a mixture of two polymers, one having $M = 62\,kg\,mol^{-1}$ and the other $M = 78\,kg\,mol^{-1}$, with their amounts (numbers of moles) in the ratio 3:2.

17D.2(a) A solution consists of solvent, 30 per cent by mass, of a dimer with $M = 30\,kg\,mol^{-1}$ and its monomer. What average molar mass would be obtained from measurement of (i) osmotic pressure, (ii) light scattering?

17D.2(b) A solution consists of 25 per cent by mass of a trimer with $M = 22\,kg\,mol^{-1}$ and its monomer. What average molar mass would be obtained from measurement of: (i) osmotic pressure, (ii) light scattering?

17D.3(a) What is the relative rate of sedimentation for two spherical particles of the same density, but which differ in radius by a factor of 10?

17D.3(b) What is the relative rate of sedimentation for two spherical particles with densities $1.10\,g\,cm^{-3}$ and $1.18\,g\,cm^{-3}$ and which differ in radius by a factor of 8.4, the former being the larger? Use $\rho = 0.794\,g\,cm^{-3}$ for the density of the solution.

17D.4(a) Human haemoglobin has a specific volume of $0.749 \times 10^{3}\,m^3\,kg^{-1}$, a sedimentation constant of 4.48 Sv, and a diffusion coefficient of $6.9 \times 10^{-11}\,m^2\,s^{-1}$. Determine its molar mass from this information.

17D.4(b) A synthetic polymer has a specific volume of $8.01 \times 10^{-4}\,m^3\,kg^{-1}$, a sedimentation constant of 7.46 Sv, and a diffusion coefficient of $7.72 \times 10^{-11}\,m^2\,s^{-1}$. Determine its molar mass from this information.

17D.5(a) Find the drift speed of a particle of radius 20 μm and density $1750\,kg\,m^3$ which is settling from suspension in water (density $1000\,kg\,m^{-3}$) under the influence of gravity alone. The viscosity of water is $8.9 \times 10^{-4}\,kg\,m^{-1}\,s^{-1}$.

17D.5(b) Find the drift speed of a particle of radius 15.5 μm and density $1250\,kg\,m^{-3}$ which is settling from suspension in water (density $1000\,kg\,m^{-3}$) under the influence of gravity alone. The viscosity of water is $8.9 \times 10^{-4}\,kg\,m^{-1}\,s^{-1}$.

17D.6(a) At 20 °C the diffusion coefficient of a macromolecule is found to be $8.3 \times 10^{-11}\,m^2\,s^{-1}$. Its sedimentation constant is 3.2 Sv in a solution of density $1.06\,g\,cm^{-3}$. The specific volume of the macromolecule is $0.656\,cm^3\,g^{-1}$. Determine the molar mass of the macromolecule.

17D.6(b) At 20 °C the diffusion coefficient of a macromolecule is found to be $7.9 \times 10^{-11}\,m^2\,s^{-1}$. Its sedimentation constant is 5.1 Sv in a solution of density $997\,kg\,m^{-1}$. The specific volume of the macromolecule is $0.721\,cm^3\,g^{-1}$. Determine the molar mass of the macromolecule.

17D.7(a) The data from a sedimentation equilibrium experiment performed at 300 K on a macromolecular solute in aqueous solution show that a graph of $\ln c$ against $(r/cm)^2$ is a straight line with a slope of 729. The rotation rate of the centrifuge was 50 000 r.p.m. The specific volume of the solute is $0.61\,cm^3\,g^{-1}$. Calculate the molar mass of the solute.

17D.7(b) The data from a sedimentation equilibrium experiment performed at 293 K on a macromolecular solute in aqueous solution show that a graph of $\ln c$ against $(r/cm)^2$ is a straight line with a slope of 821. The rotation rate of the centrifuge was 1080 cycles per second. The specific volume of the solute is $7.2 \times 10^{-4}\,m^3\,kg^{-1}$. Calculate the molar mass of the solute.

Problems

17D.1 A polymerization process produced a Gaussian distribution of polymers in the sense that the proportion of molecules having a molar mass in the range M to $M + dM$ was proportional to $e^{-(M-\bar{M})^2/2\gamma}$. What is the number average molar mass when the distribution is narrow?

17D.2 Polystyrene is a synthetic polymer with the structure $-(CH_2{-}CH(C_6H_5))_n-$. A batch of polydisperse polystyrene was prepared by initiating the polymerization with t-butyl radicals. As a result, the t-butyl group is expected to be covalently attached to the end of the final products.

A sample from this batch was embedded in an organic matrix containing silver trifluoroacetate and the resulting MALDI-TOF spectrum consisted of a large number of peaks separated by 104 g mol^{-1}, with the most intense peak at 25 578 g mol^{-1}. Comment on the purity of this sample and determine the number of (CH_2—$CH(C_6H_5)$) units in the species that gives rise to the most intense peak in the spectrum.

17D.3 Suppose that a rod-like DNA molecule of length 250 nm undergoes a conformational change to a closed-circular (cc) form. (a) Use the information in Problem 17A.8 and an incident wavelength $\lambda = 488$ nm to calculate the ratio of scattering intensities by each of these conformations, I_{rod}/I_{cc}, when $\theta = 20°$, 45°, and 90°. (b) Suppose that you wish to use light scattering as a technique for the study of conformational changes in DNA molecules. Based on your answer to part (a), at which angle would you conduct the experiments? Justify your choice.

17D.4 In a sedimentation experiment the position of the boundary as a function of time was found to be as follows:

t/min	15.5	29.1	36.4	58.2
r/cm	5.05	5.09	5.12	5.19

The rotation rate of the centrifuge was 45 000 r.p.m. Calculate the sedimentation constant of the solute.

17D.5 Calculate the speed of operation (in r.p.m.) of an ultracentrifuge needed to obtain a readily measurable concentration gradient in a sedimentation equilibrium experiment. Take that gradient to be a concentration at the bottom of the cell about five times greater than at the top. Use $r_{top} = 5.0$ cm, $r_{bot} = 7.0$ cm, $M \approx 10^5$ g mol^{-1}, $\rho v_s \approx 0.75$, $T = 298$ K.

17D.6 In an ultracentrifugation experiment at 20 °C on bovine serum albumin the following data were obtained: $\rho = 1.001$ g cm^{-3}, $v_s = 1.112$ cm^3 g^{-1}, $\omega/2\pi = 322$ Hz,

r/cm	5.0	5.1	5.2	5.3	5.4
c/(mg cm^{-3})	0.536	0.284	0.148	0.077	0.039

Evaluate the molar mass of the sample.

17D.7 Sedimentation studies on haemoglobin in water gave a sedimentation constant $S = 4.5$ Sv at 20 °C. The diffusion coefficient is 6.3×10^{-11} m^2 s^{-1} at the same temperature. Calculate the molar mass of haemoglobin using $v_s =$ 0.75 cm^3 g^{-1} for its partial specific volume and $\rho = 0.998$ g cm^{-3} for the density of the solution. Estimate the effective radius of the haemoglobin molecule given that the viscosity of the solution is 1.00×10^3 kg m^{-1} s^{-1}.

17D.8 The rate of sedimentation of a recently isolated protein was monitored at 20 °C and with a rotor speed of 50 000 r.p.m. The boundary receded as follows:

t/s	0	300	600	900	1200	1500	1800
r/cm	6.127	6.153	6.179	6.206	6.232	6.258	6.284

Calculate the sedimentation constant and the molar mass of the protein on the basis that its partial specific volume is 0.728 cm^3 g^{-1} and its diffusion coefficient is 7.62×10^{-11} m^2 s^{-1} at 20 °C, the density of the solution then being 0.9981 g cm^{-3}. Suggest a shape for the protein given that the viscosity of the solution is 1.00×10^3 kg m^{-1} s^{-1} at 20 °C.

17D.9 The concentration dependence of the viscosity of a polymer solution is found to be as follows:

c/(g dm^{-3})	1.32	2.89	5.73	9.17
η/(g m^{-1} s^{-1})	1.08	1.20	1.42	1.73

The viscosity of the solvent is 0.985 g m^{-1} s^{-1}. What is the intrinsic viscosity of the polymer?

17D.10 The times of flow of dilute solutions of polystyrene in benzene through a viscometer at 25 °C are given in the table below. From these data, calculate the molar mass of the polystyrene samples. Because the solutions are dilute, assume that the densities of the solutions are the same as those of pure benzene. η(benzene) = 0.601×10^3 kg m^{-1} s^{-1} (0.601 cP) at 25 °C.

c/(g dm^{-3})	0	2.22	5.00	8.00	10.00
t/s	208.2	248.1	303.4	371.8	421.3

17D.11 The viscosities of solutions of polyisobutylene in benzene were measured at 23 °C with the following results:

c/(g/10^2 cm^3)	0	0.2	0.4	0.6	0.8	1.0
η/(10^3 kg m^{-1} s^{-1})	0.647	0.690	0.733	0.777	0.821	0.865

Use the information in Table 17D.3 to deduce the molar mass of the polymer.

Integrated activities

17.1 In formamide as solvent, poly(γ-benzyl-L-glutamate) is found by light scattering experiments to have a radius of gyration proportional to M; in contrast, polystyrene in butanone has R_g proportional to $M^{1/2}$. Present arguments to show that the first polymer is a rigid rod whereas the second is a random coil.

17.2 Consider the thermodynamic description of stretching rubber. The observables are the tension, t, and length, l (the analogues of p and V for gases). Because $dw = tdl$, the basic equation is $dU = TdS + tdl$. If $G = U - TS - tl$, find expressions for dG and dA, and deduce the Maxwell relations

$$\left(\frac{\partial S}{\partial l}\right)_T = -\left(\frac{\partial t}{\partial T}\right)_l \quad \left(\frac{\partial S}{\partial t}\right)_T = -\left(\frac{\partial l}{\partial T}\right)_t$$

Go on to deduce the equation of state for rubber,

$$\left(\frac{\partial U}{\partial l}\right)_T = t - \left(\frac{\partial t}{\partial T}\right)_l$$

17.3 Commercial software (more specifically 'molecular mechanics' or 'conformational search' software) automate the calculations that lead to Ramachandran plots, such as those in Fig. 17A.13. In this problem our model for the protein is the dipeptide (**1**) in which the terminal methyl groups replace the rest of the polypeptide chain. (a) Draw three initial conformers of the dipeptide with R=H: one with $\phi = +75°$, $\psi = -65°$, a second with $\phi = \psi = +180°$, and a third with $\phi = +65°$, $\psi = +35°$. Use software of your instructor's choice to optimize the geometry of each conformer and find the final ϕ and ψ angles in each case. Did all the initial conformers converge to the same final conformation? If not, what do these final conformers represent? (b) Use the approach in part (a) to investigate the case R=CH_3, with the same three initial conformers as starting points for the calculations. Rationalize any similarities and differences between the final conformers of the dipeptides with R=H and R=CH_3.

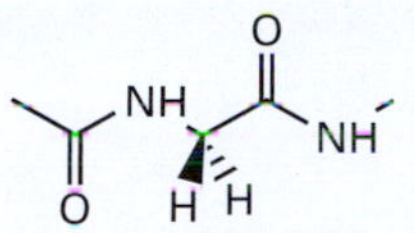

1

17.4 The effective radius, a, of a random coil is related to its radius of gyration, R_g, by $a=\gamma R_g$, with $\gamma=0.85$. Deduce an expression for the osmotic virial coefficient, B (Topic 5B), in terms of the number of chain units for (a) a freely jointed chain, (b) a chain with tetrahedral bond angles. Evaluate B for $l=154$ pm and $N=4000$. Estimate B for a randomly coiled polyethylene chain of arbitrary molar mass, M, and evaluate it for $M=56\,\text{kg mol}^{-1}$. Use $B=\frac{1}{2}N_A v_P$, where v_P is the excluded volume due to a single molecule.

17.5 A manufacturer of polystyrene beads claims that they have an average molar mass of 250 kg mol^{-1}. Solutions of these beads are studied by a physical chemistry student by dilute solution viscometry with an Ostwald viscometer in both toluene and cyclohexane. The drainage times, t_D, as a function of concentration for the two solvents are given in the table below. (a) Fit the data to the virial equation for viscosity,

$$\eta=\eta_0(1+[\eta]c+k'[\eta]^2c^2+\cdots)$$

where k' is called the *Huggins constant* and is typically in the range 0.35–0.40. From the fit, determine the intrinsic viscosity and the Huggins constant. (b) Use the empirical Mark–Kuhn–Houwink–Sakurada equation (eqn 17D.18) to determine the molar mass of polystyrene in the two solvents. For theta solvents, $a=0.5$ and $K=8.2\times10^{-5}\,\text{dm}^{-3}\,\text{g}^{-1}$ for cyclohexane; for the good solvent toluene $a=0.72$ and $K=1.15\times10^{-5}\,\text{dm}^{-3}\,\text{g}^{-1}$. (c) According to a general theory proposed by Kirkwood and Riseman, the root mean square end-to-end distance of a polymer chain in solution is related to $[\eta]$ by $\Phi\langle r^2\rangle^{3/2}/M$, where Φ is a universal constant with the value 2.84×10^{26} when $[\eta]$ is expressed in cubic decimetres per gram and the distance is in metres. Calculate this quantity for each solvent. (d) From the molar masses calculate the average number of styrene ($C_6H_5CH{=}CH_2$) monomer units, $\langle N\rangle$, (e) Calculate the length of a fully stretched, planar zigzag configuration, taking the C—C distance as 154 pm and the CCC bond angle to be 109°. (f) Use eqn 17A.6 to calculate the radius of gyration, R_g. Also calculate $\langle r^2\rangle^{1/2}=N^{1/2}l$. Compare this result with that predicted by the Kirkwood–Riseman theory: which gives the better fit? (g) Compare your values for M to the results of Problem 17D.2. Is there any reason why they should or should not agree? Is the manufacturer's claim valid?

$c/(\text{g dm}^{-3}$ toluene)	0	1.0	3.0	5.0
t_D/s	8.37	9.11	10.72	12.52
$c/(\text{g dm}^{-3}$ cyclohexane)	0	1.0	1.5	2.0
t_D/s	8.32	8.67	8.85	9.03

CHAPTER 18

Solids

The solid state includes most of the materials that make modern technology possible. It includes the wide varieties of steel that are used in architecture and engineering, the semiconductors and metallic conductors that are used in information technology and power distribution, the ceramics that increasingly are replacing metals, and the synthetic and natural polymers discussed in Chapter 17 that are used in the textile industry and in the fabrication of many of the common objects of the modern world. In this chapter we explore the structures and physical properties of solids.

18A Crystal structure

In this Topic we see how to describe the regular arrangement of atoms in crystals and the symmetry of their arrangement. Then we consider the basic principles of 'X-ray diffraction' and see how the diffraction pattern can be interpreted in terms of the distribution of electron density in a 'unit cell'.

18B Bonding in solids

X-ray diffraction leads to information about the structures of metallic, ionic, and molecular solids, and in this Topic we review some typical results and their rationalization in terms of atomic and ionic radii.

18C Mechanical, electrical and magnetic properties of solids

In this Topic we begin to see how the bulk properties of solids stem from the arrangement and properties of the constituent atoms. Here we focus on rigidity, electrical conductivity, and magnetic properties.

18D The optical properties of solids

This Topic continues the exploration of properties of solids, with a focus on optical properties that render materials as useful building blocks of devices with important technological applications.

What is the impact of this material?

The deployment of X-ray diffraction techniques for the determination of the location of all the atoms in biological macromolecules has revolutionized the study of biochemistry. In *Impact* I18.1, the power of the techniques is demonstrated by exploring the most seminal X-ray images of all: the characteristic pattern obtained from strands of DNA and used in the construction of the double-helix model of DNA. We also turn our attention to research on nanometre-sized materials, which is motivated by the possibility that they will form the basis for cheaper and smaller electronic devices. In *Impact* I18.2 we discuss the synthesis of 'nanowires', nanometre-sized atomic assemblies that conduct electricity, which is a major step in the fabrication of nanodevices.

To read more about the impact of this material, scan the QR code, or go to bcs.whfreeman.com/webpub/chemistry/pchem10e/impact/pchem-18-1.html

18A Crystal structure

Contents

➤ Why do you need to know this material?

You need to understand the structures of metallic, ionic, and molecular solids if you want to be able to account for the mechanical, electrical, optical, and magnetic properties that form the basis of new materials and new technologies. A central part of this understanding is knowing how the internal structures of solids are determined and described.

➤ What is the key idea?

The details of the regular arrangement of atoms in periodic crystals can be expressed in terms of unit cells and determined by diffraction techniques.

➤ What do you need to know already?

You need to be familiar with the wave description of electromagnetic radiation (*Foundations* C), and the significance of Fourier transforms (*Mathematical background* 7). Light use is made of the de Broglie relation (Topic 7A) and the equipartition theorem (*Foundations* B).

A crucial aspect of the link between the structure and properties of a solid is the pattern in which the atoms (and molecules) are stacked together, so here we examine how the structures of solids are described and determined. First, we see how to describe the regular arrangement of atoms in solids. Then we consider the basic principles of X-ray diffraction and see how the diffraction pattern can be interpreted in terms of the distribution of electron density in a crystal.

18A.1 Periodic crystal lattices

A **periodic crystal** is built up from regularly repeating 'structural motifs', which may be atoms, molecules, or groups of atoms, molecules, or ions. A **space lattice** is the pattern formed by points representing the locations of these motifs (Fig. 18A.1). A space lattice is, in effect, an abstract scaffolding for the crystal

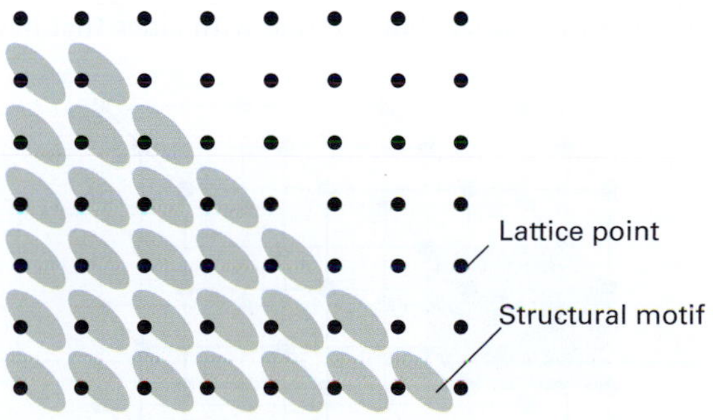

Figure 18A.1 Each lattice point specifies the location of a structural motif (for example, a molecule or a group of molecules). The crystal lattice is the array of lattice points; the crystal structure is the collection of structural motifs arranged according to the lattice.

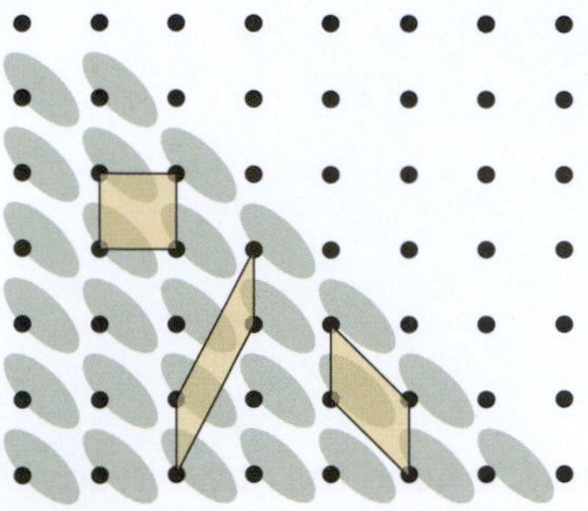

Figure 18A.2 A unit cell is a parallel-sided (but not necessarily rectangular) figure from which the entire periodic crystal structure can be constructed by using only translations (not reflections, rotations, or inversions).

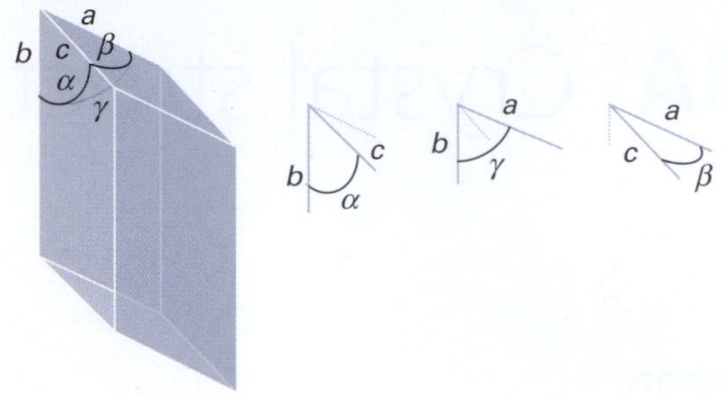

Figure 18A.4 The notation for the sides and angles of a unit cell. Note that the angle α lies in the plane (b,c) and perpendicular to the axis a.

structure. More formally, a space lattice is a three-dimensional, infinite array of points, each of which is surrounded in an identical way by its neighbours, and which defines the basic structure of the crystal. In some cases there may be a structural motif centred on each lattice point, but that is not necessary. The crystal structure itself is obtained by associating with each lattice point an identical structural motif. The solids known as **quasicrystals** are 'aperiodic', in the sense that the space lattice, though still filling space, does not have translational symmetry. Our discussion will focus on periodic crystals only and, to simplify the language, we refer to these structures simply as 'crystals'.

A **unit cell** is an imaginary parallelepiped (parallel-sided figure) that contains one unit of the translationally repeating pattern (Fig. 18A.2). It can be thought of as the fundamental region from which the entire crystal may be constructed by purely translational displacements (like bricks in a wall). A unit cell is commonly formed by joining neighbouring lattice points by straight lines (Fig. 18A.3). Such unit cells are called **primitive.** It is sometimes more convenient to draw larger **non-primitive unit cells** that also have lattice points at their centres or on pairs of opposite faces. An infinite number of different unit cells can describe the same lattice, but the one with sides that have the shortest lengths and that are most nearly perpendicular to one another is normally chosen. The lengths of the sides of a unit cell are denoted a, b, and c, and the angles between them are denoted α, β, and γ (Fig. 18A.4).

Unit cells are classified into seven **crystal systems** by noting the rotational symmetry elements they possess. A **symmetry operation** is an action (such as a rotation, reflection, or inversion) that leaves an object looking the same after it has been carried out. There is a corresponding **symmetry element** for each symmetry operation, which is the point, line, or plane with respect to which the symmetry operation is performed. For instance, an ***n*-fold rotation** (the symmetry operation) about an ***n*-fold axis of symmetry** (the corresponding symmetry element) is a rotation through $360°/n$. (See Topics 11A–11C for a more detailed discussion of symmetry.)

The following are examples of unit cells:

- A **cubic unit cell** has four threefold axes in a tetrahedral array (Fig. 18A.5).
- A **monoclinic unit cell** has one twofold axis (Fig. 18A.6).
- A **triclinic unit cell** has no rotational symmetry, and typically all three sides and angles are different (Fig. 18A.7).

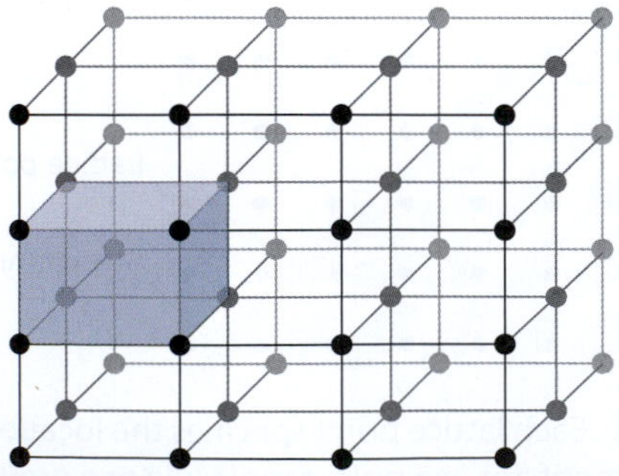

Figure 18A.3 A unit cell can be chosen in a variety of ways, as shown here. It is conventional to choose the cell that represents the full symmetry of the lattice. In this rectangular lattice, the rectangular unit cell would normally be adopted.

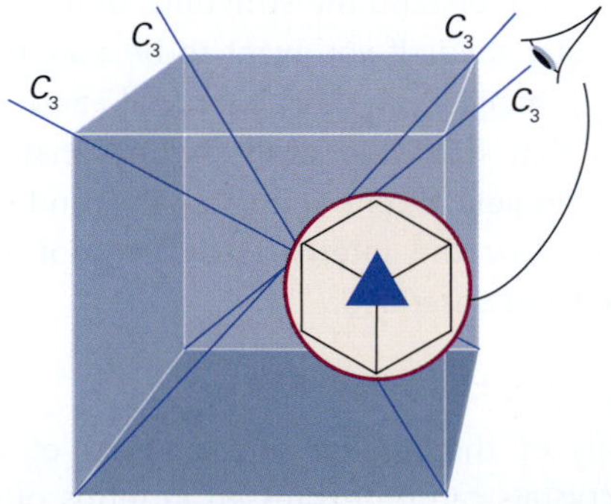

Figure 18A.5 A unit cell belonging to the cubic system has four threefold axes, denoted C_3, arranged tetrahedrally. The insert shows the threefold symmetry.

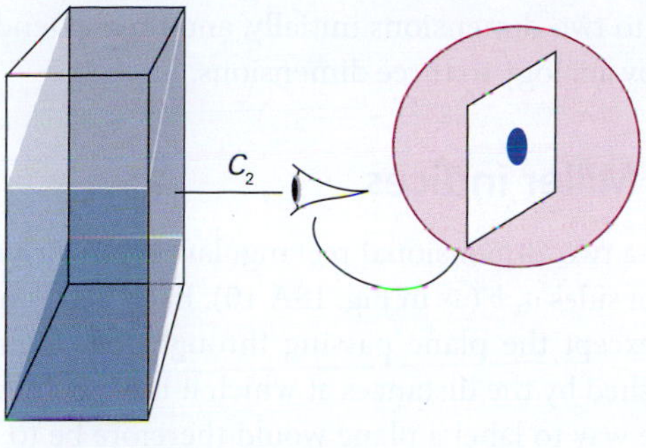

Figure 18A.6 A unit cell belonging to the monoclinic system has a twofold axis (denoted C_2 and shown in more detail in the insert).

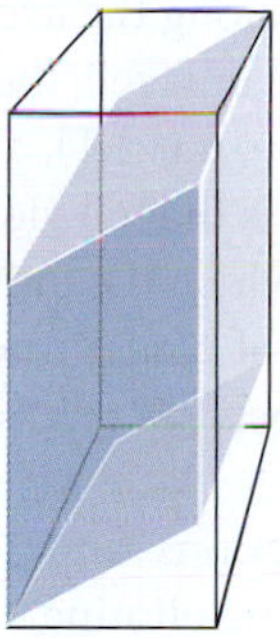

Figure 18A.7 A triclinic unit cell has no axes of rotational symmetry.

Table 18A.1 lists the **essential symmetries**, the elements that must be present for the unit cell to belong to a particular crystal system.

There are only 14 distinct space lattices in three dimensions. These **Bravais lattices** are illustrated in Fig. 18A.8. It is conventional to portray these lattices by primitive unit cells in some cases and by non-primitive unit cells in others. The following notation applies:

- A **primitive unit cell** (with lattice points only at the corners) is denoted P.

Table 18A.1 The seven crystal systems

System	Essential symmetries
Triclinic	None
Monoclinic	One C_2 axis
Orthorhombic	Three perpendicular C_2 axes
Rhombohedral	One C_3 axis
Tetragonal	One C_4 axis
Hexagonal	One C_6 axis
Cubic	Four C_3 axes in a tetrahedral arrangement

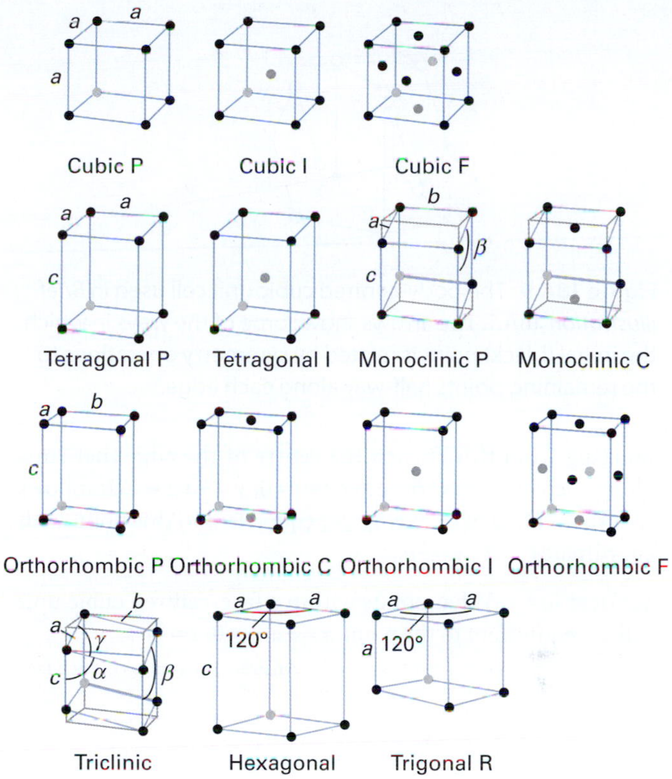

Figure 18A.8 The 14 Bravais lattices. The points are lattice points, and are not necessarily occupied by atoms. P denotes a primitive unit cell (R is used for a trigonal lattice), I a body-centred unit cell, F a face-centred unit cell, and C (or A or B) a cell with lattice points on two opposite faces. Trigonal lattices may belong to the rhombohedral or hexagonal systems (Table 18A.1).

- A **body-centred unit cell** (I) also has a lattice point at its centre.
- A **face-centred unit cell** (F) has lattice points at its corners and also at the centres of its six faces.
- A **side-centred unit cell** (A, B, or C) has lattice points at its corners and at the centres of two opposite faces.

For simple structures, it is often convenient to choose an atom belonging to the structural motif, or the centre of a molecule, as the location of a lattice point or the vertex of a unit cell, but that is not a necessary requirement. Equivalent lattice points within the unit cell of a Bravais lattice have identical surroundings.

Brief illustration 18A.1 Bravais lattices

Consider a body-centred cubic unit cell of sides a and one of its corners with coordinates $x=0$, $y=0$, $z=0$ (Fig. 18A.9).

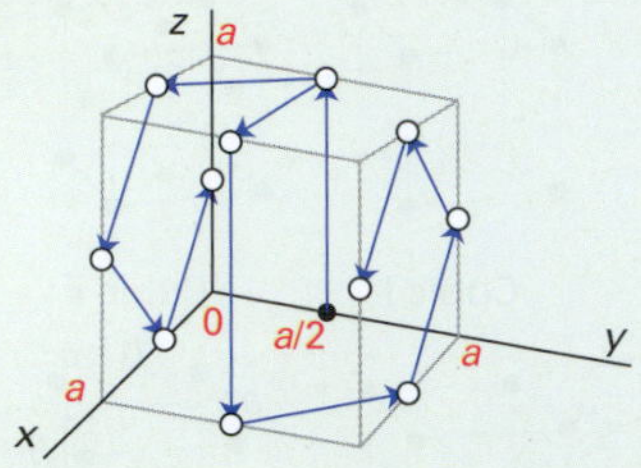

Figure 18A.9 The body-centred cubic unit cell used in *Brief illustration* 18A.1. The arrows show some of the ways in which the initial (black) point is related by symmetry operations to the remaining points half-way along each edge.

Starting from this corner, the centre of the edge that runs along the *y*-axis has coordinates $x=0$, $y=\frac{1}{2}a$, $z=0$. It follows that the centres of each edge are equivalent to this point with coordinates $x=0$, $y=\frac{1}{2}a$, $z=0$.

Self-test 18A.1 What points within a face-centred cubic unit cell are equivalent to the point $x=\frac{1}{2}a$, $y=0$, $z=\frac{1}{2}a$?

Answer: The centres of each face

18A.2 The identification of lattice planes

There are many different sets of lattice planes in a crystal (Fig. 18A.10), and we need to be able to identify them. Two-dimensional lattices are easier to visualize than three-dimensional lattices, so we shall introduce the concepts involved by referring to two dimensions initially, and then extend the conclusions by analogy to three dimensions.

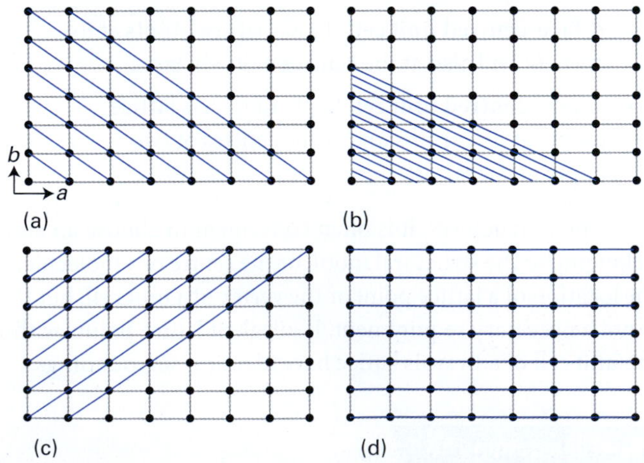

Figure 18A.10 Some of the planes that can be drawn through the points of a rectangular space lattice and their corresponding Miller indices (*hkl*): (a) (110), (b) (230), (c) ($\bar{1}$10), and (d) (010).

(a) The Miller indices

Consider a two-dimensional rectangular lattice formed from a unit cell of sides *a*, *b* (as in Fig. 18A.10). Each plane in the illustration (except the plane passing through the origin) can be distinguished by the distances at which it intersects the *a* and *b* axes. One way to label a plane would therefore be to quote the smallest intersection distances. For example, we could denote a representative plane of each type in Fig. 18A.10 as (a) $(1a,1b)$, (b) $(\frac{1}{2}a,\frac{1}{3}b)$, (c) $(-1a,1b)$, and (d) $(\infty a,1b)$, where ∞ is used to show that the planes intersect an axis at infinity. However, if we agree to quote distances along the axes as multiples of the corresponding lengths of the unit cell, then we can label the planes more simply as (1,1), $(\frac{1}{2},\frac{1}{3})$, (−1,1), and $(\infty,1)$, respectively. If the lattice in Fig. 18A.10 is the top view of a three-dimensional orthorhombic lattice in which the unit cell has a length *c* in the *z*-direction, all four sets of planes intersect the *z*-axis at infinity. Therefore, the full labels are $(1,1,\infty)$, $(\frac{1}{2},\frac{1}{3},\infty)$, $(-1,1,\infty)$, and $(\infty,1,\infty)$.

The presence of fractions and infinity in the labels is inconvenient. They can be eliminated by taking the reciprocals of the labels. As we shall see, taking reciprocals turns out to have further advantages. The **Miller indices**, (*hkl*), are the reciprocals of intersection distances. To simplify the notation while providing a great deal of information, the following rules apply:

- Negative indices are written with a bar over the number, as in ($\bar{1}$10).
- If taking the reciprocal results in a fraction, then the fraction can be cleared by multiplying through by an appropriate factor.

For example, a $(\frac{1}{3},\frac{1}{2},\infty)$ plane is denoted (230) after multiplication of all three indices by 6.

- The notation (*hkl*) denotes an *individual* plane. To specify a *set* of parallel planes we use the notation {*hkl*}.

Thus, we speak of the (110) plane in a lattice, and the set of all {110} planes that lie parallel to the (110) plane.

A helpful feature to remember is that the smaller the absolute value of *h* in {*hkl*}, the more nearly parallel the set of planes is to the *a* axis (the {*h*00} planes are an exception). The same is true of *k* and the *b* axis and *l* and the *c* axis. When $h=0$, the planes intersect the *a* axis at infinity, so the {0*kl*} planes are parallel to the *a* axis. Similarly, the {*h*0*l*} planes are parallel to the *b* axis and the {*hk*0} planes are parallel to the *c* axis.

Brief illustration 18A.2 Miller indices

The $\{1,1,\infty\}$ planes in Fig. 18A.10a are the $\{110\}$ planes in the Miller notation. Similarly, the $\{\frac{1}{3},\frac{1}{2},\infty\}$ planes are denoted $\{230\}$. Fig. 18A.10c shows the $\{\bar{1}10\}$ planes. The Miller indices for the four types of plane in Fig. 18A.10 are therefore $\{110\}$, $\{230\}$, $\{\bar{1}10\}$, and $\{010\}$. Figure 18A.11 shows a three-dimensional representation of a selection of planes, including one in a lattice with non-orthogonal axes.

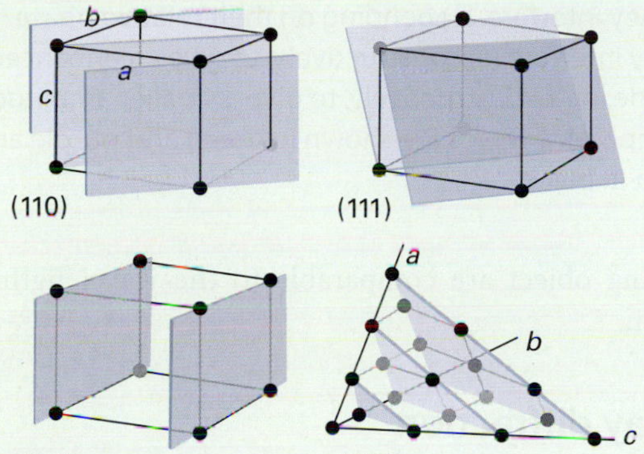

Figure 18A.11 Some representative planes in three dimensions and their Miller indices. Note that a 0 indicates that a plane is parallel to the corresponding axis, and that the indexing may also be used for unit cells with non-orthogonal axes.

Self-test 18A.2 Find the Miller indices of the planes that intersect the crystallographic axes at the distances $(3a, 2b, c)$ and $(2a, \infty b, \infty c)$

Answer: $\{236\}$ and $\{100\}$

(b) The separation of planes

The Miller indices are very useful for expressing the separation of planes. It is shown in the following *Justification* that the separation of the $\{hk0\}$ planes in the square lattice in Fig. 18A.12 is given by

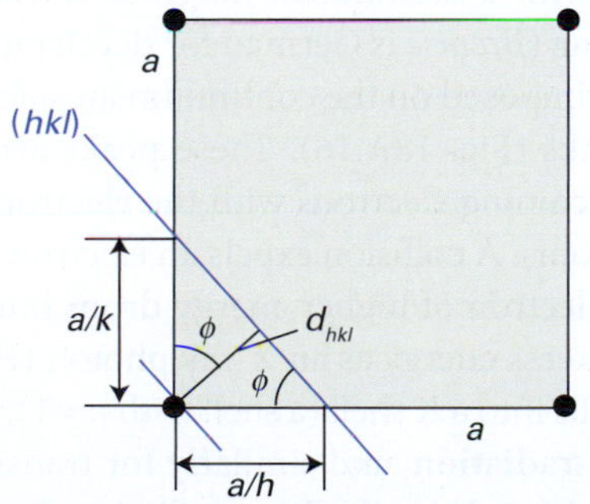

Figure 18A.12 The dimensions of a unit cell and their relation to the plane passing through the lattice points.

$$\frac{1}{d_{hk0}^2}=\frac{h^2+k^2}{a^2} \quad \text{or} \quad d_{hk0}=\frac{a}{(h^2+k^2)^{1/2}}$$ *Square lattice* Separation of planes (18A.1a)

Justification 18A.1 The separation of lattice planes

Consider the $\{hk0\}$ planes of a square lattice built from a unit cell with sides of length a (Fig. 18A.12). We can write the following trigonometric expressions for the angle ϕ shown in the illustration:

$$\sin\phi=\frac{d}{(a/h)}=\frac{hd_{hk0}}{a} \qquad \cos\phi=\frac{d}{(a/k)}=\frac{kd_{hk0}}{a}$$

Because the lattice planes intersect the horizontal axis h times and the vertical axis k times, the length of each hypotenuse is calculated by dividing a by h and a by k. Then, because $\sin^2\phi+\cos^2\phi=1$, it follows that

$$\left(\frac{hd_{hk0}}{a}\right)^2+\left(\frac{kd_{hk0}}{a}\right)^2=1$$

which we can rearrange by dividing both sides by d_{hk0}^2 into

$$\frac{1}{d_{hk0}^2}=\frac{h^2}{a^2}+\frac{k^2}{a^2}=\frac{h^2+k^2}{a^2}$$

This expression is eqn 18A.1a.

By extension to three dimensions, the separation of the $\{hkl\}$ planes of a cubic lattice is given by

$$\frac{1}{d_{hkl}^2}=\frac{h^2+k^2+l^2}{a^2} \quad \text{or} \quad d_{hkl}=\frac{a}{(h^2+k^2+l^2)^{1/2}}$$ *Cubic lattice* Separation of planes (18A.1b)

The corresponding expression for a general orthorhombic lattice (one in which the axes are mutually perpendicular) is the generalization of this expression:

$$\frac{1}{d_{hkl}^2}=\frac{h^2}{a^2}+\frac{k^2}{b^2}+\frac{l^2}{c^2}$$ *Orthorhombic lattice* Separation of planes (18A.1c)

Example 18A.1 Using the Miller indices

Calculate the separation of (a) the $\{123\}$ planes and (b) the $\{246\}$ planes of an orthorhombic unit cell with $a=0.82$ nm, $b=0.94$ nm, and $c=0.75$ nm.

Method For the first part, simply substitute the information into eqn 18A.1c. For the second part, instead of repeating the calculation, note that if all three Miller indices are multiplied by n, then their separation is reduced by that factor (Fig. 18A.13):

$$\frac{1}{d_{nh,nk,nl}^2}=\frac{(nh)^2}{a^2}+\frac{(nk)^2}{b^2}+\frac{(nl)^2}{c^2}=n^2\left(\frac{h^2}{a^2}+\frac{k^2}{b^2}+\frac{l^2}{c^2}\right)=\frac{n^2}{d_{hkl}^2}$$

which implies that

$$d_{nh,nk,nl}=\frac{d_{hkl}}{n}$$

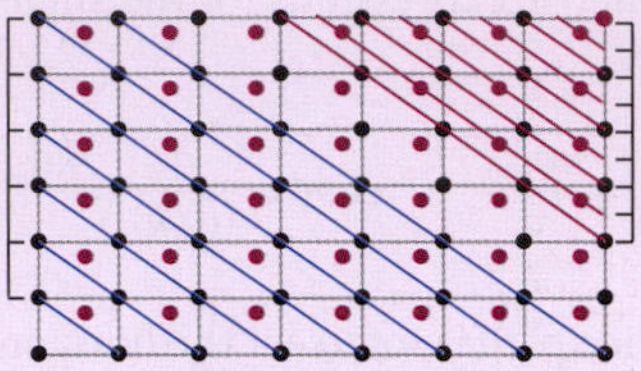

Figure 18A.13 The separation of the {220} planes is half that of the {110} planes. In general, the separation of the planes {*nh*,*nk*,*nl*} is *n* times smaller than the separation of the {*hkl*} planes.

Answer Substituting the indices into eqn 18A.1c gives

$$\frac{1}{d_{123}^2}=\frac{1^2}{(0.82\,\text{nm})^2}+\frac{2^2}{(0.94\,\text{nm})^2}+\frac{3^2}{(0.75\,\text{nm})^2}=22\,\text{nm}^{-2}$$

Hence, $d_{123}=0.21$ nm. It then follows immediately that d_{246} is one-half this value, or 0.11 nm.

A note on good practice It is always sensible to look for analytical relations between quantities rather than to evaluate expressions numerically each time, for that emphasizes the relations between quantities (and avoids unnecessary work).

Self-test 18A.3 Calculate the separation of (a) the {133} planes and (b) the {399} planes in the same lattice.

Answer: 0.19 nm, 0.063 nm

18A.3 X-ray crystallography

A characteristic property of waves is that when they are present in the same region of space they interfere with one another, giving a greater displacement where peaks or troughs coincide and a smaller displacement where peaks coincide with troughs (Fig. 18A.14 and *Foundations* C). According to classical electromagnetic theory, the intensity of electromagnetic radiation is proportional to the square of the amplitude of the waves. Therefore, the regions of constructive or destructive interference show up as regions of enhanced or diminished intensities. The phenomenon of **diffraction** is the interference caused by an object in the path of waves, and the pattern of varying intensity that results is called the **diffraction pattern**. Diffraction occurs when the dimensions of the diffracting object are comparable to the wavelength of the radiation.

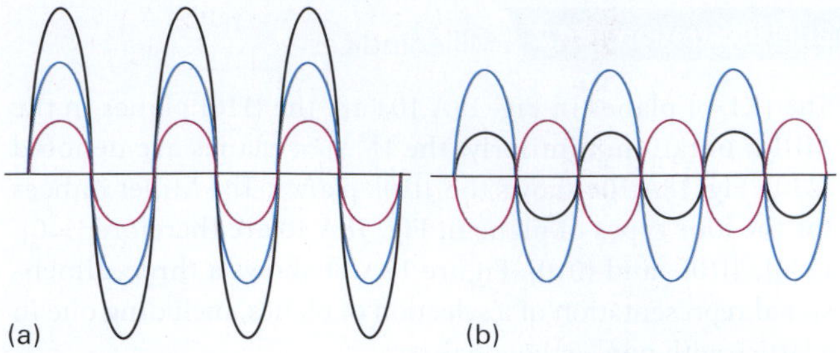

Figure 18A.14 When two waves are in the same region of space they interfere. Depending on their relative phase, they may interfere (a) constructively, to give an enhanced amplitude, or (b) destructively, to give a smaller amplitude. The component waves are shown in green and purple and the resultant in black.

(a) X-ray diffraction

Wilhelm Röntgen discovered X-rays in 1895. Seventeen years later, Max von Laue suggested that they might be diffracted when passed through a crystal, for by then he had realized that their wavelengths are comparable to the separation of lattice planes. This suggestion was confirmed almost immediately by Walter Friedrich and Paul Knipping and has grown since then into a technique of extraordinary power. The bulk of this section will deal with the determination of structures using X-ray diffraction. The mathematical procedures necessary for the determination of structure from X-ray diffraction data are enormously complex, but such is the degree of integration of computers into the experimental apparatus that the technique is almost fully automated, even for large molecules and complex solids. The analysis is aided by molecular modelling techniques, which can guide the investigation towards a plausible structure.

X-rays are electromagnetic radiation with wavelengths of the order of 10^{-10} m. They are typically generated by bombarding a metal with high-energy electrons (Fig. 18A.15). The electrons decelerate as they plunge into the metal and generate radiation with a continuous range of wavelengths called **Bremsstrahlung** (*Bremse* is German for deceleration, *Strahlung* for ray). Superimposed on the continuum are a few high-intensity, sharp peaks (Fig. 18A.16). These peaks arise from collisions of the incoming electrons with the electrons in the inner shells of the atoms. A collision expels an electron from an inner shell, and an electron of higher energy drops into the vacancy, emitting the excess energy as an X-ray photon (Fig. 18A.17). If the electron falls into a K shell (a shell with $n=1$), the X-rays are classified as **K-radiation**, and similarly for transitions into the L ($n=2$) and M ($n=3$) shells. Strong, distinct lines are labelled K_α, K_β, and so on. Increasingly, X-ray diffraction makes use of

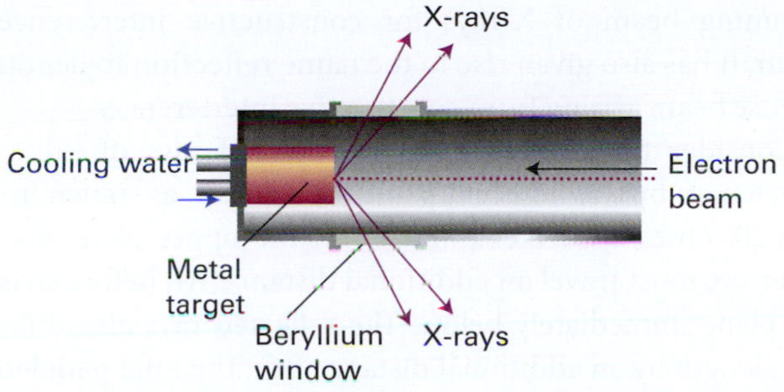

Figure 18A.15 X-rays are generated by directing an electron beam on to a cooled metal target. Beryllium is transparent to X-rays (on account of the small number of electrons in each atom) and is used for the windows.

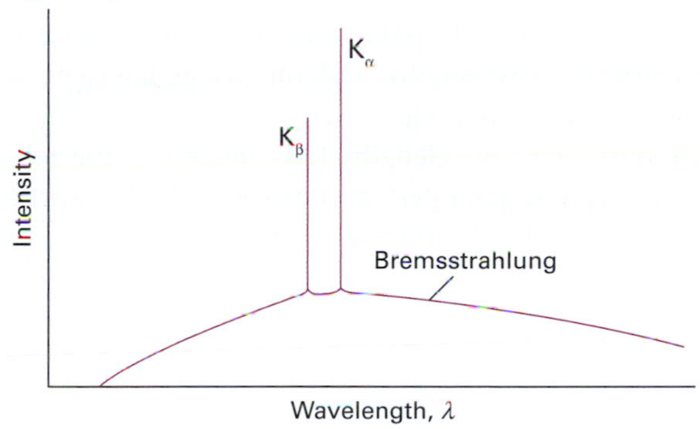

Figure 18A.16 The X-ray emission from a metal consists of a broad, featureless Bremsstrahlung background, with sharp transitions superimposed on it. The label K indicates that the radiation comes from a transition in which an electron falls into a vacancy in the K shell of the atom.

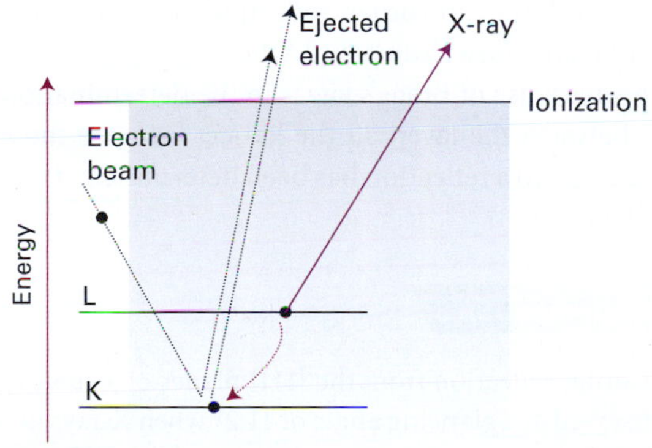

Figure 18A.17 The processes that contribute to the generation of X-rays. An incoming electron collides with an electron (in the K shell), and ejects it. Another electron (from the L shell in this illustration) falls into the vacancy and emits its excess energy as an X-ray photon.

the radiation available from synchrotron sources (Topic 12A), for its high intensity greatly enhances the sensitivity of the technique.

Von Laue's original method consisted of passing a broadband beam of X-rays into a single crystal, and recording the diffraction pattern photographically. The idea behind the approach was that a crystal might not be suitably orientated to act as a diffraction grating for a single wavelength but, whatever its orientation, diffraction would be achieved for at least one of the wavelengths if a range of wavelengths were used. There is currently a resurgence of interest in this approach because synchrotron radiation spans a range of X-ray wavelengths.

An alternative technique was developed by Peter Debye and Paul Scherrer and independently by Albert Hull. They used monochromatic radiation and a powdered sample. When the sample is a powder, at least some of the crystallites will be orientated so as to give rise to diffraction. In modern powder diffractometers the intensities of the reflections are monitored electronically as the detector is rotated around the sample in a plane containing the incident ray (Fig. 18A.18). Powder diffraction techniques are used to identify the composition of a sample of a solid substance by comparison of the positions of the diffraction lines and their intensities with diffraction patterns stored in a large data bank. Powder diffraction data are also used to help determine phase diagrams, for different crystalline phases result in different diffraction patterns, and to determine the relative amounts of each phase present in a mixture. The technique is also used for the initial determination of the dimensions and symmetries of unit cells.

The method developed by the Braggs (William and his son Lawrence, who later jointly won the Nobel Prize) is the foundation of almost all modern work in X-ray crystallography. They used a single crystal and a monochromatic beam of X-rays, and rotated the crystal until a reflection was detected. There are many different sets of planes in a crystal, so there are many angles at which a reflection occurs. The complete set of data consists of the list of angles at which reflections are observed and their intensities.

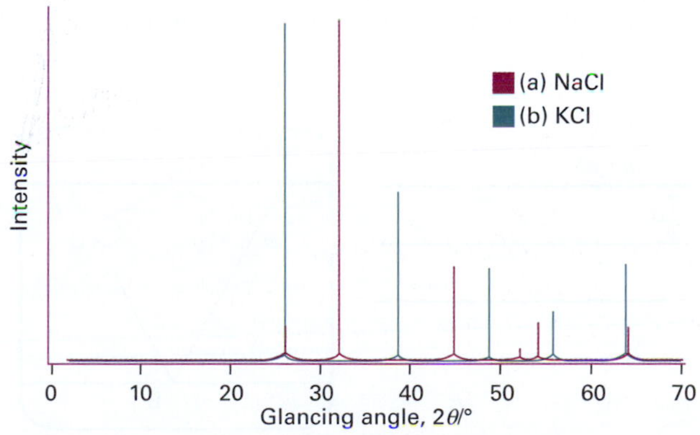

Figure 18A.18 X-ray powder diffraction patterns of (a) NaCl, (b) KCl. The smaller number of lines in (b) is a consequence of the similarity of the K^+ and Cl^- scattering factors, as discussed later in the Topic.

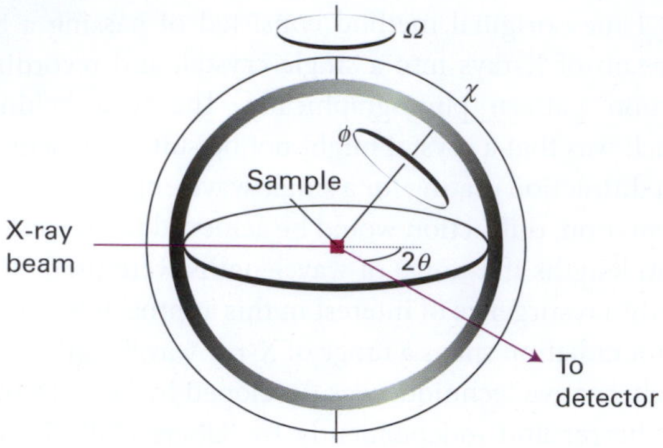

Figure 18A.19 A four-circle diffractometer. The settings of the orientations (ϕ, χ, θ, and Ω) of the components are controlled by computer; each (*hkl*) reflection is monitored in turn, and their intensities are recorded.

Single-crystal diffraction patterns are measured by using a **four-circle diffractometer** (Fig. 18A.19). An integrated computer identifies the angular settings of the diffractometer's four circles that are needed to observe any particular intensity peak in the diffraction pattern. The diffraction intensity is measured at each setting and background intensities are assessed by making measurements at slightly different settings. Computing techniques are now available that lead not only to automatic indexing but also to the automated determination of the shape, symmetry, and size of the unit cell. Moreover, several techniques are now available for sampling large amounts of data, including area detectors and image plates, which sample whole regions of diffraction patterns simultaneously.

(b) Bragg's law

An early approach to the analysis of diffraction patterns produced by crystals was to regard a lattice plane as a semi-transparent mirror and to model a crystal as a stack of reflecting lattice planes of separation d (Fig. 18A.20). The model makes it easy to calculate the angle the crystal must make to the incoming beam of X-rays for constructive interference to occur. It has also given rise to the name **reflection** to denote an intense beam arising from constructive interference.

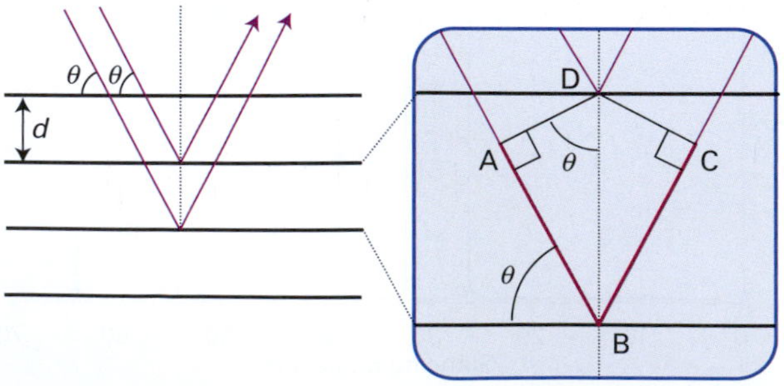

Figure 18A.20 The conventional derivation of Bragg's law treats each lattice plane as a plane reflecting the incident radiation. The path lengths differ by AB+BC, which depends on the angle θ. Constructive interference (a 'reflection') occurs when AB+BC is equal to an integer number of wavelengths.

Consider the reflection of two parallel rays of the same wavelength by two adjacent planes of a lattice, as shown in Fig. 18A.20. One ray strikes point D on the upper plane but the other ray must travel an additional distance AB before striking the plane immediately below. The reflected rays also differ in path length by an additional distance BC. The total path length difference of the two rays is then

$$\mathrm{AB}+\mathrm{BC}=2d\sin\theta$$

where 2θ is the **glancing angle** (2θ rather than θ, because the beam is deflected through 2θ from its initial direction). For many glancing angles the path-length difference is not an integer number of wavelengths, and the waves interfere largely destructively. However, when the path-length difference is an integer number of wavelengths ($\mathrm{AB}+\mathrm{BC}=n\lambda$), the reflected waves are in phase and interfere constructively. It follows that a reflection should be observed when the glancing angle satisfies **Bragg's law**:

$$n\lambda=2d\sin\theta \quad \text{Bragg's law} \quad (18\text{A}.2\text{a})$$

Reflections with $n=2, 3, \ldots$ are called *second order*, *third order*, and so on; they correspond to path-length differences of 2, 3, … wavelengths. In modern work it is normal to absorb the n into d, to write Bragg's law as

$$\lambda=2d\sin\theta \quad \textit{Alternative form} \quad \text{Bragg's law} \quad (18\text{A}.2\text{b})$$

and to regard the nth-order reflection as arising from the $\{nh,nk,nl\}$ planes (see *Example* 18A.1).

The primary use of Bragg's law is in the determination of the spacing between the layers in the lattice, for once the angle θ corresponding to a reflection has been determined, d may readily be calculated.

Brief illustration 18A.3 Bragg's law 1

A first-order reflection from the {111} planes of a cubic crystal was observed at a glancing angle of 11.2° when X-rays of wavelength 154 pm were used. According to eqn 18A.2b, the {111} planes responsible for the diffraction have separation $d_{111}=\lambda/(2\sin\theta)$. The separation of the {111} planes of a cubic lattice of side a is given by eqn 18A.1 as $d_{111}=a/3^{1/2}$. Therefore,

$$a=\frac{3^{1/2}\lambda}{2\sin\theta}=\frac{3^{1/2}\times(154\,\text{pm})}{2\sin 11.2^\circ}=687\,\text{pm}$$

Self-test 18A.4 Calculate the angle θ at which the same crystal will give a reflection from the {123} planes.

Answer: 24.8°

Brief illustration 18A.4 Bragg's law 2

Some types of unit cell give characteristic and easily recognizable patterns of lines. In a cubic lattice of unit cell dimension a the spacing is given by eqn 18A.2, so the angles at which the $\{hkl\}$ planes give first-order reflections are given by

$$\sin\theta = (h^2+k^2+l^2)^{1/2}\frac{\lambda}{2a}$$

The reflections are then predicted by substituting the values of h, k, and l:

$\{hkl\}$	{100}	{110}	{111}	{200}	{210}	{211}	{220}	{300}	{221}	{310}…
$h^2+k^2+l^2$	1	2	3	4	5	6	8	9	9	10…

Notice that 7 (and 15, …) is missing because the sum of the squares of three integers cannot equal 7 (or 15, …). Such absences from the pattern are characteristic of the cubic P lattice.

Self-test 18A.5 Normally, experimental procedures measure the glancing angle 2θ rather than θ itself. A diffraction examination of the element polonium gave lines at the following values of 2θ (in degrees) when 71.0 pm X-rays were used: 12.1, 17.1, 21.0, 24.3, 27.2, 29.9, 34.7, 36.9, 38.9, 40.9, 42.8. Identify the unit cell and determine its dimensions.

Answer: cubic P; $a = 337$ pm

(c) Scattering factors

To prepare the way to discussing methods of structural analysis we need to note that the scattering of X-rays is caused by the oscillations an incoming electromagnetic wave generates in the electrons of atoms. Heavy, electron-rich atoms give rise to stronger scattering than light atoms. This dependence on the number of electrons is expressed in terms of the **scattering factor**, f, of the element. If the scattering factor is large, then the atoms scatter X-rays strongly. An analysis that we do not repeat here concludes that the scattering factor of an atom is related to the electron density distribution in a spherically symmetrical atom, $\rho(r)$, and the angle through which the beam is scattered, 2θ, by

$$f = 4\pi\int_0^\infty \rho(r)\frac{\sin kr}{kr}r^2\mathrm{d}r \quad k = \frac{4\pi}{\lambda}\sin\theta \quad \text{Scattering factor} \quad (18A.3)$$

The value of f is greatest in the forward direction ($\theta = 0$, Fig. 18A.21). The detailed analysis of the intensities of reflections must take this dependence on direction into account. We show in the following *Justification* that, in the forward direction f is equal to the total number of electrons in the atom. For example, the scattering factors of Na^+, K^+, and Cl^- are 8, 18, and 18, respectively.

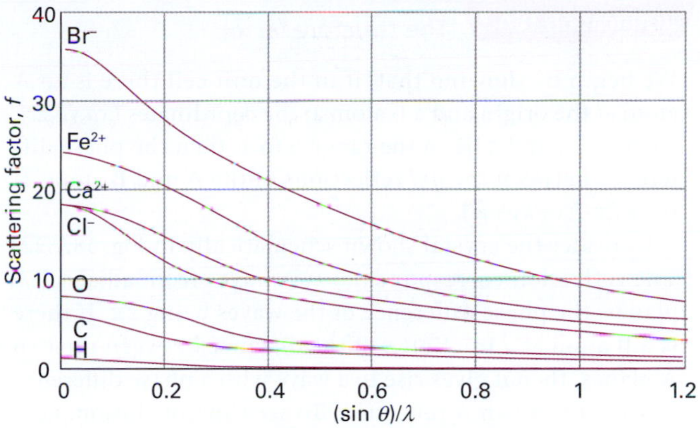

Figure 18A.21 The variation of the scattering factor of atoms and ions with atomic number and angle. The scattering factor in the forward direction (at $\theta = 0$, and hence at $(\sin\theta)/\lambda = 0$) is equal to the number of electrons present in the species.

Justification 18A.2 The forward scattering factor

As $\theta \to 0$, so $k \propto \sin\theta \to 0$. Because $\sin x = x - \frac{1}{6}x^3 + \cdots$,

$$\lim_{k\to 0}\frac{\sin kr}{kr} = \lim_{k\to 0}\frac{kr - \frac{1}{6}(kr)^3 + \cdots}{kr} = \lim_{k\to 0}(1 - \tfrac{1}{6}(kr)^2 + \cdots) = 1$$

The factor $(\sin kr)/kr$ is therefore equal to 1 for forward scattering. It follows that in the forward direction

$$f = 4\pi\int_0^\infty \rho(r)r^2\mathrm{d}r$$

The integral over the electron density ρ (the number of electrons in an infinitesimal region divided by the volume of the region) multiplied by the volume element $4\pi r^2\mathrm{d}r$, the volume of a spherical shell of radius r and thickness $\mathrm{d}r$, is the total number of electrons, N_e, in the atom. Hence, in the forward direction, $f = N_e$.

(d) The electron density

If a unit cell contains several atoms with scattering factors f_j and coordinates (x_ja, y_jb, z_jc), then we show in the following *Justification* that the overall amplitude of a wave diffracted by the $\{hkl\}$ planes is given by

$$F_{hkl} = \sum_j f_j \mathrm{e}^{\mathrm{i}\phi_{hkl}(j)} \quad \text{Structure factor} \quad (18A.4)$$
$$\text{where } \phi_{hkl}(j) = 2\pi(hx_j + ky_j + lz_j)$$

The sum is over all the atoms in the unit cell. The quantity F_{hkl} is called the **structure factor**.

Justification 18A.3 The structure factor

We begin by showing that, if in the unit cell there is an A atom at the origin and a B atom at the coordinates (xa,yb,zc), where x, y, and z lie in the range 0 to 1, then the phase difference between the hkl reflections of the A and B atoms is $\phi_{hkl}=2\pi(hx+ky+lz)$.

Consider the crystal shown schematically in Fig. 18A.22. The reflection corresponds to two waves from adjacent A planes, the phase difference of the waves being 2π. If there is a B atom at a fraction x of the distance between the two A planes, then it gives rise to a wave with a phase difference $2\pi x$ relative to an A reflection. To see this conclusion, note that, if $x=0$, there is no phase difference; if $x=\frac{1}{2}$ the phase difference is π; if $x=1$, the B atom lies where the upper A atom is and the phase difference is 2π. Now consider a (200) reflection. There is now a $2\times 2\pi$ difference between the waves from the two A layers, and if B were to lie at $x=0.5$ it would give rise to a wave that differed in phase by 2π from the wave from the lower A layer. Thus, for a general fractional position x, the phase difference for a (200) reflection is $2\times 2\pi x$. For a general $(h00)$ reflection, the phase difference is therefore $h\times 2\pi x$. For three dimensions, this result generalizes to $\phi_{hkl}=2\pi(hx+ky+lz)$.

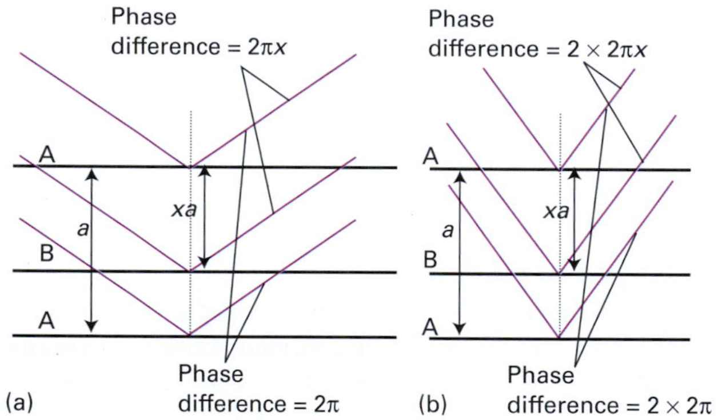

Figure 18A.22 Diffraction from a crystal containing two kinds of atoms. (a) For a (100) reflection from the A planes, there is a phase difference of 2π between waves reflected by neighbouring planes. (b) For a (200) reflection, the phase difference is 4π. The reflection from a B plane at a fractional distance xa from an A plane has a phase that is x times these phase differences.

If the amplitude of the waves scattered from A is f_A at the detector, that of the waves scattered from B is $f_B e^{i\phi_{hkl}}$. The total amplitude at the detector is therefore

$$F_{hkl}=f_A+f_B e^{i\phi_{hkl}}$$

This expression generalizes to eqn 18A.4 when there are several atoms present each with scattering factor f_i.

Example 18A.2 Calculating a structure factor

Calculate the structure factors for the unit cell in Fig. 18A.23.

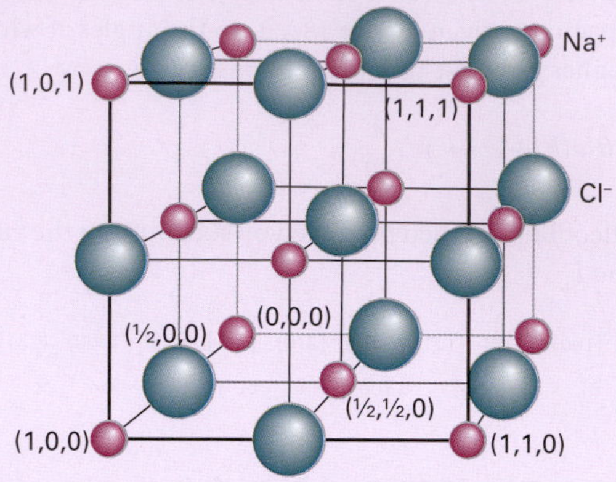

Figure 18A.23 The location of the atoms for the structure factor calculation in *Example* 18A.2. The red spheres are Na^+, the green spheres are Cl^-.

Method The structure factor is defined by eqn 18A.4. To use this equation, consider the ions at the locations specified in Fig. 18A.23. Write f^+ for the Na^+ scattering factor and f^- for the Cl^- scattering factor. Note that ions in the body of the cell contribute to the scattering with a strength f. However, ions on faces are shared between two cells (use $\frac{1}{2}f$), those on edges by four cells (use $\frac{1}{4}f$), and those at corners by eight cells (use $\frac{1}{8}f$). Two useful relations are (*Mathematical background* 3)

$$e^{i\pi}=-1 \qquad \cos\phi=\tfrac{1}{2}(e^{i\phi}+e^{-i\phi})$$

Answer From eqn 18A.4, and summing over the coordinates of all 27 atoms in the illustration:

$$\begin{aligned}F_{hkl}=&f^+(\tfrac{1}{8}+\tfrac{1}{8}e^{2\pi il}+\dots+\tfrac{1}{2}e^{2\pi i(\frac{1}{2}h+\frac{1}{2}k+l)})\\&+f^-(e^{2\pi i(\frac{1}{2}h+\frac{1}{2}k+\frac{1}{2}l)}+\tfrac{1}{4}e^{2\pi i(\frac{1}{2}h)}+\dots+\tfrac{1}{4}e^{2\pi i(\frac{1}{2}h+l)})\end{aligned}$$

To simplify this 27-term expression, we use $e^{2\pi ih}=e^{2\pi ik}=e^{2\pi il}=1$ because h, k, and l are all integers:

$$\begin{aligned}F_{hkl}=&f^+\{1+\cos(h+k)\pi+\cos(h+l)\pi+\cos(k+l)\pi\}\\&+f^-\{(-1)^{h+k+l}+\cos k\pi+\cos l\pi+\cos h\pi\}\end{aligned}$$

Then, because $\cos h\pi=(-1)^h$,

$$\begin{aligned}F_{hkl}=&f^+\{1+(-1)^{h+k}+(-1)^{h+l}+(-1)^{l+k}\}\\&+f^-\{(-1)^{h+k+l}+(-1)^h+(-1)^k+(-1)^l\}\end{aligned}$$

Now note that:

- if h, k, and l are all even, $F_{hkl}=f^+\{1+1+1+1\}+f^-\{1+1+1+1\}=4(f^++f^-)$
- if h, k, and l are all odd, $F_{hkl}=4(f^+-f^-)$
- if one index is odd and two are even, or vice versa, $F_{hkl}=0$

The *hkl* all-odd reflections are less intense than the *hkl* all-even. For $f^+=f^-$, which is the case for identical atoms in a cubic P arrangement, the *hkl* all-odd have zero intensity, corresponding to the absences that are characteristic of cubic P unit cells (see *Brief illustration* 18A.4).

Self-test 18A.6 Which reflections cannot be observed for a cubic I lattice?

Answer: for $h+k+l$ odd, $F_{hkl}=0$

Because the intensity is proportional to the square modulus of the amplitude of the wave, the intensity, I_{hkl}, at the detector is

$$I_{hkl} \propto F_{hkl}^{*}F_{hkl} = (f_A + f_B e^{-i\phi_{hkl}})(f_A + f_B e^{i\phi_{hkl}})$$

This expression expands to

$$I_{hkl} \propto f_A^2 + f_B^2 + f_A f_B (e^{i\phi_{hkl}} + e^{-i\phi_{hkl}}) = f_A^2 + f_B^2 + 2f_A f_B \cos\phi_{hkl}$$

The cosine term either adds to or subtracts from $f_A^2+f_B^2$ depending on the value of ϕ_{hkl}, which in turn depends on h, k, and l and x, y, and z. Hence, there is a variation in the intensities of the lines with different *hkl*. The A and B reflections interfere destructively when the phase difference is π, and the total intensity is zero if the atoms have the same scattering power. For example, if the unit cells are cubic I with a B atom at $x=y=z=\frac{1}{2}$, then the A,B phase difference is $(h+k+l)\pi$. Therefore, all reflections for odd values of $h+k+l$ vanish (as we saw in *Example* 18A.2) because the waves are displaced in phase by π. Hence the diffraction pattern for a cubic I lattice can be constructed from that for the cubic P lattice (a cubic lattice without points at the centre of its unit cells) by striking out all reflections with odd values of $h+k+l$. Recognition of these **systematic absences** in a powder spectrum immediately indicates a cubic I lattice (Fig. 18A.24).

The intensity of the (*hkl*) reflection is proportional to $|F_{hkl}|^2$, so in principle we can determine the structure factors experimentally by taking the square root of the corresponding intensities (but see next section). Then, once we know all the structure factors F_{hkl}, we can calculate the electron density distribution, $\rho(r)$, in the unit cell by using the expression

$$\rho(r) = \frac{1}{V}\sum_{hkl} F_{hkl} e^{-2\pi i(hx+ky+lz)} \qquad \text{Fourier synthesis} \qquad (18A.5)$$

where V is the volume of the unit cell. Equation 18A.5 is called a **Fourier synthesis** of the electron density. Fourier transforms occur throughout chemistry in a variety of guises, and are described in more detail in *Mathematical background* 7 following this chapter.

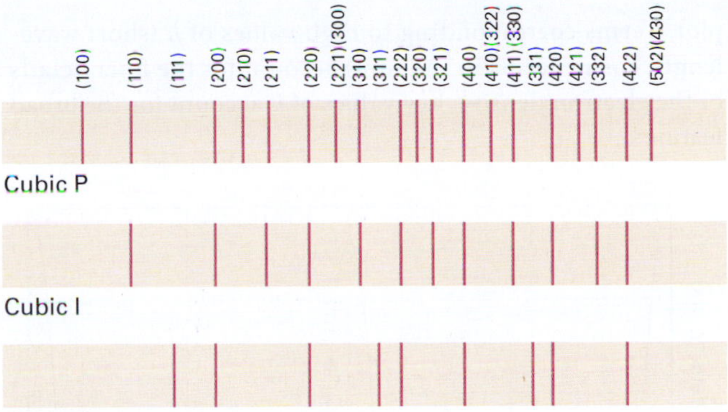

Figure 18A.24 The powder diffraction patterns and the systematic absences of three versions of a cubic cell as a function of angle: cubic F (fcc; h, k, l all even or all odd are present), cubic I (bcc; $h+k+l=$odd are absent), cubic P. Comparison of the observed pattern with patterns like these enables the unit cell to be identified. The locations of the lines give the cell dimensions.

Example 18A.3 Calculating an electron density by Fourier synthesis

Consider the $\{h00\}$ planes of a crystal extending indefinitely in the x-direction. In an X-ray analysis the structure factors were found as follows:

h:	0	1	2	3	4	5	6	7	8	9
F_h	16	−10	2	−1	7	−10	8	−3	2	−3

h:	10	11	12	13	14	15
F_h	6	−5	3	−2	2	−3

(and $F_{-h}=F_h$). Construct a plot of the electron density projected on to the x-axis of the unit cell.

Method Because $F_{-h}=F_h$, it follows from eqn 18A.5 that

$$V\rho(x) = \sum_{h=-\infty}^{\infty} F_h e^{-2\pi i hx} = F_0 + \sum_{h=1}^{\infty}(F_h e^{-2\pi i hx} + F_{-h}e^{2\pi i hx})$$

$$= F_0 + \sum_{h=1}^{\infty} F_h (e^{-2\pi i hx} + e^{2\pi i hx})$$

$$\overset{\frac{1}{2}(e^{-2\pi i hx}+e^{2\pi i hx})=\cos 2\pi hx}{=} F_0 + 2\sum_{h=1}^{\infty} F_h \cos 2\pi hx$$

and we evaluate the sum (truncated at $h=15$) for points $0 \le x \le 1$ by using mathematical software.

Answer The results are plotted in Fig. 18A.25 (green line). The positions of three atoms can be discerned very readily. The more terms there are included, the more accurate the density plot. Terms corresponding to high values of h (short wavelength cosine terms in the sum) account for the finer details of the electron density; low values of h account for the broad features.

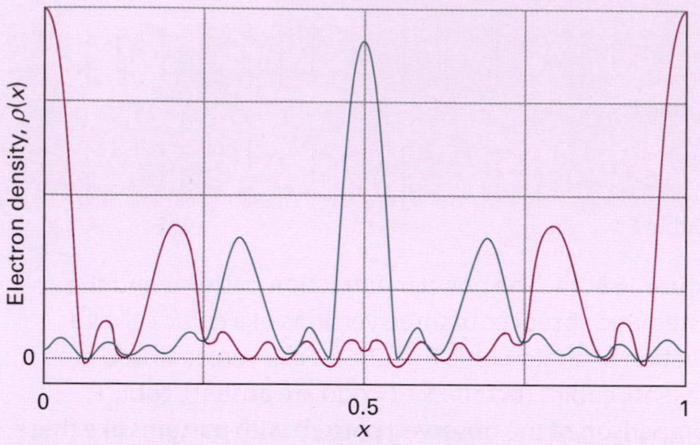

Figure 18A.25 The plot of the electron density calculated in *Example* 18A.3 (green) and *Self-test* 18A.7 (purple).

Self-test 18A.7 Use mathematical software to experiment with different structure factors (including changing signs as well as amplitudes). For example, use the same values of F_h as above, but with positive signs for $h \geq 6$.

Answer: See Fig. 18A.25 (purple line).

(e) Determination of the structure

A problem with the procedure outlined above is that the observed intensity I_{hkl} is proportional to the square modulus $|F_{hkl}|^2$, so we do not know whether to use $+|F_{hkl}|$ or $-|F_{hkl}|$ in the sum in eqn 18A.5. In fact, the difficulty is more severe for noncentrosymmetric unit cells, because if we write F_{hkl} as the complex number $|F_{hkl}|e^{i\alpha}$, where α is the phase of F_{hkl} and $|F_{hkl}|$ is its magnitude, then the intensity lets us determine $|F_{hkl}|$ but tells us nothing of its phase, which may lie anywhere from 0 to 2π. This ambiguity is called the **phase problem**; its consequences are illustrated by comparing the two plots in Fig. 18A.25. Some way must be found to assign phases to the structure factors, for otherwise the sum for ρ cannot be evaluated and the method would be useless.

The phase problem can be overcome to some extent by a variety of methods. One procedure that is widely used for inorganic materials with a reasonably small number of atoms in a unit cell, and for organic molecules with a small number of heavy atoms, is the **Patterson synthesis**. Instead of the structure factors F_{hkl}, the values of $|F_{hkl}|^2$, which can be obtained without ambiguity from the intensities, are used in an expression that resembles eqn 18A.5:

$$P(\boldsymbol{r})=\frac{1}{V}\sum_{hkl}|F_{hkl}|^2\,e^{-2\pi i(hx+ky+lz)} \quad \text{Patterson synthesis} \quad (18A.6)$$

where the $\boldsymbol{r}$ values correspond to the vector separations between the atoms in the unit cell; that is, the distances and directions between atoms. Whereas the electron density function $\rho(\boldsymbol{r})$ is the probability density of the positions of atoms, the function $P(\boldsymbol{r})$ is a map of the probability density of the separations between atoms: a peak in P at a vector separation $\boldsymbol{r}$ arises from pairs of atoms that are separated by the same separation $\boldsymbol{r}$. Thus, if atom A is at the coordinates (x_A,y_A,z_A) and atom B is at (x_B,y_B,z_B), then there will be a peak at $(x_A-x_B,\ y_A-y_B,\ z_A-z_B)$ in the Patterson map. There will also be a peak at the negative of these coordinates, because there is a separation vector from B to A as well as a separation vector from A to B. The height of the peak in the map is proportional to the product of the atomic numbers of the two atoms, Z_AZ_B.

Brief illustration 18A.5 The Patterson synthesis

If the unit cell has the structure shown in Fig. 18A.26a, the Patterson synthesis would be the map shown in Fig. 18A.26b, where the location of each spot relative to the origin gives the separation and relative orientation of each pair of atoms in the original structure.

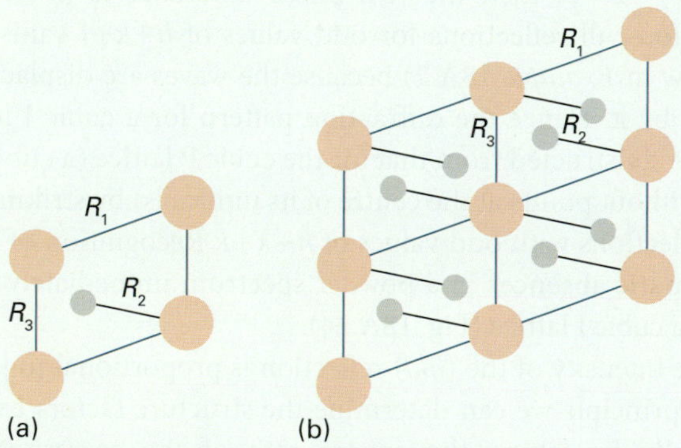

Figure 18A.26 The Patterson synthesis corresponding to the pattern in (a) is the pattern in (b). The distance and orientation of each spot from the origin gives the orientation and separation of one atom–atom separation in (a). Some of the typical distances and their contribution to (b) are shown as R_1, etc.

Self-test 18A.8 Consider the data in *Example* 18A.3. Show that $VP(x)=|F_0|^2+2\sum_{h=1}^{\infty}|F_h|^2\cos 2\pi hx$ and plot the Patterson synthesis.

Answer: See Fig. 18A.27.

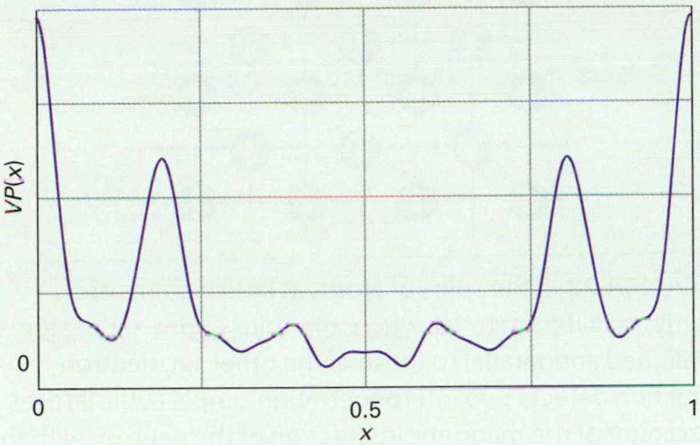

Figure 18A.27 Patterson synthesis of the data from *Example* 18A.3.

Heavy atoms dominate the scattering because their scattering factors are large, of the order of their atomic numbers, and their locations may be deduced quite readily. The sign of F_{hkl} can now be calculated from the locations of the heavy atoms in the unit cell, and to a high probability the phase calculated for them will be the same as the phase for the entire unit cell. To see why this is so, we have to note that a structure factor of a centrosymmetric cell has the form

$$F=(\pm)f_{\text{heavy}}+(\pm)f_{\text{light}}+(\pm)f_{\text{light}}+\cdots \quad (18A.7)$$

where f_{heavy} is the scattering factor of the heavy atom and f_{light} the scattering factors of the light atoms. The f_{light} are all much smaller than f_{heavy}, and their phases are more or less random if the atoms are distributed throughout the unit cell. Therefore, the net effect of the f_{light} is to change F only slightly from f_{heavy}, and we can be reasonably confident that F will have the same sign as that calculated from the location of the heavy atom. This phase can then be combined with the observed $|F|$ (from the reflection intensity) to perform a Fourier synthesis of the full electron density in the unit cell, and hence to locate the light atoms as well as the heavy atoms.

Modern structural analyses make extensive use of **direct methods**. Direct methods are based on the possibility of treating the atoms in a unit cell as being virtually randomly distributed (from the radiation's point of view), and then to use statistical techniques to compute the probabilities that the phases have a particular value. It is possible to deduce relations between some structure factors and sums (and sums of squares) of others, which have the effect of constraining the phases to particular values (with high probability, so long as the structure factors are large). For example, the **Sayre probability relation** has the form

$$\text{sign of } F_{h+h',k+k',l+l'} \text{ is probably equal to}(\text{sign of } F_{hkl})\times(\text{sign of } F_{h'k'l'})$$

Sayre probability relation (18A.8)

For example, if F_{122} and F_{232} are both large and negative, then it is highly likely that F_{354}, provided it is large, will be positive.

In the final stages of the determination of a crystal structure, the parameters describing the structure (atom positions, for instance) are adjusted systematically to give the best fit between the observed intensities and those calculated from the model of the structure deduced from the diffraction pattern. This process is called **structure refinement**. Not only does the procedure give accurate positions for all the atoms in the unit cell, but it also gives an estimate of the errors in those positions and in the bond lengths and angles derived from them. The procedure also provides information on the vibrational amplitudes of the atoms.

18A.4 Neutron and electron diffraction

According to the de Broglie relation (Topic 7A, $\lambda=h/p$), particles have wavelengths and may therefore undergo diffraction. Neutrons generated in a nuclear reactor and then slowed to thermal velocities have wavelengths similar to those of X-rays and may also be used for diffraction studies. For instance, a neutron generated in a reactor and slowed to thermal velocities by repeated collisions with a moderator (such as graphite) until it is travelling at about 4 km s^{-1} has a wavelength of about 100 pm. In practice, a range of wavelengths occurs in a neutron beam, but a monochromatic beam can be selected by diffraction from a crystal, such as a single crystal of germanium.

Example 18A.4 Calculating the typical wavelength of thermal neutrons

Calculate the typical wavelength of neutrons after reaching thermal equilibrium with their surroundings at 373 K. For simplicity, assume that the particles are travelling in one dimension.

Method We need to relate the wavelength to the temperature. There are two linking steps. First, the de Broglie relation expresses the wavelength in terms of the linear momentum. Then the linear momentum can be expressed in terms of the kinetic energy, the mean value of which is given in terms of the temperature by the equipartition theorem (*Foundations* B).

Answer From the equipartition principle, we know that the mean translational kinetic energy of a neutron at a temperature T travelling in the x-direction is $E_k=\frac{1}{2}kT$. The kinetic energy is also equal to $p^2/2m$, where p is the momentum of the neutron and m is its mass. Hence, $p=(mkT)^{1/2}$. It follows from the de Broglie relation $\lambda=h/p$ that the neutron's wavelength is

$$\lambda=\frac{h}{(mkT)^{1/2}}$$

Therefore, at 373 K,

$$\lambda = \frac{6.626\times10^{-34}\,\text{J s}}{\{(1.675\times10^{-27}\,\text{kg})\times(1.381\times10^{-23}\,\text{J K}^{-1})\times(373\,\text{K})\}^{1/2}}$$

$$\overset{1\,\text{J}=1\,\text{kg m}^2\,\text{s}^{-2}}{=} \frac{6.626\times10^{-34}}{(1.675\times10^{-27}\times1.381\times10^{-23}\times373)^{1/2}}\,\frac{\text{kg m}^2\,\text{s}^{-1}}{(\text{kg}^2\,\text{m}^2\,\text{s}^{-2})^{1/2}}$$

$$=2.26\times10^{-10}\,\text{m}=226\,\text{pm}$$

Self-test 18A.9 Calculate the temperature needed for the average wavelength of the neutrons to be 100 pm.

Answer: 1.90 kK

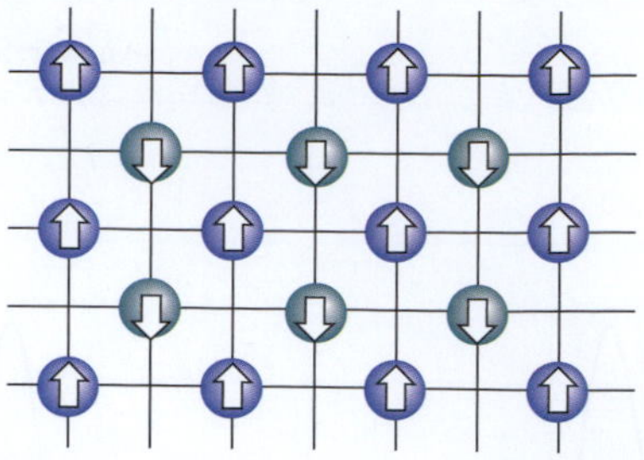

Figure 18A.28 If the spins of atoms at lattice points are orderly, as in this material, where the spins of one set of atoms are aligned antiparallel to those of the other set, neutron diffraction detects two interpenetrating simple cubic lattices on account of the magnetic interaction of the neutron with the atoms, but X-ray diffraction would see only a single bcc lattice.

Neutron diffraction differs from X-ray diffraction in two main respects. First, the scattering of neutrons is a nuclear phenomenon. Neutrons pass through the extranuclear electrons of atoms and interact with the nuclei through the 'strong force' that is responsible for binding nucleons together. As a result, the intensity with which neutrons are scattered is independent of the number of electrons and neighbouring elements in the periodic table might scatter neutrons with markedly different intensities. Neutron diffraction can be used to distinguish atoms of elements such as Ni and Co that are present in the same compound and to study order–disorder phase transitions in FeCo. A second difference is that neutrons possess a magnetic moment due to their spin. This magnetic moment can couple to the magnetic fields of atoms or ions in a crystal (if the ions have unpaired electrons) and modify the diffraction pattern. One consequence is that neutron diffraction is well suited to the investigation of magnetically ordered lattices in which neighbouring atoms may be of the same element but have different orientations of their electronic spin (Fig. 18A.28).

Electrons accelerated through a potential difference of 40 kV have wavelengths of about 6 pm, and so are also suitable for diffraction studies of molecules. Consider the scattering of electrons (or neutrons) from a pair of nuclei separated by a distance R_{ij} and orientated at a definite angle θ to an incident beam of electrons (or neutrons). When the molecule consists of a number of atoms, the scattering intensity can be calculated by summing over the contribution from all pairs. The total intensity $I(\theta)$ is given by the **Wierl equation:**

$$I(\theta)=\sum_{i,j} f_i f_j \frac{\sin sR_{ij}}{sR_{ij}} \qquad s=\frac{4\pi}{\lambda}\sin\tfrac{1}{2}\theta \qquad \text{Wierl equation} \qquad (18\text{A}.9)$$

where λ is the wavelength of the electrons in the beam, and f is the **electron scattering factor**, a measure of the electron scattering power of the atom. The main application of electron diffraction techniques is to the study of surfaces (Topic 22A), and you are invited to explore the Wierl equation in Problem 18A.17.

Checklist of concepts

☐ 1. A **space lattice** is the pattern formed by points representing the locations of structural motifs (atoms, molecules, or groups of atoms, molecules, or ions).

☐ 2. The **Bravais lattices** are the 14 distinct space lattices in three dimensions (Fig. 18A.8).

☐ 3. A **unit cell** is an imaginary parallelepiped that contains one unit of a translationally repeating pattern.

☐ 4. Unit cells are classified into seven **crystal systems** according to their rotational symmetries.

☐ 5. A crystal plane is specified by a set of **Miller indices** (hkl); sets of planes are denoted $\{hkl\}$.

☐ 6. The **scattering factor** is a measure of the ability of an atom to diffract radiation.

☐ 7. The **structure factor** is the overall amplitude of a wave diffracted by the $\{hkl\}$ planes.

☐ 8. **Fourier synthesis** is the construction of the electron density distribution from structure factors.

☐ 9. A **Patterson synthesis** is a map of interatomic vectors obtained by Fourier analysis of diffraction intensities.

☐ 10. **Structure refinement** is the adjustment of structural parameters to give the best fit between the observed intensities and those calculated from the model of the structure deduced from the diffraction pattern.

Checklist of equations

Property	Equation	Comment	Equation number
Separation of planes in a rectangular lattice	$1/d_{hkl}^2 = h^2/a^2 + k^2/b^2 + l^2/c^2$	h, k, and l are Miller indices	18A.1c
Bragg's law	$\lambda = 2d \sin\theta$	d is the lattice spacing, 2θ the glancing angle	18A.2b
Scattering factor	$f = 4\pi \int_0^\infty [\{\rho(r)\sin kr\}/kr]r^2 \mathrm{d}r,\ k = (4\pi/\lambda)\sin\theta$	Spherically symmetrical atom	18A.3
Structure factor	$F_{hkl} = \sum_j f_j \mathrm{e}^{\mathrm{i}\phi_{hkl}(j)},\quad \phi_{hkl}(j) = 2\pi(hx_j + ky_j + lz_j)$	Definition	18A.4
Fourier synthesis	$\rho(r) = (1/V)\sum_{hkl} F_{hkl}\mathrm{e}^{-2\pi\mathrm{i}(hx+ky+lz)}$	V is the volume of the unit cell	18A.5
Patterson synthesis	$P(r) = (1/V)\sum_{hkl} \lvert F_{hkl}\rvert^2\, \mathrm{e}^{-2\pi\mathrm{i}(hx+ky+lz)}$		18A.6
Wierl equation	$I(\theta) = \sum_{i,j} f_i f_j(\sin sR_{ij}/sR_{ij}),\quad s = (4\pi/\lambda)\sin\tfrac{1}{2}\theta$		18A.9

18B Bonding in solids

Contents

➤ Why do you need to know this material?

To prepare for the study of the properties of materials and the structures they adopt, you need to know how atoms and molecules interact to form metallic, covalent, ionic, and molecular solids.

➤ What is the key idea?

Four characteristic types of bonding result in metals, ionic solids, covalent solids, and molecular solids.

➤ What do you need to know already?

You need to be familiar with molecular interactions (Topic 16B) and the general features of crystal structure (Topic 18A). For the discussion of metallic bonding you should be aware of the principles of Hückel molecular orbital theory (Topic 10E). The discussion of ionic bonding makes use of the concept of enthalpy and the fact that it is a state function (Topic 2C).

The bonding within a solid may be of various kinds. Simplest of all (in concept, at least) are **metals**, where electrons are delocalized over arrays of identical cations and bind the whole together into a rigid but ductile and malleable structure. **Ionic solids** consist of cations and anions packed together by electrostatic interactions in a crystal (Topic 18A). In **covalent solids**, covalent bonds in a definite spatial orientation link the atoms in a network extending through a crystal. **Molecular solids** are bonded together by van der Waals interactions (Topic 16B).

18B.1 Metallic solids

The crystalline forms of metallic elements can be discussed in terms of a model in which their atoms are represented as identical hard spheres. Most metallic elements crystallize in one of three simple forms, two of which can be explained in terms of the hard spheres packing together in the closest possible arrangement. In this section we consider not only the geometrical arrangement of the atoms in the crystal, but also the distribution of electrons over the atoms.

(a) Close packing

Figure 18B.1 shows a **close-packed** layer of identical spheres, one with maximum utilization of space. A close-packed three-dimensional structure is obtained by stacking such close-packed layers on top of one another. However, this stacking can be done in different ways, which result in close-packed **polytypes**, or structures that are identical in two dimensions (the close-packed layers) but differ in the third dimension.

In all polytypes, the spheres of second close-packed layer lie in the depressions of the first layer (Fig. 18B.2). The third layer may be added in either of two ways. In one, the spheres are placed directly above the first layer to give an ABA pattern of layers (Fig. 18B.3a). Alternatively, the spheres may be placed over the gaps in the first layer, so giving an ABC pattern (Fig.

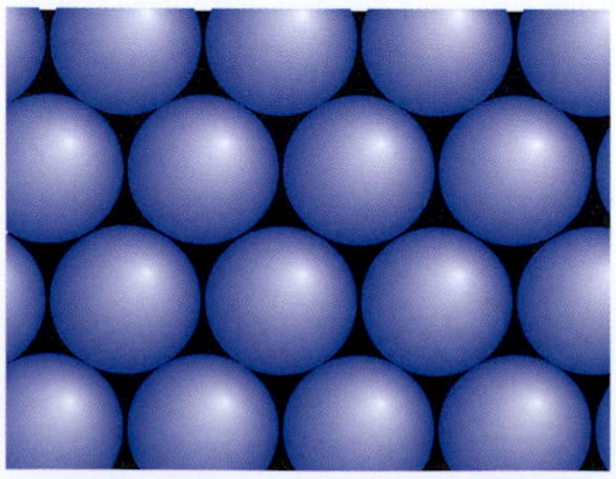

Figure 18B.1 The first layer of close-packed spheres used to build a three-dimensional close-packed structure.

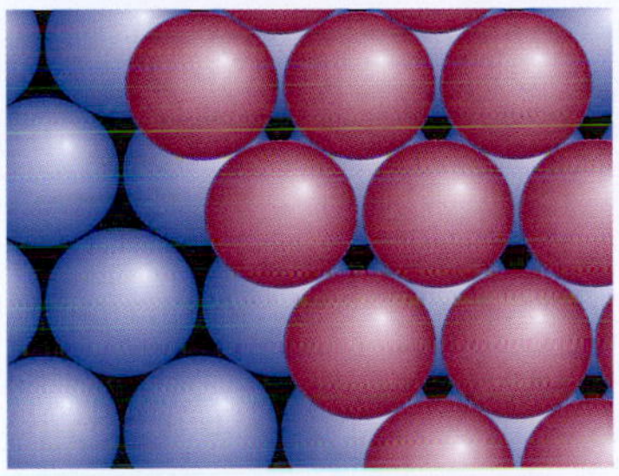

Figure 18B.2 The second layer of close-packed spheres occupies the dips of the first layer. The two layers are the AB component of the close-packed structure.

18B.3b). Two polytypes are formed if the two stacking patterns are repeated in the vertical direction:

- **Hexagonally close-packed** (hcp): the ABA pattern is repeated, to give the sequence of layers ABABAB....
- **Cubic close-packed** (ccp): the ABC pattern is repeated, to give the sequence ABCABC....

The origins of these names can be seen by referring to Fig. 18B.4. The ccp structure gives rise to a face-centred unit cell, so may also be denoted cubic F (or fcc, for face-centred cubic). It is also possible to have random sequences of layers; however, the hcp and ccp polytypes are the most important. Table 18B.1 lists some elements possessing these structures.

The compactness of close-packed structures is indicated by their **coordination number**, the number of spheres immediately surrounding any selected sphere, which is 12 in all cases. Another measure of their compactness is the **packing fraction**, the fraction of space occupied by the spheres, which is 0.740 (see *Example* 18B.1). That is, in a close-packed solid of identical hard spheres, only 26.0 per cent of the volume is empty space. The fact that many metals are close-packed accounts for their high mass densities.

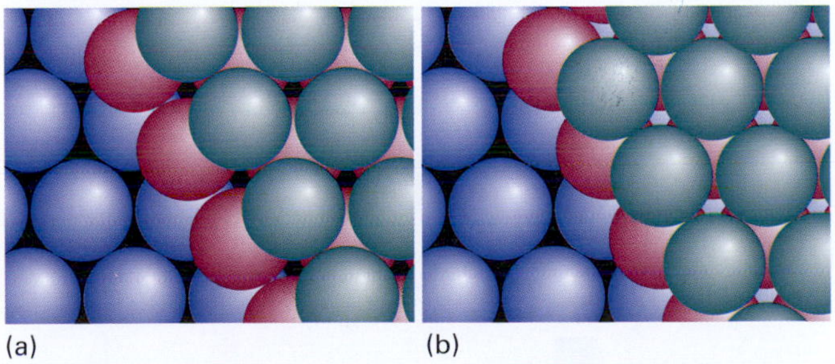

Figure 18B.3 (a) The third layer of close-packed spheres might occupy the dips lying directly above the spheres in the first layer, resulting in an ABA structure, which corresponds to hexagonal close-packing. (b) Alternatively, the third layer might lie in the dips that are not above the spheres in the first layer, resulting in an ABC structure, which corresponds to cubic close-packing.

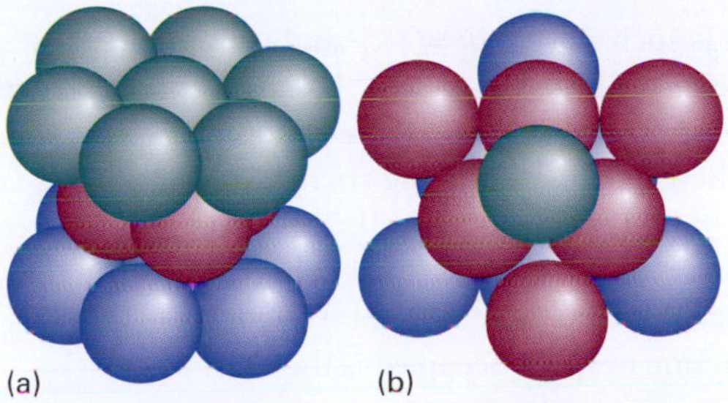

Figure 18B.4 A fragment of the structure shown in Fig. 18B.3 revealing the (a) hexagonal, (b) cubic symmetry. The colours of the spheres are the same as for the layers in Fig. 18B.3.

Table 18B.1 The crystal structures of some elements*

Structure	Element
hcp‡	Be, Cd, Co, He, Mg, Sc, Ti, Zn
fcc‡ (ccp, cubic F)	Ag, Al, Ar, Au, Ca, Cu, Kr, Ne, Ni, Pd, Pb, Pt, Rh, Rn, Sr, Xe
bcc (cubic I)	Ba, Cs, Cr, Fe, K, Li, Mn, Mo, Rb, Na, Ta, W, V
cubic P	Po

* The notation used to describe primitive unit cells is introduced in Topic 18A.
‡ Close-packed structures.

Example 18B.1 Calculating a packing fraction

Calculate the packing fraction of a ccp structure with spheres of radius R.

Method Refer to Fig. 18B.5. First calculate the volume of a unit cell, and then calculate the total volume of the spheres that occupy it fully or partially. The first part of the calculation is an exercise in geometry and the use of the Pythagorean theorem ($a^2+b^2=c^2$ in a right-angled triangle). The second part involves counting the fraction of spheres that occupy the cell.

Answer We see in Fig. 18B.5 that a diagonal of any face passes completely through one sphere and halfway through two other spheres. Therefore, the length of a diagonal is $4R$. The length of

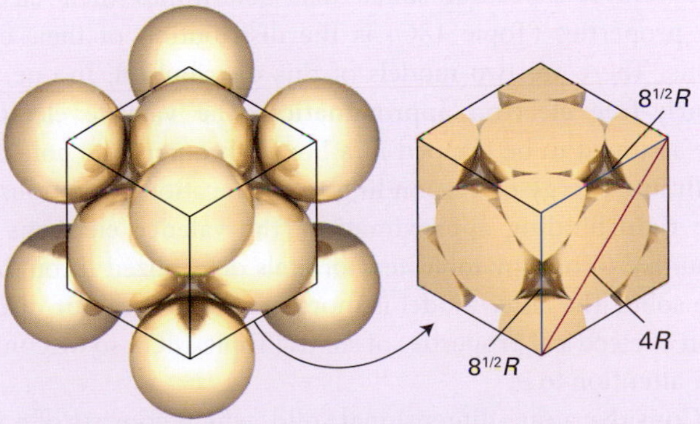

Figure 18B.5 The calculation of the packing fraction of a ccp unit cell.

a side l is such that $l^2+l^2=(4R)^2$ and is therefore $l=8^{1/2}R$. The volume of the unit cell is $l^3=8^{3/2}R^3$. As Fig 18B.5 shows, each of the eight vertices of the cube contains the equivalent of $\frac{1}{8}$ of a sphere. Also, each of the six remaining spheres contributes $\frac{1}{2}$ of its volume to the cell. Therefore, each cell contains the equivalent of $6\times\frac{1}{2}+8\times\frac{1}{8}=4$ spheres. Because the volume of each sphere is $\frac{4}{3}\pi R^3$, the total occupied volume is $\frac{16}{3}\pi R^3$. The fraction of space occupied is therefore

$$\frac{\frac{16}{3}\pi R^3}{8^{3/2}R^3}=\frac{16\pi}{3\times 8^{3/2}}=0.740$$

Because an hcp structure has the same coordination number, its packing fraction is the same.

Self-test 18B.1 The packing fractions of structures that are not close-packed are calculated similarly. Calculate the packing fraction of a structure with one sphere at the centre of a cube formed by eight others: this is a cubic I (bcc) structure.

Answer: 0.68

As shown in Table 18B.1, a number of common metals adopt structures that are less than close-packed. The departure from close packing suggests that factors such as specific covalent bonding between neighbouring atoms are beginning to influence the structure and impose a specific geometrical arrangement. One such arrangement results in a cubic I (bcc, for body-centred cubic) structure, with one sphere at the centre of a cube formed by eight others. The coordination number of a bcc structure is only 8, but there are six more atoms not much further away than the eight nearest neighbours. The packing fraction of 0.68 (*Self-test* 18B.1) is not much smaller than the value for a close-packed structure (0.74), and shows that about two-thirds of the available space is actually occupied.

(b) Electronic structure of metals

The central aspect of solids that determines their electrical properties (Topic 18C) is the distribution of their electrons. There are two models of this distribution. In one, the **nearly-free electron approximation**, the valence electrons are assumed to be trapped in a box with a periodic potential, with low energy corresponding to the locations of cations. In the **tight-binding approximation**, the valence electrons are assumed to occupy molecular orbitals delocalized throughout the solid. The latter model is more in accord with our discussion of electrical properties of solids (Topic 18C), so we confine our attention to it.

Consider a one-dimensional solid, which consists of a single, infinitely long line of atoms. At first sight, this model may seem too restrictive and unrealistic. However, not only does it give us the concepts we need to understand the structure and electrical properties of three-dimensional, macroscopic samples of metals and semiconductors, it is also the starting point for the description of long and thin structures, such carbon nanotubes.

Suppose that each atom has one s orbital available for forming molecular orbitals. We can construct the LCAO-MOs of the solid by adding N atoms in succession to a line, and then infer the electronic structure by using the building-up principle. One atom contributes one s orbital at a certain energy (Fig. 18B.6). When a second atom is brought up it overlaps the first and forms bonding and antibonding orbitals. The third atom overlaps its nearest neighbour (and only slightly the next-nearest), and from these three atomic orbitals, three molecular orbitals are formed: one is fully bonding, one fully antibonding, and the intermediate orbital is nonbonding between neighbours. The fourth atom leads to the formation of a fourth molecular orbital. At this stage, we can begin to see that the general effect of bringing up successive atoms is to spread the range of energies covered by the molecular orbitals, and also to fill in the range of energies with more and more orbitals (one more for each atom). When N atoms have been added to the line, there are N molecular orbitals covering a band of energies of finite width, and the Hückel secular determinant (Topic 10E) is

$$\begin{vmatrix} \alpha-E & \beta & 0 & 0 & 0 & \cdots & 0 \\ \beta & \alpha-E & \beta & 0 & 0 & \cdots & 0 \\ 0 & \beta & \alpha-E & \beta & 0 & \cdots & 0 \\ 0 & 0 & \beta & \alpha-E & \beta & \cdots & 0 \\ 0 & 0 & 0 & \beta & \alpha-E & \cdots & 0 \\ \vdots & \vdots & \vdots & \vdots & \vdots & \cdots & \vdots \\ 0 & 0 & 0 & 0 & 0 & \cdots & \alpha-E \end{vmatrix}=0$$

where α is the Coulomb integral and β is the (s,s) resonance integral. The theory of determinants applied to such a

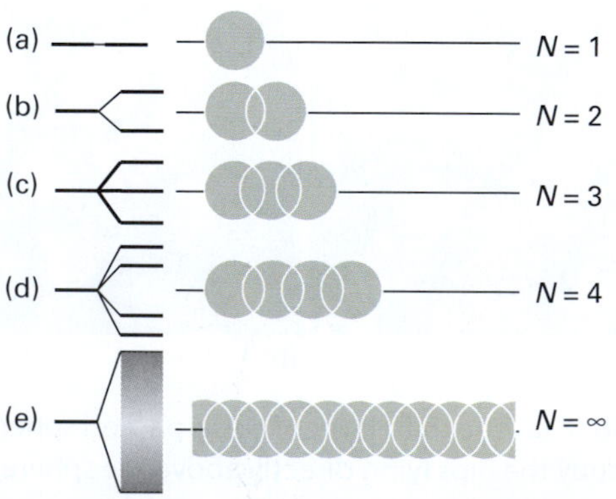

Figure 18B.6 The formation of a band of N molecular orbitals by successive addition of N atoms to a line. Note that the band remains of finite width as N becomes infinite and, although it then looks continuous, it consists of N different orbitals.

symmetrical example as this (technically a 'tridiagonal determinant') leads to the following expression for the roots:

$$E_k = \alpha + 2\beta\cos\frac{k\pi}{N+1} \qquad k=1,2,\ldots,N \qquad \textit{Linear array of s orbitals} \quad \text{Energy levels} \qquad (18B.1)$$

We show in the following *Justification* that when N is infinitely large, the separation between neighbouring levels, $E_{k+1}-E_k$, is infinitely small, but the band still has finite width overall (Fig. 18B.6):

$$E_N - E_1 \to -4\beta \quad \text{as} \quad N\to\infty \qquad \textit{Linear array of s orbitals} \quad \text{Band width} \qquad (18B.2)$$

(Note that because $\beta<0$, $-4\beta>0$.) We can think of this band as consisting of N different molecular orbitals, the lowest-energy orbital ($k=1$) being fully bonding, and the highest-energy orbital ($k=N$) being fully antibonding between adjacent atoms (Fig. 18B.7). The molecular orbitals of intermediate energy have $k-1$ nodes distributed along the chain of atoms. Similar bands form in three-dimensional solids.

Justification 18B.1 The properties of a band

From eqn 18B.1 we see that the energy separation between neighbouring energy levels k and $k+1$ is

$$E_{k+1}-E_k = \left(\alpha+2\beta\cos\frac{(k+1)\pi}{N+1}\right) - \left(\alpha+2\beta\cos\frac{k\pi}{N+1}\right)$$
$$= 2\beta\left(\cos\frac{(k+1)\pi}{N+1} - \cos\frac{k\pi}{N+1}\right)$$

By using the trigonometric identity $\cos(A+B)=\cos A\cos B - \sin A\sin B$ followed by $\cos 0=1$ and $\sin 0=0$ the first (blue) term in parentheses is

$$\cos\frac{(k+1)\pi}{N+1} = \cos\frac{k\pi}{N+1}\overbrace{\cos\frac{\pi}{N+1}}^{\to 1 \text{ as } N\to\infty} - \sin\frac{k\pi}{N+1}\overbrace{\sin\frac{\pi}{N+1}}^{\to 0 \text{ as } N\to\infty}$$

Therefore, as $N\to\infty$,

$$E_{k+1}-E_k \to 2\beta\left(\cos\frac{k\pi}{N+1} - \cos\frac{k\pi}{N+1}\right) = 0$$

It follows that when N is infinitely large, the difference between neighbouring energy levels is infinitely small.

To assess the effect of N on the width $E_N - E_1$ of a band, we proceed as follows. The energy of the level with $k=1$ is

$$E_1 = \alpha + 2\beta\cos\frac{\pi}{N+1}$$

As $N\to\infty$, the cosine approaches $\cos 0 = 1$. Therefore, in this limit

$$E_1 = \alpha + 2\beta$$

When k has its maximum value of N,

$$E_N = \alpha + 2\beta\cos\frac{N\pi}{N+1}$$

As $N\to\infty$, we can ignore the 1 in the denominator, and the cosine term becomes $\cos\pi = -1$. Therefore, in this limit $E_N = \alpha - 2\beta$, and $E_N - E_1 \to -4\beta$, as in eqn 18B.2.

Brief illustration 18B.1 Energy levels in a band

To illustrate the dependence of $E_{k+1}-E_k$ on N, we use the first equation in the *Justification* to calculate

$$N=3:\quad E_2-E_1 = 2\beta\left(\cos\frac{2\pi}{4} - \cos\frac{\pi}{4}\right) \approx -1.414\beta$$
$$N=30:\quad E_2-E_1 = 2\beta\left(\cos\frac{2\pi}{31} - \cos\frac{\pi}{31}\right) \approx -0.0307\beta$$
$$N=300:\quad E_2-E_1 = 2\beta\left(\cos\frac{2\pi}{301} - \cos\frac{\pi}{301}\right) \approx -0.00036\beta$$

We see that the energy difference decreases with increasing N.

Self-test 18B.2 For $N=300$, at which value of k would $E_{k+1}-E_k$ have its maximum value? *Hint*: Use mathematical software.

Answer: $k=150$

The band formed from overlap of s orbitals is called the **s band**. If the atoms have p orbitals available, the same procedure leads to a **p band** (as shown in the upper half of Fig. 18B.7). If the atomic p orbitals lie higher in energy than the s orbitals, then the p band lies higher than the s band, and there may be a **band gap**, a range of energies to which no orbital corresponds.

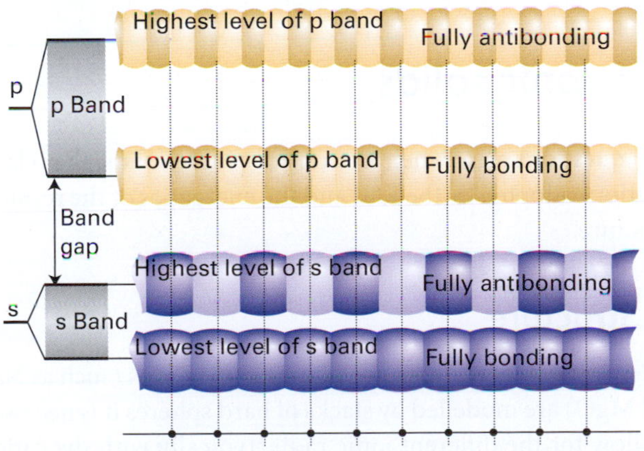

Figure 18B.7 The overlap of s orbitals gives rise to an s band and the overlap of p orbitals gives rise to a p band. In this case, the s and p orbitals of the atoms are so widely spaced in energy that there is a band gap. In many cases the separation is less and the bands overlap.

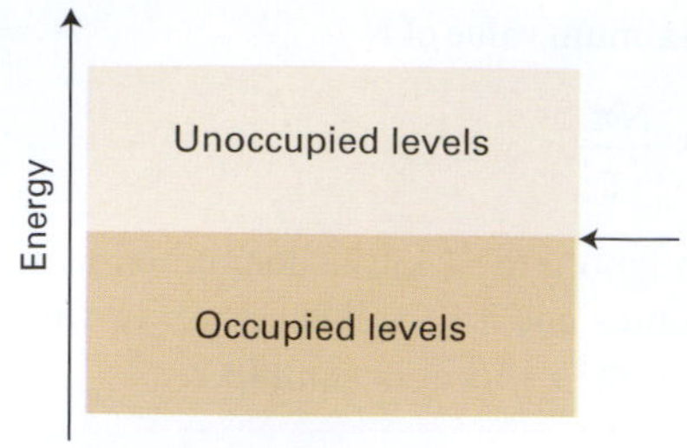

Figure 18B.8 When N electrons occupy a band of N orbitals, it is only half full and the electrons near the Fermi level (the top of the filled levels) are mobile.

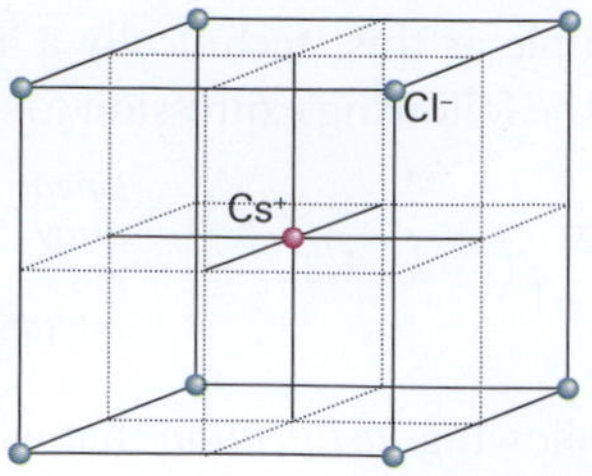

Figure 18B.9 The caesium chloride structure consists of two interpenetrating simple cubic arrays of ions, one of cations and the other of anions, so that each cube of ions of one kind has a counter-ion at its centre.

However, the s and p bands may also be contiguous with the highest orbital of the s band coincident with the lowest level of the p band or even overlap (as is the case for the 3s and 3p bands in magnesium).

Now consider the electronic structure of a solid formed from atoms each able to contribute one electron (for example, the alkali metals). There are N atomic orbitals and therefore N molecular orbitals packed into an apparently continuous band. There are N electrons to accommodate. At $T=0$, only the lowest $\frac{1}{2}N$ molecular orbitals are occupied (Fig. 18B.8), and the HOMO is called the **Fermi level**. However, unlike in molecules, there are empty orbitals very close in energy to the Fermi level, so it requires hardly any energy to excite the uppermost electrons. Some of the electrons are therefore very mobile and give rise to electrical conductivity (Topic 18C). Only the small number of electrons close to the Fermi level can undergo thermal excitation, so only these electrons contribute to the heat capacity of the metal. It is for this reason that Dulong and Petit's law for heat capacities (Topic 7A) gives reasonable agreement with experiment at normal temperatures by counting only the atoms in a sample, not the atoms plus the 'free' electrons.

18B.2 Ionic solids

Two questions arise when we consider ionic solids: the relative locations adopted by the ions and the energetics of the resulting structure.

(a) Structure

When crystals of compounds of monatomic ions (such as NaCl and MgO) are modelled by stacks of hard spheres it is necessary to allow for the different ionic radii (typically with the cations smaller than the anions) and different electrical charges. The coordination number of an ion is the number of nearest neighbours of opposite charge; the structure itself is characterized as having **(N_+,N_-) coordination**, where N_+ is the coordination number of the cation and N_- that of the anion.

Even if, by chance, the ions have the same size, the problems of ensuring that the unit cells are electrically neutral make it impossible to achieve 12-coordinate close-packed ionic structures. As a result, ionic solids are generally less dense than metals. The best packing that can be achieved is the (8,8)-coordinate **caesium chloride structure** in which each cation is surrounded by eight anions and each anion is surrounded by eight cations (Fig. 18B.9). In this structure, an ion of one charge occupies the centre of a cubic unit cell with eight counter ions at its corners. The structure is adopted by CsCl itself and also by CaS.

When the radii of the ions differ more than in CsCl, even eight-coordinate packing cannot be achieved. One common structure adopted is the (6,6)-coordinate **rock salt structure** typified by rock salt itself, NaCl (Fig. 18B.10). In this structure, each cation is surrounded by six anions and each anion is surrounded by six cations. The rock salt structure can be pictured as consisting of two interpenetrating slightly expanded cubic F (fcc) arrays, one composed of cations and the other of anions. This structure is adopted by NaCl itself and also by several other MX compounds, including KBr, AgCl, MgO, and ScN.

The switch from the caesium chloride structure to the rock salt structure is related to the value of the **radius ratio**, γ:

$$\gamma=\frac{r_{\text{smaller}}}{r_{\text{larger}}} \qquad \textit{Definition} \quad \text{Radius ratio} \qquad (18B.3)$$

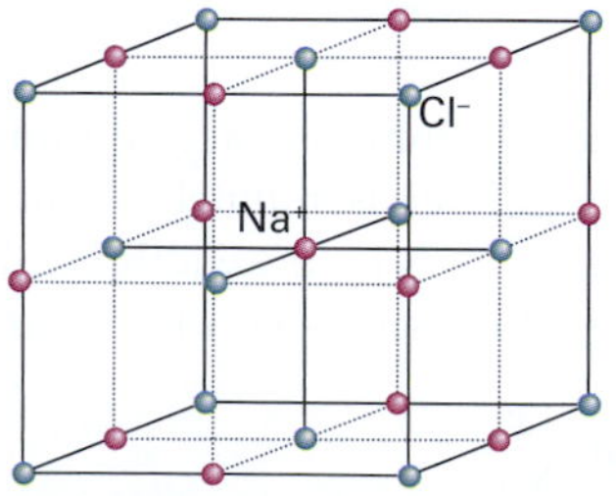

Figure 18B.10 The rock salt (NaCl) structure consists of two mutually interpenetrating slightly expanded face-centred cubic arrays of ions. The entire assembly shown here is the unit cell.

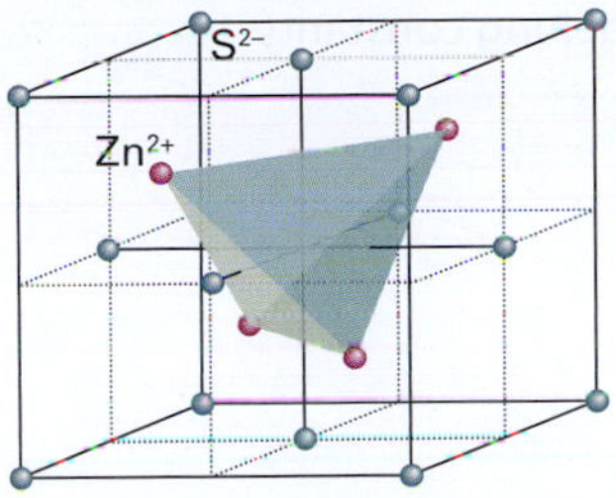

Figure 18B.11 The structure of the sphalerite form of ZnS showing the location of the Zn atoms in the tetrahedral holes formed by the array of S atoms. (There is an S atom at the centre of the cube inside the tetrahedron of Zn atoms.)

The two radii are those of the larger and smaller ions in the crystal. The **radius-ratio rule**, which is derived by considering the geometrical problem of packing the maximum number of hard spheres of one radius around a hard sphere of a different radius, can be summarized as follows:

Radius ratio	Structural type
$\gamma < 2^{1/2} - 1 = 0.414$	sphalerite (or zinc blende, Fig. 18B.11)
$0.414 < \gamma < 3^{1/2} - 1 = 0.732$	rock salt (Fig. 18B.10)
$\gamma > 0.732$	caesium chloride (Fig. 18B.9)

The deviation of a structure from that expected on the basis of this rule is often taken to be an indication of a shift from ionic towards covalent bonding. A major source of unreliability, though, is the arbitrariness of ionic radii (as we explain in a moment) and their variation with coordination number.

Ionic radii are derived from the distance between centres of adjacent ions in a crystal. However, we need to apportion the total distance between the two ions by defining the radius of one ion and then inferring the radius of the other ion. One scale that is widely used is based on the value 140 pm for the radius of the O^{2-} ion (Table 18B.2). Other scales are also available (such as one based on F^- for discussing halides), and it is essential not to mix values from different scales. Because ionic radii are so arbitrary, predictions based on them must be viewed cautiously.

Table 18B.2* Ionic radii, r/pm

Na^+	102 (6‡), 116 (8)
K^+	138 (6), 151 (8)
F^-	128 (2), 131 (4)
Cl^-	181 (close packing)

* This scale is based on a value 140 pm for the radius of the O^{2-} ion. More values are given in the *Resource section*.

‡ Coordination number.

Brief illustration 18B.2 The radius ratio

Using values of ionic radii from the *Resource section*, the radius ratio for MgO is

$$\gamma = \frac{\overbrace{72\,\text{pm}}^{\text{radius of Mg}^{2+}}}{\underbrace{140\,\text{pm}}_{\text{radius of O}^{2-}}} = 0.51$$

which is consistent with the observed rock salt structure of MgO crystals.

Self-test 18B.3 Predict the crystal structure of TlCl.

Answer: $\gamma = 0.88$; caesium chloride structure

(b) Energetics

The **lattice energy** of a solid is the difference in Coulombic potential energy of the ions packed together in a solid and widely separated as a gas. The lattice energy is always positive; a high lattice energy indicates that the ions interact strongly with one another to give a tightly bonded solid. The **lattice enthalpy**, ΔH_L, is the change in standard molar enthalpy for the process

$$MX(s) \rightarrow M^+(g) + X^-(g)$$

and its equivalent for other charge types and stoichiometries. At $T=0$ the lattice enthalpy is equal to the lattice energy; at normal temperatures they differ by only a few kilojoules per mole, and the difference is normally neglected.

Each ion in a solid experiences electrostatic attractions from all the other oppositely charged ions and repulsions from all the other like-charged ions. The total Coulomb potential energy is the sum of all the electrostatic contributions. Each cation is surrounded by anions, and there is a large negative contribution from the attraction of the opposite charges. Beyond those nearest neighbours, there are cations that contribute a positive term to the total potential energy of the central cation. There is also a negative contribution from the anions beyond those cations, a positive contribution from the cations beyond them, and so on to the edge of the solid. These repulsions and attractions become progressively weaker as the distance from the central ion increases, but the net outcome of all these contributions is dominated by the interaction between nearest neighbours and is a lowering of energy.

First, consider a simple one-dimensional model of a solid consisting of a long line of uniformly spaced alternating cations and anions, with d the distance between their centres, the sum of the ionic radii (Fig. 18B.12). If the charge numbers of the ions have the same absolute value (+1 and −1, or +2 and −2, for instance), then $z_1=+z$, $z_2=-z$, and $z_1z_2=-z^2$. The potential

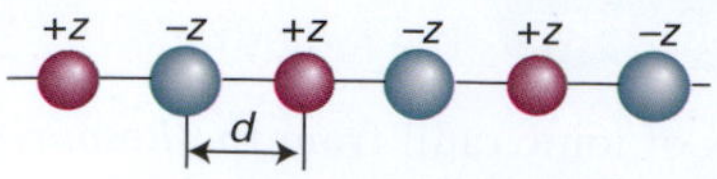

Figure 18B.12 A line of alternating cations and anions used in the calculation of the Madelung constant in one dimension.

energy of the central ion is calculated by summing all the terms, with negative terms representing attractions to oppositely charged ions and positive terms representing repulsions from like-charged ions. For the interaction with ions extending in a line to the right of the central ion, the lattice energy is

$$E_p = \frac{1}{4\pi\varepsilon_0} \times \left(-\frac{z^2e^2}{d} + \frac{z^2e^2}{2d} - \frac{z^2e^2}{3d} + \frac{z^2e^2}{4d} - \cdots \right)$$
$$= \frac{z^2e^2}{4\pi\varepsilon_0 d} \times \left(-1 + \frac{1}{2} - \frac{1}{3} + \frac{1}{4} - \cdots \right)$$
$$= -\frac{z^2e^2}{4\pi\varepsilon_0 d} \times \ln 2$$

We have used the relation $1 - \frac{1}{2} + \frac{1}{3} - \frac{1}{4} + \cdots = \ln 2$. Finally, we multiply E_p by 2 to obtain the total energy arising from interactions on each side of the ion and then multiply by Avogadro's constant, N_A, to obtain an expression for the (molar) lattice energy. The outcome is

$$E_p = -2\ln 2 \times \frac{z^2 N_A e^2}{4\pi\varepsilon_0 d}$$

with $d = r_{\text{cation}} + r_{\text{anion}}$. This energy is negative, corresponding to a net attraction. This calculation can be extended to three-dimensional arrays of ions with different charges:

$$E_p = -A \times \frac{|z_A z_B| N_A e^2}{4\pi\varepsilon_0 d} \qquad \text{(18B.4)}$$

The factor A is a positive numerical constant called the **Madelung constant**; its value depends on how the ions are arranged about one another. For a rock salt structure, $A = 1.748$. Table 18B.3 lists Madelung constants for other common structures.

There are also repulsions arising from the overlap of the filled atomic orbitals of the ions and, consequently, the role of the Pauli principle. These repulsions are taken into account by supposing that, because wavefunctions decay exponentially with distance at large distances from the nucleus, and repulsive interactions depend on the overlap of orbitals, the repulsive contribution to the potential energy has the form

$$E_p^* = N_A C' e^{-d/d^*} \qquad \text{(18B.5)}$$

Table 18B.3 Madelung constants

Structural type	A
Caesium chloride	1.763
Fluorite	2.519
Rock salt	1.748
Rutile	2.408
Sphalerite (zinc blende)	1.638
Wurtzite	1.641

with C' and d^* constants; the value of C' is not needed (it cancels in expressions that make use of this formula; see below) and that of d^* is commonly taken to be 34.5 pm. The total potential energy is the sum of E_p and E_p^*, and passes through a minimum when $\mathrm{d}(E_p + E_p^*)/\mathrm{d}d = 0$ (Fig. 18B.13). A short calculation leads to the following expression for the minimum total potential energy (see Problem 18B.9):

$$E_{p,\min} = -\frac{N_A |z_A z_B| e^2}{4\pi\varepsilon_0 d}\left(1 - \frac{d^*}{d}\right) A \qquad \text{Born–Mayer equation} \qquad \text{(18B.6)}$$

This expression is called the **Born–Mayer equation**. Provided we ignore zero-point contributions to the energy, the negative of this potential energy can be identified with the lattice energy. The important features of this equation are:

- Because $E_{p,\min} \propto |z_A z_B|$, the potential energy decreases (becomes more negative) with increasing charge number of the ions.
- Because $E_{p,\min} \propto 1/d$, the potential energy decreases (becomes more negative) with decreasing ionic radius.

Physical interpretation

The second conclusion follows from the fact that the smaller the ionic radii, the smaller the value of d. We see that large lattice energies are expected when the ions are highly charged (so $|z_A z_B|$ is large) and small (so d is small).

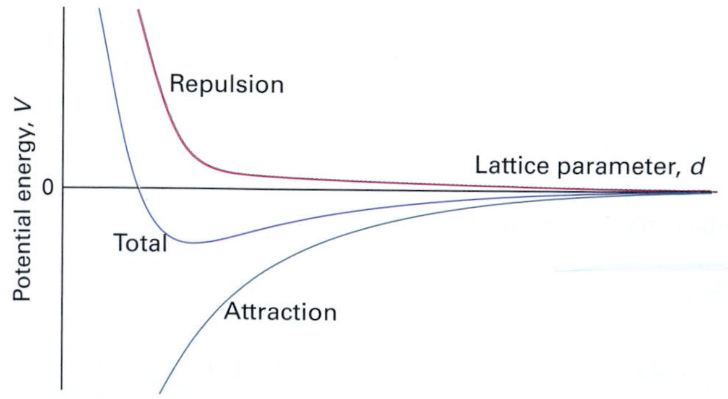

Figure 18B.13 The contributions to the total potential energy of an ionic crystal.

Brief illustration 18B.3 The Born–Mayer equation

To estimate $E_{p,min}$ for MgO, which has a rock salt structure ($A = 1.748$), we use $d = r(Mg^{2+}) + r(O^{2-}) = 72 + 140\,pm = 212\,pm$. We also use

$$\frac{N_A e^2}{4\pi\varepsilon_0} = \frac{(6.022\,14\times10^{23}\,mol^{-1})\times(1.601\,176\times10^{-19}\,C)^2}{4\pi\times(8.854\,19\times10^{-12}\,J^{-1}\,C^2\,m^{-1})}$$
$$= 1.387\,62\times10^{-4}\,J\,m\,mol^{-1}$$

and obtain

$$E_{p,min} = -\frac{\overbrace{4}^{|z_{Mg^{2+}}z_{O^{2-}}|}}{\underbrace{2.12\times10^{-10}\,m}_{d}}\times\overbrace{(1.38762\times10^{-4}\,J\,m\,mol^{-1})}^{N_A e^2/4\pi\varepsilon_0}\times \overbrace{\left(1-\frac{34.5\,pm}{212\,pm}\right)}^{1-d^*/d}\times\overbrace{1.748}^{A}$$
$$= -3.83\times10^3\,kJ\,mol^{-1}$$

Self-test 18B.4 Which can be expected to have the greater lattice energy, magnesium oxide or strontium oxide?

Answer: MgO

Experimental values of the lattice enthalpy (the enthalpy, rather than the energy) are obtained by using a **Born–Haber cycle**, a closed path of transformations starting and ending at the same point, one step of which is the formation of the solid compound from a gas of widely separated ions.

Example 18B.2 Using the Born–Haber cycle

Use the Born–Haber cycle to calculate the lattice enthalpy of KCl.

Method The Born–Haber cycle for KCl is shown in Fig 18B.14. It consists of the following steps (for convenience, starting at the elements):

	$\Delta H/(kJ\,mol^{-1})$	
1. Sublimation of K(s)	+89	[dissociation enthalpy of K(s)]
2. Dissociation of $\frac{1}{2}Cl_2(g)$	+122	[$\frac{1}{2}\times$dissociation enthalpy of $Cl_2(g)$]
3. Ionization of K(g)	+418	[ionization enthalpy of K(g)]
4. Electron attachment to Cl(g)	−349	[electron gain enthalpy of Cl(g)]
5. Formation of solid from gaseous ions	$-\Delta H_L/(kJ\,mol^{-1})$	
6. Decomposition of compound	+437	[negative of enthalpy of formation of KCl(s)]

Because this is a closed cycle, the sum of these enthalpy changes is equal to zero, and the lattice enthalpy can be inferred from the resulting equation.

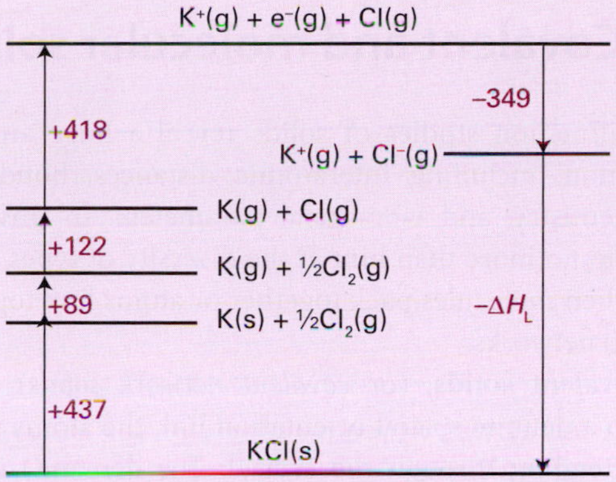

Figure 18B.14 The Born–Haber cycle for KCl at 298 K. Enthalpy changes are in kilojoules per mole.

Answer The equation associated with the cycle is

$$89 + 122 + 418 - 349 - \Delta H_L/(kJ\,mol^{-1}) + 437 = 0$$

It follows that $\Delta H_L = +717\,kJ\,mol^{-1}$.

Self-test 18B.5 Calculate the lattice enthalpy of CaO from the following data:

	$\Delta H/(kJ\,mol^{-1})$
Sublimation of Ca(s)	+178
Ionization of Ca(g) to $Ca^{2+}(g)$	+1735
Dissociation of $O_2(g)$	+249
Electron attachment to O(g)	−141
Electron attachment to $O^-(g)$	+844
Formation of CaO(s) from Ca(s) and $O_2(g)$	−635

Answer: $+3500\,kJ\,mol^{-1}$

Some lattice enthalpies obtained by the Born–Haber cycle are listed in Table 18B.4. As can be seen from the data, the trends in values are in general accord with the predictions of the Born–Mayer equation. Agreement is typically taken to imply that the ionic model of bonding is valid for the substance; disagreement

Table 18B.4* Lattice enthalpies at 298 K, $\Delta H_L/(kJ\,mol^{-1})$

NaF	787
NaBr	751
MgO	3850
MgS	3406

* More values are given in the *Resource section*.

implies that there is a covalent contribution to the bonding. It is important, though, to be cautious, because numerical agreement might be coincidental.

18B.3 Covalent and molecular solids

X-ray diffraction studies of solids reveal a huge amount of information, including interatomic distances, bond angles, stereochemistry, and vibrational parameters. In this section we can do no more than hint at the diversity of types of solids found when molecules pack together or atoms link together in extended networks.

In **covalent solids**, (or *covalent network solids*) covalent bonds in a definite spatial orientation link the atoms in a network extending through the crystal. The demands of directional bonding, which have only a small effect on the structures of many metals, now override the geometrical problem of packing spheres together, and elaborate and extensive structures may be formed.

Brief illustration 18B.4 Diamond and graphite

Diamond and graphite are two allotropes of carbon. In diamond each sp^3-hybridized carbon is bonded tetrahedrally to its four neighbours (Fig. 18B.15). The network of strong C—C bonds is repeated throughout the crystal and, as a result, diamond is very hard (in fact, the hardest known substance). In graphite, σ bonds between sp^2-hybridized carbon atoms form hexagonal rings which, when repeated throughout a plane, give rise to 'graphene' sheets (Fig. 18B.16). Because the sheets can slide against each other when impurities are present, impure graphite is used widely as a lubricant.

Self-test 18B.6 Identify the solids that form covalent networks: silicon, boron nitride, red phosphorus, and calcium carbonate.

Answer: Silicon, boron nitride, and red phosphorus are covalent networks; calcium carbonate is an ionic solid

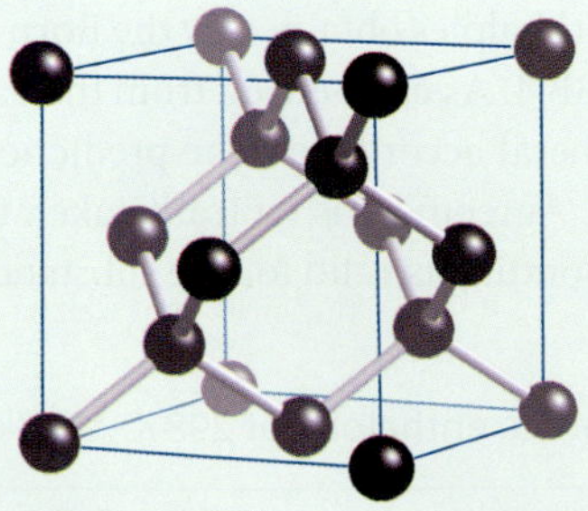

Figure 18B.15 A fragment of the structure of diamond. Each C atom is tetrahedrally bonded to four neighbours. This framework-like structure results in a rigid crystal.

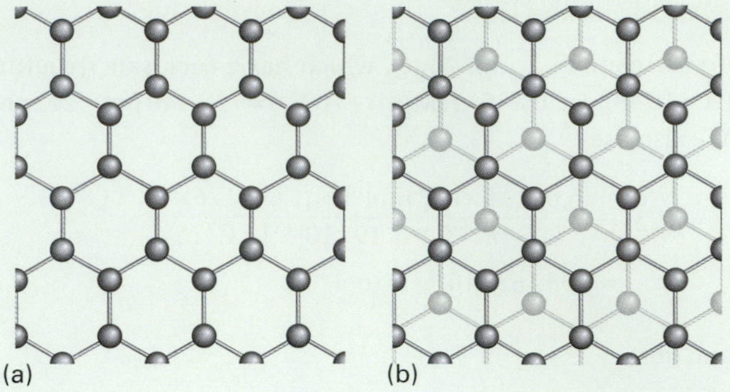

Figure 18B.16 Graphite consists of flat planes of hexagons of carbon atoms lying above one another. (a) The arrangement of carbon atoms in a 'graphene' sheet; (b) the relative arrangement of neighbouring sheets. The planes can slide over one another easily when impurities are present.

Molecular solids, which are the subject of the overwhelming majority of modern structural determinations, are held together by van der Waals interactions between the individual molecular components (Topic 16B). The observed crystal structure is nature's solution to the problem of condensing objects of various shapes into an aggregate of minimum energy (actually, for $T>0$, of minimum Gibbs energy). The prediction of the structure is difficult, but software specifically designed to

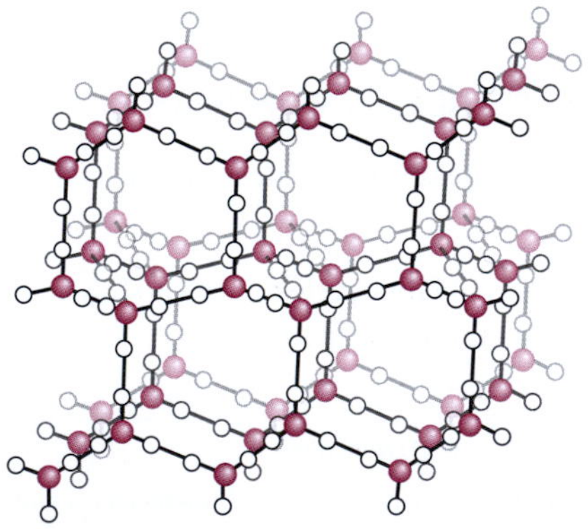

Figure 18B.17 A fragment of the crystal structure of ice (ice-I). Each O atom is at the centre of a tetrahedron of four O atoms at a distance of 276 pm. The central O atom is attached by two short O—H bonds to two H atoms and by two long hydrogen bonds to the H atoms of two of the neighbouring molecules. Both alternative H atoms locations are shown for each O—O separation. Overall, the structure consists of planes of hexagonal puckered rings of H_2O molecules (like the chair form of cyclohexane).

explore interaction energies can now make reasonably reliable predictions. The problem is made more complicated by the role of hydrogen bonds, which in some cases dominate the crystal structure, as in ice (Fig. 18B.17), but in others (for example, in phenol) distort a structure that is determined largely by the van der Waals interactions.

Checklist of concepts

- ☐ 1. The **coordination number** of an atom in a metal is the number of its nearest neighbours.
- ☐ 2. Many elemental metals have **close-packed structures** with coordination number 12.
- ☐ 3. Close-packed structures may be either cubic (ccp) or hexagonal (hcp).
- ☐ 4. The **packing fraction** is the fraction of space occupied by spheres in a crystal.
- ☐ 5. Electrons in metals occupy molecular orbitals formed from the overlap of atomic orbitals.
- ☐ 6. The **Fermi level** is the highest occupied molecular orbital at $T=0$.
- ☐ 7. Representative ionic structures include the caesium chloride, rock salt, and zinc blende structures.
- ☐ 8. The **coordination number of an ionic lattice** is denoted (N_+,N_-), with N_+ the number of nearest neighbour anions around a cation and N_- the number of nearest neighbour cations around an anion.
- ☐ 9. The **radius ratio** (see below) is a guide to the likely lattice type.
- ☐ 10. The **lattice enthalpy** is the change in enthalpy (per mole of formula units) accompanying the complete separation of the components of the solid.
- ☐ 11. A **Born–Haber cycle** is a closed path of transformations starting and ending at the same point, one step of which is the formation of the solid compound from a gas of widely separated ions.
- ☐ 12. A covalent network solid is a solid in which covalent bonds in a definite spatial orientation link the atoms in a network extending through the crystal.
- ☐ 13. A molecular solid is a solid consisting of discrete molecules held together by van der Waals interactions.

Checklist of equations

Property	Equation	Comment	Equation number
Energy levels of a linear array of orbitals	$E_k=\alpha+2\beta\cos(k\pi/(N+1)),\ k=1,2,\ldots,N$	Hückel approximation	18B.1
Band width	$E_N-E_1\to-4\beta$ as $N\to\infty$	Hückel approximation	18B.2
Radius ratio	$\gamma=r_{\text{smaller}}/r_{\text{larger}}$	For criteria, see Section 18B.2.	18B.3
Born–Mayer equation	$E_{\text{p,min}}=-\{N_A\lvert z_Az_B\rvert e^2/4\pi\varepsilon_0 d\}(1-d^*/d)A$	A is the Madelung constant	18B.6

18C Mechanical, electrical, and magnetic properties of solids

Contents

➤ Why do you need to know this material?

Careful consideration and manipulation of the physical properties of solids are needed for the development of modern materials and an understanding of their properties.

➤ What is the key idea?

The mechanical, electrical, and magnetic properties of solids stem from the properties of their constituent atoms and how they stack together.

➤ What do you need to know already?

You need to be familiar with electromagnetic fields (*Foundations* C), atomic structure (Topics 9A and 9B), and bonding arrangements in solids (Topic 18B), especially the formation of bands of orbitals. This Topic draws a little on the properties of the Boltzmann distribution (*Foundations* B and Topic 15A).

This Topic addresses the mechanical, electrical, and magnetic properties of solids. Optical properties are covered in Topic 18D.

18C.1 Mechanical properties

The fundamental concepts for the discussion of the mechanical properties of solids are stress and strain. The **stress** on an object is the applied force divided by the area to which it is applied. The **strain** is the resulting fractional distortion of the sample. The general field of the relations between stress and strain is called **rheology** from the Greek word for 'flow'.

Stress may be applied in a number of different ways (Fig. 18C.1):

- **Uniaxial stress** is a simple compression or extension in one direction.
- **Hydrostatic stress** is a stress applied simultaneously in all directions, as in a body immersed in a fluid.
- **Pure shear** is a stress that tends to push opposite faces of the sample in opposite directions.

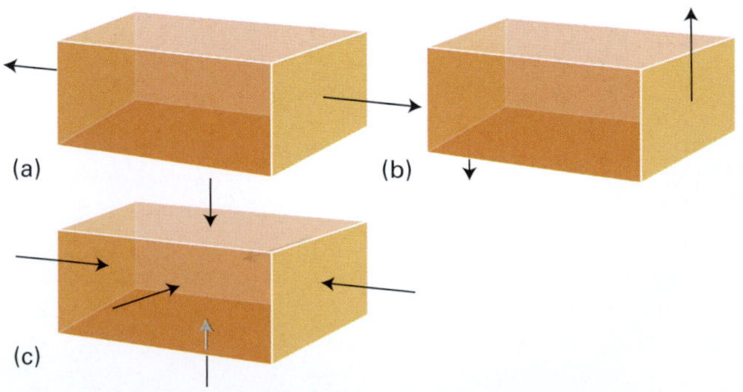

Figure 18C.1 Types of stress applied to a body. (a) Uniaxial stress, (b) shear stress, (c) hydrostatic pressure.

A sample subjected to a small stress typically undergoes **elastic deformation** in the sense that it recovers its original shape when the stress is removed. For low stresses, the strain is linearly proportional to the stress, and the stress–strain relation is a Hooke's law of force (as depicted in Fig. 17B.2 and reproduced here as Fig. 18C.2). The response becomes nonlinear at high stresses but may remain elastic. Above a certain threshold, the strain becomes **plastic** in the sense that recovery does not occur when the stress is removed. Plastic deformation occurs when bond breaking takes place and, in pure metals, typically takes place through the agency of dislocations. Brittle solids, such as ionic solids, exhibit sudden fracture as the stress focused by cracks causes them to spread catastrophically.

The response of a solid to an applied stress is commonly summarized by a number of coefficients of proportionality known as **moduli**:

$$E = \frac{\text{normal stress}}{\text{normal strain}} \quad \textit{Definition} \quad \text{Young's modulus} \quad (18C.1a)$$

$$K = \frac{\text{pressure}}{\text{fractional change in volume}} \quad \textit{Definition} \quad \text{Bulk modulus} \quad (18C.1b)$$

$$G = \frac{\text{shear stress}}{\text{shear strain}} \quad \textit{Definition} \quad \text{Shear modulus} \quad (18C.1c)$$

'Normal stress' refers to stretching and compression of the material, as shown in Fig. 18C.3a, and 'shear stress' refers to the stress depicted in Fig. 18C.3b. The fractional change in volume is $\delta V/V$, where δV is the change in volume of a sample of volume V; similarly, the strains are (dimensionless) fractional changes in dimensions. The bulk modulus is the inverse of the isothermal compressibility, κ, first encountered in Topic 2D (eqn 2D.7, $\kappa = -(\partial V/\partial p)_T/V$). A third ratio, called **Poisson's ratio**, indicates how the sample changes its shape:

$$\nu_P = \frac{\text{transverse strain}}{\text{normal strain}} \quad \textit{Definition} \quad \text{Poisson's ratio} \quad (18C.2)$$

The moduli are interrelated (see Problem 18C.1):

$$G = \frac{E}{2(1+\nu_P)} \qquad K = \frac{E}{3(1-2\nu_P)} \quad \text{Relations between moduli} \quad (18C.3)$$

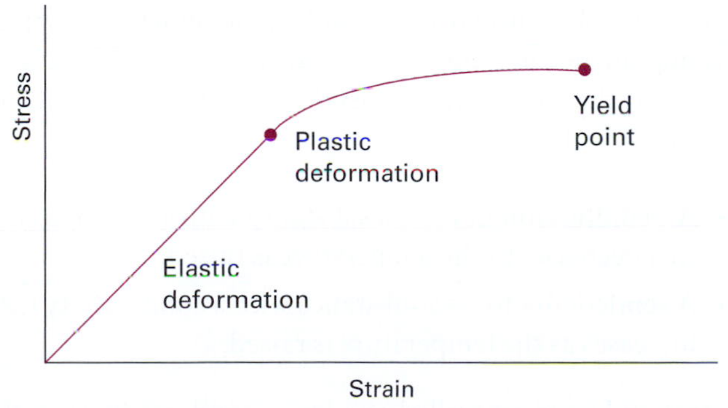

Figure 18C.2 At small strains, a body obeys Hooke's law (stress proportional to strain) and is elastic (recovers its shape when the stress is removed). At high strains, the body is no longer elastic, may yield and become plastic. At even higher strains, the solid fails (at its limiting tensile strength) and finally fractures.

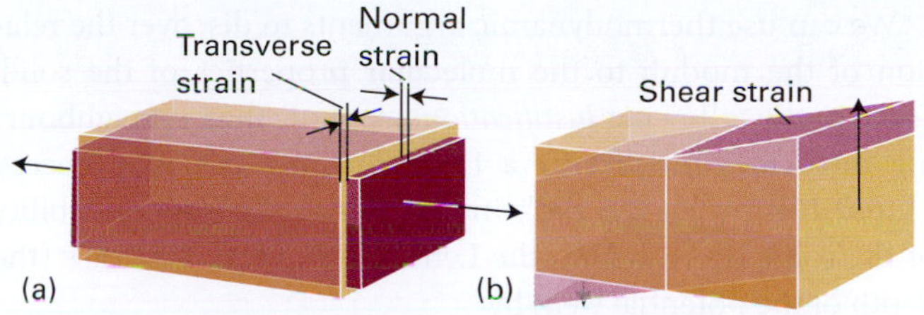

Figure 18C.3 (a) Normal stress and the resulting strain. (b) Shear stress. Poisson's ratio indicates the extent to which a body changes shape when subjected to a uniaxial stress.

Brief illustration 18C.1 Young's modulus

Young's modulus for iron at room temperature is 215 GPa. The normal strain produced when a mass of $m = 10.0\,\text{kg}$ is suspended from an iron wire of diameter $d = 0.10\,\text{mm}$ is

$$\text{normal strain} = \frac{\text{normal stress}}{E}$$

where the normal stress is the force F on the wire divided by the area on which the force acts. This area is $\pi(d/2)^2$, the area at the base of the cylindrical wire. The force is mg, where g is the acceleration of free fall. It follows that

$$\text{normal strain} = \frac{\overbrace{mg}^{F} / \overbrace{\pi(d/2)^2}^{A}}{E} = \frac{(10.0\,\text{kg}) \times \dfrac{9.81\,\text{m}\,\text{s}^{-2}}{\pi(5.0\times10^{-5}\,\text{m})^2}}{2.15\times10^{11}\,\underbrace{\text{kg}\,\text{m}^{-1}\,\text{s}^{-2}}_{\text{Pa}}} = 0.058$$

which corresponds to elongation of the wire by 5.8 per cent.

Self-test 18C.1 Young's modulus for polyethene at room temperature is 1.2 GPa. What strain will be produced when a mass of 1.0 kg is suspended from a polyethene thread of diameter 1.0 mm?

Answer: 0.010

We can use thermodynamic arguments to discover the relation of the moduli to the molecular properties of the solid. Thus, in the following *Justification*, we show that, if neighbouring molecules interact by a Lennard-Jones potential energy (Topic 16B), then the bulk modulus and the compressibility of the solid are related to the Lennard-Jones parameter ε (the depth of the potential well) by

$$K=\frac{8N_A\varepsilon}{V_m} \qquad \kappa=\frac{V_m}{8N_A\varepsilon} \tag{18C.4}$$

We see that the bulk modulus is large (the solid stiff) if the potential well represented by the Lennard-Jones potential energy is deep and the solid is dense (its molar volume small).

Justification 18C.1 The relation between compressibility and molecular interactions

First, we combine the relations $K=1/\kappa$ and $\kappa=-(\partial V/\partial p)_T/V$ with the thermodynamic relation $p=-(\partial U/\partial V)_T$ (this is eqn 3D.3 of Topic 3D), to obtain

$$K=-V\left(\frac{\partial p}{\partial V}\right)_T=V\left(\frac{\partial^2 U}{\partial V^2}\right)_T$$

This expression shows that the bulk modulus (and through eqn 18C.3, the other two moduli) depends on the curvature of a plot of the internal energy against volume. To develop this conclusion, we note that the variation of internal energy with volume can be expressed in terms of its variation with a lattice parameter, R, such as the length of the side of a unit cell:

$$\frac{\partial U}{\partial V}=\frac{\partial U}{\partial R}\frac{\partial R}{\partial V}$$

and so

$$\frac{\partial^2 U}{\partial V^2}=\frac{\partial U}{\partial R}\left(\frac{\partial^2 R}{\partial V^2}\right)+\left(\frac{\partial^2 U}{\partial V\partial R}\right)\frac{\partial R}{\partial V}$$

$$\frac{\partial^2 U}{\partial V\partial R}=\left(\frac{\partial^2 U}{\partial R^2}\right)\left(\frac{\partial R}{\partial V}\right)$$

$$\overset{\uparrow}{=}\frac{\partial U}{\partial R}\left(\frac{\partial^2 R}{\partial V^2}\right)+\frac{\partial^2 U}{\partial R^2}\left(\frac{\partial R}{\partial V}\right)^2$$

To calculate K at the equilibrium volume of the sample, we set $R=R_0$ and recognize that $\partial U/\partial R=0$ at equilibrium, so the first (blue) term on the right disappears and we are left with

$$K=V\left(\frac{\partial^2 U}{\partial R^2}\right)_{T,0}\left(\frac{\partial R}{\partial V}\right)_{T,0}^2$$

where the 0 denotes that the derivatives are evaluated at the equilibrium dimensions of the unit cell by setting $R=R_0$ after the derivative has been calculated. At this point we can write $V=aR^3$, where a is a constant that depends on the crystal structure, which implies that $\partial R/\partial V=1/3aR^2$. Then, if the internal energy is given by a pairwise Lennard-Jones (12,6)-potential energy, eqn 16B.14 of Topic 16B in the form $U=nN_AE_p$, with $E_p=4\varepsilon\{(R_0/R)^{12}-(R_0/R)^6\}$, we can write

$$\left(\frac{\partial^2 U}{\partial R^2}\right)_{T,0}=\frac{72nN_A\varepsilon}{R_0^2}$$

where n is the amount of substance in the sample of volume V_0. It then follows that

$$K=\frac{72nN_A\varepsilon}{9aR^3}=\frac{8nN_A\varepsilon}{V_0}=\frac{8N_A\varepsilon}{V_m}$$

where we have used $V_m=V_0/n$, which is the first of eqn 18C.4. The reciprocal of K is κ. To obtain the expression for $(\partial^2U/\partial R^2)_{T,0}$, we have used the fact that, at equilibrium, $R=R_0$ and $\sigma^6/R_0^6=\frac{1}{2}$ where σ is the scale parameter for the intermolecular potential energy (R_0 in the expression for E_p).

The differing rheological characteristics of metals can be traced to the presence of **slip planes**, which are planes of atoms that under stress may slip or slide relative to one another. The slip planes of a ccp structure are the close-packed planes, and careful inspection of a unit cell shows that there are eight sets of slip planes in different directions. As a result, metals with cubic close-packed structures, like copper, are malleable: they can easily be bent, flattened, or pounded into shape. In contrast, a hexagonal close-packed structure has only one set of slip planes; and metals with hexagonal close packing, like zinc or cadmium, tend to be brittle.

18C.2 Electrical properties

We confine attention to electronic conductivity, but note that some ionic solids display ionic conductivity in which complete ions migrate through the lattice. Two types of solid are distinguished by the temperature dependence of their electrical conductivity (Fig. 18C.4):

- A **metallic conductor** is a substance with a conductivity that decreases as the temperature is raised.
- A **semiconductor** is a substance with a conductivity that increases as the temperature is raised.

A semiconductor generally has a lower conductivity than that typical of metals, but the magnitude of the conductivity is not the criterion of the distinction. It is conventional to classify semiconductors with very low electrical conductivities, such as most synthetic polymers, as **insulators**. We shall use this term, but it should be appreciated that it is one of convenience rather than one of fundamental significance. A **superconductor** is a solid that conducts electricity without resistance.

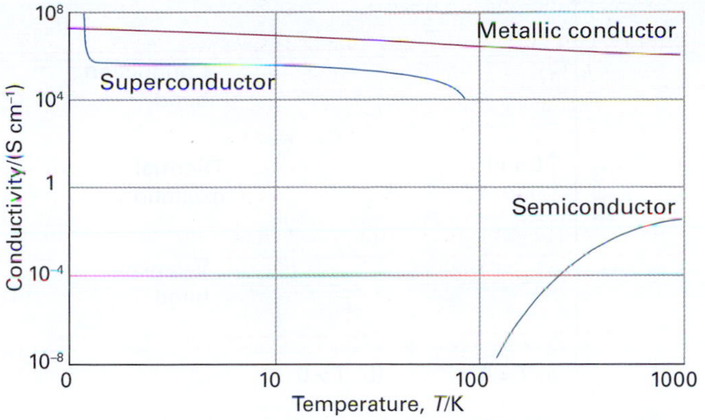

Figure 18C.4 The variation of the electrical conductivity of a substance with temperature is the basis of its classification as a metallic conductor, a semiconductor, or a superconductor. Conductivity is expressed in siemens per metre (S m^{-1} or, as here, S cm^{-1}), where 1 S = 1 Ω^{-1} (the resistance is expressed in ohms, Ω).

(a) Conductors

To understand the origins of electronic conductivity in conductors and semiconductors, we need to explore the consequences of the formation of bands in different materials (Topic 18B). Our starting point is Fig 18B.8, which is repeated here for convenience (Fig. 18C.5). It shows the electronic structure of a solid formed from atoms each able to contribute one electron (such as the alkali metals). At $T=0$, only the lowest $\frac{1}{2}N$ molecular orbitals are occupied, up to the Fermi level.

At temperatures above absolute zero, electrons are excited by the thermal motion of the atoms. The electrical conductivity of a metallic conductor decreases with increasing temperature even though more electrons are excited into empty orbitals. This apparent paradox is resolved by noting that the increase in temperature causes more vigorous thermal motion of the atoms, so collisions between the moving electrons and an atom are more likely. That is, the electrons are scattered out of their paths through the solid, and are less efficient at transporting charge.

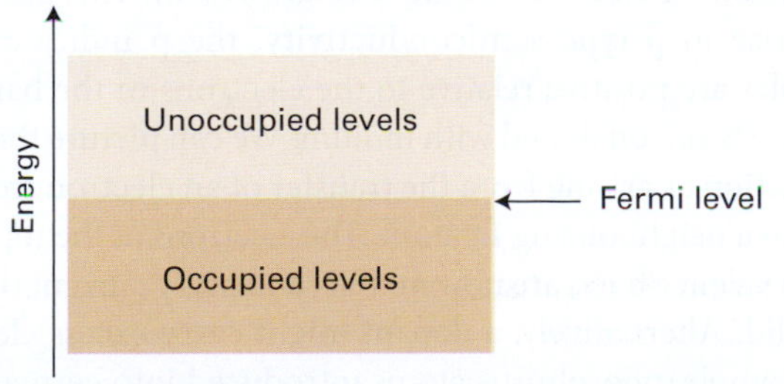

Figure 18C.5 When N electrons occupy a band of N orbitals, it is only half full and the electrons near the Fermi level (the top of the filled levels) are mobile.

A more quantitative treatment of conductivity in metals requires an expression for the variation with temperature of the distribution of electrons over the available states. We begin by considering the density of states, $\rho(E)$, at the energy E: the number of states between E and $E+dE$ divided by dE. Note that the 'state' of an electron includes its spin, so each spatial orbital counts as two states. Then it follows that $\rho(E)dE$ is the number of states between E and $E+dE$. To obtain the number of electrons $dN(E)$ that occupy states between E and $E+dE$, we multiply $\rho(E)dE$ by the probability $f(E)$ of occupation of the state with energy E. That is,

$$dN(E)=\overbrace{\rho(E)dE}^{\text{Number of states between } E \text{ and } E+dE}\times\overbrace{f(E)}^{\text{Probability of occupation of a state with energy } E} \quad (18C.5)$$

The function $f(E)$ is the **Fermi–Dirac distribution**, a version of the Boltzmann distribution that takes into account the Pauli exclusion principle that each orbital can be occupied by no more than two electrons (Fig. 18C.6):

$$f(E)=\frac{1}{e^{(E-\mu)/kT}+1} \quad \text{Fermi–Dirac distribution} \quad (18C.6a)$$

where μ is a temperature-dependent parameter known as the 'chemical potential' (it has a subtle relation to the familiar chemical potential of thermodynamics), and provided $T>0$ is the energy of the state for which $f=\frac{1}{2}$. At $T=0$, only states up to a certain energy known as the **Fermi energy**, E_F, are occupied (Fig. 18C.5). Provided the temperature is not so high that many electrons are excited to states above the Fermi energy, the chemical potential can be identified with E_F, in which case the Fermi–Dirac distribution becomes

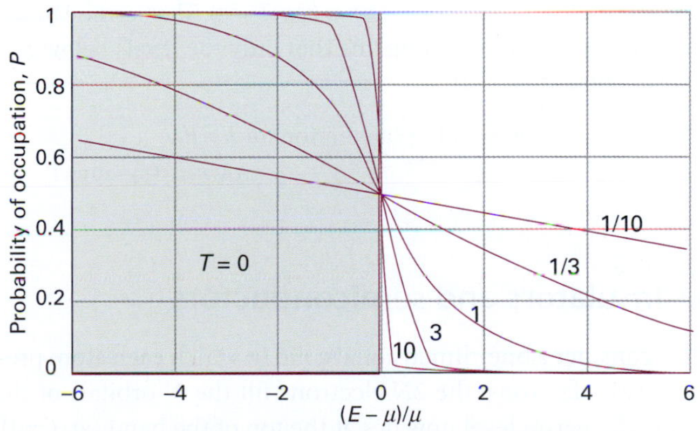

Figure 18C.6 The Fermi–Dirac distribution, which gives the probability of occupation of the state at a temperature T. The high-energy tail decays exponentially towards zero. The curves are labelled with the value of μ/kT. The tinted region shows the occupation of levels at $T=0$.

$$f(E)=\frac{1}{e^{(E-E_F)/kT}+1}$$ Fermi–Dirac distribution (18C.6b)

Moreover, for energies well above E_F, the exponential term in the denominator is so large that the 1 in the denominator can be neglected, and then

$$f(E)\approx e^{-(E-E_F)/kT}$$ *Approximate form for $E>E_F$* Fermi–Dirac distribution (18C.6c)

The function now resembles a Boltzmann distribution, decaying exponentially with increasing energy; the higher the temperature, the longer the exponential tail.

There is a distinction between the Fermi energy and the Fermi level. First, note that if $E=E_F$, then from eqn 18C.6b, $f(E_F)=\frac{1}{2}$. That is:

- The Fermi level is the uppermost occupied level at $T=0$.
- The Fermi energy is the energy level at which $f(E)=\frac{1}{2}$ at any temperature.

The Fermi energy coincides with the Fermi level as $T\rightarrow 0$.

Brief illustration 18C.2 The Fermi–Dirac distribution at $T=0$

Consider cases in which $E<E_F$. Then, as $T\rightarrow 0$ we write

$$\lim_{T\rightarrow 0}\{E-E_F\}/kT=-\infty$$

because $E_F>0$ and $E-E_F<0$. It follows that

$$\lim_{T\rightarrow 0} f(E)=\lim_{T\rightarrow 0}\frac{1}{\underbrace{e^{(E-E_F)/kT}}_{\rightarrow 0}+1}=1$$

We conclude that as $T\rightarrow 0$, $f(E)\rightarrow 1$, and all the energy levels below $E=E_F$ are populated. A similar calculation for $E>E_F$ (*Self-test* 18C.2) shows that $f(E)\rightarrow 0$ as $T\rightarrow 0$. The Fermi–Dirac distribution function confirms that only the levels below E_F are populated as $T\rightarrow 0$.

Self-test 18C.2 Repeat the calculation for $E>E_F$.

Answer: $f(E)\rightarrow 0$ as $T\rightarrow 0$

(b) Insulators and semiconductors

Now consider a one-dimensional solid in which each atom provides two electrons: the $2N$ electrons fill the N orbitals of the band. The Fermi level now lies at the top of the band (at $T=0$), and there is a gap before the next band begins (Fig. 18C.7). As the temperature is increased, the tail of the Fermi–Dirac distribution extends across the gap, and electrons leave the lower band, which is called the **valence band** and populate the empty orbitals of the upper band, which is called the **conduction band**. As a consequence of electron promotion, positively charged 'holes' are left in in the valence band. The holes and promoted electrons are now mobile, and the material is now a conductor. In fact, we call it a semiconductor, because the electrical conductivity depends on the number of electrons that are promoted across the gap, and that number increases as the temperature is raised. If the gap is large, though, very few electrons are promoted at ordinary temperatures and the conductivity remains close to zero, resulting in an insulator. Thus, the conventional distinction between an insulator and a semiconductor is related to the size of the band gap and is not an absolute distinction like that between a metal (incomplete bands at $T=0$) and a semiconductor (full bands at $T=0$).

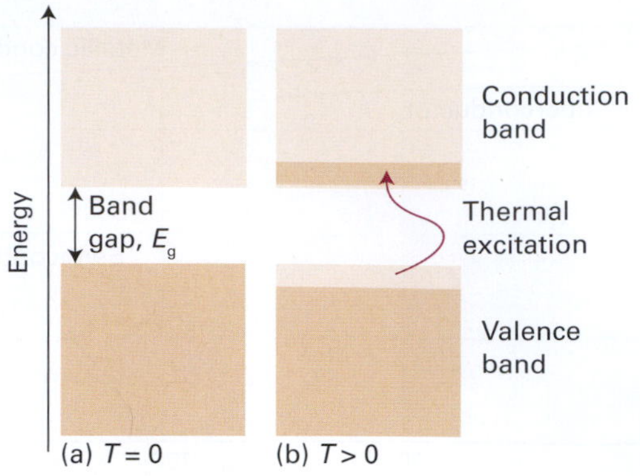

Figure 18C.7 (a) When $2N$ electrons are present, the band is full and the material is an insulator at $T=0$. (b) At temperatures above $T=0$, electrons populate the levels of the upper conduction band and the solid is a semiconductor.

Figure 18C.7 depicts conduction in an **intrinsic semiconductor**, in which semiconduction is a property of the band structure of the pure material. Examples of intrinsic semiconductors include silicon and germanium. A **compound semiconductor** is an intrinsic semiconductor that is a combination of different elements, such as GaN, CdS, and many d-metal oxides. An **extrinsic semiconductor** is one in which charge carriers are present as a result of the replacement of some atoms (to the extent of about 1 in 10^9) by **dopant** atoms, the atoms of another element. If the dopants can trap electrons, they withdraw electrons from the filled band, leaving holes which allow the remaining electrons to move (Fig. 18C.8a). This procedure gives rise to **p-type semiconductivity**, the p indicating that the holes are positive relative to the electrons in the band. An example is silicon doped with indium. We can picture the semiconduction as arising from the transfer of an electron from a Si atom to a neighbouring In atom. The electrons at the top of the silicon valence band are now mobile, and carry current through the solid. Alternatively, a dopant might carry excess electrons (for example, phosphorus atoms introduced into germanium), and these additional electrons occupy otherwise empty bands, giving **n-type semiconductivity**, where n denotes the negative charge of the carriers (Fig. 18C.8b).

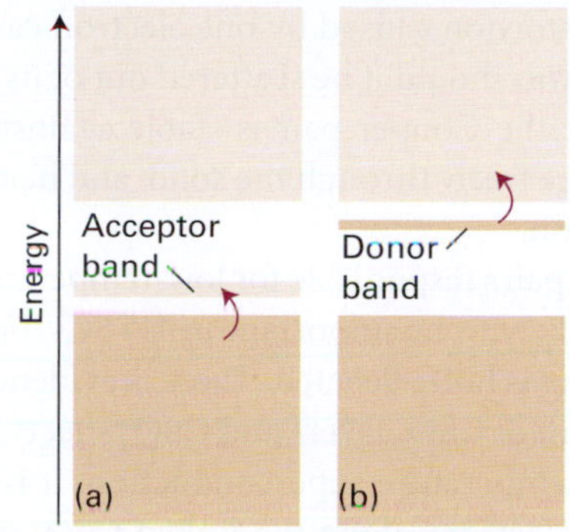

Figure 18C.8 (a) A dopant with fewer electrons than its host can form a narrow band that accepts electrons from the valence band. The holes in the band are mobile and the substance is a p-type semiconductor. (b) A dopant with more electrons than its host forms a narrow band that can supply electrons to the conduction band. The electrons it supplies are mobile and the substance is an n-type semiconductor.

Brief illustration 18C.3 The effect of doping on semiconductivity

Consider the doping of pure silicon (a Group 14 element) by arsenic (a Group 15 element). Because each Si atom has four valence electrons and each As atom has five valence electrons, the addition of arsenic increases the number of electrons in the solid. These electrons populate the empty conduction band of silicon, and the doped material is an n-type semiconductor.

Self-test 18C.3 Is gallium-doped germanium a p-type or an n-type semiconductor?

Answer: p-type semiconductor

Now consider the properties of a **p–n junction**, the interface of a p-type and an n-type semiconductor. When a 'reverse bias' is applied to the junction, in the sense that a negative electrode is attached to the p-type semiconductor and a positive electrode is attached to the n-type semiconductor, the positively charged holes in the p-type semiconductor are attracted to the negative electrode and the negatively charged electrons in the n-type semiconductor are attracted to the positive electrode (Fig. 18C.9a). As a consequence, charge does not flow across the junction. Now consider the application of a 'forward bias' to the junction, in the sense that the positive electrode is attached to the p-type semiconductor and the negative electrode is attached to the n-type semiconductor (Fig. 18C.9b). Now charge flows across the junction, with electrons in the n-type semiconductor moving towards the positive electrode and holes moving in the opposite direction. It follows that a p–n junction affords a great deal of control over the magnitude and direction of current through a material. This control is essential for the operation

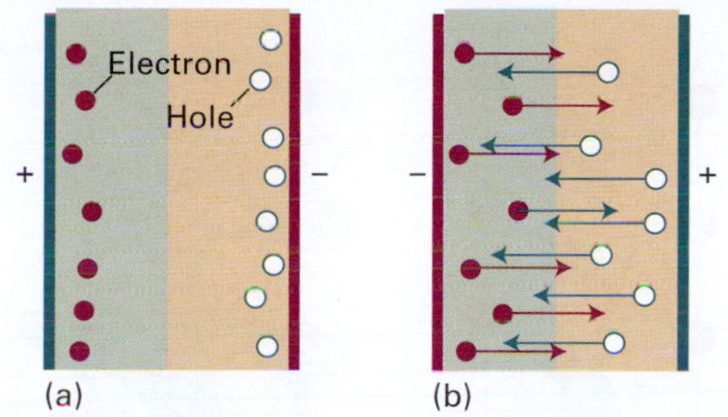

Figure 18C.9 A p–n junction under (a) reverse bias, (b) forward bias.

of transistors and diodes, which are key components of modern electronic devices.

As electrons and holes move across a p–n junction under forward bias, they recombine and release energy. However, as long as the forward bias continues to be applied, the flow of charge from the electrodes to the semiconductors replenishes them with electrons and holes, so the junction sustains a current. In some solids, the energy of electron–hole recombination is released as heat and the device becomes warm. The reason lies in the fact that the return of the electron to a hole involves a change in the electron's linear momentum. The atoms of the lattice must absorb the difference, and therefore electron-hole recombination stimulates lattice vibrations. This is the case for silicon semiconductors, and is one reason why computers need efficient cooling systems.

(c) Superconductivity

The resistance to flow of electrical current of a normal metallic conductor decreases smoothly with decreasing temperature but never vanishes. However, certain solids known as **superconductors** conduct electricity without resistance below a critical temperature, T_c. Following the discovery in 1911 that mercury is a superconductor below 4.2 K, the normal boiling point of liquid helium, physicists and chemists made slow but steady progress in the discovery of superconductors with higher values of T_c. Metals, such as tungsten, mercury, and lead, tend to have T_c values below about 10 K. Intermetallic compounds, such as Nb_3X (X=Sn, Al, or Ge), and alloys, such as Nb/Ti and Nb/Zr, have intermediate T_c values ranging between 10 K and 23 K. In 1986, **high-temperature superconductors** (HTSC) were discovered. Several *ceramics*, inorganic powders that have been fused and hardened by heating to a high temperature, containing oxocuprate motifs, Cu_mO_n, are now known with T_c values well above 77 K, the boiling point of the inexpensive refrigerant liquid nitrogen. For example, $HgBa_2Ca_2Cu_2O_8$ has $T_c = 153$ K.

There is a degree of periodicity in the elements that exhibit superconductivity. The metals iron, cobalt, nickel, copper, silver, and gold do not display superconductivity; nor do the alkali metals. One of the most widely studied oxocuprate superconductors $YBa_2Cu_3O_7$ (informally known as '123' on account of

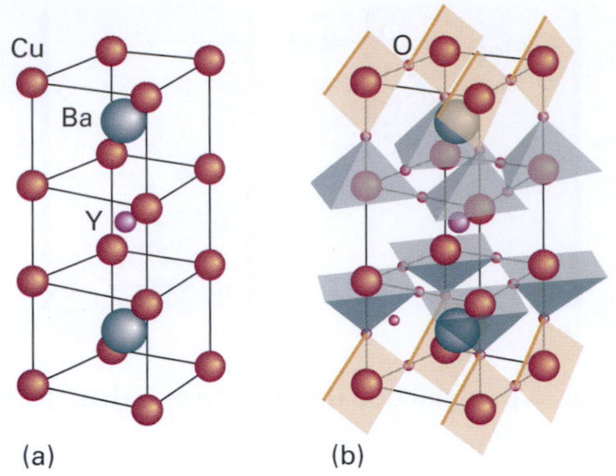

Figure 18C.10 Structure of the $YBa_2Cu_3O_7$ superconductor. (a) Metal atom positions. (b) The polyhedra show the positions of oxygen atoms and indicate that the metal ions are in square-planar and square-pyramidal coordination environments.

the proportions of the metal atoms in the compound) has the structure shown in Fig. 18C.10. The square-pyramidal CuO_5 units arranged as two-dimensional layers and the square planar CuO_4 units arranged in sheets are common structural features of oxocuprate HTSCs.

The mechanism of superconduction is well-understood for low-temperature materials, and is based on the properties of a **Cooper pair**, a pair of electrons that exists on account of the indirect electron–electron interactions mediated by the nuclei of the atoms in the lattice. Thus, if one electron is in a particular region of a solid, the nuclei there move toward it to give a distorted local structure (Fig. 18C.11). Because that local distortion is rich in positive charge, it is favourable for a second electron to join the first. Hence, there is a virtual attraction between the two electrons and they move together as a pair. The local distortion is disrupted by thermal motion of the ions in the solid, so the virtual attraction occurs only at very low temperatures. A Cooper pair undergoes less scattering than an individual electron as it travels through the solid

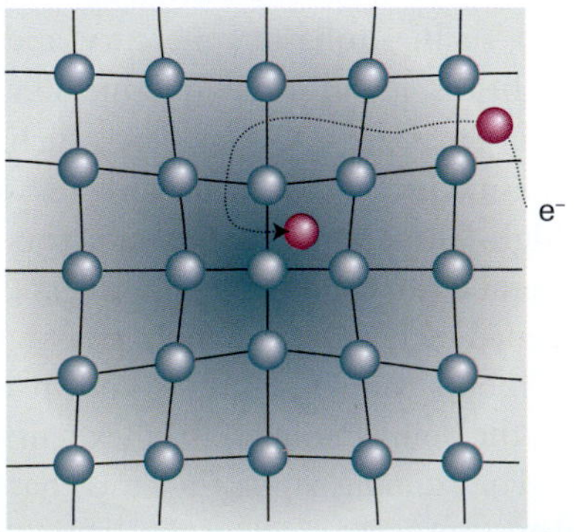

Figure 18C.11 The formation of a Cooper pair. One electron distorts the crystal lattice and the second electron has a lower energy if it goes to that region. These electron–lattice interactions effectively bind the two electrons into a pair.

because the distortion caused by one electron can attract back the other electron should it be scattered out of its path in a collision. Because the Cooper pair is stable against scattering, it can carry charge freely through the solid, and hence give rise to superconduction.

The Cooper pairs responsible for low-temperature superconductivity are likely to be important in HTSCs, but the mechanism for pairing is hotly debated. There is evidence implicating the arrangement of CuO_5 layers and CuO_4 sheets in the mechanism of high-temperature superconduction. It is believed that movement of electrons along the linked CuO_4 units accounts for superconductivity, whereas the linked CuO_5 units act as 'charge reservoirs' that maintain an appropriate number of electrons in the superconducting layers.

18C.3 Magnetic properties

The magnetic properties of metallic solids and semiconductors depend strongly on the band structures of the material. Here we confine our attention largely to magnetic properties that stem from collections of individual molecules or ions such as d-metal complexes. Much of the discussion applies to liquid and gas phase samples as well as to solids.

(a) Magnetic susceptibility

The magnetic and electric properties of molecules and solids are analogous. For instance, some molecules possess permanent magnetic dipole moments, and an applied magnetic field can induce a magnetic moment, with the result that the entire solid sample becomes magnetized. The **magnetization**, $\mathcal{M}$, is the average molecular magnetic dipole moment multiplied by the number density of molecules in the sample. The magnetization induced by a field of strength $\mathcal{H}$ is proportional to $\mathcal{H}$, and we write

$$\mathcal{M} = \chi\mathcal{H} \qquad \text{Magnetization} \qquad (18C.7)$$

where χ is the dimensionless **volume magnetic susceptibility**. A closely related quantity is the **molar magnetic susceptibility**, χ_m:

$$\chi_m = \chi V_m \qquad \text{Molar magnetic susceptibility} \qquad (18C.8)$$

where V_m is the molar volume of the substance.

We can think of the magnetization as contributing to the density of lines of force in the material (Fig. 18C.12). Materials for which $\chi > 0$ are called **paramagnetic**; they tend to move into a magnetic field and the density of lines of force within them is greater than in a vacuum. Those for which $\chi < 0$ are called **diamagnetic** and tend to move out of a magnetic field; the density

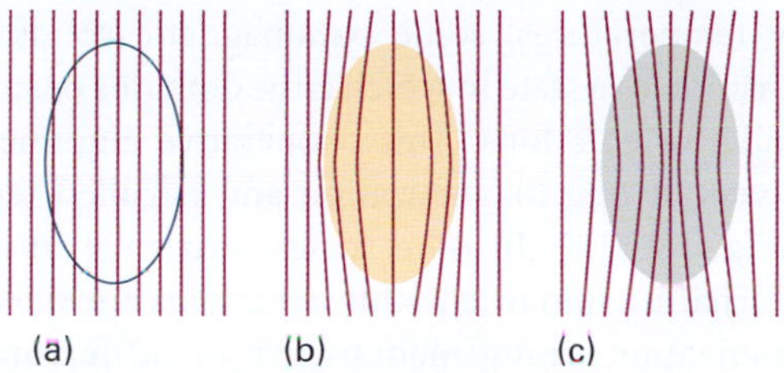

Figure 18C.12 (a) In a vacuum, the strength of a magnetic field can be represented by the density of lines of force; (b) in a diamagnetic material, the lines of force are reduced; (c) in a paramagnetic material, the lines of force are increased.

of lines of force within them is lower than in a vacuum. A paramagnetic material consists of ions or molecules with unpaired electrons, such as radicals and many d-metal complexes; a diamagnetic substance (a far more common property) is one with no unpaired electrons.

Brief illustration 18C.4 The magnetic character of metallic solids and molecules

Solid magnesium is a metal in which the two valence electrons of each Mg atom are donated to a band of orbitals constructed from 3s orbitals. From N atomic orbitals we can construct N molecular orbitals spreading through the metal. Each atom supplies two electrons, so there are $2N$ electrons to accommodate. These occupy and fill the N molecular orbitals. There are no unpaired electrons, so the metal is diamagnetic. An O_2 molecule has the electronic structure described in Topic 10C, where we see that two electrons occupy separate antibonding π orbitals with parallel spins. We conclude that O_2 is a paramagnetic gas.

Self-test 18C.4 Repeat the analysis for Zn(s) and NO(g).

Answer: Zn diamagnetic, NO paramagnetic

The magnetic susceptibility is traditionally measured with a **Gouy balance**. This instrument consists of a sensitive balance from which the sample hangs in the form of a narrow cylinder and lies between the poles of a magnet. If the sample is paramagnetic, it is drawn into the field, and its apparent weight is greater than when the field is off. A diamagnetic sample tends to be expelled from the field and appears to weigh less when the field is turned on. The balance is normally calibrated against a sample of known susceptibility. The modern version of the determination makes use of a **superconducting quantum interference device** (SQUID, Fig. 18C.13). A SQUID takes advantage of the quantization of magnetic flux and the property of current loops in superconductors that, as part of the circuit, include a weakly conducting link through which electrons must tunnel. The current that flows in the loop in a magnetic field depends on the value of the magnetic flux, and a SQUID can be exploited as a very sensitive magnetometer. Table 18C.1 lists some experimental values.

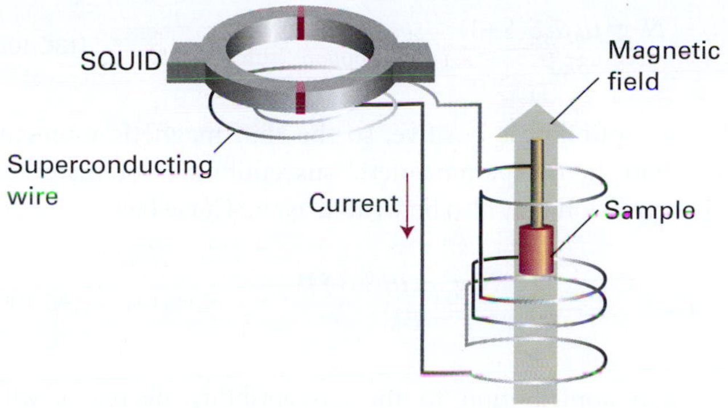

Figure 18C.13 The arrangement used to measure magnetic susceptibility with a SQUID. The sample is moved upwards in small increments and the potential difference across the SQUID is measured.

(b) Permanent and induced magnetic moments

The permanent magnetic moment of a molecule arises from any unpaired electron spins in the molecule. The magnitude of the magnetic moment of an electron is proportional to the magnitude of the spin angular momentum, $\{s(s+1)\}^{1/2}\hbar$:

$$m = g_e\{s(s+1)\}^{1/2}\mu_B \qquad \mu_B = \frac{e\hbar}{2m_e} \qquad \textit{Magnitude} \quad \text{Magnetic moment} \qquad (18C.9)$$

where $g_e = 2.0023$ and μ_B, the Bohr magneton, has the value $9.274\times10^{-24}\,\text{J T}^{-1}$. If there are several electron spins in each molecule, they combine to give a total spin S, and then $s(s+1)$ should be replaced by $S(S+1)$.

The magnetization and consequently the magnetic susceptibility depend on the temperature because the orientations of the electron spins fluctuate, whether the molecules are in fluid phases or trapped in solids: some orientations have lower energy than others, and the magnetization depends on the randomizing influence of thermal motion. Thermal averaging of the permanent magnetic moments in the presence of an applied magnetic field contributes to the magnetic susceptibility an amount proportional to $m^2/3kT$.[1] It follows that the spin contribution to the molar magnetic susceptibility is

Table 18C.1* Magnetic susceptibilities at 298 K

	$\chi/10^{-6}$	$\chi_m/(10^{-10}\,\text{m}^3\,\text{mol}^{-1})$
$H_2O(l)$	−9.02	−1.63
NaCl(s)	−16	−3.8
Cu(s)	−9.7	−0.69
$CuSO_4{\cdot}5H_2O(s)$	+167	+183

* More values are given in the *Resource section*.

[1] See our *Physical chemistry: Quanta, matter, and change* (2014) for the derivation of this contribution.

$$\chi_m = \frac{N_A g_e^2 \mu_0 \mu_B^2 S(S+1)}{3kT}$$ *Spin contribution* Molar magnetic susceptibility (18C.10a)

The susceptibility is positive, so the spin magnetic moments contribute to the paramagnetic susceptibilities of materials. This expression may also be written as the **Curie law**:

$$\chi_m = \frac{C}{T} \qquad C = \frac{N_A g_e^2 \mu_0 \mu_B^2 S(S+1)}{3k}$$ Curie law (18C.10b)

The spin contribution to the susceptibility decreases with increasing temperature because the thermal motion randomizes the spin orientations. In practice, a contribution to the paramagnetism also arises from the orbital angular momenta of electrons: we have discussed the spin-only contribution.

Example 18C.1 Calculating a molar magnetic susceptibility

Consider a complex salt with three unpaired electrons per complex cation at 298 K and molar volume 61.7 $cm^3 mol^{-1}$. Calculate the molar magnetic susceptibility and the volume magnetic susceptibility of the complex.

Method Use the data and eqn 18C.10 to calculate the molar magnetic susceptibility. Then use the values of χ_m and V_m, and eqn 18C.8 to calculate the volume magnetic susceptibility.

Answer First note that the constants can be collected into the term:

$$\frac{N_A g_e^2 \mu_0 \mu_B^2}{3k} = 6.3001\times10^{-6}\,m^3\,K^{-1}\,mol^{-1}$$

Consequently eqn 18C.10 becomes

$$\chi_m = 6.3001\times10^{-6}\times\frac{S(S+1)}{T/K}\,m^3\,mol^{-1}$$

Substitution of the data with $S=\frac{3}{2}$ gives

$$\chi_m = 6.3001\times10^{-6}\times\frac{\frac{3}{2}(\frac{3}{2}+1)}{298}\,m^3\,mol^{-1} = 7.93\times10^{-8}\,m^3\,mol^{-1}$$

If follows from eqn 18C.8 that, to obtain the volume magnetic susceptibility, the molar susceptibility is divided by the molar volume $V_m = 61.7\,cm^3\,mol^{-1} = 6.17\times10^{-5}\,m^3\,mol^{-1}$ and

$$\chi = \frac{\chi_m}{V_m} = \frac{7.93\times10^{-8}\,m^3\,mol^{-1}}{6.17\times10^{-5}\,m^3\,mol^{-1}} = 1.29\times10^{-3}$$

Self-test 18C.5 Repeat the calculation for a complex with five unpaired electrons, molar mass 322.4 g mol^{-1}, and a mass density of 2.87 g cm^{-3} at 273 K.

Answer: $\chi_m = 2.02\times10^{-7}\,m^3\,mol^{-1}$; $\chi = 1.79\times10^{-3}$

At low temperatures, some paramagnetic solids make a phase transition to a state in which large domains of spins align with parallel orientations. This cooperative alignment gives rise to a very strong magnetization and is called **ferromagnetism** (Fig. 18C.14). In other cases, exchange interactions lead to alternating spin orientations: the spins are locked into a low-magnetization arrangement to give an **antiferromagnetic phase**. The ferromagnetic phase has a nonzero magnetization in the absence of an applied field, but the antiferromagnetic phase has a zero magnetization because the spin magnetic moments cancel. The ferromagnetic transition occurs at the **Curie temperature**, and the antiferromagnetic transition occurs at the **Néel temperature**. Which type of cooperative behaviour occurs depends on the details of the band structure of the solid.

Magnetic moments can also be induced in molecules. To see how this effect arises, we need to note that the circulation of electronic currents induced by an applied field gives rise to a magnetic field which usually opposes the applied field, so the substance is diamagnetic. In these cases, the induced electron currents occur within the orbitals of the molecule that are occupied in its ground state. In the few cases in which molecules are paramagnetic despite having no unpaired electrons, the induced electron currents flow in the opposite direction because they can make use of unoccupied orbitals that lie close to the HOMO in energy. This orbital paramagnetism can be distinguished from spin paramagnetism by the fact that it is temperature independent: this is why it is called **temperature-independent paramagnetism** (TIP).

We can summarize these remarks as follows. All molecules have a diamagnetic component to their susceptibility, but it is dominated by spin paramagnetism if the molecules have unpaired electrons. In a few cases (where there are low-lying excited states) TIP is strong enough to make the molecules paramagnetic even though their electrons are paired.

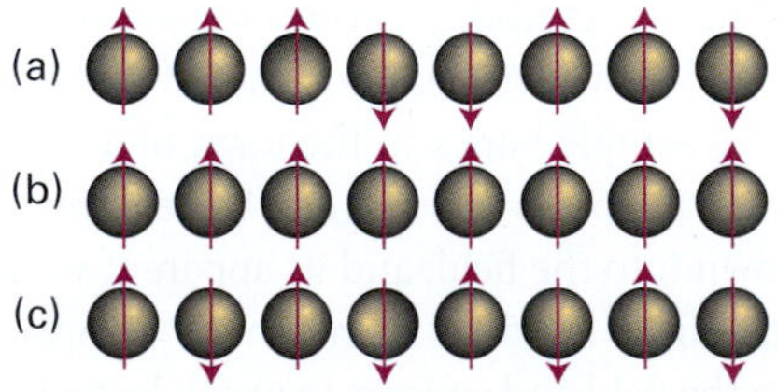

Figure 18C.14 (a) In a paramagnetic material, the electron spins are aligned at random in the absence of an applied magnetic field. (b) In a ferromagnetic material, the electron spins are locked into a parallel alignment over large domains. (c) In an antiferromagnetic material, the electron spins are locked into an antiparallel arrangement. The latter two arrangements survive even in the absence of an applied field.

(c) Magnetic properties of superconductors

Superconductors have unique magnetic properties. Some superconductors, classed as *Type I*, show abrupt loss of superconductivity when an applied magnetic field exceeds a critical value $\mathcal{H}_c$ characteristic of the material. It is observed that the value of $\mathcal{H}_c$ depends on temperature and T_c as

$$\mathcal{H}_c(T)=\mathcal{H}_c(0)\left(1-\frac{T^2}{T_c^2}\right) \qquad \text{Dependence of } \mathcal{H}_c \text{ on } T_c \qquad (18C.11)$$

where $\mathcal{H}_c(0)$ is the value of $\mathcal{H}_c$ as $T\rightarrow 0$.

Example 18C.2 Calculating the temperature at which a material becomes superconducting

Lead has $T_c=7.19\,\mathrm{K}$ and $\mathcal{H}_c(0)=63.9\,\mathrm{kA\,m^{-1}}$. At what temperature does lead become superconducting in a magnetic field of strength $20\,\mathrm{kA\,m^{-1}}$?

Method Rearrange eqn 18C.11 and use the data to calculate the temperature at which the substance becomes superconducting.

Answer Rearrangement of eqn 18C.11 gives

$$T=T_c\left(1-\frac{\mathcal{H}_c(T)}{\mathcal{H}_c(0)}\right)^{1/2}$$

and substitution of the data gives

$$T=7.19\,\mathrm{K}\times\left(1-\frac{20\,\mathrm{kA\,m^{-1}}}{63.9\,\mathrm{kA\,m^{-1}}}\right)^{1/2}=6.0\,\mathrm{K}$$

We conclude that lead becomes superconducting at temperatures below 6.0 K.

Self-test 18C.6 Tin has $T_c=3.72\,\mathrm{K}$ and $\mathcal{H}_c(0)=25\,\mathrm{kA\,m^{-1}}$. At what temperature does tin become superconducting in a magnetic field of strength $15\,\mathrm{kA\,m^{-1}}$?

Answer: 2.4 K

Type I superconductors are also completely diamagnetic below $\mathcal{H}_c$, meaning that the magnetic field does not penetrate into the material. This complete exclusion of a magnetic field from a material is known as the **Meissner effect**, which can be demonstrated by the levitation of a superconductor above a magnet. *Type II* superconductors, which include the HTSCs, show a gradual loss of superconductivity and diamagnetism with increasing magnetic field.

Checklist of concepts

- ☐ 1. **Uniaxial stress** is a simple compression or extension in one direction.
- ☐ 2. **Hydrostatic stress** is a stress applied simultaneously in all directions, as in a body immersed in a fluid.
- ☐ 3. A **pure shear** is a stress that tends to push opposite faces of the sample in opposite directions.
- ☐ 4. A sample subjected to a small stress typically undergoes **elastic deformation**.
- ☐ 5. The response of a solid to an applied stress is summarized by the **Young's modulus**, the **bulk modulus**, the **shear modulus**, and **Poisson's ratio**.
- ☐ 6. The differing rheological characteristics of metals can be traced to the presence of **slip planes**.
- ☐ 7. Electronic conductors are classified as **metallic conductors** or **semiconductors** according to the temperature dependence of their conductivities.
- ☐ 8. Semiconductors are classified as **p-type** or **n-type** according to whether conduction is due to holes in the valence band or electrons in the conduction band.
- ☐ 9. An **insulator** is a semiconductor with a very low electrical conductivity.
- ☐ 10. A **superconductor** conducts electricity without resistance below a critical temperature T_c.
- ☐ 11. A **Cooper pair** is a pair of electrons that exists on account of the indirect electron–electron interactions mediated by the nuclei of the atoms in the lattice.
- ☐ 12. A **diamagnetic material** moves out of a magnetic field; it has a negative magnetic susceptibility.
- ☐ 13. A **paramagnetic material** moves into a magnetic field; it has a positive magnetic susceptibility.
- ☐ 14. The **Curie law** describes the temperature dependence of the magnetic susceptibility.
- ☐ 15. **Ferromagnetism** is the cooperative alignment of electron spins in a material and gives rise to strong permanent magnetization.
- ☐ 16. **Antiferromagnetism** results from alternating spin orientations in a material and leads to weak magnetization.
- ☐ 17. **Temperature-independent paramagnetism** arises from induced electron currents that make use of excited states of molecules.

Checklist of equations

Property	Equation	Comment	Equation number
Young's modulus	E=normal stress/normal strain	Definition	18C.1a
Bulk modulus	K=pressure/fractional change in volume	Definition	18C.1b
Shear modulus	G=shear stress/shear strain	Definition	18C.1c
Poisson's ratio	ν_P=transverse strain/normal strain	Definition	18C.2
Fermi–Dirac distribution	$f(E)=1/\{e^{(E-\mu)/kT}+1\}$	μ is the chemical potential	18C.6
Magnetization	$\mathcal{M}=\chi\mathcal{H}$	Definition	18C.7
Molar magnetic susceptibility	$\chi_m=\chi V_m$	Definition	18C.8
Magnetic moment	$m=g_e\{s(s+1)\}^{1/2}\mu_B$	$\mu_B=e\hbar/2m_e$	18C.9
Molar magnetic susceptibility	$\chi_m=\{N_A g_e^2\mu_0\mu_B^2 S(S+1)\}/3kT$	Spin contribution	18C.10a
Curie law	$\chi_m=C/T,\quad C=N_A g_e^2\mu_0\mu_B^2 S(S+1)/3k$	Paramagnetism	18C.10b
Dependence of $\mathcal{H}_c$ on T_c	$\mathcal{H}_c(T)=\mathcal{H}_c(0)(1-T^2/T_c^2)$	Empirical	18C.11

18D The optical properties of solids

Contents

➤ Why do you need to know this material?

The optical properties of solids are of ever increasing importance in modern technology, not only for the generation of light but for the propagation and manipulation of information. You need to be aware of these properties to understand and contribute to the development of new optical technologies.

➤ What is the key idea?

The optical properties of solids stem from electronic transitions between the available orbitals in the material and the interaction between the transition dipoles.

➤ What do you need to know already?

You need to be familiar with basic properties of electromagnetic fields (*Foundations* C) and bonding arrangements in solids (Topic 18B), especially the band structure of solids. The Topic draws on the factors that determine the absorption of light by atoms and molecules (especially Topic 12A) and the operation of lasers (Topic 13C).

In this Topic we explore the consequences of interactions between electromagnetic radiation and solids. Our focus is on the origins of phenomena that are the basis of the design of a variety of useful devices, such as lasers and light-emitting diodes.

18D.1 Light absorption by excitons in molecular solids

Topic 12A explains the factors that determine the energy and intensity of light absorbed by isolated atoms and molecules in the gas phase and in solution. Here we consider the effects on their electronic absorption spectra that result from bringing the atoms or molecules together to form a solid.

Consider an electronic excitation of a molecule (or an ion) in a crystal. If the excitation corresponds to the removal of an electron from one orbital of a molecule and its elevation to an orbital of higher energy, then the excited state of the molecule can be envisaged as the coexistence of an electron and a hole. This electron–hole pair, which behaves as a particle-like **exciton**, migrates from molecule to molecule in the crystal (Fig. 18D.1). A migrating excitation of this kind is called a **Frenkel exciton**. Frenkel excitons are more common in molecular solids. The electron and hole can also be on different molecules, but in each other's vicinity. A migrating excitation of this kind, which is now spread over several molecules (more usually ions), is called a **Wannier exciton**. Exciton formation causes spectral lines to shift, split, and change intensity.

The migration of a Frenkel exciton (the only type we consider) implies that there is an interaction between the species that constitute the crystal, for otherwise the excitation on one unit could not move to another. This interaction affects the energy levels of the system. The strength of the interaction also governs the rate at which an exciton moves through the crystal: a strong interaction results in fast migration and a vanishingly

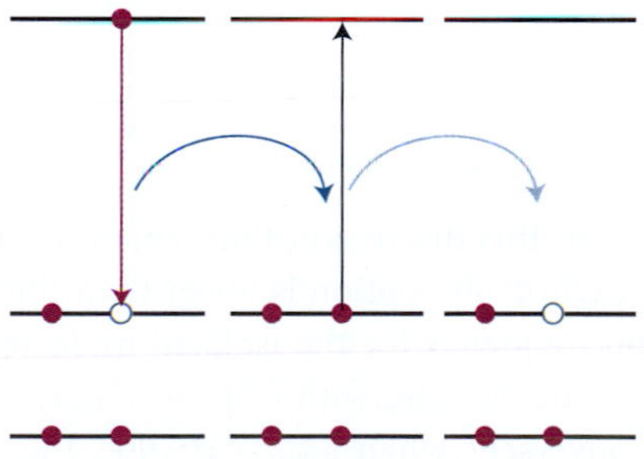

Figure 18D.1 The electron–hole pair shown on the left can migrate through a solid lattice as the excitation hops from molecule to molecule. The mobile excitation is called an exciton.

small interaction leaves the exciton localized on its original molecule. The specific mechanism of interaction that leads to exciton migration is the interaction between the transition dipole moments of the excitation (Topic 12A). Thus, an electric dipole transition in a molecule is accompanied by a shift of charge, and the transient dipole exerts a force on an adjacent molecule. The latter responds by shifting its charge. This process continues and the excitation migrates through the crystal.

The energy shift arising from the interaction between transition dipoles can be understood in terms of their electrostatic interaction. As we see in the following *Justification*, an all-parallel arrangement of the transition dipoles (Fig. 18D.2a) is energetically unfavourable, so the absorption occurs at a higher frequency than in the isolated molecule. Conversely, a head-to-tail alignment of transition dipoles (Fig. 18D.2b) is energetically favourable, and the transition occurs at a lower frequency than in the isolated molecules.

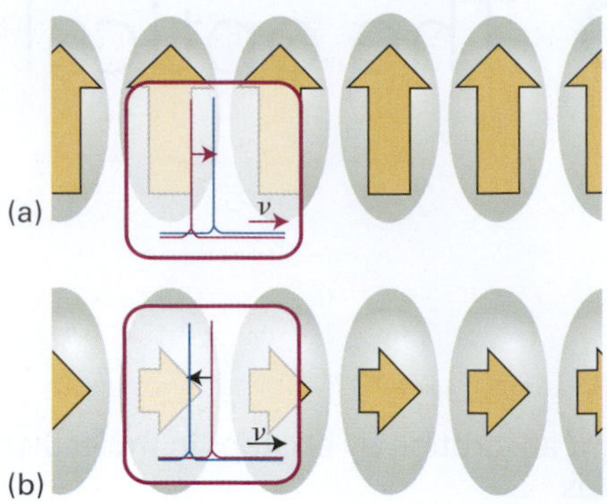

Figure 18D.2 (a) The alignment of transition dipoles (the yellow arrows) is energetically unfavourable, and the exciton absorption is shifted to higher energy (higher frequency). (b) The alignment is energetically favourable for a transition in this orientation, and the exciton band occurs at lower frequency than in the isolated molecules.

Justification 18D.1 The energy of interaction of transition dipoles

The potential energy of interaction between two parallel electric dipole moments $\boldsymbol{\mu}_1$ and $\boldsymbol{\mu}_2$ separated by a distance r is $V=\mu_1\mu_2(1-3\cos^2\theta)/4\pi\varepsilon_0 r^3$, where the angle θ is defined in **1** (Topic 16B). We see that $\theta=0°$ for a head-to-tail alignment and $\theta=90°$ for a parallel alignment. It follows that $V<0$ (an attractive interaction) for $0°\leq\theta<54.7°$, $V=0$ when $\theta=54.7°$ (for then $1-3\cos^2\theta=0$), and $V>0$ (a repulsive interaction) for $54.7°<\theta\leq 90°$. This result is expected on the basis of qualitative arguments. In a head-to-tail arrangement, the interaction between the region of partial positive charge in one molecule and the region of partial negative charge in the other molecule is attractive. By contrast, in a parallel arrangement, the molecular interaction is repulsive because of the close approach of regions of partial charge with the same sign.

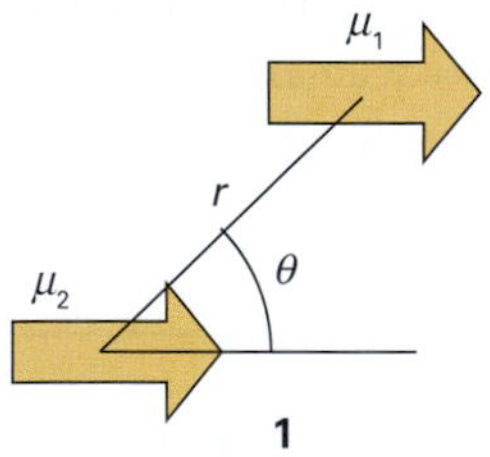

1

It follows from this discussion that, when $0°\leq\theta<54.7°$, the frequency of exciton absorption is lower than the corresponding absorption frequency for the isolated molecule (a *red shift* in the spectrum of the solid with respect to that of the isolated molecule). Conversely, when $54.7°<\theta\leq 90°$, the frequency of exciton absorption is higher than the corresponding absorption frequency for the isolated molecule (a *blue shift* in the spectrum of the solid with respect to that of the isolated molecule). In the special case $\theta=54.7°$ the solid and the isolated molecule have absorption lines at the same frequency.

If there are N molecules per unit cell, there are N **exciton bands** in the spectrum (if all of them are allowed). The splitting between the bands is the **Davydov splitting**. To understand the origin of the splitting, consider the case $N=2$ with the molecules arranged as in Fig. 18D.3 and suppose that the transition dipoles are along the length of the molecules. The radiation stimulates the collective excitation of the transition dipoles that are in-phase between neighbouring unit cells. Within each unit cell the transition dipoles may be arrayed in the two different ways shown in the illustration. The two orientations correspond to different interaction energies, with interaction being repulsive in one and attractive in the other, so the two transitions appear in the spectrum at two bands of different frequencies. The magnitude of the Davydov splitting is determined by the energy of interaction between the transition dipoles within the unit cell.

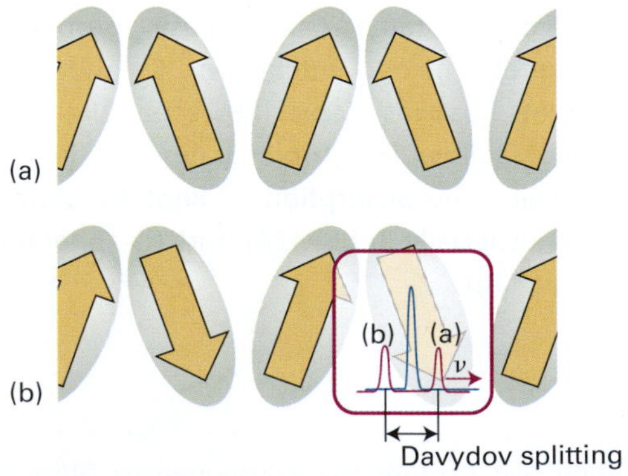

Figure 18D.3 When the transition moments within a unit cell lie in different relative directions, as depicted in (a) and (b), the energies of the transitions are shifted and give rise to the two bands labelled (a) and (b) in the spectrum. The separation of the bands is the Davydov splitting.

18D.2 Light absorption by metals and semiconductors

Now we turn our attention to metallic conductors and semiconductors. Again we need to consider the consequences of interactions between particles, in this case atoms, which are now so strong that we need to abandon arguments based primarily on van der Waals interactions in favour of a full molecular orbital treatment, the band model of Topic 18B.

Consider again Fig. 18C.5, reproduced here as Fig. 18D.4, which shows bands in an idealized metallic conductor. The absorption of a photon can excite electrons from the occupied levels to the unoccupied levels. There is a near continuum of unoccupied energy levels above the Fermi level, so we expect to observe absorption over a wide range of frequencies. In metals, the bands are sufficiently wide that radiation from the radiofrequency to the middle of the ultraviolet region of the electromagnetic spectrum is absorbed. Metals are transparent to very high-frequency electromagnetic radiation, such as X-rays and γ-rays. Because this range of absorbed frequencies includes the entire visible spectrum, we might therefore expect all metals to appear black. However, we know that metals are lustrous (that is, they reflect light) and some are coloured (that is, they absorb light of only certain wavelengths), so we need to extend our model.

To explain the lustrous appearance of a smooth metal surface, we need to realize that the absorbed energy can be re-emitted very efficiently as light, with only a small fraction of the energy being released to the surroundings as heat. Because the atoms near the surface of the material absorb most of the radiation, emission also occurs primarily from the surface. In essence, if the sample is excited with visible light, then electrons near the surface are driven into oscillation at the same frequency, and visible light will be emitted from the surface, so accounting for the lustre of the material.

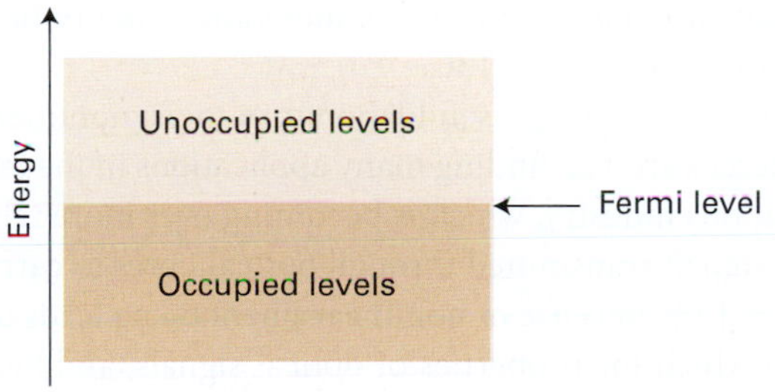

Figure 18D.4 When *N* electrons occupy a band of *N* orbitals, it is only half full and the electrons near the Fermi level (the top of the filled levels) are mobile.

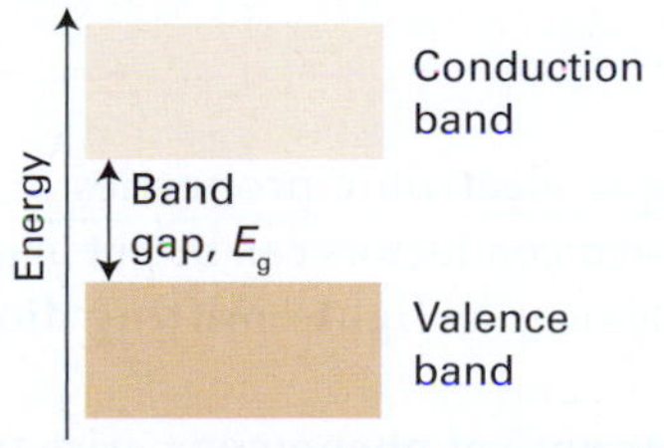

Figure 18D.5 In some materials, the band gap E_g is very large and electron promotion can occur only by excitation with electromagnetic radiation.

The perceived colour of a metal depends on the frequency range of reflected light which, in turn, depends on the frequency range of light that can be absorbed and, by extension, on the band structure. Silver reflects light with nearly equal efficiency across the visible spectrum because its band structure has many unoccupied energy levels that can be populated by absorption of, and depopulated by emission of, visible light. On the other hand, copper has its characteristic colour because it has relatively fewer unoccupied energy levels that can be excited with violet, blue, and green light. The material reflects at all wavelengths, but more light is emitted at lower frequencies (corresponding to yellow, orange, and red). Similar arguments account for the colours of other metals, such as the yellow of gold. The colour of gold, incidentally, can be accounted for only by including relativistic effects in the calculation of its band structure.

Now consider semiconductors. We have already seen that promotion of electrons from the conduction to the valence band of a semiconductor can be the result of thermal excitation, if the band gap E_g is comparable to the energy that can be supplied by heating. In some materials, the band gap is very large and electron promotion can occur only by excitation with electromagnetic radiation. However, we see from Fig. 18D.5 that there is a frequency $\nu_{min}=E_g/h$ below which light absorption cannot occur. Above this frequency threshold, a wide range of frequencies can be absorbed by the material, as in a metal.

Brief illustration 18D.1 The optical properties of a semiconductor

The energy of the band gap in semiconductor cadmium sulfide (CdS) is 2.4 eV (equivalent to 3.8×10^{-19} J). It follows that the minimum electronic absorption frequency is

$$\nu_{min}=\frac{3.8\times10^{-19}\,\text{J}}{6.626\times10^{-34}\,\text{J}}=5.8\times10^{14}\,\text{s}^{-1}$$

This frequency, of 580 THz, corresponds to a wavelength of 520 nm (green light). Lower frequencies, corresponding to yellow, orange, and red, are not absorbed and consequently CdS appears yellow-orange.

Self-test 18D.1 Predict the colours of the following materials, given their band-gap energies (in parentheses): GaAs (1.43 eV), HgS (2.1 eV), and ZnS (3.6 eV).

Answer: Black, red, and colourless

18D.3 Light-emitting diodes and diode lasers

The unique electrical properties of p–n junctions between semiconductors can be put to good use in optical devices. In some materials, most notably gallium arsenide, GaAs, energy from electron–hole recombination is released not as heat but is carried away by photons as electrons move across the junction under forward bias. Practical **light-emitting diodes** of this kind are widely used in electronic displays. The wavelength of emitted light depends on the band gap of the semiconductor. Gallium arsenide itself emits infrared light, but its band gap is widened by incorporating phosphorus, and a material of composition approximately $GaAs_{0.6}P_{0.4}$ emits light in the red region of the spectrum.

A light-emitting diode is not a laser, because neither a resonance cavity nor stimulated emission is involved. In **diode lasers**, light emission due to electron–hole recombination is employed as the basis of laser action. The population inversion can be sustained by sweeping away the electrons that fall into the holes of the p-type semiconductor, and a resonant cavity can be formed by using the high refractive index of the semiconducting material and cleaving single crystals so that the light is trapped by the abrupt variation of refractive index. One widely used material is $Ga_{1-x}Al_xAs$, which produces infrared laser radiation and is widely used in compact-disc (CD) players.

High-power diode lasers are also used to pump other lasers. One example is the pumping of Nd:YAG lasers (Topic 13C) by $Ga_{0.91}Al_{0.09}As/Ga_{0.7}Al_{0.3}As$ diode lasers. The Nd:YAG laser is often used to pump yet another laser, such as a Ti:sapphire laser (Topic 13C). As a result, it is now possible to construct a laser system for steady-state or time-resolved spectroscopy entirely out of solid-state components.

18D.4 Nonlinear optical phenomena

Nonlinear optical phenomena arise from changes in the optical properties of a material in the presence of an intense electric field from electromagnetic radiation. In **frequency doubling**, or **second harmonic generation**, an intense laser beam is converted to radiation with twice (and in general a multiple) of its initial frequency as it passes through a suitable material. It follows that frequency doubling and tripling of a Nd:YAG laser, which emits radiation at 1064 nm (Topic 13C), produce green light at 532 nm and ultraviolet radiation at 355 nm, respectively.

We can account for frequency doubling by examining how a substance responds nonlinearly to incident radiation of frequency $\omega = 2\pi\nu$. Radiation of a particular frequency arises from oscillations of an electric dipole at that frequency and the incident electric field $\mathcal{E}$ induces an electric dipole of magnitude μ, in the substance. At low light intensity, most materials respond linearly, in the sense that $\mu = \alpha\mathcal{E}$, where α is the polarizability (Topic 16A). At high light intensity, the hyperpolarizability β of the material becomes important (Topic 16B) and we write.

$$\mu = \alpha\mathcal{E} + \tfrac{1}{2}\beta\mathcal{E}^2 + \cdots \qquad \text{Induced dipole moment in terms of the hyperpolarizability} \qquad (18D.1)$$

The nonlinear term $\beta\mathcal{E}^2$ can be expanded as follows if we suppose that the incident electric field is $\mathcal{E}_0 \cos \omega t$:

$$\beta\mathcal{E}^2 = \beta\mathcal{E}_0^2 \cos^2\omega t = \tfrac{1}{2}\beta\mathcal{E}_0^2(1+\cos 2\omega t) \qquad (18D.2)$$

Hence, the nonlinear term contributes an induced electric dipole that oscillates at the frequency 2ω and that can act as a source of radiation of that frequency. Common materials that can be used for frequency doubling in laser systems include crystals of potassium dihydrogenphosphate (KH_2PO_4), lithium niobate ($LiNbO_3$), and β-barium borate (β-BaB_2O_4). Another important nonlinear optical phenomenon is the optical Kerr effect discussed in Topic 13C.

In addition to being useful laboratory tools, nonlinear optical materials are also finding many applications in the telecommunications industry, which is becoming ever more reliant on optical signals transmitted through optical fibres to carry voice and data. Judicious use of nonlinear phenomena leads to more ways in which the properties of optical signals, and hence the information they carry, can be manipulated.

Checklist of concepts

- ☐ 1. The optical properties of molecular solids can be understood in terms of the formation and migration of **excitons**.
- ☐ 2. The spectroscopic properties of metallic conductors and semiconductors can be understood in terms of the photon-induced promotion of electrons from valence bands to conduction bands.
- ☐ 3. The unique electronic properties of p–n junctions between semiconductors can be put to good use in such optical devices as **light-emitting diodes** and **diode lasers**.
- ☐ 4. **Nonlinear optical phenomena** arise from changes in the optical properties of a material in the presence of intense electromagnetic radiation.

CHAPTER 18 Solids

TOPIC 18A Crystal structure

Discussion questions

18A.1 Describe the relationship between the space lattice and unit cell.

18A.2 Explain how planes of lattice points are labelled.

18A.3 Describe the procedure for identifying the type and size of a cubic unit cell.

18A.4 Discuss what is meant by 'scattering factor'. How is it related to the number of electrons in the atoms scattering X-rays?

18A.5 What is meant by a systematic absence? How do they arise?

18A.6 Describe the consequences of the phase problem in determining structure factors and how the problem is overcome.

Exercises

18A.1(a) The orthorhombic unit cell of $NiSO_4$ has the dimensions $a = 634$ pm, $b = 784$ pm, and $c = 516$ pm, and the density of the solid is estimated as 3.9 g cm^{-3}. Determine the number of formula units per unit cell and calculate a more precise value of the density.
18A.1(b) An orthorhombic unit cell of a compound of molar mass 135.01 g mol^{-1} has the dimensions $a = 589$ pm, $b = 822$ pm, and $c = 798$ pm. The density of the solid is estimated as 2.9 g cm^{-3}. Determine the number of formula units per unit cell and calculate a more precise value of the density.

18A.2(a) Find the Miller indices of the planes that intersect the crystallographic axes at the distances ($2a$, $3b$, $2c$) and ($2a$, $2b$, ∞c).
18A.2(b) Find the Miller indices of the planes that intersect the crystallographic axes at the distances ($-a$, $2b$, $-c$) and (a, $4b$, $-4c$).

18A.3(a) Calculate the separations of the planes {112}, {110}, and {224} in a crystal in which the cubic unit cell has side 562 pm.
18A.3(b) Calculate the separations of the planes {123}, {222}, and {246} in a crystal in which the cubic unit cell has side 712 pm.

18A.4(a) The unit cells of $SbCl_3$ are orthorhombic with dimensions $a = 812$ pm, $b = 947$ pm, and $c = 637$ pm. Calculate the spacing, d, of the {321} planes.
18A.4(b) An orthorhombic unit cell has dimensions $a = 769$ pm, $b = 891$ pm, and $c = 690$ pm. Calculate the spacing, d, of the {312} planes.

18A.5(a) The angle of a Bragg reflection from a set of crystal planes separated by 99.3 pm is 20.85°. Calculate the wavelength of the X-rays.
18A.5(b) The angle of a Bragg reflection from a set of crystal planes separated by 128.2 pm is 19.76°. Calculate the wavelength of the X-rays.

18A.6(a) What are the values of the angle θ of the first three diffraction lines of bcc iron (atomic radius 126 pm) when the X-ray wavelength is 72 pm?
18A.6(b) What are the values of the angle θ of the first three diffraction lines of fcc gold (atomic radius 144 pm) when the X-ray wavelength is 129 pm?

18A.7(a) Potassium nitrate crystals have orthorhombic unit cells of dimensions $a = 542$ pm, $b = 917$ pm, and $c = 645$ pm. Calculate the values of θ for the (100), (010), and (111) reflections using radiation of wavelength 154 pm.
18A.7(b) Calcium carbonate crystals in the form of aragonite have orthorhombic unit cells of dimensions $a = 574.1$ pm, $b = 796.8$ pm, and $c = 495.9$ pm. Calculate the values of θ for the (100), (010), and (111) reflections using radiation of wavelength 83.42 pm.

18A.8(a) Radiation from an X-ray source consists of two components of wavelengths 154.433 pm and 154.051 pm. Calculate the difference in glancing angles (2θ) of the diffraction lines arising from the two components in a diffraction pattern from planes of separation 77.8 pm.
18A.8(b) Consider a source that emits X-radiation at a range of wavelengths, with two components of wavelengths 93.222 and 95.123 pm. Calculate the separation of the glancing angles (2θ) arising from the two components in a diffraction pattern from planes of separation 82.3 pm.

18A.9(a) What is the value of the scattering factor in the forward direction for Br^-?
18A.9(b) What is the value of the scattering factor in the forward direction for Mg^{2+}?

18A.10(a) The coordinates, in units of a, of the atoms in a primitive cubic unit cell are (0,0,0), (0,1,0), (0,0,1), (0,1,1), (1,0,0), (1,1,0), (1,0,1), and (1,1,1). Calculate the structure factors F_{hkl} when all the atoms are identical.
18A.10(b) The coordinates, in units of a, of the atoms in a body-centred cubic unit cell are (0,0,0), (0,1,0), (0,0,1), (0,1,1), (1,0,0), (1,1,0), (1,0,1), (1,1,1), and $(\frac{1}{2},\frac{1}{2},\frac{1}{2})$. Calculate the structure factors F_{hkl} when all the atoms are identical.

18A.11(a) Calculate the structure factors for a face-centred cubic structure (C) in which the scattering factors of the ions on the two faces are twice that of the ions at the corners of the cube.
18A.11(b) Calculate the structure factors for a body-centred cubic structure in which the scattering factor of the central ion is twice that of the ions at the corners of the cube.

18A.12(a) In an X-ray investigation, the following structure factors were determined (with $F_{-h00} = F_{h00}$):

h	0	1	2	3	4	5	6	7	8	9
F_{h00}	10	−10	8	−8	6	−6	4	−4	2	−2

Construct the electron density along the corresponding direction.
18A.12(b) In an X-ray investigation, the following structure factors were determined (with $F_{-h00} = F_{h00}$):

h	0	1	2	3	4	5	6	7	8	9
F_{h00}	10	10	4	4	6	6	8	8	10	10

Construct the electron density along the corresponding direction.
18A.13(a) Construct the Patterson synthesis from the information in Exercise 18A.12(a).
18A.13(b) Construct the Patterson synthesis from the information in Exercise 18A.12(b).

18A.14(a) In a Patterson synthesis, the spots correspond to the lengths and directions of the vectors joining the atoms in a unit cell. Sketch the pattern that would be obtained for a planar, triangular isolated BF_3 molecule.
18A.14(b) In a Patterson synthesis, the spots correspond to the lengths and directions of the vectors joining the atoms in a unit cell. Sketch the pattern that would be obtained from the C atoms in an isolated benzene molecule.

18A.15(a) What speed should neutrons have if they are to have a wavelength of 65 pm?
18A.15(b) What speed should electrons have if they are to have a wavelength of 105 pm?

Problems

18A.1 Although the crystallization of large biological molecules may not be as readily accomplished as that of small molecules, their crystal lattices are no different. Tobacco seed globulin forms face-centred cubic crystals with unit cell dimension of 12.3 nm and a mass density of 1.287 g cm^{-3}. Determine its molar mass.

18A.2 Show that the volume of a monoclinic unit cell is $V = abc \sin \beta$.

18A.3 Derive an expression for the volume of a hexagonal unit cell.

18A.4 Show that the volume of a triclinic unit cell of sides a, b, and c and angles α, β, and γ is

$$V = abc(1-\cos^2\alpha-\cos^2\beta-\cos^2\gamma+2\cos\alpha\cos\beta\cos\gamma)^{1/2}$$

Use this expression to derive expressions for monoclinic and orthorhombic unit cells. For the derivation, it may be helpful to use the result from vector analysis that $V = \boldsymbol{a}\cdot\boldsymbol{b}\times\boldsymbol{c}$ and to calculate V^2 initially. The compound Rb_3TlF_6 has a *tetragonal* unit cell with dimensions $a = 651$ pm and $c = 934$ pm. Calculate the volume of the unit cell.

18A.5 The volume of a monoclinic unit cell is $abc \sin \beta$ (see Problem 18A.2). Naphthalene has a monoclinic unit cell with two molecules per cell and sides in the ratio 1.377:1:1.436. The angle β is 122.82° and the mass density of the solid is 1.152 g cm^{-3}. Calculate the dimensions of the cell.

18A.6 Fully crystalline polyethylene has its chains aligned in an orthorhombic unit cell of dimensions 740 pm × 493 pm × 253 pm. There are two repeating CH_2CH_2 units per unit cell. Calculate the theoretical mass density of fully crystalline polyethylene. The actual density ranges from 0.92 to 0.95 g cm^{-3}.

18A.7‡ B.A. Bovenzi and G.A. Pearse, Jr. (*J. Chem. Soc. Dalton Trans.*, 2793 (1997)) synthesized coordination compounds of the tridentate ligand pyridine-2,6-diamidoxime (**1**, $C_7H_9N_5O_2$). The compound they isolated from the reaction of the ligand with $CuSO_4$(aq) did not contain a $[Cu(C_7H_9N_5O_2)_2]^{2+}$ complex cation as expected. Instead, X-ray diffraction analysis revealed a linear polymer of formula $[\{Cu(C_7H_9N_5O_2)(SO_4)\}\cdot 2H_2O]_n$, which features bridging sulfate groups. The unit cell was primitive monoclinic with $a = 1.0427$ nm, $b = 0.8876$ nm, $c = 1.3777$ nm, and $\beta = 93.254°$. The mass density of the crystals is 2.024 g cm^{-3}. How many monomer units are there in the unit cell?

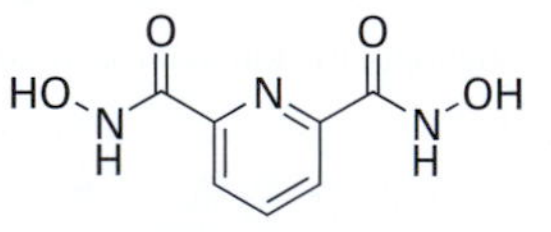

1 Pyridine-2,6-diamidoxime

18A.8‡ D. Sellmann et al. (*Inorg. Chem.* **36**, 1397 (1997)) describe the synthesis and reactivity of the ruthenium nitrido compound $[N(C_4H_9)_4][Ru(N)(S_2C_6H_4)_2]$. The ruthenium complex anion has the two 1,2-benzenedithiolate ligands (**2**) at the base of a rectangular pyramid and the nitrido ligand at the apex. Compute the mass density of the compound given that it crystallizes with an orthorhombic unit cell with $a = 3.6881$ nm, $b = 0.9402$ nm, and $c = 1.7652$ nm and eight formula units per cell. The replacement of the ruthenium with osmium results in a compound with the same crystal structure and a unit cell with a volume less than 1 per cent larger. Estimate the mass density of the osmium analogue.

‡ These problems were supplied by Charles Trapp and Carmen Giunta.

18A.16(a) Calculate the wavelength of neutrons that have reached thermal equilibrium by collision with a moderator at 350 K.
18A.16(b) Calculate the wavelength of electrons that have reached thermal equilibrium by collision with a moderator at 380 K.

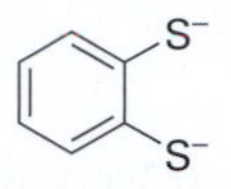

2 1,2-Benzenedithiolate ion

18A.9 Show that the separation of the $\{hkl\}$ planes in an orthorhombic crystal with sides a, b, and c is given by eqn 18A.1c.

18A.10 In the early days of X-ray crystallography there was an urgent need to know the wavelengths of X-rays. One technique was to measure the diffraction angle from a mechanically ruled grating. Another method was to estimate the separation of lattice planes from the measured density of a crystal. The mass density of NaCl is 2.17 g cm^{-3} and the (100) reflection using radiation of a certain wavelength occurred at 6.0°. Calculate the wavelength of the X-rays.

18A.11 The element polonium crystallizes in a cubic system. Bragg reflections, with X-rays of wavelength 154 pm, occur at $\sin \theta = 0.225$, 0.316, and 0.388 from the {100}, {110}, and {111} sets of planes. The separation between the sixth and seventh lines observed in the diffraction pattern is larger than between the fifth and sixth lines. Is the unit cell primitive, body centred, or face centred? Calculate the unit cell dimension.

18A.12 Elemental silver reflects X-rays of wavelength 154.18 pm at angles of 19.076°, 22.171°, and 32.256°. However, there are no other reflections at angles of less than 33°. Assuming a cubic unit cell, determine its type and dimension. Calculate the mass density of silver.

18A.13 In their book *X-rays and crystal structures* (which begins 'It is now two years since Dr. Laue conceived the idea…') the Braggs give a number of simple examples of X-ray analysis. For instance, they report that the reflection from {100} planes in KCl occurs at 5.38°, but for NaCl it occurs at 6.00° for X-rays of the same wavelength. If the side of the NaCl unit cell is 564 pm, what is the side of the KCl unit cell? The mass densities of KCl and NaCl are 1.99 g cm^{-3} and 2.17 g cm^{-3}, respectively. Do these values support the X-ray analysis?

18A.14 Use mathematical software to draw a graph of the scattering factor f against $(\sin \theta)/\lambda$ for an atom of atomic number Z for which $\rho(r) = 3Z/4\pi R^3$ for $0 \le r \le R$ and $\rho(r) = 0$ for $r > R$, with R a parameter that represents the radius of the atom. Explore how f varies with Z and R.

18A.15 The coordinates of the four I atoms in the unit cell of KIO_4 are $(0,0,0), (0,\tfrac{1}{2},\tfrac{1}{2}), (\tfrac{1}{2},\tfrac{1}{2},\tfrac{1}{2}), (\tfrac{1}{2},0,\tfrac{3}{4})$. By calculating the phase of the I reflection in the structure factor, show that the I atoms contribute no net intensity to the (114) reflection.

18A.16 The coordinates, as multiples of a, of the A atoms, with scattering factor f_A, in a cubic lattice are (0,0,0), (0,1,0), (0,0,1), (0,1,1), (1,0,0), (1,1,0), (1,0,1), and (1,1,1). There is also a B atom, with scattering factor f_B, at $(\tfrac{1}{2},\tfrac{1}{2},\tfrac{1}{2})$. Calculate the structure factors F_{hkl} and predict the form of the diffraction pattern when (a) $f_A = f, f_B = 0$, (b) $f_B = \tfrac{1}{2} f_A$, and (c) $f_A = f_B = f$.

18A.17 Here we explore electron diffraction patterns. (a) Predict from the Wierl equation, eqn 18A.9, the positions of the first maximum and first minimum in the neutron and electron diffraction patterns of a Br_2 molecule obtained with neutrons of wavelength 78 pm and electrons of wavelength 4.0 pm. (b) Use the Wierl equation to predict the appearance of the electron diffraction pattern of CCl_4 with an (as yet) undetermined C—Cl bond length but of known tetrahedral symmetry. Take $f_{Cl} = 17f$ and $f_C = 6f$ and note that $R(Cl,Cl) = (8/3)^{1/2}R(C,Cl)$. Plot I/f^2 against positions of the maxima occurred at 3.17°, 5.37°, and 7.90° and minima occurred at 1.77°, 4.10°, 6.67°, and 9.17°. What is the C–Cl bond length in CCl_4?

TOPIC 18B Bonding in solids

Discussion questions

18B.1 In what respects is the hard-sphere model of metallic solids deficient?

18B.2 Describe the caesium chloride and rock salt structures in terms of the occupation of holes in expanded close-packed lattices.

Exercises

18B.1(a) Calculate the packing fraction for close-packed cylinders. (For a generalization of this Exercise, see Problem 18B.2.)

18B.1(b) Calculate the packing fraction for equilateral triangular rods stacked as shown in 3.

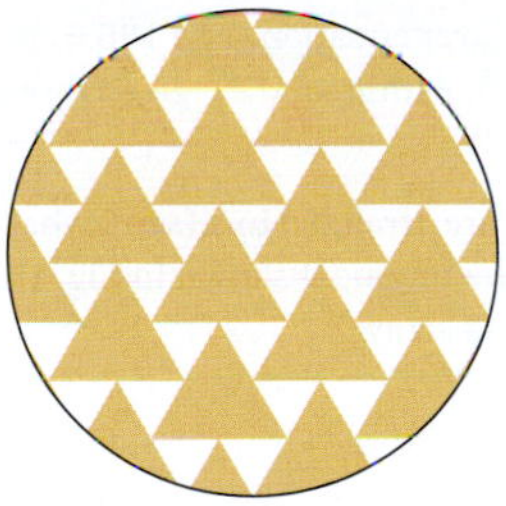

3

18B.2(a) Calculate the packing fractions of (i) a primitive cubic unit cell, (ii) a bcc unit cell, (iii) an fcc unit cell composed of identical hard spheres.

18B.2(b) Calculate the atomic packing factor for a face-centred (C) cubic unit cell.

18B.3(a) From the data in Table 18B.2 determine the radius of the smallest cation that can have (i) sixfold and (ii) eightfold coordination with the Cl^- ion.

18B.3(b) From the data in Table 18B.2 determine the radius of the smallest cation that can have (i) sixfold and (ii) eightfold coordination with the Rb^+ ion.

18B.4(a) Does titanium expand or contract as it transforms from hcp to bcc? The atomic radius of titanium is 145.8 pm in hcp but 142.5 pm in bcc.

18B.4(b) Does iron expand or contract as it transforms from hcp to bcc? The atomic radius of iron is 126 in hcp but 122 pm in bcc.

18B.5(a) Calculate the lattice enthalpy of CaO from the following data:

	$\Delta H/(\text{kJ mol}^{-1})$
Sublimation of Ca(s)	+178
Ionization of Ca(g) to Ca^{2+}(g)	+1735
Dissociation of O_2(g)	+249
Electron attachment to O(g)	−141
Electron attachment to O^-(g)	+844
Formation of CaO(s) from Ca(s) and O_2(g)	−635

18B.5(b) Calculate the lattice enthalpy of $MgBr_2$ from the following data:

	$\Delta H/(\text{kJ mol}^{-1})$
Sublimation of Mg(s)	+148
Ionization of Mg(g) to Mg^{2+}(g)	+2187
Vaporization of Br_2(l)	+31
Dissociation of Br_2(g)	+193
Electron attachment to Br(g)	−331
Formation of $MgBr_2$(s) from Mg(s) and Br_2(l)	−524

Problems

18B.1 Calculate the atomic packing factor for diamond.

18B.2 Rods of elliptical cross-section with semi-minor and major axes a and b are close-packed as shown in 4. What is the packing fraction? Draw a graph of the packing fraction against the eccentricity ε of the ellipse. For an ellipse with semi-major axis a and semi-minor axis b, $\varepsilon = (1 - b^2/a^2)^{1/2}$.

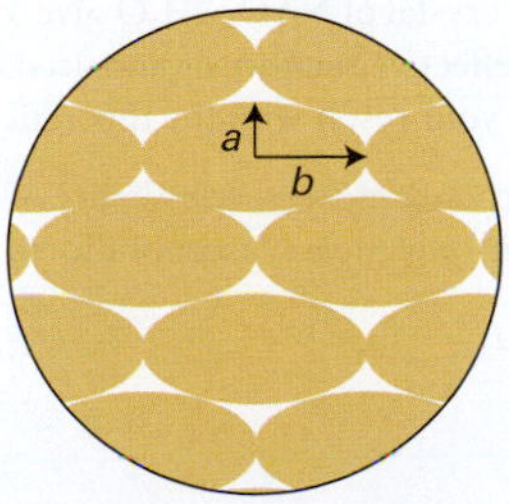

4

18B.3 The carbon–carbon bond length in diamond is 154.45 pm. If diamond were considered to be a close-packed structure of hard spheres with radii equal to half the bond length, what would be its expected mass density? The diamond lattice is face-centred cubic and its actual mass density is 3.516 g cm^{-3}. Can you explain the discrepancy?

18B.4 When energy levels in a band form a continuum, the density of states $\rho(E)$, the number of levels in an energy range divided by the width of the range, may be written as $\rho(E) = dk/dE$, where dk is the change in the quantum number k and dE is the energy change. (a) Use eqn 18B.1 to show that

$$\rho(E) = -\frac{(N+1)/2\pi\beta}{\left\{1-\left(\dfrac{E-\alpha}{2\beta}\right)^2\right\}^{1/2}}$$

where k, N, α and β have the meanings described in Topic 18B. (b) Use this expression to show that $\rho(E)$ becomes infinite as E approaches $\alpha \pm 2\beta$. That is,

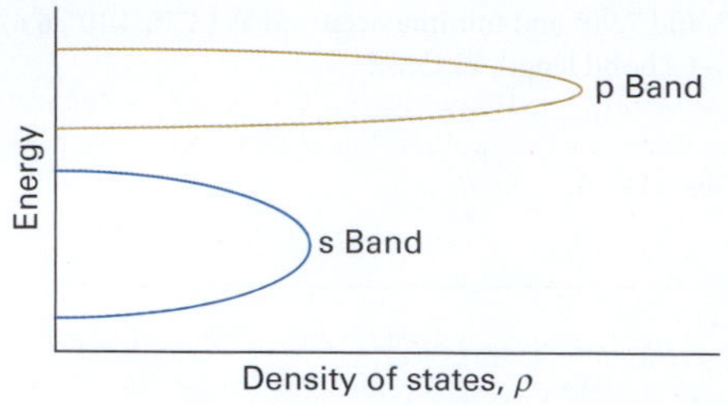

Figure 18.1 The variation of density of states in a three-dimensional solid.

show that the density of states increases towards the edges of the bands in a one-dimensional metallic conductor.

18B.5 The treatment in Problem 18B.4 applies only to one-dimensional solids. In three dimensions, the variation of density of states is more like that shown in Fig. 18.1. Account for the fact that in a three-dimensional solid the greatest density of states is near the centre of the band and the lowest density is at the edges.

18B.6 The energy levels of N atoms in the tight-binding Hückel approximation are the roots of a tridiagonal determinant (eqn 18B.1):

$$E_k = \alpha + 2\beta \cos\frac{k\pi}{N+1} \qquad k = 1,2,\ldots,N$$

If the atoms are arranged in a ring, the solutions are the roots of a 'cyclic' determinant:

$$E_k = \alpha + 2\beta \cos\frac{2k\pi}{N} \qquad k = 0,\pm1,\pm2,\ldots,\pm\tfrac{1}{2}N$$

(for N even). Discuss the consequences, if any, of joining the ends of an initially straight length of material.

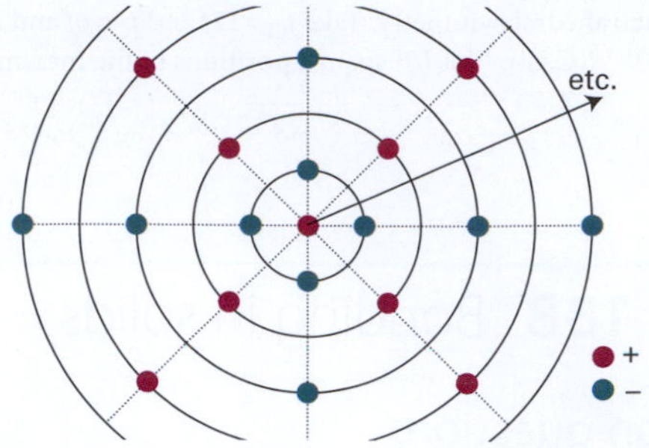

Figure 18.2 Fragment of a two-dimensional lattice used as a model in Problem 18B.10.

18B.7 Verify that the radius ratio for (a) sixfold coordination is 0.414, and (b) for eightfold coordination is 0.732.

18B.8 Use the Born–Mayer equation for the lattice enthalpy and a Born–Haber cycle to show that formation of CaCl is an exothermic process (the sublimation enthalpy of Ca(s) is 176 kJ mol^{-1}). Show that an explanation for the nonexistence of CaCl can be found in the reaction enthalpy for the reaction 2 CaCl(s) $\rightarrow$ Ca(s) + $CaCl_2$.

18B.9 Derive the Born–Mayer equation (eqn 18B.6) by calculating the energy at which $d(E_p + E_p^*)/dd = 0$, with E_p and E_p^* given by eqns 18B.4 and 18B.5, respectively.

18B.10 Suppose that ions are arranged in a (somewhat artificial) two-dimensional lattice like the fragment shown in Fig. 18.2. Calculate the Madelung constant for this array.

TOPIC 18C Mechanical, electrical, and magnetic properties of solids

Discussion questions

18C.1 Describe the characteristics of the Fermi–Dirac distribution.

18C.2 To what extent are the electric and magnetic properties of molecules analogous? How do they differ?

Exercises

18C.1(a) Poisson's ratio for polyethene is 0.45. What change in volume takes place when a cube of polyethene of volume 1.0 cm^3 is subjected to a uniaxial stress that produces a strain of 1.0 per cent?
18C.1(b) Poisson's ratio for lead is 0.41. What change in volume takes place when a cube of lead of volume 1.0 dm^3 is subjected to a uniaxial stress that produces a strain of 2.0 per cent?

18C.2(a) Is arsenic-doped germanium a p-type or n-type semiconductor?
18C.2(b) Is gallium-doped germanium a p-type or n-type semiconductor?

18C.3(a) The magnitude of the magnetic moment of $CrCl_3$ is 3.81μ_B. How many unpaired electrons does the Cr possess?
18C.3(b) The magnitude of the magnetic moment of Mn^{2+} in its complexes is typically 5.3μ_B. How many unpaired electrons does the ion possess?

18C.4(a) Calculate the molar susceptibility of benzene given that its volume susceptibility is -7.2×10^{-7} and its mass density 0.879 g cm^{-3} at 25 °C.
18C.4(b) Calculate the molar susceptibility of cyclohexane given that its volume susceptibility is -7.9×10^{-7} and its mass density 811 kg m^{-3} at 25 °C.

18C.5(a) Data on a single crystal of MnF_2 give $\chi_m = 0.1463$ cm^3 mol^{-1} at 294.53 K. Determine the effective number of unpaired electrons in this compound and compare your result with the theoretical value.
18C.5(b) Data on a single crystal of $NiSO_4\cdot7H_2O$ give $\chi_m = 6.00\times10^{-8}$ m^3 mol^{-1} at 298 K. Determine the effective number of unpaired electrons in this compound and compare your result with the theoretical value.

18C.6(a) Estimate the spin-only molar susceptibility of $CuSO_4\cdot5H_2O$ at 25 °C.
18C.6(b) Estimate the spin-only molar susceptibility of $MnSO_4\cdot4H_2O$ at 298 K.

Problems

18C.1 For an isotropic substance, the moduli and Poisson's ratio may be expressed in terms of two parameters λ and μ called the *Lamé constants*:

$$E=\frac{\mu(3\lambda+2\mu)}{\lambda+\mu} \qquad K=\frac{3\lambda+2\mu}{3} \qquad G=\mu \qquad \nu_P=\frac{\lambda}{2(\lambda+\mu)}$$

Use the Lamé constants to confirm the relations between G, K, and E given in eqn 18C.3.

18C.2 Refer to eqn 18C.6 and express $f(E)$ as a function of the variables $(E-\mu)/\mu$ and μ/kT. Then, using mathematical software, display the set of curves shown in Fig. 18C.6 as a single surface.

18C.3 In this and the following problem we explore further some of the properties of the Fermi–Dirac distribution, eqn 18C.6. For a three-dimensional solid of volume V, it turns out that $\rho(E)=CE^{1/2}$, with $C=4\pi V(2m_e/h^2)^{3/2}$. Show that at $T=0$,

$$f(E)=1 \text{ for } E<\mu \qquad f(E)=0 \text{ for } E>\mu$$

and deduce that $\mu(0)=(3\mathcal{N}/8\pi)^{2/3}(h^2/2m_e)$, where $\mathcal{N}=N/V$, the number density of electrons in the solid. Evaluate $\mu(0)$ for sodium (where each atom contributes one electron).

18C.4 By inspection of eqn 18C.6 and the expression for dN in eqn 18C.5 (and without attempting to evaluate integrals explicitly), show that in order for N to remain constant as the temperature is raised, the chemical potential must decrease in value from $\mu(0)$.

18C.5 In an intrinsic semiconductor, the band gap is so small that the Fermi–Dirac distribution results in some electrons populating the conduction band. It follows from the exponential form of the Fermi–Dirac distribution that the conductance G, the inverse of the resistance (with units of siemens, $1\,S=1\,\Omega^{-1}$), of an intrinsic semiconductor should have an Arrhenius-like temperature dependence, shown in practice to have the form $G=G_0e^{-E_g/2kT}$, where E_g is the band gap. The conductance of a sample of germanium varied with temperature as indicated below. Estimate the value of E_g.

T/K	312	354	420
G/S	0.0847	0.429	2.86

18C.6 A transistor is a semiconducting device that is commonly used either as a switch or an amplifier of electrical signals. Prepare a brief report on the design of a nanometre-sized transistor that uses a carbon nanotube as a component. A useful starting point is the work summarized by Tans, et al. (*Nature* **393**, 49 (1998)).

18C.7‡ J.J. Dannenberg et al. (*J. Phys. Chem.* **100**, 9631 (1996)) carried out theoretical studies of organic molecules consisting of chains of unsaturated four-membered rings. The calculations suggest that such compounds have large numbers of unpaired spins, and that they should therefore have unusual magnetic properties. For example, the lowest-energy state of the compound shown as **5** is computed to have $S=3$, but the energies of $S=2$ and $S=4$ structures are each predicted to be 50 kJ mol^{-1} higher in energy. Compute the molar magnetic susceptibility of these three low-lying levels at 298 K. Estimate the molar susceptibility at 298 K if each level is present in proportion to its Boltzmann factor (effectively assuming that the degeneracy is the same for all three of these levels).

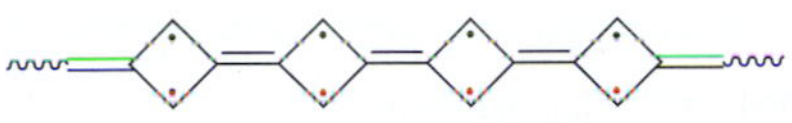

5

18C.8 An NO molecule has thermally accessible electronically excited states. It also has an unpaired electron, and so may be expected to be paramagnetic. However, its ground state is not paramagnetic because the magnetic moment of the orbital motion of the unpaired electron almost exactly cancels the spin magnetic moment. The first excited state (at 121 cm^{-1}) is paramagnetic because the orbital magnetic moment adds to, rather than cancels, the spin magnetic moment. The upper state has a magnetic moment of magnitude $2\mu_B$. Because the upper state is thermally accessible, the paramagnetic susceptibility of NO shows a pronounced temperature dependence even near room temperature. Calculate the molar paramagnetic susceptibility of NO and plot it as a function of temperature.

18C.9‡ P.G. Radaelli et al. (*Science* **265**, 380 (1994)) reported the synthesis and structure of a material that becomes superconducting at temperatures below 45 K. The compound is based on a layered compound $Hg_2Ba_2YCu_2O_{8-\delta}$, which has a tetragonal unit cell with $a=0.38606$ nm and $c=2.8915$ nm; each unit cell contains two formula units. The compound is made superconducting by partially replacing Y by Ca, accompanied by a change in unit cell volume by less than 1 per cent. Estimate the Ca content x in superconducting $Hg_2Ba_2Y_{1-x}Ca_xCu_2O_{7.55}$ given that the mass density of the compound is 7.651 g cm^{-3}.

TOPIC 18D The optical properties of solids

Discussion questions

18D.1 Explain the origin of Davydov splitting in the exciton bands of a crystal.

18D.2 Distinguish between light-emitting diodes and diode lasers.

Exercises

18D.1(a) The promotion of an electron from the valence band into the conduction band in pure TIO_2 by light absorption requires a wavelength of less than 350 nm. Calculate the energy gap in electronvolts between the valence and conduction bands.

18D.1(b) The band gap in silicon is 1.12 eV. Calculate the maximum wavelength of electromagnetic radiation that results in promotion of electrons from the valence to the conduction band.

Problems

18D.1 Here we investigate quantitatively the spectra of molecular solids. We begin by considering a dimer, with each monomer having a single transition with transition dipole moment μ_{mon} and wavenumber $\tilde{\nu}_{\text{mon}}$. We assume that the ground state wavefunctions are not perturbed as a result of dimerization, and then write the dimer excited state wavefunctions Ψ_i as linear combinations of the excited state wavefunctions ψ_1 and ψ_2 of the monomer: $\Psi_i = c_j\psi_1 + c_k\psi_2$. Now we write the hamiltonian matrix with diagonal elements set to the energy between the excited and ground state of the monomer (which, expressed as a wavenumber, is simply $\tilde{\nu}_{\text{mon}}$), and off-diagonal elements correspond to the energy of interaction between the transition dipoles. Using the arrangement discussed in **1** of *Justification* 18D.1, we write this interaction energy (as a wavenumber) as:

$$\beta = \frac{\mu_{\text{mon}}^2}{4\pi\varepsilon_0 hcr^3}(1-3\cos^2\theta)$$

It follows that the hamiltonian matrix is

$$\hat{H} = \begin{pmatrix} \tilde{\nu}_{\text{mon}} & \beta \\ \beta & \tilde{\nu}_{\text{mon}} \end{pmatrix}$$

The eigenvalues of the matrix are the dimer transition wavenumbers $\tilde{\nu}_1$ and $\tilde{\nu}_2$. The eigenvectors are the wavefunctions for the excited states of the dimer and have the form $\begin{pmatrix} c_j \\ c_k \end{pmatrix}$. (a) The intensity of absorption of incident radiation is proportional to the square of the transition dipole moment (Topic 12A). The monomer transition dipole moment is $\mu_{\text{mon}} = \int \psi_1^* \hat{\mu} \psi_0 \, d\tau = \int \psi_2^* \hat{\mu} \psi_0 \, d\tau$, where ψ_0 is the wavefunction of the monomer ground state. Assume that the dimer ground state may also be described by ψ_0 and show that the transition dipole moment μ_i of each dimer transition is given by $\mu_i = \mu_{\text{mon}}(c_j + c_k)$. (b) Consider a dimer of monomers with $\mu_{\text{mon}} = 4.00\,\text{D}$, $\tilde{\nu}_{\text{mon}} = 25\,000\,\text{cm}^{-1}$, and $r = 0.5\,\text{nm}$. How do the transition wavenumbers $\tilde{\nu}_1$ and $\tilde{\nu}_2$ vary with the angle θ? The relative intensities of the dimer transitions may be estimated by calculating the ratio μ_2^2/μ_1^2. How does this ratio vary with the angle θ? (c) Now expand the treatment given above to a chain of N monomers ($N = 5$, 10, 15, and 20), with $\mu_{\text{mon}} = 4.00\,\text{D}$, $\tilde{\nu}_{\text{mon}} = 25\,000\,\text{cm}^{-1}$, and $r = 0.5\,\text{nm}$. For simplicity, assume that $\theta = 0$ and that only nearest neighbours interact with interaction energy V. For example the hamiltonian matrix for the case $N = 4$ is

$$\hat{H} = \begin{pmatrix} \tilde{\nu}_{\text{mon}} & V & 0 & 0 \\ V & \tilde{\nu}_{\text{mon}} & V & 0 \\ 0 & V & \tilde{\nu}_{\text{mon}} & V \\ 0 & 0 & V & \tilde{\nu}_{\text{mon}} \end{pmatrix}$$

How does the wavenumber of the lowest energy transition vary with size of the chain? How does the transition dipole moment of the lowest energy transition vary with the size of the chain?

18D.2 Show that if a substance responds nonlinearly to two sources of radiation, one of frequency ω_1 and the other of frequency ω_2, then it may give rise to radiation of the sum and difference of the two frequencies. This nonlinear optical phenomenon is known as *frequency mixing* and is used to expand the wavelength range of lasers in laboratory applications, such as spectroscopy and photochemistry.

Integrated activities

18.1 Calculate the thermal expansion coefficient, $\alpha = (\partial V/\partial T)_p/V$, of diamond given that the (111) reflection shifts from 22.0403° to 21.9664° on heating a crystal from 100 K to 300 K and 154.0562 pm X-rays are used.

18.2 Calculate the scattering factor for a hydrogenic atom of atomic number Z in which the single electron occupies (a) the 1s orbital, (b) the 2s orbital. Plot f as a function of $(\sin\theta)/\lambda$. *Hint*: Interpret $4\pi\rho(r)r^2$ as the radial distribution function $P(r)$.

18.3 Explore how the scattering factor of Problem 18.2 changes when the actual 1s wavefunction of a hydrogenic atom is replaced by a Gaussian function.

Mathematical background 7 Fourier series and Fourier transforms

Some of the most versatile mathematical functions are the trigonometric functions sine and cosine. As a result, it is often very helpful to express a general function as a linear combination of these functions and then to carry out manipulations on the resulting series. Because sines and cosines have the form of waves, the linear combinations often have a straightforward physical interpretation. Throughout this discussion, the function $f(x)$ is real.

MB7.1 Fourier series

A *Fourier series* is a linear combination of sines and cosines that replicates a periodic function:

$$f(x)=\tfrac{1}{2}a_0+\sum_{n=1}^{\infty}\left\{a_n\cos\frac{n\pi x}{L}+b_n\sin\frac{n\pi x}{L}\right\} \qquad \text{(MB7.1)}$$

A periodic function is one that repeats periodically, such that $f(x+2L)=f(x)$ where $2L$ is the period. Although it is perhaps not surprising that sines and cosines can be used to replicate continuous functions, it turns out that—with certain limitations—they can also be used to replicate discontinuous functions too. The coefficients in eqn MB7.1 are found by making use of the orthogonality of the sine and cosine functions

$$\int_{-L}^{L}\sin\frac{m\pi x}{L}\cos\frac{m\pi x}{L}\,dx=0 \qquad \text{(MB7.2a)}$$

and the integrals

$$\int_{-L}^{L}\sin\frac{m\pi x}{L}\sin\frac{n\pi x}{L}\,dx=\int_{-L}^{L}\cos\frac{m\pi x}{L}\cos\frac{n\pi x}{L}\,dx=L\delta_{mn} \qquad \text{(MB7.2b)}$$

where $\delta_{mn}=1$ if $m=n$ and 0 if $m\neq n$. Thus, multiplication of both sides of eqn MB7.1 by $\cos(k\pi x/L)$ and integration from $-L$ to L gives an expression for the coefficient a_k, and multiplication by $\sin(k\pi x/L)$ and integration likewise gives an expression for b_k:

$$a_k=\frac{1}{L}\int_{-L}^{L}f(x)\cos\frac{k\pi x}{L}\,dx \quad k=0,1,2,\ldots$$
$$b_k=\frac{1}{L}\int_{-L}^{L}f(x)\sin\frac{k\pi x}{L}\,dx \quad k=1,2,\ldots \qquad \text{(MB7.3)}$$

Brief illustration MB7.1 A square wave

Figure MB7.1 shows a graph of a square wave of amplitude A that is periodic between $-L$ and L. The mathematical form of the wave is

$$f(x)=\begin{cases}-A & -L\le x<0\\ +A & 0\le x<L\end{cases}$$

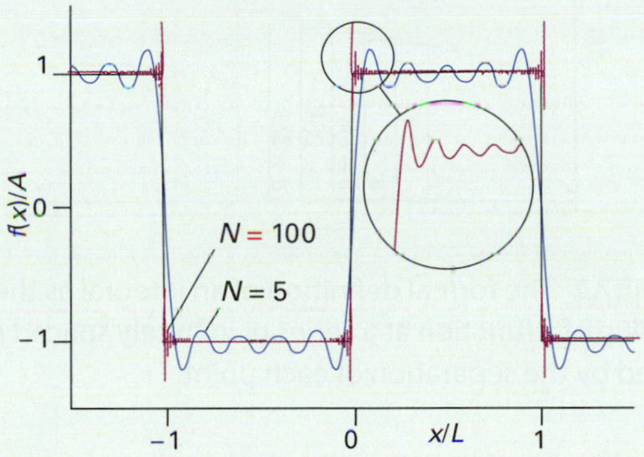

Figure MB7.1 A square wave and two successive approximations by Fourier series ($N=5$ and $N=100$). The inset shows a magnification of the $N=100$ approximation.

The coefficients a are all zero because $f(x)$ is antisymmetric ($f(-x)=-f(x)$) whereas all the cosine functions are symmetric ($\cos(-x)=\cos(x)$) and so cosine waves make no contribution to the sum. The coefficients b are obtained from

$$\begin{aligned}b_k&=\frac{1}{L}\int_{-L}^{L}f(x)\sin\frac{k\pi x}{L}\,dx\\&=\frac{1}{L}\int_{-L}^{0}(-A)\sin\frac{k\pi x}{L}\,dx+\frac{1}{L}\int_{0}^{L}A\sin\frac{k\pi x}{L}\,dx\\&=\frac{2A}{\pi}\frac{\{1-(-1)^k\}}{k}\end{aligned}$$

The final expression has been formulated to acknowledge that the two integrals cancel when k is even but add together when k is odd. Therefore,

$$f(x)=\frac{2A}{\pi}\sum_{k=1}^{N}\frac{1-(-1)^k}{k}\sin\frac{k\pi x}{L}=\frac{4A}{\pi}\sum_{n=1}^{N}\frac{1}{2n-1}\sin\frac{(2n-1)\pi x}{L}$$

with $N\to\infty$. The sum over n is the same as the sum over k; in the latter, terms with k even are all zero. This function is plotted in Fig. MB7.1 for two values of N to show how the series becomes more faithful to the original function as N increases.

MB7.2 Fourier transforms

The Fourier series in eqn MB7.1 can be expressed in a more succinct manner if we allow the coefficients to be complex numbers and make use of *de Moivre's relation*

$$e^{in\pi x/L}=\cos\left(\frac{n\pi x}{L}\right)+i\sin\left(\frac{n\pi x}{L}\right) \qquad \text{(MB7.4)}$$

for then we may write

$$f(x)=\sum_{n=-\infty}^{\infty}c_n e^{in\pi x/L} \quad c_n=\frac{1}{2L}\int_{-L}^{L}f(x)e^{-in\pi x/L}\,dx \qquad \text{(MB7.5)}$$

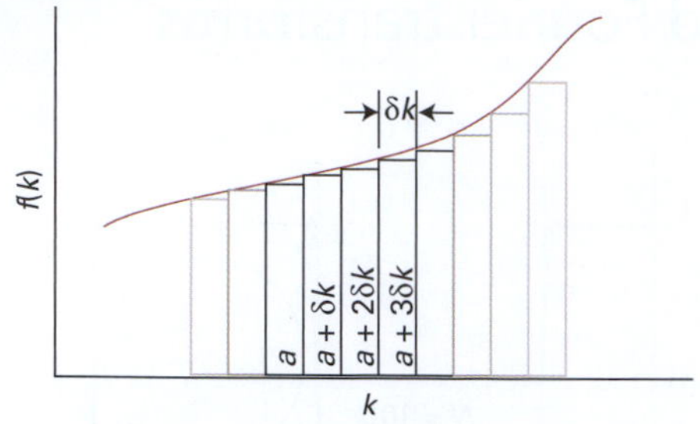

Figure MB7.2 The formal definition of an integral as the sum of the value of a function at a series of infinitely spaced points multiplied by the separation of each point.

This complex formalism is well suited to the extension of this discussion to functions with periods that become infinite. If a period is infinite, then we are effectively dealing with a non-periodic function, such as the decaying exponential function e^{-x}.

We write $\delta k=\pi/L$ and consider the limit as $L\to\infty$ and therefore $\delta k\to 0$: that is, eqn MB7.5 becomes

$$\begin{aligned} f(x)&=\lim_{L\to\infty}\sum_{n=-\infty}^{\infty}\frac{1}{2L}\left\{\int_{-L}^{L} f(x')e^{-in\pi x'/L}dx'\right\}e^{in\pi x/L}\\ &=\lim_{\delta k\to 0}\sum_{n=-\infty}^{\infty}\frac{\delta k}{2\pi}\left\{\int_{-\pi/\delta k}^{\pi/\delta k} f(x')e^{-in\delta k x'}dx'\right\}e^{in\delta k x/L}\\ &=\lim_{\delta k\to 0}\sum_{n=-\infty}^{\infty}\frac{1}{2\pi}\left\{\int_{-\infty}^{\infty} f(x')e^{-in\delta k(x'-x)}dx'\right\}\delta k \end{aligned} \quad \text{(MB7.6)}$$

In the last line we have anticipated that the limits of the integral will become infinite. At this point we should recognize that a formal definition of an integral is the sum of the value of a function at a series of infinitely spaced points multiplied by the separation of each point (Fig. MB7.2; see *Mathematical background* 1):

$$\int_a^b F(k)dk=\lim_{\delta k\to 0}\sum_{n=-\infty}^{\infty}F(n\delta k)\delta k \quad \text{(MB7.7)}$$

Exactly this form appears on the right-hand side of eqn MB7.6, so we can write that equation as

$$f(x)=\frac{1}{2\pi}\int_{-\infty}^{\infty}\tilde{f}(k)e^{ikx}dk \quad \text{where } \tilde{f}(k)=\int_{-\infty}^{\infty}f(x')e^{-ikx'}dx' \quad \text{(MB7.8)}$$

At this stage we can drop the prime on x in the expression for $\tilde{f}(k)$. We call the function $\tilde{f}(k)$ the *Fourier transform* of $f(x)$; the original function $f(x)$ is the *inverse Fourier transform* of $\tilde{f}(k)$.

Brief illustration MB7.2 A Fourier transform

The Fourier transform of the symmetrical exponential function $f(x)=e^{-a|x|}$ is

$$\begin{aligned}\tilde{f}(k)&=\int_{-\infty}^{\infty}f(x)e^{-ikx}dx=\int_{-\infty}^{\infty}e^{-a|x|-ikx}dx\\ &=\int_{-\infty}^{0}e^{ax-ikx}dx+\int_{0}^{\infty}e^{-ax-ikx}dx\\ &=\frac{1}{a-ik}+\frac{1}{a+ik}=\frac{2a}{a^2+k^2}\end{aligned}$$

The original function and its Fourier transform are drawn in Fig. MB7.3.

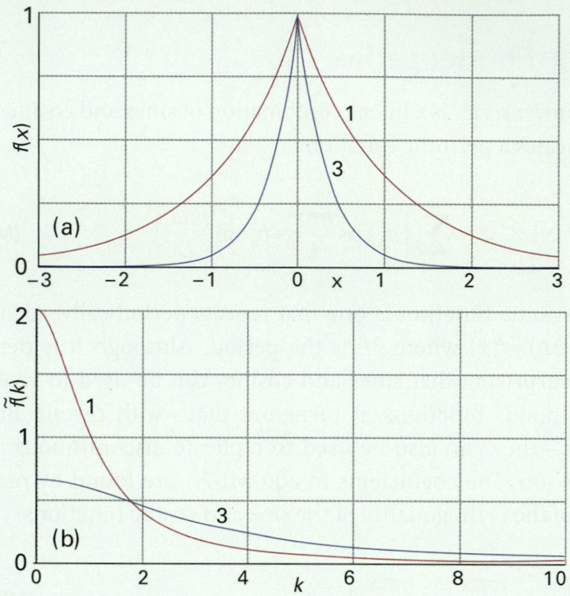

Figure MB7.3 (a) The symmetrical exponential function $f(x)=e^{-a|x|}$ and (b) its Fourier transform for two values of the decay constant a. Note how the function with the more rapid decay has a Fourier transform richer in short-wavelength (high k) components.

The physical interpretation of eqn MB7.8 is that $f(x)$ is expressed as a superposition of harmonic (sine and cosine) functions of wavelength $\lambda=2\pi/k$, and that the weight of each constituent function is given by the Fourier transform at the corresponding value of k. This interpretation is consistent with the calculation in *Brief illustration* MB7.2. As we see from Fig. MB7.3, when the exponential function falls away rapidly with time, the Fourier transform is extended to high values of k, corresponding to a significant contribution from short-wavelength waves. When the exponential function decays only slowly, the most significant contributions to the superposition come from low-frequency components, which is reflected in the Fourier transform, with its predominance of small-k contributions in this case. In general, a slowly varying function has a Fourier transform with significant contributions from small-k components.

MB7.3 The convolution theorem

A final point concerning the properties of Fourier transforms is the **convolution theorem**, which states that if a function is the 'convolution' of two other functions, that is if

$$F(x)=\int_{-\infty}^{\infty} f_1(x')f_2(x-x')\mathrm{d}x' \qquad \text{(MB7.9a)}$$

then the Fourier transform of $F(x)$ is the product of the Fourier transforms of its component functions:

$$\tilde{F}(k)=\tilde{f}_1(k)\tilde{f}_2(k) \qquad \text{(MB7.9b)}$$

Brief illustration MB7.3 Convolutions

Suppose that $F(x)$ is the convolution of two Gaussian functions:

$$F(x)=\int_{-\infty}^{\infty} \mathrm{e}^{-a^2x'^2}\mathrm{e}^{-b^2(x-x')^2}\mathrm{d}x'$$

The Fourier transform of a Gaussian function is itself a Gaussian function:

$$\tilde{f}(k)=\int_{-\infty}^{\infty} \mathrm{e}^{-c^2x^2}\mathrm{e}^{-\mathrm{i}kx}\mathrm{d}x=\left(\frac{\pi}{c^2}\right)^{1/2}\mathrm{e}^{-k^2/4c^2}$$

Therefore, the transform of $F(x)$ is the product

$$\tilde{F}(k)=\left(\frac{\pi}{a^2}\right)^{1/2}\mathrm{e}^{-k^2/4a^2}\left(\frac{\pi}{b^2}\right)^{1/2}\mathrm{e}^{-k^2/4b^2}=\frac{\pi}{ab}\mathrm{e}^{-(k^2/4)(1/a^2+1/b^2)}$$

PART THREE

Change

In Part 3 we consider the processes by which chemical change occurs and one form of matter is converted into another. We prepare the ground for a discussion of the rates of reactions by considering the motion of molecules in gases and in liquids. Then we establish the precise meaning of reaction rate and see how the overall rate, and the complex behaviour of some reactions, may be expressed in terms of elementary steps and the atomic events that take place when molecules meet. Of enormous importance in industry are reactions on solid surfaces, such as redox reactions at electrodes and various chemical transformations accelerated by solid catalysts. We discuss these processes in the final chapter of the text.

CHAPTER 19

Molecules in motion

This chapter provides techniques for discussing the motion of all kinds of particles in all kinds of fluids. It makes extensive use of the kinetic theory of gases treated in Topic 1B.

19A Transport in gases

We set the scene by showing that molecular motion in fluids (both gases and liquids) shows a number of similarities. We concentrate on the 'transport properties' of a substance, its ability to transfer matter, energy, or some other property from one place to another. These properties include diffusion, thermal conduction, viscosity, and effusion and we see that their rates are expressed in terms of the kinetic theory of gases.

19B Motion in liquids

Molecular motion in liquids is different from that in gases on account of the intermolecular forces, which now play an important role and govern, for instance the viscosity. One way to probe motion in liquids is to drag ions through a solvent by applying an electric field, and we see how the conductivities and mobilities of ions give some insight into motion in liquids.

19C Diffusion

One very important type of motion in fluids is diffusion, and it turns out that it can be discussed in a uniform way by introducing the concept of a 'thermodynamic force'. The spread of molecules can be explored by setting up and solving the 'diffusion equation', and that equation can be interpreted in terms of the molecules undergoing a random walk.

What is the impact of this material?

A great deal of chemistry, chemical engineering, and biology depends on the ability of molecules and ions to migrate through media of various kinds. In *Impact* I19.1 we see how conductivity measurements are used to analyse the motion of nutrients and other matter through biological membranes.

To read more about the impact of this material, scan the QR code, or go to bcs.whfreeman.com/webpub/chemistry/pchem10e/impact/pchem-19-1.html

19A Transport in gases

Contents

➤ Why do you need to know this material?

The transport of properties by gas molecules plays an important role in the atmosphere. The Topic also extends the approach of kinetic theory, showing how to extract quantitative expressions from simple models.

➤ What is the key idea?

A molecule carries properties through space for about the distance of its mean free path.

➤ What do you need to know already?

This Topic builds on and extends the kinetic theory of gases (Topic 1B) and you need to be familiar with the expressions from that Topic for the mean speed of molecules and with the significance of the mean free path and its pressure-dependence.

Transport properties are commonly expressed in terms of a number of equations that are empirical summaries of experimental observations. These equations apply to all kinds of properties and media. In the following sections, we introduce the equations for the general case and then show how to calculate the parameters that appear in them on the basis of a model of molecular behaviour in gases. A more general approach is taken in Topic 19C.

19A.1 The phenomenological equations

By a 'phenomenological equation', a term encountered commonly in the study of fluids, we mean an equation that summarizes empirical observations on phenomena without, initially at least, being based on an understanding of the molecular processes responsible for the property.

The rate of migration of a property is measured by its **flux**, J, the quantity of that property passing through a given area in a given time interval divided by the area and the duration of the interval. If matter is flowing (as in diffusion), we speak of a **matter flux** of so many molecules per square metre per second (number or amount $m^{-2}\ s^{-1}$); if the property is energy (as in thermal conduction), then we speak of the **energy flux** and express it in joules per square metre per second ($J\ m^{-2}\ s^{-1}$), and so on. To calculate the total quantity of each property transferred through a given area A in a given time interval Δt, we multiply the flux by the area and the time interval and form $JA\Delta t$.

Experimental observations on transport properties show that the flux of a property is usually proportional to the first derivative of some other related property. For example, the flux of matter diffusing parallel to the z-axis of a container is found to be proportional to the first derivative of the concentration:

$$J(\text{matter}) \propto \frac{d\mathcal{N}}{dz} \qquad \text{Fick's first law of diffusion} \qquad (19A.1)$$

where $\mathcal{N}$ is the number density of particles with units number per metre cubed (m^{-3}). The proportionality of the flux of matter to the concentration gradient is sometimes called **Fick's first law of diffusion**: the law implies that diffusion will be faster when the concentration varies steeply with position than when the concentration is nearly uniform. There is no net flux if the concentration is uniform ($d\mathcal{N}/dz = 0$). Similarly, the rate of thermal conduction (the flux of the energy associated with

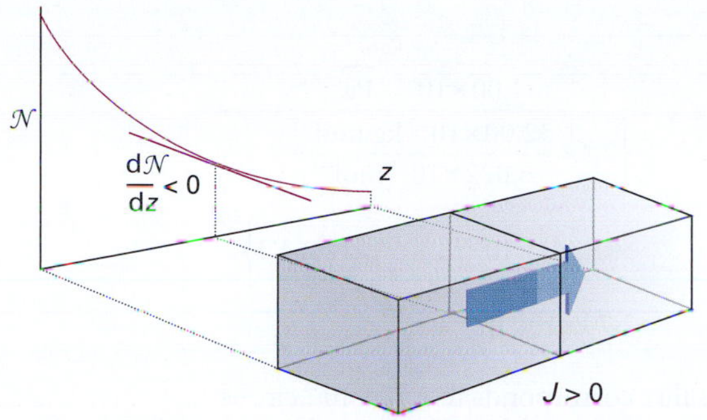

Figure 19A.1 The flux of particles down a concentration gradient. Fick's first law states that the flux of matter (the number of particles passing through an imaginary window in a given interval divided by the area of the window and the length of the interval) is proportional to the density gradient at that point.

thermal motion) is found to be proportional to the temperature gradient:

$$J(\text{energy of thermal motion}) \propto \frac{dT}{dz} \qquad \text{Flux of energy} \qquad (19A.2)$$

A positive value of J signifies a flux towards positive z; a negative value of J signifies a flux towards negative z. Because matter flows down a concentration gradient, from high concentration to low concentration, J is positive if $d\mathcal{N}/dz$ is negative (Fig. 19A.1). Therefore, the coefficient of proportionality in eqn 19A.1 must be negative, and we write it $-D$:

$$J(\text{matter}) = -D\frac{d\mathcal{N}}{dz} \qquad \text{Fick's first law in terms of the diffusion coefficient} \qquad (19A.3)$$

The constant D is the called the **diffusion coefficient**; its SI units are metre squared per second ($m^2\ s^{-1}$). Energy migrates down a temperature gradient, and the same reasoning leads to

$$J(\text{energy of thermal motion}) = -\kappa\frac{dT}{dz} \qquad \text{Flux of energy in terms of the coefficient of thermal conductivity} \qquad (19A.4)$$

where κ is the **coefficient of thermal conductivity**. The SI units of κ are joules per kelvin per metre per second ($J\ K^{-1}\ m^{-1}\ s^{-1}$) or, because $1\ J\ s^{-1} = 1\ W$, watts per kelvin per metre ($W\ K^{-1}\ m^{-1}$). Some experimental values are given in Table 19A.1.

Brief illustration 19A.1 Energy flux

Suppose that there is a temperature difference of 10 K between two metal plates that are separated by 1.0 cm in air (for which $\kappa = 24.1\ mW\ K^{-1}\ m^{-1}$). The temperature gradient is

$$\frac{dT}{dz} = -\frac{-10\ K}{1.0\times10^{-2}\ m} = -1.0\times10^{3}\ K\ m^{-1}$$

Therefore, because for air the energy flux is

$$J(\text{energy of thermal motion}) = -(24.1\ mW\ K^{-1}\ m^{-1}) \times(-1.0\times10^{3}\ K\ m^{-1}) = +24\ W\ m^{-2}$$

As a result, in 1.0 h (3600 s) the transfer of energy through an area of the opposite walls of 1.0 cm² is

$$\text{Transfer} = (24\ W\ m^{-2})\times(1.0\times10^{-4}\ m^2)\times(3600\ s) = 8.6\ J$$

Self-test 19A.1 The thermal conductivity of glass is 0.92 $W\ K^{-1}\ m^{-1}$. What is the rate of energy transfer through a window pane of thickness 0.50 cm and area 1.0 m² when the room is at 22 °C and the exterior is at 0 °C?

Answer: 2.8 Kw

Table 19A.1* Transport properties of gases at 1 atm

	$\kappa/(mW\ K^{-1}\ m^{-1})$	$\eta/\mu P^{\ddagger}$	
	273 K	273 K	293 K
Ar	16.3	210	223
CO_2	14.5	136	147
He	144.2	187	196
N_2	24.0	166	176

* More values are given in the *Resource section*.
‡ $1\ \mu P = 10^{-7}\ kg\ m^{-1}\ s^{-1}$.

To see the connection between the flux of momentum and the viscosity, consider a fluid in a state of **Newtonian flow**, which can be imagined as occurring by a series of layers moving past one another (Fig. 19A.2). The layer next to the wall

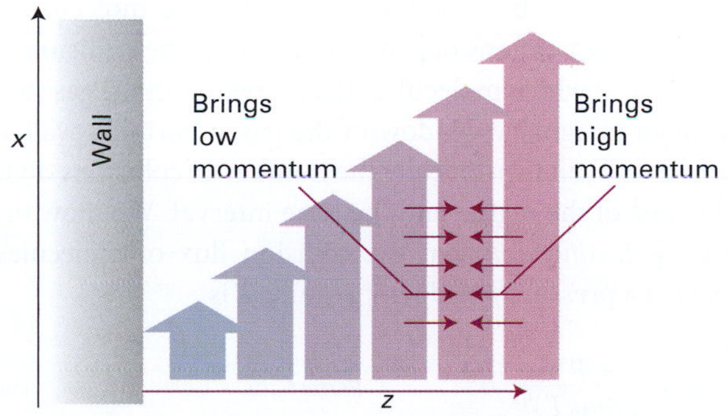

Figure 19A.2 The viscosity of a fluid arises from the transport of linear momentum. In this illustration the fluid is undergoing Newtonian (laminar) flow, and particles bring their initial momentum when they enter a new layer. If they arrive with high x-component of momentum they accelerate the layer; if with low x-component of momentum they retard the layer.

of the vessel is stationary, and the velocity of successive layers varies linearly with distance, z, from the wall. Molecules ceaselessly move between the layers and bring with them the x-component of linear momentum they possessed in their original layer. A layer is retarded by molecules arriving from a more slowly moving layer because they have a low momentum in the x-direction. A layer is accelerated by molecules arriving from a more rapidly moving layer. We interpret the net retarding effect as the fluid's viscosity.

Because the retarding effect depends on the transfer of the x-component of linear momentum into the layer of interest, the viscosity depends on the flux of this x-component in the z-direction. The flux of the x-component of momentum is proportional to $\mathrm{d}v_x/\mathrm{d}z$ because there is no net flux when all the layers move at the same velocity. We can therefore write

$$J(x\text{-component of momentum}) = -\eta\frac{\mathrm{d}v_x}{\mathrm{d}z}$$

Momentum flux in terms of the coefficient of viscosity (19A.5)

The constant of proportionality, η, is the **coefficient of viscosity** (or simply 'the viscosity'). Its units are kilograms per metre per second (kg m^{-1} s^{-1}). Viscosities are often reported in poise (P), where 1 P = 10^{-1} kg m^{-1} s^{-1}. Some experimental values are given in Table 19A.1.

Although it is not strictly a transport property, closely related to diffusion is **effusion**, the escape of matter through a small hole. The essential empirical observations on effusion are summarized by **Graham's law of effusion**, which states that the rate of effusion is inversely proportional to the square root of the molar mass, M.

19A.2 The transport parameters

Here we derive expressions for the diffusion characteristics of a perfect gas on the basis of a model, the kinetic-molecular theory. All the expressions depend on knowing the **collision flux**, Z_W, the rate at which molecules strike a region in the gas (which may be an imaginary window in the gas, a part of a wall, or a hole in a wall) and specifically the number of collisions divided by the area of the region and the time interval. We show in the following *Justification* that the collision flux of molecules of mass m at a pressure p and temperature T is

$$Z_W = \frac{p}{(2\pi mkT)^{1/2}}$$

Perfect gas Collision flux (19A.6)

Brief illustration 19A.2 The collision flux

The collision flux of O_2 molecules, with $m = M/N_A$ and $M = 32.00$ g mol^{-1}, at 25 °C and at 1.00 bar is

$$Z_W = \frac{1.00\times10^5\ \overbrace{\mathrm{Pa}}^{\mathrm{kg\,m^{-1}\,s^{-2}}}}{\left\{2\pi\times\frac{32.00\times10^{-3}\ \mathrm{kg\,mol^{-1}}}{6.022\times10^{23}\ \mathrm{mol^{-1}}}\times(1.381\times10^{-23}\ \mathrm{J\,K^{-1}})\times(298\ \mathrm{K})\right\}^{1/2}}$$
$$= 2.70\times10^{27}\ \mathrm{m^{-2}\,s^{-1}}$$

This flux corresponds to 2.70×10^{23} cm^{-2} s^{-1}.

Self-test 19A.2 Evaluate the collision flux of H_2 molecules under the same conditions.

Answer: 1.07×10^{28} m^{-2} s^{-1}

Justification 19A.1 The collision flux

Consider a wall of area A perpendicular to the x-axis (Fig. 19A.3). If a molecule has $v_x > 0$ (that is, it is travelling in the direction of positive x), then it will strike the wall within an interval Δt if it lies within a distance $v_x\Delta t$ of the wall. Therefore, all molecules in the volume $Av_x\Delta t$, and with positive x-component of velocities, will strike the wall in the interval Δt. The total number of collisions in this interval is therefore the volume $Av_x\Delta t$ multiplied by the number density, $\mathcal{N}$, of molecules. However, to take account of the presence of a range of velocities in the sample, we must sum the result over all the positive values of v_x weighted by the probability distribution of velocities given in *Justification* 1B.2:

$$f(v_x) = \left(\frac{m}{2\pi kT}\right)^{1/2} \mathrm{e}^{-mv_x^2/2kT}$$

That is,

$$\text{Number of collisions} = \mathcal{N}A\Delta t\int_0^\infty v_x f(v_x)\mathrm{d}v_x$$

The collision flux is the number of collisions divided by A and Δt, so

$$Z_W = \mathcal{N}\int_0^\infty v_x f(v_x)\mathrm{d}v_x$$

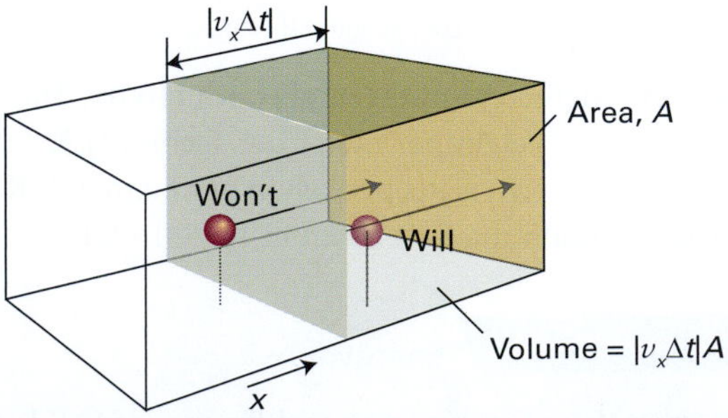

Figure 19A.3 A molecule will reach the wall on the right within an interval Δt if it is within a distance $v_x\Delta t$ of the wall and travelling to the right.

Then, because

$$\int_0^\infty v_x f(v_x)\mathrm{d}v_x = \left(\frac{m}{2\pi kT}\right)^{1/2}\int_0^\infty v_x \mathrm{e}^{-mv_x^2/2kT}\mathrm{d}v_x \overset{\text{Integral G.2}}{=} \left(\frac{kT}{2\pi m}\right)^{1/2}$$

it follows that

$$Z_\mathrm{W} = \mathcal{N}\left(\frac{kT}{2\pi m}\right)^{1/2}$$

Substitution of $\mathcal{N}=p/kT$ then gives eqn 19A.6.

According to eqn 19A.6, the collision flux increases with pressure, because the rate of collisions on the region of interest increases with pressure. The flux decreases with increasing mass of the molecules because heavy molecules move more slowly that light molecules. Caution, however, is needed with the interpretation of the role of temperature, for it is wrong to conclude that because $T^{1/2}$ appears in the denominator that the collision flux decreases with increasing temperature. If the system has constant volume, the pressure increases with temperature ($p \propto T$), so the collision flux is in fact proportional to $T/T^{1/2}=T^{1/2}$, and increases with temperature (because the molecules are moving faster).

(a) The diffusion coefficient

Consider the arrangement depicted in Fig. 19A.4. The molecules passing through the area A at $z=0$ have travelled an average of about one mean free path λ since their last collision. Therefore, the number density where they originated is $\mathcal{N}(z)$ evaluated at $z=-\lambda$. This number density is approximately

$$\mathcal{N}(-\lambda)=\mathcal{N}(0)-\lambda\left(\frac{\mathrm{d}\mathcal{N}}{\mathrm{d}z}\right)_0 \qquad (19\text{A}.7\text{a})$$

where we have used a Taylor expansion of the form $f(x)=f(0)+(\mathrm{d}f/\mathrm{d}x)_0x+\cdots$ truncated after the second term (see

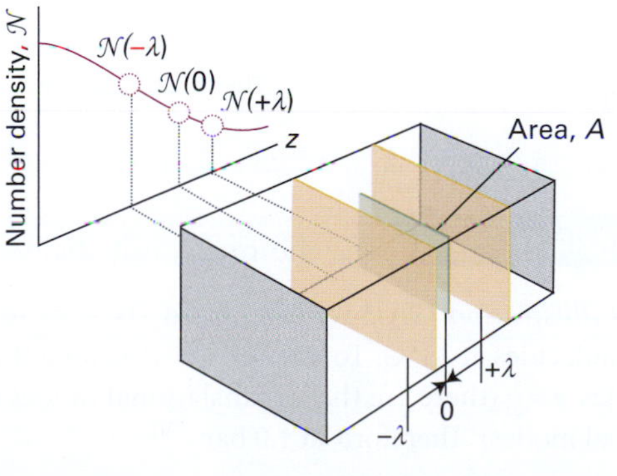

Figure 19A.4 The calculation of the rate of diffusion of a gas considers the net flux of molecules through a plane of area A as a result of arrivals from an average distance λ away in each direction, where λ is the mean free path.

Mathematical Background 1). Similarly, the number density at an equal distance on the other side of the area is

$$\mathcal{N}(\lambda)=\mathcal{N}(0)+\lambda\left(\frac{\mathrm{d}\mathcal{N}}{\mathrm{d}z}\right)_0 \qquad (19\text{A}.7\text{b})$$

The average number of impacts on the imaginary region of area A_0 during an interval Δt is $Z_\mathrm{W}A_0\Delta t$, where Z_W is the collision flux. Therefore, the flux from left to right, $J(\mathrm{L}\rightarrow\mathrm{R})$, arising from the supply of molecules on the left, is

$$J(\mathrm{L}\rightarrow\mathrm{R})=\frac{\overbrace{\tfrac{1}{4}\mathcal{N}(-\lambda)v_\mathrm{mean}}^{Z_\mathrm{W}}A_0\Delta t}{A_0\Delta t}=\tfrac{1}{4}\mathcal{N}(-\lambda)v_\mathrm{mean} \qquad (19\text{A}.8\text{a})$$

There is also a flux of molecules from right to left. On average, the molecules making the journey have originated from $z=+\lambda$ where the number density is $\mathcal{N}(\lambda)$. Therefore,

$$J(\mathrm{L}\leftarrow\mathrm{R})=\tfrac{1}{4}\mathcal{N}(\lambda)v_\mathrm{mean} \qquad (19\text{A}.8\text{b})$$

The net flux from left to right is

$$\begin{aligned}J_z &= J(\mathrm{L}\rightarrow\mathrm{R})-J(\mathrm{L}\leftarrow\mathrm{R})\\ &=\tfrac{1}{4}v_\mathrm{mean}\{\mathcal{N}(-\lambda)-\mathcal{N}(\lambda)\}\\ &=\tfrac{1}{4}v_\mathrm{mean}\left\{\left[\mathcal{N}(0)-\lambda\left(\frac{\mathrm{d}\mathcal{N}}{\mathrm{d}z}\right)_0\right]-\left[\mathcal{N}(0)+\lambda\left(\frac{\mathrm{d}\mathcal{N}}{\mathrm{d}z}\right)_0\right]\right\}\end{aligned}$$

That is,

$$J_z=-\tfrac{1}{2}v_\mathrm{mean}\lambda\left(\frac{\mathrm{d}\mathcal{N}}{\mathrm{d}z}\right)_0 \qquad (19\text{A}.9)$$

This equation shows that the flux is proportional to the first derivative of the concentration, in agreement with Fick's law, eqn 19A.1.

At this stage it looks as though we can pick out a value of the diffusion coefficient by comparing eqns 19A.9 and 19A.3, so obtaining $D=\tfrac{1}{2}\lambda v_\mathrm{mean}$. It must be remembered, however, that the calculation is quite crude, and is little more than an assessment of the order of magnitude of D. One aspect that has not been taken into account is illustrated in Fig. 19A.5, which shows that although a molecule may have begun its journey very close to the window, it could have a long flight before it gets there. Because the path is long, the molecule is likely to collide before reaching the window, so it ought to be added to the graveyard of other molecules that have collided. To take this effect into account involves a lot of work, but the end result is the appearance of a factor of $\tfrac{2}{3}$ representing the lower flux. The modification results in

$$D=\tfrac{1}{3}\lambda v_\mathrm{mean} \qquad \text{Diffusion coefficient} \qquad (19\text{A}.10)$$

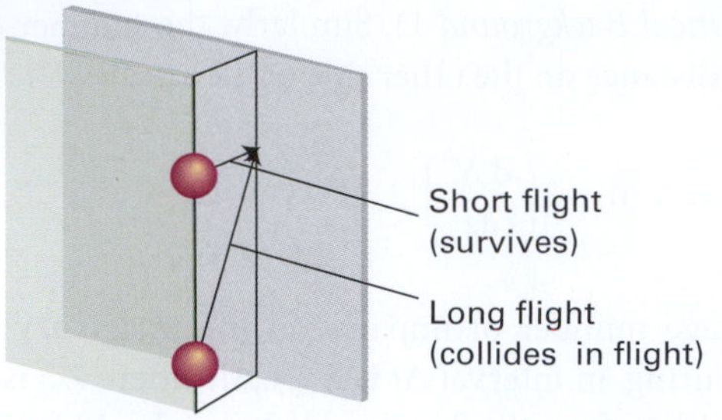

Figure 19A.5 One approximation ignored in the simple treatment is that some particles might make a long flight to the plane even though they are only a short perpendicular distance away, and therefore they have a higher chance of colliding during their journey.

Brief illustration 19A.3 The diffusion coefficient

In *Brief illustration* 1B.4 of Topic 1B it is established that the mean free path of N_2 molecules in a gas at 1.0 bar is 95 nm; in *Example* 1B.1 of the same Topic it is calculated that the mean speed of N_2 molecules at 25 °C is 475 m s^{-1}. Therefore, the diffusion coefficient for N_2 molecules under these conditions is

$$D=\tfrac{1}{3}\times(9.5\times10^{-8}\,\text{m})\times475\,\text{m s}^{-1}=1.5\times10^{-5}\,\text{m}^2\text{s}^{-1}$$

The experimental value (for N_2 in O_2) is $2.0\times10^{-5}\,\text{m}^2\,\text{s}^{-1}$.

Self-test 19A.3 Evaluate D for H_2 under the same conditions.

Answer: $9.0\times10^{-5}\,\text{m}^2\,\text{s}^{-1}$

There are three points to note about eqn 19A.10:

Physical interpretation

- The mean free path, λ, decreases as the pressure is increased (eqn 1B.13 of Topic 1B, $\lambda=kT/\sigma p$), so D decreases with increasing pressure and, as a result, the gas molecules diffuse more slowly.
- The mean speed, v_{mean}, increases with the temperature (eqn 1B.8 of Topic 1B, $v_{\text{mean}}=(8RT/\pi M)^{1/2}$), so D also increases with temperature. As a result, molecules in a hot sample diffuse more quickly than those in a cool sample (for a given concentration gradient).
- Because the mean free path increases when the collision cross-section σ of the molecules decreases (eqn 1B.13 again, $\lambda=kT/\sigma p$), the diffusion coefficient is greater for small molecules than for large molecules.

(b) Thermal conductivity

According to the equipartition theorem (*Foundations* C), each molecule carries an average energy $\varepsilon=\nu kT$, where ν is a number of the order of 1. For atoms, $\nu=\tfrac{3}{2}$. When one molecule passes through the imaginary window, it transports that average energy. We suppose that the number density is uniform but that the temperature is not. Molecules arrive from the left after travelling a mean free path from their last collision in a hotter region, and therefore with a higher energy. Molecules also arrive from the right after travelling a mean free path from a cooler region. The two opposing energy fluxes are therefore

$$J(\text{L}\leftarrow\text{R})=\overbrace{\tfrac{1}{4}\mathcal{N}v_{\text{mean}}}^{Z_W}\,\varepsilon(-\lambda)\quad J(\text{L}\leftarrow\text{R})=\overbrace{\tfrac{1}{4}\mathcal{N}v_{\text{mean}}}^{Z_W}\,\varepsilon(\lambda)\qquad(19\text{A}.11)$$

and the net flux is

$$\begin{aligned}J_z&=J(\text{L}\rightarrow\text{R})-J(\text{L}\leftarrow\text{R})\\&=\tfrac{1}{4}v_{\text{mean}}\mathcal{N}\{\varepsilon(-\lambda)-\varepsilon(\lambda)\}\\&=\tfrac{1}{4}v_{\text{mean}}\mathcal{N}\left\{\left[\varepsilon(0)-\lambda\left(\frac{\mathrm{d}\varepsilon}{\mathrm{d}z}\right)_0\right]-\left[\varepsilon(0)+\lambda\left(\frac{\mathrm{d}\varepsilon}{\mathrm{d}z}\right)_0\right]\right\}\end{aligned}$$

That is,

$$J_z=-\tfrac{1}{2}v_{\text{mean}}\lambda\mathcal{N}\left(\frac{\mathrm{d}\varepsilon}{\mathrm{d}z}\right)_0=-\tfrac{1}{2}\nu v_{\text{mean}}\lambda\mathcal{N}k\left(\frac{\mathrm{d}T}{\mathrm{d}z}\right)_0\qquad(19\text{A}.12)$$

The energy flux is proportional to the temperature gradient, as we wanted to show. As before, we multiply by $\tfrac{2}{3}$ to take long flight paths into account, and after comparison of this equation with eqn 19A.4 arrive at

$$\kappa=\tfrac{1}{3}\nu v_{\text{mean}}\lambda\mathcal{N}k\qquad\text{Thermal conductivity}\qquad(19\text{A}.13\text{a})$$

If we now identify $\mathcal{N}=nN_A/V=[\text{J}]N_A$, where [J] is the molar concentration of the carrier particles J and N_A is Avogadro's constant, and identify νkN_A as the molar constant-volume heat capacity of a perfect gas (which follows from $C_{V,\text{m}}=N_A(\partial\varepsilon/\partial T)_V$), this expression becomes

$$\kappa=\tfrac{1}{3}v_{\text{mean}}\lambda[\text{J}]C_{V,\text{m}}\qquad\text{Thermal conductivity}\qquad(19\text{A}.13\text{b})$$

Yet another form is found by recognizing that $\mathcal{N}=p/kT$ and the expression for D in eqn 19A.10, for then

$$\kappa=\frac{\nu pD}{T}\qquad\text{Thermal conductivity}\qquad(19\text{A}.13\text{c})$$

Brief illustration 19A.4 The thermal conductivity

In *Brief illustration* 19A.3 we calculated $D=1.5\times10^{-5}\,\text{m}^2\,\text{s}^{-1}$ for N_2 molecules at 25 °C. To use eqn 19A.13c note that for N_2 molecules $\nu=\tfrac{5}{2}$ (there are three translational modes and two rotational modes). Therefore, at 1.0 bar

$$\kappa=\frac{\tfrac{5}{2}\times(1.0\times10^5\,\overbrace{\text{Pa}}^{\text{Jm}^{-3}})\times(1.5\times10^{-5}\,\text{m}^2\,\text{s}^{-1})}{298\,\text{K}}=1.3\times10^{-2}\,\text{J K}^{-1}\,\text{m}^{-1}\,\text{s}^{-1}$$

or 13 mW K^{-1} m^{-1}. The experimental value is 26 mW K^{-1} m^{-1}.

Self-test 19A.4 Estimate the thermal conductivity of argon gas at 25 °C and 1.0 bar.

Answer: 8 mW K^{-1} m^{-1}

To interpret eqn 19A.13, we note that:

- Because λ is inversely proportional to the pressure (eqn 1B.13 of Topic 1B, $\lambda = kT/\sigma p$), and hence inversely proportional to the molar concentration of the gas, and $\mathcal{N}$ is proportional to the pressure ($\mathcal{N} = p/kT$), the thermal conductivity, which is proportional to the product λp, is independent of the pressure.
- The thermal conductivity is greater for gases with a high heat capacity (eqn 19A.13b) because a given temperature gradient then corresponds to a greater energy gradient.

Physical interpretation

The physical reason for the pressure independence of the thermal conductivity is that it can be expected to be large when many molecules are available to transport the energy, but the presence of so many molecules limits their mean free path and they cannot carry the energy over a great distance. These two effects balance. The thermal conductivity is indeed found experimentally to be independent of the pressure, except when the pressure is very low, when $\kappa \propto p$. At low pressures λ exceeds the dimensions of the apparatus, and the distance over which the energy is transported is determined by the size of the container and not by collisions with the other molecules present. The flux is still proportional to the number of carriers, but the length of the journey no longer depends on λ, so $\kappa \propto$ [A], which implies that $\kappa \propto p$.

(c) Viscosity

Molecules travelling from the right in Fig. 19A.6 (from a fast layer to a slower one) transport a momentum $mv_x(\lambda)$ to their new layer at $z=0$; those travelling from the left transport $mv_x(-\lambda)$ to it. If it is assumed that the density is uniform, the collision flux is $\frac{1}{4}\mathcal{N}v_{\text{mean}}$. Those arriving from the right on average carry a momentum

$$mv_x(\lambda) = mv_x(0) + m\lambda\left(\frac{dv_x}{dz}\right)_0 \qquad (19A.14a)$$

Those arriving from the left bring a momentum

$$mv_x(-\lambda) = mv_x(0) - m\lambda\left(\frac{dv_x}{dz}\right)_0 \qquad (19A.14b)$$

The net flux of x-momentum in the z-direction is therefore

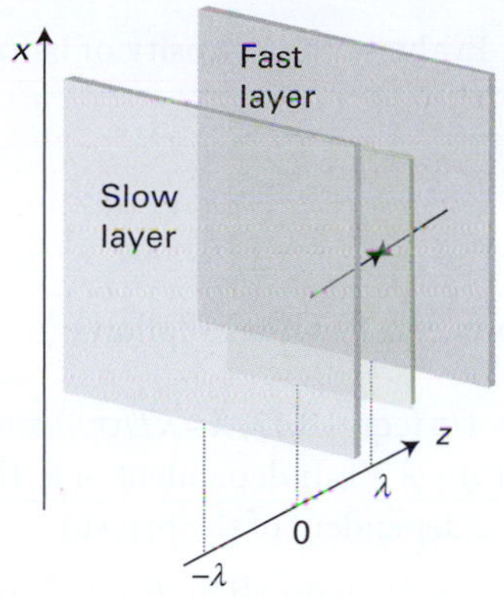

Figure 19A.6 The calculation of the viscosity of a gas examines the net x-component of momentum brought to a plane from faster and slower layers on average a mean free path away in each direction.

$$J_z = \tfrac{1}{4}v_{\text{mean}}\mathcal{N}\left\{\left[mv_x(0) - m\lambda\left(\frac{dv_x}{dz}\right)_0\right] - \left[mv_x(0) + m\lambda\left(\frac{dv_x}{dz}\right)_0\right]\right\} = -\tfrac{1}{2}v_{\text{mean}}\lambda m\mathcal{N}\left(\frac{dv_x}{dz}\right)_0 \qquad (19A.15)$$

The flux is proportional to the velocity gradient, as we wished to show. Comparison of this expression with eqn 19A.5, and multiplication by $\frac{2}{3}$ in the normal way, leads to

$$\eta = \tfrac{1}{3}v_{\text{mean}}\lambda m\mathcal{N} \qquad \text{Viscosity} \qquad (19A.16a)$$

Two alternative forms of this expression (after using $mN_A = M$) are

$$\eta = MD[\text{J}] \qquad \text{Viscosity} \qquad (19A.16b)$$

$$\eta = \frac{pMD}{RT} \qquad \text{Viscosity} \qquad (19A.16c)$$

where [J] is the molar concentration of the gas molecules and M is their molar mass.

Brief illustration 19A.5 The viscosity

We have already calculated $D = 1.5\times10^{-5}\,\text{m}^2\,\text{s}^{-1}$ for N_2 at 25 °C. Because $M = 28.02\,\text{g mol}^{-1}$, for the gas at 1.0 bar, eqn 19A.17c gives

$$\eta = \frac{(1.0\times10^5\ \overbrace{\text{Pa}}^{\text{J m}^{-3}})\times(28.02\times10^{-3}\,\text{kg mol}^{-1})\times(1.5\times10^{-5}\,\text{m}^2\,\text{s}^{-1})}{(8.3145\,\text{J K}^{-1}\,\text{mol}^{-1})\times(298\,\text{K})}$$
$$= 1.7\times10^{-5}\,\text{kg m}^{-1}\,\text{s}^{-1}$$

or 171 μP. The experimental value is 176 μP.

Self-test 19A.5 Evaluate the viscosity of benzene vapour at 0.10 bar and 25 °C.

Answer: 140 μP

We can interpret eqn 19A.16a as follows:

- Because $\lambda \propto 1/p$ (eqn 1B.13, $\lambda = kT/\sigma p$)) and $[A] \propto p$, it follows that $\eta \propto \lambda\mathcal{N}$ is independent of p. That is, the viscosity is independent of the pressure.
- Because $v_{mean} \propto T^{1/2}$ (eqn 1B.8), $\eta \propto T^{1/2}$. That is, the viscosity of a gas *increases* with temperature.

Physical interpretation

The physical reason for the pressure independence of the viscosity is the same as for the thermal conductivity: more molecules are available to transport the momentum, but they carry it less far on account of the decrease in mean free path. The increase of viscosity with temperature is explained when we remember that at high temperatures the molecules travel more quickly, so the flux of momentum is greater. By contrast, as discussed in Topic 19B, the viscosity of a liquid *decreases* with increase in temperature because intermolecular interactions must be overcome.

(d) Effusion

Because the mean speed of molecules is inversely proportional to $M^{1/2}$, the rate at which they strike the area of the hole through which they are effusing is also inversely proportional to $M^{1/2}$, in accord with Graham's law. However, by using the expression for the collision flux, we can obtain a more detailed expression for the rate of effusion and hence use effusion data more effectively.

When a gas at a pressure p and temperature T is separated from a vacuum by a small hole, the rate of escape of its molecules is equal to the rate at which they strike the area of the hole, which is the product of the collision flux and the area of the hole, A_0:

$$\text{Rate of effusion} = Z_W A_0 = \frac{pA_0}{(2\pi mkT)^{1/2}} \overset{m=M/N_A,\ k=R/N_A}{=} \frac{pA_0N_A}{(2\pi MRT)^{1/2}}$$

Rate of effusion (19A.17)

This rate is inversely proportional to $M^{1/2}$, in accord with Graham's law. Do not conclude from eqn 19A.17, however, that effusion is slower at high temperatures than at low. Because $p \propto T$, the rate is in fact proportional to $T^{1/2}$ and increases with temperature.

Equation 19A.17 is the basis of the **Knudsen method** for the determination of the vapour pressures of liquids and solids, particularly of substances with very low vapour pressures. Thus, if the vapour pressure of a sample is p, and it is enclosed in a cavity with a small hole, then the rate of loss of mass from the container is proportional to p.

Example 19A.1 Calculating the vapour pressure from a mass loss

Caesium (m.p. 29 °C, b.p. 686 °C) was introduced into a container and heated to 500 °C. When a hole of diameter 0.50 mm was opened in the container for 100 s, a mass loss of 385 mg was measured. Calculate the vapour pressure of liquid caesium at 500 K.

Method The pressure of vapour is constant inside the container despite the effusion of atoms because the hot liquid metal replenishes the vapour. The rate of effusion is therefore constant, and given by eqn 19A.17. To express the rate in terms of mass, multiply the number of atoms that escape by the mass of each atom.

Answer The mass loss Δm in an interval Δt is related to the collision flux by

$$\Delta m = Z_W A_0 m \Delta t$$

where A_0 is the area of the hole and m is the mass of one atom. It follows that

$$Z_W = \frac{\Delta m}{A_0 m \Delta t}$$

Because Z_W is related to the pressure by eqn 19A.17, we can write

$$p = \left(\frac{2\pi RT}{M}\right)^{1/2} \frac{\Delta m}{A_0 \Delta t}$$

Because $M = 132.9\ \text{g mol}^{-1}$, substitution of the data gives $p = 8.7\ \text{kPa}$ (using $1\ \text{Pa} = 1\ \text{N m}^{-2} = 1\ \text{J m}^{-1}$), or 65 Torr.

Self-test 19A.6 How long would it take 1.0 g of Cs atoms to effuse out of the oven under the same conditions?

Answer: 260 s

Checklist of concepts

☐ 1. **Flux** is the quantity of a property passing through a given area in a given time interval divided by the area and the duration of the interval.

☐ 2. **Diffusion** is the migration of matter down a concentration gradient.

☐ 3. **Fick's first law of diffusion** states that the flux of matter is proportional to the concentration gradient.
☐ 4. **Thermal conduction** is the migration of energy down a temperature gradient and the flux of energy is proportional to the temperature gradient.
☐ 5. **Viscosity** is the migration of linear momentum down a velocity gradient and the flux of momentum is proportional to the velocity gradient.
☐ 6. **Effusion** is the emergence of a gas from a container through a small hole.
☐ 7. **Graham's law of effusion** states that the rate of effusion is inversely proportional to the square root of the molar mass.
☐ 8. The coefficients of diffusion, thermal conductivity, and viscosity of a perfect gas are proportional to the product of the mean free path and mean speed.

Checklist of equations

Property	Equation	Comment	Equation number
Fick's first law of diffusion	$J = -D\mathrm{d}\mathcal{N}/\mathrm{d}z$		19A.3
Flux of energy of thermal motion	$J = -\kappa \mathrm{d}T/\mathrm{d}z$		19A.4
Flux of momentum	$J = -\eta \mathrm{d}v_x/\mathrm{d}z$		19A.5
Diffusion coefficient of a perfect gas	$D = \frac{1}{3}\lambda v_{\mathrm{mean}}$	KMT*	19A.10
Coefficient of thermal conductivity of a perfect gas	$\kappa = \frac{1}{3} v_{\mathrm{mean}} \lambda [\mathrm{J}] C_{V,\mathrm{m}}$	KMT and equipartition	19A.13b
Coefficient of viscosity of a perfect gas	$\eta = \frac{1}{3} v_{\mathrm{mean}} \lambda m \mathcal{N}$	KMT	19A.16a
Rate of effusion	Rate $\propto 1/M^{1/2}$	Graham's law	19A.17

* KMT indicates that the equation is based on the kinetic theory of gases.

19B Motion in liquids

Contents

➤ **Why do you need to know this material?**

Liquids are central to chemical reactions, and it is important to know how the mobility of their molecules and solutes in them varies with the conditions. Ionic motion is a way of exploring this motion as forces to move them can be applied electrically. From electrical measurements the properties of diffusing neutral molecules may also be inferred.

➤ **What is the key idea?**

The viscosity of a liquid decreases with increasing temperature; ions reach a terminal velocity when the electrical force on them is balanced by the drag due to the viscosity of the solvent.

➤ **What do you need to know already?**

The discussion of viscosity starts with the definition of viscosity coefficient introduced in Topic 19A. One derivation uses the same argument about flux as was used in Topic 19A. The final section quotes the relation between the drift speed and a generalized force acting on a solute particle, which is derived in Topic 19C. You need to be aware of several concepts from electrostatics, which are introduced in *Foundations* B.

In this Topic we consider two aspects of motion in liquids. First, we deal with pure liquids, and examine how the mobilities of their molecules, as measured by their viscosity, varies with temperature. Then we consider the motion of solutes. A particularly simple and to some extent controllable type of motion through a liquid, is that of an ion, and we shall see that the information that motion provides can be used to infer the behaviour of uncharged species too.

19B.1 Experimental results

The motion of molecules in liquids can be studied experimentally by a variety of methods. Relaxation time measurements in NMR (Topic 14C) and EPR can be interpreted in terms of the mobilities of the molecules, and have been used to show that big molecules in viscous fluids typically rotate in a series of small (about 5°) steps, whereas small molecules in non-viscous fluids typically jump through about 1 radian (57°) in each step. Another important technique is **inelastic neutron scattering**, in which the energy neutrons collect or discard as they pass through a sample is interpreted in terms of the motion of its particles. The same technique is used to examine the internal dynamics of macromolecules.

(a) Liquid viscosity

The coefficient of viscosity, η (eta), is introduced in Topic 19A as a phenomenological coefficient, the constant of proportionality between the flux of linear momentum and the velocity gradient in a fluid:

$$J_z(x\text{-component of momentum}) = -\eta\frac{dv_x}{dz} \quad \text{Viscosity} \quad (19B.1)$$

(This is eqn 19A.5 of Topic 19A.) Some values for liquids are given in Table 19B.1. The SI units of viscosity are kilograms per metre per second (kg m^{-1} s^{-1}), but they may also be reported in the equivalent units of pascal seconds (Pa s). The non-SI unit poise (P) and centipoise (cP) are still widely encountered: 1 P = 10^{-1} Pa s and so 1 cP = 1 mPa s.

Unlike in a gas, for a molecule to move in a liquid it must acquire at least a minimum energy (an 'activation energy' in the language of Topic 20D) to escape from its neighbours. The probability that a molecule has at least an energy E_a is proportional to $e^{-E_a/RT}$, so the mobility of the molecules in the liquid

Table 19B.1* Viscosities of liquids at 298 K, $\eta/(10^{-3}\,\mathrm{kg\,m^{-1}\,s^{-1}})$

	$\eta/(10^{-3}\,\mathrm{kg\,m^{-1}\,s^{-1}})$
Benzene	0.601
Mercury	1.55
Pentane	0.224
Water‡	0.891

* More values are given in the *Resource section*.
‡ The viscosity of water corresponds to 0.891 cP.

should follow this type of temperature dependence. Because the coefficient of viscosity is inversely proportional to the mobility of the particles, we should expect that

$$\eta = \eta_0 e^{E_a/RT} \qquad \text{Temperature dependence of viscosity (liquid)} \qquad (19B.2)$$

(Note the positive sign of the exponent, because the viscosity is *inversely* proportion to the mobility.) This expression implies that the viscosity should decrease sharply with increasing temperature. Such a variation is found experimentally, at least over reasonably small temperature ranges (Fig. 19B.1). The activation energy typical of viscosity is comparable to the mean potential energy of intermolecular interactions.

One problem with the interpretation of viscosity measurements is that the change in density of the liquid as it is heated makes a pronounced contribution to the temperature variation of the viscosity. Thus, the temperature dependence of viscosity at constant volume, when the density is constant, is much less than that at constant pressure. The intermolecular interactions between the molecules of the liquid govern the magnitude of E_a, but the problem of calculating it is immensely difficult and still largely unsolved. At low temperatures, the viscosity of water decreases as the pressure is increased. This behaviour is consistent with the need to rupture hydrogen bonds for migration to occur.

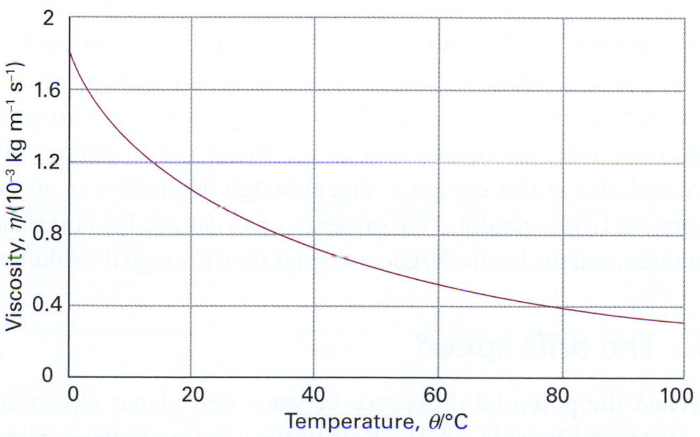

Figure 19B.1 The experimental temperature dependence of the viscosity of water. As the temperature is increased, more molecules are able to escape from the potential wells provided by their neighbours, and so the liquid becomes more fluid.

Brief illustration 19B.1 Liquid viscosity

The viscosity of water at 25 °C and 50 °C is 0.890 mPa s and 0.547 mPa s, respectively. It follows from eqn 19B.2 that the activation energy for molecular migration is the solution of

$$\frac{\eta(T_2)}{\eta(T_1)} = e^{(E_a/R)(1/T_2 - 1/T_1)}$$

which is

$$E_a = \frac{R\ln\{\eta(T_2)/\eta(T_1)\}}{1/T_2 - 1/T_1} = \frac{(8.3145\,\mathrm{J\,K^{-1}\,mol^{-1}})\ln(0.547/0.890)}{1/320\,\mathrm{K} - 1/298\,\mathrm{K}}$$

or 17.5 kJ mol^{-1}. That value is comparable to the strength of a hydrogen bond.

Self-test 19B.1 The corresponding values of the viscosity of benzene are 0.604 mPa s and 0.436 mPa s. Evaluate the activation energy for viscosity.

Answer: 11.7 kJ mol^{-1}

(b) Electrolyte solutions

Further insight into the nature of molecular motion can be obtained by studying the net transport of charged species through solution, for ions can be dragged through the solvent by the application of a potential difference between two electrodes immersed in the sample. By studying the transport of charge through electrolyte solutions it is possible to build up a picture of the events that occur in them and, in some cases, to extrapolate the conclusions to species that have zero charge; that is, to neutral molecules.

The fundamental measurement used to study the motion of ions is that of the electrical resistance, R, of the solution. The **conductance**, G, of a solution is the inverse of its resistance R: $G=1/R$. As resistance is expressed in ohms, Ω, the conductance of a sample is expressed in Ω^{-1}. The reciprocal ohm used to be called the mho, but its SI designation is now the siemens, S, and $1\,\mathrm{S} = 1\,\Omega^{-1} = 1\,\mathrm{C\,V^{-1}\,s^{-1}}$. It is found that the conductance of a sample decreases with its length l and increases with its cross-sectional area A. We therefore write

$$G = \kappa\frac{A}{l} \qquad \textit{Definition of } \kappa \quad \text{Conductivity} \qquad (19B.3)$$

where κ (kappa) is the electrical **conductivity**. With the conductance in siemens and the dimensions in metres, it follows that the SI units of κ are siemens per metre (S m^{-1}).

The conductivity of a solution depends on the number of ions present, and it is normal to introduce the **molar conductivity**, Λ_m, which is defined as

$$\Lambda_m = \frac{\kappa}{c} \qquad \textit{Definition} \quad \text{Molar conductivity} \qquad (19B.4)$$

where c is the molar concentration of the added electrolyte. The SI unit of molar conductivity is siemens metre-squared per mole (S m^2 mol^{-1}), and typical values are about 10 mS m^2 mol^{-1} (where 1 mS = 10^{-3} S).

The values of the molar conductivity as calculated by eqn 19B.4 are found to vary with the concentration. One reason for this variation is that the number of ions in the solution might not be proportional to the nominal concentration of the electrolyte. For instance, the concentration of ions in a solution of a weak electrolyte depends on the concentration of the solute in a complicated way, and doubling the concentration of the solute added does not double the number of ions. Secondly, because ions interact strongly with one another, the conductivity of a solution is not exactly proportional to the number of ions present.

In an extensive series of measurements during the nineteenth century, Friedrich Kohlrausch established the **Kohlrausch law**, that at low concentrations the molar conductivities of strong electrolytes vary linearly with the square root of the concentration:

$$\Lambda_m = \Lambda_m^\circ - \mathcal{K}c^{1/2} \qquad \text{Kohlrausch law} \qquad (19B.5)$$

He also established that Λ_m°, the **limiting molar conductivity**, the molar conductivity in the limit of zero concentration, is the sum of contributions from its individual ions. If the limiting molar conductivity of the cations is denoted λ_+ and that of the anions λ_-, then his **law of the independent migration of ions** states that

$$\Lambda_m^\circ = \nu_+\lambda_+ + \nu_-\lambda_- \qquad \textit{Limiting law} \qquad \text{Law of the independent migration of ions} \qquad (19B.6)$$

where ν_+ and ν_- are the numbers of cations and anions per formula unit of electrolyte. For example, $\nu_+ = \nu_- = 1$ for HCl, NaCl, and $CuSO_4$, but $\nu_+ = 1$, $\nu_- = 2$ for $MgCl_2$.

Example 19B.1 Determining the limiting molar conductivity

The conductivity of KCl(aq) at 25 °C is 14.668 mS m^{-1} when c = 1.0000 mmol dm^{-3} and 71.740 mS m^{-1} when c = 5.0000 mmol dm^{-3}. Determine the values of the limiting molar conductivity Λ_m° and the Kohlrausch constant $\mathcal{K}$.

Method Use eqn 19B.4 to determine the molar conductivities at the two concentrations, then use the Kohlrausch law, eqn 19B.5, in the form

$$\Lambda_m(c_2) - \Lambda_m(c_1) = \mathcal{K}(c_1^{1/2} - c_2^{1/2})$$

to determine $\mathcal{K}$. Then find Λ_m° from the law in the form

$$\Lambda_m^\circ = \Lambda_m + \mathcal{K}c^{1/2}$$

With more data available, a better procedure is to perform a linear regression.

Answer It follows that the molar conductivity of KCl when c = 1.0000 mmol dm^{-3} (which is the same as 1.0000 mol m^{-3}) is

$$\Lambda_m = \frac{14.688\,\text{mS m}^{-1}}{1.0000\,\text{mol m}^{-3}} = 14.688\,\text{mS m}^2\,\text{mol}^{-1}$$

Similarly, when c = 5.0000 mol dm^{-3} the molar conductivity is 14.348 mS m^2 mol^{-1}. It then follows that

$$\begin{aligned}\mathcal{K} &= \frac{\Lambda_m(c_2) - \Lambda_m(c_1)}{c_1^{1/2} - c_2^{1/2}} \\ &= \frac{(14.348 - 14.688)\,\text{mS m}^2\,\text{mol}^{-1}}{(0.0010000^{1/2} - 0.0050000^{1/2})\,(\text{mol dm}^{-3})^{1/2}} \\ &= 8.698\,\text{mS m}^2\,\text{mol}^{-1}/(\text{mol dm}^{-3})^{1/2}\end{aligned}$$

(It is best to keep this awkward but convenient array of units as they are rather than converting them to the equivalent $10^{-3/2}$ S m$^{7/2}$ mol$^{-3/2}$.) Now we find the limiting value from the data for c = 0.0100 mol dm^{-3}:

$$\begin{aligned}\Lambda_m^\circ &= 14.688\,\text{mS m}^2\,\text{mol}^{-1} + 8.698\frac{\text{mS m}^2\,\text{mol}^{-1}}{(\text{mol dm}^{-3})^{1/2}} \\ &\quad \times (0.001\,0000\,\text{mol dm}^{-3})^{1/2} = 14.963\,\text{mS m}^2\,\text{mol}^{-1}\end{aligned}$$

Self-test 19B.2 The conductivity of $KClO_4$(aq) at 25 °C is 13.780 mS m^{-1} when c = 1.000 mmol dm^{-3} and 67.045 mS m^{-1} when c = 5.000 mmol dm^{-3}. Determine the values of the limiting molar conductivity Λ_m° and the Kohlrausch constant $\mathcal{K}$ for this system.

Answer: $\mathcal{K}$ = 9.491 mS m^2 mol^{-1}/(mol dm^{-3})$^{1/2}$, Λ_m° = 14.08 mS m^2 mol^{-1}

19B.2 The mobilities of ions

To interpret conductivity measurements we need to know why ions move at different rates, why they have different molar conductivities, and why the molar conductivities of strong electrolytes decrease with the square root of the molar concentration. The central idea in this section is that although the motion of an ion remains largely random, the presence of an electric field biases its motion, and the ion undergoes net migration through the solution.

(a) The drift speed

When the potential difference between two planar electrodes a distance l apart is $\Delta\phi$, the ions in the solution between them experience a uniform electric field of magnitude

$$\mathcal{E} = \frac{\Delta\phi}{l} \qquad (19B.7)$$

In such a field, an ion of charge ze experiences a force of magnitude

$$\mathcal{F} = ze\mathcal{E} = \frac{ze\Delta\phi}{l}$$ Electric force (19B.8)

where here and throughout this section we disregard the sign of the charge number and so avoid notational complications.

A cation responds to the application of the field by accelerating towards the negative electrode and an anion responds by accelerating towards the positive electrode. However, this acceleration is short lived. As the ion moves through the solvent it experiences a frictional retarding force, $\mathcal{F}_{\text{fric}}$, proportional to its speed. For a spherical particle of radius a travelling at a speed s, this force is given by **Stokes' law**, which was derived by considering the hydrodynamics of the passage of a sphere through a continuous fluid:

$$\mathcal{F}_{\text{fric}} = fs \qquad f = 6\pi\eta a$$ Stokes' law (19B.9)

where η is the viscosity. In writing eqn 19B.9, we have assumed that it applies on a molecular scale, and independent evidence from magnetic resonance suggests that it often gives at least the right order of magnitude.

The two forces act in opposite directions, and the ions quickly reach a terminal speed, the **drift speed**, when the accelerating force is balanced by the viscous drag. The net force is zero when $fs = ze\mathcal{E}$, or

$$s = \frac{ze\mathcal{E}}{f}$$ Drift speed (19B.10)

It follows that the drift speed of an ion is proportional to the strength of the applied field. We write

$$s = u\mathcal{E}$$ *Definition of u* Mobility (19B.11)

where u is called the **mobility** of the ion (Table 19B.2). Comparison of the last two equations shows that

$$u = \frac{ze}{f} = \frac{ze}{6\pi\eta a}$$ Mobility (19B.12)

Table 19B.2* Ionic mobilities in water at 298 K, $u/(10^{-8}\,\text{m}^2\,\text{s}^{-1}\,\text{V}^{-1})$

	$u/(10^{-8}\,\text{m}^2\,\text{s}^{-1}\,\text{V}^{-1})$		$u/(10^{-8}\,\text{m}^2\,\text{s}^{-1}\,\text{V}^{-1})$
H^+	36.23	OH^-	20.64
Na^+	5.19	Cl^-	7.91
K^+	7.62	Br^-	8.09
Zn^{2+}	5.47	SO_4^{2-}	8.29

* More values are given in the *Resource section*.

Brief illustration 19B.2 Ion mobility

For an order of magnitude estimate we can take $z=1$ and a the radius of an ion such as Cs^+ (which might be typical of a smaller ion plus its hydration sphere), which is 170 pm. For the viscosity, we use $\eta = 1.0$ cP (1.0 mPa s, Table 19B.1). Then

$$u = \frac{1.6\times10^{-19}\,\overbrace{\text{C}}^{\text{J V}^{-1}}}{6\pi\times\left(1.0\times10^{-3}\,\underbrace{\text{Pa}}_{\text{J m}^{-3}}\,\text{s}\right)\times(170\times10^{-12}\,\text{m})}$$
$$= 5.0\times10^{-8}\,\text{m}^2\,\text{V}^{-1}\,\text{s}^{-1}$$

This value means that when there is a potential difference of 1 V across a solution of length 1 cm (so $\mathcal{E} = 100\,\text{V m}^{-1}$), the drift speed is typically about $5\,\mu\text{m s}^{-1}$. That speed might seem slow, but not when expressed on a molecular scale, for it corresponds to an ion passing about 10^4 solvent molecules per second.

Self-test 19B.3 The mobility of an SO_4^{2-} ion in water at 25 °C is $8.29\times10^{-8}\,\text{m}^2\,\text{V}^{-1}\,\text{s}^{-1}$. What is its effective radius?

Answer: 205 pm

Because the drift speed governs the rate at which charged species are transported, we might expect the conductivity to decrease with increasing solution viscosity and ion size. Experiments confirm these predictions for bulky ions (such as R_4N^+ and RCO_2^-) but not for small ions. For example, the mobilities of the alkali metal ions in water increase from Li^+ to Cs^+ (Table 19B.2) even though the ionic radii increase. The paradox is resolved when we realize that the radius a in the Stokes formula is the **hydrodynamic radius** (or 'Stokes radius') of the ion, its effective radius in the solution taking into account all the H_2O molecules it carries in its hydration shell. Small ions give rise to stronger electric fields than large ones (the electric field at the surface of a sphere of radius r is proportional to ze/r^2 and it follows that the smaller the radius the stronger the field), so small ions are more extensively solvated than big ions. Thus, an ion of small ionic radius may have a large hydrodynamic radius because it drags many solvent molecules through the solution as it migrates. The hydrating H_2O molecules are often very labile, however, and NMR and isotope studies have shown that the exchange between the coordination sphere of the ion and the bulk solvent is very rapid for ions of low charge but may be slow for ions of high charge.

The proton, although it is very small, has a very high mobility (Table 19B.2)! Proton and ^{17}O-NMR show that the times characteristic of protons hopping from one molecule to the next are about 1.5 ps, which is comparable to the time that inelastic neutron scattering shows it takes a water molecule to

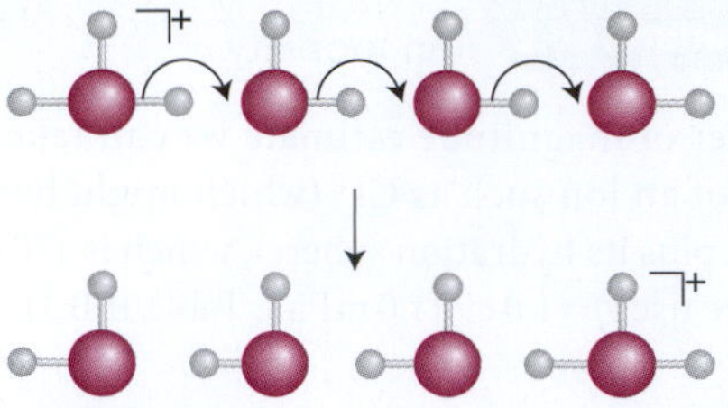

Figure 19B.2 A highly schematic diagram showing the effective motion of a proton in water.

reorient through about 1 radian (1 to 2 ps). According to the **Grotthuss mechanism**, there is an effective motion of a proton that involves the rearrangement of bonds in a group of water molecules (Fig. 19B.2). However, the actual mechanism is still highly contentious. The mobility of NH_4^+ in liquid ammonia is also anomalous and presumably occurs by an analogous mechanism.

(b) Mobility and conductivity

Ionic mobilities provide a link between measurable and theoretical quantities. As a first step we establish in the following *Justification* the relation between an ion's mobility and its molar conductivity:

$$\lambda = zuF \qquad \text{Ion conductivity} \qquad (19B.13)$$

where F is Faraday's constant ($F=N_Ae$).

Justification 19B.1 The relation between ionic mobility and molar conductivity

To keep the calculation simple, we ignore signs in the following, and concentrate on the magnitudes of quantities.

Consider a solution of a fully dissociated strong electrolyte at a molar concentration c. Let each formula unit give rise to ν_+ cations of charge z_+e and ν_- anions of charge z_-e. The molar concentration of each type of ion is therefore νc (with $\nu=\nu_+$ or ν_-), and the number density of each type is νcN_A. The number of ions of one kind that pass through an imaginary window of area A during an interval Δt is equal to the number within the distance $s\Delta t$ (Fig. 19B.3), and therefore to the number in the volume $s\Delta tA$. (The same sort of argument is used in Topic 19A in the discussion of the transport properties of gases.) The number of ions of that kind in this volume is equal to $s\Delta tA\nu cN_A$. The flux through the window (the number of this type of ion passing through the window divided by the area of the window and the duration of the interval) is therefore

$$J(\text{ions}) = \frac{s\Delta tA\nu cN_A}{\Delta tA} = s\nu cN_A$$

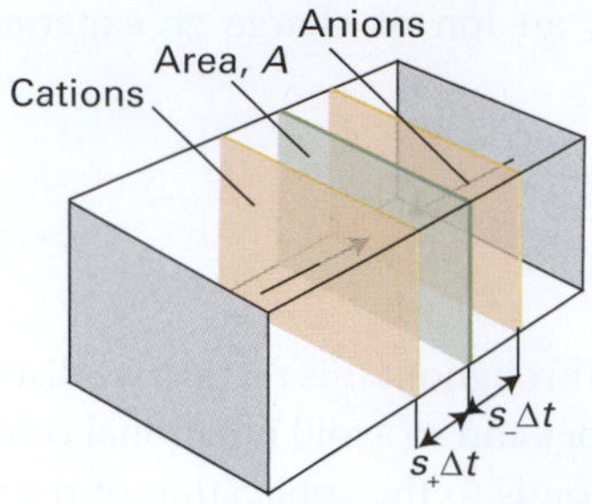

Figure 19B.3 In the calculation of the current, all the cations within a distance $s_+\Delta t$ (that is, those in the volume $s_+A\Delta t$) will pass through the area A. The anions in the corresponding volume the other side of the window will also contribute to the current similarly.

Each ion carries a charge ze, so the flux of charge is

$$J(\text{charge}) = zs\nu ceN_A = zs\nu cF$$

Because $s=u\mathcal{E}$, the flux is

$$J(\text{charge}) = zu\nu cF\mathcal{E}$$

The current, I, through the window due to the ions we are considering is the charge flux times the area:

$$I = JA = zu\nu cF\mathcal{E}A$$

Because the electric field is the potential gradient (eqn 19B.7, $\mathcal{E}=\Delta\phi/l$), we can write

$$I = \frac{zu\nu cFA\Delta\phi}{l}$$

Current and potential difference are related by Ohm's law, $\Delta\phi = IR$, so it follows that

$$I = \frac{\Delta\phi}{R} = G\Delta\phi = \frac{\kappa A\Delta\phi}{l}$$

where we have used eqn 19B.3 in the form $\kappa = Gl/A$. Comparison of the last two expressions gives $\kappa = zu\nu cF$. Division by the molar concentration of ions, νc, then results in eqn 19B.13.

Equation 19B.13 applies to the cations and to the anions. Therefore, for the solution itself in the limit of zero concentration (when there are no ionic interactions),

$$\Lambda_m^\circ = (z_+u_+\nu_+ + z_-u_-\nu_-)F \qquad (19B.14a)$$

For a symmetrical z:z electrolyte (for example, $CuSO_4$ with $z=2$), this equation simplifies to

$$\Lambda_m^\circ = z(u_+ + u_-)F \qquad (19B.14b)$$

Brief illustration 19B.3 Ionic conductivity

Earlier, we estimated the typical ionic mobility as $5.0\times10^{-8}\,m^2\,V^{-1}\,s^{-1}$; so, with $z=1$ for both the cation and anion, we can estimate that a typical limiting molar conductivity should be about

$$\lambda=(5.0\times10^{-8}\,m^2\,V^{-1}\,s^{-1})\times(9.648\times10^4\,C\,mol^{-1})$$
$$=4.8\times10^{-3}\,m^2\,V^{-1}\,s^{-1}\,C\,mol^{-1}$$

But $1\,V\,C^{-1}\,s=1\,S$ (see the remark preceding eqn 19B.3), so $\lambda\approx5\,mS\,m^2\,mol^{-1}$, and about twice that value for Λ_m°, in accord with experiment. The experimental value for KCl, for instance, is $15\,mS\,m^2\,mol^{-1}$.

Self-test 19B.4 Estimate the ionic conductivity of an SO_4^{2-} ion in water at 25 °C from its mobility (Table 19B.2).

Answer: $16\,mS\,m^2\,mol^{-1}$

(c) The Einstein relations

An important relation between the drift speed s and a force $\mathcal{F}$ of any kind acting on a particle is derived in Topic 19C:

$$s=\frac{D\mathcal{F}}{RT} \quad \text{Drift speed} \quad (19B.15)$$

where D is the diffusion coefficient for the species (Table 19B.3). We have seen that an ion in solution has a drift speed $s=u\mathcal{E}$ when it experiences a force $N_Aez\mathcal{E}$ from an electric field of strength $\mathcal{E}$. Therefore, substituting these known values into eqn 19B.15 and using $N_Ae=F$ gives $u\mathcal{E}=DFz\mathcal{E}/RT$ and hence, on cancelling the $\mathcal{E}$, we obtain the **Einstein relation**:

$$u=\frac{zDF}{RT} \quad \text{Einstein relation} \quad (19B.16)$$

The Einstein relation provides a link between the molar conductivity of an electrolyte and the diffusion coefficients of its ions. First, by using eqns 19B.13 and 19B.16 we write

$$\lambda=zuF=\frac{z^2DF^2}{RT} \quad (19B.17)$$

for each type of ion. Then, from $\Lambda_m^\circ=\nu_+\lambda_++\nu_-\lambda_-$, the limiting molar conductivity is

$$\Lambda_m^\circ=(\nu_+z_+^2D_++\nu_-z_-^2D_-)\frac{F^2}{RT} \quad \text{Nernst–Einstein equation} \quad (19B.18)$$

which is the **Nernst–Einstein equation**. An application of this equation is to the determination of ionic diffusion coefficients from conductivity measurements; another is to the prediction of conductivities using models of ionic diffusion.

Table 19B.3* Diffusion coefficients at 298 K, $D/(10^{-9}\,m^2\,s^{-1})$

Molecules in liquids		Ions in water			
I_2 in hexane	4.05	K^+	1.96	Br^-	2.08
in benzene	2.13	H^+	9.31	Cl^-	2.03
Glycine in water	1.055	Na^+	1.33	I^-	2.05
H_2O in water	2.26			OH^-	5.03
Sucrose in water	0.5216				

* More values are given in the *Resource section*.

Equations 19B.12 ($u=ze/f$) and 19B.16 ($u=zDF/RT$ in the form $u=zDe/kT$) relate the mobility of an ion to the frictional force and to the diffusion coefficient, respectively. We can combine the two expressions and cancel the ze and obtain the **Stokes–Einstein equation**:

$$D=\frac{kT}{f} \quad \text{Stokes–Einstein equation} \quad (19B.19a)$$

If the frictional force is described by Stokes' law, then we also obtain a relation between the diffusion coefficient and the viscosity of the medium:

$$D=\frac{kT}{6\pi\eta a} \quad \text{Stokes–Einstein equation} \quad (19B.19b)$$

An important feature of eqn 19B.19b is that it makes no reference to the charge of the diffusing species. Therefore, the equation also applies in the limit of vanishingly small charge; that is, it also applies to neutral molecules. This feature is taken further in Topic 19C. It must not be forgotten, however, that both equations depend on the assumption that the viscous drag is proportional to the speed.

Brief illustration 19B.4 Mobility and diffusion

From Table 19B.2, the mobility of SO_4^{2-} is $8.29\times10^{-8}\,m^2\,V^{-1}\,s^{-1}$. It follows from eqn 19B.16 in the form $D=uRT/zF$ that the diffusion coefficient for the ion in water at 25 °C is

$$D=\frac{(8.29\times10^{-8}\,m^2\,V^{-1}\,s^{-1})\times(8.3145\,J\,K^{-1}\,mol^{-1})\times(298\,K)}{2\times\left(9.649\times10^4\,\underbrace{C}_{J\,V^{-1}}\,mol^{-1}\right)}$$
$$=1.06\times10^{-9}\,m^2\,s^{-1}$$

Self-test 19B.5 Repeat the calculation for the NH_4^+ ion.

Answer: $1.96\times10^{-9}\,m^2\,s^{-1}$

Checklist of concepts

- ☐ 1. The **viscosity of a liquid** decreases with increasing temperature.
- ☐ 2. The **conductance**, G, of a solution is the inverse of its resistance.
- ☐ 3. **Kohlrausch's law** states that at low concentrations the molar conductivities of strong electrolytes vary linearly with the square root of the concentration.
- ☐ 4. The **law of the independent migration of ions** states the molar conductivity in the limit of zero concentration, is the sum of contributions from its individual ions.
- ☐ 5. An ion reaches a **drift speed** when the acceleration due to the electrical force is balanced by the viscous drag.
- ☐ 6. The **hydrodynamic radius** of an ion may be greater than its geometrical radius due to solvation.
- ☐ 7. The high mobility of a proton in water is explained by the **Grotthuss mechanism**.

Checklist of equations

Property	Equation	Comment	Equation number
Viscosity of a liquid	$\eta=\eta_0 e^{E_a/RT}$	Narrow temperature range	19B.2
Conductivity	$\kappa=Gl/A,\ G=1/R$	Definition	19B.3
Molar conductivity	$\Lambda_m=\kappa/c$	Definition	19B.4
Kohlrausch's law	$\Lambda_m=\Lambda_m^\circ-\mathcal{K}c^{1/2}$	Empirical observation	19B.5
Law of independent migration of ions	$\Lambda_m^\circ=\nu_+\lambda_+ +\nu_-\lambda_-$	Limiting law	19B.6
Stokes' law	$\mathcal{F}_{fric}=fs \quad f=6\pi\eta a$	Hydrodynamic radius	19B.9
Drift speed	$s=u\mathcal{E}$	Defines u	19B.11
Ion mobility	$u=ze/6\pi\eta a$	Asumes Stokes' law	19B.12
Conductivity and mobility	$\lambda=zuF$		19B.13
Molar conductivity and mobility	$\Lambda_m^\circ=(z_+u_+\nu_+ +z_-u_-\nu_-)F$		19B.14a
Drift speed	$s=DF/RT$		19B.15
Einstein relation	$u=zDF/RT$		19B.16
Nernst–Einstein relation	$\Lambda_m^\circ=(\nu_+z_+^2D_+ +\nu_-z_-^2D_-)(F^2/RT)$		19B.18
Stokes–Einstein relation	$D=kT/f$		19B.19a

19C Diffusion

Contents

➤ Why do you need to know this material?

Diffusion is a hugely important process both in the atmosphere and in solution, and it is important to be able to predict the spread of one material through another when discussing reactions in solution and the spread of substances into the environment. The interpretation of diffusion in terms of a random walk also gives insight into the molecular basis of the process.

➤ What is the key idea?

Particles tend to spread and achieve a uniform distribution.

➤ What do you need to know already?

This Topic draws on arguments relating to flux that are treated in Topic 19A, particularly the way to calculate the flux of particles through a window of given area. This Topic goes into more detail about the diffusion coefficient, which was introduced in Topic 19A and used in Topic 19B. It uses the concept of chemical potential (Topic 5A) to discuss the direction of spontaneous change. One of the *Justifications* develops the random walk model introduced in Topic 17A.

That solutes in gases, liquids, and solids have a tendency to spread can be discussed from three points of view. One viewpoint is from the Second Law of thermodynamics and the tendency for entropy to increase or, if the temperature and pressure are constant, for the Gibbs energy to decrease. When this law is applied to solutes it appears that there is a force acting to disperse the solute. That force is illusory, but it provides an interesting and useful approach to discussing diffusion. The second approach is to set up a differential equation for the change in concentration in a region by considering the flux of material through its boundaries. The resulting 'diffusion equation' can then be solved (in principle, at least) for various configurations of the system, including taking into account the shape of a reaction vessel. The third approach is more mechanistic, and is to imagine diffusion as taking place in a series of random small steps.

19C.1 The thermodynamic view

It is established in Topic 3C that, at constant temperature and pressure, the maximum non-expansion work that can be done per mole when a substance moves from one location to another differing in molar Gibbs energy by $\mathrm{d}G_\mathrm{m}$ is $\mathrm{d}w_\mathrm{e}=\mathrm{d}G_\mathrm{m}$. In terms of the chemical potentials of the substance in the two locations (its partial molar Gibbs energy), $\mathrm{d}w_\mathrm{e}=\mathrm{d}\mu$. In a system in which the chemical potential depends on the position x,

$$\mathrm{d}w=\mathrm{d}\mu=\left(\frac{\partial\mu}{\partial x}\right)_{T,p}\mathrm{d}x$$

It is also shown in Topic 2A that in general work can always be expressed in terms of an opposing force (which here we write $\mathcal{F}$), and that $\mathrm{d}w=-\mathcal{F}\mathrm{d}x$. By comparing these two expressions for $\mathrm{d}w$ we see that the slope of the chemical potential with respect to position can be interpreted as an effective force per mole of molecules. We write this **thermodynamic force** as

$$\mathcal{F}=-\left(\frac{\partial\mu}{\partial x}\right)_{T,p} \qquad \text{Thermodynamic force} \qquad (19C.1)$$

There is not necessarily a real force pushing the particles down the slope of the chemical potential. As we shall see, the force may represent the spontaneous tendency of the molecules to disperse as a consequence of the Second Law and the hunt for maximum entropy.

In a solution in which the activity of the solute is a, the chemical potential is $\mu=\mu^{\ominus}+RT\ln a$. If the solution is not uniform the activity depends on the position and we can write

$$\mathcal{F}=-RT\left(\frac{\partial \ln a}{\partial x}\right)_{T,p} \quad \text{(19C.2a)}$$

If the solution is ideal, a may be replaced by the molar concentration c, and then

$$\mathcal{F}=-RT\left(\frac{\partial \ln c}{\partial x}\right)_{T,p} \overset{\mathrm{d}\ln y/\mathrm{d}x=(1/y)(\mathrm{d}y/\mathrm{d}x)}{=} -\frac{RT}{c}\left(\frac{\partial c}{\partial x}\right)_{T,p} \quad \text{(19C.2b)}$$

Brief illustration 19C.1 The thermodynamic force

Suppose a linear concentration gradient is set up across a container at 25 °C, with points separated by 1.0 cm differing in concentration by 0.10 mol dm^{-3} around a mean value of 1.0 mol dm^{-3}. According to eqn 19C.2b, the solute experiences a thermodynamic force of magnitude

$$|\mathcal{F}|=\frac{(8.3145\,\mathrm{J\,K^{-1}\,mol^{-1}})\times(298\,\mathrm{K})}{1.0\,\mathrm{mol\,dm^{-3}}}\times\frac{0.10\,\mathrm{mol\,dm^{-3}}}{1.0\times10^{-2}\,\mathrm{m}}$$

$$=2.5\times10^{4}\,\overbrace{\mathrm{J\,m^{-1}}}^{\mathrm{N}}\,\mathrm{mol^{-1}}$$

or 25 kN mol^{-1}. Note that the thermodynamic force is a molar quantity.

Self-test 19C.1 Suppose that the concentration of a solute decreases exponentially as $c(x)=c_0\mathrm{e}^{-x/\lambda}$. Derive an expression for the thermodynamic force.

Answer: $\mathcal{F}=RT/\lambda$

In Topic 19A it is established that Fick's first law of diffusion, which we write here in the form

$$J(\text{number})=-D\frac{\mathrm{d}\mathcal{N}}{\mathrm{d}x} \quad \text{Fick's first law} \quad \text{(19C.3)}$$

can be deduced from the kinetic model of gases. Here we generalize that result and show that it applies to the diffusion of species in condensed phases too. To do so, we suppose that the flux of diffusing particles is a response to a thermodynamic force due to a concentration gradient. The diffusing particles reach a steady 'drift speed', s, when the thermodynamic force, $\mathcal{F}$, is matched by the drag due to the viscosity of the medium. This drift speed is proportional to the thermodynamic force, and we write $s\propto\mathcal{F}$. However, the particle flux, J, is proportional to the drift speed, and the thermodynamic force is proportional to the concentration gradient, $\mathrm{d}c/\mathrm{d}x$. The chain of proportionalities ($J\propto s$, $s\propto\mathcal{F}$, and $\mathcal{F}\propto \mathrm{d}c/\mathrm{d}x$) implies that $J\propto \mathrm{d}c/\mathrm{d}x$, which is the content of Fick's law.

If we divide both sides of eqn 19C.3 by Avogadro's constant, so converting numbers into amounts (numbers of moles), noting that $\mathcal{N}/N_\mathrm{A}=(N/V)/N_\mathrm{A}=(nN_\mathrm{A}/V)/N_\mathrm{A}=n/V=c$, the molar concentration, then Fick's law becomes

$$J(\text{amount})=-D\frac{\mathrm{d}c}{\mathrm{d}x} \quad \text{(19C.4)}$$

In this expression, D is the diffusion coefficient and $\mathrm{d}c/\mathrm{d}x$ is the slope of the molar concentration. The flux is related to the drift speed by

$$J(\text{amount})=sc \quad \text{(19C.5)}$$

This relation follows from the argument used in Topic 19A. Thus, all particles within a distance $s\Delta t$, and therefore in a volume $s\Delta tA$, can pass through a window of area A in an interval Δt. Hence, the amount of substance that can pass through the window in that interval is $s\Delta tAc$. The particle flux is this quantity divided by the area A and the time interval Δt, and is therefore simply sc.

By combining the last two equations for J(amount) and using eqn 19C.2b

$$sc=-D\frac{\mathrm{d}c}{\mathrm{d}x}=\frac{Dc\mathcal{F}}{RT} \quad \text{or} \quad s=\frac{D\mathcal{F}}{RT} \quad \text{(19C.6)}$$

Therefore, once we know the effective force and the diffusion coefficient, D, we can calculate the drift speed of the particles (and vice versa) whatever the origin of the force. This equation is used in Topic 19B, where the force is applied electrically to an ion.

Brief illustration 19C.2 The thermodynamic force and the drift speed

Laser measurements showed that a molecule has a drift speed of 1.0 μm s^{-1} in water at 25 °C, with diffusion coefficient 5.0×10^{-9} m^2 s^{-1}. The corresponding thermodynamic force from eqn 19C.6 in the form $\mathcal{F}=sRT/D$ is

$$\mathcal{F}=\frac{(1.0\times10^{-6}\,\mathrm{m\,s^{-1}})\times(8.3145\,\mathrm{J\,K^{-1}\,mol^{-1}})\times(298\,\mathrm{K})}{(5.0\times10^{-9}\,\mathrm{m^2\,s^{-1}})}$$

$$=5.0\times10^{5}\,\overbrace{\mathrm{J\,m^{-1}}}^{\mathrm{N}}\,\mathrm{mol^{-1}}$$

or about 500 kN mol^{-1}.

Self-test 19C.2 What thermodynamic force would achieve a drift speed of 10 μm s^{-1} in water at 25 °C?

Answer: 5.0 MN mol^{-1}

19C.2 The diffusion equation

We now turn to the discussion of time-dependent diffusion processes, where we are interested in the spreading of inhomogeneities with time. One example is the temperature of a metal bar that has been heated at one end: if the source of heat is removed, then the bar gradually settles down into a state of uniform temperature. When the source of heat is maintained and the bar can radiate, it settles down into a steady state of non-uniform temperature. Another example (and one more relevant to chemistry) is the concentration distribution in a solvent to which a solute is added. We shall focus on the description of the diffusion of particles, but similar arguments apply to the diffusion of physical properties, such as temperature. Our aim is to obtain an equation for the rate of change of the concentration of particles in an inhomogeneous region.

(a) Simple diffusion

The central equation of this section is the **diffusion equation**, also called 'Fick's second law of diffusion', which relates the rate of change of concentration at a point to the spatial variation of the concentration at that point:

$$\frac{\partial c}{\partial t}=D\frac{\partial^2 c}{\partial x^2} \qquad \text{Diffusion equation} \qquad (19C.7)$$

We show in the following *Justification* that the diffusion equation follows from Fick's first law of diffusion.

Justification 19C.1 The diffusion equation

Consider a thin slab of cross-sectional area A that extends from x to $x+\lambda$ (Fig. 19C.1). Let the concentration at x be c at the time t. The rate at which the amount (in moles) of particles

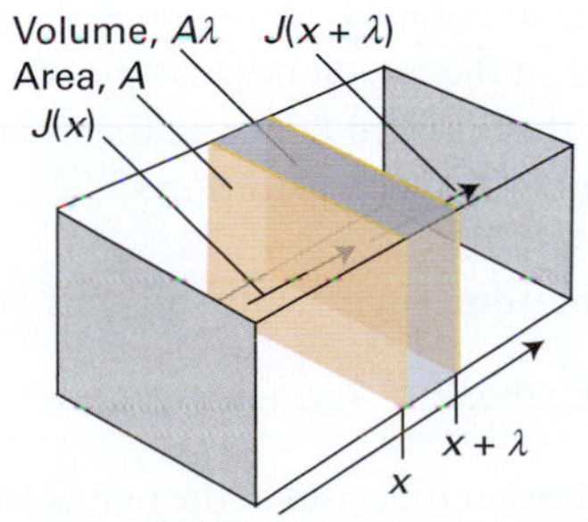

Figure 19C.1 The net flux in a region is the difference between the flux entering from the region of high concentration (on the left) and the flux leaving to the region of low concentration (on the right).

enter the slab is JA, so the rate of increase in molar concentration inside the slab (which has volume $A\lambda$) on account of the flux from the left is

$$\frac{\partial c}{\partial t}=\frac{JA}{A\lambda}=\frac{J}{\lambda}$$

There is also an outflow through the right-hand window. The flux through that window is J', and the rate of change of concentration that results is

$$\frac{\partial c}{\partial t}=-\frac{J'}{\lambda}$$

The net rate of change of concentration is therefore

$$\frac{\partial c}{\partial t}=\frac{J-J'}{\lambda}$$

Each flux is proportional to the concentration gradient at the respective window. So, by using Fick's first law, we can write

$$J-J'=-D\frac{\partial c}{\partial x}+D\frac{\partial c'}{\partial x}$$

The concentration at the right-hand window is related to that on the left by

$$c'=c+\left(\frac{\partial c}{\partial x}\right)\lambda$$

which implies that

$$J-J'=-D\frac{\partial c}{\partial x}+D\frac{\partial}{\partial x}\left\{c+\left(\frac{\partial c}{\partial x}\right)\lambda\right\}=D\lambda\frac{\partial^2 c}{\partial x^2}$$

When this relation is substituted into the expression for the rate of change of concentration in the slab, we get eqn 19C.7.

The diffusion equation shows that the rate of change of concentration is proportional to the curvature (more precisely, to the second derivative) of the concentration with respect to distance. If the concentration changes sharply from point to point (i.e. if the distribution is highly wrinkled), then the concentration changes rapidly with time. Where the curvature is positive (a dip, Fig. 19C.2), the change in concentration is positive; the dip tends to fill. Where the curvature is negative (a heap), the change in concentration is negative; the heap tends to spread. If the curvature is zero, then the concentration is constant in time. If the concentration decreases linearly with distance, then the concentration at any point is constant because the inflow of particles is exactly balanced by the outflow.

The diffusion equation can be regarded as a mathematical formulation of the intuitive notion that there is a natural tendency for the wrinkles in a distribution to disappear. More succinctly: Nature abhors a wrinkle.

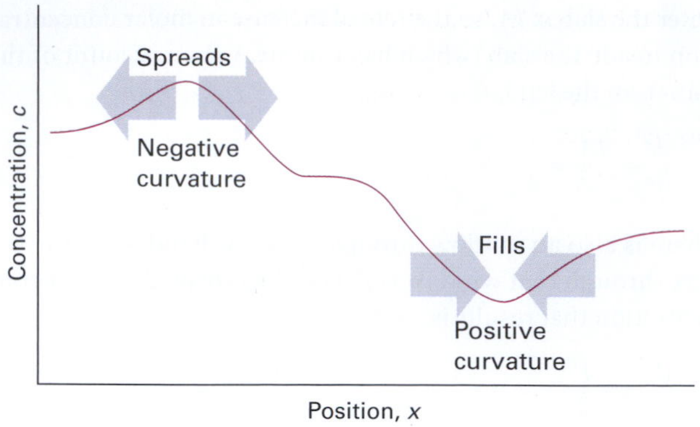

Figure 19C.2 Nature abhors a wrinkle. The diffusion equation tells us that peaks in a distribution (regions of negative curvature) spread and troughs (regions of positive curvature) fill in.

Brief illustration 19C.3 The diffusion equation

If a concentration falls linearly across a small region of space, in the sense that $c=c_0-ax$ then $\partial^2c/\partial x^2=0$ and consequently $\partial c/\partial t=0$. The concentration in the region is constant because the inward flow through one window is matched by the outward flow through the other window (Fig. 19C.3a). If the concentration varies as $c=c_0-\frac{1}{2}ax^2$ then $\partial^2c/\partial x^2=-a$ and consequently $\partial c/\partial t=-Da$. Now the concentration decreases, because there is a greater outward flow than inward flow (Fig. 19C.3b).

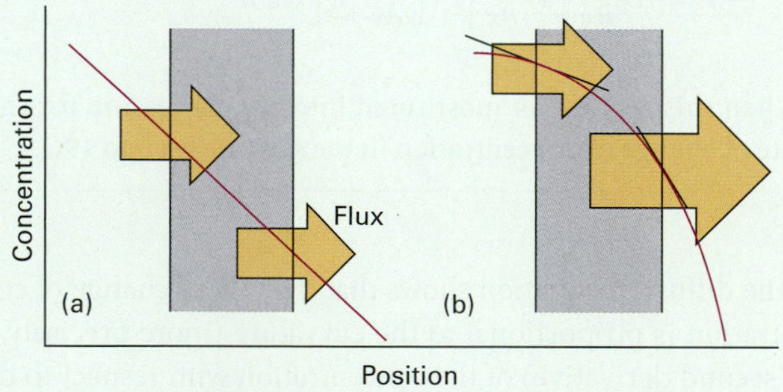

Figure 19C.3 The two instances treated in *Brief illustration* 19C.3: (a) linear concentration gradient, (b) parabolic concentration gradient.

Self-test 19C.3 What is the change in concentration when the concentration falls exponentially across a region? Take $c=c_0e^{-x/\lambda}$.

Answer: $\partial c/\partial t=-(D/\lambda^2)c$

(b) Diffusion with convection

The transport of particles arising from the motion of a streaming fluid is called **convection**. If for the moment we ignore diffusion, then the flux of particles through an area A in an interval Δt when the fluid is flowing at a velocity v can be calculated in the way we have used several times elsewhere (such as in Topic 19A, by counting the particles within a distance $v\Delta t$), and is

$$J_{\text{conv}}=\frac{cAv\Delta t}{A\Delta t}=cv \quad \text{Convective flux} \quad (19C.8)$$

This J is called the **convective flux**. The rate of change of concentration in a slab of thickness l and area A is, by the same argument as before and assuming that the velocity does not depend on the position,

$$\frac{\partial c}{\partial t}=\frac{J_{\text{conv}}-J'_{\text{conv}}}{\lambda}=\frac{cv}{\lambda}-\left\{c+\left(\frac{\partial c}{\partial x}\right)\lambda\right\}\frac{v}{\lambda}=-\left(\frac{\partial c}{\partial x}\right)v \quad \text{Convection} \quad (19C.9)$$

When both diffusion and convection occur, the total change of concentration in a region is the sum of the two effects, and the **generalized diffusion equation** is

$$\frac{\partial c}{\partial t}=D\frac{\partial^2 c}{\partial x^2}-v\frac{\partial c}{\partial x} \quad \text{Generalized diffusion equation} \quad (19C.10)$$

A further refinement, which is important in chemistry, is the possibility that the concentrations of particles may change as a result of reaction. When reactions are included in eqn 19C.10 (which we do in Topic 21B) we get a powerful differential equation for discussing the properties of reacting, diffusing, convecting systems, which is the basis of reactor design in chemical industry and of the utilization of resources in living cells.

Brief illustration 19C.4 Convection

Here we continue the discussion of the systems treated in *Brief illustration* 19C.3 and suppose that there is a convective flow v. If the concentration falls linearly across a small region of space, in the sense that $c=c_0-ax$ then $\partial c/\partial x=-a$ and the change in concentration in the region is $\partial c/\partial t=av$. There is now an increase in the region because the inward convective flow outweighs the outward flow, and there is no diffusion. If $a=0.010\,\text{mol dm}^{-3}\,\text{m}^{-1}$ and $v=+1.0\,\text{mm s}^{-1}$,

$$\begin{aligned}\frac{\partial c}{\partial t}&=(0.010\,\text{mol dm}^{-3}\,\text{m}^{-1})\times(1.0\times10^{-3}\,\text{m s}^{-1})\\&=1.0\times10^{-5}\,\text{mol dm}^{-3}\,\text{s}^{-1}\end{aligned}$$

and the concentration increases at the rate of $10\,\mu\text{mol dm}^{-3}\,\text{s}^{-1}$.

Self-test 19C.4 What rate of flow is needed to replenish the concentration when the concentration varies exponentially as $c=c_0e^{-x/\lambda}$ across the region?

Answer: $v=D/\lambda$

(c) Solutions of the diffusion equation

The diffusion equation is a second-order differential equation with respect to space and a first-order differential equation with respect to time. Therefore, we must specify two boundary conditions for the spatial dependence and a single initial condition for the time dependence (see *Mathematical background* 4 following Chapter 8).

As an illustration, consider a solvent in which the solute is initially coated on one surface of the container (for example, a layer of sugar on the bottom of a deep beaker of water). The single initial condition is that at $t=0$ all N_0 particles are concentrated on the yz-plane (of area A) at $x=0$. The two boundary conditions are derived from the requirements (1) that the concentration must everywhere be finite and (2) that the total amount (number of moles) of particles present is n_0 (with $n_0=N_0/N_A$) at all times. These requirements imply that the flux of particles is zero at the top and bottom surfaces of the system. Under these conditions it is found that

$$c(x,t)=\frac{n_0}{A(\pi Dt)^{1/2}}\mathrm{e}^{-x^2/4Dt} \quad \text{One-dimensional diffusion} \quad (19C.11)$$

as may be verified by direct substitution. Figure 19C.4 shows the shape of the concentration distribution at various times, and it is clear that the concentration spreads and tends to uniformity.

Another useful result is for a localized concentration of solute in a three-dimensional solvent (a sugar lump suspended in a large flask of water). The concentration of diffused solute is spherically symmetrical, and at a radius r is

$$c(x,t)=\frac{n_0}{8(\pi Dt)^{3/2}}\mathrm{e}^{-r^2/4Dt} \quad \text{Three-dimensional diffusion} \quad (19C.12)$$

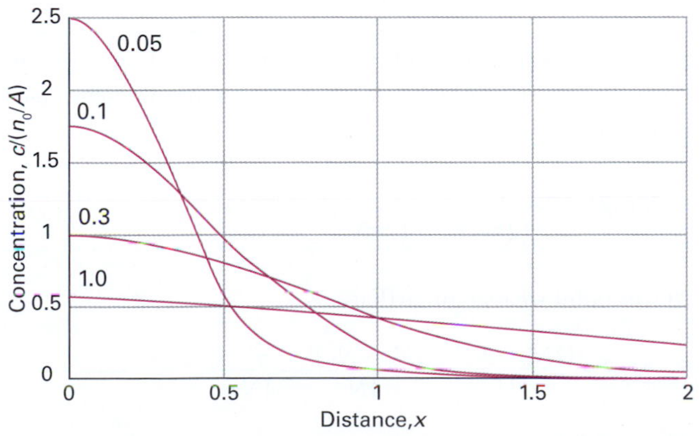

Figure 19C.4 The concentration profiles above a plane from which a solute is diffusing. The curves are plots of eqn 19C.11 and are labelled with different values of Dt. The units of Dt and x are arbitrary, but are related so that Dt/x^2 is dimensionless. For example, if x is in metres, Dt would be in metres²; so, for $D=10^{-9}$ m² s⁻¹, $Dt=0.1$ m² corresponds to $t=10^8$ s.

Other chemically (and physically) interesting arrangements, such as transport of substances across biological membranes can be treated. In many cases the solutions are more cumbersome.

The solutions of the diffusion equation are useful for experimental determinations of diffusion coefficients. In the **capillary technique**, a capillary tube, open at one end and containing a solution, is immersed in a well-stirred larger quantity of solvent, and the change of concentration in the tube is monitored. The solute diffuses from the open end of the capillary at a rate that can be calculated by solving the diffusion equation with the appropriate boundary conditions, so D may be determined. In the **diaphragm technique**, the diffusion occurs through the capillary pores of a sintered glass diaphragm separating the well-stirred solution and solvent. The concentrations are monitored and then related to the solutions of the diffusion equation corresponding to this arrangement. Diffusion coefficients may also be measured by a number of techniques, including NMR spectroscopy.

The solutions of the diffusion equation can be used to predict the concentration of particles (or the value of some other physical quantity, such as the temperature in a non-uniform system) at any location. We can also use them to calculate the average displacement of the particles in a given time.

Example 19C.1 Calculating the average displacement

Calculate the average displacement of particles in a time t in a one-dimensional system if they have a diffusion constant D.

Method We need to calculate the probability that a particle will be found at a certain distance from the origin, and then calculate the average by weighting each distance by that probability.

Answer The number of particles in a slab of thickness dx and area A at x, where the molar concentration is c, is $cAN_A\mathrm{d}x$. The probability that any of the $N_0=n_0N_A$ particles is in the slab is therefore $cAN_A\mathrm{d}x/N_0$. If the particle is in the slab, it has travelled a distance x from the origin. Therefore, the average displacement of all the particles is the sum of each x weighted by the probability of its occurrence:

$$\langle x\rangle=\int_0^\infty x\frac{c(x,t)AN_A}{\underbrace{N_0}_{n_0N_A}}\mathrm{d}x=\frac{1}{(\pi Dt)^{1/2}}\int_0^\infty x\mathrm{e}^{-x^2/4Dt}\,\mathrm{d}x \overset{\text{Integral G.2}}{=} 2\left(\frac{Dt}{\pi}\right)^{1/2}$$

The average displacement varies as the square root of the lapsed time.

Self-test 19C.5 Derive an expression for the root-mean-square distance travelled by diffusing particles in a time t in a one-dimensional system. You will need Integral G.3 from the *Resource section.*

Answer: $\langle x^2\rangle^{1/2}=(2Dt)^{1/2}$

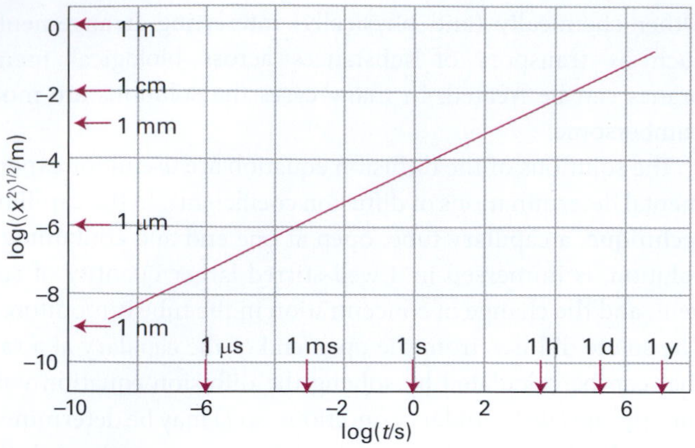

Figure 19C.5 The root-mean-square distance covered by particles with $D=5\times10^{-10}\,m^2\,s^{-1}$. Note the great slowness of diffusion.

As shown in *Example* 19C.1, the average displacement of a diffusing particle in a time t in a one-dimensional system is

$$x=2\left(\frac{Dt}{\pi}\right)^{1/2} \quad \textit{One dimension} \quad \text{Mean displacement} \quad (19C.13)$$

and the root-mean-square displacement in the same time is

$$\langle x^2\rangle^{1/2}=(2Dt)^{1/2} \quad \textit{One dimension} \quad \text{Root-mean-square displacement} \quad (19C.14)$$

The latter is a valuable measure of the spread of particles when they can diffuse in both directions from the origin (for then $\langle x\rangle=0$ at all times). The root-mean-square displacement of particles with a typical diffusion coefficient in a liquid ($D=5\times10^{-10}\,m^2\,s^{-1}$) is illustrated in Fig. 19C.5, which shows how long it takes for diffusion to increase the net distance travelled on average to about 1 cm in an unstirred solution. The graph shows that diffusion is a very slow process (which is why solutions are stirred, to encourage mixing by convection). The diffusion of pheromones in still air is also very slow, and greatly accelerated by convection.

19C.3 The statistical view

An intuitive picture of diffusion is of the particles moving in a series of small steps and gradually migrating from their original positions. We explore this idea by using a model in which the particles can jump through a distance λ in a time τ. The total distance travelled by a particle in a time t is therefore $t\lambda/\tau$. However, the particle will not necessarily be found at that distance from the origin. The direction of each step may be different, and the net distance travelled must take the changing directions into account.

If we simplify the discussion by allowing the particles to travel only along a straight line (the x-axis), and for each step (to the left or the right) to be through the same distance λ, then we obtain the **one-dimensional random walk**. The same model can be used to discuss the random coil structures of denatured polymers (Topic 17B).

We show in the following *Justification* that the probability of a particle being at a distance x from the origin after a time t is

$$P(x,t)=\left(\frac{2\tau}{\pi t}\right)^{1/2}e^{-x^2\tau/2t\lambda^2} \quad \textit{One dimension} \quad \text{Probability} \quad (19C.15)$$

Justification 19C.2 The one-dimensional random walk

The starting point for this calculation is the expression derived in *Justification* 17A.1, with steps in place of the bonds treated there, for the probability that the net distance $n\lambda$ reached from the origin with $n=N_R-N_L$ after N steps each of length λ, with N_R steps to the right and N_L to the left is

$$P(n\lambda)=\frac{N!}{(N-N_R)!N_R!2^N}$$

As in *Justification* 17A.1, this expression can be developed by making use of Stirling's approximation (Topic 15A) in the form

$$\ln x!\approx\ln(2\pi)^{1/2}+\left(x+\tfrac{1}{2}\right)\ln x-x$$

but here we use the parameter

$$\mu=\frac{N_R}{N}-\tfrac{1}{2}\ll1$$

which is small because almost half the steps are to the right. The smallness of μ allows us to use the expansion

$$\ln\left(\tfrac{1}{2}\pm\mu\right)=-\ln2\pm2\mu-2\mu^2+\cdots$$

and retain terms through $O(\mu^2)$ in the overall expression for ln $P(n\lambda)$. The final result, after quite a lot of algebra (see Problem 19C.11) is

$$P(n\lambda)=\frac{2^{N+1}e^{-2N\mu^2}}{2^N(2\pi N)^{1/2}}=\frac{2e^{-2N\mu^2}}{(2\pi N)^{1/2}}$$

At this point we recognize that

$$N\mu^2=\frac{(2N_R-N)^2}{4N}=\frac{(N_R-N_L)^2}{4N}=\frac{n^2}{4N}$$

The net distance from the origin is $x=n\lambda$ and the number of steps taken in a time t is $N=t/\tau$, so $N\mu^2=\tau x^2/4t\lambda^2$. Substitution of these quantities into the expression for P gives eqn 19C.15.

The differences of detail between eqns 19C.11 (for one-dimensional diffusion) and 19C.15 arises from the fact that in the present calculation the particles can migrate in either direction from the origin. Moreover, they can be found only at discrete points separated by λ instead of being anywhere on a continuous line. The fact that the two expressions are so similar suggests that diffusion can indeed be interpreted as the outcome of a large number of steps in random directions.

We can now relate the coefficient D to the step length λ and the rate at which the jumps occur. Thus, by comparing the two exponents in eqn 19C.11 and eqn 19C.15 we can immediately write down the **Einstein–Smoluchowski equation**:

$$D=\frac{\lambda^2}{2\tau} \qquad \text{Einstein–Smoluchowski equation} \qquad (19C.16)$$

Brief illustration 19C.5 Random walk

Suppose an SO_4^{2-} ion jumps through its own diameter each time it makes a move in an aqueous solution, then because $D=1.1\times10^{-9}\,m^2\,s^{-1}$ and $a=250$ pm (as deduced from mobility measurements, Topic 19B), it follows from $\lambda=2a$ that

$$\tau=\frac{(2a)^2}{2D}=\frac{2a^2}{D}=\frac{2\times(250\times10^{-12}\,pm)^2}{1.1\times10^{-9}\,m^2\,s^{-1}}=1.1\times10^{-10}\,s$$

or $\tau=110$ ps. Because τ is the time for one jump, the ion makes about 1×10^{10} jumps per second.

Self-test 19C.6 What would be the diffusion constant for a similar ion that is 50 per cent larger than SO_4^{2-} and leaps through its own diameter at only 30 per cent of the rate?

Answer: $2.1\times10^{-9}\,m^2\,s^{-1}$

The Einstein–Smoluchowski equation is the central connection between the microscopic details of particle motion and the macroscopic parameters relating to diffusion. It also brings us back full circle to the properties of the perfect gas treated in Topic 19A. For if we interpret λ/τ as v_{mean}, the mean speed of the molecules, and interpret λ as a mean free path, then we can recognize in the Einstein–Smoluchowski equation essentially the same expression as we obtained from the kinetic model of gases (eqn 19A.10 of Topic 19A, $D=\frac{1}{3}\lambda v_{mean}$). That is, the diffusion of a perfect gas is a random walk with an average step size equal to the mean free path.

Checklist of concepts

- ☐ 1. A **thermodynamic force** represents the spontaneous tendency of the molecules to disperse as a consequence of the Second Law and the hunt for maximum entropy.
- ☐ 2. The **diffusion equation** (Fick's second law; see below) can be regarded as a mathematical formulation of the notion that there is a natural tendency for concentration to become uniform.
- ☐ 3. **Convection** is the transport of particles arising from the motion of a streaming fluid.
- ☐ 4. An intuitive picture of diffusion is of the particles moving in a series of small steps, a **random walk**, and gradually migrating from their original positions.

Checklist of equations

Property	Equation	Comment	Equation number
Thermodynamic force	$\mathcal{F}=-(\partial\mu/\partial x)_{T,p}$	Definition	19C.1
Fick's first law	$J(\text{amount})=-Ddc/dx$		19C.4
Convective flux	$J=sc$		19C.5
Drift speed	$s=D\mathcal{F}/RT$		19C.6
Diffusion equation	$\partial c/\partial t=D\partial^2c/\partial x^2$	One dimension	19C.7
Generalized diffusion equation	$\partial c/\partial t=D\partial^2c/\partial x^2-v\partial c/\partial x$	One dimension	19C.10

Property	Equation	Comment	Equation number
Mean displacement	$\langle x\rangle=2(Dt/\pi)^{1/2}$	One-dimensional diffusion	19C.13
Root-mean-square displacement	$\langle x^2\rangle^{1/2}=(2Dt)^{1/2}$	One-dimensional diffusion	19C.14
Probability of displacement	$P(x,t)=(2\tau/\pi t)^{1/2}\,e^{-x^2t/2\tau\lambda^2}$	One-dimensional random walk	19C.15
Einstein–Smoluchowski equation	$D=\lambda^2/2\tau$	One-dimensional random walk	19C.16

CHAPTER 19 Molecules in motion

TOPIC 19A Transport properties of a perfect gas

Discussion questions

19A.1 Explain how Fick's first law arises from the concentration gradient of gas molecules.

19A.2 Provide molecular interpretations for the dependencies of the diffusion constant and the viscosity on the temperature, pressure, and size of gas molecules.

19A.3 What might be the effect of molecular interactions on the transport properties of a gas?

Exercises

19A.1(a) Calculate the thermal conductivity of argon ($C_{V,m} = 12.5\,J\,K^{-1}\,mol^{-1}$, $\sigma = 0.36\,nm^2$) at 298 K.
19A.1(b) Calculate the thermal conductivity of nitrogen ($C_{V,m} = 20.8\,J\,K^{-1}\,mol^{-1}$, $\sigma = 0.43\,nm^2$) at 298 K.

19A.2(a) Calculate the diffusion constant of argon at 20 °C and (i) 1.00 Pa, (ii) 100 kPa, (iii) 10.0 MPa. If a pressure gradient of $1.0\,bar\,m^{-1}$ is established in a pipe, what is the flow of gas due to diffusion?
19A.2(b) Calculate the diffusion constant of nitrogen at 20 °C and (i) 100.0 Pa, (ii) 100 kPa, (iii) 20.0 MPa. If a pressure gradient of $1.20\,bar\,m^{-1}$ is established in a pipe, what is the flow of gas due to diffusion?

19A.3(a) Calculate the flux of energy arising from a temperature gradient of $10.5\,K\,m^{-1}$ in a sample of argon in which the mean temperature is 280 K.
19A.3(b) Calculate the flux of energy arising from a temperature gradient of $8.5\,K\,m^{-1}$ in a sample of hydrogen in which the mean temperature is 290 K.

19A.4(a) Use the experimental value of the thermal conductivity of neon (Table 19A.2) to estimate the collision cross-section of Ne atoms at 273 K.
19A.4(b) Use the experimental value of the thermal conductivity of nitrogen (Table 19A.2) to estimate the collision cross-section of N_2 molecules at 298 K.

19A.5(a) In a double-glazed window, the panes of glass are separated by 1.0 cm. What is the rate of transfer of heat by conduction from the warm room (28 °C) to the cold exterior (−15 °C) through a window of area $1.0\,m^2$? What power of heater is required to make good the loss of heat?
19A.5(b) Two sheets of copper of area $2.00\,m^2$ are separated by 5.00 cm in air. What is the rate of transfer of heat by conduction from the warm sheet (70 °C) to the cold sheet (0 °C).

19A.6(a) Use the experimental value of the coefficient of viscosity for neon (Table 19A.1) to estimate the collision cross-section of Ne atoms at 273 K.
19A.6(b) Use the experimental value of the coefficient of viscosity for nitrogen (Table 19A.1) to estimate the collision cross-section of the molecules at 273 K.

19A.7(a) Calculate the viscosity of air at (i) 273 K, (ii) 298 K, (iii) 1000 K. Take $\sigma \approx 0.40\,nm^2$. (The experimental values are 173 μP at 273 K, 182 μP at 20 °C, and 394 μP at 600 °C.)
19A.7(b) Calculate the viscosity of benzene vapour at (i) 273 K, (ii) 298 K, (iii) 1000 K. Take $\sigma \approx 0.88\,nm^2$.

19A.8(a) A solid surface with dimensions 2.5 mm × 3.0 mm is exposed to argon gas at 90 Pa and 500 K. How many collisions do the Ar atoms make with this surface in 15 s?
19A.8(b) A solid surface with dimensions 3.5 mm × 4.0 cm is exposed to helium gas at 111 Pa and 1500 K. How many collisions do the He atoms make with this surface in 10 s?

19A.9(a) An effusion cell has a circular hole of diameter 2.50 mm. If the molar mass of the solid in the cell is $260\,g\,mol^{-1}$ and its vapour pressure is 0.835 Pa at 400 K, by how much will the mass of the solid decrease in a period of 2.00 h?
19A.9(b) An effusion cell has a circular hole of diameter 3.00 mm. If the molar mass of the solid in the cell is $300\,g\,mol^{-1}$ and its vapour pressure is 0.224 Pa at 450 K, by how much will the mass of the solid decrease in a period of 24.00 h?

19A.10(a) A solid compound of molar mass $100\,g\,mol^{-1}$ was introduced into a container and heated to 400 °C. When a hole of diameter 0.50 mm was opened in the container for 400 s, a mass loss of 285 mg was measured. Calculate the vapour pressure of the compound at 400 °C.
19A.10(b) A solid compound of molar mass $200\,g\,mol^{-1}$ was introduced into a container and heated to 300 °C. When a hole of diameter 0.50 mm was opened in the container for 500 s, a mass loss of 277 mg was measured. Calculate the vapour pressure of the compound at 300 °C.

19A.11(a) A manometer was connected to a bulb containing carbon dioxide under slight pressure. The gas was allowed to escape through a small pinhole, and the time for the manometer reading to drop from 75 cm to 50 cm was 52 s. When the experiment was repeated using nitrogen (for which $M = 28.02\,g\,mol^{-1}$) the same fall took place in 42 s. Calculate the molar mass of carbon dioxide.
19A.11(b) A manometer was connected to a bulb containing nitrogen under slight pressure. The gas was allowed to escape through a small pinhole, and the time for the manometer reading to drop from 65.1 cm to 42.1 cm was 18.5 s. When the experiment was repeated using a fluorocarbon gas, the same fall took place in 82.3 s. Calculate the molar mass of the fluorocarbon.

19A.12(a) A space vehicle of internal volume $3.0\,m^3$ is struck by a meteor and a hole of radius 0.10 mm is formed. If the oxygen pressure within the vehicle is initially 80 kPa and its temperature 298 K, how long will the pressure take to fall to 70 kPa?
19A.12(b) A container of internal volume $22.0\,m^3$ was punctured, and a hole of radius 0.050 mm was formed. If the nitrogen pressure within the container is initially 122 kPa and its temperature 293 K, how long will the pressure take to fall to 105 kPa?

Problems

19A.1‡ A. Fenghour et al. (*J. Phys. Chem. Ref. Data* **24**, 1649 (1995)) compiled an extensive table of viscosity coefficients for ammonia in the liquid and vapour phases. Deduce the effective molecular diameter of NH_3 based on each of the following vapour-phase viscosity coefficients: (a) $\eta = 9.08 \times 10^{-6}\,kg\,m^{-1}\,s^{-1}$ at 270 K and 1.00 bar; (b) $\eta = 1.749 \times 10^{-5}\,kg\,m^{-1}\,s^{-1}$ at 490 K and 10.0 bar.

‡ These problems were provided by Charles Trapp and Carmen Giunta.

19A.2 Calculate the ratio of the thermal conductivities of gaseous hydrogen at 300 K to gaseous hydrogen at 10 K. Be circumspect, and think about the modes of motion that are thermally active at the two temperatures.

19A.3 Interstellar space is quite different from the gaseous environments we commonly encounter on Earth. For instance, a typical density of the medium is about 1 atom cm^{-3} and that atom is typically H; the effective temperature due to stellar background radiation is about 10 kK. Estimate the diffusion coefficient and thermal conductivity of H under these conditions. *Comment*: Energy is in fact transferred much more effectively by radiation.

19A.4 A Knudsen cell was used to determine the vapour pressure of germanium at 1000 °C. During an interval of 7200 s the mass loss through a hole of radius 0.50 mm amounted to 43 μg. What is the vapour pressure of germanium at 1000 °C? Assume the gas to be monatomic.

19A.5 An atomic beam is designed to function with (a) cadmium, (b) mercury. The source is an oven maintained at 380 K, there being a small slit of dimensions $1.0\,cm \times 1.0 \times 10^{-3}$ cm. The vapour pressure of cadmium is 0.13 Pa and that of mercury is 12 Pa at this temperature. What is the atomic current (the number of atoms per second) in the beams?

19A.6 Derive an expression that shows how the pressure of a gas inside an effusion oven (a heated chamber with a small hole in one wall) varies with time if the oven is not replenished as the gas escapes. Then show that $t_{1/2}$, the time required for the pressure to decrease to half its initial value, is independent of the initial pressure. *Hint*. Begin by setting up a differential equation relating dp/dt to $p = NkT/V$, and then integrating it.

TOPIC 19B Motion in liquids

Discussion questions

19B.1 Discuss the difference between the hydrodynamic radius of an ion and its ionic radius and explain why a small ion can have a large hydrodynamic radius.

19B.2 Discuss the mechanism of proton conduction in water. How could the model be tested?

19B.3 Why is a proton less mobile in liquid ammonia than in water?

Exercises

19B.1(a) The viscosity of water at 20 °C is 1.002 cP and 0.7975 cP at 30 °C. What is the energy of activation for the transport process?
19B.1(b) The viscosity of mercury at 20 °C is 1.554 cP and 1.450 cP at 40 °C. What is the energy of activation for the transport process?

19B.2(a) The mobility of a chloride ion in aqueous solution at 25 °C is $7.91 \times 10^{-8}\,m^2\,s^{-1}\,V^{-1}$. Calculate the molar ionic conductivity.
19B.2(b) The mobility of an acetate ion in aqueous solution at 25 °C is $4.24 \times 10^{-8}\,m^2\,s^{-1}\,V^{-1}$. Calculate the molar ionic conductivity.

19B.3(a) The mobility of a Rb^+ ion in aqueous solution is $7.92 \times 10^{-8}\,m^2\,s^{-1}\,V^{-1}$ at 25 °C. The potential difference between two electrodes placed in the solution is 25.0 V. If the electrodes are 7.00 mm apart, what is the drift speed of the Rb^+ ion?
19B.3(b) The mobility of a Li^+ ion in aqueous solution is $4.01 \times 10^{-8}\,m^2\,s^{-1}\,V^{-1}$ at 25 °C. The potential difference between two electrodes placed in the solution is 24.0 V. If the electrodes are 5.0 mm apart, what is the drift speed of the ion?

19B.4(a) The limiting molar conductivities of NaI, $NaNO_3$, and $AgNO_3$ are $12.69\,mS\,m^2\,mol^{-1}$, $12.16\,mS\,m^2\,mol^{-1}$ and $13.34\,mS\,m^2\,mol^{-1}$, respectively (all at 25 °C). What is the limiting molar conductivity of AgI at this temperature?
19B.4(b) The limiting molar conductivities of KF, KCH_3CO_2, and $Mg(CH_3CO_2)_2$ are $12.89\,mS\,m^2\,mol^{-1}$, $11.44\,mS\,m^2\,mol^{-1}$ and $18.78\,mS\,m^2\,mol^{-1}$, respectively (all at 25 °C). What is the limiting molar conductivity of MgF_2 at this temperature?

19B.5(a) At 25 °C the molar ionic conductivities of Li^+, Na^+, and K^+ are $3.87\,mS\,m^2\,mol^{-1}$, $5.01\,mS\,m^2\,mol^{-1}$, and $7.35\,mS\,m^2\,mol^{-1}$, respectively. What are their mobilities?
19B.5(b) At 25 °C the molar ionic conductivities of F^-, Cl^-, and Br^- are $5.54\,mS\,m^2\,mol^{-1}$, $7.635\,mS\,m^2\,mol^{-1}$, and $7.81\,mS\,m^2\,mol^{-1}$, respectively. What are their mobilities?

19B.6(a) Estimate the effective radius of a sucrose molecule in water at 25 °C given that its diffusion coefficient is $5.2 \times 10^{-10}\,m^2\,s^{-1}$ and that the viscosity of water is 1.00 cP.
19B.6(b) Estimate the effective radius of a glycine molecule in water at 25 °C given that its diffusion coefficient is $1.055 \times 10^{-9}\,m^2\,s^{-1}$ and that the viscosity of water is 1.00 cP.

19B.7(a) The mobility of a NO_3^- ion in aqueous solution at 25°C is $7.40 \times 10^{-8}\,m^2\,s^{-1}\,V^{-1}$. Calculate its diffusion coefficient in water at 25 °C.
19B.7(b) The mobility of a $CH_3CO_2^-$ ion in aqueous solution at 25°C is $4.24 \times 10^{-8}\,m^2\,s^{-1}\,V^{-1}$. Calculate its diffusion coefficient in water at 25 °C.

Problems

19B.1 The viscosity of benzene varies with temperature as shown in the following table. Use the data to infer the activation energy for viscosity.

θ/°C	10	20	30	40	50	60	70
η/cP	0.758	0.652	0.564	0.503	0.442	0.392	0.358

19B.2 An empirical expression that reproduces the viscosity of water in the range 20–100 °C is

$$\log \frac{\eta}{\eta_{20}} = \frac{1.3272(20-\theta/°C) - 0.001053(20-\theta/°C)^2}{\theta/°C + 105}$$

where η_{20} is the viscosity at 20 °C. Explore (by using mathematical software) the possibility of fitting an exponential curve to this expression and hence identifying an activation energy for the viscosity.

19B.3 The conductivity of aqueous ammonium chloride at a series of concentrations is listed in the following table. Deduce the molar conductivity and determine the parameters that occur in Kohlrausch's law.

$c/(\text{mol dm}^{-3})$	1.334	1.432	1.529	1.672	1.725
$\kappa/(\text{mS cm}^{-1})$	131	139	147	156	164

19B.4 Conductivities are often measured by comparing the resistance of a cell filled with the sample to its resistance when filled with some standard solution, such as aqueous potassium chloride. The conductivity of water is $76\,\text{mS m}^{-1}$ at 25 °C and the conductivity of $0.100\,\text{mol dm}^{-3}$ KCl(aq) is $1.1639\,\text{S m}^{-1}$. A cell had a resistance of 33.21 Ω when filled with $0.100\,\text{mol dm}^{-3}$ KCl(aq) and 300.0 Ω when filled with $0.100\,\text{mol dm}^{-3}$ CH_3COOH(aq). What is the molar conductivity of acetic acid at that concentration and temperature?

19B.5 The resistances of a series of aqueous NaCl solutions, formed by successive dilution of a sample, were measured in a cell with cell constant (the constant C in the relation $\kappa = C/R$) equal to $0.2063\,\text{cm}^{-1}$. The following values were found:

$c/(\text{mol dm}^{-3})$	0.00050	0.0010	0.0050	0.010	0.020	0.050
R/Ω	3314	1669	342.1	174.1	89.08	37.14

Verify that the molar conductivity follows Kohlrausch's law and find the limiting molar conductivity. Determine the coefficient $\mathcal{K}$. Use the value of $\mathcal{K}$ (which should depend only on the nature, not the identity, of the ions) and the information that $\lambda(Na^+) = 5.01\,\text{mS m}^2\,\text{mol}^{-1}$ and $\lambda(I^-) = 7.68\,\text{mS m}^2\,\text{mol}^{-1}$ to predict (a) the molar conductivity, (b) the conductivity, (c) the resistance it would show in the cell of $0.010\,\text{mol dm}^{-3}$ NaI(aq) at 25 °C.

19B.6 What are the drift speeds of Li^+, Na^+, and K^+ in water when a potential difference of 100 V is applied across a 5.00 cm conductivity cell? How long would it take an ion to move from one electrode to the other? In conductivity measurements it is normal to use alternating current: what are the displacements of the ions in (a) centimetres, (b) solvent diameters, about 300 pm, during a half cycle of 2.0 kHz applied potential difference?

19B.7‡ G. Bakale, et al. (*J. Phys. Chem.* **100**, 12477 (1996)) measured the mobility of singly charged C_{60}^- ions in a variety of nonpolar solvents. In cyclohexane at 22 °C, the mobility is $1.1\,\text{cm}^2\,\text{V}^{-1}\,\text{s}^{-1}$. Estimate the effective radius of the C_{60}^- ion. The viscosity of the solvent is $0.93\times10^{-3}\,\text{kg m}^{-1}\,\text{s}^{-1}$. Suggest a reason why there is a substantial difference between this number and the van der Waals radius of neutral C_{60}.

19B.8 Estimate the diffusion coefficients and the effective hydrodynamic radii of the alkali metal cations in water from their mobilities at 25 °C. Estimate the approximate number of water molecules that are dragged along by the cations. Ionic radii are given Table 18B.2.

19B.9 Nuclear magnetic resonance can be used to determine the mobility of molecules in liquids. A set of measurements on methane in carbon tetrachloride showed that its diffusion coefficient is $2.05\times10^{-9}\,\text{m}^2\,\text{s}^{-1}$ at 0 °C and $2.89\times10^{-9}\,\text{m}^2\,\text{s}^{-1}$ at 25 °C. Deduce what information you can about the mobility of methane in carbon tetrachloride.

19B.10‡ A dilute solution of a weak (1,1)-electrolyte contains both neutral ion pairs and ions in equilibrium ($AB \rightleftharpoons A^+ + B^-$). Prove that molar conductivities are related to the degree of ionization by the equations:

$$\frac{1}{\Lambda_m} = \frac{1}{\Lambda_m(\alpha)} + \frac{(1-\alpha)\Lambda_m^\circ}{\alpha^2\Lambda_m(\alpha)^2}, \quad \Lambda_m(\alpha) = \Lambda_m^\circ - \mathcal{K}(\alpha c)^{1/2}$$

where Λ_m° is the molar conductivity at infinite dilution and $\mathcal{K}$ is the constant in Kohlrausch's law.

TOPIC 19C Diffusion

Discussion questions

19C.1 Describe the origin of the thermodynamic force. To what extent can it be regarded as an actual force?

19C.2 Account physically for the form of the diffusion equation.

Exercises

19C.1(a) The diffusion coefficient of glucose in water at 25 °C is $6.73\times10^{-10}\,\text{m}^2\,\text{s}^{-1}$. Estimate the time required for a glucose molecule to undergo a root-mean-square displacement of 5.0 mm.

19C.1(b) The diffusion coefficient of H_2O in water at 25 °C is $2.26\times10^{-9}\,\text{m}^2\,\text{s}^{-1}$. Estimate the time required for an H_2O molecule to undergo a root-mean-square displacement of 1.0 cm.

19C.2(a) A layer of 20.0 g of sucrose is spread uniformly over a surface of area $5.0\,\text{cm}^2$ and covered in water to a depth of 20 cm. What will be the molar concentration of sucrose molecules at 10 cm above the original layer at (i) 10 s, (ii) 24 h? Assume diffusion is the only transport process and take $D = 5.216\times10^{-9}\,\text{m}^2\,\text{s}^{-1}$.

19C.2(b) A layer of 10.0 g of iodine is spread uniformly over a surface of area $10.0\,\text{cm}^2$ and covered in hexane to a depth of 10 cm. What will be the molar concentration of sucrose molecules at 5.0 cm above the original layer at (i) 10 s, (ii) 24 h? Assume diffusion is the only transport process and take $D = 4.05\times10^{-9}\,\text{m}^2\,\text{s}^{-1}$.

19C.3(a) Suppose the concentration of a solute decays linearly along the length of a container. Calculate the thermodynamic force on the solute at 25 °C and 10 cm and 20 cm given that the concentration falls to half its value in 10 cm.

19C.3(b) Suppose the concentration of a solute increases as x^2 along the length of a container. Calculate the thermodynamic force on the solute at 25 °C and 8 cm and 16 cm given that the concentration falls to half its value in 8 cm.

19C.4(a) Suppose the concentration of a solute follows a Gaussian distribution (proportional to e^{-x^2}) along the length of a container. Calculate the thermodynamic force on the solute at 20 °C and 5.0 cm given that the concentration falls to half its value in 5.0 cm.

19C.4(b) Suppose the concentration of a solute follows a Gaussian distribution (proportional to e^{-x^2}) along the length of a container. Calculate the thermodynamic force on the solute at 18 °C and 10.0 cm given that the concentration falls to half its value in 10.0 cm.

19C.5(a) The diffusion coefficient of CCl_4 in heptane at 25 °C is $3.17\times10^{-9}\,\text{m}^2\,\text{s}^{-1}$. Estimate the time required for a CCl_4 molecule to have a root-mean-square displacement of 5.0 mm.

19C.5(b) The diffusion coefficient of I_2 in hexane at 25 °C is $4.05\times10^{-9}\,\text{m}^2\,\text{s}^{-1}$. Estimate the time required for an iodine molecule to have a root-mean-square displacement of 1.0 cm.

19C.6(a) Estimate the effective radius of a sucrose molecule in water 25 °C given that its diffusion coefficient is $5.2\times10^{-10}\,\text{m}^2\,\text{s}^{-1}$ and that the viscosity of water is 1.00 cP.

19C.6(b) Estimate the effective radius of a glycine molecule in water at 25 °C given that its diffusion coefficient is $1.055 \times 10^{-9}\,m^2\,s^{-1}$ and that the viscosity of water is 1.00 cP.

19C.7(a) The diffusion coefficient for molecular iodine in benzene is $2.13 \times 10^{-9}\,m^2\,s^{-1}$. How long does a molecule take to jump through about one molecular diameter (approximately the fundamental jump length for translational motion)?

19C.7(b) The diffusion coefficient for CCl_4 in heptane is $3.17 \times 10^{-9}\,m^2\,s^{-1}$. How long does a molecule take to jump through about one molecular diameter (approximately the fundamental jump length for translational motion)?

19C.8(a) What are the root-mean-square distances travelled by an iodine molecule in benzene and by a sucrose molecule in water at 25 °C in 1.0 s?

19C.8(b) About how long, on average, does it take for the molecules in Exercise 19C.8(a) to drift to a point (i) 1.0 mm, (ii) 1.0 cm from their starting points?

Problems

19C.1 A dilute solution of potassium permanganate in water at 25 °C was prepared. The solution was in a horizontal tube of length 10 cm, and at first there was a linear gradation of intensity of the purple solution from the left (where the concentration was $0.100\,mol\,dm^{-3}$) to the right (where the concentration was $0.050\,mol\,dm^{-3}$). What is the magnitude and sign of the thermodynamic force acting on the solute (a) close to the left face of the container, (b) in the middle, (c) close to the right face? Give the force per mole and force per molecule in each case.

19C.2 A dilute solution of potassium permanganate in water at 25 °C was prepared. The solution was in a horizontal tube of length 10 cm, and at first there was a Gaussian distribution of concentration around the centre of the tube at $x=0$, $c(x)=c_0 e^{-ax^2}$, with $c_0=0.100\,mol\,dm^{-3}$ and $a=0.10\,cm^{-2}$. Determine the thermodynamic force acting on the solute as a function of location, x, and plot the result. Give the force per mole and force per molecule in each case. What do you expect to be the consequence of the thermodynamic force?

19C.3 Instead of a Gaussian 'heap' of solute, as in Problem 19C.2, suppose that there is a Gaussian dip, a distribution of the form $c(x)=c_0(1-e^{-ax^2})$. Repeat the calculation in Problem 19C.2 and its consequences.

19C.4 A lump of sucrose of mass 10.0 g is suspended in the middle of a spherical flask of water of radius 10 cm at 25 °C. What is the concentration of sucrose at the wall of the flask after (a) 1.0 h, (b) 1.0 week? Take $D=5.22 \times 10^{-10}\,m^2\,s^{-1}$.

19C.5 Confirm that eqn 19C.11 is a solution of the diffusion equation with the correct initial value.

19C.6 Confirm that

$$c(x)=\frac{c_0}{(4\pi Dt)^{1/2}}e^{-(x-x_0-vt)^2/4Dt}$$

is a solution of the diffusion equation with convection (eqn 19C.10) with all the solute concentrated at $x=x_0$ at $t=0$ and plot the concentration profile at a series of times to show how the distribution spreads and its centroid drifts.

19C.7 Calculate the relation between $\langle x^2\rangle^{1/2}$ and $\langle x^4\rangle^{1/4}$ for diffusing particles at a time t if they have a diffusion constant D.

19C.8 The thermodynamic force has a direction as well as a magnitude, and in a three-dimensional ideal system eqn 19C.7 becomes $\mathcal{F}=-RT\Delta(\ln c)$. What is the thermodynamic force acting to bring about the diffusion summarized by eqn 19C.12 (that of a solute initially suspended at the centre of a flask of solvent)? *Hint*: Use $\nabla=\boldsymbol{i}\partial/\partial x+\boldsymbol{j}\partial/\partial y+\boldsymbol{k}\partial/\partial z$.

19C.9 The diffusion equation is valid when many elementary steps are taken in the time interval of interest, but the random walk calculation lets us discuss distributions for short times as well as for long. Use the expression $P(n\lambda)=N!/(N-N_R)!N_R!2^N$ to calculate the probability of being six paces from the origin (that is, at $x=6\lambda$) after (a) four, (b) six, (c) twelve steps.

19C.10 Use mathematical software to calculate $P(n\lambda)$ in a one-dimensional random walk, and evaluate the probability of being at $x=n\lambda$ for $n=6$, 10, 14, …, 60. Compare the numerical value with the analytical value in the limit of a large number of steps. At what value of n is the discrepancy no more than 0.1 per cent? Use $n=6$ and $N=6$, 8, ..., 180.

19C.11 Supply the intermediate mathematical steps in *Justification* 19C.2.

19C.12 The diffusion coefficient of a particular kind of t-RNA molecule is $D=1.0 \times 10^{-11}\,m^2\,s^{-1}$ in the medium of a cell interior. How long does it take molecules produced in the cell nucleus to reach the walls of the cell at a distance 1.0 μm, corresponding to the radius of the cell?

19C.13‡ In this problem, we examine a model for the transport of oxygen from air in the lungs to blood. First, show that, for the initial and boundary conditions $c(x,t)=c(x,0)=c_o$, $(0<x<\infty)$ and $c(0,t)=c_s$ $(0\le t\le\infty)$ where c_o and c_s are constants, the concentration, $c(x,t)$, of a species is given by

$$c(x,t)=c_o+(c_s-c_o)\{1-\mathrm{erf}(\xi)\}$$

where erf(ξ) is the error function (see the collection of integrals in the *Resource section*) and the concentration $c(x,t)$ evolves by diffusion from the *yz*-plane of constant concentration, such as might occur if a condensed phase is absorbing a species from a gas phase. Now draw graphs of concentration profiles at several different times of your choice for the diffusion of oxygen into water at 298 K (when $D=2.10 \times 10^{-9}\,m^2\,s^{-1}$) on a spatial scale comparable to passage of oxygen from lungs through alveoli into the blood. Use $c_o=0$ and set c_s equal to the solubility of oxygen in water. *Hint*: Use mathematical software.

Integrated activities

19.1 Use mathematical software, a spreadsheet, or the *Living graphs* on the web site of this book to generate a family of curves similar to that shown in Fig. 19C.4 but by using eqn 19C.14, which describes diffusion in three dimensions.

19.2 In Topic 20D it is shown that a general expression for the activation energy of a chemical reaction is $E_a = RT^2(\mathrm{d}\ln k/\mathrm{d}T)$. Confirm that the same expression may be used to extract the activation energy from eqn 19B.2 for the viscosity and then apply the expression to deduce the temperature-dependence of the activation energy when the viscosity of water is given by the empirical expression in Problem 19B.2. Plot this activation energy as a function of temperature. Suggest an explanation of the temperature dependence of E_a.

CHAPTER 20

Chemical kinetics

This chapter introduces the principles of 'chemical kinetics', the study of reaction rates. The rate of a chemical reaction might depend on variables under our control, such as the pressure, the temperature, and the presence of a catalyst, and we may be able to optimize the rate by the appropriate choice of conditions. Here we begin to see how such manipulations are possible. In the remaining chapters of the text we develop this material in more detail and apply it to more complicated or more specialized cases.

20A The rates of chemical reactions

This Topic discusses the definition of reaction rate and outlines the techniques for its measurement. The results of such measurements show that reaction rates depend on the concentration of reactants (and products) and 'rate constants' that are characteristic of the reaction. This dependence can be expressed in terms of differential equations known as 'rate laws'.

20B Integrated rate laws

'Integrated rate laws' are the solutions of the differential equations that describe rate laws. They are used to predict the concentrations of species at any time after the start of the reaction and to provide procedures for measuring rate constants. This Topic explores simple yet very useful integrated rate laws that appear throughout the chapter.

20C Reactions approaching equilibrium

In this Topic we see that in general the rate laws must take into account both the forward and reverse reactions and that they give rise to expressions that describe the approach to equilibrium, when the forward and reverse rates are equal. A result of the analysis is a useful relation, which can be explored experimentally, between the equilibrium constant of the overall process and the rate constants of the forward and reverse reactions in the proposed mechanism.

20D The Arrhenius equation

The rate constants of most reactions increase with increasing temperature. In this Topic we see that the 'Arrhenius equation' captures this temperature dependence by using only two parameters that can be determined experimentally.

20E Reaction mechanisms

The study of reaction rates also leads to an understanding of the 'mechanisms' of reactions, their analysis into a sequence of elementary steps. In this Topic we see how to construct rate laws from a proposed mechanism. The elementary steps themselves have simple rate laws which can be combined together by invoking the concept of the 'rate-determining step' of a reaction or making either the 'steady-state approximation' or the existence of a 'pre-equilibrium'.

20F Examples of mechanisms

This Topic develops two examples of reaction mechanisms. The first describes a special class of reactions in the gas phase that depend on the collisions between reactants. The second gives insight into the formation of polymers and shows how the kinetics of their formation affects their properties.

20G Photochemistry

'Photochemistry' is the study of reactions that are initiated by light. In this Topic we explore mechanisms of photochemical reactions, with special emphasis on electron and energy transfer processes.

20H Enzymes

In this Topic we discuss the general mechanism of action of 'enzymes', which are biological catalysts. We show how to assemble expressions for their influence on the rate of reactions and the effect of substances that inhibit their function.

What is the impact of this material?

Plants, algae, and some species of bacteria evolved apparatus that perform 'photosynthesis', the capture of visible and near-infrared radiation for the purpose of synthesizing complex molecules in the cell. In *Impact* I20.1 we explore plant photosynthesis in detail.

To read more about the impact of this material, scan the QR code, or go to bcs.whfreeman.com/webpub/chemistry/pchem10e/impact/pchem-20-1.html

20A The rates of chemical reactions

Contents

➤ **Why do you need to know this material?**

Studies of the rates of disappearance of reactants and appearance of products allow us to predict how quickly a reaction mixture approaches equilibrium. Furthermore, studies of reaction rates lead to detailed descriptions of the molecular events that transform reactants into products.

➤ **What is the key idea?**

Reaction rates can be expressed mathematically in terms of the concentrations of reactants and, in some cases, products.

➤ **What do you need to know already?**

This introductory Topic is the foundation of a sequence: all you need to be aware of initially is the significance of stoichiometric numbers (Topic 2C). For more background on the spectroscopic determination of concentration, refer to Topic 12A.

This Topic introduces the principles of **chemical kinetics**, the study of reaction rates, by showing how the rates of reactions are defined and measured. The results of such measurements show that reaction rates depend on the concentration of reactants (and products) in characteristic ways that can be expressed in terms of differential equations known as rate laws.

20A.1 Monitoring the progress of a reaction

The first steps in the kinetic analysis of reactions are to establish the stoichiometry of the reaction and identify any side reactions. The basic data of chemical kinetics are then the concentrations of the reactants and products at different times after a reaction has been initiated.

(a) General considerations

The rates of most chemical reactions are sensitive to the temperature (as described in Topic 20D), so in conventional experiments the temperature of the reaction mixture must be held constant throughout the course of the reaction. This requirement puts severe demands on the design of an experiment. Gas-phase reactions, for instance, are often carried out in a vessel held in contact with a substantial block of metal. Liquid-phase reactions, including flow reactions, must be carried out in an efficient thermostat. Special efforts have to be made to study reactions at low temperatures, as in the study of the kinds of reactions that take place in interstellar clouds. Thus, supersonic expansion of the reaction gas can be used to attain temperatures as low as 10 K. For work in the liquid-phase and the solid-phase, very low temperatures are often reached by flowing cold liquid or cold gas around the reaction vessel. Alternatively, the entire reaction vessel is immersed in a thermally insulated container filled with a cryogenic liquid, such as liquid helium (for work at around 4 K) or liquid nitrogen (for work at around 77 K). Non-isothermal conditions are sometimes employed. For instance, the shelf life of an expensive pharmaceutical may be explored by slowly raising the temperature of a single sample.

Spectroscopy is widely applicable to the study of reaction kinetics, and is especially useful when one substance in the reaction mixture has a strong characteristic absorption in a conveniently accessible region of the electromagnetic spectrum. For example, the progress of the reaction

$$H_2(g) + Br_2(g) \rightarrow 2HBr(g)$$

can be followed by measuring the absorption of visible light by bromine. A reaction that changes the number or type of ions present in a solution may be followed by monitoring the electrical conductivity of the solution. The replacement of neutral molecules by ionic products can result in dramatic changes in the conductivity, as in the reaction

$$(CH_3)_3CCl(aq)+H_2O(l) \rightarrow (CH_3)_3COH(aq)+H^+(aq)+Cl^-(aq)$$

If hydrogen ions are produced or consumed, the reaction may be followed by monitoring the pH of the solution.

Other methods of determining composition include emission spectroscopy (Topic 13B), mass spectrometry (Topic 17D), gas chromatography, nuclear magnetic resonance (Topics 14B and 14C), and electron paramagnetic resonance (for reactions involving radicals or paramagnetic d-metal ions; see Topic 14D). A reaction in which at least one component is a gas might result in an overall change in pressure in a system of constant volume, so its progress may be followed by recording the variation of pressure with time.

Example 20A.1 Monitoring the variation in pressure

Predict how the total pressure varies during the gas-phase decomposition $2\,N_2O_5(g) \rightarrow 4\,NO_2(g)+O_2(g)$ in a constant-volume container.

Method The total pressure (at constant volume and temperature and assuming perfect gas behaviour) is proportional to the number of gas-phase molecules. Therefore, because each mole of N_2O_5 gives rise to $\frac{5}{2}$ mol of gas molecules, we can expect the pressure to rise to $\frac{5}{2}$ times its initial value. To confirm this conclusion, express the progress of the reaction in terms of the fraction, α, of N_2O_5 molecules that have reacted.

Answer Let the initial pressure be p_0 and the initial amount of N_2O_5 molecules present be n. When a fraction α of the N_2O_5 molecules has decomposed, the amounts of the components in the reaction mixture are:

	N_2O_5	NO_2	O_2	Total
Amount:	$n(1-\alpha)$	$2\alpha n$	$\frac{1}{2}\alpha n$	$n(1+\frac{3}{2}\alpha)$

When $\alpha=0$ the pressure is p_0, so at any stage the total pressure is

$$p=(1+\tfrac{3}{2}\alpha)p_0$$

When the reaction is complete ($\alpha=1$), the pressure will have risen to $\frac{5}{2}$ times its initial value.

Self-test 20A.1 Repeat the calculation for $2\,NOBr(g) \rightarrow 2\,NO(g)+Br_2(g)$.

Answer: $p=(1+\frac{1}{2}\alpha)p_0$

(b) Special techniques

The method used to monitor concentrations depends on the species involved and the rapidity with which their concentrations change. Many reactions reach equilibrium over periods of minutes or hours, and several techniques may then be used to follow the changing concentrations. In a **real-time analysis** the composition of the system is analysed while the reaction is in progress. Either a small sample is withdrawn or the bulk solution is monitored. In the **flow method** the reactants are mixed as they flow together in a chamber (Fig. 20A.1). The reaction continues as the thoroughly mixed solutions flow through the outlet tube, and observation of the composition at different positions along the tube is equivalent to the observation of the reaction mixture at different times after mixing. The disadvantage of conventional flow techniques is that a large volume of reactant solution is necessary. This requirement makes the study of fast reactions particularly difficult because to spread the reaction over a length of tube the flow must be rapid. This disadvantage is avoided by the **stopped-flow technique**, in which the reagents are mixed very quickly in a small chamber fitted with a syringe instead of an outlet tube (Fig. 20A.2). The flow ceases when the plunger of the syringe reaches a stop and the reaction continues in the mixed solutions. Observations, commonly using spectroscopic techniques such as ultraviolet–visible absorption, circular dichroism, and fluorescence emission (all introduced in Topics 13A and 13B), are made on the sample as a function of time. The technique allows for the study

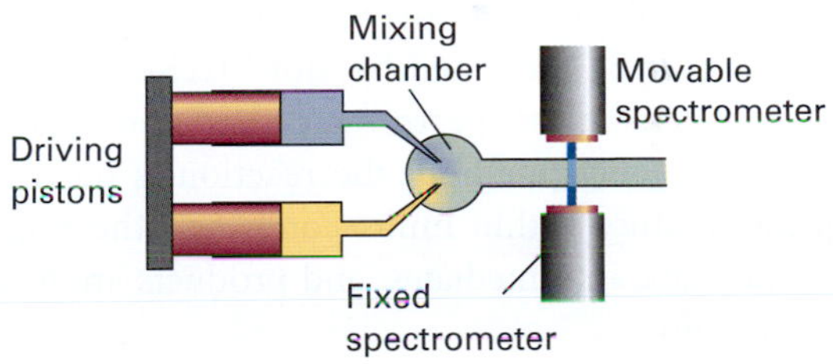

Figure 20A.1 The arrangement used in the flow technique for studying reaction rates. The reactants are injected into the mixing chamber at a steady rate. The location of the spectrometer corresponds to different times after initiation.

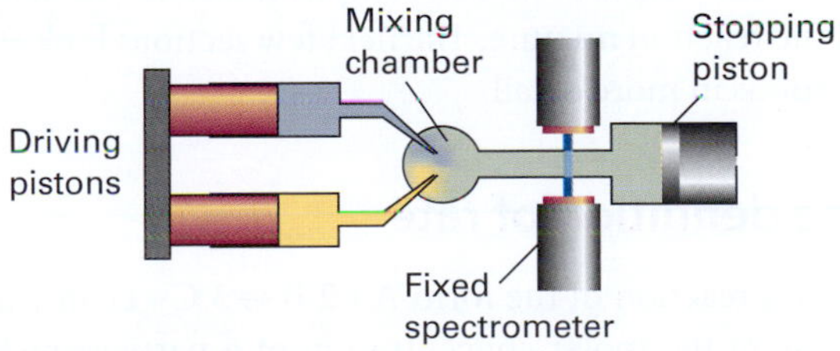

Figure 20A.2 In the stopped-flow technique the reagents are driven quickly into the mixing chamber by the driving pistons and then the time-dependence of the concentrations is monitored.

of reactions that occur on the millisecond to second timescale. The suitability of the stopped-flow method to the study of small samples means that it is appropriate for many biochemical reactions; it has been widely used to study the kinetics of protein folding and enzyme action.

Very fast reactions can be studied by **flash photolysis**, in which the sample is exposed to a brief flash of light that initiates the reaction and then the contents of the reaction chamber are monitored. The apparatus used for flash photolysis studies is based on the experimental design for time-resolved spectroscopy, in which reactions occurring on a picosecond or femtosecond timescale may be monitored by using electronic absorption or emission, infrared absorption, or Raman scattering (Topic 13C).

In contrast to real-time analysis, **quenching methods** are based on quenching, or stopping, the reaction after it has been allowed to proceed for a certain time. In this way the composition is analysed at leisure and reaction intermediates may be trapped. These methods are suitable only for reactions that are slow enough for there to be little reaction during the time it takes to quench the mixture. In the **chemical quench flow method**, the reactants are mixed in much the same way as in the flow method but the reaction is quenched by another reagent, such as solution of acid or base, after the mixture has travelled along a fixed length of the outlet tube. Different reaction times can be selected by varying the flow rate along the outlet tube. An advantage of the chemical quench flow method over the stopped-flow method is that spectroscopic fingerprints are not needed in order to measure the concentration of reactants and products. Once the reaction has been quenched, the solution may be examined by 'slow' techniques, such as gel electrophoresis, mass spectrometry, and chromatography. In the **freeze quench method**, the reaction is quenched by cooling the mixture within milliseconds and the concentrations of reactants, intermediates, and products are measured spectroscopically.

20A.2 The rates of reactions

Reaction rates depend on the composition and the temperature of the reaction mixture. The next few sections look at these observations in more detail.

(a) The definition of rate

Consider a reaction of the form A+2 B→3 C+D, in which at some instant the molar concentration of a participant J is [J] and the volume of the system is constant. The instantaneous **rate of consumption** of one of the reactants at a given time is $-\mathrm{d}[\mathrm{R}]/\mathrm{d}t$, where R is A or B. This rate is a positive quantity (Fig. 20A.3). The **rate of formation** of one of the products (C or D, which we denote P) is $\mathrm{d}[\mathrm{P}]/\mathrm{d}t$ (note the difference in sign). This rate is also positive.

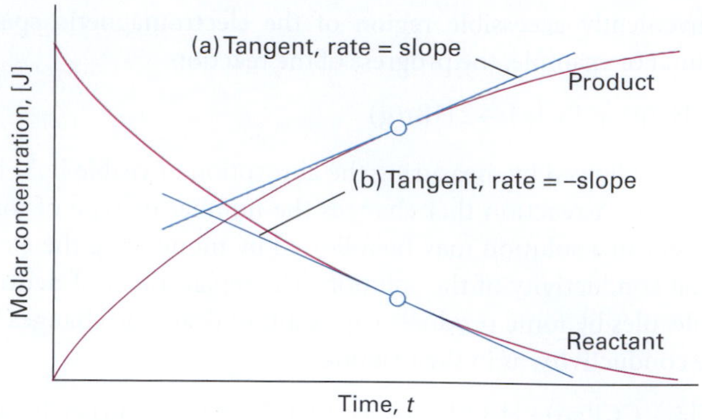

Figure 20A.3 The definition of (instantaneous) rate as the slope of the tangent drawn to the curve showing the variation of concentration of (a) products, (b) reactants with time. For negative slopes, the sign is changed when reporting the rate, so all reaction rates are positive.

It follows from the stoichiometry of the reaction A+2 B→3 C+D that

$$\frac{\mathrm{d}[\mathrm{D}]}{\mathrm{d}t}=\frac{1}{3}\frac{\mathrm{d}[\mathrm{C}]}{\mathrm{d}t}=-\frac{\mathrm{d}[\mathrm{A}]}{\mathrm{d}t}=-\frac{1}{2}\frac{\mathrm{d}[\mathrm{B}]}{\mathrm{d}t}$$

so there are several rates connected with the reaction. The undesirability of having different rates to describe the same reaction is avoided by using the extent of reaction, ξ (xi, the quantity introduced in Topic 6A):

$$\xi=\frac{n_\mathrm{J}-n_{\mathrm{J},0}}{\nu_\mathrm{J}} \qquad \textit{Definition} \quad \text{Extent of reaction} \qquad (20\text{A}.1)$$

where ν_J is the stoichiometric number of species J (Topic 2C; remember that ν_J is negative for reactants and positive for products), and defining the unique **rate of reaction**, v, as the rate of change of the extent of reaction:

$$v=\frac{1}{V}\frac{\mathrm{d}\xi}{\mathrm{d}t} \qquad \textit{Definition} \quad \text{Rate of reaction} \qquad (20\text{A}.2)$$

where V is the volume of the system. It follows that

$$v=\frac{1}{\nu_\mathrm{J}}\times\frac{1}{V}\frac{\mathrm{d}n_\mathrm{J}}{\mathrm{d}t} \qquad (20\text{A}.3\text{a})$$

For a homogeneous reaction in a constant-volume system the volume V can be taken inside the differential and we use $[\mathrm{J}]=n_\mathrm{J}/V$ to write

$$v=\frac{1}{\nu_\mathrm{J}}\frac{\mathrm{d}[\mathrm{J}]}{\mathrm{d}t} \qquad (20\text{A}.3\text{b})$$

For a heterogeneous reaction we use the (constant) surface area, A, occupied by the species in place of V and then use $\sigma_J = n_J/A$ to write

$$v = \frac{1}{\nu_J}\frac{d\sigma_J}{dt} \qquad (20A.3c)$$

In each case there is now a single rate for the entire reaction (for the chemical equation as written). With molar concentrations in moles per cubic decimetre and time in seconds, reaction rates of homogeneous reactions are reported in moles per cubic decimetre per second ($mol\,dm^{-3}\,s^{-1}$) or related units. For gas-phase reactions, such as those taking place in the atmosphere, concentrations are often expressed in molecules per cubic centimetre ($molecules\,cm^{-3}$) and rates in molecules per cubic centimetre per second ($molecules\,cm^{-3}\,s^{-1}$). For heterogeneous reactions, rates are expressed in moles per square metre per second ($mol\,m^{-2}\,s^{-1}$) or related units.

Brief illustration 20A.1 Reaction rates from balanced chemical equations

If the rate of formation of NO in the reaction $2\,NOBr(g) \rightarrow 2\,NO(g) + Br_2(g)$ is reported as $0.16\,mmol\,dm^{-3}\,s^{-1}$, we use $\nu_{NO} = +2$ to report that $v = 0.080\,mmol\,dm^{-3}\,s^{-1}$. Because $\nu_{NOBr} = -2$ it follows that $d[NOBr]/dt = -0.16\,mmol\,dm^{-3}\,s^{-1}$. The rate of consumption of NOBr is therefore $0.16\,mmol\,dm^{-3}\,s^{-1}$, or $9.6 \times 10^{16}\,molecules\,cm^{-3}\,s^{-1}$.

Self-test 20A.2 The rate of change of molar concentration of CH_3 radicals in the reaction $2\,CH_3(g) \rightarrow CH_3CH_3(g)$ was reported as $d[CH_3]/dt = -1.2\,mol\,dm^{-3}\,s^{-1}$ under particular conditions. What is (a) the rate of reaction and (b) the rate of formation of CH_3CH_3?

Answer: (a) $0.60\,mol\,dm^{-3}\,s^{-1}$, (b) $0.60\,mol\,dm^{-3}\,s^{-1}$

(b) Rate laws and rate constants

The rate of reaction is often found to be proportional to the concentrations of the reactants raised to a power. For example, the rate of a reaction may be proportional to the molar concentrations of two reactants A and B, so we write

$$v = k_r[A][B] \qquad (20A.4)$$

with each concentration raised to the first power. The coefficient k_r is called the **rate constant** for the reaction. The rate constant is independent of the concentrations but depends on the temperature. An experimentally determined equation of this kind is called the **rate law** of the reaction. More formally, a rate law is an equation that expresses the rate of reaction as a function of the concentrations of all the species present in the overall chemical equation for the reaction at the time of interest:

$$v = f([A],[B],\ldots) \qquad \text{General form} \quad \text{Rate law in terms of concentrations} \qquad (20A.5a)$$

For homogeneous gas-phase reactions, it is often more convenient to express the rate law in terms of partial pressures, which are related to molar concentrations by $p_J = RT[J]$. In this case, we write

$$v = f(p_A, p_B, \ldots) \qquad \text{General form} \quad \text{Rate law in terms of partial pressures} \qquad (20A.5b)$$

The rate law of a reaction is determined experimentally, and cannot in general be inferred from the chemical equation for the reaction. The reaction of hydrogen and bromine, for example, has a very simple stoichiometry, $H_2(g) + Br_2(g) \rightarrow 2\,HBr(g)$, but its rate law is complicated:

$$v = \frac{k_a[H_2][Br_2]^{3/2}}{[Br_2] + k_b[HBr]} \qquad (20A.6)$$

In certain cases the rate law does reflect the stoichiometry of the reaction; but that is either a coincidence or reflects a feature of the underlying reaction mechanism (see Topic 20E).

A note on good (or, at least, our) practice We denote a general rate constant k_r to distinguish it from the Boltzmann constant k. In some texts k is used for the former and k_B for the latter. When expressing the rate constants in a more complicated rate law, such as that in eqn 20A.6, we use k_a, k_b, and so on.

The units of k_r are always such as to convert the product of concentrations into a rate expressed as a change in concentration divided by time. For example, if the rate law is the one shown in eqn 20A.4, with concentrations expressed in $mol\,dm^{-3}$, then the units of k_r will be $dm^3\,mol^{-1}\,s^{-1}$ because

$$dm^3\,mol^{-1}s^{-1} \times mol\,dm^{-3} \times mol\,dm^{-3} = mol\,dm^{-3}\,s^{-1}$$

In gas-phase studies, including studies of the processes taking place in the atmosphere, concentrations are commonly expressed in $molecules\,cm^{-3}$, so the rate constant for the reaction above would be expressed in $cm^3\,molecule^{-1}\,s^{-1}$. We can use the approach just developed to determine the units of the rate constant from rate laws of any form. For example, the rate constant for a reaction with rate law of the form $k_r[A]$ is commonly expressed in s^{-1}.

Brief illustration 20A.2 The units of rate constants

The rate constant for the reaction $O(g) + O_3(g) \rightarrow 2\,O_2(g)$ is $8.0 \times 10^{-15}\,cm^3\,molecule^{-1}\,s^{-1}$ at 298 K. To express this rate constant in $dm^3\,mol^{-1}\,s^{-1}$, we make use of the two relations

1 cm = 10^{-1} dm and 1 molecule = (1 mol)/(6.022×10^{23}). It follows that

$$
\begin{aligned}
k_r &= 8.0\times10^{-15}\,\text{cm}^3\,\text{molecule}^{-1}\,\text{s}^{-1}\\
&= 8.0\times10^{-15}\left(10^{-1}\,\text{dm}\right)^3\left(\frac{1\,\text{mol}}{6.022\times10^{23}}\right)^{-1}\text{s}^{-1}\\
&= 8.0\times10^{-15}\times10^{-3}\times6.022\times10^{23}\,\text{dm}^3\,\text{mol}^{-1}\,\text{s}^{-1}\\
&= 4.8\times10^{6}\,\text{dm}^3\,\text{mol}^{-1}\,\text{s}^{-1}
\end{aligned}
$$

Self-test 20A.3 A reaction has a rate law of the form $k_r[\text{A}]^2[\text{B}]$. What are the units of the rate constant if the reaction rate is measured in mol dm^{-3} s^{-1}?

Answer: dm^6 mol^{-2} s^{-1}

A practical application of a rate law is that once we know the law and the value of the rate constant, we can predict the rate of reaction from the composition of the mixture. Moreover, as demonstrated in Topic 20B, by knowing the rate law, we can go on to predict the composition of the reaction mixture at a later stage of the reaction. Moreover, a rate law is a guide to the mechanism of the reaction, for any proposed mechanism must be consistent with the observed rate law. This application is developed in Topic 20E.

(c) Reaction order

Many reactions are found to have rate laws of the form

$$v = k_r[\text{A}]^a[\text{B}]^b\cdots \tag{20A.7}$$

The power to which the concentration of a species (a product or a reactant) is raised in a rate law of this kind is the **order** of the reaction with respect to that species. A reaction with the rate law in eqn 20A.4 is first order in A and first order in B. The **overall order** of a reaction with a rate law like that in eqn 20A.7 is the sum of the individual orders, $a+b+\cdots$. The rate law in eqn 20A.4 is therefore second order overall.

A reaction need not have an integral order, and many gas-phase reactions do not. For example, a reaction having the rate law

$$v = k_r[\text{A}]^{1/2}[\text{B}] \tag{20A.8}$$

is half order in A, first order in B, and three-halves order overall.

Brief illustration 20A.3 The interpretation of rate laws

The experimentally determined rate law for the gas-phase reaction $H_2(g) + Br_2(g) \rightarrow 2\,HBr(g)$ is given by eqn 20A.6. Although the reaction is first-order in H_2, it has an indefinite order with respect to both Br_2 and HBr and an indefinite order overall.

Self-test 20A.4 Repeat this analysis for a typical rate law for the action of an enzyme E on a substrate S: $v = k_r[\text{E}][\text{S}]/([\text{S}]+K_M)$, where K_M is a constant.

Answer: First order in E; no specific order with respect to S

Some reactions obey a zero-order rate law, and therefore have a rate that is independent of the concentration of the reactant (so long as some is present). Thus, the catalytic decomposition of phosphine (PH_3) on hot tungsten at high pressures has the rate law

$$v = k_r \tag{20A.9}$$

PH_3 decomposes at a constant rate until it has almost entirely disappeared. Zero-order reactions typically occur when there is a bottle-neck of some kind in the mechanism, as in heterogeneous reactions when the surface is saturated and the subsequent reaction slow and in a number of enzyme reactions when there is a large excess of substrate relative to the enzyme.

As we saw in *Brief illustration* 20A.3, when a rate law is not of the form in eqn 20A.7, the reaction does not have an overall order and may not even have definite orders with respect to each participant.

These remarks point to three important tasks:

- To identify the rate law and obtain the rate constant from the experimental data. We concentrate on this aspect in this Topic.
- To construct reaction mechanisms that are consistent with the rate law. We introduce the techniques for doing so in Topic 20E.
- To account for the values of the rate constants and explain their temperature dependence. This task is undertaken in Topic 20D.

(d) The determination of the rate law

The determination of a rate law is simplified by the **isolation method** in which the concentrations of all the reactants except one are in large excess. If B is in large excess in a reaction between A and B, for example, then to a good approximation its concentration is constant throughout the reaction. Although the true rate law might be $v = k_r[\text{A}][\text{B}]$, we can approximate [B] by $[\text{B}]_0$, its initial value, and write

$$v = k_r'[\text{A}] \qquad k_r' = k_r[\text{B}]_0 \tag{20A.10}$$

which has the form of a first-order rate law. Because the true rate law has been forced into first-order form by assuming that the concentration of B is constant, eqn 20A.10 is called a **pseudofirst-order rate law**. The dependence of the rate on the concentration of each of the reactants may be found by isolating them in turn (by having all the other substances present in large excess), and so constructing a picture of the overall rate law.

In the **method of initial rates**, which is often used in conjunction with the isolation method, the rate is measured at the beginning of the reaction for several different initial concentrations of reactants. We shall suppose that the rate law for a reaction with A isolated is $v = k_r'[A]^a$, then its initial rate, v_0, is given by the initial values of the concentration of A, and we write $v = k_r'[A]_0^a$. Taking logarithms gives:

$$\log v_0 = \log k_r' + a \log [A]_0 \qquad (20A.11)$$

A plot of the logarithms of the initial rates against the logarithms of a series of initial concentrations of A should be a straight line with slope a.

Example 20A.2 Using the method of initial rates

The recombination of iodine atoms in the gas phase in the presence of argon was investigated and the order of the reaction was determined by the method of initial rates. The initial rates of reaction of $2\,I(g) + Ar(g) \rightarrow I_2(g) + Ar(g)$ were as follows:

$[I]_0/(10^{-5}\,mol\,dm^{-3})$		1.0	2.0	4.0	6.0
$v_0/(mol\,dm^{-3}\,s^{-1})$	(a)	8.70×10^{-4}	3.48×10^{-3}	1.39×10^{-2}	3.13×10^{-2}
	(b)	4.35×10^{-3}	1.74×10^{-2}	6.96×10^{-2}	1.57×10^{-1}
	(c)	8.69×10^{-3}	3.47×10^{-2}	1.38×10^{-1}	3.13×10^{-1}

The Ar concentrations are (a) 1.0 mmol dm^{-3}, (b) 5.0 mmol dm^{-3}, and (c) 10.0 mmol dm^{-3}. Determine the orders of reaction with respect to the I and Ar atom concentrations and the rate constant.

Method Plot the logarithm of the initial rate, $\log v_0$, against $\log [I]_0$ for a given concentration of Ar and, separately, against $\log [Ar]_0$ for a given concentration of I. The slopes of the two lines are the orders of reaction with respect to I and Ar, respectively. The intercepts with the vertical axis give $\log k_r$.

Answer The plots are shown in Fig. 20A.4. The slopes are 2 and 1, respectively, so the (initial) rate law is $v_0 = k_r[I]_0^2[Ar]_0$. This rate law signifies that the reaction is second order in [I], first order in [Ar], and third order overall. The intercept corresponds to $k_r = 9\times10^9\,mol^{-2}\,dm^6\,s^{-1}$.

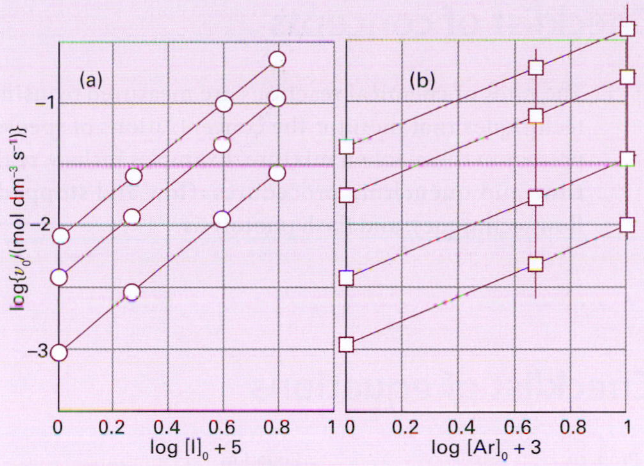

Figure 20A.4 The plot of $\log v_0$ against (a) $\log [I]_0$ for a given $[Ar]_0$, and (b) $\log [Ar]_0$ for a given $[I]_0$.

A note on good practice The units of k_r come automatically from the calculation, and are always such as to convert the product of concentrations to a rate in concentration/time (for example, mol dm^{-3} s^{-1}).

Self-test 20A.5 The initial rate of a reaction depended on concentration of a substance J as follows:

$[J]_0/(mmol\,dm^{-3})$	5.0	8.2	17	30
$v_0/(10^{-7}\,mol\,dm^{-3}\,s^{-1})$	3.6	9.6	41	130

Determine the order of the reaction with respect to J and calculate the rate constant.

Answer: 2, $1.4\times10^{-2}\,dm^3\,mol^{-1}\,s^{-1}$

The method of initial rates might not reveal the full rate law, for once the products have been generated they might participate in the reaction and affect its rate. For example, products participate in the synthesis of HBr, because eqn 20A.6 shows that the full rate law depends on the concentration of HBr. To avoid this difficulty, the rate law should be fitted to the data throughout the reaction. The fitting may be done, in simple cases at least, by using a proposed rate law to predict the concentration of any component at any time, and comparing it with the data. A rate law should also be tested by observing whether the addition of products or, for gas phase reactions, a change in the surface-to-volume ratio in the reaction chamber affects the rate.

Checklist of concepts

☐ 1. The rates of chemical reactions are measured by using techniques that monitor the concentrations of species present in the reaction mixture. Examples include **real-time** and **quenching** procedures, **flow** and **stopped-flow** techniques, and **flash photolysis**.

☐ 2. The **instantaneous rate** of a reaction is the slope of the tangent to the graph of concentration against time (expressed as a positive quantity).

☐ 3. A **rate law** is an expression for the reaction rate in terms of the concentrations of the species that occur in the overall chemical reaction.

Checklist of equations

Property	Equation	Comment	Equation number
Extent of reaction	$\xi=(n_J-n_{J,0})/\nu_J$	Definition	20A.1
Rate of a reaction	$v=(1/V)(d\xi/dt)$, $\xi=(n_J-n_{J,0})/\nu_J$	Definition	20A.2
Rate law (in some cases)	$v=k_r[A]^a[B]^b\cdots$	a, b, …: orders; $a+b+\ldots$: overall order	20A.7
Method of initial rates	$\log v_0=\log k_r'+a\log[A]_0$		20A.11

20B Integrated rate laws

Contents

➤ Why do you need to know this material?

You need the integrated rate law if you want to predict the composition of a reaction mixture as it approaches equilibrium. It is also used to determine the rate law and rate constants of a reaction, which is a necessary step in the formulation of the mechanism of the reaction.

➤ What is the key idea?

A comparison between experimental data and the integrated form of the rate law is used to verify a proposed rate law and determine the order and rate constant of a reaction.

➤ What do you need to know already?

You need to be familiar with the concepts of rate law, reaction order, and rate constant (Topic 20A). The manipulation of simple rate laws requires only elementary techniques of integration (see the *Resource section* for standard integrals).

Rate laws (Topic 20A) are differential equations. We must integrate them if we want to find the concentrations as a function of time. Even the most complex rate laws may be integrated numerically. However, in a number of simple cases analytical solutions, known as **integrated rate laws**, are easily obtained and prove to be very useful. We examine a few of these simple cases here.

20B.1 First-order reactions

As shown in the following *Justification*, the integrated form of the first-order rate law

$$\frac{d[A]}{dt} = -k_r[A] \qquad (20B.1a)$$

is

$$\ln\frac{[A]}{[A]_0} = -k_r t \qquad [A] = [A]_0 e^{-k_r t} \qquad \text{Integrated first-order rate law} \qquad (20B.1b)$$

where $[A]_0$ is the initial concentration of A (at $t=0$).

Justification 20B.1 First-order integrated rate law

First, we rearrange eqn 20B.1a into

$$\frac{d[A]}{[A]^2} = -k_r dt$$

This expression can be integrated directly because k_r is a constant independent of t. Initially (at $t=0$) the concentration of A is $[A]_0$, and at a later time t it is $[A]$, so we make these values the limits of the integrals and write

$$\int_{[A]_0}^{[A]} \frac{d[A]}{[A]} = -k_r \int_0^t dt$$

Because the integral of $1/x$ is $\ln x + \text{constant}$, eqn 20B.1b is obtained immediately.

Equation 20B.1b shows that if $\ln([A]/[A]_0)$ is plotted against t, then a first-order reaction will give a straight line of slope $-k_r$. Some rate constants determined in this way are given in Table 20B.1. The second expression in eqn 20B.1b shows that in a first-order reaction the reactant concentration decreases exponentially with time with a rate determined by k_r (Fig. 20B.1).

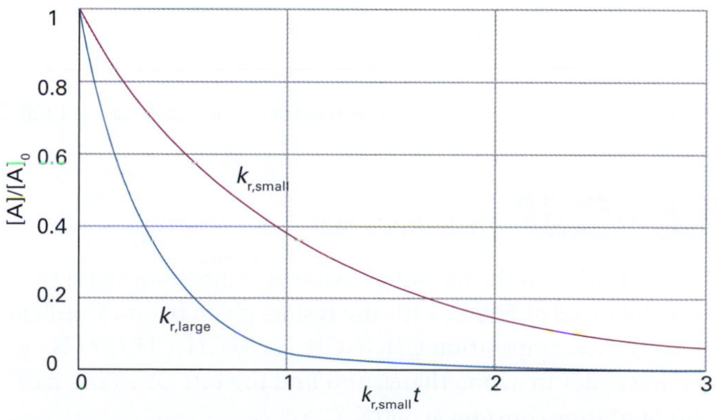

Figure 20B.1 The exponential decay of the reactant in a first-order reaction. The larger the rate constant, the more rapid is the decay: here $k_{r,large} = 3k_{r,small}$.

Table 20B.1* Kinetic data for first-order reactions

Reaction	Phase	θ/°C	k_r/s^{-1}	$t_{1/2}$
$2\,N_2O_5 \rightarrow 4\,NO_2 + O_2$	g	25	3.38×10^{-5}	5.70 h
	$Br_2(l)$	25	4.27×10^{-5}	4.51 h
$C_2H_6 \rightarrow 2\,CH_3$	g	700	5.36×10^{-4}	21.6 min

* More values are given in the *Resource section*.

A useful indication of the rate of a first-order chemical reaction is the **half-life**, $t_{1/2}$, of a substance, the time taken for the concentration of a reactant to fall to half its initial value. This quantity is readily obtained from the integrated rate law. Thus, the time for [A] to decrease from $[A]_0$ to $\frac{1}{2}[A]_0$ in a first-order reaction is given by eqn 20B.1b as

$$k_r t_{1/2} = -\ln\frac{\frac{1}{2}[A]_0}{[A]_0} = -\ln\tfrac{1}{2} = \ln 2$$

Hence

$$t_{1/2} = \frac{\ln 2}{k_r} \qquad \textit{First-order reaction} \quad \text{Half-life} \qquad (20B.2)$$

(Note that ln 2 = 0.693.) The main point to note about this result is that for a first-order reaction, the half-life of a reactant is independent of its initial concentration. Therefore, if the concentration of A at some *arbitrary* stage of the reaction is [A], then it will have fallen to $\frac{1}{2}[A]$ after a further interval of $(\ln 2)/k_r$. Some half-lives are given in Table 20B.1.

Another indication of the rate of a first-order reaction is the **time constant**, τ (tau), the time required for the concentration of a reactant to fall to 1/e of its initial value. From eqn 20B.1b it follows that

$$k_r\tau = -\ln\frac{\frac{1}{e}[A]_0}{[A]_0} = -\ln\tfrac{1}{e} = 1$$

That is, the time constant of a first-order reaction is the reciprocal of the rate constant:

$$\tau = \frac{1}{k_r} \qquad \textit{First-order reaction} \quad \text{Time constant} \qquad (20B.3)$$

Example 20B.1 Analysing a first-order reaction

The variation in the partial pressure of azomethane with time was followed at 600 K, with the results given below. Confirm that the decomposition $CH_3N_2CH_3(g) \rightarrow CH_3CH_3(g) + N_2(g)$ is first order in azomethane, and find the rate constant, half-life, and time constant at 600 K.

t/s	0	1000	2000	3000	4000
p/Pa	10.9	7.63	5.32	3.71	2.59

Method As indicated in the text, to confirm that a reaction is first order, plot $\ln([A]/[A]_0)$ against time and expect a straight line. Because the partial pressure of a gas is proportional to its concentration, an equivalent procedure is to plot $\ln(p/p_0)$ against t. If a straight line is obtained, its slope can be identified with $-k_r$. The half-life and time constant are then calculated from k_r by using eqns 20B.2 and 20B.3, respectively.

Answer We draw up the following table:

t/s	0	1000	2000	3000	4000
$\ln(p/p_0)$	1	−0.357	−0.717	−1.078	−1.437

Figure 20B.2 shows the plot of $\ln(p/p_0)$ against t. The plot is straight, confirming a first-order reaction, and its slope is -3.6×10^{-4}. Therefore, $k_r = 3.6\times10^{-4}\,s^{-1}$.

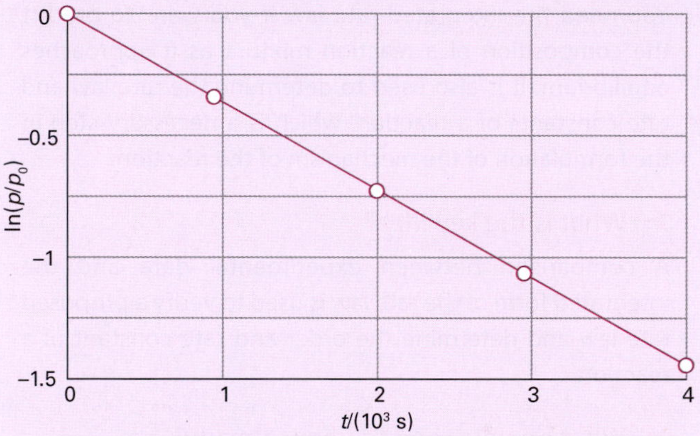

Figure 20B.2 The determination of the rate constant of a first-order reaction: a straight line is obtained when ln [A] (or, as here, ln p/p_0) is plotted against t; the slope gives k_r.

A note on good practice Because the horizontal and vertical axes of graphs are labelled with pure numbers, the slope of a graph is always dimensionless. For a graph of the form $y = b + mx$ we can write $y = b + (m \text{ units})(x/\text{units})$, where 'units' are the units of x, and identify the (dimensionless) slope with 'm units'. Then m = slope/units. In the present case, because the graph shown here is a plot of $\ln(p/p_0)$ against t/s (with 'units' = s) and k_r is the negative value of the slope of $\ln(p/p_0)$ against t itself, k_r = −slope/s.

It follows from eqns 5.2 and 5.3 that the half-life and time constant are, respectively,

$$t_{1/2} = \frac{\ln 2}{3.6\times10^{4}\,s^{-1}} = 1.9\times10^{-5}\,s \qquad \tau = \frac{1}{3.6\times10^{4}\,s^{-1}} = 2.8\times10^{-5}\,s$$

Self-test 20B.1 In a particular experiment, it was found that the concentration of N_2O_5 in liquid bromine varied with time as follows:

t/s	0	200	400	600	1000
$[N_2O_5]$/(mol dm^{-3})	0.110	0.073	0.048	0.032	0.014

Confirm that the reaction is first order in N_2O_5 and determine the rate constant.

Answer: $k_r = 2.1\times10^{-3}\,s^{-1}$

20B.2 Second-order reactions

We show in the following *Justification* that the integrated form of the second-order rate law

$$\frac{d[A]}{dt} = -k_r[A]^2 \qquad (20B.4a)$$

is either of the following two forms:

$$\frac{1}{[A]} - \frac{1}{[A]_0} = k_r t \qquad \textit{Second-order reaction} \quad \text{Integrated rate law} \qquad (20B.4b)$$

$$[A] = \frac{[A]_0}{1 + k_r t[A]_0} \qquad \textit{Second-order reaction; alternative form} \quad \text{Integrated rate law} \qquad (20B.4c)$$

where $[A]_0$ is the initial concentration of A (at $t=0$).

Justification 20B.2 Second-order integrated rate law

To integrate eqn 20B.4a we rearrange it into

$$\frac{d[A]}{[A]^2} = -k_r dt$$

The concentration is $[A]_0$ at $t=0$ and $[A]$ at a general time t later. Therefore,

$$-\int_{[A]_0}^{[A]} \frac{d[A]}{[A]^2} = k_r \int_0^t dt$$

Because the integral of $1/x^2$ is $-1/x+$constant, we obtain eqn 20B.4b by substitution of the limits

$$\left.\frac{1}{[A]} + \text{constant}\right|_{[A]_0}^{[A]} = \frac{1}{[A]} - \frac{1}{[A]_0} = k_r t$$

We can then rearrange this expression into eqn 20B.4c.

Equation 20B.4b shows that to test for a second-order reaction we should plot $1/[A]$ against t and expect a straight line.

Table 20B.2* Kinetic data for second-order reactions

Reaction	Phase	θ/°C	k_r/(dm^3 mol^{-1} s^{-1})
$2\,NOBr \rightarrow 2\,NO + Br_2$	g	10	0.80
$2\,I \rightarrow I_2$	g	23	7×10^9
$CH_3Cl + CH_3O^-$	$CH_3OH(l)$	20	2.29×10^{-6}

* More values are given in the *Resource section*.

The slope of the graph is k_r. Some rate constants determined in this way are given in Table 20B.2. The rearranged form, eqn 20B.4c, lets us predict the concentration of A at any time after the start of the reaction. It shows that the concentration of A approaches zero more slowly than in a first-order reaction with the same initial rate (Fig. 20B.3).

It follows from eqn 20B.4b by substituting $t=t_{1/2}$ and $[A]=\frac{1}{2}[A]_0$ that the half-life of a species A that is consumed in a second-order reaction is

$$t_{1/2} = \frac{1}{k_r[A]_0} \qquad \textit{Second-order reaction} \quad \text{Half-life} \qquad (20B.5)$$

Therefore, unlike a first-order reaction, the half-life of a substance in a second-order reaction varies with the initial concentration. A practical consequence of this dependence is that species that decay by second-order reactions (which includes some environmentally harmful substances) may persist in low concentrations for long periods because their half-lives are long when their concentrations are low. In general, for an nth-order reaction (with n neither 0 nor 1) of the form A→products, the half-life is related to the rate constant and the initial concentration of A by (see Problem 20B.16)

$$t_{1/2} = \frac{2^{n-1}-1}{(n-1)k_r[A]_0^{n-1}} \qquad \textit{nth-order reaction} \quad \text{Half-life} \qquad (20B.6)$$

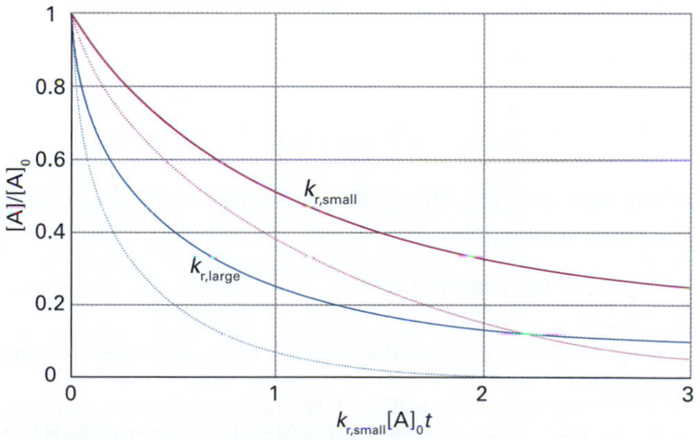

Figure 20B.3 The variation with time of the concentration of a reactant in a second-order reaction. The dotted lines are the corresponding decays in a first-order reaction with the same initial rate. For this illustration, $k_{r,large} = 3k_{r,small}$.

Another type of second-order reaction is one that is first order in each of two reactants A and B:

$$\frac{d[A]}{dt} = -k_r[A][B] \qquad (20B.7)$$

To integrate this rate law we need to know how the concentration of B is related to that of A. For example, if the reaction is $A+B \rightarrow P$, where P denotes products, and the initial concentrations are $[A]_0$ and $[B]_0$, then it is shown in the following *Justification* that at a time t after the start of the reaction, the concentrations satisfy the relation

$$\ln\frac{[B]/[B]_0}{[A]/[A]_0} = ([B]_0 - [A]_0)k_r t \qquad (20B.8)$$

Second-order reaction of the type $A+B\rightarrow P$ — Integrated rate law

Therefore, a plot of the expression on the left against t should be a straight line from which k_r can be obtained. As shown in the following *Brief illustration*, the rate constant may be estimated quickly by using data from only two measurements.

Brief illustration 20B.1 Second-order reactions

Consider a second-order reaction of the type $A+B\rightarrow P$ carried out in a solution. Initially, the concentrations of reactants were $[A]_0 = 0.075\,\text{mol dm}^{-3}$ and $[B]_0 = 0.050\,\text{mol dm}^{-3}$. After 1.0 h the concentration of B fell to $[B] = 0.020\,\text{mol dm}^{-3}$. Because $\Delta[B] = \Delta[A]$, it follows that during this time interval

$$\Delta[B] = (0.020 - 0.050)\,\text{mol dm}^{-3} = -0.030\,\text{mol dm}^{-3}$$

$$\Delta[A] = -0.030\,\text{mol dm}^{-3}$$

Therefore, the concentrations of A and B after 1.0 h are

$$[A] = \Delta[A] + [A]_0 = (-0.030 + 0.075)\,\text{mol dm}^{-3} = 0.045\,\text{mol dm}^{-3}$$

$$[B] = 0.020\,\text{mol dm}^{-3}$$

It follows from rearrangement of eqn 20B.8 that

$$k_r(3600\,\text{s}) = \frac{1}{(0.050-0.075)\,\text{mol dm}^{-3}}\ln\frac{0.020/0.050}{0.045/0.075}$$

where we have used 1 hr = 3600 s. Solving this expression for the rate constant gives

$$k_r = 4.5\times10^{-3}\,\text{dm}^3\,\text{mol}^{-1}\,\text{s}^{-1}$$

Self-test 20B.2 Calculate the half-life of the reactants for the reaction.

Answer: $t_{1/2}(A) = 5.1\times10^3\,\text{s}$, $t_{1/2}(B) = 2.1\times10^3\,\text{s}$

Justification 20B.3 Overall second-order rate law

It follows from the reaction stoichiometry that when the concentration of A has fallen to $[A]_0 - x$, the concentration of B will have fallen to $[B]_0 - x$ (because each A that disappears entails the disappearance of one B). It follows that

$$\frac{d[A]}{dt} = -k_r([A]_0 - x)([B]_0 - x)$$

Because $[A] = [A]_0 - x$, it follows that $d[A]/dt = -dx/dt$ and the rate law may be written as

$$\frac{dx}{dt} = k_r([A]_0 - x)([B]_0 - x)$$

The initial condition is that $x=0$ when $t=0$; so the integration required is

$$\int_0^x \frac{dx}{([A]_0 - x)([B]_0 - x)} = k_r\int_0^t dt$$

The integral on the right is simply $k_r t$. The integral on the left is evaluated by using the method of partial fractions (see *The chemist's toolkit* 20B.1):

$$\int_0^x \frac{dx}{([A]_0 - x)([B]_0 - x)} = \frac{1}{[B]_0 - [A]_0}\left\{\ln\frac{[A]_0}{[A]_0 - x} - \ln\frac{[B]_0}{[B]_0 - x}\right\}$$

The two logarithms can be combined as follows:

$$\ln\frac{[A]_0}{\underbrace{[A]_0 - x}_{[A]}} - \ln\frac{[B]_0}{\underbrace{[B]_0 - x}_{[B]}} = \ln\frac{[A]_0}{[A]} - \ln\frac{[B]_0}{[B]}$$

$$= \ln\frac{1}{[A]/[A]_0} - \ln\frac{1}{[B]/[B]_0}$$

$$= \ln\frac{[B]/[B]_0}{[A]/[A]_0}$$

Combining all the results so far gives eqn 20B.8. Similar calculations may be carried out to find the integrated rate laws for other orders, and some are listed in Table 20B.3.

The chemist's toolkit 20B.1 Integration by the method of partial fractions

To solve an integral of the form

$$I = \int \frac{1}{(a-x)(b-x)}dx$$

where a and b are constants, we use the method of partial fractions in which a fraction that is the product of terms (as in the denominator of this integrand) is written as a sum of fractions. To implement this procedure we write the integrand as

$$\frac{1}{(a-x)(b-x)}=\frac{1}{b-a}\left(\frac{1}{a-x}-\frac{1}{b-x}\right)$$

Then we integrate each term on the right. It follows that

$$I=\frac{1}{b-a}\left(\int\frac{\mathrm{d}x}{a-x}-\int\frac{\mathrm{d}x}{b-x}\right)\overset{\text{Integral A.2}}{=}\frac{1}{b-a}\left(\ln\frac{1}{a-x}-\ln\frac{1}{b-x}\right)a$$
$$+\text{constant}$$

Table 20B.3 Integrated rate laws

Order	Reaction	Rate law*	$t_{1/2}$
0	$A\rightarrow P$	$v=k_r$ $k_rt=x$ for $0\le x\le[A]_0$	$[A]_0/2k_r$
1	$A\rightarrow P$	$v=k_r[A]$ $k_rt=\ln\frac{[A]_0}{[A]_0-x}$	$(\ln 2)/k_r$
2	$A\rightarrow P$	$v=k_r[A]^2$ $k_rt=\frac{x}{[A]_0([A]_0-x)}$	$1/k_r[A]_0$
	$A+B\rightarrow P$	$v=k_r[A][B]$ $k_rt=\frac{1}{[B]_0-[A]_0}\ln\frac{[A]_0([B]_0-x)}{([A]_0-x)[B]_0}$	
	$A+2\,B\rightarrow P$	$v=k_r[A][B]$ $k_rt=\frac{1}{[B]_0-2[A]_0}\ln\frac{[A]_0([B]_0-2x)}{([A]_0-x)[B]_0}$	
	$A\rightarrow P$ with autocatalysis	$v=k_r[A][P]$ $k_rt=\frac{1}{[A]_0+[P]_0}\ln\frac{[A]_0([P]_0+x)}{([A]_0-x)[P]_0}$	
3	$A+2\,B\rightarrow P$	$v=k_r[A][B]^2$ $k_rt=\frac{2x}{(2[A]_0-[B]_0)([B]_0-2x)[B]_0}+\frac{1}{(2[A]_0-[B]_0)^2}\ln\frac{[A]_0([B]_0-2x)}{([A]_0-x)[B]_0}$	
$n\ge 2$	$A\rightarrow P$	$v=k_r[A]^n$ $k_rt=\frac{1}{n-1}\left\{\frac{1}{([A]_0-x)^{n-1}}-\frac{1}{[A]_0^{n-1}}\right\}$	$\frac{2^{n-1}-1}{(n-1)k_r[A]_0^{n-1}}$

* $x=[P]$ and $v=\mathrm{d}x/\mathrm{d}t$

Checklist of concepts

- ☐ 1. An **integrated rate law** is an expression for the concentration of a reactant or product as a function of time (Table 20B.3).
- ☐ 2. The **half-life** of a reactant is the time it takes for its concentration to fall to half its initial value.
- ☐ 3. Analysis of experimental data using integrated rate laws allow for the prediction of the composition of a reaction system at any stage, the verification of the rate law, and the determination of the rate constant.

Checklist of equations

Property	Equation	Comment	Equation number
Integrated rate law	$\ln([A]/[A]_0) = -k_r t$ or $[A] = [A]_0 e^{-k_r t}$	First order, $A \rightarrow P$	20B.1b
Half-life	$t_{1/2} = (\ln 2)/k_r$	First order, $A \rightarrow P$	20B.2
Time constant	$\tau = 1/k_r$	First order	20B.3
Integrated rate law	$1/[A] - 1/[A]_0 = k_r t$ or $[A] = [A]_0/(1 + k_r t[A]_0)$	Second order, $A \rightarrow P$	20B.4b,c
Half-life	$t_{1/2} = 1/k_r[A]_0$	Second order, $A \rightarrow P$	20B.5
	$t_{1/2} = (2^{n-1} - 1)/(n-1)k_r[A]_0^{n-1}$	nth order, $n \neq 0,1$	20B.6
Integrated rate law	$\ln\{([B]/[B]_0)/([A]/[A]_0)\} = ([B]_0 - [A]_0)k_r t$	Second order, $A + B \rightarrow P$	20B.8

20C Reactions approaching equilibrium

Contents

➤ Why do you need to know this material?

All reactions approach equilibrium, so it is important to be able to describe the changing composition as they approach this composition. Analysis of the time dependence shows that there is an important relation between the rate constants and the equilibrium constant.

➤ What is the key idea?

Both forward and reverse reactions must be incorporated into a reaction scheme to account for the approach to equilibrium.

➤ What do you need to know already?

You need to be familiar with the concepts of rate law, reaction order, and rate constant (Topic 20A), integrated rate laws (Topic 20B), and equilibrium constants (Topic 6A). As in Topic 20B, the manipulation of simple rate laws requires only elementary techniques of integration.

In practice, most kinetic studies are made on reactions that are far from equilibrium and if products are in low concentration the reverse reactions are unimportant. Close to equilibrium the products may be so abundant that the reverse reaction must be taken into account.

20C.1 First-order reactions approaching equilibrium

We can explore the variation of the composition with time as a reaction approaches equilibrium by considering a reaction in which A forms B and both forward and reverse reactions are first order (as in some isomerizations):

$$\begin{aligned} &\mathrm{A} \rightarrow \mathrm{B} \quad v = k_\mathrm{r}[\mathrm{A}] \\ &\mathrm{B} \rightarrow \mathrm{A} \quad v = k_\mathrm{r}'[\mathrm{B}] \end{aligned} \tag{20C.1}$$

The concentration of A is reduced by the forward reaction (at a rate $k_\mathrm{r}[\mathrm{A}]$) but it is increased by the reverse reaction (at a rate $k_\mathrm{r}'[\mathrm{B}]$). The net rate of change at any stage is therefore

$$\frac{\mathrm{d}[\mathrm{A}]}{\mathrm{d}t} = -k_\mathrm{r}[\mathrm{A}] + k_\mathrm{r}'[\mathrm{B}] \tag{20C.2}$$

If the initial concentration of A is $[\mathrm{A}]_0$, and no B is present initially, then at all times $[\mathrm{A}] + [\mathrm{B}] = [\mathrm{A}]_0$. Therefore,

$$\begin{aligned} \frac{\mathrm{d}[\mathrm{A}]}{\mathrm{d}t} &= -k_\mathrm{r}[\mathrm{A}] + k_\mathrm{r}'([\mathrm{A}]_0 - [\mathrm{A}]) \\ &= -(k_\mathrm{r} + k_\mathrm{r}')[\mathrm{A}] + k_\mathrm{r}'[\mathrm{A}]_0 \end{aligned} \tag{20C.3}$$

The solution of this first-order differential equation (as may be checked by differentiation, Problem 20C.1) is

$$[\mathrm{A}] = \frac{k_\mathrm{r}' + k_\mathrm{r}\mathrm{e}^{-(k_\mathrm{r}+k_\mathrm{r}')t}}{k_\mathrm{r} + k_\mathrm{r}'}[\mathrm{A}]_0 \tag{20C.4}$$

Figure 20C.1 shows the time dependence predicted by this equation, with $[\mathrm{B}] = [\mathrm{A}]_0 - [\mathrm{A}]$.

As $t \rightarrow \infty$, the concentrations reach their equilibrium values, which are given by eqn 20C.4 as:

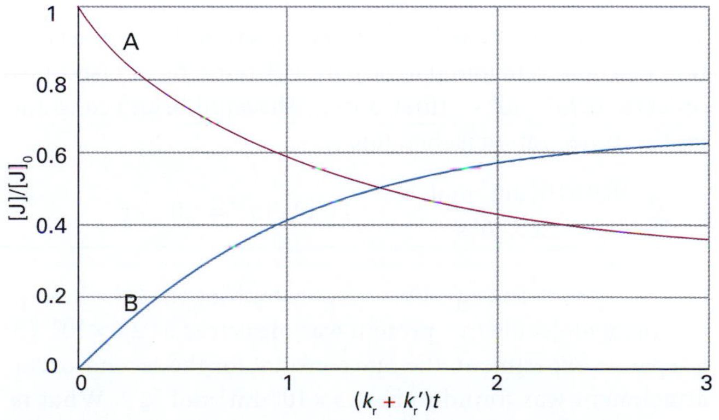

Figure 20C.1 The approach of concentrations to their equilibrium values as predicted by eqn 20C.4 for a reaction $\mathrm{A} \rightleftharpoons \mathrm{B}$ that is first order in each direction, and for which $k_\mathrm{r} = 2k_\mathrm{r}'$.

$$[A]_{eq}=\frac{k_r'[A]_0}{k_r+k_r'} \qquad [B]_{eq}=[A]_0-[A]_{eq}=\frac{k_r[A]_0}{k_r+k_r'} \qquad (20C.5)$$

It follows that the equilibrium constant of the reaction is

$$K=\frac{[B]_{eq}}{[A]_{eq}}=\frac{k_r}{k_r'} \qquad (20C.6)$$

(As explained in Topic 5E, we are justified in replacing activities with the numerical values of molar concentrations if the system is treated as ideal.) Exactly the same conclusion can be reached—more simply, in fact—by noting that, at equilibrium, the forward and reverse rates must be the same, so

$$k_r[A]=k_r'[B] \qquad (20C.7)$$

This relation rearranges into eqn 20C.6. The theoretical importance of eqn 20C.6 is that it relates a thermodynamic quantity, the equilibrium constant, to quantities relating to rates. Its practical importance is that if one of the rate constants can be measured, then the other may be obtained if the equilibrium constant is known.

Equation 20C.6 is valid even if the forward and reverse reactions have different orders, but in that case we need to be careful with units. For instance, if the reaction A + B → C is second order forward and first order in reverse, then the condition for equilibrium is $k_r[A]_{eq}[B]_{eq}=k_r'[C]_{eq}$ and the dimensionless equilibrium constant in full dress is

$$K=\frac{[C]_{eq}/c^{\ominus}}{([A]_{eq}/c^{\ominus})([B]_{eq}/c^{\ominus})}=\left(\frac{[C]}{[A][B]}\right)_{eq}c^{\ominus}=\frac{k_r}{k_r'}\times c^{\ominus}$$

The presence of $c^{\ominus}=1\,\text{mol dm}^{-3}$ in the last term ensures that the ratio of second-order to first-order rate constants, with their different units, is turned into a dimensionless quantity.

Brief illustration 20C.1 The equilibrium constant from rate constants

The rates of the forward and reverse reactions for a dimerization reaction were found to be $8.0\times10^8\,\text{dm}^3\,\text{mol}^{-1}\,\text{s}^{-1}$ (second-order) and $2.0\times10^6\,\text{s}^{-1}$ (first-order). The equilibrium constant for the dimerization is therefore

$$K=\frac{8.0\times10^8\,\text{dm}^3\,\text{mol}^{-1}\,\text{s}^{-1}}{2.0\times10^6\,\text{s}^{-1}}\times1\,\text{mol dm}^{-3}=4.0\times10^2$$

Self-test 20C.1 The equilibrium constant for the attachment of a drug molecule to a protein was measured as 2.0×10^2. In a separate experiment, the rate constant for the second-order attachment was found to be $1.5\times10^8\,\text{dm}^3\,\text{mol}^{-1}\,\text{s}^{-1}$. What is the rate constant for the loss of the drug molecule from the protein?

Answer: $7.5\times10^5\,\text{s}^{-1}$

For a more general reaction, the overall equilibrium constant can be expressed in terms of the rate constants for all the intermediate stages of the reaction mechanism (see Problem 20C.4):

$$K=\frac{k_a}{k_a'}\times\frac{k_b}{k_b'}\times\cdots \qquad \text{Equilibrium constant in terms of the rate constants} \qquad (20C.8)$$

where the k_r are the rate constants for the individual steps and the k_r' are those for the corresponding reverse steps.

20C.2 Relaxation methods

The term **relaxation** denotes the return of a system to equilibrium. It is used in chemical kinetics to indicate that an externally applied influence has shifted the equilibrium position of a reaction, normally suddenly, and that the reaction is adjusting to the equilibrium composition characteristic of the new conditions (Fig. 20C.2). We shall consider the response of reaction rates to a **temperature jump**, a sudden change in temperature. We know from Topic 6B that the equilibrium composition of a reaction depends on the temperature (provided $\Delta_r H^{\ominus}$ is nonzero), so a shift in temperature acts as a perturbation on the system. One way of achieving a temperature jump is to discharge a capacitor through a sample that has been made conducting by the addition of ions, but laser or microwave discharges can also be used. Temperature jumps of between 5 and 10 K can be achieved in about 1 μs with electrical discharges. The high energy output of pulsed lasers is sufficient to generate temperature jumps of between 10 and 30 K within nanoseconds in aqueous samples. Some equilibria are also sensitive to pressure, and **pressure-jump techniques** may then also be used.

We show in the following *Justification* that when a sudden temperature increase is applied to a simple A ⇌ B equilibrium that is first order in each direction, the composition relaxes exponentially to the new equilibrium composition:

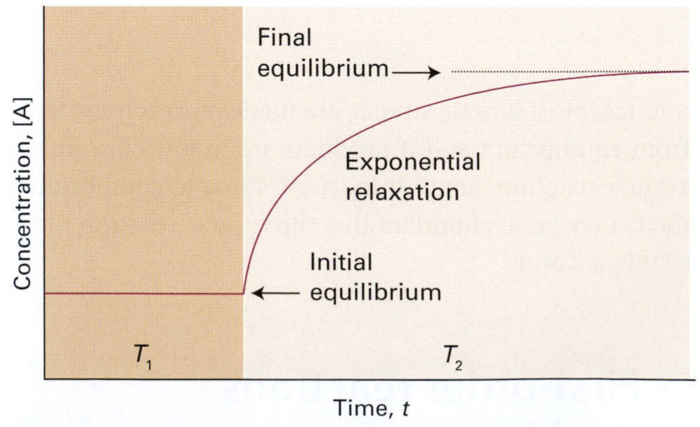

Figure 20C.2 The relaxation to the new equilibrium composition when a reaction initially at equilibrium at a temperature T_1 is subjected to a sudden change of temperature, which takes it to T_2.

$$x = x_0 e^{-t/\tau} \qquad \tau = \frac{1}{k_r + k_r'}$$ First-order reaction — Relaxation after a temperature jump (20C.9)

where x_0 is the departure from equilibrium immediately after the temperature jump, x is the departure from equilibrium at the new temperature after a time t, and k_r and k_r' are the forward and reverse rate constants, respectively, at the new temperature.

Justification 20C.1 Relaxation to equilibrium

When the temperature of a system at equilibrium is increased suddenly, the rate constants change from their earlier values to the new values k_r and k_r' characteristic of that temperature, but the concentrations of A and B remain for an instant at their old equilibrium values. As the system is no longer at equilibrium, it readjusts to the new equilibrium concentrations, which are now given by

$$k_r[\mathrm{A}]_{eq} = k_r'[\mathrm{B}]_{eq}$$

and it does so at a rate that depends on the new rate constants. We write the deviation of [A] from its new equilibrium value as x, so $[\mathrm{A}] = [\mathrm{A}]_{eq} + x$ and $[\mathrm{B}] = [\mathrm{B}]_{eq} - x$. The concentration of A then changes as follows:

$$\begin{aligned}\frac{d[\mathrm{A}]}{dt} &= -k_r[\mathrm{A}] + k_r'[\mathrm{B}] \\ &= -k_r([\mathrm{A}]_{eq} + x) + k_r'([\mathrm{B}]_{eq} - x) \\ &= -(k_r + k_r')x\end{aligned}$$

because the two terms involving the equilibrium concentrations cancel. Because $d[\mathrm{A}]/dt = dx/dt$, this equation is a first-order differential equation with the solution that resembles eqn 20A.1b and is given in eqn 20C.9.

Equation 20C.9 shows that the concentrations of A and B relax into the new equilibrium at a rate determined by the sum of the two new rate constants. Because the equilibrium constant under the new conditions is $K \approx k_r/k_r'$, its value may be combined with the relaxation time measurement to find the individual k_r and k_r'.

Example 20C.1 Analysing a temperature-jump experiment

The equilibrium constant for the autoprotolysis of water, $H_2O(l) \rightleftharpoons H^+(aq) + OH^-(aq)$, is $K_w = a(H^+)a(OH^-) = 1.008 \times 10^{-14}$ at 298 K, where we have used the exact expression in terms of activities. After a temperature-jump, the reaction returns to equilibrium with a relaxation time of 37 μs at 298 K and pH ≈ 7. Given that the forward reaction is first order and the reverse is second order overall, calculate the rate constants for the forward and reverse reactions.

Method We need to derive an expression for the relaxation time, τ (the time constant for return to equilibrium), in terms of k_r (forward, first-order reaction) and k_r' (reverse, second-order reaction). We can proceed as above, but it will be necessary to make the assumption that the deviation from equilibrium (x) is so small that terms in x^2 can be neglected. Relate k_r and k_r' through the equilibrium constant, but be careful with units because K_w is dimensionless.

Answer The forward rate at the final temperature is $k_r[H_2O]$ and the reverse rate is $k_r'[H^+][OH^-]$. The net rate of deprotonation of H_2O is

$$\frac{d[H_2O]}{dt} = -k_r[H_2O] + k_r'[H^+][OH^-]$$

We write $[H_2O] = [H_2O]_{eq} + x$, $[H^+] = [H^+]_{eq} - x$, and $[OH^-] = [OH^-]_{eq} - x$, and obtain

$$\begin{aligned}\frac{dx}{dt} &= -\{k_r + k_r'([H^+]_{eq} + [OH^-]_{eq})\}x \\ &\quad -k_r[H_2O]_{eq} + k_r'[H^+]_{eq}[OH^-]_{eq} + k_r'x^2 \\ &\approx -\{k_r + k_r'([H^+]_{eq} + [OH^-]_{eq})\}x\end{aligned}$$

where we have neglected the term in x^2 because it is so small and have used the equilibrium condition $k_r[H_2O]_{eq} = k_r'[H^+]_{eq}[OH^-]_{eq}$ to eliminate the terms (in blue) that are independent of x. It follows that

$$\frac{1}{\tau} = k_r + k_r'([H^+]_{eq} + [OH^-]_{eq})$$

At this point we note that

$$\begin{aligned}K_w &= a(H^+)a(OH^-) \approx ([H^+]_{eq}/c^\ominus)([OH^-]_{eq}/c^\ominus) \\ &= [H^+]_{eq}[OH^-]_{eq}/c^{\ominus 2}\end{aligned}$$

with $c^\ominus = 1\ \mathrm{mol\ dm^{-3}}$. For this electrically neutral system, $[H^+] = [OH^-]$, so the concentration of each type of ion is $K_w^{1/2}c^\ominus$, and hence

$$\frac{1}{\tau} = k_r + k_r'(K_w^{1/2}c^\ominus + K_w^{1/2}c^\ominus) = k_r'\left\{\frac{k_r}{k_r'} + 2K_w^{1/2}c^\ominus\right\}$$

At this point we note that

$$\frac{k_r}{k_r'} = \frac{[H^+]_{eq}[OH^-]_{eq}}{[H_2O]_{eq}} = \frac{K_w c^{\ominus 2}}{[H_2O]_{eq}}$$

and therefore

$$\frac{1}{\tau} = 2k_r'\left(1 + \overbrace{\frac{K_w^{1/2}c^\ominus}{2[H_2O]_{eq}}}^{K}\right)K_w^{1/2}c^\ominus = 2k_r'(1+K)K_w^{1/2}c^\ominus$$

The molar concentration of pure water is 55.6 mol dm^{-3}, so $[H_2O]_{eq}/c^\ominus = 55.6$ and

$$K = \frac{(1.008 \times 10^{-14})^{1/2}}{2 \times 55.6} = 9.03 \times 10^{-10}$$

which implies that $1+K$ may be replaced by 1 and therefore that

$$k_r' \approx \frac{1}{2\tau K_w^{1/2}c^{\ominus}}$$
$$= \frac{1}{2(3.7\times10^{-5}\,\text{s})\times(1.008\times10^{-14})^{1/2}\times(1\,\text{mol dm}^{-3})}$$
$$= 1.4\times10^{11}\,\text{dm}^3\,\text{mol}^{-1}\,\text{s}^{-1}$$

It follows from the expression for k_r/k_r' that

$$k_r = \frac{K_w c^{\ominus 2} k_r'}{[H_2O]_{eq}}$$
$$= \frac{(1.008\times10^{-14})\times(1\,\text{mol dm}^{-3})^2\times(1.4\times10^{11}\,\text{dm}^3\,\text{mol}^{-1}\,\text{s}^{-1})}{55.6\,\text{mol dm}^{-3}}$$
$$= 2.5\times10^{-5}\,\text{s}^{-1}$$

The reaction is faster in ice, where $k_r' = 8.6\times10^{12}\,\text{dm}^3\,\text{mol}^{-1}\,\text{s}^{-1}$.

A note on good practice Notice how we keep track of units through the use of $c^{\ominus}$: K and K_w are dimensionless; k_r' is expressed in $\text{dm}^3\,\text{mol}^{-1}\,\text{s}^{-1}$ and k_r is expressed in s^{-1}.

Self-test 20C.2 Derive an expression for the relaxation time of a concentration when the reaction $A+B \rightleftharpoons C+D$ is second order in both directions.

Answer: $1/\tau = k_r([A]+[B])_{eq} + k_r'([C]+[D])_{eq}$

Checklist of concepts

☐ 1. There is a relation between the equilibrium constant, a thermodynamic quantity, and the rate constants of the forward and reverse reactions (see Checklist of equations).

☐ 2. In **relaxation methods** of kinetic analysis, the equilibrium position of a reaction is first shifted suddenly and then allowed to readjust to the equilibrium composition characteristic of the new conditions.

Checklist of equations

Property	Equation	Comment	Equation number
Equilibrium constant in terms of rate constants	$K = k_a/k_a' \times k_b/k_b' \times \cdots$	include $c^{\ominus}$ as appropriate	20C.8
Relaxation of an equilibrium $A \rightleftharpoons B$ after a temperature jump	$x = x_0 e^{-t/\tau}$ $\tau = 1/(k_r + k_r')$	First order in each direction	20C.9

20D The Arrhenius equation

Contents

➤ Why do you need to know this material?

The rates of reactions depend on the temperature. Exploration of this dependence leads to the formulation of theories that can help you understand the details of the processes that occur when reactant molecules meet and why a collection of reactants under specific conditions leads to certain products but not others.

➤ What is the key idea?

The temperature dependence of the rate of a reaction is summarized by the activation energy, the minimum energy needed for reaction to occur in an encounter between reactants.

➤ What do you need to know already?

You need to know that the rate of a chemical reaction is expressed by a rate constant (Topic 20A).

In this Topic we interpret the common experimental observation that chemical reactions usually go faster as the temperature is raised. We also begin to see how exploration of the temperature dependence of reaction rates can reveal some details of the energy requirements for molecular encounters that lead to the formation of products.

20D.1 The temperature dependence of reaction rates

It is found experimentally for many reactions that a plot of $\ln k_r$ against $1/T$ gives a straight line with a negative slope, indicating that an increase in $\ln k_r$ (and therefore an increase in k_r) results from a decrease in $1/T$ (that is, an increase in T). This behaviour is normally expressed mathematically by introducing two parameters, one representing the intercept and the other the slope of the straight line, and writing the **Arrhenius equation**

$$\ln k_r = \ln A - \frac{E_a}{RT} \quad \text{Arrhenius equation} \quad (20D.1)$$

The parameter A, which corresponds to the intercept of the line at $1/T=0$ (at infinite temperature, Fig. 20D.1), is called the **pre-exponential factor** or the 'frequency factor'. The parameter E_a, which is obtained from the slope of the line ($-E_a/R$), is called the **activation energy**. Collectively the two quantities are called the **Arrhenius parameters** (Table 20D.1).

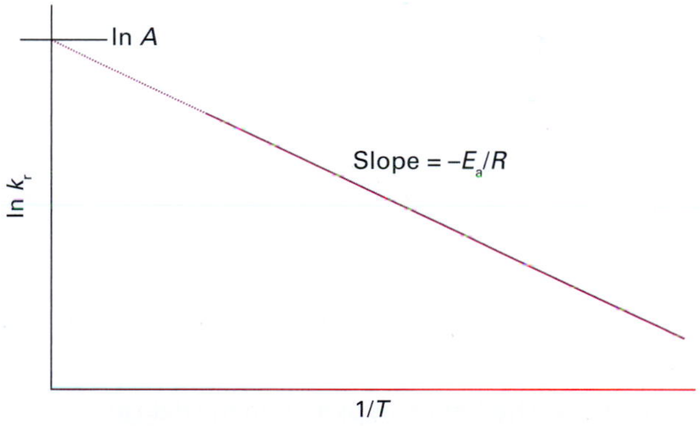

Figure 20D.1 A plot of $\ln k_r$ against $1/T$ is a straight line when the reaction follows the behaviour described by the Arrhenius equation (eqn 20D.1). The slope gives $-E_a/R$ and the intercept at $1/T=0$ gives $\ln A$.

Example 20D.1 Determining the Arrhenius parameters

The rate of the second-order decomposition of acetaldehyde (ethanal, CH_3CHO) was measured over the temperature range 700–1000 K, and the rate constants are reported below. Find E_a and A.

T/K	700	730	760	790	810	840	910	1000
k_r/(dm^3 mol^{-1} s^{-1})	0.011	0.035	0.105	0.343	0.789	2.17	20.0	145

Method According to eqn 20D.1, the data can be analysed by plotting $\ln(k_r/\text{dm}^3\,\text{mol}^{-1}\,\text{s}^{-1})$ against $1/(T/\text{K})$, or more conveniently $(10^3\,\text{K})/T$, and getting a straight line. Obtain the activation energy from the dimensionless slope by writing $-E_a/R = \text{slope/units}$, where in this case 'units' $= 1/(10^3\,\text{K})$, so $E_a = -\text{slope}\times R\times 10^3\,\text{K}$. The intercept at $1/T = 0$ is $\ln(A/\text{dm}^3\,\text{mol}^{-1}\,\text{s}^{-1})$. Use a least-squares procedure to determine the plot parameters.

Answer We draw up the following table:

$(10^3\,\text{K})/T$	1.43	1.37	1.32	1.27	1.23	1.19	1.10	1.00
$\ln(k_r$/dm^3 mol^{-1} s^{-1})	−4.51	−3.35	−2.25	−1.07	−0.24	0.77	3.00	4.98

Now plot $\ln k_r$ against $1/T$ (Fig. 20D.2). The least-squares fit results in a line with slope −22.7 and intercept 27.7. Therefore,

$$E_a = 22.7\times(8.3145\,\text{J K}^{-1}\,\text{mol}^{-1})\times(10^3\,\text{K}) = 189\,\text{kJ mol}^{-1}$$
$$A = e^{27.7}\,\text{dm}^3\,\text{mol}^{-1}\,\text{s}^{-1} = 1.1\times10^{12}\,\text{dm}^3\,\text{mol}^{-1}\,\text{s}^{-1}$$

Note that A has the same units as k_r.

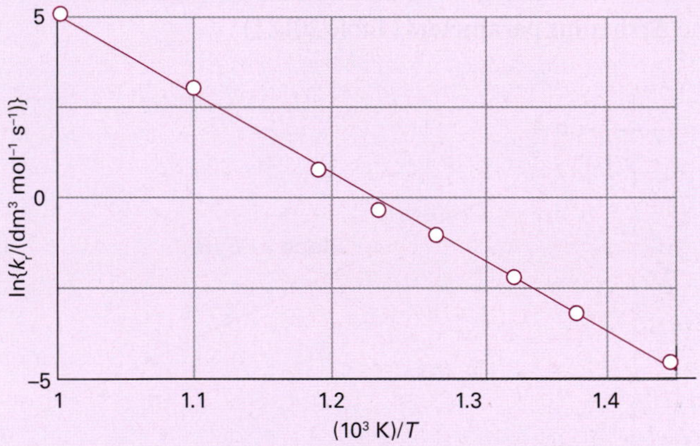

Figure 20D.2 The Arrhenius plot using the data in *Example* 20D.1.

Self-test 20D.1 Determine A and E_a from the following data:

T/K	300	350	400	450	500
k_r/(dm^3 mol^{-1} s^{-1})	7.9×10^6	3.0×10^7	7.9×10^7	1.7×10^8	3.2×10^8

Answer: 8×10^{10} dm^3 mol^{-1} s^{-1}, 23 kJ mol^{-1}

Once the activation energy of a reaction is known, it is a simple matter to predict the value of a rate constant $k_{r,2}$ at a temperature T_2 from its value $k_{r,1}$ at another temperature T_1. To do so, we write

Table 20D.1* Arrhenius parameters

(1) First-order reactions	A/s^{-1}	E_a/(kJ mol^{-1})
$CH_3NC \rightarrow CH_3CN$	3.98×10^{13}	160
$2\,N_2O_5 \rightarrow 4\,NO_2 + O_2$	4.94×10^{13}	103.4
(2) Second-order reactions	**A/(dm^3 mol^{-1} s^{-1})**	**E_a/(kJ mol^{-1})**
$OH + H_2 \rightarrow H_2O + H$	8.0×10^{10}	42
$NaC_2H_5O + CH_3I$ in ethanol	2.42×10^{11}	81.6

* More values are given in the *Resource section*.

$$\ln k_{r,2} = \ln A - \frac{E_a}{RT_2}$$

and then subtract eqn 20D.1 (with T identified as T_1 and k_r as $k_{r,1}$), so obtaining

$$\ln k_{r,2} - \ln k_{r,1} = -\frac{E_a}{RT_2} + \frac{E_a}{RT_1}$$

We can rearrange this expression to

$$\ln\frac{k_{r,2}}{k_{r,1}} = \frac{E_a}{R}\left(\frac{1}{T_1} - \frac{1}{T_2}\right) \qquad \text{Temperature dependence of the rate constant} \qquad (20D.2)$$

Brief illustration 20D.1 The Arrhenius equation

For a reaction with an activation energy of 50 kJ mol^{-1}, an increase in the temperature from 25 °C to 37 °C (body temperature) corresponds to

$$\ln\frac{k_{r,2}}{k_{r,1}} = \frac{50\times10^3\,\text{J mol}^{-1}}{8.3145\,\text{J K}^{-1}\,\text{mol}^{-1}}\left(\frac{1}{298\,\text{K}} - \frac{1}{310\,\text{K}}\right) = \frac{50\times10^3}{8.3145}\left(\frac{1}{298} - \frac{1}{310}\right) = 0.781\ldots$$

By taking natural antilogarithms (that is, by forming e^x), $k_{r,2} = 2.18k_{r,1}$. This result corresponds to slightly more than a doubling of the rate constant as the temperature is increased from 298 K to 310 K.

Self-test 20D.2 The activation energy of one of the reactions in a biochemical process is 87 kJ mol^{-1}. What is the change in rate constant when the temperature falls from 37 °C to 15 °C?

Answer: $k_r(15\,°\text{C}) = 0.076k_r(37\,°\text{C})$

The fact that E_a is given by the slope of the plot of $\ln k_r$ against $1/T$ leads to the following conclusions:

- The stronger the temperature dependence of the rate constant (that is, the steeper the slope), the higher the activation energy.

- A high activation energy signifies that the rate constant depends strongly on temperature.
- If a reaction has zero activation energy, its rate is independent of temperature.
- A negative activation energy indicates that the rate decreases as the temperature is raised.

Physical interpretation

The temperature dependence of some reactions is 'non-Arrhenius' in the sense that a straight line is not obtained when $\ln k_r$ is plotted against $1/T$. However, it is still possible to define an activation energy at any temperature as

$$E_a = RT^2\left(\frac{d\ln k_r}{dT}\right) \quad \textit{Definition} \quad \text{Activation energy} \quad (20D.3)$$

This definition reduces to the earlier one (as the slope of a straight line) for a temperature-independent activation energy (see Problem 20D.1). However, the definition in eqn 20D.3 is more general than that in eqn 20D.1, because it allows E_a to be obtained from the slope (at the temperature of interest) of a plot of $\ln k_r$ against $1/T$ even if the Arrhenius plot is not a straight line. Non-Arrhenius behaviour is sometimes a sign that quantum mechanical tunnelling (Topic 8A) is playing a significant role in the reaction. In biological reactions it might signal that an enzyme has undergone a structural change and has become less efficient.

20D.2 The interpretation of the Arrhenius parameters

For the present Topic we shall regard the Arrhenius parameters as purely empirical quantities that enable us to summarize the variation of rate constants with temperature. However, it is useful to have an interpretation in mind. Topics 21A–21F provide a more elaborate interpretation.

(a) A first look at the energy requirements of reactions

To interpret E_a we consider how the molecular potential energy changes in the course of a chemical reaction that begins with a collision between molecules of A and molecules of B (Fig. 20D.3). In the gas phase that is an actual collision; in solution it is best regarded as a close encounter, possibly with excess energy, and might involve the solvent too. As the reaction event proceeds, A and B come into contact, distort, and begin to exchange or discard atoms. The **reaction coordinate** summarizes the collection of motions, such as changes in interatomic distances and bond angles, that are directly involved in the formation of products from reactants. (The reaction coordinate is essentially a geometrical concept and quite distinct from the extent of reaction.) The potential energy rises to a maximum and the cluster of atoms that corresponds to the region close to the maximum is called the **activated complex**.

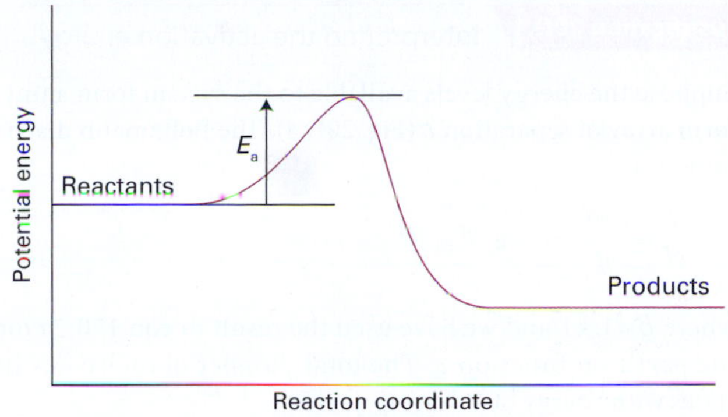

Figure 20D.3 A potential energy profile for an exothermic reaction. The height of the barrier between the reactants and products is the activation energy of the reaction.

After the maximum, the potential energy falls as the atoms rearrange in the cluster and reaches a value characteristic of the products. The climax of the reaction is at the peak of the potential energy, which corresponds to the activation energy E_a. Here two reactant molecules have come to such a degree of closeness and distortion that a small further distortion will send them in the direction of products. This crucial configuration is called the **transition state** of the reaction. Although some molecules entering the transition state might revert to reactants, if they pass through this configuration then it is inevitable that products will emerge from the encounter. (The terms 'activated complex' and 'transition state' are often used as synonyms; however, we shall preserve a distinction.)

We conclude from the preceding discussion that *the activation energy is the minimum energy reactants must have in order to form products*. For example, in a reaction mixture there are numerous molecular encounters each second, but only very few are sufficiently energetic to lead to reaction. The fraction of close encounters between reactants with energy in excess of E_a is given by the Boltzmann distribution (*Foundations* B and Topic 15A) as $e^{-E_a/RT}$. This interpretation is confirmed by comparing this expression with the Arrhenius equation written in the form

$$k_r = Ae^{-E_a/RT} \quad \textit{Alternative form} \quad \text{Arrhenius equation} \quad (20D.4)$$

which is obtained by taking antilogarithms of both sides of eqn 20D.1. We show in the following *Justification* that the exponential factor in eqn 20D.4 can be interpreted as the fraction of encounters that have enough energy to lead to reaction. This point is explored further for gas-phase reactions in Topic 21A and for reactions in solution in Topic 21C.

Justification 20D.1 Interpreting the activation energy

Suppose the energy levels available to the system form a uniform array of separation ε (Fig. 20D.4). The Boltzmann distribution is

$$\frac{N_i}{N}=\frac{e^{-i\varepsilon\beta}}{q}=(1-e^{-\varepsilon\beta})e^{-i\varepsilon\beta}$$

where $\beta=1/kT$ and we have used the result in eqn 15B.2a for the partition function q. The total number of molecules in states with energy of at least $i_{\min}\varepsilon$ is

$$\sum_{i=i_{\min}}^{\infty}N_i=\sum_{i=0}^{\infty}N_i-\sum_{i=0}^{i_{\min}-1}N_i=N-\frac{N}{q}\sum_{i=0}^{i_{\min}-1}e^{-i\varepsilon\beta}$$

The sum of the (blue) finite geometrical series is

$$\sum_{i=0}^{i_{\min}-1}e^{-i\varepsilon\beta}=\frac{1-e^{-i_{\min}\varepsilon\beta}}{1-e^{-\varepsilon\beta}}=q(1-e^{-i_{\min}\varepsilon\beta})$$

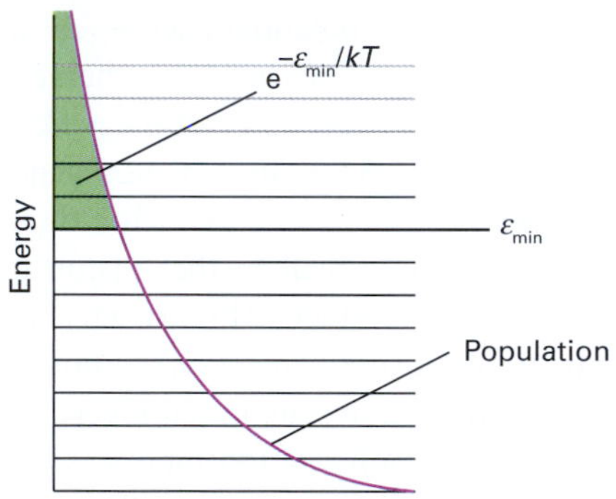

Figure 20D.4 Equally spaced energy levels of an idealized system. As shown in *Justification* 20D.1, the fraction of molecules with energy of at least $e^{-\varepsilon_{\min}/kT}$.

Therefore, the fraction of molecules in states with energy of at least $\varepsilon_{\min}=i_{\min}\varepsilon$ is

$$\frac{1}{N}\sum_{i=i_{\min}}^{\infty}N_i=1-(1-e^{-i_{\min}\varepsilon\beta})=e^{-i_{\min}\varepsilon\beta}=e^{-\varepsilon_{\min}/kT}$$

which has the form of eqn 20D.4.

Brief illustration 20D.2 The fraction of reactive collisions

From *Justification* 20D.1 the fraction of molecules with energy at least $\varepsilon_{\min}$ is $e^{-\varepsilon_{\min}/kT}$, By multiplying $\varepsilon_{\min}$ and k by N_A, Avogadro's constant, and identifying $N_A\varepsilon_{\min}$ with E_a, then the fraction f of molecular collisions that occur with a kinetic energy E_a becomes $f=e^{-E_a/RT}$. With $E_a=50$ kJ mol^{-1} $=5.0\times10^4$ J mol^{-1} and $T=298$ K, we calculate

$$f=e^{-(5.0\times10^4\,\mathrm{J\,mol^{-1}})/(8.3145\,\mathrm{J\,K^{-1}\,mol^{-1}}\times298\,\mathrm{K})}=1.7\times10^{-9}$$

or about 1 in a billion.

Self-test 20D.3 At what temperature would $f=0.10$ if $E_a=50$ kJ mol^{-1}?

Answer: $T=2612$ K

The pre-exponential factor is a measure of the rate at which collisions occur irrespective of their energy. Hence, the product of A and the exponential factor, $e^{-E_a/RT}$ gives the rate of *successful* collisions. We develop these remarks in Topics 21A and 21C, and see that they have their analogues for reactions that take place in liquids.

(b) The effect of a catalyst on the activation energy

The Arrhenius equation tells us that the rate constant of a reaction can be increased by increasing the temperature or by decreasing the energy of activation. Changing the temperature of a reaction mixture is an easy strategy. Reducing the energy of activation is more challenging, but is possible if a reaction takes place in the presence of a suitable **catalyst,** a substance that accelerates a reaction but undergoes no net chemical change. The catalyst lowers the activation energy of the reaction by providing an alternative path that avoids the slow, rate-determining step of the uncatalysed reaction (Fig 20D.5).

Heterogeneous catalysts, which are discussed in Topic 22C, function in a different phase from the reaction mixture. For example, some gas-phase reactions are accelerated in the presence of a solid catalyst. **Homogeneous catalysts** function in the same phase as the reaction mixture. For example, the OH^- ion is a catalyst for a number of organic and inorganic transformations in solution.

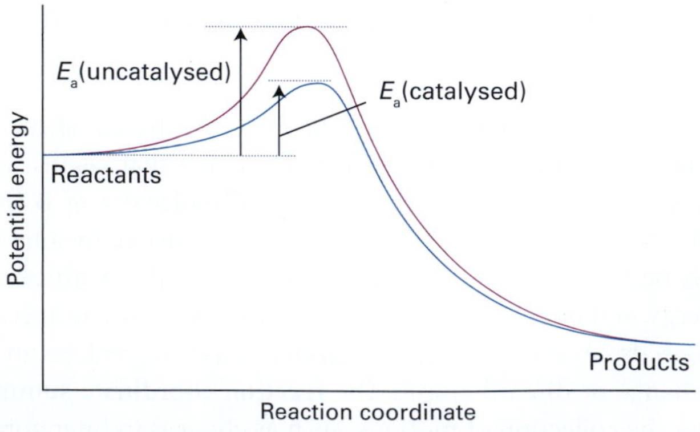

Figure 20D.5 A catalyst provides a different path with a lower activation energy. The result is an increase in the rate of formation of products.

Brief illustration 20D.3 The effect of a catalyst on the rate constant

The enzyme catalase reduces the activation energy for the decomposition of hydrogen peroxide from 76 kJ mol^{-1} to 8 kJ mol^{-1}. From eqn 20D.4 and assuming that the exponential factor is the same in both cases, it follows that the ratio of rate constants is:

$$\frac{k_{\mathrm{r,catalysed}}}{k_{\mathrm{r,uncatalysed}}}=\frac{A\mathrm{e}^{-E_{\mathrm{a,catalysed}}/RT}}{A\mathrm{e}^{-E_{\mathrm{a,uncatalysed}}/RT}}=\mathrm{e}^{-(E_{\mathrm{a,catalysed}}-E_{\mathrm{a,uncatalysed}})/RT}$$
$$=\mathrm{e}^{(68\times10^{3}\,\mathrm{J\,mol^{-1}})/(8.3145\,\mathrm{J\,K^{-1}\,mol^{-1}})\times(298\,\mathrm{K})}=8.3\times10^{11}$$

Self-test 20D.4 Consider the decomposition of hydrogen peroxide, which can be catalysed in solution by iodide ion. By how much is the activation energy of the reaction reduced if the rate constant of reaction increases by a factor of 2000 at 298 K upon addition of the catalyst?

Answer: 25 per cent

Checklist of concepts

☐ 1. The **activation energy**, the parameter E_a in the **Arrhenius equation**, is the minimum energy of close molecular encounters able to result in reaction.

☐ 2. The larger the activation energy, the more sensitive the rate constant is to the temperature.

☐ 3. The **pre-exponential factor** is a measure of the rate at which encounters occur irrespective of their energy.

☐ 4. A **catalyst** lowers the activation energy of a reaction.

Checklist of equations

Property	Equation	Comment	Equation number
Arrhenius equation	$\ln k_r = \ln A - E_a/RT$		20D.1
Activation energy	$E_a = RT^2(\mathrm{d}\ln k_r/\mathrm{d}T)$	Definition	20D.3

20E Reaction mechanisms

Contents

➤ Why do you need to know this material?

You need to know how to construct the rate law for a reaction that takes place by a sequence of steps partly because that gives insight into the atomic processes going on when reactions take place, but also because it indicates how the yield of desired products can be optimized.

➤ What is the key idea?

Many chemical reactions occur as a sequence of simpler steps, with corresponding rate laws that can be combined together by applying one or more approximations.

➤ What do you need to know already?

You need to be familiar with the concept of rate laws (Topic 20A) and how to integrate them (Topics 20B and 20C). You also need to be familiar with the Arrhenius equation for the effect of temperature on reaction rate (Topic 20D).

The study of reaction rates leads to an understanding of the **mechanisms** of reactions, their analysis into a sequence of elementary steps. Simple elementary steps have simple rate laws, which can be combined together by invoking one or more approximations. These approximations include the concept of the rate-determining step of a reaction, the steady-state concentration of a reaction intermediate, and the existence of a pre-equilibrium.

20E.1 Elementary reactions

Most reactions occur in a sequence of steps called **elementary reactions**, each of which involves only a small number of molecules or ions. A typical elementary reaction is

$$\mathrm{H + Br_2 \rightarrow HBr + Br}$$

Note that the phase of the species is not specified in the chemical equation for an elementary reaction and the equation represents the specific process occurring to individual molecules. This equation, for instance, signifies that an H atom attacks a Br_2 molecule to produce an HBr molecule and a Br atom. The **molecularity** of an elementary reaction is the number of molecules coming together to react in an elementary reaction. In a **unimolecular reaction**, a single molecule shakes itself apart or its atoms into a new arrangement, as in the isomerization of cyclopropane to propene. In a **bimolecular reaction**, a pair of molecules collide and exchange energy, atoms, or groups of atoms, or undergo some other kind of change. It is most important to distinguish molecularity from order:

- *reaction order* is an empirical quantity, and obtained from the experimentally determined rate law;
- *molecularity* refers to an elementary reaction proposed as an individual step in a mechanism.

The rate law of a unimolecular elementary reaction is first-order in the reactant:

$$\mathrm{A \rightarrow P} \qquad \frac{\mathrm{d[A]}}{\mathrm{d}t} = -k_\mathrm{r}[\mathrm{A}] \quad \text{Unimolecular elementary reaction} \qquad (20E.1)$$

where P denotes products (several different species may be formed). A unimolecular reaction is first order because the number of A molecules that decay in a short interval is proportional to the number available to decay. For instance, ten times as many decay in the same interval when there are initially 1000 A molecules as when there are only 100 present. Therefore, the rate of decomposition of A is proportional to its molar concentration at any moment during the reaction.

An elementary bimolecular reaction has a second-order rate law:

$$A+B \rightarrow P \qquad \frac{d[A]}{dt}=-k_r[A][B] \qquad \text{Bimolecular elementary reaction} \qquad (20E.2)$$

A bimolecular reaction is second-order because its rate is proportional to the rate at which the reactant species meet, which in turn is proportional to both their concentrations. Therefore, if we have evidence that a reaction is a single-step, bimolecular process, we can write down the rate law (and then go on to test it).

Brief illustration 20E.1 The rate laws of elementary steps

Bimolecular elementary reactions are believed to account for many homogeneous reactions, such as the dimerizations of alkenes and dienes and reactions such as

$$CH_3I(alc)+CH_3CH_2O^-(alc) \rightarrow CH_3OCH_2CH_3(alc)+I^-(alc)$$

(where 'alc' signifies alcohol solution). There is evidence that the mechanism of this reaction is a single elementary step:

$$CH_3I+CH_3CH_2O^- \rightarrow CH_3OCH_2CH_3+I^-$$

This mechanism is consistent with the observed rate law

$$v=k_r[CH_3I][CH_3CH_2O^-]$$

Self-test 20E.1 The following are elementary processes: (a) the dimerization of NO(g) to form N_2O_2(g), and (b) the decomposition of the N_2O_2(g) dimer into NO(g) molecules. Write the rate laws for these processes.

Answer: (a) bimolecular process: $k_r[NO]^2$, (b) unimolecular process: $k_r[N_2O_2]$

We shall see in the following sections how to combine a series of simple steps together into a mechanism and how to arrive at the corresponding overall rate law. For the present we emphasize that, *if the reaction is an elementary bimolecular process, then it has second-order kinetics, but if the kinetics is second order, then the reaction might be complex.* The postulated mechanism can be explored only by detailed detective work on the system and by investigating whether side products or intermediates appear during the course of the reaction. Detailed analysis of this kind was one of the ways, for example, in which the reaction $H_2(g)+I_2(g) \rightarrow 2\,HI(g)$ was shown to proceed by a complex mechanism. For many years the reaction had been accepted on good but insufficiently meticulous evidence as a fine example of a simple bimolecular reaction, $H_2+I_2 \rightarrow HI+HI$, in which atoms exchanged partners during a collision.

20E.2 Consecutive elementary reactions

Some reactions proceed through the formation of an intermediate (I), as in the consecutive unimolecular reactions

$$A \xrightarrow{k_a} I \xrightarrow{k_b} P$$

Note that the intermediate occurs in the reaction steps but does not appear in the overall reaction, which in this case is $A \rightarrow P$. We are ignoring any reverse reactions, so the reaction proceeds from all A to all P, not to an equilibrium mixture of the two. An example of this type of mechanism is the decay of a radioactive family, such as

$$^{239}U \xrightarrow{23.5\ \text{min}} {}^{239}Np \xrightarrow{2.35\ \text{days}} {}^{239}Pu$$

(The times are half-lives.) The characteristics of this type of reaction are discovered by setting up the rate laws for the net rate of change of the concentration of each substance and then combining them in the appropriate manner.

The rate of unimolecular decomposition of A is

$$\frac{d[A]}{dt}=-k_a[A] \qquad (20E.3a)$$

and A is not replenished. The intermediate I is formed from A (at a rate $k_a[A]$) but decays to P (at a rate $k_b[I]$). The net rate of formation of I is therefore

$$\frac{d[I]}{dt}=k_a[A]-k_b[I] \qquad (20E.3b)$$

The product P is formed by the unimolecular decay of I:

$$\frac{d[P]}{dt}=k_b[I] \qquad (20E.3c)$$

We suppose that initially only A is present and that its concentration is then $[A]_0$.

The first of the rate laws, eqn 20E.3a, is an ordinary first-order decay, so we can write

$$[A]=[A]_0 e^{-k_a t} \qquad (20E.4a)$$

When this equation is substituted into eqn 20E.3b, we obtain after rearrangement

$$\frac{d[I]}{dt}+k_b[I]=k_a[A]_0 e^{-k_a t}$$

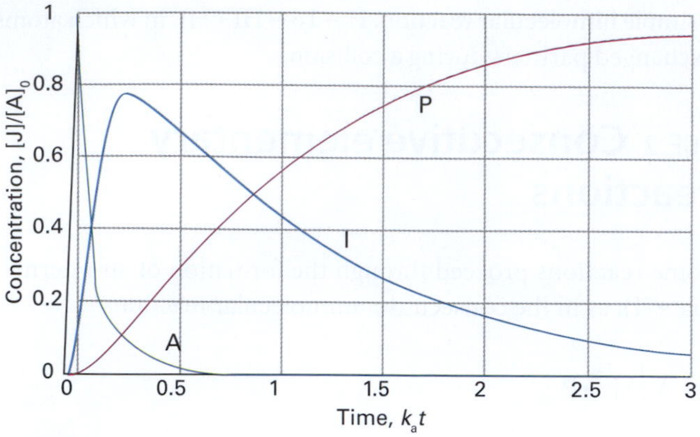

Figure 20E.1 The concentrations of A, I, and P in the consecutive reaction scheme A→I→P. The curves are plots of eqns 20E.4a–c with $k_a = 10k_b$. If the intermediate I is in fact the desired product, it is important to be able to predict when its concentration is greatest; see *Example* 20E.1.

This differential equation has a standard form (see *Mathematical background* 4) and, after setting $[I]_0 = 0$ (no intermediate present initially), the solution is

$$[I] = \frac{k_a}{k_b - k_a}(e^{-k_a t} - e^{-k_b t})[A]_0 \qquad (20E.4b)$$

At all times $[A]+[I]+[P]=[A]_0$, so it follows that

$$[P] = \left\{1 + \frac{k_a e^{-k_b t} - k_b e^{-k_a t}}{k_b - k_a}\right\}[A]_0 \qquad (20E.4c)$$

The concentration of the intermediate I rises to a maximum and then falls to zero (Fig. 20E.1). The concentration of the product P rises from zero towards $[A]_0$, when all A has been converted to P.

Example 20E.1 Analysing consecutive reactions

Suppose that in an industrial batch process a substance A produces the desired compound I which goes on to decay to a worthless product C, each step of the reaction being first order. At what time will I be present in greatest concentration?

Method The time dependence of the concentration of I is given by eqn 20E.4b. We can find the time at which [I] passes through a maximum, t_{max}, by calculating d[I]/dt and setting the resulting rate equal to zero.

Answer It follows from eqn 20E.4b that

$$\frac{d[I]}{dt} = -\frac{k_a(k_a e^{-k_a t} - k_b e^{-k_b t})[A]_0}{k_b - k_a}$$

This rate is equal to zero when $k_a e^{-k_a t} = k_b e^{-k_b t}$. Therefore,

$$t_{max} = \frac{1}{k_a - k_b}\ln\frac{k_a}{k_b}$$

For a given value of k_a, as k_b increases both the time at which [I] is a maximum and the yield of I decrease.

Self-test 20E.2 Calculate the maximum concentration of I and justify the last remark.

Answer: $[I]_{max}/[A]_0 = (k_a/k_b)^c$, $c = k_b/(k_b - k_a)$

20E.3 The steady-state approximation

One feature of the calculation so far has probably not gone unnoticed: there is a considerable increase in mathematical complexity as soon as the reaction mechanism has more than a couple of steps and reverse reactions are taken into account. A reaction scheme involving many steps is nearly always unsolvable analytically, and alternative methods of solution are necessary. One approach is to integrate the rate laws numerically. An alternative approach, which continues to be widely used because it leads to convenient expressions and more readily digestible results, is to make an approximation.

The **steady-state approximation** (which is also widely called the **quasi-steady-state approximation**, QSSA, to distinguish it from a true steady state) assumes that the intermediate, I, is in a low, constant concentration. More specifically, after an initial **induction period**, an interval during which the concentrations of intermediates rise from zero, and during the major part of the reaction, the rates of change of concentrations of all reaction intermediates are negligibly small (Fig. 20E.2):

$$\frac{d[I]}{dt} \approx 0 \qquad \text{Steady-state approximation} \qquad (20E.5)$$

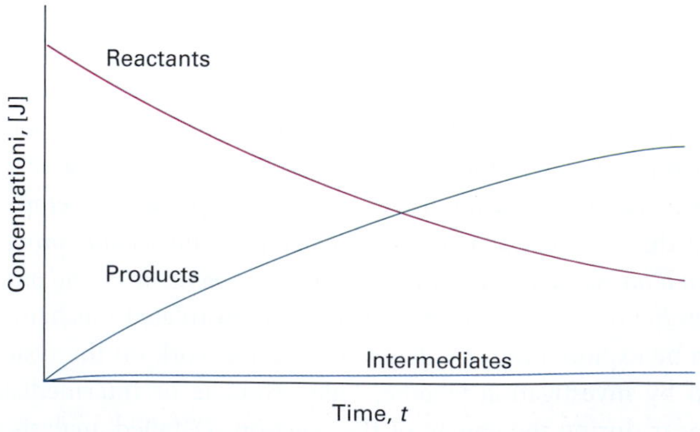

Figure 20E.2 The basis of the steady-state approximation. It is supposed that the concentrations of intermediates remain small and hardly change during most of the course of the reaction.

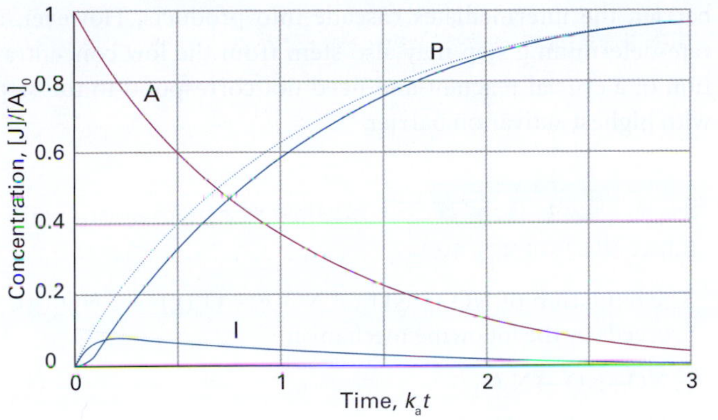

Figure 20E.3 A comparison of the exact result for the concentrations of a consecutive reaction and the concentrations obtained by using the steady-state approximation (dotted lines) for $k_b = 20k_a$. (The curve for [A] is unchanged.)

This approximation greatly simplifies the discussion of reaction schemes. For example, when we apply the approximation to the consecutive first-order mechanism, we set $d[I]/dt = 0$ in eqn 20E.3b, which then becomes $k_a[A] - k_b[I] = 0$. Then

$$[I] = (k_a/k_b)[A] \qquad (20E.6)$$

For this expression to be consistent with eqn 20E.5, we require $k_a/k_b \ll 1$ (so that, even though [A] does depend on the time, the dependence of [I] on the time is negligible). On substituting this value of [I] into eqn 20E.3c, that equation becomes

$$\frac{d[P]}{dt} = k_b[I] \approx k_a[A] \qquad (20E.7)$$

and we see that P is formed by a first-order decay of A, with a rate constant k_a, the rate-constant of the slower, rate-determining, step. We can write down the solution of this equation at once by substituting the solution for [A], eqn 20E.4a, and integrating:

$$[P] = k_a[A]_0 \int_0^t e^{-k_a t} dt = (1 - e^{-k_a t})[A]_0 \qquad (20E.8)$$

This is the same (approximate) result as before, eqn 20E.4c (when $k_b \gg k_a$), but much more quickly obtained. Figure 20E.3 compares the approximate solutions found here with the exact solutions found earlier: k_b does not have to be very much bigger than k_a for the approach to be reasonably accurate.

Example 20E.2 Using the steady-state approximation

Devise the rate law for the decomposition of N_2O_5, $2\,N_2O_5(g) \rightarrow 4\,NO_2(g) + O_2(g)$ on the basis of the following mechanism:

$$N_2O_5 \rightarrow NO_2 + NO_3 \qquad k_a$$
$$NO_2 + NO_3 \rightarrow N_2O_5 \qquad k_a'$$
$$NO_2 + NO_3 \rightarrow NO_2 + O_2 + NO \qquad k_b$$
$$NO + N_2O_5 \rightarrow NO_2 + NO_2 + NO_2 \qquad k_c$$

A note on good practice Note that when writing the equation for an elementary reaction all the species are displayed individually; so we write $A \rightarrow B + B$, for instance, not $A \rightarrow 2\,B$.

Method First identify the intermediates and write expressions for their net rates of formation. Then, all net rates of change of the concentrations of intermediates are set equal to zero and the resulting equations are solved algebraically.

Answer The intermediates are NO and NO_3; the net rates of change of their concentrations are

$$\frac{d[NO]}{dt} = k_b[NO_2][NO_3] - k_c[NO][N_2O_5] \approx 0$$
$$\frac{d[NO_3]}{dt} = k_a[N_2O_5] - k_a'[NO_2][NO_3] - k_b[NO_2][NO_3] \approx 0$$

The solutions of these two simultaneous equations (in blue) are

$$[NO_3] = \frac{k_a[N_2O_5]}{(k_a' + k_b)[NO_2]} \qquad [NO] = \frac{k_b[NO_2][NO_3]}{k_c[N_2O_5]} = \frac{k_a k_b}{(k_a' + k_b)k_c}$$

The net rate of change of concentration of N_2O_5 is then

$$\frac{d[N_2O_5]}{dt} = -k_a[N_2O_5] + k_a'[NO_2][NO_3] - k_c[NO][N_2O_5]$$
$$= -k_a[N_2O_5] + \frac{k_a k_a'[N_2O_5]}{k_a' + k_b} - \frac{k_a k_b}{k_a' + k_b}[N_2O_5]$$
$$= -\frac{2k_a k_b[N_2O_5]}{k_a' + k_b}$$

That is, N_2O_5 decays with a first-order rate law with a rate constant that depends on k_a, k_a' and k_b but not on k_c.

Self-test 20E.3 Derive the rate law for the decomposition of ozone in the reaction $2\,O_3(g) \rightarrow 3\,O_2(g)$ on the basis of the (incomplete) mechanism

$$O_3 \rightarrow O_2 + O \qquad k_a$$
$$O_2 + O \rightarrow O_3 \qquad k_a'$$
$$O + O_3 \rightarrow O_2 + O_2 \qquad k_b$$

Answer: $d[O_3]/dt = -2k_a k_b[O_3]^2/(k_a'[O_2] + k_b[O_3])$

20E.4 The rate-determining step

Equation 20E.8 shows that when $k_b \gg k_a$ the formation of the final product P depends on only the *smaller* of the two rate constants. That is, the rate of formation of P depends on the rate at

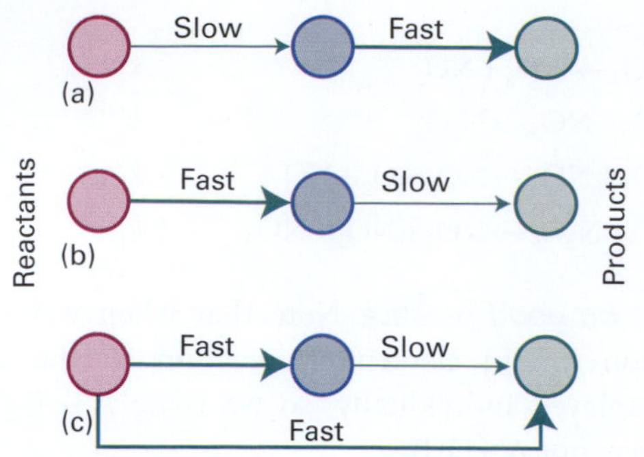

Figure 20E.4 In these diagrams of reaction schemes, heavy arrows represent fast steps and light arrows represent slow steps. (a) The first step is rate-determining; (b) the second step is rate-determining; (c) although one step is slow, it is not rate-determining because there is a fast route that circumvents it.

which I is formed, not on the rate at which I changes into P. For this reason, the step $A \rightarrow I$ is called the 'rate-determining step' of the reaction. Its existence has been likened to building a six-lane highway up to a single-lane bridge: the traffic flow is governed by the rate of crossing the bridge. Similar remarks apply to more complicated reaction mechanisms. In general, the **rate-determining step** (RDS) is the slowest step in a mechanism and controls the overall rate of the reaction. The rate-determining step is not just the slowest step: it must be slow *and* be a crucial gateway for the formation of products. If a faster reaction can also lead to products, then the slowest step is irrelevant because the slow reaction can then be sidestepped (Fig. 20E.4).

The rate law of a reaction that has a rate-determining step can often—but certainly not always—be written down almost by inspection. If the first step in a mechanism is rate-determining, then the rate of the overall reaction is equal to the rate of the first step because all subsequent steps are so fast that once the first intermediate is formed it results immediately in the formation of products. Figure 20E.5 shows the reaction profile for a mechanism of this kind in which the slowest step is the one with the highest activation energy. Once over the initial barrier, the intermediates cascade into products. However, a rate-determining step may also stem from the low concentration of a crucial reactant and need not correspond to the step with highest activation barrier.

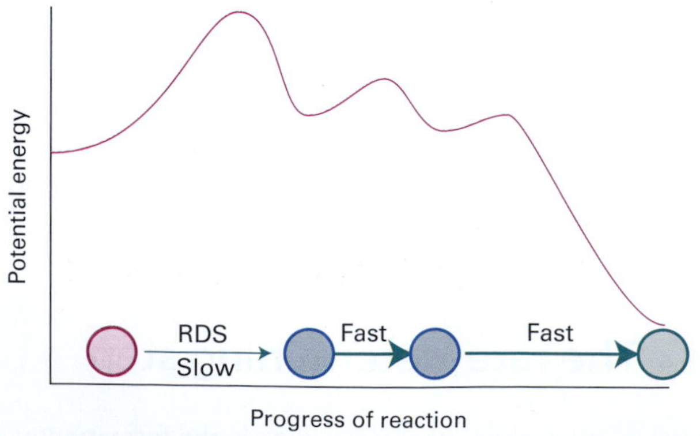

Figure 20E.5 The reaction profile for a mechanism in which the first step (RDS) is rate-determining.

Brief illustration 20E.2 The rate law of a mechanism with a rate-determining step

The oxidation of NO to NO_2, $2\,NO(g) + O_2(g) \rightarrow 2\,NO_2(g)$, proceeds by the following mechanism:

$$NO + NO \rightarrow N_2O_2 \qquad k_a$$

$$N_2O_2 \rightarrow NO + NO \qquad k_a'$$

$$N_2O_2 + O_2 \rightarrow NO_2 + NO_2 \qquad k_b$$

with rate law (see the *Self-test*)

$$\frac{d[NO_2]}{dt} = \frac{2k_a k_b [NO]^2[O_2]}{k_a' + k_b[O_2]}$$

When the concentration of O_2 in the reaction mixture is so large that the third step is very fast in the sense that $[O_2]k_b \gg k_a'$, then the rate law simplifies to

$$\frac{d[NO_2]}{dt} = 2k_a[NO]^2$$

and the formation of N_2O_2 in the first step is rate-determining. We could have written the rate law by inspection of the mechanism, because the rate law for the overall reaction is simply the rate law of that rate-determining step.

Self-test 20E.4 Verify that application of the steady-state approximation to the intermediate N_2O_2 results in the rate law.

20E.5 Pre-equilibria

From a simple sequence of consecutive reactions we now turn to a slightly more complicated mechanism in which an intermediate I reaches an equilibrium with the reactants A and B:

$$A + B \rightleftharpoons I \rightarrow P \qquad \text{Pre-equilibrium} \qquad (20E.9)$$

The rate constants are k_a and k_a' for the forward and reverse reactions of the equilibrium and k_b for the final step. This scheme involves a **pre-equilibrium**, in which an intermediate is in equilibrium with the reactants. A pre-equilibrium can arise when the rate of decay of the intermediate back into reactants is much faster than the rate at which it forms products; thus, the condition is possible when $k_a' \gg k_b$ but not when $k_b \gg k_a'$. Because we assume that A, B, and I are in equilibrium, we can write

$$K = \frac{[I]}{[A][B]} \quad \text{with} \quad K = \frac{k_a}{k_a'} \qquad (20E.10)$$

In writing these equations, we are presuming that the rate of reaction of I to form P is too slow to affect the maintenance of the pre-equilibrium (see the following *Example*). We are also ignoring the fact, as is commonly done, that the standard concentration $c^{\ominus}$ should appear in the expression for K to ensure that it is dimensionless. The rate of formation of P may now be written:

$$\frac{d[P]}{dt}=k_b[I]=k_bK[A][B] \tag{20E.11}$$

This rate law has the form of a second-order rate law with a composite rate constant:

$$\frac{d[P]}{dt}=k_r[A][B] \quad \text{with} \quad k_r=k_bK=\frac{k_ak_b}{k_a'} \tag{20E.12}$$

Example 20E.3 Analysing a pre-equilibrium

Repeat the pre-equilibrium calculation but without ignoring the fact that I is slowly leaking away as it forms P.

Method Begin by writing the net rates of change of the concentrations of the substances and then invoke the steady-state approximation for the intermediate I. Use the resulting expression to obtain the rate of change of the concentration of P.

Answer The net rates of change of P and I are

$$\frac{d[P]}{dt}=k_b[I]$$

$$\frac{d[I]}{dt}=k_a[A][B]-k_a'[I]-k_b[I]\approx 0$$

The second equation solves to

$$[I]\approx\frac{k_a[A][B]}{k_a'+k_b}$$

When we substitute this result into the expression for the rate of formation of P, we obtain

$$\frac{d[P]}{dt}\approx k_r[A][B] \quad \text{with} \quad k_r=\frac{k_ak_b}{k_a'+k_b}$$

This expression reduces to that in eqn 20E.12 when the rate constant for the decay of I into products is much smaller than that for its decay into reactants, $k_b \ll k_a'$.

Self-test 20E.5 Show that the pre-equilibrium mechanism in which $2\,A \rightleftharpoons I$ (K) followed by $I+B \rightarrow P$ (k_b) results in an overall third-order reaction.

Answer: $d[P]/dt=k_bK[A]^2[B]$

One feature to note is that although each of the rate constants in eqn 20E.12 increases with temperature, that might not

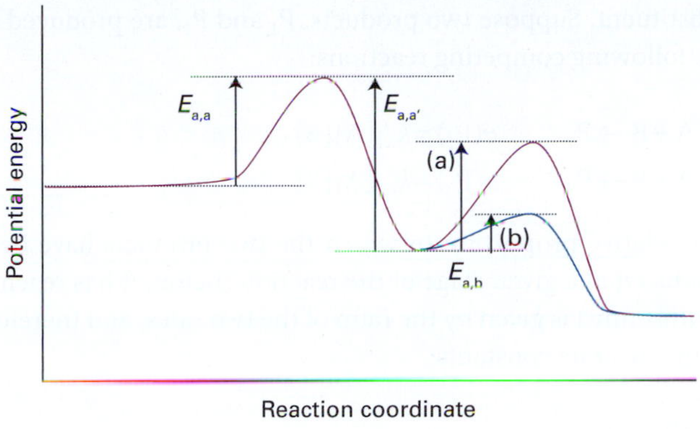

Figure 20E.6 For a reaction with a pre-equilibrium, there are three activation energies to take into account: two referring to the reversible steps of the pre-equilibrium and one for the final step. The relative magnitudes of the activation energies determine whether the overall activation energy is (a) positive or (b) negative.

be true of k_r itself. Thus, if the rate constant k_a' increases more rapidly than the product k_ak_b increases, then $k_r=k_ak_b/k_a'$ will decrease with increasing temperature and the reaction will go more slowly as the temperature is raised. Mathematically, we would say that the composite reaction had a 'negative activation energy'. For example, suppose that each rate constant in eqn 20E.12 exhibits an Arrhenius temperature dependence (Topic 20D). It follows from the Arrhenius equation (eqn 20D.4, $k_r=Ae^{-E_a/RT}$) that

$e^{x+y}=e^xe^y$
$e^{x+y}=e^x/e^y$

$$k_r=\frac{(A_ae^{-E_{a,a}/RT})(A_be^{-E_{a,b}/RT})}{A_{a'}e^{-E_{a,a'}/RT}} \stackrel{\cdot}{=} \frac{A_aA_b}{A_{a'}}e^{-(E_{a,a}+E_{a,b}-E_{a,a'})/RT}$$

The effective activation energy of the reaction is therefore

$$E_a=E_{a,a}+E_{a,b}-E_{a,a'} \tag{20E.13}$$

This activation energy is positive if $E_{a,a}+E_{a,b}>E_{a,a'}$ (Fig. 20E.6a) but negative if $E_{a,a'}>E_{a,a}+E_{a,b}$ (Fig. 20E.6b). An important consequence of this discussion is that we have to be very cautious about making predictions about the effect of temperature on reactions that are the outcome of several steps.

20E.6 Kinetic and thermodynamic control of reactions

In some cases reactants can give rise to a variety of products, as in nitrations of mono-substituted benzene, when various proportions of the *ortho*-, *meta*-, and *para*-substituted products are obtained, depending on the directing power of the original substituent. Suppose two products, P_1 and P_2, are produced by the following competing reactions:

$$A+B \rightarrow P_1 \qquad v(P_1)=k_{r,1}[A][B]$$
$$A+B \rightarrow P_2 \qquad v(P_2)=k_{r,2}[A][B]$$

The relative proportion in which the two products have been produced at a given stage of the reaction (before it has reached equilibrium) is given by the ratio of the two rates, and therefore of the two rate constants:

$$\frac{[P_2]}{[P_1]}=\frac{k_{r,2}}{k_{r,1}} \quad \text{Kinetic control} \quad (20E.14)$$

This ratio represents the **kinetic control** over the proportions of products, and is a common feature of the reactions encountered in organic chemistry where reactants are chosen that facilitate pathways favouring the formation of a desired product. If a reaction is allowed to reach equilibrium, then the proportion of products is determined by thermodynamic rather than kinetic considerations, and the ratio of concentrations is controlled by considerations of the standard Gibbs energies of all the reactants and products.

Brief illustration 20E.3 The outcome of kinetic control

Consider two products formed from reactant R in reactions for which: (a) product P_1 is thermodynamically more stable than product P_2; and (b) the activation energy E_a for the reaction leading to P_2 is greater than that leading to P_1. It follows from eqn 20E.14 and the Arrhenius equation ($k_r = Ae^{-E_a/RT}$, eqn 20D.4) that the ratio of products is

$$\frac{[P_2]}{[P_1]}=\frac{k_2}{k_1}=\frac{A_2e^{-E_{a,2}/RT}}{A_1e^{-E_{a,1}/RT}}=\frac{A_2}{A_1}e^{-(E_{a,2}-E_{a,1})/RT}=\frac{A_2}{A_1}e^{-\Delta E_a/RT}$$

Because $\Delta E_a = E_{a,2} - E_{a,1} > 0$, as T increases,

- the term $\Delta E_a/RT$ decreases, and
- the term $e^{-\Delta E_a/RT}$ increases.

Consequently, the ratio $[P_2]/[P_1]$ increases with increasing temperature before equilibrium is reached.

Self-test 20E.6 Consider the reactions from *Brief illustration* 20E.3. Derive an expression for the ratio $[P_2]/[P_1]$ when the reaction is under thermodynamic control. State your assumptions.

Answer: $[P_2]/[P_1]=e^{-(\Delta_r G_2^{\ominus}-\Delta_r G_1^{\ominus})/RT}$, assuming that activities can be replaced by concentrations

Checklist of concepts

- ☐ 1. The **mechanism** of reaction is the sequence of elementary steps that leads from reactants to products.
- ☐ 2. The **molecularity** of an elementary reaction is the number of molecules coming together to react.
- ☐ 3. An elementary unimolecular reaction has first-order kinetics; an elementary bimolecular reaction has second-order kinetics.
- ☐ 4. The **rate-determining step** is the slowest step in a reaction mechanism that controls the rate of the overall reaction.
- ☐ 5. In the **steady-state approximation**, it is assumed that the concentrations of all reaction intermediates remain constant and small throughout the reaction.
- ☐ 6. **Pre-equilibrium** is a state in which an intermediate is in equilibrium with the reactants and which arises when the rates of formation of the intermediate and its decay back into reactants are much faster than its rate of formation of products.
- ☐ 7. Provided a reaction has not reached equilibrium, the products of competing reactions are controlled by kinetics.

Checklist of equations

Property	Equation	Comment	Equation number
Unimolecular reaction	$d[A]/dt=-k_r[A]$	$A \rightarrow P$	20E.1
Bimolecular reaction	$d[A]/dt=-k_r[A][B]$	$A+B \rightarrow P$	20E.2
Consecutive reactions	$[A]=[A]_0e^{-k_at}$ $[I]=(k_a/(k_b-k_a))(e^{-k_at}-e^{-k_bt})[A]_0$ $[P]=\{1+(k_ae^{-k_bt}-k_be^{-k_at})/(k_b-k_a)\}[A]_0$	$A \xrightarrow{k_a} I \xrightarrow{k_b} P$	20E.4
Steady-state approximation	$d[I]/dt \approx 0$	I is an intermediate	20E.5

20F Examples of reaction mechanisms

Contents

➤ **Why do you need to know this material?**

Some important reactions have complex mechanisms and need special treatment, so you need to see how to make and implement assumptions about the relative rates of the steps in a mechanism.

➤ **What is the key idea?**

The steady-state approximation can often be used to derive rate laws for proposed mechanisms.

➤ **What do you need to know already?**

You need to be familiar with the concept of rate laws (Topic 20A) and the steady-state approximation (Topic 20E).

Many reactions take place by mechanisms that involve several elementary steps. We focus here on the kinetic analysis of a special class of reactions in the gas phase and polymerization kinetics. Photochemical processes are treated in Topic 20G and the role of catalysis in Topics 20H and 22C.

20F.1 Unimolecular reactions

A number of gas-phase reactions follow first-order kinetics, as in the isomerization of cyclopropane:

$$cyclo\text{-}C_3H_6(g) \rightarrow CH_3CH{=}CH_2(g) \quad v = k_r[cyclo\text{-}C_3H_6]$$

The problem with the interpretation of first-order rate laws is that presumably a molecule acquires enough energy to react as a result of its collisions with other molecules. However, collisions are simple bimolecular events, so how can they result in a first-order rate law? First-order gas-phase reactions are widely called 'unimolecular reactions' because they also involve an elementary unimolecular step in which the reactant molecule changes into the product. This term must be used with caution, however, because the overall mechanism has bimolecular as well as unimolecular steps.

The first successful explanation of unimolecular reactions was provided by Frederick Lindemann in 1921 and then elaborated by Cyril Hinshelwood. In the **Lindemann–Hinshelwood mechanism** it is supposed that a reactant molecule A becomes energetically excited by collision with another A molecule in a bimolecular step (Fig. 20F.1):

$$A + A \rightarrow A^* + A \qquad \frac{d[A^*]}{dt} = k_a[A]^2 \qquad (20F.1a)$$

The energized molecule (A*) might lose its excess energy by collision with another molecule:

$$A + A^* \rightarrow A + A \qquad \frac{d[A^*]}{dt} = -k_a'[A][A^*] \qquad (20F.1b)$$

Alternatively, the excited molecule might shake itself apart and form products P. That is, it might undergo the unimolecular decay

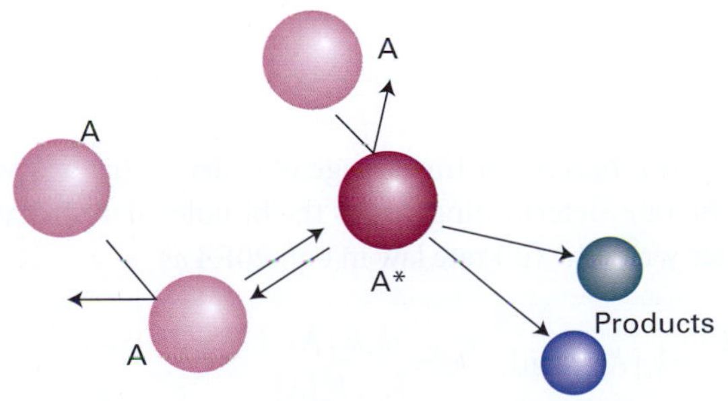

Figure 20F.1 A representation of the Lindemann–Hinshelwood mechanism of unimolecular reactions. The species A is excited by collision with A, and the excited A molecule (A*) may either be deactivated by a collision with A or go on to decay by a unimolecular process to form products.

$$A^* \to P \qquad \frac{d[A^*]}{dt} = -k_b[A^*] \qquad (20F.1c)$$

If the unimolecular step is slow enough to be the rate-determining step, the overall reaction will have first-order kinetics, as observed. This conclusion can be demonstrated explicitly by applying the steady-state approximation to the net rate of formation of A*:

$$\frac{d[A^*]}{dt} = k_a[A]^2 - k_a'[A][A^*] - k_b[A^*] \approx 0 \qquad (20F.2)$$

This equation solves to

$$[A^*] = \frac{k_a[A]^2}{k_b + k_a'[A]} \qquad (20F.3)$$

so the rate law for the formation of P is

$$\frac{d[P]}{dt} = k_b[A^*] = \frac{k_a k_b [A]^2}{k_b + k_a'[A]} \qquad (20F.4)$$

At this stage the rate law is not first-order. However, if the rate of deactivation by (A*,A) collisions is much greater than the rate of unimolecular decay, in the sense that $k_a'[A][A^*] \gg k_b[A^*]$, or (after cancelling the $[A^*]$), $k_a'[A] \gg k_b$, then we can neglect k_b in the denominator and obtain

$$\frac{d[P]}{dt} = k_r[A] \quad \text{with} \quad k_r = \frac{k_a k_b}{k_a'} \qquad \text{Lindemann–Hinshelwood rate law} \qquad (20F.5)$$

Equation 20F.5 is a first-order rate law, as we set out to show.

The Lindemann–Hinshelwood mechanism can be tested because it predicts that, as the concentration (and therefore the partial pressure) of A is reduced, the reaction should switch to overall second-order kinetics. Thus, when $k_a'[A] \ll k_b$, the rate law in eqn 20F.4 becomes

$$\frac{d[P]}{dt} = k_a[A]^2 \qquad (20F.6)$$

The physical reason for the change of order is that at low pressures the rate-determining step is the bimolecular formation of A*. If we write the full rate law in eqn 20F.4 as

$$\frac{d[P]}{dt} = k_r[A] \quad \text{with} \quad k_r = \frac{k_a k_b [A]}{k_b + k_a'[A]} \qquad (20F.7)$$

then the expression for the effective rate constant, k_r, can be rearranged (by inverting each side) to

$$\frac{1}{k_r} = \frac{k_a'}{k_a k_b} + \frac{1}{k_a[A]} \qquad \textit{Lindemann–Hinshelwood mechanism} \qquad \text{Effective rate constant} \qquad (20F.8)$$

Hence, a test of the theory is to plot $1/k_r$ against $1/[A]$, and to expect a straight line. This behaviour is observed often at low concentrations but deviations are common at high concentrations. In Topic 21A we develop the description of the mechanism to take into account experimental results over a range of concentrations and pressures.

Example 20F.1 Analysing the Lindemann–Hinshelwood mechanism

At 300 K the effective rate constant for a gaseous reaction A → P, which has a Lindemann–Hinshelwood mechanism, is $k_{r,1} = 2.50 \times 10^{-4}\,s^{-1}$ at $[A]_1 = 5.21 \times 10^{-4}\,mol\,dm^{-3}$ and $k_{r,2} = 2.10 \times 10^{-5}\,s^{-1}$ at $[A]_2 = 4.81 \times 10^{-6}\,mol\,dm^{-3}$. Calculate the rate constant for the activation step in the mechanism.

Method Use eqn 20F.8 to write an expression for the difference $1/k_{r,2} - 1/k_{r,1}$ and then use the data to solve for k_a, the rate constant for the activation step.

Answer It follows from eqn 20F.8 that

$$\frac{1}{k_{r,2}} - \frac{1}{k_{r,1}} = \frac{1}{k_a}\left(\frac{1}{[A]_2} - \frac{1}{[A]_1}\right)$$

and so

$$k_a = \frac{1/[A]_2 - 1/[A]_1}{1/k_{r,2} - 1/k_{r,1}}$$

$$= \frac{1/(4.81\times10^{-6}\,mol\,dm^{-3}) - 1/(5.21\times10^{-4}\,mol\,dm^{-3})}{1/(2.10\times10^{-5}\,s^{-1}) - 1/(2.50\times10^{-4}\,s^{-1})}$$

$$= 4.72\,dm^3\,mol^{-1}\,s^{-1}$$

Self-test 20F.1 The effective rate constants for a gaseous reaction A → P, which has a Lindemann–Hinshelwood mechanism, are $1.70 \times 10^{-3}\,s^{-1}$ and $2.20 \times 10^{-4}\,s^{-1}$ at $[A] = 4.37 \times 10^{-4}\,mol\,dm^{-3}$ and $1.00 \times 10^{-5}\,mol\,dm^{-3}$, respectively. Calculate the rate constant for the activation step in the mechanism.

Answer: $24.6\,dm^3\,mol^{-1}\,s^{-1}$

20F.2 Polymerization kinetics

There are two major classes of polymerization processes and the average molar mass of the product varies with time in distinctive ways. In **stepwise polymerization** any two monomers present in the reaction mixture can link together at any time and growth of the polymer is not confined to chains that are already forming (Fig. 20F.2). As a result, monomers are consumed early in the reaction and, as we shall see, the average molar mass of the product grows linearly with time. In **chain polymerization** an activated monomer, M, attacks another monomer, links to it, then that unit attacks another monomer, and so on. The monomer is used up as it becomes linked to

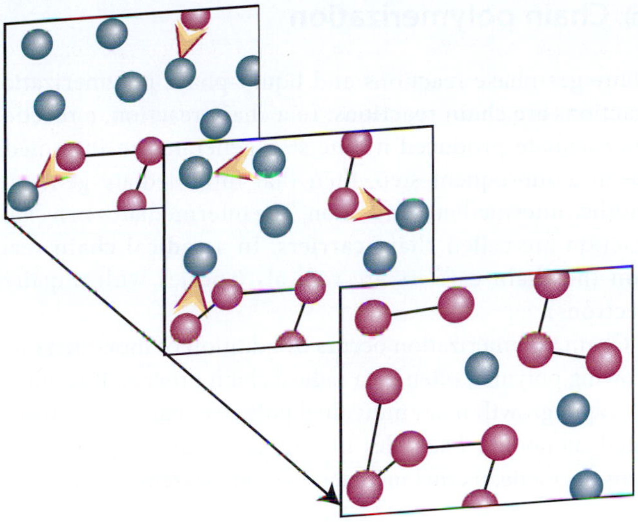

Figure 20F.2 In stepwise polymerization, growth can start at any pair of monomers (in green), and so new chains (in purple) begin to form throughout the reaction.

the growing chains (Fig. 20F.3). High polymers are formed rapidly and only the yield, not the average molar mass, of the polymer is increased by allowing long reaction times.

(a) Stepwise polymerization

Stepwise polymerization commonly proceeds by a **condensation reaction**, in which a small molecule (typically H_2O) is eliminated in each step. Stepwise polymerization is the mechanism of production of polyamides, as in the formation of nylon-66:

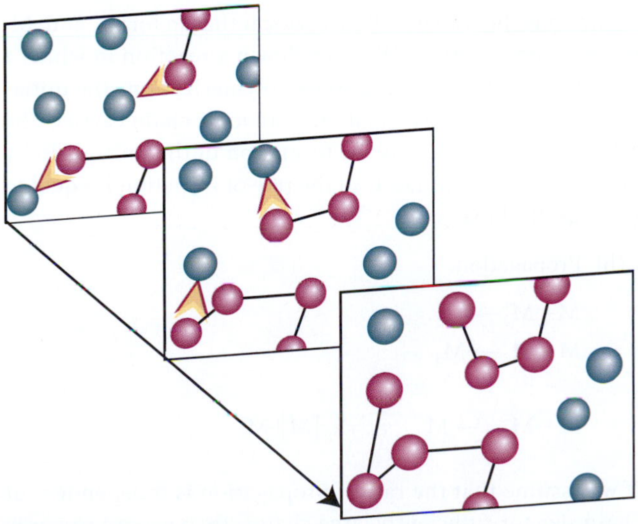

Figure 20F.3 The process of chain polymerization. Chains (in purple) grow as each chain acquires additional monomers (in green).

$$
\begin{aligned}
&H_2N(CH_2)_6NH_2 + HOOC(CH_2)_4COOH \rightarrow \\
&\quad H_2N(CH_2)_6NHCO(CH_2)_4COOH + H_2O \\
&\xrightarrow{\text{continuing to}} H{-}[HN(CH_2)_6NHCO(CH_2)_4CO]_n{-}OH
\end{aligned}
$$

Polyesters and polyurethanes are formed similarly (the latter without elimination). A polyester, for example, can be regarded as the outcome of the stepwise condensation of a hydroxyacid HO–R–COOH. We shall consider the formation of a polyester from such a monomer, and measure its progress in terms of the concentration of the –COOH groups in the sample (which we denote A), for these groups gradually disappear as the condensation proceeds. Because the condensation reaction can occur between molecules containing any number of monomer units, chains of many different lengths can grow in the reaction mixture.

In the absence of a catalyst, we can expect the condensation to be overall second-order in the concentration of the –OH and –COOH (or A) groups, and write

$$\frac{d[A]}{dt} = -k_r[OH][A] \tag{20F.9a}$$

However, because there is one –OH group for each –COOH group, this equation is the same as

$$\frac{d[A]}{dt} = -k_r[A]^2 \tag{20F.9b}$$

If we assume that the rate constant for the condensation is independent of the chain length, then k_r remains constant throughout the reaction. The solution of this rate law is given by eqn 20B.4, and is

$$[A] = \frac{[A]_0}{1 + k_r t[A]_0} \tag{20F.10}$$

The fraction, p, of –COOH groups that have condensed at time t is, after application of eqn 20F.10:

$$p = \frac{[A]_0 - [A]}{[A]_0} = \frac{k_r t[A]_0}{1 + k_r t[A]_0} \quad \textit{Stepwise polymerization} \quad \text{Fraction of condensed groups} \tag{20F.11}$$

Next, we calculate the **degree of polymerization**, which is defined as the average number of monomer residues per polymer molecule. This quantity is the ratio of the initial concentration of A, $[A]_0$, to the concentration of end groups, [A], at the time of interest, because there is one A group per polymer molecule. For example, if there were initially 1000 A groups and there are now only 10, each polymer must be 100 units long on average. Because we can express [A] in terms of p (the first part of eqn 20F.11), the average number of monomers per polymer molecule, $\langle N \rangle$, is

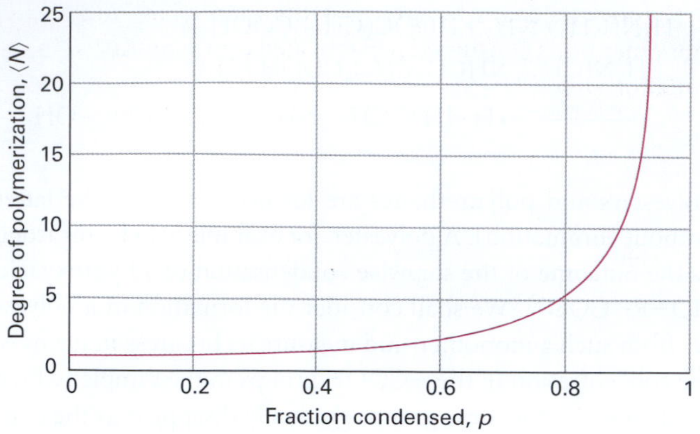

Figure 20F.4 The average chain length of a polymer as a function of the fraction of reacted monomers, p. Note that p must be very close to 1 for the chains to be long.

$$\langle N\rangle=\frac{[A]_0}{[A]}=\frac{1}{1-p} \qquad \textit{Stepwise polymerization} \qquad \text{Degree of polymerization} \qquad (20F.12a)$$

This result is illustrated in Fig. 20F.4. When we express p in terms of the rate constant k_r (the second part of eqn 20F.11), we find

$$\langle N\rangle=1+k_r t[A]_0 \qquad \textit{Stepwise polymerization} \qquad \text{Degree of polymerization in terms of the rate constant} \qquad (20F.12b)$$

The average length grows linearly with time. Therefore, the longer a stepwise polymerization proceeds, the higher the average molar mass of the product.

Brief illustration 20F.1 The degree of polymerization

Consider a polymer formed by a stepwise process with $k_r=1.00\,dm^3\,mol^{-1}\,s^{-1}$ and an initial monomer concentration of $[A]_0=4.00\times10^{-3}\,mol\,dm^{-3}$. From eqn 20F.12b, the degree of polymerization at $t=1.5\times10^4\,s$ is

$$\langle N\rangle=1+(1.00\,dm^3\,mol^{-1}\,s^{-1})\times(1.5\times10^4\,s)\times(4.00\times10^{-3}\,mol\,dm^{-3})=61$$

From eqn 20F.12a, the fraction condensed, p, is

$$p=\frac{\langle N\rangle-1}{\langle N\rangle}=\frac{61-1}{61}=0.98$$

Self-test 20F.2 Calculate the fraction condensed and the degree of polymerization at $t=1.0\,h$ of a polymer formed by a stepwise process with $k_r=1.80\times10^{-2}\,dm^3\,mol^{-1}\,s^{-1}$ and an initial monomer concentration of $3.00\times10^{-2}\,mol\,dm^{-3}$.

Answer: $\langle N\rangle=2.9$; $p=0.66$

(b) Chain polymerization

Many gas-phase reactions and liquid-phase polymerization reactions are **chain reactions**. In a chain reaction, a reaction intermediate produced in one step generates an intermediate in a subsequent step, then that intermediate generates another intermediate, and so on. The intermediates in a chain reaction are called **chain carriers.** In a **radical chain reaction** the chain carriers are radicals (species with unpaired electrons).

Chain polymerization occurs by addition of monomers to a growing polymer, often by a radical chain process. It results in the rapid growth of an individual polymer chain for each activated monomer. Examples include the addition polymerizations of ethene, methyl methacrylate, and styrene, as in

$$-CH_2CH_2X\cdot+CH_2{=}CHX\rightarrow-CH_2CHXCH_2CHX\cdot$$

and subsequent reactions. The central feature of the kinetic analysis (which is summarized in the following *Justification*) is that the rate of polymerization is proportional to the square root of the initiator, In, concentration:

$$v=k_r[In]^{1/2}[M] \qquad \textit{Chain polymerization} \qquad \text{Rate of polymerization} \qquad (20F.13)$$

Justification 20F.1 The rate of chain polymerization

There are three basic types of reaction step in a chain polymerization process:

(a) Initiation:

$$In\rightarrow R\cdot+R\cdot \qquad v_i=k_i[In]$$
$$M+R\cdot\rightarrow\cdot M_1 \qquad \text{(fast)}$$

where In is the initiator, R· the radical that In forms, and $\cdot M_1$ is a monomer radical. We have shown a reaction in which a radical is produced, but in some polymerizations the initiation step leads to the formation of an ionic chain carrier. The rate-determining step is the formation of the radicals R· by homolysis of the initiator, so the rate of initiation is equal to the v_i given above.

(b) Propagation:

$$M+\cdot M_1\rightarrow\cdot M_2$$
$$M+\cdot M_2\rightarrow\cdot M_3$$
$$\vdots$$
$$M+\cdot M_{n-1}\rightarrow\cdot M_n \qquad v_p=k_p[M][\cdot M]$$

If we assume that the rate of propagation is independent of chain size for sufficiently large chains, then we can use only the equation given above to describe the propagation process. Consequently, for sufficiently large chains, the rate of propagation is equal to the overall rate of polymerization.

Because this chain of reactions propagates quickly, the rate at which the total concentration of radicals grows is equal to the rate of the rate-determining initiation step. It follows that

$$\left(\frac{d[\cdot M]}{dt}\right)_{\text{production}} = 2f\,k_i[\text{In}]$$

where f is the fraction of radicals R· that successfully initiate a chain.

(c) Termination:

mutual termination: $\cdot M_n + \cdot M_m \rightarrow M_{n+m}$

disproportionation: $\cdot M_n + \cdot M_m \rightarrow M_n + M_m$

chain transfer: $M + \cdot M_n \rightarrow \cdot M + M_n$

In **mutual termination** two growing radical chains combine. In termination by **disproportionation** a hydrogen atom transfers from one chain to another, corresponding to the oxidation of the donor and the reduction of the acceptor. In **chain transfer**, a new chain initiates at the expense of the one currently growing.

Here we suppose that only mutual termination occurs. If we assume that the rate of termination is independent of the length of the chain, the rate law for termination is

$$v_t = k_t[\cdot M]^2$$

and the rate of change of radical concentration by this process is

$$\left(\frac{d[\cdot M]}{dt}\right)_{\text{depletion}} = -2k_t[\cdot M]^2$$

The steady-state approximation gives:

$$\frac{d[\cdot M]}{dt} = 2f\,k_i[\text{In}] - 2k_t[\cdot M]^2 \approx 0$$

The steady-state concentration of radical chains is therefore

$$[\cdot M] = \left(\frac{f\,k_i}{k_t}\right)^{1/2}[\text{In}]^{1/2}$$

Because the rate of propagation of the chains is the negative of the rate at which the monomer is consumed, we can write $v_p = -d[M]/dt$ and

$$v_p = k_p[\cdot M][M] = k_p\left(\frac{f\,k_i}{k_t}\right)^{1/2}[\text{In}]^{1/2}[M]$$

This rate is also the rate of polymerization, which has the form of eqn 20F.13.

The **kinetic chain length**, λ, is the ratio of the number of monomer units consumed per activated centre produced in the initiation step:

$$\lambda = \frac{\text{number of monomer units consumed}}{\text{number of activated centres produced}}$$

Definition Kinetic chain length (20F.14a)

The kinetic chain length can be expressed in terms of the rate expressions in *Justification* 20F.1. To do so, we recognize that monomers are consumed at the rate that chains propagate. Then,

$$\lambda = \frac{\text{rate of propagation of chains}}{\text{rate of production of radicals}}$$

Kinetic chain length in terms of reaction rates (20F.14b)

By making the steady-state approximation, we set the rate of production of radicals equal to the termination rate. Therefore, we can write the expression for the kinetic chain length as

$$\lambda = \frac{k_p[\cdot M][M]}{2k_t[\cdot M]^2} = \frac{k_p[M]}{2k_t[\cdot M]}$$

When we substitute the steady-state expression, $[\cdot M] = (fk_i/k_t)^{1/2}[\text{In}]^{1/2}$, for the radical concentration, we obtain

$$\lambda = k_r[M][\text{In}]^{-1/2} \quad \text{with } k_r = k_p(fk_ik_t)^{-1/2}$$

Chain polymerization Kinetic chain length (20F.15)

Consider a polymer produced by a chain mechanism with mutual termination. In this case, the average number of monomers in a polymer molecule, $\langle N\rangle$, produced by the reaction is the sum of the numbers in the two combining polymer chains. The average number of units in each chain is λ. Therefore,

$$\langle N\rangle = 2\lambda = 2k_r[M][\text{In}]^{-1/2}$$

Chain polymerization Degree of polymerization (20F.16)

with k_r given in eqn 20F.15. We see that, the slower the initiation of the chain (the smaller the initiator concentration and the smaller the initiation rate constant), the greater is the kinetic chain length, and therefore the higher is the average molar mass of the polymer. Some of the consequences of molar mass for polymers are explored in Topic 17D: here we have seen how we can exercise kinetic control over them.

Checklist of concepts

☐ 1. The **Lindemann–Hinshelwood mechanism** of 'unimolecular' reactions account for the first-order kinetics of some gas-phase reactions.

☐ 2. In **stepwise polymerization** any two monomers in the reaction mixture can link together at any time.

☐ 3. The longer a stepwise polymerization proceeds, the higher the average molar mass of the product.

☐ 4. In **chain polymerization** an activated monomer attacks another monomer and links to it.

☐ 5. The slower the initiation of the chain, the higher the average molar mass of the polymer.

☐ 6. The **kinetic chain length** is the ratio of the number of monomer units consumed per activated centre produced in the initiation step.

Checklist of equations

Property	Equation	Comment	Equation number
Lindemann–Hinshelwood rate law	$d[P]/dt = k_r[A]$ with $k_r = k_a k_b / k_a'$	$k_a'[A] \gg k_b$	20F.5
Effective rate constant	$1/k_r = k_a'/k_a k_b + 1/k_a[A]$	Lindemann–Hinshelwood mechanism	20F.8
Fraction of condensed groups	$p = k_r t[A]_0/(1 + k_r t[A]_0)$	Stepwise polymerization	20F.11
Degree of polymerization	$\langle N \rangle = 1/(1-p) = 1 + k_r t[A]_0$	Stepwise polymerization	20F.12
Rate of polymerization	$v = k_r[In]^{1/2}[M]$	Chain polymerization	20F.13
Kinetic chain length	$\lambda = k_r[M][In]^{-1/2}$, $k_r = k_p(fk_i k_t)^{-1/2}$	Chain polymerization	20F.15
Degree of polymerization	$\langle N \rangle = 2k_r[M][In]^{-1/2}$	Chain polymerization	20F.16

20G Photochemistry

Contents

➤ Why do you need to know this material?

Many chemical and biological processes, including photosynthesis and vision, can be initiated by the absorption of electromagnetic radiation, so you need to know how to include its effect in rate laws. You also need to see how to obtain insight into these processes by the quantitative analysis of their mechanisms.

➤ What is the key idea?

The mechanisms of many photochemical reactions lead to relatively simple rate laws that yield rate constants and quantitative measures of the efficiency with which radiant energy induces reactions.

➤ What do you need to know already?

You need to be familiar with the concepts of singlet and triplet states (Topics 9B and 13B), modes of radiative decay (fluorescence and phosphorescence, Topic 13B), concepts of electronic spectroscopy (Topic 13A), and the formulation of a rate law from a proposed mechanism (Topic 20E).

Photochemical processes are initiated by the absorption of electromagnetic radiation. Among the most important of these processes are those that capture the radiant energy of the Sun. Some of these reactions lead to the heating of the atmosphere during the daytime by absorption of ultraviolet radiation. Others include the absorption of visible radiation during photosynthesis. Without photochemical processes, the Earth would be simply a warm, sterile, rock.

20G.1 Photochemical processes

Table 20G.1 summarizes common photochemical reactions. Photochemical processes are initiated by the absorption of radiation by at least one component of a reaction mixture. In a **primary process**, products are formed directly from the excited state of a reactant. Examples include fluorescence (Topic 13B) and the *cis–trans* photoisomerization of retinal. Products of a **secondary process** originate from intermediates that are

Table 20G.1 Examples of photochemical processes

Process	General form	Example
Ionization	$A^* \rightarrow A^+ + e^-$	$NO^* \xrightarrow{134\,nm} NO^+ + e^-$
Electron transfer	$A^* + B \rightarrow A^+ + B^-$ or $A^- + B^+$	$[Ru(bpy)_3^{2+}]^* + Fe^{3+} \xrightarrow{452\,nm} Ru(bpy)_3^{3+} + Fe^{2+}$
Dissociation	$A^* \rightarrow B + C$	$O_3^* \xrightarrow{1180\,nm} O^2 + O$
	$A^* + B{-}C \rightarrow A + B + C$	$Hg^* + CH_4 \xrightarrow{254\,nm} Hg + CH_3 + H$
Addition	$2\,A^* \rightarrow B$	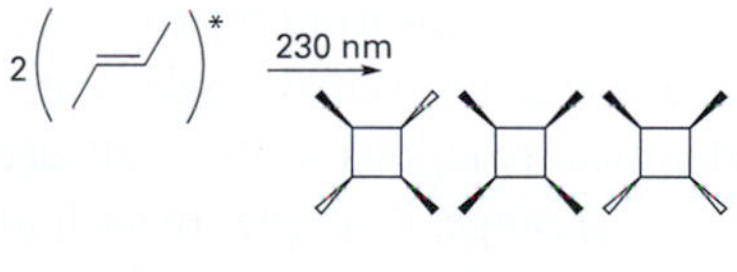
	$A^* + B \rightarrow AB$	
Abstraction	$A^* + B{-}C \rightarrow A{-}B + C$	$Hg^* + H_2 \xrightarrow{254\,nm} HgH + H$
Isomerization or rearrangement	$A^* \rightarrow A'$	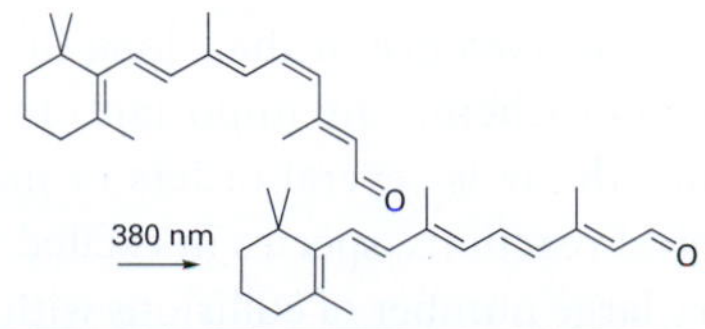

* Excited state.

Table 20G.2 Common photophysical processes

Primary absorption	$S+h\nu \rightarrow S^*$
Excited-state absorption	$S^*+h\nu \rightarrow S^{**}$
	$T^*+h\nu \rightarrow T^{**}$
Fluorescence	$S^* \rightarrow S+h\nu$
Stimulated emission	$S^*+h\nu \rightarrow S+2\,h\nu$
Intersystem crossing (ISC)	$S^* \rightarrow T^*$
Phosphorescence	$T^* \rightarrow S+h\nu$
Internal conversion (IC)	$S^* \rightarrow S$
Collision-induced emission	$S^*+M \rightarrow S+M+h\nu$
Collisional deactivation	$S^*+M \rightarrow S+M$
	$T^*+M \rightarrow S+M$
Electronic energy transfer:	
Singlet–singlet	$S^*+S \rightarrow S+S^*$
Triplet–triplet	$T^*+T \rightarrow T+T^*$
Excimer formation	$S^*+S \rightarrow (SS)^*$
Energy pooling	
Singlet–singlet	$S^*+S^* \rightarrow S^{**}+S$
Triplet–triplet	$T^*+T^* \rightarrow S^{**}+S$

* Denotes an excited state, S a singlet state, T a triplet state, and M is a third-body.

formed directly from the excited state of a reactant, such as oxidative processes initiated by the oxygen atom formed by ozone photodissociation.

Competing with the formation of photochemical products are numerous primary photophysical processes that can deactivate the excited state (Table 20G.2). Therefore, it is important to consider the timescales of the formation and decay of excited states before describing the mechanisms of photochemical reactions.

Electronic transitions caused by absorption of ultraviolet and visible radiation occur within 10^{-16}–10^{-15} s. We expect, then, the upper limit for the rate constant of a first-order photochemical reaction to be about $10^{16}\,s^{-1}$. Fluorescence is slower than absorption, with typical lifetimes of 10^{-12}–10^{-6} s. Therefore, the excited singlet state can initiate very fast photochemical reactions in the femtosecond (10^{-15} s) to picosecond (10^{-12} s) range. Examples of such ultrafast reactions are the initial events of vision and of photosynthesis. Typical intersystem crossing (ISC, Topic 13B) and phosphorescence times for large organic molecules are 10^{-12}–10^{-4} s and 10^{-6}–10^{-1} s, respectively. As a consequence of their long lifetimes, excited triplet states are photochemically important. Indeed, because phosphorescence decay is several orders of magnitude slower than most typical reactions, species in excited triplet states can undergo a very large number of collisions with other reactants before they lose their energy radiatively.

Brief illustration 20G.1 The nature of the excited state

To judge whether the excited singlet or triplet state of the reactant is a suitable product precursor, we compare the emission lifetimes with the time constant for chemical reaction of the reactant, τ (Topic 20B). Consider a unimolecular photochemical reaction with rate constant $k_r = 1.7\times10^4\,s^{-1}$ and therefore time constant $\tau = 1/(1.7\times10^4\,s^{-1}) = 59\,\mu s$ that involves a reactant with an observed fluorescence lifetime of 1.0 ns and an observed phosphorescence lifetime of 1.0 ms. The excited singlet state is too short-lived to be a major source of product in this reaction. On the other hand, the relatively long-lived excited triplet state is a good candidate for a precursor.

Self-test 20G.1 Consider a molecule with a fluorescence lifetime of 10.0 ns that undergoes unimolecular photoisomerization. What approximate value of the half-life would be consistent with the excited singlet state being the product precursor?

Answer: The value of $t_{1/2}$ should be less than about 7 ns

20G.2 The primary quantum yield

The rates of deactivation of the excited state by radiative, non-radiative, and chemical processes determine the yield of product in a photochemical reaction. The **primary quantum yield**, ϕ, is defined as the number of photophysical or photochemical events that lead to primary products divided by the number of photons absorbed by the molecule in the same interval:

$$\phi = \frac{\text{number of events}}{\text{number of photons absorbed}} \quad \textit{Definition} \quad \text{Primary quantum yield} \qquad (20G.1a)$$

When both the numerator and denominator of this expression are divided by the time interval over which the events occur, we see that the primary quantum yield is also the rate of radiation-induced primary events divided by the rate of photon absorption, I_{abs}:

$$\phi = \frac{\text{rate of process}}{\text{rate of photon absorption}} = \frac{v}{I_{abs}} \quad \text{Primary quantum yield in terms of rates of processes} \qquad (20G.1b)$$

Example 20G.1 Calculating a primary quantum yield

In an experiment to determine the quantum yield of a photochemical reaction, the absorbing substance was exposed to 490 nm light from a 100 W source for 2700 s, with 60 per cent of the incident light being absorbed. As a result of irradiation,

0.344 mol of the absorbing substance decomposed. Determine the primary quantum yield.

Method We need to calculate the terms used in eqn 20G.1a. To calculate the number of absorbed photons N_{abs}, which is the denominator of the expression on the right-hand side of eqn 20G.1a, we note that:

- The energy absorbed by the substance is $E_{abs}=fPt$, where P is the incident power, t is the time of exposure, and the factor f (in this case $f=0.60$) is the proportion of incident light that is absorbed.
- E_{abs} is also related to the number N_{abs} of absorbed photons through $E_{abs}=N_{abs}hc/\lambda$, where hc/λ is the energy of a single photon of wavelength λ (eqn 7A.5).

By combining both expressions for the absorbed energy, the value of N_{abs} follows readily. The number of photochemical events, and hence the numerator of the expression on the right-hand side of eqn 20G.1a, is simply the number of decomposed molecules $N_{decomposed}$. The primary quantum yield follows from $\phi=N_{decomposed}/N_{abs}$.

Answer From the expressions for the absorbed energy, it follows that

$$E_{abs}=fPt=N_{abs}\left(\frac{hc}{\lambda}\right)$$

and therefore that $N_{abs}=fPt\lambda/hc$. Now we use eqn 20G.1a to write

$$\phi=\frac{N_{decomposed}}{N_{abs}}=\frac{N_{decomposed}hc}{fPt\lambda}$$

With $N_{decomposed}=(0.344\,\text{mol})\times(6.022\times10^{23}\,\text{mol}^{-1})$, $P=100\,\text{W}=100\,\text{J s}^{-1}$, $t=2700\,\text{s}$, $\lambda=490\,\text{nm}=4.90\times10^{-7}\,\text{m}$, and $f=0.60$ it follows that

$$\phi=\frac{(0.344\,\text{mol})\times(6.022\times10^{23}\,\text{mol}^{-1})\times(6.626\times10^{-34}\,\text{J s})\times(2.998\times10^{8}\,\text{m s}^{-1})}{0.60\times(100\,\text{J s}^{-1})\times(2700\,\text{s})\times(4.90\times10^{-7}\,\text{m})}$$
$$=0.52$$

Self-test 20G.2 In an experiment to measure the quantum yield of a photochemical reaction, the absorbing substance was exposed to 320 nm radiation from a 87.5 W source for 38 min. The intensity of the transmitted light was 0.35 that of the incident light. As a result of irradiation, 0.324 mol of the absorbing substance decomposed. Determine the primary quantum yield.

Answer: $\phi=0.93$

A molecule in an excited state must either decay to the ground state or form a photochemical product. Therefore, the total number of molecules deactivated by radiative processes, non-radiative processes, and photochemical reactions must be equal to the number of excited species produced by absorption of the incident radiation. We conclude that the sum of primary quantum yields ϕ_i for *all* photophysical and photochemical events i must be equal to 1, regardless of the number of reactions involving the excited state:

$$\sum_i \phi_i=\sum_i \frac{v_i}{I_{abs}}=1 \qquad (20G.2)$$

It follows that for an excited singlet state that decays to the ground state only by the photophysical processes described in Section 20G.1 (and without reacting), we write

$$\phi_F+\phi_{IC}+\phi_P=1$$

where ϕ_F, ϕ_{IC}, and ϕ_P are the quantum yields of fluorescence, internal conversion, and phosphorescence, respectively (intersystem crossing from the singlet to the triplet state is taken into account by the presence of ϕ_P). The quantum yield of photon emission by fluorescence and phosphorescence is $\phi_{emission}=\phi_F+\phi_P$, which is less than 1. If the excited singlet state also participates in a primary photochemical reaction with quantum yield ϕ_r, we write

$$\phi_F+\phi_{IC}+\phi_P+\phi_r=1$$

We can now strengthen the link between reaction rates and primary quantum yield already established by eqns 20G.1 and 20G.2. By taking the constant I_{abs} out of the sum in eqn 20G.2 and rearranging, we obtain $I_{abs}=\Sigma_i v_i$. Substituting this result into eqn 20G.2 gives the general result

$$\phi_i=\frac{v_i}{\sum_i v_i} \qquad (20G.3)$$

Therefore, the primary quantum yield may be determined directly from the experimental rates of *all* photophysical and photochemical processes that deactivate the excited state.

20G.3 Mechanism of decay of excited singlet states

Consider the formation and decay of an excited singlet state in the absence of a chemical reaction:

Absorption:	$S+h\nu_i\rightarrow S^*$	$v_{abs}=I_{abs}$
Fluorescence:	$S^*\rightarrow S+h\nu_f$	$v_F=k_F[S^*]$
Internal conversion:	$S^*\rightarrow S$	$v_{IC}=k_{IC}[S^*]$
Intersystem crossing:	$S^*\rightarrow T^*$	$v_{ISC}=k_{ISC}[S^*]$

in which S is an absorbing singlet-state species, S* an excited singlet state, T* an excited triplet state, and $h\nu_i$ and $h\nu_f$ are the energies of the incident and fluorescent photons, respectively. From the methods presented in Topic 20E and the rates of the steps that form and destroy the excited singlet state S*, we write the rate of formation and decay of S* as:

$$\text{Rate of formation of } S^* = I_{abs}$$

$$\begin{aligned}\text{Rate of disappearance of } S^* &= k_F[S^*]+k_{ISC}[S^*]+k_{IC}[S^*]\\ &=(k_F+k_{ISC}+k_{IC})[S^*]\end{aligned}$$

It follows that the excited state decays by a first-order process so, when the light is turned off, the concentration of S* varies with time t as:

$$[S^*](t)=[S^*]_0 e^{-t/\tau_0} \qquad (20G.4)$$

where the **observed lifetime**, τ_0, of the first excited singlet state is defined as:

$$\tau_0 = \frac{1}{k_F+k_{ISC}+k_{IC}} \qquad \textit{Definition} \quad \text{Observed lifetime of the excited singlet state} \qquad (20G.5)$$

We show in the following *Justification* that the quantum yield of fluorescence is

$$\phi_{F,0} = \frac{k_F}{k_F+k_{ISC}+k_{IC}} \qquad \text{Quantum yield of fluorescence} \qquad (20G.6)$$

Justification 20G.1 The quantum yield of fluorescence

Most fluorescence measurements are conducted by illuminating a relatively dilute sample with a continuous and intense beam of light or ultraviolet radiation. It follows that [S*] is small and constant, so we may invoke the steady-state approximation (Topic 20E) and write:

$$\begin{aligned}\frac{d[S^*]}{dt} &= I_{abs}-k_F[S^*]-k_{ISC}[S^*]-k_{IC}[S^*]\\ &= I_{abs}-(k_F+k_{ISC}+k_{IC})[S^*]\approx 0\end{aligned}$$

Consequently,

$$I_{abs}=(k_F+k_{ISC}+k_{IC})[S^*]$$

By using this expression and eqn 20G.1b, we write the quantum yield of fluorescence as:

$$\phi_{F,0} = \frac{v_F}{I_{abs}} = \frac{k_F[S^*]}{(k_F+k_{ISC}+k_{IC})[S^*]}$$

which, by cancelling the [S*], simplifies to eqn 20G.6.

The observed fluorescence lifetime can be measured by using a pulsed laser technique. First, the sample is excited with a short light pulse from a laser using a wavelength at which S absorbs strongly. Then, the exponential decay of the fluorescence intensity after the pulse is monitored. From eqns 20G.5 and 20G.6, it follows that

$$\tau_0 = \frac{1}{k_F+k_{ISC}+k_{IC}} = \frac{k_F}{k_F+k_{ISC}+k_{IC}} \times \frac{1}{k_F} = \frac{\phi_{F,0}}{k_F} \qquad (20G.7)$$

Brief illustration 20G.2 The fluorescence rate constant

The fluorescence quantum yield and observed fluorescence lifetime of tryptophan in water are $\phi_{F,0}=0.20$ and $\tau_0=2.6$ ns, respectively. It follows from eqn 20G.7 that the fluorescence rate constant k_F is

$$k_F = \frac{\phi_{F,0}}{\tau_0} = \frac{0.20}{2.6\times10^{-9}\,\text{s}} = 7.7\times10^7\,\text{s}^{-1}$$

Self-test 20G.3 A substance has a fluorescence quantum yield of $\phi_{F,0}=0.35$. In an experiment to measure the fluorescence lifetime of this substance, it was observed that the fluorescence emission decayed with a half-life of 5.6 ns. Determine the fluorescence rate constant of this substance.

Answer: $k_F=4.3\times10^7\,\text{s}^{-1}$

20G.4 Quenching

The shortening of the lifetime of the excited state by the presence of another species is called **quenching**. Quenching may be either a desired process, such as in energy or electron transfer, or an undesired side reaction that can decrease the quantum yield of a desired photochemical process. Quenching effects may be studied by monitoring the emission from the excited state that is involved in the photochemical reaction.

The addition of a quencher, Q, opens an additional channel for deactivation of S*:

$$\text{Quenching: } S^*+Q\rightarrow S+Q \qquad v_Q = k_Q[Q][S^*]$$

The **Stern–Volmer equation**, which is derived in the following *Justification*, relates the fluorescence quantum yields $\phi_{F,0}$ and ϕ_F measured in the absence and presence, respectively, of a quencher Q at a molar concentration [Q]:

$$\frac{\phi_{F,0}}{\phi_F} = 1+\tau_0 k_Q[Q] \qquad \text{Stern–Volmer equation} \qquad (20G.8)$$

This equation tells us that a plot of $\phi_{F,0}/\phi_F$ against [Q] should be a straight line with slope $\tau_0 k_Q$. Such a plot is called a **Stern–Volmer plot** (Fig. 20G.1). The method may also be applied to the quenching of phosphorescence.

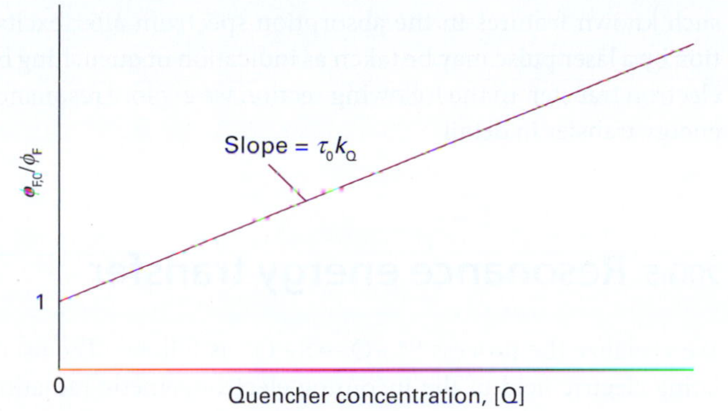

Figure 20G.1 The format of a Stern–Volmer plot and the interpretation of the slope in terms of the rate constant for quenching and the observed fluorescence lifetime in the absence of quenching.

Justification 20G.2 The Stern–Volmer equation

With the addition of quenching, the steady-state approximation for $[S^*]$ now gives:

$$\frac{d[S^*]}{dt} = I_{abs} - (k_F + k_{ISC} + k_{IC} + k_Q[Q])[S^*] \approx 0$$

and the fluorescence quantum yield in the presence of the quencher is:

$$\phi_F = \frac{k_F}{k_F + k_{ISC} + k_{IC} + k_Q[Q]}$$

It follows that

$$\frac{\phi_{F,0}}{\phi_F} = \frac{k_F}{k_F + k_{ISC} + k_{IC}} \times \frac{k_F + k_{ISC} + k_{IC} + k_Q[Q]}{k_F}$$
$$= \frac{k_F + k_{ISC} + k_{IC} + k_Q[Q]}{k_F + k_{ISC} + k_{IC}}$$
$$= 1 + \frac{k_Q}{k_F + k_{ISC} + k_{IC}}[Q]$$

By using eqn 20G.7, this expression simplifies to eqn 20G.8.

Because the fluorescence intensity and lifetime are both proportional to the fluorescence quantum yield (specifically, from eqn 20G.7, $\tau = \phi_F/k_F$), plots of $I_{F,0}/I_F$ and τ_0/τ (where the subscript 0 indicates a measurement in the absence of quencher) against [Q] should also be linear with the same slope and intercept as those shown for eqn 20G.8.

Example 20G.2 Determining the quenching rate constant

The molecule 2,2′-bipyridine (**1**, bpy) forms a complex with the Ru^{2+} ion. Tris-(2,2′-bipyridyl)ruthenium(II), $Ru(bpy)_3^{2+}$ (**2**), has a strong metal-to-ligand charge transfer (MLCT) transition (Topic 13A) at 450 nm.

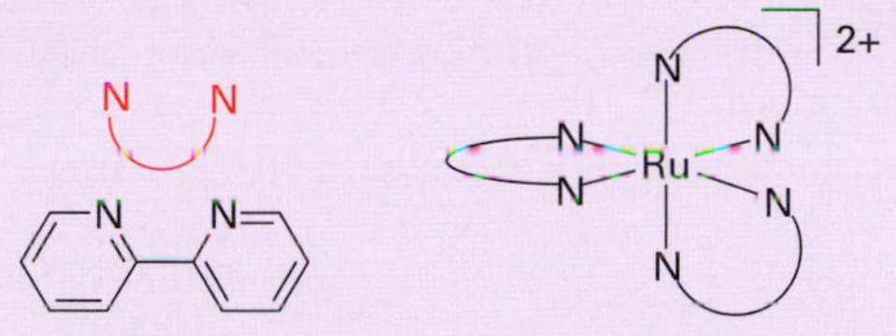

1 2,2′-Bipyridine (bpy) **2** $[Ru(bpy)_3]^{2+}$

The quenching of the $^*Ru(bpy)_3^{2+}$ excited state by Fe^{3+} (present as the complex ion $Fe(OH_2)_6^{3+}$) in acidic solution was monitored by measuring emission lifetimes at 600 nm. Determine the quenching rate constant for this reaction from the following data:

$[Fe(OH_2)_6^{3+}]/(10^{-4}\,mol\,dm^{-3})$	0	1.6	4.7	7	9.4
$\tau/(10^{-7}\,s)$	6	4.05	3.37	2.96	2.17

Method Rewrite the Stern–Volmer equation (eqn 20G.8) for use with lifetime data; then fit the data to a straight-line.

Answer Upon substitution of τ_0/τ for $\phi_{F,0}/\phi_F$ in eqn 20G.8 and after rearrangement, we obtain:

$$\frac{1}{\tau} = \frac{1}{\tau_0} + k_Q[Q]$$

Figure 20G.2 shows a plot of $1/\tau$ against $[Fe^{3+}]$ and the results of a fit to this expression. The slope of the line is 2.8×10^9, so $k_Q = 2.8 \times 10^9\,dm^3\,mol^{-1}\,s^{-1}$. This example shows that measurements of emission lifetimes are preferred because they yield the value of k_Q directly. To determine the value of k_Q from intensity or quantum yield measurements, it is necessary to make an independent measurement of τ_0.

Self-test 20G.4 The quenching of tryptophan fluorescence by dissolved O_2 gas was monitored by measuring

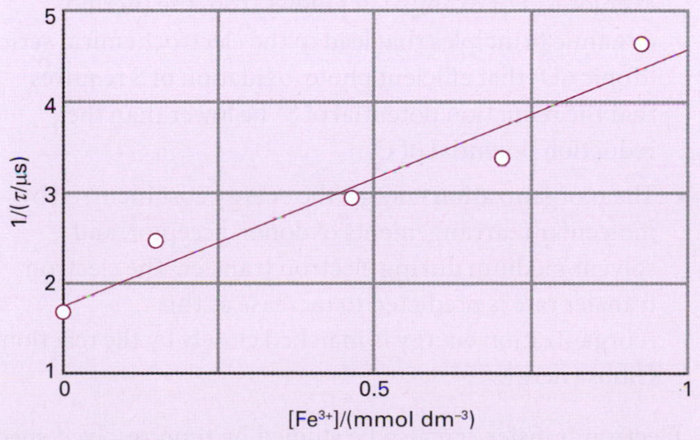

Figure 20G.2 The Stern–Volmer plot of the data for *Example* 20G.2.

emission lifetimes at 348 nm in aqueous solutions. Determine the quenching rate constant for this process from the following data:

$[O_2]/(10^{-2}\,mol\,dm^{-3})$	0	2.3	5.5	8	10.8
$\tau/(10^{-9}\,s)$	2.6	1.5	0.92	0.71	0.57

Answer: $1.3\times10^{10}\,dm^3\,mol^{-1}\,s^{-1}$

Three common mechanisms for bimolecular quenching of an excited singlet (or triplet) state are:

Collisional deactivation:	$S^*+Q\rightarrow S+Q$
Resonance energy transfer:	$S^*+Q\rightarrow S+Q^*$
Electron ransfer:	$S^*+Q\rightarrow S^{+/-}+Q^{-/+}$

The quenching rate constant itself does not give much insight into the mechanism of quenching. For the system of *Example* 20G.2, it is known that the quenching of the excited state of $Ru(bpy)_3^{2+}$ is a result of electron transfer to Fe^{3+}, but the quenching data do not allow us to prove the mechanism.

There are, however, some criteria that govern the relative efficiencies of collisional quenching, resonance energy transfer, and electron transfer. Collisional quenching is particularly efficient when Q is a species, such as iodide ion, which receives energy from S* and then decays to the ground state primarily by releasing energy as heat. As we show in detail in Topic 21E, according to the **Marcus theory** of electron transfer, which was proposed by R.A. Marcus in 1965, the rates of electron transfer (from ground or excited states) depend on:

- The distance between the donor and acceptor, with electron transfer becoming more efficient as the distance between donor and acceptor decreases.
- The reaction Gibbs energy, $\Delta_r G$, with electron transfer becoming more efficient as the reaction becomes more exergonic. For example, it follows from the thermodynamic principles that lead to the electrochemical series (Topic 6D) that efficient photo-oxidation of S requires that the reduction potential of S* be lower than the reduction potential of Q.
- The reorganization energy, the energy cost incurred by molecular rearrangements of donor, acceptor, and solvent medium during electron transfer. The electron transfer rate is predicted to increase as this reorganization energy is matched closely by the reaction Gibbs energy.

Electron transfer can also be studied by time-resolved spectroscopy (Topic 13C). The oxidized and reduced products often have electronic absorption spectra distinct from those of their neutral parent compounds. Therefore, the rapid appearance of such known features in the absorption spectrum after excitation by a laser pulse may be taken as indication of quenching by electron transfer. In the following section we explore resonance energy transfer in detail.

20G.5 Resonance energy transfer

We visualize the process $S^*+Q\rightarrow S+Q^*$ as follows. The oscillating electric field of the incoming electromagnetic radiation induces an oscillating electric dipole moment in S. Energy is absorbed by S if the frequency of the incident radiation, ν, is such that $\nu=\Delta E_S/h$, where ΔE_S is the energy separation between the ground and excited electronic states of S and h is Planck's constant. This is the 'resonance condition' for absorption of radiation (essentially the Bohr frequency condition, eqn 7A.12). The oscillating dipole on S can now affect electrons bound to a nearby Q molecule by inducing an oscillating dipole moment in them. If the frequency of oscillation of the electric dipole moment in S is such that $\nu=\Delta E_Q/h$ then Q will absorb energy from S.

The efficiency, η_T, of resonance energy transfer is defined as

$$\eta_T=1-\frac{\phi_F}{\phi_{F,0}} \quad \textit{Definition} \quad \text{Efficiency of resonance energy transfer} \quad (20G.9)$$

According to the **Förster theory** of resonance energy transfer, energy transfer is efficient when:

- The energy donor and acceptor are separated by a short distance (of the order of nanometres).
- Photons emitted by the excited state of the donor can be absorbed directly by the acceptor.

For donor–acceptor systems held rigidly either by covalent bonds or by a protein 'scaffold', η_T increases with decreasing distance, R, according to

$$\eta_T=\frac{R_0^6}{R_0^6+R^6} \quad \text{Efficiency of energy transfer in terms of the donor–acceptor distance} \quad (20G.10)$$

where R_0 is a parameter (with dimensions of distance) that is characteristic of each donor–acceptor pair. It can be regarded as the distance at which energy transfer is 50 per cent efficient for a given donor–acceptor pair. (You can verify this assertion by using $R=R_0$ in eqn 20G.10.) Equation 20G.10 has been verified experimentally and values of R_0 are available for a number of donor–acceptor pairs (Table 20G.3).

The emission and absorption spectra of molecules span a range of wavelengths, so the second requirement of the Förster theory is met when the emission spectrum of the donor molecule overlaps significantly with the absorption spectrum of the

Table 20G.3 Values of R_0 for some donor–acceptor pairs*

Donor‡	Acceptor‡	R_0/nm
Naphthalene	Dansyl	2.2
Dansyl	ODR	4.3
Pyrene	Coumarin	3.9
1.5-I AEDANS	FITC	4.9
Tryptophan	1.5-I AEDANS	2.2
Tryptophan	Haem (heme)	2.9

*Additional values may be found in J.R. Lacowicz *Principles of fluorescence spectroscopy*, Kluwer Academic/Plenum, New York (1999).

‡Abbreviations:

Dansyl: 5-dimethylamino-1-naphthalenesulfonic acid

FITC: fluorescein 5-isothiocyanate

1.5-I AEDANS: 5-((((2-iodoacetyl)amino)ethyl)amino)naphthalene-1-sulfonic acid

ODR: octadecyl-rhodamine

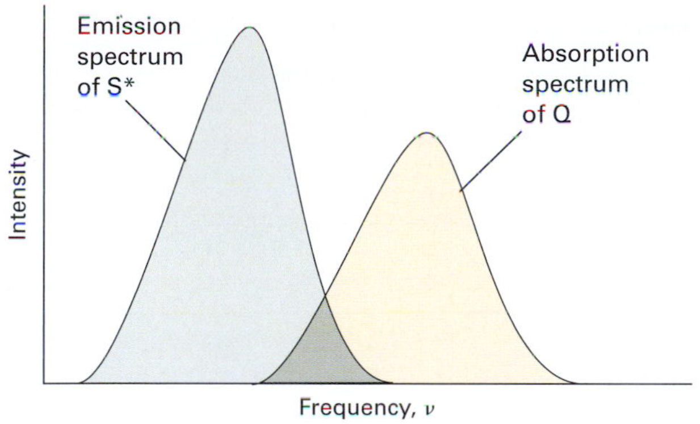

Figure 20G.3 According to the Förster theory, the rate of energy transfer from a molecule S* in an excited state to a quencher molecule Q is optimized at radiation frequencies in which the emission spectrum of S* overlaps with the absorption spectrum of Q, as shown in the (dark green) shaded region.

acceptor. In the overlap region, photons emitted by the donor have the appropriate energy to be absorbed by the acceptor (Fig. 20G.3).

Equation 20G.10 forms the basis of **fluorescence resonance energy transfer** (FRET), in which the dependence of the energy transfer efficiency, η_T, on the distance, R, between energy donor and acceptor is used to measure distances in biological systems. In a typical FRET experiment, a site on a biopolymer or membrane is labelled covalently with an energy donor and another site is labelled covalently with an energy acceptor. In certain cases, the donor or acceptor may be natural constituents of the system, such as amino acid groups, cofactors, or enzyme substrates. The distance between the labels is then calculated from the known value of R_0 and eqn 20G.10. Several tests have shown that the FRET technique is useful for measuring distances ranging from 1 to 9 nm.

Brief illustration 20G.3 The FRET technique

As an illustration of the FRET technique, consider a study of the protein rhodopsin. When an amino acid on the surface of rhodopsin was labelled covalently with the energy donor 1.5-I AEDANS (**3**), the fluorescence quantum yield of the label decreased from 0.75 to 0.68 due to quenching by the visual pigment 11-*cis*-retinal (**4**). From eqn 20G.10, we calculate $\eta_T = 1 - 0.68/0.75 = 0.093$ and from eqn 20G.10 and the known value of $R_0 = 5.4$ nm for the 1.5-I AEDANS/11-*cis*-retinal pair we calculate $R = 7.9$ nm. Therefore, we take 7.9 nm to be the distance between the surface of the protein and 11-*cis*-retinal.

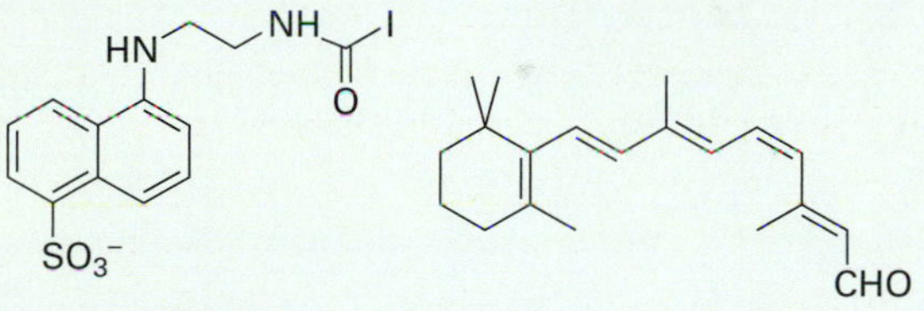

3 1.5-I AEDANS **4** 11-*cis*-Retinal

Self-test 20G.5 An amino acid on the surface of a protein was labelled covalently with 1.5-I AEDANS and another was labelled covalently with FITC. The fluorescence quantum yield of 1.5-I AEDANS decreased by 10 per cent due to quenching by FITC. What is the distance between the amino acids?

Answer: 7.1 nm

If donor and acceptor molecules diffuse in solution or in the gas phase, Förster theory predicts that the efficiency of quenching by energy transfer increases as the average distance travelled between collisions of donor and acceptor decreases. That is, the quenching efficiency increases with concentration of quencher, as predicted by the Stern–Volmer equation.

Checklist of concepts

☐ 1. The **primary quantum yield** of a photochemical reaction is the number of reactant molecules producing specified primary products for each photon absorbed.

☐ 2. The **observed lifetime** of an excited state is related to the quantum yield and rate constant of emission.

☐ 3. A **Stern–Volmer plot** is used to analyse the kinetics of fluorescence quenching in solution.

☐ 4. Collisional deactivation, electron transfer, and resonance energy transfer are common fluorescence quenching processes.

☐ 5. The efficiency of resonance energy transfer decreases with increasing separation between donor and acceptor molecules.

Checklist of equations

Property	Equation	Comment	Equation number
Primary quantum yield	$\phi = v/I_{abs}$		20G.1b
Excited state lifetime	$\tau_0 = 1/(k_F + k_{ISC} + k_{IC})$	No quencher present	20G.5
Quantum yield of fluorescence	$\phi_{F,0} = k_F/(k_F + k_{ISC} + k_{IC})$	Without quencher present	20G.6
Observed excited state lifetime	$\tau_0 = \phi_{F,0}/k_F$		20G.7
Stern–Volmer equation	$\phi_{F,0}/\phi_F = 1 + \tau_0 k_Q[Q]$		20G.8
Efficiency of resonance energy transfer	$\eta_T = 1 - \phi_F/\phi_{F,0}$	Definition	20G.9
	$\eta_T = R_0^6/(R_0^6 + R^6)$	Förster theory	20G.10

20H Enzymes

Contents

➤ Why do you need to know this material?

The role of enzymes in controlling chemical reactions is central to biology and the maintenance of life. It is at the centre of attention of much of the application of physical chemistry to biology.

➤ What is the key idea?

Enzymes are homogeneous catalysts that can have a dramatic effect on the rates of the reactions they control but are subject to inhibition.

➤ What do you need to know already?

You need to be familiar with the analysis of reaction mechanisms in terms of the steady-state approximation (Topic 20E).

A catalyst is a substance that accelerates a reaction but undergoes no net chemical change (Topic 20D): the catalyst lowers the activation energy of the reaction by providing an alternative path that avoids the slow, rate-determining step of the uncatalysed reaction (Fig. 20H.1). **Enzymes**, which are homogeneous biological catalysts, are very specific and can have a dramatic effect on the reactions they control. The enzyme catalase reduces the activation energy from 76 kJ mol^{-1} to 8 kJ mol^{-1}, corresponding to an acceleration of the reaction by a factor of 10^{15} at 298 K.

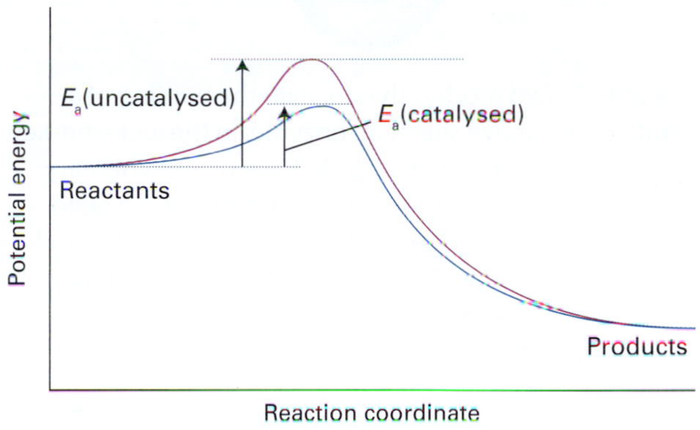

Figure 20H.1 A catalyst provides a different path with a lower activation energy. The result is an increase in the rate of formation of product.

20H.1 Features of enzymes

Enzymes act in the aqueous environment of cells. These biologically ubiquitous compounds are special proteins or nucleic acids that contain an **active site**, which is responsible for binding the **substrates**, the reactants, and processing them into products. As is true of any catalyst, the active site returns to its original state after the products are released. Many enzymes consist primarily of proteins, some featuring organic or inorganic co-factors in their active sites. However, certain RNA molecules can also be biological catalysts, forming *ribozymes*. A very important example of a ribozyme is the *ribosome*, a large assembly of proteins and catalytically active RNA molecules responsible for the synthesis of proteins in the cell.

The structure of the active site is specific to the reaction that it catalyses, with groups in the substrate interacting with groups in the active site by intermolecular interactions, such as hydrogen bonding, electrostatic forces, and van der Waals interactions. Figure 20H.2 shows two models that explain the binding of a substrate to the active site of an enzyme. In the **lock-and-key model**, the active site and substrate have complementary three-dimensional structures and dock without the need for major atomic rearrangements. Experimental evidence favours the **induced fit model**, in which binding of the substrate induces a conformational change in the active site. Only after the change does the substrate fit snugly in the active site.

Enzyme-catalysed reactions are prone to inhibition by molecules that interfere with the formation of product. Many drugs for the treatment of disease function by inhibiting

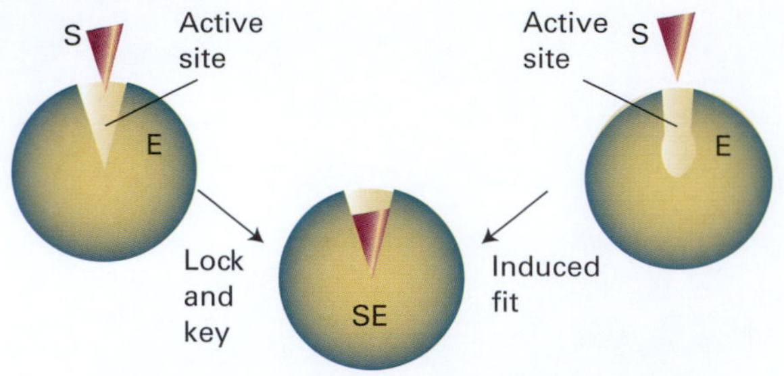

Figure 20H.2 Two models that explain the binding of a substrate to the active site of an enzyme. In the lock-and-key model, the active site and substrate have complementary three-dimensional structures and dock without the need for major atomic rearrangements. In the induced fit model, binding of the substrate induces a conformational change in the active site. The substrate fits well in the active site after the conformational change has taken place.

enzymes. For example, an important strategy in the treatment of acquired immune deficiency syndrome (AIDS) involves the steady administration of a specially designed protease inhibitor. The drug inhibits an enzyme that is key to the formation of the protein envelope surrounding the genetic material of the human immunodeficiency virus (HIV). Without a properly formed envelope, HIV cannot replicate in the host organism.

20H.2 The Michaelis–Menten mechanism

Experimental studies of enzyme kinetics are typically conducted by monitoring the initial rate of product formation in a solution in which the enzyme is present at very low concentration. Indeed, enzymes are such efficient catalysts that significant accelerations may be observed even when their concentration is more than three orders of magnitude smaller than that of the substrate.

The principal features of many enzyme-catalysed reactions are as follows:

- For a given initial concentration of substrate, $[S]_0$, the initial rate of product formation is proportional to the total concentration of enzyme, $[E]_0$.
- For a given $[E]_0$ and low values of $[S]_0$, the rate of product formation is proportional to $[S]_0$.
- For a given $[E]_0$ and high values of $[S]_0$, the rate of product formation becomes independent of $[S]_0$, reaching a maximum value known as the **maximum velocity**, v_{max}.

The **Michaelis–Menten mechanism** accounts for these features. According to this mechanism, an enzyme–substrate complex is formed in the first step and either the substrate is released unchanged or after modification to form products:

$$\mathrm{E+S \rightleftarrows ES} \quad k_a, k_a'$$
$$\mathrm{ES \rightarrow P+E} \quad k_b$$

Michaelis–Menten mechanism

We show in the following *Justification* that this mechanism leads to the **Michaelis–Menten equation** for the rate of product formation

$$v = \frac{k_b[\mathrm{E}]_0}{1+K_M/[\mathrm{S}]_0} \qquad \text{Michaelis–Menten equation} \quad (20\mathrm{H}.1)$$

where $K_M = (k_a' + k_b)/k_a$ is the **Michaelis constant**, characteristic of a given enzyme acting on a given substrate and having the dimensions of a molar concentration.

Justification 20H.1 The Michaelis–Menten equation

The rate of product formation according to the Michaelis–Menten mechanism is

$$v = k_b[\mathrm{ES}]$$

We can obtain the concentration of the enzyme–substrate complex by invoking the steady-state approximation and writing

$$\frac{d[\mathrm{ES}]}{dt} = k_a[\mathrm{E}][\mathrm{S}] - k_a'[\mathrm{ES}] - k_b[\mathrm{ES}] \approx 0$$

It follows that

$$[\mathrm{ES}] = \frac{k_a[\mathrm{E}][\mathrm{S}]}{k_a' + k_b}$$

where [E] and [S] are the concentrations of *free* enzyme and substrate, respectively. Now we define the Michaelis constant as

$$K_M = \frac{k_a' + k_b}{k_a}$$

To express the rate law in terms of the concentrations of enzyme and substrate added, we note that $[\mathrm{E}]_0 = [\mathrm{E}] + [\mathrm{ES}]$ and

$$[\mathrm{E}]_0 = \overbrace{\frac{K_M[\mathrm{ES}]}{[\mathrm{S}]}}^{[\mathrm{E}]} + [\mathrm{ES}] = [\mathrm{ES}]\{1 + K_M/[\mathrm{S}]\}$$

Moreover, because the substrate is typically in large excess relative to the enzyme, the free substrate concentration is approximately equal to the initial substrate concentration and we can write $[\mathrm{S}] \approx [\mathrm{S}]_0$. It then follows that:

$$[\mathrm{ES}] = \frac{[\mathrm{E}]_0}{1 + K_M/[\mathrm{S}]_0}$$

Equation 20H.1 is obtained when this expression for [ES] is substituted into that for the rate of product formation ($v = k_b[\mathrm{ES}]$).

Equation 20H.1 shows that, in accord with experimental observations (Fig. 20H.3):

- When $[S]_0 \ll K_M$, the rate is proportional to $[S]_0$:

$$v = \frac{k_b}{K_M}[S]_0[E]_0 \qquad (20H.2a)$$

- When, $[S]_0 \gg K_M$, the rate reaches its maximum value and is independent of $[S]_0$:

$$v = v_{max} = k_b[E]_0 \qquad (20H.2b)$$

Substitution of the definition of v_{max} into eqn 20H.1 gives

$$v = \frac{v_{max}}{1 + K_M/[S]_0} \qquad (20H.3a)$$

which can be rearranged into a form suitable for data analysis by linear regression by taking reciprocals of both sides:

$$\frac{1}{v} = \frac{1}{v_{max}} + \left(\frac{K_M}{v_{max}}\right)\frac{1}{[S]_0} \qquad \text{Lineweaver–Burk plot} \qquad (20H.3b)$$

A **Lineweaver–Burk plot** is a plot of $1/v$ against $1/[S]_0$, and according to eqn 20H.3b it should yield a straight line with slope of K_M/v_{max}, a y-intercept at $1/v_{max}$, and an x-intercept at $-1/K_M$ (Fig. 20H.4). The value of k_b is then calculated from the y-intercept and eqn 20H.2b. However, the plot cannot give the individual rate constants k_a and k_a' that appear in the expression for K_M. The stopped-flow technique described in Topic 20A can give the additional data needed, because the rate of formation of the enzyme–substrate complex can be found by monitoring the concentration after mixing the enzyme and substrate. This procedure gives a value for k_a, and k_a' is then found by combining this result with the values of k_b and K_M.

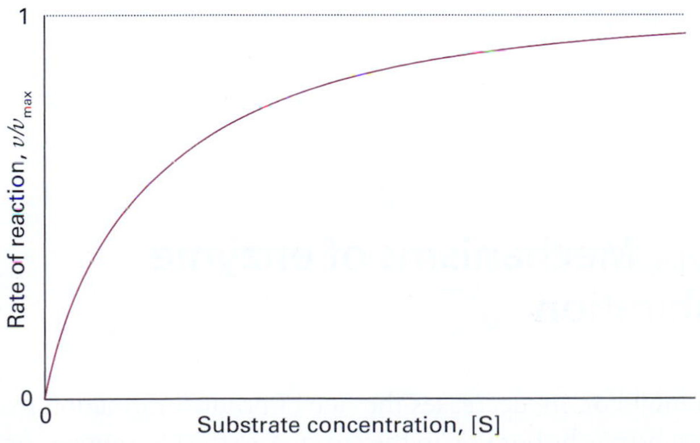

Figure 20H.3 The variation of the rate of an enzyme-catalysed reaction with substrate concentration. The approach to a maximum rate, v_{max}, for large [S] is explained by the Michaelis–Menten mechanism.

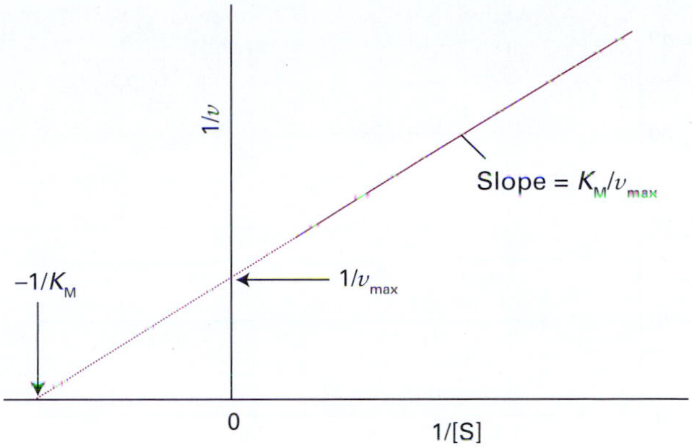

Figure 20H.4 A Lineweaver–Burk plot for the analysis of an enzyme-catalysed reaction that proceeds by a Michaelis–Menten mechanism and the significance of the intercepts and the slope.

Example 20H.1 Analysing a Lineweaver–Burk plot

The enzyme carbonic anhydrase catalyses the hydration of CO_2 in red blood cells to give bicarbonate (hydrogencarbonate) ion: $CO_2(g) + H_2O(l) \rightarrow HCO_3^-(aq) + H^+(aq)$. The following data were obtained for the reaction at pH = 7.1, 273.5 K, and an enzyme concentration of $2.3\,nmol\,dm^{-3}$:

$[CO_2]/(mmol\,dm^{-3})$	1.25	2.5	5	20
$v/(mmol\,dm^{-3}\,s^{-1})$	2.78×10^{-2}	5.00×10^{-2}	8.33×10^{-2}	1.67×10^{-1}

Determine the maximum velocity and the Michaelis constant for the reaction at 273.5 K.

Method Prepare a Lineweaver–Burk plot and determine the values of K_M and v_{max} by linear regression analysis.

Answer We draw up the following table:

$1/([CO_2]/(mmol\,dm^{-3}))$	0.800	0.400	0.200	0.0500
$1/(v/(mmol\,dm^{-3}\,s^{-1}))$	36.0	20.0	12.0	6.0

Figure 20H.5 shows the Lineweaver–Burk plot for the data. The slope is 40.0 and the y-intercept is 4.00. Hence,

$$v_{max}/(mmol\,dm^{-3}\,s^{-1}) = \frac{1}{\text{intercept}} = \frac{1}{4.00} = 0.250$$

and

$$K_M/(mmol\,dm^{-3}) = \frac{\text{slope}}{\text{intercept}} = \frac{40.00}{4.00} = 10.0$$

A note on good practice The slope and the intercept are unit-less: all graphs should be plotted as pure numbers.

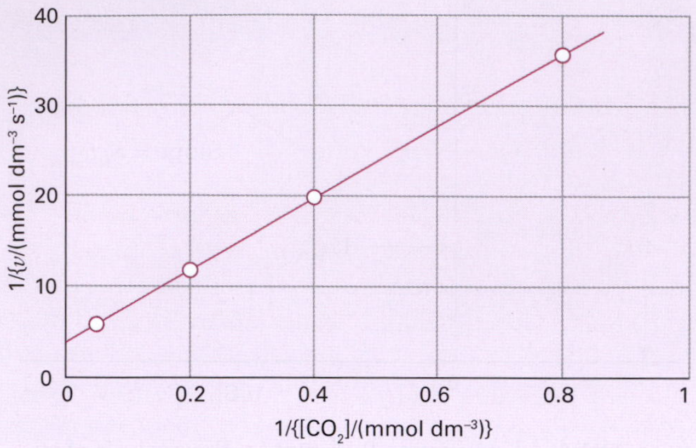

Figure 20H.5 The Lineweaver–Burk plot of the data for *Example* 20H.1.

Self-test 20H.1 The enzyme α-chymotrypsin is secreted in the pancreas of mammals and cleaves peptide bonds made between certain amino acids. Several solutions containing the small peptide *N*-glutaryl-l-phenylalanine-*p*-nitroanilide at different concentrations were prepared and the same small amount of α-chymotrypsin was added to each one. The following data were obtained on the initial rates of the formation of product:

[S]/(mmol dm^{-3})	0.334	0.450	0.667	1.00	1.33	1.67
v/(mmol dm^{-3} s^{-1})	0.152	0.201	0.269	0.417	0.505	0.667

Determine the maximum velocity and the Michaelis constant for the reaction.

Answer: $v_{max}=2.80$ mmol dm^{-3} s^{-1}, $K_M=5.89$ mmol dm^{-3}

20H.3 The catalytic efficiency of enzymes

The **turnover frequency**, or **catalytic constant**, of an enzyme, k_{cat}, is the number of catalytic cycles (turnovers) performed by the active site in a given interval divided by the duration of the interval. This quantity has units of a first-order rate constant and, in terms of the Michaelis–Menten mechanism, is numerically equivalent to k_b, the rate constant for release of product from the enzyme–substrate complex. It follows from the identification of k_{cat} with k_b and from eqn 20H.2b that

$$k_{cat}=k_b=\frac{v_{max}}{[E]_0} \qquad \text{Turnover frequency} \qquad (20H.4)$$

The **catalytic efficiency**, η (eta), of an enzyme is the ratio k_{cat}/K_M. The higher the value of η, the more efficient is the enzyme. We can think of the catalytic efficiency as the effective rate constant of the enzymatic reaction. From $K_M=(k_a'+k_b)/k_a$ and eqn 20H.4, it follows that

$$\eta=\frac{k_{cat}}{K_M}=\frac{k_ak_b}{k_a'+k_b} \qquad \text{Catalytic efficiency} \qquad (20H.5)$$

The efficiency reaches its maximum value of k_a when $k_b \gg k_a'$. Because k_a is the rate constant for the formation of a complex from two species that are diffusing freely in solution, the maximum efficiency is related to the maximum rate of diffusion of E and S in solution. This limit (which is discussed further in Topic 21B) leads to rate constants of about 10^8–10^9 dm^3 mol^{-1} s^{-1} for molecules as large as enzymes at room temperature. The enzyme catalase has $\eta=4.0\times10^8$ dm^3 mol^{-1} s^{-1} and is said to have attained 'catalytic perfection', in the sense that the rate of the reaction it catalyses is controlled only by diffusion: it acts as soon as a substrate makes contact.

Brief illustration 20H.1 The catalytic efficiency of an enzyme

To determine the catalytic efficiency of carbonic anhydrase at 273.5 K from the results from *Example* 20H.1, we begin by using eqn 20H.4 to calculate k_{cat}:

$$k_{cat}=\frac{v_{max}}{[E]_0}=\frac{2.5\times10^{-4}\ \text{mol dm}^{-3}\ \text{s}^{-1}}{2.3\times10^{-9}\ \text{mol dm}^{-3}}=1.1\times10^5\ \text{s}^{-1}$$

The catalytic efficiency follows from eqn 20H.5:

$$\eta=\frac{k_{cat}}{K_M}=\frac{1.1\times10^5\ \text{s}^{-1}}{2.3\times10^{-9}\ \text{mol dm}^{-3}}=1.1\times10^7\ \text{dm}^3\ \text{mol}^{-1}\ \text{s}^{-1}$$

Self-test 20H.2 The enzyme-catalysed conversion of a substrate at 298 K has $K_M=0.032$ mol dm^{-3} and $v_{max}=4.25\times10^{-4}$ mol dm^{-3} s^{-1} when the enzyme concentration is 3.60×10^{-9} mol dm^{-3}. Calculate k_{cat} and η. Is the enzyme 'catalytically perfect'?

Answer: $k_{cat}=1.18\times10^5$ s^{-1}, $\eta=7.9\times10^6$ dm^3 mol^{-1} s^{-1}; the enzyme is not 'catalytically perfect'

20H.4 Mechanisms of enzyme inhibition

An inhibitor, In, decreases the rate of product formation from the substrate by binding to the enzyme, to the ES complex, or to the enzyme and ES complex simultaneously. The most general kinetic scheme for enzyme inhibition is then:

$$E+S \rightleftarrows ES \qquad k_a, k_a'$$
$$ES \rightarrow P+E \qquad k_b$$

$$\text{EIn} \rightleftharpoons \text{E} + \text{In} \qquad K_I = \frac{[\text{E}][\text{In}]}{[\text{EI}]} \tag{20H.6a}$$

$$\text{ESIn} \rightleftharpoons \text{ES} + \text{In} \qquad K_I' = \frac{[\text{ES}][\text{In}]}{[\text{ESIn}]} \tag{20H.6b}$$

The lower the values of K_I and K_I' the more efficient are the inhibitors. The rate of product formation is always given by $v = k_b[\text{ES}]$, because only ES leads to product. As shown in the following *Justification*, the rate of reaction in the presence of an inhibitor is

$$v = \frac{v_{\max}}{\alpha' + \alpha K_M/[\text{S}]_0} \qquad \text{Effect of inhibition on the rate} \tag{20H.7}$$

where $\alpha = 1 + [\text{In}]/K_I$ and $\alpha' = 1 + [\text{In}]/K_I'$. This equation is very similar to the Michaelis–Menten equation for the uninhibited enzyme (eqn 20H.1) and is also amenable to analysis by a Lineweaver–Burk plot:

$$\frac{1}{v} = \frac{\alpha'}{v_{\max}} + \left(\frac{\alpha K_M}{v_{\max}}\right)\frac{1}{[\text{S}]_0} \tag{20H.8}$$

Justification 20H.2 Enzyme inhibition

By mass balance, the total concentration of enzyme is:

$$[\text{E}]_0 = [\text{E}] + [\text{EIn}] + [\text{ES}] + [\text{ESIn}]$$

By using eqns 20H.6a and 20H.6b and the definitions

$$\alpha = 1 + \frac{[\text{In}]}{K_I} \quad \text{and} \quad \alpha' = 1 + \frac{[\text{In}]}{K_I'}$$

it follows that

$$[\text{E}]_0 = [\text{E}]\alpha + [\text{ES}]\alpha'$$

By using $K_M = [\text{E}][\text{S}]/[\text{ES}]$ and replacing [S] with $[\text{S}]_0$ we can write

$$[\text{E}]_0 = \frac{K_M[\text{ES}]}{[\text{S}]_0}\alpha + [\text{ES}]\alpha' = [\text{ES}]\left(\frac{\alpha K_M}{[\text{S}]_0} + \alpha'\right)$$

The expression for the rate of product formation is then:

$$v = k_b[\text{ES}] = \frac{k_b[\text{E}]_0}{\alpha K_M/[\text{S}]_0 + \alpha'}$$

which, upon replacement of $k_b[\text{E}]_0$ with $v_{\max}$, gives eqn 20H.7.

There are three major modes of inhibition that give rise to distinctly different kinetic behaviour (Fig. 20H.6). In **competitive inhibition** the inhibitor binds only to the active site of the enzyme and thereby inhibits the attachment of the substrate.

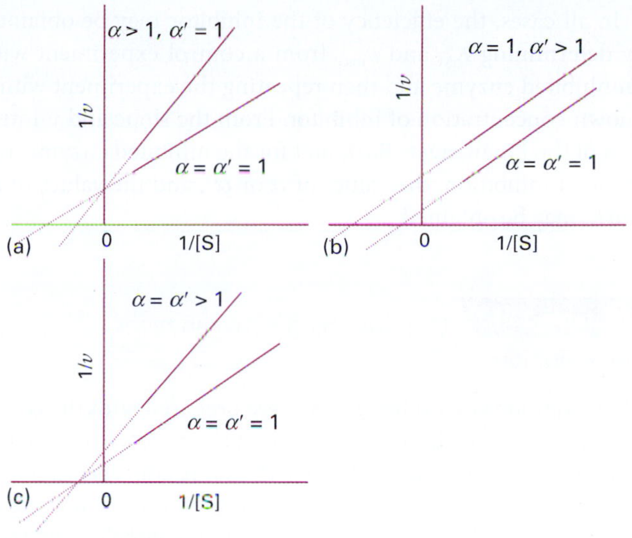

Figure 20H.6 Lineweaver–Burk plots characteristic of the three major modes of enzyme inhibition: (a) competitive inhibition, (b) uncompetitive inhibition, and (c) non-competitive inhibition, showing the special case $\alpha = \alpha' > 1$.

This condition corresponds to $\alpha > 1$ and $\alpha' = 1$ (because ESIn does not form). In this limit, eqn 20H.8 becomes

$$\frac{1}{v} = \frac{1}{v_{\max}} + \left(\frac{\alpha K_M}{v_{\max}}\right)\frac{1}{[\text{S}]_0} \qquad \text{Competitive inhibition}$$

The y-intercept is unchanged but the slope of the Lineweaver–Burk plot increases by a factor of α relative to the slope for data on the uninhibited enzyme (Fig. 20H.6a).

In **uncompetitive inhibition** the inhibitor binds to a site of the enzyme that is removed from the active site, but only if the substrate is already present. The inhibition occurs because ESI reduces the concentration of ES, the active type of complex. In this case $\alpha = 1$ (because EI does not form) and $\alpha' > 1$ and eqn 20H.8 becomes

$$\frac{1}{v} = \frac{\alpha'}{v_{\max}} + \left(\frac{K_M}{v_{\max}}\right)\frac{1}{[\text{S}]_0} \qquad \text{Uncompetitive inhibition}$$

The y-intercept of the Lineweaver–Burk plot increases by a factor of α' relative to the y-intercept for data on the uninhibited enzyme but the slope does not change (Fig. 20H.6b).

In **non-competitive inhibition** (also called **mixed inhibition**) the inhibitor binds to a site other than the active site, and its presence reduces the ability of the substrate to bind to the active site. Inhibition occurs at both the E and ES sites. This condition corresponds to $\alpha > 1$ and $\alpha' > 1$. Both the slope and y-intercept of the Lineweaver–Burk plot increase upon addition of the inhibitor. Figure 20H.6c shows the special case of $K_I = K_I'$ and $\alpha = \alpha'$, which results in intersection of the lines at the x-axis.

In all cases, the efficiency of the inhibitor may be obtained by determining K_M and v_{max} from a control experiment with uninhibited enzyme and then repeating the experiment with a known concentration of inhibitor. From the slope and y-intercept of the Lineweaver–Burk plot for the inhibited enzyme, the mode of inhibition, the values of α or α', and the values of K_I and K_I' may be obtained.

Example 20H.2 Distinguishing between types of inhibition

Five solutions of a substrate, S, were prepared with the concentrations given in the first column below and each one was divided into five equal volumes. The same concentration of enzyme was present in each one. An inhibitor, In, was then added in four different concentrations to the samples, and the initial rate of formation of product was determined with the results given below. Does the inhibitor act competitively or noncompetitively? Determine K_I and K_M.

$[S]_0/(mmol\,dm^{-3})$	$[In]/(mmol\,dm^{-3})$				
	0	0.20	0.40	0.60	0.80
0.050	0.033	0.026	0.021	0.018	0.016
0.10	0.055	0.045	0.038	0.033	0.029
0.20	0.083	0.071	0.062	0.055	0.050
0.40	0.111	0.100	0.091	0.084	0.077
0.60	0.116	0.116	0.108	0.101	0.094

$v/(\mu mol\,dm^{-3}\,s^{-1})$

Method Draw a series of Lineweaver–Burk plots for different inhibitor concentrations. If the plots resemble those in Fig. 20H.6a, then the inhibition is competitive. On the other hand, if the plots resemble those in Fig. 20H.6b, then the inhibition is non-competitive. To find K_I, we need to determine the slope at each value of [In], which is equal to $\alpha K_M/v_{max}$, or $K_M/v_{max}+K_M[In]/K_I v_{max}$, then plot this slope against [In]: the intercept at $[In]=0$ is the value of K_M/v_{max} and the slope is $K_M/K_I v_{max}$.

Answer First we draw up a table of $1/[S]_0$ and $1/v$ for each value of [I]:

$1/([S]_0/(mmol\,dm^{-3}))$	$[In]/(mmol\,dm^{-3})$				
	0	0.20	0.40	0.60	0.80
20	30	38	48	56	62
10	18	22	26	30	34
5.0	12	14	16	18	20
2.5	9.01	10.0	11.0	11.9	13.0
1.7	7.94	8.62	9.26	9.90	10.6

$1/(v/(\mu mol\,dm^{-3}\,s^{-1}))$

The five plots (one for each [In]) are given in Fig. 20H.7. We see that they pass through the same intercept on the vertical axis, so the inhibition is competitive.

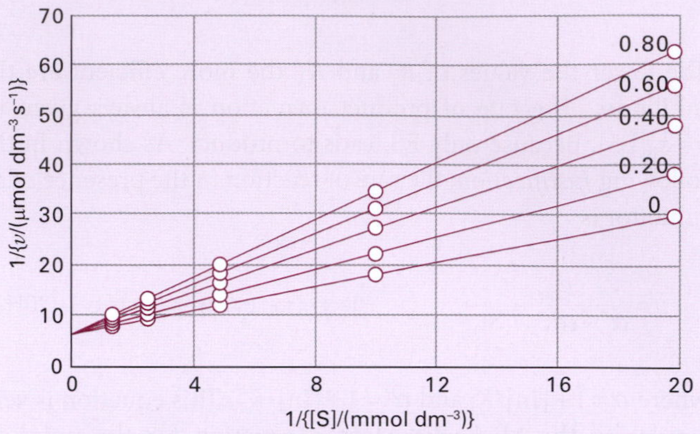

Figure 20H.7 Lineweaver–Burk plots for the data in *Example* 20H.2. Each line corresponds to a different concentration of inhibitor.

The mean of the (least squares) intercepts is 5.83, so $v_{max}=0.172\,\mu mol\,dm^{-3}\,s^{-1}$ (note how it picks up the units for v in the data). The (least squares) slopes of the lines are as follows:

$[In]/(mmol\,dm^{-3})$	0	0.20	0.40	0.60	0.80
Slope	1.219	1.627	2.090	2.489	2.832

These values are plotted in Fig. 20H.8. The intercept at $[In]=0$ is 1.234, so $K_M=0.212\,mmol\,dm^{-3}$. The (least squares) slope of the line is 2.045, so

$$K_I/(mmol\,dm^{-3})=\frac{K_M}{slope\times v_{max}}=\frac{0.212}{2.045\times 0.172}$$

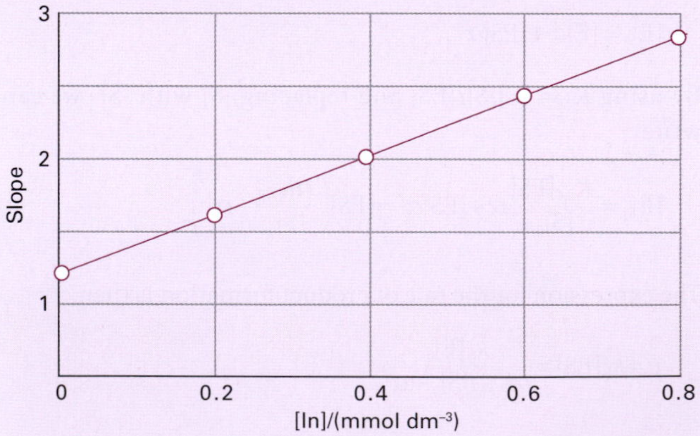

Figure 20H.8 Plot of the slopes of the plots in Fig. 20H.7 against [In] based on the data in *Example* 20H.2.

Self-test 20H.3 Repeat the question using the following data:

$[S]_0/(mmol\ dm^{-3})$	$[In]/(mmol\ dm^{-3})$				
	0	0.20	0.40	0.60	0.80
0.050	0.020	0.015	0.012	0.0098	0.0084
0.10	0.035	0.026	0.021	0.017	0.015
0.20	0.056	0.042	0.033	0.028	0.024
0.40	0.080	0.059	0.047	0.039	0.034
0.60	0.093	0.069	0.055	0.046	0.039

$v/(\mu mol\ dm^{-3}\ s^{-1})$

Answer: Non-competitive, $K_M = 0.30\ mmol\ dm^{-3}$, $K_I = 0.57\ mmol\ dm^{-3}$

Checklist of concepts

- ☐ 1. **Enzymes** are homogeneous biological catalysts.
- ☐ 2. The **Michaelis–Menten mechanism** of enzyme kinetics accounts for the dependence of rate on the concentration of the substrate and the enzyme.
- ☐ 3. A **Lineweaver–Burk plot** is used to determine the parameters that occur in the Michaelis–Menten mechanism.
- ☐ 4. In **competitive inhibition** of an enzyme, the inhibitor binds only to the active site of the enzyme.
- ☐ 5. In **uncompetitive inhibition** the inhibitor binds to a site of the enzyme that is removed from the active site, but only if the substrate is already present.
- ☐ 6. In **non-competitive inhibition**, the inhibitor binds to a site other than the active site.

Checklist of equations

Property	Equation	Comment	Equation number
Michaelis–Menten equation	$v = v_{max}/(1 + K_M/[S]_0)$		20H.3a
Lineweaver–Burk plot	$1/v = 1/v_{max} + (K_M/v_{max})(1/[S]_0)$		20H.3b
Turnover frequency	$k_{cat} = v_{max}/[E]_0$	Definition	20H.4
Catalytic efficiency	$\eta = k_{cat}/K_M$	Definition	20H.5
Effect of inhibition	$v = v_{max}/(\alpha' + \alpha K_M/[S]_0)$	Assumes Michaelis–Menten mechanism	20H.7

CHAPTER 20 Chemical kinetics

TOPIC 20A The rates of chemical reactions

Discussion question

20A.1 Summarize the characteristic of zeroth-order, first-order, second-order, and pseudofirst-order reactions.

20A.2 When can a reaction order not be ascribed?

20A.3 What are the advantages of ascribing an order to a reaction?

20A.4 Summarize the experimental procedures that can be used to monitor the composition of a reaction system.

Exercises

20A.1(a) Predict how the total pressure varies during the gas-phase reaction $2\,ICl(g)+H_2(g)\rightarrow I_2(g)+2\,HCl(g)$ in a constant-volume container.

20A.1(b) Predict how the total pressure varies during the gas-phase reaction $N_2(g)+3\,H_2(g)\rightarrow 2\,NH_3(g)$ in a constant-volume container.

20A.2(a) The rate of the reaction $A+2\,B\rightarrow 3\,C+D$ was reported as $2.7\,mol\,dm^{-3}\,s^{-1}$. State the rates of formation and consumption of the participants.

20A.2(b) The rate of the reaction $A+3\,B\rightarrow C+2\,D$ was reported as $2.7\,mol\,dm^{-3}\,s^{-1}$. State the rates of formation and consumption of the participants.

20A.3(a) The rate of formation of C in the reaction $2\,A+B\rightarrow 2\,C+3\,D$ is $2.7\,mol\,dm^{-3}\,s^{-1}$. State the reaction rate, and the rates of formation or consumption of A, B, and D.

20A.3(b) The rate of consumption of B in the reaction $A+3\,B\rightarrow C+2\,D$ is $2.7\,mol\,dm^{-3}\,s^{-1}$. State the reaction rate, and the rates of formation or consumption of A, C, and D.

20A.4(a) The rate law for the reaction in Exercise 20A.2(a) was found to be $v=k_r[A][B]$. What are the units of k_r when the concentrations are in moles per cubic decimetre? Express the rate law in terms of the rates of formation and consumption of (i) A, (ii) C.

20A.4(b) The rate law for the reaction in Exercise 20A.2(b) was found to be $v=k_r[A][B]^2$. What are the units of k_r when the concentrations are in moles per cubic decimetre? Express the rate law in terms of the rates of formation and consumption of (i) A, (ii) C.

20A.5(a) The rate law for the reaction in Exercise 20A.3(a) was reported as $d[C]/dt=k_r[A][B][C]$. Express the rate law in terms of the reaction rate v; what are the units for k_r in each case when the concentrations are in moles per cubic decimetre?

20A.5(b) The rate law for the reaction in Exercise 20A.3(b) was reported as $d[C]/dt=k_r[A][B][C]^{-1}$. Express the rate law in terms of the reaction rate v; what are the units for k_r in each case when the concentrations are in moles per cubic decimetre?

20A.6(a) If the rate laws are expressed with (i) concentrations in moles per cubic decimetre, (ii) pressures in kilopascals, what are the units of the second-order and third-order rate constants?

20A.6(b) If the rate laws are expressed with (i) concentrations in molecules per cubic metre, (ii) pressures in pascals, what are the units of the second-order and third-order rate constants?

Problems

20A.1 At 400 K, the rate of decomposition of a gaseous compound initially at a pressure of 12.6 kPa, was $9.71\,Pa\,s^{-1}$ when 10.0 per cent had reacted and $7.67\,Pa\,s^{-1}$ when 20.0 per cent had reacted. Determine the order of the reaction.

20A.2 The following initial-rate data were obtained on the rate of binding of glucose with the enzyme hexokinase present at a concentration of $1.34\,mmol\,dm^{-3}$. What is (a) the order of reaction with respect to glucose, (b) the rate constant?

$[C_6H_{12}O_6]/(mmol\,dm^{-3})$	1.00	1.54	3.12	4.02
$v_0/(mol\,dm^{-3}\,s^{-1})$	5.0	7.6	15.5	20.0

20A.3 The following data were obtained on the initial rates of a reaction of a d-metal complex with a reactant Y in aqueous solution. What is (a) the order of reaction with respect to the complex and Y, (b) the rate constant? For the experiments (i), $[Y]=2.7\,mmol\,dm^{-3}$ and for experiments (ii) $[Y]=6.1\,mmol\,dm^{-3}$.

$[complex]/(mmol\,dm^{-3})$		8.01	9.22	12.11
$v_0/(mol\,dm^{-3}\,s^{-1})$	(i)	125	144	190
	(ii)	640	730	960

20A.4 The following kinetic data (v_0 is the initial rate) were obtained for the reaction $2\,ICl(g)+H_2(g)\rightarrow I_2(g)+2\,HCl(g)$:

Experiment	$[ICl]_0/(mmol\,dm^{-3})$	$[H_2]_0/(mmol\,dm^{-3})$	$v_0/(mol\,dm^{-3}\,s^{-1})$
1	1.5	1.5	3.7×10^{-7}
2	3.0	1.5	7.4×10^{-7}
3	3.0	4.5	22×10^{-7}
4	4.7	2.7	?

(a) Write the rate law for the reaction. (b) From the data, determine the value of the rate constant. (c) Use the data to predict the reaction rate for experiment 4.

TOPIC 20B Integrated rate laws

Discussion questions

20B.1 Describe the main features, including advantages and disadvantages, of the following experimental methods for determining the rate law of a reaction: the isolation method, the method of initial rates, and fitting data to integrated rate law expressions.

20B.2 Write the rate law that corresponds to each of the following expressions: (a) $[A]=[A]_0-k_rt$, (b) $\ln([A]/[A]_0)=-k_rt$, and (c) $[A]=[A]_0/(1+k_rt[A]_0)$.

Exercises

20B.1(a) At 518 °C, the half-life for the decomposition of a sample of gaseous acetaldehyde (ethanal) initially at 363 Torr was 410 s. When the pressure was 169 Torr, the half-life was 880 s. Determine the order of the reaction.

20B.1(b) At 400 K, the half-life for the decomposition of a sample of a gaseous compound initially at 55.5 kPa was 340 s. When the pressure was 28.9 kPa, the half-life was 178 s. Determine the order of the reaction.

20B.2(a) The rate constant for the first-order decomposition of N_2O_5 in the reaction $2\,N_2O_5(g)\rightarrow 4\,NO_2(g)+O_2(g)$ is $k_r=3.38\times10^{-5}\,s^{-1}$ at 25 °C. What is the half-life of N_2O_5? What will be the pressure, initially 500 Torr, after (i) 50 s, (ii) 20 min after initiation of the reaction?

20B.2(b) The rate constant for the first-order decomposition of a compound A in the reaction $2\,A\rightarrow P$ is $k_r=3.56\times10^{-7}\,s^{-1}$ at 25 °C. What is the half-life of A? What will be the pressure, initially 33.0 kPa after (i) 50 s, (ii) 20 min after initiation of the reaction?

20B.3(a) The second-order rate constant for the reaction $CH_3COOC_2H_5(aq)+OH^-(aq)\rightarrow CH_3CO_2^-(aq)+CH_3CH_2OH(aq)$ is $0.11\,dm^3\,mol^{-1}\,s^{-1}$. What is the concentration of ester ($CH_3COOC_2H_5$) after (i) 20 s, (ii) 15 min when ethyl acetate is added to sodium hydroxide so that the initial concentrations are $[NaOH]=0.060\,mol\,dm^{-3}$ and $[CH_3COOC_2H_5]=0.110\,mol\,dm^{-3}$?

20B.3(b) The second-order rate constant for the reaction $A+2\,B\rightarrow C+D$ is $0.34\,dm^3\,mol^{-1}\,s^{-1}$. What is the concentration of C after (i) 20 s, (ii) 15 min when the reactants are mixed with initial concentrations of $[A]=0.027\,mol\,dm^{-1}$ and $[B]=0.130\,mol\,dm^{-3}$?

20B.4(a) A reaction $2\,A\rightarrow P$ has a second-order rate law with $k_r=4.30\times10^{-4}\,dm^3\,mol^{-1}\,s^{-1}$. Calculate the time required for the concentration of A to change from $0.210\,mol\,dm^{-3}$ to $0.010\,mol\,dm^{-3}$.

20B.4(b) A reaction $2\,A\rightarrow P$ has a third-order rate law with $k_r=6.50\times10^{-4}\,dm^6\,mol^{-2}\,s^{-1}$. Calculate the time required for the concentration of A to change from $0.067\,mol\,dm^{-3}$ to $0.015\,mol\,dm^{-3}$.

Problems

20B.1 For a first-order reaction of the form $A\rightarrow n\,B$ (with n possibly fractional), the concentration of the product varies with time as $[B]=n[B]_0(1-e^{-k_rt})$. Plot the time dependence of [A] and [B] for the cases $n=\frac{1}{2}$, 1, and 2.

20B.2 For a second-order reaction of the form $A\rightarrow n\,B$ (with n possibly fractional), the concentration of the product varies with time as $[B]=nk_rt[A]_0^2/(1+k_rt[A]_0)$. Plot the time dependence of [A] and [B] for the cases $n=\frac{1}{2}$, 1, and 2.

20B.3 The data below apply to the formation of urea from ammonium cyanate, $NH_4CNO\rightarrow NH_2CONH_2$. Initially 22.9 g of ammonium cyanate was dissolved in enough water to prepare $1.00\,dm^3$ of solution. Determine the order of the reaction, the rate constant, and the mass of ammonium cyanate left after 300 min.

t/min	0	20.0	50.0	65.0	150
m(urea)/g	0	7.0	12.1	13.8	17.7

20B.4 The data below apply to the reaction, $(CH_3)_3CBr+H_2O\rightarrow(CH_3)_3COH+HBr$. Determine the order of the reaction, the rate constant, and the molar concentration of $(CH_3)_3CBr$ after 43.8 h.

t/h	0	3.15	6.20	10.00	18.30	30.80
$[(CH_3)_3CBr]/(10^{-2}\,mol\,dm^{-3})$	10.39	8.96	7.76	6.39	3.53	2.07

20B.5 The thermal decomposition of an organic nitrile produced the following data:

$t/(10^3\,s)$	0	2.00	4.00	6.00	8.00	10.00	12.00	∞
[nitrile]/($mol\,dm^{-3}$)	1.50	1.26	1.07	0.92	0.81	0.72	0.65	0.40

Determine the order of the reaction and the rate constant.

20B.6 A second-order reaction of the type $A+2\,B\rightarrow P$ was carried out in a solution that was initially $0.050\,mol\,dm^{-3}$ in A and $0.030\,mol\,dm^{-3}$ in B. After 1.0 h the concentration of A had fallen to $0.010\,mol\,dm^{-3}$. (a) Calculate the rate constant. (b) What is the half-life of the reactants?

20B.7‡ The oxidation of HSO_3^- by O_2 in aqueous solution is a reaction of importance to the processes of acid rain formation and flue gas desulfurization. R.E. Connick et al. (*Inorg. Chem.* 34, 4543 (1995)) report that the reaction $2\,HSO_3^-(aq)+O_2(g)\rightarrow 2\,SO_4^{2-}(aq)+2\,H^+(aq)$ follows the rate law $v=k_r[HSO_4^-]^2[H^+]^2$. Given pH = 5.6 and an oxygen molar concentration of $0.24\,mmol\,dm^{-3}$ (both presumed constant), an initial HSO_3^- molar concentration of $50\,\mu mol\,dm^{-3}$, and a rate constant of $3.6\times10^6\,dm^9\,mol^{-3}\,s^{-1}$, what is the initial rate of reaction? How long would it take for HSO_3^- to reach half its initial concentration?

20B.8 Pharmacokinetics is the study of the rates of absorption and elimination of drugs by organisms. In most cases, elimination is slower than absorption and is a more important determinant of availability of a drug for binding to its target. A drug can be eliminated by many mechanisms, such as metabolism in the liver, intestine, or kidney followed by excretion of breakdown products through urine or faeces. As an example of pharmacokinetic analysis, consider the elimination of beta adrenergic blocking agents (beta blockers), drugs used in the treatment of hypertension. After intravenous administration of a beta blocker, the blood plasma of a patient was analysed for remaining drug and the data are shown below, where c is the drug concentration measured at a time t after the injection.

t/min	30	60	120	150	240	360	480
c/(ng cm^{-3})	699	622	413	292	152	60	24

(a) Is removal of the drug a first- or second-order process? (b) Calculate the rate constant and half-life of the process. *Comment*: An essential

‡ These problems were supplied by Charles Trapp and Carmen Giunta.

aspect of drug development is the optimization of the half-life of elimination, which needs to be long enough to allow the drug to find and act on its target organ but not so long that harmful side effects become important.

20B.9 The following data have been obtained for the decomposition of $N_2O_5(g)$ at 67 °C according to the reaction $2\,N_2O_5(g) \rightarrow 4\,NO_2(g) + O_2(g)$. Determine the order of the reaction, the rate constant, and the half-life. It is not necessary to obtain the result graphically; you may do a calculation using estimates of the rates of change of concentration.

t/min	0	1	2	3	4	5
$[N_2O_5]/(\text{mol dm}^{-3})$	1.000	0.705	0.497	0.349	0.246	0.173

20B.10 The gas phase decomposition of acetic acid at 1189 K proceeds by way of two parallel reactions:

(1) $CH_3COOH \rightarrow CH_4 + CO_2$ $k_1 = 3.74\ \text{s}^{-1}$

(2) $CH_3COOH \rightarrow CH_2CO + H_2O$ $k_2 = 4.65\ \text{s}^{-1}$

What is the maximum percentage yield of the ketene CH_2CO obtainable at this temperature?

20B.11 Sucrose is readily hydrolysed to glucose and fructose in acidic solution. The hydrolysis is often monitored by measuring the angle of rotation of plane-polarized light passing through the solution. From the angle of rotation the concentration of sucrose can be determined. An experiment on the hydrolysis of sucrose in 0.50 M HCl(aq) produced the following data:

t/min	0	14	39	60	80	110	140	170	210
[sucrose]/(mol dm^{-3})	0.316	0.300	0.274	0.256	0.238	0.211	0.190	0.170	0.146

Determine the rate constant of the reaction and the half-life of a sucrose molecule.

20B.12 The composition of a liquid phase reaction $2\,A \rightarrow B$ was followed by a spectrophotometric method with the following results:

t/min	0	10	20	30	40	∞
$[B]/(\text{mol dm}^{-3})$	0	0.089	0.153	0.200	0.230	0.312

Determine the order of the reaction and its rate constant.

20B.13 The ClO radical decays rapidly by way of the reaction, $2\,ClO \rightarrow Cl_2 + O_2$. The following data have been obtained:

t/ms	0.12	0.62	0.96	1.60	3.20	4.00	5.75
$[ClO]/(\mu\text{mol dm}^{-3})$	8.49	8.09	7.10	5.79	5.20	4.77	3.95

Determine the rate constant of the reaction and the half-life of a ClO radical.

20B.14 Cyclopropane isomerizes into propene when heated to 500 °C in the gas phase. The extent of conversion for various initial pressures has been followed by gas chromatography by allowing the reaction to proceed for a time with various initial pressures:

p_0/Torr	200	200	400	400	600	600
t/s	100	200	100	200	100	200
p/Torr	186	173	373	347	559	520

where p_0 is the initial pressure and p is the final pressure of cyclopropane. What is the order and rate constant for the reaction under these conditions?

20B.15 The addition of hydrogen halides to alkenes has played a fundamental role in the investigation of organic reaction mechanisms. In one study (M.J. Haugh and D.R. Dalton, *J. Amer. Chem. Soc.* **97**, 5674 (1975)), high pressures of hydrogen chloride (up to 25 atm) and propene (up to 5 atm) were examined over a range of temperatures and the amount of 2-chloropropane formed was determined by NMR. Show that if the reaction $A + B \rightarrow P$ proceeds for a short time δt, the concentration of product follows $[P]/[A] = k_r[A]^{m-1}[B]^n\delta t$ if the reaction is mth-order in A and nth-order in B. In a series of runs the ratio of [chloropropane] to [propene] was independent of [propene] but the ratio of [chloropropane] to [HCl] for constant amounts of propene depended on [HCl]. For $\delta t \approx 100$ h (which is short on the time scale of the reaction) the latter ratio rose from zero to 0.05, 0.03, 0.01 for $p(HCl) = 10$ atm, 7.5 atm, 5.0 atm. What are the orders of the reaction with respect to each reactant?

20B.16 Show that $t_{1/2}$ is given by eqn 20B.6 for a reaction that is nth order in A. Then deduce an expression for the time it takes for the concentration of a substance to fall to one-third the initial value in an nth-order reaction.

20B.17 Derive an integrated expression for a second-order rate law $v = k_r[A][B]$ for a reaction of stoichiometry $2\,A + 3\,B \rightarrow P$.

20B.18 Derive the integrated form of a third-order rate law $v = k_r[A]^2[B]$ in which the stoichiometry is $2\,A + B \rightarrow P$ and the reactants are initially present in (a) their stoichiometric proportions, (b) with B present initially in twice the amount.

20B.19 Show that the ratio $t_{1/2}/t_{3/4}$, where $t_{1/2}$ is the half-life and $t_{3/4}$ is the time for the concentration of A to decrease to $^3/_4$ of its initial value (implying that $t_{3/4} < t_{1/2}$), can be written as a function of n alone, and can therefore be used as a rapid assessment of the order of a reaction.

TOPIC 20C Reactions approaching equilibrium

Discussion questions

20C.1 Describe the strategy of a temperature-jump experiment. What parameters of a reaction are accessible from this technique?

20C.2 What feature of a reaction would ensure that its rate would respond to a pressure jump?

Exercises

20C.1(a) The equilibrium $NH_3(aq) + H_2O(l) \rightleftharpoons NH_4^+(aq) + OH^-(aq)$ at 25 °C is subjected to a temperature jump which slightly increased the concentration of $NH_4^+(aq)$ and $OH^-(aq)$. The measured relaxation time is 7.61 ns. The equilibrium constant for the system is 1.78×10^{-5} at 25 °C, and the equilibrium concentration of $NH_3(aq)$ is 0.15 mol dm^{-3}. Calculate the rate constants for the forward and reverse steps.

20C.1(b) The equilibrium $A \rightleftharpoons B + C$ at 25 °C is subjected to a temperature jump which slightly increases the concentrations of B and C. The measured relaxation time is 3.0 μs. The equilibrium constant for the system is 2.0×10^{-16} at 25 °C, and the equilibrium concentrations of B and C at 25 °C are both 0.20 mmol dm^{-3}. Calculate the rate constants for the forward and reverse steps.

Problems

20C.1 Show by differentiation that eqn 20C.4 is a solution of eqn 20C.3.

20C.2 Set up the rate equations and plot the corresponding graphs for the approach to an equilibrium of the form $A \rightleftharpoons 2B$.

20C.3 The reaction $A \rightleftharpoons 2B$ is first-order in both directions. Derive an expression for the concentration of A as a function of time when the initial molar concentrations of A and B are $[A]_0$ and $[B]_0$. What is the final composition of the system?

20C.4 Show that eqn 20C.8 is an expression for the overall equilibrium constant in terms of the rate constants for the intermediate steps of a reaction mechanism. To facilitate the task, begin with a mechanism containing three steps, and then argue that your expression may be generalized for any number of steps.

20C.5 Consider the dimerization $2A \rightleftarrows A_2$, with forward rate constant k_a and backward rate constant k_a'. (a) Derive the following expression for the relaxation time in terms of the total concentration of protein, $[A]_{tot}=[A]+2[A_2]$:

$$\frac{1}{\tau^2}=k_a'^2+8k_ak_a'[A]_{tot}$$

(b) Describe the computational procedures that lead to the determination of the rate constants k_a and k_a' from measurements of τ for different values of $[A]_{tot}$. (c) Use the data provided below and the procedure you outlined in part (b) to calculate the rate constants k_a and k_a', and the equilibrium constant K for formation of hydrogen-bonded dimers of 2-pyridone:

$[P]/(\text{mol dm}^{-3})$	0.500	0.352	0.251	0.151	0.101
τ/ns	2.3	2.7	3.3	4.0	5.3

20C.6 Consider the dimerization $2A \rightleftarrows A_2$ with forward rate constant k_r and backward rate constant k_r'. Show that the relaxation time is $\tau=1/(k_r'+4k_r[A]_{eq})$.

TOPIC 20D The Arrhenius equation

Discussion question

20D.1 Define the terms in and discuss the validity of the expression $\ln k_r=\ln A-E_a/RT$.

20D.2 What might account for the failure of the Arrhenius equation at low temperatures?

Exercises

20D.1(a) The rate constant for the decomposition of a certain substance is 3.80×10^{-3} dm^3 mol^{-1} s^{-1} at 35 °C and 2.67×10^{-2} dm^3 mol^{-1} s^{-1} at 50 °C. Evaluate the Arrhenius parameters of the reaction.

20D.1(b) The rate constant for the decomposition of a certain substance is 2.25×10^{-2} dm^3 mol^{-1} s^{-1} at 29 °C and 4.01×10^{-2} dm^3 mol^{-1} s^{-1} at 37 °C. Evaluate the Arrhenius parameters of the reaction.

20D.2(a) The rate of a chemical reaction is found to triple when the temperature is raised from 24 °C to 49 °C. Determine the activation energy.

20D.2(b) The rate of a chemical reaction is found to double when the temperature is raised from 25 °C to 35 °C. Determine the activation energy.

Problems

20D.1 Show that the definition of E_a given in eqn 20D.3 reduces to eqn 20D.1 for a temperature-independent activation energy.

20D.2 A first-order decomposition reaction is observed to have the following rate constants at the indicated temperatures. Estimate the activation energy.

$k_r/(10^{-3}\ \text{s}^{-1})$	2.46	45.1	576
$\theta/°\text{C}$	0	20.0	40.0

20D.3 The second-order rate constants for the reaction of oxygen atoms with aromatic hydrocarbons have been measured (R. Atkinson and J.N. Pitts, *J. Phys. Chem.* **79**, 295 (1975)). In the reaction with benzene the rate constants are 1.44×10^7 dm^3 mol^{-1} s^{-1} at 300.3 K, 3.03×10^7 dm^3 mol^{-1} s^{-1} at 341.2 K, and 6.9×10^7 dm^3 mol^{-1} s^{-1} at 392.2 K. Find the pre-exponential factor and activation energy of the reaction.

20D.4‡ Methane is a by-product of a number of natural processes (such as digestion of cellulose in ruminant animals, anaerobic decomposition of organic waste matter), and industrial processes (such as food production and fossil fuel use). Reaction with the hydroxyl radical OH is the main path by which CH_4 is removed from the lower atmosphere. T. Gierczak et al. (*J. Phys. Chem. A* **101**, 3125 (1997)) measured the rate constants for the elementary bimolecular gas-phase reaction of methane with the hydroxyl radical over a range of temperatures of importance to atmospheric chemistry. Deduce the Arrhenius parameters A and E_a from the following measurements:

T/K	295	223	218	213	206	200	195
$k_r/(10^6\ \text{dm}^3\ \text{mol}^{-1}\ \text{s}^{-1})$	3.55	0.494	0.452	0.379	0.295	0.241	0.217

20D.5‡ As we saw in Problem 20D.4, reaction with the hydroxyl radical OH is the main path by which CH_4, a by-product of many natural and industrial processes, is removed from the lower atmosphere. T. Gierczak et al. (*J. Phys. Chem. A* **101**, 3125 (1997)) measured the rate constants for the bimolecular gas-phase reaction $CH_4(g)+OH(g)\rightarrow CH_3(g)+H_2O(g)$ and found $A=1.13\times10^9\,dm^3\,mol^{-1}\,s^{-1}$ and $E_a=14.1\,kJ\,mol^{-1}$ for the Arrhenius parameters. (a) Estimate the rate of consumption of CH_4. Take the average OH concentration to be $1.5\times10^{-21}\,mol\,dm^{-3}$, that of CH_4 to be 40 nmol dm^{-3}, and the temperature to be −10 °C. (b) Estimate the global annual mass of CH_4 consumed by this reaction (which is slightly less than the amount introduced to the atmosphere) given an effective volume for the Earth's lower atmosphere of $4\times10^{21}\,dm^3$.

TOPIC 20E Reaction mechanisms

Discussion questions

20E.1 Distinguish between reaction order and molecularity.

20E.2 Assess the validity of the statement that the rate-determining step is the slowest step in a reaction mechanism.

20E.3 Distinguish between a pre-equilibrium approximation and a steady-state approximation. Why might they lead to different conclusions?

20E.4 Explain and illustrate how reaction orders may change under different circumstances.

20E.5 Distinguish between kinetic and thermodynamic control of a reaction. Suggest criteria for expecting one rather than the other.

20E.6 Explain how it is possible for the activation energy of a reaction to be negative.

Exercises

20E.1(a) The reaction mechanism for the decomposition of A_2

$$A_2 \rightleftharpoons A + A \text{ (fast)}$$
$$A + B \rightarrow P \text{ (slow)}$$

involves an intermediate A. Deduce the rate law for the reaction in two ways by (i) assuming a pre-equilibrium and (ii) making a steady-state approximation.

20E.1(b) The reaction mechanism for renaturation of a double helix from its strands A and B:

$$A + B \rightleftharpoons \text{unstable helix (fast)}$$
$$\text{Unstable helix} \rightarrow \text{stable double helix (slow)}$$

involves an intermediate. Deduce the rate law for the reaction in two ways by (i) assuming a pre-equilibrium and (ii) making a steady-state approximation.

20E.2(a) The mechanism of a composite reaction consists of a fast pre-equilibrium step with forward and reverse activation energies of $25\,kJ\,mol^{-1}$ and $38\,kJ\,mol^{-1}$, respectively, followed by an elementary step of activation energy $10\,kJ\,mol^{-1}$. What is the activation energy of the composite reaction?

20E.2(b) The mechanism of a composite reaction consists of a fast pre-equilibrium step with forward and reverse activation energies of $27\,kJ\,mol^{-1}$ and $35\,kJ\,mol^{-1}$, respectively, followed by an elementary step of activation energy $15\,kJ\,mol^{-1}$. What is the activation energy of the composite reaction?

Problems

20E.1 Use mathematical software or a spreadsheet to examine the time dependence of [I] in the reaction mechanism $A \xrightarrow{k_a} I \xrightarrow{k_b} P$. In all of the following calculations, use $[A]_0=1\,mol\,dm^{-3}$ and a time range of 0 to 5 s. (a) Plot [In] against t for $k_a=10\,s^{-1}$ and $k_b=1\,s^{-1}$. (b) Increase the ratio k_b/k_a steadily by decreasing the value of k_a and examine the plot of [I] against t at each turn. What approximation about $d[I]/dt$ becomes increasingly valid?

20E.2 Use mathematical software or a spreadsheet to investigate the effects on [A], [I], [P], and t_{max} of decreasing the ratio k_a/k_b from 10 (as in Fig. 20E.1) to 0.01. Compare your results with those shown in Fig. 20E.3.

20E.3 Set up the rate equations for the reaction mechanism:

$$A \underset{k_a'}{\overset{k_a}{\rightleftarrows}} B \underset{k_b'}{\overset{k_b}{\rightleftarrows}} C$$

Show that, under specific circumstances, the mechanism is equivalent to

$$A \underset{k_r'}{\overset{k_r}{\rightleftarrows}} C$$

20E.4 Derive an equation for the steady state rate of the sequence of reactions $A \rightleftarrows B \rightleftarrows C \rightleftarrows D$, with [A] maintained at a fixed value and the product D removed as soon as it is formed.

$HCl+HCl \rightleftharpoons (HCl)_2$	K_1
$HCl+CH_3CH{=}CH_2 \rightleftharpoons \text{complex}$	K_2
$(HCl)_2+\text{complex} \rightarrow CH_3CHClCH_3+HCl+HCl$	k_r (slow)

20E.5 Show that the following mechanism can account for the rate law of the reaction in Problem 20B.15:

What further tests could you apply to verify this mechanism?

20E.6 Polypeptides are polymers of amino acids. Suppose that a long polypeptide chain can undergo a transition from a helical conformation to a random coil. Consider a mechanism for a helix–coil transition that begins in the middle of the chain:

hhhh... ⇄ *hchh...*

hchh... ⇄ *cccc...*

in which h and c label, respectively, an amino acid in a helical or coil part of the chain. The first conversion from h to c, also called a nucleation step, is relatively slow, so neither step may be rate determining. (a) Set up the rate equations for this mechanism. (b) Apply the steady-state approximation and show that, under these circumstances, the mechanism is equivalent to *hhhh...* ⇄ *cccc...*

TOPIC 20F Examples of reaction mechanisms

Discussion questions

20F.1 Discuss the range of validity of the expression $k_r = k_a k_b[A]/(k_b + k_a'[A])$ for the effective rate constant of a unimolecular reaction according to the Lindemann–Hinshelwood mechanism.

20F.2 Bearing in mind distinctions between the mechanisms of stepwise and chain polymerization, describe ways in which it is possible to control the molar mass of a polymer by manipulating the kinetic parameters of polymerization.

Exercises

20F.1(a) The effective rate constant for a gaseous reaction which has a Lindemann–Hinshelwood mechanism is $2.50\times10^{-4}\,s^{-1}$ at 1.30 kPa and $2.10\times10^{-5}\,s^{-1}$ at 12 Pa. Calculate the rate constant for the activation step in the mechanism.
20F.1(b) The effective rate constant for a gaseous reaction which has a Lindemann–Hinshelwood mechanism is $1.7\times10^{-3}\,s^{-1}$ at 1.09 kPa and $2.2\times10^{-4}\,s^{-1}$ at 25 Pa. Calculate the rate constant for the activation step in the mechanism.

20F.2(a) Calculate the fraction condensed and the degree of polymerization at $t = 5.00$ h of a polymer formed by a stepwise process with $k_r = 1.39\,dm^3\,mol^{-1}\,s^{-1}$ and an initial monomer concentration of $10.0\,mmol\,dm^{-3}$.

20F.2(b) Calculate the fraction condensed and the degree of polymerization at $t = 10.00$ h of a polymer formed by a stepwise process with $k_r = 2.80\times10^{-2}\,dm^3\,mol^{-1}\,s^{-1}$ and an initial monomer concentration of $50.0\,mmol\,dm^{-3}$.

20F.3(a) Consider a polymer formed by a chain process. By how much does the kinetic chain length change if the concentration of initiator increases by a factor of 3.6 and the concentration of monomer decreases by a factor of 4.2?
20F.3(b) Consider a polymer formed by a chain process. By how much does the kinetic chain length change if the concentration of initiator decreases by a factor of 10.0 and the concentration of increases by a factor of 5.0?

Problems

20F.1 The isomerization of cyclopropane over a limited pressure range was examined in Problem 20B.14. If the Lindemann mechanism of unimolecular reactions is to be tested we also need data at low pressures. These have been obtained (H.O. Pritchard et al., *Proc. R. Soc. A* **217**, 563 (1953)):

p/Torr	84.1	11.0	2.89	0.569	0.120	0.067
$10^4\,k_r/s^{-1}$	2.98	2.23	1.54	0.857	0.392	0.303

Test the Lindemann–Hinshelwood theory with these data.

20F.2 Calculate the average polymer length in a polymer produced by a chain mechanism in which termination occurs by a disproportionation reaction of the form $M\cdot + \cdot M \rightarrow M + {:}M$.

20F.3 Derive an expression for the time dependence of the degree of polymerization for a stepwise polymerization in which the reaction is acid-catalysed by the –COOH acid functional group. The rate law is $d[A]/dt = -k_r[A]^2[OH]$.

TOPIC 20G Photochemistry

Discussion questions

20G.1 Consult literature sources and list the observed ranges of timescales during which the following processes occur: radiative decay of excited electronic states, molecular rotational motion, molecular vibrational motion, proton transfer reactions, energy transfer between fluorescent molecules used in FRET analysis, electron transfer events between complex ions in solution, and collisions in liquids.

20G.2 Discuss experimental procedures that make it possible to differentiate between quenching by energy transfer, collisions, and electron transfer.

Exercises

20G.1(a) In a photochemical reaction $A \rightarrow 2\,B + C$, the quantum yield with 500 nm light is $210\,mol\,einstein^{-1}$ (1 einstein = 1 mol photons). After exposure of 300 mmol of A to the light, 2.28 mmol of B is formed. How many photons were absorbed by A?
20G.1(b) In a photochemical reaction $A \rightarrow B + C$, the quantum yield with 500 nm light is $120\,mol\,einstein^{-1}$ (1 einstein = 1 mol photons). After exposure of 200 mmol A to the light, 1.77 mmol B is formed. How many photons were absorbed by A?

20G.2(a) Consider the quenching of an organic fluorescent species with $\tau_0 = 6.0$ ns by a d-metal ion with $k_Q = 3.0\times10^8\,dm^3\,mol^{-1}\,s^{-1}$. Predict the concentration of quencher required to decrease the fluorescence intensity of the organic species to 50 per cent of the unquenched value.

20G.2(b) Consider the quenching of an organic fluorescent species with $\tau_0 = 3.5$ ns by a d-metal ion with $k_Q = 2.5 \times 10^9\,dm^3\,mol^{-1}\,s^{-1}$. Predict the concentration of quencher required to decrease the fluorescence intensity of the organic species to 75 per cent of the unquenched value.

Problems

20G.1 In an experiment to measure the quantum yield of a photochemical reaction, the absorbing substance was exposed to 320 nm radiation from a 87.5 W source for 28.0 min. The intensity of the transmitted radiation was 0.257 that of the incident radiation. As a result of irradiation, 0.324 mol of the absorbing substance decomposed. Determine the quantum yield.

20G.2‡ Ultraviolet radiation photolyses O_3 to O_2 and O. Determine the rate at which ozone is consumed by 305 nm radiation in a layer of the stratosphere of thickness 1.0 km. The quantum yield is 0.94 at 220 K, the concentration about 8 nmol dm^{-3}, the molar absorption coefficient 260 dm^3 mol^{-1} cm^{-1}, and the flux of 305 nm radiation about 1×10^{14} photons cm^{-2} s^{-1}. Data from W.B. DeMore et al., *Chemical kinetics and photochemical data for use in stratospheric modeling: Evaluation Number 11*, JPL Publication 94–26 (1994).

20G.3 Dansyl chloride, which absorbs maximally at 330 nm and fluoresces maximally at 510 nm, can be used to label amino acids in fluorescence microscopy and FRET studies. Tabulated below is the variation of the fluorescence intensity of an aqueous solution of dansyl chloride with time after excitation by a short laser pulse (with I_0 the initial fluorescence intensity). The ratio of intensities is equal to the ratio of the rates of photon emission.

t/ns	5.0	10.0	15.0	20.0
I_f/I_0	0.45	0.21	0.11	0.05

(a) Calculate the observed fluorescence lifetime of dansyl chloride in water.
(b) The fluorescence quantum yield of dansyl chloride in water is 0.70. What is the fluorescence rate constant?

20G.4 When benzophenone is exposed to ultraviolet radiation it is excited into a singlet state. This singlet changes rapidly into a triplet, which phosphoresces. Triethylamine acts as a quencher for the triplet. In an experiment in the solvent methanol, the phosphorescence intensity varied with amine concentration as shown below. A time-resolved laser spectroscopy experiment had also shown that the half-life of the fluorescence in the absence of quencher is 29 μs. What is the value of k_Q?

[Q]/(mmol dm^{-3})	1.0	5.0	10.0
I_f/(arbitrary units)	0.41	0.25	0.16

20G.5 An electronically excited state of Hg can be quenched by N_2 according to $Hg^*(g) + N_2(g, v=0) \rightarrow Hg(g) + N_2(g, v=1)$ in which energy transfer from Hg* excites N_2 vibrationally. Fluorescence lifetime measurements of samples of Hg with and without N_2 present are summarized below (for $T = 300$ K):

$p_{N_2} = 0.0$ atm

Relative fluorescence intensity	1.000	0.606	0.360	0.22	0.135
t/μs	0.0	5.0	10.0	15.0	20.0

$p_{N_2} = 9.74 \times 10^{-4}$ atm

Relative fluorescence intensity	1.000	0.585	0.342	0.200	0.117
t/μs	0.0	3.0	6.0	9.0	12.0

You may assume that all gases are perfect. Determine the rate constant for the energy transfer process.

20G.6 An amino acid on the surface of an enzyme was labelled covalently with 1.5-I AEDANS and it is known that the active site contains a tryptophan residue. The fluorescence quantum yield of tryptophan decreased by 15 per cent due to quenching by 1.5-I AEDANS. What is the distance between the active site and the surface of the enzyme?

20G.7 The Förster theory of resonance energy transfer and the basis for the FRET technique can be tested by performing fluorescence measurements on a series of compounds in which an energy donor and an energy acceptor are covalently linked by a rigid molecular linker of variable and known length. L. Stryer and R.P. Haugland (*Proc. Natl. Acad. Sci. USA* **58**, 719 (1967)) collected the following data on a family of compounds with the general composition dansyl-(l-prolyl)$_n$-naphthyl, in which the distance R between the naphthyl donor and the dansyl acceptor was varied from 1.2 nm to 4.6 nm by increasing the number of prolyl units in the linker:

R/nm	1.2	1.5	1.8	2.8	3.1	3.4	3.7	4.0	4.3	4.6
η_T	0.99	0.94	0.97	0.82	0.74	0.65	0.40	0.28	0.24	0.16

Are the data described adequately by eqn 20G.10? If so, what is the value of R_0 for the naphthyl–dansyl pair?

20G.8 The first step in plant photosynthesis is absorption of light by chlorophyll molecules bound to proteins known as 'light-harvesting complexes', where the fluorescence of a chlorophyll molecule is quenched by nearby chlorophyll molecules. Given that for a pair of chlorophyll *a* molecules $R_0 = 5.6$ nm, by what distance should two chlorophyll *a* molecules be separated to shorten the fluorescence lifetime from 1 ns (a typical value for monomeric chlorophyll *a* in organic solvents) to 10 ps?

TOPIC 20H Enzymes

Discussion questions

20H.1 Discuss the features, advantages, and limitations of the Michaelis–Menten mechanism of enzyme action.

20H.2 A plot of the rate of an enzyme-catalysed reaction against temperature has a maximum, in an apparent deviation from the behaviour predicted by the Arrhenius equation (Topic 20D). Suggest an interpretation.

20H.3 Distinguish between competitive, non-competitive, and uncompetitive inhibition of enzymes. Discuss how these modes of inhibition may be detected experimentally.

20H.4 Some enzymes are inhibited by high concentrations of their own products. Sketch a plot of reaction rate against concentration of substrate for an enzyme that is prone to product inhibition.

Exercises

20H.1(a) Consider the base-catalysed reaction

$$(1)\ \mathrm{AH}+\mathrm{B} \underset{k_a'}{\overset{k_a}{\rightleftharpoons}} \mathrm{BH}^+ + \mathrm{A}^- \quad \text{(both fast)}$$

$$(2)\ \mathrm{A}^- + \mathrm{AH} \xrightarrow{k_b} \text{product} \quad \text{(slow)}$$

Deduce the rate law.

20H.1(b) Consider the acid-catalysed reaction

$$(1)\ \mathrm{HA}+\mathrm{H}^+ \underset{k_a'}{\overset{k_a}{\rightleftharpoons}} \mathrm{HAH}^+ \quad \text{(both fast)}$$

$$(2)\ \mathrm{HAH}^+ + \mathrm{B} \xrightarrow{k_b} \mathrm{BH}^+ + \mathrm{AH} \quad \text{(slow)}$$

Deduce the rate law.

20H.2(a) The enzyme-catalysed conversion of a substrate at 25 °C has a Michaelis constant of 0.046 mol dm^{-3}. The rate of the reaction is 1.04 mmol dm^{-3} s^{-1} when the substrate concentration is 0.105 mol dm^{-3}. What is the maximum velocity of this reaction?

20H.2(b) The enzyme-catalysed conversion of a substrate at 25 °C has a Michaelis constant of 0.032 mol dm^{-3}. The rate of the reaction is 0.205 mmol dm^{-3} s^{-1} when the substrate concentration is 0.875 mol dm^{-3}. What is the maximum velocity of this reaction?

20H.3(a) Consider an enzyme-catalysed reaction that follows Michaelis–Menten kinetics with $K_M=3.0$ mmol dm^{-3}. What concentration of a competitive inhibitor characterized by $K_I=20$ μmol dm^{-3} will reduce the rate of formation of product by 50 per cent when the substrate concentration is held at 0.10 mmol dm^{-3}?

20H.3(b) Consider an enzyme-catalysed reaction that follows Michaelis–Menten kinetics with $K_M=0.75$ mmol dm^{-3}. What concentration of a competitive inhibitor characterized by $K_I=0.56$ mmol dm^{-3} will reduce the rate of formation of product by 75 per cent when the substrate concentration is held at 0.10 mmol dm^{-3}?

Problems

20H.1 Michaelis and Menten derived their rate law by assuming a rapid pre-equilibrium of E, S, and ES. Derive the rate law in this manner, and identify the conditions under which it becomes the same as that based on the steady-state approximation (eqn 20H.1).

20H.2 (a) Use the Michaelis–Menten equation (eqn 20H.1) to generate two families of curves showing the dependence of v on [S]: one in which K_M varies but v_{max} is constant, and another in which v_{max} varies but K_M is constant. (b) Use eqn 20H.7 to explore the effect of competitive, uncompetitive, and non-competitive inhibition on the shapes of the plots of v against [S] for constant K_M and v_{max}. Use mathematical software, a spreadsheet, or the *Living graphs* on the web site of this book.

20H.3 For many enzymes, the mechanism of action involves the formation of two intermediates:

$\mathrm{E}+\mathrm{S}\rightarrow\mathrm{ES}$	$v=k_a[\mathrm{E}][\mathrm{S}]$
$\mathrm{ES}\rightarrow\mathrm{E}+\mathrm{S}$	$v=k_a'[\mathrm{ES}]$
$\mathrm{ES}\rightarrow\mathrm{ES}'$	$v=k_b[\mathrm{ES}]$
$\mathrm{ES}'\rightarrow\mathrm{E}+\mathrm{P}$	$v=k_c[\mathrm{ES}']$

Show that the rate of formation of product has the same form as that shown in eqn 20H.1, but with v_{max} and K_M given by

$$v_{max}=\frac{k_b k_c[\mathrm{E}]_0}{k_b+k_c} \quad \text{and} \quad K_M=\frac{k_c(k_a'+k_b)}{k_a(k_b+k_c)}$$

20H.4 The enzyme-catalysed conversion of a substrate at 25 °C has a Michaelis constant of 90 μmol dm^{-3} and a maximum velocity of 22.4 μmol dm^{-3} s^{-1} when the enzyme concentration is 1.60 nmol dm^{-3}. (a) Calculate k_{cat} and η. (b) Is the enzyme 'catalytically perfect'?

20H.5 The following results were obtained for the action of an ATPase on ATP at 20 °C, when the concentration of the ATPase was 20 nmol dm^{-3}:

[ATP]/(μmol dm^{-3})	0.60	0.80	1.4	2.0	3.0
v/(μmol dm^{-3} s^{-1})	0.81	0.97	1.30	1.47	1.69

Determine the Michaelis constant, the maximum velocity of the reaction, the turnover number, and the catalytic efficiency of the enzyme.

20H.6 Some enzymes are inhibited by high concentrations of their own substrates. (a) Show that when substrate inhibition is important the reaction rate v is given by

$$v=\frac{v_{max}}{1+K_M/[\mathrm{S}]_0+[\mathrm{S}]_0/K_I}$$

where K_I is the equilibrium constant for dissociation of the inhibited enzyme–substrate complex. (b) What effect does substrate inhibition have on a plot of $1/v$ against $1/[\mathrm{S}]_0$?

Integrated activities

20.1 Autocatalysis is the catalysis of a reaction by the products. For example, for a reaction $\mathrm{A}\rightarrow\mathrm{P}$ it may be found that the rate law is $v=k_r[\mathrm{A}][\mathrm{P}]$ and the reaction rate is proportional to the concentration of P. The reaction gets started because there are usually other reaction routes for the formation of some P initially, which then takes part in the autocatalytic reaction proper. (a) Integrate the rate equation for an autocatalytic reaction of the form $\mathrm{A}\rightarrow\mathrm{P}$, with rate law $v=k_r[\mathrm{A}][\mathrm{P}]$, and show that

$$\frac{[\mathrm{P}]}{[\mathrm{P}]_0}=\frac{(1+b)e^{at}}{1+be^{at}}$$

where $a=([\mathrm{A}]_0+[\mathrm{P}]_0)k_r$ and $b=[\mathrm{P}]_0/[\mathrm{A}]_0$. *Hint*: Starting with the expression $v=-\mathrm{d}[\mathrm{A}]/\mathrm{d}t=k_r[\mathrm{A}][\mathrm{P}]$, write $[\mathrm{A}]=[\mathrm{A}]_0-x$, $[\mathrm{P}]=[\mathrm{P}]_0+x$ and then write the expression for the rate of change of either species in terms of x. To integrate the resulting expression, use integration by the method of partial fractions

(*The chemist's toolbox* 20B.1). (b) Plot $[P]/[P]_0$ against at for several values of b. Discuss the effect of autocatalysis on the shape of a plot of $[P]/[P]_0$ against t by comparing your results with those for a first-order process, in which $[P]/[P]_0=1-e^{-k_rt}$. (c) Show that for the autocatalytic process discussed in parts (a) and (b), the reaction rate reaches a maximum at $t_{max}=-(1/a)\ln b$. (d) An autocatalytic reaction A→P is observed to have the rate law $d[P]/dt=k_r[A]^2[P]$. Solve the rate law for initial concentrations $[A]_0$ and $[P]_0$. Calculate the time at which the rate reaches a maximum. (e) Another reaction with the stoichiometry A→P has the rate law $d[P]/dt=k_r[A][P]^2$; integrate the rate law for initial concentrations $[A]_0$ and $[P]_0$. Calculate the time at which the rate reaches a maximum.

20.2 Many biological and biochemical processes involve autocatalytic steps (Problem 20.1). In the SIR model of the spread and decline of infectious diseases the population is divided into three classes; the 'susceptibles', S, who can catch the disease, the 'infectives', I, who have the disease and can transmit it, and the 'removed class', R, who have either had the disease and recovered, are dead, are immune, or isolated. The model mechanism for this process implies the following rate laws:

$$\frac{dS}{dt}=-rSI \qquad \frac{dI}{dt}=rSI-aI \qquad \frac{dR}{dt}=aI$$

Which are the autocatalytic steps of this mechanism? Find the conditions on the ratio a/r that decide whether the disease will spread (an epidemic) or die out. Show that a constant population is built into this system, namely that S+I+R=N, meaning that the time scales of births, deaths by other causes, and migration are assumed large compared to that of the spread of the disease.

20.3‡ J. Czarnowski and H.J. Schuhmacher (*Chem. Phys. Lett.* **17**, 235 (1972)) suggested the following mechanism for the thermal decomposition of F_2O in the reaction $2\,F_2O(g) \rightarrow 2\,F_2(g)+O_2(g)$:

(1) $F_2O+F_2O \rightarrow F+OF+F_2O$	k_a
(2) $F+F_2O \rightarrow F_2+OF$	k_b
(3) $OF+OF \rightarrow O_2+F+F$	k_c
(4) $F+F+F_2O \rightarrow F_2+F_2O$	k_d

(a) Using the steady-state approximation, show that this mechanism is consistent with the experimental rate law $-d[F_2O]/dt=k_r[F_2O]^2+k_r'[F_2O]^{3/2}$. (b) The experimentally determined Arrhenius parameters in the range 501–583 K are $A=7.8\times10^{13}\,dm^3\,mol^{-1}\,s^{-1}$, $E_a/R=1.935\times10^4$ K for k_r and $A=2.3\times10^{10}\,dm^3\,mol^{-1}\,s^{-1}$, $E_a/R=1.691\times10^4$ K for k_r'. At 540 K, $\Delta_fH^{\ominus}(F_2O)=+24.41\,kJ\,mol^{-1}$, $D(F\text{–}F)=160.6\,kJ\,mol^{-1}$, and $D(O\text{–}O)=498.2\,kJ\,mol^{-1}$. Estimate the bond dissociation energies of the first and second F–O bonds in F_2O and the Arrhenius activation energy of reaction 2.

20.4 Express the root mean square deviation $\{\langle M^2\rangle-\langle M\rangle^2\}^{1/2}$ of the molar mass of a condensation polymer in terms of the fraction p, and deduce its time-dependence.

20.5 Calculate the ratio of the mean cube molar mass to the mean square molar mass in terms of (a) the fraction p, (b) the chain length.

20.6 Conventional equilibrium considerations do not apply when a reaction is being driven by light absorption. Thus the steady-state concentration of products and reactants might differ significantly from equilibrium values. For instance, suppose the reaction A→B is driven by light absorption, and that its rate is I_a, but that the reverse reaction B→A is bimolecular and second order with a rate $k_r[B]^2$. What is the stationary state concentration of B? Why does this 'photostationary state' differ from the equilibrium state?

20.7 The photochemical chlorination of chloroform in the gas phase has been found to follow the rate law $d[CCl_4]/dt=k_r[Cl_2]^{1/2}I_a^{1/2}$. Devise a mechanism that leads to this rate law when the chlorine pressure is high.

CHAPTER 21

Reaction dynamics

Now we are at the heart of chemistry. In this chapter we examine the details of what happens to molecules at the climax of reactions. Extensive changes of structure are taking place and energies the size of dissociation energies are being redistributed among bonds: old bonds are being ripped apart and new bonds are being formed.

As may be imagined, the calculation of the rates of such processes from first principles is very difficult. Nevertheless, like so many intricate problems, the broad features can be established quite simply. Only when we enquire more deeply do the complications emerge. Here we look at several approaches to the calculation of a rate constant for elementary bimolecular processes, ranging from electron transfer to chemical reactions involving bond breakage and formation. Although a great deal of information can be obtained from gas-phase reactions, many reactions of interest take place in condensed phases, and we also see to what extent their rates can be predicted.

21A Collision theory

This Topic explores 'collision theory', the simplest quantitative account of reaction rates. The treatment can be used only for the discussion of reactions between simple species in the gas phase.

21B Diffusion-controlled reactions

In this Topic we see that reactions in solution are classified into two types: 'diffusion-controlled' and 'activation-controlled'. The former can be expressed quantitatively in terms of the diffusion equation.

21C Transition-state theory

This Topic discusses 'transition-state theory', in which it is assumed that the reactant molecules form a complex that can be discussed in terms of the population of its energy levels. The theory inspires a thermodynamic approach to reaction rates, in which the rate constant is expressed in terms of thermodynamic parameters. This approach is useful for parameterizing the rates of reactions in solution.

21D The dynamics of molecular collisions

The highest level of sophistication in the theoretical study of chemical reactions is in terms of potential energy surfaces and the motion of molecules on these surfaces. As we see in this Topic, such an approach gives an intimate picture of the events that occur when reactions occur and is open to experimental study.

21E Electron transfer in homogeneous systems

In this Topic we use transition-state theory to examine the transfer of electrons in homogeneous systems, including those involving proteins.

21F Processes at electrodes

Electron transfer processes on the surface of electrodes are difficult to describe theoretically, but in this Topic we develop a useful phenomenological approach that lends insight into useful experimental techniques and technological applications of electrochemistry.

What is the impact of this material?

The economic consequences of electron transfer reactions are almost incalculable. Most of the modern methods of generating electricity are inefficient, and in *Impact* I21.1 we see how the development of special electrochemical cells known as ‘fuel cells’ could revolutionize our production and deployment of energy.

To read more about the impact of this material, scan the QR code, or go to bcs.whfreeman.com/webpub/chemistry/pchem10e/impact/pchem-21-1.html

21A Collision theory

Contents

➤ Why do you need to know this material?

A major component of chemistry is the study of the mechanisms of chemical reactions. One of the earliest approaches, which continues to give insight into the details of mechanisms, is collision theory.

➤ What is the key idea?

According to collision theory, in a bimolecular gas-phase reaction, a reaction takes place on the collision of reactants provided their relative kinetic energy exceeds a threshold value and certain steric requirements are fulfilled.

➤ What do you need to know already?

This Topic draws on the kinetic theory of gases (Topic 1B) and extends the account of unimolecular reactions (Topic 20F). The latter uses combinatorial arguments like those described in Topic 15A.

In this Topic we consider the bimolecular elementary reaction

$$A+B \rightarrow P \qquad v=k_r[A][B] \tag{21A.1a}$$

where P denotes products. Our aim is to calculate the second-order rate constant k_r and to justify the form of the Arrhenius expression (Topic 20D):

$$k_r = Ae^{-E_a/RT} \qquad \text{Arrhenius expression} \tag{21A.1b}$$

where A is the 'pre-exponential factor' and E_a is the 'activation energy'. The model is then improved by examining how the energy of a collision is distributed over all the bonds in the reactant molecule. This improvement helps to account for the value of the rate constant k_b that appears in the Lindemann theory of unimolecular reactions (Topic 20F).

21A.1 Reactive encounters

We can anticipate the general form of the expression for k_r in eqn 21A.1a by considering the physical requirements for reaction. We can expect the rate v to be proportional to the rate of collisions, and therefore to the mean speed of the molecules, $v_{mean} \propto (T/M)^{1/2}$ where M is some combination of the molar masses of A and B; we also expect the rate to be proportional to their collision cross-section, σ, (Topic 1B) and to the number densities $\mathcal{N}_A$ and $\mathcal{N}_B$ of A and B:

$$v \propto \sigma(T/M)^{1/2}\mathcal{N}_A\mathcal{N}_B \propto \sigma(T/M)^{1/2}[A][B]$$

However, a collision will be successful only if the kinetic energy exceeds a minimum value which we denote E'. This requirement suggests that the rate should also be proportional to a Boltzmann factor of the form $e^{-E'/RT}$ representing the fraction of collisions with at least the minimum required energy E'. Therefore,

$$v \propto \sigma(T/M)^{1/2} e^{-E'/RT}[A][B]$$

and we can anticipate, by writing the reaction rate in the form given in eqn 21A.1, that

$$k_r \propto \sigma(T/M)^{1/2} e^{-E'/RT}$$

At this point, we begin to recognize the form of the Arrhenius equation, eqn 21A.1b, and identify the minimum kinetic energy E' with the activation energy E_a of the reaction. This identification, however, should not be regarded as precise, since collision theory is only a rudimentary model of chemical reactivity.

Not every collision will lead to reaction even if the energy requirement is satisfied, because the reactants may need to collide in a certain relative orientation. This 'steric requirement' suggests that a further factor, P, should be introduced, and that

$$k_r \propto P\sigma(T/M)^{1/2}\mathrm{e}^{-E'/RT} \qquad (21A.2)$$

As we shall see in detail below, this expression has the form predicted by collision theory. It reflects three aspects of a successful collision:

$$k_r \propto \overbrace{P}^{\text{Steric requirement}}\ \overbrace{\sigma(T/M)^{1/2}}^{\text{Encounter rate}}\ \overbrace{\mathrm{e}^{-E'/RT}}^{\text{Minimum energy requirement}}$$

(a) Collision rates in gases

We have anticipated that the reaction rate, and hence k_r, depends on the frequency with which molecules collide. The **collision density**, Z_{AB}, is the number of (A,B) collisions in a region of the sample in an interval of time divided by the volume of the region and the duration of the interval. The frequency of collisions of a single molecule in a gas was calculated in Topic 1B (eqn 1B.11a, $z=\sigma v\mathcal{N}_A$). As shown in the following *Justification*, that result can be adapted to deduce that

$$Z_{AB}=\sigma\left(\frac{8kT}{\pi\mu}\right)^{1/2}N_A^2[\mathrm{A}][\mathrm{B}] \quad \textit{KMT} \quad \text{Collision density} \qquad (21A.3a)$$

where σ is the collision cross-section (Fig. 21A.1)

$$\sigma=\pi d^2 \qquad d=\tfrac{1}{2}(d_A+d_B) \qquad \text{Collision cross-section} \qquad (21A.3b)$$

d_A and d_B are the diameters of A and B, respectively, and μ is the reduced mass,

$$\mu=\frac{m_A m_B}{m_A+m_B} \qquad \text{Reduced mass} \qquad (21A.3c)$$

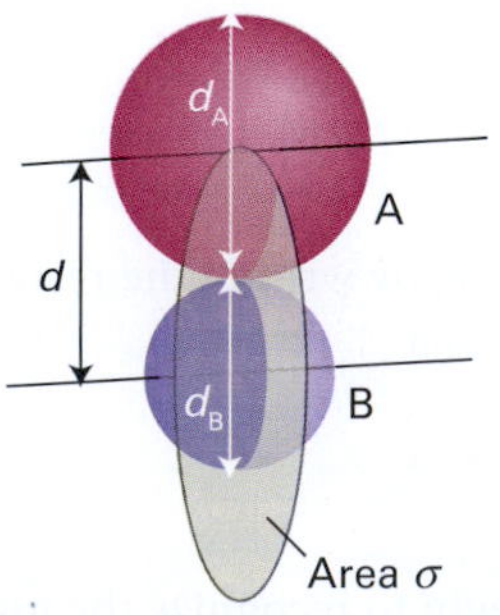

Figure 21A.1 The collision cross-section for two molecules can be regarded to be the area within which the projectile molecule (A) must enter around the target molecule (B) in order for a collision to occur. If the diameters of the two molecules are d_A and d_B, the radius of the target area is $d=\frac{1}{2}(d_A+d_B)$ and the cross-section is πd^2.

For like molecules $\mu=\frac{1}{2}m_A$ and at a molar concentration [A]

$$Z_{AA}=\tfrac{1}{2}\sigma\left(\frac{16kT}{\pi m_A}\right)^{1/2}N_A^2[\mathrm{A}]^2=\sigma\left(\frac{4kT}{\pi m_A}\right)^{1/2}N_A^2[\mathrm{A}]^2 \qquad (21A.3d)$$

The (blue) factor of $\frac{1}{2}$ is included to avoid double counting of collisions in this instance. If the collision density is required in terms of the pressure of each gas J, then we use $[\mathrm{J}]=n_J/V=p_J/RT$.

Brief illustration 21A.1 Collision density

Collision densities may be very large. For example, in nitrogen at 25 °C and 1.0 bar, when $[\mathrm{N_2}]\approx 40\,\mathrm{mol\,m^{-3}}$, with $\sigma=0.43\,\mathrm{nm^2}$ and $m_{N_2}=28.02m_u$ the collision density is

$$Z_{N_2N_2}=(4.3\times10^{-19}\,\mathrm{m^2})\times\left(\frac{4\times(1.381\times10^{-23}\,\mathrm{J\,K^{-1}})\times(298\,\mathrm{K})}{\pi\times28.02\times(1.661\times10^{-27}\,\mathrm{kg})}\right)^{1/2}$$
$$\times(6.022\times10^{23}\,\mathrm{mol^{-1}})^2\times(40\,\mathrm{mol\,m^{-3}})^2=8.4\times10^{34}\,\mathrm{m^{-3}\,s^{-1}}$$

Even in 1 cm³, there are over 8×10^{16} collisions in each picosecond.

Self-test 21A.1 Calculate the collision density in molecular hydrogen under the same conditions.

Answer: $Z_{H_2H_2}=2.0\times10^{35}\,\mathrm{m^{-3}\,s^{-1}}$

Justification 21A.1 The collision density

It follows from Topic 1B that the collision frequency, z, for a single A molecule of mass m_A in a gas of other A molecules is $z=\sigma v_{rel}\mathcal{N}_A$, where $\mathcal{N}_A$ is the number density of A molecules and v_{rel} is their relative mean speed. As indicated in Topic 1B, $v_{rel}=2^{1/2}v_{mean}$ with $v_{mean}=(8kT/\pi m)^{1/2}$. For future convenience, it is sensible to introduce $\mu=\frac{1}{2}m$ (for like molecules of mass m), and then to write $v_{rel}=(8kT/\pi\mu)^{1/2}$. This expression also applies to the mean relative speed of dissimilar molecules provided that μ is interpreted as their reduced mass.

The total collision density is the collision frequency multiplied by the number density of A molecules:

$$Z_{AA}=\tfrac{1}{2}z\mathcal{N}_A=\tfrac{1}{2}\sigma v_{rel}\mathcal{N}_A^2$$

The factor of $\frac{1}{2}$ has been introduced to avoid double counting of the collisions (so one A molecule colliding with another A molecule is counted as one collision regardless of their actual identities). For collisions of A and B molecules present at number densities $\mathcal{N}_A$ and $\mathcal{N}_B$, the collision density is

$$Z_{AB}=\sigma v_{rel}\mathcal{N}_A\mathcal{N}_B$$

The factor of $\frac{1}{2}$ has been discarded because now we are considering an A molecule colliding with any of the B molecules as a collision. The number density of a species J is $\mathcal{N}_J=N_A[\mathrm{J}]$, where [J] is their molar concentration and N_A is Avogadro's constant. Equation 21A.3 then follows.

(b) The energy requirement

According to collision theory, the rate of change in the number density, $\mathcal{N}_A$, of A molecules is the product of the collision density and the probability that a collision occurs with sufficient energy. The latter condition can be incorporated by writing the collision cross-section σ as a function of the kinetic energy ε of approach of the two colliding species, and setting the cross-section, $\sigma(\varepsilon)$, equal to zero if the kinetic energy of approach is below a certain threshold value, ε_a. Later, we shall identify $N_A\varepsilon_a$ as E_a, the (molar) activation energy of the reaction. Then, for a collision between A and B with a specific relative speed of approach v_{rel} (not, at this stage, a mean value),

$$\frac{d\mathcal{N}_A}{dt} = -\sigma(\varepsilon)v_{rel}\mathcal{N}_A\mathcal{N}_B \qquad (21A.4a)$$

or, in terms of molar concentrations,

$$\frac{d[A]}{dt} = -\sigma(\varepsilon)v_{rel}N_A[A][B] \qquad (21A.4b)$$

The kinetic energy associated with the relative motion of the two particles takes the form $\varepsilon = \frac{1}{2}\mu v_{rel}^2$ when the centre-of-mass coordinates are separated from the internal coordinates of each particle. Therefore the relative speed is given by $v_{rel} = (2\varepsilon/\mu)^{1/2}$. At this point we recognize that a wide range of approach energies ε is present in a sample, so we should average the expression just derived over a Boltzmann distribution of energies $f(\varepsilon)$, and write

$$\frac{d[A]}{dt} = -\left\{\int_0^\infty \sigma(\varepsilon)v_{rel}f(\varepsilon)d\varepsilon\right\}N_A[A][B] \qquad (21A.5)$$

and hence recognize the rate constant as

$$k_r = N_A\int_0^\infty \sigma(\varepsilon)v_{rel}f(\varepsilon)d\varepsilon \qquad \text{Rate constant} \qquad (21A.6)$$

Now suppose that the reactive collision cross-section is zero below ε_a. We show in the following *Justification* that, above ε_a, $\sigma(\varepsilon)$ varies as

$$\sigma(\varepsilon) = \left(1 - \frac{\varepsilon_a}{\varepsilon}\right)\sigma \qquad \text{Energy dependence of } \sigma \qquad (21A.7)$$

with the energy-independent σ given by eqn 21A.3b. This form of the energy-dependence for $\sigma(\varepsilon)$ is broadly consistent with experimental determinations of the reaction between H and D_2 as determined by molecular beam measurements of the kind described in Topic 21D (Fig. 21A.2).

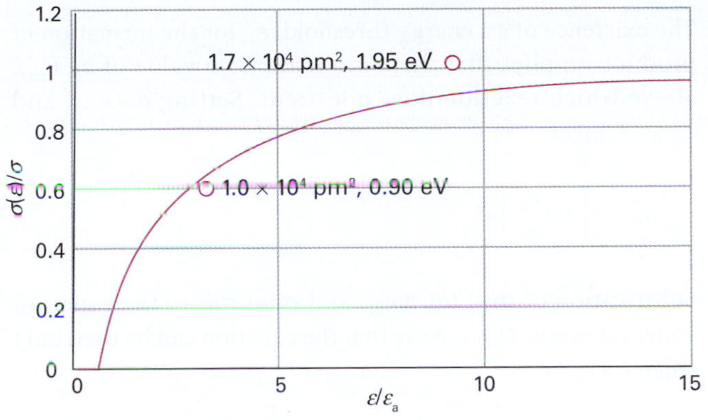

Figure 21A.2 The variation of the reactive cross-section with energy as expressed by eqn 21A.7. The data points are from experiments on the reaction $H + D_2 \rightarrow HD + D$ (K. Tsukiyama et al., *J. Chem. Phys.* **84**, 1934 (1986)).

Justification 21A.2 The collision cross-section

Consider two colliding molecules A and B with relative speed v_{rel} and relative kinetic energy $\varepsilon = \frac{1}{2}\mu v_{rel}^2$ (Fig. 21A.3). Intuitively, we expect that a head-on collision between A and B will be most effective in bringing about a chemical reaction. Therefore, $v_{rel,A-B}$, the magnitude of the relative velocity component parallel to an axis that contains the vector connecting the centres of A and B, must be large. From trigonometry and the definitions of the distances a and d and the angle θ given in Fig. 21A.3, it follows that

$$v_{rel,A-B} = v_{rel}\cos\theta = v_{rel}\left(\frac{d^2 - a^2}{d^2}\right)^{1/2}$$

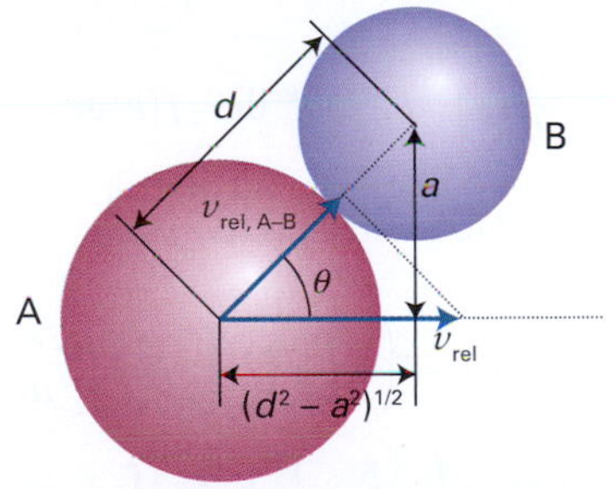

Figure 21A.3 The parameters used in the calculation of the dependence of the collision cross-section on the relative kinetic energy of two molecules A and B.

We assume that only the kinetic energy associated with the head-on component of the collision, ε_{A-B}, can lead to a chemical reaction. After squaring both sides of this equation and multiplying by $\frac{1}{2}\mu$, it follows that

$$\varepsilon_{A-B} = \varepsilon \times \frac{d^2 - a^2}{d^2}$$

The existence of an energy threshold, ε_a, for the formation of products implies that there is a maximum value of a, a_{max}, above which reaction does not occur. Setting $a=a_{max}$ and $\varepsilon_{A\text{-}B}=\varepsilon_a$ gives

$$a_{max}^2=\left(1-\frac{\varepsilon_a}{\varepsilon}\right)d^2$$

Substitution of $\sigma(\varepsilon)$ for πa_{max}^2 and σ for πd^2 in the equation above gives eqn 21A.7. Note that the equation can be used only when $\varepsilon>\varepsilon_a$.

With the energy dependence of the collision cross-section established, we can evaluate the integral in eqn 21A.6. In the following *Justification* we show that

$$k_r=\sigma N_A v_{rel} e^{-E_a/RT} \quad \textit{Collision theory} \quad \text{Rate constant} \quad (21A.8)$$

Justification 21A.3 The rate constant

The Maxwell–Boltzmann distribution of molecular speeds is eqn 1B.4 of Topic 1B:

$$f(v)dv=4\pi\left(\frac{\mu}{2\pi kT}\right)^{3/2} v^2 e^{-\mu v^2/2kT}dv$$

(We have replaced M/R by μ/k.) This expression may be written in terms of the kinetic energy, ε, by writing $\varepsilon=\frac{1}{2}\mu v^2$; then $dv=d\varepsilon/(2\mu\varepsilon)^{1/2}$, when it becomes

$$f(v)dv=4\pi\left(\frac{\mu}{2\pi kT}\right)^{3/2}\left(\frac{2\varepsilon}{\mu}\right)e^{-\varepsilon/kT}\frac{d\varepsilon}{(2\mu\varepsilon)^{1/2}}$$
$$=2\pi\left(\frac{1}{\pi kT}\right)^{3/2}\varepsilon^{1/2}e^{-\varepsilon/kT}d\varepsilon=f(\varepsilon)d\varepsilon$$

The integral we need to evaluate is therefore

$$\int_0^\infty \sigma(\varepsilon)\overbrace{v_{rel}}^{(2\varepsilon/\mu)^{1/2}} f(\varepsilon)d\varepsilon=2\pi\left(\frac{1}{\pi kT}\right)^{3/2}\int_0^\infty \sigma(\varepsilon)\left(\frac{2\varepsilon}{\mu}\right)^{1/2}\varepsilon^{1/2}e^{-\varepsilon/kT}d\varepsilon$$
$$=\left(\frac{8}{\pi\mu kT}\right)^{1/2}\left(\frac{1}{kT}\right)\int_0^\infty \varepsilon\sigma(\varepsilon)e^{-\varepsilon/kT}d\varepsilon$$

To proceed, we introduce the approximation for $\sigma(\varepsilon)$ in eqn 21A.7 and evaluate

$$\int_0^\infty \varepsilon\sigma(\varepsilon)e^{-\varepsilon/kT}d\varepsilon \overset{\sigma=0\text{ for }\varepsilon<\varepsilon_a}{=} \sigma\int_{\varepsilon_a}^\infty \varepsilon\left(1-\frac{\varepsilon_a}{\varepsilon}\right)e^{-\varepsilon/kT}d\varepsilon$$
$$=\sigma\left\{\int_{\varepsilon_a}^\infty \varepsilon e^{-\varepsilon/kT}d\varepsilon-\int_{\varepsilon_a}^\infty \varepsilon_a e^{-\varepsilon/kT}d\varepsilon\right\}$$
$$\overset{\text{Integral E.1}}{=} (kT)^2\sigma e^{-\varepsilon_a/kT}$$

It follows that

$$\int_0^\infty \sigma(\varepsilon)v_{rel}f(\varepsilon)d\varepsilon=\sigma\left(\frac{8kT}{\pi\mu}\right)^{1/2}e^{-\varepsilon_a/kT}$$

as in eqn 21A.8 (with $\varepsilon_a/kT=E_a/RT$).

Equation 21A.8 has the Arrhenius form $k_r=Ae^{-E_a/RT}$ provided the exponential temperature dependence dominates the weak square-root temperature dependence of the pre-exponential factor. It follows that we can identify (within the constraints of collision theory) the activation energy, E_a, with the minimum kinetic energy along the line of approach that is needed for reaction, and that the pre-exponential factor is a measure of the rate at which collisions occur in the gas.

The simplest procedure for calculating k_r is to use for σ the values obtained for non-reactive collisions (for example, typically those obtained from viscosity measurements) or from tables of molecular radii. If the collision cross-sections of A and B are $\sigma_A=\pi d_A^2$ and $\sigma_B=\pi d_B^2$, then an approximate value of the AB cross-section can be estimated from $\sigma=\pi d^2$, with $d=\frac{1}{2}(d_A+d_B)$. That is,

$$\sigma\approx\tfrac{1}{4}(\sigma_A^{1/2}+\sigma_B^{1/2})^2$$

Brief illustration 21A.2 The rate constant

To estimate the rate constant for the reaction $H_2+C_2H_4\rightarrow C_2H_6$ at 628 K we first calculate the reduced mass using $m(H_2)=2.016m_u$ and $m(C_2H_4)=28.05m_u$. A straightforward calculation gives $\mu=3.123\times10^{-27}$ kg. It then follows that

$$\left(\frac{8kT}{\pi\mu}\right)^{1/2}=\left(\frac{8\times(1.381\times10^{-23}\,\mathrm{J\,K^{-1}})\times(628\,\mathrm{K})}{\pi\times(3.123\times10^{-27}\,\mathrm{kg})}\right)^{1/2}=2.65\ldots\,\mathrm{km\,s^{-1}}$$

From Table 1B.1, $\sigma(H_2)=0.27\,\mathrm{nm^2}$ and $\sigma(C_2H_4)=0.64\,\mathrm{nm^2}$, giving $\sigma(H_2,C_2H_4)\approx0.44\,\mathrm{nm^2}$. The activation energy, Table 20D.1, is large: 180 kJ mol⁻¹. Therefore,

$$k_r=(4.4\times10^{-19}\,\mathrm{m^2})\times(2.65\ldots\times10^3\,\mathrm{m\,s^{-1}})\times(6.022\times10^{23}\,\mathrm{mol^{-1}})\times e^{-(1.80\times10^5\,\mathrm{J\,mol^{-1}})/(8.3145\,\mathrm{J\,K^{-1}\,mol^{-1}})\times(628\,\mathrm{K})}$$
$$=\overbrace{7.04\ldots\times10^8\,\mathrm{m^3\,mol^{-1}\,s^{-1}}}^{A}\times e^{-34.4\ldots}=7.5\times10^{-7}\,\mathrm{m^3\,mol^{-1}\,s^{-1}}$$

or $7.5\times10^{-4}\,\mathrm{dm^3\,mol^{-1}\,s^{-1}}$.

Self-test 21A.2 Evaluate the rate constant for the reaction $NO+Cl_2\rightarrow NOCl+Cl$ at 298 K from $\sigma(NO)=0.42\,\mathrm{nm^2}$ and $\sigma(Cl_2)=0.93\,\mathrm{nm^2}$ and data in Table 1B.1.

Answer: $2.7\times10^{-4}\,\mathrm{dm^3\,mol^{-1}\,s^{-1}}$

(c) The steric requirement

Table 21A.1 compares some values of the pre-exponential factor calculated from the collisional data in Table 1B.1 with values obtained from Arrhenius plots. One of the reactions shows fair agreement between theory and experiment, but for others there are major discrepancies. In some cases the experimental values are orders of magnitude smaller than those calculated, which suggests that the collision energy is not the only criterion for reaction and that some other feature, such as the relative orientation of the colliding species, is important. Moreover, one reaction in the table has a pre-exponential factor larger than theory, which seems to indicate that the reaction occurs more quickly than the particles collide!

The disagreement between experiment and theory can be eliminated by introducing a **steric factor**, P, and expressing the **reactive cross-section**, σ^*, as a multiple of the collision cross-section, $\sigma^* = P\sigma$ (Fig. 21A.4). Then the rate constant becomes

$$k_r = P\sigma N_A \left(\frac{8kT}{\pi\mu}\right)^{1/2} e^{-E_a/RT} \qquad (21A.9)$$

This expression has the form we anticipated in eqn 21A.2. The steric factor is normally found to be several orders of magnitude smaller than 1.

Table 21A.1* Arrhenius parameters for gas-phase reactions

	$A/(\mathrm{dm^3\,mol^{-1}\,s^{-1}})$		$E_a/(\mathrm{kJ\,mol^{-1}})$	P
	Experiment	Theory		
$2\,NOCl \rightarrow 2\,NO + 2\,Cl$	9.4×10^{9}	5.9×10^{10}	102	0.16
$2\,ClO \rightarrow Cl_2 + O_2$	6.3×10^{7}	2.5×10^{10}	0	2.5×10^{-3}
$H_2 + C_2H_4 \rightarrow C_2H_6$	1.24×10^{6}	7.4×10^{11}	180	1.7×10^{-6}
$K + Br_2 \rightarrow KBr + Br$	1.0×10^{12}	2.1×10^{11}	0	4.8

* More values are given in the *Resource section*.

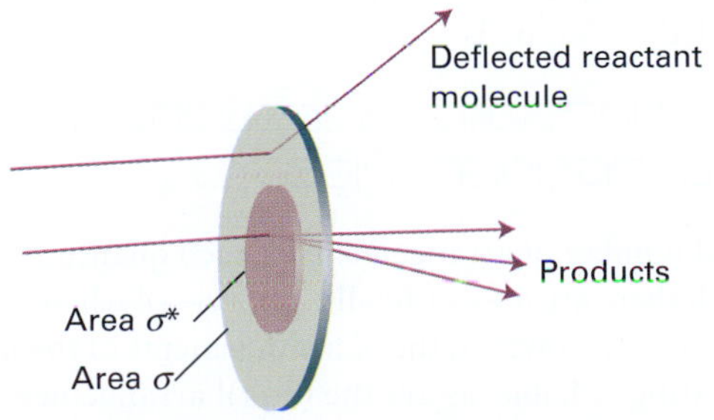

Figure 21A.4 The collision cross-section is the target area that results in simple deflection of the projectile molecule; the reactive cross-section is the corresponding area for chemical change to occur on collision.

Brief illustration 21A.3 The steric factor

It is found experimentally that the pre-exponential factor for the reaction $H_2 + C_2H_4 \rightarrow C_2H_6$ at 628 K is $1.24\times10^{6}\,\mathrm{dm^3\,mol^{-3}\,s^{-1}}$. In *Brief illustration* 21A.2 we calculated the result that can be expressed as $A = 7.04\ldots\times10^{11}\,\mathrm{dm^3\,mol^{-1}\,s^{-1}}$. It follows that the steric factor for this reaction is

$$P = \frac{A_{\text{experimental}}}{A_{\text{calculated}}} = \frac{1.24\times10^{6}\,\mathrm{dm^3\,mol^{-1}\,s^{-1}}}{7.04\ldots\times10^{11}\,\mathrm{dm^3\,mol^{-1}\,s^{-1}}} \approx 1.8\times10^{-6}$$

The very small value of P is one reason why catalysts are needed to bring this reaction about at a reasonable rate. As a general guide, the more complex the reactant molecules, the smaller the value of P.

Self-test 21A.3 It is found for the reaction $NO + Cl_2 \rightarrow NOCl + Cl$ that $A = 4.0\times10^{9}\,\mathrm{dm^3\,mol^{-1}\,s^{-1}}$ at 298 K. Estimate the P factor for the reaction (see *Self-test* 21A.2).

Answer: 0.019

An example of a reaction for which it is possible to estimate the steric factor is $K + Br_2 \rightarrow KBr + Br$, with the experimental value $P = 4.8$. In this reaction, the distance of approach at which reaction occurs appears to be considerably larger than the distance needed for deflection of the path of the approaching molecules in a non-reactive collision. It has been proposed that the reaction proceeds by a **harpoon mechanism**. This brilliant name is based on a model of the reaction which pictures the K atom as approaching a Br_2 molecule, and when the two are close enough an electron (the harpoon) flips across from K to Br_2. In place of two neutral particles there are now two ions, so there is a Coulombic attraction between them: this attraction is the line on the harpoon. Under its influence the ions move together (the line is wound in), the reaction takes place, and KBr + Br emerge. The harpoon extends the cross-section for the reactive encounter, and the reaction rate is significantly underestimated by taking for the collision cross-section the value for simple mechanical contact between K and Br_2.

Example 21A.1 Estimating a steric factor

Estimate the value of P for the harpoon mechanism by calculating the distance at which it becomes energetically favourable for the electron to leap from K to Br_2. Take the sum of the radii of the reactants (treating them as spherical) to be 400 pm.

Method Begin by identifying all the contributions to the energy of interaction between the colliding species. There are three contributions to the energy of the process $K + Br_2 \rightarrow K^+ + Br_2^-$. The first is the ionization energy, I, of K. The second is the electron affinity, E_{ea}, of Br_2. The third is the

Coulombic interaction energy between the ions when they have been formed: when their separation is R, this energy is $-e^2/4\pi\varepsilon_0 R$. The electron flips across when the sum of these three contributions changes from positive to negative (that is, when the sum is zero) and becomes energetically favourable.

Answer The net change in energy when the transfer occurs at a separation R is

$$E = I - E_{ea} - \frac{e^2}{4\pi\varepsilon_0 R}$$

The ionization energy I is larger than E_{ea}, so E becomes negative only when R has decreased to less than some critical value R^* given by

$$R = \frac{e^2}{4\pi\varepsilon_0 (I - E_{ea})}$$

When the particles are at this separation, the harpoon shoots across from K to Br_2, so we can identify the reactive cross-section as $\sigma^* = \pi R^{*2}$. This value of σ^* implies that the steric factor is

$$P = \frac{\sigma^*}{\sigma} = \frac{R^{*2}}{d^2} = \left(\frac{e^2}{4\pi\varepsilon_0 d(I - E_{ea})}\right)^2$$

where $d = R(\mathrm{K}) + R(\mathrm{Br_2})$, the sum of the radii of the spherical reactants. With $I = 420$ kJ mol^{-1} (corresponding to 0.70 aJ), $E_{ea} \approx 250$ kJ mol^{-1} (corresponding to 0.42 aJ), and $d = 400$ pm, we find $P = 4.2$, in good agreement with the experimental value (4.8).

Self-test 21A.4 Estimate the value of P for the harpoon reaction between Na and Cl_2 for which $d \approx 350$ pm; take $E_{ea} \approx 230$ kJ mol^{-1}.

Answer: 2.2

Example 21A.1 illustrates two points about steric factors. First, the concept of a steric factor is not wholly useless because in some cases its numerical value can be estimated. Second, and more pessimistically, most reactions are much more complex than K + Br_2, and we cannot expect to obtain P so easily.

21A.2 The RRK model

The steric factor P can also be estimated for unimolecular gas-phase reactions (Topic 20F), by a calculation based on the **Rice–Ramsperger–Kassel model** (RRK model), which was proposed in 1926 by O.K. Rice and H.C. Ramsperger and almost simultaneously by L.S. Kassel. The model has been elaborated, largely by R.A. Marcus, into the 'RRKM model'. Here we outline Kassel's original approach to the RRK model; the details are set out in the following *Justification*. The essential feature of the model is that although a molecule might have enough energy to react, that energy is distributed over all the modes of motion of the molecule, and reaction will occur only when enough of that energy has migrated into a particular location (such as a particular bond) in the molecule. This distribution effect leads to a P factor of the form

$$P = \left(1 - \frac{E^*}{E}\right)^{s-1} \qquad \text{RRK theory} \qquad (21A.10a)$$

where s is the number of modes of motion over which the energy E may be dissipated and E^* is the energy required for the bond of interest to break. The resulting **Kassel form** of the unimolecular rate constant for the decay of A^* to products is

$$k_b(E) = \left(1 - \frac{E^*}{E}\right)^{s-1} k_b \qquad \text{for } E \geq E^* \qquad \text{Kassel form} \qquad (21A.10b)$$

where k_b is the rate constant used in the original Lindemann theory for the decomposition of the activated intermediate (eqn 20F.8 of Topic 20F).

Justification 21A.4 The RRK model of unimolecular reactions

To set up the RRK model, we suppose that a molecule consists of s identical harmonic oscillators, each of which has frequency ν. In practice, of course, the vibrational modes of a molecule have different frequencies, but assuming that they are all the same is a reasonable first approximation. Next, we suppose that the vibrations are excited to a total energy $E = nh\nu$ and then set out to calculate the number of ways N in which the energy can be distributed over the oscillators.

We can represent the n quanta as follows:

□□□□□□□□□□□□□□□□□□□□□□□□□□
□□□□□□…□□□

These quanta must be put in s containers (the s oscillators), which can be represented by inserting $s-1$ walls, denoted by . One such distribution is

□□|□□□□|□□||□□□|□□□□□□□□|□□□□|||
□□□□□|□□□□…□|□□

The total number of arrangements of each quantum and wall (of which there are $n+s-1$ in all) is $(n+s-1)!$ where, as usual, $x! = x(x-1)\ldots1$. However, the $n!$ arrangements of the n quanta are indistinguishable, as are the $(s-1)!$ arrangements of the $s-1$ walls. Therefore, to find N we must divide $(n+s-1)!$ by these two factorials. It follows that

$$N = \frac{(n+s-1)!}{n!(s-1)!}$$

The distribution of the energy throughout the molecule means that it is too sparsely spread over all the modes for any particular bond to be sufficiently highly excited to undergo dissociation. We suppose that a bond will break only if it is excited to at least an energy $E^* = n^*h\nu$. Therefore, we isolate one critical oscillator as the one that undergoes dissociation if it has *at least* n^* of the quanta, leaving up to $n - n^*$ quanta to be accommodated in the remaining $s-1$ oscillators (and therefore with $s-2$ walls in the partition in place of the $s-1$ walls we used above). For example, consider 28 quanta distributed over six oscillators, with excitation by at least six quanta required for dissociation. Then all the following partitions will result in dissociation:

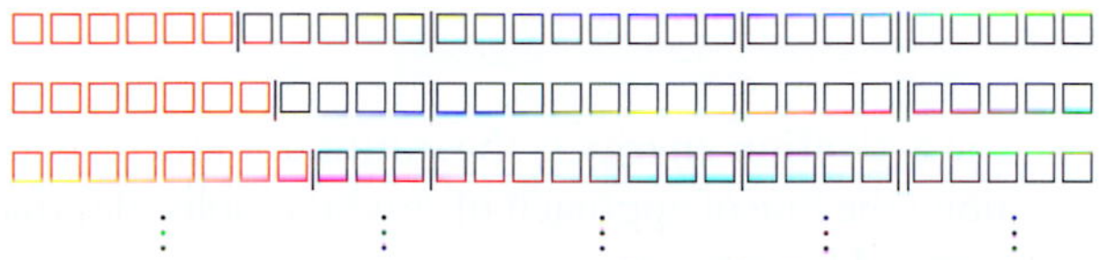

(The leftmost partition is the critical oscillator.) However, these partitions are equivalent to

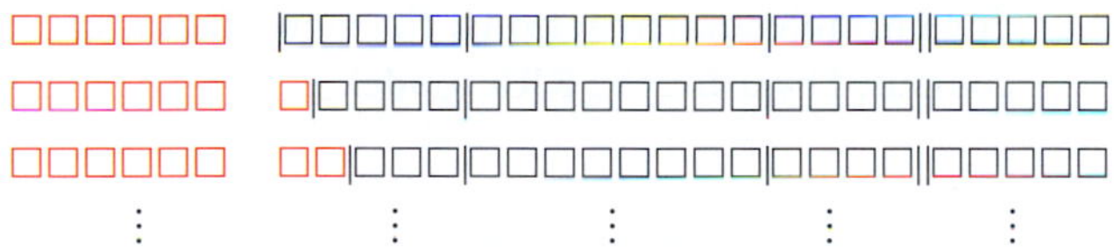

and we see that we have the problem of permuting $28-6=22$ (in general, $n-n^*$) quanta and five (in general, $s-1$) walls, and therefore a total of 27 (in general, $n-n^*+s-1$ objects). Therefore, the calculation is exactly like the one above for N, except that we have to find the number of distinguishable permutations of $n-n^*$ quanta in s containers (and therefore $s-1$ walls). The number N^* is therefore obtained from the expression for N by replacing n by $n-n^*$ and is

$$N^* = \frac{(n-n^*+s-1)!}{(n-n^*)!(s-1)!}$$

From the preceding discussion we conclude that the probability that one specific oscillator will have undergone sufficient excitation to dissociate is the ratio N^*/N, which is

$$P = \frac{N^*}{N} = \frac{n!(n-n^*+s-1)!}{(n-n^*)!(n+s-1)!}$$

This equation is still awkward to use, even when written out in terms of its factors:

$$\begin{aligned} P &= \frac{n(n-1)(n-2)\ldots 1}{(n-n^*)(n-n^*-1)\ldots 1} \times \frac{(n-n^*+s-1)(n-n^*+s-2)\ldots 1}{(n+s-1)(n+s-2)\ldots 1} \\ &= \frac{(n-n^*+s-1)(n-n^*+s-2)\ldots(n-n^*+1)}{(n+s-1)(n+s-2)\ldots(n+2)(n+1)} \end{aligned}$$

However, because $s-1$ is small (in the sense $s-1 \ll n-n^*$), we can approximate this expression by

$$P = \frac{\overbrace{(n-n^*)(n-n^*)\ldots(n-n^*)}^{s-1\text{ factors}}}{\underbrace{(n)(n)\ldots(n)}_{s-1\text{ factors}}} = \left(\frac{n-n^*}{n}\right)^{s-1}$$

An alternative derivation of this expression for P is developed in Problem 21A.7. Because the energy of the excited molecule is $E = nh\nu$ and the critical energy is $E^* = n^*h\nu$, this expression may be written

$$P = \left(1 - \frac{E^*}{E}\right)^{s-1}$$

as in eqn 21A.10.

The energy dependence of the rate constant given by eqn 21A.10b is shown in Fig. 21A.5 for various values of s. We see that the rate constant is smaller at a given excitation energy if s is large, as it takes longer for the excitation energy to migrate through all the oscillators of a large molecule and accumulate in the critical mode. As E becomes very large, however, the term in parentheses approaches 1, and $k_b(E)$ becomes independent of the energy and the number of oscillators in the molecule, as there is now enough energy to accumulate immediately in the critical mode regardless of the size of the molecule.

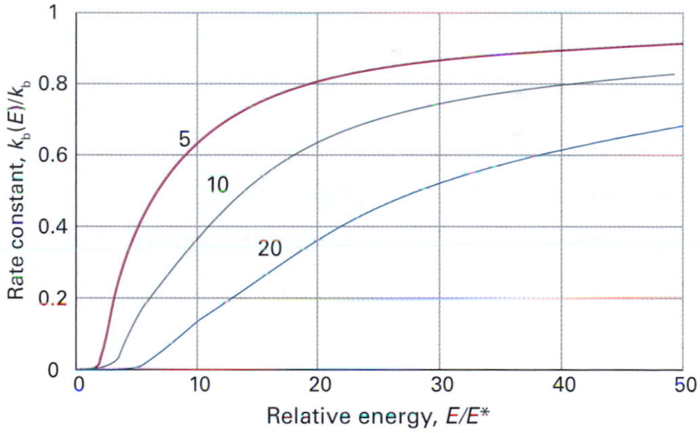

Figure 21A.5 The energy dependence of the rate constant given by eqn 21A.10b for three values of s.

Brief illustration 21A.4 The RRK model

In *Brief illustration* 21A.3 we calculated a value of $P = 1.8 \times 10^{-6}$ for the reaction $H_2 + C_2H_4 \rightarrow C_2H_6$. Although this is not a unimolecular process, it is interesting to analyse it on the basis of the RRK theory because in some sense the collision energy must accumulate in a region where bonds are broken and formed. Thus, C_2H_4 has six atoms and therefore $s=12$ vibrational modes. We can estimate the ratio E^*/E by solving

$$\left(1-\frac{E^*}{E}\right)^{11}=1.8\times10^{-6} \quad \text{or} \quad \frac{E^*}{E}=1-(1.8\times10^{-6})^{1/11}=0.70$$

This result suggests in one interpretation that the energy needed to proceed in the reaction (identified here with the energy to break the carbon-carbon bond in C_2H_4) is typically 70 per cent of the energy of a typical collision. If all eight atoms are taken to be involved in sharing the energy of the collision, the ratio works out as 0.54.

Self-test 21A.5 Apply the same analysis to the reaction in Self-test 21A.3, where it is found that $P=0.019$ for $NO+Cl_2 \rightarrow NOCl+Cl$. Take the number of atoms in the complex to be 4, so $s=6$.

Answer: 0.55

Checklist of concepts

- ☐ 1. In **collision theory**, it is supposed that the rate is proportional to the collision frequency, a steric factor, and the fraction of collisions that occur with at least the kinetic energy E_a along their lines of centres.
- ☐ 2. The **collision density** is the number of collisions in a region of the sample in an interval of time divided by the volume of the region and the duration of the interval.
- ☐ 3. The **activation energy** is the minimum kinetic energy along the line of approach of reactant molecules that is required for reaction.
- ☐ 4. The **steric factor** is an adjustment that takes into account the orientational requirements for a successful collision.
- ☐ 5. For unimolecular reactions, the steric factor can be computed by using the **RRK model**.

Checklist of equations

Property	Equation	Comment	Equation number
Collision density	$Z_{AB}=\sigma(8kT/\pi\mu)^{1/2}N_A^2[A][B]$	Unlike molecules, KMT (kinetic molecular theory)	21A.3a
	$Z_{AA}=\sigma(4kT/\pi m_A)^{1/2}N_A^2[A]^2$	Like molecules, KMT	21A.3d
Energy-dependence of σ	$\sigma(\varepsilon)=(1-\varepsilon_a/\varepsilon)\sigma$	$\varepsilon\geq\varepsilon_a$, 0 otherwise	21A.7
Rate constant	$k_r=P\sigma N_A(8kT/\pi\mu)^{1/2}e^{-E_a/RT}$	KMT, collision theory	21A.9
Steric factor	$P=(1-E^*/E)^{s-1}$	RRK theory	21A.10a

21B Diffusion-controlled reactions

Contents

➤ Why do you need to know this material?

Most chemical reactions take place in solution and for a thorough grasp of chemistry it is important to understand what controls their rates and how those rates can be modified.

➤ What is the key idea?

There are two limiting types of chemical reaction mechanism in solution: diffusion control and activation control.

➤ What do you need to know already?

This Topic makes use of the steady-state approximation (Topic 20E) and draws on the formulation and solution of the diffusion equation (Topic 19C). At one point it uses the Stokes–Einstein relation (Topic 19B).

To consider reactions in solution we have to imagine processes that are entirely different from those in gases. No longer are there collisions of molecules hurtling through space; now there is the jostling of one molecule through a dense but mobile collection of molecules making up the fluid environment.

21B.1 Reactions in solution

Encounters between reactants in solution occur in a very different manner from encounters in gases. The encounters of reactant molecules dissolved in solvent are considerably less frequent than in a gas. However, because a molecule also migrates only slowly away from a location, two reactant molecules that encounter each other stay near each other for much longer than in a gas. This lingering of one molecule near another on account of the hindering presence of solvent molecules is called the **cage effect**. Such an **encounter pair** may accumulate enough energy to react even though it does not have enough energy to do so when it first forms. The activation energy of a reaction is a much more complicated quantity in solution than in a gas because the encounter pair is surrounded by solvent and we need to consider the energy of the entire local assembly of reactant and solvent molecules.

(a) Classes of reaction

The complicated overall process can be divided into simpler parts by setting up a simple kinetic scheme. We suppose that the rate of formation of an encounter pair AB is first order in each of the reactants A and B:

$$\mathrm{A+B \rightarrow AB} \qquad v = k_d[\mathrm{A}][\mathrm{B}]$$

As we shall see, k_d (where the d signifies diffusion) is determined by the diffusional characteristics of A and B. The encounter pair can break up without reaction or it can go on to form products P. If we suppose that both processes are pseudo-first-order reactions (with the solvent perhaps playing a role), then we can write

$$\mathrm{AB \rightarrow A+B} \qquad v = k_d'[\mathrm{AB}]$$

$$\mathrm{AB \rightarrow P} \qquad v = k_a[\mathrm{AB}]$$

The concentration of AB can now be found from the equation for the net rate of change of concentration of AB:

$$\frac{d[\mathrm{AB}]}{dt} = k_d[\mathrm{A}][\mathrm{B}] - k_d'[\mathrm{AB}] - k_a[\mathrm{AB}] = 0$$

where we have applied the steady-state approximation (Topic 20E). This expression solves to

$$[\mathrm{AB}] = \frac{k_d[\mathrm{A}][\mathrm{B}]}{k_a + k_d'}$$

The rate of formation of products is therefore

$$\frac{d[\mathrm{P}]}{dt} = k_a[\mathrm{AB}] = k_r[\mathrm{A}][\mathrm{B}] \qquad k_r = \frac{k_a k_d}{k_a + k_d'} \qquad (21B.1)$$

Two limits can now be distinguished. If the rate of separation of the unreacted encounter pair is much slower than the rate at which it forms products, then $k_d' \ll k_a$ and the effective rate constant is

$$k_r \approx \frac{k_a k_d}{k_a} = k_d \qquad \text{Diffusion-controlled limit} \qquad (21B.2a)$$

In this **diffusion-controlled limit**, the rate of reaction is governed by the rate at which the reactant molecules diffuse through the solvent. Because the combination of radicals involves very little activation energy, radical and atom recombination reactions are often diffusion-controlled. An **activation-controlled reaction** arises when a substantial activation energy is involved in the reaction AB → P. Then $k_a \ll k_d'$ and

$$k_r \approx \frac{k_a k_d}{k_d'} = k_a K \qquad \text{Activation-controlled limit} \qquad (21B.2b)$$

where K is the equilibrium constant for $A+B \rightleftharpoons AB$. In this limit, the reaction proceeds at the rate at which energy accumulates in the encounter pair from the surrounding solvent. Some experimental data are given in Table 21B.1.

Table 21B.1* Arrhenius parameters for solvolysis reactions in solution

	Solvent	$A/(\mathrm{dm^3\,mol^{-1}\,s^{-1}})$	$E_a/(\mathrm{kJ\,mol^{-1}})$
$(CH_3)_3CCl$	Water	7.1×10^{16}	100
	Ethanol	3.0×10^{13}	112
	Chloroform	1.4×10^{4}	45
CH_3CH_2Br	Ethanol	4.3×10^{11}	90

* More values are given in the *Resource section*.

(b) Diffusion and reaction

The rate of a diffusion-controlled reaction is calculated by considering the rate at which the reactants diffuse together. As shown in the following *Justification*, the rate constant for a reaction in which the two molecules react if they come within a distance R^* of one another is

$$k_d = 4\pi R^* D N_A \qquad (21B.3)$$

where D is the sum of the diffusion coefficients of the two reactant species in the solution.

Brief illustration 21B.1 Diffusion control 1

The order of magnitude of R^* is 10^{-7} m (100 nm) and that of D for a species in water is $10^{-9}\,\mathrm{m^2\,s^{-1}}$. It follows from eqn 21B.3 that

$$k_d \approx 4\pi\times(10^{-7}\,\mathrm{m})\times(10^{-9}\,\mathrm{m^2\,s^{-1}})\times(6.022\times10^{23}) \approx 10^{9}\,\mathrm{m^3 mol^{-1} s^{-1}}$$

which corresponds to $10^{12}\,\mathrm{dm^3\,mol^{-1}\,s^{-1}}$. An indication that a reaction is diffusion-controlled is therefore that its rate constant is of the order of $10^{12}\,\mathrm{dm^3\,mol^{-1}\,s^{-1}}$.

Self-test 21B.1 Estimate the rate constant for a diffusion-controlled reaction in benzene ($D \approx 2\times10^{-9}\,\mathrm{m^2\,s^{-1}}$), taking $R^* \approx 100$ nm.

Answer: $1.5\times10^{12}\,\mathrm{dm^3\,mol^{-1}\,s^{-1}}$

Justification 21B.1 Solution of the radial diffusion equation

The general form of the diffusion equation (Topic 19A) corresponding to motion in three dimensions is $D_B\nabla^2[B](r,t) = \partial[B](r,t)/\partial t$; therefore, the concentration of B when the system has reached a steady state ($\partial[B](r,t)]/\partial t = 0$) satisfies $\nabla^2[B](r) = 0$, with the concentration of B now depending only on location not time. For a spherically symmetrical system, ∇^2 can be replaced by radial derivatives alone (see Table 7B.1), so the equation satisfied by $[B](r)$, as $[B](r)$ can now be written, is

$$\frac{d^2[B](r)}{dr^2} + \frac{2}{r}\frac{d[B](r)}{dr} = 0$$

The general solution of this equation is

$$[B](r) = a + \frac{b}{r}$$

as may be verified by substitution. We need two boundary conditions to pin down the values of the two constants (a and b). One condition is that $[B](r)$ has its bulk value $[B]$ as $r\to\infty$. The second condition is that the concentration of B is zero at $r = R^*$, the distance at which reaction occurs. It follows that $a = [B]$ and $b = -R^*[B]$, and hence that (for $r \ge R^*$)

$$[B](r) = \left(1 - \frac{R^*}{r}\right)[B]$$

Figure 21B.1 illustrates the variation of concentration expressed by this equation.

The rate of reaction is the (molar) flux, J, of the reactant B towards A multiplied by the area of the spherical surface of radius R^* through which B must pass:

$$\text{Rate of reaction} = 4\pi R^{*2} J$$

From Fick's first law (eqn 19C.3 of Topic 19C, $J = -D\partial[J]/\partial x$), the flux of B towards A is proportional to the concentration gradient, so at a radius R^*:

$$J = D_B\left(\frac{d[B](r)}{dr}\right)_{r=R^*} = -D_B[B]R^*\left(-\frac{1}{r^2}\right)_{r=R^*} = \frac{D_B[B]}{R^*}$$

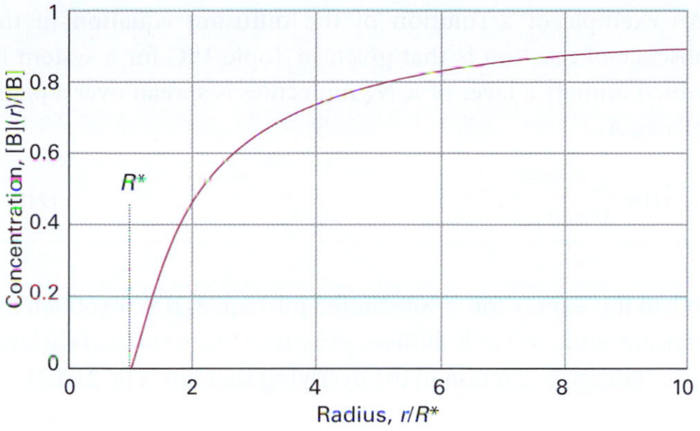

Figure 21B.1 The concentration profile for reaction in solution when a molecule B diffuses towards another reactant molecule and reacts if it reaches R^*.

(A sign change has been introduced because we are interested in the flux towards decreasing values of r.) It follows that

$$\text{Rate of reaction} = 4\pi R^* D_B[\text{B}]$$

The rate of the diffusion-controlled reaction is equal to the average flow of B molecules to all the A molecules in the sample. If the bulk concentration of A is [A], the number of A molecules in the sample of volume V is $N_A[\text{A}]V$; the global flow of all B to all A is therefore $4\pi R^* D_B N_A[\text{A}][\text{B}]V$. Because it is unrealistic to suppose that all A molecules are stationary; we replace D_B by the sum of the diffusion coefficients of the two species and write $D = D_A + D_B$. Then the rate of change of concentration of AB is

$$\frac{d[\text{AB}]}{dt} = 4\pi R^* D N_A[\text{A}][\text{B}]$$

Hence, the diffusion-controlled rate constant is as given in eqn 21B.3.

We can take eqn 21B.3 further by incorporating the Stokes–Einstein equation (eqn 19B.19 of Topic 19B, $D_J = kT/6\pi\eta R_J$) relating the diffusion constant and the hydrodynamic radius R_A and R_B of each molecule in a medium of viscosity η. As this relation is approximate, little extra error is introduced if we write $R_A = R_B = \frac{1}{2}R^*$, which leads to

$$k_d = \frac{8RT}{3\eta} \qquad \text{Diffusion-controlled rate constant} \qquad (21B.4)$$

(The R in this equation is the gas constant.) The radii have cancelled because, although the diffusion constants are smaller when the radii are large, the reactive collision radius is larger and the particles need travel a shorter distance to meet. In this approximation, the rate constant is independent of the identities of the reactants, and depends only on the temperature and the viscosity of the solvent.

Brief illustration 21B.2 Diffusion control 2

The rate constant for the recombination of I atoms in hexane at 298 K, when the viscosity of the solvent is 0.326 cP (with $1\,\text{P} = 10^{-1}\,\text{kg m}^{-1}\,\text{s}^{-1}$) is

$$k_d = \frac{8\times(8.3145\,\text{J K}^{-1}\,\text{mol}^{-1})\times(298\,\text{K})}{3\times(3.26\times10^{-4}\,\text{kg m}^{-1}\,\text{s}^{-1})} = 2.0\times10^7\,\text{m}^3\,\text{mol}^{-1}\,\text{s}^{-1}$$

where we have used $1\,\text{J} = 1\,\text{kg m}^2\,\text{s}^{-2}$. Because $1\,\text{m}^3 = 10^3\,\text{dm}^3$, this result corresponds to $2.0\times10^{10}\,\text{dm}^3\,\text{mol}^{-1}\,\text{s}^{-1}$. The experimental value is $1.3\times10^{10}\,\text{dm}^3\,\text{mol}^{-1}\,\text{s}^{-1}$, so the agreement is very good considering the approximations involved.

Self-test 21B.2 Evaluate a typical rate constant for a reaction taking place in ethanol at 20 °C, for which the viscosity is 1.06 cP.

Answer: $6.1\times10^9\,\text{dm}^3\,\text{mol}^{-1}\,\text{s}^{-1}$

21B.2 The material-balance equation

The diffusion of reactants plays an important role in many chemical processes, such as the diffusion of O_2 molecules into red blood corpuscles and the diffusion of a gas towards a catalyst. We can catch a glimpse of the kinds of calculations involved by considering the diffusion equation (Topic 19C) generalized to take into account the possibility that the diffusing, convecting molecules are also reacting.

(a) The formulation of the equation

Consider a small volume element in a chemical reactor (or a biological cell). The net rate at which J molecules enter the region by diffusion and convection is given by eqn 19C.10 of Topic 19C, which we repeat here:

$$\frac{\partial[\text{J}]}{\partial t} = D\frac{\partial^2[\text{J}]}{\partial x^2} - v\frac{\partial[\text{J}]}{\partial x} \qquad \text{Diffusion equation} \qquad (21B.5)$$

where v is the velocity of the convective flow of J and [J] in general depends on both position and time. The net rate of change of molar concentration due to chemical reaction is

$$\frac{\partial[\text{J}]}{\partial t} = -k_r[\text{J}]$$

if we suppose that J disappears by a pseudofirst-order reaction. Therefore, the overall rate of change of the concentration of J is

$$\frac{\partial[\text{J}]}{\partial t} = D\overbrace{\frac{\partial^2[\text{J}]}{\partial x^2}}^{\text{Spread due to non-uniform distribution}} - v\overbrace{\frac{\partial[\text{J}]}{\partial x}}^{\text{Change due to convection}} - \overbrace{k_r[\text{J}]}^{\text{Loss due to reaction}} \qquad \text{Material-balance equation} \qquad (21B.6)$$

Equation 21B.6 is called the **material-balance equation**. If the rate constant is large, then [J] will decline rapidly. However, if the diffusion constant is large, then the decline can be replenished as J diffuses rapidly into the region. The convection term, which may represent the effects of stirring, can sweep material either into or out of the region according to the signs of v and the concentration gradient $\partial[\mathrm{J}]/\partial x$.

(b) Solutions of the equation

The material-balance equation is a second-order partial differential equation and is far from easy to solve in general. Some idea of how it is solved can be obtained by considering the special case in which there is no convective motion (as in an unstirred reaction vessel):

$$\frac{\partial[\mathrm{J}]}{\partial t}=D\frac{\partial^2[\mathrm{J}]}{\partial x^2}-k_\mathrm{r}[\mathrm{J}] \qquad (21\mathrm{B}.7)$$

As may be verified by substitution (Problem 21B.1), if the solution of this equation in the absence of reaction (that is, for $k_\mathrm{r}=0$) is [J], then the solution $[\mathrm{J}]^*$ in the presence of reaction ($k_\mathrm{r}>0$) is

$$[\mathrm{J}]^*=[\mathrm{J}]\mathrm{e}^{-k_\mathrm{r}t} \qquad \text{Diffusion with reaction} \qquad (21\mathrm{B}.8)$$

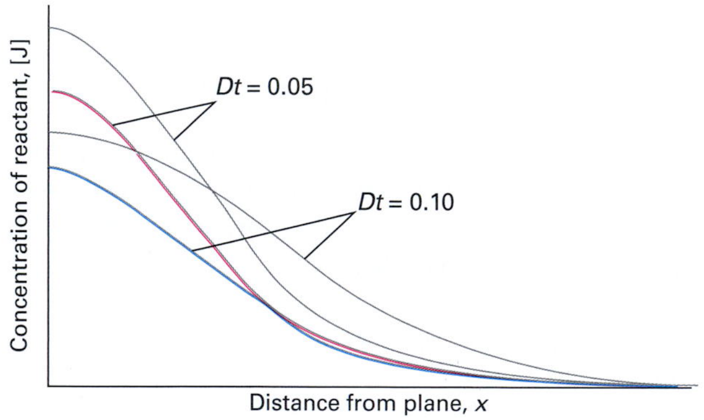

Figure 21B.2 The concentration profiles for a diffusing, reacting system (for example, a column of solution) in which one reactant is initially in a layer at $x=0$. In the absence of reaction (grey lines), the concentration profiles are the same as in Fig. 19C.3.

An example of a solution of the diffusion equation in the absence of reaction is that given in Topic 19C for a system in which initially a layer of n_0N_A molecules is spread over a plane of area A:

$$[\mathrm{J}]=\frac{n_0\mathrm{e}^{-x^2/4Dt}}{A(\pi Dt)^{1/2}} \qquad (21\mathrm{B}.9)$$

When this expression is substituted into eqn 21B.8, we obtain the concentration of J as it diffuses away from its initial surface layer and undergoes reaction in the overlying solution (Fig. 21B.2).

Brief illustration 21B.3 Reaction with diffusion

Suppose 1.0 g of iodine (3.9 mmol I_2) is spread over a surface of area 5.0 cm² under a column of hexane ($D=4.1\times10^{-9}\,\mathrm{m^2\,s^{-1}}$). As it diffuses upwards it reacts with a pseudofirst-order rate constant $k_\mathrm{r}=4.0\times10^{-5}\,\mathrm{s^{-1}}$. By substituting these values into

$$[\mathrm{J}]^*=\frac{n_0\mathrm{e}^{-x^2/4Dt-k_\mathrm{r}t}}{A(\pi Dt)^{1/2}}$$

we can construct the following table:

$[\mathrm{J}]^*/(\mathrm{mmol\,dm^{-3}})$		x	
T	1 mm	5 mm	1 cm
100 s	3.72	0	0
1000 s	1.96	0.45	0.005
10 000 s	0.46	0.40	0.25

Self-test 21B.3 What is the value of [J] at 15 000 s at the same three locations?

Answer: 0.31, 0.28, 0.21 mmol dm⁻³

Even this relatively simple example has led to an equation that is difficult to solve, and only in some special cases can the full material balance equation be solved analytically. Most modern work on reactor design and cell kinetics uses numerical methods to solve the equation, and detailed solutions for realistic environments, such as vessels of different shapes (which influence the boundary conditions on the solutions) and with a variety of inhomogeneously distributed reactants, can be obtained reasonably easily.

Checklist of concepts

☐ 1. A reaction in solution may be **diffusion-controlled** if its rate is controlled by the rate at which reactant molecules encounter each other in solution.

☐ 2. The rate of an **activation-controlled reaction** is controlled by the rate at which the encounter pair accumulates sufficient energy.

☐ 3. The **material-balance equation** relates the overall rate of change of the concentration of a species to its rates of diffusion, convection and reaction.

☐ 4. The **cage effect**, the lingering of one reactant molecule near another due to the hindering presence of solvent molecules, results in the formation of an **encounter pair** of reactant molecules.

Checklist of equations

Property	Equation	Comment	Equation number
Diffusion-controlled limit	$k_r = k_d$	$v = k_d[A][B]$ for the encounter rate	21B.2a
Activation-controlled limit	$k_r = k_d K$	K for $A+B \rightleftharpoons AB$, k_a for the decomposition of AB	21B.2b
Diffusion-controlled rate constant	$k_d = 4\pi R^* D N_A$	$D = D_A + D_B$	21B.3
	$k_d = 8RT/3\eta$	Assumes Stokes–Einstein relation	21B.4
Material-balance equation	$\partial[J]/\partial t = D\partial^2[J]/\partial x^2 - v\partial[J]/\partial x - k_r[J]$	First-order reaction	21B.6

21C Transition-state theory

Contents

➤ Why do you need to know this material?

Transition-state theory provides a way to relate the rate constant of reactions to models of the cluster of atoms that is proposed to be formed when reactants come together. It provides a link between information about the structures of reactants and the rate constant for their reaction.

➤ What is the key idea?

Reactants come together to form an activated complex that decays into products.

➤ What do you need to know already?

This Topic makes use of two strands: one is the relation between equilibrium constants and partition functions (Topic 15F); the other is the relation between equilibrium constants and thermodynamic functions, such as the Gibbs energy, enthalpy, and entropy of reaction (Topic 6A). You need to be aware of the Arrhenius equation for the temperature dependence of the rate constant (Topic 20D).

In **transition-state theory** (which is also widely referred to as *activated complex theory*), the notion of the transition state is used in conjunction with concepts of statistical thermodynamics to provide a more detailed calculation of rate constants than that presented by collision theory (Topic 21A). Transition-state theory has the advantage that a quantity corresponding to the steric factor appears automatically, and P does not need to be grafted on to an equation as an afterthought; it is an attempt to identify the principal features governing the size of a rate constant in terms of a model of the events that take place during the reaction. There are several approaches to the formulation of transition-state theory; here we present the simplest.

21C.1 The Eyring equation

In the course of a chemical reaction that begins with a collision between molecules of A and molecules of B, the potential energy of the system typically changes in a manner shown in Fig. 21C.1. Although the illustration displays an exothermic reaction, a potential barrier is also common for endothermic reactions. As the reaction event proceeds, A and B come into contact, distort, and begin to exchange or discard atoms.

(a) The formulation of the equation

The **reaction coordinate** is a representation of the atomic displacements, such as changes in interatomic distances and bond angles, that are directly involved in the formation of products from reactants. The potential energy rises to a maximum and the cluster of atoms that corresponds to the region close to the maximum is called the **activated complex**. After the maximum, the potential energy falls as the atoms rearrange in the cluster and reaches a value characteristic of the products. The climax of the reaction is at the peak of the potential energy, which can be identified with the activation energy E_a; however, as in collision theory, this identification should be regarded as approximate and we clarify it later. Here two reactant molecules have come to such a degree of closeness and distortion that a small further distortion will send them in the direction of products. This crucial configuration is called the **transition state** of the reaction. Although some molecules entering the transition state might revert to reactants, if they pass through this configuration then it is inevitable that products will emerge from the encounter.

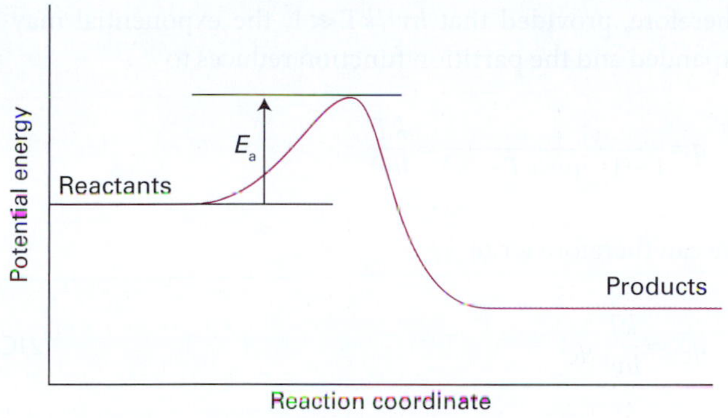

Figure 21C.1 A potential energy profile for an exothermic reaction. The height of the barrier between the reactants and products is the activation energy of the reaction.

A note on good practice The terms *activated complex* and *transition state* are often used as synonyms; however, it is best to preserve the distinction, with the former referring to the cluster of atoms and the latter to their critical configuration.

Transition state theory pictures a reaction between A and B as proceeding through the formation of an activated complex, $C^{\ddagger}$, in a rapid pre-equilibrium (Fig. 21C.2):

$$A+B \rightleftharpoons C^{\ddagger} \qquad K^{\ddagger}=\frac{p_{C^{\ddagger}}p^{\ominus}}{p_A p_B} \qquad (21C.1)$$

where we have replaced the activity of each species by $p/p^{\ominus}$. When we express the partial pressures, p_J, in terms of the molar concentrations, [J], by using $p_J = RT[J]$, the concentration of activated complex is related to the (dimensionless) equilibrium constant by

$$[C^{\ddagger}]=\frac{RT}{p^{\ominus}}K^{\ddagger}[A][B] \qquad (21C.2)$$

The activated complex falls apart by unimolecular decay into products, P, with a rate constant $k^{\ddagger}$:

$$C^{\ddagger} \rightarrow P \qquad v=k^{\ddagger}[C^{\ddagger}] \qquad (21C.3)$$

It follows that

$$v=k_r[A][B] \qquad k_r=\frac{RT}{p^{\ominus}}k^{\ddagger}K^{\ddagger} \qquad (21C.4)$$

Our task is to calculate the unimolecular rate constant $k^{\ddagger}$ and the equilibrium constant $K^{\ddagger}$.

(b) The rate of decay of the activated complex

An activated complex can form products if it passes through the transition state. As the reactant molecules approach the

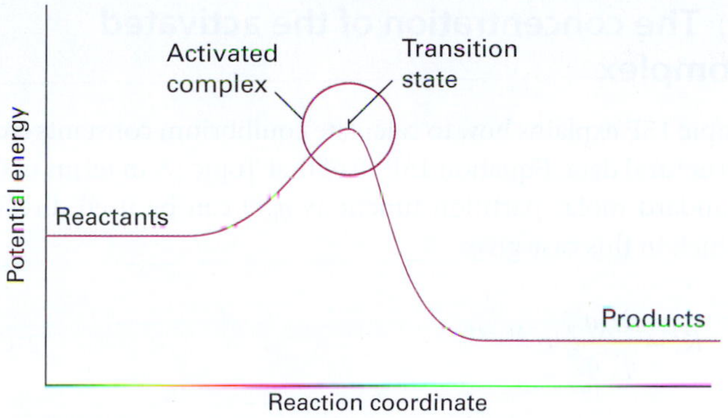

Figure 21C.2 A reaction profile (for an exothermic reaction). The horizontal axis is the reaction coordinate, and the vertical axis is potential energy. The activated complex is the region near the potential maximum, and the transition state corresponds to the maximum itself.

activated complex region, some bonds are forming and shortening while others are lengthening and breaking; therefore, along the reaction coordinate, there is a vibration-like motion of the atoms in the activated complex. If this motion occurs with a frequency $\nu^{\ddagger}$, then the frequency with which the cluster of atoms forming the complex approaches the transition state is also $\nu^{\ddagger}$. However, it is possible that not every oscillation along the reaction coordinate takes the complex through the transition state. For instance, the centrifugal effect of rotations might also be an important contribution to the break-up of the complex, and in some cases the complex might be rotating too slowly or rotating rapidly but about the wrong axis. Therefore, we suppose that the rate of passage of the complex through the transition state is proportional to the vibrational frequency along the reaction coordinate, and write

$$k^{\ddagger}=\kappa\nu^{\ddagger} \qquad (21C.5)$$

where κ (kappa) is the **transmission coefficient**. In the absence of information to the contrary, κ is assumed to be about 1.

Brief illustration 21C.1 The transmission coefficient

Typical molecular vibration wavenumbers of small molecules occur at wavenumbers of the order of $10^3\,cm^{-1}$ (C–H bends, for example, occur in the range 1340–1465 cm^{-1}) and therefore occur at frequencies of the order of 10^{13} Hz. If we suppose that the loosely bound cluster vibrates at one or two orders of magnitude lower frequency, then $\nu^{\ddagger} \approx 10^{11}$–$10^{12}$ Hz. These figures suggest that $\nu^{\ddagger} \approx 10^{11}$–$10^{12}\,s^{-1}$, with κ perhaps reducing that value further.

Self-test 21C.1 Estimate the change in $\nu^{\ddagger}$ that would occur if ^1H is replaced by ^2H in a C–H group at the site of reaction. Assume that the C atom is immobile.

Answer: $\nu^{\ddagger} \rightarrow \nu^{\ddagger}/2^{1/2}$

(c) The concentration of the activated complex

Topic 15F explains how to calculate equilibrium constants from structural data. Equation 15F.10 of that Topic (K in terms of the standard molar partition functions $q_J^{\ominus}$) can be used directly, which in this case gives

$$K^{\ddagger}=\frac{N_A q_{C^{\ddagger}}^{\ominus}}{q_A^{\ominus} q_B^{\ominus}}e^{-\Delta E_0/RT} \qquad (21C.6)$$

with

$$\Delta E_0=E_0(C^{\ddagger})-E_0(A)-E_0(B) \qquad (21C.7)$$

Note that the units of N_A and the $q_J^{\ominus}$ are mol^{-1}, so $K^{\ddagger}$ is dimensionless (as is appropriate for an equilibrium constant).

In the final step of this part of the calculation, we focus attention on the partition function of the activated complex. We have already assumed that a vibration of the activated complex $C^{\ddagger}$ tips it through the transition state. The partition function for this vibration is (see eqn 15B.15 of Topic 15B, which is essentially the following):

$$q=\frac{1}{1-e^{-h\nu^{\ddagger}/kT}}$$

where $\nu^{\ddagger}$ is its frequency (the same frequency that determines $k^{\ddagger}$). This frequency is much lower than for an ordinary molecular vibration because the oscillation corresponds to the complex falling apart (Fig. 21C.3), so the force constant is very low.

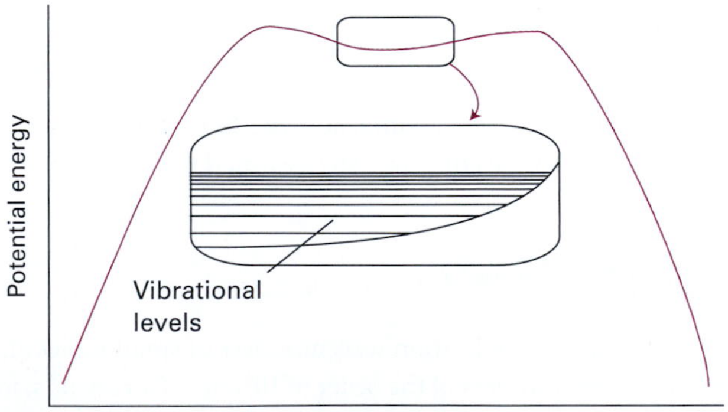

Figure 21C.3 In an elementary depiction of the activated complex close to the transition state, there is a broad, shallow dip in the potential energy surface along the reaction coordinate. The complex vibrates harmonically and almost classically in this well. However, this depiction is an oversimplification, for in many cases there is no dip at the top of the barrier, and the curvature of the potential energy, and therefore the force constant, is negative. Formally, the vibrational frequency is then imaginary. We ignore this problem here.

Therefore, provided that $h\nu^{\ddagger}/kT \ll 1$, the exponential may be expanded and the partition function reduces to

$$q=\frac{1}{1-(1-h\nu^{\ddagger}/kT+\cdots)}\approx\frac{kT}{h\nu^{\ddagger}}$$

We can therefore write

$$q_{C^{\ddagger}}^{\ominus}=\frac{kT}{h\nu^{\ddagger}}\bar{q}_{C^{\ddagger}} \qquad (21C.8)$$

where $\bar{q}_{C^{\ddagger}}$ denotes the partition function for all the other modes of the complex. The constant $K^{\ddagger}$ is therefore

$$K^{\ddagger}=\frac{kT}{h\nu^{\ddagger}}\bar{K}^{\ddagger} \qquad \bar{K}^{\ddagger}=\frac{N_A\bar{q}_{C^{\ddagger}}^{\ominus}}{q_A^{\ominus} q_B^{\ominus}}e^{-\Delta E_0/RT} \qquad (21C.9)$$

with $\bar{K}^{\ddagger}$ a kind of equilibrium constant, but with one vibrational mode of $C^{\ddagger}$ discarded.

Brief illustration 21C.2 The discarded mode

Consider the case of two structureless particles A and B colliding to give an activated complex that resembles a diatomic molecule. The activated complex is a diatomic cluster. It has one vibrational mode, but that mode corresponds to motion along the reaction coordinate and therefore does not appear in $\bar{q}_{C^{\ddagger}}^{\ominus}$. It follows that the standard molar partition function of the activated complex has only rotational and translational contributions.

Self-test 21C.2 Which mode would be discarded for a reaction in which the activated complex is modelled as a linear triatomic cluster?

Answer: Antisymmetric stretch

(d) The rate constant

We can now combine all the parts of the calculation into

$$k_r=\frac{RT}{p^{\ominus}}k^{\ddagger}K^{\ddagger}=\kappa\nu^{\ddagger}\frac{kT}{h\nu^{\ddagger}}\frac{RT}{p^{\ominus}}\bar{K}^{\ddagger}$$

At this stage the unknown frequencies $\nu^{\ddagger}$ (in blue) cancel, and after writing $\bar{K}_c^{\ddagger}=(RT/p^{\ominus})\bar{K}^{\ddagger}$, we obtain the **Eyring equation**:

$$k_r=\kappa\frac{kT}{h}\bar{K}_c^{\ddagger} \qquad \text{Eyring equation} \qquad (21C.10)$$

The factor $\bar{K}_c^{\ddagger}$ is given by eqn 21C.9 and the definition $\bar{K}_c^{\ddagger}=(RT/p^{\ominus})\bar{K}^{\ddagger}$ in terms of the partition functions of A, B, and $C^{\ddagger}$, so in principle we now have an explicit expression for calculating the second-order rate constant for a bimolecular reaction in terms of the molecular parameters for the reactants and the activated complex and the quantity κ.

The partition functions for the reactants can normally be calculated quite readily by using either spectroscopic information about their energy levels or the approximate expressions set out in Table 15C.1. The difficulty with the Eyring equation, however, lies in the calculation of the partition function of the activated complex: $C^‡$ is difficult to investigate spectroscopically (but see the following section), and in general we need to make assumptions about its size, shape, and structure. We shall illustrate what is involved in one simple but significant case.

Example 21C.1 Analysing the collision of structureless particles

Consider the case of two structureless (and different) particles A and B colliding to give an activated complex that resembles a diatomic molecule and deduce an expression for the rate constant of the reaction $A+B \rightarrow$ Products.

Method Because the reactants J = A, B are structureless 'atoms', the only contributions to their partition functions are the translational terms. The activated complex is a diatomic cluster of mass $m_{C^‡}=m_A+m_B$ and moment of inertia I. It has one vibrational mode but, as explained in *Brief illustration* 21C.2, that mode corresponds to motion along the reaction coordinate. It follows that the standard molar partition function of the activated complex has only rotational and translational contributions. Expressions for the relevant partition functions are given in Topic 15B.

Answer The translational partition functions are

$$q_J^{\ominus}=\frac{V_m^{\ominus}}{\Lambda_J^3} \qquad \Lambda_J=\frac{h}{(2\pi m_J kT)^{1/2}} \qquad V_m^{\ominus}=\frac{RT}{p^{\ominus}}$$

with J = A, B, and $C^‡$ and with $m_{C^‡}=m_A+m_B$. The expression for the partition function of the activated complex is

$$\bar{q}_{C^‡}^{\ominus}=\frac{2IkT}{\hbar^2}\frac{V_m^{\ominus}}{\Lambda_{C^‡}^3}$$

where we have used the high-temperature form of the rotational partition function (Topic 15B). By substituting these expressions into the Eyring equation, we find that the rate constant is

$$k_r=\kappa\frac{kT}{h}\frac{RT}{p^{\ominus}}\left(\frac{N_A\Lambda_A^3\Lambda_B^3}{\Lambda_{C^‡}^3 V_m^{\ominus}}\right)\frac{2IkT}{\hbar^2}e^{-\Delta E_0/RT}$$

$$=\kappa\frac{kT}{h}N_A\left(\frac{\Lambda_A\Lambda_B}{\Lambda_{C^‡}}\right)^3\frac{2IkT}{\hbar^2}e^{-\Delta E_0/RT}$$

The moment of inertia of a diatomic molecule of bond length r is μr^2, where $\mu=m_Am_B/(m_A+m_B)$, so after introducing the expressions for the thermal wavelengths and cancelling common terms, we find (Problem 21C.3)

$$k_r=\kappa N_A\left(\frac{8kT}{\pi\mu}\right)^{1/2}\pi r^2 e^{-\Delta E_0/RT}$$

Finally, by identifying $\kappa\pi r^2$ as the reactive cross-section σ^*, we arrive at precisely the same expression as that obtained from simple collision theory (eqn 21A.9):

$$k_r=N_A\left(\frac{8kT}{\pi\mu}\right)^{1/2}\sigma^* e^{-\Delta E_0/RT}$$

Self-test 21C.3 What additional factors would be present if the reaction were $AB+C \rightarrow$ Products through a linear activated complex?

Answer: Rotation and vibration of AB, bends and symmetric stretch of the activated complex

(e) Observation and manipulation of the activated complex

The development of femtosecond pulsed lasers has made it possible to make observations on species that have such short lifetimes that in a number of respects they resemble an activated complex, which often survive for only a few picoseconds. In a typical experiment designed to detect an activated complex, a femtosecond laser pulse is used to excite a molecule to a dissociative state, and then the system is exposed to a second femtosecond pulse at an interval after the dissociating pulse. The frequency of the second pulse is set at an absorption of one of the free fragmentation products, so the intensity of its absorption is a measure of the abundance of the dissociation product. For example, when ICN is dissociated by the first pulse, the emergence of CN from the photoactivated state can be monitored by watching the growth of the free CN absorption (or, more commonly, its laser-induced fluorescence). In this way it has been found that the CN signal remains zero until the fragments have separated by about 600 pm, which takes about 205 fs.

Some sense of the progress that has been made in the study of the intimate mechanism of chemical reactions can be obtained by considering the decay of the ion pair Na^+I^-. As shown in Fig. 21C.4, excitation of the ionic species with a femtosecond laser pulse forms an excited state that corresponds to a covalently bonded NaI molecule. The system can be described with two potential energy surfaces, one largely 'ionic' and another 'covalent', which cross at an internuclear separation of 693 pm. A short laser pulse is composed of a wide range of frequencies, which excite many vibrational states of NaI simultaneously. Consequently, the electronically excited complex exists as a superposition of states, or a localized wavepacket, which oscillates between the 'covalent' and 'ionic' potential energy

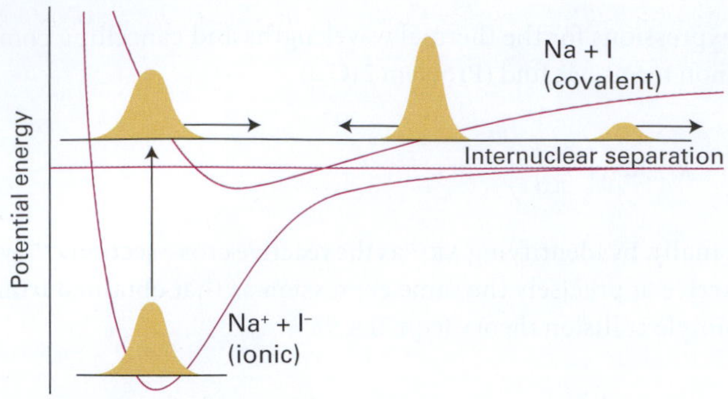

Figure 21C.4 Excitation of the ion pair Na^+I^- forms an excited state with covalent character. Also shown is migration between a 'covalent' surface (upper curve) and an 'ionic' surface (lower curve) of the wavepacket formed by laser excitation.

surfaces, as shown in Fig. 21C.4. The complex can also dissociate, shown as movement of the wavepacket towards very long internuclear separation along the dissociative surface. However, not every outward-going swing leads to dissociation because there is a chance that the I atom can be harpooned again, in which case it fails to make good its escape. The dynamics of the system is probed by a second laser pulse with a frequency that corresponds to the absorption frequency of the free Na product or to the frequency at which Na absorbs when it is a part of the complex. The latter frequency depends on the Na···I distance, so an absorption (in practice, a laser-induced fluorescence) is obtained each time the wavepacket returns to that separation.

Brief illustration 21C.3 Femtosecond analysis

A typical set of results is shown in Fig. 21C.5. The bound Na absorption intensity shows up as a series of pulses that recur in about 1 ps, showing that the wavepacket oscillates with about that period. The decline in intensity shows the rate at which the complex can dissociate as the two atoms swing away from each other. The free Na absorption also grows in an oscillating manner, showing the periodicity of wavepacket oscillation, each swing of which gives it a chance to dissociate. The precise period of the oscillation in NaI is 1.25 ps. The complex survives for about ten oscillations. In contrast, although the oscillation frequency of NaBr is similar, it barely survives one oscillation.

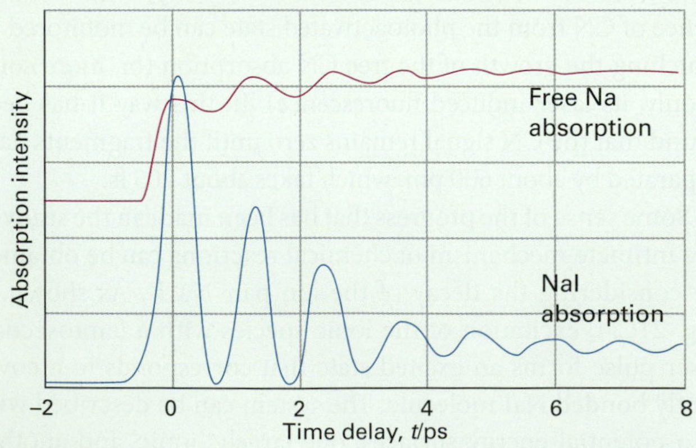

Figure 21C.5 Femtosecond spectroscopic results for the reaction in which sodium iodide separates into Na and I. The lower curve is the absorption of the electronically excited complex and the upper curve is the absorption of free Na atoms. (Adapted from A.H. Zewail, *Science* **242**, 1645 (1988).)

Self-test 21C.4 Confirm the assumption in transition-state theory that the vibrational frequency of the dissociative mode of the activated complex is very low by calculating the vibrational wavenumber corresponding to the 1.25 ps period of oscillation in NaI.

Answer: $27\ cm^{-1}$

Femtosecond spectroscopy has also been used to examine analogues of the activated complex involved in bimolecular reactions. Thus, a molecular beam can be used to produce a complex held together by van der Waals interactions (a 'van der Waals molecule'), such as IH···OCO. The HI bond can be dissociated by a femtosecond pulse, and the H atom is ejected towards the O atom of the neighbouring CO_2 molecule to form HOCO. Hence, the van der Waals molecule is a source of a species that resembles the activated complex of the reaction

$$H + CO_2 \rightarrow [HOCO]^{\ddagger} \rightarrow HO + CO$$

The probe pulse is tuned to the OH radical, which enables the evolution of $[HOCO]^{\ddagger}$ to be studied in real time.

The techniques used for the spectroscopic detection of transition states can also be used to control the outcome of a chemical reaction by direct manipulation of the activated complex. Consider the reaction $I_2 + Xe \rightarrow XeI^* + I$, which occurs by a harpoon mechanism with an activated complex denoted as $[Xe^+ \cdots I^- \cdots I]$. The reaction can be initiated by exciting I_2 to an electronic state at least 52 460 cm^{-1} above the ground state and then followed by measuring the time dependence of the chemiluminescence of XeI^*. To exert control over the yield of the product, a pair of femtosecond pulses can be used to induce the reaction. The first pulse excites the I_2 molecule to a low energy and unreactive electronic state. We already know that excitation by a femtosecond pulse generates a wavepacket that can be treated as a particle travelling across the potential energy surface. In this case, the wavepacket does not have enough energy to react, but excitation by another laser pulse with the appropriate wavelength can provide the necessary additional energy. It follows that activated complexes with different geometries can be prepared by varying the time delay between the two pulses, as the partially localized wavepacket will be at different locations on the potential energy surface as it evolves after being formed by the first pulse. Because the reaction occurs by the harpoon mechanism, the product yield is expected to be optimal if the second pulse is applied when the wavepacket is at a point where the Xe···I_2 distance is just right for electron transfer from Xe to I_2 to occur. This type of control of the I_2+Xe reaction has been demonstrated.

21C.2 Thermodynamic aspects

The statistical thermodynamic version of transition state theory rapidly runs into difficulties because only in some cases is anything known about the structure of the activated complex. However, the concepts that it introduces, principally that of an equilibrium between the reactants and the activated complex, have motivated a more general, empirical approach in which the activation process is expressed in terms of thermodynamic functions.

(a) Activation parameters

If we accept that $\bar{K}^{\ddagger}$ is an equilibrium constant (despite one mode of $C^{\ddagger}$ having been discarded), we can express it in terms of a **Gibbs energy of activation**, $\Delta^{\ddagger}G$, through the definition

$$\Delta^{\ddagger}G=-RT\ln\bar{K}^{\ddagger} \qquad \textit{Definition} \quad \text{Gibbs energy of activation} \qquad (21C.11)$$

All the $\Delta^{\ddagger}X$ in this section are *standard* thermodynamic quantities, $\Delta^{\ddagger}X^{\ominus}$, but we shall omit the standard state sign to avoid overburdening the notation. Then the rate constant becomes

$$k_r=\kappa\frac{kT}{h}\frac{RT}{p^{\ominus}}e^{-\Delta^{\ddagger}G/RT} \qquad (21C.12)$$

Because $G=H-TS$, the Gibbs energy of activation can be divided into an **entropy of activation**, $\Delta^{\ddagger}S$, and an **enthalpy of activation**, $\Delta^{\ddagger}H$, by writing

$$\Delta^{\ddagger}G=\Delta^{\ddagger}H-T\Delta^{\ddagger}S \qquad \textit{Definition} \quad \text{Entropy and enthalpy of activation} \qquad (21C.13)$$

When eqn 21C.13 is used in eqn 21C.12 and κ is absorbed into the entropy term, we obtain

$$k_r=Be^{\Delta^{\ddagger}S/R}e^{-\Delta^{\ddagger}H/RT} \qquad B=\frac{kT}{h}\frac{RT}{p^{\ominus}} \qquad (21C.14)$$

The formal definition of activation energy (eqn 20D.2 of Topic 20D, $E_a=RT^2(\partial\ln k_r/\partial T)$), then gives $E_a=\Delta^{\ddagger}H+2RT$, so[1]

$$k_r=e^2Be^{\Delta^{\ddagger}S/R}e^{-E_a/RT} \qquad (21C.15a)$$

from which it follows that the Arrhenius factor A can be identified as

$$A=e^2Be^{\Delta^{\ddagger}S/R} \qquad \textit{Transition-state theory} \quad \textit{A-factor} \qquad (21C.15b)$$

The entropy of activation is negative because throughout the system reactant species are combining to form reactive pairs. However, if there is a reduction in entropy below what would be expected for the simple encounter of A and B, then the Arrhenius factor A will be smaller than that expected on the basis of simple collision theory. Indeed, we can identify that *additional* reduction in entropy, $\Delta^{\ddagger}S_{\text{steric}}$, as the origin of the steric factor of collision theory, and write

$$P=e^{\Delta^{\ddagger}S_{\text{steric}}/R} \qquad \textit{Transition-state theory} \quad \textit{P-factor} \qquad (21C.15c)$$

Thus, the more complex the steric requirements of the encounter, the more negative the value of $\Delta^{\ddagger}S_{\text{steric}}$, and the smaller the value of P.

Brief illustration 21C.4 Activation parameters

The reaction of propylxanthate ion in acetic acid buffer solutions can be represented by the equation $A^-+H^+\rightarrow P$. Near 30 °C, $A=2.05\times10^{13}\,dm^3\,mol^{-1}\,s^{-1}$. To evaluate the entropy of activation at 30 °C we first note that because the reaction is in solution the e^2 of eqn 21C.15 should be replaced by e (see footnote 1), and then use eqn 21C.15b in the form

$$\Delta^{\ddagger}S=R\ln\frac{A}{eB} \quad \text{with } B=\frac{kT}{h}\frac{RT}{p^{\ominus}}=1.592\times10^{14}\,dm^3\,mol^{-1}\,s^{-1}$$

Therefore,

$$\Delta^{\ddagger}S=R\ln\frac{2.05\times10^{13}\,dm^3\,mol^{-1}\,s^{-1}}{e\times(1.592\times10^{14}\,dm^3\,mol^{-1}\,s^{-1})}=R\ln0.047\ldots$$
$$=-25.4\,J\,K^{-1}\,mol^{-1}$$

Self-test 21C.5 The reaction $A^-+H^+\rightarrow P$ in solution has $A=6.92\times10^{12}\,dm^3\,mol^{-1}\,s^{-1}$. Evaluate the entropy of activation at 25 °C.

Answer: $-34.1\,J\,K^{-1}\,mol^{-1}$

Gibbs energies, enthalpies, and entropies of activation (and volumes and heat capacities of activation) are widely used to report experimental reaction rates, especially for organic reactions in solution. They are encountered when relationships between equilibrium constants and rates of reaction are explored using **correlation analysis**, in which ln K (which is equal to $-\Delta_rG^{\ominus}/RT$) is plotted against ln k_r (which is proportional to $-\Delta^{\ddagger}G/RT$). In many cases the correlation is linear, signifying that as the reaction becomes thermodynamically more

[1] For reactions of the type $A+B\rightarrow P$ in the gas phase, $E_a=\Delta^{\ddagger}H+2RT$. For such reactions in solution, $E_a=\Delta^{\ddagger}H+RT$.

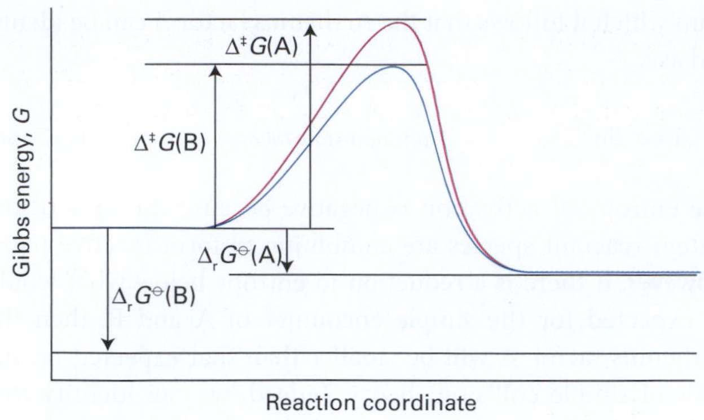

Figure 21C.6 For a related series of reactions, as the magnitude of the standard reaction Gibbs energy increases, so the activation barrier decreases and the rate constant increases. The approximate linear correlation between $\Delta^{\ddagger}G$ and $\Delta_r G^{\ominus}$ is the origin of linear free energy relations.

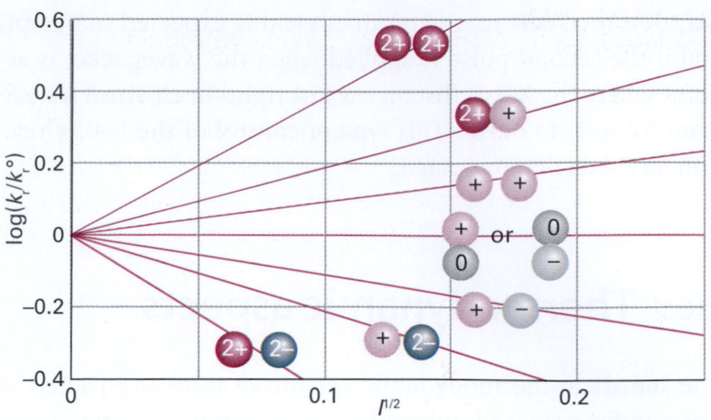

Figure 21C.7 Experimental tests of the kinetic salt effect for reactions in water at 298 K. The ion types are shown as spheres, and the slopes of the lines are those given by the Debye–Hückel limiting law and eqn 21C.18.

favourable, its rate constant increases (Fig. 21C.6). This linear correlation is the origin of the alternative name **linear free energy relation** (LFER).

(b) Reactions between ions

The full statistical thermodynamic theory is very complicated to apply because the solvent plays a role in the activated complex. The thermodynamic version of transition-state theory simplifies the discussion of reactions in solution and is applicable to non-ideal systems. In the thermodynamic approach we combine the rate law

$$\frac{d[P]}{dt}=k^{\ddagger}[C^{\ddagger}]$$

with the thermodynamic equilibrium constant (Topic 6A)

$$K=\frac{a_{C^{\ddagger}}}{a_A a_B}=K_{\gamma}\frac{[C^{\ddagger}]c^{\ominus}}{[A][B]} \qquad K_{\gamma}=\frac{\gamma_{C^{\ddagger}}}{\gamma_A\gamma_B}$$

Then

$$\frac{d[P]}{dt}=k_r[A][B] \qquad k_r=\frac{k^{\ddagger}K}{K_{\gamma}c^{\ominus}} \qquad \text{(21C.16a)}$$

If k_r° is the rate constant when the activity coefficients are 1 (that is, $k_r^{\circ}=k^{\ddagger}K/c^{\ominus}$),we can write

$$k_r=\frac{k_r^{\circ}}{K_{\gamma}} \qquad \log k_r=\log k_r^{\circ}-\log K_{\gamma} \qquad \text{(21C.16b)}$$

At low concentrations the activity coefficients can be expressed in terms of the ionic strength, I, of the solution by using the Debye–Hückel limiting law (Topic 5F, particularly eqn 5F.8, $\log \gamma_{\pm}=-A|z_+z_-|I^{1/2}$). However, we need the expressions for the individual ions rather than the mean value, and so write $\log \gamma_J=-Az_J^2I^{1/2}$ and

$$\log\gamma_A=-Az_A^2I^{1/2} \qquad \log\gamma_B=-Az_B^2I^{1/2} \qquad \text{(21C.17a)}$$

with $A=0.509$ in aqueous solution at 298 K and z_A and z_B the (signed) charge numbers of A and B, respectively. Because the activated complex forms from reaction of one of the ions of A with one of the ions of B, the charge number of the activated complex is z_A+z_B where z_J is positive for cations and negative for anions. Therefore

$$\log\gamma_{C^{\ddagger}}=-A(z_A+z_B)^2I^{1/2} \qquad \text{(21C.17b)}$$

Inserting these relations into eqn 21C.16b results in

$$\begin{aligned}\log k_r&=\log k_r^{\circ}-A\{z_A^2+z_B^2-(z_A+z_B)^2\}I^{1/2}\\&=\log k_r^{\circ}+2Az_Az_BI^{1/2}\end{aligned} \qquad \text{(21C.18)}$$

Equation 21C.18 expresses the **kinetic salt effect**, the variation of the rate constant of a reaction between ions with the ionic strength of the solution (Fig. 21C.7). If the reactant ions have the same sign (as in a reaction between cations or between anions), then increasing the ionic strength by the addition of inert ions increases the rate constant. The formation of a single, highly charged ionic complex from two less highly charged ions is favoured by a high ionic strength because the new ion has a denser ionic atmosphere and interacts with that atmosphere more strongly. Conversely, ions of opposite charge react more slowly in solutions of high ionic strength. Now the charges cancel and the complex has a less favourable interaction with its atmosphere than the separated ions.

Example 21C.2 Analysing the kinetic salt effect

The rate constant for the base (OH^-) hydrolysis of $[CoBr(NH_3)_5]^{2+}$ varies with ionic strength as tabulated below. What can be deduced about the charge of the activated complex in the rate-determining stage? We cannot assume without more evidence that it is a bimolecular process with an activated complex of charge +1.

I	0.0050	0.0100	0.0150	0.0200	0.0250	0.0300
k_r/k_r°	0.718	0.631	0.562	0.515	0.475	0.447

Method According to eqn 21C.18, a plot of $\log(k_r/k_r^\circ)$ against $I^{1/2}$ will have a slope of $1.02z_Az_B$, from which we can infer the charges of the ions involved in the formation of the activated complex.

Answer Form the following table:

I	0.0050	0.0100	0.0150	0.0200	0.0250	0.0300
$I^{1/2}$	0.071	0.100	0.122	0.141	0.158	0.173
$\log(k_r/k_r^\circ)$	−0.14	−0.20	−0.25	−0.29	−0.32	−0.35

These points are plotted in Fig. 21C.8. The slope of the (least squares) straight line is −2.04, indicating that $z_Az_B=-2$. Because $z_A=-1$ for the OH^- ion, if that ion is involved in the formation of the activated complex, then the charge number of the second ion is +2. This analysis suggests that the pentaamminebromidocobalt(III) cation $[CoBr(NH_3)_5]^{2+}$ participates in the formation of the activated complex and that the charge of the activated complex is $-1+2=+1$. Although we do not pursue the point here, you should be aware that the rate constant is also influenced by the relative permittivity of the medium.

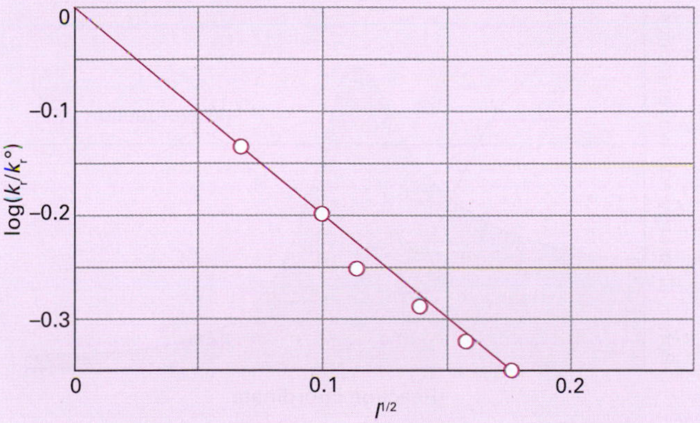

Figure 21C.8 The experimental ionic strength dependence of the rate constant of a hydrolysis reaction: the slope gives information about the charge types involved in the activated complex of the rate determining step. See *Example* 21C.2.

Self-test 21C.6 An ion of charge number +1 is known to be involved in the activated complex of a reaction. Deduce the charge number of the other ion from the following data:

I	0.0050	0.0100	0.0150	0.0200	0.0250	0.0300
k_r/k_r°	0.930	0.902	0.884	0.867	0.853	0.841

Answer: −1

21C.3 The kinetic isotope effect

The postulation of a plausible reaction mechanism requires careful analysis of many experiments designed to determine the fate of atoms during the formation of products. Observation of the **kinetic isotope effect**, a decrease in the rate of a chemical reaction upon replacement of one atom in a reactant by a heavier isotope, facilitates the identification of bond-breaking events in the rate-determining step. A **primary kinetic isotope effect** is observed when the rate-determining step requires the scission of a bond involving the isotope. A **secondary kinetic isotope effect** is the reduction in reaction rate even though the bond involving the isotope is not broken to form product. In both cases, the effect arises from the change in activation energy that accompanies the replacement of an atom by a heavier isotope on account of changes in the zero-point vibrational energies. We now explore the primary kinetic isotope effect in some detail.

Consider a reaction in which a C–H bond is cleaved. If scission of this bond is the rate-determining step (Topic 20E), then the reaction coordinate corresponds to the stretching of the C–H bond and the potential energy profile is shown in Fig. 21C.9. On deuteration, the dominant change is the reduction of the zero-point energy of the bond (because the deuterium atom is heavier). The whole reaction profile is not lowered, however, because the relevant vibration in the activated complex has a very low force constant, so there is little zero-point energy associated with the reaction coordinate in either form of the activated complex. We show in the following *Justification*, that, as a consequence of this reduction, the activation energy change upon deuteration is

$$E_a(\text{C–D})-E_a(\text{C–H})=\tfrac{1}{2}N_A\hbar\omega(\text{C–H})\left\{1-\left(\frac{\mu_{\text{CH}}}{\mu_{\text{CD}}}\right)^{1/2}\right\} \quad (21C.19)$$

where ω is the relevant vibrational frequency (in radians per second), μ is the relevant effective mass, and

$$\frac{k_r(\text{C–D})}{k_r(\text{C–H})}=e^{-\zeta} \quad \text{with} \quad \zeta=\frac{\hbar\omega(\text{C–H})}{2kT}\left\{1-\left(\frac{\mu_{\text{CH}}}{\mu_{\text{CD}}}\right)^{1/2}\right\} \quad (21C.20)$$

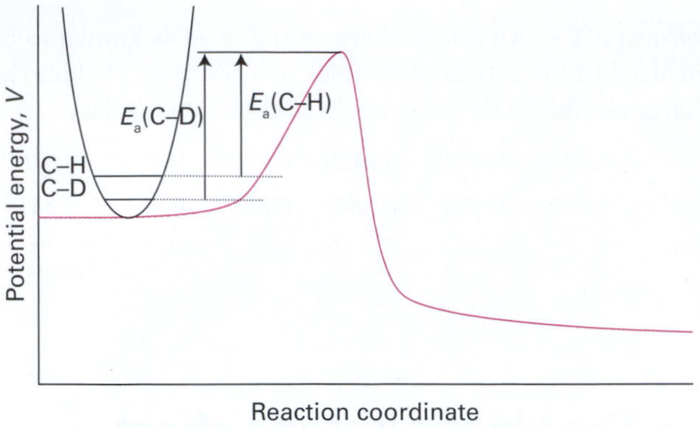

Figure 21C.9 Changes in the reaction profile when a C–H bond undergoing cleavage is deuterated. In this figure the C–H and C–D bonds are modelled as harmonic oscillators. The only significant change is in the zero-point energy of the reactants, which is lower for C–D than for C–H. As a result, the activation energy is greater for C–D cleavage than for C–H cleavage.

Note that $\zeta>0$ (ζ is zeta) because $\mu_{CD}>\mu_{CH}$ and that $k_r(\text{C–D})/k_r(\text{C–H})<1$, meaning that, as expected from Fig. 21C.9, the rate constant decreases upon deuteration. We also conclude that $k_r(\text{C–D})/k_r(\text{C–H})$ decreases with decreasing temperature.

Justification 21C.1 The primary kinetic isotope effect

We assume that, to a good approximation, a change in the activation energy arises only from the change in zero-point energy of the stretching vibration, so, from Fig. 21C.9,

$$E_a(\text{C–D})-E_a(\text{C–H})=N_A\{\tfrac{1}{2}\hbar\omega(\text{C–H})-\tfrac{1}{2}\hbar\omega(\text{C–D})\}$$
$$=\tfrac{1}{2}N_A\hbar\{\omega(\text{C–H})-\omega(\text{C–D})\}$$

where ω is the relevant vibrational frequency. From Topic 12D, we know that $\omega(\text{C–D})=(\mu_{CH}/\mu_{CD})^{1/2}\omega(\text{C–H})$, where μ is the relevant effective mass. Making this substitution in the equation above gives eqn 21C.19.

If we assume further that the pre-exponential factor does not change upon deuteration, then the rate constants for the two species should be in the ratio

$$\frac{k_r(\text{C–D})}{k_r(\text{C–H})}=e^{-\{E_a(\text{C–D})-E_a(\text{C–H})\}/RT}=e^{-\{E_a(\text{C–D})-E_a(\text{C–H})\}/N_AkT}$$

where we used $R=N_Ak$. Equation 21C.20 follows after using eqn 21C.19 for $E_a(\text{C–D})-E_a(\text{C–H})$ in this expression.

Brief illustration 21C.5 The primary kinetic isotope effect

From infrared spectra, the fundamental vibrational wavenumber $\tilde{\nu}$ for stretching of a C–H bond is about 3000 cm^{-1}. To convert this wavenumber to an angular frequency, $\omega=2\pi\nu$, we use $\omega=2\pi c\tilde{\nu}$ and it follows that

$$\omega=2\pi\times(2.998\times10^{10}\,\text{cm s}^{-1})\times(3000\,\text{cm}^{-1})$$
$$=5.65\ldots\times10^{14}\,\text{s}^{-1}$$

The ratio of effective masses is

$$\frac{\mu_{CH}}{\mu_{CD}}=\left(\frac{m_Cm_H}{m_C+m_H}\right)\times\left(\frac{m_C+m_D}{m_Cm_D}\right)$$
$$=\left(\frac{12.01\times1.0078}{12.01+1.0078}\right)\times\left(\frac{12.01+2.0140}{12.01\times2.0140}\right)=0.539\ldots$$

Now we can use eqn 21C.20, to calculate

$$\zeta=\frac{(1.055\times10^{-34}\,\text{J s})\times(5.65\ldots\times10^{14}\,\text{s}^{-1})}{2\times(1.381\times10^{-23}\,\text{J K}^{-1})\times(298\,\text{K})}\times(1-0.539\ldots^{1/2})$$
$$=1.92\ldots$$

and

$$\frac{k_r(\text{C–D})}{k_r(\text{C–H})}=e^{-1.92\ldots}=0.146$$

We conclude that at room temperature C–H cleavage should be about seven times faster than C–D cleavage, other conditions being equal. However, experimental values of $k_r(\text{C–D})/k_r(\text{C–H})$ can differ significantly from those predicted by eqn 21C.20 on account of the severity of the assumptions in the model.

Self-test 21C.7 The bromination of a deuterated hydrocarbon at 298 K proceeds 6.4 times more slowly than the bromination of the undeuterated material. What value of the force constant for the cleaved bond can account for this difference?

$k_f=450\,\text{N m}^{-1}$, which is consistent with k_f(C–H)

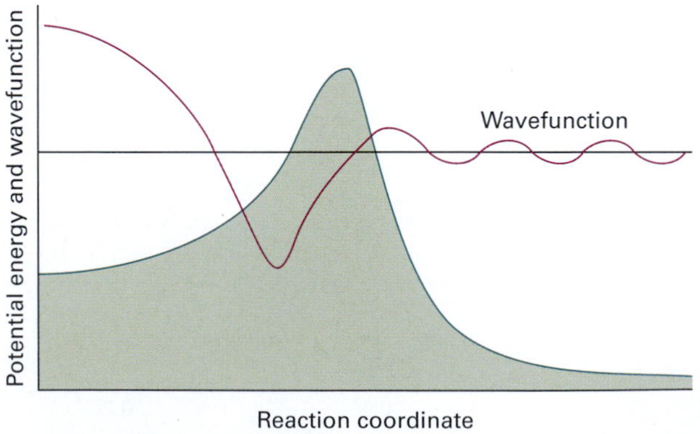

Figure 21C.10 A proton can tunnel through the activation energy barrier that separates reactants from products, so the effective height of the barrier is reduced and the rate of the proton transfer reaction increases. The effect is represented by drawing the wavefunction of the proton near the barrier. Proton tunnelling is important only at low temperatures, when most of the reactants are trapped on the left of the barrier.

In some cases, substitution of deuterium for hydrogen results in values of $k_r(\text{C–D})/k_r(\text{C–H})$ that are too low to be accounted for by eqn 21C.20, even when more complete models are used to predict ratios of rate constants. Such abnormal kinetic isotope effects are evidence for a path in which quantum mechanical tunnelling of hydrogen atoms takes place through the activation barrier (Fig 21C.10). The probability of tunnelling through a barrier decreases as the mass of the particle increases (Topic 8A), so deuterium tunnels less efficiently through a barrier than hydrogen and its reactions are correspondingly slower. Quantum mechanical tunnelling can be the dominant process in reactions involving hydrogen atom or proton transfer when the temperature is so low that very few reactant molecules can overcome the activation energy barrier. We see in Topic 21E that because m_e is so small, tunnelling is also a very important contributor to the rates of electron transfer reactions.

Checklist of concepts

- ☐ 1. In transition-state theory, it is supposed that an **activated complex** is in equilibrium with the reactants.
- ☐ 2. The rate at which the activated complex forms products depends on the rate at which it passes through a **transition state**.
- ☐ 3. The rate constant may be parameterized in terms of the **Gibbs energy, entropy, and enthalpy of activation**.
- ☐ 4. The **kinetic salt effect** is the effect of an added inert salt on the rate of a reaction between ions.
- ☐ 5. The **kinetic isotope effect** is the decrease in the rate constant of a chemical reaction upon replacement of one atom in a reactant by a heavier isotope.

Checklist of equations

Property	Equation	Comment	Equation number
'Equilibrium constant' for activated complex formation	$K^{\ddagger}=\left(N_A\bar{q}_{C^{\ddagger}}^{\ominus}/q_A^{\ominus}q_B^{\ominus}\right)e^{-\Delta E_0/RT}$	Assume equilibrium; one vibrational mode of $C^{\ddagger}$ discarded	21C.9
Eyring equation	$k_r=\kappa(kT/h)\bar{K}_c^{\ddagger}$	Transition-state theory	21C.10
Gibbs energy of activation	$\Delta^{\ddagger}G=-RT\ln\bar{K}^{\ddagger}$	Definition	21C.11
Enthalpy and entropy of activation	$\Delta^{\ddagger}G=\Delta^{\ddagger}H-T\Delta^{\ddagger}S$	Definition	21C.13
Parameterization	$k_r=e^nBe^{\Delta^{\ddagger}S/R}e^{-E_a/RT}$	$n=2$ for gas-phase reactions; $n=1$ for solution	21C.15a
A-factor	$A=e^nBe^{\Delta^{\ddagger}S/R}$		21C.15b
P-factor	$P=e^{\Delta^{\ddagger}S_{\text{steric}}/R}$		21C.15c
Kinetic salt effect	$\log k_r=\log k_r^{\circ}+2Az_Az_BI^{1/2}$	Assumes Debye–Hückel limiting law valid	21C.18
Primary kinetic isotope effect	$k_r(\text{C–D})/k_r(\text{C–H})=e^{-\zeta}$, $\zeta=(\hbar\omega(\text{C–H})/2kT)\times\{1-(\mu_{\text{CH}}/\mu_{\text{CD}})^{1/2}\}$	Cleavage of a C–H/D bond in the rate-determining step	21C.20

21D The dynamics of molecular collisions

Contents

➤ **Why do you need to know this material?**

Chemists need to be interested in the details of chemical reactions, and there is no more detailed approach than that involved in the study of the dynamics of reactive encounters, when one molecule collides with another and atoms exchange partners.

➤ **What is the key idea?**

The rates of reactions in the gas phase can be investigated by exploring the trajectories of molecules on potential energy surfaces.

➤ **What do you need to know already?**

This Topic builds on the concept of rate constant (Topic 20A) and in one part of the discussion uses the concept of partition function (Topic 15B). The discussion of potential energy surfaces is qualitative, but the underlying calculations are those of self-consistent field theory (Topic 10E).

The investigation of the dynamics of the collisions between reactant molecules is the most detailed level of the examination of the factors that govern the rates of reactions. There are two approaches: an experimental one that uses molecular beams and a theoretical one that uses the results of computations. In this Topic we describe both approaches and the link between them.

21D.1 Molecular beams

Molecular beams, which consist of collimated, narrow streams of molecules travelling through an evacuated vessel, allow collisions between molecules in preselected states (for example, specific rotational and vibrational states) to be studied, and can be used to identify the states of the products of a reactive collision. Information of this kind is essential if a full picture of the reaction is to be built, because the rate constant is an average over events in which reactants in different initial states evolve into products in their final states.

(a) Techniques

The basic arrangement of a molecular beam experiment is shown in Fig. 21D.1. If the pressure of vapour in the source is increased so that the mean free path of the molecules in the emerging beam is much shorter than the diameter of the pinhole, many collisions take place even outside the source. The net effect of these collisions, which give rise to **hydrodynamic**

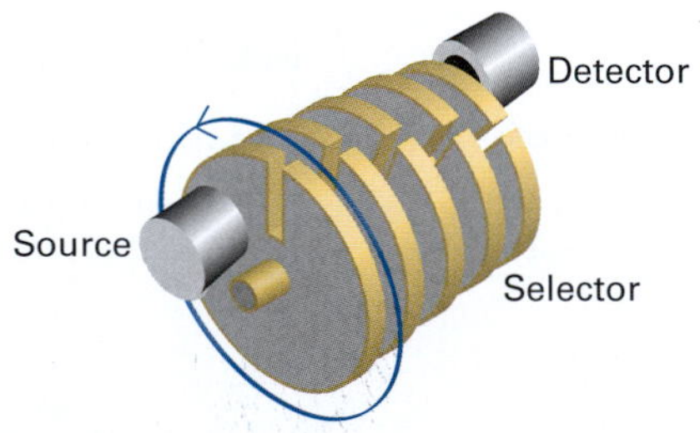

Figure 21D.1 The basic arrangement of a molecular beam apparatus. The atoms or molecules emerge from a heated source, and pass through the velocity selector, a rotating series of slotted discs, such as that discussed in Topic 1B. The scattering occurs from the target gas (which might take the form of another beam), and the flux of particles entering the detector set at some angle is recorded.

flow, is to transfer momentum into the direction of the beam. The molecules in the beam then travel with very similar speeds, so further downstream few collisions take place between them. This condition is called **molecular flow**. Because the spread in speeds is so small, the molecules are effectively in a state of very low translational temperature (Fig. 21D.2). The translational temperature may reach as low as 1 K. Such jets are called **supersonic** because the average speed of the molecules in the jet is much greater than the speed of sound in the jet.

A supersonic jet can be converted into a more parallel **supersonic beam** if it is 'skimmed' in the region of hydrodynamic flow and the excess gas pumped away. A skimmer consists of a conical nozzle shaped to avoid any supersonic shock waves spreading back into the gas and so increasing the translational temperature (Fig. 21D.3). A jet or beam may also be formed by using helium or neon as the principal gas, and injecting molecules of interest into it in the hydrodynamic region of flow.

The low translational temperature of the molecules is reflected in their low rotational and vibrational temperatures. In this context, a rotational or vibrational temperature means the temperature that should be used in the Boltzmann distribution to reproduce the observed populations of the states. However, as rotational states equilibrate more slowly than translational states, and vibrational states equilibrate even more slowly, the rotational and vibrational populations of the species correspond to somewhat higher temperatures, of the order of 10 K for rotation and 100 K for vibrations.

The target gas may be either a bulk sample or another molecular beam. The detectors may consist of a chamber fitted with a sensitive pressure gauge, a bolometer (a detector that responds to the incident energy by making use of the temperature-dependence of resistance), or an ionization detector, in which the incoming molecule is first ionized and then detected electronically. The rotational and vibrational state of the scattered molecules may also be determined spectroscopically.

(b) Experimental results

The primary experimental information from a molecular beam experiment is the fraction of the molecules in the incident beam that are scattered into a particular direction. The fraction is normally expressed in terms of dI, the rate at which molecules are scattered into a cone (described by a solid angle dΩ) that represents the area covered by the 'eye' of the detector (Fig. 21D.4). This rate is reported as the **differential scattering cross-section**, σ, the constant of proportionality between the value of dI and the intensity, I, of the incident beam, the number density of target molecules, $\mathcal{N}$, and the infinitesimal path length dx through the sample:

$$\mathrm{d}I = \sigma I \mathcal{N}\,\mathrm{d}x \quad \textit{Definition} \quad \text{Differential scattering cross-section} \quad (21D.1)$$

The value of σ (which has the dimensions of area) depends on the **impact parameter**, b, the initial perpendicular separation of the paths of the colliding molecules (Fig. 21D.5), and the details of the intermolecular potential.

The role of the impact parameter is most easily seen by considering the impact of two hard spheres (Fig. 21D.6). If $b=0$,

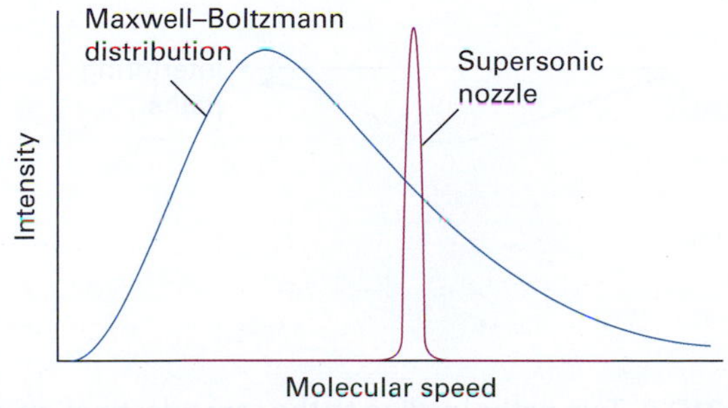

Figure 21D.2 The shift in the mean speed and the width of the distribution brought about by use of a supersonic nozzle.

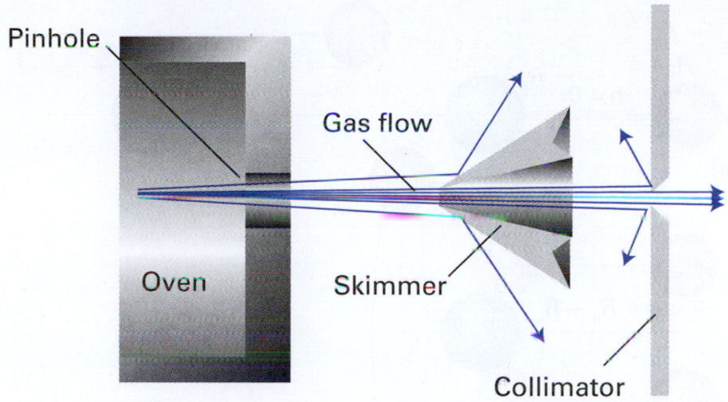

Figure 21D.3 A supersonic nozzle skims off some of the molecules of the beam and leads to a beam with well-defined velocity.

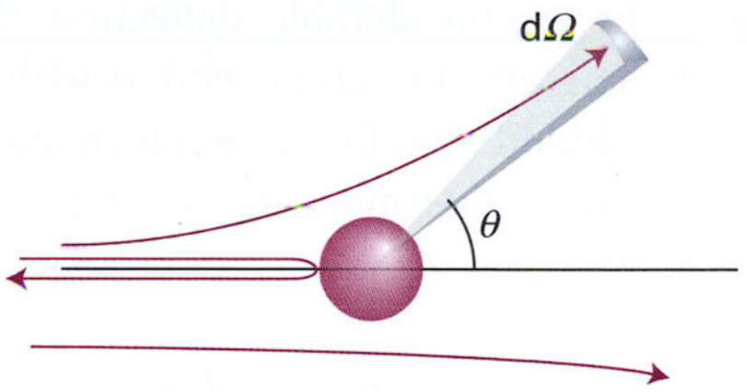

Figure 21D.4 The definition of the solid angle, dΩ, for scattering.

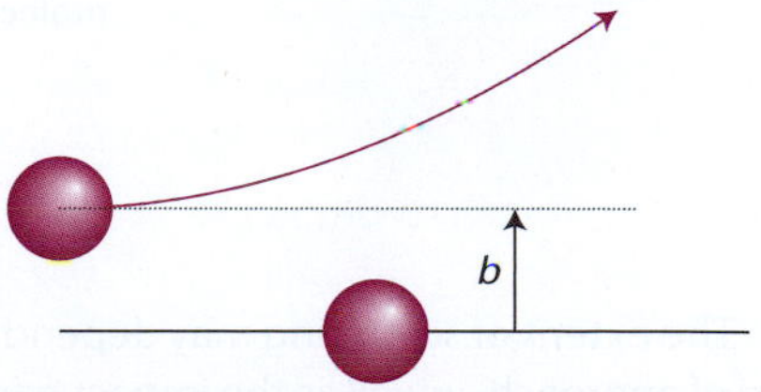

Figure 21D.5 The definition of the impact parameter, b, as the perpendicular separation of the initial paths of the particles.

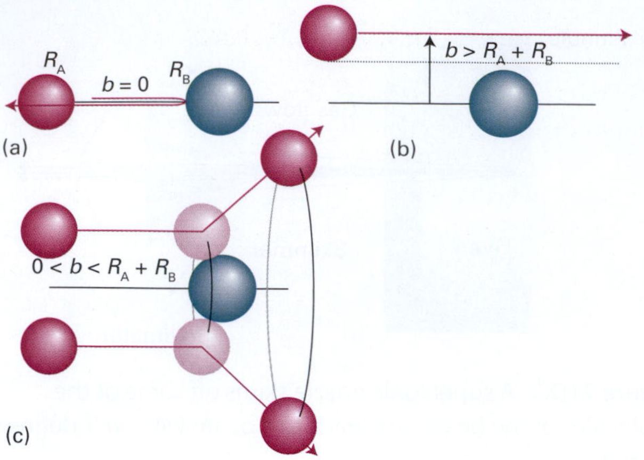

Figure 21D.6 Three typical cases for the collisions of two hard spheres: (a) $b=0$, giving backward scattering; (b) $b>R_A+R_B$, giving forward scattering; (c) $0<b<R_A+R_B$, leading to scattering into one direction on a ring of possibilities. (The target molecule is taken to be so heavy that it remains virtually stationary.)

the projectile is on a trajectory that leads to a head-on collision, so the only scattering intensity is detected when the detector is at $\theta=\pi$. When the impact parameter is so great that the spheres do not make contact ($b>R_A+R_B$), there is no scattering and the scattering cross-section is zero at all angles except $\theta=0$. Glancing blows, with $0<b\leq R_A+R_B$, lead to scattering intensity in cones around the forward direction

The scattering pattern of real molecules, which are not hard spheres, depends on the details of the intermolecular potential, including the anisotropy that is present when the molecules are non-spherical. The scattering also depends on the relative speed of approach of the two particles: a very fast particle might pass through the interaction region without much deflection, whereas a slower one on the same path might be temporarily captured and undergo considerable deflection (Fig. 21D.7). The variation of the scattering cross-section with the relative speed of approach should therefore give information about the strength and range of the intermolecular potential.

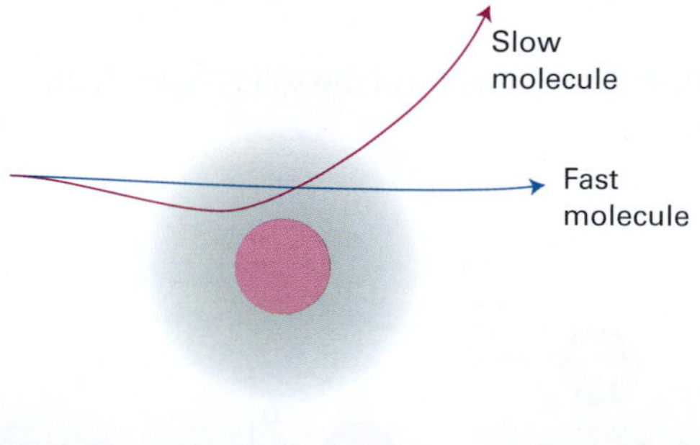

Figure 21D.7 The extent of scattering may depend on the relative speed of approach as well as the impact parameter. The dark central zone represents the repulsive core; the fuzzy outer zone represents the long-range attractive potential.

A further point is that the outcome of collisions is determined by quantum, not classical, mechanics. The wave nature of the particles can be taken into account, at least to some extent, by drawing all classical trajectories that take the projectile particle from source to detector, and then considering the effects of interference between them.

Two quantum mechanical effects are of great importance. A particle with a certain impact parameter might approach the attractive region of the potential in such a way that the particle is deflected towards the repulsive core (Fig. 21D.8), which then repels it out through the attractive region to continue its flight in the forward direction. Some molecules, however, also travel in the forward direction because they have impact parameters so large that they are undeflected. The wavefunctions of the particles that take the two types of path interfere, and the intensity in the forward direction is modified. The effect is called **quantum oscillation**. The same phenomenon accounts for the optical 'glory effect', in which a bright halo can sometimes be seen surrounding an illuminated object. (The coloured rings around the shadow of an aircraft cast on clouds by the Sun, and often seen in flight, are an example of an optical glory.)

The second quantum effect we need consider is the observation of a strongly enhanced scattering in a non-forward direction. This effect is called **rainbow scattering** because the same mechanism accounts for the appearance of an optical rainbow. The origin of the phenomenon is illustrated in Fig. 21D.9. As the impact parameter decreases, there comes a stage at which the scattering angle passes through a maximum and the interference between the paths results in a strongly scattered beam. The **rainbow angle**, θ_r, is the angle for which $d\theta/db=0$ and the scattering is strong.

Another phenomenon that can occur in certain beams is the capturing of one species by another. The vibrational temperature in supersonic beams is so low that **van der Waals molecules** may be formed, which are complexes of the form AB in which A and B are held together by van der Waals forces or hydrogen bonds. Large numbers of such molecules have been studied spectroscopically, including ArHCl, $(HCl)_2$, $ArCO_2$, and $(H_2O)_2$. More recently, van der Waals clusters of water molecules have been pursued as far as $(H_2O)_6$. The study of their

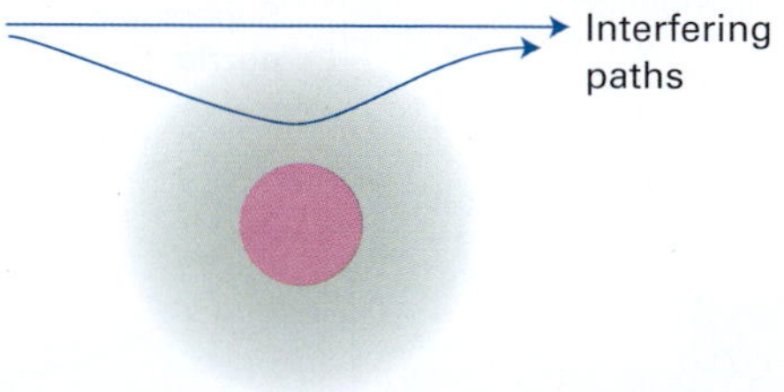

Figure 21D.8 Two paths leading to the same destination will interfere quantum mechanically; in this case they give rise to quantum oscillations in the forward direction.

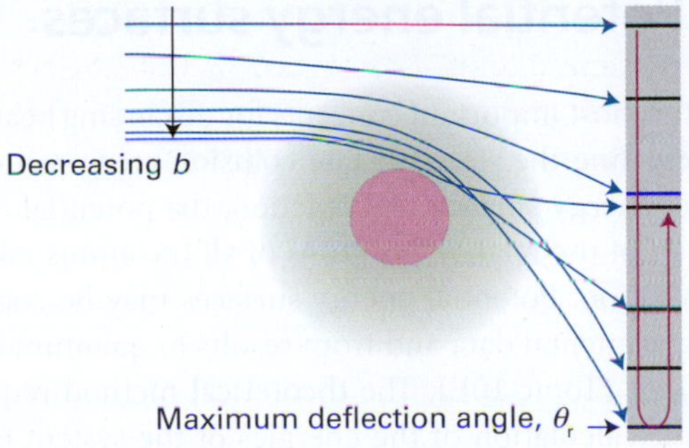

Figure 21D.9 The interference of paths leading to rainbow scattering. The rainbow angle, θ_r, is the maximum scattering angle reached as b is decreased. Interference between the numerous paths at that angle modifies the scattering intensity markedly.

spectroscopic properties gives detailed information about the intermolecular potentials involved.

21D.2 Reactive collisions

Detailed experimental information about the intimate processes that occur during reactive encounters comes from molecular beams, especially **crossed molecular beams** (Fig. 21D.10). The detector for the products of the collision of molecules in the two beams can be moved to different angles to observe so the angular distribution of the products. Because the molecules in the incoming beams can be prepared with different energies (for example, with different translational energies by using rotating sectors and supersonic nozzles, with different vibrational energies by using selective excitation with lasers, and with different orientations by using electric fields), it is possible to study the dependence of the success of collisions on these variables and to study how they affect the properties of the product molecules.

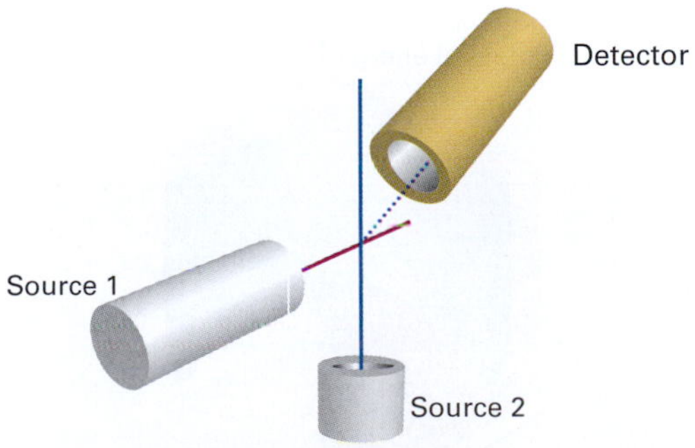

Figure 21D.10 In a crossed-beam experiment, state-selected molecules are generated in two separate sources, and are directed perpendicular to one another. The detector responds to molecules (which may be product molecules if a chemical reaction occurs) scattered into a chosen direction.

(a) Probes of reactive collisions

One method for examining the energy distribution in the products is **infrared chemiluminescence**, in which vibrationally excited molecules emit infrared radiation as they return to their ground states. By studying the intensities of the infrared emission spectrum, the populations of the vibrational states of the products may be determined (Fig. 21D.11). Another method makes use of **laser-induced fluorescence**. In this technique, a laser is used to excite a product molecule from a specific vibration–rotation level; the intensity of the fluorescence from the upper state is monitored and interpreted in terms of the population of the initial vibration–rotation state. When the molecules being studied do not fluoresce efficiently, versions of Raman spectroscopy (Topic 12A) can be used to monitor the progress of reaction. **Multiphoton ionization** (MPI) techniques are also good alternatives for the study of weakly fluorescing molecules. In MPI, the absorption by a molecule of several photons from one or more pulsed lasers results in ionization if the total photon energy is greater than the ionization energy of the molecule.

The angular distribution of products can be determined by **reaction product imaging**. In this technique, product ions are accelerated by an electric field towards a phosphorescent screen and the light emitted from specific spots where the ions struck the screen is imaged by a charge-coupled device (CCD). An important variant of MPI is **resonant multiphoton ionization** (REMPI), in which one or more photons promote a molecule to an electronically excited state and then additional photons are used to generate ions from the excited state. The power of REMPI lies in the fact that the experimenter can choose which reactant or product to study by tuning the laser frequency to the electronic absorption band of a specific molecule.

(b) State-to-state reaction dynamics

The concept of collision cross-section is introduced in connection with collision theory in Topic 21A, where it is shown

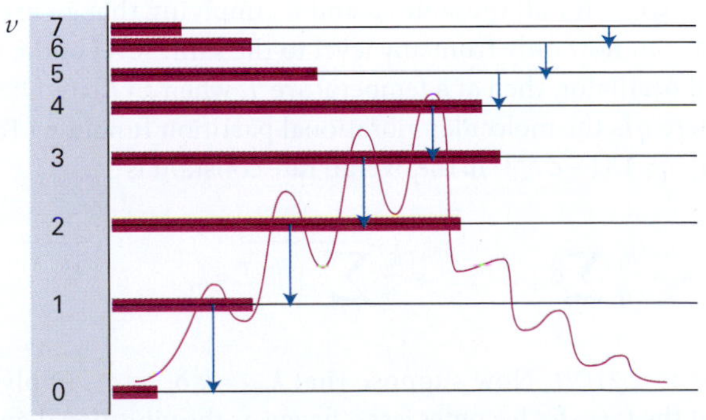

Figure 21D.11 Infrared chemiluminescence from CO produced in the reaction $O + CS \rightarrow CO + S$ arises from the non-equilibrium populations of the vibrational states of CO and the radiative relaxation to equilibrium.

that the second-order rate constant, k_r, can be expressed as a Boltzmann-weighted average of the reactive cross-section and the relative speed of approach of the colliding reactant molecules. We shall write eqn 21A.6 of that Topic ($k_r = N_A\int_0^\infty \sigma(\varepsilon)v_{rel}f(\varepsilon)d\varepsilon$) as

$$k_r = \langle \sigma v_{rel} \rangle N_A \qquad (21D.2)$$

where the angle brackets denote a Boltzmann average. Molecular beam studies provide a more sophisticated version of this quantity, for they provide the **state-to-state cross-section**, $\sigma_{nn'}$, and hence the **state-to-state rate constant**, $k_{nn'}$ for the reactive transition from initial state n of the reactants to final state n' of the products:

$$k_{nn'} = \langle \sigma_{nn'} v_{rel} \rangle N_A \qquad \text{State-to-state rate constant} \qquad (21D.3)$$

The rate constant k_r is the sum of the state-to-state rate constants over all final states (because a reaction is successful whatever the final state of the products) and over a Boltzmann-weighted sum of initial states (because the reactants are initially present with a characteristic distribution of populations at a temperature T):

$$k_r = \sum_{n,n'} k_{nn'}(T) f_n(T) \qquad (21D.4)$$

where $f_n(T)$ is the Boltzmann factor at a temperature T. It follows that if we can determine or calculate the state-to-state cross-sections for a wide range of approach speeds and initial and final states, then we have a route to the calculation of the rate constant for the reaction.

Brief illustration 21D.1 The state-to-state rate constant

Suppose a harmonic oscillator collides with another oscillator of the same effective mass and force constant. If the state-to-state rate constant for the excitation of the latter's vibration is $k_{vv'} = k_r^\circ \delta_{vv'}$ for all the states v and v', implying that an excitation can flow only from any level to the same level of the second oscillator, then at a temperature T, when $f_v(T) = e^{-vh\nu/kT}/q$, where q is the molecular vibrational partition function (Topic 15B, $q = 1/(1-e^{-h\nu/kT})$), the overall rate constant is

$$k_r = \frac{k_r^\circ}{q}\sum_{v,v'} \delta_{vv'} e^{-vh\nu/kT} = \frac{k_r^\circ}{q}\overbrace{\sum_{v'} e^{-v'h\nu/kT}}^{q} = k_r^\circ$$

Self-test 21D.1 Now suppose that $k_{vv'} = k_r^\circ \delta_{vv'} e^{-\lambda v}$, implying that the transfer becomes less efficient as the vibrational quantum number increases. Evaluate k_r.

Answer: $k_r = k_r^\circ (1-e^{-h\nu/kT})/(1-e^{-(\lambda + h\nu/kT)})$

21D.3 Potential energy surfaces

One of the most important concepts for discussing beam results and calculating the state-to-state collision cross-section is the **potential energy surface** of a reaction, the potential energy as a function of the relative positions of all the atoms taking part in the reaction. Potential energy surfaces may be constructed from experimental data and from results of quantum chemical calculations (Topic 10E). The theoretical method requires the systematic calculation of the energies of the system in a large number of geometrical arrangements. Special computational techniques, such as those described in Topic 10E, are used to take into account electron correlation, which arises from interactions between electrons as they move closer to and farther from each other in a molecule or molecular cluster. Techniques that incorporate electron correlation accurately are very time consuming and, consequently, only reactions between relatively simple particles, such as the reactions $H + H_2 \rightarrow H_2 + H$ and $H + H_2O \rightarrow OH + H_2$, currently are amenable to this type of theoretical treatment. An alternative is to use semi-empirical methods, in which results of calculations and experimental parameters are used to construct the potential energy surface.

To illustrate the features of a potential energy surface, consider the collision between an H atom and an H_2 molecule. Detailed calculations show that the approach of an atom H_A along the H_B–H_C axis requires less energy for reaction than any other approach, so initially we confine our attention to that collinear approach. Two parameters are required to define the nuclear separations: the H_A–H_B separation R_{AB} and the H_B–H_C separation R_{BC}.

At the start of the encounter R_{AB} is effectively infinite and R_{BC} is the H_2 equilibrium bond length. At the end of a successful reactive encounter R_{AB} is equal to the equilibrium bond length and R_{BC} is infinite. The total energy of the three-atom system depends on their relative separations, and can be found by doing an electronic structure calculation. The plot of the total energy of the system against R_{AB} and R_{BC} gives the potential energy surface of this collinear reaction (Fig. 21D.12). This surface is normally depicted as a contour diagram (Fig. 21D.13).

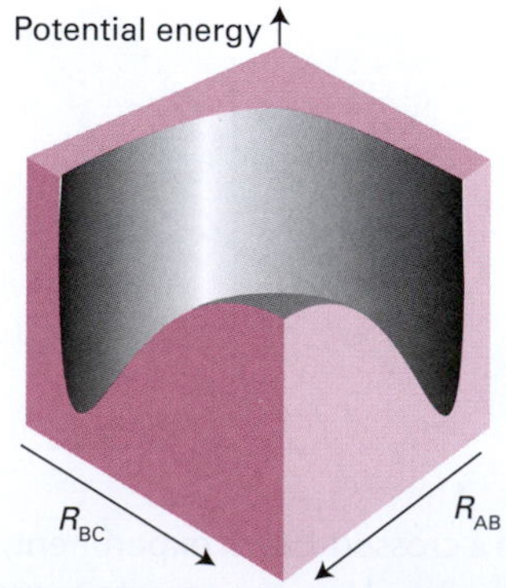

Figure 21D.12 The potential energy surface for the $H + H_2 \rightarrow H_2 + H$ reaction when the atoms are constrained to be collinear.

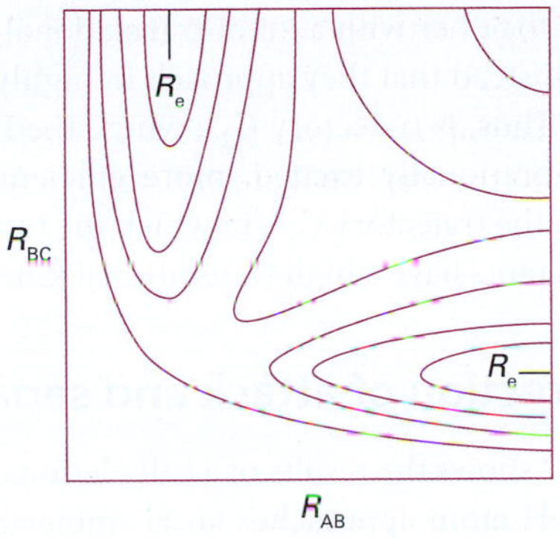

Figure 21D.13 The contour diagram (with contours of equal potential energy) corresponding to the surface in Fig. 21D.12. R_e marks the equilibrium bond length of an H_2 molecule (strictly, it relates to the arrangement when the third atom is at infinity).

When R_{AB} is very large, the variation in potential energy represented by the surface as R_{BC} changes is that of an isolated H_2 molecule as its bond length is altered. A section through the surface at $R_{AB}=\infty$, for example, is the same as the H_2 bonding potential energy curve. At the edge of the diagram where R_{BC} is very large, a section through the surface is the molecular potential energy curve of an isolated H_AH_B molecule.

Brief illustration 21D.2 A potential energy surface

The bimolecular reaction $H+O_2\rightarrow OH+O$ plays an important role in combustion processes. The reaction can be characterized in terms of the HO_2 potential energy surface and the two distances for collinear approach R_{HO_A} and $R_{O_AO_B}$. When R_{HO_A} is very large, the variation of the HO_2 potential energy with $R_{O_AO_B}$ is that of an isolated dioxygen molecule as its bond length is changed. Similarly, when $R_{O_AO_B}$ is very large, a section through the potential energy surface is the molecular potential energy curve of an isolated OH radical.

Self-test 21D.2 Repeat the analysis for $H+OD\rightarrow OH+D$.

Answer: R_{HO} at infinity: OD potential energy curve; R_{OD} at infinity: OH potential energy curve

The actual path of the atoms in the course of the encounter depends on their total energy, the sum of their kinetic and potential energies. However, we can obtain an initial idea of the paths available to the system by identifying paths that correspond to least potential energy. For example, consider the changes in potential energy as H_A approaches H_BH_C. If the H_B–H_C bond length is constant during the initial approach of H_A, then the potential energy of the H_3 cluster rises along the path marked A in Fig. 21D.14. We see that the potential energy reaches a high value as H_A is pushed into the molecule and then decreases sharply as H_C breaks off and separates to a great distance. An alternative reaction path can be imagined (B) in which the H_B–H_C bond length increases while H_A is still far away. Both paths, although feasible if the molecules have sufficient initial kinetic energy, take the three atoms to regions of high potential energy in the course of the encounter.

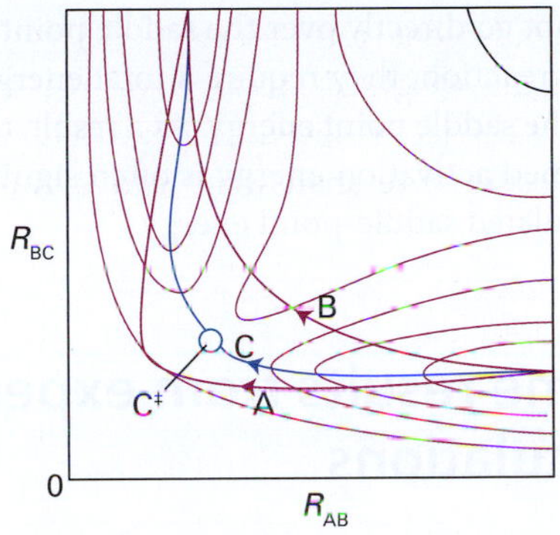

Figure 21D.14 Various trajectories through the potential energy surface shown in Fig. 21D.13. Path A corresponds to a route in which R_{BC} is held constant as H_A approaches; path B corresponds to a route in which R_{BC} lengthens at an early stage during the approach of H_A; path C is the route along the floor of the potential valley.

The path of least potential energy is the one marked C, corresponding to R_{BC} lengthening as H_A approaches and begins to form a bond with H_B. The H_B–H_C bond relaxes at the demand of the incoming atom, and the potential energy climbs only as far as the saddle-shaped region of the surface, to the **saddle point** marked $C^{\ddagger}$. The encounter of least potential energy is one in which the atoms take route C up the floor of the valley, through the saddle point, and down the floor of the other valley as H_C recedes and the new H_A–H_B bond achieves its equilibrium length. This path is the reaction coordinate.

We can now make contact with the transition-state theory of reaction rates (Topic 21C). In terms of trajectories on potential surfaces with a total energy close to the saddle point energy, the transition state can be identified with a critical geometry such that every trajectory that goes through this geometry goes on to react (Fig. 21D.15). Most trajectories on potential energy

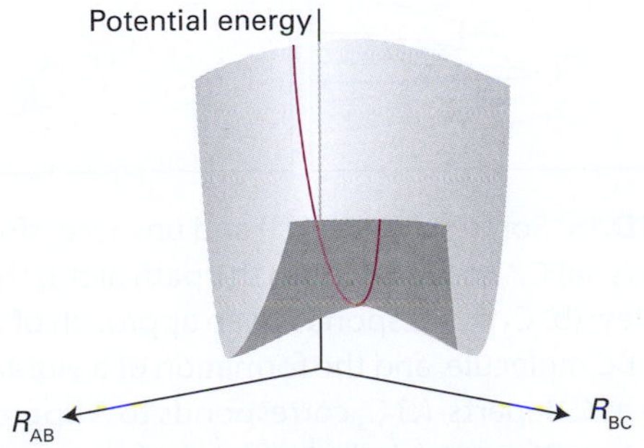

Figure 21D.15 The transition state is a set of configurations (here, marked by the purple line across the saddle point) through which successful reactive trajectories must pass.

surfaces do not go directly over the saddle point and therefore, to result in a reaction, they require a total energy significantly higher than the saddle point energy. As a result, the experimentally determined activation energy is often significantly higher than the calculated saddle-point energy.

21D.4 Some results from experiments and calculations

Although quantum mechanical tunnelling can play an important role in reactivity, particularly in hydrogen atom and electron transfer reactions, initially we can consider the classical trajectories of particles over surfaces. From this viewpoint, to travel successfully from reactants to products, the incoming molecules must possess enough kinetic energy to be able to climb to the saddle point of the potential surface. Therefore, the shape of the surface can be explored experimentally by changing the relative speed of approach (by selecting the beam velocity) and the degree of vibrational excitation and observing whether reaction occurs and whether the products emerge in a vibrationally excited state (Fig. 21D.16). For example, one question that can be answered is whether it is better to smash the reactants together with a lot of translational kinetic energy or to ensure instead that they approach in highly excited vibrational states. Thus, is trajectory C_2*, where the H_BH_C molecule is initially vibrationally excited, more efficient at leading to reaction than the trajectory C_1*, in which the total energy is the same but reactants have a high translational kinetic energy?

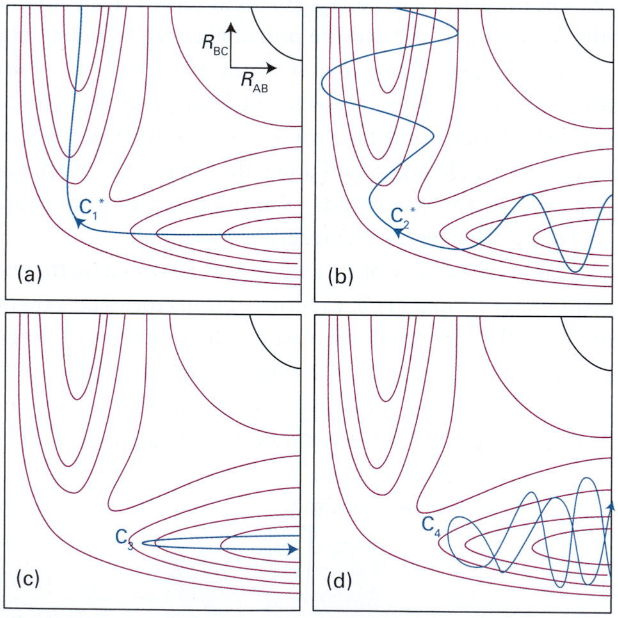

Figure 21D.16 Some successful (*) and unsuccessful encounters. (a) C_1* corresponds to the path along the foot of the valley; (b) C_2* corresponds to an approach of A to a vibrating BC molecule, and the formation of a vibrating AB molecule as C departs. (c) C_3 corresponds to A approaching a non-vibrating BC molecule, but with insufficient translational kinetic energy; (d) C_4 corresponds to A approaching a vibrating BC molecule, but still the energy, and the phase of the vibration, is insufficient for reaction.

(a) The direction of attack and separation

Figure 21D.17 shows the results of a calculation of the potential energy as an H atom approaches an H_2 molecule from different angles, the H_2 bond being allowed to relax to the optimum length in each case. The potential barrier is least for collinear attack, as we assumed earlier. (But we must be aware that other lines of attack are feasible and contribute to the overall rate.) In contrast, Fig. 21D.18 shows the potential energy changes that occur as a Cl atom approaches an HI molecule. The lowest barrier occurs for approaches within a cone of half-angle 30° surrounding the H atom. The relevance of this result to the calculation of the steric factor of collision theory should be noted: not every collision is successful, because they do not all lie within the reactive cone.

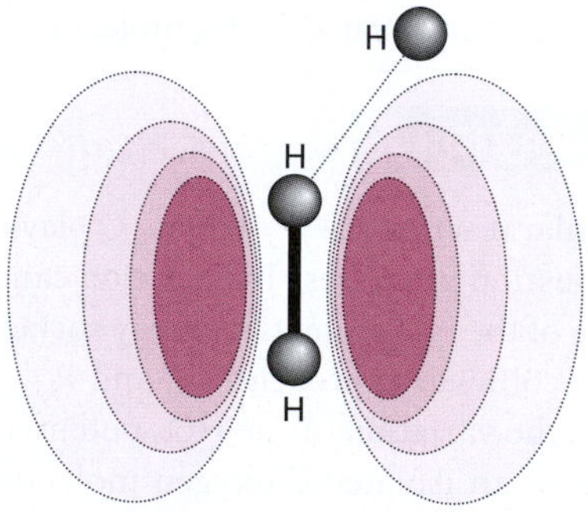

Figure 21D.17 An indication of how the anisotropy of the potential energy changes as H approaches H_2 with different angles of attack. The collinear attack has the lowest potential barrier to reaction. The surface indicates the potential energy profile along the reaction coordinate for each configuration.

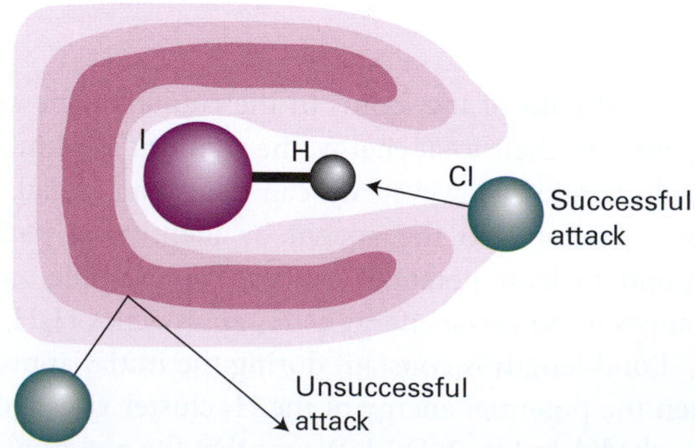

Figure 21D.18 The potential energy barrier for the approach of Cl to HI. In this case, successful encounters occur only when Cl approaches within a cone surrounding the H atom.

If the collision is sticky, so that when the reactants collide they orbit around each other, the products can be expected to emerge in random directions because all memory of the approach direction has been lost. A rotation takes about 1 ps, so if the collision is over in less than that time the complex will not have had time to rotate and the products will be thrown off in a specific direction. In the collision of K and I_2, for example, most of the products are thrown off in the forward direction (forward and backward directions refer to directions in a centre-of-mass coordinate system with the origin at the centre of mass of the colliding reactants and collision occurring when molecules are at the origin.) This product distribution is consistent with the harpoon mechanism (Topic 20A) because the transition takes place at long range. In contrast, the collision of K with CH_3I leads to reaction only if the molecules approach each other very closely. In this mechanism, K effectively bumps into a brick wall, and the KI product bounces out in the backward direction. The detection of this anisotropy in the angular distribution of products gives an indication of the distance and orientation of approach needed for reaction, as well as showing that the event is complete in less than about 1 ps.

(b) Attractive and repulsive surfaces

Some reactions are very sensitive to whether the energy has been pre-digested into a vibrational mode or left as the relative translational kinetic energy of the colliding molecules. For example, if two HI molecules are hurled together with more than twice the activation energy of the reaction, then no reaction occurs if all the energy is solely translational. For $F+HCl \rightarrow Cl+HF$, for example, the reaction is about five times more efficient when the HCl is in its first vibrational excited state than when, although HCl has the same total energy, it is in its vibrational ground state.

The origin of these requirements can be found by examining the potential energy surface. Figure 21D.19 shows an **attractive surface** in which the saddle point occurs early in the reaction coordinate. Figure 21D.20 shows a **repulsive surface** in which the saddle point occurs late. A surface that is attractive in one direction is repulsive in the reverse direction.

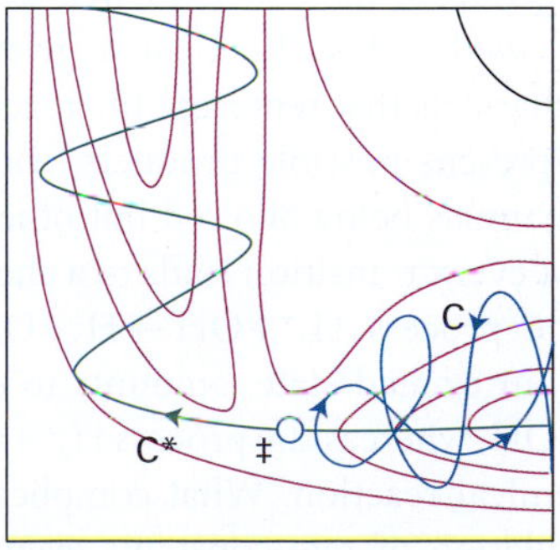

Figure 21D.19 An attractive potential energy surface. A successful encounter (C*) involves high translational kinetic energy and results in a vibrationally excited product.

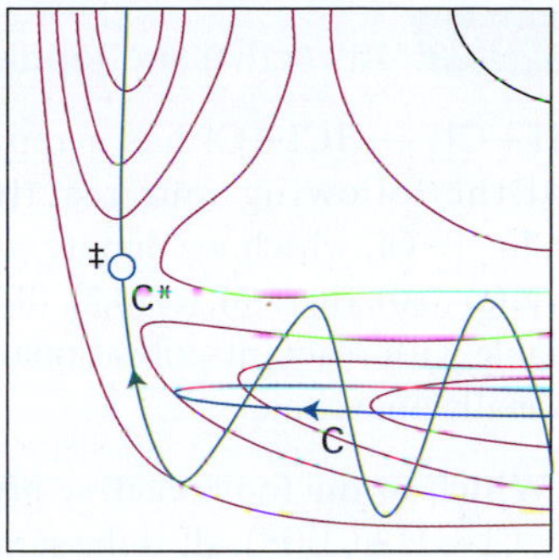

Figure 21D.20 A repulsive potential energy surface. A successful encounter (C*) involves initial vibrational excitation and the products have high translational kinetic energy. A reaction that is attractive in one direction is repulsive in the reverse direction.

Consider first the attractive surface. If the original molecule is vibrationally excited, then a collision with an incoming molecule takes the system along C. This path is bottled up in the region of the reactants, and does not take the system to the saddle point. If, however, the same amount of energy is present solely as translational kinetic energy, then the system moves along C* and travels smoothly over the saddle point into products. We can therefore conclude that reactions with attractive potential energy surfaces proceed more efficiently if the energy is in relative translational motion. Moreover, the potential surface shows that once past the saddle point the trajectory runs up the steep wall of the product valley, and then rolls from side to side as it falls to the foot of the valley as the products separate. In other words, the products emerge in a vibrationally excited state.

Now consider the repulsive surface. On trajectory C the collisional energy is largely in translation. As the reactants approach, the potential energy rises. Their path takes them up the opposing face of the valley, and they are reflected back into the reactant region. This path corresponds to an unsuccessful encounter, even though the energy is sufficient for reaction. On C* some of the energy is in the vibration of the reactant molecule and the motion causes the trajectory to weave from side to side up the valley as it approaches the saddle point. This motion may be sufficient to tip the system round the corner to the saddle point and then on to products. In this case, the product molecule is expected to be in an unexcited vibrational state. Reactions with repulsive potential surfaces can therefore be expected to proceed more efficiently if the excess energy is present as vibrations. This is the case with the $H+Cl_2 \rightarrow HCl+Cl$ reaction, for instance.

Brief illustration 21D.3 Attractive and repulsive surfaces

The reaction $H+Cl_2 \rightarrow HCl+Cl$ has a repulsive potential surface. Of the following four reactive processes, $H+Cl_2(v) \rightarrow HCl(v')+Cl$, which we denote (v,v'), all at the same total energy, (a) (0,0), (b) (2,0), (c) (0,2), (d) (2,2), reaction (b) is most probable with reactants vibrationally excited and products vibrationally unexcited.

Self-test 21D.3 Which of the four reactive processes of the reaction $HCl(v)+Cl \rightarrow H+Cl_2(v')$, all at the same total energy, (a) (0,0), (b) (2,0), (c) (0,2), (d) (2,2), is most probable?

Answer: (0,2); attractive surface

(c) Classical trajectories

A clear picture of the reaction event can be obtained by using classical mechanics to calculate the trajectories of the atoms taking place in a reaction from a set of initial conditions, such as velocities, relative orientations, and internal energies of the reacting particles. The initial values used for the internal energy reflect the quantization of electronic, vibrational, and rotational energies in molecules but the features of quantum mechanics are not used explicitly in the calculation of the trajectory.

Figure 21D.21 shows the result of such a calculation of the positions of the three atoms in the reaction $H+H_2 \rightarrow H_2+H$, the horizontal coordinate now being time and the vertical coordinate the separations. This illustration shows clearly the vibration of the original molecule and the approach of the attacking atom. The reaction itself, the switch of partners, takes place very rapidly and is an example of a **direct mode process**. The newly formed molecule shakes, but quickly settles down to steady, harmonic vibration as the expelled atom departs. In contrast, Fig. 21D.22 shows an example of a **complex mode process**, in which the activated complex survives for an extended period. The reaction in the figure is the exchange reaction $KCl+NaBr \rightarrow KBr+NaCl$. The tetratomic activated complex survives for about 5 ps, during which time the atoms make about 15 oscillations before dissociating into products.

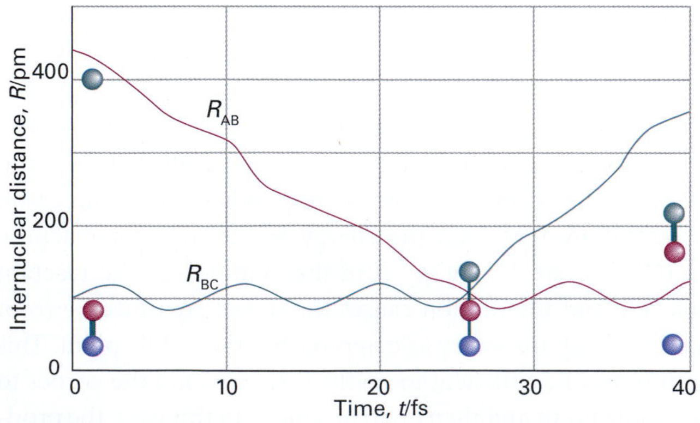

Figure 21D.21 The calculated trajectories for a reactive encounter between A and a vibrating BC molecule leading to the formation of a vibrating AB molecule. This direct-mode reaction is between H and H_2 (M. Karplus et al., *J. Chem. Phys.* **43**, 3258 (1965)).

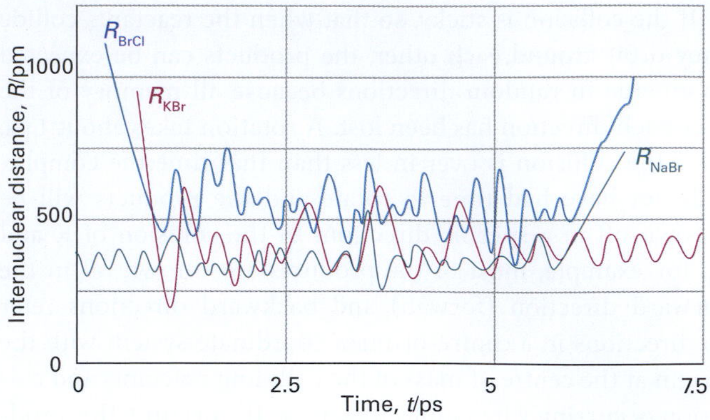

Figure 21D.22 An example of the trajectories calculated for a complex-mode reaction, $KCl+NaBr \rightarrow KBr+NaCl$, in which the collision cluster has a long lifetime (P. Brumer and M. Karplus, *Faraday Disc. Chem. Soc.* **55**, 80 (1973)).

(d) Quantum mechanical scattering theory

Classical trajectory calculations do not recognize the fact that the motion of atoms, electrons, and nuclei is governed by quantum mechanics. The concept of trajectory then fades and is replaced by the unfolding of a wavefunction that represents initially the reactants and finally products.

Complete quantum mechanical calculations of trajectories and rate constants are very onerous because it is necessary to take into account all the allowed electronic, vibrational, and rotational states populated by each atom and molecule in the system at a given temperature. It is common to define a 'channel' as a group of molecules in well-defined quantum mechanically allowed states. Then, at a given temperature, there are many channels that represent the reactants and many channels that represent possible products, with some transitions between channels being allowed but others not allowed. Furthermore, not every transition leads to a chemical reaction. For example, the process $H_2^*+OH \rightarrow H_2+OH^*$, where the asterisk denotes an excited state, amounts to energy transfer between H_2 and OH, whereas the process $H_2^*+OH \rightarrow H_2O+H$ represents a chemical reaction. What complicates a quantum mechanical calculation of rate constants even in this simple four-atom system is that many reacting channels present at a given temperature can lead to the desired products H_2O+H, which themselves may be formed as many distinct channels.

The **cumulative reaction probability**, $\bar{P}(E)$, at a fixed total energy E is then written as

$$\bar{P}(E)=\sum_{i,j} P_{ij}(E) \qquad \text{Cumulative reaction probability} \qquad (21D.5)$$

where $P_{i,j}(E)$ is the probability for a transition between a reacting channel i and a product channel j and the summation is over all possible transitions that lead to product. It is then possible to show that the rate constant is given by

$$k_r(T)=\frac{\int_0^\infty \bar{P}(E)e^{-E/kT}dE}{hQ_R(T)} \qquad \text{Rate constant} \qquad (21D.6)$$

where $Q_R(T)$ is the partition function density (the partition function divided by the volume) of the reactants at the temperature T. The significance of eqn 21D.6 is that it provides a direct connection between an experimental quantity, the rate constant, and a theoretical quantity, $\bar{P}(E)$.

Checklist of concepts

- ☐ 1. A **molecular beam** is a collimated, narrow stream of molecules travelling through an evacuated vessel.
- ☐ 2. In a molecular beam, the scattering pattern of real molecules depends on quantum mechanical effects and the details of the intermolecular potential.
- ☐ 3. A **van der Waals molecule** is a complex of the form AB in which A and B are held together by van der Waals forces or hydrogen bonds.
- ☐ 4. Techniques for the study of reactive collisions include **infrared chemiluminescence**, **laser-induced fluorescence**, **multiphoton ionization** (MPI), **reaction product imaging**, and **resonant multiphoton ionization** (REMPI).
- ☐ 5. A **potential energy surface** maps the potential energy as a function of the relative positions of all the atoms taking part in a reaction.
- ☐ 6. In an **attractive surface**, the saddle point (the highest point) occurs early on the reaction coordinate.
- ☐ 7. In a **repulsive surface**, the saddle point occurs late on the reaction coordinate.

Checklist of equations

Property	Equation	Comment	Equation number
Rate of molecular scattering	$dI=\sigma I\mathcal{N}dx$	σ is the differential scattering cross section	21D.1
Rate constant	$k_r=\langle\sigma v_{rel}\rangle N_A$		21D.2
State-to-state rate constant	$k_{nn'}=\langle\sigma_{nn'}v_{rel}\rangle N_A$		21D.3
Overall rate constant	$k_r=\sum_{n,n'} k_{nn'}(T)f_n(T)$		21D.4
Cumulative reaction probability	$\bar{P}(E)=\sum_{i,j} P_{ij}(E)$		21D.5
Rate constant	$k_r(T)=\int_0^\infty \bar{P}(E)e^{-E/kT}dE/hQ_R(T)$	$Q_R(T)$ is the partition function density	21D.6

21E Electron transfer in homogeneous systems

Contents

➤ Why do you need to know this material?

Electron transfer reactions between protein-bound cofactors or between proteins play an important role in a variety of biological processes. Electron transfer is also important in homogeneous, non-biological catalysis (especially biomimetic systems).

➤ What is the key idea?

The rate constant of electron transfer in a donor–acceptor complex depends on the distance between electron donor and acceptor, the standard reaction Gibbs energy, and the energy needed to reach a particular arrangement of atoms.

➤ What do you need to know already?

This Topic makes use of transition-state theory (Topic 21C). It also uses the concept of tunnelling (Topic 8A), the steady-state approximation (Topic 20E), and the Franck–Condon principle (Topic 13A).

Here we apply the concepts of transition state theory and quantum theory to the study of a deceptively simple process, electron transfer between molecules in homogeneous systems. We describe a theoretical approach to the calculation of rate constants and discuss the theory in the light of experimental results on a variety of systems, including protein complexes. We shall see that relatively simple expressions can be used to predict the rates of electron transfer with reasonable accuracy.

21E.1 The electron transfer rate law

Consider electron transfer from a donor species D to an acceptor species A in solution. The overall reaction is

$$\mathrm{D+A\rightarrow D^{+}+A^{-}} \qquad v=k_{\mathrm{r}}[\mathrm{D}][\mathrm{A}] \tag{21E.1}$$

In the first step of the mechanism, D and A must diffuse through the solution and on meeting form a complex DA:

$$\mathrm{D+A}\underset{k_{\mathrm{a}}'}{\overset{k_{\mathrm{a}}}{\rightleftarrows}}\mathrm{DA} \tag{21E.2a}$$

We suppose that in the complex D and A are separated by d, the distance between their outer surfaces. Next, electron transfer occurs within the DA complex to yield $\mathrm{D^+A^-}$:

$$\mathrm{DA}\underset{k_{\mathrm{et}}'}{\overset{k_{\mathrm{et}}}{\rightleftarrows}}\mathrm{D^{+}A^{-}} \tag{21E.2b}$$

The complex $\mathrm{D^+A^-}$ can also break apart and the ions diffuse through the solution:

$$\mathrm{D^{+}A^{-}}\xrightarrow{k_{\mathrm{d}}}\mathrm{D^{+}+A^{-}} \tag{21E.2c}$$

We show in the following *Justification* that on the basis of this model

$$\frac{1}{k_{\mathrm{r}}}=\frac{1}{k_{\mathrm{a}}}+\frac{k_{\mathrm{a}}'}{k_{\mathrm{a}}k_{\mathrm{et}}}\left(1+\frac{k_{\mathrm{et}}'}{k_{\mathrm{d}}}\right) \qquad \text{Electron transfer rate constant} \tag{21E.3}$$

Justification 21E.1 The rate constant for electron transfer in solution

We begin by identifying the rate of the overall reaction (eqn 21E.1) with the rate of formation of separated ions:

$$v = k_r[D][A] = k_d[D^+A^-]$$

There are two reaction intermediates, DA and D^+A^-, and we apply the steady-state approximation (Topic 20E) to both. From

$$\frac{d[D^+A^-]}{dt} = k_{et}[DA] - k'_{et}[D^+A^-] - k_d[D^+A^-] = 0$$

it follows that

$$[DA] = \frac{k'_{et} + k_d}{k_{et}}[D^+A^-]$$

and from

$$\frac{d[DA]}{dt} = k_a[D][A] - k'_a[DA] - k_{et}[DA] + k'_{et}[D^+A^-] = 0$$

it follows that

$$k_a[D][A] \overbrace{-k'_a[DA] - k_{et}\underbrace{[DA]}_{}}^{-(k_a + k_{et})\ \ [DA]\ =\ ((k'_{et}+k_d)/k_{et})[D^+A^-]} + k'_{et}[D^+A^-]$$

$$= k_a[D][A] - \left\{\frac{(k'_a + k_{et})(k'_{et} + k_d)}{k_{et}} - k'_{et}\right\}[D^+A^-]$$

$$= k_a[D][A] - \frac{1}{k_{et}}(k'_a k'_{et} + k'_a k_d + k_d k_{et})[D^+A^-] = 0$$

therefore

$$[D^+A^-] = \frac{k_a k_{et}}{k'_a k'_{et} + k'_a k_d + k_d k_{et}}[D][A]$$

When this expression is multiplied by k_d, the resulting equation has the form of the rate of electron transfer, $v = k_r[D][A]$, with k_r given by

$$k_r = \frac{k_a k_{et} k_d}{k'_a k'_{et} + k'_a k_d + k_d k_{et}}$$

To obtain eqn 21E.3, divide the numerator and denominator on the right-hand side of this expression by $k_d k_{et}$ and solve for the reciprocal of k_r.

To gain insight into eqn 21E.3 and the factors that determine the rate of electron transfer reactions in solution, we assume that the main decay route for D^+A^- is dissociation of the complex into separated ions, and therefore that $k_d \gg k'_{et}$. It follows that

$$\frac{1}{k_r} \approx \frac{1}{k_a}\left(1 + \frac{k'_a}{k_{et}}\right)$$

Physical interpretation

- When $k_{et} \gg k'_a$, $k_r \approx k_a$ and the rate of product formation is controlled by diffusion of D and A in solution, which favours formation of the DA complex.
- When $k_{et} \ll k'_a$, $k_r \approx (k_a/k'_a)k_{et} = Kk_{et}$, where K is the equilibrium constant for the diffusive encounter. The process is controlled by k_{et} and therefore the activation energy of electron transfer in the DA complex.

21E.2 The rate constant

This analysis can be taken further by introducing the implication from transition-state theory (Topic 21C) that, at a given temperature, $k_{et} \propto e^{-\Delta^{\ddagger}G/RT}$, where $\Delta^{\ddagger}G$ is the Gibbs energy of activation. Our remaining task, therefore, is to find expressions for the proportionality constant and $\Delta^{\ddagger}G$.

Our discussion concentrates on the following two key aspects of the theory of electron transfer processes, which was developed independently by R.A. Marcus, N.S. Hush, V.G. Levich, and R.R. Dogonadze:

- Electrons are transferred by tunnelling through a potential energy barrier, the height of which is partly determined by the ionization energies of the DA and D^+A^- complexes. Electron tunnelling influences the magnitude of the proportionality constant in the expression for k_{et}.
- The complex DA and the solvent molecules surrounding it undergo structural rearrangements prior to electron transfer. The energy associated with these rearrangements and the standard reaction Gibbs energy determine $\Delta^{\ddagger}G$.

According to the Franck–Condon principle (Topic 13A), electronic transitions are so fast that they can be regarded as taking place in a stationary nuclear framework. This principle also applies to an electron transfer process in which an electron migrates from one energy surface, representing the dependence of the energy of DA on its geometry, to another representing the energy of D^+A^-. We can represent the potential energy (and the Gibbs energy) surfaces of the two complexes (the reactant complex, DA, and the product complex, D^+A^-) by the parabolas characteristic of harmonic oscillators, with the displacement coordinate corresponding to the changing geometries (Fig. 21E.1). This coordinate represents a collective mode of the donor, acceptor, and solvent.

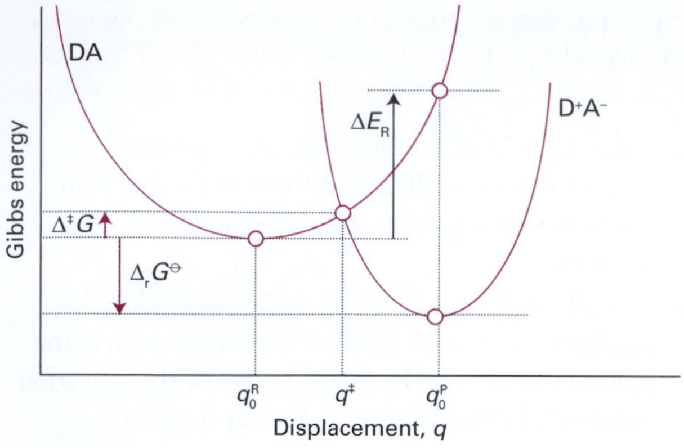

Figure 21E.1 The Gibbs energy surfaces of the complexes DA and D^+A^- involved in an electron transfer process are represented by parabolas characteristic of harmonic oscillators, with the displacement coordinate q corresponding to the changing geometries of the system.

According to the Franck–Condon principle, the nuclei do not have time to move when the system passes from the reactant to the product surface as a result of the transfer of an electron. Therefore, electron transfer can occur only after thermal fluctuations bring the geometry of DA to $q^‡$ in Fig 21E.1, the value of the nuclear coordinate at which the two parabolas intersect.

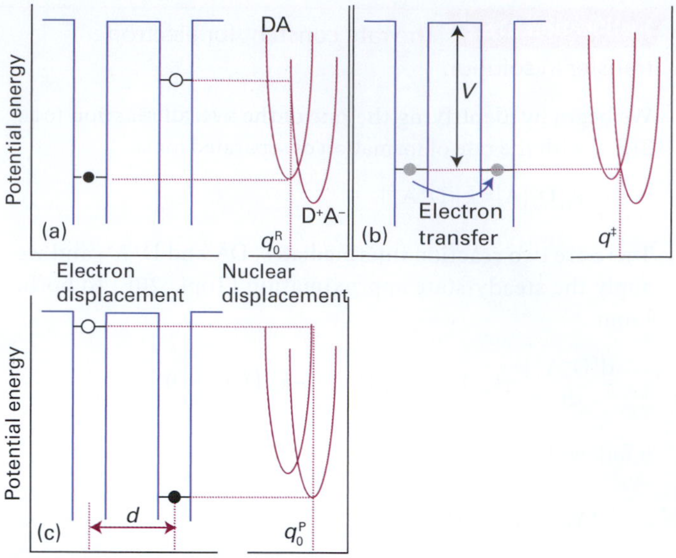

Figure 21E.2 (a) At the nuclear configuration denoted by q_0^R, the electron to be transferred in DA is in an occupied electronic energy level and the lowest unoccupied energy level of D^+A^- is of too high an energy to be a good electron acceptor. (b) As the nuclei rearrange to a configuration represented by $q^‡$, DA and D^+A^- become degenerate and electron transfer occurs by tunnelling. (c) The system relaxes to the equilibrium nuclear configuration of D^+A^- denoted by q_0^P, in which the lowest unoccupied electronic level of DA is higher in energy than the highest occupied electronic level of D^+A^-. Adapted from R.A. Marcus and N. Sutin, *Biochim. Biophys. Acta* **811**, 265 (1985).

(a) The role of electron tunnelling

The proportionality constant in the expression for k_{et} is a measure of the rate at which the system will convert from reactants (DA) to products (D^+A^-) at $q^‡$ by electron transfer within the thermally excited DA complex. To understand the process, we must turn our attention to the effect that the rearrangement of nuclear coordinates has on electronic energy levels of DA and D^+A^- for a given distance d between D and A (Fig. 21E.2). Initially, the overall energy of DA is lower than that of D^+A^- (Fig 21E.2a). As the nuclei rearrange to a configuration represented by $q^‡$ in Fig. 21E.2b, the HOMO of DA and the LUMO of D^+A^- become degenerate and electron transfer becomes energetically feasible. Over reasonably short distances d, the main mechanism of electron transfer is tunnelling through the potential energy barrier depicted in Fig 21E.2b. After an electron moves between the two frontier orbitals, the system relaxes to the configuration represented by q_0^P in Fig 21E.2c. As shown in the illustration, now the energy of D^+A^- is lower than that of DA, reflecting the thermodynamic tendency for A to remain reduced (as A^-) and for D to remain oxidized (as D^+).

The tunnelling event responsible for electron transfer is similar to that described in Topic 8A, except that in this case the electron tunnels from an electronic level of DA, with wavefunction ψ_{DA}, to an electronic level of D^+A^-, with wavefunction $\psi_{D^+A^-}$. The rate of an electronic transition from a level described by the wavefunction ψ_{DA} to a level described by $\psi_{D^+A^-}$ is proportional to the square of the integral

$$H_{et} = \int \psi_{DA}\hat{h}\psi_{D^+A^-}\,d\tau$$

where $\hat{h}$ is a hamiltonian that describes the coupling of the electronic wavefunctions. The probability of tunnelling through a potential barrier typically has an exponential dependence on the width of the barrier (Topic 8A), which suggests that we should write

$$H_{et}(d)^2 = H_{et}^{\circ 2}e^{-\beta d} \qquad (21E.4)$$

where d is the edge-to-edge distance between D and A, β is a parameter that measures the sensitivity of the electronic coupling matrix element to distance, and H_{et}° is the value of the electronic coupling matrix element when DA and D^+A^- are in contact ($d=0$).

Brief illustration 21E.1 The distance dependence of the coupling

The value of β depends on the medium through which the electron must travel from donor to acceptor. In a vacuum, $28\,\mathrm{nm}^{-1}<\beta<35\,\mathrm{nm}^{-1}$, whereas $\beta\approx 9\,\mathrm{nm}^{-1}$ when the intervening medium is a molecular link between donor and acceptor. Electron transfer between protein-bound cofactors can occur at distances of up to about 2.0 nm, a long distance on a molecular scale, corresponding to about 20 carbon atoms, with the protein providing an intervening medium between donor and acceptor.

Self-test 21E.1 By how much does H_{et}^2 change when d is increased from 1.0 nm to 2.0 nm, with $\beta\approx 9\,\mathrm{nm}^{-1}$?

Answer: Decrease by a factor of 8100

(b) The reorganization energy

The proportionality constant in $k_{\mathrm{et}}\propto \mathrm{e}^{-\Delta^{\ddagger}G/RT}$ is proportional to $H_{\mathrm{et}}(d)^2$, as expressed by eqn 21E.4. Therefore, we can expect the full expression for k_{et} to have the form

$$k_{\mathrm{et}}=CH_{\mathrm{et}}(d)^2\mathrm{e}^{-\Delta^{\ddagger}G/RT} \qquad (21\mathrm{E}.5)$$

with C a constant of proportionality and $H_{\mathrm{et}}(d)^2$ given by eqn 21E.5. We show in the following *Justification* that the Gibbs energy of activation $\Delta^{\ddagger}G$ is

$$\Delta^{\ddagger}G=\frac{(\Delta_{\mathrm{r}}G^{\ominus}+\Delta E_{\mathrm{R}})^2}{4\Delta E_{\mathrm{R}}} \qquad \text{Gibbs energy of activation} \qquad (21\mathrm{E}.6)$$

where $\Delta_{\mathrm{r}}G^{\ominus}$ is the standard reaction Gibbs energy for the electron transfer process $\mathrm{DA}\rightarrow\mathrm{D}^+\mathrm{A}^-$, and ΔE_{R} is the **reorganization energy**, the energy change associated with molecular rearrangements that must take place so that DA can take on the equilibrium geometry of $\mathrm{D}^+\mathrm{A}^-$. These molecular rearrangements include the relative reorientation of the D and A molecules in DA and the relative reorientation of the solvent molecules surrounding DA. Equation 21E.6 shows that $\Delta^{\ddagger}G=0$, with the implication that the reaction is not slowed down by an activation barrier, when $\Delta_{\mathrm{r}}G^{\ominus}=-\Delta E_{\mathrm{R}}$, corresponding to the cancellation of the reorganization energy term by the standard reaction Gibbs energy.

Justification 21E.2 The Gibbs energy of activation of electron transfer

The simplest way to derive an expression for the Gibbs energy of activation of electron transfer processes is to construct a model in which the surfaces for DA (the 'reactant complex', denoted R) and for $\mathrm{D}^+\mathrm{A}^-$ (the 'product complex', denoted P) are described by classical harmonic oscillators with identical reduced masses μ and angular frequencies ω, but displaced minima, as shown in Fig. 21E.3. The molar Gibbs energies $G_{\mathrm{m,R}}(q)$ and $G_{\mathrm{m,P}}(q)$ of the reactant and product complexes, respectively, may then be written

$$G_{\mathrm{m,R}}(q)=\tfrac{1}{2}N_{\mathrm{A}}\mu\omega^2(q-q_0^{\mathrm{R}})^2+G_{\mathrm{m,R}}(q_0^{\mathrm{R}})$$
$$G_{\mathrm{m,P}}(q)=\tfrac{1}{2}N_{\mathrm{A}}\mu\omega^2(q-q_0^{\mathrm{P}})^2+G_{\mathrm{m,P}}(q_0^{\mathrm{P}})$$

where q_0^{R} and q_0^{P} are the values of q at which the minima of the reactant and product parabolas occur, respectively. The standard reaction Gibbs energy for the electron transfer process $\mathrm{R}\rightarrow\mathrm{P}$ is $\Delta_{\mathrm{r}}G^{\ominus}=G_{\mathrm{m,P}}(q_0^{\mathrm{P}})-G_{\mathrm{m,R}}(q_0^{\mathrm{R}})$, the difference in standard molar Gibbs energy between the minima of the parabolas. In Fig. 21E.3, $\Delta_{\mathrm{r}}G^{\ominus}<0$.

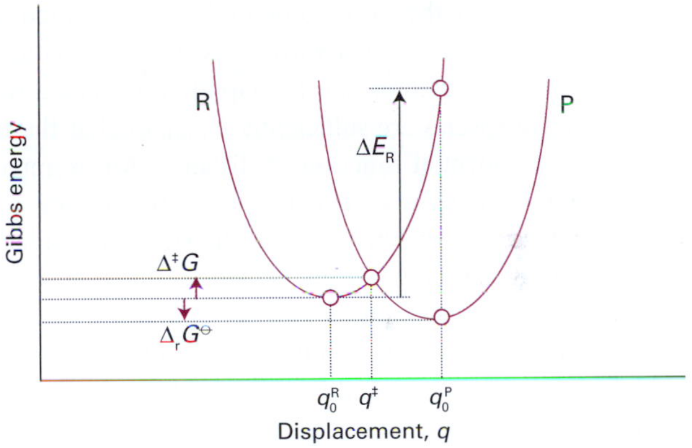

Figure 21E.3 The model system used in *Justification* 21E.2.

The value of q corresponding to the transition state of the complex, $q^{\ddagger}$, may be written in terms of the parameter α, the fractional change in q:

$$q^{\ddagger}=q_0^{\mathrm{R}}+\alpha(q_0^{\mathrm{P}}-q_0^{\mathrm{R}})$$

We see from Fig. 21E.3 that $\Delta^{\ddagger}G=G_{\mathrm{m,R}}(q^{\ddagger})-G_{\mathrm{m,R}}(q_0^{\mathrm{R}})$. It then follows that

$$\Delta^{\ddagger}G=\tfrac{1}{2}N_{\mathrm{A}}\mu\omega^2(q^{\ddagger}-q_0^{\mathrm{R}})^2=\tfrac{1}{2}N_{\mathrm{A}}\mu\omega^2\{\alpha(q_0^{\mathrm{P}}-q_0^{\mathrm{R}})^2\}$$

We now define the reorganization energy, ΔE_{R}, as

$$\Delta E_{\mathrm{R}}=\tfrac{1}{2}N_{\mathrm{A}}\mu\omega^2(q_0^{\mathrm{P}}-q_0^{\mathrm{R}})^2$$

which can be interpreted as $G_{\mathrm{m,R}}(q_0^{\mathrm{P}})-G_{\mathrm{m,R}}(q_0^{\mathrm{R}})$ and, consequently, as the (Gibbs) energy required to deform the equilibrium configuration of R to the equilibrium configuration of P (as shown in Fig. 21E.3). Then $\Delta^{\ddagger}G=\alpha^2\Delta E_{\mathrm{R}}$. Because $G_{\mathrm{m,R}}(q^{\ddagger})=G_{\mathrm{m,P}}(q^{\ddagger})$, it follows that

$$\alpha^2\Delta E_{\mathrm{R}}=\tfrac{1}{2}N_{\mathrm{A}}\mu\omega^2\{(\alpha-1)(q_0^{\mathrm{P}}-q_0^{\mathrm{R}})\}^2+\Delta_{\mathrm{r}}G^{\ominus}$$
$$=(\alpha-1)\Delta E_{\mathrm{R}}+\Delta_{\mathrm{r}}G^{\ominus}$$

which implies that

$$\alpha=\frac{1}{2}\left(\frac{\Delta_{\mathrm{r}}G^{\ominus}}{\Delta E_{\mathrm{R}}}+1\right)$$

By inserting this equation into $\Delta^{\ddagger}G=\alpha^2\Delta E_R$, we obtain eqn 21E.6. We can obtain an identical relation if we allow the harmonic oscillators to have different angular frequencies and hence different curvatures.

The only missing piece of the expression for k_{et} is the value of the constant of proportionality C. Detailed calculation, which we do not repeat here, gives

$$C=\frac{1}{h}\left(\frac{\pi^3}{RT\Delta E_R}\right)^{1/2} \qquad (21E.7)$$

Equation 21E.6 has some limitations as might be expected because perturbation theory arguments have been used. For instance, it describes processes with weak electronic coupling between donor and acceptor. Weak coupling is observed when the electroactive species are sufficiently far apart that the tunnelling is an exponential function of distance. An example of a weakly coupled system is the complex of the proteins cytochrome c and cytochrome b_5, in which the electroactive haem-bound iron ions shuttle between oxidation states Fe(II) and Fe(III) during electron transfer and are about 1.7 nm apart. Strong coupling is observed when the wavefunctions ψ_A and ψ_D overlap very extensively and, as well as other complications, the tunnelling rate is no longer a simple exponential function of distance. Examples of strongly coupled systems are mixed-valence, binuclear d-metal complexes with the general structure L_mM^{n+}–B–$M^{p+}L_m$, in which the electroactive metal ions are separated by a bridging ligand B. In these systems, $d<1.0$ nm. The weak coupling limit applies to a large number of electron transfer reactions, including those between proteins during metabolism.

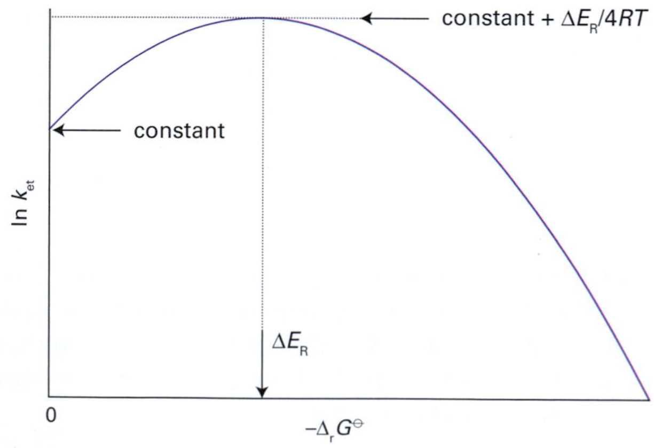

Figure 21E.4 The parabolic dependence of ln k_{et} on $-\Delta_r G^{\ominus}$ predicted by eqn 21E.9.

The most meaningful experimental tests of the dependence of k_{et} on d are those in which the same donor and acceptor are positioned at a variety of distances, perhaps by covalent attachment to molecular linkers (see **1** in *Brief illustration* 21E.2 for an example). Under these conditions, the term $e^{-\Delta^{\ddagger}G/RT}$ becomes a constant and, after taking the natural logarithm of eqn 21E.5 and using eqn 21E.4, we obtain

$$\ln k_{et}=-\beta d+\text{constant} \qquad (21E.8)$$

which implies that a plot of ln k_{et} against d should be a straight line of slope $-\beta$.

The dependence of k_{et} on the standard reaction Gibbs energy has been investigated in systems where the edge-to-edge distance and the reorganization energy are constant for a series of reactions. Then, by using eqn 21E.6 for $\Delta^{\ddagger}G$, eqn 21E.5 becomes

$$\ln k_{et}=-\frac{RT}{4\Delta E_R}\left(\frac{\Delta_r G^{\ominus}}{RT}\right)^2-\frac{1}{2}\left(\frac{\Delta_r G^{\ominus}}{RT}\right)+\text{constant} \qquad (21E.9)$$

and a plot of ln k_{et} (or log k_{et}=ln k_{et}/ln 10) against $\Delta_r G^{\ominus}$ (or $-\Delta_r G^{\ominus}$) is predicted to be shaped like a downward parabola (Fig. 21E.4). Equation 21E.9 implies that the rate constant increases as $\Delta_r G^{\ominus}$ decreases but only up to $-\Delta_r G^{\ominus}=\Delta E_R$. Beyond that, the reaction enters the **inverted region**, in which the rate constant decreases as the reaction becomes more exergonic ($\Delta_r G^{\ominus}$ becomes more negative). The inverted region has been observed in a series of special compounds in which the electron donor and acceptor are linked covalently to a molecular spacer of known and fixed size (Fig. 21E.5).

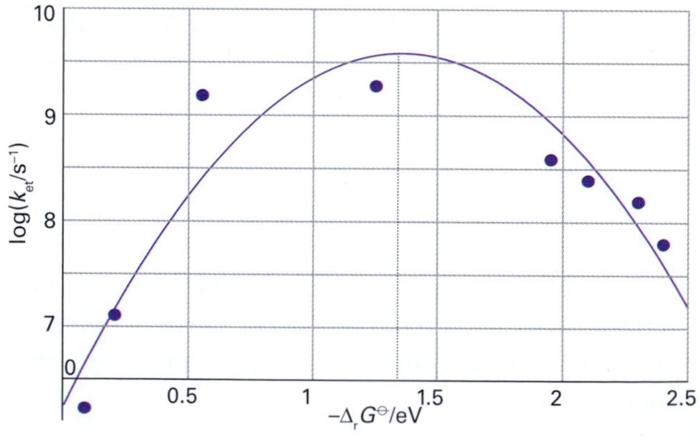

Figure 21E.5 Variation of log k_{et} with $-\Delta_r G^{\ominus}$ for a series of compounds with the structures given in **1** and as described in *Brief illustration* 21E.2. Based on J.R. Miller, et al., *J. Am. Chem. Soc.* **106**, 3047 (1984).

Brief illustration 21E.2 The determination of the reorganization energy

Kinetic measurements were conducted in 2-methyltetrahydrofuran and at 296 K for a series of compounds with the structures given in 1. The distance between donor (the reduced biphenyl group) and the acceptor is constant for all compounds in the series because the molecular linker remains the same. Each acceptor has a characteristic standard potential, so it follows that the standard Gibbs energy for the electron transfer process is different for each compound in the series. The line in Fig. 21E.5 is a fit to a version of eqn 21E.9 and the maximum of the parabola occurs at $-\Delta_r G^\ominus = \Delta E_R = 1.4\,\text{eV} = 1.4\times10^2\,\text{kJ mol}^{-1}$.

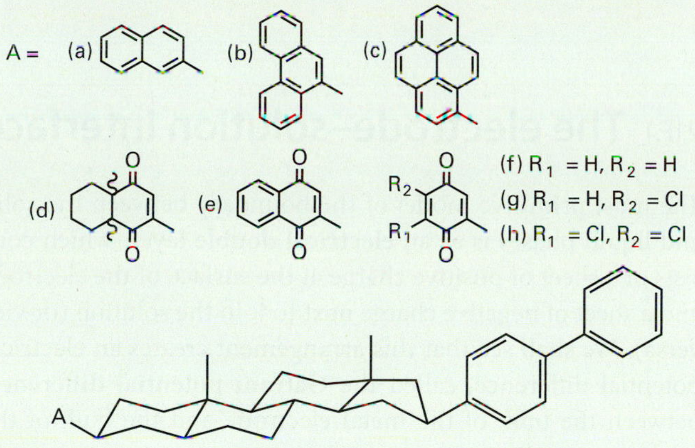

1 An electron donor–acceptor system

Self-test 21E.2 Some (invented) data on a series of complexes are as follows:

$-\Delta_r G^\ominus$/eV	0.20	0.60	1.0	1.3	1.6	2.0	2.4
$\log k_{et}$	8.2	9.7	10.2	10.1	9.4	7.7	5.1

Determine the reorganization energy.

Answer: 1.05 eV

Checklist of concepts

- ☐ 1. Electron transfer can occur only after thermal fluctuations bring the nuclear coordinate to the point at which the donor and acceptor have the same configuration.
- ☐ 2. The tunnelling rate is supposed to depend exponentially on the separation of the donor and acceptor.
- ☐ 3. The **reorganization energy** is the energy change associated with molecular rearrangements that must take place so that DA can acquire the equilibrium geometry of D^+A^-.
- ☐ 4. In the **inverted region**, the rate constant k_{et} decreases as the reaction becomes more exergonic ($\Delta_r G^\ominus$ becomes more negative).

Checklist of equations

Property	Equation	Comment	Equation number
Electron transfer rate constant	$1/k_r = 1/k_a + (k_a'/k_a k_{et})(1 + k_{et}'/k_d)$	Steady-state assumption	21E.3
Tunnelling probability	$H_{et}(d)^2 = H_{et}^{\circ 2} e^{-\beta d}$	Assumed	21E.4
Rate constant	$k_{et} = CH_{et}(d)^2 e^{-\Delta^\ddagger G/RT}$	Transition-state theory	21E.5
Gibbs energy of activation	$\Delta^\ddagger G = (\Delta_r G^\ominus + \Delta E_R)^2 / 4\Delta E_R$	Assumes parabolic potential energy	21E.6
Dependence on separation	$\ln k_{et} = -\beta d + \text{constant}$		21E.8
Dependence on $\Delta_r G^\ominus$	$\ln k_{et} = a\Delta_r G^{\ominus 2} + b\Delta_r G^\ominus + c$	$a = -1/4\Delta E_R RT$, $b = -1/2RT$, $c = \text{constant}$	21E.9

21F Processes at electrodes

Contents

➤ Why do you need to know this material?

A knowledge of the factors that determine the rate of electron transfer at electrodes leads to a better understanding of power production in batteries, and of electron conduction in metals, semiconductors, and nanometre-sized electronic devices, all of which are highly important in modern technology.

➤ What is the key idea?

Transition-state theory can be applied to the description of electron transfer processes at the surface of electrodes.

➤ What do you need to know already?

You need to be familiar with electrochemical cells (Topic 6C), electrode potentials (Topic 6D), and the thermodynamic version of transition-state theory (Topic 21C), particularly the activation Gibbs energy.

As for homogeneous systems (Topic 21E), electron transfer at the surface of an electrode involves electron tunnelling. However, the electrode possesses a nearly infinite number of closely spaced electronic energy levels rather than the small number of discrete levels of a typical complex. Furthermore, specific interactions with the electrode surface give the solute and solvent special properties that can be very different from those observed in the bulk of the solution. For this reason, we begin with a description of the electrode–solution interface. Then we describe the kinetics of electrode processes that draws on the thermodynamic language inspired by transition-state theory.

21F.1 The electrode–solution interface

The most primitive model of the boundary between the solid and liquid phases is as an **electrical double layer**, which consists of a sheet of positive charge at the surface of the electrode and a sheet of negative charge next to it in the solution (or vice versa). We shall see that this arrangement creates an electrical potential difference, called the **Galvani potential difference**, between the bulk of the metal electrode and the bulk of the solution.

More sophisticated models for the electrode–solution interface attempt to describe the gradual changes in the structure of the solution between two extremes, one the charged electrode surface and the other the bulk solution. In the **Helmholtz layer model** of the interface the solvated ions arrange themselves along the surface of the electrode but are held away from it by their hydration spheres (Fig. 21F.1). The location of the sheet

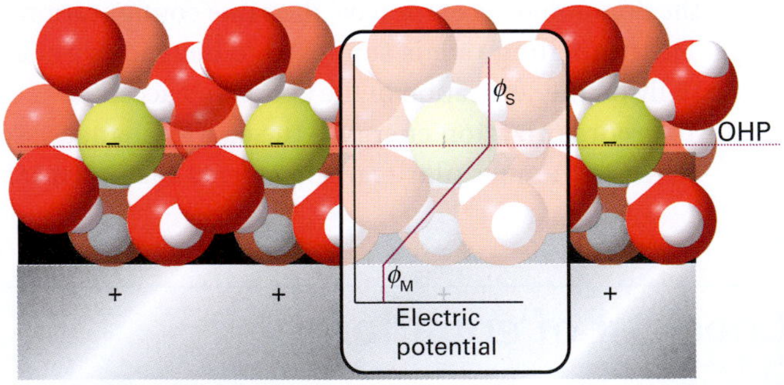

Figure 21F.1 A simple model of the electrode–solution interface treats it as two rigid planes of charge. One plane, the outer Helmholtz plane (OHP), is due to the ions with their solvating molecules and the other plane is that of the electrode itself. The plot shows the dependence of the electric potential with distance from the electrode surface according to this model. Between the electrode surface and the OHP, the potential varies linearly from ϕ_M, the value in the metal, to ϕ_S, the value in the bulk of the solution.

of ionic charge, which is called the **outer Helmholtz plane** (OHP), is identified as the plane running through the solvated ions. In this simple model, the electrical potential changes linearly within the layer bounded by the electrode surface on one side and the OHP on the other. In a refinement of this model, ions that have discarded their solvating molecules and have become attached to the electrode surface by chemical bonds are regarded as forming the **inner Helmholtz plane** (IHP). The Helmholtz layer model ignores the disrupting effect of thermal motion, which tends to break up and disperse the rigid outer plane of charge. In the **Gouy–Chapman model** of the **diffuse double layer**, the disordering effect of thermal motion is taken into account in much the same way as the Debye–Hückel model describes the ionic atmosphere of an ion (Topic 5F) with the latter's single central ion replaced by an infinite, plane electrode.

Figure 21F.2 shows how the local concentrations of cations and anions differ in the Gouy–Chapman model from their bulk concentrations. Ions of opposite charge cluster close to the electrode and ions of the same charge are repelled from it. The modification of the local concentrations near an electrode implies that it might be misleading to use activity coefficients characteristic of the bulk to discuss the thermodynamic properties of ions near the interface. This is one of the reasons why measurements of the dynamics of electrode processes are almost always done using a large excess of supporting electrolyte (for example, a 1 M solution of a salt, an acid, or a base). Under such conditions, the activity coefficients are almost constant because the inert ions dominate the effects of local changes caused by any reactions taking place. The use of a concentrated solution also minimizes ion migration effects.

Neither the Helmholtz nor the Gouy–Chapman model is a very good representation of the structure of the double layer. The former overemphasizes the rigidity of the local solution; the latter underemphasizes its structure. The two are combined in the

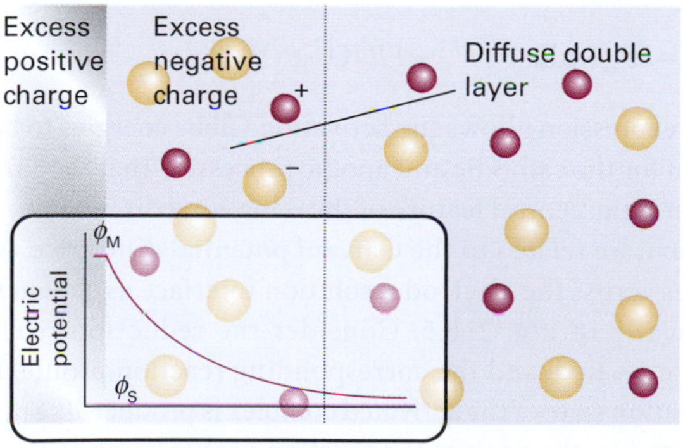

Figure 21F.2 The Gouy–Chapman model of the electrical double layer treats the outer region as an atmosphere of counter-charge, similar to the Debye–Hückel theory of ion atmospheres. The plot of electrical potential against distance from the electrode surface shows the meaning of the diffuse double layer (see text for details).

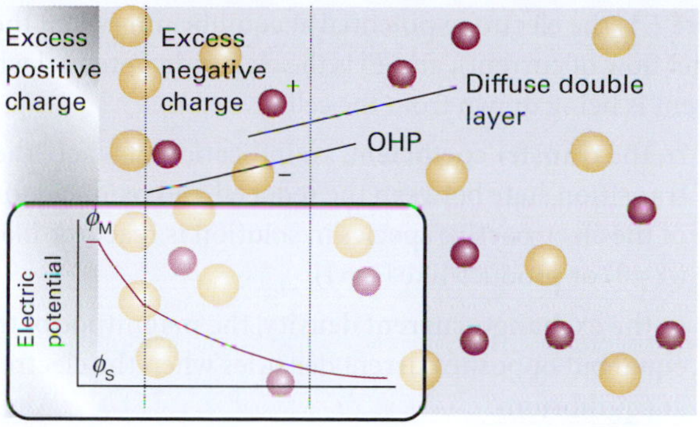

Figure 21F.3 A representation of the Stern model of the electrode–solution interface. The model incorporates the idea of an outer Helmholtz plane near the electrode surface and of a diffuse double layer further away from the surface.

Stern model, in which the ions closest to the electrode are constrained into a rigid Helmholtz plane while outside that plane the ions are dispersed as in the Gouy–Chapman model (Fig. 21F.3). Yet another level of sophistication is found in the **Grahame model**, which adds an inner Helmholtz plane to the Stern model.

The potential difference between points in the bulk metal and the bulk solution is the **Galvani potential difference**, $\Delta\phi$. Apart from a constant, this Galvani potential difference is the electrode potential that was discussed in Topic 6D. We shall ignore the constant, which cannot be measured anyway, and identify changes in $\Delta\phi$ with changes in electrode potential.

21F.2 The rate of electron transfer

We shall consider a reaction at the electrode in which an ion is reduced by the transfer of a single electron in the rate-determining step. We focus on the **current density**, j, the electric current flowing through a region of an electrode divided by the area of the region.

(a) The Butler–Volmer equation

We show in the following *Justification* that an analysis of the effect of the Galvani potential difference at the electrode on the current density leads to the **Butler–Volmer equation**:

$$j = j_0\{e^{(1-\alpha)f\eta} - e^{-\alpha f\eta}\} \qquad \text{Butler–Volmer equation} \qquad (21F.1)$$

where we have written $f = F/RT$, with F as Faraday's constant. The equation contains the following parameters:

- η (eta), the **overpotential:**

$$\eta = E' - E \qquad \text{Definition} \quad \text{Overpotential} \qquad (21F.2)$$

where E is the electrode potential at equilibrium (when there is no net flow of current), and E' is the electrode potential when a current is being drawn from the cell.

- α, the **transfer coefficient**, an indication of where the transition state between the reduced and oxidized forms of the electroactive species in solution is reactant-like ($\alpha=0$) or product-like ($\alpha=1$).
- j_0, the **exchange-current density**, the magnitude of the equal but opposite current densities when the electrode is at equilibrium.

Justification 21F.1 The Butler–Volmer equation

Because an electrode reaction is heterogeneous, we express the rate of charge transfer as the flux of products, the amount of material produced over a region of the electrode surface in an interval of time divided by the area of the region and the duration of the interval.

A first-order heterogeneous rate law has the form

$$\text{Product flux} = k_r[\text{species}]$$

where [species] is the molar concentration of the relevant electroactive species in solution close to the electrode, just outside the double layer. The rate constant has dimensions of length/time (with units, for example, of centimetres per second, cm s^{-1}). If the molar concentrations of the oxidized and reduced materials outside the double layer are [Ox] and [Red], respectively, then the rate of reduction of Ox, v_{Ox}, is $v_{\text{Ox}}=k_c[\text{Ox}]$ and the rate of oxidation of Red, v_{Red}, is $v_{\text{Red}}=k_a[\text{Red}]$. (The notation k_c and k_a is justified below.)

Now consider a reaction at the electrode in which an ion is reduced by the transfer of a single electron in the rate-determining step. The net current density at the electrode is the difference between the current densities arising from the reduction of Ox and the oxidation of Red. Because the redox processes at the electrode involve the transfer of one electron per reaction event, the current densities, j, arising from the redox processes are the rates v_{Ox} and v_{Red} multiplied by the charge transferred per mole of reaction, which is given by Faraday's constant. Therefore, there is a **cathodic current density** of magnitude

$$j_c = Fk_c[\text{Ox}] \qquad \text{for Ox}+e^- \rightarrow \text{Red}$$

arising from the reduction (because, as defined in Topic 6C, the cathode is the site of reduction). There is also an opposing **anodic current density** of magnitude

$$j_a = Fk_a[\text{Red}] \qquad \text{for Red} \rightarrow \text{Ox}+e^-$$

arising from the oxidation (because the anode is the site of oxidation). The net current density at the electrode is the difference

$$j = j_a - j_c = Fk_a[\text{Red}] - Fk_c[\text{Ox}]$$

Note that, when $j_a > j_c$, so that $j > 0$, the current is anodic (Fig. 21F.4a); when $j_c > j_a$, so that $j > 0$, the current is cathodic (Fig. 21F.4b).

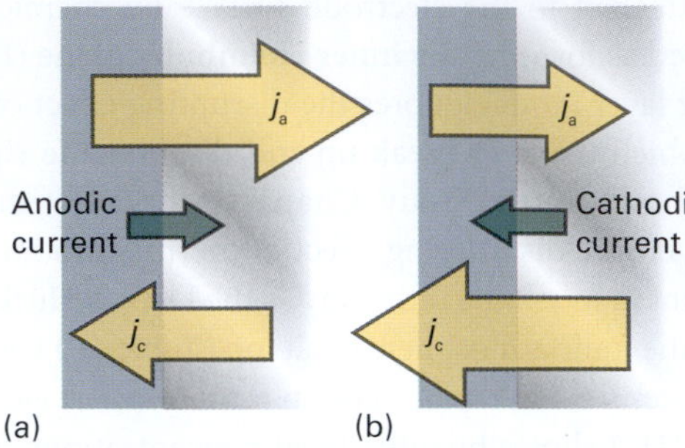

Figure 21F.4 The net current density is defined as the difference between the cathodic and anodic contributions. (a) When $j_a > j_c$, the net current is anodic, and there is a net oxidation of the species in solution. (b) When $j_c > j_a$, the net current is cathodic, and the net process is reduction.

If a species is to participate in reduction or oxidation at an electrode, it must discard any solvating molecules, migrate through the electrode–solution interface, and adjust its hydration sphere as it receives or discards electrons. Likewise, a species already at the inner plane must be detached and migrate into the bulk. Because both processes are activated, we can expect to write their rate constants in the form suggested by transition-state theory (Topic 21C) as

$$k_r = Be^{-\Delta^{\ddagger}G/RT}$$

where $\Delta^{\ddagger}G$ is the activation Gibbs energy and B is a constant with the same dimensions as k_r.

When the expressions for k_r, specifically k_c and k_a, are inserted, we obtain

$$j = FB_a[\text{Red}]e^{-\Delta^{\ddagger}G_a/RT} - FB_c[\text{Ox}]e^{-\Delta^{\ddagger}G_c/RT}$$

This expression allows the activation Gibbs energies to be different for the cathodic and anodic processes. That they are different is the central feature of the remaining discussion.

Next, we relate j to the Galvani potential difference, which varies across the electrode–solution interface as shown schematically in Fig. 21F.5. Consider the reduction reaction, $\text{Ox}+e^- \rightarrow \text{Red}$, and the corresponding reaction profile. If the transition state of the activated complex is product-like (as represented by the peak of the reaction profile being close to the electrode in Fig. 21F.6), the activation Gibbs energy is changed from $\Delta^{\ddagger}G_c(0)$, the value it has in the absence of a potential difference across the double layer, to $\Delta^{\ddagger}G_c = \Delta^{\ddagger}G_c(0) + F\Delta\phi$. Thus, if the electrode is more positive than the solution, $\Delta\phi > 0$, then more work has to be done to form an activated complex from Ox; in this case the activation Gibbs energy is increased.

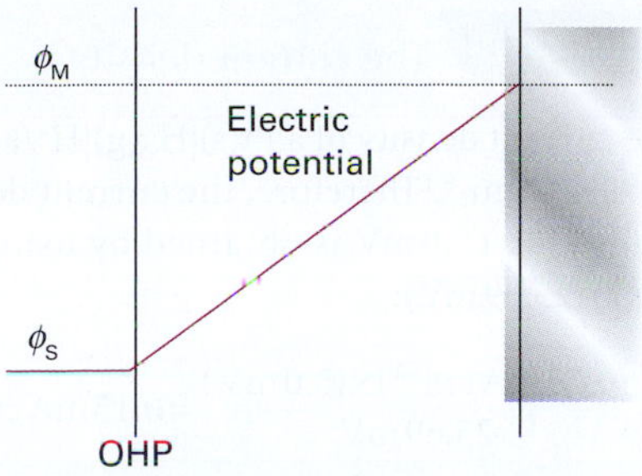

Figure 21F.5 The potential, ϕ, varies linearly between two plane parallel sheets of charge, and its effect on the Gibbs energy of the transition state depends on the extent to which the latter resembles the species at the inner or outer planes.

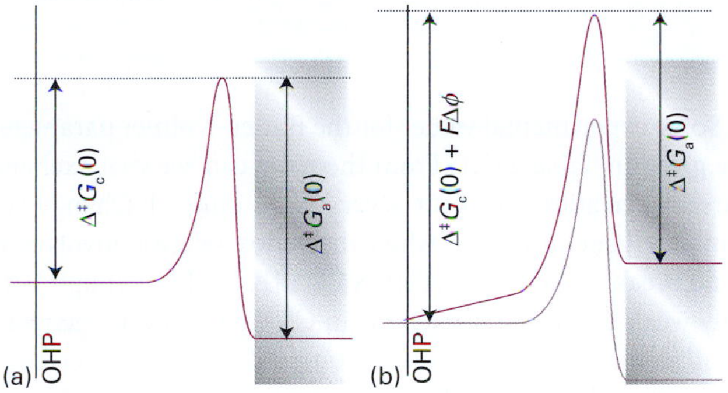

Figure 21F.6 When the transition state resembles a species that has undergone reduction, the activation Gibbs energy for the anodic current is almost unchanged, but the full effect applies to the cathodic current. (a) Zero potential difference; (b) nonzero potential difference.

If the transition state is reactant-like (represented by the peak of the reaction profile being close to the outer plane of the double-layer in Fig. 21F.7), then $\Delta^‡G_c$ is independent of $\Delta\phi$. In a real system, the transition state has an intermediate resemblance to these extremes (Fig. 21F.8) and the activation Gibbs energy for reduction may be written as

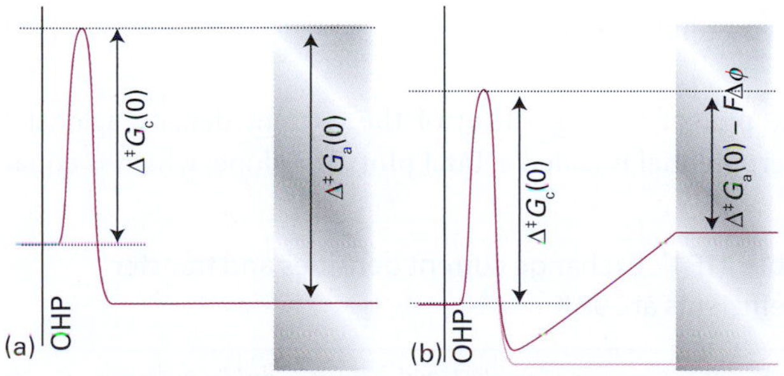

Figure 21F.7 When the transition state resembles a species that has undergone oxidation, the activation Gibbs energy for the cathodic current is almost unchanged but the activation Gibbs energy for the anodic current is strongly affected. (a) Zero potential difference; (b) nonzero potential difference.

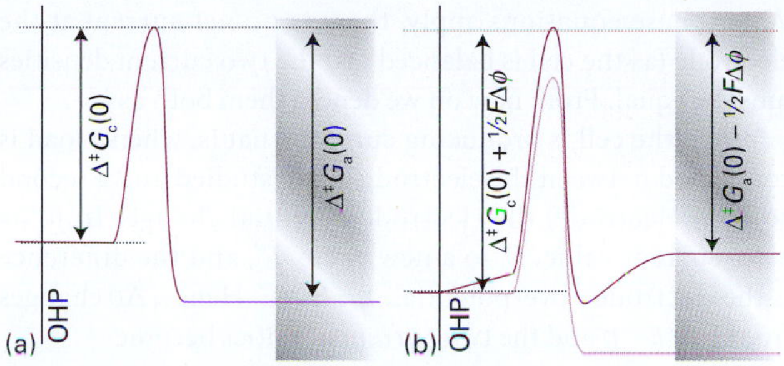

Figure 21F.8 When the transition state is intermediate in its resemblance to reduced and oxidized species, as represented here by a peak located at an intermediate position as measured by α (with $0<\alpha<1$), both activation Gibbs energies are affected; here, $\alpha \approx 0.5$. (a) Zero potential difference; (b) nonzero potential difference.

$$\Delta^‡G_c = \Delta^‡G_c(0) + \alpha F\Delta\phi$$

The parameter α lies in the range 0 to 1. Experimentally, α is often found to be about 0.5.

Now consider the oxidation reaction, $\text{Red} + e^- \rightarrow \text{Ox}$ and its reaction profile. Similar remarks apply. In this case, Red discards an electron to the electrode, so the extra work is zero if the transition state is reactant-like (represented by a peak close to the electrode). The extra work is the full $-F\Delta\phi$ if it resembles the product (the peak close to the outer plane). In general, the activation Gibbs energy for this anodic process is

$$\Delta^‡G_a = \Delta^‡G_a(0) - (1-\alpha)F\Delta\phi$$

The two activation Gibbs energies can now be inserted in the expression for j, with the result that

$$j = FB_a[\text{Red}]e^{-\Delta^‡G_a(0)/RT}e^{(1-\alpha)F\Delta\phi/RT} - FB_c[\text{Ox}]e^{-\Delta^‡G_c(0)/RT}e^{-\alpha F\Delta\phi/RT}$$

This is an explicit, if complicated, expression for the net current density in terms of the potential difference.

The appearance of the new expression for j can be simplified. First, in a purely cosmetic step we write $f = F/RT$. Next, we identify the individual cathodic and anodic current densities:

$$j = \overbrace{FB_a[\text{Red}]e^{-\Delta^‡G_a(0)/RT}e^{(1-\alpha)f\Delta\phi}}^{j_a} - \overbrace{FB_c[\text{Ox}]e^{-\Delta^‡G_c(0)/RT}e^{-\alpha f\Delta\phi}}^{j_c}$$

If the cell is balanced against an external source, the Galvani potential difference, $\Delta\phi$, can be identified as the (zero-current) electrode potential, E, and we can write

$$j_a = FB_a[\text{Red}]e^{-\Delta^‡G_a(0)/RT}e^{(1-\alpha)fE}$$
$$j_c = FB_c[\text{Ox}]e^{-\Delta^‡G_c(0)/RT}e^{-\alpha fE}$$

When these equations apply, there is no net current at the electrode (as the cell is balanced), so the two current densities must be equal. From now on we denote them both as j_0.

When the cell is producing current (that is, when a load is connected between the electrode being studied and a second counter electrode) the electrode potential changes from its zero-current value, E, to a new value, E', and the difference is the electrode's overpotential, $\eta = E' - E$. Hence, $\Delta\phi$ changes from E to $E + \eta$ and the two current densities become

$$j_a = j_0 e^{(1-\alpha)f\eta} \qquad j_c = j_0 e^{-\alpha f\eta}$$

Then from $j = j_a - j_c$ we obtain the Butler–Volmer equation, eqn 21F.1.

Figure 21F.9 shows how eqn 21F.1 predicts the current density to depend on the overpotential for different values of the transfer coefficient. When the overpotential is so small that $f\eta \ll 1$ (in practice, η less than about 10 mV) the exponentials in eqn 21F.1 can be expanded by using $e^x = 1 + x + \cdots$ to give

$$j = j_0\{\overbrace{1+(1-\alpha)f\eta+\cdots}^{e^{(1-\alpha)f\eta}} - \overbrace{(1-\alpha f\eta+\cdots)}^{e^{-\alpha f\eta}}\} \approx j_0 f\eta \qquad (21F.3)$$

This equation shows that the current density is proportional to the overpotential, so at low overpotentials the interface behaves like a conductor that obeys Ohm's law. When there is a small positive overpotential the current is anodic ($j > 0$ when $\eta > 0$), and when the overpotential is small and negative the current is cathodic ($j < 0$ when $\eta < 0$). The relation can also be reversed to calculate the potential difference that must exist if a current density j has been established by some external circuit:

$$\eta = \frac{RTj}{Fj_0} \qquad (21F.4)$$

The importance of this interpretation will become clear shortly.

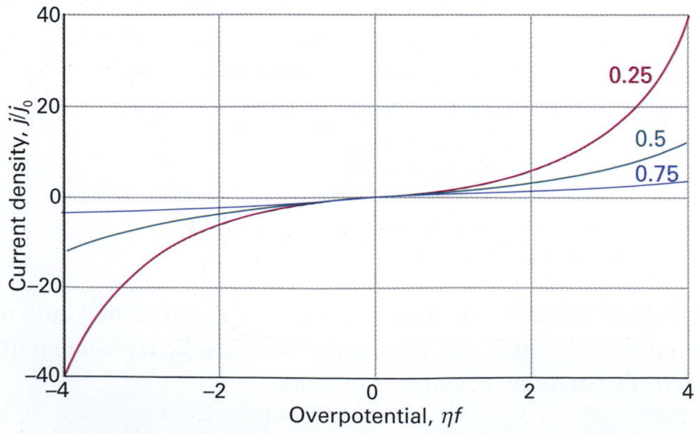

Figure 21F.9 The dependence of the current density on the overpotential for different values of the transfer coefficient.

Brief illustration 21F.1 The current density

The exchange current density of a $Pt(s)|H_2(g)|H^+(aq)$ electrode at 298 K is 0.79 mA cm^{-2}. Therefore, the current density when the overpotential is +5.0 mV is obtained by using eqn 21F.4 and $f = F/RT = 1/(25.69\,\text{mV})$:

$$j = j_0 f\eta = \frac{(0.79\,\text{mA cm}^{-2})\times(5.0\,\text{mV})}{25.69\,\text{mV}} = 0.15\,\text{mA cm}^{-2}$$

The current through an electrode of total area 5.0 cm^2 is therefore 0.75 mA.

Self-test 21F.1 What would be the current at pH = 2.0, the other conditions being the same?

Answer: −18 mA (cathodic)

Some experimental values for the Butler–Volmer parameters are given in Table 21F.1. From them we can see that exchange current densities vary over a very wide range. Exchange currents are generally large when the redox process involves no bond breaking (as in the $[Fe(CN)_6]^{3-}$, $[Fe(CN)_6]^{4-}$ couple) or if only weak bonds are broken (as in Cl_2,Cl^-). They are generally small when more than one electron needs to be transferred, or when multiple or strong bonds are broken, as in the N_2,N_3^- couple and in redox reactions of organic compounds.

(b) Tafel plots

When the overpotential is large and positive (in practice, $\eta \geq 0.12$ V), corresponding to the electrode being the anode in electrolysis, the second exponential in eqn 21F.1 is much smaller than the first, and may be neglected. Then

$$j = j_0 e^{(1-\alpha)f\eta} \qquad (21F.5a)$$

so

$$\ln j = \ln j_0 + (1-\alpha)f\eta \qquad (21F.5b)$$

The plot of the logarithm of the current density against the overpotential is called a **Tafel plot**. The slope, which is equal to

Table 21F.1* Exchange current densities and transfer coefficients at 298 K

Reaction	Electrode	$j_0/(\text{A cm}^{-2})$	α
$2\,H^+ + 2\,e^- \rightarrow H_2$	Pt	7.9×10^{-4}	
	Ni	6.3×10^{-6}	0.58
	Pb	5.0×10^{-12}	
$Fe^{3+} + e^- \rightarrow Fe^{2+}$	Pt	2.5×10^{-3}	0.58

* More values are given in the *Resource section*.

$(1-\alpha)f$, gives the value of α and the intercept at $\eta=0$ gives the exchange-current density. If instead the overpotential is large but negative (in practice, $\eta \leq -0.12$ V), the first exponential in eqn 21F.1 may be neglected. Then

$$j = j_0 e^{-\alpha f \eta} \quad (21F.6a)$$

so

$$\ln j = \ln j_0 - \alpha f \eta \quad (21F.6b)$$

In this case the slope of the Tafel plot is $-\alpha f$.

Example 21F.1 Interpreting a Tafel plot

The data below refer to the anodic current through a platinum electrode of area 2.0 cm^2 in contact with an Fe^{3+}, Fe^{2+} aqueous solution at 298 K. Calculate the exchange-current density and the transfer coefficient for the electrode process.

η/mV	50	100	150	200	250
I/mA	8.8	25.0	58.0	131	298

Method The anodic process is the oxidation $Fe^{2+}(aq) \rightarrow Fe^{3+}(aq) + e^-$. To analyse the data, we make a Tafel plot (of $\ln j$ against η) using the anodic form (eqn 21F.5b). The intercept at $\eta=0$ is $\ln j_0$ and the slope is $(1-\alpha)f$.

Answer Draw up the following table:

η/mV	50	100	150	200	250
j/(mA cm^{-2})	4.4	12.5	29.0	65.5	149
$\ln(j/(\text{mA cm}^{-2}))$	1.48	2.53	3.37	4.18	5.00

The points are plotted in Fig. 21F.10. The high overpotential region gives a straight line of intercept 0.88 and slope 0.0165. From the former it follows that $\ln(j_0/(\text{mA cm}^{-2})) = 0.88$, so $j_0 = 2.4$ mA cm^{-2}. From the latter,

$$\underbrace{(1-\alpha)f}_{\text{}} = \overbrace{0.0165}^{\text{Slope}} \overbrace{\text{mV}^{-1}}^{\text{Units of } \ln j/\eta \;(1/\text{mV})}$$

so

$$\alpha = 1 - \frac{\overbrace{0.0165\,\text{mV}^{-1}}^{16.5\,\text{V}^{-1}}}{\underbrace{f}_{38.9\,\text{V}^{-1}}} = 1 - 0.42\ldots = 0.58$$

Note that the Tafel plot is nonlinear for $\eta < 100$ mV; in this region $\alpha f \eta = 2.3$ and the approximation that $\alpha f \eta \gg 1$ fails.

Self-test 21F.2 Repeat the analysis using the following cathodic current data:

η/mV	−50	−100	−150	−200	−250	−300
I/mA	0.3	1.5	6.4	27.6	118.6	510

Answer: $\alpha = 0.75$, $j_0 = 0.041$ mA cm^{-2}

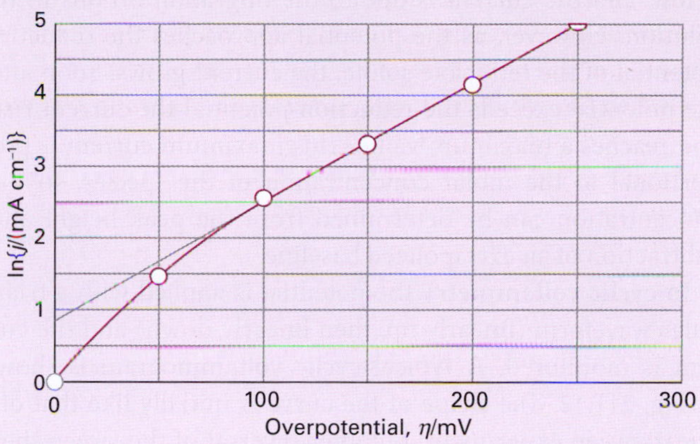

Figure 21F.10 A Tafel plot is used to measure the exchange current density (given by the extrapolated intercept at $\eta=0$) and the transfer coefficient (from the slope). The data are from *Example* 21F.1.

21F.3 Voltammetry

One of the assumptions in the derivation of the Butler–Volmer equation is the negligible conversion of the electroactive species at low current densities, resulting in uniformity of concentration near the electrode. This assumption fails at high current densities because the consumption of electroactive species close to the electrode results in a concentration gradient. The diffusion of the species towards the electrode from the bulk is slow and may become rate determining; a larger overpotential is then needed to produce a given current. This effect is called **concentration polarization**. Concentration polarization is important in the interpretation of **voltammetry**, the study of the current through an electrode as a function of the applied potential difference.

The kind of output from **linear-sweep voltammetry** is illustrated in Fig. 21F.11. Initially, the absolute value of the potential

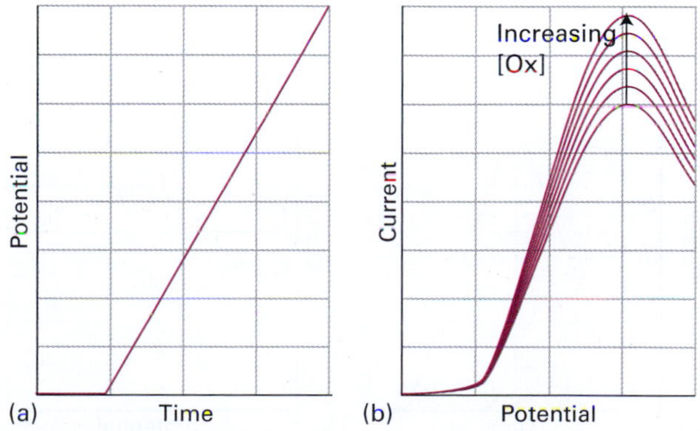

Figure 21F.11 (a) The change of potential with time and (b) the resulting current/potential curve in a voltammetry experiment. The peak value of the current density is proportional to the concentration of electroactive species (for instance, [Ox]) in solution.

is low, and the current is due to the migration of ions in the solution. However, as the potential approaches the reduction potential of the reducible solute, the current grows. Soon after the potential exceeds the reduction potential the current rises and reaches a maximum value. This maximum current is proportional to the molar concentration of the species, so that concentration can be determined from the peak height after subtraction of an extrapolated baseline.

In **cyclic voltammetry** the potential is applied with a triangular waveform (linearly up, then linearly down) and the current is monitored. A typical cyclic voltammogram is shown in Fig. 21F.12. The shape of the curve is initially like that of a linear-sweep experiment, but after reversal of the sweep there is a rapid change in current on account of the high concentration of oxidizable species close to the electrode that was generated on the reductive sweep. When the potential is close to the value required to oxidize the reduced species, there is a substantial current until all the oxidation is complete, and the current returns to zero. Cyclic voltammetry data are obtained at scan rates of about 50 mV s^{-1}, so a scan over a range of 2 V takes about 80 s.

When the reduction reaction at the electrode can be reversed, as in the case of the $[Fe(CN)_6]^{3-}/[Fe(CN)_6]^{4-}$ couple, the cyclic voltammogram is broadly symmetric about the standard potential of the couple (as in Fig. 21F.12). The scan is initiated with $[Fe(CN)_6]^{3-}$ present in solution, and as the potential approaches $E^{\ominus}$ for the couple, the $[Fe(CN)_6]^{3-}$ near the electrode is reduced and current begins to flow. As the potential continues to change, the current begins to decline again because all the $[Fe(CN)_6]^{3-}$ near the electrode has been reduced and the current reaches its limiting value. The potential is now returned linearly to its initial value, and the reverse series of events occurs with the $[Fe(CN)_6]^{4-}$ produced during the forward scan now undergoing oxidation. The peak of current lies on the other side of $E^{\ominus}$, so the species present and its standard potential can be identified, as indicated in the illustration, by noting the locations of the two peaks.

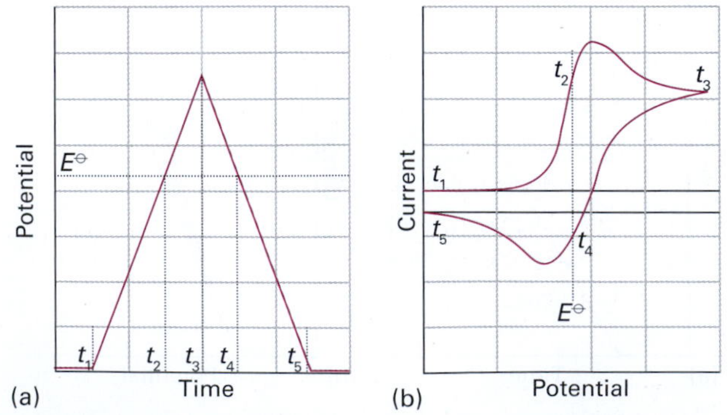

Figure 21F.12 (a) The change of potential with time and (b) the resulting current/potential curve in a cyclic voltammetry experiment.

The overall shape of the curve gives details of the kinetics of the electrode process and the change in shape as the rate of change of potential is altered gives information on the rates of the processes involved. For example, the matching peak on the return phase of the potential sweep may be missing, which indicates that the oxidation (or reduction) is irreversible. The appearance of the curve may also depend on the timescale of the sweep, for if the sweep is too fast some processes might not have time to occur. This style of analysis is illustrated in the following example.

Example 21F.2 Analysing a cyclic voltammetry experiment

The electroreduction of *p*-bromonitrobenzene in liquid ammonia is believed to occur by the following mechanism:

$$BrC_6H_4NO_2 + e^- \rightarrow BrC_6H_4NO_2^-$$
$$BrC_6H_4NO_2^- \rightarrow \cdot C_6H_4NO_2 + Br^-$$
$$\cdot C_6H_4NO_2 + e^- \rightarrow C_6H_4NO_2^-$$
$$C_6H_4NO_2^- + H^+ \rightarrow C_6H_5NO_2$$

Suggest the likely form of the cyclic voltammogram expected on the basis of this mechanism.

Method Decide which steps are likely to be reversible on the timescale of the potential sweep: such processes will give symmetrical voltammograms. Irreversible processes will give unsymmetrical shapes as reduction (or oxidation) might not occur. However, at fast sweep rates, an intermediate might not have time to react, and a reversible shape will be observed.

Answer At slow sweep rates, the second reaction has time to occur, and a curve typical of a two-electron reduction will be

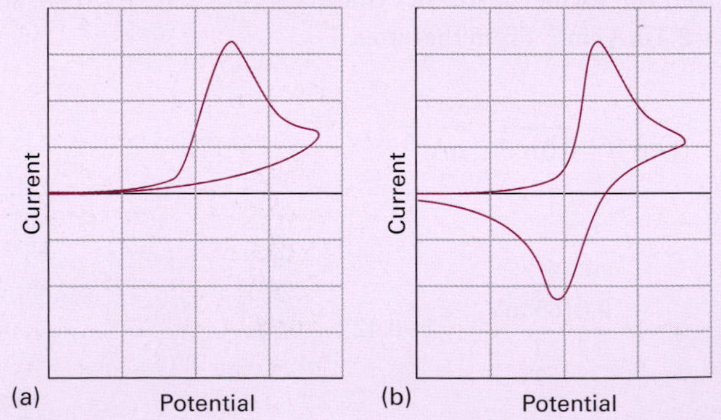

Figure 21F.13 (a) When a non-reversible step in a reaction mechanism has time to occur, the cyclic voltammogram may not show the reverse oxidation or reduction peak. (b) However, if the rate of sweep is increased, the return step may be caused to occur before the irreversible step has had time to intervene, and a typical 'reversible' voltammogram is obtained.

observed, but there will be no oxidation peak on the second half of the cycle because the product, $C_6H_5NO_2$, cannot be oxidized (Fig. 21F.13a). At fast sweep rates, the second reaction does not have time to take place before oxidation of the $BrC_6H_4NO_2^-$ intermediate starts to occur during the reverse scan, so the voltammogram will be typical of a reversible one-electron reduction (Fig. 21F.13b).

Self-test 21F.3 Suggest an interpretation of the cyclic voltammogram shown in Fig. 21F.14.The electroactive material is ClC_6H_4CN in acid solution; after reduction to $ClC_6H_4CN^-$ the radical anion may form C_6H_5CN irreversibly.

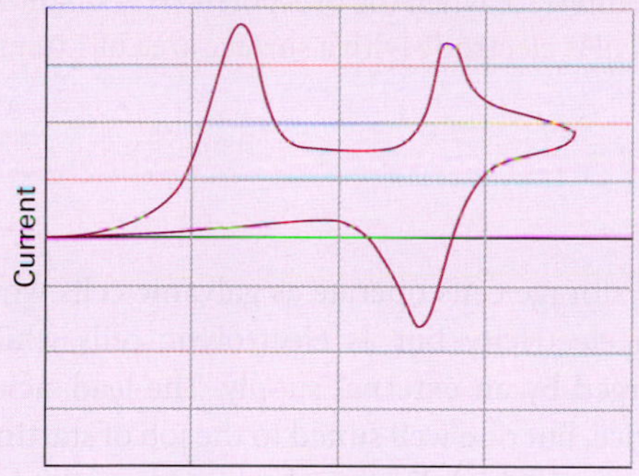

Figure 21F.14 The cyclic voltammogram referred to in *Self-test* 21F.3.

Answer: $ClC_6H_4CN+e^- \rightleftharpoons ClC_6H_4CN^-$, $ClC_6H_4CN^- + H^+ + e^- \rightarrow C_6H_5CN + Cl^-$, $C_6H_5CN + e^- \rightleftharpoons C_6H_5CN^-$

21F.4 Electrolysis

To induce current to flow through an electrolytic cell and bring about a nonspontaneous cell reaction, the applied potential difference must exceed the zero-current potential by at least the **cell overpotential**, the sum of the overpotentials at the two electrodes and the ohmic drop (IR_s, where R_s is the internal resistance of the cell) due to the current through the electrolyte. The additional potential needed to achieve a detectable rate of reaction may need to be large when the exchange current density at the electrodes is small. For similar reasons, a working galvanic cell generates a smaller potential than under zero-current conditions. In this section we see how to cope with both aspects of the overpotential.

The relative rates of gas evolution or metal deposition during electrolysis can be estimated from the Butler–Volmer equation and tables of exchange current densities. From eqn 21F.6a and assuming equal transfer coefficients, we write the ratio of the cathodic currents as

$$\frac{j'}{j}=\frac{j_0'}{j_0}e^{(\eta-\eta')\alpha f} \qquad \text{(21F.7)}$$

where j' is the current density for electrodeposition and j is that for gas evolution, and j_0' and j_0 are the corresponding exchange current densities. This equation shows that metal deposition is favoured by a large exchange current density and relatively high gas evolution overpotential (so $\eta - \eta'$ is positive and large). Note that $\eta<0$ for a cathodic process, so $-\eta'>0$. The exchange current density depends strongly on the nature of the electrode surface, and changes in the course of the electrodeposition of one metal on another. A very crude criterion is that significant evolution or deposition occurs only if the overpotential exceeds about 0.6 V.

A glance at Table 21F.1 shows the wide range of exchange current densities for a metal/hydrogen electrode. The most sluggish exchange currents occur for lead and mercury: 1 pA cm^{-2} corresponds to a monolayer of atoms being replaced in about 5 years. For such systems, a high overpotential is needed to induce significant hydrogen evolution. In contrast, the value for platinum (1 mA cm^{-2}) corresponds to a monolayer being replaced in 0.1 s, so significant gas evolution occurs for a much lower overpotential.

The exchange current density also depends on the crystal face exposed. For the deposition of copper on copper, the (100) face has $j_0=1$ mA cm^{-2}, so for the same overpotential the (100) face grows at 2.5 times the rate of the (111) face, for which $j_0=0.4$ mA cm^{-2}.

21F.5 Working galvanic cells

In working galvanic cells (those not balanced against an external potential), the overpotential leads to a smaller potential than under zero-current conditions. Furthermore, we expect the cell potential to decrease as current is generated because it is then no longer working reversibly and can therefore do less than maximum work.

We shall consider the cell $M|M^+(aq)||M'^+(aq)|M'$ and ignore all the complications arising from liquid junctions. The potential of the cell is $E'=\Delta\phi_R-\Delta\phi_L$. Because the cell potential differences differ from their zero-current values by overpotentials, we can write $\Delta\phi_X=E_X+\eta_X$ where X is L or R for the left or right electrode, respectively. The cell potential is therefore

$$E'=E+\eta_R-\eta_L \qquad \text{(21F.8a)}$$

To avoid confusion about signs (η_R is negative, η_L is positive) and to emphasize that a working cell has a lower potential than a zero-current cell, we shall write this expression as

$$E' = E - |\eta_R| - |\eta_L| \quad (21F.8b)$$

with E the cell potential. We should also subtract the ohmic potential difference IR_s, where R_s is the cell's internal resistance:

$$E' = E - |\eta_R| - |h_L| - IR_s \quad (21F.8c)$$

The ohmic term is a contribution to the cell's irreversibility—it is a thermal dissipation term—so the sign of IR_s is always such as to reduce the potential in the direction of zero.

The overpotentials in eqn 21F.8 can be calculated from the Butler–Volmer equation for a given current, I, being drawn. We shall simplify the equations by supposing that the areas, A, of the electrodes are the same, that only one electron is transferred in the rate-determining steps at the electrodes, that the transfer coefficients are both $\frac{1}{2}$, and that the high-overpotential limit of the Butler–Volmer equation may be used. Then from eqns 21F.6a and 21F.8c we find

$$E' = E - IR_s - \frac{4RT}{F}\ln\left(\frac{I}{A\bar{j}}\right) \qquad \bar{j} = (j_{0L}j_{0R})^{1/2} \quad (21F.9)$$

where j_{0L} and j_{0R} are the exchange current densities for the two electrodes.

Brief illustration 21F.2 The working potential

Suppose that a cell consists of two electrodes each of area 10 cm² with exchange current densities 5 μA cm^{-2} and has internal resistance 10 Ω. At 298 K $RT/F = 25.7$ mV. The zero-current cell potential is 1.5 V. If the cell is producing a current of 10 mA, its working potential will be

$$E' = 1.5\,\text{V} - \overbrace{(10\,\text{mA}) \times (10\,\Omega)}^{0.10\,V} - \overbrace{4(25.7\,\text{mV})\ln\left(\frac{10\,\text{mA}}{(10\,\text{cm}^2)\times(5\,\mu\text{A}\,\text{cm}^{-2})}\right)}^{0.54\,V\ldots}$$
$$= 0.9\,\text{V}$$

We have used 1 A Ω = 1 V. Note that we have ignored various other factors that reduce the cell potential, such as the inability of reactants to diffuse rapidly enough to the electrodes.

Self-test 21F.4 What is the effective resistance at 25 °C of an electrode interface when the overpotential is small? Evaluate it for a Pt,H_2|H^+ electrode with a surface area of 1.0 cm².

Answer: 33 Ω

Electric storage cells operate as galvanic cells while they are producing electricity but as electrolytic cells while they are being charged by an external supply. The lead–acid battery is an old device, but one well suited to the job of starting cars (and the only one available). During charging the cathode reaction is the reduction of Pb^{2+} and its deposition as lead on the lead electrode. Deposition occurs instead of the reduction of the acid to hydrogen because the latter has a low exchange current density on lead. The anode reaction during charging is the oxidation of Pb(II) to Pb(IV), which is deposited as the oxide PbO_2. On discharge, the two reactions run in reverse. Because they have such high exchange current densities the discharge can occur rapidly, which is why the lead battery can produce large currents on demand.

Checklist of concepts

- ☐ 1. An **electrical double layer** consists of sheets of opposite charge at the surface of the electrode and next to it in the solution.
- ☐ 2. Models of the double layer include the **Helmholtz layer model** and the **Gouy-Chapman** model.
- ☐ 3. The **Galvani potential difference** is the potential difference between the bulk of the metal electrode and the bulk of the solution.
- ☐ 4. The current density at an electrode is expressed by the **Butler–Volmer equation**.
- ☐ 5. A **Tafel plot** is the plot of the logarithm of the current density against the overpotential (see below).
- ☐ 6. **Voltammetry** is the study of the current through an electrode as a function of the applied potential difference.
- ☐ 7. To induce current to flow through an electrolytic cell and bring about a nonspontaneous cell reaction, the applied potential difference must exceed the cell potential by at least the **cell overpotential**.
- ☐ 8. In working galvanic cells the overpotential leads to a smaller potential than under zero-current conditions and the cell potential decreases as current is generated.

Checklist of equations

Property	Equation	Comment	Equation number
Butler–Volmer equation	$j = j_0\{e^{(1-\alpha)f\eta} - e^{-\alpha f\eta}\}$		21F.1
Tafel plots	$\ln j = \ln j_0 + (1-\alpha)f\eta$	Anodic current density	21F.5b
	$\ln j = \ln j_0 - \alpha f\eta$	Cathodic current density	21F.6b
Potential of a working galvanic cell	$E' = E - IR_s - (4RT/F)\ln(I/A\bar{j})$	$\bar{j} = (j_{0L}j_{0R})^{1/2}$	21F.9

CHAPTER 21 Reaction dynamics

TOPIC 21A Collision theory

Discussion questions

21A.1 Discuss how the collision theory of gases builds on the kinetic–molecular theory.

21A.2 How might collision theory change for real gases?

21A.3 Describe the essential features of the harpoon mechanism.

21A.4 Discuss the significance of the steric *P*-factor in the RRK model.

Exercises

21A.1(a) Calculate the collision frequency, z, and the collision density, Z, in ammonia, $R = 190$ pm, at 30 °C and 120 kPa. What is the percentage increase when the temperature is raised by 10 K at constant volume?
21A.1(b) Calculate the collision frequency, z, and the collision density, Z, in carbon monoxide, $R = 180$ pm at 30 °C and 120 kPa. What is the percentage increase when the temperature is raised by 10 K at constant volume?

21A.2(a) Collision theory depends on knowing the fraction of molecular collisions having at least the kinetic energy E_a along the line of flight. What is this fraction when (i) $E_a = 20\,\text{kJ mol}^{-1}$, (ii) $E_a = 100\,\text{kJ mol}^{-1}$ at (1) 350 K and (2) 900 K?
21A.2(b) Collision theory depends on knowing the fraction of molecular collisions having at least the kinetic energy E_a along the line of flight. What is this fraction when (i) $E_a = 15\,\text{kJ mol}^{-1}$, (ii) $E_a = 150\,\text{kJ mol}^{-1}$ at (1) 300 K and (2) 800 K?

21A.3(a) Calculate the percentage increase in the fractions in Exercise 21A.2(a) when the temperature is raised by 10 K.
21A.3(b) Calculate the percentage increase in the fractions in Exercise 21A.2(b) when the temperature is raised by 10 K.

21A.4(a) Use the collision theory of gas-phase reactions to calculate the theoretical value of the second-order rate constant for the reaction $H_2(g) + I_2(g) \rightarrow 2\,HI(g)$ at 650 K, assuming that it is elementary and bimolecular. The collision cross section is 0.36 nm^2, the reduced mass is 3.32×10^{-27} kg, and the activation energy is 171 kJ mol^{-1}. (Assume a steric factor of 1.)
21A.4(b) Use the collision theory of gas-phase reactions to calculate the theoretical value of the second-order rate constant for the reaction $D_2(g) + Br_2(g) \rightarrow 2\,HI(g)$ at 450 K, assuming that it is elementary and bimolecular. Take the collision cross section as 0.30 nm^2, the reduced mass as $3.930m_u$, and the activation energy as 200 kJ mol^{-1}. (Assume a steric factor of 1.)

21A.5(a) For the gaseous reaction $A + B \rightarrow P$, the reactive cross-section obtained from the experimental value of the pre-exponential factor is 9.2×10^{-22} m^2. The collision cross-sections of A and B estimated from the transport properties are 0.95 and 0.65 nm^2 respectively. Calculate the *P*-factor for the reaction.
21A.5(b) For the gaseous reaction $A + B \rightarrow P$, the reactive cross-section obtained from the experimental value of the pre-exponential factor is 8.7×10^{-22} m^2. The collision cross-sections of A and B estimated from the transport properties are 0.88 and 0.40 nm^2, respectively. Calculate the *P*-factor for the reaction.

21A.6(a) Consider the unimolecular decomposition of a nonlinear molecule containing five atoms according to RRK theory. If $P = 3.0 \times 10^{-5}$, what is the value of E^*/E?
21A.6(b) Consider the unimolecular decomposition of a linear molecule containing four atoms according to RRK theory. If $P = 0.025$, what is the value of E^*/E?

21A.7(a) Suppose that an energy of 250 kJ mol^{-1} is available in a collision but 200 kJ mol^{-1} is needed to break a particular bond in a molecule with $s = 10$. Use the RRK model to calculate the steric *P*-factor.
21.A7(b) Suppose that an energy of 500 kJ mol^{-1} is available in a collision but 300 kJ mol^{-1} is needed to break a particular bond in a molecule with $s = 12$. Use the RRK model to calculate the steric *P*-factor.

Problems

21A.1 In the dimerization of methyl radicals at 25 °C, the experimental pre-exponential factor is $2.4 \times 10^{10}\,\text{dm}^3\,\text{mol}^{-1}\,\text{s}^{-1}$. What are (a) the reactive cross-section, (b) the *P*-factor for the reaction if the C–H bond length is 154 pm?

21A.2 Nitrogen dioxide reacts bimolecularly in the gas phase: $NO_2 + NO_2 \rightarrow NO + NO + O_2$. The temperature dependence of the second-order rate constant for the rate law $d[P]/dt = k_r[NO_2]^2$ is given in the following table. What are the *P*-factor and the reactive cross-section for the reaction?

T/K	600	700	800	1000
k_r/(cm^3 mol^{-1} s^{-1})	4.6×10^2	9.7×10^3	1.3×10^5	3.1×10^6

Take $\sigma = 0.60$ nm^2.

21A.3 The diameter of the methyl radical is about 308 pm. What is the maximum rate constant in the expression $d[C_2H_6]/dt = k_r[CH_3]^2$ for second-order recombination of radicals at room temperature? 10 per cent of a sample of ethane of volume 1.0 dm^3 at 298 K and 100 kPa is dissociated into methyl radicals. What is the minimum time for 90 per cent recombination?

21A.4 The total cross-sections for reactions between alkali metal atoms and halogen molecules are given in the following table (R.D. Levine and R.B. Bernstein, *Molecular reaction dynamics*, Clarendon Press, Oxford, 72 (1974)). Assess the data in terms of the harpoon mechanism.

σ^*/nm^2	Cl_2	Br_2	I_2
Na	1.24	1.16	0.97
K	1.54	1.51	1.27
Rb	1.90	1.97	1.67
Cs	1.96	2.04	1.95

Electron affinities are approximately 1.3 eV (Cl_2), 1.2 eV (Br_2), and 1.7 eV (I_2), and ionization energies are 5.1 eV (Na), 4.3 eV (K), 4.2 eV (Rb), and 3.9 eV (Cs).

21A.5‡ One of the most historically significant studies of chemical reaction rates was that by M. Bodenstein (*Z. physik. Chem.* **29**, 295 (1899)) of the gas-phase reaction 2 HI(g)→H_2(g)+I_2(g) and its reverse, with rate constants k_r and k_r', respectively. The measured rate constants as a function of temperature are

T/K	647	666	683	700	716	781
k_r/(22.4 dm^3 mol^{-1} min^{-1})	0.230	0.588	1.37	3.10	6.70	105.9
k_r'/(22.4 dm^3 mol^{-1} min^{-1})	0.0140	0.0379	0.0659	0.172	0.375	3.58

Demonstrate that these data are consistent with the collision theory of bimolecular gas-phase reactions.

21A.6‡ R. Atkinson (*J. Phys. Chem. Ref. Data* **26**, 215 (1997)) has reviewed a large set of rate constants relevant to the atmospheric chemistry of volatile organic compounds. The recommended rate constant for the bimolecular association of O_2 with an alkyl radical R at 298 K is 4.7×10^9 dm^3 mol^{-1} s^{-1} for R=C_2H_5 and 8.4×10^9 dm^3 mol^{-1} s^{-1} for R=cyclohexyl. Assuming no energy barrier, compute the steric factor, P, for each reaction. *Hint*: Obtain collision diameters from collision cross-sections of similar molecules in the *Resource section*.

21A.7 According to the RRK model (see *Justification* 21A.1)

$$P=\frac{n!(n-n^*+s-1)!}{(n-n^*)!(n+s-1)!}$$

Use Stirling's approximation of the form $\ln x!\approx x\ln x-x$ to deduce that $P\approx(n-n^*/n)^{s-1}$ when $s-1\ll n-n^*$. *Hint*: replace terms of the form $n-n^*+s-1$ by $n-n^*$ inside logarithms but retain $n-n^*+s-1$ when it is a factor of a logarithm.

TOPIC 21B Diffusion-controlled reactions

Discussion questions

21B.1 Distinguish between a diffusion-controlled reaction and an activation-controlled reaction. Do both have activation energies?

21B.2 Describe the role of the encounter pair in the cage effect.

Exercises

21B.1(a) A typical diffusion coefficient for small molecules in aqueous solution at 25 °C is 6×10^{-9} m^2 s^{-1}. If the critical reaction distance is 0.5 nm, what value is expected for the second-order rate constant for a diffusion-controlled reaction?

21B.1(b) Suppose that the typical diffusion coefficient for a reactant in aqueous solution at 25 °C is 5.2×10^{-9} m^2 s^{-1}. If the critical reaction distance is 0.4 nm, what value is expected for the second-order rate constant for the diffusion-controlled reaction?

21B.2(a) Calculate the magnitude of the diffusion-controlled rate constant at 298 K for a species in (i) water, (ii) pentane. The viscosities are 1.00×10^{-3} kg m^{-1} s^{-1}, and 2.2×10^{-4} kg m^{-1} s^{-1}, respectively.

21B.2(b) Calculate the magnitude of the diffusion-controlled rate constant at 298 K for a species in (i) decylbenzene, (ii) concentrated sulfuric acid. The viscosities are 3.36 cP and 27 cP, respectively.

21B.3(a) Calculate the magnitude of the diffusion-controlled rate constant at 320 K for the recombination of two atoms in water, for which $\eta=0.89$ cP. Assuming the concentration of the reacting species is 1.5 mmol dm^{-3} initially, how long does it take for the concentration of the atoms to fall to half that value? Assume the reaction is elementary.

21B.3(b) Calculate the magnitude of the diffusion-controlled rate constant at 320 K for the recombination of two atoms in benzene, for which $\eta=0.601$ cP. Assuming the concentration of the reacting species is 2.0 mmol dm^{-3} initially, how long does it take for the concentration of the atoms to fall to half that value? Assume the reaction is elementary.

21B.4(a) Two neutral species, A and B, with diameters 655 pm and 1820 pm, respectively, undergo the diffusion-controlled reaction A+B→P in a solvent of viscosity 2.93×10^{-3} kg m^{-1} s^{-1} at 40 °C. Calculate the initial rate d[P]/dt if the initial concentrations of A and B are 0.170 mol dm^{-3} and 0.350 mol dm^{-3}, respectively.

21B.4(b) Two neutral species, A and B, with diameters 421 pm and 945 pm, respectively, undergo the diffusion-controlled reaction A+B→P in a solvent of viscosity 1.35 cP at 20 °C. Calculate the initial rate d[P]/dt if the initial concentrations of A and B are 0.155 mol dm^{-3} and 0.195 mol dm^{-3}, respectively.

Problems

21B.1 Confirm that eqn 21B.8 is a solution of eqn 21B.7, where [J] is a solution of the same equation but with $k_r=0$ and for the same initial conditions.

21B.2 Use mathematical software, a spreadsheet, or the *Living graphs* on the web site of this book to explore the effect of varying the value of the rate constant k_r on the spatial variation of [J]* (see eqn 21B.8 with [J] given in eqn 21B.9) for a constant value of the diffusion constant D.

21B.3 Confirm that if the initial condition is [J]=0 at $t=0$ everywhere, and the boundary condition is $[J]=[J]_0$ at $t>0$ at all points on a surface, then the solutions [J]* in the presence of a first-order reaction that removed J are related to those in the absence of reaction, [J], by

$$[J]^*=k_r\int_0^t [J]e^{-k_rt}dt+[J]e^{-k_rt}$$

Base your answer on eqn 21B.5.

21B.4‡ The compound α-tocopherol, a form of vitamin E, is a powerful antioxidant that may help to maintain the integrity of biological membranes. R.H. Bisby and A.W. Parker (*J. Amer. Chem. Soc.* **117**, 5664 (1995)) studied

‡ These problems were supplied by Charles Trapp and Carmen Giunta.

the reaction of photochemically excited duroquinone with the antioxidant in ethanol. Once the duroquinone was photochemically excited, a bimolecular reaction took place at a rate described as diffusion limited. (a) Estimate the rate constant for a diffusion-limited reaction in ethanol. (b) The reported rate constant was $2.77\times10^{9}\,\mathrm{dm^3\,mol^{-1}\,s^{-1}}$; estimate the critical reaction distance if the sum of diffusion constants is $1\times10^{-9}\,\mathrm{m^2\,s^{-1}}$.

TOPIC 21C Transition-state theory

Discussion questions

21C.1 Describe in outline the formulation of the Eyring equation.

21C.2 How is femtosecond spectroscopy used to examine the structures of activated complexes?

21C.3 Explain the physical origin of the kinetic salt effect. What might be the effect of the relative permittivity of the medium?

21C.4 How do kinetic isotope effects provide insight into the mechanism of a reaction?

Exercises

21C.1(a) The reaction of propylxanthate ion in acetic acid buffer solutions has the mechanism $A^- + H^+ \rightarrow P$. Near 30 °C the rate constant is given by the empirical expression $k_r = (2.05\times10^{13})\,e^{-(8681\,K)/T}\,\mathrm{dm^3\,mol^{-1}\,s^{-1}}$. Evaluate the energy and entropy of activation at 30 °C.

21C.1(b) The reaction $A^- + H^+ \rightarrow P$ has a rate constant given by the empirical expression $k_r = (6.92\times10^{12})e^{-(5925\,K)/T}\,\mathrm{dm^3\,mol^{-1}\,s^{-1}}$. Evaluate the energy and entropy of activation at 25 °C.

21C.2(a) When the reaction in Exercise 21C.1(a) occurs in a dioxane/water mixture which is 30 per cent dioxane by mass, the rate constant fits $k_r = (7.78\times10^{14})e^{-(9134\,K)/T}\,\mathrm{dm^3\,mol^{-1}\,s^{-1}}$ near 30 °C. Calculate $\Delta^{\ddagger}G$ for the reaction at 30 °C.

21C.2(b) A rate constant is found to fit the expression $k_r = (4.98\times10^{13})e^{-(4972\,K)/T}\,\mathrm{dm^3\,mol^{-1}\,s^{-1}}$ near 25 °C. Calculate $\Delta^{\ddagger}G$ for the reaction at 25 °C.

21C.3(a) The gas phase association reaction between F_2 and IF_5 is first order in each of the reactants. The energy of activation for the reaction is 58.6 kJ mol^{-1}. At 65 °C the rate constant is $7.84\times10^{-3}\,\mathrm{kPa^{-1}\,s^{-1}}$. Calculate the entropy of activation at 65 °C.

21C.3(b) A gas-phase recombination reaction is first order in each of the reactants. The energy of activation for the reaction is 39.7 kJ mol^{-1}. At 65 °C the rate constant is $0.35\,\mathrm{m^3\,s^{-1}}$. Calculate the entropy of activation at 65 °C.

21C.4(a) Calculate the entropy of activation for a collision between two structureless particles at 300 K, taking $M = 65\,\mathrm{g\,mol^{-1}}$ and $\sigma = 0.35\,\mathrm{nm^2}$.

21C.4(b) Calculate the entropy of activation for a collision between two structureless particles at 450 K, taking $M = 92\,\mathrm{g\,mol^{-1}}$ and $\sigma = 0.45\,\mathrm{nm^2}$.

21C.5(a) The pre-exponential factor for the gas-phase decomposition of ozone at low pressures is $4.6\times10^{12}\,\mathrm{dm^3\,mol^{-1}\,s^{-1}}$ and its activation energy is 10.0 kJ mol^{-1}. What are (i) the entropy of activation, (ii) the enthalpy of activation, (iii) the Gibbs energy of activation at 298 K?

21C.5(b) The pre-exponential factor for a gas-phase decomposition of a gas at low pressures is $2.3\times10^{13}\,\mathrm{dm^3\,mol^{-1}\,s^{-1}}$ and its activation energy is 30.0 kJ mol^{-1}. What are (i) the entropy of activation, (ii) the enthalpy of activation, (iii) the Gibbs energy of activation at 298 K?

21C.6(a) The rate constant of the reaction $H_2O_2(aq) + I^-(aq) + H^+(aq) \rightarrow H_2O(l) + HIO(aq)$ is sensitive to the ionic strength of the aqueous solution in which the reaction occurs. At 25 °C, $k_r = 12.2\,\mathrm{dm^6\,mol^{-2}\,min^{-1}}$ at an ionic strength of 0.0525. Use the Debye–Hückel limiting law to estimate the rate constant at zero ionic strength.

21C.6(b) At 25 °C, $k_r = 1.55\,\mathrm{dm^6\,mol^{-2}\,min^{-1}}$ at an ionic strength of 0.0241 for a reaction in which the rate-determining step involves the encounter of two singly charged cations. Use the Debye–Hückel limiting law to estimate the rate constant at zero ionic strength.

Problems

21C.1 The rates of thermolysis of a variety of *cis*- and *trans*-azoalkanes have been measured over a range of temperatures in order to settle a controversy concerning the mechanism of the reaction. In ethanol an unstable *cis*-azoalkane decomposed at a rate that was followed by observing the N_2 evolution, and this led to the rate constants given in the following table (P.S. Engel and D.J. Bishop, *J. Amer. Chem. Soc.* **97**, 6754 (1975)). Calculate the enthalpy, entropy, energy, and Gibbs energy of activation at −20 °C.

θ/°C	−24.82	−20.73	−17.02	−13.00	−8.95
$10^4\times k_r/\mathrm{s^{-1}}$	1.22	2.31	4.39	8.50	14.3

21C.2 In an experimental study of a bimolecular reaction in aqueous solution, the second-order rate constant was measured at 25 °C and at a variety of ionic strengths and the results are tabulated in the following table. It is known that a singly charged ion is involved in the rate-determining step. What is the charge on the other ion involved?

I/(mol kg^{-1})	0.0025	0.0037	0.0045	0.0065	0.0085
k_r/(dm^3 mol^{-1} s^{-1})	1.05	1.12	1.16	1.18	1.26

21C.3 Derive the expression for k_r given in *Example* 21C.1 by introducing the equations for the thermal wavelengths.

21C.4 The rate constant of the reaction $I^-(aq) + H_2O_2(aq) \rightarrow H_2O(l) + IO^-(aq)$ varies slowly with ionic strength, even though the Debye–Hückel limiting law predicts no effect. Use the following data from 25 °C to find the dependence of $\log k_r$ on the ionic strength:

I/(mol kg^{-1})	0.0207	0.0525	0.0925	0.1575
k_r/(dm^3 mol^{-1} min^{-1})	0.663	0.670	0.679	0.694

Evaluate the limiting value of k_r at zero ionic strength. What does the result suggest for the dependence of $\log\gamma$ on ionic strength for a neutral molecule in an electrolyte solution?

21C.5‡ For the gas phase reaction $A + A \rightarrow A_2$, the experimental rate constant, k_r, has been fitted to the Arrhenius equation with the pre-exponential factor $A = 4.07 \times 10^5\,dm^3\,mol^{-1}\,s^{-1}$ at 300 K and an activation energy of $65.43\,kJ\,mol^{-1}$. Calculate $\Delta^{\ddagger}S$, $\Delta^{\ddagger}H$, $\Delta^{\ddagger}U$, and $\Delta^{\ddagger}G$ for the reaction.

21C.6 Use the Debye–Hückel limiting law to show that changes in ionic strength can affect the rate of reaction catalysed by H^+ from the deprotonation of a weak acid. Consider the mechanism: $H^+ + B \rightarrow P$, where H^+ comes from the deprotonation of the weak acid, HA. The weak acid has a fixed concentration. First show that log $[H^+]$, derived from the ionization of HA, depends on the activity coefficients of ions and thus depends on the ionic strength. Then find the relationship between log(rate) and log $[H^+]$ to show that the rate also depends on the ionic strength.

21C.7‡ Show that bimolecular reactions between nonlinear molecules are much slower than between atoms even when the activation energies of both reactions are equal. Use transition-state theory and make the following assumptions. (1) All vibrational partition functions are close to unity; (2) all rotational partition functions are approximately $1 \times 10^{1.5}$, which is a reasonable order of magnitude number; (3) the translational partition function for each species is 1×10^{26}.

21C.8 This exercise gives some familiarity with the difficulties involved in predicting the structure of activated complexes. It also demonstrates the importance of femtosecond spectroscopy to our understanding of chemical dynamics because direct experimental observation of the activated complex removes much of the ambiguity of theoretical predictions. Consider the attack of H on D_2, which is one step in the $H_2 + D_2$ reaction. (a) Suppose that the H approaches D_2 from the side and forms a complex in the form of an isosceles triangle. Take the H–D distance as 30 per cent greater than in H_2 (74 pm) and the D–D distance as 20 per cent greater than in H_2. Let the critical coordinate be the antisymmetric stretching vibration in which one H–D bond stretches as the other shortens. Let all the vibrations be at about $1000\,cm^{-1}$. Estimate k_r for this reaction at 400 K using the experimental activation energy of about $35\,kJ\,mol^{-1}$. (b) Now change the model of the activated complex in part (a) and make it linear. Use the same estimated molecular bond lengths and vibrational frequencies to calculate k_r for this choice of model. (c) Clearly, there is much scope for modifying the parameters of the models of the activated complex. Use mathematical software or write and run a program that allows you to vary the structure of the complex and the parameters in a plausible way, and look for a model (or more than one model) that gives a value of k_r close to the experimental value, $4 \times 10^5\,dm^3\,mol^{-1}\,s^{-1}$.

21C.9‡ M. Cyfert et al. (*Int. J. Chem. Kinet.* **28**, 103 (1996)) examined the oxidation of tris(1,10-phenanthroline)iron(II) by periodate in aqueous solution, a reaction which shows autocatalytic behaviour. To assess the kinetic salt effect, they measured rate constants at a variety of concentrations of Na_2SO_4 far in excess of reactant concentrations and reported the following data:

$[Na_2SO_4]/(mol\,kg^{-1})$	0.2	0.15	0.1	0.05	0.025	0.0125	0.005
$k_r/(dm^{3/2}\,mol^{-1/2}\,s^{-1})$	0.462	0.430	0.390	0.321	0.283	0.252	0.224

What can be inferred about the charge of the activated complex of the rate-determining step?

21C.10 The study of conditions that optimize the association of proteins in solution guides the design of protocols for formation of large crystals that are amenable to analysis by X-ray diffraction techniques. It is important to characterize protein dimerization because the process is considered to be the rate-determining step in the growth of crystals of many proteins. Consider the variation with ionic strength of the rate constant of dimerization in aqueous solution of a cationic protein P:

$I/(mol\,kg^{-1})$	0.0100	0.0150	0.0200	0.0250	0.0300	0.0350
k_r/k_r°	8.10	13.30	20.50	27.80	38.10	52.00

What can be deduced about the charge of P?

21C.11 Predict the order of magnitude of the primary isotope effect on the relative rates of displacement of (a) ^1H and ^3H in a C–H bond, (b) ^{16}O and ^{18}O in a C–O bond. Will raising the temperature enhance the difference? Take $k_f(C–H) = 450\,N\,m^{-1}$, $k_f(C–O) = 1750\,N\,m^{-1}$.

TOPIC 21D The dynamics of molecular collisions

Discussion questions

21D.1 Describe how the following techniques are used in the study of chemical dynamics: infrared chemiluminescence, laser-induced fluorescence, multiphoton ionization, resonant multiphoton ionization, and reaction product imaging.

21D.2 Discuss the relationship between the saddle-point energy and the activation energy of a reaction.

21D.3 A method for directing the outcome of a chemical reaction consists of using molecular beams to control the relative orientations of reactants during a collision. Consider the reaction $Rb + CH_3I \rightarrow RbI + CH_3$. How should CH_3I molecules and Rb atoms be oriented to maximize the production of RbI?

21D.4 Consider a reaction with an attractive potential energy surface. Discuss how the initial distribution of reactant energy affects how efficiently the reaction proceeds. Repeat for a repulsive potential energy surface.

21D.5 Describe how molecular beams are used to investigate intermolecular potentials.

Exercises

21D.1(a) The interaction between two diatomic molecules is described by an attractive potential energy surface. What distribution of vibrational and translational energies among reactants and products is most likely to lead to a successful reaction?

21D.1(b) The interaction between two diatomic molecules has a repulsive potential energy surface. What distribution of vibrational and translational energies among reactants and products is most likely to lead to a successful reaction?

21D.2(a) If the cumulative reaction probability were independent of energy, what is the temperature dependence of the rate constant predicted by the numerator of eqn 21D.6?

21D.2(b) If the cumulative reaction probability equalled 1 for energies less than a barrier height *V* and vanished for higher energies, what is the temperature dependence of the rate constant predicted by the numerator of eqn 21D.6?

Problems

21D.1 Show that the intensities of a molecular beam before and after passing through a chamber of length L containing inert scattering atoms are related by $I=I_0e^{-\mathcal{N}\sigma L}$, where σ is the collision cross-section and $\mathcal{N}$ the number density of scattering atoms.

21D.2 In a molecular beam experiment to measure collision cross-sections it was found that the intensity of a CsCl beam was reduced to 60 per cent of its intensity on passage through CH_2F_2 at 10 μTorr, but that when the target was Ar at the same pressure the intensity was reduced only by 10 per cent. What are the relative cross-sections of the two types of collision? Why is one much larger than the other?

21D.3 Consider the collision between a hard-sphere molecule of radius R_1 and mass m, and an infinitely massive impenetrable sphere of radius R_2. Plot the scattering angle θ as a function of the impact parameter b. Carry out the calculation using simple geometrical considerations.

21D.4 The dependence of the scattering characteristics of atoms on the energy of the collision can be modelled as follows. We suppose that the two colliding atoms behave as impenetrable spheres, as in Problem 21D.3, but that the effective radius of the heavy atoms depends on the speed v of the light atom. Suppose its effective radius depends on v as R_2e^{-v/v^*}, where v^* is a constant. Take $R_1=\frac{1}{2}R_2$ for simplicity and an impact parameter $b=\frac{1}{2}R_2$, and plot the scattering angle as a function of (a) speed, (b) kinetic energy of approach.

TOPIC 21E Electron transfer in homogeneous systems

Discussion questions

21E.1 Discuss how the following factors determine the rate of electron transfer in homogeneous systems: the distance between electron donor and acceptor, the standard Gibbs energy of the process, and the reorganization energy of the redox active species and the surrounding medium.

21E.2 What role does tunnelling play in electron transfer?

21E.3 Explain why the rate constant decreases as the reaction becomes more exergonic in the inverted region.

Exercises

21E.1(a) For a pair of electron donor and acceptor at 298 K, $H_{et}(d)=0.04\,cm^{-1}$, $\Delta_rG^{\ominus}=-0.185$ eV and $k_{et}=37.5\,s^{-1}$. Estimate the value of the reorganization energy.

21E.1(b) For a pair of electron donor and acceptor at 298 K, $k_{et}=2.02\times10^5\,s^{-1}$ for $\Delta_rG^{\ominus}=-0.665$ eV. The standard reaction Gibbs energy changes to $\Delta_rG^{\ominus}=-0.975$ eV when a substituent is added to the electron acceptor and the rate constant for electron transfer changes to $k_{et}=3.33\times10^6\,s^{-1}$. Assuming that the distance between donor and acceptor is the same in both experiments, estimate the values of $H_{et}(d)$ and ΔE_R.

21E.2(a) For a pair of electron donor and acceptor, $k_{et}=2.02\times10^5\,s^{-1}$ when $d=1.11$ nm and $k_{et}=4.51\times10^4\,s^{-1}$ when $r=1.23$ nm. Assuming that $\Delta_rG^{\ominus}$ and ΔE_R are the same in both experiments, estimate the value of β.

21E.2(b) Refer to Exercise 21E.2(a). Estimate the value of k_{et} when $d=1.59$ nm.

Problems

21E.1 Consider the reaction $D+A\rightarrow D^++A^-$. The rate constant k_r may be determined experimentally or may be predicted by the *Marcus cross-relation* $k_r=(k_{DD}k_{AA}K)^{1/2}f$, where k_{DD} and k_{AA} are the experimental rate constants for the electron self-exchange processes $^*D+D^+\rightarrow{}^*D^++D$ and $^*A+A^+\rightarrow{}^*A^++A$, respectively, and f is a function of $K=[D^+][A^-]/[D][A]$, k_{DD}, k_{AA}, and the collision frequencies. Derive the approximate form of the Marcus cross-relation by following these steps. (a) Use eqn 21E.7 to write expressions for $\Delta^{\ddagger}G$, $\Delta^{\ddagger}G_{DD}$, and $\Delta^{\ddagger}G_{AA}$, keeping in mind that $\Delta_rG^{\ominus}=0$ for the electron self-exchange reactions. (b) Assume that the reorganization energy $\Delta E_{R,DA}$ for the reaction $D+A\rightarrow D^++A^-$ is the average of the reorganization energies $\Delta E_{R,DD}$ and $\Delta E_{R,AA}$ of the electron self-exchange reactions. Then show that in the limit of small magnitude of $\Delta_rG^{\ominus}$, or $|\Delta_rG^{\ominus}|\ll\Delta E_{R,DA}$, $\Delta^{\ddagger}G=\frac{1}{2}(\Delta^{\ddagger}G_{DD}+\Delta^{\ddagger}G_{AA}+\Delta_rG^{\ominus})$, where $\Delta_rG^{\ominus}$ is the standard Gibbs energy for the reaction $D+A\rightarrow D^++A^-$. (c) Use an equation of the form of eqn 21E.4 to write expressions for k_{DD} and k_{AA}. (d) Use eqn 21E.4 and the result above to write an expression for k_r. (e) Complete the derivation by using the results from part (c), the relation $K=e^{-\Delta_rG^{\ominus}/RT}$, and assuming that all $\kappa v^{\dagger}$ terms, which may be interpreted as collision frequencies, are identical.

21E.2 Consider the reaction $D+A\rightarrow D^++A^-$. The rate constant k_r may be determined experimentally or may be predicted by the Marcus cross-relation (see Problem 21E.1). It is common to make the assumption that $f\approx1$. Use the approximate form of the Marcus relation to estimate the rate constant for the reaction $Ru(bpy)_3^{3+}+Fe(H_2O)_6^{2+}\rightarrow Ru(bpy)_3^{2+}+Fe(H_2O)_6^{3+}$, where bpy stands for 4,4′-bipyridine. The following data will be useful:

$Ru(bpy)_3^{3+}+e^-\rightarrow Ru(bpy)_3^{2+}$	$E^{\ominus}=1.26$ V
$Fe(H_2O)_6^{3+}+e^-\rightarrow Fe(H_2O)_6^{2+}$	$E^{\ominus}=0.77$ V
$^*Ru(bpy)_3^{3+}+Ru(bpy)_3^{2+}\rightarrow{}^*Ru(bpy)_3^{2+}+Ru(bpy)_3^{3+}$	$k_{Ru}=4.0\times10^8\,dm^3\,mol^{-1}\,s^{-1}$
$^*Fe(H_2O)_6^{3+}+Fe(H_2O)_6^{2+}\rightarrow{}^*Fe(H_2O)_6^{2+}+Fe(H_2O)_6^{3+}$	$k_{Fe}=4.2\,dm^3\,mol^{-1}\,s^{-1}$

21E.3 A useful strategy for the study of electron transfer in proteins consists of attaching an electroactive species to the protein's surface and then measuring k_{et} between the attached species and an electroactive protein cofactor. J.W. Winkler and H.B. Gray (*Chem. Rev.* **92**, 369 (1992)) summarize data for cytochrome *c* modified by replacement of the haem iron by a zinc ion, resulting in a zinc–porphyrin (ZnP) group in the interior of the protein, and by attachment of a ruthenium ion complex to a surface histidine amino acid. The edge-to-edge distance between the electroactive species was thus fixed at 1.23 nm. A variety of ruthenium ion complexes with different standard potentials was used. For each ruthenium-modified protein, either the $Ru^{2+} \rightarrow ZnP^+$ or the $ZnP^* \rightarrow Ru^{3+}$, in which the electron donor is an electronically excited state of the zinc–porphyrin group formed by laser excitation, was monitored. This arrangement leads to different standard reaction Gibbs energies because the redox couples ZnP^+/ZnP and ZnP^+/ZnP^* have different standard potentials, with the electronically excited porphyrin being a more powerful reductant. Use the following data to estimate the reorganization energy for this system:

$\Delta_r G^{\ominus}$/eV	0.665	0.705	0.745	0.975	1.015	1.055
$k_{et}/(10^6\ s^{-1})$	0.657	1.52	1.12	8.99	5.76	10.1

21E.4 The photosynthetic reaction centre of the purple photosynthetic bacterium *Rhodopseudomonas viridis* contains a number of bound co-factors that participate in electron transfer reactions. The following table shows data compiled by Moser et al. (*Nature* **355**, 796 (1992)) on the rate constants for electron transfer between different co-factors and their edge-to-edge distances:

Reaction	$BChl^- \rightarrow BPh$	$BPh^- \rightarrow BChl_2^+$	$BPh^- \rightarrow Q_A$	cyt $c_{559} \rightarrow BChl_2$	$Q_A^- \rightarrow Q_B$	$Q_A^- \rightarrow BChl_2^+$
d/nm	0.48	0.95	0.96	1.23	1.35	2.24
k_{et}/s^{-1}	1.58×10^{12}	3.98×10^{9}	1.00×10^{9}	1.58×10^{8}	3.98×10^{7}	63.1

(BChl, bacteriochlorophyll; $BChl_2$, bacteriochlorophyll dimer, functionally distinct from BChl; BPh, bacteriophaeophytin; Q_A and Q_B, quinone molecules bound to two distinct sites; cyt c_{559}, a cytochrome bound to the reaction centre complex). Are these data in agreement with the behaviour predicted by eqn 21E.9? If so, evaluate the value of β.

21E.5 The rate constant for electron transfer between a cytochrome c and the bacteriochlorophyll dimer of the reaction centre of the purple bacterium *Rhodobacter sphaeroides* (Problem 21E.4) decreases with decreasing temperature in the range 300 K to 130 K. Below 130 K, the rate constant becomes independent of temperature. Account for these results.

TOPIC 21F Processes at electrodes

Discussion questions

21F.1 Describe the various models of the electrode–electrolyte interface.

21F.2 In what sense is electron transfer at an electrode an activated process?

21F.3 Discuss the technique of cyclic voltammetry and account for the characteristic shape of a cyclic voltammogram, such as those shown in Figs. 21F.13 and 21F.14.

Exercises

21F.1(a) The transfer coefficient of a certain electrode in contact with M^{3+} and M^{4+} in aqueous solution at 25 °C is 0.39. The current density is found to be 55.0 mA cm^{-2} when the overpotential is 125 mV. What is the overpotential required for a current density of 75 mA cm^{-2}?

21F.1(b) The transfer coefficient of a certain electrode in contact with M^{2+} and M^{3+} in aqueous solution at 25°C is 0.42. The current density is found to be 17.0 mA cm^{-2} when the overpotential is 105 mV. What is the overpotential required for a current density of 72 mA cm^{-2}?

21F.2(a) Determine the exchange current density from the information given in Exercise 21F.1(a).

21F.2(b) Determine the exchange current density from the information given in Exercise 21F.1(b).

21F.3(a) To a first approximation, significant evolution or deposition occurs in electrolysis only if the overpotential exceeds about 0.6 V. To illustrate this criterion determine the effect that increasing the overpotential from 0.40 V to 0.60 V has on the current density in the electrolysis of 1.0 M NaOH(aq), which is 1.0 mA cm^{-2} at 0.4 V and 25 °C. Take $\alpha = 0.5$.

21F.3(b) Determine the effect that increasing the overpotential from 0.50 V to 0.60 V has on the current density in the electrolysis of 1.0 M NaOH(aq), which is 1.22 mA cm^{-2} at 0.50 V and 25 °C. Take $\alpha = 0.50$.

21F.4(a) Use the data in Table 21F.1 for the exchange current density and transfer coefficient for the reaction $2\,H^+ + 2\,e^- \rightarrow H_2$ on nickel at 25 °C to determine what current density would be needed to obtain an overpotential of 0.20 V as calculated from (i) the Butler–Volmer equation, and (ii) the Tafel equation. Is the validity of the Tafel approximation affected at higher overpotentials (of 0.4 V and more)?

21F.4(b) Use the data in Table 21F.1 for the exchange current density and transfer coefficient for the reaction $Fe^{3+} + e^- \rightarrow Fe^{2+}$ on platinum at 25 °C to determine what current density would be needed to obtain an overpotential of 0.30 V as calculated from (i) the Butler–Volmer equation, and (ii) the Tafel equation. Is the validity of the Tafel approximation affected at higher overpotentials (of 0.4 V and more)?

21F.5(a) A typical exchange current density, that for H^+ discharge at platinum, is 0.79 mA cm^{-2} at 25 °C. What is the current density at an electrode when its overpotential is (i) 10 mV, (ii) 100 mV, (iii) −5.0 V? Take $\alpha = 0.5$.

21F.5(b) The exchange current density for a Pt|Fe^{3+},Fe^{2+} electrode is 2.5 mA cm^{-2}. The standard potential of the electrode is +0.77 V. Calculate the current flowing through an electrode of surface area 1.0 cm^2 as a function of the potential of the electrode. Take unit activity for both ions.

21F.6(a) How many electrons or protons are transported through the double layer in each second when the Pt,H_2|H^+, Pt|Fe^{3+},Fe^{2+}, and Pb,H_2|H^+ electrodes are at equilibrium at 25 °C? Take the area as 1.0 cm^2 in each case. Estimate the number of times each second a single atom on the surface takes part in an electron transfer event, assuming an electrode atom occupies about (280 pm)^2 of the surface.

21F.6(b) How many electrons or protons are transported through the double layer in each second when the Cu,H_2|H^+ and Pt|Ce^{4+},Ce^{3+} electrodes are at equilibrium at 25 °C? Take the area as 1.0 cm^2 in each case. Estimate the number of times each second a single atom on the surface takes part in an electron transfer event, assuming an electrode atom occupies about (260 pm)^2 of the surface.

21F.7(a) What is the effective resistance at 25 °C of an electrode interface when the overpotential is small? Evaluate it for 1.0 cm^2 (i) Pt,H_2|H^+, (ii) Hg,H_2|H^+ electrodes.

21F.7(b) Evaluate the effective resistance at 25 °C of an electrode interface for 1.0 cm^2 (i) Pb,H_2|H^+, (ii) Pt|Fe^{2+},Fe^{3+} electrodes.

21F.8(a) The exchange current density for H^+ discharge at zinc is about 50 pA cm^{-2}. Can zinc be deposited from a unit activity aqueous solution of a zinc salt?

21F.8(b) The standard potential of the Zn^{2+}|Zn electrode is −0.76 V at 25 °C. The exchange current density for H^+ discharge at platinum is 0.79 mA cm^{-2}. Can zinc be plated on to platinum at that temperature? (Take unit activities.)

Problems

21F.1 In an experiment on the Pt|H_2|H^+ electrode in dilute H_2SO_4 the following current densities were observed at 25 °C. Evaluate α and j_0 for the electrode.

η/mV	50	100	150	200	250
j/(mA cm^{-2})	2.66	8.91	29.9	100	335

How would the current density at this electrode depend on the overpotential of the same set of magnitudes but of opposite sign?

21F.2 The standard potentials of lead and tin are −126 mV and −136 mV, respectively, at 25 °C, and the overpotential for their deposition are close to zero. What should their relative activities be in order to ensure simultaneous deposition from a mixture?

21F.3‡ The rate of deposition of iron, v, on the surface of an iron electrode from an aqueous solution of Fe^{2+} has been studied as a function of potential, E, relative to the standard hydrogen electrode, by J. Kanya (*J. Electroanal. Chem.* **84**, 83 (1977)). The values in the table below are based on the data obtained with an electrode of surface area 9.1 cm^2 in contact with a solution of concentration 1.70 μmol dm^{-3} in Fe^{2+}. (a) Assuming unit activity coefficients, calculate the zero current potential of the Fe^{2+}/Fe cathode and the overpotential at each value of the working potential. (b) Calculate the cathodic current density, j_c, from the rate of deposition of Fe^{2+} for each value of E. (c) Examine the extent to which the data fit the Tafel equation and calculate the exchange current density.

v/(pmol s^{-1})	1.47	2.18	3.11	7.26
$-E$/mV	702	727	752	812

21F.4‡ V.V. Losev and A.P. Pchel'nikov (*Soviet Electrochem.* **6**, 34 (1970)) obtained the following current–voltage data for an indium anode relative to a standard hydrogen electrode at 293 K:

$-E$/V	0.388	0.365	0.350	0.335
j/(A m^{-2})	0	0.590	1.438	3.507

Use these data to calculate the transfer coefficient and the exchange current density. What is the cathodic current density when the potential is 0.365 V?

21F.5‡ An early study of the hydrogen overpotential is that of H. Bowden and T. Rideal (*Proc. Roy. Soc.* **A120**, 59 (1928)), who measured the overpotential for H_2 evolution with a mercury electrode in dilute aqueous solutions of H_2SO_4 at 25 °C. Determine the exchange current density and transfer coefficient, α, from their data:

j/(mA m^{-2})	2.9	6.3	28	100	250	630	1650	3300
η/V	0.60	0.65	0.73	0.79	0.84	0.89	0.93	0.96

Explain any deviations from the result expected from the Tafel equation.

21F.6 If $\alpha=\frac{1}{2}$, an electrode interface is unable to rectify alternating current because the current density curve is symmetrical about $\eta=0$. When $\alpha\neq\frac{1}{2}$, the magnitude of the current density depends on the sign of the overpotential, and so some degree of 'faradaic rectification' may be obtained. Suppose that the overpotential varies as $\eta=\eta_0\cos\omega t$. Derive an expression for the mean flow of current (averaged over a cycle) for general α, and confirm that the mean current is zero when $\alpha=\frac{1}{2}$. In each case work in the limit of small η_0 but to second order in η_0F/RT. Calculate the mean direct current at 25 °C for a 1.0 cm^2 hydrogen–platinum electrode with $\alpha=0.38$ when the overpotential varies between ±10 mV at 50 Hz.

21F.7 Now suppose that the overpotential is in the high overpotential region at all times even though it is oscillating. What waveform will the current across the interface show if it varies linearly and periodically (as a sawtooth waveform) between η_- and η_+ around η_0? Take $\alpha=\frac{1}{2}$.

21F.8 Figure 21.1 shows four different examples of voltammograms. Identify the processes occurring in each system. In each case the vertical axis is the current and the horizontal axis is the (negative) electrode potential.

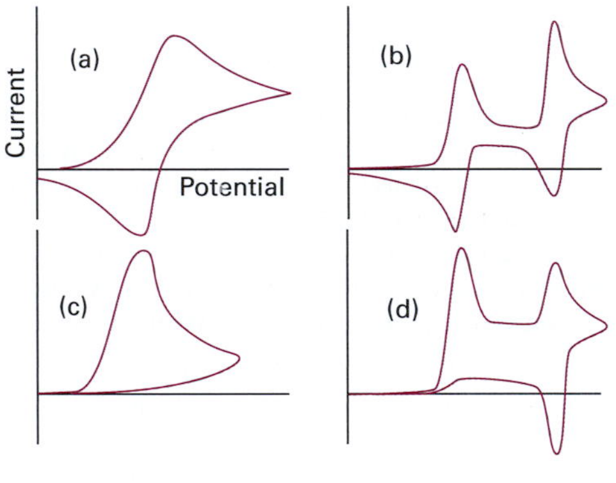

Figure 21.1 The voltammograms used in Problem 21F.8.

Integrated activities

21.1 Estimate the orders of magnitude of the partition functions involved in a rate expression. State the order of magnitude of $q^{\mathrm{T}}_{\mathrm{m}}/N_{\mathrm{A}}$, q^{R}, q^{V}, q^{E} for typical molecules. Check that in the collision of two structureless molecules the order of magnitude of the pre-exponential factor is of the same order as that predicted by collision theory. Go on to estimate the P-factor for a reaction in which A + B→P, and A and B are nonlinear triatomic molecules.

21.2 Discuss the factors that govern the rates of photo-induced electron transfer according to Marcus theory and that govern the rates of resonance energy transfer according to Förster theory (Topic 20G). Can you find similarities between the two theories?

21.3 Calculate the thermodynamic limit to the zero-current potential of fuel cells operating on (a) hydrogen and oxygen, (b) methane and air, and (c) propane and air. Use the Gibbs energy information in the *Resource section*, and take the species to be in their standard states at 25 °C.

CHAPTER 22

Processes on solid surfaces

Processes at solid surfaces govern the viability of industry constructively, as in catalysis, and the permanence of its products destructively, as in corrosion. Chemical reactions at solid surfaces may differ sharply from reactions in the bulk, for reaction pathways of much lower activation energy may be provided by the surface, and hence result in catalysis. This chapter extends the material introduced in Chapters 20 and 21 by showing how to deal with processes on solid surfaces.

22A An introduction to solid surfaces

We begin by exploring the structure of solid surfaces. This Topic also describes a number of experimental techniques commonly used in surface science.

22B Adsorption and desorption

Although we began with a discussion of clean surfaces, for chemists the important aspects of a surface are the attachment of substances to it and the reactions that take place there. In this Topic we discuss the extent to which a solid surface is covered and the variation of the extent of coverage with pressure and temperature.

22C Heterogeneous catalysis

This Topic discusses chemical reactions on solid surfaces. We focus on how surfaces affect the rate and course of chemical change by acting as the site of catalysis.

What is the impact of this material?

Almost the whole of modern chemical industry depends on the development, selection, and application of catalysts, with heterogeneous catalysts being particularly important. All we can hope to do in *Impact* I22.1 is to give a brief indication of some of the problems involved. Other than the ones we consider, these problems include the danger of the catalyst being poisoned by by-products or impurities, and economic considerations relating to cost and lifetime.

To read more about the impact of this material, scan the QR code, or go to bcs.whfreeman.com/webpub/chemistry/pchem10e/impact/pchem-22-1.html

22A An introduction to solid surfaces

Contents

➤ Why do you need to know this material?

To understand the thermodynamics and kinetics of chemical reactions occurring on solid surfaces, which underlie much of catalysis and therefore the chemical industry, you need to understand surface structure, composition, and growth.

➤ What is the key idea?

Structural features, including defects, play important roles in physical and chemical processes occurring on solid surfaces.

➤ What do you need to know already?

You need to be aware of the structure of solids (Topic 18A), but not in detail. This Topic draws on results from the kinetic theory of gases (Topic 1B).

A great deal of chemistry occurs at solid surfaces. Heterogeneous catalysis (Topic 22C) is just one example, with the surface providing reactive sites where reactants can attach, be torn apart, and react with other reactants. Even as simple an act as dissolving is intrinsically a surface phenomenon, with the solid gradually escaping into the solvent from sites on the surface. Surface deposition, in which atoms are laid down on a surface to create layers, is crucial to the semiconductor industry, as it is the way in which integrated circuits are created. Electrodes are essentially surfaces at which electron transfer occurs, and their efficiency depends crucially on an understanding of the events there (Topic 21F).

22A.1 Surface growth

Adsorption is the attachment of particles to a solid surface; **desorption** is the reverse process. The substance that adsorbs is the **adsorbate** and the material to which it adsorbs is the **adsorbent** or **substrate**.

A simple picture of a perfect crystal surface is as a tray of oranges in a grocery store (Fig. 22A.1). A gas molecule that collides with the surface can be imagined as a ping-pong ball bouncing erratically over the oranges. The molecule loses energy as it bounces, but it is likely to escape from the surface before it has lost enough kinetic energy to be trapped. The same is true, to some extent, of an ionic crystal in contact with a solution. There is little energy advantage for an ion in solution to discard some of its solvating molecules and stick at an exposed position on the surface.

The picture changes when the surface has defects, for then there are ridges of incomplete layers of atoms or ions. A common type of surface defect is a **step** between two otherwise flat layers of atoms called **terraces** (Fig. 22A.2). A step defect might itself have defects, for it might have kinks. When an atom settles on a terrace it bounces across it under the influence of the intermolecular potential, and might come to a step or a corner formed by a kink. Instead of interacting with a single terrace

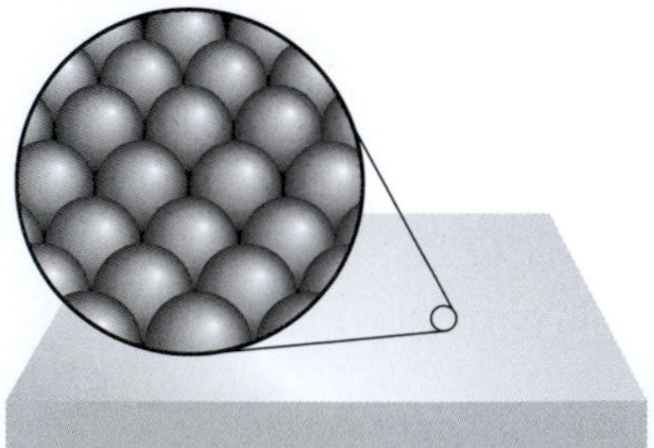

Figure 22A.1 A schematic diagram of the flat surface of a solid. This primitive model is largely supported by scanning tunnelling microscope images.

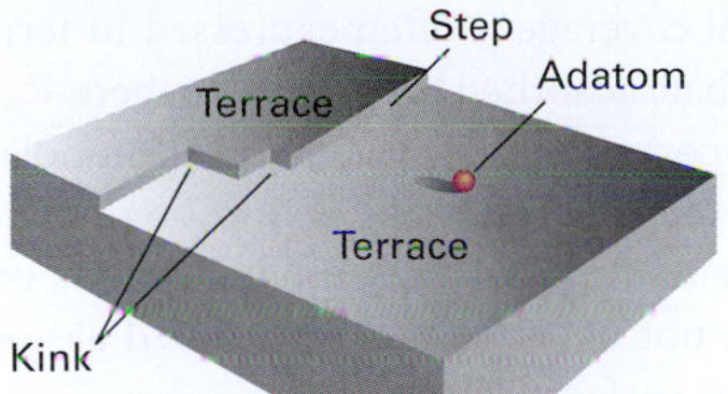

Figure 22A.2 Some of the kinds of defects that may occur on otherwise perfect terraces. Defects play an important role in surface growth and catalysis.

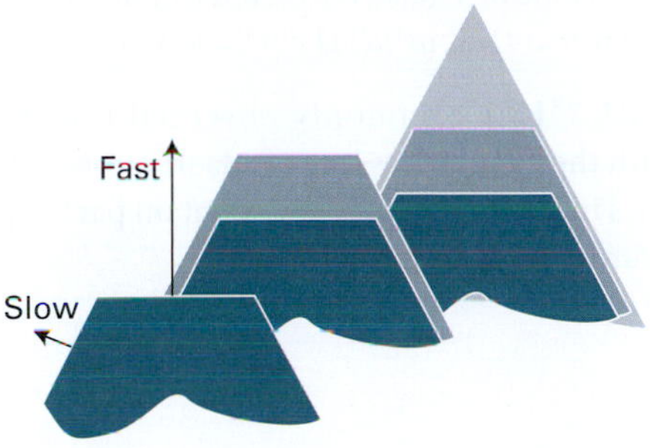

Figure 22A.3 The slower-growing faces of a crystal dominate its final external appearance. Three successive stages of the growth are shown.

atom, the molecule now interacts with several, and the interaction may be strong enough to trap it. Likewise, when ions deposit from solution, the loss of the solvation interaction is offset by a strong Coulombic interaction between the arriving ions and several ions at the surface defect.

The rapidity of growth depends on the crystal plane concerned, and the slowest growing faces dominate the appearance of the crystal. This feature is explained in Fig. 22A.3, where we see that, although the horizontal face grows forward most rapidly, it grows itself out of existence, and the slower-growing faces survive.

Under normal conditions, a surface exposed to a gas is constantly bombarded with molecules and a freshly prepared surface is covered very quickly. Just how quickly can be estimated using the kinetic model of gases and the following expression for the collision flux (eqn 19A.6):

$$Z_W = \frac{p}{(2\pi mkT)^{1/2}} \qquad \text{Collision flux} \qquad (22A.1)$$

Brief illustration 22A.1 The collision flux

If we write $m = M/N_A$, where M is the molar mass of the gas, eqn 22A.1 becomes

$$Z_W = \frac{\overbrace{(N_A/2\pi k)^{1/2}}^{Z_0} p}{(TM)^{1/2}}$$

After inserting numerical values for the constants and selecting units for the variables, the practical form of this expression is:

$$Z_W = \frac{Z_0(p/\text{Pa})}{\{(T/\text{K})(M/(\text{g mol}^{-1}))\}^{1/2}} \quad \text{with } Z_0 = 2.63\times10^{24}\,\text{m}^{-2}\,\text{s}^{-1}$$

For air, with $M \approx 29\,\text{g mol}^{-1}$, at $p = 1\,\text{atm} = 1.013\,25\times10^5\,\text{Pa}$ and $T = 298\,\text{K}$, we obtain $Z_W = 2.9\times10^{27}\,\text{m}^{-2}\,\text{s}^{-1}$. Because $1\,\text{m}^2$ of metal surface consists of about 10^{19} atoms, each atom is struck about 10^8 times each second. Even if only a few collisions leave a molecule adsorbed to the surface, the time for which a freshly prepared surface remains clean is very short.

Self-test 22A.1 Calculate the collision flux with a surface of a vessel containing propane at 25 °C when the pressure is 100 Pa.

Answer: $Z_W = 2.30\times10^{20}\,\text{cm}^{-2}\,\text{s}^{-1}$

22A.2 Physisorption and chemisorption

Molecules and atoms can attach to surfaces in two ways. In **physisorption** (a contraction of 'physical adsorption'), there is a van der Waals interaction (for example, a dispersion or a dipolar interaction, Topic 16B) between the adsorbate and the substrate; van der Waals interactions have a long range but are weak, and the energy released when a particle is physisorbed is of the same order of magnitude as the enthalpy of condensation. Such small energies can be absorbed as vibrations of the lattice and dissipated as thermal motion, and a molecule bouncing across the surface will gradually lose its energy and finally adsorb to it in the process called **accommodation**.

The enthalpy of physisorption can be measured by monitoring the rise in temperature of a sample of known heat capacity, and typical values are in the region of $-20\,\text{kJ mol}^{-1}$ (Table 22A.1). This small enthalpy change is insufficient to lead to bond breaking, so a physisorbed molecule retains its identity, although it might be distorted by the presence of the surface.

In **chemisorption** (a contraction of 'chemical adsorption'), the molecules (or atoms) stick to the surface by forming a chemical (usually covalent) bond, and tend to find sites that maximize their coordination number with the substrate. The

Table 22A.1* Maximum observed standard enthalpies of physisorption, $\Delta_{ad}H^{\ominus}/(\text{kJ mol}^{-1})$, at 298 K

Adsorbate	$\Delta_{ad}H^{\ominus}/(\text{kJ mol}^{-1})$
CH_4	−21
H_2	−84
H_2O	−59
N_2	−21

* More values are given in the *Resource section*.

Table 22A.2* Standard enthalpies of chemisorption, $\Delta_{ad}H^{\ominus}$/(kJ mol^{-1}), at 298 K

Adsorbate	Adsorbent (substrate)		
	Cr	Fe	Ni
C_2H_4	−427	−285	−243
CO		−192	
H_2	−188	−134	
NH_3		−188	−155

* More values are given in the *Resource section*.

enthalpy of chemisorption is very much greater than that for physisorption, and typical values are in the region of −200 kJ mol^{-1} (Table 22A.2). The distance between the surface and the closest adsorbate atom is also typically shorter for chemisorption than for physisorption. A chemisorbed molecule may be torn apart at the demand of the unsatisfied valencies of the surface atoms, and the existence of molecular fragments on the surface as a result of chemisorption is one reason why solid surfaces catalyse reactions (Topic 22C).

Except in special cases, chemisorption must be exothermic. A spontaneous process requires $\Delta G<0$ at constant pressure and temperature. Because the translational freedom of the adsorbate is reduced when it is adsorbed, ΔS is negative. Therefore, in order for $\Delta G=\Delta H-T\Delta S$ to be negative, ΔH must be negative (that is, the process must be exothermic). Exceptions may occur if the adsorbate dissociates and has high translational mobility on the surface. For example, H_2 adsorbs endothermically on glass because there is a large increase of translational entropy accompanying the dissociation of the molecules into atoms that move quite freely over the surface. In this case, the entropy change in the process $H_2(g)\rightarrow 2\,H(glass)$ is sufficiently positive to overcome the small positive enthalpy change.

The enthalpy of adsorption depends on the extent of surface coverage, mainly because the adsorbate particles interact with each other. If the particles repel each other (as for CO on palladium) the adsorption becomes less exothermic (the enthalpy of adsorption less negative) as coverage increases. Moreover, studies show that such species settle on the surface in a disordered way until packing requirements demand order. If the adsorbate particles attract one another (as for O_2 on tungsten), then they tend to cluster together in islands, and growth occurs at the borders. These adsorbates also show order–disorder transitions when they are heated enough for thermal motion to overcome the particle–particle interactions, but not so much that they are desorbed.

Whether a result of physisorption or chemisorption, the extent of surface coverage is normally expressed as the **fractional coverage**, θ:

$$\theta=\frac{\text{number of adsorption sites occupied}}{\text{number of adsorption sites available}}$$

Definition Fractional coverage (22A.2)

The fractional coverage is often expressed in terms of the volume of adsorbate adsorbed by $\theta=V/V_\infty$, where V_∞ is the volume of adsorbate corresponding to complete monolayer coverage. In each case, the volumes in the definition of θ are those of the free gas measured under the same conditions of temperature and pressure, not the volume the adsorbed gas occupies when attached to the surface.

Brief illustration 22A.2 Fractional coverage

For the adsorption of CO on charcoal at 273 K, $V_\infty=111\,\text{cm}^3$, a value corrected to 1 atm. When the partial pressure of CO is 80.0 kPa, the value of V (also corrected to 1 atm) is 41.6 cm^3, so it follows that $\theta=(41.6\,\text{cm}^3)/(111\,\text{cm}^3)=0.375$.

Self-test 22A.2 It is commonly observed that θ increases sharply with the partial pressure of adsorbate at low pressures, but becomes increasingly less dependent on partial pressure at high pressures. Explain this behaviour.

Answer: See Topic 22B

22A.3 Experimental techniques

A vast array of experimental techniques are used to study the composition and structure of solid surfaces at the atomic level. Many of the arrangements allow for direct visualization of changes in the surface as adsorption and chemical reactions take place there.

Experimental procedures must begin with a clean surface. The obvious way to retain cleanliness of a surface is to reduce the pressure and reduce the number of impacts on the surface. When the pressure is reduced to 0.1 mPa (as in a simple vacuum system) the collision flux falls to about $10^{18}\,\text{m}^{-2}\,\text{s}^{-1}$, corresponding to one hit per surface atom in each 0.1 s. Even that is too frequent in most experiments, and in **ultrahigh vacuum** (UHV) techniques pressures as low as 0.1 μPa (when $Z_W=10^{15}\,\text{m}^{-2}\,\text{s}^{-1}$) are reached on a routine basis and as low as 1 nPa (when $Z_W=10^{13}\,\text{m}^{-2}\,\text{s}^{-1}$) are reached with special care. These collision fluxes correspond to each surface atom being hit once every 10^5 to 10^6 s, or about once a day.

(a) Microscopy

The basic approach of illuminating a small area of a sample and collecting light with a microscope has been used for many years to image small specimens. However, the resolution of a microscope, the minimum distance between two objects that leads to two distinct images, is on the order of the wavelength of the light being used. Therefore, conventional microscopes employing visible light have resolutions in the micrometre range and are blind to features on a scale of nanometres.

One technique that is often used to image nanometre-sized objects is **electron microscopy**, in which a beam of electrons with a well-defined de Broglie wavelength (Topic 7A) replaces the lamp found in traditional light microscopes. Instead of glass or quartz lenses, magnetic fields are used to focus the beam. In **transmission electron microscopy** (TEM), the electron beam passes through the specimen and the image is collected on a screen. In **scanning electron microscopy** (SEM), electrons scattered back from a small irradiated area of the sample are detected and the electrical signal is sent to a video screen. An image of the surface is then obtained by scanning the electron beam across the sample.

As in traditional light microscopy, the wavelength of and the ability to focus the incident beam—in this case a beam of electrons focused by magnetic fields—govern the resolution. It is now possible to achieve atomic resolution with TEM instruments. Resolution on the order of a few nanometres is possible with SEM instruments.

Scanning probe microscopy (SPM) is a collection of techniques that can be used to make visible and manipulate objects as small as atoms on surfaces. One version is **scanning tunnelling microscopy** (STM), in which a platinum–rhodium or tungsten needle is scanned across the surface of a conducting solid. When the tip of the needle is brought very close to the surface, electrons tunnel across the intervening space (Fig. 22A.4). In the 'constant-current mode' of operation, the stylus moves up and down according to the form of the surface, and the topography of the surface, including any adsorbates, can be mapped on an atomic scale. The vertical motion of the stylus is achieved by fixing it to a piezoelectric cylinder, which contracts or expands according to the potential difference it experiences. In the 'constant-*z* mode', the vertical position of the stylus is held constant and the current is monitored. Because the tunnelling probability is very sensitive to the size of the gap, the microscope can detect tiny, atom-scale variations in the height of the surface.

Figure 22A.5 shows an example of the kind of image obtained with a surface, in this case of gallium arsenide that has been modified by addition of caesium atoms. Each 'bump' on the surface corresponds to an atom. In a further variation of the STM technique, the tip may be used to nudge single atoms around on the surface, making possible the fabrication of complex and yet very tiny nanometre-sized materials and devices.

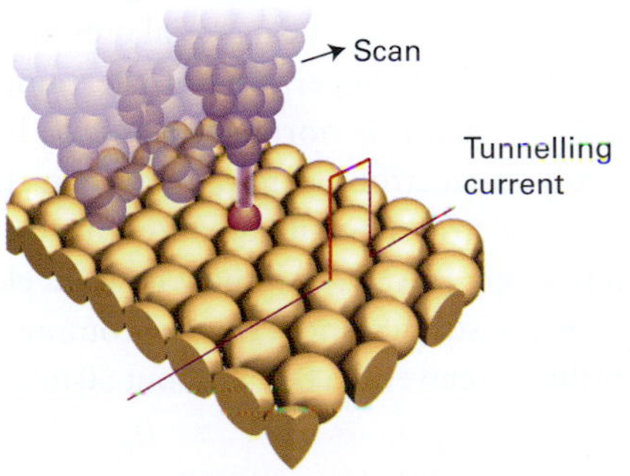

Figure 22A.4 A scanning tunnelling microscope makes use of the current of electrons that tunnel between the surface and the tip. That current is very sensitive to the distance of the tip above the surface.

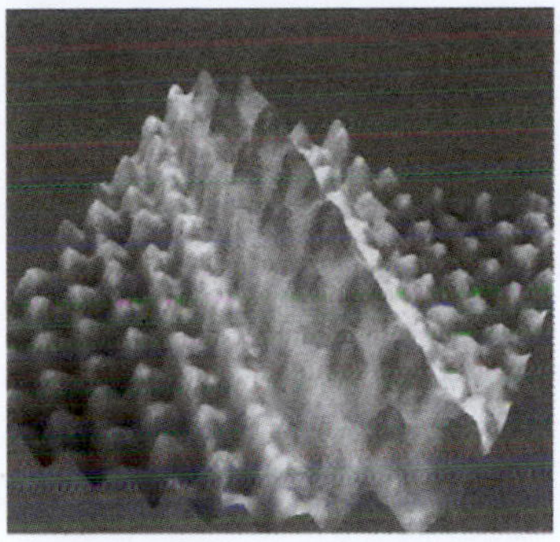

Figure 22A.5 An STM image of caesium atoms on a gallium arsenide surface.

Diffusion characteristics of an adsorbate can be examined by using STM to follow the change in surface characteristics. An adsorbed atom makes a random walk across the surface, and the diffusion coefficient, D, can be inferred from the mean distance, d, travelled in an interval τ by using the two-dimensional random walk expression $d=(D\tau)^{1/2}$. The value of D for different crystal planes at different temperatures can be determined directly in this way, and the activation energy for migration over each plane obtained from the Arrhenius-like expression

$$D=D_0\mathrm{e}^{-E_{\mathrm{a,diff}}/RT} \qquad \text{Temperature dependence of the diffusion coefficient} \qquad (22A.3)$$

where $E_{\mathrm{a,diff}}$ is the activation energy for diffusion and D_0 is the diffusion coefficient in the limit of infinite temperature.

Brief illustration 22A.3 Diffusion coefficients

Typical values for W atoms on tungsten have $E_{\mathrm{a,diff}}$ in the range 57–87 kJ mol^{-1} and $D_0\approx3.8\times10^{-11}\,\mathrm{m^2\,s^{-1}}$. It follows from eqn 22A.3 that at 800 K the diffusion coefficient varies approximately from

$$D=(3.8\times10^{-11}\,\mathrm{m^2\,s^{-1}})\times\mathrm{e}^{-5.7\times10^{-2}\,\mathrm{J\,mol^{-1}}/(8.3145\,\mathrm{J\,K^{-1}\,mol^{-1}}\times800\,\mathrm{K})}$$
$$=7.2\times10^{-15}\,\mathrm{m^2\,s^{-1}}$$

to

$$D=(3.8\times10^{-11}\,\mathrm{m^2\,s^{-1}})\times\mathrm{e}^{-8.7\times10^{-2}\,\mathrm{J\,mol^{-1}}/(8.3145\,\mathrm{J\,K^{-1}\,mol^{-1}}\times800\,\mathrm{K})}$$
$$=7.9\times10^{-17}\,\mathrm{m^2\,s^{-1}}$$

Self-test 22A.3 For CO on tungsten, the activation energy falls from 144 kJ mol^{-1} at low surface coverage to 88 kJ mol^{-1} when the coverage is high. Calculate the ratio $D_{\mathrm{high}}/D_{\mathrm{low}}$ of diffusion coefficients at 800 K.

Answer: 4.5×10^3

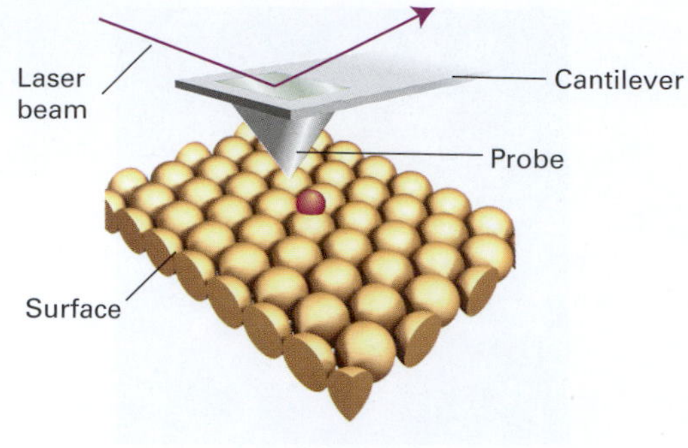

Figure 22A.6 In atomic force microscopy, a laser beam is used to monitor the tiny changes in position of a probe as it is attracted to or repelled by atoms on a surface.

In **atomic force microscopy** (AFM), a sharpened tip attached to a cantilever is scanned across the surface. The force exerted by the surface and any molecules attached to it pushes or pulls on the tip and deflects the cantilever (Fig. 22A.6). The deflection is monitored by using a laser beam. Because no current needs to pass between the sample and the probe, the technique can be applied to non-conducting surfaces and to liquid samples.

Two modes of operation of AFM are common. In 'contact mode', or 'constant-force mode', the force between the tip and surface is held constant and the tip makes contact with the surface. This mode of operation can damage fragile samples on the surface. In 'non-contact', or 'tapping mode', the tip bounces up and down with a specified frequency and never quite touches the surface. The amplitude of the tip's oscillation changes when it passes over a species adsorbed on the surface.

(b) Ionization techniques

The chemical composition of a surface can be determined by a variety of ionization techniques. The same techniques can be used to detect any remaining contamination after cleaning and to detect layers of material adsorbed later in the experiment.

One technique is **photoemission spectroscopy**, a derivative of the photoelectric effect (Topic 7A), in which X-rays (for XPS) or hard (short wavelength) ultraviolet (for UPS) ionizing radiation is used, giving rise to ejected electrons from adsorbed species. The kinetic energies of the electrons ejected from their orbitals are measured and the pattern of energies is a fingerprint of the material present (Fig. 22A.7). UPS, which examines electrons ejected from valence shells, is also used to establish the bonding characteristics and the details of valence shell electronic structures of substances on the surface. Its usefulness is its ability to reveal which orbitals of the adsorbate are involved in the bond to the substrate.

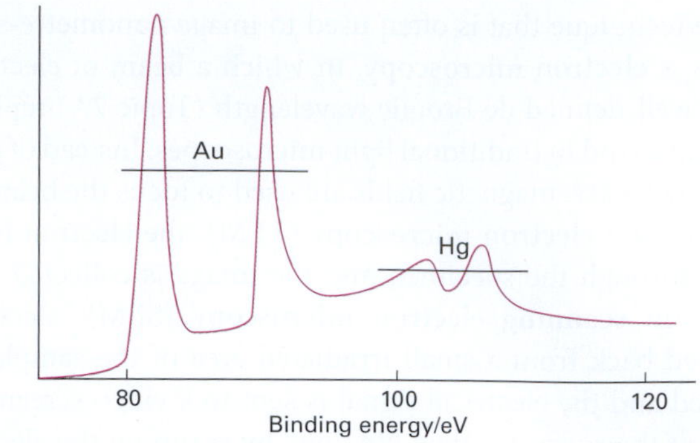

Figure 22A.7 The X-ray photoelectron emission spectrum of a sample of gold contaminated with a surface layer of mercury. (M.W. Roberts and C.S. McKee, *Chemistry of the metal–gas interface*, Oxford (1978).)

Brief illustration 22A.4 A UPS spectrum

The principal difference between the photoemission results on free benzene and benzene adsorbed on palladium is in the energies of the π electrons. This difference is interpreted as meaning that the C_6H_6 molecules lie parallel to the surface and are attached to it by their π orbitals.

Self-test 22A.4 When adsorbed to palladium, pyridine (C_5H_5N) stands almost perpendicular to the surface. Suggest a mode of attachment of the molecule to palladium atoms on the surface.

Answer: Data are consistent with a σ bond formed by the nitrogen lone pair.

A very important technique, which is widely used in the microelectronics industry, is **Auger electron spectroscopy** (AES). The **Auger effect** (pronounced oh-zhey) is the emission of a second electron after high energy radiation has expelled another. The first electron to depart leaves a hole in a low-lying orbital, and an upper electron falls into it. The energy this transition releases may result either in the generation of radiation, which is called **X-ray fluorescence** (Fig. 22A.8a) or in the ejection of another electron (Fig. 22A.8b). The latter is the 'secondary electron' of the Auger effect. The energies of the secondary electrons are characteristic of the material present, so the Auger effect effectively takes a fingerprint of the sample. In practice, the Auger spectrum is normally obtained by irradiating the sample with an electron beam of energy in the range 1–5 keV rather than electromagnetic radiation. In **scanning Auger electron microscopy** (SAM), the finely focused electron beam is scanned over the surface and a map of composition is compiled; the resolution can reach below about 50 nm.

(c) Diffraction techniques

A useful technique for determining the arrangement of the atoms close to the surface is **low energy electron diffraction**

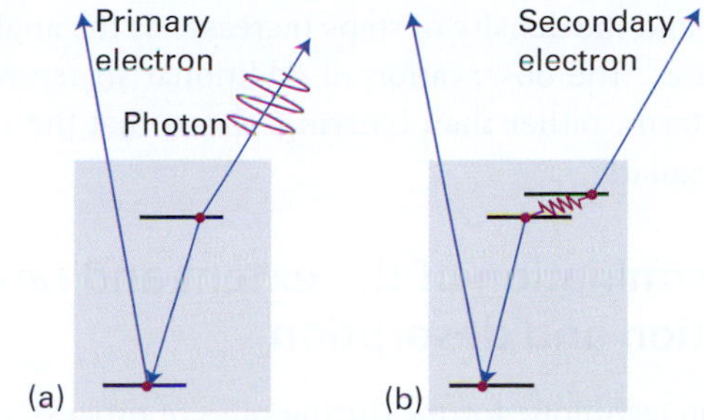

Figure 22A.8 When an electron is expelled from a solid (a) an electron of higher energy may fall into the vacated orbital and emit an X-ray photon to produce X-ray fluorescence. Alternatively (b) the electron falling into the orbital may give up its energy to another electron, which is ejected in the Auger effect.

(LEED). This technique is like X-ray diffraction (Topic 18A) but uses the wave character of electrons, and the sample is now the surface of a solid. The use of low energy electrons (with energies in the range 10–200 eV, corresponding to wavelengths in the range 100–400 pm) ensures that the diffraction is caused only by atoms on and close to the surface. The experimental arrangement is shown in Fig. 22A.9, and typical LEED patterns, obtained by photographing the fluorescent screen through the viewing port, are shown in Fig. 22A.10.

Observations using LEED show that the surface of a crystal rarely has exactly the same form as a slice through the bulk because surface and bulk atoms experience different forces. **Reconstruction** refers to processes by which atoms on the surface achieve their equilibrium structures. As a general rule, it is found that metal surfaces are simply truncations of the bulk lattice, but the distance between the top layer of atoms and the one below is contracted by around 5 per cent. Semiconductors generally have surfaces reconstructed to a depth of several layers. Reconstruction occurs in ionic solids. For example, in

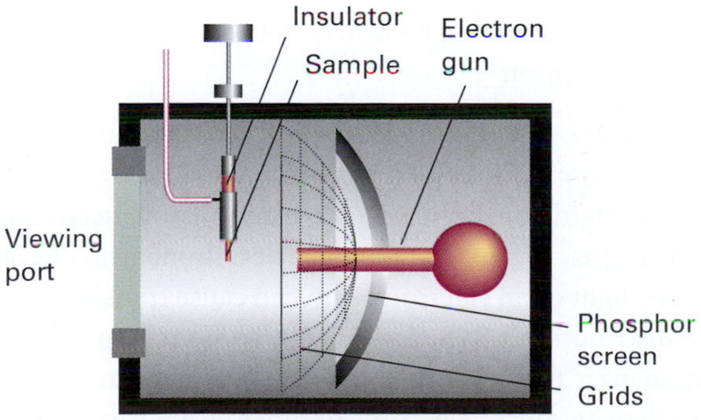

Figure 22A.9 A schematic diagram of the apparatus used for a LEED experiment. The electrons diffracted by the surface layers are detected by the fluorescence they cause on the phosphor screen.

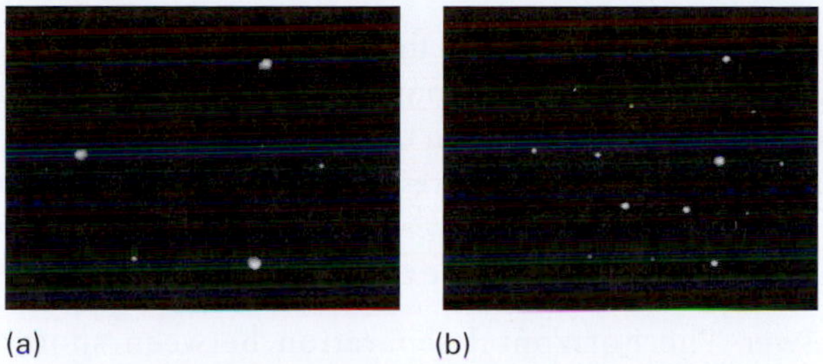

Figure 22A.10 LEED photographs of (a) a clean platinum surface and (b) after its exposure to propyne, $CH_3C{\equiv}CH$. (Photographs provided by Professor G.A. Somorjai.)

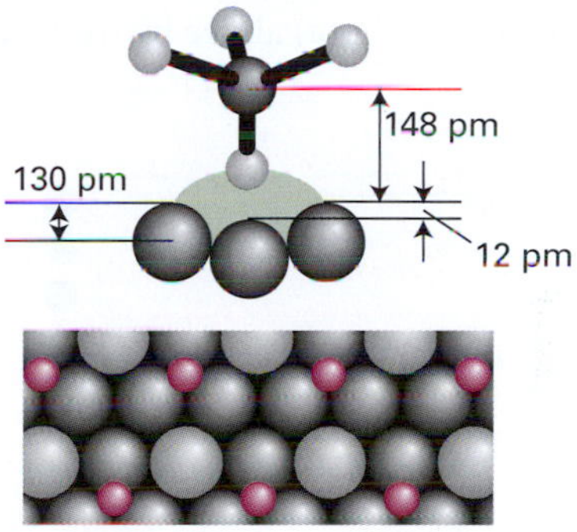

Figure 22A.11 The structure of a surface close to the point of attachment of CH_3C– to the (110) surface of rhodium at 300 K and the changes in positions of the metal atoms that accompany chemisorption.

lithium fluoride the Li^+ and F^- ions close to the surface apparently lie on slightly different planes. An actual example of the detail that can now be obtained from refined LEED techniques is shown in Fig. 22A.11 for CH_3C– adsorbed on a (111) plane of rhodium.

Example 22A.1 Interpreting a LEED pattern

The LEED pattern from a clean (110) face of palladium is shown in (a) below. The reconstructed surface gives a LEED pattern shown as (b). What can be inferred about the structure of the reconstructed surface?

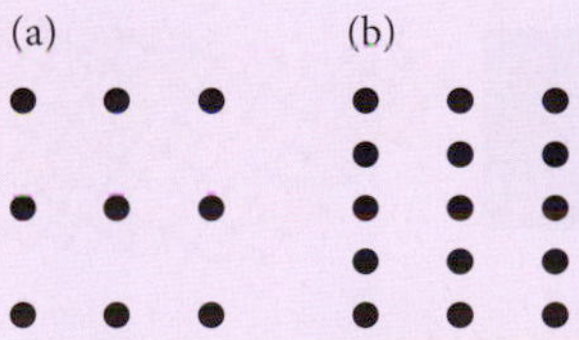

Method Recall from Bragg's law (Topic 18A), $\lambda = 2d \sin\theta$, that for a given wavelength, the greater the separation d of the layers, the smaller is the scattering angle (so that $2d \sin\theta$

remains constant). It follows that, in terms of the LEED pattern, the farther apart the atoms responsible for the pattern, the closer the spots appear in the pattern. Twice the separation between the atoms corresponds to half the separation between the spots, and vice versa. Therefore, inspect the two patterns and identify how the new pattern relates to the old.

Answer The horizontal separation between spots is unchanged, which indicates that the atoms remain in the same position in that dimension when reconstruction occurs. However, the vertical spacing is halved, which suggests that the atoms are twice as far apart in that direction as they are in the unreconstructed surface.

Self-test 22A.5 Sketch the LEED pattern for a surface that differs from that shown in (a) above by tripling the vertical separation.

Answer:

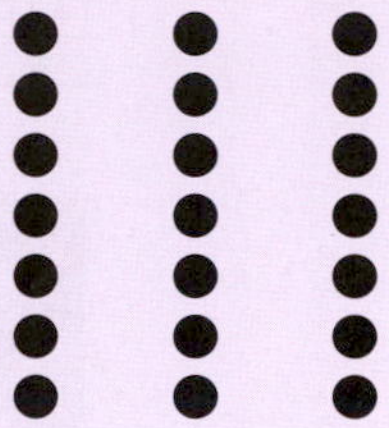

The presence of terraces, steps, and kinks in a surface shows up in LEED patterns, and their surface density (the number of defects in a region divided by the area of the region) can be estimated. The importance of this type of measurement will emerge later. Three examples of how steps and kinks affect the pattern are shown in Fig. 22A.12. The samples used were obtained by cleaving a crystal at different angles to a plane of atoms. Only terraces are produced when the cut is parallel to the plane, and the density of steps increases as the angle of the cut increases. The observation of additional structure in the LEED patterns, rather than blurring, shows that the steps are arrayed regularly.

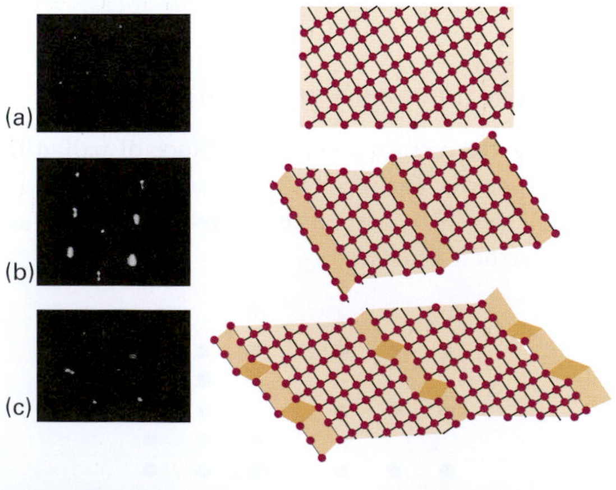

Figure 22A.12 LEED patterns may be used to assess the defect density of a surface. The photographs correspond to a platinum surface with (a) low defect density, (b) regular steps separated by about six atoms, and (c) regular steps with kinks. (Photographs provided by Professor G.A. Samorjai.)

(d) Determination of the extent and rates of adsorption and desorption

A common technique for measuring rates of processes on surfaces is to monitor the rates of flow of gas into and out of the system: the difference is the rate of gas uptake by the sample. Integration of this rate then gives the fractional coverage at any stage.

- **Gravimetry**, in which the sample is weighed on a microbalance during the experiment, is used to determine the extent and kinetics of adsorption and desorption.

The technique commonly uses a **quartz crystal microbalance** (QCM), in which the mass of a sample adsorbed on the surface of a quartz crystal is related to changes in the characteristic vibrational frequency of the crystal. In this way, masses as small as a few nanograms can be measured reliably.

- **Second harmonic generation** (SHG), the conversion of an intense, pulsed laser beam to radiation with twice its initial frequency is very important for the study of all types of surfaces, including thin films and liquid–gas interfaces.

For example, adsorption of gas molecules on to a surface alters the intensity of the SHG signal, allowing for determination of the rates of surface processes and the fractional coverage. Because pulsed lasers are the excitation sources, time-resolved measurements of the kinetics and dynamics of surface processes are possible on timescales as short as femtoseconds.

- **Surface plasmon resonance** (SPR), the absorption of energy from an incident beam of electromagnetic radiation by surface 'plasmons', is a very sensitive technique now used routinely in the study of adsorption and desorption.

To understand the technique we need to examine the terms 'surface plasmon' and 'resonance' in its name.

The mobile delocalized valence electrons of metals form a **plasma**, a dense gas of charged particles. Bombardment of this plasma by light or an electron beam can cause transient changes in the distribution of electrons, with some regions becoming slightly denser than others. Coulomb repulsion in the regions of high density causes electrons to move away from each other, so lowering their density. The resulting oscillations in electron density, the **plasmons**, can be excited both in the bulk and on the surface of a metal. A surface plasmon propagates away from the surface, but the amplitude of the wave, also called an

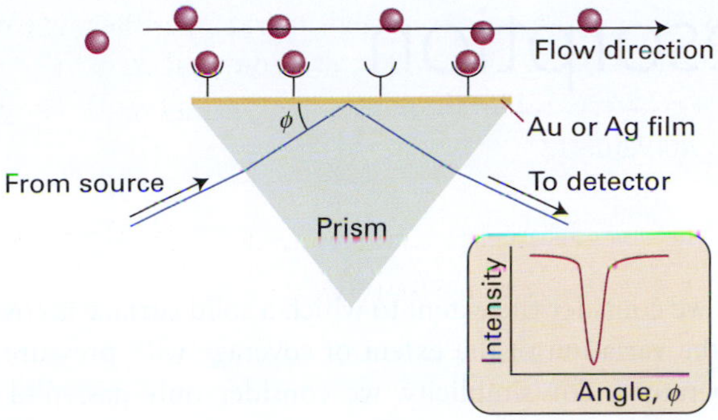

Figure 22A.13 The experimental arrangement for the observation of surface plasmon resonance, as explained in the text.

evanescent wave, decreases sharply with distance from the surface. The 'resonance' in the name refers to the absorption that can be observed with appropriate choice of the wavelength and angle of incidence of the excitation beam.

It is common practice to use a monochromatic beam and to vary the angle of incidence (the ϕ in Fig. 22A.13). The beam passes through a prism that strikes one side of a thin film of gold or silver. Because the evanescent wave interacts with material a short distance away from the surface, the angle at which resonant absorption occurs depends on the refractive index of the medium on the opposite side of the metallic film. Thus, changing the identity and quantity of material on the surface changes the resonance angle.

The SPR technique can be used in the study of the binding of molecules to a surface or binding of ligands to a biopolymer attached to the surface; this interaction mimics the biological recognition processes that occur in cells. Examples of complexes amenable to analysis include antibody–antigen and protein–DNA interactions. The most important advantage of SPR is its sensitivity: it is possible to measure the deposition of nanograms of material on to a surface. The main disadvantage of the technique is its requirement for immobilization of at least one of the components of the system under study.

Checklist of concepts

☐ 1. **Adsorption** is the attachment of molecules to a surface; the substance that adsorbs is the adsorbate and the underlying material is the adsorbent or substrate. The reverse of adsorption is desorption.

☐ 2. Surface defects play an important role in surface growth and catalysis.

☐ 3. **Reconstruction** refers to processes by which atoms on the surface achieve their equilibrium structures.

☐ 4. Techniques for studying surfaces include **scanning electron microscopy** (SEM), **transmission electron microscopy** (TEM), **scanning probe microscopy** (SPM), **photoemission spectroscopy**, **Auger electron spectroscopy** (AES), **low energy electron diffraction** (LEED), **gravimetry**, **second harmonic generation** (SHG), and **surface plasmon resonance** (SPR).

Checklist of equations

Property	Equation	Comment	Equation number
Collision flux	$Z_W = p/(2\pi mkT)^{1/2}$	KMT	22A.1
Fractional coverage	θ = (number of adsorption sites occupied)/(number of adsorption sites available)	Definition	22A.2

22B Adsorption and desorption

Contents

➤ **Why do you need to know this material?**

To understand how surfaces can affect the rates of chemical reactions, you need to know how to assess the extent of surface coverage and the factors that determine the rates at which molecules attach to and detach from solid surfaces.

➤ **What is the key idea?**

The extent of surface coverage can be expressed in terms of isotherms derived on the basis of dynamic equilibria between adsorbed and free material.

➤ **What do you need to know already?**

This Topic extends the discussion of adsorption in Topic 22A. You need to be familiar with the basic ideas of chemical kinetics (Topics 20A–20C), the Arrhenius equation (Topic 20D), and the expression of reaction mechanisms as rate laws (Topic 20E).

Here we consider the extent to which a solid surface is covered and the variation of the extent of coverage with pressure and temperature. For simplicity, we consider only gas/solid systems. We use this material in Topic 22C to discuss how surfaces affect the rate and course of chemical change by acting as the site of catalysis.

22B.1 Adsorption isotherms

In adsorption (Topic 22A) the free gas and the adsorbed gas are in dynamic equilibrium, and the fractional coverage, θ, of the surface (eqn 22A.2) depends on the pressure of the overlying gas. The variation of θ with pressure at a chosen temperature is called the **adsorption isotherm**.

Many of the techniques discussed in Topic 22A can be used to measure θ. Another is **flash desorption**, in which the sample is suddenly heated (electrically) and the resulting rise of pressure is interpreted in terms of the amount of adsorbate originally on the sample.

(a) The Langmuir isotherm

The simplest physically plausible isotherm is based on three assumptions:

- Adsorption cannot proceed beyond monolayer coverage.
- All sites are equivalent.
- The ability of a molecule to adsorb at a given site is independent of the occupation of neighbouring sites (that is, there are no interactions between adsorbed molecules).

The dynamic equilibrium is

$$\mathrm{A(g)+M(surface)} \rightleftharpoons \mathrm{AM(surface)}$$

with rate constants k_a for adsorption and k_d for desorption. The rate of change of the surface coverage, $\mathrm{d}\theta/\mathrm{d}t$, due to adsorption is proportional to the partial pressure p of A and the number of vacant sites $N(1-\theta)$, where N is the total number of sites:

$$\frac{\mathrm{d}\theta}{\mathrm{d}t}=k_a pN(1-\theta) \qquad \text{Rate of adsorption} \qquad (22B.1a)$$

The rate of change of θ due to desorption is proportional to the number of adsorbed species, $N\theta$:

$$\frac{d\theta}{dt}=-k_dN\theta \quad \text{Rate of desorption} \quad (22B.1b)$$

At equilibrium there is no net change (that is, the sum of these two rates is zero), and solving $k_apN(1-\theta)-k_dN\theta=0$ for θ gives the **Langmuir isotherm**:

$$\theta=\frac{\alpha p}{1+\alpha p} \quad \alpha=\frac{k_a}{k_d} \quad \text{Langmuir isotherm} \quad (22B.2)$$

The dimensions of α are 1/pressure.

Example 22B.1 Using the Langmuir isotherm

The data given below are for the adsorption of CO on charcoal at 273 K. Confirm that they fit the Langmuir isotherm, and find the constant α and the volume corresponding to complete coverage. In each case V has been corrected to 1 atm (101.325 kPa).

p/kPa	13.3	26.7	40.0	53.3	66.7	80.0	93.3
V/cm^3	10.2	18.6	25.5	31.5	36.9	41.6	46.1

Method From eqn 22B.2, $\alpha p\theta+\theta=\alpha p$. With $\theta=V/V_\infty$ (eqn 22A.2), where V_∞ is the volume corresponding to complete coverage, this expression can be rearranged into

$$\frac{p}{V}=\frac{p}{V_\infty}+\frac{1}{\alpha V_\infty}$$

Hence, a plot of p/V against p should give a straight line of slope $1/V_\infty$ and intercept $1/\alpha V_\infty$.

Answer The data for the plot are as follows:

p/kPa	13.3	26.7	40.0	53.3	66.7	80.0	93.3
$(p$/kPa$)/(V$/cm$^3)$	1.30	1.44	1.57	1.69	1.81	1.92	2.02

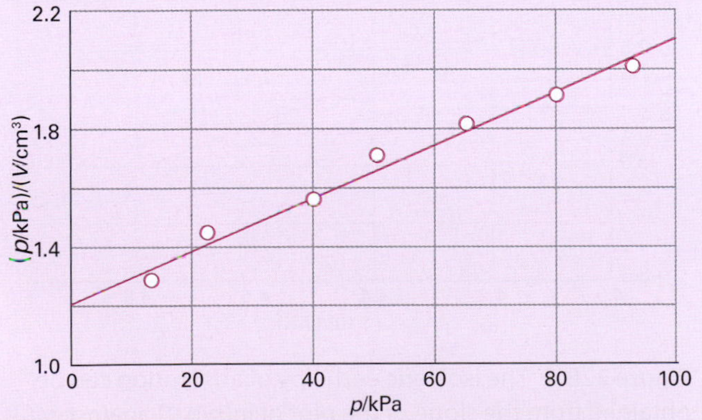

Figure 22B.1 The plot of the data in Example 22B.1. As illustrated here, the Langmuir isotherm predicts that a straight line should be obtained when p/V is plotted against p.

The points are plotted in Fig. 22B.1.The (least squares) slope is 0.009 00, so $V_\infty=111\,\text{cm}^3$. The intercept at $p=0$ is 1.20, so

$$\alpha=\frac{1}{(111\,\text{cm}^3)\times(1.20\,\text{kPa cm}^3)}=7.51\times10^{-3}\,\text{kPa}^{-1}$$

Self-test 22B.1 Repeat the calculation for the following data:

p/kPa	13.3	26.7	40.0	53.3	66.7	80.0	93.3
V/cm^3	10.3	19.3	27.3	34.1	40.0	45.5	48.0

Answer: 128 cm^3, 6.69×10^{-3} kPa^{-1}

For adsorption with dissociation, when A_2 adsorbs as 2 A, the rate of adsorption is proportional to the pressure and to the probability that both fragments A will find sites. The latter is now proportional to the *square* of the number of vacant sites:

$$\frac{d\theta}{dt}=k_a p\{N(1-\theta)\}^2 \quad (22B.3a)$$

The rate of desorption is proportional to the frequency of encounters of the fragments on the surface, and is therefore second order in the number of fragments present:

$$\frac{d\theta}{dt}=-k_d(N\theta)^2 \quad (22B.3b)$$

The condition for no net change leads to the isotherm

$$\theta=\frac{(\alpha p)^{1/2}}{1+(\alpha p)^{1/2}} \quad \text{Langmuir isotherm for adsorption with dissociation} \quad (22B.4)$$

The surface coverage now depends more weakly on pressure than for non-dissociative adsorption.

The shapes of the Langmuir isotherms with and without dissociation are shown in Figs. 22B.2 and 22B.3. The fractional

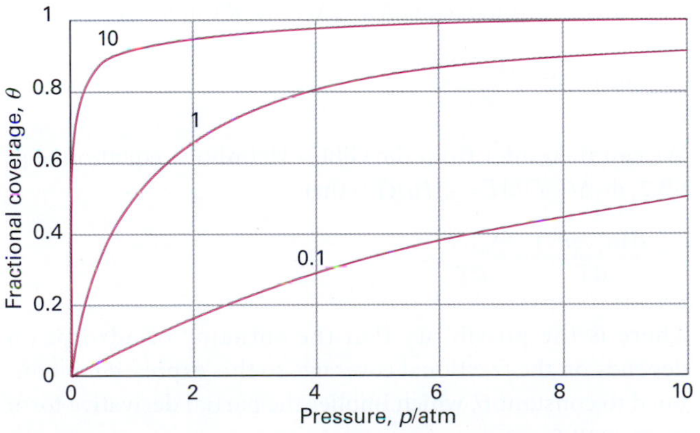

Figure 22B.2 The Langmuir isotherm for dissociative adsorption, $A_2(g)\rightarrow 2\,A(\text{surface})$, for different values of α.

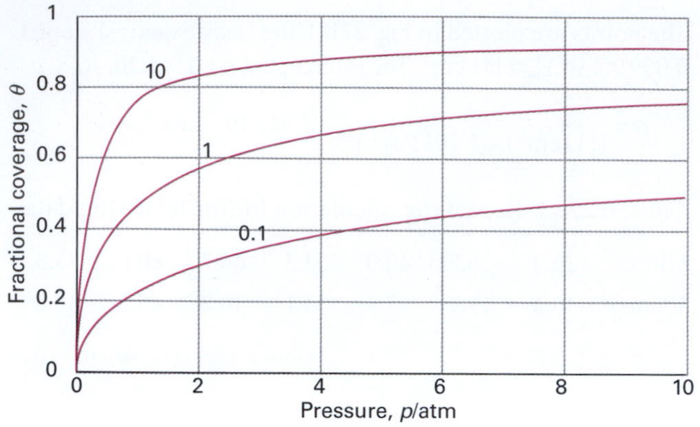

Figure 22B.3 The Langmuir isotherm for non-dissociative adsorption for different values of α.

coverage increases with increasing pressure, and approaches 1 only at very high pressure, when the gas is forced on to every available site of the surface.

(b) The isosteric enthalpy of adsorption

The Langmuir isotherm depends on the value of $\alpha = k_a/k_d$, which depends on the temperature. As we show in the following *Justification*, the temperature dependence of α can be used to determine the **isosteric enthalpy of adsorption**, $\Delta_{ad}H^{\ominus}$, the standard enthalpy of adsorption at a fixed surface coverage, by using the relation

$$\left(\frac{\partial \ln(\alpha p^{\ominus})}{\partial T}\right)_{\theta} = \frac{\Delta_{ad}H^{\ominus}}{RT^2} \quad \text{Isosteric enthalpy of adsorption} \quad (22B.5)$$

Justification 22B.1 The isosteric enthalpy of adsorption

It follows from the treatment in Topic 20C that the quantity $\alpha p^{\ominus} = (k_a/k_d) \times p^{\ominus}$ is an equilibrium constant for the process $A(g) + M(surface) \rightleftharpoons AM(surface)$, and can therefore be expressed in terms of the standard Gibbs energy of adsorption, $\Delta_{ad}G^{\ominus}$ through eqn 6A.14 ($\Delta G^{\ominus} = -RT \ln K$) as

$$\ln(\alpha p^{\ominus}) = -\frac{\Delta_{ad}G^{\ominus}}{RT}$$

We can then infer from the Gibbs–Helmholtz equation (eqn 6B.2, $d((\Delta G/T)/dT = -\Delta H/RT^2$) that

$$\frac{d\ln(\alpha p^{\ominus})}{dT} = \frac{\Delta_{ad}H^{\ominus}}{RT^2}$$

There is the possibility that the enthalpy of adsorption depends on the fractional coverage, so this expression is confined to constant θ, which implies the partial derivative form in eqn 22B.5.

Example 22B.2 Measuring the isosteric enthalpy of adsorption

The data below show the pressures of CO needed for the volume of adsorption (corrected to 1 atm and 0 °C) to be 10.0 cm^3 using the same sample as in *Example* 22B.1. In this case, there is no dissociation. Calculate the adsorption enthalpy at this surface coverage.

T/K	200	210	220	230	240	250
p/kPa	4.00	4.95	6.03	7.20	8.47	9.85

Method The Langmuir isotherm for adsorption without dissociation (eqn 22B.2), can be rearranged to $\alpha p = \theta/(1-\theta)$, a constant when θ is constant. We need to guard against problems with units as we manipulate expressions, and in this case it will prove useful to write αp = constant as $(\alpha p^{\ominus}) \times (p/p^{\ominus})$ = constant. It then follows that

$$\ln\{(\alpha p^{\ominus})(p/p^{\ominus})\} = \ln(\alpha p^{\ominus}) + \ln(p/p^{\ominus}) = \text{constant}$$

and from eqn 22B.5 that

$$\left(\frac{\partial \ln(p/p^{\ominus})}{\partial T}\right)_{\theta} = -\left(\frac{\partial \ln(\alpha p^{\ominus})}{\partial T}\right)_{\theta} = -\frac{\Delta_{ad}H^{\ominus}}{RT^2}$$

With $d(1/T)/dT = -1/T^2$, and therefore $dT = -T^2 d(1/T)$, this expression becomes

$$\left(\frac{\partial \ln(p/p^{\ominus})}{\partial(1/T)}\right)_{\theta} = \frac{\Delta_{ad}H^{\ominus}}{R}$$

Therefore, a plot of $\ln(p/p^{\ominus})$ against $1/T$ should be a straight line of slope $\Delta_{ad}H^{\ominus}/R$.

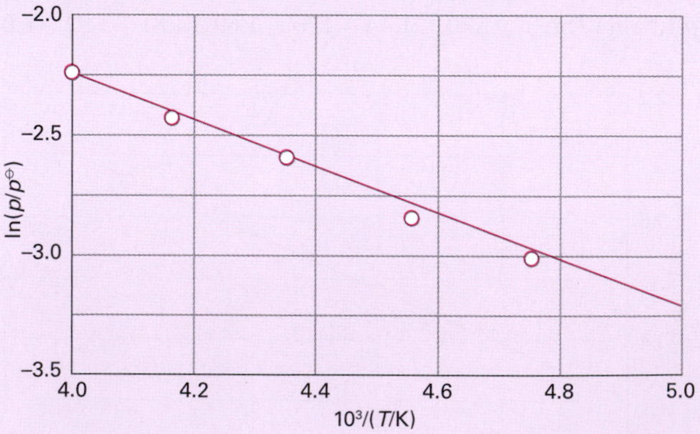

Figure 22B.4 The isosteric enthalpy of adsorption can be obtained from the slope of the plot of $\ln(p/p^{\ominus})$ against $1/T$, where p is the pressure needed to achieve the specified surface coverage. The data used are from *Example* 22B.2.

Answer With $p^\ominus = 1\ \text{bar} = 10^2\ \text{kPa}$, we draw up the following table:

T/K	200	210	220	230	240	250
$10^3/(T/\text{K})$	5.00	4.76	4.55	4.35	4.17	4.00
$(p/p^\ominus)\times 10^2$	4.00	4.95	6.03	7.20	8.47	9.85
$\ln(p/p^\ominus)$	−3.22	−3.01	−2.81	−2.63	−2.47	−2.32

The points are plotted in Fig. 22B.4. The slope (of the least squares fitted line) is −0.904, so

$$\Delta_{\text{ad}}H^\ominus = -(0.904\times 10^3\ \text{K})\times R = -7.52\ \text{kJ mol}^{-1}$$

Self-test 22B.2 Repeat the calculation using the following data:

T/K	200	210	220	230	240	250
p/kPa	4.32	5.59	7.07	8.80	10.67	12.80

Answer: −9.0 kJ mol⁻¹

Two assumptions of the Langmuir isotherm are the independence and equivalence of the adsorption sites. Deviations from the isotherm can often be traced to the failure of these assumptions. For example, the enthalpy of adsorption often becomes less negative as θ increases, which suggests that the energetically most favourable sites are occupied first. Also, substrate–substrate interactions on the surface can be important. A number of isotherms have been developed to deal with cases where deviations from the Langmuir isotherm are important.

(c) The BET isotherm

If the initial adsorbed layer can act as a substrate for further (for example, physical) adsorption, then, instead of the isotherm levelling off to some saturated value at high pressures, it can be expected to rise indefinitely. The most widely used isotherm dealing with multilayer adsorption was derived by Stephen Brunauer, Paul Emmett, and Edward Teller and is called the **BET isotherm**:

$$\frac{V}{V_{\text{mon}}} = \frac{cz}{(1-z)\{1-(1-c)z\}} \quad \text{with} \quad z = \frac{p}{p^*} \qquad \text{BET isotherm} \qquad (22\text{B}.6)$$

In this expression, which is obtained in the following *Justification*, p^* is the vapour pressure above a layer of adsorbate that is more than one molecule thick and which resembles a pure bulk liquid, V_{mon} is the volume corresponding to monolayer coverage, and c is a constant which is large when the enthalpy of desorption from a monolayer is large compared with the enthalpy of vaporization of the liquid adsorbate:

$$c = e^{(\Delta_{\text{des}}H^\ominus - \Delta_{\text{vap}}H^\ominus)/RT} \qquad (22\text{B}.7)$$

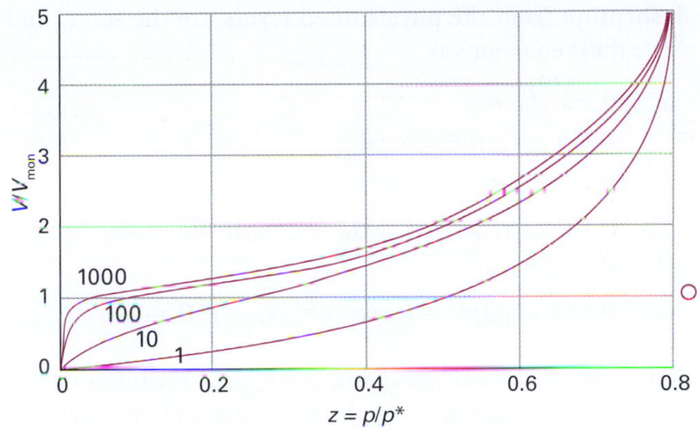

Figure 22B.5 Plots of the BET isotherm for different values of c. The value of V/V_{mon} rises indefinitely because the adsorbate may condense on the covered substrate surface.

Figure 22B.5 illustrates the shapes of BET isotherms. They rise indefinitely as the pressure is increased because there is no limit to the amount of material that may condense when multilayer coverage is possible. A BET isotherm is not accurate at all pressures, but it is widely used in industry to determine the surface areas of solids.

Justification 22B.2 The BET isotherm

We suppose that at equilibrium a fraction θ_0 of the surface sites are unoccupied, a fraction θ_1 is covered by a monolayer, a fraction θ_2 is covered by a bilayer, and so on. The number of adsorbed molecules is therefore

$$N = N_{\text{sites}}(\theta_1 + 2\theta_2 + 3\theta_3 + \cdots)$$

where N_{sites} is the total number of sites. We now follow the derivation that led to the Langmuir isotherm (eqn 22B.2) but allow for different rates of desorption from the substrate and the various layers:

First layer:	Rate of adsorption $= Nk_{\text{a},0}p\theta_0$
	Rate of desorption $= Nk_{\text{d},0}\theta_1$
	At equilibrium, $k_{\text{a},0}p\theta_0 = k_{\text{d},0}\theta_1$
Second layer:	Rate of adsorption $= Nk_{\text{a},1}p\theta_1$
	Rate of desorption $= Nk_{\text{d},1}\theta_2$
	At equilibrium, $k_{\text{a},1}p\theta_1 = k_{\text{d},1}\theta_2$
Third layer:	Rate of adsorption $= Nk_{\text{a},2}p\theta_2$
	Rate of desorption $= Nk_{\text{d},2}\theta_3$
	At equilibrium, $k_{\text{a},2}p\theta_2 = k_{\text{d},2}\theta_3$

and so on. We now suppose that once a monolayer has been formed, all the rate constants involving adsorption and desorption from the physisorbed layers are the same, and write these equations as

$$k_{a,0}p\theta_0 = k_{d,0}\theta_1, \text{ so}$$
$$\theta_1 = (k_{a,0}/k_{d,0})p\theta_0 = \alpha_0 p\theta_0$$
$$k_{a,1}p\theta_1 = k_{d,1}\theta_2, \text{ so}$$
$$\theta_2 = (k_{a,1}/k_{d,1})p\theta_1 = (k_{a,0}/k_{d,0})(k_{a,1}/k_{d,1})p^2\theta_0 = \alpha_0\alpha_1 p^2\theta_0$$
$$k_{a,1}p\theta_2 = k_{d,1}\theta_3, \text{ so}$$
$$\theta_3 = (k_{a,1}/k_{d,1})p\theta_2 = (k_{a,0}/k_{d,0})(k_{a,1}/k_{d,1})^2 p^3\theta_0 = \alpha_0\alpha_1^2 p^3\theta_0$$

and so on, with $\alpha_0 = k_{a,0}/k_{d,0}$ and $\alpha_1 = k_{a,1}/k_{d,1}$ the ratios of rate constants for adsorption to the substrate and an overlayer, respectively. Now, because $\theta_0 + \theta_1 + \theta_2 + \cdots = 1$, it follows that

$$\theta_0 + \alpha_0 p\theta_0 + \alpha_0\alpha_1 p^2\theta_0 + \alpha_0\alpha_1^2 p^3\theta_0 + \cdots$$
$$= \theta_0 + \alpha_0 p\theta_0\{1 + \alpha_1 p + \alpha_1^2 p^2 + \cdots\}$$
$$\overset{1+x+x^2+\cdots=1/(1-x)}{\triangleq} \left\{1 + \frac{\alpha_0 p}{1-\alpha_1 p}\right\}\theta_0 = \left\{\frac{1-\alpha_1 p + \alpha_0 p}{1-\alpha_1 p}\right\}\theta_0$$

Then, because this expression is equal to 1,

$$\theta_0 = \frac{1-\alpha_1 p}{1-(\alpha_1-\alpha_0)p}$$

In a similar way, we can write the number of adsorbed species as

$$N = N_{\text{sites}}\alpha_0 p\theta_0 + 2N_{\text{sites}}\alpha_0\alpha_1 p^2\theta_0 + \cdots$$
$$= N_{\text{sites}}\alpha_0 p\theta_0(1 + 2\alpha_1 p + 3\alpha_1^2 p^2 + \cdots)$$
$$\overset{1+2x+3x^2+\cdots=1/(1\ \ x)^2}{\triangleq} \frac{N_{\text{sites}}\alpha_0 p\theta_0}{(1-\alpha_1 p)^2}$$

By combining the last two expressions, we obtain

$$N = \frac{N_{\text{sites}}\alpha_0 p}{(1-\alpha_1 p)^2} \times \frac{1-\alpha_1 p}{1-(\alpha_1-\alpha_0)p} = \frac{N_{\text{sites}}\alpha_0 p}{(1-\alpha_1 p)\{1-(\alpha_1-\alpha_0)p\}}$$

The ratio N/N_{sites} is equal to the ratio V/V_{mon}, where V is the total volume adsorbed and V_{mon} the volume adsorbed had there been complete monolayer coverage. The equilibrium of the adsorption and desorption from the overlayers is equivalent to the vaporization $A(l) \rightleftharpoons A(g)$ of the pure adsorbate, with matching forward and reverse rates: $k_d = k_a p^*$, where p^* is the vapour pressure of the liquid adsorbate. Therefore, $\alpha_1 = k_{a,1}/k_{d,1} = 1/p^*$. Then, with $z = p/p^*$ and $c = \alpha_0/\alpha_1$, the last equation becomes

$$\frac{V}{V_{\text{mon}}} = \frac{\alpha_0 p}{(1-p/p^*)\{1-(1-\alpha_0/\alpha_1)p/p^*\}} = \frac{cz}{(1-z)\{1-(1-c)z\}}$$

as in eqn 22B.6.

As in *Justification* 22B.1, α_0 and α_1 are related to the Gibbs energy changes accompanying adsorption to the substrate and condensation on the adsorbed layers, $\Delta_{\text{ad}}G^\ominus$ and $\Delta_{\text{con}}G^\ominus$, which in turn can be related to the Gibbs energies for the opposite processes, desorption from the substrate and vaporization from the overlayer, by $\Delta_{\text{des}}G^\ominus = -\Delta_{\text{ad}}G^\ominus$ and $\Delta_{\text{vap}}G^\ominus = -\Delta_{\text{con}}G^\ominus$. Therefore, from $\ln(\alpha p^\ominus) = -\Delta G^\ominus/RT$ in each case,

$$\alpha_0 p^\ominus = e^{-\Delta_{\text{ad}}G^\ominus/RT} = e^{\Delta_{\text{des}}G^\ominus/RT} \text{ and } \alpha_1 p^\ominus = e^{-\Delta_{\text{con}}G^\ominus/RT} = e^{\Delta_{\text{vap}}G^\ominus/RT}$$

The ratio c then becomes (after cancelling the $p^\ominus$ and writing $\Delta G^\ominus = \Delta H^\ominus - T\Delta S^\ominus$ in each case)

$$c = \frac{\alpha_0}{\alpha_1} = \frac{e^{\Delta_{\text{des}}G^\ominus/RT}}{e^{\Delta_{\text{vap}}G^\ominus/RT}} = \frac{e^{\Delta_{\text{des}}H^\ominus/RT}e^{-\Delta_{\text{des}}S^\ominus/R}}{e^{\Delta_{\text{vap}}H/RT}e^{-\Delta_{\text{vap}}S^\ominus/R}}$$

If the entropies of desorption and vaporization are assumed to be the same because they correspond to very similar processes in terms of the escape of the condensed adsorbate to the gas phase, this ratio becomes

$$c = \frac{e^{\Delta_{\text{des}}H^\ominus/RT}}{e^{\Delta_{\text{vap}}H^\ominus/RT}} = e^{(\Delta_{\text{des}}H^\ominus - \Delta_{\text{vap}}H^\ominus)/RT}$$

as in eqn 21B.7.

Example 22B.3 Using the BET isotherm

The data below relate to the adsorption of N_2 on rutile (TiO_2) at 75 K. Confirm that they fit a BET isotherm in the range of pressures reported, and determine V_{mon} and c.

p/kPa	0.160	1.87	6.11	11.67	17.02	21.92	27.29
V/mm^3	601	720	822	935	1046	1146	1254

At 75 K, $p^* = 76.0$ kPa. The volumes have been corrected to 1.00 atm and 273 K and refer to 1.00 g of substrate.

Method Equation 22B.6 can be reorganized into

$$\frac{z}{(1-z)V} = \frac{1}{cV_{\text{mon}}} + \frac{(c-1)z}{cV_{\text{mon}}}$$

It follows that $(c-1)/cV_{\text{mon}}$ can be obtained from the slope of a plot of the expression on the left against z, and cV_{mon} can be found from the intercept at $z = 0$. The results can then be combined to give c and V_{mon}.

Answer We draw up the following table:

p/kPa	0.160	1.87	6.11	11.67	17.02	21.92	27.29
$10^3 z$	2.11	24.6	80.4	154	224	288	359
$10^4 z/(1-z)$ (V/mm^3)	0.035	0.350	1.06	1.95	2.76	3.53	4.47

These points are plotted in Fig. 22B.6. The least squares best line has an intercept at 0.0398, so

$$\frac{1}{cV_{\text{mon}}} = 3.98 \times 10^{-6} \text{ mm}^{-3}$$

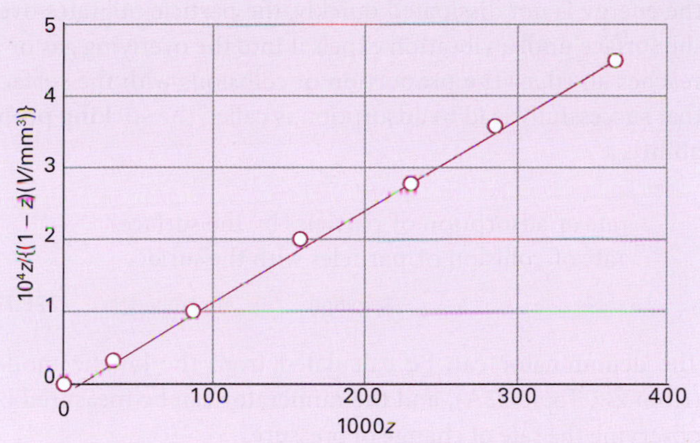

Figure 22B.6 The BET isotherm can be tested, and the parameters determined, by plotting $z/(1-z)V$ against $z=p/p^*$. The data are from *Example* 22B.3.

The slope of the line is 1.23×10^{-2}, so

$$\frac{c-1}{cV_{\text{mon}}}=(1.23\times10^{-2})\times10^{3}\times10^{-4}\,\text{mm}^{-3}=1.23\times10^{-3}\,\text{mm}^{-3}$$

The solutions of these equations are $c=310$ and $V_{\text{mon}}=811\,\text{mm}^3$. At 1.00 atm and 273 K, 811 mm^3 corresponds to 3.6×10^{-5} mol, or 2.2×10^{19} atoms. Because each atom occupies an area of about 0.16 nm^2, the surface area of the sample is about 3.5 m^2.

Self-test 22B.3 Repeat the calculation for the following data:

p/kPa	0.160	1.87	6.11	11.67	17.02	21.92	27.29
V/cm^3	235	559	649	719	790	860	950

Answer: 370, 615 cm^3

When $c\gg1$, the BET isotherm takes the simpler form

$$\frac{V}{V_{\text{mon}}}=\frac{1}{1-z} \qquad \text{BET isotherm when } c\gg1 \qquad (22\text{B}.8)$$

This expression is applicable to unreactive gases on polar surfaces, for which $c\approx10^2$ because $\Delta_{\text{des}}H^{\ominus}$ is then significantly greater than $\Delta_{\text{vap}}H^{\ominus}$ (eqn 22B.7). The BET isotherm fits experimental observations moderately well over restricted pressure ranges, but it errs by underestimating the extent of adsorption at low pressures and by overestimating it at high pressures.

(d) The Temkin and Freundlich isotherms

An assumption of the Langmuir isotherm is the independence and equivalence of the adsorption sites. Deviations from the isotherm can often be traced to the failure of these assumptions. For example, the enthalpy of adsorption often becomes less negative as θ increases, which suggests that the energetically most favourable sites are occupied first. Various attempts have been made to take these variations into account. The **Temkin isotherm**,

$$\theta=c_1\ln(c_2p) \qquad \text{Temkin isotherm} \qquad (22\text{B}.9)$$

where c_1 and c_2 are constants, corresponds to supposing that the adsorption enthalpy changes linearly with pressure. The **Freundlich isotherm**

$$\theta=c_1p^{1/c_2} \qquad \text{Freundlich isotherm} \qquad (22\text{B}.10)$$

corresponds to a logarithmic change. This isotherm attempts to incorporate the role of substrate–substrate interactions on the surface.

Different isotherms agree with experiment more or less well over restricted ranges of pressure, but they remain largely empirical. Empirical, however, does not mean useless for, if the parameters of a reasonably reliable isotherm are known, reasonably reliable results can be obtained for the extent of surface coverage under various conditions. This kind of information is essential for any discussion of heterogeneous catalysis (Topic 22C).

22B.2 The rates of adsorption and desorption

We have noted that adsorption and desorption are activated processes, in the sense that they have an activation energy and follow Arrhenius behaviour. Now we are ready to look more closely at the origin of the activation energy in these processes, with a special focus on chemisorption.

(a) The precursor state

Figure 22B.7 shows how the potential energy of a molecule varies with its distance from the substrate surface. As the molecule approaches the surface its energy falls as it becomes physisorbed into the **precursor state** for chemisorption (see Topic 22A). Dissociation into fragments often takes place as a molecule moves into its chemisorbed state, and after an initial increase of energy as the bonds stretch there is a sharp decrease as the adsorbate–substrate bonds reach their full strength. Even if the molecule does not fragment, there is likely to be an initial increase of potential energy as the molecule approaches the surface and the bonds adjust.

In most cases, therefore, we can expect there to be a potential energy barrier separating the precursor and chemisorbed states. This barrier, though, might be low, and might not rise above the energy of a distant, stationary particle (as in Fig. 22B.7a). In this case, chemisorption is not an activated process and can be expected to be rapid. Many gas adsorptions on clean metals appear to be non-activated. In some cases, however, the

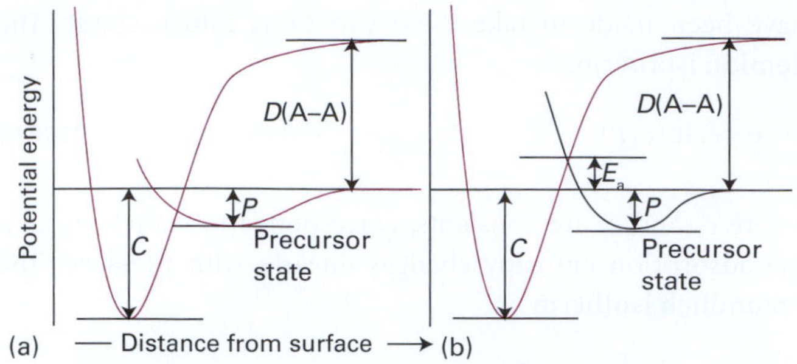

Figure 22B.7 The potential energy profiles for the dissociative chemisorption of an A_2 molecule. In each case, *P* is the enthalpy of (non-dissociative) physisorption and *C* that for chemisorption (at $T=0$). The relative locations of the curves determine whether the chemisorption is (a) not activated or (b) activated.

barrier rises above the zero axis (as in Fig. 22B.7b); such chemisorptions are activated and slower than the non-activated kind. An example is H_2 on copper, which has an activation energy in the region of 20–40 kJ mol^{-1}.

One point that emerges from this discussion is that rates are not good criteria for distinguishing between physisorption and chemisorption. Chemisorption can be fast if the activation energy is small or zero, but it may be slow if the activation energy is large. Physisorption is usually fast, but it can appear to be slow if adsorption is taking place on a porous medium.

Brief illustration 22B.1 The rate of activated adsorption

Consider two adsorption experiments for hydrogen on different faces of a copper crystal. In one, Face 1, the activation energy is 28 kJ mol^{-1} and on the other, Face 2, the activation energy is 33 kJ mol^{-1}. The ratio of the rates of chemisorption on equal areas of the two faces at 250 K is

$$\frac{\text{Rate}(1)}{\text{Rate}(2)}=\frac{Ae^{-E_{a,ads}(1)/RT}}{Ae^{-E_{a,ads}(2)/RT}}=e^{-\{E_{a,ads}(1)-E_{a,ads}(2)\}/RT}$$
$$=e^{5\times10^3\,\text{J mol}^{-1}/(8.3145\,\text{J K}^{-1}\,\text{mol}^{-1})\times(250\,\text{K})}=11$$

We have assumed that the *A* factor is the same for each face.

Self-test 22B.4 What are the relative rates when the temperature is increased to 300 K?

Answer: 7

(b) Adsorption and desorption at the molecular level

The rate at which a surface is covered by adsorbate depends on the ability of the substrate to dissipate the energy of the incoming particle as thermal motion as it crashes on to the surface. If the energy is not dissipated quickly, the particle migrates over the surface until a vibration expels it into the overlying gas or it reaches an edge. The proportion of collisions with the surface that successfully lead to adsorption is called the **sticking probability**, *s*:

$$s=\frac{\text{rate of adsorption of particles by the surface}}{\text{rate of collision of particles with the surface}}$$

Definition Sticking probability (22B.11)

The denominator can be calculated from the kinetic model (from Z_W, Topic 22A), and the numerator can be measured by observing the rate of change of pressure.

Values of *s* vary widely. For example, at room temperature CO has *s* in the range 0.1–1.0 for several d-metal surfaces, but for N_2 on rhenium $s<10^{-2}$, indicating that more than a hundred collisions are needed before one molecule sticks successfully. Beam studies on specific crystal planes show a pronounced specificity: for N_2 on tungsten, *s* ranges from 0.74 on the (320) faces down to less than 0.01 on the (110) faces at room temperature. The sticking probability decreases as the surface coverage increases (Fig. 22B.8). A simple assumption is that *s* is proportional to $1-\theta$, the fraction uncovered, and it is common to write

$$s=(1-\theta)s_0$$

Commonly used form of the sticking probability (22B.12)

where s_0 is the sticking probability on a perfectly clean surface. The results in the illustration do not fit this expression because they show that *s* remains close to s_0 until the coverage has risen to about 6×10^{13} molecules cm^{-2}, and then falls steeply. The explanation is probably that the colliding molecule does not enter the chemisorbed state at once, but moves over the surface until it encounters an empty site.

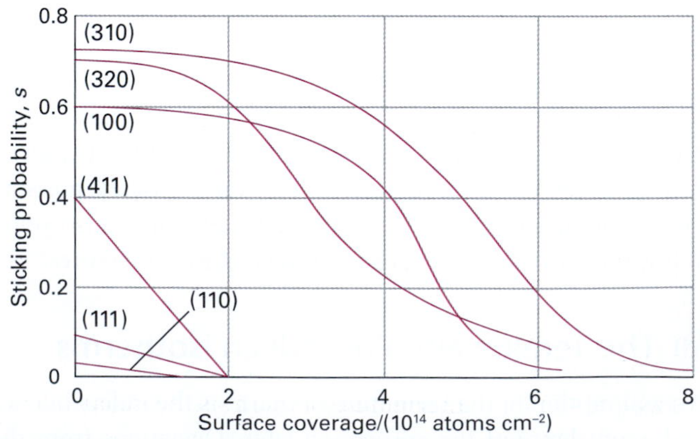

Figure 22B.8 The sticking probability of N_2 on various faces of a tungsten crystal and its dependence on surface coverage. Note the very low sticking probability for the (110) and (111) faces. (Data provided by Professor D.A. King.)

Desorption is always activated because the particles have to be lifted from the foot of a potential well. A physisorbed particle vibrates in its shallow potential well, and might shake itself off the surface after a short time. If the temperature dependence of the first-order rate of departure follows Arrhenius behaviour, then $k_d = Ae^{-E_{a,des}/RT}$, with $E_{a,des}$ the activation energy for desorption. Therefore, the half-life for remaining on the surface has a temperature dependence

$$t_{1/2} = \frac{\ln 2}{k_d} = \tau_0 e^{E_{a,des}/RT} \quad \tau_0 = \frac{\ln 2}{A} \qquad \text{Residence half-life} \qquad (22B.13)$$

Note the positive sign in the exponent: the greater the activation energy for desorption, the larger the residence half-life.

Brief illustration 22B.2 Residence half-lives

If we suppose that $1/\tau_0$ is approximately the same as the vibrational frequency of the weak particle–surface bond (about 10^{12} Hz) and $E_d \approx 25$ kJ mol^{-1}, then residence half-lives of around 10 ns are predicted at room temperature. Lifetimes close to 1 s are obtained only by lowering the temperature to about 100 K. For chemisorption, with $E_d = 100$ kJ mol^{-1} and guessing that $\tau_0 = 10^{-14}$ s (because the adsorbate–substrate bond is quite stiff), we expect a residence half-life of about 3×10^3 s (about an hour) at room temperature, decreasing to 1 s at about 350 K.

Self-test 22B.5 For how long on average would an atom remain on a surface at 800 K if its desorption activation energy is 200 kJ mol^{-1}? Take $\tau_0 = 0.10$ ps.

Answer: $t_{1/2} = 1.3$ s

The desorption activation energy can be measured in several ways. However, we must be guarded in its interpretation because it often depends on the fractional coverage, and so might change as desorption proceeds. Moreover, the transfer of concepts such as 'reaction order' and 'rate constant' from bulk studies to surfaces is hazardous, and there are few examples of strictly first-order or second-order desorption kinetics (just as there are few integral-order reactions in the gas phase too).

If we disregard these complications, one way of measuring the desorption activation energy is to monitor the rate of increase in pressure when the sample is maintained at a series of temperatures, and to attempt to make an Arrhenius plot. A more sophisticated technique is **temperature programmed desorption** (TPD) or **thermal desorption spectroscopy** (TDS). The basic observation is a surge in desorption rate (as monitored by a mass spectrometer) when the temperature is raised linearly to the temperature at which desorption occurs rapidly, but once the desorption has occurred there is no more adsorbate to escape from the surface, so the desorption flux falls again

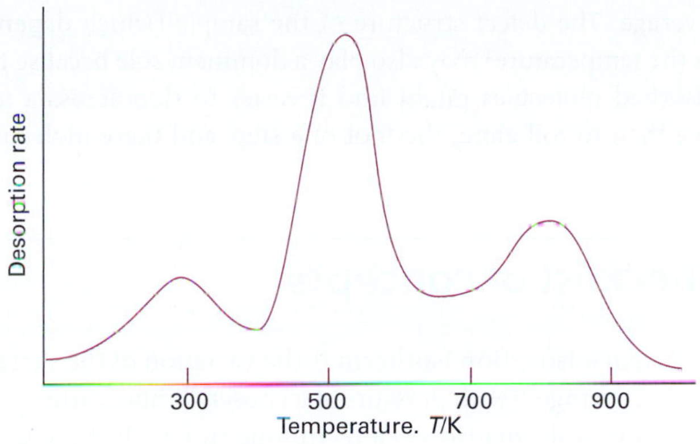

Figure 22B.9 The TPD spectrum of H_2 on the (100) face of tungsten. The three peaks indicate the presence of three sites with different adsorption enthalpies and therefore different desorption activation energies (P.W. Tamm and L.D. Schmidt, *J. Chem. Phys.* **51**, 5352 (1969)).

as the temperature continues to rise. The TPD spectrum, the plot of desorption flux against temperature, therefore shows a peak, the location of which depends on the desorption activation energy. There are three maxima in the example shown in Fig. 22B.9, indicating the presence of three sites with different activation energies.

In many cases only a single activation energy (and a single peak in the TPD spectrum) is observed. When several peaks are observed they might correspond to adsorption on different crystal planes or to multilayer adsorption. For instance, Cd atoms on tungsten show two activation energies, one of 18 kJ mol^{-1} and the other of 90 kJ mol^{-1}. The explanation is that the more tightly bound Cd atoms are attached directly to the substrate, and the less strongly bound are in a layer (or layers) above the primary overlayer. Another example of a system showing two desorption activation energies is CO on tungsten, the values being 120 kJ mol^{-1} and 300 kJ mol^{-1}. The explanation is believed to be the existence of two types of metal–adsorbate binding site, one involving a simple M-CO bond, the other adsorption with dissociation into individually adsorbed C and O atoms.

(c) Mobility on surfaces

A further aspect of the strength of the interactions between adsorbate and substrate is the mobility of the adsorbate. Mobility is often a vital feature of a catalyst's activity, because a catalyst might be impotent if the reactant molecules adsorb so strongly that they cannot migrate.

The activation energy for diffusion over a surface need not be the same as for desorption because the particles may be able to move through valleys between potential peaks without leaving the surface completely. In general, the activation energy for migration is about 10–20 per cent of the energy of the surface–adsorbate bond, but the actual value depends on the extent of coverage. The defect structure of the sample (which depends on the temperature) may also play a dominant role because the adsorbed molecules might find it easier to skip across a terrace than to roll along the foot of a step, and these molecules might become trapped in vacancies in an otherwise flat terrace. Diffusion may also be easier across one crystal face than another, and so the surface mobility depends on which lattice planes are exposed.

Checklist of concepts

- ☐ 1. An **adsorption isotherm** is the variation of the surface coverage θ with pressure at a chosen temperature.
- ☐ 2. **Flash desorption** is a technique in which the sample is suddenly heated (electrically) and the resulting rise of pressure is interpreted in terms of the amount of adsorbate originally on the substrate.
- ☐ 3. Examples of adsorption isotherms include the **Langmuir, BET, Temkin,** and **Freundlich** isotherms.
- ☐ 4. The **sticking probability** is the proportion of collisions with the surface that successfully lead to adsorption.
- ☐ 5. Desorption is an activated process; the desorption activation energy is measured by **temperature-programmed desorption** or **thermal desorption spectroscopy**.
- ☐ 6. The mobility of adsorbates on a surface is dominated by diffusion.

Checklist of equations

Property	Equation	Comment	Equation number
Langmuir isotherm:		Independent and equivalent sites, monolayer coverage	
(a) without dissociation	$\theta = \alpha p/(1+\alpha p)$		22B.2
(b) with dissociation	$\theta = (\alpha p)^{1/2}/\{1+(\alpha p)^{1/2}\}$		22B.4
Isosteric enthalpy of adsorption	$(\partial \ln(\alpha p^{\ominus})/\partial T)_{\theta} = \Delta_{\text{ad}}H^{\ominus}/RT^2$		22B.5
BET isotherm	$V/V_{\text{mon}} = cz/(1-z)\{1-(1-c)z\}$, $z = p/p^*$, $c = e^{(\Delta_{\text{des}}H^{\ominus}-\Delta_{\text{vap}}H^{\ominus})/RT}$	Multilayer adsorption	22B.6–7
Temkin isotherm	$\theta = c_1 \ln(c_2 p)$	Enthalpy of adsorption varies with θ	22B.9
Freundlich isotherm	$\theta = c_1 p^{1/c_2}$	Substrate–substrate interactions	22B.10
Sticking probability	$s = (1-\theta)s_0$	Approximate form	22B.12

22C Heterogeneous catalysis

Contents

➤ Why do we need to know this material?

Catalysis is at the heart of the chemical industry, and an understanding of the concepts is essential for developing new catalysts.

➤ What is the key idea?

In heterogeneous catalysis, the pathway for lowering the activation energy of a reaction commonly involves chemisorption of one or more reactants.

➤ What do we need to know already?

Catalysis is introduced in Topic 20H. This Topic builds on the discussion of reaction mechanisms (Topic 20E), the Arrhenius equation (Topic 20D), and adsorption isotherms (Topic 22B).

A **heterogeneous catalyst** is a catalyst in a different phase from the reaction mixture. For example, the hydrogenation of ethene to ethane, a gas-phase reaction, is accelerated in the presence of a solid catalyst such as palladium, platinum, or nickel. The metal provides a surface to which the reactants bind; this binding facilitates encounters between reactants and increases the rate of the reaction. This Topic is an exploration of catalytic activity on surfaces, building on the concepts developed in Topic 22B.

22C.1 Mechanisms of heterogeneous catalysis

Many catalysts depend on **co-adsorption**, the adsorption of two or more species. One consequence of the presence of a second species may be the modification of the electronic structure at the surface of a metal. For instance, partial coverage of d-metal surfaces by alkali metals has a pronounced effect on the electron distribution at the surface and reduces the work function of the metal (the energy needed to remove an electron; see Topic 7A). Such modifiers can act as promoters (to enhance the action of catalysts) or as poisons (to inhibit catalytic action).

Figure 22C.1 shows the potential energy curve for a reaction influenced by the action of a heterogeneous catalyst. Differences between Fig. 22C.1 and 20H.1 arise from the fact that heterogeneous catalysis normally depends on at least one reactant being adsorbed (usually chemisorbed) and modified to an **active phase** in which it readily undergoes reaction, and desorption of products. Modification of the reactant often takes the form of a fragmentation of the reactant molecules. In practice, the active phase is dispersed as very small particles of linear dimension less than 2 nm on a porous oxide support. **Shape-selective catalysts**, such as the zeolites, which have a pore size that can distinguish shapes and sizes at a molecular scale, have high internal specific surface areas, in the range of $100–500\,m^2\,g^{-1}$.

Mechanisms of reactions catalysed by surfaces can be treated quantitatively by using the techniques of Topic 20E (on the

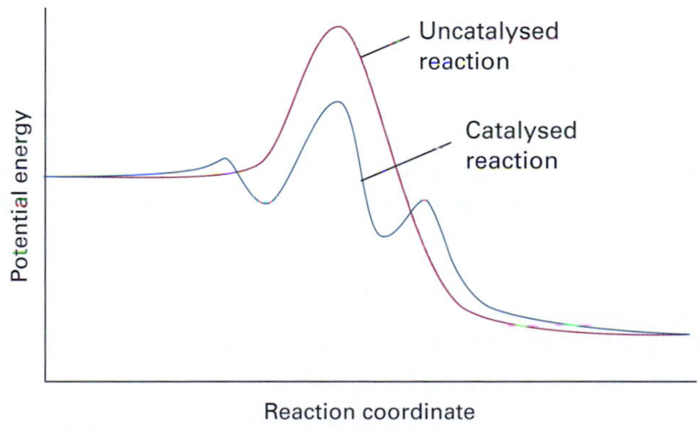

Figure 22C.1 The reaction profile for catalysed and uncatalysed reactions. The catalysed reaction path includes activation energies for adsorption and desorption as well as an overall lower activation energy for the process.

development of rate laws based on proposed reaction mechanisms) and the adsorption isotherms developed in Topic 22B. Here we explore some simple mechanisms that can give significant insight into surface-catalysed reactions.

(a) Unimolecular reactions

The rate law of a surface-catalysed unimolecular reaction, such as the decomposition of a substance on a surface, can be written in terms of an adsorption isotherm if the rate is supposed to be proportional to the extent of surface coverage. For example, if the fractional coverage θ is given by the Langmuir isotherm (eqn 22B.2, $\theta = \alpha p/(1+\alpha p)$), we would write

$$v = k_r\theta = \frac{k_r \alpha p}{1+\alpha p} \qquad (22C.1)$$

where p is the pressure of the adsorbing substance.

Brief illustration 22C.1 Surface-catalysed unimolecular decomposition

Consider the decomposition of phosphine (PH_3) on tungsten, which is first order at low pressures. We can use eqn 22C.1 to account for this observation. When the pressure is so low that $\alpha p \ll 1$, we can neglect αp in the denominator of eqn 22C.1 and obtain $v = k_r\alpha p$. The decomposition is predicted to be first order, as observed experimentally.

Self-test 22C.1 Write a rate law for the decomposition of PH_3 on tungsten at high pressures

Answer: $v = k_r$; the reaction is zeroth order at high pressures

(b) The Langmuir–Hinshelwood mechanism

In the **Langmuir–Hinshelwood mechanism** (LH mechanism) of surface-catalysed reactions, the reaction takes place by encounters between molecular fragments and atoms adsorbed on the surface. We therefore expect the rate law to be second order in the extent of surface coverage:

$$A+B \rightarrow P \quad v = k_r\theta_A\theta_B \qquad \text{Langmuir–Hinshelwood rate law} \qquad (22C.2)$$

Insertion of the appropriate isotherms for A and B then gives the reaction rate in terms of the partial pressures of the reactants.

Example 22C.1 Writing a rate law based on the Langmuir–Hinshelwood mechanism

Consider a reaction $A+B \rightarrow P$ in which A and B follow Langmuir isotherms and adsorb without dissociation. Devise a rate law that is consistent with the Langmuir–Hinshelwood mechanism.

Method Begin by following the procedures outlined in Topic 22B for the derivation of the Langmuir isotherm to write expressions for θ_A and θ_B, the fractional coverages of A and B, respectively. However, note that, unlike the simple situation in Topic 22B, two species compete for the same sites on the surface. Then, use eqn 22C.2 to express the rate law.

Answer Because two species compete for sites on the surface, the number of vacant sites is equal to $N(1-\theta_A-\theta_B)$, where N is the total number of sites. It follows from eqns 22B.1a and 22B.1b that the rates of adsorption and desorption are given by

Rate of adsorption of A $= k_{a,A}p_A N(1-\theta_A-\theta_B)$	Rate of desorption of A $= k_{d,A}N\theta_A$
Rate of adsorption of B $= k_{a,B}p_B N(1-\theta_A-\theta_B)$	Rate of desorption of B $= k_{d,B}N\theta_B$

At equilibrium, the rates of adsorption and desorption for each species are equal, and, with $\alpha_A = k_{a,A}/k_{d,A}$ and $\alpha_B = k_{a,B}/k_{d,B}$, it follows that

$$\alpha_A p_A(1-\theta_A-\theta_B) = \theta_A$$
$$\alpha_B p_B(1-\theta_A-\theta_B) = \theta_B$$

The solutions of this pair of simultaneous equations (see *Self-test* 22C.2) are

$$\theta_A = \frac{\alpha_A p_A}{1+\alpha_A p_A+\alpha_B p_B} \qquad \theta_B = \frac{\alpha_B p_B}{1+\alpha_A p_A+\alpha_B p_B}$$

It follows from eqn 22C.2 that the rate law is

$$v = \frac{k_r\alpha_A\alpha_B p_A p_B}{(1+\alpha_A p_A+\alpha_B p_B)^2}$$

The parameters α in the isotherms and the rate constant k_r are all temperature-dependent, so the overall temperature dependence of the rate may be strongly non-Arrhenius (in the sense that the reaction rate is unlikely to be proportional to $e^{-E_a/RT}$). The LH mechanism is dominant for the catalytic oxidation of CO to CO_2.

Self-test 22C.2 Provide the missing steps in the derivation of the expression for v.

(c) The Eley–Rideal mechanism

In the **Eley–Rideal mechanism** (ER mechanism) of a surface-catalysed reaction, a gas-phase molecule collides with another molecule already adsorbed on the surface. The rate of formation of product is expected to be proportional to the partial pressure, p_B, of the non-adsorbed gas B and the extent of surface coverage, θ_A, of the adsorbed gas A. It follows that the rate law should be

$$A+B \rightarrow P \quad v = k_r p_B\theta_A \qquad \text{Eley–Rideal rate law} \qquad (22C.3)$$

The rate constant, k_r, might be much larger than for the uncatalysed gas-phase reaction because the reaction on the surface has a low activation energy and the adsorption itself is often not activated.

If we know the adsorption isotherm for A, we can express the rate law in terms of its partial pressure, p_A. For example, the adsorption of A follows a Langmuir isotherm in the pressure range of interest, then the rate law would be

$$v = \frac{k_r \alpha p_A p_B}{1+\alpha p_A} \qquad (22C.4)$$

Brief illustration 22C.2 The Eley–Rideal mechanism

According to eqn 22C.4, when the partial pressure of A is high (in the sense $\alpha p_A \gg 1$) there is almost complete surface coverage, and the rate is equal to $k_r p_B$. Now the rate-determining step is the collision of B with the adsorbed fragments. When the pressure of A is low ($\alpha p_A \ll 1$), perhaps because of its reaction, the rate is equal to $k_r \alpha p_A p_B$; now the extent of surface coverage is important in the determination of the rate.

Self-test 22C.3 Rewrite eqn 22C.4 for cases where A is a diatomic molecule that adsorbs as atoms.

Answer: $v = k_r p_B (\alpha p_A)^{1/2}/(1+(\alpha p_A)^{1/2})$

Almost all thermal surface-catalysed reactions are thought to take place by the LH mechanism but a number of reactions with an ER mechanism have also been identified from molecular beam investigations. For example, the reaction between H(g) and D(ad) to form HD(g) is thought to be by an ER mechanism involving the direct collision and pick-up of the adsorbed D atom by the incident H atom. However, the two mechanisms should really be thought of as ideal limits with all reactions lying somewhere between the two and showing features of each one.

22C.2 Catalytic activity at surfaces

It has become possible to investigate how the catalytic activity of a surface depends on its structure as well as its composition. For instance, the cleavage of C–H and H–H bonds appears to depend on the presence of steps and kinks, and a terrace often has only minimal catalytic activity.

The reaction $H_2+D_2 \rightarrow 2HD$ has been studied in detail. For this reaction, terrace sites are inactive but one molecule in ten reacts when it strikes a step. Although the step itself might be the important feature, it may be that the presence of the step merely exposes a more reactive crystal face (the step face itself). Likewise, the dehydrogenation of hexane to hexene depends strongly on the kink density, and it appears that kinks are needed to cleave C–C bonds. These observations suggest a reason why even small amounts of impurities may poison a catalyst: they are likely to attach to step and kink sites, and so impair the activity of the catalyst entirely. A constructive outcome is that the extent of dehydrogenation may be controlled relative to other types of reactions by seeking impurities that adsorb at kinks and act as specific poisons.

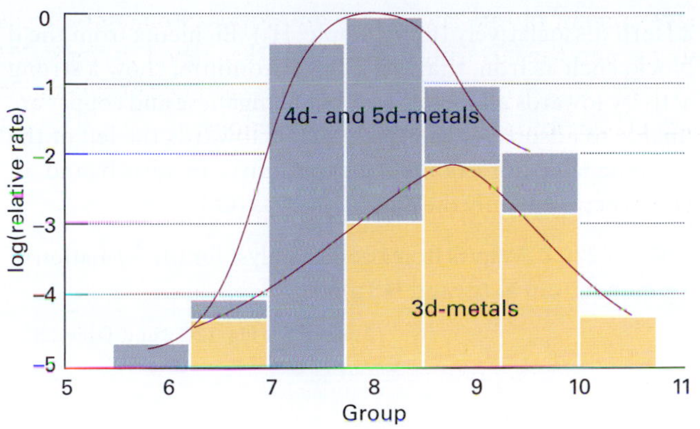

Figure 22C.2 A volcano curve of catalytic activity arises because although the reactants must adsorb reasonably strongly, they must not adsorb so strongly that they are immobilized. The lower curve refers to the first series of d-block metals, the upper curve to the second and third series of d-block metals. The group numbers relate to the periodic table inside the back cover.

The activity of a catalyst depends on the strength of chemisorption as indicated by the 'volcano' curve in Fig. 22C.2 (which is so-called on account of its general shape). To be active, the catalyst should be extensively covered by adsorbate, which is the case if chemisorption is strong. On the other hand, if the strength of the substrate–adsorbate bond becomes too great, the activity declines either because the other reactant molecules cannot react with the adsorbate or because the adsorbate molecules are immobilized on the surface. This pattern of behaviour suggests that the activity of a catalyst should initially increase with strength of adsorption (as measured, for instance, by the enthalpy of adsorption) and then decline, and that the most active catalysts should be those lying near the summit of the volcano. Most active metals are those that lie close to the middle of the d block. Many metals are suitable for adsorbing gases, and some trends are summarized in Table 22C.1.

Brief illustration 22C.3 Trends in chemisorption abilities

We see from Table 22C.1 that for a number of metals the general order of adsorption strengths decreases along the series O_2, C_2H_2, C_2H_4, CO, H_2, CO_2, N_2. Some of these molecules adsorb dissociatively (for example, H_2). Elements from the d block, such as iron, titanium, and chromium, show a strong activity towards all these gases, but manganese and copper are unable to adsorb N_2 and CO_2. Metals towards the left of the periodic table (for example, magnesium) can adsorb (and, in fact, react with) only the most active gas (O_2).

Self-test 22C.4 Why is iron a good catalyst for the formation of ammonia from N_2(g) and H_2(g)?

Answer: See Fig. 22C.2 and Table 22C.1

Table 22C.1 Chemisorption abilities*

	O_2	C_2H_2	C_2H_4	CO	H_2	CO_2	N_2
Ti, Cr, Mo, Fe	+	+	+	+	+	+	+
Ni, Co	+	+	+	+	+	+	–
Pd, Pt	+	+	+	+	+	–	–
Mn, Cu	+	+	+	+	±	–	–
Al, Au	+	+	+	–	–	–	–
Li, Na, K	+	+	–	–	–	–	–
Mg, Ag, Zn, Pb	+	–	–	–	–	–	–

* +, Strong chemisorption; ±, chemisorption; –, no chemisorption.

Checklist of concepts

☐ 1. A **catalyst** is a substance that accelerates a reaction but undergoes no net chemical change.

☐ 2. A **heterogeneous catalyst** is a catalyst in a different phase from the reaction mixture.

☐ 3. In the **Langmuir–Hinshelwood mechanism** of surface-catalysed reactions, the reaction takes place by encounters between molecular fragments and atoms adsorbed on the surface.

☐ 4. In the **Eley–Rideal mechanism** of a surface-catalysed reaction, a gas-phase molecule collides with another molecule already adsorbed on the surface.

☐ 5. The activity of a catalyst depends on the strength of chemisorption.

Checklist of equations

Property	Equation	Comment	Equation number
Langmuir–Hinshelwood mechanism	$v = k_r\theta_A\theta_B$	Competitive adsorption	22C.2
Eley–Rideal mechanism	$v = k_r p_B \theta_A$	Adsorption of A	22C.3

CHAPTER 22 Processes on solid surfaces

TOPIC 22A An introduction to solid surfaces

Discussion questions

22A.1 (a) What topographical features are found on clean surfaces? (b) Describe how steps and terraces can be formed by dislocations.

22A.2 Drawing from knowledge you have acquired through the text, describe the advantages and limitations of each of the microscopy, diffraction, and ionizations techniques designated by the acronyms AFM, LEED, SAM, SEM, STM, and TEM.

Exercises

22A.1(a) Calculate the frequency of molecular collisions per square centimetre of surface in a vessel containing (i) hydrogen, (ii) propane at 25 °C when the pressure is 0.10 μTorr.
22A.1(b) Calculate the frequency of molecular collisions per square centimetre of surface in a vessel containing (i) nitrogen, (ii) methane at 25 °C when the pressure is 10.0 Pa. Repeat the calculations for a pressure of 0.150 μTorr.

22A.2(a) What pressure of argon gas is required to produce a collision rate of $4.5\times10^{20}\,s^{-1}$ at 425 K on a circular patch of surface of diameter 1.5 mm?
22A.2(b) What pressure of nitrogen gas is required to produce a collision rate of $5.00\times10^{19}\,s^{-1}$ at 525 K on a circular patch of surface of diameter 2.0 mm?

Problems

22A.1 The movement of atoms and ions on a surface depends on their ability to leave one position and stick to another, and therefore on the energy changes that occur. As an illustration, consider a two-dimensional square lattice of singly charged positive and negative ions separated by 200 pm, and consider a cation on the upper terrace of this array. Calculate, by direct summation, its Coulombic interaction when it is in an empty lattice point directly above an anion. Now consider a high step in the same lattice, and let the cation move into the corner formed by the step and the terrace. Calculate the Coulombic energy for this position, and decide on the likely settling point for the cation.

22A.2 In a study of the catalytic properties of a titanium surface it was necessary to maintain the surface free from contamination. Calculate the collision frequency per square centimetre of surface made by O_2 molecules at (a) 100 kPa, (b) 1.00 Pa and 300 K. Estimate the number of collisions made with a single surface atom in each second. The conclusions underline the importance of working at very low pressures (much lower than 1 Pa, in fact) in order to study the properties of uncontaminated surfaces. Take the nearest neighbour distance as 291 pm.

22A.3 Nickel is face-centred cubic with a unit cell of side 352 pm. What is the number of atoms per square centimetre exposed on a surface formed by (a) (100), (b) (110), (c) (111) planes? Calculate the frequency of molecular collisions with a single atom in a vessel containing (i) hydrogen, (ii) propane at 25 °C when the pressure is 100 Pa and 0.10 μTorr.

22A.4 The LEED pattern from a clean unreconstructed (110) face of a metal is shown below. Sketch the LEED pattern for a surface that was reconstructed by tripling the horizontal separation between the atoms.

TOPIC 22B Adsorption and desorption

Discussion questions

22B.1 Distinguish between the following adsorption isotherms: Langmuir, BET, Temkin, and Freundlich. Indicate when and why each is likely to be appropriate.

22B.2 What approximations underlie the formulation of the Langmuir and BET isotherms?

Exercises

22B.1(a) The volume of oxygen gas at 0 °C and 104 kPa adsorbed on the surface of 1.00 g of a sample of silica at 0 °C was 0.286 cm^3 at 145.4 Torr and 1.443 cm^3 at 760 Torr. What is the value of V_{mon}?
22B.1(b) The volume of gas at 20 °C and 1.00 bar adsorbed on the surface of 1.50 g of a sample of silica at 0 °C was 1.52 cm^3 at 56.4 kPa and 2.77 cm^3 at 108 kPa. What is the value of V_{mon}?

22B.2(a) The enthalpy of adsorption of CO on a surface is found to be $-120\,kJ\,mol^{-1}$. Estimate the mean lifetime of a CO molecule on the surface at 400 K.
22B.2(b) The enthalpy of adsorption of ammonia on a nickel surface is found to be $-155\,kJ\,mol^{-1}$. Estimate the mean lifetime of an NH_3 molecule on the surface at 500 K.

22B.3(a) A certain solid sample adsorbs 0.44 mg of CO when the pressure of the gas is 26.0 kPa and the temperature is 300 K. The mass of gas adsorbed when the pressure is 3.0 kPa and the temperature is 300 K is 0.19 mg. The Langmuir isotherm is known to describe the adsorption. Find the fractional coverage of the surface at the two pressures.

22B.3(b) A certain solid sample adsorbs 0.63 mg of CO when the pressure of the gas is 36.0 kPa and the temperature is 300 K. The mass of gas adsorbed when the pressure is 4.0 kPa and the temperature is 300 K is 0.21 mg. The Langmuir isotherm is known to describe the adsorption. Find the fractional coverage of the surface at the two pressures.

22B.4(a) The adsorption of a gas is described by the Langmuir isotherm with $\alpha = 0.75\,\text{kPa}^{-1}$ at 25 °C. Calculate the pressure at which the fractional surface coverage is (i) 0.15, (ii) 0.95.

22B.4(b) The adsorption of a gas is described by the Langmuir isotherm with $\alpha = 0.548\,\text{kPa}^{-1}$ at 25 °C. Calculate the pressure at which the fractional surface coverage is (i) 0.20, (ii) 0.75.

22B.5(a) A solid in contact with a gas at 12 kPa and 25 °C adsorbs 2.5 mg of the gas and obeys the Langmuir isotherm. The enthalpy change when 1.00 mmol of the adsorbed gas is desorbed is +10.2 J. What is the equilibrium pressure for the adsorption of 2.5 mg of gas at 40 °C?

22B.5(b) A solid in contact with a gas at 8.86 kPa and 25 °C adsorbs 4.67 mg of the gas and obeys the Langmuir isotherm. The enthalpy change when 1.00 mmol of the adsorbed gas is desorbed is +12.2 J. What is the equilibrium pressure for the adsorption of the same mass of gas at 45 °C?

22B.6(a) Nitrogen gas adsorbed on charcoal to the extent of $0.921\,\text{cm}^3\,\text{g}^{-1}$ at 490 kPa and 190 K, but at 250 K the same amount of adsorption was achieved only when the pressure was increased to 3.2 MPa. What is the enthalpy of adsorption of nitrogen on charcoal?

22B.6(b) Nitrogen gas adsorbed on a surface to the extent of $1.242\,\text{cm}^3\,\text{g}^{-1}$ at 350 kPa and 180 K, but at 240 K the same amount of adsorption was achieved only when the pressure was increased to 1.02 MPa. What is the enthalpy of adsorption of nitrogen on the surface?

22B.7(a) In an experiment on the adsorption of oxygen on tungsten it was found that the same volume of oxygen was desorbed in 27 min at 1856 K and 2.0 min at 1978 K. What is the activation energy of desorption? How long would it take for the same amount to desorb at (i) 298 K, (ii) 3000 K?

22B.7(b) In an experiment on the adsorption of ethene on iron it was found that the same volume of the gas was desorbed in 1856 s at 873 K and 8.44 s at 1012 K. What is the activation energy of desorption? How long would it take for the same amount of ethene to desorb at (i) 298 K, (ii) 1500 K?

22B.8(a) The average time for which an oxygen atom remains adsorbed to a tungsten surface is 0.36 s at 2548 K and 3.49 s at 2362 K. What is the activation energy for chemisorption?

22B.8(b) The average time for which a hydrogen atom remains adsorbed on a manganese surface is 35 per cent shorter at 1000 K than at 600 K. What is the activation energy for chemisorption?

22B.9(a) For how long on average would an H atom remain on a surface at 400 K if its desorption activation energy is (i) $15\,\text{kJ}\,\text{mol}^{-1}$, (ii) $150\,\text{kJ}\,\text{mol}^{-1}$? Take $\tau_0 = 0.10\,\text{ps}$. For how long on average would the same atoms remain at 1000 K?

22B.9(b) For how long on average would an atom remain on a surface at 298 K if its desorption activation energy is (i) $20\,\text{kJ}\,\text{mol}^{-1}$, (ii) $200\,\text{kJ}\,\text{mol}^{-1}$? Take $\tau_0 = 0.12\,\text{ps}$. For how long on average would the same atoms remain at 800 K?

22B.10(a) Hydrogen iodide is very strongly adsorbed on gold but only slightly adsorbed on platinum. Assume the adsorption follows the Langmuir isotherm and predict the order of the HI decomposition reaction on each of the two metal surfaces.

22B.10(b) Suppose it is known that ozone adsorbs on a particular surface in accord with a Langmuir isotherm. How could you use the pressure dependence of the fractional coverage to distinguish between adsorption (i) without dissociation, (ii) with dissociation into $O + O_2$, (iii) with dissociation into $O + O + O$?

Problems

22B.1 Use mathematical software, a spreadsheet, or the *Living graphs* on the web site of this book to perform the following calculations: (a) Use eqn 22B.2 to generate a family of curves showing the dependence of $1/\theta$ on $1/p$ for several values of α. (b) Use eqn 22B.4 to generate a family of curves showing the dependence of $1/\theta$ on $1/p$ for several values of α. On the basis of your results from parts (a) and (b), discuss how plots of $1/\theta$ against $1/p$ can be used to distinguish between adsorption with and without dissociation. (c) Use eqn 22B.6 to generate a family of curves showing the dependence of $zV_{\text{mon}}/(1-z)V$ on z for different values of c.

22B.2 The following data are for the chemisorption of hydrogen on copper powder at 25 °C. Confirm that they fit the Langmuir isotherm at low coverages. Then find the value of α for the adsorption equilibrium and the adsorption volume corresponding to complete coverage.

p/Pa	25	129	253	540	1000	1593
V/cm^3	0.042	0.163	0.221	0.321	0.411	0.471

22B.3 The data for the adsorption of ammonia on barium fluoride are reported in the following tables. Confirm that they fit a BET isotherm and find values of c and V_{mon}.

(a) $\theta = 0\,°\text{C}$, $p^* = 429.6\,\text{kPa}$:

p/kPa	14.0	37.6	65.6	79.2	82.7	100.7	106.4
V/cm^3	11.1	13.5	14.9	16.0	15.5	17.3	16.5

(b) $\theta = 18.6\,°\text{C}$, $p^* = 819.7\,\text{kPa}$:

p/kPa	5.3	8.4	14.4	29.2	62.1	74.0	80.1	102.0
V/cm^3	9.2	9.8	10.3	11.3	12.9	13.1	13.4	14.1

22B.4 The following data have been obtained for the adsorption of H_2 on the surface of 1.00 g of copper at 0 °C. The volume of H_2 below is the volume that the gas would occupy at STP (0 °C and 1 atm).

p/atm	0.050	0.100	0.150	0.200	0.250
V/cm^3	23.8	13.3	8.70	6.80	5.71

Determine the volume of H_2 necessary to form a monolayer and estimate the surface area of the copper sample. The density of liquid hydrogen is $0.708\,\text{g}\,\text{cm}^{-3}$.

22B.5‡ M.-G. Olivier and R. Jadot (*J. Chem. Eng. Data* **42**, 230 (1997)) studied the adsorption of butane on silica gel. They report the following amounts of absorption (in moles per kilogram of silica gel) at 303 K:

p/kPa	31.00	38.22	53.03	76.38	101.97
n/(mol kg^{-1})	1.00	1.17	1.54	2.04	2.49

p/kPa	130.47	165.06	182.41	205.75	219.91
n/(mol kg^{-1})	2.90	3.22	3.30	3.35	3.36

Fit these data to a Langmuir isotherm, and determine the value of n that corresponds to complete coverage and the constant K.

‡ These problems were supplied by Charles Trapp and Carmen Giunta.

22B.6 The designers of a new industrial plant wanted to use a catalyst code-named CR-1 in a step involving the fluorination of butadiene. As a first step in the investigation they determined the form of the adsorption isotherm. The volume of butadiene adsorbed per gram of CR-1 at 15 °C varied with pressure as given below. Is the Langmuir isotherm suitable at this pressure?

p/kPa	13.3	26.7	40.0	53.3	66.7	80.0
V/cm³	17.9	33.0	47.0	60.8	75.3	91.3

Investigate whether the BET isotherm gives a better description of the adsorption of butadiene on CR-1. At 15 °C, p^*(butadiene) = 200 kPa. Find V_{mon} and c.

22B.7‡ C. Huang and W.P. Cheng (*J. Colloid Interface Sci.* **188**, 270 (1997)) examined the adsorption of the hexacyanoferrate(III) ion, $[Fe(CN)_6]^{3-}$, on γ-Al_2O_3 from aqueous solution. They modelled the adsorption with a modified Langmuir isotherm, obtaining the following values of α at pH = 6.5:

T/K	283	298	308	318
$10^{-11}\alpha$	2.642	2.078	1.286	1.085

Determine the isosteric enthalpy of adsorption, $\Delta_{ads}H^{\ominus}$, at this pH. The researchers also reported $\Delta_{ads}S^{\ominus} = +146\ \mathrm{J\ mol^{-1}\ K^{-1}}$ under these conditions. Determine $\Delta_{ads}G^{\ominus}$.

22B.8‡ In a study relevant to automobile catalytic converters, C.E. Wartnaby et al. (*J. Phys. Chem.* **100**, 12483 (1996)) measured the enthalpy of adsorption of CO, NO, and O_2 on initially clean platinum (110) surfaces. They report $\Delta_{ads}H^{\ominus}$ for NO to be −160 kJ mol⁻¹. How much more strongly adsorbed is NO at 500 °C than at 400 °C?

22B.9‡ The removal or recovery of volatile organic compounds (VOCs) from exhaust gas streams is an important process in environmental engineering. Activated carbon has long been used as an adsorbent in this process, but the presence of moisture in the stream reduces its effectiveness. M.-S. Chou and J.-H. Chiou (*J. Envir. Engrg. ASCE* **123**, 437(1997)) have studied the effect of moisture content on the adsorption capacities of granular activated carbon (GAC) for normal hexane and cyclohexane in air streams. From their data for dry streams containing cyclohexane, shown in the following table, they conclude that GAC obeys a Langmuir type model in which $q_{VOC,RH=0} = abc_{VOC}/(1 + bc_{VOC})$, where $q = m_{VOC}/m_{GAC}$, RH denotes relative humidity, a the maximum adsorption capacity, b is an affinity parameter, and p is the abundance in parts per million (ppm). The following table gives values of $q_{VOC,RH=0}$ for cyclohexane:

c/ppm	33.6 °C	41.5 °C	57.4 °C	76.4 °C	99 °C
200	0.080	0.069	0.052	0.042	0.027
500	0.093	0.083	0.072	0.056	0.042
1000	0.101	0.088	0.076	0.063	0.045
2000	0.105	0.092	0.083	0.068	0.052
3000	0.112	0.102	0.087	0.072	0.058

(a) By linear regression of $1/q_{VOC,RH=0}$ against $1/c_{VOC}$, test the goodness of fit and determine values of a and b. (b) The parameters a and b can be related to $\Delta_{ads}H$, the enthalpy of adsorption, and Δ_bH, the difference in activation energy for adsorption and desorption of the VOC molecules, through Arrhenius type equations of the form $a = k_a e^{-\Delta_{ads}H/RT}$ and $b = k_b e^{-\Delta_b H/RT}$. Test the goodness of fit of the data to these equations and obtain values for k_a, k_b, $\Delta_{ads}H$, and Δ_bH. (c) What interpretation might you give to k_a and k_b?

22B.10 The adsorption of solutes on solids from liquids often follows a Freundlich isotherm. Check the applicability of this isotherm to the following data for the adsorption of acetic acid on charcoal at 25 °C and find the values of the parameters c_1 and c_2.

[acid]/(mol dm⁻³)	0.05	0.10	0.50	1.0	1.5
w_a/g	0.04	0.06	0.12	0.16	0.19

where w_a is the mass adsorbed per gram of charcoal.

22B.11‡ A. Akgerman and M. Zardkoohi (*J. Chem. Eng. Data* **41**, 185 (1996)) examined the adsorption of phenol from aqueous solution on to fly ash at 20 °C. They fitted their observations to a Freundlich isotherm of the form $c_{ads} = Kc_{sol}^{1/n}$, where c_{ads} is the concentration of adsorbed phenol and c_{sol} is the concentration of aqueous phenol. Among the data reported are the following:

c_{sol}/(mg g⁻¹)	8.26	15.65	25.43	31.74	40.00
c_{ads}/(mg g⁻¹)	4.41	9.2	35.2	52.0	67.2

Determine the constants K and n. What further information would be necessary in order to express the data in terms of fractional coverage, θ?

22B.12‡ The following data were obtained for the extent of adsorption, s, of acetone on charcoal from an aqueous solution of molar concentration, c, at 18 °C:

c/(mmol dm⁻³)	15.0	23.0	42.0	84.0	165	390	800
s/(mmol acetone/g charcoal)	0.60	0.75	1.05	1.50	2.15	3.50	5.10

Which isotherm fits this data best, Langmuir, Freundlich, or Temkin?

22B.13‡ M.-S. Chou and J.-H. Chiou (*J. Envir. Engrg. ASCE* **123**, 437(1997)) have studied the effect of moisture content on the adsorption capacities of granular activated carbon (GAC, Norit PK 1-3) for the volatile organic compounds (VOCs) normal hexane and cyclohexane in air streams. The following table shows the adsorption capacities ($q_{water} = m_{water}/m_{GAC}$) of GAC for pure water from moist air streams as a function of relative humidity (RH) in the absence of VOCs at 41.5 °C:

RH	0.00	0.26	0.49	0.57	0.80	1.00
q_{water}	0.00	0.026	0.072	0.091	0.161	0.229

The authors conclude that the data at this and other temperatures obey a Freundlich type isotherm, $q_{water} = k(\mathrm{RH})^{1/n}$. (a) Test this hypothesis for their data at 41.5 °C and determine the constants k and n. (b) Why might VOCs obey the Langmuir model, but water the Freundlich model? (c) When both water vapour and cyclohexane were present in the stream the values given in the table below were determined for the ratio $r_{VOC} = q_{VOC}/q_{VOC,RH=0}$ at 41.5 °C:

The authors propose that these data fit the equation $r_{VOC} = 1 - q_{water}$. Test their

RH	0.00	0.10	0.25	0.40	0.53	0.76	0.81
r_{VOC}	1.00	0.98	0.91	0.84	0.79	0.67	0.61

proposal and determine values for k and n and compare to those obtained in part (b) for pure water. Suggest reasons for any differences.

22B.14‡ The release of petroleum products by leaky underground storage tanks is a serious threat to clean ground water. BTEX compounds (benzene, toluene, ethylbenzene, and xylenes) are of primary concern due to their ability to cause health problems at low concentrations. D.S. Kershaw et al. (*J. Geotech. Geoenvir. Engrg.* **123**, 324 (1997)) have studied the ability of ground tyre rubber to sorb (adsorb and absorb) benzene and o-xylene. Though sorption involves more than surface interactions, sorption data is usually found to fit one of the adsorption isotherms. In this study, the authors have tested how well their data fit the linear ($q = Kc_{eq}$), Freundlich ($q = K_F c_{eq}^{1/n}$), and Langmuir ($q = K_L M c_{eq}/(1 + K_L c_{eq})$) type isotherms, where q is the mass of solvent sorbed per gram of ground rubber (in milligrams per gram), the Ks and M are empirical constants, and c_{eq} the equilibrium concentration of contaminant in solution (in milligrams per litre). (a) Determine the units of the empirical constants. (b) Determine which of the isotherms best fits the data in the following table for the sorption of benzene on ground rubber:

c_{eq}/(mg dm⁻³)	97.10	36.10	10.40	6.51	6.21	2.48
q/(mg g⁻¹)	7.13	4.60	1.80	1.10	0.55	0.31

(c) Compare the sorption efficiency of ground rubber to that of granulated activated charcoal which for benzene has been shown to obey the Freundlich isotherm in the form $q = 1.0c_{eq}^{1.6}$ with coefficient of determination $R^2 = 0.94$.

TOPIC 22C Heterogeneous catalysis

Discussion questions

22C.1 Describe the essential features of the Langmuir–Hinshelwood and Eley–Rideal mechanisms for surface-catalysed reactions.

22C.2 Account for the dependence of catalytic activity of a surface on the strength of chemisorption, as shown in Fig. 22B.8.

Exercises

22C.1(a) A monolayer of N_2 molecules is adsorbed on the surface of 1.00 g of an Fe/Al_2O_3 catalyst at 77 K, the boiling point of liquid nitrogen. Upon warming, the nitrogen occupies 3.86 cm^3 at 0 °C and 760 Torr. What is the surface area of the catalyst?

22C.1(b) A monolayer of CO molecules is adsorbed on the surface of 1.00 g of an Fe/Al_2O_3 catalyst at 77 K, the boiling point of liquid nitrogen. Upon warming, the carbon monoxide occupies 3.75 cm^3 at 0 °C and 1.00 bar. What is the surface area of the catalyst?

Problem

22C.1 In some catalytic reactions the products adsorb more strongly than the reacting gas. This is the case, for instance, in the catalytic decomposition of ammonia on platinum at 1000 °C. As a first step in examining the kinetics of this type of process, show that the rate of ammonia decomposition should follow

$$\frac{\mathrm{d}p_{\mathrm{NH_3}}}{\mathrm{d}t} = -k_c \frac{p_{\mathrm{NH_3}}}{p_{\mathrm{H_2}}}$$

in the limit of very strong adsorption of hydrogen. Start by showing that when a gas J adsorbs very strongly, and its pressure is p_J, that the fraction of uncovered sites is approximately $1/Kp_J$. Solve the rate equation for the catalytic decomposition of NH_3 on platinum and show that a plot of $F(t) = (1/t) \times \ln(p/p_0)$ against $G(t) = (p - p_0)/t$, where p is the pressure of ammonia, should give a straight line from which k_c can be determined. Check the rate law on the basis of the following data, and find k_c for the reaction.

t/s	0	30	60	100	160	200	250
p/kPa	13.3	11.7	11.2	10.7	10.3	9.9	9.6

Integrated activities

22.1 Although the attractive van der Waals interaction between individual molecules varies as R^{-6} the interaction of a molecule with a nearby solid (a homogeneous collection of molecules) varies as R^{-3}, where R is its vertical distance above the surface. Confirm this assertion. Calculate the interaction energy between an Ar atom and the surface of solid argon on the basis of a Lennard-Jones (6,12)-potential. Estimate the equilibrium distance of an atom above the surface.

22.2 Electron microscopes can obtain images with much higher resolution than optical microscopes because of the short wavelength obtainable from a beam of electrons. For electrons moving at speeds close to c, the speed of light, the expression for the de Broglie wavelength (eqn 7A.14, $\lambda = h/p$) needs to be corrected for relativistic effects:

$$\lambda = \frac{h}{\left\{2m_e e\Delta\phi\left(1 + \frac{e\Delta\phi}{2m_e c^2}\right)\right\}^{1/2}}$$

where c is the speed of light in vacuum and $\Delta\phi$ is the potential difference through which the electrons are accelerated. (a) Use this expression to calculate the de Broglie wavelength of electrons accelerated through 50 kV. (b) Is the relativistic correction important?

22.3 The forces measured by AFM arise primarily from interactions between electrons of the stylus and on the surface. To get an idea of the magnitudes of these forces, calculate the force acting between two electrons separated by 2.0 nm. To calculate the force between the electrons, use $F = -\mathrm{d}V/\mathrm{d}r$ where V is their mutual Coulombic potential energy and r is their separation.

22.4 To appreciate the distance dependence of the tunnelling current in scanning tunnelling microscopy, suppose that the electron in the gap between sample and needle has an energy 2.0 eV smaller than the barrier height. By what factor would the current drop if the needle is moved from $L_1 = 0.50$ nm to $L_2 = 0.60$ nm from the surface?

RESOURCE SECTION

Contents

PART 1 Common integrals

Algebraic functions

A.1 $\int x^n \mathrm{d}x = \frac{x^{n+1}}{n+1} + \text{constant},\ n \neq -1$

A.2 $\int \frac{1}{x} \mathrm{d}x = \ln x + \text{constant}$

Exponential functions

E.1 $\int_0^\infty x^n \mathrm{e}^{-ax} \mathrm{d}x = \frac{n!}{a^{n+1}},\quad n! = n(n-1)\ldots 1;\ 0! \equiv 1$

E.2 $\int_0^\infty \frac{x^4 \mathrm{e}^x}{(\mathrm{e}^x - 1)^2} \mathrm{d}x = \frac{\pi^4}{15}$

Gaussian functions

G.1 $\int_0^\infty \mathrm{e}^{-ax^2} \mathrm{d}x = \frac{1}{2}\left(\frac{\pi}{a}\right)^{1/2}$

G.2 $\int_0^\infty x\mathrm{e}^{-ax^2} \mathrm{d}x = \frac{1}{2a}$

G.3 $\int_0^\infty x^2 \mathrm{e}^{-ax^2} \mathrm{d}x = \frac{1}{4}\left(\frac{\pi}{a^3}\right)^{1/2}$

G.4 $\int_0^\infty x^3 \mathrm{e}^{-ax^2} \mathrm{d}x = \frac{1}{2a^2}$

G.5 $\int_0^\infty x^4 \mathrm{e}^{-ax^2} \mathrm{d}x = \frac{3}{8a^2}\left(\frac{\pi}{a}\right)^{1/2}$

G.6 $\operatorname{erf} z = \frac{2}{\pi^{1/2}} \int_0^z \mathrm{e}^{-x^2} \mathrm{d}x \qquad \operatorname{erfc} z = 1 - \operatorname{erf} z$

G.7 $\int_0^\infty x^{2m+1} \mathrm{e}^{-ax^2} \mathrm{d}x = \frac{m!}{2a^{m+1}}$

G.8 $\int_0^\infty x^{2m} \mathrm{e}^{-ax^2} \mathrm{d}x = \frac{(2m-1)!!}{2^{m+1} a^m}\left(\frac{\pi}{a}\right)^{1/2}$

$(2m-1)!! = 1 \times 3 \times 5 \cdots \times (2m-1)$

Trigonometric functions

T.1 $\int \sin ax \, \mathrm{d}x = -\frac{1}{a} \cos ax + \text{constant}$

T.2 $\int \sin^2 ax \, \mathrm{d}x = \frac{1}{2}x - \frac{\sin 2ax}{4a} + \text{constant}$

T.3 $\int \sin^3 ax \, \mathrm{d}x = -\frac{(\sin^2 ax + 2)\cos ax}{3a} + \text{constant}$

T.4 $\int \sin^4 ax \, \mathrm{d}x = \frac{3x}{8} - \frac{3}{8a} \sin ax \cos ax - \frac{1}{4a} \sin^3 ax \cos ax + \text{constant}$

T.5 $\int \sin ax \sin bx \, \mathrm{d}x = \frac{\sin(a-b)x}{2(a-b)} - \frac{\sin(a+b)x}{2(a+b)} + \text{constant},\ a^2 \neq b^2$

T.6 $\int_0^L \sin nax \sin^2 ax \, \mathrm{d}x = -\frac{1}{2a}\left\{\frac{1}{n} - \frac{1}{2(n+2)} - \frac{1}{2(n-2)}\right\} \times \{(-1)^n - 1\}$

T.7 $\int \sin ax \cos ax \, \mathrm{d}x = \frac{1}{2a} \sin^2 ax + \text{constant}$

T.8 $\int \sin bx \cos ax \, \mathrm{d}x = \frac{\cos(a-b)x}{2(a-b)} - \frac{\cos(a+b)x}{2(a+b)} + \text{constant},\ a^2 \neq b^2$

T.9 $\int x \sin ax \sin bx \, \mathrm{d}x = -\frac{\mathrm{d}}{\mathrm{d}a} \int \sin bx \cos ax \, \mathrm{d}x$

T.10 $\int \cos^2 ax \sin ax \, \mathrm{d}x = -\frac{1}{3a} \cos^3 ax + \text{constant}$

T.11 $\int x \sin^2 ax \, \mathrm{d}x = \frac{x^2}{4} - \frac{x \sin 2ax}{4a} - \frac{\cos 2ax}{8a^2} + \text{constant}$

T.12 $\int x^2 \sin^2 ax \, \mathrm{d}x = \frac{x^3}{6} - \left(\frac{x^2}{4a} - \frac{1}{8a^3}\right) \sin 2ax - \frac{x \cos 2ax}{4a^2} + \text{constant}$

T.13 $\int x \cos ax \, \mathrm{d}x = \frac{1}{a^2} \cos ax + \frac{x}{a} \sin ax + \text{constant}$

PART 2 Units

Table A.1 Some common units

Physical quantity	Name of unit	Symbol for unit	Value*
Time	minute	min	60 s
	hour	h	3600 s
	day	d	86 400 s
	year	a	31 556 952 s
Length	ångström	Å	10^{-10} m
Volume	litre	L, l	1 dm^3
Mass	tonne	t	10^3 kg
Pressure	bar	bar	10^5 Pa
	atmosphere	atm	101.325 kPa
Energy	electronvolt	eV	$1.602\ 177\ 33 \times 10^{-19}$ J 96.485 31 kJ mol^{-1}

* All values are exact, except for the definition of 1 eV, which depends on the measured value of *e*, and the year, which is not a constant and depends on a variety of astronomical assumptions.

Table A.2 Common SI prefixes

Prefix	y	z	a	f	p	n	μ	m	c	d
Name	yocto	zepto	atto	femto	pico	nano	micro	milli	centi	deci
Factor	10^{-24}	10^{-21}	10^{-18}	10^{-15}	10^{-12}	10^{-9}	10^{-6}	10^{-3}	10^{-2}	10^{-1}
Prefix	da	h	k	M	G	T	P	E	Z	Y
Name	deca	hecto	kilo	mega	giga	tera	peta	exa	zeta	yotta
Factor	10	10^2	10^3	10^6	10^9	10^{12}	10^{15}	10^{18}	10^{21}	10^{24}

Table A.3 The SI base units

Physical quantity	Symbol for quantity	Base unit
Length	l	metre, m
Mass	m	kilogram, kg
Time	t	second, s
Electric current	I	ampere, A
Thermodynamic temperature	T	kelvin, K
Amount of substance	n	mole, mol
Luminous intensity	I_v	candela, cd

Table A.4 A selection of derived units

Physical quantity	Derived unit*	Name of derived unit
Force	1 kg m s^{-2}	newton, N
Pressure	1 kg m^{-1} s^{-2} 1 N m^{-2}	pascal, Pa
Energy	1 kg m^2 s^{-2} 1 N m 1 Pa m^3	joule, J
Power	1 kg m^2 s^{-3} 1 J s^{-1}	watt, W

* Equivalent definitions in terms of derived units are given following the definition in terms of base units.

PART 3 Data

The following is a directory of all tables in the text; those included in this *Resource section* are marked with an asterisk. The remainder will be found on the pages indicated. These tables reproduce and expand the data given in the short tables in the text, and follow their numbering. Standard states refer to a pressure of $p^{\ominus} = 1$ bar. The general references are as follows:

AIP: D.E. Gray (ed.), *American Institute of Physics handbook*. McGraw-Hill, New York (1972).

E: J. Emsley, *The elements*. Oxford University Press, Oxford (1991).

HCP: D.R. Lide (ed.), *Handbook of chemistry and physics*. CRC Press, Boca Raton (2000).

JL: A.M. James and M.P. Lord, *Macmillan's chemical and physical data*. Macmillan, London (1992).

KL: G.W.C. Kaye and T.H. Laby (ed.), *Tables of physical and chemical constants*. Longman, London (1973).

LR: G.N. Lewis and M. Randall, revised by K.S. Pitzer and L. Brewer, *Thermodynamics*. McGraw-Hill, New York (1961).

NBS: *NBS tables of chemical thermodynamic properties*, published as *J. Phys. Chem. Reference Data*, **11**, Supplement 2 (1982).

RS: R.A. Robinson and R.H. Stokes, *Electrolyte solutions*, Butterworth, London (1959).

TDOC: J.B. Pedley, J.D. Naylor, and S.P. Kirby, *Thermochemical data of organic compounds*. Chapman & Hall, London (1986).

Table 0.1 Physical properties of selected materials

	$\rho/(\text{g cm}^{-3})$ at 293 K†	T_f/K	T_b/K		$\rho/(\text{g cm}^{-3})$ at 293 K†	T_f/K	T_b/K
Elements				**Inorganic compounds**			
Aluminium(s)	2.698	933.5	2740	$CaCO_3$(s, calcite)	2.71	1612	1171^{d}
Argon(g)	1.381	83.8	87.3	$CuSO_4{\cdot}5H_2O$(s)	2.284	383($-H_2O$)	423($-5H_2O$)
Boron(s)	2.340	2573	3931	HBr(g)	2.77	184.3	206.4
Bromine(l)	3.123	265.9	331.9	HCl(g)	1.187	159.0	191.1
Carbon(s, gr)	2.260	3700^{s}		HI(g)	2.85	222.4	237.8
Carbon(s, d)	3.513			H_2O(l)	0.997	273.2	373.2

(*Continued*)

Table 0.1 (Continued)

	$\rho/(\mathrm{g\ cm^{-3}})$ at 293 K†	T_f/K	T_b/K		$\rho/(\mathrm{g\ cm^{-3}})$ at 293 K†	T_f/K	T_b/K
Elements (Continued)				**Inorganic compounds (continued)**			
Chlorine(g)	1.507	172.2	239.2	D_2O(l)	1.104	277.0	374.6
Copper(s)	8.960	1357	2840	NH_3(g)	0.817	195.4	238.8
Fluorine(g)	1.108	53.5	85.0	KBr(s)	2.750	1003	1708
Gold(s)	19.320	1338	3080	KCl(s)	1.984	1049	1773[s]
Helium(g)	0.125		4.22	NaCl(s)	2.165	1074	1686
Hydrogen(g)	0.071	14.0	20.3	H_2SO_4(l)	1.841	283.5	611.2
Iodine(s)	4.930	386.7	457.5				
Iron(s)	7.874	1808	3023	**Organic compounds**			
Krypton(g)	2.413	116.6	120.8	Acetaldehyde, CH_3CHO(l)	0.788	152	293
Lead(s)	11.350	600.6	2013	Acetic acid, CH_3COOH(l)	1.049	289.8	391
Lithium(s)	0.534	453.7	1620	Acetone, $(CH_3)_2CO$(l)	0.787	178	329
Magnesium(s)	1.738	922.0	1363	Aniline, $C_6H_5NH_2$(l)	1.026	267	457
Mercury(l)	13.546	234.3	629.7	Anthracene, $C_{14}H_{10}$(s)	1.243	490	615
Neon(g)	1.207	24.5	27.1	Benzene, C_6H_6(l)	0.879	278.6	353.2
Nitrogen(g)	0.880	63.3	77.4	Carbon tetrachloride, CCl_4(l)	1.63	250	349.9
Oxygen(g)	1.140	54.8	90.2	Chloroform, $CHCl_3$(l)	1.499	209.6	334
Phosphorus(s, wh)	1.820	317.3	553	Ethanol, C_2H_5OH(l)	0.789	156	351.4
Potassium(s)	0.862	336.8	1047	Formaldehyde, HCHO(g)		181	254.0
Silver(s)	10.500	1235	2485	Glucose, $C_6H_{12}O_6$(s)	1.544	415	
Sodium(s)	0.971	371.0	1156	Methane, CH_4(g)		90.6	111.6
Sulfur(s, α)	2.070	386.0	717.8	Methanol, CH_3OH(l)	0.791	179.2	337.6
Uranium(s)	18.950	1406	4018	Naphthalene, $C_{10}H_8$(s)	1.145	353.4	491
Xenon(g)	2.939	161.3	166.1	Octane, C_8H_{18}(l)	0.703	216.4	398.8
Zinc(s)	7.133	692.7	1180	Phenol, C_6H_5OH(s)	1.073	314.1	455.0
				Sucrose, $C_{12}H_{22}O_{11}$(s)	1.588	457[d]	

d: decomposes; s: sublimes; Data: AIP, E, HCP, KL. † For gases, at their boiling points.

Table 0.2 Masses and natural abundances of selected nuclides

Nuclide		m/m_u	Abundance/%
H	^{1}H	1.0078	99.985
	^{2}H	2.0140	0.015
He	^{3}He	3.0160	0.000 13
	^{4}He	4.0026	100
Li	^{6}Li	6.0151	7.42
	^{7}Li	7.0160	92.58
B	^{10}B	10.0129	19.78
	^{11}B	11.0093	80.22
C	^{12}C	12*	98.89
	^{13}C	13.0034	1.11
N	^{14}N	14.0031	99.63
	^{15}N	15.0001	0.37
O	^{16}O	15.9949	99.76
	^{17}O	16.9991	0.037
	^{18}O	17.9992	0.204
F	^{19}F	18.9984	100
P	^{31}P	30.9738	100
S	^{32}S	31.9721	95.0
	^{33}S	32.9715	0.76

Table 0.2 (Continued)

	Nuclide	m/m_u	Abundance/%
	^{34}S	33.9679	4.22
Cl	^{35}Cl	34.9688	75.53
	^{37}Cl	36.9651	24.4
Br	^{79}Br	78.9183	50.54
	^{81}Br	80.9163	49.46
I	^{127}I	126.9045	100

* Exact value.

Table 1B.1 Collision cross-sections, σ/nm^2

Ar	0.36
C_2H_4	0.64
C_6H_6	0.88
CH_4	0.46
Cl_2	0.93
CO_2	0.52
H_2	0.27
He	0.21
N_2	0.43
Ne	0.24
O_2	0.40
SO_2	0.58

Data: KL.

Table 1C.1 Second virial coefficients, $B/(cm^3\ mol^{-1})$

	100 K	273 K	373 K	600 K
Air	−167.3	−13.5	3.4	19.0
Ar	−187.0	−21.7	−4.2	11.9
CH_4		−53.6	−21.2	8.1
CO_2		−142	−72.2	−12.4
H_2	−2.0	13.7	15.6	
He	11.4	12.0	11.3	10.4
Kr		−62.9	−28.7	1.7
N_2	−160.0	−10.5	6.2	21.7
Ne	−6.0	10.4	12.3	13.8
O_2	−197.5	−22.0	−3.7	12.9
Xe		−153.7	−81.7	−19.6

Data: AIP, JL. The values relate to the expansion in eqn 1C.3 of Topic 1C; convert to eqn 1C.3 using $B'=B/RT$.
For Ar at 273 K, $C = 1200\ cm^6\ mol^{-1}$.

Table 1C.2 Critical constants of gases

	p_c/atm	$V_c/(cm^3\ mol^{-1})$	T_c/K	Z_c	T_B/K
Ar	48.0	75.3	150.7	0.292	411.5
Br_2	102	135	584	0.287	
C_2H_4	50.50	124	283.1	0.270	
C_2H_6	48.20	148	305.4	0.285	
C_6H_6	48.6	260	562.7	0.274	
CH_4	45.6	98.7	190.6	0.288	510.0
Cl_2	76.1	124	417.2	0.276	
CO_2	72.9	94.0	304.2	0.274	714.8
F_2	55	144			
H_2	12.8	34.99	33.23	0.305	110.0
H_2O	218.3	55.3	647.4	0.227	
HBr	84.0	363.0			
HCl	81.5	81.0	324.7	0.248	
He	2.26	57.8	5.2	0.305	22.64
HI	80.8	423.2			
Kr	54.27	92.24	209.39	0.291	575.0
N_2	33.54	90.10	126.3	0.292	327.2
Ne	26.86	41.74	44.44	0.307	122.1

(Continued)

Table 1C.2 (Continued)

	p_c/atm	V_c/(cm^3 mol^{-1})	T_c/K	Z_c	T_B/K
NH_3	111.3	72.5	405.5	0.242	
O_2	50.14	78.0	154.8	0.308	405.9
Xe	58.0	118.8	289.75	0.290	768.0

Data: AIP, KL.

Table 1C.3 van der Waals coefficients

	a/(atm dm^6 mol^{-2})	b/(10^{-2} dm^3 mol^{-1})		a/(atm dm^6 mol^{-2})	b/(10^{-2} dm^3 mol^{-1})
Ar	1.337	3.20	H_2S	4.484	4.34
C_2H_4	4.552	5.82	He	0.0341	2.38
C_2H_6	5.507	6.51	Kr	5.125	1.06
C_6H_6	18.57	11.93	N_2	1.352	3.87
CH_4	14.61	4.31	Ne	0.205	1.67
Cl_2	6.260	5.42	NH_3	4.169	3.71
CO	1.453	3.95	O_2	1.364	3.19
CO_2	3.610	4.29	SO_2	6.775	5.68
H_2	0.2420	2.65	Xe	4.137	5.16
H_2O	5.464	3.05			

Data: HCP.

Table 2B.1 Temperature variation of molar heat capacities, $C_{p,m}$/(J K^{-1} mol^{-1}) = $a + bT + c/T^2$

	a	b/(10^{-3} K^{-1})	c/(10^5 K^2)
Monatomic gases			
	20.78	0	0
Other gases			
Br_2	37.32	0.50	−1.26
Cl_2	37.03	0.67	−2.85
CO_2	44.22	8.79	−8.62
F_2	34.56	2.51	−3.51
H_2	27.28	3.26	0.50
I_2	37.40	0.59	−0.71
N_2	28.58	3.77	−0.50
NH_3	29.75	25.1	−1.55
O_2	29.96	4.18	−1.67
Liquids (from melting to boiling)			
$C_{10}H_8$, naphthalene	79.5	0.4075	0
I_2	80.33	0	0
H_2O	75.29	0	0
Solids			
Al	20.67	12.38	0
C (graphite)	16.86	4.77	−8.54
$C_{10}H_8$, naphthalene	−110	936	0
Cu	22.64	6.28	0
I_2	40.12	49.79	0
NaCl	45.94	16.32	0
Pb	22.13	11.72	0.96

Source: Mostly LR.

Table 2C.1 Standard enthalpies of fusion and vaporization at the transition temperature, $\Delta_{trs}H^{\ominus}/(\mathrm{kJ\ mol^{-1}})$

	T_f/K	Fusion	T_b/K	Vaporization		T_f/K	Fusion	T_b/K	Vaporization
Elements					**Inorganic compounds**				
Ag	1234	11.30	2436	250.6	CO_2	217.0	8.33	194.6	25.23^s
Ar	83.81	1.188	87.29	6.506	CS_2	161.2	4.39	319.4	26.74
Br_2	265.9	10.57	332.4	29.45	H_2O	273.15	6.008	373.15	40.656
Cl_2	172.1	6.41	239.1	20.41					44.016 at 298 K
F_2	53.6	0.26	85.0	3.16	H_2S	187.6	2.377	212.8	18.67
H_2	13.96	0.117	20.38	0.916	H_2SO_4	283.5	2.56		
He	3.5	0.021	4.22	0.084	NH_3	195.4	5.652	239.7	23.35
Hg	234.3	2.292	629.7	59.30	**Organic compounds**				
I_2	386.8	15.52	458.4	41.80	CH_4	90.68	0.941	111.7	8.18
N_2	63.15	0.719	77.35	5.586	CCl_4	250.3	2.47	349.9	30.00
Na	371.0	2.601	1156	98.01	C_2H_6	89.85	2.86	184.6	14.7
O_2	54.36	0.444	90.18	6.820	C_6H_6	278.61	10.59	353.2	30.8
Xe	161	2.30	165	12.6	C_6H_{14}	178	13.08	342.1	28.85
K	336.4	2.35	1031	80.23	$C_{10}H_8$	354	18.80	490.9	51.51
					CH_3OH	175.2	3.16	337.2	35.27
									37.99 at 298 K
					C_2H_5OH	158.7	4.60	352	43.5

Data: AIP; s denotes sublimation.

Table 2C.3 Lattice enthalpies at 298 K, $\Delta H_L/(\mathrm{kJ\ mol^{-1}})$. See Table 18B.4.

Table 2C.4 Thermodynamic data for organic compounds at 298 K

	$M/(\mathrm{g\ mol^{-1}})$	$\Delta_f H^{\ominus}/(\mathrm{kJ\ mol^{-1}})$	$\Delta_f G^{\ominus}/(\mathrm{kJ\ mol^{-1}})$	$S_m^{\ominus}/(\mathrm{J\ K^{-1}\ mol^{-1}})$	$C_{p,m}^{\ominus}/(\mathrm{J\ K^{-1}\ mol^{-1}})$	$\Delta_c H^{\ominus}/(\mathrm{kJ\ mol^{-1}})$
C(s) (graphite)	12.011	0	0	5.740	8.527	−393.51
C(s) (diamond)	12.011	+1.895	+2.900	2.377	6.113	−395.40
CO_2(g)	44.040	−393.51	−394.36	213.74	37.11	
Hydrocarbons						
CH_4(g), methane	16.04	−74.81	−50.72	186.26	35.31	−890
CH_3(g), methyl	15.04	+145.69	+147.92	194.2	38.70	
C_2H_2(g), ethyne	26.04	+226.73	+209.20	200.94	43.93	−1300
C_2H_4(g), ethene	28.05	+52.26	+68.15	219.56	43.56	−1411
C_2H_6(g), ethane	30.07	−84.68	−32.82	229.60	52.63	−1560
C_3H_6(g), propene	42.08	+20.42	+62.78	267.05	63.89	−2058
C_3H_6(g), cyclopropane	42.08	+53.30	+104.45	237.55	55.94	−2091
C_3H_8(g), propane	44.10	−103.85	−23.49	269.91	73.5	−2220
C_4H_8(g), 1-butene	56.11	−0.13	+71.39	305.71	85.65	−2717
C_4H_8(g), *cis*-2-butene	56.11	−6.99	+65.95	300.94	78.91	−2710
C_4H_8(g), *trans*-2-butene	56.11	−11.17	+63.06	296.59	87.82	−2707
C_4H_{10}(g), butane	58.13	−126.15	−17.03	310.23	97.45	−2878

(Continued)

Table 2C.4 (Continued)

	M/(g mol^{-1})	$\Delta_f H^\ominus$/(kJ mol^{-1})	$\Delta_f G^\ominus$/(kJ mol^{-1})	$S_m^\ominus$/(J K^{-1} mol^{-1})†	$C_{p,m}^\ominus$/(J K^{-1} mol^{-1})	$\Delta_c H^\ominus$/(kJ mol^{-1})
C_5H_{12}(g), pentane	72.15	−146.44	−8.20	348.40	120.2	−3537
C_5H_{12}(l)	72.15	−173.1				
C_6H_6(l), benzene	78.12	+49.0	+124.3	173.3	136.1	−3268
C_6H_6(g)	78.12	+82.93	+129.72	269.31	81.67	−3302
C_6H_{12}(l), cyclohexane	84.16	−156	+26.8	204.4	156.5	−3920
C_6H_{14}(l), hexane	86.18	−198.7		204.3		−4163
$C_6H_5CH_3$(g), methylbenzene (toluene)	92.14	+50.0	+122.0	320.7	103.6	−3953
C_7H_{16}(l), heptane	100.21	−224.4	+1.0	328.6	224.3	
C_8H_{18}(l), octane	114.23	−249.9	+6.4	361.1		−5471
C_8H_{18}(l), iso-octane	114.23	−255.1				−5461
$C_{10}H_8$(s), naphthalene	128.18	+78.53				−5157
Alcohols and phenols						
CH_3OH(l), methanol	32.04	−238.66	−166.27	126.8	81.6	−726
CH_3OH(g)	32.04	−200.66	−161.96	239.81	43.89	−764
C_2H_5OH(l), ethanol	46.07	−277.69	−174.78	160.7	111.46	−1368
C_2H_5OH(g)	46.07	−235.10	−168.49	282.70	65.44	−1409
C_6H_5OH(s), phenol	94.12	−165.0	−50.9	146.0		−3054
Carboxylic acids, hydroxy acids, and esters						
HCOOH(l), formic	46.03	−424.72	−361.35	128.95	99.04	−255
CH_3COOH(l), acetic	60.05	−484.5	−389.9	159.8	124.3	−875
CH_3COOH(aq)	60.05	−485.76	−396.46	178.7		
$CH_3CO_2^-$(aq)	59.05	−486.01	−369.31	+86.6	−6.3	
$(COOH)_2$(s), oxalic	90.04	−827.2			117	−254
C_6H_5COOH(s), benzoic	122.13	−385.1	−245.3	167.6	146.8	−3227
$CH_3CH(OH)COOH$(s), lactic	90.08	−694.0				−1344
$CH_3COOC_2H_5$(l), ethyl acetate	88.11	−479.0	−332.7	259.4	170.1	−2231
Alkanals and alkanones						
HCHO(g), methanal	30.03	−108.57	−102.53	218.77	35.40	−571
CH_3CHO(l), ethanal	44.05	−192.30	−128.12	160.2		−1166
CH_3CHO(g)	44.05	−166.19	−128.86	250.3	57.3	−1192
CH_3COCH_3(l), propanone	58.08	−248.1	−155.4	200.4	124.7	−1790
Sugars						
$C_6H_{12}O_6$(s), α-D-glucose	180.16	−1274				−2808
$C_6H_{12}O_6$(s), β-D-glucose	180.16	−1268	−910	212		
$C_6H_{12}O_6$(s), β-D-fructose	180.16	−1266				−2810
$C_{12}H_{22}O_{11}$(s), sucrose	342.30	−2222	−1543	360.2		−5645
Nitrogen compounds						
$CO(NH_2)_2$(s), urea	60.06	−333.51	−197.33	104.60	93.14	−632
CH_3NH_2(g), methylamine	31.06	−22.97	+32.16	243.41	53.1	−1085
$C_6H_5NH_2$(l), aniline	93.13	+31.1				−3393
$CH_2(NH_2)COOH$(s), glycine	75.07	−532.9	−373.4	103.5	99.2	−969

Data: NBS, TDOC. † Standard entropies of ions may be either positive or negative because the values are relative to the entropy of the hydrogen ion.

Table 2C.5 Thermodynamic data for elements and inorganic compounds at 298 K

	$M/(\text{g mol}^{-1})$	$\Delta_f H^\ominus/(\text{kJ mol}^{-1})$	$\Delta_f G^\ominus/(\text{kJ mol}^{-1})$	$S_m^\ominus/(\text{J K}^{-1}\text{ mol}^{-1})^\dagger$	$C_{p,m}^\ominus/(\text{J K}^{-1}\text{ mol}^{-1})$
Aluminium (aluminum)					
Al(s)	26.98	0	0	28.33	24.35
Al(l)	26.98	+10.56	+7.20	39.55	24.21
Al(g)	26.98	+326.4	+285.7	164.54	21.38
Al^{3+}(g)	26.98	+5483.17			
Al^{3+}(aq)	26.98	−531	−485	−321.7	
Al_2O_3(s, α)	101.96	−1675.7	−1582.3	50.92	79.04
$AlCl_3$(s)	133.24	−704.2	−628.8	110.67	91.84
Argon					
Ar(g)	39.95	0	0	154.84	20.786
Antimony					
Sb(s)	121.75	0	0	45.69	25.23
SbH_3(g)	124.77	+145.11	+147.75	232.78	41.05
Arsenic					
As(s, α)	74.92	0	0	35.1	24.64
As(g)	74.92	+302.5	+261.0	174.21	20.79
As_4(g)	299.69	+143.9	+92.4	314	
AsH_3(g)	77.95	+66.44	+68.93	222.78	38.07
Barium					
Ba(s)	137.34	0	0	62.8	28.07
Ba(g)	137.34	+180	+146	170.24	20.79
Ba^{2+}(aq)	137.34	−537.64	−560.77	+9.6	
BaO(s)	153.34	−553.5	−525.1	70.43	47.78
$BaCl_2$(s)	208.25	−858.6	−810.4	123.68	75.14
Beryllium					
Be(s)	9.01	0	0	9.50	16.44
Be(g)	9.01	+324.3	+286.6	136.27	20.79
Bismuth					
Bi(s)	208.98	0	0	56.74	25.52
Bi(g)	208.98	+207.1	+168.2	187.00	20.79
Bromine					
Br_2(l)	159.82	0	0	152.23	75.689
Br_2(g)	159.82	+30.907	+3.110	245.46	36.02
Br(g)	79.91	+111.88	+82.396	175.02	20.786
Br^-(g)	79.91	−219.07			
Br^-(aq)	79.91	−121.55	−103.96	+82.4	−141.8
HBr(g)	90.92	−36.40	−53.45	198.70	29.142
Cadmium					
Cd(s, γ)	112.40	0	0	51.76	25.98
Cd(g)	112.40	+112.01	+77.41	167.75	20.79
Cd^{2+}(aq)	112.40	−75.90	−77.612	−73.2	
CdO(s)	128.40	−258.2	−228.4	54.8	43.43
$CdCO_3$(s)	172.41	−750.6	−669.4	92.5	

(Continued)

Table 2C.5 (Continued)

	M/(g mol^{-1})	$\Delta_f H^\ominus$/(kJ mol^{-1})	$\Delta_f G^\ominus$/(kJ mol^{-1})	$S_m^\ominus$/(J K^{-1} mol^{-1})†	$C_{p,m}^\ominus$/(J K^{-1} mol^{-1})
Caesium (cesium)					
Cs(s)	132.91	0	0	85.23	32.17
Cs(g)	132.91	+76.06	+49.12	175.60	20.79
Cs^+(aq)	132.91	−258.28	−292.02	+133.05	−10.5
Calcium					
Ca(s)	40.08	0	0	41.42	25.31
Ca(g)	40.08	+178.2	+144.3	154.88	20.786
Ca^{2+}(aq)	40.08	−542.83	−553.58	−53.1	
CaO(s)	56.08	−635.09	−604.03	39.75	42.80
$CaCO_3$(s) (calcite)	100.09	−1206.9	−1128.8	92.9	81.88
$CaCO_3$(s) (aragonite)	100.09	−1207.1	−1127.8	88.7	81.25
CaF_2(s)	78.08	−1219.6	−1167.3	68.87	67.03
$CaCl_2$(s)	110.99	−795.8	−748.1	104.6	72.59
$CaBr_2$(s)	199.90	−682.8	−663.6	130	
Carbon (for 'organic' compounds of carbon, see Table 2C.4)					
C(s) (graphite)	12.011	0	0	5.740	8.527
C(s) (diamond)	12.011	+1.895	+2.900	2.377	6.113
C(g)	12.011	+716.68	+671.26	158.10	20.838
C_2(g)	24.022	+831.90	+775.89	199.42	43.21
CO(g)	28.011	−110.53	−137.17	197.67	29.14
CO_2(g)	44.010	−393.51	−394.36	213.74	37.11
CO_2(aq)	44.010	−413.80	−385.98	117.6	
H_2CO_3(aq)	62.03	−699.65	−623.08	187.4	
HCO_3^-(aq)	61.02	−691.99	−586.77	+91.2	
CO_3^{2-}(aq)	60.01	−677.14	−527.81	−56.9	
CCl_4(l)	153.82	−135.44	−65.21	216.40	131.75
CS_2(l)	76.14	+89.70	+65.27	151.34	75.7
HCN(g)	27.03	+135.1	+124.7	201.78	35.86
HCN(l)	27.03	+108.87	+124.97	112.84	70.63
CN^-(aq)	26.02	+150.6	+172.4	+94.1	
Chlorine					
Cl_2(g)	70.91	0	0	223.07	33.91
Cl(g)	35.45	+121.68	+105.68	165.20	21.840
Cl^-(g)	34.45	−233.13			
Cl^-(aq)	35.45	−167.16	−131.23	+56.5	−136.4
HCl(g)	36.46	−92.31	−95.30	186.91	29.12
HCl(aq)	36.46	−167.16	−131.23	56.5	−136.4
Chromium					
Cr(s)	52.00	0	0	23.77	23.35
Cr(g)	52.00	+396.6	+351.8	174.50	20.79
CrO_4^{2-}(aq)	115.99	−881.15	−727.75	+50.21	
$Cr_2O_7^{2-}$(aq)	215.99	−1490.3	−1301.1	+261.9	

Table 2C.5 (Continued)

	M/(g mol^{-1})	$\Delta_f H^\ominus$/(kJ mol^{-1})	$\Delta_f G^\ominus$/(kJ mol^{-1})	$S_m^\ominus$/(J K^{-1} mol^{-1})†	$C_{p,m}^\ominus$/(J K^{-1} mol^{-1})
Copper					
Cu(s)	63.54	0	0	33.150	24.44
Cu(g)	63.54	+338.32	+298.58	166.38	20.79
Cu^+(aq)	63.54	+71.67	+49.98	+40.6	
Cu^{2+}(aq)	63.54	+64.77	+65.49	−99.6	
Cu_2O(s)	143.08	−168.6	−146.0	93.14	63.64
CuO(s)	79.54	−157.3	−129.7	42.63	42.30
$CuSO_4$(s)	159.60	−771.36	−661.8	109	100.0
$CuSO_4\cdot H_2O$(s)	177.62	−1085.8	−918.11	146.0	134
$CuSO_4\cdot 5H_2O$(s)	249.68	−2279.7	−1879.7	300.4	280
Deuterium					
D_2(g)	4.028	0	0	144.96	29.20
HD(g)	3.022	+0.318	−1.464	143.80	29.196
D_2O(g)	20.028	−249.20	−234.54	198.34	34.27
D_2O(l)	20.028	−294.60	−243.44	75.94	84.35
HDO(g)	19.022	−245.30	−233.11	199.51	33.81
HDO(l)	19.022	−289.89	−241.86	79.29	
Fluorine					
F_2(g)	38.00	0	0	202.78	31.30
F(g)	19.00	+78.99	+61.91	158.75	22.74
F^-(aq)	19.00	−332.63	−278.79	−13.8	−106.7
HF(g)	20.01	−271.1	−273.2	173.78	29.13
Gold					
Au(s)	196.97	0	0	47.40	25.42
Au(g)	196.97	+366.1	+326.3	180.50	20.79
Helium					
He(g)	4.003	0	0	126.15	20.786
Hydrogen (see also deuterium)					
H_2(g)	2.016	0	0	130.684	28.824
H(g)	1.008	+217.97	+203.25	114.71	20.784
H^+(aq)	1.008	0	0	0	0
H^+(g)	1.008	+1536.20			
H_2O(s)	18.015			37.99	
H_2O(l)	18.015	−285.83	−237.13	69.91	75.291
H_2O(g)	18.015	−241.82	−228.57	188.83	33.58
H_2O_2(l)	34.015	−187.78	−120.35	109.6	89.1
Iodine					
I_2(s)	253.81	0	0	116.135	54.44
I_2(g)	253.81	+62.44	+19.33	260.69	36.90
I(g)	126.90	+106.84	+70.25	180.79	20.786
I^-(aq)	126.90	−55.19	−51.57	+111.3	−142.3
HI(g)	127.91	+26.48	+1.70	206.59	29.158

(Continued)

Table 2C.5 (Continued)

	M/(g mol^{-1})	$\Delta_f H^\ominus$/(kJ mol^{-1})	$\Delta_f G^\ominus$/(kJ mol^{-1})	$S_m^\ominus$/(J K^{-1} mol^{-1})†	$C_{p,m}^\ominus$/(J K^{-1} mol^{-1})
Iron					
Fe(s)	55.85	0	0	27.28	25.10
Fe(g)	55.85	+416.3	+370.7	180.49	25.68
Fe^{2+}(aq)	55.85	−89.1	−78.90	−137.7	
Fe^{3+}(aq)	55.85	−48.5	−4.7	−315.9	
Fe_3O_4(s) (magnetite)	231.54	−1118.4	−1015.4	146.4	143.43
Fe_2O_3(s) (haematite)	159.69	−824.2	−742.2	87.40	103.85
FeS(s, α)	87.91	−100.0	−100.4	60.29	50.54
FeS_2(s)	119.98	−178.2	−166.9	52.93	62.17
Krypton					
Kr(g)	83.80	0	0	164.08	20.786
Lead					
Pb(s)	207.19	0	0	64.81	26.44
Pb(g)	207.19	+195.0	+161.9	175.37	20.79
Pb^{2+}(aq)	207.19	−1.7	−24.43	+10.5	
PbO(s, yellow)	223.19	−217.32	−187.89	68.70	45.77
PbO(s, red)	223.19	−218.99	−188.93	66.5	45.81
PbO_2(s)	239.19	−277.4	−217.33	68.6	64.64
Lithium					
Li(s)	6.94	0	0	29.12	24.77
Li(g)	6.94	+159.37	+126.66	138.77	20.79
Li^+(aq)	6.94	−278.49	−293.31	+13.4	68.6
Magnesium					
Mg(s)	24.31	0	0	32.68	24.89
Mg(g)	24.31	+147.70	+113.10	148.65	20.786
Mg^{2+}(aq)	24.31	−466.85	−454.8	−138.1	
MgO(s)	40.31	−601.70	−569.43	26.94	37.15
$MgCO_3$(s)	84.32	−1095.8	−1012.1	65.7	75.52
$MgCl_2$(s)	95.22	−641.32	−591.79	89.62	71.38
Mercury					
Hg(l)	200.59	0	0	76.02	27.983
Hg(g)	200.59	+61.32	+31.82	174.96	20.786
Hg^{2+}(aq)	200.59	+171.1	+164.40	−32.2	
Hg_2^{2+}(aq)	401.18	+172.4	+153.52	+84.5	
HgO(s)	216.59	−90.83	−58.54	70.29	44.06
Hg_2Cl_2(s)	472.09	−265.22	−210.75	192.5	102
$HgCl_2$(s)	271.50	−224.3	−178.6	146.0	
HgS(s, black)	232.65	−53.6	−47.7	88.3	
Neon					
Ne(g)	20.18	0	0	146.33	20.786
Nitrogen					
N_2(g)	28.013	0	0	191.61	29.125
N(g)	14.007	+472.70	+455.56	153.30	20.786

Table 2C.5 (Continued)

	$M/(\mathrm{g\,mol^{-1}})$	$\Delta_f H^{\ominus}/(\mathrm{kJ\,mol^{-1}})$	$\Delta_f G^{\ominus}/(\mathrm{kJ\,mol^{-1}})$	$S_m^{\ominus}/(\mathrm{J\,K^{-1}\,mol^{-1}})^{\dagger}$	$C_{p,m}^{\ominus}/(\mathrm{J\,K^{-1}\,mol^{-1}})$
$NO(g)$	30.01	+90.25	+86.55	210.76	29.844
$N_2O(g)$	44.01	+82.05	+104.20	219.85	38.45
$NO_2(g)$	46.01	+33.18	+51.31	240.06	37.20
$N_2O_4(g)$	92.1	+9.16	+97.89	304.29	77.28
$N_2O_5(s)$	108.01	−43.1	+113.9	178.2	143.1
$N_2O_5(g)$	108.01	+11.3	+115.1	355.7	84.5
$HNO_3(l)$	63.01	−174.10	−80.71	155.60	109.87
$HNO_3(aq)$	63.01	−207.36	−111.25	146.4	−86.6
$NO_3^-(aq)$	62.01	−205.0	−108.74	+146.4	−86.6
$NH_3(g)$	17.03	−46.11	−16.45	192.45	35.06
$NH_3(aq)$	17.03	−80.29	−26.50	111.3	
$NH_4^+(aq)$	18.04	−132.51	−79.31	+113.4	79.9
$NH_2OH(s)$	33.03	−114.2			
$HN_3(l)$	43.03	+264.0	+327.3	140.6	43.68
$HN_3(g)$	43.03	+294.1	+328.1	238.97	98.87
$N_2H_4(l)$	32.05	+50.63	+149.43	121.21	139.3
$NH_4NO_3(s)$	80.04	−365.56	−183.87	151.08	84.1
$NH_4Cl(s)$	53.49	−314.43	−202.87	94.6	
Oxygen					
$O_2(g)$	31.999	0	0	205.138	29.355
$O(g)$	15.999	+249.17	+231.73	161.06	21.912
$O_3(g)$	47.998	+142.7	+163.2	238.93	39.20
$OH^-(aq)$	17.007	−229.99	−157.24	−10.75	−148.5
Phosphorus					
P(s, wh)	30.97	0	0	41.09	23.840
$P(g)$	30.97	+314.64	+278.25	163.19	20.786
$P_2(g)$	61.95	+144.3	+103.7	218.13	32.05
$P_4(g)$	123.90	+58.91	+24.44	279.98	67.15
$PH_3(g)$	34.00	+5.4	+13.4	210.23	37.11
$PCl_3(g)$	137.33	−287.0	−267.8	311.78	71.84
$PCl_3(l)$	137.33	−319.7	−272.3	217.1	
$PCl_5(g)$	208.24	−374.9	−305.0	364.6	112.8
$PCl_5(s)$	208.24	−443.5			
$H_3PO_3(s)$	82.00	−964.4			
$H_3PO_3(aq)$	82.00	−964.8			
$H_3PO_4(s)$	94.97	−1279.0	−1119.1	110.50	106.06
$H_3PO_4(l)$	94.97	−1266.9			
$H_3PO_4(aq)$	94.97	−1277.4	−1018.7	−222	
$PO_4^{3-}(aq)$	94.97	−1277.4	−1018.7	−221.8	
$P_4O_{10}(s)$	283.89	−2984.0	−2697.0	228.86	211.71
$P_4O_6(s)$	219.89	−1640.1			

(Continued)

Table 2C.5 (Continued)

	$M/(\text{g mol}^{-1})$	$\Delta_f H^\ominus/(\text{kJ mol}^{-1})$	$\Delta_f G^\ominus/(\text{kJ mol}^{-1})$	$S_m^\ominus/(\text{J K}^{-1}\text{ mol}^{-1})^\dagger$	$C_{p,m}^\ominus/(\text{J K}^{-1}\text{ mol}^{-1})$
Potassium					
K(s)	39.10	0	0	64.18	29.58
K(g)	39.10	+89.24	+60.59	160.336	20.786
K^+(g)	39.10	+514.26			
K^+(aq)	39.10	−252.38	−283.27	+102.5	21.8
KOH(s)	56.11	−424.76	−379.08	78.9	64.9
KF(s)	58.10	−576.27	−537.75	66.57	49.04
KCl(s)	74.56	−436.75	−409.14	82.59	51.30
KBr(s)	119.01	−393.80	−380.66	95.90	52.30
KI(s)	166.01	−327.90	−324.89	106.32	52.93
Silicon					
Si(s)	28.09	0	0	18.83	20.00
Si(g)	28.09	+455.6	+411.3	167.97	22.25
SiO_2(s, α)	60.09	−910.94	−856.64	41.84	44.43
Silver					
Ag(s)	107.87	0	0	42.55	25.351
Ag(g)	107.87	+284.55	+245.65	173.00	20.79
Ag^+(aq)	107.87	+105.58	+77.11	+72.68	21.8
AgBr(s)	187.78	−100.37	−96.90	107.1	52.38
AgCl(s)	143.32	−127.07	−109.79	96.2	50.79
Ag_2O(s)	231.74	−31.05	−11.20	121.3	65.86
$AgNO_3$(s)	169.88	−129.39	−33.41	140.92	93.05
Sodium					
Na(s)	22.99	0	0	51.21	28.24
Na(g)	22.99	+107.32	+76.76	153.71	20.79
Na^+(aq)	22.99	−240.12	−261.91	+59.0	46.4
NaOH(s)	40.00	−425.61	−379.49	64.46	59.54
NaCl(s)	58.44	−411.15	−384.14	72.13	50.50
NaBr(s)	102.90	−361.06	−348.98	86.82	51.38
NaI(s)	149.89	−287.78	−286.06	98.53	52.09
Sulfur					
S(s, α) (rhombic)	32.06	0	0	31.80	22.64
S(s, β) (monoclinic)	32.06	+0.33	+0.1	32.6	23.6
S(g)	32.06	+278.81	+238.25	167.82	23.673
S_2(g)	64.13	+128.37	+79.30	228.18	32.47
S^{2-}(aq)	32.06	+33.1	+85.8	−14.6	
SO_2(g)	64.06	−296.83	−300.19	248.22	39.87
SO_3(g)	80.06	−395.72	−371.06	256.76	50.67
H_2SO_4(l)	98.08	−813.99	−690.00	156.90	138.9
H_2SO_4(aq)	98.08	−909.27	−744.53	20.1	−293
SO_4^{2-}(aq)	96.06	−909.27	−744.53	+20.1	−293
HSO_4^-(aq)	97.07	−887.34	−755.91	+131.8	−84
H_2S(g)	34.08	−20.63	−33.56	205.79	34.23

Table 2C.5 (Continued)

	$M/(\mathrm{g\ mol^{-1}})$	$\Delta_f H^\ominus/(\mathrm{kJ\ mol^{-1}})$	$\Delta_f G^\ominus/(\mathrm{kJ\ mol^{-1}})$	$S_m^\ominus/(\mathrm{J\,K^{-1}\,mol^{-1}})$†	$C_{p,m}^\ominus/(\mathrm{J\,K^{-1}\,mol^{-1}})$
$H_2S(aq)$	34.08	−39.7	−27.83	121	
$HS^-(aq)$	33.072	−17.6	+12.08	+62.08	
$SF_6(g)$	146.05	−1209	−1105.3	291.82	97.28
Tin					
Sn(s, β)	118.69	0	0	51.55	26.99
Sn(g)	118.69	+302.1	+267.3	168.49	20.26
$Sn^{2+}(aq)$	118.69	−8.8	−27.2	−17	
SnO(s)	134.69	−285.8	−256.9	56.5	44.31
$SnO_2(s)$	150.69	−580.7	−519.6	52.3	52.59
Xenon					
Xe(g)	131.30	0	0	169.68	20.786
Zinc					
Zn(s)	65.37	0	0	41.63	25.40
Zn(g)	65.37	+130.73	+95.14	160.98	20.79
$Zn^{2+}(aq)$	65.37	−153.89	−147.06	−112.1	46
ZnO(s)	81.37	−348.28	−318.30	43.64	40.25

Source: NBS. † Standard entropies of ions may be either positive or negative because the values are relative to the entropy of the hydrogen ion.

Table 2C.6 Standard enthalpies of formation of organic compounds at 298 K, $\Delta_f H^\ominus/(\mathrm{kJ\ mol^{-1}})$. See Table 2C.4.

Table 2D.1 Expansion coefficients (α) and isothermal compressibilities (κ_T) at 298 K

	$\alpha/(10^{-4}\ \mathrm{K^{-1}})$	$\kappa_T/(10^{-6}\ \mathrm{atm^{-1}})$
Liquids		
Benzene	12.4	92.1
Carbon tetrachloride	12.4	90.5
Ethanol	11.2	76.8
Mercury	1.82	38.7
Water	2.1	49.6
Solids		
Copper	0.501	0.735
Diamond	0.030	0.187
Iron	0.354	0.589
Lead	0.861	2.21

The values refer to 20 °C.
Data: AIP(α), KL(κ_T).

Table 2D.2 Inversion temperatures (T_I), normal freezing (T_f) and boiling points (T_b), and Joule–Thomson coefficients (μ) at 1 atm and 298 K

	T_I/K	T_f/K	T_b/K	$\mu/(\mathrm{K\ atm^{-1}})$
Air	603			0.189 at 50 °C
Argon	723	83.8	87.3	
Carbon dioxide	1500	194.7s		1.11 at 300 K
Helium	40		4.22	−0.062
Hydrogen	202	14.0	20.3	−0.03
Krypton	1090	116.6	120.8	
Methane	968	90.6	111.6	
Neon	231	24.5	27.1	
Nitrogen	621	63.3	77.4	0.27
Oxygen	764	54.8	90.2	0.31

s: sublimes.
Data: AIP, JL, and M.W. Zemansky, *Heat and thermodynamics*. McGraw-Hill, New York (1957).

Table 3A.1 Standard entropies (and temperatures) of phase transitions, $\Delta_{trs}S^{\ominus}/(J\,K^{-1}\,mol^{-1})$

	Fusion (at T_f)	Vaporization (at T_b)
Ar	14.17 (at 83.8 K)	74.53 (at 87.3 K)
Br_2	39.76 (at 265.9 K)	88.61 (at 332.4 K)
C_6H_6	38.00 (at 278.6 K)	87.19 (at 353.2 K)
CH_3COOH	40.4 (at 289.8 K)	61.9 (at 391.4 K)
CH_3OH	18.03 (at 175.2 K)	104.6 (at 337.2 K)
Cl_2	37.22 (at 172.1 K)	85.38 (at 239.0 K)
H_2	8.38 (at 14.0 K)	44.96 (at 20.38 K)
H_2O	22.00 (at 273.2 K)	109.1 (at 373.2 K)
H_2S	12.67 (at 187.6 K)	87.75 (at 212.0 K)
He	4.8 (at 1.8 K and 30 bar)	19.9 (at 4.22 K)
N_2	11.39 (at 63.2 K)	75.22 (at 77.4 K)
NH_3	28.93 (at 195.4 K)	97.41 (at 239.73 K)
O_2	8.17 (at 54.4 K)	75.63 (at 90.2 K)

Data: AIP.

Table 3A.2 The standard enthalpies and entropies of vaporization of liquids at their normal boiling point

	$\Delta_{vap}H^{\ominus}/(kJ\,mol^{-1})$	$\theta_b/°C$	$\Delta_{vap}S^{\ominus}/(J\,K^{-1}\,mol^{-1})$
Benzene	30.8	80.1	+87.2
Carbon disulfide	26.74	46.25	+83.7
Carbon tetrachloride	30.00	76.7	+85.8
Cyclohexane	30.1	80.7	+85.1
Decane	38.75	174	+86.7
Dimethyl ether	21.51	−23	+86
Ethanol	38.6	78.3	+110.0
Hydrogen sulfide	18.7	−60.4	+87.9
Mercury	59.3	356.6	+94.2
Methane	8.18	−161.5	+73.2
Methanol	35.21	65.0	+104.1
Water	40.7	100.0	+109.1

Data: JL.

Table 3B.1 Standard Third-Law entropies at 298 K, $S_m^{\ominus}/(J\,K^{-1}\,mol^{-1})$. See Tables 2C.4 and 2C.5

Table 3C.1 Standard Gibbs energies of formation at 298 K, $\Delta_f G^{\ominus}/(kJ\,mol^{-1})$. See Tables 2C.4 and 2C.5

Table 3D.2 The fugacity coefficients of nitrogen at 273 K, ϕ

p/atm	ϕ	p/atm	ϕ
1	0.999 55	300	1.0055
10	0.9956	400	1.062
50	0.9912	600	1.239
100	0.9703	800	1.495
150	0.9672	1000	1.839
200	0.9721		

To convert to fugacities, use $f=\phi p$

Data: LR.

Table 5A.1 Henry's law constants for gases at 298 K, $K/(kPa\,kg\,mol^{-1})$

	Water	Benzene
CH_4	7.55×10^4	44.4×10^3
CO_2	3.01×10^3	8.90×10^2
H_2	1.28×10^5	2.79×10^4
N_2	1.56×10^5	1.87×10^4
O_2	7.92×10^4	

Data: converted from R.J. Silbey and R.A. Alberty, *Physical chemistry*. Wiley, New York (2001).

Table 5B.1 Freezing-point (K_f) and boiling-point (K_b) constants

	$K_f/(K\,kg\,mol^{-1})$	$K_b/(K\,kg\,mol^{-1})$
Acetic acid	3.90	3.07
Benzene	5.12	2.53
Camphor	40	
Carbon disulfide	3.8	2.37
Carbon tetrachloride	30	4.95
Naphthalene	6.94	5.8
Phenol	7.27	3.04
Water	1.86	0.51

Data: KL.

Table 5F.2 Mean activity coefficients in water at 298 K

$b/b^{\ominus}$	HCl	KCl	$CaCl_2$	H_2SO_4	$LaCl_3$	$In_2(SO_4)_3$
0.001	0.966	0.966	0.888	0.830	0.790	
0.005	0.929	0.927	0.789	0.639	0.636	0.16
0.01	0.905	0.902	0.732	0.544	0.560	0.11
0.05	0.830	0.816	0.584	0.340	0.388	0.035
0.10	0.798	0.770	0.524	0.266	0.356	0.025
0.50	0.769	0.652	0.510	0.155	0.303	0.014
1.00	0.811	0.607	0.725	0.131	0.387	
2.00	1.011	0.577	1.554	0.125	0.954	

Data: RS, HCP, and S. Glasstone, *Introduction to electrochemistry*. Van Nostrand (1942).

Table 6D.1 Standard potentials at 298 K, $E^{\ominus}$/V. (a) In electrochemical order

Reduction half-reaction	$E^{\ominus}$/V	Reduction half-reaction	$E^{\ominus}$/V
Strongly oxidizing		$Cu^+ + e^- \rightarrow Cu$	+0.52
$H_4XeO_6 + 2\,H^+ + 2\,e^- \rightarrow XeO_3 + 3\,H_2O$	+3.0	$NiOOH + H_2O + e^- \rightarrow Ni(OH)_2 + OH^-$	+0.49
$F_2 + 2\,e^- \rightarrow 2\,F^-$	+2.87	$Ag_2CrO_4 + 2e^- \rightarrow 2\,Ag + CrO_4^{2-}$	+0.45
$O_3 + 2\,H^+ + 2\,e^- \rightarrow O_2 + H_2O$	+2.07	$O_2 + 2\,H_2O + 4\,e^- \rightarrow 4\,OH^-$	+0.40
$S_2O_8^{2-} + 2e^- \rightarrow 2\,SO_4^{2-}$	+2.05	$ClO_4^- + H_2O + 2e^- \rightarrow ClO_3^- + 2OH^-$	+0.36
$Ag^{2+} + e^- \rightarrow Ag^+$	+1.98	$[Fe(CN)_6]^{3-} + e^- \rightarrow [Fe(CN)_6]^{4-}$	+0.36
$Co^{3+} + e^- \rightarrow Co^{2+}$	+1.81	$Cu^{2+} + 2\,e^- \rightarrow Cu$	+0.34
$H_2O_2 + 2\,H^+ + 2\,e^- \rightarrow 2\,H_2O$	+1.78	$Hg_2Cl_2 + 2\,e^- \rightarrow 2\,Hg + 2\,Cl^-$	+0.27
$Au^+ + e^- \rightarrow Au$	+1.69	$AgCl + e^- \rightarrow Ag + Cl^-$	+0.22
$Pb^{4+} + 2\,e^- \rightarrow Pb^{2+}$	+1.67	$Bi^{3+} + 3\,e^- \rightarrow Bi$	+0.20
$2\,HClO + 2\,H^+ + 2\,e^- \rightarrow Cl_2 + 2\,H_2O$	+1.63	$Cu^{2+} + e^- \rightarrow Cu^+$	+0.16
$Ce^{4+} + e^- \rightarrow Ce^{3+}$	+1.61	$Sn^{4+} + 2\,e^- \rightarrow Sn^{2+}$	+0.15
$2\,HBrO + 2\,H^+ + 2\,e^- \rightarrow Br_2 + 2\,H_2O$	+1.60	$NO_3^- + H_2O + 2e^- \rightarrow NO_2^- + 2OH^-$	+0.10
$MnO_4^- + 8H^+ + 5\,e^- \rightarrow Mn^{2+} + 4H_2O$	+1.51	$AgBr + e^- \rightarrow Ag + Br^-$	+0.0713
$Mn^{3+} + e^- \rightarrow Mn^{2+}$	+1.51	$Ti^{4+} + e^- \rightarrow Ti^{3+}$	0.00
$Au^{3+} + 3\,e^- \rightarrow Au$	+1.40	$2\,H^+ + 2\,e^- \rightarrow H_2$	0, by definition
$Cl_2 + 2\,e^- \rightarrow 2\,Cl^-$	+1.36	$Fe^{3+} + 3\,e^- \rightarrow Fe$	−0.04
$Cr_2O_7^{2-} + 14H^+ + 6e^- \rightarrow 2Cr^{3+} + 7H_2O$	+1.33	$O_2 + H_2O + 2\,e^- \rightarrow HO_2^- + OH^-$	−0.08
$O_3 + H_2O + 2\,e^- \rightarrow O_2 + 2\,OH^-$	+1.24	$Pb^{2+} + 2\,e^- \rightarrow Pb$	−0.13
$O_2 + 4\,H^+ + 4\,e^- \rightarrow 2\,H_2O$	+1.23	$In^+ + e^- \rightarrow In$	−0.14
$ClO_4^- + 2H^+ + 2e^- \rightarrow ClO_3^- + H_2O$	+1.23	$Sn^{2+} + 2\,e^- \rightarrow Sn$	−0.14
$MnO_2 + 4\,H^+ + 2\,e^- \rightarrow Mn^{2+} + 2\,H_2O$	+1.23	$AgI + e^- \rightarrow Ag + I^-$	−0.15
$Pt^{2+} + 2\,e^- \rightarrow Pt$	+1.20	$Ni^{2+} + 2\,e^- \rightarrow Ni$	−0.23
$Br_2 + 2\,e^- \rightarrow 2Br^-$	+1.09	$V^{3+} + e^- \rightarrow V^{2+}$	−0.26
$Pu^{4+} + e^- \rightarrow Pu^{3+}$	+0.97	$Co^{2+} + 2\,e^- \rightarrow Co$	−0.28
$NO_3^- + 4H^+ + 3e^- \rightarrow NO + 2H_2O$	+0.96	$In^{3+} + 3\,e^- \rightarrow In$	−0.34
$2Hg^{2+} + 2e^- \rightarrow Hg_2^{2+}$	+0.92	$Tl^+ + e^- \rightarrow Tl$	−0.34
$ClO^- + H_2O + 2\,e^- \rightarrow Cl^- + 2\,OH^-$	+0.89	$PbSO_4 + 2\,e^- \rightarrow Pb + SO_4^{2-}$	−0.36
$Hg^{2+} + 2\,e^- \rightarrow Hg$	+0.86	$Ti^{3+} + e^- \rightarrow Ti^{2+}$	−0.37
$NO_3^- + 2H^+ + e^- \rightarrow NO_2 + H_2O$	+0.80	$Cd^{2+} + 2\,e^- \rightarrow Cd$	−0.40
$Ag^+ + e^- \rightarrow Ag$	+0.80	$In^{2+} + e^- \rightarrow In^+$	−0.40
$Hg_2^{2+} + 2e^- \rightarrow 2Hg$	+0.79	$Cr^{3+} + e^- \rightarrow Cr^{2+}$	−0.41
$AgF + e^- \rightarrow Ag + F^-$	+0.78	$Fe^{2+} + 2\,e^- \rightarrow Fe$	−0.44
$Fe^{3+} + e^- \rightarrow Fe^{2+}$	+0.77	$In^{3+} + 2\,e^- \rightarrow In^+$	−0.44
$BrO^- + H_2O + 2\,e^- \rightarrow Br^- + 2\,OH^-$	+0.76	$S + 2\,e^- \rightarrow S^{2-}$	−0.48
$Hg_2SO_4 + 2e^- \rightarrow 2Hg + SO_4^{2-}$	+0.62	$In^{3+} + e^- \rightarrow In^{2+}$	−0.49
$MnO_4^{2-} + 2H_2O + 2e^- \rightarrow MnO_2 + 4OH^-$	+0.60	$O_2 + e^- \rightarrow O_2^-$	−0.56
$MnO_4^- + e^- \rightarrow MnO_4^{2-}$	+0.56	$U^{4+} + e^- \rightarrow U^{3+}$	−0.61
$I_2 + 2\,e^- \rightarrow 2\,I^-$	+0.54	$Cr^{3+} + 3\,e^- \rightarrow Cr$	−0.74
$I_3^- + 2e^- \rightarrow 3I^-$	+0.53	$Zn^{2+} + 2\,e^- \rightarrow Zn$	−0.76

(Continued)

Table 6D.1 (Continued)

Reduction half-reaction	$E^\ominus$/V	Reduction half-reaction	$E^\ominus$/V
$Cd(OH)_2 + 2e^- \rightarrow Cd + 2OH^-$	−0.81	$Ce^{3+} + 3e^- \rightarrow Ce$	−2.48
$2H_2O + 2e^- \rightarrow H_2 + 2OH^-$	−0.83	$La^{3+} + 3e^- \rightarrow La$	−2.52
$Cr^{2+} + 2e^- \rightarrow Cr$	−0.91	$Na^+ + e^- \rightarrow Na$	−2.71
$Mn^{2+} + 2e^- \rightarrow Mn$	−1.18	$Ca^{2+} + 2e^- \rightarrow Ca$	−2.87
$V^{2+} + 2e^- \rightarrow V$	−1.19	$Sr^{2+} + 2e^- \rightarrow Sr$	−2.89
$Ti^{2+} + 2e^- \rightarrow Ti$	−1.63	$Ba^{2+} + 2e^- \rightarrow Ba$	−2.91
$Al^{3+} + 3e^- \rightarrow Al$	−1.66	$Ra^{2+} + 2e^- \rightarrow Ra$	−2.92
$U^{3+} + 3e^- \rightarrow U$	−1.79	$Cs^+ + e^- \rightarrow Cs$	−2.92
$Be^{2+} + 2e^- \rightarrow Be$	−1.85	$Rb^+ + e^- \rightarrow Rb$	−2.93
$Sc^{3+} + 3e^- \rightarrow Sc$	−2.09	$K^+ + e^- \rightarrow K$	−2.93
$Mg^{2+} + 2e^- \rightarrow Mg$	−2.36	$Li^+ + e^- \rightarrow Li$	−3.05

Table 6D.1 Standard potentials at 298 K, $E^\ominus$/V. (b) In alphabetical order

Reduction half-reaction	$E^\ominus$/V	Reduction half-reaction	$E^\ominus$/V
$Ag^+ + e^- \rightarrow Ag$	+0.80	$Cr^{2+} + 2e^- \rightarrow Cr$	−0.91
$Ag^{2+} + e^- \rightarrow Ag^+$	+1.98	$Cr_2O_7^{2-} + 14H^+ + 6e^- \rightarrow 2Cr^{3+} + 7H_2O$	+1.33
$AgBr + e^- \rightarrow Ag + Br^-$	+0.0713	$Cr^{3+} + 3e^- \rightarrow Cr$	−0.74
$AgCl + e^- \rightarrow Ag + Cl^-$	+0.22	$Cr^{3+} + e^- \rightarrow Cr^{2+}$	−0.41
$Ag_2CrO_4 + 2e^- \rightarrow 2Ag + CrO_4^{2-}$	+0.45	$Cs^+ + e^- \rightarrow Cs$	−2.92
$AgF + e^- \rightarrow Ag + F^-$	+0.78	$Cu^+ + e^- \rightarrow Cu$	+0.52
$AgI + e^- \rightarrow Ag + I^-$	−0.15	$Cu^{2+} + 2e^- \rightarrow Cu$	+0.34
$Al^{3+} + 3e^- \rightarrow Al$	−1.66	$Cu^{2+} + e^- \rightarrow Cu^+$	+0.16
$Au^+ + e^- \rightarrow Au$	+1.69	$F_2 + 2e^- \rightarrow 2F^-$	+2.87
$Au^{3+} + 3e^- \rightarrow Au$	+1.40	$Fe^{2+} + 2e^- \rightarrow Fe$	−0.44
$Ba^{2+} + 2e^- \rightarrow Ba$	−2.91	$Fe^{3+} + 3e^- \rightarrow Fe$	−0.04
$Be^{2+} + 2e^- \rightarrow Be$	−1.85	$Fe^{3+} + e^- \rightarrow Fe^{2+}$	+0.77
$Bi^{3+} + 3e^- \rightarrow Bi$	+0.20	$[Fe(CN)_6]^{3-} + e^- \rightarrow [Fe(CN)_6]^{4-}$	+0.36
$Br_2 + 2e^- \rightarrow 2Br^-$	+1.09	$2H^+ + 2e^- \rightarrow H_2$	0, by definition
$BrO^- + H_2O + 2e^- \rightarrow Br^- + 2OH^-$	+0.76	$2H_2O + 2e^- \rightarrow H_2 + 2OH^-$	−0.83
$Ca^{2+} + 2e^- \rightarrow Ca$	−2.87	$2HBrO + 2H^+ + 2e^- \rightarrow Br_2 + 2H_2O$	+1.60
$Cd(OH)_2 + 2e^- \rightarrow Cd + 2OH^-$	−0.81	$2HClO + 2H^+ + 2e^- \rightarrow Cl_2 + 2H_2O$	+1.63
$Cd^{2+} + 2e^- \rightarrow Cd$	−0.40	$H_2O_2 + 2H^+ + 2e^- \rightarrow 2H_2O$	+1.78
$Ce^{3+} + 3e^- \rightarrow Ce$	−2.48	$H_4XeO_6 + 2H^+ + 2e^- \rightarrow XeO_3 + 3H_2O$	+3.0
$Ce^{4+} + e^- \rightarrow Ce^{3+}$	+1.61	$Hg_2^{2+} + 2e^- \rightarrow 2Hg$	+0.79
$Cl_2 + 2e^- \rightarrow 2Cl^-$	+1.36	$Hg_2Cl_2 + 2e^- \rightarrow 2Hg + 2Cl^-$	+0.27
$ClO^- + H_2O + 2e^- \rightarrow Cl^- + 2OH^-$	+0.89	$Hg^{2+} + 2e^- \rightarrow Hg$	+0.86
$ClO_4^- + 2H^+ + 2e^- \rightarrow ClO_3^- + H_2O$	+1.23	$2Hg^{2+} + 2e^- \rightarrow Hg_2^{2+}$	+0.92
$ClO_4^- + H_2O + 2e^- \rightarrow ClO_3^- + 2OH^-$	+0.36	$Hg_2SO_4 + 2e^- \rightarrow 2Hg + SO_4^{2-}$	+0.62
$Co^{2+} + 2e^- \rightarrow Co$	−0.28	$I_2 + 2e^- \rightarrow 2I^-$	+0.54
$Co^{3+} + e^- \rightarrow Co^{2+}$	+1.81	$I_3^- + 2e^- \rightarrow 3I^-$	+0.53

Table 6D.1a (Continued)

Reduction half-reaction	$E^{\ominus}$/V	Reduction half-reaction	$E^{\ominus}$/V
$In^+ + e^- \rightarrow In$	−0.14	$O_3 + 2H^+ + 2e^- \rightarrow O_2 + H_2O$	+2.07
$In^{2+} + e^- \rightarrow In^+$	−0.40	$O_3 + H_2O + 2e^- \rightarrow O_2 + 2OH^-$	+1.24
$In^{3+} + 2e^- \rightarrow In^+$	−0.44	$Pb^{2+} + 2e^- \rightarrow Pb$	−0.13
$In^{3+} + 3e^- \rightarrow In$	−0.34	$Pb^{4+} + 2e^- \rightarrow Pb^{2+}$	+1.67
$In^{3+} + e^- \rightarrow In^{2+}$	−0.49	$PbSO_4 + 2e^- \rightarrow Pb + SO_4^{2-}$	−0.36
$K^+ + e^- \rightarrow K$	−2.93	$Pt^{2+} + 2e^- \rightarrow Pt$	+1.20
$La^{3+} + 3e^- \rightarrow La$	−2.52	$Pu^{4+} + e^- \rightarrow Pu^{3+}$	+0.97
$Li^+ + e^- \rightarrow Li$	−3.05	$Ra^{2+} + 2e^- \rightarrow Ra$	−2.92
$Mg^{2+} + 2e^- \rightarrow Mg$	−2.36	$Rb^+ + e^- \rightarrow Rb$	−2.93
$Mn^{2+} + 2e^- \rightarrow Mn$	−1.18	$S + 2e^- \rightarrow S^{2-}$	−0.48
$Mn^{3+} + e^- \rightarrow Mn^{2+}$	+1.51	$S_2O_8^{2-} + 2e^- \rightarrow 2SO_4^{2-}$	+2.05
$MnO_2 + 4H^+ + 2e^- \rightarrow Mn^{2+} + 2H_2O$	+1.23	$Sc^{3+} + 3e^- \rightarrow Sc$	−2.09
$MnO_4^{2-} + 8H^+ + 5e^- \rightarrow Mn^{2+} + 4H_2O$	+1.51	$Sn^{2+} + 2e^- \rightarrow Sn$	−0.14
$MnO_4^- + e^- \rightarrow MnO_4^{2-}$	+0.56	$Sn^{4+} + 2e^- \rightarrow Sn^{2+}$	+0.15
$MnO_4^{2-} + 2H_2O + 2e^- \rightarrow MnO_2 + 4OH^-$	+0.60	$Sr^{2+} + 2e^- \rightarrow Sr$	−2.89
$Na^+ + e^- \rightarrow Na$	−2.71	$Ti^{2+} + 2e^- \rightarrow Ti$	−1.63
$Ni^{2+} + 2e^- \rightarrow Ni$	−0.23	$Ti^{3+} + e^- \rightarrow Ti^{2+}$	−0.37
$NiOOH + H_2O + e^- \rightarrow Ni(OH)_2 + OH^-$	+0.49	$Ti^{4+} + e^- \rightarrow Ti^{3+}$	0.00
$NO_3^- + 2H^+ + e^- \rightarrow NO_2 + H_2O$	+0.80	$Tl^+ + e^- \rightarrow Tl$	−0.34
$NO_3^- + 4H^+ + 3e^- \rightarrow NO + 2H_2O$	+0.96	$U^{3+} + 3e^- \rightarrow U$	−1.79
$NO_3^- + H_2O + 2e^- \rightarrow + NO_2^- + 2OH^-$	+0.10	$U^{4+} + e^- \rightarrow U^{3+}$	−0.61
$O_2 + 2H_2O + 4e^- \rightarrow 4OH^-$	+0.40	$V^{2+} + 2e^- \rightarrow V$	−1.19
$O_2 + 4H^+ + 4e^- \rightarrow 2H_2O$	+1.23	$V^{3+} + e^- \rightarrow V^{2+}$	−0.26
$O_2 + e^- \rightarrow O_2^-$	−0.56	$Zn^{2+} + 2e^- \rightarrow Zn$	−0.76
$O_2 + H_2O + 2e^- \rightarrow HO_2^- + OH^-$	−0.08		

Table 9B.1 Effective nuclear charge, $Z_{eff} = Z - \sigma$*

	H							He
1s	1							1.6875
	Li	Be	B	C	N	O	F	Ne
1s	2.6906	3.6848	4.6795	5.6727	6.6651	7.6579	8.6501	9.6421
2s	1.2792	1.9120	2.5762	3.2166	3.8474	4.4916	5.1276	5.7584
2p			2.4214	3.1358	3.8340	4.4532	5.1000	5.7584
	Na	Mg	Al	Si	P	S	Cl	Ar
1s	10.6259	11.6089	12.5910	13.5745	14.5578	15.5409	16.5239	17.5075
2s	6.5714	7.3920	8.3736	9.0200	9.8250	10.6288	11.4304	12.2304
2p	6.8018	7.8258	8.9634	9.9450	10.9612	11.9770	12.9932	14.0082
3s	2.5074	3.3075	4.1172	4.9032	5.6418	6.3669	7.0683	7.7568
3p			4.0656	4.2852	4.8864	5.4819	6.1161	6.7641

* The actual charge is $Z_{eff}e$.

Data: E. Clementi and D.L. Raimondi, *Atomic screening constants from SCF functions*. IBM Res. Note NJ-27 (1963). *J. Chem. Phys.* **38**, 2686 (1963).

Table 9B.2 First and subsequent ionization energies, I/(kJ mol^{-1})

H							He
1312.0							2372.3
							5250.4
Li	Be	B	C	N	O	F	Ne
513.3	899.4	800.6	1086.2	1402.3	1313.9	1681	2080.6
7298.0	1757.1	2427	2352	2856.1	3388.2	3374	3952.2
Na	Mg	Al	Si	P	S	Cl	Ar
495.8	737.7	577.4	786.5	1011.7	999.6	1251.1	1520.4
4562.4	1450.7	1816.6	1577.1	1903.2	2251	2297	2665.2
		2744.6		2912			
K	Ca	Ga	Ge	As	Se	Br	Kr
418.8	589.7	578.8	762.1	947.0	940.9	1139.9	1350.7
3051.4	1145	1979	1537	1798	2044	2104	2350
		2963	2735				
Rb	Sr	In	Sn	Sb	Te	I	Xe
403.0	549.5	558.3	708.6	833.7	869.2	1008.4	1170.4
2632	1064.2	1820.6	1411.8	1794	1795	1845.9	2046
		2704	2943.0	2443			
Cs	Ba	Tl	Pb	Bi	Po	At	Rn
375.5	502.8	589.3	715.5	703.2	812	930	1037
2420	965.1	1971.0	1450.4	1610			
		2878	3081.5	2466			

Data: E.

Table 9B.3 Electron affinities, E_{ea}/(kJ mol^{-1})

H							He
72.8							−21
Li	Be	B	C	N	O	F	Ne
59.8	≤0	23	122.5	−7	141	322	−29
					−844		
Na	Mg	Al	Si	P	S	Cl	Ar
52.9	≤0	44	133.6	71.7	200.4	348.7	−35
					−532		
K	Ca	Ga	Ge	As	Se	Br	Kr
48.3	2.37	36	116	77	195.0	324.5	−39
Rb	Sr	In	Sn	Sb	Te	I	Xe
46.9	5.03	34	121	101	190.2	295.3	−41
Cs	Ba	Tl	Pb	Bi	Po	At	Rn
45.5	13.95	30	35.2	101	186	270	−41

Data: E.

Table 10C.1 Bond lengths, R_e/pm

(a) Bond lengths in specific molecules

Br_2	228.3
Cl_2	198.75
CO	112.81
F_2	141.78
H_2^+	106
H_2	74.138
HBr	141.44
HCl	127.45
HF	91.680
HI	160.92
N_2	109.76
O_2	120.75

(b) Mean bond lengths from covalent radii*

H	37						
C	77(1)	N	74(1)	O	66(1)	F	64
	67(2)		65(2)		57(2)		
	60(3)						
Si	118	P	110	S	104(1)	Cl	99
					95(2)		
Ge	122	As	121	Se	104	Br	114
		Sb	141	Te	137	I	133

* Values are for single bonds except where indicated otherwise (values in parentheses). The length of an A–B covalent bond (of given order) is the sum of the corresponding covalent radii.

Table 10C.2a Bond dissociation enthalpies, $\Delta H^{\ominus}$(A–B)/(kJ mol^{-1}) at 298 K*

Diatomic molecules

H–H	436	F–F	155	Cl–Cl	242	Br–Br	193	I–I	151
O=O	497	C=O	1076	N≡N	945				
H–O	428	H–F	565	H–Cl	431	H–Br	366	H–I	299

Polyatomic molecules

H–CH_3	435	H–NH_2	460	H–OH	492	H–C_6H_5	469
H_3C–CH_3	368	H_2C=CH_2	720	HC≡CH	962		
HO–CH_3	377	Cl–CH_3	352	Br–CH_3	293	I–CH_3	237
O=CO	531	HO–OH	213	O_2N–NO_2	54		

* To a good approximation bond dissociation enthalpies and dissociation energies are related by $\Delta H^{\ominus} = D_e + \frac{3}{2}RT$ with $D_e = D_0 + \frac{1}{2}\hbar\omega$. For precise values of D_0 for diatomic molecules, see Table 12D.1.
Data: HCP, KL.

Table 10C.2b Mean bond enthalpies, $\Delta H^{\ominus}(\text{A–B})/(\text{kJ mol}^{-1})$*

	H	C	N	O	F	Cl	Br	I	S	P	Si
H	436										
C	412	348(i)									
		612(ii)									
		838(iii)									
		518(a)									
N	388	305(i)	163(i)								
		613(ii)	409(ii)								
		890(iii)	946(iii)								
O	463	360(i)	157	146(i)							
		743(ii)		497(ii)							
F	565	484	270	185	155						
Cl	431	338	200	203	254	242					
Br	366	276				219	193				
I	299	238				210	178	151			
S	338	259			496	250	212		264		
P	322									201	
Si	318		374	466							226

* Mean bond enthalpies are such a crude measure of bond strength that they need not be distinguished from dissociation energies.
(i) Single bond, (ii) double bond, (iii) triple bond, (a) aromatic.
Data: HCP and L. Pauling, *The nature of the chemical bond*. Cornell University Press (1960).

Table 10D.1 Pauling (*italics*) and Mulliken electronegativities

H							He
2.20							
3.06							
Li	Be	B	C	N	O	F	Ne
0.98	*1.57*	*2.04*	*2.55*	*3.04*	*3.44*	*3.98*	
1.28	1.99	1.83	2.67	3.08	3.22	4.43	4.60
Na	Mg	Al	Si	P	S	Cl	Ar
0.93	*1.31*	*1.61*	*1.90*	*2.19*	*2.58*	*3.16*	
1.21	1.63	1.37	2.03	2.39	2.65	3.54	3.36
K	Ca	Ga	Ge	As	Se	Br	Kr
0.82	*1.00*	*1.81*	*2.01*	*2.18*	*2.55*	*2.96*	*3.0*
1.03	1.30	1.34	1.95	2.26	2.51	3.24	2.98
Rb	Sr	In	Sn	Sb	Te	I	Xe
0.82	*0.95*	*1.78*	*1.96*	*2.05*	*2.10*	*2.66*	*2.6*
0.99	1.21	1.30	1.83	2.06	2.34	2.88	2.59
Cs	Ba	Tl	Pb	Bi			
0.79	*0.89*	*2.04*	*2.33*	*2.02*			

Data: Pauling values: A.L. Allred, *J. Inorg. Nucl. Chem.* **17**, 215 (1961); L.C. Allen and J.E. Huheey, ibid., **42**, 1523 (1980). Mulliken values: L.C. Allen, *J. Am. Chem. Soc.* **111**, 9003 (1989). The Mulliken values have been scaled to the range of the Pauling values.

Table 11B.1 The C_{3v} character table; see Part 4

Table 11B.2 The C_{2v} character table; see Part 4

Table 12D.1 Properties of diatomic molecules

	$\tilde{\nu}/cm^{-1}$	θ^V/K	$\tilde{B}/cm^{-1}$	θ^R/K	R_e/pm	$k_f/(N\,m^{-1})$	$hc\tilde{D}_0/(kJ\,mol^{-1})$	σ
$^1H_2^+$	2321.8	3341	29.8	42.9	106	160	255.8	2
$^1H_2$	4400.39	6332	60.864	87.6	74.138	574.9	432.1	2
$^2H_2$	3118.46	4487	30.442	43.8	74.154	577.0	439.6	2
$^1H^{19}F$	4138.32	5955	20.956	30.2	91.680	965.7	564.4	1
$^1H^{35}Cl$	2990.95	4304	10.593	15.2	127.45	516.3	427.7	1
$^1H^{81}Br$	2648.98	3812	8.465	12.2	141.44	411.5	362.7	1
$^1H^{127}I$	2308.09	3321	6.511	9.37	160.92	313.8	294.9	1
$^{14}N_2$	2358.07	3393	1.9987	2.88	109.76	2293.8	941.7	2
$^{16}O_2$	1580.36	2274	1.4457	2.08	120.75	1176.8	493.5	2
$^{19}F_2$	891.8	1283	0.8828	1.27	141.78	445.1	154.4	2
$^{35}Cl_2$	559.71	805	0.2441	0.351	198.75	322.7	239.3	2
$^{12}C^{16}O$	2170.21	3122	1.9313	2.78	112.81	1903.17	1071.8	1
$^{79}Br^{81}Br$	323.2	465	0.0809	10.116	283.3	245.9	190.2	1

Data: AIP.

Table 12E.1 Typical vibrational wavenumbers, $\tilde{\nu}/cm^{-1}$

C–H stretch	2850–2960
C–H bend	1340–1465
C–C stretch, bend	700–1250
C=C stretch	1620–1680
C≡C stretch	2100–2260
O–H stretch	3590–3650
H-bonds	3200–3570
C=O stretch	1640–1780
C≡N stretch	2215–2275
N–H stretch	3200–3500
C–F stretch	1000–1400
C–Cl stretch	600–800
C–Br stretch	500–600
C–I stretch	500
CO_3^{2-}	1410–1450
NO_3^-	1350–1420
NO_2^-	1230–1250
SO_4^{2-}	1080–1130
Silicates	900–1100

Data: L.J. Bellamy, *The infrared spectra of complex molecules* and *Advances in infrared group frequencies*. Chapman and Hall.

Table 13A.1 Colour, wavelength, frequency, and energy of light

Colour	λ/nm	$\nu/(10^{14}$ Hz)	$\tilde{\nu}/(10^4$ cm^{-1})	E/eV	E/(kJ mol^{-1})
Infrared	>1000	<3.00	<1.00	<1.24	<120
Red	700	4.28	1.43	1.77	171
Orange	620	4.84	1.61	2.00	193
Yellow	580	5.17	1.72	2.14	206
Green	530	5.66	1.89	2.34	226
Blue	470	6.38	2.13	2.64	254
Violet	420	7.14	2.38	2.95	285
Ultraviolet	<400	>7.5	>2.5	>3.10	>300

Data: J.G. Calvert and J.N. Pitts, *Photochemistry*. Wiley, New York (1966).

Table 13A.2 Absorption characteristics of some groups and molecules

Group	$\tilde{\nu}_{max}/(10^4\ cm^{-1})$	λ_{max}/nm	$\varepsilon_{max}/(dm^3\ mol^{-1}\ cm^{-1})$
C=C ($\pi^* \leftarrow \pi$)	6.10	163	1.5×10^4
	5.73	174	5.5×10^3
C=O ($\pi^* \leftarrow n$)	3.7–3.5	270–290	10–20
–N=N–	2.9	350	15
	>3.9	<260	Strong
$-NO_2$	3.6	280	10
	4.8	210	1.0×10^4
C_6H_5-	3.9	255	200
	5.0	200	6.3×10^3
	5.5	180	1.0×10^5
$[Cu(OH_2)_6]^{2+}$(aq)	1.2	810	10
$[Cu(NH_3)_4]^{2+}$(aq)	1.7	600	50
H_2O ($\pi^* \leftarrow n$)	6.0	167	7.0×10^3

Table 14A.2 Nuclear spin properties

Nuclide	Natural abundance, %	Spin, I	Magnetic Moment, μ/μ_N	g-value	$\gamma/(10^7\ T^{-1}s^{-1})$	NMR frequency at 1 T, ν/MHz
^1n*		$\frac{1}{2}$	−1.9130	−3.8260	−18.324	29.164
^1H	99.9844	$\frac{1}{2}$	2.792 85	5.5857	26.752	42.576
^2H	0.0156	1	0.857 44	0.857 44	4.1067	6.536
^3H*		$\frac{1}{2}$	2.978 96	−4.2553	−20.380	45.414
^{10}B	19.6	3	1.8006	0.6002	2.875	4.575
^{11}B	80.4	$\frac{3}{2}$	2.6886	1.7923	8.5841	13.663
^{13}C	1.108	$\frac{1}{2}$	0.7024	1.4046	6.7272	10.708
^{14}N	99.635	1	0.403 56	0.403 56	1.9328	3.078
^{17}O	0.037	$\frac{5}{2}$	−1.893 79	−0.7572	−3.627	5.774
^{19}F	100	$\frac{1}{2}$	2.628 87	5.2567	25.177	40.077
^{31}P	100	$\frac{1}{2}$	1.1316	2.2634	10.840	17.251
^{33}S	0.74	$\frac{3}{2}$	0.6438	0.4289	2.054	3.272
^{35}Cl	75.4	$\frac{3}{2}$	0.8219	0.5479	2.624	4.176
^{37}Cl	24.6	$\frac{3}{2}$	0.6841	0.4561	2.184	3.476

* Radioactive.

μ is the magnetic moment of the spin state with the largest value of m_I: $\mu = g_I\mu_N I$ and μ_N is the nuclear magneton (see inside front cover).

Data: KL and HCP.

Table 14D.1 Hyperfine coupling constants for atoms, a/mT

Nuclide	Spin	Isotropic coupling	Anisotropic coupling
^{1}H	$\frac{1}{2}$	50.8(1s)	
^{2}H	1	7.8(1s)	
^{13}C	$\frac{1}{2}$	113.0(2s)	6.6(2p)
^{14}N	1	55.2(2s)	4.8(2p)
^{19}F	$\frac{1}{2}$	1720(2s)	108.4(2p)
^{31}P	$\frac{1}{2}$	364(3s)	20.6(3p)
^{35}Cl	$\frac{3}{2}$	168(3s)	10.0(3p)
^{37}Cl	$\frac{3}{2}$	140(3s)	8.4(3p)

Data: P.W. Atkins and M.C.R. Symons, *The structure of inorganic radicals*. Elsevier, Amsterdam (1967).

Table 16A.1 Magnitudes of dipole moments (μ), polarizabilities (α), and polarizability volumes (α')

	$\mu/(10^{-30}$ C m)	μ/D	$\alpha'/(10^{-30}$ m^3)	$\alpha/(10^{-40}$ J^{-1} C^2 m^2)
Ar	0	0	1.66	1.85
C_2H_5OH	5.64	1.69		
$C_6H_5CH_3$	1.20	0.36		
C_6H_6	0	0	10.4	11.6
CCl_4	0	0	10.3	11.7
CH_2Cl_2	5.24	1.57	6.80	7.57
CH_3Cl	6.24	1.87	4.53	5.04
CH_3OH	5.70	1.71	3.23	3.59
CH_4	0	0	2.60	2.89
$CHCl_3$	3.37	1.01	8.50	9.46
CO	0.390	0.117	1.98	2.20
CO_2	0	0	2.63	2.93
H_2	0	0	0.819	0.911
H_2O	6.17	1.85	1.48	1.65
HBr	2.67	0.80	3.61	4.01
HCl	3.60	1.08	2.63	2.93
He	0	0	0.20	0.22
HF	6.37	1.91	0.51	0.57
HI	1.40	0.42	5.45	6.06
N_2	0	0	1.77	1.97
NH_3	4.90	1.47	2.22	2.47
1,2-$C_6H_4(CH_3)_2$	2.07	0.62		

Data: HCP and C.J.F. Böttcher and P. Bordewijk, *Theory of electric polarization*. Elsevier, Amsterdam (1978).

Table 16B.2 Lennard-Jones parameters for the (12,6) potential

	(ε/k)/K	r_0/pm
Ar	111.84	362.3
C_2H_2	209.11	463.5
C_2H_4	200.78	458.9
C_2H_6	216.12	478.2
C_6H_6	377.46	617.4
CCl_4	378.86	624.1
Cl_2	296.27	448.5
CO_2	201.71	444.4
F_2	104.29	357.1
Kr	154.87	389.5
N_2	91.85	391.9
O_2	113.27	365.4
Xe	213.96	426.0

Source: F. Cuadros, I. Cachadiña, and W. Ahamuda, *Molec. Engineering* **6**, 319 (1996).

Table 16C.1 Surface tensions of liquids at 293 K, γ/(mN m^{-1})

	γ/(mN m^{-})
Benzene	28.88
Carbon tetrachloride	27.0
Ethanol	22.8
Hexane	18.4
Mercury	472
Methanol	22.6
Water	72.75
	72.0 at 25 °C
	58.0 at 100 °C

Data: KL.

Table 17D.1 Radius of gyration

	M/(kg mol^{-1})	R_g/nm
Serum albumin	66	2.98
Myosin	493	46.8
Polystyrene	3.2×10^3	50†
DNA	4×10^3	117
Tobacco mosaic virus	3.9×10^4	92.4

† In a poor solvent.

Table 17D.2 Frictional coefficients and molecular geometry

a/b	Prolate	Oblate
2	1.04	1.04
3	1.18	1.17
4	1.18	1.17
5	1.25	1.22
6	1.31	1.28
7	1.38	1.33
8	1.43	1.37
9	149	1.42
10	1.54	1.46
50	2.95	2.38
100	4.07	2.97

Data: K.E. Van Holde, *Physical biochemistry*. Prentice-Hall, Englewood Cliffs (1971)

Sphere; radius a, $c=af_0$

Prolate ellipsoid; major axis $2a$, minor axis $2b$, $c=(ab)^{1/3}$

$$f=\left\{\frac{(1-b^2/a^2)^{1/2}}{(b/a)^{2/3}\ln\{\left[1+(1-b^2/a^2)^{1/2}\right]/(b/a)\}}\right\}f_0$$

Oblate ellipsoid; major axis $2a$, minor axis $2b$, $c=(a^2b)^{1/3}$

$$f=\left\{\frac{(a^2/b^2-1)^{1/2}}{(a/b)^{2/3}\arctan[(a^2/b^2-1)^{1/2}]}\right\}f_0$$

Long rod; length l, radius a, $c=(3a^2/4)^{1/3}$

$$f=\left\{\frac{(1/2a)^{2/3}}{(3/2)^{1/3}\{2\ln(l/a)-0.11\}}\right\}f_0$$

In each $f_0=6\pi\eta c$ with the appropriate value of c.

Table 17D.3 Intrinsic viscosity

Macromolecule	Solvent	θ/°C	K/(10^{-3} cm^3 g^{-1})	a
Polystyrene	Benzene	25	9.5	0.74
	Cyclobutane	34†	81	0.50
Polyisobutylene	Benzene	23†	83	0.50
	Cyclohexane	30	26	0.70
Amylose	0.33 M KCl(aq)	25†	113	0.50
Various proteins‡	Guanidine hydrochloride + $HSCH_2CH_2OH$		7.16	0.66

† The θ temperature.
‡ Use $[\eta] = KN^a$; N is the number of amino acid residues.
Data: K.E. Van Holde, *Physical biochemistry*. Prentice-Hall, Englewood Cliffs (1971).

Table 18B.2 Ionic radii, r/pm*

Li^+(4)	Be^{2+}(4)	B^{3+}(4)	N^{3-}	O^{2-}(6)	F^-(6)		
59	27	12	171	140	133		
Na^+(6)	Mg^{2+}(6)	Al^{3+}(6)	P^{3-}	S^{2-}(6)	Cl^-(6)		
102	72	53	212	184	181		
K^+(6)	Ca^{2+}(6)	Ga^{3+}(6)	As^{3-}(6)	Se^{2-}(6)	Br^-(6)		
138	100	62	222	198	196		
Rb^+(6)	Sr^{2+}(6)	In^{3+}(6)		Te^{2-}(6)	I^-(6)		
149	116	79		221	220		
Cs^+(6)	Ba^{2+}(6)	Tl^{3+}(6)					
167	136	88					
d-block elements (high-spin ions)							
Sc^{3+}(6)	Ti^{4+}(6)	Cr^{3+}(6)	Mn^{3+}(6)	Fe^{2+}(6)	Co^{3+}(6)	Cu^{2+}(6)	Zn^{2+}(6)
73	60	61	65	63	61	73	75

* Numbers in parentheses are the coordination numbers of the ions. Values for ions without a coordination number stated are estimates.
Data: R.D. Shannon and C.T. Prewitt, *Acta Cryst.* **B25**, 925 (1969).

Table 18B.4 Lattice enthalpies at 298 K, ΔH_L/(kJ mol^{-1})

	F	Cl	Br	I
Halides				
Li	1037	852	815	761
Na	926	787	752	705
K	821	717	689	649
Rb	789	695	668	632
Cs	750	676	654	620
Ag	969	912	900	886
Be		3017		
Mg		2524		
Ca		2255		
Sr		2153		

Oxides							
MgO	3850	CaO	3461	SrO	3283	BaO	3114
Sulfides							
MgS	3406	CaS	3119	SrS	2974	BaS	2832

Entries refer to $MX(s) \rightarrow M^+(g) + X^-(g)$.
Data: Principally D. Cubicciotti et al., *J. Chem. Phys.* **31**, 1646 (1959).

Table 18C.1 Magnetic susceptibilities at 298 K

	$\chi/10^{-6}$	$\chi_m/(10^{-10}\ m^3\ mol^{-1})$
$H_2O(l)$	−9.02	−1.63
$C_6H_6(l)$	−8.8	−7.8
$C_6H_{12}(l)$	−10.2	−11.1
$CCl_4(l)$	−5.4	−5.2
NaCl(s)	−16	−3.8
Cu(s)	−9.7	−0.69
S(rhombic)	−12.6	−1.95
Hg(l)	−28.4	−4.21
Al(s)	+20.7	+2.07
Pt(s)	+267.3	+24.25
Na(s)	+8.48	+2.01
K(s)	+5.94	+2.61
$CuSO_4\cdot 5H_2O(s)$	+167	+183
$MnSO_4\cdot 4H_2O(s)$	+1859	+1835
$NiSO_4\cdot 7H_2O(s)$	+355	+503
$FeSO_4(s)$	+3743	+1558

Source: Principally HCP, with $\chi_m = \chi V_m = \chi\rho/M$.

Table 19A.1 Transport properties of gases at 1 atm

	$\kappa/(mW\ K^{-1}\ m^{-1})$	$\eta/\mu P$	
	273 K	273 K	293 K
Air	24.1	173	182
Ar	16.3	210	223
C_2H_4	16.4	97	103
CH_4	30.2	103	110
Cl_2	7.9	123	132
CO_2	14.5	136	147
H_2	168.2	84	88
He	144.2	187	196
Kr	8.7	234	250
N_2	24.0	166	176
Ne	46.5	298	313
O_2	24.5	195	204
Xe	5.2	212	228

Data: KL.

Table 19B.1 Viscosities of liquids at 298 K, $\eta/(10^{-3}\ kg\ m^{-1}\ s^{-1})$

Benzene	0.601
Carbon tetrachloride	0.880
Ethanol	1.06
Mercury	1.55
Methanol	0.553
Pentane	0.224
Sulfuric acid	27
Water†	0.891

† The viscosity of water over its entire liquid range is represented with less than 1 per cent error by the expression $\log(\eta_{20}/\eta) = A/B$,
$A = 1.370\ 23(t-20) + 8.36\times10^{-4}(t-20)^2$
$B = 109 + t \quad t = \theta/°C$
Convert $kg\ m^{-1}\ s^{-1}$ to centipoise (cP) by multiplying by 10^3 (so $\eta \approx 1$ cP for water).
Data: AIP, KL.

Table 19B.2 Ionic mobilities in water at 298 K, $u/(10^{-8}\ m^2\ s^{-1}\ V^{-1})$

Cations		Anions	
Ag^+	6.24	Br^-	8.09
Ca^{2+}	6.17	$CH_3CO_2^-$	4.24
Cu^{2+}	5.56	Cl^-	7.91
H^+	36.23	CO_3^{2-}	7.46
K^+	7.62	F^-	5.70
Li^+	4.01	$[Fe(CN)_6]^{3-}$	10.5
Na^+	5.19	$[Fe(CN)_6]^{4-}$	11.4
NH_4^+	7.63	I^-	7.96
$[N(CH_3)_4]^+$	4.65	NO_3^-	7.40
Rb^+	7.92	OH^-	20.64
Zn^{2+}	5.47	SO_4^{2-}	8.29

Data: Principally Table 19B.2 and $u = \lambda/zF$.

Table 19B.3 Diffusion coefficients at 298 K, $D/(10^{-9}\ m^2\ s^{-1})$

Molecules in liquids				Ions in water			
I_2 in hexane	4.05	H_2 in $CCl_4(l)$	9.75	K^+	1.96	Br^-	2.08
in benzene	2.13	N_2 in $CCl_4(l)$	3.42	H^+	9.31	Cl^-	2.03
CCl_4 in heptane	3.17	O_2 in $CCl_4(l)$	3.82	Li^+	1.03	F^-	1.46
Glycine in water	1.055	Ar in $CCl_4(l)$	3.63	Na^+	1.33	I^-	2.05
Dextrose in water	0.673	CH_4 in $CCl_4(l)$	2.89			OH^-	5.03
Sucrose in water	0.5216	H_2O in water	2.26				
		CH_3OH in water	1.58				
		C_2H_5OH in water	1.24				

Data: AIP.

Table 20B.1 Kinetic data for first-order reactions

	Phase	θ/°C	k_r/s^{-1}	$t_{1/2}$
$2\,N_2O_5 \rightarrow 4\,NO_2 + O_2$	g	25	3.38×10^{-5}	5.70 h
	$HNO_3(l)$	25	1.47×10^{-6}	131 h
	$Br_2(l)$	25	4.27×10^{-5}	4.51 h
$C_2H_6 \rightarrow 2\,CH_3$	g	700	5.36×10^{-4}	21.6 min
Cyclopropane → propene	g	500	6.71×10^{-4}	17.2 min
$CH_3N_2CH_3 \rightarrow C_2H_6 + N_2$	g	327	3.4×10^{-4}	34 min
Sucrose → glucose + fructose	$aq(H^+)$	25	6.0×10^{-5}	3.2 h

g: High pressure gas-phase limit.
Data: Principally K.J. Laidler, *Chemical kinetics*. Harper & Row, New York (1987); M.J. Pilling and P.W. Seakins, *Reaction kinetics*. Oxford University Press (1995); J. Nicholas, *Chemical kinetics*. Harper & Row, New York (1976). See also JL.

Table 20B.2 Kinetic data for second-order reactions

	Phase	θ/°C	$k_r/(dm^3\,mol^{-1}\,s^{-1})$
$2\,NOBr \rightarrow 2\,NO + Br_2$	g	10	0.80
$2\,NO_2 \rightarrow 2\,NO + O_2$	g	300	0.54
$H_2 + I_2 \rightarrow 2\,HI$	g	400	2.42×10^{-2}
$D_2 + HCl \rightarrow DH + DCl$	g	600	0.141
$2\,I \rightarrow I_2$	g	23	7×10^{9}
	hexane	50	1.8×10^{10}
$CH_3Cl + CH_3O^-$	methanol	20	2.29×10^{-6}
$CH_3Br + CH_3O^-$	methanol	20	9.23×10^{-6}
$H^+ + OH^- \rightarrow H_2O$	water	25	1.35×10^{11}
	ice	−10	8.6×10^{12}

Data: Principally K.J. Laidler, *Chemical kinetics*. Harper & Row, New York (1987); M.J. Pilling and P.W. Seakins, *Reaction kinetics*. Oxford University Press, (1995); J. Nicholas, *Chemical kinetics*. Harper & Row, New York (1976).

Table 20D.1 Arrhenius parameters

First-order reactions	A/s^{-1}	$E_a/(kJ\,mol^{-1})$
Cyclopropane → propene	1.58×10^{15}	272
$CH_3NC \rightarrow CH_3CN$	3.98×10^{13}	160
cis-CHD=CHD → *trans*-CHD=CHD	3.16×10^{12}	256
Cyclobutane → $2\,C_2H_4$	3.98×10^{13}	261
$C_2H_5I \rightarrow C_2H_4 + HI$	2.51×10^{17}	209
$C_2H_6 \rightarrow 2\,CH_3$	2.51×10^{7}	384
$2\,N_2O_5 \rightarrow 4\,NO_2 + O_2$	4.94×10^{13}	103.4
$N_2O \rightarrow N_2 + O$	7.94×10^{11}	250
$C_2H_5 \rightarrow C_2H_4 + H$	1.0×10^{13}	167

(*Continued*)

Table 20D.1 (Continued)

Second-order, gas-phase	$A/(\mathrm{dm^3\ mol^{-1}\ s^{-1}})$	$E_a/(\mathrm{kJ\ mol^{-1}})$
$O+N_2 \rightarrow NO+N$	1×10^{11}	315
$OH+H_2 \rightarrow H_2O+H$	8×10^{10}	42
$Cl+H_2 \rightarrow HCl+H$	8×10^{10}	23
$2\ CH_3 \rightarrow C_2H_6$	2×10^{10}	*ca.*0
$NO+Cl_2 \rightarrow NOCl+Cl$	4.0×10^{9}	85
$SO+O_2 \rightarrow SO_2+O$	3×10^{8}	27
$CH_3+C_2H_6 \rightarrow CH_4+C_2H_5$	2×10^{8}	44
$C_6H_5+H_2 \rightarrow C_6H_6+H$	1×10^{8}	*ca.*25
Second-order, solution	$A/(\mathrm{dm^3\ mol^{-1}\ s^{-1}})$	$E_a/(\mathrm{kJ\ mol^{-1}})$
$C_2H_5ONa+CH_3I$ in ethanol	2.42×10^{11}	81.6
$C_2H_5Br+OH^-$ in water	4.30×10^{11}	89.5
$C_2H_5I+C_2H_5O^-$ in ethanol	1.49×10^{11}	86.6
$C_2H_5Br+OH^-$ in ethanol	4.30×10^{11}	89.5
CO_2+OH^- in water	1.5×10^{10}	38
$CH_3I+S_2O_3^{2-}$ in water	2.19×10^{12}	78.7
Sucrose $+H_2O$ in acidic water	1.50×10^{15}	107.9
$(CH_3)_3CCl$ solvolysis		
in water	7.1×10^{16}	100
in methanol	2.3×10^{13}	107
in ethanol	3.0×10^{13}	112
in acetic acid	4.3×10^{13}	111
in chloroform	1.4×10^{4}	45
$C_6H_5NH_2 + C_6H_5COCH_2Br$ in benzene	91	34

Data: Principally J. Nicholas, *Chemical kinetics*. Harper & Row, New York (1976) and A.A. Frost and R.G. Pearson, *Kinetics and mechanism*. Wiley, New York (1961).

Table 21A.1 Arrhenius parameters for gas-phase reactions

	$A/(\mathrm{dm^3\ mol^{-1}\ s^{-1}})$			
	Experiment	Theory	$E_a/(\mathrm{kJ\ mol^{-1}})$	P
$2\ NOCl \rightarrow 2\ NO+Cl_2$	9.4×10^{9}	5.9×10^{10}	102.0	0.16
$2\ NO_2 \rightarrow 2\ NO+O_2$	2.0×10^{9}	4.0×10^{10}	111.0	5.0×10^{-2}
$2\ ClO \rightarrow Cl_2+O_2$	6.3×10^{7}	2.5×10^{10}	0.0	2.5×10^{-3}
$H_2+C_2H_4 \rightarrow C_2H_6$	1.24×10^{6}	7.4×10^{11}	180	1.7×10^{-6}
$K+Br_2 \rightarrow KBr+Br$	1.0×10^{12}	2.1×10^{11}	0.0	4.8

Data: Principally M.J. Pilling and P.W. Seakins, *Reaction kinetics*. Oxford University Press (1995).

Table 21B.1 Arrhenius parameters for reactions in solution. See Table 20D.1.

Table 21F.1 Exchange current densities (j_0) and transfer coefficients (α) at 298 K

Reaction	Electrode	j_0/(A cm^{-2})	α
$2\,H^+ + 2\,e^- \rightarrow H_2$	Pt	7.9×10^{-4}	
	Cu	1×10^{-6}	
	Ni	6.3×10^{-6}	0.58
	Hg	7.9×10^{-13}	0.50
	Pb	5.0×10^{-12}	
$Fe^{3+} + e^- \rightarrow Fe^{2+}$	Pt	2.5×10^{-3}	0.58
$Ce^{4+} + e^- \rightarrow Ce^{3+}$	Pt	4.0×10^{-5}	0.75

Data: Principally J.O'M. Bockris and A.K.N. Reddy, *Modern electrochemistry*. Pleanum, New York (1970).

Table 22A.1 Maximum observed standard enthalpies of physisorption, $\Delta_{ad}H^{\ominus}$/(kJ mol^{-1}) at 298 K

C_2H_2	−38	H_2	−84
C_2H_4	−34	H_2O	−59
CH_4	−21	N_2	−21
Cl_2	−36	NH_3	−38
CO	−25	O_2	−21
CO_2	−25		

Data: D.O. Haywood and B.M.W. Trapnell, *Chemisorption*. Butterworth (1964).

Table 22A.2 Standard enthalpies of chemisorption, $\Delta_{ad}H^{\ominus}$/(kJ mol^{-1}) at 298 K

Adsorbate	Adsorbent (substrate)											
	Ti	Ta	Nb	W	Cr	Mo	Mn	Fe	Co	Ni	Rh	Pt
H_2		−188			−188	−167	−71	−134			−117	
N_2		−586						−293				
O_2						−720					−494	−293
CO	−640							−192	−176			
CO_2	−682	−703	−552	−456	−339	−372	−222	−225	−146	−184		
NH_3				−301				−188		−155		
C_2H_4		−577		−427	−427			−285		−243	−209	

Data: D.O. Haywood and B.M.W. Trapnell, *Chemisorption*. Butterworth (1964).

PART 4 Character tables

The groups C_1, C_s, C_i

C_1 (1)	E	$h=1$
A	1	

$C_s = C_h$ m	E	σ_h	$h=2$	
A′	1	1	x, y, R_z	x^2, y^2, z^2, xy
A″	1	−1	z, R_x, R_y	yz, zx

$C_i = S_2$ $\bar{1}$	E	i	$h=2$	
A_g	1	1	R_x, R_y, R_z	$x^2, y^2, z^2, xy, yz, zx,$
A_u	1	−1	x, y, z	

The groups C_{nv}

C_{2v}, $2mm$	E	C_2	σ_v	σ_v'	$h=4$	
A_1	1	1	1	1	z, z^2, x^2, y^2	
A_2	1	1	−1	−1	xy	R_z
B_1	1	−1	1	−1	x, zx	R_y
B_2	1	−1	−1	1	y, yz	R_x

C_{3v}, $3m$	E	$2C_3$	$3\sigma_v$	$h=6$	
A_1	1	1	1	z, z^2, x^2+y^2	
A_2	1	1	−1		R_z
E	2	−1	0	$(x, y), (xy, x^2-y^2)$ (yz, zx)	(R_x, R_y)

C_{4v}, $4mm$	E	C_2	$2C_4$	$2\sigma_v$	$2\sigma_d$	$h=8$	
A_1	1	1	1	1	1	z, z^2, x^2+y^2	
A_2	1	1	1	−1	−1		R_z
B_1	1	1	−1	1	−1	x^2-y^2	
B_2	1	1	−1	−1	1	xy	
E	2	−2	0	0	0	$(x, y), (yz, zx)$	(R_x, R_y)

C_{5v}	E	$2C_5$	$2C_5^2$	$5\sigma_v$	$h=10$, $\alpha=72°$	
A_1	1	1	1	1	z, z^2, x^2+y^2	
A_2	1	1	1	−1		R_z
E_1	2	$2\cos\alpha$	$2\cos 2\alpha$	0	$(x, y), (yz, zx)$	(R_x, R_y)
E_2	2	$2\cos 2\alpha$	$2\cos\alpha$	0	(xy, x^2-y^2)	

C_{6v}, $6mm$	E	C_2	$2C_3$	$2C_6$	$3\sigma_d$	$3\sigma_v$	$h=12$	
A_1	1	1	1	1	1	1	z, z^2, x^2+y^2	
A_2	1	1	1	1	−1	−1		R_z
B_1	1	−1	1	−1	−1	1		
B_2	1	−1	1	−1	1	−1		
E_1	2	−2	−1	1	0	0	$(x, y), (yz, zx)$	(R_x, R_y)
E_2	2	2	−1	−1	0	0	(xy, x^2-y^2)	

$C_{\infty v}$	E	$2C_\phi$†	$\infty\sigma_v$	$h=\infty$	
$A_1(\Sigma^+)$	1	1	1	z, z^2, x^2+y^2	
$A_2(\Sigma^-)$	1	1	−1		R_z
$E_1(\Pi)$	2	$2\cos\phi$	0	$(x, y), (yz, zx)$	(R_x, R_y)
$E_2(\Delta)$	2	$2\cos 2\phi$	0	(xy, x^2-y^2)	

† There is only one member of this class if $\phi=\pi$.

The groups D_n

D_2, 222	E	C_2^z	C_2^y	C_2^x	$h=4$	
A_1	1	1	1	1	x^2, y^2, z^2	
B_1	1	1	−1	−1	z, xy	R_z
B_2	1	−1	1	−1	y, zx	R_y
B_3	1	−1	−1	1	x, yz	R_x

D_3, 32	E	$2C_3$	$3C_2'$	$h=6$	
A_1	1	1	1	z^2, x^2+y^2	
A_2	1	1	−1	z	R_z
E	2	−1	0	$(x, y), (yz, zx), (xy, x^2-y^2)$	(R_x, R_y)

D_4, 422	E	C_2	$2C_4$	$2C_2'$	$2C_2''$	$h=8$	
A_1	1	1	1	1	1	z^2, x^2+y^2	
A_2	1	1	1	−1	−1	z	R_z
B_1	1	1	−1	1	−1	x^2-y^2	
B_2	1	1	−1	−1	1	xy	
E	2	−2	0	0	0	$(x, y), (yz, zx)$	(R_x, R_y)

The groups D_{nh}

D_{3h}, $\bar{6}2m$	E	σ_h	$2C_3$	$2S_3$	$3C_2'$	$3\sigma_v$	$h=12$	
A_1'	1	1	1	1	1	1	z^2, x^2+y^2	
A_2'	1	1	1	1	−1	−1		R_z
A_1''	1	−1	1	−1	1	−1		
A_2''	1	−1	1	−1	−1	1	z	
E'	2	2	−1	−1	0	0	$(x, y), (xy, x^2-y^2)$	
E''	2	−2	−1	1	0	0	(yz, zx)	(R_x, R_y)

D_{4h}, $4/mmm$	E	$2C_4$	C_2	$2C_2'$	$2C_2''$	i	$2S_4$	σ_h	$2\sigma_v$	$2\sigma_d$	$h=16$	
A_{1g}	1	1	1	1	1	1	1	1	1	1	x^2+y^2, z^2	
A_{2g}	1	1	1	−1	−1	1	1	1	−1	−1		R_z
B_{1g}	1	−1	1	1	−1	1	−1	1	1	−1	x^2-y^2	
B_{2g}	1	−1	1	−1	1	1	−1	1	−1	1	xy	
E_g	2	0	−2	0	0	2	0	−2	0	0	(yz, zx)	(R_x, R_y)
A_{1u}	1	1	1	1	1	−1	−1	−1	−1	−1		
A_{2u}	1	1	1	−1	−1	−1	−1	−1	1	1	z	
B_{1u}	1	−1	1	1	−1	−1	1	−1	−1	1		
B_{2u}	1	−1	1	−1	1	−1	1	−1	1	−1		
E_u	2	0	−2	0	0	−2	0	2	0	0	(x, y)	

D_{5h}	E	$2C_5$	$2C_5^2$	$5C_2$	σ_h	$2S_5$	$2S_5^3$	$5\sigma_v$	$h=20$	$\alpha=72°$
A_1'	1	1	1	1	1	1	1	1	x^2+y^2, z^2	
A_2'	1	1	1	−1	1	1	1	−1		R_z
E_1'	2	$2\cos\alpha$	$2\cos 2\alpha$	0	2	$2\cos\alpha$	$2\cos 2\alpha$	0	(x, y)	
E_2'	2	$2\cos 2\alpha$	$2\cos\alpha$	0	2	$2\cos 2\alpha$	$2\cos\alpha$	0	(x^2-y^2, xy)	
A_1''	1	1	1	1	−1	−1	−1	−1		
A_2''	1	1	1	−1	−1	−1	−1	1	z	
E_1''	2	$2\cos\alpha$	$2\cos 2\alpha$	0	−2	$-2\cos\alpha$	$-2\cos 2\alpha$	0	(yz, zx)	(R_x, R_y)
E_2''	2	$2\cos 2\alpha$	$2\cos\alpha$	0	−2	$-2\cos 2\alpha$	$-2\cos\alpha$	0		

$D_{\infty h}$	E	$2C_\phi$	…	$\infty\sigma_v$	i	$2S_\infty$	…	$\infty C_2'$	$h=\infty$	
$A_{1g}(\Sigma_g^+)$	1	1	…	1	1	1	…	1	z^2, x^2+y^2	
$A_{1u}(\Sigma_u^+)$	1	1	…	1	−1	−1	…	−1	z	
$A_{2g}(\Sigma_g^-)$	1	1	…	−1	1	1	…	−1		R_z
$A_{2u}(\Sigma_u^-)$	1	1	…	−1	−1	−1	…	1		
$E_{1g}(\Pi_g)$	2	$2\cos\phi$	…	0	2	$-2\cos\phi$	…	0	(yz, zx)	(R_x, R_y)
$E_{1u}(\Pi_u)$	2	$2\cos\phi$	…	0	−2	$2\cos\phi$	…	0	(x, y)	
$E_{2g}(\Delta_g)$	2	$2\cos 2\phi$	…	0	2	$2\cos 2\phi$	…	0	(xy, x^2-y^2)	
$E_{2u}(\Delta_u)$	2	$2\cos 2\phi$	…	0	−2	$-2\cos 2\phi$	…	0		
⋮	⋮	⋮	…	⋮	⋮	⋮	…	⋮		

The cubic groups

T_d, $\bar{4}3m$	E	$8C_3$	$3C_2$	$6\sigma_d$	$6S_4$	$h=24$	
A_1	1	1	1	1	1	$x^2+y^2+z^2$	
A_2	1	1	1	−1	−1		
E	2	−1	2	0	0	$(3z^2-r^2, x^2-y^2)$	
T_1	3	0	−1	−1	1		(R_x, R_y, R_z)
T_2	3	0	−1	1	−1	(x, y, z), (xy, yz, zx)	

O_h, $m3m$	E	$8C_3$	$6C_2$	$6C_4$	$3C_2(=C_4^2)$	i	$6S_4$	$8S_6$	$3\sigma_h$	$6\sigma_d$	$h=48$	
A_{1g}	1	1	1	1	1	1	1	1	1	1	$x^2+y^2+z^2$	
A_{2g}	1	1	−1	−1	1	1	−1	1	1	−1		
E_g	2	−1	0	0	2	2	0	−1	2	0	$(2z^2-x^2-y^2, x^2-y^2)$	
T_{1g}	3	0	−1	1	−1	3	1	0	−1	−1		(R_x, R_y, R_z)
T_{2g}	3	0	1	−1	−1	3	−1	0	−1	1	(xy, yz, zx)	
A_{1u}	1	1	1	1	1	−1	−1	−1	−1	−1		
A_{2u}	1	1	−1	−1	1	−1	1	−1	−1	1		
E_u	2	−1	0	0	2	−2	0	1	−2	0		
T_{1u}	3	0	−1	1	−1	−3	−1	0	1	1	(x, y, z)	
T_{2u}	3	0	1	−1	−1	−3	1	0	1	−1		

The icosahedral group

I	E	$12C_5$	$12C_5^2$	$20C_3$	$15C_2$	$h=60$	
A	1	1	1	1	1	$x^2+y^2+z^2$	
T_1	3	$\frac{1}{2}(1+5^{1/2})$	$\frac{1}{2}(1-5^{1/2})$	0	−1	(x, y, z)	(R_x, R_y, R_z)
T_2	3	$\frac{1}{2}(1-5^{1/2})$	$\frac{1}{2}(1+5^{1/2})$	0	−1		
G	4	−1	−1	1	0		
H	5	0	0	−1	1	$(2z^2-x^2-y^2, x^2-y^2, xy, yz, zx)$	

Further information: P.W. Atkins, M.S. Child, and C.S.G. Phillips, *Tables for group theory.* Oxford University Press, (1970). In this source, which is available on the web (see p. *x* for more details), other character tables such as D_2, D_4, D_{2d}, D_{3d}, and D_{5d} can be found.

INDEX

D

R

S

T

U